OF THE ELEMENTS

Atomic weights are based on ^{12}C. Atomic weights in parentheses indicate the most stable or best-known isotope. Names and symbols for elements 104, 105, and 106 are unofficial.

	IB	IIB	IIIA	IVA	VA	VIA	VIIA	Noble gases
								2 Helium **He** 4.00260
			5 Boron **B** 10.81	6 Carbon **C** 12.011	7 Nitrogen **N** 14.0067	8 Oxygen **O** 15.9994	9 Fluorine **F** 18.99840	10 Neon **Ne** 20.179
			13 Aluminum **Al** 26.98154	14 Silicon **Si** 28.086	15 Phosphorus **P** 30.97376	16 Sulfur **S** 32.06	17 Chlorine **Cl** 35.453	18 Argon **Ar** 39.948
28 Nickel **Ni** 58.71	29 Copper **Cu** 63.546	30 Zinc **Zn** 65.38	31 Gallium **Ga** 69.72	32 Germanium **Ge** 72.59	33 Arsenic **As** 74.9216	34 Selenium **Se** 78.96	35 Bromine **Br** 79.904	36 Krypton **Kr** 83.80
46 Palladium **Pd** 106.4	47 Silver **Ag** 107.868	48 Cadmium **Cd** 112.40	49 Indium **In** 114.82	50 Tin **Sn** 118.69	51 Antimony **Sb** 121.75	52 Tellurium **Te** 127.60	53 Iodine **I** 126.9045	54 Xenon **Xe** 131.30
78 Platinum **Pt** 195.09	79 Gold **Au** 196.9665	80 Mercury **Hg** 200.59	81 Thallium **Tl** 204.37	82 Lead **Pb** 207.2	83 Bismuth **Bi** 208.9804	84 Polonium **Po** (210)[a]	85 Astatine **At** (210)[a]	86 Radon **Rn** (222)[a]

Inner transition elements

63 Europium **Eu** 151.96	64 Gadolinium **Gd** 157.25	65 Terbium **Tb** 158.9254	66 Dysprosium **Dy** 162.50	67 Holmium **Ho** 164.9304	68 Erbium **Er** 167.26	69 Thulium **Tm** 168.9342	70 Ytterbium **Yb** 173.04	71 Lutetium **Lu** 174.97
95 Americium **Am** (243)[a]	96 Curium **Cm** (247)[a]	97 Berkelium **Bk** (249)[a]	98 Californium **Cf** (251)[a]	99 Einsteinium **Es** (254)[a]	100 Fermium **Fm** (253)[a]	101 Mendelevium **Md** (256)[a]	102 Nobelium **No** (254)[a]	103 Lawrencium **Lr** (257)[a]

8/12/91

ORGANIC CHEMISTRY

Organic Chemistry

Douglas J. Raber
Nancy K. Raber

University of South Florida

West Publishing Company

St. Paul   New York   Los Angeles   San Francisco

A STUDENT SOLUTIONS MANUAL AND STUDY GUIDE

A solutions manual that contains complete, detailed answers to all exercises and problems is available to all students from the local bookstore under the title *Student Solutions Manual and Study Guide to Accompany Organic Chemistry*, prepared by the authors. An examination copy is available to instructors by contacting West Publishing Company.

Copyediting: Ernestine R. Daniels

Interior design: Signature Design, Inc.

Interior illustration: York Production Services, Inc.

Composition: York Graphics Services, Inc.

Cover design: Designteam/Paul Konsterlie

Cover image: Molecular image created with SYBYL/MENDYL Modeling Software developed by Tripos Associates, Inc., St. Louis, Missouri.

Printed in the United States of America

Library of Congress Cataloging-in-Publication Data

Raber, Douglas.
Organic chemistry/Douglas Raber, Nancy Raber.
p. cm.
Includes index.
ISBN 0-314-28508-3
1. Chemistry, Organic. I. Raber, Nancy. II. Title.
QD251.2.R33 1988
547–dc19

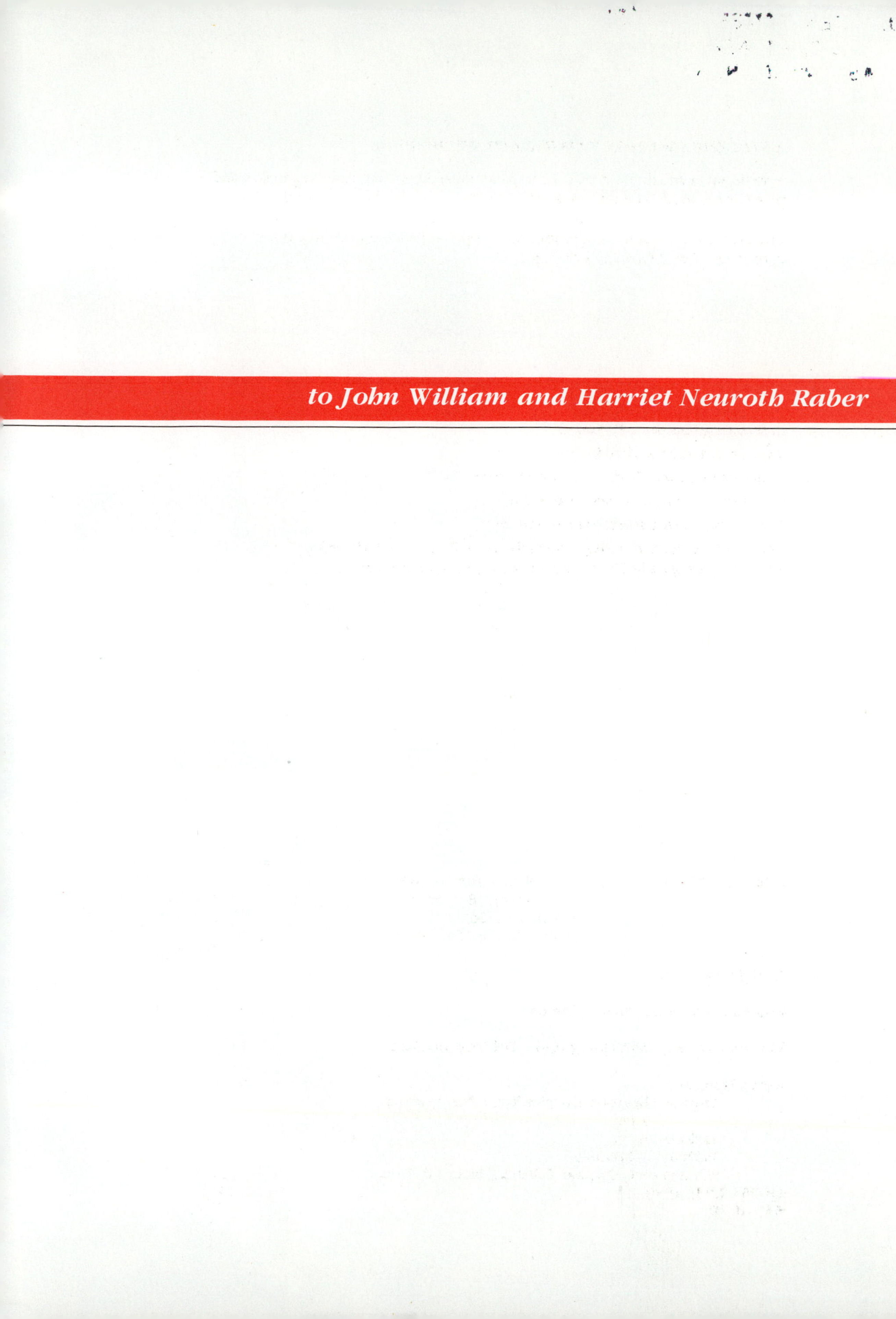

to John William and Harriet Neuroth Raber

CONTENTS

CHAPTER 1 ALKANES

CHAPTER 2 STRUCTURE, STABILITY, AND REACTIVITY

CHAPTER 3 CYCLOALKANES

CHAPTER 6 STEREOCHEMISTRY

CHAPTER 7 ALKYNES

CHAPTER 8 COMPOUNDS WITH MORE THAN ONE REACTIVE CENTER

CHAPTER 16 CARBOXYLIC ACIDS AND THEIR DERIVATIVES

CHAPTER 17 AMINES

CHAPTER 18 AROMATIC SUBSTITUTION REACTIONS

This text has been written using the *functional group approach.* The organic chemistry begins early in the book, and a complete discussion of all the major functional groups is completed by Chapter 17. At the same time, we have carefully included the necessary attention to mechanistic and synthetic organic chemistry as well as to spectroscopy.

This text has been designed with *flexibility* in mind. Once the instructor has finished the first 17 chapters, all the functional group chemistry has been covered. This allows the instructor to design the overall orientation of the course by targeting specific chapters among the last ten. One instructor might aim toward a highly *synthetic* emphasis, using the chapters on Dicarbonyl Compounds and Polyfunctional Synthesis as the cornerstones; another might use the chapters on Aromatic Substitution and Molecular Orbital Theory (together with the supplementary chapter on Molecular Rearrangements in the Solutions Manual) to emphasize topics in *physical organic chemistry.*

Our approach is one that consciously teaches *problem solving,* not only in the text where we present worked examples, but also in the Solutions Manual. There, in addition to the answers to the questions, the student will find detailed explanations of the reasoning needed to solve them. The many problems at the end of each chapter always encompass the full range of complexity, from straightforward nomenclature and basic reactions, to word problems and multi-step synthetic conversions, to spectroscopic problems (where appropriate), and finally to the intricate roadmap problems that will provide an intellectual challenge to the best of students.

An emphasis on biologically and medically important topics is found throughout the book, and the appropriate pages are marked with a segment of double helix that serves as a *bioorganic logo.* This emphasis is in part a response to the fact that organic chemistry courses serve a wide range of students who intend to specialize in health-related fields, but equally it recognizes the origins of organic chemistry. We hope that students using this text will find it unnecessary to turn away from organic chemistry in order to discover topics that are exciting to them. Much of the intellectual excitement that students hope to find in the health-related professional fields can be discovered more quickly in the area of organic chemistry. For this reason, we have introduced topics such as insect pheromones as early as Chapter 11, and sections are devoted to such topics in many subsequent chapters. The last ten chapters include several (Carbohydrates, Amino Acids and Peptides, Natural Products, and Heterocyclic Compounds) that provide very strong coverage of biologically important organic chemistry.

We have strived to keep this text modern and accurate, yet we have also avoided "fad" reactions. This is an introductory text, so only organic reactions that have received wide acceptance by active workers in the research community have been included in our coverage. Every reaction shown in the text is an actual example taken from the chemical literature, and percent yields are shown. The examples are representative, so the yields often provide a good illustration of the reliability or generality of a particular method. Many of the examples are taken from *Organic Syntheses* (usually recognizable from the range in reported percent yield), so they represent reliable and useful preparative methods. The full literature citation for each reaction is provided in the Instructor's Manual, enabling the instructor to conveniently locate additional information on a particular topic.

Industrial Chemistry and *Polymers* are presented as separate chapters, and other economically important aspects of modern organic chemistry are also included.

For example, discussions of *pharmaceuticals* are presented in the chapters on Carbohydrates, Amino Acids and Peptides, Natural Products, and Heterocyclic Compounds.

Spectroscopy is first introduced in Chapter 9 and is subsequently integrated throughout the text. Full, up-to-date coverage of both ^{13}C and ^{1}H NMR spectroscopy is provided without immersing the student in the incredible complexity of modern instrumental techniques. Infrared and Ultraviolet-Visible spectroscopy are presented at a basic but accurate (and useful) level. Spectroscopic data is employed regularly as part of our approach to problem solving. The introductory discussion of Mass Spectrometry in Chapter 9 is complemented by a complete Supplementary Chapter in the Solutions Manual.

In recognition of a new generation of organic chemistry, we have integrated *molecular orbital theory* throughout the text. Instructors who do not wish to emphasize this topic will have no difficulty, because the treatment is at a convenient, qualitative level. On the other hand, an instructor who wishes to stress the importance of molecular orbital theory in modern organic chemistry can take advantage of the more complete coverage that is presented in Chapter 27.

ACKNOWLEDGMENTS

This book was created with the assistance and support of a variety of people, and we give them our warm thanks. Wayne C. Guida listened patiently during the formative stages of the manuscript and provided many helpful ideas. The entire book was proofread by Anne Richards (Tripos Associates, Inc.). Our two coauthors on the Solutions Manual, James W. Leahy and Ellen M. Salcines, provided invaluable assistance in working, drawing, and checking the Exercises and Problems. We thank Ronald F. Federspiel for his assistance in recording several NMR spectra. Many others reviewed one or more chapters during the early stages of writing, and their comments and criticisms helped tremendously as the early manuscripts evolved into a finished text:

Guy Mattson
University of Central Florida

Patty O'Hara-Mays
Arizona State University

David Macaulay
William Rainey Harper College

Harry E. Peery
Tompkins Cortland Community College

Eugene J. Agnello
Hofstra University

Frank L. Setliff
University of Arkansas—Little Rock

Edwin Kaiser
University of Missouri—Columbia

James Herschberger
Miami University

Jim Hagen
University of Nebraska—Omaha

Gerald Selter
San Jose State University

James M. Garrett
Stephen F. Austin State University

Kenneth Raymond
Eastern Washington University

Mike Rathke
Michigan State University

John C. Shelton
California State University—Hayward

Alan Morgan
University of Toledo

Gary Crowther
James Madison University

J. E. Gurst
University of West Florida

Tom Green
Western Kentucky University

Edward Leete
University of Minnesota

John Huffman
Clemson University

Martin Saltzman
Providence College

James MacMillan
University of North Iowa

John Swenton
Ohio State University

David Todd
Worcester Polytechnic

Frank Cartledge
Louisiana State University

Robert Gadwood
Northwestern University

Leroy Nyquist
California State University—Northridge

Thomas Cantrell
American University

D. Howard Miles
Mississippi State University

Charles Spangler
Northern Illinois State University

Roland Flynn
Montclair State College

Berton C. Weberg
Mankato State University

Paul Kropp
University of North Carolina—Chapel Hill

Larry Singer
University of Southern California

David Boykin
Georgia State University

Word processing came of age during the writing of this book, and we are indebted to the staff of the University Computing Center at the University of South Florida for the assistance that they have provided. In particular we thank Gary Grandon, Ron Hickman and Richard Wain. We also thank our colleagues in the Chemistry Department at the University of South Florida for their support, especially during some of the more difficult parts of the project. Jefferson C. Davis, Jr. and Stewart W. Schneller have been particularly helpful.

We appreciate the help that has been provided by the editors at West: Gary Woodruff, Pete Marshall, Pam McClanahan, and Laura Mezner. The quality of this book reflects their efforts.

The NMR and infrared spectra in this text were supplied by the Aldrich Chemical Company, Inc., and we thank Charles J. Pouchert and Jacqlynn Behnke for their assistance. The NMR spectra concerned with lanthanide shift reagents were recorded at the University of South Florida by Ronald F. Federspiel. All mass spectra are from the EPA/NIH Database, and we thank Sharon Lias (National Bureau of Standards) and William L. Budde (United States Environmental Protection Agency) for their assistance.

Finally (and most importantly), we thank our daughters, Wendy Elizabeth Raber and Harriet Jessica Raber, for their love and patience during this project. One is now a college student and can barely remember times before "the book." The other, is in the second grade and has never known a time when we were not working on the book. Now it is their turn.

CHAPTER 1

ALKANES

Organic chemistry is a young area of science, and its origins can be traced to the same exciting period of history that saw the birth of the United States, together with political and intellectual revolution in Europe. In those times there were no "organic chemists" as such, and science was not a field in which one could make a career. Education was limited to but a fraction of the population, and the world of science and research was very small. What limited scientific experimentation did take place resulted largely from the intellectual curiosity of individuals who were primarily physicians or apothecaries. Others may have had the inclination, but only these few had the necessary training and materials at their disposal.

Most investigations during the early stages of organic chemistry were concerned with the isolation and characterization of compounds obtained from animals or plants. Of particular historical significance was the isolation in the late 1700s of *urea* (CH_4N_2O) from human urine, because this compound would play a key role in a philosophical debate.

The first use of the word **organic** to describe compounds derived from living *organisms* is attributed to the Swedish chemist Berzelius in the early 1800s. Prevalent at that time was a philosophy which held that organic compounds were a unique class of chemicals that could only be synthesized inside living organisms. No chemist had ever prepared an organic compound except by employing some other organic compound as the starting material. Consequently, the German chemist Wöhler achieved a dramatic breakthrough in the development of science in 1828 when he carried out the preparation of urea from *inorganic* starting materials (lead cyanate and aqueous ammonia). The possibility of laboratory preparation of the same compounds which are produced by living organisms removed the distinction between naturally occurring and synthetic compounds. The definition of *organic chemistry* has slowly evolved, and the term is now generally accepted as referring to the *chemistry of compounds of carbon* because carbon atoms provide the basic building blocks of organic molecules.

In the remainder of the 1800s and the first part of the present century chemists were largely concerned with the isolation, structure determination, reactions, and eventually synthesis of *natural products,* i.e., naturally occurring organic compounds obtained from animals and plants. Even today the role of natural products in organic chemistry remains important; for example, many antibiotics and other drugs are obtained from plant or microbiological sources. But a major difference in *modern* organic chemistry is the parallel use of *synthetic* procedures to produce structurally related compounds. In many cases these *analogs* are superior to the natural products.

Synthetic organic chemistry is not restricted to natural products and pharmaceuticals, however; the influence of organic synthesis pervades all aspects of our modern lives. Various "synthetics" now complement naturally occurring materials, and for many uses they have largely replaced such materials as natural rubber, silk, and cotton. In some instances even structural materials such as wood and steel can be replaced by synthetic organic materials.

Organic chemistry is an exciting and vibrant area of modern science, and it is found everywhere in today's world. To meet the challenges of our times both an understanding and an appreciation of this area of science are indispensable. Only with such a background will you be prepared to face (and perhaps solve) the problems of the coming decades in such diverse areas as medicine, biochemistry, ecology, pollution control, environmental safety, petrochemical management, and energy conservation.

1.1 THE STRUCTURE OF ALKANES

The class of compounds called *alkanes* provides a useful starting point for the study of organic chemistry because we can use these compounds to introduce the basic principles of organic chemistry. Moreover, the chemistry of alkanes is fairly straightforward since alkanes are quite unreactive in comparison with other types of organic compounds. The most important reaction that they undergo is not even a typical laboratory reaction at all but is the high-temperature oxidation reaction known as *combustion.* In fact, you are already familiar with alkanes because they are the major components of all the petroleum-derived fuels that have become so much a part of our lives in the twentieth century.

Alkanes are a specific class of compounds within a more general group called **hydrocarbons,** compounds composed entirely of carbon and hydrogen. Alkanes are characterized by a specific relationship between the numbers of carbon and hydrogen atoms in each molecule, which can be expressed by the general molecular formula, C_nH_{2n+2}. In other words, an alkane molecule has two more than twice as many hydrogen atoms as carbon atoms.

Methane is the simplest of the alkanes. It is the major component of natural gas, an important fuel that is often found in conjunction with petroleum deposits. Methane has a single carbon in its molecular formula, CH_4. The next two members of the series are *ethane*, C_2H_6, and *propane,* C_3H_8. Propane is the major constituent of liquefied petroleum gas (LPG), another commercially important fuel that is obtained directly from crude petroleum. Figure 1.1 illustrates the **structural formulas** together with the molecular formulas of these three alkanes.

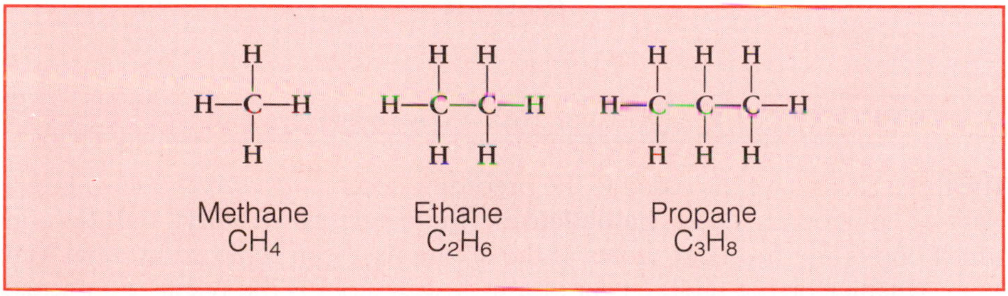

```
   H           H  H           H  H  H
   |           |  |           |  |  |
H--C--H     H--C--C--H     H--C--C--C--H
   |           |  |           |  |  |
   H           H  H           H  H  H
```

Methane CH_4 — Ethane C_2H_6 — Propane C_3H_8

Figure 1.1
The three simplest alkanes: structural formulas and molecular formulas for methane, ethane and propane.

As suggested by Figure 1.1, there is more than one way in which a formula can be used to designate a particular compound, and selection of the appropriate type depends upon the information to be conveyed. The molecular formulas—CH_4, C_2H_6, and C_3H_8—tell the number of atoms of each element in the molecule, but they provide no information about which of the various atoms are bonded to each other. On the other hand, the structural formulas show these bonding relationships explicitly, thereby describing not only the number of each type of atom but also the **constitution** of the molecules, that is, which atoms are bonded to which other atoms. Clearly the structural formula conveys more information, but it is used at the expense of the additional effort required to draw that structure. To draw all the bonds of a large molecule is very time-consuming, and for this reason we frequently use an abbreviated structural formula. For example, we can represent the structure of propane by the formula $CH_3—CH_2—CH_3$, in which we have drawn the bonds between carbon atoms but have not shown any of the bonds between hydrogen and carbon. By convention it is *understood* that those atoms immediately following a carbon atom are bonded to it. Sometimes these formulas are abbreviated still further and even the carbon-carbon bonds are not shown explicitly. Propane would then be drawn as $CH_3CH_2CH_3$, and as long as you understand that there is a chain of carbon atoms, all the necessary structural information is present.

The structures of the three alkanes in Figure 1.1 illustrate a general structural rule in organic chemistry: *In neutral organic compounds, each carbon atom has a total of four bonds.* (Recognize too, that each hydrogen atom has only one bond.) The structural formulas in Figure 1.1 use *lines* to denote the electrons of a bond. This is a convenient variation of the drawings known as Lewis structures, in which each pair of electrons is drawn as a pair of dots. For example, methane could be drawn in the following way:

```
   H
   ..
H : C : H
   ..
   H
```

When drawing organic structures, it is more convenient to use lines (rather than a pair of dots) to represent bonds, and we will employ structures such as those in Figure 1.1. On the other hand, we will use pairs of dots to denote *nonbonded* electrons such as the unshared pairs of electrons on the oxygen atom of water.

```
    ..
H—O—H
    ..
```

1.2 DERIVATIVES OF ALKANES: HALOALKANES

The three alkanes we discussed in the preceding section also serve to introduce a class of compounds called **haloalkanes** (frequently called **alkyl halides**), in which one of the hydrogen atoms of the alkane has been replaced by a halogen atom. Whether or not replacement of a hydrogen atom is the result of an actual reaction, this type of relationship between two compounds or classes of compounds allows us to consider the idea that one is derived from the other. On this basis the haloalkanes are frequently referred to as **derivatives** of alkanes. Similarly, the idea of replacement or *substitution* of a halogen for a hydrogen allows us to designate such atoms as **substituents** on a carbon atom. This notion is used throughout organic chemistry, and any atom or group attached to a carbon atom can be referred to as a substituent on that carbon atom.

Replacement of any of the four hydrogens of methane with chlorine would afford the chloro derivative, *chloromethane* (*methyl chloride*). Likewise, if any one of the six hydrogens of ethane were replaced by a single chlorine, the resulting derivative would be *chloroethane* (*ethyl chloride*).

$CH_3—Cl$	$CH_3CH_2—Cl$
Chloromethane (methyl chloride)	Chloroethane (ethyl chloride)

Fluorine, bromine, and iodine atoms can also occur in place of hydrogen atoms in alkanes. Substitution with these halogens would produce the corresponding fluoro-, bromo-, and iodoalkanes. The Greek prefixes **mono-, di-, tri-,** etc. are used to indicate how many substituents are present. The preceding chloroalkanes are both monochloro derivatives because each has one chloro substituent. An example of a dihalide is the dibromoethane known as EDB (from the common name, *ethylene dibromide*).

$$Br—CH_2—CH_2—Br$$

This compound has been used extensively as a pesticide, but its use has been sharply curtailed in recent years as a result of environmental concerns.

1.3 ISOMERISM

A new situation arises when we consider derivatives of propane. Methane has only a single carbon atom, and only a single monochloromethane can be derived from it. Ethane has two carbon atoms but they are equivalent, as suggested by the symmetrical formula $CH_3—CH_3$. Consequently only a single monochloroethane is possible. Propane, however, has two kinds of carbon atoms. The carbon atoms of the CH_3 groups at each end of the propane molecule are equivalent, as suggested by the symmetrical formula $CH_3—CH_2—CH_3$, and each is attached to only one other carbon atom. The central carbon, however, is connected to *two* other carbons, and it is clearly different from the terminal carbons. As a result the alkyl chloride derived from replacement of one of the hydrogens of the central CH_2 group of propane is not the same as the alkyl chloride derived from replacement of one of the hydrogens of a CH_3 group. These two chloropropanes are shown in Figure 1.2.

The two different chloropropanes shown in Figure 1.2 illustrate a very important type of relationship in organic chemistry, that of **isomerism.** Two compounds are called **isomers** if they have the *same* molecular formula but nevertheless are *different* compounds. Such is the situation with 1-chloropropane and 2-chloropropane; each has the same molecular formula, C_3H_7Cl, but they clearly do not have the same structure. As a result we can conclude that 1-chloropropane and 2-chloropropane are isomers.

$$\underset{\text{Propane}}{\overset{3}{C}H_3—\overset{2}{C}H_2—\overset{1}{C}H_3} \longrightarrow \underset{\text{1-Chloropropane}}{\overset{3}{C}H_3—\overset{2}{C}H_2—\overset{1}{C}H_2—Cl} + \underset{\text{2-Chloropropane}}{\overset{3}{C}H_3—\underset{\substack{|\\Cl}}{\overset{2}{C}H}—\overset{1}{C}H_3}$$

Figure 1.2
Two different chloropropanes can be formed from propane. If the carbons of the three-carbon chain are numbered from the right side of the structure, these two compounds have the chloro substituent on C-1 and on C-2, respectively.

We will introduce a variety of different kinds of isomerism in subsequent chapters, but some general comments about isomerism are appropriate here. First, *isomerism is a relationship between two (or more) compounds.* It is never appropriate to ask whether any single structure is "an isomer." Instead you must compare two structures and ask whether or not they are isomers. Notice that we find it necessary to compare *structures,* but we are actually testing a relationship between the *compounds* which they represent. Second, *isomerism always requires two aspects of the relationship;* one way in which the two structures are the *same* and another way in which they are *different* must be identified. The way in which the two structures are the *same* must always be that they have identical molecular formulas. There are many ways in which they can be *different,* but ultimately the difference must be in their molecular structures. For the case illustrated in Figure 1.2, the isomeric chloropropanes have different constitutions, that is, the sequences in which the atoms are bonded are different. Accordingly, these are called **constitutional isomers.** The idea of *constitution* can be difficult to grasp at first, and it is useful to think of it in terms of **connectivity.** For example, we can describe the connectivity of the chlorine and all the carbon atoms of 1-chloropropane (Figure 1.2) by writing them in sequence: Cl, C-1, C-2, C-3. However, we cannot write a single sequence (without doubling back) to describe the connectivity of 2-chloropropane. Instead, we must write two lists: C-1, C-2, C-3 and C-2, Cl. Such connectivity lists provide ways of explicitly defining the constitutions of organic molecules.

Constitutional isomers have frequently been described by another name, *structural isomers.* But all kinds of isomerism require some structural differences between compounds that are isomers. Consequently, the term "structural isomers" can lead to confusion, and we will not use it in this text.

1.4 STEREOCHEMICAL CONVENTIONS

The structural drawings which we have used up to this point convey varying amounts of structural information. At one extreme the notation C_2H_6 for ethane simply gives the molecular formula but provides no information about the structure of the ethane molecule. The notation

```
    H  H
    |  |
 H—C—C—H
    |  |
    H  H
```

completely describes the connectivity of all the atoms in ethane (i.e., it fully describes the constitution), but it still provides no information regarding the three-dimensional shape of the molecule.

In order to reasonably depict a three-dimensional structure on a printed page (which is clearly only two-dimensional), we must adopt certain *conventions* which are widely accepted by organic chemists. The representation of molecules in three dimensions together with the study of the chemical consequences of three-dimensional structures falls in the category of **stereochemistry.** The prefix *stereo* is derived from the Greek *stereos,* meaning "solid" and is used to indicate a quality that has more than one dimension (for example, *stereo*phonic sound). We will present a detailed discussion of stereochemistry in Chapter 6. For the time being we will only be concerned with ways of *drawing* structures so as to convey three-

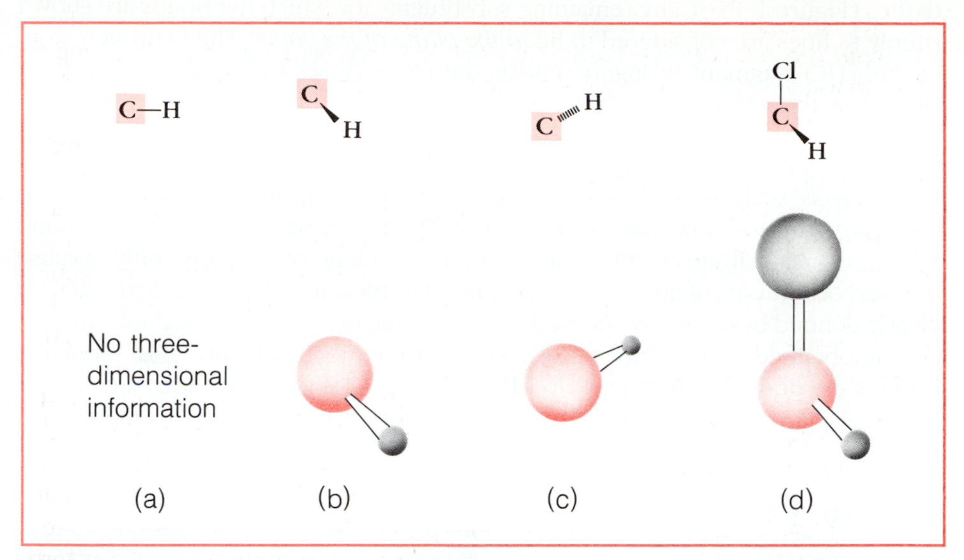

Figure 1.3 Stereochemical conventions.
(a) A C—H fragment with no stereochemical information; (b) a C—H fragment with the hydrogen closer to you than the carbon; (c) a C—H fragment with the hydrogen farther from you than the carbon; (d) a CH—Cl fragment with the chlorine directly above the carbon and the hydrogen closer to you than the carbon.

dimensional information. Figure 1.3 illustrates the basic stereochemical conventions that we will employ to show the relative positions of two atoms or groups. In accord with the central position of carbon in organic chemistry, we will *designate the orientation of an atom or group relative to the carbon to which it is attached.*

Figure 1.3b shows a C—H fragment in which the hydrogen is closer to you than the carbon atom. It is particularly useful to think of the carbon atom as lying in the plane of the paper so that the hydrogen atom would actually be above the plane of the paper. Note that the wedge used to indicate the relative orientation of the hydrogen atom is similar to the perspective drawing (with "balls and sticks") shown beneath it in Figure 1.3b. Ball-and-stick drawings (which are true perspective drawings) can also be used to show the three-dimensional features of a molecule, but they are considerably more difficult and inconvenient to draw properly, particularly for complex organic molecules.

A second orientation of a C—H fragment, with the hydrogen atom farther away from the viewer than the carbon, is shown in Figure 1.3c. This convention consists of a series of dashes between the H and C, where the dashes are *perpendicular* to the C—H bond, that is, C⦀H, not C---H. The latter convention, in which the dashes are *parallel* to the bond, is used to indicate a partial or incomplete bond and has no stereochemical implications. Although there is no clear resemblance to the perspective drawing below it in Figure 1.3c, this representation is nevertheless widely accepted as a convenient means of indicating three-dimensional structure. You may use both wedges and dashes in a single drawing and use them as often as necessary when drawing a complicated molecule.

One additional convention is shown in Figure 1.3d. When the stereochemistry of one or more substituents on a carbon atom is depicted by use of the wedge or dashes (Figure 1.3b,c), any remaining substituents for which the bonds are shown simply as lines are considered to lie *in the plane of the paper*. The hydrogen atom of the CHCl fragment in Figure 1.3d is therefore closer to you than the carbon atom, but the carbon and chlorine atoms are both in the plane of the paper.

One final, very important point must be emphasized about the stereochemical conventions illustrated in Figure 1.3. In the absence of specific three-dimensional designations such as wedges or dashed lines, a structural formula conveys *no three-dimensional information whatsoever*. The line designating the C—Cl bond in Figure 1.3d indicates that both atoms lie in the plane of the paper only because the stereochemistry of another group bonded to the carbon (i.e., the hydrogen) is clearly defined by the use of the wedge. In contrast, no stereochemical information can be obtained by the drawing of the C—H fragment in Figure 1.3a. Similarly, none of the drawings shown previously in Figures 1.1 and 1.2 provided any stereochemical information because none of them were three-dimensional representations.

EXERCISE 1.1

Draw a C(H)(Br) fragment (a) with the carbon and hydrogen atoms in the plane of the paper and the bromine atom behind the plane of the paper; and (b) with the carbon in the plane of the paper, the bromine behind the plane of the paper, and the hydrogen closer to you than the carbon atom.

1.5 THE TETRAHEDRAL GEOMETRY OF CARBON COMPOUNDS

Many years of experimental studies have firmly established the structure of methane as one in which all C—H bond lengths are equal and all H—C—H bond angles are equal. With these restrictions only a single structure is possible, and this is illustrated in Figure 1.4 from several different perspectives.

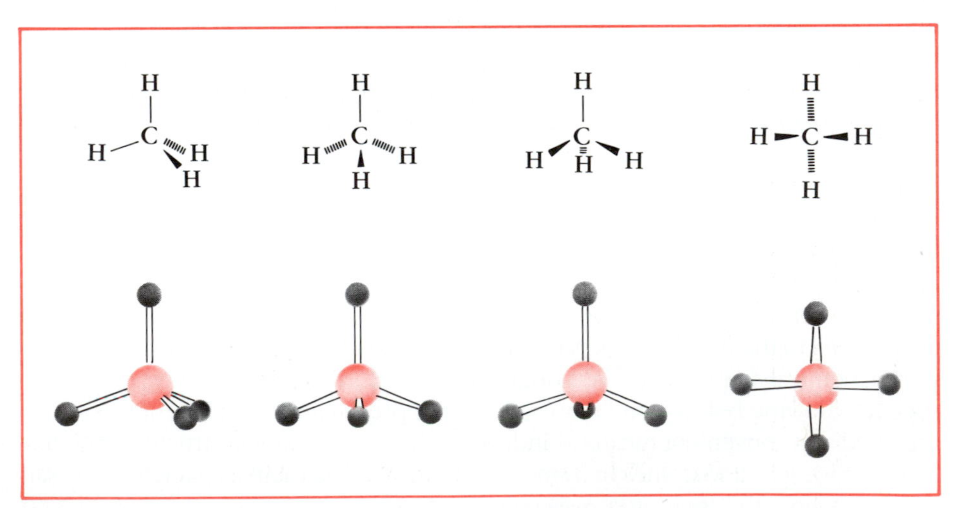

Figure 1.4 Three-dimensional structure of methane from various perspectives.

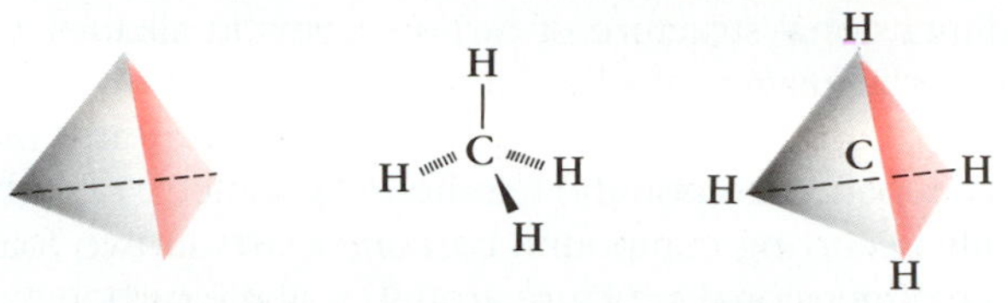

C—H = 1.09 Å, H—C—H = 109.5°

Figure 1.5 The tetrahedral geometry of methane.
Each hydrogen atom defines one apex of a tetrahedron, and the carbon atom is located at the exact center. All C—H bonds are 1.09 Å, and all H—C—H angles are 109.5°.

The carbon-hydrogen bond length for methane is approximately 1.09 Å (1.09×10^{-1}nm), and this value also applies to the C—H bonds of alkanes in general. The H—C—H angle of methane is 109.5°, a value which is frequently referred to as the *tetrahedral angle.* A tetrahedron is a regular geometric solid having four sides, each of which is an equilateral triangle of the same size. The molecular structure of the highly symmetrical methane molecule defines a tetrahedron, as illustrated in Figure 1.5. Each hydrogen corresponds to one apex of the tetrahedron, and the carbon atom lies at the exact center.

The three-dimensional structures of other alkanes are very similar to that of methane. In Figure 1.6 we summarize the structural details of ethane and propane, focusing in each case on individual carbon atoms. The C—H bond lengths are slightly longer than the 1.09 Å for methane, but for our purposes the difference is not significant. Similarly, the C—C bond length of 1.54 Å for ethane is a value that we will employ as typical of the carbon-carbon bond length in alkanes and related compounds. There is greater variability in the H—C—H and C—C—H angles of ethane and propane, but they are all in the range of 109° ± 3°. This range of 6° may at first seem quite large to you, but stop to consider the minute hand of a clock. During the time interval from noon to 12:01, the minute hand sweeps through one-sixtieth of a circle (360°), which is an angle of 6°. In other words the variation of H—C—H angles in methane, ethane, and propane is about the same as the change in the minute hand of a clock over a time interval of 1 min. This variation is

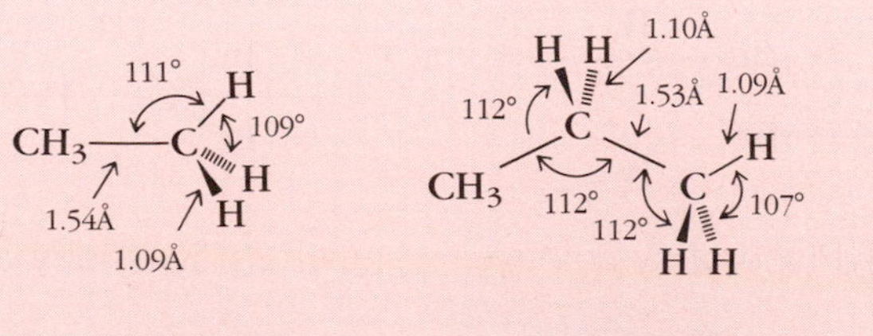

Ethane CH_3CH_3 Propane $CH_3CH_2CH_3$

Figure 1.6 Three-dimensional representations of ethane and propane.

sufficiently small that we can refer to the tetrahedral angle of methane as a typical bond angle in other alkanes.

The three-dimensional structure of carbon atoms in alkanes and other organic compounds is not at matter of idle curiosity. As you proceed in your study of organic chemistry, you will discover that many chemical and biochemical processes are highly dependent upon the tetrahedral geometry of carbon. As an example, the naturally occurring compound *carvone* exists in two forms, which differ only in the three-dimensional structure around a single carbon atom. One form of carvone is responsible for the flavor that you know as spearmint, while the other form is found in caraway seeds and provides the characteristic flavor of rye bread.

1.6 SYMMETRY

We have already made several references in this chapter to structures or shapes that are *symmetrical.* The concept of symmetry in organic chemistry is extremely important, and we will stop at this point to carefully define some terms. We will present a more detailed analysis of symmetry in Chapter 6, when we discuss stereochemistry.

Symmetry is a property of a structure such that a plane or line can be used to divide the structure into parts having the same size, shape, and relative orientation. There are two types of symmetry and you must be aware of both of them. The first type is **reflection symmetry,** involving a **plane of symmetry** which passes through the center of the molecule. A molecule has reflection symmetry when a plane passing through the center of the molecule divides it into two equivalent halves. (More precisely, as you will see in Chapter 6, the two halves must actually be mirror images.) The second type of symmetry is **rotational symmetry,** involving a **symmetry axis** which passes through the center of the molecule. If a molecule has rotational symmetry, rotation about that axis (by some angle less than 360°) will generate a structure that is *indistinguishable* from the original. Figure 1.7 illustrates both types of symmetry for dibromomethane.

In Figure 1.7a we have illustrated a plane of symmetry for the CH_2Br_2 molecule. Three of the atoms (the carbon and bromines) lie in the symmetry plane (the shaded plane), and the two hydrogen atoms lie at equal distances above and below this plane. Consequently, the symmetry plane divides the molecule into two identical parts, each of which contains half of the carbon atom, half of each of the two

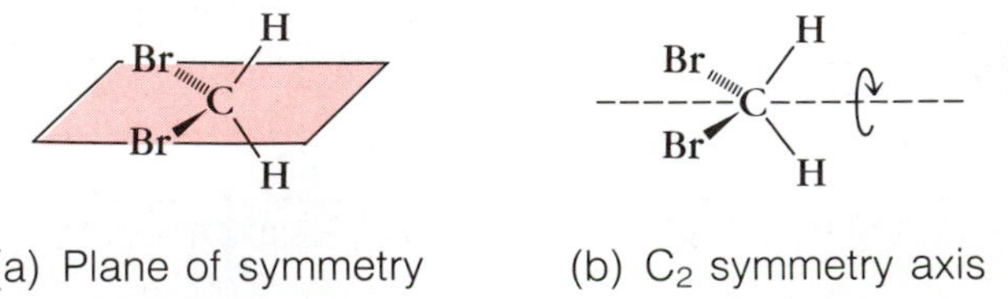

(a) Plane of symmetry (b) C_2 symmetry axis

Figure 1.7 Symmetry elements in the CH_2Br_2 molecule.
(a) The shaded plane is a plane of symmetry. (b) A horizontal line through the carbon atom is a twofold axis of rotational symmetry.

bromines, and a hydrogen atom. The two hydrogen atoms are *symmetry-related* by this plane of symmetry, and this means that they are **equivalent.**

Figure 1.7b illustrates a symmetry axis for the same molecule. The horizontal line through the carbon atom bisects the H—C—H angle and lies in the plane defined by those three atoms. When the entire molecule is rotated by 180° about this axis, the resulting structure is indistinguishable from the original drawing. The two hydrogens would have been interchanged, as would the two bromine atoms, but an observer would not know the difference. This tells you that the two hydrogens are *equivalent,* as are the two bromines. Because only one-half of a complete 360° rotation was required, the symmetry axis is defined as a C_2 (two-fold) axis. (If one-third of a rotation had been required to produce a structure indistinguishable from the original, we would have called it a C_3 axis.)

The symmetry plane and symmetry axis depicted in Figure 1.7 are called **symmetry elements.** Identification of symmetry elements in a molecule can be very useful because symmetry provides information about chemical reactivity. Whenever two atoms or groups are shown to be equivalent by symmetry, you can conclude that their chemical reactivity will be the same. Moreover, reaction at either of two equivalent sites will lead to the same product.

The *absence* of symmetry can also be useful in evaluating chemical structure and reactivity. We showed in the preceding section that the bond lengths and angles for methane, ethane, and propane exhibit small differences. That is not surprising, since the structures themselves are quite different. There is no reason that we should expect the length of a CH_3—H bond to be exactly the same as that of a CH_3CH_2—H bond. Even within a single molecule the various bond lengths and angles can differ, and this was seen for propane in Figure 1.6. On the basis of symmetry we can say that only when all four substituents on a carbon atom are identical (as in methane) will all bond lengths and angles be the same. For other alkanes the structural features should be similar although not identical. Nonetheless the description of the carbon atoms of alkanes as tetrahedral is common even if it is only approximate. You can anticipate that bond angles in alkanes will be within a few degrees of 109° and that C—H and C—C bond lengths will be within several tenths of an angstrom of 1.09 and 1.54 Å, respectively.

EXERCISE 1.2

Dibromomethane has a second plane of symmetry in addition to that shown in Figure 1.7. Show this second symmetry plane.

EXERCISE 1.3

Chloromethane has a threefold axis of rotational symmetry (a C_3 axis, one-third of a rotation about which produces a structure that is indistinguishable from the original). Show this axis.

1.7 A BRIEF INTRODUCTION TO MOLECULAR ORBITAL THEORY

Organic compounds are characterized by **covalent bonding.** Bonds between carbon atoms and other atoms involve pairs of electrons that are *shared* by the two attached atoms, and we usually indicate each of these electron pair bonds by a line connecting the atoms. But the use of a line to represent a bond is only a convenience; the actual location of the electrons must be described in terms of *orbitals.* As you work to develop a good understanding of organic chemistry, your efforts will be greatly assisted by a familiarity with the atomic and molecular orbitals that are used to describe organic structures. Both molecular reactivity and molecular structure are determined by the various interactions between negatively charged electrons and positively charged atomic nuclei in the molecule. A complete mathematical analysis would require calculation of all electron-electron repulsions, all nuclear-nuclear repulsions, and all electron-nuclear attractions, and this becomes extremely difficult when the motion of the electrons in their orbitals is taken into consideration. However, we do not need to carry out such mathematical calculations to take advantage of the descriptive aspects of molecular orbital theory. In this section we will present a highly simplified view of the way that atomic orbitals interact to form molecular orbitals, and this in turn will give you the basis for understanding the structure and reactivity of the different kinds of organic molecules that we will discuss in this text.

Molecular orbitals result from the interaction of atomic orbitals, and our first concern must be to identify the appropriate atomic orbitals. In the case of hydrogen the atomic orbital involved in bonding interactions with other atoms is the 1*s* orbital. All other orbitals are of much higher energy and we will neglect them. The situation is quite different for elements in the second row (lithium, beryllium, boron, carbon, nitrogen, oxygen, fluorine, and neon). For these atoms the 1*s* orbital is always filled and it is not usually considered in terms of bonding with other atoms. The important atomic orbitals for carbon and the other second-row elements are the 2*s* and 2*p* orbitals, the **valence atomic orbitals.** These orbitals are depicted in Figure 1.8.

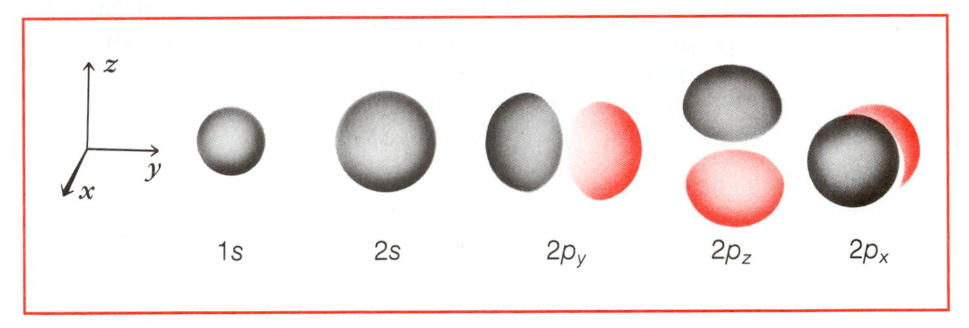

Figure 1.8
The atomic orbitals that are involved in forming bonds between the atoms of organic compounds. For hydrogen only the 1*s* orbital is involved. For carbon and the other second-row elements four atomic orbitals are important, namely, the 2*s* and the three different 2*p* orbitals.

The orbital drawings in Figure 1.8 serve to illustrate several important features of atomic orbitals (and of molecular orbitals as well):

1. *These orbital drawings are very great simplifications of extremely complex mathematical functions* (called wave functions) that describe the electron distribution in an atom or molecule. If you consider an electron as a particle,

these orbital representations correspond to the region in space within which the electron would usually be found. In other words, we might say that the electron resides within this region more than 90 percent of the time. Alternatively, we could say that there is a 90 percent probability of finding the electron within this region at any given instant.

2. *Both the orientation and the shading of the orbitals is arbitrary.* We have selected a particular orientation with respect to a set of axes (x, y, z) simply to demonstrate that the three p orbitals are mutually perpendicular. Real atoms and molecules can, of course, assume all possible orientations in space.

The shading of the orbitals is intended only to show that the two **lobes** of the p orbitals are different. Lobes are spatial regions of an orbital that are separated by **nodes,** regions where the probability of finding an electron is zero. The specific difference between lobes of an orbital is in the sign (**phase**) of the wave function, the mathematical description of the orbital. The phase of the wave function is sometimes denoted as (+) or (−) in an orbital drawing, but this can be confused with the use of (+) and (−) to indicate *ionic charge.* Consequently, we will use shading instead to distinguish between orbital lobes of the same or opposite phase. The phase of an orbital or orbital lobe becomes important only when we consider interactions between orbitals, so that for an isolated orbital it makes no difference which lobe is shown as shaded or unshaded.

The orbital drawings of Figure 1.8 are pictorial representations of complex mathematical functions, and mathematical manipulations can be carried out on orbitals. We can even do this qualitatively without knowing the mathematical equations that define the orbitals. As an example, we can combine two orbitals and obtain two new orbitals, as illustrated in Figure 1.9 for two atomic orbitals on the same atom. This process can be viewed most simply in terms of addition and subtraction of the orbitals (wave functions). For the case of a 2*s* and a 2*p* orbital on a carbon atom, this leads to two new **hybrid** orbitals that are designated *sp* to indicate their origin.

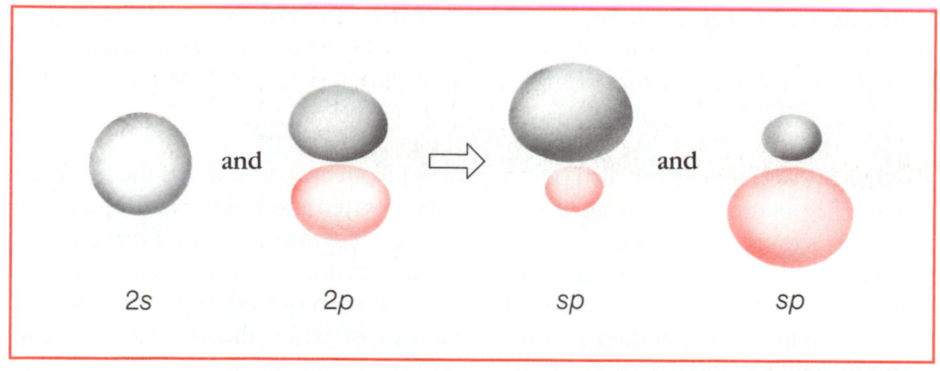

Figure 1.9 Combination of a 2*s* and a 2*p* orbital on the same atom to generate two equivalent *sp* orbitals.
Addition of the original atomic orbitals results in an increase in the magnitude of the unshaded lobe and a decrease in the magnitude of the shaded lobe. *Subtraction* of the original orbitals produces the opposite result. The two resulting *sp* orbitals are equivalent, differing only in spatial orientation.

Addition of the two starting orbitals in Figure 1.9 produces the first of the two *sp* orbitals shown. The upper lobe of the *p* orbital increases in magnitude on combination with the 2*s* orbital of the *same* phase, and the lower lobe of the *p* orbital decreases in magnitude because the 2*s* orbital and lower lobe of the 2*p* orbital have *opposite* phases. (Remember that phase corresponds to the sign of the mathematical function, so addition of two lobes with opposite phase results in a decrease in magnitude.) The second *sp* orbital corresponds to the *subtraction* of one orbital from the other, and this is mathematically equivalent to addition using the 2*s* orbital with the opposite (i.e., shaded) phase. The two resulting *sp* orbitals are fully equivalent except that they differ in directionality. Even the shading is arbitrary, and two equivalent *sp* orbitals would have resulted if the subtraction had been carried out by adding the *p* orbital of opposite phase to the unshaded 2*s* orbital. The results depicted in Figure 1.9 illustrate the important point that *combination of any two orbitals always yields two new orbitals.*

In a similar way it is possible to mathematically combine the four valence atomic orbitals on carbon, *one* 2*s* orbital and *three* 2*p* orbitals, to generate *four equivalent* sp^3 *hybrid orbitals* (where the designation sp^3 describes their origin). The use of such hybrid orbitals is helpful in understanding the highly symmetrical tetrahedral structure of methane, with its identical bond lengths and angles. Figure 1.10 shows one way of drawing four sp^3 orbitals, with the orientations corresponding to those of a methane molecule.

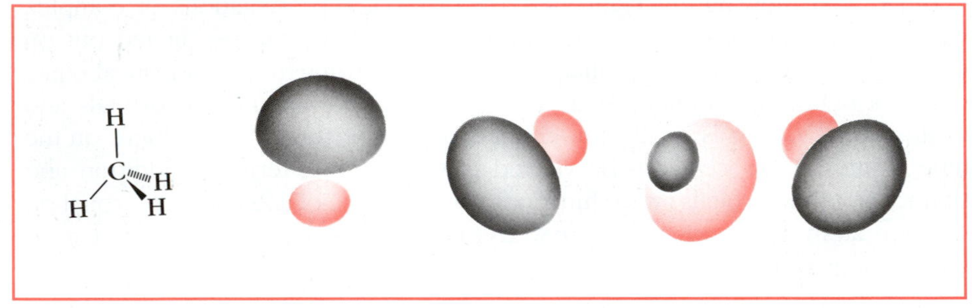

Figure 1.10 The four equivalent sp^3 orbitals generated by combination of the 2*s* and three 2*p* orbitals of a second-row element.
Each of the orbitals is equivalent, except for directionality, and their orientation corresponds precisely to that of the corresponding C—H bonds of methane.

Other hybrid orbitals can also be constructed from a 2*s* orbital and the three 2*p* orbitals. To obtain four equivalent sp^3 orbitals with tetrahedral geometry, the hybrid orbitals must result from interaction of exactly one-fourth of the 2*s* orbital with the appropriate combination of 2*p* orbitals. Interaction of very similar but not identical combinations of 2*s* and 2*p* orbitals would be required to produce sets of hybrid orbitals corresponding to the geometries of other alkanes that are only approximately tetrahedral.

The preceding discussion has been concerned mainly with interactions of orbitals on a single atom, but bonding interactions between two atoms are a result of interactions between atomic orbitals on *different* atoms. This is conveniently illustrated in Figure 1.11 for the case of two hydrogen atoms and their 1*s* valence atomic orbitals. If the two atoms are sufficiently close in space, their orbitals can **overlap.**

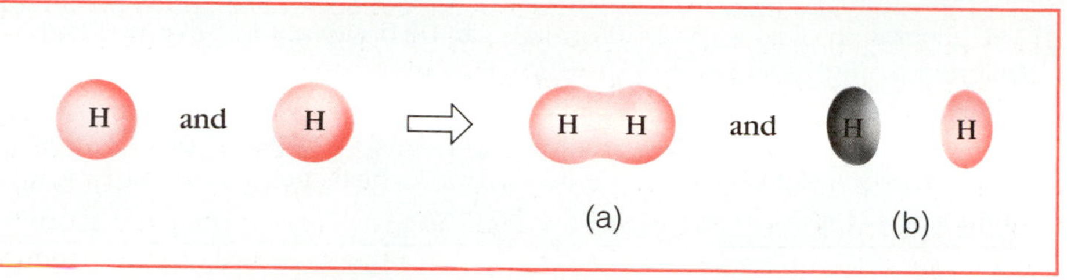

Figure 1.11 Interaction of 1*s* atomic orbitals on two hydrogen atoms to give (a) bonding (in-phase combination) and (b) antibonding (out-of-phase combination) molecular orbitals.

Just as was shown for the interactions of two atomic orbitals on the same atom, this interaction can be either additive (i.e, the interacting orbitals are **in phase**) or subtractive (i.e, the interacting orbitals are **out of phase**). The interaction of two atomic orbitals in this way produces two **molecular orbitals.** The in-phase combination (Figure 1.11a) is more favorable because the resulting molecular orbital corresponds to a bonding interaction between the two hydrogen atoms. This molecular orbital is called a **bonding** molecular orbital. (A bond, of course, must contain a pair of electrons in the appropriate molecular orbital.) The out-of-phase combination (Figure 1.11b) corresponds to an **antibonding** molecular orbital. When a pair of electrons is placed in such an antibonding molecular orbital, the interaction between the atoms is repulsive rather than attractive.

We present a more detailed description of molecular orbital theory in Chapter 27, but even the highly simplified picture presented here serves as a basis for understanding molecular structure and bonding. Figure 1.11 illustrates several points that you should understand. As stated previously for the interaction of orbitals on the same atom (Figure 1.9), the choice of phase (i.e., shading) on the starting orbitals is arbitrary. Figure 1.11 could equally well have been drawn with both starting orbitals unshaded, and the resulting molecular orbitals would still be equivalent to those shown—one combination would be in phase (bonding) and the other would be out of phase (antibonding). Also notice that the interaction of two atomic orbitals results in two molecular orbitals; the *total number of orbitals* is preserved.

What about the molecular orbitals of methane? A complete analysis is beyond the scope of this chapter, but certain features should be clear from the preceding discussion. Interaction of the four equivalent sp^3 orbitals of carbon (Figure 1.10) with the 1*s* orbitals of four different hydrogen atoms allows us to describe the highly symmetrical structure of CH_4, with its equal bond lengths and angles. The eight valence atomic orbitals (four from carbon and one from each hydrogen) must yield eight molecular orbitals; four of these will be bonding and four will be antibonding molecular orbitals. The total of eight valence electrons in the CH_4 molecule, four from carbon (the 1*s* core electrons of carbon are ignored) and one from each hydrogen, is precisely the right number to fill each of the four bonding molecular orbitals. While the individual molecular orbitals of methane each involve valence atomic orbitals from more than two constituent atoms, the overall result is equivalent to the traditional picture of methane with a total of four individual C—H bonds, each involving only two atoms.

EXERCISE 1.4

The generation of two *sp* hybrid orbitals can be done in a way that is slightly different from that shown in Figure 1.9. First generate one hybrid orbital by adding the 2*s* and 2*p* orbitals shown at the left of the figure. Then subtract the *p* orbital from the *s* orbital (i.e., add the *p* orbital of opposite phase). What differences are seen between the two hybrid orbitals generated in this way and those depicted in Figure 1.9?

1.8 NOMENCLATURE

Virtually any activity in the field of chemistry requires the transmission of information. Students, teachers, and research scientists all need to communicate their ideas about molecular structure, and it is essential that this communication be clear and unambiguous. The use of structural formulas is often satisfactory but it becomes extremely inconvenient for large molecules. Moreover, two individuals might draw the identical compound in two different orientations, with the result that they would appear to be different. To avoid such difficulties a standard system for naming organic compounds has been developed by chemists around the world. The *International Union of Pure and Applied Chemistry* (IUPAC) has developed an extensive set of rules for naming even the most complex organic compounds. We will present only the basic applications of IUPAC nomenclature in this text, but this will allow you to go back and forth efficiently between structural formulas and names of compounds. As you should expect from a reliable system of nomenclature, each different compound will have a unique name. When used properly, the IUPAC nomenclature will always produce only one name for a particular compound. Conversely, only a single compound can be drawn for any particular name. This unique one-to-one relationship between name and structure will provide you with a valuable aid in differentiating between compounds that are different even if they appear to be similar. Moreover, application of systematic nomenclature can also help you to identify two structural drawings of the same compound, even if they are drawn in quite different orientations.

The simplest alkanes are the **straight-chain alkanes,** and the nomenclature for these compounds provides the basis for naming all other organic compounds as well. In Table 1.1 we show the first 10 members of the **homologous series** (i.e., the series in which each successive member differs by a single CH_2 group). With the exception of the first four entries, the names are simply based on the prefixes from the Greek numbers with the ending *-ane.* You are probably already familiar with the numerical prefixes from the names of common geometric figures such as pentagons, hexagons, and octagons.

You should be careful to understand the term "straight-chain" as it is used here. The carbon chain is certainly not "straight" in the sense of bond angles. In Section 1.5 we emphasized that the bond angles at the carbon atoms of alkanes are approximately 109°, not the 180° that would correspond to a linear arrangement. Instead the term straight-chain indicates that all the carbon atoms are connected in a single continuous chain. The connectivity is such that by starting at one end of the molecule you can proceed from one carbon to the next until you arrive at the other end of the chain, having encountered every carbon atom in the molecule without ever

TABLE 1.1 Straight-Chain Alkanes

Number of Carbons	Name	Molecular Formula	Structure	Melting Point (°C)	Boiling Point (°C)
1	Methane	CH_4	CH_4	−182	−161
2	Ethane	C_2H_6	CH_3CH_3	−183	−89
3	Propane	C_3H_8	$CH_3CH_2CH_3$	−190	−45
4	Butane	C_4H_{10}	$CH_3(CH_2)_2CH_3$	−138	−1
5	Pentane	C_5H_{12}	$CH_3(CH_2)_3CH_3$	−130	36
6	Hexane	C_6H_{14}	$CH_3(CH_2)_4CH_3$	−95	68
7	Heptane	C_7H_{16}	$CH_3(CH_2)_5CH_3$	−91	98
8	Octane	C_8H_{18}	$CH_3(CH_2)_6CH_3$	−56	125
9	Nonane	C_9H_{20}	$CH_3(CH_2)_7CH_3$	−51	151
10	Decane	$C_{10}H_{22}$	$CH_3(CH_2)_8CH_3$	−30	174

backtracking. This is illustrated for octane by numbering each of the eight carbon atoms consecutively.

$$\overset{1}{CH_3}-\overset{2}{CH_2}-\overset{3}{CH_2}-\overset{4}{CH_2}-\overset{5}{CH_2}-\overset{6}{CH_2}-\overset{7}{CH_2}-\overset{8}{CH_3}$$

Keep in mind that structural formulas such as the preceding formula for octane impart no three-dimensional information, so it is not necessary to draw the carbon atoms in a straight line. For example, propane could be drawn equally well in either of the following ways:

$$\begin{array}{l} CH_3 \\ | \\ CH_2-CH_2 \end{array} \qquad CH_3-CH_2-CH_3$$

It is possible to name a very large number of organic compounds as derivatives of the straight-chain alkanes. The following rules for naming alkanes will also help you to name many of their derivatives.

Nomenclature of Alkanes and Their Derivatives

1. *Locate the principal carbon chain.* For alkanes this is defined as the longest carbon chain in the molecule. *The straight-chain alkane of the same number of carbons provides the basis of the name.*

2. *Name and number the substituents* on the principal carbon chain.
   a. The principal chain is numbered so that the lower of two possible numbers results at the occasion of the first difference (i.e., the first point of substitution).
   b. When there is more than one substituent, they are listed in alphabetical order. If several of the substituents are identical, this is indicated by the numerical prefixes *di-*, *tri-*, *tetra-*, *penta-*, *hexa-*, etc.

These rules allow you to name a variety of compounds quite easily. For the following two halo derivatives of butane, the carbon chains are numbered so that the carbon bearing the halogen has the lower of the two possible numbers.

$$CH_3{-}CH_2{-}CH_2{-}CH_2{-}Cl \qquad CH_3{-}CH_2{-}\overset{\overset{\displaystyle Br}{|}}{C}H{-}CH_3$$

1-Chlorobutane
(*not* 4-Chlorobutane)

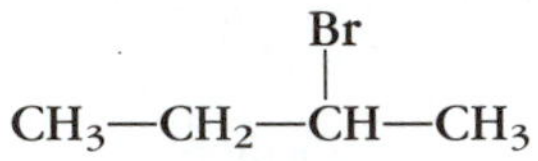

2-Bromobutane
(*not* 3-Bromobutane)

EXERCISE 1.5

Draw and name all possible constitutional isomers that could result from the substitution of a fluorine atom for a single hydrogen atom of octane.

EXERCISE 1.6

Draw and name all possible constitutional isomers that could result from the substitution of an iodine atom for a single hydrogen atom of decane.

The presence of several substituents on the carbon chain requires the use of rules 2a and 2b. The presence of two or three identical substituents is illustrated in the following structures:

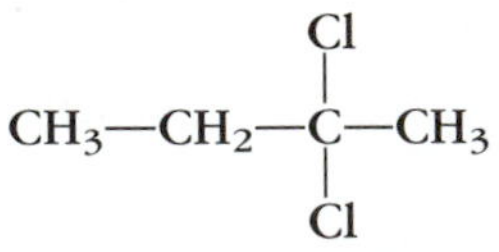

$$CH_3{-}CH_2{-}\underset{\underset{\displaystyle Cl}{|}}{\overset{\overset{\displaystyle Cl}{|}}{C}}{-}CH_3 \qquad CH_3{-}\overset{\overset{\displaystyle Br}{|}}{C}H{-}CHBr_2$$

2,2-Dichlorobutane

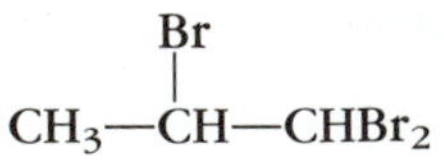

1,1,2-Tribromopropane

In the next example, numbering is begun at the left side of the molecule so that the second carbon will have a halo substituent, but the bromo substituent on C-3 is named first because of its higher alphabetical priority.

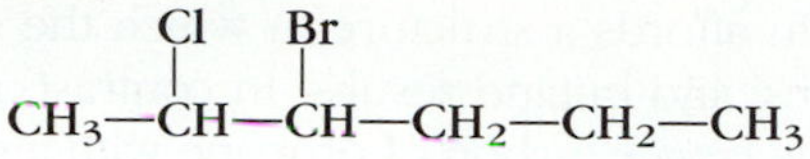

$$\begin{array}{c} ClBr \\ CH_3-CH-CH-CH_2-CH_2-CH_3 \end{array}$$

3-Bromo-2-chlorohexane
(*not* 2-Chloro-3-bromohexane)

And what if a compound has multiple substituents that not all are the same?

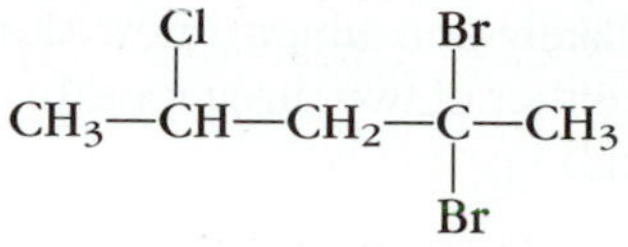

$$\begin{array}{c} Cl \qquad\quad Br \\ CH_3-CH-CH_2-C-CH_3 \\ \qquad\qquad\quad Br \end{array}$$

2,2-Dibromo-4-chloropentane

Notice that the bromo substituents are named before the chloro substituent. The letters *b* of bromo and *c* of chloro (and not the *d* of the numerical prefix di-) are used in deciding the order.

Branched-Chain Alkanes

The process of naming an organic compound becomes slightly more complex in the case of molecules with carbon skeletons that are not those of straight-chain alkanes. Before we discuss the specific procedure for naming these compounds, however, we will consider some of their structural features.

Recall that substitution of a chlorine for a hydrogen of propane (Figure 1.2) could, in principle, lead to either 1-chloropropane or 2-chloropropane. In the same manner Figure 1.12 shows that the hypothetical replacement of one of the hydrogens of propane with a CH_3 substituent (called a *methyl group*) can also afford two isomeric structures. Replacement of one of the hydrogens at either *end*

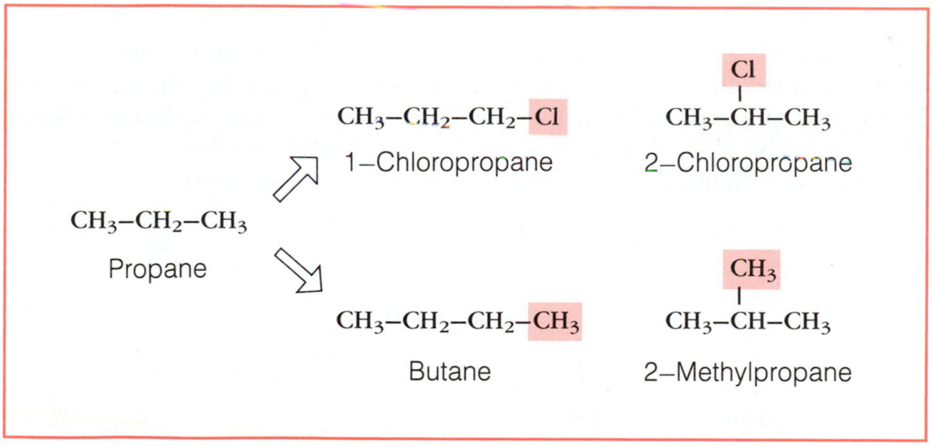

$$CH_3-CH_2-CH_3 \text{ (Propane)} \nearrow CH_3-CH_2-CH_2-Cl \text{ (1-Chloropropane)}, \quad CH_3-CH(Cl)-CH_3 \text{ (2-Chloropropane)}$$

$$CH_3-CH_2-CH_3 \text{ (Propane)} \searrow CH_3-CH_2-CH_2-CH_3 \text{ (Butane)}, \quad CH_3-CH(CH_3)-CH_3 \text{ (2-Methylpropane)}$$

Figure 1.12
Replacement of a hydrogen atom of propane with a chlorine atom (top) produces either of two isomeric chloropropanes. Replacement of the same hydrogen atoms with a methyl group produces two isomeric hydrocarbons.

of the three-carbon chain affords a structure in which the chain length has been increased to four carbons, and butane results. In contrast, replacement of one of the two hydrogens on the *center* carbon of propane with a methyl group does not increase the chain length. For the resulting structure, 2-methylpropane, it is not possible to proceed from one carbon to the next and encounter all four carbons without backtracking. The longest carbon chain is still three carbons, but we can identify several *different* three-carbon chains in this structure. This is a **branched-chain** alkane. Beginning at one of the CH_3 (methyl group) carbons and proceeding along the chain, we immediately encounter a new situation at the second carbon atom. We could continue in either of two directions. In other words, this represents a *branch point* in the chain.

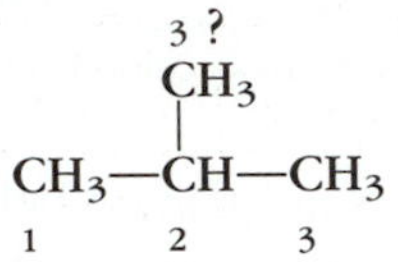

```
       3 ?
       CH3
       |
CH3—CH—CH3
 1     2     3
```

In contrast to the straight-chain alkanes, which contain only CH_2 and CH_3 groups, branched-chain alkanes can have carbons which are attached in turn to three or even four other carbon atoms. An example of the latter type is provided by 2,2-dimethylpropane:

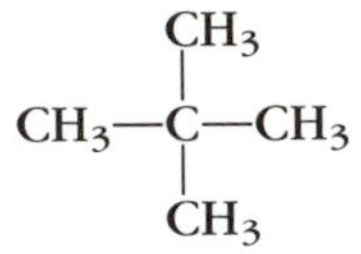

```
       CH3
       |
CH3—C—CH3
       |
       CH3
```

The number of other carbons to which a carbon is directly bonded plays an important role in chemical reactivity, and names (derived from Latin numbers) have consequently been assigned to these different types of carbon. When a carbon is attached to only *one* other carbon (e.g, the CH_3 group of propane) it is called **primary.** When bonded to *two* other carbons, as is the CH_2 group of propane, it is designated **secondary.** Carbons such as the central CH group of 2-methylpropane, which are bonded to *three* other carbons, are called **tertiary,** and a carbon attached to *four* other carbons (e.g., the central carbon of 2,2-dimethylpropane) is described as **quaternary.** Note that methane is unique because it has but a single carbon atom and none of these labels apply to it. Primary, secondary, tertiary, and quaternary carbon atoms are illustrated in the following structure.

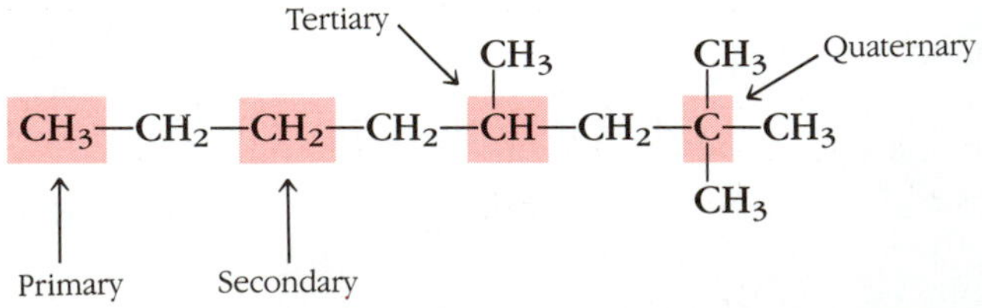

```
                          Tertiary
                               ↘  CH3          CH3  Quaternary
                                   |           | ↙
CH3—CH2—CH2—CH2—CH—CH2—C—CH3
 ↑             ↑                               |
 |             |                               CH3
Primary     Secondary
```

The terms primary, secondary, and tertiary refer specifically to a carbon atom, but they are sometimes used to designate a substituent on that type of a carbon atom. For example, the hydrogens of the CH_2 group of propane can be called secondary hydrogens. Similarly, 1-chloropropane ($CH_3CH_2CH_2Cl$) is often referred to as a primary alkyl halide because the chlorine is attached to a primary carbon.

The words primary, secondary, and tertiary are sometimes symbolized by 1°, 2°, and 3°, respectively.

As seen for the cases of 2-methylpropane and 2,2-dimethylpropane, the nomenclature of the branched-chain alkanes requires that alkyl groups (e.g., the methyl group, CH_3) be named as substituents on the principal alkyl chain. Table 1.2 presents the various alkyl groups and the alkanes from which they are derived. It is important for you to recognize that the alkyl groups under discussion are only parts of molecules and not separate chemical species. Although it is possible to generate a highly reactive chemical species with the formula CH_3, the term "methyl group" refers specifically to a CH_3 that is bonded to some other atom.

TABLE 1.2 Straight-Chain Alkyl Groups

Alkane	Molecular Formula	Alkyl Group	Molecular Formula	Structure
Methane	CH_4	Methyl	CH_3	$CH_3—$
Ethane	C_2H_6	Ethyl	C_2H_5	$CH_3—CH_2—$
Propane	C_3H_8	Propyl	C_3H_7	$CH_3—CH_2—CH_2—$
Butane	C_4H_{10}	Butyl	C_4H_9	$CH_3—(CH_2)_2—CH_2—$
Pentane	C_5H_{12}	Pentyl	C_5H_{11}	$CH_3—(CH_2)_3—CH_2—$
Hexane	C_6H_{14}	Hexyl	C_6H_{13}	$CH_3—(CH_2)_4—CH_2—$
Heptane	C_7H_{16}	Heptyl	C_7H_{15}	$CH_3—(CH_2)_5—CH_2—$
Octane	C_8H_{18}	Octyl	C_8H_{17}	$CH_3—(CH_2)_6—CH_2—$
Nonane	C_9H_{20}	Nonyl	C_9H_{19}	$CH_3—(CH_2)_7—CH_2—$
Decane	$C_{10}H_{22}$	Decyl	$C_{10}H_{21}$	$CH_3—(CH_2)_8—CH_2—$

Now that we have introduced the names of alkyl groups (Table 1.2), we can proceed to the naming of a much larger number of alkanes and their derivatives. This is illustrated by the following examples.

$$\begin{array}{ccccccccc} & & & & CH_3 & & & & \\ & & & & | & & & & \\ & & & & CH_2 & & & & \\ & & & & | & & & & \\ CH_3 & — & CH_2 & — & CH & — & CH_2 & — & CH_3 \\ 1 & & 2 & & 3 & & 4 & & 5 \end{array}$$

3-Ethylpentane

$$\begin{array}{ccccccccccccccc} & & & & & & & & CH_2—CH_2—CH_3 & & & & & & \\ & & & & & & & & | & & & & & & \\ CH_3 & — & CH_2 & — & CH_2 & — & CH_2 & — & CH & — & CH_2 & — & CH_2 & — & CH_3 \\ 8 & & 7 & & 6 & & 5 & & 4 & & 3 & & 2 & & 1 \end{array}$$

4-Propyloctane
(*not* 5-propyloctane)

The longest carbon chain is not necessarily that which is drawn in a straight line, so the first step in naming the following two alkanes is identification of the longest chain. In neither structure does the horizontal row of carbons correspond to the longest chain, which is highlighted in the drawings.

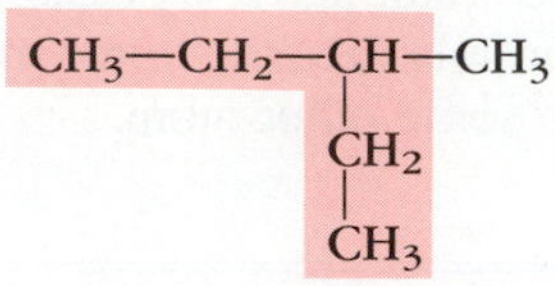

$$CH_3{-}CH_2{-}\underset{\underset{\displaystyle CH_3}{|}}{\underset{\displaystyle CH_2}{\underset{|}{CH}}}{-}CH_3$$

3-Methylpentane

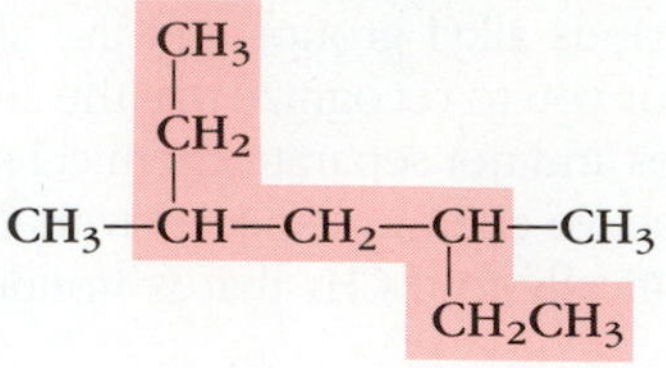

$$CH_3{-}\overset{\overset{\displaystyle CH_3}{|}}{\overset{\displaystyle CH_2}{\overset{|}{CH}}}{-}CH_2{-}\underset{\underset{\displaystyle CH_2CH_3}{|}}{CH}{-}CH_3$$

3,5-Dimethylheptane

The two compounds will be named as derivatives of pentane and heptane, respectively, and the full names are 3-methylpentane and 3,5-dimethylheptane. Of course, we could also draw these two compounds with the longest chain horizontal. None of these drawings conveys any three-dimensional information, so it makes absolutely no difference which drawing is used.

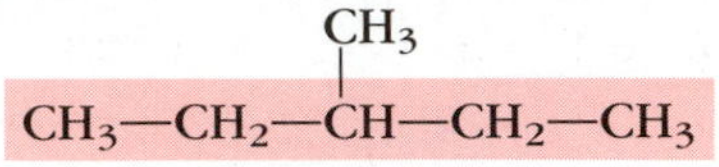

$$CH_3{-}CH_2{-}\overset{\overset{\displaystyle CH_3}{|}}{CH}{-}CH_2{-}CH_3$$

3-Methylpentane

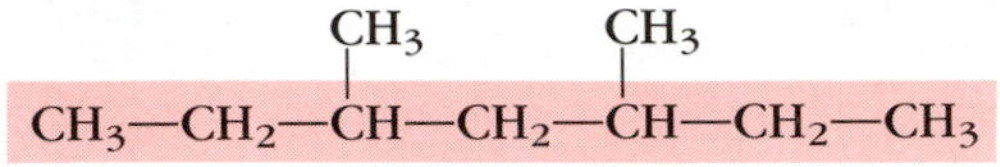

$$CH_3{-}CH_2{-}\overset{\overset{\displaystyle CH_3}{|}}{CH}{-}CH_2{-}\overset{\overset{\displaystyle CH_3}{|}}{CH}{-}CH_2{-}CH_3$$

3,5-Dimethylheptane

The next two examples show that the naming of branched alkyl halides is done in the same manner.

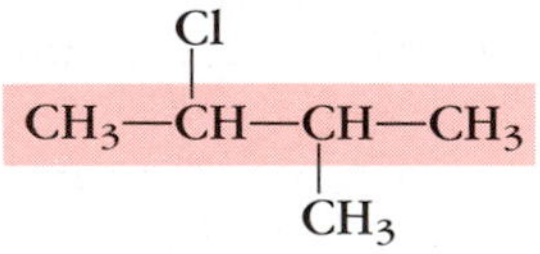

$$CH_3{-}\overset{\overset{\displaystyle Cl}{|}}{CH}{-}\underset{\underset{\displaystyle CH_3}{|}}{CH}{-}CH_3$$

2-Chloro-3-methylbutane

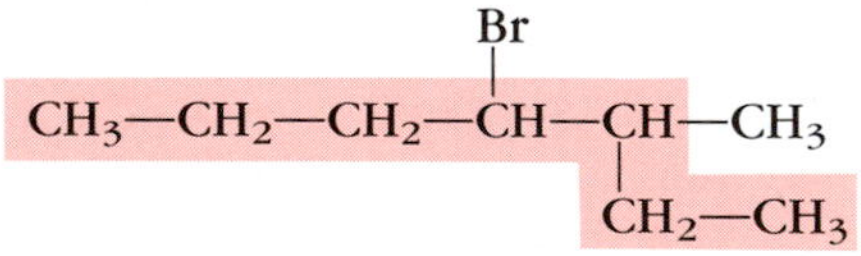

$$CH_3{-}CH_2{-}CH_2{-}\overset{\overset{\displaystyle Br}{|}}{CH}{-}\underset{\underset{\displaystyle CH_2{-}CH_3}{|}}{CH}{-}CH_3$$

4-Bromo-3-methylheptane

The substituents in the first structure are on C-2 and C-3 regardless of which end of the chain is designated as C-1. Which name is preferable? Either possibility would fully define the structure, but 2-chloro-3-methylbutane is preferred over 3-chloro-2-methylbutane because the first-named substituent has a lower number. The second of the preceding examples is easily named according to the IUPAC rules as 4-bromo-3-methylheptane. The alternative numbering, from the other end of the parent seven-carbon chain, would have placed the substituents on C-4 and C-5 instead of C-3 and C-4.

The final degree of complexity that we will consider here results from the presence of alkyl substituents (or *side chains*) which are themselves substituted. This situation is illustrated by the following alkyl halide:

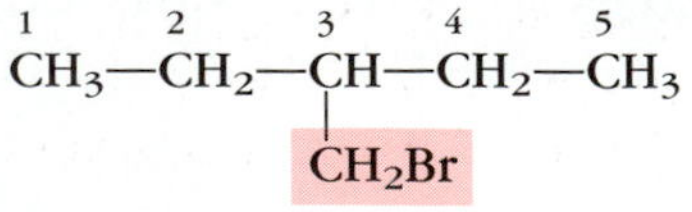

$$\overset{1}{CH_3}{-}\overset{2}{CH_2}{-}\underset{\underset{\displaystyle CH_2Br}{|}}{\overset{3}{CH}}{-}\overset{4}{CH_2}{-}\overset{5}{CH_3}$$

This is correctly named as a derivative of pentane because the longest carbon chain in the molecule has five carbons. We must therefore assign a name to the $—CH_2Br$ group. It is in fact a bromo-substituted methyl group, so it is called a *bromomethyl* group, and the correct name for the entire molecule is 3-(bromomethyl)pentane. Parentheses are needed to indicate the presence of a bromomethyl group rather than individual bromo and methyl groups. A similar situation arises when an alkyl substituent is itself branched, as in the following structure:

$$\overset{1}{CH_3}—\overset{2}{CH_2}—\overset{3}{CH_2}—\overset{4}{CH}(—CH(CH_3)—CH_3)—\overset{5}{CH_2}—\overset{6}{CH_2}—\overset{7}{CH_2}—\overset{8}{CH_2}—\overset{9}{CH_2}—\overset{10}{CH_3}$$

Clearly, this must be named as a derivative of decane because the longest chain has 10 carbons. However, what name is to be given to the three-carbon side chain? It is not a *propyl* group (since propyl refers specifically to $CH_3CH_2CH_2—$), and therefore some other name must be used. It is possible to consider this three-carbon substituent as a substituted ethyl group. An **alkyl substituent is always numbered from the point of attachment to the principal chain,** so the three-carbon side chain would have a methyl at its 1-position. The complete name of the alkane is therefore 4-(1-methylethyl)decane, where the branched side chain is placed in parentheses to minimize confusion with the numbering of the parent chain. Despite the hyphens, numbers, and parentheses used here, there are no spaces in the name. If you look at other examples, both in this chapter and in subsequent ones, you will find that this is typical of organic nomenclature.

Common Names

Branched side chains are encountered frequently, and the nomenclature we presented in the previous example is cumbersome. A more typical way of naming such compounds takes advantage of *common names* for a series of branched alkyl groups. For example, a three-carbon alkyl group attached via the central carbon (i.e., $CH_3—\overset{|}{C}H—CH_3$) is called an *isopropyl* group, and the compound in the preceding example can therefore be named *4-isopropyldecane.*

The use of common names could itself become overwhelming if every compound or group had a common name that we had to remember. Fortunately, only a few such common names are recognized for IUPAC nomenclature, and Table 1.3 presents the most important alkyl groups having accepted common names. In the table we also show the parent alkane and any other alkyl groups derived from the parent. Notice that the prefixes *sec-* and *tert-* indicate the type of carbon (secondary or tertiary) from which a hydrogen is "missing" relative to the parent alkane. For two of the parent alkanes in Table 1.3, 2-methylpropane and 2,2-dimethylpropane, we have also listed common names. Ordinarily, systematic names are preferable to common names because they allow you to draw the correct structure using the simple rules of nomenclature.

How did common names arise? In the early days of organic chemistry many compounds were isolated and purified even though their structures were unknown. A systematic name based on molecular structure could not be employed in

TABLE 1.3 Names and Common Names* for Alkanes and Alkyl Groups

Alkane Name	Alkane Structure	Alkyl Group Structure	Alkyl Group Name
Propane	$CH_3-CH_2-CH_3$	$CH_3-CH_2-CH_2-$	Propyl
		$CH_3-\underset{\vert}{C}H-CH_3$	Isopropyl*
Butane	$CH_3-CH_2-CH_2-CH_3$	$CH_3-CH_2-CH_2-CH_2-$	Butyl
		$CH_3-CH_2-\underset{\vert}{C}H-CH_3$	*sec*-Butyl*
2-Methylpropane (isobutane*)	$CH_3-\overset{CH_3}{\overset{\vert}{C}}H-CH_3$	$CH_3-\overset{CH_3}{\overset{\vert}{C}}H-CH_2-$	Isobutyl*
		$CH_3-\overset{CH_3}{\underset{CH_3}{\overset{\vert}{\underset{\vert}{C}}}}-$	*tert*-Butyl* (*t*-butyl*)
2,2-Dimethyl-propane (neopentane*)	$CH_3-\overset{CH_3}{\underset{CH_3}{\overset{\vert}{\underset{\vert}{C}}}}-CH_3$	$CH_3-\overset{CH_3}{\underset{CH_3}{\overset{\vert}{\underset{\vert}{C}}}}-CH_2-$	Neopentyl*

such cases, and the early chemists resorted to common names so that they could catalog and describe the compounds they were studying. Only much later, when structures of organic compounds could be reliably determined, did a systematic method for naming compounds become practical. Modern organic chemistry employs systematic names as much as possible, but the few common names shown in Table 1.3 will actually make it much easier for you to go back and forth between organic structures and their names.

The common names of the alkyl groups shown in Table 1.3 now make it possible for you to assign correct names to a very large number of alkanes and their derivatives. In the following examples we will show the step-by-step procedure for naming compounds according to the IUPAC rules for nomenclature.

Example 1.1 Name the following compound:

$$CH_3-CH_2-CH_2-\underset{CH_3-\underset{CH_3}{\underset{|}{C}}-CH_3}{\underset{|}{CH}}-CH_2-CH_2-CH_3$$

Solution *(a)* Find the longest chain:

```
CH3—CH2—CH2—CH—CH2—CH2—CH3
             |
       CH3—C—CH3
             |
            CH3
```

(b) Name and number the substituents.

```
CH3—CH2—CH2—CH—CH2—CH2—CH3
             |
       CH3—C—CH3
             |
            CH3 ↖
                 tert-butyl
```

The name is 4-*tert*-butylheptane.

Example 1.2 Name the following compound.

```
     CH3
     |
CH3—CH—CH2—CH—CH2—CH2—CH2—CH3
             |
             CH—CH2—CH3
             |
             CH3
```

Solution *(a)* Find the longest chain.

```
     CH3
     |
CH3—CH—CH2—CH—CH2—CH2—CH2—CH3
             |
             CH—CH2—CH3
             |
             CH3
```

(b) Name and number the substituents.

```
     CH3 ←—— methyl
     |
CH3—CH—CH2—CH—CH2—CH2—CH2—CH3
             |
             CH—CH2—CH3
             |     ↖
             CH3     sec-butyl
```

The name is 4-*sec*-butyl-2-methyloctane. As illustrated in this example, the prefixes *sec*- and *tert*- are not used in determining alphabetical priority. The priorities in this example are based on the *b* of butyl and the *m* of methyl.

Example 1.3 Name the following compound.

```
      CH3              Cl
       |               |
CH3—CH—CH—CH2—CH—CH—CH2—CH3
            |                |
      CH3—CH           CH2
            |                |
           CH3    CH3—C—CH3
                             |
                            CH3
```

Solution The longest chain has nine carbons, and there are several such chains. Because the two isopropyl groups at the left side of the structure are equivalent, any of these chains could be selected as the principal chain. For example:

```
      CH3              Cl
       |               |
CH3—CH—CH—CH2—CH—CH—CH2—CH3
            |                |
      CH3—CH           CH2
            |                |
           CH3    CH3—C—CH3
                             |
                            CH3
```

or

```
      CH3              Cl
       |               |
CH3—CH—CH—CH2—CH—CH—CH2—CH3
            |                |
      CH3—CH           CH2
            |                |
           CH3    CH3—C—CH3
                             |
                            CH3
```

Since the isopropyl groups are equivalent, selection of either chain as the principal chain produces the same final result. The substituents on the chain (in alphabetical order) are a chloro, an ethyl, an isopropyl, and three methyl groups. Numbering from either end of the nine-carbon chain would place the lowest-numbered substituent at the 2-position, but the preferred numbering also places the second-named substituent on C-2 (rather than on C-3). The correct name is 5-chloro-4-ethyl-7-isopropyl-2,2,8-trimethylnonane (rather than 5-chloro-6-ethyl-3-isopropyl-2,8,8-trimethylnonane).

1.9 CONSTITUTIONAL ISOMERS

We showed previously for the case of propane that two isomeric chloropropanes could be formed, and in Exercises 1.5 and 1.6 you found that an even greater number of isomeric halo derivatives is possible for octane and decane. We further showed in the preceding section that two isomeric hydrocarbons with the formula C_4H_{10} could result from replacement of a hydrogen atom of propane with a methyl group. By considering branched-chain alkanes as derivatives of straight-chain alkanes, we can systematically analyze the number of possible isomers for various alkanes.

Only a single structure is possible for each of the first three alkanes (methane, ethane, and propane), but there are two constitutional isomers with the formula C_4H_{10}.

$$CH_3{-}CH_2{-}CH_2{-}CH_3 \qquad CH_3{-}\underset{}{\overset{\displaystyle CH_3}{\overset{|}{C}}H}{-}CH_3$$

Butane — 2-Methylpropane (isobutane)

You can demonstrate that only two isomeric butanes are possible by replacing the hydrogens of propane with methyl groups, as we showed with chlorine atoms in Figure 1.12.

There are three isomers having the formula C_5H_{12}. Once again you can convince yourself of this by replacing the different hydrogens of the two butane isomers with methyl groups. Notice, however, that the same structure (2-methylbutane) is produced by replacement of one of the secondary (CH_2) hydrogens of butane and by replacement of one of the primary (CH_3) hydrogens of isobutane.

$$CH_3{-}CH_2{-}CH_2{-}CH_2{-}CH_3 \qquad CH_3{-}\overset{\displaystyle CH_3}{\overset{|}{C}}H{-}CH_2{-}CH_3 \qquad CH_3{-}\underset{\underset{\displaystyle CH_3}{|}}{\overset{\displaystyle CH_3}{\overset{|}{C}}}{-}CH_3$$

Pentane — 2-Methylbutane — 2,2-Dimethylpropane (neopentane)

The preceding analysis of C_4H_{10} and C_5H_{12} isomers shows clearly that the number of possible isomeric alkanes increases with the number of carbon atoms. Unfortunately, there is no simple relationship by which you could calculate the number of possible isomers for a particular molecular formula. There are five six-carbon alkanes that are constitutional isomers and nine constitutional isomers corresponding to the formula C_7H_{16}. For the $C_{10}H_{22}$ alkanes there are 75 possible constitutional isomers, and this increases to over 300,000 for $C_{20}H_{42}$!

EXERCISE 1.7

Draw and name all isomeric alkanes with the molecular formula C_6H_{14}.

Alkyl halides are a convenient group of compounds for learning about isomers and how they might be formed in chemical reactions. The halogenation of alkanes is not usually a useful reaction in the laboratory because complex mixtures of products are typically produced. Nevertheless, halogenation does provide us with a very useful device for teaching you how to *predict* the products of an organic reaction. Development of such a predictive ability is one of our major goals for you in this course.

We have already illustrated the two possible isomeric monochloro products from propane in Figure 1.12. That example was relatively straightforward because you can easily see that there are only two kinds of carbon atoms in the propane molecule. For larger alkanes prediction of the number of different chlorination products is often more difficult, so we will show you how to evaluate such a problem logically. Hexane, for example, has six carbons on which chlorine substitution might occur. Does this mean that six isomeric chlorohexanes could be formed? The answer is no, because some of the carbons of hexane are equivalent and substitution at either of two equivalent positions would afford the same product. How can you determine whether two groups are equivalent? One way is to evaluate the molecule to see if it has any symmetry.

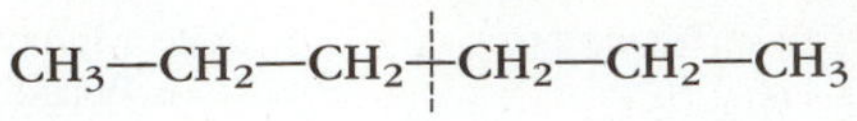

$$CH_3—CH_2—CH_2 \mid CH_2—CH_2—CH_3$$

The hexane molecule is symmetric about the midpoint, and we could draw either a symmetry plane or a C_2 axis. This means that there are only three different kinds of carbon atoms in the molecule (each of which is equivalent to another carbon by symmetry). We can therefore conclude that only three constitutional isomers could be formed by monochlorination, and this would correspond to substitution at each of the three unique types of carbon atom in the molecule. In the case of molecules for which no such symmetry is present (or is perhaps overlooked), a second approach is available. You can consider substitution at each carbon in the molecule and then evaluate the resulting structures to see if any are the same. For hexane this would afford the following six structural formulas:

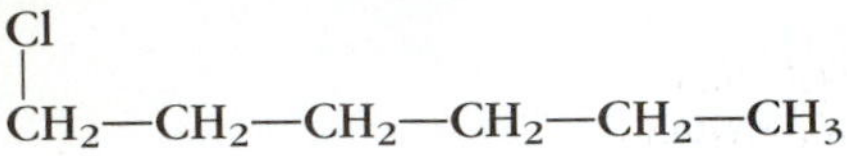

$$\underset{\mathbf{1}}{\overset{Cl}{\overset{|}{CH_2}}—CH_2—CH_2—CH_2—CH_2—CH_3} \qquad \underset{\mathbf{2}}{CH_3—\overset{Cl}{\overset{|}{CH}}—CH_2—CH_2—CH_2—CH_3} \qquad \underset{\mathbf{3}}{CH_3—CH_2—\overset{Cl}{\overset{|}{CH}}—CH_2—CH_2—CH_3}$$

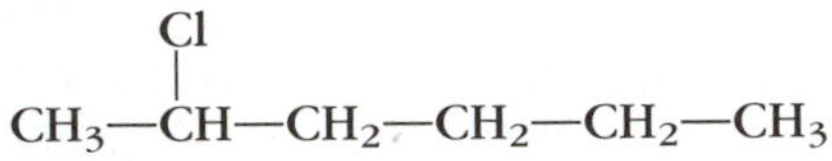

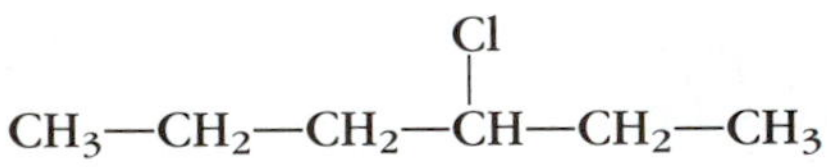

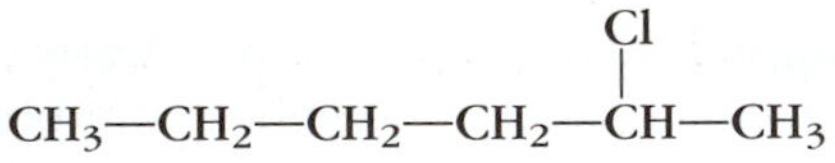

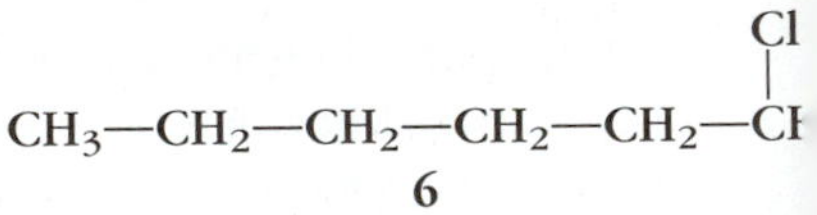

$$\underset{\mathbf{4}}{CH_3—CH_2—CH_2—\overset{Cl}{\overset{|}{CH}}—CH_2—CH_3} \qquad \underset{\mathbf{5}}{CH_3—CH_2—CH_2—CH_2—\overset{Cl}{\overset{|}{CH}}—CH_3} \qquad \underset{\mathbf{6}}{CH_3—CH_2—CH_2—CH_2—CH_2—\overset{Cl}{\overset{|}{CH_2}}}$$

How should you compare these structures to see if they represent the same or different compounds? With sufficient experience you will be able to do this by a visual inspection, and you would conclude that structures **1** and **6** are two drawings of the same compound, as are structures **2** and **5** and structures **3** and **4.** Naming the compounds according to the IUPAC rules gives you another reliable way to evaluate whether or not two structural formulas represent the same compound or constitutional isomers. Two drawings of the *same compound* will necessarily yield the same name, and, conversely, constitutional isomers must have different names. The names of the six structures drawn above are: **1** = 1-chlorohexane, **2** = 2-chlorohexane, **3** = 3-chlorohexane, **4** = 3-chlorohexane, **5** = 2-chlorohexane, and **6** = 1-chlorohexane.

The use of the systematic names demonstrates that **1** and **6, 2** and **5,** and **3** and **4** are in each case two drawings of the same compound. Again you can conclude that only three constitutional isomers could be formed by substitution of a chlorine for one of the hydrogens in hexane.

Many molecules do lack elements of symmetry which would cause some of the carbons to be equivalent. In the case of 3-methylhexane, for example, no two of the carbons are equivalent, and seven constitutional isomers could be produced by monochlorination.

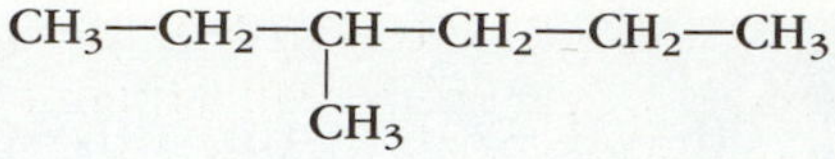

$$\begin{array}{l} CH_3{-}CH_2{-}\underset{\displaystyle CH_3}{\underset{|}{CH}}{-}CH_2{-}CH_2{-}CH_3 \end{array}$$

EXERCISE 1.8

Draw and name the seven constitutional isomers that could be produced by monochlorination of 3-methylhexane.

Sometimes it is difficult to recognize equivalent groups because a molecule is drawn in a way which makes the groups look different. For example, the equivalence of the two methyl groups attached to the tertiary carbon of 2-methylbutane is not obvious in the following drawing:

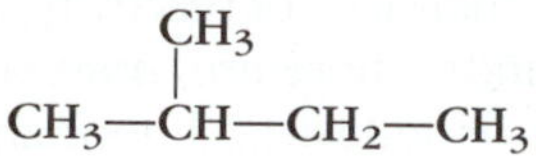

$$CH_3{-}\overset{\displaystyle CH_3}{\overset{|}{CH}}{-}CH_2{-}CH_3$$

Recall that such drawings convey no three-dimensional information, so the alkane could equally well be drawn as:

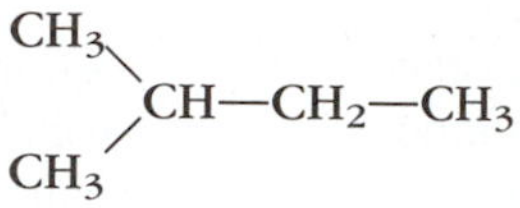

$$\begin{matrix} CH_3\diagdown \\ \quad CH{-}CH_2{-}CH_3 \\ CH_3\diagup \end{matrix}$$

In this drawing the symmetry of the molecule and hence the equivalence of the two methyl groups on its left side are easily seen. Recognize that these two methyl groups are not equivalent to the single methyl group at the other end of the carbon chain.

Symmetry is an extremely useful tool in organic chemistry and we will consider it in greater detail in subsequent chapters. In the meantime we will continue to employ the concept of groups that are equivalent because they are symmetry-related. Recall that the *constitution* of a molecule is defined only by the sequence in which atoms are attached to each other and not by any spatial arrangement. Consequently, any identical groups (e.g., methyl groups) attached to the same atom in a molecule are **constitutionally equivalent.** This means that the two methyls of an isopropyl group must be constitutionally equivalent, as must the three methyl groups of a *tert*-butyl group. Reaction at any one of several equivalent groups will always yield the same products (i.e., products having the same constitution), and this means that *constitutional equivalence* can be very useful in helping you predict the products of a reaction.

EXERCISE 1.9

(a) Use the letters **a, b, c** . . . to indicate each type of hydrogen atom in the following alkanes. Constitutionally equivalent hydrogen atoms should be designated by the same letter. The answer is provided for the first structure.

(i) 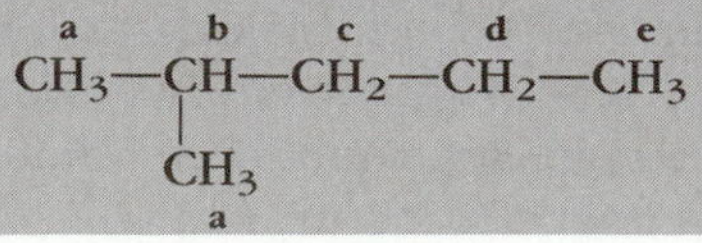 $\overset{a}{CH_3}{-}\overset{b}{CH}(\underset{a}{CH_3}){-}\overset{c}{CH_2}{-}\overset{d}{CH_2}{-}\overset{e}{CH_3}$

(ii) $CH_3-CH_2-CH(CH_3)-CH_2-CH_3$ (iii) $CH_3-C(CH_3)_2-CH_2-CH_3$

(b) Draw and name all the monochloro derivatives which could be formed by chlorination of the alkanes shown in (a).

1.10 REACTIONS OF ALKANES

Alkanes are characterized by their lack of reactivity, particularly in comparison with most other organic compounds. There are, however, several reactions of alkanes that have great commercial or industrial importance, and we will discuss these briefly.

Oxidation

The high-temperature oxidation of alkanes is certainly the most common reaction of this class of organic compounds, and it is probably more familiar to you by the name of *combustion.* Combustion of alkanes obtained from petroleum and natural gas has become the dominant source of energy in our society. The oxidation of alkanes is exothermic, and the heat released from this process can be used directly. For example, combustion of natural gas (primarily methane) is used extensively for home heating, as well as for gas stoves and clothes dryers.

Gasoline and other fuels derived from petroleum are mixtures of alkanes, and the heat produced by their combustion can be converted to different forms of energy. For example, the heat of combustion can be converted to electricity by boiling water to operate a steam turbine. In automobiles the mechanical energy of the hot, expanding gases produced by combustion is used directly to move the pistons of an internal combustion engine. The general reaction (not balanced) can be written:

$$C_nH_{2n+2} + O_2 \longrightarrow CO_2 + H_2O$$

If insufficient oxygen is present, carbon monoxide is produced, together with a variety of compounds in which the alkane molecule has been only partially oxidized. At the high temperatures of the combustion reaction in a gasoline engine, oxides of nitrogen are also produced. These are not derived from the fuel itself but from the nitrogen present in the air that is used as the source of oxygen. Considerable effort has been expended in recent years to reduce the quantity of these byproducts released into the atmosphere.

Halogenation

While relatively unimportant as a laboratory reaction for alkanes, the direct reaction of alkanes with halogens is an important industrial process. Bromination is sometimes used, but chlorination is of far greater importance. (Fluorine is too reactive to be used for this purpose and iodine is too unreactive.)

The *overall* chlorination reaction of an alkane is highly exothermic, but the *initiation* of the reaction requires heat or light. This is also the case for oxidation of alkanes, which are quite stable in the presence of oxygen (i.e., air) until the high temperature of a flame or spark causes combustion to begin. Step-by-step details of the chlorination reaction and the significance of the initiation step will be discussed in Chapter 2. The net result, as illustrated here for methane, involves the replacement of a hydrogen by a chlorine to afford a chloroalkane, together with the by-product, hydrogen chloride.

$$CH_4 + Cl_2 \xrightarrow{\text{heat or light}} CH_3Cl + HCl$$

As we showed in the preceding section, a large number of isomeric products may be possible in the halogenation of an alkane. This restricts the usefulness of the reaction to compounds having only a few different types of hydrogen that could be replaced by halogen. Methane affords a good example of such a compound, and indeed the chlorination of methane is an important industrial process.

EXERCISE 1.10

(a) Draw the structures of two alkanes other than methane that have only a single kind of hydrogen.
(b) Draw the structures of three alkanes that have only two kinds of hydrogen.

Even in the case of a compound such as methane, which has only a single type of hydrogen, the chlorination reaction produces a complex mixture of products. The initial product, chloromethane, still has three hydrogens and is capable of undergoing further chlorination. Moreover, the process can continue until all four hydrogens have been replaced. This is summarized in Figure 1.13, which also shows the common names for the various chlorinated methanes. In the case of chloroform and carbon tetrachloride, the common names are used much more frequently than the systematic names.

The chlorination of methane proceeds very easily past the stage of monochlorination (in fact, the second step of Figure 1.13 proceeds more readily than the first). This makes it quite difficult to obtain a high yield of a single product. (The term **yield** refers to the quantity of material actually obtained from a reaction.) The

$$CH_4 \longrightarrow CH_3Cl \longrightarrow CH_2Cl_2 \longrightarrow CHCl_3 \longrightarrow CCl_4$$

Methane	Chloromethane (Methyl chloride)	Dichloromethane (Methylene chloride)	(Trichloromethane) Chloroform	(Tetrachloromethane) Carbon tetrachloride

Figure 1.13 The chlorination of methane.
Chlorination can proceed to replace one, two, three, or all four hydrogens of methane. The trivial names of the chloro derivatives are shown below the IUPAC names. The names in parentheses are used less frequently.

following discussion will focus on one method by which we might regulate the chlorination reaction in order to obtain a product mixture that would be enriched in the desired product. We have oversimplified the mathematical analysis of competing reactions, but this will enable you to see the results more easily. For the hypothetical situation in which an equimolar mixture of chlorine and methane reacted to the extent of 50 percent, the following mixture would be present:

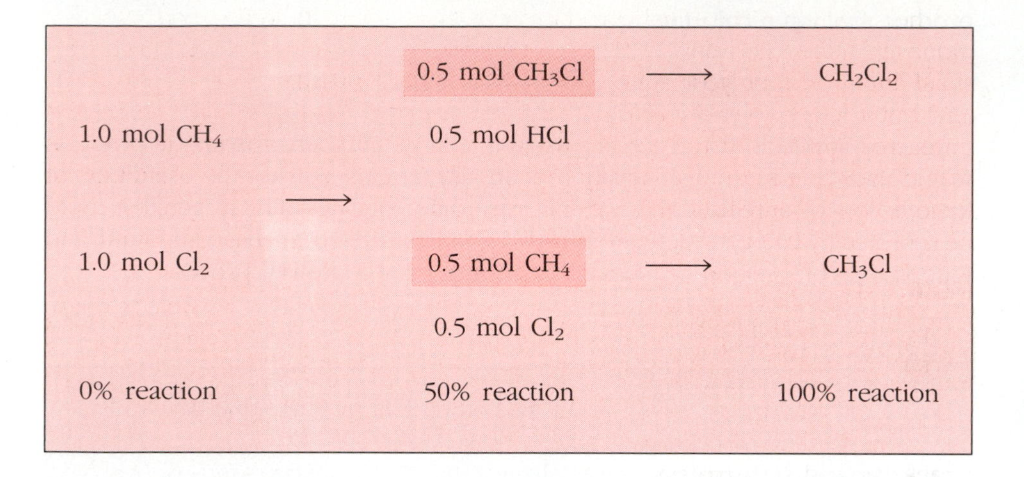

The oversimplification, of course, is that only monochlorination will have occurred up to this point. In actuality, formation of dichloromethane will begin to take place as soon as the first molecules of chloromethane have been formed. Nevertheless, if we assume that CH_4 and CH_3Cl are present in equal amounts, we might expect that further reaction would involve chlorination of each to a comparable extent. This would result in a 50-50 mixture of chloromethane and dichloromethane when the reaction reached completion.

The preceding example shows how the reaction of methane with an equimolar quantity of chlorine would yield a product mixture in which the monochloro derivative would not even be the predominant product. How might we circumvent this difficulty if we needed to obtain just one of the chlorinated derivatives? A partial solution could be achieved by using a large excess of one of the starting materials.

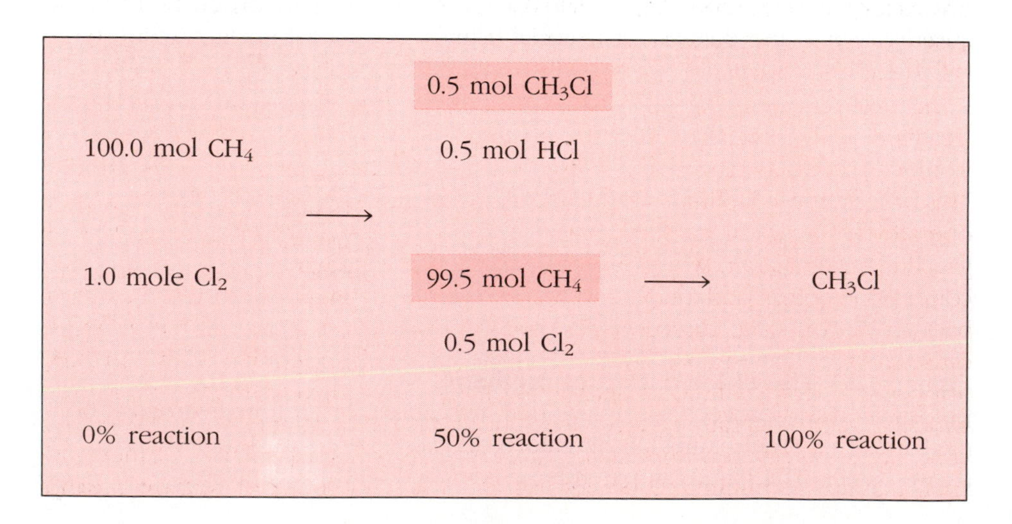

Consider another oversimplified picture of the same chlorination reaction, but this time with the use of a large excess of methane. If we assume that only monochlorination takes place at first, there would still be a very large excess of methane relative to chloromethane when 50 percent of the starting chlorine had been consumed.

Because there are so many more molecules of CH_4 than of CH_3Cl, we have shown further reaction of CH_4 as the predominant pathway during the second half of the reaction. The final product mixture would therefore contain the monosubstituted derivative, CH_3Cl, as the major chlorination product, with only small amounts of polychlorinated derivatives. This approach to solving the problem would not be without its disadvantages, however, because a large quantity of unreacted methane would remain when all the chlorine had been consumed. It would then be necessary to purify the product (CH_3Cl) by separating it from the large excess of unreacted starting material (CH_4).

EXERCISE 1.11

Suggest a procedure by which CCl_4 could be prepared from methane in high yield.

Cracking and Reforming

A substantial portion of crude petroleum consists of alkanes which have high molecular weights (and therefore high boiling points). Alkanes with molecular structures in the range of 20 to 70 carbons are valuable as *lubricating oils,* but their low volatility greatly decreases their use as fuels. An industrial process, called *cracking,* converts these high molecular weight alkanes into smaller molecules, and the resulting hydrocarbons are used as raw materials for other industrial processes (Chapter 26). Cracking involves cleavage of the alkane molecules at very high temperatures, usually 450°C or above. (To gain a feeling for how high these temperatures are, note that the oven of a typical kitchen stove has a maximum setting of about 550°F, which is less than 300°C.) The cracking process is generally carried out in the presence of metal catalysts, which assist the reactions. When *catalytic cracking* is carried out in the presence of hydrogen gas (*hydrocracking*), the resulting products are lower molecular weight alkanes which can be used in gasoline. In the absence of hydrogen gas, cracking produces lower molecular weight hydrocarbons called alkenes (discussed in Chapter 5) rather than alkanes.

In a related industrial process, *catalytic reforming,* alkanes are converted by treatment with a metal catalyst at high temperature into more highly branched isomers, as well as into aromatic hydrocarbons (to be discussed in Chapter 10). This process is used to improve the *antiknock* properties of gasoline. In the proper operation of a typical gasoline engine, the piston first compresses a mixture of gasoline vapor and air. When the compressed mixture is ignited by the spark plug, combustion occurs, and this generates hot, expanding gases, which drive the piston back in the opposite direction. The initial compression of the air-fuel mixture causes a large temperature increase, which can result in ignition of the mixture prior to the desired time of spark plug firing. This phenomenon is known as *knocking,* and it greatly reduces the efficiency of the engine. The problem of knocking is less severe with branched alkanes than with straight-chain alkanes, and knocking can be minimized by the use of a fuel with a higher **octane rating.**

Octane rating is based on an arbitrary scale in which the straight-chain alkane heptane is assigned a value of zero and the highly branched alkane 2,2,4-trimethylpentane is assigned a value of 100.

$$CH_3{-}CH_2{-}CH_2{-}CH_2{-}CH_2{-}CH_2{-}CH_3$$

Heptane:
octane rating 0

$$CH_3{-}\underset{\displaystyle CH_3}{\overset{\displaystyle CH_3}{\overset{|}{\underset{|}{C}}}}{-}CH_2{-}\overset{\displaystyle CH_3}{\overset{|}{C}H}{-}CH_3$$

2,2,4-Trimethylpentane:
octane rating 100

2,2,4-Trimethylpentane has the common (and confusing) name *isooctane,* and this provides the origin of the term "octane rating." Octane itself, however, is a straight-chain alkane and has a much lower octane rating.

Gasoline blends containing aromatic hydrocarbons such as toluene (Chapter 10) also have higher octane ratings, and increasing amounts of these aromatic hydrocarbons are being added to gasoline as antipollution regulations require the use of *unleaded* gasoline. *Leaded* gasoline contains small amounts of *antiknock compounds* such as tetraethyllead, which cannot be used in automobiles equipped with *catalytic converters* in their exhaust systems. The catalytic converters bring about complete reaction of partially oxidized hydrocarbons that are produced as by-products in a gasoline engine, but their activity is destroyed by lead compounds. Consequently, unleaded gasoline with larger quantities of branched alkanes and aromatic hydrocarbons is being produced in increasing quantities, and its use is required in new automobiles.

1.11 PREPARATION OF ALKANES

Petroleum Technology

Alkanes are formed by a variety of chemical processes used in petroleum refining. These include hydrocracking and catalytic reforming (discussed in the previous section) and *alkylation,* in which small hydrocarbon units are joined together to produce branched alkanes with good octane ratings. Alkanes with very high molecular weights can be produced by polymerization, which we discuss in Section 25.4.

Reduction of Haloalkanes

The reactions of petroleum technology are only used as industrial processes, and we will now turn to reactions that are used by chemists to prepare alkanes from haloalkanes in the laboratory. This process is a *reduction* of the alkyl halide. Oxidation states are not usually assigned to organic compounds, and more operational definitions of oxidation and reduction are used instead. Most reactions of organic compounds are centered on either a single carbon atom or two adjacent carbon atoms in a molecule, and we can conveniently evaluate oxidation-reduction reactions by looking at only those carbon atoms undergoing changes. Sometimes the oxidation-reduction reactions of organic compounds do not involve oxygen, and only a change in the number of hydrogen atoms occurs. As a result we must consider changes in the number of both carbon-hydrogen and carbon-oxygen bonds.

An increase in the number of bonds to hydrogen atoms corresponds to *reduction*, whereas a decrease in the number of bonds to hydrogen atoms is classified as *oxidation*. Conversely, an increase in the number of bonds to oxygen atoms (or other electronegative atoms) corresponds to *oxidation*, but a decrease in the number of bonds to electronegative atoms would be classified as *reduction*.

This definition tells you that halogenation of an alkane is an oxidation, while the reverse process would be a reduction.

$$CH_4 \underset{\text{reduction}}{\overset{\text{oxidation}}{\rightleftharpoons}} CH_3Cl$$

Another example of a reduction reaction would be hydrocracking, in which a carbon-carbon bond is cleaved and two new carbon-hydrogen bonds are formed:

$$—CH_2—CH_2—CH_2—CH_2— \xrightarrow{H_2} —CH_2—CH_3 + CH_3—CH_2—$$

Sometimes two adjacent carbon atoms might undergo changes in opposite directions, but the overall reaction would be neither an oxidation nor a reduction. We will present reactions of that type in Chapter 5.

The reduction of an alkyl halide to an alkane can be accomplished with a reactive metal in the presence of an acid. You will encounter additional ways of carrying out this reaction in subsequent chapters, but the most common method for reducing an alkyl halide to an alkane is the use of zinc and hydrochloric acid, as illustrated in the following example:

$$CH_3—(CH_2)_{14}—CH_2—I \xrightarrow{\text{Zn, HCl}} CH_3—(CH_2)_{14}—CH_3 \qquad 85\%$$

This reduction reaction is the first of many examples in this book of actual reactions reported in the chemical literature, all of which are easily recognized by the appearance of a percent yield after the equation. A percent yield describes the quantity of material obtained in the reaction relative to the amount predicted for complete reaction according to the stoichiometry of the balanced equation. Notice that only 85 percent of the theoretical quantity of alkane was obtained by the chemists who carried out the preceding reaction. Their results illustrate the important point that *chemical reactions very rarely afford products in 100 percent yield.* Even when only a single compound is formed in a reaction, the physical manipulations of isolating and purifying the product invariably result in the loss of some material. The preceding reaction also illustrates a second general feature of the reactions presented in this book. In most cases the equations are not balanced, and often we do not show all products. We have done this intentionally in order to focus attention on the structural changes taking place in the reaction rather than on the stoichiometry. On the basis of your training in general chemistry, you should be capable of balancing these equations.

Now that we have begun to show you reactions of specific compounds, we will introduce several common abbreviations that will be useful in your study of organic chemistry. The letter R is widely used to designate an alkyl group. For example, the general reaction for chlorination of an alkane might be written as:

$$R—H \xrightarrow{Cl_2} R—Cl$$

In a similar manner the letter X is commonly used to indicate an electronegative group, typically a halogen. The general reduction of alkyl halides (not just chlorides) can be written in the following way:

$$\text{R—X} \xrightarrow{\text{Zn, HCl}} \text{R—H}$$

Different alkyl groups in a single structure may be indicated with a superscript, for example:

$$\text{R—}\underset{\displaystyle\text{Cl}}{\underset{|}{\text{CH}}}\text{—R}'$$

Coupling of Alkyl Halides

By using the appropriate reagents it is possible to join the alkyl fragments of two haloalkanes and generate a new alkane with a larger number of carbon atoms.

$$\text{R—X} + \text{R}'\text{—X} \longrightarrow \text{R—R}'$$

This introduces you to a new and extremely important concept in organic chemistry, the fact that reactions can result in the *formation of new carbon-carbon bonds.* Whether in a university research laboratory or in the manufacturing plant of a large chemical or pharmaceutical company, one of the most fundamental requirements of modern organic chemistry is the ability to carry out the *synthesis* of complex organic compounds from less complex starting materials. Typically, the reactants are smaller organic molecules, and building up the desired carbon skeleton invariably requires reactions which form new carbon-carbon bonds. Only a limited number of reaction types are available for carbon-carbon bond formation, and we will emphasize these appropriately as we introduce them to you.

The classical method for coupling of alkyl halides was introduced by the French chemist C.A. Wurtz over 100 years ago, and it involves heating a mixture of the alkyl halide and sodium metal:

$$\text{R—X} \xrightarrow{\text{Na}} \text{R—R}$$

Unfortunately, the reaction does not work very well. The yields are often quite low, and they are very dependent upon the precise *reaction conditions* (time, temperature, stirring, etc.). The variability of yield can be seen from the following results for the coupling of 1-bromobutane.

$$CH_3\text{—}CH_2\text{—}CH_2\text{—}CH_2\text{—Br} \xrightarrow{\text{Na}} CH_3\text{—}CH_2\text{—}CH_2\text{—}CH_2\text{—}CH_2\text{—}CH_2\text{—}CH_2\text{—}CH_3 \quad 15\text{–}70\%$$

In addition, the reaction affords only symmetrical alkanes R—R, where both R groups are the same, and attempts to couple two different alkyl halides have been unsuccessful. Because of these limitations the Wurtz reaction is not widely used in the organic laboratory.

A more recent coupling reaction, initially developed by E. J. Corey (Harvard University) and H.O. House (Massachusetts Institute of Technology, Georgia Institute of Technology), has proved to be quite valuable. The reaction involves several steps, and it permits the coupling of two different alkyl halides. The reaction sequence begins with preparation of an **organometallic** reagent, i.e., a species that has a carbon-metal bond. Thus the reaction of a haloalkane with lithium metal produces an alkyllithium. (Because we want to focus on the organic species, the by-product of the reaction, a lithium halide salt, is not shown.)

$$R{-}X \xrightarrow{Li} R{-}Li \qquad (1)$$

Organometallic compounds are usually quite reactive and not easily isolated, so the solution of alkyllithium formed in step 1 is allowed to react with cuprous iodide to form an organocopper reagent:

$$R{-}Li \xrightarrow{CuI} R_2CuLi \qquad (2)$$

In this organometallic compound the alkyl groups are bonded to the copper, and the reagent is named as a *lithium dialkylcuprate.* The organocopper reagent, like the alkyllithium formed in the previous step, is not isolated but is allowed to react with an excess quantity of a second alkyl halide.

$$R_2CuLi + R'CH_2{-}X \longrightarrow R'CH_2{-}R \qquad (3)$$

As long as the organocopper reagent is prepared carefully, there are few restrictions on the nature of its alkyl groups, but there are definite structural limitations on the alkyl halide with which it reacts. The latter must be a *primary alkyl halide* (typically a bromide or iodide), and for this reason we have abbreviated it as $R'CH_2X$. The reaction of lithium dialkylcuprates with secondary alkyl halides is not dependable and frequently gives poor yields; the reaction with a tertiary alkyl halide typically affords none of the coupling product. As long as these limitations are recognized, the reaction is quite versatile, as illustrated by the following examples. Only the final step, the actual coupling reaction, is shown for these reactions, but in each case the organocopper reagent is prepared according to the procedure we have just described.

$$(CH_3{-}CH_2{-}CH_2{-}CH_2)_2CuLi \xrightarrow{Cl{-}CH_2{-}CH_2{-}CH_2{-}CH_2{-}CH_3} CH_3{-}CH_2{-}CH_2{-}CH_2{-}CH_2{-}CH_2{-}CH_2{-}CH_2{-}CH_3 \quad 80\%$$

Nonane

$$(CH_3)_2CuLi \xrightarrow{I{-}CH_2{-}CH_2{-}CH_2{-}CH_2{-}CH_2{-}CH_2{-}CH_2{-}CH_3} CH_3{-}CH_2{-}CH_2{-}CH_2{-}CH_2{-}CH_2{-}CH_2{-}CH_2{-}CH_3 \quad 88\%$$

Nonane

$$(CH_3{-}CH_2{-}CH_2{-}CH_2)_2CuLi \xrightarrow{Br{-}(CH_2)_7{-}CH_3} CH_3{-}(CH_2)_{10}{-}CH_3 \quad 93\%$$

Dodecane

$$(CH_3—CH_2—\overset{\overset{\large CH_3}{|}}{CH})_2CuLi \xrightarrow{Br—(CH_2)_7—CH_3} CH_3—CH_2—\overset{\overset{\large CH_3}{|}}{CH}—(CH_2)_7—CH_3 \quad 77\%$$

3-Methylundecane

$$\left(CH_3—\underset{\underset{\large CH_3}{|}}{\overset{\overset{\large CH_3}{|}}{C}}—\right)_2CuLi \xrightarrow{Br—(CH_2)_7—CH_3} CH_3—\underset{\underset{\large CH_3}{|}}{\overset{\overset{\large CH_3}{|}}{C}}—(CH_2)_7—CH_3 \quad 93\% \quad 93\%$$

2,2-Dimethyldecane

The preceding examples show you that a wide variety of alkyl halides can be used in these coupling reactions. The halide can be Cl, Br or I, and the alkyl group in the organometallic reagent can be methyl, primary, secondary or tertiary. As long as a primary alkyl halide is used for reaction with the organometallic compound, you should expect to obtain the desired product in good yield.

EXERCISE 1.12

There are two sequences of reactions that would allow you to couple 1-bromobutane with 1-bromopropane. Write both of these sequences.

EXERCISE 1.13

Write the preferred sequence of reactions that would allow you to couple 1-bromobutane with 2-bromobutane.

1.12 SYNTHETIC PLANNING

When a chemist needs to prepare a particular organic compound, it is necessary to design a synthesis that meets two requirements: (1) *The proposed scheme must consist of a logical sequence of reactions that can be expected to work;* and (2) *The proposed starting materials must be available.* A professional organic chemist has considerable experience and many reference books available to determine the validity of a proposed reaction, but you, as a beginning student, must rely on this textbook and the lectures of your instructor. In a similar way, the availability of starting materials to a professional chemist is restricted only by the requirement that they be sold commercially by chemical suppliers. For you as a student, however, commercial availability is frequently not an appropriate criterion. In the first place, you have little idea what compounds could be purchased commercially. Moreover, many of the less complex molecules that we will present as synthetic targets are themselves commercially available. Consequently, the criteria for availability of starting materials will always be given in the instructions for problems in this book.

Example 1.4 Propose a method of synthesis of pentane starting with alkyl halides containing four or fewer carbon atoms.

Solution This example illustrates a common feature of synthesis problems in organic chemistry: *There is more than one correct answer.* Since the product contains five carbons and each of the starting materials may contain no more than four carbons, it is clear that a reaction must be used which results in carbon-carbon bond formation. Pentane has an odd number of carbon atoms, so it is not composed of two identical alkyl groups, and the Wurtz reaction cannot be used. The only alternative method for synthesizing alkanes that we have discussed so far is the coupling of an alkyl halide with a lithium dialkylcuprate. There are two possible combinations of alkyl halides which fit the stated requirements:

$$CH_3{-}CH_2{-}CH_2{-}CH_2{-}CH_3 \Leftarrow \begin{cases} CH_3{-}CH_2{-}X \quad \text{and} \quad X{-}CH_2{-}CH_2{-}CH_3 \\ CH_3{-}X \quad \text{and} \quad X{-}CH_2{-}CH_2{-}CH_2{-}CH_3 \end{cases}$$

Both alternatives would be satisfactory in this instance. In addition, *either* of the alkyl halides could be used to form the organocopper reagent in each case, so at least four distinct answers are possible:

$$(CH_3{-}CH_2)_2CuLi \xrightarrow{CH_3-CH_2-CH_2-X} CH_3{-}CH_2{-}CH_2{-}CH_2{-}CH_3$$

$$(CH_3{-}CH_2{-}CH_2)_2CuLi \xrightarrow{CH_3-CH_2-X} CH_3{-}CH_2{-}CH_2{-}CH_2{-}CH_3$$

$$(CH_3)_2CuLi \xrightarrow{CH_3-CH_2-CH_2-CH_2-X} CH_3{-}CH_2{-}CH_2{-}CH_2{-}CH_3$$

$$(CH_3{-}CH_2{-}CH_2{-}CH_2)_2CuLi \xrightarrow{CH_3-X} CH_3{-}CH_2{-}CH_2{-}CH_2{-}CH_3$$

Moreover, the halogen for each starting alkyl halide could be either chlorine, bromine, or iodine. As a result, a very large number of complete answers could be written, and we have shown only one of these:

$$CH_3{-}Br \xrightarrow{Li} CH_3Li \xrightarrow{CuI} (CH_3)_2CuLi \xrightarrow{Cl-CH_2-CH_2-CH_2-CH_3} CH_3{-}CH_2{-}CH_2{-}CH_2{-}CH_3$$

Example 1.5 Suggest a method of synthesizing 3-methylpentane starting with alkyl halides containing four or fewer carbon atoms.

Solution The goal is 3-methylpentane, which has a total of six carbons. A maximum of four carbons is permitted in any starting compound, so you must evaluate the possibilities of joining either a C_4 and a C_2 fragment or two C_3 fragments. Because of the branching of the carbon chain, there is no way to form it by joining two three-carbon fragments. Therefore you must join a C_2 and a C_4 fragment, and this corresponds to forming a bond between C-2 and C-3.

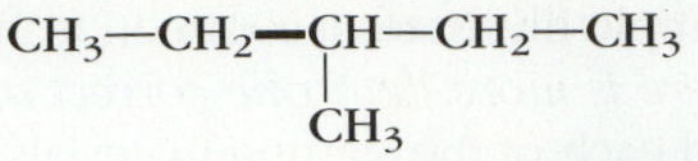

$$CH_3{-}CH_2{-}\underset{\displaystyle CH_3}{\underset{|}{CH}}{-}CH_2{-}CH_3$$

Notice that the symmetry of the target molecule makes the bond between C-3 and C-4 equivalent to that between C-2 and C-3.

This analysis means that the *final* step of the synthesis must be formation of the bond between C-2 and C-3. It also requires that the original alkyl halides be chloroethane and 2-chlorobutane. (The choice of chlorine is arbitrary; the bromo or iodo compounds would work equally well.) Which of these two halides should be used to form the organometallic derivative? This is best answered by considering both alternatives for the final step:

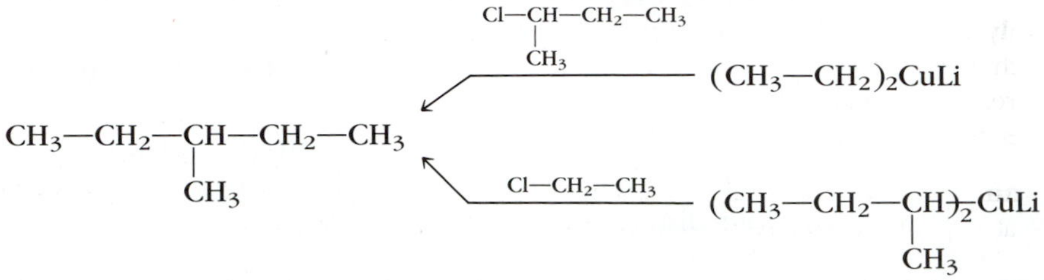

$$CH_3{-}CH_2{-}\underset{\displaystyle CH_3}{\underset{|}{CH}}{-}CH_2{-}CH_3 \xleftarrow{Cl{-}\underset{CH_3}{CH}{-}CH_2{-}CH_3} (CH_3{-}CH_2)_2CuLi$$

$$CH_3{-}CH_2{-}\underset{\displaystyle CH_3}{\underset{|}{CH}}{-}CH_2{-}CH_3 \xleftarrow{Cl{-}CH_2{-}CH_3} (CH_3{-}CH_2{-}\underset{\displaystyle CH_3}{\underset{|}{CH}})_2CuLi$$

Only the second possibility would work, because the organocopper reagent must react with a *primary* alkyl halide.

The complete proposed synthesis can now be written out. Observe that we have written the reactions in reverse sequence to be consistent with the reasoning process used in planning a synthesis. Frequently, you will find this sort of "working backwards" to be of tremendous assistance in working out synthesis problems. It is a standard technique in synthetic organic chemistry, and organic chemists frequently refer to the approach as *retrosynthetic* chemistry.

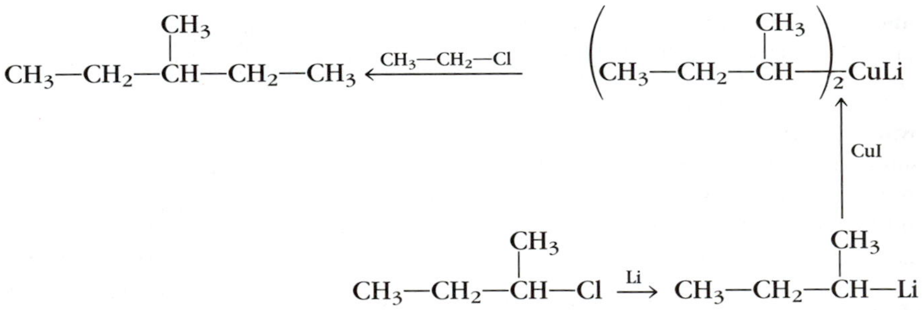

$$CH_3{-}CH_2{-}\overset{\displaystyle CH_3}{\overset{|}{CH}}{-}CH_2{-}CH_3 \xleftarrow{CH_3{-}CH_2{-}Cl} \left(CH_3{-}CH_2{-}\overset{\displaystyle CH_3}{\overset{|}{CH}}{-}\right)_2CuLi$$

$$\uparrow \text{CuI}$$

$$CH_3{-}CH_2{-}\overset{\displaystyle CH_3}{\overset{|}{CH}}{-}Cl \xrightarrow{Li} CH_3{-}CH_2{-}\overset{\displaystyle CH_3}{\overset{|}{CH}}{-}Li$$

1.13 TERMS AND DEFINITIONS

Alkane. A hydrocarbon with the general formula C_nH_{2n+2}, in which each carbon atom is bonded to four other atoms (either carbon or hydrogen).

Alkyl group. A hydrocarbon portion of a molecule corresponding to an alkane from which a hydrogen atom has been removed.

Antibonding. A description of a molecular orbital produced by the out-of-phase combination of two atomic orbitals. When occupied by electrons, this corresponds to a repulsive interaction between the two atoms.

Atomic orbital. An allowed energy state for an electron in an atom; the orbital is described mathematically by a wave function. The three-dimensional distribution of electron density is described by the square of the wave function.

Bonding. The attractive interaction (i.e., a bond) between two atoms that results when electrons occupy a molecular orbital formed by the in-phase interaction between two atomic orbitals.

Branched-chain. A term describing a molecule in which all the carbon atoms are not connected sequentially, i.e., in which it is not possible to encounter all atoms of the carbon chain by proceeding consecutively from one to the next without backtracking.

Catalyst. A material that acts to facilitate a reaction but which is not consumed in the reaction. A catalyst causes the reaction to proceed more rapidly or to occur at a lower temperature.

Common name. A name that is based only partially (or not at all) on the molecular structure of a compound.

Connectivity. The specific sequence in which atoms in a molecule are joined (connected) by bonds.

Constitution. The features of a molecular structure that result from the connectivity of the individual atoms. The constitution can be defined either by drawing a structural formula or by stating the IUPAC name.

Constitutional equivalence. The relationship between two atoms or groups that cannot be differentiated according to their connectivity.

Constitutional isomers. Two molecules (compounds) with the same molecular formula that have different connectivities (i.e., different constitutions).

Covalent bonding. Bonding between two atoms that results from sharing of electrons by the two atoms in a bonding molecular orbital. (This contrasts with ionic bonding, which is an electrostatic attraction between two ions of opposite charge.)

Derivative. A compound that is formed or derived from another compound, either in an actual reaction or in a hypothetical sense by changing a structural formula on paper.

Equivalent. The same. A term describing two or more groups that are indistinguishable on the basis of symmetry or connectivity.

Haloalkane. A derivative of an alkane in which a hydrogen atom has been replaced by a halogen atom; also known as an alkyl halide.

Homolog. A molecule that differs from a reference compound by a single CH_2 group.

Hybrid orbital. Atomic orbital for which the wave function has been generated by mathematical combination of other wave functions (usually for *s* and *p* atomic orbitals) on a single atom.

Hydrocarbon. A compound composed exclusively of carbon and hydrogen atoms.

In phase. Term describing the interaction between atomic orbitals of two different atoms that leads to a bonding molecular orbital. This results from overlap of lobes for which the wave function has the same sign (phase).

Isomers. Different compounds (molecules) having the same molecular formula but differing in their molecular structures.

IUPAC name. The name of a compound that is systematically derived from its molecular structure according to the rules set down by the International Union of Pure and Applied Chemistry.

Lobe. A region of an orbital for which the wave function describing the orbital has a single phase.

Molecular formula. A chemical formula denoting how many atoms of each element are present in a single molecule.

Molecular orbital. An orbital derived from interaction of two or more atomic orbitals. The molecular orbital can be either bonding or antibonding, but normally only the bonding molecular orbitals are occupied by electrons.

Node. A region (usually a plane or a spherical surface) of an atomic (or molecular) orbital that separates lobes of opposite phase. At the nodal surface the electron density for that orbital is zero.

Organometallic compound. A compound having a metal-carbon bond.

Out of phase. A term describing the interaction between atomic orbitals of two different atoms that leads to an antibonding molecular orbital. This results from overlap of lobes for which the wave function has the opposite sign (phase).

Overlap. The interaction (either favorable or unfavorable) between orbitals of two or more atoms that results from their occupying the same region in space.

Oxidation. A reaction that causes the oxidation state of a compound to increase. For organic compounds this usually involves an increase in the number of bonds between carbon and oxygen or other electronegative atoms, or a decrease in the number of bonds between carbon and hydrogen.

Phase. The sign (positive or negative) of the wave function that describes an orbital. We have used shaded and unshaded drawings to distinguish between orbitals or lobes of opposite sign in this text.

Primary. A term describing a carbon atom that is bonded to only one other carbon atom. When used to describe an alkyl group, it refers to the carbon from which a hydrogen has been removed to generate the group (i.e., the carbon that is the point of attachment for the group).

Quaternary. A term describing a carbon atom that is bonded to four other carbon atoms.

Reduction. A reaction that causes the oxidation state of a compound to decrease. For organic compounds this usually involves a decrease in the number of bonds between carbon and oxygen or other electronegative atoms, or an increase in the number of bonds between carbon and hydrogen.

Reflection symmetry. A property of a molecule such that a plane passing through its center divides it into two equivalent halves.

Rotational symmetry. A property of a molecule such that rotation (by some angle of less than 360°) about an axis passing through the center of the molecule generates a structure that is indistinguishable from the original.

Secondary. A term describing a carbon atom that is bonded to two other carbon atoms. When used to describe an alkyl group, it refers to the carbon from which a hydrogen has been removed to generate the group (i.e., the carbon that is the point of attachment for the group).

Stereochemistry. The study of the three-dimensional aspects of molecular structures and chemical reactions; also a term used to describe the three-dimensional structure of a molecule.

Straight-chain. A term describing a molecule in which all the carbon atoms are connected sequentially, i.e., in which it is possible to encounter all atoms of the carbon chain by proceeding consecutively from one to the next without backtracking.

Structural formula. A representation of a molecule that denotes the constitution (and sometimes the three-dimensional structure as well).

Substituent. Any atom or group that is bonded to a particular (carbon) atom is a substituent of that (carbon) atom.

Symmetry axis. An axis passing through the center of a molecule such that rotation about the axis (by some angle of less than 360°) generates a structure indistinguishable from the original.

Tertiary. A term describing a carbon atom that is bonded to three other carbon atoms. When used to describe an alkyl group, it refers to the carbon from which a hydrogen has been removed to generate the group (i.e., the carbon that is the point of attachment for the group).

Tetrahedral. A term describing the (approximate) geometry of carbon atoms in an alkane. In methane the four hydrogen substituents on the carbon atom define the corners of a tetrahedron, and the H—C—H angle is 109.5° (called the tetrahedral angle).

Valence atomic orbital. An atomic orbital of the level that is occupied by electrons involved in bonding (i.e., 1*s* for hydrogen and 2*s*, 2*p* for carbon and other second-row elements).

Yield. The quantity of material actually obtained from a reaction; usually expressed as a percentage of the theoretical yield (which is the quantity predicted for complete reaction according to the stoichiometry of the balanced equation).

1.14 SUMMARY OF REACTIONS

Table 1.4 summarizes the reactions of alkanes, and Table 1.5 summarizes the reactions used to prepare them.

TABLE 1.4 Reactions of Alkanes

Reaction	Comments
1. Combustion (oxidation) $C_nH_{2n+2} \xrightarrow{O_2} CO_2 + H_2O$	Section 1.10. Used as an energy source; not applicable to synthesis.

TABLE 1.4 Reactions of Alkanes (continued)

Reaction	Comments
2. Halogenation $C_nH_{2n+2} \xrightarrow[\text{heat or light}]{X_2} C_nH_{2n+1}X$ isomers	Section 1.10. X may be Cl or Br. Products typically consist of the various possible constitutional isomers together with polyhalo derivatives. In most cases not a useful reaction for laboratory synthesis.
3. Catalytic cracking, reforming (a) Hydrocracking $C_nH_{2n+2} \xrightarrow[\text{catalyst}]{H_2}$ smaller alkanes (b) Reforming $C_nH_{2n+2} \xrightarrow[\text{catalyst}]{} C_nH_{2n+2}$ isomers	Section 1.10. High-temperature reactions used in petroleum refining; not applicable to laboratory synthesis.

TABLE 1.5 Preparation of Alkanes

Reaction	Comments
1. Catalytic cracking, reforming	Section 1.10. Industrial processes (see Reactions of Alkanes)
2. Reduction of alkyl halides $R{-}X \xrightarrow{Zn/HCl} R{-}H$	Section 1.11. X may be Cl, Br, or I. A useful procedure for replacement of halogen by hydrogen. (Other metals can sometimes be used.)
3. Coupling of alkyl halides	Section 1.11.
(a) Wurtz reaction $R{-}X \xrightarrow{Na} R{-}R$	Affords alkanes derived from two identical alkyl groups only. Not very useful for organic synthesis.

TABLE 1.5 Preparation of Alkanes (continued)

Reaction	Comments
(b) With organocopper reagents 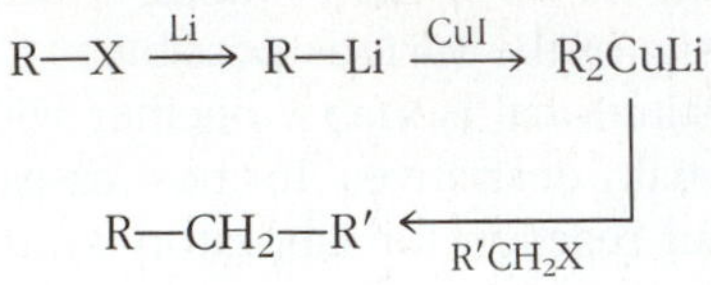 $R{-}X \xrightarrow{Li} R{-}Li \xrightarrow{CuI} R_2CuLi \xrightarrow{R'CH_2X} R{-}CH_2{-}R'$	Can be used to prepare a wide variety of alkanes. Must use primary alkyl halide in the reaction with the organocopper reagent. A good synthetic reaction.
4. Reduction of alkenes	Section 5.5

1.15 PROBLEMS

1.1 Draw the structures of the following molecules:

(a) 1,2-Dibromoethane
(b) Carbon tetrachloride
(c) 1-Chloro-2,4,4-trimethylhexane
(d) 2,2,3,3-Tetramethylbutane
(e) 1-Bromo-3-methylbutane
(f) 5-*sec*-Butyl-2,7-dimethylnonane
(g) 5-*tert*-Butyl-3-methyloctane
(h) Isobutyl chloride
(i) Neopentyl bromide
(j) 3-Ethyl-1-iodo-4-methylpentane

1.2 Name the following structures:

(a) 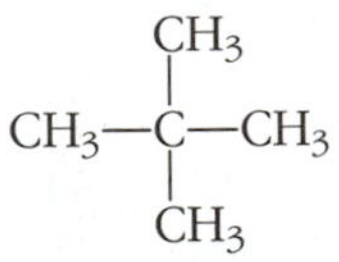 $CH_3{-}C(CH_3)_2{-}CH_3$ (central C bearing CH$_3$ above and below)

(b) $CHCl_3$

(c) 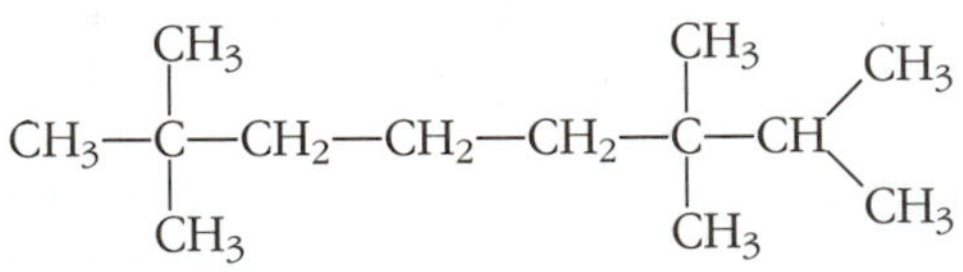 $CH_3{-}C(CH_3)_2{-}CH_2{-}CH_2{-}CH_2{-}C(CH_3)_2{-}CH(CH_3)_2$

(d) 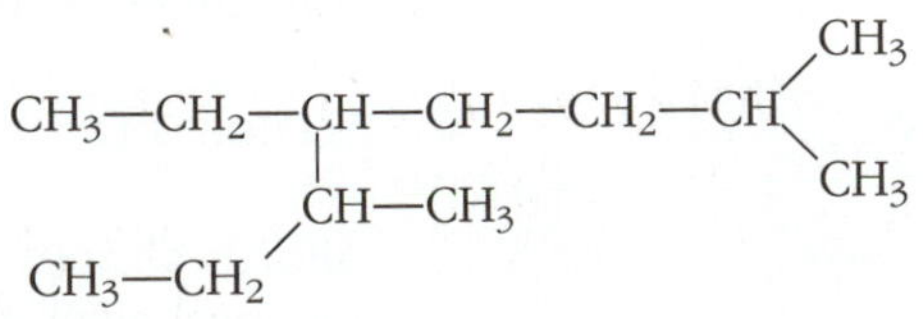 $CH_3{-}CH_2{-}CH(CH(CH_3){-}CH_2{-}CH_3){-}CH_2{-}CH_2{-}CH(CH_3)_2$

(e)
CH_3
$CH_3—CH—CH_2—CH—CH_2—CH—CH_3$
CH_3 CH_2
CH_2
$CH_3—CH—CH_3$

(f)
CH_3
$CH_3—CH—Br$

(g)
CH_3 Br
$CH_3—C—CH_2—C—CH_2—CH—CH_3$
Br Br CH_2CH_3

(h) CH_3I

(i) $CF_3—CF_3$

(j)
CH_3
$CH_3—CH_2—CH_2—CH—CH_2—C—CH_3$
CH_2 CH_3
CH_2
$CH_2—CCl_3$

1.3 Draw and name all constitutional isomers that could be produced upon monochlorination of the following alkanes:

(a)
CH_3
$CH_3—C—CH_2—CH_3$
CH_3

(b)
CH_3
$CH—CH_2CH—CH_3$
CH_3 CH_3

(c) $CH_3—CH_2—CH—CH_2—CH_3$
CH_3

(d) $CH_3—CH_2—CH—CH_2—CH_3$
CH_2
CH_3

(e)
CH_3 CH_3
$CH_3—CH—CH—CH_2—CH_2—CH—CH_3$
$CH_2—CH_3$

1.4 Draw and name all constitutional isomers that could be produced upon monobromination of each of the following alkyl halides. (Note that the product in each case will be a *dihalide*.)

(a)
CH_3
$CH_3—CH$
CH_2Br

(b)
Br
$CH_3—C—CH_2—CH—CH_3$
CH_3 CH_3

(c) $CH_3—CH_2—CH—CH_3$
Br

(d) $CH_3—CH_2—CH—CH_2Cl$
CH_3

(e)
CH_3 CH_3
$CH_3—C—CH_2—C—CH_2Br$
CH_3 CH_3

1.5 Draw three-dimensional representations (using wedges and dashed lines) of 2-chloro-2-fluoropropane in the following orientations. Use the carbon atom at position 2 as the central atom in each case (see Figure 1.4).

(a) One methyl group is to the left of the central carbon and in the plane of the paper; the other methyl group is behind the plane of the paper; the chlorine atom is above the plane of the paper; and the fluorine atom lies in the plane of the paper directly above the central carbon atom.

(b) The chlorine atom is above the plane of the paper; one methyl group is in the plane of the paper and directly above the central carbon atom; and both the fluorine atom and the other methyl group are behind the plane of the paper.

(c) One methyl group is in the plane of the paper; the other methyl group is directly behind the central carbon; and the chlorine and fluorine atoms are in front of the central carbon atom.

(d) Both methyl groups are in front of the central carbon atom and the fluorine and chlorine atoms are behind the central carbon atom.

1.6 What is the relationship between each of the following pairs of structures? Are they totally *different* molecules (i.e., which do not have the same molecular formula), are they *constitutional isomers,* or are they two drawings of the *same* compound (i.e., having the same molecular formula and the same constitution)?

(a) C_4H_{10} $\qquad$ $CH_3—CH(CH_3)—CH_2—CH_3$

(b) $CH_3—CH(CH_3)—CH_3$ (drawn with CH_3 above and below CH) $\qquad$ $CH_3—CH_2—CH_2—CH_3$

(c) C with CH_3 (up), H, H (dashed), CH_3 (wedge) $\qquad$ $CH_3—CH_2—CH_3$

(d) H—C—H with CH_2Cl (up, dashed) and Br (down, dashed), H atoms wedged $\qquad$ $CH_3—CHClBr$

(e) H (dashed)—C (dashed)—$CH_2—CH_3$, with CH_3 (wedge, up) and CH_3 (wedge, down) $\qquad$ $CH_3—CH(CH_3)—C$ with H (up), CH_3 (dashed), H (wedge)

(f) C with CH_3 (up), Br (dashed), CH_3 (dashed), Br (wedge) $\qquad$ $BrCH_2—C$ with Br (up), H (dashed), CH_3 (wedge)

(g) $BrCH_2—CH(Br)—CH(CH_3)—CH_2—CH_3$ $\qquad$ $Br—CH(CH_2—CH_3)—C(CH_3)(H)—CH_2—CH_2Br$

(h) $ClCH_2—CH_2—CH(Cl)—CH_3$ $\qquad$ $CH_3—C(CH_3)(CH_2Cl)—CH_2Cl$

(i) $Br—C(H)(Br)—CH_2—CH_3$ $\qquad$ $BrCH_2\blacktriangleright C(H)(CH_3)\blacktriangleleft Br$

(j) $(CH_3)_2C(H)Br$ $\qquad$ $CH_3—CH_2—CH_2—CH_2Br$

1.7 Consider the molecule $CH_3—CH_2—CH_3$ in terms of its valence level atomic and molecular orbitals.

(a) What is the total number of valence level atomic orbitals?

(b) What is the total number of valence level molecular orbitals?

(c) How many of these molecular orbitals are bonding orbitals? How many are antibonding?

(d) How many valence electrons are there?

(e) Compare the number of valence level electron pairs with the number of bonding molecular orbitals.

(f) What is the number of filled (doubly occupied) bonding molecular orbitals in propane? Compare this with the number of bonds in the structural formula.

1.8 Consider the preparation of lithium dimethylcuprate from methyl iodide. Write the *balanced* equation for each step in the complete reaction sequence.

1.9 Suggest methods of synthesizing the indicated compounds. All carbon atoms of the final product must be derived from the indicated starting materials.

(a) $CH_3—CH_2—CH(CH_3)—CH_2—CH_2—CH(CH_2CH_3)—CH_3$ $\quad$ from alkyl halides containing five or fewer carbon atoms

(b) $CH_3—CH_2—CH(CH_3)—CH_2—CH_3$ $\quad$ from alkyl halides containing four or fewer carbon atoms

(c) $CH_3—CH_2—CH_2—C(CH_3)_2—CH_3$ $\quad$ from alkyl halides containing four or fewer carbon atoms

1.10 Suggest methods for carrying out the following conversions. More than one step may be necessary in some cases.

(a) $CH_3-CH_2-CH(Br)-CH_3 \longrightarrow (CH_3-CH_2-CH(CH_3))_2CuLi$

(b) $CH_3-CH_2-CH_2Br \longrightarrow CH_3-CH_2-CH_3$

(c) $CH_3-C(CH_3)_2-Cl \longrightarrow CH_3-CH(CH_3)-CH_3$

(d) $CH_3-CH(CH_3)-CH_2Br \longrightarrow CH_3-CH(CH_3)-CH_2-CH_2-CH(CH_3)-CH_3$

(e) $CH_3-CH_3 \longrightarrow CH_3-CH_2Cl$

1.11 Compounds **A–H** are all different. On the basis of the reactions shown, deduce the structures of compounds **A–G**.

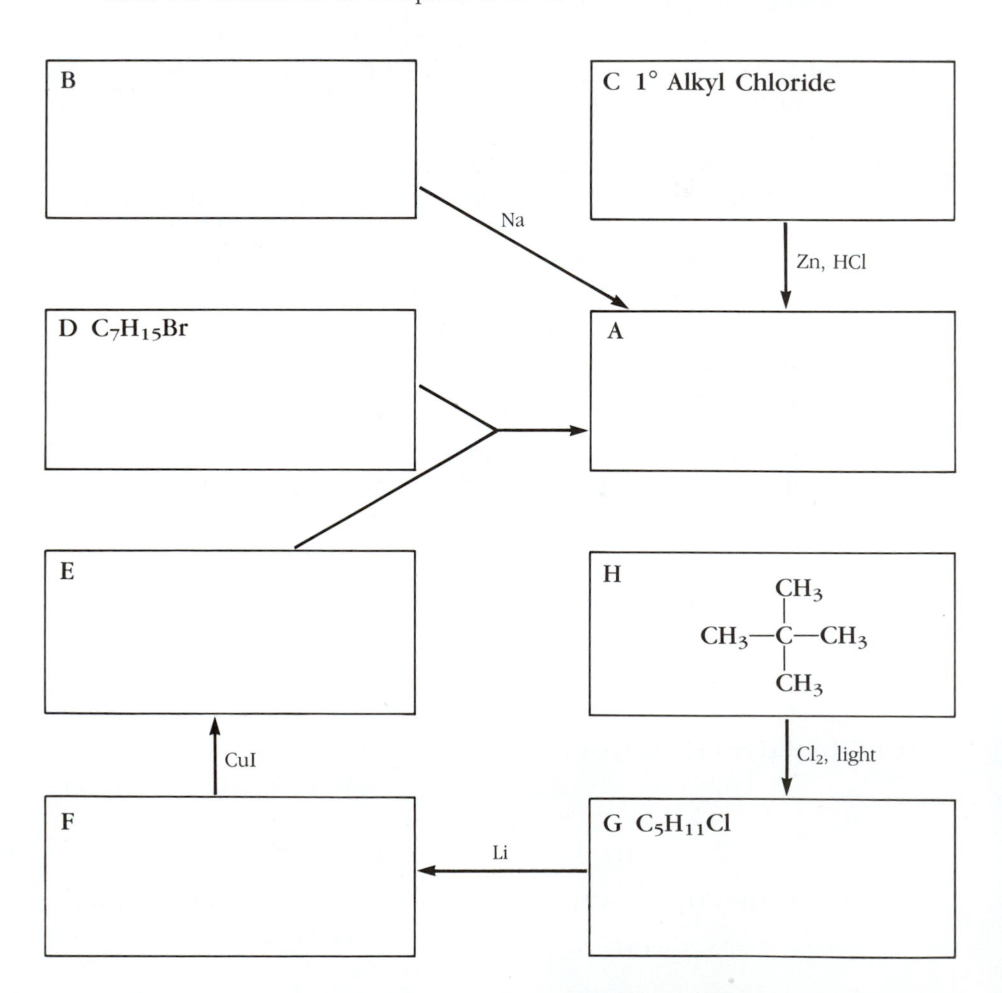

1.12 Compounds **A–G** are all different. On the basis of the reactions shown, deduce the structures of compounds **A–F**.

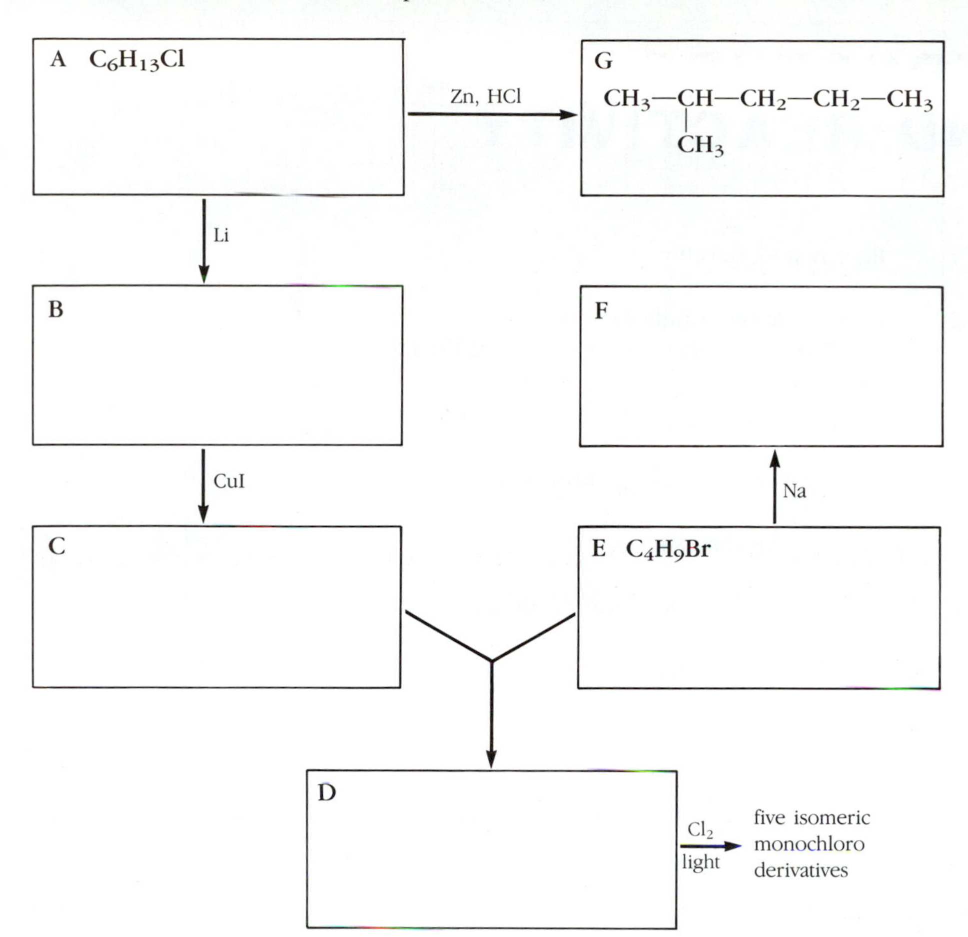

CHAPTER 2

STRUCTURE, STABILITY, AND REACTIVITY

If you mix an organic compound with some other reagent, will it react? Will it react very slowly over a period of weeks or months? Or will it react very rapidly, perhaps even explosively? Certainly you would want to be able to predict the answers to these questions *before* you mixed the reagents together, and in this chapter we will provide you with the background needed to do so.

2.1 ENERGY AND STABILITY

Your first step in understanding chemical reactivity will be to develop an understanding of the concepts of **energy** and **stability.** Despite the common use of these words, they refer to concepts which are often poorly understood. The idea of *stability* from a chemical viewpoint is basically the same as in everyday usage: it is the tendency of something to remain in its present state. Statesmen discussing an international crisis might say that the situation is *stable,* and the same term might be used by a physician evaluating an accident victim. Chemists describe a molecule or some other chemical species as stable when it tends to remain in its present state without undergoing changes. In other words, a *stable* species corresponds to one that is *unreactive* under a particular set of conditions, whereas an *unstable* species is *reactive* under those same conditions.

Sometimes we regard stability as a long-term property, as in many modern synthetic materials (for example, paints and fibers), which can last for years without fading or becoming physically weak and brittle. In other situations we will be much more concerned with a short-term view of stability. If we are carrying out a reaction, will the expected product be sufficiently stable that it can be isolated? Or will it rapidly decompose within seconds after it is formed? The preceding statements make it clear that you must view stability as a *relative* property. A molecule may be stable under one set of conditions but not others, and you are already familiar with examples of this from your experience with general chemistry. Sodium metal, for instance, can be stored for long periods of time if properly protected from the atmosphere, so under those conditions we would describe sodium as stable. But if the sodium were dropped into a pool of water, the resulting reaction would be violent. Clearly, in the presence of water sodium should be regarded as unstable. Another example is provided by the organic compound nitroglycerin, which is well known as an explosive and would usually be considered *unstable.* On the other hand it can be stored without decomposition when handled carefully, so under the appropriate conditions we might call it stable. Because stability is relative, the term is most useful when we use it to *compare* two species or two situations. We could certainly agree that sodium metal is more stable when it is immersed in oil than in water, and we would also agree that nitroglycerin is less stable than water.

As indicated by the preceding discussion, the meaning of *stability* is often rather vague. For quantitative evaluations we have much better terms, *energy* and (for comparison of two systems) *energy differences.* It is not possible to determine the total energy of any system, so we rely on energy differences because they can be measured experimentally. The word "energy" has many meanings and connotations (heat, light, electricity, nuclear energy, solar energy, etc.), but in chemistry its meaning is more restricted. The energy change of a system for some event is the algebraic sum of the heat evolved and the work done by the system. Ordinarily, experimental measurements are carried out at constant pressure (atmospheric pressure), and the heat evolved or absorbed under these conditions is defined as the change in the thermodynamic quantity **enthalpy** (ΔH). For reactions that are allowed to reach equilibrium, experimental measurements provide the difference in **free energy** (ΔG) between reactants and products. Changes in enthalpy and free energy differ by $T\Delta S$, where ΔS is the change in **entropy,** a quantity related to probability. Changes in free energy, enthalpy, and entropy are related according to the following equation:

$$\Delta G = \Delta H - T\Delta S$$

The free energy difference between reactants and products in a reaction can easily be calculated from the equilibrium constant K:

$$\Delta G = -2.3\ RT \log K$$

where R is the ideal gas constant (2.0×10^{-3} kcal mol^{-1} deg^{-1}) and T is the absolute temperature (degrees K).

At various places in this book we will use both free energy changes (ΔG) and enthalpy changes (ΔH), but frequently we will evaluate changes in the *total energy,* which is the sum of the *kinetic energy* and the *potential energy.* The division of total energy into kinetic and potential energy components is often a major point of confusion for students, so we will attempt to clarify the meanings of these terms for you.

The **kinetic energy** results from motion, and the kinetic energy of a molecule is equal to one-half the product of the mass of the molecule and the square of its velocity.

$$\text{Kinetic energy} = \tfrac{1}{2}mv^2$$

You will recall from your general chemistry studies of ideal gases that the square of the average velocity of the component molecules is directly related to the temperature. This in turn means that the average kinetic energy of the molecules varies according to the temperature, and it is convenient to remember that the molecules in a particular reaction mixture will have *more kinetic energy* (and therefore more total energy) *if the temperature is increased.* This kinetic energy need not be simply motion of the entire molecule; it can also be motion of one part of a molecule relative to the remainder (**rotational** and **vibrational** motions).

Potential energy represents the energy that is stored in a molecule as a function of its elemental composition and molecular geometry. An analogy is shown in Figure 2.1. In Figure 2.1a a large boulder is poised, *motionless,* on the top of the hill, and we have an observer standing at the bottom of the hill. Clearly the boulder has no kinetic energy at this time. However, it does have a substantial amount of potential energy because it is not in a stable position at the top of the hill. In Figure 2.1b the boulder has begun to roll down the hill, and the observer can clearly see the kinetic energy which has resulted from partial conversion of the original potential energy. No outside forces were used to move the boulder, so its *total energy* remains unchanged.

In Figure 2.1c the boulder has come to a stop. All its potential energy had been converted to kinetic energy by the time it reached the level ground at the foot of the hill, and the kinetic energy was dissipated when the boulder demolished the brick wall at the foot of the hill. For the purposes of this discussion, the boulder now has neither kinetic nor potential energy, because it is motionless and has

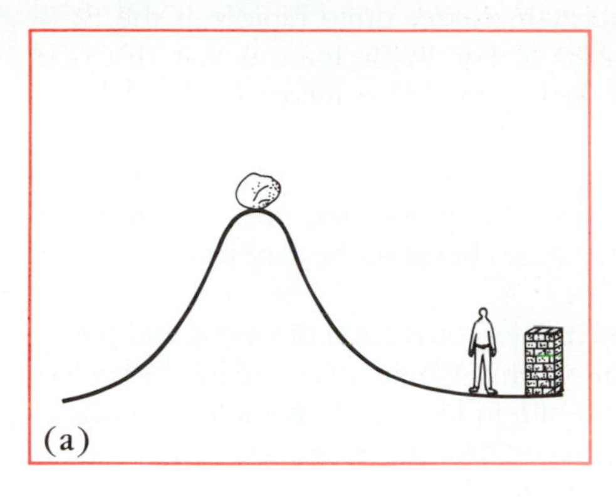
(a)

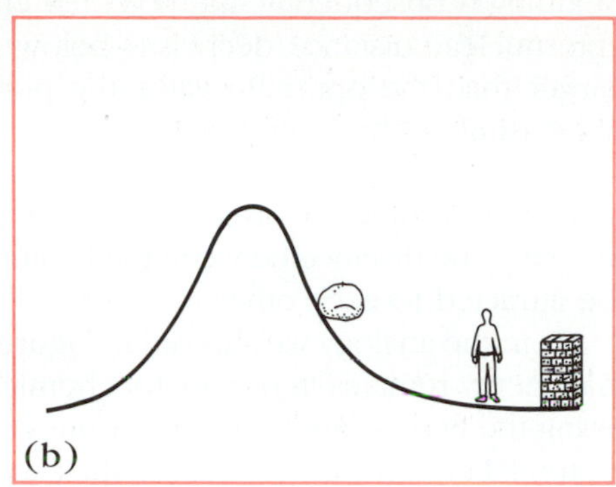
(b)

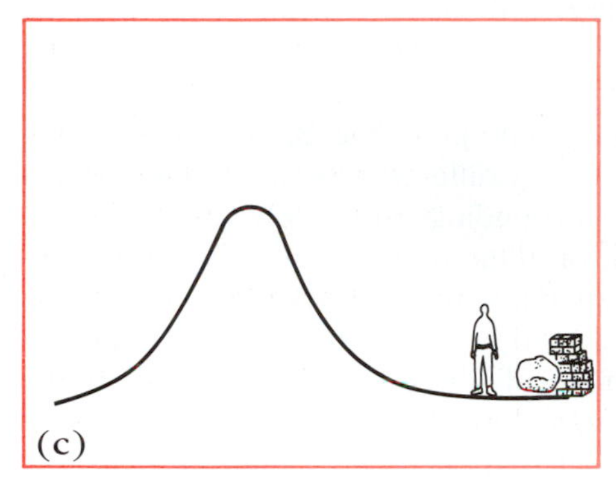
(c)

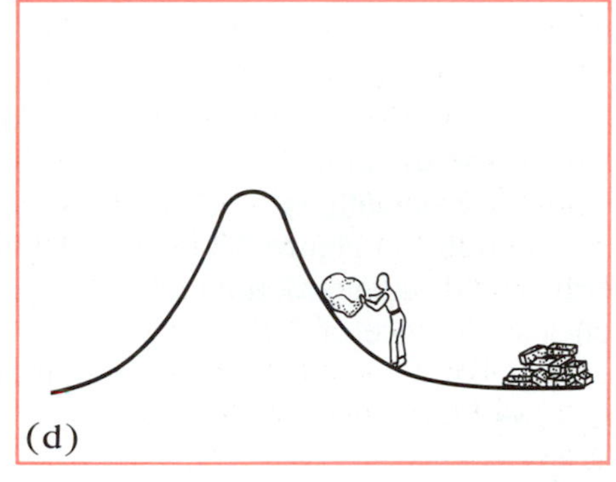
(d)

Figure 2.1 Potential Energy and Kinetic Energy.
(a) A motionless boulder (no kinetic energy) sits atop a hill. (b) The boulder rolls down the hill, and some of its original potential energy is converted to kinetic energy. (c) All the boulder's energy has been dissipated through the work of demolishing the brick wall at the foot of the hill. (d) Work is being done to move the boulder to its original position atop the hill, and this gives it potential energy again.

already rolled to the lowest possible location. In Figure 2.1d the observer is attempting to restore the boulder to its original unstable position atop the hill. The boulder is being moved very slowly, so there is little kinetic energy at this point. However, the potential energy of the boulder is increasing in proportion to the work being done by the observer to push the boulder up the hill, presumably a rather substantial effort.

The chemical description of potential energy is related to that depicted in Figure 2.1. The potential energy of a molecule depends on the relative positions of its constituent atoms, and it also reflects the amount of energy that might be released (or absorbed) if some chemical reaction were to occur. In Figure 2.2 we show how the potential energy of a bromine molecule depends on the Br-Br internuclear distance. The lowest potential energy is at a bromine-bromine interatomic distance of 2.28 Å, and it would require energy to distort that bond length in either direction. This **energy minimum** corresponds to the normal, *optimum* bond length, which is the *most stable* geometry for two bromine atoms. If the two atoms were moved closer together, unfavorable, repulsive interactions would begin to develop between the two positive nuclei and between the various electrons not involved in bonding. Consequently, the potential energy increases quite rapidly as the Br-Br internuclear distance decreases below 2.28 Å. For Br-Br internuclear distances larger than the optimum value the potential energy also increases, but it stops increasing as the interatomic distance becomes still larger. Why is this? A large increase in the Br-Br internuclear distance corresponds to stretching of the bond until it is completely broken. After that point it no longer requires any energy to increase the distance between the bromine atoms because they are too far apart to be attracted to each other.

As in the analogy we showed in Figure 2.1, interconversion of kinetic and potential energy remains important for chemical systems. We can illustrate this by considering the two isolated bromine atoms at point *b* in Figure 2.2. Point *b* has a higher potential energy than point *a*, so the two isolated bromine atoms are in an unstable arrangement relative to a molecule of Br_2. If we allow the two atoms to approach each other more closely, the potential energy will decrease steadily until the most stable geometry (a Br-Br internuclear distance of 2.28 Å) is reached. The total energy must not change, however, so the original potential energy of the two bromine atoms will be converted to kinetic energy of the Br_2 molecule. (This kinetic energy, which corresponds to the energy difference between points *a* and *b*, might eventually be released into the surroundings in the form of heat.)

According to Figure 2.2, a hypothetical bromine molecule with no kinetic energy would remain motionless at the energy minimum corresponding to an internuclear distance of 2.28 Å. If the molecule did have kinetic energy, it could be exhibited either as motion of the entire molecule through space or as vibrational motion. The latter can be described as alternating motion of the two bromine atoms toward and away from each other, much as one might expect for two balls

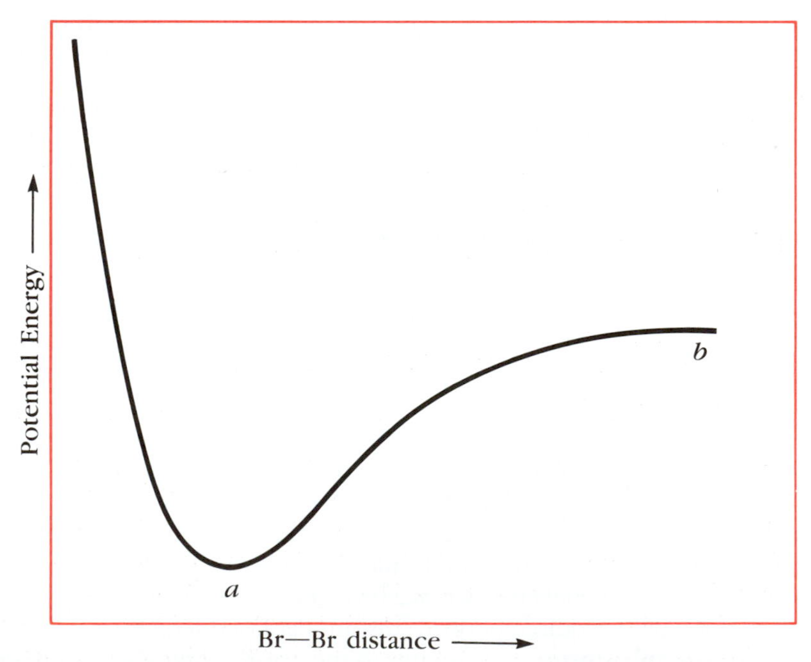

Figure 2.2 Potential Energy Diagram of the Bromine Molecule. Point *a* denotes the optimum bond length of 2.28 Å for a Br_2 molecule, and point *b* corresponds to two isolated bromine atoms which do not interact.

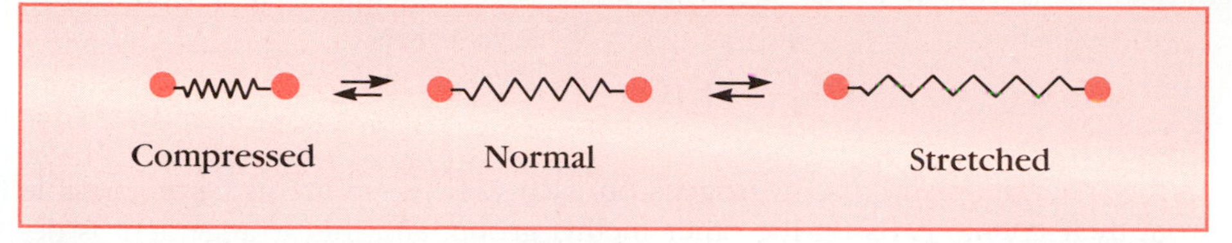

Figure 2.3 Vibrational Motion of Two Objects Connected by a Spring.

attached to the ends of a spring as illustrated in Figure 2.3. This type of vibrational motion involves distortion from the optimum bond length by stretching and compressing the bond, and it is the only kind of vibration possible for a diatomic molecule. For larger molecules a second type of vibration is possible which involves distortions of bond angles from their optimum values. You will see later that both types of vibrations are important for the study of molecular structure by infrared spectroscopy (Section 9.8).

Now consider a collision between two bromine molecules which previously had a total kinetic energy slightly larger than the energy difference between points *a* and *b* in Figure 2.2. The collision could give one of the Br_2 molecules sufficient vibrational energy to dissociate into two bromine atoms. Indeed, at sufficiently high temperatures, some bromine molecules do have enough kinetic energy for this cleavage process to occur. The dissociated bromine atoms are too unstable to isolate (and are usually produced only in extremely low concentrations), but they can nevertheless play an important part in chemical reactions in the role of **reactive intermediates.** Reactive intermediates are chemical species, produced during the intermediate stages of chemical reactions, which are too reactive (too unstable) to be isolated.

EXERCISE 2.1

Sketch a curve depicting the variation in potential energy of the dichloromethane molecule as a function of the Cl—C—Cl bond angle. Consider the optimum bond angle to be tetrahedral. Draw structures that illustrate the changes in geometry in your energy graph.

2.2 ROTATION ABOUT SINGLE BONDS: THE THREE-DIMENSIONAL STRUCTURE OF ETHANE

Our evaluation of the relationship between kinetic and potential energy does not stop with bond lengths and bond angles; we must also consider other ways in which one part of a molecule can move with respect to the remainder of the molecule. In this section we will analyze such internal motions in the ethane molecule. An infinite number of intermediate geometries could be drawn, but we will restrict our analysis of the structure of ethane to two extremes which differ by the twisting of the CH_3 groups about the central C—C bond. In the first geometry the hydrogens on each of the CH_3 groups are oriented as closely as possible to their counterparts on the other methyl group. This arrangement is described as **eclipsed:**

Eclipsed geometry of ethane

In the other geometry, the hydrogens on each CH_3 group are as far as possible from their counterparts on the other methyl group, and this arrangement is described as **staggered:**

Staggered geometry of ethane

The differences between the eclipsed and staggered arrangements are better understood if you examine a variety of different drawings, and Figure 2.4 shows three types of drawings used to display complete three-dimensional information for two tetrahedral atoms that are connected to each other.

The first drawing in Figure 2.4 (a ball-and-stick model) is probably the most easily understood but is difficult and time-consuming to draw properly. The second

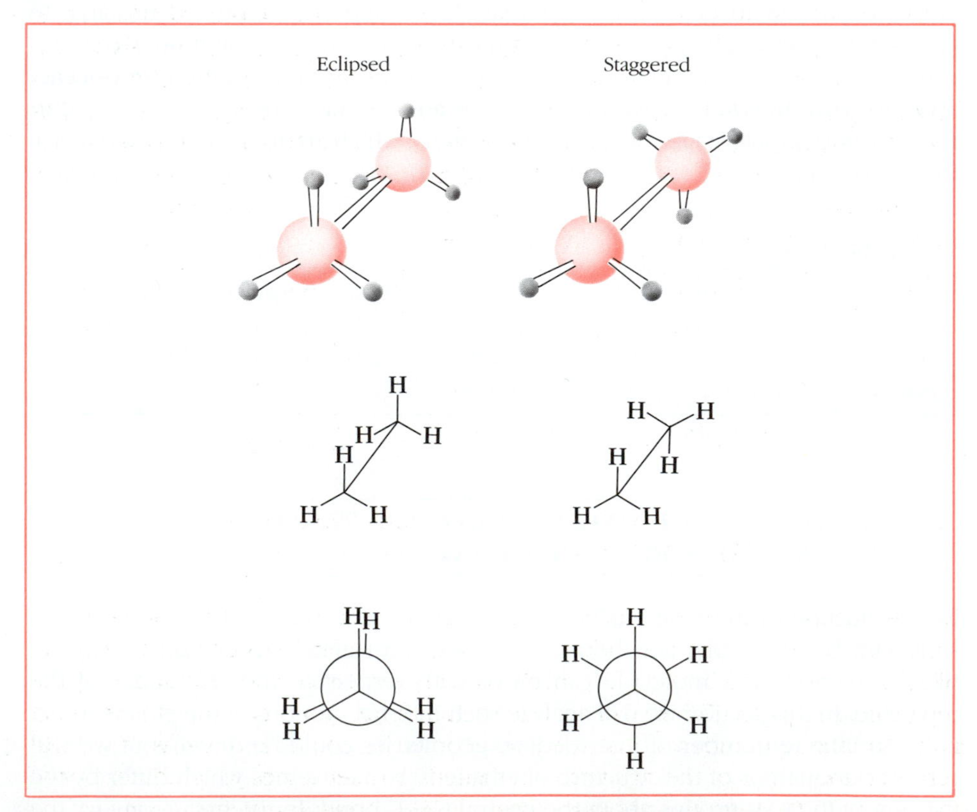

Figure 2.4 Eclipsed (*left*) and Staggered (*right*) Geometries of Ethane. (*Top*) Ball-and-stick models; (*center*) sawhorse representations; (*bottom*) Newman projections.

type of drawing, commonly referred to as a **sawhorse** representation, provides approximately the same perspective as the ball-and-stick drawing, but it is quite easy to draw.

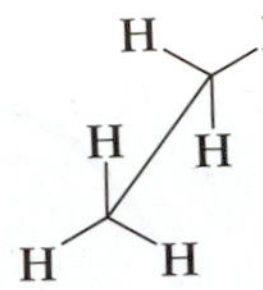

The carbon atoms are not shown explicitly in these representations but correspond to those points at which the bonds (i.e., the *lines*) to the various hydrogens intersect. Sawhorse drawings employ a convention that you must remember: *the end of the central bond that is lower on the page is closer to the viewer.* According to this convention, for both of the following two sawhorse drawings of 1,1,1-trichloroethane the CCl_3 group is closer to you than the CH_3 group.

The third type of drawing in Figure 2.4 is called a **Newman projection.** It is named after M. S. Newman of Ohio State University, who developed its use. A Newman projection places the molecule in an orientation such that the viewer's line of sight is directly along the bond between the two central atoms, and this partially obscures the rear portion of the molecule. Newman projections are sometimes more difficult to "visualize" in three dimensions, but they have the advantage of clearly defining spatial relationships between the substituents on the two central atoms. In particular the precise meaning of the terms "eclipsed" and "staggered" is readily apparent with these drawings:

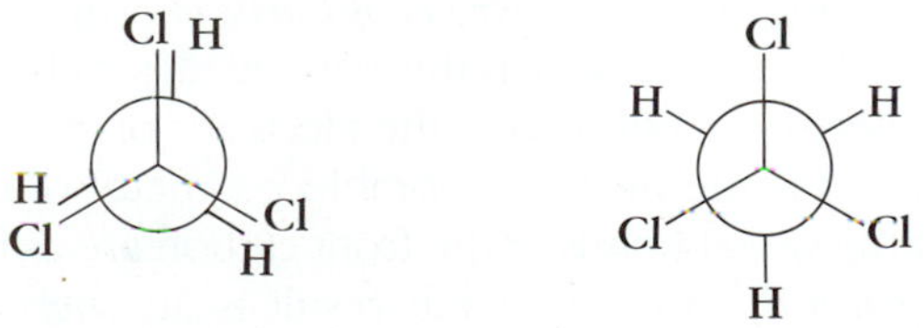

Eclipsed geometry of 1,1,1-trichloroethane

Staggered geometry of 1,1,1-trichloroethane

Notice that the two central atoms are not drawn but are implied by the bonds to substituents (i.e., to the three chlorines or three hydrogens in the preceding example). The "front" and "rear" of the molecule (relative to the two central atoms) are separated by a circle. When you draw Newman projections, the lines representing bonds to the rear carbon must not extend into this circle. In contrast, the lines representing bonds to the front carbon should all meet at a point in the center of the circle. In this way the drawings have precise three-dimensional meaning.

Since there is an infinite number of *possible* three-dimensional arrangements, ranging between the extremes of eclipsed and staggered, the next logical question is: What is the *actual* three-dimensional arrangement of ethane? The answer to

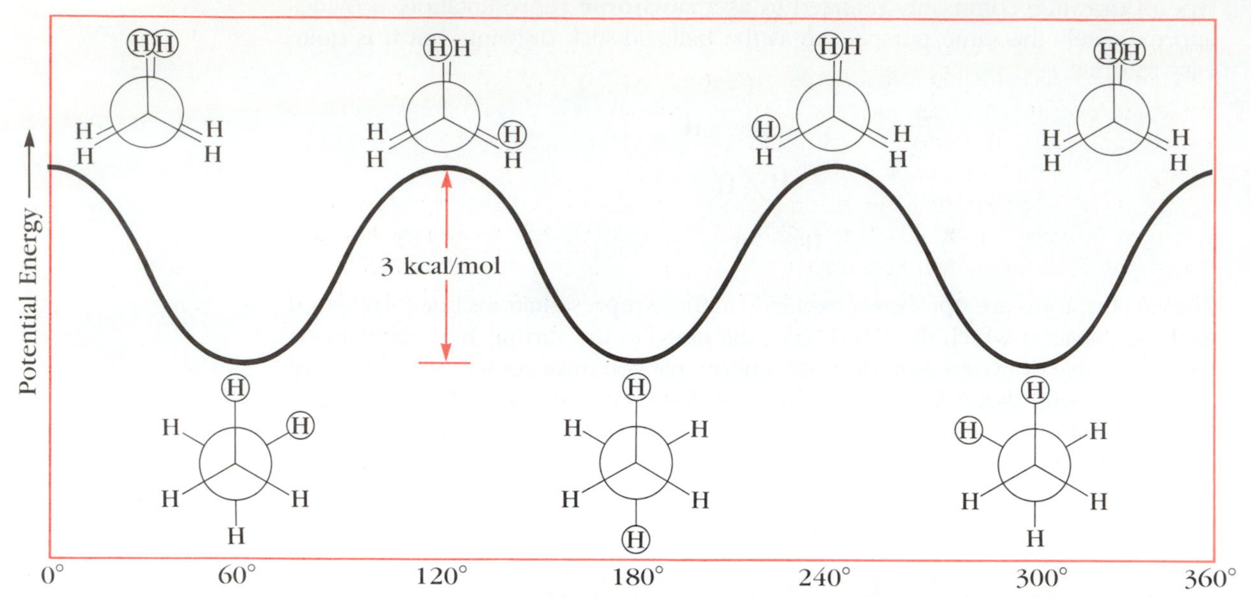

Figure 2.5 Potential Energy of Ethane as a Function of Rotation about the Central Carbon-Carbon Bond.
Two of the hydrogen atoms are circled in order to show the rotation.

such a question is best approached by means of an energy diagram, and Figure 2.5 shows a plot of the potential energy of an ethane molecule as one methyl group is rotated relative to the other.

Although the individual hydrogens on ethane cannot actually be distinguished from each other, we have labeled one hydrogen on each carbon (by drawing a circle around it) to show the rotation. The methyl group at the back rotates while the methyl group at the front remains stationary. Inspection of the energy curve clearly shows that the *staggered geometry is of lower energy and is therefore more stable than the eclipsed geometry*. Why is this so? The major effect appears to be the increased electron-electron repulsions of the electrons of the C—H bonds for the eclipsed form, and this is the most unfavorable geometry of ethane. In the staggered arrangement the C—H bonds of the front carbon are as far as possible from those on the rear carbon, and the overall result is an *energy minimum* in the potential energy curve of Figure 2.5. In contrast, the eclipsed geometry corresponds to an *energy maximum*.

Energy minima and energy maxima are very important in the study of organic chemistry. *The actual structure of any molecule will always correspond to an energy minimum.* As shown in Figure 2.5, there are several energy minima for ethane, but they are all equivalent structures with the same energy. Other compounds can have several different energy minima corresponding to structures that are not equivalent, but the geometry of any single molecule will correspond to one of those energy minima. This view is slightly oversimplified because there will be some vibrational motion of the molecular structure, but the average geometry of such a vibrating molecule will correspond to a minimum on the potential energy curve. How does this relate to the three-dimensional structure of ethane? Each of the energy minima corresponds to the staggered geometry, so we know that this is the preferred structure. Any experimental measurement we might make on ethane would afford results corresponding to the preferred, staggered geometry. None of the other points on a potential energy curve represent observable molecular structures; instead, they simply indicate the potential energy of a molecule which is proceeding along the energy curve (e.g., by rotation as in Figure 2.5) from one energy minimum to another.

The *energy difference* between an energy minimum and an energy maximum can also be very important. This difference, referred to as an **energy barrier,** corresponds to the amount of kinetic energy required for a molecule to proceed from one energy minimum to another, passing over an energy maximum in the process. Analysis of Figure 2.5 reveals that this energy barrier is about 3 kcal/mol for ethane. If an ethane molecule in the staggered arrangement had vibrational kinetic energy equivalent to 3 kcal/mol, this would be sufficient for it to proceed from one energy minimum to the next. We will discuss this more extensively in the following section.

2.3 ROTATION ABOUT SINGLE BONDS: FREE ROTATION

In the preceding section we made several statements that may appear to be in conflict. Does ethane have a static, staggered structure? Or does it have a dynamic structure, with an equilibrium between different geometries which differ by rotation about the central carbon-carbon bond? At any given instant essentially all the ethane molecules are in the minimum-energy, staggered arrangement. However, molecules are constantly undergoing collisions, which result in the exchange of kinetic energy. This in turn results in the rapid conversion of ethane molecules from one energy minimum to another by rotation about the carbon-carbon bond. An important point with regard to such rotation involves the meaning of the term "rapid." Rapid relative to what? The answer is rapid relative to experimental measurements and, of particular importance, rapid relative to typical chemical reactions.

At room temperature the *average* total energy of a molecule is approximately 1 kcal/mol, but not all molecules correspond to the average. The statistical variation of energy for a large number of molecules is known as the **Boltzmann distribution,** named after Ludwig Boltzmann, who studied the problem during the late 1800s. The energy of some molecules is greater than the average value while that of others is less. Even though the average energy of 1 kcal/mol is insufficient to pass over the rotational energy barrier of 3 kcal/mol, the Boltzmann distribution results in a small but significant fraction of ethane molecules having the necessary energy. Since molecular collisions are extremely fast (on the order of 10^{10} collisions per second for an ideal gas at STP), this kinetic energy is transferred to other ethane molecules at an extremely rapid rate. The net result, then, is that relative to most experimental observations the rate of interconversion between various staggered arrangements is fast. Because it takes place so rapidly, this process (which involves rotation about a carbon-carbon bond) has frequently been described as **free rotation.** You should be careful to avoid the trap of thinking that the rotation is "free" of energy requirements; the 3 kcal/mol requirement for ethane is quite explicit. Instead, you should remember that the interconversion of the different staggered arrangements occurs rapidly, that is, freely in the sense of time. In general, such rapid or free rotation is to be expected for nearly all carbon-carbon single bonds.

The energy barriers are usually higher than that of ethane, but the interconversion between different staggered arrangements is still very rapid.

Rotation about single bonds is obviously of great importance in studying the three-dimensional structure of organic compounds, and there is a precise word which should be used in that context, **conformation.** Conformations of a molecule are *three-dimensional arrangements which differ only by rotation about single bonds.* On this basis we can describe the two molecular geometries that we have been discussing as the *eclipsed and staggered conformations of ethane.*

2.4 CONFORMATIONAL ANALYSIS: BUTANE

As in the case for ethane, the most stable geometry about the central carbon-carbon bond of butane is a staggered arrangement. However, for butane there are two distinct possibilities: the methyl groups may be oriented in opposite directions or may be in much closer proximity. These are the two minimum energy geometries which butane can adopt, and they are known as the *anti* and *gauche* conformations, respectively.

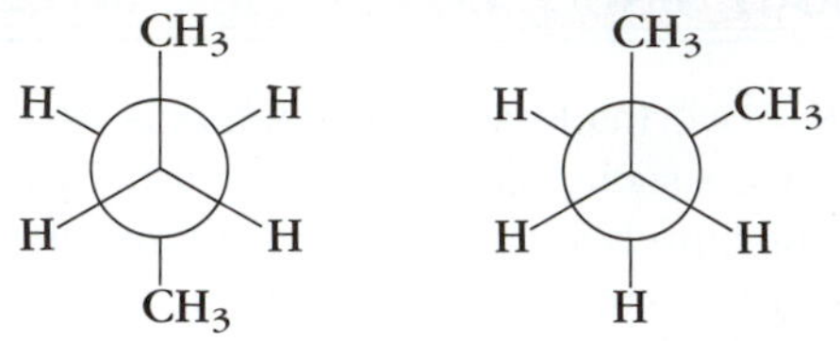

Anti

Gauche

To properly analyze the various conformations of a molecule, we must introduce a new term, **dihedral angle.** The relative orientation of two substituents, X and Y, on adjacent carbons of an alkane is described by the dihedral angle X–C–C–Y. This angle is evaluated by sighting down the carbon-carbon bond and measuring the angle between the C–X and C–Y bonds. This process corresponds quite simply to drawing a Newman projection and evaluating the angular difference between X and Y. According to this definition, we can specify the dihedral angle between the methyl groups of butane as 180° in the *anti* conformation and 60° in the *gauche* conformation:

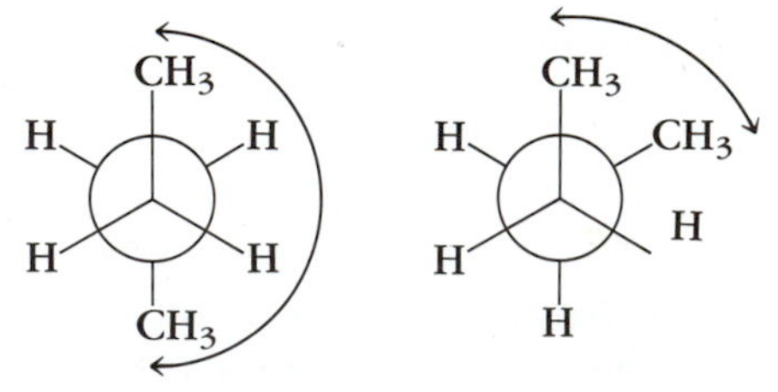

CH_3–C–C–CH_3 = 180°

CH_3–C–C–CH_3 = 60°

Both the *anti* and *gauche* conformations of butane are more stable than any of the other geometries that could result from rotation about the central carbon-carbon bond. Butane is therefore an example of a molecule which can exist in more than one stable conformation. This provides us with an opportunity to introduce *conformational analysis,* the detailed study of energetics and stability in relation to single-bond rotation. Conformational analysis will focus on the relationships between different conformations of a molecule: Which conformations correspond to energy minima? Which conformation is most stable? What are the energy differences between the different conformations? In which conformation or conformations does the molecule actually exist? How rapidly are the molecules converted from one conformation to another? Figure 2.6 shows the potential energy of butane as a function of rotation about the central carbon-carbon bond, and the information in this plot will allow us to answer all the preceding questions.

Both the *anti* and *gauche* conformations of butane are energy minima, but as we would have expected for two different geometries, their potential energies are not the same. Figure 2.6 shows that the *anti* conformation is more stable than the *gauche* conformation, and it will therefore be the predominant geometry of butane molecules. The energy difference between the two conformations is approximately 1 kcal/mol.

As with any equilibrium reaction, the equilibrium constant, *K*, is always defined so that the products appear in the numerator. If we write the equilibrium as:

$$\text{Butane}_{gauche} \overset{K}{\rightleftarrows} \text{Butane}_{anti}$$

then the equilibrium constant is defined as:

$$K = \frac{[\text{butane}_{anti}]}{[\text{butane}_{gauche}]}$$

and the energy difference is:

$$\Delta G = G_{anti} - G_{gauche}$$

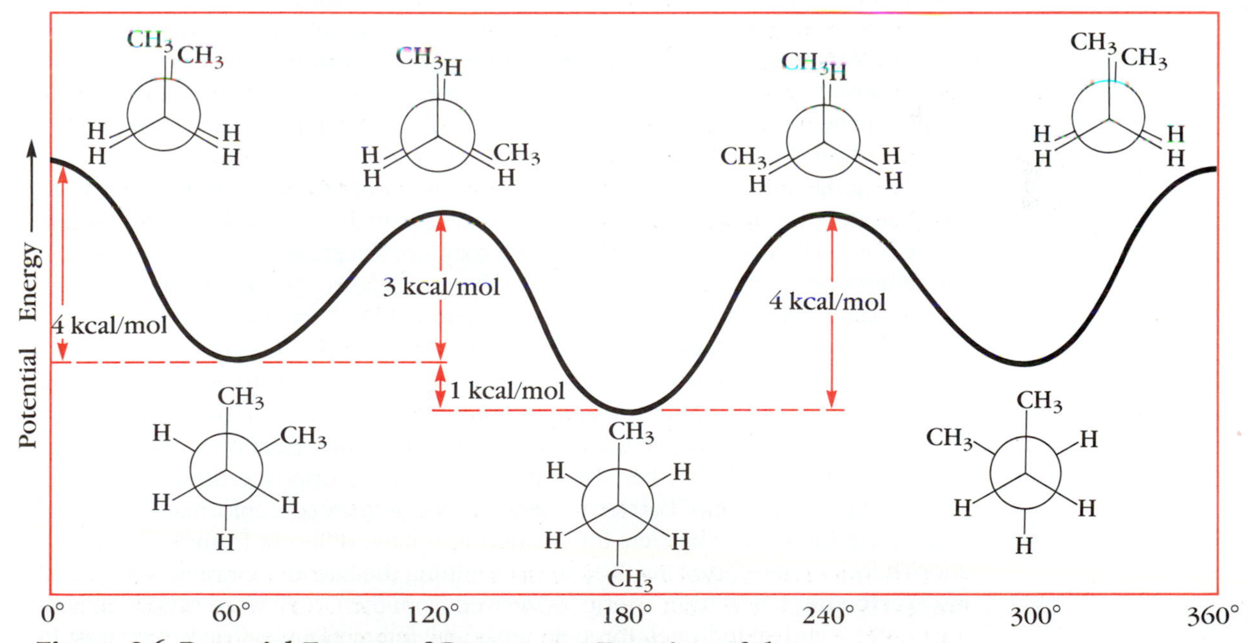

Figure 2.6 Potential Energy of Butane as a Function of the CH_3—C—C—CH_3 Dihedral Angle.

These conventions are followed throughout chemistry. Equilibrium constants are always defined in terms of [products]/[reactants], and energy differences are always defined in terms of $E_{products} - E_{reactants}$ (where E is simply a more general abbreviation for energy). For the present discussion we can also assume that energy changes correspond to changes in heat that is evolved or absorbed in a chemical process. (In other words, we are assuming that the entropy contribution to the free energy change is small.) These conventions therefore define the sign associated with exothermic and endothermic processes. When a molecule is converted to a product with less potential energy, the difference in potential energy will be converted to kinetic energy; in other words, energy will be *released* in the form of heat, and the process will be **exothermic.** The products will have less energy than the reactants, so that $E_{products} - E_{reactants}$ will have a negative sign. (A process that occurs with absorption of heat is described as **endothermic.**)

Referring back to Figure 2.6 allows you to evaluate these terms for butane. The *anti* conformation, which we have arbitrarily defined as the "product," is more stable than the *gauche* conformation, so the reaction as drawn

$$\text{Butane}_{gauche} \overset{K}{\rightleftharpoons} \text{Butane}_{anti}$$

is exothermic, and ΔE (or ΔG) is -1 kcal/mol. Notice that we can calculate the actual value of ΔE from Figure 2.6 even though we only know *relative* values of the potential energy. The sign is determined by the higher energy of the "reactant" compared with that of the "product." We can now calculate the equilibrium constant K for the interconversion of the *gauche* and *anti* conformations of butane at 25°C (298 K) by substituting into $\Delta G = -2.3\ RT \log K$.

$$-1 \text{ kcal/mol} = -2.3(2.0 \times 10^{-3} \text{ kcal/mol}^{-1} \text{ deg}^{-1})(298 \text{ deg}) \log K$$

Solution of this equation affords $\log K = 0.7$, so $K = 5$, for an *anti/gauche* ratio of 5:1. Notice that Figure 2.6 actually shows *two gauche* conformations, at dihedral angles of 60° and 300°, so the ratio of conformations (*anti*/$gauche_{60}$/$gauche_{300}$) is fully described as 5:1:1. This means that five of every seven, or approximately 70%, of the butane molecules exist in the *anti* conformation, while about 15% are found in each of the two *gauche* conformations.

The equilibrium between the *gauche* and *anti* conformations of butane is very rapid at room temperature, and consequently (as with nearly all rotations about single bonds) it falls in the category of free rotation. Nevertheless, we must evaluate the differences between the energetics of butane (Figure 2.6) and ethane (Figure 2.5). Ethane has only a single rotational barrier, but there are several different energy barriers for rotation about the central carbon-carbon bond of butane. Conversion of the *anti* conformation to either of the *gauche* conformations involves a barrier of 4 kcal/mol, but for the *gauche* conformations there are two possibilities. Conversion of the $gauche_{60}$ conformation to the *anti* conformation would require passage over a 3 kcal/mol barrier, but rotation in the opposite direction would proceed via a 4 kcal/mol barrier to generate the $gauche_{300}$ conformation.

Why are the energy barriers for rotation in butane different from those of ethane? There is a new factor involved in determining the barriers for butane, and this is a **steric effect.** The term "steric" is derived from the Greek word *stereos,* meaning "solid," and it indicates three-dimensional interactions. Steric effects result from interactions between atoms that are not bonded to each other; they are also known as **nonbonded interactions.** In contrast to the unfavorable eclipsing interactions in ethane, which result from proximity of the electrons in the eclipsed *bonds,* unfavorable nonbonded effects result from repulsions between the electron clouds surrounding *atoms* that are forced into close proximity despite being separated by at least three or four intervening bonds.

We can best evaluate the influence of steric effects in butane by comparing two of the structures in Figure 2.6, and we will focus on interactions between the two methyl groups. In the *anti* conformation, the methyl groups are oriented in opposite directions, and any interaction between them should be at a minimum. But in the eclipsed conformation at 0° (or 360°, which is identical) the interactions between the methyl groups will be at a maximum. As seen in Figure 2.7, two of the hydrogens on these methyl groups are quite close to each other. In fact, the distance between these two hydrogens is less than the sum of their van der Waals radii, and there is a repulsive interaction between them. Whenever two atoms are forced into close proximity in this manner, the repulsive interaction of their electron clouds results in an unfavorable steric interaction. Put in a different way, the unfavorable steric interaction in the conformation of butane with the methyl groups eclipsed can be likened to a physical "bumping" of the two methyl groups.

If nonbonded interactions are so important, why is the energy barrier for conversion of the *gauche*$_{60}$ conformation of butane to the *anti* conformation only 3 kcal/mol? This is the same as the ΔE for rotation about the carbon-carbon bond in ethane. Comparison of the 120° eclipsed conformation of ethane drawn in Figure 2.5 with that of butane in Figure 2.6 suggests that the methyl-hydrogen eclipsing interactions in butane should be energetically more unfavorable than are the hydrogen-hydrogen eclipsing interactions for ethane rotation. However, there is also an unfavorable methyl-methyl steric interaction in the *gauche*$_{60}$ conformation, so the minimum energy conformation (*gauche*$_{60}$) and the maximum energy geometry at 120° are *both* destabilized by unfavorable steric interactions. This results in a barrier height that is very similar to that in ethane, where these nonbonded interactions are absent.

The example of the conformational analysis of butane provides you with the basic principles that you need for conformational analysis of many other systems as well. The key features that you should remember are, *first,* that staggered confor-

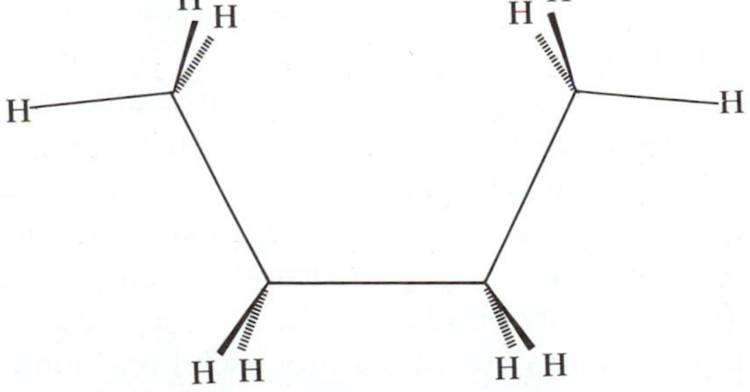

H H H H
H H
H H H H

Figure 2.7 The Conformation of Butane in Which the Methyl Groups are Eclipsed.
The close proximity of the hydrogens on the methyl groups produces an unfavorable steric interaction. This nonbonded interaction is in addition to the unfavorable interactions that exist for any eclipsed geometry.

mations are always more stable than eclipsed geometries, and *second,* unfavorable steric interactions are minimized for conformations in which large substituents are far apart.

EXERCISE 2.2

Sketch an approximate potential energy curve for 2-methylbutane. For each energy maximum and minimum, draw the corresponding structure. Use the carbon atoms designated by * to construct the central bond.

$$CH_3-\overset{*}{C}H(CH_3)-\overset{*}{C}H_2-CH_3$$

EXERCISE 2.3

Draw (using Newman projections) each of the possible staggered conformations of the following compounds. Where applicable, indicate which is the most stable and which is the least stable. Use the carbon atoms designated by * to construct the central bond.

(a) $\overset{*}{C}H_3-\overset{*}{C}(CH_3)_2-CH_3$ (b) $CH_3-\overset{*}{C}H(CH_3)-\underset{*}{C}H(CH_3)-CH_3$ (c) $CH_3-\overset{*}{C}H(CH_3)-\overset{*}{C}H_3$

2.5 REACTION MECHANISM

Two of your major goals in studying organic chemistry should be to gain an *understanding* of organic reactions and to achieve a *predictive* ability with regard to reactions that you have not previously encountered. This second goal is particularly important because you cannot possibly memorize all the different examples of organic reactions. Instead you must learn how to recognize certain relationships between structure and reactivity. This will let you predict the results of a reaction even if they are not identical to any of the examples you have already seen.

One of the best ways to reach these goals is to study reactions in terms of **reaction mechanisms.** A reaction mechanism is a step-by-step, detailed accounting of a reaction at the molecular level. The reaction mechanism provides the answers to such questions as: Which bonds are broken and formed, and in what sequence? If these points are all well understood, then you can apply them to similar reactions of other compounds.

The halogenation of alkanes provides a very useful reaction for learning about reaction mechanisms. This is certainly not a result of the synthetic value of this reaction, which is minimal, but simply because the mechanism in this instance is quite well understood. Halogenation is a fairly complex multistep process, but the individual steps are fairly straightforward. As a convenient example, we will analyze the chlorination of ethane. We will make a major simplification in our discussion by ignoring the possibility of polychlorination, which occurs readily in the case of all alkanes. On this basis our discussion can be restricted to the replacement of just one of the six equivalent hydrogen atoms of ethane. The mechanism of any subsequent reaction (i.e., polychlorination) would be the same in all essential features.

A mixture of chlorine and ethane at room temperature in the dark undergoes no reaction, but rapid chlorination takes place either at higher temperatures (a range of 300–600°C is used industrially) or in the presence of light. At room temperature virtually none of the molecules have sufficient energy for the chlorination to occur, and additional energy (in the form of heat or light) is required for *initiation* of the reaction. The light needed for the initiation must have a wavelength shorter than approximately 490 nm (the green region of the visible spectrum); this corresponds to a minimum energy of 58 kcal/mol, which is the bond dissociation energy of the chlorine molecule. These data correctly suggest that the initiation process involves cleavage of a chlorine molecule into two chlorine atoms:

$$:\ddot{\underset{..}{Cl}}—\ddot{\underset{..}{Cl}}: \longrightarrow :\ddot{\underset{..}{Cl}}\cdot + \cdot\ddot{\underset{..}{Cl}}:$$

Energy is absorbed by Cl_2, so the total energy (kinetic and potential) of the resulting chlorine atoms must be greater than that of the original chlorine molecule by a total of at least 58 kcal/mol.

The higher energy of chlorine atoms relative to chlorine molecules causes them to be highly reactive. Consideration of the electronic structure of a chlorine atom reveals that there are only seven electrons in the valence level

$$:\ddot{\underset{..}{Cl}}\cdot$$

and it is convenient to associate the instability (or high reactivity) of such a species with this **electron deficiency.** In principle a chlorine atom could achieve an electron configuration corresponding to the stable octet of the inert gas, neon, in either of two ways. Addition of an electron would yield the chloride ion

$$:\ddot{\underset{..}{Cl}}\cdot + e^- \longrightarrow \left[:\ddot{\underset{..}{Cl}}:\right]^-$$

but there are no free electrons available under the conditions of the halogenation reaction. Instead an alternative process occurs, the transfer of a *hydrogen atom* (with a single electron) from a molecule of ethane.

$$:\ddot{\underset{..}{Cl}}\cdot + H—\underset{H}{\overset{H}{|}}\!\!\!\!C—\underset{H}{\overset{H}{|}}\!\!\!\!C—H \longrightarrow :\ddot{\underset{..}{Cl}}—H + \cdot\underset{H}{\overset{H}{|}}\!\!\!\!C—\underset{H}{\overset{H}{|}}\!\!\!\!C—H$$

This hydrogen transfer produces the stable chlorine compound HCl (in which chlorine now has the stable octet of electrons in its valence shell), but there is simultaneous generation of a *carbon* species which is electron-deficient:

$$\cdot\underset{H}{\overset{H}{|}}\!\!\!\!C—\underset{H}{\overset{H}{|}}\!\!\!\!C—H$$

This species, $C_2H_5\cdot$, is an example of a **free radical,** which has as its definitive characteristic the presence of an *odd number of electrons.* In stable organic compounds, each carbon atom has four two-electron bonds, which constitute a filled valence shell with eight electrons. In contrast, the electron-deficient carbon atom of the *ethyl radical* from which the hydrogen was transferred has only seven valence-level electrons. You will recognize that the chlorine atom, with its seven valence-level electrons, is also a free radical. With a few rare exceptions free radicals are highly unstable species that cannot be isolated. In other words, they are examples of reactive intermediates (Section 2.1).

In some older chemistry books you might encounter the use of the term "radical" to indicate a portion of a stable molecule. For example, a methyl group might have been referred to as a "methyl radical," but this terminology is no longer common.

For most reactions of free radicals you can focus on the atom which has an odd number of electrons. You can expect the radical to form a product in which this atom has an even number of electrons, typically corresponding to the stable inert gas configuration. The most common way in which this happens is by transfer of an atom together with only one electron of a bonding pair. This is precisely what happens in the reaction of a chlorine atom with ethane. Transfer of a hydrogen atom from ethane to the chlorine radical produces a stable molecule of HCl and a new free radical, an ethyl radical. What is the fate of this ethyl radical? Just as we suggested for a chlorine radical, you should expect transfer of some atom to the electron-deficient carbon of the ethyl radical. One possibility would be reaction of the ethyl radical with a molecule of ethane (in exact analogy with the reaction of the chlorine atom):

$$\begin{array}{ccccc} & \mathrm{H} & & \mathrm{H} & \\ & | & & | & \\ \mathrm{H}- & \mathrm{C} & - & \mathrm{C}^{\cdot} & \\ & | & & | & \\ & \mathrm{H} & & \mathrm{H} & \end{array} + \begin{array}{ccccc} & \mathrm{H} & & \mathrm{H} & \\ & | & & | & \\ \mathrm{H}- & \mathrm{C} & - & \mathrm{C} & -\mathrm{H} \\ & | & & | & \\ & \mathrm{H} & & \mathrm{H} & \end{array} \longrightarrow \begin{array}{ccccc} & \mathrm{H} & & \mathrm{H} & \\ & | & & | & \\ \mathrm{H}- & \mathrm{C} & - & \mathrm{C} & -\mathrm{H} \\ & | & & | & \\ & \mathrm{H} & & \mathrm{H} & \end{array} + \begin{array}{ccccc} & \mathrm{H} & & \mathrm{H} & \\ & | & & | & \\ {}^{\cdot} & \mathrm{C} & - & \mathrm{C} & -\mathrm{H} \\ & | & & | & \\ & \mathrm{H} & & \mathrm{H} & \end{array}$$

But this reaction produces no net change, and is normally undetectable because the products are identical to the reactants. The final product in the monochlorination of ethane is ethyl chloride, so a carbon-chlorine bond must be formed. This suggests two possibilities for reaction of the ethyl radical, the reaction with either H—Cl or Cl—Cl. Transfer of Cl from hydrogen chloride is very unlikely because it would produce a much more unstable (i.e., higher-energy) radical, H·. The actual process which generates chloroethane is transfer of a chlorine atom from Cl_2 to the ethyl radical:

$$\begin{array}{ccccc} & \mathrm{H} & & \mathrm{H} & \\ & | & & | & \\ \mathrm{H}- & \mathrm{C} & - & \mathrm{C}^{\cdot} & \\ & | & & | & \\ & \mathrm{H} & & \mathrm{H} & \end{array} + :\ddot{\underset{..}{\mathrm{Cl}}}-\ddot{\underset{..}{\mathrm{Cl}}}: \longrightarrow \begin{array}{ccccc} & \mathrm{H} & & \mathrm{H} & \\ & | & & | & \\ \mathrm{H}- & \mathrm{C} & - & \mathrm{C} & -\mathrm{Cl} \\ & | & & | & \\ & \mathrm{H} & & \mathrm{H} & \end{array} + :\ddot{\underset{..}{\mathrm{Cl}}}^{\cdot}$$

Notice that the other product of this reaction is a chlorine atom. Although the net change has been the conversion of an ethane and a chlorine molecule into a chloroethane and a hydrogen chloride molecule, a chlorine atom is still available to continue the process.

$$\mathrm{Cl}^{\cdot} + \mathrm{C_2H_6} \longrightarrow \mathrm{HCl} + \mathrm{C_2H_5}^{\cdot}$$

$$\mathrm{C_2H_5}^{\cdot} + \mathrm{Cl_2} \longrightarrow \mathrm{C_2H_5Cl} + \mathrm{Cl}^{\cdot}$$

Such a reaction, which can continue to operate in a cyclic manner (after an initiation step), is called a **chain reaction.** The two preceding reaction steps, which define the actual chain process, are called *propagation* steps. Experimental data shows that a single initiation step (production of two chlorine atoms) can lead to the formation of a very large number of molecules of chloroethane. Would the production of only two chlorine atoms be sufficient to cause the reaction to continue until one of the reactants is completely consumed? If not, what causes the chain reaction to end? Two chlorine atoms would not be sufficient to cause complete reaction because there are processes that can interrupt, or *terminate,* the chain reaction. In the halogenation of an alkane, free radicals propagate the chain, and any process that removes a free radical from the reaction mixture without generating another free radical will terminate the chain.

The only process which *increases* the total number of free radicals is the initiation step, in which two radicals are produced from a single chlorine molecule. The logical processes which could *decrease* the total number of free radicals therefore involve combination of two radicals to give a stable molecule. In the monochlorination of ethane two radicals are generated, so three different termination steps are possible:

$$Cl^{\cdot} + Cl^{\cdot} \longrightarrow Cl_2$$

$$CH_3{-}CH_2^{\cdot} + {}^{\cdot}CH_2{-}CH_3 \longrightarrow CH_3{-}CH_2{-}CH_2{-}CH_3$$

$$CH_3{-}CH_2^{\cdot} + Cl^{\cdot} \longrightarrow CH_3{-}CH_2{-}Cl$$

The termination step involving an ethyl radical and a chlorine radical generates ethyl chloride, the same product that results from chain propagation. How much do the termination steps contribute to the total products of the overall reaction? They contribute only to a very minor extent. Reactive intermediates such as free radicals are high-energy, unstable species, which tend to react very quickly after they are formed. This means that they are present in only very low concentrations, and it is uncommon for two such intermediates to encounter each other. Consequently, termination steps occur infrequently, and the chain reaction can go through many cycles before it is finally interrupted.

EXERCISE 2.4

For the free radical chlorination of ethane, write equations for all reaction steps involved in (a) initiation, (b) propagation, and (c) termination of the reaction.

2.6 FREE RADICAL HALOGENATION OF ALKANES

Free radical halogenation is not restricted to the chlorination of ethane but is a general reaction of alkanes. Despite the severe limitations of halogenation as a preparative method (i.e., polyhalogenation and formation of a mixture of isomers) both chlorination and bromination of alkanes are important industrial processes. Direct halogenations using fluorine and iodine are not important because fluorine is too reactive (frequently reacting explosively with organic compounds) and io-

Initiation:

$$X_2 \xrightarrow{\text{heat or light}} 2\,X^{\bullet}$$

Propagation:

$$X^{\bullet} + R\text{—}H \longrightarrow H\text{—}X + R^{\bullet}$$

$$R^{\bullet} + X_2 \longrightarrow R\text{—}X + X^{\bullet}$$

Termination:

$$X^{\bullet} + X^{\bullet} \longrightarrow X\text{—}X$$

$$R^{\bullet} + R^{\bullet} \longrightarrow R\text{—}R$$

$$R^{\bullet} + X^{\bullet} \longrightarrow R\text{—}X$$

Figure 2.8 Free Radical Halogenation of Alkanes (X = Cl, Br). This is a chain reaction, and it proceeds by initiation, propagation, and termination steps.

dine is so unreactive that the reaction does not work. Even chlorine and bromine exhibit differences in their reactivities, and we will discuss this later.

The reaction mechanism for either bromination or chlorination of an alkane is basically the same as that shown for chlorination of ethane. The overall reaction of an alkane (R—H) to give the alkyl halide (R—X) proceeds via a chain reaction, as summarized in Figure 2.8. A halogen atom (X·), produced in the initiation step by the action of heat or light, abstracts a hydrogen from the alkane to produce an alkyl radical (R·). Subsequent reaction of the alkyl free radical with a molecule of the halogen (X_2) produces the alkyl halide and regenerates a halogen atom. The chain reaction continues until two radicals encounter each other and combine to form a stable product.

Up to this point we have restricted our discussion of free radical halogenation to ethane and the general structure R—H, both of which have only a single kind of hydrogen. What should you expect if there are several kinds of hydrogen on the original alkane? Certainly several monohalogenated products *could* be formed, but would each of these be produced in significant amounts? The following chlorination reactions of some simple alkanes show that substantial amounts of each possible isomer are indeed produced. In each case we have only shown the monochloro products, and the percentages refer not to *yield* (i.e., the percentage of the reactant that is converted to product) but to **product distribution.** The product distribution tells you the relative amounts of each of the products that you are comparing (regardless of the actual yield), and by definition it must total 100 percent.

$$\underset{\text{Propane}}{CH_3\text{—}CH_2\text{—}CH_3} \xrightarrow[300°C]{Cl_2} \underset{48\%}{CH_3\text{—}CH_2\text{—}\underset{|}{\overset{}{CH_2}}\text{—}Cl} + \underset{52\%}{CH_3\text{—}\underset{\substack{|\\Cl}}{CH}\text{—}CH_3}$$

$$\underset{\text{Butane}}{CH_3-CH_2-CH_2-CH_3} \xrightarrow[300°C]{Cl_2} \underset{30\%}{CH_3-CH_2-CH_2-\underset{Cl}{\underset{|}{CH_2}}} + \underset{70\%}{CH_3-CH_2-\underset{Cl}{\underset{|}{CH}}-CH_3}$$

$$\underset{\text{2-Methylpropane}}{CH_3-\overset{CH_3}{\overset{|}{CH}}-CH_3} \xrightarrow[300°C]{Cl_2} \underset{70\%}{CH_3-\overset{CH_3}{\overset{|}{CH}}-\underset{Cl}{\underset{|}{CH_2}}} + \underset{30\%}{CH_3-\overset{CH_3}{\overset{|}{\underset{Cl}{\underset{|}{C}}}}-CH_3}$$

The preceding reactions demonstrate that substantial amounts of both possible isomers are formed in each case, but you should find the relative amounts to be surprising. On a simple statistical basis the yield of primary alkyl halide always seems to be "too low." For example, propane has two methyl groups and only a single CH_2 (methylene) group, so you might have expected a 2:1 ratio of primary to secondary alkyl chloride. Alternatively, you might have predicted a 3:1 ratio, because propane has six primary but only two secondary hydrogen atoms that could be abstracted by Cl· to generate the intermediate alkyl radical. Yet less than half of the actual product mixture is the primary product, 1-chloropropane.

Similar behavior is observed in the chlorination of butane, where the secondary product predominates. The low preference for chlorination at a primary carbon is even more dramatic in the case of 2-methylpropane. Only twice as much primary as tertiary alkyl chloride is formed despite a 9:1 predominance of primary hydrogens in the starting alkane. Clearly some factor other than the *number* of available sites for reaction plays an important role in determining the amounts of the different products formed.

Studies have shown that the different alkyl halides do not interconvert under the conditions of free radical halogenation, so once a particular molecule is produced it does not **isomerize** (interconvert from one isomer to another). This has an important consequence: the faster a particular isomer is generated, the greater will be the percentage of that isomer in the final product mixture. In other words, the amount of each product obtained in the reaction is governed by the *rate* at which it is produced. We describe such reactions as being under **kinetic control.** In subsequent chapters you will encounter other reactions in which an equilibrium is established between the reaction products. We will describe that type of reaction as being under **thermodynamic control.**

Returning to the chlorination of alkanes, we can conclude that the small fraction of primary halides in the products is a direct consequence of a lower rate of formation. This poses the new question: What is responsible for the different rates of formation? We will present a more detailed discussion of reaction rates in Section 2.9, but for the present we will argue qualitatively that the rate at which a particular product is formed depends upon the *ease* with which it is formed. Certainly it makes sense to say that the product which can be formed most easily will be obtained in greatest yield. Putting this into chemical terminology, "ease" must refer to an *energy* requirement. More specifically, in the case of halogenation we are concerned with the differences in potential energy of the different free radicals that could be formed from the starting alkane. The higher the potential energy of a particular radical, the more difficult will be its formation. In other words, the more stable of two alternative radical intermediates will be formed at a greater rate.

We can discuss this problem conveniently by returning to the chlorination reaction of butane. The key step (the energetically difficult step) is abstraction of a hydrogen atom by a chlorine atom, and either the 1-butyl radical or the 2-butyl radical could be formed:

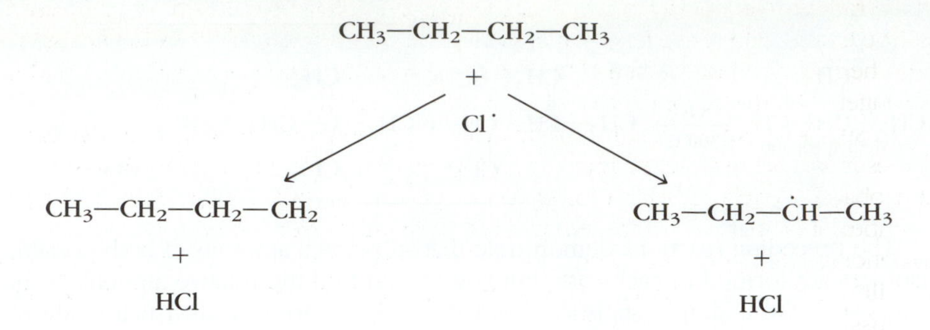

Once either of these radicals is generated, the structure of the final product has been completely determined. If the radical center is C-1, then the product will be 1-chlorobutane; if the radical center is C-2, then the product will be 2-chlorobutane.

From the predominance of 2-chlorobutane in the reaction products, we know that the abstraction of a secondary hydrogen atom is preferred over the abstraction of a primary hydrogen atom. Now we can conclude that starting from butane, formation of the secondary 2-butyl radical requires less energy (i.e., is easier) than formation of the primary 1-butyl radical. This is illustrated in Figure 2.9, which

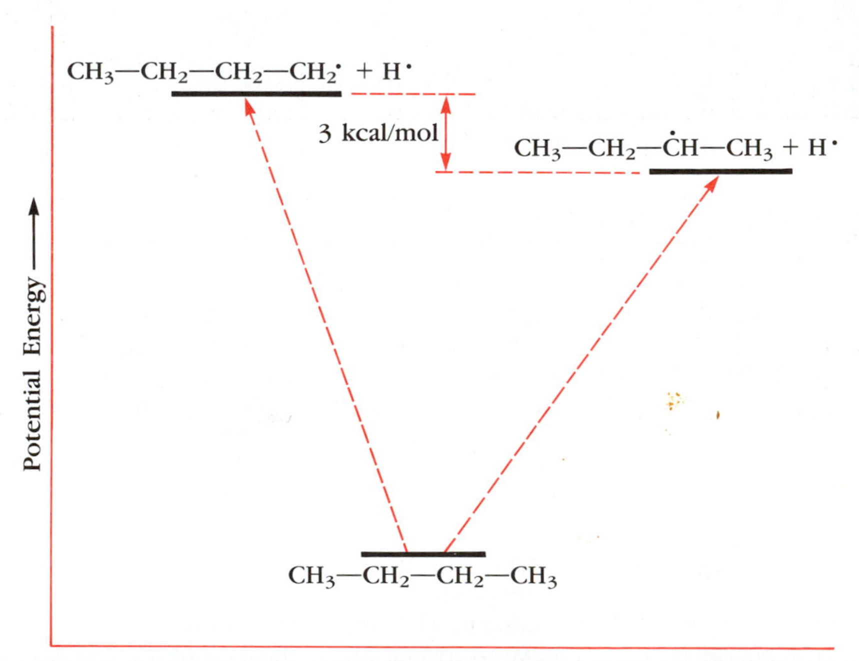

Figure 2.9 Comparison of the Stabilities of the Primary and Secondary Free Radicals Derived from Butane.
In the hypothetical dissociation of butane into a hydrogen atom and a butyl radical, less energy is required to produce the secondary radical. The energy contribution of the hydrogen atom is the same in both cases, so 3 kcal/mol is the difference in stability between the 1-butyl and 2-butyl radicals.

compares the purely hypothetical processes of breaking either the primary or the secondary carbon-hydrogen bond to form a hydrogen atom and the corresponding butyl radical. The energy of a hydrogen atom, H·, must be identical in both cases, so the energy difference between the primary and secondary radicals is 3 kcal/mol. Another way to view the situation would be to say that the primary C—H bond is stronger than the secondary C—H bond by 3 kcal/mol.

In summary, any prediction of product distributions in the halogenation of alkanes must take two factors into account. Using butane as an example, first there is a probability effect, in which formation of 1-chlorobutane is favored by the larger number of primary as compared with secondary hydrogen atoms. Second, there is an energy effect, in which formation of 2-chlorobutane is favored by the greater stability of the 2-butyl radical. The relative importance of these two factors will be discussed further in Section 2.10.

EXERCISE 2.5

Chlorination of propane gives almost equal amounts of the two possible monochloro products. Account for the *difference* in the product distributions observed for chlorination of propane and of butane in terms of the reasoning used in this section.

EXERCISE 2.6

Chlorination of 2-methylpropane affords a mixture of products that is 30% 2-chloro-2-methylpropane and 70% 1-chloro-2-methylpropane. On the basis of reasoning analogous to that used for the case of butane, what can you conclude about the relative stability of primary and tertiary free radicals?

2.7 REACTIVE INTERMEDIATES

Free radicals are not the only kind of unstable organic species that can be formed as intermediates in chemical reactions. Altogether there are four important types of reactive intermediates that you will encounter in your study of organic chemistry, and they are illustrated in Figure 2.10 for hypothetical reactions of methane. These are *not* the actual reactions that would be used to generate the various species; they merely provide a common basis for comparison.

$$CH_4 \longrightarrow H^{\cdot} + {}^{\cdot}CH_3 \quad \textit{a free radical}$$

$$CH_4 \longrightarrow H{:}^{-} + CH_3^{+} \quad \textit{a carbocation}$$

$$CH_4 \longrightarrow H^{+} + {:}CH_3^{-} \quad \textit{a carbanion}$$

$$CH_4 \longrightarrow H_2 + {:}CH_2 \quad \textit{a carbene (methylene)}$$

Figure 2.10 Hypothetical Reactions of Methane Illustrating Different Types of Reactive Intermediates.

The first two species, the methyl radical and the methyl cation, are both *electron deficient:* neither has an octet of electrons in the valence shell of the carbon atom. The third species, the methyl anion, is *electron rich:* the carbon atom has the stable inert gas configuration but also carries a negative charge. The fourth species, a carbene, shares electronic properties of both carbocations and carbanions. A carbene has only six electrons in the valence shell of carbon, and it is therefore electron deficient; it is, however, an uncharged species. The carbon atom also carries an unshared pair of electrons, a feature it shares with the electron-rich carbanion.

Carbocations, free radicals, and carbanions differ successively by a single electron. Thus addition of a single electron to the methyl cation would afford the methyl radical, and addition of another electron would yield the methyl anion:

$$CH_3^+ \xrightarrow{e^-} CH_3\cdot \xrightarrow{e^-} :CH_3^-$$

We used simple methane derivatives in Figure 2.10 to illustrate the different kinds of reactive intermediates, but most of our discussions in this text will involve more complex structures. How does the stability of a reactive intermediate depend on its structure? This important question will play an important role in your study of organic chemistry. The stability of any of these reactive intermediates is highly dependent upon the nature of the substituents on the **reactive center,** the carbon atom which is either electron-deficient or electron-rich. For the present time we will only consider *alkyl substitution,* i.e., replacement by an alkyl group of one or more of the hydrogen atoms of the intermediates shown in Figure 2.10. In addition, we will restrict our discussion to free radicals and carbocations.

The electronegativities of carbon and hydrogen are about the same, but carbon is surrounded by a larger number of electrons. The higher electron density around the carbon atom of an alkyl substituent makes it better able than a hydrogen to stabilize an electron-deficient carbon. Each replacement of a hydrogen in CH_3^+ or $CH_3\cdot$ by an alkyl group results in greater stability for that intermediate; this is summarized in Figure 2.11.

As long as there is no equilibrium between the different reactive intermediates (or their products), we can extend this relationship between structure and stability of electron-deficient intermediates so that you can predict the products of a reaction. If a reactant could give several different intermediates, we can predict which will be formed on the basis of their relative stabilities. Whichever intermediate is

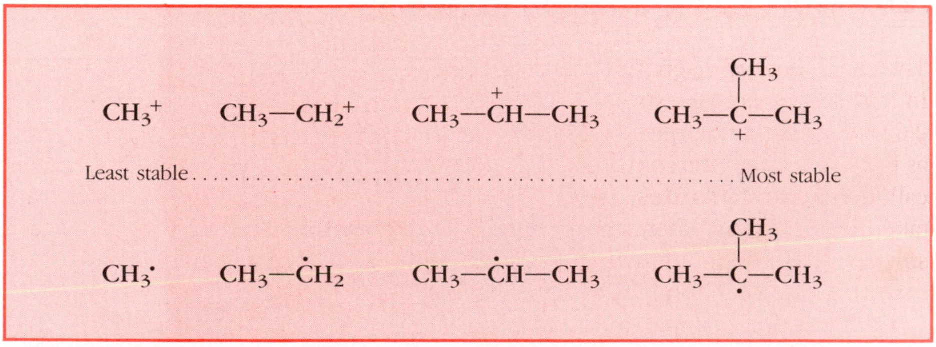

$$CH_3^+ \qquad CH_3{-}CH_2^+ \qquad CH_3{-}\overset{+}{C}H{-}CH_3 \qquad CH_3{-}\underset{+}{\overset{CH_3}{\overset{|}{C}}}{-}CH_3$$

Least stable .... Most stable

$$CH_3\cdot \qquad CH_3{-}\dot{C}H_2 \qquad CH_3{-}\dot{C}H{-}CH_3 \qquad CH_3{-}\underset{\bullet}{\overset{CH_3}{\overset{|}{C}}}{-}CH_3$$

Figure 2.11 The Effect of Alkyl Group Substitution on the Stability of Free Radicals and Carbocations.

Least stable			Most stable
CH_3^+ or $CH_3^{\cdot}$	1° R^+ or $R^{\cdot}$	2° R^+ or $R^{\cdot}$	3° R^+ or $R^{\cdot}$
Least easily formed			Most easily formed

Figure 2.12 Correlation of Stability and Rate (Ease) of Formation of Electron-Deficient Intermediates.
As long as the intermediates and their products do not interconvert, the major product of a reaction will result from the intermediate that is formed most rapidly (normally, the most stable intermediate).

most stable will be formed most easily (i.e., most rapidly), and the major product of the reaction will be the one formed from this intermediate. This now completes the picture for the chlorination reactions that we showed in the preceding section. In general, *the ease of formation of a particular radical corresponds directly to its stability*. Hence the chlorination of 2-methylpropane proceeds by abstraction of a tertiary hydrogen atom to the extent of 30% (despite a 9:1 statistical advantage for attack on a primary hydrogen) to give the more stable tertiary free radical and subsequently the tertiary alkyl halide. The correlation between stability and rate of formation of reactive intermediates is summarized in Figure 2.12.

In Figure 2.12 primary, secondary, and tertiary carbons are labeled 1°, 2°, and 3°, respectively. Despite the presence of the degree sign, you should read these labels as "primary, secondary, tertiary," not "first degree, second degree, third degree."

EXERCISE 2.7

Two different radicals could be formed by hydrogen abstraction in the chlorination of 3,3-diethylpentane. Which radical would be more stable? Which product should be formed in higher yield?

2.8 THE REACTION COORDINATE

The use of energy diagrams to illustrate changes in energy as a function of changes in a molecule can be extremely useful. We used this technique in Figures 2.5 and 2.6 to discuss the energies of different conformations of ethane and butane as well as the barriers to interconversion of the different conformations. Similar diagrams, called **reaction profiles,** are very useful in describing the energy changes that take place during a chemical reaction. The total potential energy of the reacting species is plotted as a function of the **reaction coordinate,** which is some measure (e.g., an interatomic distance) of the reaction progress.

Consider, for example, the step in the chlorination of methane in which a chlorine atom abstracts a hydrogen atom.

$$Cl^{\cdot} + CH_4 \longrightarrow CH_3^{\cdot} + HCl$$

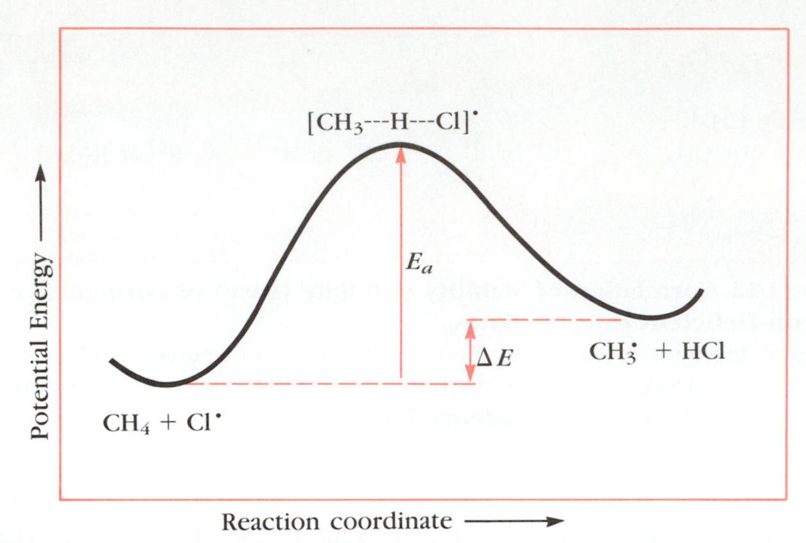

Figure 2.13 Reaction Profile for the Abstraction of a Hydrogen Atom from Methane by a Chlorine Atom.
For this process E_a = 4 kcal/mol and ΔE = 1 kcal/mol.

The energetics of this process are known to proceed as shown in Figure 2.13.

The overall process shown in the figure is endothermic, and energy must be supplied to the reactants if the reaction is to proceed. The total energy change during this reaction step (shown in Figure 2.13 as ΔE) is 1 kcal/mol, but the reactants need a greater energy than ΔE in order to proceed over the energy barrier (4 kcal/mol) between them and the products. This is designated as E_a in the figure, and it is called the **activation energy.** *The activation energy of a reaction is that amount of energy required by the reactants in order for reaction to occur.*

The activation energy of a reaction step is equal to the difference between the energy of the reactants and that of the **transition state.** *The transition state of a reaction is a point on the reaction coordinate which is an energy maximum.* As an energy maximum it has only a fleeting existence, simply representing a location (the point of highest energy) along the reaction coordinate through which the reactants must pass as they are converted to products. Because the transition state has an extremely short lifetime, its structure cannot be observed experimentally and cannot be determined exactly. However, for the abstraction of a hydrogen atom from methane by a chlorine atom we know that the transition state must have the structural features shown in Figure 2.13. The bonds to be broken in the reaction step must be *partially* broken and those to be formed must be *partially* formed:

$$[CH_3\text{----}H\text{----}Cl]^{\cdot}$$

Virtually all chemical reactions have some activation energy, regardless of whether the overall process is endothermic or exothermic. We see this again in Figure 2.14, which shows the abstraction by a chlorine atom of the tertiary hydrogen of 2-methylpropane. The overall reaction step is exothermic, but there is still a small activation energy. Be careful not to think that the *tert*-butyl radical is more

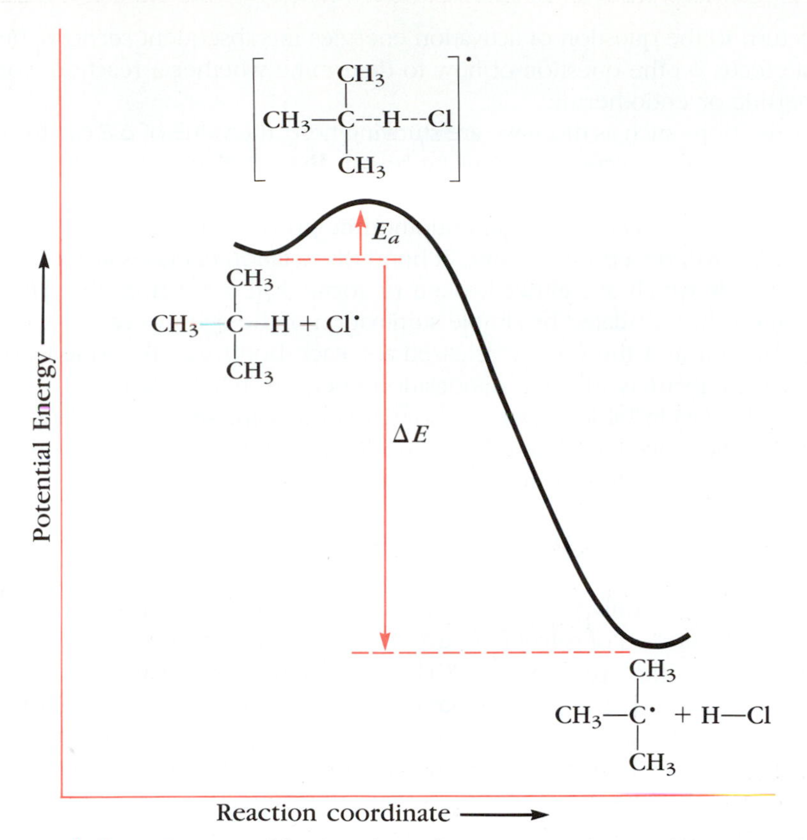

Figure 2.14 Reaction Profile for the Abstraction of a Hydrogen from 2-Methylpropane by a Chlorine Atom.
For this process $E_a = 0.4$ kcal/mol and $\Delta E = -11$ kcal/mol.

stable than the starting alkane, as Figure 2.14 merely shows that the *total* energy of a *tert*-butyl radical and an HCl molecule is less than the *total* energy of a chlorine radical and a 2-methylpropane molecule.

The origin of the activation energy is not always readily apparent, but we can analyze it conveniently in terms of the structure of the transition state. For Figures 2.13 and 2.14 the transition states are closely related, and, using the abbreviation R for the alkyl group, we can draw them as:

$$[\text{R----H----Cl}]^{\cdot}$$

The combination of a partially broken bond (R---H) and a partially formed bond (H---Cl) somehow results in a higher energy for the transition state than for *either* the reactants or the products. In any case, the energy profiles shown in Figures 2.13 and 2.14 are typical of most one-step processes.

Now that we have shown you how the reaction profile summarizes the energy changes that occur during a reaction, there are some important questions to be asked. How can you predict whether a reaction step will be exothermic or endothermic? How can you predict the activation energy of a particular reaction step? What structural features govern the variation in these different energy changes? We will return to the question of activation energies in subsequent sections, but now we can focus on the question of how to determine whether a reaction would be exothermic or endothermic.

For reactions such as those we are studying here, the value of ΔE can frequently be estimated quite accurately by using **bond dissociation energies.** The bond dissociation energy is the energy cost of cleaving a particular bond to produce two free radicals. Such cleavage is an endothermic process, so the bond dissociation energy, ΔE, will be a positive value. If bond dissociation energies are available for all the bonds which are either broken or formed in a reaction, the ΔE for the reaction can be calculated by simple summation of the energy cost for each bond that is broken and the energy released for each bond that is formed. We have summarized a variety of bond dissociation energies in Table 2.1.

Using the data in Table 2.1, we can now carry out some simple calculations of ΔE for the reactions we have been discussing. What is the energy change for abstraction of a hydrogen atom from methane by a chlorine atom?

$$CH_4 + Cl^{\cdot} \longrightarrow CH_3^{\cdot} + H{-}Cl$$

We are concerned only with the energy difference between the initial and final states, so this method of calculating ΔE does not require that you know the actual mechanism of the reaction. A CH_3—H bond has been broken ($\Delta E =$ +104 kcal/mol), and a Cl—H bond has been formed ($\Delta E = -103$ kcal/mol), so the net change is +1 kcal/mol (and the process is endothermic). The greater stability of the tertiary *tert*-butyl radical in comparison with the methyl radical is evident from the analogous hydrogen abstraction:

$$(CH_3)_3C{-}H + Cl^{\cdot} \longrightarrow (CH_3)_3C^{\cdot} + H{-}Cl$$

TABLE 2.1 Bond Dissociation Energies, kcal/mol ($X{-}Y \rightarrow X^{\cdot} + Y^{\cdot}$).

Bond	ΔE	Bond	ΔE	Bond	ΔE	Bond	ΔE
CH_3—H	104	1° R—H	98	2° R—H	95	3° R—H	92
CH_3—F	109	1° R—F	107	2° R—F	106	3° R—F	108
CH_3—Cl	84	1° R—Cl	81	2° R—Cl	80	3° R—Cl	80
CH_3—Br	70	1° R—Br	68	2° R—Br	68	3° R—Br	66
CH_3—I	56	1° R—I	53	2° R—I	53	3° R—I	50
F—F	38	Cl—Cl	58	Br—Br	46	I—I	36
H—F	136	H—Cl	103	H—Br	88	H—I	71

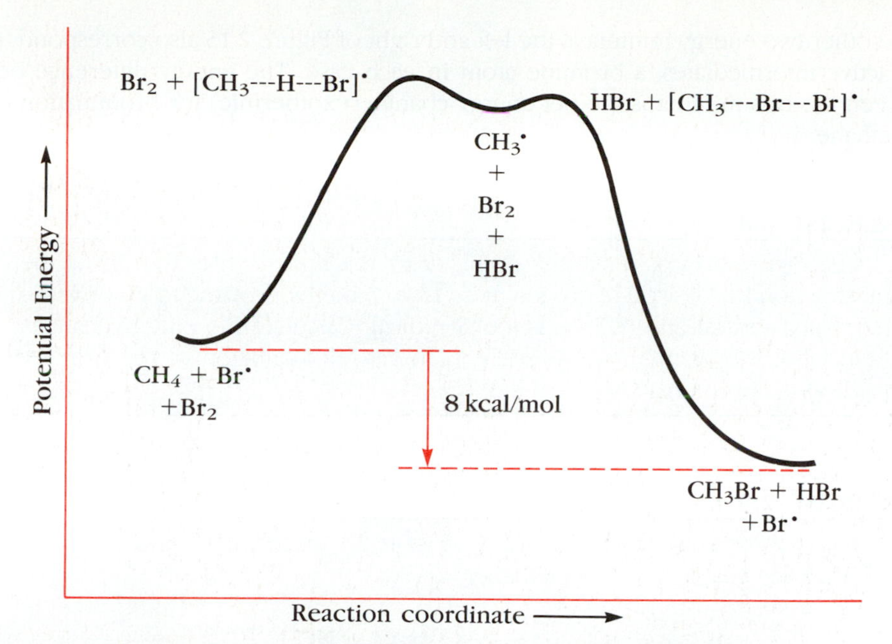

Figure 2.15 Reaction Profile for the Bromination of Methane. Only that part of the total reaction which involves the actual replacement of a hydrogen atom with a bromine atom is shown.

Again a C—H bond has been broken, but now it is a tertiary C—H bond (ΔE = +92 kcal/mol). A Cl—H bond has also been formed (ΔE = −103 kcal/mol), so the net change is −11 kcal/mol, and this process is exothermic. These values of ΔE were shown graphically in Figures 2.13 and 2.14.

Many reactions involve several steps, and this is clearly the case with free radical halogenation of alkanes, in which formation of the alkyl radical is only the first step. The reaction profiles for such multistep processes typically exhibit several energy minima and maxima, as illustrated in Figure 2.15 for the chain reaction process in the bromination of methane:

$$Br^{\cdot} + CH_4 + Br_2 \longrightarrow CH_3Br + HBr + Br^{\cdot}$$

The first transition state in Figure 2.15 involves abstraction of a hydrogen atom from the alkane by a bromine atom, although we have also included a molecule of Br_2 in the figure to maintain the same total number of atoms throughout the reaction.

$$[CH_3\text{----}H\text{----}Br]^{\cdot}$$

The second transition state is that for transfer of a bromine atom from Br_2 to the methyl radical.

$$HBr + [CH_3\text{----}Br\text{----}Br]^{\cdot}$$

The small energy minimum between these two transition states is characteristic of a *reactive intermediate,* in this case the methyl radical. Actually, you will notice that the other two energy minima at the left and right of Figure 2.15 also correspond to reactive intermediates, a bromine atom in each case. The energy difference between the two minima is the net energy change (exothermic) for bromination of methane.

EXERCISE 2.8

Using Figure 2.14 as a starting point, draw reaction profiles that show the abstraction by Cl· of a hydrogen atom from 2-methylpropane to give (a) the *tert*-butyl radical and (b) the isobutyl radical. Calculate the energy changes and label the graphs appropriately. (The E_a for formation of the primary radical is approximately 1 kcal/mol.)

EXERCISE 2.9

Using the bond dissociation energies in Table 2.1, calculate the overall energy change for chlorination of methane:

$$Cl_2 + CH_4 \rightarrow CH_3Cl + HCl$$

2.9 RATE OF REACTION

All chemists must be concerned with **reaction rate,** the speed at which a reaction takes place. If the rate is very slow, the reaction may not be finished within a convenient period of time. Organic chemistry students, for example, typically have laboratory classes which last three or four hours, and a reaction which required six hours would certainly be unsatisfactory for that situation. At the other extreme a reaction could also proceed too rapidly, with the disastrous result of an explosion. Clearly there is a limited range of reaction rates which would be satisfactory for any particular situation. Fortunately, we can regulate the rate of a reaction by controlling the experimental conditions.

The rate of any reaction depends on several variables, and we can analyze rates by considering just three factors: *energy, collision frequency,* and *probability.* Consider the first factor, energy. As we illustrated in Figures 2.13 and 2.14, all reactions have an activation energy (labeled E_a in the figures). Only collisions between reactants having a combined kinetic energy equal to or greater than E_a can result in reaction; other collisions are simply ineffectual. As illustrated by Figure 2.16, the kinetic energy of molecules in a reaction system is characterized by a Boltzmann distribution about the average value, and only a tiny fraction of the molecules in a typical reaction mixture have enough energy at any given instant to be able to undergo reaction. Only as a consequence of the rapid exchange of kinetic energy through molecular collisions do other molecules obtain sufficient energy to pass over the activation barrier.

It is not possible to change E_a for a reaction, but we can regulate the number of molecules which have enough kinetic energy to react. The average kinetic energy of the molecules in a system varies directly with the absolute temperature, so *temperature* provides the key to the energy factor in reaction rate. An increase in temperature will result in an increase in reaction rate, and a decrease in tempera-

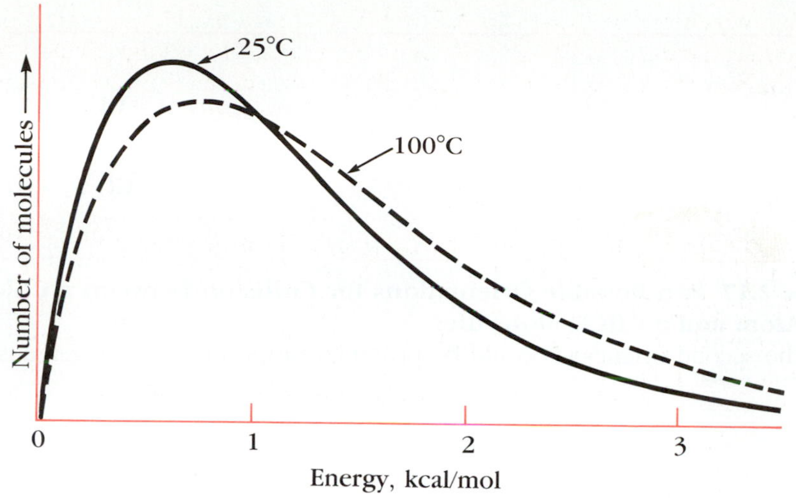

Figure 2.16 Boltzmann Distribution of Energy for a Large Number of Molecules.
The distribution shown here is for 25°C (where the average energy is 0.9 kcal/mol) and for 100°C (where the average energy is 1.1 kcal/mol).

ture will cause a reaction to proceed more slowly. Refrigeration is a common way of using decreased temperature to retard the rates of various chemical reactions.

The second factor governing reaction rate is the collision frequency. Reaction can occur only when the reactant molecules collide, so the greater the number of collisions in a given time, the greater will be the rate of reaction. The collision frequency is determined by the temperature and by the concentration of reactants. The temperature determines the average kinetic energy and therefore the average velocity of the molecules in a reacting system, since kinetic energy is defined as $\frac{1}{2}mv^2$. At higher temperatures the greater average velocity will result in more frequent collisions between molecules, and this will increase the rate of reaction. The frequency of collisions is also regulated by concentration, and the greater the concentration of the reactant molecules, the more frequently they will collide. An increase in concentration of reactants will therefore result in a greater rate of reaction.

The third parameter that regulates reaction rate is the probability factor, and the relationship between probability and rate of reaction concerns the *orientation* of two molecules when they collide. If the orientation is unfavorable, even molecules with sufficient kinetic energy will not react when they collide. We have illustrated this problem of orientation in Figure 2.17 for the abstraction of a hydrogen atom from a $CHCl_3$ molecule by a chlorine atom. The collision on the right of the figure appears to be ideally oriented for hydrogen abstraction, but that on the left could not be productive since there would be no Cl—H interaction. Like the activation energy, the probability factor is an inherent property of a particular reaction and cannot be regulated by the chemist.

The overall reaction rate can be summarized in terms of the factors discussed above by using the following equation:

$$\text{Rate} = \begin{pmatrix}\text{collision}\\\text{frequency}\end{pmatrix} \times \begin{pmatrix}\text{energy}\\\text{factor}\end{pmatrix} \times \begin{pmatrix}\text{probability}\\\text{factor}\end{pmatrix}$$

Cl• ---→ Cl, Cl, Cl C—H   Cl• ---→ H—C Cl, Cl, Cl

Figure 2.17 Two Possible Orientations for Collision between a Chlorine Atom and a $CHCl_3$ Molecule.
Only the second orientation could be productive in terms of hydrogen abstraction.

We will not attempt to assign numerical values to the terms in this equation, but you will find it very useful for *qualitative* evaluation of the ways that changes in molecular structure or in reaction conditions can alter the rate of a reaction.

A potentially confusing situation arises when the effect of a catalyst on a reaction is considered. As we indicated in Chapter 1, a catalyst facilitates a reaction by increasing the rate at a particular temperature or by allowing the reaction to be carried out at a lower temperature. This is frequently explained by stating that a catalyst decreases the activation energy of the reaction, but this statement is not strictly true. As we have illustrated in Figure 2.18, a catalyst can act in either of two ways to modify the profile of the uncatalyzed reaction (Figure 2.18a). First, the catalyst can change the nature of the transition state, thereby decreasing E_a, as shown in Figure 2.18b. This is the way that metal catalysts frequently function in organic reactions, with formation of metal-organic species rather than generation of free radicals or other high-energy reactive intermediates. Second, the catalyst can act to form a complex with the reactants, as shown in Figure 2.18c. This breaks up the reaction into two steps, each of which has an activation energy that is smaller

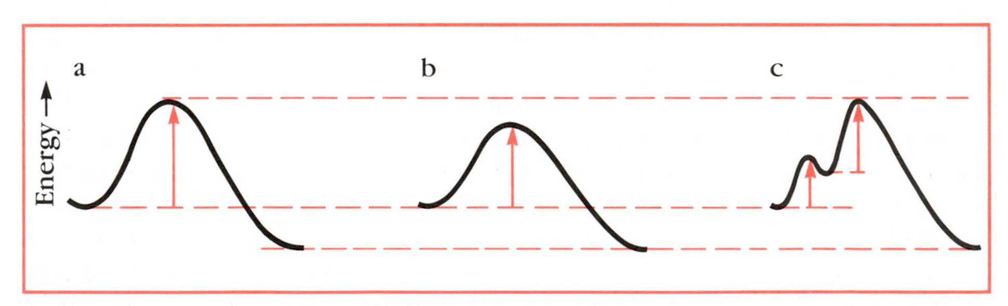

Figure 2.18 Effect of a Catalyst on a Reaction.
(a) The uncatalyzed reaction. (b) A reaction in which the catalyst interacts with the reacting species to produce a transition state that has a lower energy (and a different structure) than that of the uncatalyzed reaction. (c) A reaction in which the catalyst forms a complex with the reacting species. The overall transition state is not changed, but the reaction is broken down into two steps, each of which has an activation energy smaller than that of the uncatalyzed reaction.

than that of the uncatalyzed reaction. Each individual step can occur more rapidly, and the overall rate of reaction increases as a result. Catalysis of biochemical reactions by enzymes usually occurs in this way. Notice that in all cases the energies of the reactants and products are unaffected by the catalyst.

EXERCISE 2.10

For which collision would you expect a larger probability factor, hydrogen abstraction resulting from collision of a chlorine atom with (a) CH_4 or (b) $CHCl_3$? Why?

2.10 SELECTIVITY AND STABILITY

Reactions which exhibit **selectivity,** that is, *preference for one out of several possible reaction pathways,* are very important in organic chemistry. If you were attempting to prepare a particular compound, a reaction which affords a mixture of products would clearly be undesirable. On the other hand, you would almost certainly be satisfied with a reaction that provided a single product in high yield. In this section we will introduce you to a very important relationship between *selectivity* and *stability.* This relationship is only approximate, but it can be extremely useful when you want to predict or interpret chemical reactivity.

Free radical chlorination of alkanes is a reaction which shows little selectivity, and this is largely responsible for its minimal value in synthetic processes. The selectivity of free radical halogenation is quite different for the various halogens, and this is illustrated by the product distribution for free radical halogenation of butane. Although not used frequently, fluorination is included for comparison. The percentages are product distributions, not yields, of the monohalo products.

$$CH_3{-}CH_2{-}CH_2{-}CH_3 \xrightarrow{X_2} CH_3{-}CH_2{-}CH_2{-}\underset{\displaystyle X}{\underset{|}{C}H_2} + CH_3{-}CH_2{-}\underset{\displaystyle X}{\underset{|}{C}H}{-}CH_3$$

F_2	56 ± 2%	44 ± 2%
Cl_2	31 ± 1%	69 ± 1%
Br_2	2 ± 0.5%	98 ± 0.5%

As we indicated previously, halogenation reactions are governed by *kinetic control,* that is, the *yields* of products are proportional to the *rates at which they are formed.* This means that we can analyze the product ratios in terms of factors which govern the rates of the different reaction paths. We are comparing the rates of several processes that occur simultaneously in a reaction mixture, all at the same temperature. The effect of temperature, both on average kinetic energy and on collision frequency, will be the same for all species in the reaction mixture, so these factors can be disregarded. The key parameters that determine the relative rates of formation of the isomeric products are the activation energy (energy factor) and the probability factor for each pathway.

Butane has six primary and four secondary hydrogen atoms; thus the probability factors favor formation of the *primary* product by a ratio of 3:2. This is almost exactly the observed product ratio for fluorination, so no contribution to the observed selectivity from the energy factor can be seen. In contrast, chlorination affords the *secondary* product in a yield that is more than twice that of the primary product. Instead of a 3:2 ratio, a 3:7 ratio is observed, so the energy factor must enhance the rate at which secondary product is formed by a factor of 3.5. This is certainly a detectable preference for reaction at the secondary position, but because substantial amounts of both products are formed, we must still consider the reaction to be *nonselective.* Only for the third reaction, bromination of butane, is substantial selectivity observed. The primary to secondary product ratio is 2:98 (or 3:150), so the energy factor must enhance the formation rate of secondary product by a factor of nearly 80.

The selectivity of halogenation is even greater for abstraction of *tertiary* hydrogen atoms, at least for chlorination and bromination. This is illustrated by the following product distributions for the halogenation of 2-methylpropane:

$$CH_3-\underset{CH_3}{\overset{CH_3}{C}}-H \xrightarrow{X_2} CH_3-\underset{CH_2-X}{\overset{CH_3}{C}}-H + CH_3-\underset{CH_3}{\overset{CH_3}{C}}-X$$

F_2	90%	10%
Cl_2	70%	30%
Br_2	1%	99%

Once again, there is no observable selectivity for fluorination, but a substantial fraction of the chlorination product is tertiary, and bromination affords the tertiary product almost exclusively.

We can use the data in these halogenation reactions to calculate a set of *selectivities,* and these are summarized in Table 2.2. These represent the differences in energy factor of the rates for competing pathways, and you can use them to predict the approximate product distributions in halogenations. We can return to the equation presented in Section 2.9 for reaction rate, and by employing a *ratio* of rates for competing pathways 1 and 2, the unknown parameters will cancel out:

$$\frac{\text{Product 1}}{\text{Product 2}} = \frac{\text{rate 1}}{\text{rate 2}} = \frac{\text{probability factor 1}}{\text{probability factor 2}} \times \frac{\text{energy factor 1}}{\text{energy factor 2}}$$

TABLE 2.2 Selectivity of Halogenation: Contributions to Energy Factors for Reaction Rates (Relative to a value of 1 for abstraction of 1° hydrogen in each case)

Halogen	1° R-H	2° R-H	3° R-H
F_2	1	1	1
Cl_2	1	3.5	4
Br_2	1	80	1600

We can illustrate the use of this equation by using the data in Table 2.2 to "predict" the product distribution for chlorination of butane. Using pathways 1 and 2 for primary and secondary products, the probability factors are equal to the number of primary and secondary hydrogens, respectively. The selectivities in Table 2.2 provide the relative contributions to the energy factor, and this leads to the following:

$$\frac{\text{1-Clorobutane}}{\text{2-Clorobutane}} = \frac{\text{rate 1}}{\text{rate 2}} = \frac{6}{4} \times \frac{1}{3.5} = \frac{0.43}{1}$$

Solving this equation gives a product ratio of 0.43 to 1, which is the same as the experimental results of 30% primary product and 70% secondary product.

How can we account for the variation in the energy factors for abstraction of primary, secondary and tertiary hydrogens? We have previously indicated that the transition state is intermediate in structure between reactant and product, but it will not always be exactly "half-way" between the two. Ordinarily, *the transition state for a reaction tends to resemble more closely whichever is higher in energy, the reactant or the product.* This simple rule, originally stated in different form by G.S. Hammond (then at Iowa State University) is known as the **Hammond postulate,** and it allows us to fully interpret the variation in selectivity for the reactions of the different halogens with butane. First of all, we must recognize that the identity of the product in the halogenation of an alkane is determined by the initial hydrogen abstraction. If the 1-butyl radical is produced, then the product will be the 1-halobutane, but the 2-butyl radical will yield a 2-halobutane. This means that we can focus on the energetics of the hydrogen abstraction reactions and calculate ΔE for each of these processes using the data in Table 2.1.

$$X^{\cdot} + CH_3CH_2CH_2CH_3 \longrightarrow H{-}X + CH_3CH_2CH_2CH_2^{\cdot}$$

X = F	$\Delta E = -38$ kcal/mol
X = Cl	$\Delta E = -5$ kcal/mol
X = Br	$\Delta E = +10$ kcal/mol

$$X^{\cdot} + CH_3CH_2CH_2CH_3 \longrightarrow H{-}X + CH_3CH_2\dot{C}HCH_3$$

X = F	$\Delta E = -41$ kcal/mol
X = Cl	$\Delta E = -8$ kcal/mol
X = Br	$\Delta E = +7$ kcal/mol

The abstraction of a hydrogen by a fluorine atom is highly exothermic in both cases, and we would therefore expect the transition state to resemble the *fluorine radical.* This is the same species in both reactions, so we would expect the energies of the transition states to be very similar. This is exactly what is found in the experimental results, and Table 2.2 shows no selectivity for fluorination. The abstraction of hydrogen by a bromine atom lies at the other extreme. It is endothermic for both the processes we are considering, and we would therefore expect the transition state to resemble the *alkyl radical.* We know that a secondary alkyl radical is more stable than a primary radical, so we would expect the transition state for the formation of the secondary radical to be of lower energy. The rate of formation of secondary radical should therefore be enhanced, and this again is exactly what is found experimentally (Table 2.2). The energy changes calculated for hydrogen abstraction by a chlorine atom are intermediate between those for the other two halogens, and this leads to a selectivity which also lies between those for fluorine and bromine.

In the preceding discussion we have established a correlation between selectivity and transition state energy, together with a relationship between transition state energy and the structures of the reactive intermediates that are involved. We can summarize all this information in terms of a relatively simple relationship between *stability* and *selectivity*. As a general rule, *the selectivity of a reactive intermediate increases with its stability.* Conversely, highly reactive intermediates tend to exhibit less selectivity among alternative reaction pathways.

2.11 TERMS AND DEFINITIONS

Activation energy. The amount of energy that the reactants must have in order for a reaction to take place; the energy difference (usually abbreviated as E_a) between the reactants and the transition state for the reaction.

Anti. A term describing the orientation of two substituents (on adjacent carbons) in which they are at a dihedral angle of 180°; for alkanes this is a stable arrangement corresponding to an energy minimum.

Boltzmann distribution. The statistical distribution of energy for a large number of molecules. The average kinetic energy varies with the temperature, but most individual molecules have either more or less energy than this average.

Bond dissociation energy. The energy required to cleave a bond and form two free radicals.

Chain reaction. A reaction that operates in a cyclic manner because the reactive intermediate is regenerated. Chain reactions are characterized by three stages: *initiation,* in which a reactive intermediate is formed; *propagation,* in which the intermediate reacts to form product with regeneration of the intermediate; and *termination,* in which the reactive intermediate is consumed.

Conformation. A three-dimensional arrangement of a molecule that differs from other arrangements only by rotation about single bonds.

Dihedral angle. The angle between two substituents on adjacent carbons that is perceived by viewing along the C—C bond. The dihedral angle X—C—C—Y is equal to the angle between the C—X and C—Y bonds when the structure is viewed along the C—C bond (i.e., with a Newman projection).

Eclipsed. A term describing a three-dimensional arrangement in which the dihedral angle between substituents on adjacent carbons is 0°. When viewed as a Newman projection, the substituent on the rear carbon is directly behind (i.e., is eclipsed by) the substituent on the front carbon.

Electron deficient. A term describing an atom lacking the eight electrons of the inert gas configuration in its valence shell (e.g., the reactive center in a free radical or carbocation).

Energy. Ability of a system to liberate heat and to do work. Experimental measurement gives an energy change, and this is the sum of the heat evolved and the work done by the system.

Energy barrier. The activation energy for a reaction or conformational change. Molecules with less than this amount of energy cannot undergo the change.

Energy minimum. A position on a reaction coordinate for which any change results in an increase in the energy. This corresponds to the optimum geometry (or at least to one of the stable conformations) of the species.

Enthalpy. A thermodynamic quantity for which the change, ΔH, is the heat evolved or absorbed in a process occurring at constant pressure.

Entropy. A thermodynamic quantity related to probability. Experimental measurements provide the change in entropy (ΔS) for a process, which is related to the enthalpy and free energy changes by $\Delta G = \Delta H - T\Delta S$

Exothermic. Description of a process that liberates heat (i.e., $\Delta H < 0$). In contrast, an **endothermic** process occurs with absorption of heat (i.e., $\Delta H > 0$).

Free energy. A thermodynamic quantity for which the difference (ΔG) describes the combined enthalpy and entropy differences between two species. The free energy difference between reactants and products of a reaction is related to the equilibrium constant (K) by: $\Delta G = -2.3\, RT \log K$.

Free radical. An electron-deficient species having an odd number of electrons (i.e., having at least one unpaired electron).

Free rotation. The situation in which interconversion of conformations occurs rapidly (i.e., freely) at room temperature—in other words the energy barrier for rotation about single bonds is small.

Gauche. The orientation of two substituents (on adjacent carbons) in which they are at a dihedral angle of about 60°. This corresponds to a staggered conformation of the molecule, but the two substituents may have unfavorable steric interactions.

Hammond postulate. A general rule about a reaction profile: The transition state for a reaction tends to resemble more closely whichever is higher in energy, the reactant or the product.

Intermediate. A reactive intermediate; also a compound that is obtained at an intermediate stage in a synthesis.

Kinetic control. Describes a reaction for which the amount of each product obtained is governed by the rate at which it is produced.

Kinetic energy. The energy of a molecule that results from motion; it equals $\frac{1}{2}mv^2$, where m is the mass of the molecule and v is its velocity.

Newman projection. A type of drawing used to convey three-dimensional information about two adjacent carbon atoms in an organic structure.

Nonbonded interaction. The interaction between the electron clouds of atoms that are not bonded to each other (or to the same central atom); when two atoms or groups approach each other closely, the interaction is repulsive.

Potential energy. The energy a molecule posesses that results from the relative positions of its constituent atoms; potential energy reflects the chemical energy that might be released or absorbed in a reaction.

Reaction coordinate. The horizontal axis of a reaction profile. Describes the progress of the reaction (for example, the distance X---Y between two atoms, X and Y).

Reaction mechanism. A detailed step-by-step description of the events of a reaction at the molecular level.

Reaction profile. A plot of the potential energy of the molecules involved in a reaction during the course of the reaction (as specified by the reaction coordinate).

Reaction rate. The rate at which reactants are converted into products.

Reactive center. The atom (or group of atoms) in a molecule at which reaction occurs.

Reactive intermediate. A high-energy species that is a minimum-energy structure, but which cannot be isolated (i.e., a free radical, carbocation, carbene, or carbanion).

Rotational energy. The kinetic energy of a molecule that results from rotational motion of all or part of the molecule relative to its surroundings.

Sawhorse representation. A type of drawing used to convey three-dimensional information about two adjacent carbon atoms in an organic structure.

Selectivity. Preferential reaction at a particular site in a molecule.

Stability. Used to describe relative potential energy; the more stable a species, the lower is its energy. Also used to denote reactivity; the more stable a species, the lower is its reactivity.

Staggered. The orientation of the substituents on adjacent carbons in a stable conformation of an alkane; the dihedral angles between such substituents are either 60° or 180°.

Steric effect. A chemical effect on reactivity resulting from nonbonded interactions.

Thermodynamic control. Reaction conditions such that the amount of each product obtained results from establishment of an equilibrium between them. The equilibrium constant is related to the change in the thermodynamic quantity, free energy, by the relationship: $\Delta G = -2.3\ RT \log K$.

Transition state. An energy-maximum on a reaction profile; the energy difference between the transition state and the reactant defines the activation energy, E_a.

Vibrational energy. The kinetic energy of a molecule that results from vibrational motion, i.e., oscillating motion of one part of the molecule relative to the remainder.

2.12 PROBLEMS

2.1 Consider the following reaction profile, in which starting material *A* can react to give either *B* or *C*. (Assume that the reactions are not reversible, i.e, that *B* and *C* do not interconvert.)

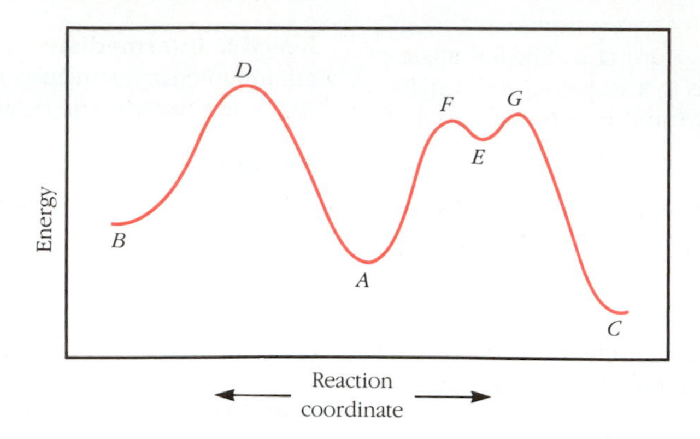

(a) Is the formation of *B* from *A* exothermic or endothermic?
(b) Is the formation of *C* from *A* exothermic or endothermic?
(c) Which is more stable, *B* or *C*?
(d) What term is used to describe *D*?
(e) What term is used to describe *E*?
(f) What term is used to describe *F*?
(g) What term is used to describe *G*?
(h) Which will be the major product, *B* or *C*?

2.2 Free radical chlorination of 2-chloro-2,4-dimethylpentane gives more than one dichloro product.

$$CH_3{-}\underset{}{\overset{CH_3}{\overset{|}{C}H}}{-}CH_2{-}\underset{\underset{Cl}{|}}{\overset{CH_3}{\overset{|}{C}}}{-}CH_3 \xrightarrow[\text{light or heat}]{Cl_2} C_7H_{14}Cl_2 \text{ isomers}$$

Using the selectivities presented in Table 2.2, calculate the approximate percentage expected for each of the isomers produced by this reaction.

2.3 For each of the products in the preceding question, use the bond dissociation energies in Table 2.1 to calculate the overall energy change for that reaction pathway.

2.4 Draw each of the following compounds in the most stable staggered conformation, utilizing Newman projections. Use the atoms indicated by * to form the central bond. (In analyzing this problem you should consider a bromine to be larger than a methyl group.)

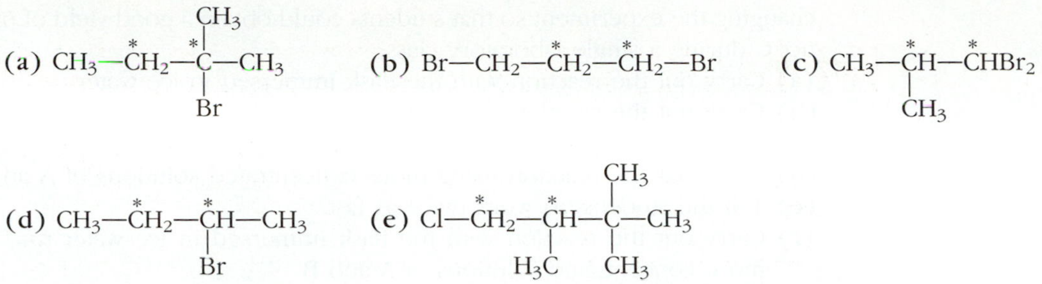

(a) $CH_3—\overset{*}{C}H_2—\overset{*}{C}(CH_3)(Br)—CH_3$ (b) $Br—CH_2—\overset{*}{C}H_2—\overset{*}{C}H_2—Br$ (c) $CH_3—\overset{*}{C}H(CH_3)—\overset{*}{C}HBr_2$

(d) $CH_3—\overset{*}{C}H_2—\overset{*}{C}H(Br)—CH_3$ (e) $Cl—\overset{*}{C}H_2—\overset{*}{C}H(H_3C)—C(CH_3)_2—CH_3$

2.5 Draw the answers for Problem 2.4 using the corresponding sawhorse representations.

2.6 Draw each of the compounds in Problem 2.4 in the *least* stable staggered conformation, utilizing Newman projections. Use the atoms indicated by * to form the central bond.

2.7 Draw the answers for Problem 2.6 using the corresponding sawhorse representations.

2.8 Draw each of compounds in Problem 2.4 in the *least* stable eclipsed conformation, utilizing Newman projections. Use the atoms indicated by * to form the central bond.

2.9 Draw the answers for Problem 2.8 using the corresponding sawhorse representations.

2.10 What is the dihedral angle between the chlorine atoms in each of the following structures?

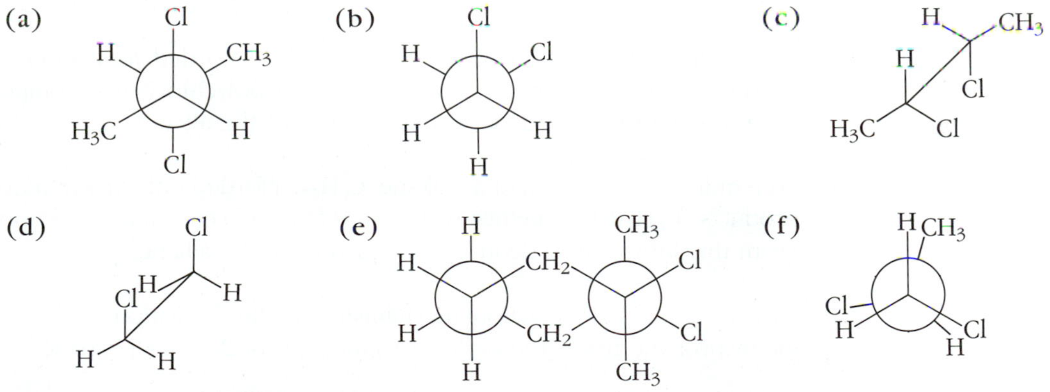

(a) Cl, H, CH_3, H_3C, H, Cl
(b) Cl, H, Cl, H, H, H
(c) H, CH_3, H, Cl, H_3C, Cl
(d) Cl, Cl, H, H, H, H
(e) H, H, CH_2, H, CH_2, H, CH_3, Cl, Cl, CH_3
(f) H, CH_3, Cl, H, Cl, H

2.11 An instructor for an organic chemistry laboratory course was trying out a new experiment before assigning it to students. The instructor mixed 100 mL of a 1 *M* solution of A and 100 mL of a 1 *M* solution of B in a 500-mL flask at room temperature. The reaction took place at a rate such that only 50 percent of the desired product *C* was produced after 6 h. Unfortunately, the laboratory classes last 3 h. Which of the following would be reasonable approaches to changing the experiment so that students could obtain a good yield of product C during a single laboratory class?

(a) Carry out the reaction with the flask immersed in ice water.
(b) Carry out the reaction using more dilute solutions of A and B.
(c) Use a 250-mL flask.
(d) Carry out the reaction using more concentrated solutions of A and B.
(e) Tell the students to work twice as fast.
(f) Carry out the reaction with the flask immersed in ice water *and* use more concentrated solutions of A and B.
(g) Use a 1-L flask.
(h) Carry out the reaction with the flask immersed in boiling water.
(i) Add powdered bat wings.

2.12 Consider a hypothetical molecule that could exist in two possible conformations, A and B, where A is more stable. Calculate the fraction of each conformation that would be expected at 25°C if the energy difference between A and B were:

(a) 0.1 kcal/mol
(b) 0.5 kcal/mol
(c) 1.0 kcal/mol
(d) 2.0 kcal/mol
(e) 5.0 kcal/mol
(f) 10.0 kcal/mol

2.13 Which of the following radicals would be most stable? Which would be least stable?

(a) $CH_3-\dot{C}(CH_3)-CH_2-CH(CH_3)-CH_3$

(b) $CH_3-CH(CH_3)-\dot{C}H-CH(CH_3)-CH_3$

(c) $CH_3-CH(CH_3)-CH_2-CH(CH_3)-CH_2^{\cdot}$

(d) $CH_3-CH(CH_3)-CH_2-CH(CH_2^{\cdot})-CH_3$

2.14 Free radical chlorination of an alkane, C_5H_{12}, afforded *four* monochloro products ($C_5H_{11}Cl$), together with a variety of polychlorinated compounds. From this information deduce the structure of the alkane.

2.15 Free radical chlorination of an alkane, C_6H_{14}, afforded only *two* monochloro products ($C_6H_{13}Cl$), together with a variety of polychlorinated compounds. From this information deduce the structure of the alkane.

2.16 Write the complete reaction mechanism for the formation of 2-bromo-2-methylpropane from free radical bromination of 2-methylpropane.

2.17 An alternative mechanism, in which a halogen atom reacts with an alkane to produce the alkyl halide directly, could be written for the propagation steps of free radical halogenation. This is illustrated for the chlorination of methane in the following equations:

$$Cl^{\cdot} + CH_4 \longrightarrow Cl-CH_3 + H^{\cdot}$$
$$H^{\cdot} + Cl_2 \longrightarrow HCl + Cl^{\cdot}$$

Use the data in Table 2.1 to calculate ΔE for each of these reaction steps, and compare these values with those for the mechanism we presented in Section 2.6. Would the alternative mechanism presented in this problem be expected to compete with the one described earlier?

2.18 Experimental measurements have shown that 2-chloro-2-methylbutane is more stable than the isomeric 1-chloro-2-methylbutane by 9.5 kcal/mol. Assuming that this is the free energy difference, calculate the ratio of the two compounds that would be present at equilibrium.

2.19 Use the bond dissociation energies in Table 2.1 to calculate the energy difference between 2-chloro-2-methylbutane and the isomeric 1-chloro-2-methylbutane. How well does the result of this approximate method agree with the experimental data cited in the preceding problem?

2.20 The experimental energy difference between butane and isobutane is 2.0 kcal/mol. Assuming that this is the free energy difference, what ratio of these two compounds would you predict for an industrial process that created an equilibrium mixture of the two isomers?

2.21 Using the selectivity data of Table 2.2, calculate the expected product distribution for the chlorination of 2-methylpentane.

2.22 Using the selectivity data of Table 2.2, calculate the expected product distribution for the bromination of 2-methylpentane.

2.23 A chemist found that an equilibrium mixture of two isomeric compounds, A and B, consisted of 92% A and 8% B. Which is more stable, A or B? Calculate the free energy difference between the two compounds.

CHAPTER 3

CYCLOALKANES

In Chapter 1 we discussed a variety of alkanes, all of which were described as either *straight-chain* or *branched-chain* alkanes. There is, however, still another type of chain that is frequently found in alkanes, and that is one in which the carbon chain is *cyclic*. Cyclic compounds are very widespread, both in nature and in synthetic materials, and in this chapter we will introduce you to the structure and reactivity of cyclic alkanes.

3.1 CYCLIC ALKANES

A straight-chain alkane such as hexane consists of a series of CH_2 groups with a methyl group at each end of the chain. In contrast, the analogous cyclic alkane, *cyclohexane,* consists entirely of a sequence of connected CH_2 groups, and the chain has no ends.

$CH_3CH_2CH_2CH_2CH_2CH_3$

Hexane

Cyclohexane (ring of six CH_2 groups)

As this example suggests, cyclic alkanes are known as *cycloalkanes,* and the cyclic chains are called *rings.* In general, compounds that contain rings are called **cyclic,** and those that do not contain any rings are known as **acyclic.**

Cycloalkanes are generally quite similar to their acyclic analogs, and important differences in chemical reactivity exist for only a few special cases, which we will discuss later in this chapter. The major differences between cyclic and acyclic compounds are *structural,* and we will devote much of this chapter to discussion of the three-dimensional shapes and conformations of cycloalkanes. The simplest and most consistent point of difference between cyclic and acyclic alkanes is found in their molecular formulas. The general molecular formula for a cyclic alkane with n carbon atoms can be expressed as C_nH_{2n}, whereas that for the corresponding acyclic alkane is C_nH_{2n+2}. Looking back at the structural formulas of hexane (C_6H_{14}) and cyclohexane (C_6H_{12}), you will see that these hexanes differ by two hydrogen atoms. Why are two hydrogens "missing" in the cyclic compound? You can see this easily from a simple comparison between cyclopentane and pentane.

Cyclopentane (ring of five CH_2 groups)
C_5H_{10}

Pentane ($CH_3CH_2CH_2CH_2CH_3$)
C_5H_{12}

Each carbon atom can have only four bonds, so each carbon of the cycloalkane has two bonds to other carbon atoms and two bonds to hydrogens. The extra carbon-carbon bond needed to complete the ring of cyclopentane exists at the expense of a C—H bond on each of the terminal methyl groups of pentane. As you will see in

Section 3.4, it is sometimes possible to convert a cyclic compound to the acyclic analog by *addition* of two hydrogen atoms.

In Chapter 1 we showed you several ways of drawing structural formulas for alkanes. As we proceed to more complex molecular structures, we will introduce additional types of structural formulas, which provide the maximum structural information for the minimum effort in drawing them. A particularly convenient way to draw cycloalkanes is to use a regular polygon to define the carbon skeleton. In this way cyclohexane and cyclopentane can be drawn respectively, as:

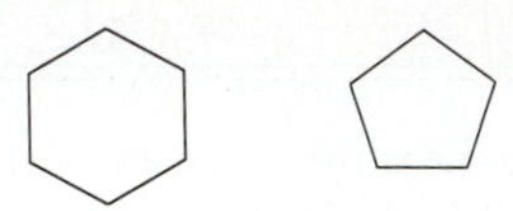

In this type of drawing each corner of the polygon corresponds to a carbon atom, and the lines represent carbon-carbon bonds. All carbon atoms of stable compounds have four bonds, so *any bonds not explicitly drawn are understood to be to hydrogen atoms.*

The nomenclature of cyclic alkanes is only slightly different from that of the acyclic compounds, and the rules presented in Section 1.8 apply to cyclic compounds as well. The prefix "cyclo" is used to indicate the cyclic nature of both cycloalkanes and the corresponding cycloalkyl groups. A new difficulty arises in identifying the *principal chain* of a cycloalkane. Once again, this is the longest carbon chain, but now there is the additional restriction that the principal chain must be located either entirely within a ring or entirely outside of it. The following two compounds illustrate both possibilities. In the first case the ring has only three carbons, so the longer pentyl group is the principal chain, and the compound is named as cyclopropyl*pentane.* In the second case the six-membered ring has a greater number of carbons, so it becomes the principal chain, and the name of the compound is pentyl*cyclohexane.*

$CH_2CH_2CH_2CH_2CH_3$ $CH_2CH_2CH_2CH_2CH_3$

Cyclopropylpentane Pentylcyclohexane

As illustrated by the preceding examples, the position of attachment of either an alkyl group or a cycloalkyl group is, by definition, C-1 of that group. Only when the principal chain is cyclic and contains more than one substituent must you decide among different numbering possibilities for the ring. As we described in Chapter 1 for alkanes, C-1 is then selected so that the lowest numbers are obtained for the positions of the substituents at the first point of difference. The remaining carbon atoms of the ring are numbered consecutively starting from C-1. In the case of a monosubstituted derivative, the ring carbon bearing the substituent is always understood to be C-1 (regardless of the orientation in which it is drawn). A number at the beginning of the name of a compound refers to the principal chain; in the following example this denotes C-2 of the six-carbon chain of 2-cyclobutylhexane.

$CH_3—CH—CH_2—CH_2—CH_2—CH_3$

2-Cyclobutylhexane

The following examples illustrate the nomenclature for several substituted cycloalkanes, and the first two help to emphasize that the name is independent of the orientation in which we draw the structure.

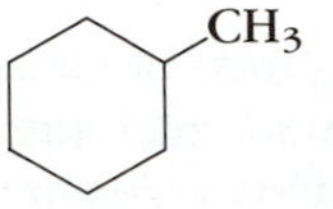

CH_3 CH_3

Methylcyclohexane 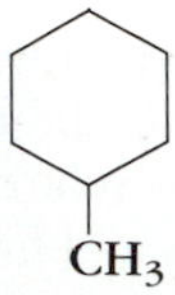 Methylcyclohexane

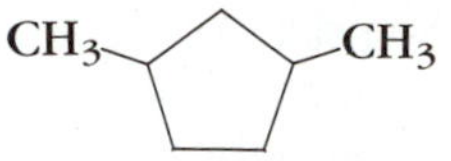

CH_3 CH_3 CH_3 CH_3

1,3-Dimethylcyclopentane 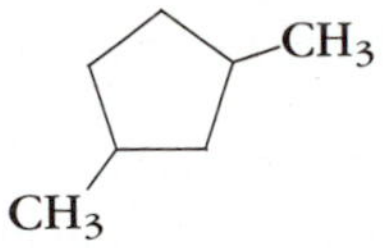 1,3-Dimethylcyclopentane

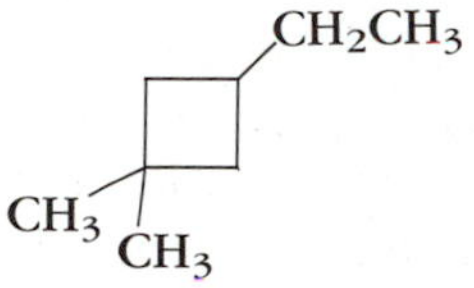

CH_2CH_3

CH_3 CH_3

3-Ethyl-1,1-dimethylcyclobutane

In the last case the four-membered ring was numbered so that C-1 has the two methyl groups. This leads to 3-ethyl-1,1-dimethylcyclobutane, whereas the alternative would have been 1-ethyl-3,3-dimethylcyclobutane. For each name there would be a substituent on C-1 and C-3, but the preferred name has two substituents on the carbon atom with the lower number. (In other words "1,1,3-" is better than "1,3,3-".) Notice also that the ethyl group is cited before the methyl groups in the complete name, because the "e" of ethyl takes alphabetical precedence over the "m" of di*methyl*. The complete names for cycloalkanes and their derivatives sometimes include descriptions of their three-dimensional structures, and we will introduce the necessary terminology in Section 3.5.

The selection of C-1 on a ring is much simpler when the ring itself is a substituent on another chain, because C-1 by definition is the point of attachment:

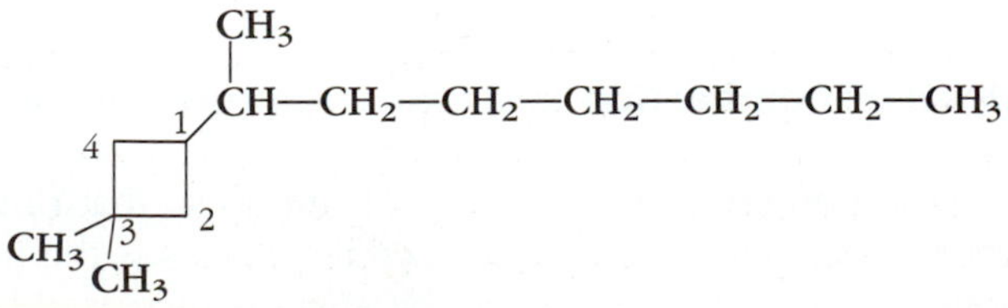

CH_3

$CH—CH_2—CH_2—CH_2—CH_2—CH_2—CH_3$

1 2 3 4

CH_3 CH_3

2-(3,3-Dimethylcyclobutyl)octane

The preceding examples show the basic method for naming cycloalkanes, and their derivatives are named similarly. Additional examples will be presented throughout the text.

3.2 BAEYER STRAIN THEORY

You will better understand some of the unique aspects of cyclic compounds if we describe some key historical developments. Approximately 100 years ago organic chemistry was still largely concerned with the study of naturally occurring compounds, and a large number of cyclic compounds had been isolated, primarily from plant materials. Of the many cyclic compounds which had been studied, there was a great predominance of six- and five-membered ring structures. In addition, cyclopropane and its derivatives appeared to be unstable relative to similar compounds with larger rings. This led the German chemist Adolf von Baeyer (who received the Nobel Prize in chemistry in 1905) to propose an explanation based on the idea that the stability of cyclic compounds should vary with ring size. At the heart of his proposal was the premise that ring sizes other than five and six would be unstable and would therefore not be readily produced in nature.

Baeyer considered the representation of each cycloalkane as a regular polygon, as shown in Figure 3.1. (We have denoted the angle between adjacent sides of the polygon for each ring size.) Baeyer observed that this angle was quite far from the

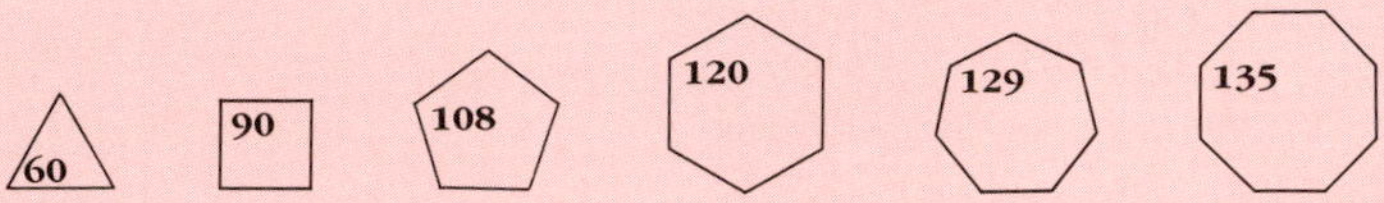

	60 (triangle)	90 (square)	108 (pentagon)	120 (hexagon)	129 (heptagon)	135 (octagon)	Acyclic $-(CH_2)_n-$
Heat of combustion per CH_2 group, kcal/mol	167	164	159	157	158	158	157
Strain per CH_2 group, kcal/mol	10	7	2	0	1	1	0
Total strain, kcal/mol	30	28	10	0	7	8	0

Figure 3.1 Heats of Combustion and Strain Energies of Various Cycloalkanes.
The heats of combustion are presented per CH_2 group so that ring sizes can be compared directly. The value given for the acyclic alkanes is the increase in heat of combustion that is observed when the chain is increased by a single CH_2 group. The interior angle for each regular polygon is shown in order to allow comparison with the tretrahedral angle of 109.5°. The deviation from the acyclic value of 157 kcal/mol provides the strain per CH_2 for each cycloalkane, and multiplication by the number of carbon atoms in the ring yields the total strain.

tetrahedral angle of 109.5° for the first two cases, and he concluded that the necessary distortion from the ideal geometry would cause cyclopropane and cyclobutane to be quite strained and unstable relative to five- and six-membered rings. Indeed, only the five-sided polygon in Figure 3.1 has an angle that is almost the same as the tetrahedral angle, and Baeyer argued that even cyclohexane should be slightly strained relative to cyclopentane. Although they were not considered explicitly by Baeyer, we have also shown the seven- and eight-membered rings in Figure 3.1 for comparison; the interior angles of a heptagon and an octagon also show large deviations from 109.5°.

Some evidence was already available to support the idea that cyclopropane and cyclobutane were rather strained, and Baeyer's theory provided an explanation for the scarcity of known structures with rings larger than six-membered. However, this was not conclusive, and Baeyer proposed a test that could be used to prove or disprove his theory. His proposed test was based on the heat of combustion of the cycloalkanes, that is, the energy evolved when the hydrocarbon is completely oxidized to H_2O and CO_2. Baeyer suggested that the heat of combustion of cyclohexane would be twice that for cyclopropane if there were no strain effects. On the other hand, if cyclopropane were highly strained (and therefore of higher energy), its combustion would release more energy in the form of heat.

$$C_6H_{12} + 9\,O_2 \longrightarrow 6\,CO_2 + 6\,H_2O + \text{heat}$$

$$C_3H_6 + 4.5\,O_2 \longrightarrow 3\,CO_2 + 3\,H_2O + \text{heat}$$

Results for the test proposed by Baeyer are summarized in Figure 3.1. Each of the cycloalkanes is composed only of CH_2 groups, so we can make direct comparisons between them if we consider the heat of combustion per CH_2 group. We have also shown the *change* in the heat of combustion (157 kcal/mol) that results when the chain length of an acyclic alkane is increased by a single CH_2 group, and this provides a reference point for the heat of combustion per CH_2 group in the absence of any effects caused by the presence of a ring. If there were no ring strain, all the cycloalkanes would be expected to yield this same value of 157 kcal/mol.

In accord with Baeyer's original prediction, the heats of combustion per CH_2 group for cyclopropane and cyclobutane are much larger than those of the other cycloalkanes as well as that of the acyclic model. This is in agreement with considerable modern evidence supporting the notion that three- and four-membered ring compounds are strained and are therefore less stable than their acyclic counterparts. However, the data in Figure 3.1 show that the original Baeyer theory breaks down for the larger ring sizes. In fact, it is cyclohexane rather than cyclopentane that corresponds most closely to the acyclic value of 157 kcal/mole per CH_2 group. The flaw in the proposal put forth by Baeyer was the implicit assumption that the carbon skeletons of the cyclic compounds were all planar. While this is necessarily true for the three-membered ring, *all the other cycloalkanes have nonplanar conformations* and are consequently able to avoid the angle distortions that would have been found for planar molecules with the shapes of regular polygons.

Even though cyclobutane is not a planar molecule, the four-membered ring cannot achieve bond angles of 109.5°. Consequently, both cyclopropane and cyclobutane have substantial ring strain. For all the other cycloalkanes the total strain energy is much smaller, especially if we look at the value per CH_2 group. There is, nevertheless, some strain for all the compounds except cyclohexane.

What is the origin of this strain? While each of the molecules with ring sizes larger than four-membered can achieve conformations in which the bond angles of the carbon skeleton are approximately tetrahedral, this frequently results in other interactions that are unfavorable. The five-membered ring of cyclopentane, for example, would have C—C—C bond angles of 108° if the structure were planar, but this would produce a highly unfavorable consequence: all the hydrogens on the ring would be eclipsed. Therefore cyclopentane cannot adopt a conformation which is free of *both* angle strain and other unfavorable interactions. As you will see in the following section, only cyclohexane can achieve a conformation that is free of strain from all these sources.

3.3 THE ACTUAL SHAPES OF CYCLIC COMPOUNDS

The ring carbons of cyclopropane and its derivatives are always planar, a necessary consequence of the geometric property that three points define a plane. The planarity of the cyclopropane ring requires that substituents on adjacent carbon atoms be in the eclipsed arrangement, and this certainly contributes to the strain of three-membered rings. The eclipsing interactions that must be present in a planar ring result in a distinct preference for cycloalkanes of other ring sizes to adopt nonplanar conformations. Even cyclobutane exists in a slightly puckered shape, despite the fact that this must require a further distortion (although only a small one) of the C—C—C bond angle away from the tetrahedral angle of 109.5°. Cyclopentane similarly exists in a conformation that is slightly distorted from planarity in order to alleviate eclipsing interactions. The minimum-energy conformations of each of these cycloalkanes is shown in Figure 3.2. In each case we have drawn the ring with and without the hydrogens.

Cyclohexane presents a quite different situation, and alternation of successive atoms in a zigzag fashion allows the molecule to achieve a conformation in which all bond angles are essentially tetrahedral and all substituents on adjacent carbon atoms of the ring are almost perfectly staggered. No distortion from tetrahedral geometry is required, and no eclipsing interactions are present. On this basis cyclohexane is considered to be essentially *strain free.* This is in good agreement with the combustion data of Figure 3.1, which shows no difference between cyclohexane and the acyclic alkanes.

Bond angles depend upon the atoms bonded to a central carbon, so a precisely tetrahedral geometry will be found only for a carbon atom bearing four identical substituents. The bond angles for a CH_2 group in a carbon chain are expected to exhibit small but significant deviations from 109.5°. For example, the C—C—C angle for cyclohexane is 111.5°, and such small distortions are typical for the bond angles of carbon atoms for which all four substituents are not identical. Keep in mind that a distortion of just a few degrees is quite small. When the minute hand of a clock changes by 1 min, a nearly imperceptible change, it has moved through an angle of 6°. Therefore a distortion of only 2° is almost undetectable to the human eye.

The most stable conformation for cyclohexane, shown in Figure 3.3, is known as the **chair conformation.** The conformations of cyclohexane derivatives play an important role in organic chemistry, and they have been studied in great detail. We will discuss the conformational analysis of cyclohexanes more fully in Section 3.6.

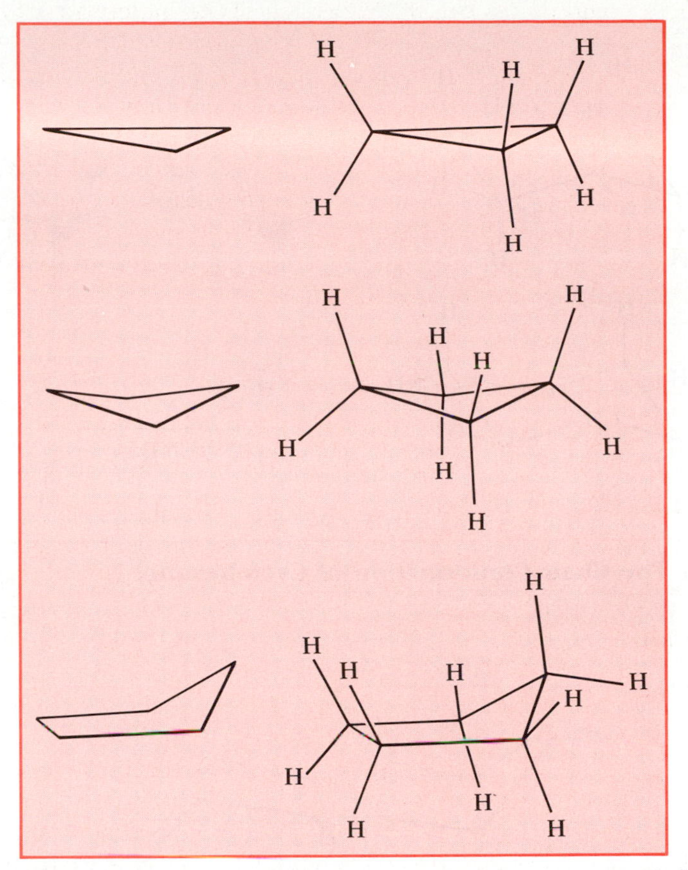

Figure 3.2 Three-dimensional Shapes of Cyclopropane, Cyclobutane, and Cyclopentane.
Only in the case of cyclopropane is a planar geometry required. The other cycloalkanes adopt nonplanar conformations.

The conformations of cycloalkanes with seven or more carbon atoms in the ring are also nonplanar. However, they tend to have relatively large numbers of similar conformations which do not vary substantially in energy. It is nevertheless possible to identify the most stable geometry, and we have shown the most stable conformations of cycloheptane and cyclooctane in Figure 3.4.

If you carefully analyze the structures in Figure 3.4 and compare them with those in Figures 3.2 and 3.3, you will see that only in cyclohexane is the staggered arrangement achieved for all adjacent carbon atoms. On that basis you can see clearly why the heat of combustion data in Figure 3.1 indicate that cyclohexane is the only cycloalkane that is free of strain.

We have carefully emphasized in this section the nonplanar structures of cycloalkanes. Nevertheless, we will continue to use regular polygons in the structural formulas of these compounds. Only when we are depicting actual conformations will we use the more complex drawings such as those in Figures 3.2–3.4. You must remember that the use of the polygons in drawings of cyclobutane and larger cycloalkanes is merely a matter of convenience, and the actual structures are distinctly nonplanar.

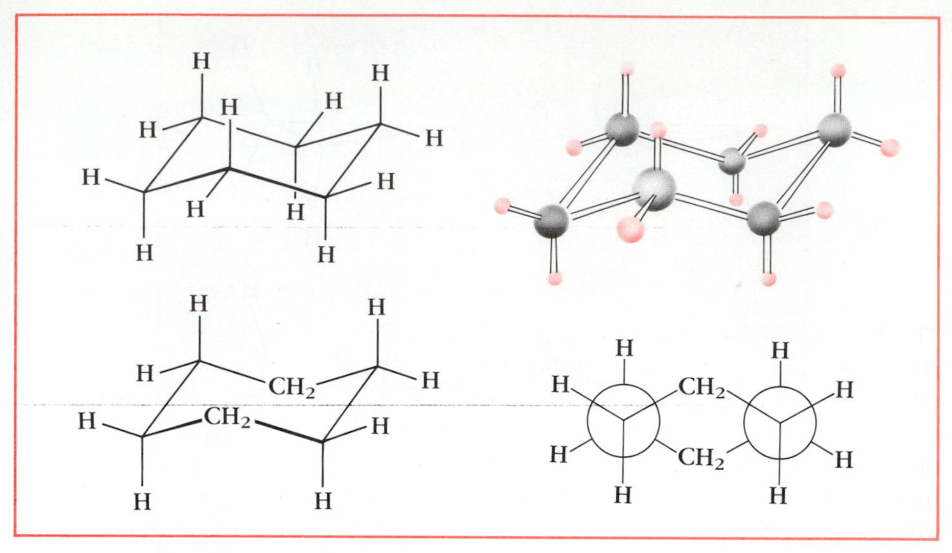

Figure 3.3 The Chair Conformation of Cyclohexane.

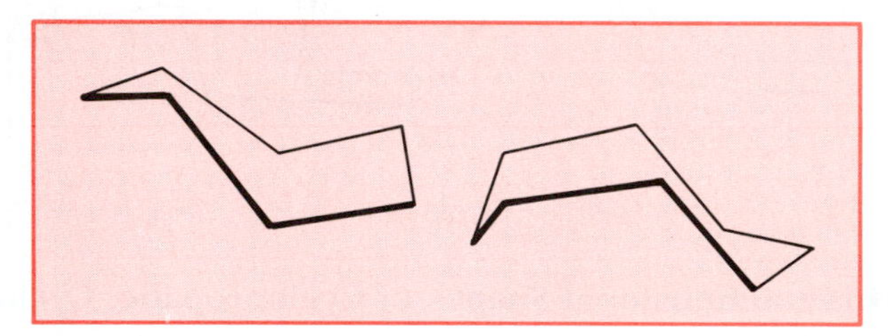

Figure 3.4 The Most Stable Conformations of Cycloheptane and Cyclooctane.

3.4 REACTIVITY OF CYCLOALKANES

As we indicated previously, the reactivity of cycloalkanes is generally no different than that of their acyclic counterparts. Only the two strained ring compounds, cyclopropane and cyclobutane, and their derivatives exhibit unusual reactivity, and this is usually seen in reactions that result in cleavage of the ring. Under very severe reaction conditions, such as those of catalytic cracking and reforming in petroleum refining (Section 1.11), normal carbon-carbon bonds can be cleaved. But laboratory reactions do not employ such stringent conditions, and the carbon-carbon bonds of most alkanes are not readily cleaved. Cyclopropanes and cyclobutanes provide an exception to this behavior, because cleavage of one of the carbon-carbon bonds of the ring relieves the very substantial strain, about 30 kcal/mol, that is present in these compounds.

The reaction of cyclopropanes with hydrogen and a metal catalyst under fairly vigorous conditions results in the rupture of the three-membered ring and the addition of two hydrogen atoms. This is illustrated by the following examples:

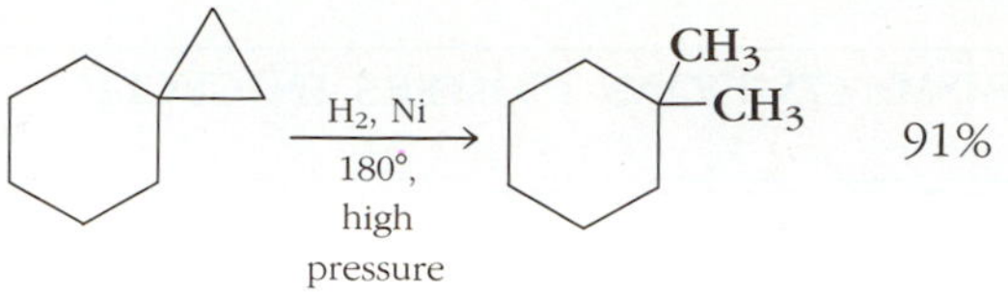

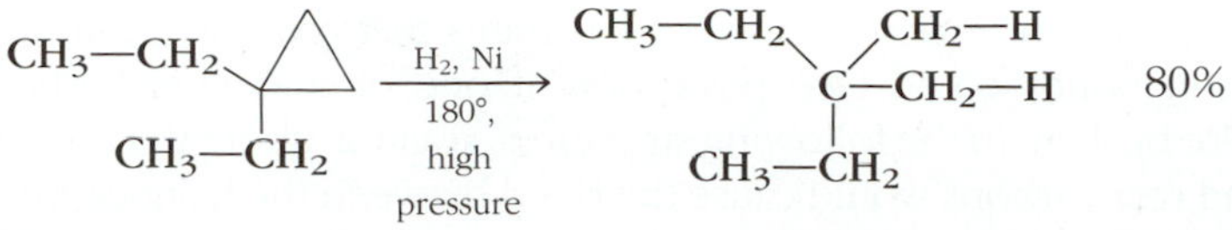

$$\text{spiro[2.5]octane} \xrightarrow[\text{180°, high pressure}]{H_2,\ Ni} \text{1,1-dimethylcyclohexane} \quad 91\%$$

$$CH_3{-}CH_2\text{(1,1-diethylcyclopropane)}\xrightarrow[\text{180°, high pressure}]{H_2,\ Ni} (CH_3{-}CH_2)_2C(CH_2{-}H)(CH_2{-}H) \quad 80\%$$

The high temperature and pressure (approximately 130 atm) of the preceding reactions are necessary to effect the cleavage. Notice, however, that the *cyclohexane* ring of the first example remains intact. Cyclobutane rings can also be cleaved by hydrogen and a metal catalyst, but higher temperatures and pressures are needed than for the cleavage of cyclopropanes. While both three- and four-membered rings are cleaved under milder conditions than are other cycloalkanes, only in the case of cyclopropanes is this a common laboratory procedure.

Cyclopropane rings are also cleaved by strong acids or halogens at room temperature, as illustrated by the following reactions:

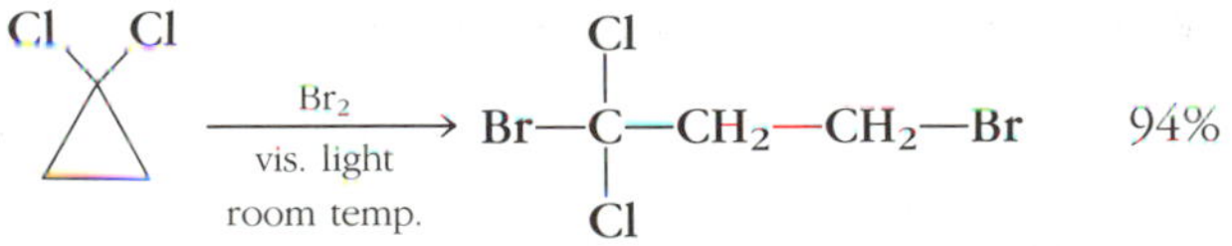

$$\triangle \xrightarrow{HBr} H{-}CH_2{-}CH_2{-}CH_2{-}Br \quad 86\%$$

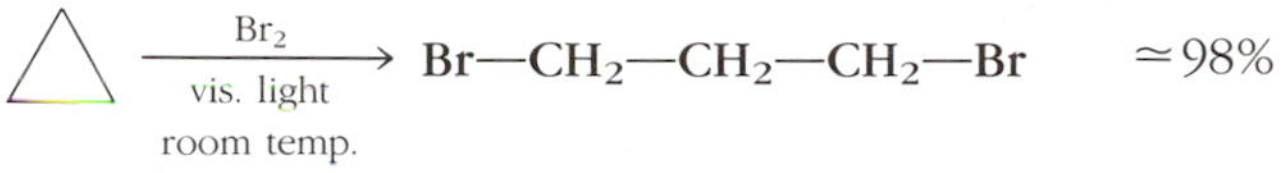

$$\triangle \xrightarrow[\text{vis. light, room temp.}]{Br_2} Br{-}CH_2{-}CH_2{-}CH_2{-}Br \quad \simeq 98\%$$

$$\text{1,1-dichlorocyclopropane} \xrightarrow[\text{vis. light, room temp.}]{Br_2} Br{-}CCl_2{-}CH_2{-}CH_2{-}Br \quad 94\%$$

The last two reactions are both free-radical chain processes, in which bromine atoms cause the actual cleavage. Under the mild conditions of the preceding reactions, the rings of other cycloalkanes are unaffected.

EXERCISE 3.1

Irradiation with visible light of a mixture of 1,1-dimethylcyclopropane and Br_2 could yield more than one ring-opened product.

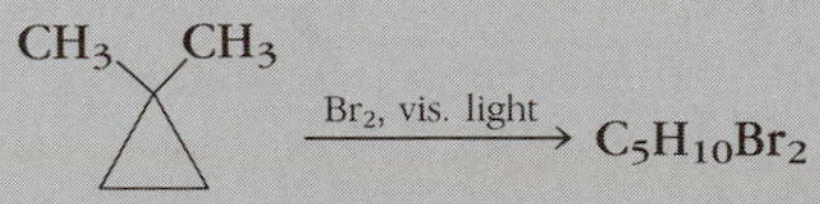

$$\text{1,1-dimethylcyclopropane} \xrightarrow{Br_2,\ \text{vis. light}} C_5H_{10}Br_2$$

Draw the possible products on the basis of the intermediates formed in the free-radical reaction. Which product should predominate? Why?

3.5 STEREOISOMERISM: *CIS-TRANS* ISOMERS IN CYCLIC SYSTEMS

The presence of a ring in cyclic compounds limits the conformational possibilities by restricting the rotation about the single bonds of the ring. Examination of a Newman projection of cyclopropane demonstrates that rotation about one of the carbon-carbon bonds could take place only if one of the other carbon-carbon linkages were broken. In the following structures, rotation about the bond between the front and rear carbons would cause the bond between the front carbon and the CH_2 group at the right to break. In the last structure (after rotation by 60°) these two atoms are no longer joined together, and we have indicated this with "bonds" to nonexistent substituents designated by question marks.

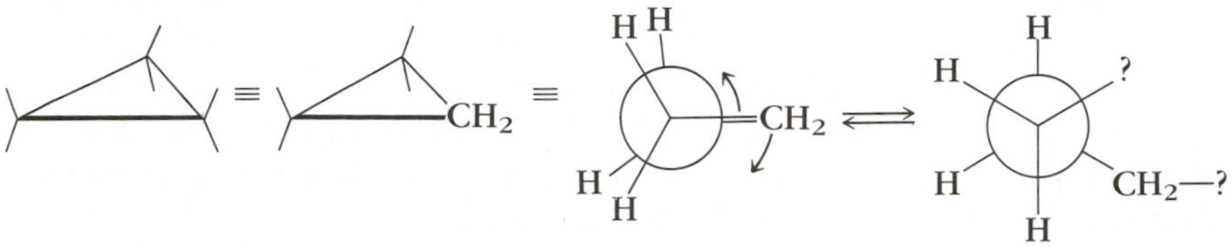

Larger rings do have some flexibility, and we illustrate this for two possible conformations of cyclopentane in the following drawings. These depict a conformational equilibrium which occurs through rotation of approximately 60° about the C—C single bond between the front and back carbons of the Newman projection.

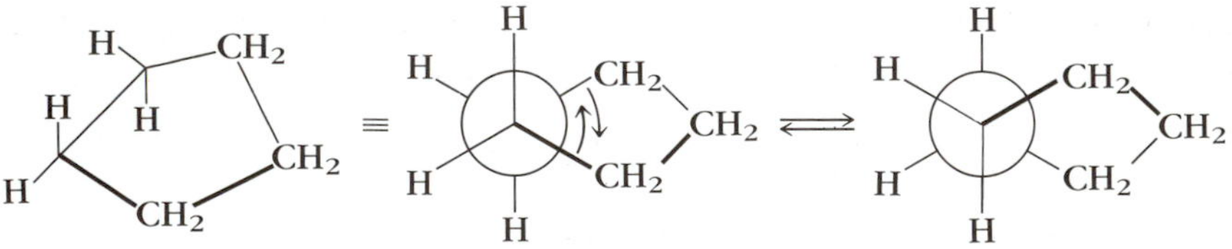

Nevertheless, rotational possibilities are still restricted, and the arrangement corresponding to a Newman projection with the CH_2 groups *anti* (resulting from a 120° rotation of the front carbon) is clearly impossible if carbon-carbon bond lengths and angles are to retain their normal values. In the following drawings this is again indicated by "bonds" to nonexistent substituents. The cycloalkanes we have been considering, in distinct contrast to acyclic compounds, are characterized by **restricted rotation** about the carbon-carbon bonds of the ring. Only for much larger rings is free rotation possible.

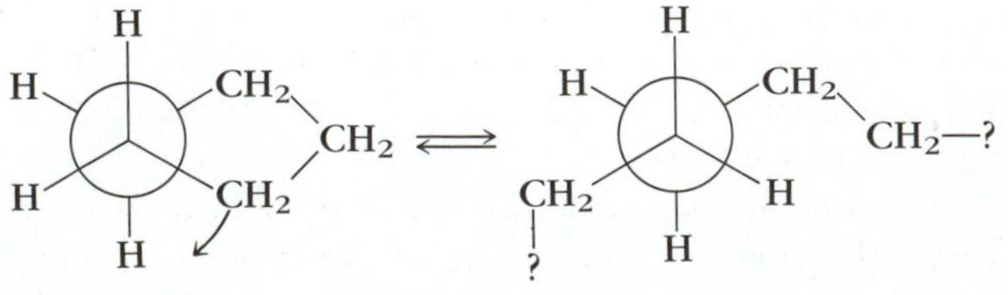

The restricted rotation of cyclic compounds allows us to show you a new kind of isomerism in which the *difference* between two isomers is in their three-dimensional structures. (As is always the case, two isomeric structures must have the *same*

molecular formula.) This isomerism is called **stereoisomerism,** where the prefix *stereo* indicates the spatial or three-dimensional aspects of a structure. Do not confuse this with the relationship between two or more *conformations* of a molecule. Isomers are distinct compounds with different physical properties, and they can be separated and isolated individually. This is not the case with conformations, which are different spatial arrangements of the *same* compound. These are ordinarily in rapid equilibrium and cannot be separated or isolated individually.

We can illustrate the concept of stereoisomers with dimethylcyclopropane, a hydrocarbon for which three different isomeric forms have been isolated, each with unique physical properties. Clearly all three are not constitutional isomers, because only two constitutions are possible for dimethylcyclopropane. The methyl groups must be on either the same or adjacent carbon atoms of the ring, leading to 1,1-dimethylcyclopropane and 1,2-dimethylcyclopropane, respectively.

CH_2 / CH_2—C(CH_3)(CH_3)

1,1-Dimethylcyclopropane

CH_2 / CH_3—CH—CH—CH_3

1,2-Dimethylcyclopropane

However, two different three-dimensional structures are possible for the latter. Both stereoisomers have the same constitution (the connectivity of atoms which is denoted by the name), yet they differ in their three-dimensional structures. In order to show this stereoisomerism, we must use structural formulas that have three-dimensional significance. The following structures clearly show the two different spatial arrangements of the methyl groups.

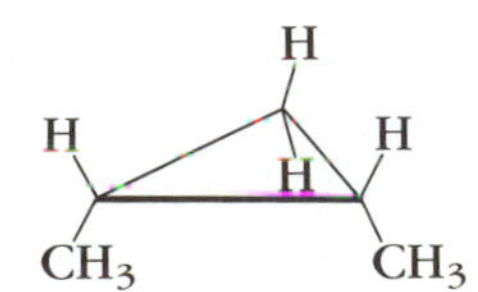

cis-1,2-Dimethylcyclopropane

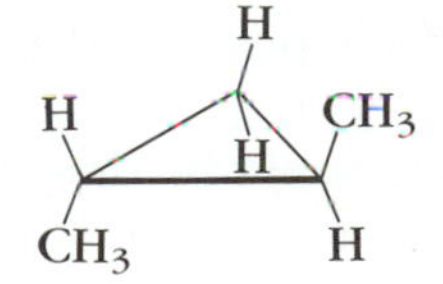

trans-1,2-Dimethylcyclopropane

In the first structure both methyl groups lie below the plane of the three-membered ring. This isomer is designated as ***cis*** (from the Latin word meaning "on this side of"), because both substituents are on the same side of the ring. The second isomer has one methyl group above the plane of the ring and one below the plane. This isomer is designated as ***trans*** (from the Latin word meaning "across") since the substituents are on opposite sides of the three-membered ring. Stereoisomers of this type are properly known as **diastereomers,** and we will present a detailed discussion of this in Chapter 6. For the time being we will simply refer to them as *cis-trans isomers.*

A second type of drawing is commonly used to show *cis-trans* isomerism with cyclic compounds. The ring is simply drawn as the appropriate polygon, and the *relative* orientation of the appropriate substituents is shown with wedges and dashed lines. Keeping in mind that we are drawing a "planar" ring for convenience, we use a wedge to indicate that a substituent lies above the ring (i.e., above the plane of the paper), whereas a dashed line denotes a substituent that is below the ring (i.e., behind the plane of the paper). The two isomeric 1,2-dimethylcyclopropanes can be drawn in the following manner:

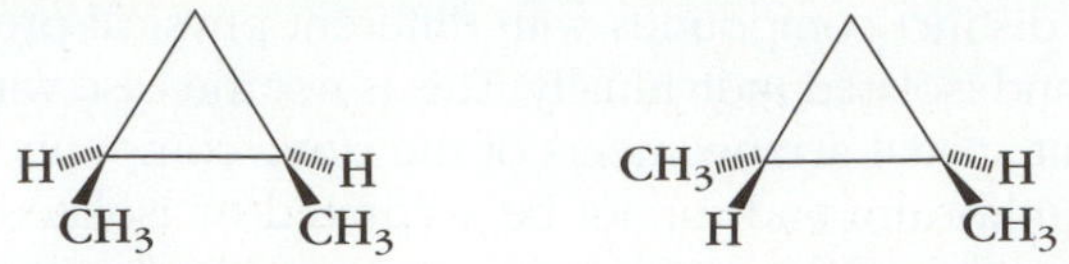

cis-1,2-Dimethylcyclopropane *trans*-1,2-Dimethylcyclopropane

Similarly, this type of drawing can conveniently be used to show the *cis*- and *trans* isomers of 1,3-dichlorocyclopentane.

cis-1,3-Dichlorocyclopentane *trans*-1,3-Dichlorocyclopentane

When should you look for stereoisomerism in cyclic compounds? The necessary conditions can be specified quite easily:

***Cis-trans* isomerism is possible whenever two or more carbon atoms of a ring each have two different substituents.**

Of course, each carbon atom has a total of four substituents and the other two substituents are carbon atoms *of the ring*. In Chapter 6 you will encounter stereoisomers for which you will need to evaluate all four substituents of a ring carbon, including the two that are part of the ring, but for the purpose of deciding *cis-trans* isomerism it is only necessary to consider the substituents that are attached to the ring.

A variety of other representations can also be used to show the orientations of substituents on a ring. For example, an "edge on" view of *trans*-1,2-dibromocyclobutane might be drawn in any of the following ways:

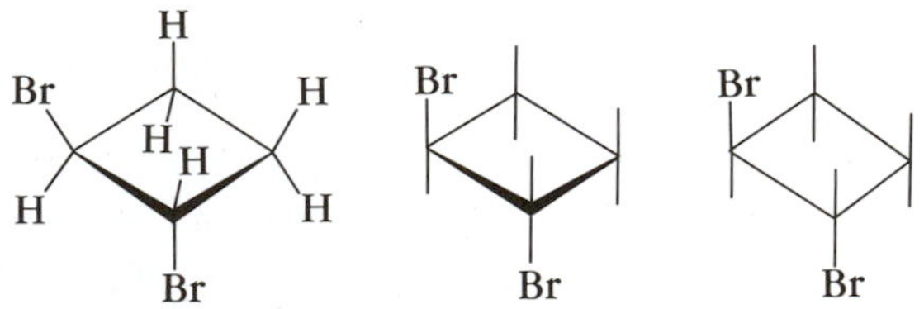

However, such drawings frequently lead to mistakes, and therefore we will not ordinarily use them in this text. When a three-dimensional representation of the *entire* molecule is needed, drawings such as those in Figures 3.2–3.4 (which show the actual shape of the ring) are preferable.

How do you distinguish between *cis* and *trans* isomers when you name a compound? As we have illustrated in the preceding examples, you merely begin the name of the compound with the appropriate prefix, *cis*- or *trans*-. Even when there are several substituents on the ring, you only need to specify the relationships to a *reference* position, which is the lowest-numbered carbon in the ring that is appropriately substituted. This is denoted by the prefix *r,* as shown in the following example (where C-1 is the reference position):

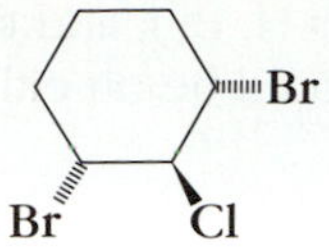

Br Br Cl

r-1-*cis*-3-Dibromo-*trans*-2-chlorocyclohexane

The possibility that substituents on a ring could have either a *cis* or a *trans* relationship means that you must evaluate the possible products of a reaction not only in terms of *constitutional isomers* but also with respect to *stereoisomers*. We will illustrate this problem by looking at a specific reaction, the monochlorination of bromocyclobutane.

$$C_4H_7Br \xrightarrow[\text{vis. light}]{Cl_2} C_4H_6BrCl \text{ isomers} + \text{polychlorinated products}$$

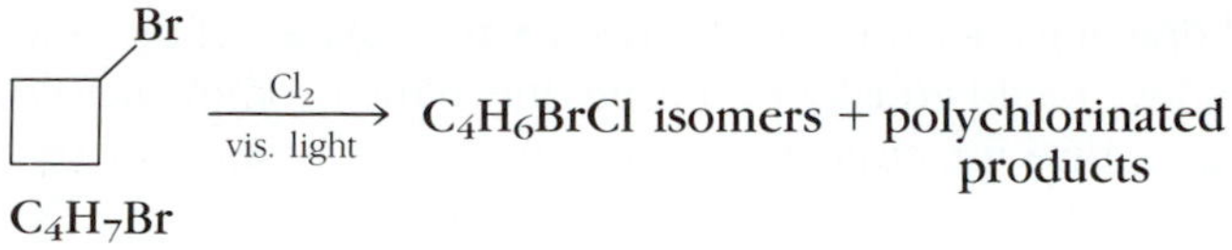

In evaluating this reaction you should keep in mind that the two CH_2 groups adjacent to the CHBr of bromocyclobutane are equivalent. This is easily seen when the molecule is drawn in an orientation to show its plane of symmetry,

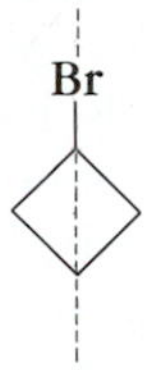

Br

The equivalence of the two CH_2 groups is also demonstrated by naming the following structures:

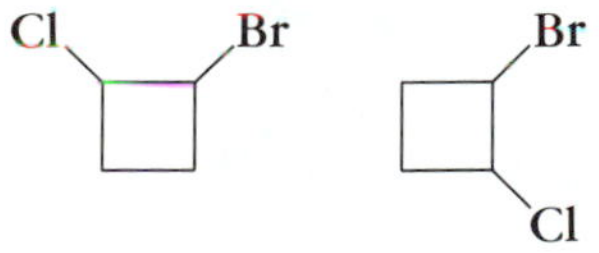

Cl Br Br Cl

Both would be named 1-bromo-2-chlorocyclobutane, so they have the same constitution and must have been formed by replacement of hydrogen at equivalent positions.

The chlorination reaction of bromocyclobutane could give monochloro products with three different constitutions:

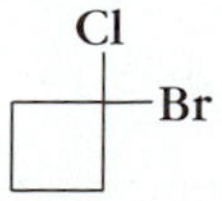

1-Bromo-1-chloro-cyclobutane	1-Bromo-2-chloro-cyclobutane	1-Bromo-3-chloro-cyclobutane

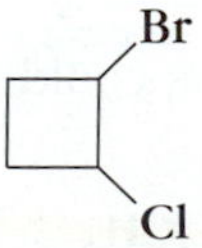

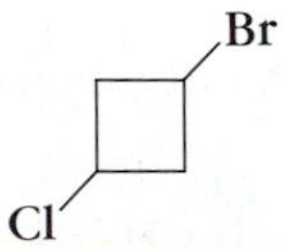

Both the second and third structures possess the structural requirements necessary for *cis-trans* isomerism. In each case two carbon atoms of the ring have two different substituents (either H, Cl or H, Br), and the two halogens (or the two hydrogens) on these carbon atoms could be on either the *same* or on *opposite* sides of the ring.

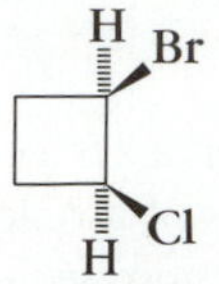

cis-1-Bromo-2-chlorocyclobutane 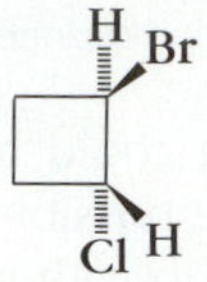 | *trans*-1-Bromo-2-chlorocyclobutane 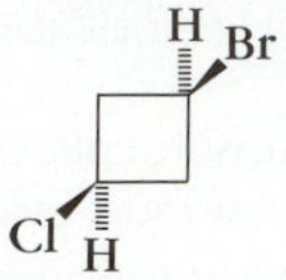 | *cis*-1-Bromo-3-chlorocyclobutane 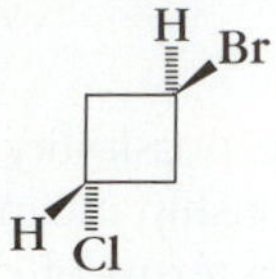 | *trans*-1-Bromo-3-chlorocyclobutane

The choice of drawing the bromine atom *above* the "plane" of the ring in each case is arbitrary, and we could equally well have drawn the bromine *below* the "plane." The important feature for *cis-trans* isomerism is the location of one substituent *relative to the other.*

For the third constitutional isomer, 1-bromo-1-chlorocyclobutane, only one of the ring carbons has two different substituents (Cl, Br). Each of the other three carbon atoms has two identical substituents (H, H), so stereoisomerism is not possible for this compound.

To avoid confusion you should recognize that the terms *cis* and *trans* are used to describe *relationships between atoms (or groups) in a single molecule.* Sometimes these terms are used even when stereoisomerism is not possible. For example, the hydrogen atom of 1-bromo-1-chlorocyclobutane that is designated with the label *a* in the following drawing, can be described as being *cis* to the chlorine atom and *trans* to the bromine atom. The ability to distinguish between two nonequivalent atoms such as H_a and H_b can be very useful for you in predicting and interpreting the reactions of such molecules.

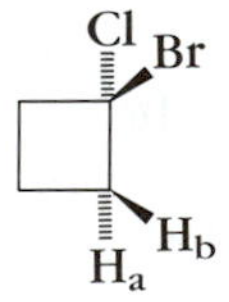

Example 3.1 Chlorination of methylcyclohexane would yield a variety of isomeric monochloro derivatives, and *cis-trans* isomerism is possible for several of the constitutional isomers that could be formed. In each such case draw both possible stereoisomers.

Solution The following constitutional isomers could be *formed*:

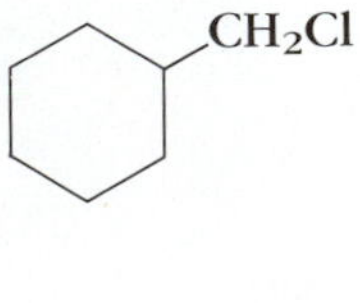

1 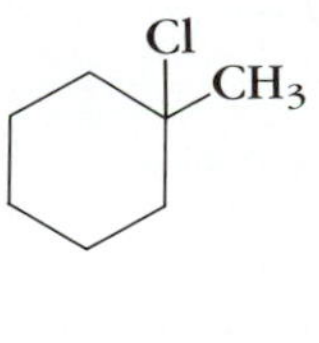 2 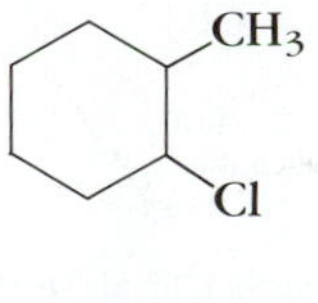 3 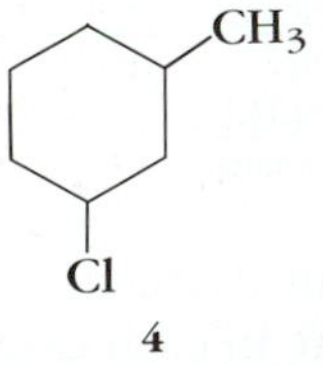 4 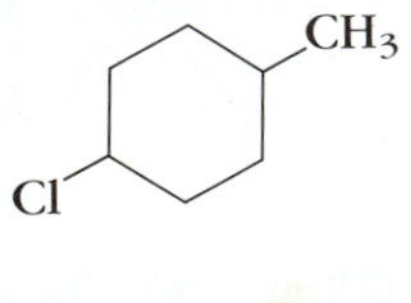 5

Cis-trans isomerism is not possible for structures **1** and **2** but is possible for structures **3, 4,** and **5.** You should recognize that we have made an arbitrary choice in drawing the methyl group *forward* in these structures. If we had drawn the methyl group to the *rear,* then the chlorine atom would have been drawn *back* in the *cis* structures and *forward* in the *trans* structures.

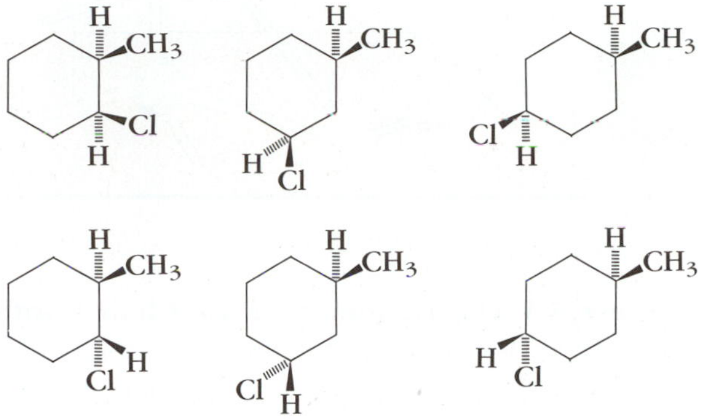

H CH3 Cl H   H CH3 H Cl   H CH3 Cl H

H CH3 H Cl   H CH3 Cl H   H CH3 H Cl

3   4   5

3.6 CONFORMATIONAL ANALYSIS: CYCLOHEXANE

Cyclohexane derivatives are ubiquitous in nature, and many of these compounds have been found to possess valuable medical or biochemical activity. As modern techniques of organic chemistry developed during the first half of the twentieth century, scientists devoted considerable effort to understanding the structure of cyclohexane and its derivatives. Two of these scientists, Odd Hassel (Norway) and Derek H.R. Barton (Great Britain) shared the 1969 Nobel Prize in chemistry for their work on the conformational analysis of cyclohexanes.

If you closely inspect the chair conformation of cyclohexane in Figure 3.3, you will see that there are two kinds of hydrogen substituents. Taking the orientation of the carbon framework to be essentially horizontal, there are C—H bonds that are vertical and others that are more or less horizontal. The vertically oriented bonds are designated **axial,** and those which are approximately horizontal are called **equatorial.** The chair form of cyclohexane has a vertical C_3 axis through the center of the ring, and the *axial* bonds are so named because they are all parallel to this axis. The C_3 axis also provides the name for the other set of substituents. By analogy to a rotating planet, a plane perpendicular to the axis defines the equator, and bonds in this region are called *equatorial.* We have labeled the axial and equatorial bonds as "a" and "e," respectively, in the following structure.

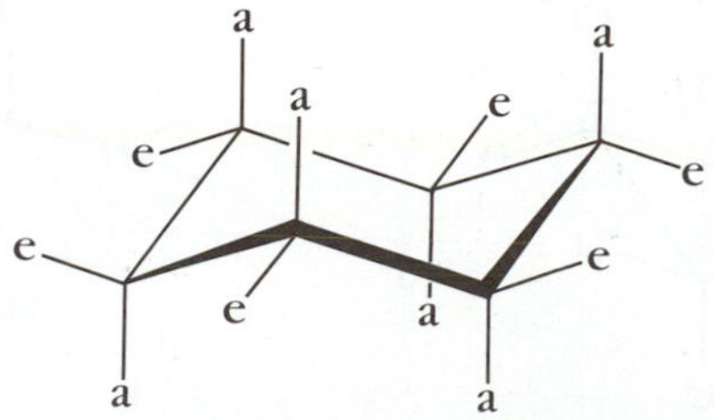

a a a e e e e e e e a a a

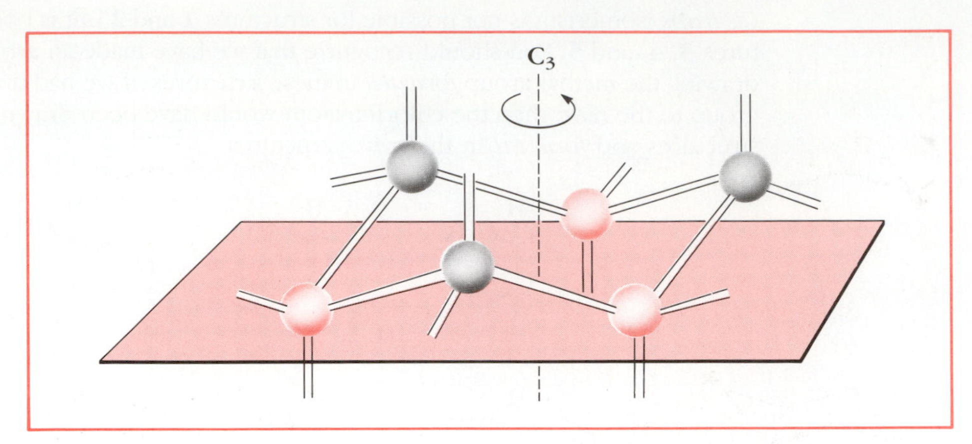

Figure 3.5 The Geometry of the Chair Conformation of Cyclohexane. The three "lower" carbon atoms (light color) define a plane, shown as the shaded surface on which they are resting. A second plane, parallel to the first, but not depicted, is defined by the three "upper" carbon atoms (dark color). The axial bonds are all perpendicular to these two planes; the axial bonds on the upper and lower carbon atoms lie above and below the respective planes. This is readily seen for the lower three carbon atoms, for which the axial bonds lie below the shaded plane. The equatorial bonds, on the other hand, lie between the two planes. The chair conformation of cyclohexane has a C_3 axis, and each pair of carbon-carbon bonds on opposite sides of the ring forms a set of parallel line segments. The carbon-carbon bonds of the ring also define the orientation of the equatorial substituents; if you take any ring carbon-carbon bond, the equatorial bonds on *adjacent* carbon atoms will be parallel to that carbon-carbon bond.

A useful way of visualizing the three-dimensional aspects of a cyclohexane ring is illustrated in Figure 3.5. The three "lower" carbons of the ring define a plane, and the three "upper" carbons define a second, parallel plane (not shown). All the axial bonds are perpendicular to these two planes. The equatorial bonds, although approximately horizontal, all lie *between* these two planes. You can help visualize these relationships by constructing a molecular model (see Section 3.7) and placing the model on a flat surface after removing the axial substituents from the three lower carbon atoms.

The preceding description of the cyclohexane ring might lead you to guess that a substituted derivative such as chlorocyclohexane could exist in two forms, one with the chlorine in an axial orientation and another in which the chlorine is equatorial.

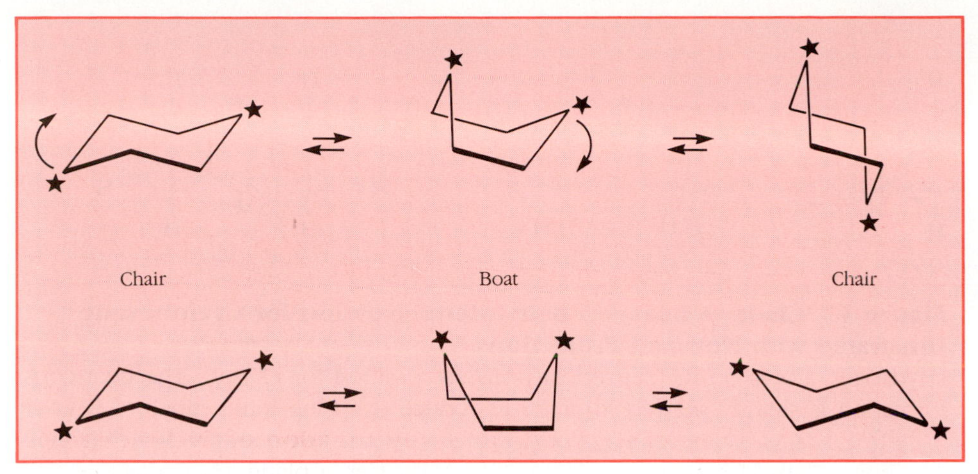

Figure 3.6 Chair → Boat → Chair Interconversion for Cyclohexane. Two of the carbon atoms are starred for reference, and the hydrogens are omitted for simplicity. In the lower equation the structures are rotated slightly in order to show the usual perspective.

Your guess would be correct, and these two structures are indeed possible. They are not isomers, however, because the individual structures cannot be isolated. Instead, they are merely two conformations of a single compound that are interconverting in a rapid equilibrium by means of rotation about carbon-carbon single bonds. The process proceeds via a different conformation called a *boat* and is illustrated for cyclohexane itself in Figure 3.6. As we have depicted in the figure, movement of one of the carbon atoms relative to the other five produces the boat conformation. Subsequent movement of the other starred carbon atom in the opposite direction then leads to the second possible chair conformation. The boat conformation is considerably less stable (by about 6 kcal/mol) than the chair, and the equilibrium mixture of conformers contains only an extremely small fraction of the boat. Nevertheless the boat does play an important role in the conformational analysis of cyclohexane, because the chair-chair interconversion must proceed by way of a boat conformation.

The chair-boat-chair equilibrium can also be analyzed with Newman projections, and we have shown this in Figure 3.7. These drawings do not show the overall shape of the carbon skeleton as well as the drawings in Figure 3.6, but other features are much more evident. Consider, for example, the upward movement of the starred carbon atom in the structure on the left to produce the boat. You can now see that this is simply a simultaneous rotation of the two carbon-carbon bonds that join the front and back carbon atoms of the double Newman projections.

The drawings in Figure 3.7 also reveal why the chair is considerably more stable than the boat. All bonds between adjacent carbon atoms are completely staggered in the chair, but the situation for the boat is quite different. The unskewed boat geometry is unfavorable, and only by adopting a slightly twisted geometry can it reduce the eclipsing interactions. The correct name for this minimum-energy conformation is the *twist boat,* and this is the structure that is being considered when chemists discuss the boat conformation of a cyclohexane derivative.

Figure 3.7 Chair → Boat → Chair Interconversion for Cyclohexane Illustrated with Newman Projections.
(Hydrogen atoms on the starred carbons are omitted for clarity.) The boat conformation is slightly skewed in order to reduce eclipsing interactions.

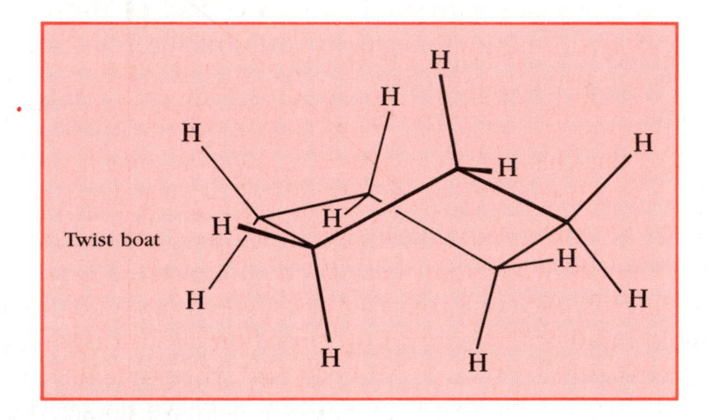

The transition state for interconversion of the chair and boat conformations corresponds closely to the following drawing with five of the carbon atoms in one plane.

Transition state for
chair-boat interconversion

The substituents on three of these carbons are forced to be eclipsed, and this results in an activation barrier that is somewhat higher than the rotational barriers observed for acyclic alkanes. A barrier of about 11 kcal/mol is found for the chair → boat interconversion. Because this is the difficult step in the overall chair → chair interconversion, the overall barrier for that process is approximately the same. Even though this energy barrier is larger than those for acyclic alkanes, the facile transfer of kinetic energy among molecules still results in rapid equilibration of the two chair conformations of cyclohexane at normal temperatures.

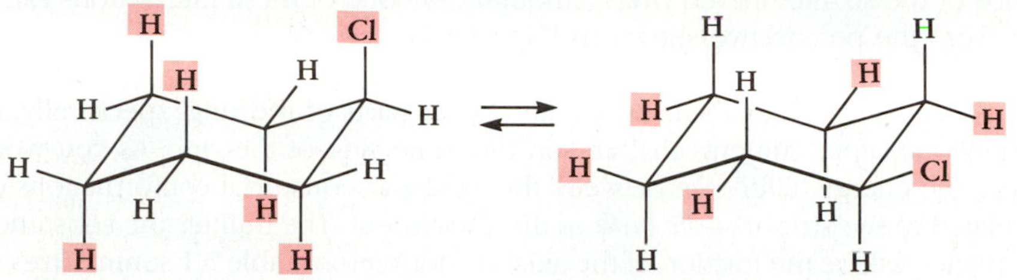

Figure 3.8 Chair-Chair Equilibrium of Chlorocyclohexane.
All the axial substituents (highlighted) of the structure on the left become equatorial in the second conformation.

Figure 3.8 depicts the chair-chair equilibrium for chlorocyclohexane, and these drawings show you that the axial and equatorial forms of a monosubstituted cyclohexane are indeed merely two conformations of a single molecule. An important consequence of chair-chair interconversion is also illustrated in this figure. We have highlighted each of the axial substituents in the first drawing, and *all* these axial substituents become equatorial in the second chair conformation. Simultaneously, all equatorial substituents become axial. This complete interchange of axial and equatorial substituents is *always* observed when one chair form of a cyclohexane is converted to the other.

The two chair forms of a monosubstituted cyclohexane are clearly not equivalent, and we conclude that they must have different energies. One of them will therefore predominate in the equilibrium mixture, but which one? Your knowledge of conformational analysis from Chapter 2 actually allows you to answer this question if you first draw the molecules appropriately, and we have done this in Figure 3.9. The more stable (and therefore predominant) conformation is that in which the substituent is equatorial. In the equatorial conformation the substituent

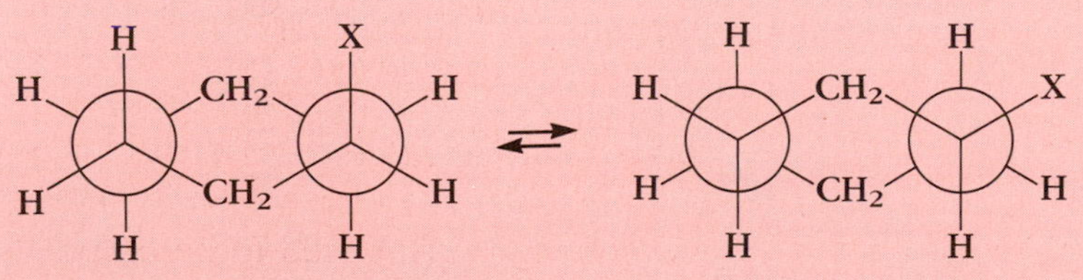

Figure 3.9 Chair-Chair Equilibrium for a Monosubstituted Cyclohexane.
In the structure on the left, the axial substituent, X, is gauche to the CH_2 group of the ring. For the more stable conformation (right) in which X is equatorial, the relationship between X and the CH_2 group is *anti*.

has the more favorable *anti* relationship to the CH_2 group on the adjacent carbon. In contrast, an axial substituent suffers *gauche* interactions with two of the CH_2 groups of the six-membered ring (although only one of these interactions can be seen from the perspective shown in Figure 3.9).

The decreased stability of the axial conformation results from *nonbonded* interactions between the axial substituent and other parts of the ring, specifically, the other two axial substituents that are on the same side of the ring. As you would expect, the energy difference between the axial and equatorial conformations can be related to the size or *steric bulk* of the substituent. The bulkier the substituent, the smaller will be the fraction of the axial conformation. Table 3.1 summarizes the preference for the equatorial conformer of a simple series of alkyl- and halocyclohexanes. For both series the predominance of the equatorial form increases as the size of the group increases. These values will give you a good idea of the relative "size" or "bulk" of various substituents. Notice that the fraction of the equatorial conformation for 1-bromocyclohexane is smaller than that for 1-chlorocyclohexane, even though a bromine atom, with a larger number of electrons, has a larger diameter. Why is this? The carbon-bromine bond is longer than the carbon-chlorine bond, which increases its distance from the other axial substituents. This offsets any increase in steric interactions that would result from the larger diameter of the bromine atom, and the *effective* steric bulk is approximately the same for Br and Cl.

Conformational analysis becomes more complex for cyclohexanes with more than one substituent. For example, in neither conformation of *cis*-1-chloro-4-methylcyclohexane can both substituents be equatorial.

TABLE 3.1 Axial-Equatorial Preferences of Some Monosubstituted Cyclohexanes at Room Temperature*

Substituent	ΔG kcal/mol	Equatorial/axial ratio	Percent equatorial
F	0.2	1.3	56
Cl	0.4	2.1	67
Br	0.4	1.9	65
CH_3	1.7	17.3	94.5
C_2H_5	1.8	18.8	95
iso-C_3H_7	2.2	36.9	97
tert-C_4H_9	4.7	3.5×10^3	>99.9

*$\Delta G = -2.3\,RT \log K$, where $K = [Ax]/[Eq]$.

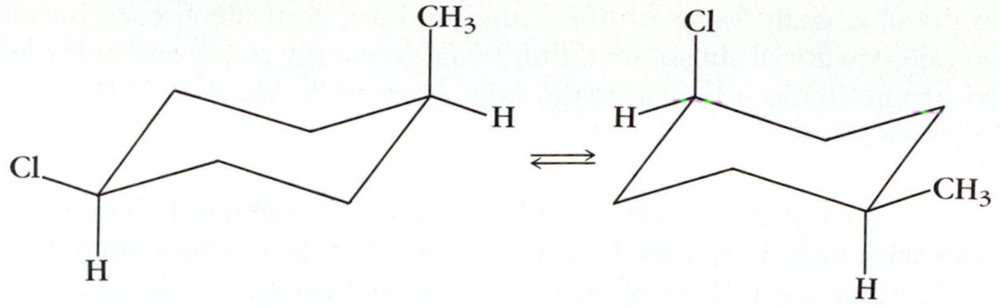

You should be careful to understand that the substituents do have a *cis* relationship in these drawings. To recognize the relationship, you should consider both substituents on each of the carbon atoms. On C-1 the chlorine is *above* the hydrogen atom, and on C-4 the methyl group is also *above* the hydrogen atom. Therefore the chloro and methyl groups are *cis* to each other. You might anticipate that the second conformation would be more stable, because the data of Table 3.1 indicate that the methyl group has a greater preference for being equatorial than does the chlorine atom. Indeed, *cis*-1-chloro-4-methylcyclohexane has been found experimentally to exist predominantly in the conformation in which the methyl group is equatorial.

The interaction between two axial substituents (other than hydrogen) on the same side of the ring is particularly unfavorable.

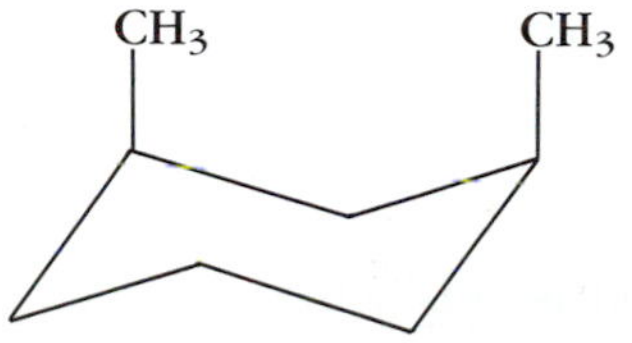

The energy associated with such *1,3-diaxial interactions* is in the range of 4 kcal/mol for two methyl groups. Calculation of the energy difference using the relationship $\Delta G = -2.3\,RT \log K$ will convince you that a compound such as *cis*-1,3-dimethylcyclohexane would exist almost entirely in the *other* chair conformation.

We stated earlier that the boat conformation is much less stable than the chair conformation. In most instances the only role played by the boat is that of a rather unstable intermediate in the interconversion of the more stable chair conformations. But there are unusual circumstances in which a cyclohexane derivative can exist to a large degree in a boat conformation, and *cis*-1,4-di-*tert*-butylcyclohexane is an example.

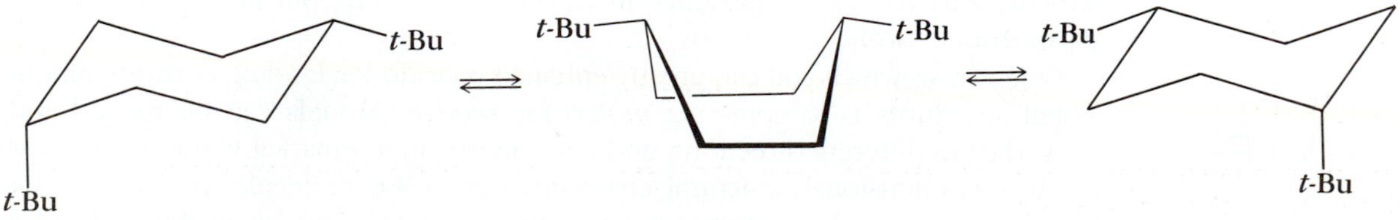

The two possible chair forms of this compound are equivalent, each having one axial and one equatorial substituent. But the large energy requirement for an axial *tert*-butyl group (Table 3.1) causes the boat form to be more favorable than the chair. Such cases are quite rare, and ordinarily you only need to consider chair conformations of cyclohexane derivatives.

Although chair representations are necessary for describing the conformations of cyclohexanes, they do not always provide the best way to show other structural features. Consider the following drawing of *trans*-1-bromo-2-chlorocyclohexane.

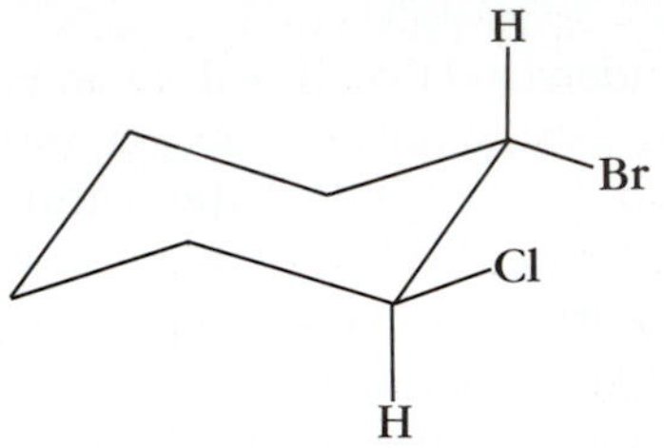

H
Br
Cl
H

It is obvious that the axial hydrogen atoms on C-1 and C-2 are *trans* to each other, but the *trans* relationship between the halogen atoms is less apparent. Unless it is important to show the conformation, this compound can be drawn much more conveniently as:

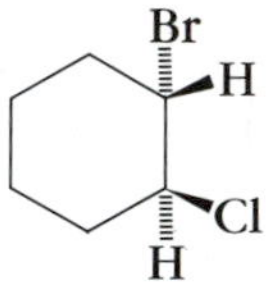

Br
H
Cl
H

This drawing fully defines all aspects of the molecular structure (other than conformation), and the *cis-trans* relationships are actually easier to recognize. Unless we need to show the actual conformations, we will continue to draw cyclohexane rings as hexagons throughout this text.

3.7 MOLECULAR MODELS

To a very large extent organic chemistry is a three-dimensional study, yet communication via printed words and drawings is limited to only two dimensions. The problem is alleviated to a great extent by use of the stereochemical conventions that we introduced in Chapters 1 and 2, but these three-dimensional representations must still be drawn on two-dimensional paper. The third dimension is sometimes difficult to visualize. You will have less difficulty with three-dimensional structures as you gain experience in working with them, but the problem never disappears entirely.

One way in which you can greatly enhance your understanding of three-dimensional structures is to construct *molecular models*. Models can be moved and viewed from different directions, and you can obtain a remarkably good "feel" for the three-dimensional structural characteristics of the molecule in question. Of course, you must always remember that these are only *models* of the molecular structure. The hardness and elasticity of the metal, plastic, or wood cannot accurately represent the properties of the nuclei and electrons of a real molecule. But within these limitations molecular models can provide a great deal of information in terms of spatial characteristics of molecular structures. Molecular models are particularly valuable for illustrating differences between isomers or between two or more conformations of a single molecule. Most importantly, the use of molecular models in combination with a pencil and paper provides one of the best possible ways for you to to develop skill in drawing and understanding the various three-dimensional representations that must be used in organic chemistry.

Two main types of molecular models are commonly employed in the study of organic chemistry. One type delineates and emphasizes the connectivity (i.e., the bonding relationships) between the various atoms, and the other type emphasizes the size and shape of the various atoms and groups in a molecule. Models of the first type are called *skeletal* models because the bonding skeleton is the primary feature that is illustrated in three dimensions. Many types of skeletal models are available commercially, and these range from inexpensive models intended for student use to rather expensive sets that are designed for research applications. In Figures 3.10 and 3.11 we show two types of inexpensive skeletal models of chloromethane.

Figure 3.10 shows ball-and-stick models in which spheres of various colors represent different atoms. The "bonds" of the model shown in Figure 3.10 are metal springs, which fit into holes drilled in the wooden spheres. Models of this type are usually fabricated from wood or plastic, and plastic or wooden rods (rather than metal springs) are often used to connect the spheres.

A second type of inexpensive skeletal model employs metal "atoms" (Figure 3.11, far left) and "bonds" made of plastic tubes, into which the prongs of the atom are inserted. With this type of skeletal model there are no spheres to denote the atomic centers, and plastic tubes of various colors are used to designate different substituents.

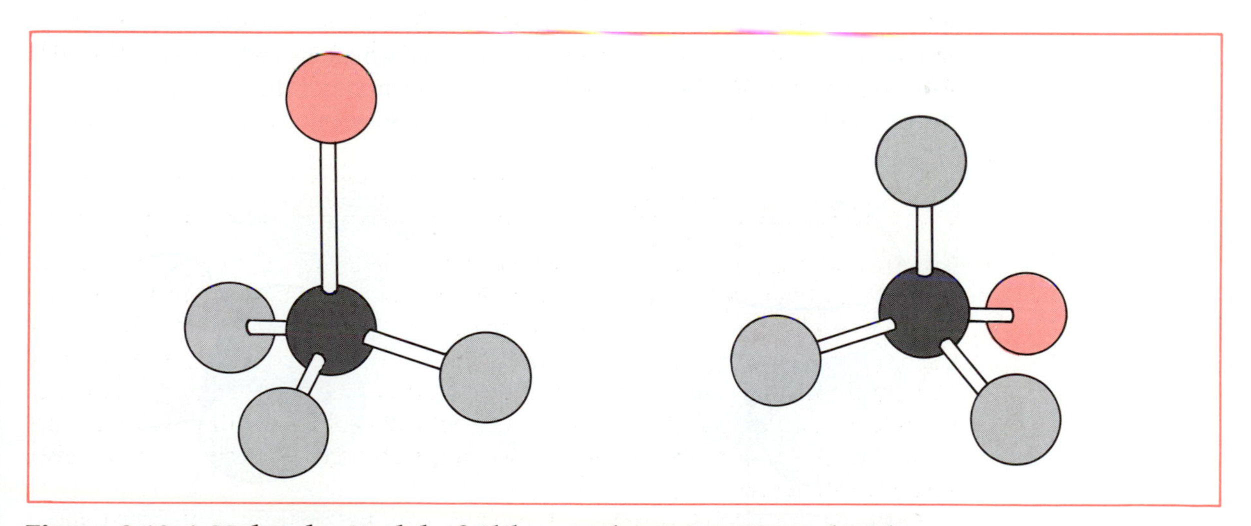

Figure 3.10 A Molecular Model of Chloromethane Constructed with Skeletal Molecular Models.
With this type of skeletal model the atoms are represented by spheres that are connected by "bonds." Different atom types are designated by using spheres of different colors.

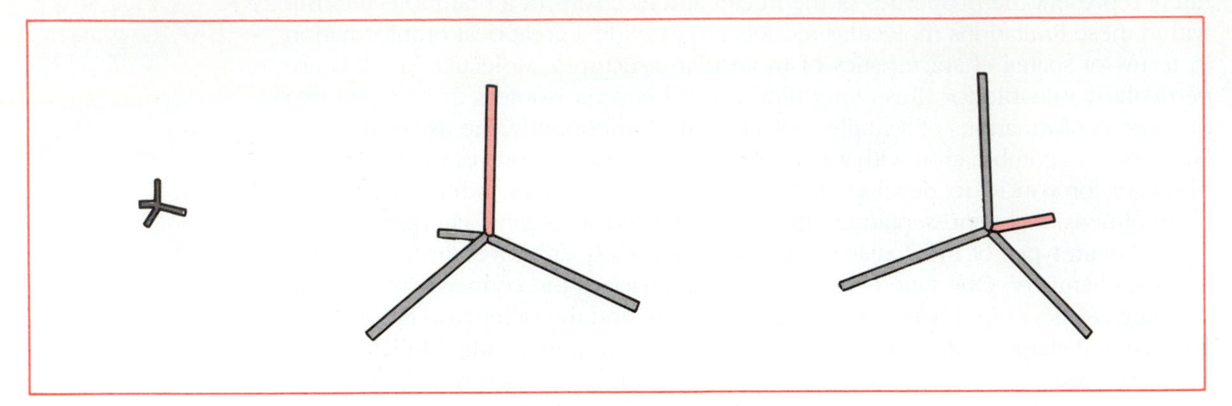

Figure 3.11 A Molecular Model of Chloromethane Constructed with a Second Type of Skeletal Molecular Model.
The small metal "atoms" (far left) are connected with plastic tubes. Plastic tubes of various colors are used to denote different substituents.

The second basic type of molecular model is a *space-filling model.* In contrast to the skeletal models, which do not differentiate between the sizes of various atoms, space-filling models are designed to show the three-dimensional size of the electron cloud around each of the atoms in a molecule. This allows you to analyze the steric interactions between different atoms and groups, and even the interactions between two different molecules can be evaluated. A space-filling model of chloromethane is shown in Figure 3.12. While space-filling models provide information about the relative sizes of different atoms, the bonding relationships are much more difficult to see than with skeletal models. You should carefully compare the various models of chloromethane in Figures 3.10–3.12.

Skeletal and space-filling models each have their advantages and disadvantages, and you can see some of these in Figure 3.13. This figure shows both space-filling and skeletal models of the *gauche* and *anti* conformations of 1,2-dibromoethane. The bonding relationships as well as the orientations of the various bonds are clearly illustrated by the skeletal models, but these relationships are much more

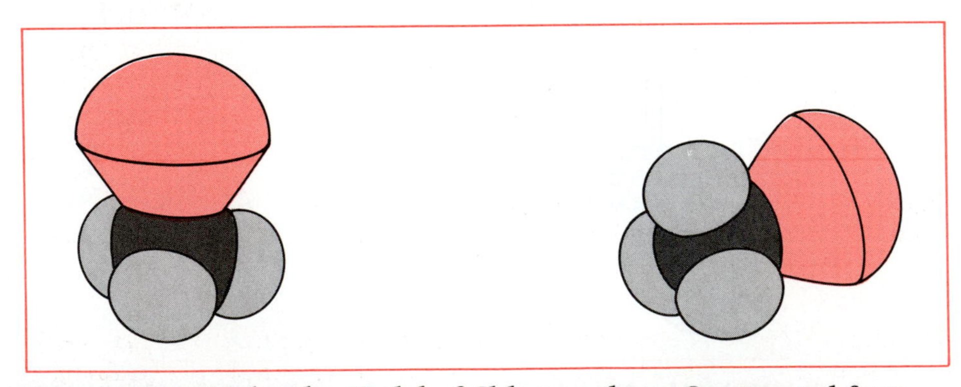

Figure 3.12 A Molecular Model of Chloromethane Constructed from Space-Filling Models.
The partial spheres represent the electron clouds surrounding the individual atoms, and they provide an indication of the "size" of each of the various groups in a molecule. Different atoms are denoted by different colors.

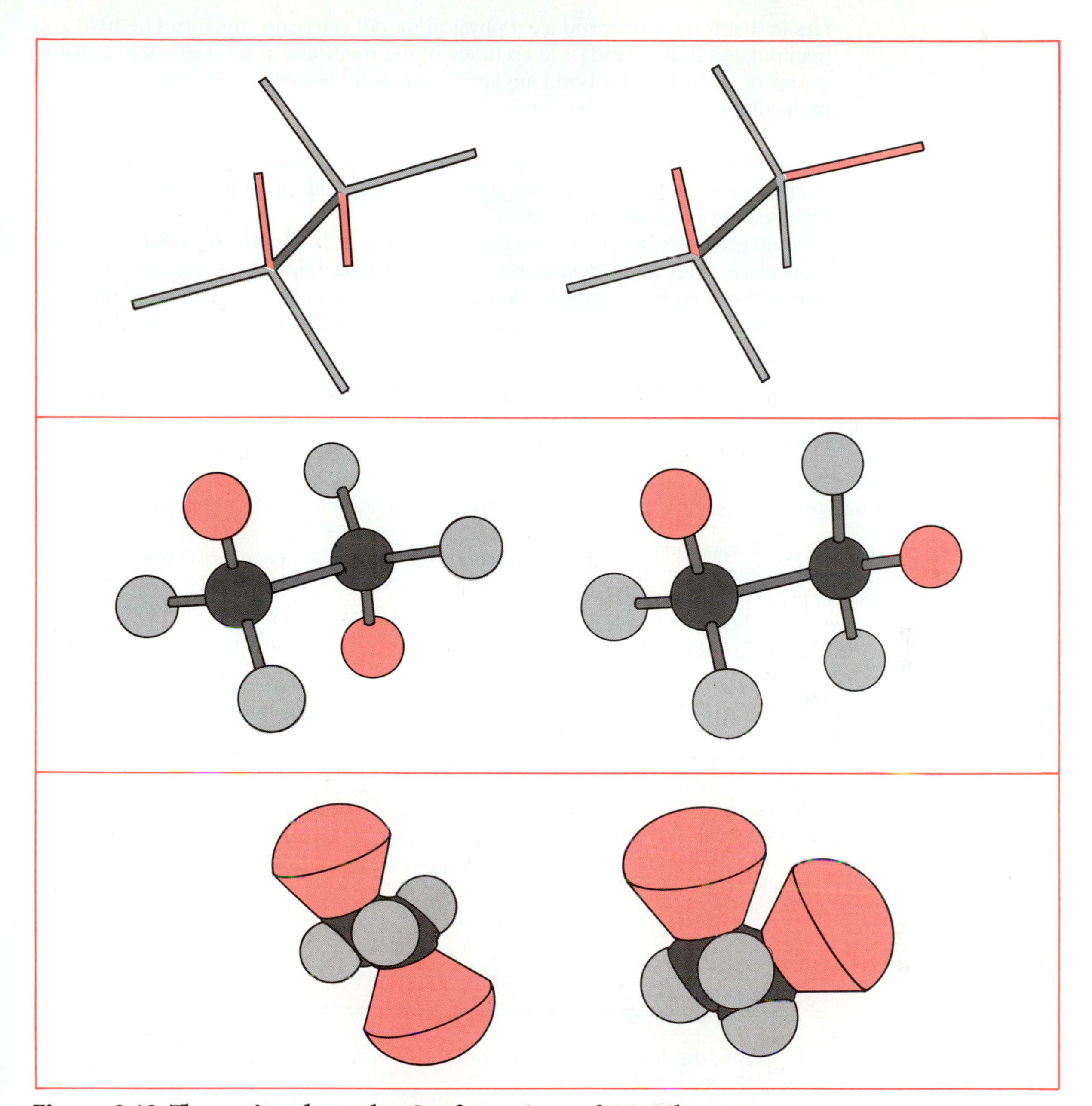

Figure 3.13 The anti and gauche Conformations of 1,2-Dibromoethane Constructed with Several Different Kinds of Molecular Models.

difficult to see with the space-filling models. On the other hand, the large steric interaction between the two bromine substituents in the gauche conformation is evident only with the space-filling models.

For construction of molecular models of very high accuracy, none of the above types of model offers the ideal solution. The rapid advances in computing and computer graphics in recent years have led to an entirely different sort of "molecular model," one which is entirely mathematical. Although computer modeling suffers in that it does not produce a physical model (i.e., one which can be held and manipulated in the hand), the accuracy of the method can be very good. Reliable values of bond lengths, bond angles, interatomic distances, and other features of molecular structure can be obtained from relatively easy computations. New techniques in computer graphics even allow you to rapidly manipulate the display (i.e., to rotate it in various directions), and this provides a sense of the three-dimensional nature of the structure. Programs for molecular graphics are now available for a variety of home computers.

Simple molecular models such as those shown in Figures 3.10–3.13 still have their place. They are portable and convenient, and they can be obtained inexpensively. They can be invaluable either to a research chemist or to a beginning student such as yourself, who is learning to use a pencil and paper to draw and manipulate three-dimensional organic structures.

3.8 TERMS AND DEFINITIONS

Acyclic. Noncyclic; containing no rings.

Boat. A conformation of the cyclohexane ring system. It is almost always much higher in energy than the chair conformations, and only an extremely small fraction of the total molecules will exist in the boat form at any one time.

Chair. A conformation of the cyclohexane ring system, which is almost always the optimum structure. Two chair conformations are always possible, differing by interchange of axial and equatorial positions of the substituents.

Cis. A term describing that relationship between two substituents on a ring in which the substituents (which must be on different carbon atoms) are on the same side of the ring.

Cyclic. Having atoms that are connected to form a ring.

Restricted rotation. The absence of free rotation about a bond. Restricted rotation results from a structural feature that makes rotation about the bond difficult or impossible.

Stereoismerism. The relationship between two compounds having the same constitution but differing in some aspect of three-dimensional molecular structure.

Trans. A term describing that relationship between two substituents on a ring in which the substituents (which must be on different carbon atoms) are on opposite sides of the ring.

3.9 PROBLEMS

3.1 Name the following compounds:

(a) 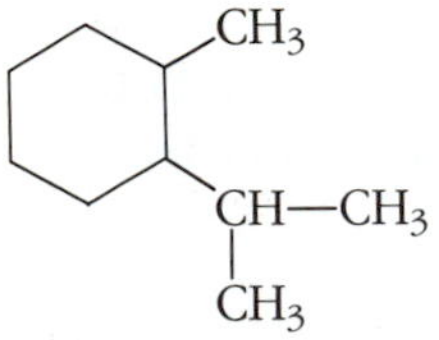 CH_3 / $CH—CH_3$ / CH_3

(b)

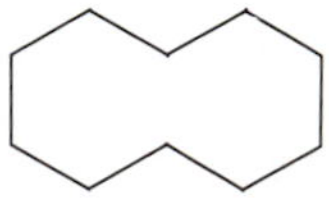

(c) 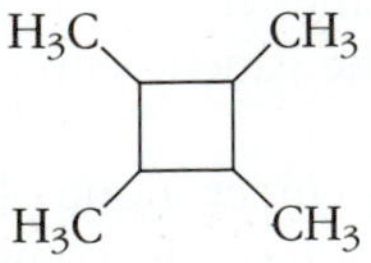 H_3C CH_3 / H_3C CH_3

(d) 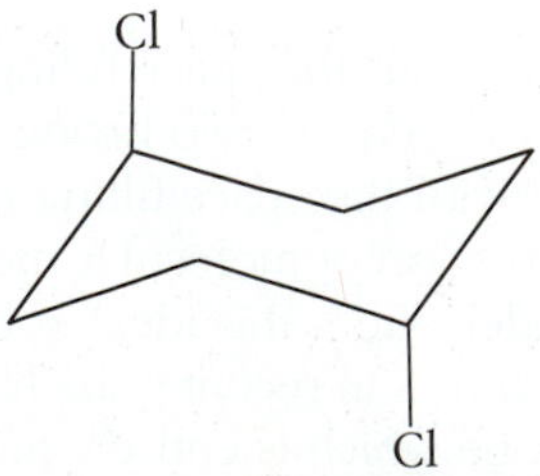 Cl / Cl

(e) 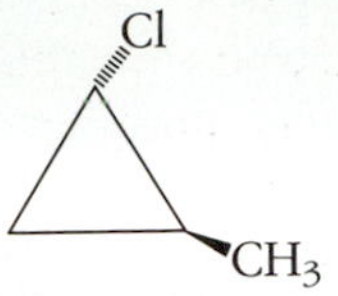 (f) 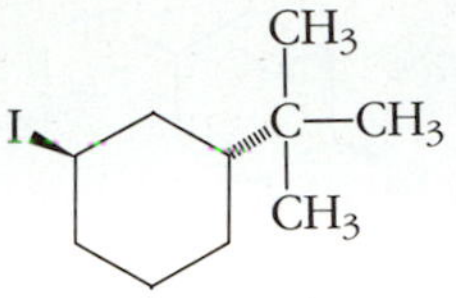 (g) 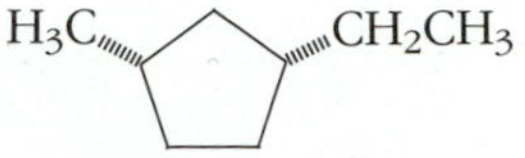 (h) 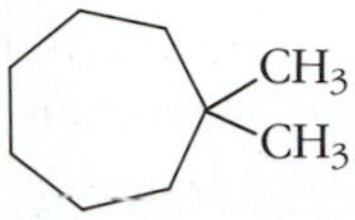 (i) 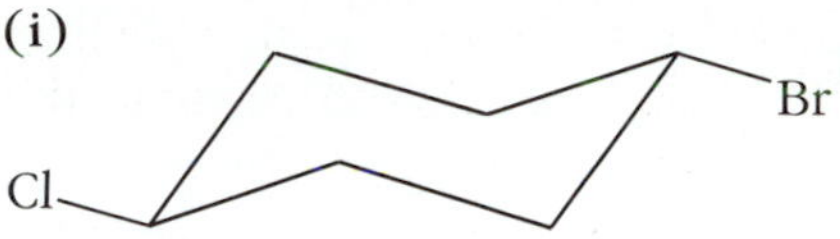 (j)

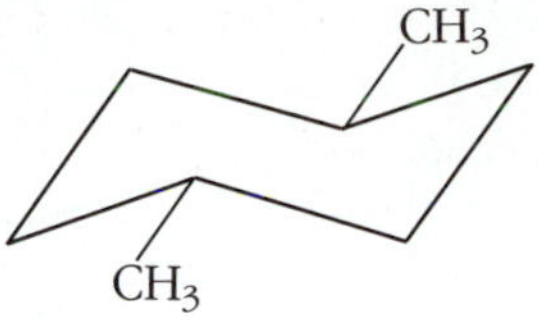

3.2 Draw the following compounds.

(a) 1-Bromo-1-methylcyclopropane
(b) 3-*tert*-Butyl-1,1-dimethylcyclopentane
(c) *trans*-1,3-Dibromocyclobutane
(d) *trans*-1-Chloro-4-propylcyclooctane
(e) *cis*-1,3-Difluorocyclohexane
(f) *trans*-4-Bromo-*r*-1-*cis*-3-dichlorocyclohexane

3.3 Chlorination of 1-methyl-1-isopropylcyclobutane would yield a variety of isomeric monochloro derivatives.

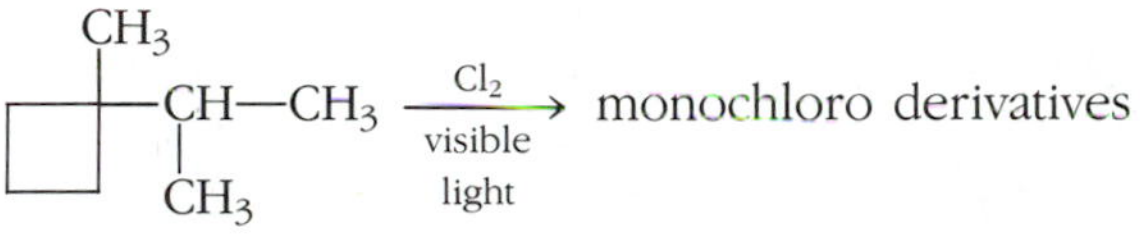

$\xrightarrow[\text{visible light}]{Cl_2}$ monochloro derivatives

Draw and name each of the constitutional isomers that could be formed.

3.4 *Cis-trans* isomerism is possible for more than one of the constitutional isomers formed as products in Problem 3.3. For each such case draw and name both isomers.

3.5 Chlorination of 1-(chloromethyl)-1-methylcyclopentane would yield a variety of isomeric dichloro derivatives.

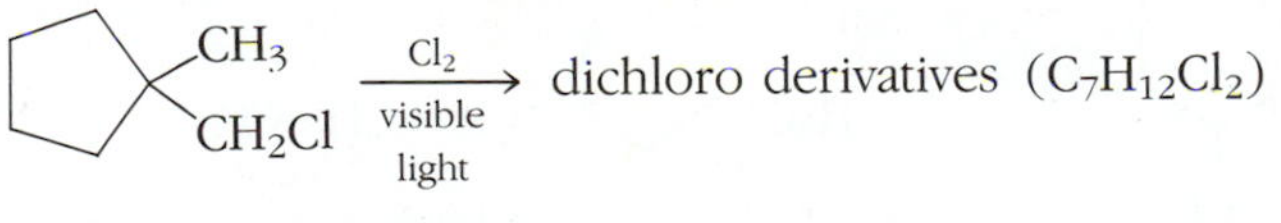

$\xrightarrow[\text{visible light}]{Cl_2}$ dichloro derivatives ($C_7H_{12}Cl_2$)

Draw and name each of the constitutional isomers that could be formed.

3.6 *Cis-trans* isomerism is possible for more than one of the constitutional isomers formed as products in Problem 3.5. For each such case draw and name both isomers.

3.7 Describe the relationship between each pair of structural formulas as that of *constitutional* isomers, *stereoisomers (cis-trans* isomers), two drawings of the *same* compound (although perhaps in different conformations), or *different* compounds that are not isomers.

(a)

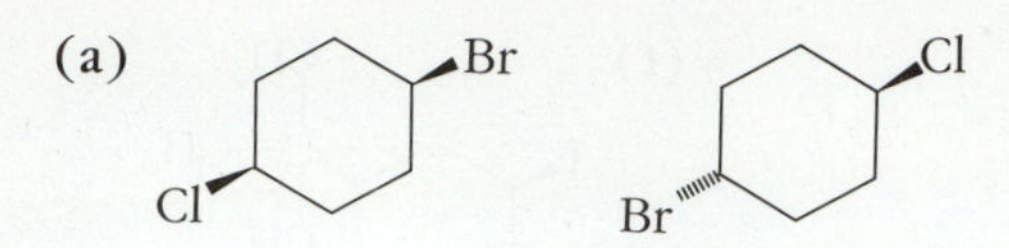

(b)

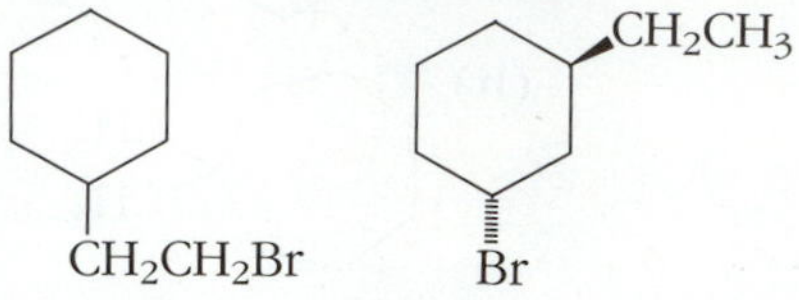

CH_2CH_2Br

CH_2CH_3

Br

(c)

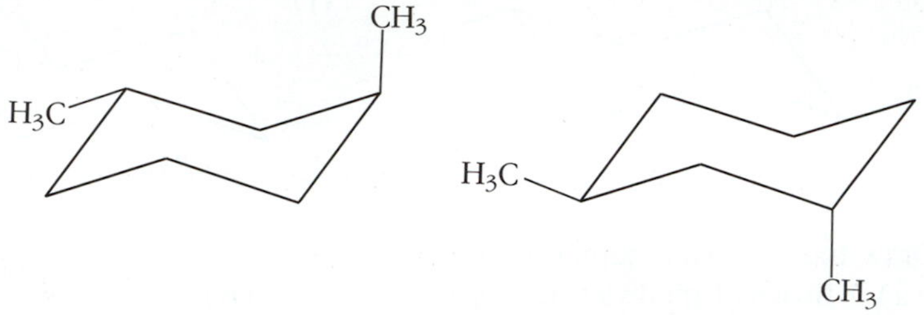

CH_3

H_3C

H_3C

CH_3

(d)

Br Br Br Br

(e)

Br

CH_2Br

(f)

Cl

CH_2Cl

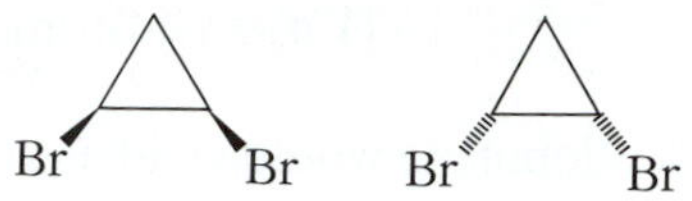

Cl

Cl

(g)

CH_2CH_3

CH_2CH_3

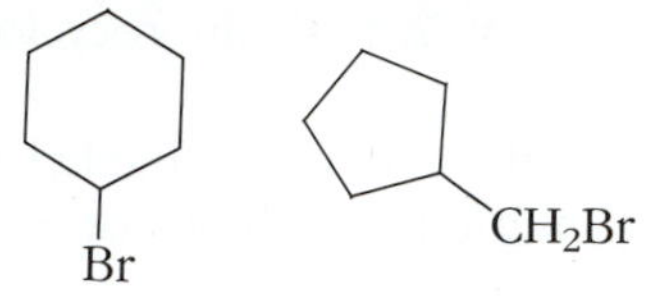

CH_2CH_3

CH_2CH_3

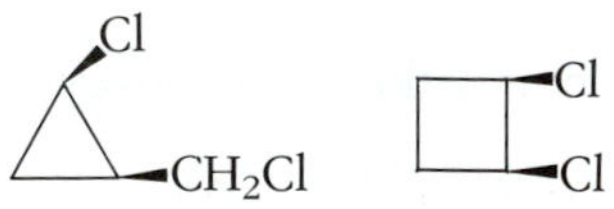

(h)

CH_2Br

Cl

H_3C — $CH—CH_2—Br$

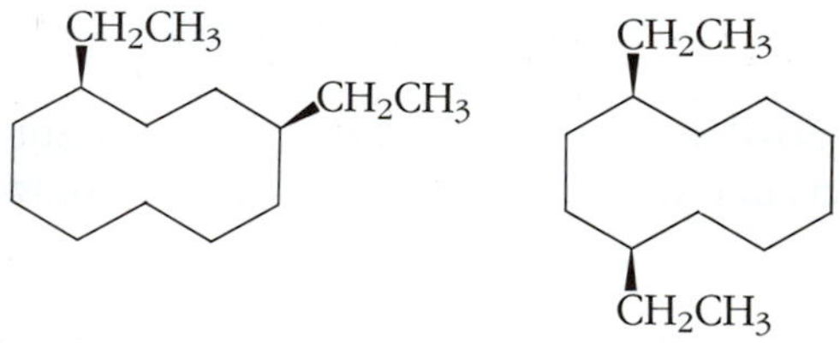

Cl

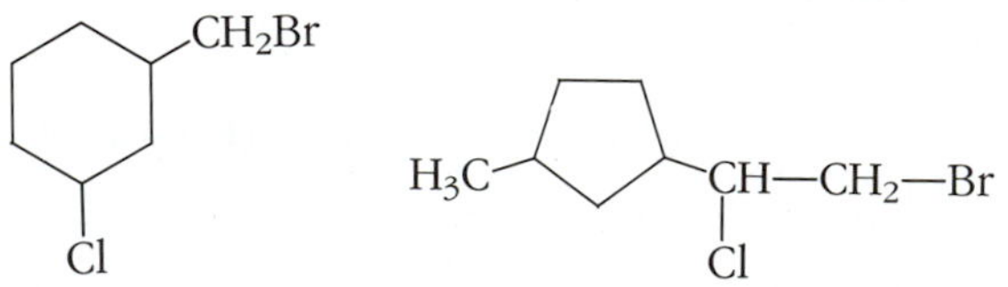

(i) H_3C

H

H

CH_3

H

H_3C

CH_3

H

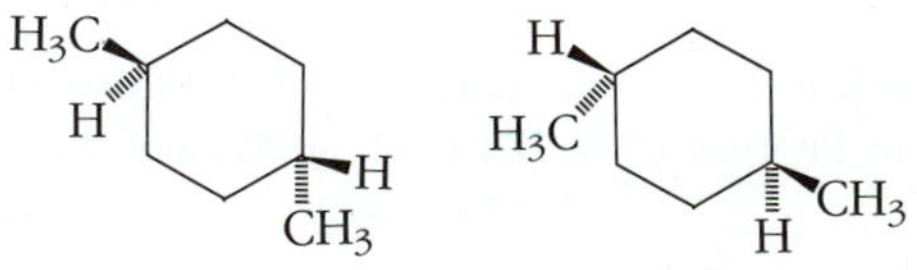

(j)

CH_3

CH_3

CH_2CH_3

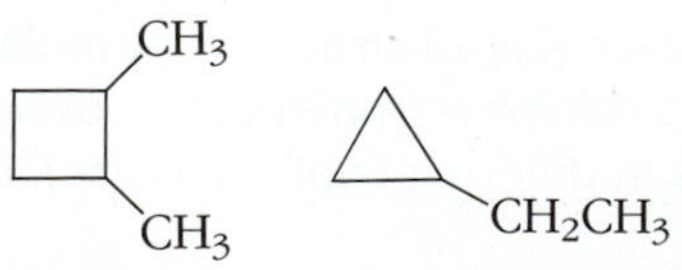

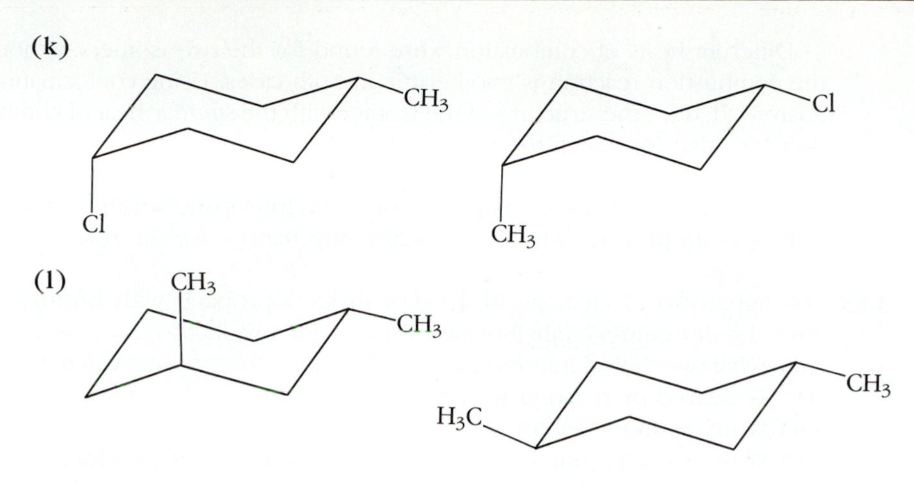

3.8 Draw the most stable conformation for each of the following structures:

(a) F, Cl

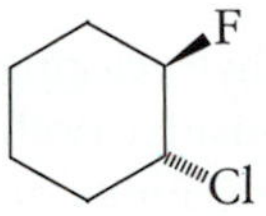

(b) CH_3, Br

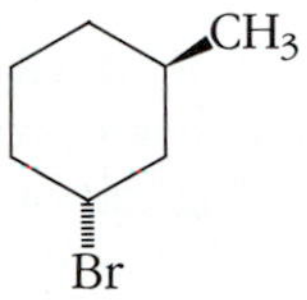

(c) CH_2CH_3, CH_3CH_2

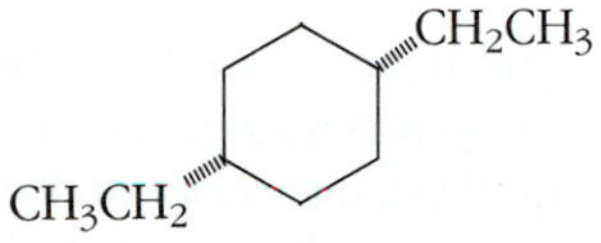

(d) CH_3, $CH_3—C$, CH_3, CH_3

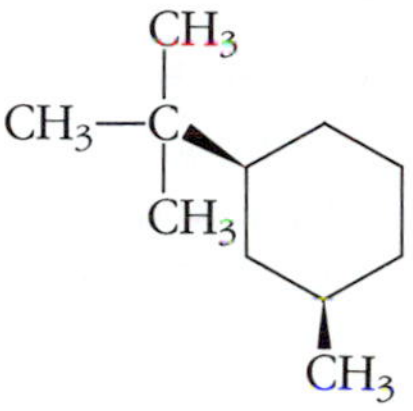

(e) Br, $H_3C—CH—CH_3$

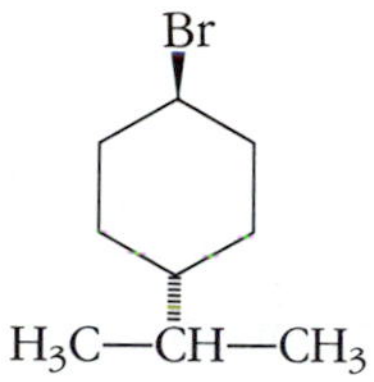

3.9 Suggest a method of synthesizing 2-cyclobutylheptane. All carbon atoms of the final product must be derived from starting materials containing six or fewer carbon atoms.

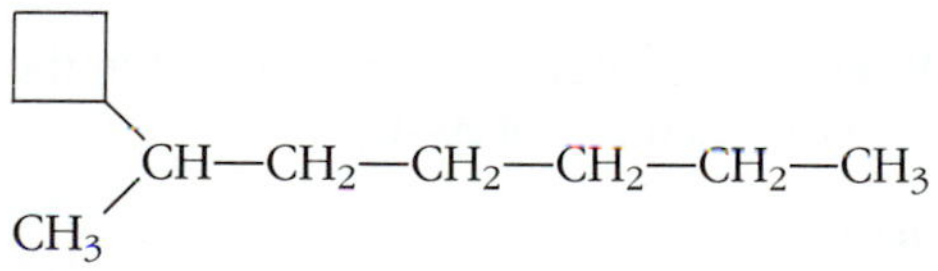

$CH—CH_2—CH_2—CH_2—CH_2—CH_3$, CH_3

3.10 Suggest a method of preparing 3-methylpentane in *one* step. All carbon atoms of the product must be derived from a single cycloalkane.

3.11 The heat of combustion has been measured for both *cis-* and *trans-* 3-methylcyclohexanol.

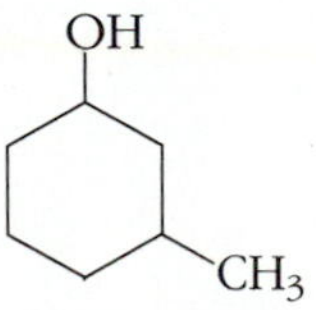

OH, CH_3

3-Methylcyclohexanol

Different heats of combustion were found for the two isomers, although the combustion reaction is exothermic in both cases. Using conformational analysis, deduce the structure of the isomer with the *smaller* heat of combustion (i.e., that for which the reaction is less exothermic).

3.12 The light-induced ring opening reaction of cyclopropane with Br_2 is a free-radical chain process. Write a reasonable mechanism for the reaction.

3.13 The light-induced cleavage of 1,1-dimethylcyclopropane with bromine affords 1,3-dibromo-3-methylbutane as the major product.

(a) What two radical intermediates could lead to this product when the ring is cleaved by reaction with a bromine atom?

(b) Which is more stable?

(c) Which would require more unfavorable steric interactions for attack on the ring by the bromine atom?

(d) Which is the actual intermediate in the reaction?

3.14 The light-induced cleavage reaction of 1,1,2,2-tetramethylcyclopropane with Br_2 yields a single compound as the major product. Consider both the steric hindrance to attack at the different ring carbons and the stability of the different free radicals that could be formed upon cleavage of the ring. On that basis deduce the structure of the product.

3.15 Both the *cis* and *trans* isomers of 1,4-dimethylcyclohexane have been studied, and the *cis* isomer is less stable by a free energy difference of 1.9 kcal/mol. Draw the possible chair conformations for each isomer, and use these drawings to explain the observed free energy difference. How does this energy difference compare with the data in Table 3.1?

3.16 Draw the two possible chair conformations of *cis*-1-bromo-4-methylcyclohexane. Using the data in Table 3.1, predict which should be more stable. By how much? Use your estimated free energy difference to calculate the approximate fraction of each conformation.

3.17 Compounds **A–H** are all different compounds. On the basis of the reactions shown, deduce the structures of **A–G.**

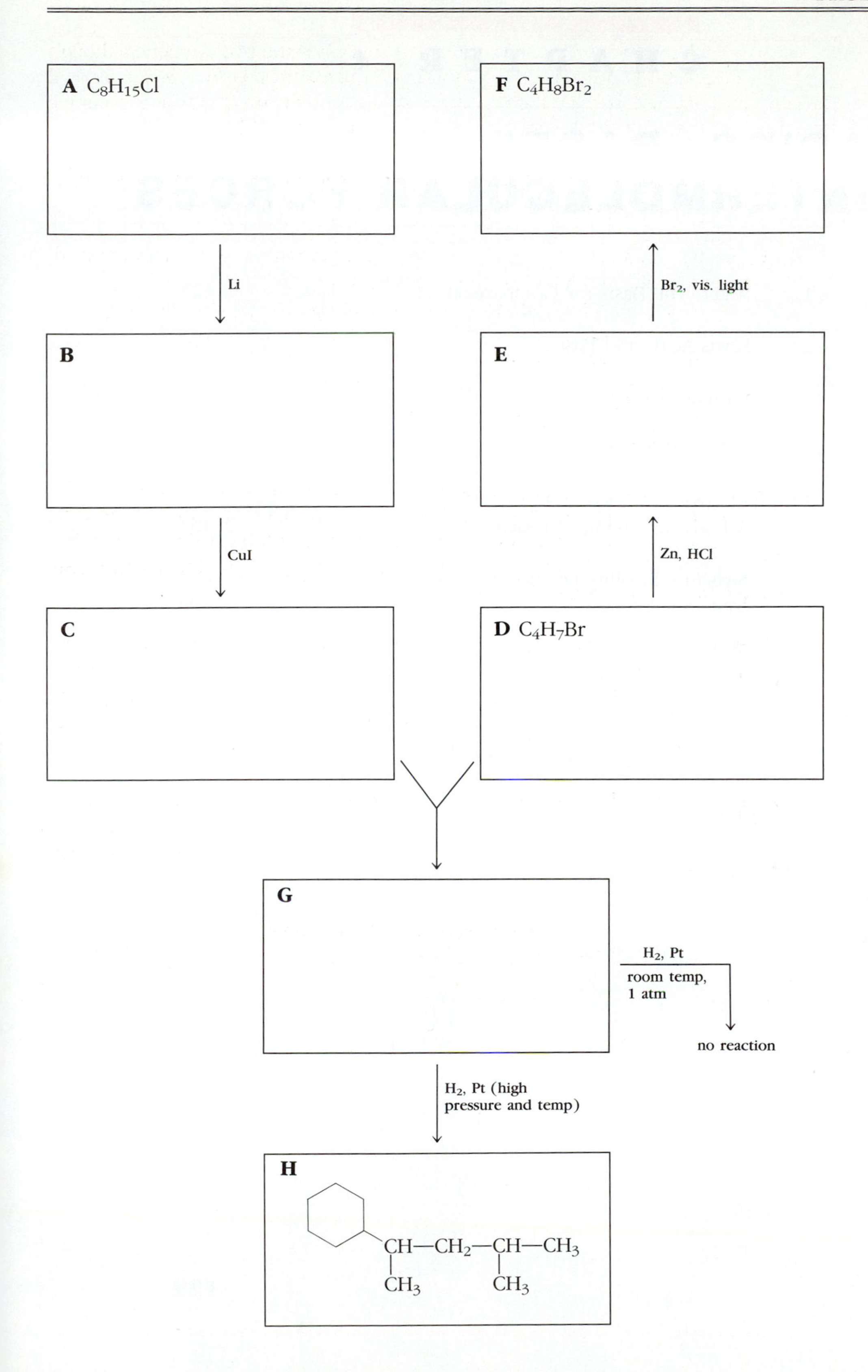
A $C_8H_{15}Cl$
Li
B
CuI
C
F $C_4H_8Br_2$
Br_2, vis. light
E
Zn, HCl
D C_4H_7Br
G
H_2, Pt
room temp,
1 atm
no reaction
H_2, Pt (high
pressure and temp)
H
CH—CH$_2$—CH—CH$_3$
CH$_3$
CH$_3$

CHAPTER 4

ACIDS, BASES, AND INTERMOLECULAR FORCES

During your study of general chemistry you received an extensive exposure to the fundamentals of acid-base chemistry. Now that you have begun to study organic chemistry, there is a tremendous temptation to assume that everything is "different." One of our major aims in this chapter is to show you that organic chemistry is based on the same fundamental principles that are used in all other areas of chemistry.

Without a doubt the chemical substance most familiar to you is water. You have been drinking and washing with it your entire life, and it is the liquid component in a wide variety of foods, drinks, and household products. You have also been exposed to the chemistry of water, and most of the reactions you encountered in general chemistry employed water as a solvent. Water is not the typical solvent for organic chemistry, but there are remarkable similarities between organic reactions and those occurring in aqueous solution. The similarities are particularly striking for acid-base chemistry, and we will begin with a review of these reactions in aqueous systems.

4.1 WATER: THE BASIS FOR COMPARISON

A hydrogen atom consists of a single electron in combination with a nucleus that has only a single proton. Removal of the electron would afford a hydrogen ion, H^+, consisting simply of a single proton. In aqueous solution ions do not exist as isolated species; they are surrounded by and coordinated to water molecules. A hydrogen ion, for example, exists as a *hydronium ion,* H_3O^+, or as an aggregate with a larger number of water molecules. In the gas phase a hydronium ion can be formed by an acid-base reaction of a proton with water:

$$H^+ + H_2O \rightleftharpoons [H_3O]^+$$

In aqueous solution, however, the proton would be supplied by an acid such as sulfuric acid.

$$H_2SO_4 + H_2O \rightleftharpoons [H_3O]^+ + [HSO_4]^-$$

As illustrated for the following acid-base reactions, in which $[OH]^-$ and NH_3 act as bases, the hydronium ion serves as the actual acid in aqueous solution:

$$[H_3O]^+ + [OH]^- \rightleftharpoons 2\,H_2O$$

$$[H_3O]^+ + NH_3 \rightleftharpoons H_2O + [NH_4]^+$$

Organic compounds also undergo acid-base reactions, as illustrated by the following reaction of methanol (CH_3OH):

$$H_2SO_4 + CH_3{-}OH \rightleftharpoons [CH_3{-}OH_2]^+ + [HSO_4]^-$$

Methanol is an example of a class of compounds called **alcohols** (Chapter 11) with the general structure R—OH, in which an alkyl group is covalently bonded to the oxygen of a **hydroxyl** (—OH) group. Alternatively, we can view an alcohol as a structure in which one of the two hydrogens of water has been replaced by an alkyl group. This structural relationship to water is accompanied by a similarity in reactivity. The preceding reactions of methanol and of water with sulfuric acid illustrate just one of the many similarities between the reactivities of alcohols and of water. As you continue your study of organic chemistry, you will encounter many more reactions of organic compounds that are analogous to those of water.

4.2 LEWIS ACIDS AND BASES

For each of the acid-base reactions that we showed in the previous section, the *acid* is either H^+ or some species that can transfer H^+ to the molecule or ion that acts as the *base.* This classification of acids and bases as H^+ donors and H^+ acceptors, respectively, was formulated in 1923 by J. N. Brønsted in Denmark and T. M. Lowry in England. The **Brønsted-Lowry** classification further labels the products of these acid-base reactions as **conjugate acids** and **conjugate bases.** Transfer of an H^+ ion to a base produces the conjugate acid of that base, while loss of a proton from an acid generates the conjugate base. The Brønsted-Lowry description serves well for acid-base reactions in water and in other hydroxylic solvents such as alcohols, but there are a number of situations in the study of organic chemistry for which a broader concept of acid-base reactions is needed. Such a broad definition of acid-base reactions was introduced in 1923 by the American chemist G. N. Lewis. A few years earlier Lewis had helped to establish the concept of *electron pairs* as key elements in molecular structure and bonding, particularly the idea that covalent bonds resulted from the sharing of electron pairs by two atoms. Lewis extended the significance of electron pairs to a new definition of acids and bases, and his name is commonly used with this definition. A **Lewis base** is a species having a *pair of electrons available for bonding to the corresponding acid* in some reaction. Conversely, a **Lewis acid** is a species capable of reacting with the electron pair of a base because it has an *electron-deficient* atom, that is, an atom that does not have the electron configuration of an inert gas. In the simplest terms we can define a Lewis base as an *electron pair donor* and a Lewis acid as an *electron pair acceptor.*

How do the Lewis and Brønsted-Lowry definitions compare? In many cases they are equivalent, and we can illustrate this by looking at the gas-phase reaction of a proton with a chloride ion.

$$H^+ + Cl^- \rightleftharpoons HCl$$

The acid, H^+, in this case fits both definitions. Obviously it is an H^+ donor, and it also fits the Lewis definition because it is electron-deficient, lacking the two electrons in its valence shell that would correspond to the electron configuration of helium. Similarly, the chloride ion is a Brønsted-Lowry base because it acts as a proton acceptor in this reaction. It is also a Lewis base, because it has an available pair of electrons for reaction with the H^+. In fact there is little difference between the two definitions of a base, because an electron pair donor will always be capable of acting as a proton acceptor as well. But the two definitions for an acid are not always equivalent. For the protonation reaction of chloride ion in aqueous solution, the acid is a hydronium ion, H_3O^+.

$$[H_3O]^+ + Cl^- \rightleftharpoons H_2O + HCl$$

The hydronium ion acts as a Brønsted-Lowry acid by transferring a proton, but it is not a Lewis acid.

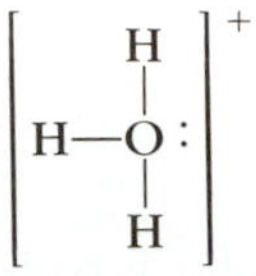

$$\left[\begin{array}{c} \mathrm{H} \\ | \\ \mathrm{H{-}\ddot{O}{:}} \\ | \\ \mathrm{H} \end{array}\right]^{+}$$

Each of the hydrogen atoms shares two electrons in a bond with oxygen, giving them the inert gas configuration of helium. Those three electron pairs, together with the unshared pair, give oxygen the stable octet corresponding to the electron configuration of neon.

Only by *transferring* an electron-deficient hydrogen ion can the hydronium ion act as an acid, and you can begin to see that there are limitations to any set of definitions that we may choose to place on acids and bases. You must remember that although free hydrogen ions are not found in solution, it is nevertheless an H^+ that is transferred in a great many acid-base reactions. As a matter of convenience, we will circumvent some of the difficulty by using the abbreviation H^+ throughout this text, even when the actual proton donor is a hydronium ion or some other species.

In acid-base chemistry, as in most other areas of chemistry, we need to identify those structural features which govern the reactivity of the molecules being studied. As we showed previously in Section 2.10, relationships between *structure* and *reactivity* give us the capability of both understanding reactions and predicting how other compounds will react. The Lewis definitions of acids and bases provide a useful framework for evaluating structural effects on acidity and basicity, and the relationship between ionic charge and basicity provides a convenient example. As a general rule, *a negatively charged species will be a stronger base than a structurally similar species that is uncharged.* In accord with this statement, ammonia (NH_3) is a much weaker base than the amide ion (NH_2^-):

$$:NH_3 + H_2O \rightleftharpoons [NH_4]^+ + [OH]^- \qquad K \simeq 10^{-5}$$

$$[NH_2]^- + H_2O \rightleftharpoons NH_3 + [OH]^- \qquad K \simeq 10^{21}$$

Why is the amide ion a stronger base by a factor of 10^{26}? The only structural difference between ammonia and the amide ion is a proton ($NH_3 = NH_2^- + H^+$), so the much greater electron-donating ability of NH_2^- is clearly associated with the negative charge. The ability of NH_3 to function as a base results from the presence of an **unshared pair of electrons** on the nitrogen atom. The electron configuration of the nitrogen atom of ammonia corresponds to the stable octet of neon, but there are only three bonds. This leaves two electrons (drawn as a pair of dots) that are localized on the nitrogen atom as an unshared pair of electrons. Such **nonbonding electrons** (also described as a **lone pair** of electrons) are readily available for the formation of a bond to an H^+ or to some other electron acceptor:

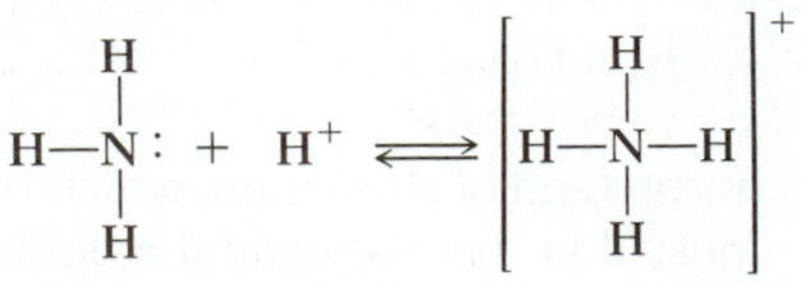

$$\begin{array}{c} \mathrm{H} \\ | \\ \mathrm{H{-}\ddot{N}{:}} \\ | \\ \mathrm{H} \end{array} + H^+ \rightleftharpoons \left[\begin{array}{c} \mathrm{H} \\ | \\ \mathrm{H{-}N{-}H} \\ | \\ \mathrm{H} \end{array}\right]^{+}$$

The amide ion (NH_2^-) is structurally similar, although it has two unshared pairs of electrons:

$$\left[\mathrm{H{-}\ddot{N}{:}}\atop{|\atop \mathrm{H}}\right]^- + \mathrm{H^+} \rightleftharpoons \mathrm{H{-}\ddot{N}{-}H}\atop{|\atop \mathrm{H}}$$

A preference for electrical neutrality is typical behavior for organic compounds and most other covalent compounds, and this provides a clear answer to our question about the relative basicities of ammonia and the amide ion. In the first of the two preceding equations, the product and one of the reactants both carry a positive charge, so there is no overall change in charge. In contrast, in the second equation the ionic charges of the two reactants are neutralized in the reaction. The instability associated with ionic charge is relieved by the formation of a neutral product in the second reaction, and that reaction is energetically much more favorable, as indicated by its larger equilibrium constant.

A molecule can also function as a base by donation of *bonding electrons,* and you will discover the importance of this reactivity when you study the reactions of alkenes (Chapter 5) and other molecules that have multiple bonds. However, the use of bonding electrons for the reaction with an acid requires breaking (or at least greatly weakening) the original bond. As a consequence, bases that react via donation of bonding electrons are typically much weaker than those using nonbonded electrons.

We have indicated that the structural requirement for a species to function as a Lewis base is an available pair of electrons that can react with the acid. What structural features are necessary for a species to act as a Lewis acid? The species must have an empty orbital that the electrons of the base could occupy. If there is to be a significant interaction between the acid and base, this must be an unoccupied orbital of relatively low energy. All molecules and atoms have many unoccupied high-energy orbitals, corresponding to atomic orbitals at levels higher than the valence shell, but these high-energy orbitals do not interact significantly with the electron pairs of Lewis bases. A Lewis acid is ordinarily characterized by the presence of an *unoccupied valence shell orbital.* Clearly the hydrogen ion meets this requirement, because it has no electrons in its 1*s* valence orbital. An acid such as HCl does not itself have any vacant orbitals of low energy, and it does not meet the criteria of a Lewis acid. The acidity of such *protic acids* results solely from their ability to donate H^+ ions.

Most metal cations formed by the loss of one or more electrons from low-energy valence shell orbitals can function as Lewis acids. Neutral molecules can also function as Lewis acids, as illustrated by the following reaction of boron trifluoride (BF_3) with diethyl ether. Ethers are another class of organic compounds that are structurally related to water. *Both* hydrogen atoms of water have been replaced by alkyl groups, but the oxygen atom of an ether molecule has lone pairs of electrons that permit it to function as a base.

$$\mathrm{BF_3} + \underset{\text{Diethyl ether}}{\mathrm{C_2H_5{-}\ddot{\underset{..}{O}}{-}C_2H_5}} \rightleftharpoons \mathrm{(C_2H_5)_2\ddot{O}^+BF_3}$$

Trivalent compounds of boron (and of aluminum, which is directly below boron in the periodic table) are unusual in that the central atom is electron-deficient. In a molecule such as BF_3 the boron has only six electrons in its valence shell, and a 2*p*

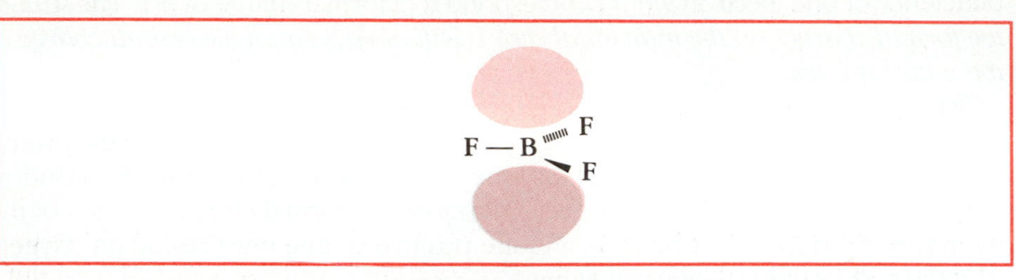

Figure 4.1 The Structure of Boron Trifluoride.
The total of 24 valence-level electrons is distributed so that each fluorine atom has an octet of electrons. The boron atom has only the six electrons that it shares with the fluorines in the three B—F bonds. The optimum geometry is that with the three substituents at the maximum possible distance from each other, that is, with all atoms in a single plane and F—B—F angles of 120°. The empty *p* orbital, which is perpendicular to the plane of the molecule, allows BF_3 to function as a Lewis acid.

orbital remains unoccupied (Figure 4.1). Donation of a pair of electrons from the ether oxygen proceeds with formation of an oxygen-boron bond, and boron achieves the stable octet of electrons in its valence shell.

4.3 FORMAL CHARGE

In the preceding sections of this chapter we have drawn a number of structures with ionic charges. We are now ready to address an extremely important question: Where are the charges located? Consider the hydronium ion, H_3O^+. Is the positive charge centered on one specific hydrogen, is it shared by several atoms, or is it located on the oxygen? It is essential that you be able to answer such questions if you are to develop the ability to *explain and predict chemical reactivity on the basis of molecular structure.*

Exact answers to these questions are difficult. According to molecular orbital theory, we should expect that an ionic charge would be distributed over several atoms. Putting this a different way, a molecule consists of positively charged nuclei surrounded by a delocalized cloud of electron density. How can we decide where to put the "dividing line" between nuclei, in order to determine how much negative charge should be assigned to each nucleus? This can be done by quantum mechanical calculations, but is there a method that you can employ using only pencil and paper? The answer is yes, if we adopt the appropriate *formalism.* In the standard representations of molecular structure, the positively charged nuclei are denoted by the appropriate atomic symbols (C, H, etc.). The negatively charged electrons are denoted by lines between nuclei (to indicate the pair of electrons in a bond) or by dots associated with a specific nucleus (to indicate nonbonding electrons). Chemists have developed a procedure for calculating the **formal charge** on any atomic center using the formalism that a pair of electrons in a bond is always divided equally between the two atomic centers.

If the number of protons in the nucleus of an atom is equal to the number of electrons surrounding the atom, then we consider the atom to be neutral. An excess of a single electron would result in a formal charge of −1 on the atom, and a deficiency of one electron would correspond to a formal charge of +1. The *sum of the formal charges on the individual atoms will always equal the overall charge of the entire species.*

We can now return to the hydronium ion and calculate the formal charge for each of the constituent atoms. Each of the three hydrogen atoms has a single proton in its nucleus (atomic number 1), so one of the electrons used in the bond to oxygen must be counted in each case. Therefore the formal charge on each of the hydrogens of H_3O^+ must be zero, and the positive charge must reside on oxygen, as indicated in the following drawing:

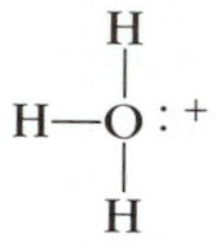

```
      H
      |
   H—O:+
      |
      H
```

The oxygen atom is bonded to three hydrogen atoms, so three of the six bonding electrons are allocated to oxygen, and the lone pair of electrons adds two more to the total. Finally, we must count the two additional electrons that are not part of the valence shell. These two **core electrons** (in the 1*s* level) bring the total number of electrons on oxygen to *seven*. Taken together with the *eight* protons in the nucleus (atomic number 8), this clearly shows that the positive charge of the hydronium ion formally resides on oxygen.

For most anions the formal charge is associated with a lone pair of electrons on the atom bearing the formal negative charge. Moreover, the chemical reactivity of the ion can usually be attributed to that particular pair of electrons. The importance of such nonbonding electron pairs and the relationship to negative charge requires that we discuss some of the different ways that chemists draw pairs of electrons. Some chemists always draw electron pairs as two dots, and they indicate negative charge with a minus sign. Other chemists use lines to indicate electron pairs, and the following structures illustrate how water can be drawn in these two ways.

```
   ..
 H:O:H      H—O—H
   ..
```

In this book the electron pairs of bonds will be represented as lines, and non-bonded pairs of electrons will be denoted by a pair of dots. A minus sign will be used to denote the formal charge of an anion, but in most cases this charge will be associated with a nonbonding pair of electrons on the anion. This is highly convenient, because it allows the reactivity of an anion to be depicted without explicitly drawing all the nonbonding electrons. According to this procedure the hydroxide ion can be drawn in either of the following ways:

```
    ..
 H—O:⁻     or just     H—O⁻
    ..
```

The following two equations show how the electron pair (denoted explicitly as a pair of dots or implicitly as a minus sign) can be used to show the reactivity of the hydroxide ion.

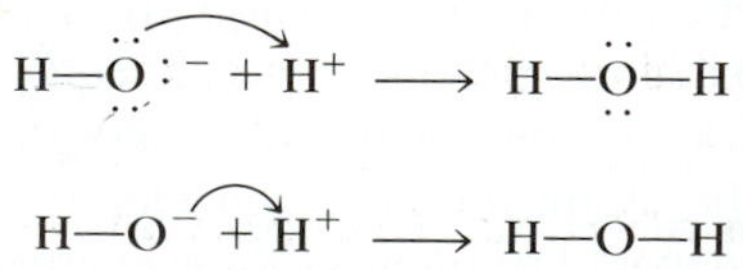

```
    ..                    ..
 H—O:⁻ + H⁺  ⟶   H—O—H
    ..                    ..

 H—O⁻ + H⁺  ⟶   H—O—H
```

In these two equations the reactants and products are drawn as similarly as possible, and attention is focused upon the negative charge that makes a hydroxide ion more reactive than a neutral water molecule.

Occasionally you will encounter an anion that does not have a pair of nonbonding electrons associated with the negative charge. In such cases you must be careful that you do not use a minus sign to denote a pair of electrons in a way that might be misunderstood. This situation arises for tetracoordinate boron and aluminum. For example, in the BF_4^- anion the formal negative charge is situated on boron, but it does not correspond to a lone pair of electrons. All eight electrons of the octet for boron are involved in B—F bonds.

```
[     F     ] -
[     |     ]
[  F—B—F   ]
[     |     ]
[     F     ]
```

By enclosing the anionic species within brackets we have correctly indicated the net charge on the species, but there is no suggestion that the minus sign would denote a nonbonded pair of electrons.

We have summarized the calculation of formal charge in Figure 4.2. As you gain experience in working with organic reactions, you will find that the formal charges on various atoms will be apparent to you without having to calculate them explicitly. In the meantime, however, you should be certain that you can calculate formal charges quickly and reliably for a variety of chemical species.

You may already have recognized one important pattern in the changes in formal charge associated with acid-base reactions. When a Lewis base reacts with an acid, it uses a lone pair of electrons to form a bond to the acid. Prior to reaction this

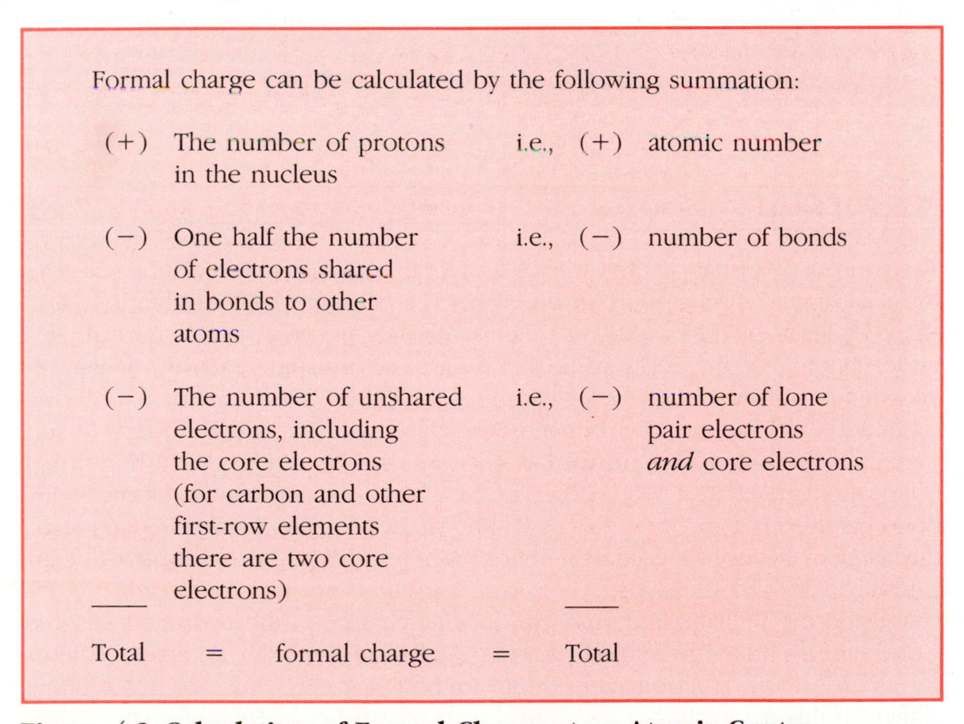

Formal charge can be calculated by the following summation:

(+)	The number of protons in the nucleus	i.e.,	(+)	atomic number
(−)	One half the number of electrons shared in bonds to other atoms	i.e.,	(−)	number of bonds
(−)	The number of unshared electrons, including the core electrons (for carbon and other first-row elements there are two core electrons)	i.e.,	(−)	number of lone pair electrons *and* core electrons
Total	= formal charge =		Total	

Figure 4.2 Calculation of Formal Charge at an Atomic Center.

nonbonding pair of electrons is associated with a single atom, but that atom undergoes a decrease in the number of electrons formally assigned to it. When reaction of these electrons produces a covalent bond, they are counted as being shared by the two bonded atoms. The net result is a change to a less negative (or more positive) formal charge for the atom that donates a pair of electrons in an acid-base reaction. This generalization allows you to easily recognize the positively charged "onium ions" in which oxygen has three bonds or nitrogen has four bonds.

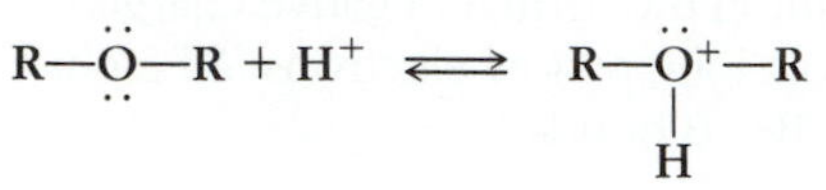

$$\mathrm{R{-}\ddot{\underset{..}{O}}{-}R} + \mathrm{H^+} \rightleftharpoons \mathrm{R{-}\ddot{O}^+{-}R}\ (\text{with H on O})$$

An ether (a derivative of water) — An oxonium ion

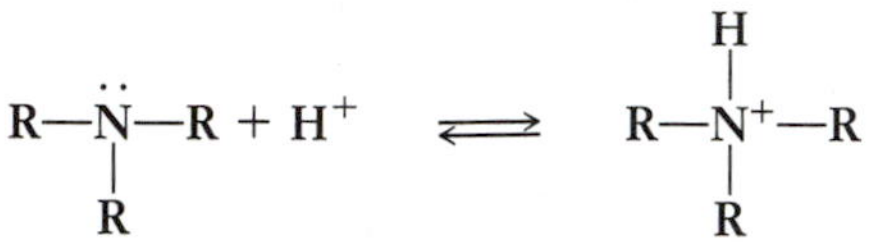

$$\mathrm{R{-}\ddot{N}(R){-}R} + \mathrm{H^+} \rightleftharpoons \mathrm{R{-}N^+(H)(R){-}R}$$

An amine (a derivative of ammonia) — An ammonium ion

EXERCISE 4.1

(a) What is the formal charge on the oxygen atom of water?
(b) Which atom carries the formal negative charge in the hydroxide ion, hydrogen or oxygen?

EXERCISE 4.2

Calculate formal charges on boron and oxygen in the adduct formed when the Lewis acid BF_3 reacts with the Lewis base diethyl ether ($C_2H_5OC_2H_5$).

4.4 POLARIZED BONDS

The concept of formal charge is very useful for explaining and predicting reactions, but it is limited by the approximation that the electrons of a covalent bond are shared equally by the two atoms. This is certainly not true, and a more reliable understanding of the relationship between structure and reactivity demands a knowledge of how bonding electrons are distributed. In symmetrical molecules such as H_2 and Br_2 the electrons must be shared equally, and the bonds in such molecules are classified as **nonpolar.** The same can be said for the carbon-carbon bond of ethane ($H_3C{-}CH_3$). However, we cannot expect that the electrons in a bond between two different groups will be shared equally. In the H—F molecule the bonding electrons are attracted much more toward the electronegative fluorine atom than toward the hydrogen atom. This effect produces a *partial* negative charge on the fluorine and a *partial* positive charge on the hydrogen. In other words, the bond is **polarized.** We can symbolize this with an arrow pointing toward the more electronegative atom (although such bonds are not routinely drawn in this manner).

H↔F

The other end of the arrow is crossed to designate the more positive of the two atoms. Remember that the *formal charge* on both hydrogen and fluorine is zero; here we are discussing *partial charges* (i.e., charges less than that of a single electron). Such partial charge is frequently denoted by the symbol δ (the lowercase Greek letter delta):

$$\overset{\delta+}{H}—\overset{\delta-}{F}$$

Bond polarities can be predicted on the basis of electronegativities, and these follow systematic patterns in the periodic table. In the study of organic chemistry we are concerned with bonds to carbon, and some simple trends can be summarized:

1. *Bonds between two atoms of the same element are nonpolar.* This means that nearly all carbon-carbon bonds will be nonpolar. Exceptions will arise when the substituents on the carbons are quite different, as in the case of $CH_3—CCl_3$, which has three electronegative chlorine atoms on one of the carbon atoms.

2. *Carbon and hydrogen have approximately the same electronegativities.* Therefore C—H bonds are generally considered to be nonpolar.

3. *Elements to the right of carbon in the periodic table* (most commonly nitrogen, oxygen, and the halogens) *are more electronegative than carbon.* For most cases in which a carbon atom is bonded to one of these other atoms, the carbon atom will have a partial positive charge.

4. *Elements to the left of carbon in the periodic table* (in particular the metallic elements) *are less electronegative than carbon.* For most cases in which carbon is bonded to a metal atom, the carbon atom will have a partial negative charge.

As you will see in subsequent sections, a knowledge of partial positive or negative charges in a molecule enables you to understand and predict chemical reactivity.

EXERCISE 4.3

Using the crossed arrow pointing toward the negative end of the dipole, show the bond polarity (if any) for each of the bonds between the atoms indicated in the following structures.

(a) $CH_3—Cl$ (b) $CH_3—CH_2—CH_2—O—CH_3$

(c) $CH_3—\underset{\displaystyle CH_3}{\underset{|}{CH}}—CH_2—CH_3$ (d) $CH_3—CH_2—\overset{\displaystyle NH_2}{\overset{|}{CH}}—CH_3$

(e) $CH_3-C(H)(CH_3)-CH_3$ (f) $CH_3-CH(CH_3)-OH$ (g) $CH_3-C(CH_3)_2-Li$

(h) CH_3-CH_2-MgBr (i) $CH_3-N(CH_3)-CH_3$

4.5 ELECTRONIC ATTRACTION AND REPULSION: A BASIS FOR CHEMICAL REACTIVITY

If you are like most science students, you learned at an early stage that like charges repel and opposite charges attract. This rule can be of considerable value in organic chemistry, particularly if it is restated in terms of energy considerations. A negatively charged ion is attracted to a positively charged ion, so a *decrease in potential energy* will result as the two species are brought closer together. (There is of course a limit to how close the two species may approach each other; they cannot occupy the same space at the same time.) In Figure 4.3 we have illustrated this attractive interaction between methyl and fluoride ions.

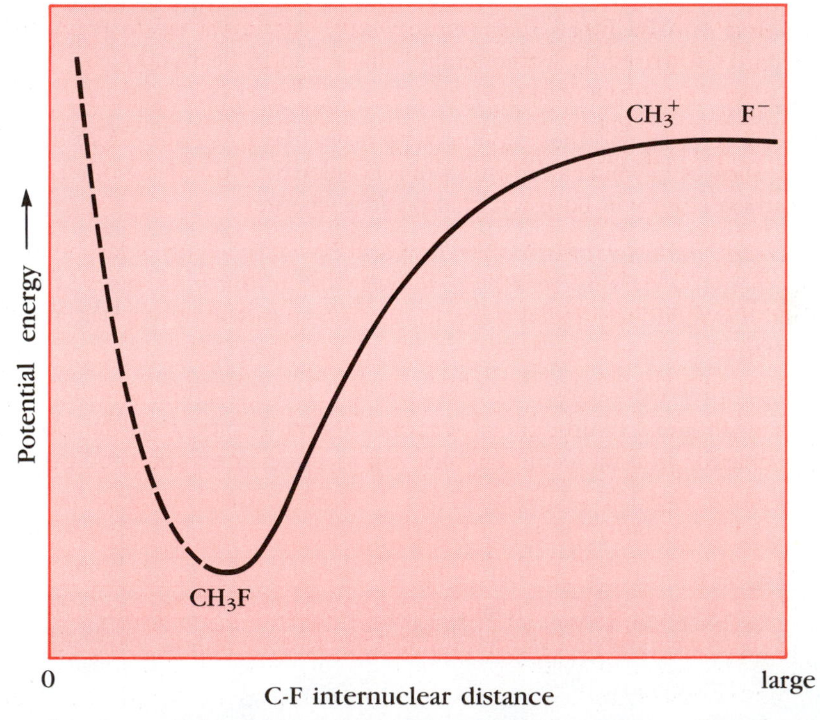

Figure 4.3 Potential Energy Curve for a Methyl Cation and a Fluoride Ion as a Function of the Carbon-Fluorine Internuclear Distance. When the two ions are very far apart (right), they do not interact, but as they are brought closer together, the potential energy decreases. The energy continues to decrease uniformly until the optimum distance of 1.33 Å (the covalent bond length) is reached. Any further decrease in the C—F distance would require these atoms to occupy the same space at the same time, and this would result in the rapid increase in energy that is indicated by the dotted line.

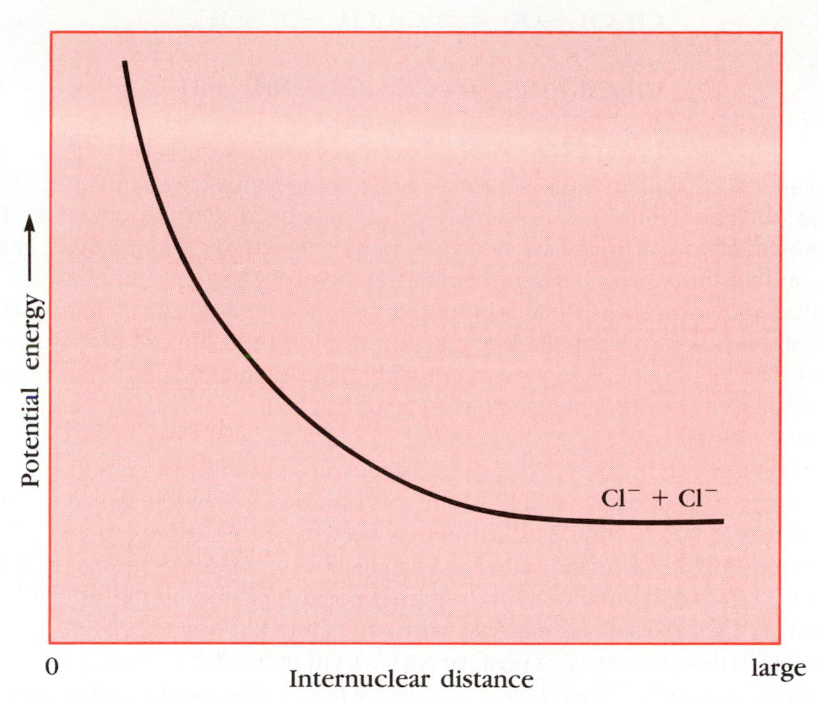

Figure 4.4 Potential Energy Curve for the Interaction of Two Chloride Ions as a Function of the Internuclear Distance.
As the two ions are brought closer together (i.e., movement from right to left in the figure), the potential energy increases steadily. The most stable arrangement is that in which the ions of like charge are very far apart.

A similar "experiment" can be considered for two ions having the *same* charge, as shown in Figure 4.4. Two chloride ions will not interact when they are extremely far apart, but as they approach each other, the charge repulsion causes a continuous increase in potential energy.

For a molecule in which all atoms have electron configurations corresponding to the inert gases (i.e., the second-row atoms each have an *octet* of electrons in their valence shell), another generalization can be made: *The neutral species will be more stable than its ionic fragments.* This provides us with a basis for evaluating relationships between structure and reactivity of organic compounds: *The energetics of organic reactions that proceed via charged intermediates or give charged products can be evaluated by focusing on the ions.* When uncharged organic reactants produce ionic products or intermediates, the transition state will also have ionic character. The rate of reaction will be determined by the energy difference between reactants and transition state, and the position of equilibrium will be determined by the energy difference between reactants and products. In either case it is the energy of the higher energy ionic species that will be most sensitive to structural effects, and these structural effects will in turn govern rates and equilibria.

We can illustrate this point by considering the relative acidities of an alcohol and an amine. Just as alcohols can be considered derivatives of water, amines are derived from ammonia by replacement of one or more hydrogens with alkyl groups. Both alcohols and amines are very weak acids, but we can still compare their acidities.

$$CH_3CH_2—OH \rightleftharpoons CH_3CH_2—O^- + H^+$$

$$CH_3CH_2—NH_2 \rightleftharpoons CH_3CH_2—NH^- + H^+$$

Each reaction produces H^+ (or its solvated equivalent), so this provides no basis for comparison. Similarly, the neutral organic compounds have similar structures and no obvious differences in stability can be assigned. On the other hand, the anions of the two reactions have negative charges on *oxygen* and *nitrogen,* respectively, and this provides a basis for comparing the two. Oxygen is more electronegative than nitrogen, so it is better able to accommodate a negative charge. Therefore you should predict that the first reaction would proceed more readily than the second. This prediction is in agreement with experimental results, which confirm that the alcohol is better able to act as an acid than is the amine (although both are very weak acids).

The concept of attraction of opposite charges also permits us to rationalize many other organic reactions. We will illustrate this idea by briefly looking at an example of *nucleophilic substitution,* a reaction that we will discuss extensively in Chapter 12. A **nucleophile** (from the Latin for "nucleus-loving") is a species that will attack a carbon atom that has either a full or partial positive charge. In the strictest sense the term applies to a species reacting with *any* cation, but organic chemists usually use it only to describe attack at positive *carbon.* (In particular, a reagent that attacks a positively charged *hydrogen atom* is called a *base.*) The counterpart to a nucleophile is an **electrophile** (from the Latin for "electron-loving"), which reacts with a negatively charged or electron-rich species. This of course describes any Lewis acid, but the term "electrophile" is also used in a restricted sense, specifically to describe a species that reacts at an electron-rich *carbon atom.*

Consider the reaction of hydroxide ion, a nucleophile, with bromomethane:

$$HO^- + \overset{\delta+}{CH_3}—\overset{\delta-}{Br} \longrightarrow ?$$

On the basis of the bond polarity of the carbon-bromine bond, the negatively charged nucleophile might be expected to attack the carbon atom of the methyl group. This is precisely what occurs, and the formation of a carbon-oxygen bond is accompanied by expulsion of the bromine atom (with its bonding electrons) as a bromine ion:

$$H—O^- + CH_3—\ddot{\underset{..}{Br}}: \longrightarrow HO—CH_3 + {}^-:\ddot{\underset{..}{Br}}:$$

Such reactivity patterns will continue to appear throughout your study of organic chemistry. You may find it surprising, but it is nevertheless extraordinarily useful that the course of complex organic reactions can be predicted in such a simple way.

4.6 STRUCTURE-STABILITY RELATIONSHIPS OF IONIC SPECIES

In the preceding section we introduced a very useful approximation for analyzing the energetics of organic reactions that involve ionic intermediates or products: *The energy difference between neutral reactants and ionic products (or an ionic transition state) is determined predominantly by the stability of the ionic species.*

This allows us to focus on the ions, but what governs that stability of an ionic species? In the next three sections we will consider this problem in detail. First, in this section we will discuss the dependence of stability on the atom which formally carries the ionic charge. Then we will turn to **substituent effects,** changes in stability or reactivity that result when a hydrogen of an organic species is replaced by some other group. We will discuss *resonance effects* in Section 4.7 and *inductive effects* in Section 4.8.

How does the stability of an organic ion depend on the atom that bears the charge? We will begin by looking at the formation of some cations produced by reaction of neutral bases with hydrogen ions. Differences in basicity of various compounds are conveniently evaluated by comparing them with respect to **proton affinity,** the energy released when the base combines with an H^+ ion. Proton affinities are measured in the gas phase, so there are no solvation effects. Table 4.1 shows the reactions and lists the proton affinities for water, ammonia, and hydrogen fluoride.

The trend in Table 4.1 is clear and easily understood. Ammonia is the strongest base because it releases the greatest amount of energy upon protonation. Water and hydrogen fluoride are successively weaker bases. This pattern should not be surprising to you because fluorine is the most electronegative element and can donate its electrons to form a bond much less easily than other atoms. On the other hand, nitrogen is the least electronegative of the three elements and can most easily use its unshared pair of electrons for bond formation. In aqueous solution the much lower basicity of HF relative to H_2O effectively precludes the action of HF as a proton acceptor, and you should only expect to observe reactions in which it acts as an acid.

The proton affinity data in Table 4.1 can be extended to the alkyl substituted analogs. This allows us to explain the acid-base properties of the alkyl derivatives of ammonia (amines) and the alkyl derivatives of water (alcohols and ethers). As you would expect from Table 4.1, amines are stronger bases than alcohols and ethers because *ammonium* ions are more stable than *oxonium* ions (although both are formed quite easily).

The same kinds of reasoning that we just used to analyze base strengths can also be applied to acid strength. The acidity of organic compounds is highly dependent upon the atom which donates the H^+ because this atom bears the formal negative charge in the corresponding anion:

TABLE 4.1 Proton Affinities of Some Simple Bases

Reaction	Proton Affinity, kcal/mol
$\ddot{N}H_3 + H^+ \longrightarrow NH_4^+$	202
$H_2\ddot{O}: + H^+ \longrightarrow H_3O:^+$	170
$H{-}\ddot{\underset{..}{F}}: + H^+ \longrightarrow H{-}\ddot{\underset{..}{F}}^+{-}H$	113

$$HA \overset{K_a}{\rightleftharpoons} H^+ + A^-$$

The dissociation constant of an acid, HA, is defined by the following expression:

$$K_a = \frac{[H^+][A^-]}{[HA]}$$

We have summarized in Table 4.2 the reactions and dissociation constants for compounds of fluorine, oxygen, nitrogen, and carbon. Each compound consists of the central atom with hydrogen atoms as the only substituents. For purposes of comparison the dissociation constants correspond to reaction in aqueous solution, but the last two reactions (for NH_3 and CH_4) have such small equilibrium constants that they cannot be measured in aqueous solution. The values for those reactions are extrapolated from other solvent systems, and we have written H^+ (rather than H_3O^+) as one of the products in each case.

The dissociation constant listed for water in Table 4.2 raises a minor point, but one which could be the source of considerable confusion. The commonly used value of 10^{-14} is only the product of the two ion concentrations:

$$K_w = [H^+][OH^-] = 10^{-14}$$

The dissociation constants in Table 4.2 all include the concentration of the reactant, which in the case of H_2O is large (55.5 *M*) because it is also the solvent. The resulting equilibrium constant is therefore smaller than K_w by two orders of magnitude:

$$K_a = \frac{[H^+][OH^-]}{[H_2O]} = 10^{-16}$$

The trend in Table 4.2 is again quite clear. The more electronegative the atom, the better it can accommodate a negative charge. If other structural features are the same, you should always expect that organic compounds with an OH group will be more acidic than the corresponding NH derivatives. Similarly, a comparably substituted CH group will be much less acidic than either the OH or NH analogs.

TABLE 4.2 Dissociation Constants for Some Simple Acids

Reaction	Dissociation Constant, K_a
$HF \rightleftharpoons H^+ + F^-$	10^{-4}
$H_2O \rightleftharpoons H^+ + HO^-$	10^{-16}
$NH_3 \rightleftharpoons H^+ + NH_2^-$	10^{-32}
$CH_4 \rightleftharpoons H^+ + CH_3^-$	10^{-40}

In the first section of this chapter we stated that water is the implicit standard for comparison in acid-base reactions. This allows you to use Tables 4.1 and 4.2 as a basis for predicting the chemical behavior of a wide variety of organic compounds. You only need to use a simple approximation: The replacement of a hydrogen atom on a reactive center by an alkyl group does not cause large changes in acid-base reactivity at that center. From the data in Table 4.1 you could correctly predict that *amines* (e.g., R—NH_2), like ammonia, are *bases.* In a similar way the data in Table 4.2 would allow you to predict that amines are much less acidic than water.

We can use electronegativity differences between oxygen and nitrogen to predict that amines will be less acidic and more basic than water. But the *absence* of any difference in electronegativity can also be useful. The structural similarities of ethers (R—O—R) and alcohols (R—OH) suggest that these derivatives should have basicities similar to that of water, and this is indeed the case. The presence of a hydroxyl group also allows alcohols to act as acids, and their K_a values are very close to that of water. (The oxygen atom of an ether has no hydrogen substituent, so a comparison of acidity is not appropriate.) These comparisons are only qualitative, but they do give you a good idea of the kinds of reactivity that you should expect for different classes of organic compounds.

In this section we have introduced the relationship between stability of reaction products and the position of the equilibrium. Throughout the remainder of this book we will continue to relate structure and reactivity, and we will stop here to summarize and review some of the key relationships. If reactants A and B yield product C, we can write the reaction and the equilibrium constant in the following way:

$$A + B \rightleftharpoons C \qquad K = \frac{[C]}{[A][B]}$$

The energy change is defined as,

$$\Delta E = E_{\text{products}} - E_{\text{reactants}}$$

and (when expressed as the free energy, ΔG) the energy difference is related to the equilibrium constant by the following equation:

$$\Delta G = -2.3\ RT\ (\log K)$$

In an equilibrium reaction any effect that stabilizes the reactants (either A or B) will decrease their tendency to produce product. Stabilization of the reactants would increase the energy difference, ΔG, for an endothermic reaction (i.e., make it more positive), and this would decrease the value of K. The net result would be a larger amount of A and B in the equilibrium mixture. Any effect which stabilizes the product C will *increase* the equilibrium constant and produce a corresponding increase in the amount of C that is present at equilibrium. We can also express these relationships in the opposite manner. Any effect which increases the energy of A or B (i.e., makes it more reactive) will result in an increased value of K and a shift of the equilibrium to the right. A decrease in stability of the product, C, would make it less likely to form, and you would expect a reduction in the magnitude of the equilibrium constant.

EXERCISE 4.4

Predict the relative order of basicity of $(CH_3CH_2)_3N$, $(CH_3CH_2)_2O$, CH_3CH_2F.

EXERCISE 4.5

Predict the relative order of acidity of $CH_3NH_3^+$, $CH_3OH_2^+$, and CH_3FH^+.

EXERCISE 4.6

Predict the relative order of acidities of $CH_3CH_2—CH_3$, $CH_3CH_2—NH_2$, and $CH_3CH_2—OH$.

EXERCISE 4.7

Predict the relative order of basicities of the ions: $(CH_3)_3C^-$, $(CH_3)_2N^-$, CH_3O^-.

4.7 RESONANCE STABILIZATION OF IONS

The presence of multiple bonds, particularly of double bonds to oxygen, can lead to striking changes in acidity for certain hydrogens in a molecule. From the preceding section you know that the K_a for the hydroxyl group of water and of alcohols is approximately 10^{-16}, and you might have expected all hydroxyl groups to be comparable. But this is not true, and the following inorganic acids have OH groups that are more acidic than that of water by as much as a factor of 10^{20}.

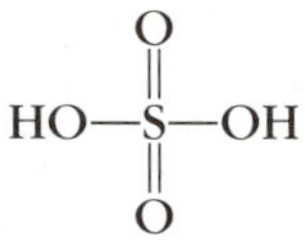

HO—S(=O)₂—OH | HO—⁺N(=O)—O⁻ | HO—P(=O)(OH)—OH

Sulfuric acid
$K_a = 10^3$

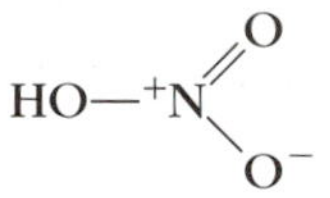

Nitric acid
$K_a = 10$

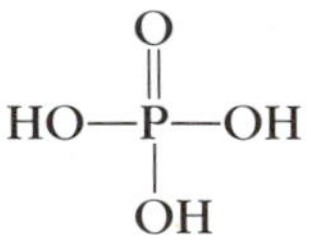

Phosphoric acid
$K_a = 10^{-2}$

The OH group in organic compounds can be highly acidic even when it is attached to carbon:

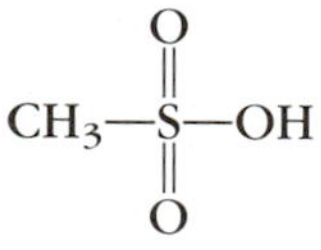

$CH_3—S(=O)_2—OH$ | HO—C(=O)—OH | $CH_3—C(=O)—OH$

Methanesulfonic acid
$K_a \simeq 1$

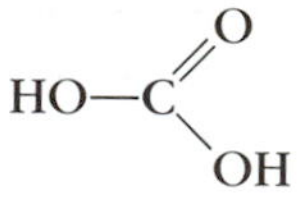

Carbonic acid
$K_a = 10^{-6}$

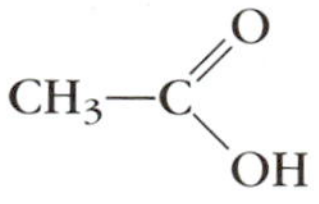

Acetic acid
$K_a = 10^{-5}$

What structural features make these compounds so acidic? In order to analyze the relationship between acidity and structure for these compounds, we must first discuss the way that the structural formulas are drawn for sulfuric and phosphoric acids. In both cases we have drawn more than eight electrons for the valence shell of the central atom. These atoms, sulfur and phosphorus, are third-row elements, and we should expect some differences in chemical and electronic properties as compared with those of the second-row elements. The third-row atoms have *d* orbitals of relatively low energy, which can interact with the *s* and *p* orbitals of the attached oxygens. We will not attempt to analyze the molecular orbitals, but the *d* orbitals afford a convenient rationale for the "expanded" valence shell of the third-row elements.

Some chemists object to the notion of more than eight electrons in the valence shell, even for third-row elements, and prefer the following way of drawing the structural formulas for sulfuric and phosphoric acids:

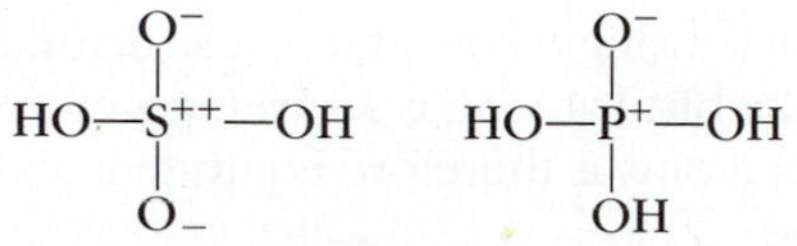

```
      O⁻              O⁻
      |               |
HO—S⁺⁺—OH       HO—P⁺—OH
      |               |
      O⁻              OH
```

When drawn in this way, the central atoms have only eight electrons in their valence shells, but formal charges must be placed on some of the atoms. For a variety of reasons, particularly to aid comparisons with the carbon-oxygen double bonds of organic compounds, we will ordinarily use the type of drawings that have expanded valence shells for third-row atoms.

What structural feature is responsible for the very high acidity of sulfuric acid and the other strong acids we are considering here? In each case the acidic hydrogen is bonded to oxygen, but the OH group is also present in very weak acids such as water and alcohols. The unique structural feature of the strong acids is attachment of the hydroxyl group to an atom which also has a double bond to oxygen. The question, of course, is how does this simple structural effect, the presence of an S=O, P=O, N=O, or C=O group, produce such a remarkable enhancement in acidity relative to the K_a for water and the alcohols?

For convenience, we will center our discussion on sulfuric acid, and this allows us to rephrase the question. Why is sulfuric acid so much more acidic than water? Acidity is concerned with a *reaction,* and our analysis must include *both* the acid and its conjugate base. Moreover, we are comparing the acidity of two different acids, so we must compare two reactions:

$$H_2SO_4 \rightleftharpoons H^+ + HSO_4^- \qquad K_a = 10^3$$

$$H_2O \rightleftharpoons H^+ + HO^- \qquad K_a = 10^{-16}$$

As we indicated in the previous section, you can focus your attention on the ionic species when you evaluate the energetics of such reactions. Both equilibria involve a single H^+ ion, so the important differences in this case must be with the two anions:

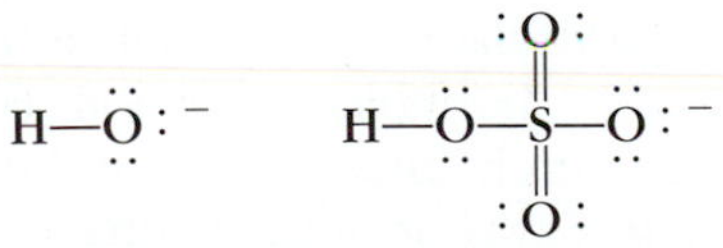

```
                         :O:
                          ‖
H—Ö:⁻            H—Ö—S—Ö:⁻
                          ‖
                         :O:
```

Hydroxide ion   Bisulfate ion

Because the acid dissociation constant for sulfuric acid is so much larger than that for water, we can conclude that the bisulfate ion is very stable in comparison with the hydroxide ion. This brings us to a new question: Which oxygen of the bisulfate ion carries the negative charge? The answer to the question is that the bisulfate ion can be drawn in any of three equivalent ways:

$$\underset{\textbf{I}}{HO{-}\overset{\overset{O}{\|}}{\underset{\underset{O}{\|}}{S}}{-}O^-} \qquad \underset{\textbf{II}}{HO{-}\overset{\overset{O^-}{|}}{\underset{\underset{O}{\|}}{S}}{=}O} \qquad \underset{\textbf{III}}{HO{-}\overset{\overset{O}{\|}}{\underset{\underset{O^-}{|}}{S}}{=}O}$$

This is a situation in which a single conventional structural formula is inadequate. The "true" structure of the bisulfate ion is an average, or *hybrid*, of structures I–III. Each of the three oxygen atoms is therefore equivalent and each carries only one-third of the ionic charge.

Finally, we can answer our questions. As we indicated previously, neutral species are ordinarily more stable than ions. To state this in a different way, an energetically unfavorable condition exists when the number of negatively charged electrons on an atomic center exceeds the number of positively charged protons in the nucleus. This means that anything which *decreases* the electron density surrounding an anionic center will have a stabilizing effect. Even though the bisulfate ion still has a total charge of −1, the charge is divided among three different oxygens, and the overall electron-electron repulsions are reduced. This is a general phenomenon for cations as well as anions; *any effect that delocalizes ionic charge over several atomic centers will stabilize the ion.*

We can now summarize the increased acidity of sulfuric acid relative to water by comparing the anions in the following way:

$$H_2SO_4 \rightleftharpoons H^+ + HSO_4^-$$

More stable than HO^-; reaction has larger equilibrium constant

$$H_2O \rightleftharpoons H^+ + HO^-$$

Less stable than HSO_4^-; reaction has smaller equilibrium constant

When the actual structure of a molecule or ion can only be described as an average, or hybrid, of several structural formulas, the mathematical description of the structure exhibits a similarity to the classical equations for *resonance.* We will present a more detailed description of the development of resonance theory in Section 10.5, but for the time being you only need to understand the modern interpretation. Individual structural formulas such as I to III for the bisulfate ion are called **resonance contributors** or **resonance forms** of the actual structure (which in turn is often called a **resonance hybrid**). The conventional structural formulas that we use to draw organic molecules can only describe bonding interactions that involve exactly two electrons, and a particular electron or electronic charge must be associated with one specific atom in the molecule or ion. For many compounds the conventional structures are entirely satisfactory, but the delocalization of charge usually requires the drawing of two or more resonance contributors in order to fully describe the structure. Because charge delocalization and resonance go hand in hand, the increased stability of charge-delocalized ions is often called **resonance stabilization.**

The concept of resonance provides us with a powerful tool for interpreting and predicting reactivity for certain types of compounds and ions, and there are several important conventions and rules that govern the use of resonance. The first is the use of a double-headed arrow (↔) to denote the resonance forms of a structure.

The three structures which contribute to the overall structure of the bisulfate ion are therefore drawn in the following fashion:

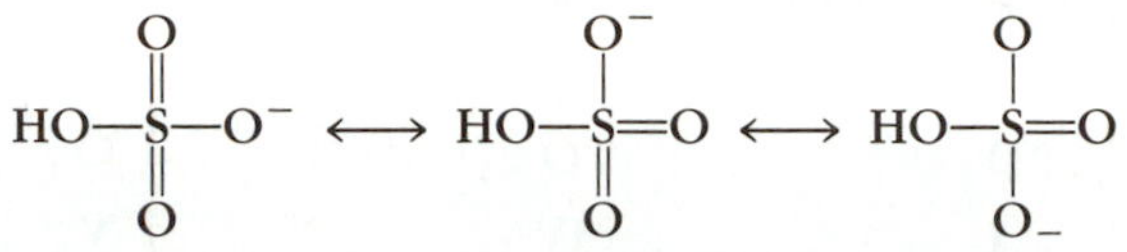

$$\underset{\textbf{I}}{HO{-}S({=}O)_2{-}O^-} \longleftrightarrow \underset{\textbf{II}}{HO{-}S(O^-)({=}O)_2} \longleftrightarrow \underset{\textbf{III}}{HO{-}S({=}O)_2{-}O^-}$$

The double-headed arrow specifically denotes two resonance contributors to a particular structure. It is particularly important that you do not confuse this with the double arrows (⇄) that are used to designate an equilibrium between two different structures. The *double-headed arrow* (↔) in no way signifies a reaction; rather it denotes resonance forms that by definition are two drawings of the same species. Although a single resonance form does not adequately describe the actual structure of a charge-delocalized ion, we often draw only one such structure for convenience. It will be assumed that you are aware of the other resonance forms that could be drawn.

There are some very strict limitations on the structural formulas that can be drawn to represent resonance contributors. Resonance contributors to a structure must have the same constitution and the same net charge. The only ways in which individual resonance forms can differ are in the location of the *multiple* bonds and the location of formal charges. If you drew structural formulas that differed in other ways, you would be describing *isomers* rather than *resonance contributors.*

The relationship between resonance contributors is frequently shown with curved arrows, as in the following drawings:

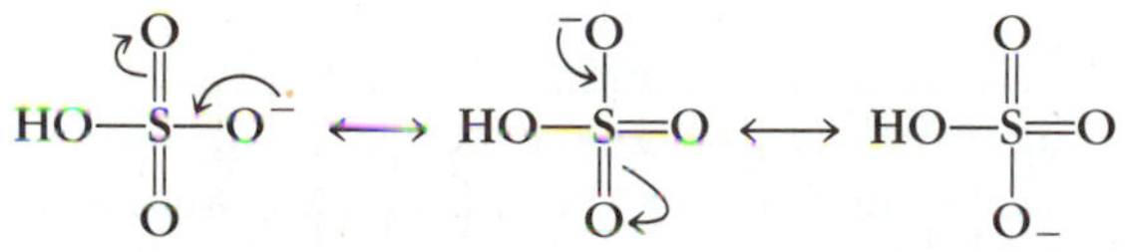

$$\underset{\textbf{I}}{HO{-}S({=}O)_2{-}O^-} \longleftrightarrow \underset{\textbf{II}}{HO{-}S(O^-)({=}O)_2} \longleftrightarrow \underset{\textbf{III}}{HO{-}S({=}O)_2{-}O^-}$$

These arrows *do not* imply any reaction of the molecule or ion. They merely provide us with a convenient way of showing the relationships between the various resonance forms once any single contributor has been drawn.

The rationale is as follows: The lone pair of electrons corresponding to the negative charge in structure I could also be involved in a bond to sulfur. An arrow shows this pair of electrons "moving" toward sulfur while also continuing to be shared by the oxygen. The result is a "new" sulfur-oxygen bond (structure II). But the total number of bonds to sulfur cannot increase if the sulfur is to maintain electrical neutrality. Therefore, when we use the pair of electrons from oxygen to form a new double bond to sulfur, one of the existing double bonds must be "broken." This is shown as migration of one of the electron pairs of a double bond in structure I from sulfur, the electron pair becoming completely localized on another oxygen. It is essential for you to remember that the electrons are not actually moving in this fashion. Nevertheless, this technique of "moving" the electrons provides you with a very useful device for drawing the different resonance forms of an ion.

It is possible to draw additional resonance forms for the bisulfate ion:

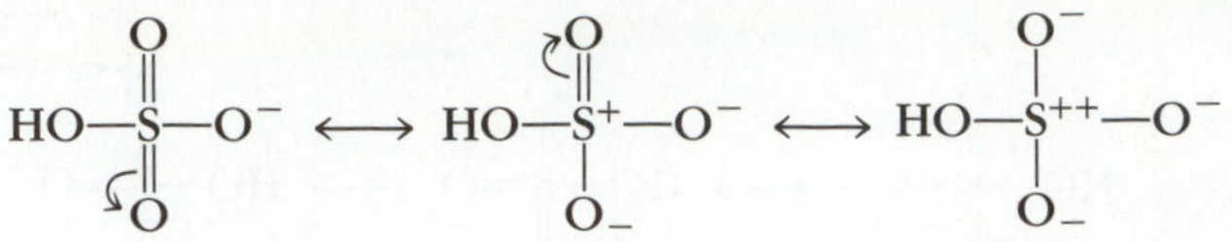

$$\underset{\textbf{I}}{HO{-}S(=O)_2{-}O^-} \longleftrightarrow \underset{\textbf{IV}}{HO{-}S^+(=O)(O^-){-}O^-} \longleftrightarrow \underset{\textbf{V}}{HO{-}S^{++}(O^-)(O^-){-}O^-}$$

Two of these structures, IV and V, have additional formal charges, however, and this brings up another important feature of resonance theory. *An individual resonance form contributes to the actual structure only to the extent that the contributor itself corresponds to a stable structure.* You saw in Figure 4.3 that *separation* of positive and negative charges is energetically unfavorable (except at very large distances). We can therefore conclude that structures such as IV and V would contribute much less than I, II, and III to the resonance hybrid for the bisulfate ion. This argument further explains why we prefer the use of an expanded valence shell to draw structures for sulfuric acid and other derivatives of third-row elements. You will now recognize that our preferred representation for sulfuric acid and the drawing we showed earlier are actually two *resonance forms* of the same structure.

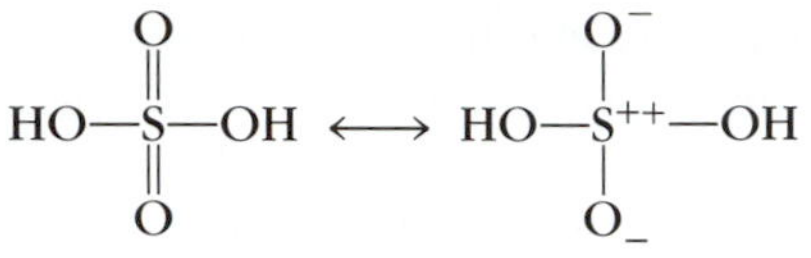

$$HO{-}S(=O)_2{-}OH \longleftrightarrow HO{-}S^{++}(O^-)_2{-}OH$$

The charge separation of the second structure would correspond to a higher-energy arrangement, so it would contribute *less* to the actual resonance hybrid for sulfuric acid.

We have discussed extensively the large enhancement in acidity of an OH group that is bonded to an S=O group. Enhancement of acidity is also found with carbon-oxygen double bonds. The C=O group is called a **carbonyl group,** and it plays an important role in much of organic chemistry. Organic compounds that have an OH group bonded directly to a carbonyl group are known as **carboxylic acids,** and the resulting —CO_2H group is known as a **carboxyl group.** Although they are less acidic than the inorganic acids such as sulfuric, nitric, and phosphoric acid, carboxylic acids are still highly acidic in comparison with water. The relatively minor structural change from the alcohols (in which the hydroxyl group is coordinated directly to an alkyl group) to the carboxylic acids (which have a carbonyl group interposed between the hydroxyl and alkyl fragments) results in a remarkable increase in acidity. The acid dissociation constant of a carboxylic acid is greater by a factor of about 10^{13}.

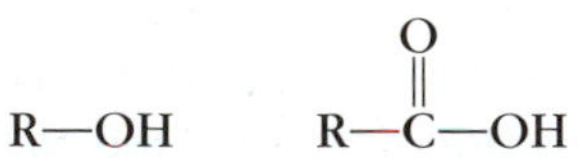

R—OH	R—C(=O)—OH
An alcohol	A carboxylic acid
$K_a \simeq 10^{-18}$	$K_a \simeq 10^{-5}$

Once again we can use resonance to explain the change in acidity. When an alcohol loses an H^+, the resulting negative charge must be localized on a single oxygen. In contrast, the loss of a proton from a carboxylic acid generates an anion

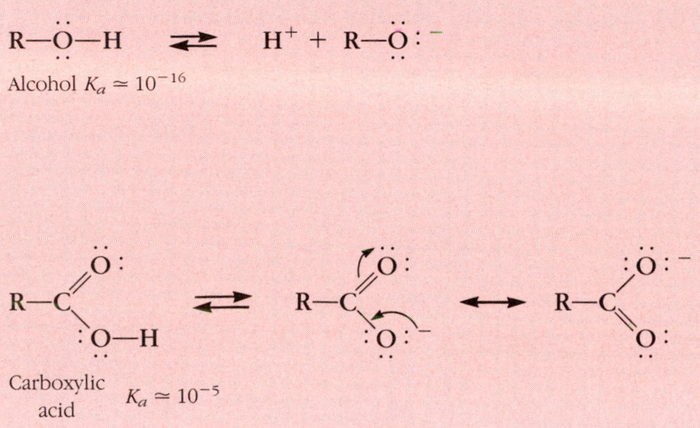

$$R{-}\ddot{\underset{..}{O}}{-}H \rightleftharpoons H^+ + R{-}\ddot{\underset{..}{O}}{:}^-$$

Alcohol $K_a \simeq 10^{-16}$

Negative charge localized on a single oxygen. Dissociation constant is normal (relative to that of water).

Carboxylic acid $K_a \simeq 10^{-5}$

Negative charge delocalized over two equivalent oxygen atoms. Dissociation constant is very large (relative to that of water).

Figure 4.5
Enhanced acidity of an OH group results when the OH is bonded directly to a carbonyl group. Loss of an H^+ produces an ion in which the negative charge is delocalized over two oxygen atoms. The resulting resonance stabilization causes carboxylic acids to be much more acidic than alcohols.

in which the charge is delocalized over two *equivalent* oxygen atoms (Figure 4.5).

In order to show the electron delocalization, the carboxylate anion is sometimes drawn as:

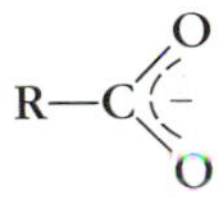

While such a drawing does indicate the equivalence of the two oxygen atoms, it fails to illustrate the key feature of the resonance argument that the negative charge is distributed over the two *oxygen* atoms. The use of the dotted line in this structure incorrectly suggests that the reactive charge is also delocalized onto the central *carbon* atom. In order to avoid any confusion, we will not employ this type of drawing for the carboxylate anion.

The remarkable ability of a carbonyl group to increase the acidity of hydrogen atoms is not restricted to the OH group. As you can see by inspection of Table 4.3, the acidity of a hydrogen on any atom is greatly increased when that atom is also bonded to a carbonyl group. We have compared the acidities of a series of compounds having the substituent $CH_3{-}CH_2$ with similar compounds in which the CH_2 has been replaced by C=O. In each case this replacement produces an extremely large increase in the acidity of the hydrogens on the adjacent atom, ranging from 11 to 24 powers of 10. Clearly, resonance stabilization of the negative charge in the conjugate bases is very substantial for compounds containing a carbonyl group.

As you study the data in Table 4.3, you will notice that the magnitude of the effect of a carbonyl group is strongly dependent upon the atom to which the acidic hydrogen is bonded. In the case of oxygen (alcohol → carboxylic acid), the acid

TABLE 4.3 The Effect of a Carbonyl Group on the Acidity of Organic Compounds

Compound	Dissociation Constant, K_a	Compound	Dissociation Constant, K_a
$CH_3{-}CH_2{-}OH$ Alcohol	10^{-17}	$CH_3{-}C(=O){-}OH$ Carboxylic acid	10^{-5}
$CH_3{-}CH_2{-}NH_2$ Amine	10^{-34}	$CH_3{-}C(=O){-}NH_2$ Amide	10^{-17}
$CH_3{-}CH_2{-}CH_3$ Alkane	$<10^{-40}$	$CH_3{-}C(=O){-}CH_3$ Ketone	10^{-20}

dissociation constant increases by approximately 10^{12}. For nitrogen the effect is larger, and an increase by a factor of 10^{17} is found for the acidity of an amide compared with that of an amine. The effect is largest for ionization of hydrogen bound to carbon. The K_a of a **ketone** (in which the carbonyl group is coordinated to two alkyl groups) is larger than that of the corresponding alkane by a factor of more than 10^{20}. Once again, we can interpret these changes in equilibria by looking at the ionic species, and for this series of compounds we can make a further generalization: *The more unstable the anion, the more it can benefit from the resonance stabilization of an adjacent carbonyl group.* This is the case for a hydrogen bound to carbon, which yields the least stable of the anions in Table 4.3; the carbonyl group exerts its largest effect on acidity for a C—H group. On the other hand, the most stable of the anions under consideration is a negatively charged *oxygen,* and it is for an OH group that the carbonyl group produces the smallest effect.

The data in Table 4.3 allow us to make another very important observation. The presence of the carbonyl groups in amides and ketones causes both these classes of compounds to be comparable in acidity with the alcohols. This has extremely important consequences for many organic reactions, because alcohols are often used as solvents and their salts are commonly employed as bases. This means that the sodium salt of ethanol will undergo the following acid-base reaction with acetone (a ketone):

$$CH_3CH_2O^- \ {}^+Na + \underset{\text{Acetone}}{CH_3{-}C(=O){-}CH_3} \rightleftharpoons CH_3CH_2OH + \left[CH_3{-}C(=O){-}CH_2^- \longleftrightarrow CH_3{-}C(O^-){=}CH_2 \right] Na^+$$

The equilibrium lies to the left in this reaction, and you can calculate an equilibrium constant of 10^{-3} from the data in Table 4.3. A small but significant amount of the salt of the ketone is therefore present in solution, and such anions are important intermediates in the reactions of ketones. Even though the concentration at any one time is small, the anion is in equilibrium with the ketone, and the equilibrium concentration can be maintained throughout a reaction. You will encounter many examples of such reactions in subsequent chapters of this book.

What is the effect of a *second* carbonyl group on the acidity of a hydrogen bonded to a carbon? As you might expect, a further increase in acidity results.

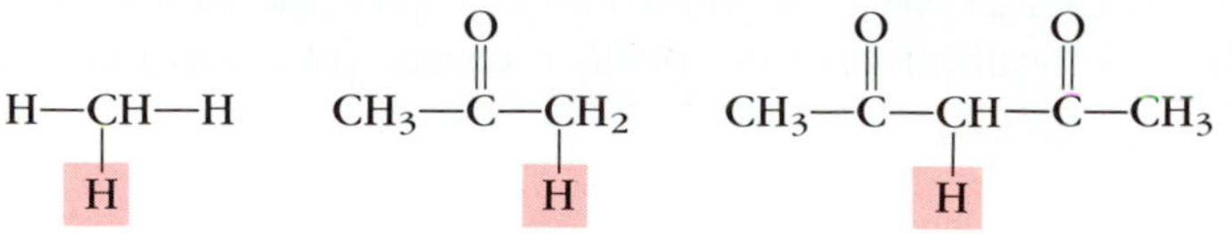

$$H{-}CH(H){-}H \qquad CH_3{-}C(=O){-}CH_2(H) \qquad CH_3{-}C(=O){-}CH(H){-}C(=O){-}CH_3$$

$$K_a \simeq 10^{-40} \qquad K_a \simeq 10^{-20} \qquad K_a \simeq 10^{-9}$$

You should recognize two important features in this trend in acidity. First, the effect of the second carbonyl group is not as large as that of the initial one, and this is yet another example of the pattern we have mentioned previously. The more unstable the anion, the more it benefits from resonance stabilization. The presence of one carbonyl group results in a very large resonance stabilization of the corresponding anion, so the additional stabilization provided by a second carbonyl group is somewhat smaller. Second, the combined influence of two carbonyl groups enhances the acidity of the adjacent C—H to such an extent that the compound is *more acidic than water and the alcohols.* The reaction of the dicarbonyl compound with the sodium salt of ethanol would result in essentially complete conversion of that diketone to its anion. This remarkable acidity of carbon-bound hydrogen atoms has allowed such dicarbonyl compounds to become valuable reagents in organic reactions, and we will discuss the reactions of these compounds at length in Chapter 21.

Resonance stabilization can also be observed for cations, although examples of this are less common. We can illustrate this with the following two compounds, which contain basic nitrogen atoms:

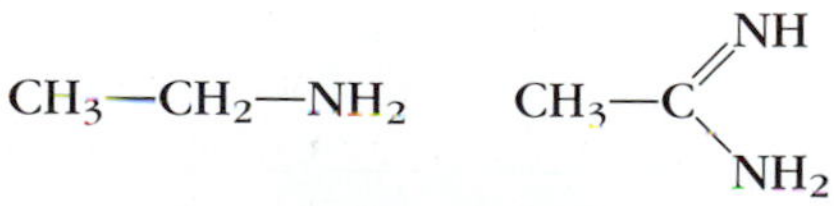

$$CH_3{-}CH_2{-}NH_2 \qquad CH_3{-}C(=NH){-}NH_2$$

The K_a values of the conjugate acids of these bases are shown, together with the equations for dissociation of these acids (in aqueous solution).

$$CH_3{-}CH_2{-}NH_3^+ \rightleftharpoons CH_3{-}CH_2{-}NH_2 + H^+ \qquad K_a \simeq 10^{-10}$$

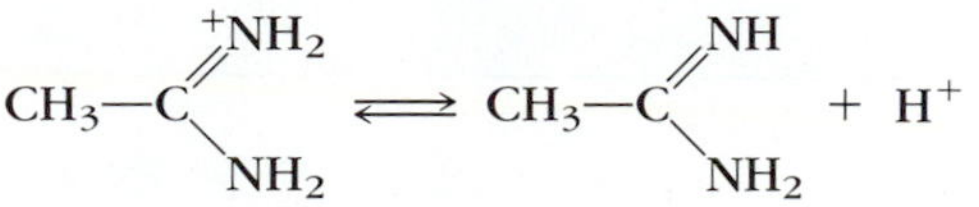

$$CH_3{-}C(={}^+NH_2){-}NH_2 \rightleftharpoons CH_3{-}C(=NH){-}NH_2 + H^+ \qquad K_a \simeq 10^{-13}$$

We can explain the smaller acid dissociation constant of the second compound on the basis of resonance stabilization of the *cation,* which therefore has a decreased tendency to undergo dissociation.

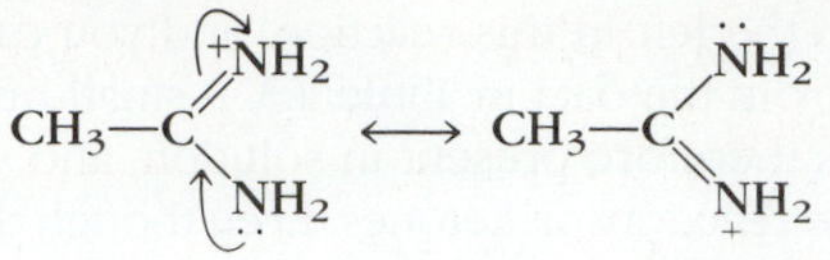

$$CH_3-C(=\overset{+}{N}H_2)(-\ddot{N}H_2) \longleftrightarrow CH_3-C(-\ddot{N}H_2)(=\overset{+}{N}H_2)$$

The overall effect is relatively small, and the equilibrium constants for the two reactions differ by a factor of only 10^3. This is still another illustration of the pattern of resonance stabilization being large for very unstable ions but less for ions that are more stable. Ammonium ions are quite stable, even in the absence of resonance, and you can see this from the relatively small magnitude of the acid dissociation constant for $CH_3CH_2NH_3^+$. As you would anticipate, the additional stabilization resulting from delocalization of the positive charge over two nitrogen atoms is fairly small.

EXERCISE 4.8

(a) Draw the anions resulting from loss of the indicated hydrogen in each of the following compounds:

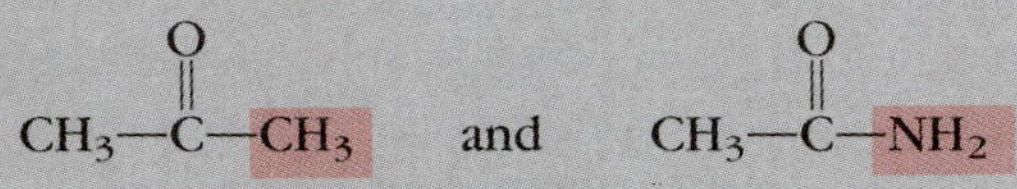

$$CH_3-\overset{O}{\overset{\|}{C}}-CH_3 \quad \text{and} \quad CH_3-\overset{O}{\overset{\|}{C}}-NH_2$$

(b) Draw both resonance forms for each anion. Which resonance form contributes most to the overall structure in each case?

EXERCISE 4.9

Write the full equilibrium for the reaction between

$$CH_3CH_2O^-\ Na^+ \quad \text{and} \quad CH_3-\overset{O}{\overset{\|}{C}}-CH_2-\overset{O}{\overset{\|}{C}}-CH_3$$

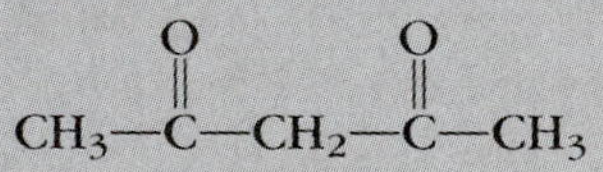

Draw all resonance forms of the anion, and indicate the relative contribution of each to the overall structure.

EXERCISE 4.10

In the presence of a strong acid such as HCl, a carboxylic acid can be protonated. On the basis of resonance stabilization, which oxygen of acetic acid would be protonated more readily? Draw appropriate structures (including resonance forms) to support your answer.

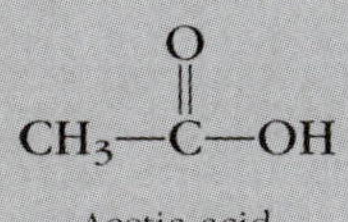

$$CH_3-\overset{O}{\overset{\|}{C}}-OH$$

Acetic acid

4.8 INDUCTIVE EFFECTS

Resonance interactions are not the only structural effects governing the stability of ions; we will now consider **inductive effects.** Resonance interactions result from charge delocalization through multiple bonds, but inductive effects operate strictly through single bonds. Inductive effects result from the *polarization* of individual bonds (Section 4.4) and are simply related to the electronegativity of a substituent. We can illustrate this nicely by comparing the acidity of ethanol with that of 2,2,2-trifluoroethanol, in which three hydrogen atoms have been replaced by highly electronegative fluorine atoms. This change results in a substantial increase in acidity for the fluorinated alcohol.

$$CH_3—CH_2—OH \qquad\qquad CF_3—CH_2—OH$$

Ethanol $K_a \simeq 10^{-16}$ 2,2,2-Trifluoroethanol $K_a \simeq 10^{-12}$

Focusing once again on the *charged* species in the equilibrium, we can conclude that the increased acidity results from a greater stability of the conjugate base of the fluorinated alcohol.

$$CF_3CH_2OH \rightleftharpoons CF_3CH_2O^- + H^+$$

How do the fluorine atoms stabilize the anion on oxygen? Fluorine is strongly electronegative, and each of the three fluorine-carbon bonds is highly polarized so that electron density is moved toward the fluorine atoms. (The use of $\delta+$ and $\delta-$ in the following drawing only indicates *partial* charge, and no *magnitude* is implied. Clearly the net charge on the entire ion remains -1.)

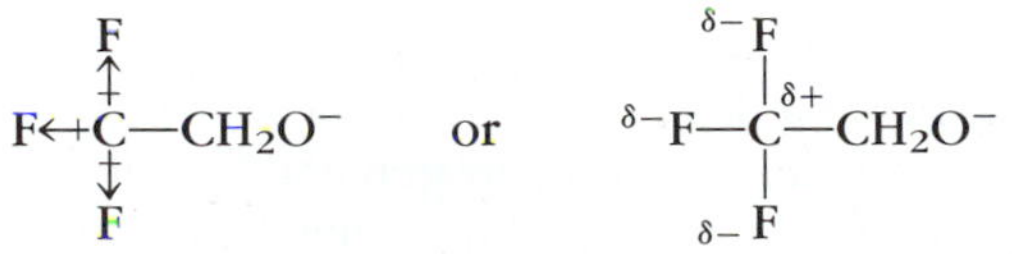

$$F \leftarrow\!\!\!+ C(\uparrow F)(\downarrow F)—CH_2O^- \quad \text{or} \quad {}^{\delta-}F—\overset{\delta+}{C}({}^{\delta-}F)({}^{\delta-}F)—CH_2O^-$$

This **electron-withdrawing** effect of the fluorine atoms causes the carbon atom of a CF_3 group to have a partial positive charge, and the CF_3 group in turn behaves as an electronegative substituent.

$$CF_3 \leftarrow\!\!\!+ CH_2—O^-$$

The polarization is transmitted through yet another single bond, that between the CH_2 group and oxygen, and this reduces the net electron density on the negatively charged oxygen.

$$CF_3—CH_2 \leftarrow\!\!\!+ O^-$$

The final result of all these interactions is dispersal of some electron density from the negatively charged oxygen atom onto the atoms of the CF_3CH_2 group. The delocalization from inductive effects is not as extensive as that from resonance interaction, but it can nevertheless greatly affect the stability of an ion.

You can obtain a better feeling for the magnitude of inductive effects by comparing the effects of successively replacing the hydrogen atoms of the CH_3 group of acetic acid by fluorine atoms.

Acetic acid	Fluoroacetic acid	Difluoroacetic acid	Trifluoroacetic acid
$K_a = 1.8 \times 10^{-5}$	$K_a = 200 \times 10^{-5}$	$K_a = 6000 \times 10^{-5}$	$K_a = 50{,}000 \times 10^{-5}$

The total of three fluorines results in a remarkably strong acid, and the dissociation constant for trifluoroacetic acid is comparable with that for inorganic acids such as nitric and phosphoric acid (Section 4.7). The increased acidity results from stabilization of the negatively charged conjugate base, but as in the case of resonance interactions, the stabilizing effect diminishes as the anion becomes more stable. The introduction of a single fluorine substituent causes K_a to increase by a factor of 50 for fluoroacetic acid relative to acetic acid. The second fluorine substitution increases the K_a by a factor of 30, and the third substitution results in only an eightfold increase.

Inductive effects operate by polarization of individual bonds, and you might anticipate that the effect should depend upon the number of bonds through which the effect must be transmitted. In other words, the magnitude of an inductive effect should depend on the "distance" between the two interacting groups. This is indeed the case, and you can see the pattern in the dissociation constants for a series of carboxylic acids of varying chain length (Figure 4.6).

The three fluorine atoms of trifluoroacetic acid (CF_3CO_2H) produce an increase in the dissociation constant by more than a factor of 10^4 relative to acetic acid (CH_3CO_2H). However, the effect is much smaller for the other examples in Figure 4.6. With a single CH_2 group interposed between the CF_3 and CO_2H groups, the effect decreases to a factor of less than 10^2, and a second CH_2 group reduces the ratio of dissociation constants for the hydrogen and fluorine derivatives to only 4. In this last example a total of five single bonds separate the electronegative fluorine atoms from the negatively charged oxygen atom,

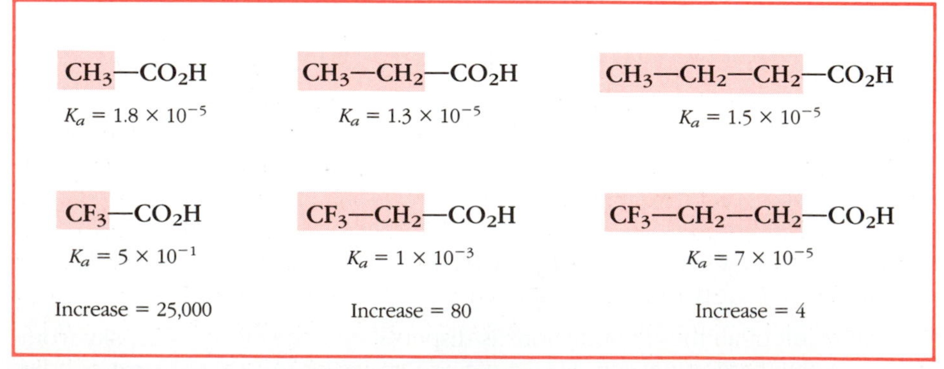

Figure 4.6 Inductive Effects on the Acidity of Carboxylic Acids. The highly electronegative fluorine substituents can produce a large increase in the acid dissociation constant, but the effect decreases as the number of intervening bonds between the fluorine atoms and the carboxyl group is increased.

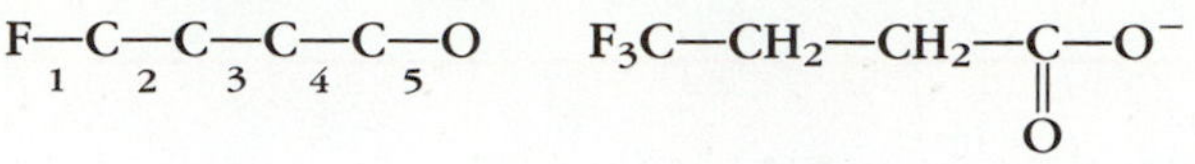

$$\underset{1}{\text{F}}\text{—}\underset{2}{\text{C}}\text{—}\underset{3}{\text{C}}\text{—}\underset{4}{\text{C}}\text{—}\underset{5}{\text{C}}\text{—O} \qquad F_3C\text{—}CH_2\text{—}CH_2\text{—}C(=O)\text{—}O^-$$

and this corresponds to the maximum distance for which an inductive effect will be observed.

Other electronegative substituents will have similar effects on acidities of organic compounds. The inductive effects of other elements are usually somewhat smaller in magnitude than that of fluorine, which is the most electronegative element. The electron-withdrawing inductive effect of oxygen and halogens, other than fluorine, is reflected by the increase in acidity that they produce in the following substituted acetic acids:

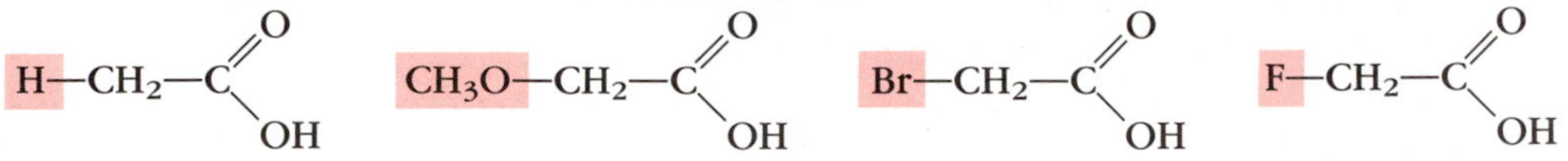

$$H\text{—}CH_2\text{—}C(=O)OH \qquad CH_3O\text{—}CH_2\text{—}C(=O)OH \qquad Br\text{—}CH_2\text{—}C(=O)OH \qquad F\text{—}CH_2\text{—}C(=O)OH$$

$$K_a = 1.7 \times 10^{-5} \qquad K_a = 29 \times 10^{-5} \qquad K_a = 130 \times 10^{-5} \qquad K_a = 200 \times 10^{-5}$$

Hydrogen and carbon have approximately the same electronegativity, and as you would expect for nonpolar substituents, simple alkyl groups have almost no effect on the acidity of a carboxylic acid:

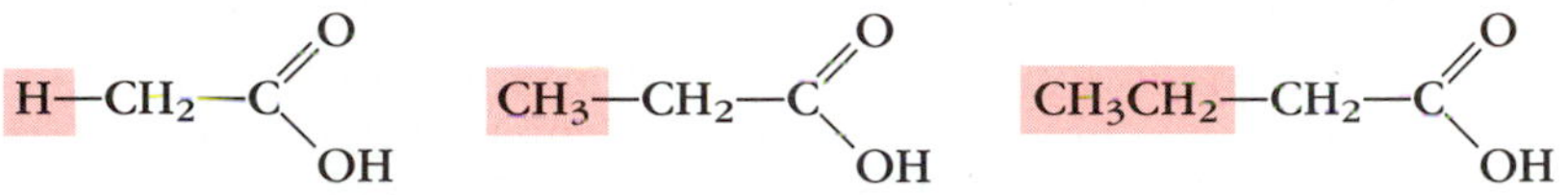

$$H\text{—}CH_2\text{—}C(=O)OH \qquad CH_3\text{—}CH_2\text{—}C(=O)OH \qquad CH_3CH_2\text{—}CH_2\text{—}C(=O)OH$$

$$K_a = 1.7 \times 10^{-5} \qquad K_a = 1.3 \times 10^{-5} \qquad K_a = 1.5 \times 10^{-5}$$

The stabilizing effect of an alkyl group relative to a hydrogen atom on a carbocation (see Section 2.7) might appear to conflict with this statement, but it is not appropriate to compare the two situations. A carbocation is *electron-deficient,* and an alkyl substituent can provide stabilization simply because a carbon atom has more electrons in its valence shell than does hydrogen. This allows the electrons of a $C\text{—}C^+$ bond to become more highly polarized than the electrons of an $H\text{—}C^+$ bond. However, you should not confuse this with our present discussion of electronegativity effects of atoms separated from the charged site by one or more single bonds.

The fluorinated derivatives of ethanol and acetic acid also permit us to compare resonance and inductive effects, and we have done this in Figure 4.7. We have used ethanol as the reference compound, and this allows us to directly compare the effects of a carbonyl group with those of fluorine substituents. The replacement of CH_3 by CF_3 increases the acidity of both the alcohol and the carboxylic acid by approximately the same amount; an increase in K_a by a factor of 10^4 is observed in each case. Replacement of the CH_2 group of either alcohol by a carbonyl group increases K_a by a factor of 10^{11} to 10^{12}. These results clearly demonstrate that resonance effects are considerably larger than inductive effects for the stabilization of anions. Nevertheless, you cannot neglect inductive effects because they can also bring about significant changes in chemical reactivity.

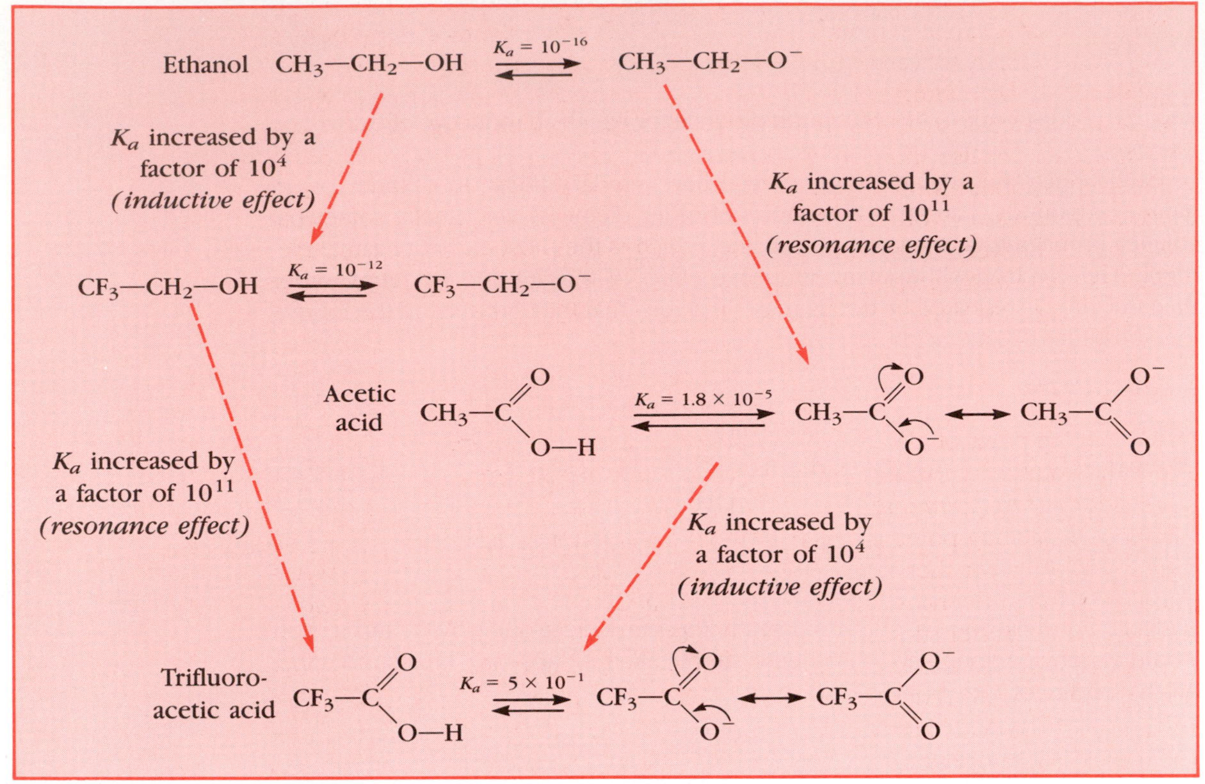

Figure 4.7 Influence of Resonance and Inductive Effects on Acidity. Both effects result in an increase in acidity by stabilizing the negative charge of the conjugate base, but the magnitude of the resonance effect is considerably larger.

EXERCISE 4.11

Carbocations are usually very unstable, but an oxygen substituent can stabilize them greatly. For example, it is possible to isolate certain salts of ions such as $[CH_3—CH—OCH_3]^+$.

(a) On the basis of resonance indicate why $HO—CH_2^+$ is more stable than CH_3^+.
(b) Which resonance contributor is most important?
(c) In the resonance hybrid, which atom bears most of the positive charge?
(d) On the basis of inductive effects *only* (i.e., if resonance interactions did not exist) which would be more stable, $HO—CH_2^+$ or CH_3^+?
(e) Which effect appears to have greater effect on stability, induction or resonance?

4.9 STRUCTURE AND PHYSICAL PROPERTIES

Just as the chemical reactivity of a compound is dependent upon its molecular structure, so are various physical properties. Melting point, boiling point, and solubility can all be correlated with features of molecular structure. We cannot simply examine the structure of a molecule and then predict either the melting or boiling point, but frequently we can predict which of two similar structures would have the higher melting or boiling point. In this section we will show you how different features of molecular structure are related to physical properties, and we will show you how you can use these to make predictions about melting points, boiling points, and solubilities of organic compounds. These physical properties are all related to processes involving changes in relative orientations of molecules and in the distances between them. For this reason physical properties are determined to a large extent by *intermolecular forces,* and much of our discussion in this section will center on how these are dependent upon molecular structure.

Melting Point

At the melting point of a compound its solid and liquid states are in equilibrium. The same equilibrium is involved in freezing, and we will use the terms *melting point* and *freezing point* interchangeably. To evaluate the effect of some structural variation on melting point, you must determine whether the change in structural features would favor the liquid or the the solid. This in turn requires an understanding of the differences between solids and liquids, and we will briefly discuss some of the characteristics of these two states.

A crystalline solid is composed of molecules that are arranged in a very regular pattern. This typically requires that all the molecules adopt the *same conformation* and that each molecule have a *specific orientation* relative to its neighbors. We have illustrated this schematically in Figure 4.8a. These restrictions are removed in the liquid state, as depicted in Figure 4.8b. In a liquid the individual molecules will be distributed among the energetically accessible conformations, and the relative orientations will be more random. These differences between solid and liquid tell us what kinds of structural features will influence the preference of a compound for either the solid or liquid state.

Intermolecular forces play a dominant role in determining the melting point of a compound, and sufficient energy to overcome these forces must be provided to

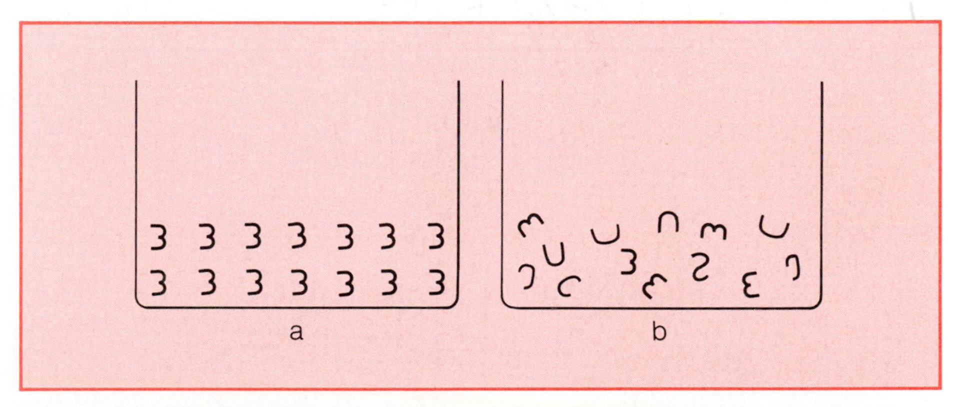

Figure 4.8 Idealized Views of the Solid and Liquid States.
(a) The solid state. Molecules in the crystal are ordered in a very regular fashion and have the same conformation. (b) The liquid state. The orientation of molecules in a liquid is more random, and the molecules can adopt different conformations.

melt a solid. In order to melt a solid you must increase the average kinetic energy of the individual molecules, and this is done by raising the temperature. Ionic compounds have particularly high melting points because the strong attraction between ions of opposite charge favors the well-defined arrangement of the solid. Only at very high temperatures do the ions have sufficient kinetic energy to overcome these powerful forces and move freely in a liquid. For example, sodium chloride melts at 801°C, a much higher melting point than you will find for organic compounds, even organic salts. This is illustrated by a comparison of acetic acid and its sodium salt; the melting point of acetic acid is 15°C; the sodium salt melts at 324°C.

We will now turn our attention to nonpolar organic compounds and evaluate the effects of structural variations on melting points. If you refer back to Table 1.1, you will notice that there is a tendency for melting points to increase with molecular weight in a series of structurally related compounds. In order to avoid the complications that would result from changes in molecular weight, we will restrict our discussion to the comparison of isomeric structures. You will be able to apply the arguments we develop to nonisomeric structures only if they have very similar molecular weights.

Consider the isomeric alkanes, pentane and neopentane (2,2-dimethylpropane). The unbranched alkane has a variety of conformations with fairly similar energies, and we have drawn these in Figure 4.9. In the liquid we would expect each of these conformations of pentane to be present, although to different extents according to their relative stabilities. In the crystal, however, all molecules of pentane should have the same conformation. If we consider the freezing process, molecules having the "wrong" conformation would not easily fit into a growing crystal, and this would favor the liquid state over the solid. When a molecule can exist in a variety of conformations, this is usually reflected by a low melting point.

In contrast to the conformational flexibility of pentane, the isomeric 2,2-dimethylpropane has a single stable conformation.

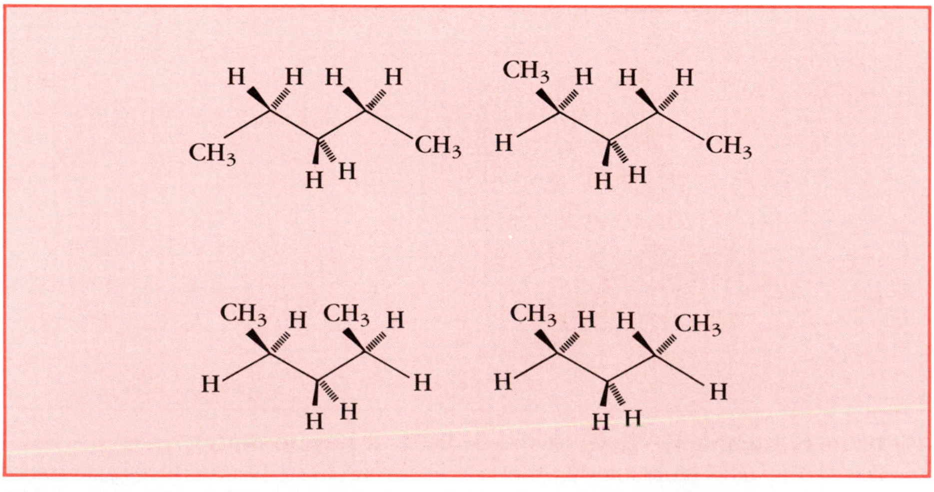

Figure 4.9 The Different Staggered Conformations of Pentane. Each of these conformations has a different but similar potential energy.

H H H H
H C C C H
H_3C CH_3

You will also notice that all hydrogen atoms are equivalent in this compound, and the high symmetry of the structure increases the probability that a molecule will have the correct orientation to fit into a "vacancy" in a growing crystal. Both of these effects favor the solid in the liquid-solid equilibrium, and higher temperatures should be required to convert it to the liquid. Our arguments are fully supported by the observed melting points for pentane and neopentane:

$CH_3CH_2CH_2CH_2CH_3$

Pentane

mp −130°C

$$CH_3-\underset{CH_3}{\overset{CH_3}{C}}-CH_3$$

Neopentane
(2,2-dimethylpropane)
mp −20°C

Our somewhat oversimplified explanation of factors which influence the melting points of nonpolar organic compounds is illustrated in Figure 4.10. Figure 4.10 a and b illustrate how a molecule in the wrong conformation or wrong orientation

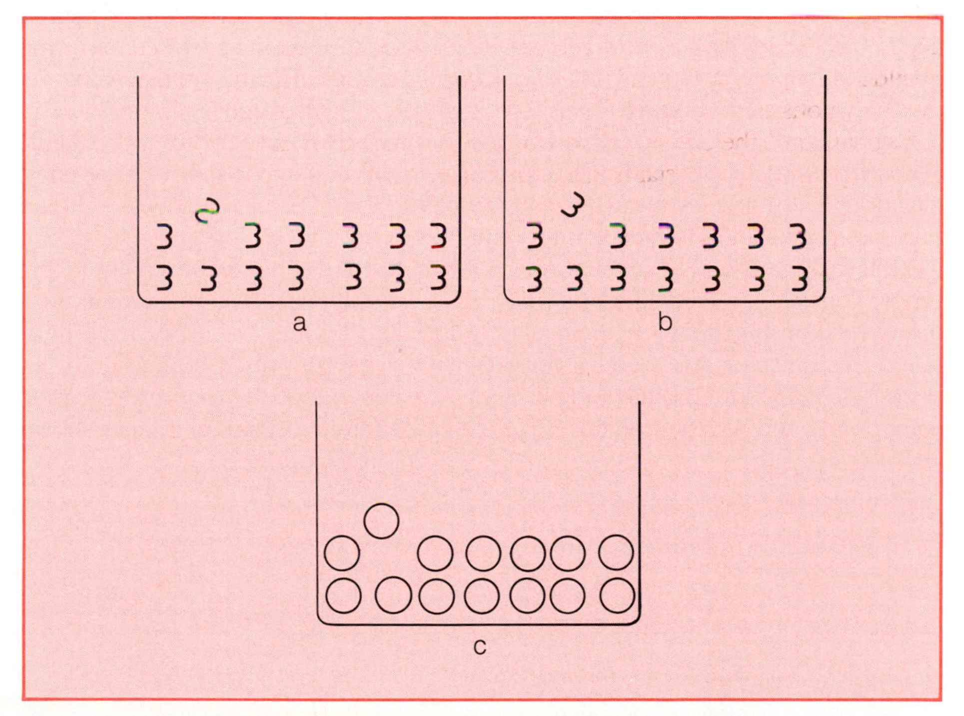

Figure 4.10 Idealized View of the Crystallization Process.
(a) The molecule adjacent to the vacancy in the crystal is in the wrong conformation for crystallization. (b) The molecule adjacent to the vacancy has the wrong orientation for crystallization. (c) In the case of a symmetrical or spherical molecule, all orientations will allow it to fit into the vacancy in the crystal.

might not fit into a growing crystal. Figure 4.10c shows how a symmetrical, highly branched molecule that is relatively spherical could easily fit into a growing crystal. A molecule with extensive branching or high symmetry has fewer available conformations, and nearly any orientation would be appropriate for fitting into the vacancy.

We can summarize the factors that favor high and low melting points in the following way. *Low melting points* are characteristic of nonpolar compounds with unbranched chains that can exist in a variety of conformations. *High melting points,* in contrast, are typical of compounds that are symmetrical and have highly branched carbon skeletons. Strong intermolecular forces such as ionic interactions can result in extremely high melting points, and weaker intermolecular forces such as hydrogen bonding (which we will discuss later in this section) can also be reflected by higher melting points.

Boiling Point

The boiling point of a substance is defined as the temperature at which the vapor pressure of the liquid is equal to the atmospheric pressure. When we discuss boiling points, we are therefore comparing the liquid and vapor phases. To some extent this is easier than comparing solid and liquid because the molecules in a gas are quite far apart. To a first approximation there are no interactions between molecules in the gas phase, so we can focus on the interactions that are present in the liquid. The ordering and orientation of molecules in a liquid are far more random than in a solid, but interactions between molecules remain important in the liquid phase. Attractive forces between molecules must be overcome in the process of vaporization, and *intermolecular forces* govern the relationships between structure and boiling point. Intermolecular forces of major importance in this context are *van der Waals forces, dipolar attraction,* and *hydrogen bonding.* The energies associated with these forces are small in comparison with the energies of chemical bonds, but when taken together for all the molecules in a liquid they can significantly influence the boiling point.

Molecular weight plays an important role in the equilibrium between liquid and vapor, volatility decreasing as the molecular weight increases. This results in a direct relationship between molecular weight and boiling point. We have illustrated this behavior in Table 4.4 for the inert gases, for which boiling point increases steadily with atomic weight. A similar dependence is found for organic compounds, and you can see this by referring back to the series of alkanes shown

TABLE 4.4 Boiling Points of the Inert Gases

	He	Ne	Ar	Kr	Xe	Rn
Atomic weight	4	20	40	84	131	222
Boiling point, °C	−269	−246	−186	−152	−137	−71

in Table 1.1. The effect is smaller for the alkanes of higher molecular weight, but a fairly regular increase of 20–30° in the boiling point is observed when the chain length is increased by a single CH_2 group. To avoid the influence of variation in molecular weight, our discussion of the dependence of boiling point on molecular structure will be restricted to compounds that are isomeric or else have very similar molecular weights.

Van der Waals forces are weak intermolecular forces that are always present, but they do not lend themselves to a simple definition because they are the net result of the complex attractive and repulsive interactions between all the electrons and nuclei of the two molecules. At very small intermolecular distances, that is, when the two molecules are "bumping" into each other, the overall interaction is repulsive, but at larger distances there will be a weak attraction between the molecules. We have illustrated this in Figure 4.11, which shows the potential energy of two molecules as a function of the distance between them. At very large intermolecular distances there is *no interaction* between the molecules (dotted line), but as the two molecules approach each other more closely, a favorable interaction results. The optimum distance between the molecules depends on the substituents that are interacting, but it typically corresponds to an internuclear distance of 2–4 Å for the atoms that are in closest proximity. At shorter distances the potential energy increases again, and very strong repulsions develop when the molecules are forced together. The attractive forces that develop at the optimum intermolecular distance are quite small, typically on the order of 0.1 kcal/mol for the interaction between any two atoms. However, the net stabilization will be somewhat larger when several pairs of atoms have such favorable attractions.

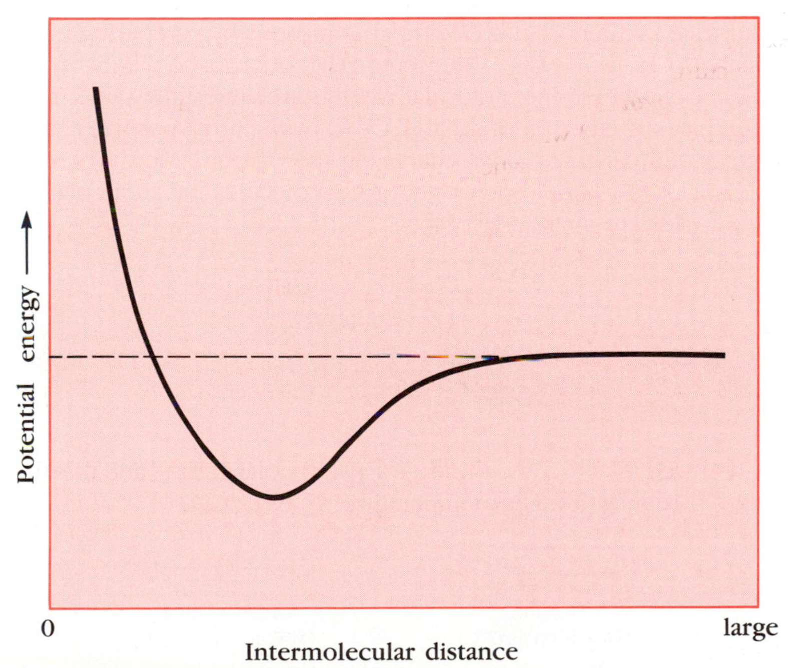

Figure 4.11 Van der Waals Forces between Molecules.
At large intermolecular distances there is no interaction, but the potential energy decreases slightly as the molecules approach each other. At very small intermolecular distances strong repulsive effects develop, and the energy increases substantially.

The distances between molecules in a liquid fall in the region in which van der Waals forces are attractive, and these forces must be overcome when a liquid vaporizes. Once again, larger intermolecular forces will require greater kinetic energy and higher temperatures for boiling to occur. The dependence of these forces on molecular structure is largely based on the *shape* of the molecules. Highly spherical molecules have minimal surface area, and this lessens the van der Waals interactions. Highly branched or symmetrical molecules have somewhat spherical shapes, and they tend to have lower boiling points than their unbranched or unsymmetrical isomers. Van der Waals interactions are maximized for molecules with unbranched skeletons, and these compounds exhibit higher boiling points. The trends in boiling point that can be attributed to van der Waals forces are evident for the following octane isomers:

$CH_3—CH_2—CH_2—CH_2—CH_2—CH_2—CH_2—CH_3$

Octane
bp 126°C

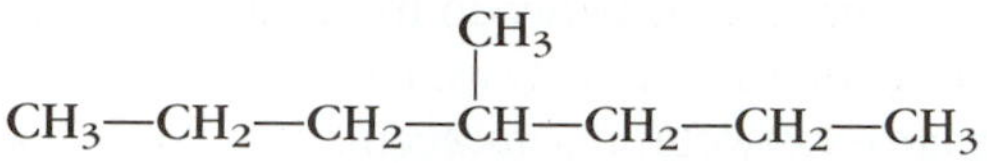

$CH_3—CH_2—CH_2—CH(CH_3)—CH_2—CH_2—CH_3$

4-Methylheptane
bp 118°C

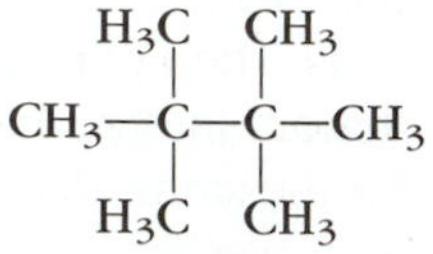

$(CH_3)_3C—C(CH_3)_3$

2,2,3,3-tetra-
methylbutane
bp 106°C

Dipolar attractions between molecules can occur when the molecules have electronegative substituents. Charge polarization in such a bond produces a *dipole*. We can illustrate the importance of dipolar attractions by comparing propane and dimethyl ether ($CH_3—O—CH_3$). The central CH_2 of propane differs by only 2 amu from the ether oxygen, so the molecular weights are almost the same. The carbon-oxygen bonds of the ether act as dipoles, C↔O, with partial positive charge on the carbon atoms and partial negative charge on the oxygen atoms. There are many nearby molecules in a liquid, and neighboring molecules can adopt orientations in which the dipoles are aligned favorably.

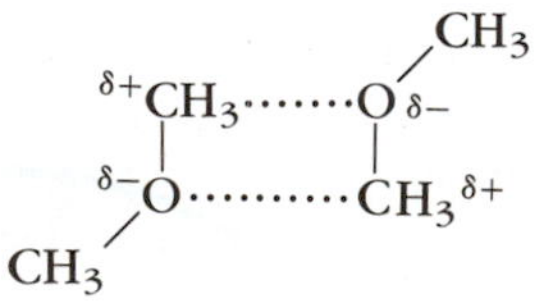

$$\begin{array}{ccc} & & CH_3 \\ {}^{\delta+}CH_3 & \cdots\cdots & O^{\delta-} \\ | & & | \\ {}^{\delta-}O & \cdots\cdots & CH_3^{\delta+} \\ CH_3 & & \end{array}$$

Energy is required to overcome these attractive dipolar forces, and this means that boiling should require a higher temperature.

$CH_3—CH_2—CH_3$

Propane
(MW 44; bp −42°C)

$CH_3—O—CH_3$

Dimethyl ether
(MW 46; bp −23°C)

As you would expect, the more polar ether exhibits a boiling point that is higher by approximately 20°C.

Hydrogen bonding is an additional type of intermolecular interaction, and it plays an important role in many chemical and biochemical processes. It is a relatively weak interaction, although it involves energies somewhat larger than those associated with van der Waals and dipolar interactions. *Hydrogen bonds are typically formed between the hydrogen atom of an NH or OH group and another oxygen or nitrogen atom.* The positive end of an N↔H or O↔H dipole interacts favorably with the nonbonding electrons of some other nitrogen or oxygen. Hydrogen bonding can also occur for other combinations and is very important for hydrogen fluoride, but we are primarily concerned with organic compounds. The hydrogen bonds involving other elements such as sulfur and chlorine are usually much weaker, and we will restrict our discussion to those involving nitrogen and oxygen.

The hydrogen bond can be considered a very weak or *partial bond,* and the energies of hydrogen bonds of nitrogen and oxygen derivatives usually fall in the range of 3–6 kcal/mol. A hydrogen bond is an interaction over and above the single bond between that hydrogen and the oxygen or nitrogen to which it is attached. One of the most important cases of hydrogen bonding involves water, and we have shown the formation of a hydrogen bond, represented by a dashed line, between two water molecules in Figure 4.12.

The effect of hydrogen bonding on boiling point can be seen by a comparison of dimethyl ether and the isomeric compound ethanol:

$CH_3—O—CH_3$
Dimethyl ether
(bp −23°C)

$CH_3—CH_2—OH$
Ethanol
(bp 78°C)

Clearly the boiling point of ethanol is substantially higher than that of dimethyl ether, and we can attribute this increase to hydrogen bonding in the liquid phase of ethanol.

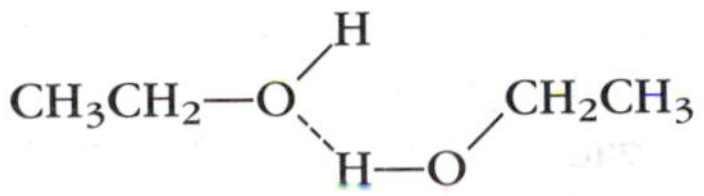

$CH_3CH_2—O(H)\cdots H—O—CH_2CH_3$

Dimethyl ether has no hydrogen atoms attached to oxygen, so in contrast to ethanol hydrogen bonding is not possible for this compound. As suggested by the large difference in the boiling points of these two isomers, intermolecular hydrogen bonding in the liquid probably has the largest effect on boiling point of the various factors considered in this section.

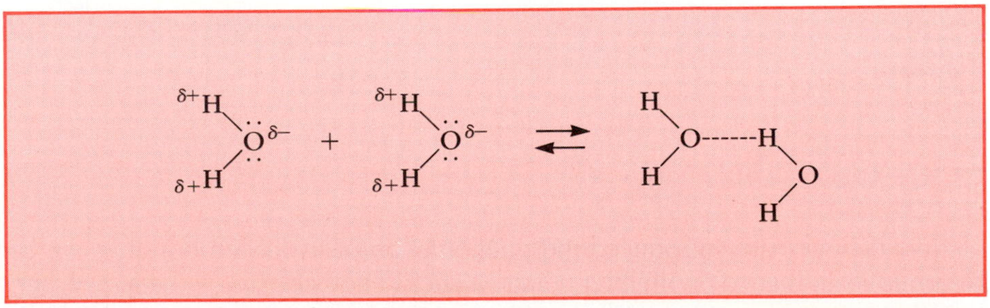

H_2O ($\delta+$H, $\delta+$H, $\ddot{\text{O}}^{\delta-}$) + H_2O ($\delta+$H, $\delta+$H, $\ddot{\text{O}}^{\delta-}$) ⇌ $H_2O\text{------}H—O—H$

Figure 4.12 Hydrogen Bonding.
Two isolated water molecules are shown approaching each other. A weak association results from the attraction of a hydrogen of one molecule to the oxygen of the other.

The presence of two groups in a molecule that are each capable of hydrogen bonding interactions sometimes leads to unusual behavior as a result of *intramolecular hydrogen bonding.* Intramolecular hydrogen bonding is very important if it can occur via a *five- or six-membered ring.* This is illustrated by the boiling points of the following series of isomeric compounds that contain both carbonyl and hydroxyl groups:

$CH_3CH_2CH_2{-}C({=}O){-}CH_2OH$ — bp 152°C

$CH_3CH_2CH(OH){-}C({=}O){-}CH_3$ — bp 147°C

$CH_3CH(OH)CH_2{-}C({=}O){-}CH_3$ — bp 177°C

$CH_2(OH)CH_2CH_2{-}C({=}O){-}CH_3$ — bp > 208°C

Why does the last of these compounds have a higher boiling point than the other three? We can explain this behavior on the basis of *intramolecular* hydrogen bonding in the first three compounds.

$CH_3{-}CH_2{-}CH_2{-}C({=}O{\cdots}H{-}O){-}CH_2$ (five-membered ring)

$CH_3{-}CH_2{-}CH(O{-}H{\cdots}O{=}){-}C{-}CH_3$ (five-membered ring)

$CH_3{-}CH(O{-}H{\cdots}O{=}){-}CH_2{-}C{-}CH_3$ (six-membered ring)

Each of the preceding compounds can form an intramolecular hydrogen bond via either a five- or six-membered ring, and you will recall from Chapter 3 that these are precisely the ring sizes which are stable and easily formed. The preferential formation of five- and six-membered rings is a pattern that you will encounter repeatedly in your study of organic chemistry. Since these compounds can form *intra-* rather than *inter*molecular hydrogen bonds, there are no strong intermolecular forces which must be overcome upon vaporization, and their boiling points are not elevated. In contrast, an intramolecular hydrogen bond for the fourth compound could occur only via a less likely seven-membered ring. The hydrogen bonding interactions for this compound are therefore predominantly intermolecular rather than intramolecular, and the intermolecular forces result in an elevated boiling point.

$O{-}H{\cdots}O{=}C(CH_3){-}CH_2{-}CH_2{-}CH_2{-}$ (seven-membered ring)

$HO{-}CH_2{-}CH_2{-}CH_2{-}C({=}O{\cdots}HO{-}CH_2{-}CH_2{-}CH_2{-}C({=}O){-}CH_3){-}CH_3$ (intermolecular)

Now that we have introduced the phenomenon of hydrogen bonding, we can return to its influence on melting points. It should not surprise you that hydrogen bonding can produce large effects on melting points, because transition from the solid to the liquid state requires that some of the favorable intermolecular interactions be overcome. Hydrogen bonding interactions should be optimal in the crystal, so melting would require more energy for a compound with intermolecular hydrogen bonding. Therefore you would predict that such intermolecular hydrogen bonding would result in higher melting points as well as higher boiling points. This prediction is borne out by the following examples:

$CH_3—CH_2—OH$	$CH_3—O—CH_3$	$HO—\overset{\displaystyle O}{\overset{\|}{C}}—CH_2—CH_2—\overset{\displaystyle O}{\overset{\|}{C}}—OH$	$CH_3—O—\overset{\displaystyle O}{\overset{\|}{C}}—\overset{\displaystyle O}{\overset{\|}{C}}—O—CH_3$
Ethanol (bp 78°C; mp −117°C)	Dimethyl ether (bp −23°C; mp −138°C)	Succinic acid (bp > 235°C; mp 188°C)	Dimethyl oxalate (bp 165°C; mp 54°C)

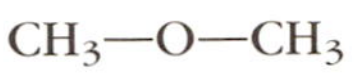

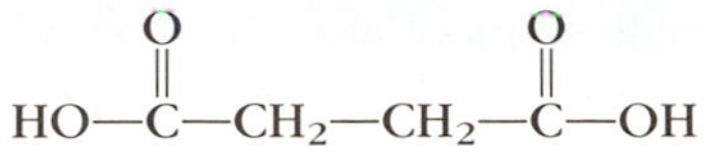

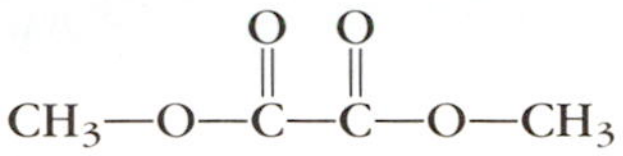

For each of the preceding pairs of isomers, the structure on the left has OH groups that would lead to intermolecular hydrogen bonding. This is reflected by higher melting and boiling points than for the isomeric structures on the right.

Solubility

Nearly all chemical reactions are carried out in solution. Unless the reactants are dissolved in a single phase, they will not mix evenly, and reactants that are in different phases may have so little contact with each other that the reaction will be extremely slow or even nonexistent. When all reactants are present in a single phase, this guarantees uniform mixing, permits concentrations of the reactive species to be controlled, and facilitates control of reaction conditions such as temperature.

The extent to which a compound (solute) is soluble in any particular solvent will be a matter of some concern to a chemist, and correlations between molecular structure and solubility properties are both useful and important. How is solubility related to structure? We can state a simple rule that is remarkably general: *A compound will dissolve most easily in a solvent when the solvent and solute are structurally similar.* By "structurally similar" we mean that polar groups or groups that can undergo hydrogen bonding are either present in both compounds or absent in both. This pattern is conveniently illustrated by using an example that is probably familiar to you. Neither oil nor gasoline will dissolve in water, and if either of these nonpolar organic liquids (each is a mixture of alkanes) is poured into water, it will not dissolve in the water but will float on the surface.

Why is solubility greater when the solvent and solute are structurally similar? When a substance dissolves, a variety of intermolecular forces are affected. The substance which is dissolving loses virtually all the favorable intermolecular attractions between molecules that it originally had in the solid (or liquid). Similarly, the solute molecules will occupy positions between solvent molecules, and this requires disruption of some of the favorable interactions that originally existed between solvent molecules. Only if these original intermolecular attractions are replaced by new, favorable *solvent-solute interactions* will there be substantial solubility of one compound in the other.

Returning to the lack of solubility of hydrocarbons in water, water is highly polar and it has little affinity for the distinctly nonpolar hydrocarbon molecules. To the extent that any of the hydrocarbon does dissolve, it will be at the expense of favorable intermolecular interactions for both water and the hydrocarbon, yet no strong solvent-solute attractions will replace them. This will not be an energetically favorable process, and the solubility is indeed very low. On the other hand, water is a rather good solvent for highly polar substances, and ionic compounds such as sodium chloride are typically quite soluble in water. The loss of favorable ionic interactions in the solid sodium chloride and of the intermolecular hydrogen bonding between some water molecules is more than offset by the favorable interactions of sodium and chloride ions with water molecules.

We normally think of water as an excellent solvent for inorganic salts but not for organic compounds. However, organic compounds do exhibit substantial water solubility if they have the appropriate structural features. The general requirement for an organic compound to be soluble in water is a substantial number of polar oxygen or nitrogen substituents relative to the total number of carbon atoms. A major factor in the solubility of such compounds is the capability of forming hydrogen bonds to water molecules. Sucrose (table sugar), for example, is highly soluble in water, even though each molecule contains 12 carbon atoms. The presence of 12 oxygen atoms, including 8 hydroxyl groups, allows it to function as both a donor and an acceptor in hydrogen bonding, so the solvent-solute interactions with water are very strong.

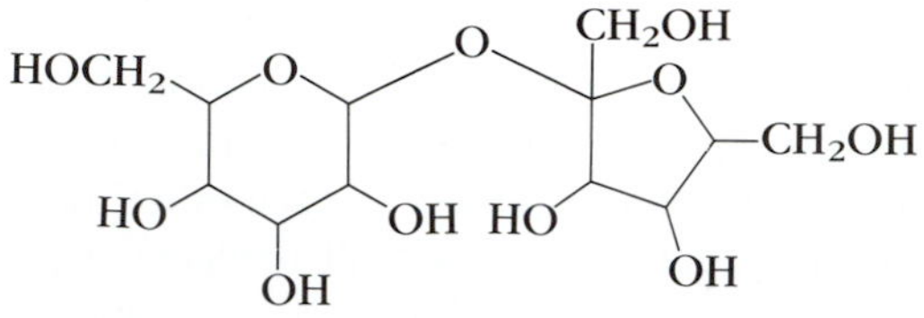

$HOCH_2$ O O CH_2OH O CH_2OH
HO OH HO
OH OH

Sucrose

Sucrose also illustrates how highly polar substituents will greatly reduce the solubility of an organic compound in many organic solvents, since it is almost completely insoluble in a nonpolar organic solvent such as hexane.

The typical water solubility of ionic compounds is also observed for organic salts. For example, stearic acid is water-insoluble, but the sodium salt of this carboxylic acid will dissolve in water.

$CH_3—(CH_2)_{16}—CO_2H$

Stearic acid
(water insoluble)

$CH_3—(CH_2)_{16}—CO_2^-Na^+$

Sodium stearate
(water soluble)

Compounds such as sodium stearate provide the basis for soaps and detergents, since one end of the molecule is highly polar and the other end is very nonpolar. The high polarity of the ionic end of the molecule imparts water solubility, and the interaction of the nonpolar hydrocarbon chain with organic compounds permits the breakup and washing away of dirt and grease.

In Table 4.5 we have illustrated some fairly typical trends in solubility for several common solvents: water, ethanol, diethyl ether, and hexane. These solvents decrease in polarity from left to right across the table, as you would expect on the basis of their structure. Water is the most polar of these solvents, and replacement of just one of the hydrogen atoms by an alkyl group causes a substantial decrease in polarity. Nevertheless, ethanol is still moderately polar, and it can act both as a donor and as an acceptor in hydrogen bonding. These properties make ethanol a good solvent for many organic reactions because it will dissolve a large variety of both organic and inorganic reagents.

Replacement of the second hydrogen atom of water with another ethyl group affords diethyl ether, and this solvent is only weakly polar. The electronegative oxygen atom can act only as an acceptor in hydrogen bonding, because there is no

TABLE 4.5 Solubility Trends

Solute	Solvent: Water H_2O	Solvent: Ethanol CH_3CH_2OH	Solvent: Diethyl Ether $CH_3CH_2OCH_2CH_3$	Solvent: Hexane $CH_3(CH_2)_4CH_3$
$CH_3-C(=O)-CH_3$ Acetone	Soluble	Soluble	Soluble	Soluble
$CH_3-C(=O)-(CH_2)_5CH_3$ 2-Octanone	Insoluble	Soluble	Soluble	Soluble
CH_3Cl Chloromethane	Slightly soluble	Soluble	Soluble	Soluble
$CH_3CH_2CH_2CH_2Cl$ 1-Chlorobutane	Insoluble	Soluble	Soluble	Soluble
$CH_3(CH_2)_6Cl$ 1-Chloroheptane	Insoluble	Soluble	Soluble	Soluble
$CH_3CH_2-CH(OH)-CH_2CH_3$ 3-Pentanol	Slightly soluble	Soluble	Soluble	Soluble
$HO-CH_2-CH(OH)-CH_2-OH$ Glycerol (1,2,3-trihydroxypropane)	Soluble	Soluble	Slightly soluble	Slightly soluble

OH group to act as a hydrogen donor. Finally, hexane is a very nonpolar solvent since it contains only carbon and hydrogen.

The first two entries in Table 4.5, acetone and 2-octanone, illustrate the variation in solubility as a function of the relative numbers of carbon and oxygen atoms in a molecule. Acetone, which has a carbonyl group and two methyl groups, is *miscible* (i.e., completely soluble in all proportions) with each of the four solvents. Replacement of one of the methyl groups of acetone by a hexyl group does not decrease the solubility in organic solvents, but the water solubility decreases substantially, and 2-octanone is insoluble in water.

The three alkyl chlorides in Table 4.5 show similar trends. Chloromethane is soluble in both water and ethanol. Although hydrogen bonding to this solute is much less important than to solutes containing oxygen or nitrogen, the carbon-chlorine bond is polar, and this accounts for the slight water solubility. As the chain length is increased from methyl to butyl to heptyl, the water solubility decreases accordingly. All three of these compounds are soluble in the organic solvents.

The last two entries in Table 4.5, 3-pentanol and glycerol, clearly demonstrate how the solubility in *organic solvents* is influenced by the relative number of polar groups per carbon atom. 3-Pentanol and glycerol have comparable molecular weights, and glycerol differs by the replacement of two methyl groups by hydroxyl groups. Hydrogen bonding is much more important for glycerol; as you would expect, its three polar hydroxyl groups lead to better water solubility than that for 3-pentanol. On the other hand, the solubility of glycerol in the less polar organic solvents is substantially lower than that of 3-pentanol, which has only a single hydroxyl group and five carbon atoms.

While we cannot easily make numerical predictions of physical properties on the basis of molecular structure, the discussion in this section should make it clear that you can anticipate on a qualitative basis how physical properties will vary as a function of structural differences between compounds. A fairly simple evaluation of molecular structure in terms of size, shape, symmetry, and relative number of polar and nonpolar substituents (especially those which can form hydrogen bonds) will allow you to make surprisingly accurate predictions of physical properties.

EXERCISE 4.12

Two bottles were known to contain *r*-1,*cis*-3,*cis*-5-trimethylcyclohexane and 1-propylcyclohexane, but the labels had fallen off. The material in bottle A solidified at −95°C and the material in bottle B solidified at −50°C. Which substance was in each bottle?

EXERCISE 4.13

Hexane (MW 86) boils at 69°C, and the analog in which two CH_2 groups have been replaced by oxygen atoms (i.e., $CH_3—O—CH_2—CH_2—O—CH_3$) has a boiling point which differs from that of hexane by 14°C. What is the boiling point of this latter compound?

EXERCISE 4.14

Trimethylamine actually boils at a slightly lower temperature than dimethylamine, despite the fact that it has a higher molecular weight. Suggest a reason for this.

$CH_3—NH(CH_3)$ $CH_3—N(CH_3)_2$

Dimethylamine
MW 45
bp 7°C

Trimethylamine
MW 59
bp 3°C

4.10 TERMS AND DEFINITIONS

Alcohol. A member of the class of organic compounds with the general structure R—OH, in which an alkyl group is covalently bonded to the oxygen of a hydroxyl group.

Ammonium ion. An ion characterized by a nitrogen atom that has four bonds (and therefore a formal positive charge).

Brønsted-Lowry classification. Classification of acids as H^+ donors and bases as H^+ acceptors.

Carbonyl group. The C=O group, which characterizes carboxylic acids (RCO_2H, where the substituents are R and OH) and ketones ($R_2C{=}O$, where both substituents are alkyl groups).

Carboxyl group. The CO_2H group, which characterizes carboxylic acids; this consists of a carbonyl group for which one substituent is a hydroxyl group.

Carboxylic acid. An organic compound having a carboxyl group; the general structure is $R{-}CO_2H$. The K_a is usually about 10^{-5}.

Conjugate acid. The species produced by reaction of a base with H^+.

Conjugate base. The species produced by loss of an H^+ from an acid.

Core electrons. The electrons on an atom that are not part of the valence level, i.e., the 1*s* electrons for carbon and other second-row elements.

Dipolar attraction. The attractive interaction between the oppositely charged ends of two polarized bonds; one of the main types of intermolecular forces.

Electron-withdrawing. The inductive or resonance effect of an electronegative substituent. Electron density is shifted to the substituent from the atom to which it is bonded.

Formal charge. The charge that is assigned to an atomic center according to the formalism that electrons in covalent bonds are shared equally by the bonded atoms.

Hydrogen bonding. The favorable interaction between the hydrogen of an N—H or O—H group and the nonbonding electrons of another electronegative atom (usually nitrogen or oxygen).

Inductive effect. The influence on structure or reactivity produced by a substituent as a result of polarization of individual bonds.

Ketone. An organic compound having a carbonyl group with two alkyl groups as substituents; the general formula is $R_2C{=}O$.

Lewis acid. A species that can act as an electron pair acceptor; it must have an atomic center with a low-energy unoccupied orbital.

Lewis base. A species that can act as an electron pair donor; it can also act as an H^+ acceptor.

Lone pair. An unshared pair of valence-level electrons on an atomic center; a pair of nonbonding electrons.

Nonbonding electrons. Valence-level electrons that are not involved in bonding; lone pair electrons; an unshared pair of electrons.

Nonpolar. A term describing a covalent bond in which the electrons are shared equally; a bond between two atoms with the same electronegativity. The term *nonpolar* is also used to describe a compound containing no polar substituents.

Nucleophile. A species with nonbonding electrons that can attack a carbon atom bearing full or partial positive charge.

Oxonium ion. An ion characterized by an oxygen that has three bonds (and therefore a formal positive charge).

Polarized bond. A bond between two atoms with different electronegativities. This forms a dipole, placing partial positive charge on one atom and partial negative charge on the other.

Proton affinity. The energy released when a base combines with an H^+ ion in the gas phase.

Resonance contributor. One of the individual structures that is drawn when a single conventional drawing does not adequately represent a molecule or ion; also called a *resonance form.*

Resonance form. One of the individual structures that is drawn when a single conventional drawing does not adequately represent a molecule or ion; also called a *resonance contributor.* The individual resonance forms differ only in the positions of multiple bonds and formal charges; there can be no difference in constitution.

Resonance hybrid. The overall, "average" structure corresponding to a group of individual resonance forms.

Resonance stabilization. The stabilization of a molecule or ion for which several resonance forms can be drawn; the stabilization is associated with electron delocalization.

Substituent effect. The change in structure or reactivity that results when one substituent is replaced by another.

Unshared pair. A pair of valence-level electrons that is not involved in bonding; nonbonding electrons; a lone pair of electrons.

Van der Waals forces. Weak interactions between molecules (or between atoms in the same molecule); the net result of attractions and repulsions for all electrons and nuclei.

4.11 PROBLEMS

4.1 Complete each of the following reactions. Write the formal charge for all atoms having a formal charge other than zero. If more than one resonance form is possible, show formal charges for each.

(a) $CH_3NH_2 + H^+ \longrightarrow$

(b) $H_2SO_3 \longrightarrow H^+ +$ (both hydrogen atoms of H_2SO_3 are bound to oxygen)

(c) $BF_3 + (CH_3)_3N \longrightarrow$

(d) $CH_3CH_2CH_2OH + H^+ \longrightarrow$

(e) $(CH_3)_2NH \longrightarrow H^+ +$

(f) $CH_3CH_2CO_2H \longrightarrow H^+ +$

(g) $H-\overset{O}{\overset{\|}{C}}-NH_2 + CH_3O^-Na^+ \longrightarrow$

4.2 Indicate the most acidic hydrogen atom(s) in each compound:

(a) $CH_3-CH_2-\overset{OH}{\overset{|}{C}H}-CH_3$

(b) $NH_2-CH_2-CH_2-NH-\overset{O}{\overset{\|}{C}}-CH_3$

(c) $H-\overset{O}{\overset{\|}{C}}-CH_2-CH_3$

(d) $CH_3-O-\overset{O}{\overset{\|}{C}}-CH_2-\overset{O}{\overset{\|}{C}}-CH_2-CH_3$

(e) $CH_3-\overset{O}{\overset{\|}{C}}-\underset{NH_2}{\underset{|}{C}H}-CH_3$

(f) $CH_3-\overset{OH}{\overset{|}{C}H}-CH_2-NH_2$

(g) $CH_3-\overset{HO}{\overset{|}{C}H}-\overset{O}{\overset{\|}{C}}-OH$

(h) $CH_3-\overset{HO}{\overset{|}{C}H}-\overset{O}{\overset{\|}{C}}-O-CH_3$

4.3 For each of the following pairs of compounds, indicate which should be more acidic. Why?

(a) $CH_3-CH_2-CH_2-NH-CH_3$ $\quad$ $CH_3-CH_2-CH_2-CH_2-OH$

(b) $CH_3-CH_2-CH_2-CH_3$ $\quad$ $CH_3-CH_2-\overset{O}{\overset{\|}{C}}-CH_3$

(c) $CH_3-\overset{O}{\overset{\|}{C}}-O-CH_2-O-\overset{O}{\overset{\|}{C}}-CH_3$ $\quad$ $CH_3-O-\overset{O}{\overset{\|}{C}}-CH_2-\overset{O}{\overset{\|}{C}}-O-CH_3$

(d) $CH_3-CH_2-C(=O)-NH-CH_3$ $\quad$ $CH_3-C(=O)-N(CH_3)_2$

(e) $CH_3-CH_2-C(=O)-NH-CH_3$ $\quad$ $CH_3-CH_2-CH_2-NH-CH_3$

(f) $CH_3-C(=O)-CH_2-CH_3$ $\quad$ $CH_3-CH(OH)-CH_2-CH_3$

(g) $CH_3-CH(CH_3)-CO_2H$ $\quad$ $CH_3-C(Cl)(CH_3)-CO_2H$

(h) $CH_3-CH_2-CH_2-NH-CH_3$ $\quad$ $CH_3-CH_2-CH_2-O-CH_3$

4.4 For the compounds in the preceding question, locate all polar bonds and indicate the direction of the bond polarity.

4.5 For each of the following pairs of compounds, indicate which should be more basic. Why?

(a) $(CH_3CH_2CH_2)_3N$ $\quad$ $(CF_3CF_2CF_2CF_2)_3N$

(b) $CH_3CH_2O^-Na^+$ $\quad$ $CH_3-C(=O)-O^-Na^+$

(c) $CH_3-CH_2-NH_2$ $\quad$ $CH_3-C(=O)-CH_3$

(d) $CH_3-NH-CH_3$ $\quad$ CH_3-O-CH_3

(e) $CF_3CH_2O^-K^+$ $\quad$ $CF_3CH_2CH_2O^-K^+$

4.6 All the atoms of boron trifluoride lie in a plane, but the same is not true for ammonia. On the basis of electron pair repulsions, explain this difference in the structures of NH_3 and BF_3.

4.7 On the basis of electron pair repulsions, draw the three-dimensional structures expected for the products of the following reactions:

(a) $H_2O + H^+ \longrightarrow [H_3O]^+$

(b) $BF_3 + F^- \longrightarrow [BF_4]^-$

(c) $CH_3NH_2 + H^+ \longrightarrow [CH_3NH_3]^+$

4.8 Using arguments based on electron pair repulsion to predict geometry, draw the three-dimensional structures of

(a) The methyl cation, CH_3^+

(b) The methyl anion, CH_3^-

4.9 For each of the following pairs of compounds, indicate which should have the higher melting point. Why?

(a) $CH_3-CH_2-CH_2-CH_2-CH_2-CH(CH_3)-CH_3$ $\quad$ $CH_3-C(CH_3)_2-C(CH_3)_2-CH_3$

(b) $CH_3-C(CH_3)_2-C(=O)-CH_3$ $\quad$ $CH_3-CH(CH_3)-CH_2-C(=O)-CH_3$

(c) $\quad$ $CH_3-(CH_2)_4-CH_3$

(d) $CH_3-CH_2-NH_3^+Cl^-$ $\quad$ $CH_3-CH_2-CH_2-Cl$

(e) $CH_3-CH_2-CH_2-CH_2-C(=O)-OH$ $\quad$ $CH_3-C(CH_3)_2-CO_2H$

4.10 For each of the following pairs of compounds, indicate which should have the higher boiling point. Why?

(a) $CH_3-CH(CH_3)-NH_2$ $\quad$ $CH_3-N(CH_3)-CH_3$

(b) $CH_3CH_2CH_2CH_2OH$ $\quad$ $CH_3CH_2CH_2CH_2CH_2OH$

(c) $CH_3-CH(OH)-CH_2-CH_2OH$ $\quad$ $HOCH_2-CH_2-CH_2-CH_2OH$

(d) $H-C(=O)-N(CH_3)-CH_3$ $\quad$ $CH_3-CH_2-C(=O)-NH_2$

(e) $CH_3CH_2-O-CH_2CH_3$ $\quad$ $CH_3CH_2CH_2CH_2OH$

4.11 For each of the following pairs of compounds, indicate which should be more soluble in water. Why?

(a)

(b) $CH_3-C(=O)-OH$ $\quad$ $CH_3-CH_2-CH_2-CH_2-C(=O)-OH$

4.12 For each of the following pairs of compounds, indicate which should be more soluble in hexane. Why?

(a) CCl_4 $CH_3-\overset{O}{\overset{\|}{C}}-NH_2$

(b) 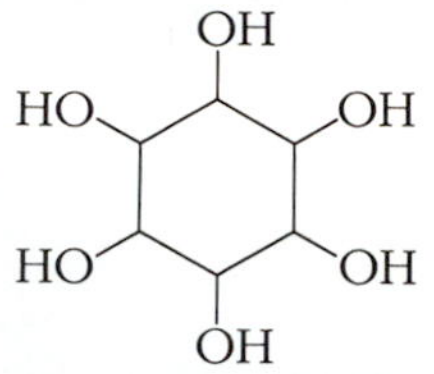  [1,2,3,4,5,6-hexahydroxycyclohexane: ring with OH, HO, OH, HO, OH, OH] [cyclohexane ring]

(c) $CH_3-CH_2-NH_3^+\ Cl^-$ $Cl-CH_2-CH_2-NH_2$

(d) $CH_3CH_2CH_2-\overset{O}{\overset{\|}{C}}-OH$ $CH_3CH_2CH_2-\overset{O}{\overset{\|}{C}}-O^-\ K^+$

4.13 For each pair of compounds in the preceding problem, predict which should be more soluble in water. Why?

4.14 A bromination experiment with propionic acid ($CH_3CH_2CO_2H$) produced a variety of products. Two of these were determined to be dibromopropionic acid isomers ($C_3H_4Br_2O_2$), and they exhibited acid dissociation constants of 3×10^{-2} and 7×10^{-3}. Using the information in Figure 4.6 and your knowledge of inductive effects, suggest possible structures for these two acids.

4.15 A cyano substituent ($C{\equiv}N$) results in a substantial increase in the acidity of hydrogen atoms on the carbon to which it is attached. For example:

$$CH_4 \quad K_a \simeq 10^{-40} \qquad \text{and} \qquad CH_3-C{\equiv}N \quad K_a = 10^{-25}$$

A second cyano group results in still greater acidity:

$$N{\equiv}C-CH_2-C{\equiv}N \quad K_a = 10^{-11}$$

Explain these increases in acidity, using appropriate structural formulas.

4.16 Two isomers of 1,4-dibromocyclohexane are known. One of them has a melting point of 113°C, and the other is a liquid at room temperature. On the basis of these physical properties, what are the complete structures of the two compounds? (*Suggestion*: You will solve the problem more easily if you draw the chair conformations of the cyclohexane rings.)

4.17 Salts are generally characterized by very high boiling points. Nevertheless, when some ammonium salts are heated, they appear to evaporate quite readily. This is observed for salts such as $RNH_3^+\ Cl^-$ but not for those in which all four substituents on nitrogen are alkyl groups (e.g., $R_4N^+\ Cl^-$). Suggest an explanation. (Remember that salts such as $RNH_3^+\ Cl^-$ can be considered as the products of an acid-base reactions, and that such reactions are equilibrium processes.)

4.18 In Section 4.7 we showed you that a carbonyl group greatly increases the acidity of hydrogens on an adjacent atom. This effect should increase the acidity of the hydrogens in *both* the CH_3 and NH_2 groups in the following compound:

$$CH_3-\overset{\overset{\Large O}{\|}}{C}-NH_2$$

(a) Using the data in Table 4.3 for the K_a values, use the relationship $\Delta G = -2.3\ RT \log K$ to estimate the free energy difference between the neutral reactant and the products of acid dissociation for each of the two ionization reactions.

(b) Use the free energy differences estimated in (a) to calculate the free energy difference between the two possible anions.

(c) Use the free energy equation written in (a) to calculate the ratio of the two anions that would be present at equilibrium.

CHAPTER 5

ALKENES

In Chapter 1 we emphasized that *alkanes* are characterized by their lack of chemical reactivity. But not all *hydrocarbons* are unreactive. For example, ethylene (C_2H_4) undergoes a wide variety of chemical reactions under very mild conditions. It is a major product of petroleum refining, and it serves as the raw material for the industrial synthesis of many important organic compounds. Ethylene also plays an important biological role in plant metabolism, inducing the ripening process in fruits and vegetables. This phenomenon is exploited commercially, and it allows farm produce to be picked before it is ripe. The produce can be shipped in special vehicles, in which exposure to ethylene causes it to ripen during transit. Ethylene is the simplest member of the alkenes, and in this chapter we will introduce you to the important features of structure and reactivity for members of this class of hydrocarbons.

5.1 THE CARBON-CARBON DOUBLE BOND

The **alkenes** form a class of hydrocarbons that differ from the corresponding alkanes by the absence of two hydrogen atoms. Open-chain alkanes are represented by the molecular formula C_nH_{2n+2}, but the analogous alkenes have the formula C_nH_{2n}. An alkene contains two adjacent carbon atoms having only *three substituents,* but each of these two carbons still has the four *bonds* that are always found for carbon in stable carbon compounds. We have just described the structural features of a *carbon-carbon double bond,* and this is the structural subunit that defines an alkene. You can see this by comparing the C_3 hydrocarbons propane and propene.

$$CH_3{-}CH_2{-}CH_3 \qquad\qquad CH_2{=}CH{-}CH_3$$

Propane (an alkane) — Propene (an alkene)

The carbon atoms of an alkane (and of other compounds), which have the maximum *four* substituents, are sometimes called **saturated** carbon atoms. In contrast, we can describe alkenes as **unsaturated** because the carbon atoms of the double bond have only three substituents. Notice that the general molecular formula, C_nH_{2n}, for an alkene is the same as that for a cycloalkane. Do not let this similarity in the molecular formulas confuse you into thinking that cycloalkanes are unsaturated. The "absence" of two hydrogens in a cycloalkane results simply from the closure of the ring, and all carbon atoms of a cycloalkane are saturated. You may occasionally encounter the term *olefin* in your studies of chemistry, and this is simply another word for alkene. In modern organic chemistry alkene is used more commonly.

In marked contrast to alkanes, the alkenes are a fairly reactive class of compounds, which undergo a large number of chemical reactions under very mild conditions. The carbon-carbon double bond provides the basis for this reactivity, and we will refer to the double bond as the *reactive center* in these reactions. In this chapter we will show that alkenes of a wide range of structure all undergo similar reactions. Because of this consistent reactivity, alkenes are sometimes called a **functional class** of organic compounds, and the carbon-carbon double bond is a **functional group.** During your study of organic chemistry you will find that nearly all organic reactions occur at a specific functional group, so the double bond is merely the first of a variety of functional groups that we will discuss.

Why are the reactions of alkenes governed by the carbon-carbon double bond? To answer this question, you must first understand the geometry and electronic structure of the double bond. When we first discussed the basis of molecular geometry in Chapter 1, we argued that electron-electron repulsions caused the substituents on a carbon atom to lie as far apart from each other as possible. For a saturated carbon atom this results in a geometry that is tetrahedral, so that the bond angles are approximately 109°. But alkenes have only *three* substituents on the carbon atoms of the double bond, and their maximum separation is found at a bond angle of about 120°, with the central carbon atom and all three substituents lying in a single plane. This geometry is known as **trigonal,** in contrast to the *tetrahedral* geometry of a saturated carbon atom. As we indicated in Chapter 1, a carbon atom will be exactly tetrahedral only when the four substituents are the same. Similarly, the bond angles at a trigonal center will be exactly 120° only if all three substituents are identical.

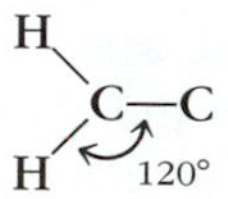

H
C—C
H 120°

The double bond further requires that both carbon atoms and all atoms bonded to them lie in the same plane. This is referred to as the *plane of the double bond.* A molecule such as ethene (normally called by its common name, ethylene) is therefore completely planar.

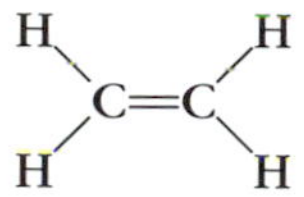

H H
C=C
H H

Ethylene

Larger alkenes will have three-dimensional substituents (e.g., the methyl group of propene), and only the local region of the double bond will be planar. We have used propene to illustrate the molecular geometry of alkenes in Figure 5.1, and this figure also shows the geometry for the corresponding saturated compound, propane.

The major differences between propane and propene are found at the unsaturated carbon atoms of the latter. The carbon-carbon bond length of the double bond (1.34 Å) of propene is *shorter* than that of the carbon-carbon single bond of propane (1.53 Å), and this shorter C—C distance is typical for double bonds. The C—C—C bond angle for the alkene is 124°, while that for the saturated compound is 112°. In each case the angle is slightly distorted (from the ideal geometry of 120° or 109.5°) in a way that increases the distance between the "large" carbon atoms. The remaining bond angles and bond lengths of propene are not significantly affected by the double bond, and their values fall in the same range previously shown for alkanes.

Why are alkenes more reactive than alkanes? In the following section you will see that there are two doubly occupied molecular orbitals associated with the double bond, and one of these is a fairly high-energy orbital. Consequently, the electrons of a carbon-carbon double bond can more easily interact with a Lewis

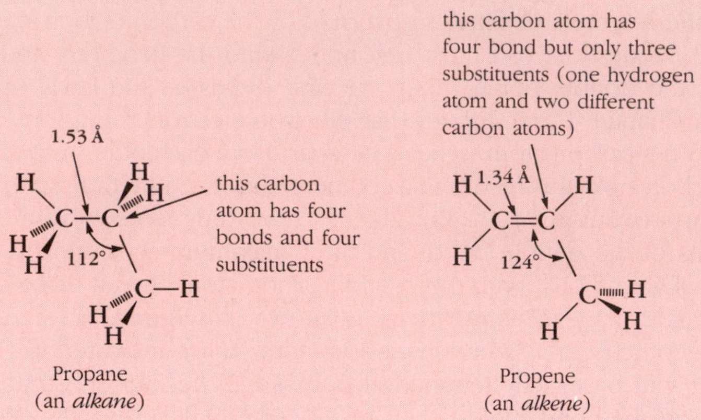

1.53 Å

H H H C C H H H 112° C—H H H

this carbon atom has four bonds and four substituents

Propane
(an *alkane*)

this carbon atom has four bond but only three substituents (one hydrogen atom and two different carbon atoms)

H 1.34 Å H C=C H 124° C H H H

Propene
(an *alkene*)

Figure 5.1 The Carbon-Carbon Double Bond: A Comparison of the Molecular Geometries of Propane and Propene.
Each doubly bonded carbon atom of the alkene has four bonds but only three substituents. These substituents are arranged in such a way that repulsive interactions are minimized, lying in a plane with a separation of approximately 120°. (In contrast, the C—C—C angle of propane is approximately tetrahedral.) The length of the carbon-carbon double bond in propane is 1.34 Å, and this is significantly shorter than that of the corresponding single bond of propane (1.53 Å). Other bond lengths and angles in these two compounds are very similar. The two carbon atoms of the double bond of propene and all four atoms attached to the doubly bonded carbons lie in a single plane.

acid, permitting the carbon-carbon double bond to act as a Lewis base. Most reactions of alkenes are initiated by a step in which the double bond interacts with a Lewis acid, and this is often described as **electrophilic** attack on the double bond.

5.2 MOLECULAR ORBITAL DESCRIPTION OF THE CARBON-CARBON DOUBLE BOND

You can gain a better understanding of both the reactivity and electronic structure of the carbon-carbon double bond through molecular orbital theory. Before considering the actual bonding interactions, we will first look at the atomic orbitals on one of the carbon atoms of a double bond. The interaction of the atomic orbitals on a single atom is governed by local symmetry, and here the carbon atom together with the three atoms attached to it all lie in a plane. Only two of the three *p* atomic orbitals on carbon can lie in such a plane; the remaining *p* orbital must be perpendicular to it. Interaction of the 2*s* orbital and two of the three 2*p* orbitals leads to three equivalent sp^2 orbitals on carbon, and we have drawn these in Figure 5.2.

Each of the sp^2 hybrid atomic orbitals is capable of interacting with a comparable orbital on another atom to form molecular orbitals. When the orbitals on each of the atoms are directed toward the other atom, a **sigma (σ) molecular orbital** is formed. This is the type of molecular orbital we discussed in Section 1.7 for sp^3

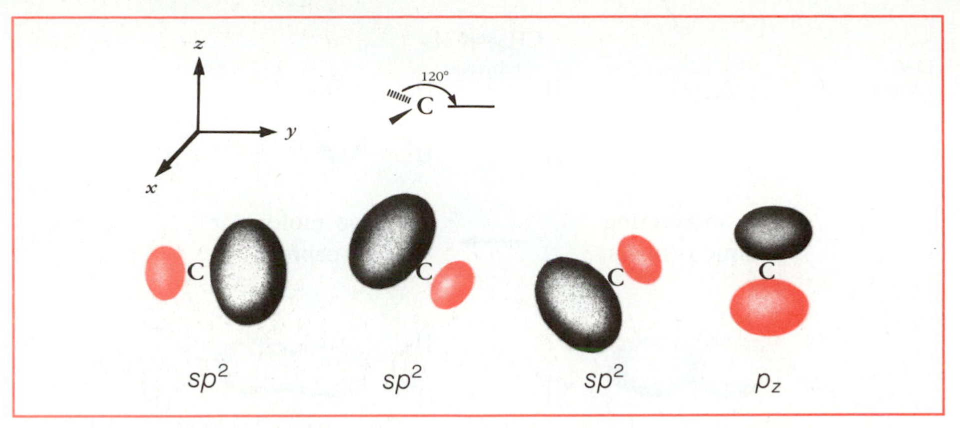

Figure 5.2 Hybrid Atomic Orbitals for the Carbon Atom of a Carbon-Carbon Double Bond.
The sp^2 hybrid orbitals are separated by 120° and lie in a single plane (in this drawing, the *x-y* plane, which is perpendicular to the page). These three hybrid orbitals result from combining the 2*s* orbital and two of the three 2*p* orbitals on carbon. The p_z orbital (which is perpendicular to the plane of the trigonal center) is "left over."

orbitals, and it leads to **sigma bonds,** which ordinarily are normal single bonds. The molecular orbitals corresponding to sigma bonds are characterized by rotational symmetry about the bonding axis, the imaginary line connecting the two nuclei. In Figure 5.3 we have illustrated the interaction between the sp^2 orbitals of the two carbon atoms of ethylene. Only the sp^2 orbital directed toward the other carbon atom is shown in each case; the remaining two sp^2 orbitals on each of these carbon atoms of ethylene are involved in C—H bonding.

The bonding interaction shown in Figure 5.3 is for σ bonding, but this is normally associated with **single bonds.** What is the orbital interaction that results in the **double bond** of an alkene? The double bond of an alkene results from a second type of bonding. If you refer back to Figure 5.2, you will note that we described the p_z orbital as "left over." This orbital is left over only in the sense that it does not mix with the other *p* orbitals in forming the hybrid atomic orbitals. When two trigonal carbon atoms are adjacent to each other, as in a carbon-carbon double bond, the p_z orbitals on these carbon atoms interact strongly to form another type of molecular orbital, called a π molecular orbital (where *pi* is the Greek letter corresponding to *p*). The orientation of these *p* orbitals is *perpendicular* to the C—C bonding axis, and the region in which they overlap is not along the bonding axis but above and below the plane of the double bond. We have illustrated these orbital interactions in Figure 5.4.

Figures 5.3 and 5.4 show that the two bonds of a carbon-carbon double bond differ considerably from each other. The σ bond is very similar to other carbon-carbon single bonds, but the π bond is quite different. The p_z atomic orbitals which combine to make up the bonding and antibonding π molecular orbitals are both parallel to the *z* axis, which is perpendicular to the bonding axis. As a result these orbitals do not *overlap* as effectively as do the sp^2 orbitals that form the σ molecular orbitals. Because the overlap of the atomic orbitals is less, the π bond is *weaker* than the σ bond. This means that the occupied π orbital is of higher energy than

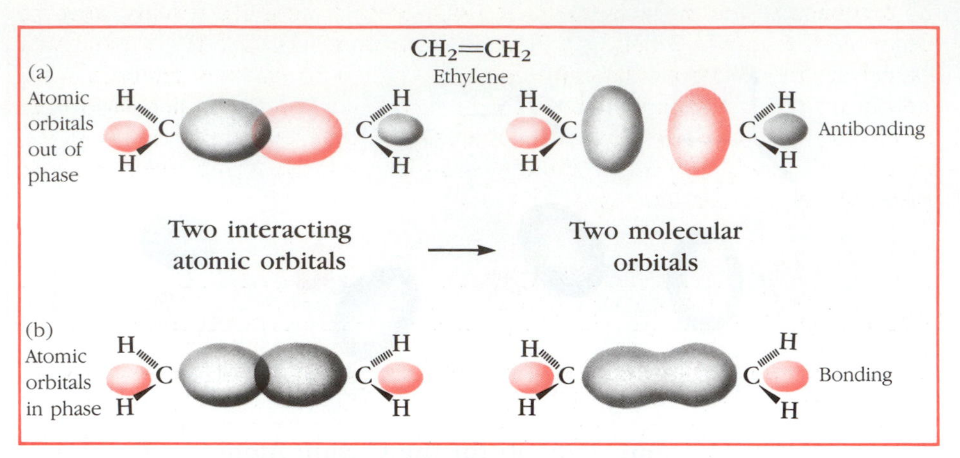

Figure 5.3 Sigma Bonding in Ethylene.
Interaction of the sp^2 orbitals that lie along the bonding axis between the two carbon atoms can be (a) *out of phase,* leading to an *antibonding* molecular orbital; or (b) *in phase,* leading to a *bonding* molecular orbital. In a ground-state ethylene molecule the bonding molecular orbital is doubly occupied, and the antibonding orbital is unoccupied.

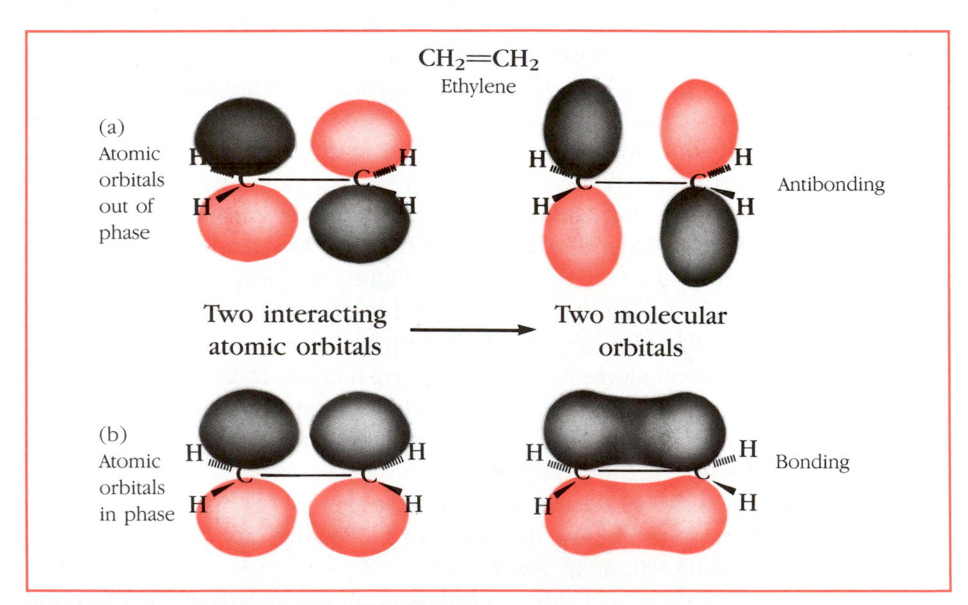

Figure 5.4 π-Bonding in Ethylene.
Interaction of two *p* orbitals oriented perpendicular to the bonding axis can be either: (a) *out of phase,* leading to an *antibonding* molecular orbital; or (b) *in phase,* leading to a *bonding* molecular orbital. The bonding molecular orbital is doubly occupied in a ground state ethylene molecule, and the antibonding orbital is unoccupied.

the σ orbital, so the π electrons of the double bond are more reactive than the electrons of σ bonds. Finally, we have the answer to our questions about why alkenes are more reactive than alkanes and why their reactivity is centered at the carbon-carbon double bond. Even though they are in a bonding molecular orbital, the electrons of the π bond are sufficiently reactive that the double bond is able to act as a weak Lewis base. You will find that this characteristic will surface over and over again as you continue to learn about alkenes.

5.3 STEREOISOMERISM IN ALKENES

In distinct contrast to the rapid rotation about single bonds that we discussed for alkanes in Section 2.3, rotation about the carbon-carbon double bond is *restricted*. The fixed geometry of the double bond creates a situation similar to that which we discussed for cycloalkanes in Section 3.5; two alkenes with the *same constitution* can nevertheless have different three-dimensional structures. In other words, the carbon-carbon double bond can result in *stereoisomers*.

Consider, for example, the molecular formula $C_2H_2Cl_2$, which corresponds to dichloroethylene. As we have shown in Figure 5.5, the two chlorine atoms could be attached either to the same carbon atom or to different carbon atoms. The two structures at the top of the figure are clearly *constitutional isomers,* because the chlorines are on the *same end* of the double bond in the first case but on *different ends* of the double bond in the second.

What of the two structures in Figure 5.5 that have a chlorine atom at *each end* of the double bond? These must have the same constitution because constitution is

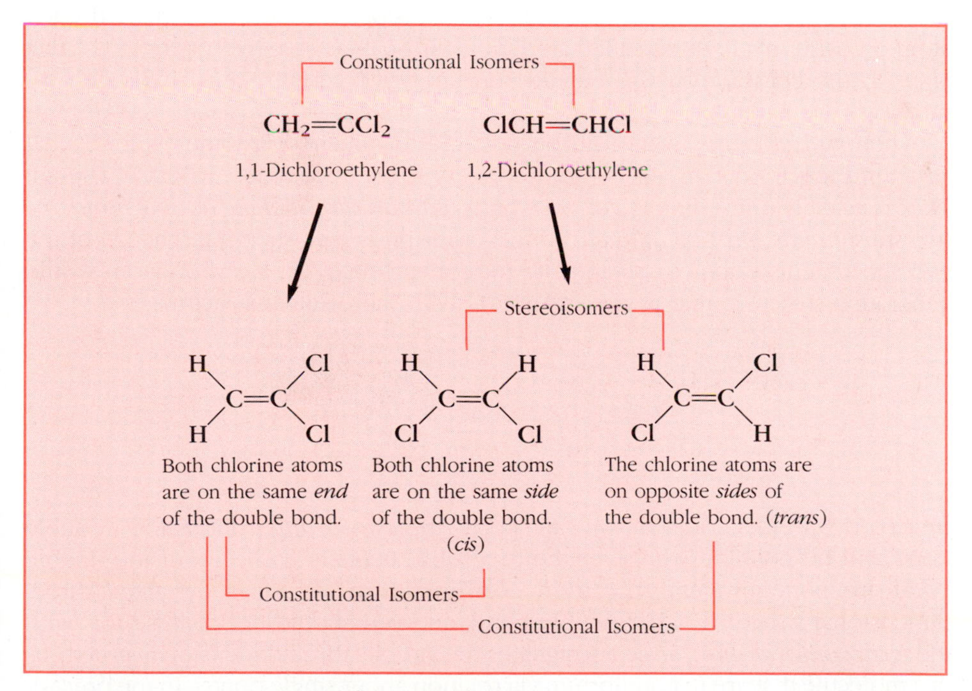

Figure 5.5 Isomeric Dichloroethylenes.

defined only by the sequence in which the atoms are connected. Nevertheless, the two structures clearly differ in a three-dimensional sense, and the chlorine atoms are on the *same side* of the double bond in one case but *opposite sides* of the double bond in the other. Note the convenient distinction between *sides* and *ends* of double bonds that we have illustrated in Figure 5.5. When the two chlorine atoms are on the *same side* of the double bond, the relationship between them is specified as ***cis,*** and if they are on *opposite sides,* the relationship is ***trans.***

These two isomeric 1,2-dichloroethylenes are frequently referred to as the *cis*- and the *trans*- isomers, where the prefix *cis* or *trans* is understood to describe the relationship between the substituents (i.e., the chlorine atoms) on the double bond. This terminology is only used for convenience, however, and we will discuss the IUPAC nomenclature of alkenes in Section 5.5

The terms *cis* and *trans* can always be used to describe relationships between substituents on a double bond, even in cases in which isomerism is not possible. For example, we might discuss the two different hydrogen atoms on C-2 of chloroethylene in terms of their relationship to the chlorine substituent.

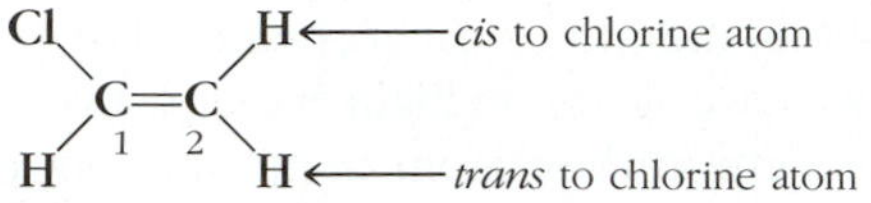

Cl, H (C-1) C=C (C-2) H ← *cis* to chlorine atom; H ← *trans* to chlorine atom

One of these hydrogens is *cis* to the chlorine atom and the other is *trans* to it. *It is essential for you to remember that these terms describe a relationship between two substituents, so both substituents must be specified.* Accordingly, we could not properly refer to either the "*cis* hydrogen" or the "*trans* hydrogen" without specifying the other half of the relationship. *Cis* or *trans* to what? The hydrogen atom that is *cis* to the chlorine atom on C-1 is at the same time *trans* to the hydrogen atom on C-1. Only if we specified a relationship such as "the hydrogen on C-2 that is *cis* to the chlorine on C-1," could you know for certain which hydrogen we meant.

When will *cis-trans* isomerism be possible for alkenes? In other words, what structural features of an alkene are necessary for stereoisomers to exist? The answer is actually quite simple. *For each of the carbon atoms of the double bond, the two substituents must be different.* If this criterion is met, then interchange of the two substituents (labeled "a" and "b" for convenience) on a carbon atom would generate a structure that is stereoisomeric with the original structure.

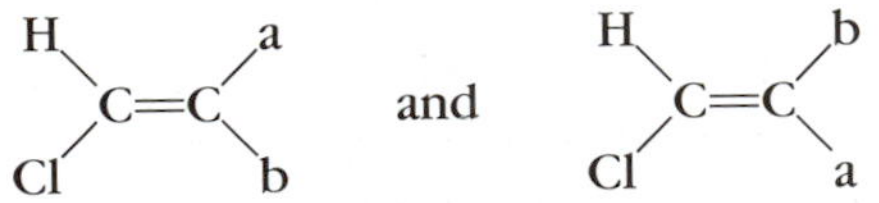

(H)(Cl)C=C(a)(b) and (H)(Cl)C=C(b)(a)

In accord with this criterion, we can specify that two *configurations* for a double bond will be possible whenever the two substituents at each end of the double bond are nonequivalent. The term **configuration** describes the *fixed* three-dimensional structure of the double bond, and you should not confuse this with the term *conformation,* which designates the various three-dimensional structures of a molecule that are in equilibrium via rotation about single bonds. In the preceding example, substituent *a* could either be *cis* or *trans* to the chlorine, and only those two configurations of the double bond are possible.

Cis-trans isomerism is *not possible* if the two substituents are the same at *either end* of the double bond.

$$(Cl)_2C{=}C(a)(b) \quad \text{and} \quad (Cl)_2C{=}C(b)(a)$$

These are two drawings of the same structure; they differ only in their orientations, where the second is drawn upside down relative to the first. Whenever the two substituents at either end of the double bond are equivalent, only a single configuration is possible for the double bond.

EXERCISE 5.1

For which of the following alkenes would stereoisomerism be possible? For each of those cases draw both stereoisomers.

(a) $CH_3{-}CH{=}CH{-}CH_2CH_3$
(b) $CH_3{-}CH{=}CH_2$
(c) $CH_3{-}CH{=}C(CH_3)_2$
(d) $Cl{-}CH{=}CH{-}CH_2Cl$

5.4 *CIS-TRANS* ISOMERIZATION OF ALKENES

As we stated in the previous section, the existence of stereoisomerism in appropriately substituted alkenes is a consequence of the fixed configuration of the double bond. If alkenes could rotate about the double bond, we would observe *cis-trans* isomerization. The following equation shows that such rotation would interconvert the two possible stereoisomers of an alkene.

$$\underset{\text{(CH}_3\text{, H)}}{} CH_3(H)C{=}C(CH_3)(H) \overset{?}{\rightleftharpoons} CH_3(H)C{=}C(H)(CH_3)$$

However, direct *cis-trans* isomerization does not occur readily. Both stereoisomers are stable, and they can be separated and isolated in the same way as other isomeric compounds.

The configurational stability of alkenes is a direct consequence of the electronic structure of the carbon-carbon double bond. Rotation about the double bond would have to pass through a transition state in which the π bond would be completely broken, and we have illustrated this process in Figure 5.6.

The key point that you should learn from Figure 5.6 is that the p atomic orbitals are *parallel* (the orientation necessary for π bonding) in the initial and final geometries. The center structure is highly unstable because the p orbitals are perpendicular to each other and cannot overlap. High-temperature and photochemical studies of the *cis-trans* isomerization of alkenes have shown that rotation about the double bond of an alkene requires energy on the order of 50 kcal/mol. This is much lower than the bond dissociation energy for carbon-carbon σ bonds, which are in the range of 80–90 kcal/mol. As we indicated in Section 5.2, π bonds are

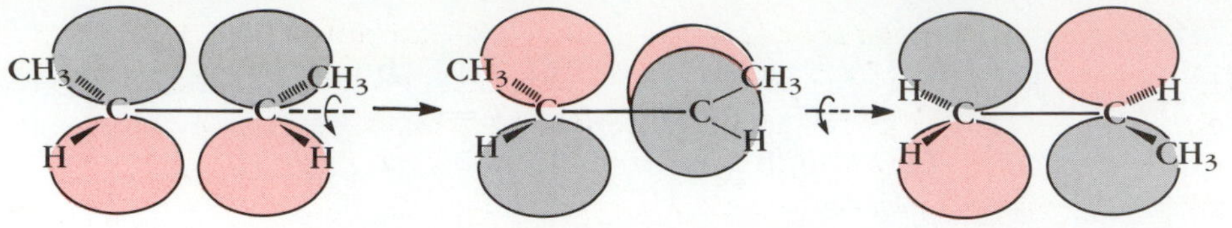

Transition state:
p Orbitals perpendicular
to each other;
no overlap

Figure 5.6 Rotation about the Carbon-Carbon Double Bond of an Alkene.
Interconversion of the *cis* and *trans* isomers would require a transition state in which the p orbitals of the π bond are perpendicular to each other. There could no net overlap in this orientation and the π bond would be completely broken. (The shading of the orbitals in the figure is arbitrary and we have used it only to show the rotation. It is *not intended to indicate an antibonding orbital.)*

weaker than σ bonds because the overlap of the parallel p atomic orbitals is less effective. While free rotation about a double bond does not occur, the weaker π bond is energetically more susceptible to cleavage by chemical reagents than are the σ bonds of saturated compounds.

5.5 NOMENCLATURE

The nomenclature of alkenes employs the same IUPAC system that we introduced for alkanes in Section 1.8. Only a few changes are needed to name alkenes. First, the **principal chain** is the longest carbon chain *that contains the double bond.* Second, the suffix *-ane* that would be used for the alkane is replaced by *-ene*. Third, the configuration of the double bond must be specified.

We will begin by disregarding the problem of stereochemistry, and after we cover the other aspects of the name, will turn to the procedure for specifying the configuration. The similarity between the names for alkanes and alkenes is illustrated by the following example:

$$CH_3{-}CH_2{-}CH_3 \qquad CH_3{-}CH{=}CH_2$$

Propane Propene

The two-carbon alkene, $CH_2{=}CH_2$, is called by its common name, ethylene, rather than its IUPAC name, ethene. Common names exist for some of the other alkenes as well, but the IUPAC names are preferred. The overall procedure for naming alkenes is summarized as follows:

Nomenclature of Alkenes and Their Derivatives

1. *Locate the principal carbon chain.* For alkenes this is defined as the longest carbon chain in the molecule *that contains the double bond.* The basic name is then formed by adding the suffix *-ene* to the root of the name for the alkane chain having the same number of carbon atoms.

2. *Name and number the substituents* on the principal chain.
   (a) The principal chain is numbered so that the lower of two possible numbers results for *the first carbon of the double bond.* This number is prefixed to the name of the principal chain with a hyphen. If the location of the double bond does not distinguish between two different ways of numbering the principal chain, the substituents are evaluated so that the lower number results at the first point of difference.
   (b) When there is more than one substituent, they are listed in alphabetical order. If several of the substituents are identical, this is indicated by the numerical prefixes *di-, tri-, tetra-, penta-, hexa-,* etc.

3. *Specify the stereochemistry* for any center in the molecule that could exist in two alternative configurations. In the case of alkenes the geometry of the double bond is specified by a prefix, either (*E*) or (*Z*), if two configurations are possible.

The following examples show you how the principal chain is numbered:

$$CH_3—CH_2—CH═CH—CH_3$$

2-Pentene
(not 3-pentene)

$$CH_2═CH—CH_2—CH_3$$

1-Butene
(not 3-butene)

Numbering begins at the end of the chain that leads to the lower number for the *first carbon of the double bond,* and the number of this carbon immediately precedes the name of the parent chain. (Clearly the double bond *starts* at this carbon and *ends* at that with the next higher number.) The lower possible number is given to the first carbon atom of the double bond even if that results in higher numbers for substituents on the chain.

$$CH_2═CH—CH_2—CH(CH_3)—CH(CH_3)—CH_3$$

4,5-Dimethyl-1-hexene
(not 2,3-dimethyl-5-hexene)

Only if the first carbon atom has the same number starting from either end of the principal chain do you use the substituents as the basis for numbering.

$CH_3—CH_2—CH=CH—CH_2—CH_2Cl$

1-Chloro-3-hexene
(not 6-chloro-3-hexene)

The next step in naming an alkene is specifying the configuration of the double bond, i.e., designating of the specific stereoisomer in cases where *cis-trans* isomerism is possible. In common practice the descriptors *cis* and *trans* are frequently used.

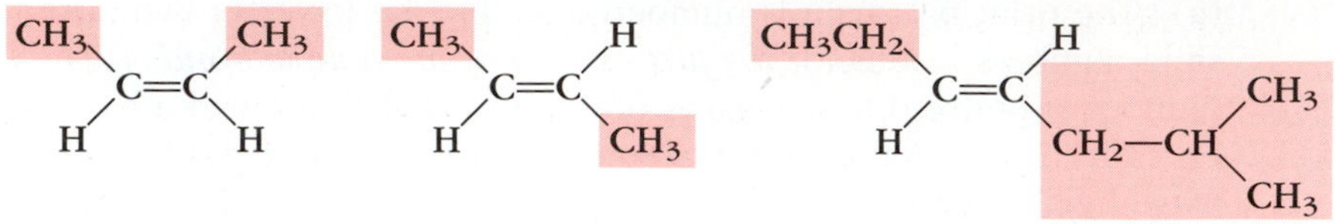

cis-2-Butene

trans-2-Butene

6-Methyl-*trans*-3-heptene

In such cases as the preceding three examples, this provides a convenient description of the *carbon chain*, and we could refer to the double bond in *cis*-2-butene as a *cis* double bond. But this terminology breaks down for tri- or tetrasubstituted double bonds, because it would not be clear which two substituents were being described as *cis* or *trans* to each other.

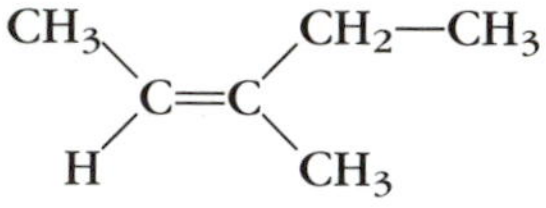

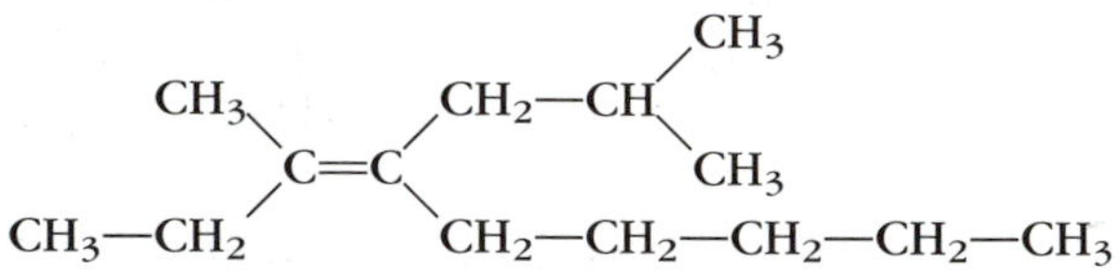

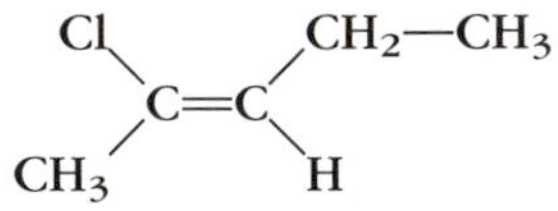

For example, the first of the preceding three compounds might be named 3-methyl-*cis*-2-pentene to indicate a *cis* relationship for the carbons of the principal chain or between the *larger* groups (CH_3 and CH_2CH_3) at each end of the double bond. Alternatively, you might name this compound as 3-methyl-*trans*-2-pentene to indicate the relationship between the *equivalent* methyl groups. Similar uncertainties would arise if you tried to use *cis* and *trans* to name the other two examples.

Because *cis* and *trans* cannot always define the configuration of a double bond unambiguously, alternative descriptors, *E* and *Z*, were introduced into IUPAC nomenclature. To assign the configuration of a double bond, you must follow two steps:

1. At each end of the double bond, identify the substituent atom which has the higher priority. *The atom with the higher atomic number has the higher priority.*

chlorine has higher priority than carbon

carbon has higher priority than hydrogen

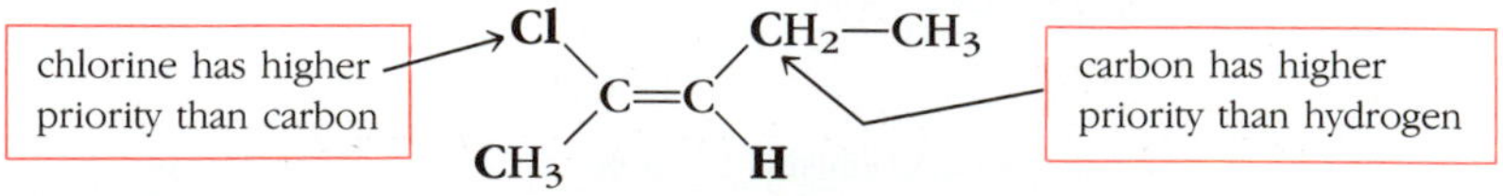

2. If the two higher-priority groups are on the *same side* of the double bond (i.e., if they are *cis* to each other), the configuration of the double bond is **Z** (German *zusammen,* "together"). If the two higher-priority groups are on

opposite sides of the double bond (i.e., if they are *trans* to each other), the configuration of the double bond is **E** (German *entgegen,* "opposite").

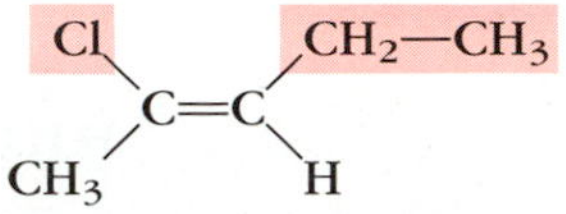

Cl $CH_2{-}CH_3$
$C{=}C$
CH_3 H

(*Z*)-2-Chloro-2-pentene

The higher priority groups at each end of the double bond are on the same side of the double bond.

In some cases the simple comparison of atomic numbers proves to be insufficient for determining priorities in the first of the two preceding steps. This is the case when two different alkyl groups are attached to one end of a double bond. Both groups are attached to the double bond via carbon atoms, so the atomic number is the same.

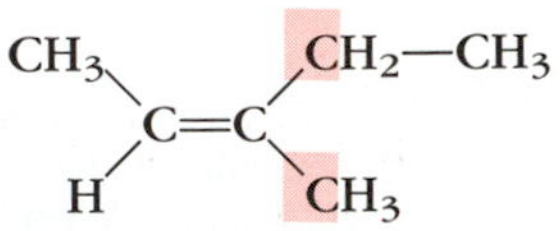

CH_3 $CH_2{-}CH_3$
$C{=}C$
H CH_3

Both substituents are attached to the double bond by carbon atoms—no basis for assigning different priorities.

As with earlier problems in nomenclature, you must resolve the problem by examining the other atoms in the substituent until you encounter the *first point of difference.* This is done sequentially along the carbon chain, beginning with the three substituents on each of the two carbon atoms that you originally compared.

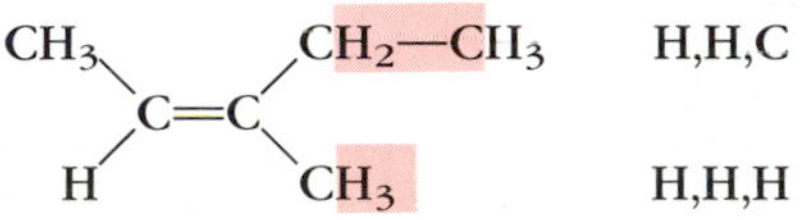

CH_3 $CH_2{-}CH_3$ H,H,C
$C{=}C$
H CH_3 H,H,H

An atom (in this example, C) is found which has higher atomic number than any of those for the other set.

Two of the comparisons (H vs H, and H vs H) still afford no distinction between the two alkyl groups, but the third comparison is between a hydrogen and a carbon, and finally we have a difference in atomic number. Carbon has the higher atomic number, so the ethyl group has a higher priority than the methyl group and the compound is named (*Z*)-3-methyl-2-pentene.

Sometimes you must proceed even further along the carbon chains of two different substituents to find a basis for assigning relative priorities. The distinction between ethyl and methyl was analyzed in the preceding example, and it presents no problem for the following alkene:

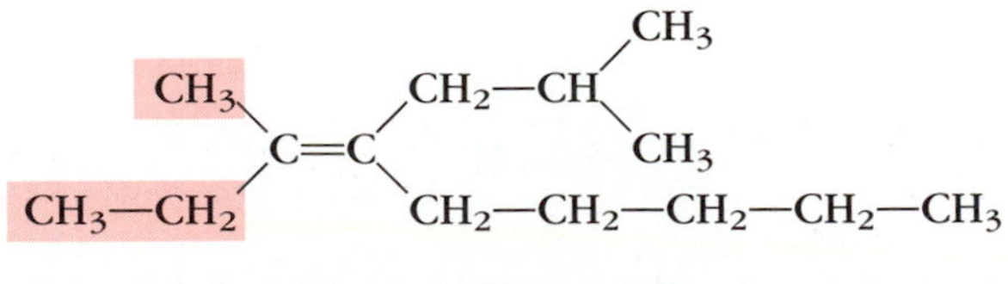

CH_3
CH_3 $CH_2{-}CH$
$C{=}C$ CH_3
$CH_3{-}CH_2$ $CH_2{-}CH_2{-}CH_2{-}CH_2{-}CH_3$

However, the distinction between the pentyl and isobutyl groups on the other end of the double bond is not so obvious. The relative priority *is not* simply a consequence of the overall size of the substituent.

Proceeding as before, neither the *atoms attached directly to the double bond* nor *their substituents* provide a basis for differentiation in this compound.

$$\mathrm{(CH_3)(CH_3CH_2)C{=}C(CH_2{-}CH(CH_3)_2)(CH_2{-}CH_2{-}CH_2{-}CH_2{-}CH_3)}$$

H,H,C

H,H,C

Only when you come to the next set of substituent atoms along each of the carbon chains do you encounter the point of difference which allows you to assign relative priorities.

$$\mathrm{(CH_3)(CH_3CH_2)C{=}C(CH_2{-}CH(CH_3)_2)(CH_2{-}CH_2{-}CH_2{-}CH_2{-}CH_3)}$$

H,C,C

H,H,C

Carbon has the highest atomic number and is present in both sets, but the upper set has *two* carbon atoms.

Neither set of three atoms contains an atom with a higher atomic number than any of the three atoms in the other set, so the atoms are again paired off. We first discount the substituents which are the same (H vs H, and C vs C), and this leaves us with a point of difference, H vs C. The higher atomic number of carbon therefore allows assignment of higher priority to the isobutyl group. The correct IUPAC name for this compound is (*E*)-4-isobutyl-3-methyl-3-nonene.

$$\mathrm{(CH_3)(CH_3CH_2)C{=}C(CH_2{-}CH(CH_3)_2)(CH_2{-}CH_2{-}CH_2{-}CH_2{-}CH_3)}$$

(*E*)-4-Isobutyl-3-methyl-3-nonene

One additional rule for assigning priority may sometimes be necessary. For an atom which is itself multiply bonded, each of the bonds counts as a substituent. For example, the substituents on the carbon atom in the group

$$-\overset{\overset{\displaystyle O}{\|}}{C}-H$$

would be H,O,O, since there are two bonds to oxygen. The configuration of the double bond in the following example would therefore be specified as *Z*. (The complete nomenclature of compounds containing this functional group will be covered in Chapter 13.)

CH_3, H, C=C, $\overset{O}{\|}$CH, CH—CH_3, OH

H,O,O

H,C,O

Oxygen has the highest atomic number of the six substituent atoms, but the upper set has *two* and therefore the $—\overset{O}{\overset{\|}{C}}—H$ group has higher priority than the $—\underset{OH}{\underset{|}{CH}}—CH_3$ group.

Cyclic Alkenes

The procedure for naming a cycloalkene is essentially the same as that which we have just discussed, except that the principal chain is that of the ring and the prefix *cyclo* is used. The term *cycloalkene* refers specifically to compounds in which the double bond is contained within the ring. If the double bond is not in a ring, then the compound would be named as usual, with the ring listed as a substituent on the principal alkene chain. As we have shown in the following three structures, a cycloalkene is always numbered so that the double bond is between C-1 and C-2. The direction in which the ring is numbered is selected so that the lower number is obtained at the first point of difference.

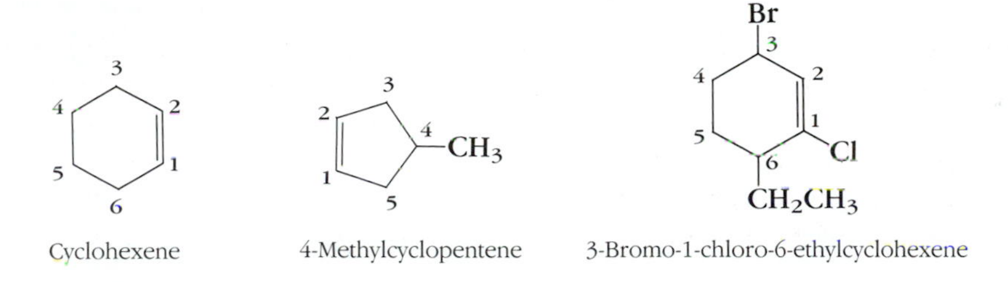

Cyclohexene   4-Methylcyclopentene   3-Bromo-1-chloro-6-ethylcyclohexene

The naming of cyclic compounds present a few unique problems because the basic rules assume that the principal chain must be entirely within or entirely outside of the ring. But what do you do when the carbon-carbon double bond is positioned between one carbon that is part of the ring and a second which is part of a substituent on the ring? In these cases the suffix *-ylidene* is used to name the group as a substituent of a cycloal*kane*. For example, $=CH—CH_3$ is *ethylidene,* $=C(CH_3)_2$ is *isopropylidene,* and so on. However, the one-carbon residue CH_2 is named *methylene* rather than "methylidene."

CH_2 CH_3, CH—CH_3 =CH—CH_3

Methylenecyclohexane   Isopropylidenecyclopentane   Ethylidenecyclopropane

Notice that these names are closely related to those of the *saturated* cycloalkanes that would result if two hydrogens were added to the double bond (i.e., methylcyclohexane, isopropylcyclopentane, etc.).

We will introduce one final note about naming alkenes here. Sometimes the double bond must be considered as part of a *substituent.* Usually this is straightforward, and you derive the name of an *alkenyl* substituent from the name of the parent alkene in the same way that we showed for *alkyl* and *alkane* in Section 1.8. The point of attachment will be C-1 of the substituent, as shown in the following example:

$$CH_3—CH=CH—CH_2—CH_2—$$

3-Pentenyl

Several alkenyl groups have common names; the most important of these are *vinyl* and *allyl*.

$$CH_2=CH— \qquad CH_2=CH—CH_2—$$

Vinyl   Allyl

EXERCISE 5.2

Name the following alkenes:

(a) 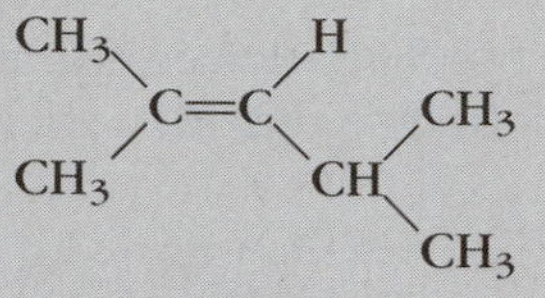 $(CH_3)_2C=CH—CH(CH_3)_2$

(b) 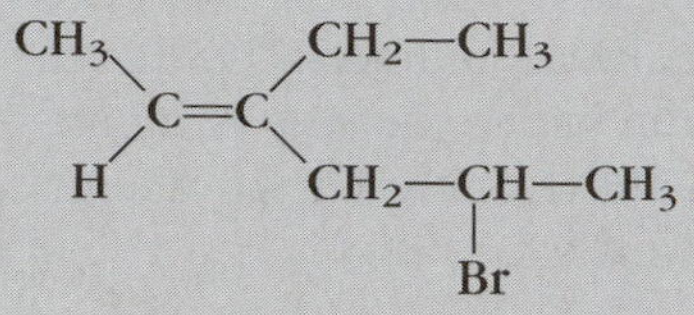 $CH_3(H)C=C(CH_2—CH_3)(CH_2—CH(Br)—CH_3)$

(c) $CH_3—CH_2(H)C=C(H)CH_2—CH_3$ (both H on the same side)

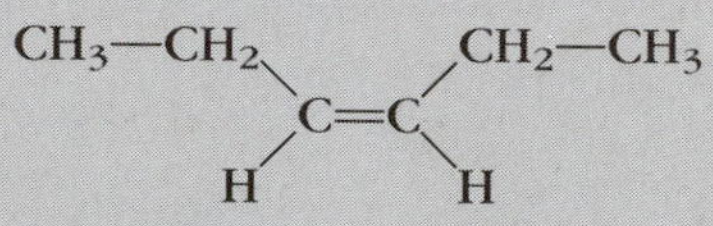

(d) 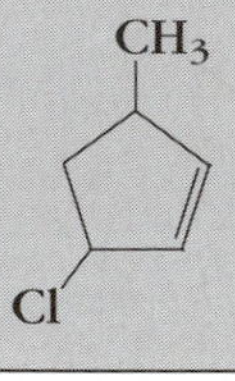 Cyclopentene ring bearing CH_3 and Cl

5.6 REDUCTION OF ALKENES: RELATIVE STABILITIES OF ISOMERIC ALKENES

If two or more isomeric alkenes could be produced in a reaction, how could you predict which of these you would expect to obtain? The answer to this question is frequently quite simple: the *more stable isomer is expected to predominate.* This leads us to another question that must be answered if you want to fully understand reactions involving alkenes. What determines the relative stabilities of various alkenes?

An experimental comparison of the stability of two compounds must be based on some reaction which they both undergo. This is most reliable when both compounds yield the same products. We employed this type of reaction in our discussion of ring strain in Section 3.2, where the heat of combustion was used to compare the stabilities of cycloalkanes. It was necessary to use the heat of combustion per CH_2 group when comparing cycloalkanes with different numbers of carbons, but that approach is only applicable to a homologous series of compounds. For example, not only do ethylene and 2,3-dimethyl-2-butene have different molecular weights, but the ratio of carbon atoms to hydrogen atoms in the two compounds is also different. The relative quantities of CO_2 and H_2O produced by the oxidation of these two alkenes would differ, so a direct comparison of their heats of combustion would not be meaningful. There is, however, a simple reaction of alkenes that allows a direct comparison of the heats of reaction, and that is the addition of two hydrogen atoms to the double bond to form an alkane. This process falls into the category of *reduction,* and before we analyze its energetics, we will stop to evaluate the meaning of oxidation and reduction in organic chemistry.

Reactions involving the addition of oxygen clearly fall into the category of oxidation, and reactions which remove oxygen atoms from organic compounds must be reductions. A unique aspect of organic reactions is an alternative way of defining *reduction.* The *addition* of hydrogen atoms to an unsaturated compound corresponds to *reduction,* while the *removal* of hydrogen atoms corresponds to *oxidation.* These are convenient operational definitions, and you can readily calculate the oxidation state of a carbon atom according to rules we will discuss in Section 11.3. The reaction of an alkene to generate an alkane is a reduction.

$$R{-}CH_2{-}CH{=}CH_2 \longrightarrow R{-}CH_2{-}CH_2{-}CH_3$$

Addition of hydrogen to an alkene is accomplished by treating the alkene with hydrogen gas in the presence of a metal catalyst, typically finely powdered platinum or palladium metal coated on particles of carbon. This is illustrated by the conversion of 1-octene to octane.

$$C_6H_{13}{-}CH{=}CH_2 \xrightarrow{H_2,\ Pt} C_6H_{13}{-}CH_2{-}CH_3 \qquad \sim 100\%$$

The reaction is known as *catalytic hydrogenation,* and it is typically exothermic.

To obtain legitimate energy comparisons, you must employ reactions that yield the *same product.* We have shown such a comparison in Figure 5.7 for the isomeric 2-butenes, each of which affords butane. The energy evolved in the catalytic hydrogenation of (Z)-2-butene is greater than that for the E isomer by 1.1 kcal/mol. This value provides a direct measure of the greater stability of the E-isomer.

Figure 5.7 illustrates a general property of 1,2-disubstituted alkenes (where "1,2-" indicates that there is one substituent on each of the two carbon atoms of the double bond). The isomer in which the two alkyl groups are *trans* to each other is more stable than the isomer in which they are *cis* to each other, and the energy difference between them is on the order of 1 kcal/mol.

$$\underset{\text{Less stable}}{(R)(H)C{=}C(R')(H)\ \text{(cis)}} \qquad \underset{\text{More stable}}{(R)(H)C{=}C(H)(R')\ \text{(trans)}}$$

This preference for the alkyl groups to be far apart in space is another example of *nonbonded* or *steric* effects, and it is entirely analogous to the conformational preferences of alkanes discussed in Section 2.4. The alkyl groups are bulkier than hydrogen atoms, and this results in unfavorable steric interactions when they are in close proximity. These steric effects are alleviated when the alkyl groups are in a *trans* configuration on a double bond.

Heats of hydrogenation can be used to compare the relative stabilities in any series of isomeric compounds. In Figure 5.8 we have done this for a series of

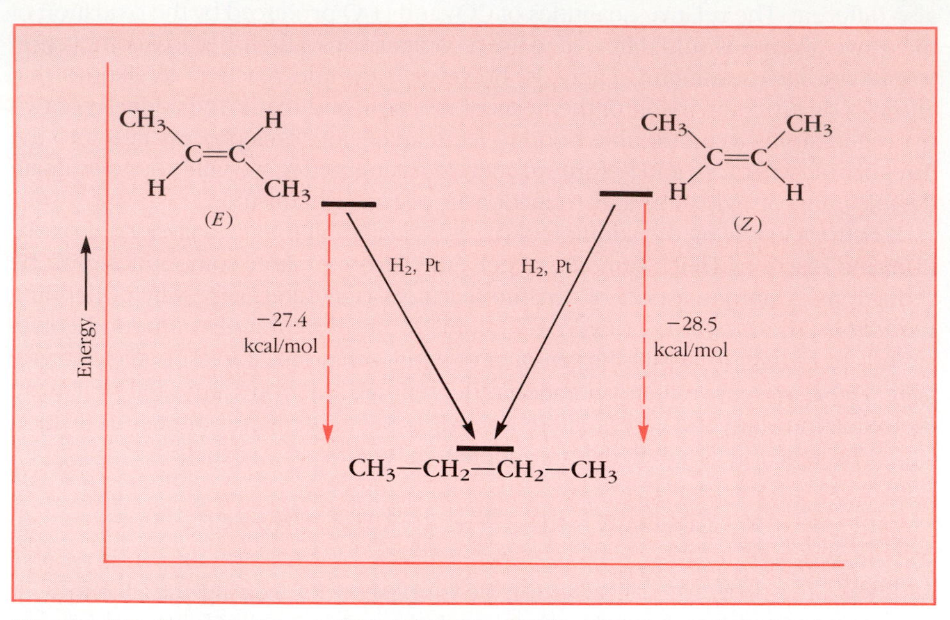

Figure 5.7 Comparison of Stabilities of (*E*)- and (*Z*)-2-butene by Catalytic Hydrogenation to Produce Butane.
The heat of hydrogenation (ΔH) is shown in each case.

alkenes which all have the same carbon skeleton. The double bond is located between different pairs of carbon atoms of this skeleton, and in one case there are two configurations of the double bond.

You can see from Figure 5.8 that there is a clear trend toward lower energy (greater stability) with an increase in the number of alkyl groups attached directly to the double bond. The least stable isomer is that with a *monosubstituted* double bond, while the most stable alkene has a *tetrasubstituted* double bond. The energies for those isomers with di- and trisubstituted double bonds are rather similar, so we cannot make any "safe" generalization about their stabilities. The only clear patterns are that tetrasubstituted alkenes will be most stable, monosubstituted isomers will be least stable, and the *E* isomer (alkyl groups *trans*) will be more stable than the *Z* isomer for a disubstituted double bond.

The preference for the *E* configuration of disubstituted double bonds cannot be extended to *cyclo*alkenes. The atoms attached directly to a carbon-carbon double bond all lie in a plane, and the distance between two carbon atoms having a *trans* relationship is nearly 4 Å.

C—C=C—C Interatomic distance is approximately 3.9 Å

The bond length of a carbon-carbon single bond is only 1.5 Å, and a minimum of two additional CH_2 groups (three additional C—C bonds) would be necessary to bridge the 4 Å gap. Even then, we are neglecting angle distortions, and *E,* or *trans,* cycloalkenes are highly distorted unless the ring size is large. (*E*)-Cyclooctene has the smallest ring size of any *trans* cycloalkene that is stable enough to be isolated at normal temperatures.

Alkene	Number of Alkyl Groups on the Double Bond	Relative Energy, kcal/mol
$(CH_3)_2CH{-}CH(CH_3){-}CH_2{-}CH{=}CH_2$	1	4.8
$CH_2{=}C(CH_3){-}CH(CH_3){-}CH_2{-}CH_2{-}CH_3$	2	2.8
$(CH_3)_2CH{-}CH(CH_3){-}CH{=}CH{-}CH_3$ (H, H on the same side)	2	2.8
$(CH_3)_2CH{-}CH(CH_3){-}CH{=}CH{-}CH_3$ (H, CH_3 on the same side)	2	1.4
$(CH_3)_2CH{-}C(CH_3){=}CH{-}CH_2{-}CH_3$	3	2.2
$(CH_3)_2C{=}C(CH_3){-}CH_2{-}CH_2{-}CH_3$	4	0.0 most stable

cis-trans difference 1.4 kcal/mol

Figure 5.8 Relative Stability of Isomeric Alkenes.
The stability increases with the number of alkyl groups bonded directly to the carbon atoms of the double bond. The energies are relative to an arbitrary value of zero for the most stable isomer.

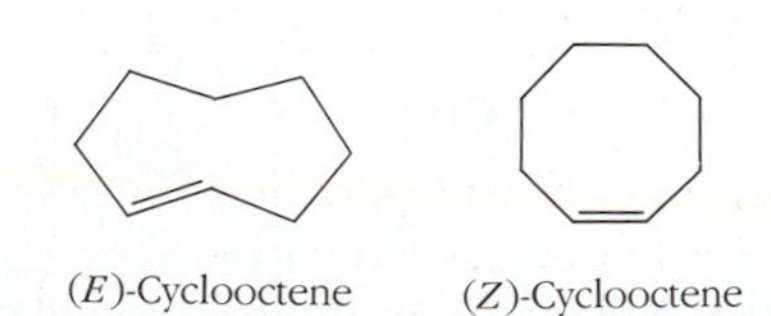

(*E*)-Cyclooctene (*Z*)-Cyclooctene

(*E*)-Cyclooctene is less stable than (*Z*)-cyclooctene, but it can still be prepared and isolated quite easily. As you can see from Figure 5.9, the carbon chain of (*E*)-cyclooctene must twist around in order to accommodate the *trans* double bond.

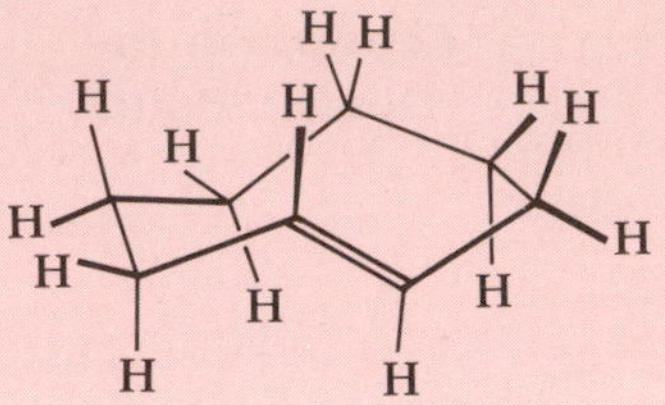

Figure 5.9 **The Three-Dimensional Structure of (*E*)-Cyclooctene.**

EXERCISE 5.3

(a) Name each of the alkenes in Figure 5.8.
(b) Draw and name the compound which would be produced by catalytic hydrogenation of each of these alkenes.

EXERCISE 5.4

For most alkenes of the general formula R—CH═CH—R, the *E* isomer is more stable by about 1 kcal/mol. In the case of 4,4-dimethyl-2-pentene, however, the energy difference is much larger, 4 kcal/mol. Why? What does this suggest about the *steric bulk* of the *tert*-butyl group? (Show complete stereochemistry in the drawings you use in your answer.)

5.7 ADDITION OF CARBENES TO ALKENES: PREPARATION OF CYCLOPROPANES

Now that we have explained the basic structural aspects of alkenes, we can proceed to their reactions. The typical reaction of an alkene is *addition,* and we have already introduced you to one such reaction, catalytic hydrogenation. Catalytic hydrogenation of an alkene proceeds by the addition of two hydrogen atoms to the double bond to give an alkane. The common feature of addition reactions is attachment of a new substituent to each end of the double bond to form a product in which both of these carbon atoms are *saturated.*

In this section we will show you a very simple addition reaction in which a *carbene* adds to the double bond of an alkene to generate a cyclopropane.

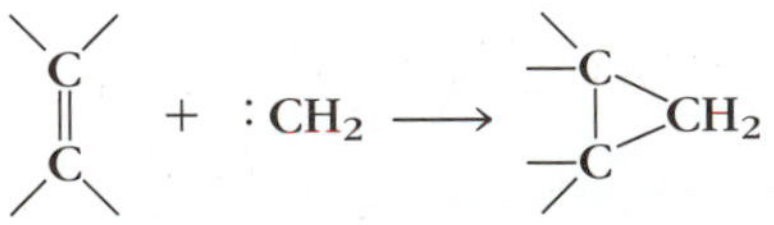

$$\text{>C=C<} + :CH_2 \longrightarrow \text{cyclopropane (C–C–}CH_2\text{ ring)}$$

In most of the reactions we will consider, the addition is **concerted,** which means that the reactants proceed over a single transition state to give the product directly, without the formation of any intermediate. We will discuss concerted reactions in much greater detail in later chapters, particularly Chapter 27.

Methylene itself, $:CH_2$, can be generated by decomposition of diazomethane, CH_2N_2, but diazomethane is highly toxic and can decompose explosively. For most laboratory purposes a preferred method is the use of methylene iodide (CH_2I_2) with a zinc catalyst such as a zinc-copper couple (i.e., a zinc-copper alloy). This does not actually generate the free carbene but, instead forms a *carbenoid* intermediate, which is a zinc complex of the carbene.

$$CH_2I_2 \xrightarrow{Zn-Cu} [CH_2I_2Zn]$$

This method for generating a carbenoid intermediate was discovered by H. E. Simmons and R. D. Smith (E. I. du Pont de Nemours and Co.) in the late 1950s and has since become a widely used method for the synthesis of cyclopropanes. The following examples illustrate the Simmons-Smith reaction with alkenes.

$$CH_3CH_2CH{=}CHCH_2CH_3 \xrightarrow[Zn-Cu]{CH_2I_2} CH_3CH_2-\triangle-CH_2CH_3 \qquad 35\%$$

$$\text{cyclohexene} \xrightarrow[Zn-Cu]{CH_2I_2} \text{bicyclo[4.1.0]heptane} \qquad 58\%$$

Notice in the second equation that reaction of a cycloalkene produces a *bicyclic* alkane, that is, one which has *two* rings.

A related way of forming cyclopropanes employs the addition of a dihalocarbene to an alkene. The carbene intermediate can be generated by the reaction of chloroform or bromoform with bases such as NaOH, KOH, or the potassium salt of *tert*-butyl alcohol (abbreviated as KOtBu). The following equation illustrates the generation of dichlorocarbene from chloroform by using NaOH.

$$CHCl_3 \xrightarrow{NaOH} [:CCl_2]$$

The reaction of dihalocarbenes with alkenes produces dihalocyclopropanes, as illustrated by the following two equations:

$$\text{cyclohexene} \xrightarrow[KOtBu]{CHBr_3} \text{7,7-dibromobicyclo[4.1.0]heptane (Br, Br)} \qquad 75\%$$

$$(CH_3)_2C{=}CH_2 \xrightarrow[KOtBu]{CHCl_3} \text{1,1-dichloro-2,2-dimethylcyclopropane (Cl, Cl; } CH_3, CH_3) \qquad 65\%$$

EXERCISE 5.5

Draw the product of the reaction of methylene iodide and a zinc-copper couple with (a) 1-heptene and (b) 2-methyl-2-pentene.

EXERCISE 5.6

Draw the product of the reaction of chloroform and sodium hydroxide with (a) 1-heptene and (b) 3-methyl-2-pentene.

5.8 POLAR ADDITION TO ALKENES

Addition to the double bond is the characteristic reaction of alkenes, and it involves bonding of an additional atom or group to each of the carbon atoms of the original double bond. This has two important consequences: first, the carbon atoms of the double bond will become *saturated,* and second, the two new substituents will bear a 1,2-relationship to each other, i.e., they will be on adjacent carbon atoms.

$$\rangle C{=}C\langle \xrightarrow{X-Y} X-\overset{|}{\underset{|}{C}}-\overset{|}{\underset{|}{C}}-Y$$

General equation for addition of reagent X—Y to an alkene

Addition reactions to alkenes can be classified according to their reaction mechanisms, and both of the additions we have discussed so far are *nonpolar* additions. As we described in the previous section, the reaction of a carbene with an alkene normally proceeds in a concerted manner to give a cyclopropane without the formation of any intermediate. Catalytic hydrogenation takes place stepwise, and as we will describe more fully in Section 6.8, the hydrogen atoms are transferred one at a time from the catalyst surface to the carbon atoms of the alkene.

In contrast to the reactions that we have already discussed, *polar* additions are found for many other additions to alkenes. We call a reaction *polar* when ionic charge develops at the transition state of the reaction. At the transition state the charge is only *partial,* but the reaction can continue via the formation of an intermediate *carbocation.* We have summarized this for the stepwise ionic addition of the hypothetical reagent X—Y in the following equation:

$$\rangle C{=}C\langle \xrightarrow{X-Y} \left[\rangle C-\overset{|}{\underset{|}{C}}-Y\right]^{+} + X^{-} \longrightarrow X-\overset{|}{\underset{|}{C}}-\overset{|}{\underset{|}{C}}-Y$$

Ionic addition of reagent X—Y to an alkene

In the first step of this reaction the π electrons of the alkene are used to form a bond to Y, and it is useful to think of the alkene reacting with Y^+. This produces the anion X^-, together with a carbocation. In a subsequent step the highly reactive carbocation undergoes nucleophilic attack by X^- to complete the overall addition.

In many polar addition reactions, attack by X^- or some other nucleophile occurs before the carbocation is fully formed, but for the present we will write the free carbocation as an intermediate because it makes it easier to understand the reactions. For example, what would happen if our hypothetical reagent X—Y were to react with an *unsymmetrical* alkene? Two different products could be formed, as we have illustrated for 2-methylbutene.*

*Remember that drawings such as those used to show the products in this equation convey *no three-dimensional information.* For example, the following three structural formulas are all equivalent:

$$Y-\overset{|}{\underset{|}{C}}-\overset{|}{\underset{|}{C}}-X \qquad -\overset{Y}{\underset{|}{C}}-\overset{|}{\underset{|}{C}}-X \qquad -\overset{Y}{\underset{|}{C}}-\overset{X}{\underset{|}{C}}-$$

$$(CH_3)(H)C{=}C(CH_3)_2 \xrightarrow{XY} CH_3{-}CH(X){-}C(CH_3)_2{-}Y \quad \text{or} \quad CH_3{-}CH(Y){-}C(CH_3)_2{-}X$$

In most addition reactions one of the two possible products will predominate. There is usually a specific correlation between the *structure* of the alkene and the *orientation* of the addition, and you can predict the outcome in advance. Such reactions are called **regiospecific,** (from the Latin *regio,* "direction, boundary") because the fragments (X and Y) add predictably to the double bond in a specific direction. The preference for one of the two possible products is a direct result of the difference in stability between the carbocations that could be formed in the first step:

$$(CH_3)(H)C{=}C(CH_3)_2 \xrightarrow{XY} (CH_3)(H)\overset{+}{C}{-}C(CH_3)_2{-}Y \quad \text{or} \quad CH_3{-}CH(Y){-}\overset{+}{C}(CH_3)_2$$

A secondary carbocation — A tertiary carbocation

You will recall from Section 2.7 that a tertiary carbocation is more stable than a secondary carbocation, and it is also more rapidly formed. As we indicated for free radicals in Sections 2.6 and 2.10, the energies for two alternative transition states in a reaction reflect the energies of the intermediates that would be formed in each case. As you would expect, the transition state leading to the tertiary cation would be of lower energy, and that cation would be generated preferentially in the first step of the addition reaction. The final product that you would expect from our hypothetical reaction is:

$$CH_3{-}CH(Y){-}C(CH_3)_2{-}X$$

Sometimes the two alternative carbocations that could be produced in a polar addition reaction will have similar energies, and in such cases you would expect to obtain a mixture of products. For example, attack of the hypothetical "Y^+" of X—Y at either carbon of the double bond of 2-pentene would generate a secondary carbocation. You would therefore expect both pathways to be energetically similar, and both products should be formed.

$$CH_3{-}CH{=}CH{-}CH_2{-}CH_3 \xrightarrow{X-Y} CH_3{-}CH(X){-}CH(Y){-}CH_2{-}CH_3 \quad \text{and} \quad CH_3{-}CH(Y){-}CH(X){-}CH_2{-}CH_3$$

Whenever the alternative pathways in a polar addition reaction would generate positive charge on carbon atoms that are substituted differently, you can predict the products of the reaction on the basis of a very simple rule: *The initial step of the reaction will proceed in the orientation that generates the more stable carbocation.*

Acid-Mediated Addition Reactions

A major subgroup of polar addition reactions is the *acid-mediated* addition of H—X, where X can be any of a variety of electronegative substituents.

$$\mathrm{>C{=}C<} \xrightarrow[\mathrm{H^+}]{\mathrm{H{-}X}} \mathrm{H{-}\overset{|}{\underset{|}{C}}{-}\overset{|}{\underset{|}{C}}{-}X}$$

H—X can be HCl, H—OH, H—OR, etc.

Recall from our discussions in Chapter 4 that we will employ H^+ as a convenient symbol in proton transfer reactions even though free H^+ ions are not present in solution. The actual proton donors will be acids such as H—X and H_3O^+. If the molecule H—X is itself a strong acid, no additional acid is needed, and the reaction proceeds in two simple steps: protonation to give a carbocation, followed by nucleophilic attack of X^- on the cation.

$$\mathrm{>C{=}C<} \xrightarrow{\mathrm{H{-}X}} \mathrm{>\overset{+}{C}{-}\overset{|}{\underset{|}{C}}{-}H} + \mathrm{X^-} \longrightarrow \mathrm{{-}\overset{X}{\underset{|}{C}}{-}\overset{|}{\underset{|}{C}}{-}H}$$

The preceding equation introduces you to a new convention that we will use to illustrate reaction mechanisms in this book. *Curved arrows are used to show the fate of electron pairs that are involved in bond making and bond breaking,* just as they were used earlier (Sec. 4.7) to denote resonance stabilization of ions. In this equation the arrows in the first step show that the π electrons of the double bond form a new bond to the hydrogen atom of H—X, and the electron pair of the H—X bond remains on atom X to form the anion X^-. In the second step the nonbonding electrons of the anion are shown interacting with the carbocation center to form a C—X bond. You will find that this technique of using curved arrows to describe reaction mechanisms is a valuable device for understanding and interpreting organic reactions.

The direct addition of H—X is the expected pathway for additions involving mineral acids such as HCl, HBr, HI, and H_2SO_4. The reactions typically proceed readily at room temperature.

1-methylcyclopentene $\xrightarrow{\mathrm{HCl}}$ 1-chloro-1-methylcyclopentane (CH₃, Cl on same carbon) ~100%

$$\mathrm{CH_3{-}CH_2{-}CH{=}CH{-}CH_2{-}CH_3} \xrightarrow[\mathrm{CHCl_3}]{\mathrm{HBr}} \mathrm{CH_3{-}CH_2{-}\overset{Br}{\overset{|}{C}H}{-}CH_2{-}CH_2{-}CH_3} \quad 76\%$$

The addition of sulfuric acid to alkenes is an important industrial process. It is not used for preparative purposes in the laboratory, but it provides a useful way to distinguish between alkenes and alkanes. *Alkanes* are insoluble in sulfuric acid, but *alkenes* undergo addition of cold concentrated H_2SO_4 to produce alkyl hydrogen sulfates, which are soluble.

$$\mathrm{>C{=}C<} \xrightarrow{\text{cold conc. } \mathrm{H_2SO_4}} \mathrm{H{-}\overset{|}{\underset{|}{C}}{-}\overset{|}{\underset{|}{C}}{-}OSO_3H}$$

An alkyl hydrogen sulfate

(The name of this reaction product is analogous to that of the monosodium salt of sulfuric acid; $Na^{+}\,^{-}OSO_3H$ is called sodium bisulfate or sodium hydrogen sulfate.) Alkyl hydrogen sulfates are the intermediates in an industrial preparation of alcohols, and they are converted to alcohols in a reaction with water called *hydrolysis* (a name derived from the Greek *hudor,* "water" and *lysis,* "loosening" and meaning "cleavage by water").

$$H-\overset{|}{\underset{|}{C}}-\overset{|}{\underset{|}{C}}-OSO_3H \xrightarrow[H_2SO_4]{H_2O} H-\overset{|}{\underset{|}{C}}-\overset{|}{\underset{|}{C}}-OH$$

Ethyl alcohol (ethanol) is an important example of the manufacture of alcohols from alkenes. Prior to 1930 ethanol was prepared entirely by fermentation processes, the method that is still exclusively used for the production of alcoholic beverages. The rapid growth of the chemical industry in the first half of the twentieth century demanded an increased supply of ethanol for use as a reagent and as a solvent. Ethanol became available from ethylene, which in turn is a major product of petroleum refining processes such as cracking. When a stream of ethylene is passed through concentrated sulfuric acid, 99% of the ethylene gas is absorbed, and the alkyl hydrogen sulfate is formed. The reaction mixture is then diluted with water and heated, resulting in hydrolysis to the alcohol (which is removed by distillation).

$$CH_2{=}CH_2 \xrightarrow{\text{conc. } H_2SO_4} CH_3-CH_2-OSO_3H \xrightarrow[100°C]{H_2O} CH_3-CH_2-OH$$

All the preceding examples involved the addition of strong acids to the double bonds of alkenes. When H—X itself is not highly acidic, a strong acid such as H_2SO_4 or H_3PO_4 is added to the reaction mixture as a *catalyst.* This procedure is employed for the addition of water (H—OH), alcohols (H—OR), and even weak organic acids. Again the first step is protonation of the double bond (although in this case by the acid catalyst). The subsequent reaction is slightly different, however, because H—X is a weak acid and little or no X^- is present. In the case of water the concentration of ^{-}OH is only 10^{-7} at neutral pH. The ion product $[H^+][^{-}OH]$ must equal 10^{-14}, so it is clear that the ^{-}OH ion concentration must be far smaller under acidic conditions.

How does the reaction proceed if there is not a significant concentration of X^-? In all the cases we are considering, the oxygen atom in HX has nonbonding electrons that allow it to function as a nucleophile. As we show in the following equation, nucleophilic attack by water produces an oxonium ion, but this is a strong acid (analogous to H_3O^+). Subsequent transfer of a proton from the oxonium ion to some other species in solution (such as H_2O) produces the neutral addition product and regenerates the acid catalyst.

$$\rangle C{=}C\langle \xrightarrow{H^+} \rangle\overset{+}{C}-\overset{|}{\underset{|}{C}}-H \xrightarrow{H-\ddot{O}-H} -\overset{H-\overset{+}{O}-H}{\underset{|}{\overset{|}{C}}}-\overset{|}{\underset{|}{C}}-H \xrightarrow[H^+ \text{ transfer}]{:OH_2} -\overset{OH}{\underset{|}{\overset{|}{C}}}-\overset{|}{\underset{|}{C}}-H$$

The acid-catalyzed addition to alkenes of water and of organic compounds containing a hydroxyl group is illustrated by the following examples:

$$(CH_3)_2C{=}CHCH_3 \xrightarrow[H_2SO_4]{H_2O} CH_3{-}C(CH_3)(OH){-}CH_2{-}CH_3 \quad 74\%$$

$$(CH_3)_2C{=}CH_2 \xrightarrow[HF,\ BF_3]{CH_3OH} CH_3{-}C(CH_3)(OCH_3){-}CH_3 \quad 86\%$$

$$CH_3CH{=}C(CH_3)_2 \xrightarrow[H_2SO_4]{CH_3CH_2OH} CH_3{-}CH_2{-}C(CH_3)_2{-}O{-}CH_2{-}CH_3 \quad 40\%$$

In each of these reactions the major product is the isomer that results from protonation of the alkene to give the more stable cation.

The acid-catalyzed addition of water and alcohols to alkenes is most successful when fairly stable tertiary carbocations are formed. The laboratory conversion of alkenes to alcohols and ethers is generally carried out by other methods (Section 5.11), but the acid-catalyzed addition of water *(hydration)* has achieved considerable success in the industrial synthesis of low-molecular-weight alcohols. The production of 2-propanol, for example, can be effected (at elevated temperatures and pressures) by passing a mixture of propylene (propene) and water over an acid catalyst such as phosphoric acid coated on a solid mineral support. High yields can be achieved by recycling unreacted alkene after dissolving the alcohol product in water.

$$CH_3{-}CH{=}CH_2 \xrightarrow[\substack{H_3PO_4\ (\text{catalyst}) \\ 180\text{–}260°,\ 25\text{–}45\ \text{atm}}]{H_2O} CH_3{-}CH(OH){-}CH_3$$

This direct hydration process has become increasingly important in recent years because the indirect process (via the alkyl hydrogen sulfate) generates large quantities of diluted sulfuric acid that must be recycled. The recycling process is expensive, and it can also lead to problems of environmental pollution because the recycling is not 100 percent efficient.

Example 5.1 An unknown alkene undergoes reaction with HBr to produce 3-bromohexane in 76% yield. Deduce the structure of the alkene.

Solution Addition of HBr places a hydrogen and a bromine on adjacent carbon atoms. In the product, there is a hydrogen atom on *each* of the carbon atoms adjacent to that carbon atom which has the bromo substituent

$$CH_3{-}CH_2{-}CH(Br){-}CH_2{-}CH_2{-}CH_3$$

so *two* alkenes must be considered:

$$CH_3—CH{=}CH—CH_2—CH_2—CH_3 \quad \text{and} \quad CH_3—CH_2—CH{=}CH—CH_2—CH_3$$

The second structure (3-hexene) is symmetrical, and it would afford only 3-bromohexane on reaction with HBr, but the first alkene (2-hexene) could add HBr in two ways. Each pathway would proceed via a secondary carbocation, and a 50–50 mixture (approximately) of 2- and 3-bromohexanes would be expected. This is not consistent with the formation of a single product in greater than 50% yield. Consequently the unknown alkene must be 3-hexene.

Reaction with Halogens

The addition of a halogen to an alkene takes place readily at room temperature or below, and it is typically carried out in an inert solvent (i.e, one that does not enter into the reaction) such as CCl_4. For laboratory purposes the procedure is generally used with Br_2 and Cl_2. Fluorine is so reactive that the reaction with an alkene cannot be easily controlled, and the diiodo compounds formed from addition of I_2 tend to be unstable and to revert to the alkene. The mechanism can again be written as a two-step process involving the transient formation of a carbocation. In the first step the π electrons of the double bond displace a bromide ion from Br_2, and this bromide ion subsequently attacks the cationic intermediate. The net result is the very smooth formation of a **vicinal** dihalide (where the term *vicinal* indicates that the halogens are on adjacent carbons).

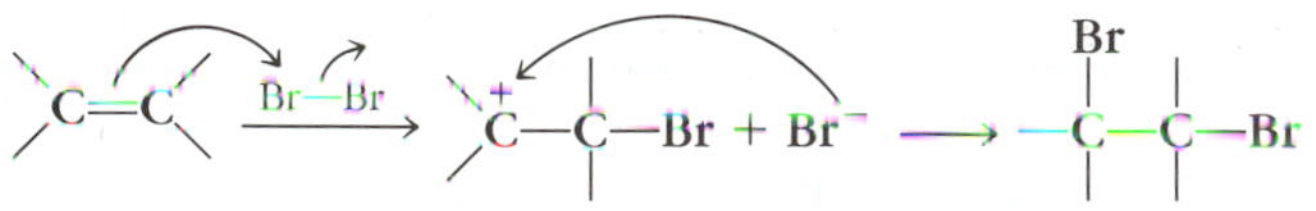

$$\text{>C=C<} \xrightarrow{Br—Br} \text{>}\overset{+}{C}—C—Br + Br^- \longrightarrow —\underset{|}{\overset{Br}{\overset{|}{C}}}—\underset{|}{\overset{|}{C}}—Br$$

There is considerable evidence (particularly stereochemical results, Section 6.8) which suggests that the intermediate cation has the halogen atom partially bonded to both carbon atoms of the original double bond:

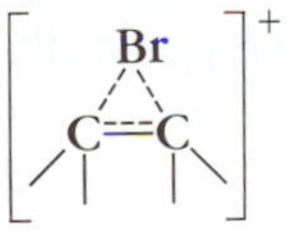

$$\left[\begin{array}{c} Br \\ / \ \backslash \\ C{=}C \end{array}\right]^+$$

This structure would be stabilized relative to a free carbocation by distribution of the positive charge over all three atoms. The bonding interaction between the halogen and the two carbon atoms in the *halonium ion* cannot be fully symmetrical, however, because other mechanistic evidence shows that the intermediate behaves in the manner expected for a *carbocation.* For this reason we will continue to draw the positive charge on carbon, but you should remember that there is an interaction between the carbocation center and the halogen atom.

The addition of halogen to an alkene is illustrated by the following two additions of bromine:

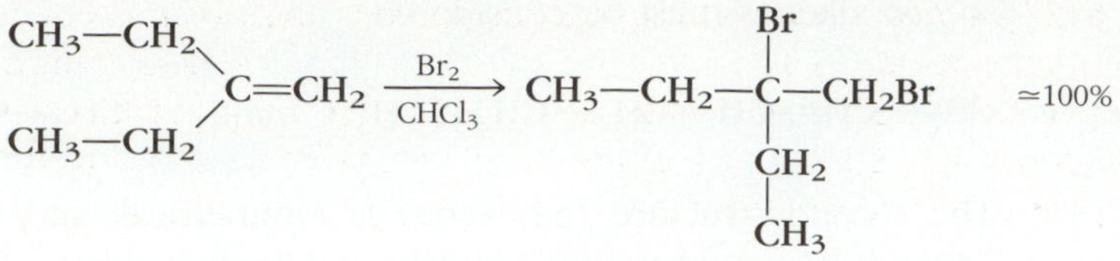

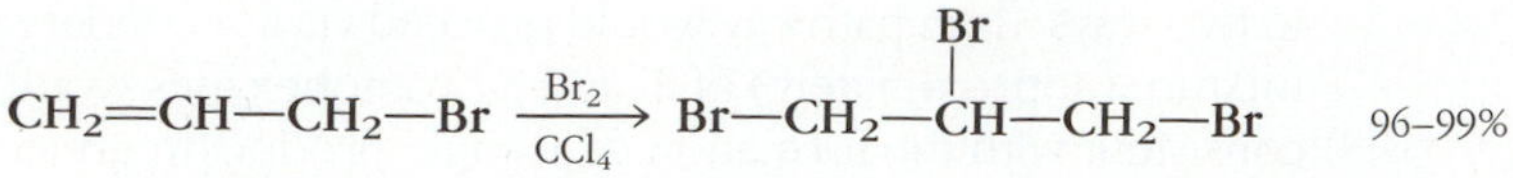

$$(CH_3CH_2)_2C{=}CH_2 \xrightarrow[CHCl_3]{Br_2} CH_3{-}CH_2{-}C(Br)(CH_2CH_3){-}CH_2Br \qquad \approx 100\%$$

$$CH_2{=}CH{-}CH_2{-}Br \xrightarrow[CCl_4]{Br_2} Br{-}CH_2{-}CH(Br){-}CH_2{-}Br \qquad 96\text{–}99\%$$

The reaction of an alkene with a halogen can also be carried out by using a *reactive* solvent. Typically a **hydroxylic** solvent (i.e., one with an OH group) is used, and it acts a nucleophile because of the nonbonding electrons on the oxygen atom. Abbreviating such a solvent as SOH, we can summarize this modification in the following way:

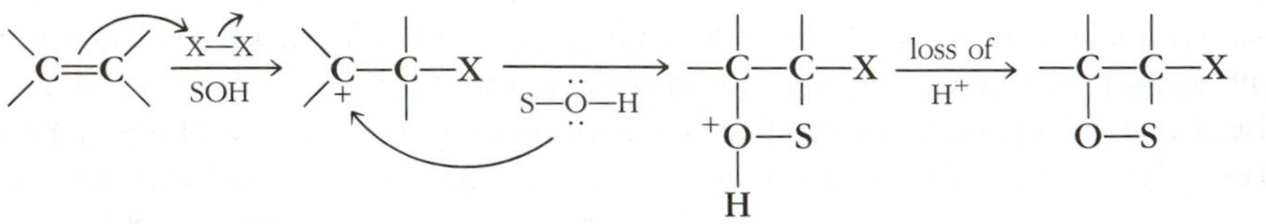

$$>C{=}C< \xrightarrow[SOH]{X{-}X} >\overset{+}{C}{-}C{-}X \xrightarrow{S{-}\ddot{O}{-}H} {-}C({}^{+}O(H){-}S){-}C{-}X \xrightarrow[H^+]{\text{loss of}} {-}C(O{-}S){-}C{-}X$$

The overall process is mechanistically quite similar to the acid-catalyzed additions we discussed previously, except that the first step corresponds to reaction with "X^+" rather than "H^+." The actual reaction is with X_2, and an X^- ion is displaced by the π electrons of the alkene. Solvent attack on the intermediate cation generates an oxonium ion, and proton transfer generates the final product. In some cases, the solvent can first react with the halogen to form a *hypohalite,* e.g.,

$$SO{-}H + Cl_2 \longrightarrow HCl + SO{-}Cl$$

The reaction could then be viewed as simple addition of the latter compound to the alkene, but the reaction would still proceed via attack of *solvent* on the intermediate cation.

The reaction can be carried out by using such solvents as water, alcohols, and acetic acid. The following reaction illustrates the formation of a *bromohydrin* from the reaction of an alkene with bromine in water. (Compounds with adjacent halo and hydroxy substituents are known generally as **halohydrins.**)

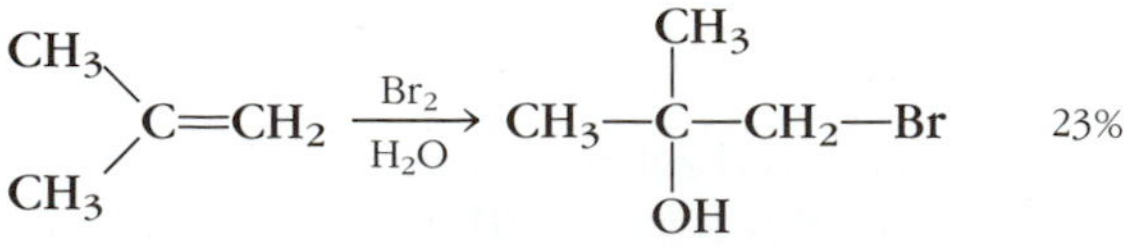

$$(CH_3)_2C{=}CH_2 \xrightarrow[H_2O]{Br_2} CH_3{-}C(CH_3)(OH){-}CH_2{-}Br \qquad 23\%$$

Only one of the two possible constitutional isomers is produced in this overall addition of Br and OH groups to the alkene. This example shows why you must understand the reaction mechanism if you want to predict the products of a reaction. The initial reaction with Br_2 produces a tertiary carbocation, with a bromine attached to the *less* substituted carbon atom of the double bond. Subsequent reaction of the cation with water results in attachment of the hydroxyl substituent at the tertiary center.

As you would expect, there is competition between the two nucleophiles that are present in these reactions, namely the solvent and the halide ion produced in the first step. Usually the higher concentration of the solvent favors its reaction with the carbocation, but the following example shows that the halide ion can sometimes compete very well.

$$CH_3-CH=CH-CH_2-CH_3 \xrightarrow[CH_3-C(=O)-OH]{Cl_2} CH_3-CH(Cl)-CH(O-C(=O)-CH_3)-CH_2-CH_3 + CH_3-CH(O-C(=O)-CH_3)-CH(Cl)-CH_2-CH_3 \quad (46\%)$$

$$+ \; CH_3-CH(Cl)-CH(Cl)-CH_2-CH_3 \quad 52\%$$

For the laboratory synthesis of bromo derivatives, this problem can be alleviated by using the special reagent *N*-bromosuccinimide (NBS). *N*-Bromosuccinimide reacts with the alkene to give the appropriate carbocation, but no bromide ion is formed. The initially formed nitrogen anion becomes protonated and is thereby rendered nonnucleophilic.

$$>C=C< \; + \; Br-N(\text{succinimide}) \longrightarrow >\overset{+}{C}-C(-Br)< \; + \; {}^{-}N(\text{succinimide})$$

Subsequent reaction of the intermediate cation then occurs exclusively with the nucleophilic solvent, and the bromohydrins are obtained in good yield.

$$(CH_3)_2C=CH(CH_3) \xrightarrow[H_2O]{NBS} CH_3-C(CH_3)(OH)-CH(Br)-CH_3 \quad 76\%$$

$$\text{cyclohexene} \xrightarrow[H_2O]{NBS} \text{2-bromocyclohexanol (Br, OH)} \quad 79\%$$

EXERCISE 5.7

In this section we showed specific examples of the acid-catalyzed additions of water and hydroxyl substituted organic compounds to alkenes. For each of these reactions draw the two isomeric carbocations that could be formed by protonation of the alkene. Which carbocation is more stable? Is the product shown in the text derived from the more stable cation?

EXERCISE 5.8

(a) Draw the other constitutional isomer which might have been formed in the reaction of 2-methylpropene with Br_2/H_2O.
(b) Draw the intermediate carbocation which would lead to the compound drawn in (a).
(c) Draw the intermediate carbocation which would lead to the observed product.
(d) Is the observed reaction consistent with the prediction that the reaction will proceed via the more stable cation?

EXERCISE 5.9

Write the reaction mechanism for the addition to propene of:

(a) Br_2 in CCl_4
(b) HCl
(c) Cl_2 in H_2O

5.9 CARBOCATION REARRANGEMENTS

Prior to the 1930s the reaction mechanisms for additions to alkenes were not known. The intermediacy of carbocations was proposed by F. C. Whitmore (Pennsylvania State University), and a key piece of evidence was obtained from the reaction of 3-methyl-1-butene with hydrogen chloride gas:

$$CH_3-CH(CH_3)-CH{=}CH_2 \xrightarrow{HCl} \underset{45\%}{CH_3-CH(CH_3)-CH(Cl)-CH_3} + \underset{45\%}{CH_3-C(CH_3)(Cl)-CH_2-CH_3}$$

In addition to the expected 3-chloro-2-methylbutane, the product mixture contained an equal amount of the isomeric 2-chloro-2-methylbutane. This observation of a *rearranged* product (the chlorine is not bonded to *either* of the carbon atoms of the original double bond) is now explained in the following way: (1) Initial protonation of the double bond (to yield the more stable secondary cation) takes place in the normal manner. However, secondary cations are not very stable, and (2) a *rearrangement of the cation occurs to give a more stable cation.* There is evidence that the rearrangement step may occur prior to complete formation of the carbocation, but the overall result is the same. Finally, (3) the more stable tertiary cation undergoes nucleophilic attack by chloride ion to form the rearranged product.

$$CH_3-CH(CH_3)-CH{=}CH_2 \xrightarrow[\text{protonation}]{H^+} CH_3-CH(CH_3)-\overset{+}{C}H-CH_2-H \xrightarrow{\text{rearrangement}}$$

$$CH_3-\overset{+}{C}(CH_3)-CH_2-CH_3 \xrightarrow[\text{nucleophilic attack}]{Cl^-} CH_3-C(CH_3)(Cl)-CH_2-CH_3$$

Three very important aspects of this (and other) rearrangements should be emphasized. First, the rearrangement itself involves a *1,2-shift,* that is some group (in this case a hydrogen) migrates to the electron-deficient carbon from an *adjacent* position. Second, the group migrates *with its bonding pair of electrons,* thereby eliminating the electron deficiency at the carbon which originally had the positive charge. Of course, a positive charge is generated on the carbon atom from which the migrating group has departed. Third, the rearrangement has resulted in the formation of a *more stable carbocation.*

The migrating group need not be *hydride* (so designated because the hydrogen shifts *with its bonding electrons*). You can see this from the following reaction of HCl with 3,3-dimethyl-1-butene (the nitromethane merely serves as an inert solvent). Only a small part of the product is formed by addition of HCl without rearrangement.

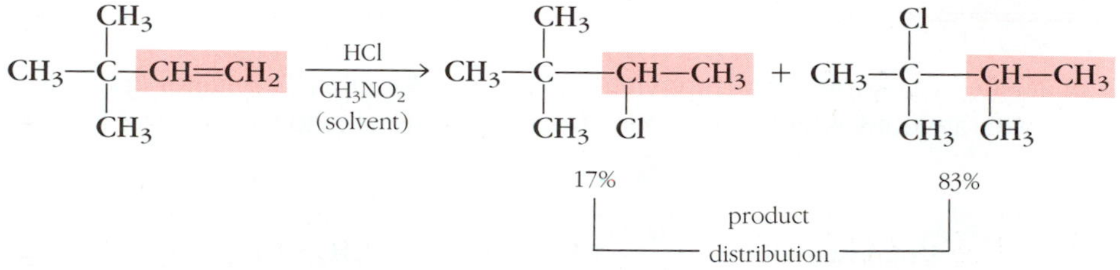

$$(CH_3)_3C\text{—}CH\text{=}CH_2 \xrightarrow[CH_3NO_2\ (\text{solvent})]{HCl} \underset{17\%}{(CH_3)_3C\text{—}CHCl\text{—}CH_3} + \underset{83\%}{(CH_3)_2CCl\text{—}CH(CH_3)\text{—}CH_3}$$

product distribution

Initial protonation again gives the more stable (secondary rather than primary) of the two possible cations, but attack by chloride ion is in competition with rearrangement to a still more stable (tertiary) cation. This rearrangement occurs via a 1,2-methyl shift, and the resulting tertiary cation then undergoes nucleophilic attack by chloride ion.

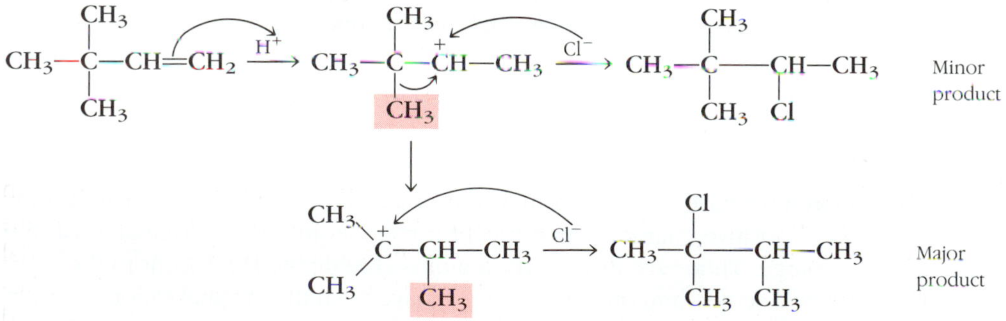

$$(CH_3)_3C\text{—}CH\text{=}CH_2 \xrightarrow{H^+} (CH_3)_3C\text{—}\overset{+}{C}H\text{—}CH_3 \xrightarrow{Cl^-} (CH_3)_3C\text{—}CHCl\text{—}CH_3 \quad \text{Minor product}$$

$$\downarrow$$

$$(CH_3)_2\overset{+}{C}\text{—}CH(CH_3)\text{—}CH_3 \xrightarrow{Cl^-} (CH_3)_2CCl\text{—}CH(CH_3)\text{—}CH_3 \quad \text{Major product}$$

The rearrangement process is not restricted to the migration of hydride or a methyl group but can also involve a 1,2-shift of larger alkyl groups. Even an alkyl group which is part of a ring could migrate:

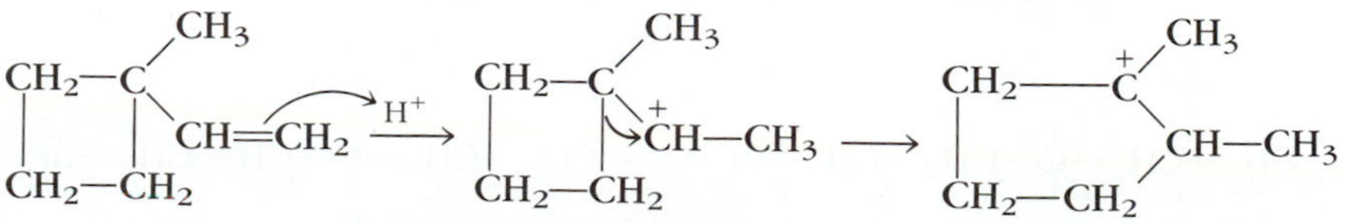

$$\text{1-methyl-1-vinylcyclobutane} \xrightarrow{H^+} \text{1-methylcyclobutyl–}\overset{+}{C}H\text{—}CH_3 \longrightarrow \text{1,2-dimethylcyclopentyl cation } (\overset{+}{C}\text{—}CH_3 \text{ ring carbon, adjacent } CH\text{—}CH_3)$$

In this example the 1,2-alkyl shift not only produces a more stable cation but also converts a four-membered ring to a less strained five-membered ring. Notice that, in sharp contrast to a 1,2-hydride shift, the migration of an alkyl group generates a cation with a *different carbon skeleton.*

It is not always easy to predict when or to what extent carbocation rearrangement will occur; the overall reaction pathway may vary with changes in solvent and other reaction conditions. When will a carbocation rearrange? A useful rule of thumb, and one that you should follow in this study of organic chemistry, is that *a carbocation will rearrange whenever rearrangement will yield a more stable cation.* For most reaction conditions the rearrangement of one carbocation to another of comparable stability is not observed. Remember that certain restrictions still govern the rearrangement process: it must be a 1,2-shift, and the choice of the group that migrates will be determined by the relative stabilities of the cations that would be formed as a result of the rearrangement. In predicting the products of a reaction in which such a rearrangement is possible, it is safest for you to expect a mixture of *both* rearranged *and* unrearranged products.

Example 5.2 An alkene of unknown structure yielded 1-bromo-3-methyl-3-pentanol upon reaction with aqueous NBS. Deduce the structure of the alkene.

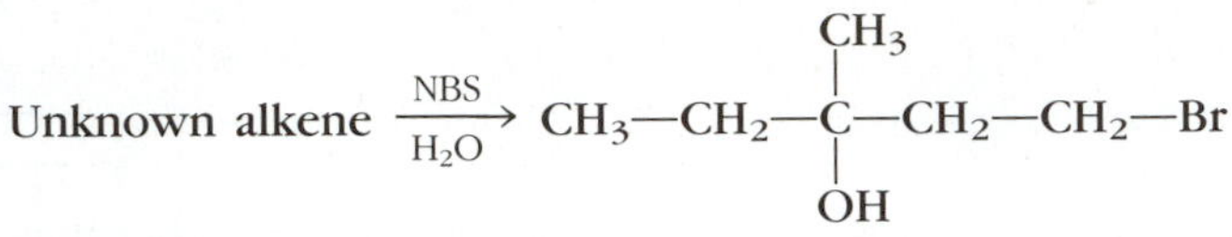

$$\text{Unknown alkene} \xrightarrow[H_2O]{NBS} CH_3-CH_2-\underset{OH}{\overset{CH_3}{C}}-CH_2-CH_2-Br$$

1-Bromo-3-methyl-3-pentanol

Solution The reaction of NBS/H_2O with an alkene would ordinarily result in the introduction of Br and OH groups on *adjacent* carbon atoms. Consequently a rearrangement must have occurred. The final product is derived from attack by water on a carbocation, so the structure of that cation must have been:

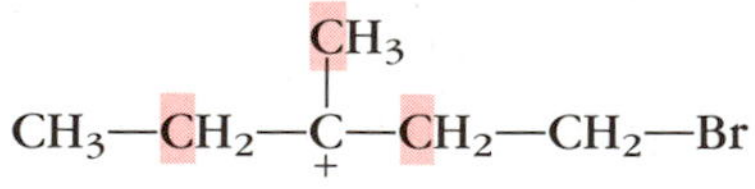

$$CH_3-CH_2-\underset{+}{\overset{CH_3}{C}}-CH_2-CH_2-Br$$

This cation must have resulted from a shift of a group *from* the carbon atom now bearing the positive charge *to* an adjacent carbon atom, which originally had the positive charge. There are three carbon atoms (highlighted) adjacent to the positively charged center, and one of these must have been the original cationic center. Moreover, the original cationic center must have also been adjacent to the carbon atom that is now bonded to the bromine, because the cation was generated by addition of "Br^+" to one end of a carbon-carbon double bond. The only carbon atom which fits these requirements has two hydrogen atoms as its remaining substituents, and one of them must have been the migrating group.

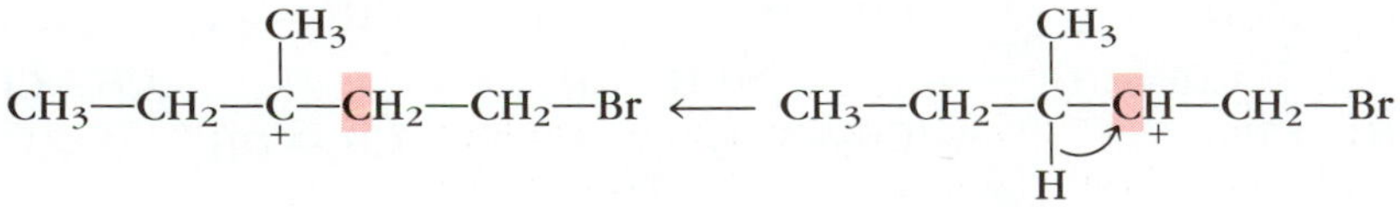

$$CH_3-CH_2-\underset{+}{\overset{CH_3}{C}}-CH_2-CH_2-Br \longleftarrow CH_3-CH_2-\underset{H}{\overset{CH_3}{C}}-\overset{+}{C}H-CH_2-Br$$

You can therefore conclude that the overall reaction must have involved a hydride shift after initial attack on 3-methyl-1-pentene:

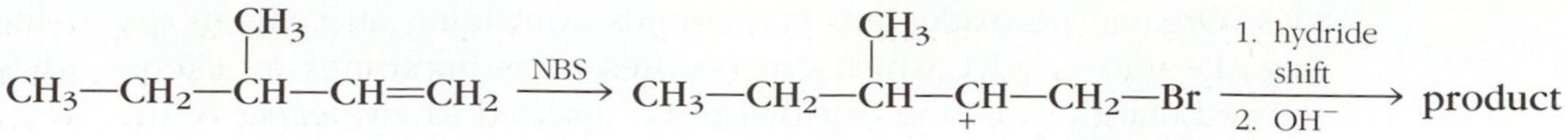

$$CH_3-CH_2-\underset{}{\overset{CH_3}{\overset{|}{C}}H}-CH=CH_2 \xrightarrow{NBS} CH_3-CH_2-\overset{CH_3}{\overset{|}{C}H}-\underset{+}{C}H-CH_2-Br \xrightarrow[2.\ OH^-]{1.\ \text{hydride shift}} \text{product}$$

EXERCISE 5.10

Each of the following cations could conceivably undergo rearrangement by any of three pathways:

(a) a 1,2-hydride shift
(b) a 1,2- methyl shift
(c) a 1,2-ethyl shift

(i)

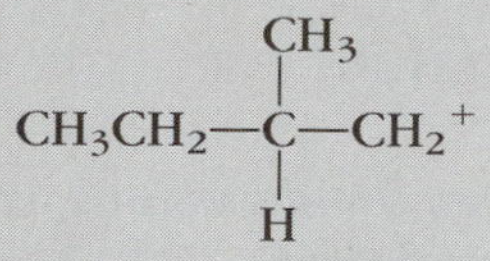

$$CH_3CH_2-\underset{H}{\overset{CH_3}{C}}-CH_2^+$$

(ii)

$$CH_3CH_2-\underset{CH_2CH_3}{\overset{CH_3-CH_2}{C}}-\underset{+}{C}H-\underset{CH_3}{\overset{H}{C}}-H$$

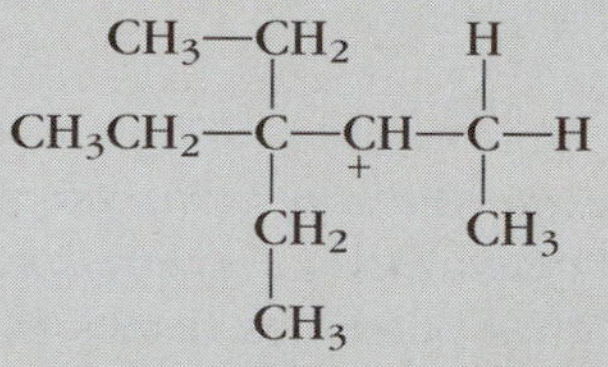

Draw each of the three rearranged cations in each case, and indicate which one of the three processes would be favored.

5.10 FREE RADICAL ADDITION TO ALKENES

In the period prior to the 1930s when the mechanisms of addition reactions of alkenes were not understood, chemists were forced to rely on empirical rules for predicting products of these reactions. The most common rule was that formulated by the Russian chemist V. Markovnikov in 1870. This rule, known as Markovnikov's rule summarizes the regiospecificity of hydrogen halide addition to an alkene in the following way: The hydrogen will become attached to the *less* substituted carbon atom and the halogen will become bonded to the *more* substituted carbon atom of the double bond. Now that we understand the behavior of carbocation intermediates, such empirical rules are not needed, but they were valuable aids to organic chemists for many years.

Development of a satisfactory mechanistic theory for polar addition reactions of alkenes was at first hindered by the occasional observation of "anti-Markovnikov" additions such as the following:

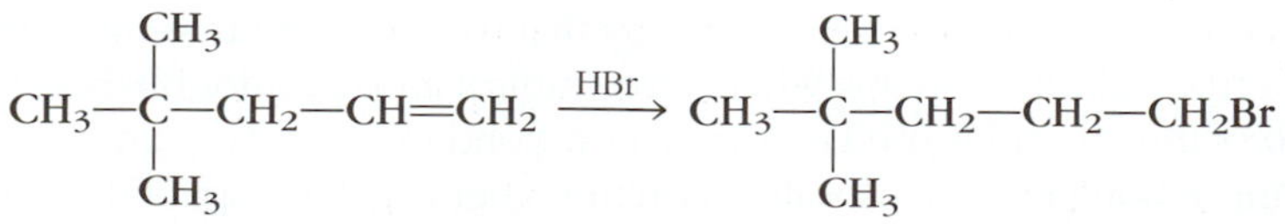

$$CH_3-\underset{CH_3}{\overset{CH_3}{C}}-CH_2-CH=CH_2 \xrightarrow{HBr} CH_3-\underset{CH_3}{\overset{CH_3}{C}}-CH_2-CH_2-CH_2Br \quad 88\%$$

Only the isomer with bromine at the terminal position was observed in this reaction, and none of the "expected" product was formed.

The confusion was resolved by N. Kharasch at the University of Chicago. He demonstrated that this anti-Markovnikov addition was the consequence of a free-radical process caused by exposure to light or by the presence of organic peroxides. Organic peroxides are compounds containing an oxygen-oxygen linkage (e.g., R—O—O—R), which can be present as impurities in alkenes. Kharasch showed that the addition of HBr can be directed to give *either* of the alternative products if the purity of the reagents and the reaction conditions are carefully controlled.

$$CH_3{-}C(CH_3)_2{-}CH_2{-}CH{=}CH_2 \xrightarrow[\text{peroxides}]{HBr} CH_3{-}C(CH_3)_2{-}CH_2{-}CH(H){-}CH_2{-}Br \quad \sim 100\%$$

$$CH_3{-}C(CH_3)_2{-}CH_2{-}CH{=}CH_2 \xrightarrow{HBr} CH_3{-}C(CH_3)_2{-}CH_2{-}CH(Br){-}CH_2{-}H \quad \sim 100\%$$

For all subsequent examples in this textbook you should assume that there are no peroxides present unless they are specifically indicated.

The effect of peroxides results from their decomposition to produce oxygen free radicals, which react with hydrogen bromide to give bromine radicals.

$$R{-}O{-}O{-}R \longrightarrow RO\cdot + \cdot OR$$

$$RO\cdot + HBr \longrightarrow Br\cdot + ROH$$

This initiates a free radical chain reaction, which takes place much more rapidly than the "normal" ionic addition of HBr.

$$Br\cdot + \;\rangle C{=}C\langle \; \longrightarrow Br{-}\overset{|}{\underset{|}{C}}{-}\dot{C}\langle$$

$$Br{-}\overset{|}{\underset{|}{C}}{-}\dot{C}\langle \xrightarrow{H{-}Br} Br{-}\overset{|}{\underset{|}{C}}{-}\overset{|}{\underset{|}{C}}{-}H + Br\cdot$$

The addition mechanism is actually quite similar to that for the ionic addition of HBr except that initial attack is by Br· rather than H^+. Once again, the addition proceeds in the orientation which gives the *more stable* reactive intermediate (in this case a free radical rather than a carbocation). In contrast to reactions of carbocations, however, *free radicals do not undergo rearrangement* to more stable radicals.

There is an important difference in the way that we write reaction mechanisms for ionic and free radical reactions. When a free radical adds to a double bond, only *one electron* of the π bond is used to form a new bond to the radical, and the *other* electron of the π bond becomes localized on the other carbon atom of the original double bond to produce a carbon free radical. When we write reaction mechanisms for processes involving only a single electron, we will use single-barbed arrows (⌒) rather than normal arrows (⌒). This is standard practice in organic chemistry.

Of all the polar addition reactions that we have discussed in this section, *only the reaction with hydrogen bromide is affected by peroxides.* For all the other reactions there is at least one step in the free radical process that would be energetically unfavorable, so it does not compete with ionic addition. Free radical additions to alkenes are important for other reagents, however, particularly in certain industrial processes. We will discuss some of these in Chapter 25.

EXERCISE 5.11

For the addition of hydrogen bromide to 4,4-dimethyl-1-pentene, (a) in the presence of peroxides and (b) in the absence of peroxides, draw the two possible reactive intermediates that could be formed in each case. Which of the two is more stable? Which would actually be generated? What product would be formed?

5.11 HYDROBORATION AND OXYMERCURATION: LABORATORY METHODS FOR THE HYDRATION OF ALKENES

A major role of alkenes, both in the laboratory and in industry, is that of **synthetic intermediates.** Alkenes are used to make many polymers (Chapter 25), and they are also important starting materials for the industrial preparation of alcohols by the acid-catalyzed addition of water. Alcohols are also versatile synthetic intermediates, and organic chemists have expended a great deal of effort to develop laboratory methods to prepare them from alkenes in high yields without rearrangements. The two most successful methods are *hydroboration* and *oxymercuration,* and these have been extensively investigated by H. C. Brown (Purdue University), who received the Nobel prize for chemistry in 1979.

Hydroboration

Hydroboration is a actually a two-step process for conversion of an alkene into an alcohol, and the complete process is often called *hydroboration-oxidation.* First the alkene reacts with borane (BH_3) to form an alkylborane.

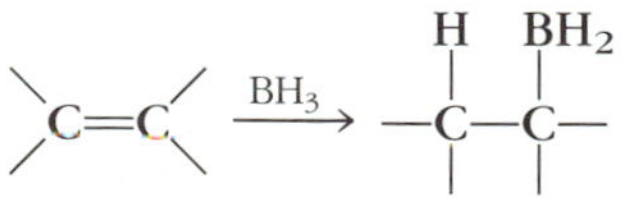

$$\text{>C=C<} \xrightarrow{BH_3} \text{—}\underset{|}{\overset{\text{H}}{\overset{|}{\text{C}}}}\text{—}\underset{|}{\overset{\text{BH}_2}{\overset{|}{\text{C}}}}\text{—}$$

Although this is not shown in this equation, all three hydrogens of the BH_3 are reactive. If an excess of the alkene is present, two more additions occur to produce a trialkylborane:

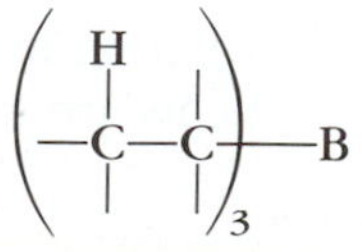

$$\left(\text{—}\underset{|}{\overset{\text{H}}{\overset{|}{\text{C}}}}\text{—}\underset{|}{\overset{|}{\text{C}}}\right)_3\text{—B}$$

In the gas phase borane exists as the dimeric species diborane (B_2H_6), but hydroboration is carried out in solution using an ether (R—O—R) as the solvent. The most common solvent for hydroboration is tetrahydrofuran (THF), and this forms a 1:1 complex with the monomer, BH_3. As you would expect from our discussion of

BF_3 in Section 4.2, BH_3 is a Lewis acid and it can interact with the nonbonding electrons of the ether oxygen.

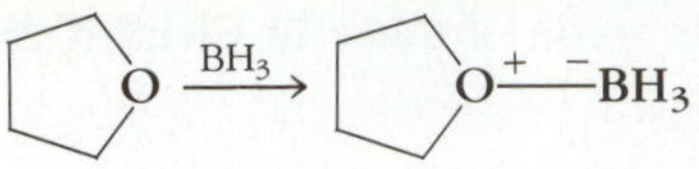

$$\text{THF} \xrightarrow{BH_3} \text{THF}\overset{+}{O}-\overset{-}{B}H_3$$

THF

Hydroboration does not involve carbocation intermediates, but BH_3 interacts with the alkene to generate *partial* positive charge on the *more* substituted carbon. Hydrogen is then transferred (with its bonding electrons) from boron to that carbon. (This process is then repeated two more times to form the trialkylborane.)

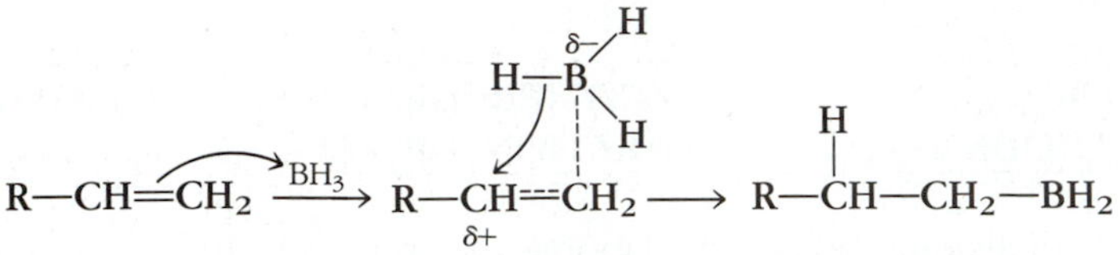

$$R-CH=CH_2 \xrightarrow{BH_3} R-\overset{\delta+}{CH}\cdots CH_2(\cdots BH_3^{\delta-}) \longrightarrow R-CH(H)-CH_2-BH_2$$

In a second step alkaline hydrogen peroxide is added to the trialkylborane. Oxidation of the carbon-boron bond occurs, and that oxidation reaction results in the replacement of boron by an OH group.

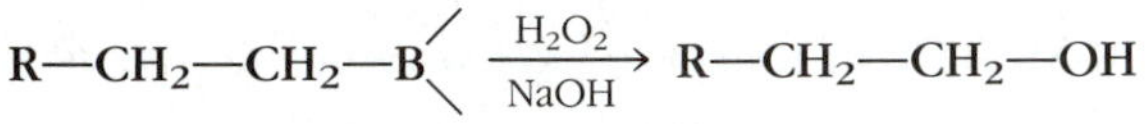

$$R-CH_2-CH_2-B\langle \xrightarrow[NaOH]{H_2O_2} R-CH_2-CH_2-OH$$

In complete contrast to the direct, acid-catalyzed addition of water (Section 5.8), this hydroboration-oxidation sequence affords the *less* substituted alcohol as the major product.

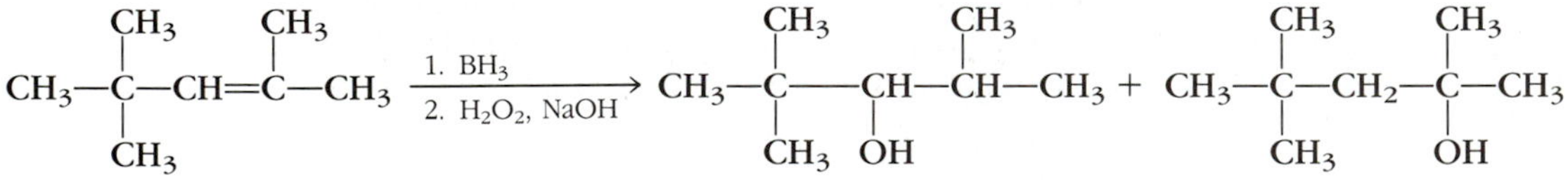

$$CH_3-C(CH_3)_2-CH=C(CH_3)-CH_3 \xrightarrow[2.\ H_2O_2,\ NaOH]{1.\ BH_3} CH_3-C(CH_3)_2-CH(OH)-CH(CH_3)-CH_3 + CH_3-C(CH_3)_2-CH_2-C(CH_3)(OH)-CH_3$$

77% 2%

$$\text{1-methylcyclopentene} \xrightarrow[2.\ H_2O_2,\ NaOH]{1.\ BH_3} \text{2-methylcyclopentanol}$$

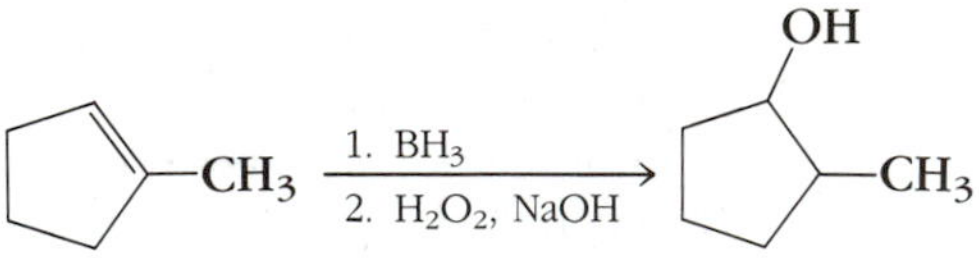

86%

As you would expect for an addition reaction in which the transition state is only slightly polar, carbocation rearrangements are not observed.

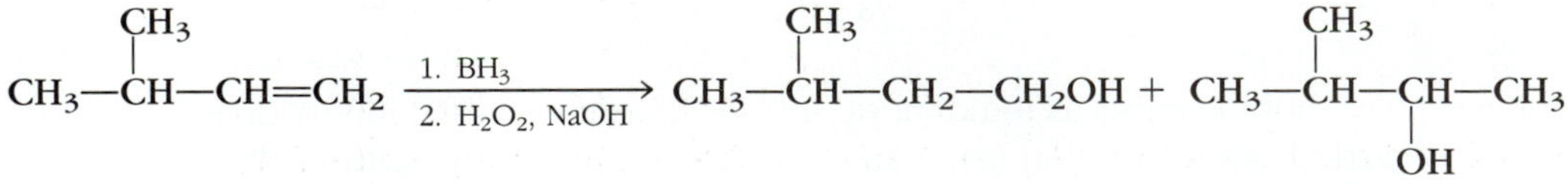

$$CH_3-CH(CH_3)-CH=CH_2 \xrightarrow[2.\ H_2O_2,\ NaOH]{1.\ BH_3} CH_3-CH(CH_3)-CH_2-CH_2OH + CH_3-CH(CH_3)-CH(OH)-CH_3$$

94% 1%

$$(CH_3)_3C{-}CH{=}CH{-}C(CH_3)_3 \xrightarrow[2.\ H_2O_2,\ NaOH]{1.\ BH_3} (CH_3)_3C{-}CH(OH){-}CH_2{-}C(CH_3)_3$$

82%

Oxymercuration

The second common laboratory method for adding water to a carbon-carbon double bond is *oxymercuration,* which also proceeds in two stages. A mercury substituent that is introduced in the first stage is removed in the second, and the overall process is frequently called *oxymercuration-demercuration.* In the first stage the alkene reacts with mercuric acetate, and the cationic center is rapidly attacked by the solvent.

$$\text{>C=C<} \xrightarrow{Hg(OCCH_3)_2} \text{>}\overset{+}{C}{-}C{-}HgOCCH_3 \xrightarrow{H_2\ddot{O}} {-}C(\overset{+}{O}H_2){-}C{-}HgOCCH_3 \xrightarrow{-H^+} {-}C(OH){-}C{-}HgOCCH_3$$

Carbocation rearrangements are not ordinarily observed for oxymercuration, and this suggests that nucleophilic attack by solvent occurs at the developing cationic center while it still has only a *partial* positive charge.

The intermediate organomercury compound is not isolated, and the reaction mixture is treated directly with *sodium borohydride,* $NaBH_4$. This useful *reducing agent* (also developed by H. C. Brown) brings about replacement of the mercury substituent with a hydrogen.

$$-C(OH){-}C{-}HgOCCH_3 \xrightarrow{NaBH_4} -C(OH){-}C{-}H + \text{Hg metal}$$

The net result of the two-step oxymercuration-demercuration process is addition of water to the original double bond. Initial attack by mercuric acetate places partial positive charge on the more substituted carbon atom, so this is the carbon atom to which the OH group becomes attached. Oxymercuration therefore produces alcohols with the regiospecificity that is *opposite* to that found with hydroboration.

The following examples show that the yields of alcohol tend to be quite good, and only small amounts of the isomeric less substituted alcohol are formed.

$$CH_3{-}CH_2{-}CH_2{-}CH{=}CH_2 \xrightarrow[2.\ NaBH_4]{1.\ Hg(OCCH_3)_2,\ H_2O} CH_3{-}CH_2{-}CH_2{-}CH(OH){-}CH_3$$

97%

$$\underset{CH_3CH_2}{\overset{CH_3}{\diagdown}}C{=}C\underset{H}{\overset{H}{\diagup}} \xrightarrow[\text{2. NaBH}_4]{\text{1. Hg}(\overset{O}{\overset{\|}{O}}CCH_3)_2,\ H_2O} \underset{90\%}{CH_3{-}CH_2{-}\underset{OH}{\overset{CH_3}{C}}{-}CH_3} + \underset{<1\%}{CH_3{-}CH_2{-}\underset{H}{\overset{CH_3}{C}}{-}CH_2OH}$$

$$\text{1-methylcyclohexene} \xrightarrow[\text{2. NaBH}_4]{\text{1. Hg}(OCCH_3)_2,\ H_2O} \underset{\sim100\%}{\text{1-methylcyclohexanol (OH, } CH_3\text{)}}$$

The last example also confirms that the reaction normally proceeds without rearrangements.

Solvents other than water can also be used in the reaction with mercuric acetate (and the general reaction with solvent is sometimes called *solvomercuration*).

$$>C{=}C< \xrightarrow{Hg(OCCH_3)_2} >\overset{+}{C}{-}C{-}HgOCCH_3 \xrightarrow{\ddot{R}OH} {-}\underset{|}{C}{-}\underset{|}{C}{-}HgOCCH_3 \text{ (with } \overset{+}{O}(H)R \text{ on C)} \xrightarrow{H^+ \text{ transfer}} {-}\overset{OR}{\underset{|}{C}}{-}\underset{|}{C}{-}HgOCCH_3$$

The use of alcohols as solvents is therefore a useful method for preparing the corresponding ethers.

$$CH_3CH_2CH_2CH_2CH{=}CH_2 \xrightarrow[\text{2. NaBH}_4]{\text{1. Hg}(OCCH_3)_2,\ CH_3{-}\overset{OH}{CH}{-}CH_3} \underset{99\%}{CH_3CH_2CH_2CH_2\overset{CH_3}{CH}{-}O{-}CH(CH_3)_2}$$

$$\underset{CH_3{-}CH_2}{\overset{CH_3}{\diagdown}}C{=}CH_2 \xrightarrow[\text{2. NaBH}_4]{\text{1. Hg}(OCCH_3)_2,\ CH_3OH} \underset{100\%}{CH_3{-}CH_2{-}\underset{OCH_3}{\overset{CH_3}{C}}{-}CH_3}$$

Hydroboration versus Oxymercuration

The hydroboration and oxymercuration sequences both afford alcohols in good yields. Which one is preferable? That depends on the alcohol you want to prepare, i.e., the **target molecule.** If the carbon-carbon double bond is symmetrical, then both procedures will afford the same product. However, each of these hydration procedures is regiospecific; moreover, they are *complementary*. Hydroboration yields the *less* substituted alcohol, while oxymercuration gives the *more* substituted alcohol. Both procedures avoid the carbocation rearrangements that commonly occur in the acid-catalyzed addition of water, and these methods combine to provide chemists with a very powerful method of organic synthesis using alkenes.

EXERCISE 5.12

We showed in this section that hydroboration of 3-methyl-1-butene and of 2,2,5,5,-tetramethyl-3-hexene yields unrearranged alcohols. What products would result from the acid-catalyzed addition of water to these two alkenes?

EXERCISE 5.13

Predict the major product from hydroboration-oxidation of each of the following alkenes:

(a) $CH_3{-}(CH_2)_6{-}CH{=}CH_2$

(b) $CH_3{-}\underset{\displaystyle CH_3}{\overset{}{CH}}{-}CH{=}CH_2$ (with CH_3 on the second carbon)

(c) $CH_3{-}CH{=}CH{-}CH_2{-}CH_3$

EXERCISE 5.14

(a) Not all alkenes afford a single product in hydroboration and oxymercuration reactions. What products would be anticipated for hydroboration-oxidation and for oxymercuration-demercuration of (*Z*)-2-pentene?

(b) What two isomeric alkenes would be expected to produce 3-hexanol

$$CH_3{-}CH_2{-}\underset{\displaystyle OH}{\underset{|}{CH}}{-}CH_2{-}CH_2{-}CH_3$$

as the only product of *both* hydroboration and oxymercuration?

EXERCISE 5.15

What alkene would best yield the following products using the specified method? Write the reactions, including the necessary reagents.

(a) $CH_3{-}\overset{\displaystyle CH_3}{\overset{|}{CH}}{-}\overset{\displaystyle OH}{\overset{|}{CH}}{-}CH_2{-}CH_3$ via hydroboration

(b) $CH_3{-}\overset{\displaystyle CH_3}{\overset{|}{CH}}{-}\underset{\displaystyle OCH_3}{\underset{|}{CH}}{-}CH_3$ via solvomercuration

5.12 OXIDATION OF ALKENES

Alkenes are quite susceptible to oxidation, and they will even react slowly with molecular oxygen if they are exposed to air over a period of time, particularly in the presence of sunlight. Such air oxidation produces peroxides, and that in turn creates a safety hazard because the peroxides are potentially explosive. For this reason alkenes should be stored in the dark in tightly closed containers to exclude air.

In contrast to the air oxidation of alkenes, which is an undesirable event, alkenes undergo a variety of other oxidation reactions to yield useful products. We will first discuss reactions that proceed by *addition* of oxygen to each of the carbon atoms of the double bond, and then we will turn to oxidations that result in *cleavage* of the carbon skeleton.

There are two ways that oxidation of an alkene can occur with addition of an oxygen substituent to each carbon of the double bond. Depending on the reagents employed, the product can be either a 1,2-diol, called a **glycol**

$$\text{>C=C<} \longrightarrow \text{HO—}\underset{|}{\overset{|}{\text{C}}}\text{—}\underset{|}{\overset{|}{\text{C}}}\text{—OH}$$

or an **epoxide,** which is a three-membered ring containing an oxygen atom.

$$\text{>C=C<} \longrightarrow \text{—C(—O—)C—}$$

The strained three-membered ring causes epoxides to be relatively reactive, and they can be hydrolyzed to the 1,2-diols by aqueous acid.

$$\text{—C(—O—)C—} \xrightarrow[H_2O]{H^+} \text{HO—}\underset{|}{\overset{|}{\text{C}}}\text{—}\underset{|}{\overset{|}{\text{C}}}\text{—OH}$$

The direct formation of 1,2-diols can be carried out with catalytic amounts of osmium tetroxide (OsO_4) in the presence of excess chlorate ion or with alkaline potassium permanganate.

$$\text{cyclohexene} \xrightarrow[NaClO_3]{OsO_4} \text{cyclohexane-1,2-diol} \quad 76\%$$

$$\text{cyclooctene} \xrightarrow[NaOH]{KMnO_4} \text{cyclooctane-1,2-diol} \quad 50\%$$

Osmium tetroxide is both expensive and toxic, so potassium permanganate is a superior reagent for addition of two OH groups to a double bond.

The epoxidation of alkenes proceeds smoothly with many organic peroxyacids,

$$R{-}\overset{\overset{\Large O}{\|}}{C}{-}O{-}OH$$

and peracetic acid (in which R = CH_3) has been used successfully, as shown in the reaction of cyclooctene:

$$\text{cyclooctene} \xrightarrow{CH_3\overset{O}{\overset{\|}{C}}{-}OOH} \text{cyclooctene oxide} \qquad 48\%$$

The hydrolysis of epoxides to the diols is illustrated by the following examples:

$$CH_3{-}\underset{CH_3}{\underset{|}{C}}\overset{O}{\frown}CH{-}CH_2Cl \xrightarrow[H_2SO_4]{H_2O} CH_3{-}\underset{CH_3}{\underset{|}{\overset{OH}{\overset{|}{C}}}}{-}\overset{OH}{\overset{|}{C}}{-}CH_2Cl \qquad 95\%$$

$$CH_3{-}CH\overset{O}{\frown}CH{-}CH_2CH_3 \xrightarrow[H_2O]{H_2SO_4} CH_3{-}\overset{OH}{\overset{|}{CH}}{-}\overset{OH}{\overset{|}{CH}}{-}CH_2{-}CH_3 \qquad 81\%$$

We will return to the mechanisms of these two oxidation reactions in Section 6.8.

If the reaction of potassium permanganate with an alkene is not carried out under alkaline conditions, oxidation proceeds past the stage of the diol and cleavage of the carbon chain results. This reaction is difficult to control, so it is not used extensively in preparative organic chemistry. However, a related process, *ozonolysis* (the cleavage of a carbon-carbon double bond with ozone), can be a useful reaction. Ozonolysis proceeds in stepwise fashion, with the initial formation of an adduct of the alkene and ozone called an ozonide, which contains a five-membered ring with three consecutive oxygen atoms. This unstable compound rapidly rearranges to an isomeric ozonide in which there is no longer a bond between the carbon atoms of the original double bond.

$$\rangle C{=}C\langle \xrightarrow{O_3} \text{(C–C bonded ring: C–O–O–O–C)} \longrightarrow \text{(ring: C–O–O–C–O)}$$

No attempt is made to isolate the ozonide, and the reaction mixture is treated directly with a reducing agent such as zinc. The overall result is cleavage of the original alkene into two carbonyl fragments, and for each of these the original double bond to carbon has been replaced by a double bond to oxygen.

$$\rangle C{=}C\langle \xrightarrow{O_3} [\text{ozonide}] \xrightarrow{Zn} \rangle C{=}O + O{=}C\langle$$

The reaction is illustrated by the ozonolysis of 6-methyl-1-heptene.

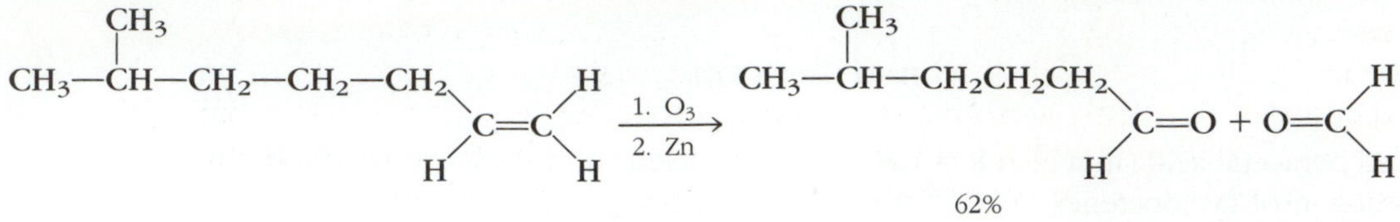

$$CH_3-\underset{\displaystyle CH_3}{CH}-CH_2-CH_2-CH_2-CH{=}CH_2 \xrightarrow[2.\ Zn]{1.\ O_3} CH_3-\underset{\displaystyle CH_3}{CH}-CH_2CH_2CH_2-CH{=}O + O{=}CH_2$$

62%

Ozonolysis has played an important role in helping chemists to determine the structures of many alkenes. Cleavage of the alkene generates two smaller molecules, which can be more easily identified. Once the carbonyl fragments have been identified, you can reconstruct the original alkene on paper by removing the two oxygens and connecting the unsaturated carbon atoms by a double bond.

EXERCISE 5.16

An alkene of unknown structure was subjected to ozonolysis, and the following two carbonyl compounds were obtained.

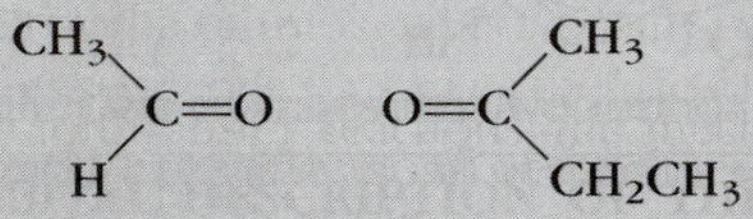

$$CH_3(H)C{=}O \qquad O{=}C(CH_3)CH_2CH_3$$

Deduce the structure of the original alkene.

5.13 PREPARATION OF ALKENES VIA ELIMINATION REACTIONS

Reactions that yield alkenes as products can be divided into two general classes: (1) those which involve *modification of functional groups* without changing the total number of carbon atoms; and (2) those which join two fragments to give a larger molecule via *formation of a carbon-carbon bond.* Reactions of the second type play a particularly important role in the synthesis of organic compounds from commonly available starting compounds.

The most common methods for preparation of alkenes are *elimination reactions,* in which two substituents are "removed" from adjacent positions on a saturated carbon chain.

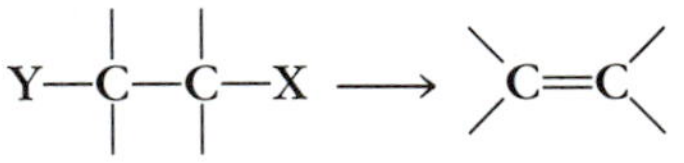

$$Y-\overset{|}{\underset{|}{C}}-\overset{|}{\underset{|}{C}}-X \longrightarrow \;\rangle C{=}C\langle$$

Many variations of this reaction (where X and Y represent different atoms and functional groups) have been used to prepare alkenes. Others will be presented in subsequent chapters, but in this chapter we will discuss only three types of elimination reactions:

1. *Dehydrohalogenation,* where Y and X are hydrogen and halogen
2. *Dehydration* (removal of water), where Y and X are H and OH
3. *Dehalogenation,* where Y and X are both halogen

Dehydrohalogenation

The elimination of hydrogen and halogen from an alkyl halide takes place in the presence of a *base*. The reaction is typically a one-step process, and we can describe the mechanism in the following way. The base abstracts a proton from a carbon *adjacent* to the carbon atom with the halogen substituent, and the electrons from the C—H bond are used to form a carbon-carbon double bond, with concurrent expulsion of a halide ion. We have shown this in the following equation for the case of hydroxide ion as the base and bromine as the halogen:

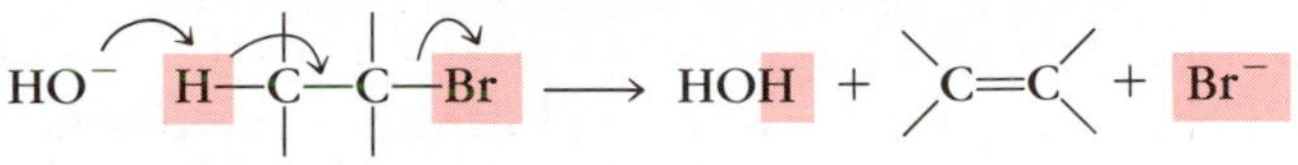

$$HO^- \; H-\overset{|}{\underset{|}{C}}-\overset{|}{\underset{|}{C}}-Br \longrightarrow HOH + \;\rangle C=C \langle\; + Br^-$$

The reaction specifically involves the loss of H and X from adjacent carbon atoms of the starting alkyl halide.

In order to dissolve both the nonpolar alkyl halide and the ionic base, alcohols have commonly been employed as solvents in these reactions, although other polar solvents such as dimethyl sulfoxide (DMSO)* are also useful. The most common bases are KOH, NaOH, and the corresponding conjugate bases of the solvent. The potassium salt of *tert*-butyl alcohol is a particularly useful reagent in these elimination reactions. Although we normally avoid the use of abbreviations in this text, the use of tBuOH and KOtBu for *tert*-butyl alcohol and its salt, potassium *tert*-butoxide, are exceptions. These reagents are used frequently, and it would be cumbersome to draw the full structures in every case.

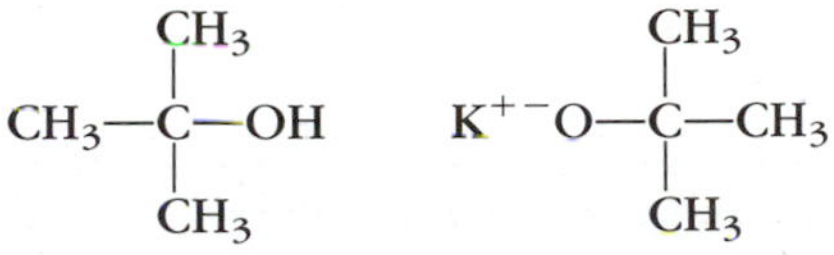

$$CH_3-\overset{CH_3}{\underset{CH_3}{C}}-OH \qquad K^{+\,-}O-\overset{CH_3}{\underset{CH_3}{C}}-CH_3$$

tert-Butyl alcohol — Potassium *tert*-butoxide

The following three examples illustrate the formation of alkenes by reaction of haloalkanes with base.

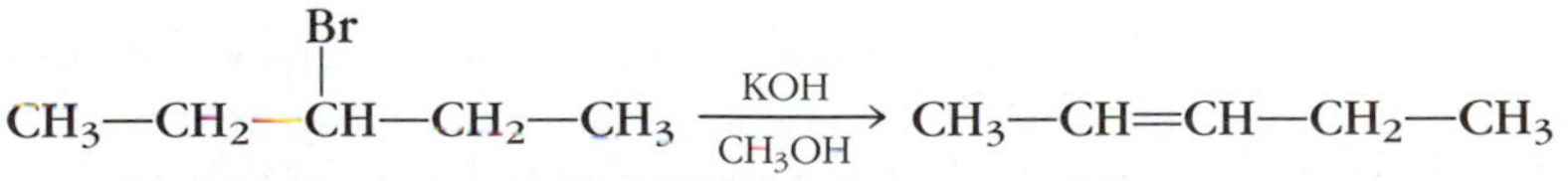

$$CH_3-CH_2-\overset{Br}{\overset{|}{C}H}-CH_2-CH_3 \xrightarrow[CH_3OH]{KOH} CH_3-CH=CH-CH_2-CH_3 \qquad 90\%$$

$$CH_3-(CH_2)_{15}-CH_2-CH_2Br \xrightarrow[tBuOH]{KOtBu} CH_3-(CH_2)_{15}-CH=CH_2 \qquad 85\%$$

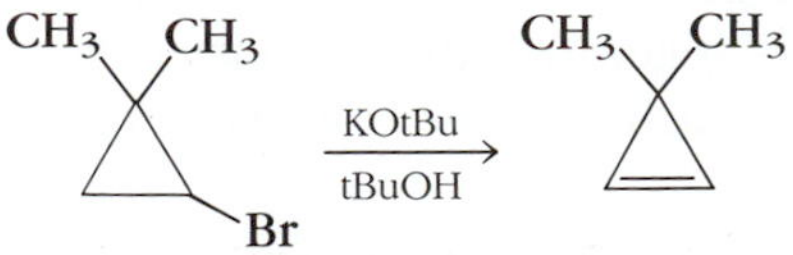

1,1-Dimethyl-2-bromocyclopropane (ring carbon bearing two CH_3 groups; Br on adjacent ring carbon) $\xrightarrow[tBuOH]{KOtBu}$ 3,3-dimethylcyclopropene (ring carbon bearing two CH_3 groups) 81%

---

*DMSO is a polar solvent that has no OH groups. Its structure is quite simple:

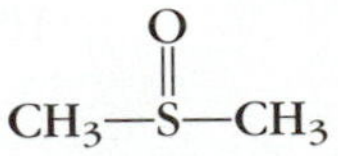

$$CH_3-\overset{O}{\overset{\|}{S}}-CH_3$$

DMSO is a useful solvent for many organic reactions that must be carried out under aprotic conditions (i.e., not even weak acids such as water or an alcohol may be present). It also has some medical uses, although these are somewhat controversial.

When the starting alkyl halide has two or more different C—H groups adjacent to the carbon atom bearing the halogen substituent, a mixture of isomeric products could be formed, but usually the most stable product predominates:

$$CH_3—CH_2—CH_2—CH(Br)—CH_3 \xrightarrow[C_2H_5OH]{KOEt} \underset{38\%}{(CH_3CH_2)HC{=}CH(CH_3)} + \underset{13\%}{(CH_3CH_2)HC{=}CH(CH_3)} + \underset{23\%}{CH_3CH_2CH_2CH{=}CH_2}$$

As you can see from this example, the preference for the most stable isomer is not very large, and variation in experimental conditions such as solvent and temperature, as well as in the base used, can change the product distribution. We will discuss these influences on the product distribution more thoroughly in Chapter 12. In the meantime, when isomeric products could be formed in an elimination reaction, you should assume that the most stable alkene will predominate.

Dehydration

When an alcohol is heated with a strong acid such as sulfuric or phosphoric acid, the elimination of water occurs and the corresponding alkene is formed. The need for acid catalysis is typical in reactions involving the loss of water, because the OH group itself is a poor **leaving group.** By definition, in organic chemistry a leaving group, X, leaves from a carbon atom *with its bonding pair of electrons.*

$$R—X \longrightarrow R^+ + X^-$$

The leaving ability of a group is a directly parallel to its stability. Hydroxide ion is a poor leaving group because it is the conjugate base of the weak acid water. In contrast, a halide ion (the conjugate base of a strong acid) is a good leaving group, as you saw for dehydrohalogenation reactions.

The hydroxide ion may be a poor leaving group, but when the OH group of an alcohol reacts with a strong acid, a protonated alcohol is formed. This contains the exceptionally good leaving group *water.* Loss of water from the protonated alcohol produces a *carbonation,* and even a weak base such as HSO_4^- can act as a proton acceptor to generate the alkene.

$$H—C—C—\ddot{O}H \xrightleftharpoons{H_2SO_4} H—C—C—\overset{+}{\ddot{O}}H_2 \rightleftharpoons H—C—C^+ \;(HSO_4^-) \rightleftharpoons \;\; C{=}C$$

In some cases the protonated alcohol may react with a base by a mechanism similar to that for dehydrohalogenation:

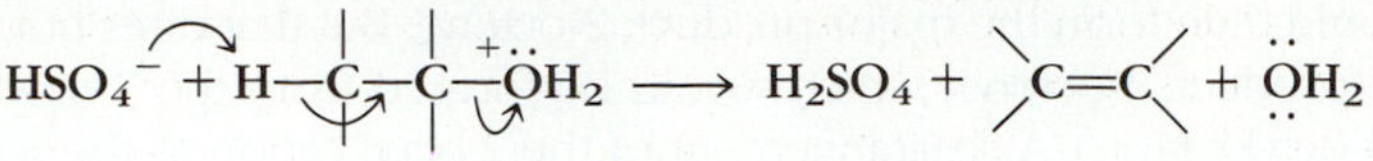

$$HSO_4^- + H{-}\overset{|}{\underset{|}{C}}{-}\overset{|}{\underset{|}{C}}{-}\overset{+}{O}H_2 \longrightarrow H_2SO_4 + {>}C{=}C{<} + OH_2$$

But we will assume that carbocations are formed as intermediates in acid-catalyzed dehydration reactions.

The dehydration of an alcohol is actually a *reversible reaction,* and you will recall that we discussed acid-catalyzed *addition* of water to an alkene in Section 5.8. Because the reaction is reversible and because acid can cause unwanted reactions, the alkene produced in an acid-catalyzed dehydration typically is removed from the reaction mixture by distillation to prevent further reaction. The following examples show that acid-catalyzed dehydration of alcohols can afford alkenes in good yield.

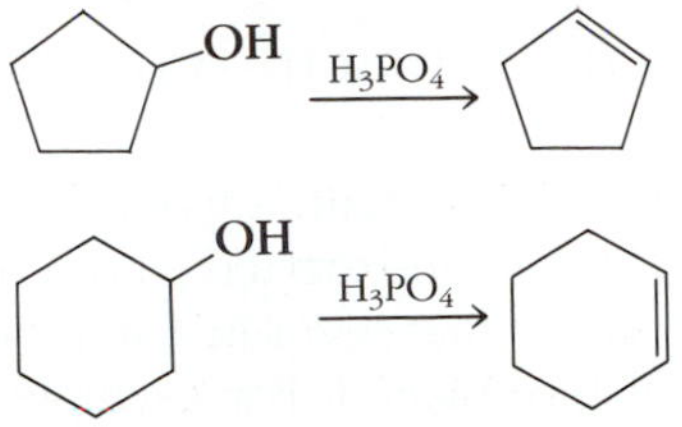

Cyclopentanol $\xrightarrow{H_3PO_4}$ cyclopentene 90%

Cyclohexanol $\xrightarrow{H_3PO_4}$ cyclohexene 79–84%

Carbocations are formed as intermediates in dehydration reactions, so you should expect to observe rearrangements when they are possible. Treatment of 3,3-dimethyl-2-butanol with acid affords a product mixture containing only a small fraction of unrearranged product; the other two alkenes both result from a 1,2-methyl shift. As we indicated earlier, the isomer formed in greatest yield is the more stable tetrasubstituted alkene.

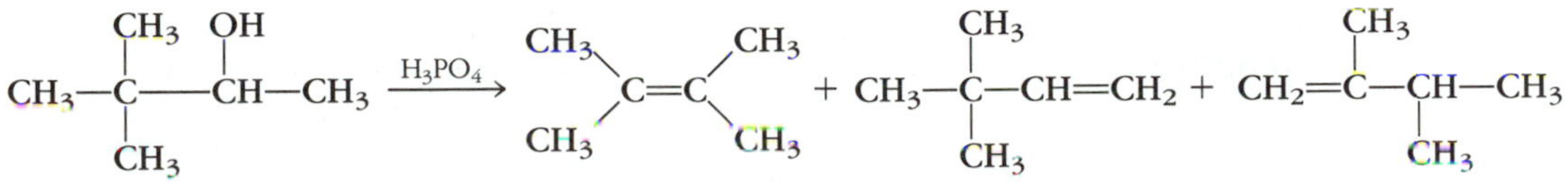

$$CH_3{-}C(CH_3)_2{-}CH(OH){-}CH_3 \xrightarrow{H_3PO_4} (CH_3)_2C{=}C(CH_3)_2 + CH_3{-}C(CH_3)_2{-}CH{=}CH_2 + CH_2{=}C(CH_3){-}CH(CH_3){-}CH_3$$

48% 2% 24%

The reversibility of acid-catalyzed dehydration introduces the additional complication that isomerization of the initial product can occur. For example, the dehydration of 1-octanol affords 2-octene as the major product, along with substantial amounts of 3-octene.

$$CH_3{-}CH_2{-}CH_2{-}CH_2{-}CH_2{-}CH_2{-}CH_2{-}CH_2OH \xrightarrow{H_3PO_4}$$

$$CH_3{-}CH_2{-}CH_2{-}CH_2{-}CH_2{-}CH{=}CH{-}CH_3$$

(major product)

$$+\ CH_3{-}CH_2{-}CH_2{-}CH_2{-}CH{=}CH{-}CH_2{-}CH_3$$

(minor product)

How are these products formed? Protonation of the alcohol followed by loss of water would generate the 1-octyl cation, and you would expect that to rearrange to the more stable 2-octyl cation via a 1,2-hydride shift. Loss of H^+ from the adjacent carbon would then form the major product, 2-octene. But that does not account for the minor product, 3-octene, which would be formed from the 3-octyl cation (or even the 4-octyl cation). A rearrangement of the 2-octyl cation to the 3-octyl cation is unlikely because they are both secondary.

A more reasonable explanation for the formation of 3-octene involves the reversibility of the overall reaction. Any 2-octene that is formed could be protonated to regenerate the 2-octyl cation from which it was formed, but a second alternative exists as well. Actually, protonation of 2-octene could yield *either* of two isomeric secondary cations.

$$CH_3-CH_2-CH_2-CH_2-CH_2-CH{=}CH-CH_3 \underset{}{\overset{H^+}{\rightleftharpoons}} CH_3-CH_2-CH_2-CH_2-CH_2-CH_2-\overset{+}{C}H-CH_3$$

$$CH_3-CH_2-CH_2-CH_2-CH_2-CH{=}CH-CH_3 \underset{}{\overset{H^+}{\rightleftharpoons}} CH_3-CH_2-CH_2-CH_2-CH_2-\overset{+}{C}H-CH_2-CH_3$$

Loss of H^+ from C-2 of the second of these cations would again yield 2-octene, but loss of H^+ from C-4 would afford the "unexpected" 3-octene.

In order to avoid the isomerizations that occur in acid-catalyzed dehydrations, a variety of alternatives have been investigated. For example, dehydration can be carried out by high-temperature treatment with aluminum oxide, but this reaction is limited to relatively low-boiling alcohols. None of the other methods for direct dehydration of alcohols is completely free of problems, and elimination under alkaline conditions (i.e., dehydrohalogenation, in which no carbocations are formed) remains the method of choice for preparing an alkene. (In Section 11.9 we will show you how to convert an alcohol into the corresponding halide so that such an elimination reaction could be carried out.)

Dehalogenation

The treatment of a vicinal dibromide with a reducing agent, particularly zinc metal, results in the formation of the alkene.

$$Br-\overset{|}{\underset{|}{C}}-\overset{|}{\underset{|}{C}}-Br \xrightarrow{:Zn} \rangle C{=}C\langle + ZnBr_2$$

This reaction is of limited synthetic value because preparation of the dibromide requires the alkene as a starting material. The reaction has been used in the purification of unsaturated compounds by *temporary* conversion to the dibromide, followed by regeneration of the double bond by treatment with zinc. Sometimes the dibromide can be purified more easily than the alkene. In such a case the pure alkene could be obtained by the indirect procedure of addition of bromine, purification of the dibromide, and subsequent debromination with zinc. Such a two-step bromination-debromination sequence is shown in the following equation.

$$CH_3-\underset{CH_3}{\overset{CH_3}{C}}-CH_2-CH{=}CH_2 \xrightarrow{Br_2} \underset{85\%}{CH_3-\underset{CH_3}{\overset{CH_3}{C}}-CH_2-\underset{Br}{CH}-CH_2Br} \xrightarrow{Zn} \underset{91\%}{CH_3-\underset{CH_3}{\overset{CH_3}{C}}-CH_2-CH{=}CH_2}$$

EXERCISE 5.17

Draw the carbocation that would be formed by loss of water after protonation of 3,3-dimethyl-2-butanol. What rearranged cation would result from a 1,2-methyl migration? What alkenes would be formed from this cation? Are these the same alkenes that are actually formed in the reaction with H_3PO_4, which we showed in this section?

EXERCISE 5.18

What isomeric alkenes would be formed from the treatment of 2-bromopentane with KOH in ethanol? On the basis of their stabilities, which alkene should be the major product?

EXERCISE 5.19

Write the mechanism for the elimination reaction of 2-bromopropane with: (a) KOtBu, tBuOH; and (b) KOH, ethanol.

5.14 PREPARATION OF ALKENES VIA CARBON-CARBON BOND-FORMING REACTIONS

The preparation of an organic compound, whether in industry or in the research laboratory, must usually be carried out by building up the desired molecule from smaller, readily available fragments. This almost always requires the formation of one or more carbon-carbon bonds, so carbon-carbon bond-forming reactions are extremely important in synthetic chemistry. As you proceed through this text, you will also discover that the number of available reactions for carbon-carbon bond formation is quite limited. We will therefore place special emphasis on each new reaction of this type when we first introduce it.

There are several carbon-carbon bond-forming reactions that afford alkenes directly, and they generally involve the treatment of an alkyl halide with an organometallic reagent. The most versatile reaction employs the organocuprate reagents that we first discussed in Section 1.11, and the coupling reaction works very well even if one of the groups to be coupled contains a double bond. In the following example the reactive centers of both the iodide and the organocuprate are saturated carbons, but the organometallic reagent also contains a double bond. The product of the reaction is therefore an *alkene.*

$$CH_3{-}CH_2{-}CH_2{-}CH_2{-}CH_2{-}I \xrightarrow{(CH_2{=}CH{-}CH_2)_2CuLi} CH_3{-}CH_2{-}CH_2{-}CH_2{-}CH_2{-}CH_2{-}CH{=}CH_2 \quad 98\%$$

Moreover, either of the two reactants can contain an *alkenyl* group, i.e., a group in which the substituent is bonded directly to an sp^2 carbon atom.

$$\text{4-methylcyclohexyl bromide} \xrightarrow{(H_2C{=}C(CH_3){-})_2CuLi} H_2C{=}C(CH_3){-}(\text{4-methylcyclohexyl}) \quad 80\%$$

$$\text{1-bromocyclohexene} \xrightarrow{(CH_3-CH_2-CH_2-CH_2)_2CuLi} \text{1-butylcyclohexene } (CH_2-CH_2-CH_2-CH_3) \quad 80\%$$

Although we indicated in Section 1.11 that the reaction of organocuprates with secondary alkyl halides is usually unreliable, *cyclic* alkyl halides such as 1-bromo-4-methylcyclohexane (in the first of the preceding examples) appear to give satisfactory results. When *alkenyl* halides are employed in the organocuprate coupling reaction, there is no restriction on the degree of substitution of the sp^2 carbon to which the halogen is attached.

The preparation of an organocopper reagent from an alkenyl halide is carried out in the same way that we described in Section 1.11 for alkyl halides. The alkenyl halide is treated successively with lithium metal and with cuprous iodide, and the resulting organometallic reagent is allowed to react with an alkyl halide.

$$(E)\text{-}CH_3CH{=}CHCl \xrightarrow[2.\ CuI]{1.\ Li} \left((E)\text{-}CH_3CH{=}CH\right)_2CuLi \xrightarrow{CH_3-(CH_2)_6-CH_2-I}$$

$$(E)\text{-}CH_3CH{=}CH-CH_2-(CH_2)_6-CH_3 \quad + 4\%\ (Z)\text{-2-undecene}$$

(*E*)-2-undecene
86–89%

This example further demonstrates that the reaction of a lithium dialkenylcuprate and an alkyl halide proceeds with *retention of configuration* at the carbon-carbon double bond. The stereochemical relationship *(trans)* between the chlorine and the methyl group of the starting alkene is maintained for the methyl group and the new alkyl group that is introduced in the reaction. In the example we are considering here, (*E*)-1-chloropropene affords almost exclusively (*E*)-2-undecene. (You should be careful not to jump to the incorrect conclusion that the *E*/*Z* descriptor for nomenclature will always remain unchanged, because the relative priorities of groups on the double bond could change in the course of the reaction. (See Problem 5.15 at the end of the chapter.)

The reactions we have presented in this section greatly expand the synthetic procedures that you learned in Chapter 1. When a carbon-carbon bond-forming reaction is used to prepare an alkane, this usually represents the *end* of a synthetic sequence because the alkane has no functional groups and is basically unreactive. In contrast, the synthesis of an alkene by means of an organocopper reagent provides a molecule which still contains a reactive functional group, a carbon-carbon double bond. You could add a hydrogen halide to the double bond to produce an alkyl halide, or you could add Cl_2 or Br_2 to give a dihalide. Dehydrohalogenation of the latter with base would yield an alkenyl halide. Either an alkyl halide or an alkenyl halide would be a useful reagent for further synthetic operations.

Example 5.3 Suggest a method for the preparation of 3-butyl-2-methyl-2-heptene using starting materials that contain four or fewer carbon atoms.

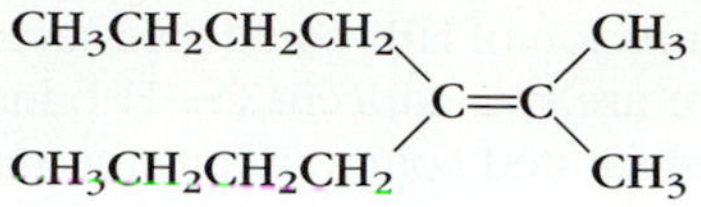

$$(CH_3CH_2CH_2CH_2)_2C{=}C(CH_3)_2$$

Solution Using the restraint that any starting compound can have no more than four carbon atoms, we can visualize the target molecule as consisting of three four-carbon fragments which were joined in some reaction sequence.

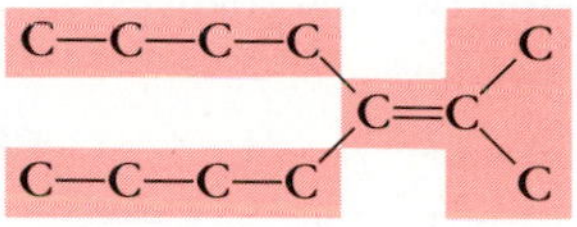

C—C—C—C, C—C—C—C, C=C, C, C

If we attempt to use these three four-carbon fragments as our starting materials, the bonds which join them in the final product would have to be the carbon-carbon bonds formed during the synthesis. Both bonds are to butyl groups, so the last reaction must have resulted from a coupling reaction starting from 1-bromobutane and 3-bromo-2-methyl-2-heptene. (The choice of bromine is arbitrary.)

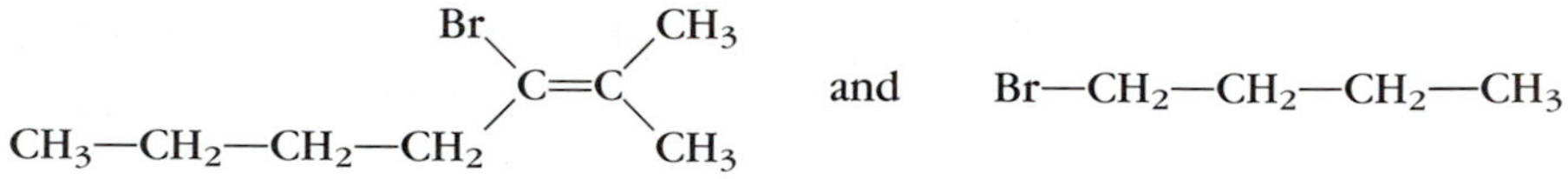

$$CH_3{-}CH_2{-}CH_2{-}CH_2(Br)C{=}C(CH_3)_2 \quad \text{and} \quad Br{-}CH_2{-}CH_2{-}CH_2{-}CH_3$$

In principle either of these halides could be used to prepare the organocopper reagent. In practice, however, an excess of the organocopper reagent is ordinarily employed in these reactions, and it would be wasteful to use an excess of the eight-carbon compound that must be prepared in another synthetic sequence. A more efficient approach would be to use an excess of a four-carbon reagent which you can obtain at no cost (according to the rules of this problem). The following reactions would therefore be appropriate for the final stages of the synthesis:

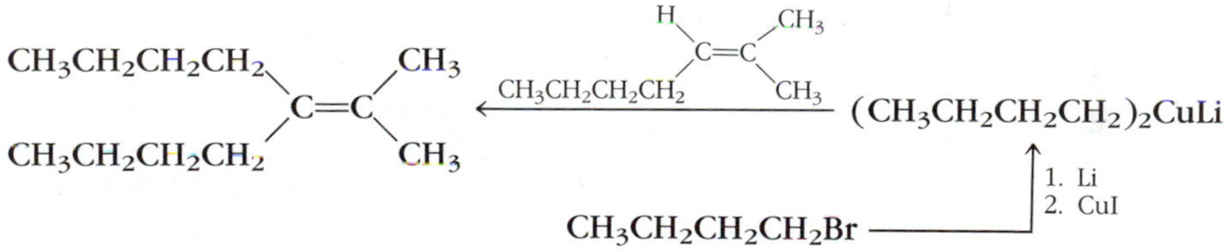

$$CH_3CH_2CH_2CH_2Br \xrightarrow{1.\ Li;\ 2.\ CuI} (CH_3CH_2CH_2CH_2)_2CuLi \xrightarrow{CH_3CH_2CH_2CH_2(Br)C{=}C(CH_3)_2} (CH_3CH_2CH_2CH_2)_2C{=}C(CH_3)_2$$

Completion of the problem still requires the preparation of 3-bromo-2-methyl-2-heptene. This may be difficult for you to see at first, but the alkenyl bromide could be prepared by elimination of HBr following addition of Br_2 to the parent alkene.

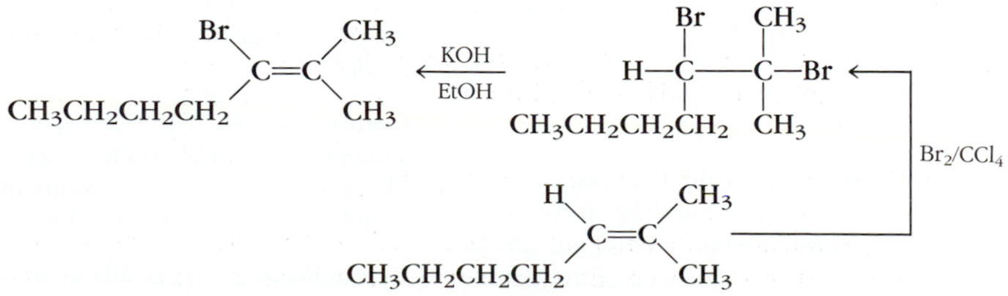

$$CH_3CH_2CH_2CH_2(H)C{=}C(CH_3)_2 \xrightarrow{Br_2/CCl_4} CH_3CH_2CH_2CH_2CH(Br){-}C(Br)(CH_3)_2 \xrightarrow{KOH,\ EtOH} CH_3CH_2CH_2CH_2(Br)C{=}C(CH_3)_2$$

The elimination of HBr could give several products (there are two bromine atoms and there are two adjacent C—H bonds for each), but we have drawn the most highly substituted compound as the major product. (You will see in Sections 7.8 and 7.9 why the second bromine is not eliminated as well.)

Preparation of 2-methyl-2-heptene could be accomplished by starting with the two four-carbon halides depicted earlier (again, we have arbitrarily chosen the bromides). *Either* of them could be converted to the copper derivative, and a coupling reaction would yield the necessary alkene. The following equation shows just one of the two alternatives:

$$CH_3—CH_2—CH_2—CH_2(H)C{=}C(CH_3)_2 \xleftarrow{(H)(Br)C{=}C(CH_3)_2} (CH_3CH_2CH_2CH_2)_2CuLi \xleftarrow{1.\ Li;\ 2.\ CuI} CH_3CH_2CH_2CH_2Br$$

Example 5.3 demonstrates that you could design fairly complex synthetic sequences using only the few reactions that we have introduced in Chapters 1 and 5. As you continue in your study of organic chemistry, you will learn additional ways to form carbon-carbon bonds and to modify existing functional groups so that a large variety of organic compounds could be synthesized.

EXERCISE 5.20

In Example 5.3 we showed only a single product for the elimination of HBr from 2,3-dibromo-2-methylheptane.

(a) Write the reaction mechanism for formation of this compound.
(b) Draw the other products which could have been formed by elimination of a single molecule of HBr. Rank all these in order of stability.

EXERCISE 5.21

In Example 5.3 we showed only one of the two ways to prepare 2-methyl-2-heptene from the four-carbon bromides. Show the second way.

5.15 TERMS AND DEFINITIONS

Alkene. A hydrocarbon with the general formula C_nH_{2n}, characterized by the presence of a carbon-carbon double bond.

Bonding axis. The imaginary line connecting two nuclei that are bonded to each other.

Cis. A term describing a relationship between two substituents on a ring or double bond. The substituents (which must be on different carbon atoms) are on the same side of the ring or on the same side of the double bond.

Concerted. A term describing a process in which the reactants proceed through a single transition state to give the product directly, without the formation of any intermediate.

Configuration. The fixed spatial arrangement of substituents on a double bond; two configurations are possible whenever the two substituents at each end of the double bond are nonequivalent.

Cycloalkene. A cyclic alkene in which the double bond is specifically contained within the ring.

Double bond. The combination of a π bond and a σ bond between two atoms; the carbon-carbon double bond is the characteristic functional group of an alkene.

E. Entgegen (the German word for "opposite"). This term defines the configuration of a double bond in which the two substituents of highest priority are on opposite sides of the double bond.

Electrophilic. Electron seeking; the interaction of a Lewis acid with the double bond of an alkene (or with some other electron donor) is described as electrophilic attack on the double bond.

Epoxide. A three-membered ring with two sp^3 carbon atoms and an oxygen atom.

Functional class. A class of compounds characterized by a particular type of reactivity; the members of a functional class all contain a specific functional group.

Functional group. A structural feature, such as a carbon-carbon double bond, that imparts a uniform type of reactivity to molecules of a particular functional class.

Halohydrin. A compound with adjacent halo and hydroxy substituents.

Hydroxylic. A term describing a solvent that contains OH (i.e., hydroxyl) groups.

Leaving group. A group that departs from a carbon atom with its bonding pair of electrons during the course of a reaction.

π Molecular orbital. A molecular orbital formed by the interaction of two parallel atomic *p* orbitals.

σ Molecular orbital. A molecular orbital formed by the interaction of two atomic *s* orbitals or two hybrid orbitals (*s* and *p*) that are directed toward each other.

Principal chain. The longest carbon chain in a molecule that contains the functional group used to name the compound; for an alkene this is the longest carbon chain that contains the double bond.

Regiospecific. Describes a reaction, such as addition to a double bond, for which there is a specific correlation between the structure of the reactant and the site of the addition.

Saturated. Describes an sp^3 carbon that is bonded to the maximum number of possible substituents; all four substituents are connected via single bonds.

Sigma bond. A single bond, formed when electrons occupy a bonding σ orbital.

Single bond. A sigma bond; a normal two-electron bond between two atoms.

Target molecule. The compound that is the immediate goal in a synthetic sequence.

Trans. A term describing a relationship between two substituents on a ring or on a double bond in which the substituents (which must be on different carbon atoms) are on opposite sides of the ring or double bond.

Trigonal. The geometry of a trisubstituted atom (such as a doubly bonded carbon) in which the three substituents and the central atom all lie in a plane; the three substituents are separated by angles of approximately 120°.

Unsaturated. Describes a molecule that contains atoms with less than the maximum number of substituents; for example, an alkene (in which the doubly bonded carbon atoms have only three substituents).

Vicinal. A term describing the relationship between two substituents that are bonded to adjacent carbon atoms.

Z. Zusammen (the German word for "together"); defines the configuration of a double bond in which the two substituents of highest priority are on the same side of the double bond.

5.16 SUMMARY OF REACTIONS

Table 5.1 summarizes the reactions of alkenes, and Table 5.2 summarizes the reactions used to prepare them.

TABLE 5.1 Reactions of Alkenes

Reaction	Comments
1. Reduction: catalytic hydrogenation $\mathrm{>C{=}C<} \xrightarrow[\text{cat.}]{H_2} \mathrm{H{-}\overset{|}{\underset{|}{C}}{-}\overset{|}{\underset{|}{C}}{-}H}$	Section 5.6. Either Pt or Pd/C is a satisfactory catalyst.

TABLE 5.1 Reactions of Alkenes (continued)

Reaction	Comments
2. Addition of carbenes	Section 5.7.
$>C=C< \xrightarrow{CH_2I_2,\ ZnCu}$ cyclopropane (C with H, H bridging C—C)	Simmons-Smith reaction.
$>C=C< \xrightarrow{CHX_3,\ base}$ cyclopropane (C with X, X bridging C—C)	X is usually Cl or Br; KOH and KOtBu are common bases.
3. Polar addition reactions	Section 5.8.
(a) Acid-mediated additions $>C=C< \xrightarrow[H^+]{HX} H—C—C—X$	Carbocation rearrangements may occur. No H^+ catalyst needed if HX itself a strong acid.
(b) Halogenation: inert solvent $>C=C< \xrightarrow{X_2} X—C—C—X$	Carbocation rearrangements may occur.
(c) Halogenation: nucleophilic solvent $>C=C< \xrightarrow[SOH]{X_2} X—C—C—OS$	Carbocation rearrangements may occur. NBS/H_2O is a good synthetic procedure for adding Br, OH.
4. Other addition reactions	
(a) Hydration via hydroboration $>C=C< \xrightarrow[2.\ NaOH,\ H_2O_2]{1.\ BH_3} H—C—C—OH$	Section 5.11. Yields the less substituted alcohol.
(b) Hydration via oxymercuration $>C=C< \xrightarrow[2.\ NaBH_4]{1.\ Hg(OCOCH_3)_2,\ H_2O} H—C—C—OH$	Section 5.11. Yields the more substituted alcohol. Use of alcohols (ROH) as solvent in place of water allows synthesis of ether by addition of H, OR.
(c) Free-radical addition of HBr $>C=C< \xrightarrow[peroxides]{HBr} H—C—C—Br$	Section 5.10. Affords the less substituted bromide. Not applicable to other hydrogen halides.

TABLE 5.1 Reactions of Alkenes (continued)

Reaction	Comments
5. Oxidation (a) Addition of OH groups $\rangle C{=}C\langle \xrightarrow[\text{or alkaline } KMnO_4]{OsO_4,\ NaClO_3} HO-\overset{\vert}{\underset{\vert}{C}}-\overset{\vert}{\underset{\vert}{C}}-OH$	Section 5.12. Formation of glycols. The high toxicity of osmium compounds makes potassium permanganate the reagent of choice.
(b) Epoxidation $\rangle C{=}C\langle \xrightarrow{CH_3-\overset{O}{\overset{\Vert}{C}}-OOH}$ epoxide ($C-C$ bridged by O)	Hydrolysis of the epoxide with aqueous acid provides an alternative preparation of glycols.
(c) Ozonolysis $\rangle C{=}C\langle \xrightarrow[2.\ Zn]{1.\ O_3} \rangle C{=}O + O{=}C\langle$	Oxidative cleavage yields aldehydes and ketones with carbon skeletons corresponding to the two "halves" of the alkene. Good for structure proof.
(d) Allylic oxidation $\rangle C{=}C\langle_{CH\langle} \xrightarrow{NBS} \rangle C{=}C\langle_{C(Br)\langle}$	Section 8.4.

TABLE 5.2 Preparation of Alkenes

Reaction	Comments
Functional Group Modification: No Change in Carbon Skeleton	
1. Base-mediated elimination $H-\overset{\vert}{\underset{\vert}{C}}-\overset{\vert}{\underset{\vert}{C}}-X \xrightarrow{\text{base}} \rangle C{=}C\langle$	Section 5.13. X must be a good leaving group, i.e., HX must be a strong acid. Common bases are sodium and potassium salts of water and of alcohols.
2. Dehydration of alcohols $H-\overset{\vert}{\underset{\vert}{C}}-\overset{\vert}{\underset{\vert}{C}}-OH \xrightarrow{\text{acid}} \rangle C{=}C\langle$	Section 5.13. Normally use conc. sulfuric or phosphoric acid. Rearrangements should be expected.

TABLE 5.2 Preparation of Alkenes (continued)

Reaction	Comments
Functional Group Modification: No Change in Carbon Skeleton	
3. Dehalogenation 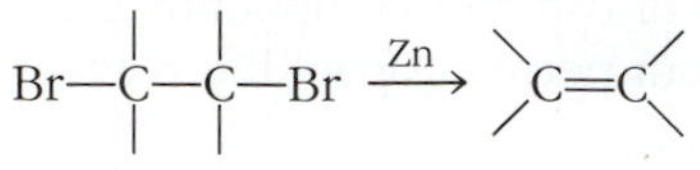 Br—C—C—Br $\xrightarrow{Zn}$ >C=C<	Section 5.13. The dibromide is prepared from the alkene; the two-step sequence can be useful in purification.
4. Pyrolysis of esters 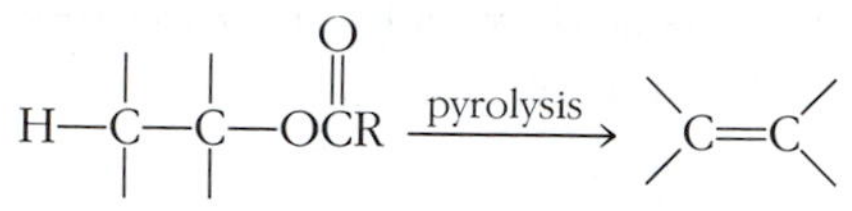 H—C—C—OC(=O)R $\xrightarrow{pyrolysis}$ >C=C<	Section 6.9.
5. Pyrolysis of amine oxides 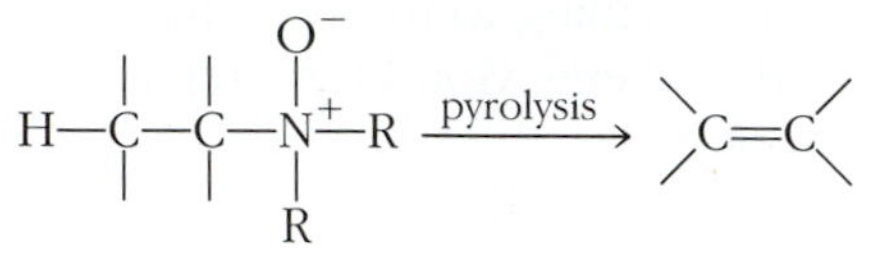 H—C—C—N⁺(O⁻)(R)—R $\xrightarrow{pyrolysis}$ >C=C<	Sections 6.9, 17.5.
6. Reduction of alkynes —C≡C— ⟶ —CH=CH—	Section 7.6.
Reactions Yielding Alkenes via Carbon-Carbon Bond Formation	
7. Reactions of organocuprates	Section 5.14 (Cf. Section 1.11)
(a) With a *remote* double bond R′—X $\xrightarrow{R_2CuLi}$ R′—R	Either R or R′ can contain a remote double bond, i.e., neither Cu nor X is attached directly to unsaturated carbon. See Section 1.11 for other limitations.
(b) With an *alkenyl* derivative 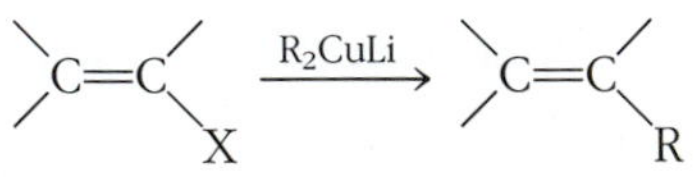 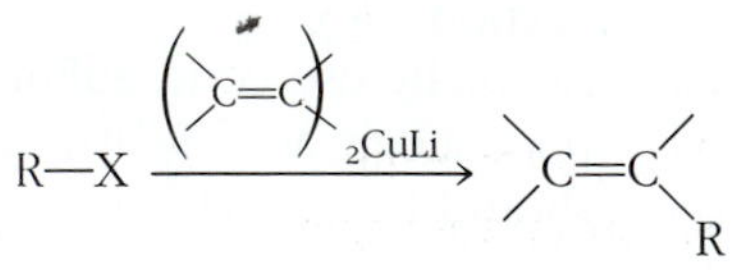 >C=C(X) $\xrightarrow{R_2CuLi}$ >C=C(R) or R—X $\xrightarrow{(>C=C<)_2CuLi}$ >C=C(R)	The reaction proceeds with retention of stereochemistry of the double bond.

TABLE 5.2 Preparation of Alkenes (continued)

Reaction	Comments
Functional Group Modification: No Change in Carbon Skeleton	
8. Wittig reaction 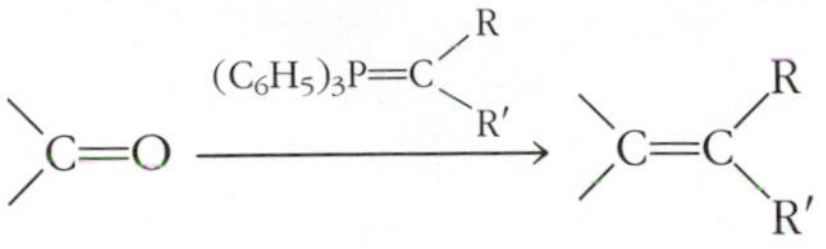 $\rangle C{=}O \xrightarrow{(C_6H_5)_3P{=}CRR'} \rangle C{=}CRR'$	Section 13.9.

5.17 PROBLEMS

5.1 Name the following compounds:

(a) 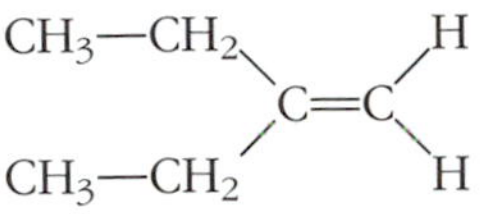 $(CH_3{-}CH_2)_2C{=}CH_2$ — $CH_3{-}CH_2$ and $CH_3{-}CH_2$ on one carbon; H and H on the other

(b) 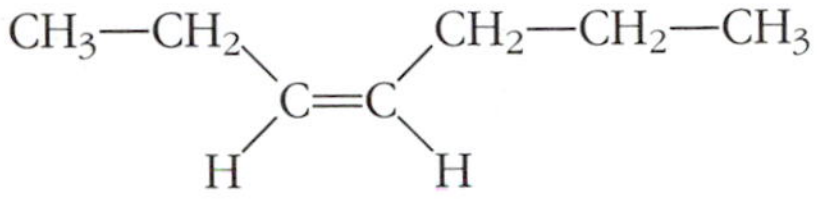 $CH_3{-}CH_2$, H / $CH_2{-}CH_2{-}CH_3$, H on $C{=}C$

(c) $CH_3{-}CH_2{-}CH_2$, H / $CH_2{-}CH_3$, CH_3 on $C{=}C$

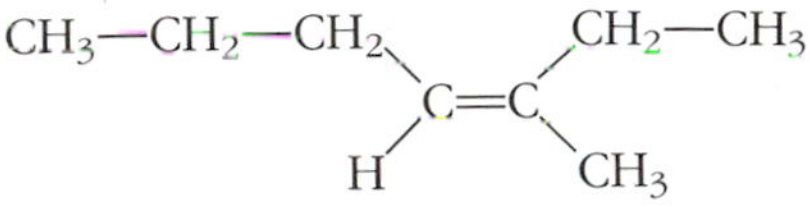

(d) 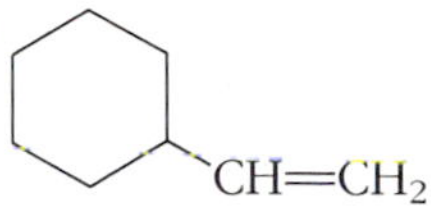 cyclohexane ring with $CH{=}CH_2$

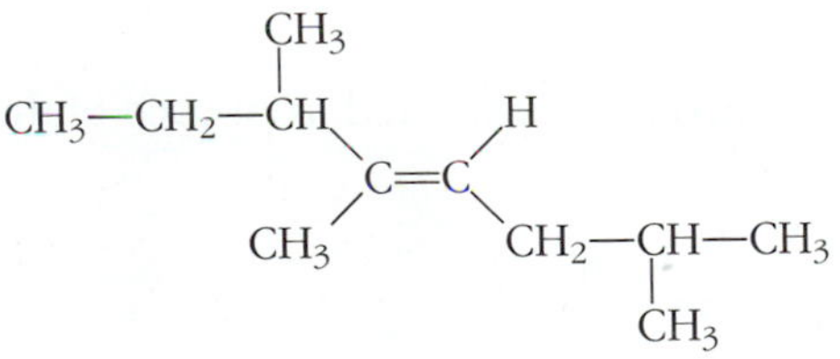

(e) 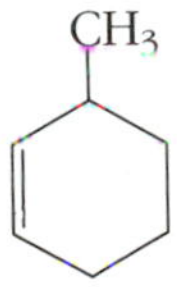 cyclohexene ring with CH_3

(f) 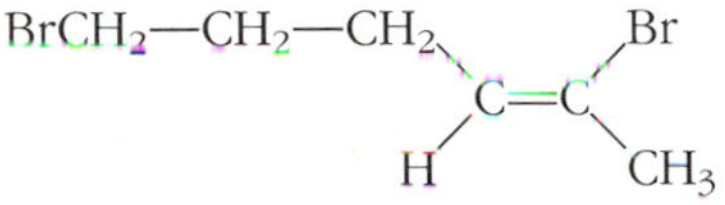 $BrCH_2{-}CH_2{-}CH_2$, H / Br, CH_3 on $C{=}C$

(g) $CH_3{-}CH_2{-}CH(CH_3)$, CH_3 / H, $CH_2{-}CH(CH_3){-}CH_3$ on $C{=}C$

(h) cyclopentane ring with $=CH_2$

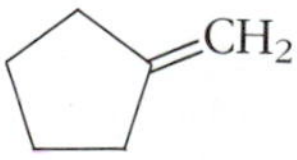

(i) 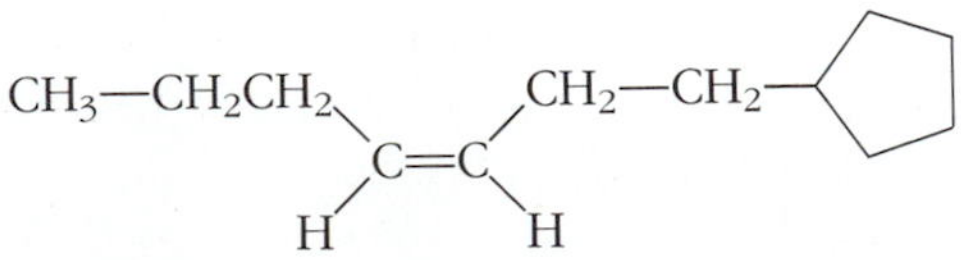 $CH_3{-}CH_2CH_2$, H / $CH_2{-}CH_2{-}$(cyclopentyl), H on $C{=}C$

(j) 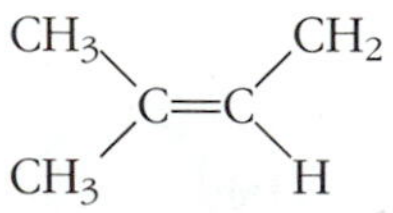 CH_3, CH_3 / CH_2, H on $C{=}C$

5.2 Draw the following structures.

(a) (*E*)-1,3-Difluoro-2-methyl-1-butene
(b) 1-Pentene
(c) 1,3-Dimethylcyclopentene
(d) Methylenecyclohexane
(e) (*Z*)-6-Bromo-3-propyl-2-hexene
(f) 1,2-Dichlorocyclopropene
(g) 2,3-Dimethyl-2-butene
(h) (*Z*)-4-*tert*-Butyl-3-heptene
(i) (*E*)-4,5-Dichloro-4-octene
(j) (*Z*)-1,2-Dibromoethene

5.3 The ethyl hydrogen sulfate that is formed in the reaction of concentrated H_2SO_4 and ethylene (Section 5.8) is capable of undergoing further reaction with ethylene. What is the product of this reaction? What would this compound yield in the subsequent hydrolysis step?

5.4 Describe the relationship between each pair of structural formulas as that of *constitutional* isomers, *stereoisomers (cis-trans* isomers), two drawings of the *same* compound (although perhaps in different conformations), or *different* compounds that are not isomers.

(a) 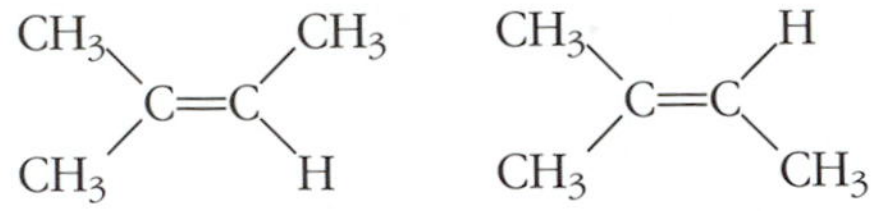 $(CH_3)_2C{=}C(CH_3)H$ and $(CH_3)_2C{=}C(H)CH_3$

(b)

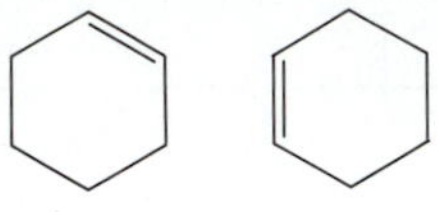

(c) 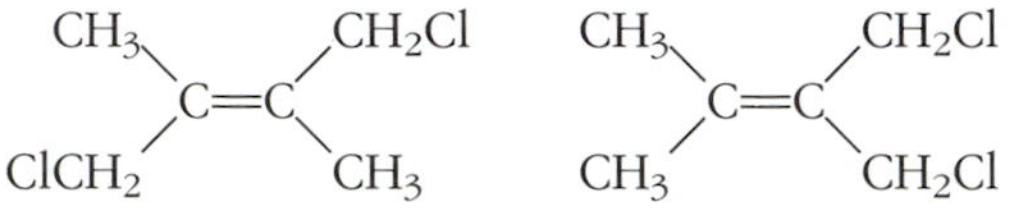 $CH_3(ClCH_2)C{=}C(CH_2Cl)CH_3$ and $CH_3(CH_3)C{=}C(CH_2Cl)CH_2Cl$

(d) 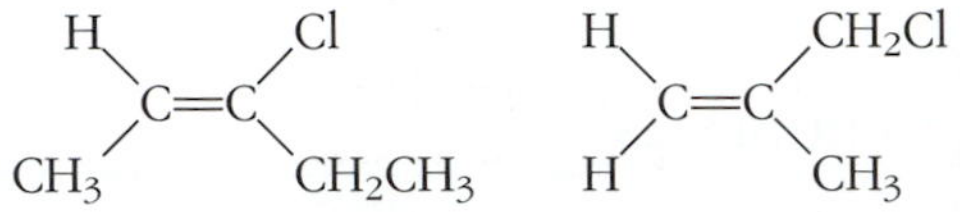 $H(CH_3)C{=}C(Cl)CH_2CH_3$ and $H(H)C{=}C(CH_2Cl)CH_3$

(e) 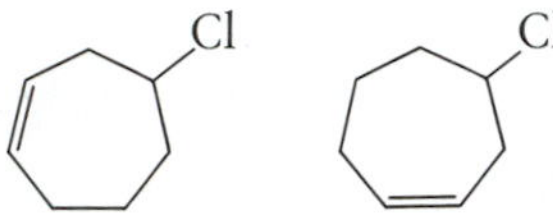 Cl Cl

(f) 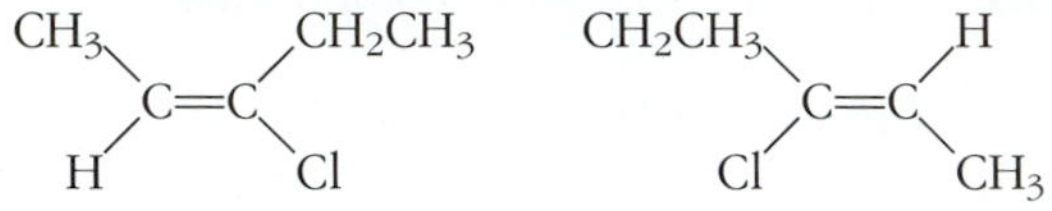 $CH_3(H)C{=}C(CH_2CH_3)Cl$ and $CH_2CH_3(Cl)C{=}C(H)CH_3$

(g) 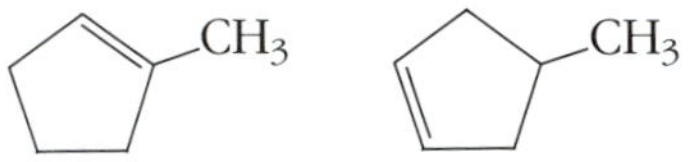 CH_3 CH_3

(h)

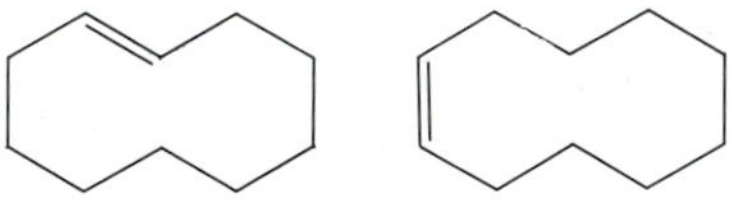

(i) 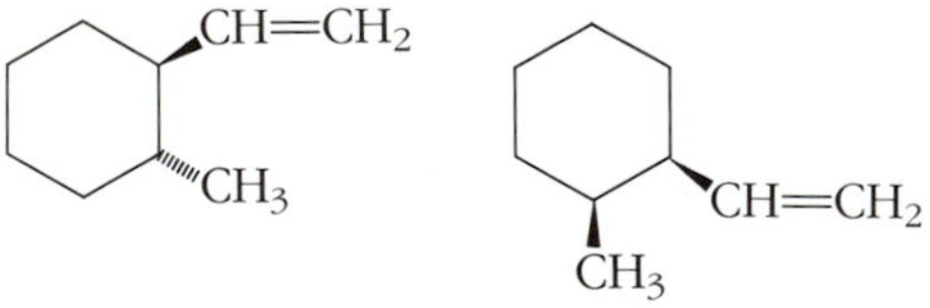 $CH{=}CH_2$, CH_3 and $CH{=}CH_2$, CH_3

(j) 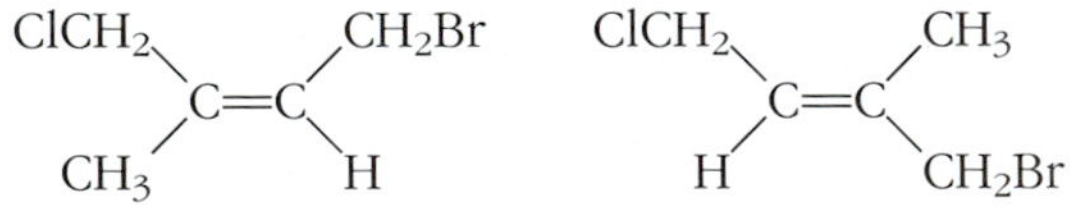 $ClCH_2(CH_3)C{=}C(CH_2Br)H$ and $ClCH_2(H)C{=}C(CH_3)CH_2Br$

5.5 Draw the major product (or products) expected from each of the following reactions:

(a) $\xrightarrow{H_2,\ Pt}$?

(b) CH_3, $CH{-}CH_3$ $\xrightarrow[2.\ NaBH_4]{1.\ Hg(OCCH_3)_2\ (C{=}O),\ H_2O}$?

(c) $\xrightarrow{H_2SO_4,\ C_2H_5OH}$?

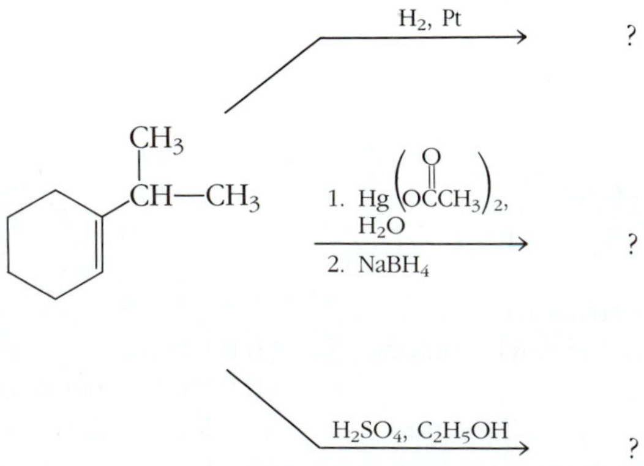

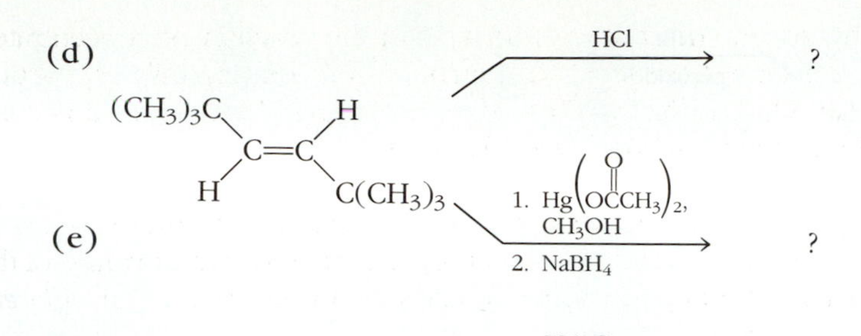

(d) $\xrightarrow{HCl}$?

(e) $\xrightarrow[2.\ NaBH_4]{1.\ Hg\left(O\overset{O}{\overset{\|}{C}}CH_3\right)_2,\ CH_3OH}$?

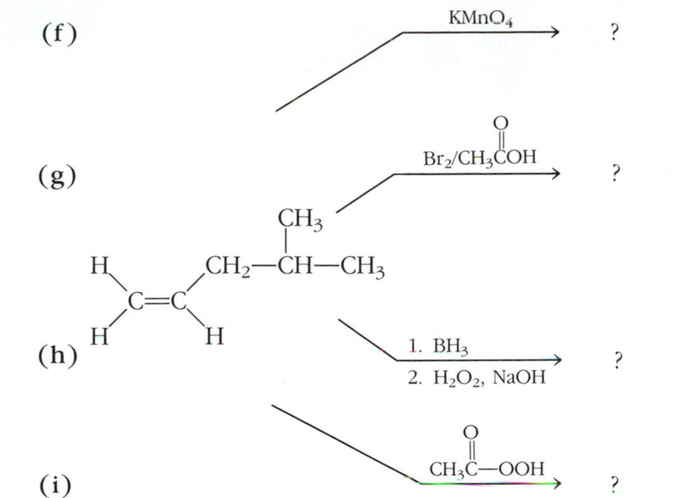

(f) $\xrightarrow{KMnO_4}$?

(g) $\xrightarrow{Br_2/CH_3\overset{O}{\overset{\|}{C}}OH}$?

(h) $\xrightarrow[2.\ H_2O_2,\ NaOH]{1.\ BH_3}$?

(i) $\xrightarrow{CH_3\overset{O}{\overset{\|}{C}}—OOH}$?

(j) 2,2-dimethylcyclopentanol (CH_3, CH_3, OH) $\xrightarrow{H_3PO_4}$?

(k) 1,2-dimethylcyclopentene (CH_3, CH_3) $\xrightarrow{Cl_2,\ CH_3OH}$?

(l) methylenecyclopentane (=CH_2) $\xrightarrow{NBS/H_2O}$?

(m) $\xrightarrow[2.\ Zn]{1.\ O_3}$?

(n) $\xrightarrow[\text{peroxides}]{\text{HBr}}$?

(o) CH_3, CH_3 $\xrightarrow{\text{HCl}}$?

(p) $\xrightarrow[\text{2. } H_2O_2\text{, NaOH}]{\text{1. } BH_3}$?

(q) Cl, C, H $\xrightarrow{(CH_3)_2CuLi}$?

(r) $\xrightarrow{NaOCH_3/CH_3OH}$?

$CH_3-C(CH_3)_2-CH(Br)-CH_2-Br$

(s) $\xrightarrow{\text{Zn}}$?

(t) $CH_3-CH(CH_3)-CH_2-CH(Br)-CH_3 \xrightarrow{\text{NaOEt/EtOH}}$?

5.6 An alkene of unknown structure afforded a high yield (87%) of a single alcohol when subjected to hydroboration followed by oxidation with alkaline hydrogen peroxide. Treatment with aqueous mercuric acetate followed by reduction with sodium borohydride afforded an isomeric alcohol in 96% yield. Reduction of the alkene with hydrogen and a platinum catalyst yielded hexane. Deduce the structure of the alkene.

5.7 Suggest methods for carrying out the following conversions. More than one step may be necessary in each case.

(a) CH_2 $\xrightarrow{?}$ $CH_2CH_2CH_3$

(b) $CH_3-CH(CH_3)-CH(OH)-CH_3 \xrightarrow{?} CH_3-C(Cl)(CH_3)-CH(Cl)-CH_3$

(c)

$$\begin{array}{l} \quad\;\; CH_3 \qquad\qquad\qquad\qquad\qquad\qquad\qquad\quad CH_3 \\ CH_3{-}CH{-}CH{-}CH_2Br \xrightarrow{?} CH_3{-}CH{-}CH_2{-}CH_2Br \\ \qquad\qquad\;\; Br \end{array}$$

(d)

$$\begin{array}{l} H \qquad\quad CH_3 \qquad\qquad\qquad\qquad\qquad OCH_3 \\ \quad C{=}C \qquad\xrightarrow{?} CH_3{-}CH_2{-}C{-}CH_3 \\ CH_3 \qquad\; CH_3 \qquad\qquad\qquad\qquad\quad CH_3 \end{array}$$

(e) cyclopentene $\xrightarrow{?}$ cyclopentane epoxide (O)

(f)

$$\begin{array}{l} \qquad\qquad\qquad\quad Br \quad Br \qquad\qquad\qquad\qquad\qquad H \qquad\quad CH_3 \\ CH_3{-}CH_2{-}CH{-}C{-}CH_3 \xrightarrow{?} \qquad\qquad\quad C{=}C \\ \qquad\qquad\qquad\qquad\quad CH_3 \qquad\qquad\qquad CH_3CH_2 \qquad\; CH_3 \end{array}$$

(g)

$$\begin{array}{l} \qquad\qquad\qquad\quad OH \\ CH_3{-}CH_2{-}CH{-}CH_3 \xrightarrow{?} CH_3{-}CH_2{-}CH_2{-}CH_3 \end{array}$$

(h)

$$\begin{array}{l} CH_3 \qquad\quad CH_2CH_3 \qquad\qquad\qquad\qquad CH_3 \;\; OH \\ \quad C{=}C \qquad\qquad\xrightarrow{?} CH_3{-}CH{-}CH{-}CH_2{-}CH_3 \\ CH_3 \qquad\quad H \end{array}$$

(i)

$$\begin{array}{l} \qquad\quad CH_3 \quad Br \qquad\qquad\qquad\qquad CH_3 \;\; CH_3 \\ CH_3{-}C{=}C \qquad\xrightarrow{?} CH_3{-}CH{-}CH{-}CH_2{-}CH_3 \\ \qquad\qquad\qquad CH_3 \end{array}$$

(j)

$$\begin{array}{l} \qquad\qquad\qquad\qquad\qquad\qquad\qquad\qquad\qquad\qquad\qquad\quad Cl \\ CH_3CH_2CH_2CH_2CH_2Cl \xrightarrow{?} CH_3CH_2CH_2CH{-}CH_2Cl \end{array}$$

(k) 1-methylcyclohexanol (CH_3, OH) $\xrightarrow{?}$ cyclohexane bearing CH_3 and OH on adjacent carbons

(l) 1-methylcyclohexanol (CH_3, OH) $\xrightarrow{?}$ cyclohexane bearing CH_3 and OCH_3 on the same carbon

5.8 Deduce the structures of compounds A–C.

$$\begin{array}{l} \qquad\qquad\qquad\qquad\qquad\qquad\quad CH_3 \\ \boxed{A} \xrightarrow{Br_2/CCl_4} CH_3{-}CH_2{-}C{-}CH_2{-}CH_2{-}Br \\ \qquad\qquad\qquad\qquad\qquad\qquad\quad Br \end{array}$$

B $\xrightarrow{NBS/H_2O}$ $CH_3-C(OH)(CH_3)-CH(Br)-CH_3$

C $C_7H_{13}Br$ $\xrightarrow[\text{peroxides}]{HBr,}$ $CH_3-C(CH_3)(Br)-CH_2-CH(CH_3)-CH_2-Br$

5.9 Oxymercuration-demercuration of an unknown alkene afforded an alcohol in 94% yield. Hydroboration-oxidation of the alkene gave an isomeric alcohol in 96% yield. Reduction of the alkene afforded an alkane with the molecular formula C_5H_{12}, while ozonolysis yielded acetaldehyde ($O{=}CH{-}CH_3$) as one of the products. Deduce the structure of the alkene.

5.10 The presence of a highly electronegative substituent such as the CF_3 group can greatly reduce the reactivity of an alkene toward polar addition reactions because an intermediate carbocation would be less stable. For example, 3,3,3-trifluoropropene ($CF_3{-}CH{=}CH_2$) reacts slowly with HCl (in the presence of $AlCl_3$ as a Lewis acid catalyst). What effect does a CF_3 group have on the orientation of a polar addition reaction? Why?

5.11 An unknown alkene reacted with HI to give 2-iodohexane in 93% yield. What is the structure of the alkene?

5.12 In Section 5.10 we stated that HCl does not add to alkenes by a free-radical mechanism in the same way that HBr does. Write the two propagation steps for free-radical addition of HBr to propene and calculate the ΔH for each step using the data in Table 2.1 (and assuming that the bond energy for the π bond of the double bond is 50 kcal/mol). Carry out the same procedure for the hypothetical free-radical addition of HCl to propene. Why is the free-radical addition of HCl to alkenes not observed?

5.13 Consider the addition of HCl to 1-butene. Calculate the oxidation state for C-1 and C-2 both before and after the addition. Is there any change in oxidation state for either of these two carbon atoms? Is the addition of HCl to an alkene an overall oxidation, an overall reduction, or neither?

5.14 Consider the addition of bromine to 1-butene. Calculate the oxidation state for C-1 and C-2 both before and after the addition. Is there any change in oxidation state for either of these two carbon atoms? Is the addition of bromine to an alkene an overall oxidation, an overall reduction, or neither?

5.15 The coupling of an alkenyl halide to a lithium dialkylcuprate is stereospecific. Draw and name the hydrocarbon that would be formed by reaction of (*E*)-3-chloro-2-hexene with lithium dimethylcuprate. Can you predict whether an *E*-alkenyl halide will always yield either an *E* or a *Z* product?

5.16 Write the reaction mechanism for the addition of HBr to 1-hexene (a) in the absence of peroxides and (b) in the presence of peroxides.

5.17 Write the reaction mechanism for the addition of HBr to 3-methyl-1-hexene (a) in the absence of peroxides and (b) in the presence of peroxides.

5.18 In the acid-catalyzed addition of water to 3-ethyl-3-methyl-1-hexene, a carbocation rearrangement would be expected. In this case three different 1,2-alkyl shifts could occur. Draw the products that would be formed in each case, and write the complete mechanism for each of the pathways.

5.19 When the substituents at the two ends of a double bond are not equivalent, protonation of an alkene could give two isomeric carbocations, but the more stable cation is formed preferentially. A tertiary cation is favored over a secondary cation, and a secondary cation is favored over a primary cation. Draw the structure of the only alkene for which protonation would yield a primary carbocation.

5.20 Write the mechanism for the reaction of 1-methylcyclohexene with: (a) dilute H_2SO_4; (b) HBr; (c) HBr, peroxides; (d) NBS, H_2O; (e) Br_2, CH_3CO_2H.

5.21 Write the mechanism for the formation of cyclohexene from:
(a) Cyclohexanol and conc. sulfuric acid
(b) *trans*-1,2-Dibromocyclohexane and zinc
(c) Bromocyclohexane and KOtBu, tBuOH

5.22 Suggest a method of synthesis for each of the following compounds. All the carbon atoms of the final product must be derived from the specified starting materials, but any other organic or inorganic reagents may be used.

(a) $(CH_3)_3C{-}CH{=}C(CH_3)_2$ from compounds containing four or fewer carbon atoms

(b) cyclopentyl$-CH{=}CH_2$ from compounds containing five or fewer carbon atoms

(c) $CH_3{-}CH(CH_3){-}CH_2{-}CH_2{-}CH_3$ from *alkenes* containing three or fewer carbon atoms

(d) cyclobutyl$-CH_2{-}CH(CH_3){-}CH_2OH$ from compounds containing five or fewer carbon atoms

(e) 1-ethylcyclohexanol (cyclohexane ring bearing $CH_2{-}CH_3$ and OH on the same carbon) from compounds containing six or fewer carbon atoms

5.23 Compounds **A–C** are all different compounds. From the reactions shown deduce their structures.

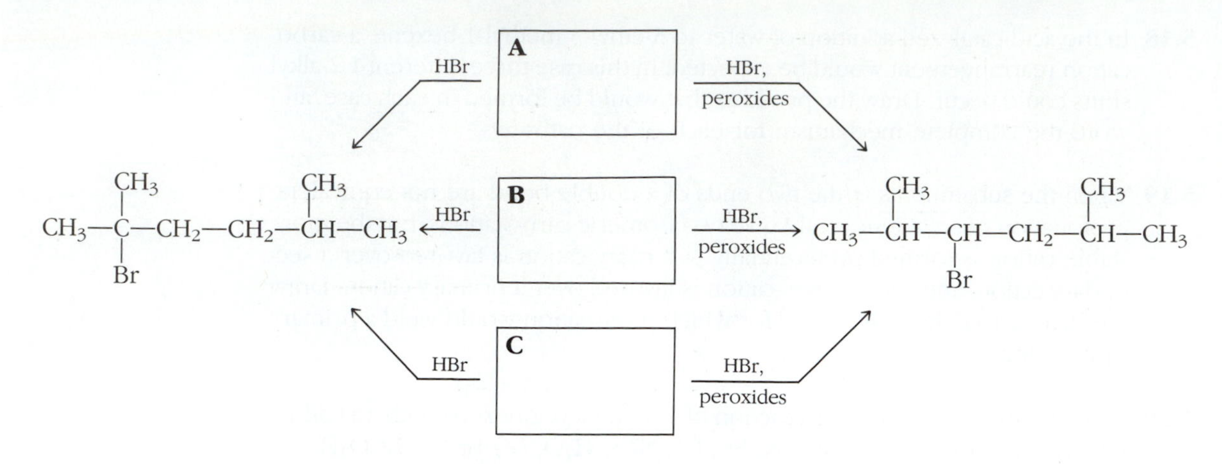

5.24 Write the complete reaction mechanism for each of the reactions in the preceding question.

5.25 Compounds **A–F** are all different compounds, but A and C have the *same carbon skeleton.* Deduce the structures of **A–D.**

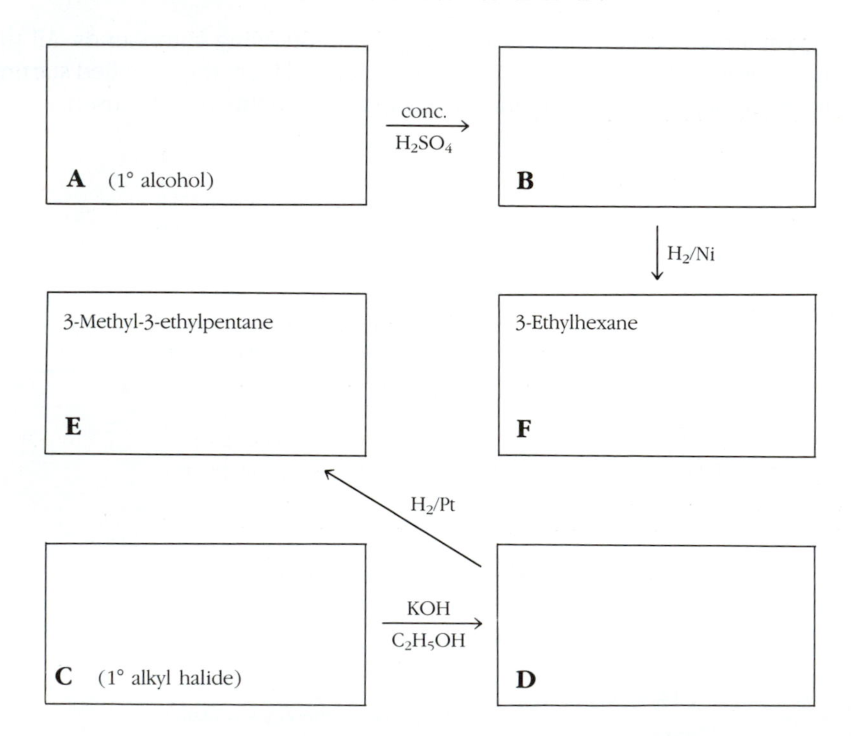

5.26 Compounds **A–H** are all different compounds. On the basis of the information provided, deduce the structures of compounds **A–E.**

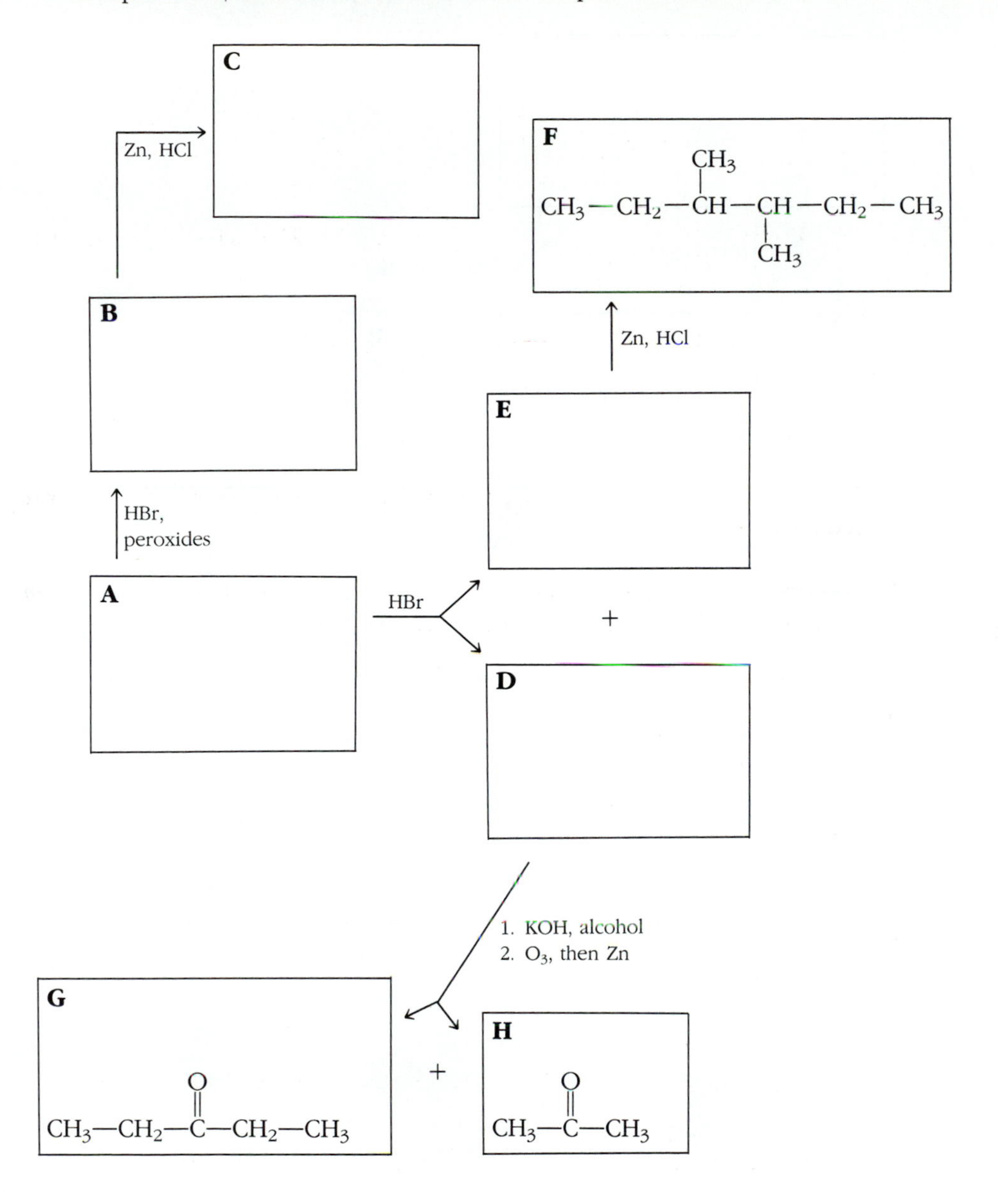

5.27 Compounds **A–E** are all different compounds. On the basis of the information provided, deduce the structures of compounds **A–D.**

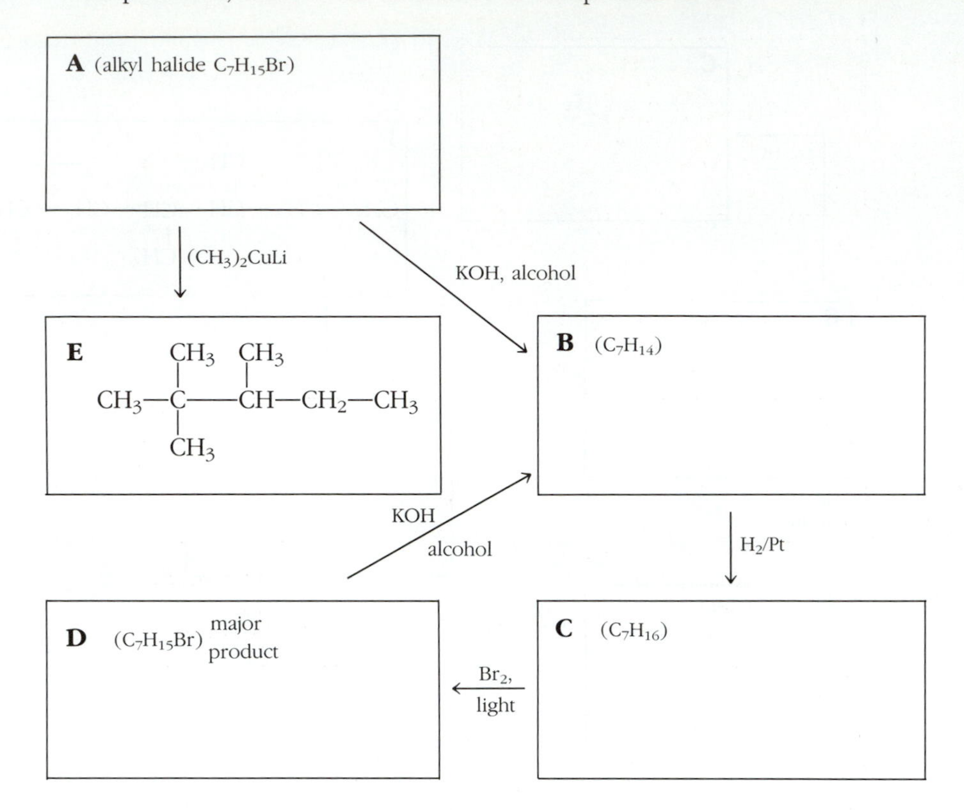

5.28 Compounds **A–G** are all different compounds. On the basis of the information provided, deduce the structures of compounds **B–G.**

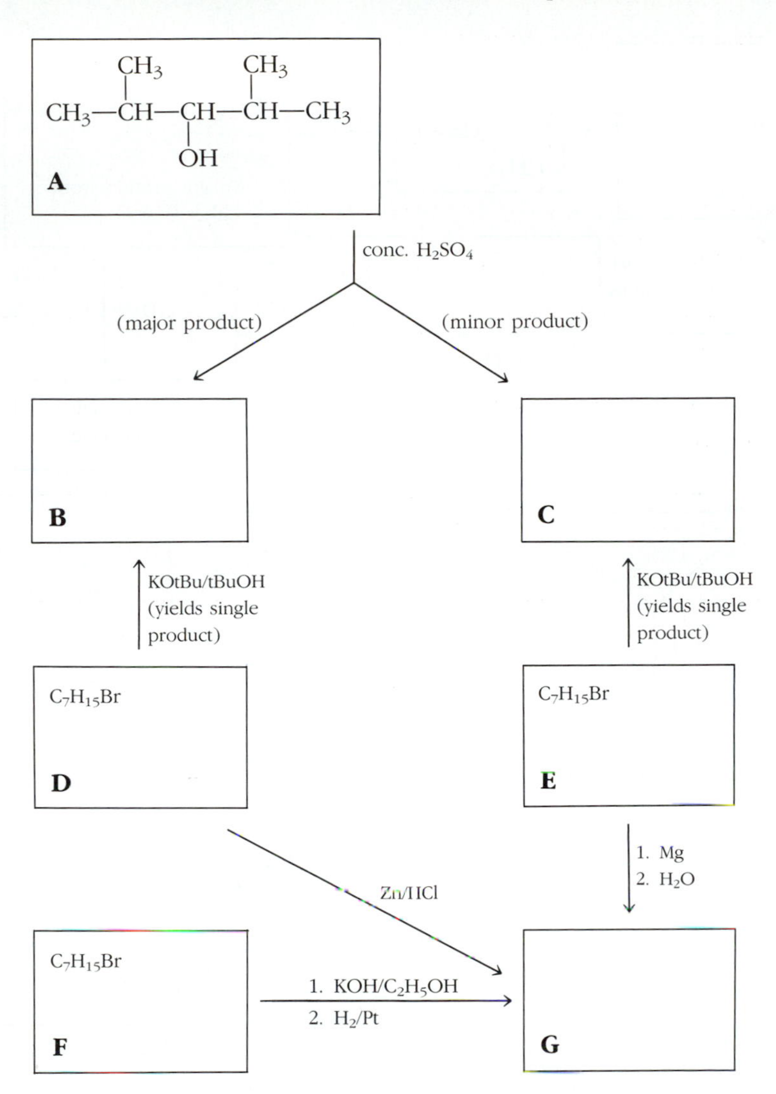

5.29 Compounds **A–F** are all different compounds. On the basis of the information provided, deduce their structures.

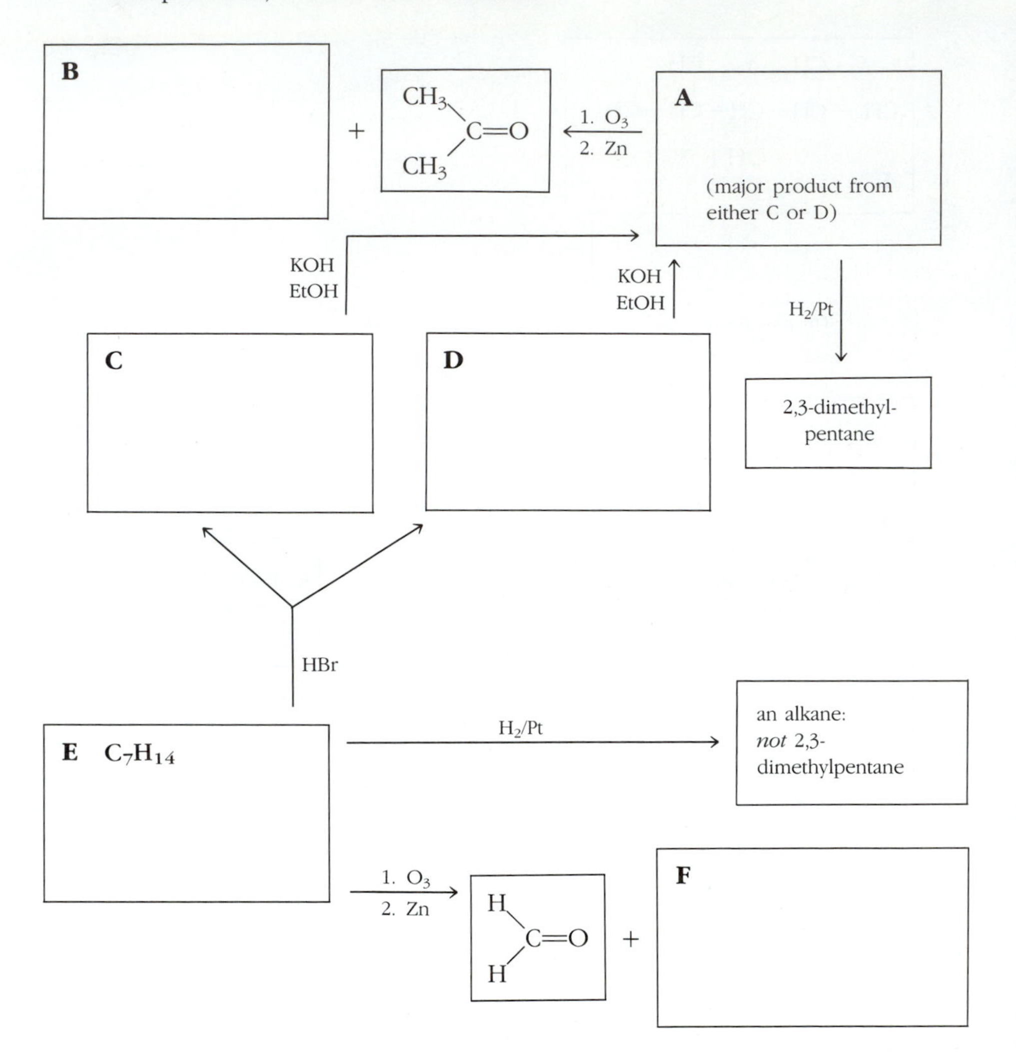

5.30 Draw the major product (or products) expected from each of the following reactions:

Starting alkene for (a)–(c): $CH_3CH{=}CHCH_3$ (trans: CH_3 and H on each carbon, CH_3 groups on opposite sides)

(a) 1. $CH_3-\overset{O}{\overset{\|}{C}}-OOH$ 2. H_2SO_4/H_2O $\longrightarrow$?

(b) $\xrightarrow{H_2/Pd\text{-}C}$?

(c) $\xrightarrow{\text{cold conc. } H_2SO_4}$?

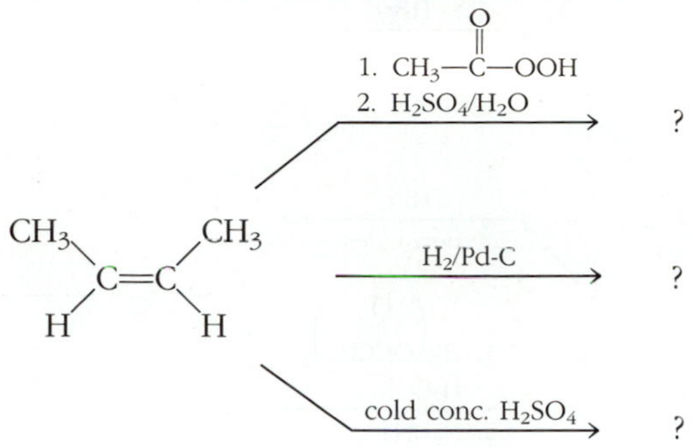

(d) $CH_3-CH_2-\underset{Cl}{\overset{CH_3}{C}}-CH_2-CH_3 \xrightarrow{KOtBu/tBuOH}$?

(e) $CH_3-\overset{CH_3}{CH}-\underset{CH_3}{CH}-CH_2OH \xrightarrow{H_3PO_4}$?

Starting alkene for (f)–(i): $(CH_3)_3C-CH{=}CH_2$

(f) 1. $Hg\left(O\overset{O}{\overset{\|}{C}}CH_3\right)_2$ 2. $NaBH_4$ $\longrightarrow$?

(g) $\xrightarrow{Cl_2,\ H_2O}$?

(h) $\xrightarrow[\text{peroxides}]{HBr}$?

(i) 1. O_3 2. Zn, H_2O $\longrightarrow$?

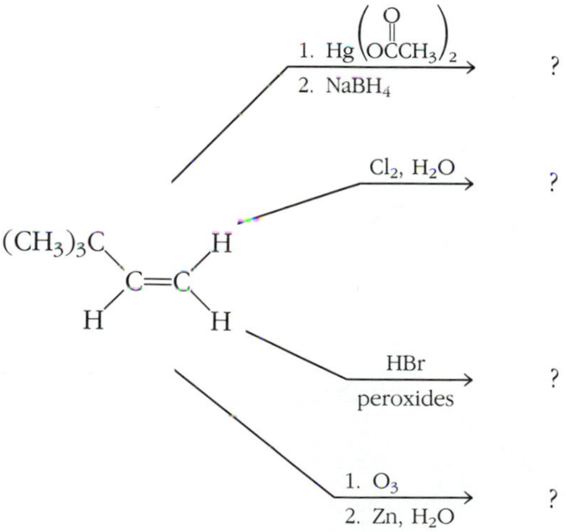

(j) Cyclohexyl–CH_2OH $\xrightarrow{\text{conc. } H_2SO_4}$?

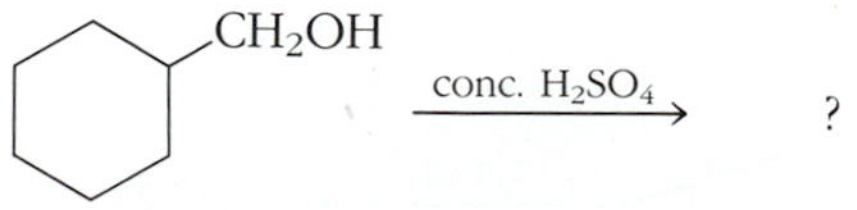

(k) 1-Bromo-1-methylcyclohexane (CH_3, Br on the same ring carbon) $\xrightarrow{KOH,\ EtOH}$?

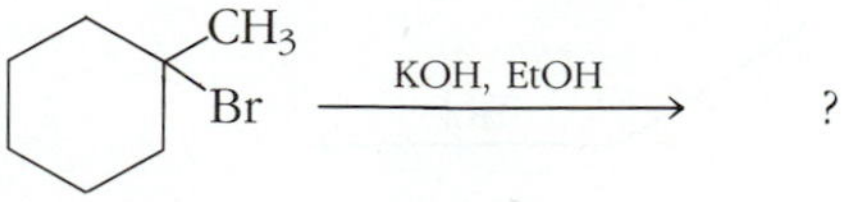

5.31 Draw the major product (or products) expected from each of the following reactions:

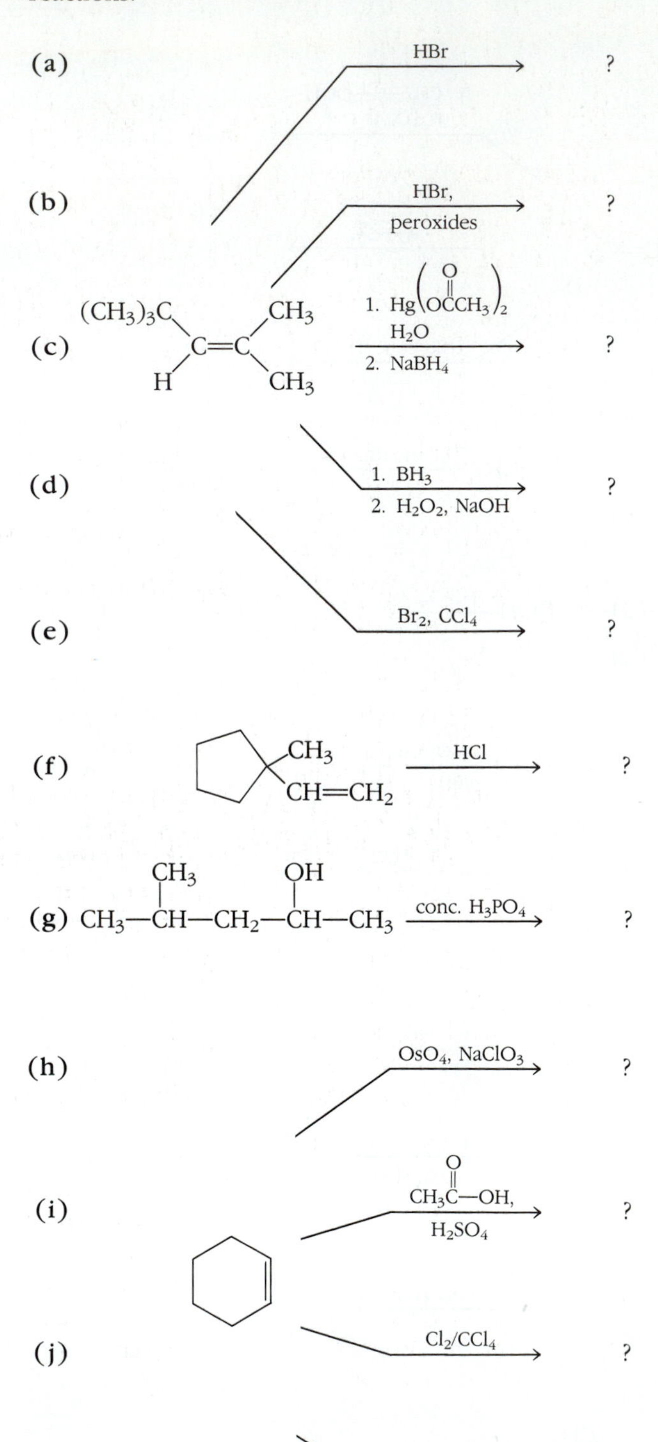

CHAPTER 6

STEREOCHEMISTRY

What do molecules actually look like in three dimensions? This question has intrigued chemists for more than 100 years, and we have come a long way since 1884 when two young scientists, J. H. van't Hoff in the Netherlands and J. A. LeBel in France, independently suggested that carbon atoms are tetrahedral. Their idea was met with scorn and ridicule by other scientists, and the influential German chemist, Herman Kolbe, wrote an article attacking van't Hoff's suggestions as "fanciful nonsense" and "supernatural explanations."

Only slowly did chemists begin to accept the notion of tetrahedral carbon, but this ultimately provided the foundation upon which modern organic chemistry was built. In particular, the concept of tetrahedral carbon allowed the development of **stereochemistry,** the study of three-dimensional molecular structure and of the relationships between the three-dimensional structures of reactants and products in chemical reactions. In this chapter we will present the basic principles of stereochemistry. We will begin with an analysis of stereoisomers, molecules that are isomeric as a result of differences in their three-dimensional structures. We will then turn to the stereochemistry of organic reactions, and we will show how you can use the structures of the reactants to predict the three-dimensional structures of the products for a variety of organic reactions.

6.1 CLASSIFICATION OF ISOMERS

In the previous chapters we have discussed various types of isomers: *constitutional isomers* (compounds with the same molecular formula but different connectivities), *cis-trans* isomers of cycloalkanes, and *E-Z* isomers of alkenes. The last two types are both examples of *stereoisomers* because the differences are in the three-dimensional bonding arrangements of the atoms rather than in the connectivities. The relationships that can exist between molecules sometimes appear bewildering, but we can greatly simplify the problem with a reliable classification system. Figure 6.1 shows a classification of the possible relationships between any two molecules.

Isomerism always indicates a relationship between two or more molecules, and you can determine the precise relationship by answering a brief series of questions. The goal of the process is to rule out all categories but one, and it is not necessary to answer the questions in a particular sequence. However, we have found it very convenient to start at the top of Figure 6.1 and work downward until a final decision is reached.

1. *Do the two structures have different molecular formulas?* If so, they cannot be isomers and are classified as *different, nonisomeric compounds.* On the other hand, if they *do* have the *same* molecular formula, either they are the same compound or they are isomers.

2. *Are they two drawings of the same molecule* (although perhaps in different conformations)? This question is sometimes difficult to answer, and you may need to draw the structures in a different way. The use of the term *isomers* in this book is restricted to the experimental criterion that two compounds that are isomers can each be *isolated* under normal conditions. Ordinarily the individual conformations of a compound rapidly interconvert by rotation about single bonds and cannot be isolated. Hence different conformations of a molecule are classified as the *same compound* and not as *isomers.* As an

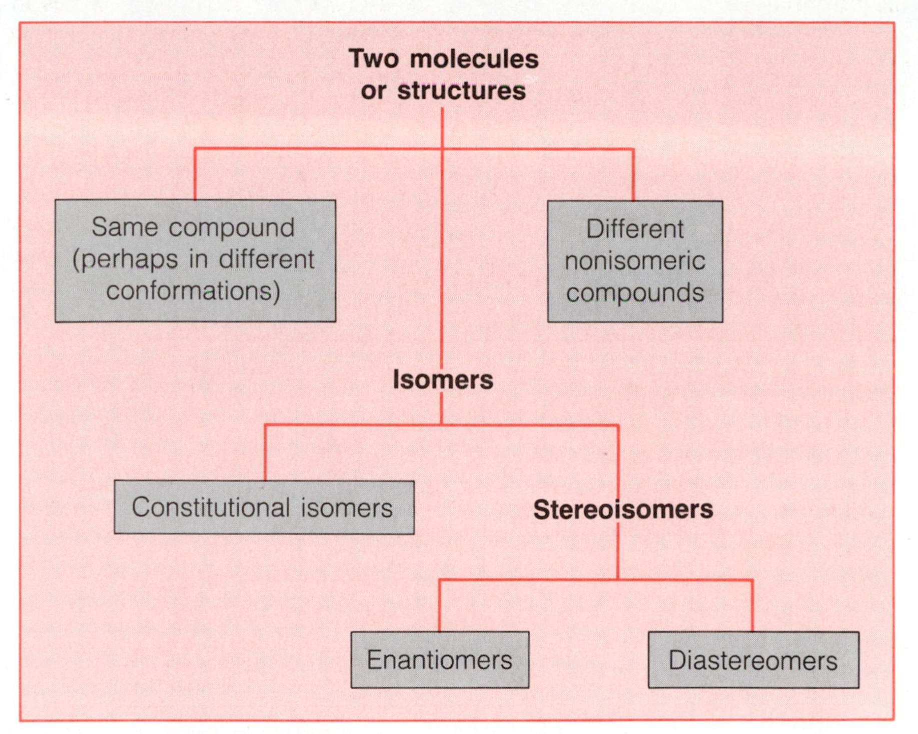

Figure 6.1 Relationships Between Molecular Structures.
The relationship between any two structures can be specified according to this classification scheme.

alternative to drawing both structures, you could write the complete names, including specification of stereochemistry (Section 6.2). If both compounds have the same name, they are two drawings of the *same* compound; otherwise they are isomers.

3. *Do they have different constitutions?* An answer of no to each of the first two questions has established an isomeric relationship. To see if both molecules have all their atoms connected in the same sequence, it may be useful to either name them or draw them similarly (but *not* necessarily using three-dimensional representations). If the connectivities are different, then the structures are *constitutional isomers;* otherwise they are *stereoisomers.*

4. *Are the two structures mirror images of each other?* An answer of no to each of the three preceding questions has established that the two structures are stereoisomeric. If they are also mirror images, they are **enantiomers;** otherwise they must be **diastereomers,** the category which includes all other stereoisomeric structures (*cis-trans* isomers among them).

EXERCISE 6.1

Using Figure 6.1 classify each of the following pairs of structures as either the same compound, different nonisomeric compounds, constitutional isomers, or stereoisomers. (You will learn how to distinguish between the two types of stereoisomers later in this chapter.)

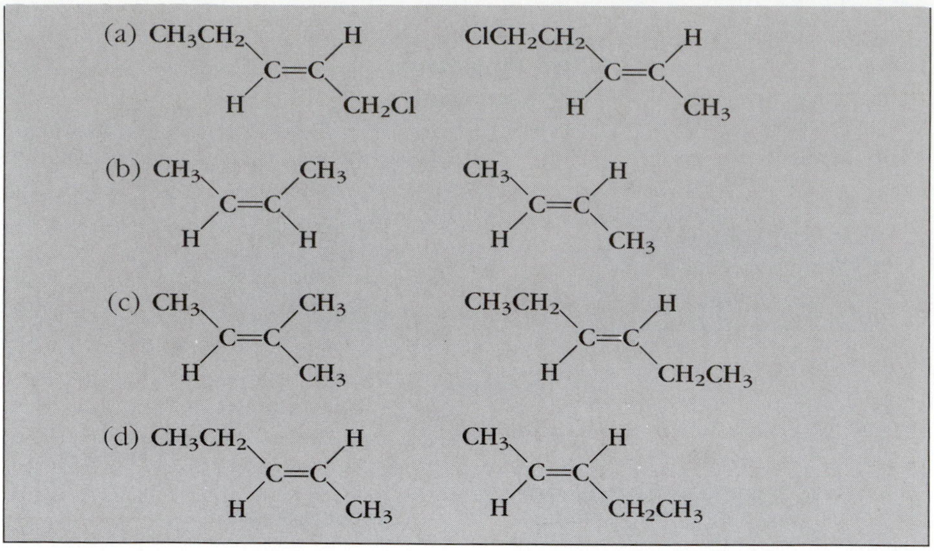

(a) CH_3CH_2 H $C{=}C$ H CH_2Cl | $ClCH_2CH_2$ H $C{=}C$ H CH_3

(b) CH_3 CH_3 $C{=}C$ H H | CH_3 H $C{=}C$ H CH_3

(c) CH_3 CH_3 $C{=}C$ H CH_3 | CH_3CH_2 H $C{=}C$ H CH_2CH_3

(d) CH_3CH_2 H $C{=}C$ H CH_3 | CH_3 H $C{=}C$ H CH_2CH_3

6.2 MIRROR IMAGES AND CHIRALITY

Mirror Image Relationships

In the preceding section we stated that the difference between the two types of stereoisomers is determined by whether or not the structures under consideration are *mirror images*. What do we mean by this term? While you may not have encountered this concept from a chemical viewpoint, you have almost certainly benefited from the regular use of a mirror in your everyday life. What you see when you view some object in a mirror is the **mirror image** of that object. There is very little difference between an object and its mirror image, but that difference is real. You could easily convince yourself of this by holding a book up to a mirror and trying to read its mirror image.

We learn to work with mirror images easily, and we tend to forget how difficult it can be for the inexperienced. Perhaps you can recall an early (and possibly humorous) experience such as learning to shave or to put on makeup in front of a mirror. Or maybe you remember your early difficulties as a new driver in mastering the use of a rearview mirror. The use of mirror images in chemistry tends to be similar; it is difficult at first but becomes easier with practice.

The idea of mirror image relationships can be clarified with familiar objects. Figure 6.2a shows a pair of hands, one left hand and one right hand. Figure 6.2b shows a right hand and its image in a mirror. The image in the mirror is indistinguishable from the left hand in Figure 6.2a. Consequently we say the left and right hands of Figure 6.2a exhibit a *mirror image relationship.* A person's left hand is certainly different from the right, so Figure 6.2 also demonstrates that *an object and its mirror image can be different.* On the other hand, a mirror image relationship does not always mean that two objects are different. A Ping-Pong ball and its mirror image would be indistinguishable.

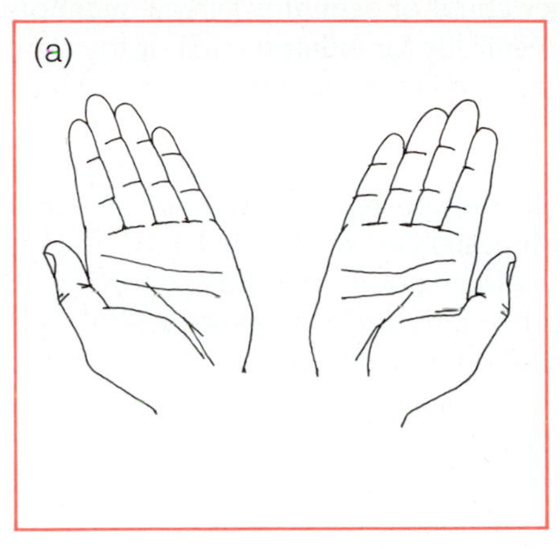

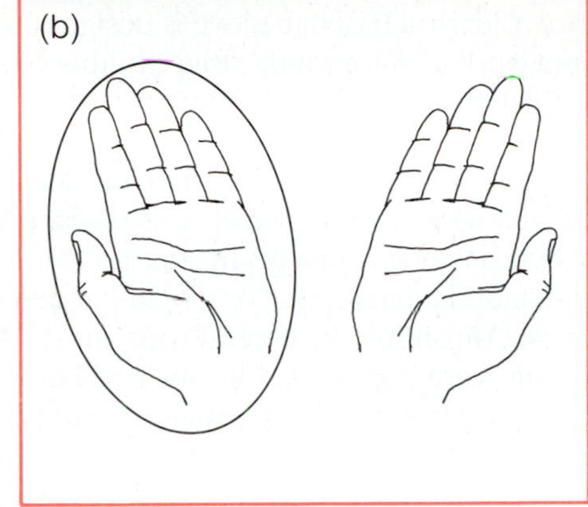

Figure 6.2 Mirror Image Relationships.
(a) A right hand and a left hand. (b) A right hand and its image in a mirror.

Chirality

When two objects are indistinguishable, we say that they are the *same,* and it does not matter that they might also have a mirror image relationship. However, *if an object and its mirror image are different,* each is described as **chiral** (pronounced "ki′ ral," where the "ki" has the same sound as in the word, kite). The term, chiral, is derived from the Greek *kheir,* "hand" and it refers to the properties of a left or right hand, that is, objects that are mirror images but are nevertheless different. In the opposite situation, an object that is indistinguishable from its mirror image is called **achiral** because it lacks the property of chirality. The commonplace objects shown in the illustration allow us to discuss and evaluate the property of chirality.

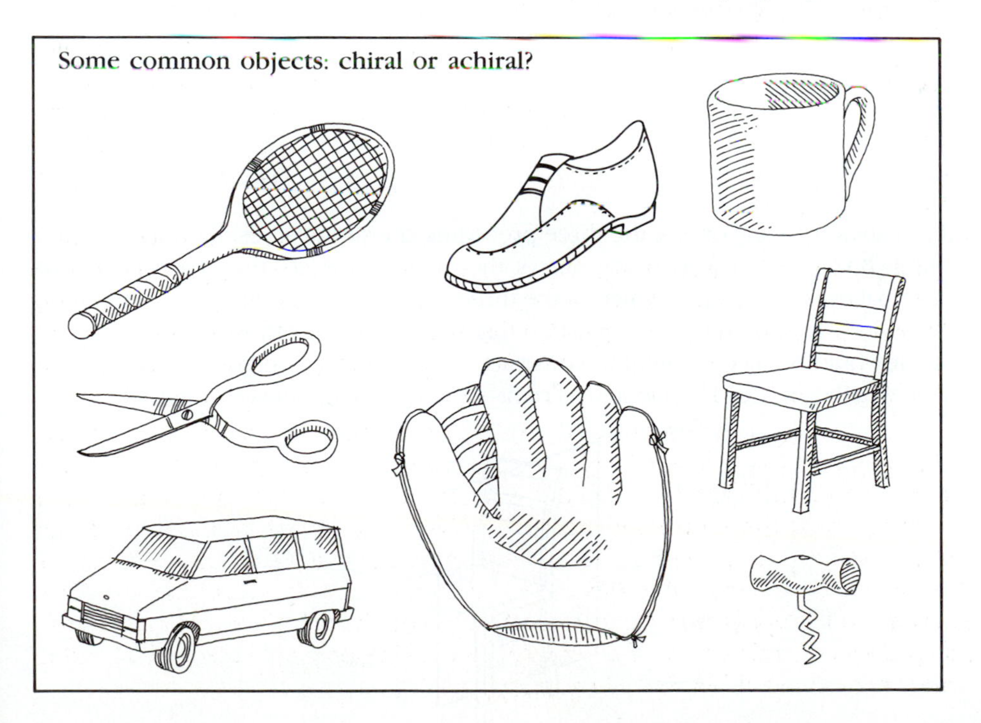

The classification of these objects as either chiral or achiral is fairly straightforward. Clearly a baseball glove is designed specifically for either the left or the right hand, and an automobile driver in this country expects to find the steering wheel on the left side of the car. These two objects are chiral, as are the corkscrew (which must be turned clockwise to insert it into a cork) and the pair of scissors, which is designed for a right-handed person. Most people are right-handed and may not realize that a normal pair of scissors cannot be used easily with the left hand. (Try this to convince yourself.) In contrast to a pair of scissors, you would not purchase a left-handed tennis racket or a right-handed coffee mug; those items are achiral, as is a chair. We should be careful with our decision on the coffee mug, however. If the writing were considered in our evaluation, we would have to classify the mug as chiral because its mirror image would be different:

How can you determine whether or not an object is chiral? You will be able to make a decision if you can answer any one of the three questions in the following box.

Determination of Chirality

1. *Is the object different from its mirror image?* You might need to redraw the structure or construct its mirror image to answer this question.

2. *Can the object be labeled as either right-handed or left-handed?*

3. *Does the object lack reflection symmetry?* The most common type of reflection symmetry is a plane of symmetry (Section 1.6), and a plane of symmetry divides an object into two "halves" that are mirror images of each other.

If the answer to any one of the three preceding questions is yes, then the object is chiral. But if the answer to any one of the questions is no, the object is *achiral*.

It makes no difference which of the three questions is used to reach a decision; therefore you should use the question that you can answer most easily. For example, it would be inconvenient to construct a chair, and you might not see immediately whether it could be labeled as right- or left-handed, but you can easily draw a

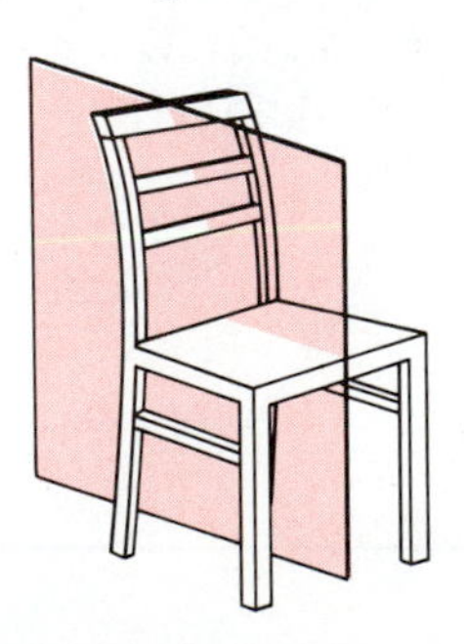

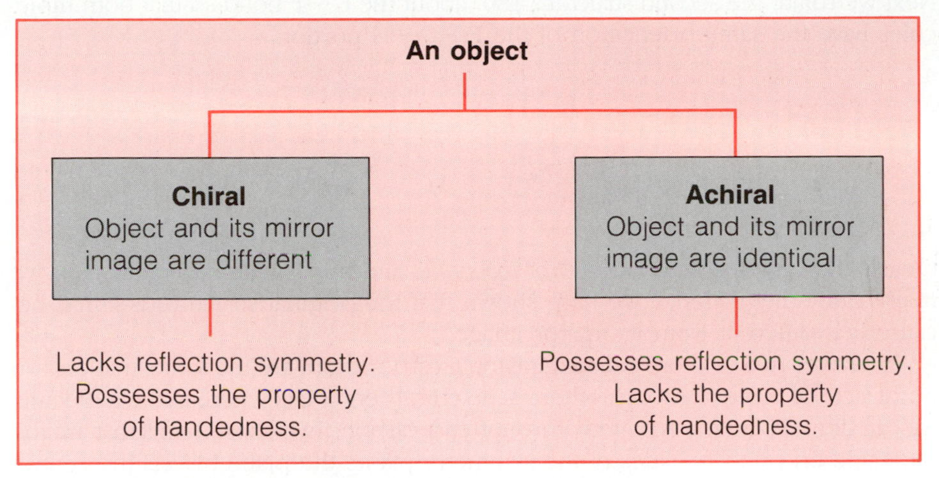
An object
Chiral
Object and its mirror image are different
Achiral
Object and its mirror image are identical
Lacks reflection symmetry. Possesses the property of handedness.
Possesses reflection symmetry. Lacks the property of handedness.

Figure 6.3 The Criteria That Determine Chirality.

plane of symmetry through the center of the chair shown previously. The half of the chair on one side of this symmetry plane is the mirror image of the other half, so the answer to the third question is *no,* and the chair is achiral.

In Figure 6.3 we have summarized schematically the criteria that determine whether or not an object is chiral. If you study this figure carefully, you will see that an answer to any one of the three questions about chirality is sufficient to clearly define an object as either chiral or achiral.

Chirality: A Property of Molecules

Molecules also can possess the property of chirality. Most commonly, a molecule will be chiral when it contains a tetrahedral atom with *four different substituents.* We describe such an atom as a **chiral center;** frequently this is a carbon atom and we call it a *chiral carbon.* Each of the following structures contains one chiral carbon and is a drawing of a chiral molecule:

F, H, C, Br, Cl CH_2OH, CH_3, C, H, Cl CH_2Cl, H, C, HO, CH_2Br

How can you determine whether or not a molecule is chiral? A decision can be reached by answering any of the three questions that we just discussed. Although an individual molecule is not called right- or left-handed, we will describe a naming procedure in Section 6.3 that does distinguish between the two mirror images of a chiral molecule. We can conveniently demonstrate that the first of the three structures we have just shown is chiral by drawing both the molecule and its mirror image (where the dotted line represents an imaginary mirror).

F, H, C, Br, Cl | F, Br, Cl, C, H

Next we rotate the second structure 180° about the C—F bond so that both molecules have the same orientation of the F—C—H portion.

Clearly these two structures *are not the same,* and this is true no matter how we might draw them. Hence we have shown that the original structure is chiral because it is different from its mirror image.

Whenever two or more substituents on a carbon atom are the same, that carbon is not a chiral center. As long as there is no chiral center elsewhere in the molecule, we can draw a plane of symmetry through that carbon. Two substituents that are the same* lie on opposite sides of the symmetry plane; the plane bisects the central carbon and each of the two remaining substituents.

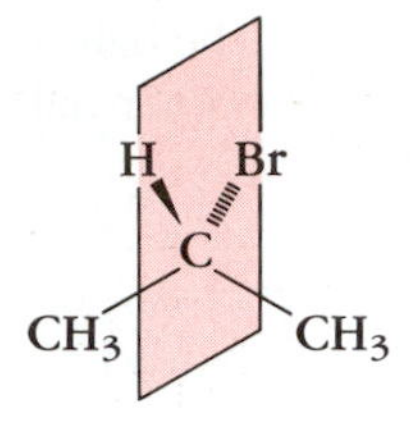

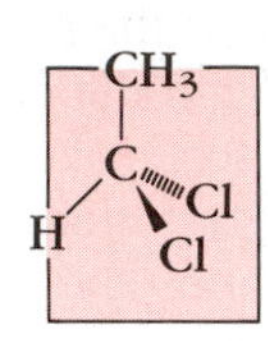

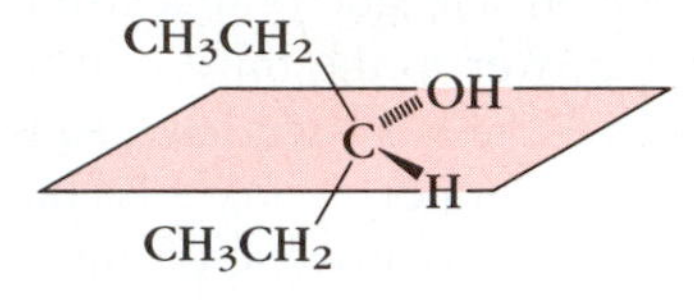

The symmetry plane is perpendicular to the plane of the paper.

The symmetry plane is equivalent to the plane of the paper.

The symmetry plane is perpendicular to the plane of the paper.

The definition of a chiral center provides us with a fourth way to determine if a molecule is chiral: *If a structure contains a single chiral center, the molecule is chiral.* It is also usually true that most structures that lack a chiral center are achiral. The few instances in which a chiral molecule has no chiral center result from unusual functional groups (for example, allenes, Section 7.9). We will discuss these situations individually as they arise in the course of this book.

EXERCISE 6.2

For the objects below answer the following questions: Which are chiral? Which are achiral? Which have a plane of symmetry?

(a) (b) A cube (c) A dinner fork

(d) A solid-color T-shirt (e) A button-front overcoat

*Strictly speaking these two substituents must be *mirror images* of each other because they are on opposite sides of the symmetry plane.

(f) CH_3CH_2, CH_3CH_2, Cl, H on C

(g) $CH_3CH_2-CH(Br)-CH(CH_3)-CH_3$

(h) CH_3, CH_3, CH_3, Cl on C

(i) $CH_2CH_2CH_3$, H, CH_3, $CH_3-CH(CH_3)$ on C

(j) An individual ball used to play pocket billiards (pool).

6.3 ENANTIOMERISM: A RELATIONSHIP BETWEEN TWO MOLECULES

In Section 6.1 we introduced enantiomers as a new type of stereoisomer. Now that you have also been introduced to chirality we can fully define **enantiomers** as two isomeric molecules that are *chiral* and are *mirror images* of each other. Unlike chirality, which is a property of an individual molecule, *enantiomerism* is a relationship between molecules. In most ways enantiomers are indistinguishable: they have the *same molecular formula,* the *same constitution,* and the *same energy.* Most of their physical and chemical properties are also the same, and we will discuss the few exceptions in Sections 6.5 and 6.6. In this section we will show you how to name enantiomers by specifying the configuration at chiral centers, and we will discuss ways to determine whether or not two structures are enantiomers.

Configuration

The spatial arrangement of the four substituents on a chiral carbon atom is called the **configuration** of that chiral center. Only two configurations are possible for a chiral center, and they are mirror images of each other. Unless a carbon atom is chiral, only a single configuration is possible. We can now define the only structural difference between enantiomers: they have opposite configurations at the chiral center. (We will show in Section 6.4 that enantiomers with more than one chiral center have the opposite configuration at *each and every chiral center.*)

What is the difference between *configuration* and *conformation*? As we stated for alkenes in Section 5.3, both terms describe the three-dimensional structure of a molecule, but different conformations can be interconverted by rotation about single bonds. In marked contrast, the *configuration* of a chiral center can be changed only by *breaking* one or more bonds in order to interchange two of the substituents. We therefore regard the configuration of a chiral center as *fixed,* whereas conformational changes occur easily and rapidly. This is precisely the situation that we described for the fixed configuration of alkenes that results in *cis-trans* isomerism.

Nomenclature of Chiral Centers

The only structural difference between enantiomers is the configuration at the chiral center (or centers), so a complete system of nomenclature must include a way of denoting the configuration at a chiral center. Early methods for doing this were awkward and ambiguous. In the 1950s however, the English chemists R. S. Cahn and C. K. Ingold and the Swiss chemist V. Prelog proposed a system, which is now widely used. Their system allows us to name the configuration of any chiral center as *R* or *S* (from the Latin *rectus,* "right," and *sinister* "left"). Figure 6.4 illustrates the use of this system to assign configuration.

Example: Br, C, F, H, Cl

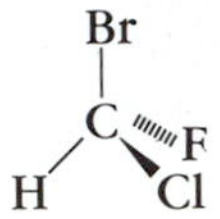

1. Assign relative priorities to each of the four different substituents according to the atomic numbers of the four atoms bonded to the chiral center. (The complete rules for assigning priorities are given in Section 5.5.)

Substituent	—Br	—Cl	—F	—H
Relative priority	3	2	1	0

2. Reorient the molecule (either mentally or by drawing it in a different orientation) so that the lowest-priority substituent is "directly behind" the central carbon atom. (A detailed discussion of how to reorient three-dimensional drawings is found in the Solution's Manual.)

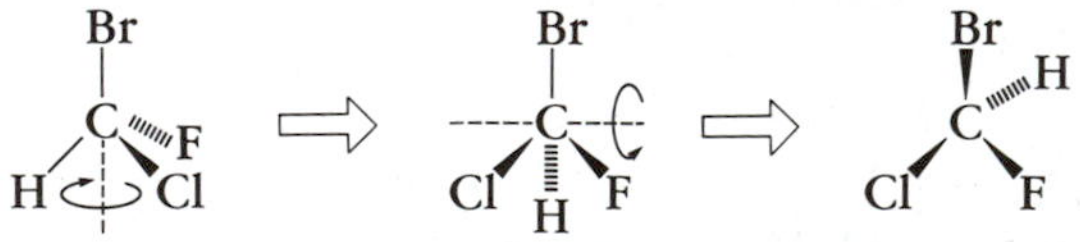

Br, C, F, H, Cl ⇨ Br, C, Cl, H, F ⇨ Br, H, C, Cl, F

3. Move successively from the highest-priority group (3) to the next two substituents (2,1). A clockwise motion defines an *R* configuration and a counterclockwise motion defines an *S* configuration.

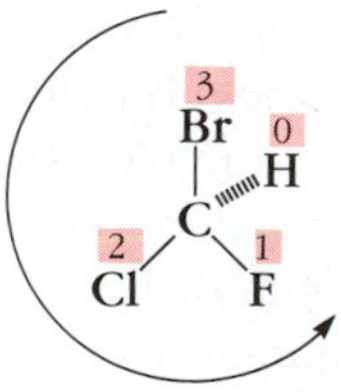

3 Br, 0 H, C, 2 Cl, 1 F — Counterclockwise, *S* configuration

Figure 6.4 Assignment of Configuration as *R* or *S* at a Chiral Center.

The procedure consists of three steps. *First,* assign priorities to the four substituents on the chiral center. These priorities are assigned according to the atomic numbers of the atoms bonded directly to the chiral center in the way that we showed previously for designating the *E*/*Z* configuration of alkenes (Section 5.5). If any two atoms bonded directly to the chiral center are the same, you must next compare the priorities of the substituents on those atoms. (If two atoms are *isotopes* of the same element, the higher atomic *mass* has the higher priority.) *Second,* orient the molecule so that the substituent with lowest priority is directly behind the chiral carbon (i.e., the bond from carbon to this substituent is pointing away from you). A thorough discussion of how to redraw molecular structures in different orientations is presented in the Solutions Manual. *Third,* trace a path along the remaining three substituents, starting at the substituent with the highest priority and proceeding toward that with the lowest priority. This will describe either a clockwise (*R*) or counterclockwise (*S*) motion.

The three-dimensional drawings that we have used in Figure 6.4 are particularly useful in assigning *R*/*S* configuration. If you place one substituent "directly behind" the central carbon atom, the remaining three substituents must all be projecting slightly toward you. By convention these three groups can be connected to the central carbon atom with lines rather than wedges, so this type of three-dimensional structure can be drawn quickly and easily:

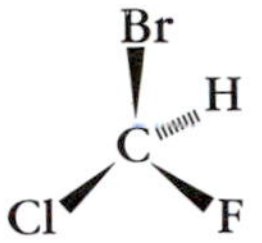

Br H C Cl F is the same as Br H C Cl F

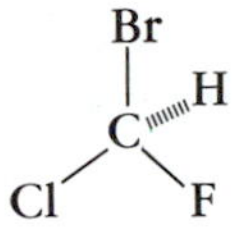

EXERCISE 6.3

Designate as *R* or *S* the configuration of each chiral center in the following structures.

(a) 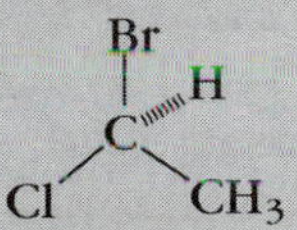 Br H C Cl CH_3

(b) 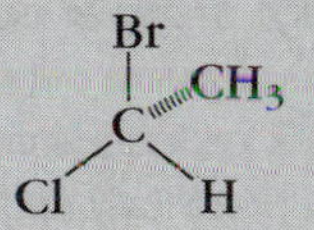 Br CH_3 C Cl H

(c) OH H C CH_3 CH_2OH

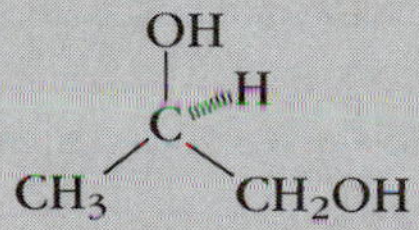

(d) 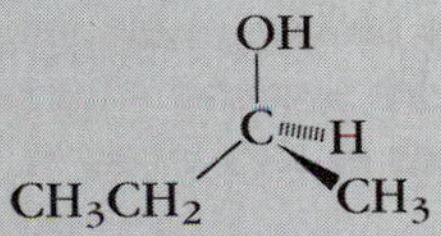 OH C H CH_3CH_2 CH_3

(e) 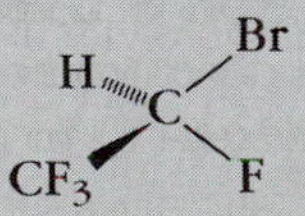 H C Br CF_3 F

(f) $BrCH_2$ CH_3 C FCH_2 OH

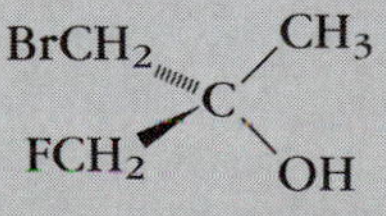

(g) 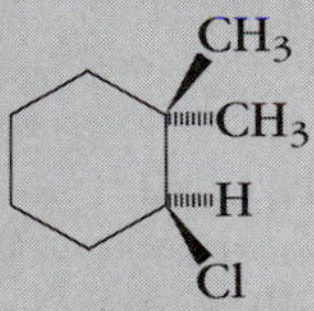 CH_3 CH_3 H Cl

Identification of Enantiomeric Relationships

The similarity between enantiomers can often make it difficult for you to determine whether two structures with the same constitution are enantiomers or are actually the same. For example, what is the relationship between the following two structures?

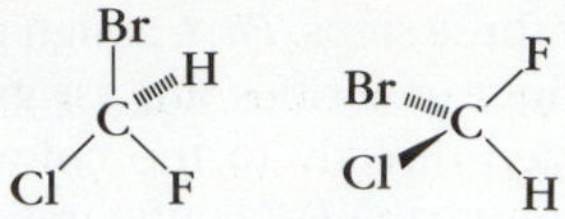

Clearly they have the same constitution because both are drawings of bromochlorofluoromethane. But are they *enantiomers* or *two drawings of the same molecule?* You could answer this question by several alternative methods, and these are summarized in Figure 6.5 below.

Example:

(a) and (b)

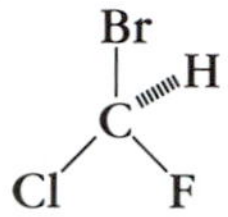

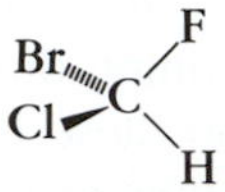

Method 1: Determine the *R*/*S* configuration of each chiral carbon.

If the two structures are the same, the configuration of the chiral center will be the same. If the structures are enantiomers, the configurations of the chiral carbon atoms will be opposite.

In this example, it is necessary to reorient structure (b):

(a) Configuration is *S*

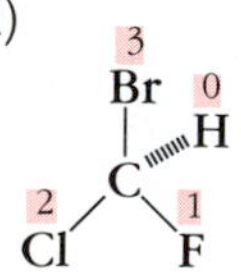

(b) Configuration is *S*

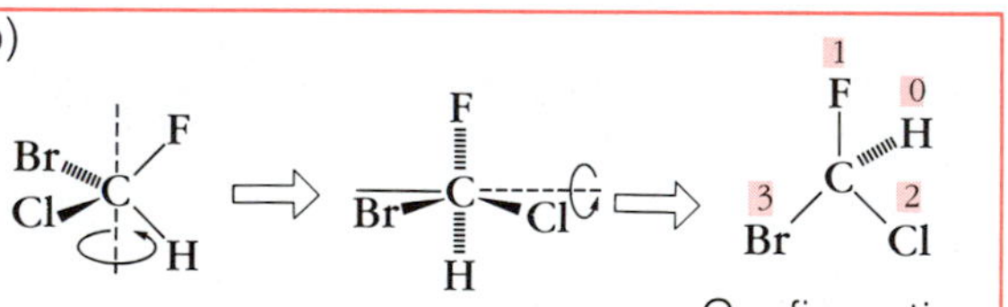

Method 2: Test to see if both are structural drawings of the same compound.

This typically requires the redrawing of one or both structures so that they are in the *same orientation* with regard to the chiral carbon atom and two of its substituents.

One possibility in the present example might be to select an orientation with the Br and the Cl both to the left and Br behind the plane of the paper, as in (b).

(a) (b)

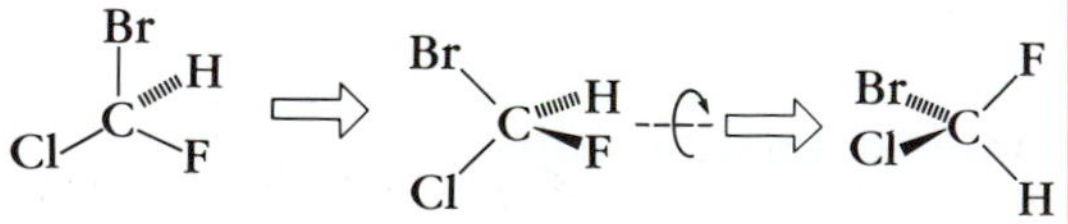

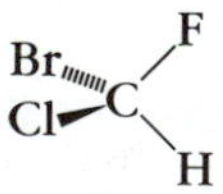

these are the same

Method 3: Test to see if the structures bear a mirror image relationship to each other.

When drawn in the same conformation and in the proper orientation, enantiomers *must* exhibit this relationship. It is typically necessary to redraw one or both structures so that they have the same orientation relative to the (imaginary) mirror for the chiral center and two of its substituents.

For example, the two structures in this case could be drawn so that the fluorine, carbon, and hydrogen are in the plane of the paper, with the fluorine above the carbon and toward the mirror and the hydrogen below the carbon and toward the mirror. Leaving structure (b) unchanged, we see:

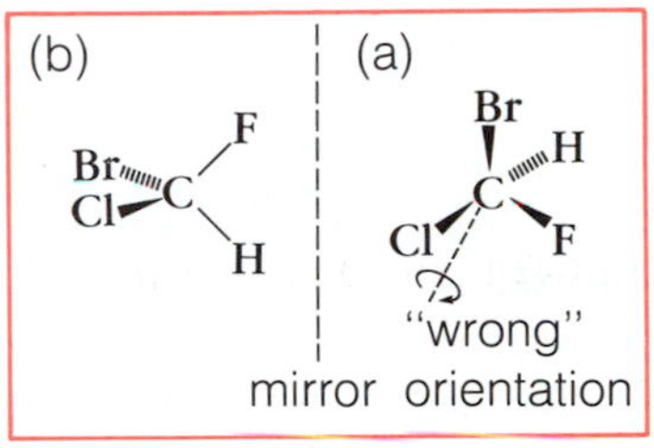

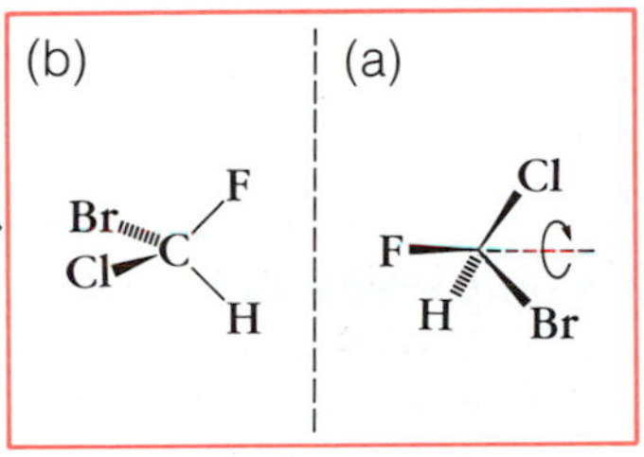

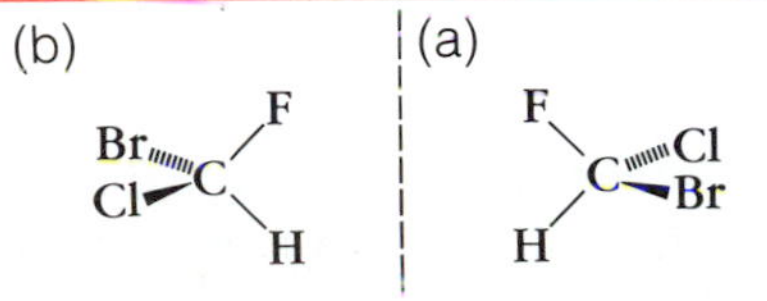

(b) | (a)

"wrong" orientation

mirror

(b) | (a)

(b) | (a)

The chlorine and bromine atoms are not in a mirror image relationship. The structures are not mirror images; they must be two drawings of the same compound.

Method 4: Construct molecular models of each of the structures and compare them.

Figure 6.5 Comparison of Two Structures That Have the Same Constitution and a Single Chiral Center to Determine Whether They are Enantiomers or the Same.

In the first method stated in Figure 6.5 you assign each chiral center as *R* or *S*. If the configuration is different for the two structures, they must be enantiomers, but if they are both *R* or both *S*, they are two drawings of the same compound. In the second method you redraw one or both molecules to test if they are *superim-*

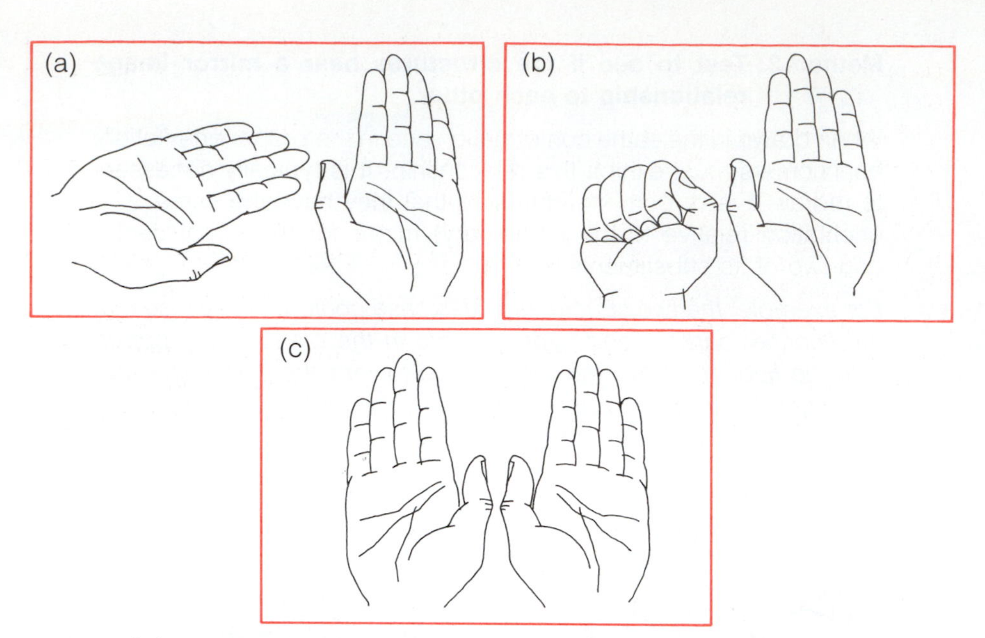

Figure 6.6 Comparison of Two Chiral Objects.
A left and right hand are mirror images, but this is not apparent in (a) because the two objects are not oriented appropriately. The two hands have the correct orientation in (b) but are in different "conformations" which obscures the mirror image relationship. Only in (c) where both objects are drawn with the appropriate orientations and in the same "conformation" is the mirror image relationship obvious.

*posable** *(i.e., the same)*. The chiral center is the key to this test, and you must draw that carbon together with any two of its substituents in the same orientation; the two remaining groups on each structure will then be oriented either in the same way or differently. If they are oriented differently, the two structures are not the same and must be enantiomers.

The third procedure shown in Figure 6.5 is a test for the mirror image relationship that enantiomers must always exhibit. You must orient the two structures appropriately with respect to an imaginary mirror; the molecules are enantiomers only if all groups and atoms show a mirror image relationship. First, redraw at least one of the structures so that the chiral carbon and any two of its substituents are placed in a mirror image relationship to the corresponding parts of the other

---

*The idea of two things being superimposable is straightforward in two dimensions. Two drawings can be superimposed, as shown for the following three-dimensional representations of a chiral molecule.

F, Br, Cl, H on C and F, Br, Cl, H on C ⟹ (superimposed drawings) ⟹ F, Br, Cl, H on C

But this cannot be done physically in three dimensions and must be a mental process. The word *superimposable* simply indicates that two species are identical in all respects; it has the same meaning as the mathematical term *congruent*. For example, two congruent triangles have the same angles, and the lengths of their corresponding sides are the same. Thus they match up point for point; they are superimposable.

structure. The two structures are now in the proper relative orientations. Next you compare them. If they exhibit a mirror image relationship they are enantiomers; otherwise they are the same compound.

Finally, the fourth method stated in Figure 6.5 employs molecular models. If you construct a molecular model of each of the structures under consideration, you can manipulate the models in order to test whether the two structures are the same or have a mirror image relationship. In contrast to drawings on paper, molecular models can be turned around, viewed from the side or back, etc., without any concern that you might have accidentally changed the configuration.

One important precaution must be taken when using any of the last three methods for comparing two chiral structures. Not only must you orient the structures appropriately, but you must draw both in the *same conformation.* As we have shown in Figures 6.6 and 6.7, two objects cannot be properly compared if they are in different conformations.

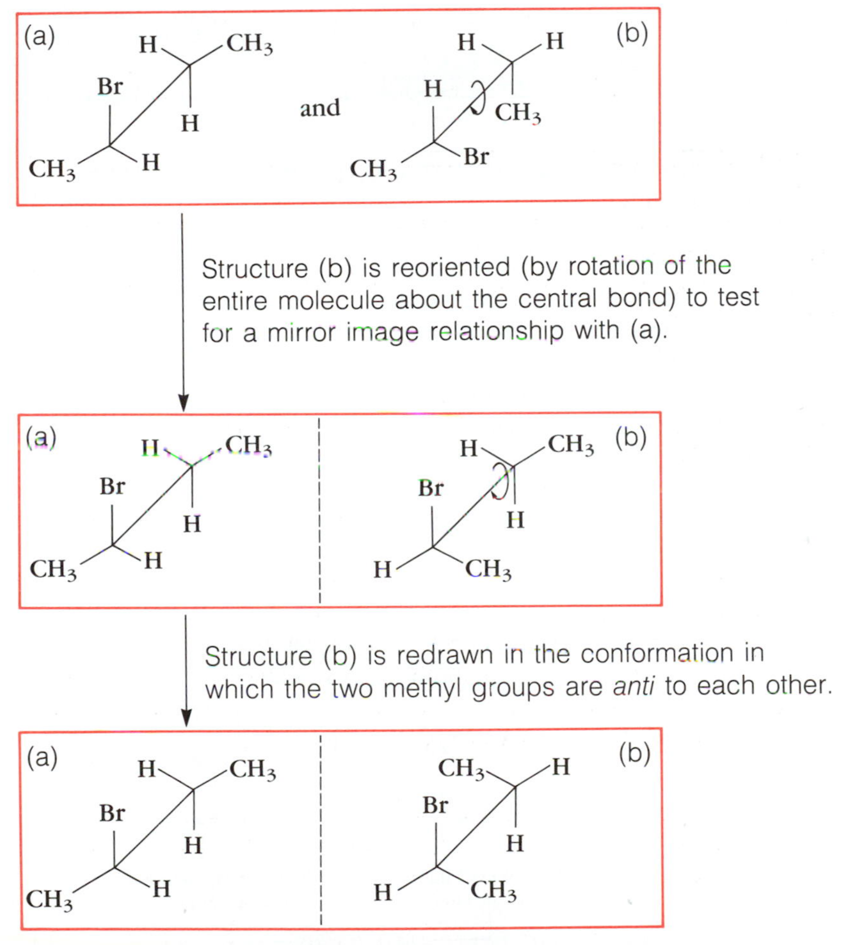

Figure 6.7 Comparison of Two Drawings of 2-Bromobutane.
Are they the same compound or enantiomers? The molecules are not drawn appropriately at the top of the figure. At the center of the figure the structures have the appropriate orientation at the chiral center to test for a mirror image relationship, but they are in different conformations. At the bottom of the figure the structures have both the appropriate orientation and conformation, so the mirror image relationship is readily apparent.

EXERCISE 6.4

Describe each pair of structures as either the *same compound* or *enantiomers.*

(a) (b)

(c)

(d)

6.4 DIASTEREOMERS AND COMPOUNDS WITH MULTIPLE CHIRAL CENTERS

What are diastereomers? As with all types of isomerism, we are concerned with a relationship between two structures, but we define diastereomers in a negative way. **Diastereomers** *are stereoisomers that are not enantiomeric,* that is, they differ in their three-dimensional structures but are not mirror images. We have already seen two types of diastereomers in previous chapters; *E-Z* isomers of alkenes (Chapter 5) and *cis-trans* isomers of cyclic compounds (Chapter 3). The following pairs of structures are examples of diastereomers:

(*E*)-2-Chloro-2-butene (*Z*)-2-Chloro-2-butene

Diastereomers

cis-1-Chloro-2-fluorocyclohexane *trans*-1-Chloro-2-fluorocyclohexane

Diastereomers

The preceding examples illustrate the two common types of diastereomers: alkenes for which two configurations of the double bond are possible; and compounds containing more than one chiral center. As you can see for the second pair of structures, diastereomers result when compounds with multiple chiral centers differ in configuration at *some but not all* chiral centers. The two cyclohexane derivatives shown above have the *same* configuration at C-1 but *opposite* configurations at C-2, and are therefore diastereomers.

The presence of two chiral centers can also lead to diastereomers in acyclic compounds:

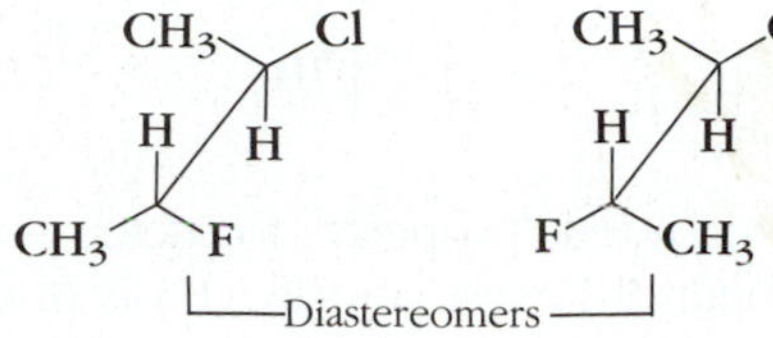

Diastereomers

Visual inspection clearly shows that these two chlorofluorobutanes are not identical, yet they both have the same molecular formula and the same constitution, so they must be stereoisomers. Since they are not mirror images, they can only be diastereomers. They have the same configuration at the "back" carbon but different configurations at the "front" carbon.

Nomenclature of Molecules with More than One Chiral Center

How are molecules named when they contain more than one chiral center? As long as the complete three-dimensional structure is known, we simply specify the configuration of each chiral center in turn. Consider the following 2-chloro-3-fluorobutane:

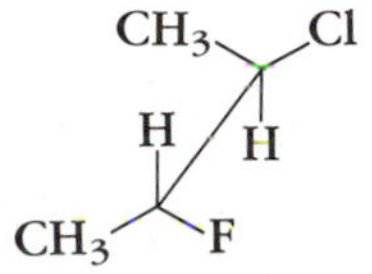

We can determine the configuration of C-2 by assigning priorities to each of the substituents at this position and reorienting the molecule with the lowest-priority group pointed away from us. The substituent with lowest priority is hydrogen, so we can orient the molecule in the following way:

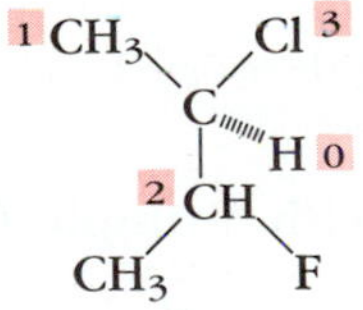

The configuration at C-2 is *R*

The configuration at C-3 is determined similarly by reorienting the molecule so that the substituent of lowest priority on C-3 is pointed away from us.

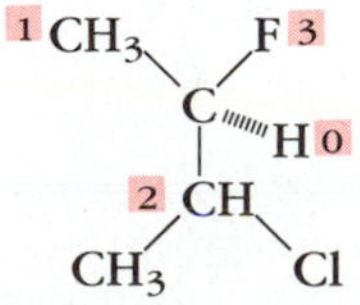

The configuration at C-3 is also *R*

Therefore the full name of the compound is (2*R*,3*R*)-2-chloro-3-fluorobutane.

Cyclic compounds with multiple chiral centers are analyzed in the same way, and we will illustrate this by naming the following enantiomer of *trans*-1-chloro-2-fluorocyclohexane.

Cl
H
H
F

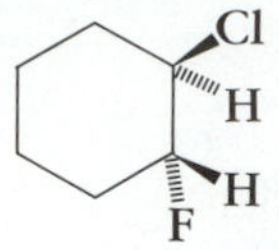

The molecule is already oriented "properly" for determining the configuration at C-1, because the substituent of lowest priority (H) is oriented away from us.

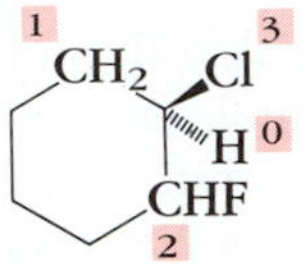

1 CH_2 3 Cl H 0 CHF 2

The configuration at C-1 is *R*

The molecule is not oriented properly for evaluation of C-2, however, because the substituent of lowest priority (again a hydrogen) is pointed toward, rather than away from us. Therefore we must turn the structure over.

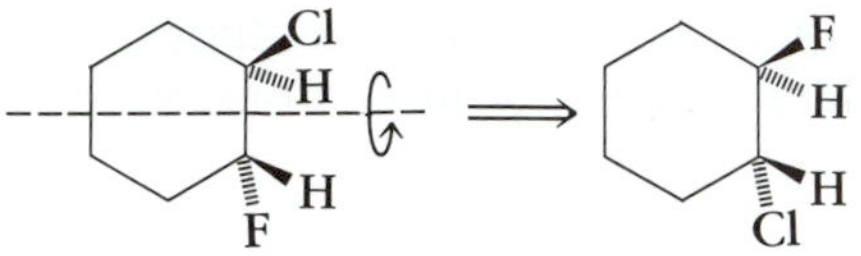

Cl H H F ⟹ F H H Cl

After priorities have been assigned to the substituents, you can see that C-2 also has the *R* configuration.*

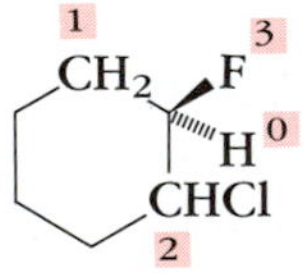

1 CH_2 3 F H 0 CHCl 2

The configuration at C-2 is *R*

Therefore the enantiomer of *trans*-1-chloro-2-fluorocyclohexane that we have drawn here can be named (1*R*,2*R*)-1-chloro-2-fluorocyclohexane.

Stereochemical Drawings of Molecules with Multiple Chiral Centers

In Section 1.4 we introduced several types of three-dimensional representations that depict the stereochemical features of an individual carbon atom, and these representations can also be used to draw structures with more than one chiral center. For example, dashed lines and wedges clearly define the stereochemistry in cyclic compounds:

*You may have noticed that when a hydrogen (the lowest-priority group) on a chiral center is oriented *toward* the viewer, the clockwise/counterclockwise arrangement of the remaining three substituents is precisely opposite to that needed to assign *R/S* configuration. If great care is used you could *predict the "wrong answer"* and thereby know the correct configuration.

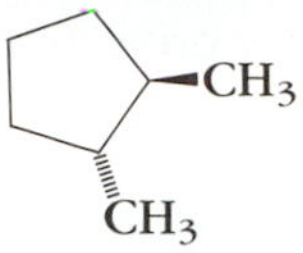
CH3
CH3

trans-1,2-Dimethylcyclopentane

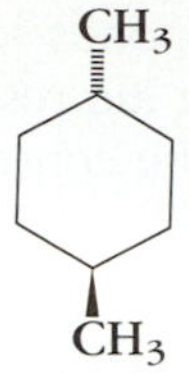
CH3
CH3

trans-1,4-Dimethylcylohexane

However, this type of drawing can be cumbersome for acyclic molecules:

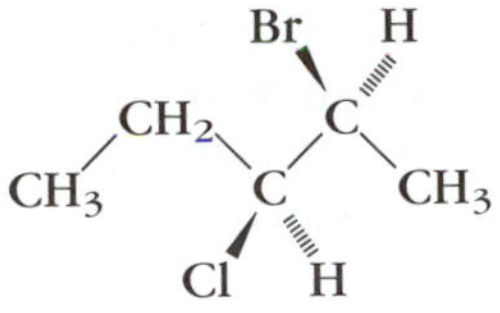
Br H
CH2 C
CH3 C CH3
Cl H

(2*S*,3*S*)-2-Bromo-3-chloropentane

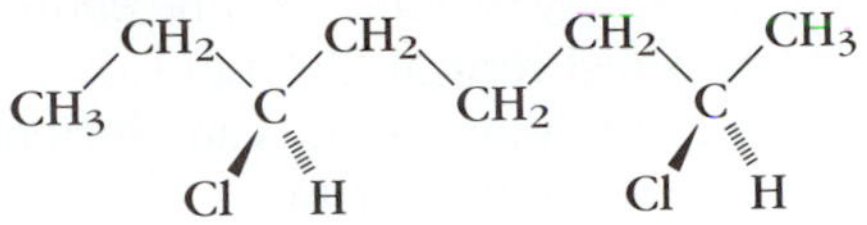
CH2 CH2 CH2 CH3
CH3 C CH2 C
Cl H Cl H

(2*R*,6*S*)-2,6-Dichlorooctane

When there are only two chiral centers and they are on adjacent atoms, Newman projections and sawhorse drawings are more convenient:

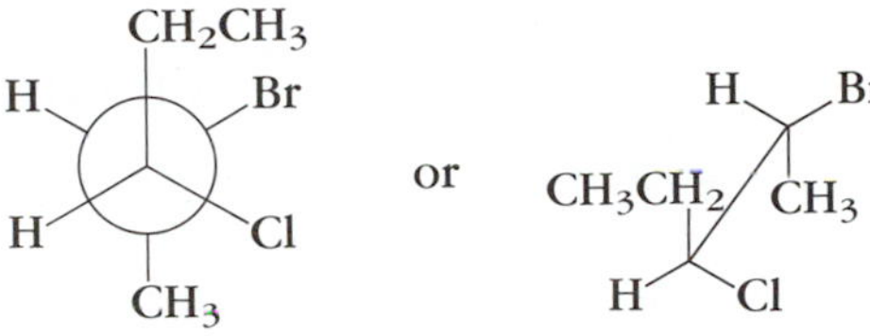
CH2CH3
H Br
H Cl
CH3
or
H Br
CH3CH2 CH3
H Cl

(2*S*,3*S*)-2-Bromo-3-chloropentane

When a molecule has three or more chiral centers or when two chiral centers are not on adjacent atoms, a **Fischer projection** is a particularly useful drawing. This type of three-dimensional representation was devised by the German chemist Emil Fischer, whose stereochemical studies of sugars earned him the Nobel Prize for chemistry in 1902. Fischer projections fully define the *configuration* of each chiral center, but as you will see in the following discussion, they provide no information whatsoever about the *conformation* of a molecule.

In order to explain the stereochemical significance of Fischer projections, we will first consider molecules containing only a single chiral center. A chiral carbon is not explicitly drawn in a Fischer projection but is defined by the intersection of the lines representing bonds to its four substituents, as illustrated for (*R*)-2-chlorobutane:

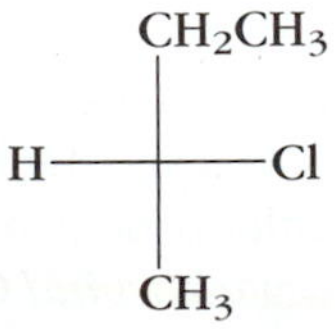
CH2CH3
H Cl
CH3

The orientation of these intersecting lines has a very specific meaning. *The substituents on the vertical line are further away from you than the chiral center, and the substituents on the horizontal line are closer to you.* The preceding Fischer projection therefore corresponds to the following drawing (and some students have found it useful to use the resemblance of the horizontal wedges to a bow tie in order to remember the conventions for a Fischer projection).

CH_2CH_3
H—C—Cl
CH_3

The specific orientation of the "vertical" and "horizontal" substituents in a Fischer projection requires that you use considerable care when manipulating this type of drawing. Any operation on the entire molecule which does not maintain the original horizontal-vertical directions of the substituents is invalid. Consequently, the only legitimate way that you can reorient an entire Fischer projection is to turn it 180° in the plane of the paper:

CH_2CH_3 / H—+—Cl / CH_3 ⟹ CH_3 / Cl—+—H / CH_2CH_3

(*R*)-2-Chlorobutane (*R*)-2-Chlorobutane

When drawing Fischer projections for compounds with more than one chiral center, we draw the carbon chain that includes these centers as a single vertical line. As shown by the following examples, it is not necessary for the chiral centers to be adjacent to one another nor must each carbon on the vertical line be chiral.

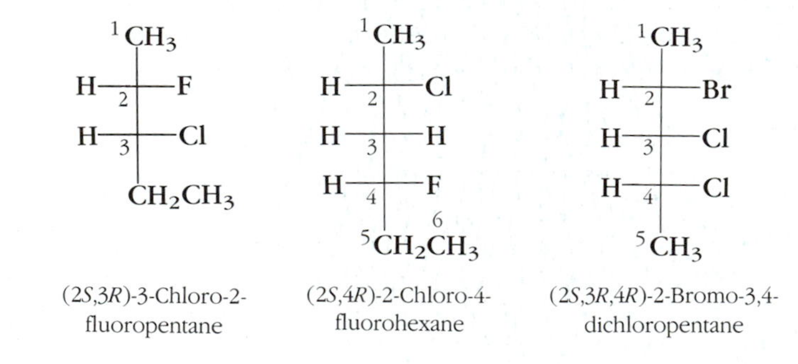

(2*S*,3*R*)-3-Chloro-2-fluoropentane (2*S*,4*R*)-2-Chloro-4-fluorohexane (2*S*,3*R*,4*R*)-2-Bromo-3,4-dichloropentane

As in the case of the Fischer projection for a molecule having only a single chiral center, horizontal substituents are always closer to the viewer than the chiral center to which they are attached.

Problems may arise, however, in viewing the vertical substituents. In the preceding drawing of (2*S*,3*R*)-3-chloro-2-fluoropentane, the configuration of C-2 is defined as having both C-1 and C-3 located *behind* C-2. Yet the configuration indicated for C-3 requires C-2 to be located behind C-3. This apparent conflict arises because the drawing is a *projection,* that is, all the carbon atoms of the vertical line have been "squashed" into the plane of the paper. As a result, each chiral center must be considered individually in order to say that the two vertical substituents are farther from the viewer than the chiral center.

The first of the three preceding examples, (2*S*,3*R*)-3-chloro-2-fluoropentane, corresponds to a sawhorse drawing in the eclipsed conformation viewed from above:

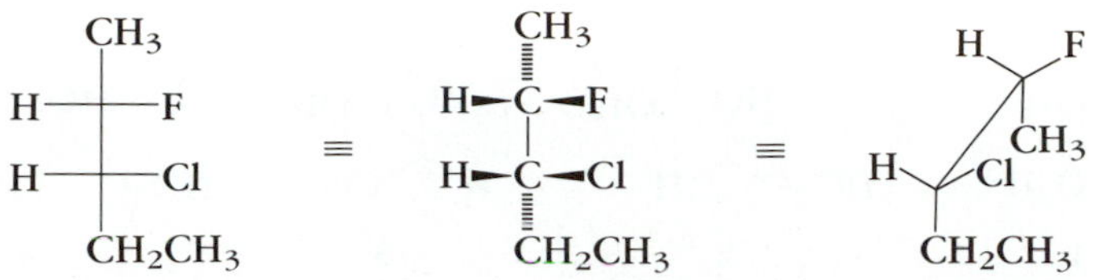

This is not a minimum-energy conformation, but corresponds to a transition state between two staggered conformations. The same situation is found for all Fischer projections, and they simply do not correspond to the actual conformations of molecules. Their sole purpose is to define the configuration of each chiral center.

Chirality and Symmetry in Molecules Having Multiple Chiral Centers

We have previously shown that a molecule will be chiral when it contains a single chiral center. Now we will turn to molecules containing more than one chiral center. Will such molecules always be chiral or will they sometimes be achiral? How can you evaluate whether or not a molecule is chiral when it contains multiple chiral centers?

Since a chiral center has two possible configurations, we would expect a molecule with n chiral centers to have a total of 2^n stereoisomers. This is indeed the case for 2-bromo-3-chlorobutane. This molecule has two chiral centers, and four stereoisomers exist—two pairs of enantiomers.

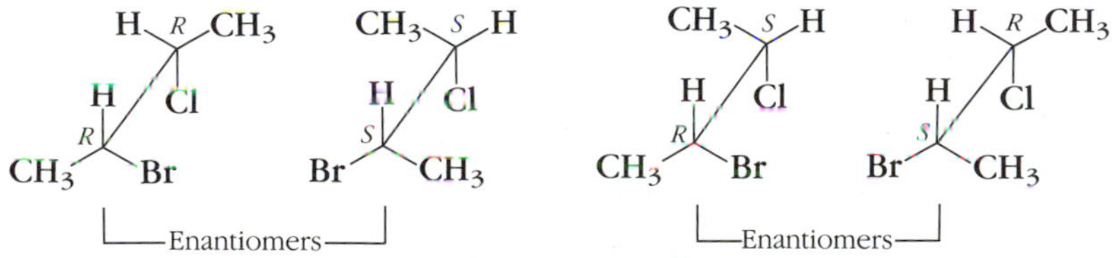

The enantiomeric relationships are labeled in the preceding drawings; the relationship between any other two of these four stereoisomers is that of diastereomers. Notice that the enantiomers differ in configuration at *each* and *every* chiral center.

As the number of chiral centers increases, so should the number of stereoisomeric structures. 2-Bromo-3-chloro-4-fluoropentane, for example, has three chiral centers, and eight stereoisomers exist. (Draw them.) However, *the number of possible stereoisomers will be less than 2^n when the molecule has chiral centers which are constitutionally equivalent,* that is, when they have the same four substituents. An example of a molecule containing constitutionally equivalent chiral centers is tartaric acid, a compound which played a pivotal role in the development of organic stereochemistry (Section 6.6).

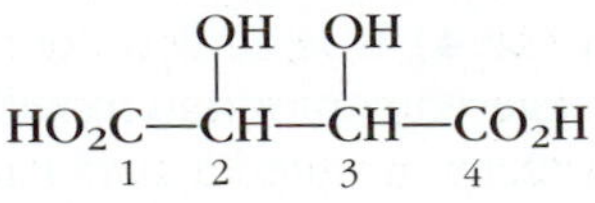

$$HO_2C-CH(OH)-CH(OH)-CO_2H$$

Tartaric acid

Each of the two chiral centers (C-2 and C-3) in tartaric acid has the same four substituents: —H, —OH, —CO_2H, and —$CH(OH)CO_2H$. By drawing the *R* and *S* configurations at both the front and back carbons, we can generate four structures:

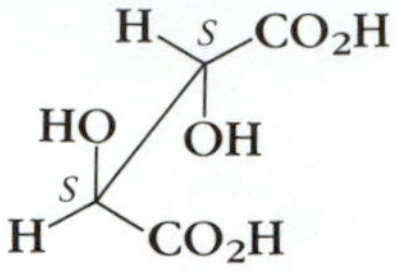

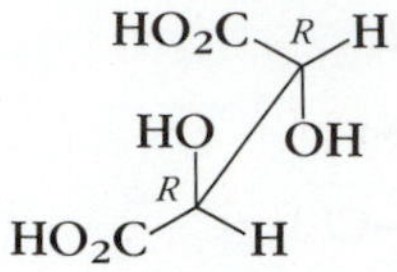

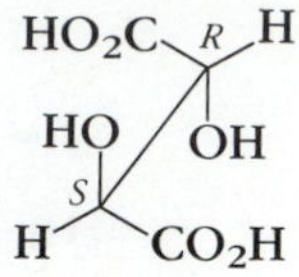

Structures **1** and **2** are clearly mirror images, as are **3** and **4**. But are the mirror images different in each case? Earlier in this chapter we showed that if a molecule posesses a plane of symmetry, it is achiral. Therefore we can search for reflection symmetry in these structures. By drawing **1** with the —OH groups eclipsed, we can demonstrate the absence of reflection symmetry.

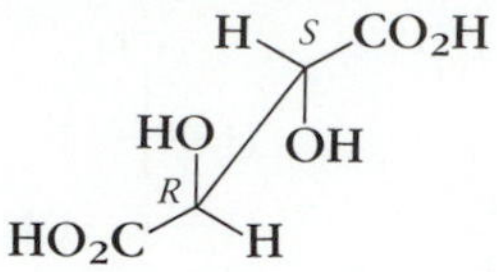

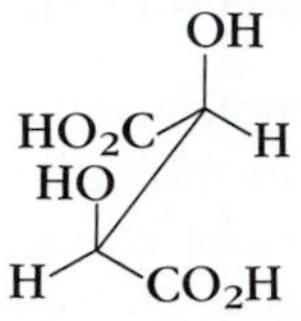

Therefore **1** is chiral and it must be different from its mirror image, **2**; in other words, **1** and **2** are enantiomers. Applying the same test to **3**, we find that this structure does possess reflection symmetry, with a plane of symmetry corresponding to the plane of the paper. Each substituent on the front carbon atom "sees" its mirror image on the rear carbon atom.

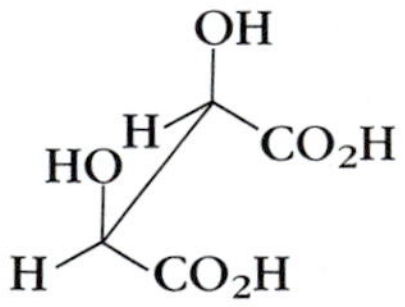

Structure **3** is therefore achiral and is identical to its mirror image, **4** despite the presence of two chiral centers; in other words, **3** and **4** are two drawings of the *same molecule.* We can show this in another way by redrawing **4**.

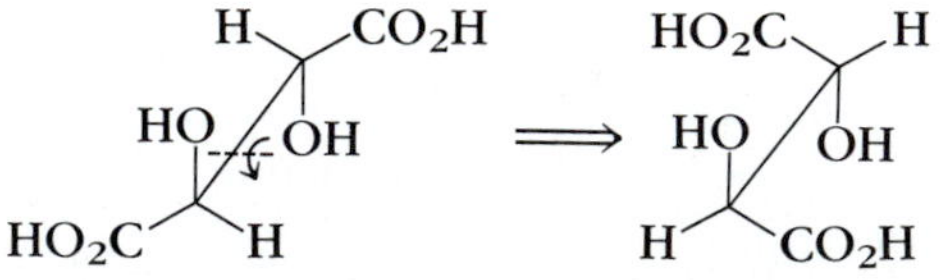

We could also show that **3** (or **4**) is achiral by demonstrating another type of reflection symmetry. The staggered conformation with the OH groups in an *anti* relationship has a symmetry element called a *center of symmetry.* A line from the center of symmetry to any atom in the molecule will reach an equivalent atom when extended an equal distance in the opposite direction. The following drawing is reoriented to show this more clearly, a dark circle being used to denote the center of symmetry.

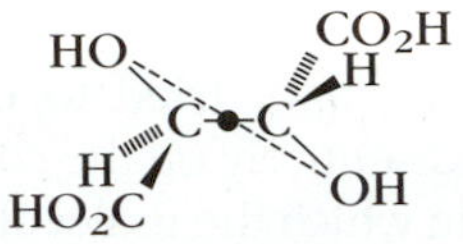

HO, CO_2H, H, C•C, H, HO_2C, OH

In most cases a plane of symmetry is easier to visualize than a center of symmetry, but the presence of either one proves that the molecule is achiral.

We can now conclude that there are only three stereoisomeric tartaric acids, and the following drawings illustrate the stereoisomeric relationships between them. Structure **3** is an example of a *meso compound, a molecule that is achiral despite the presence of chiral centers.*

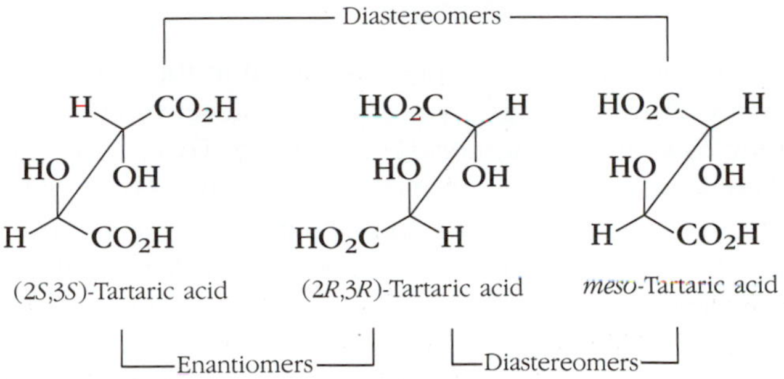

Diastereomers

(2*S*,3*S*)-Tartaric acid (2*R*,3*R*)-Tartaric acid *meso*-Tartaric acid

Enantiomers Diastereomers

Fischer projections can be very convenient for showing a plane of symmetry in a compound because the stereochemical convention used to draw them corresponds to the necessary eclipsed conformation of the carbon skeleton. The three isomeric tartaric acids all have constitutionally equivalent chiral centers, but only in the case of *meso*-tartaric acid is reflection symmetry present (dotted line).

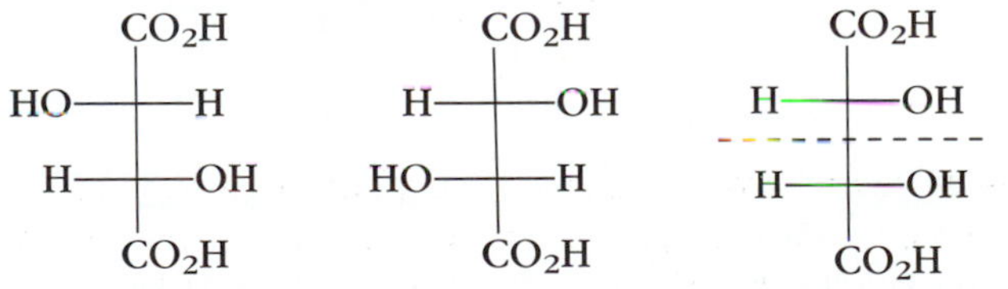

(2*S*,3*S*)-Tartaric acid (2*R*,3*R*)-Tartaric acid *meso*-Tartaric acid

When will a molecule with multiple chiral centers be a *meso* compound? It must have an even number of chiral atoms, and the molecule as a whole must possess reflection symmetry. In other words, *every chiral center in the molecule must be matched by a second chiral center that is constitutionally equivalent but has the opposite R/S* configuration. As with other problems of chirality and symmetry, it is often helpful to refer to a familiar example. In this case an analogy can be made with the human body, which to a first approximation has a vertical plane of symmetry. This symmetry exists *despite* such "chiral centers" as hands and feet. The symmetry plane is possible only because you have one right hand and one left hand, not two of the same. The identical requirement holds for a plane of symmetry in a molecule containing several chiral centers: each such center must be matched by another that is constitutionally equivalent but has the opposite configuration.

Cyclic Compounds with Multiple Chiral Centers

In the preceding discussion we have tried to emphasize that the property of chirality for a molecule depends entirely on the *configuration* at the chiral centers and not on the *conformation* in which the molecule is drawn. This remains true for cyclic compounds, and you must be careful not to reach incorrect conclusions about chirality based on the evaluation of individual conformations. The potential problems are illustrated by the case of *cis*-1,2-dichlorocyclohexane.

If the second compound is reoriented as shown in the third structure, you can see that the two conformations are mirror images. Moreover, they are not superimposable, so they have an enantiomeric relationship. This is not unusual behavior for individual conformers, and if you refer back to Figure 2.6 you will find the same relationship between the two *gauche* conformations of butane. But for either butane or for *cis*-1,2-dichlorocyclohexane the conformers are rapidly interconverting, and we are concerned with the "average" structure of a molecule over a period of time; this, in turn is equivalent to the "average structure" of a large number of molecules at any instant. Since we are concerned with configuration, there is no need to utilize structures which show the conformation of a cyclic molecule when determining whether or not it is chiral. It is usually preferable to draw cyclic compounds as polygons, using wedges and dashed lines to indicate stereochemistry. Such drawings clearly show that *cis*-1,2-dichlorocyclohexane is a *meso* compound and that the *trans* isomer is chiral.

Enantiomers

Meso
cis-1,2-Dichlorocyclohexane

trans-1,2-Dichlorocyclohexane

A note of caution: For these three-dimensional representations the substituent connected to a carbon atom by a dashed line is in reality *directly behind* the other substituent; only for legibility is it drawn next to the other. Therefore, a molecule such as chlorocyclopentane has a plane of symmetry (dotted line) that bisects not only the carbon atom at the top of the ring but also the chlorine and carbon atoms attached to it.

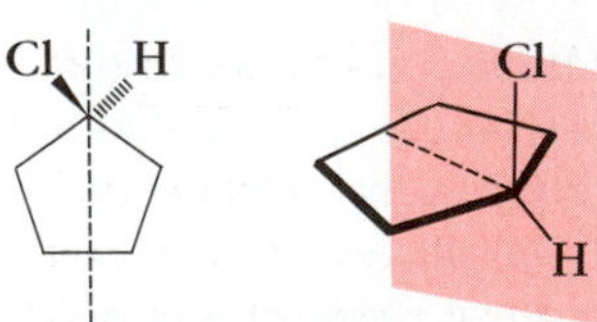

Chlorocyclopentane (achiral)

EXERCISE 6.5

Redraw the following compounds using sawhorse representations:

(a) CH_3; H—C—Cl; H—C—Cl; CH_2CH_3

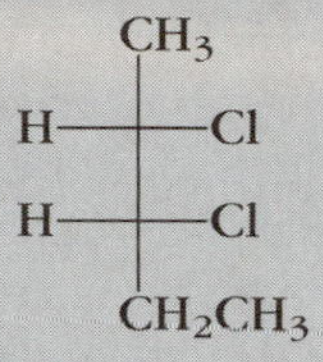

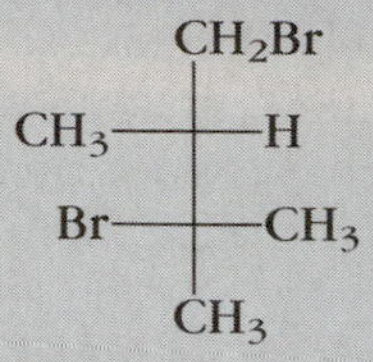

(b) CH_2Br; CH_3—C—H; Br—C—CH_3; CH_3

(c) CH_3; H—C—H; H—C—Br; Br—C—H; CH_3

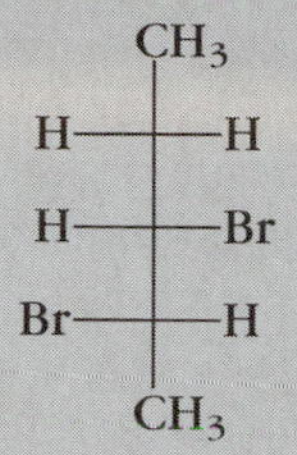

EXERCISE 6.6

Name the compounds in Exercise 6.5, using full stereochemical designations.

EXERCISE 6.7

Determine if the following sets of compounds are *diastereomers, enantiomers,* or two drawings of the *same compound.*

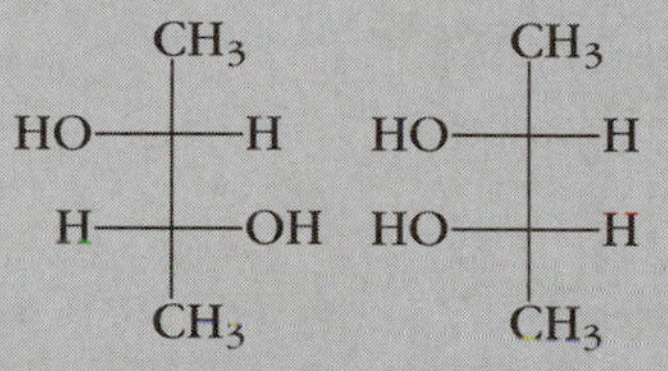

(a) CH_3; C_2H_5—C—Br; H   and C_2H_5; CH_3—C—H; Br

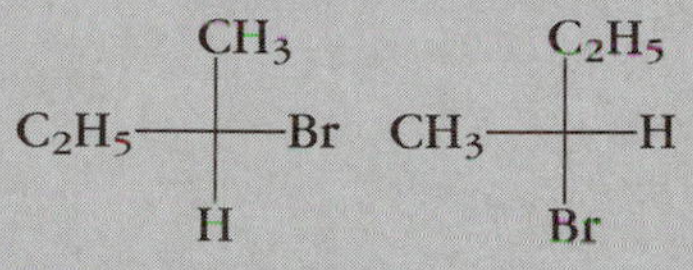

(b) CH_3; HO—C—H; H—C—OH; CH_3 and CH_3; HO—C—H; HO—C—H; CH_3

(c) CH_3; H—C—H; Br—C—H; CH_3 and CH_3; H—C—H; H—C—Br; CH_3

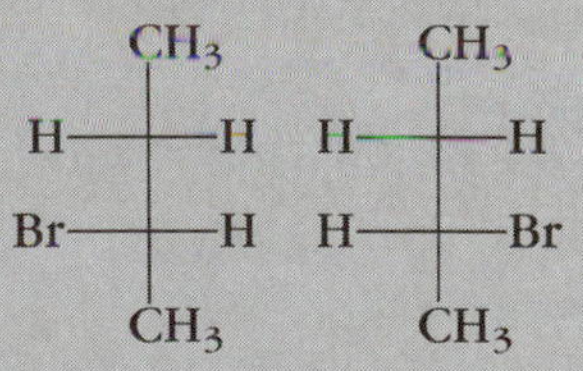

(d) CH_3 / CH_3 and CH_3 / CH_3

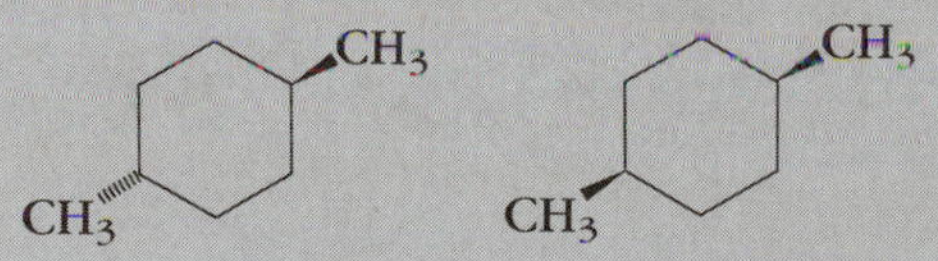

EXERCISE 6.8

Two of the staggered conformers of 1-bromo-2-chloroethane are nonsuperimposable mirror images. They cannot be isolated because they interconvert rapidly, and therefore they are not considered enantiomers. Draw these mirror image conformations.

6.5 PHYSICAL PROPERTIES OF STEREOISOMERS

Chemical reactions rarely afford a single product in 100 percent yield. In fact yields may vary over a wide range, and in nearly every case a *purification* of the product is necessary. Even if a reaction proceeds in very high yield we must usually remove the product from the solvent and any inorganic reagents that were used. Any purification procedure requires the *separation* of two or more compounds, and this is possible only if the compounds have different physical or chemical properties.

The most common separation methods rely on differences in physical properties. Two compounds with different boiling points can be separated by distillation, and compounds with different solubilities can be separated by techniques such as recrystallization. Many organic reactions afford isomeric products, and the isolation, separation, and purification of such products rely upon differences in physical properties. It is therefore imperative that you understand when to expect two isomeric compounds to share the same properties and when to expect them to exhibit different properties.

Physical Properties

Two isomeric compounds have the same molecular formula but differ structurally in some way, and this structural difference leads to distinctive physical and chemical properties for each of the compounds. As an example, the melting point of pentane is more than 100° lower than that of 2,2-dimethylpropane (Section 4.9). Similarly, the boiling points of these compounds are about 25° apart. But these two alkanes are *constitutional* isomers. Do *stereoisomers* exhibit comparable differences in physical properties? The answer to this question is a qualified yes, as long as we restrict our discussion to *diastereomers*. You can see this for the diastereomeric 2-octenes, which differ in both melting point and boiling point (although the boiling points are very similar).

H, H, CH_3, $(CH_2)_4CH_3$ — C=C

(*Z*)-2-Octene
(mp = −100°C;
bp = 126°C)

CH_3, H, H, $(CH_2)_4CH_3$ — C=C

(*E*)-2-Octene
(mp = −88°C;
bp = 125°C)

Different physical properties are similarly observed for the diastereomeric tartaric acids.

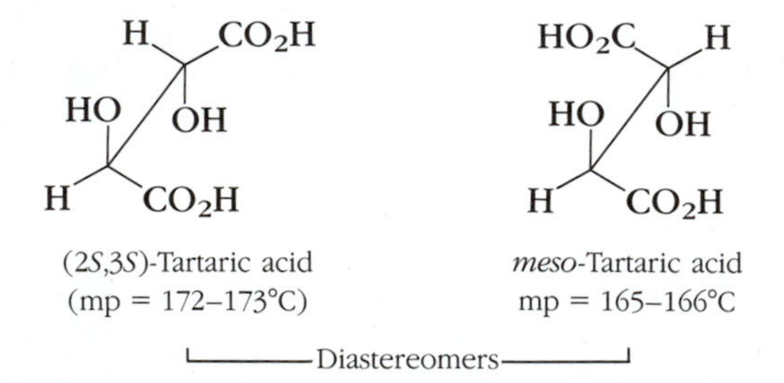

(2*S*,3*S*)-Tartaric acid
(mp = 172–173°C)

meso-Tartaric acid
mp = 165–166°C

Constitutional isomers and diastereomers all have different molecular shapes, and it should not surprise you that they exhibit different physical properties. But what about enantiomers? Enantiomers differ only in chirality and they have the same shape, so it should not surprise you to learn that nearly all their physical and chemical properties are identical. For example, the enantiomeric tartaric acids exhibit the same melting point. The melting points (obtained with samples of slightly different purity) were reported as 172–173°C for (2*S*,3*S*)-tartaric acid and 170°C for (2*R*,3*R*)-tartaric acid. For accurate determinations with samples of the same purity, enantiomers will have *exactly* the same melting point, boiling point, and spectroscopic behavior (Chapter 9). The only physical property which differs for two enantiomers is *optical activity* (Section 6.6), and the only chemical property in which they differ is reactivity toward a chiral reagent. The *only* way that we can separate a mixture of two enantiomers is by their interaction with a *chiral agent;* the *only* physical way to distinguish between two enantiomers is through their different interactions with *polarized light.* We shall examine these two types of interactions in the following discussions of *resolution* and *optical activity.*

Resolution

Resolution is the process of resolving or separating a mixture of two enantiomers. As we indicated previously, this requires interaction of the enantiomers with a chiral reagent. Figure 6.8 illustrates one possible method for resolving a pair of enantiomers.

A mixture of enantiomeric carboxylic acids is treated with only one of the enantiomeric forms of a chiral amine. Each of the enantiomeric carboxylic acids forms a salt that has two chiral centers, but the products differ at only one of these centers, so these salts are *diastereomeric.* Since diastereomers have different physical properties, this mixture can be separated by standard laboratory methods into the *R,R* and *S,R* salts. As shown by the following equation, acidification of one of these salts would regenerate just one of the enantiomeric carboxylic acids.

$$RCO_2^- \; R'{-}NH_3^+ \xrightarrow{HCl} RCO_2H + R'{-}NH_3^+ \; Cl^-$$

R,R salt — *R*-carboxylic acid — *R*-ammonium salt

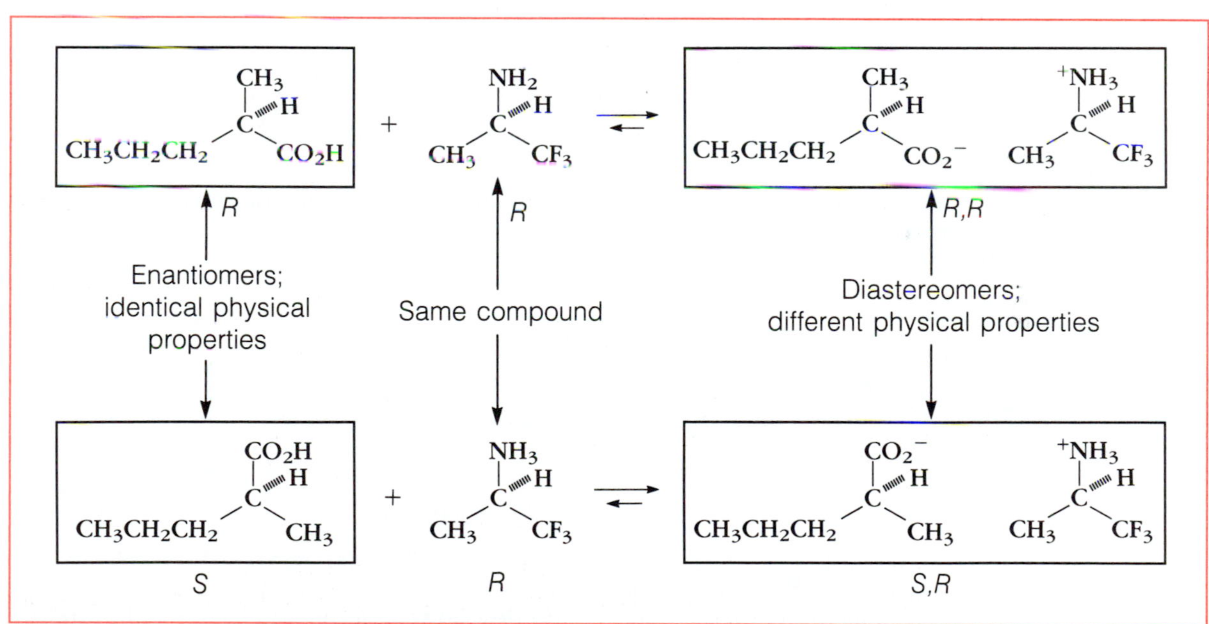

Figure 6.8 A Method for Resolution of Enantiomers.
Reaction of two enantiomeric acids with a chiral reagent produces a pair of diastereomeric salts. Although each of the enantiomers has the same physical properties, the diastereomeric products have different physical properties and can be separated by a technique such as crystallization. (In a reaction not shown, the individual enantiomers of the carboxylic acid may then be obtained from the salts by an acid-base reaction.)

Because each of the diastereomeric salts would yield only one of the enantiomeric carboxylic acids, the overall process would result in the *resolution* of the original mixture of enantiomers.

6.6 OPTICAL ACTIVITY

We have indicated that enantiomeric compounds interact differently with polarized light, and this phenomenon played an important role in the historical development of stereochemistry. During the early part of the nineteenth century a number of European scientists were striving to understand the nature of light. They discovered that light which was reflected from a surface or which had passed through a prism made of the mineral Iceland spar became *plane-polarized* (Figure 6.9).

Subsequent work by nineteenth century scientists showed that various substances can cause *rotation* of the plane of the polarized light, a phenomenon known as **optical activity** (Figure 6.10), which is measured with a *polarimeter*.

When the optical activity of a substance is measured, the plane of the polarized light undergoes rotation through some angle, α, and this is the *optical rotation* of the sample. When the optical rotation is in a clockwise direction, the sample is *dextrorotatory* (from the Latin *dexter,* "right") and is designated as (+). When rotation is in a counterclockwise direction, the sample is *levorotatory* (from the Latin *laevus,* "left") and is designated as (−). (The designations *d* and *l* were formerly used instead of (+) and (−), and you may still encounter these symbols in other books.) A 50–50 mixture of enantiomers exhibits no net optical rotation. Such a mixture is called **racemic** and is often designated as (±).

Some of the aspects of a polarimeter that we have illustrated in Figure 6.10 can be easily simulated with the lenses from a defunct pair of Polaroid sunglasses, which are made of a polarizing material. If you place one of the lenses behind the other and then rotate it, the effect shown in parts (a) and (b) of the figure can be demonstrated. When the lenses are oriented in the *same* way, the transmission of light is maximized, but if one of them is turned by 90°, the transmission decreases to a minimum. If you experiment with a single polarizing lens, you can demon-

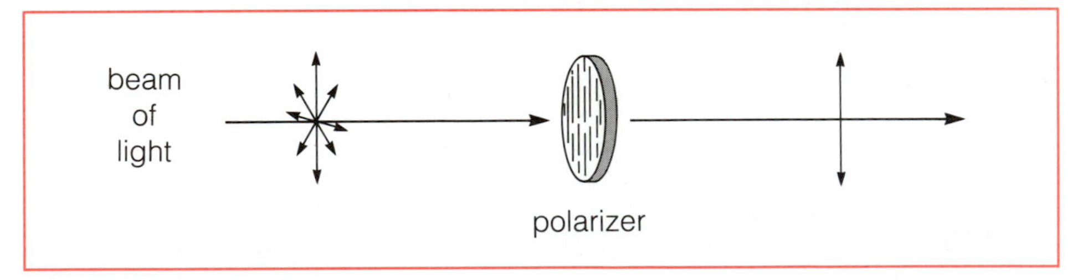

Figure 6.9 Polarization of Light.
The wave properties of light can be described in terms of electric and magnetic fields. A beam of ordinary light traveling from left to right in the plane of the paper has electric field components in all directions perpendicular to the direction of the beam. After passing through a polarizer (such as a prism of Iceland spar) the electric field components are still perpendicular to the beam, but they lie in a single plane, in this case, the plane of the paper. Such light is called *plane-polarized.*

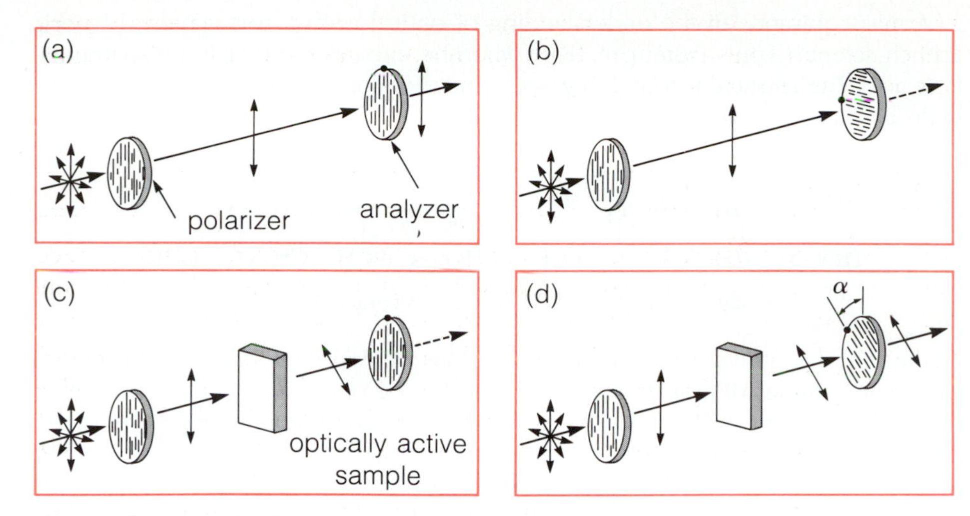

Figure 6.10 Optical Activity.
(a) An idealized polarimeter. A light beam becomes plane-polarized by passing through a polarizer. The beam then passes through a second polarizer, called an *analyzer.* The intensity of the beam is at its maximum level after it passes through the analyzer because the analyzer has the same orientation as the polarizer. (b) When the orientation of the analyzer is perpendicular to that of the polarizer (note position of mark on analyzer), the intensity of the beam is minimized. (c) The orientation of the analyzer is the same as that of the polarizer, but the presence of an optically active sample causes rotation of the plane of the polarized light. Consequently the intensity of the beam emerging from the analyzer is less than maximum. (d) The analyzer has been rotated by an angle α to maximize the intensity of the light beam. The sample has an *optical rotation* equal to α.

strate that *reflected* light (for example, sunlight reflected from the hood of an automobile) is partially plane-polarized. When the lens is rotated, an orientation can be found which minimizes the transmission of the reflected light.

Historical Development

The history of optical activity provides a fascinating glimpse into the evolution of modern stereochemistry. During the early 1800s individual crystals of quartz had been found to rotate polarized light, and sometimes two crystals would rotate light in opposite directions. In these cases the two crystals were always chiral and had a mirror image relationship to each other. This was the first indication of a correlation between optical activity and chirality. In 1815 Jean Baptiste Biot (France) reported that some liquid organic compounds and some *solutions* of organic compounds also exhibited optical activity. The individual molecules of a liquid are rapidly tumbling and are not ordered in a specific way, so these observations require that optical activity be a consequence of the structure of *individual molecules.* However, the possibility that optical activity might be related to *molecular structure* was not at all obvious in the early 1800s, when an understanding of structural chemistry was in its infancy.

A major advance in the understanding of optical activity was achieved by the French scientist Louis Pasteur in 1848, when he unraveled a puzzle involving tartaric acid. Tartaric acid is found in grapes, and the monopotassium salt (potassium hydrogen tartrate, called "tartar") is formed as a by-product in winemaking. (The German name for tartaric acid is *Weinsäure,* "wine acid.")

```
   O   OH   OH   O                  O   OH   OH   O
   ‖   |    |    ‖                  ‖   |    |    ‖
HO—C—CH—CH—C—OH             HO—C—CH—CH—C—O⁻K⁺
```

Tartaric acid — Potassium hydrogen tartrate

Ordinarily, the tartaric acid isolated from grapes exhibits a *positive* optical rotation, but some samples of tartaric acid were found to be optically inactive. In order to learn why, Pasteur carefully purified a sample of the sodium ammonium salt of the optically inactive tartaric acid.

```
          O   OH   OH   O
          ‖   |    |    ‖
Na⁺ ⁻O—C—CH—CH—C—O⁻NH₄⁺
```

He found that there were two types of crystals, which differed *only* in that they had a mirror image relationship. He undertook the difficult and tedious task of separating the small crystals, using a magnifying glass and a pair of tweezers. After this task was completed, he converted the salts back to the corresponding acids and obtained two different samples of tartaric acid. One was the well-known (+)-tartaric acid and the other exhibited an optical rotation which was *equal in magnitude* but had the *opposite sign.* In other words, the second sample was (−)-tartaric acid. Thus the optically inactive tartaric acid was actually a 50–50 mixture of enantiomers,* and Pasteur had succeeded in resolving it into its individual components. He recognized that molecules of an optically active compound must be chiral (although he did not use that term), but the existing structural theories of chemistry were not sufficiently advanced for him to be able to write three-dimensional representations of these molecules.

Not until 1874 was a structural theory put forth that could adequately account for chiral organic molecules. In that year J. H. van't Hoff and J. A. LeBel independently proposed the idea of the tetrahedral structure of carbon. This was the first structural hypothesis which could explain mirror image isomers. Despite the elegance of the idea, other chemists of the day were reluctant to accept it, and some of van't Hoff and LeBel's contemporaries reacted to their proposals with considerable hostility. Slowly, however, the resistance to this structural theory waned, and the concept of tetrahedral carbon became the cornerstone of modern organic stereochemistry. In 1901 van't Hoff was awarded the first Nobel Prize in Chemistry.

Specific Rotation and Optical Purity

The optical activity of a substance is one of its characteristic physical properties, but the magnitude of the optical rotation of a particular sample depends on how many

*The French chemist Gay-Lussac had previously named this optically inactive acid using the Latin term for a bunch of grapes, *racemus.* He called it *racemic* acid and thereby provided the name for an entire class of chemical substances.

chiral molecules actually interact with the beam of polarized light. When corrected for the length of the sample tube and the concentration of the sample, optical rotation is normally expressed as $[\alpha]$, the *specific rotation* of the substance. The specific rotation is calculated by the following relationship:

$$[\alpha] = \frac{\alpha}{l \times c}$$

where α is the experimentally measured optical rotation of a particular sample, l is the length of sample tube which the beam of light must transverse, and c is the concentration of the sample.

The length of the sample tube is ordinarily expressed in decimeters (1 dm = 10 cm), and the unit for concentration of the sample is grams per milliliter (g/mL) of solution. (For a pure liquid the concentration is also expressed as grams per milliliter.) In order to fully define an optical rotation, the temperature and the wavelength of light used in the measurement are usually specified, along with the solvent. A typical specific rotation would be reported as $[\alpha]_D^{25} = -13°(CHCl_3)$, which means that the measurement was made at 25°C with chloroform as the solvent, with the sodium D line (589 nm, a common wavelength for polarimetry) as the light source.

The specific rotation of a sample depends upon its purity. Clearly, if half of the weight of a sample were an achiral impurity, you would measure an optical rotation of only half the true value. We also need to consider the effect on optical activity of a second kind of contamination. Many optically active samples are prepared by resolution of racemic material, and frequently a sample will still contain some of the second enantiomer even though it is completely free of other impurities. The presence of the second enantiomer would result in a decrease in the optical rotation that you would measure. We can describe such a sample in the following way: it is *chemically pure,* but it is not *optically pure.* Only when a sample consists entirely of one of the enantiomers can you describe it as having 100 percent optical purity. The optical purity is defined as the ratio of the observed optical rotation to the maximum value that would be obtained if only a single enantiomer were present. At the other extreme, an equal mixture of the two (i.e., a racemic sample) will be optically inactive: any effect on a beam of polarized light that is caused by interaction with molecules of one enantiomer will be exactly counterbalanced by interaction with molecules of the other enantiomer. A racemic sample is optically inactive and therefore has an optical purity of zero.

What does optical purity tell you about the composition of a sample? Imagine that you had a sample of 2-chlorobutane with a specific rotation of +4.8°. The reported value for pure (−)-2-chlorobutane is −8.5, so first of all you know that your sample consists mainly of the (+) enantiomer. The specific rotation for the pure dextrorotatory enantiomer must be equal and opposite to that of the levorotatory isomer, so that value is +8.5°. The optical purity of the sample is therefore (4.8/8.5) × 100 = 56%. This means that 56% of the sample is excess (+)-2-chlorobutane; the remaining 44% is optically *inactive* and is *racemic* 2-chlorobutane. Some chemists prefer to use an equivalent term, *enantiomeric excess,* instead of optical purity, and would describe this sample as having an enantiomeric excess of (+)-2-chlorobutane equal to 56%. Stated in another way, 78% of the entire sample is (+)-2-chlorobutane (56% + half of 44%), and 22% is (−)-2-chlorobutane (i.e., half of 44%).

Absolute and Relative Configuration

The term *absolute configuration* denotes the precise three-dimensional structure of the chiral center (or centers) of a molecule and is used to distinguish between enantiomers. The absolute configuration is usually specified by labeling the chiral centers as *R* or *S*, but prior to 1951 it was not possible to unambiguously determine the absolute configuration of the molecules in a laboratory sample.

During the first half of the twentieth century the configuration of a variety of compounds had been related to that of glyceraldehyde, and in order to bring consistency to the stereochemical studies the *R* configuration* was arbitrarily assigned to (+)-glyceraldehyde. This arbitrary assignment had a 50 percent chance of being wrong, and not until 1951 did x-ray crystallographic studies of a salt of (+)-tartaric acid prove that it was indeed correct. It had previously been demonstrated that (+)-tartaric acid and (+)-glyceraldehyde had the same *relative configuration* because both had been converted to (−)-glyceric acid by reactions that do not affect the chiral centers. The x-ray studies showed that (+)-tartaric acid is the enantiomer having the *R* configuration at both chiral centers, and this means that (+)-glyceraldehyde also has the *R* configuration at its chiral center.

O
‖
COH
H—(R)—OH
HO—(R)—H
CO_2H
(+)-Tartaric acid

⟶

O
‖
COH
H—(R)—OH
CH_2OH
(−)-Glyceric acid

⟵

O
‖
CH
H—(R)—OH
CH_2OH
(+)-Glyceraldehyde

Optical Rotation and Absolute Configuration

The preceding discussion has shown that the enantiomers of tartaric acid and glyceraldehyde which have the *R* configuration both exhibit a positive optical rotation. Is this part of a general trend? What is the exact relationship between *R/S* configuration of a chiral center and the direction of optical rotation? The answer is extremely simple: *There is none.* The direction of optical rotation is a complex function of the nature of the substituents surrounding a chiral center, and there is no simple method which allows us to predict the direction of optical rotation. In fact, the magnitude and direction of optical rotation of a particular chiral molecule will typically change as a function of experimental variables such as the wavelength of the light or the solvent. Remember, the *R/S* designation of configuration is merely an arbitrary system of nomenclature, so you should not be surprised that that it does not correlate with optical rotation. The following examples clearly demonstrate that the two are quite independent.

CO_2H	CO_2H	CO_2H	CO_2H	CO_2H	CO_2H
H—C—NH_2	H—C—OH	H—C—OH	H—C—OH	H—C—OH	H—C—NH_2
CH_2OH	CH_2OH	CH_2NH_2	CH_2Br	CH_3	CH_3
R, (+)	*R*, (−)	*R*, (+)	*S*, (−)	*R*, (−)	*R*, (−)

*The *R/S* naming system was not introduced until the 1950s; a different system, which employed the symbols D and L, was used previously. That system is still used in sugar chemistry (Chapter 19).

All the preceding compounds are structurally similar, but both the sign of the optical rotation and the *R/S* configuration vary. Even if we consider only structures with the *R* configuration, a structural change such as replacement of −OH with $-NH_2$ sometimes results in a change in the sign of the optical rotation (first two structures) but other times does not (last two structures).

Optical Activity, Chirality, and Symmetry

When will a compound be optically active? This question is deceptively simple. We could give the obvious answer that a compound will be optically active when it consists of chiral molecules. But this is true only when just one of the enantiomers is present, (or at least is present in excess of the other). Sometimes it is easier to answer the question: When will the compound be optically *inactive*? Optical inactivity is always the result of one of two conditions: either the compound in question is *achiral* or it is *racemic*. You can test for each of these situations with reflection symmetry. An achiral molecule has reflection symmetry, usually a plane of symmetry passing through its center. In the case of a racemic compound there are equal

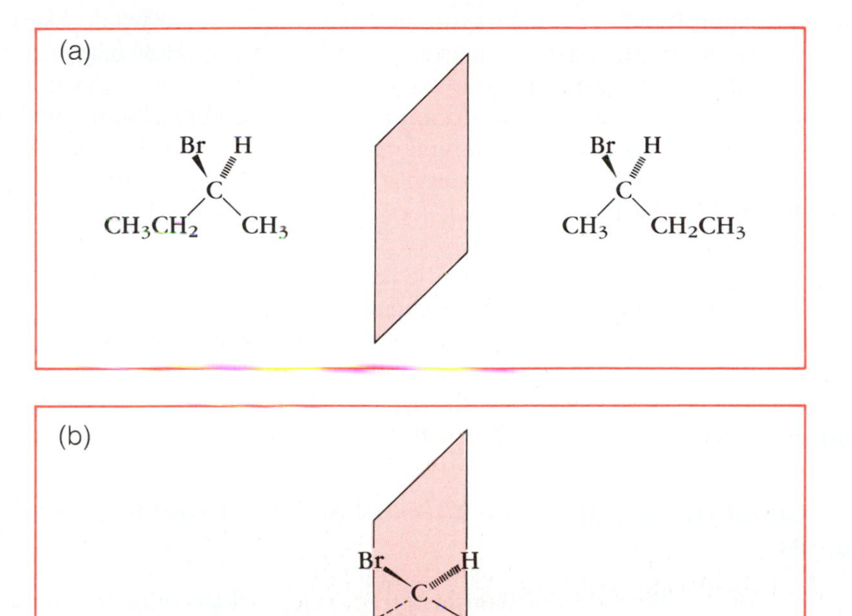

Figure 6.11 Reflection Symmetry as a Criterion for the Absence of Optical Activity.
(a) A racemic compound (50–50 mixture of enantiomers) has reflection symmetry in the sense that each molecule can be paired with its enantiomer.
(b) Any single achiral molecule exhibits reflection symmetry if drawn in the appropriate conformation. The type of reflection symmetry shown here is a *plane of symmetry*. The chiral center and two of its substituents (Br and H) lie in the symmetry plane, and the remaining substituents (two methyl groups) have a mirror image relationship.

amounts of the two enantiomers, and each molecule is accompanied by its mirror image. As a result, we can consider each such *pair* of molecules to exhibit reflection symmetry—one-half of the pair is the mirror image of the other half (Figure 6.11a).

As we have stated previously, an achiral molecule will exhibit reflection symmetry, and we have illustrated this in Figure 6.11b. Whenever you can demonstrate such reflection symmetry, the compound in question must be optically inactive.

6.7 STEREOCHEMISTRY OF REACTIONS: FORMATION AND DESTRUCTION OF CHIRAL CENTERS IN FREE RADICAL SUBSTITUTION

In the preceding sections we have discussed the structural aspects of stereochemistry. We will now turn our attention to *dynamic stereochemistry,* the stereochemical features of reactions. What are the stereochemical relationships between the reactants and products in a reaction? What three-dimensional features are necessary to fully describe the details of the reaction mechanism? As always we are concerned with the relationships between structure and reactivity, and we would like to be able to answer a variety of specific questions about a reaction: How many products are expected? Will the products be obtained in optically active or inactive form (only physical methods such as distillation being used to purify them)? Will each compound be made up of chiral or achiral molecules? If an optically inactive product is obtained, will it be inactive because it is racemic or because the individual molecules are achiral? In the following sections we will show you how to answer these questions.

The *reaction mechanism* provides the key to any structural relationship between the reactants and products of a chemical reaction. If you understand the mechanism, you can predict the products of a reaction, and if you know the products of a reaction, you can utilize the mechanism to interpret their formation. In the next several sections we will analyze the dynamic stereochemistry of a wide variety of organic reactions, beginning here with free-radical halogenation.

Free Radical Halogenation: The Three-Dimensional Structure of Free Radicals

We previously described the mechanism of free radical halogenation in Section 2.6, but we will review the important features so that we can consider the stereochemical aspects of the different steps. This is shown in the following scheme for chlorination:

$$Cl_2 \xrightarrow{\text{light}} 2\,Cl\cdot$$

$$R{-}H + Cl\cdot \longrightarrow R\cdot + HCl$$

$$R\cdot + Cl_2 \longrightarrow R{-}Cl + Cl\cdot$$

The light-induced cleavage of molecular chlorine generates chlorine atoms, which abstract hydrogens from organic molecules. This produces organic free radicals, which react further with Cl_2 to yield an organic halide.

The orbital structure of the intermediate free radical is similar to that of a carbon atom which is part of a double bond. Repulsion between the bonding electrons causes the three substituents on the radical center to lie as far apart as possible, that is, approximately in a plane and separated by angles of about 120°. The unpaired electron occupies a *p* orbital perpendicular to the plane of the three substituents, as shown for the 2-propyl radical.

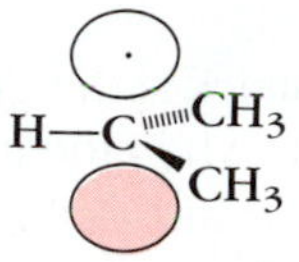

H—C(CH$_3$)(CH$_3$)

The dot indicating an unpaired electron is arbitrarily drawn in one of the lobes of the *p* orbital, but at any given instant it could reside in either lobe, and the net electron density corresponding to the unpaired electron is divided equally between the two. Reaction could take place at either lobe, that is, at either *face* of the planar free radical.

The Fate of Optically Inactive Starting Materials

When optically inactive reactants undergo a chemical reaction, will the products also be optically inactive? Or could they be optically active? As long as each of the reactants is optically inactive, we can state as a general rule that *optically inactive reactants produce optically inactive products.* We will illustrate this pattern with the free radical chlorination of 2-methylbutane (ignoring the polychloro products that could be formed and focusing on the monochloro isomers, $C_5H_{11}Cl$). First we will draw each of the *constitutional* isomers that would result, and then we will determine which of these would be produced in stereoisomeric forms. The two methyl groups on C-2 are equivalent, but each of the other carbon atoms is unique. Consequently the reaction would yield four constitutional isomers.

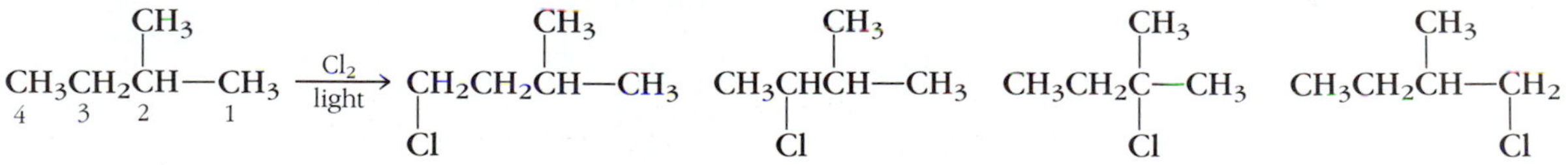

$$\underset{4\ \ 3\ \ \ 2\ \ \ \ \ 1}{CH_3CH_2CH(CH_3){-}CH_3} \xrightarrow[\text{light}]{Cl_2} \underset{\mathbf{1}}{CH_2(Cl)CH_2CH(CH_3){-}CH_3} \quad \underset{\mathbf{2}}{CH_3CH(Cl)CH(CH_3){-}CH_3} \quad \underset{\mathbf{3}}{CH_3CH_2C(CH_3)(Cl){-}CH_3} \quad \underset{\mathbf{4}}{CH_3CH_2CH(CH_3){-}CH_2Cl}$$

Neither **1** nor **3** contains a chiral center, and these achiral molecules would necessarily be optically inactive. But **2** and **4** each do contain a chiral center. Why must these products also be optically inactive?

We shall first examine the formation of **4**. The two enantiomers of **4** would be produced by substitution at the two *equivalent* methyl groups on C-2 of the reactant. Because these two methyl groups are equivalent, attack by chlorine atoms would proceed at identical rates to form equal amounts of the enantiomeric products. In other words the 1-chloro-2-methylbutane (**4**) formed in this reaction would be racemic.

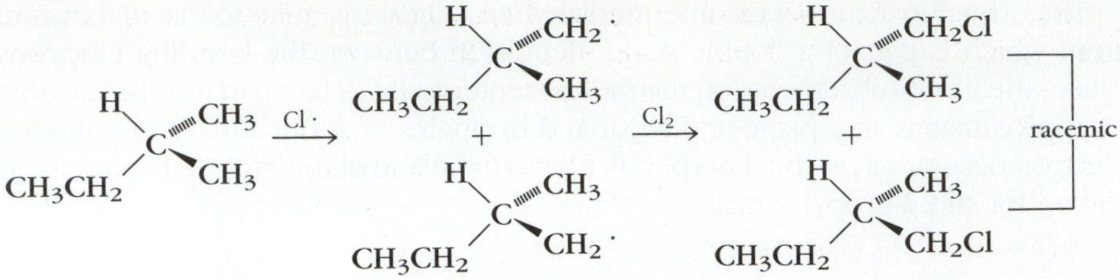

The formation of **2** begins with attack at the CH_2 group of the reactant. The chiral center of the product is formed only when the intermediate free radical reacts with Cl_2, and it is this second step that we must evaluate. The free radical is achiral, and a plane of symmetry can be drawn through the three substituents on the radical center. Consequently collisions with Cl_2 from above and below this symmetry plane will occur at identical rates to yield **2** as a racemic product.

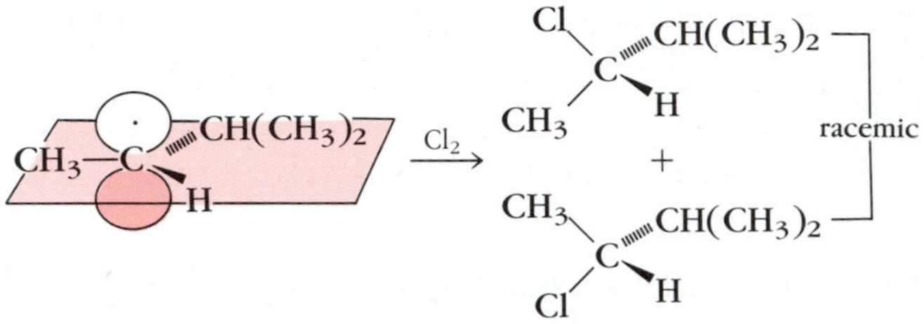

There are actually a total of four ways that inactive reactants yield optically inactive products, and we have only shown you two of these. In the preceding example the reactant was *achiral,* and the products could be either *achiral* or *racemic.* The remaining two ways that optically inactive reactants form optically inactive products are found when the reactants are *racemic.*

We will show you why optically inactive products are formed from racemic starting materials by evaluating the hypothetical chlorination of (±)-2-chloro-1,1,1-trifluoropropane.

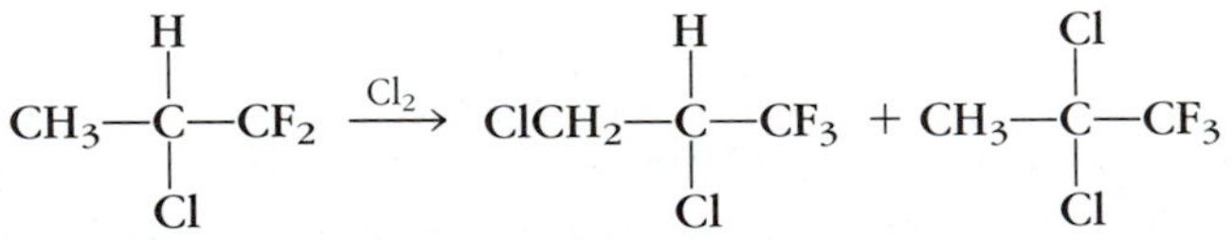

$$CH_3-CHCl-CF_2 \xrightarrow{Cl_2} ClCH_2-CHCl-CF_3 + CH_3-CCl_2-CF_3$$

(±)-2-Chloro-1,1,1-trifluoropropane; (±)-1,2-Dichloro-3,3,3-trifluoropropane; 2,2-Dichloro-1,1,1-trifluoropropane

The first product, 1,2-dichloro-3,3,3-trifluoropropane, is still chiral, but at all stages of the reaction every molecule has been accompanied by its enantiomer. Consequently, this product is also racemic (and is therefore optically inactive). The second product, 2,2-dichloro-1,1,1-trifluoropropane, is achiral because the original chiral center was destroyed by replacement of a hydrogen with a second chlorine.

The Fate of Optically Active Starting Materials

When the starting material in a reaction is *optically active,* the reaction products are frequently (but not always) optically active. We will illustrate this with the free radical chlorination of (*R*)-2-chlorobutane, again restricting our discussion to the products in which a single chlorine atom has been introduced. As in previous examples, we will first draw the constitutional isomers that would be formed.

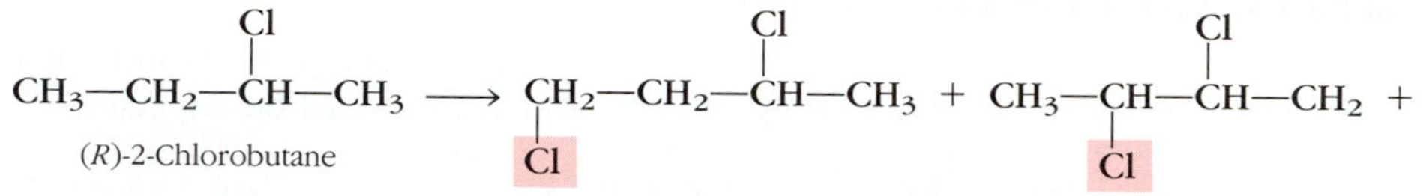

$$CH_3{-}CH_2{-}CH(Cl){-}CH_3 \longrightarrow CH_2(Cl){-}CH_2{-}CH(Cl){-}CH_3 + CH_3{-}CH(Cl){-}CH(Cl){-}CH_2 + \ldots$$

(*R*)-2-Chlorobutane; **1**; **2**

$$CH_3{-}CH_2{-}C(Cl)_2{-}CH_3 + CH_3{-}CH_2{-}CH(Cl){-}CH_2Cl$$

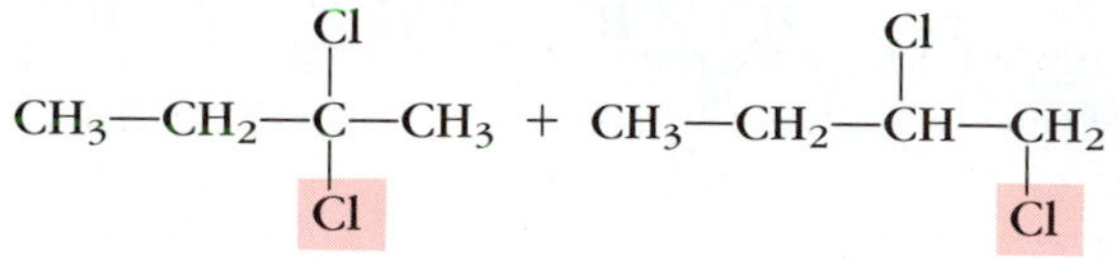

3; **4**

Of the four constitutional isomers, only compound **3** contains no chiral center. The original chiral center of the reactant has been destroyed in **3** by replacing a hydrogen with a second chlorine. Each of the other products has at least one chiral center, so we must ascertain how many stereoisomers would be produced in each case.

The formation of **1** and **4** does not involve breaking any of the bonds to the original chiral center, and configuration at the original center must remain unchanged. Both **1** and **4** would be produced as a single stereoisomer and would be obtained in optically active form.

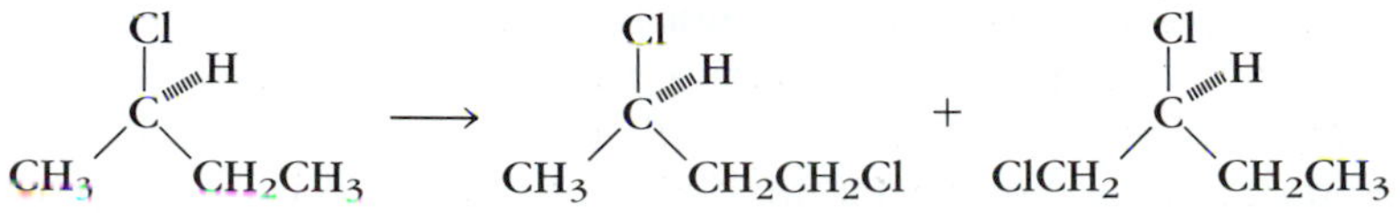

Cl, H, C, CH_3, CH_2CH_3 → Cl, H, C, CH_3, CH_2CH_2Cl + Cl, H, C, $ClCH_2$, CH_2CH_3

(*R*)-2-Chlorobutane — (*R*)-1,3-Dichlorobutane **1** — (*S*)-1,2-Dichlorobutane **4**

Note that both the reactant and the product, **1**, have the *R* configuration, but the configuration of **4** is *S*, even though all the original bonds to the chiral center have remained intact throughout the reaction. How did this happen? The answer is really quite simple. Replacement of a hydrogen on the methyl group by a chlorine changes the relative priority of that group. In the starting compound, $-CH_2CH_3$ has a higher priority than $-CH_3$, but in the product $-CH_2Cl$ has a higher priority than $-CH_2CH_3$. The actual configuration of the chiral center has not changed; only the *R*/*S* label is different.

Structure **2** contains two chiral centers, and we must consider the possibility that several diastereomers would be formed. The two chiral centers are adjacent, so a sawhorse drawing can be used to show the stereochemistry. Abstraction of either hydrogen atom from C-3 of the reactant would yield the same free radical:

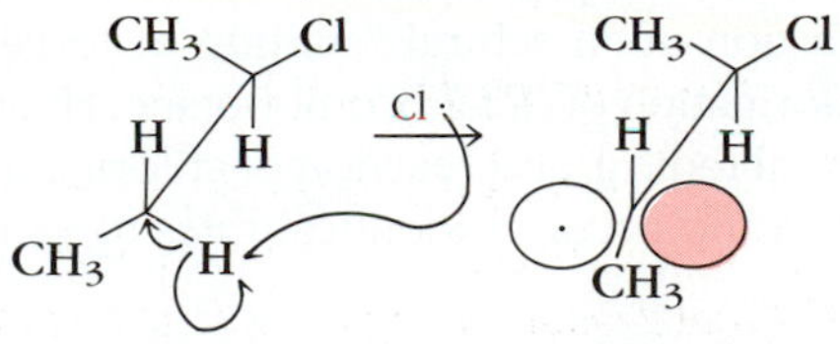

CH_3, Cl, H, H, CH_3, H —Cl·→ CH_3, Cl, H, H, CH_3

This intermediate is *chiral,* and the two faces of the radical center are nonequivalent. Reaction with Cl_2 could take place at either face, and two diastereomers would result. Different interactions between Cl_2 and the radical would result from attack at the two faces, so the activation energies for the two pathways would be different and the diastereomers **2a** and **2b** would be formed in unequal amounts. It is difficult to predict which would predominate, and both should be expected.

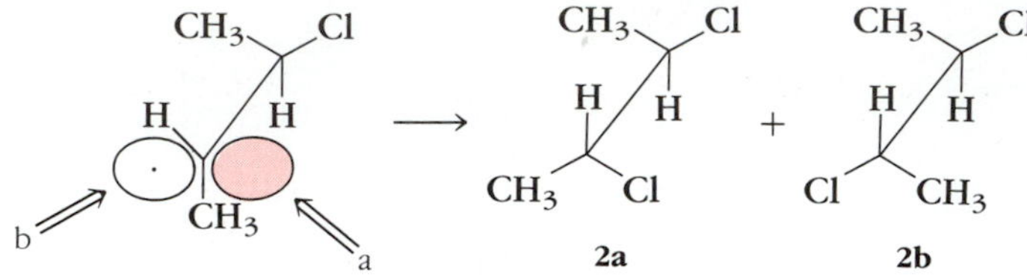

2a **2b**

In both **2a** and **2b** the configuration of the original chiral center (at the back of the molecule) remains unchanged. Nevertheless, while **2a** would be produced in optically active form, **2b** would be optically inactive because it is a *meso* compound.

The chlorination reaction of (*R*)-2-chlorobutane would therefore yield a total of five isomeric dichloro products. The complete results are summarized in Figure 6.12.

The preceding example demonstrates that we cannot state any simple rule to describe the stereochemical outcome of all reactions involving optically active starting materials. But we *can* emphasize the two circumstances under which optically active reactants yield optically *inactive* product. These are the cases in which the product is *achiral* or in which it is *racemic.* Now that you know what to look for, your task is greatly simplified.

There are two ways in which an optically active reactant could generate an *achiral* product. First, the generation of a second chiral center could produce a *meso* compound, as in the formation of **2b** in Figure 6.12. Second, the original chiral center could be destroyed if two of the four substituents became equivalent, as in the formation of **3** in Figure 6.12. Breaking a bond to the chiral center is not necessary, however, and you can see this by considering the chlorination of (+)-1-chloro-3,3,3-trifluoro-2-methylpropane. The reaction would produce a variety of products, but the product resulting from chlorination of the methyl group would generate a second CH_2Cl group, so that product would not have a chiral center.

$$(+)\ CF_3{-}\underset{}{\overset{CH_3}{\overset{|}{C}}}H{-}CH_2Cl \longrightarrow CF_3{-}\overset{CH_2Cl}{\overset{|}{C}}H{-}CH_2Cl$$

chiral;
optically active

achiral;
optically inactive

There is only one way that an optically active reactant could yield a *racemic* product, and that is for the reaction to proceed through an *achiral intermediate* (or transition state). If at any point along a reaction coordinate the structure becomes achiral, then the product cannot be optically active; for all intents and purposes this would have become a reaction of an achiral "starting material." We can illustrate this by considering the chlorination of (*R*)-2-bromobutane. Hydrogen abstraction at C-2 would produce an achiral radical, and reaction of chlorine at the two equivalent faces of this intermediate would occur at identical rates to yield racemic product.

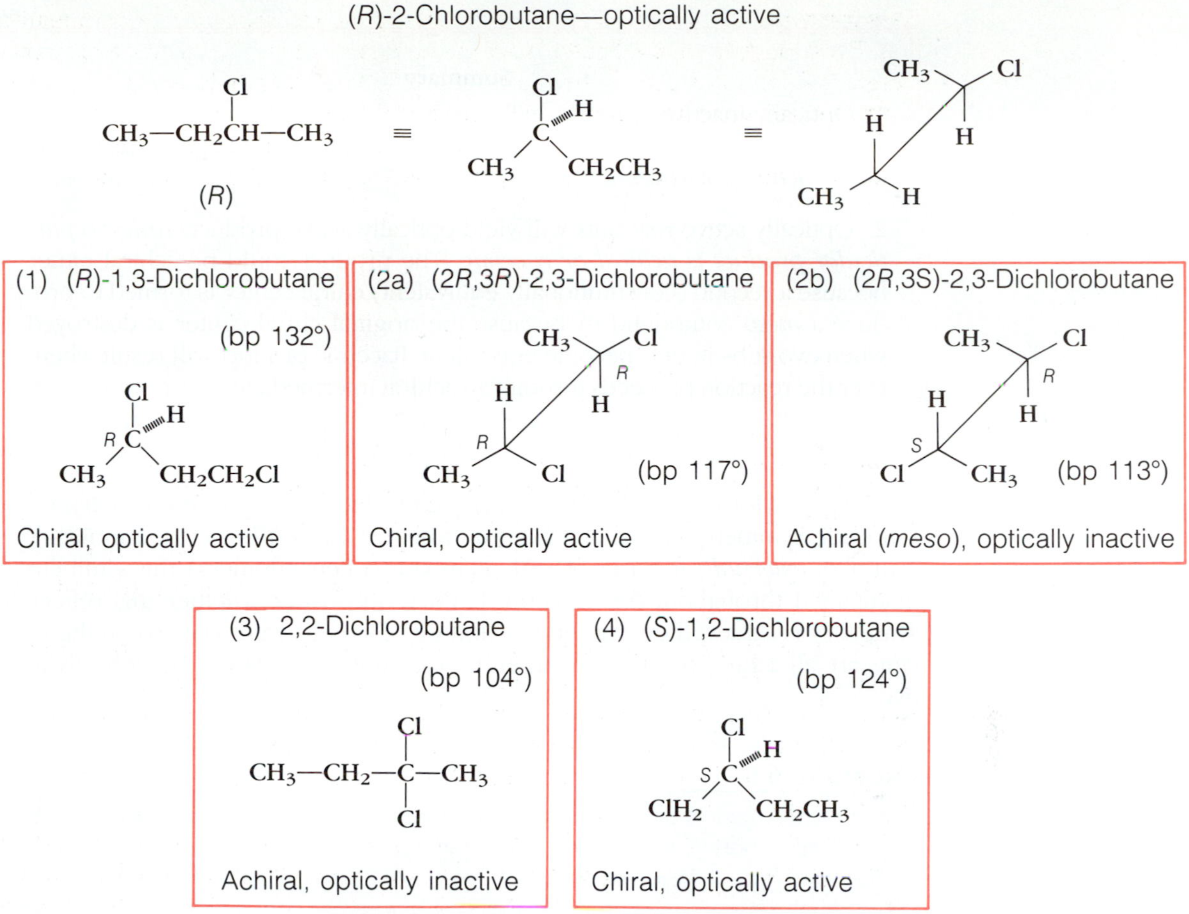

(*R*)-2-Chlorobutane—optically active

$CH_3—CH_2CH(Cl)—CH_3$ (*R*) ≡ ≡

(1) (*R*)-1,3-Dichlorobutane (bp 132°)
Chiral, optically active

(2a) (2*R*,3*R*)-2,3-Dichlorobutane (bp 117°)
Chiral, optically active

(2b) (2*R*,3S)-2,3-Dichlorobutane (bp 113°)
Achiral (*meso*), optically inactive

(3) 2,2-Dichlorobutane (bp 104°)
$CH_3—CH_2—CCl_2—CH_3$
Achiral, optically inactive

(4) (*S*)-1,2-Dichlorobutane (bp 124°)
Chiral, optically active

Figure 6.12 Chlorination of (*R*)-2-Chlorobutane.
Five dichloro derivatives would be formed which could be separated by distillation. (Boiling points are shown to illustrate that these isomers have different physical properties.)

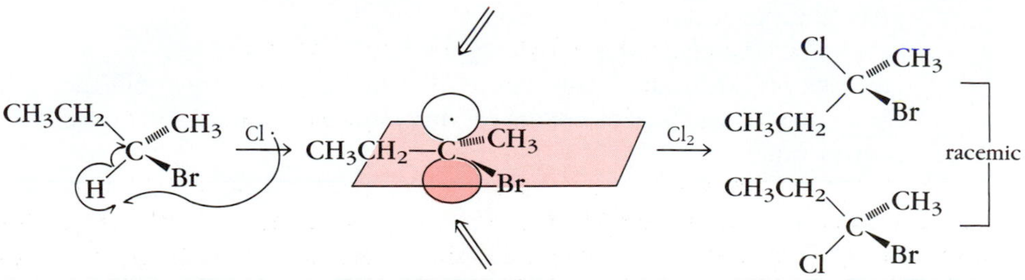

$CH_3CH_2C(CH_3)(H)Br \xrightarrow{Cl\cdot} CH_3CH_2—\dot{C}(CH_3)Br \xrightarrow{Cl_2}$ racemic

In the preceding discussion we utilized free radical substitution to illustrate several principles of dynamic stereochemistry. These principles apply to all organic reactions, and they are summarized in the following box.

Summary

1. Optically inactive reactants will always yield optically inactive products. If each of the reactants is either achiral or racemic, each of the products will be either *achiral* or *racemic*.

2. Optically active reactants will yield optically active products *unless a particular product is achiral or racemic*. The product could be achiral either because a second (constitutionally equivalent) chiral center is formed to produce a *meso* compound or because the original chiral center is destroyed when two substituents become equivalent. Racemic product will result whenever the reaction proceeds through an achiral intermediate or transition state.

The key to formation of optically inactive products is *reflection symmetry*. Once reflection symmetry has been achieved (whether *internally* so that a molecule is achiral or *externally* in terms of a 50–50 mixture of enantiomers), this symmetry is maintained throughout the reaction. It does not matter whether this reflection symmetry is achieved during the reaction or before it begins (i.e., when the reactants are all achiral or racemic.) In either case the products will be optically inactive.

EXERCISE 6.9

Monochlorination of (*S*)-1-chloro-2,3-dimethylbutane would yield a variety of dichloro products which could be separated by distillation. Show all stereochemistry where appropriate.

$$Cl{-}CH_2{-}CH(CH_3){-}CH(CH_3){-}CH_3 \xrightarrow[\text{light}]{Cl_2} C_6H_{12}Cl_2 \text{ isomers}$$

(a) Draw the starting compound.
(b) Draw the products of the reaction.
(c) Label all chiral centers as *R* or *S*.
(d) For each product drawn state clearly: (1) whether or not the molecule is *chiral*, and (2) whether or not the product would be obtained in optically active form.

6.8 STEREOCHEMISTRY OF ADDITION REACTIONS OF ALKENES

Addition to the double bond of an alkene brings up a new set of stereochemical problems. The reactive center consists of two carbon atoms, and this might lead to complex mixtures of stereoisomers. For example, the addition of chlorine to (*E*)-2-butene produces two chiral centers.

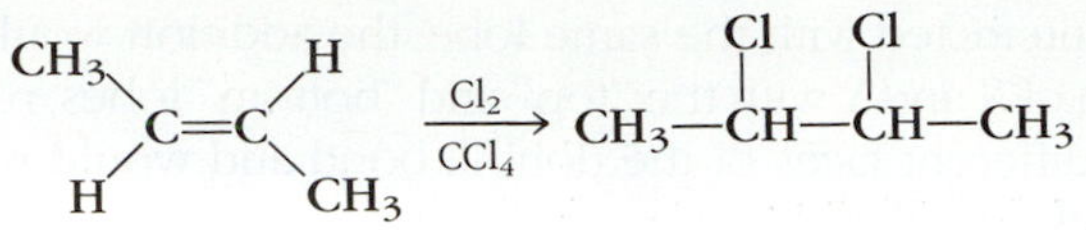

$$(E)\text{-}CH_3CH{=}CHCH_3 \xrightarrow[CCl_4]{Cl_2} CH_3\text{—}CHCl\text{—}CHCl\text{—}CH_3$$

How many of the three possible stereoisomers (one pair of enantiomers and a *meso* compound) would actually be formed? The answer is remarkable: only a single dichloro compound is produced, *meso*-2,3-dichlorobutane.

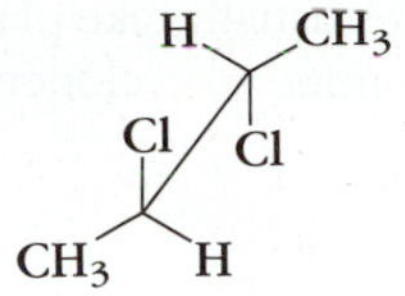

This behavior is typical for addition reactions of alkenes, and a knowledge of the reaction mechanism allows us to *predict* the stereochemistry of the products. We describe such reactions as **stereospecific.** *A stereospecific reaction is defined as one in which there is a specific relationship between the configuration of the reactant and the configuration of the product.* This means that the E isomer of an alkene would give a different product (i.e., different stereoisomer) than the Z isomer. More importantly, we can predict the precise stereochemistry of the reaction products on the basis of the structure of the starting material.

These results differ markedly from those of the free radical substitution reactions discussed in the previous section. When a chiral center is converted to a radical, the chirality is destroyed. Subsequent regeneration of the chiral center will result in the formation of the *same* mixture of stereoisomers regardless of whether the original chiral center was R or S, so there is no correlation between the configurations of the reactant and the products. Such reactions are not stereospecific.

How is stereospecificity achieved? What brings about the specific relationship between the configuration of an alkene and the configuration the product of an addition reaction? In order to answer these questions, you must understand the *reaction mechanism* for the particular reaction you are looking at. In this section we will discuss the different reactions mechanisms that are observed for additions to alkenes and show you how you can use these to predict the stereochemistry of the addition products.

Syn and *Anti* Modes of Addition

The two carbon atoms of a double bond and the four atoms attached directly to the double bond all lie in a plane. We frequently refer to this as the *plane of the double bond,* and we can further use this plane to define two *faces* of the double bond, one above the plane and one below it. The addition of two atoms (or two groups) to an alkene could, in principle, take place in either of two modes. If both groups add to *same face* it is called a *syn addition,* but if they add to *opposite faces* it is called an *anti addition.* As we discussed in Section 5.2, the π electrons occupy an orbital which has two lobes, one above and one below the plane of the double bond, and it is with this orbital that a reagent X—Y will interact.

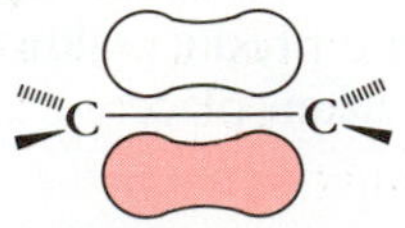

If X and Y both interacted with the same lobe, the addition would be *syn*. Alternatively, interaction of X and Y with the "top" and "bottom" lobes, respectively, would involve the two different faces of the double bond and would result in *anti* addition. The stereochemical consequences of each of these modes of addition are summarized in Figure 6.13.

Anti Additions: The Common Addition Mode

Polar addition reactions of alkenes usually take place via the *anti* mode of addition. For example, the addition of bromine to cyclopentene would yield racemic *trans*-1,2-dibromocyclopentane.

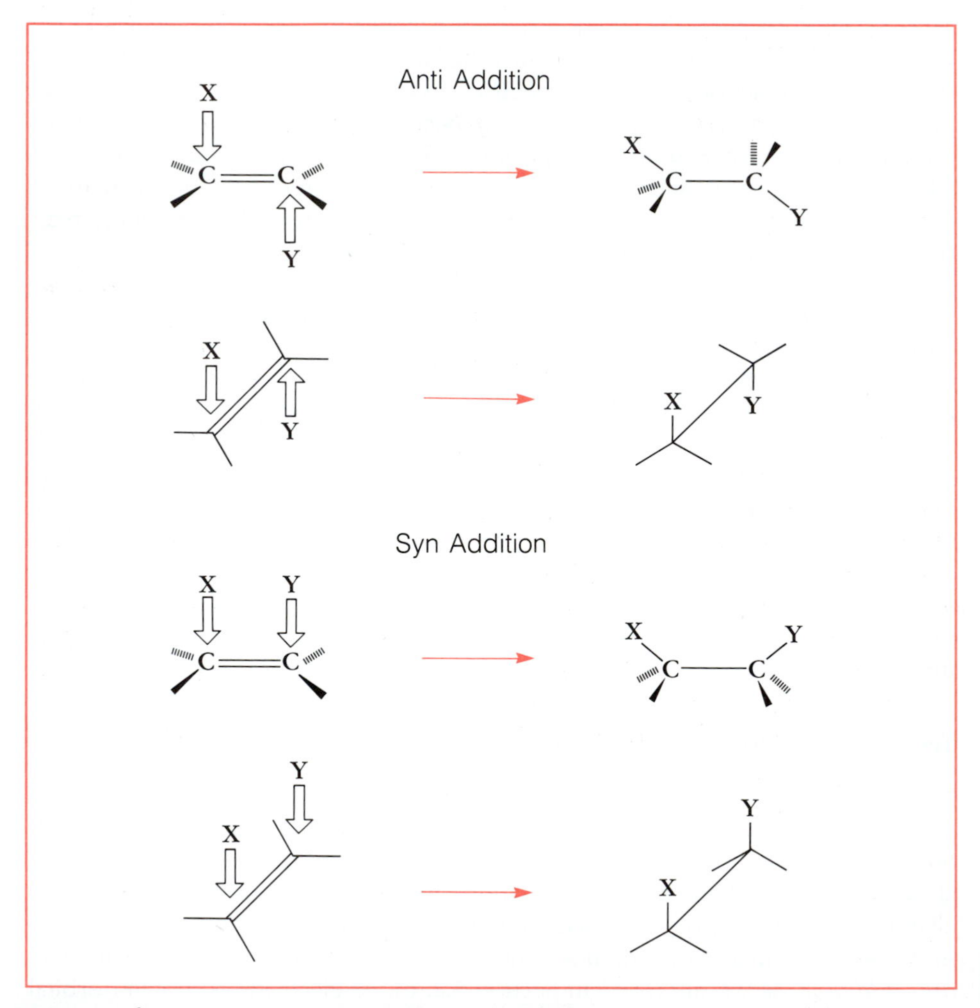

Figure 6.13 *Syn* and *Anti* Modes of Addition.
In the second drawing for each addition mode, the alkene is oriented so that the product can be drawn with a sawhorse representation. The addition mode defines the *configuration* of the product, although the *conformation* may change once the addition has taken place.

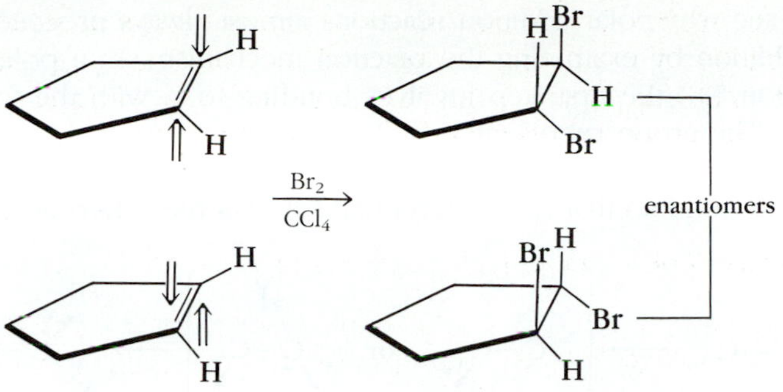

H H $\xrightarrow[CCl_4]{Br_2}$ H Br H Br — Br H Br H — enantiomers

The two faces of the cyclopentene double bond are equivalent, so both enantiomers would be formed in equal amounts.

But what of the reaction of an acyclic alkene? The situation is quite similar, as shown for the reaction of (*E*)-2-pentene with bromine.

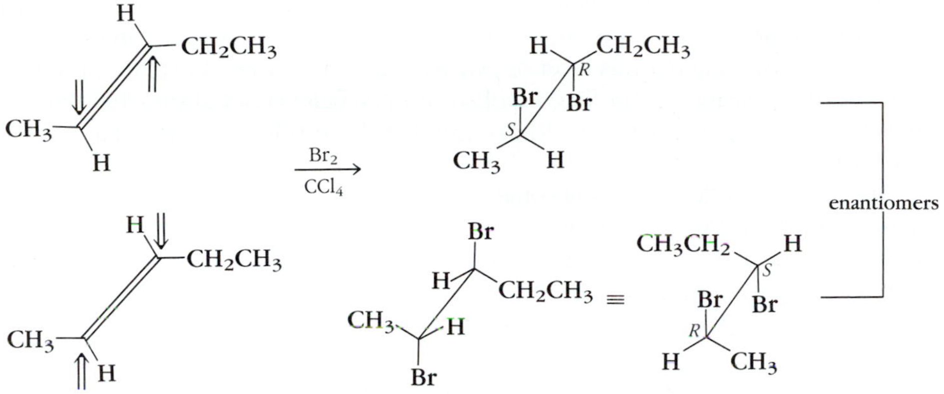

H CH_2CH_3 CH_3 H $\xrightarrow[CCl_4]{Br_2}$ H CH_2CH_3 *R* Br Br *S* CH_3 H

H CH_2CH_3 CH_3 H — Br H CH_2CH_3 CH_3 H Br $\equiv$ CH_3CH_2 H *S* Br Br *R* H CH_3 — enantiomers

The major product from the reaction is again a mixture of enantiomers. *Cis-trans* labels are not applicable to an acyclic compound, so the systematic name must indicate the configuration at each chiral center. The racemic product would contain equal amounts of (2*S*,3*R*)-2,3-dibromopentane and its enantiomer, (2*R*,3*S*)-2,3-dibromopentane. The other pair of enantiomers (2*R*,3*R*) and (2*S*,3*S*)-2,3-dibromopentane, would not be formed in detectable amounts in the reaction of (*E*)-2-pentene because their formation corresponds to *syn* addition. However, they would be produced in the reaction of (*Z*)-2-pentene:

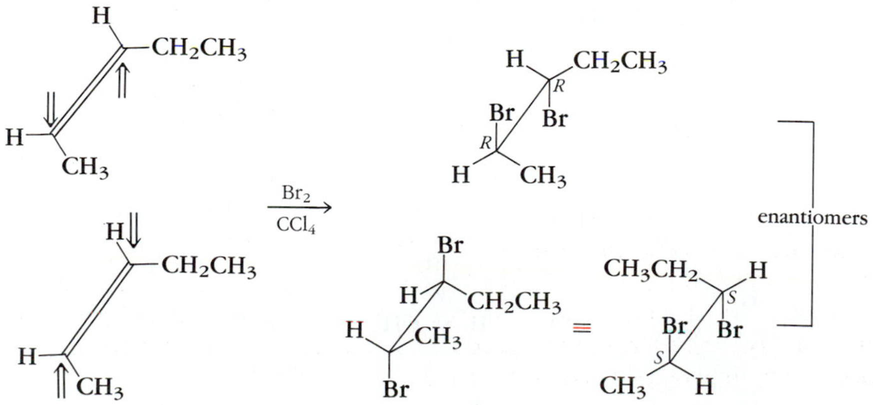

H CH_2CH_3 H CH_3 $\xrightarrow[CCl_4]{Br_2}$ H CH_2CH_3 *R* Br Br *R* H CH_3

H CH_2CH_3 H CH_3 — Br H CH_2CH_3 H CH_3 Br $\equiv$ CH_3CH_2 H *S* Br Br *S* CH_3 H — enantiomers

You can see why polar addition reactions almost always proceed via the *anti* mode of addition by examining the reaction mechanism. In a polar addition of X—Y (Section 5.8) the first step involves bonding to Y, with the formation of a carbocation. The group represented by Y is sometimes partially bonded to both carbons of the original double bond, but in either case it physically blocks one face of the double bond so that X^- preferentially attacks the other face.

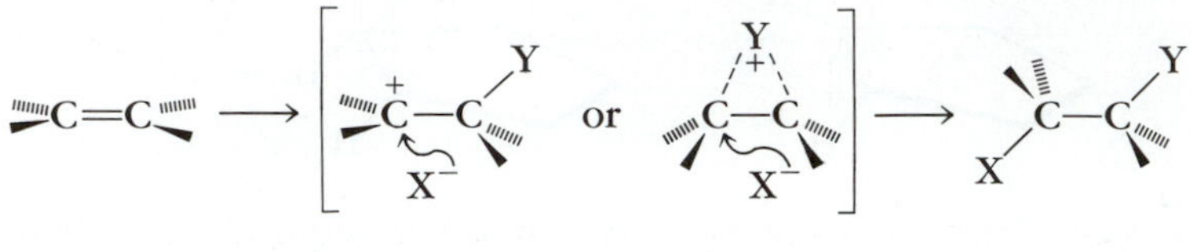

In many reactions attack by X^- occurs when the carbon has only a *partial* positive charge, that is, before a carbocation is fully developed, but this does not change the preference for *anti* addition. You can see from the preceding equation that *anti* addition generates the product in a stable staggered conformation. In contrast, *syn* addition would initially form a structure with unfavorable eclipsed interactions.

For the preceding reasons most stepwise addition reactions take place preferentially via the *anti* mode. The degree of stereospecificity is not always 100 percent but if you must guess, you should assume that the product of *anti* addition will predominate.

The stereochemical results of some typical stepwise addition reactions are summarized in the following examples, all of which proceed by *anti* addition. Note that in the second equation the stereochemistry of the reaction is observable only because of an *isotopic* difference in substituents: ^{1}H, *protium* (commonly abbreviated simply as H) and ^{2}H, *deuterium* (abbreviated as D).

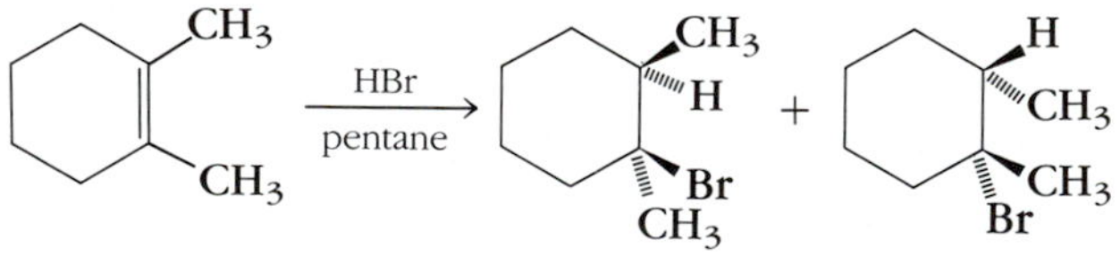

HBr, pentane

(racemic)
only product observed

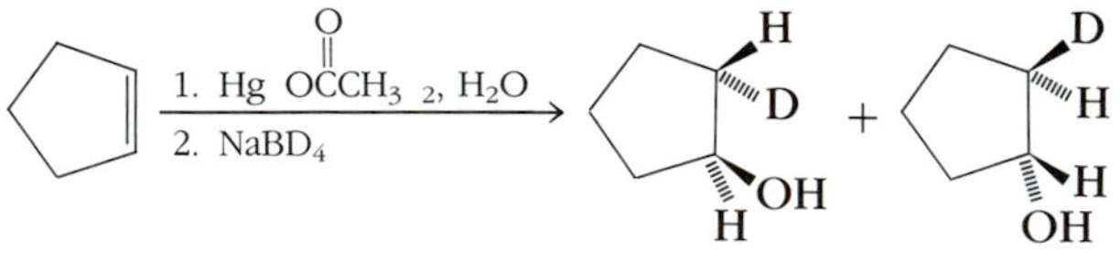

1. Hg(OCCH$_3$)$_2$, H$_2$O
2. NaBD$_4$

(racemic)
> 95% *anti* addition

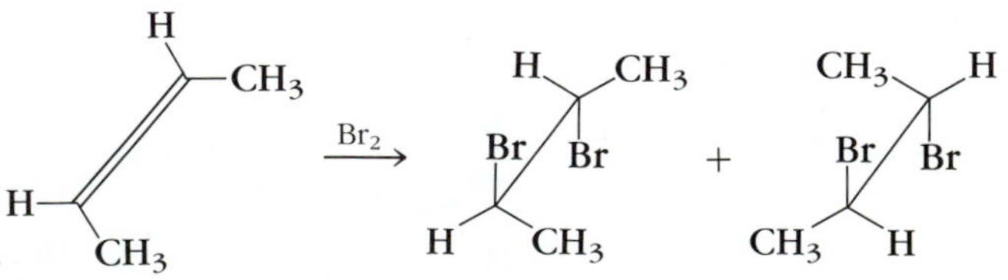

Br$_2$

(racemic)
96%

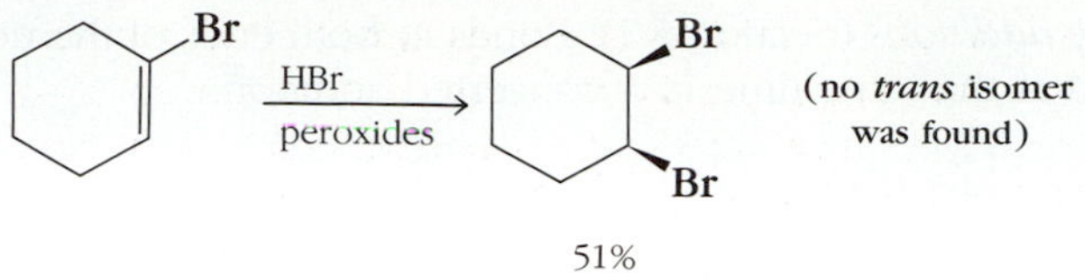

HBr / peroxides → (no *trans* isomer was found)

51%

The last of these examples proceeds by a free radical rather than ionic mechanism, but it shows that there is still a preference for *anti* addition in a stepwise reaction.

Syn Addition

A limited number of addition reactions to alkenes proceed via the *syn* addition mode. In contrast to the stepwise reactions that we considered previously, the two groups of X—Y add to the double bond *at the same time* (or nearly so) in most syn additions. Such processes are called *synchronous* or *concerted* (see Section 5.7), where the latter description further specifies that the product is generated directly without the formation of an intermediate. Because X and Y are initially joined by a bond, synchronous addition to a double bond requires that they add to the same face of the double bond. We have illustrated this for cyclopentene.

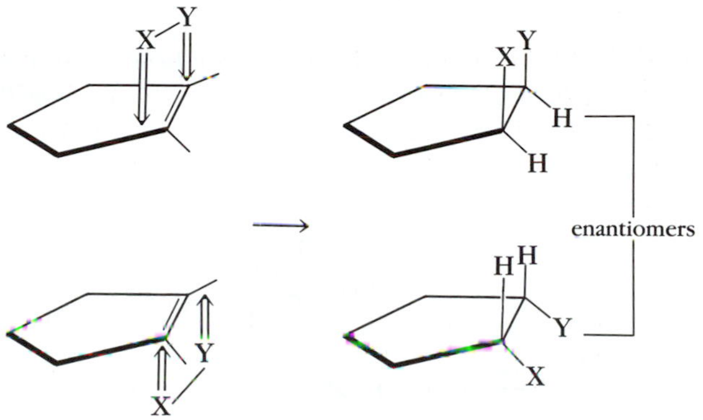

enantiomers

In this example the product is racemic. Note that if X and Y are the same, the product will be a *meso* compound.

The following list summarizes those reactions discussed in the text to this point that proceed via *syn* addition. The comments in each case help to explain why the *syn* addition takes place.

1. *Hydroboration-oxidation* (Section 5.11): The hydrogen is transferred (with its bonding electrons) directly from boron to the adjacent carbon atom in a cyclic transition state.

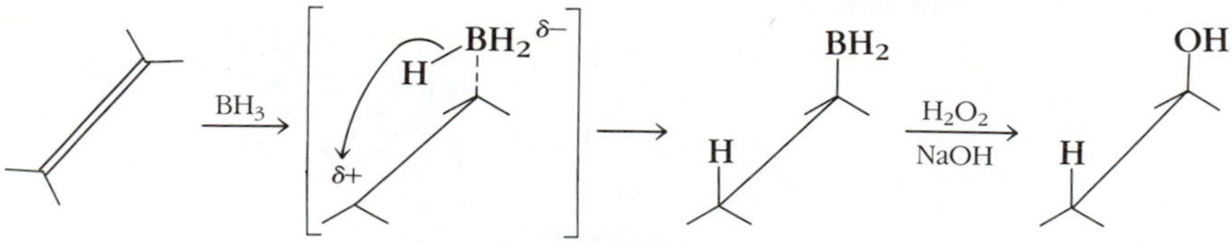

BH_3 → [$BH_2^{\delta-}$ H $\delta+$] → BH_2 H $\xrightarrow[NaOH]{H_2O_2}$ OH H

2. *Carbene additions* (Section 5.7): Bonds to both ends of the double bond are formed at the same time in a concerted addition.

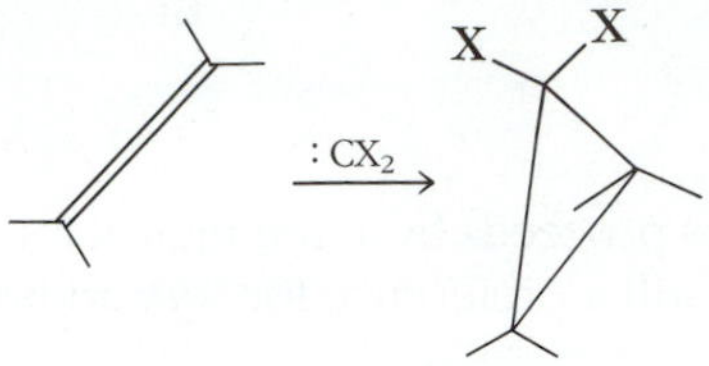

$: CX_2$

3. *Catalytic hydrogenation* (Section 5.6): The alkene binds to the catalyst, and both hydrogen atoms are delivered to the alkene from the surface of the metal catalyst.

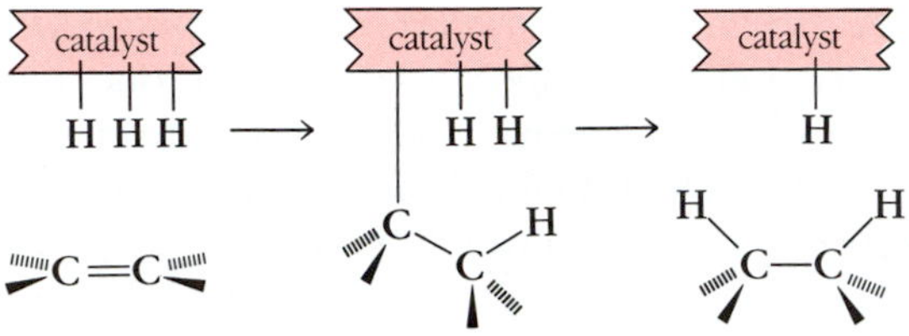

catalyst catalyst catalyst

4. *Oxidation with OsO_4 or $KMnO_4$* (Section 5.12): Both these reactions proceed via cyclic intermediates that are formed by synchronous addition of the reagent to the double bond. Subsequent hydrolysis of the metal-oxygen bonds yields the diol.

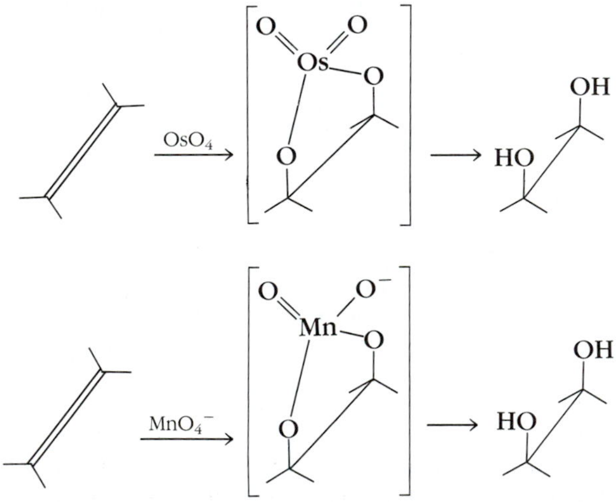

OsO_4

MnO_4^-

5. *Epoxidation* (Section 5.12). A single oxygen atom coordinates to both ends of the double bond.

CH_3COOH (peroxy acid, $CH_3C(=O)OOH$)

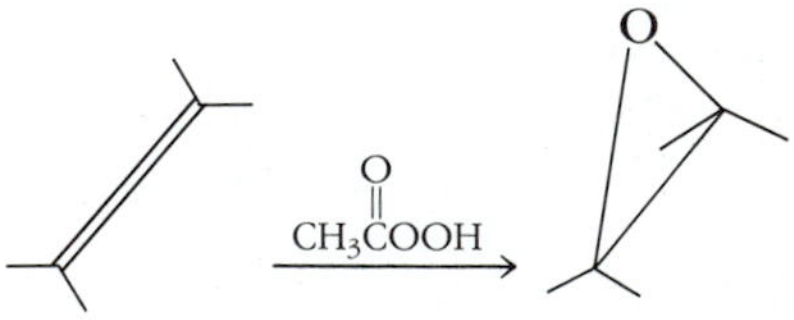

The following examples illustrate the stereochemical results of reactions that proceed by the *syn* addition mode:

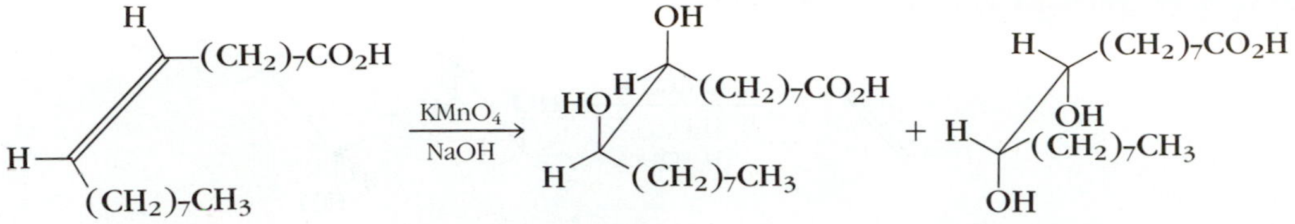

H, $(CH_2)_7CO_2H$, H, $(CH_2)_7CH_3$ $\xrightarrow[NaOH]{KMnO_4}$ OH, HO, H, $(CH_2)_7CO_2H$, H, $(CH_2)_7CH_3$ + H, $(CH_2)_7CO_2H$, OH, H, $(CH_2)_7CH_3$, OH

94% (racemic)

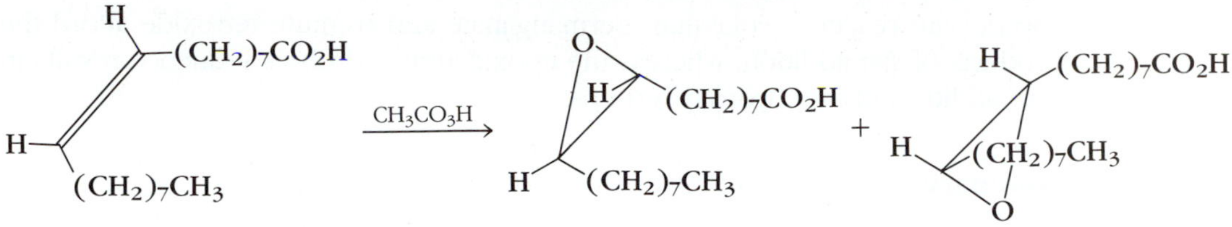

H, $(CH_2)_7CO_2H$, H, $(CH_2)_7CH_3$ $\xrightarrow{CH_3CO_3H}$ O, H, $(CH_2)_7CO_2H$, H, $(CH_2)_7CH_3$ + H, $(CH_2)_7CO_2H$, H, $(CH_2)_7CH_3$, O

82% (racemic)

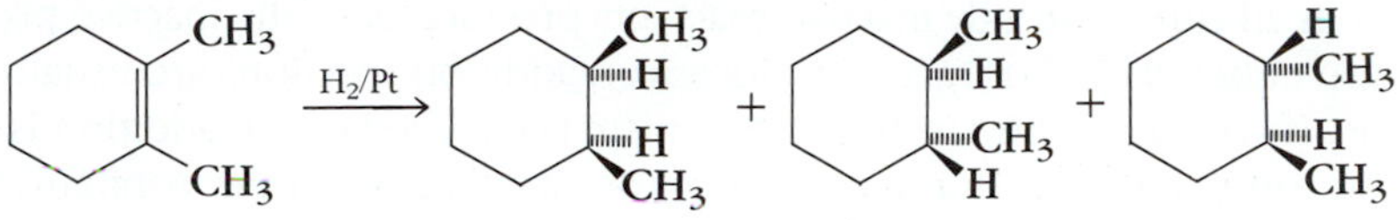

CH_3, CH_3 $\xrightarrow{H_2/Pt}$ CH_3, H, H, CH_3 + CH_3, H, CH_3, H + H, CH_3, H, CH_3

82–95%

5–18% (racemic)

H, H, C=C, CH_3, CH_3 $\xrightarrow[KOtBu]{CHBr_3}$ Br, Br, CH_3, CH_3

78%

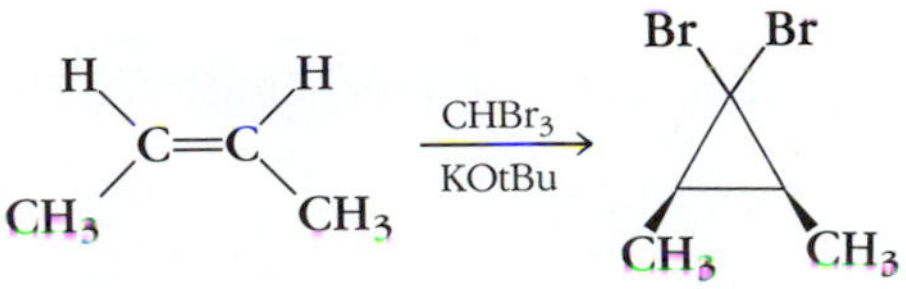

(no detectable *trans* isomer)

CH_3, CH_3 $\xrightarrow[2.\ NaOH/H_2O_2]{1.\ BH_3}$ CH_3, OH, H, CH_3 ; OH, CH_3, CH_3, H

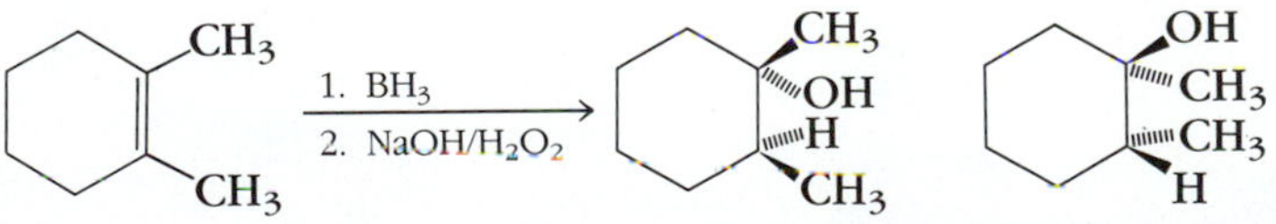

82% (racemic)

Only in the catalytic hydrogenation of 1,2-dimethylcyclohexene was any product of *anti* addition observed. Products of *anti* addition are sometimes found in the catalytic hydrogenation of alkenes because the initial step, binding of the alkene to the catalyst, is reversible. This can lead to isomeric alkenes, which in turn yield different stereoisomers upon reduction.

Although the epoxidation of an alkene proceeds by a *syn* addition, the hydrolysis of the epoxide takes place with *inversion of configuration* (Chapter 12).

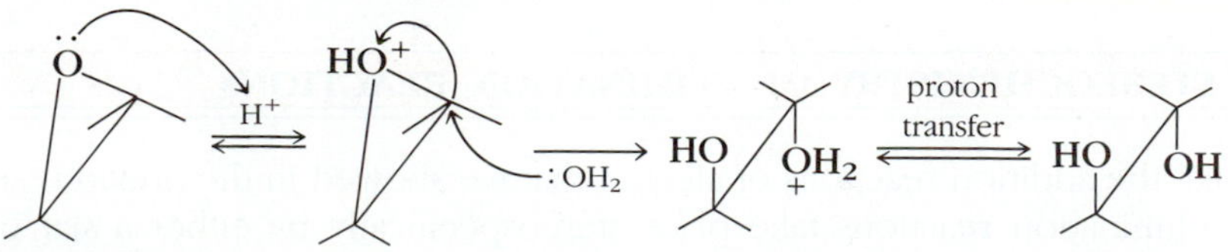

O $\overset{H^+}{\rightleftharpoons}$ HO^+ $\xrightarrow{:OH_2}$ HO, OH_2^+ $\overset{\text{proton transfer}}{\rightleftharpoons}$ HO, OH

Therefore the net result of an epoxidation-hydrolysis sequence is formation of a 1,2-diol by the overall equivalent of an *anti* addition:

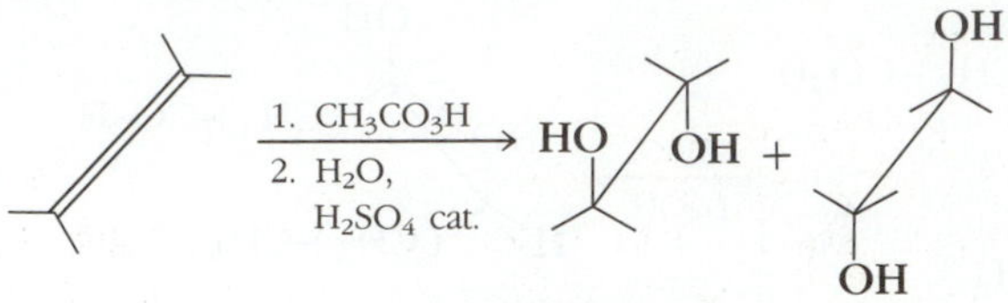

$$\xrightarrow[\text{2. } H_2O,\ H_2SO_4 \text{ cat.}]{\text{1. } CH_3CO_3H}$$

HO OH + OH OH

This means that we can obtain the products of either addition mode by using the appropriate reagents. Potassium permanganate and osmium tetroxide afford the products of *syn* addition, whereas the epoxidation-hydrolysis sequence results in *anti* addition of two hydroxyl groups.

Summary

As shown by the examples presented in this section, the stereochemistry of addition reactions to alkenes follows the same general rules that we presented in Section 6.7. In all cases optically inactive reactants produce optically inactive products, which are either achiral or racemic. Moreover, addition reactions are usually stereospecific. If you know the reaction mechanism (i.e., whether the addition is *syn* or *anti*, you can predict the configuration of the products from the configuration of the starting alkene.

EXERCISE 6.11

Complete the following reactions. Indicate the stereochemistry of the products.

(a) 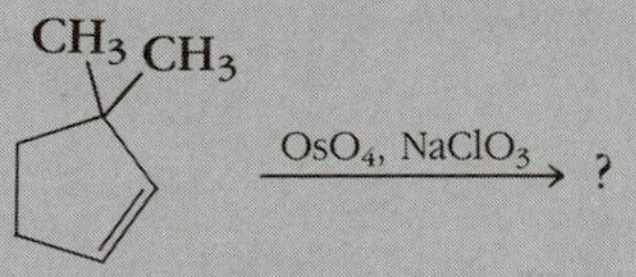 CH_3 CH_3 $\xrightarrow{OsO_4,\ NaClO_3}$?

(b) 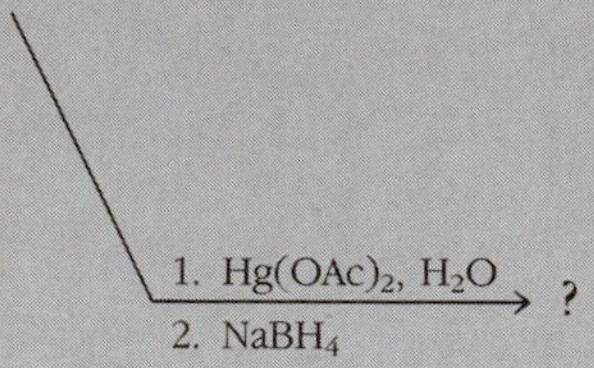 $\xrightarrow[\text{2. } NaBH_4]{\text{1. } Hg(OAc)_2,\ H_2O}$?

(c) 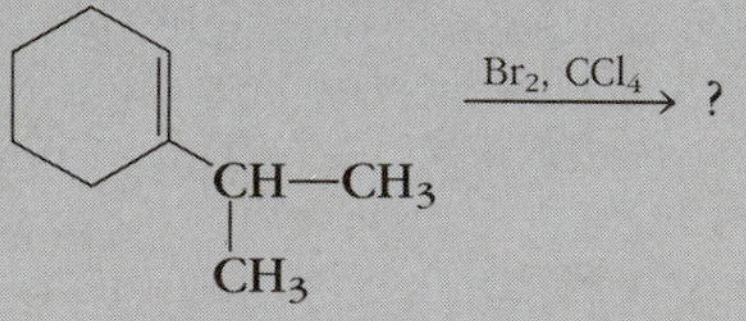 $\xrightarrow{Br_2,\ CCl_4}$?

$CH—CH_3$
CH_3

6.9 STEREOCHEMISTRY OF ELIMINATION REACTIONS

Just like the addition reactions of alkenes that we showed in the previous section, most elimination reactions take place stereospecifically by either a *syn* or *anti* mode. In this section we will present the reaction mechanisms for a variety of elimination reactions and show you how to predict the *E/Z* configuration of the products from the configuration of the reactants.

Anti Eliminations

The *anti* mode of elimination is characterized by a geometry in which the two departing groups are arranged at a dihedral angle of 180°. For any *bimolecular* elimination reaction, in which the alkene is formed in a single step by the collision of two molecules (such as an alkyl halide and a base), the *anti* elimination is usually preferred.

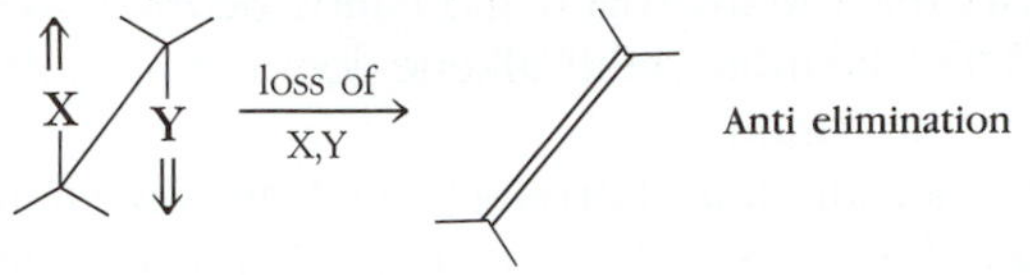

X Y loss of X,Y → Anti elimination

You can understand the preference for *anti* elimination by examining the atomic orbitals involved in the reaction. The *p* orbital contributions to the hybrid orbitals used for σ bonding to X and Y in the reactant are subsequently used in the π bond of the alkene. For the alkene these *p* orbitals must be in the same plane, and this geometric requirement is already satisfied in the stable *anti* conformation of the starting material. The relationship is summarized in the following equation. (Bear in mind that we are only showing spatial relationships between *atomic* orbitals; these are not *molecular* orbitals).

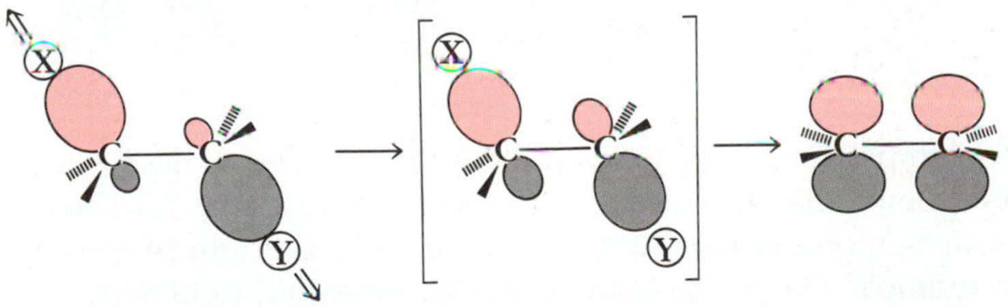

The preferred *anti* mode of elimination is illustrated by the following examples:

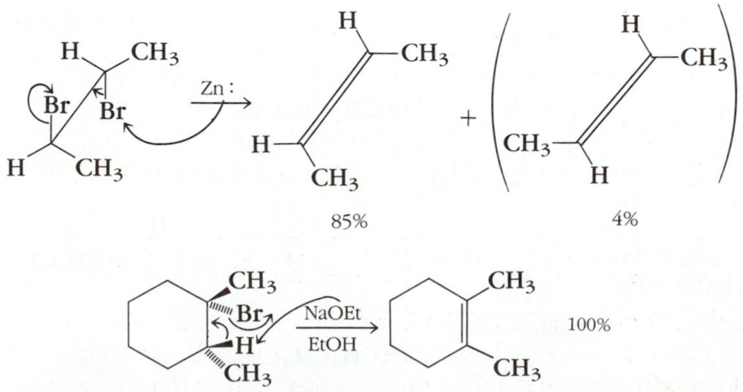

85% 4%

NaOEt / EtOH 100%

CH₃ ... NaOEt / EtOH → CH₃ + (CH₃)

CH_3 Cl CH_3 CH_3 — NaOEt, EtOH → 98% + 2%

The second reaction illustrates the preferential formation of the most highly substituted alkene (Section 5.13), even though *anti* elimination could also have led to other isomeric products. For a cyclic halide to undergo elimination by the *anti* mode, it is necessary that the hydrogen and halide be *trans* to each other, and the last reaction shows that the more stable alkene does not predominate if it cannot be formed by an *anti* elimination.

What about the small amounts of product which are shown in parentheses in the first and third examples? These appear to result from *syn* elimination. For many years organic chemists believed that all bimolecular eliminations proceeded by the *anti* mode, but research beginning in the 1960s demonstrated that small amounts of *syn* elimination can sometimes contribute to an overall reaction. Although the eclipsed conformation (which gives rise to *syn* elimination) is higher in energy than the staggered conformation for an *anti* elimination, the orientation is correct for π bond formation. This is illustrated in the following equation (which shows the appropriate *atomic* orbitals).

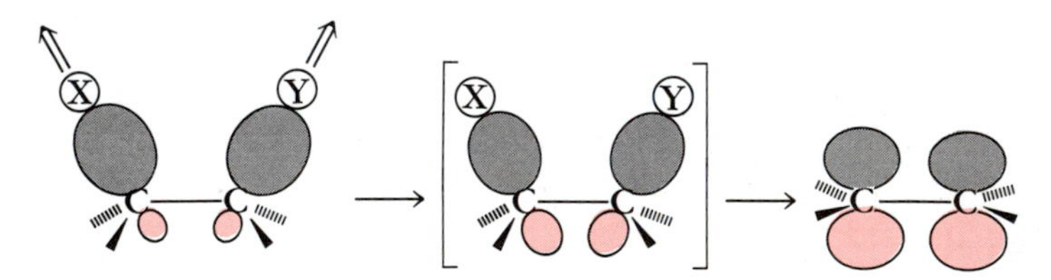

A direct comparison of *syn* and *anti* elimination pathways has been carried out for several compounds by means of isotopic labeling. The formation of (*E*)-5-decene from 5-chlorodecane was studied by using deuterium-labeled starting material. For example, (5*S*, 6*S*)-5-chloro[6-$^2H_1$]decane would yield deuterium labeled (*E*)-5-decene by *anti* elimination. *Syn* elimination, however, would produce an alkene containing no deuterium.

D, $CH_2CH_2CH_2CH_3$, Cl, H, $CH_3CH_2CH_2CH_2$, H —KOtBu, *anti* elimination→ D, $CH_2CH_2CH_2CH_3$, $CH_3CH_2CH_2CH_2$, H

⇅

D, H, Cl, $CH_2CH_2CH_2CH_3$, $CH_3CH_2CH_2CH_2$, H —KOtBu, *syn* elimination→ H, $CH_2CH_2CH_2CH_3$, $CH_3CH_2CH_2CH_2$, H

The elimination of HCl from 5-chlorodecane was found to proceed by about 7% *syn* elimination and 93% *anti* elimination when carried out with potassium *tert*-butoxide with dimethyl sulfoxide as the solvent.

The net contribution of the *syn* pathway to an elimination reaction depends upon a variety of factors, including the solvent used for the reaction. For the hydroxylic solvents typically used for elimination reactions we would expect even less *syn* elimination than the 7% observed in the previous example. This evidence supports our previous generalization that *bimolecular elimination reactions proceed by the anti mode.* These reactions are therefore stereospecific and you can predict the configuration of the alkene if you know the stereochemistry of the reactant.*

Syn Eliminations

A limited number of elimination reactions proceed stereospecifically by the *syn* pathway. These *syn* eliminations are *intramolecular* (only a single molecule is involved in the actual elimination) and proceed by a cyclic transition state in which the leaving group also acts as a base for removal of a hydrogen on the adjacent carbon.

We will present two examples of *syn* eliminations here, and both are *pyrolysis* (from the Greek *pyr,* "fire," and *lysis,* "loosening cleavage") reactions: ester pyrolysis and amine oxide pyrolysis. As suggested by the names, these reactions are carried out at relatively high temperatures. Ester pyrolysis takes place when an ester (Chapter 16) is heated at 300–500°C and yields the alkene and a carboxylic acid as products.

Amine oxide pyrolysis (see Section 17.5) takes place at a somewhat lower temperature (100–200°C) and yields the alkene together with a hydroxylamine:

An amine oxide

The following examples illustrate the stereochemistry of these pyrolysis reactions:

*Acid catalyzed dehydrations constitute an exception, however. These eliminations are reversible (Section 5.13), and the stereochemical correlation between the configurations of reactants and products is frequently destroyed by isomerization of the alkene.

pyrolysis

36% 64%

These reactions proceed readily in both cyclic and acyclic compounds, although the preceding examples all happen to involve cyclic compounds. For a *syn* elimination to occur in a cyclic system, the leaving group must be *cis* to the hydrogen which is removed. The more highly substituted alkene can be formed as a major product only when there is a hydrogen *cis* to the leaving group. The high stereospecificity of pyrolytic *syn* eliminations makes them useful synthetic reactions.

6.10 FORMATION OF OPTICALLY ACTIVE PRODUCT FROM AN ACHIRAL REACTANT

Can an optically active molecule ever be produced from an achiral or racemic organic compound? Under certain circumstances the answer is yes. Most reactions involve two or more reactants, and *one of them* must be optically active if the product is to be optically active. We have illustrated such a situation (Section 6.6) in the resolution of enantiomers. Reaction of a racemic carboxylic acid with an optically active amine produced diastereomeric products, which could be separated by physical methods.

Two enantiomers —chiral reagent (optically active)→ two diastereomers —separation→ optically active products

Enzymatic Reactions

Optically active products can be formed from an achiral starting material, particularly in biological reactions which are catalyzed by enzymes. The first step in an enzymatic reaction is the formation of an *enzyme-substrate complex*. Reaction then converts this to an *enzyme-product complex,* which subsequently dissociates to liberate the product and regenerate the free enzyme. Enzymes are chiral, and only a single enantiomeric form is present in an organism. Consequently when the enzymatic reaction of an achiral molecule produces a new chiral center in the organic substrate, one of two diastereomeric enzyme complexes must be produced, enzyme–*R* product or enzyme–*S* product. Diastereomers have different physical and chemical properties, including different stabilities, so the activation energies for the alternative pathways will be different. Therefore, the rate of formation of *R* product will be different from the rate of formation of *S* product. For many enzymatic reactions this rate difference is sufficiently large that only a single enantiomer of the chiral product is formed, and the overall process results in the conversion of an achiral starting material into an optically active product.

One important biochemical reaction which illustrates this process is the hydration of the double bond of fumaric acid to produce malic acid.

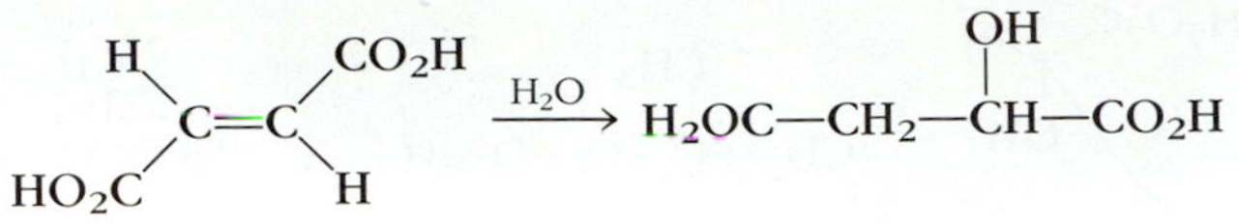

$$\underset{\text{Fumaric acid}}{(H)(HO_2C)C{=}C(CO_2H)(H)} \xrightarrow{H_2O} \underset{\text{Malic acid}}{H_2OC{-}CH_2{-}\overset{\displaystyle OH}{\overset{|}{C}}H{-}CO_2H}$$

Despite its high symmetry, the two faces of the double bond in fumaric acid are not totally equivalent. The following equation shows that an addition reaction to fumaric acid can generate a chiral center, and the configuration will depend upon which face of the double bond undergoes reaction.

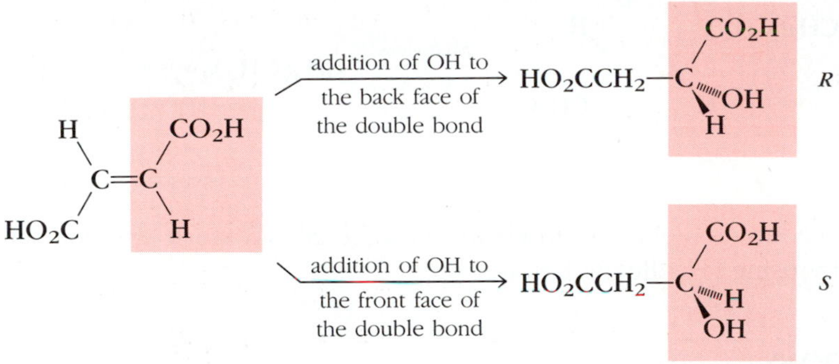

H, CO_2H, C=C, HO_2C, H

addition of OH to the back face of the double bond → $HO_2CCH_2{-}C(CO_2H)(OH)(H)$ *R*

addition of OH to the front face of the double bond → $HO_2CCH_2{-}C(CO_2H)(H)(OH)$ *S*

The two faces of the double bond of fumaric acid are called **prochiral.** The term *prochiral* indicates a specific difference between the two faces; reaction at one of the faces generates a product that is the enantiomer of the product formed by reaction at the other face.

The addition of water to fumaric acid is part of the biochemical pathway of *respiration,* in which cells utilize molecular oxygen to produce energy, and the reaction is catalyzed by the enzyme *fumarase.* Binding of fumaric acid to fumarase takes place in an orientation such that the hydroxyl group is added specifically to just one of the prochiral faces of the double bond, and only the *S* enantiomer of malic acid is formed.

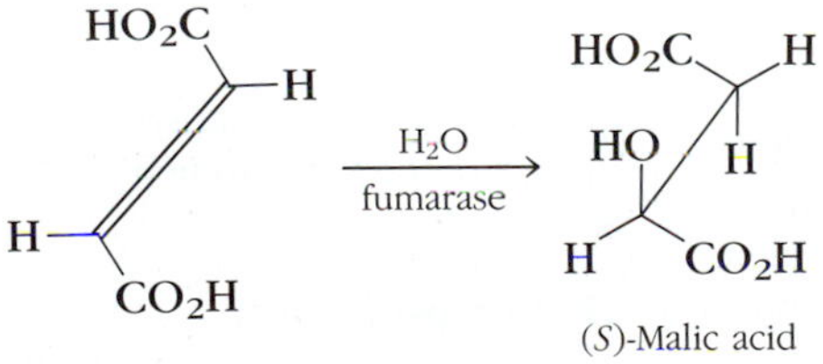

HO_2C, H, H, CO_2H $\xrightarrow[\text{fumarase}]{H_2O}$ HO_2C, H, HO, H, H, CO_2H

(*S*)-Malic acid

Laboratory Reactions

Although the biochemical conversion of achiral reactants into optically active products is widespread, there are relatively few examples of such reactions that can be carried out in the absence of an enzyme. One reaction which has been found to work well in the laboratory is the epoxidation of certain achiral alkenes to yield optically active epoxides. In the absence of an enzyme some other chiral reagent must be used, and some of the most successful results have been obtained with a derivative of tartaric acid, the same compound which originally revealed the relationship between molecular structure and optical activity. The chiral oxidizing agent is a mixture of (+)-diethyl tartrate, *tert*-butyl hydroperoxide, and titanium tetra-(2-propoxide).

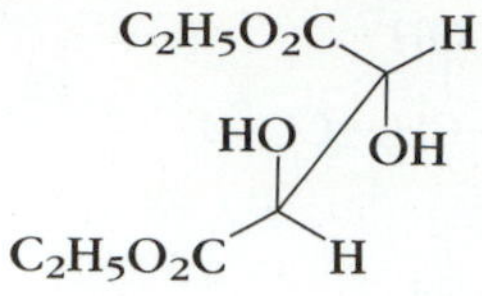

$C_2H_5O_2C$ H HO OH $C_2H_5O_2C$ H
(+)-Diethyl tartrate

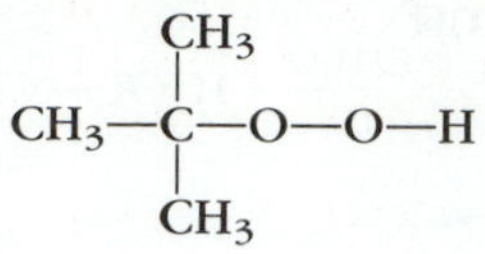

CH_3—C(CH_3)(CH_3)—O—O—H
t-Butyl hydroperoxide

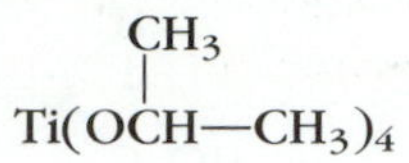

Ti(OCH(CH_3)—CH_3)$_4$
Titanium tetra-(2-propoxide)

The reagent was developed by K. B. Sharpless at the Massachusetts Institute of Technology. Its reaction with an alkene often leads to a single enantiomer in high yield, as illustrated by the following example:

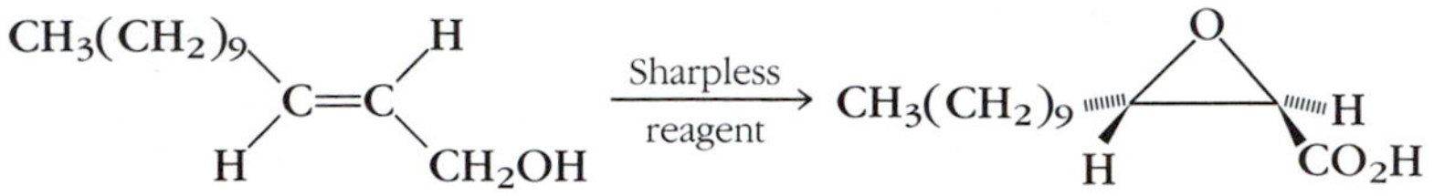

$$CH_3(CH_2)_9(H)C{=}C(H)CH_2OH \xrightarrow[\text{reagent}]{\text{Sharpless}} CH_3(CH_2)_9 \text{—(epoxide, O)—} CO_2H$$

75% yield
(> 95% optical purity)

The procedure is highly stereospecific, and the enantiomeric epoxide can be obtained by using (−)-diethyl tartrate.

6.11 TERMS AND DEFINITIONS

Achiral. Possessing reflection symmetry. Describes an object or molecule that is identical to its mirror image.

Chiral. Lacking reflection symmetry. Describes an object or molecule that is different from its mirror image.

Chiral center. A site in a molecule that is characterized by the absence of reflection symmetry. For organic compounds this is usually a tetrahedral carbon atom with four different substituents. A molecule with a single chiral center will always be a chiral molecule.

Configuration. The fixed spatial arrangement of substituents on an atom or at a double bond. A chiral center has two possible configurations, as does a double bond if neither carbon has two equivalent substituents. (If two substituents on a tetrahedral carbon atom are equivalent, only a single configuration is possible for that carbon.) Configuration at a chiral center is specified by *R/S* nomenclature (Section 6.3), and configuration at a double bond is specified by *E/Z* nomenclature (Section 5.5).

Conformation. A spatial arrangement of the atoms of a molecule that can be converted to another arrangement by rotation about single bonds. Under normal conditions the individual conformations of a molecule cannot be isolated; therefore they are not considered isomers.

Constitution. The connectivity sequence by which atoms in a molecule are bonded to each other. The constitution can be defined either by drawing a structural formula or by stating the IUPAC name.

Diastereomers. Stereoisomers that are not mirror images of each other. Usually two diastereomers differ either in the configuration at a carbon-carbon double bond or in the configurations at some but not all chiral centers.

Enantiomers. Two stereoisomers that are nonsuperimposable mirror images. Each enantiomer is necessarily chiral.

Equivalent. The same. This term is used to describe two or more groups or parts in a molecule or to describe the relationship between groups in several molecules. *Constitutional equivalence* of groups in a molecule requires either rotational symmetry (the groups are identical) or reflection symmetry (the groups are mirror images).

Fischer projection. A three-dimensional representation used to show the orientation of substituents on a carbon atom, in which the carbon is denoted by the intersection of a horizontal and a vertical line. The substituents on the vertical line are farther away from you than the chiral center, and the substituents on the horizontal line are closer to you.

Meso compound. A molecule that is achiral despite having more than one chiral center. It has reflection symmetry, and it contains an even number of constitutionally equivalent chiral centers.

Mirror image. The image that would be seen if an object or molecule were viewed in a mirror.

Optical activity. The rotation of polarized light by chiral molecules. One enantiomer must be present in excess of the other for optical activity to be observed.

Orientation. The position of a molecule or drawing relative to its surroundings. Reorienting a molecule does not affect the constitution, configuration, or conformation; only its position relative to its environment is affected.

Prochiral. Capable of becoming chiral by a simple reaction: by *adding* a fourth substituent to an sp^2 carbon or *modifying* one of two equivalent substituents on an sp^3 carbon.

Racemic. Consisting of exactly equal amounts of two enantiomers. A racemic substance is optically inactive and behaves as a pure compound since the two enantiomers cannot be separated by physical methods such as distillation.

Reflection symmetry. A property of an achiral object or molecule, wherein a structure is symmetric about its midpoint and the two "halves" are mirror images of each other. The most common types of reflection symmetry are a plane of symmetry and a center of symmetry.

Relative configuration. Specification of the configuration of a chiral center (or of an entire molecule) in relation to the chiral center of another compound. If one compound can be converted to the other, the relative configuration can be discussed even if the absolute configuration is not known for either.

Resolution. The process by which a mixture of two enantiomers is separated (resolved) into the two optically active, enantiomeric forms. Since the enantiomers have identical physical properties, this is accomplished by reaction with a chiral (optically active) reagent.

Stereochemistry. The study of the three-dimensional aspects of molecular structures and chemical reactions. The term is also used to describe the three-dimensional structure of a molecule.

Stereoisomers. Isomers that have the same constitution but differ in some aspect of their three-dimensional structure. Stereoisomers are either enantiomers or diastereomers.

Stereospecific. A term describing a reaction in which there is a specific relationship between the configuration of the reactants and the configuration of the products.

6.12 PROBLEMS

6.1 Label each chiral center as either *R* or *S* in the following structures:

(a)

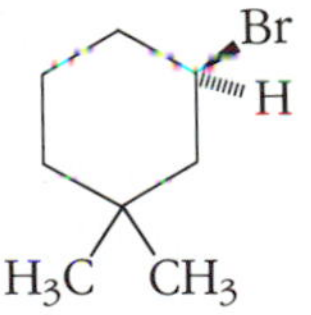

(b)

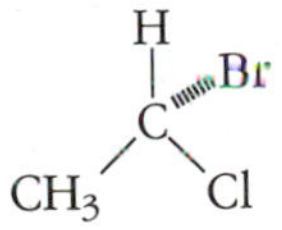

(c)

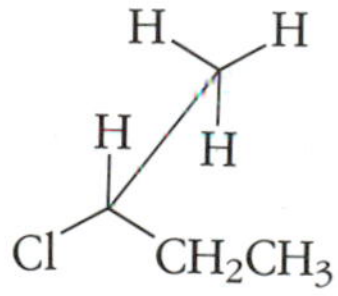

(d)

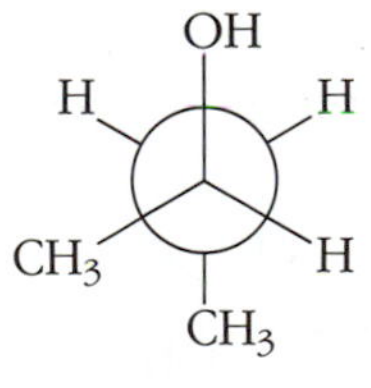

(e)

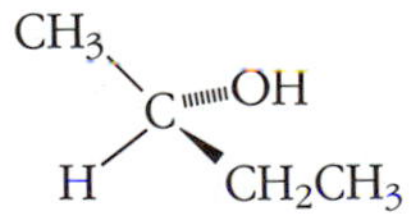

(f)

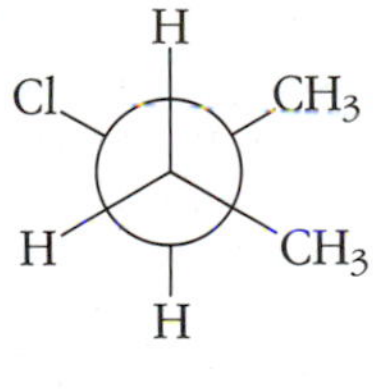

(g)

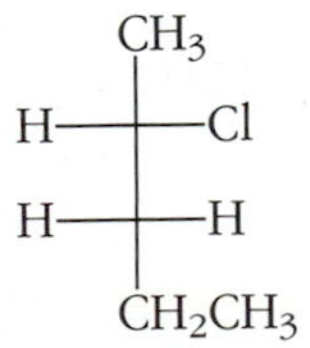

(h)

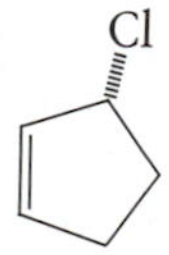

(i)

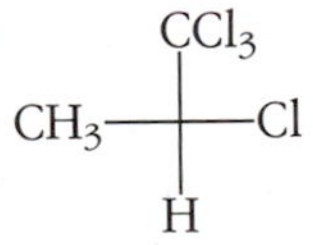

(j)

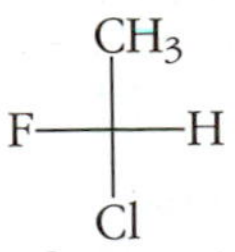

(k)

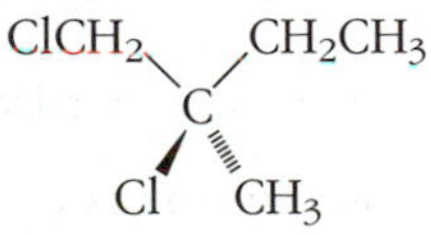

(l)

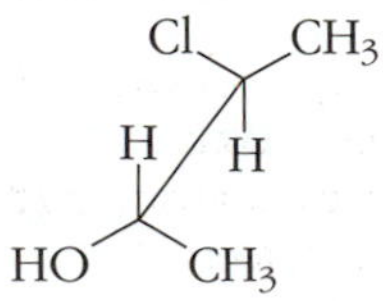

6.2 Describe the relationship between each pair of structural formulas as that of *constitutional* isomers, *enantiomers,* two drawings of the *same* compound (although perhaps in different conformations), *diastereomers,* or completely *different* compounds that are not isomers.

(a)

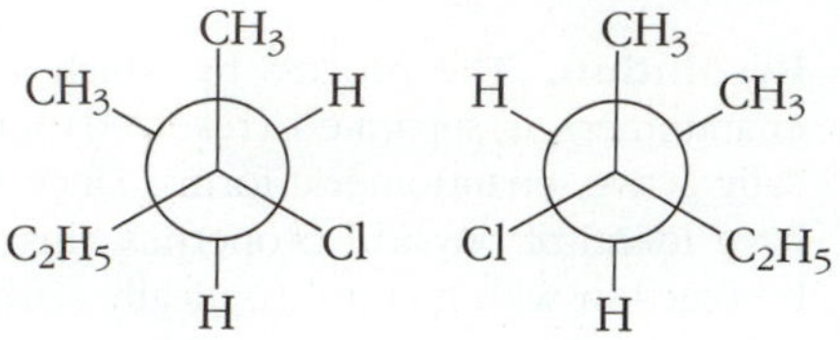

(b)

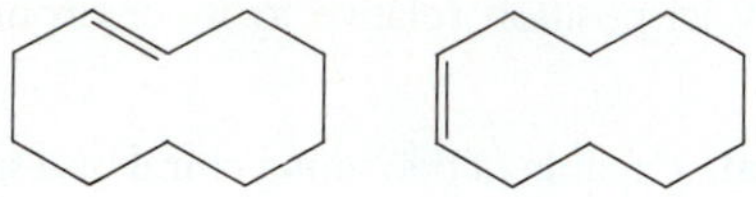

(c)

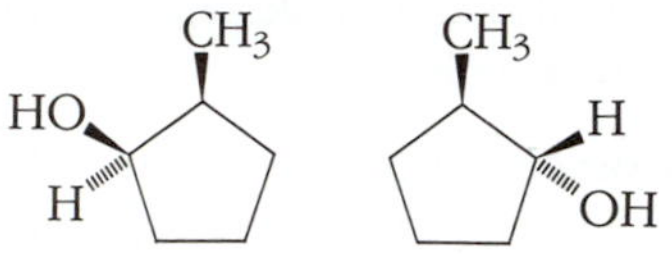

(d)

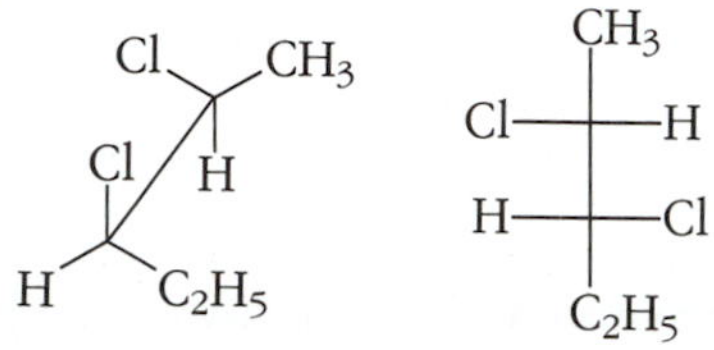

(e)

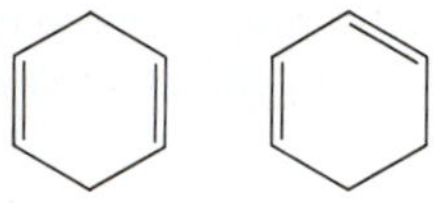

(f)

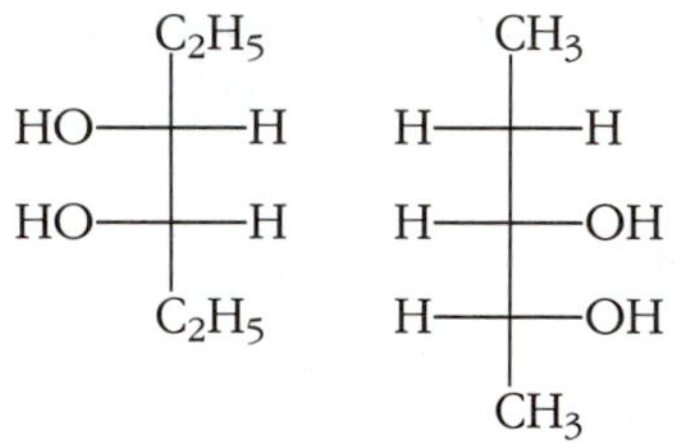

(g)

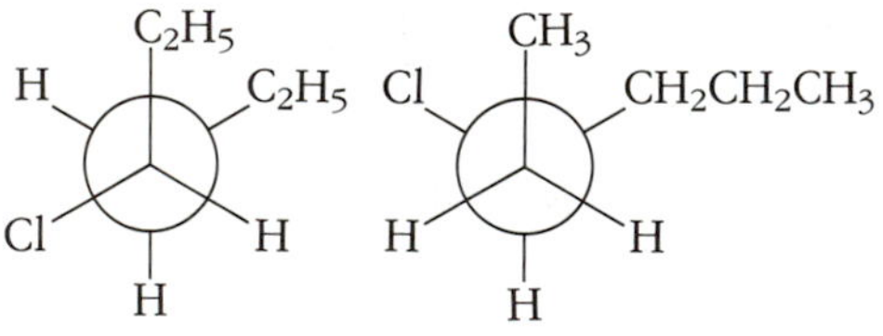

(h)

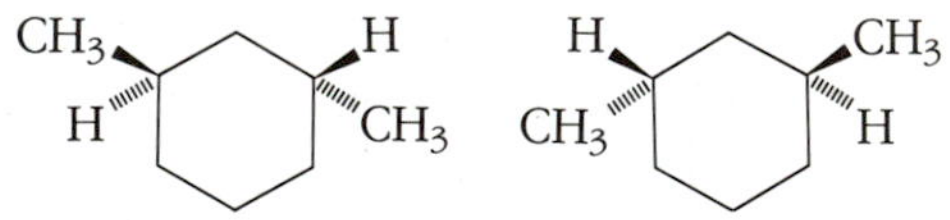

(i)

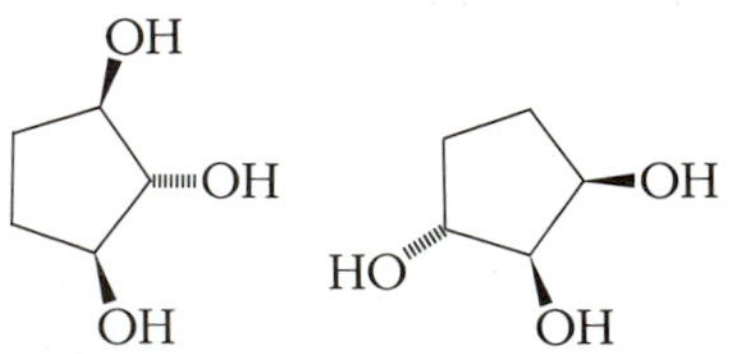

(j)

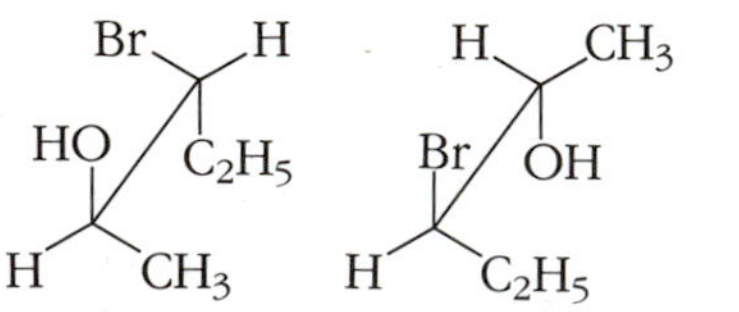

(k)

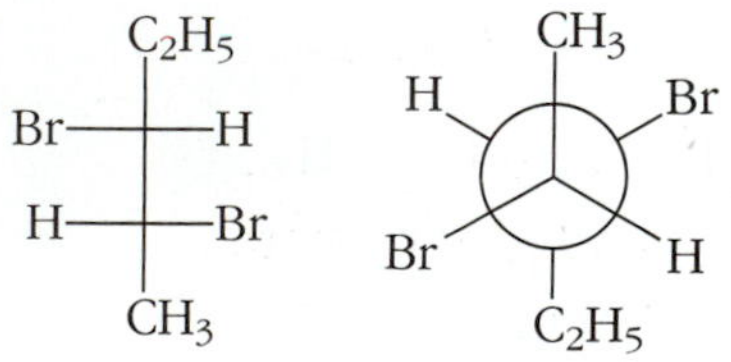

(l)

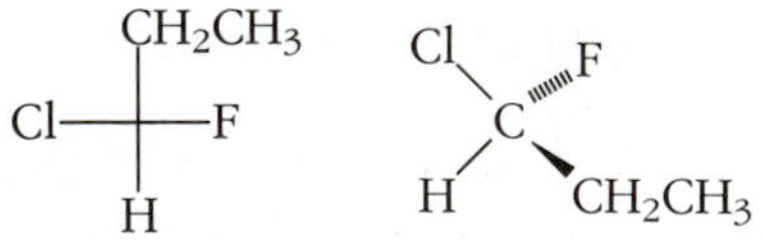

(m)

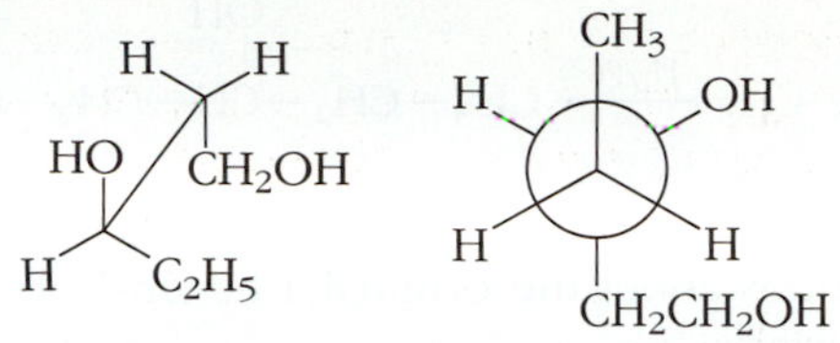

(n)

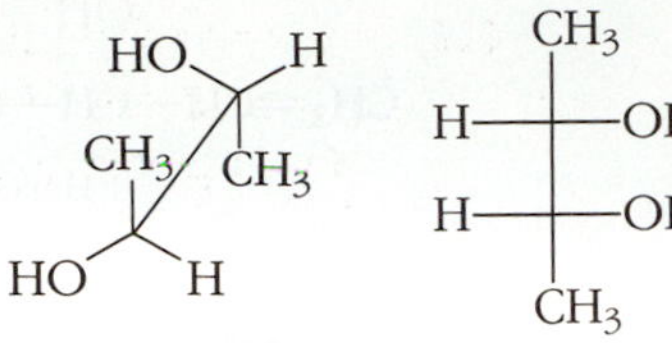

(o)

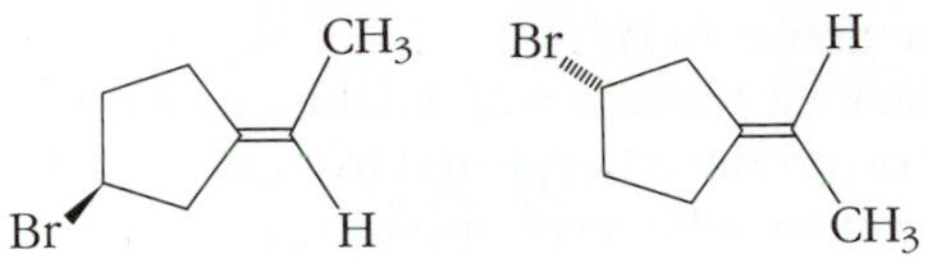

(p)

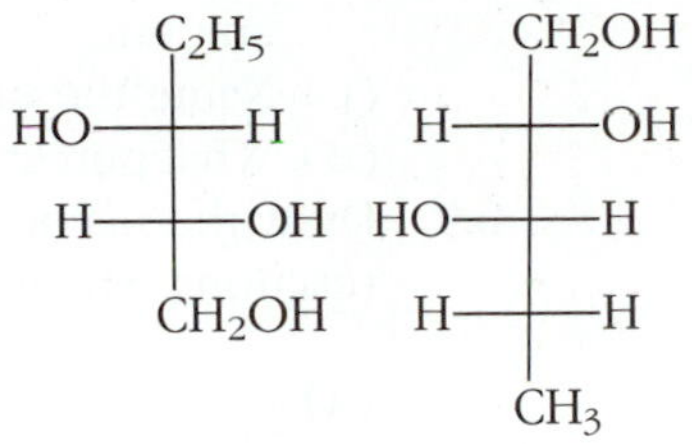

(q)

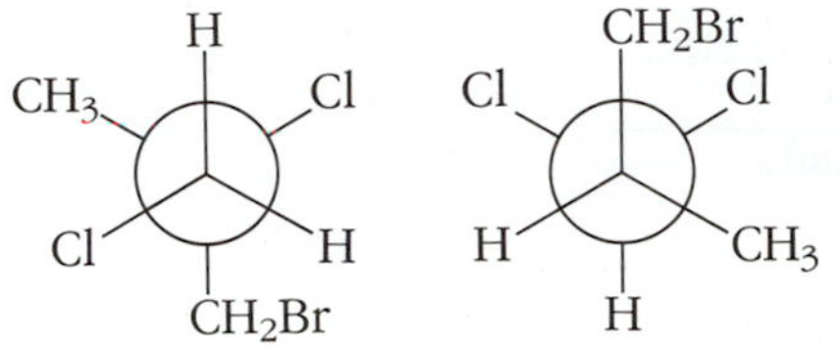

(r)

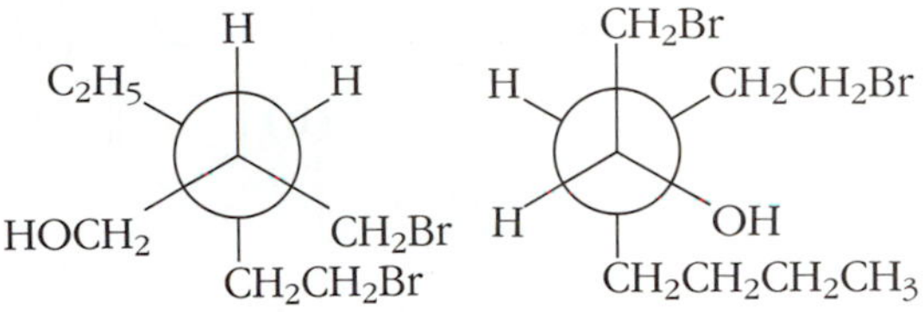

(s)

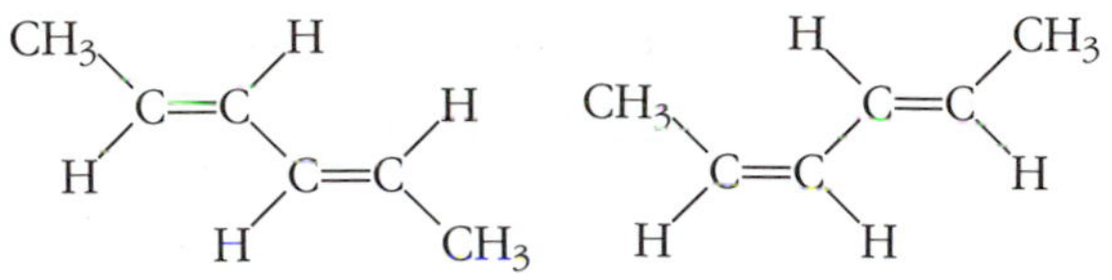

(t)

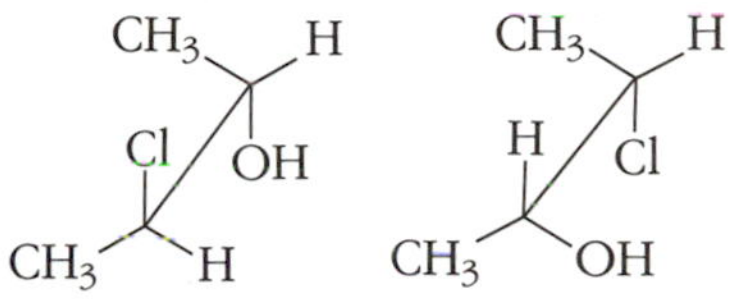

6.3 Monochlorination of (*S*)-1-chloro-2,5-dimethylhexane would yield a variety of dichloro products ($C_8H_{16}Cl_2$), which could be separated by distillation.
(a) Draw the reactant. *Show all stereochemistry.*
(b) Draw the products of the reaction. *Show all stereochemistry.*
(c) Label all chiral centers in your drawings as *R* or *S*.
(d) Name each product.
(e) For each product you draw, state clearly: (1) whether or not the molecule is chiral; and (2) whether or not the product would be obtained in optically active form.

6.4 Answer Problem 6.3 for the chlorination of (*R*)-1-bromo-3-chloro-2-methylpropane to yield $C_4H_7BrCl_2$ isomers.

6.5 Answer Problem 6.3 for the chlorination of (*S*)-1-chloro-2,4,4,6-tetramethylheptane to yield $C_{11}H_{22}Cl_2$ isomers.

6.6 Explain why it is not possible to assign the configuration of a carbon atom as either *R* or *S* if the carbon atom is not chiral.

6.7 What was the chiral "reagent" in Pasteur's resolution of potassium ammonium tartrate?

6.8 Pure (*S*)-3-hexanol was found to have a specific rotation of −5.20°. However, when synthesized by hydrogenation of a sample of 1-hexen-3-ol, the 3-hexanol had a specific rotation of +3.12°.

$$CH_2{=}CH{-}\underset{|}{\overset{OH}{CH}}{-}CH_2{-}CH_2{-}CH_3 \xrightarrow{H_2/Pt} CH_3{-}CH_2{-}\underset{|}{\overset{OH}{CH}}{-}CH_2{-}CH_2{-}CH_3$$

1-Hexen-3-ol 3-Hexanol

Answer the following questions about the original 1-hexen-3-ol:

(a) What was its optical purity?

(b) Draw the structure of the enantiomer that was present in greater amount.

(c) Name the enantiomer drawn in (b).

(d) What percentage of the total 1-hexen-3-ol did this enantiomer comprise?

6.9 Draw the major product (or products) expected from each of the following reactions. Show stereochemistry whenever appropriate.

(a) (isopropylbenzene, $C_6H_5{-}CH(CH_3){-}CH_3$) $\xrightarrow{D_2,\ Pt}$?

(b) $C_6H_5{-}CH(CH_3){-}CH_3$ $\xrightarrow[\text{2. } NaBH_4]{\text{1. } Hg(OCCH_3)_2,\ H_2O}$? (with $\overset{O}{\|}$ on the acetate C)

(c) $C_6H_5{-}CH(CH_3){-}CH_3$ $\xrightarrow{H_2SO_4,\ C_2H_5OH}$?

(d) $(CH_3)_3C(H)C{=}C(H)C(CH_3)_3$ $\xrightarrow{HCl}$?

(e) $(CH_3)_3C(H)C{=}C(H)C(CH_3)_3$ $\xrightarrow[\text{2. } NaBD_4]{\text{1. } Hg(OCCH_3)_2,\ CH_3OH}$? (with $\overset{O}{\|}$ on the acetate C)

(f) 1-methylcyclohexene (CH_3) $\xrightarrow{KMnO_4}$?

(g) 1-methylcyclohexene $\xrightarrow{Br,\ CH_3\overset{O}{\overset{\|}{C}}OH}$?

(h) 1-methylcyclohexene $\xrightarrow[\text{2. } H_2O_2,\ NaOH]{\text{1. } BH_3}$?

(i) 1-methylcyclohexene $\xrightarrow{CH_3{-}\overset{O}{\overset{\|}{C}}{-}OOH}$?

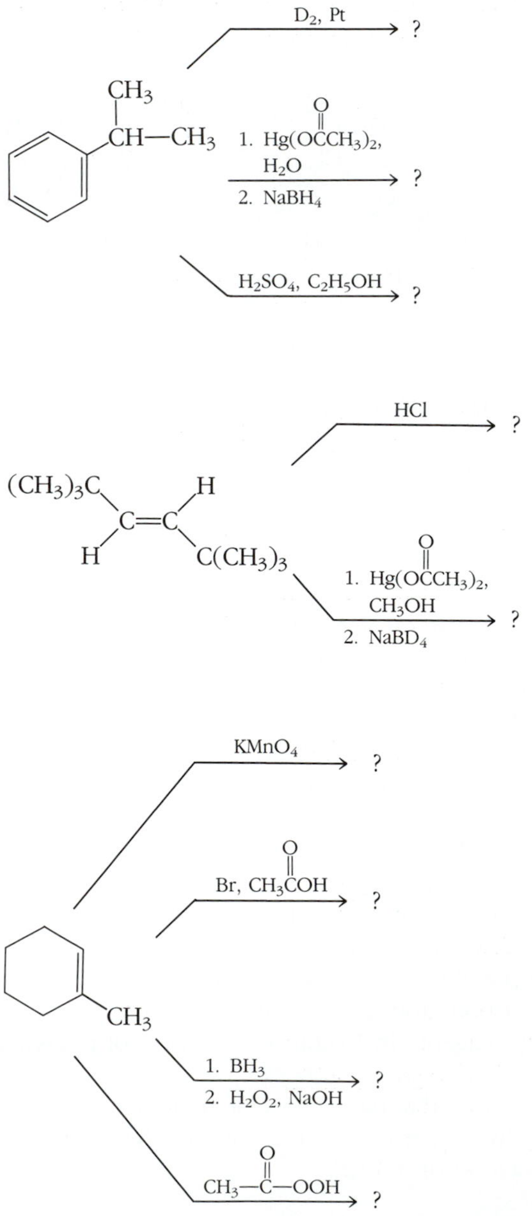

(j) 1-isopropylcyclohexene (ring with $CH(CH_3)CH_3$ substituent) $\xrightarrow{CH_3-\overset{O}{\overset{\|}{C}}-OOH}$?

(k) 1,2-dimethylcyclopentene (CH_3, CH_3) $\xrightarrow{Cl_2,\ CH_3OH}$?

6.10 Draw the major product (or products) expected from each of the following reactions. Show stereochemistry whenever appropriate.

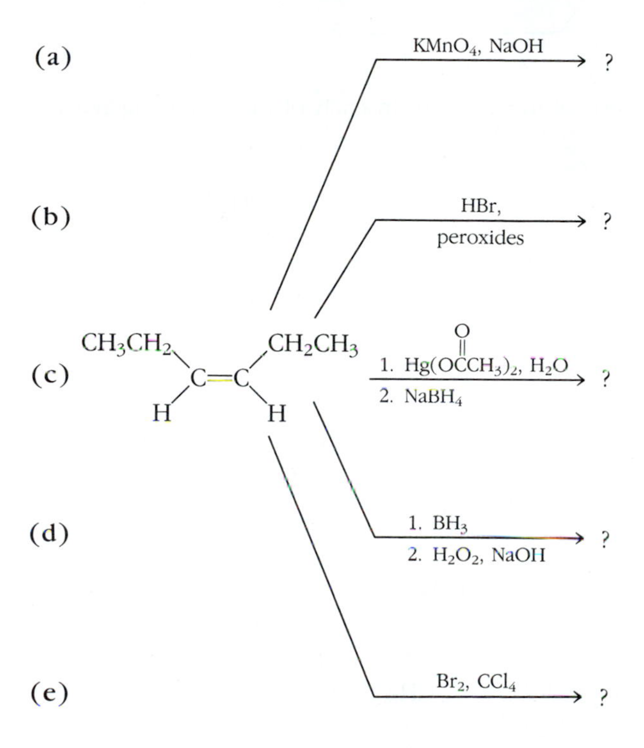

(f) $CH_3(H)C{=}C(CH_3)(CH_2CH_3)$ $\xrightarrow{Br_2,\ CCl_4}$?

(g) 1-isopropylcyclohexene (ring with $CH(CH_3)CH_3$ substituent) $\xrightarrow[2.\ NaBH_4]{1.\ Hg(OAc)_2,\ CH_3OH}$?

(h) $\xrightarrow{OsO_4,\ NaClO_3}$?

(i) $\xrightarrow[H_2SO_4]{CH_3-C(=O)-OH,}$?

(j) $\xrightarrow{Cl_2,\ CCl_4}$?

(k) $\xrightarrow{DCl}$?

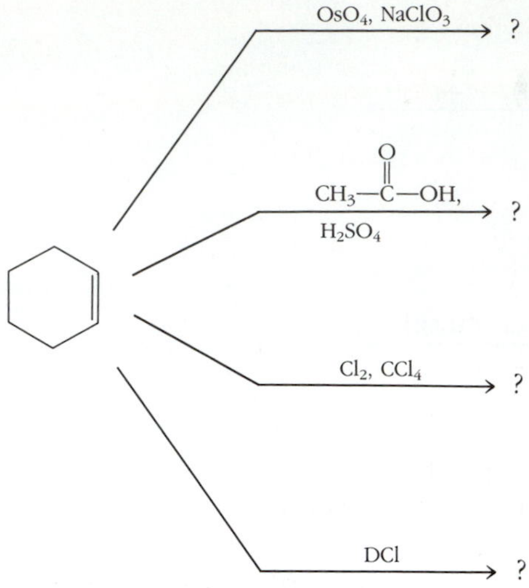

6.11 Deduce the structure of the reactant in each of the following reactions:

(a) $C_{10}H_{21}Cl$ (optically active) $\xrightarrow{KOH}$ $(CH_3)_2CH(H)C{=}C(CH_3)CH_2{-}CH(CH_3)_2$

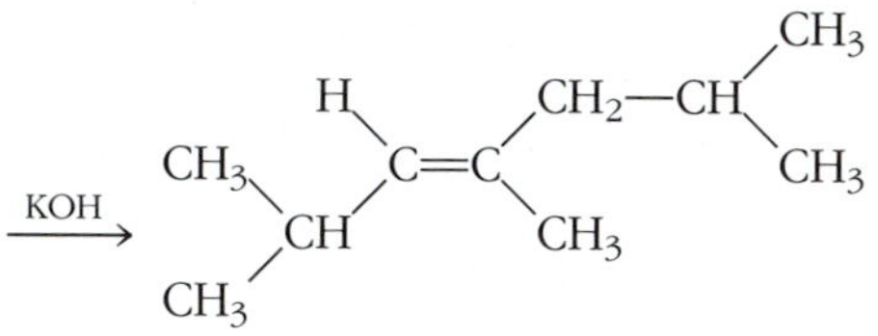

(b) $\xrightarrow{Br_2,\ CCl_4}$ [Br, CH_3, H / Br, H, $(CH_3)_2CH$] + [CH_3, Br, H / Br, H, $CH(CH_3)_2$]

(c) $\xrightarrow{Cl_2,\ CCl_4}$ [CH_3, H, Cl / CH_3, Cl, H]

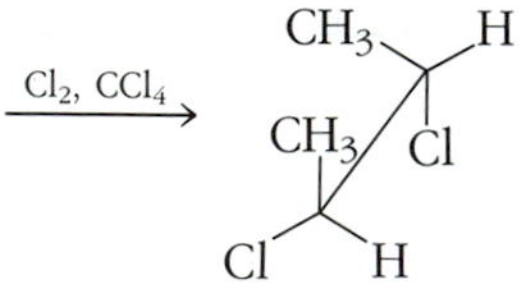

(d) $\xrightarrow{KMnO_4}$ [H, CH_2CH_3, OH / HO, H, CH_2CH_3] + [CH_3CH_2, H, OH / HO, CH_3CH_2, H]

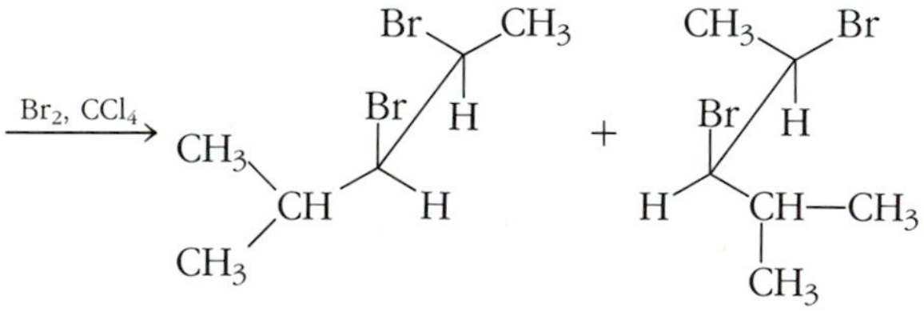

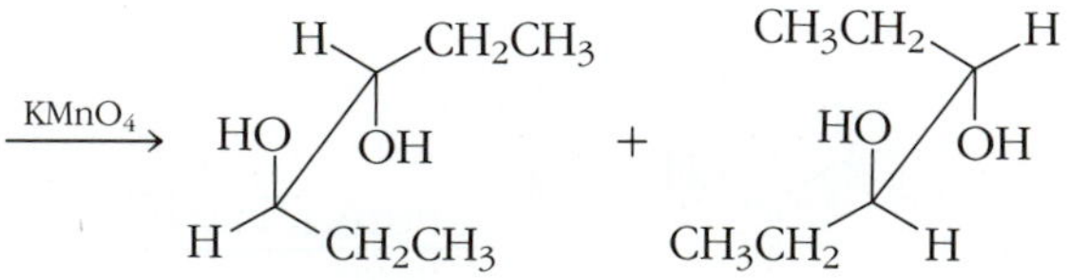

(e)

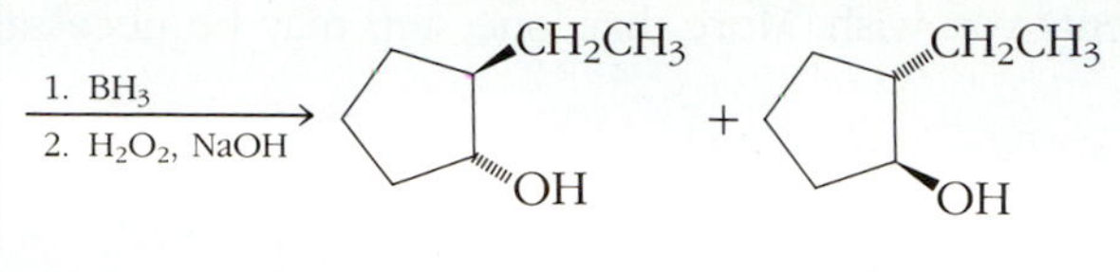

1. BH_3
2. H_2O_2, NaOH

CH_2CH_3 OH + CH_2CH_3 OH

(f)

HCl

Cl H CH_3 C CH_3 C—Cl CH_3

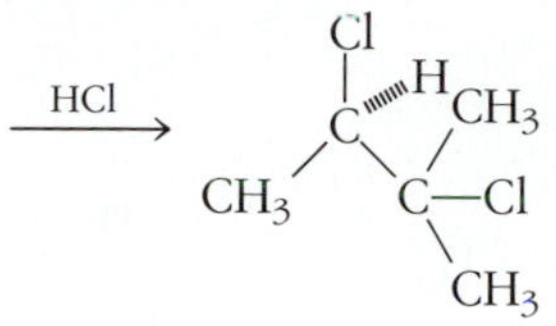

6.12 Draw the major product (or products) expected from each of the following reactions. Show stereochemistry whenever appropriate.

(a) CH_3 O—C(=O)—CH_3 CH_3 H C_2H_5 H $\xrightarrow{\text{pyrolysis}}$?

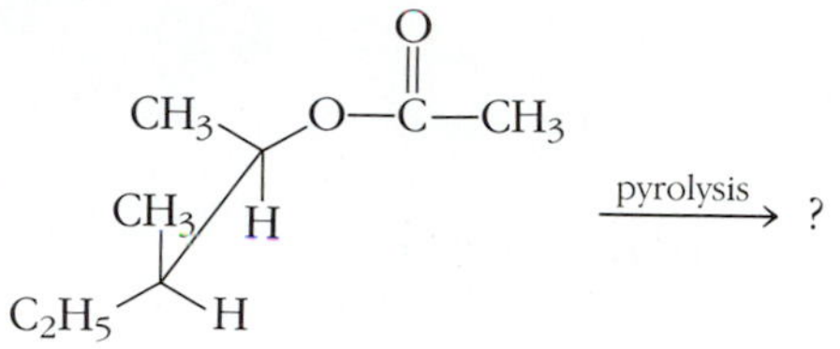

(b) CH_3 CH_2CH_3 $\xrightarrow{\text{NBS/H}_2\text{O}}$?

(c) CH_3 O—C(=O)—CH_3 $\xrightarrow{\text{pyrolysis}}$?

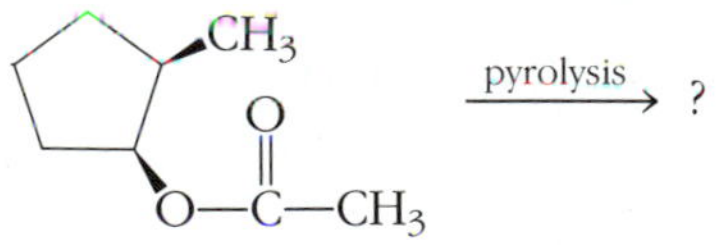

(d) CH_2CH_3 $\xrightarrow[\text{peroxides}]{\text{HBr}}$?

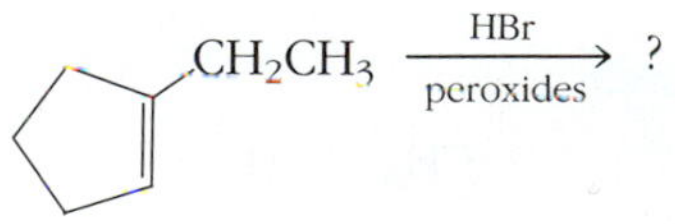

(e) CH_3 $\overset{+}{N}(CH_3)_2$ CH_3 O_- $\xrightarrow{\text{pyrolysis}}$?

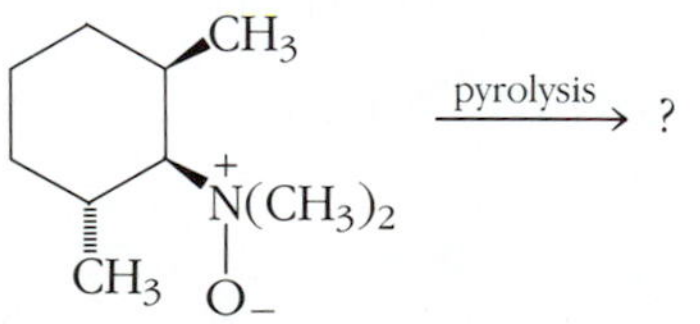

(f) CH_3 CH_3 C=C H CH—CH_3 CH_3 $\xrightarrow{CH_3CO_3H}$?

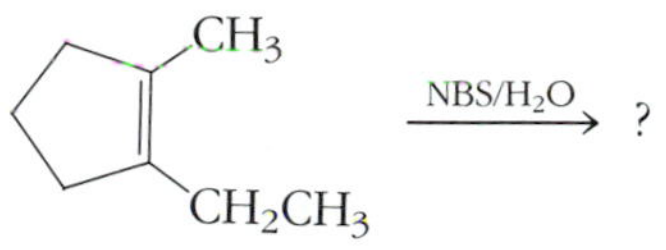

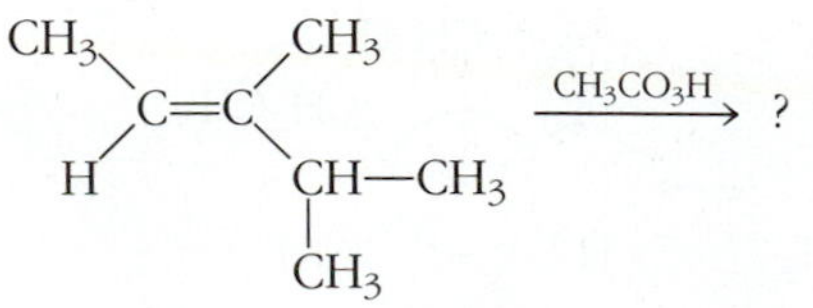

6.13 Suggest methods for carrying out the following conversions, using any reagents you wish. More than one step may be necessary in each case.

(a) 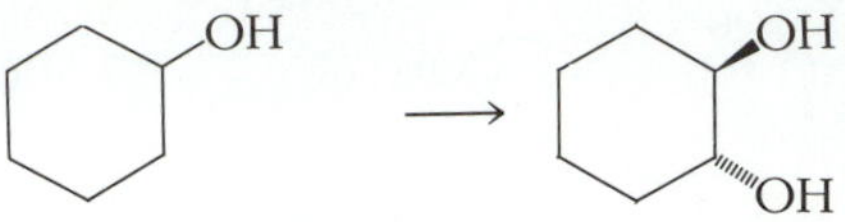 OH → OH, OH (racemic)

(b) 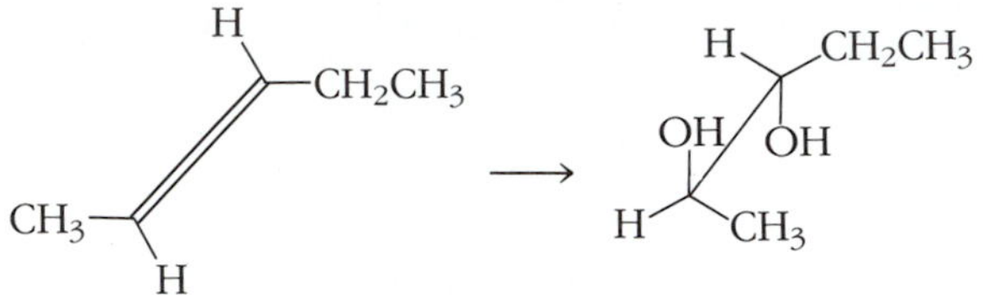 H, CH_2CH_3, CH_3, H → H, CH_2CH_3, OH, OH, H, CH_3 (racemic)

(c) 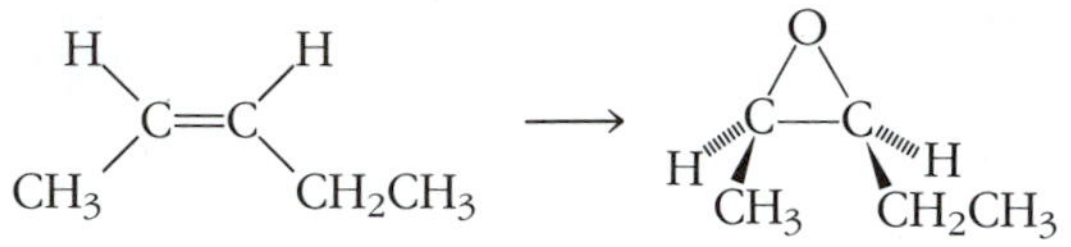 H, H, C=C, CH_3, CH_2CH_3 → O, H, C—C, H, CH_3, CH_2CH_3 (racemic)

(d)  Br → Br, OH (racemic)

(e) 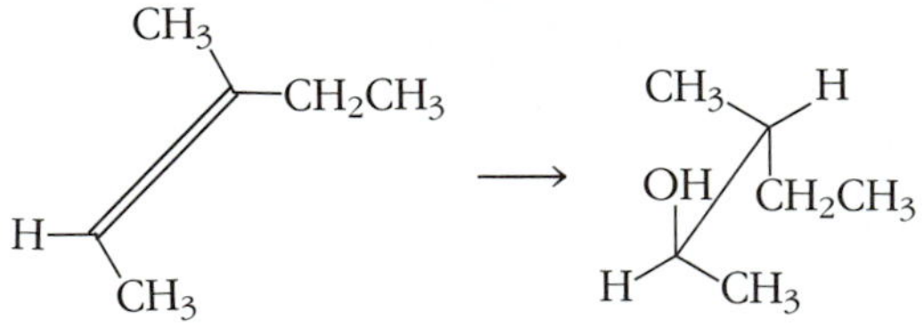 CH_3, CH_2CH_3, H, CH_3 → CH_3, H, OH, CH_2CH_3, H, CH_3 (racemic)

(f) 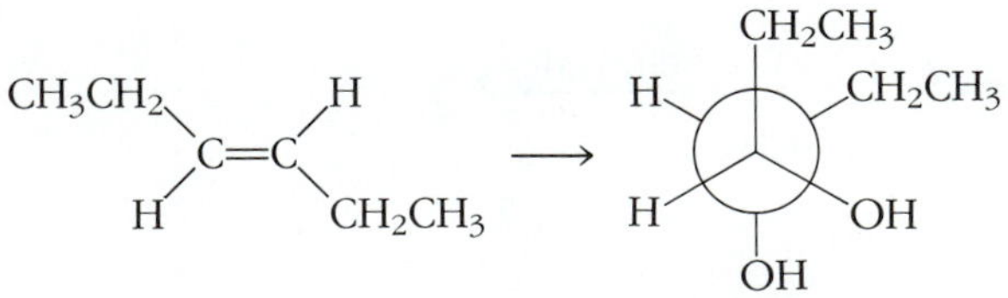 CH_3CH_2, H, C=C, H, CH_2CH_3 → CH_2CH_3, H, CH_2CH_3, H, OH, OH

6.14 Structures **A–H** are all different compounds, which can be separated by distillation. On the basis of the reactions shown, deduce the structures of **A–F.** *Show stereochemistry.*

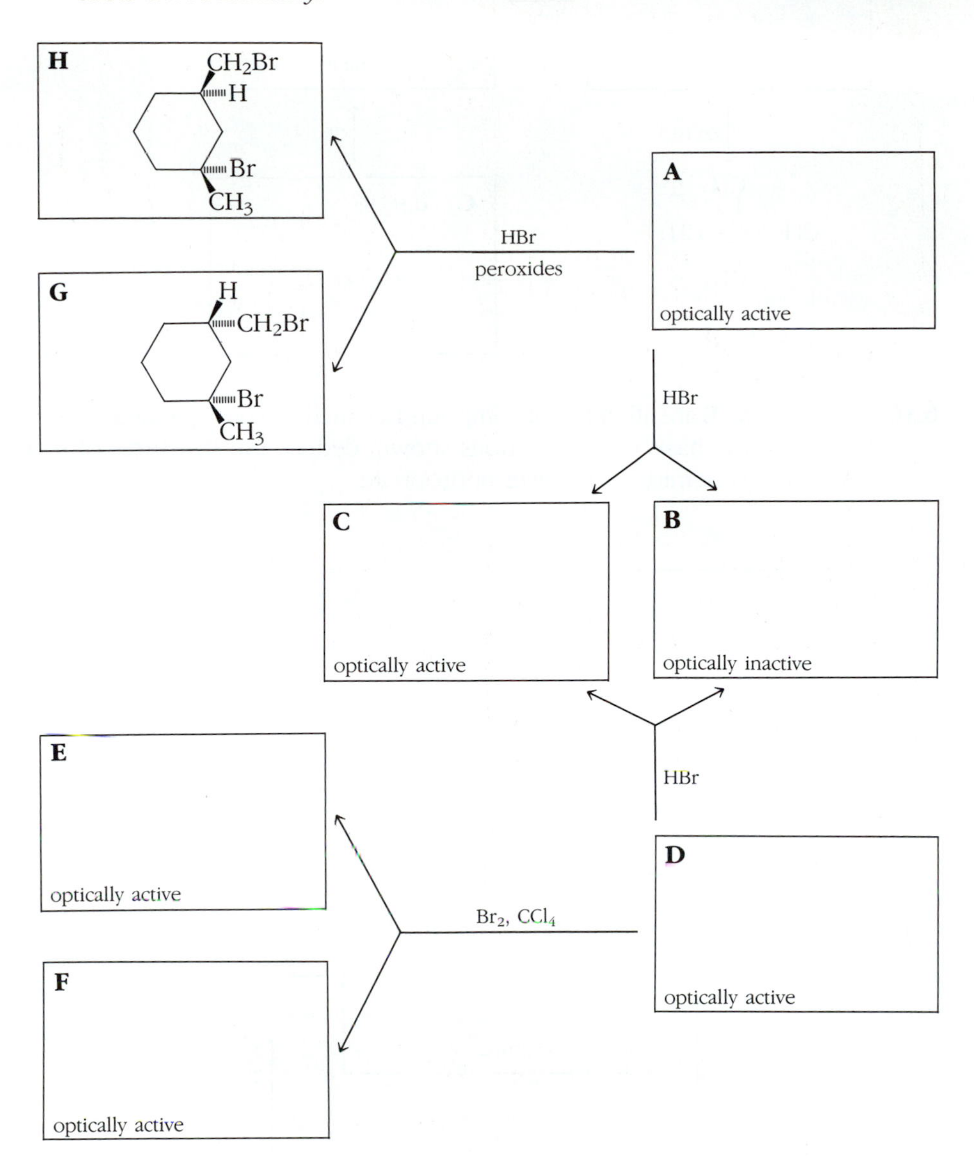

6.15 Structures **A–C** are all different compounds. On the basis of the reactions shown, deduce the structures of **A–C.**

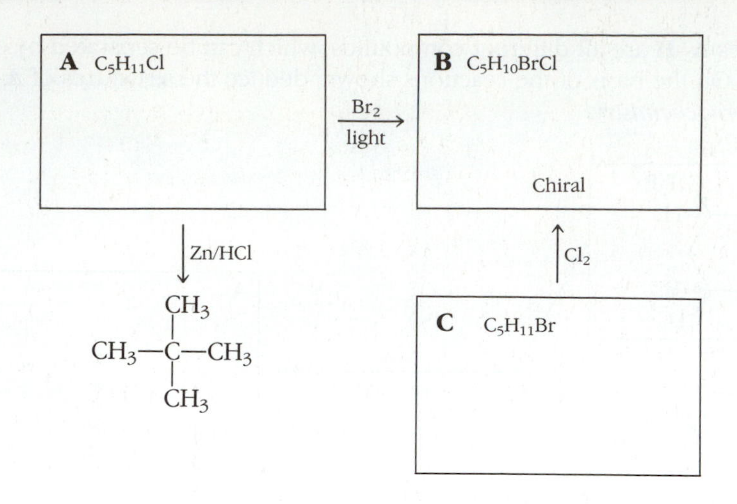

6.16 Structures **A–E** are all different compounds, which can be separated by distillation. On the basis of the reactions shown, deduce the structures of **A–E.** Show stereochemistry whenever appropriate.

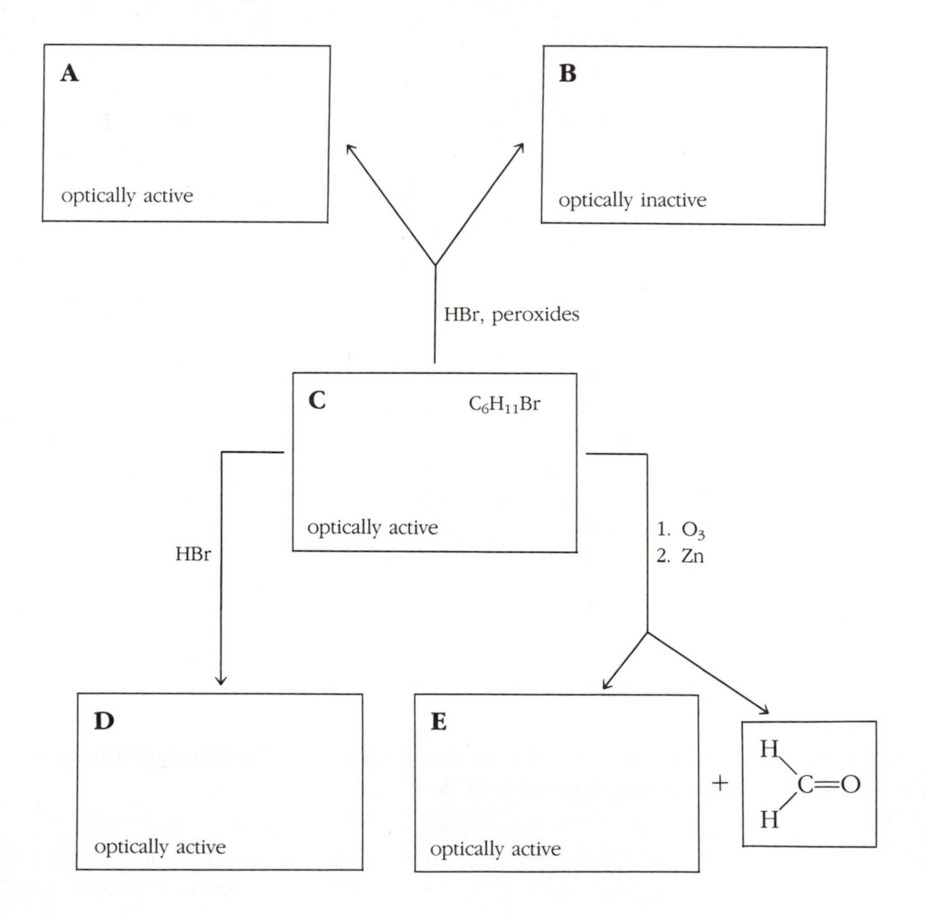

6.17 Studies on the enzymatic addition of D_2O to fumaric acid (Section 6.10) have shown that this is an *anti* addition. Draw the complete structure of the deuterated malic acid that is produced in this way, and label each of the chiral centers as *R* or *S*.

6.18 On the basis of the reactions shown, deduce the structures of **A** and **B.**

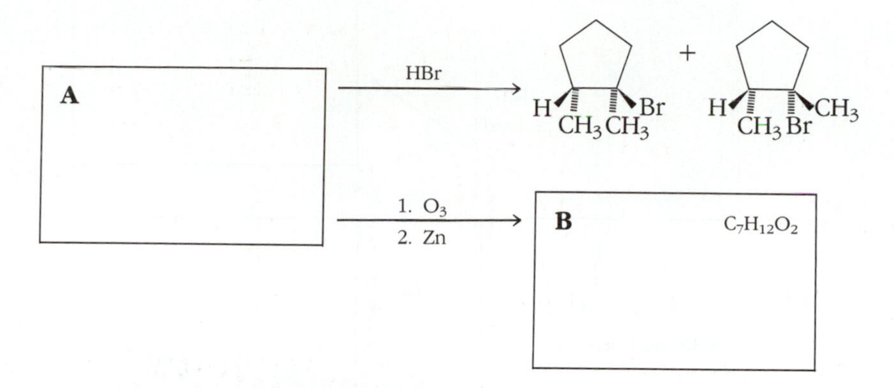

6.19 On the basis of the reactions shown, deduce the structures of compounds **A–C.**

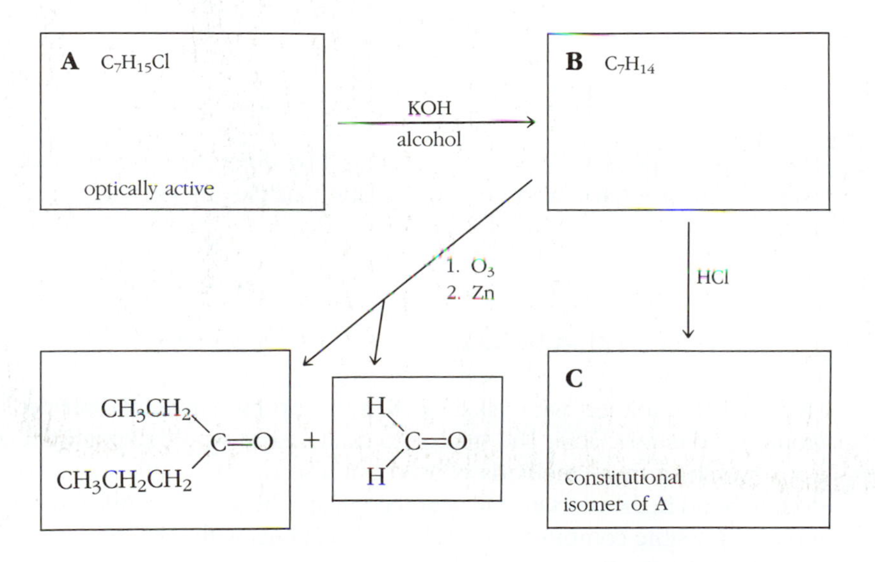

6.20 Complete the following reactions. *Show stereochemistry.* If isomeric products would be produced, specify the relationship between the products (e.g., constitutional isomers). Indicate for each product whether it would be obtained in optically active or optically inactive form.

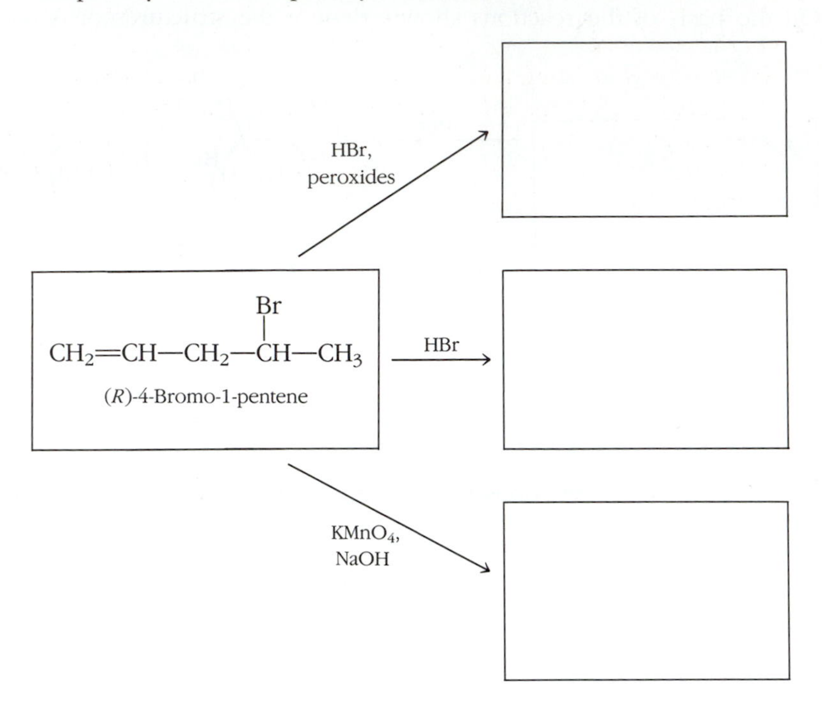

6.21 The same compound can often be prepared by different methods, as in the preparation of tartaric acid by hydroxylation of the unsaturated carboxylic acid.

$$HO_2C{-}CH{=}CH{-}CO_2H \longrightarrow HO_2C{-}\underset{}{CH(OH)}{-}CH(OH){-}CO_2H$$

Hydroxylation either with alkaline $KMnO_4$ or by epoxidation-hydrolysis would yield tartaric acid. The stereochemical results would depend not only upon which of these methods is used but also on the configuration of the double bond in the reactant. Draw and compare the stereochemical results of the four possible combinations of the two reagents with the *E* and *Z* stereoisomers of the reactant.

6.22 In contrast to other elimination reactions, the acid catalyzed dehydration of an alcohol to an alkene is frequently not stereospecific. Use the reaction mechanism to suggest a reason for this behavior.

6.23 Intramolecular elimination reactions such as ester pyrolysis proceed by a *syn* pathway, and the ester group acts as an internal base to remove a proton from the adjacent carbon.

(a) Use appropriate drawings to show that the reactive conformation must have the ester group and the hydrogen in a gauche (or even eclipsed) relationship. Show why this would preclude elimination from a cyclohexane derivative with the ester and hydrogen in a *trans* diaxial relationship.

(b) For ester pyrolysis of a cyclohexane derivative, the ester group and the hydrogen that are eliminated must be *cis*. One of these substituents would therefore be equatorial and one would be axial, and they would be separated by a 60° dihedral angle. But the substituents would also be separated by a 60° dihedral angle in the *trans* isomer with both substituents equatorial. Why does *syn* elimination not occur in this way? (*Hint:* Draw the structure of the product that would be produced and show its complete stereochemical structure).

CHAPTER 7

ALKYNES

In Chapter 5 we showed you that unsaturated hydrocarbons exhibit reactivity that is very different from that of the alkanes. Now we will turn to another class of unsaturated hydrocarbons, the **alkynes,** which are characterized by a carbon-carbon *triple bond,* —C≡C—. The simplest alkyne, acetylene (C_2H_2), may be familiar to you as the gaseous fuel that is used in welding torches. The "cannons" that are sometimes shot off after a touchdown in a football game act by the ignition of an explosive mixture of acetylene and air. The small quantities of acetylene that are needed for these reactions are generated by mixing CaC_2 with water.

As you will see in this chapter, alkynes share many features of the reactivity exhibited by alkenes, but in some cases they can also react in a way that is very different. These patterns of reactivity make alkynes a valuable class of synthetic intermediates, and we will show you how they can be used to prepare a variety of organic compounds. We will also show you how the reactions of alkynes can be understood and predicted on the basis of their molecular structures.

7.1 THE CARBON-CARBON TRIPLE BOND

The alkynes form a class of compounds which differ from the alkanes by the absence of four hydrogen atoms on two adjacent carbon atoms. While the simple open-chain alkanes are represented by the molecular formula C_nH_{2n+2}, the analogous alkynes have the formula C_nH_{2n-2}. Alkynes are therefore more unsaturated than the alkenes (C_nH_{2n}). Whereas an alkene is characterized by a carbon-carbon double bond, the unique structural feature of an alkyne is the carbon-carbon *triple bond.*

$CH_3—CH_2—CH_3$	$CH_3—CH{=}CH_2$	$CH_3—C{\equiv}CH$
Prop*ane* (an alkane)	Prop*ene* (an alkene)	Prop*yne* (an alkyne)

Each of the triply bonded carbons has only two substituents (but, of course, a total of four bonds). As you would expect on the basis of electron repulsion arguments, the substituents lie as far apart as possible, that is, the C—C—H angle of the following fragment is 180°.

$$C{\equiv}C—H$$

This applies to *both* carbon atoms of the triple bond, so the geometry of a triple bond is characterized by a linear array of four atoms, as illustrated in Figure 7.1.

The reactivity of alkynes, just as you saw for alkenes, in Chapter 5, is directly related to the high electron density in the vicinity of the multiple bond. The carbon-carbon triple bond can function as a Lewis base, and reactions of alkynes are commonly initiated by electrophilic attack on the triple bond. Not surprisingly, addition reactions similar to those of alkenes are one of the two major kinds of reactions which alkynes undergo. The other major type of reaction is found only with **terminal alkynes,** i.e., those for which the triple bond is at the end of the carbon chain. In contrast to the hydrogens of alkanes and alkenes, the hydrogen of a terminal alkyne (R—C≡CH) is relatively acidic. Removal of this hydrogen atom by a strong base generates an anion, which undergoes a variety of important reactions.

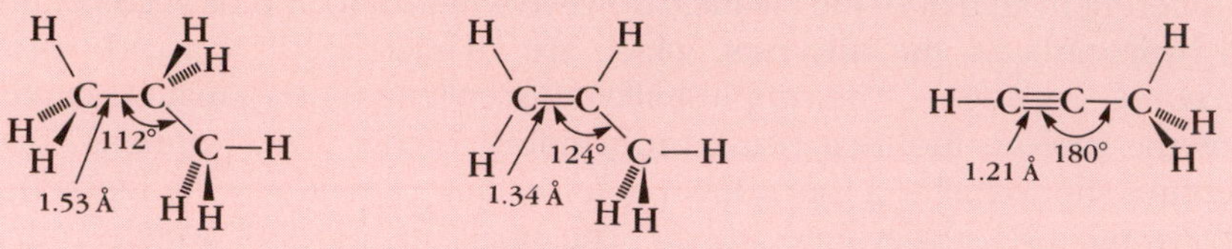

Propane
(an *alkane*)

Propene
(an *alkene*)

Propyne
(an *alkyne*)

Figure 7.1 The Carbon-Carbon Triple Bond.
In comparison with an alkane, the corresponding alkyne differs by the absence of four hydrogens on two adjacent carbon atoms; these two carbon atoms are joined by a triple bond. There is a regular decrease in bond length in proceeding from a single bond to a double bond to a triple bond, as shown by the three structures in the figure. The C—C—C bond angle of propyne is 180°, and the triple bond, together with the two atoms attached to it, constitutes a linear array of four atoms.

7.2 MOLECULAR ORBITAL DESCRIPTION OF THE CARBON-CARBON TRIPLE BOND

The electronic structure of the triple bond is conveniently analyzed in terms of molecular orbital theory. The combination of the 2*s* orbital and one of the 2*p* orbitals on carbon leads to two equivalent *sp* orbitals, as shown in Figure 7.2. Each of these is available for bond formation by overlap with a corresponding orbital

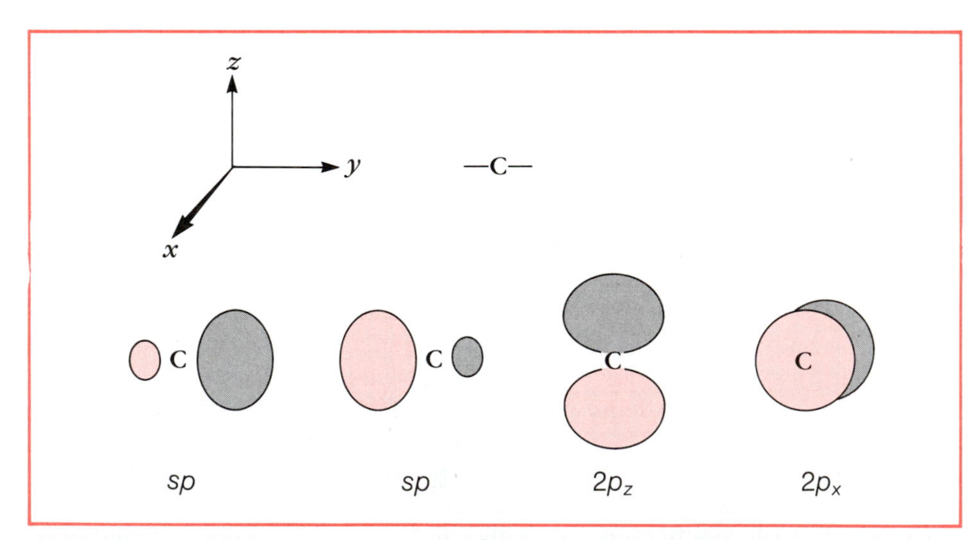

Figure 7.2 A Simplified Description of the Atomic Orbitals for One of the Carbon Atoms of a Triple Bond.
This carbon and the two atoms to which it is bonded all lie on a line (as drawn here, the *y* axis). Combination of the 2*s* orbital and the $2p_y$ orbital results in the two equivalent *sp* hybrid orbitals needed to form sigma bonds to the substituents. Two orbitals, p_x and p_z, are "left over."

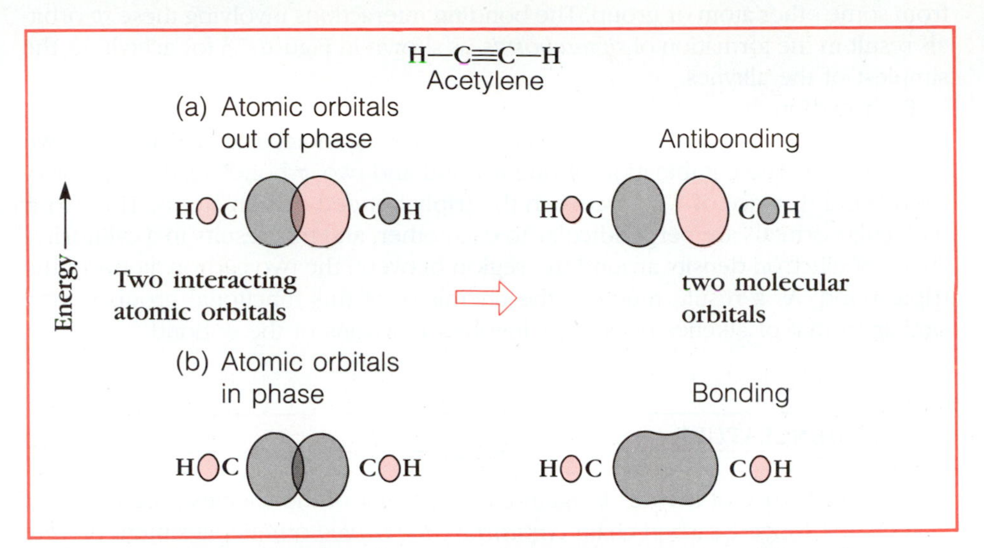

Figure 7.3 Sigma Bonding between the Carbon Atoms of Acetylene.
Two *sp* orbitals that are directed toward each other along the bonding axis of the molecule can interact in two ways. (a) The *out-of-phase* combination leads to an *antibonding* molecular orbital, and the *in-phase* combination leads to a *bonding* molecular orbital. In a ground-state alkyne only the bonding orbital is occupied by a pair of electrons, and this constitutes the σ bond portion of the carbon-carbon double bond.

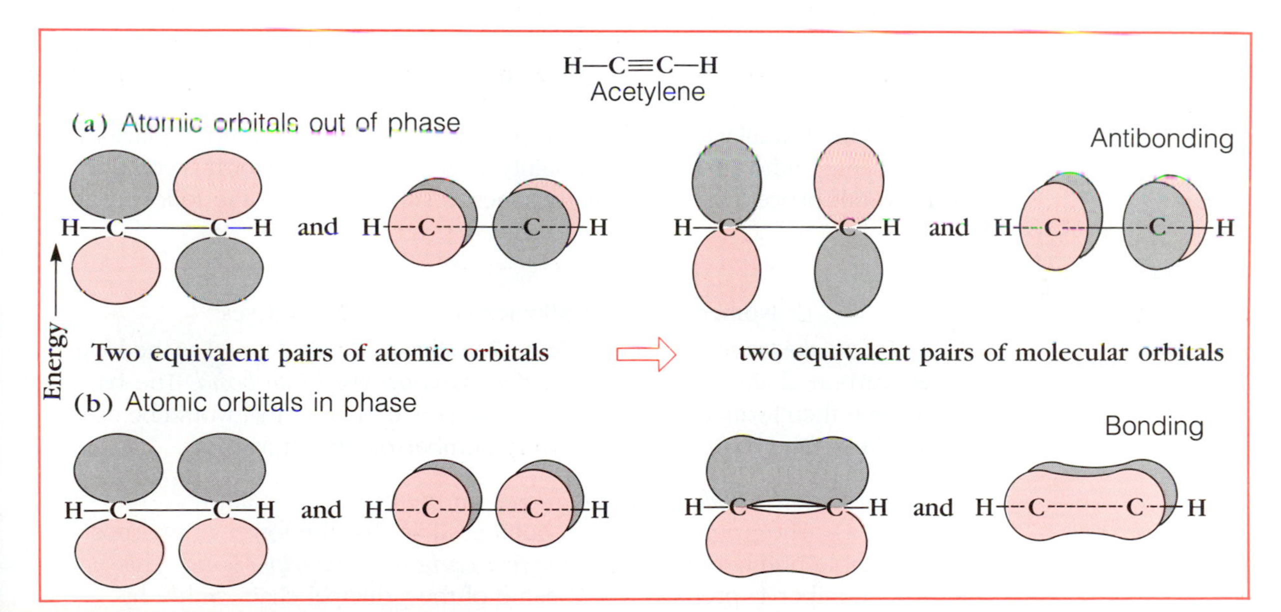

Figure 7.4 π Bonding in Acetylene.
Interactions of two equivalent pairs of *p* orbitals oriented perpendicular to the bonding axis can be either (a) *out-of-phase* or (b) *in-phase*. The out-of-phase interactions lead to two equivalent *antibonding* molecular orbitals, and the in-phase interactions lead to two equivalent *bonding* molecular orbitals. In a ground state alkyne only the bonding molecular orbitals are occupied, and these constitute the π bonding portion of the carbon-carbon triple bond.

from some other atom or group. The bonding interactions involving these *sp* orbitals result in the formation of *sigma bonds,* as shown in Figure 7.3 for acetylene, the simplest of the alkynes.

Each carbon atom of the triple bond has *two p* orbitals "left over," and this results in the formation of two equivalent π bonding molecular orbitals as shown in Figure 7.4. The combination of one σ bond and two π bonds leads to the very short bond distance of 1.2 Å between the triply bonded carbon atoms. The two π molecular orbitals are perpendicular to each other, and this results in a cylindrical "belt" of electron density around the region between the two carbon atoms of the triple bond. As a result, much of the chemistry of this functional group is very similar to that of alkenes because it involves reactions of the π bond.

7.3 NOMENCLATURE

The nomenclature of alkynes is again based on that of the alkanes. The principal chain is the longest carbon chain *containing the triple bond.*Replacement of *-ane* by *-yne* in the name of the principal chain affords the basic name, as illustrated by the three-carbon alkyne, which is called *propyne.*

$CH_3CH_2CH_3$ — Prop*ane*

$CH_3C{\equiv}CH$ — Prop*yne*

The two-carbon alkyne acetylene

$$H{-}C{\equiv}C{-}H$$

Acetylene

is one of the few alkynes for which a trivial name is normally used in place of the IUPAC name (although *ethyne* is certainly an acceptable name for this compound). The overall procedure for naming alkynes is summarized in the following box.

Nomenclature of Alkynes and Their Derivatives

1. *Locate the principal carbon chain.* For alkynes this is defined as the longest carbon chain in the molecule *that contains the triple bond.* The basic name is then formed by adding the suffix *-yne* to the root that would be used for the alkane chain having the same number of carbon atoms.

2. *Name and number the substituents* on the principal chain.
   (a) The principal chain is numbered so that the lower of two possible numbers results for *the first carbon of the triple bond.* This number is prefixed to the name of the principal chain with a hyphen. If the location of the triple bond does not distinguish between two different ways of numbering the principal chain, the substituents are evaluated so that the lower number results at the occasion of the first difference.
   (b) When there is more than one substituent, they are listed in alphabetical order.

The following two examples show how the chain is numbered to give the lower number for the first carbon of the triple bond.

$$CH_3CH_2C{\equiv}C{-}CH_3 \qquad\qquad CH_3CH_2C{\equiv}CH$$

2-Pentyne 1-Butyne

Since the carbon atoms of a triple bond and the two atoms which are attached directly to the triple bond form a linear array, stereoisomerism at the triple bond is not possible, and the only remaining step in the naming process is to name and number any substituents on the parent chain. In those few cases in which the triple bond is at the "center" of the parent chain, the direction of the numbering is that which leads to the lower number for the first substituent encountered. This is illustrated by the following example:

$$\underset{1}{CH_3}{-}\underset{2}{\overset{\overset{\displaystyle CH_3}{|}}{CH}}{-}\underset{3}{CH_2}{-}\underset{4}{C}{\equiv}\underset{5}{C}{-}\underset{6}{\overset{\overset{\displaystyle CH_2CH_3}{|}}{CH}}{-}\underset{7}{CH_2}{-}\underset{8}{CH_3}$$

6-Ethyl-2-methyl-4-octyne

Cycloalkynes present no unusual nomenclature problems, but they are not very common. As you might expect from the ideal linear geometry of the triple bond and the two atoms bonded to it, only cycloalkynes having fairly large rings are stable. The situation is quite similar to that for the *trans* cycloalkenes (Section 5.5). Cycloalkynes with rings of six or fewer atoms are apparently so strained (and therefore so reactive) that they cannot be isolated. Even those with slightly larger rings still exhibit high reactivity as a result of strain. The following two hydrocarbons are examples of cycloalkynes that have been synthesized.

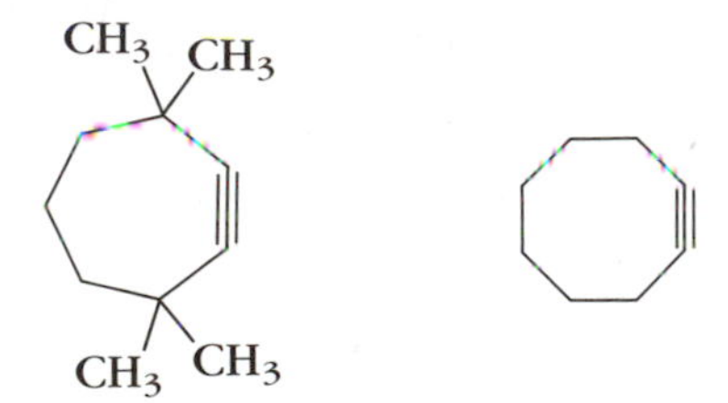

3,3,7,7-Tetramethylcycloheptyne Cyclooctyne

EXERCISE 7.1

Draw the following compounds:

(a) 1,2-Dichloroacetylene
(b) 4-Methyl-2-pentyne
(c) 1-Bromo-5-methyl-2-hexyne
(d) 6-Bromo-4-methylcyclooctyne
(e) 5-Ethyl-7-fluoro-6-methyl-3-heptyne

7.4 ACIDITY OF TERMINAL ALKYNES

The chemical reactivities of alkynes and alkenes are very similar in that both undergo addition reactions involving a π bond. However, alkynes exhibit one very important chemical property which makes them unique among the simple hydrocarbons. In marked contrast to the behavior of alkanes and alkenes, the —C≡C—H group of a terminal alkyne can act as an acid. Terminal alkynes are very weak acids, but salts of the conjugate bases can be readily prepared in the laboratory.

$$\text{R—C≡C—H} \xrightarrow[\text{base}]{\text{strong}} \text{R—C≡C:}^-$$

Table 7.1 provides a comparison of the acidities of acetylene and a variety of other organic compounds, of which some are weaker acids and some are stronger.

The data in Table 7.1 show clearly that acetylene, which is representative of terminal alkynes, is far less acidic than compounds having an OH group. On the other hand, the alkynes ($K_a \sim 10^{-25}$) are considerably more acidic than ammonia ($K_a \sim 10^{-32}$). As a consequence, the sodium salt of ammonia (in a solvent such as liquid ammonia) can convert an alkyne to the corresponding salt:

$$\text{R—C≡C—H} + Na^+\ NH_2^- \rightleftarrows \text{R—C≡C}^-\ Na^+ + NH_3$$

In Section 7.10 we will show you that such salts are very useful reagents.

Why are alkynes so much more acidic than either alkanes or alkenes? The generally accepted explanation is based on the type of orbital occupied by the electrons of the C—H bond because this corresponds to the orbital occupied by the un-

TABLE 7.1 Acidity of Terminal Alkynes in Comparison with Other Compounds

Compound	K_a
CH_3—C(=O)—O**H**	10^{-5}
H—O—**H**	10^{-16}
CH_3—CH_2—O**H**	10^{-16}
H—C≡C—**H**	10^{-25}
NH_2—**H** (H—N(H)—**H**)	10^{-32}
H_2C=CH—**H** (H₂C=C(H)**H**)	10^{-36}
CH_3—CH_2—**H** (H₃C—C(H)(H)—**H**)	10^{-42}

shared pair of electrons in the conjugate base (—C≡C:$^-$). The bonding C—H orbital of a hydrocarbon results from overlap of the 1*s* atomic orbital on hydrogen with a hybrid of 2*s* and 2*p* atomic orbitals on the carbon atom. After ionization, the hydrogen atom no longer plays a role, and the unshared electron pair of the anion must occupy an orbital that is a hybrid of the 2*s* and 2*p* atomic orbitals on carbon. A 2*s* orbital is of lower energy than a 2*p* orbital, so the anion will be more stable if the unshared electron pair occupies an orbital having substantial "*s* character." This is basically a statement of the *Aufbau principle,* which describes the energy levels of orbitals in a free atom. An electron in a 2*s* orbital is, on the average, closer to the nucleus than an electron in a corresponding 2*p* orbital. It therefore experiences a stronger attraction to the nucleus, and at the same time the repulsive interaction with other electrons is decreased.

We expect an increase in acidity of the hydrocarbon when the conjugate base is stabilized, and Figure 7.5 shows that there is a good correlation between the percent *s* character of the hybrid atomic orbital on carbon and the acidity of the corresponding C—H bond. Acidity is greatest for the C—H bond of an alkyne, in which *sp* hybridization corresponds to 50% *s* (and 50% *p*) character of the hybrid atomic orbital. An alkane on the other hand, is least acidic because the hybrid orbital (sp^3) has 75% *p* character and only 25% *s* character.

In contrast to a *terminal* alkyne, an **internal alkyne,** has no hydrogens on the *sp* carbon atoms, and its acidity will be similar to that of an alkane. This dependence

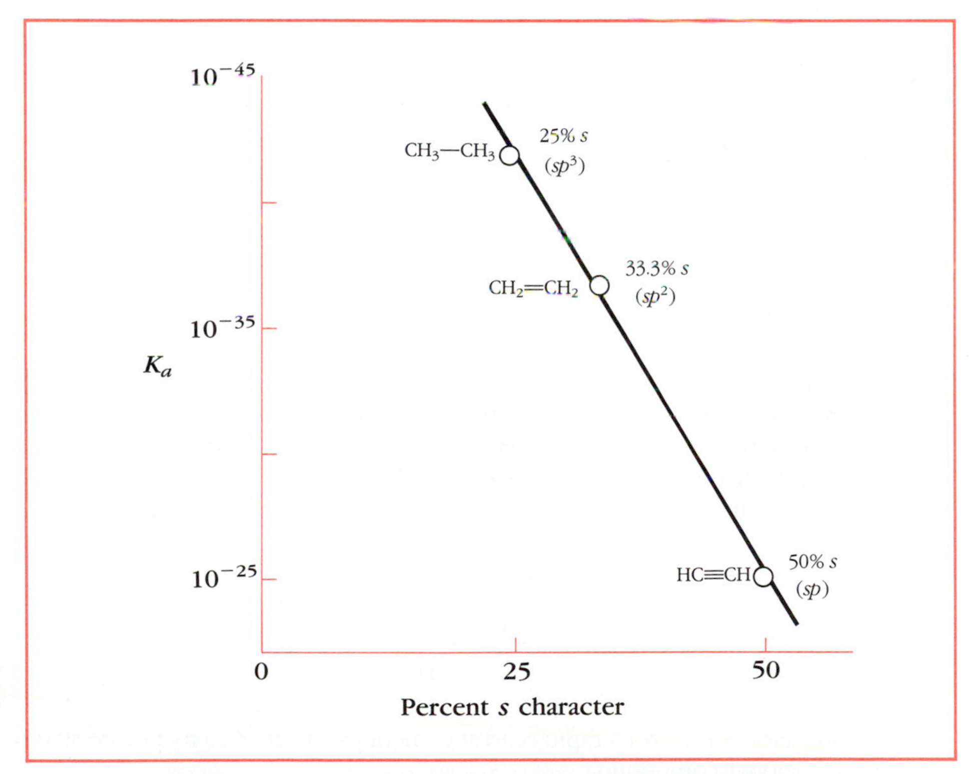

Figure 7.5 Acidity of a C—H Bond is a Function of the Percent *s* Character of the Hybrid Orbital on the Carbon Atom.
The conjugate base is more stable and acidity is greater when the lone pair of electrons occupies an orbital of lower energy, that is, with more *s* character.

of acid-base behavior upon the location of the triple bond permits a simple chemical test for a —C≡CH group. Reaction of a terminal alkyne with cuprous chloride in aqueous ammonia results in the formation of an insoluble *cuprous acetylide.*

$$\text{R—C=C—H} \xrightarrow[\text{H}_2\text{O, NH}_3]{\text{CuCl}} \text{R—C≡C—Cu}$$

The cuprous acetylide is a covalent compound rather than a salt, and the reaction probably proceeds via initial formation of a copper complex of the alkyne followed by loss of H^+ rather than by formation of the conjugate base of the alkyne, since there is no strong base present. The reaction cannot occur unless there is a hydrogen atom bonded directly to the triple bond, and the procedure clearly distinguishes between an internal alkyne and a terminal alkyne.

$$\text{R—C≡C—H} \xrightarrow[\text{H}_2\text{O, NH}_3]{\text{CuCl}} \text{R—C≡C—Cu} \quad \text{(precipitate)}$$

$$\text{R—C≡C—R} \xrightarrow[\text{H}_2\text{O, NH}_3]{\text{CuCl}} \text{R—C≡C—R} \quad \text{(no reaction, no precipitate)}$$

The treatment of an alkyne with alcoholic silver nitrate can also be used to distinguish between terminal and internal alkynes. A terminal alkyne affords a precipitate of the corresponding silver acetylide, but no reaction is observed with an internal alkyne.

$$\text{R—C≡CH} \xrightarrow[\text{C}_2\text{H}_5\text{OH}]{\text{AgNO}_3} \text{R—C≡C—Ag} \quad \text{(precipitate)}$$

The silver and copper acetylides can decompose explosively when they are dry, so they are rarely isolated. Nevertheless the formation of either compound as a precipitate provides a very simple and convenient test for the presence of a terminal alkyne.

7.5 HYDRATION OF ALKYNES

In this and the next two sections we will present a variety of addition reactions of alkynes. We will begin in this section with the addition of water to the carbon-carbon triple bond. The addition of a single molecule of water to an alkyne would yield an alkenyl alcohol (called an **enol,** from the endings *-ene* and *-ol*), in which an OH group is attached directly to one of the carbon atoms of the carbon-carbon double bond.

$$\text{—C≡C—} \xrightarrow[\text{of H}_2\text{O}]{\text{addition}} (\text{H})\text{C=C}(\text{OH})$$

An enol ordinarily undergoes rapid rearrangement (Section 13.9) to a more stable isomer, a carbonyl compound.

$$(\text{H})\text{C=C}(\text{OH}) \underset{}{\overset{\text{acid or base}}{\rightleftharpoons}} \text{—CH}_2\text{—C}(=\text{O})$$

Consequently, the hydration of alkynes offers a potential method for the synthesis of aldehydes and ketones.

$$\underset{\text{Aldehyde}}{R-C(=O)-H} \qquad \underset{\text{Ketone}}{R-C(=O)-R'}$$

Direct Hydration of Alkynes

The direct sulfuric acid-catalyzed addition of water to an alkyne occurs readily, although the addition of mercuric sulfate (a Lewis acid) as a catalyst is necessary.

$$CH_3-(CH_2)_4-C\equiv CH \xrightarrow[H_2SO_4,\ HgSO_4]{H_2O} CH_3-(CH_2)_4-\overset{O}{\overset{\|}{C}}-CH_3 \quad 91\%$$

$$CH_3CH_2CH_2CH_2-C\equiv CH \xrightarrow[H_2SO_4,\ HgSO_4]{H_2O} CH_3CH_2CH_2CH_2-\overset{O}{\overset{\|}{C}}-CH_3 \quad 80\%$$

This reaction is a polar addition. For an internal alkyne both ends of the triple bond are comparably substituted, but a terminal alkyne reacts in the orientation that produces positive charge on the *more substituted* carbon atom. As shown by the following simplified mechanism (which ignores the catalytic role of the mercuric ion), the carbon atom that is attacked by water subsequently becomes the carbon atom of the carbonyl group.

$$R-C\equiv CH \xrightarrow{H^+} [R-\overset{+}{C}=CH_2] \xrightarrow{:OH_2} R-\overset{^+OH_2}{\overset{|}{C}}=CH_2 \underset{\text{transfer}}{\overset{\text{proton}}{\rightleftharpoons}} R-\overset{OH}{\overset{|}{C}}=CH_2 \rightleftharpoons R-\overset{O}{\overset{\|}{C}}-CH_3$$

As indicated by this mechanism, the hydration product is a ketone rather than an aldehyde. Acetylene itself, which has no alkyl substituents, is the only alkyne which would yield an aldehyde by this method.

Unless the two substituents on the triple bond are identical, an internal alkyne yields a mixture of products (constitutional isomers) because the two possible carbocation intermediates would have approximately the same stability. The formation of such mixtures is illustrated by the following examples.

$$CH_3-(CH_2)_4-C\equiv C-CH_3 \xrightarrow[H_2SO_4,\ HgSO_4]{H_2O} \underset{33\%}{CH_3-(CH_2)_5-\overset{O}{\overset{\|}{C}}-CH_3} + \underset{37\%}{CH_3-(CH_2)_4-\overset{O}{\overset{\|}{C}}-CH_2CH_3}$$

$$CH_3-\overset{CH_3}{\overset{|}{C}H}-C\equiv C-CH_3 \xrightarrow[H_2SO_4,\ HgSO_4]{H_2O} \underset{54\%}{CH_3-\overset{CH_3}{\overset{|}{C}H}-CH_2-\overset{O}{\overset{\|}{C}}-CH_3} + \underset{46\%}{CH_3-\overset{CH_3}{\overset{|}{C}H}-\overset{O}{\overset{\|}{C}}-CH_2-CH_3}$$

product distribution

The preceding reactions show that the direct hydration of alkynes with sulfuric acid and mercuric sulfate is of somewhat limited usefulness. Good yields of a single product can be expected in only two cases:

1. Hydration of *terminal* alkynes to produce methyl ketones (ketones in which one of the two alkyl substituents on the carbonyl is a methyl group).

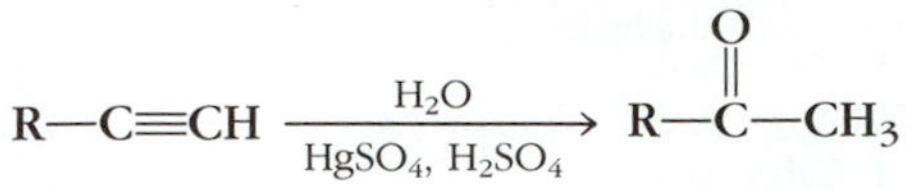

$$R{-}C{\equiv}CH \xrightarrow[HgSO_4,\ H_2SO_4]{H_2O} R{-}\overset{O}{\overset{\|}{C}}{-}CH_3$$

2. Hydration of *symmetrical* internal alkynes to yield ketones.

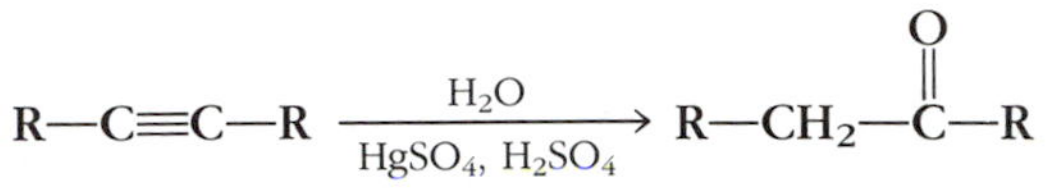

$$R{-}C{\equiv}C{-}R \xrightarrow[HgSO_4,\ H_2SO_4]{H_2O} R{-}CH_2{-}\overset{O}{\overset{\|}{C}}{-}R$$

Hydration of Alkynes via Borane Intermediates

In a reaction analogous to that with alkenes (Section 5.11), borane reacts with an alkyne via a *syn* addition process to form an alkenylborane.

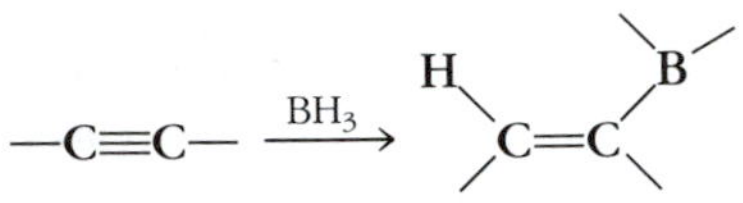

$$-C{\equiv}C- \xrightarrow{BH_3} \overset{H}{\diagdown} C{=}C \overset{B}{\diagup}$$

In principle this product could be converted to the enol (and hence to the corresponding carbonyl compound), but serious difficulties are encountered. For an internal alkyne two constitutional isomers are expected unless the alkyne is symmetrical. Moreover, the alkenylborane can react a second time to give a derivative containing two boron atoms; this is a particularly important problem in the reaction of terminal alkynes.

$$-C{\equiv}C{-}H \xrightarrow{BH_3} [-CH{=}\overset{B}{\overset{|}{C}}H] \longrightarrow -CH_2{-}CH\begin{matrix}B\\B\end{matrix}$$

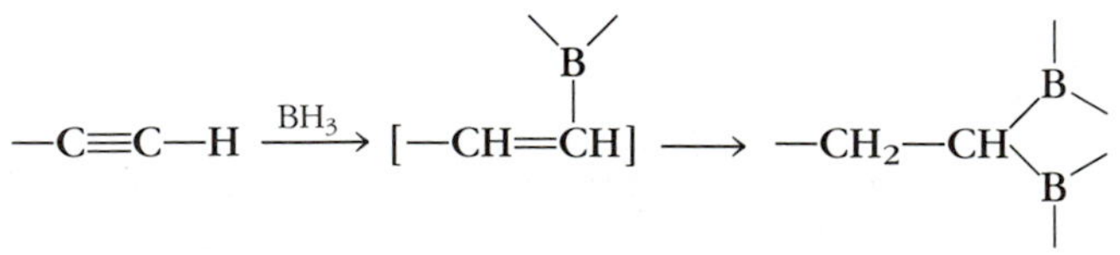

H.C. Brown (Purdue University) showed that these difficulties can be greatly alleviated by the use of *substituted boranes,* and two of the most useful are compounds known by the trivial names *disiamylborane* and *catecholborane.*

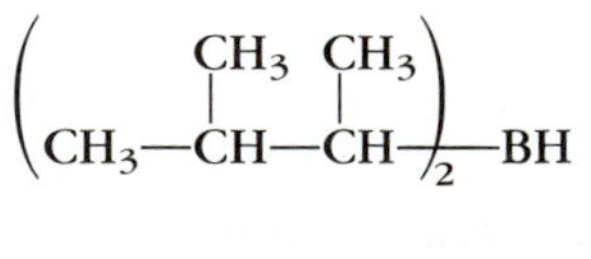

$$\left(CH_3{-}\overset{CH_3}{\overset{|}{C}H}{-}\overset{CH_3}{\overset{|}{C}H}\right)_2 BH$$

Disiamylborane

Catecholborane (benzene ring fused to O–B(H)–O ring: $C_6H_4O_2B{-}H$)

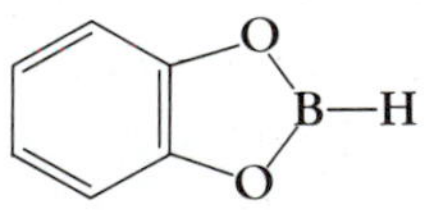

Both these reagents react smoothly with alkynes, and the reaction stops after addition of a single molecule of the borane derivative. The much greater steric bulk of these borane derivatives prevents further reaction of the alkenylborane because two bulky BR_2 groups on the same carbon atom would produce highly unfavorable steric interactions.

The addition of a borane to an alkyne proceeds by the orientation which places partial positive charge on the more substituted carbon (Section 5.11), so the boron atom becomes attached to the less substituted carbon atom. This orientation also minimizes unfavorable steric interactions between the bulky substituents on boron and substituents on the alkyne.

$$R{-}C{\equiv}C{-}H \xrightarrow{HB\left(CH(CH_3){-}CH(CH_3){-}CH_3\right)_2} \left[R{-}\overset{\delta+}{C}{\equiv}C(H) \cdots \overset{\delta-}{H}{-}B\left(CH(CH_3){-}CH(CH_3){-}CH_3\right)_2 \right] \longrightarrow (H)(R)C{=}C(H){-}B\left(CH(CH_3){-}CH(CH_3){-}CH_3\right)_2$$

Oxidation of the organoborane with alkaline hydrogen peroxide then generates a carbonyl compound. For an internal alkyne the overall process results in a ketone, but a terminal alkyne is converted to the corresponding *aldehyde*.

$$(H)(R)C{=}C(H){-}B\left(CH(CH_3){-}CH(CH_3){-}CH_3\right)_2 \xrightarrow[NaOH]{H_2O_2} R{-}CH_2{-}\overset{O}{\overset{\|}{C}}{-}H$$

The overall reaction is illustrated by the following example:

$$CH_3{-}(CH_2)_5{-}C{\equiv}CH \xrightarrow[2.\ H_2O_2,\ NaOH]{1.\ HB\left(CH(CH_3){-}CH(CH_3){-}CH_3\right)_2} CH_3{-}(CH_2)_5{-}CH_2{-}\overset{O}{\overset{\|}{C}}{-}H$$

70%

The hydroboration-oxidation sequence is complementary to the direct hydration catalyzed by sulfuric acid and mercuric sulfate. A terminal alkyne can therefore be selectively converted to *either* the methyl ketone (via acid-catalyzed hydration) *or* to the aldehyde (via hydroboration-oxidation).

The substituted boranes, disiamylborane and catecholborane, offer an additional advantage as synthetic reagents. Even when the two ends of the triple bond are substituted, these reagents sometimes exhibit substantial regioselectivity (Section 5.8). The selectivity again results from steric effects, and the bulky borane derivative bonds to that end of the triple bond which has the less bulky alkyl group (a tertiary alkyl group being the bulkiest). In other words, the boron becomes attached to the carbon atom that is less *sterically hindered*. This is illustrated by the following example, which shows the relative amounts of the two alkenylboranes that are produced.

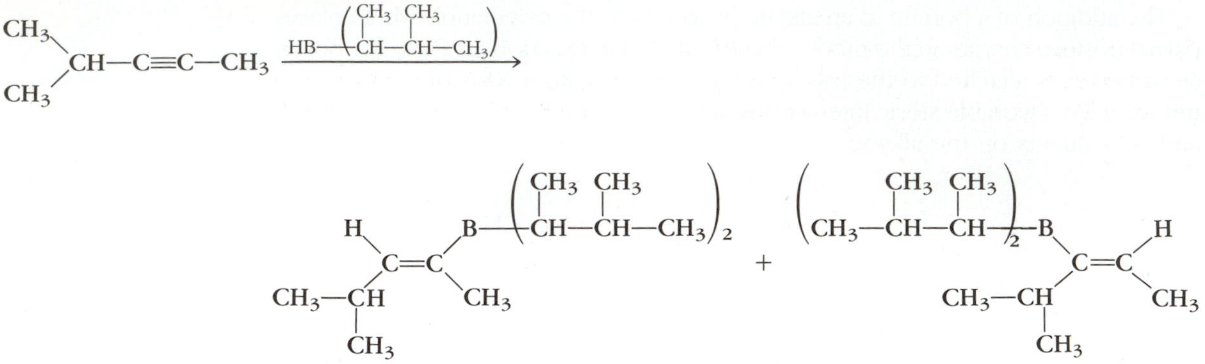

$$(CH_3)_2CH{-}C{\equiv}C{-}CH_3 \xrightarrow{HB\left(\overset{CH_3}{\underset{|}{}}\overset{CH_3}{\underset{|}{}}CH{-}CH{-}CH_3\right)_2}$$

$$\underset{93\%}{(CH_3)_2CH(H)C{=}C(CH_3)B(CH(CH_3)CH(CH_3)CH_3)_2} + \underset{7\%}{(CH_3CH(CH_3)CH(CH_3))_2B(H)C{=}C(CH_3)CH(CH_3)_2}$$

The high regioselectivity observed in the preceding hydroboration reaction is found only when there is a substantial difference in the steric bulk of the two substituents on the triple bond. If one group is primary and the other tertiary or if one is methyl and the other is secondary, the difference is usually sufficient to allow the preparation of a single ketone in good yield.

EXERCISE 7.2

Predict the major products of hydration for each of the following alkynes using (i) H_2O, $HgSO_4$, H_2SO_4; and (ii) either disiamylborane or catecholborane followed by treatment with alkaline hydrogen peroxide.

(a) $CH_3CH_2CH_2C{\equiv}C{-}H$

(b) $CH_3CH_2CH_2C{\equiv}C{-}CH_2CH_3$

(c) cyclohexyl$-C{\equiv}C{-}H$

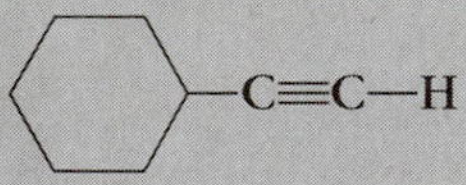

(d) cyclohexyl$-C{\equiv}C{-}CH_3$

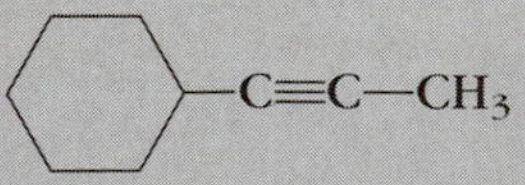

(e) cyclopentyl$-C{\equiv}C-$cyclopentyl

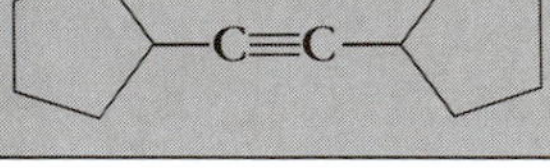

7.6 REDUCTION OF ALKYNES

Syn Addition of Hydrogen

The catalytic hydrogenation of alkynes is a straightforward reaction, and a total of two molar equivalents of hydrogen can be added to the triple bond. With typical catalysts the intermediate alkene reacts rapidly, and the reaction cannot be stopped before complete reduction to the alkane.

$$-C\equiv C- \xrightarrow[\text{Pt or Pd}]{H_2} \left[\begin{matrix} H & & H \\ & C=C & \end{matrix}\right] \longrightarrow -CH_2-CH_2-$$

However, the use of a *deactivated* catalyst such as palladium on barium sulfate permits the reaction to be stopped after the first step. Hydrogenation proceeds by a *syn* addition (Section 6.8), and the corresponding Z alkenes can be obtained.

$$CH_3-C\equiv C-CH_2CH_3 \xrightarrow[\text{Pd-BaSO}_4]{H_2} \begin{matrix} H & & H \\ & C=C & \\ CH_3 & & CH_2CH_3 \end{matrix} \quad 50\%$$

An alternative method for carrying out the *syn* addition of two hydrogen atoms has been developed in recent years. When the intermediate formed by hydroboration of an alkyne is treated with *acid* (rather than with an oxidizing agent), the boron is replaced by a hydrogen atom to yield the alkene. Acetic acid (CH_3CO_2H) has frequently been used as the acid in this step.

$$-C\equiv C- \xrightarrow{\text{hydroboration}} \begin{matrix} H & & B< \\ & C=C & \end{matrix} \xrightarrow[\text{NaOH}]{H_2O_2} \begin{matrix} H & & OH \\ & C=C & \end{matrix} \rightleftarrows -CH_2-\overset{O}{\overset{\|}{C}}-$$

$$\begin{matrix} H & & B< \\ & C=C & \end{matrix} \xrightarrow{CH_3CO_2H} \begin{matrix} H & & H \\ & C=C & \end{matrix}$$

Either disiamylborane or catecholborane can be used in this scheme, and treatment of the resulting alkenylborane derivative with acetic acid causes *protolysis* ("cleavage" by a "proton") to yield the alkene.

$$C_6H_{11}-C\equiv C-CH_3 \xrightarrow[\text{2. } CH_3CO_2H]{\text{1. catecholborane (B–H)}} \begin{matrix} H & & H \\ & C=C & \\ C_6H_{11} & & CH_3 \end{matrix} \quad 82\%$$

$$CH_3CH_2-C\equiv C-CH_2CH_3 \xrightarrow[\text{2. } CH_3CO_2H]{\text{1. } HB\left(CH(CH_3)-CH(CH_3)-CH_3\right)_2} \begin{matrix} H & & H \\ & C=C & \\ CH_3CH_2 & & CH_2CH_3 \end{matrix} \quad 82\%$$

Anti Addition of Hydrogen

It is possible to obtain the alkene of opposite configuration if the alkyne is reduced with sodium in liquid ammonia. In this reaction the metal acts as a reducing agent by supplying electrons and the ammonia functions as a proton donor (with the formation of NH_2^- as the by-product). Notice that addition of an electron followed by a proton corresponds to addition of a hydrogen, in accord with our earlier definition of reduction (Section 5.6).

$$-C\equiv C- \xrightarrow[NH_3]{Na} \text{(trans) } HC{=}CH$$

Liquid ammonia (bp −34°) is an excellent solvent for sodium metal. The sodium dissolves to give a highly reactive, deep blue solution, which acts as a reducing agent.* Lithium metal can been used in place of sodium, and other solvents such as ethylamine ($CH_3CH_2NH_2$) have sometimes been employed, but the general result is the same.

$$CH_3CH_2{-}C\equiv C{-}CH_2CH_3 \xrightarrow[NH_3]{Na} \text{trans-}CH_3CH_2CH{=}CHCH_2CH_3 \quad 87\%$$

$$CH_3CH_2CH_2CH_2{-}C\equiv C{-}CH_2CH_2CH_2CH_3 \xrightarrow[CH_3CH_2NH_2]{Li} \text{trans-}CH_3CH_2CH_2CH_2CH{=}CHCH_2CH_2CH_2CH_3 \quad 55\%$$

$$\text{cyclodecyne} \xrightarrow[NH_3]{Na} \text{trans-cyclodecene (H, H)}$$

This procedure also works for terminal alkynes, except that the weak acid, ammonium sulfate, must be added to the reaction mixture, in order to neutralize the NH_2^- ions that are produced when ammonia acts as a proton donor in the reduction. Unless the ammonium sulfate is added, the NH_2^- ions will generate the conjugate base of the terminal alkyne, and this negatively charged species is not susceptible to reduction by addition of another electron.

$$CH_3{-}(CH_2)_5{-}C\equiv CH \xrightarrow[NH_3,\ (NH_4)_2SO_4]{Na} CH_3{-}(CH_2)_5{-}CH{=}CH_2 \quad 90\%$$

EXERCISE 7.3

What reagents could be used to carry out the following reductions? If more than one of the procedures discussed in this section would yield the indicated compound as the major product, show all such possibilities.

(a) $CH_3{-}(CH_2)_5{-}C\equiv C{-}CH_2CH_3 \longrightarrow$ cis-$CH_3(CH_2)_5CH{=}CHCH_2CH_3$ (H, H on same side)

*The solution of metal in ammonia slowly decomposes in an oxidation-reduction reaction to yield the *base* NH_2^-, but this is ordinarily only a relatively unimportant side reaction:

$$Na + NH_3 \longrightarrow NaNH_2 + H_2\uparrow$$

(b) $CH_3-CH(CH_3)-CH_2-C{\equiv}CH \longrightarrow CH_3-CH(CH_3)-CH_2-CH{=}CH_2$

(c) $CH_3CH_2CH_2-C{\equiv}CH \longrightarrow CH_3CH_2CH_2CH_2CH_3$

(d) $CH_3CH_2-C{\equiv}C-CH_2-CH(CH_3)-CH_3 \longrightarrow$

$$\underset{H}{\overset{CH_3CH_2}{}}\!\!>C{=}C<\!\!\underset{CH_2-CH(CH_3)_2}{\overset{H}{}}$$

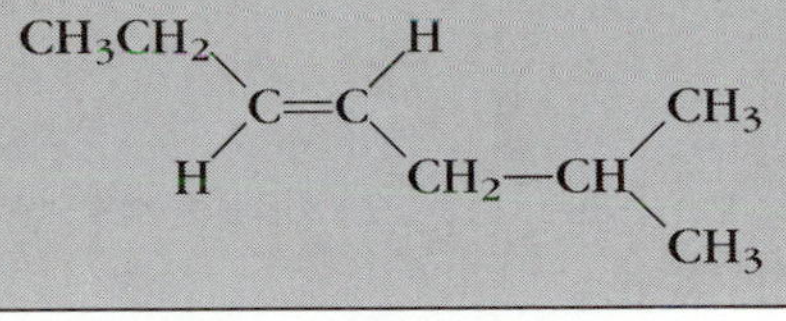

7.7 OTHER ADDITION REACTIONS TO ALKYNES

The addition reactions of alkynes are in many respects quite similar to those of alkenes. For alkynes, however, either one or two equivalents of a reagent can add to the triple bond. Clearly, it is very important to know what products will be obtained for a particular addition reaction. What orientation (i.e., regiochemistry) will be followed? What stoichiometry will be observed? Will two equivalents of the reagent add to the triple bond, or will the reaction stop after the addition of just one equivalent? In Section 7.5 we showed how you could predict the orientation of addition for hydration reactions of alkynes, and in the preceding section we showed how the use of a deactivated catalyst can stop the reduction of an alkyne after the addition of a single equivalent of hydrogen. In this section we will show you how the products can be regulated and predicted for a variety of addition reactions of alkynes. In addition to the regiochemistry and the stoichiometry, we will discuss the stereochemistry of these reactions.

Addition of Hydrogen Halides

The reaction of alkynes with hydrogen halides (HX) is a polar addition reaction, and like the analogous reaction of alkenes (Section 6.8), it is an *anti* addition. The orientation of the addition is that expected for a carbocation mechanism.

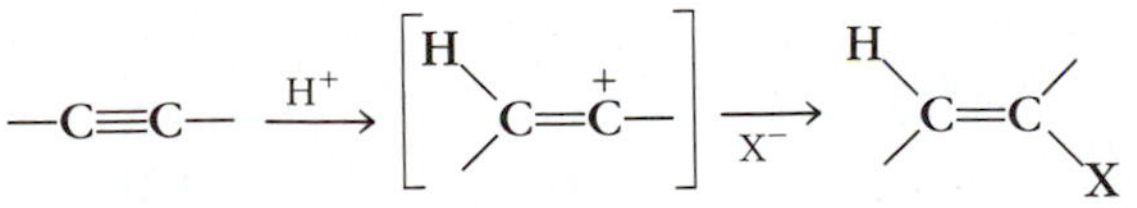

$$-C{\equiv}C- \xrightarrow{H^+} \left[\overset{H}{}\!\!>C{=}\overset{+}{C}- \right] \xrightarrow[X^-]{} \overset{H}{}\!\!>C{=}C<\!\!\underset{X}{}$$

Further reaction of the halo-substituted alkene proceeds readily if a second equivalent of the reagent is present.

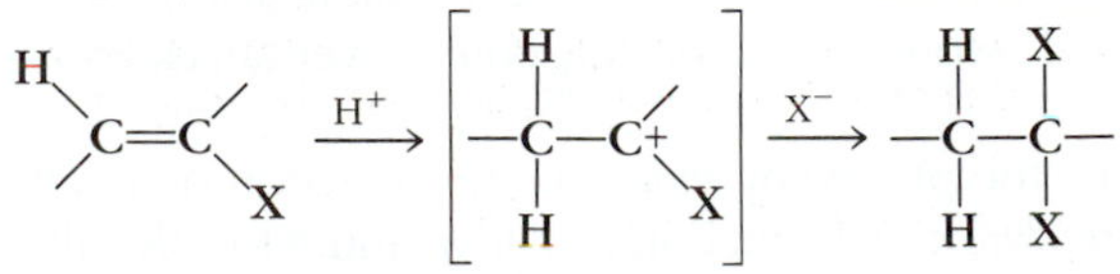

$$\overset{H}{}\!\!>C{=}C<\!\!\underset{X}{} \xrightarrow{H^+} \left[-\underset{H}{\overset{H}{C}}-\overset{+}{C}<\!\!\underset{X}{} \right] \xrightarrow{X^-} -\underset{H}{\overset{H}{C}}-\underset{X}{\overset{X}{C}}-$$

This second step proceeds by protonation to form the more stable of the two possible cations, that which has the halogen atom bonded to the positive center. The electron-withdrawing *inductive* effect of a halogen is unfavorable, but this is outweighed by the ability of the unshared pairs of electrons on the halogen atom to stabilize the cation by resonance.

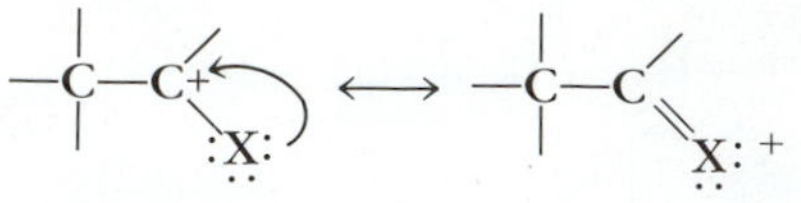

$$-\overset{|}{\underset{|}{C}}-\overset{+}{C}\!-\!\ddot{X}: \longleftrightarrow -\overset{|}{\underset{|}{C}}-C\!=\!\ddot{X}:^{+}$$

This results in bonding of both halogens to the same carbon atom, as illustrated by the reaction of 2-butyne with hydrogen bromide.

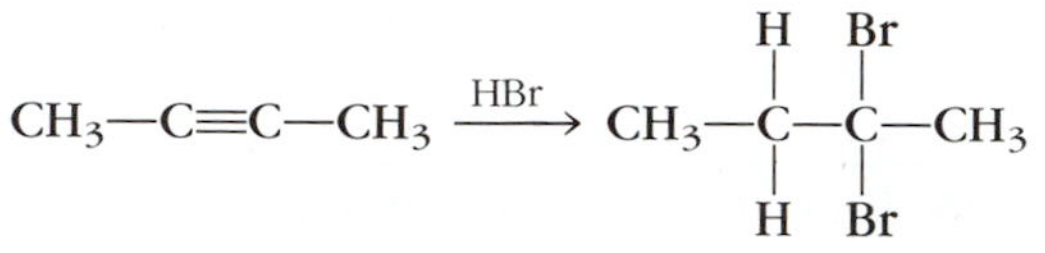

$$CH_3-C\equiv C-CH_3 \xrightarrow{HBr} CH_3-CH_2-CBr_2-CH_3 \quad 95\%$$

In the case of a terminal alkyne the initial protonation takes place at the less substituted carbon of the triple bond. This produces the more substituted carbocation, which is then attacked by bromide ion. This is shown for the addition of HBr to propyne.

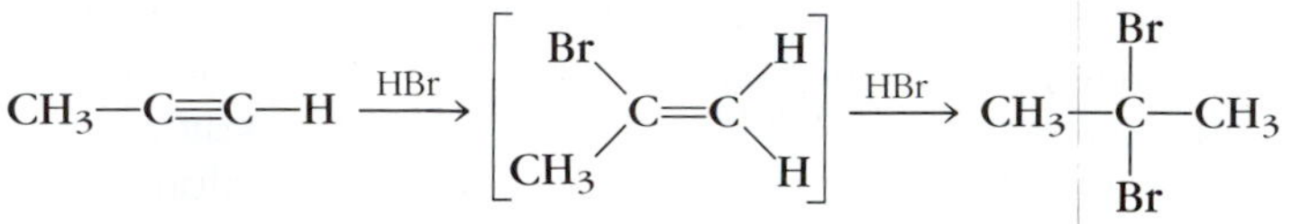

$$CH_3-C\equiv C-H \xrightarrow{HBr} \left[(Br)(CH_3)C=CH_2 \right] \xrightarrow{HBr} CH_3-CBr_2-CH_3 \quad 100\%$$

The addition of the first equivalent of HBr to propyne yields a single product because reaction proceeds via the more stable cation, and only a single product is obtained in the reaction of 2-butyne because the alkyne is symmetrical. But in the reaction of an unsymmetrical internal alkyne a mixture of isomers is expected, because the two possible cations are of approximately equal stability. This is illustrated by the following reaction of 2-hexyne with hydrogen chloride. Less than a single molar equivalent of HCl was used, so the reaction proceeded only to the alkene stage to yield approximately equal amounts of (*Z*)-2-chloro-2-hexene and (*Z*)-3-chloro-2-hexene.

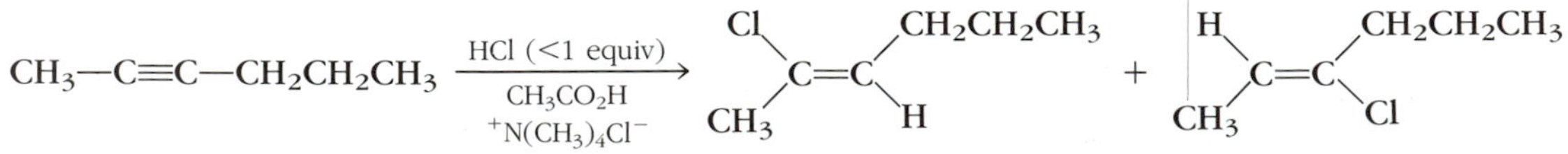

$$CH_3-C\equiv C-CH_2CH_2CH_3 \xrightarrow[\substack{CH_3CO_2H \\ ^{+}N(CH_3)_4Cl^{-}}]{HCl\ (<1\ equiv)} (Cl)(CH_3)C=C(H)(CH_2CH_2CH_3) + (H)(CH_3)C=C(Cl)(CH_2CH_2CH_3)$$

36% 36%

The reaction was carried out in a nucleophilic solvent (CH_3CO_2H), and tetramethylammonium chloride was added so that attack on the intermediate carbocation by chloride would occur preferentially over attack by solvent (cf. Section 5.8).

As illustrated by the preceding example, the reaction of an alkyne can often be limited to *monoaddition* by using equimolar quantities of the alkyne and the other reagent. This is not always possible, because the rate of reaction of the alkene intermediate is sometimes greater than that of the starting alkyne. For other reactions mixtures of products can result even when an excess of reagent is used. The latter situation is illustrated by the reaction of 1-hexyne with HBr.

$$CH_3CH_2CH_2CH_2{-}C{\equiv}CH \xrightarrow[\text{(4 equiv)}]{\text{HBr}}$$

$$\underset{44\%}{CH_3CH_2CH_2CH_2(Br)C{=}CH_2} + \underset{2\%}{CH_3CH_2CH_2CH_2CH{=}CHBr} + \underset{8\%}{CH_3CH_2CH_2CH_2{-}CBr_2{-}CH_3} + \underset{27\%}{CH_3CH_2CH_2CH_2{-}C{\equiv}CH}$$

The isolation of a small amount of 1-bromo-1-hexene in the preceding experiment should remind you of the free-radical addition of HBr to alkenes, which affords the less substituted bromide (Section 5.10). Indeed if 1-hexyne is treated with HBr in the presence of peroxides, a good yield of 1-bromo-1-hexene is obtained.

$$CH_3CH_2CH_2CH_2{-}C{\equiv}CH \xrightarrow[\text{peroxides}]{\text{HBr (1.5 equiv)}} \underset{74\%}{CH_3CH_2CH_2CH_2CH{=}CHBr} + \underset{1\%}{CH_3CH_2CH_2CH_2{-}CH(Br){-}CH_2Br}$$

The orientation of this addition is precisely that expected for a free-radical addition of HBr. The first step is reaction of the alkyne with a bromine atom, in which the more stable (more substituted) radical is formed.

$$CH_3CH_2CH_2CH_2{-}C{\equiv}CH \xrightarrow{\cdot Br} CH_3CH_2CH_2CH_2{-}\dot{C}{=}CHBr \xrightarrow{H{-}Br} CH_3CH_2CH_2CH_2{-}CH{=}CHBr + Br\cdot$$

Notice that the minor product in this reaction (formed by addition of two equivalents of HBr) proceeds via generation of a free radical on the carbon that already has a bromo substituent, even though it is a primary carbon atom. Clearly a bromine atom can stabilize a free radical on the carbon atom to which it is attached (just as it can stabilize a cation).

$$CH_3CH_2CH_2CH_2{-}CH{=}CHBr \xrightarrow{Br\cdot} CH_3CH_2CH_2CH_2{-}CH(Br){-}\dot{C}HBr \xrightarrow{H{-}Br} CH_3CH_2CH_2CH_2{-}CH(Br){-}CH_2Br + Br\cdot$$

Addition of Halogens

The addition of halogens to alkynes could produce either dihalides or tetrahalides.

$$-C{\equiv}C- \xrightarrow{X_2} \begin{matrix} X \\ \diagdown \\ \diagup \end{matrix} C{=}C \begin{matrix} \diagup \\ \diagdown \\ X \end{matrix} \xrightarrow{X_2} -\overset{\displaystyle X}{\underset{\displaystyle X}{C}}-\overset{\displaystyle X}{\underset{\displaystyle X}{C}}-$$

The reaction has been studied most carefully for the addition of bromine, and it can be stopped at the dibromide stage by the use of just one equivalent of Br_2.

$$CH_3-\overset{\displaystyle CH_3}{\underset{\displaystyle CH_3}{C}}-C{\equiv}CH \xrightarrow[CH_3CO_2H,\ LiBr]{Br_2\ (1\ equiv)} \begin{matrix} Br \\ \diagdown \\ \diagup \\ (CH_3)_3C \end{matrix} C{=}C \begin{matrix} H \\ \diagup \\ \diagdown \\ Br \end{matrix} \quad 78\%$$

$$CH_3CH_2C{\equiv}C-CH_2CH_3 \xrightarrow[CH_3CO_2H]{Br_2\ (1\ equiv)} \begin{matrix} CH_3CH_2 \\ \diagdown \\ \diagup \\ Br \end{matrix} C{=}C \begin{matrix} Br \\ \diagup \\ \diagdown \\ CH_2CH_3 \end{matrix} \quad 90\%$$

As illustrated by the preceding reactions the addition of Br_2 ordinarily proceeds via the *anti* mode (as with alkenes, Section 5.8), although some of the isomeric product may also be formed.

$$CH_3CH_2CH_2CH_2C{\equiv}CH \xrightarrow[CH_3CO_2H]{Br_2\ (1\ equiv)} \underset{57\%}{\begin{matrix} Br \\ \diagdown \\ \diagup \\ CH_3CH_2CH_2CH_2 \end{matrix} C{=}C \begin{matrix} H \\ \diagup \\ \diagdown \\ Br \end{matrix}} + \underset{22\%}{\begin{matrix} Br \\ \diagdown \\ \diagup \\ CH_3CH_2CH_2CH_2 \end{matrix} C{=}C \begin{matrix} Br \\ \diagup \\ \diagdown \\ H \end{matrix}}$$

Alkenyl Halides via Hydroboration

The direct addition of halogens or hydrogen halides to alkynes is of very limited synthetic utility. The dihaloalkenes and tetrahaloalkanes that result from addition of halogen to the alkynes are not very useful for the preparation of other organic compounds. On the other hand, the alkenyl halides formed by addition of a single equivalent of hydrogen halide to an alkyne are useful synthetic intermediates, particularly in organocuprate reactions (Section 5.14). Unfortunately, as we showed on the preceding pages, direct addition of a hydrogen halide to an alkyne frequently continues past the alkene stage, and poor yields of the alkenyl halides are obtained.

Alkenylboranes have recently been found to offer a solution to this problem, and they allow many alkenyl halides to be prepared in good yield. When an alkyne is allowed to react with either disiamylborane or catecholborane, an alkenylborane is formed. Addition of bromine affords a dibromoorganoborane, and subsequent treatment of this with base yields the alkenyl bromide.

$$-C{\equiv}C- \xrightarrow[\text{or catecholborane (}C_6H_4O_2B{-}H\text{)}]{\left(CH_3-\overset{CH_3}{CH}-\overset{CH_3}{CH}\right)_2BH} \begin{matrix} H \\ \diagdown \\ \diagup \end{matrix} C{=}C \begin{matrix} B \\ \diagup \\ \diagdown \end{matrix} \xrightarrow{Br_2} Br-\overset{\displaystyle H}{\underset{|}{C}}-\overset{\displaystyle B}{\underset{|}{C}}-Br \xrightarrow{NaOCH_3} -CH{=}C \begin{matrix} Br \\ \diagup \\ \diagdown \end{matrix}$$

This sequence proceeds to give overall *anti* addition of HBr via a remarkable combination of stereospecific reactions. The initial *syn* hydroboration is followed by an *anti* addition of bromine to the alkenylborane. This in turn is followed by an *anti* elimination of one bromine and the boron substituent. (The loss of boron and bromine proceeds rapidly under mild conditions, and subsequent elimination of HBr from the alkenyl bromide is not a problem.)

Br_2, anti addition → ⇌ → anti elimination

The following examples show that the reaction works well with both terminal and internal alkynes. (Recall from Section 7.5 that catecholborane reacts with internal alkynes by attachment of boron to the less sterically hindered position.)

$CH_3-(CH_2)_5-C{\equiv}CH$ → 1. catecholborane (B—H); 2. Br_2; 3. $NaOCH_3$ → (Z)-1-Bromo-1-octene, 82%

$(CH_3)_3C-C{\equiv}C-CH_3$ → 1. catecholborane (B—H); 2. Br_2; 3. $NaOCH_3$ → (Z)-2-Bromo-4,4-dimethyl-2-pentene, 96%

Alkenyl iodides can be generated in a related sequence with the opposite stereochemistry, although the sequence is limited to terminal alkynes. The intermediate alkenylborane is first treated with water to cleave the catechol group and form the alkenylboronic acid.

$\xrightarrow{H_2O}$ Alkenylboronic acid

The alkenylboronic acid is not isolated, and reaction with iodine and base results in *anti* addition of I_2 followed by intramolecular *syn* elimination of iodo and $B(OH)_2$ groups to generate the alkenyl iodide. The reaction sequence corresponds to an overall *syn* addition of HI and yields (*E*)-1-iodo-1-alkenes.

Cyclohexyl—C≡CH → 1. catecholborane (B—H); 2. H_2O; 3. I_2, NaOH → (E)-cyclohexyl–CH=CHI, 93%

EXERCISE 7.4

What reactions of alkynes could be used to prepare the following compounds? If more than one method would work, write all such possibilities.

(a) $CH_3CH_2CH_2(H)C{=}C(I)H$ — H and $CH_3CH_2CH_2$ on one carbon; I and H on the other

(b) $CH_3CH_2CH_2(H)C{=}C(CH_2CH_2CH_3)Br$ — H and $CH_3CH_2CH_2$ on one carbon; $CH_2CH_2CH_3$ and Br on the other

(c) $CH_3CH_2CH_2(Br)C{=}C(CH_2CH_2CH_3)Br$ — Br and $CH_3CH_2CH_2$ on one carbon; $CH_2CH_2CH_3$ and Br on the other

7.8 PREPARATION OF ALKYNES

The reactions that yield alkynes can be grouped into two general categories that we will discuss in this section: (1) reactions involving **functional group modification,** in which the carbon skeleton does not change; and (2) reactions that produce an alkyne with a different number of carbon atoms by way of **carbon-carbon bond formation.** Any synthetic sequence which affords a "large" molecule from smaller reactants must at some stage involve carbon-carbon bond formation, so the second category is a particularly important reaction type.

Functional Group Modification: Dehydrohalogenation

The carbon-carbon triple bond is produced from alkyl or alkenyl compounds by removal of substituents from two adjacent carbon atoms, that is, by elimination reactions. The most frequently used elimination reaction for the preparation of alkynes is the dehydrohalogenation of dihalides. The halogens may be located on either the same carbon (a **geminal** dihalide) or on adjacent carbon atoms (a **vicinal** dihalide), and the elimination proceeds via the alkenyl halide.

$$\text{—CH}_2\text{—CX}_2\text{—} \xrightarrow[\text{(elimination of HX)}]{\text{base}} \text{—CH=CX—} \xrightarrow[\text{(elimination of HX)}]{\text{base}} \text{—C}{\equiv}\text{C—}$$

$$\text{—CHX—CHX—} \xrightarrow[\text{(elimination of HX)}]{\text{base}} \text{—CH=CX—}$$

The reaction can sometimes be stopped at the alkenyl halide if a single equivalent of base is used (and other variables such as reaction temperature are controlled), but direct formation of the alkyne is more common.

The starting 1,2-dihalides are readily prepared by addition of chlorine or bromine to the corresponding alkene (Section 5.8).

$$\text{—CH=CH—} \xrightarrow[\text{CCl}_4]{\text{X}_2} \text{—CXH—CXH—}$$

The 1,1-dihalides can be made by treatment of the corresponding aldehyde or ketone with a phosphorus halide such as phosphorus pentachloride, PCl_5.

$$\text{—CH}_2\text{—C(=O)—} \xrightarrow{\text{PCl}_5} \text{—CH}_2\text{—CCl}_2\text{—}$$

Many bases and solvents have been used for elimination reactions in the preparation of alkynes. The most general reagent is sodium amide ($NaNH_2$) in liquid ammonia, but $NaNH_2$ has also been used with other solvents such as mineral oil (a mixture of high-boiling alkanes) and DMSO (Section 5.13). Alkali metal hydroxides and alkoxides such as KOH and KOtBu have also been utilized successfully in ethanol and other solvents.

$$\text{CH}_3\text{—(CH}_2)_7\text{—CHBr—CH}_2\text{Br} \xrightarrow[\text{NH}_3]{\text{NaNH}_2} \text{CH}_3\text{—(CH}_2)_7\text{—C}\equiv\text{CH} \quad 54\%$$

$$\text{CH}_3\text{—CH}_2\text{—CHBr—CHBr—CH}_3 \xrightarrow[\text{CH}_3\text{CH}_2\text{OH}]{\text{KOH}} \text{CH}_3\text{—CH}_2\text{—C}\equiv\text{C—CH}_3 \quad 70\%$$

$$\text{CH}_3\text{—(CH}_2)_{11}\text{—CHBr—CH}_2\text{Br} \xrightarrow[\text{DMSO}]{\text{NaNH}_2} \text{CH}_3\text{—(CH}_2)_{11}\text{—C}\equiv\text{CH} \quad 94\%$$

In some instances the alkenyl halide rather than the dihaloalkane may be available, and the same treatment with base can be used to convert it directly to the alkyne.

$$\text{C}_6\text{H}_{11}\text{—CH}_2\text{—CBr=CH}_2 \xrightarrow[\text{mineral oil}]{\text{NaNH}_2} \text{C}_6\text{H}_{11}\text{—CH}_2\text{—C}\equiv\text{CH} \quad 87\%$$

There are some dihalides for which elimination cannot proceed past the alkenyl halide stage even if an excess of base is used. 1,2-Dibromo-2-methylpropane, for example, affords 1-bromo-2-methylpropene. This alkene has no hydrogen on the carbon atom adjacent to that with the bromo substituent, so the reaction stops with formation of the alkenyl bromide.

$$\mathrm{CH_3{-}C(CH_3)(Br){-}CH_2Br \xrightarrow[CH_3CH_2OH]{KOH} (CH_3)_2C{=}CHBr} \quad 27\%$$

The overall conversion of a ketone, aldehyde, or alkene to the corresponding alkyne can be effected quite simply and in good yield* by conversion to the corresponding geminal or vicinal dihalide followed by dehydrohalogenation, as illustrated by the following examples:

$$\mathrm{CH_3{-}C(CH_3)_2{-}C({=}O){-}CH_3 \xrightarrow[(79\%)]{PCl_5} CH_3{-}C(CH_3)_2{-}CCl_2{-}CH_3 \xrightarrow[DMSO\ (95\%)]{KOtBu} CH_3{-}C(CH_3)_2{-}C{\equiv}CH}$$

75% overall yield

$$\mathrm{CH_3{-}(CH_2)_7{-}CH{=}CH{-}(CH_2)_7{-}CO_2H \xrightarrow{Br_2} CH_3{-}(CH_2)_7{-}CHBr{-}CHBr{-}(CH_2)_7{-}CO_2H \xrightarrow[NH_3]{NaNH_2} CH_3{-}(CH_2)_7{-}C{\equiv}C{-}(CH_2)_7{-}CO_2H}$$

52–62% overall yield

Carbon-Carbon Bond Formation—Alkylation of Sodium Acetylides

The acidity of acetylene or a terminal alkyne permits its conversion to the corresponding sodium salt by reaction with sodium amide in liquid ammonia.

$$\mathrm{R{-}C{\equiv}C{-}H \overset{NaNH_2}{\rightleftarrows} R{-}C{\equiv}C^-Na^+}$$

Likewise, the analogous reaction with lithium amide can be used to form the lithium salt. These sodium and lithium *acetylides* are quite reactive and undergo facile reaction with many alkyl halides. The negatively charged carbon atom of the acetylide ion attacks the (partially positive) carbon atom to which the halogen is bonded.

$$\mathrm{R{-}C{\equiv}C^- + {-}\overset{|}{\underset{|}{C}}{-}X\ (\delta+\ \delta-) \longrightarrow R{-}C{\equiv}C{-}\overset{|}{\underset{|}{C}}{-} + X^-}$$

This is an example of *nucleophilic substitution* or *nucleophilic displacement* (Chapter 12), in which the acetylide ion acts as a nucleophile and displaces a halide ion from the alkyl halide. The net result of the process is *alkylation* of the original alkyne via formation of a new carbon-carbon bond.

*The concept of a "good" yield can be misleading. While a 75% yield might be considered to fall in this category, a two-step sequence in which each step proceeds in 75% yield would afford only a 56% overall yield. Even with 90% yields, a six-step sequence would result in an overall yield of only 53%.

Since terminal alkynes are very weak acids ($K_a \sim 10^{-25}$), the sodium acetylides are rather strong bases. Consequently, an important side reaction, the action of the sodium acetylide as a base to cause dehydrohalogenation, often occurs in these alkylations.

$$R{-}C{\equiv}C^- + {-}\overset{|}{\underset{|}{C}}(H){-}\overset{|}{\underset{|}{C}}{-}X \longrightarrow R{-}C{\equiv}CH + \;\rangle C{=}C\langle\; + X^-$$

This elimination reaction takes place readily with secondary and tertiary alkyl halides, so these halides cannot be used in the alkylation reaction. Successful alkylation requires the use of a primary (or methyl) halide, and it has been most commonly employed for the construction of long-chain linear (i.e., unbranched) organic molecules.

The overall synthetic process can then be summarized in the following equation, which emphasizes the requirement for a primary alkyl halide.

$$-C{\equiv}CH \xrightarrow{NaNH_2} -C{\equiv}C^-Na^+ \xrightarrow{R-CH_2-X} -C{\equiv}C{-}CH_2R$$

The following examples illustrate the use of the reaction, both for the conversion of acetylene itself to a terminal alkyne and for the conversion of terminal alkynes to internal alkynes.

$$HC{\equiv}CH \xrightarrow[2.\ CH_3CH_2CH_2CH_2Br]{1.\ NaNH_2} CH_3CH_2CH_2CH_2C{\equiv}CH \quad 89\%$$

$$HC{\equiv}CH \xrightarrow[2.\ CH_3-(CH_2)_{17}-Br]{1.\ NaNH_2} CH_3{-}(CH_2)_{17}{-}C{\equiv}CH \quad 61\%$$

$$CH_3{-}\underset{CH_3}{\overset{CH_3}{C}}{-}C{\equiv}CH \xrightarrow[2.\ CH_3Cl]{1.\ NaNH_2} CH_3{-}\underset{CH_3}{\overset{CH_3}{C}}{-}C{\equiv}C{-}CH_3 \quad 55\%$$

$$HC{\equiv}CH \xrightarrow[2.\ CH_3CH_2CH_2CH_2Br]{1.\ NaNH_2} CH_3CH_2CH_2CH_2{-}C{\equiv}CH \xrightarrow[2.\ CH_3CH_2Br]{1.\ NaNH_2} CH_3CH_2CH_2CH_2{-}C{\equiv}C{-}CH_2CH_3$$

64% overall yield

Despite the limitation that a primary alkyl halide must be used, this method for building up a carbon chain has an important advantage. The product of the alkylation still contains a useful functional group, the carbon-carbon triple bond. As a result, the alkylation of sodium acetylides is a valuable synthetic process, because the products can in turn be transformed into a variety of other compounds.

EXERCISE 7.5

Predict the major product of each of the following reactions.

(a) $HC{\equiv}CH \xrightarrow[2.\ CH_3I]{1.\ NaNH_2}$

(b) $CH_3CH_2{-}C{\equiv}CH \xrightarrow[2.\ CH_3CH_2Br]{1.\ LiNH_2}$

(c) $CH_3C{\equiv}CH \xrightarrow[2.\ (CH_3)_3C{-}Br]{1.\ NaNH_2}$

EXERCISE 7.6

(a) Suggest a method for preparing the following compound from an aldehyde, ketone, or alkene. If more than one method would work, write all such methods.

$$CH_3CH_2{-}C(CH_3)_2{-}C{\equiv}CH$$

(b) Suggest a method of preparing 2-octyne using a terminal alkyne and an alkyl halide. If more than one method would work, write all such methods.

7.9 ALLENES

In a number of instances the dehydrohalogenation of a dihaloalkane could conceivably yield several isomeric products. For example, 2,3-dibromopentane could undergo elimination to give 1,3-pentadiene* (Equation 1), 1,2-pentadiene (Equation 2), 2,3-pentadiene (Equation 3), or 2-pentyne (Equation 4).

$$\underset{H}{CH_2}{-}\overset{Br}{CH}{-}\overset{Br}{CH}{-}\underset{H}{CH}{-}CH_3 \xrightarrow{\text{base}} CH_2{=}CH{-}CH{=}CH{-}CH_3 \quad (1)$$

$$\underset{H}{CH_2}{-}\underset{H}{\overset{Br}{C}}{-}\overset{Br}{CH}{-}CH_2CH_3 \xrightarrow{\text{base}} CH_2{=}C{=}CH{-}CH_2{-}CH_3 \quad (2)$$

*The suffix *-diene* designates a carbon chain containing two double bonds (i.e., *di-* and *-ene*). Similarly, a carbon chain containing three double bonds is designated as a *triene,* and so forth.

$$CH_3-\underset{H}{\overset{Br}{C}}H-\underset{H}{\overset{Br}{C}}-CH-CH_3 \xrightarrow{\text{base}} CH_3-CH{=}C{=}CH-CH_3 \quad (3)$$

$$CH_3-\underset{H}{\overset{Br}{C}}-\underset{H}{\overset{Br}{C}}-CH_2CH_3 \xrightarrow{\text{base}} CH_3-C{\equiv}C-CH_2CH_3 \quad (4)$$

In the previous section we showed that the reaction of 2,3-dibromopentane with alcoholic KOH produces 2-pentyne (Equation 4) in 70% yield. Why are the products of Equations 1–3 not observed?

The first step in this elimination is formation of an alkenyl halide. Removal of a proton from one of the carbon atoms that has a bromo substituent is favored by the inductive effect of the halogen, so a mixture of 3-bromo-2-pentene and 2-bromo-2-pentene is formed.

$$CH_3-\underset{H}{\overset{Br}{C}}-\underset{H}{\overset{Br}{C}}-CH_2CH_3 \longrightarrow CH_3-CH{=}\overset{Br}{C}-CH_2CH_3 \quad \text{and} \quad CH_3-\overset{Br}{C}{=}CH-CH_2CH_3$$

Neither of these two intermediates could directly yield 1,3-pentadiene, so it is evident why the product of Equation 1 is not observed. But the other three products (Equations 2, 3, and 4) could still be formed from these two alkenyl halides, and the question remains as to why only the product of Equation 4 is observed. The answer is simply that the alkyne is more stable than the other two isomeric products, and the activation energy for alkyne formation is typically lower than that for formation of a diene. Moreover, as we will show later in this section, the 1,2- and 2,3-pentadienes are converted to 2-pentyne by the action of strong base. For these reasons alkynes rather than dienes are expected as the major products from elimination reactions of vicinal and geminal dihalides.

The group of dienes in which the double bonds lie between three consecutive carbon atoms exhibit unusual structural and chemical properties. These 1,2-dienes are called **allenes,** and they can be prepared by methods that do not cause rearrangement to an alkyne. For example, the parent compound, allene itself, is produced in high yield by the reaction of zinc with 2,3-dibromopropene.

$$CH_2{=}C\begin{matrix}Br\\CH_2Br\end{matrix} \xrightarrow{\text{Zn}} \underset{\text{Allene}}{CH_2{=}C{=}CH_2} \quad 95\text{–}98\%$$

A Brief Molecular Orbital Description of Allene

The orbital structure of allene and other 1,2-dienes is unusual because the central carbon atom is part of both double bonds. The two terminal carbon atoms of allene have three substituents (two hydrogen atoms and one carbon atom) and are sp^2 hybridized like the carbon atoms of ordinary double bonds, but the central carbon has only two substituents and is *sp* hybridized like a carbon of a triple bond. The central carbon atom has two *p* orbitals available for π bonding, but they are perpendicular to each other. The two π orbitals of the double bonds will therefore be

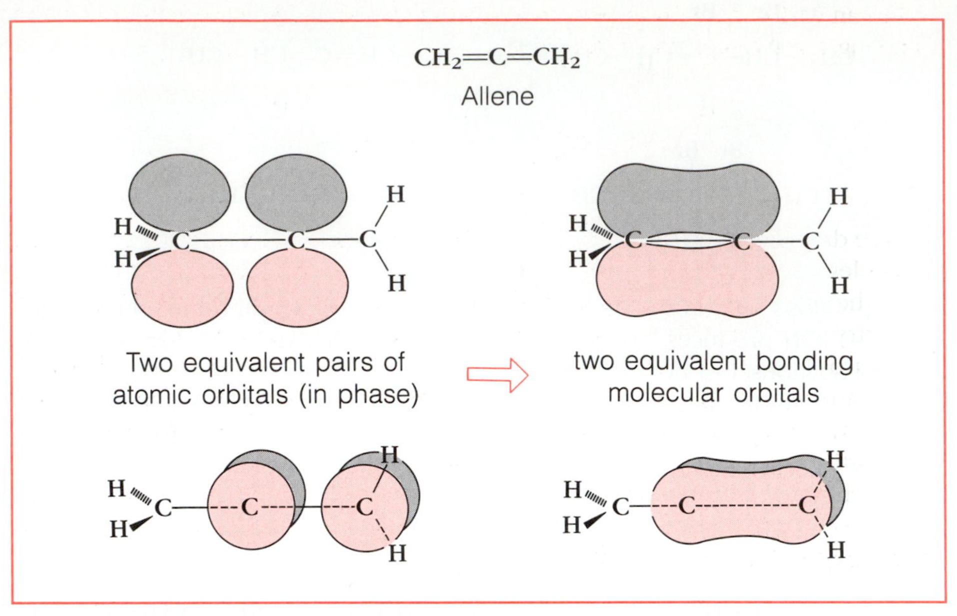

Figure 7.6 The Two π Bonds in Allene.
The combination of four atomic *p* orbitals leads to four π molecular orbitals (although we have only drawn the in-phase combinations of the atomic orbitals that lead to the bonding molecular orbitals). C-2 is *sp* hybridized and has two *p* orbitals available for π bonding. One of these interacts with a *p* orbital on C-1 and the other with a *p* orbital on C-3. The two *p* orbitals on C-2 are perpendicular to each other, so the π molecular orbitals must also be perpendicular to each other. As a consequence the molecule as a whole is not planar, and the substituents at the two ends of the allene unit lie in perpendicular planes.

perpendicular to each other, as well, and we have illustrated this in Figure 7.6. Consequently, the allene molecule is not planar, and the two CH_2 groups at the ends of the molecule lie in perpendicular planes:

$$H_2C{=}C{=}CH_2$$

Chirality in Allenes

The perpendicular relationship of the two end groupings of the allene unit leads to an unusual stereochemical property. Despite the absence of a chiral carbon atom, an allene will be chiral as long as neither end has two identical substituents. Allene itself is achiral because each end has two identical substituents. In contrast, 2,3-pentadiene is a chiral molecule because each end of the allene unit has one methyl group and one hydrogen atom. 2-Methyl-2,3-pentadiene is again achiral, because one end of the allene unit has two identical (methyl) substituents.

$CH_2{=}C{=}CH_2$ — Allene (achiral)

$CH_3{-}CH{=}C{=}CH{-}CH_3$ — 2,3-Pentadiene (chiral)

$(CH_3)_2C{=}C{=}CH{-}CH_3$ — 2-Methyl-2,3-pentadiene (achiral)

We can easily show that allene and 2-methyl-2,3-pentadiene are achiral with the following three-dimensional drawings.

In these drawings the plane of the paper is also a plane of symmetry for each of the molecules. Most of the atoms in each molecule lie *in* this plane, and those lying above the plane have mirror image counterparts below it. Allene itself has a second symmetry plane (perpendicular to the plane of the paper and passing through all three carbon atoms). In the case of 2-methyl-2,3-pentadiene the two methyl groups on C-2 are equally disposed in front of and behind the symmetry plane; the two nonequivalent substituents (H and CH_3) on C-4 both lie *in the symmetry plane* (i.e., the substituents themselves are bisected by this plane).

The chirality of 2,3-pentadiene can be demonstrated by the absence of a symmetry plane or by the nonequivalence (nonsuperimposability) of its mirror image, as shown by the following structures:

enantiomers

In the third drawing, the second structure has been rotated so that the methyl group at the left side of the structure lies in the plane of the paper and is higher on the page than the remainder of the molecule. This is the same orientation as the first of the three drawings, but the structures are not the same. Hence they are nonsuperimposable mirror images and 2,3-pentadiene is chiral.

Chiral molecules which do not have a chiral atom are relatively uncommon, and allenes are among the few organic compounds that can fit this description. The three-carbon allene unit can be considered as the chiral center, and such compounds obey all the rules of stereochemistry that we discussed in Chapter 6.

Alkyne-Allene Equilibria

Alkynes are ordinarily more stable than allenes, and they will predominate in any equilibrium mixture. The equilibrium can be established by prolonged treatment of either isomer with strong base, as illustrated for the equilibrium between 2-pentyne and 1,2-pentadiene.

$$\underset{96\%}{CH_3{-}C{\equiv}C{-}CH_2CH_3} \rightleftharpoons \underset{4\%}{CH_2{=}C{=}CH{-}CH_2CH_3}$$

An unusual effect has been observed in the case of cyclic systems. Whereas an alkyne has a total of four atoms which are linear, this number is reduced to only three in the case of an allene. Consequently, there is less strain in a cyclic allene than in the isomeric alkyne. Although the alkyne is still more stable, the 11-membered ring system shows an increase in the fraction of allene (26%) at equilibrium relative to the 4% found in the preceding example for an acyclic hydrocarbon.

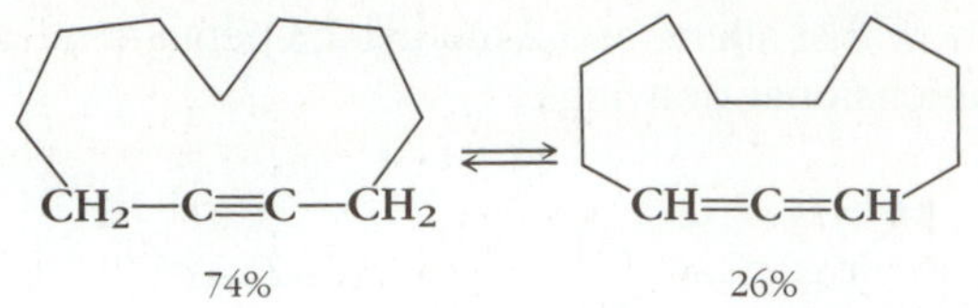

$CH_2—C≡C—CH_2$ ⇌ $CH=C=CH$

74% 26%

For a ring with only nine carbons, there is considerably greater strain for the alkyne, and the equilibrium in this case actually favors the allene.

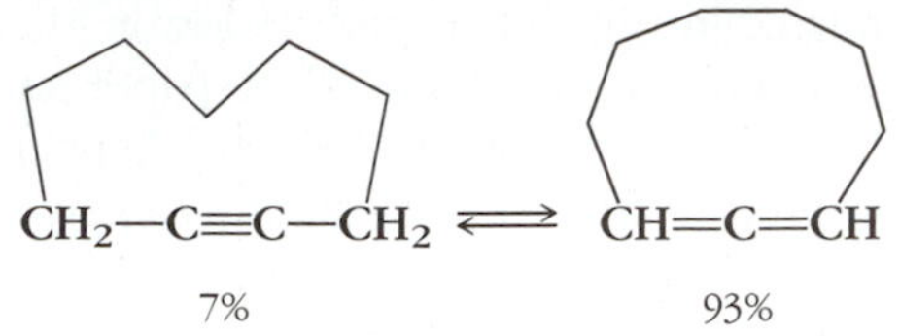

$CH_2—C≡C—CH_2$ ⇌ $CH=C=CH$

7% 93%

EXERCISE 7.7

Which of the following molecules are chiral?

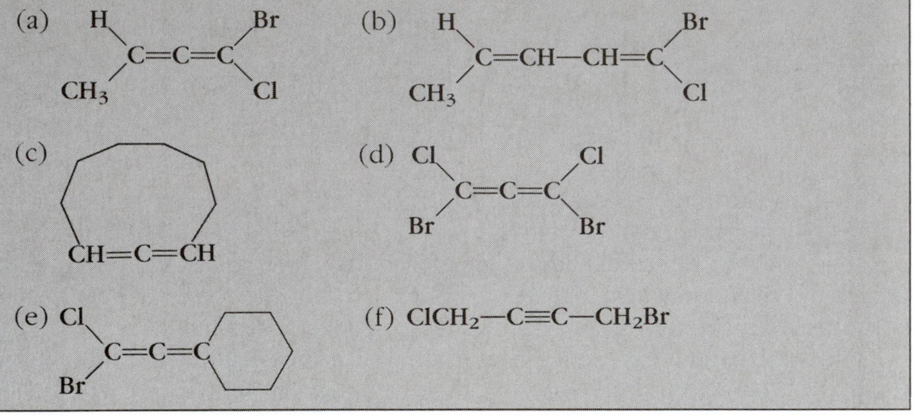

(a) (H)(CH_3)C=C=C(Br)(Cl)

(b) (H)(CH_3)C=CH—CH=C(Br)(Cl)

(c) cyclic CH=C=CH

(d) (Cl)(Br)C=C=C(Cl)(Br)

(e) (Cl)(Br)C=C=C(cyclohexylidene)

(f) $ClCH_2—C≡C—CH_2Br$

7.10 SYNTHESIS OF ORGANIC COMPOUNDS VIA ALKYNES

Because they can undergo a variety of different reactions, particularly those which form carbon-carbon bonds, alkynes are valuable intermediates in organic synthesis. In this section we will consider some of the various ways in which alkynes can be used for the efficient construction of organic molecules. We will evaluate a series of synthetic sequences as potential solutions to problems of preparing specific compounds, and we will show you how to select the best method in each case. We will place particular emphasis on the technique of "working backwards" by using the structural information that is present in the target molecule to help you design reactants that would yield the desired product.

In each synthesis problem your first goal is to find an immediate **precursor** of the target molecule, that is, a compound that will afford the target molecule when treated with the appropriate reagent. This in turn creates a new problem with a new target molecule, and you must seek another compound that can be converted to this new target molecule. The process is repeated until your current target molecule can be prepared by using starting compounds that are specified as allowable in the problem.

In working backwards in a synthetic problem, your first step should always be to find the *key functional group,* because this will provide clues to possible precursors. We define the key functional group as the functionality in the molecule that was produced in the previous reaction of the synthetic sequence we are trying to deduce. Therefore, if you know the reactions which afford that particular functional group, you will be able to select a reasonable precursor for the target molecule. The compounds we consider in this section will contain only one functional group, so by default this must be the key functional group. As an example, the key functional group of 3-hexyne is the carbon-carbon triple bond.

$$CH_3{-}CH_2{-}C{\equiv}C{-}CH_2{-}CH_3$$

If you consider all the reactions of this chapter, you will realize that the triple bond might have been introduced by an elimination reaction

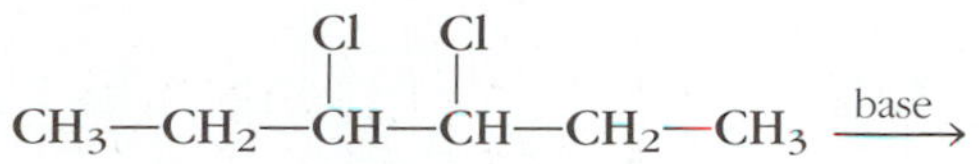

$$CH_3{-}CH_2{-}\underset{}{\overset{Cl}{\overset{|}{C}H}}{-}\overset{Cl}{\overset{|}{C}H}{-}CH_2{-}CH_3 \xrightarrow{\text{base}}$$

or by alkylation of a sodium acetylide.

$$CH_3{-}CH_2{-}C{\equiv}C^{-}\,{}^{+}Na \xrightarrow{CH_3CH_2Br}$$

Which of these two possibilities would correspond to the "better" precursor would depend on the restrictions of the particular synthetic problem.

To further illustrate the concept of the key functional group, we have drawn below a variety of compounds, and for each structure we have highlighted the key functional group.

$$CH_3{-}CH_2{-}\overset{Br}{\overset{|}{C}H}{-}CH_3$$

$$CH_3{-}CH_2{-}CH{-}CH{-}CH_3 \quad (\text{CH and CH bridged by } CH_2 \text{ in a three-membered ring})$$

$$\overset{H}{}\;\; C{=}C \;\;\overset{H}{} \quad (CH_3{-}CH_2 \text{ and } CH_3 \text{ on the two carbons; } H \text{ on each})$$

$$CH_3{-}CH_2{-}CH_2{-}\underset{Cl}{\underset{|}{\overset{Cl}{\overset{|}{C}}}}{-}CH_3$$

$$CH_3{-}CH_2{-}CH_2(H)C{=}C(I)(H)$$

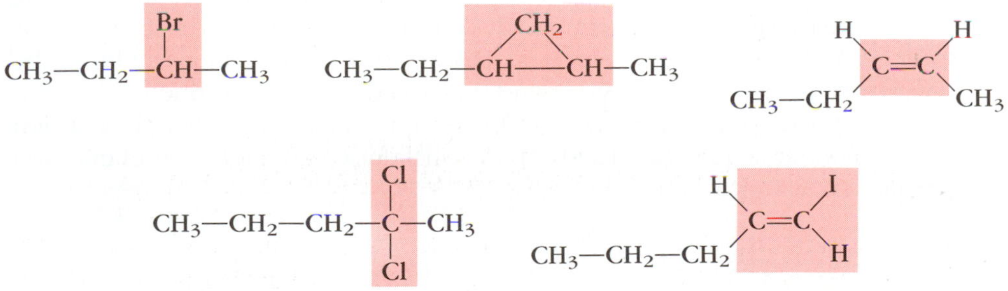

EXERCISE 7.8

Write the logical precursor of each of the preceding compounds. What reagents would be required to convert each precursor to the indicated compound? If more than one method is possible, write all such possibilities.

Example 7.1 Suggest a method for synthesizing 2,2-dimethyl-3-octyne from compounds with six or fewer carbon atoms that contain no metal atoms. Use any other reagents you wish.

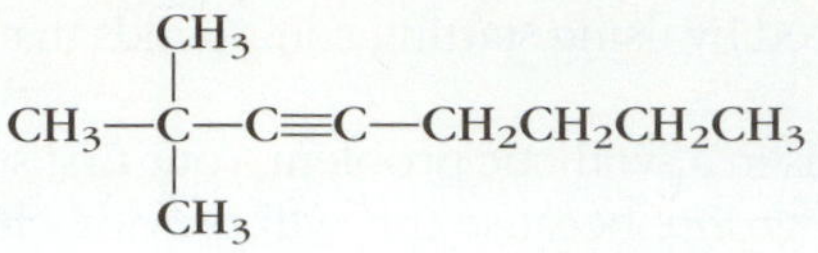

$$CH_3-C(CH_3)_2-C\equiv C-CH_2CH_2CH_2CH_3$$

Solution The key functional group can readily be identified as the triple bond.

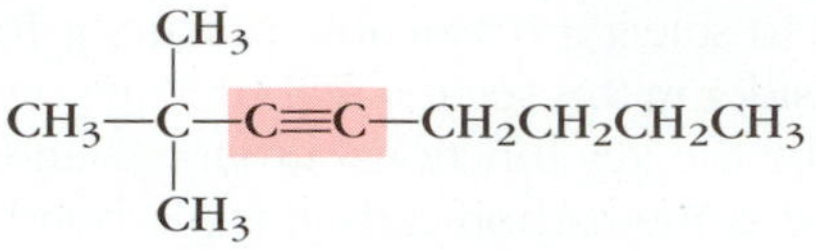

$$CH_3-C(CH_3)_2-\boxed{C\equiv C}-CH_2CH_2CH_2CH_3$$

The target molecule contains 10 carbon atoms, so a carbon-carbon bond-forming reaction must be used at some stage in the synthesis. If a synthesis can be devised which requires only a single such reaction, then the target molecule will be formed by combination either of two C_5 fragments or of a C_4 and a C_6 fragment. The question is which combination.

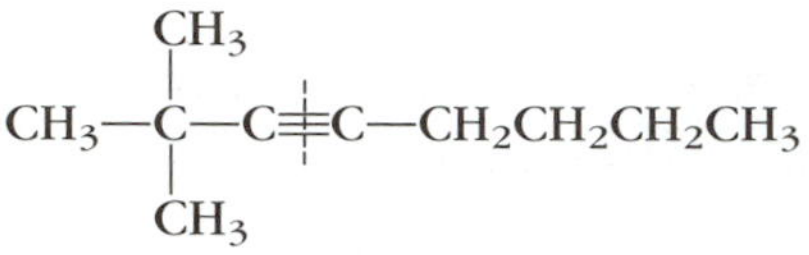

$$CH_3-C(CH_3)_2-C\,\vdots\!\equiv\!\vdots\,C-CH_2CH_2CH_2CH_3 \quad \textbf{(i)}$$

$$CH_3-C(CH_3)_2\,\vdots\,-C\equiv C-CH_2CH_2CH_2CH_3 \quad \textbf{(ii)}$$

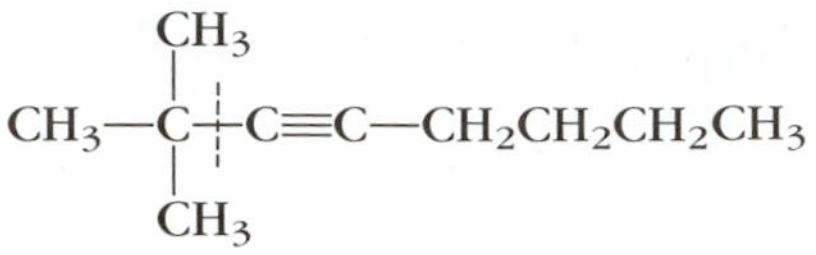

$$CH_3-C(CH_3)_2-C\equiv C\,\vdots\,-CH_2CH_2CH_2CH_3 \quad \textbf{(iii)}$$

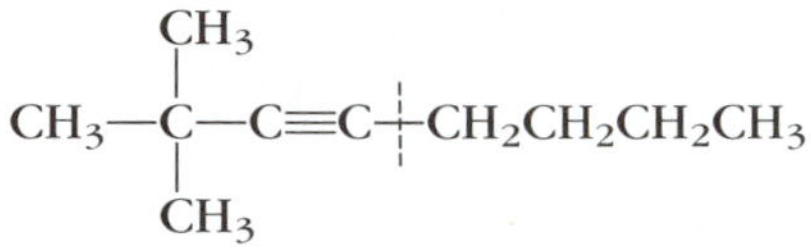

The first possibility **(i)** is not reasonable, because you know of no way to join two fragments with simultaneous formation of a triple bond between them. On the other hand, you have learned how to form a new carbon-carbon bond by joining two fragments when one of them already contains a triple bond. This corresponds to either of the situations depicted by **(ii)** and **(iii)**. The former would not be satisfactory, however, because the bond that would have to be formed is between the triple bond and a *tertiary* alkyl group. As we indicated previously, the reaction of a sodium acetylide with a tertiary alkyl halide would afford the elimination product exclusively.

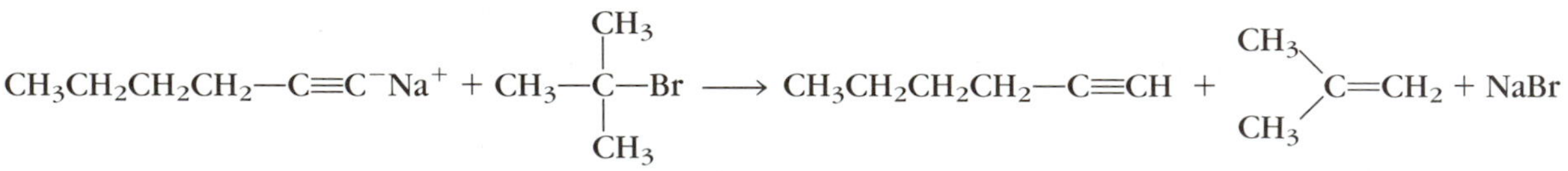

$$CH_3CH_2CH_2CH_2-C\equiv C^-Na^+ + CH_3-C(CH_3)_2-Br \longrightarrow CH_3CH_2CH_2CH_2-C\equiv CH + (CH_3)_2C{=}CH_2 + NaBr$$

This difficulty does not arise in the third alternative, so choice **(iii)** provides an adequate answer to the problem.

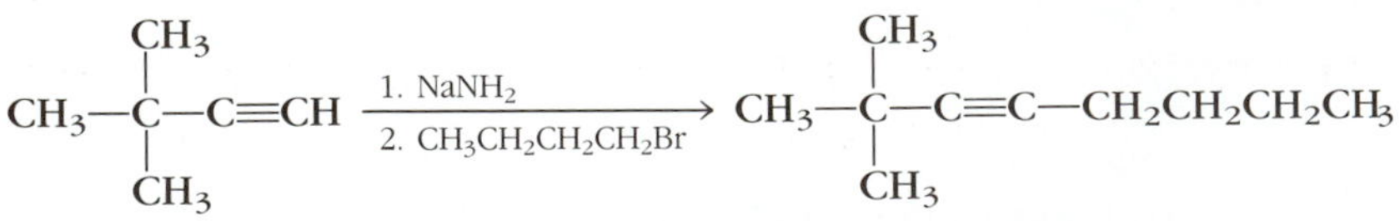

$$CH_3-C(CH_3)_2-C\equiv CH \xrightarrow[\text{2. } CH_3CH_2CH_2CH_2Br]{\text{1. } NaNH_2} CH_3-C(CH_3)_2-C\equiv C-CH_2CH_2CH_2CH_3$$

Example 7.2 Suggest a method of synthesizing (*E*)-4-octene starting from hydrocarbons containing four or fewer carbon atoms.

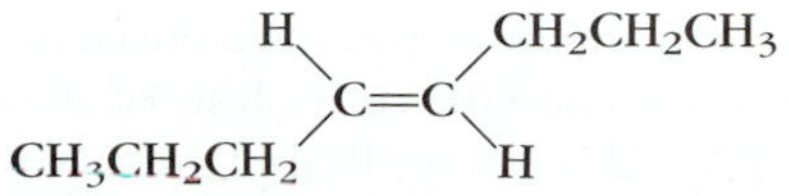

$$\underset{CH_3CH_2CH_2}{\overset{H}{}}\!\!>C{=}C<\!\!\underset{H}{\overset{CH_2CH_2CH_3}{}}$$

Solution The key functional group is the carbon-carbon double bond.

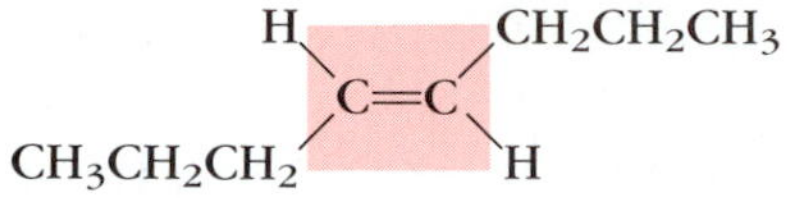

$$\underset{CH_3CH_2CH_2}{\overset{H}{}}\!\!>C{=}C<\!\!\underset{H}{\overset{CH_2CH_2CH_3}{}}$$

Although two C_4 fragments might be joined directly, none of the reactions that we have presented permit the coupling of two groups with simultaneous formation of a double bond between them. Consequently, two C_4 groups cannot be used and a larger number of smaller fragments must be joined. Recall that a disubstituted alkene can easily be prepared by reduction of the corresponding alkyne.

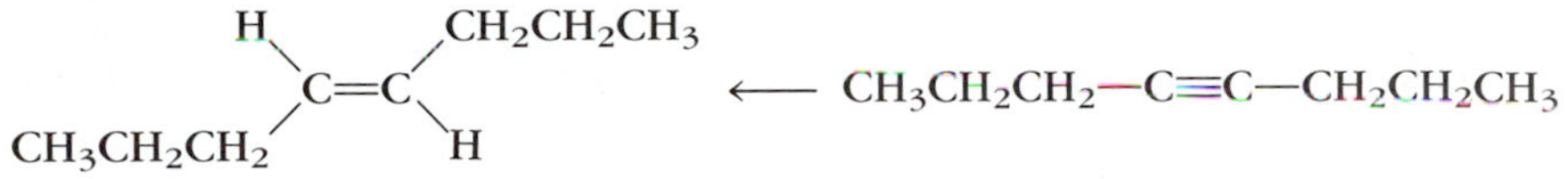

$$\underset{CH_3CH_2CH_2}{\overset{H}{}}\!\!>C{=}C<\!\!\underset{H}{\overset{CH_2CH_2CH_3}{}} \longleftarrow CH_3CH_2CH_2{-}C{\equiv}C{-}CH_2CH_2CH_3$$

The alkyne itself could then be prepared by sodium acetylide alkylation of alkyl halides. 4-Octyne thus becomes the new target molecule and the key functional group is the carbon-carbon triple bond.

$$CH_3CH_2CH_2{-}C{\equiv}C{-}CH_2CH_2CH_3$$

The two substituents on the triple bond are both propyl groups, so two successive alkylations of acetylene with 1-bromopropane would afford the required alkyne.

$$HC{\equiv}CH \xrightarrow[\text{2. } CH_3CH_2CH_2Br]{\text{1. } NaNH_2} CH_3CH_2CH_2{-}C{\equiv}CH \xrightarrow[\text{2. } CH_3CH_2CH_2Br]{\text{1. } NaNH_2} CH_3CH_2CH_2{-}C{\equiv}C{-}CH_2CH_2CH_3$$

Completion of the synthesis could then be accomplished by reduction with sodium in liquid ammonia to give the alkene with the desired *E* stereochemistry.

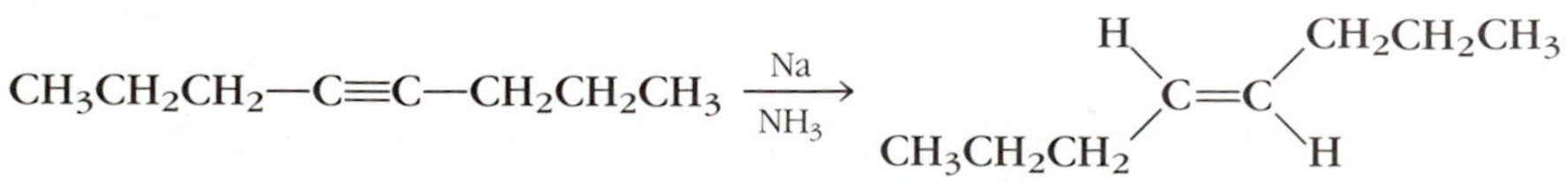

$$CH_3CH_2CH_2{-}C{\equiv}C{-}CH_2CH_2CH_3 \xrightarrow[NH_3]{Na} \underset{CH_3CH_2CH_2}{\overset{H}{}}\!\!>C{=}C<\!\!\underset{H}{\overset{CH_2CH_2CH_3}{}}$$

Example 7.3 Suggest a method of synthesizing 3-hexanone from starting compounds containing three or fewer carbon atoms.

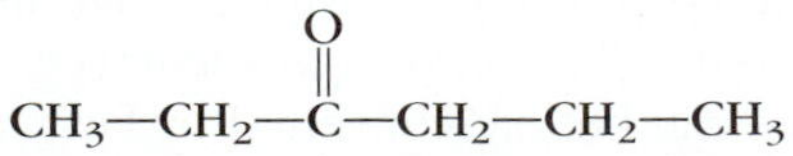

$$CH_3{-}CH_2{-}\overset{\overset{\displaystyle O}{\|}}{C}{-}CH_2{-}CH_2{-}CH_3$$

Solution The key functional group is the carbonyl group

$$CH_3{-}CH_2{-}\overset{\overset{\displaystyle O}{\|}}{C}{-}CH_2{-}CH_2{-}CH_3$$

and the only methods you have encountered for preparing ketones are the hydration of alkynes and the ozonolysis of alkenes. The latter involves cleavage of larger molecules and is not applicable in this case. Therefore the appropriate precursor is an alkyne. However, this problem is deceptively simple. The six-carbon alkyne 2-hexyne, derived from two C_3 fragments, is a logical candidate for the precursor, but it is unsatisfactory because hydration would yield a mixture of products.

$$CH_3{-}C{\equiv}C{-}CH_2CH_2CH_3 \longrightarrow CH_3{-}CH_2{-}\overset{\overset{\displaystyle O}{\|}}{C}{-}CH_2{-}CH_2{-}CH_3 + CH_3{-}\overset{\overset{\displaystyle O}{\|}}{C}{-}CH_2{-}CH_2{-}CH_2{-}CH_3$$

In addition to the desired 3-hexanone, a comparable amount of 2-hexanone would be formed from 2-hexyne. But there is a second alkyne, 3-hexyne, that would yield 3-hexanone as a hydration product. This alkyne is symmetrical, and addition of water would yield a single ketone.

$$CH_3{-}CH_2{-}C{\equiv}C{-}CH_2{-}CH_3 \xrightarrow[\text{HgSO}_4,\ \text{H}_2\text{SO}_4]{\text{H}_2\text{O}} CH_3{-}CH_2{-}\overset{\overset{\displaystyle O}{\|}}{C}{-}CH_2{-}CH_2{-}CH_3$$

Clearly, 3-hexyne is the alkyne needed to carry out the synthesis, and it could in turn be prepared by successive alkylations of acetylene with ethyl iodide or ethyl bromide.

$$H{-}C{\equiv}C{-}H \xrightarrow[\text{2. CH}_3\text{CH}_2\text{I}]{\text{1. NaNH}_2} CH_3CH_2C{\equiv}CH \xrightarrow[\text{2. CH}_3\text{CH}_2\text{I}]{\text{1. NaNH}_2} CH_3CH_2{-}C{\equiv}C{-}CH_2CH_3$$

Example 7.4 Suggest a method of synthesizing 2,6-dimethylnonane from starting materials containing no more than six carbon atoms and having no elements of atomic number greater than 17.

$$CH_3{-}\underset{}{\overset{\overset{\displaystyle CH_3}{|}}{C}H}{-}CH_2CH_2CH_2{-}\overset{\overset{\displaystyle CH_3}{|}}{C}H{-}CH_2CH_2CH_3$$

Solution We presented strategies for the synthesis of alkanes in Section 1.12, but the restriction on atomic number in this problem precludes the obvious answer involving a lithium dialkylcuprate reagent. The only other way that you know to form a carbon-carbon bond is by alkylation of a terminal alkyne, so you must devise an answer using that reaction. But what is the key functional group in the target molecule that will tell you how to synthesize it? Remember that an alkyne can be successively reduced to an alkene and then to an alkane, and the following groups can be considered as *functional equivalents* (i.e., compounds with the same carbon skeletons that can be interconverted by fairly simple functional group modifications).

$$-C{\equiv}C- \longrightarrow -CH{=}CH- \longrightarrow -CH_2-CH_2-$$

Moreover, the alkylation of a sodium acetylide requires a primary alkyl halide, so the product will have a $-CH_2-$ group attached to the triple bond. Again considering successive reduction steps, the functional equivalents are:

$$-C{\equiv}C-CH_2- \longrightarrow -CH{=}CH-CH_2- \longrightarrow -CH_2-CH_2-CH_2-$$

In other words, complete reduction of the product of a sodium acetylide alkylation will produce a compound containing three adjacent CH_2 groups. The key functional group of the target molecule can now be identified.

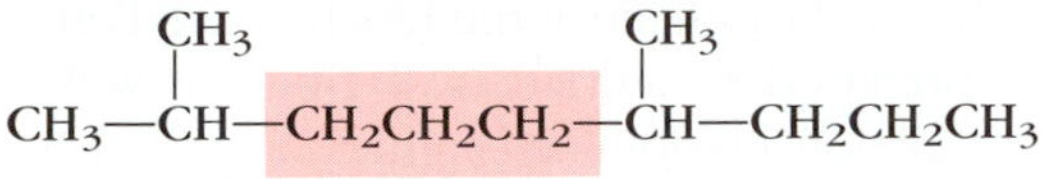

$$CH_3-\underset{}{\overset{CH_3}{\overset{|}{C}H}}-\boxed{CH_2CH_2CH_2}-\overset{CH_3}{\overset{|}{C}H}-CH_2CH_2CH_3$$

This compound would be produced by the reduction of either of the following alkynes, where the bond which would have been formed in the preceding alkylation step has been indicated by an arrow.

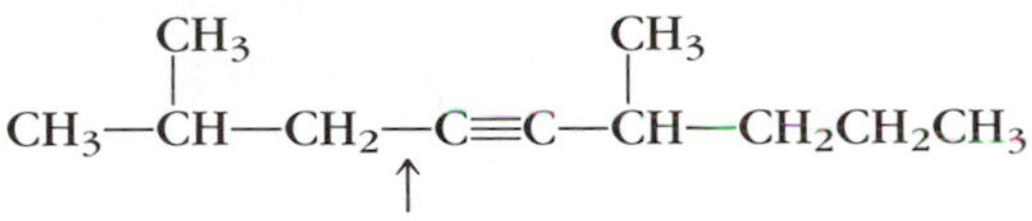

$$CH_3-\overset{CH_3}{\overset{|}{C}H}-CH_2\underset{\uparrow}{-}C{\equiv}C-\overset{CH_3}{\overset{|}{C}H}-CH_2CH_2CH_3 \quad \text{or} \quad CH_3-\overset{CH_3}{\overset{|}{C}H}-C{\equiv}C\underset{\uparrow}{-}CH_2-\overset{CH_3}{\overset{|}{C}H}-CH_2CH_2CH_3$$

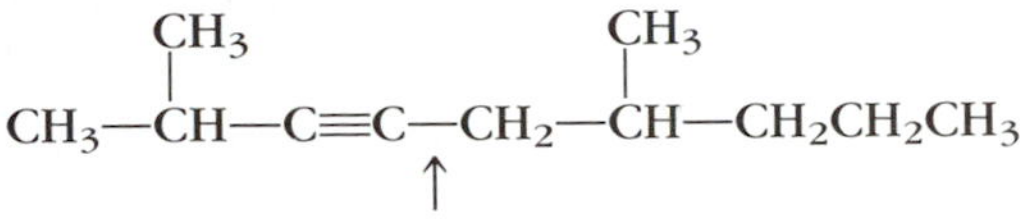

Only for the second of these two compounds would both fragments fall within the six-carbon limit stated in the instructions for this problem, so that must be the alkyne formed by alkylation.

A complete synthetic proposal can now be outlined:

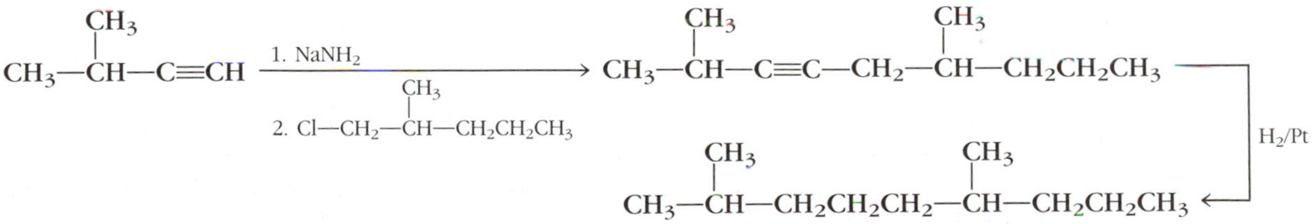

$$CH_3-\overset{CH_3}{\overset{|}{C}H}-C{\equiv}CH \xrightarrow[\text{2. } Cl-CH_2-\overset{CH_3}{\overset{|}{C}H}-CH_2CH_2CH_3]{\text{1. } NaNH_2} CH_3-\overset{CH_3}{\overset{|}{C}H}-C{\equiv}C-CH_2-\overset{CH_3}{\overset{|}{C}H}-CH_2CH_2CH_3$$

$$\xrightarrow{H_2/Pt} CH_3-\overset{CH_3}{\overset{|}{C}H}-CH_2CH_2CH_2-\overset{CH_3}{\overset{|}{C}H}-CH_2CH_2CH_3$$

Example 7.5 Suggest a method for synthesizing 1-cyclopentyl-2-butanone from starting materials containing six or fewer carbon atoms.

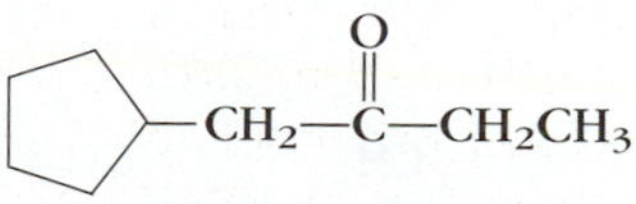

$$\text{(cyclopentyl)}-CH_2-\overset{O}{\overset{\|}{C}}-CH_2CH_3$$

Solution The key functional group is the carbonyl group, and this could be introduced by the addition of water to either of two isomeric alkynes.

$$\text{C}_5\text{H}_9\text{—CH}_2\text{—}\overset{\text{O}}{\overset{\|}{\text{C}}}\text{—CH}_2\text{CH}_3 \quad \text{from either} \quad \text{C}_5\text{H}_9\text{—C}\equiv\text{C—CH}_2\text{CH}_3 \ \textbf{(i)} \quad \text{or} \quad \text{C}_5\text{H}_9\text{—CH}_2\text{—C}\equiv\text{C—CH}_3 \ \textbf{(ii)}$$

The second alkyne **(ii)** could be prepared via the alkylation of propyne, but it would yield a mixture of two ketones upon hydration because neither of the carbon atoms attached to the triple bond is sterically hindered. On the other hand, the alkyl substituents on alkyne **(i)** are secondary and primary, so steric effects would probably result in formation of the desired ketone by hydroboration-oxidation with catecholborane at the less hindered end of the triple bond. Yet **(i)** could not be prepared by alkylation of 1-butyne with the secondary alkyl halide, because elimination predominates unless the alkyl halide is primary.

$$\text{C}_5\text{H}_9\text{—Br} + \text{Na}^{+}\,{}^{-}\text{C}\equiv\text{C—CH}_2\text{CH}_3 \longrightarrow \textbf{mainly elimination}$$

To avoid these difficulties, you could utilize the functional equivalent of the *alkene,* and employ an initial coupling reaction with an organocuprate to generate the required carbon skeleton.

$$\text{C}_5\text{H}_9\text{—Br} \xrightarrow[\text{2. CuI}]{\text{1. Li}} (\text{C}_5\text{H}_9)_2\text{CuLi} \xrightarrow{\text{ICH=CH—CH}_2\text{CH}_3} \text{C}_5\text{H}_9\text{—CH=CH—CH}_2\text{CH}_3$$

The synthesis could then be completed by the following sequence, in which the triple bond is introduced by addition of Br_2 followed by elimination of two equivalents of HBr.

$$\text{C}_5\text{H}_9\text{—CH=CH—CH}_2\text{CH}_3 \xrightarrow[\text{2. NaNH}_2\text{, NH}_3]{\text{1. Br}_2\text{, CCl}_4} \text{C}_5\text{H}_9\text{—C}\equiv\text{C—CH}_2\text{CH}_3 \xrightarrow[\text{2. H}_2\text{O}_2\text{/NaOH}]{\text{1. catecholborane (BH)}} \text{C}_5\text{H}_9\text{—CH}_2\text{—}\overset{\text{O}}{\overset{\|}{\text{C}}}\text{—CH}_2\text{CH}_3$$

EXERCISE 7.9

Suggest methods of synthesizing the following compounds. All atoms of the final product must be derived from the indicated starting materials.

(a) $\text{CH}_3\text{—}\overset{\text{CH}_3}{\overset{|}{\text{CH}}}\text{—CH}_2\text{(H)C=C(H)CH}_2\text{CH}_3$ (cis) from starting materials containing four or fewer carbon atoms

(b) $CH_3CH_2CH_2(H)C{=}C(Br)CH_2CH_2CH_3$ from starting materials containing three or fewer carbon atoms

(c) $CH_3CH_2CH_2{-}C({=}O){-}CH_3$ from starting materials containing four or fewer carbon atoms

(d) $CH_3{-}CH(CH_3){-}CH_2CH_2CH_2{-}CH({=}O)$ from starting materials containing six or fewer carbon atoms

7.11 STRUCTURE DETERMINATION AND MOLECULAR FORMULA

Organic chemists frequently obtain compounds for which the structure is either unknown or uncertain, and the assignment of "unknowns" to students is also common in organic laboratory courses. Sometimes characterizing the product of a reaction can cause problems. You may know the *desired* reaction product, but how can you be sure that the material isolated from the reaction actually has that structure? Now that we have introduced a substantial variety of functional groups, this is an important problem to consider.

Many spectroscopic techniques have been developed in recent years, and these have enabled chemists to solve extremely complicated structural problems. We will discuss some of these techniques in Chapter 9, but in this section we will show you how a series of relatively simple chemical reactions can be used to obtain useful information about the structure of an organic compound.

Molecular Formula and Degrees of Unsaturation

Elemental analysis can provide the percent composition of a compound, and it allows you to calculate the empirical formula. If you know (even approximately) the molecular weight, you can then calculate the actual molecular formula. The determination of how many bromine atoms, oxygen atoms, etc. are contained in the molecule is of great help in determining the structure, but even in the case of a hydrocarbon the molecular formula provides valuable information.

An acyclic alkane has a molecular formula C_nH_{2n+2}, whereas an analogous alkene is described by C_nH_{2n}. The difference between them, a "missing" pair of hydrogen atoms, can be conveniently described as a **degree of unsaturation.** The total number of such missing pairs of hydrogens corresponds to the total degrees of unsaturation, and it can be calculated as shown in the following box.

Determining Degrees of Unsaturation

1. For a hydrocarbon with n carbon atoms, calculate the number of hydrogen atoms that would be present in the corresponding acyclic alkane, C_nH_{2n+2}.

2. Subtract the actual number of hydrogens. This difference corresponds to the total number of missing hydrogen atoms.

3. The number of *pairs* of such atoms is one-half of this quantity, so multiplication by 1/2 affords the degrees of unsaturation.

The following analysis shows that 1-pentene (C_5H_{10}) has one degree of unsaturation:

1. 2_{n+2} for $n = 5$ is 12
2. $12H - 10H = 2H$
3. $2 \times 1/2 = 1$ degree of unsaturation

Similarly, 1-pentyne (C_5H_8) has two degrees of unsaturation.

1. 2_{n+2} for $n = 5$ is 12
2. $12H - 8H = 4H$
3. $4 \times 1/2 = 2$ degrees of unsaturation

Next consider cyclopentane. This compound has the molecular formula, C_5H_{10}, which is the same as that for 1-pentene. This means that cyclopentane must also have one degree of unsaturation even though it contains no multiple bonds. The concept of degrees of unsaturation is dependent only on the molecular formula and not the details of molecular structure, so we can write the following general rule:

Each ring and each double bond correspond to one degree of unsaturation, and each triple bond corresponds to two degrees of unsaturation.

The preceding discussion has been limited strictly to hydrocarbons, but what about other classes of organic compounds? Compounds containing other functional groups can also be analyzed in terms of unsaturation. A multiple bond or a ring will always correspond to a degree of unsaturation, but how do you calculate the total degrees of unsaturation when the molecule contains atoms other than carbon and hydrogen? For the time being, we will only consider the halogens and oxygen. 2,2-Dichloropropane contains no rings and no double bonds, so a halogen atom can be considered as *replacing* a hydrogen atom:

$CH_3CH_2CH_3$

Propane (C_3H_8); no degrees of unsaturation

$CH_3{-}C(Cl)_2{-}CH_3$

2,2-Dichloropropane ($C_3H_6Cl_2$); no degrees of unsaturation

Halogen atoms are counted as hydrogen atoms when determining degrees of unsaturation.

On the other hand, the following comparison of propane and 2-propanol (which also has no rings and no multiple bonds) demonstrates that an oxygen atom has no effect on the ratio of carbon and hydrogen atoms, so it is ignored in calculating the degrees of unsaturation.

$CH_3CH_2CH_3$

Propane (C_3H_8); no degrees of unsaturation

$CH_3—CH(OH)—CH_3$

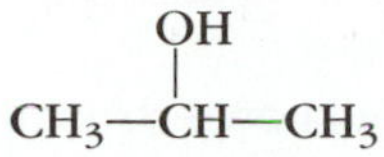

2-Propanol (C_3H_8O); no degrees of unsaturation

Oxygen atoms in the molecular formula are ignored when calculating the number of degrees of unsaturation.

If you should ever forget these rules, you can easily derive the pattern by comparing CH_4, CH_3Cl, and CH_3OH. None of these has either a ring or a multiple bond so they each have zero degrees of unsaturation. Hence halogens count as hydrogens and oxygens are ignored.

The molecular formula tells you how many degrees of unsaturation are present, but it does not tell you what form they have. For example, the following isomeric structures each have one degree of unsaturation.

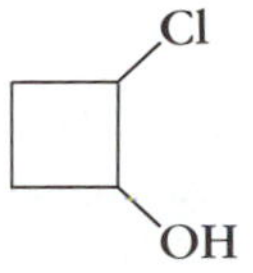

(cyclobutane with Cl and OH) $ClCH_2—CH{=}CH—CH_2OH$ $CH_3CH_2—C(=O)—CH_2Cl$ (tetrahydrofuran ring with Cl)

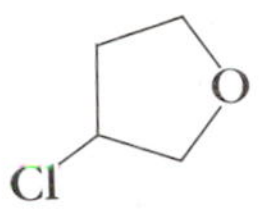

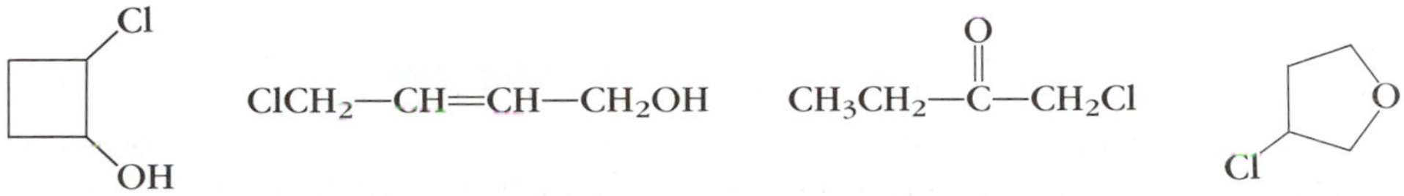

It does not matter whether the double bond is between two carbons or between a carbon and an oxygen. Nor does it matter whether a ring consists entirely of carbon atoms or if an oxygen atom is part of the ring. The number of degrees of unsaturation simply provides a total count of rings and multiple bonds (a triple bond counting as two degrees of unsaturation), regardless of what atoms may be present in the molecule. As you will see in the following discussion, the molecular formula together with other simple chemical information can tell you a great deal about the structure of an unknown compound.

Catalytic Hydrogenation: Determination of Number of Rings

Multiple bonds are readily reduced by catalytic hydrogenation while most rings remain intact under these conditions. Consequently, reduction with hydrogen and a metal catalyst provides a distinction between degrees of unsaturation resulting from multiple bonds and those resulting from rings. For example, 1-cyclopentyl-3-hexyne ($C_{11}H_{18}$) has three degrees of unsaturation, but catalytic hydrogenation would afford a compound with the molecular formula $C_{11}H_{22}$. The remaining degree of unsaturation, which is resistant to hydrogenation, could therefore have been designated as a ring even if you had not known the structure.

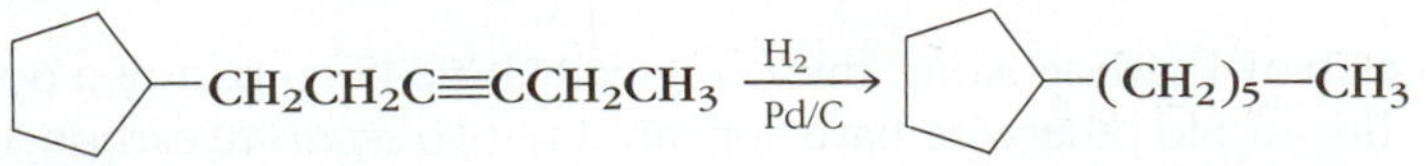

$$\text{(cyclopentyl)}—CH_2CH_2C{\equiv}CCH_2CH_3 \xrightarrow[\text{Pd/C}]{H_2} \text{(cyclopentyl)}—(CH_2)_5—CH_3$$

Cyclopropyl and cyclobutyl rings can be cleaved by hydrogen and a metal catalyst at high temperatures and pressures (Section 3.4), and this can provide some information about ring size. Failure to undergo reduction under such vigorous conditions must lead to the conclusion that a ring was neither a three- nor a four-membered ring.

Example 7.6 A compound of unknown structure having the molecular formula $C_{10}H_{14}$ was reduced by hydrogenation with platinum to $C_{10}H_{18}$. When the reduction was carried out at high temperature and pressure, the product had the molecular formula $C_{10}H_{20}$. What structural conclusions can be reached from this information?

Solution The original compound has four degrees of unsaturation. Two of them ($C_{10}H_{14} \rightarrow C_{10}H_{18}$) are easily reduced by H_2/Pt and must be multiple bonds. *Hence the compound has either two double bonds or one triple bond.*

The remaining two degrees of unsaturation which resisted hydrogenation must be rings. The absorption of 1 mol of H_2 ($C_{10}H_{18} \rightarrow C_{10}H_{20}$) at high temperature and pressure indicates that *one of these is either a three- or four-membered ring,* and the other must have a ring size of five or larger.

Ozonolysis of Cycloalkenes: Evidence for Ring Size

Ozonolysis of an acyclic alkene affords two separate carbonyl products, but a cyclic alkene would yield a single molecule containing two carbonyl groups.

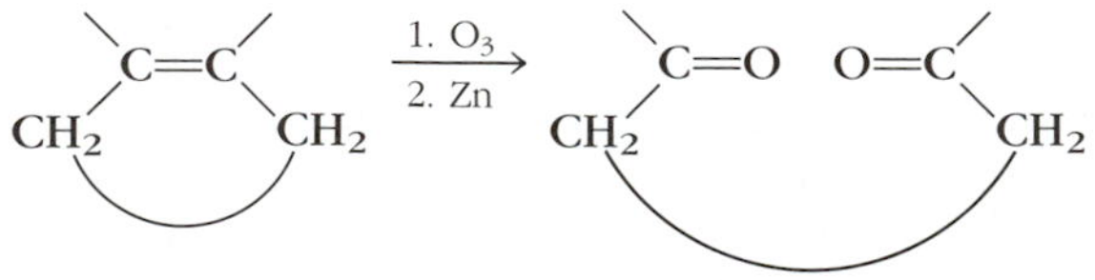

$$\text{CH}_2\text{—C=C—CH}_2 \text{ (ring)} \xrightarrow[\text{2. Zn}]{\text{1. O}_3} \text{CH}_2\text{—C=O} \quad \text{O=C—CH}_2 \text{ (chain)}$$

The number of intervening carbon atoms provides information regarding the size of the original ring.

Example 7.7 Ozonolysis of a compound of unknown structure with the molecular formula C_8H_{14} afforded the following dicarbonyl compound:

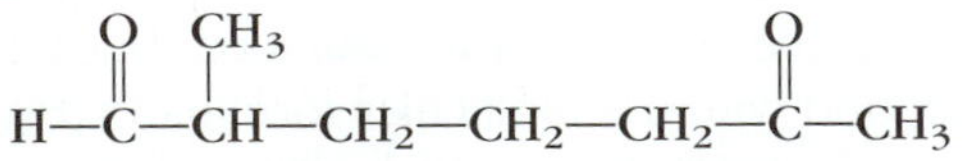

$$\text{H—C(=O)—CH(CH}_3\text{)—CH}_2\text{—CH}_2\text{—CH}_2\text{—C(=O)—CH}_3$$

What was the structure of the unknown compound?

Solution The two carbonyl carbon atoms must originally have been part of a double bond, because they would otherwise have appeared as two *separate* carbonyl fragments.

Four additional carbon atoms intervene, so the original double bond must have been in a six-membered ring:

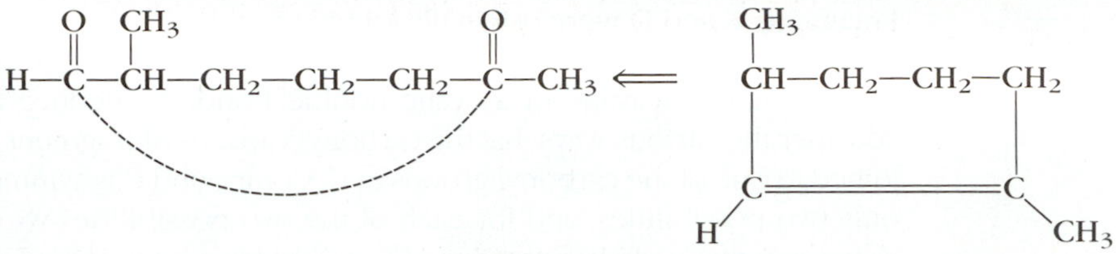

$$H{-}\overset{\overset{\displaystyle O}{\|}}{C}{-}\overset{\overset{\displaystyle CH_3}{|}}{CH}{-}CH_2{-}CH_2{-}CH_2{-}\overset{\overset{\displaystyle O}{\|}}{C}{-}CH_3 \Longleftarrow$$

(ring: $CH(CH_3){-}CH_2{-}CH_2{-}CH_2{-}C(CH_3){=}C(H){-}$, closing back to the CH)

The compound was therefore 1,3-dimethylcyclohexene, ordinarily drawn in the following manner.

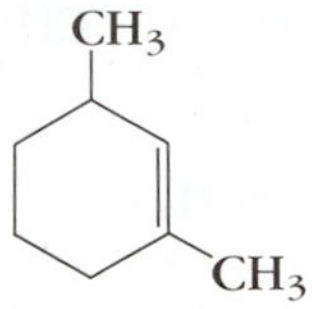

CH_3

CH_3

Structure Determination From Interrelated Pieces of Evidence

By combining various kinds of evidence you can often completely determine the structure of even a complex organic molecule. However, the various clues and pieces of evidence may not always fall into place by themselves; frequently you will find it necessary to join pieces of evidence, much as you might fit together the pieces of a jigsaw puzzle. By making sure that the pieces do fit, and with occasional use of a trial guess, you can employ the same techniques to solve both kinds of puzzle.

Example 7.8 Compounds A–D are all different compounds. On the basis of the information given, deduce the structures of **A** and **B**.

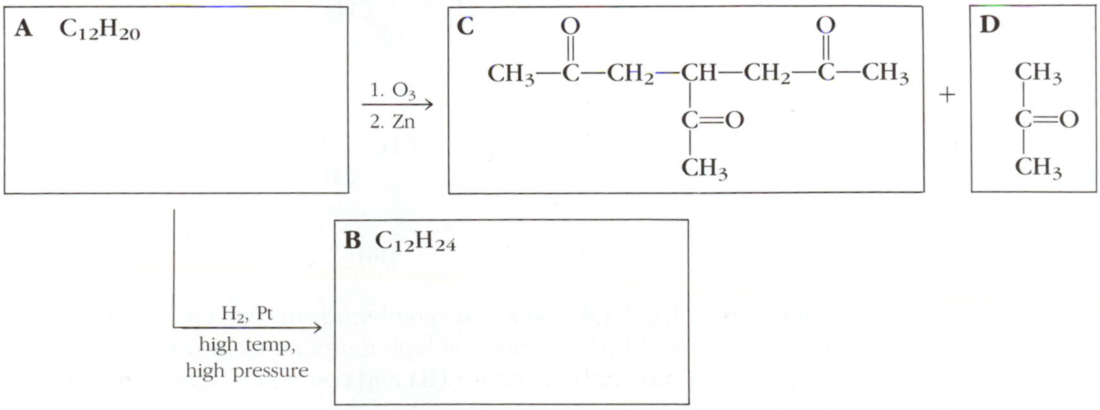

A $C_{12}H_{20}$ $\xrightarrow[\text{2. Zn}]{\text{1. } O_3}$ **C** $CH_3{-}\overset{\overset{\displaystyle O}{\|}}{C}{-}CH_2{-}\underset{\underset{\displaystyle CH_3}{|}}{\underset{\displaystyle C{=}O}{\underset{|}{CH}}}{-}CH_2{-}\overset{\overset{\displaystyle O}{\|}}{C}{-}CH_3$ + **D** $CH_3{-}C({=}O){-}CH_3$

A $\xrightarrow{H_2,\ Pt}$ high temp, high pressure → **B** $C_{12}H_{24}$

Solution Compound A has three degrees of unsaturation and one of these is a ring, as shown by the reduction to B ($C_{12}H_{24}$), which still has one degree of unsaturation. The ozonolysis of A yields two fragments, one of which has three carbonyl groups. Fragments C and D were originally joined via a double bond, and this accounts for *one* of the carbonyl groups of C. The other two carbonyl groups of C must have resulted from cleavage of a cyclic double bond. To deduce the structure of A, consider the various ways that the carbonyl carbons of fragment D might have been joined to one of the carbonyl groups of C. Compound C is symmetrical, so there are only two possibilities, and for each of the two possibilities we can show how the remaining two carbonyl carbons of C would have been joined as part of the cyclic double bond.

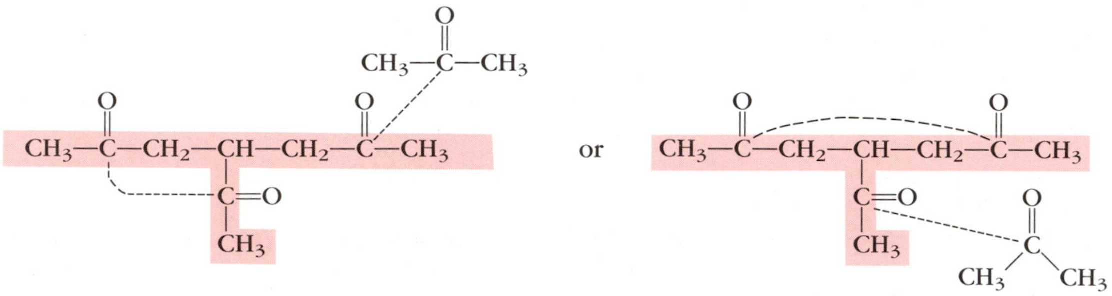

$CH_3-C(=O)-CH_2-CH(C(=O)CH_3)-CH_2-C(=O)-CH_3$ + $CH_3-C(=O)-CH_3$ or $CH_3-C(=O)-CH_2-CH(C(=O)CH_3)-CH_2-C(=O)-CH_3$ + $CH_3-C(=O)-CH_3$

Compound A must therefore have either structure **(i)** or structure **(ii):**

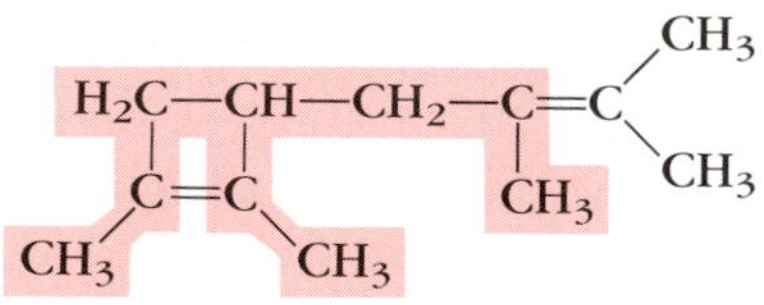

(i) 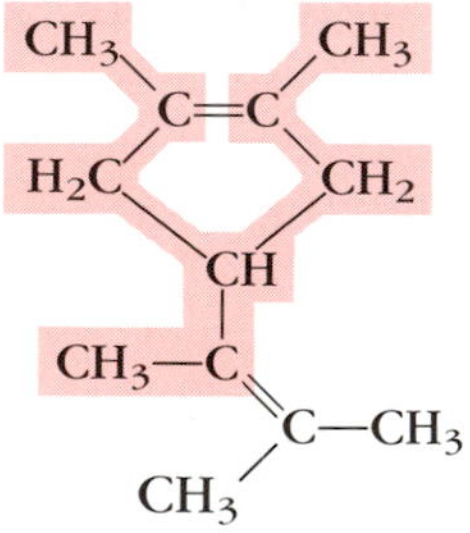 **(ii)**

The reduction of **(ii)** would lead to **(iii),** which has the correct molecular formula ($C_{12}H_{24}$) for compound B.

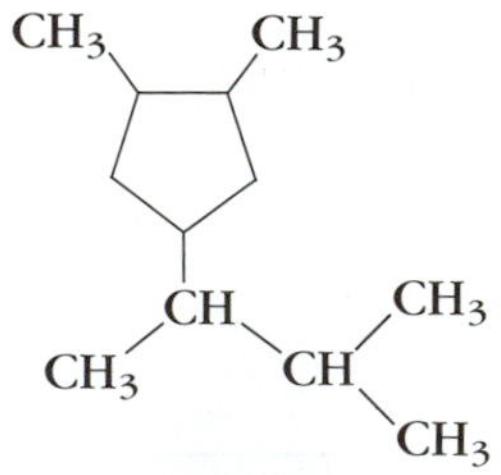

(iii)

On the other hand, **(ii)** has a four-membered ring, which would be cleaved under the conditions of hydrogenation at high temperature and pressure. Consequently, compound A must have structure **(ii)** and compound B must have structure **(iii).**

7.12 TERMS AND DEFINITIONS

Alkyne. A hydrocarbon with the general formula C_nH_{2n-2}, characterized by the presence of a carbon-carbon triple bond.

Allene. A hydrocarbon with the general formula C_nH_{2n-2}, characterized by the presence of two carbon-carbon double bonds between three contiguous carbon atoms (i.e., >C=C=C<).

Carbon-carbon bond formation. A class of reactions in which a new carbon-carbon bond is formed. Such reactions are the key synthetic steps in building up a new carbon skeleton.

Degree of unsaturation. The number of "missing" pairs of hydrogen atoms in a molecule that correspond to multiple bonds or rings. Each ring or double bond corresponds to one degree of unsaturation; each triple bond corresponds to two degrees of unsaturation.

Enol. A compound containing a hydroxyl group bonded directly to a carbon-carbon double bond. An enol ordinarily undergoes rapid rearrangement to an isomeric carbonyl compound.

Functional group modification. A class of reactions in which the carbon skeleton is unchanged and only the substituents on the carbon skeleton are affected.

Geminal. The relationship between two substituents that are bonded to the same carbon atom.

Internal alkyne. An alkyne in which the triple bond is located in the interior part of the carbon chain (i.e., —C—C≡C—C—).

Precursor. A compound that precedes another in a synthetic sequence; the compound that will afford the target molecule when subjected to the appropriate sequence of reactions.

Terminal alkyne. An alkyne in which the triple bond is located at the end of the carbon chain (i.e., —C—C≡C—H).

Vicinal. The relationship between two substituents that are bonded to adjacent carbon atoms.

7.13 SUMMARY OF REACTIONS

The reactions of alkynes are summarized in Table 7.2, and various methods for the preparation of alkynes are summarized in Table 7.3.

TABLE 7.2 Reactions of Alkynes

Reactions	Comments
1. Salt formation	Section 7.4. Terminal alkynes only.
(a) Alkali metal salts	
$R{-}C{\equiv}C{-}H \xrightarrow{NaNH_2} R{-}C{\equiv}C^-Na^+$	Used for alkylation to form new carbon-carbon bonds.
$R{-}C{\equiv}C{-}H \xrightarrow{LiNH_2} R{-}C{\equiv}C^-Li^+$	

TABLE 7.2 Reactions of Alkynes (continued)

Reactions	Comments
(b) Copper and silver salts	
$R{-}C{\equiv}C{-}H \xrightarrow[\text{aqueous } NH_3]{CuCl} R{-}C{\equiv}C{-}Cu$ $R{-}C{\equiv}C{-}H \xrightarrow[\text{alcohol}]{AgNO_3} R{-}C{\equiv}C{-}Ag$	Used to identify terminal alkynes. Precipitate indicates a terminal alkyne; internal alkynes give no reaction.
2. Hydration	Section 7.5.
(a) Direct hydration	
$-C{\equiv}C- \xrightarrow[H_2SO_4,\ HgSO_4]{H_2O} -\overset{O}{\overset{\Vert}{C}}-CH_2-$	Terminal alkynes yield ketones; unsymmetrical internal alkynes yield mixtures of the two possible ketones.
(b) Via hydroboration	
$-C{\equiv}C- \xrightarrow[\text{2. } H_2O_2,\ NaOH]{\text{1. } (CH_3{-}\overset{CH_3}{\overset{\vert}{CH}}{-}\overset{CH_3}{\overset{\vert}{CH}}{-})_2BH} -\overset{O}{\overset{\Vert}{C}}-CH_2-$ $-C{\equiv}C- \xrightarrow[\text{2. } H_2O_2,\ NaOH]{\text{1. catecholborane (benzo-1,3,2-dioxaborole, BH)}} -\overset{O}{\overset{\Vert}{C}}-CH_2-$	Terminal alkynes yield aldehydes; internal alkynes yield a predominance of the ketone with oxygen on the less hindered carbon.
3. Reduction	Section 7.6.
(a) Catalytic hydrogenation	
$-C{\equiv}C- \xrightarrow[\text{catalyst}]{H_2} -CH_2-CH_2-$	Reduction proceeds to the alkane unless a deactivated catalyst is used.
$-C{\equiv}C- \xrightarrow[Pd{-}BaSO_4]{H_2}$ (H)C=C(H) (cis)	Reaction stops at the alkene with a deactivated catalyst. *Syn* addition.

TABLE 7.2 Reactions of Alkynes (continued)

Reactions	Comments
(b) Via hydroboration	
$-C{\equiv}C- \xrightarrow[\text{2. } CH_3CO_2H]{\text{1. } (CH_3{-}CH(CH_3){-}CH(CH_3){-})_2BH}$ H(H)C=C	A good synthetic method for laboratory use. *Syn* addition.
$-C{\equiv}C- \xrightarrow[\text{2. } CH_3CO_2H]{\text{1. catecholborane (C}_6H_4O_2BH)}$ H(H)C=C	
(c) Metal reduction	
$-C{\equiv}C- \xrightarrow[NH_3]{\text{Na or Li}}$ H–C=C–H (trans)	A good synthetic method for laboratory use. *Anti* addition.
4. Addition of hydrogen halides	Section 7.7.
(a) Direct addition	
$-C{\equiv}C- \xrightarrow{HX}$ H–C=C–X $\longrightarrow$ $-CHX-CHX-$ (H, X on each carbon: $-C(H)(H)-C(X)(X)-$)	Reaction proceeds via the more stable cation. (Addition of HBr proceeds by a radical mechanism if peroxides are present.) Difficult to stop at the alkenyl halide.
(b) Via hydroboration with catecholborane	
$R{-}C{\equiv}C{-}H \xrightarrow[\text{2. } H_2O \text{ 3. } I_2, NaOH]{\text{1. catecholborane}}$ (H)(R)C=C(I)(H)	Terminal alkynes only. Overall *syn* addition of HI.
$-C{\equiv}C- \xrightarrow[\text{2. } Br_2 \text{ 3. } NaOCH_3, CH_3OH]{\text{1. catecholborane}}$ H–C=C–Br	Terminal alkynes yield 1-bromo-1-alkenes. Internal alkynes yield predominantly the product with bromine on the less hindered carbon atom. Overall *anti* addition of HBr.

TABLE 7.3 Preparation of Alkynes

Reaction	Comments
Functional Group Modification: No Change in Carbon Skeleton	
1. Elimination of HX	Section 7.8. Many bases will work: $NaNH_2/NH_3$, KOH/alcohol, KOtBu/DMSO, etc.
(a) From 1,2-dihalides 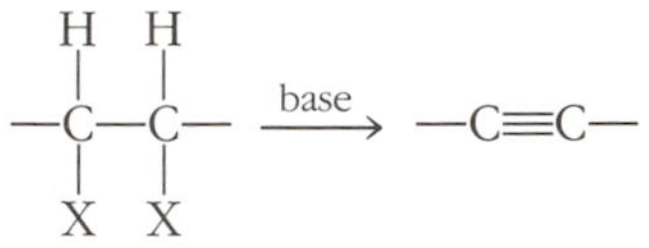 $$\text{—CH(X)—CH(X)—} \xrightarrow{\text{base}} \text{—C}\equiv\text{C—}$$	The halides are available from addition of X_2 to the alkene.
(b) From 1,1-dichlorides 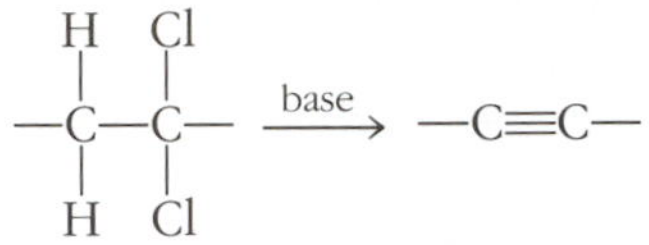 $$\text{—CH}_2\text{—CCl}_2\text{—} \xrightarrow{\text{base}} \text{—C}\equiv\text{C—}$$	The dichlorides are available from treatment of the aldehyde or ketone with PCl_5.
Reactions Yielding Alkynes via Carbon-Carbon Bond Formation	
2. Alkylation of sodium acetylides	Section 7.10.
$$\text{—C}\equiv\text{C—H} \xrightarrow{\text{NaNH}_2} \text{—C}\equiv\text{C}^-\text{Na}^+ \xrightarrow{\text{RX}} \text{—C}\equiv\text{C—R}$$	RX must be a primary alkyl halide.

7.14 PROBLEMS

7.1 Name the following compounds:

(a) 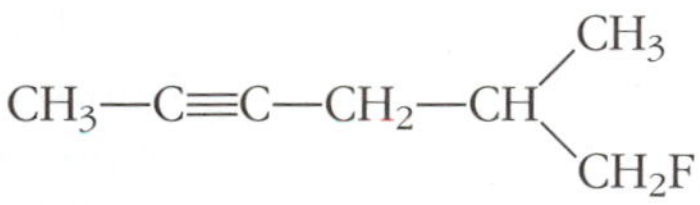 $CH_3\text{—C}\equiv\text{C—}CH_2\text{—}CH(CH_3)(CH_2F)$

(b) $CH_3\text{—C}(CH_3)(Br)\text{—C}\equiv\text{C—}CH_2CH_2Br$

(c) $(H)(CH_3)C\text{=C=}C(Cl)(H)$

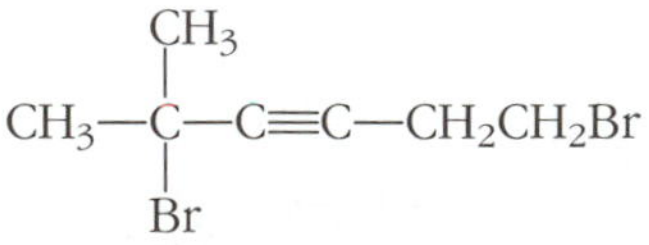

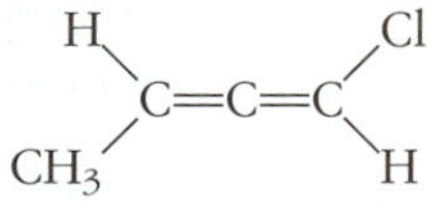

(d) $Br\text{—C}\equiv\text{C—}Br$

(e) $BrCH_2{-}C{\equiv}C{-}CH_2{-}CH(CH_2CH_3){-}CH_3$

7.2 Draw the following compounds:
(a) 2,7-Dimethyl-4-octyne
(b) 1,1-Dichloro-2,3-hexadiene
(c) 3-Ethyl-4-methyl-1-pentyne
(d) Cyclodecyne
(e) 1-Cyclohexyl-1-octyne

7.3 Write the products for the reaction of 2-pentyne with each of the following reagents:
(a) Disiamylborane followed by H_2O_2, NaOH
(b) H_2, Pt
(c) Li, NH_3
(d) Catecholborane followed by H_2O_2, NaOH
(e) H_2, Pd-$BaSO_4$
(f) H_2O, H_2SO_4, $HgSO_4$
(g) Excess HCl
(h) HBr (1 equivalent)
(i) Catecholborane, then Br_2, followed by $NaOCH_3$, CH_3OH
(j) Na, NH_3

7.4 Write the products for the reaction of 1-pentyne with each of reagents listed in the preceding question.

7.5 What reagents are necessary for converting each of the following compounds into 2-pentyne?
(a) 2,3-Dibromopentane
(b) 3,3-Dichloropentane
(c) Propyne
(d) 2-Bromo-(*Z*)-2-pentene

7.6 Draw the major product (or products) expected for each of the following reactions. Show stereochemistry whenever appropriate.

(a)  $\xrightarrow[H_2SO_4,\ HgSO_4]{H_2O}$?

(b) $CH_3{-}CH_2{-}CH(CH_3){-}CH_2{-}C{\equiv}CH \xrightarrow[NH_3,\ H_2O]{CuCl}$?

(c)  $\xrightarrow[2.\ CH_3I]{1.\ NaNH_2}$?

(d) $CH_3{-}CH_2{-}CH(CH_3){-}CH_2{-}CH(Cl){-}Cl \xrightarrow[NH_3]{NaNH_2}$?

(e) $CH_3{-}CH_2{-}C{\equiv}C{-}CH_2{-}CH_3 \xrightarrow[\text{2. } H_2O_2,\ NaOH]{\text{1. catecholborane (benzo-1,3,2-dioxaborole, BH)}} ?$

(f) $CH_3{-}CH_2{-}C{\equiv}C{-}CH_2{-}CH_3 \xrightarrow{Li,\ NH_3} ?$

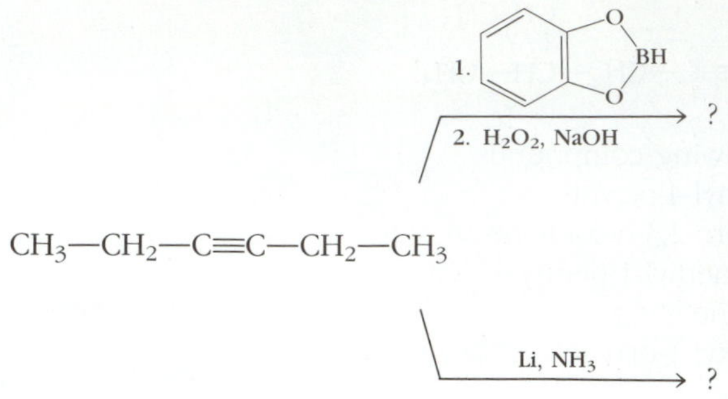

7.7 Draw the major product (or products) expected for each of the following reactions. Show stereochemistry whenever appropriate.

(a) $CH_3{-}CH_2{-}CH(CH_3){-}CH_2{-}C(=O){-}CH_3 \xrightarrow{PCl_5} ?$

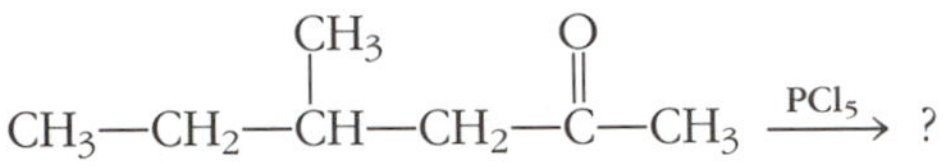

(b) $\text{cyclopentyl}{-}C{\equiv}C{-}CH_3 \xrightarrow[\text{(1 equiv)}]{Br_2,\ CCl_4} ?$

(c) $\text{cyclopentyl}{-}C{\equiv}C{-}CH_3 \xrightarrow{H_2,\ Pd{-}BaSO_4} ?$

(d) $\text{cyclopentyl}{-}C{\equiv}C{-}CH_3 \xrightarrow[\text{2. } Br_2 \text{; 3. } NaOCH_3,\ CH_3OH]{\text{1. } (CH_3CH(CH_3){-}CH(CH_3){-})_2BH} ?$

(e) $CH_3{-}CH(CH_3){-}CH_2{-}C{\equiv}C{-}CH_2{-}CH(CH_3){-}CH_3 \xrightarrow{HBr} ?$

(f) $CH_3{-}CH(CH_3){-}CH_2{-}C{\equiv}C{-}CH_2{-}CH(CH_3){-}CH_3 \xrightarrow[\text{2. } CH_3CO_2H]{\text{1. } (CH_3{-}CH(CH_3){-}CH(CH_3){-})_2BH} ?$

(g) $\text{cyclohexyl}{-}CH_2{-}C{\equiv}CH \xrightarrow[\text{2. } I_2,\ NaOH]{\text{1. catecholborane (benzo-1,3,2-dioxaborole, BH)}} ?$

(h) $\text{cyclohexyl}{-}CH_2{-}C{\equiv}CH \xrightarrow[\text{2. } H_2O_2,\ NaOH]{\text{1. } (CH_3{-}CH(CH_3){-}CH(CH_3){-})_2BH} ?$

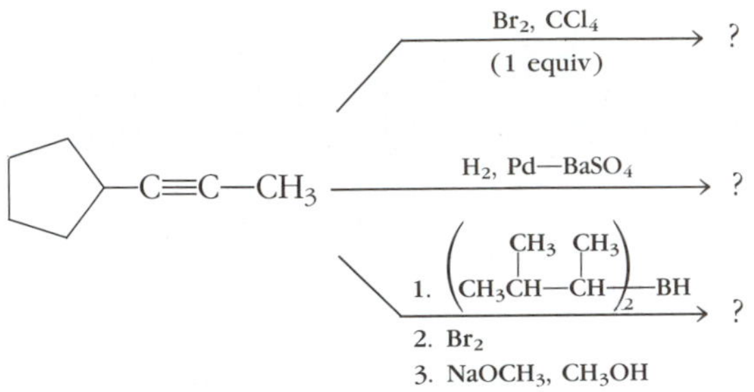

(i) $\text{cyclohexyl}{-}CH_2{-}C{\equiv}CH \xrightarrow{H_2,\ Pt} ?$

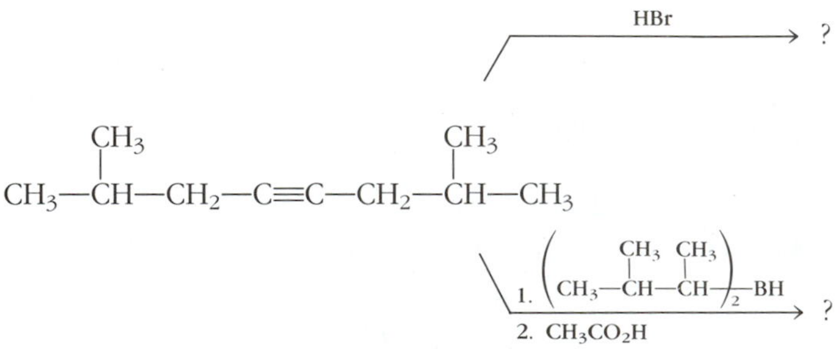

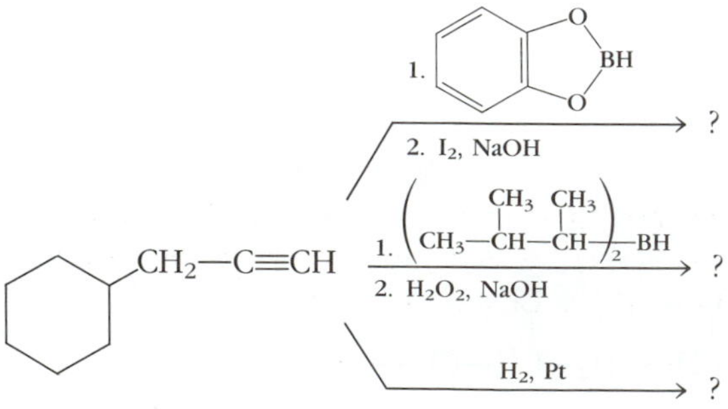

(j) 1,1-dichloroethylcyclohexane: cyclohexyl–C(Cl)(Cl)–CH_3 $\xrightarrow[\text{DMSO}]{\text{KOtBu}}$?

(k) $CH_3—CH_2—CH(CH_3)—C\equiv CH$ $\xrightarrow[\text{alcohol}]{AgNO_3}$?

(l) $CH_3—CH_2—CH(CH_3)—C\equiv CH$ $\xrightarrow[(NH_4)_2SO_4]{Na,\ NH_3}$?

(m) $CH_3—CH_2—CH(Br)—CH(Br)—CH(CH_3)—CH_3$ $\xrightarrow[NH_3]{NaNH_2}$?

7.8 Suggest a method of preparation of each of the following compounds starting from an aldehyde, a ketone, or an alkene with the same number of carbon atoms. If more than one method would work, write all such methods.

(a) $CH_3CH_2CH_2C\equiv CH$ (b) $CH_3CH_2C\equiv CCH_2CH_2CH_3$

(c) 1-ethynyl-1-methylcyclohexane: cyclohexane ring bearing CH_3 and $—C\equiv CH$ on the same carbon

7.9 Suggest methods of preparation for each of the following compounds. In each case use a terminal alkyne and an alkyl halide. If more than one combination would work, write all such combinations.

(a) $CH_3CH_2CH_2C\equiv CCH_2CH_2CH_3$

(b) $CH_3CH_2CH_2—C\equiv C—CH(CH_3)—CH_3$

(c) $CH_3CH_2—C\equiv CH$

(d) $CH_3CH_2CH_2—C\equiv C—CH_2CH_3$

7.10 What reactions of alkynes could be used to prepare the following compounds from starting materials with the same number of carbon atoms? If more than one method would work, write all such methods.

(a) $(Br)(CH_3CH_2CH_2)C=CH_2$

(b) $\underset{CH_3CH_2CH_2}{\overset{H}{}}\!\!>C{=}C<\!\!\underset{Br}{\overset{H}{}}$

(c) $\underset{CH_3CH_2CH_2}{\overset{H}{}}\!\!>C{=}C<\!\!\underset{Cl}{\overset{CH_2CH_2CH_3}{}}$

(d) $\underset{\text{cyclohexyl}}{\overset{H}{}}\!\!>C{=}C<\!\!\underset{Br}{\overset{CH_2CH_3}{}}$

(e) $\underset{\text{cyclopentyl}-CH_2}{\overset{H}{}}\!\!>C{=}C<\!\!\underset{H}{\overset{I}{}}$

(f) $\underset{H}{\overset{CH_3CH_2}{}}\!\!>C{=}C<\!\!\underset{CH_2CH_3}{\overset{Br}{}}$

(g) $\underset{Br}{\overset{CH_3CH_2}{}}\!\!>C{=}C<\!\!\underset{CH_2CH_3}{\overset{Br}{}}$

7.11 Structures **A–D** are all different compounds. On the basis of the reactions shown below, deduce the structures of compounds **A–C.** Show all stereochemistry.

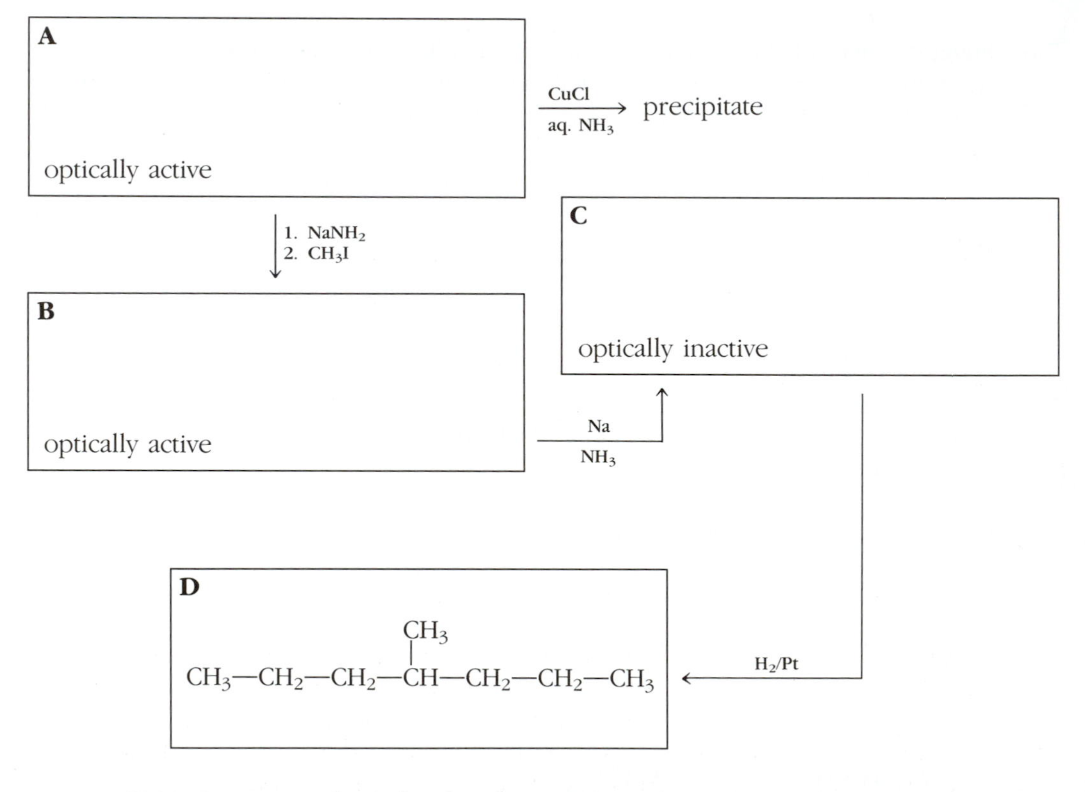

7.12 Suggest methods for carrying out the indicated conversions. More than one step may be necessary in each case.

(a) $CH_3{-}CH_2{-}C{\equiv}CH \longrightarrow CH_3{-}CH_2{-}CH_2{-}\overset{\displaystyle O}{\overset{\|}{C}}H$

(b) $CH_3{-}CH{=}CH_2 \longrightarrow CH_3{-}C{\equiv}CH$

(c) 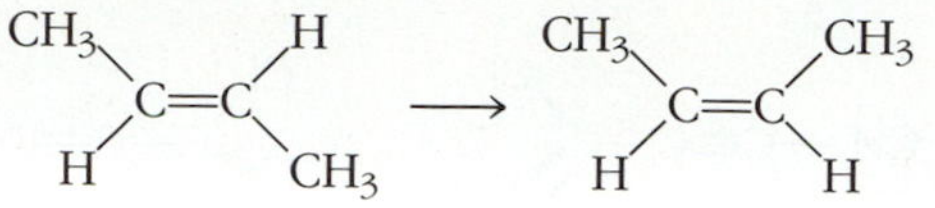 $CH_3(H)C{=}C(H)CH_3 \longrightarrow CH_3(H)C{=}C(H)CH_3$ (CH₃ and CH₃ cis)

(d) 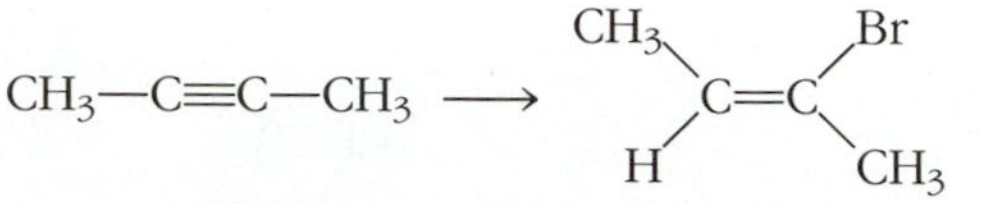 $CH_3{-}C{\equiv}C{-}CH_3 \longrightarrow$ $CH_3(H)C{=}C(Br)CH_3$

(e) $CH_3{-}CH_2{-}C{\equiv}CH \longrightarrow CH_3{-}CH_2{-}\overset{O}{\overset{\|}{C}}{-}CH_3$

(f) 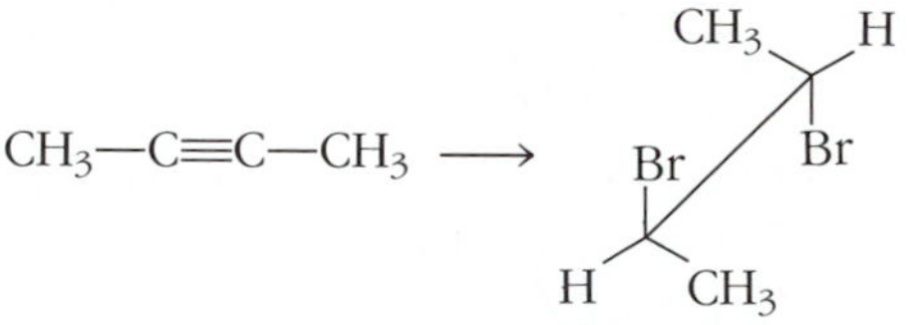 $CH_3{-}C{\equiv}C{-}CH_3 \longrightarrow$ [structure: CH₃, H, Br on upper carbon; Br, H, CH₃ on lower carbon]

(g) 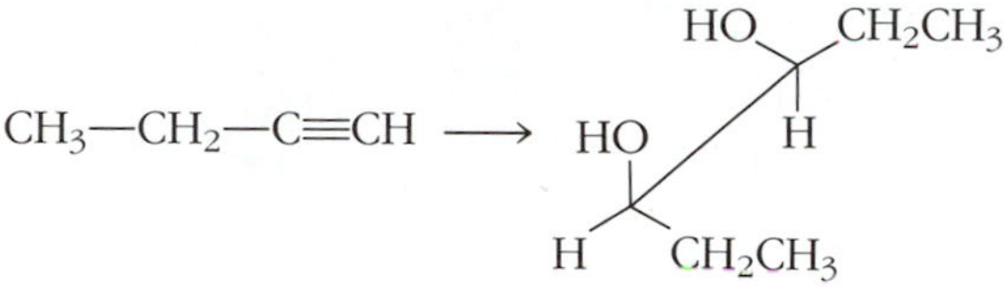 $CH_3{-}CH_2{-}C{\equiv}CH \longrightarrow$ [structure: HO, CH₂CH₃, H on upper carbon; HO, H, CH₂CH₃ on lower carbon]

(h) $CH_3CH_2{-}C{\equiv}C{-}CH_2CH_3 \longrightarrow CH_3CH_2CH_2{-}\overset{O}{\overset{\|}{C}}{-}CH_2CH_3$

(i) 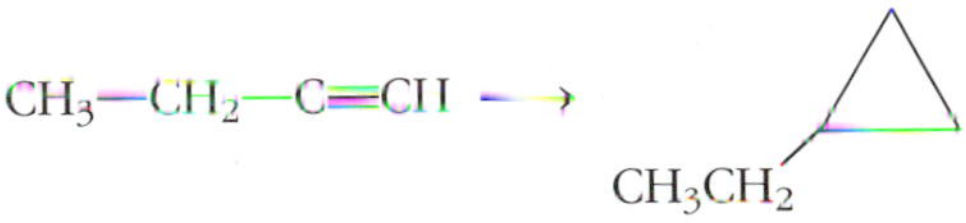 $CH_3{-}CH_2{-}C{\equiv}CH \longrightarrow$ [cyclopropane with CH_3CH_2 substituent]

(j) 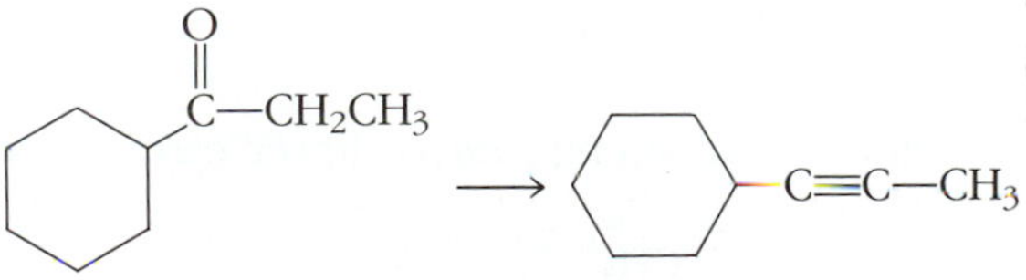 [cyclohexyl]$-\overset{O}{\overset{\|}{C}}{-}CH_2CH_3 \longrightarrow$ [cyclohexyl]$-C{\equiv}C{-}CH_3$

7.13 The reaction of 2-butyne with excess HBr in the presence of peroxides yields (±)-2,3-dibromobutane. Under the same conditions (*E*)-2-bromo-2-butene yields *meso*-2,3-dibromobutane whereas (*Z*)-2-bromo-2-butene yields (±)-2,3-dibromobutane. Show how these experiments prove that the free radical addition of HBr to 2-butyne proceeds by the *anti* addition mode.

7.14 Describe the relationship between each pair of structural formulas as that of *constitutional* isomers, *enantiomers,* two drawings of the *same* compound (although perhaps in different conformations), *diastereomers,* or completely *different* compounds that are not isomers.

(a)

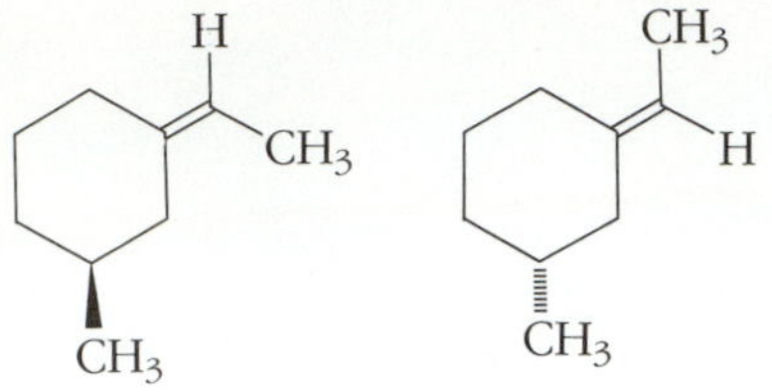

(b) $CH_3-C\equiv C-CH(Cl)CH_3$ $CH_3CH(Cl)-C\equiv C-CH_3$

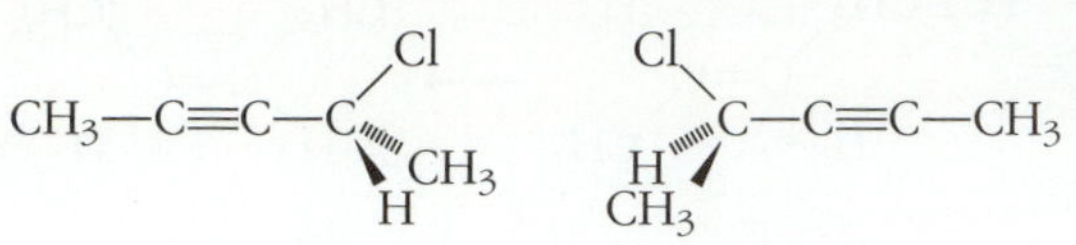

(c) $HClC=C=CHCl$ $ClHC=C=CHCl$

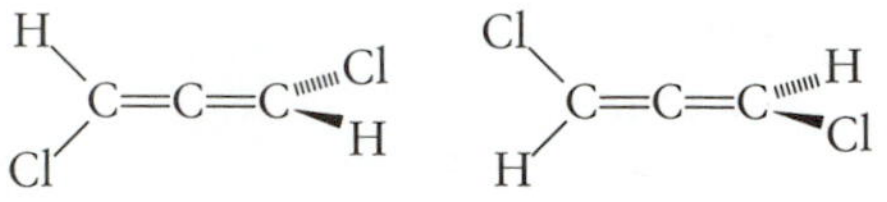

(d)

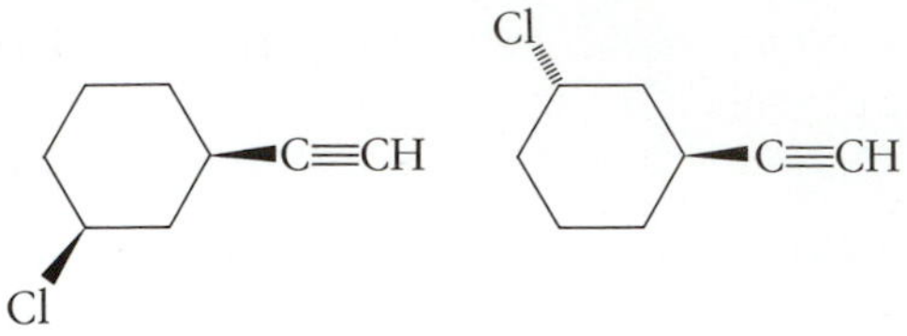

(e)

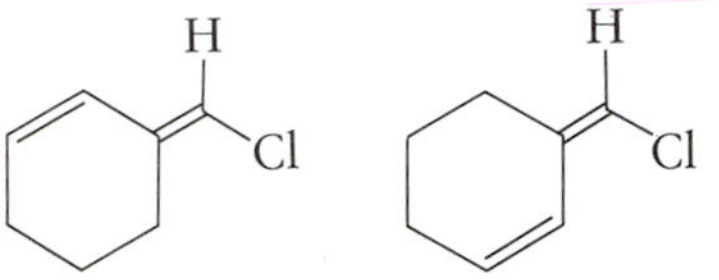

(f) $CH_3-C\equiv C-CH=CH-CH_3$ $CH_3-C\equiv C-CH=CH-CH_3$

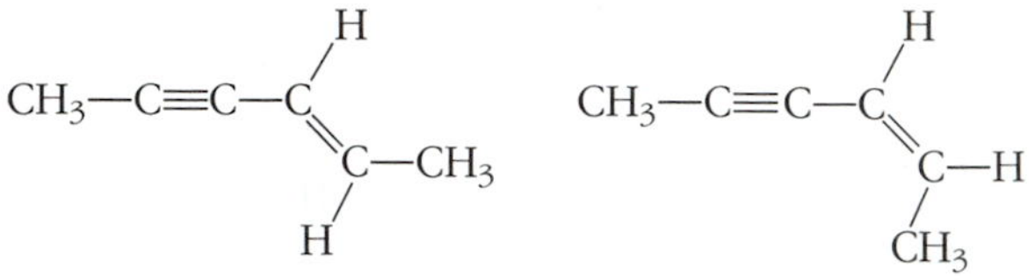

(g) $Cl_2C=C=CHCH_3$ $HCH_3C=C=CCl_2$

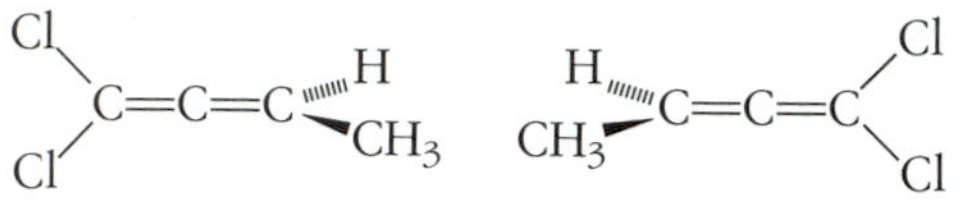

(h)

7.15 Suggest a method of synthesis for each of the following compounds. You may use any other reagents you wish, but all the carbon atoms of the product must be derived from reactants with the indicated number of carbon atoms.

(a) From starting materials containing *five* or fewer carbon atoms:

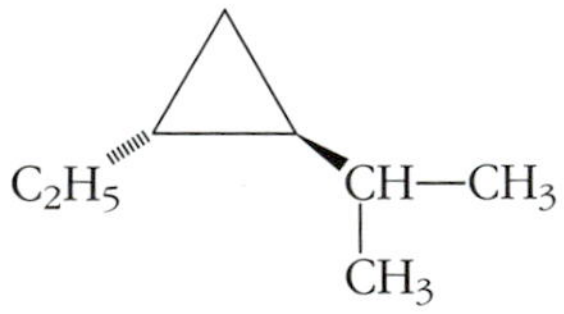

(racemic)

(b) From starting materials containing *four* or fewer carbon atoms:

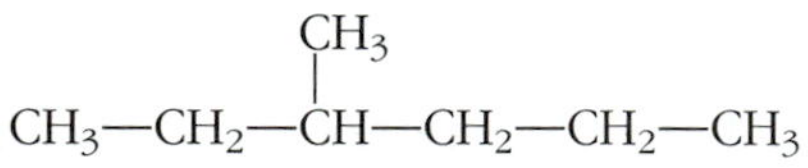

$$CH_3-CH_2-CH(CH_3)-CH_2-CH_2-CH_3$$

(c) From starting materials containing *three* or fewer carbon atoms:

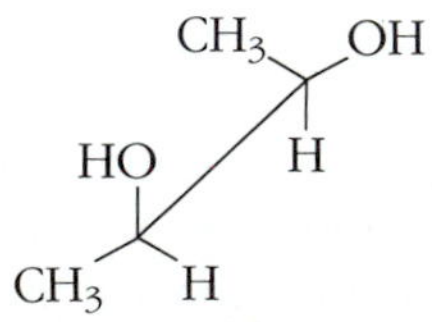

(d) From starting materials containing *five* or fewer carbon atoms:

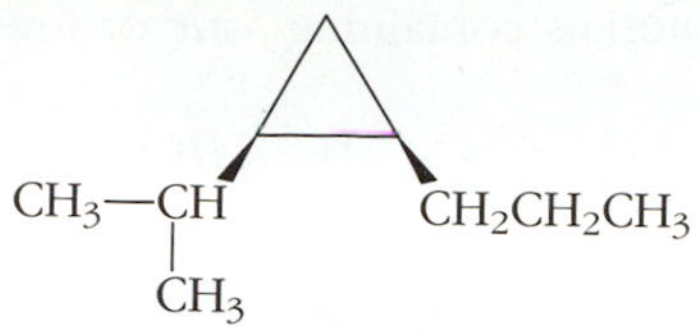

$$\text{cyclopropane with } CH_3-\underset{\underset{CH_3}{|}}{CH}- \text{ and } -CH_2CH_2CH_3 \text{ (wedged, cis)}$$

(racemic)

7.16 Suggest a method of synthesis for each of the following compounds. You may use any other reagents you wish, but all the carbon atoms of the product must be derived from reactants with the indicated number of carbon atoms.

(a) From starting materials containing *four* or fewer carbon atoms:

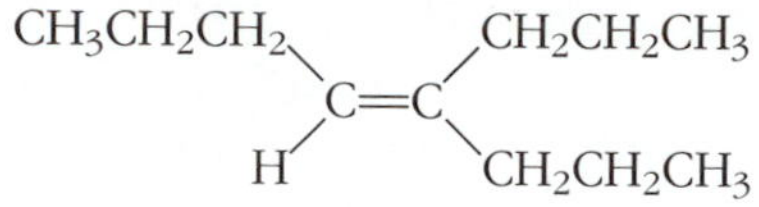

$$(CH_3CH_2CH_2)(H)C{=}C(CH_2CH_2CH_3)_2$$

(b) From starting materials containing *five* or fewer carbon atoms:

$$CH_3-CH_2-CH_2-CH_2-\underset{\underset{Cl}{|}}{\overset{\overset{Cl}{|}}{C}}-CH_2-CH_2-CH_3$$

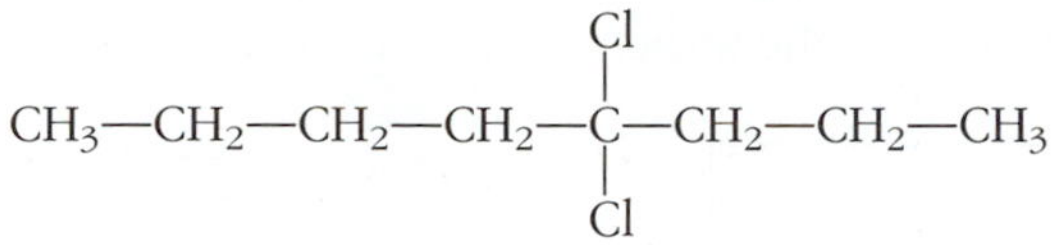

(c) From starting materials containing *four* or fewer carbon atoms:

$$CH_3-\underset{\underset{CH_3}{|}}{\overset{\overset{CH_3}{|}}{C}}-CH_2-\overset{\overset{O}{\|}}{C}-CH_3$$

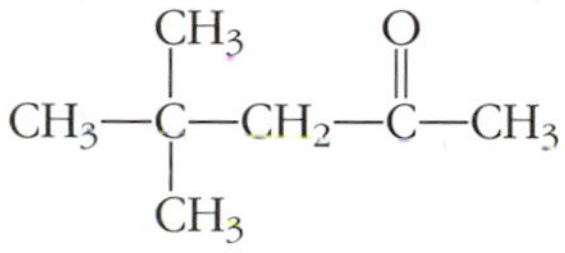

(d) From starting materials containing *three* or fewer carbon atoms:

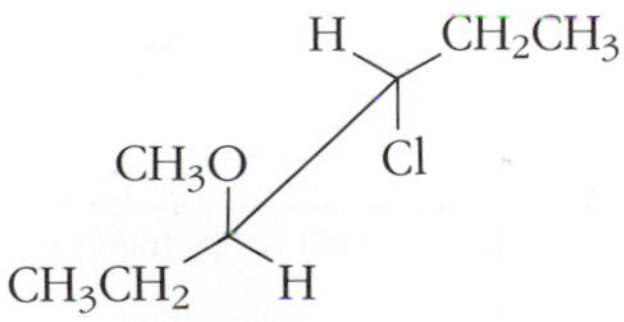

$$\text{(sawhorse) } CH_3O,\ CH_3CH_2,\ H \text{ on front carbon; } H,\ CH_2CH_3,\ Cl \text{ on back carbon}$$

7.17 Suggest a method of synthesis for each of the following compounds. You may use any other reagents you wish, but all the carbon atoms of the product must be derived from reactants with the indicated number of carbon atoms.

(a) From starting materials containing *seven* or fewer carbon atoms:

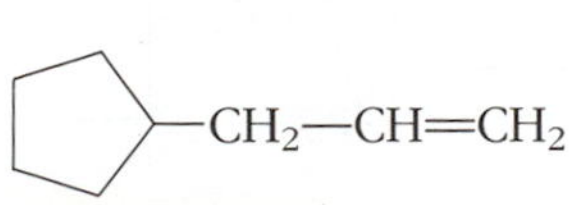

$$\text{cyclopentyl}-CH_2-CH{=}CH_2$$

(b) From starting materials containing *five* or fewer carbon atoms:

$$CH_3-\overset{\overset{CH_3}{|}}{CH}-CH_2CH_2-\overset{\overset{CH_3}{|}}{CH}-CH_3$$

(c) From starting materials containing *four* or fewer carbon atoms:

H, Br, Br, CH_3, H, $CH_2CH_2CH_3$

(racemic)

(d) From starting materials containing *six* or fewer carbon atoms:

H, H, C=C, CH_2

7.18 Structures **A–F** are all different compounds. On the basis of the information provided, deduce the structures of compounds **A–D.**

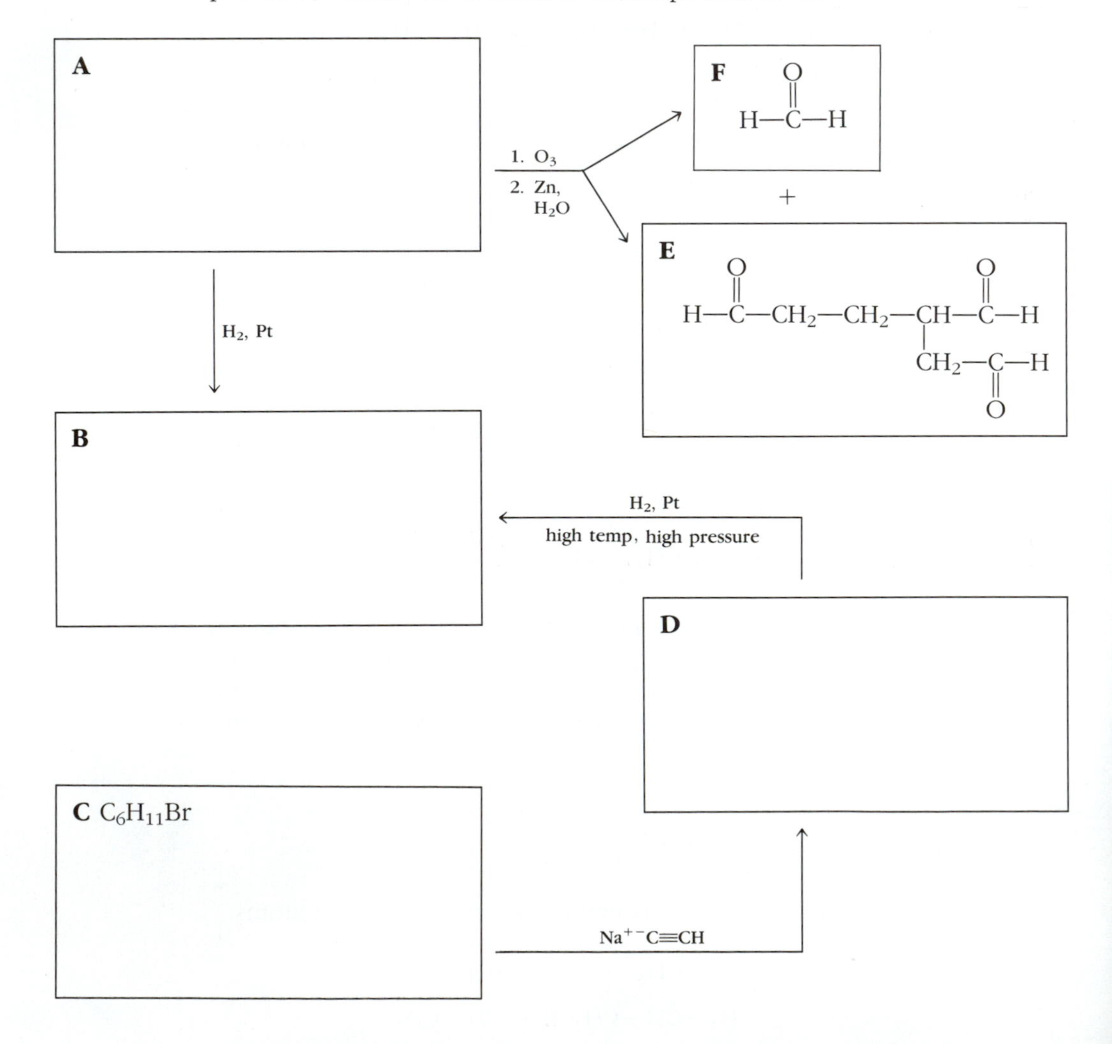

7.19 Compounds **A–F** are all different. On the basis of the information provided, deduce their structures.

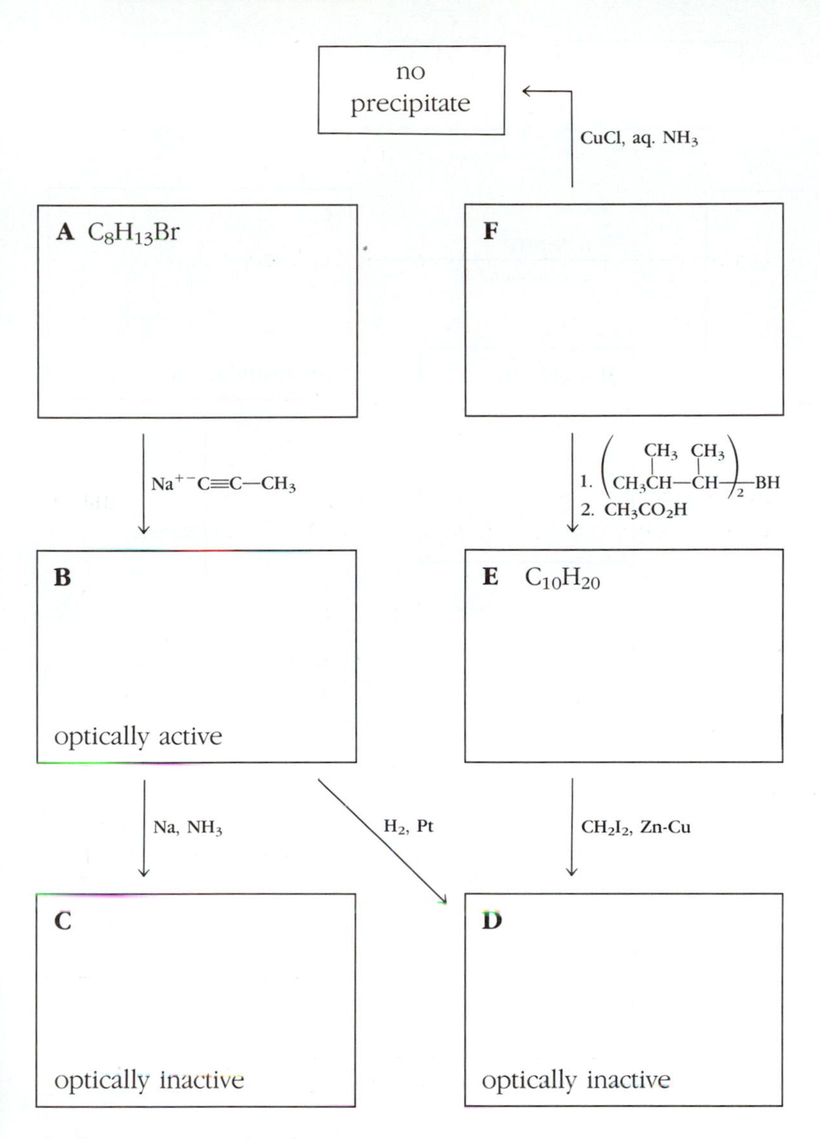

7.20 Structures **A–E** are all different compounds. On the basis of the information provided, deduce the structures of compounds **A–D.**

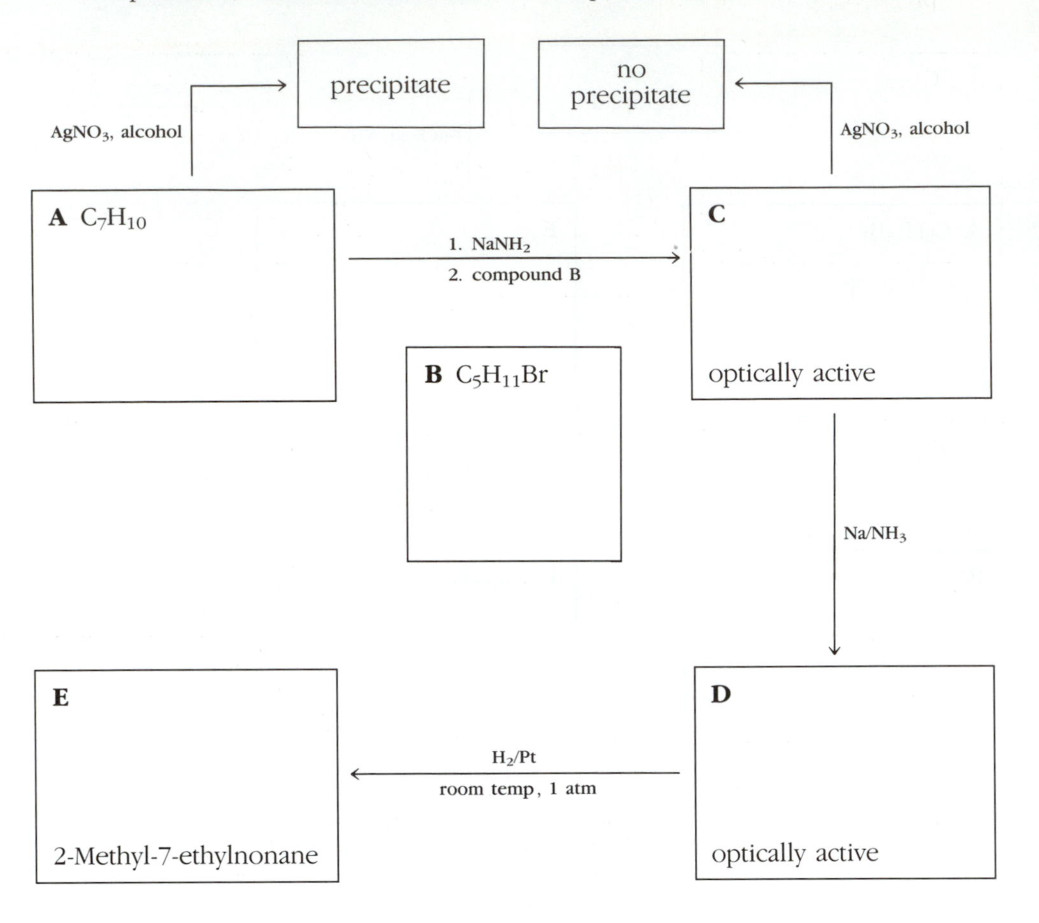

7.21 Structures **A–K** are all different compounds. On the basis of the information provided, deduce the structures of compounds **A–J**.

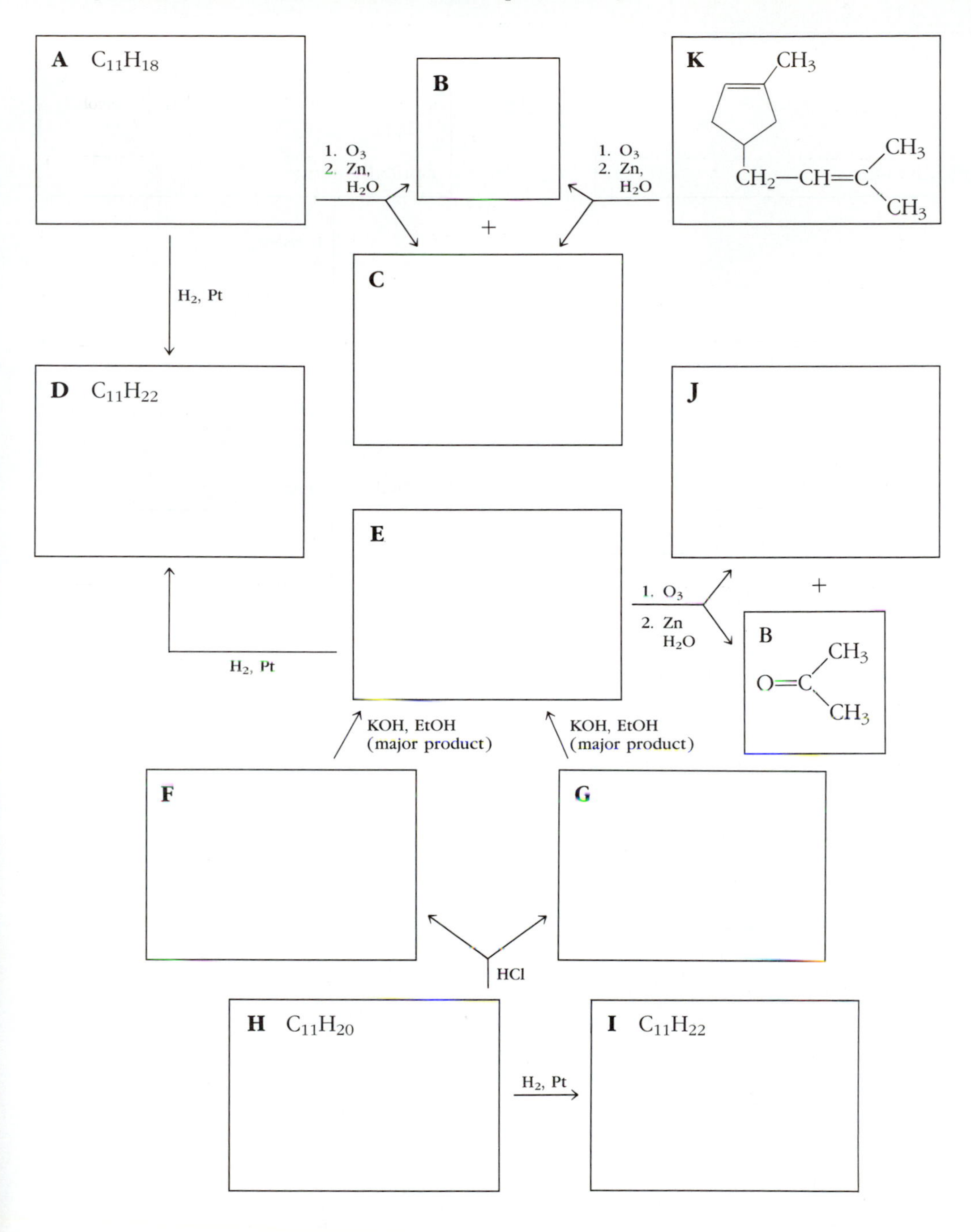

7.22 Structures **A–G** are all different compounds. On the basis of the information provided, deduce the structures of compounds **A–F.**

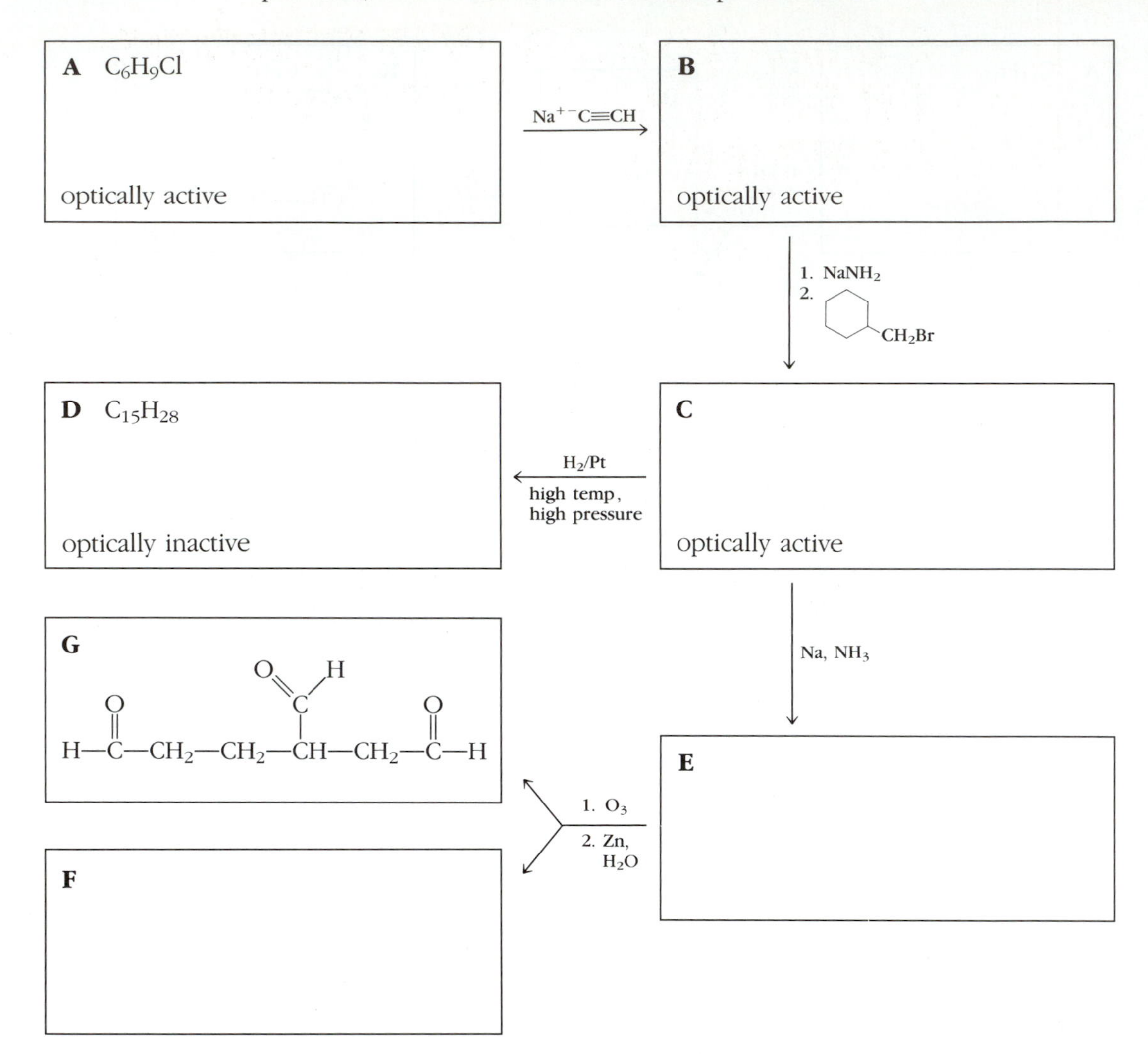

CHAPTER 8

COMPOUNDS WITH MORE THAN ONE REACTIVE CENTER

In Section 5.1 we introduced the term *functional group* to describe the carbon-carbon double bond because its consistent pattern of chemical reactivity allowed us to describe alkenes as a *functional class* of organic compounds. We also described the carbon-carbon double bond as the *reactive center* in an alkene, because that is the site of all the chemical reactions that we presented for alkenes. Nearly all the compounds that we have discussed so far in this book have contained only a single reactive center. Yet a very large number of important organic compounds, particularly those of biological importance, contain more than one reactive center. Consequently, it is very important for you to understand the structure and reactivity of such compounds.

8.1 INTERACTIONS BETWEEN REACTIVE CENTERS

Frequently, two reactive centers in a molecule will be situated in two distinct locations as parts of two different functional groups; such compounds are called **polyfunctional.** In other cases two reactive centers can simply be two atoms that are part of the same functional group. We will classify the chemical reactions of compounds with more than one reactive center into three general categories, depending on the interaction between the reactive centers.

1. *Two reactive centers can interact in such a way that their reactivities are substantially altered.* This usually results when the centers are in close proximity. In an extreme case the interaction can be so strong that we would describe the compound as a member of a separate functional class. Recall from Section 4.7 the effect of a carbonyl group on the acidity of an OH group.

$$\underset{\text{Alcohol, } K_a \simeq 10^{-18}}{R{-}CH_2{-}OH} \qquad \underset{\text{Carboxylic acid, } K_a \simeq 10^{-5}}{R{-}\overset{\overset{\displaystyle O}{\|}}{C}{-}OH}$$

The presence of a carbon-oxygen double bond on the same carbon to which the OH group is attached enhances the acidity by 10^{13}, and the resulting compound belongs to a new functional class, the *carboxylic acids.*

A strong interaction between two reactive centers does not always correspond to a new functional group, however. In this chapter we will show you how a carbon-carbon double bond can activate an adjacent carbon atom. Formation of a cation, radical, or anion on that adjacent carbon affords an intermediate with properties that are very different than those that would be observed if the two reactive centers did not interact. For example

$$\text{>C=C<} \text{ and } \text{—}\dot{C}\text{<} \qquad \text{vs.} \qquad \text{>C=C—}\dot{C}\text{<}$$

2. *Two noninteracting reactive centers can differ sufficiently in reactivity that reaction will occur selectively at just one of them.* This is depicted schematically by the following equation, in which the reactive center with substit-

uent X is unaffected by the reaction, but Y is converted to some other group, Z.

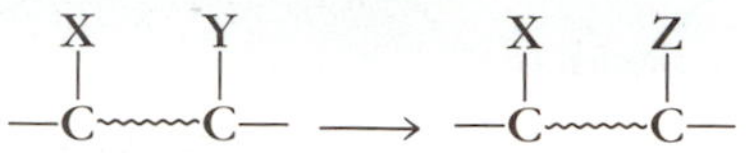

$$-\underset{|}{\overset{X}{C}}\sim\sim\underset{|}{\overset{Y}{C}}- \longrightarrow -\underset{|}{\overset{X}{C}}\sim\sim\underset{|}{\overset{Z}{C}}-$$

This selectivity should remind you of the differences in rates of free radical bromination at primary and tertiary centers (Section 2.10).

3. *Two reactive centers can interact in such a way that a bond-forming reaction takes place to yield a cyclic product.* This is shown symbolically by the following reaction, in which a bond is formed between two atoms that originally were attached to substituents X and Y.

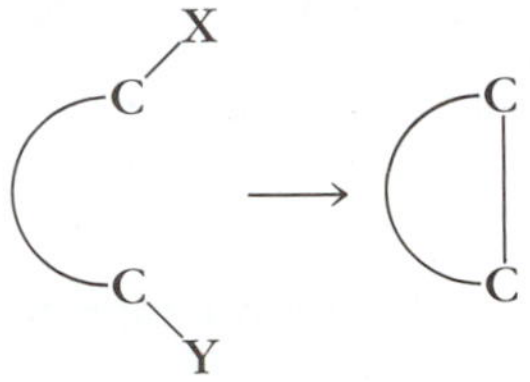

In a competing pathway the analogous *intermolecular* reaction can occur.

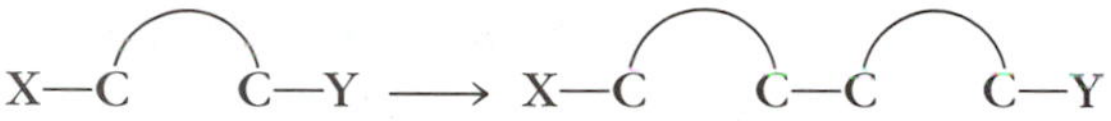

$$X{-}C\frown C{-}Y \longrightarrow X{-}C\frown C{-}C\frown C{-}Y$$

Repetition of this intermolecular bond-forming process can lead to the formation of a **polymer,** that is, a large molecule consisting of many repeating subunits.

The three categories that we have listed here are somewhat arbitrary, but they can be extremely useful when you are evaluating the potential reactivity of a compound with more than one reactive center. For such a compound you must be able to decide whether the two centers will interact with each other, and if not, which would undergo reaction preferentially. In this chapter we will introduce you to some of the reactions that fall into each of these categories.

8.2 ALLYLIC SUBSTITUTION

In this section we will show you how a carbon-carbon double bond can interact strongly with an adjacent free radical center. In most circumstances we would consider alkenes as monofunctional compounds with only a single reactive center, but a different view would be appropriate for free radical halogenation because the carbon-carbon double bond greatly affects the chemical reactivity at adjacent carbon atoms. Before proceeding with our discussion of this reactivity, we will introduce some new terms.

A carbon atom that is attached directly to a carbon-carbon double bond is described as *allylic,* and the same term is applied to substituents on these carbon atoms.

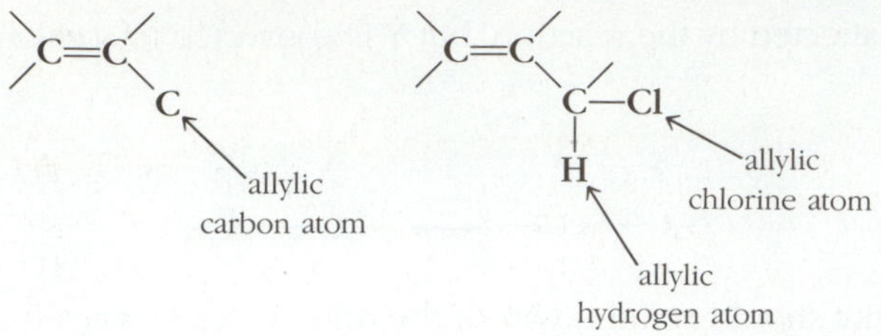

allylic carbon atom; allylic chlorine atom; allylic hydrogen atom

A carbon atom that is part of the double bond and substituents attached directly to such a carbon atom are described as *vinyl* or *vinylic*.

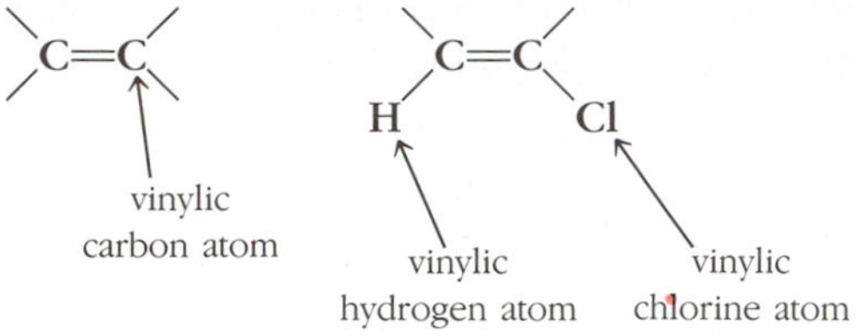

vinylic carbon atom; vinylic hydrogen atom; vinylic chlorine atom

The terms *vinyl* and *allyl* are the common names for the ethenyl and 2-propenyl groups, respectively.

$$CH_2{=}CH{-} \qquad \overset{3}{C}H_2{=}\overset{2}{C}H{-}\overset{1}{C}H_2{-}$$

Vinyl group (ethenyl) — Allyl group (2-propenyl)

A double bond activates the allylic position toward certain substitution reactions, and these reactions are called *allylic substitutions*.

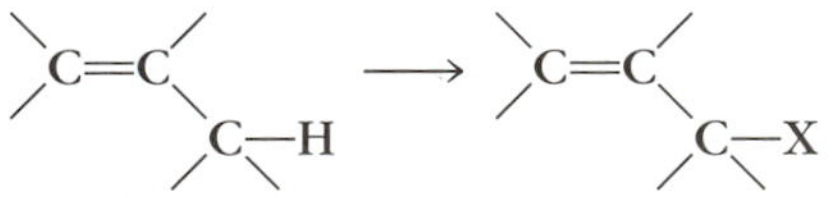

$$>C{=}C<\!\!\!\!\!-C{-}H \longrightarrow >C{=}C<\!\!\!\!\!-C{-}X$$

Allylic substitution

A particularly important reaction of this type is allylic bromination with *N*-bromosuccinimide (NBS). It is a free radical reaction that is usually initiated by addition of an organic peroxide, and it is carried out in a nonpolar solvent such as CCl_4. Allylic bromination is therefore quite different from the ionic addition (of Br and OH) that results from treatment of an alkene with aqueous NBS (Section 5.7). The reaction of 1-octene illustrates this reaction.

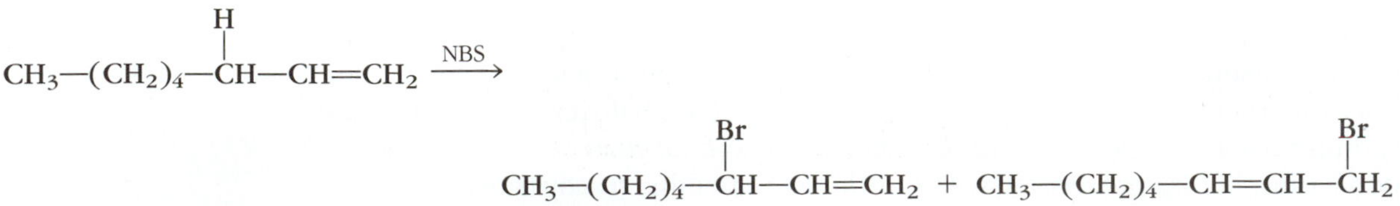

$$CH_3{-}(CH_2)_4{-}\underset{}{\overset{H}{C}}H{-}CH{=}CH_2 \xrightarrow{NBS}$$

$$CH_3{-}(CH_2)_4{-}\overset{Br}{C}H{-}CH{=}CH_2 + CH_3{-}(CH_2)_4{-}CH{=}CH{-}\overset{Br}{C}H_2$$

3-Bromo-1-octene 17% — 1-Bromo-2-octene 73%

Two features of this reaction are noteworthy. First, the reaction is highly selective, and only an allylic hydrogen atom is abstracted. This shows that the reaction intermediate, *an allylic free radical, is more stable than any of the other radicals* that could have been formed in the reaction of 1-octene. Second, two constitutionally isomeric allylic bromides are formed despite the fact that a hydrogen atom is abstracted from only a single position. This demonstrates that the reaction intermediate, *an allylic free radical, can undergo reaction at either end of the three-carbon allylic unit.* In our previous discussion of the free radical halogenation of alkanes (Section 2.6), the halogen atom was invariably attached to the same carbon atom from which the hydrogen atom was abstracted. Yet the major product in the reaction of 1-octene with NBS has the double bond between C-2 and C-3 with the bromine bonded to C-1, despite the fact that the hydrogen abstraction took place at C-3. Obviously, allylic free radicals are unusual, and their properties are explained in the following section.

8.3 ALLYLIC INTERMEDIATES

As we showed you in the previous section, the free radical halogenation of an alkene with NBS proceeds by the selective abstraction of allylic hydrogen atoms. This selectivity demonstrates that an allylic radical is more stable than the radicals which could be formed by abstraction of other hydrogen atoms in the molecule. Allylic radicals in general are more stable than their saturated counterparts, and the same is true for the relative stabilities of the corresponding cations and anions. In each case the allylic intermediate is more stable.

$$>C{=}C<\!\!-C^{+}< \quad \text{is more stable than} \quad >CH{-}CH<\!\!-C^{+}<$$

$$>C{=}C<\!\!-C^{\cdot}< \quad \text{is more stable than} \quad >CH{-}CH<\!\!-C^{\cdot}<$$

$$>C{=}C<\!\!-C{:}^{-}< \quad \text{is more stable than} \quad >CH{-}CH<\!\!-C{:}^{-}<$$

A second important property of allylic intermediates is their ability to undergo reaction at either end of the three-carbon unit. In the case of allylic bromination, for example, two isomeric bromides can be formed.

$$>C{=}C<\!\!-\underset{|}{C}{-}H \longrightarrow >C{=}C<\!\!-\underset{|}{C}{-}Br \;+\; -\overset{Br}{\underset{|}{C}}{-}C{=}\underset{|}{C}{-}$$

Similar reactivity is observed for allylic anions and cations.

How can you account for these unusual features of allylic intermediates? We will present two alternative explanations, the first in terms of *molecular orbital theory* and the second in terms of *resonance theory*. Molecular orbital theory and resonance theory both lead to the same conclusions about the stability and reactivity of allylic intermediates, but the two theories have quite different origins. Each has its advantages and disadvantages in explaining chemical concepts, and we will use both theories throughout this text. In this section we will use both theories to analyze allylic intermediates.

Allylic Intermediates: A Molecular Orbital Description

For any of the three allylic intermediates (cation, radical, or anion) the sigma bonding framework remains essentially unchanged. The key differences are in the π bonding interactions between the three carbon atoms of the allylic unit. In Figure 8.1 we have illustrated the interaction of three atomic p orbitals to form three molecular orbitals. There is one bonding π orbital (π_1) and one antibonding π

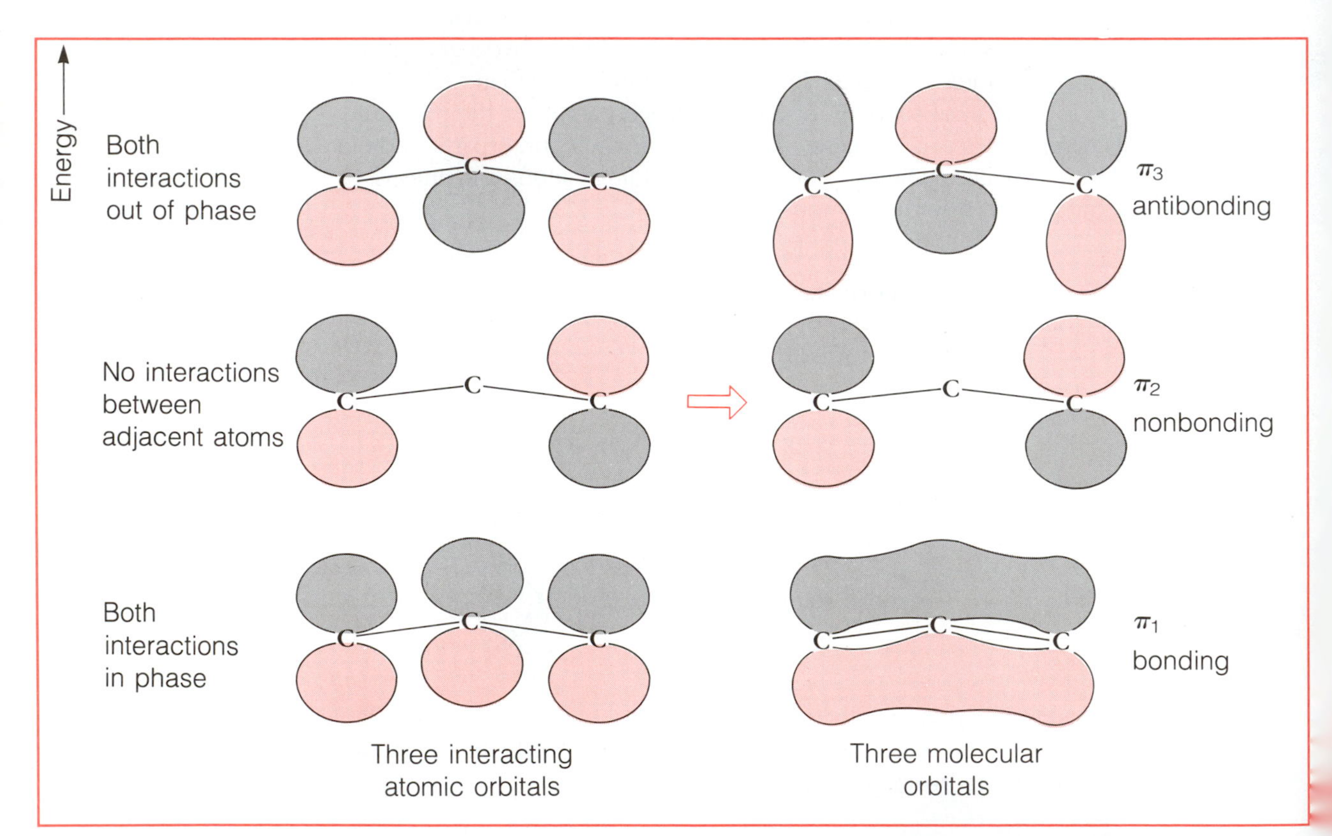

Figure 8.1 π Bonding in the Allyl System: A Molecular Orbital Description.
Three interacting p orbitals yield three molecular orbitals. In-phase interaction of all three atomic p orbitals leads to the *bonding* orbital, π_1, and out-of-phase interaction of the three orbitals leads to the *antibonding* orbital, π_3. The remaining molecular orbital (π_2) has no contribution from the p orbital on the central carbon atom and is essentially *nonbonding*.

orbital (π_3). The third π orbital (π_2) is essentially *nonbonding*. This molecular orbital has no π interaction between adjacent carbon atoms because it results exclusively from contributions of the p orbitals on the terminal carbons. Therefore it is almost the same as two isolated (nonbonding) p orbitals.

The properties of these π molecular orbitals lead to three important consequences:

1. *The π_1 orbital is substantially lower in energy than the bonding π orbital of an isolated double bond,* which involves the interaction of only two p orbitals (Figure 5.4). The energy increases for π_2, and the highest-energy orbital is π_3. There are no interactions between p orbitals on adjacent carbon atoms for π_2, so the energy for this orbital is very close to that for an isolated (or nonbonding) p orbital.* On the other hand, the in-phase interactions of a total of three p orbitals on three adjacent carbon atoms result in a very low energy for π_1. Conversely, the out-of-phase interactions of three p orbitals result in a correspondingly high energy for π_3.

2. *The greater stability of allyl intermediates can be directly attributed to the lower energy of the doubly occupied bonding molecular orbital,* π_1. The energy of any chemical species depends on the energies of *all occupied molecular orbitals,* but comparisons between two structures can be made by considering only those orbitals for which we expect substantial differences. For our discussion these are the π orbitals of the allylic intermediates and the p orbital at the reactive center of an ordinary cation, radical, or anion. First consider the occupancy of the π orbitals of an allylic intermediate. By referring to conventional structural formulas, you will see that the allyl cation has only two electrons in the π system, the radical has three, and the anion has four.

*The absence of any contribution from the p orbital on C-2 in this molecular orbital results from a requirement that molecular orbitals must have the same symmetry as the molecular species. (See Section 27.2 for a more complete explanation.) Any contribution from the p orbital on C-2 would generate a molecular orbital such as the following:

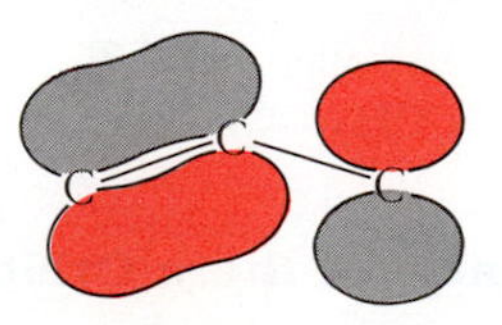

This lacks the plane of symmetry (a vertical plane bisecting C-2) that characterizes the allyl group.

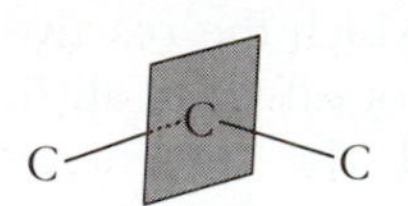

As shown in Figure 8.2, the changes in orbital occupancy among allylic cations, radicals, and anions are in the *nonbonding orbital,* and this is also true for nonallylic intermediates. Because it is nonbonding, π_2 is comparable in energy with the nonbonding p orbital of a nonallylic intermediate, and this orbital does not provide any basis for comparison of energies. Rigorous comparison of stability would require the use of isomeric structures. In Figure 8.3 we have presented a simplified comparison of an allylic radical with an isomeric unsaturated radical in which the double bond is separated from the radical center by several CH_2 groups.

Inspection of Figure 8.3 will show you that the partially occupied orbital (π_2 versus p) is of comparable energy for each of the intermediates. The important difference in orbital energies, and therefore in the overall energies of the radicals, is in the bonding π orbital (π_1 for the allyl radical, π for the other).

3. *The molecular orbital description allows a full interpretation of the chemical reactivity of allylic intermediates in terms of the nonbonding orbital,* π_2. Although the *stability* of the allyl system can be evaluated in terms of π_1, the *reactivity* of allylic intermediates is described in terms of π_2.

The reactivity of anions and radicals can be interpreted almost entirely in terms of the *highest occupied molecular orbital* (**HOMO**). The orbitals of lower energy are involved in bonding interactions, so it is the HOMO in which the nonbonding electrons of the anion or radical reside. The situation for a cation is slightly different. The chemical reactivity is again associated with the same nonbonding orbital, but the orbital is *unoccupied.* As a result we describe it as the *lowest unoccupied molecular orbital* (**LUMO**).

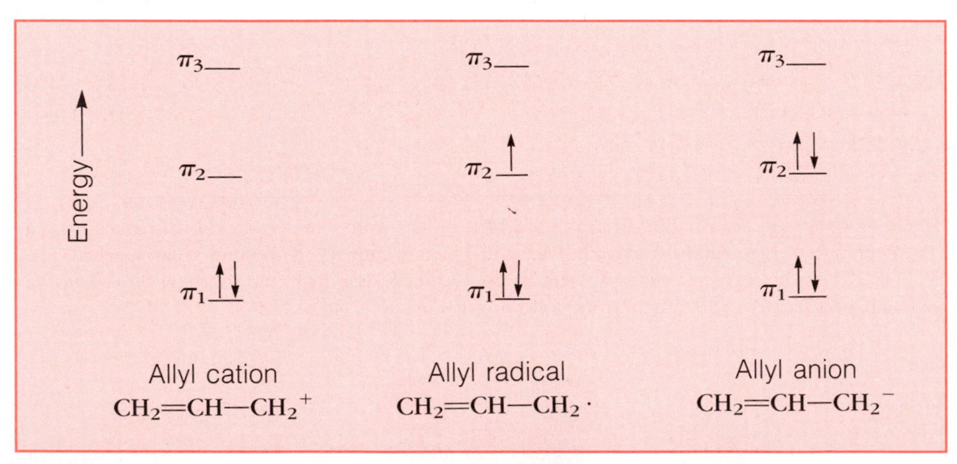

Figure 8.2 Orbital Occupancy for the π Orbitals of Allylic Intermediates.
The nonbonding orbital, π_2, is unoccupied in the cation, singly occupied in the radical, and doubly occupied in the anion. This is also typical of nonallylic cations, radicals, and anions, for which the reactive center is characterized by a single, nonbonding p orbital. For all three allylic intermediates, the bonding π orbital (π_1) is doubly occupied. The low energy of this orbital is the origin of the greater stability of allylic intermediates.

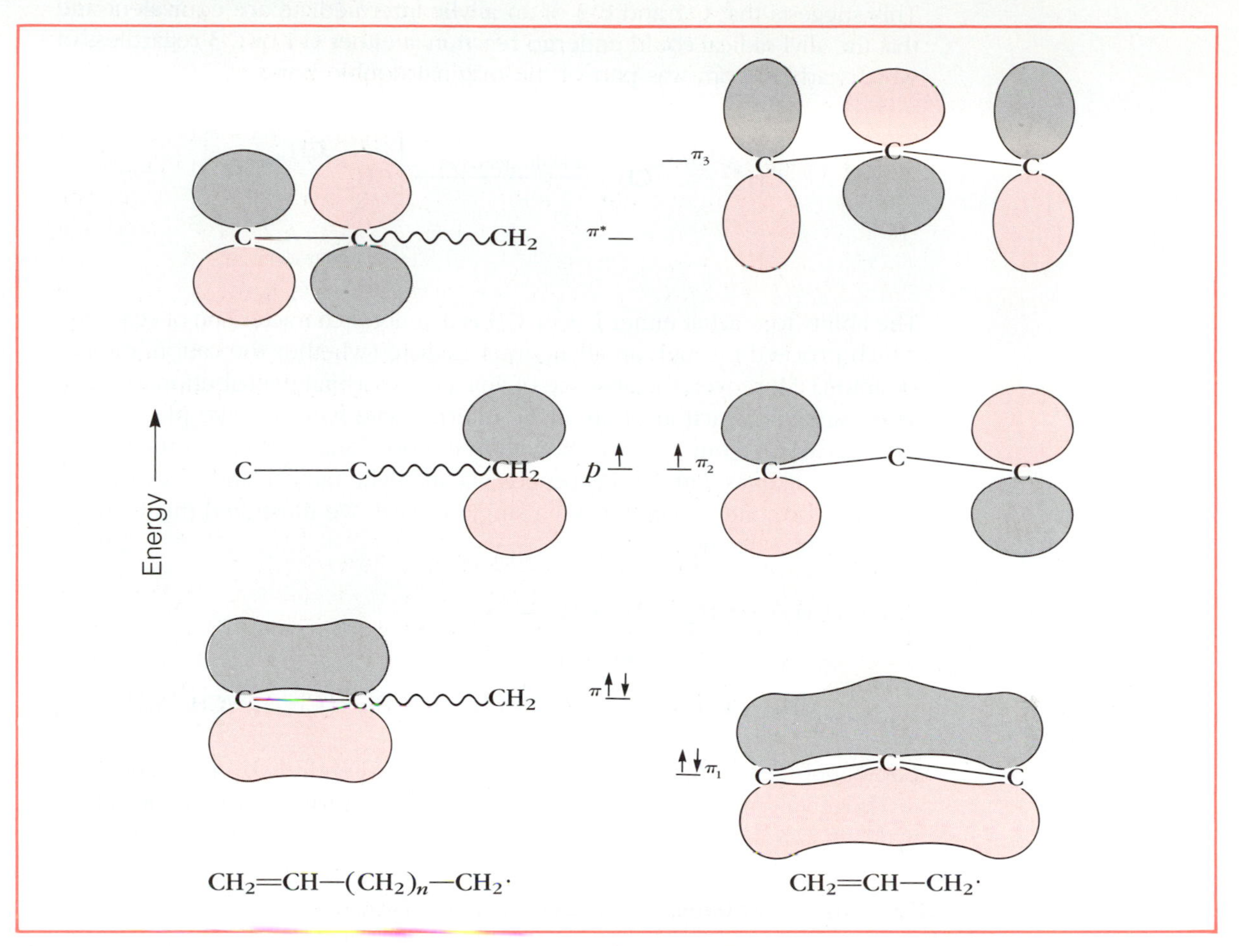

Figure 8.3 Comparison of Energies of an Allylic Radical with Those of an Isomeric Unsaturated Radical in Which There Is No Interaction Between the Double Bond and the Radical Center.
The orbitals are arranged in order of increasing energy so that the lowest-energy orbitals are at the bottom of the figure. The allylic intermediate is more stable, not because of any differences in the partially occupied molecular orbitals but because the bonding π molecular orbital of the allyl system is significantly lower in energy than that of an isolated double bond.

In every case the orbital of interest for an allylic intermediate is π_2 of Figure 8.2; for the allyl anion and radical this orbital is the HOMO, and for the allyl cation it is the LUMO. For the parent three-carbon allyl system π_2 is totally equivalent with respect to C-1 and C-3 but has no orbital contribution at C-2.

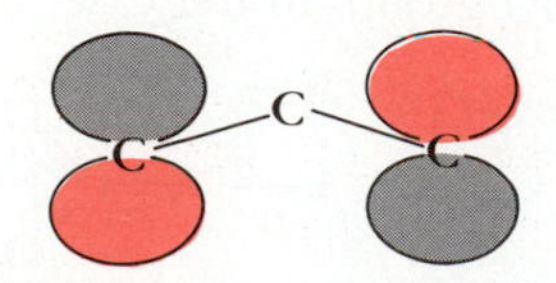

This suggests that C-1 and C-3 of an allylic intermediate are equivalent and that the allyl radical could undergo reaction at either C-1 or C-3 regardless of which carbon atom was part of the original double bond.

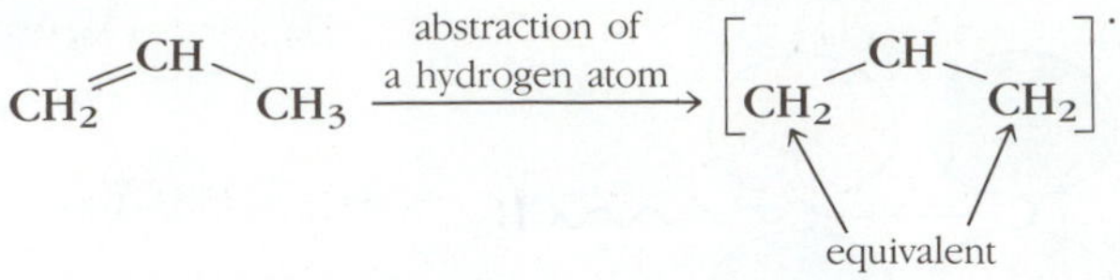

$$CH_2{=}CH{-}CH_3 \xrightarrow[\text{a hydrogen atom}]{\text{abstraction of}} \left[CH_2{-}CH{-}CH_2\right]^{\cdot}$$

(C-1 CH_2 and C-3 CH_2: equivalent)

The ability to react at either C-1 or C-3 is a general characteristic of reactions which proceed through an allylic intermediate (whether the cation, radical, or anion). Moreover, the absence of any atomic orbital contribution at C-2 in π_2 means that reaction of an allylic intermediate will *not* take place at the central carbon atom, C-2. We showed in the previous section that the bromination of alkenes with NBS proceeds via an allylic radical, and two isomeric products are often formed from a single radical. We illustrated this with the reaction of 1-octene:

$$CH_3{-}(CH_2)_4{-}CH_2{-}CH{=}CH_2 \longrightarrow$$

$$CH_3{-}(CH_2)_4{-}\underset{}{\overset{Br}{\overset{|}{C}}}H{-}CH{=}CH_2 + CH_3{-}(CH_2)_4{-}CH{=}CH{-}\overset{Br}{\overset{|}{C}}H_2$$

Although the intermediate free radical is derived entirely by abstraction of a hydrogen atom from C-3, the products are clearly formed by reaction at both C-1 and C-3.

Allylic Intermediates: A Resonance Description

In the preceding discussion we emphasized two unusual characteristics of allylic intermediates: *they are more stable than their saturated counterparts* and *they are capable of undergoing reaction at either end of the allylic unit.* Yet neither of these characteristics is evident from a normal structural drawing, such as that for the allyl anion.

$$CH_2{=}CH{-}CH_2^{-}$$

Our analysis of such a species by molecular orbital theory provided satisfactory explanations of both stability and reactivity, but it would certainly be inconvenient to draw all the molecular orbitals every time you wanted to evaluate a reactive intermediate. A highly convenient alternative to molecular orbital theory is provided by resonance theory.

The situation in which a single conventional structure does not adequately describe a chemical species is precisely the situation in which resonance theory excels. In Section 4.7 we used resonance to describe anions that were centered on an atom adjacent to a carbonyl group. In the case of a carboxylate anion

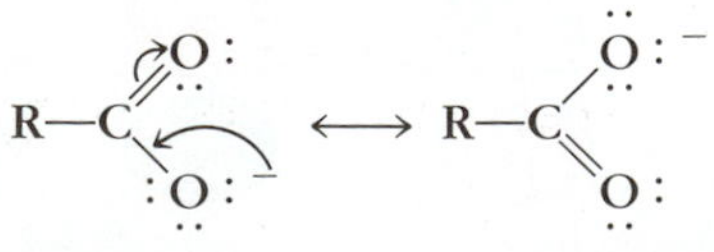

$$R{-}C({=}\ddot{O}{:}){-}\ddot{O}{:}^{-} \longleftrightarrow R{-}C({-}\ddot{O}{:}^{-}){=}\ddot{O}{:}$$

we used resonance to explain not only the high stability of the anion (and hence the high acidity of the carboxylic acid), but also the equivalence of the two oxygen atoms in the anion. A similar explanation is applicable to the allyl anion.

The allyl anion is not accurately described by any single conventional structural formula, but it can be adequately represented as a resonance hybrid of two contributing structures:

$$CH_2{=}CH{-}CH_2{:}^- \longleftrightarrow {}^-{:}CH_2{-}CH{=}CH_2$$

Allyl anion

This simple device provides a convenient method for drawing the allyl anion, and it also provides us with an explanation for the unusual characteristics of the allyl anion. It clearly illustrates that C-1 and C-3 of the allyl unit are equivalent and that both carry negative charge. It also shows the absence of negative charge on the center carbon of the allyl unit. The *delocalization* of the negative charge over two atoms (C-1 and C-3) then accounts for the greater stability of the allyl anion relative to a saturated analog, in which the negative charge would be *localized* on a single carbon atom.

Similar explanations also apply to the allyl cation and allyl radical. In each case there is no single Lewis structure which adequately describes the species, but a satisfactory description is provided by a pair of resonance contributors.

$$CH_2{=}CH{-}CH_2^+ \longleftrightarrow {}^+CH_2{-}CH{=}CH_2$$

Allyl cation

$$CH_2{=}CH{-}CH_2\cdot \longleftrightarrow \cdot CH_2{-}CH{=}CH_2$$

Allyl radical

Resonance theory provides an adequate description of all three types of allylic intermediates using a single pair of structural formulas in each case. These resonance contributors show that the reactive centers are the terminal carbon atoms of the allyl group, while the central carbon is essentially normal. *Delocalization* of either positive charge, unpaired electron density, or negative charge results in stabilization of each of the three allylic intermediates (cation, radical, and anion) and we therefore describe them as *resonance-stabilized.*

We can illustrate the strength of resonance theory by returning to the reaction of 1-octene with NBS. Only one kind of allylic hydrogen is present in the molecule

$$CH_3{-}(CH_2)_4{-}CH_2{-}CH{=}CH_2 \xrightarrow{NBS}$$

and abstraction of one of these hydrogen atoms will lead to the allylic radical.

$$CH_3{-}(CH_2)_4{-}\dot{C}H{-}CH{=}CH_2 \longleftrightarrow CH_3{-}(CH_2)_4{-}CH{=}CH{-}CH_2\cdot$$

The two resonance contributors are not equivalent, so reaction at the two ends of the allylic system will yield isomeric products. Moreover, since free radical character is located on secondary and primary carbon atoms in the two resonance contributors, the two products should not be formed in equal amounts.

$$\longrightarrow \underset{17\%}{CH_3{-}(CH_2)_4{-}\underset{|}{\overset{Br}{C}}H{-}CH{=}CH_2} + \underset{73\%}{CH_3{-}(CH_2)_4{-}CH{=}CH{-}CH_2Br}$$

The relative amounts of isomeric products are not easily predicted in such reactions. Steric effects, interaction with solvent, stability of the products, and other less obvious factors can all play a role in determining product ratios. However, you can predict when more than one product is *possible,* even if you cannot anticipate the yields of each.

EXERCISE 8.1

Draw the resonance forms of the radical formed by abstraction of the indicated hydrogen atom:

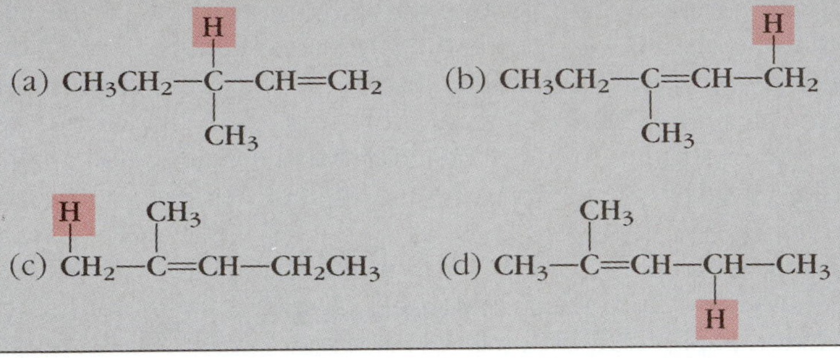

(a) $CH_3CH_2-C(H)(CH_3)-CH=CH_2$ (the H on the C bearing CH_3 is indicated)

(b) $CH_3CH_2-C(CH_3)=CH-CH_2-H$ (the H on the terminal CH_2 is indicated)

(c) $H-CH_2-C(CH_3)=CH-CH_2CH_3$ (the H on the first CH_2 is indicated)

(d) $CH_3-C(CH_3)=CH-CH(H)-CH_3$ (the H on the fourth carbon is indicated)

EXERCISE 8.2

Draw the second resonance contributor for each of the following allylic intermediates:

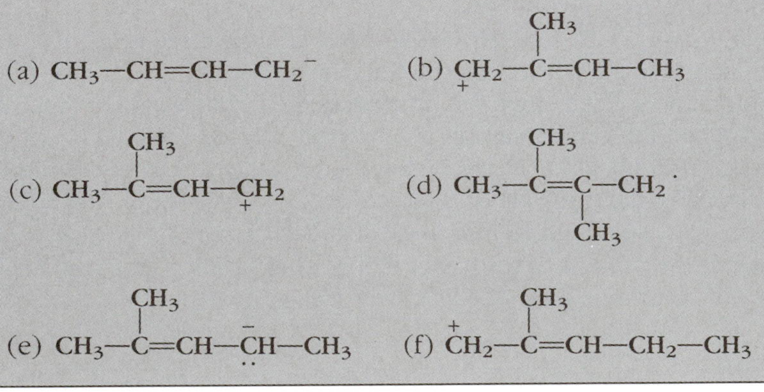

(a) $CH_3-CH=CH-CH_2^-$

(b) $\overset{+}{C}H_2-C(CH_3)=CH-CH_3$

(c) $CH_3-C(CH_3)=CH-\overset{+}{C}H_2$

(d) $CH_3-C(CH_3)=C(CH_3)-CH_2\cdot$

(e) $CH_3-C(CH_3)=CH-\overset{-}{\ddot{C}}H-CH_3$

(f) $\overset{+}{C}H_2-C(CH_3)=CH-CH_2-CH_3$

8.4 ALLYLIC BROMINATION AND OXIDATION

A variety of allylic substitution reactions are possible, but two of these are particularly important as synthetic procedures. These are *allylic bromination* with *N*-bromosuccinimide

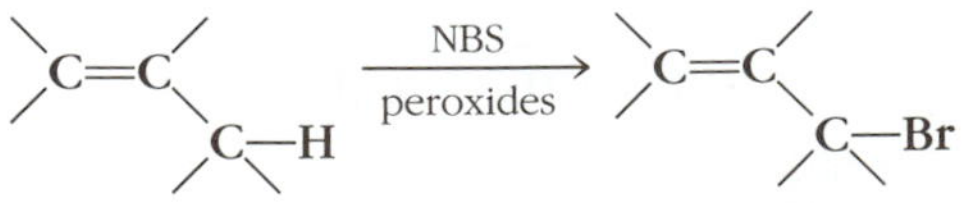

$$\text{>C=C<}\text{ (allylic C–H) }\xrightarrow[\text{peroxides}]{\text{NBS}}\text{>C=C<}\text{ (allylic C–Br)}$$

and *allylic oxidation* with selenium dioxide (which has been used extensively despite the toxicity of the reagent).

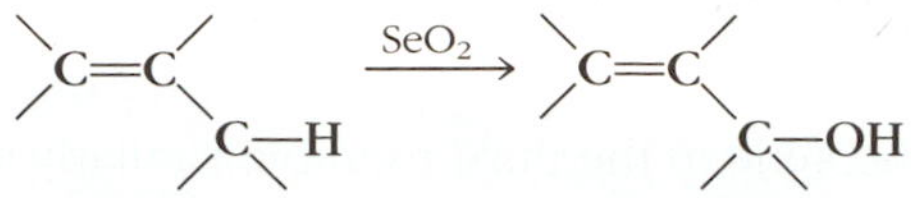

$$\text{C=C–C–H} \xrightarrow{SeO_2} \text{C=C–C–OH}$$

Allylic bromination is a free radical halogenation reaction, and it proceeds via the mechanism that we discussed in Section 2.6. Under the reaction conditions NBS decomposes slowly to produce Br_2, and this enters into the chain reaction with the alkene.

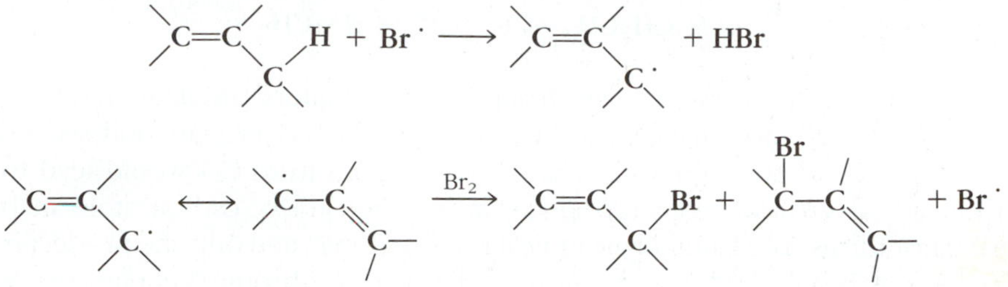

$$\text{C=C–C–H} + Br\cdot \longrightarrow \text{C=C–C}\cdot + HBr$$

$$\text{C=C–C}\cdot \longleftrightarrow \cdot\text{C–C=C–} \xrightarrow{Br_2} \text{C=C–C–Br} + \text{Br–C–C=C–} + Br\cdot$$

The reaction is initiated by an organic peroxide or by irradiation with light. The greater stability of the allylic free radical results in the selective abstraction of allylic hydrogen atoms, with subsequent formation of allylic bromides. Under the conditions used for allylic substitution, ionic addition of Br_2 to the double bond does not usually occur to a significant extent.

The following examples illustrate the use of NBS in allylic bromination.

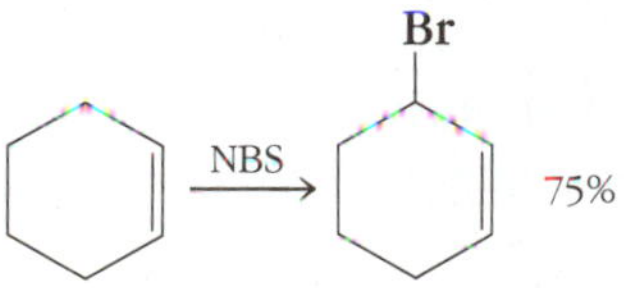

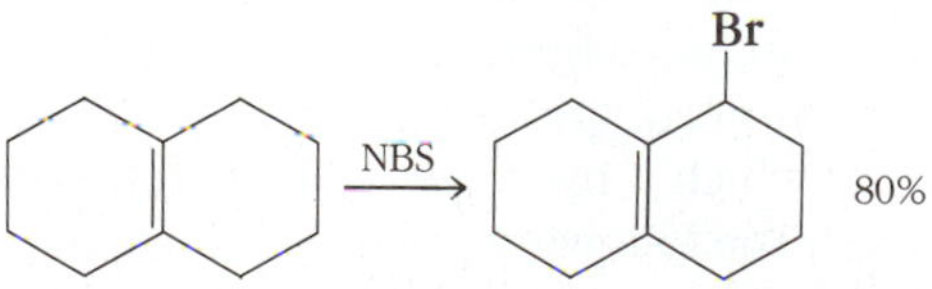

NBS 75%

NBS 80%

$$CH_3CH_2CH_2CH_2\text{—}CH\text{=}CH\text{—}CH_3 \xrightarrow{NBS} CH_3CH_2CH_2\text{—}CH(Br)\text{—}CH\text{=}CH\text{—}CH_3 \quad 58\text{–}64\%$$

In each of these cases only a single allylic bromide was isolated. This should not surprise you in the case of cyclohexene, because only one allylic bromination product is possible. In the second example all four allylic CH_2 groups are equivalent by symmetry, so only one allylic radical could be formed. This radical might have yielded a second product as well:

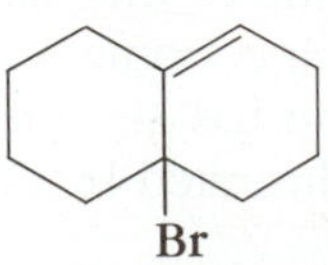

Br

None of this second product was formed, however. (Can you think of any reasons why the observed product would predominate?) While you might have predicted both isomers as possible products, only by carrying out the reaction in the laboratory could you determine which products would actually be formed in such a reaction.

The reaction of 2-heptene in the third example similarly affords a single major product, despite the fact that four constitutionally isomeric allylic bromides could have been formed. To some extent the selectivity can be predicted in this case. There are two types of allylic hydrogen atoms which could be abstracted in the case of 2-heptene:

$$CH_3CH_2CH_2—\overset{\displaystyle H}{\overset{|}{C}}H—CH{=}CH—\overset{\displaystyle H}{\overset{|}{C}}H_2$$

Abstraction of a hydrogen atom from C-1 would afford an allylic radical with resonance contributors having the unpaired electron on primary and secondary carbon atoms, whereas removal of a hydrogen atom from C-4 would lead to an allylic radical with the unpaired electron on a secondary carbon atom in both resonance forms. The latter allylic radical is more stable, and only the product from this pathway is observed. The absence of the isomeric 2-bromo-3-heptene is somewhat surprising, however. A prediction of products for the reaction (if the answer were not already known) would involve both isomers:

$$[CH_3CH_2CH_2—\dot{C}H—CH{=}CH—CH_3 \longleftrightarrow CH_3CH_2CH_2—CH{=}CH—\dot{C}H—CH_3] \longrightarrow$$

$$CH_3CH_2CH_2—\overset{\displaystyle Br}{\overset{|}{C}}H—CH{=}CH—CH_3 + CH_3CH_2CH_2—CH{=}CH—\overset{\displaystyle Br}{\overset{|}{C}}H—CH_3$$

Undoubtedly both isomers are formed in this reaction, but 4-bromo-2-heptene predominates, and only that compound was isolated after purification of the reaction mixture.

Allylic oxidation with selenium dioxide proceeds by a relatively complex mechanism, which will not be discussed here. A free (uncomplexed) allylic intermediate is not formed, however. Instead, the OH group is introduced specifically on the same carbon atom from which a hydrogen atom is removed, and no shift of the double bond is observed. The following examples illustrate the use of this reaction.

$$CH_3—(CH_2)_6—CH_2—CH{=}CH_2 \xrightarrow{SeO_2} CH_3—(CH_2)_6—\overset{\displaystyle OH}{\overset{|}{C}}H—CH{=}CH_2 \quad 56\%$$

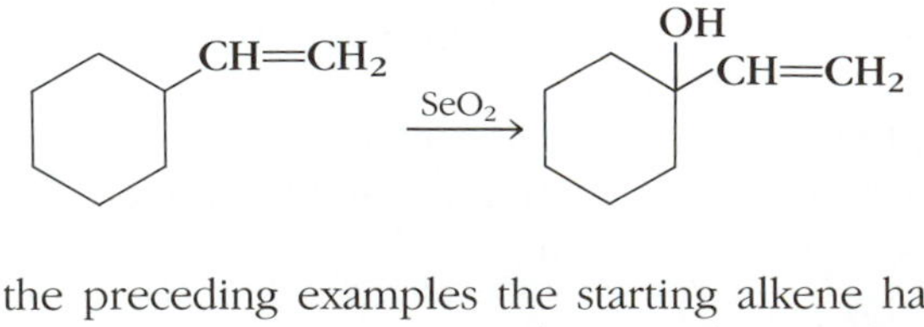

$$\text{cyclohexyl}—CH{=}CH_2 \xrightarrow{SeO_2} \text{1-hydroxycyclohexyl}—CH{=}CH_2 \quad 44\%$$

In each of the preceding examples the starting alkene has only one kind of allylic hydrogen atom, so you expect only a single constitutional isomer to be produced. However, if the alkene has several nonequivalent allylic CH groups, several products can result, as illustrated by the reaction of (*Z*)-2-octene.

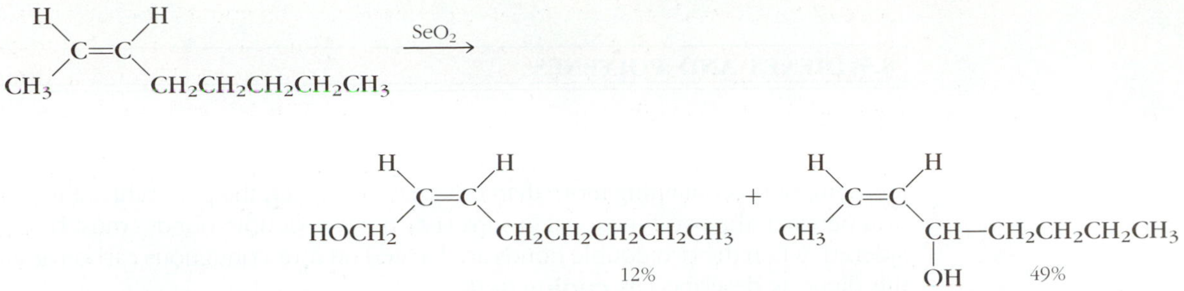

$$\underset{CH_3}{\overset{H}{}}\!\!\diagdown C{=}C \diagup^{H}_{\diagdown CH_2CH_2CH_2CH_2CH_3} \xrightarrow{SeO_2}$$

$$HOCH_2CH{=}CHCH_2CH_2CH_2CH_2CH_3 \;(12\%) \quad + \quad CH_3CH{=}CH{-}\underset{OH}{\underset{|}{CH}}{-}CH_2CH_2CH_2CH_3 \;(49\%)$$

On the other hand, a single isomer is sometimes formed in high yield, even when several products might be expected.

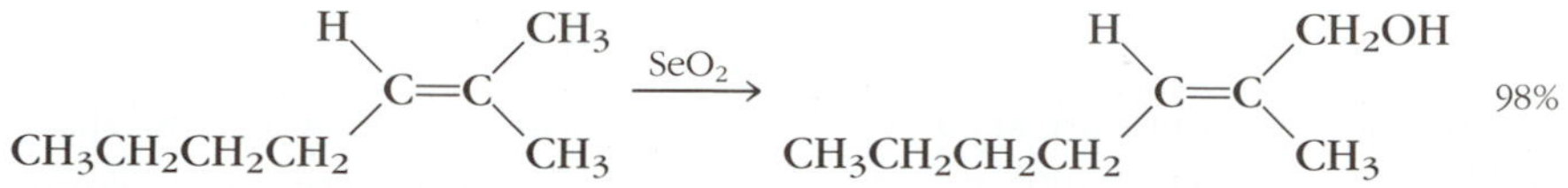

$$CH_3CH_2CH_2CH_2CH{=}C(CH_3)_2 \xrightarrow{SeO_2} CH_3CH_2CH_2CH_2CH{=}C(CH_3)CH_2OH \quad 98\%$$

Allylic bromination with NBS and allylic oxidation with SeO_2 both result in the formation of a second reactive center (either $-\overset{|}{\underset{|}{C}}-Br$ or $-\overset{|}{\underset{|}{C}}-OH$) adjacent to the double bond. These are difunctional compounds: in some reactions they will behave as the alkene, while in others they will act as the alkyl halide or the alcohol. In still other reactions the two groups will interact strongly, and allylic intermediates will be formed. If you select the proper reagents, it is possible to select which of these reactions will occur. As a result allylic substitution reactions and their products are quite useful in organic synthesis.

EXERCISE 8.3

For each of the allylic bromination reactions that we presented in this section draw the allylic free radicals that could be formed. Draw both resonance contributors, and for each of them show the product that would result from attachment of a bromine to the carbon on which you have drawn the unpaired electron.

EXERCISE 8.4

Predict all the expected products of allylic substitution for each of the following alkenes using (1) NBS and (2) SeO_2.

(a) 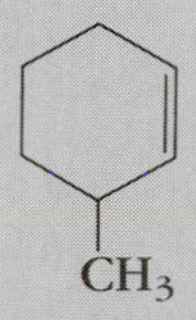 4-methylcyclohexene (ring with CH_3)

(b) 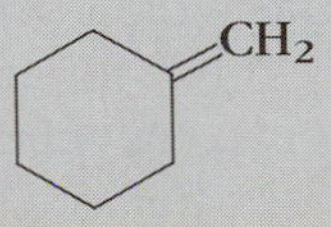 methylenecyclohexane (ring $=CH_2$)

(c) $CH_3CH_2{-}CH{=}CH{-}CH_2CH_3$

8.5 DIENES AND POLYENES

Dienes are hydrocarbons that contain two (*di-*) carbon-carbon double bonds (*-ene*). In most respects their chemistry is similar to that of alkenes. However with any compound containing more than one reactive center, the possibility of interaction between the two functional groups (i.e., the two double bonds) must be considered. When the two double bonds are located on four contiguous carbon atoms, the diene is described as **conjugated.**

$$\text{>C=C-C=C<}$$

a Conjugated diene

When the double bonds are separated by one or more carbon atoms, the diene is described as nonconjugated, and the double bonds are described as **isolated.**

$$\text{>C=C-C-C=C<}$$

a Nonconjugated diene

Allenes are also dienes, but they constitute a unique class of compounds and are not included in either of the preceding categories.

$$\text{>C=C=C<}$$

an Allene

Conjugated Dienes: A Molecular Orbital Description

Conjugated dienes are characterized by substantial interaction between the double bonds. The central bond of 1,3-butadiene, for example, has a bond length of only 1.46 Å, in contrast to the typical 1.53 Å for a carbon-carbon single bond in an alkane.

$$CH_2{=}CH\overset{1.46\,\text{Å}}{-}CH{=}CH_2$$

This shortened bond length suggests some *multiple bond* character for this "single" bond, and in the following discussion we will show that this is indeed the case.

The orbital interactions in a conjugated diene lead to a net stabilization in comparison with a nonconjugated isomer. This is illustrated in Figure 8.4 for 1,3-cyclohexadiene and 1,4-cyclohexadiene. Both compounds can be converted to cyclohexane by catalytic hydrogenation, but the conjugated diene would evolve less energy because it is more stable.

The cyclohexadienes provide a good comparison, not only because they are isomeric compounds but also because both double bonds are disubstituted in each case. The net stabilization of the conjugated isomer is about 1 kcal/mol, a small but significant difference.

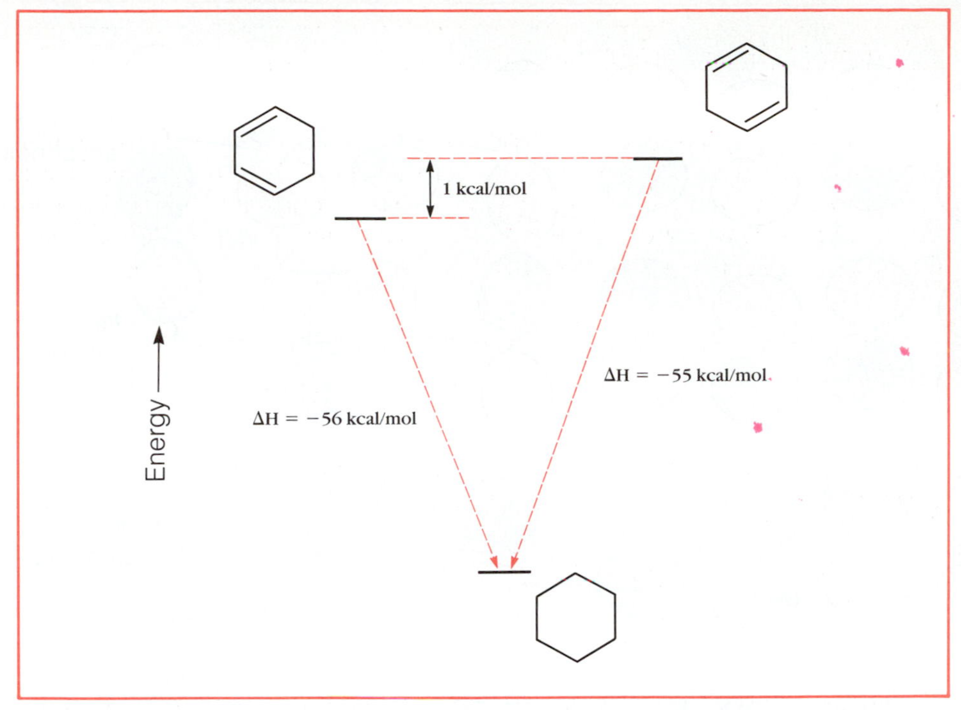

Figure 8.4 Stabilization of Conjugated Dienes, as Illustrated for the Isomeric Cyclohexadienes.
Both 1,3- and 1,4-cyclohexadiene can be reduced to cyclohexane, but the reaction of the conjugated isomer is less exothermic because this isomer is more stable.

The orbital interactions that lead to stabilization of a conjugated diene are maximized in conformations having the four carbon atoms of the diene in a plane. Two such planar conformations are possible for an acyclic diene, and these are shown for 1,3-butadiene.

Anti ⇌ *Syn*

The *anti* conformer predominates for most conjugated dienes because there are fewer unfavorable steric interactions. On the other hand, cyclic dienes (except for very large rings) can achieve only the *syn* conformation.

In both planar conformations, each of the four carbon atoms of the diene unit has a p orbital perpendicular to the plane. Consequently, these p orbitals are all parallel, the precise orientation required for maximum π interaction. Combination of four atomic p orbitals leads to four π molecular orbitals, and these are depicted in Figure 8.5.

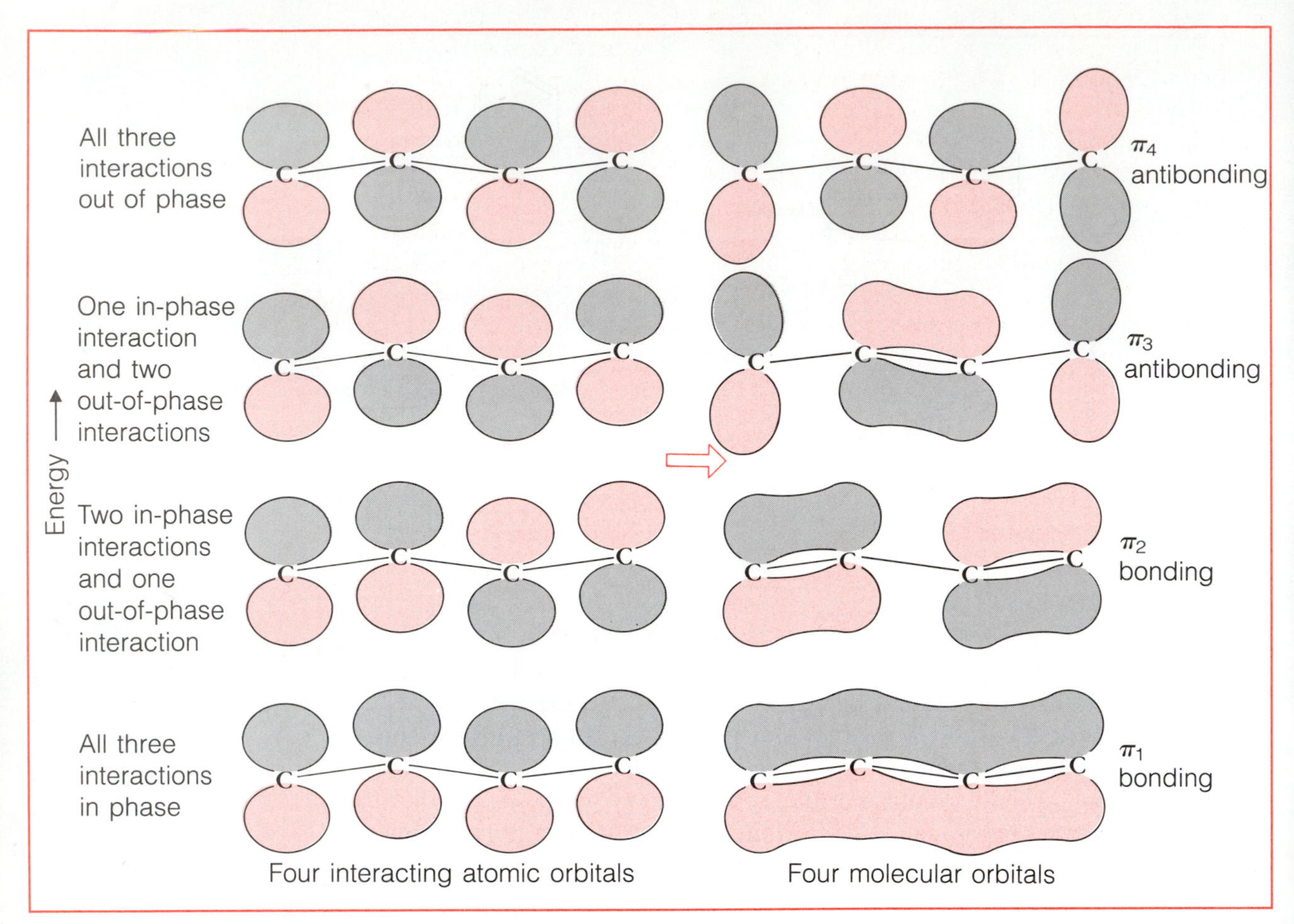

Figure 8.5 The π Molecular Orbitals of a Conjugated Diene. Four atomic *p* orbitals combine to yield four molecular orbitals. As the number of out-of-phase interactions (between orbitals on adjacent carbon atoms) increases, so does the orbital energy. Consequently, π_1 and π_2 are bonding, but π_3 and π_4 are antibonding. In a normal diene the π system has a total of four electrons, and only the two bonding π orbitals are occupied.

In Section 27.1 we will show you how these molecular orbitals are formed by combination of the appropriate atomic orbitals. The present discussion will focus on the properties of the molecular orbitals rather than on their formation. For the lowest-energy π orbital in Figure 8.5, all three interactions between *p* orbitals on adjacent carbon atoms are in phase, i.e., bonding. For π_2 there are two bonding interactions (between C-1 and C-2 and between C-3 and C-4), so the interaction between C-2 and C-3 is antibonding. There are more bonding than antibonding interactions, so the net result is still a bonding molecular orbital. In π_3, however, the antibonding interactions (C-1–C-2 and C-3–C-4) predominate over the one bonding interaction between C-2 and C-3. Similarly, all interactions in π_4 are antibonding. In the ground state a diene has a total of four electrons in the π system (corresponding to one electron pair for each of the double bonds). Therefore only two of the π molecular orbitals, the bonding orbitals π_1 and π_2, are occupied.

The structural properties of a molecule are governed by the occupied molecular orbitals, and some of the unusual properties of conjugated dienes are readily interpreted in terms of π_1, the lower-energy occupied π molecular orbital. The bonding interaction between C-2 and C-3 in π_1 explains the relatively short (1.46 Å) bond length between these two carbon atoms. We can also account for the preferred planar conformations of a conjugated diene in terms of π_1, because the required overlap of atomic *p* orbitals is achieved only when all four carbon atoms lie in a single plane. Finally, the lower energy of this orbital, resulting from bonding interactions between all carbon atoms of the diene unit, is responsible for the net stabilization of a conjugated diene relative to a nonconjugated isomer.

We can explain the stability of a conjugated diene in terms of π_1, but the chemical reactivity (at least in reactions with electrophiles) is governed largely by π_2, the highest occupied molecular orbital. This orbital resembles two individual double bonds, and as a result the reactions of such conjugated dienes are quite similar to those that we presented for alkenes in Chapter 5.

Addition to Conjugated Dienes

Except for some minor differences, the reactions of conjugated dienes are the same as those of alkenes. As an example, 1,3-butadiene reacts with HCl to yield 3-chloro-1-butene and 1-chloro-2-butene. This corresponds to addition of H and Cl not only to one of the original double bonds of the butadiene (a 1,2 addition), but also to the terminal carbons of the diene system (a 1,4 addition).

$$CH_2{=}CH{-}CH{=}CH_2 \xrightarrow{HCl} \underset{49\%}{CH_2{=}CH{-}\overset{Cl}{\overset{|}{C}}H{-}\overset{H}{\overset{|}{C}}H_2} + \underset{26\%}{\overset{Cl}{\overset{|}{C}}H_2{-}CH{=}CH{-}\overset{H}{\overset{|}{C}}H_2}$$

The reaction takes place by the mechanism we outlined in Section 5.8, and it proceeds stepwise by addition of H^+ to give the more stable carbocation, which in this case is not only secondary but also allylic.

$$CH_2{=}CH{-}CH{=}CH_2 \xrightarrow{H^+} CH_2{=}CH{-}\overset{+}{C}H{-}\overset{H}{\overset{|}{C}}H_2$$

This structural formula shows how the 1,2 addition product arises, and if you consider the second resonance contributor, you can see how the 1,4 addition takes place:

$$CH_2{=}CH{-}\overset{+}{C}H{-}\overset{H}{\overset{|}{C}}H_2 \longleftrightarrow \overset{+}{C}H_2{-}CH{=}CH{-}\overset{H}{\overset{|}{C}}H_2 \xrightarrow{Cl^-} \overset{Cl}{\overset{|}{C}}H_2{-}CH{=}CH{-}\overset{H}{\overset{|}{C}}H_2$$

The reaction of 1,3-butadiene via an allylic intermediate illustrates an important and unique feature of conjugated dienes. Although they are *more stable* than their nonconjugated isomers, they are also *more reactive.* Ordinarily the opposite trend is observed, and a compound that is particularly stable is also relatively unreactive. What is different about a conjugated diene? The key to its high reactivity is formation of an intermediate that is also highly stabilized.

The high reactivity of a conjugated diene is explained by Figure 8.6, which

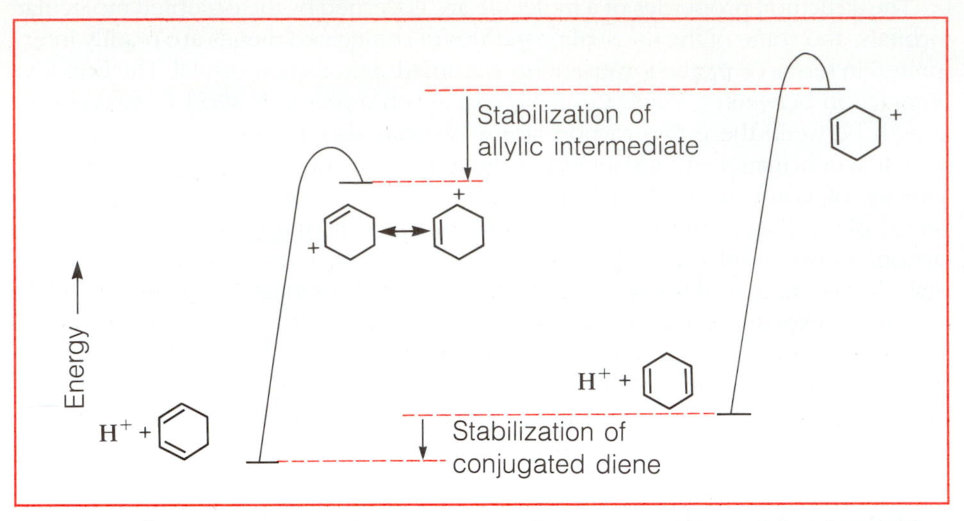

Figure 8.6 Comparison of Energies in the Protonation of Conjugated and Nonconjugated Dienes.
(In order to emphasize the comparison, the vertical axis is not drawn to scale.) The conjugated isomer (1,3-cyclohexadiene) is slightly more stable than 1,4-cyclohexadiene, but the energy difference between the cations is even greater. Consequently, the activation energy for protonation of 1,3-cyclohexadiene is lower. The conjugated diene is both more stable and more reactive.

compares the addition of H^+ (the rate-limiting step in acid-mediated additions to alkenes) of 1,3- and 1,4-cyclohexadiene. Although the conjugated diene is more stable, it yields an allylic cation when protonated. As we indicated in Section 4.6, stabilization of an ionic intermediate is usually greater than stabilization of a neutral intermediate. This results in an activation energy for reaction of the conjugated diene which is less than that for the nonconjugated isomer, and the conjugated diene reacts more readily.

Other addition reactions of conjugated dienes are illustrated by the following examples:

$$CH_2{=}CH{-}C(CH_3){=}CH_2 \xrightarrow{HCl} CH_2{=}CH{-}C(CH_3)_2{-}Cl \quad 66\%$$

$$CH_2{=}CH{-}C(CH_3){=}CH_2 \xrightarrow[CCl_4]{Br_2} BrCH_2{-}CH{=}C(CH_3){-}CH_2Br \quad 87\%$$

The first example is actually a 1,2 addition, but the mechanism is the same. Protonation initially generates the allylic cation with resonance contributors that are tertiary and primary rather than the alternative in which the contributors are secondary and primary.

$$CH_2{=}CH{-}C(CH_3){=}CH_2 \xrightarrow{H^+} CH_2{=}CH{-}\overset{+}{C}(CH_3)_2 \longleftrightarrow \overset{+}{C}H_2{-}CH{=}C(CH_3)_2$$

$$CH_2{=}CH{-}C(CH_3){=}CH_2 \xrightarrow{H^+} CH_3{-}\overset{+}{C}H{-}C(CH_3){=}CH_2 \longleftrightarrow CH_3{-}CH{=}C(CH_3){-}CH_2^+$$

The observation of 1,2 addition in this case is largely a function of reaction conditions; the observed product results from *kinetic control.* At higher temperature, the products of 1,2 and 1,4 addition are in equilibrium (i.e., the reaction is under *thermodynamic control*), and the 1,4-addition product predominates.

Additional examples show that both 1,2- and 1,4-addition products are frequently formed in the reactions of conjugated dienes. It is difficult to predict which product will be observed in any particular reaction, and you should expect both. However, you can always predict the orientation of the addition for each of these pathways because an allylic intermediate will be generated in the first step.

$$CH_3{-}CH{=}CH{-}CH{=}CH{-}CH_3 \xrightarrow[DMSO]{NBS,\ H_2O} CH_3{-}\underset{}{CH(Br)}{-}CH(OH){-}CH{=}CH{-}CH_3 \quad 93\%$$

$$CH_2{=}CH{-}CH{=}CH{-}CH_3 \xrightarrow{HCl} CH_3{-}CH{=}CH{-}CH(Cl){-}CH_3 \quad 70\%$$

EXERCISE 8.5

(a) For the preceding examples draw the allylic intermediates that would be generated by attack of "Br^+" and H^+, respectively. Only one of the two possible allylic cations is involved in the reaction of 1,3-pentadiene with HCl. Why? Illustrate your answer with appropriate structural formulas.

(b) In the reaction of 1,3-pentadiene with HCl it cannot be determined whether the product was formed by a 1,2 or 1,4 addition of HCl. Show by means of resonance contributors to the intermediate cation that both processes would yield the same product.

Additions to Nonconjugated Dienes

The products formed by addition to nonconjugated dienes are not as easily predicted as those for conjugated dienes. The double bonds react independently of each other, and mixtures of products are frequently obtained. For some compounds reaction occurs at both double bonds, but other compounds react exclusively at just one of the two sites (Section 8.8). In still other cases an initial reaction at one of the double bonds produces an intermediate that undergoes intramolecular reaction with the other double bond. We will show you examples of this in Section 8.7.

Polyenes

Polyenes are hydrocarbons that contain multiple (*poly-*) double bonds (*-ene*). Whether conjugated or nonconjugated, polyenes are similar in reactivity to the corresponding dienes. A larger number of isomeric products can often be formed in the reactions of polyenes, but their basic reactions are the same as those of alkenes and dienes. The following compounds are examples of biologically and industrially important polyenes.

Squalene

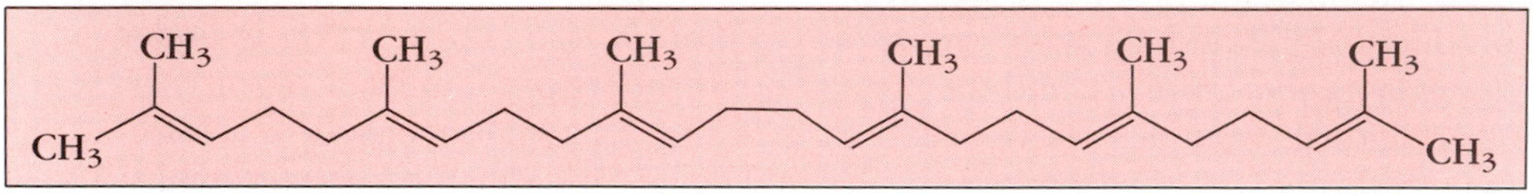

Squalene is a symmetrical hydrocarbon ($C_{30}H_{50}$) with six nonconjugated double bonds. It serves as the biological precursor to cholesterol and other steroids (Section 23.3).

β-Carotene

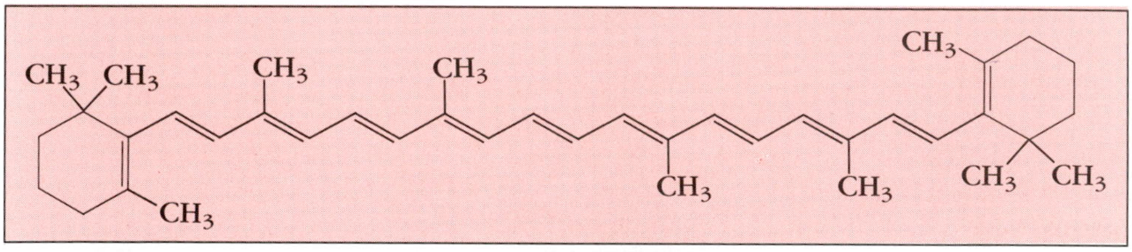

β-Carotene is a highly conjugated, reddish polyene ($C_{40}H_{56}$), which serves as a biological precursor to vitamin A.

Natural Rubber

$$\left(CH_2-\overset{\displaystyle CH_3}{\overset{|}{C}}=CH-CH_2 \right)_n$$

Natural rubber is a **polymer** (from the Greek *poly,* "many" and *meros,* "part") with a molecular weight in the range of 500,000, composed of thousands of repeating C_5 subunits.

The biochemical formation of natural rubber is closely related to that of squalene. The carbon skeleton is built up by a series of reactions beginning with five-carbon phosphate derivatives, and we will present the mechanism of squalene formation in Section 12.9. Natural rubber was at one time considered to be derived from *isoprene* (2-methyl-1,3-butadiene) by a series of carbocation reactions. We now know that the biochemical process is quite different, but isoprene can undergo polymerization in the laboratory by the following acid-catalyzed mechanism:

$$CH_2{=}C(CH_3){-}CH{=}CH_2 \xrightarrow{H^+} H{-}CH_2{-}\overset{+}{C}(CH_3){-}CH{=}CH_2 \longleftrightarrow H{-}CH_2{-}C(CH_3){=}CH{-}\overset{+}{C}H_2$$

$$\xrightarrow{CH_2=C(CH_3)-CH=CH_2} H{-}CH_2{-}C(CH_3){=}CH{-}CH_2{-}CH_2{-}\overset{+}{C}(CH_3){-}CH{=}CH_2 \longleftrightarrow H{-}CH_2{-}C(CH_3){=}CH{-}CH_2{-}CH_2{-}C(CH_3){=}CH{-}CH_2^+$$

$\longrightarrow$ etc., etc.

This reaction pathway illustrates that in addition to nucleophilic attack by a molecule such as water, a carbocation can also undergo nucleophilic attack by the electrons of an *alkene* to generate a new carbocation.

8.6 BENZENE AND BENZYLIC INTERMEDIATES

The reactivity of most polyenes is very similar to that of simple alkenes and dienes, but benzene and related *aromatic* compounds are important exceptions. The term **aromatic** originally referred to the characteristic odors of these compounds, but the meaning has evolved and now signifies certain unusual chemical properties that are characteristic of benzene and related compounds. Aromatic compounds are discussed extensively in Chapter 10, and here we will only look at the reactions that occur on the side chains of some benzene derivatives.

Benzene has the molecular formula C_6H_6, and it is usually drawn as a six-membered ring with three double bonds. Frequently, the hydrogen substituents are not drawn.

The benzene molecule is planar, and the atomic *p* orbitals of all six carbon atoms can interact. This orbital interaction (discussed in Chapters 10 and 27) causes benzene to be an exceptionally stable compound.

In Section 8.1 we listed three ways in which two reactive centers in a molecule could interact, and two of these ways are seen with benzene derivatives that have a second functional group. First, the aromatic ring can greatly affect the reactivity of any substituent. Second, as a consequence of the exceptional stability of the benzene ring, reaction tends to take place preferentially at the *other* functional group. The aromatic ring of benzene and its derivatives are essentially inert to nearly all the reactions that we have presented so far in this text. On the other hand, while the double bonds of the benzene ring do not undergo reaction easily, the π orbitals are capable of interacting with substituents. In this respect a **phenyl group** (C_6H_5—) is analogous to a *vinyl* group (CH_2=CH—) and can stabilize a cation, radical, or anion on an adjacent carbon. The resulting intermediates, in which charge or radical character is located on the CH_2 of a **benzyl group** ($C_6H_5CH_2$), are analogous to allylic intermediates but are described as **benzylic.**

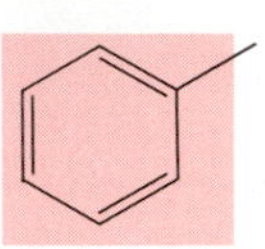

Phenyl group

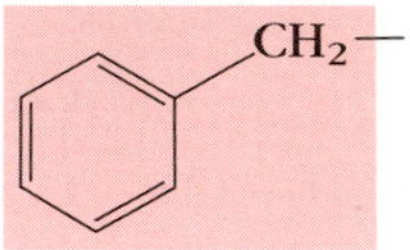

CH_2—

Benzyl group

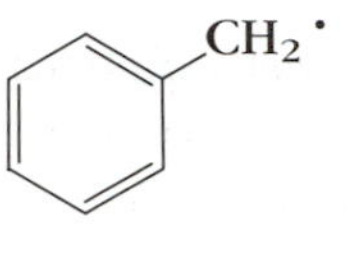

$CH_2^{\bullet}$ CH_2^{+} CH_2^{-}

Benzyl radical 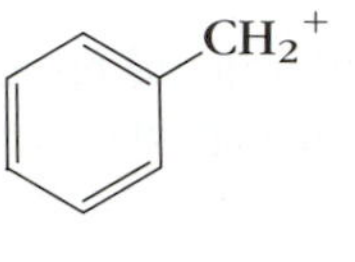 Benzyl cation 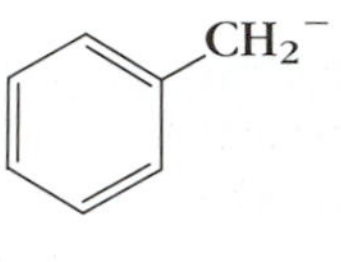 Benzyl anion

There is one major difference between benzylic and allylic intermediates, however. Allylic intermediates undergo reaction at both ends of the allyl group, but the great stability of the aromatic ring normally limits reactions of benzylic intermediates to the position adjacent to the aromatic ring.

The following examples illustrate the selective reactions of a series of benzene derivatives. In all cases reaction occurs on the side chain rather than at the phenyl group. These examples also demonstrate the directive effect of an aromatic ring. Development of ionic charge or free radical character takes place preferentially at a benzylic position.

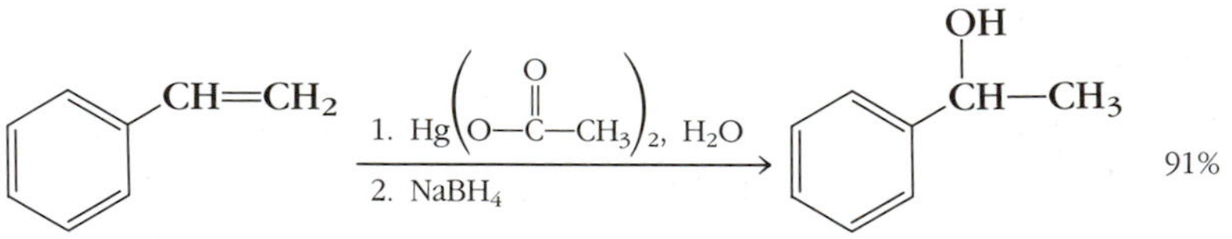

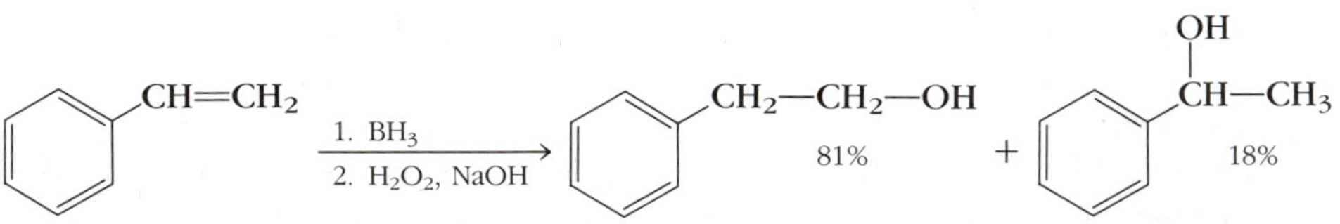

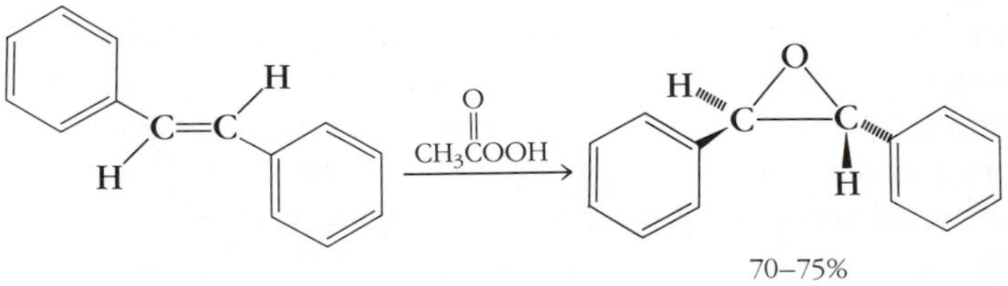

C_6H_5–CH=CH_2 —(1. $Hg(O–\overset{O}{\overset{\|}{C}}–CH_3)_2$, H_2O; 2. $NaBH_4$)→ C_6H_5–CH(OH)–CH_3 91%

C_6H_5–CH=CH_2 —(1. BH_3; 2. H_2O_2, NaOH)→ C_6H_5–CH_2–CH_2–OH 81% + C_6H_5–CH(OH)–CH_3 18%

C_6H_5(H)C=C(H)C_6H_5 —($CH_3\overset{O}{\overset{\|}{C}}OOH$)→ trans-$C_6H_5$(H)C–C(H)$C_6H_5$ epoxide (O bridge) 70–75%

$C_6H_5CH_2{-}CH_3 \xrightarrow{NBS} C_6H_5CH(Br){-}CH_3$ 80%

$C_6H_5C{\equiv}CC_6H_5 \xrightarrow[2.\ CH_3CO_2H]{1.\ (CH_3{-}CH(CH_3){-}CH(CH_3){-})_2BH}$ (Z)-$C_6H_5CH{=}CHC_6H_5$ 69%

$C_6H_5CH{=}CH{-}CH_3 \xrightarrow[CH_2Cl_2]{Cl_2} C_6H_5CH(Cl){-}CH(Cl){-}CH_3$ ~84%

The stereochemistry of the last reaction highlights an important feature of reactions involving benzylic cations. Addition of halogens to a double bond ordinarily occurs by the *anti* mode, but both the *E* and *Z* isomers of 1-phenyl-1-propene afford nearly the same mixture of products.

(E)- or (Z)-$C_6H_5CH{=}CHCH_3 \xrightarrow[CCl_4]{Cl_2}$ 55–63% + 21–29%

The reaction exhibits no stereospecificity (i.e., the configuration of the product is *not* related to the configuration of the reactant). The formation of the same products from either of the stereoisomeric alkenes indicates that both reactions proceed via the same intermediate (or mixture of intermediates). The addition of chlorine proceeds by an ionic addition to produce a cation, in this case a benzylic cation. As compared with ordinary carbocations, the more stable benzylic cation survives for a longer time before undergoing attack by chloride ion, long enough to undergo rotation about the central bond. Both alkenes therefore yield the same conformational mixture of carbocations.

Such a loss of stereospecificity is found for a variety of reactions which proceed via a benzylic cation.

Under vigorous conditions a benzene ring can be reduced, but catalytic hydrogenation takes place preferentially at an unsaturated side chain, as shown by the following reactions.

$$C_6H_5-OCH_2CH_3 \xrightarrow{H_2,\ Ni} C_6H_{11}-OCH_2CH_3 \quad (85\%)$$

$$C_6H_5-CH{=}CH-C(=O)-C_6H_5 \xrightarrow{H_2,\ Pt} C_6H_5-CH_2-CH_2-C(=O)-C_6H_5 \quad (81\text{–}95\%)$$

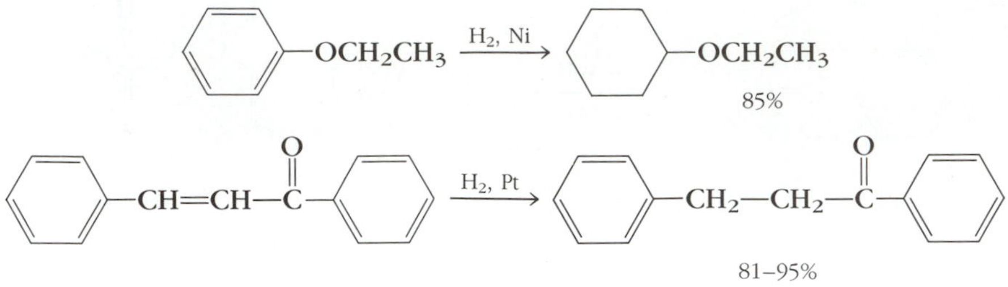

The stability of benzylic radicals sometimes results in the *hydrogenolysis* (cleavage by hydrogen) of oxygen and halogen substituents at a benzylic position. The following example illustrates the hydrogenolysis of a benzyl ether.

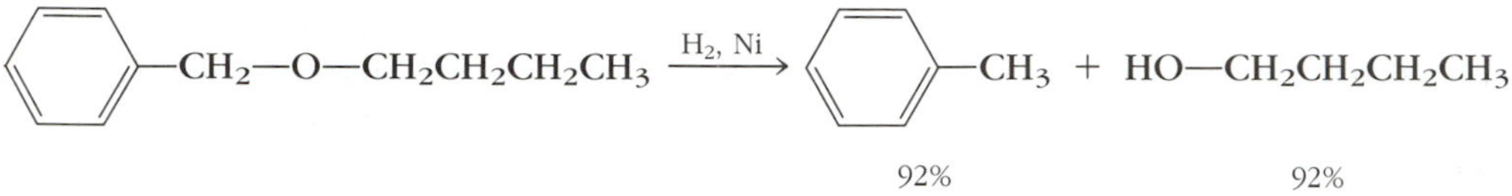

$$C_6H_5-CH_2-O-CH_2CH_2CH_2CH_3 \xrightarrow{H_2,\ Ni} C_6H_5-CH_3 + HO-CH_2CH_2CH_2CH_3$$

92% 92%

EXERCISE 8.6

Draw the major product (or products) expected from each of the following reactions. Show stereochemistry whenever appropriate.

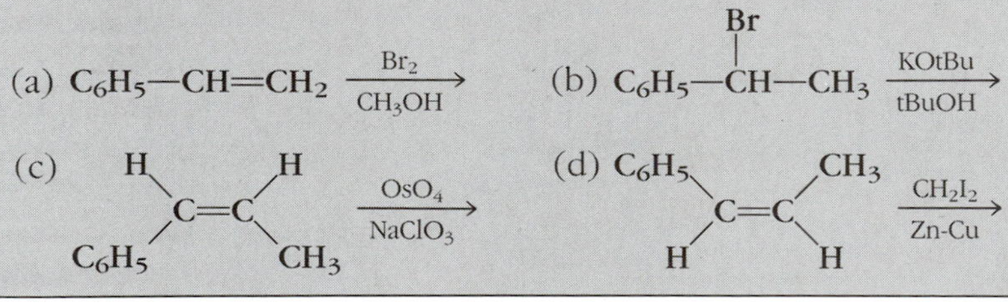

(a) $C_6H_5-CH{=}CH_2 \xrightarrow[CH_3OH]{Br_2}$

(b) $C_6H_5-CH(Br)-CH_3 \xrightarrow[tBuOH]{KOtBu}$

(c) $(H)(C_6H_5)C{=}C(H)(CH_3) \xrightarrow[NaClO_3]{OsO_4}$

(d) $(C_6H_5)(H)C{=}C(CH_3)(H) \xrightarrow[Zn\text{-}Cu]{CH_2I_2}$

8.7 CYCLIZATION REACTIONS

Many of the reactions that take place between two molecules can also be carried out *intramolecularly* with a molecule that has two functional groups. For reactions that form new bonds, the intramolecular process will generate a cyclic compound.

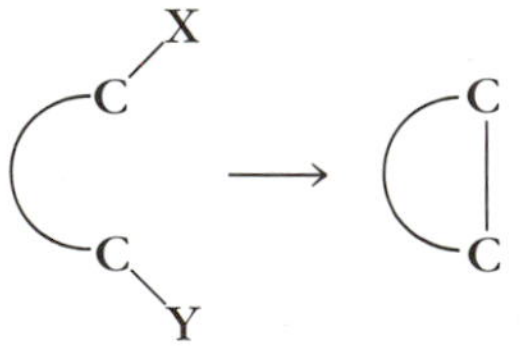

$$\text{C–X} \ \cdots \ \text{C–Y} \longrightarrow \text{C–C (ring)}$$

Frequently an intramolecular reaction will proceed more easily, because the two interacting functional groups are held in close proximity.

As we discussed previously (Sections 3.2, 3.4) three- and four-membered rings are highly strained, and rings of these sizes are produced only by using specialized reactions, such as epoxidation (Section 5.12) and cyclopropanation (Section 5.7) of alkenes. Small-ring compounds can sometimes be formed by coupling reactions of dihalides with reactive metals in a process analogous to the Wurtz reaction (Section 1.11):

$$CH_2(CH_2\text{—}Cl)_2 \xrightarrow{Zn} \text{cyclopropane} \quad 95\%$$

$$ClCH_2\text{—}CH_2\text{—}Br / CH_2\text{—}CH_2\text{—}Br \xrightarrow{Li\text{—}Hg} \text{cyclobutane} \quad 70\%$$

$$\text{PhCHBr—CH}_2\text{—CH}_2\text{Br} \xrightarrow{Zn\text{—}Cu} \text{phenylcyclopropane} \quad 75\text{–}85\%$$

These reactions are successful in the synthesis of three- and four-membered rings, where the two carbon atoms having halogen substituents are always in close proximity, but the procedure cannot be used to prepare larger rings.

Other cyclization reactions are most useful for the generation of five- and six-membered rings. Formation of a five- or six-membered ring is often favored in a cyclization reaction for two reasons. First, five- and six-membered rings are not strained, and the activation energies for their formation are comparable with those of intermolecular reactions. A much higher activation energy is expected for closure of a strained three- or four-membered ring. Second, the probability of forming five- or six-membered rings is relatively high. With only three or four atoms intervening between the two atoms to be joined, there is still a good probability that the molecule will be in a conformation that is appropriate for the cyclization (i.e., with C—X and C—Y in close proximity).

1 2 3 C—X C Y   1 2 3 4 C C

The likelihood that the two reactive centers will be in close proximity decreases for a larger number of intervening atoms, and closure of seven-membered and larger rings is often unfavorable. Cyclization to generate five- and six-membered rings is by far the most common mode of ring formation, and *intramolecular reactions that occur via a five- or six-membered ring are the basis for many important synthetic, industrial, and biological processes.* This applies not only to the formation of cyclic compounds, but also to reactions which proceed via a cyclic intermediate or transition state, such as hydrogen bond formation (Section 4.9) and pyrolytic eliminations (Section 6.9).

Reactions that generate five- and six-membered rings are illustrated by the examples that we will present in the remainder of this section. The first of these examples is an acid-catalyzed cyclization of a 1,5-diene to produce a six-membered ring.

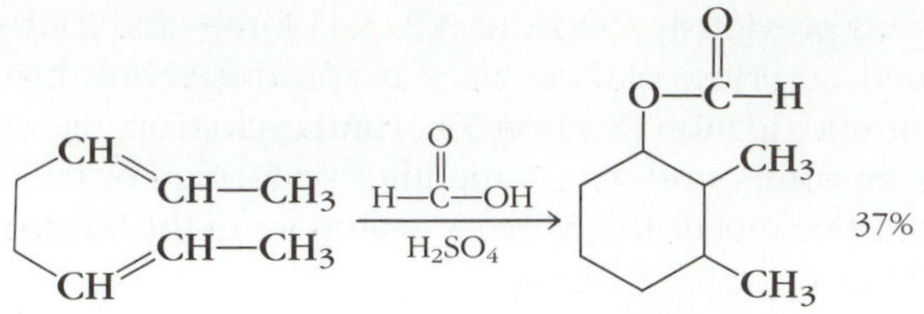

If you consider the other reactions that could take place between a strong acid and a diene (such as simple addition to one or both of the double bonds), the 37% yield is surprisingly high. The reaction proceeds by initial protonation of one of the double bonds to generate a secondary cation. This relatively unstable cation then undergoes intramolecular attack by the π electrons of the remaining double bond to produce another carbocation. In the final steps of the reaction this new cation reacts with formic acid (HCO_2H) by the same steps of nucleophilic attack and proton transfer that we discussed for polar additions to alkenes (Section 5.8). The reversibility of carbocation reactions in strongly acidic conditions favors the formation of a product with a six-membered ring because six-membered rings are more stable than analogous five-membered ring compounds (Section 3.2).

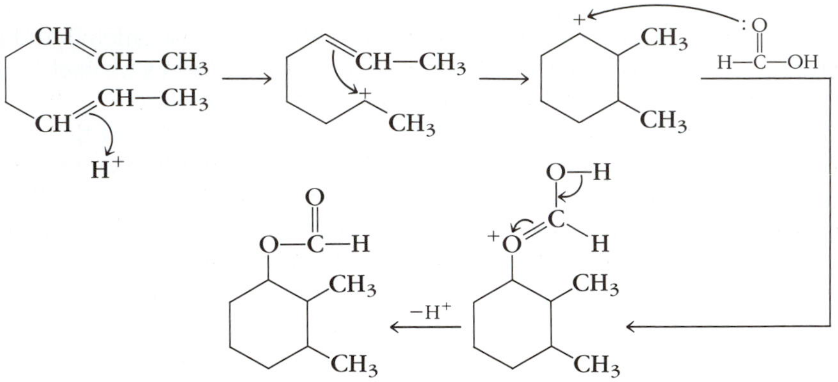

The second example is another carbocation cyclization, and the carbocation is generated by initial reaction of the double bond with halogen.

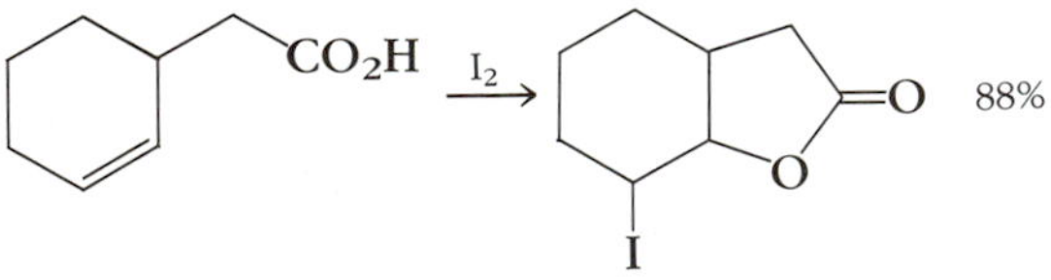

Initial attack by iodine generates an iodonium ion (Section 5.8), which then undergoes intramolecular attack by the carboxylic acid group.

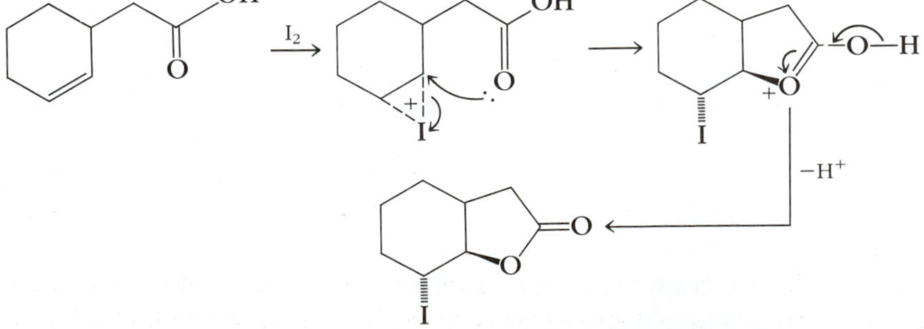

These steps are not reversible, and the five-membered ring is formed preferentially. The acid-catalyzed reaction of the previous example is thermodynamically controlled, but this example is a case of kinetic control. Notice that the polar addition proceeds by the expected *anti* mode, so the oxygen and iodine atoms are *trans*. As you will see in Sections 23.3 and 23.4, this type of cyclization reaction can be quite valuable in organic synthesis.

The third example is once again an acid-catalyzed cyclization, but this time two successive cyclizations occur to generate two six-membered rings.

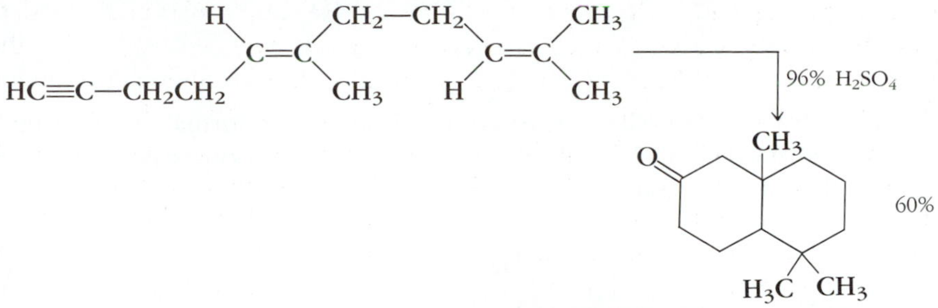

$HC{\equiv}C{-}CH_2CH_2(H)C{=}C(CH_3)CH_2{-}CH_2(H)C{=}C(CH_3)CH_3$ $\xrightarrow{96\%\ H_2SO_4}$ 60%

The details of the reaction are summarized in the following scheme. Initial protonation occurs to generate carbocation **(i)**, and this undergoes attack by the π electrons of the other double bond to form a six-membered ring. A new tertiary carbocation **(ii)** is produced in this step, and intramolecular attack by the π electrons of the triple bond closes the second six-membered ring. Cation **(iii)** is a *vinyl* cation, and this corresponds to the intermediate formed during the hydration of an alkyne (Section 7.5). Vinyl cations are highly reactive (Section 12.7), and rapid nucleophilic attack by water leads to the ketone via the enol.

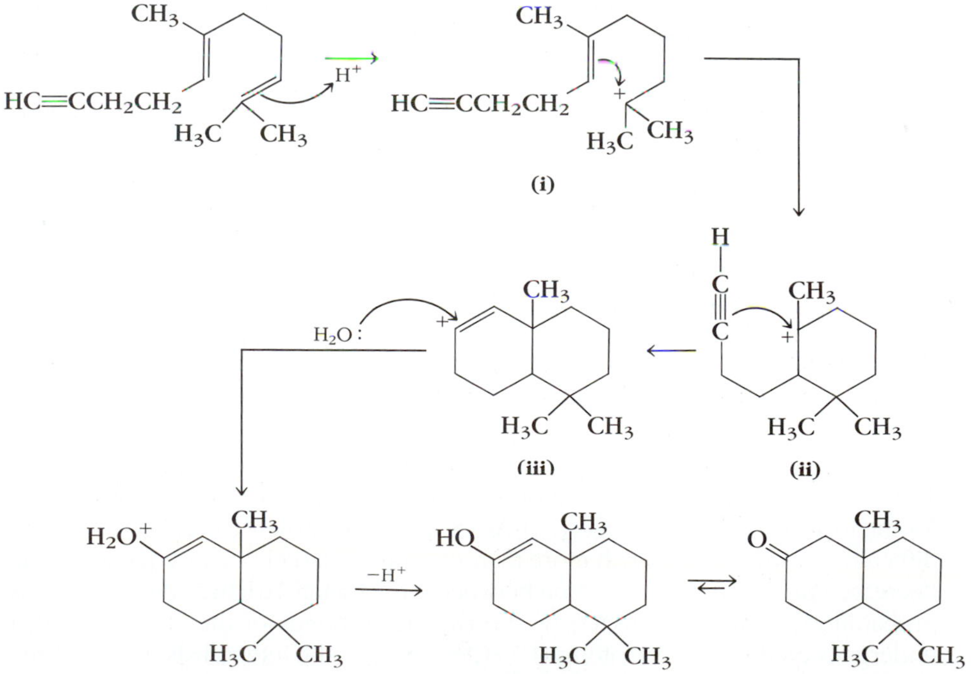

$HC{\equiv}CCH_2CH_2$, CH_3, H_3C, CH_3 $\xrightarrow{H^+}$ **(i)** → **(ii)** → **(iii)** + H_2O: → H_2O^+ $\xrightarrow{-H^+}$ HO (enol) ⇌ O (ketone)

Several different cations could be formed by protonation of the starting compound, but protonation of an alkene is reversible and only the formation of the tertiary cation **(i)** leads to the observed product. The other tertiary cation **(iv)** that might be generated by protonation of the starting material could cyclize to a five-membered ring with a secondary cation or to a four-membered ring with a tertiary cation. Neither pathway would be as favorable as that for the observed reaction.

CH_3, +, $HC{\equiv}CCH_2CH_2$, H_3C, CH_3

(iv)

Our final example of cyclization reactions illustrates the formation of a ring by intramolecular alkylation of an alkyne. A total of four carbon-carbon bonds are formed in this two-reaction sequence:

$$Br{-}CH_2{-}(CH_2)_3{-}CH_2{-}Br \xrightarrow{Na^{+\,-}C{\equiv}CH} HC{\equiv}C{-}CH_2{-}(CH_2)_3{-}CH_2{-}C{\equiv}CH$$

1. $NaNH_2$
2. $BrCH_2{-}(CH_2)_3{-}CH_2Br$

→ cyclic product: $CH_2{-}C{\equiv}C{-}CH_2$ / $(CH_2)_3$ $(CH_2)_3$ / $CH_2{-}C{\equiv}C{-}CH_2$

57%

In the first reaction 1,5-dibromopentane reacts with two equivalents of sodium acetylide to form 1,8-nonadiyne, and this is then treated with sodium amide to form the disodium salt. The second reaction proceeds by two distinct alkylation steps.

$$HC{\equiv}C{-}CH_2{-}(CH_2)_3{-}CH_2{-}C{\equiv}CH \xrightarrow{NaNH_2} Na^{+\,-}C{\equiv}C{-}CH_2{-}(CH_2)_3{-}CH_2{-}C{\equiv}C^{-}Na^{+}$$

+ $Br{-}CH_2{-}(CH_2)_3{-}CH_2Br$ → $(CH_2)_3$-linked intermediate: $CH_2{-}C{\equiv}C{-}CH_2$ / $(CH_2)_3$ $(CH_2)_3$ / $H_2C{-}C{\equiv}C^{-}$ Na^{+} $CH_2{-}Br$ → $CH_2{-}C{\equiv}C{-}CH_2$ / $(CH_2)_3$ $(CH_2)_3$ / $CH_2{-}C{\equiv}C{-}CH_2$

The overall yield for the two reaction sequence is 57%, and this is unusually high for the formation of a large ring (C_{14}). As we indicated earlier in this discussion, the difficulty in forming rings with more than six atoms is largely a consequence of the decreased probability of interaction between the "distant" reactive centers. But this probability is markedly affected by the two triple bonds of the compound that undergoes cyclization. Four atoms (C—C≡C—C) of the internal alkyne and three atoms (C—C≡C) of the terminal alkyne are held as linear arrays. This greatly reduces the conformational flexibility of the reacting species, and cyclization becomes more likely despite the large ring that is formed.

EXERCISE 8.7

Draw the cyclic product that would be formed in each of the following reactions:

(a) $CH_2{=}CH{-}(CH_2)_2{-}CO_2H \xrightarrow{I_2}$ (b) $Br{-}(CH_2)_{10}{-}C{\equiv}CH \xrightarrow{NaNH_2}$

(c) $CH_2{=}C(CH_3){-}(CH_2)_3{-}C(CH_3){=}CH_2 \xrightarrow{H_2SO_4,\ H_2O}$

(d) $Br{-}CH_2{-}C(CH_3)_2{-}CH_2Br \xrightarrow{Na}$

8.8 SELECTIVE REACTIONS OF COMPOUNDS WITH TWO REACTIVE CENTERS

In Section 8.1 we suggested three categories for the reactions of difunctional compounds. In Sections 8.4–8.6 we showed you how a double bond or a benzene ring could greatly affect the reactivity of the respective allylic or benzylic positions, and in Section 8.7 we illustrated the interaction of two reactive centers to form a cyclic product. Now we will turn to the third category and introduce you to compounds containing two reactive centers with different reactivities. Unless the two reactive centers are identical, they cannot have identical reactivity, and with an appropriate reagent reaction will occur selectively at just one of them.

What sort of a difference in reactivity is necessary if you are to observe a "satisfactory" ratio of products? A 51:49 ratio would not be judged adequate, but you would almost certainly be satisfied with a 99:1 ratio. The criteria for acceptability in any individual reaction would depend on a variety of factors such as the difficulty of carrying out the reaction, the cost of the reagents, and the ease of purification to remove unwanted components. In the end, however, your goal would be to isolate just one of the several products that might conceivably be formed. For the examples presented in this section, we will usually show only the major product, and the percentage yield indicates how much of that specific compound was actually isolated in reasonably pure form. Many reactions of compounds with two reactive centers are not selective, but we have chosen examples of useful organic processes which do exhibit high selectivity.

Differences in reactivity between two reactive centers are ordinarily a result of two factors, *steric effects* and *electronic effects.* These do not always operate in the same direction, and a prediction regarding selectivity may also require you to guess whether steric or electronic effects will predominate in a particular case.

Most of the reactions that we will discuss involve carbon-carbon double and triple bonds as the functional groups, and you will be able to evaluate steric effects by the number of alkyl substituents and their bulkiness. In the following sequence of alkenes, steric effects would increase from **(i)** to **(iv).**

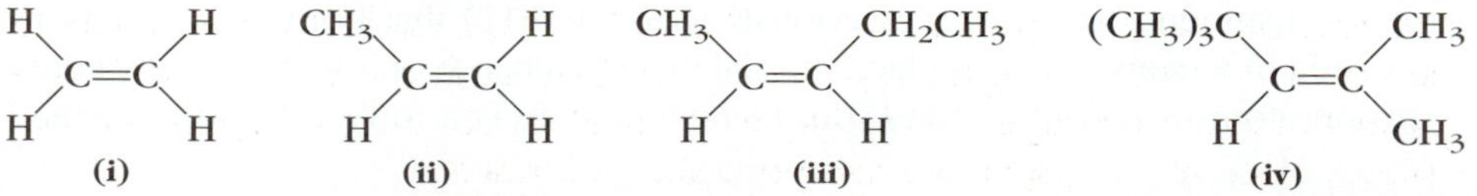

$H_2C{=}CH_2$ **(i)** $\quad CH_3CH{=}CH_2$ **(ii)** $\quad CH_3CH{=}CHCH_2CH_3$ **(iii)** $\quad (CH_3)_3CCH{=}C(CH_3)_2$ **(iv)**

A reaction governed largely by steric effects would proceed most rapidly with ethylene but progressively more slowly for **(ii), (iii)** and **(iv).** On the other hand, electronic effects on reactivity would show the opposite trend. Alkyl groups can stabilize a free radical or a carbocation via inductive effects, and **(iv)** would be the most reactive of these alkenes in a reaction governed only by electronic effects. Compounds **(iii), (ii)** and **(i)** would react progressively more slowly.

How can you anticipate whether steric or electronic effects will predominate in a particular reaction? Sometimes you can make a prediction if you know the reaction mechanism. Ionic addition reactions of alkenes, for example, tend to be governed by electronic effects, particularly when the attacking reagent is small. These reactions proceed via cationic intermediates, and faster reaction is expected for a more highly substituted double bond because it will afford a more stable cation. Similar preferences are observed in the case of free radical additions, although steric effects become increasingly important as the size of the attacking species is increased. For nonpolar or concerted addition reactions, steric effects tend to predominate, and more rapid reaction is expected at a less substituted double bond.

Some of these trends are illustrated in the reactions of 4-vinylcyclohexene. The double bond in the ring is disubstituted while that of the side chain is monosubstituted. Therefore the double bond of the ring should be more electron-rich and should be the primary site for an ionic reaction (even though reaction of either double bond could afford a secondary cation). In contrast, the double bond of the side chain is less hindered and it should be the preferential site for relatively nonpolar reactions. The following examples substantiate these expectations.

$CH{=}CH_2$ (4-vinylcyclohexene) $\xrightarrow[Ni_2B]{H_2}$ CH_2CH_3 (4-ethylcyclohexene) 98%

$CH{=}CH_2$ (4-vinylcyclohexene) $\xrightarrow[2.\ H_2O_2,\ NaOH]{1.\ (CH_3CH(CH_3){-}CH(CH_3))_2BH}$ CH_2CH_2OH (cyclohexenyl) 72%

$CH{=}CH_2$ (4-vinylcyclohexene) $\xrightarrow[2.\ NaBH_4]{1.\ Hg(OCCH_3)_2\ (O\ double-bonded),\ H_2O}$ $CH(OH){-}CH_3$ (cyclohexenyl) 100%

$CH{=}CH_2$ (4-vinylcyclohexene) $\xrightarrow{CH_3COOH\ (C{=}O)}$ epoxide (O bridging ring) with $CH{=}CH_2$ 86%

The first of these reactions, catalytic hydrogenation of 4-vinylcyclohexene, is a nonpolar addition, and it occurs preferentially at the less substituted double bond. The nickel boride catalyst shown in this example has been found quite useful in selective reductions, and it can also be employed for the partial reduction of alkynes to alkenes (Section 7.6). The second and third reactions, hydration via hydroboration and oxymercuration, respectively, also take place preferentially at the less hindered vinyl substituent. We showed previously (Section 5.11) that both these reactions proceed in a manner which places partial positive charge on the more substituted carbon atom of a double bond. But in neither case is a full carbocation formed (recall the absence of rearrangements for these two reactions). As a result, steric effects play a dominant role, and reaction occurs selectively at the less substituted double bond. The final reaction shown for 4-vinylcyclohexene, epoxidation with peracetic acid, occurs at the double bond of the ring. Epoxidation of an alkene is a reaction for which electronic effects predominate, and it occurs more rapidly at the more substituted double bond.

A similar pattern of reactivity is observed for limonene (4-isopropenyl-1-methylcyclohexene), a naturally occurring diene that is the major volatile constituent in citrus peels. The two double bonds in this compound differ sterically and electronically. The double bond of the ring is more hindered because it is trisubstituted while that in the side chain is only disubstituted. Steric effects will therefore favor reaction at the isopropenyl group, but electronic effects will favor reaction at the more highly substituted double bond in the ring.

1. $Hg(OCCH_3)_2$, H_2O; 2. $NaBH_4$ → 70% + 7% + unreacted limonene 19%

CH_2I_2, Zn—Cu → 50% + unreacted limonene 29%

CH_3COOH (peracetic acid) → 63%

As we showed for the reaction of 4-vinylcyclohexene, oxymercuration is governed largely by steric factors, and the less substituted double bond undergoes preferential hydration. The nonpolar cyclopropanation reaction similarly takes place at the less hindered double bond. In contrast, reaction with peracetic acid occurs at the more substituted double bond. As we showed for the corresponding reaction of 4-vinylcyclohexene, epoxidation is governed largely by electronic effects.

The selective reaction of a polyfunctional compound requires reactivity differences at the two reactive centers, and the attacking reagent must also be selective. A selective reagent is usually somewhat less reactive than a related reagent that lacks selectivity. You would expect a nonselective reagent to attack both reactive centers of a difunctional compound, and this is illustrated by the following reaction of limonene with a monoalkylborane to form a cyclic product.

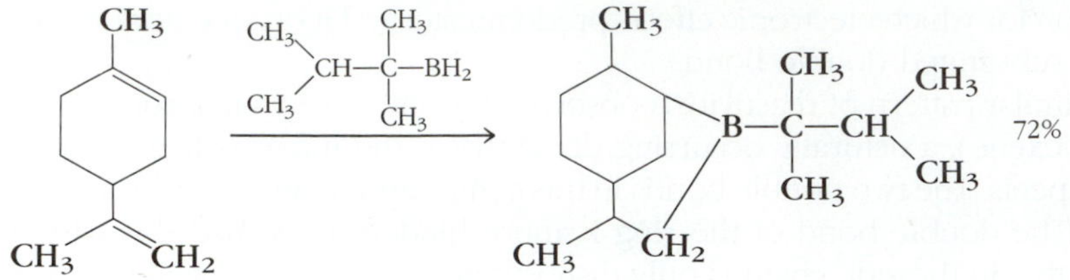

72%

In this reaction the boron reagent has *two* active hydrogens. The initial reaction still occurs more rapidly at the less substituted double bond, but the resulting intermediate has another reactive hydrogen atom on boron. An intramolecular reaction results, with favorable six-membered ring formation.

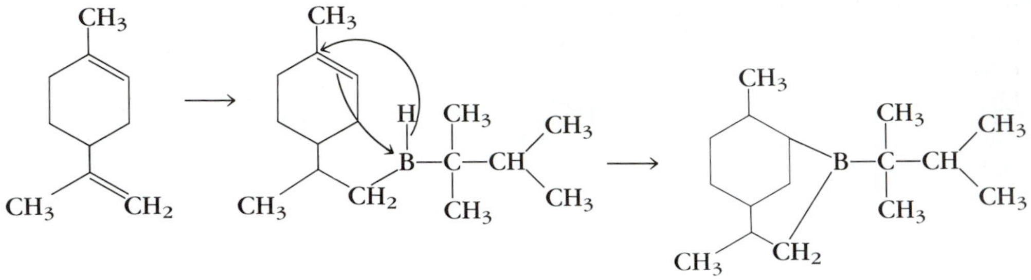

Both functional groups are attacked by the reagent, so this is an example of a reaction that is not selective. You should contrast this with the reaction we showed earlier between 4-vinylcyclohexene and disiamylborane. Disiamylborane has only a single active hydrogen and it reacted selectively with just one of the double bonds.

Most functional groups containing multiple bonds are susceptible to reduction, and the selective reduction of one such group is frequently an important problem in organic synthesis. In many cases it is possible to reduce only the desired functional group if the proper combination of reducing agent and reaction conditions is chosen. The following reaction illustrates the reduction of a carbon-carbon triple bond in the presence of a carbon-carbon double bond.

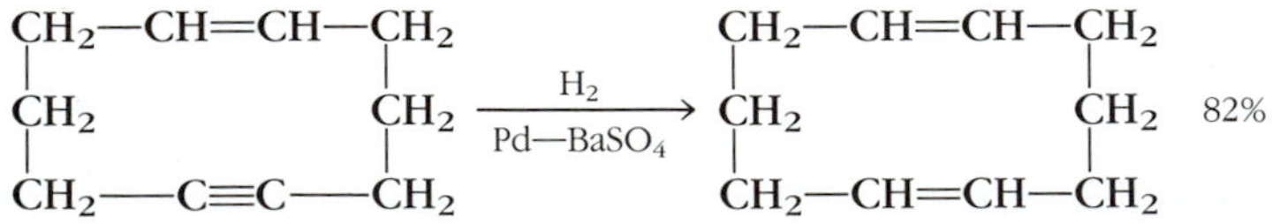

H_2, Pd—$BaSO_4$ — 82%

With many metal catalysts the reaction would not be selective and both multiple bonds would be reduced. The palladium-barium sulfate catalyst (Section 7.6) is a selective reagent that can be used to reduce alkynes to alkenes, so it clearly reduces the triple bond of an alkyne more readily than the double bond of an alkene. The present example, formation of 1,6-cyclodecadiene, is merely a situation in which both a triple bond and a double bond are contained in the same molecule.

Another way to selectively attack just one of two functional groups is to temporarily modify the other in such a way that it becomes unreactive. The simplest way to do this is by means of acid-base reactions, as shown by the following reaction, in which a triple bond is reduced in the presence of a carbonyl group.

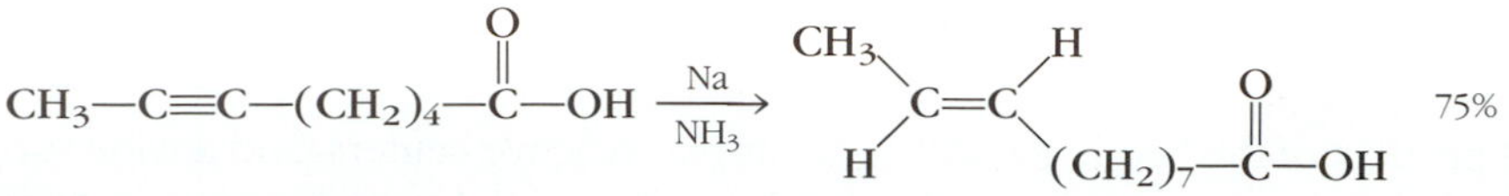

$$CH_3-C\equiv C-(CH_2)_4-\overset{\overset{\large O}{\|}}{C}-OH \xrightarrow[NH_3]{Na} \underset{H}{\overset{CH_3}{}}\!\!\!\!\!\!\;\;C=C\!\!\!\!\!\!\;\;\underset{(CH_2)_7-\overset{\overset{\large O}{\|}}{C}-OH}{\overset{H}{}} \quad 75\%$$

In this reaction ammonia not only serves as solvent but also reacts with the carboxyl group to produce a salt.

$$CH_3-C\equiv C-(CH_2)_7-\overset{\overset{\large O}{\|}}{C}-OH + NH_3 \longrightarrow CH_3-C\equiv C-(CH_2)_7-\overset{\overset{\large O}{\|}}{C}-O^-\ NH_4^+$$

A neutral carboxyl group would react readily with a reducing agent (Section 16.5), but the *anion* is quite unreactive because transfer of an electron to a negatively charged group would be energetically unfavorable. The reduction of this difunctional compound therefore takes place selectively at the carbon-carbon triple bond. The carboxyl group, which has been *protected* by reaction with ammonia, remains as an unreactive carboxylate anion until the final stage of the reaction, when the crude reaction mixture is acidified.

A similar protection process has been used to reduce an internal triple bond in the presence of a terminal alkyne.

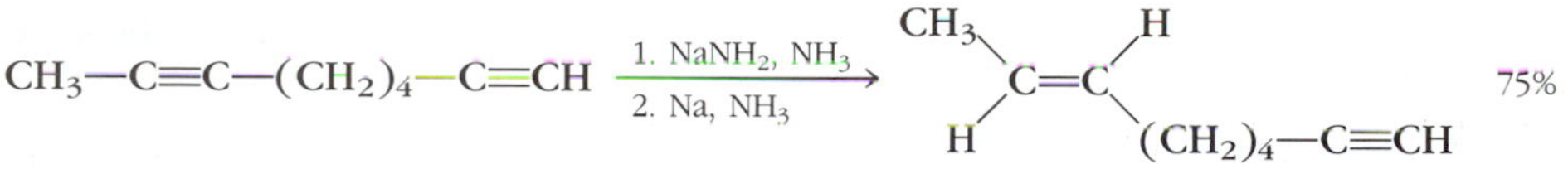

$$CH_3-C\equiv C-(CH_2)_4-C\equiv CH \xrightarrow[2.\ Na,\ NH_3]{1.\ NaNH_2,\ NH_3} \underset{H}{\overset{CH_3}{}}\!\!\!\!\!\!\;\;C=C\!\!\!\!\!\!\;\;\underset{(CH_2)_4-C\equiv CH}{\overset{H}{}} \quad 75\%$$

By first treating the starting material with sodium amide, the terminal triple bond is converted to its sodium salt.

$$CH_3-C\equiv C-(CH_2)_4-C\equiv C^-Na^+$$

The anionic group is resistant to electron transfer, so only the internal triple bond undergoes reduction. As in the previous example, addition of acid after the reaction is complete regenerates the original functional group.

Another type of selective reaction that you can predict easily involves alkylation of alkyl halides. As long as two alkyl halides have the same level of substitution (i.e., both are primary, both are secondary, etc.), the reactivity will decrease from the iodide to the bromide to the chloride. For example, sodium acetylide reacts preferentially at position 6 of 1-chloro-6-iodohexane to displace the more reactive iodide.

$$Cl-(CH_2)_6-I \xrightarrow{NaC\equiv CH} Cl-(CH_2)_6-C\equiv CH \quad 75\%$$

This sort of selective reaction of dihalides has been put to great advantage by organic chemists in the synthesis of complex molecules, and the following sequence shows how several subunits can be combined in a specific and predictable way.

$$CH_3CH_2CH_2I \xrightarrow{NaC\equiv CH} CH_3CH_2CH_2C\equiv CH \xrightarrow{NaNH_2} CH_3CH_2CH_2C\equiv C^-Na^+ \xrightarrow{Br-(CH_2)_4-Cl} \underset{60\%}{CH_3CH_2CH_2-C\equiv C-(CH_2)_4-Cl}$$

The product of this sequence still contains two reactive centers, and a wide variety of further reactions are still possible for this compound. As you continue your study of organic chemistry, you will recognize how important it is to be able to construct and manipulate molecules that contain more than one functional group.

8.9 TERMS AND DEFINITIONS

Aromatic. Describes compounds related to benzene that exhibit unusual properties, including low reactivity in comparison with similar nonaromatic compounds.

Benzyl group. The group $C_6H_5CH_2$, where C_6H_5 is a phenyl group, and the C of the CH_2 is the benzylic carbon.

Benzylic. A term describing the carbon atom of a benzyl (or a substituted benzyl) group that is bonded to the aromatic ring; also used to describe a substituent on a benzylic carbon or to describe a reactive intermediate in which the reactive center is a benzylic carbon.

Conjugated. Describes two multiple bonds on four contiguous carbon atoms (i.e., C=C—C=C for conjugated double bonds), which form a continuous π system over the four carbons. Conjugation can also extend over a larger number of carbon atoms.

Diene. A hydrocarbon that contains two carbon-carbon double bonds.

HOMO. The highest occupied molecular orbital of a species.

Isolated multiple bond. A multiple bond that is not in conjugation with another multiple bond; all carbons attached directly to the multiple bond are saturated (sp^3) carbons.

LUMO. The lowest unoccupied molecular orbital of a species.

Phenyl group. The C_6H_5 group; the group that would be produced by removal of a hydrogen from a benzene molecule.

Polyene. A hydrocarbon that contains multiple carbon-carbon double bonds.

Polyfunctional. A molecule that contains two (or more) independent functional groups.

Polymer. A molecule that consists of repeating subunits; the number of subunits can be very large.

8.10 PROBLEMS

8.1 Draw the major product or products expected for each of the following reactions:

(a) 1,2-dimethylcyclohexene (CH_3, CH_3) $\xrightarrow{NBS}$?

(b) 1,2-dimethylcyclohexene (CH_3, CH_3) $\xrightarrow{SeO_2}$?

(c) $\xrightarrow[CCl_4]{Br_2}$?

(d) $\xrightarrow[\text{peroxides}]{HBr}$?

(e) $\xrightarrow{HCl}$?

CH_3 H
C=C
CH_3

(f) $\xrightarrow{Br_2}$?

8.2 Draw the major product or products expected for each of the following reactions:

(a) $\xrightarrow{CH_3\overset{O}{\overset{\|}{C}}OOH}$?

(b) $\xrightarrow{Br_2,\ CH_3OH}$?

(c) $\xrightarrow{NBS}$?

(d) $\xrightarrow{SeO_2}$?

(e) C≡C—H $\xrightarrow[H_2SO_4,\ HgSO_4]{H_2O,}$?

(f) CH_3 $\xrightarrow{NBS}$?

CH_3 CH_2

8.3 Draw the major product or products expected for each of the following reactions:

(a) $\xrightarrow{H_2,\ Pt}$?

(b) CH_3 $\xrightarrow{HCl}$?

(c) $\xrightarrow[CCl_4]{Cl_2}$?

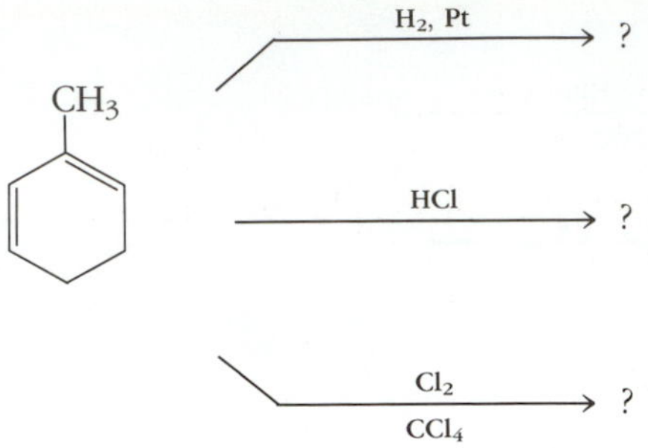

(d) $CH{=}CH{-}CH_3$ $\xrightarrow[\text{2. } NaBH_4]{\text{1. } Hg(OCCH_3)_2,\ CH_3OH}$?

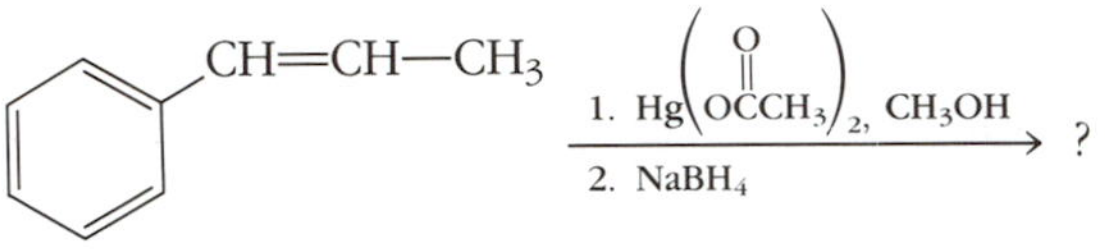

(e) $CH_2CH_2CO_2H$ $\xrightarrow{I_2}$?

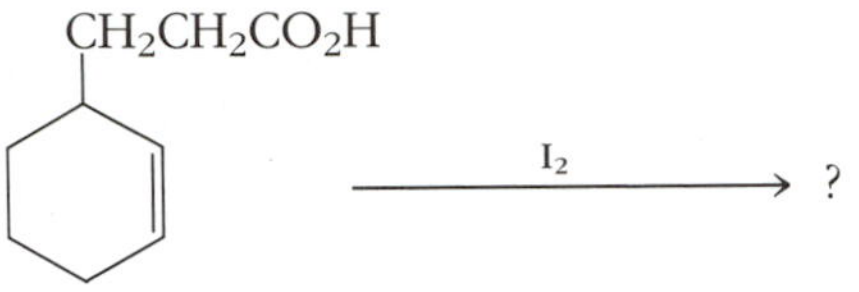

(f) CH_3, CH_3, H, C=C, H, C=C, H, H $\xrightarrow{HBr}$?

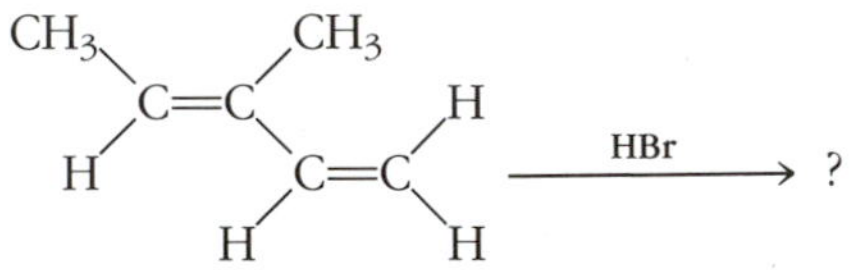

8.4 Draw the major product or products expected for each of the following reactions:

(a) $\xrightarrow[\text{2. } NaOH,\ H_2O_2]{\text{1. } (CH_3{-}CH(CH_3){-}CH(CH_3){-})_2BH}$?

(b) CH_3, CH_3, C=C, H, H, CH_2CH_2, C=C, H, H $\xrightarrow{CH_3COOH\ (\text{with } C{=}O)}$?

(c) $\xrightarrow{CH_3{-}CH(CH_3){-}C(CH_3)_2{-}BH_2}$?

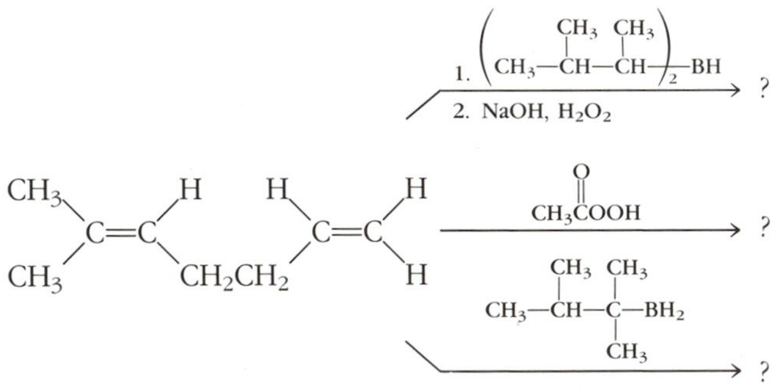

(d) CH_3, CH_3, CH_2 $\xrightarrow[\text{2. } NaOH,\ H_2O_2]{\text{1. } (CH_3{-}CH(CH_3){-}CH(CH_3){-})_2BH}$?

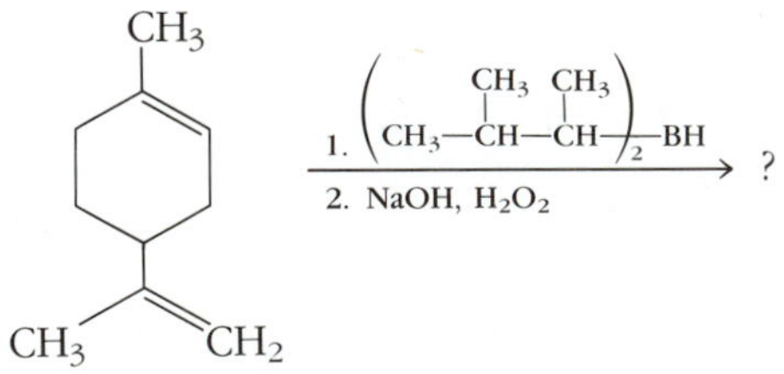

(e) C_6H_5–CH_2—CH=CH_2 $\xrightarrow{\text{NBS}}$?

(f) $\xrightarrow{\text{SeO}_2}$?

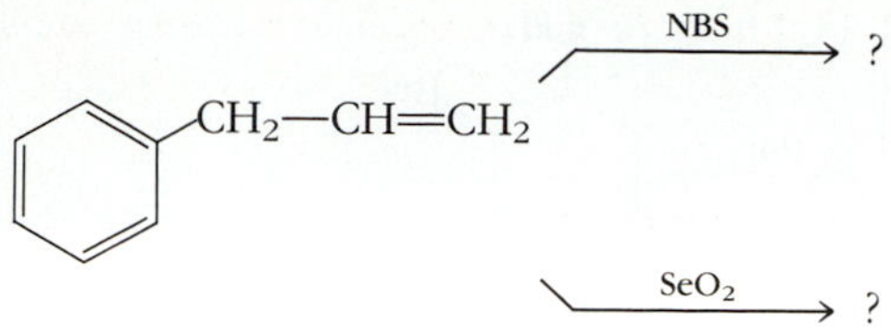

8.5 Suggest methods for carrying out the indicated conversions. More than one step may be necessary in each case.

(a) cyclopentene ⟶ cyclopentadiene

(b) $C_6H_5CH_2CH_2CH_3$ ⟶ $C_6H_5CH(Br)$—CH_2CH_3

(c) CH_3—C≡C—CH_2CH_2—CH=CH_2 ⟶ CH_3(H)C=C(H)CH_2CH_2CH=CH_2

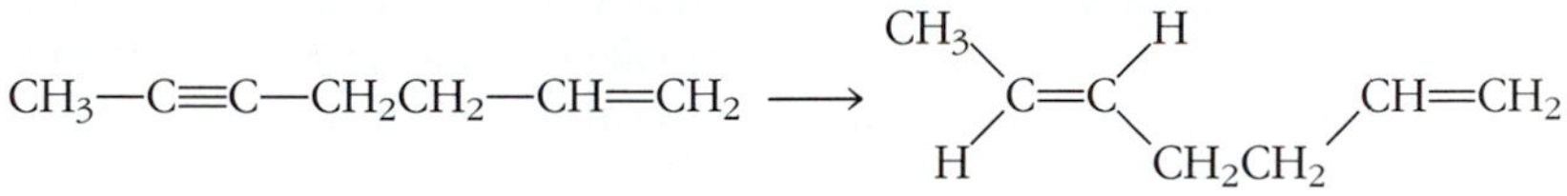

8.6 Deduce the structures of compounds **A–D.**

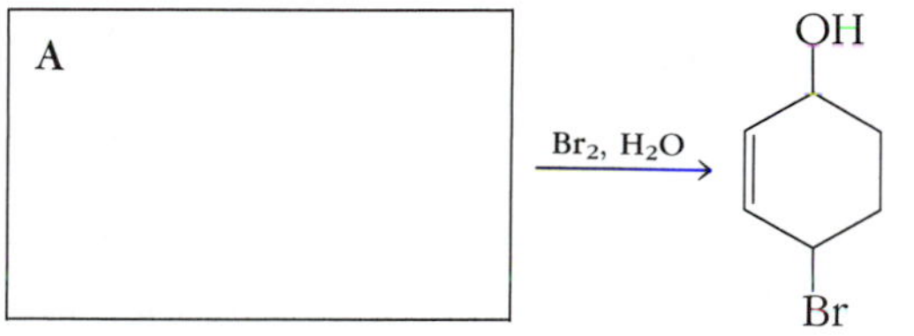

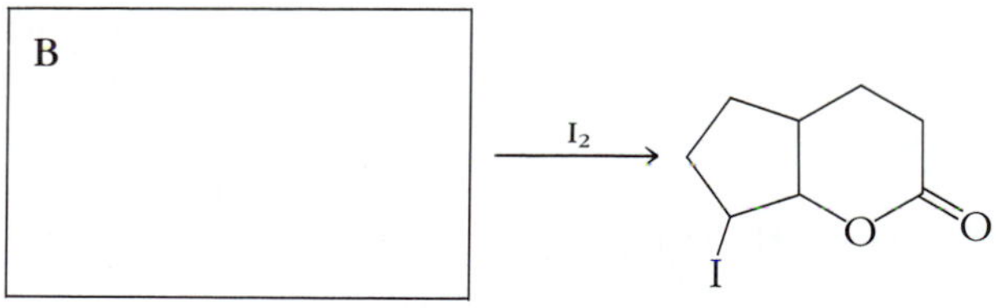

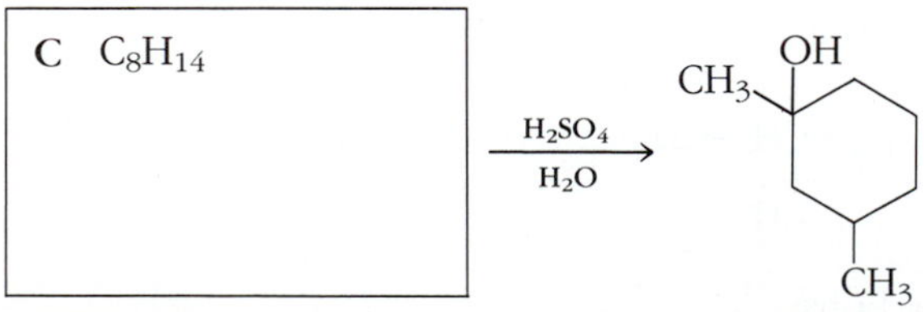

A $\xrightarrow{\text{Br}_2,\ \text{H}_2\text{O}}$ (OH, Br substituted cyclohexene)

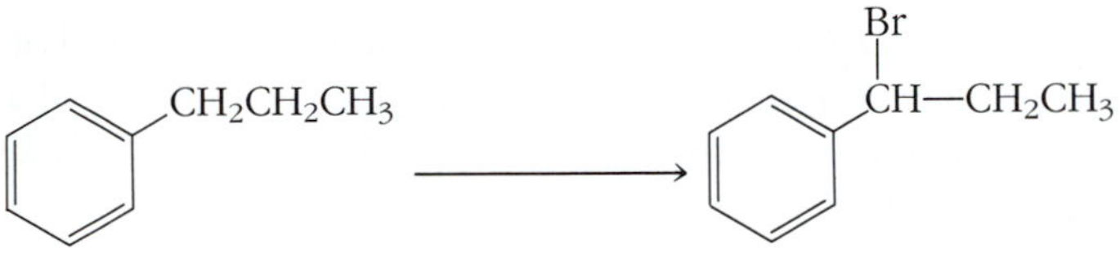

B $\xrightarrow{\text{I}_2}$ (iodo bicyclic lactone, I, O, O)

C C_8H_{14} $\xrightarrow[\text{H}_2\text{O}]{\text{H}_2\text{SO}_4}$ (CH_3, OH, CH_3 substituted cyclohexane)

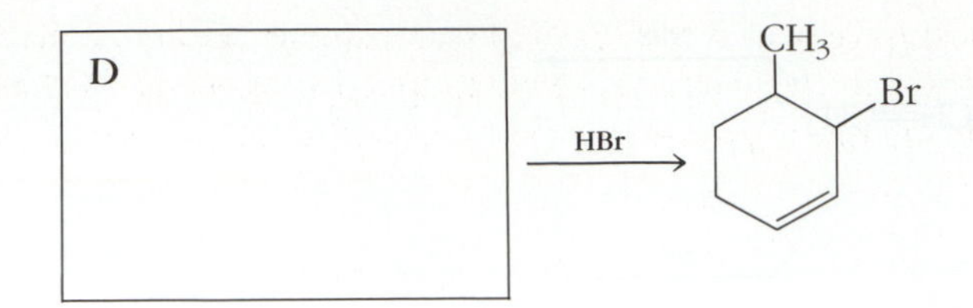

8.7 Write the reaction mechanism for formation of the product in each of the reactions shown in the preceding problem.

8.8 The reaction of 1,5-cyclooctadiene with borane affords a good yield of a mixture of two compounds with the molecular formula $C_8H_{15}B$. Suggest a reasonable structure for each of these compounds.

8.9 The reaction of 1,3-cyclooctadiene with NBS affords a 53% yield of monobromination product. The product consists of approximately equal amounts of two isomeric compounds having the molecular formula $C_8H_{11}Br$. Propose structures for each of these two products.

8.10 In Section 8.8 we showed that limonene affords a cyclic boron-containing product when it reacts with a monoalkylborane. However when limonene is first treated with disiamylborane (Section 7.5) and then oxidized with alkaline hydrogen peroxide, a single unsaturated alcohol is isolated in good yield. Deduce its structure.

8.11 When 1,3,5-tribromopentane is treated with zinc metal, a compound with the molecular formula C_5H_9Br is obtained in 72% yield. This compound undergoes catalytic hydrogenation at room temperature and atmospheric pressure to give another compound, $C_5H_{11}Br$. Deduce the structures of both compounds.

8.12 Remembering the difference in reactivity between alkyl chlorides and alkyl iodides, suggest methods of synthesizing the following compounds. All carbon atoms of the final product must be derived from reactants with the indicated number of carbons. Use any other reagents you wish.

(a) *Three* or fewer carbons:

$$CH_3{-}CH_2{-}CH_2{-}CH_2{-}\underset{\displaystyle CH_3}{\underset{|}{CH}}{-}CH_3$$

(b) *Four* or fewer carbons:

$$CH_3{-}\underset{\displaystyle CH_3}{\underset{|}{CH}}{-}CH_2{-}CH_2{-}\underset{\displaystyle CH_2CH_3}{\underset{|}{CH}}{-}CH_3$$

(c) *Five* or fewer carbons:

$$CH_3{-}CH_2{-}CH_2{-}\overset{\displaystyle CH_3}{\overset{|}{\underset{\displaystyle CH_3}{\underset{|}{C}}}}{-}CH_2{-}\underset{\displaystyle CH_3}{\underset{|}{CH}}{-}CH_2{-}CH_3$$

8.13 Describe the relationship between each pair of structural formulas as that of *constitutional* isomers, two drawings of the *same* species (although perhaps in two different conformations), *diastereomers,* or completely *different* species that are not isomeric.

(a)

(b)

(c)

(d)

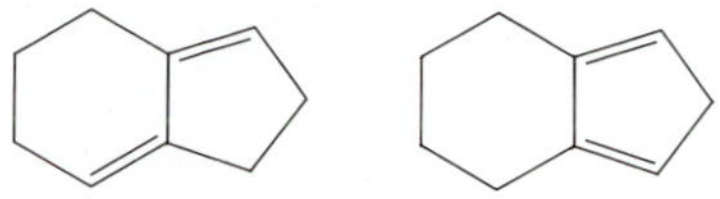

(e)

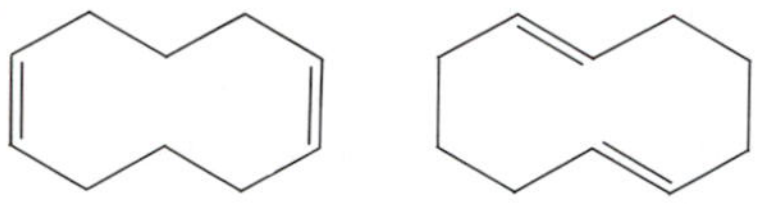

(f) + +

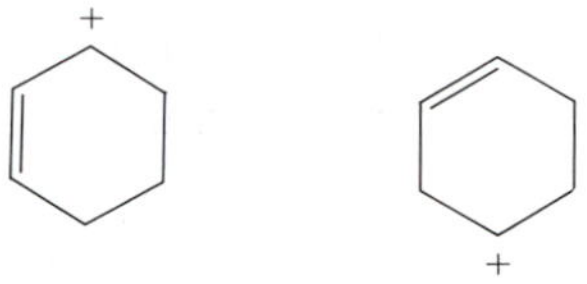

(g) CH_3 + CH_3 +

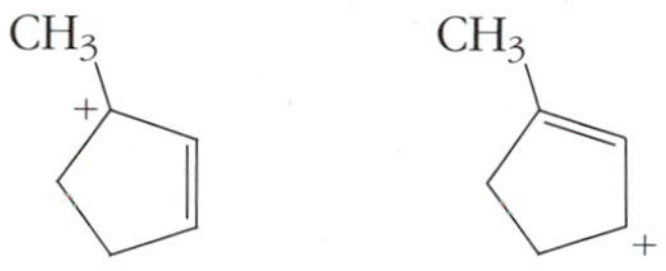

(h) $CH_2\cdot$ CH_2 ·

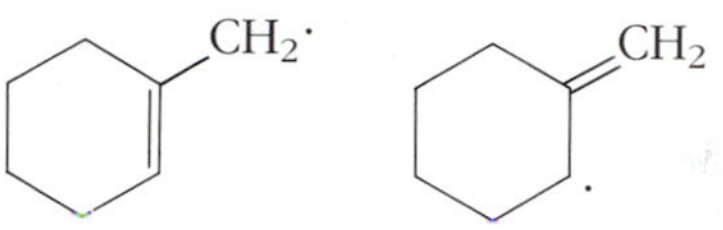

(i) $CH{=}CH_2$ $^+CH{-}CH_3$

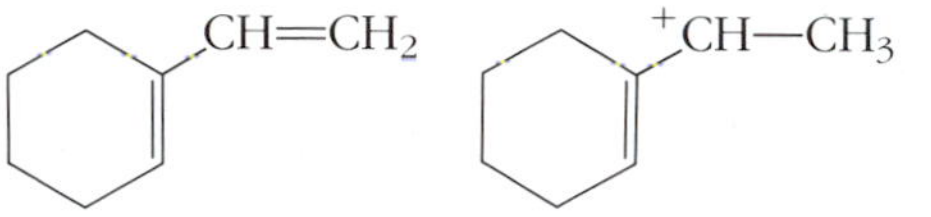

(j) CO_2CH_3 CH_3O_2C CO_2CH_3 CO_2CH_3

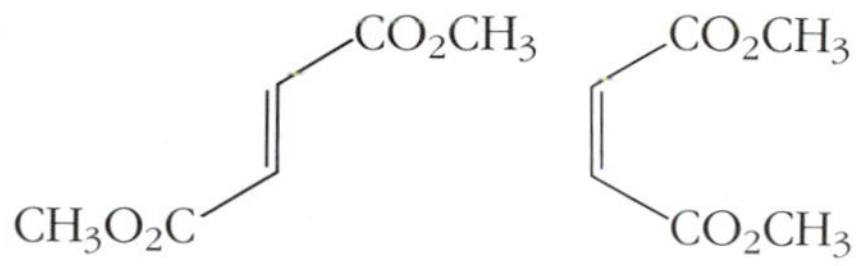

8.14 Suggest a method of synthesis for each of the following compounds. All carbon atoms of the final product must be derived from reactants with the indicated number of carbons. Use any other reagents you wish.

(a) *Four* or fewer carbons:

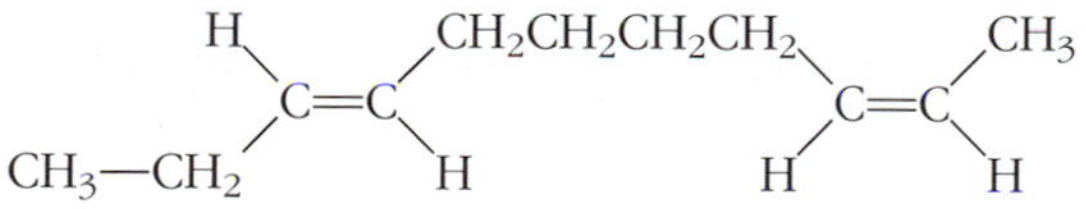

$$\mathrm{CH_3{-}CH_2(H)C{=}C(H)CH_2CH_2CH_2CH_2(H)C{=}C(H)CH_3}$$

(b) *Six* or fewer carbons:

$$\mathrm{CH_3CH_2CH_2(H)C{=}C(H)(CH_2)_6{-}C{\equiv}C{-}CH_2CH_2CH_3}$$

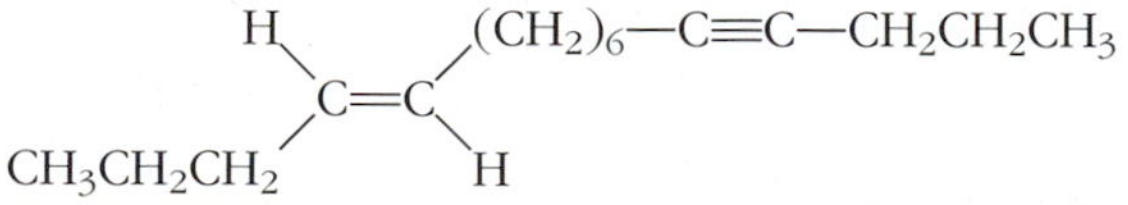

CHAPTER 9

SPECTROSCOPY AND MOLECULAR STRUCTURE

At any instant a molecule, an atom, or even a reactive intermediate must exist in just one of many possible energy states. These various energy states are *quantized,* so that each unique state is associated with a specific energy. Consequently, a species in one energy state can absorb or emit energy (a photon) and thereby proceed to a new energy state. Because the energy states themselves are quantized, the differences between them also correspond to discrete, well-defined energy values. Studies of such absorption or emission of electromagnetic radiation constitute the field of *spectroscopy,* and we will show you in this chapter how you can use spectroscopic data to obtain information about molecular structure.

9.1 THE ELECTROMAGNETIC SPECTRUM

The term **spectrum** is often used to describe the different colors of visible light that can be seen when a beam of light passes through a prism, but chemists use the term in a more general sense to describe the different wavelengths of electromagnetic radiation. The field of **spectroscopy** involves studies of the absorption or emission of the different wavelengths of electromagnetic radiation by chemical compounds. It is an extremely important field because it provides information about the molecular structure of the compounds absorbing or emitting the radiation. The combination of various possible energy states tends to be unique for that species, and this means that you can positively identify a compound by matching the experimental results with those previously observed for a known structure. If the spectroscopic results are the same, the structures will be the same. Conversely, if the spectroscopic results are different, you will know that the two species are not the same. The possible energy states of a particular species are related to the molecular structure in a way that makes it possible to *predict* the spectroscopic results for a possible structure. This means that you can compare the experimental and predicted results, as illustrated in Figure 9.1.

The task of determining the structure of an organic compound can be formidable. Fortunately, a chemist usually has some ideas about the molecular structure. For example, if you isolate a compound from a reaction, you have only three possible choices: (1) the compound is the desired product; (2) the compound is the product of a side reaction; or (3) the compound is unreacted starting material. You could easily evaluate the last possibility by simple comparison of the spectroscopic data for the compound isolated with those for the reactant; identical spectra would show that no reaction had taken place. To determine if the compound is the desired product or one of the possible by-products, the approach outlined in Figure 9.1 could be employed. You could predict the spectroscopic results for each possible structure, and whichever prediction agreed most closely with the observed data would indicate the correct structure. If none of your predictions agreed well with the experimental results, you would need to propose yet another structure.

We have shown in Figure 9.2 that the electromagnetic spectrum spans a very wide range of energy, and almost all parts of the spectrum have been used for spectroscopic studies in chemistry. Those regions which have proved to be particularly useful in organic spectroscopy are shown as the shaded areas in the figure. Recall that electromagnetic radiation can be specified in terms of *energy,* E (using any of a variety of units), *frequency,* ν ($E = h\nu$), or *wavelength,* λ ($E = hc/\lambda$). There is an inverse relationship between energy and wavelength; therefore very short wavelength radiation corresponds to high energy.

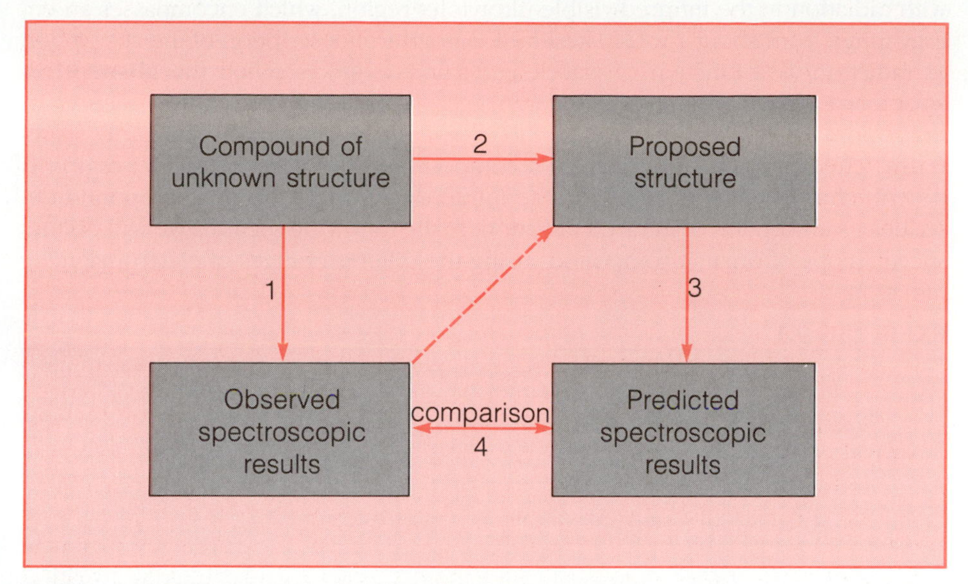

Figure 9.1 Structure Evaluation by Spectroscopy.
(1) A compound of unknown or uncertain structure is studied by spectroscopic methods. (2) A structure is proposed on the basis of other information (such as knowledge of the starting material in a reaction). (3) Spectroscopic results are predicted for the proposed structure. (4) The observed and predicted spectroscopic results are compared; agreement of the two supports the proposed structure. Sometimes the experimental results provide hints that help you suggest the proposed structure, as indicated by the dotted line.

The very low energy radiation of the radio frequency (rf) portion of the spectrum has found wide application to studies of molecules in the presence of strong magnetic fields. Its specific application to energy states of atomic nuclei in organic compounds, *nuclear magnetic resonance,* is discussed at length in the following sections. Radiation from other parts of the electromagnetic spectrum can be absorbed directly by many organic molecules in the absence of a magnetic field. For such direct absorption the greatest amount of structural information is obtained

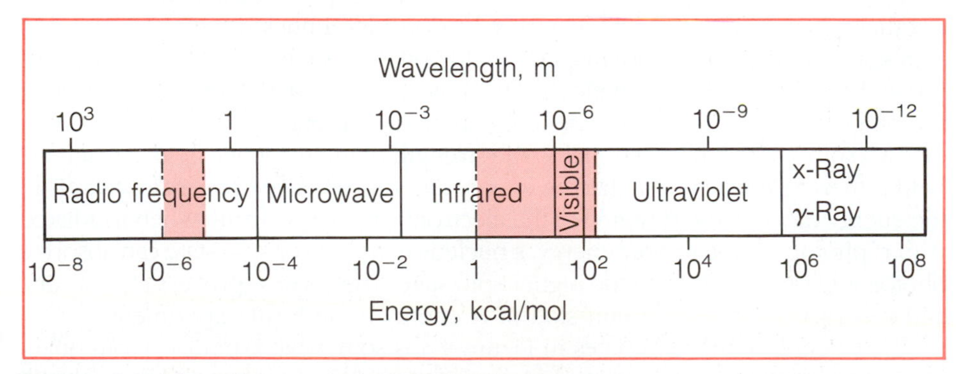

Figure 9.2 The Electromagnetic Spectrum.
In several regions of the spectrum the radiation is particularly useful for organic spectroscopy, and we have indicated those regions by shading.

with radiation in the infrared-visible-ultraviolet region, which encompasses an energy range from about 1 to 240 kcal/mol. A graphical description of the absorption of radiation as a function of wavelength (or energy) is called the **absorption spectrum** of that compound. Similarly, a graphical description of the emission of radiation as a function of wavelength (or energy) is called the **emission spectrum.** The absorption spectrum of a compound in the ultraviolet (UV) region is described as its UV spectrum, and the infrared (IR) spectrum of a compound describes its absorption of infrared radiation. In the following sections we will discuss spectroscopy of organic compounds using the different ranges of electromagnetic radiation.

EXERCISE 9.1

(a) Using Figure 9.2, estimate the energy (in kilocalories per mole) at each of the following wavelengths: $\lambda = 100$ m, 0.1 m, 240×10^{-9} m.
(b) For each of the following energy values, estimate the value of λ: $E = 10^7$ kcal/mol, 50 kcal/mol, 10^{-5} kcal/mol. In what region of the electromagnetic spectrum (infrared, ultraviolet, etc.) is each?

9.2 NUCLEAR MAGNETIC RESONANCE: THE BASIC EXPERIMENT

Certain nuclei can exist in spin states that have different energies in the presence of a magnetic field. This is the case for all nuclei which possess an odd number of either neutrons or protons (or both). The study of nuclear spin states in the presence of a strong magnetic field by absorption and emission of electromagnetic radiation is called **nuclear magnetic resonance spectroscopy,** frequently abbreviated as *NMR spectroscopy.* This name results from the classical description of a spinning nucleus in a magnetic field, which is similar to the description of a spinning top. Not only does a top spin about its axis, but the axis itself frequently "wobbles" or *precesses,* particularly just before the top falls over and comes to a stop. In the classical NMR description, *resonance* is said to occur when the frequency of precession of the spinning nucleus is equal to the frequency of electromagnetic radiation used in the experiment. Such resonance is considered to result in the absorption of energy, with a simultaneous "spin flip" of the nucleus (equivalent to a top flipping to spin on the other end).*

A more convenient explanation of the NMR experiment is based on its similarity to other types of spectroscopy (Figure 9.3). Consider a nucleus with two possible spin states. In the absence of a magnetic field both spin states have the same energy, but in the presence of a magnetic field the energy of one state increases and that of the other decreases by a corresponding amount. The energy difference between the two spin states is proportional to the strength of the magnetic field; a magnetic field is therefore selected so that the energy difference between the spin states is in the energy range of the rf region of the electromagnetic spectrum. When irradiated with rf photons of the proper energy, a nucleus in the lower spin state can absorb a photon and be converted to the higher spin state. The reverse process also occurs, and it is possible to detect emission of rf energy in an NMR experiment.

The labeling of the spin states in Figure 9.3 is somewhat arbitrary. From quantum mechanical requirements the spin states must differ by unity and be centered about zero, which leads to values of $+\frac{1}{2}$ and $-\frac{1}{2}$. These labels are sometimes

*The resonance in NMR spectroscopy is unrelated to the concept of resonance that we have employed (Sections 4.7, 8.3) to interpret chemical stability and reactivity for π systems.

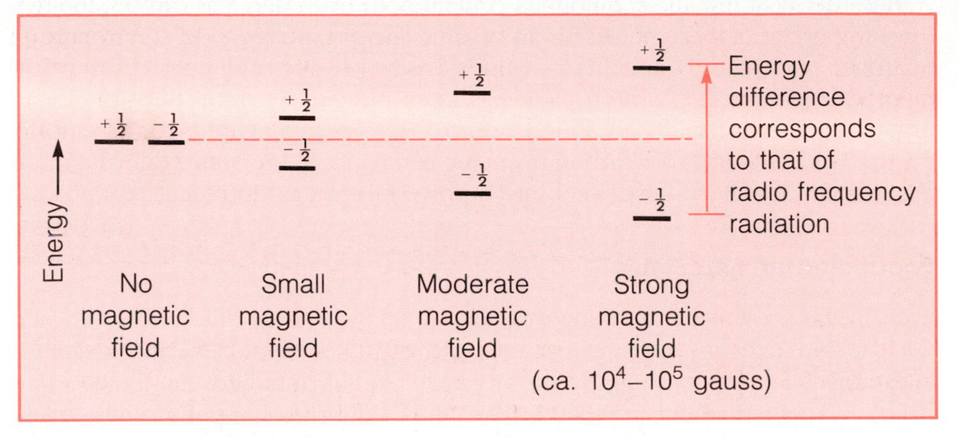

Figure 9.3 Variation of the Energy Difference Between Two Nuclear Spin States as a Function of the Applied Magnetic Field.
Under the conditions of an NMR experiment (a magnetic field of 10^4 to 10^5 gauss) a nucleus in the lower spin state (−½) can absorb a photon of the appropriate rf radiation and be converted to the higher spin state (+½).

simplified to just (+) and (−) or α and β. Different textbooks may use different labels, but in all cases the labels distinguish between two spin states of different energy.

Nuclei with Magnetic Spin

As long as a nucleus has either an odd atomic number or an odd mass number, it will have more than one spin state. For reasons that are beyond the scope of this chapter, nuclei with a total of only *two* possible spin states have been found most useful for NMR studies. Some of these nuclei are listed in Table 9.1, together with their natural isotopic abundance, their inherent sensitivity to detection by NMR, and the frequency of radiation corresponding to the energy difference between spin states in a magnetic field of about 23,000 gauss.

TABLE 9.1 Magnetic Properties of Selected Nuclei Having Two Spin States

Isotope	Natural Abundance, %	Inherent Sensitivity Relative to ^1H	Radio Frequency at 23,000 Gauss, MHz
^1H	99.98	1.00	100
^{13}C	1.1	0.02	25
^{15}N	0.4	0.1	10
^{19}F	100	0.8	94
^{31}P	100	0.06	40

Because most organic compounds contain both hydrogen and carbon, the two most important of these nuclei for an organic chemist are ^{1}H and ^{13}C. Virtually all hydrogen nuclei are ^{1}H, and this is the most easily detected of all nuclei by the NMR method. Consequently, *proton NMR* has been extensively developed for use in organic chemistry. The lower abundance of ^{13}C, combined with its lower sensitivity, has made ^{13}C NMR (or carbon magnetic resonance, *CMR*) more difficult, but it has nevertheless been developed into a powerful method in recent years.

Population of Spin States

Even in the presence of a strong magnetic field, the energy differences between nuclear spin states are very small, on the order of 10^{-6} kcal/mol. Energy transfer via intermolecular collisions rapidly establishes an equilibrium between the two spin states, and the ratio of the populations (i.e., the equilibrium constant) can be calculated by $\Delta G = -RT \ln K$. For a very small energy difference of about 10^{-6} kcal/mol, the equilibrium constant is nearly unity, so the populations of the two spin states must be nearly equal. (Of course in the *absence* of a magnetic field they have the same energy, and the populations will be *exactly* equal.)

Although these nuclei in the lower state are only slightly in excess, their absorption of radiation can be detected by modern instrumental methods. This absorption of rf radiation is the basis of all NMR studies. One of the effects of absorption of rf radiation is to decrease the population of the lower state and increase the population of the upper state, that is, to make the populations more nearly equal. If the rf radiation is increased in intensity, the two populations do approach equality, and no further net absorption of the rf radiation is observed. This condition, known as **saturation,** has consequences that we will discuss later.

The Chemical Shift

If the frequency of rf radiation required for a nuclear transition were always exactly the same (e.g., 100 MHz for protons under the conditions of Table 9.1), NMR spectroscopy would be of no value. Fortunately, most nuclei of *chemically nonequivalent* atoms do not absorb at precisely the same radio frequency. The energy differences between spin states are usually not the same for chemically nonequivalent nuclei, and these nuclei are also *magnetically nonequivalent.*

The variation in rf energy that is absorbed by nonequivalent nuclei is known as the **chemical shift,** and we have illustrated the origin of this effect in Figure 9.4. The electron density in the immediate vicinity of the nucleus modifies the effects of the magnetic field, and this perturbation is different for nonequivalent atoms. The energy difference between spin states differs for nonequivalent nuclei, and the rf energy absorbed by the nuclei is shifted as a function of the chemical environment.

The NMR experiment can be carried out in either of two ways. For many years the standard procedure employed continuous irradiation with a single frequency of rf radiation, and the magnetic field was varied until the energy separation for each different nucleus matched the energy of the rf radiation. Newer spectrometers employ a constant magnetic field, and the sample is irradiated with pulses that include a wide range of radio frequencies. The newer *pulsed NMR* techniques differ primarily in instrumental design and in computer analysis of the data, but the basic results are the same for both methods. For simplicity we will consider only the second experimental method when we describe NMR experiments in this chapter.

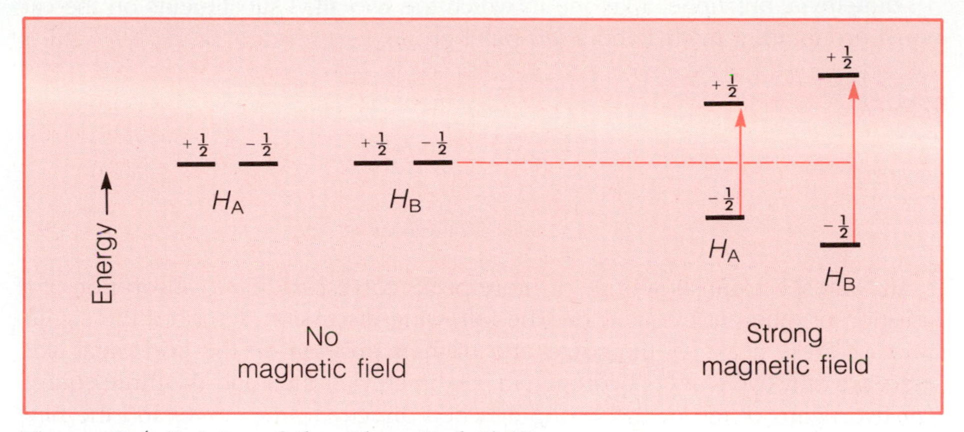

Figure 9.4 Origin of the Chemical Shift.
The nuclei of two different hydrogens (H_A and H_B) in a molecule are in different chemical environments. The different electron densities in the regions surrounding the two nuclei cause different perturbations in the magnetic field. Consequently, the energy differences between spin states are not the same and H_B would absorb at higher-energy rf than H_A. The shift in energy at which the nucleus absorbs rf radiation results from variation in the chemical environment, so it is called the *chemical shift.* (For purposes of clarity we have exaggerated the variation in the energy differences between spin states.)

The NMR Spectrum

We can best describe the experimental results obtained with NMR spectroscopy by considering an actual example. Figure 9.5 shows the proton NMR spectrum of

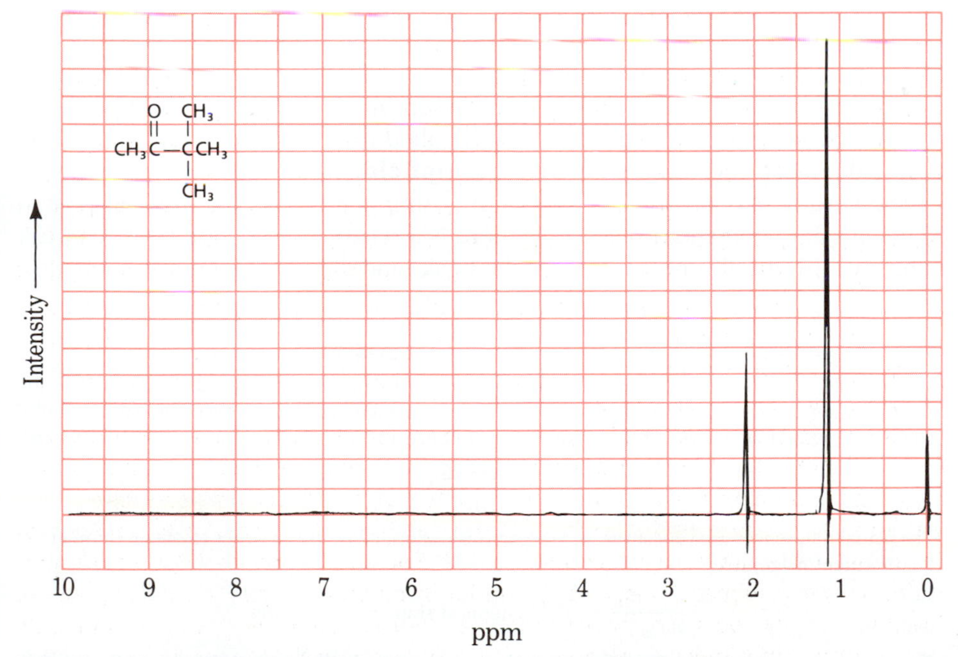

Figure 9.5 Proton NMR Spectrum of 3,3-Dimethyl-2-butanone.

3,3-dimethyl-2-butanone, a ketone in which the two alkyl substituents on the carbonyl group are a methyl and a *tert*-butyl group.

$$CH_3-C(CH_3)_2-C(=O)-CH_3$$

3,3-Dimethyl-2-butanone

An NMR spectrum shows one or more peaks corresponding to absorption of rf radiation of different frequencies. The following discussion describes the significance of these peaks, of their size, and of their position on the horizontal axis. There are only two types of hydrogen in 3,3-dimethyl-2-butanone, the three equivalent hydrogens of the methyl group attached directly to the C=O and the nine equivalent hydrogens of the *tert*-butyl group. As you would expect on this basis, the NMR spectrum shows two peaks, one for each type of hydrogen. A third signal (i.e., peak), located at the position marked zero on the horizontal axis, results from the protons of tetramethylsilane (TMS).

$$(CH_3)_4Si$$

Tetramethylsilane (TMS)

Values of chemical shifts in NMR spectra are always expressed relative to the chemical shift of a *reference compound.* In most instances (and for all the examples in this book) the reference compound is TMS. Chemical shifts are expressed in parts per million (ppm) relative to TMS, which has a chemical shift of 0 ppm by definition.* At a specific magnetic field, peaks on the left side of an NMR spectrum correspond to lower-energy transitions than those on the right side. With the earlier type of spectrometer, in which the magnetic field is varied, the same spectrum corresponds to a magnetic field that increases from left to right. For this reason the left side of an NMR spectrum is frequently called the **downfield** region, and the right side of the spectrum is described as **upfield.**

The vertical axis on an NMR spectrum is simply the peak intensity. In proton NMR (but frequently not in ^{13}C NMR) there is a reliable correlation between peak intensity (specifically, the *area* under the peak) and the *relative* number of hydro-

*Consider the absorption of rf radiation by the protons of TMS and by protons designated H_X in the sample. At a specific magnetic field these protons will absorb at different frequencies, and the chemical shift is expressed as a *ratio* of the frequency difference to the frequency at which TMS absorbs.

$$\frac{rf_{TMS} - rf_{H_x}}{rf_{TMS}}$$

The actual differences are very small, and the ratio is multiplied by 10^6 in order to obtain numbers of a convenient magnitude.

$$\frac{rf_{TMS} - rf_{H_x}}{rf_{TMS}} \times 10^6 = \text{chemical shift } (H_X) \qquad \text{ppm}$$

This is analogous to the use of percentages, where a ratio is multiplied by 10^2. A percentage is simply *parts per hundred.*

gens which give rise to the peak. The peak at 1 ppm in Figure 9.5 corresponds to the *tert*-butyl group and is three times as intense as the peak at 2 ppm for the methyl group. We will return to the measurement of peak intensities in Section 9.4, but in the meantime you should notice that there is a rough correlation between peak intensity and peak height. Note, however, that the absolute magnitude of the peaks has no particular significance; only their *relative* intensities are important. The NMR spectrometer is ordinarily adjusted so that the tallest peak has a convenient height, typically the full height of the chart paper.

9.3 ^{13}C NUCLEAR MAGNETIC RESONANCE

The predominant isotope of carbon is ^{12}C, but this has only a single nuclear spin state, so it cannot be studied by NMR. Only 1% of all carbon atoms are ^{13}C, and the low natural abundance of this isotope, together with its low sensitivity (ability to absorb rf radiation), made development of ^{13}C NMR difficult. However, tremendous advances in the capabilities and availability of ^{13}C NMR spectrometers took place during the 1970s, and ^{13}C NMR spectroscopy has become a common technique in the research laboratory.

The range of ^{13}C chemical shifts for typical organic compounds is considerably wider than that for proton signals, and a typical organic compound might exhibit

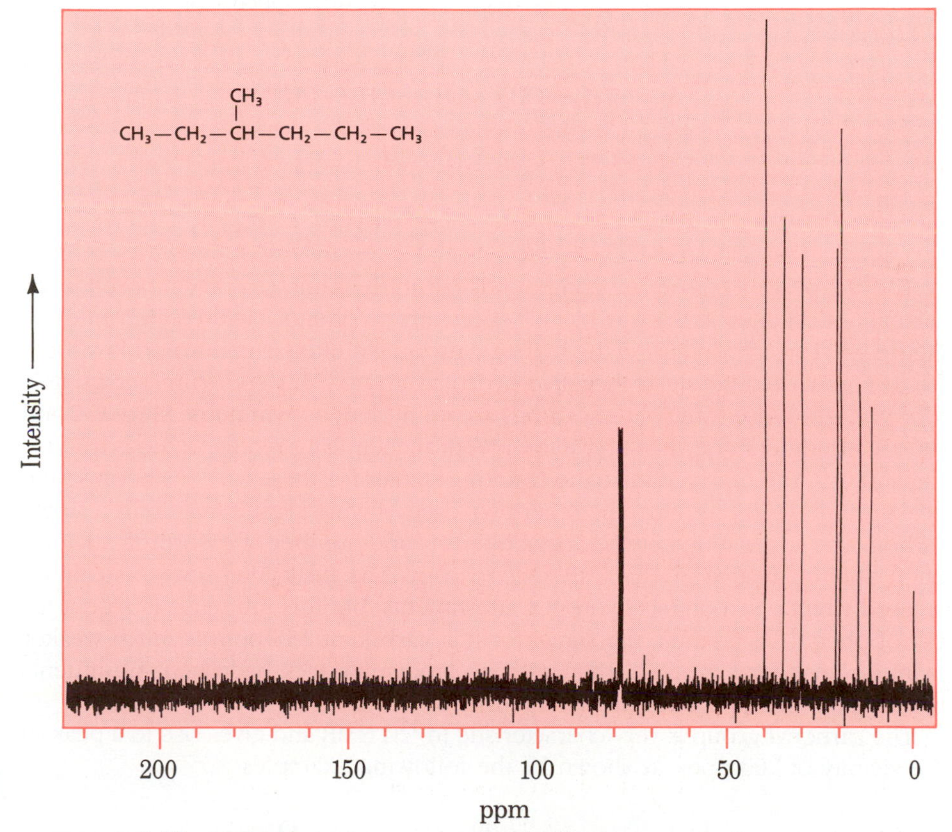

Figure 9.6 ^{13}C NMR Spectrum of 3-Methylhexane.

shifts of 0 to 200 ppm (relative to TMS). Because ^{13}C NMR spectra extend over such a wide range, chemically nonequivalent carbon atoms almost always afford individual and distinct NMR peaks. You can see this in the ^{13}C NMR spectrum of 3-methylhexane (Figure 9.6).

The spectrum exhibits seven distinct peaks between 10 and 40 ppm, the typical range of chemical shifts for an alkane. Two other important signals also appear in this spectrum. The peak at 0 ppm results from TMS, which is present as the reference compound. A second signal, consisting of a group of three peaks at 77 ppm, results from the $CDCl_3$ that was used as the solvent. All of the ^{13}C NMR spectra in this text will show these characteristic peaks for TMS and $CDCl_3$.

A polar substituent can produce substantial changes in the chemical shifts, as seen in the ^{13}C NMR spectrum of 2-methyl-1-butanol (Figure 9.7). All five carbon atoms appear as distinct peaks, but the carbon atom with the electronegative OH substituent is shifted downfield to 68 ppm.

The assignments of the peaks in Figures 9.6 and 9.7 show that large downfield shifts are produced by an electronegative substituent such as OH, and smaller but significant shifts are produced by alkyl substituents. The peaks for saturated hydrocarbons fall in the range of 10–40 ppm, but we can subdivide this range according to substitution patterns. Methyl groups on an alkyl chain give rise to peaks near 10 ppm, whereas secondary and tertiary carbon atoms afford peaks at the downfield end of the range.

The ^{13}C chemical shifts of unsaturated carbon atoms are far downfield from their saturated analogs, and such peaks allow you to identify carbon atoms that are part of multiple bonds. You can see this from the ^{13}C NMR spectra of the following hydrocarbons:

$$\underset{14}{CH_3}-\underset{23}{CH_2}-\underset{34}{CH_2}-CH_2-CH_2-CH_3$$

$$\underset{14}{CH_3}-\underset{21}{CH_2}-\underset{131}{CH}=CH-CH_2-CH_3$$

$$\underset{14}{CH_3}-\underset{12}{CH_2}-\underset{80}{C}\equiv C-CH_2-CH_3$$

The ^{13}C peaks for aromatic compounds fall in the same region as those for sp^2 carbons of alkenes as shown by the spectrum of phenol (hydroxybenzene) in Figure 9.8. The carbon having the hydroxyl group is shifted even further downfield (to 155 ppm) by the electronegative hydroxyl group. This spectrum shows only four peaks because some of the carbons are equivalent by symmetry. Since C-2 and C-6 are equivalent, they must have the identical chemical shift; the same is true for C-3 and C-5. You can recognize the resulting signals for these carbons because the peaks are more intense than the other two, each of which result from only a single carbon. The correlation between peak intensity and number of absorbing nuclei is often only qualitative in ^{13}C NMR spectroscopy. The peak intensity is usually increased when a carbon has hydrogen substituents, but this influence is absent for C-1 of phenol. As a result, the signal for that carbon at 155 ppm is much weaker than the peak for C-4 at 121 ppm. Carbon atoms that lack hydrogen substituents usually give rise to signals of low intensity.

The carbonyl group is very characteristic in ^{13}C NMR and gives rise to a peak in the vicinity of 200 ppm, as shown by the following examples:

$$CH_3-\underset{204}{\overset{\overset{\displaystyle O}{\|}}{C}}-CH_3 \qquad CH_3-\underset{200}{\overset{\overset{\displaystyle O}{\|}}{C}}-H \qquad CH_3-\underset{174}{\overset{\overset{\displaystyle O}{\|}}{C}}-OCH_3$$

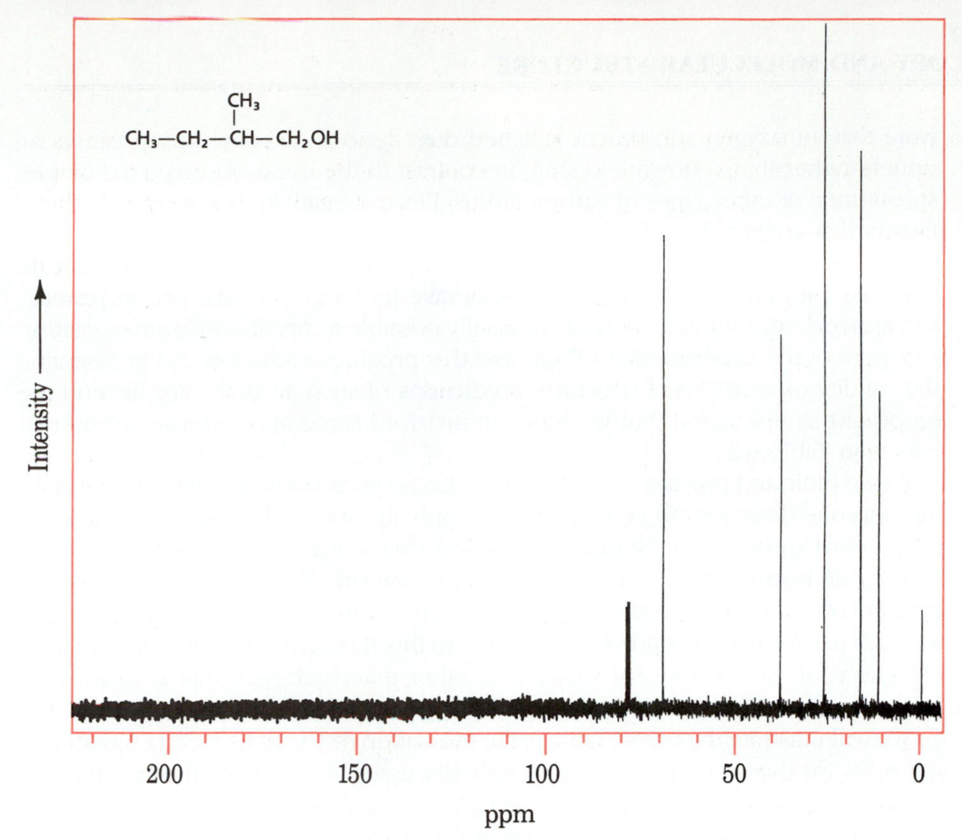

Figure 9.7 ¹³C NMR Spectrum of 2-Methyl-1-butanol.

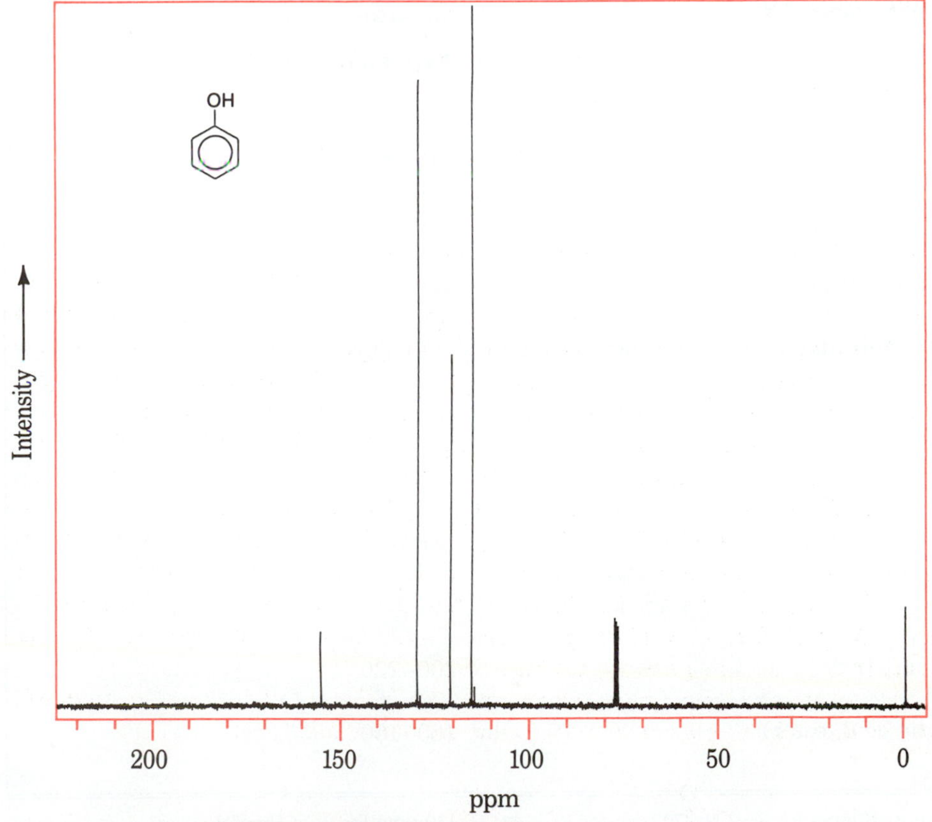

Figure 9.8 ¹³C NMR Spectrum of Phenol.

Note that an oxygen substituent attached directly to a carbonyl group causes an upfield rather than a downfield shift, in contrast to the trend observed for oxygen substitution at other types of carbon atoms. Electronegativity is just one of several factors that govern the influence of a substituent.

Accurate correlations of ^{13}C chemical shifts with molecular structure can be obtained with empirical relationships that take into account substitution patterns and stereochemical interactions. It is usually possible to predict ^{13}C shifts to within 1–2 ppm of the experimental values, and this provides a valuable aid in assessing the validity of a proposed structure. Predictions of such accuracy are beyond the scope of this discussion, but we have summarized some approximate substituent effects in Table 9.2.

As we indicated previously, substituent effects on carbonyl groups tend to be in the opposite direction, so you should not apply the substituent effects in the table to carbonyl carbons. To estimate a chemical shift using these substituent effects, you should begin with the chemical shift at the low end of the range for that type of carbon (i.e., 10 ppm for a saturated carbon). This is the approximate chemical shift for such a carbon at the end of a chain, and to this base value you add the substituent effect for each additional substituent other than hydrogen that is directly attached to the carbon atom you are evaluating.

We will illustrate the use of Table 9.2 in evaluation of ^{13}C NMR spectra by estimating shifts for those compounds previously discussed. For 3-methylhexane the ter-

TABLE 9.2 Approximate ^{13}C Chemical Shifts and Substituent Effects

Type of Carbon Atom		Expected Chemical Shift, ppm
Alkane (*sp*^{3} carbon)		10–40
Alkyne (*sp* carbon)		70–90
Alkene, aromatic (*sp*^{2} carbon)		110–150
Substituent effects for alkanes, alkenes, and alkynes*		
	Oxygen	+40
	Cl	+20
	Br	~0
	Carbon	+10
Carbonyl (ketone, aldehyde)		200–220
Carbonyl (carboxylic acid, ester)		160–180

*Substituent effects are relative to 10 ppm for R—CH_3, 70 ppm for R—C≡CH, 110 ppm for R—CH=CH_2.

minal methyl groups (primary carbon atoms) would all be predicted to yield peaks at about 10 ppm. The four CH_2 groups (secondary carbon atoms) each have one additional carbon substituent (relative to a methyl group), and the predicted chemical shift for each is 20 ppm. The single tertiary carbon atom (C-3) has *two* more alkyl substituents than a CH_3, so its predicted chemical shift is 30 ppm. The agreement with experiment is remarkably good (within 10–20 ppm) in view of the simplified method that we have used to calculate the predicted values. The greatest approximation in this method is neglect of substituents on neighboring atoms, and the largest discrepancies are for the carbons *adjacent* to the tertiary center.*

Predicted

$$\underset{10}{CH_3}{-}\underset{20}{CH_2}{-}\underset{30}{\overset{\overset{10}{CH_3}}{\overset{|}{CH}}}{-}\underset{20}{CH_2}{-}\underset{20}{CH_2}{-}\underset{10}{CH_3}$$

Observed

$$\underset{11}{CH_3}{-}\underset{30}{CH_2}{-}\underset{34}{\overset{\overset{19}{CH_3}}{\overset{|}{CH}}}{-}\underset{39}{CH_2}{-}\underset{20}{CH_2}{-}\underset{14}{CH_3}$$

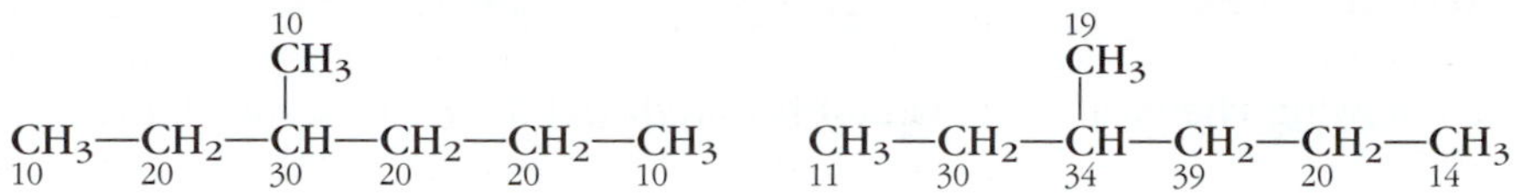

Similar prediction of the chemical shifts for 2-methyl-1-butanol also leads to agreement with the experimental values within 10–20 ppm, and the greatest discrepancies are again seen for carbon atoms adjacent to the tertiary center. Prediction of the chemical shifts is straightforward, and we will illustrate this for the CH_2OH group. This group differs from a CH_3 group at the end of a chain by the replacement of a hydrogen with a hydroxyl group. Therefore the substituent effect of +40 for OH is added to the base value of 10 ppm for a CH_3 to give the predicted shift of 50 ppm.

Predicted

$$\underset{10}{CH_3}{-}\underset{20}{CH_2}{-}\underset{30}{\overset{\overset{10}{CH_3}}{\overset{|}{CH}}}{-}\underset{50}{CH_2OH}$$

Observed

$$\underset{11}{CH_3}{-}\underset{26}{CH_2}{-}\underset{38}{\overset{\overset{14}{CH_3}}{\overset{|}{CH}}}{-}\underset{68}{CH_2OH}$$

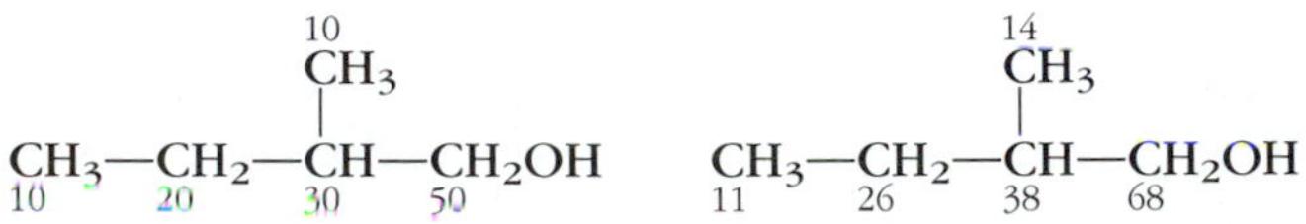

In the absence of branching the predictions are slightly more accurate, as you can see from the following examples:

Predicted

$$\underset{10}{CH_3}{-}\underset{20}{CH_2}{-}\underset{20}{CH_2}{-}CH_2{-}CH_2{-}CH_3$$

$$\underset{10}{CH_3}{-}\underset{20}{CH_2}{-}\underset{120}{CH}{=}CH{-}CH_2{-}CH_3$$

$$\underset{10}{CH_3}{-}\underset{20}{CH_2}{-}\underset{80}{C}{\equiv}C{-}CH_2{-}CH_3$$

Observed

$$\underset{14}{CH_3}{-}\underset{23}{CH_2}{-}\underset{34}{CH_2}{-}CH_2{-}CH_2{-}CH_3$$

$$\underset{14}{CH_3}{-}\underset{21}{CH_2}{-}\underset{131}{CH}{=}CH{-}CH_2{-}CH_3$$

$$\underset{14}{CH_3}{-}\underset{12}{CH_2}{-}\underset{80}{C}{\equiv}C{-}CH_2{-}CH_3$$

The predicted ^{13}C chemical shifts that are obtained by using Table 9.2 are only approximate, but they can help you deduce the correct structure of a compound from its ^{13}C NMR spectrum. A distinct peak is usually observed for each carbon atom in a compound, with the exception of those that are equivalent by symmetry. Moreover, the region of the spectrum in which a peak appears helps to define what type of carbon atom gives rise to the peak. If you have any information about the chemical origin of a compound, the NMR data frequently allow you to suggest

*Branching at an adjacent position typically increases the chemical shift of a carbon atom by 5–10 ppm.

several *possible* structures. If you predict the chemical shifts for each of these according to Table 9.2, you will often find that the correct structure is that for which the predicted shifts best match the experimental NMR data.

Example 9.1 A sample is known to be one of the following isomers:

A $CH_2{=}CH{-}CH_2{-}CH_2OH$ or **B** $CH_3{-}CH_2{-}CH_2{-}CHO$

Its ^{13}C NMR spectrum shows four lines, all between 0 and 150 ppm. Which is the correct structure?

Solution The following chemical shifts would be predicted for each isomer by using Table 9.2:

$$\textbf{A}\ \underset{110}{CH_2}{=}\underset{120}{CH}{-}\underset{20}{CH_2}{-}\underset{50}{CH_2}OH \qquad \textbf{B}\ \underset{10}{CH_3}{-}\underset{20}{CH_2}{-}\underset{20}{CH_2}{-}\underset{200\text{–}220}{\overset{\overset{\displaystyle O}{\|}}{C}}{-}H$$

A much better agreement between predicted and observed shifts is found for compound **A,** and this is the correct structure.

EXERCISE 9.2

How many peaks would be expected in the ^{13}C NMR spectrum of each of the following compounds?

(a) $CH_3{-}CH_2{-}CH_2{-}CH_2OH$ (b) 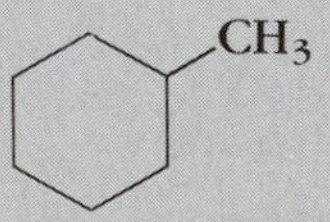 methylcyclohexane (CH_3 on a cyclohexane ring) (c) 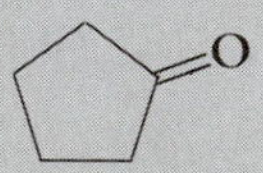 cyclopentanone (ring C=O)

(d) 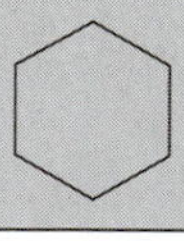 cyclohexane (e) $CH_3{-}CH_2{-}\underset{\displaystyle OH}{\underset{|}{CH}}{-}CH_2{-}CH_3$

EXERCISE 9.3

An alkene was subjected to hydroboration-oxidation in an effort to obtain the corresponding alcohol. The ^{13}C NMR spectrum of the material isolated from the reaction showed a total of seven peaks, all of which were within the range of 10 to 60 ppm. Does this indicate that the reaction was successful or unsuccessful?

9.4 PROTON NUCLEAR MAGNETIC RESONANCE

Despite the recent advances in ^{13}C NMR, proton NMR remains the single most important spectroscopic method in organic chemistry. As we discussed in the previous section, two types of information are readily available from ^{13}C NMR spectra:

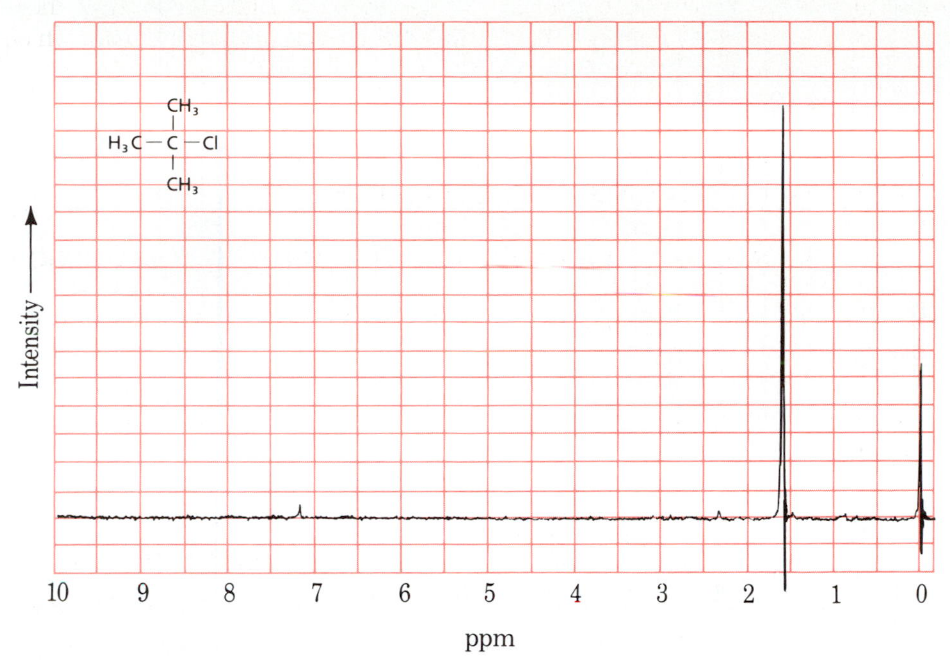

Figure 9.9 Proton NMR Spectrum of 2-Chloro-2-methylpropane.

(1) the *number* of signals and (2) their *chemical shifts*. Even more information can be obtained from the spectra, and for proton NMR we will also consider (3) the *intensity* of the peaks and (4) the *multiplicity* of the signal (i.e., the number of lines or peaks that result from each nucleus). All four kinds of information will be discussed in this section.

Number of Signals

In principle all chemically nonequivalent atoms should give rise to separate signals in the NMR. This is frequently the case with ^{13}C NMR, as we showed in the preceding section. In proton NMR, however, hydrogen atoms that are chemically *similar* (but not chemically equivalent) tend to give peaks that have very similar chemical shifts. Often these peaks overlap, and it can be difficult to determine the total number of signals unambiguously. Nevertheless, the total number of signals provides important structural information. If the total number of NMR signals is less than the total number of hydrogen atoms in the molecule, then some of these hydrogens must be equivalent. Chemical equivalence can result from symmetry of a group or symmetry of the entire molecule, and you can see this in the following spectra. The proton NMR spectrum of 2-chloro-2-methylpropane (Figure 9.9) shows only a single peak because all nine hydrogens are equivalent. The molecule has *rotational symmetry,* a C_3 axis (Section 1.6), which is an extension of the C—Cl bond. Rotation about this axis by one-third of a revolution generates a structure that is indistinguishable from the original.

CH_3 / CH_3 / CH_3 —C—Cl $\xrightarrow{\text{1/3 revolution}}$ CH_3 / CH_3 / CH_3 —C—Cl

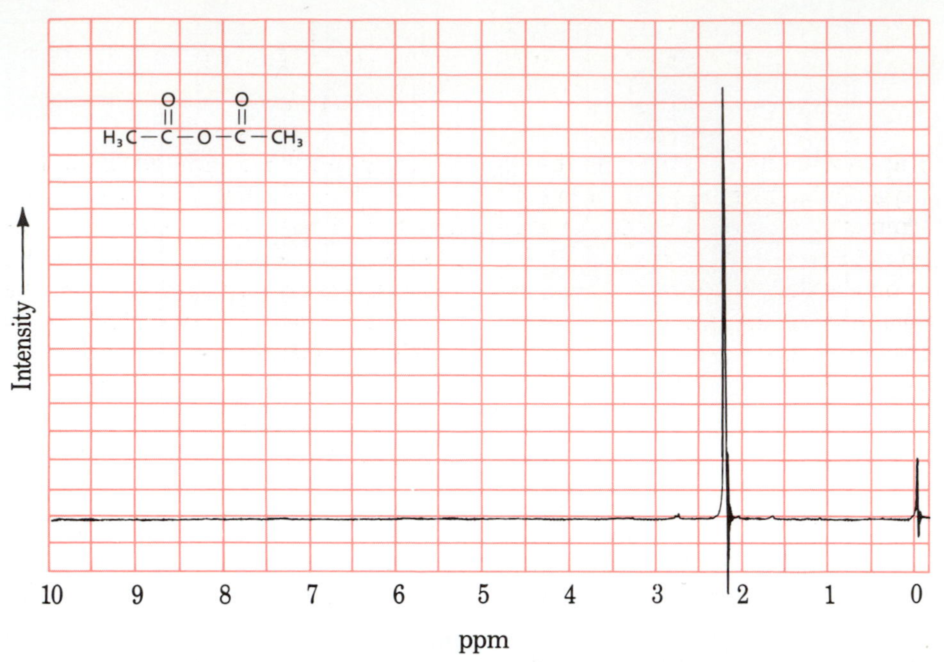

Figure 9.10 Proton NMR Spectrum of Acetic Anhydride.

Note that a methyl group has the same kind of local symmetry (a C_3 axis), and its three hydrogens are completely equivalent.

The proton NMR spectra of acetic anhydride (Figure 9.10) and methyl pyruvate (Figure 9.11) are spectra of two isomeric compounds with the molecular formula $C_4H_6O_3$.

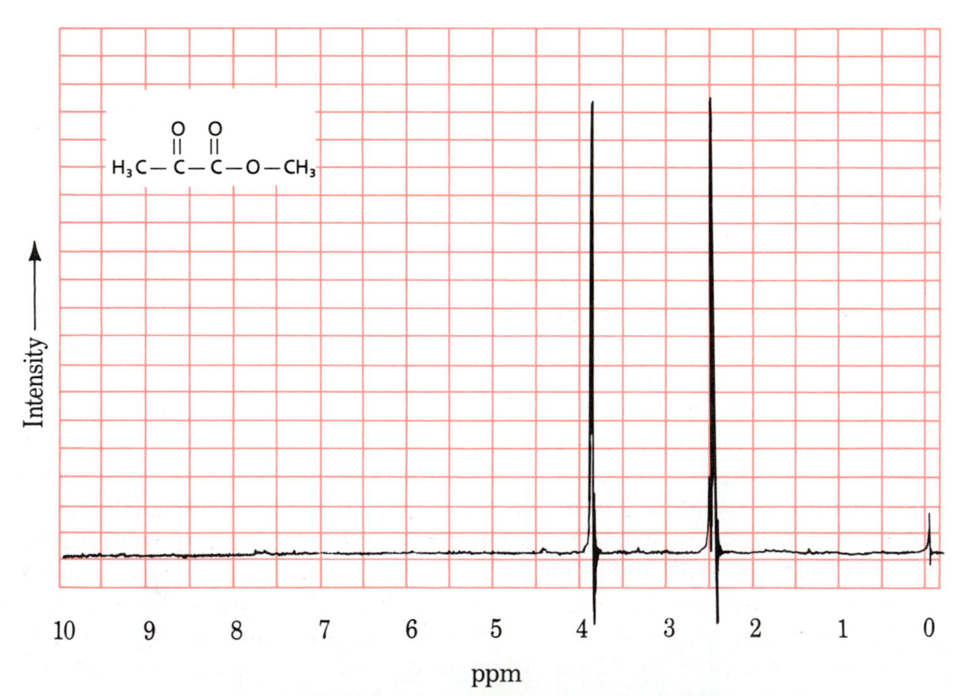

Figure 9.11 Proton NMR Spectrum of Methyl Pyruvate.

$$CH_3{-}\overset{\overset{\large O}{\|}}{C}{-}O{-}\overset{\overset{\large O}{\|}}{C}{-}CH_3 \qquad CH_3{-}\overset{\overset{\large O}{\|}}{C}{-}\overset{\overset{\large O}{\|}}{C}{-}O{-}CH_3$$

Acetic anhydride   Methyl pyruvate

The two structures differ only by the interchange of a carbonyl group and an oxygen atom, but the resulting spectra are quite different. Methyl pyruvate has no symmetry, and two distinct peaks are seen in its NMR spectrum. Acetic anhydride, on the other hand, is symmetrical, and the two equivalent methyl groups give rise to only a single peak in the NMR spectrum. The type of symmetry depends on the conformation in which the molecule is drawn. The following conformation actually has both reflection (mirror) symmetry *and* rotational symmetry.

Plane of symmetry   Axis of symmetry

As long as the conformational equilibrium includes any structure with either rotational or reflection symmetry (either as a minimum-energy conformation or as a transition state between conformations), the two groups will be equivalent.

Chemical Shift

The position of a peak in the proton NMR spectrum is related to the chemical environment of the proton that gives rise to the signal, and this is the origin of the term *chemical shift.* In the previous section we showed that ^{13}C chemical shifts are correlated with molecular structure and similar correlations are found with proton NMR. Figure 9.12 summarizes the regions of an NMR spectrum in which various kinds of hydrogen atoms typically absorb.

Certain patterns can be seen easily in Figure 9.12. Hydrogen atoms situated on sp^3 carbon atoms with no polar substituents give signals near 1 ppm. This location is shifted downfield by the presence of an adjacent double bond or carbonyl group, and even larger downfield shifts result from halogen and oxygen substituents. Vinyl hydrogens (i.e, those attached directly to a carbon-carbon double bond) show a still larger chemical shift, and they yield peaks in the 5–6 ppm region. Hydrogen atoms attached directly to an aromatic ring (designated Ar—H) have chemical shifts between 6 and 8 ppm. Finally, at the far left of the figure are the characteristic chemical shifts of aldehydes (9.5–11 ppm) and carboxylic acids (near 12 ppm). The proton NMR spectra that we have included in this text normally extend only to 10 ppm. Any peaks with chemical shifts greater than 10 are shown with a second tracing on the same chart, and you will encounter examples of this subsequently.

The normal range of chemical shifts in proton NMR (about 12 ppm) is much narrower than that for ^{13}C NMR (200 ppm), but the correlation of chemical shift with structure can still be quite accurate and reliable. In Table 9.3 we have summarized characteristic chemical shifts for various kinds of hydrogen. We have also presented substituent effects for hydrogens attached to saturated carbon atoms, and you can predict the chemical shifts quite accurately with these substituent effects. These substituent effects are not applicable to hydrogens on unsaturated carbons,

TABLE 9.3 Typical Chemical Shifts in Proton NMR

Type of Hydrogen	Chemical Shift, ppm
1. Hydrogen attached to saturated carbon	**1–5**
a. Substituent effects relative to a value of 0.5 ppm for CH_4*	

Substituent	Shift	Substituent	Shift
$—CH_3$	0.4	$—Ar$	1.7
$—CH_2Ar$	0.7	$—C(=O)—Ar$	2.0
†$—CH_2X$	0.8	$—I$	2.3
$—CH{=}CH_2$	1.1	$—Br$	2.5
$—C{\equiv}CR$	1.3	$—Cl$	2.6
$—C{\equiv}N$	1.5	$—OR$	2.6
$—C(=O)—OR$	1.5	$—OH$	2.7
$—C(=O)—R$	1.6	$—O—C(=O)—R$	3.2
$—SR$	1.6	$—O—Ar$	3.3
$—NR_2$	1.7		

If carbon is in a ring, add +0.2 ppm.

Type of Hydrogen		Chemical Shift, ppm
b. Cyclic compounds: CH_2 groups	Cyclopropane	0.2
	Other cycloalkanes	1.5–1.6
2. Hydrogen attached to unsaturated carbon		**5–11**
a. Alkynes, $—C{\equiv}C—H$		2.3–2.6
b. Alkenes		4.6–6.2

$C{=}CH_2$	$—CH{=}CH—$	$R_2C{=}CHR$	$—CH{=}CH_2$
4.6	5.4	5.1–5.2	5.7–4.9

c. Aldehydes — 9.4–11

$C{=}C—C(=O)—H$	$R—C(=O)—H$	$Ar—C(=O)—H$
α,β-Unsaturated	Saturated	Aromatic
9.4–9.5	9.5–9.8	9.9–11.0

Type of Hydrogen	Chemical Shift, ppm
d. Aromatic compounds	6–8
(with nonpolar alkyl substituents only)	6.9–7.2
3. Hydrogen attached to electronegative atoms	**0.5–12.5**
a. Amines, alcohols, thiols (RNH_2, ROH, RSH)	variable 0.5–5
b. Phenols (ArOH)	4–8
c. Carboxylic acids (RCO_2H)	11.5–12.5

*For each hydrogen of methane replaced by a substituent, the appropriate substituent constant is added to the base value of 0.5 ppm. (This is not the actual chemical shift of methane but is a value that gives good predictions for most compounds.)

†X indicates an electronegative group such as —OH, Cl, etc.

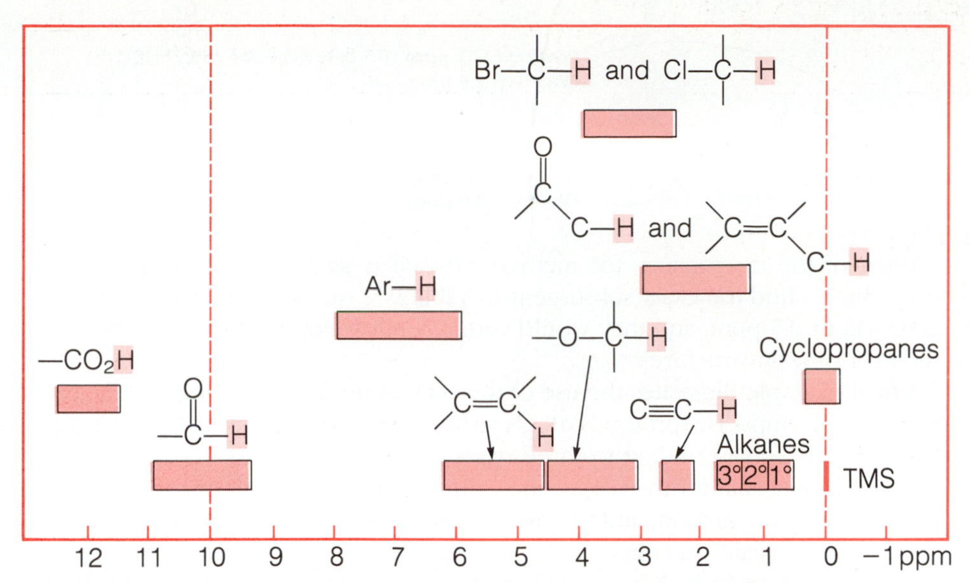

Figure 9.12 Characteristic Chemical Shifts in Proton NMR Spectra. The region from 0–10 ppm is that normally shown in NMR spectra.

and the chemical shifts listed for those types of hydrogen are only approximate. For example, the vinyl hydrogens of an alkene with polar substituents might well afford peaks outside the range specified in the table.

The chemical shifts that you can predict with the substituent effects in Table 9.3 are quite accurate for CH_3 and CH_2 groups. They usually agree with experimental values within 0.3 ppm (and almost always within 0.5 ppm). The values that you would predict for trisubstituted carbon atoms are much less reliable, however, and you should only employ them as rough estimates.

We will illustrate the use of these substituent effects with the proton NMR spectra of Figures 9.9 to 9.11. For 2-chloro-2-methylpropane there is only one kind of hydrogen, so any one of the three methyl groups may be chosen for evaluation. The "substituent" is then a 2-chloro-2-propyl group. This exact group is not listed in the table, and it is therefore necessary to use the substituent which is structurally most similar, $—CH_2X$.

$$(CH_3)_3C—Cl$$

Predicted: **1.2 ppm** (0.5 base + 0.7 for CH_2X)
Observed: **1.6 ppm**

The prediction in this case differs from the observed value by 0.4 ppm. Such large deviations are frequently observed when the hydrogen atom in question is attached to a carbon atom adjacent to a tetrasubstituted carbon.

The shifts of the two equivalent methyl groups of acetic anhydride (Figure 9.10) are predicted within 0.2 ppm of the experimental value; the corresponding discrepancies for methyl pyruvate (Figure 9.11) are 0.4 and 0.2 ppm.

$$CH_3—C(=O)—O—C(=O)—CH_3$$

Acetic anhydride

Predicted: **2.0 ppm** (0.5 base + 1.5 for $—C(=O)—OR$)
Observed: **2.2 ppm**

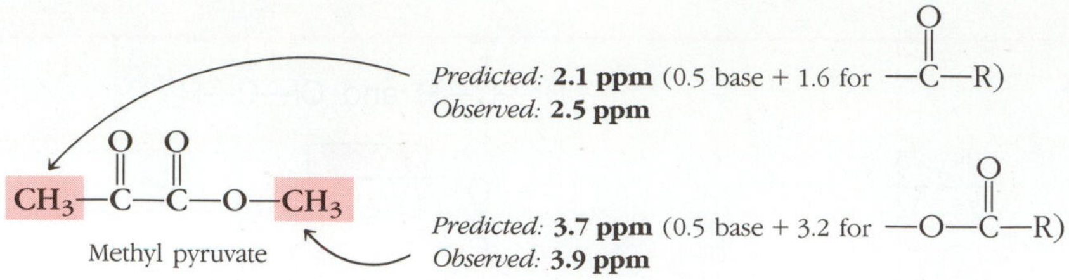

CH_3—C(=O)—C(=O)—O—CH_3

Methyl pyruvate

Predicted: **2.1 ppm** (0.5 base + 1.6 for —C(=O)—R)
Observed: **2.5 ppm**

Predicted: **3.7 ppm** (0.5 base + 3.2 for —O—C(=O)—R)
Observed: **3.9 ppm**

Much of the discrepancy for methyl pyruvate is again a consequence of not being able to find the exact substituent in Table 9.3. Nevertheless, the predictions agree within 0.5 ppm, and they would certainly allow you to distinguish between the two isomeric structures.

A final example illustrates the use of this technique for predicting the chemical shift of CH_2 groups. The proton NMR spectrum of methyl acetoacetate (Figure 9.13) shows signals corresponding to three types of hydrogen atoms.

The chemical shift for the CH_2 group is predicted on the basis of two successive replacements of hydrogen, and the overall agreement between observed and predicted shifts is again quite good.

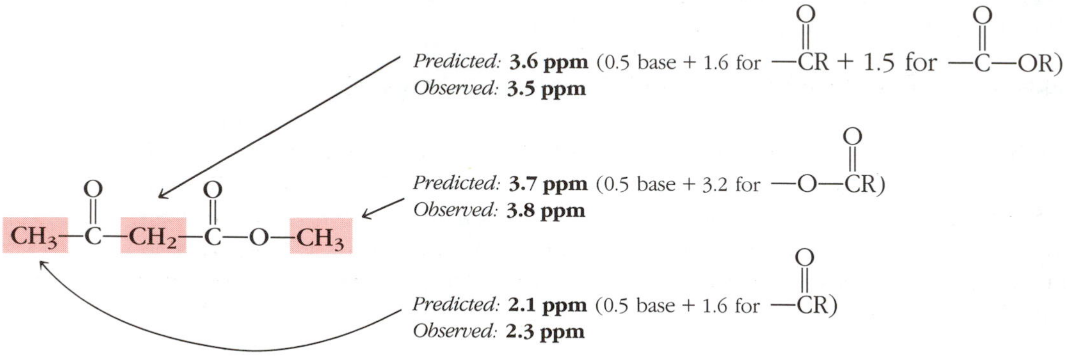

CH_3—C(=O)—CH_2—C(=O)—O—CH_3

Predicted: **3.6 ppm** (0.5 base + 1.6 for —C(=O)R + 1.5 for —C(=O)—OR)
Observed: **3.5 ppm**

Predicted: **3.7 ppm** (0.5 base + 3.2 for —O—C(=O)R)
Observed: **3.8 ppm**

Predicted: **2.1 ppm** (0.5 base + 1.6 for —C(=O)R)
Observed: **2.3 ppm**

Peak Intensities

The absolute peak intensity on an NMR spectrum has no particular significance because it varies with the instrument settings, but the *relative* intensities of signals provide important structural information, particularly in proton NMR. The relative intensities of the peaks in the spectrum are directly proportional to the number of equivalent hydrogens that give rise to each signal. The intensity is not simply the peak height but the area under the peak, and this is measured by **integration** of the curve. Integration is ordinarily done electronically and is frequently recorded as a second tracing on the actual spectrum, as we have illustrated in Figure 9.14, which again shows the spectrum of methyl acetoacetate. The second tracing, called the **integral,** has the appearance of a step function, and vertical changes in the tracing of the integral are proportional to the area under the corresponding NMR signal.

The actual number of hydrogens corresponding to each signal can be calculated if you know the total number of hydrogens in the molecule. Using the "squares" of the grid in Figure 9.14 (or a ruler) for your measurements, you can determine the changes in height of the integral tracing as 5, 3.5, and 5 squares for the peaks at 3.8, 3.5, and 2.3 ppm, respectively. These values (5, 3.5, and 5) are the relative areas of

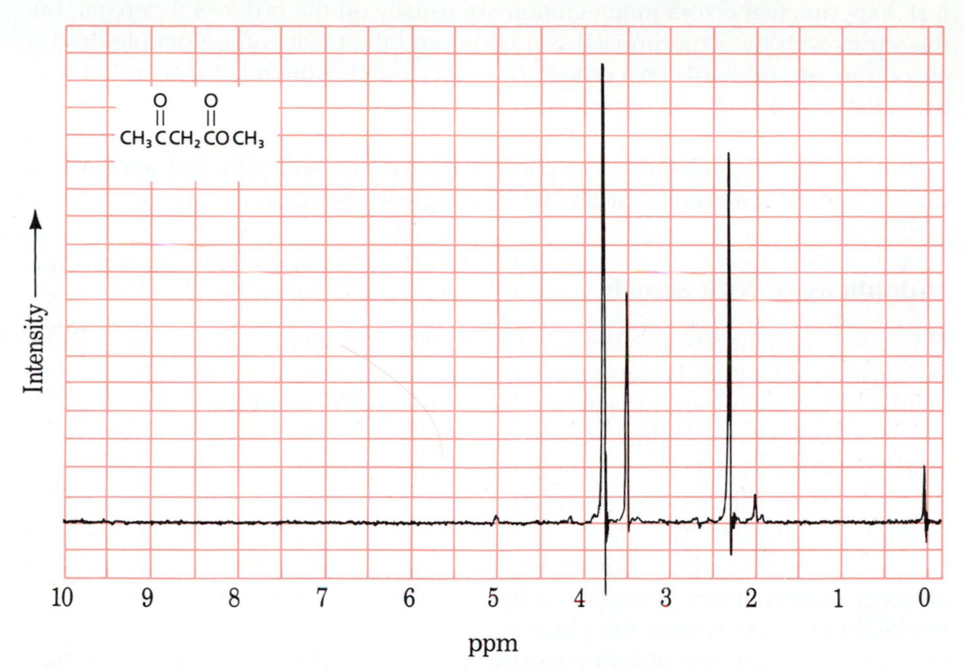

Figure 9.13 Proton NMR Spectrum of Methyl Acetoacetate.

the three peaks. The total change in height of 13.5 squares for the entire integral corresponds to the total of eight hydrogens in the molecule. Division of this total of 13.5 by 8 tells you that the area for each hydrogen is 1.7 squares. You can next divide the relative area of each peak by this value (1.7 squares per hydrogen), and this would tell you how many hydrogens correspond to each peak. The results are 5.2/1.7 = 3.1, 3.0/1.7 = 1.8, and 5.3/1.7 = 3.1, which you round off to 3 H, 2 H, and

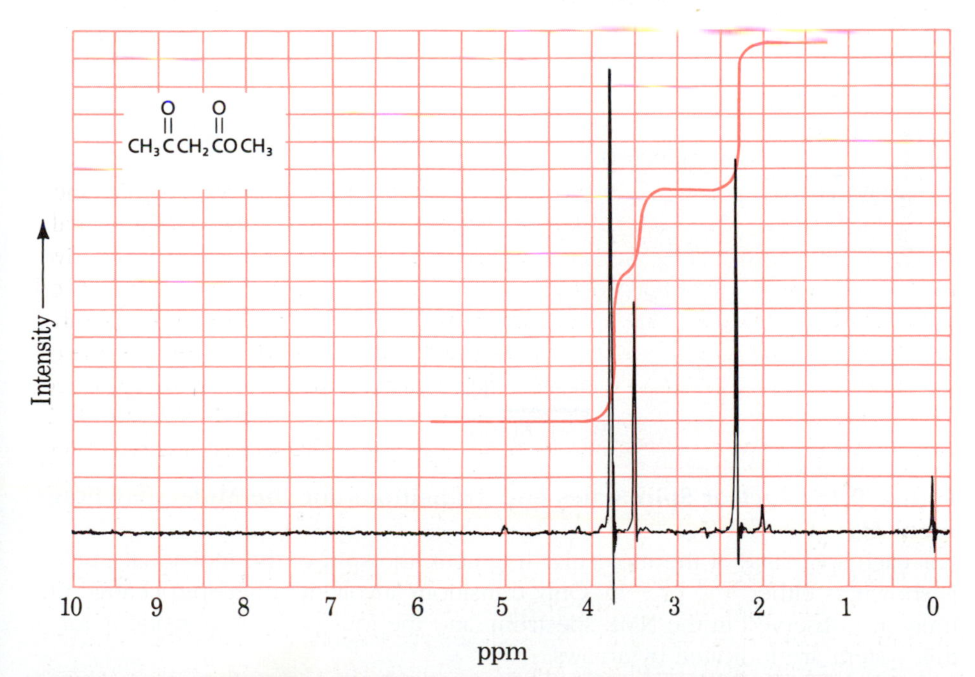

Figure 9.14 Integrated Proton NMR Spectrum of Methyl Acetoacetate.

3 H. Experimental errors in integration are usually on the order of 5 percent, but this varies with the experimental conditions and the purity of the sample. In this chapter we are primarily concerned not with calculating the relative areas but with using them to deduce the correct structures, so we will not usually show the integral tracings on the spectra. When it is important for you to know the number of hydrogen atoms that give rise to a particular signal, we will print that information directly on the spectrum, typically by labeling each peak as 3H, 2H, etc.

Multiplicity of NMR Signals

For complete analysis of an NMR spectrum the the spin state of each nucleus cannot be considered by itself because there may be interactions with the nuclei of other nearby atoms. Such spin-spin interaction, or **spin-spin coupling,** between nuclei is transmitted via the bonding electrons in a molecule, and it determines the **multiplicity** of the signal (i.e., the number of lines or peaks that result from each nucleus). As a result of spin interactions NMR signals frequently are not single peaks but *multiplets.* The effect diminishes with the number of intervening σ bonds, but it is still quite important for up to three bonds, that is, for hydrogens on adjacent carbon atoms. Spin-spin coupling can occur between various types of nuclei, but we will restrict the present discussion to spin coupling between protons. (A brief discussion of spin coupling in ^{13}C NMR will be found in Section 9.6.)

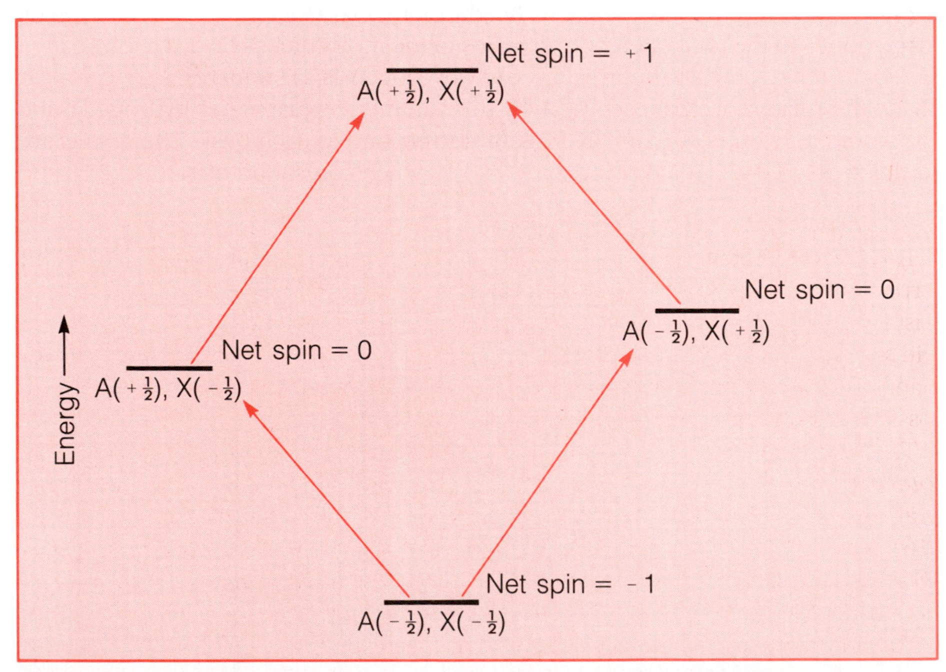

Figure 9.15 Nuclear Spin States and Transitions for the Molecular Fragment H_A—C—C—H_X.
For each spin state of the molecular fragment, the spin of the individual nuclei is shown as either +½ or −½. Only transitions involving a net spin change of unity are observed in the NMR spectrum, and the four possible transitions for this system are indicated by arrows.

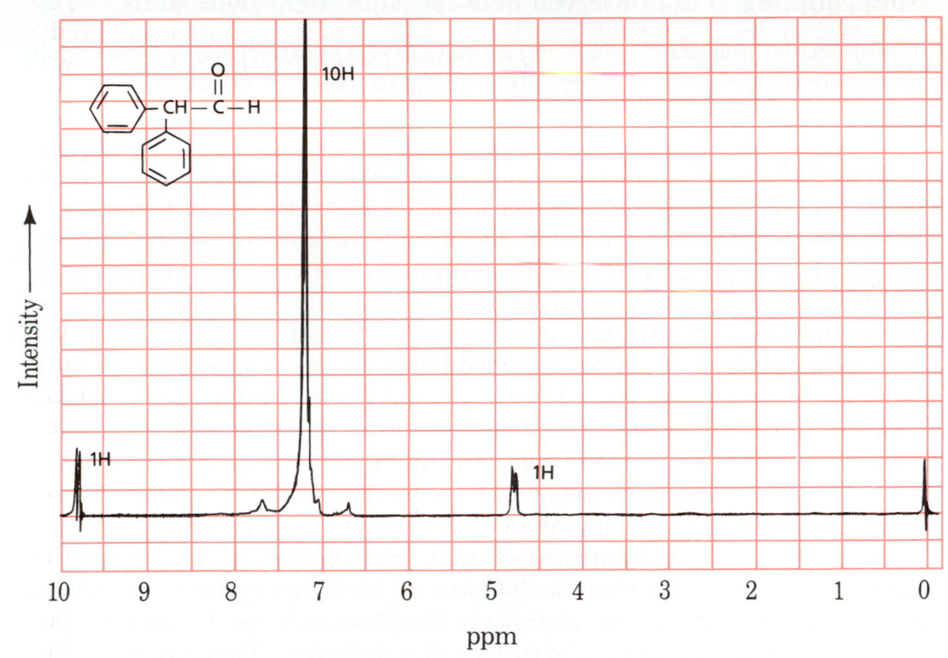

Figure 9.16 Proton NMR Spectrum of Diphenylacetaldehyde.

Consider a molecular fragment with a hydrogen atom on each of two adjacent carbon atoms:

$$\begin{array}{cc} | & | \\ -\mathrm{C}- & -\mathrm{C}- \\ | & | \\ \mathrm{H_A} & \mathrm{H_X} \end{array}$$

If H_A is not equivalent to H_X, you might expect to observe a total of two peaks in the corresponding proton NMR spectrum. However, four different nuclear spin states exist for this fragment, and four energetically different transitions are possible. The spin state for each nucleus is either $+\frac{1}{2}$ or $-\frac{1}{2}$, and only transitions involving a net change of 1 in total spin are observed. We have illustrated these spin states and the possible transitions in Figure 9.15.

The four transitions shown in Figure 9.15 appear as two pairs of peaks (two *doublets*) in the proton NMR spectra of molecules containing this isolated structural unit. An example is seen in the spectrum of diphenylacetaldehyde (Figure 9.16). The aldehyde peak appears as a doublet at 9.8 ppm and the tertiary CH (which is shifted by the carbonyl as well as by the two phenyl groups) appears as a second doublet at 4.75 ppm. The phenyl groups give rise to the peak at 7.2 ppm.

You could ask an important question about the diagram of spin states and transitions in Figure 9.15. *What if H_A and H_X were equivalent?* This would be the case if the overall molecule were symmetrical, and symmetry would lead to two consequences for the diagram. First, two of the transitions would have zero probability because of quantum mechanical symmetry restrictions, and second, the other two transitions would have equal energy. This would result in only a single peak for the two equivalent hydrogens in the NMR spectrum. This is an example of a general rule regarding spin interactions:

Spin coupling is not observed between equivalent hydrogens.

For molecular fragments containing more than two hydrogens, diagrams of spin states become increasingly complex and difficult to interpret. Fortunately an alternative way of analyzing spin coupling is available. This approach involves some simplifications, but it works well as long as it is applied only to interactions of hydrogens that have quite different chemical shifts. We will return to the case of the fragment $H_A—C—C—H_X$ and consider the influence of H_X on the transitions of H_A. The spin property of a nucleus is also a magnetic property, so the magnetic moment of H_X can affect the transition energy of H_A. A transition for H_A can occur under the influence of either of two different spin states for H_X, and the energies of these two transitions for H_A will be different. As a result, the signal for H_A will appear as two peaks, only slightly separated, that we describe as a *doublet.* The same type of analysis correctly predicts a second doublet for H_X, caused by the presence of two different spin states of H_A. Some molecules give one transition and other molecules give the second type of transition, but an NMR spectrum results from the experimental observation of many molecules in solution. Therefore, both transitions will be observed and the signal will appear as a doublet.

Other spin systems can be treated easily by this method. The ethyl group, for example, affords a very characteristic NMR pattern. We can analyze it by considering the $—CH_2—$ and $—CH_3$ groups separately. The absorptions of the methyl group can occur under the influence of three different spin states for the neighboring CH_2 group. Moreover, one of these spin states can be formed in two ways, so it has a higher probability. These spin states can be summarized in the following way, where $+½$ and $-½$ are abbreviated as + and −, respectively.

$—CH_2—$	+ +
Three spin states	+ − , − +
	− −

The net effect of the CH_2 group is the appearance of the signal for the methyl group as a *triplet,* in which the center line has twice the intensity of the other two.

The signal arising from the CH_2 of an ethyl group can be predicted similarly. The adjacent methyl has four spin states with probabilities in the ratio 1:3:3:1.

$—CH_3$	+ + +
Four spin states	+ + − , + − + , − + +
	− − + , − + − , + − −
	− − −

These four spin states of the methyl group cause the NMR signal for the CH_2 group to appear as a *quartet* in the NMR spectrum, with the two center lines having an intensity about three times that of the outer lines. The characteristic appearance of an ethyl group (actually, two equivalent ethyl groups) is seen in Figure 9.17, and

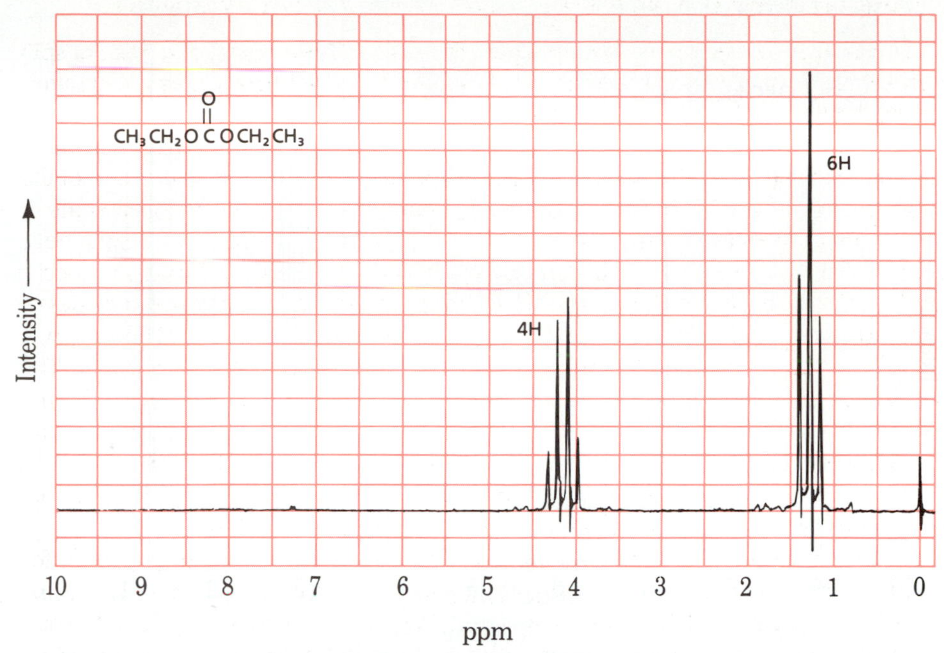

Figure 9.17 Proton NMR Spectrum of Diethyl Carbonate.

the quartet for the CH_2 group invariably appears downfield from the methyl triplet. The spacing between each of the four peaks of the quartet is equal, and it is called the **coupling constant** (usually denoted by the letter J). This defines the magnitude of the coupling between the hydrogens of the CH_2 and CH_3 groups, and the *identical* coupling constant (i.e., spacing between peaks) is observed for the three peaks of the methyl triplet.

As you can see in Figure 9.17, splitting patterns are not perfectly symmetrical. In fact the multiplets of hydrogens that are coupled tend to "point" toward each other, and the signals for an ethyl group usually exhibit the following pattern:

The cases just discussed are but two examples that illustrate a second general rule for spin coupling interactions:

A hydrogen atom that is coupled to a total of *n* equivalent neighboring hydrogens will give rise to a multiplet with *n* + 1 peaks.

This is further illustrated in Figure 9.18, which shows the characteristic pattern for an isopropyl group. The methyl group shows up as a *doublet* (split by a single adjacent CH), and the CH appears as a *septet* (split by the six hydrogens of two equivalent methyl groups).

Figure 9.18 illustrates a relatively common occurrence in proton NMR spectra, the accidental equivalence of protons in a molecule. The ring hydrogens of isopropylbenzene are not all chemically equivalent. Nevertheless they are in very

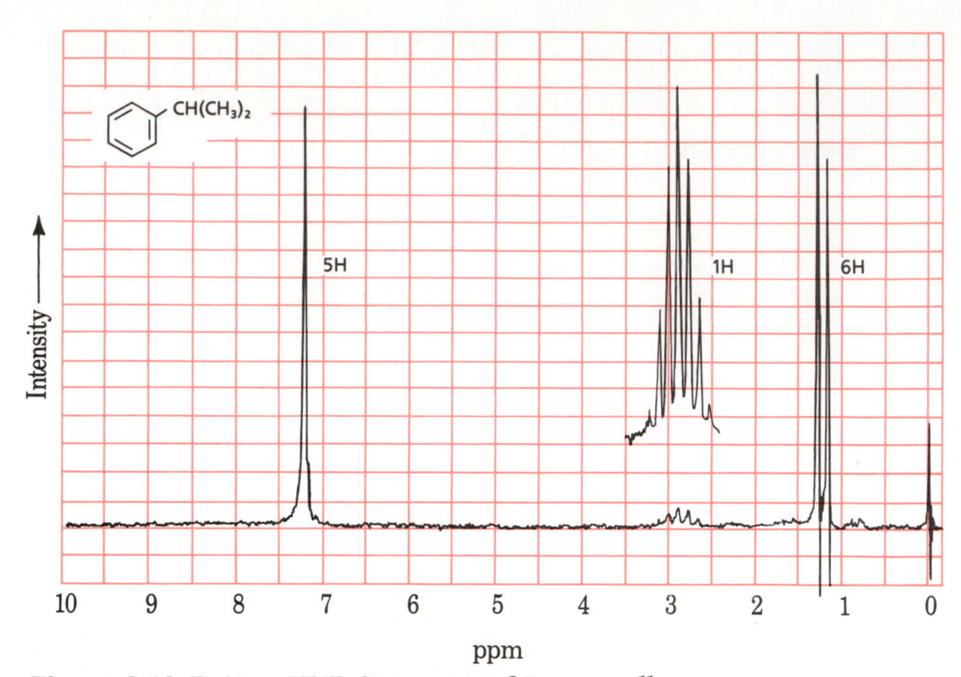

Figure 9.18 Proton NMR Spectrum of Isopropylbenzene.
An enlargement of the CH septet is shown directly above the regular spectrum.

similar chemical (and magnetic) environments, so they have essentially the same chemical shift. Hydrogens that are not necessarily equivalent by symmetry but have the same chemical shift are described as *magnetically equivalent,* and you will not observe spin coupling between them.

When a benzene ring has a polar substituent, the ring hydrogens tend to have different chemical shifts, and spin coupling between them is observed. The π system of an aromatic ring transmits nuclear spin quite effectively, and all the ring hydrogens are coupled to each other even when they are not on adjacent carbons. This frequently results in complicated splitting patterns, as seen in Figure 9.19, where the aromatic protons yield a complex multiplet extending from 6.5 to 7.2 ppm.

This spectrum also illustrates a third way in which hydrogen atoms can be equivalent in the NMR spectrum, namely, as a consequence of *chemical exchange.* Acid-base reactions involving proton exchange are extremely fast, whether intermolecular or intramolecular, and this results in rapid *exchange* of hydrogens on OH groups. This fast exchange causes the protons to absorb energy corresponding to their *average* environment, and only a single peak is seen at about 10 ppm for the two different —OH protons. Notice that the chemical shift is intermediate between values expected for carboxylic acids and alcohols. The averaging effects of protons undergoing rapid chemical exchange has one additional result: spin coupling to other nuclei is not observed. For this reason protons bonded to oxygen almost always appear as singlets, although these may sometimes be rather broad, as in Figure 9.19.

The concept of an *average* chemical environment is quite important and it explains why a proton normally gives only a single NMR signal (although perhaps a multiplet because of spin coupling) even when the molecule has more than one

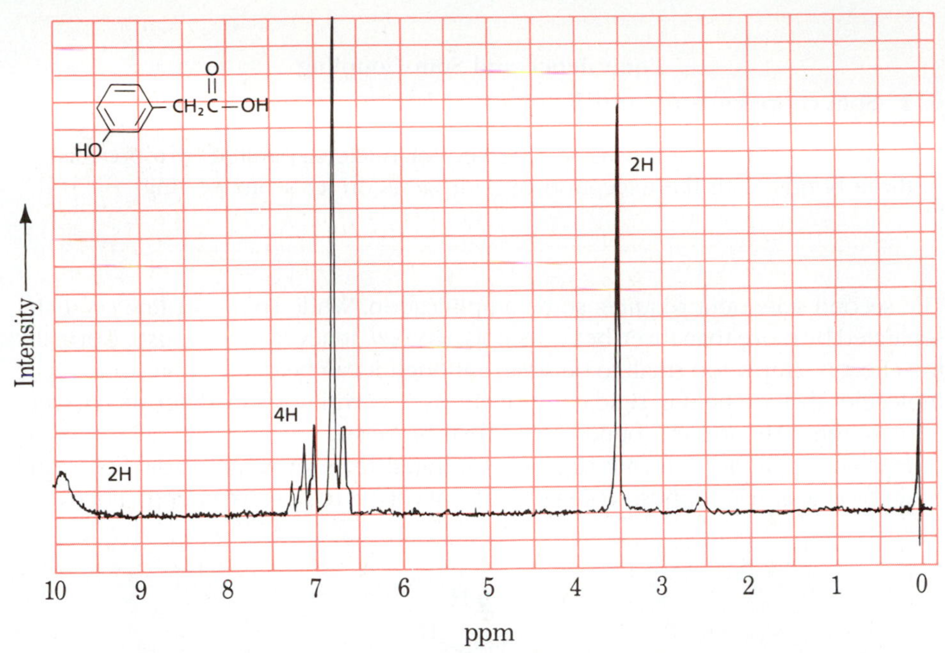

Figure 9.19 Proton NMR Spectrum of 3-Hydroxyphenylacetic Acid.

conformation. Conformational equilibria are ordinarily quite rapid at room temperature, and the observed chemical shift corresponds to the average for the individual conformers. This should not be confused with chemical exchange, however, since conformational equilibrium does not involve the making or breaking of any bonds.

Because the concept is rather important for the interpretation of NMR spectra, we have summarized in the box below the ways in which two or more nuclei can become equivalent in NMR spectroscopy.

Equivalence of Nuclei in NMR Spectroscopy

1. Nuclei that are equivalent by symmetry (either reflection symmetry or rotational symmetry) will have the same chemical shift.

2. Nuclei that are interchanged by rapid chemical equilibrium (chemical exchange) will behave as equivalent nuclei.

3. Nuclei that are structurally nonequivalent but accidentally have the same chemical shift will behave as equivalent nuclei.

These requirements for equivalence in NMR spectroscopy also make it possible to establish when you will observe spin coupling. Actually, it is easier to define the situations for which coupling will not be observed; these are listed in the following box.

Equivalence and Spin Coupling

1. Spin coupling is not usually observed between equivalent nuclei.

2. Spin coupling is not usually observed for nuclei separated by more than three bonds, with the exception of hydrogens on an aromatic ring.

The second statement requires some amplification. While spin coupling of nuclei separated by more than three bonds is not *observed,* it nevertheless *exists.* Normally the effect is simply too small to be seen, although it can be detected in some cyclic molecules that are held in rigid conformations. You will not encounter such exceptions in the NMR spectra found in this text, however.

Hydrogens on an aromatic ring are separated from hydrogen atoms of a side chain by at least four bonds, so you should not expect to observe coupling between them.

You can see this in Figure 9.19, where the CH_2 group appears as a singlet and no coupling to the hydrogens of the aromatic ring is detected.

What happens when hydrogens on adjacent carbon atoms (i.e., hydrogens separated by three bonds) are not equivalent but have very similar chemical shifts? Very

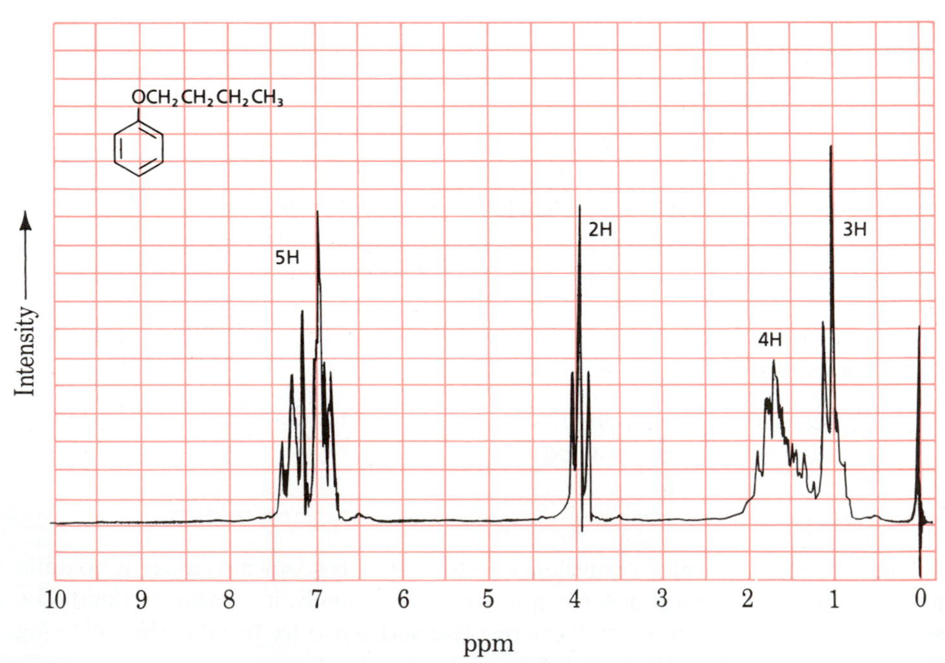

Figure 9.20 Proton NMR Spectrum of Butyl Phenyl Ether.

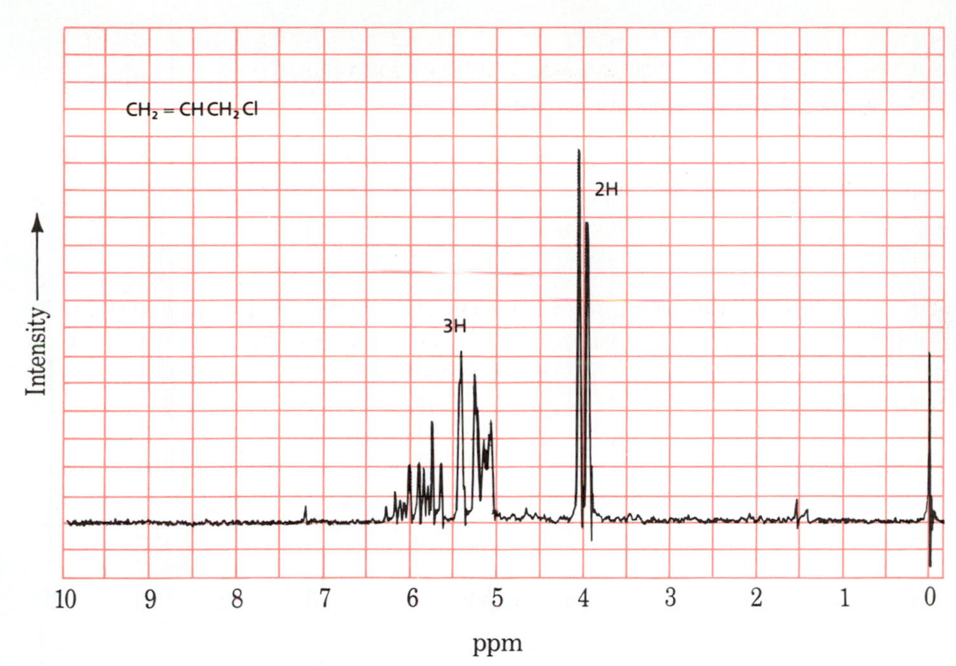

Figure 9.21 Proton NMR Spectrum of 3-Chloro-1-propene.

complex spin coupling interactions often result. An example is provided by the proton NMR spectrum of butyl phenyl ether (Figure 9.20). The hydrogens of the CH_2 group bonded directly to oxygen (triplet, 3.9 ppm) are quite different from the other protons of the butyl group, but the chemical shifts of the other two CH_2 groups and of the CH_3 group are all rather similar, and their spin coupling leads to complex multiplets. The "triplet" for the methyl group at 1.0 ppm is just barely recognizable, and the peak at 1.7 ppm for the other two CH_2 groups (which are very close in chemical shift) is virtually indecipherable.

Figure 9.21 shows one additional example of a complex spin coupling pattern that you will often encounter, that of the vinyl group, $—CH{=}CH_2$. The spectrum of 3-chloro-1-propene shows the expected doublet for the saturated CH_2 group at 4.0 ppm, but the three alkenyl hydrogens afford a complex pattern between 5.0 and 6.3 ppm.

The coupling of two hydrogens attached to the *same* carbon can sometimes greatly complicate an NMR spectrum. The three hydrogen atoms of a methyl group are always equivalent, and coupling is not observed between them. But the two hydrogens of a CH_2 group are not always equivalent. In fact, the two hydrogens of a CH_2 group in a cyclic compound are not equivalent if the ring is monosubstituted. For example, in cyclopentyl chloride the two hydrogens on C-2 are clearly nonequivalent; one is *cis* to the chlorine and the other is *trans*.

Cl
H
H H
Nonequivalent

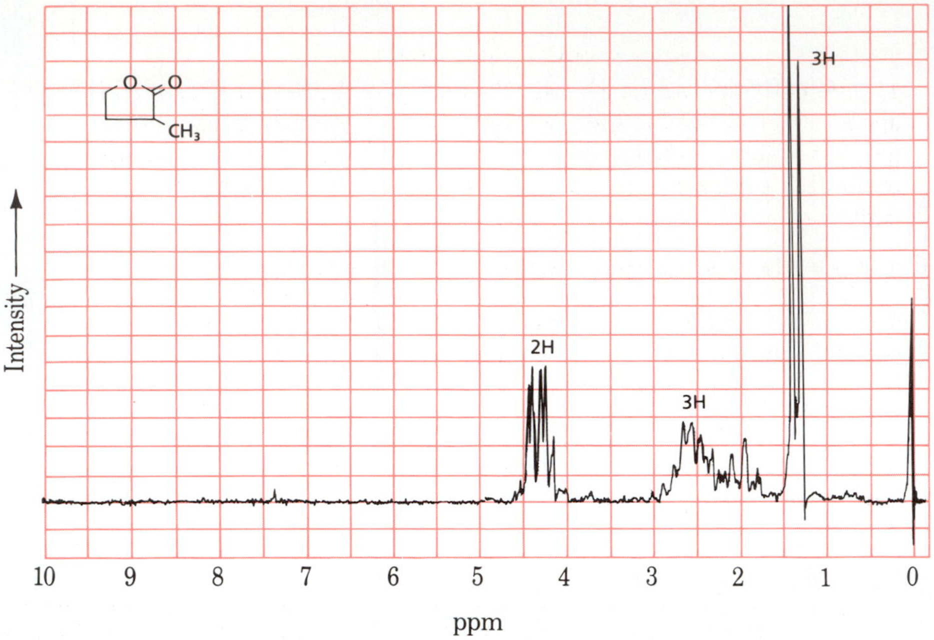

Figure 9.22 Proton NMR Spectrum of 2-Methyl-γ-butyrolactone.

A similar situation exists for the methylene groups of 2-methyl-γ-butyrolactone (Figure 9.22), in which each proton is either *cis* or *trans* to the methyl group. The methyl group appears as a clear doublet, but the remaining hydrogens give rise to complex multiplets. This sort of complexity is common for cyclic compounds, and their proton NMR spectra are often difficult to interpret.

The nonequivalence of the two hydrogens of a CH_2 group in an *acyclic* compound is less common, but you will sometimes observe this for compounds containing a chiral center. In such cases the CH_2 group has neither rotational nor reflection symmetry. You can easily verify this by consideration of a molecule such as (*S*)-2-hydroxybutyric acid.

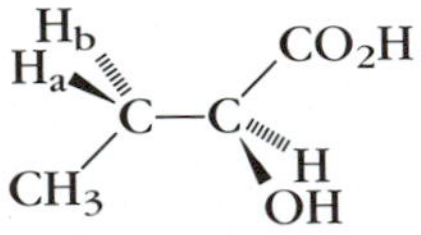

(*S*)-2-Hydroxybutyric acid

If you were to specifically replace H_a or H_b with another substituent such as a chlorine, the respective products would be diastereomers. For this reason hydrogens such as H_a and H_b are called **diastereotopic.** Fortunately, diastereotopic hydrogens in acyclic compounds often have the same chemical shift, so complex splitting patterns are not observed.

Our discussion of spin coupling interactions is summarized in the following box.

Spin Coupling Interactions

1. Spin coupling is a reciprocal relationship. If the hydrogens on carbon A appear as a multiplet in the NMR spectrum as a result of coupling to hydrogens on carbon B, then the hydrogens on carbon B must also appear as a multiplet because of their spin coupling interaction with the hydrogens on carbon A.

2. Coupling is not observed between equivalent nuclei.

3. Coupling is usually restricted to interactions between hydrogens on adjacent *carbon* atoms (and in some cases on the same carbon atom) or between hydrogens on the same aromatic ring.

4. A multiplet resulting from spin coupling tends to "point" toward the signal of the hydrogen to which it is coupled.

EXERCISE 9.4

For each of the following compounds predict the approximate proton NMR spectrum. The predictions should include the four kinds of information listed below:

1. Number of signals
2. Approximate chemical shift
3. Relative intensity of each signal
4. Multiplicity of each signal

(a) $CH_3-C(=O)-CH_3$

(b) $CH_3CH_2-O-C(CH_3)_2-CH_3$

(c) $HO-CH_2-CH_2-CH_3$

(d) $Cl-CH(CH_3)-CH_3$

(e) $CH_3CH_2-C(=O)-CH_2-C(=O)-OCH_2CH_3$

9.5 STRUCTURE DETERMINATION WITH NMR

A chemist in the laboratory usually has information about the chemical origin of a compound (i.e., about the starting material from which it was made), but there are also times when such information is not available. For example, a sample container might have been mislabeled, or the compound might have been isolated by extraction from plant or animal tissues rather than from a reaction mixture. In such cases the structure of the compound is truly unknown.

The determination of molecular structure for an unknown compound can be extremely difficult, but when the proton NMR spectrum is relatively simple, it is frequently possible to deduce the correct structure solely on the basis of proton NMR spectroscopy and knowledge of the molecular formula. The molecular formula can generally be determined from elemental analysis or mass spectrometry, and additional pieces of evidence from IR and UV spectroscopy sometimes allow even fairly complex structures to be deduced rather easily. In this section we will present examples that illustrate how a molecular structure can be deduced from spectroscopic information.

Most of the examples in this section will actually have more than enough information to deduce the structure. If you refer back to Figure 9.1, this means that once a structure has been *proposed* on the basis of some of the spectroscopic data, it can be *evaluated* by comparison of the remaining data with values predicted for that proposed structure. It is important for you to use all the available data when determining a structure, and you should be sure that all the data are consistent with the proposed structure. Any single piece of information that is inconsistent with a proposed structure would provide you with a sufficient reason for rejecting that structural possibility. The types of information that you should utilize are summarized below. Information will not always be available in all categories, and the examples in this section will utilize only the molecular formula and NMR spectrum.

1. *Molecular formula* and *degrees of unsaturation* (see Section 7.11). How many rings, double bonds, etc.?

2. *Chemical information.* What functional groups are present? What are the likely products for the reaction which was carried out?

3. *Infrared and Ultraviolet Spectroscopy.* What functional groups are present?

4. *NMR spectroscopy*
   a. *Number of signals.* How many nonequivalent nuclei are present? Does the molecule have symmetry?
   b. *Chemical shift.* What is the chemical environment of each of the nuclei?
   c. *Signal intensity (proton NMR).* To how many hydrogens in the molecule does each signal correspond?
   d. *Spin coupling (proton NMR).* What is the substitution pattern of adjacent carbon atoms?

At this stage we will introduce a useful device for interpreting proton NMR spectra. Certain structural subunits can only be found at the *end* of a carbon chain, while others must be situated in the *interior* portion of the carbon chain. We will designate such groups as *"ends"* and *"middles"* of organic structures.

Ends: $—CH_3$, $—C_6H_5$ (phenyl), —H, —Cl, etc.

Middles: —O—, $—CH_2—$, —C(=O)—, —CH—, etc.

This device is simplistic but useful. Note that a molecule which is branched will have more than two "ends." In addition, you can modify the preceding categories. For example, if a $—CH_3$ group is split into a triplet, you could redefine the "end" group as an ethyl group, $—CH_2CH_3$.

Example 9.2 Deduce the structure of the compound with the molecular formula, $C_3H_6Cl_2$, from its proton NMR spectrum (Figure 9.23).

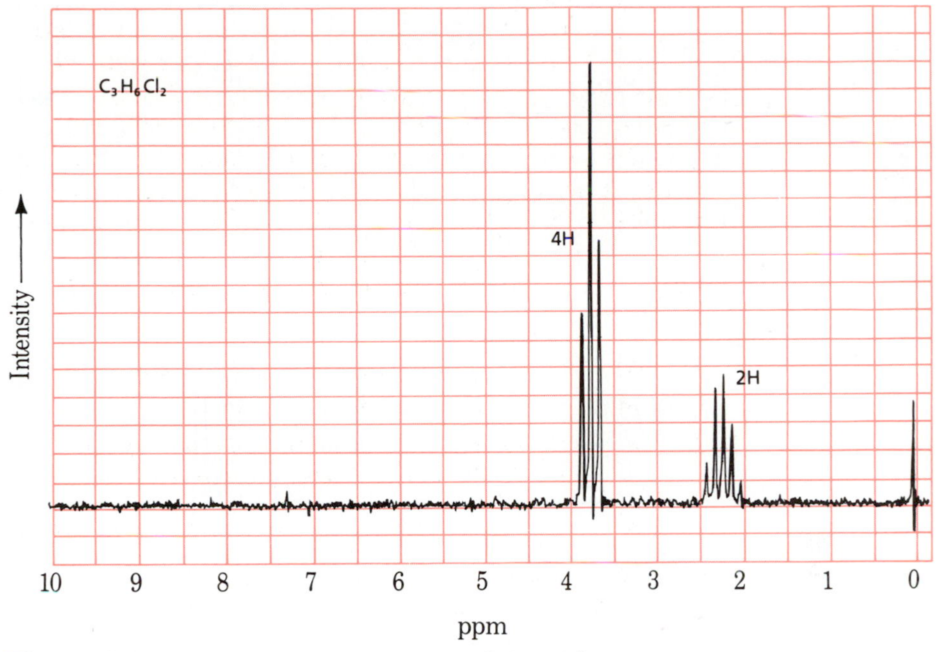

Figure 9.23 Proton NMR Spectrum of $C_3H_6Cl_2$.

Solution There are no degrees of unsaturation, and there are only two signals. There is no such thing as a "CH_4 group," so the triplet at 3.7 ppm must result from two equivalent CH_2 groups. This leaves only one carbon atom, so the two remaining hydrogens must be on a CH_2 group that yields the signal centered at 2.2 ppm. The signal is a quintet, so this CH_2 group must be coupled to *both* the other two methylene groups. We can therefore draw a partial structure:

$$—CH_2—CH_2—CH_2— \qquad (C_3H_6)$$

Only two chlorines remain, so the complete structure must be:

$$Cl—CH_2—CH_2—CH_2—Cl$$

This proposed structure can be checked by using the remaining pieces of information. The two CH_2Cl groups are each split by two equivalent hydrogens and should appear as a triplet, as is indeed the case. The predicted chemical shifts for 1,3-dichloropropane (from Table 9.3) are 3.5 ppm (0.5 base + 0.4 for alkyl + 2.6 for Cl) for the terminal CH_2Cl groups and 2.1 ppm (0.5 base + 0.8 for each of two CH_2X groups) for the central methylene group. Each of these predictions is within 0.2 ppm of the observed values, a satisfactory result for the proposed structure.

Example 9.3 Deduce the structure of the compound with the molecular formula $C_4H_8O_2$ from its proton NMR spectrum (Figure 9.24).

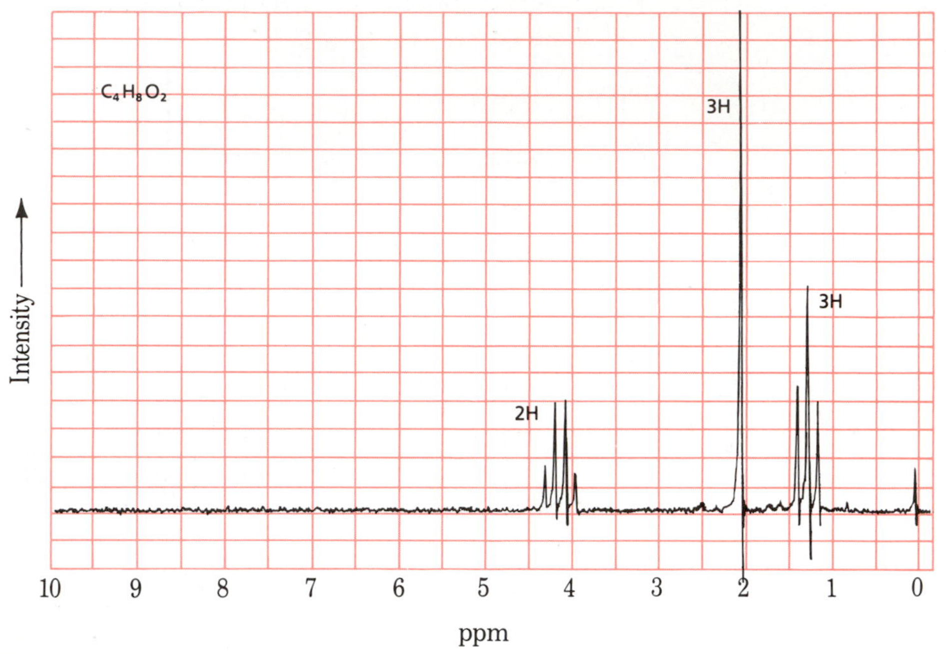

Figure 9.24 Proton NMR Spectrum of $C_4H_8O_2$.

Solution The molecular formula indicates that the molecule has one degree of unsaturation, either a double bond or a ring. There are three signals in the spectrum, so the molecule contains three kinds of hydrogen. The integration indicates two CH_3 groups and one CH_2 group. The spin coupling patterns indicate an ethyl group, and the downfield shift of the CH_2 group strongly suggests that the ethyl group is attached to oxygen. A partial structure can now be drawn which consists of two "ends," although the "middle" is missing:

$$CH_3CH_2O— \qquad —CH_3 \qquad (C_3H_8O)$$

This leaves only one carbon, one oxygen, and one degree of unsaturation, so this must be a carbonyl group, $\rangle C{=}O$. Hence the complete structure must be:

$$CH_3CH_2—O—\overset{\overset{\displaystyle O}{\|}}{C}—CH_3$$

The proposed structure is consistent with the presence of an isolated (in the sense that it is not spin coupled to any other hydrogens) ethyl group and an isolated methyl group. The predicted chemical shifts are: CH_2, 4.1 ppm (0.5 base + 0.4 for CH_3 and 3.2 for OC(O)R); CH_3 of ethyl group, 1.3 ppm (0.5 base + 0.8 for $—CH_2X$); and $—O_2CCH_3$, 2.0 ppm (0.5 base + 1.5 for $—CO_2R$). These are all in excellent agreement with the respective experimental values of 4.1, 1.2, and 2.0 ppm.

Example 9.4 Deduce the structure of the compound with the molecular formula $C_9H_{10}O_2$ from its proton NMR spectrum (Figure 9.25).

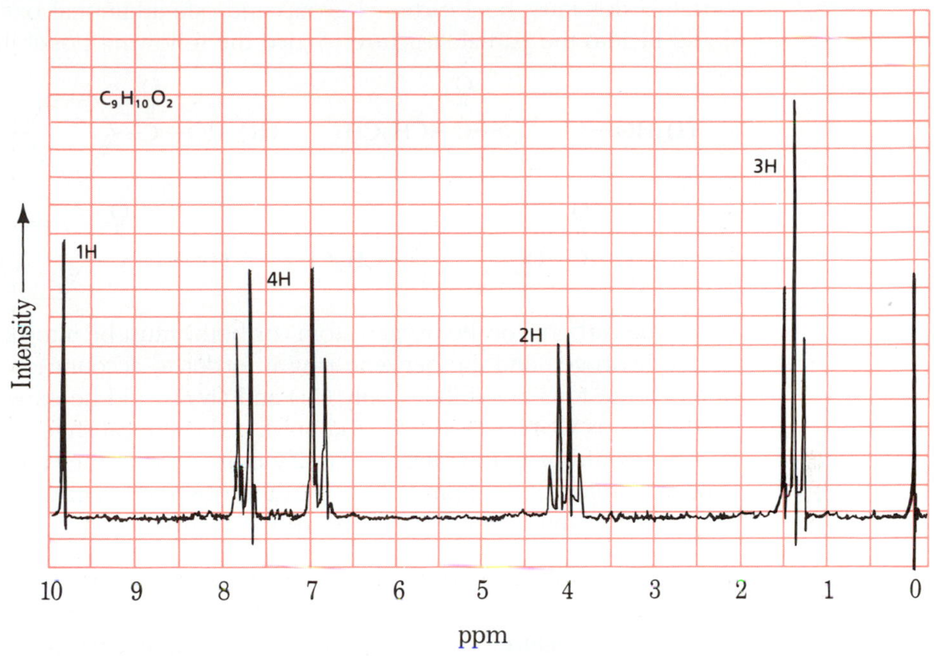

Figure 9.25 Proton NMR Spectrum of $C_9H_{10}O_2$.

Solution The compound has five degrees of unsaturation, and the peaks in the region from 6 to 8 ppm strongly suggest an aromatic ring, which would account for four degrees of unsaturation (one ring plus three double bonds). Moreover, the splitting pattern of the four-hydrogen multiplet centered at 7.4 ppm is highly symmetrical, a pattern that is characteristic of a 1,4-disubstituted benzene ring*:

*The location of the double bonds in a benzene ring is not relevant when determining equivalent positions. As you will see in Chapter 10, two resonance forms of a benzene ring can be drawn

and the bonds between the carbon atoms of the ring are neither double bonds nor single bonds but are intermediate between the two.

The remaining signals show the presence of an ethyl group and of one additional hydrogen, which gives rise to the singlet at 9.9 ppm. These last two groups correspond to ends of the molecule, and the partial structure can be revised to

H— (benzene ring, para-disubstituted) —CH_2CH_3 (C_8H_{10})

This leaves one carbon, two oxygens, and one degree of unsaturation. Almost certainly this must be a carbonyl group and one additional oxygen atom, which could fit into the partial structure to give the following possibilities:

(i) HO—C_6H_4—C(=O)—CH_2CH_3 **(ii)** HO—C(=O)—C_6H_4—CH_2CH_3

(iii) H—C(=O)—O—C_6H_4—CH_2CH_3 **(iv)** H—C(=O)—C_6H_4—OCH_2CH_3

Either the carbonyl or the oxygen atom (or both) must be interposed between the simple hydrogen and the benzene ring in order to account for the singlet at 9.9 ppm. Of the four possibilities only (iii) and (iv) would be expected to show the singlet in this vicinity. Moreover, the predicted chemical shift of the CH_2 group in (i)–(iii) would be in the range of 2.6–2.9 ppm. Only in the case of (iv) would the predicted value of 4.2 ppm be close to the observed chemical shift of 4.1 ppm. Structure (iv) must represent the correct structure.

Example 9.5 Deduce the structure of the compound with the molecular formula $C_3H_4O_3$ from its proton NMR spectrum (Figure 9.26).

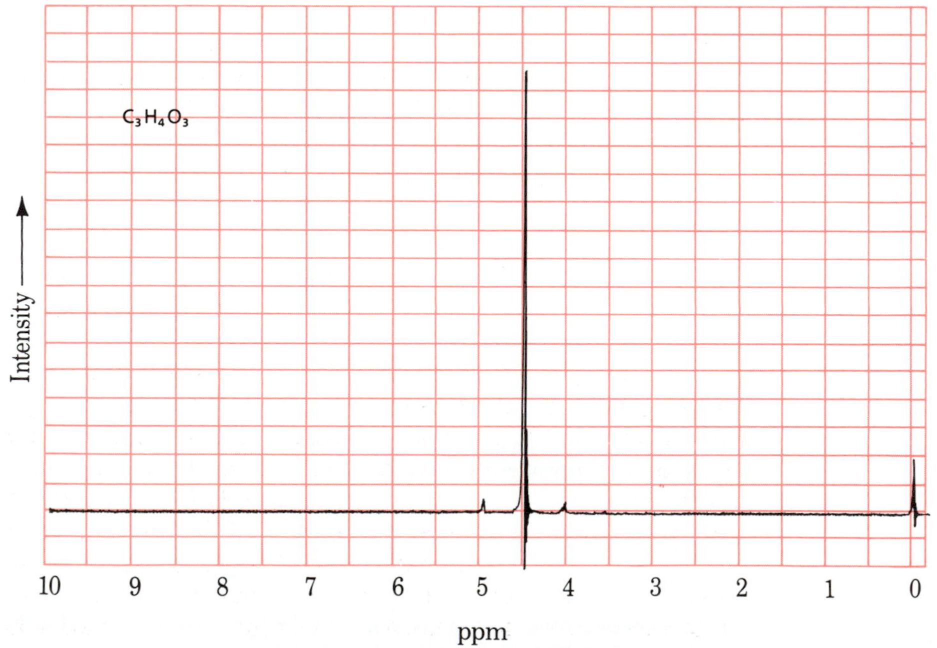

Figure 9.26 Proton NMR Spectrum of $C_3H_4O_3$.

Solution This spectrum is so simple that it is difficult to interpret. There is either molecular symmetry or accidental equivalence because only a single peak is observed. The molecular formula requires two degrees of unsaturation, and this could result from two rings, two double bonds, one triple bond, or one ring and one double bond.

Assuming that all four hydrogen atoms are symmetry-equivalent, this requires two equivalent CH_2 groups. Using a dotted line to denote symmetry, these must be equally distant from the dotted line in a partial structure:

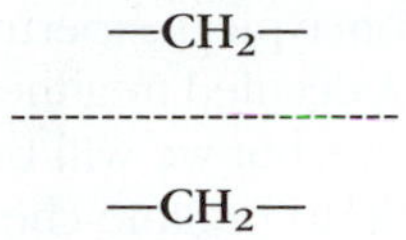

—CH_2—
- - - - - - - - - - - - (${C_2H_4}$)
—CH_2—

An odd number of oxygen atoms remain, so one must lie *on* the symmetry line. The chemical shift of the CH_2 groups (4.5 ppm) strongly suggests that they are bonded to oxygen, either both to the single oxygen that lies on the symmetry line or each to one of the remaining oxygens.

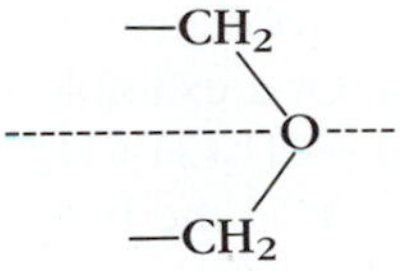

(i) C_2H_4O or (ii) $C_2H_4O_3$

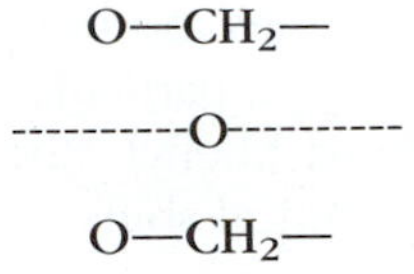

The only way in which the two remaining oxygens could be incorporated into structure (i) while maintaining symmetry would lead to a molecule with a divalent carbon atom, an unsatisfactory answer.

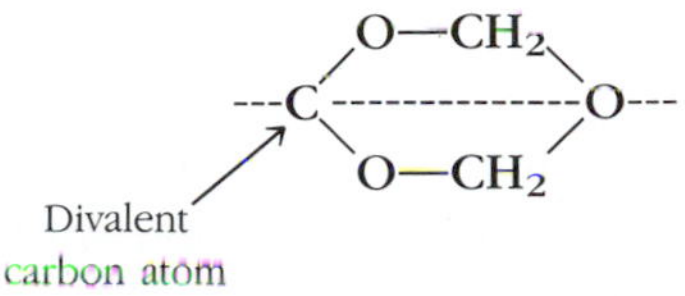

Divalent carbon atom

Consequently, partial structure (ii) must correspond to the correct structure. This differs from the complete structure by only a carbon atom, which could be incorporated into the partial structure in several ways.

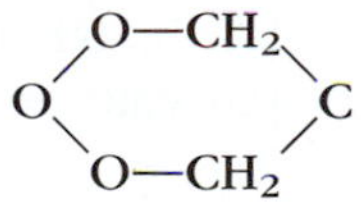

ii(a) 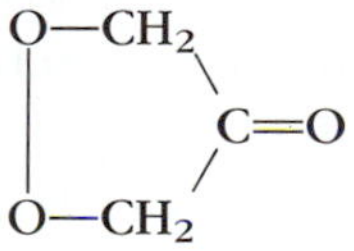 ii(b) 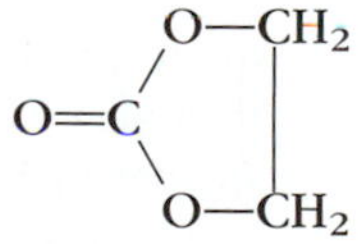 ii(c)

Structure ii(a) is totally unacceptable because it has a divalent carbon atom. Structure ii(b) has a peroxide linkage and might be unstable (explosive) but could reasonably exist. The predicted chemical shift of 4.7 ppm (0.5 base + 2.6 for —OR + 1.6 for —C(O)R) is close to the observed value of 4.5 ppm. Structure ii(c) has no unreasonable features and the predicted shift is 4.5 (base 0.5 + 3.2 for O—C(O)R + 0.8 for —CH_2X). This agrees better with the observed value than that calculated for ii(b), but by itself is not sufficient to allow a choice between ii(b) and ii(c). The correct structure is actually ii(c), but you would need additional evidence (chemical or spectroscopic) to definitively rule out ii(b) as a possibility.

9.6 SPECIALIZED NMR TECHNIQUES: SPIN DECOUPLING, DEUTERIUM EXCHANGE, AND LANTHANIDE SHIFT REAGENTS

Many specialized techniques have been developed in NMR spectroscopy that further increase its value to chemists. Some of these techniques yield additional information not available from typical experiments, while others facilitate the interpretation of typical spectra. A detailed treatment of these techniques is beyond the scope of an introductory course, but we will briefly discuss several of them because they are particularly valuable to organic chemists.

Spin Decoupling

The spin coupling patterns in a proton NMR spectrum often provide definitive information regarding substitution patterns in an organic molecule. But sometimes complex coupling patterns can obscure virtually all the important information about a particular NMR signal. One example of this was seen in the spectrum of butyl phenyl ether, $C_6H_5—O—CH_2CH_2CH_2CH_3$ (Figure 9.20), in which similar chemical shifts of the hydrogens on the butyl group led to complex and overlapping multiplets. Additional information can be obtained from such complicated spectra by using the technique of **spin decoupling** (also called *double irradiation* or *double resonance*).

We can explain the technique by evaluating a hypothetical example with the partial structure

$$H_A—\overset{|}{\underset{|}{C}}—\overset{|}{\underset{|}{C}}—H_X$$

In a normal spectrum, H_A would appear as a doublet because of spin splitting by H_X. When H_A is absorbs radiation, H_X could be in either of two possible spin states, and this results in two energetically different transitions for H_A. What happens if H_X is irradiated with the appropriate rf for its transition *at the same time?* The frequencies for H_A and H_X will be different, and if a very intense rf signal is used to irradiate H_X, this would result in the very fast interconversion of the two possible spin states of H_X. This is the phenomenon of *saturation* of H_X described earlier, and only an *average* effect would be exerted on H_A. In other words, H_X would behave as though it were in a single, "average" spin state, and *H_A would give rise to only a singlet in the NMR spectrum.*

Such a decoupling experiment provides two additional kinds of information: (1) It reveals precisely which hydrogens are coupled to which other hydrogens. (2) It often converts a complex multiplet that is obscured by other peaks in the spectrum to a reasonably sharp signal, for which the chemical shift and coupling patterns (to other nuclei) can be measured.

Spin Decoupling in ^{13}C NMR

We did not discuss this in Section 9.3, but most ^{13}C spectra are recorded with extensive use of decoupling; otherwise the spin coupling to *hydrogen* substituents would complicate the *carbon* spectrum and make it difficult to interpret. The low natural abundance (1%) of ^{13}C yields a very low probability (0.01%) that two specific carbon atoms in a single molecule will *both* be ^{13}C. Consequently, ^{13}C–^{13}C coupling is not observed in ordinary ^{13}C NMR spectroscopy.

Normal ^{13}C spectra such as those we showed in Section 9.3 are fully *proton-decoupled,* so each carbon signal appears as a singlet. However, with modern spectrometers it is also possible to obtain proton–^{13}C spin coupling information without making the spectra unacceptably complex. For example, a fully proton-decoupled spectrum could be depicted normally, with proton–^{13}C coupling information noted adjacent to each peak (s = singlet, d = doublet, t = triplet, q = quartet). The couplings of interest here are *one-bond couplings,* that is, coupling between a ^{13}C atom and any hydrogens attached directly to it. A peak marked "s" would have no hydrogen substituents, and a doublet would have a single hydrogen. A CH_2 group would be designated as a triplet and a CH_3 group as a quartet. Figure 9.27 illustrates this technique for the ^{13}C NMR spectrum of vinyl acetate.

As we stated in Section 9.3 and illustrated with the preceding spectrum, the signal intensities in ^{13}C NMR spectra are not ordinarily directly proportional to the number of carbon atoms that give rise to the signal. However, recent advances in ^{13}C NMR pulse techniques have made it possible to obtain spectra in which integration of the peaks tells you the relative numbers of each type of carbon atom in the molecule.

Still another advance that has become possible with computer-interfaced NMR spectrometers is *two-dimensional NMR spectroscopy.* This powerful technique makes it possible to record a complete map of all coupling interactions in the

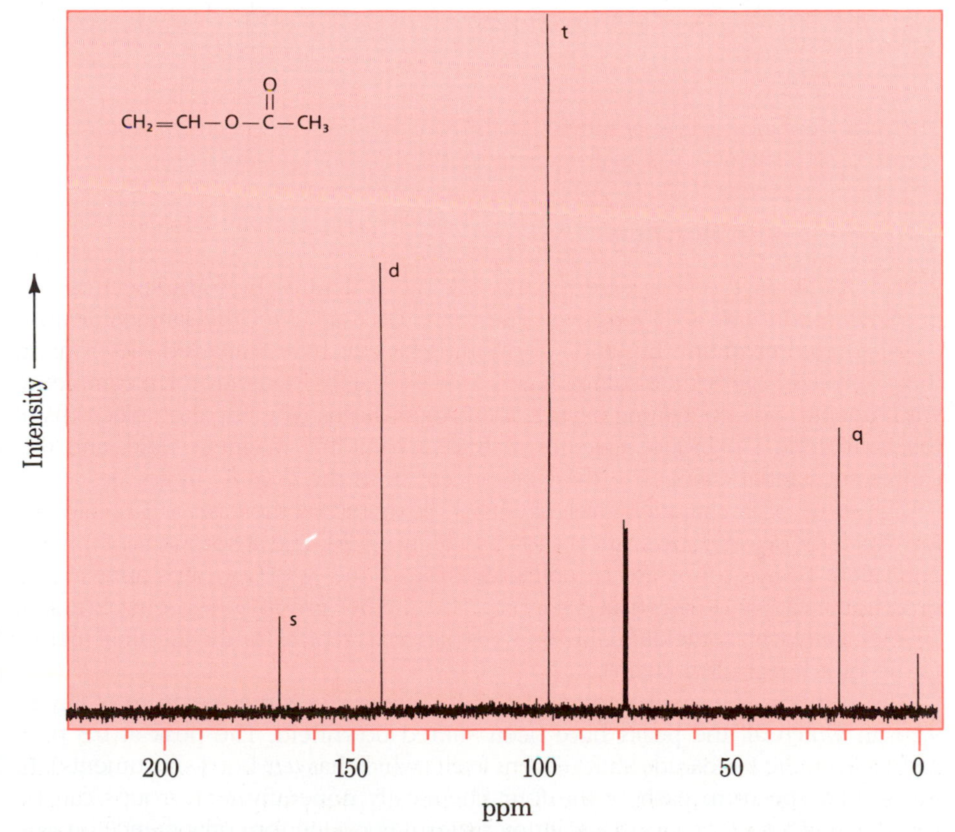

Figure 9.27 ^{13}C NMR Spectrum of Vinyl Acetate.

molecule. Because spin coupling interactions between nuclei decrease with the number of intervening bonds, two-dimensional spectroscopy allows connectivity relationships to be deduced.

Deuterium Exchange

The isotope of hydrogen with an atomic mass of 2, *deuterium,* (abbreviated as D or ^{2}H), has energetically different spin states when it is placed in a magnetic field. But ^{1}H and ^{2}H do not absorb rf radiation at the same frequency, and a deuterium signal would be found about 800,000 ppm downfield from the proton signal of TMS. As a result, the replacement of ^{1}H by ^{2}H would cause the proton NMR signal to "disappear." Chemical exchange of protons bound to oxygen (in alcohols, carboxylic acids, etc.) occurs rapidly with a hydroxylic solvent such as water or an alcohol, and simple treatment of a sample with excess D_2O (i.e., ^2H_2O) will convert the —OH groups to —OD groups. By this chemical method an OH peak that obscured other important signals could simply be removed from the NMR spectrum.

Deuterium exchange can also be used with carbonyl compounds. Recall from Section 4.7 that the hydrogens α to a carbonyl group are far more acidic than other hydrogens bonded to carbon. A base-catalyzed equilibrium in deuterated solvent will therefore produce exchange of the acidic hydrogens.

$$\text{—C(H)—C(=O)—} \xrightleftharpoons{\text{NaOD/D}_2\text{O}} \left[\text{—}\ddot{\text{C}}\text{—C(=O)—}\right]^- \rightleftharpoons \text{—C(D)—C(=O)—}$$

The disappearance of one or more signals from the the NMR spectrum would allow you to assign them to the hydrogens α to the carbonyl group.

Lanthanide Shift Reagents

A very useful method for determining organic structures by NMR spectroscopy employs *lanthanide shift reagents.* These are derivatives of the lanthanide ions (particularly europium, Eu, and ytterbium, Yb) that have unpaired electrons in their *f* orbitals. Lanthanide shift reagents are Lewis acids, and they form complexes with organic acids containing oxygen or nitrogen atoms. The unpaired electrons of the lanthanide ion induce changes in the surrounding magnetic field, and this causes substantial changes in the NMR spectrum of the organic molecule.

The value of lanthanide shift reagents is illustrated by the case of 5-nonanone, for which the normal spectrum is shown in Figure 9.28. The absorptions of the two equivalent butyl groups are complex, and the β and γ CH_2 groups appear as a broad absorption centered at 1.4 ppm. The methyl group is recognizable as a "triplet" only with some difficulty (and this pattern is typical for the terminal methyl group of a linear alkyl chain).

The addition of a small amount of a shift reagent yields the spectrum of Figure 9.29, in which all the peaks have been shifted downfield. The peak at 0.8 ppm results from the lanthanide shift reagent itself (which has *tert*-butyl substituents). In the shifted spectrum each of the four chemically nonequivalent groups can be recognized by the appropriate splitting pattern. Focusing for convenience on one of the two equivalent butyl groups, you can see that the α CH_2 group yields a triplet

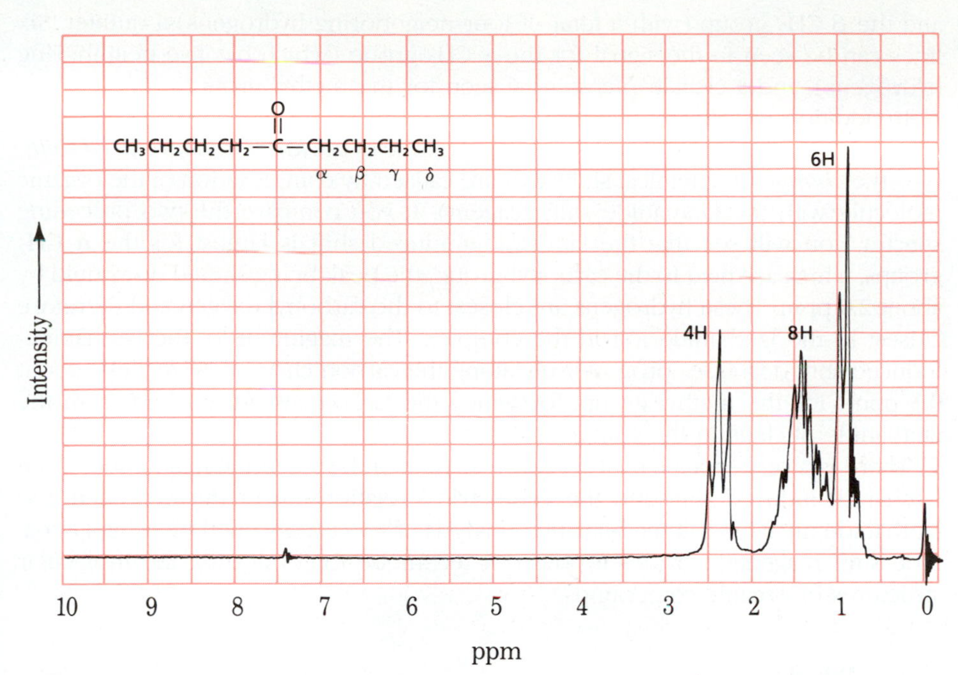

Figure 9.28 Proton NMR Spectrum of 5-Nonanone.

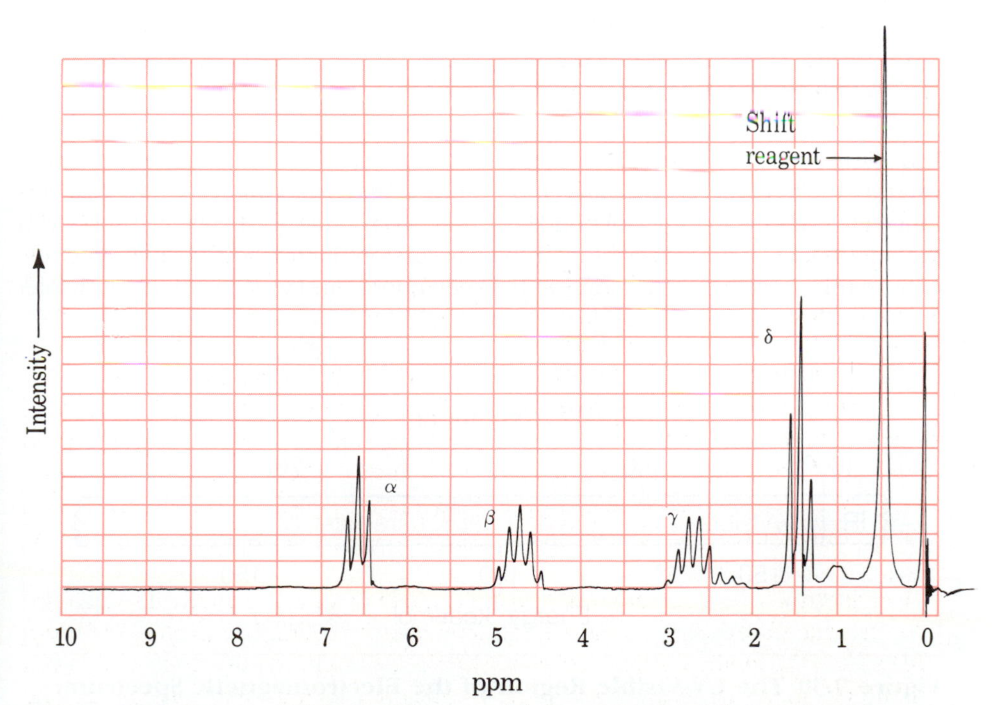

Figure 9.29 Proton NMR Spectrum of 5-Nonanone with Lanthanide Shift Reagent.

and the β CH_2 group (with a total of four neighboring hydrogens) a quintet. Six lines can be seen in the signal for the γ CH_2 group (which has five neighboring hydrogens), and a clear triplet is now seen for the methyl group.

In addition to the simplification that can be provided by shift reagents, additional information is available from the magnitude of the *lanthanide-induced shifts,* (i.e., the *changes* in chemical shifts that are caused by complexation of the organic molecule with the lanthanide shift reagent). If you compare the two preceding spectra, you will see that the lanthanide-induced shift is largest for the α CH_2 groups, which are next to the carbonyl group, the peak being shifted downfield by about 2.0 ppm. These hydrogens are closest to the carbonyl oxygen and therefore closest to the lanthanide ion in the complex. The magnitude of the lanthanide-induced shift decreases progressively along the carbon chain. It is smallest (about 0.3 ppm) for the methyl group, for which the hydrogens are furthest from the lanthanide nucleus in the complex.

There is a very specific mathematical correlation between the magnitude of the lanthanide-induced shift and the geometrical relationship of the hydrogen and lanthanide nuclei. A full mathematical analysis of spectra obtained by using lanthanide shift reagents provides a powerful technique for evaluating the molecular structures of organic compounds.

EXERCISE 9.5

What chemical shifts would you predict for each of the signals for vinyl acetate in Figure 9.27 using the substituent effects in Table 9.2? Compare the assignments you make on the basis of the predicted shifts with the multiplicities noted on the spectrum for proton coupling.

9.7 ULTRAVIOLET-VISIBLE SPECTROSCOPY

Radiation in the ultraviolet and visible regions of the electromagnetic spectrum extends from about 35 to nearly 150 kcal/mol, and you will recognize that this range includes the energies of bonds between atoms (Table 2.1). We will show you the importance of this observation in our discussion of ultraviolet-visible spectroscopy in this section. The relationship between energy and wavelength of ultraviolet

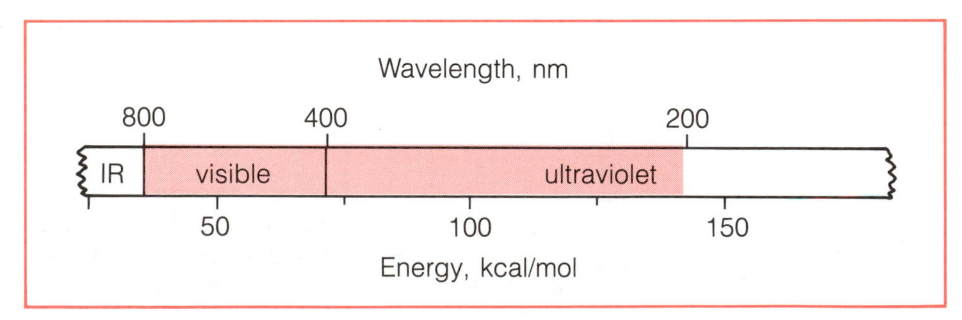

Figure 9.30 The UV-Visible Region of the Electromagnetic Spectrum. The shaded portion of this region indicates the wavelengths commonly used in UV-visible spectroscopy (1 nm = 10^{-9} m).

and visible light is summarized in Figure 9.30. The wavelength in the figure is shown in nanometers (nm), the units commonly used for wavelength in this portion of the electromagnetic spectrum.

Some organic compounds absorb radiation in the UV-visible region of the electromagnetic spectrum while many others do not. What structural features of an organic molecule can be correlated with the ability to absorb either ultraviolet or visible light? Ultraviolet and visible light are adjacent to each other in a continuous spectrum, and the structural features required for absorption of ultraviolet and visible radiation are closely related. It is easy for you to determine whether or not a substance is colored, so we will first focus on one part of the question: What structural features result in the absorption of **visible light**? First, think of the

TABLE 9.4 Some Colorless Organic Compounds

Structure	Name (Functional Class)	Common Uses
CH_3CH_2OH	Ethanol (an alcohol)	Alcoholic beverages
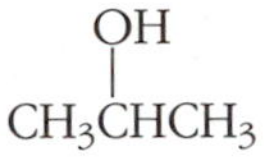 $CH_3CH(OH)CH_3$	2-Propanol (an alcohol)	Rubbing alcohol
$CH_3—(CH_2)_5—CH_3$	Heptane (an alkane)	Component of gasoline
CCl_4	Carbon tetrachloride (a halogenated alkane)	Dry cleaning solvent
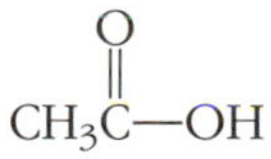 $CH_3C(=O)—OH$	Acetic acid (a carboxylic acid)	Vinegar is a dilute solution of acetic acid in water
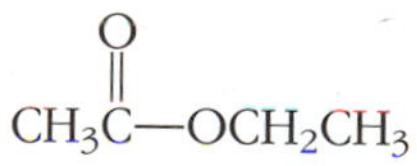 $CH_3C(=O)—OCH_2CH_3$	Ethyl acetate (an ester)	Solvent; nail polish remover
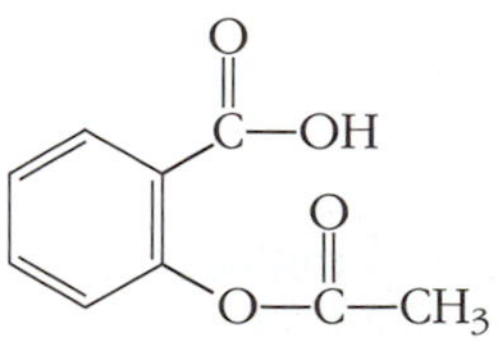 $C_6H_4(C(=O)—OH)(O—C(=O)—CH_3)$	Acetylsalicylic acid (aspirin) (carboxylic acid, ester)	Analgesic
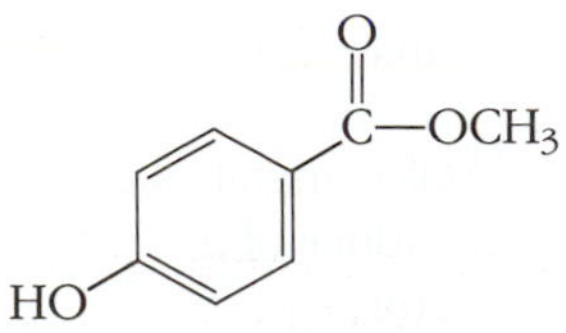 $HO—C_6H_4—C(=O)—OCH_3$	Methyl 4-hydroxybenzoate, methylparaben (carboxylic acid, phenol)	Ingredient in suntan lotion

definition of color. A substance that transmits or reflects all wavelengths of white light (such as that of sunlight) is perceived as being white (colorless). But when certain wavelengths (i.e., certain energies) of light are selectively absorbed, the light that is transmitted or reflected no longer corresponds to the white light of sunlight. Certain wavelengths of the white light have been removed, and only those wavelengths that remain can reach your eye. You would then perceive the substance as having color.

Now that we have established a working definition of color, consider some common organic compounds (Table 9.4) that do not absorb in the visible region of the spectrum and are therefore colorless. Then contrast their structures with the

TABLE 9.5 Some Colored Organic Compounds

Structure	Name (Color)	Common Uses
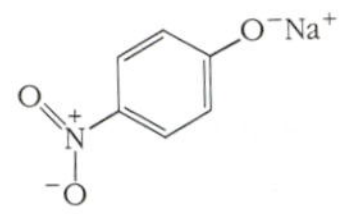	Sodium 4-nitrophenolate (yellow)	—
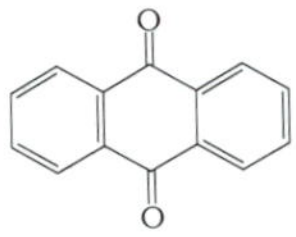	Anthraquinone (yellow)	Manufacture of dyes
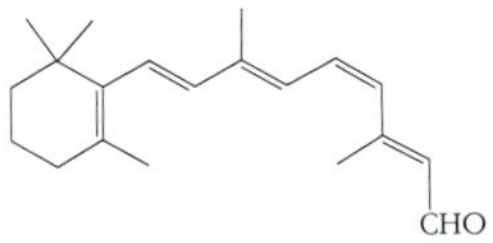	Retinal (red)	Visual pigment in the human eye: the light-sensitive portion of rhodopsin "visual purple"
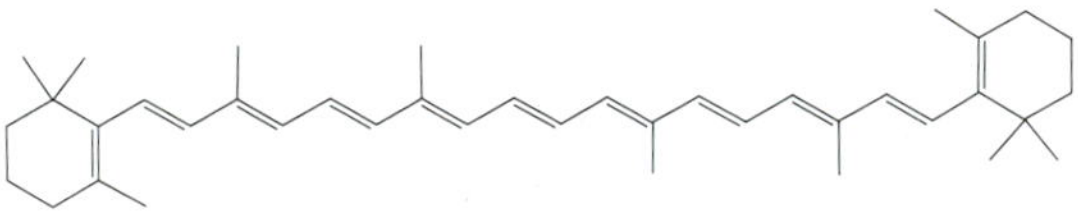	*β*-Carotene (red)	Plant pigment: precursor of vitamin A
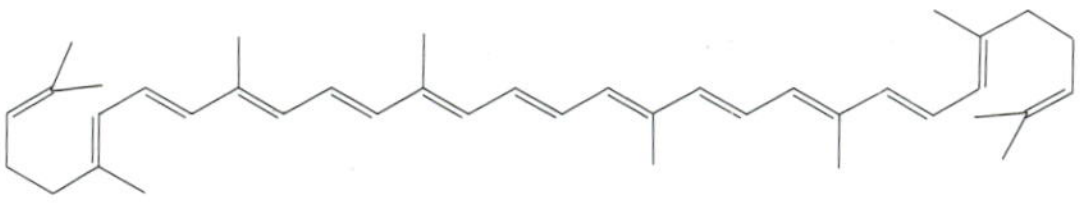	Lycopene (red)	Plant pigment (characteristic coloring of tomatoes)
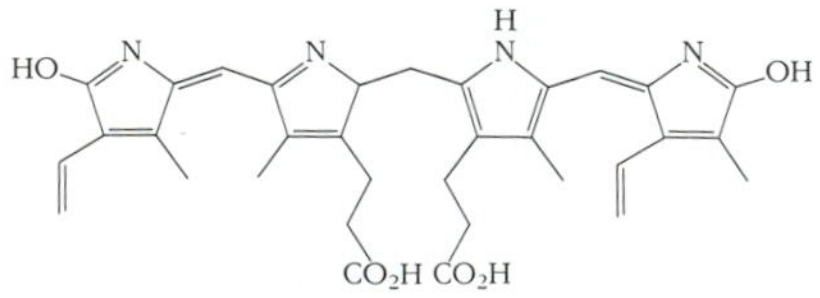	Bilirubin (orange)	Major metabolic product of heme

structures of a variety of organic compounds which do absorb light in the visible region (Table 9.5).*

As you can see from Tables 9.4 and 9.5, both colorless and colored compounds can contain a wide variety of functional groups. However, those organic compounds which absorb light in the visible region of the spectrum all share one common feature, a large number of *conjugated multiple bonds.* This feature is often described as *extended conjugation.* Simple alkanes and alkenes are therefore colorless, but compounds such as lycopene that have a large number of conjugated double bonds absorb in the visible region and are intensely colored. We cannot tell you any specific extent of conjugation that is necessary for a compound to absorb visible light, but the following compounds provide qualitative limits.

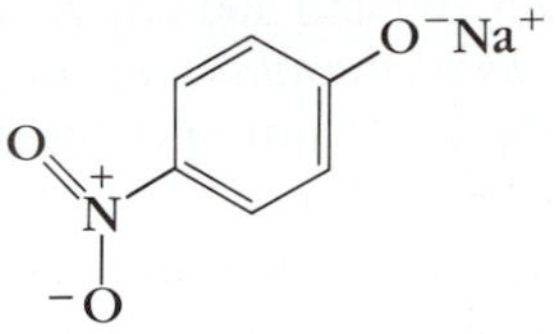

Sodium 4-nitrophenolate (yellow) 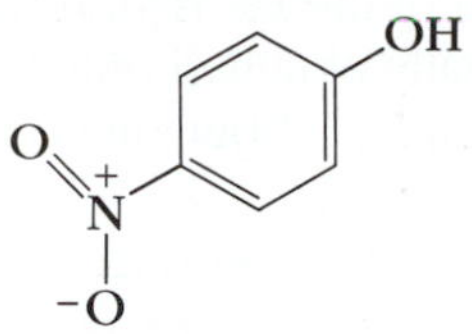 — 4-Nitrophenol (faint yellow) 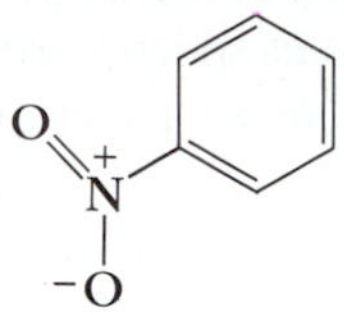 — Nitrobenzene (colorless) 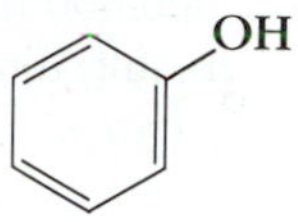 — Phenol (colorless)

4-Nitrophenol exhibits only a pale yellow color, and its sodium salt is much more intensely colored. But there is no absorption in the visible region when either the nitro group or the hydroxyl group is absent, and so both nitrobenzene and phenol are colorless. Thus 4-nitrophenol, with the six unsaturated carbon atoms of the benzene ring and two additional groups that can interact with this π system, represents the lower limit of extended conjugation needed for absorption in the visible region of the spectrum.

Absorption of ultraviolet light also requires unsaturation, but extended conjugation is not required. In practice it is necessary for a compound to have two or more conjugated multiple bonds in order for UV absorption to be observed with a typical spectrometer. Because the structural requirements are similar, many compounds that absorb in the visible region also absorb in the ultraviolet region of the electromagnetic spectrum.

Electronic Transitions: The Mechanism of Absorption of UV and Visible Light

Visible and ultraviolet absorption both result in changes in the electronic configurations of molecules, particularly with respect to the π molecular orbitals. We will

*We have introduced a very useful convention in Table 9.5 that we will continue to use in this text: methyl groups and methylene groups at the end of a carbon chain are represented by lines without writing the CH_3 or CH_2. 3-Methyl-1-butene is therefore drawn as

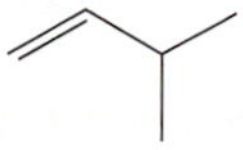

rather than as

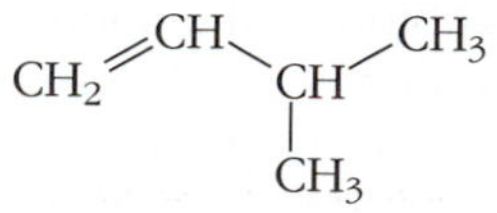

$CH_2{=}CH{-}CH(CH_3){-}CH_3$

illustrate the overall process with a simple series of unsaturated hydrocarbons, the simplest case being ethylene (Figure 9.31).

In a **ground-state** ethylene molecule the electrons occupy the lowest-energy molecular orbitals. Absorption of light produces an **excited state,** in which an electron occupies an antibonding orbital such as the π^* molecular orbital. The energy difference between the π and π^* orbitals of ethylene is approximately 159 kcal/mol, and this corresponds to UV radiation with a wavelength of 180 nm. When an ethylene molecule absorbs a photon of this wavelength, the resulting *electronic transition* is called a $\pi \rightarrow \pi^*$ transition.

The 180 nm absorption of ethylene is outside the range of normal UV-visible spectrometers, which usually permit observation of the entire visible region of 400-800 nm but only that portion of the UV region between 200 and 400 nm. As a result, ultraviolet spectra are usually obtained only for *molecules containing conjugated multiple bonds.* For example, 1,3-butadiene absorbs at 217 nm, and this is well inside the range typically studied.

The UV absorption of butadiene again produces a change in the electronic configuration of the π system. Other transitions are possible at higher energy (shorter wavelength), but the 217 nm absorption is the lowest-energy transition. The lowest-energy electronic transition must always occur from the highest *occupied* molecular orbital to the lowest *unoccupied* molecular orbital, and the energy of the transition is therefore governed by the HOMO-LUMO energy difference. Both these orbitals are again π orbitals for butadiene, and the transition shown in Figure 9.32 for butadiene is again described as a $\pi \rightarrow \pi^*$ transition.

The difference in the wavelength of light absorbed by ethylene and by 1,3-butadiene illustrates an important trend. As the number of conjugated multiple bonds is increased, the electronic absorption is shifted to longer wavelength (*lower energy*). Ethylene absorbs at 180 nm (159 kcal/mol), 1,3-butadiene absorbs at 217 nm (131

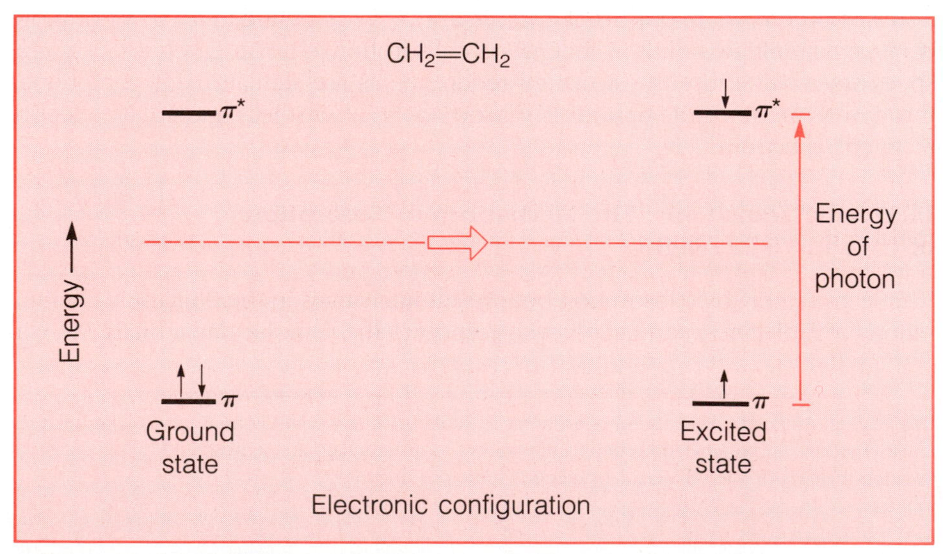

Figure 9.31 Absorption of Ultraviolet Light by Ethylene.
In the ground state both electrons of the π system occupy the bonding molecular orbital. Absorption of a photon of the appropriate energy (wavelength 180 nm) produces an excited state in which one electron now occupies the antibonding π^* orbital. Transitions involving other orbitals are also possible.

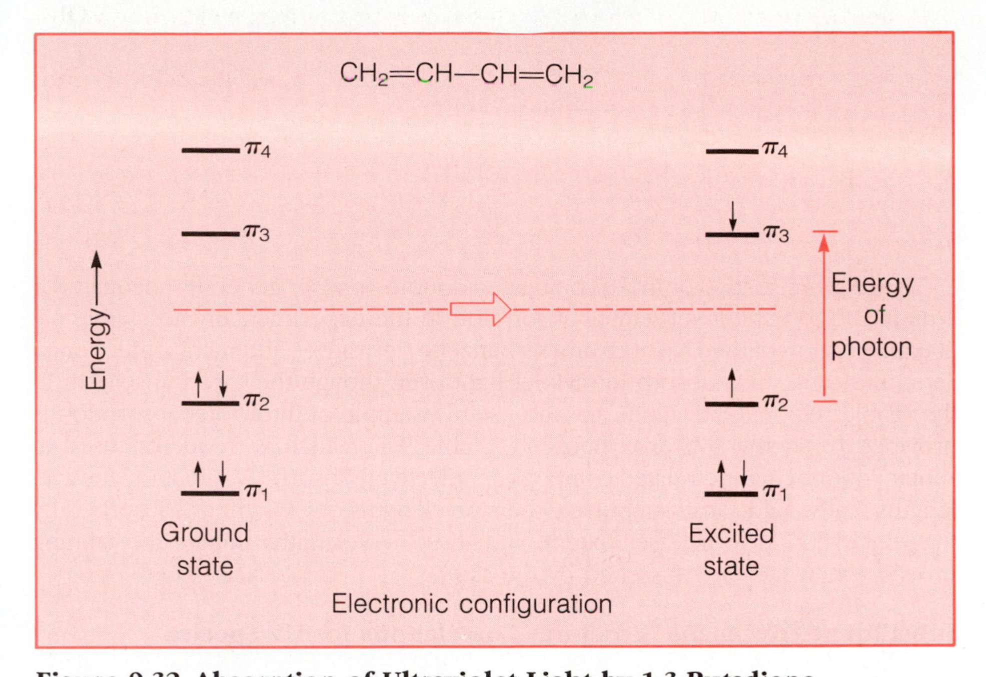

Figure 9.32 Absorption of Ultraviolet Light by 1,3-Butadiene. In the ground state both the bonding π molecular orbitals are doubly occupied. Absorption of a photon of the appropriate energy (wavelength 217 nm) produces an excited state in which an electron occupies an antibonding π orbital.

kcal/mol), and 1,3,5-hexatriene, a molecule with three conjugated double bonds, absorbs at 258 nm (111 kcal/mol). β-Carotene, with a total of 11 conjugated double bonds, exhibits a strong absorption at 478 nm (60 kcal/mol), a shift all the way into the visible region of the spectrum. What is the origin of this shift to longer wavelength? We are concerned with the lowest-energy transition, and that corresponds to the HOMO-LUMO energy difference.

For a simple alkene the bonding and antibonding π orbitals lie above and below a *nonbonding* reference level (shown as a dotted line in Figure 9.31). When two double bonds are conjugated, the interactions of additional p atomic orbitals result in a lower energy for the lowest (bonding) π orbital and a higher energy for the highest (antibonding) π orbitals. However, the intermediate π orbitals lie closer in energy to the nonbonding level (e.g., the dotted line in Figure 9.32). The net result is a smaller HOMO-LUMO energy difference as the extent of conjugation is increased, and this is reflected in the longer wavelengths at which conjugated molecules absorb light.

You should recognize that there is no particular chemical or physical distinction between the ways that ultraviolet and visible radiation interact with an organic molecule; absorption in either region produces a change in the electronic state of a molecule. The region labeled "visible" simply corresponds to those wavelengths that you can detect because they are absorbed by specific organic compounds present in the human eye. For example, human vision partly results from absorption of light by molecules of retinal that are bound to proteins in the rod cells of the eye. *Cis-trans* isomerization of the E double bond occurs in the excited state, and this change triggers a nerve impulse.

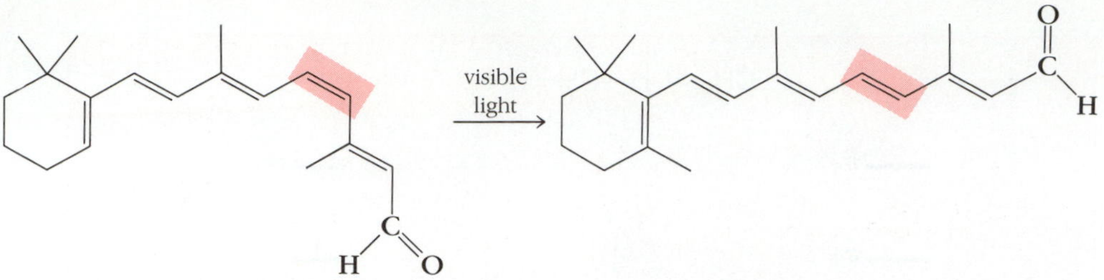

visible light →

The presence of two or more conjugated double bonds is generally adequate for a molecule to exhibit substantial absorption in the experimentally accessible (> 200 nm) region of the UV spectrum, so benzene derivatives and many other conjugated molecules will absorb ultraviolet light even though they are transparent in the visible region. One significant and useful example of ultraviolet absorption is provided by methyl 4-hydroxybenzoate (Table 9.4), which is frequently used in suntan lotion. This and related compounds selectively absorb UV radiation at wavelengths that would cause sunburn. Other wavelengths of UV light are absorbed by the suntan lotion to a lesser extent, and these wavelengths initiate the tanning process when they are absorbed by the skin.

Substituent Effects and Structural Correlations for UV Spectra

When the absorption of light is plotted against wavelength for a typical unsaturated compound, one or more peaks is observed. The maximum value for each peak, known as λ_{max}, is the wavelength at which the absorption of light is greatest. Remarkably consistent patterns in the values of λ_{max} as a function of molecular structure have been found for both conjugated dienes and α,β-unsaturated carbonyl compounds.* These patterns were analyzed by R.B. Woodward (Harvard) in the early 1940s, when UV-visible spectroscopy was the only type of spectroscopy readily available to organic chemists. We have presented a simplified scheme in Figure 9.33, which shows you how these patterns can be used to predict the wavelength of absorption for a wide variety of conjugated dienes and α,β-unsaturated carbonyl compounds. (Even more reliable predictions can be made by using the full analysis of Woodward.)

The predictions that you can make according to Figure 9.33 frequently allow you to decide between one or more possible structures, and we have illustrated this in the following examples.

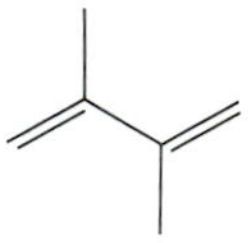

Predicted: **225 nm** (215 base; +5 for each of two alkyl substituents)
Observed: **226 nm**

---

*The designation "α,β-unsaturated" employs the first two letters of the Greek alphabet to indicate that the double bond is between the first two carbon atoms adjacent to the carbonyl group, i.e., that the double bond is conjugated with the carbonyl group. Positions near a functional group are commonly designated with letters of the Greek alphabet, as illustrated by the following structure, in which the α carbon is the carbon to which the functional group is attached:

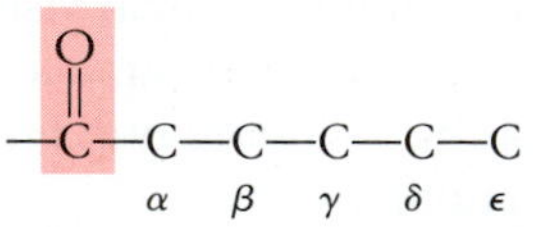

$$-\overset{O}{\overset{\|}{C}}-\underset{\alpha}{C}-\underset{\beta}{C}-\underset{\gamma}{C}-\underset{\delta}{C}-\underset{\epsilon}{C}$$

	C=C–C=C	C=C–C(R)=O
Base value:	215 nm	215 nm
Substituent effect for each alkyl group on the carbon-carbon double bond:	+5 nm	+10 nm

Figure 9.33 Substituent Effects on the λ_{max} of UV Absorptions for Conjugated Dienes and α,β-Unsaturated Ketones.

Predicted: **225 nm** (215 base; +5 for each of two alkyl substituents, i.e., the CH_2 groups of the six-membered ring)
Observed: **230 nm**

Predicted: **235 nm** (215 base; +10 for each of two alkyl substituents on the double bond)
Observed: **237 nm**

Predicted: **235 nm** (215 base; +10 for each of two alkyl substituents on the double bond)
Observed: **235 nm**

Predicted: **245 nm** (215 base; +10 for each of three alkyl substituents on the double bond)
Observed: **256 nm**

Except for the last example, all the predictions are within 5 nm of the observed value. Part of the discrepancy for this last example results from the presence of an *exocyclic double bond,* that is, a double bond in which the two unsaturated carbon atoms are not both part of the same ring. Two important structural features that result in a systematic change in the absorption wavelength are found with cyclopentenones, which absorb at wavelengths about 10 nm less than the analogous cyclohexenones, and with dienes having both double bonds in a single ring, for which λ_{max} will be larger by about 30 nm than for a comparably substituted acyclic diene.

Predicted: **255 nm** (215 base; +30 for two double bonds in the same ring; +5 for each of two alkyl substituents, which are CH_2 groups of the ring)
Observed: **256 nm**

EXERCISE 9.6

Predict the λ_{max} for each of the following compounds:

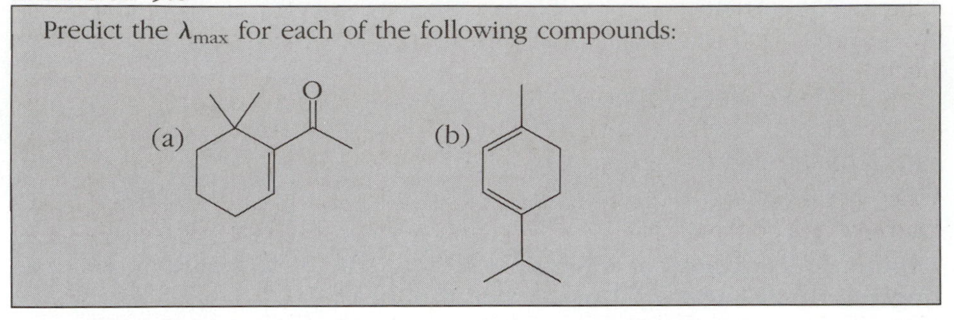

EXERCISE 9.7

A compound was known to have either structure (i) or structure (ii). On the basis of the observed λ_{max} of 245 nm, which structure is correct?

(i) (ii)

9.8 INFRARED SPECTROSCOPY

The infrared (IR) region of the electromagnetic spectrum is adjacent to the visible region but at longer wavelength (lower energy). Infrared radiation corresponds to the "heat" radiated by a hot object, although it is distinct from any *visible* glow because it is not detected by the human eye. However, certain wavelengths can be "felt" as a consequence of absorption by the skin. The absorption of infrared radiation by a molecule depends on the types of functional groups present, and it is therefore correlated with molecular structure. Infrared spectroscopy was extensively developed during the 1940s and provided a powerful tool for organic chemists because it permitted identification of different functional groups in a compound of unknown or uncertain structure.

In contrast to the absorption of UV or visible radiation, which produces an *electronically* excited state, absorption of infrared radiation generates a *vibrationally* excited state. All molecules have characteristic vibrational modes corresponding to stretching (small changes in the bond length between two atoms),

bending (small changes in the bond angle of two atoms bonded to a central atom),

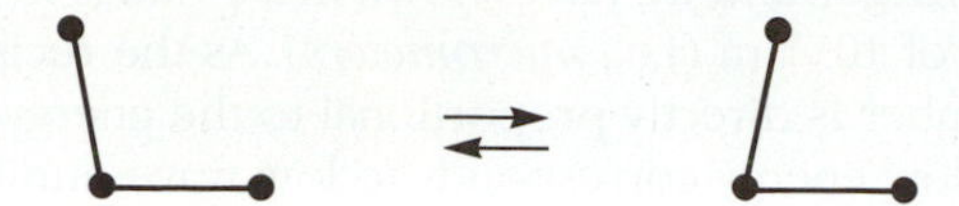

and *combinations* of stretching and bending. Vibrational states are quantized, and if the change in stretching or bending produces a change in the dipole moment of the functional group, absorption of an infrared photon can occur to produce a higher energy vibrational state.

The infrared spectra of most molecules are quite complex because the various bonds and groups absorb radiation at slightly different wavelengths. This makes it difficult to identify each absorption peak in an infrared spectrum, but you can obtain considerable information by concentrating on a few absorptions that are characteristic for certain functional groups. We have summarized some of the most useful infrared stretching absorptions in Table 9.6.

The unit commonly used for infrared radiation is *wave number,* which has units of cm^{-1}. The wave number is the reciprocal of the wavelength of the radiation (when specified in centimeters), and it is usually described as *frequency.* (Actually, frequency has units of s^{-1}; it is obtained by multiplying the wave number by the velocity of light, which has units of cm/s). Sometimes infrared spectra are reported

TABLE 9.6 Characteristic IR Stretching Frequencies*

Structural Unit		Wave Number, cm^{-1}		Wavelength, 10^{-6} m
C—H		2800–3100		3.6–3.2
O—H	Alcohol (conc. solutions)	3200–3400		3.1–2.9
O—H	Carboxylic acid	2500–3400	very broad	4.0–2.9
N—H	Amine, amide	3300–3500		3.0–2.9
C≡C	Unsymmetrical alkyne	2100–2250	may be very weak	4.8–4.4
C=C	Alkene	1620–1680	may be very weak	6.2–6.0
C=O	Ketone	1700–1725		5.9–5.8
C—O	Ether, alcohol	1000–1250		9.5–8.0

*Approximate ranges are shown for common functional groups.

as a function of wavelength, and we have shown these values in the second column in Table 9.6 in units of 10^{-6} m (i.e., *micrometers*). As the reciprocal of the wavelength, the wave number is directly proportional to the energy of the radiation ($E = hc/\lambda$). Therefore, low energy corresponds to low wave number and long wavelength.

As in the case of UV-visible spectra, a peak is identified by the wave number (or wavelength) at which absorption of radiation is at a maximum. The ranges reported in Table 9.6 indicate the regions in which particular absorption maxima should be found. Several patterns are readily discernible from Table 9.6. For example, all the stretching vibrations in the region 2500–3650 cm^{-1} involve hydrogen, and this range is further subdivided according to the element to which the hydrogen atom is attached. Thus, O—H stretching is normally found from 3200–3600 cm^{-1}, N—H stretching results in absorption between 3300 and 3500 cm^{-1}, and the characteristic absorptions of C—H groups fall between 2800 and 3100 cm^{-1}. These regions can be further classified according to structural types. For C—H stretching we can specify ranges for alkanes (2850–2950 cm^{-1}), aromatic C—H (near 3030 cm^{-1}), and vinyl C=C—H (3000–3100 cm^{-1}). Terminal alkynes are an exception, and the C≡C—H absorption occurs at about 3300 cm^{-1}. Most organic compounds have a variety of different kinds of hydrogens, and this results in a complex pattern of absorption peaks in the 2800–3100 cm^{-1} region.

The O—H stretching frequency is particularly sensitive to certain structural changes, and hydrogen bonding can cause shifts of 200 cm^{-1} or more. The range of 3200–3400 cm^{-1} given in Table 9.6 is that expected for concentrated solutions, in which there is substantial intermolecular hydrogen bonding. At higher dilution a shift to the range 3400–3600 cm^{-1} is found (unless there is intramolecular hydrogen bonding). The OH group of a carboxylic acid presents an important exception, with a very broad and characteristic absorption between 2500 and 3400 cm^{-1}.

Infrared spectra exhibit a direct relationship between bond strength and the energy of the radiation that is absorbed. For example, the carbon-oxygen double bond of a ketone absorbs in the vicinity of 1700 cm^{-1}, but the weaker C—O single bond absorbs at much lower energy (near 1100 cm^{-1}). Similarly, a carbon-carbon double bond absorbs in the region of 1650 cm^{-1}, but the stronger triple bond absorbs at higher energy in the neighborhood of 2200 cm^{-1}. Absorption in this region almost always indicates the presence of a triple bond (either C≡C or C≡N). A symmetrical alkyne lacks this absorption, because stretching of the triple bond does not produce the change in dipole moment that is needed for infrared absorption. A symmetrical C≡C has no bond moment regardless of changes in bond length.

We have shown some typical infrared spectra of organic compounds in Figures 9.34–9.52. These spectra are relatively complex, but you can quickly obtain important structural information by focusing on absorptions for the functional groups presented in Table 9.6. The horizontal axis of each spectrum is labeled in wave number (although wavelength is also noted at the top); the high-energy end of the spectrum is at the left side in each case. The vertical axis corresponds to absorption of infrared radiation. The spectra shown here are recorded as *percent transmittance* or as *absorbance.** If there were *no absorption,* the spectrum would consist

*The absorbance, A, is given by $A = \log(I_o/I)$, where I_o is the intensity of incident light and I is the intensity of transmitted light. Transmittance, T, is reported as a percentage of transmitted light and is represented by $\%T = (I/I_o) \times 100$.

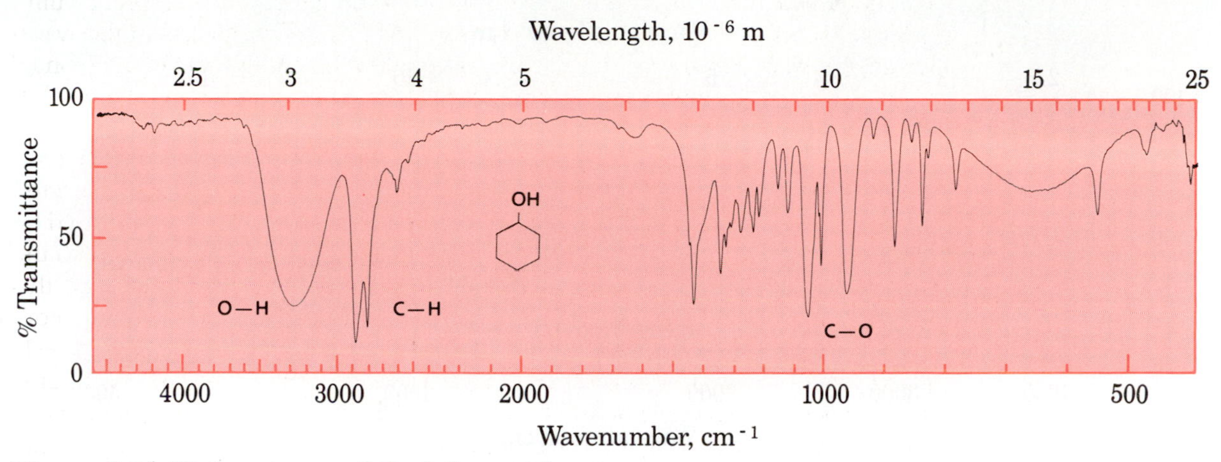

Figure 9.34 IR Spectrum of Cyclohexanol.

of a horizontal line at the *top* of the spectrum (i.e., 100 percent of the radiation would be transmitted by the sample and none would be absorbed). On the other hand, *absorption* of radiation (i.e., less than 100 percent transmittance) results in a downward deflection, called a *peak,* in the plot.

Figure 9.34 shows the IR spectrum of cyclohexanol, in which the peaks corresponding to O—H, C—H, and C—O appear at 3300, 2850–2950, and 1050 cm^{-1}, as expected.

The IR spectrum of vinylcyclohexane (Figure 9.35) demonstrates that the C═C stretching absorption near 1650 cm^{-1} can sometimes be quite weak.

The strong and characteristic absorption of a carbonyl group is illustrated in the spectrum of 2,6-dimethyl-4-heptanone (Figure 9.36).

The carbonyl group of a carboxylic acid absorbs in the same region as that of a ketone, as illustrated by the peak at 1700 cm^{-1} for butyric acid (Figure 9.37). Note the very broad O—H absorption which characterizes carboxylic acids, and compare it with the narrower O—H absorption shown for cyclohexanol in Figure 9.34.

The weak but sharp peak for the C≡C stretching absorption is seen at 2120 cm^{-1} in Figure 9.38, which shows the spectrum of 1-hexyne. The characteristic

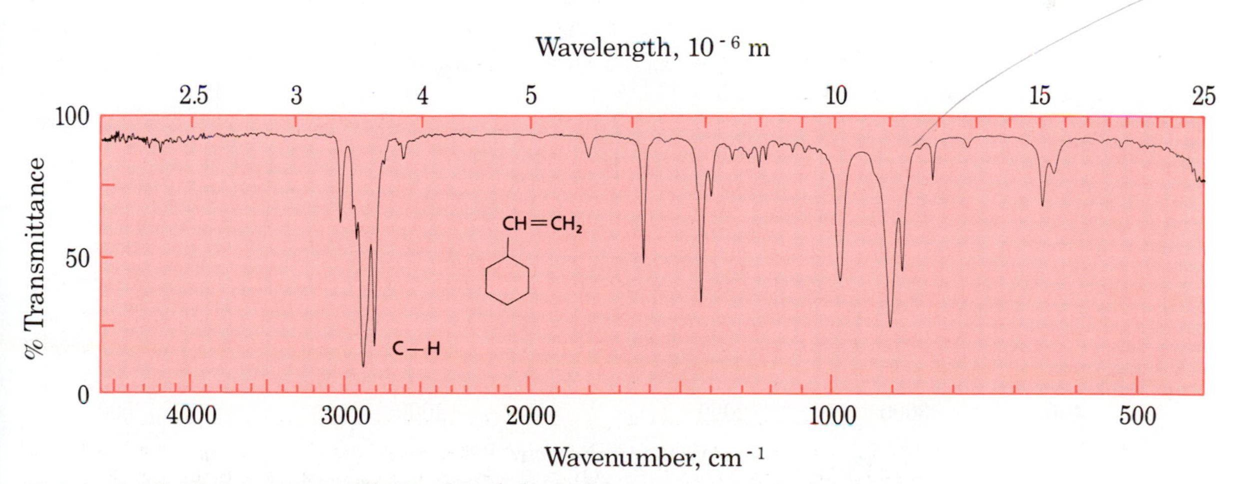

Figure 9.35 IR Spectrum of Vinylcyclohexane.

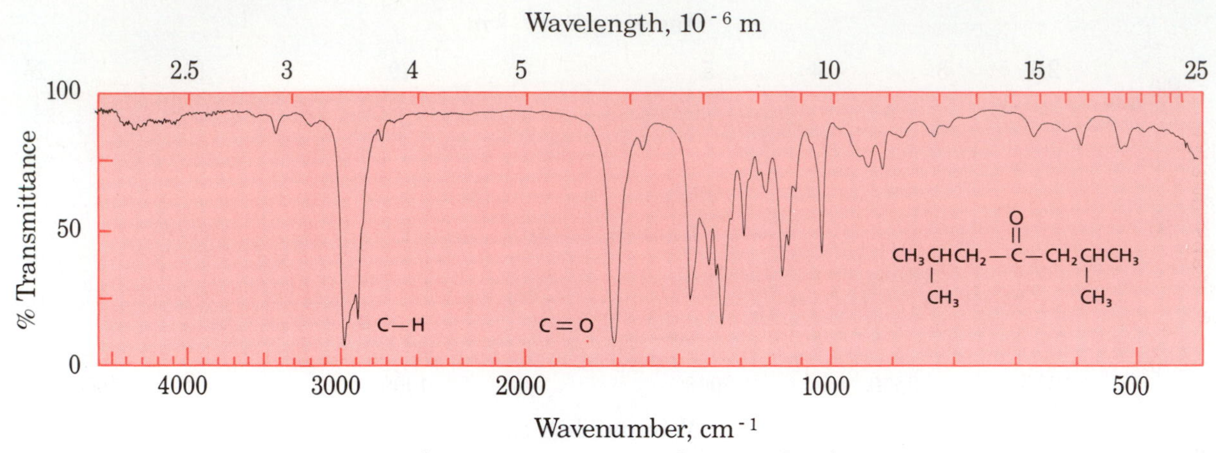

Figure 9.36 IR Spectrum of 2,6-Dimethyl-4-heptanone.

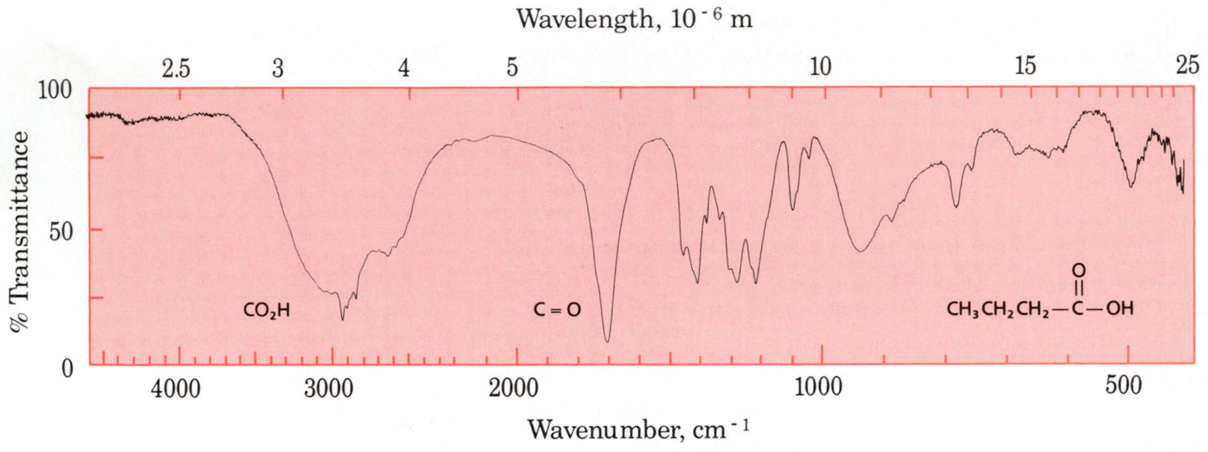

Figure 9.37 IR Spectrum of Butyric Acid.

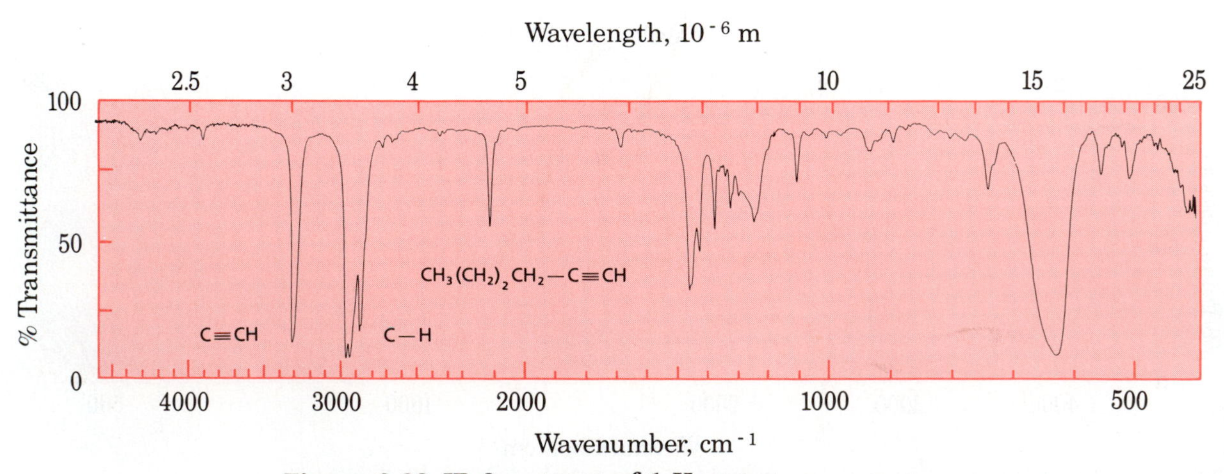

Figure 9.38 IR Spectrum of 1-Hexyne.

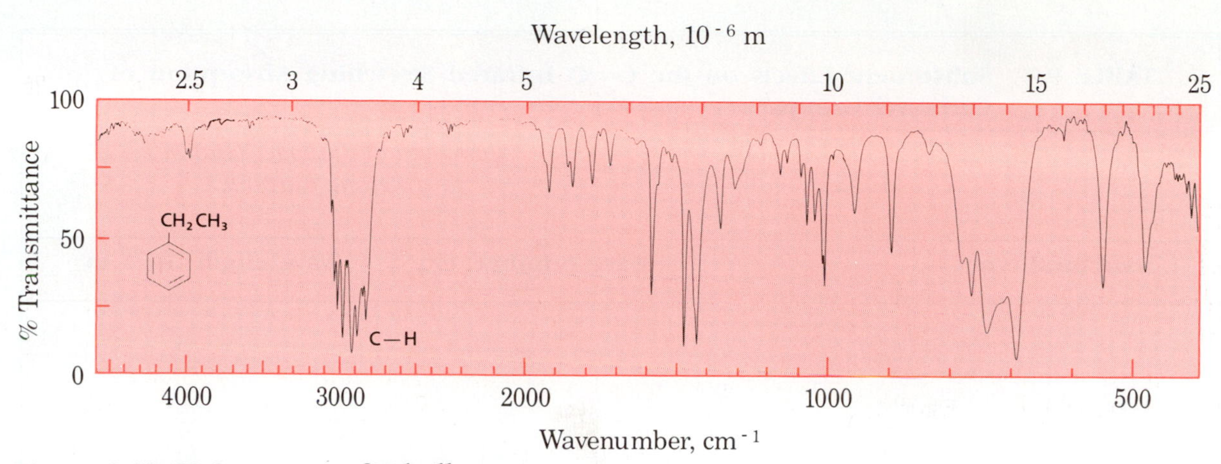

Figure 9.39 IR Spectrum of Ethylbenzene.

C—H absorption of a terminal alkyne (3320 cm^{-1}) is also seen in this example.

As you will see in Chapter 10, a benzene ring cannot be adequately described in terms of three individual double bonds. Accordingly, the IR spectra of benzene derivatives do not exhibit the specific C═C peaks of alkenes, and a variety of absorption peaks between 1500 and 2000 cm^{-1} are observed instead. The number of peaks and their locations are characteristic of the substitution pattern of the benzene ring, but we will not discuss these patterns in detail. Figure 9.39 shows one such example (ethylbenzene); the peaks corresponding to the aromatic ring (although weak) are seen between 1600 and 2000 cm^{-1}.

Carbonyl groups provide some of the most valuable structural information that you can obtain from infrared spectra because the various types of carbonyl groups (ketone, ester, etc.) do not all absorb at the same frequency. In Table 9.7 we have summarized some of the substituent effects on carbonyl groups relative to the 1700–1725 cm^{-1} absorption expected for a ketone.

The substituent effects summarized in Table 9.7 are illustrated in Figures 9.40–9.44. We have already shown you examples of a ketone and of a carboxylic acid in Figures 9.36 and 9.37, respectively, and the spectrum of ethyl propionate (Figure 9.40) illustrates the shift to higher wave number for an ester carbonyl (1700–1725

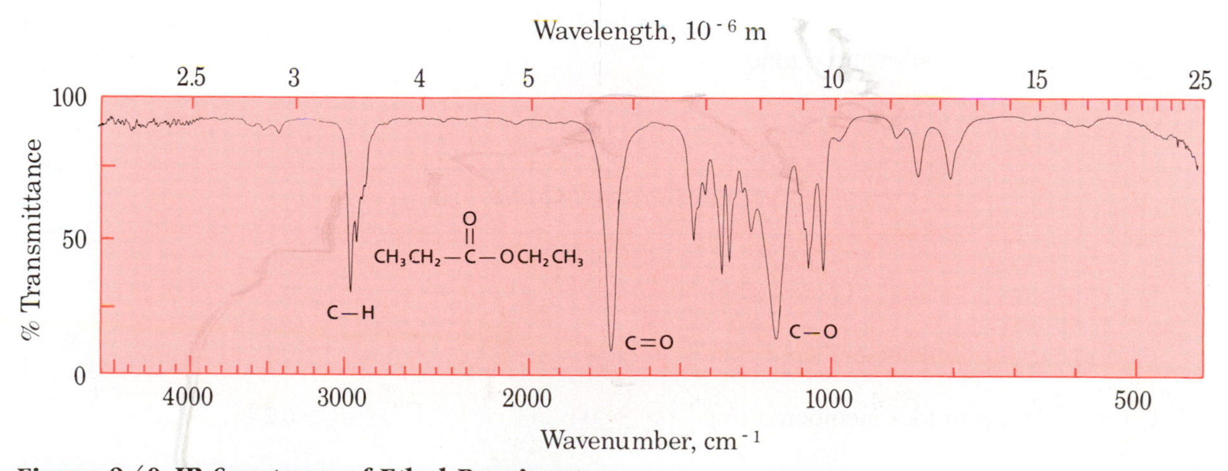

Figure 9.40 IR Spectrum of Ethyl Propionate.

TABLE 9.7 Substituent Effects on the C=O Infrared Stretching Absorption of Carbonyl Groups*

Structural type		*Approximate Change Relative to a Ketone Carbonyl* Wave number, cm^{-1}	Wavelength, 10^{-6} m
R—C(=O)—OR	Ester	+30	−0.1
R—C(=O)—H	Aldehyde	+20	−0.1
R—C(=O)—OH	Carboxylic acid	0	0
R—C(=O)—R	Ketone	0	0
R—C(=O)—N<	Amide	~−50	~+0.2
	Effects of Conjugation		
C=C—C(=O)—R	α,β Unsaturation	−40	+0.1
C_6H_5—C(=O)—R	Aromatic ketone	−25	+0.1
	Cyclic Carbonyl Group		
Carbonyl group in six-membered ring		0	0
Carbonyl group in five-membered ring		+30	−0.1
Carbonyl group in four-membered ring		+60	−0.2

*Changes are relative to the 1700–1725 cm^{-1} absorption of ketones.

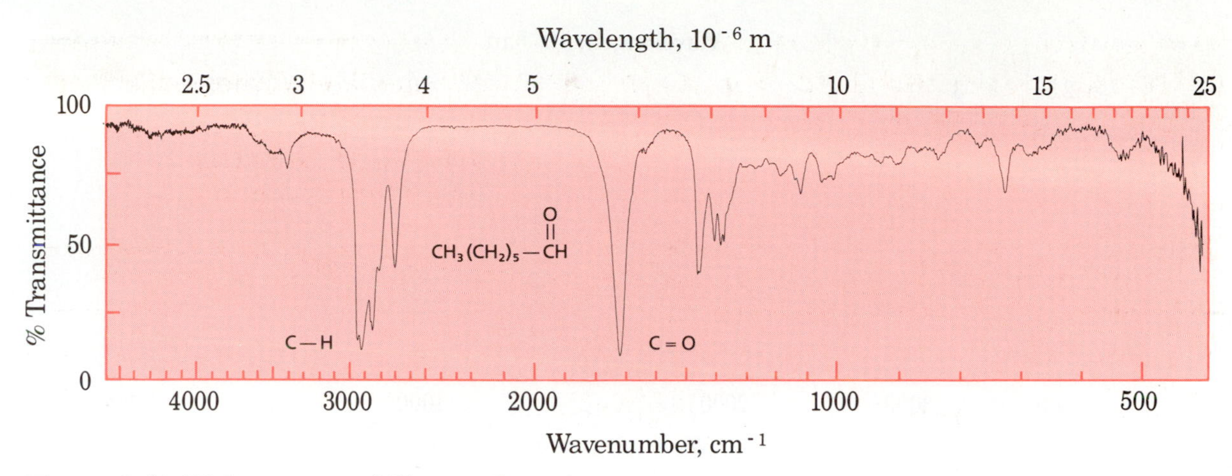

Figure 9.41 IR Spectrum of Heptanal.

cm^{-1} is increased to 1730–1755 cm^{-1}, and the value of 1730–1740 cm^{-1} for ethyl propionate is within this expected range).

The shifts of the carbonyl group to slightly higher wave number for aldehydes (an increase of 20 cm^{-1} to 1720–1745 cm^{-1}) and to lower wave number for amides (a decrease of 50 cm^{-1} to 1650 to 1675 cm^{-1}) are illustrated by Figures 9.41 and 9.42, respectively. Both the aldehyde carbonyl (1725 cm^{-1}) and the amide carbonyl (1650 cm^{-1}) fall within the expected ranges. The carbonyl stretching frequency is more variable for amides than for other carbonyl compounds, however, and it may not always fall within the narrow range predicted by Table 9.7. Figure 9.41 also illustrates the unique C—H stretching absorption that is almost always found near 2720 cm^{-1} for a CHO group. This peak gives you a very useful way to identify an aldehyde.

The effects of conjugation and of ring size on carbonyl stretching frequency are illustrated by Figures 9.43 and 9.44, respectively. Six-membered rings are "normal" (Sections 3.2 and 3.3), so only the conjugated double bond should affect the carbonyl absorption of 2-cyclohexen-1-one. The expected range for this α,β-unsaturated ketone is therefore shifted from 1700–1725 cm^{-1} to 1660–1685 cm^{-1} (Table 9.7), and this agrees with the observed peak (Figure 9.43) at 1675 cm^{-1}.

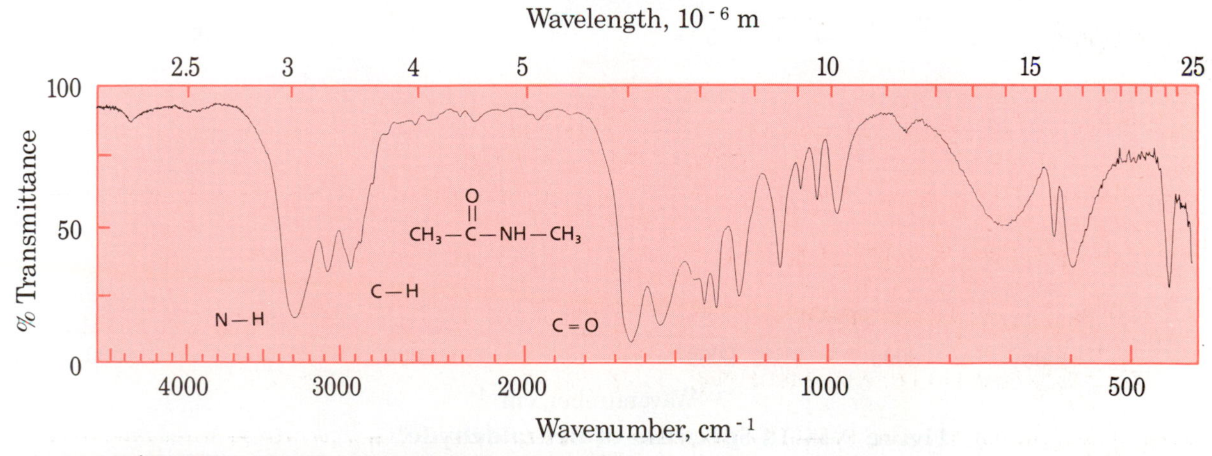

Figure 9.42 IR Spectrum of *N*-Methylacetamide.

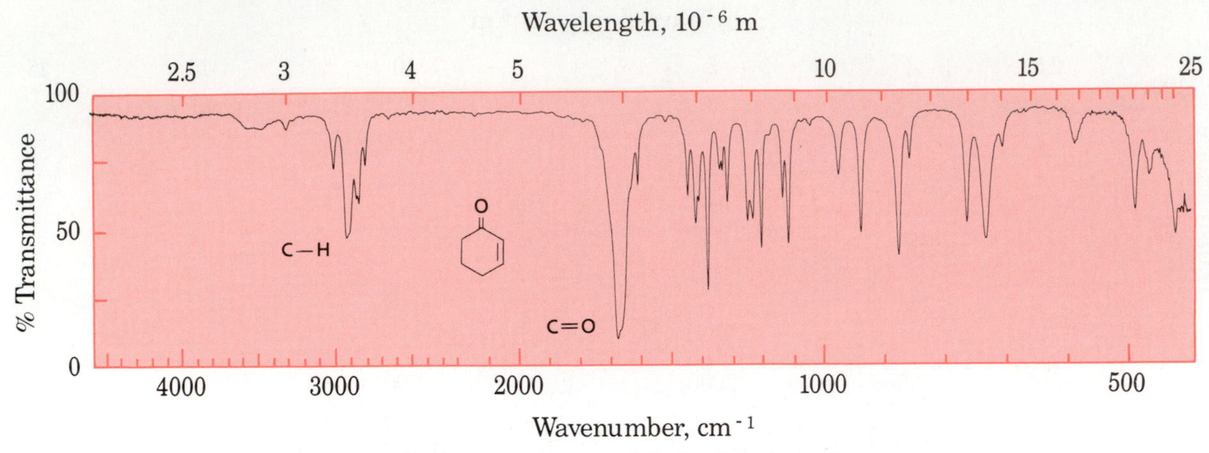

Figure 9.43 IR Spectrum of 2-Cyclohexen-1-one.

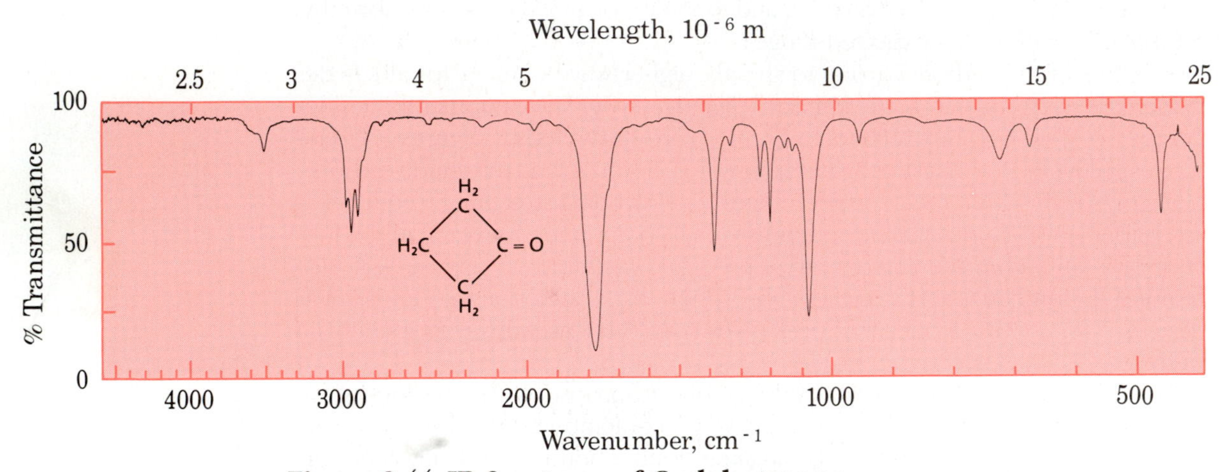

Figure 9.44 IR Spectrum of Cyclobutanone.

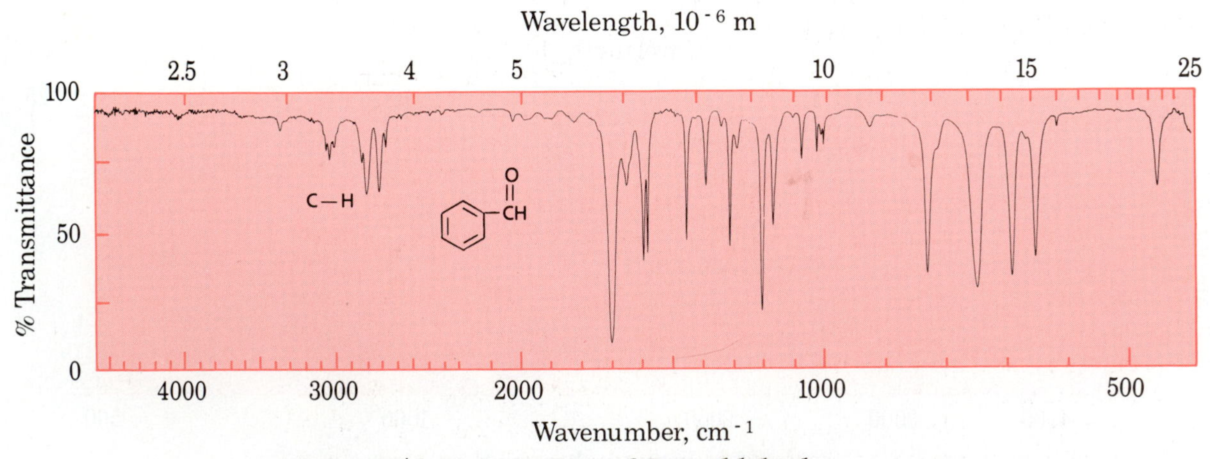

Figure 9.45 IR Spectrum of Benzaldehyde.

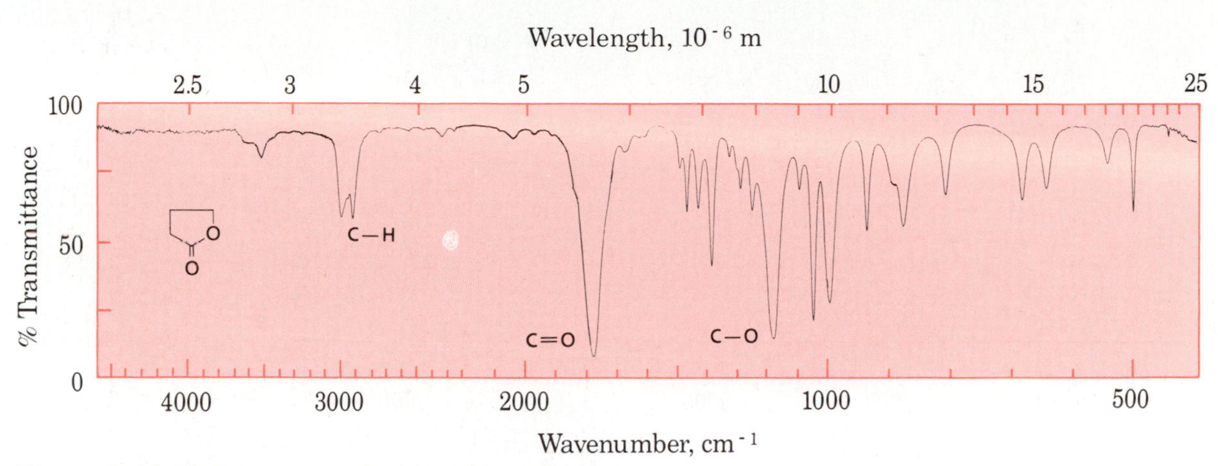

Figure 9.46 IR Spectrum of Butyrolactone.

Five- and four-membered rings are strained relative to acyclic compounds (Section 3.2), and a shift of the carbonyl stretching absorption to higher wave number is expected when the carbonyl group is part of a strained ring. Cyclobutanone (Figure 9.44) absorbs at 1785 cm^{-1}, and this is in agreement with the shift of $+60$ cm^{-1} (i.e., from 1700–1725 cm^{-1} to 1760–1785 cm^{-1}) that is predicted by Table 9.7.

The substituent effects presented in Table 9.7 are *additive,* and you can see this in Figures 9.45 and 9.46. The carbonyl absorption of benzaldehyde (Figure 9.45) should differ in two ways from that of a simple ketone: the expected shift for an aldehyde is $+20$ cm^{-1} and that for aromatic conjugation is -25 cm^{-1}. The net result is only -5 cm^{-1}, and you would therefore expect an aromatic aldehyde to show a carbonyl absorption between 1695 and 1720 cm^{-1}. The observed peak for benzaldehyde falls within this range at 1700 cm^{-1}.

The substituent effects for butyrolactone, a five-membered ring cyclic ester, are both in the same direction. A shift of $+30$ cm^{-1} is expected for a carbonyl group in a five-membered ring, and the shift for an ester is also $+30$ cm^{-1}. The net result is a predicted shift of $+60$ cm^{-1}, and the observed peak for butyrolactone (Figure 9.46) falls within the expected range of 1760–1785 cm^{-1}.

Infrared Spectra of Polyfunctional Compounds

Absorption of infrared radiation is associated with individual functional groups and bonds, so the infrared spectra of polyfunctional compounds can provide information about each of the functional groups present. As long as the individual groups do not interact substantially, the wave number for absorption of each functional group is usually found within the range expected for the corresponding monofunctional compound. This is illustrated in Figures 9.47 and 9.48, which are the spectra of a hydroxy ester and a keto ester, respectively. For each compound the peaks corresponding to the individual functional groups are clearly observed within the normal range specified by Tables 9.6 and 9.7. Figure 9.47 shows both OH (3400 cm^{-1}) and ester carbonyl (1725 cm^{-1}) stretching absorptions.

Similarly, the presence of two different carbonyl groups is readily apparent in the spectrum of 2-acetylbutyrolactone (Figure 9.48). The peak at higher wave number (1740 cm^{-1}) corresponds to the cyclic ester carbonyl, while the ketone carbonyl gives rise to the peak at 1690 cm^{-1}.

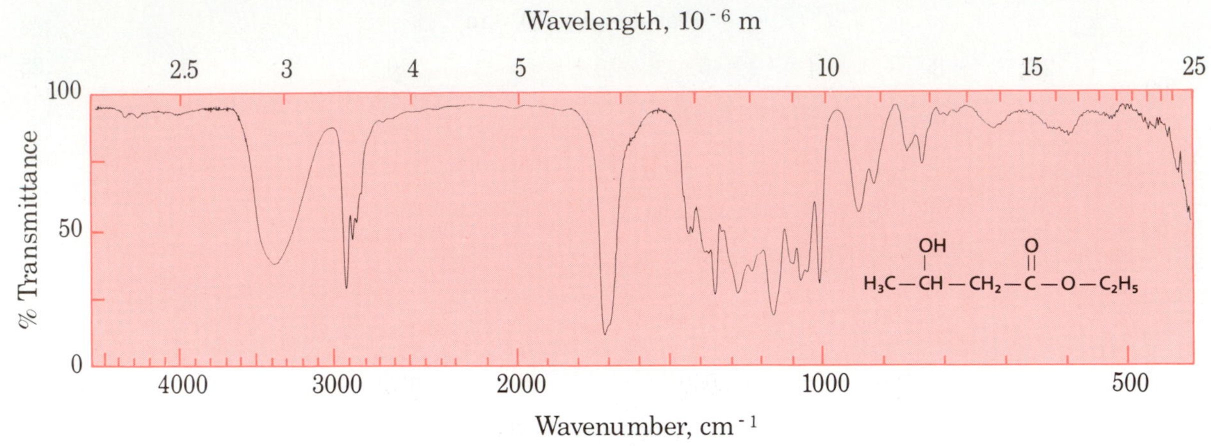

Figure 9.47 IR Spectrum of Ethyl 3-Hydroxybutyrate, a Hydroxy Ester.

Evaluation of Reaction Progress with IR Spectroscopy

One of the most useful applications of spectroscopy in the laboratory is in evaluating the results of a chemical reaction. As a chemist, you would always have at least two questions: (1) Did any reaction occur, or is the apparent product merely unreacted starting material? (2) If reaction did occur, was the expected product formed or did some unanticipated side reaction predominate? We have shown previously (Section 9.1) that you could easily demonstrate the recovery of unreacted starting material because the spectrum of your "product" would be identical to that of the starting compound. However, the situation would be more complicated if incomplete reaction had occurred. The crude product would then be a mixture of compounds, which you could separate and analyze separately. Alternatively, the IR spectrum of the mixture would exhibit the peaks for the starting compound *together with* those for the product.

Use of IR spectroscopy to evaluate reaction progress is generally easier than attempting to deduce the structure of an unknown compound. You would already have considerable structural information, and you could make reasonable guesses about the spectrum of the expected product from its structural features. Much of a

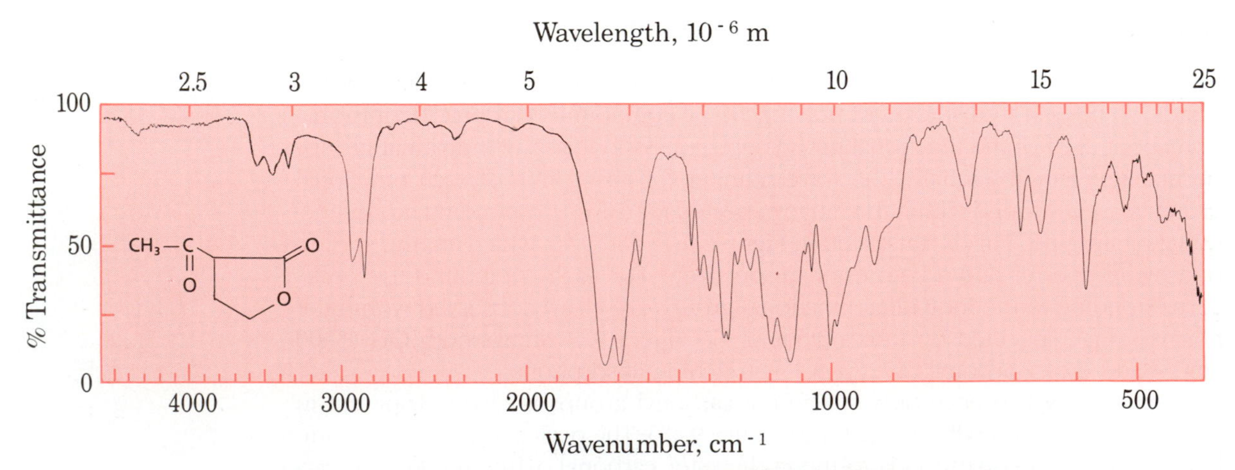

Figure 9.48 IR Spectrum of 2-Acetylbutyrolactone, a Keto Ester.

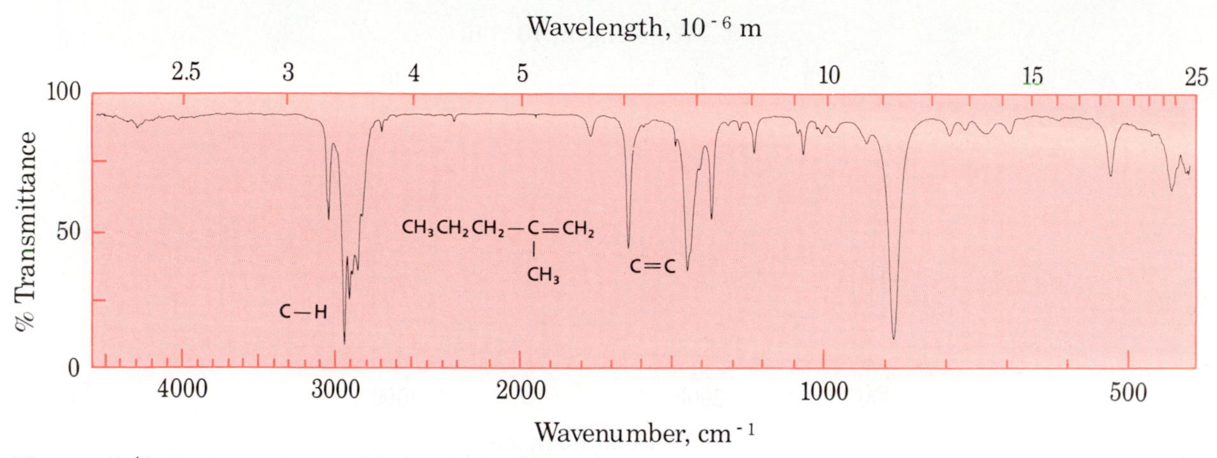

Figure 9.49 IR Spectrum of 2-Methyl-1-butene.

molecule remains unchanged during any particular reaction, so it is possible to focus your attention on those parts of the IR spectrum that *change.* These points are illustrated in the following examples, which correspond to reactions that we have already discussed in Chapters 5 through 8.

The hydration of 2-methyl-1-butene via hydroboration is expected to yield 2-methyl-1-butanol.

$$(CH_3)(CH_3CH_2)C{=}CH_2 \xrightarrow[2.\ H_2O_2,\ NaOH]{1.\ BH_3} CH_3CH_2{-}\underset{}{\overset{CH_3}{|}}{CH}{-}CH_2OH$$

The spectra for both starting material (Figure 9.49) and product (Figure 9.50) are shown below. Clearly, if you obtained a spectrum equivalent to that of Figure 9.50, you would have demonstrated that the reaction had been successful. Although there are many similarities to the spectrum of the starting alkene, the C=C stretching absorption at 1650 cm^{-1} is absent (as is the vinyl CH absorption near 3100 cm^{-1}). In addition, the strong absorption centered at 3350 cm^{-1} confirms the

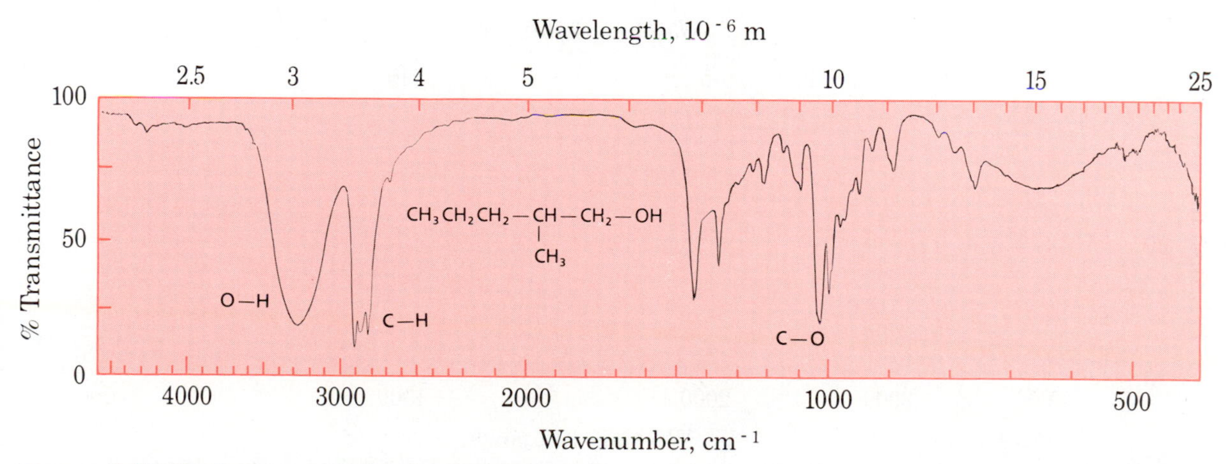

Figure 9.50 IR Spectrum of 2-Methyl-1-butanol.

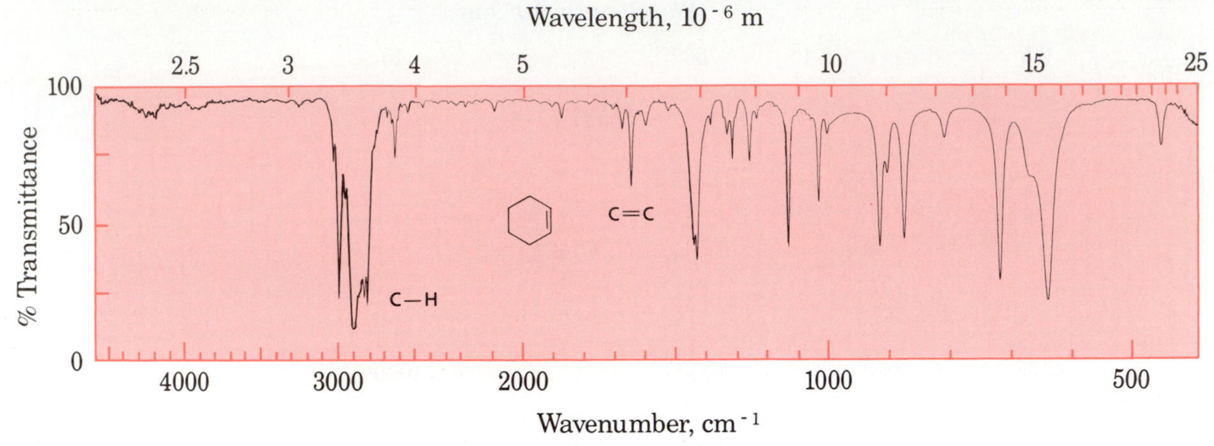

Figure 9.51 IR Spectrum of Cyclohexene.

presence of an OH group, and the corresponding C—O stretching results in a peak near 1030 cm^{-1}.

The selenium dioxide oxidation of cyclohexene provides another example of the value of IR spectroscopy in evaluating the products of a reaction. One of the major side reactions in such oxidations is the formation of an α,β-unsaturated ketone in addition to the desired allylic alcohol.

$$\text{cyclohexene} \xrightarrow{SeO_2} \text{2-cyclohexen-1-ol (OH)} + \text{2-cyclohexen-1-one (O)}$$

The spectra of the reactant (Figure 9.51) and desired allylic alcohol (Figure 9.52) are shown below. Again there are many similarities in the spectra, and the peak for C=C stretching is present in both cases at 1650 cm^{-1}. However, the presence of an OH group in the product is easily recognized from the peaks at 3350 cm^{-1} (OH) and 1060 cm^{-1} (C—O) in Figure 9.52. If any cyclohexenone were produced, you could see this easily by the presence of a peak at 1675 cm^{-1} (see Figure 9.43).

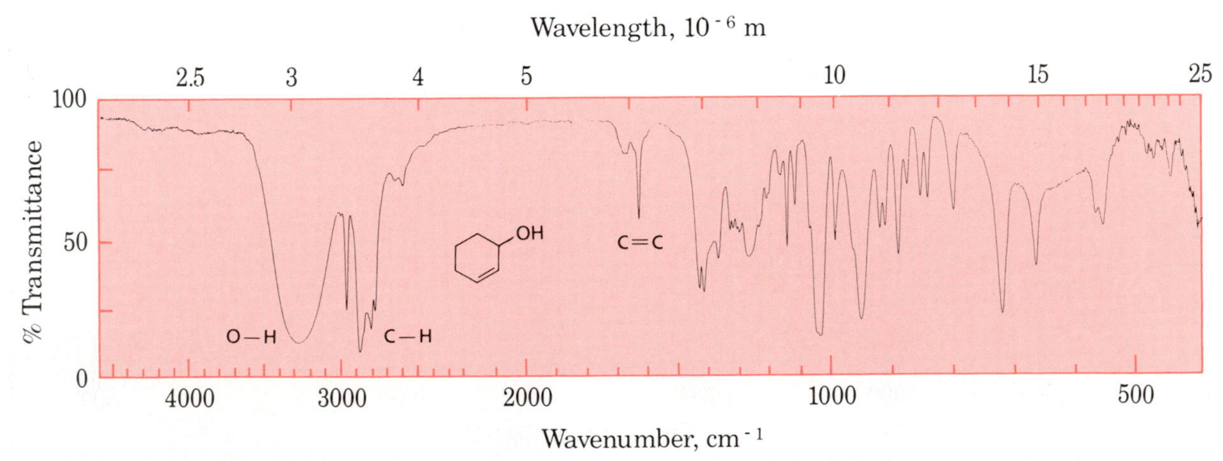

Figure 9.52 IR Spectrum of 2-Cyclohexen-1-ol.

EXERCISE 9.8

A compound of uncertain structure was determined to be one of the following:

(a) $C_6H_5—CO_2H$ (b) $CH_3—CH_2CH(OCH_3)—CH_2CH_2CH_2CH_3$

(c) $CH_3—CH(OH)—(CH_2)_6CH_3$ (d) $CH_3—CH(CH_3)—(CH_2)_5CH_3$

(e) $CH_3CH_2CH{=}CHCH_2CH_3$

Deduce the correct structure from the IR spectrum of the compound (Figure 9.53).

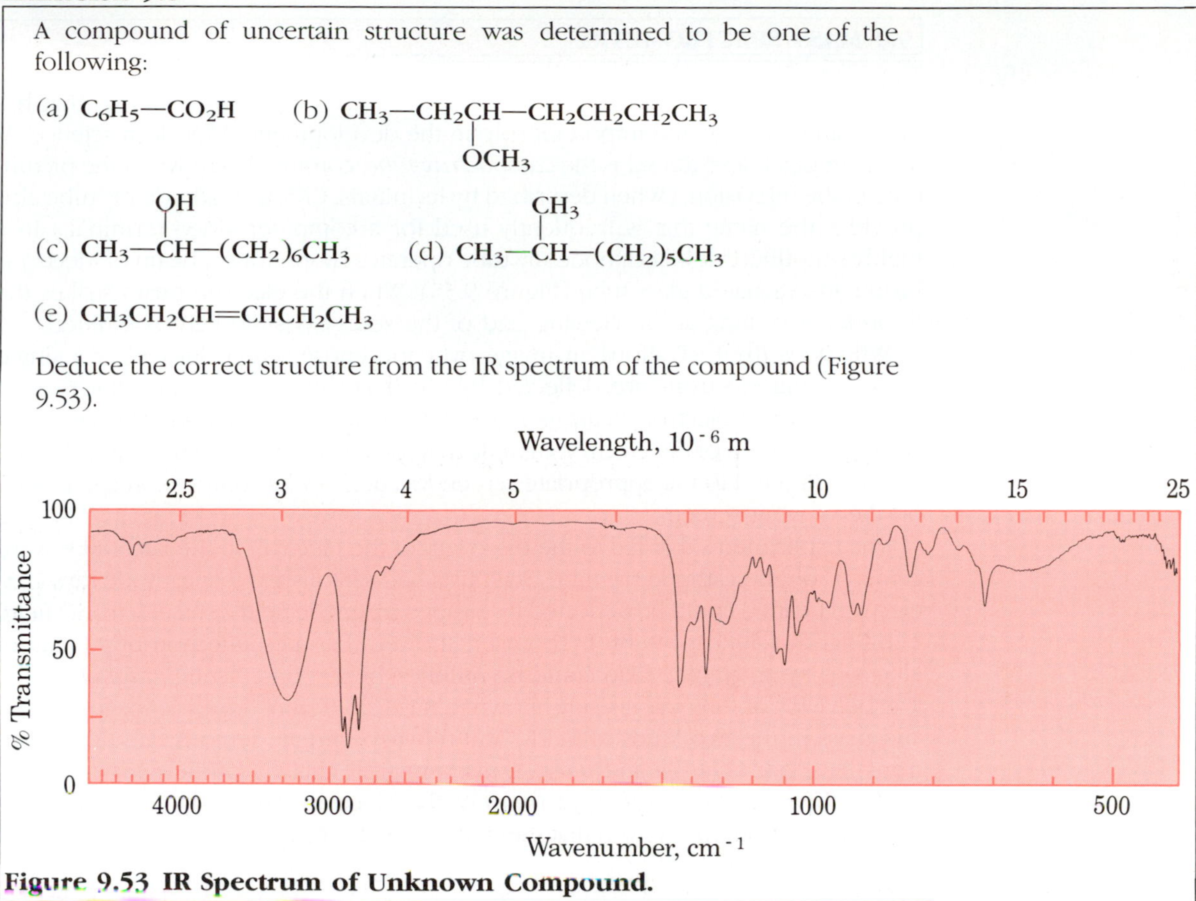

Figure 9.53 IR Spectrum of Unknown Compound.

EXERCISE 9.9

A compound of unknown structure was found to have the molecular formula C_5H_8O. Deduce its structure from its infrared spectrum (Figure 9.54). (Note the number of degrees of unsaturation.)

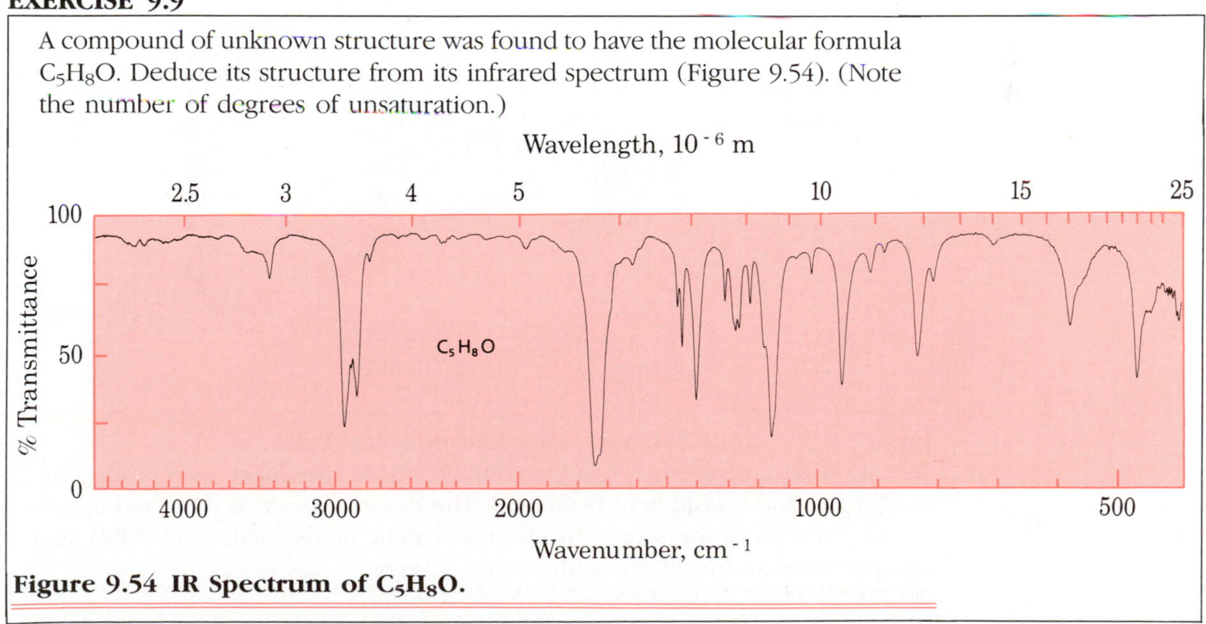

Figure 9.54 IR Spectrum of C_5H_8O.

9.9 MASS SPECTROMETRY

Every time you turn on your television set you are employing a device that has played an extensive and important role in the development of modern science. At the heart of a television set is the *cathode ray tube,* commonly known as the picture tube of the television. (When described by its initials, CRT, the cathode ray tube also provides the name that is frequently used for a computer video terminal.) In a highly simplified view, a cathode ray tube operates by emitting a beam of electrons inside an evacuated glass tube (Figure 9.55). When the electron beam strikes the fluorescent coating at the viewing end of the screen, visible light is emitted.

Why does the CRT afford an image over the entire screen instead of a single spot? Moving electrons are deflected by electromagnetic fields, and the precise direction of the electron beam is controlled by a *deflection yoke.* By rapid and accurate variation of the electromagnetic field produced by the deflection yoke, the beam is directed to the appropriate screen location, and a complete image is seen on the viewing screen.

The experiments that led to the discovery of the electron in the late 1800s were centered on the cathode ray tube. Scientists had established that cathode rays (i.e., electron beams) could be deflected by either a magnetic field or an electronic field. The English scientist Joseph J. Thomson showed that the deflection of a beam of electrons by an *electric* field could be nullified by a counteracting *magnetic* field. The behavior of ions was already known to have a different dependence for charge and mass in the two kinds of fields, and this experiment demonstrated that electrons have both *charge* and *mass.* A mathematical analysis of the experiment afforded the *mass to charge ratio (m/e)* of the electron. The early experiments of Thomson and others showed that the deflection of ions in electromagnetic fields is

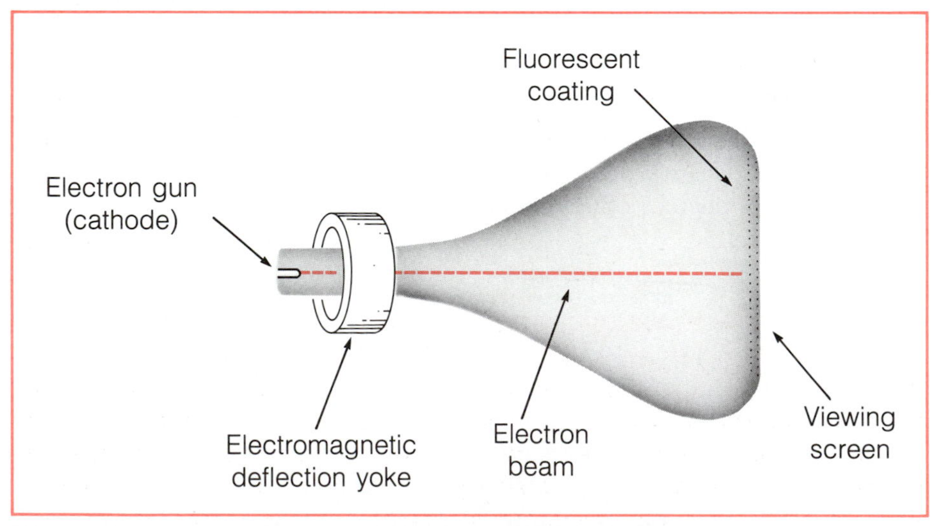

Figure 9.55 Simplified View of a Cathode Ray Tube.
A beam of electrons emitted at the cathode strikes the fluorescent coating of the screen, and visible light is emitted. The electron beam is deflected to different positions of the screen by electromagnetic fields (deflection yoke), and this produces an image across the entire screen.

a function of the mass to charge ratio. This provides the basis of modern **mass spectrometry,** separation of charged particles on the basis of mass.

Mass spectrometry has evolved into a powerful method for analyzing molecular structure, both as a research tool and as a standard analytical technique. But it has also seen some other, less common uses. For example, during the second World War, when Allied scientists with the Manhattan Project were racing to complete the first atomic bomb before it was built by enemy forces, mass spectrometry was used *preparatively* to isolate the needed isotope of uranium, ^{235}U. A curious aspect of this episode involved problems in construction of large electromagnets for the mass spectrometers at Oak Ridge, Tennessee. Copper wire was in very short supply, much of the available supply going to munitions manufacture and other war-related production. Another metal with good electrical conductivity was needed, so arrangements were made to borrow 15,000 tons of silver from Fort Knox to use as windings for the electromagnets. At today's prices this would be worth about $3 billion!

The Mass Spectrometer

There are many kinds of mass spectrometers, differing in both design and complexity. The costs of such instruments range from only a few thousand dollars to hundreds of thousands of dollars or more for some specialized research instruments. A discussion of the different types of mass spectrometer is beyond the scope of this chapter, and we will only illustrate the basic principles by which all these instruments operate.

A simplified view of a mass spectrometer is shown in Figure 9.56. In many ways this instrument is quite similar to a cathode ray tube. Positive ions from the sample are produced in the region of the spectrometer called the *ion source,* and these ions can undergo a series of decomposition reactions to give new ions with different masses. The resulting mixture of positively charged ions emerges through a small slit as an *ion beam.*

The ion source depicted in Figure 9.56 is an *electron impact* source, which is the most common type. A beam of high-energy electrons passes through the ionization chamber; when one of these electrons collides with a sample molecule, a positive ion is produced by ejection of a second electron:

$$:\ddot{\underset{\cdot\cdot}{A}}: + e^- \longrightarrow [:\dot{\underset{\cdot\cdot}{A}}:]^+ + 2e^-$$

The ion beam is accelerated in the direction of the ion collector by an electric potential, and the accelerated beam passes through the *mass analyzer.*

There are several types of mass analyzers, but in all cases they rely on the deflection of the ion beam by electric and/or magnetic fields. The magnitude of the deflection of a particular ion varies inversely with its mass to charge ratio, so the heaviest ions are deflected least. When ions strike the *ion collector,* this registers as an electric current, and it is possible to measure the current corresponding to each different mass to charge ratio. Consequently a record is obtained of each value of the ratio for which ions are produced. Moreover, the intensity of the signal at each such value is proportional to the relative number of ions with that mass to charge ratio. Hence the mass spectrum tells you not only the mass to charge ratios of the different ions but also the extent to which each contributes to the total ion beam.

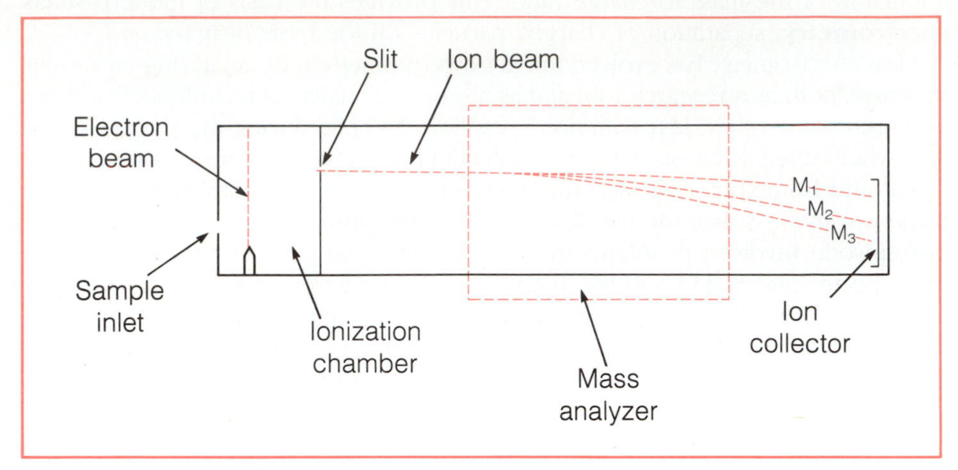

Figure 9.56 Simplified View of a Mass Spectrometer.

Under normal conditions of mass spectrometry, most of the ions formed have unit positive charge. Therefore we will frequently refer simply to the mass of an ion rather than to its mass to charge ratio.

The Mass Spectrum

Ejection of an electron from a sample molecule (M) in the ion source of a mass spectrometer produces a positively charged species called the **molecular ion.** The molecular ion is generated by loss of a single electron from a doubly occupied orbital, so it is actually a *cation radical* ($M^{+}\cdot$).

$$\underset{\text{Sample molecule}}{M} \xrightarrow{\text{ionization}} \underset{\text{Molecular ion}}{M^{+\cdot}}$$

In most cases the molecular ions decompose into ions with different masses, and the masses of these ions (together with their relative amounts) provide valuable information about the molecular structure of the sample. Here we will discuss the origin of ions with different mass, and we will show how the resulting information is recorded for subsequent interpretation.

Two general methods are available for recording mass spectral data, graphical display and numerical tabulation, but we will use primarily the first method, plotting the ion intensity as a function of the mass to charge ratio. The intensity of a peak in the mass spectrum is a measure of the number of ions with that mass-to-charge ratio, and it is recorded as the *relative abundance*. The presentation of data is illustrated in Figure 9.57, the mass spectrum for methane.

Notice that *the most intense peak in the spectrum is assigned a relative abundance of 100.* This is standard practice in mass spectrometry, and it allows us to conveniently compare the relative abundance of any ion in the spectrum with that of the most abundant ion. In the case of methane this is the molecular ion, $CH_4^{+}\cdot$, but frequently an ion of different mass will be most abundant. In fact, the relative abundance of the molecular ion is sometimes quite low, which makes interpretation of the mass spectrum much more difficult.

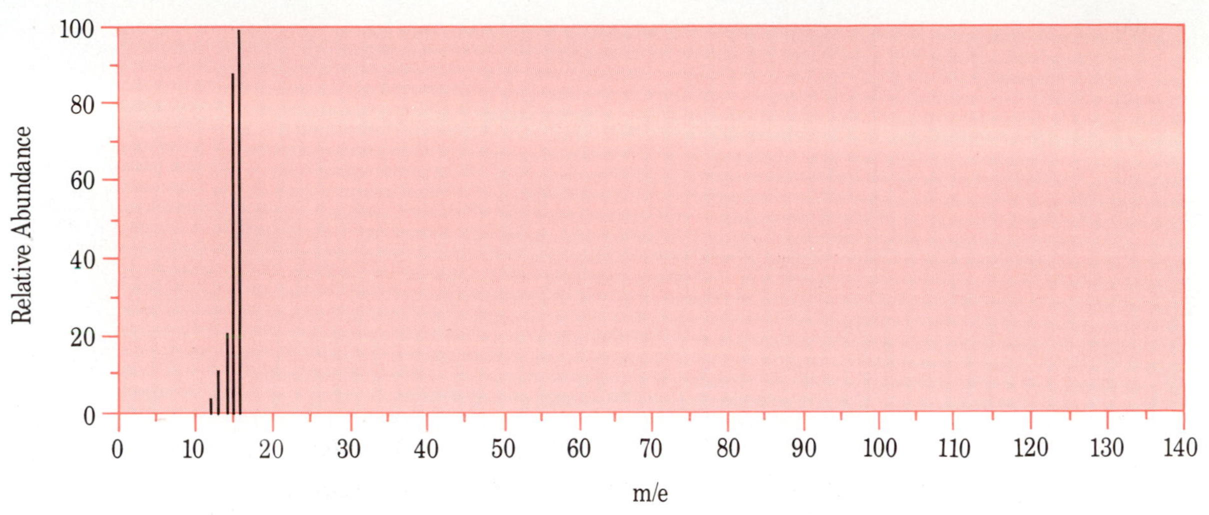

Figure 9.57 Mass Spectrum of Methane.

At this point we have accounted for only one of the peaks in the mass spectrum of methane, the molecular ion (*m*/*e* 16), which is formed by loss of a single electron from CH_4 (MW 16). What are the other peaks in the spectrum? For example, what is the origin of the peak at *m*/*e* 15? Starting from the molecular ion, $CH_4^{+\cdot}$, the only way to form a species with a mass of 15 amu is through the loss of one hydrogen. Moreover, since the 15 amu species is positively charged, the 1 amu species that was lost must have been electrically neutral. In other words, some of the molecular ions must undergo *fragmentation* into a methyl cation and a hydrogen radical.

$$CH_4^{+\cdot} \xrightarrow{\text{fragmentation}} CH_3^+ + H^\cdot$$

The peaks at *m*/*e* 14, 13, and 12 must result from further loss of one, two, and three hydrogen atoms, respectively.

Why do such fragmentations occur? The reactions of ions in a mass spectrometer are quite different from ionic reactions that occur in solution. When they are generated by collisions with high-energy electrons, the molecular ions usually have considerable potential energy. Unlike ions that are formed in solution, these ions are in the gas phase at very low pressures, so they do not undergo collisions with other molecules. Consequently, dissipation of the potential energy can only occur through a *unimolecular* reaction such as decomposition into two fragments. There are certain predictable patterns which fragmentations follow, and these provide information about molecular structure.

There are still two peaks for which we have not accounted in the mass spectrum of methane, namely, the peaks at *m*/*e* 1 and 17. Clearly, the former can only correspond to H^+, and this must result from an alternative mode of decomposition of one of the ions to generate H^+ and a methyl radical. The low relative abundance of the *m*/*e* 1 peak shows that fragmentation to a hydrogen radical and a carbocation is much more favorable. The molecular weight of methane is 16 amu, and you might suspect that *m*/*e* 17 corresponds to CH_5^+. But such a species could only result from a *bimolecular* reaction, and this is usually precluded by the very low pressures

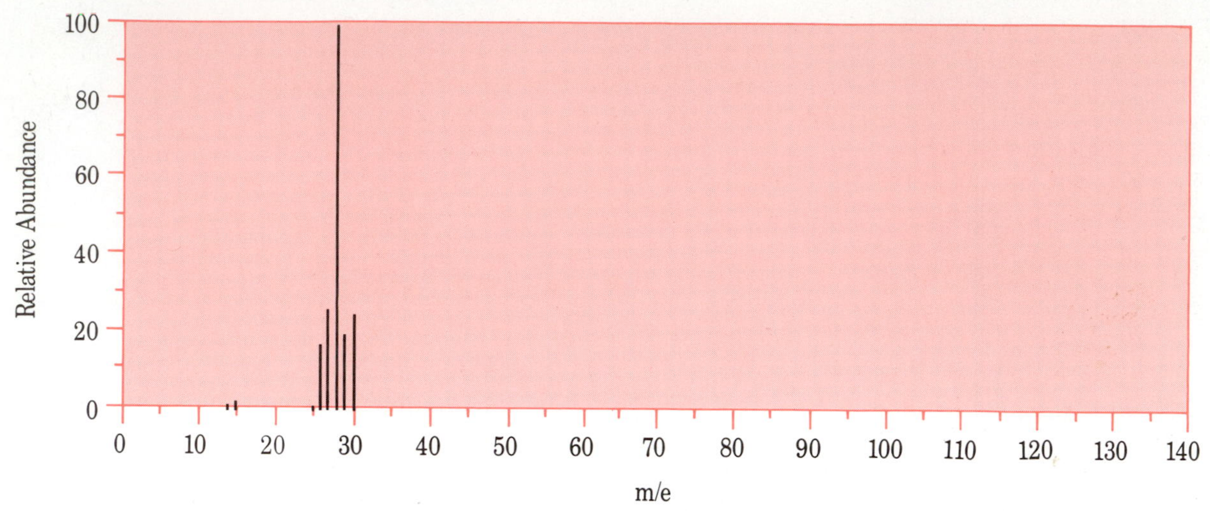

Figure 9.58 Mass Spectrum of Ethane.

inside the mass spectrometer; the concentrations of various species at these pressures are so low that bimolecular collisions are highly improbable.

To understand the origin of the *m/e* 17 peak you must recognize a unique aspect of mass spectrometry. Most chemical measurements provide information on the *average* behavior of a large assembly of molecules (6×10^{23} for one mole of a compound). Mass spectrometry is quite different because the data result from detection of *individual ions,* although a large number of ions is counted during the experiment. This means that *isotopically substituted* ions will be detected at different values of *m/e*. Approximately 1 percent of all carbon atoms are ^{13}C rather than ^{12}C, and in a normal chemical experiment this is reflected by an effective atomic weight of 12.01 amu. By contrast in mass spectrometry, where we see the results of individual ions and not an average result, $CH_4^{+\cdot}$ ions with ^{12}C will have *m/e* 16, but those with ^{13}C will have *m/e* 17. The natural abundance of ^{13}C, 1.1%, corresponds precisely to the 100:1 intensity ratio for the peaks at *m/e* 16 and 17.

Molecular Formulas and Isotopic Contribution

The peak from a molecular ion is conveniently abbreviated as M, and the peak at the next integral value of *m/e* will be called the M+1 peak. The intensity ratio of these two peaks provides direct information about the molecular formulas of the ions that produce these peaks. In the case of methane the 1.1% natural abundance of ^{13}C results in a M/(M+1) ratio of 100:1, but now consider the case of a two-carbon compound, ethane (Figure 9.58).

The molecular ion for C_2H_6 appears at *m/e* 30 and the M+1 peak at *m/e* 31. But now there are *two* carbon atoms, and *each* of them has a 1.1 percent probability of being ^{13}C rather than ^{12}C. Consequently the probability of finding a ^{13}C in ethane is twice that for methane,* and the M/(M+1) ratio is 100:2. (The actual intensities are 30 and 0.6, relative to the most intense peak at *m/e* 28.)

*We have oversimplified this analysis somewhat, because there will also be small contributions to the M+1 peak from the presence of ^{2}H in the sample (Table 9.8). In addition, the presence of more than one ^{13}C atom in a large molecule can reduce the expected intensity of the M+1 peak; two ^{13}C atoms in a single molecule will increase the M+2 peak at the expense of the M+1 peak. For highly accurate determinations, such factors must be considered explicitly, but for our purposes they can be ignored.

The correlation between molecular formula and the M/(M+1) ratio continues for larger molecules, and it can be used to determine the number of carbon atoms in an ion. If the two peaks in question are designated M and M+1, their intensity ratio can be expressed as 100:x. The number x will be a multiple of 1.1 because each carbon in the molecule will have a 1.1% probability of being ^{13}C so division of x by 1.1 will yield the number of carbons in the ion M. In fact the analysis can be applied to *any* ion, not just to the molecular ion. It is frequently possible to determine the molecular formulas of many ions in a mass spectrum by carrying out such an analysis.

Other elements than carbon can make isotopic contributions to the mass spectrum, and Table 9.8 presents the natural isotopic abundances of the elements you are most likely to encounter. We have only shown isotopes that have a natural abundance large enough for a contribution to be observed by mass spectrometry. (For example, the radioactive isotopes ^{3}H and ^{14}C have extremely low natural abundances and are not listed.)

The data in Table 9.8 are presented in two ways. In the third column the abundance of each isotope is shown as a percentage of the total. In the fourth column, the same data are presented as relative natural abundance; the most abundant isotope is given as 100, and the values for the other isotopes are scaled accordingly. The isotopic contributions to the ions might at first seem to greatly complicate the

TABLE 9.8 Natural Abundances for Isotopes of Selected Elements

Element	Isotope	Natural Abundance, %	Relative Abundance
Hydrogen	^{1}H	99.98	100.
	^{2}H	0.02	0.02
Carbon	^{12}C	98.9	100.
	^{13}C	1.1	1.1
Oxygen	^{16}O	99.8	100.
	^{17}O	0.04	0.04
	^{18}O	0.2	0.2
Chlorine	^{35}Cl	75.5	100.
	^{37}Cl	24.5	32.
Bromine	^{79}Br	50.5	100.
	^{81}Br	49.5	98.

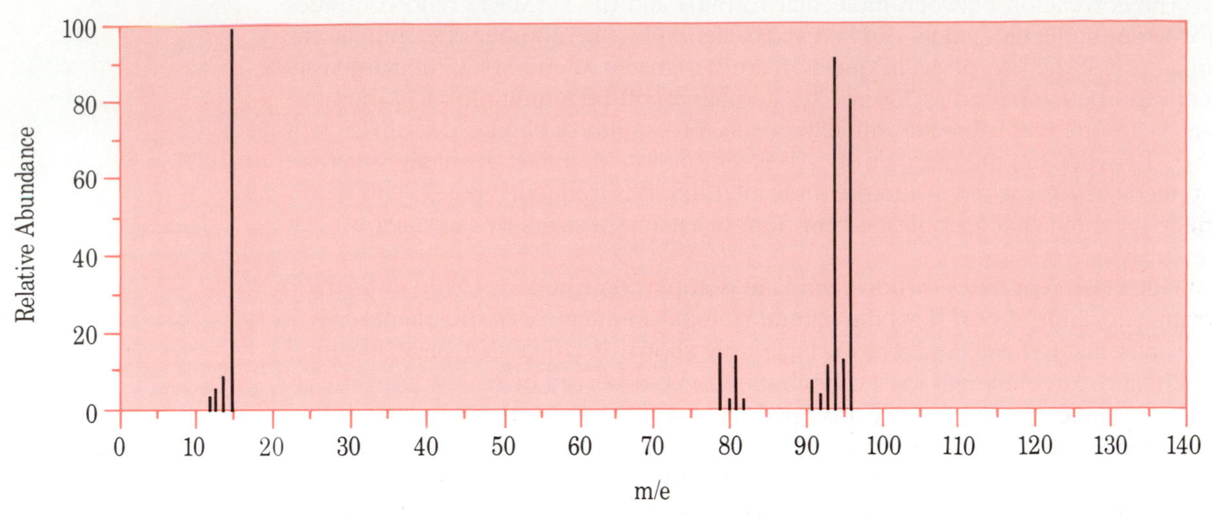

Figure 9.59 Mass Spectrum of Methyl Bromide.

interpretation of mass spectra, but they actually provide highly useful information in many cases. Several elements have characteristic isotopic abundances that produce easily recognizable patterns in the mass spectrum. Bromine, for example, exists as two isotopes, ^{79}Br and ^{81}Br, in almost equal quantities. Hence, any ion that contains a single bromine atom will give rise to a doublet of peaks in the mass spectrum, the individual peaks differing by two mass units. This can be seen in the case of methyl bromide in Figure 9.59.

The two peaks of nearly equal intensity at *m/e* 94 and 96 correspond to the two isotopic forms of the molecular ion, $CH_3{}^{79}Br^{+}\cdot$ and $CH_3{}^{81}Br^{+}\cdot$. A second doublet of peaks can also be found at *m/e* 79 and 81, and these must correspond to the two isotopic forms of Br^+ (formed by fragmentation of the molecular ion into Br^+ and $CH_3\cdot$).

Characteristic patterns are also observed when an ion contains more than one bromine atom. Consider the case of dibromomethane. What should we expect for its molecular ion, $CH_2Br_2{}^{+}\cdot$? There are three possible molecules (if we ignore the minor contributions from ^{13}C): $CH_2{}^{79}Br_2$, $CH_2{}^{79}Br^{81}Br$ and $CH_2{}^{81}Br_2$. The probability that any single bromine atom will be ^{79}Br is 50% or one-half, so the probability that both bromines of CH_2Br_2 will be ^{79}Br is $0.5 \times 0.5 = 0.25$, or 25%. The same argument leads to the prediction that 25% of all molecular ions will have two ^{81}Br atoms, and the remaining 50% will have one ^{79}Br together with one ^{81}Br. The net result is that an ion with two bromines will give rise to a 1:2:1 triplet, each peak separated by *m/e* 2.

These expectations for dibromomethanes are confirmed by Figure 9.60, as the molecular ion appears as a triplet at *m/e* 172, 174, and 176. Notice also that fragmentation of the molecular ion generates ions that appear at lower *m/e* as the characteristic doublets for ions containing only a single bromine.

The two isotopes of chlorine, ^{35}Cl and ^{37}Cl, lead to similar patterns in the mass spectrum. However, they are not as easily recognized because the natural abundances are 75 and 25%, so the peaks of the doublet for an ion with a single chlorine differ in intensity by a factor of 3. The expected pattern for an ion with two chlorine

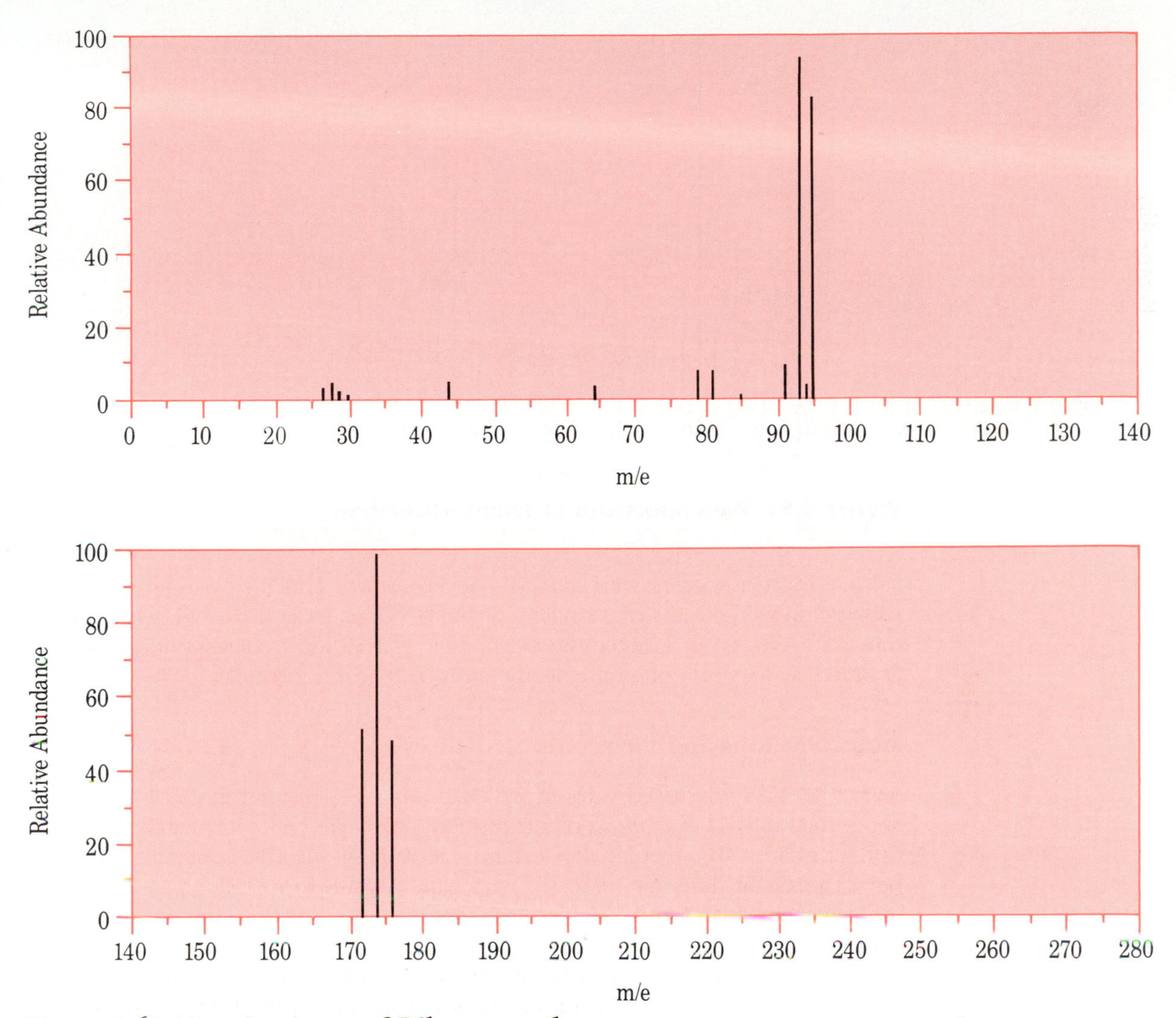

Figure 9.60 Mass Spectrum of Dibromomethane.

atoms will be a triplet with a ratio of 100:65:11.* Both these patterns are seen in the mass spectrum of dichloromethane (Figure 9.61).

The dichloromethane molecular ion, $CH_2Cl_2^{+\cdot}$, affords the expected three peaks at *m/e* 84, 86, and 88 with intensity ratios of 86:55:9 (i.e., 100:64:10, which are quite close to the predicted relative intensities). A second pattern is easily recognized at *m/e* 49 and 51, and here the ratio for the two peaks is 100:31, which is very close to the 100:32 ratio predicted for a species containing a single chlorine

*Consider the mass of a Cl_2 molecule. The probability that a chlorine will be mass 35 is 75.5%, or 0.755. The probability that both will have mass 35 is $(0.755)^2$, or 0.570. Similarly, the probability that both chlorines will have mass 37 is $(0.245)^2 = 0.060$. Therefore 57% of all Cl_2 molecules will have mass 70 and 6% will have mass 74. The remainder (37%) will have one atom of each isotope, with a total mass of 72. Hence the three species will be present in a ratio of 57:37:6 (i.e., 100:65:11).

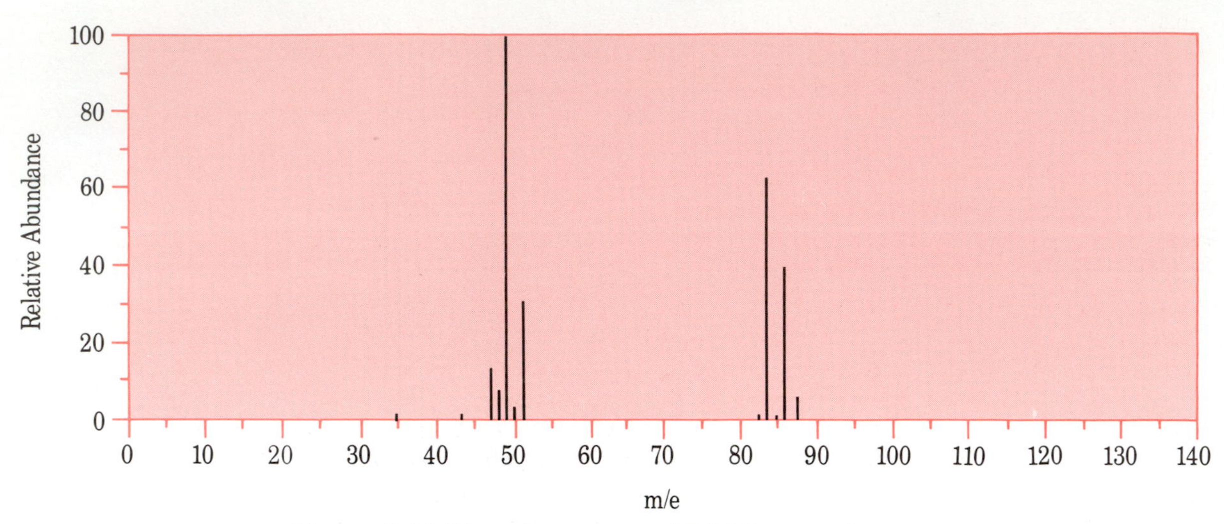

Figure 9.61 Mass Spectrum of Dichloromethane.

atom. In fact a species with mass 49 or 51 *cannot* contain two chlorines, and subtraction of the mass of a single chlorine (^{35}Cl or ^{37}Cl) leaves 14, which is the mass of a CH_2 group. Hence the peaks near *m/e* 50 must correspond to CH_2Cl^+ (formed by loss of a neutral chlorine atom from the molecular ion).

Molecular Ions and Electronic Structures

So far we have discussed ways in which the mass spectrum affords information about the *molecular formula* of a compound. Now we will consider the information it provides about *molecular structure.* Most mass spectra contain a large number of peaks at different values of *m/e,* which result from fragmentation of the molecular ion into ions of lower mass. The ways in which an ion fragments are strongly correlated with its molecular structure. For this reason an understanding of fragmentation pathways is the first step in obtaining structural information from a mass spectrum. Space precludes a complete analysis of the fragmentation pathways of organic ions, but we will present a brief overview of some of the more important ones.

The molecular ion is a cation radical, so the unpaired electron and the positive charge must both appear in the fragments. But only *one* of the two fragments can be positively charged, and only *one* of the two fragments can have an unpaired electron. One major fragmentation pathway involves decomposition of the molecular ion into a cation plus a free radical:

$$M^{+}\cdot \longrightarrow A^{+} + B\cdot$$

In the other major pathway a neutral molecule is lost, with the formation of a new cation radical (a process that usually involves a cyclic rearrangement).

$$M^{+}\cdot \longrightarrow C^{+}\cdot + D$$

Only positively charged species can be detected in the mass spectrometer, so no direct information is available about radical B· or molecule D in the preceding equations. But we would know their masses *indirectly* from the difference in *m/e* between $M^{+}\cdot$ and A^{+} or between $M^{+}\cdot$ and $C^{+}\cdot$. Consequently the *masses* of each of the individual fragments could be deduced. Secondary fragmentations of the ions formed by initial decomposition of the molecular ion are also important, but these are beyond the scope of the present discussion.

The molecular ion formed in the ion source of a mass spectrometer is symbolized by $M^{+}\cdot$, but this symbol provides no information about which electron was lost in the ionization. The formation of the molecular ion must occur by loss of an electron from one of the *occupied molecular orbitals* of the neutral molecule. We can then ask the question: Which occupied orbital loses an electron? The answer is quite reasonable: The most probable ionization pathway is that which requires the least energy. In other words, the most likely orbital from which an electron could be removed is the *highest occupied molecular orbital.* This may not be the *only* ionization pathway, but it is certainly the *major* pathway in most cases. Loss of an electron from the HOMO is important because it allows us to make correlations between the electronic structure of the molecular ion and the reaction pathways for fragmentation. To do this we must first identify the HOMO in typical organic molecules. In a molecule that contains a heteroatom the HOMO should be that associated with a *nonbonding* pair of electrons. The equations in the following box illustrate the most common fragmentation pathways for a molecule containing a heteroatom.

Common Fragmentation Pathways for Molecules with a Heteroatom

$$-\overset{|}{\underset{|}{C}}-\overset{|}{\underset{|}{C}}-\dot{X}^{+} \longrightarrow -C\cdot + \;C=\overset{+}{X}$$ α-Cleavage

$$-\overset{|}{\underset{|}{C}}-\overset{|}{\underset{|}{C}}-X^{+} \longrightarrow -\overset{|}{\underset{|}{C}}-C^{+} + \cdot\ddot{X}$$ Charged site cleavage

$$H-R-\dot{X}^{+} \longrightarrow [R]^{+\cdot} + HX$$ Elimination

The first type of fragmentation is called *α cleavage* because the bond cleavage occurs between the carbon atoms that are α, β to the charged site. This produces a rather stable cation in which the positive charge is localized on a heteroatom with eight electrons in its valence shell. For example, an alcohol (or ether) would generate an *oxonium ion* by this fragmentation pathway. The other fragment would be a carbon free radical.

$$-\overset{|}{\underset{|}{C}}-\overset{|}{\underset{|}{C}}-\overset{+}{\underset{\cdot\cdot}{O}}H \longrightarrow -C\cdot + \;C=\overset{+}{\underset{\cdot\cdot}{O}}H$$

In the second fragmentation mode, which we will call *charged–site cleavage,* a bond to the electron-deficient heteroatom is broken. This produces a carbocation and a heteroatom radical, as illustrated again for an alcohol. (Both electrons from the C—O bond remain with the oxygen to produce a neutral oxygen fragment but one that still has an unpaired electron.) The other fragment is simply a carbocation.

$$-\overset{|}{\underset{|}{C}}-\overset{|}{\underset{|}{C}}-\overset{\cdot +}{\ddot{O}}H \longrightarrow -\overset{|}{\underset{|}{C}}-C^{+} + :\dot{\ddot{O}}H$$

The third fragmentation mode involves elimination of a neutral molecule, and this generates a new cation radical as the other fragment. The elimination typically involves loss of a remote hydrogen (via a five- or six-membered ring transition state). For example, the cation radical from an alcohol could lose water in the following way:

$$\text{H–C–C–C–}\overset{\cdot +}{\ddot{O}}H \longrightarrow \cdot C\text{–C–C–}\overset{+}{O}H_2 \longrightarrow \cdot C\text{–C–C}^{+} + H_2O$$

These three fragmentation modes are illustrated by the mass spectrum of diisobutyl ether (Figure 9.62). The molecular ion (*m/e* 130) is very weak for this compound because several very efficient fragmentation pathways are available. The molecular ion decomposes predominantly via charged-site cleavage to form an isobutyl cation (*m/e* 57), and the peak at *m/e* 87 corresponding to α cleavage has a low relative intensity of only 8%. Elimination (of isobutyl alcohol) occurs to an even lesser extent, and there is only a very small peak at *m/e* 56.

$$(CH_3)_2CH-CH_2-\overset{\cdot +}{\ddot{O}}-CH_2-CH(CH_3)_2 \xrightarrow[\text{cleavage}]{\text{charged site}} (CH_3)_2CH-CH_2-\ddot{O}\cdot + \overset{+}{C}H_2CH(CH_3)_2$$

m/e 57

$$(CH_3)_2CH-CH_2-\overset{\cdot +}{\ddot{O}}-CH_2CH(CH_3)_2 \xrightarrow{\alpha \text{ cleavage}} (CH_3)_2CH\cdot + CH_2=\overset{+}{\ddot{O}}-CH_2-CH(CH_3)_2$$

m/e 87

$$(CH_3)_2CH-CH_2-\overset{\cdot +}{\ddot{O}}-CH_2CH(CH_3)_2 \xrightarrow{\text{elimination}} [(CH_3)_2C=CH_2]^{\cdot +} + HOCH_2CH(CH_3)_2$$

m/e 56

Fragmentation pathways for a molecular ion are highly dependent on structure, but such structure-reactivity patterns are beyond the scope of this introductory chapter. We have shown you some of the ways that ions can decompose, and even this limited discussion will enable you to deduce useful structural information from the mass spectra of organic compounds.

EXERCISE 9.11

Calculate the mass of the molecular ion for each of the indicated compounds.

(a) $CH_3CH_2CH_2CH(CH_3)_2$
(b) $CH_3-O-CH_2CH_2C(CH_3)_3$
(c) $ClCH_2CH_2CH(CH_3)_2$
(d) $CH_3CH_2CH_2CHBr_2$

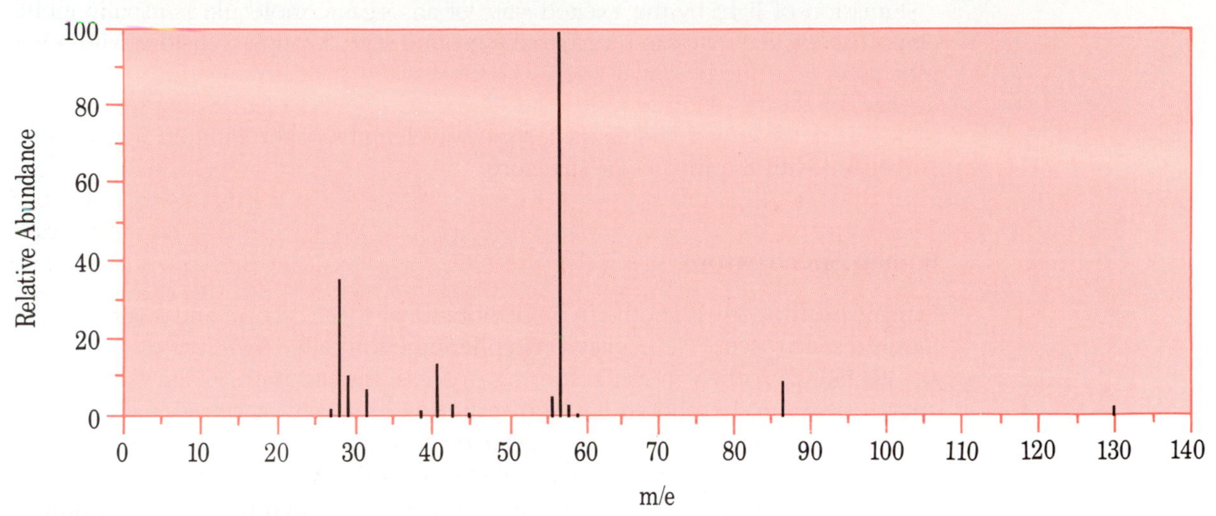

Figure 9.62 Mass Spectrum of Diisobutyl Ether.

EXERCISE 9.12

For each of the compounds in the previous question calculate the ratio of the M and M+1 peaks expected from the natural abundance of ^{13}C.

EXERCISE 9.13

The molecular ion for an unknown compound appears at *m/e* 210 with a relative intensity of 100. The intensity of the M+1 peak is 16.5.

(a) How many carbon atoms are present in the compound?
(b) Chemical tests for the compound indicate that it is a hydrocarbon. What is its molecular formula?

9.10 OTHER SPECTROSCOPIC METHODS USED FOR STRUCTURE EVALUATION

A variety of spectroscopic techniques can provide valuable information about the structures of organic compounds. We have already discussed NMR, ultraviolet-visible, and infrared spectroscopy and mass spectrometry, but the other techniques are more specialized, and only brief descriptions will be presented here.

Emission Spectroscopy

Just as a molecule can absorb a photon of electromagnetic radiation to generate an excited state species, a molecule in an excited state can *emit* a photon and return to a lower energy state. The resulting emission of light is familiar to anyone who has ever seen a firefly on a summer evening. The category of emission can be further subdivided into *fluorescence* and *phosphorescence*. The distinction between the two is beyond the scope of this discussion, but it depends on the electronic configuration of the excited state that emits a photon.

Emission of light by the excited state of an organic molecule is usually inefficient. The excited state can return to the ground state through collisions with other molecules, and the potential energy of the excited state is converted into kinetic energy for the resulting ground-state molecules. Nevertheless, when emission of light can be observed, the characteristic wavelengths of the emitted light can be correlated with the molecular structure.

Raman Spectroscopy

Raman spectroscopy is another type of vibrational spectroscopy, and it is related to infrared spectroscopy. It results from a phenomenon called *light scattering*. Light is not actually absorbed in the scattering process; instead there is an exchange of energy between the photon and the scattering molecule. A vibrationally excited molecule is produced, and this is detected from a *change* in wavelength of the incident radiation. In contrast to the absorptions in IR spectroscopy these transitions do not require a change in dipole moment. Consequently, the two techniques are complementary, and Raman spectroscopy is particularly valuable for the study of highly symmetrical molecules and functional groups, for which IR spectroscopy provides little information.

ORD and CD Spectroscopy

Optical rotatory dispersion (ORD) and *circular dichroism (CD)* utilize radiation in the UV-visible region. Both techniques are applicable only to optically active compounds, and can provide valuable information about the stereochemistry of such molecules. An ORD spectrum shows the optical rotation as a function of wavelength, and a CD spectrum yields the difference in absorption of left- and right-handed circularly polarized light as a function of wavelength. The results of such studies depend on which enantiomer of a chiral substance is being studied, and empirical rules have been developed to determine the stereochemistry of substituents near functional groups that absorb UV-visible radiation.

Electron Paramagnetic Resonance

One of the properties of an electron that is specified by quantum mechanics is that of *spin*. In the classical model you can think of an electron as a tiny sphere that is spinning in the same sense that the earth rotates about its axis. Clockwise or counterclockwise rotation about this axis corresponds to the two possible spin states. Ordinarily the two spin states of an unpaired electron have the same energy, but they become different when the electron is in a magnetic field. This provides the basis for studying organic free radicals by *electron spin resonance (ESR),* also called *electron paramagnetic resonance (EPR).* The energy differences are rather small, and EPR studies utilize radiation in the microwave region (see Figure 9.2).

In a strong magnetic field the two possible spin states for the unpaired electron of a free radical have different energies, and absorption of a photon of appropriate energy can convert a free radical from the lower- to the higher-energy spin state. The unpaired electron of a free radical must reside in a molecular orbital, and the EPR spectrum provides information about the various atoms of the molecule which interact via that particular molecular orbital.

Microwave Spectroscopy

Absorption in the microwave region of the electromagnetic spectrum by ordinary molecules produces changes in *rotational* energy levels. This phenomenon provides the basis for cooking food with microwave radiation. The radiation generated in microwave ovens is preferentially absorbed by the large amounts of water that are present in nearly all foods. The rotationally excited water molecules collide with neighboring molecules, and there is an increase in the overall kinetic energy. Kinetic energy is directly related to temperature, so the food is heated and becomes cooked (while the container remains relatively cool).

The rotational levels of a molecule are quantized and are related both to the masses of individual nuclei in the molecule and to their internuclear distances. Microwave spectra therefore yield information about bond lengths and bond angles in a molecule.

Diffraction Methods

In *x-ray diffraction* a crystalline sample is bombarded with x-radiation, and x-ray photons are deflected by the nuclei in the sample. The radiation has wave properties, so constructive and destructive interference produce a diffraction pattern that contains information about the geometric relationships between the various nuclei in the sample. Extensive mathematical analysis of the diffraction data provides a complete structural description (bond lengths and angles), even for large and complex molecules. Diffraction patterns can also be obtained with beams of atomic particles. Consequently, *neutron diffraction* and *electron diffraction* are also useful for structure determination. Electron diffraction is generally carried out in the gas phase.

Electron Spectroscopy for Chemical Analysis (ESCA)

High-energy photons (typically x-rays or very short wavelength UV radiation) can cause the ejection of electrons from a molecule. x-Rays cause the ejection of core electrons and UV radiation causes ejection of valence electrons. In either case the kinetic energy of the ejected electrons can be measured. This kinetic energy is related to how strongly the electron was held in the original molecule, that is, to its *binding energy*. The binding energy of a core electron is related to the chemical environment of that particular atom. For example, a core electron on an atom in a +3 oxidation state would be bound more strongly than if the atom were in only a +1 oxidation state. The binding energy of valence-level electrons can be correlated with calculated energies for molecular orbitals.

9.11 TERMS AND DEFINITIONS

Absorption spectrum. A graphical description of the absorption of radiation as a function of wavelength (or energy) of the radiation.

Chemical shift. The variation in radio frequency energy absorbed by nonequivalent nuclei in an NMR experiment as a function of their chemical environments.

Coupling constant. The magnitude of the NMR coupling interaction between two nuclei, usually denoted by the letter J.

Diastereotopic. A term describing two atoms or groups in a molecule that are different in the sense that replacement by another substituent would lead to diastereomeric products, depending on which of the two is replaced.

Downfield. A term describing transitions of lower energy in NMR spectroscopy; it indicates peaks to the left of some point of reference.

Emission spectrum. A graphical description of the emission of radiation as a function of wavelength (or energy) of the radiation.

Excited state. A state of a species that has absorbed radiation. The absorbed energy can be liberated by emission of a photon or by transfer to the surroundings in the form of heat.

Ground state. The lowest-energy state of a species. For electronic configurations this is the energy state normally occupied at room temperature.

Integral. A second tracing on an NMR spectrum that indicates the relative area under each peak.

Integration. In NMR spectroscopy, measurement of the area under the peaks of a spectrum.

Mass spectrometry. An instrumental technique for studying molecular structure, which is based on the separation of charged particles according to mass.

Molecular ion. The cation radical that is produced in a mass spectrometer by ejection of an electron from a neutral molecule.

Multiplicity. The number of lines in an NMR signal that are observed for a particular nucleus as a result of spin coupling to other nuclei.

Nuclear magnetic resonance (NMR) spectroscopy. A technique for studying molecular structure based on transitions between energetically different spin states of nuclei in a strong magnetic field.

Saturation. The result in NMR spectroscopy of irradiation with a strong rf signal. The populations become nearly equal for the two spin states of the irradiated nucleus, and there is no net absorption or emission. Coupling interactions with other nuclei are thereby removed.

Spectroscopy. The study of absorption or emission of different wavelengths of electromagnetic radiation in order to gain information about molecular structure.

Spectrum. The different colors of visible light that can be seen when a beam of light passes through a prism; more generally, the different wavelengths of electromagnetic radiation. The term is also used to describe a graphical representation of the absorption (or emission) of radiation by a sample as a function of wavelength.

Spin decoupling. Elimination of spin coupling between nuclei by double irradiation of one of them to produce saturation.

Spin-spin coupling. The interaction between nuclei in different spin states; it can result in multiple peaks for a particular nucleus in an NMR spectrum. The term is often shortened to *spin coupling.*

Upfield. A term describing transitions of higher energy in NMR spectroscopy; it indicates peaks to the right of some point of reference.

Visible light. That region of the electromagnetic spectrum that can be detected by the human eye; the region with wavelengths shorter than those of the infrared region but longer than those of the ultraviolet region.

9.12 PROBLEMS

9.1 How many peaks would be expected in the ^{13}C NMR spectrum of each of the following compounds?

(a)

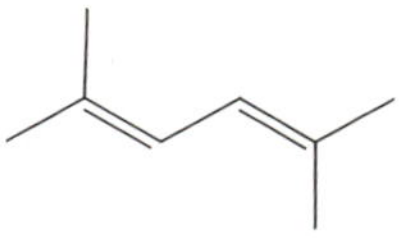

(b) 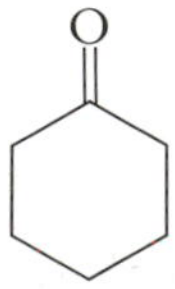 O

(c) 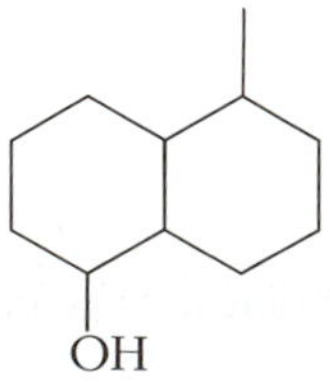 OH

(d) 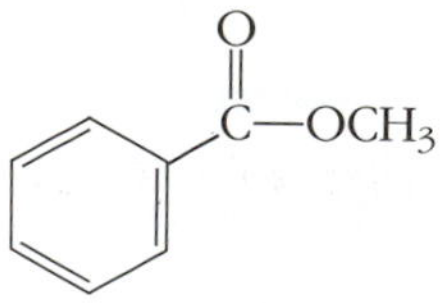 C—OCH_3 O

(e) 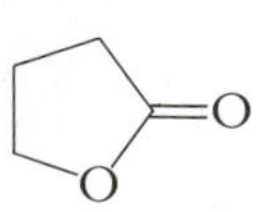 O O

9.2 Predict the approximate ^{13}C chemical shift of each peak expected for the compounds in Problem 9.1.

9.3 Predict the approximate ^{13}C chemical shift of each peak expected for the following compounds:

(a) $CH_3CH_2{-}C({=}O){-}CH_2CH_2{-}O{-}CH_2CH_2{-}Br$

(b) $CH_3{-}C(CH_3)_2{-}O{-}CH_2{-}C({=}O){-}OCH_3$ (drawn with CH₃ above and below the central C)

(c) $CH_3{-}CCl_2{-}CH_2CH_2Br$ (drawn with Cl above and below the central C)

(d) six-membered ring lactone (ring O adjacent to C=O) bearing two CH_3 groups on one ring carbon

(e) $CH_3CH_2{-}O{-}CH_2Cl$

(f) cyclopentanone bearing CH_3 and CH_2OH on the same ring carbon

9.4 A compound labeled simply "dimethylcyclohexene" was subjected to catalytic hydrogenation, and the product exhibited the NMR spectrum shown in Figure 9.63. Ozonolysis of the dimethylcyclohexene afforded a dialdehyde in which the CHO groups were not equivalent. Assuming that the name on the label was correct (but incomplete), deduce the structure of the dimethylcyclohexene. (You have sufficient information to determine the constitution but not the stereochemistry.)

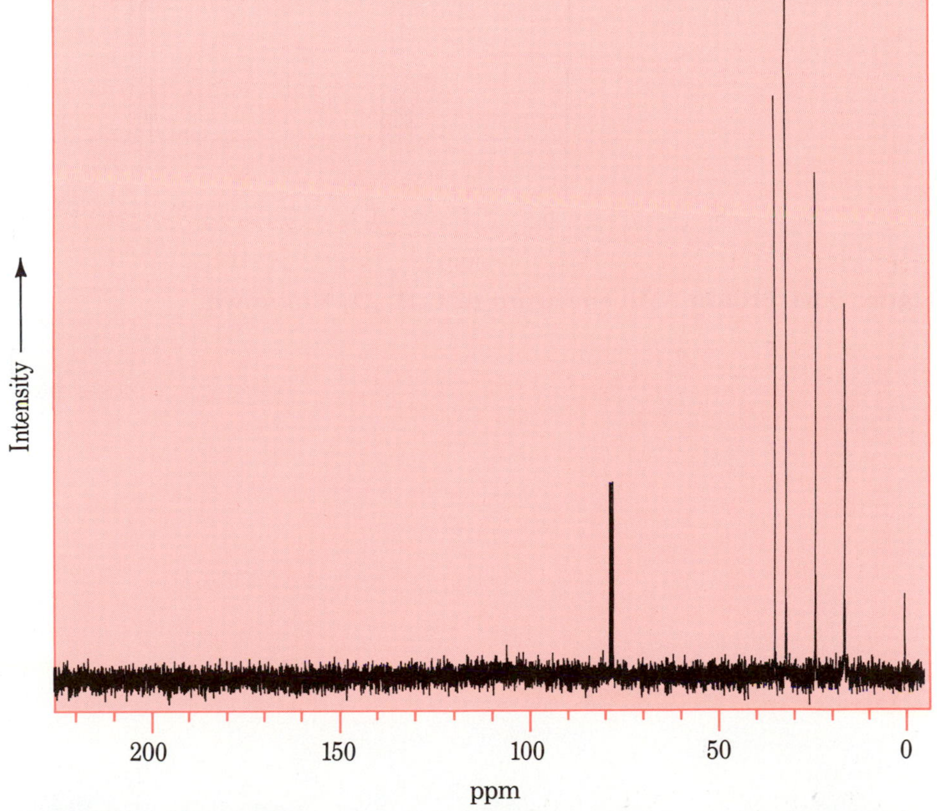

Figure 9.63 ^{13}C NMR Spectrum of Dimethylcyclohexene Hydrogenation Product.

9.5 The following are ^{13}C NMR data for a series of isomeric compounds, each of which has the molecular formula C_4H_8O. What structures are possible in each case?

(a) 7.9, 29.4, 36.9, 209.2 ppm
(b) 25.7, 68.0 ppm
(c) 13.7, 15.7, 45.8, 202.6 ppm
(d) 15.5, 41.0, 204.9 ppm

9.6 Deduce the structure of the unknown compound from its molecular formula and proton NMR spectrum, shown in Figure 9.64.

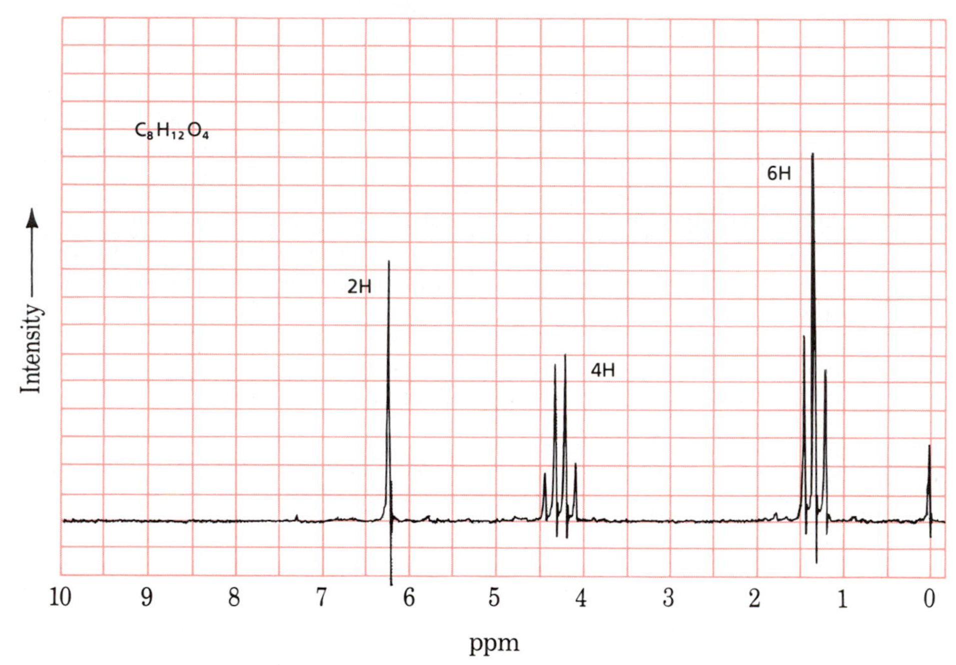

Figure 9.64 Proton NMR Spectrum of $C_8H_{12}O_4$ Unknown.

9.7 Deduce the structure of the unknown compound from its molecular formula and proton NMR spectrum, shown in Figure 9.65.

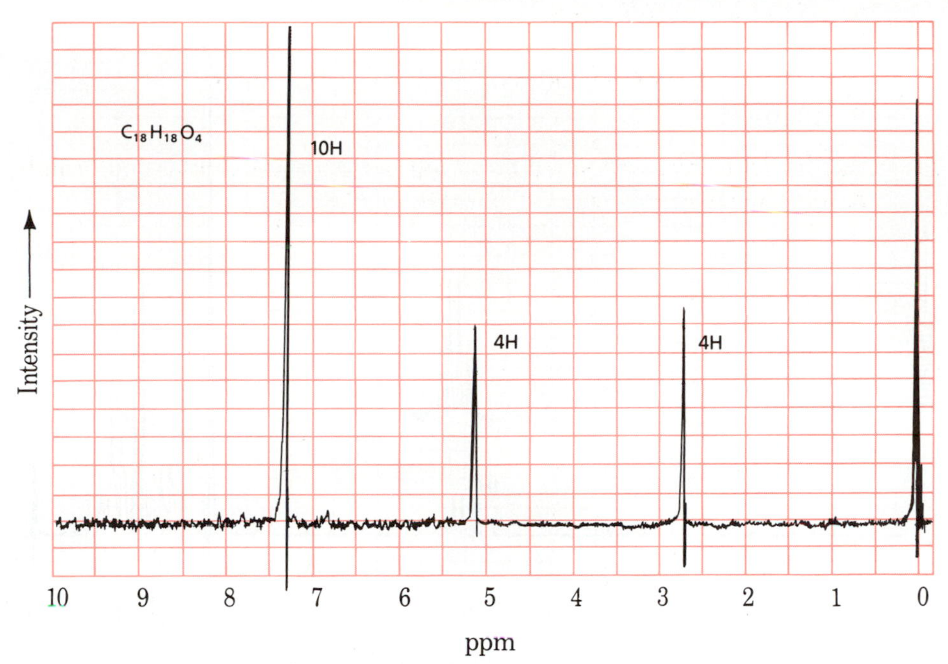

Figure 9.65 Proton NMR Spectrum of $C_{18}H_{18}O_4$ Unknown.

9.8 Deduce the structure of the unknown compound from its molecular formula and proton NMR spectrum, shown in Figure 9.66.

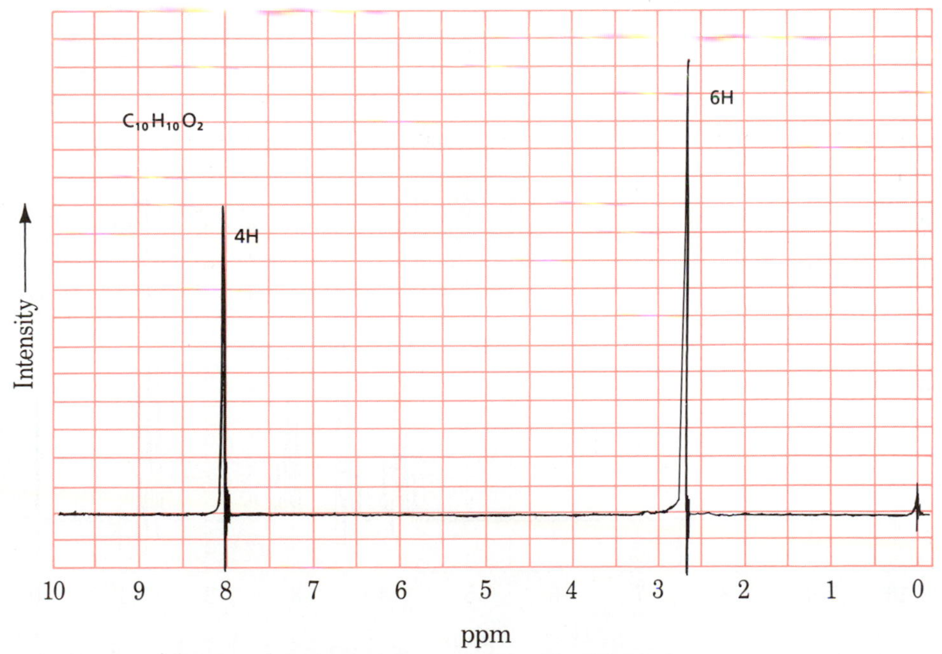

Figure 9.66 Proton NMR Spectrum of $C_{10}H_{10}O_2$ Unknown.

9.9 Deduce the structure of the unknown compound from its molecular formula and its proton NMR spectrum, shown in Figure 9.67.

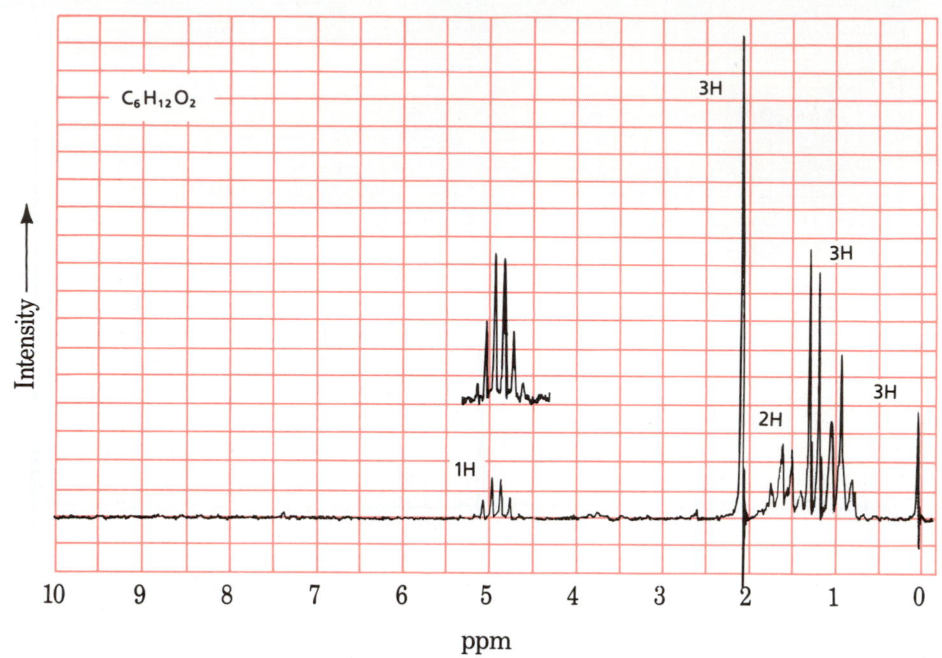

Figure 9.67 Proton NMR Spectrum of $C_6H_{12}O_2$ Unknown.

9.10 Deduce the structure of the unknown compound from its molecular formula and proton NMR spectrum, shown in Figure 9.68.

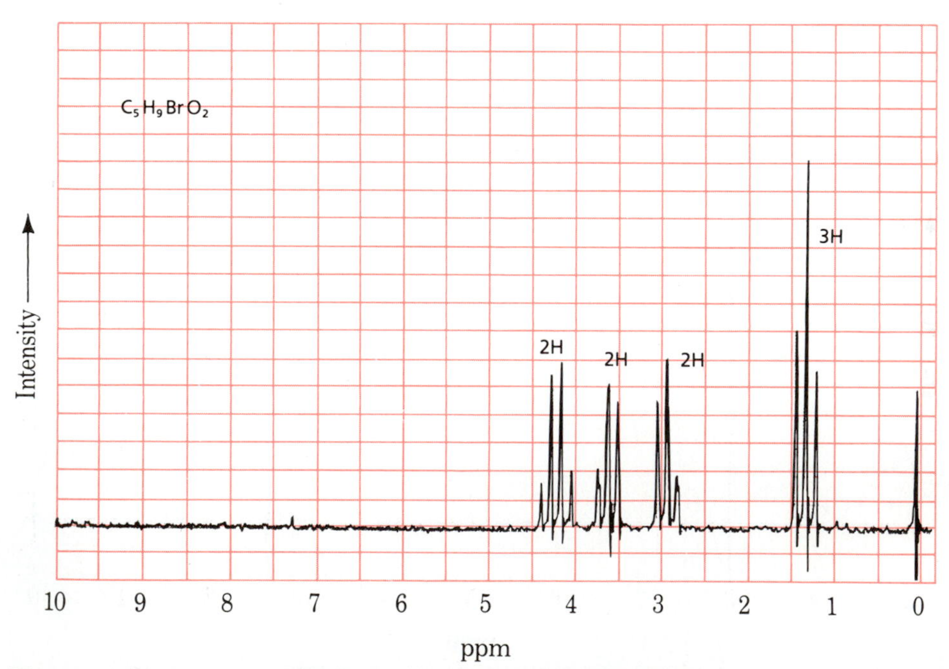

Figure 9.68 Proton NMR Spectrum of $C_5H_9BrO_2$ Unknown.

9.11 Deduce the structure of the unknown compound from its molecular formula and proton NMR spectrum, shown in Figure 9.69.

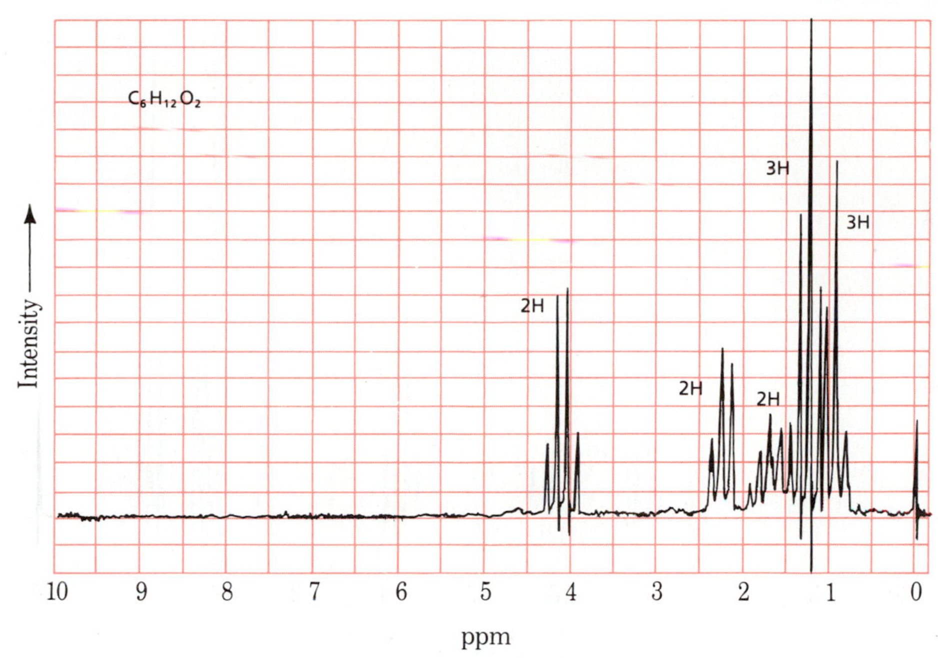

Figure 9.69 Proton NMR Spectrum of $C_6H_{12}O_2$ Unknown.

9.12 Deduce the structure of the unknown compound from its molecular formula and proton NMR spectrum, shown in Figure 9.70.

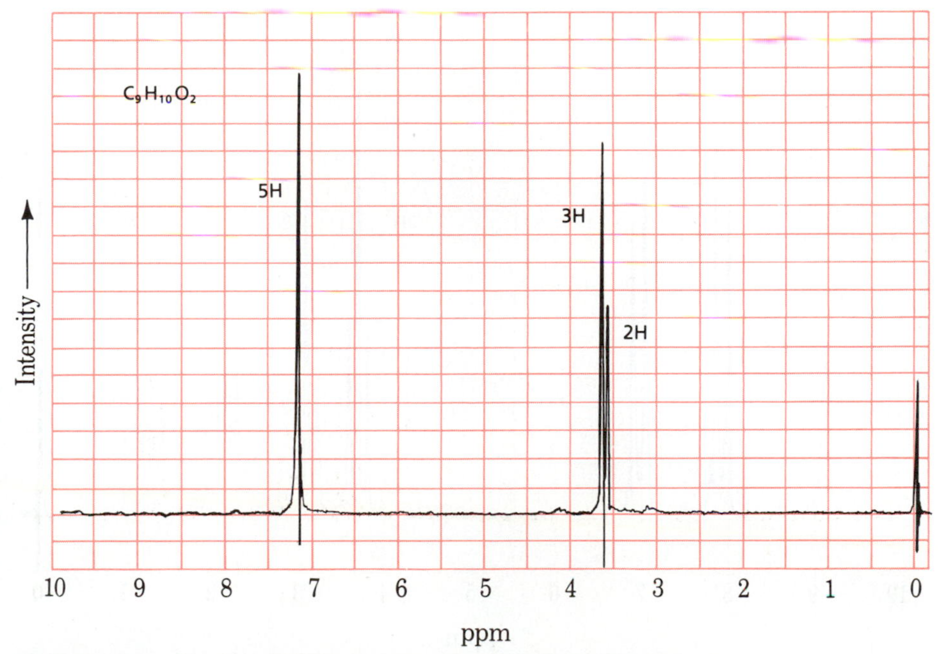

Figure 9.70 Proton NMR Spectrum of $C_9H_{10}O_2$ Unknown.

9.13 Deduce the structure of the unknown compound from its molecular formula and proton NMR spectrum, shown in Figure 9.71.

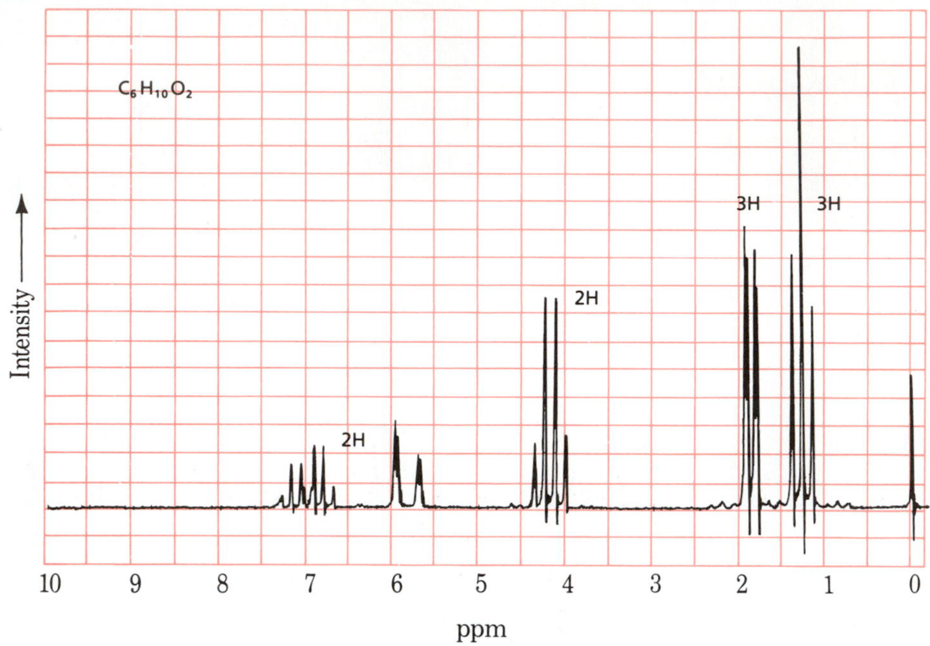

Figure 9.71 Proton NMR Spectrum of $C_6H_{10}O_2$ Unknown.

9.14 Deduce the structure of the unknown compound from its molecular formula and proton NMR spectrum, shown in Figure 9.72.

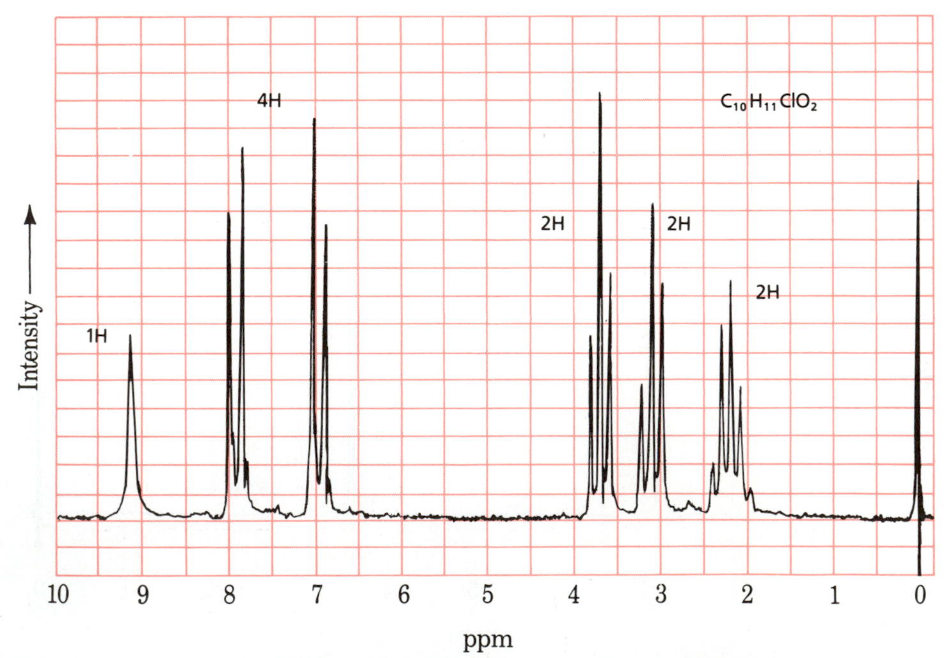

Figure 9.72 Proton NMR Spectrum of $C_{10}H_{11}ClO_2$ Unknown.

9.15 Deduce the structure of the unknown compound from its molecular formula and proton NMR spectrum, shown in Figure 9.73.

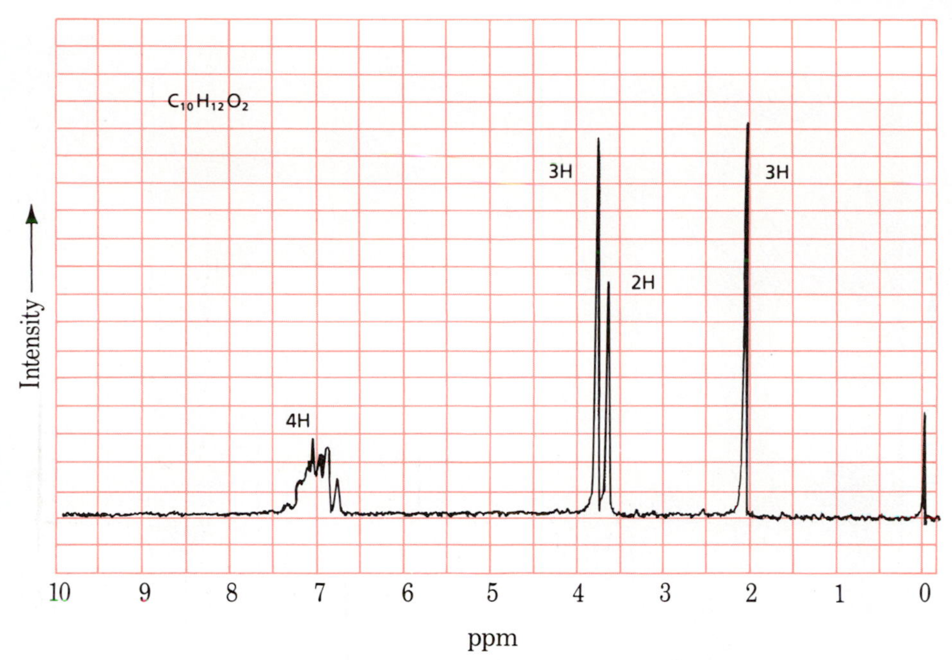

Figure 9.73 Proton NMR Spectrum of $C_{10}H_{12}O_2$ Unknown.

9.16 Deduce the structure of the unknown compound from its molecular formula and proton NMR spectrum, shown in Figure 9.74, and its ^{13}C NMR spectrum,

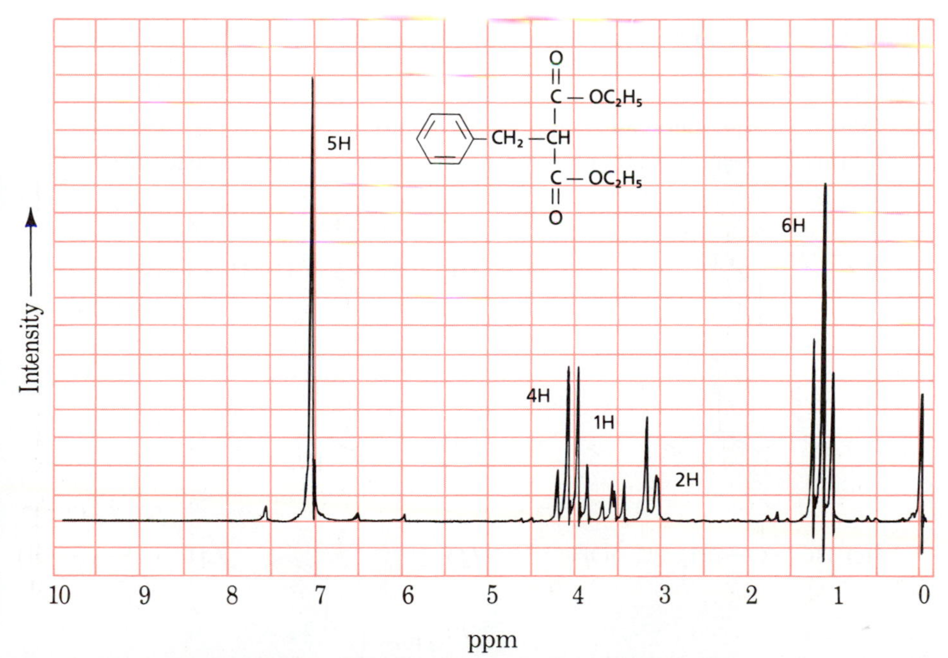

Figure 9.74 Proton NMR Spectrum of $C_{14}H_{18}O_4$ Unknown.

shown in Figure 9.75. (*Hint:* Remember that carbon atoms with no hydrogen substituents often give rise to very weak signals.)

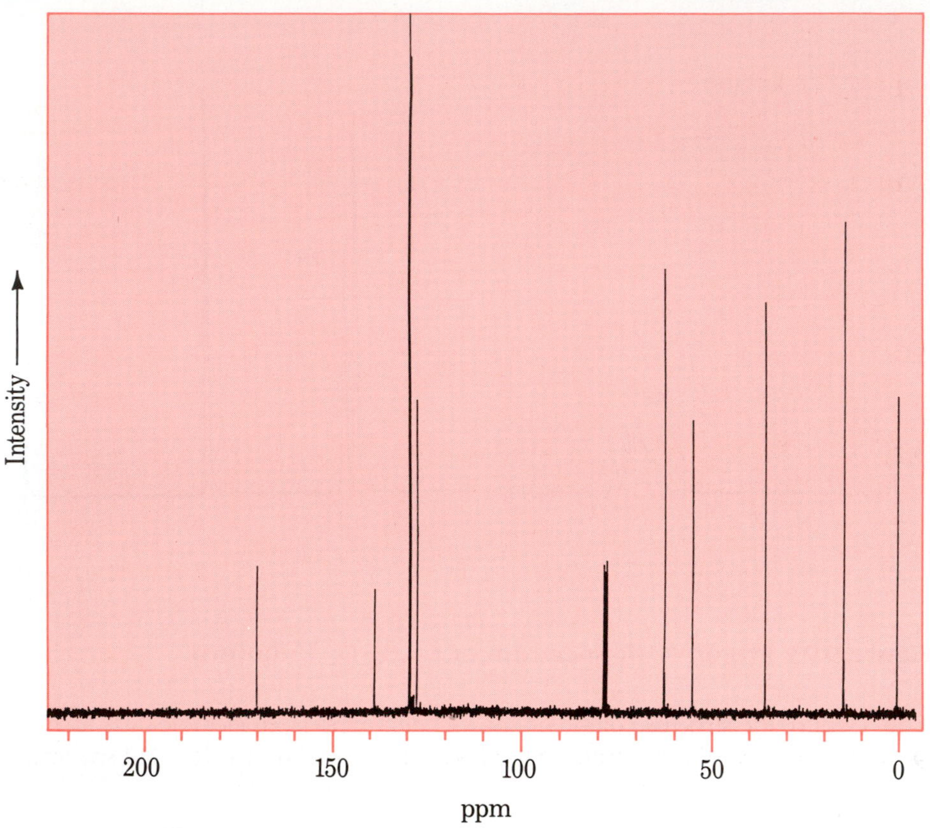

Figure 9.75 ^{13}C NMR Spectrum of $C_{14}H_{18}O_4$ Unknown.

9.17 Which of the following compounds would be expected to absorb in the ultraviolet region (200–400 nm) of the electromagnetic spectrum?

(a) 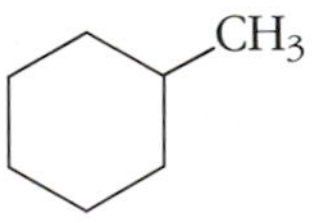 CH_3

(b) 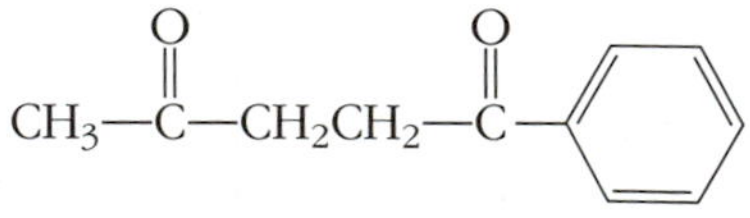 $CH_3-\overset{O}{\overset{\|}{C}}-CH_2CH_2-\overset{O}{\overset{\|}{C}}-$

(c)

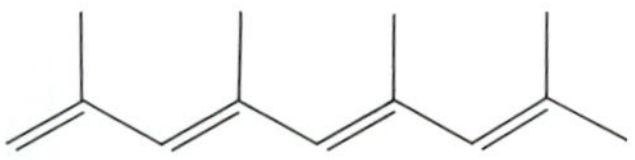

(d)

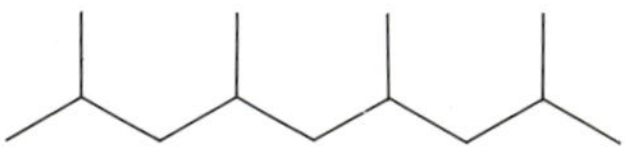

(e) 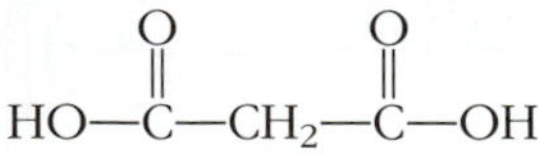 $HO-\overset{O}{\overset{\|}{C}}-CH_2-\overset{O}{\overset{\|}{C}}-OH$

(f) 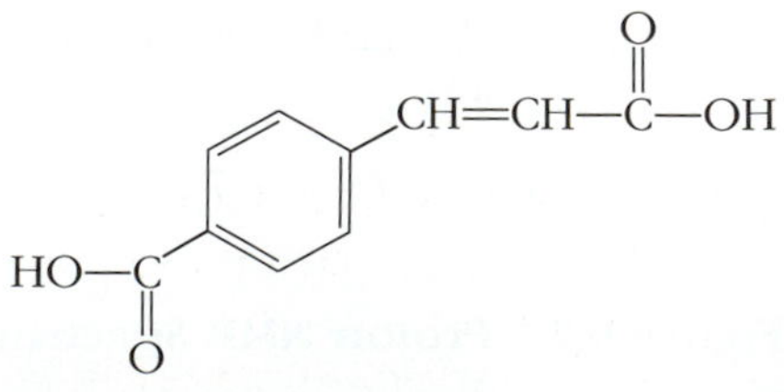 $CH=CH-\overset{O}{\overset{\|}{C}}-OH$

$HO-\underset{O}{\underset{\|}{C}}$

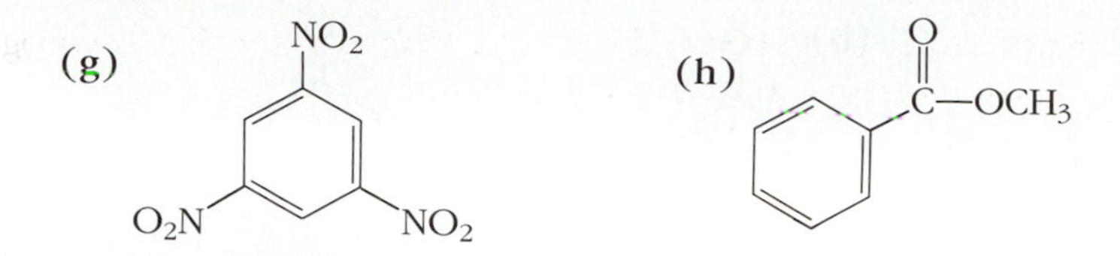

9.18 Which of the compounds in Problem 9.17 would you expect to be colored? Which would you expect to be colorless?

9.19 Predict the λ_{max} for UV absorption of each of the following compounds, using the substituent effects in Figure 9.33.

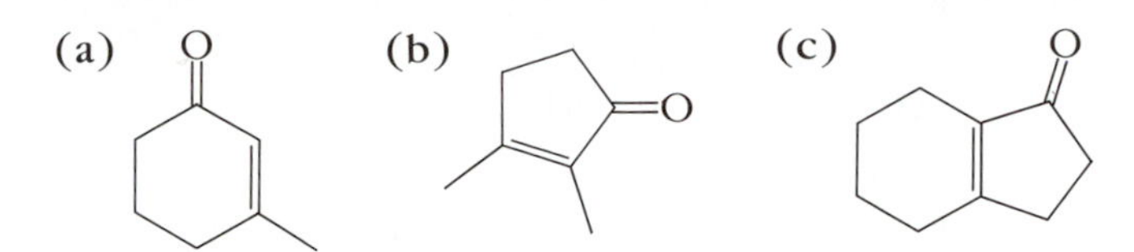

9.20 A UV spectrum for a compound of uncertain structure exhibited a λ_{max} at 228 nm. Which structure agrees best with the UV spectrum?

(a) (b) (c)

9.21 A UV spectrum for a compound of uncertain structure exhibited a λ_{max} at 249 nm. Which structure agrees best with the UV spectrum?

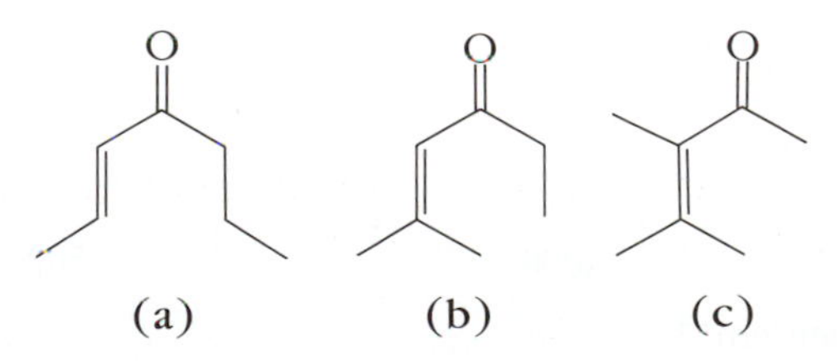

9.22 A UV spectrum for a compound of uncertain structure exhibited a λ_{max} at 221 nm. Which structure agrees best with the UV spectrum?

(a) (b) (c)

9.23 For each of the following compounds predict the approximate frequency of the C=O stretching absorption in the infrared region of the electromagnetic spectrum.

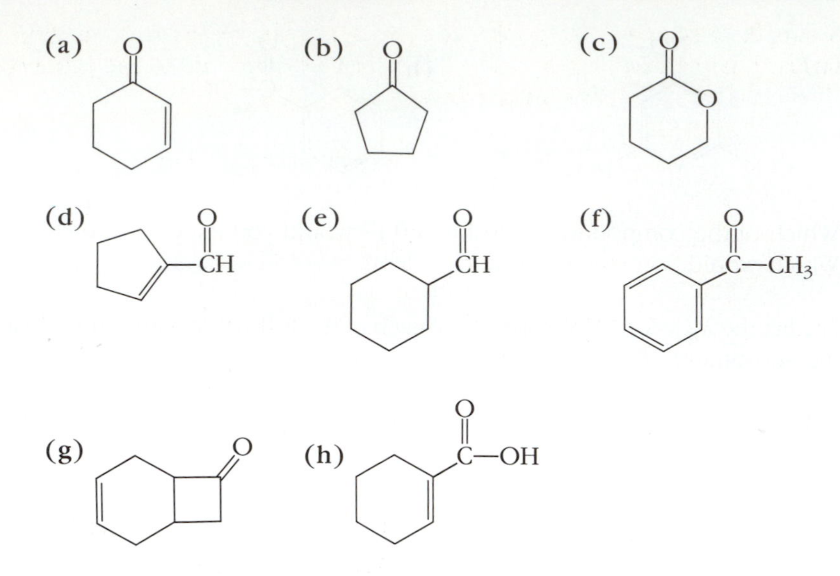

9.24 A sample of 4-methyl-1-pentyne was subjected to hydroboration-oxidation using disiamylborane. The infrared spectrum of one of the compounds isolated from the reaction mixture is shown below. What is the structure of the compound?

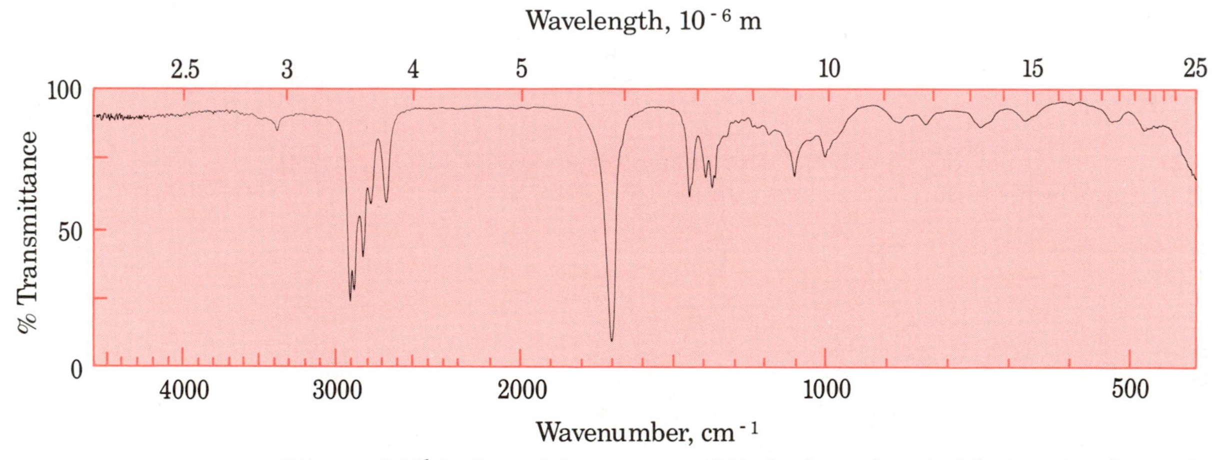

Figure 9.76 Infrared Spectrum of Hydroboration-Oxidation Product of 4-Methyl-1-pentyne.

9.25 A sample of 1-nonyne was subjected to catalytic hydrogenation in an effort to prepare the corresponding alkene. The infrared spectrum of the product is shown below. What was the product?

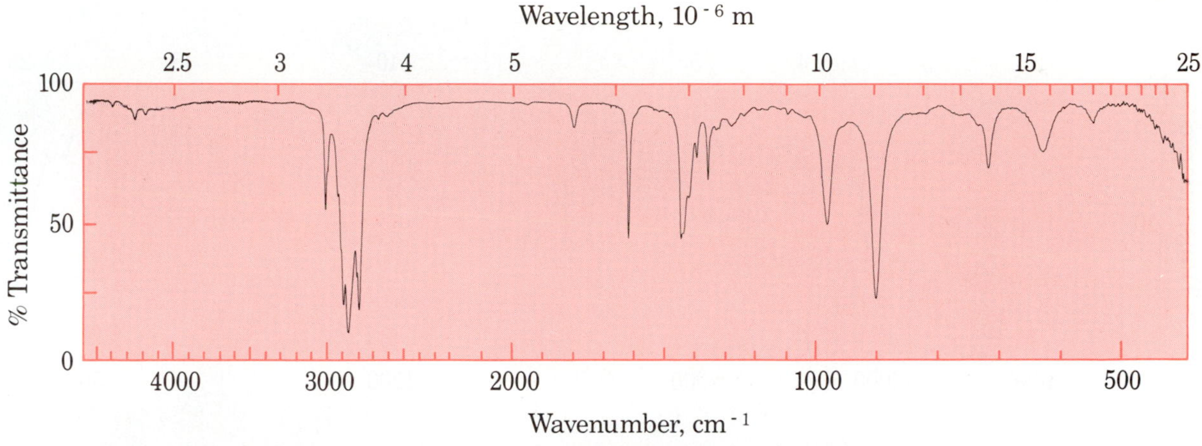

Figure 9.77 Infrared Spectrum of 1-Nonyne Hydrogenation Product.

9.26 A sample of 3-hexyne was subjected to catalytic hydrogenation in an effort to prepare (*Z*)-3-hexene. The infrared spectrum of the product is shown below. What was the product?

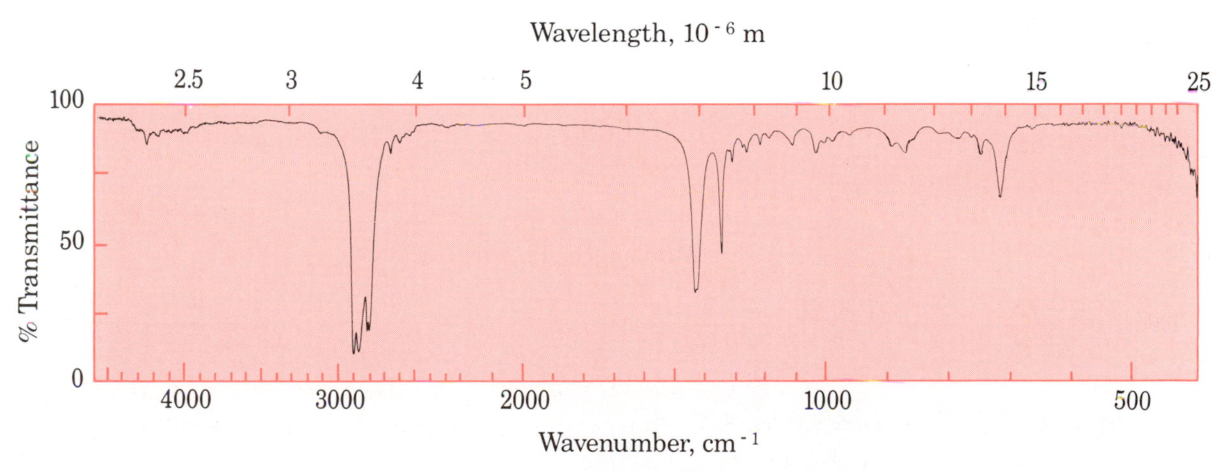

Figure 9.78 Infrared Spectrum of 3-Hexyne Hydrogenation Product.

9.27 A compound with the molecular formula $C_9H_{10}O$ was found to be a benzene derivative. Its proton NMR spectrum indicated the presence of an ethyl group. From this information and the IR spectrum below, deduce the structure.

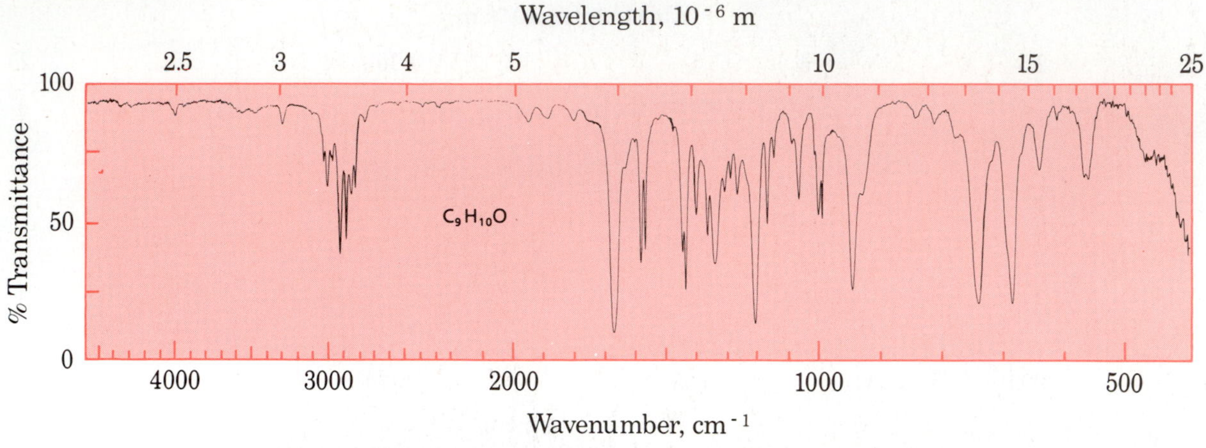

Figure 9.79 Infrared Spectrum of $C_9H_{10}O$ Unknown.

9.28 Compound **A** ($C_{16}H_{28}$) afforded compound **D** ($C_{16}H_{30}$) upon catalytic hydrogenation. Ozonolysis of A yielded two products, **B** and **C,** both of which have the same molecular weight. Deduce the structure of **A.** The infrared spectrum of compound **B** is shown below, and compound **D** has the following structure:

CH_3CH_2—(cyclohexane)—$(CH_2)_3$—(cyclopentane)

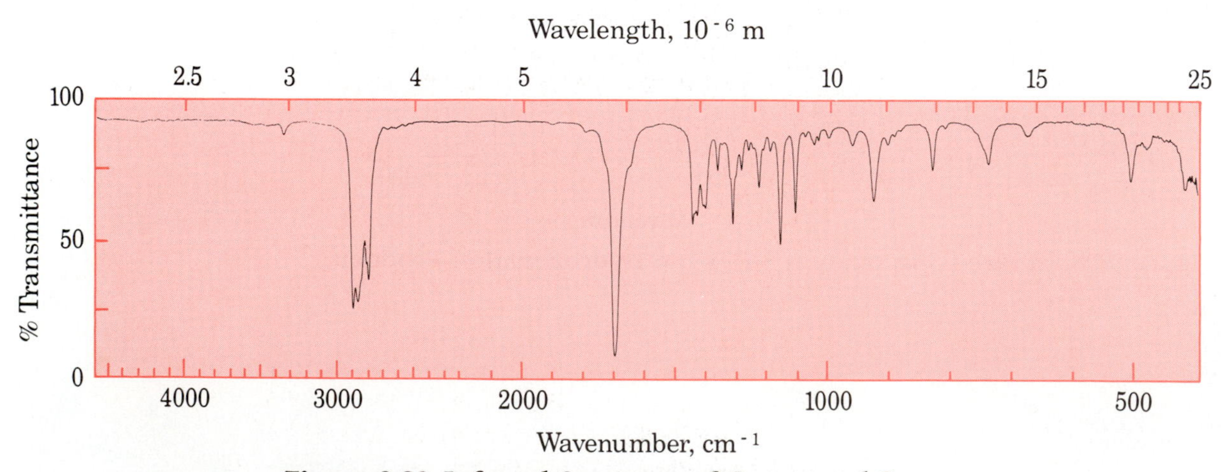

Figure 9.80 Infrared Spectrum of Compound B.

CHAPTER 10

AROMATIC COMPOUNDS

Aromatic compounds are a class of unsaturated organic compounds most of which are derivatives of benzene, and for this reason benzene will play a central role in our discussion of this class. We will refer to the structural subunit that consists of a fully conjugated six-membered ring as a *benzene ring* or an *aromatic ring*.

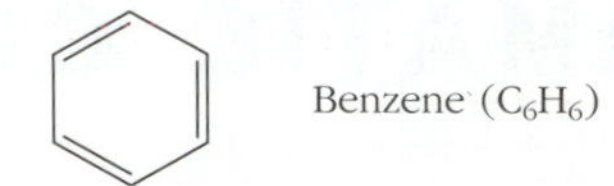

Some highly conjugated molecules that are not structurally related to benzene can also be described as aromatic on the basis of molecular orbital theory, and we will discuss these compounds in Section 10.5.

The primary source of a wide variety of aromatic compounds is coal, although petroleum refining is the major source of simple benzene derivatives. Coal itself is a mixture of high-molecular-weight structures containing many benzene rings. When coal is heated to temperatures above 1000°C in the absence of oxygen, these complex molecules decompose to produce a variety of volatile materials, along with a solid residue. The solid residue is called *coke,* and the major volatile components are *coal gas* (mainly CH_4 and H_2) and *coal tar.*

Coal tar is a high-boiling mixture of aromatic compounds, many of which have several aromatic rings and contain nitrogen and oxygen (in addition to carbon and hydrogen). For a long time coke and coal gas were important fuels, while the coal tar was considered a waste product. But the rapid development of organic chemistry and the emergence of the dye industry in the late 1800s caused coal tar to become a valuable source of aromatic compounds.

A substantial shift from coal to petroleum as the primary source of organic chemicals occurred in the first half of the twentieth century. However, the impending depletion of the world's petroleum reserves has recently given impetus to an increasing use of coal for the production of organic compounds.

In this chapter we will introduce you to the structure and reactivity of aromatic compounds. We will address a series of fundamental questions about aromatic compounds: How are they named? How are they classified? By what mechanisms do they undergo reaction? This chapter provides only an introduction to the reactivity of aromatic compounds, and we will present a more complete analysis of aromatic substitution reactions in Chapter 18.

10.1 NOMENCLATURE OF BENZENE DERIVATIVES

The general procedures for naming benzene derivatives are similar to those we have discussed for other compounds in previous chapters. However, the nomenclature of aromatic structures is complicated by the use of a large number of common names. Consequently, much of this section will simply be an introduction to the common names for a variety of benzenoid compounds.

Monosubstituted Benzenes

The key to naming a simple benzene derivative is based on whether or not the substituent contains a functional carbon atom (i.e., C—OH, C=O, etc.). If the substituent lacks such a functional carbon, the compound is named as a substituted *benzene.*

$C_6H_5—Cl$ — Chlorobenzene

$C_6H_5—CH_2CH_3$ — Ethylbenzene

$C_6H_5—NO_2$ — Nitrobenzene

The nomenclature changes, however, when a carbon-containing substituent, also called a *side chain,* includes a functional group. Such a substituent on a benzene ring is considered the *principal chain,* and we name the compound as a *phenyl* derivative of the parent structure.

$CH_3CH_2CH_2CO_2H$

Butyric acid

2-Bromocyclopentanone

$C_6H_5—CH(CH_3)—CH_2—CO_2H$

3-Phenylbutyric acid

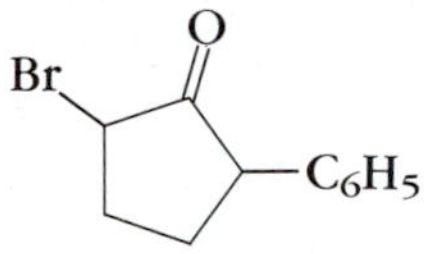

2-Bromo-5-phenylcyclopentanone

In a few cases common names are used for a group which includes the phenyl substituent and one or more carbons of the side chain. We have listed these in Table 10.1.

The following compounds illustrate the use of these common names.

$C_6H_5—CH_2—OC(=O)CH_3$

Benzyl acetate

$C_6H_5—CH(NH_2)—CH_3$

α-Phenethylamine

$C_6H_5—CH_2CH_2—NH_2$

β-Phenethylamine

TABLE 10.1 Common Names for Some Phenyl-Substituted Groups

Group	Common Name
$C_6H_5—CH_2—$	Benzyl
$C_6H_5—CH(—)—CH_3$	α-Phenethyl
$C_6H_5—CH_2—CH_2—$	β-Phenethyl
$C_6H_5—CH=CH—$	Styryl
$C_6H_5—CH=CH—CH_2—$	Cinnamyl

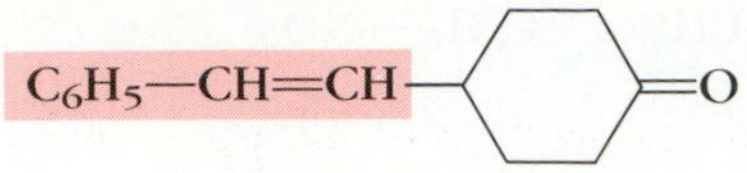

$C_6H_5—CH=CH—C_6H_9=O$ (cyclohexanone ring)

4-Styrylcyclohexanone

$C_6H_5—CH=CH—CH_2—O—C_6H_5$

Cinnamyl phenyl ether

Table 10.2 lists a number of important monosubstituted benzenes for which common names are used almost exclusively. These common names are also used as parent names, and we will show you examples of this nomenclature on the following pages.

TABLE 10.2 Some Monosubstituted Benzenes for which Common Names are Used Preferentially

Compound	Common Name	IUPAC Name
$C_6H_5—CH_3$	Toluene	Methylbenzene
$C_6H_5—CH(CH_3)—CH_3$	Cumene	Isopropylbenzene
$C_6H_5CH=CH_2$	Styrene	Vinylbenzene
$C_6H_5—CH=CH—C_6H_5$	Stilbene	1,2-Diphenylethylene
$C_6H_5—C_6H_5$	Biphenyl	Phenylbenzene
$C_6H_5—OCH_3$	Anisole	Methoxybenzene
$C_6H_5—OC_2H_5$	Phenetole	Ethoxybenzene
$C_6H_5—NH_2$	Aniline	Phenylamine
$C_6H_5—OH$	Phenol	Hydroxybenzene
$C_6H_5—C(=O)—CH_3$	Acetophenone*	Methyl phenyl ketone
$C_6H_5—CHO$	Benzaldehyde	Benzenecarbaldehyde
$C_6H_5—CH=CH—CHO$	Cinnamaldehyde	3-Phenylpropenal
$C_6H_5CO_2H$	Benzoic acid†	Benzenecarboxylic acid
$C_6H_5CH=CH—CO_2H$	Cinnamic acid	3-Phenylpropenoic acid

*Other alkyl phenyl ketones are named similarly (Section 13.1).

†Carboxylic acid derivatives (esters, amides, etc.) are named according to the procedures in Section 16.1.

Disubstituted Benzenes

Just as we have shown for the case of monosubstitution, disubstituted compounds are named as derivatives of benzene if the substituents contain no functional groups. In agreement with the nomenclature of other cyclic compounds (Section 3.1), the ring carbon bearing the substituent with the lowest alphabetical ranking is designated as C-1.

1,2-Dichlorobenzene (*o*-dichlorobenzene) | 1-Bromo-3-chlorobenzene (*m*-bromochlorobenzene) | 1,4-Diethylbenzene (*p*-diethylbenzene)

The preceding examples also illustrate an alternative to numbering the substituents.

> **In disubstituted benzenes the prefixes *o*- (for *ortho*), *m*- (for *meta*), and *p*- (for *para*) indicate 1,2-, 1,3-, and 1,4- disubstitution, respectively.**

These prefixes are commonly used to indicate the positions of substituents on *benzene* derivatives, but they are *never* used for naming other cyclic compounds. When the principal group is located on a side chain, a disubstituted benzene can be named as the appropriate (*ortho, meta,* or *para*) phenyl derivative. The name of the substituted phenyl group (also called an *aryl* group to indicate an *aromatic* substituent) is often enclosed in parentheses to clearly differentiate it from the rest of the structure.

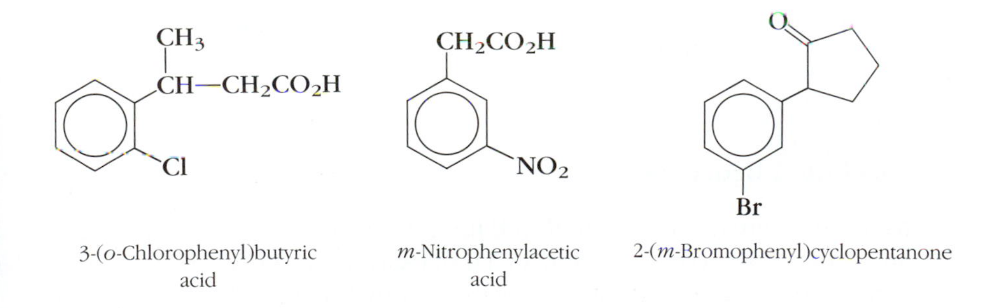

3-(*o*-Chlorophenyl)butyric acid | *m*-Nitrophenylacetic acid | 2-(*m*-Bromophenyl)cyclopentanone

When one of the two substituents is a methyl group, the compound may be named as a *tolyl* derivative (i.e., as a derivative of toluene).

CH_3 — O—$CH_2CH_2CH_3$

Propyl *m*-tolyl ether

A disubstituted benzene can also be named as a derivative of a monosubstituted benzene when the latter has a common name (see Table 10.2).

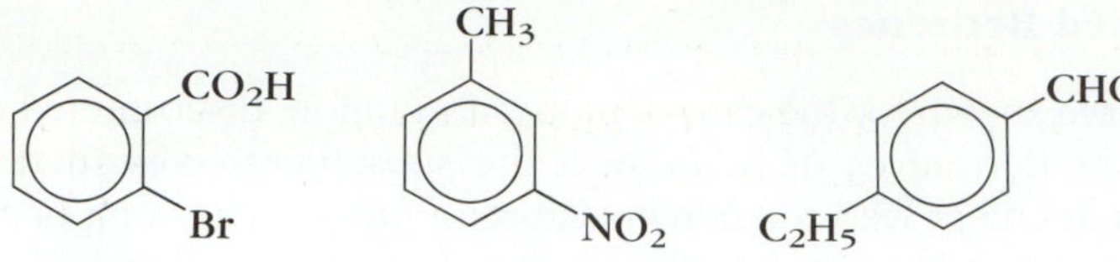

o-Bromobenzoic acid *m*-Nitrotoluene *p*-Ethylbenzaldehyde

If both substituents correspond to monosubstituted benzenes with common names, the parent name corresponds to the substituent with the higher-priority functional group according to the ranking shown in the following box:

Functional Group Priorities for Nomenclature

Carboxylic acids > carboxylic acid derivatives (esters, etc.) > aldehydes > ketones > alcohols and phenols > amines > ethers

We will discuss these priorities more fully in Section 16.1, but for the time being they allow us to name the following compound as a derivative of an acid rather than as a derivative of an amine.

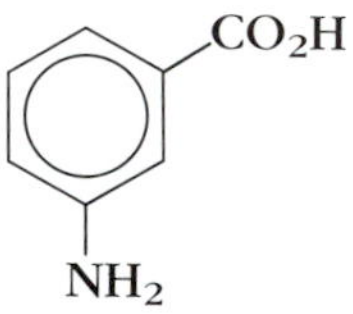

m-Aminobenzoic acid

Many disubstituted benzenes have common names; some of the more frequently encountered ones are summarized in Table 10.3.

Polysubstituted Benzenes

Systematic names are ordinarily employed for benzene derivatives having three or more substituents. The common name of a monosubstituted benzene is frequently used as the parent name, as illustrated by the first two of the following examples.

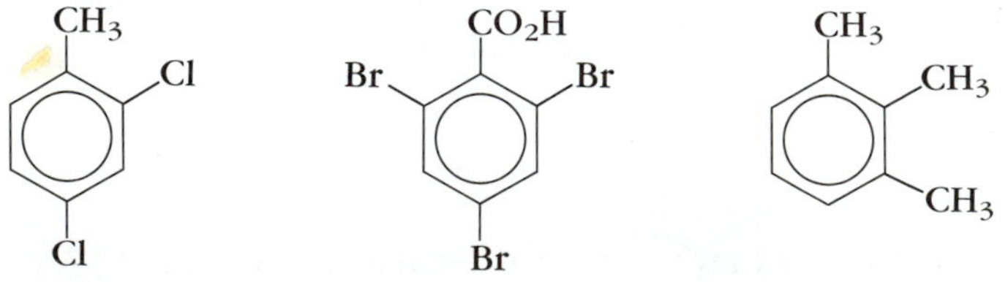

2,4-Dichlorotoluene 2,4,6-Tribromobenzoic acid 1,2,3-Trimethylbenzene

Common names are encountered less frequently for polysubstituted benzenes than for mono- and disubstituted derivatives, and we will not consider them here.

TABLE 10.3 Some Disubstituted Benzenes for which Common Names are Frequently Used*

CH_3, CH_3 *o*-Xylene (also *m*-, *p*-)	CO_2H, CH_3 *o*-Toluic acid (also *m*-, *p*-)	CO_2H, OCH_3 *o*-Anisic acid (also *m*-, *p*-)
CH_3, NH_2 *o*-Toluidine (also *m*-, *p*-)	OCH_3, NH_2 *o*-Anisidine (also *m*-, *p*-)	CH_3, OH *o*-Cresol (also *m*-, *p*-)
CHO, OCH_3 *o*-Anisaldehyde (also *m*-, *p*-)	CO_2H, OH Salicylic acid	
CO_2H, CO_2H Phthalic acid	CO_2H, CO_2H Isophthalic acid	HO_2C, CO_2H Terephthalic acid
OH, OH Pyrocatechol	OH, OH Resorcinol	HO, OH Hydroquinone

*When all three isomers use the same common name, only the *ortho* is shown.

Fused Polycyclic Aromatic Hydrocarbons

When two adjacent atoms of one ring are shared by a second ring, the two rings are described as **fused.** A variety of fused polycyclic aromatics are known, and many of them can be isolated from coal tar. Some important examples are listed in Table 10.4.

TABLE 10.4 Representative Fused Polycyclic Aromatic Hydrocarbons

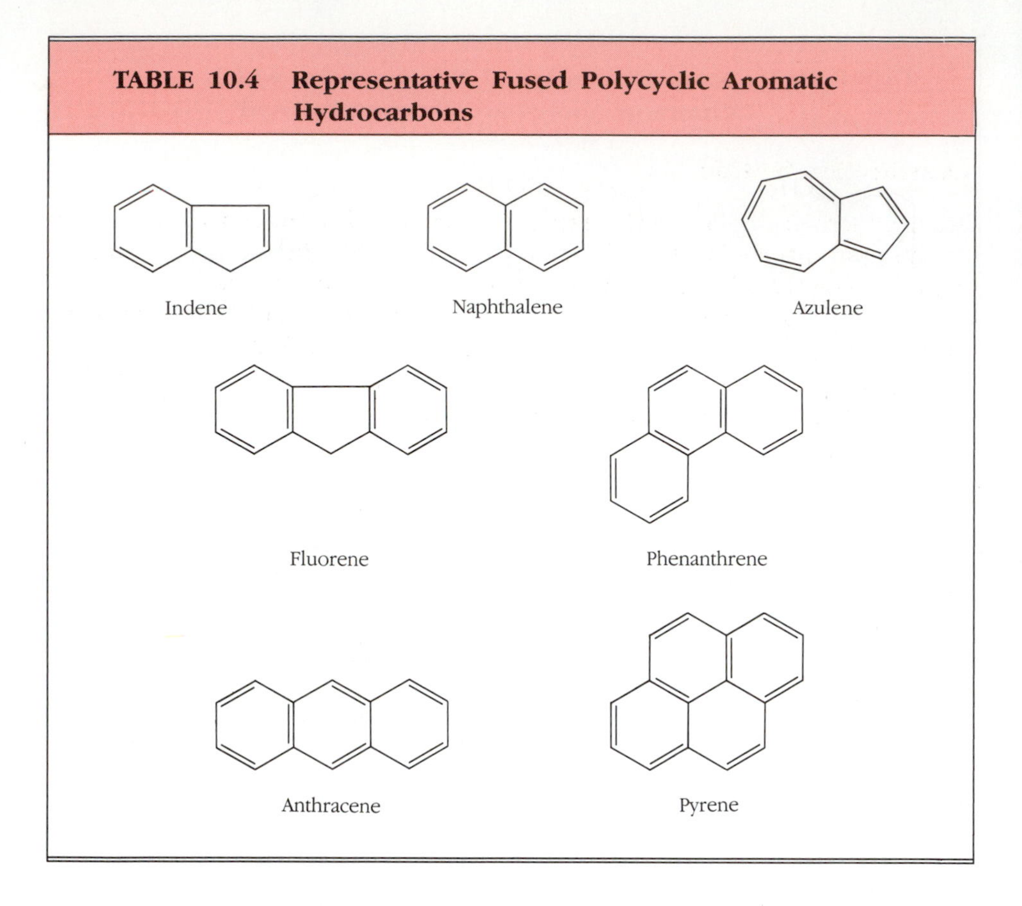

Although the mechanism of their action is not yet fully established, a number of the fused aromatic systems with more than four rings are carcinogenic, that is, they induce carcinoma (cancer). Laboratory and industrial workers who might come in contact with such compounds must take special precautions to avoid skin contact and accidental ingestion or inhalation of these aromatic hydrocarbons. Small quantities of these compounds are formed in high-temperature pyrolysis reactions as by-products of the combustion of organic materials, and some of them are prime suspects as possible carcinogens in cigarette smoke.

EXERCISE 10.1

Provide systematic names for the compounds in Table 10.3.

EXERCISE 10.2

Draw the following compounds:

(a) *m*-Xylene
(b) *p*-Toluic acid
(c) *m*-Anisic acid
(d) *m*-Toluidine
(e) *p*-Anisidine
(f) *p*-Cresol
(g) *m*-Anisaldehyde

10.2 AROMATICITY

Classification by Odor

Aromatic compounds, as the name suggests, were originally placed in this category on the basis of their *odor*. In the early days of organic chemistry only naturally occurring compounds were available for study, and those with characteristic (and often quite pleasant) aromas were designated as aromatic. Figure 10.1 illustrates a variety of fairly simple organic structures that fit this classification.

However, not all the pleasant-smelling compounds known to early organic chemists contain a benzene ring. Carvone and menthol, for example, were at first considered "aromatic" because of their odors.

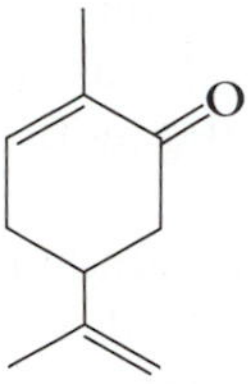

O

Carvone
(spearmint)

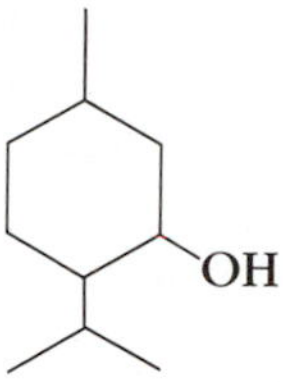

OH

Menthol
(peppermint)

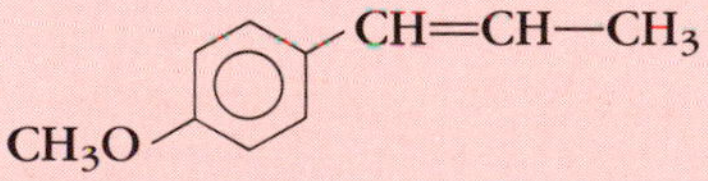

CH_3O CH=CH—CH_3

Anethole
(anise)

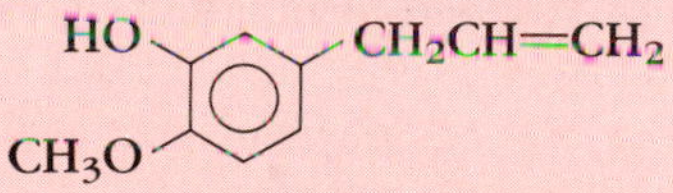

HO CH_2CH=CH_2
CH_3O

Eugenol
(cloves)

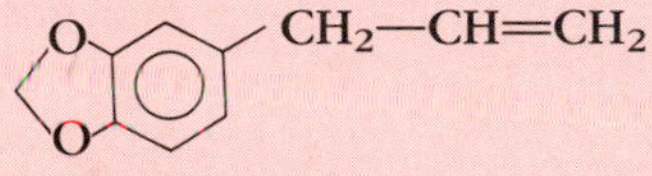

O O CH_2—CH=CH_2

Safrole
(sassafras)

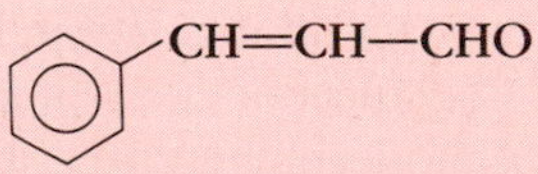

CH=CH—CHO

Cinnamaldehyde
(cinnamon)

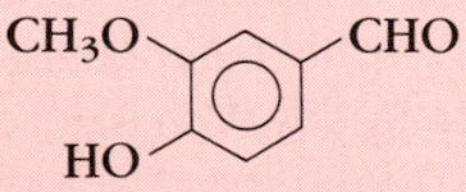

CH_3O CHO
HO

Vanillin
(vanilla)

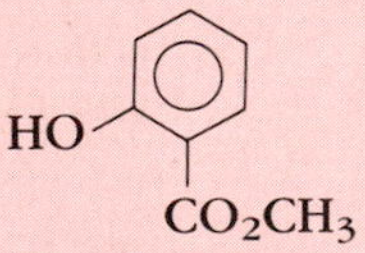

HO CO_2CH_3

Methyl salicylate
(wintergreen)

Figure 10.1 Some Naturally Occurring Organic Compounds with Pleasant Aromas.
Each of the structures shown contains a benzene ring. The characteristic spice or flavor associated with each compound is shown in parentheses.

Classification by Composition

In the early 1800s some chemists began to recognize the limitations of classifying different compounds by odor. Improved methods for combustion analysis led to classification based on *composition*. "Aromatic compounds" were found to have low hydrogen-carbon ratios, and this served as a useful, if somewhat vague, criterion for aromaticity.

Classification by Structure

The next stage in classification of organic compounds was based on relationships in *molecular structure*. Benzene was first obtained in 1815 from the pyrolysis of whale oil by the English chemist Michael Faraday. By the 1830s the molecular formula C_6H_6 had been determined, and benzene was classified as aromatic because of its low hydrogen-carbon ratio. Benzene was subsequently obtained as a product in other chemical reactions, and by the 1840s chemists had learned how to obtain substantial amounts of benzene from coal tar. It was not until 1865, however, that Friedrich August Kekulé (Germany) showed the fundamental importance of benzene.

Kekulé observed that all known aromatic compounds not only had low hydrogen-carbon ratios but also had *homologs* (i.e., related compounds with molecular formulas differing by CH_2, as in the case of alkanes). A CH_2 unit does not have a low hydrogen-carbon ratio, so this indicated that these molecules contained some *other* structural unit which caused them to be aromatic. The idea of an aromatic subunit was supported by the observation that reactions of aromatic compounds nearly always yielded products that were also aromatic. Kekulé then provided the key to the puzzle with the observation that all known aromatic compounds had a molecular formula with *at least six carbons*. From that point he was able to propose a six-carbon subunit as the *aromatic nucleus*. Over a period of 40 years, therefore, benzene had progressed from a mere curiosity to the parent structure for all aromatic compounds.

Classification by Reactivity

Almost as soon as the structural relationship between aromatic compounds was recognized by Kekulé, another German chemist, Emil Erlenmeyer, proposed that compounds could be classified according to their chemical reactivity. In marked contrast to other compounds with somewhat low hydrogen-carbon ratios (particularly alkenes), aromatic compounds undergo *substitution* rather than *addition* reactions. Similarly, aromatic compounds are resistant to oxidation by $KMnO_4$, whereas alkenes react readily with that reagent. Classification according to reactivity was an important step in the development of modern chemistry, and you will see in Chapter 18 that aromatic compounds do exhibit unique reactivity.

Modern Classification

Classification of aromaticity by both structure and reactivity has continued in modern chemistry. In most cases the two systems agree, and compounds that contain the six-carbon aromatic nucleus exhibit the chemical reactivity expected for aromatic compounds. But scientists found new questions to answer as modern chemical theory developed in the twentieth century. *Why* do aromatic compounds exhibit unusual reactivity? And *how* is this reactivity related to molecular structure? Such questions could not be answered prior to the development of quantum mechanics and molecular orbital theory in the 1920s and 1930s. The answers provided by these theories (see Sections 10.5 and 10.6) resulted in yet another system of classifying compounds as aromatic according to the properties of their molecular orbitals. Once again this new system usually (but not always) results in the same classification as the earlier methods based on structure and on reactivity.

What, then, is the current status of classification? We have no simple answer. An international symposium on aromaticity was held in 1970 and there appeared to be as many definitions of "aromaticity" as there were participants in the conference. But these arguments over the fine points of a definition ultimately make no difference to you as a student. Our goal in this book is to help you learn the *relationships* between structure and reactivity. In nearly all our discussions the concept of the benzene ring as the aromatic nucleus will provide a satisfactory basis for correlating molecular structure with chemical reactivity. Exceptions to this pattern are infrequent, and we will treat them as special cases (Section 10.5).

10.3 THE STRUCTURE OF BENZENE

Despite the ease with which we now draw a hexagon and three double bonds, the structure of benzene posed a formidable problem to chemists for over 100 years after its discovery in 1825. When Kekulé proposed benzene as the basic subunit of aromatic compounds, no satisfactory structure could be drawn. Chemists at that time were only beginning to consider "bonding" between atoms. Remember that the discovery of the electron did not take place until the late 1890s, and theories about the role of electrons in chemical bonds were not developed until the beginning of the twentieth century. Consequently, the structural drawings of the 1800s did not have the same meaning that we give to modern drawings.

By the mid 1830s the molecular formula of benzene had been established as C_6H_6. According to the notation we developed in Section 7.11, this corresponds to four degrees of unsaturation, and many isomeric structures could be drawn that have various combinations of double bonds and triple bonds (Figure 10.2).

Some important clues came from substitution reactions of benzene. *Mono*substitution always yielded a single compound (in other words, no *isomeric* monosubstituted benzenes could be detected), and this strongly suggested a very

$CH_3{-}C{\equiv}C{-}C{\equiv}C{-}CH_3$ $CH_3{-}C{\equiv}C{-}CH_2{-}C{\equiv}CH$

$CH_3{-}CH_2{-}C{\equiv}C{-}C{\equiv}CH$ $HC{\equiv}C{-}CH_2{-}CH_2{-}C{\equiv}CH$

$CH_3CH{=}C{=}CH{-}C{\equiv}CH$ $CH_2{=}CH{-}CH{=}CH{-}C{\equiv}CH$

$CH_2{=}C{=}CH{-}CH_2{-}C{\equiv}CH$ $CH_2{=}CH{-}C{\equiv}C{-}CH{=}CH_2$

$CH_2{=}C{=}CH{-}C{\equiv}C{-}CH_3$

Figure 10.2 Some C_6H_6 Unsaturated Hydrocarbons Once Considered Candidates for the Structure of Benzene.

symmetrical structure. The same product was apparently formed no matter which of the six hydrogens of benzene was replaced. Similarly, *di*substitution of benzene never afforded more than three isomeric products. This evidence led Kekulé to propose in 1865 that benzene had a *cyclic* structure. A vigorous debate about the structure took place among chemists during the following decade. Conflicting and occasionally erroneous experiments prevented an early resolution of the problem, and even Kekulé, who favored structure **1,** shown in Figure 10.3, did not rule out the other structures depicted in that figure.

Kekulé advocated structure **1** because all six C—H groups are equivalent, and this structure would afford only a single monosubstitution product. The three possible disubstitution products are those in which the substituents have 1,2-, 1,3-, and 1,4-relationships.

Several of the structures in Figure 10.3 might appear to be a bit strange to you. From our viewpoint of modern bonding theory, it is not clear what all the "bonds" mean in **2** and **3.** Apparently such structures were merely attempts to depict the six carbon atoms in an arrangement that permitted each to have four bonds and was at the same time highly symmetrical. Bonding theory was in its infancy, and the chemists of the time saw no problems in drawing very long or short bonds between atoms.

Despite the fact that Kekulé's proposal for the structure of benzene has since been shown to be correct, several pieces of evidence suggested in the 1860s that it was not. Another German chemist, A. Ladenburg, objected that two different isomers should be possible if the substituents were on adjacent carbons. The disubstituted isomers are shown on page 505. In **1a** the two adjacent substituents are on carbons joined by a *double* bond, whereas in **1d** they are on carbons joined by a *single* bond.

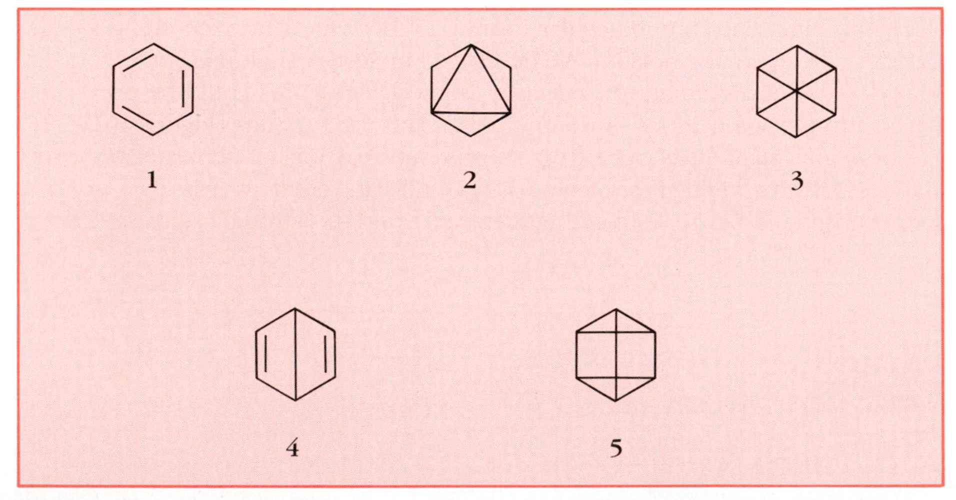

Figure 10.3 Possible Structures for Benzene Considered by Kekulé in 1869.
He considered **1, 2,** and **3** most likely and favored **1** on the basis of the available evidence. He felt that **4** and **5** were less likely candidates because they were less symmetrical.

X Y X Y X Y Y X

1 1a 1b 1c 1d

Ladenburg argued vigorously that formula **5** in Figure 10.3 was a better candidate for the structure of benzene. He did not use the planar structure suggested in Figure 10.3 but instead drew a three-dimensional "prism formula."

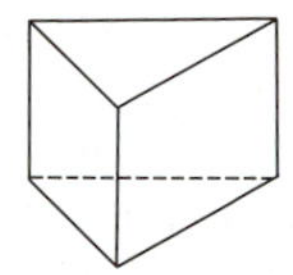

Ladenburg's proposal for the structure of benzene

Structure **5** is highly symmetrical, and all six CH groups are equivalent. Consequently only one monosubstituted derivative is possible, which is in agreement with experimental observation. In contrast to the situation with structure **1,** only *three* disubstitution products can be drawn for **5.**

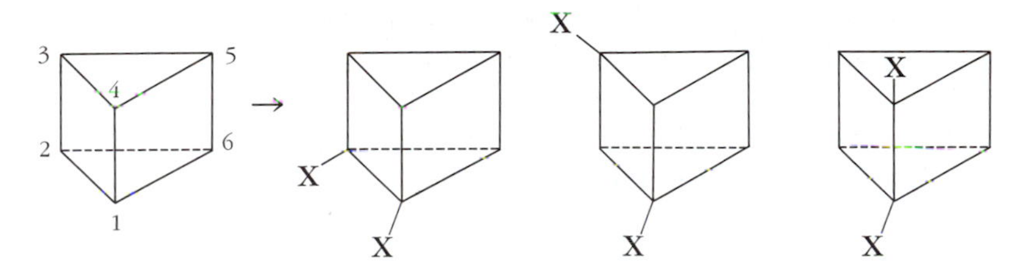

However, the Ladenburg formula was not without its flaws. In 1876 van't Hoff published an article showing that 1,2 disubstitution of **5** would afford *two* isomeric structures if the substituents are not identical. (By modern terminology these are enantiomeric.)

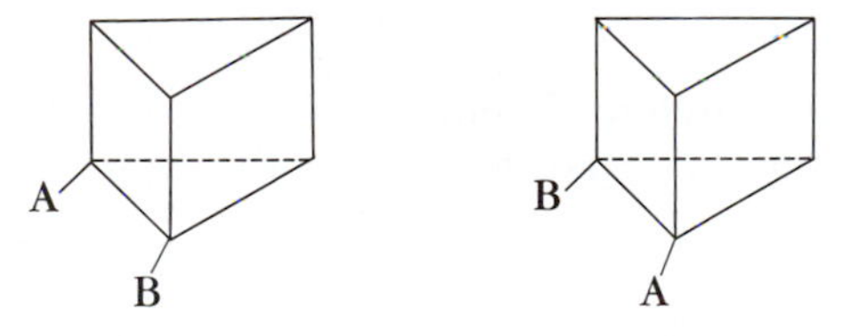

At the time van't Hoff's ideas were still new and radical. His revolutionary proposal for tetrahedral carbon was only two years old, and his arguments did not end the debate about the structure of benzene.

Kekulé responded to the problem of disubstitution products by clarifying the meaning of structure **1**. He argued that his drawing was not intended to represent a different kind of bonding between C-1 and C-2 (single bond) as compared with C-1 and C-6 (double bond). Kekulé was merely trying to indicate the idea that each carbon should have a total of four bonds, but he believed that the interaction between each pair of adjacent carbon atoms was the same.

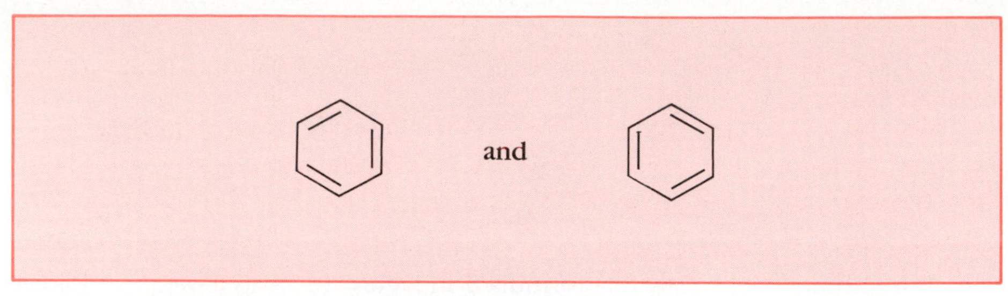

Figure 10.4 Kekulé's Two Formulas for Benzene.
He considered the actual structure to be a composite of the two drawings.

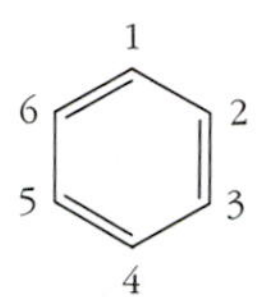

Consequently, he proposed the use of *two* formulas, with the true structure of benzene corresponding to a composite, or average, of the two.

Kekulé's interpretation of the structure of benzene is remarkably similar to our modern description using two resonance contributors (see Section 10.5), but his explanation predated the theory of resonance by more than 50 years. Nevertheless, we should not underestimate the importance of the concept that was introduced by Kekulé, the use of two structural formulas when a single drawing cannot adequately depict the true structure of a molecule.

Contemporary organic chemists still use the two Kekulé formulas to represent benzene, but a double-headed arrow is drawn between them to indicate that each structure is only one resonance contributor to the actual structure.

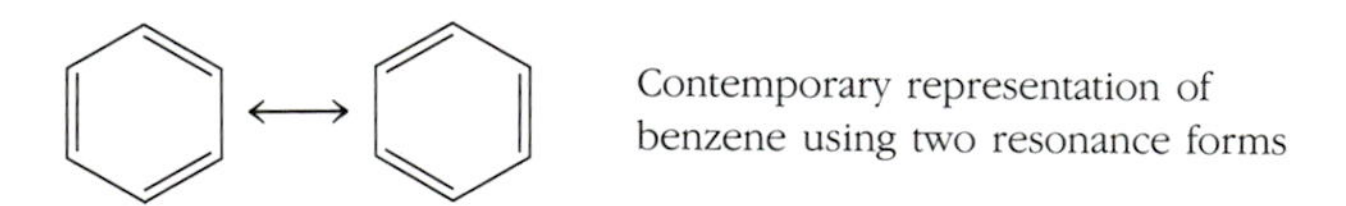

Contemporary representation of
benzene using two resonance forms

Modern chemists also frequently employ another structural formula for benzene that has a circle in the center to indicate the equivalence of all six carbon atoms:

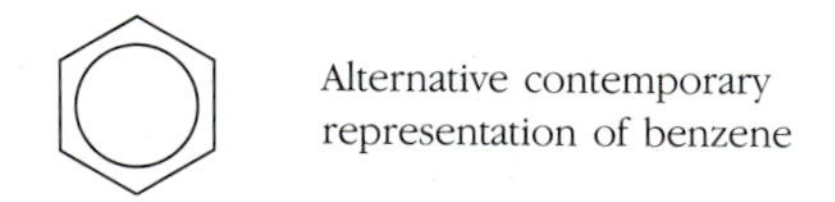

We will sometimes utilize the latter type of drawing, but it is often inadequate when interactions between substituents must be analyzed. In those situations and in the analysis of reaction mechanisms we will use the Kekulé formulas. Since you are now well aware that the true structure of benzene is a composite of the two formulas, we will usually not draw both of them. When we do use a single drawing, however, you must bear in mind that we have shown (for convenience) only one of the two possible resonance contributors.

Chemists of the 1800s had great difficulty relating the structure of benzene to its unusual reactivity. If structure **1** (with its three double bonds) were correct, why did benzene not undergo the typical addition reactions of alkenes? If **5** (or one of the other structures in Figure 10.3) were the correct alternative, why was benzene so stable? Baeyer did not postulate his strain theory (Section 3.2) until 1885, but it was certainly clear to chemists of the time that three- and four-membered rings were not common.

Despite the controversy, Kekulé's proposal gained increasing acceptance by other chemists. Final resolution of the question occurred in the 1930s, when X-ray and electron diffraction studies demonstrated beyond any doubt that benzene has a planar, hexagonal structure in which all six carbon-carbon bond lengths are the same. Figure 10.5 summarizes results obtained from a highly accurate diffraction study carried out in 1956.

As a footnote to our story about the structure of benzene, we can return to the alternatives that were considered by Kekulé in 1869 (Figure 10.3). Formulas **2** and **3** do not correspond to any reasonable modern structures composed of six C—H groups, but compounds with the three other structures in the figure have all been prepared. Structure **1** of course, is benzene. Structure **4** (commonly called *Dewar benzene* because it had been discussed by the British chemist J. Dewar in the 1860s) was first prepared by E. Van Tamelen and S. Pappas at the University of Wisconsin in the early 1960s. *Prismane* (structure **5**), named according to its shape, was synthesized by T.J. Katz and N. Acton at Columbia University in 1973. Both these isomers are considerably less stable than benzene, in part because they are not aromatic and in part as a consequence of the strained rings they contain.

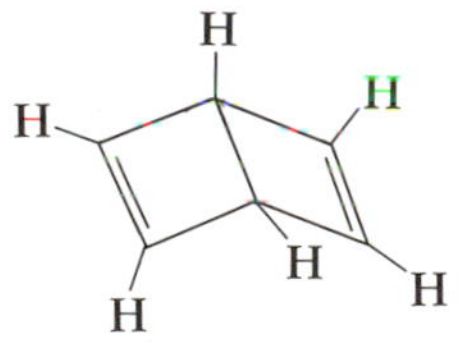

Dewar benzene (**4**)

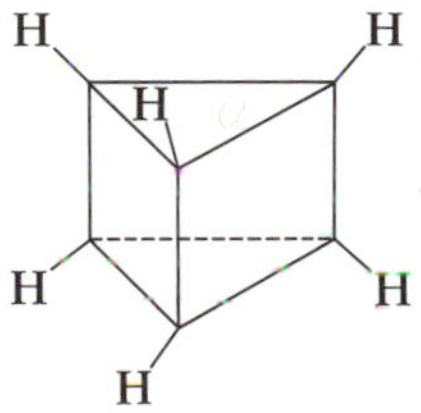

Prismane (**5**)

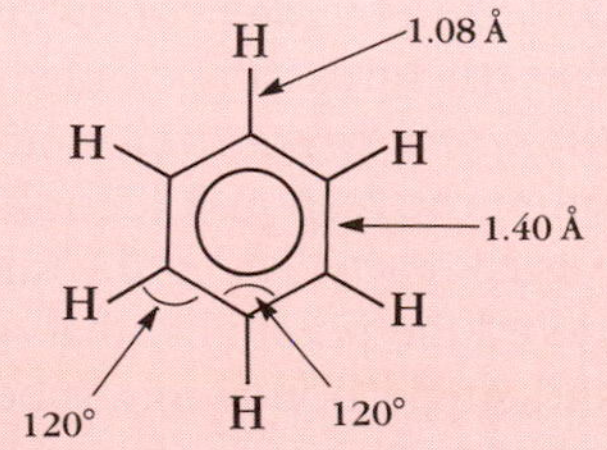

1.08 Å

1.40 Å

120°

120°

Figure 10.5 Structural Parameters for Benzene as Determined by x-Ray Crystallography.

EXERCISE 10.3

An erroneous report in the 1860s of two isomeric forms of pentachlorobenzene was consistent with structure **4** (Figure 10.3) for benzene.

(a) How many dichloro isomers would be possible for **4**?
(b) Is structure **4** consistent with the experimental observation of three disubstituted benzenes?

EXERCISE 10.4

(a) How many different dichloro products could be formed from substitutions of hydrogens in prismane (structure **5**)? Draw them.
(b) Draw the isomers that could be formed by introduction of one bromine and one chlorine.
(c) Explain the different results for (a) and (b).

EXERCISE 10.5

How many isomers could be formed by introduction of one chlorine and one bromine in Dewar benzene (structure **4**)? Draw them.

EXERCISE 10.6

A total of five hydrocarbons with the formula C_6H_6 have been prepared which contain six C—H groups. Three of these (**1, 4,** and **5**) are shown in Figure 10.3. Deduce the structures of the other two isomers.

10.4 ALTERNATING SINGLE AND DOUBLE BONDS: A CRITERION FOR AROMATICITY?

Toward the end of the nineteenth century Kekulé's proposed structure for benzene was gaining general acceptance. Having established the structure for the parent aromatic compound, chemists started considering *why* compounds are aromatic. Several features of the benzene structure are unusual, and these provided chemists with possible criteria for aromaticity. The molecule is cyclic and the carbon atoms lie at the corners of a regular hexagon, so all six C—H groups are equivalent. In addition the carbons are connected by *alternating* single and double bonds, and this means that two equivalent structures can be drawn. Since each carbon is part of a double bond, there is *continuous unsaturation* around the ring.

Could these structural features be responsible for benzene's *aromatic character* (i.e., its high stability and unusual reactivity)? At the turn of the century the German chemist R. Willstätter considered the possibility that *all* cyclic compounds consisting of alternating double and single bonds might be aromatic. He therefore set about synthesizing the series of compounds shown in Figure 10.6. Two formulas analogous to those drawn by Kekulé for benzene can be written for each of the compounds.

Willstätter was unable to prepare cyclobutadiene. When he attempted to carry out a double elimination of HBr from 1,2-dibromocyclobutane, no product was

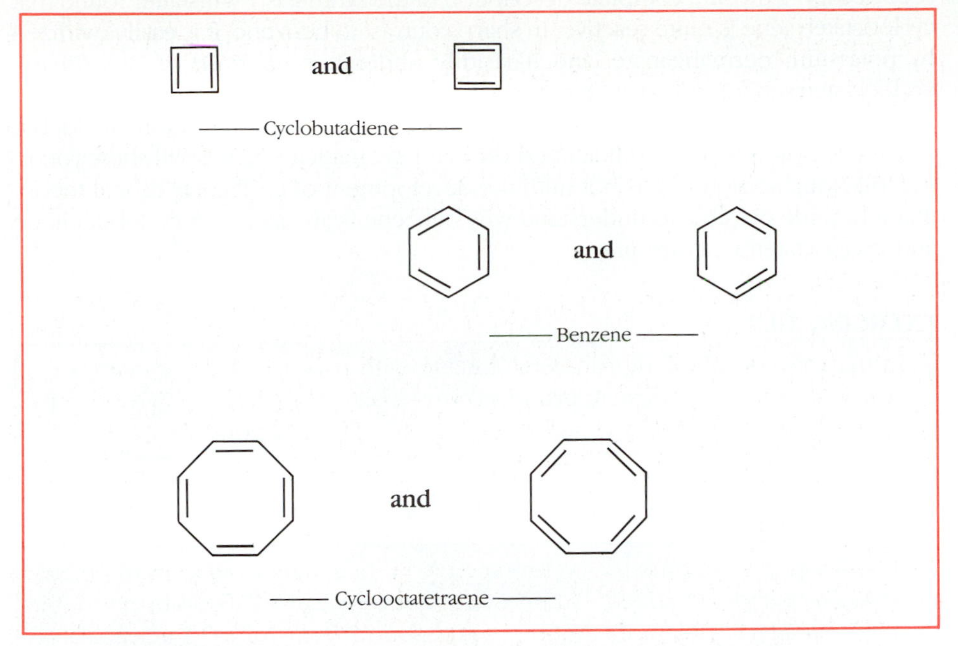

Figure 10.6 Possible Aromatic Compounds?
Cyclic compounds with alternating single and double bonds. For each compound two formulas could be drawn that differ only in the single-double bond sequence.

isolated with the molecular formula C_4H_4.* Willstätter's failure to prepare cyclobutadiene provided no direct evidence about its aromatic properties, but his experiments certainly suggested that the compound does not exhibit any unusual stability.

Conclusive evidence on the question of alternating single and double bonds was obtained when Willstätter prepared cyclooctatetraene. His synthesis began with the alkaloid pseudopelletierine, which was known to have an eight-membered ring with a nitrogen bridge. A series of elimination reactions was employed to introduce the four double bonds.

O, NCH_3 ⟶ ⟶

Pseudopelletierine Cyclooctatetraene

*Willstätter obtained a small amount of acetylene in this reaction. This suggests the possibility that some cyclobutadiene had been formed but subsequently fragmented to acetylene.

Br, Br $\xrightarrow{KOH}$ [ ] ⟶ 2 HC≡CH (low yield)

(See Section 17.5 for a complete description of the synthesis.) Willstätter found that cyclooctatetraene is quite reactive. In sharp contrast to benzene, it is easily oxidized by potassium permanganate, and it readily undergoes *addition* of Br_2. Clearly cyclooctatetraene is not aromatic.

Willstätters's results with cyclooctatetraene proved that alternation of double and single bonds is not a sufficient criterion for aromaticity. As we will show you in the following section, it was not until the development of molecular orbital theory that it became possible to understand why benzene is aromatic, yet cyclobutadiene and cyclooctatetraene are not.

EXERCISE 10.7

In the reaction of 1,2-dibromocyclobutane with base, several pathways are available which could lead to products other than cyclobutadiene. What other compounds would be expected as products in this reaction?

EXERCISE 10.8

In Willstätter's synthesis of cyclooctatetraene, he carried out a reaction in which bromine was added to 1,3,5-cyclooctatriene to yield 5,8-dibromo-1,3-cyclooctadiene. Explain this mode of addition in terms of the intermediate that is generated in the reaction.

10.5 QUANTUM MECHANICS, MOLECULAR ORBITAL THEORY, AND AROMATICITY

The emergence of quantum mechanics and molecular orbital theory in the first third of the twentieth century had a major impact on our understanding of aromatic compounds. In this section we will present a brief overview of the ways in which these theoretical developments influenced organic chemistry.

Resonance

As quantum mechanics advanced beyond the case of the hydrogen atom to more complex species, scientists had to find ways of analyzing the interactions between three or more particles. The helium atom, for example, has a nucleus and two electrons, and it is extremely difficult to solve the equations which describe the behavior of all three particles simultaneously. In 1926 the German theoretician Werner Heisenberg discovered that a good mathematical description of the helium atom could be achieved with a *combination* of two mathematical functions (wavefunctions) even though neither of the individual wavefunctions corresponded to a correct description by itself. He noticed a strong similarity between his equations for the helium atom and the classical mathematics that described the resonance interaction between two tuning forks vibrating at the same frequency. Heisenberg applied the term resonance to the quantum mechanical situation in which two "structures" (neither of which is correct by itself) are combined to give a good description of the atom or molecule being considered.

Heisenberg's idea of resonance was then adopted by organic chemists as a way to describe molecules and ions that cannot be adequately represented by a single conventional drawing. We use two or more *resonance contributors* to describe the molecule, and the true structure is a sort of average (a *resonance hybrid*) of the individual contributors.

A second consequence of Heisenberg's idea of resonance is *resonance stabilization.* Heisenberg found that a lower (and more accurate) energy was calculated for the helium atom when he used two wavefunctions than when he used either of the individual wavefunctions by itself. This energy difference was called the **resonance energy.** The idea of resonance energy was also adopted by organic chemists to explain the unusual stability of a molecule such as benzene. This is illustrated in Figure 10.7.

If the benzene molecule were simply a six-membered ring containing three noninteracting double bonds, its heat of hydrogenation (to form cyclohexane) would be about three times that of cyclohexene. In 1935 George Kistiakowski (Harvard) carried out the appropriate experiments and found that the heat of hydrogenation of benzene is only 50 kcal/mol, much less than three times the value of almost 29 kcal/mol found for cyclohexene. The difference of 36 kcal/mol is the resonance energy of benzene.

We have used resonance extensively throughout this book and we shall continue to do so because it provides a remarkably reliable basis for predicting and interpreting chemical reactions. Yet the attempted use of simple resonance arguments to predict aromatic character is a failure. We draw benzene by writing two resonance forms that are related by a double-headed arrow, so it would seem that we should be able to do the same with other unsaturated hydrocarbons such as cyclo-

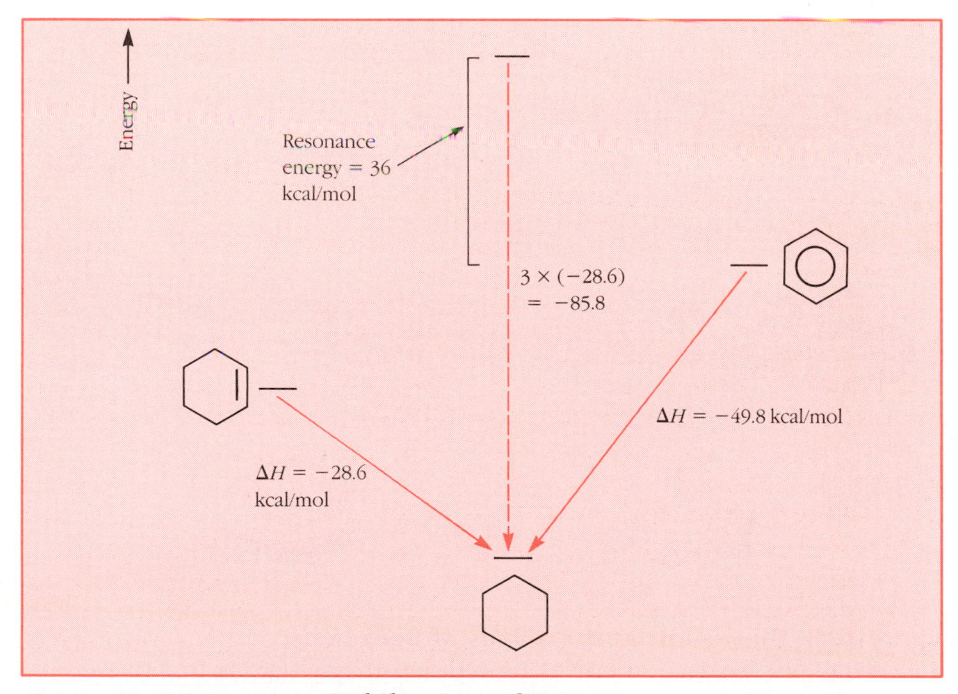

Figure 10.7 Resonance Stabilization of Benzene.
Hydrogenation of cyclohexene releases 28.6 kcal/mol of heat, but hydrogenation of benzene liberates less than three times this quantity of heat. The difference corresponds to the "resonance energy" of benzene.

butadiene and cyclooctatetraene. The latter two compounds, however, are *not* aromatic; they exhibit no unusual stabilization, and it has been shown that cyclooctatetraene has four distinct single bonds. Resonance provides an excellent device for understanding and interpreting chemical reactions, but an understanding of *aromaticity* requires that we turn to molecular orbital theory.

Molecular Orbital Theory

After 100 years of uncertainty about the structure of benzene and the reasons for its unusual stability, molecular orbital theory finally provided the answer. Figure 10.8 shows the six molecular orbitals of the π system of benzene. (The remaining molecular orbitals are not relevant to our discussion, so we will not draw them.)

A total of six electrons reside in the π orbitals of benzene (one for each carbon in the ring, or two for each of the three "double bonds"). Consequently, only the three lowest-energy orbitals are occupied. These three orbitals are constructed from atomic orbitals with mostly in-phase (bonding) interactions between the atomic orbitals on adjacent carbons. In π_1 and π_3 there are no out-of-phase (antibonding) interactions, and π_2 has only two. Hence the occupied π molecular orbit-

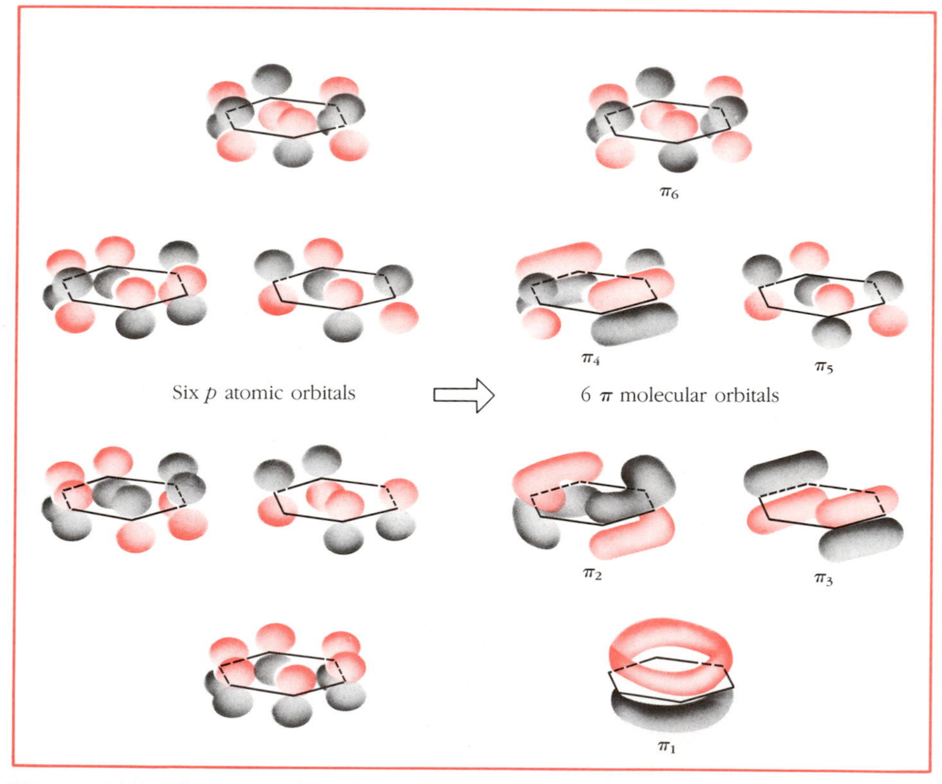

Figure 10.8 The π Molecular Orbitals of Benzene.
The orbitals are arranged in order of increasing energy starting from the lowest-energy obital, for which all interactions between atomic p orbitals on adjacent carbons are in phase. Only the three lowest orbitals are occupied, and the great stability of benzene results from the low energy of these three bonding molecular orbitals (π_1–π_3).

als are very low in energy, and this is responsible for the great stability of benzene. In other words, aromaticity results from a π system in which only low-energy molecular orbitals are occupied.

Does molecular orbital theory also explain why cyclobutadiene and cyclooctatetraene are *not* aromatic? The answer is a definite yes. Figures 10.9 and 10.10 show the combinations of *p* atomic orbitals that would produce the π molecular orbitals for these two compounds if they were planar and symmetrical.

Just as you saw in the case of benzene, the lowest π molecular orbital for each compound results from combinations of *p* orbitals that are all in phase (i.e., bonding). But orbitals that are essentially *nonbonding* (an equal number of in-phase and out-of-phase interactions) would also be occupied for these two compounds. Cyclobutadiene has four electrons in the π system and only two can occupy the lowest-energy orbital π_1. Consequently, two electrons would have to be in the nonbonding orbitals π_2 and/or π_3. Similarly, the π system of cyclooctatetraene has eight electrons. Each of the low-energy orbitals, π_1–π_3, would be doubly occupied, but the last two electrons would have to occupy the nonbonding level of π_4 and π_5.

With π electrons occupying orbitals at the *nonbonding* level, we would certainly *not* have an aromatic system. In fact the out-of-phase interactions of these orbitals are sufficiently unfavorable that neither cyclobutadiene nor cyclooctatetraene has a structure corresponding to a regular polygon. For each of these

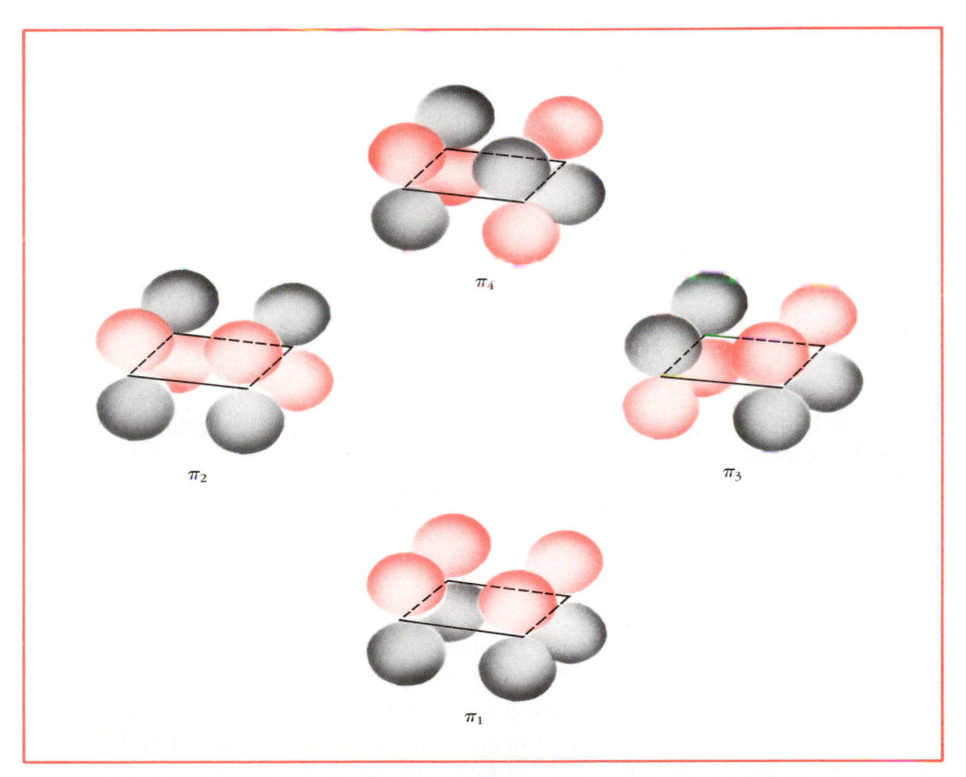

Figure 10.9 The π Molecular Orbitals for a Hypothetical Planar, Square Cyclobutadiene.
The drawings show the various combinations of atomic *p* orbitals corresponding to the π molecular orbitals. There are four electrons in the π system, and two electrons must occupy the nonbonding level of π_2 and π_3.

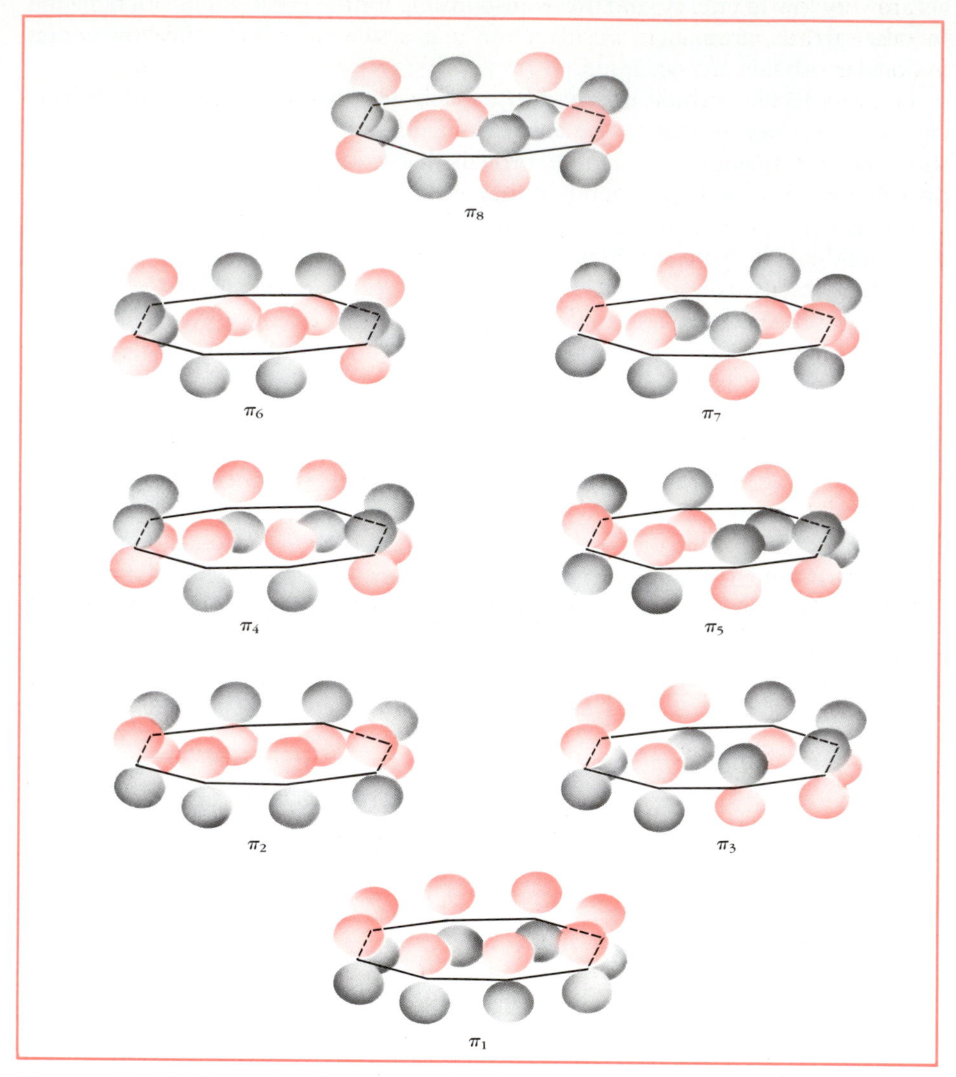

Figure 10.10 The π Molecular Orbitals for a Hypothetical Planar, Octagonal Cyclooctatetraene.
The drawings show the various combinations of atomic p orbitals corresponding to the π molecular orbitals. There are eight electrons in the π system, and two electrons must occupy the nonbonding level of π_4 and π_5.

two compounds the unfavorable orbital interactions are alleviated by distortion of the molecular geometry.

Cyclobutadiene was finally prepared in 1965 by R. Pettit at the University of Texas. As you would expect, it is a highly unstable molecule. The two double bonds greatly increase the strain that is present in any four-membered ring, so cyclobutadiene is extremely reactive and difficult to study. It appears to have a *rectangular* (rather than square) geometry. This geometric distortion causes π_2 and π_3 to have different energies; they are bonding and antibonding, respectively. Therefore the four electrons of the π system reside in π_1 and π_2, and only bonding molecular orbitals are occupied.

X-ray crystallographic studies have clearly shown that cyclooctatetraene is *nonplanar*. The actual structure is tub-shaped, and there are two distinct types of carbon-carbon bond. This distortion from planarity permits excellent overlap between the *p* atomic orbitals of the adjacent carbons that are joined by *double* bonds.

1.33 Å

1.46 Å

Cyclooctatetraene: the actual molecule is tub-shaped and has fixed single and double bonds

In contrast, the *p* orbitals on carbons joined by only a single bond are nearly perpendicular. As a result the actual π system of cyclooctatetraene corresponds very closely to four noninteracting double bonds. You can therefore see why cyclooctatetraene lacks the exceptional stability of an aromatic compound and instead behaves as an "ordinary" alkene.

EXERCISE 10.9

If cyclooctatetraene were a planar aromatic species, how many dimethyl isomers would be possible? Draw them.

EXERCISE 10.10

Cyclooctatetraene is in fact tub-shaped. How many constitutionally isomeric dimethyl derivatives should be possible? Draw them.

10.6 NONBENZENOID AROMATICITY AND HÜCKEL AROMATICITY

The development of molecular orbital theory finally provided a consistent explanation for why benzene is aromatic and why cyclobutadiene and cyclooctatetraene are not. But there were still compounds other than benzene derivatives that exhibited reactivity corresponding to an aromatic compound. Pyrrole, for example, appeared to be aromatic. It was suggested that this results from the pair of nonbonding electrons on nitrogen to give a total of *six* electrons in the π system, the same as in benzene. There are two electrons on nitrogen and two π electrons in each of the "double bonds" of pyrrole.

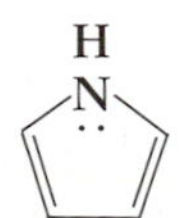

An explanation for the aromaticity of pyrrole and other nonbenzenoid structures was provided by E. Hückel (Germany) in the 1930s. Using molecular orbital theory, he was able to develop criteria for aromaticity that include benzene derivatives and a variety of other species as well.

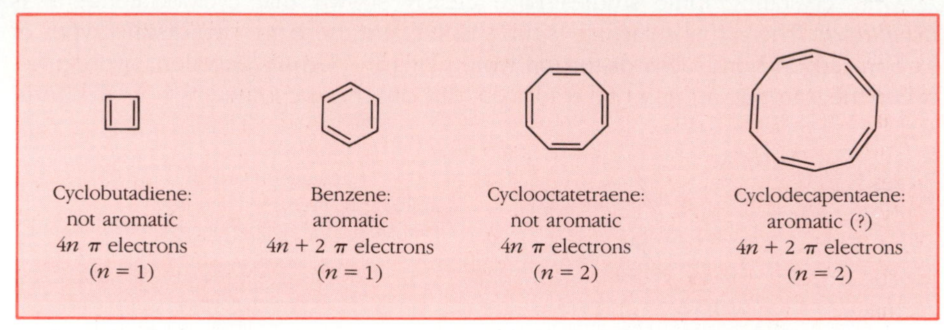

Figure 10.11 Hückel's Predictions for Aromaticity of Cyclic Compounds.

Hückel observed that molecular orbital diagrams similar to those of Figures 10.9 and 10.10 show a repeating pattern. The highest occupied π molecular orbital for cyclobutadiene is nonbonding, but that for benzene is strongly bonding. For cyclooctatetraene this orbital is again nonbonding, but for the next member in the series, cyclodecapentaene, it is once again strongly bonding (Figure 10.11).

Hückel concluded that these cyclic polyenes alternate between being aromatic and nonaromatic as the number of double bonds increases, and he characterized them according to the number of electrons in the π system. The nonaromatic compounds each have $4n$ π electrons, where n is some integer: 1, 2, 3, ... etc. The aromatic hydrocarbons are characterized by a total of $4n + 2$ π electrons, so cyclic polyenes having 6, 10, 14, ... etc. π electrons should be aromatic. (The alternating pattern of aromaticity could also be expressed in terms of the number of double bonds; the nonaromatic compounds in Figure 10.11 have an *even* number of double bonds whereas those predicted to be aromatic have an *odd* number of double bonds.)

Hückel's original analysis was for cyclic polyenes, but it has since been extended to include other compounds. We can summarize the criteria for Hückel aromaticity by the rules listed in the following box:

Criteria for Hückel Aromaticity

1. The structure must be monocyclic and capable of planarity.

2. Each atom in the ring must have an available p orbital perpendicular to the ring.

3. If the π system has a total of $4n + 2$ π electrons it is aromatic according to the Hückel criterion.

We will now consider these rules in detail and examine some specific examples that show how they can be applied. If a structure fits the first two rules, we will call it a Hückel system, and only then will we apply the third rule to see if it is aromatic. The first rule requires that the structure be capable of planarity, and the reason for this is quite simple. In order for the molecule to have π bonding, the atomic p

orbitals that contribute to the π system must be parallel, and this can occur only when all the atoms of the ring lie in the same plane. The actual molecule will invariably adopt the most stable shape, as we discussed previously for cyclooctatetraene. The question here is merely to be sure that planarity could be achieved so that we can determine if such a planar structure would be aromatic.

The second rule requires an available *p* orbital on each atom. This requirement serves a similar purpose, to ensure that the molecule is a Hückel system before we apply the $4n + 2$ rule. A structure such as cycloheptatriene has six π electrons ($4n + 2$, where $n = 1$), but C-7 has two hydrogens and therefore lacks an available *p* orbital. As a result, cycloheptatriene is not a Hückel system and cannot be aromatic.

H
H
1,3,5-Cycloheptatriene

Next consider pyrrole, which we described as an aromatic compound at the beginning of this section. At first glance it would seem that this is not a Hückel system either. Ordinarily the nitrogen of an amine is sp^3 hybridized (Section 17.4), and this would mean that the nitrogen would not have an available *p* orbital:

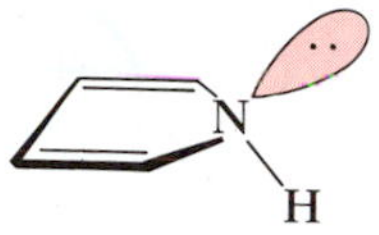

N
H

But recall the statement of our first rule, namely, that the structure must be *capable* of planarity. If a small distortion produces a large gain in stability by establishing an aromatic structure, the molecule will surely undergo the distortion. Therefore we must evaluate the following planar structure:

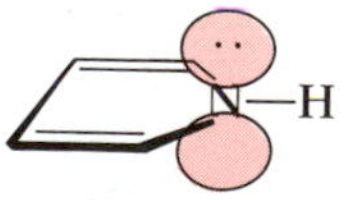

N—H

Now we have a Hückel system and the electrons can be counted. Each carbon contributes one electron to the π system (i.e., there are two π electrons for each double bond in the structure) and nitrogen contributes two electrons, so a total of *six* electrons occupy the π system. Therefore pyrrole is aromatic by the Hückel criteria, a prediction that is in good agreement with its chemical properties. A dramatic illustration of the stability of pyrrole is provided by its weak basicity.

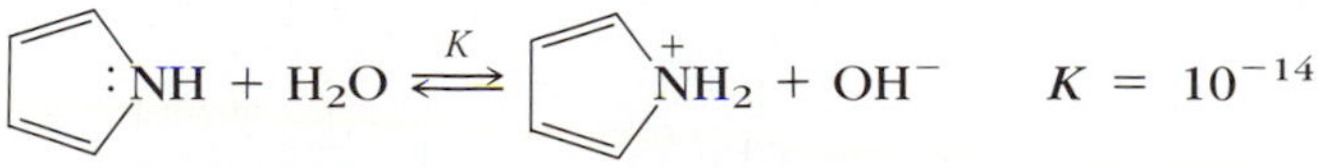

$$:\text{NH} + H_2O \overset{K}{\rightleftharpoons} \overset{+}{\text{N}}H_2 + OH^- \qquad K = 10^{-14}$$

The corresponding equilibrium constants for ammonia and simple aliphatic amines are in the region of 10^{-5}, so the basicity of pyrrole has been reduced by a factor of 10^9. Why is this? When pyrrole is protonated, the resulting cation has no available atomic *p* orbital on nitrogen and is no longer a Hückel system:

As a result, protonation of pyrrole can occur only at the expense of its aromaticity, so it is an extremely weak base.

Two further examples of aromatic heterocycles help to show how we use the criteria for Hückel aromaticity. Both pyridine and furan are planar and have available atomic p orbitals on every atom.

Pyridine Furan

Your initial reaction might be that both systems have *eight* π electrons, but that would be erroneous. We can only count the electrons that are *in* the π system; electrons in nonbonding orbitals *perpendicular* to the π system are ignored.

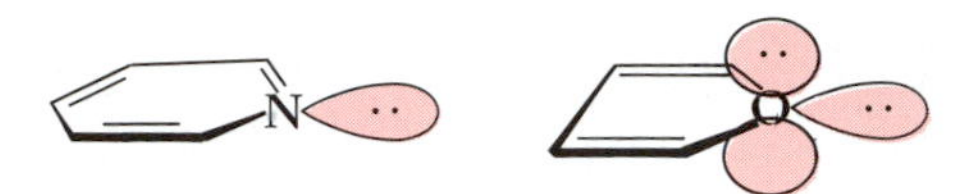

Therefore the two electrons that occupy the nonbonding orbital in the plane of the ring are not counted, so each of these molecules has six π electrons and is aromatic.

In contrast to the preceding examples, the seven-membered heterocycle oxepin is not aromatic. The planar form would be a Hückel system, but it would have eight π electrons (six from the carbons and two from the oxygen), and this falls in the category of $4n$ rather than $4n + 2$ π electrons. Such molecules are classified as *nonaromatic.*

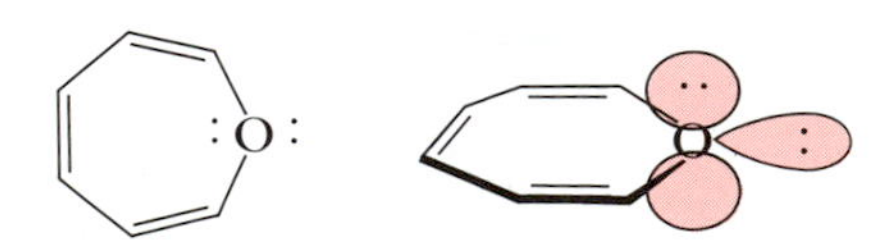

Ionic Species

A valuable feature of Hückel aromaticity is its ability to help us interpret the reactivity of certain ionic species. Consider, for example, the reactivity of 3-chlorocyclopropene and 5-iodo-1,3-cyclopentadiene:

If these compounds were to undergo reaction via a carbocation mechanism (Section 12.2), the following cations would be formed as intermediates:

Both cations are Hückel systems, but only the cyclopropenyl cation is aromatic. (It has two π electrons, which corresponds to $4n + 2$ with $n = 0$.) In the presence of silver ion the two compounds exhibit a dramatic difference in reactivity. 3-Chlorocyclopropene ionizes rapidly to generate the cation as a stable salt.

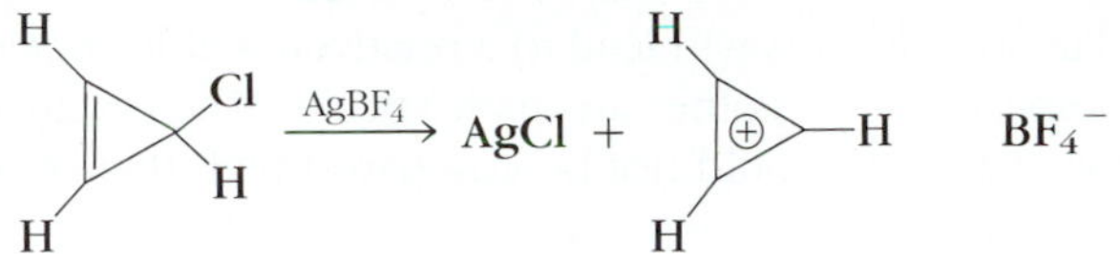

$$\xrightarrow{AgBF_4} AgCl + \quad BF_4^-$$

In contrast, the cyclopentadienyl compound is extremely resistant to ionization.

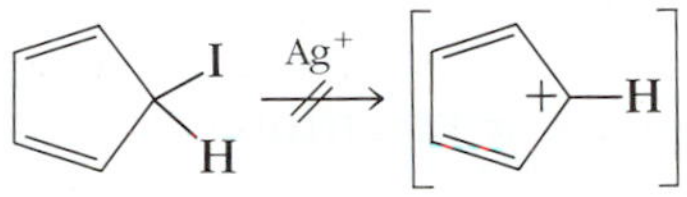

The available data indicate that ionization of 3-chlorocyclopropene must occur at least 1 million times more rapidly than ionization of the corresponding cyclopentadienyl halide.

Another example of abnormal reactivity is provided by cyclopentadiene. In most reactions it behaves as a normal diene, but its acid-base reactivity is unusual. Aliphatic hydrocarbons are very weak acids, with dissociation constants in the range of $10^{-40} - 10^{-50}$, and even allylic and benzylic hydrogens have K_a's less than 10^{-30}. But cyclopentadiene has a dissociation constant comparable with that of water.

$R—CH_2—R$	$CH_3CH{=}CH_2$	$C_6H_5—CH_2—C_6H_5$	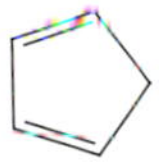 cyclopentadiene
$Ka = 10^{-40}–10^{-50}$	10^{-36}	10^{-34}	10^{-16}

Why is cyclopentadiene so much more acidic than other hydrocarbons? The answer lies in the anion that is formed. In contrast to the very unstable anions produced by ionization of other hydrocarbons, the cyclopentadienyl anion is a six-π-electron aromatic system; as a result it is easily formed in an equilibrium reaction. Treatment of cyclopentadiene with a base such as KOH produces an equilibrium mixture in which most of the hydrocarbon has been converted to its potassium salt.

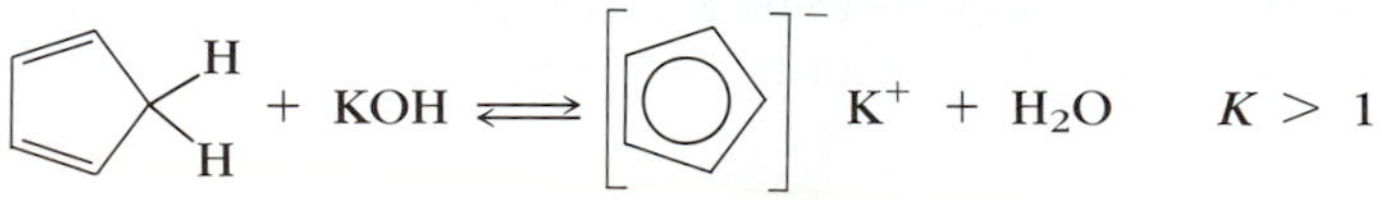

$$+ \ KOH \rightleftharpoons [\ ]^- \ K^+ + H_2O \qquad K > 1$$

Despite its remarkable stability, the cyclopentadienyl anion is a good nucleophile. For example, cyclopentadiene can be alkylated in good yield in a reaction analogous to the alkylation of terminal acetylenes.

$\xrightarrow[\text{2. } CH_3Br]{\text{1. } NaNH_2}$ —CH_3 85%

Notice that the use of a strong base in this reaction ensures complete conversion to the anion before the alkylating agent is added.

The concept of Hückel aromaticity is useful in understanding the reactivity of highly unsaturated cyclic ions, but it is important that you recognize certain limitations. The notion of aromaticity is somewhat vague, and the term Hückel aromaticity merely denotes a situation in which the π electrons all occupy relatively low energy molecular orbitals. This will lead to a relatively stable species, but you must be careful in comparisons with other aromatic species. The reactivity of cyclopentadienyl anion, for example, should not be compared with that of neutral benzene. The former is an anion, and it is far more reactive than benzene; its aromatic character is instead reflected by its remarkably weak basicity in contrast to other *carbanions.* A similar argument applies to the cyclopropenyl cation; while extremely stable in comparison with other *carbocations,* it is still a very reactive species.

Systems with More Than Six π Electrons

Up to this point we have only considered aromaticity for ring sizes of 10 or fewer atoms, but Hückel systems have been prepared with much larger rings. The 18-membered ring polyene shown below (structure **1**) has 18 π electrons and should be aromatic ($4n + 2$, with $n = 4$) by the Hückel criteria. This hydrocarbon has two types of carbon atoms (6 that lie near the center of the ring and 12 that lie on the periphery), so there are two types of carbon-hydrogen bond. Experimental studies have shown that there is no alternation between single and double bonds, and all C—C bond lengths fall in the narrow range of 1.38–1.42 Å. This is intermediate between single and double bonds (1.54 and 1.34 Å, respectively), as you would expect for an aromatic system.

1

Despite their many successes, the Hückel criteria for aromaticity must be used with care. If there are structural features in a molecule that disfavor a planar geometry, a compound might not be aromatic even if it were a Hückel system with $4n + 2$ π electrons. This appears to be the case with the $C_{10}H_{10}$ cyclic polyene, **2,** for which we questioned the aromaticity in Figure 10.11.

2

This compound does not exhibit the chemical and physical properties that would be expected for an aromatic compound. If you consider that the interior angle of a regular decagon is 144°, an explanation for the lack of aromaticity of **2** is apparent. A planar, symmetrical structure would have bond angles of 144°, a substantial distortion from the idealized value of 120° for an sp^2 hybridized carbon. The calculated "resonance energy" for **2** is only about 10 kcal/mol, but the total angle strain for all 10 bond angles of a planar structure would be substantially larger. As a result, a "normal," nonplanar polyene structure is more stable than the planar geometry that would be aromatic according to the Hückel criteria. Structure **2** shows that you must be careful when using Hückel aromaticity to predict chemical stability and reactivity. Hückel aromaticity is concerned only with the π system, but the overall stability of a species is also affected by factors such as angle strain and nonbonded (steric) interactions.

NMR Chemical Shifts as a Criterion for Aromaticity

When NMR spectroscopy was being developed as a structural tool in the 1950s, chemists noticed an unusual effect of aromatic rings: chemical shifts of nearby nuclei in the plane of the ring are shifted *downfield.* This is illustrated by the absorbance of benzene protons at 7.4 ppm, in contrast to the absorptions of ordinary alkenes in the region 5.5–6.0 ppm.

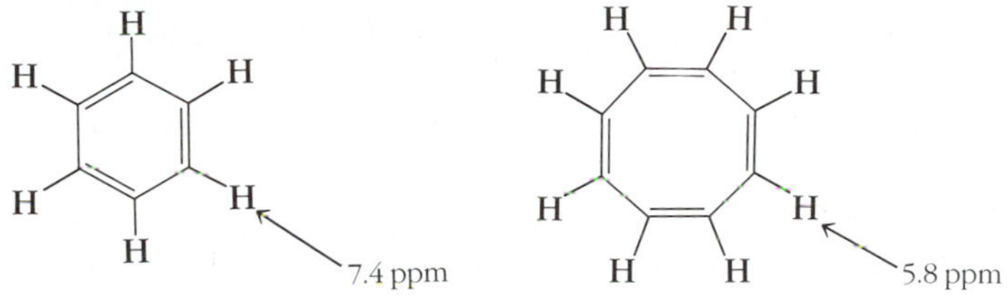

A mathematical description of this phenomenon is analogous to the explanation of the electromagnetic effects produced when an electric current flows in a circular path in the presence of a magnetic field. The mathematical similarity has resulted in the use of the term *ring current* to describe the effect of aromatic rings. In addition to the downfield shift observed for nuclei in the plane of the ring, an upfield shift is predicted for nuclei that lie above the ring. The NMR spectrum of the following benzene derivative demonstrates this effect.

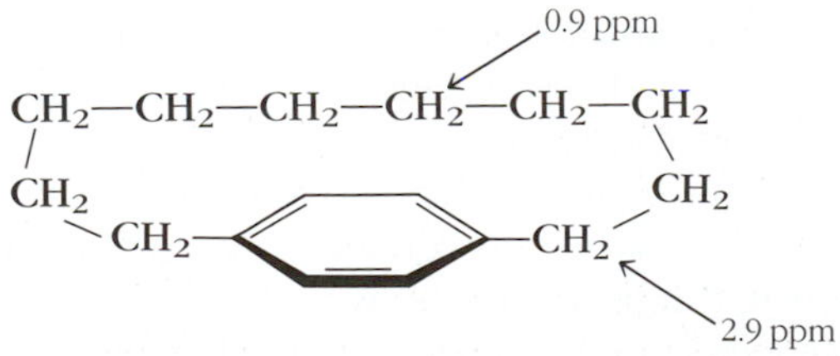

Alkyl hydrogens in a typical cycloalkane absorb at about 1.5 ppm (Section 9.8), but the CH_2 hydrogens that lie (approximately) in the plane of the ring are shifted *downfield* to about 2.9 ppm. In contrast, the CH_2 groups that lie directly above the aromatic ring produce a peak that is shifted *upfield* to 0.9 ppm.

The upfield shift should actually be greatest for a substituent that lies at the exact center of the ring, but this is not possible for a benzene derivative. Some of the larger cyclic polyenes do have hydrogens oriented toward the center of the ring, and aromatic ring currents have been observed. The inner hydrogens in such compounds show substantial upfield shifts, whereas the outer hydrogens exhibit a downfield shift analogous to that observed for the hydrogens of benzene.

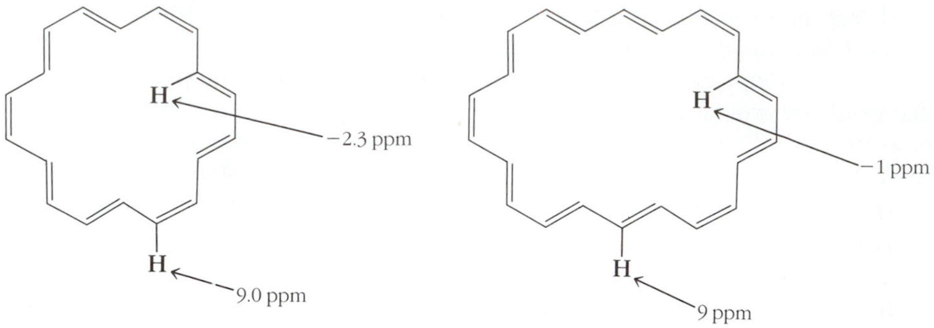

The presence or absence of such ring current effects has been used extensively by chemists to help evaluate whether or not a system is aromatic.

EXERCISE 10.11

(a) Which of the following species are Hückel systems?
(b) Which of these should be aromatic by the Hückel criteria?

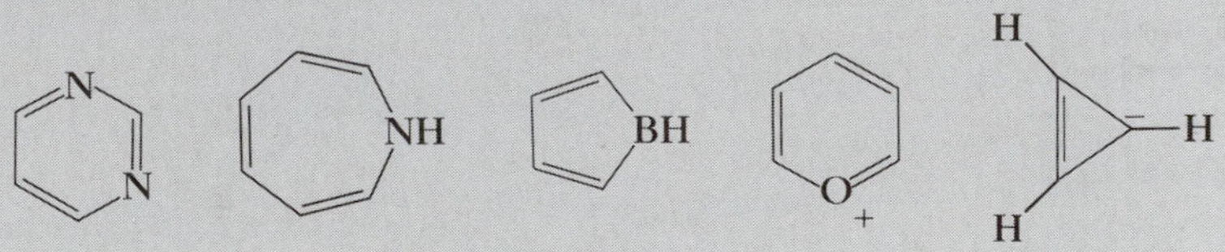

10.7 THE MECHANISM OF ELECTROPHILIC SUBSTITUTION

Substitution reactions of aromatic compounds had been extensively developed by the 1800s, but it was not until the 1930s and 1940s that any real progress was made in understanding how these processes occur at the molecular level. It turned out that all the major aromatic substitution reactions were found to proceed by essentially the same mechanism, and this provides the general basis for our understanding of the reactivity of aromatic compounds.

Benzene and its derivatives are well known to undergo substitution rather than addition reactions, but their structures certainly suggest that there should be similarities between their reactions and those of alkenes. For example, if an alkene is treated with acid, we expect (Section 5.8) that it will protonate to form a carbocation as a reactive intermediate.

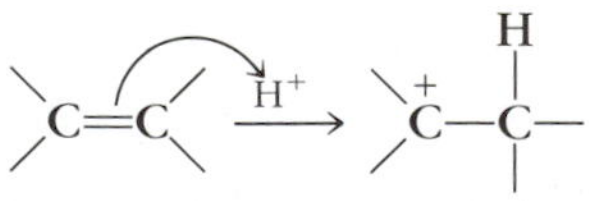

Therefore, should not benzene react in the same manner to generate a carbocation?

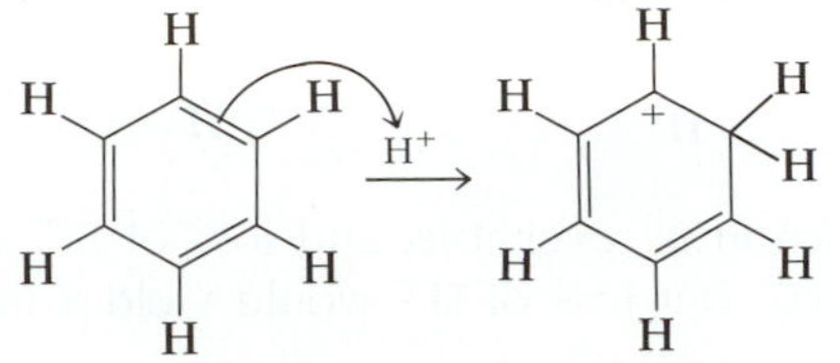

The answer most certainly is yes (although somewhat more slowly). But the overall difference in reactivity between alkenes and aromatic compounds does not lie in their ability to react with H^+ and other electrophiles. Instead the difference arises in the subsequent reactions of the carbocations. In reactions of alkenes the intermediate cation tends to react with a nucleophile, X^-, to give overall *addition* of HX.

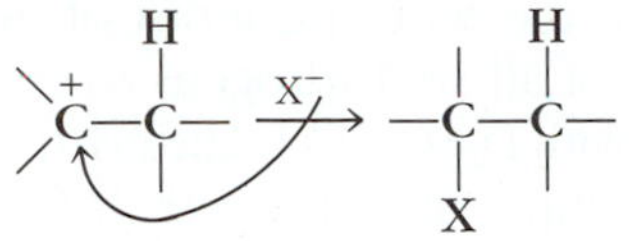

In contrast, the cation derived from an aromatic compound undergoes an acid-base reaction to regenerate a stable aromatic species.

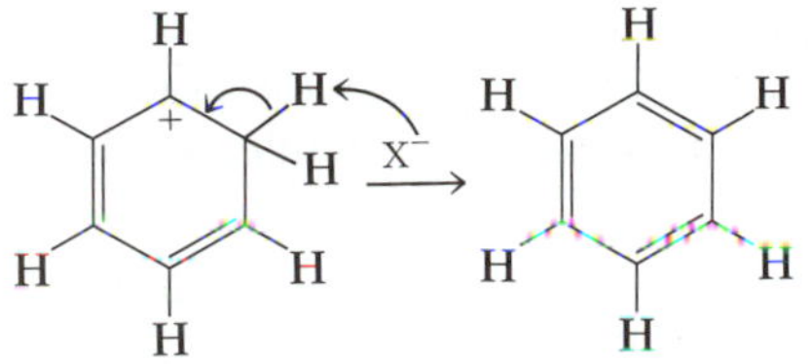

If treatment of benzene with an acid merely affords benzene as the product of the "reaction," how do we know that carbocation formation ever occurred at all? The answer comes from the use of *isotopically labeled* acid. When benzene is treated with deuterated sulfuric acid, D_2SO_4, exchange occurs to generate ($^2H_6$)benzene.

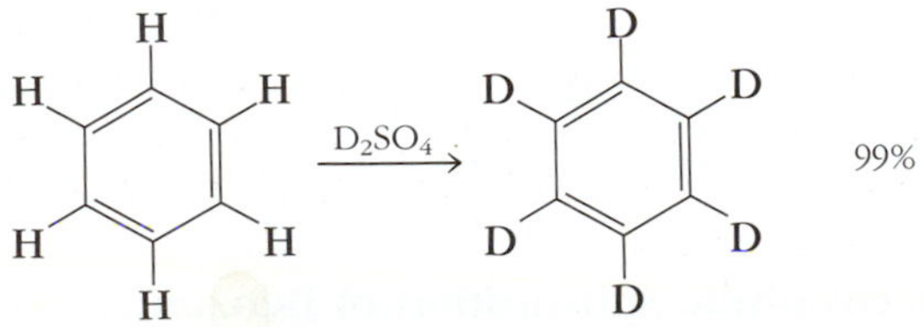

D_2SO_4 99%

The reaction mechanism for this isotope exchange is straightforward. Initial "protonation" of benzene by D^+ generates a carbocation containing a single deuterium:

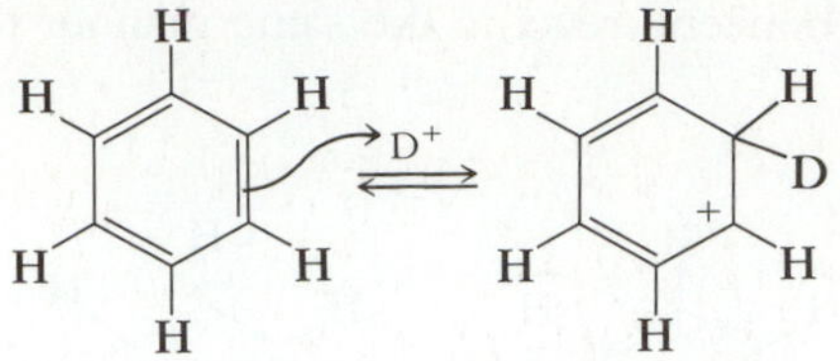

This "protonation" reaction is reversible, and loss of D^+ would regenerate the unlabeled benzene, C_6H_6. But loss of H^+ would yield a monodeuteriobenzene, C_6H_5D:

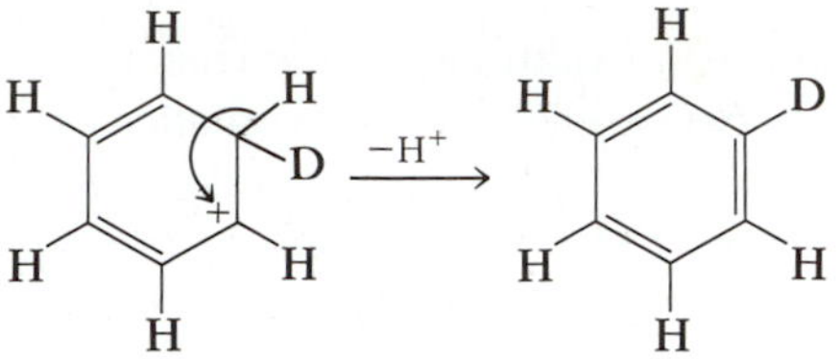

Repetition of this process at the other positions of the benzene ring eventually yields C_6D_6, with exchange of all six hydrogens.

The reaction of benzene with D_2SO_4 is characteristic of all electrophilic substitution reactions of aromatic compounds. However, in most other cases the reaction can be stopped after a single substitution. Electrophilic substitution proceeds via attack on the ring by some electrophile, E^+, to generate an intermediate carbocation, and subsequent loss of H^+ from the *same carbon* that was attacked by E^+ yields a substitution product. This is summarized in Figure 10.12 for a generalized electrophile, E^+. The designation of this process as *electrophilic* substitution is to some extent arbitrary, since benzene, after all, is behaving as a *nucleophile* in the reaction. The classification refers to the carbon atom that is undergoing substitution; the carbon is attacked by an electrophile, so the process is described as electrophilic substitution.

The intermediate carbocation formed by electrophilic attack on a benzene ring is relatively stable, because it is conjugated with two double bonds. Although we have been drawing a single structural formula for convenience, the actual structure of the cation formed by attack of D^+ on benzene is a hybrid of three resonance forms.

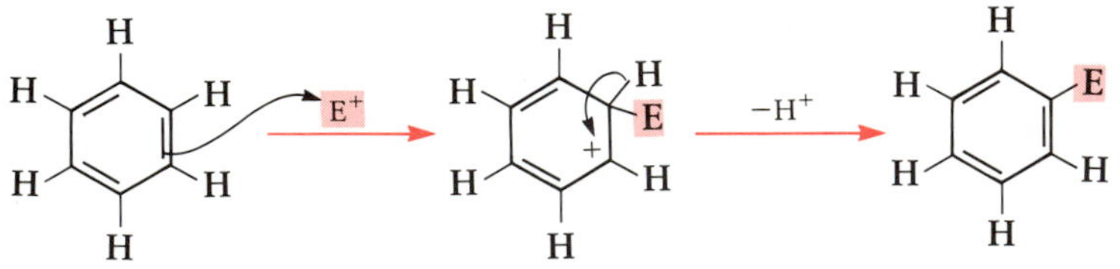

Figure 10.12 Electrophilic Substitution of Benzene.
A hypothetical electrophile, E^+, attacks the aromatic ring to form an intermediate carbocation. Loss of a proton (from the *same carbon*) that was attacked by E^+ then regenerates the aromatic system. In the overall process the substituent E has replaced an H.

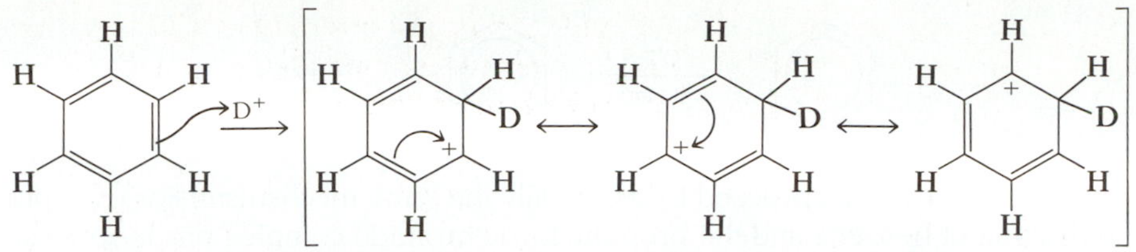

These resonance forms illustrate a very important feature of electrophilic substitution that we will use later to explain substituent effects: *The positive charge is specifically delocalized to the three carbons of the ring that are ortho and para to the site of attack by the electrophile.*

10.8 ELECTROPHILIC SUBSTITUTION REACTIONS OF BENZENE: HALOGENATION, NITRATION, AND SULFONATION

We have already illustrated electrophilic substitution with isotopic labeling. Now we will turn our attention to some synthetically important reactions that introduce potentially reactive functional groups as substituents on the aromatic ring.

Halogenation

Unlike simple alkenes, benzene does not react with chlorine or bromine at an appreciable rate at room temperature. In order to carry out the halogenation of benzene under reasonable conditions, a Lewis acid is normally used to catalyze the reaction. The most common catalysts are halides of metals such as iron, aluminum, and tin. Iron halides ($FeCl_3$, $FeBr_3$) are used most frequently in laboratory reactions. Generally, the metal halide is generated in the reaction vessel by mixing the free metal and the halogen:

$$2Fe + 3Br_2 \longrightarrow 2FeBr_3$$

Consequently, the reagents for a halogenation reaction may sometimes be written as the *halogen* and the *metal,* but the catalyst is still the *metal halide.* The Lewis acid can form a complex with the halogen, and this is a much more powerful electrophile than the free halogen. The complex behaves as a donor of "Br^+" or "Cl^+."

$$:\ddot{B}r—\ddot{B}r: + FeBr_3 \rightleftharpoons \overset{\delta+}{Br—Br}\text{---}\overset{\delta-}{FeBr_3}$$

The reaction of benzene with bromine and ferric bromide at 65–70°C affords a good yield of bromobenzene, along with some disubstitution product.

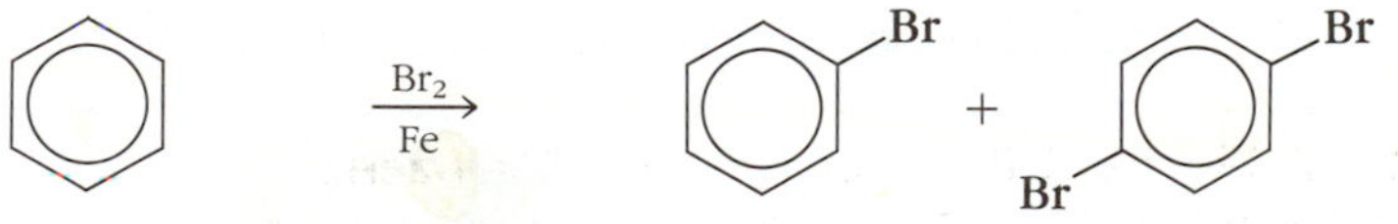

60% 10%

Chlorobenzene has been prepared similarly. The following example illustrates the use of aluminum chloride (prepared from aluminum amalgam) as a catalyst.

$$C_6H_6 \xrightarrow[\text{Al-Hg}]{Cl_2} C_6H_5Cl \quad 49\%$$

Both these reactions proceed by essentially the same mechanism. For example, the reaction of benzene and the bromine-ferric bromide complex produces a carbocation and the $FeBr_4^-$ anion. Dissociation of the $FeBr_4^-$ ion regenerates the ferric bromide, and transfer of a proton to a bromide ion yields the HBr that is formed as a by-product in the reaction.

$$C_6H_6 \xrightarrow{\overset{\delta+}{Br}-\overset{\delta-}{Br}\text{---}FeBr_3} [C_6H_6Br]^+ \xrightarrow{Br^-} C_6H_5Br + HBr$$

Electrophilic substitution with iodine can sometimes be carried out by using similar procedures, but direct fluorination of benzene and its derivatives is precluded by the very high reactivity of F_2. Ordinarily, both fluoro and iodo substituents are commonly introduced by indirect procedures, which we will discuss in Section 18.5.

Nitration

The electrophilic species that brings about nitration of aromatic rings is the *nitronium ion,* $^+NO_2$. This is a relatively stable species, and it is electronically similar to carbon dioxide. (In fact the only difference between the two species is the presence of an additional proton and two neutrons in the nucleus of nitrogen.)

$$O{=}\overset{+}{N}{=}O \quad \text{Nitronium ion}$$

Stable salts such as $NO_2^+BF_4^-$ are known, and these have been used extensively as nitrating agents. The most common reagent for nitration is a mixture of concentrated nitric and sulfuric acids. In the presence of concentrated H_2SO_4 nitric acid undergoes protonation and dehydration to form the nitronium ion.

$$HO-\overset{+}{N}(=O)O^- \xrightleftharpoons{H_2SO_4} H_2\overset{+}{O}-N(=O)O^- \rightleftharpoons O{=}N^+{=}O + H_2O$$

The nitronium ion is a fairly active electrophile and it can attack the π electron system of an aromatic ring. This results in replacement of a hydrogen by an NO_2 group, as shown in the following two examples of the conversion of benzene to nitrobenzene.

$$C_6H_6 \xrightarrow[H_2SO_4]{HNO_3} C_6H_5NO_2 \quad 85\%$$

$$C_6H_6 \xrightarrow{NO_2^+BF_4^-} C_6H_5NO_2 \quad 93\%$$

Nitrobenzene has a characteristic odor, which is often associated with shoe polish. However, it is fairly toxic and its use in household products has decreased substantially in recent years.

Nitration proceeds by the same mechanism as other electrophilic substitutions. Reaction of the aromatic ring with a nitronium ion generates a carbocation, and subsequent transfer of a proton to some other species in solution (H_2O, HSO_4^-, HNO_3) affords the nitro product.

$$C_6H_6 \xrightarrow{O{=}\overset{+}{N}{=}O} [C_6H_6NO_2]^+ \longrightarrow C_6H_5NO_2 + H^+$$

Sulfonation

Sulfonation of aromatic compounds produces sulfonic acids, $ArSO_3H$, and these are used extensively in the formation of detergents and other commercially important products. The active electrophile in sulfonation is sulfur trioxide, a powerful electrophile which reacts rapidly, even violently, with water to generate H_2SO_4.

$$H_2\ddot{O}: + SO_3 \longrightarrow HO{-}SO_2{-}OH$$

Just as it reacts rapidly with water, SO_3 can react with the π electrons of an aromatic ring to form a carbon-sulfur bond. A solution of SO_3 in H_2SO_4 is typically used for this reaction. This reagent is known as *fuming sulfuric acid,* because of the "smoke" that results when it is exposed to atmospheric moisture. The treatment of benzene with fuming sulfuric acid produces benzenesulfonic acid.

$$C_6H_6 \xrightarrow{H_2SO_4,\ SO_3} C_6H_5SO_3H \xrightarrow[H_2O]{NaCl} C_6H_5SO_3^-\ Na^+ \quad 43\%$$

The product in the preceding reaction was isolated as the sodium salt, and this procedure is frequently used because the sulfonic acids themselves are quite soluble in aqueous sulfuric acid. When sodium chloride is added to such a solution, exchange of Na^+ for H^+ produces the sodium salt, which precipitates from solution. In general, however, we will not be concerned with the method of isolation in these reactions, and we will simply show the formation of the sulfonic acid.

Sulfonation is somewhat different than aromatic halogenation and nitration because each step of the mechanism is *fully reversible.* Electrophilic attack of SO_3 on the aromatic ring and proton transfer to oxygen in the highly acidic medium generates the intermediate carbocation that is typical of electrophilic substitution reactions.

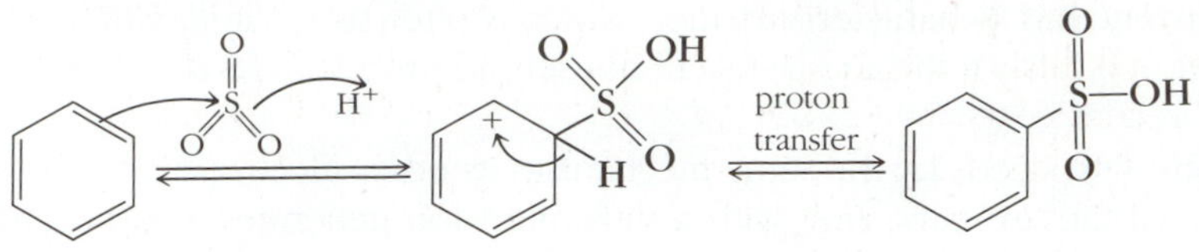

Transfer of a proton (from carbon to some proton acceptor in solution) converts the carbocation to the sulfonic acid. But a second pathway, loss of the sulfonic acid group, is also open to this cation.

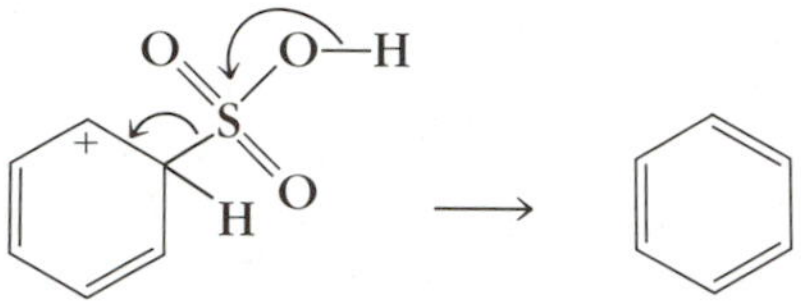

In principle the individual steps of other electrophilic substitutions should also be reversible. What feature of the sulfonation reaction makes it unusual? The answer quite simply is that the products and reactants in the reaction are comparable in stability. It is sometimes convenient to think of this reversibility as the consequence of a relatively *weak bond* between carbon and sulfur. You will encounter another example of reversible formation of sulfonic acids in Section 13.7, when we discuss the bisulfite addition products of aldehydes and ketones. In contrast to sulfonation, halogenation and nitration reactions are highly exothermic, so the products are much lower in energy than the reactants. The reverse processes in these reactions would very much be energetically "uphill," so they are not observed.

The reversibility of the sulfonation reaction means that a second kind of electrophilic substitution reaction, *desulfonation,* can be carried out. When an aromatic sulfonic acid is treated with strong acid in the *absence* of SO_3, equilibrium favors the unsubstituted compound. The reverse process for benzenesulfonic acid would occur by initial protonation of the carbon bearing the $-SO_3H$ group to generate the same intermediate carbocation that is produced in the formation of the sulfonic acid. Loss of the SO_3H group would then yield benzene.

As you will see in Chapter 18, desulfonation has been exploited by organic chemists for synthetic purposes. The $-SO_3H$ group can be employed as a *blocking group,* because it alters the reactivity of the aromatic ring and can subsequently be removed.

EXERCISE 10.12

In Section 10.7 we drew several resonance forms for the carbocation generated by protonation of benzene. Draw resonance forms of the intermediate cations produced by chlorination, nitration, and sulfonation of benzene. Is there any pattern for the relationship between the carbon that undergoes electrophilic attack and the carbons of the ring that bear positive charge?

EXERCISE 10.13

In Section 10.7 we showed all the hydrogen substituents on the carbocation produced by protonation of benzene. In this section we have frequently omitted the hydrogens, but this can sometimes be confusing when you first encounter these reactive intermediates. Consider the three resonance forms for the intermediate carbocation formed in the bromination of benzene. Draw them (a) showing all hydrogens and (b) showing only the hydrogens at the reaction center. Make sure all carbons in every structure have either four bonds or else three bonds and a positive charge.

EXERCISE 10.14

The desulfonation reaction of benzenesulfonic acid is illustrated in this section as proceeding via initial protonation of the same carbon that bears the $-SO_3H$ group. Draw the carbocations (including all resonance forms) that would result from protonation *ortho, meta,* or *para* to the SO_3H group. Could any of these yield benzene directly upon loss of the $-SO_3H$ group?

10.9 RESONANCE AND INDUCTIVE EFFECTS IN SUBSTITUTED BENZENES

Despite the inability of simple resonance theory to predict and explain the *phenomenon* of aromaticity, resonance can be extremely useful for predicting and explaining the *reactivity* of aromatic compounds. In this section we will briefly illustrate how you can use resonance to predict substituent effects in reactions of benzene derivatives.

Acidity

We will first consider the acidity of some phenols. Phenol itself ($K_a = 10^{-10}$) is considerably more acidic than simple alcohols ($K_a \cong 10^{-16} - 10^{-18}$).

$$C_6H_5OH \rightleftharpoons C_6H_5O^- + H^+ \qquad K_a = 1 \times 10^{-10}$$

In Section 8.6 we indicated that benzylic ions were stabilized by orbital interactions with the π system of the ring. Similar interactions stabilize the phenolate ion and result in increased acidity of the phenol. This orbital interaction, however, is difficult to illustrate, and we can give you an alternative explanation using resonance theory. (Despite the fact that these arguments may seem quite different, the resonance and orbital explanations are basically two ways of saying the same thing.) Five resonance forms can be drawn for the phenolate ion, the two Kekulé forms, **1a** and **1b**, and three additional structures having the negative charge on *carbon*.

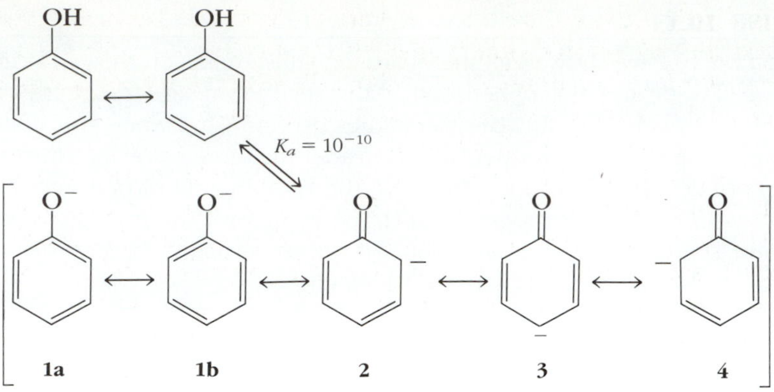

$K_a = 10^{-10}$

1a 1b 2 3 4

The negative charge resides mainly on oxygen, but as structures **2–4** show, it is also delocalized to the carbons of the ring. This delocalization stabilizes the phenolate ion and is responsible for the increased acidity of phenol (in comparison with alcohols).

Structures **2–4** show one feature of resonance that is common to all situations involving resonance stabilization of an ion by a benzene ring: *Charge can be delocalized only onto the carbons that are ortho and para to the charged substituent.*

Next compare the acidity of phenol with that of its *m*- and *p*-nitro derivatives. What effect should you expect a nitro group to exert on the acidity of a phenol? And how should the effect depend on the location of the nitro group? The experimentally measured acid dissociation constants are shown below the structures of the three phenols.

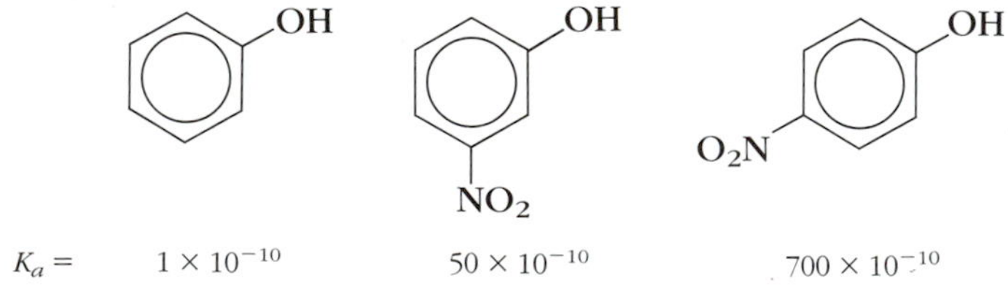

$K_a =$ 1×10^{-10} 50×10^{-10} 700×10^{-10}

Both *m*- and *p*-nitrophenol are substantially more acidic than phenol itself. In part this must reflect the *inductive* effect of a nitro group, which is strongly electron-withdrawing. But the large difference in acidity between the *meta* and *para* isomers results from a *resonance* effect. The nitro group has a nitrogen-oxygen double bond and can act as an electron-withdrawing group by resonance. We have shown this with the following resonance forms, particularly the last structure, which shows delocalization of the negative charge onto the nitro group.

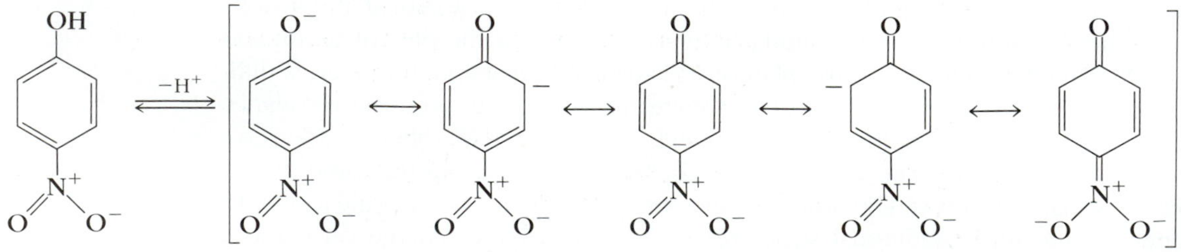

$-H^+$

Notice that we have not drawn both Kekulé forms of the "aromatic" resonance forms because they have the same charge distribution. In general we will draw only a single Kekulé form for aromatic compounds. By now you are well aware that the true structure is a resonance hybrid, so it is no longer necessary to draw the individual contributors in each instance.

The *p*-nitrophenolate ion is stabilized by a resonance interaction with the nitro group, but such an interaction is absent for the *meta* compound. Only structures having charge on the phenolate oxygen and on the ring carbons can be drawn. While there is charge delocalization in the anion from *m*-nitrophenol there is no direct resonance interaction between the oxygen and the nitro group. Consequently the nitro group stabilizes the anion only through a (weaker) inductive effect.

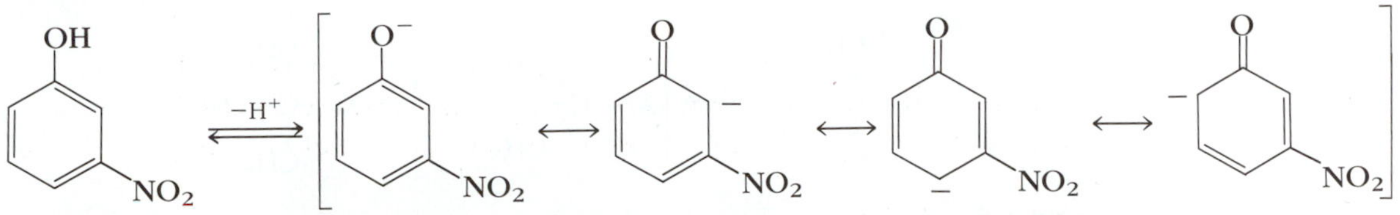

OH, NO_2 $\xrightleftharpoons{-H^+}$ [O^-, NO_2 ⟷ O, NO_2 ⟷ O, NO_2 ⟷ O, NO_2]

This reaffirms our previous statement that charge delocalization can occur only at the ring positions *ortho* and *para* to the charged substituent. We can extend this idea to the following generalization:

> **Direct resonance interaction between substituents on a benzene ring can occur only when the substituents are *ortho* or *para* to each other.**

The magnitude of resonance stabilization of an anion by a nitro group is further illustrated by a pair of nitrophenols that also have two methyl groups:

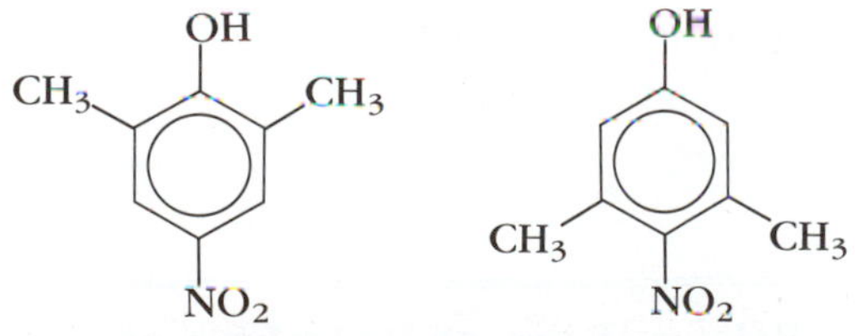

OH, CH_3, CH_3, NO_2 — $K_a = 700 \times 10^{-10}$

OH, CH_3, CH_3, NO_2 — $K_a = 60 \times 10^{-10}$

In the first case (2,6-dimethyl-4-nitrophenol) the methyl substituents have virtually no effect, and the acid dissociation constant is the same as that for *p*-nitrophenol. But the acidity of the second compound is reduced by a factor of 10. How do the methyl groups exert this influence? The answer is found in steric effects. In order for a resonance interaction to exist between a substituent and the ring, the *p* orbitals constituting the π system must be parallel. This in turn requires that the nitro group lie in the plane of the ring, a geometry that is not possible for the second compound because of steric interactions with the two methyl groups.

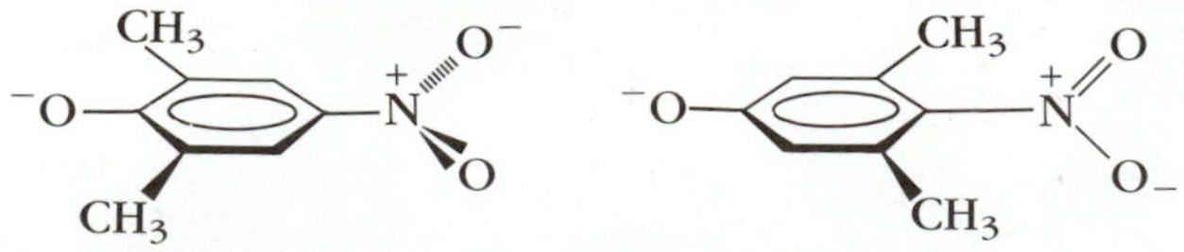

CH_3, ^-O, CH_3, $\overset{+}{N}$, O^-, O CH_3, ^-O, CH_3, $\overset{+}{N}$, O, O^-

Consequently, the nitro group must twist out of the plane of the aromatic ring, so there is no resonance stabilization of the anion and only inductive effects are operating. The dissociation constant of 60×10^{-10} is nearly the same as that for *m*-nitrophenol, which also has no resonance interaction between the oxygen and the nitro group.

Reaction Rates

Resonance interactions are not limited to stabilization of anions; resonance effects on carbocation reactions, for example, can be quite large. The hydrolysis reactions of 2-aryl-2-chloropropanes proceed via carbocations (Section 12.2), and the relative rate of reaction for each compound (Table 10.5) is directly related to the stability of the intermediate cation.

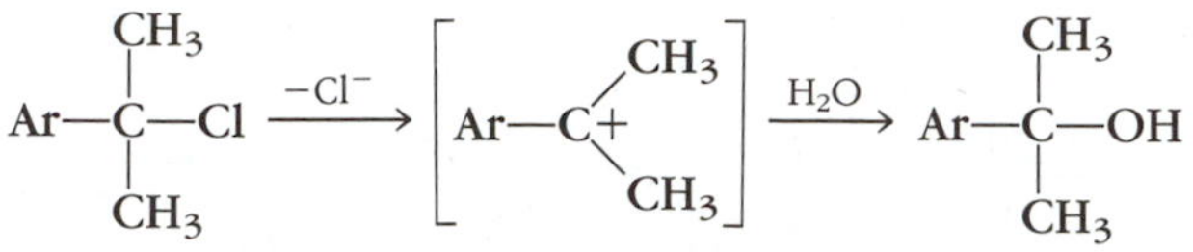

$$\text{Ar—C(CH}_3\text{)}_2\text{—Cl} \xrightarrow{-\text{Cl}^-} \left[\text{Ar—C}^+\text{(CH}_3\text{)}_2\right] \xrightarrow{\text{H}_2\text{O}} \text{Ar—C(CH}_3\text{)}_2\text{—OH}$$

Consider first the two methoxy (—OCH_3) compounds. The *meta*-methoxy derivative reacts more slowly than the parent phenyl compound, but the *para* isomer reacts much more rapidly. The *para* isomer affords a carbocation with a direct, stabilizing resonance interaction between the cationic center and the methoxy group (see structure **5**).

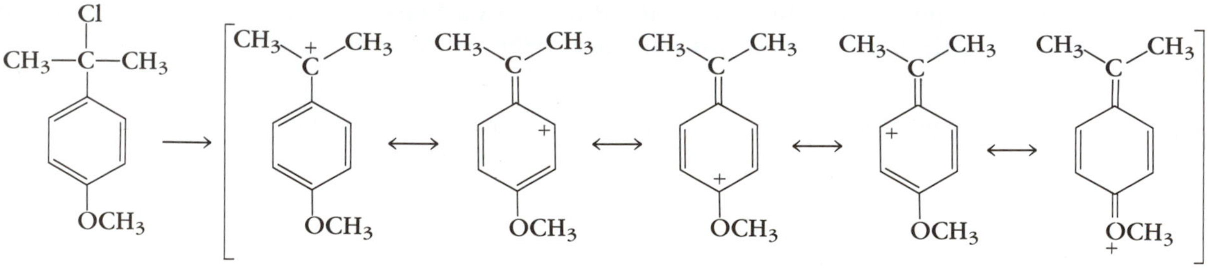

CH_3—C(Cl)—CH_3 on a benzene ring with para OCH_3 → [resonance structures **1** ↔ **2** ↔ **3** ↔ **4** ↔ **5** of the $(CH_3)_2C^+$-C_6H_4-OCH_3 cation; in **5**, $C=C_6H_4=\overset{+}{O}CH_3$]

1 **2** **3** **4** **5**

TABLE 10.5 Relative Rates of Hydrolysis of Some 2-Aryl-2-Chloropropanes, Ar—$C(CH_3)_2Cl$

Aryl Group	Relative Rate
C_6H_5	1
m-OCH_3—C_6H_4	0.6
p-OCH_3—C_6H_4	340
m-Cl—C_6H_4	0.02
p-Cl—C_6H_4	0.30

No such interaction between the methoxy group and the cationic center is possible for the *meta* isomer, however.

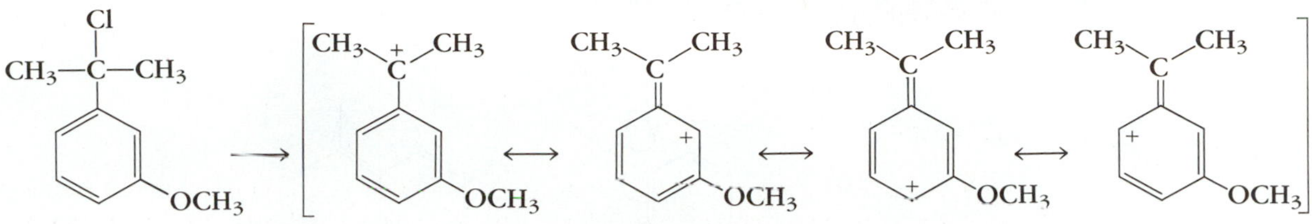

The slight decrease in reactivity of the *m*-methoxy derivative relative to the parent phenyl compound in Table 10.5 results from the electron-withdrawing inductive effect of the methoxy group. This small destabilizing inductive effect is completely overwhelmed by resonance stabilization of the cation in the *para* isomer, which reacts about 500 times more rapidly.

A comparison of inductive and resonance effects can also be made with the chloro compounds. Both isomers ionize to the corresponding carbocation more slowly than does the parent compound, 2-phenyl-2-chloropropane. This reflects the strong electron-withdrawing inductive effect of the electro negative chloro substituent. The inductive effect destabilizes the carbocation and retards the rate of ionization. However, the *para* isomer is more reactive than the *meta*-chloro compound by a factor of 15 because the strongly destabilizing inductive effect is partially compensated by a stabilizing resonance effect. This is shown by the following two resonance contributors:

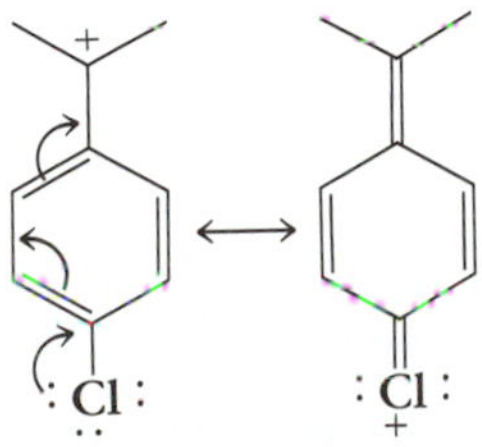

The ability of substituents *ortho* or *para* to a reaction center to stabilize an ionic intermediate by resonance can be illustrated without showing all the possible resonance forms. If you indicate the "flow" of electrons, it is possible to draw directly the resonance form of the interaction in which you are interested (i.e., the interaction with the other substituent). In the following equations we have used this technique to show how each reaction yields a product that is resonance-stabilized by a substituent on the aromatic ring.

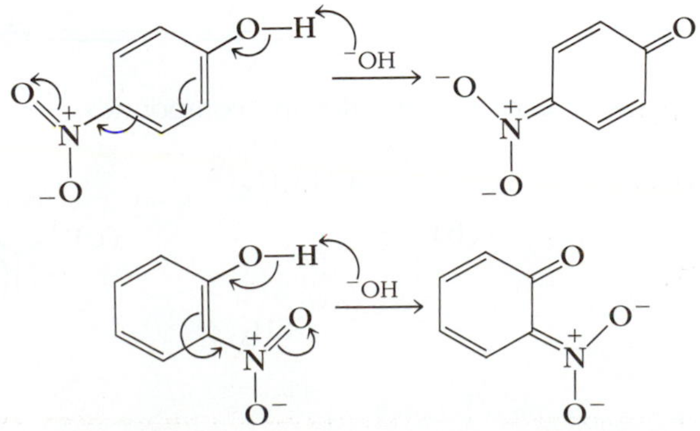

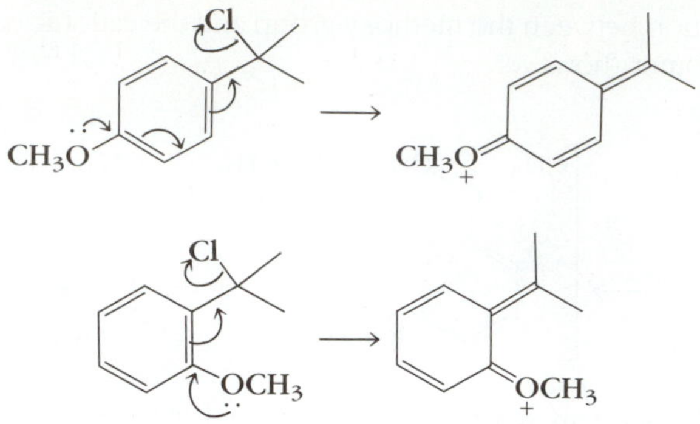

CH_3O → $CH_3\overset{+}{O}$

Cl → OCH₃ +

Resonance stabilization of intermediates is very important in the reactions of aromatic compounds. In Chapter 18 you will see that these same interactions can be used to predict the products and relative rates for a variety of aromatic substitution reactions.

EXERCISE 10.15

(a) Draw all the resonance forms for the ion produced by ionization of *o*-nitrophenol.
(b) Draw all the resonance forms for the ion produced by ionization of 2-(*o*-methoxyphenyl)-2-chloropropane.

10.10 TERMS AND DEFINITIONS

Aromatic. Belonging to a class of conjugated unsaturated compounds with characteristic features of structure and reactivity, usually related to those of benzene.

Fused ring system. A type of cyclic structure in which two adjacent atoms of one ring are also part of a second ring.

Hückel aromaticity. A criterion for aromaticity of fully conjugated cyclic molecules with $4n + 2$ π electrons.

Meta. A term describing the relative positions of two substituents in a 1,3 relationship on a benzene ring.

Ortho. A term describing the relative positions of two substituents in a 1,2 relationship on a benzene ring.

Para. A term describing the relative positions of two substituents in a 1,4 relationship on a benzene ring.

Resonance energy. The energy by which a conjugated molecule is stabilized through resonance effects.

10.11 PROBLEMS

10.1 Write the name of each of the following compounds:

(a) OH, CH_2CH_3

(b) OCH_2CH_3, CH_3

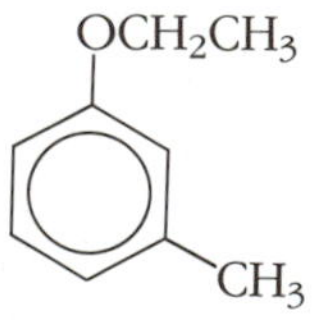

(c) I, $NHCH_3$, Br

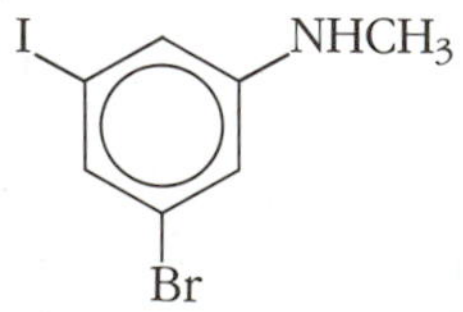

(d) 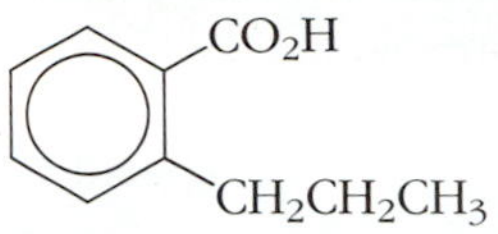 CO_2H, $CH_2CH_2CH_3$ (e) 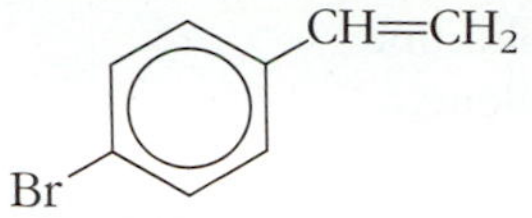 Br, $CH{=}CH_2$ (f) HO, Cl, F

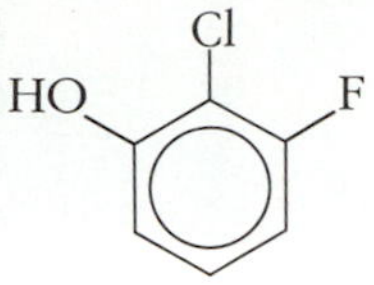

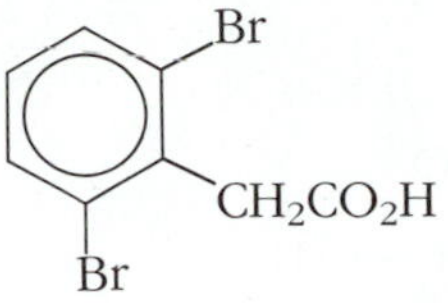

(g) 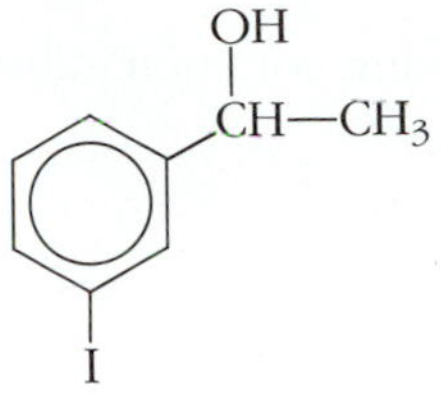 OH, $CH{-}CH_3$, I (h) Br, CH_2CO_2H, Br (i) CH_3, CO_2H, CH_3

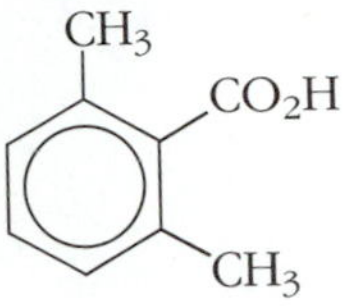

(j) 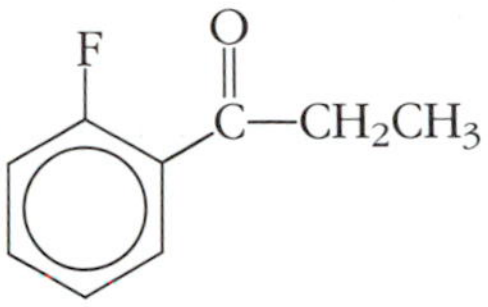 F, O, $C{-}CH_2CH_3$ (k) 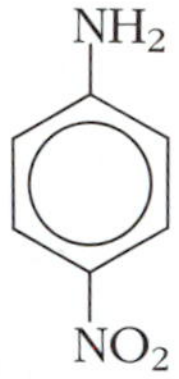 NH_2, NO_2 (l) 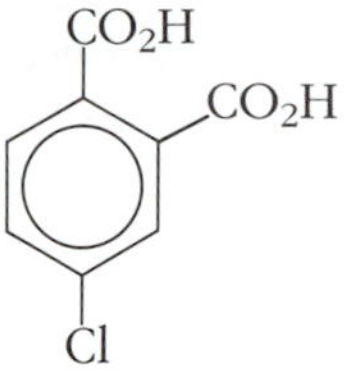 CO_2H, CO_2H, Cl

10.2 Draw the following compounds:

(a) Benzyl alcohol (b) Cinnamyl acetate
(c) Styrylcyclononane (d) Methyl α-phenethyl ether
(e) β-Phenethyl benzoate (f) Octyl phenyl ether

10.3 Draw the following compounds:

(a) 4-Bromo-3-phenylbutyric acid (b) 3-(4-Bromophenyl)butyric acid
(c) Ethyl phenylacetate (d) 3-Ethylphenyl acetate
(e) 2-Chloro-3-ethylaniline (f) *m*-(2-Chloroethyl)aniline
(g) 3,5-Diethylaniline (h) *N,N*-Diethylaniline

10.4 Draw and name all possible constitutional isomers with the molecular formula $C_8H_{10}O$ that have a benzene ring.

10.5 (a) Draw and name all structures that could be formed by replacement of the different hydrogens of *toluene* by a *chlorine*. (Restrict your answer to the isomeric compounds corresponding to a single replacement.)
(b) Draw and name all structures that could be formed by replacement of any *two* hydrogens of toluene by chlorines.

10.6 (a) Draw and name all structures that could be formed by replacement of the different hydrogens of *chlorobenzene* by a *chlorine*. (Restrict your answer to the isomeric compounds corresponding to a single replacement.)
(b) Draw and name all structures that could be formed by replacement of any *two* hydrogens of chlorobenzene by chlorines.

10.7 Draw the following compounds:

(a) 6-Chloro-2-nitrophenol (b) *N*-Methylanisidine
(c) Dimethyl phthalate (d) 3,5-Dinitrophenol
(e) *o*-Chloroaniline (f) *m*-Isopropylbenzoic acid
(g) Butyl anisate (h) *m*-Aminophenol

(i) *p*-Bromobenzoic acid
(j) 3-Bromo-4-chlorotoluene
(k) *p*-Toluidine
(l) *m*-Cresol
(m) *o*-Xylene
(n) 2,3-Diphenylphenol
(o) *o*-Benzylbenzoic acid
(p) *p*-Benzoylbenzoic acid
(q) *m*-Phenylbenzoic acid

10.8 Provide the systematic names for the following common substituted benzenes:

(a) $NH{-}C(=O){-}CH_3$ / OH — Acetaminophen (an analgesic)

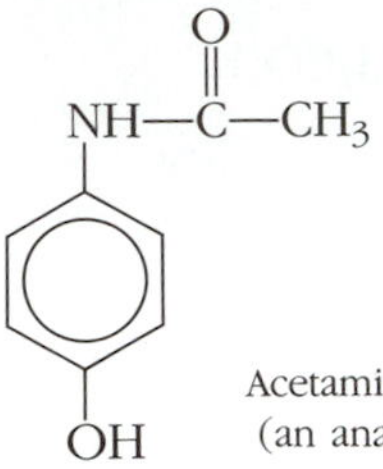

(b) $CO_2C_2H_5$ / NH_2 — Benzocaine

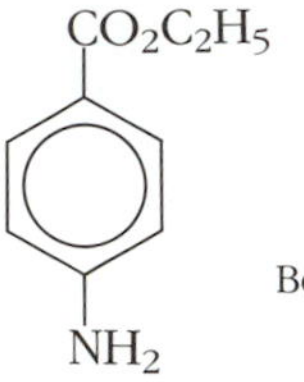

(c) CO_2H / NH_2 — PABA

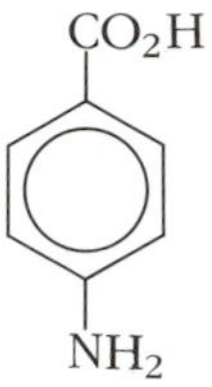

(d) CO_2H, $O{-}C(=O){-}CH_3$ — Aspirin

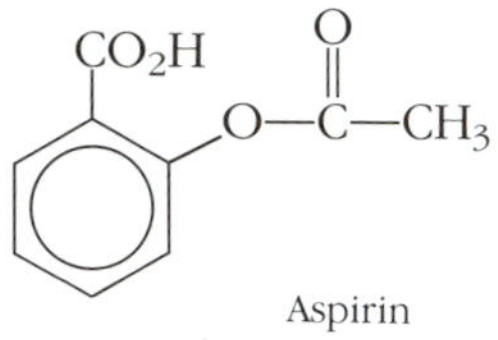

(e) CH_3, O_2N, NO_2, NO_2 — TNT

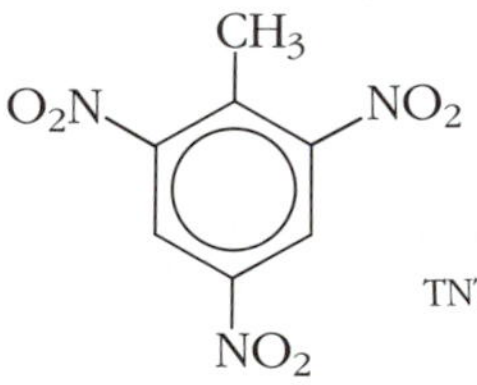

(f) OH, O_2N, NO_2, NO_2 — Picric acid

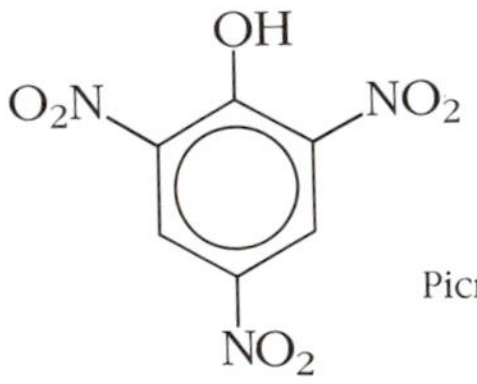

10.9 Draw all monocyclic isomers with the molecular formula C_6H_7N that would be aromatic.

10.10 Which would be more acidic, cycloheptatriene or cyclononatetraene? Why?

10.11 1,4-Di-*tert*-butylcyclooctatetraene has been shown to exist in two distinct forms. Suggest reasonable structures.

10.12 The isomer of cyclodecapentaene with two *E* double bonds shows none of the stability that would be expected for an aromatic compound. What structural feature would account for this?

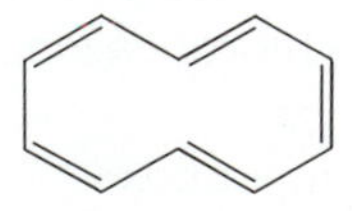

10.13 Answer the question for each pair of compounds. Justify your answer.
(a) Which would ionize to a carbocation more readily?

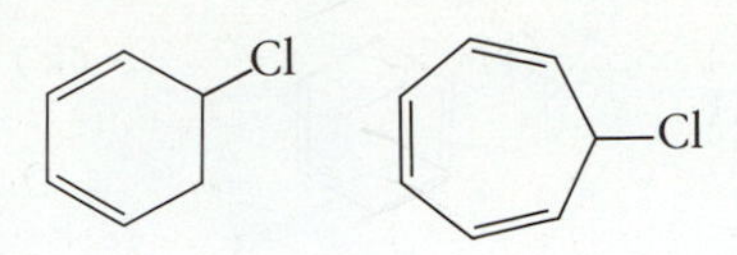

(b) Which would react more rapidly with water to yield an alcohol?

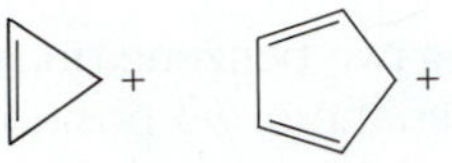

(c) Which is more basic?

(d) Which is more stable?

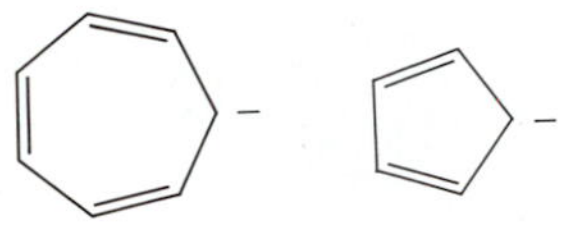

(e) Which is nonplanar?

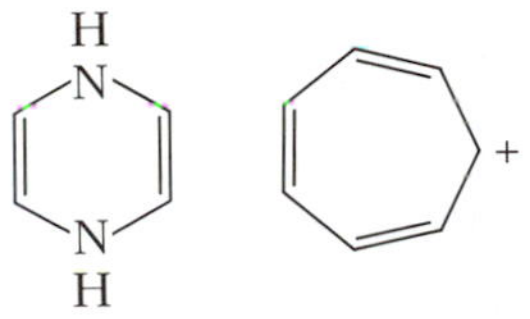

(f) Which is more acidic?

10.14 For each of the following structures, indicate: (1) whether or not the compound is a Hückel system; and (2) whether or not it is aromatic.

(a) 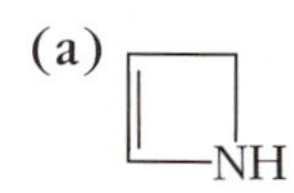

(b) 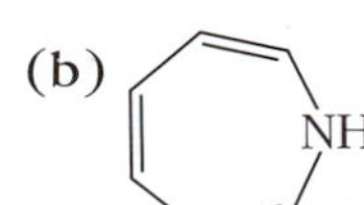

(c) 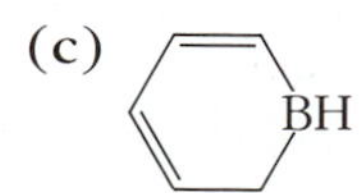

(d) 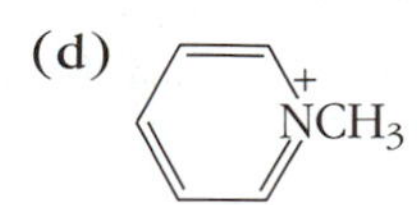

(e) 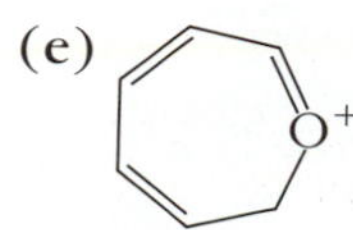

(f) 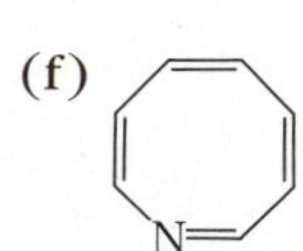

(g) 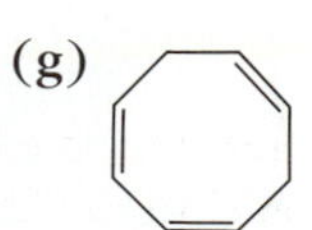

(h) 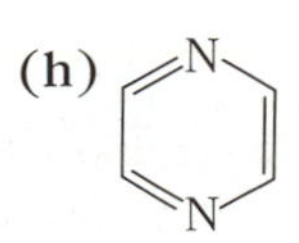

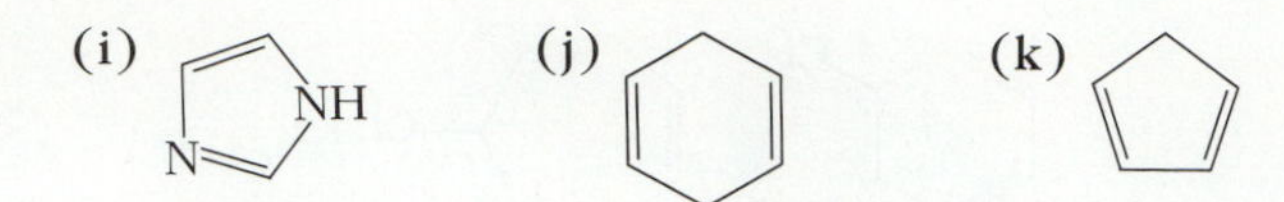

10.15 Write a detailed reaction mechanism for the nitration of benzene. Draw all resonance forms for the intermediate cation.

10.16 Compounds having two benzene rings joined by a single bond are known as *biphenyls* (A). When the *ortho* positions are substituted with bulky groups, certain biphenyls such as 2,2′-biphenyldisulfonic acid (B) can be obtained in optically active form. What does this tell about the preferred geometry of biphenyls? Draw three-dimensional representations of the two enantiomers of B.

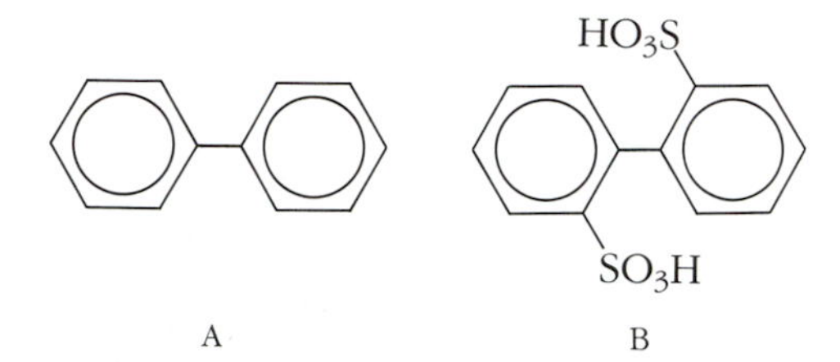

10.17 Write a detailed reaction mechanism for the nitration of toluene. Draw all resonance forms for the intermediate cations that are formed for *ortho, meta,* and *para* substitution. For which pathways does the methyl group stabilize the intermediate? What products would you expect to isolate from this reaction?

10.18 Write a detailed mechanism for the desulfonation reaction of benzenesulfonic acid.

10.19 Compound A was found to contain bromine as well as carbon and hydrogen. Deduce its structure from its ^{13}C NMR spectrum, shown in Figure 10.13.

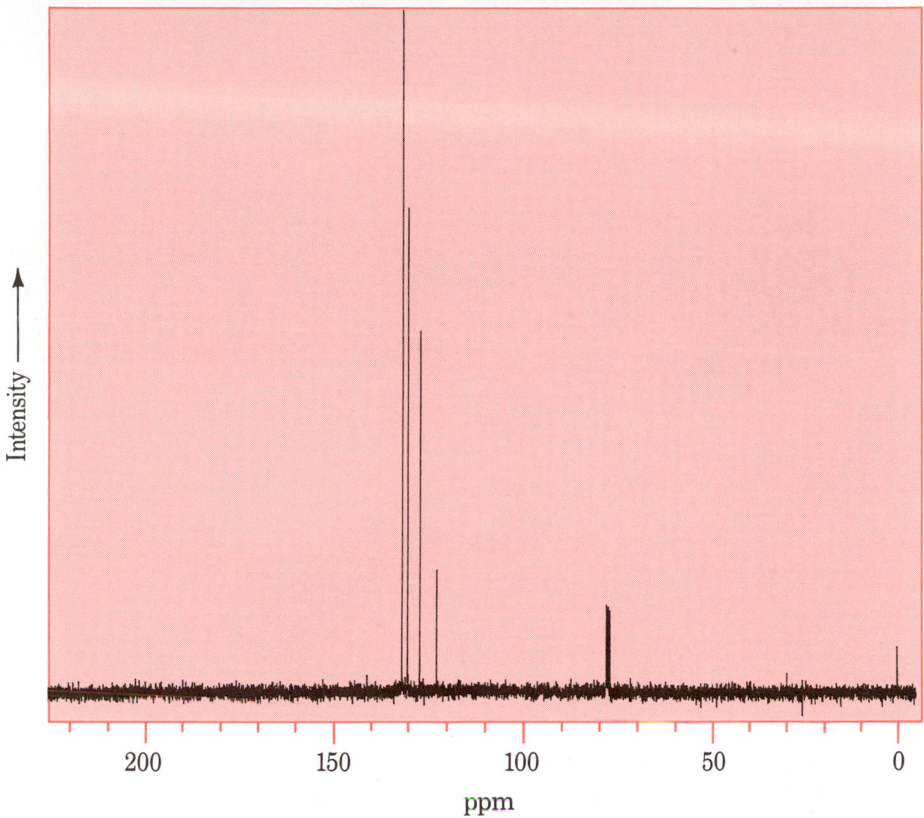

Figure 10.13 ^{13}C **NMR Spectrum of Compound A**

10.20 Benzoic acid has a dissociation constant of 6×10^{-5}, which is a typical value for a carboxylic acid (see Sections 4.7 and 4.8). Substitution of a nitro group on the ring increases the dissociation constant substantially; the dissociation constant for *m*-nitrobenzoic acid is 34×10^{-5}, and that for *p*-nitrobenzoic acid is 39×10^{-5}. In neither case is it possible to draw a resonance form in which the negative charge of the carboxylate ion is delocalized onto the nitro group, so the acidity-enhancing effect must be predominantly inductive in origin. Suggest a reason why *p*-nitrobenzoic acid is not a weaker acid than *m*-nitrobenzoic acid, even though the nitro and carboxyl groups are separated by a larger number of bonds in the former compound.

10.21 T.J. Katz (Columbia University) showed that cyclooctatetraene reacts with potassium metal to form a dianion ($C_8H_8^{2-}$) that is unusually stable. Experimental evidence indicates that this dianion is planar and symmetrical. Explain.

CHAPTER 11

ALCOHOLS

Alcohols are a class of organic compounds characterized by the presence of the hydroxyl group; they have the general formula R—OH. Alcohols play many important roles in industrial chemistry, synthetic chemistry, and biochemistry, and we will introduce you to their most important uses and reactions in this chapter.

One particular alcohol is well known to almost everyone, and that is the two-carbon compound called *ethanol.* Ethanol, frequently just called *alcohol,* is produced through the biological degradation of sugars by certain microorganisms, a chemical reaction that has been well known for many centuries. As the active constituent of intoxicating beverages, ethanol is probably the one organic compound that is most widely known (in name, taste, and biological properties) by all those who do not study organic chemistry.

11.1 THE HYDROXYL GROUP: ALCOHOLS AS DERIVATIVES OF WATER

Simple **alcohols** are hydrocarbon derivatives that carry a hydroxyl group (i.e., an OH group) as a substituent.

R—H	R—OH
Alkane	Alcohol

Except for their involvement in oxidation-reduction processes (Section 11.3), the reactivity of alcohols is very similar to that of water. For this reason it is often useful to consider an alcohol as an alkyl derivative of water.

$$H{-}OH \qquad R{-}OH$$

As we have already discussed in Chapter 4, alcohols can act as acids*

$$CH_3CH_2OH \rightleftharpoons CH_3CH_2O^- + H^+ \qquad K_a = 10^{-17}$$

and as bases

$$CH_3CH_2{-}\ddot{\underset{\cdot\cdot}{O}}H \overset{H^+}{\rightleftharpoons} CH_3CH_2{-}\overset{+}{\underset{\cdot\cdot}{O}}H_2$$

The latter reaction is the first step in the acid-catalyzed dehydration of alcohols (Section 5.13).

Still another similarity to water is illustrated by the reaction of alcohols with metals such as sodium to produce salts and generate hydrogen gas.

$$CH_3CH_2OH + Na \longrightarrow CH_3CH_2O^- + H_2\uparrow$$

In many organic reactions the metal salts of alcohols ($NaOCH_3$, KOtBu, etc.) are

*You should keep in mind that these acid-base reactions are all proton transfer reactions. Free H^+ ions are not formed in solution, and the ionization of an acid is actually a proton transfer, e.g.,

$$ROH + H_2O \rightleftharpoons RO^- + H_3O^+$$

Nevertheless, we will continue to use H^+ as a convenient notation.

used as a base in place of the corresponding salts of water (NaOH, KOH). This is done for two reasons:

1. Salts of alcohols tend to be more soluble in organic solvents (see Section 4.9). NaOH and NaOR are comparable in base strength, but the nonpolar alkyl group of NaOR increases its solubility in an organic solvent.
2. Some side reactions caused by hydroxide ion can be avoided by using salts of alcohols.

Water and hydroxide ion can both function as nucleophiles (Section 4.5), as shown in the following reactions with methyl bromide:

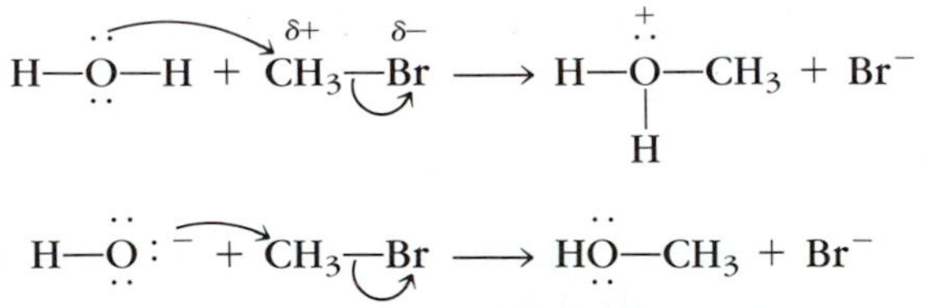

$$H{-}\ddot{\underset{\cdot\cdot}{O}}{-}H + \overset{\delta+}{CH_3}{-}\overset{\delta-}{Br} \longrightarrow H{-}\overset{+}{\underset{H}{\ddot{O}}}{-}CH_3 + Br^-$$

$$H{-}\ddot{\underset{\cdot\cdot}{O}}{:}^- + CH_3{-}Br \longrightarrow H\ddot{\underset{\cdot\cdot}{O}}{-}CH_3 + Br^-$$

In the same manner, alcohols and their conjugate bases can act as nucleophiles:

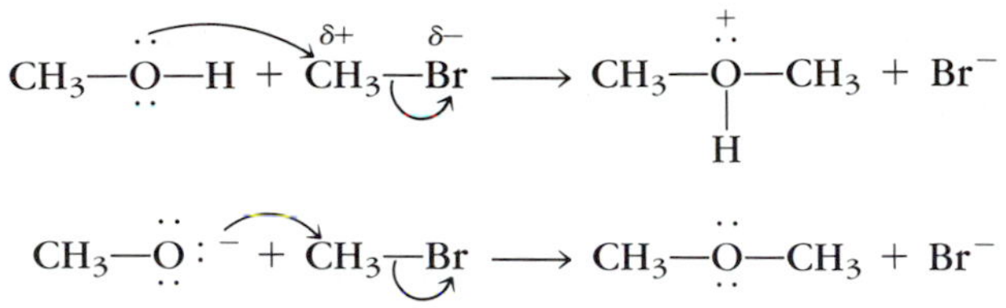

$$CH_3{-}\ddot{\underset{\cdot\cdot}{O}}{-}H + \overset{\delta+}{CH_3}{-}\overset{\delta-}{Br} \longrightarrow CH_3{-}\overset{+}{\underset{H}{\ddot{O}}}{-}CH_3 + Br^-$$

$$CH_3{-}\ddot{\underset{\cdot\cdot}{O}}{:}^- + CH_3{-}Br \longrightarrow CH_3{-}\ddot{\underset{\cdot\cdot}{O}}{-}CH_3 + Br^-$$

The latter reaction is discussed briefly in Section 11.8, and we will present a detailed discussion of nucleophilic substitution in Chapter 12. Although the chemical reactivity of alcohols is often analogous to that of water, you should recognize that there are also important differences. In this chapter we will examine the chemistry of alcohols, focusing particularly on those reactions that are unique to this functional group.

11.2 NOMENCLATURE OF ALCOHOLS

Alcohols are named by the same procedures that we previously introduced for alkanes, alkenes, and alkynes (see box).

Nomenclature of Alcohols and Their Derivatives

1. *Locate the principal carbon chain.* For an alcohol this is the longest carbon chain in the molecule *which contains the hydroxyl group.* The basic name is then formed by replacing the final *-e* in the name of the corresponding alkane by the suffix *-ol.*

2. *Name and number the substituents* on the principal chain.
   (a) The principal chain is numbered so that the lower of two possible numbers results for *the carbon bearing the hydroxyl group.* This number is prefixed to the name of the principal chain with a hyphen. If the location of the hydroxyl group does not distinguish between two different ways of numbering the principal chain, the substituents are evaluated so that the lower number results at the first point of difference.
   (b) When there is more than one substituent, they are listed in alphabetical order. If several of the substituents are identical, this is indicated by the numerical prefixes *di-, tri-, tetra-, penta-, hexa-,* etc.

3. *Specify the stereochemistry* for any center in the molecule that could exist in two alternative configurations.

The following examples show the relationship between the names of alkanes and alcohols with the same carbon skeleton:

$CH_3—CH_3$

Ethan*e*

$CH_3—CH_2—OH$

Ethan*ol*

$CH_3—CH_2—CH_2—CH_3$

Butan*e*

$CH_3—CH_2—CH_2—CH_2—OH$

1-Butan*ol*

The correct numbering of the principal chain is illustrated by the case of 6-methyl-3-heptanol:

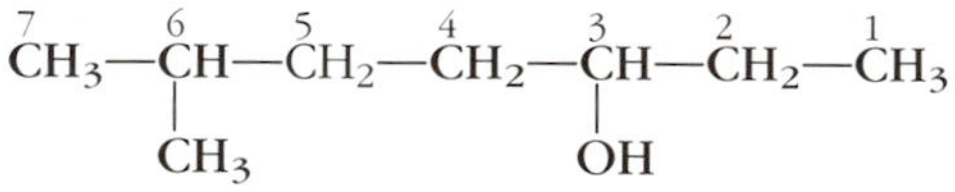

$$\overset{7}{CH_3}—\overset{6}{\underset{\underset{CH_3}{|}}{CH}}—\overset{5}{CH_2}—\overset{4}{CH_2}—\overset{3}{\underset{\underset{OH}{|}}{CH}}—\overset{2}{CH_2}—\overset{1}{CH_3}$$

6-Methyl-3-heptanol
(not 2-methyl-5-heptanol)

The following three examples show how substituted alcohols are named:

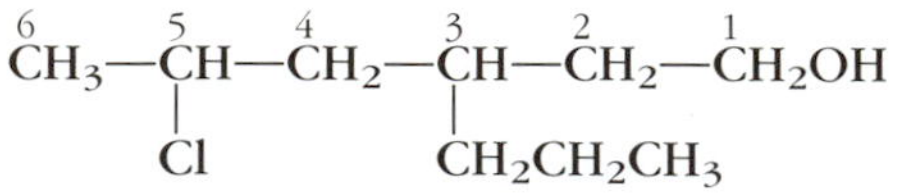

$$\overset{6}{CH_3}—\overset{5}{\underset{\underset{Cl}{|}}{CH}}—\overset{4}{CH_2}—\overset{3}{\underset{\underset{CH_2CH_2CH_3}{|}}{CH}}—\overset{2}{CH_2}—\overset{1}{CH_2OH}$$

5-Chloro-3-propyl-1-hexanol

(Note that the presence of the longer seven-carbon chain in the molecule has no relevance in naming the alcohol because it does not have the OH group as a substituent.)

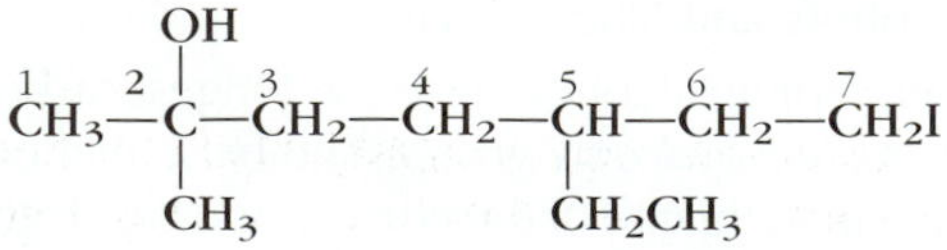

$$\overset{1}{CH_3}—\overset{\overset{OH}{|}}{\underset{\underset{CH_3}{|}}{\overset{2}{C}}}—\overset{3}{CH_2}—\overset{4}{CH_2}—\overset{5}{\underset{\underset{CH_2CH_3}{|}}{CH}}—\overset{6}{CH_2}—\overset{7}{CH_2I}$$

5-Ethyl-7-iodo-2-methyl-2-heptanol

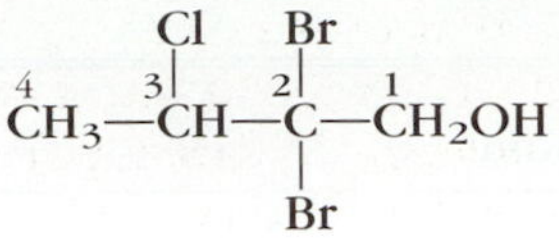

$$\underset{4}{CH_3}-\underset{3}{\overset{Cl}{CH}}-\underset{\underset{Br}{|}}{\overset{Br}{\overset{2}{C}}}-\underset{1}{CH_2OH}$$

2,2-Dibromo-3-chloro-1-butanol

(Remember that the numerical prefix *di* is not used in determining alphabetical order.)

Diols are usually named so that both hydroxyl groups are part of the principal chain, and the suffix *-diol* is simply added to the final *-ane* of the name for the corresponding alkane.

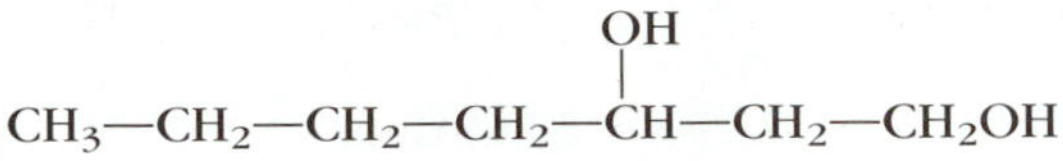

$$CH_3-CH_2-CH_2-CH_2-\overset{OH}{\overset{|}{CH}}-CH_2-CH_2OH$$

1,3-Heptanediol

$$CH_3-CH_2-CH_2-\underset{\underset{CH_2OH}{|}}{CH}-CH_2-CH_2-CH_2OH$$

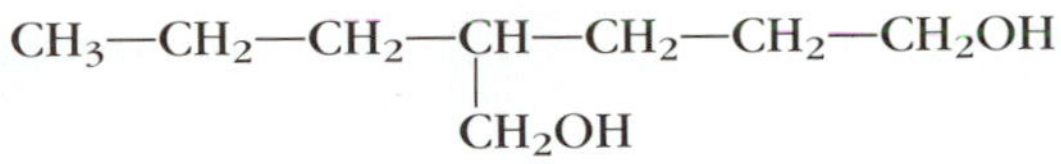

2-Propyl-1,5-pentanediol

Common Names of Alcohols

The IUPAC nomenclature just discussed can be applied to any structure, but nonsystematic names are also acceptable for a number of alcohols. These common names are frequently used in the chemical literature, and we have listed the more important examples in Table 11.1. Most of them are a combination of the name of the alkyl group and the word *alcohol*.

Unsaturated Alcohols

An alcohol that has a carbon-carbon double or triple bond as a second functional group is named by a slightly modified procedure, in which the principal chain is that which contains *both* the principal group (the OH group) and the multiple bond. The number of the first carbon atom of the multiple bond precedes the name of the principal chain, and the number of the carbon atom bearing the hydroxyl group precedes the *-ol* ending. Any other substituents on the principal chain are cited normally.

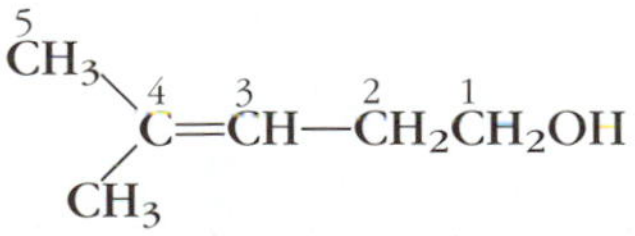

$$(\overset{5}{CH_3})(CH_3)\overset{4}{C}=\overset{3}{C}H-\overset{2}{C}H_2\overset{1}{C}H_2OH$$

4-Methyl-3-penten-1-ol

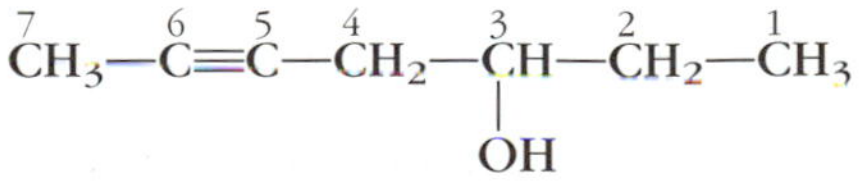

$$\overset{7}{C}H_3-\overset{6}{C}\equiv\overset{5}{C}-\overset{4}{C}H_2-\underset{\underset{OH}{|}}{\overset{3}{C}H}-\overset{2}{C}H_2-\overset{1}{C}H_3$$

5-Heptyn-3-ol

$$CH_3-CH_2-CH_2-\underset{\underset{\underset{5\quad 6}{CH=CH_2}}{|}}{\overset{4}{C}H}-\overset{3}{C}H_2-\overset{2}{C}H_2-\overset{1}{C}H_2OH$$

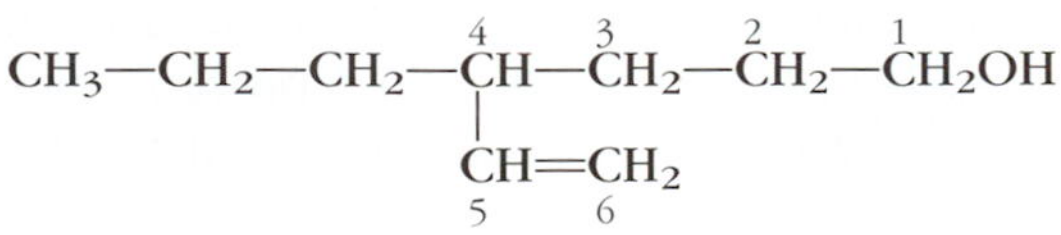

4-Propyl-5-hexen-1-ol

Metal Salts

The anion derived from an alcohol can always be named by adding the suffix *-ate* to the final *-ol* for the name of the alcohol. By this procedure the sodium salt of

TABLE 11.1 Common Names of Some Common Alcohols

Alcohol	Common Name	IUPAC Name
CH_3OH	Methyl alcohol	Methanol
CH_3CH_2OH	Ethyl alcohol	Ethanol
$CH_3-CH(OH)-CH_3$	Isopropyl alcohol	2-Propanol
$CH_3-C(CH_3)_2-OH$	*tert*-Butyl alcohol	2-Methyl-2-propanol
$CH_2{=}CH-CH_2OH$	Allyl alcohol	2-Propen-1-ol
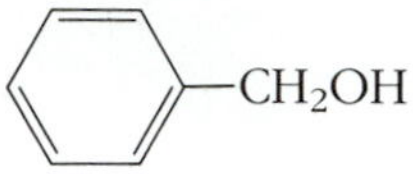 $C_6H_5-CH_2OH$	Benzyl alcohol	Phenylmethanol
$HO-CH_2CH_2-OH$	Ethylene glycol	1,2-Ethanediol
$CH_2(OH)-CH(OH)-CH_2(OH)$	Glycerol	1,2,3-Propanetriol

2-propanol is called sodium 2-propanolate, and the potassium salt of benzyl alcohol is called potassium benzyl alcoholate.

The frequently encountered anions of unsubstituted alcohols with up to four carbon atoms are usually named by an alternative procedure in which the *-yl* ending of the alkyl group is replaced by *-oxide*. This is illustrated by the following examples:

$NaOCH_3$ Sodium methoxide

$NaOCH_2CH_3$ Sodium ethoxide

$Mg(OCH(CH_3)-CH_3)_2$ Magnesium isopropoxide

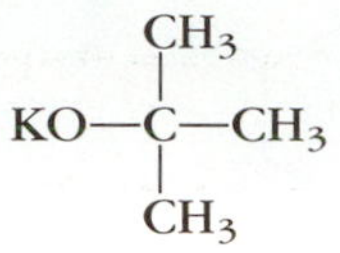

$$KO-C(CH_3)_2-CH_3$$

Potassium *tert*-butoxide (KOtBu)

EXERCISE 11.1

Name the following compounds:

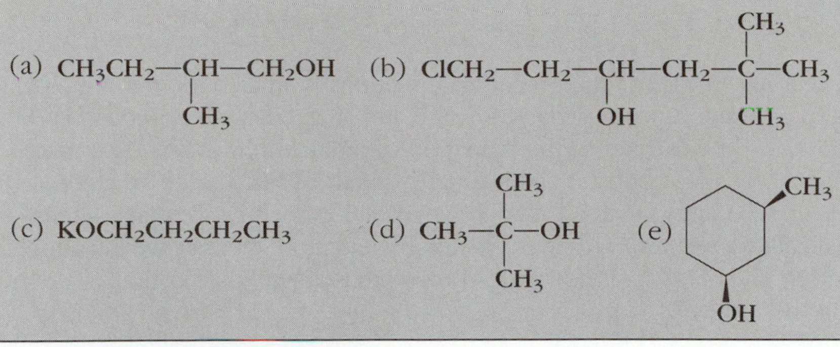

(a) $CH_3CH_2-CH(CH_3)-CH_2OH$

(b) $ClCH_2-CH_2-CH(OH)-CH_2-C(CH_3)_2-CH_3$

(c) $KOCH_2CH_2CH_2CH_3$

(d) $CH_3-C(CH_3)_2-OH$

(e) [cyclohexane ring bearing CH_3 (wedge bond) and OH (wedge bond) in a 1,3 relationship]

EXERCISE 11.2

Draw the following structures:

(a) 4-Ethyl-2-hexanol
(b) Calcium ethoxide
(c) (*Z*)-3-Bromo-4-methyl-3-hexen-1-ol
(d) Cyclobutanol
(e) 7-Octyn-4-ol

11.3 OXIDATION OF ALCOHOLS

The interconversion of functional groups is very important in organic chemistry, and oxidation-reduction reactions provide the basis for many such transformations. For example, alcohols are formed as the products of reduction reactions of carbonyl compounds. Conversely, many alcohols undergo facile oxidation to various carbonyl compounds by inorganic oxidizing agents. Carbonyl compounds, that is, compounds that contain the **carbonyl group,** $>C=O$, include: **aldehydes,** in which one of the two substituents on the carbonyl group is a hydrogen and the other is an alkyl (or aryl) group; **ketones,** in which both substituents are alkyl (or aryl) groups; and **carboxylic acids,** in which one of the substituents on the carbonyl group is an OH. The following reactions illustrate the oxidation of alcohols to carbonyl compounds:

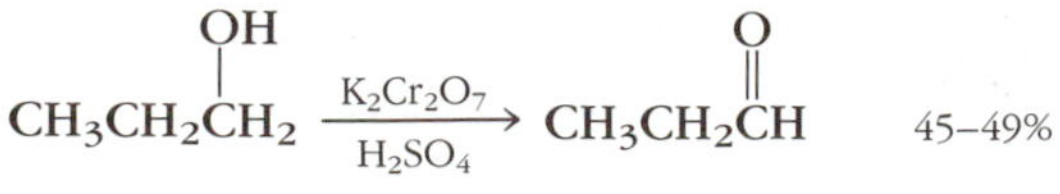

$$CH_3CH_2CH_2OH \xrightarrow[H_2SO_4]{K_2Cr_2O_7} CH_3CH_2CH{=}O \qquad 45\text{–}49\%$$

An aldehyde

$$\text{4-ethylcyclohexanol} \xrightarrow[H_2SO_4]{Na_2Cr_2O_7} \text{4-ethylcyclohexanone} \quad 90\%$$

A ketone

$$Cl_3C{-}CH_2CH_2CH_2{-}CH_2OH \xrightarrow{KMnO_4} Cl_3C{-}CH_2CH_2CH_2{-}\overset{O}{\overset{\|}{C}}{-}OH \quad 92\%$$

A carboxylic acid

The reactant in each of the preceding equations is an alcohol, but the products belong to different functional classes. How can you predict which type of carbonyl compound will be formed in the oxidation of an alcohol? This is determined by two factors: (1) the oxidizing agent employed, and (2) the type of alcohol (i.e., primary or secondary) that is undergoing oxidation. Before we discuss these factors in detail, we will return to the question of oxidation state that we introduced in Section 5.6. The oxidation state of a carbon atom is assigned according to the rules listed in the following box.

Calculation of Oxidation State

1. Calculate the oxidation state of only one carbon at a time, using a one-carbon fragment. The sum of the oxidation states of the carbon atom and of its substituents will equal the formal charge on the fragment.

2. The one-carbon fragment that you evaluate consists of the carbon atom and the atoms attached directly to it.
   (a) Carbon substituents are *ignored*.
   (b) Singly bonded oxygen substituents (OR, etc.) are counted as OH groups.

3. The oxidation state of oxygen is always −2, and the oxidation state of hydrogen is always taken as +1. Substituents other than carbon have oxidation states that are the same as the charge that they would have as free ions (i.e., OH is −1, Cl is −1, Li is +1, Mg is +2, etc.)

We will illustrate the use of this formalism by assigning oxidation states to the carbon atoms of methane and carbon dioxide, the extremes in oxidation state of carbon. In methane, the carbon atom has four hydrogens as substituents, each of which is in the +1 oxidation state. The formal charge on the carbon atom is zero, so its oxidation state must be −4. In the case of CO_2, the substituents on carbon are two doubly bonded oxygens. Again, the carbon atom has a formal charge of zero, so its oxidation state must be +4.

CH_4 (each H: +1)

Sum of oxidation states of substituents	+4
Oxidation state of carbon	−4
+	
Formal charge on carbon atom	0

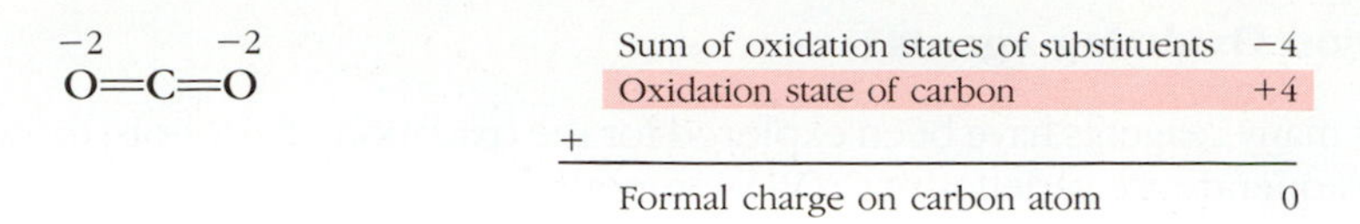

When the oxidation state of a carbon atom increases in a reaction, that carbon has been oxidized. Conversely, if its oxidation state decreases, it has been reduced. We will illustrate changes in oxidation state by evaluating several specific reactions. The increase in oxidation state at the reactive center (i.e., the carbon atom of the reactant that bears the hydroxyl group) shows that the conversion of 1-propanol to propionic acid is an oxidation.

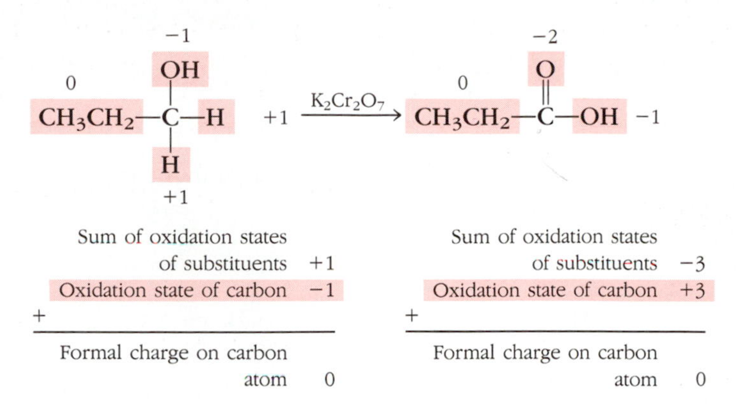

The overall reaction involves an increase in oxidation state of the carbon atom from −1 to +3, *a four-electron oxidation*. This is the only carbon atom in the molecule that is modified in the reaction, and it is the *change in oxidation state* that provides the key to evaluating the overall oxidation reaction.

The formalism of oxidation state is not used frequently in organic chemistry, but it can clearly demonstrate whether or not a net oxidation or reduction has taken place. Thus, oxidation of an alkene to a diol with $KMnO_4$ involves an increase in oxidation state at each of two carbon atoms of the reactive center.

$$\mathrm{R{-}CH_2{-}\underset{-1}{CH}{=}\underset{-2}{CH_2} \xrightarrow{KMnO_4} R{-}CH_2{-}\underset{0}{CH(OH)}{-}\underset{-1}{CH_2}{-}OH}$$

Change in oxidation state

On the other hand, there is no *net change* in oxidation state as a result of addition of water to an alkene. The oxidation state of one carbon atom increases while that of the other decreases.

$$\mathrm{R{-}CH_2{-}\underset{-1}{CH}{=}\underset{-2}{CH_2} \xrightarrow[H^+]{H_2O} R{-}CH_2{-}\underset{0}{CH(OH)}{-}\underset{-3}{CH_3}}$$

Reactions such as this, in which there is no overall change in oxidation state, are not considered oxidation-reduction reactions.

Common Oxidizing Agents

A great many reagents have been explored for the oxidation of alcohols; those used most frequently are varieties of Cr(VI) compounds (although $KMnO_4$ is used occasionally). The common Cr(VI) reagents are CrO_3, H_2CrO_4, $H_2Cr_2O_7$, $Na_2Cr_2O_7$, and $K_2Cr_2O_7$, and they are generally interchangeable. The chromium reagents are the reagents of choice for most laboratory oxidations of alcohols.

Industrial processes require inexpensive reagents, so most large-scale commercial oxidations are *catalytic processes* in which atmospheric oxygen is the oxidant. Formaldehyde, for example, can be prepared by catalytic oxidation of methanol using a silver catalyst.

$$CH_3OH \xrightarrow[O_2,\ 450\text{–}650°C]{Ag} H_2C{=}O \qquad 83\text{–}92\%$$

Similarly, a copper-silver alloy can be used as a catalyst in the oxidation of ethanol to the corresponding aldehyde.

$$CH_3CH_2OH \xrightarrow[O_2,\ >500°C]{Cu\text{-}Ag} CH_3\overset{O}{\overset{\|}{C}}H \qquad 76\%$$

These catalytic processes are not generally applicable to most oxidations carried out under laboratory conditions; the oxidizing agents most commonly used in the laboratory are the chromium reagents discussed previously.

One type of chromium reagent merits special comment, complexes of CrO_3 with pyridine. Most common procedures for Cr(VI) oxidations employ acidic conditions, but acidic conditions (which may be undesirable) can be avoided by using the basic solvent pyridine. A more convenient procedure employs complexes of pyridine with CrO_3 in an inert solvent. The dipyridine complex and the monopyridine–hydrogen chloride complex are very useful oxidizing agents.

$$[CrO_3 \cdot 2\,C_5H_5N] \qquad [CrO_3 \cdot C_5H_5N \cdot HCl]$$

For convenience we will designate both these complexes (as well as the oxidizing agent in which pyridine is used as solvent) as CrO_3-pyridine. These reagents allow oxidation under nonacidic conditions, and they also permit the reliable oxidation of a primary alcohol to an aldehyde (rather than to a carboxylic acid).

Oxidation of Primary, Secondary, and Tertiary Alcohols

The oxidation products of an alcohol are directly related to the degree of substitution of the carbon atom to which the OH group is attached. Primary alcohols, with two hydrogen atoms attached to that carbon atom, can undergo either a two-electron oxidation to an aldehyde or a four-electron oxidation to a carboxylic acid. The actual product formed depends upon which oxidizing agent is used. Most oxidizing agents yield the carboxylic acid, but the aldehydes can be prepared by utilizing one of the CrO_3-pyridine reagents.

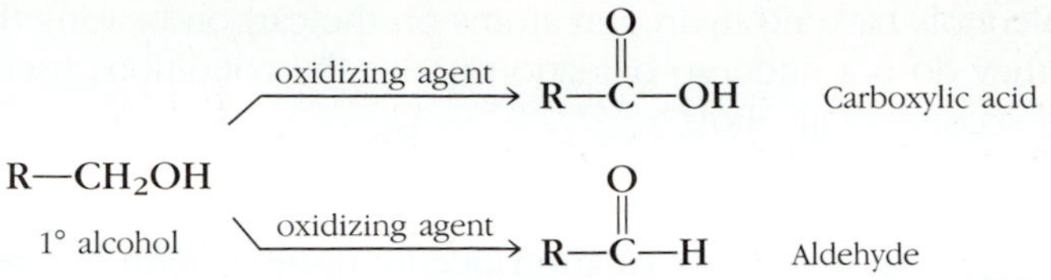

$$R{-}CH_2OH \xrightarrow{\text{oxidizing agent}} R{-}\overset{O}{\overset{\|}{C}}{-}OH \quad \text{Carboxylic acid}$$

$$R{-}CH_2OH \xrightarrow{\text{oxidizing agent}} R{-}\overset{O}{\overset{\|}{C}}{-}H \quad \text{Aldehyde}$$

1° alcohol

The following examples show how the choice of reagent governs whether an aldehyde or a carboxylic acid is formed in the oxidation of a primary alcohol.

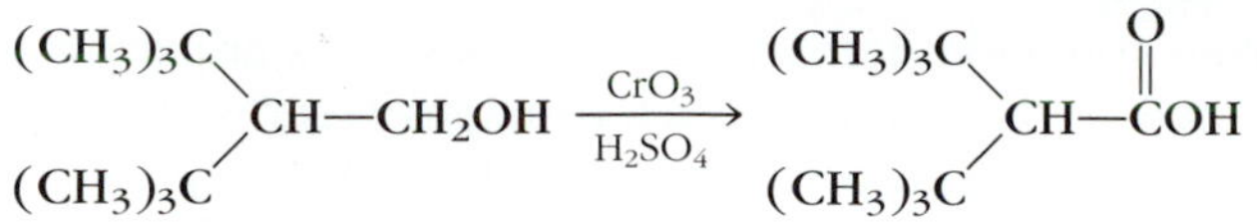

$$((CH_3)_3C)_2CH{-}CH_2OH \xrightarrow[H_2SO_4]{CrO_3} ((CH_3)_3C)_2CH{-}\overset{O}{\overset{\|}{C}}OH \quad 82\%$$

$$CH_3{-}(CH_2)_5{-}CH_2OH \xrightarrow{KMnO_4} CH_3{-}(CH_2)_5{-}\overset{O}{\overset{\|}{C}}OH \quad 70\%$$

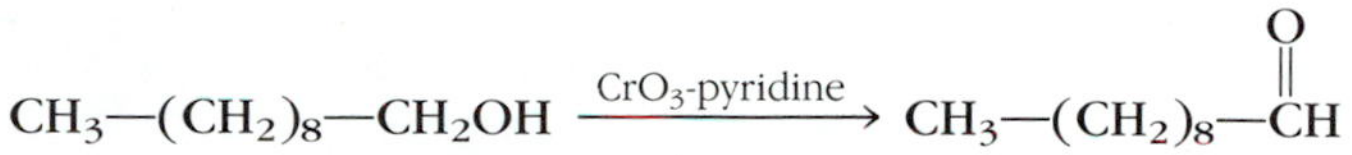

$$CH_3{-}(CH_2)_8{-}CH_2OH \xrightarrow{CrO_3\text{-pyridine}} CH_3{-}(CH_2)_8{-}\overset{O}{\overset{\|}{C}}H \quad 92\%$$

$$\text{CH}_2\text{OH} \xrightarrow{CrO_3\text{-pyridine}} \overset{O}{\overset{\|}{C}}{-}H \quad 82\%$$

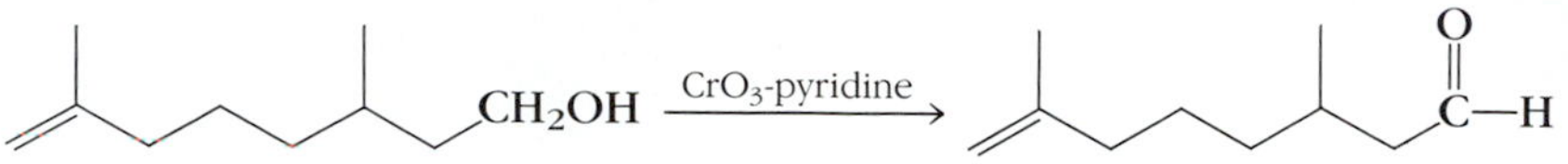

Secondary alcohols have a single hydrogen atom on the carbon to which the OH is attached. Therefore only a two-electron oxidation is expected, and a ketone is the resulting product. The choice of oxidizing agent usually makes little difference in the oxidation of secondary alcohols.

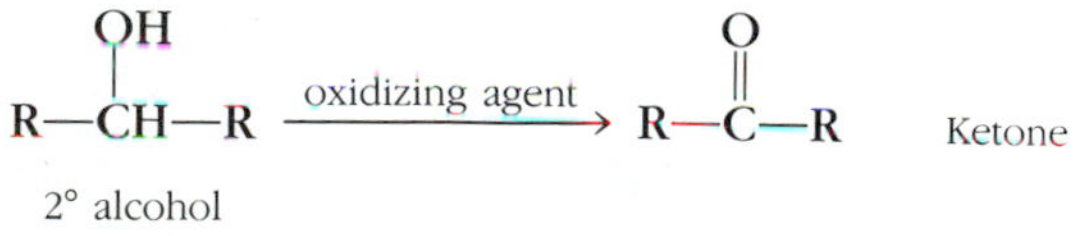

$$R{-}\underset{}{\overset{OH}{\overset{|}{C}H}}{-}R \xrightarrow{\text{oxidizing agent}} R{-}\overset{O}{\overset{\|}{C}}{-}R \quad \text{Ketone}$$

2° alcohol

The following examples illustrate the oxidation of secondary alcohols to ketones using Cr(VI) reagents.

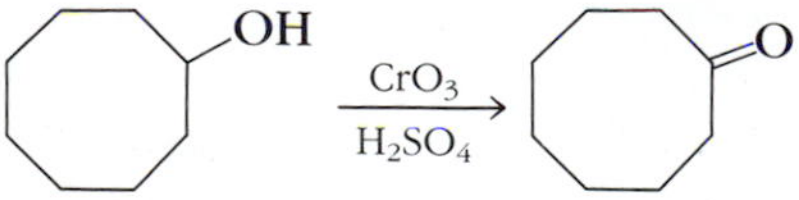

$$\text{OH} \xrightarrow[H_2SO_4]{CrO_3} \text{O} \quad 92\text{–}96\%$$

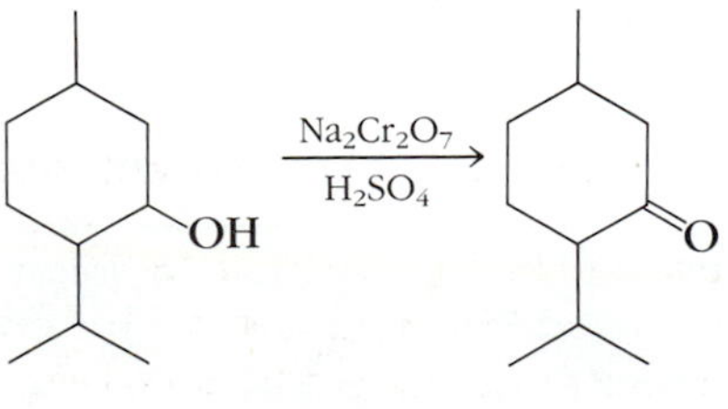

$$\text{OH} \xrightarrow[H_2SO_4]{Na_2Cr_2O_7} \text{O} \quad 94\%$$

$$(CH_3)_3C \text{ — OH} \xrightarrow{CrO_3\text{-pyridine}} (CH_3)_3C \text{ — O} \quad 97\%$$

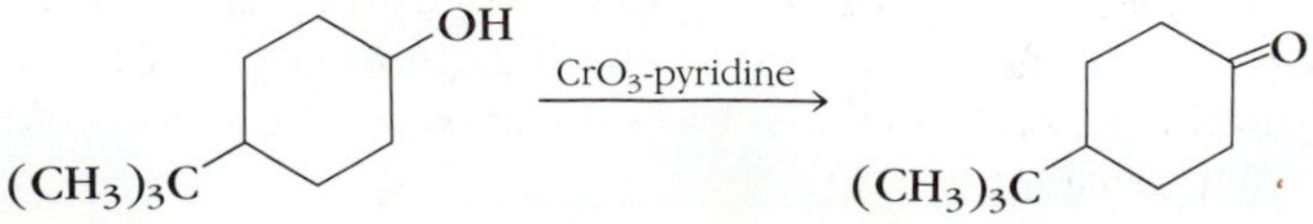

Tertiary alcohols have no hydrogen atoms on the carbon bearing the hydroxyl group, and they do not undergo oxidation under the conditions used to oxidize primary and secondary alcohols.

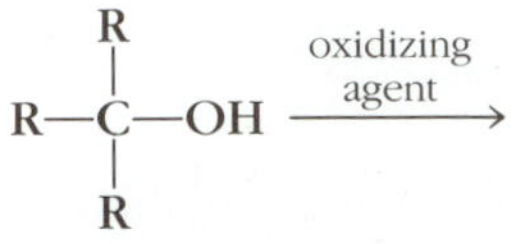

$$\underset{3^\circ \text{ alcohol}}{R_3C{-}OH} \xrightarrow[\text{agent}]{\text{oxidizing}}$$ **No oxidation occurs under conditions used to oxidize primary and secondary alcohols**

Under more vigorous conditions tertiary alcohols (as well as primary and secondary alcohols) will undergo oxidation, but such reactions occur with the cleavage of carbon-carbon bonds. All these processes usually fall into the category of unwanted side reactions.

EXERCISE 11.3

Draw the product expected from each of the following reactions.

(a) Cyclohexyl$-CH_2OH$ $\xrightarrow{KMnO_4}$; $\xrightarrow{CrO_3\text{-pyridine}}$

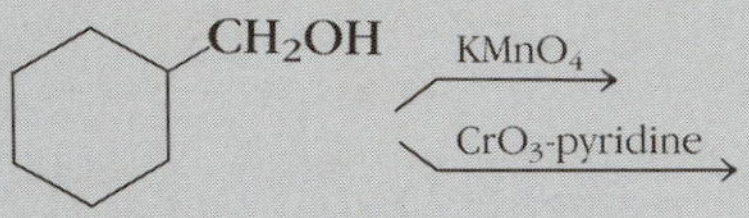

(b) $(CH_3)_2CH{-}CH(OH){-}CH_3 \xrightarrow[H_2SO_4]{CrO_3}$

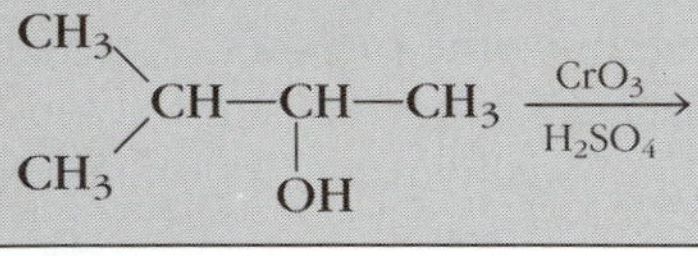

EXERCISE 11.4

What oxidizing agent would be needed to carry out each of the following conversions?

(a) $C_6H_5{-}CH_2{-}CH(CH_3){-}CH_2OH \xrightarrow{?} C_6H_5{-}CH_2{-}CH(CH_3){-}CHO$

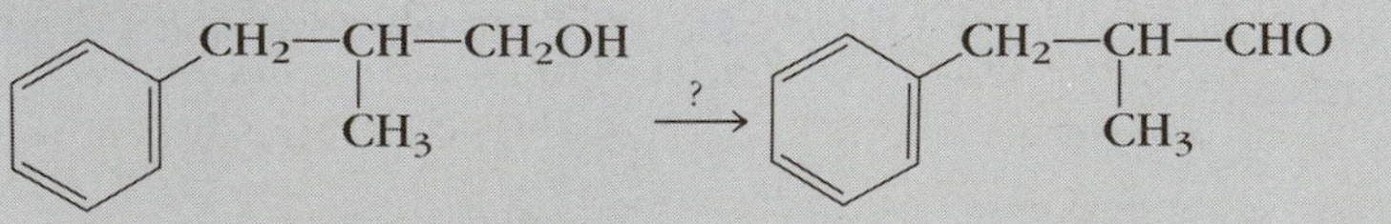

(b) $CH_3CH_2{-}CH(OH){-}CH_2C{\equiv}CH \xrightarrow{?} CH_3CH_2{-}C({=}O){-}CH_2{-}C{\equiv}CH$

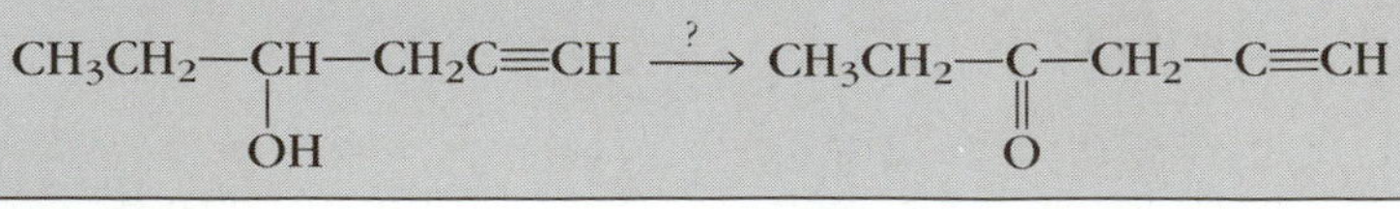

EXERCISE 11.5

Deduce the structure of the starting material for the following reaction:

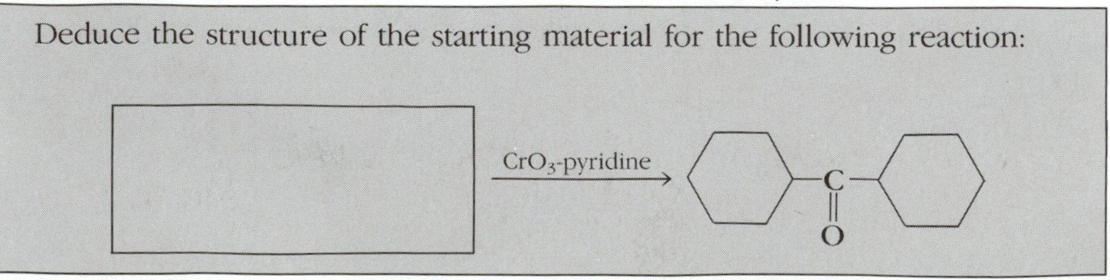

11.4 FORMATION OF ALCOHOLS BY REDUCTION OF CARBONYL COMPOUNDS

Just as alcohols can be oxidized to carbonyl compounds, so can carbonyl compounds be reduced to alcohols. The relationship between functional groups is the same as for oxidation: primary alcohols result from the reduction of aldehydes and carboxylic acids, and secondary alcohols are formed by the reduction of ketones.

$$R-\overset{O}{\overset{\|}{C}}-H \longrightarrow R-CH_2OH$$

$$R-\overset{O}{\overset{\|}{C}}-OH \longrightarrow R-CH_2OH$$

$$R-\overset{O}{\overset{\|}{C}}-R \longrightarrow R-\overset{OH}{\overset{|}{C}H}-R$$

Esters, carboxylic acid derivatives in which the substituents on the carbonyl group are one alkyl (or aryl) and one alkoxy group, also yield primary alcohols upon reduction. A second alcohol, corresponding to the alkoxy group of the starting ester, is generated as well.

$$R-\overset{O}{\overset{\|}{C}}-OR' \xrightarrow{\text{reduction}} R-CH_2OH + HOR'$$

There has been no reduction in the fragment designated as R′, and the alcohol R′OH is ordinarily just a by-product of the reduction reaction. All these reductions of carbonyl groups involve addition of hydrogen to carbon-oxygen double bonds or replacement of oxygen substituents by hydrogen; bonds between carbons atoms are unaffected.

A wide variety of reducing agents are known to cause these transformations, but some of the most useful show high selectivity; they can be used to reduce just one group in a polyfunctional compound. (Such preferential reaction at one of several different reactive centers is known as **chemoselectivity**). In this section we will discuss some of the important methods for reduction of carbonyl compounds to alcohols.

Catalytic Hydrogenation

We previously discussed catalytic hydrogenation as a method for reducing unsaturated hydrocarbons, but it can also be used for the reduction of carbonyl compounds to the corresponding alcohols.

$$H-\overset{O}{\overset{\|}{C}}-CH_2-\overset{CH_3}{\overset{|}{C}H}-CH_2\overset{O}{\overset{\|}{C}}-H \xrightarrow[100\text{ atm}]{H_2,\ Ni} HO-CH_2CH_2-\overset{CH_3}{\overset{|}{C}H}-CH_2CH_2-OH \quad 81\text{–}83\%$$

$$C_6H_{11}-\overset{O}{\overset{\|}{C}}-O-CH_2CH_3 \xrightarrow[\text{copper chromite catalyst}]{H_2,\ 4000\text{ psi}} C_6H_{11}-CH_2OH \quad (+\ HOCH_2CH_3) \quad 97\%$$

These reactions frequently require much higher pressures than are needed for reduction of carbon-carbon multiple bonds, so the latter are likely to undergo preferential reduction. As a result catalytic hydrogenation is not the most common laboratory method for reduction of carbonyl compounds, although it is used in a number of industrial processes.

Dissolving Metal Reductions

Dissolving metal reductions, in which a reactive metal is used as the reducing agent, have also been used extensively for reduction of carbonyl compounds to alcohols. The process is called a *dissolving metal reduction,* because the metal is converted to a soluble metal salt during the reaction. The reduction proceeds via two successive transfers of an electron from the metal to the carbonyl group, each of which is followed by a proton transfer from the solvent (or from some added proton source).

$$R{-}\overset{\overset{\displaystyle :\ddot{O}:}{\|}}{C}{-}R \xrightarrow{e^-} \left[R{-}\overset{\overset{\displaystyle :\dot{O}:}{|}}{\underset{\ddot{}^{-}}{C}}{-}R \longleftrightarrow R{-}\overset{\overset{\displaystyle :\ddot{O}:^-}{|}}{\underset{\cdot}{C}}{-}R \right] \xrightarrow[\text{transfer}]{\text{proton}} R{-}\overset{\overset{\displaystyle :\dot{O}:}{|}}{CH}{-}R \xrightarrow{e^-} R{-}\overset{\overset{\displaystyle :\ddot{O}:^-}{|}}{CH}{-}R \xrightarrow[\text{transfer}]{\text{proton}} R{-}\overset{\overset{\displaystyle :\ddot{O}H}{|}}{CH}{-}R$$

The reduction of carbonyl compounds by dissolving metals is illustrated by the following examples:

$$CH_3{-}(CH_2)_4{-}\overset{\overset{\displaystyle O}{\|}}{C}{-}CH_3 \xrightarrow[CH_3CH_2OH]{Na} CH_3{-}(CH_2)_4{-}\overset{\overset{\displaystyle OH}{|}}{CH}{-}CH_3 \quad 62\text{–}65\%$$

$$CH_3{-}(CH_2)_7{-}CH{=}CH{-}(CH_2)_7{-}\overset{\overset{\displaystyle O}{\|}}{C}OC_2H_5 \xrightarrow[CH_3CH_2OH]{Na}$$

$$CH_3{-}(CH_2)_7{-}CH{=}CH{-}(CH_2)_7CH_2OH \quad 49\text{–}51\%$$

The second example shows that such reactions can be carried out on some difunctional compounds without affecting the second functional group. Sodium metal in alcohol is a powerful reducing agent, but it does not normally affect an isolated carbon-carbon double bond.

Metal Hydride Reductions

Metal hydride reductions are highly useful for the selective reduction of carbonyl groups. Two reagents have gained wide use in the organic laboratory; sodium borohydride ($NaBH_4$) and lithium aluminum hydride ($LiAlH_4$).

$$Li^+ \left[H{-}\overset{\overset{\displaystyle H}{|}}{\underset{\underset{\displaystyle H}{|}}{Al}}{-}H \right]^- \qquad Na^+ \left[H{-}\overset{\overset{\displaystyle H}{|}}{\underset{\underset{\displaystyle H}{|}}{B}}{-}H \right]^-$$

Lithium aluminum hydride   Sodium borohydride

The active species in these cases are the borohydride ion (BH_4^-) and aluminohydride ion (AlH_4^-), and they react by transferring a hydride ion (H^-) to the carbonyl group. The carbonyl group is polarized in a manner such that the carbon atom is highly susceptible to nucleophilic attack, and the metal ions can act as Lewis acids to further enhance the reactivity of the carbonyl group.

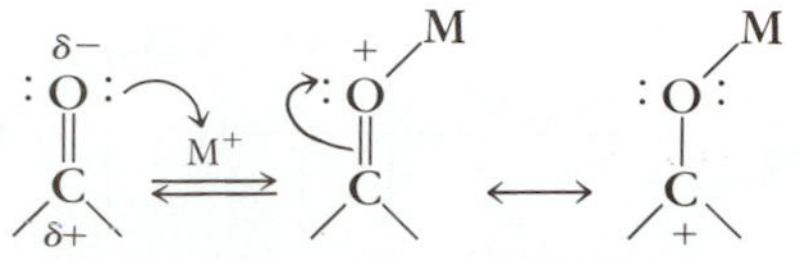

$$:\overset{\delta-}{O}:=\overset{\delta+}{C}< \;\underset{}{\overset{M^+}{\rightleftharpoons}}\; \overset{+}{O}(M)=C< \;\longleftrightarrow\; :O(M)-\overset{+}{C}<$$

The hydrogen is transferred *with its bonding pair of electrons* from the metal hydride ion to the carbonyl compound, as illustrated for the borohydride ion:

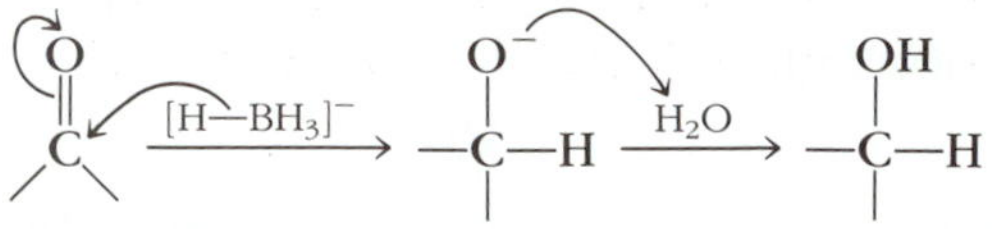

$$>C=O \xrightarrow{[H-BH_3]^-} -\overset{O^-}{\underset{|}{C}}-H \xrightarrow{H_2O} -\overset{OH}{\underset{|}{C}}-H$$

Sodium borohydride is a very selective reducing agent that readily reduces aldehydes and ketones but not carboxylic acids or esters. It is typically used in either aqueous or alcoholic solvent, and the anion produced by hydride attack is protonated by the solvent. The following reactions illustrate the reduction of aldehydes and ketones with sodium borohydride:

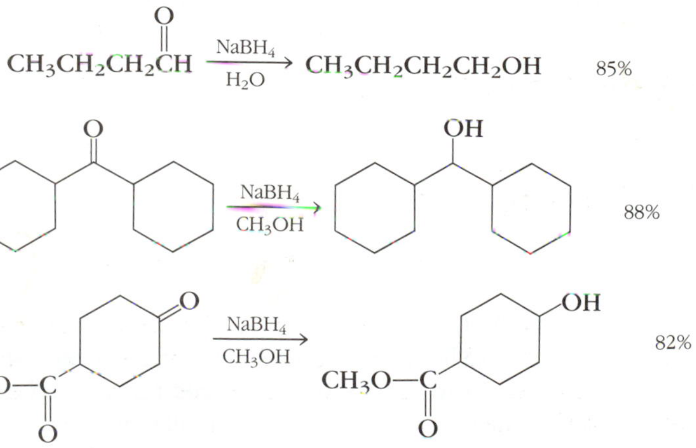

$$CH_3CH_2CH_2\overset{O}{\overset{\|}{C}}H \xrightarrow[H_2O]{NaBH_4} CH_3CH_2CH_2CH_2OH \quad 85\%$$

$$C_6H_{11}-\overset{O}{\overset{\|}{C}}-C_6H_{11} \xrightarrow[CH_3OH]{NaBH_4} C_6H_{11}-\overset{OH}{\overset{|}{C}H}-C_6H_{11} \quad 88\%$$

$$CH_3O-\overset{O}{\underset{\|}{C}}-C_6H_9{=}O \xrightarrow[CH_3OH]{NaBH_4} CH_3O-\overset{O}{\underset{\|}{C}}-C_6H_{10}-OH \quad 82\%$$

The last example clearly demonstrates the selective nature of sodium borohydride. The less reactive ester carbonyl group remains unaffected, but the ketone carbonyl is cleanly reduced.

Lithium aluminum hydride is a much more powerful reducing agent than sodium borohydride, and it reduces carboxylic acids and esters as well as aldehydes and ketones. Because it is so highly reactive, lithium aluminum hydride must be used in an **aprotic** solvent (i.e., one that cannot act as a proton donor), such as diethyl ether, $CH_3CH_2—O—CH_2CH_3$. Any solvent that contains OH or NH groups would destroy the reagent by acting as an acid (proton donor) in a reaction with "hydride."

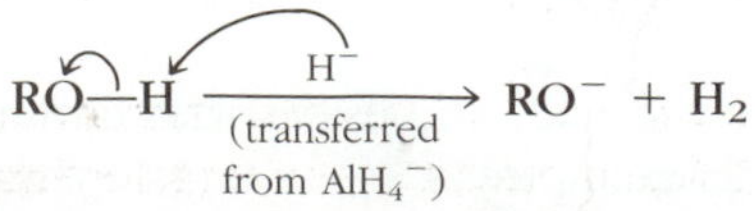

$$RO-H \xrightarrow[\text{(transferred from } AlH_4^-)]{H^-} RO^- + H_2$$

Since aprotic solvents must be used for $LiAlH_4$ reductions, the oxygen anions are generated as their aluminum salts. All four hydrides can be transferred, although we have shown only the first step in the following equation.

$$\mathrm{>C{=}O} \xrightarrow{[H{-}AlH_3]^- Li^+} \left[\mathrm{-\overset{O-AlH_3}{\underset{|}{C}}-H} \right]^- \quad Li^+$$

In a subsequent procedure, called **workup,** water or some other proton donor is added when the reaction is complete. This produces the corresponding alcohol, together with lithium and aluminum salts.

$$\mathrm{-\overset{O-{}^-Al(OR)_3Li^+}{\underset{|}{C}}-H} \xrightarrow{H_2O} \mathrm{-\overset{OH}{\underset{|}{C}}-H} + LiOH + Al(OH)_3$$

The following equations illustrate the formation of alcohols by the reduction of aldehydes and ketones with lithium alumninum hydride.

$$CH_3{-}(CH_2)_5{-}\overset{O}{\overset{\|}{C}}H \xrightarrow[2.\ H_2O]{1.\ LiAlH_4} CH_3{-}(CH_2)_5{-}CH_2OH \qquad 86\%$$

$$\text{cyclobutanone} \xrightarrow[2.\ H_2O]{1.\ LiAlH_4} \text{cyclobutanol} \qquad 90\%$$

$$\text{dicyclopropyl ketone} \xrightarrow[2.\ H_2O]{1.\ LiAlH_4} \text{dicyclopropylmethanol} \qquad 89\%$$

The reductions of carboxylic acids and esters occur by several steps, in which the *aldehyde* is an intermediate. As shown for the reduction of an ester, the addition of one hydride produces a salt, from which an alkoxide ion can be lost. This affords the aldehyde, which reacts rapidly with a second hydride. Finally, workup yields the primary alcohol.

$$R{-}\overset{O}{\overset{\|}{C}}{-}OCH_3 \xrightarrow{LiAlH_4} R{-}\overset{O-{}^-Al\langle}{\underset{H}{C}}{-}OCH_3 \longrightarrow R{-}\overset{O}{\overset{\|}{C}}{-}H \xrightarrow{LiAlH_4} R{-}\overset{O-{}^-Al\langle}{\underset{H}{C}}{-}H$$

$$\xrightarrow[\text{(workup)}]{H_2O} R{-}CH_2OH$$

The reduction of carboxylic acids and esters with lithium aluminum hydride is illustrated by the following examples:

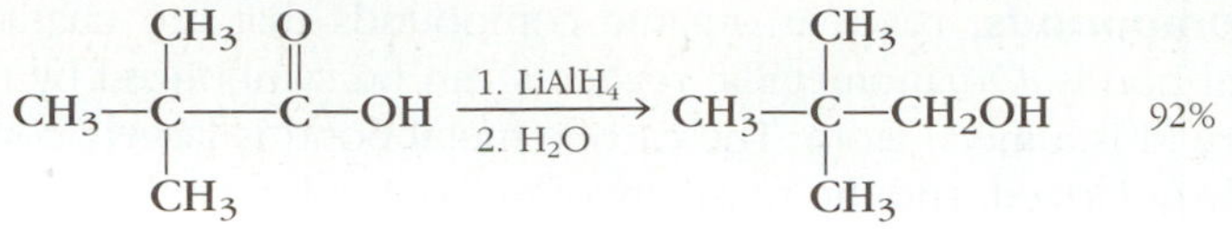

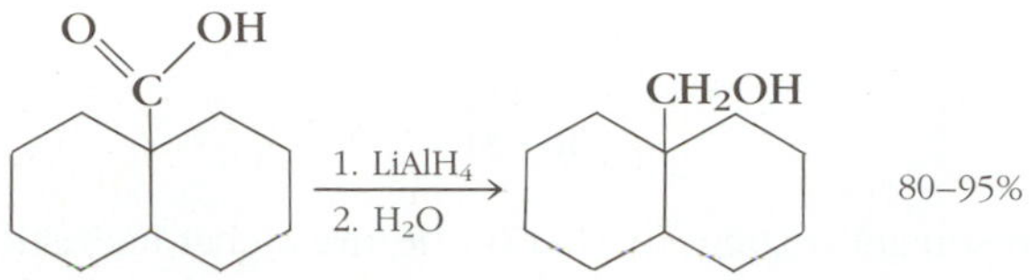

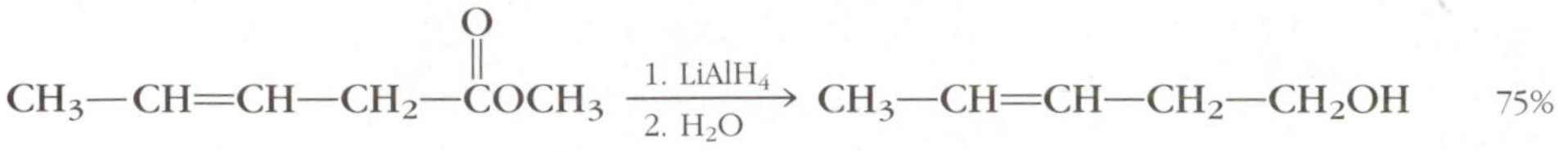

$$CH_3-C(CH_3)_2-C(=O)-OH \xrightarrow[2.\ H_2O]{1.\ LiAlH_4} CH_3-C(CH_3)_2-CH_2OH \quad 92\%$$

$$\text{(decalin-COOH)} \xrightarrow[2.\ H_2O]{1.\ LiAlH_4} \text{(decalin-}CH_2OH) \quad 80\text{–}95\%$$

$$CH_3-CH{=}CH-CH_2-\overset{O}{\overset{\|}{C}}OCH_3 \xrightarrow[2.\ H_2O]{1.\ LiAlH_4} CH_3-CH{=}CH-CH_2-CH_2OH \quad 75\%$$

Lithium aluminum hydride readily attacks all types of carbonyl groups. Carbon-carbon double bonds are not usually reduced, but virtually any *polar* multiple bond (particularly multiple bonds to oxygen or nitrogen) will be reduced by this reagent. Even alkyl halides, epoxides, and alkynes are attacked by $LiAlH_4$, and its use in the reduction of polyfunctional compounds is therefore limited by other groups that might be present in the molecule. On the other hand, sodium borohydride permits the reduction of ketone and aldehyde carbonyl groups in the presence of a large variety of functional groups. Borane (BH_3) will also reduce carbonyl groups, but this reagent must be prepared from sodium borohydride, so $NaBH_4$ is a more economical and convenient reducing agent for aldehydes and ketones.

EXERCISE 11.6

Draw the product expected from each of the following reactions:

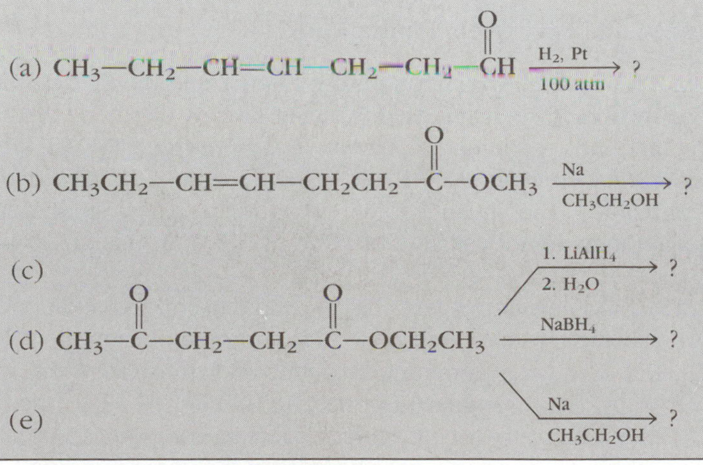

(a) $CH_3-CH_2-CH{=}CH-CH_2-CH_2-\overset{O}{\overset{\|}{C}}H \xrightarrow[100\ atm]{H_2,\ Pt} ?$

(b) $CH_3CH_2-CH{=}CH-CH_2CH_2-\overset{O}{\overset{\|}{C}}-OCH_3 \xrightarrow[CH_3CH_2OH]{Na} ?$

$CH_3-\overset{O}{\overset{\|}{C}}-CH_2-CH_2-\overset{O}{\overset{\|}{C}}-OCH_2CH_3$

(c) $\xrightarrow[2.\ H_2O]{1.\ LiAlH_4} ?$

(d) $\xrightarrow{NaBH_4} ?$

(e) $\xrightarrow[CH_3CH_2OH]{Na} ?$

11.5 PREPARATION OF ALCOHOLS VIA REACTION OF CARBONYL COMPOUNDS WITH ORGANOMETALLIC REAGENTS

In the previous section we showed you how carbonyl groups are susceptible to nucleophilic attack by the "hydride ion" of a reagent such as lithium aluminum hydride. Carbonyl groups are also susceptible to nucleophilic attack by **organo-**

metallic compounds, reactive organic compounds that are characterized by carbon-metal bonds. Organometallic reagents can be symbolized by the formula R—M, where M is a metal atom. The carbon-metal bond is largely covalent but it is also highly polarized, and the reactivity of most organometallic compounds can be predicted in terms of a carbon atom with partial negative charge:

$$\overset{\delta-}{R}—\overset{\delta+}{M}$$

This simple representation suggests two of the most characteristic and important reactions of organometallic compounds: (1) nucleophilic attack on a carbonyl group

$$\overset{\delta+}{C}=\overset{\delta-}{O} \xrightarrow{\overset{\delta-}{R}—\overset{\delta+}{M}} —C(O^-M^+)—R$$

and (2) reaction as a base by abstraction of a proton from a hydroxylic solvent

$$R'O—H + \overset{\delta-}{R}—\overset{\delta+}{M} \longrightarrow R—H + R'—O^-M^+$$

The second of these reactions is not typically used in a preparative way; instead, it is observed as an undesired side reaction with hydroxylic impurities in reactions of organometallic reagents with carbonyl compounds. Because they are highly reactive, organometallic reagents must be prepared and used in an aprotic solvent. Any hydroxylic impurity, particularly water, will destroy the reagent and convert it to the corresponding hydrocarbon, R—H.

Organometallic Reagents and Their Limitations

A variety of different organometallic compounds are known, and their reactivities differ markedly as a function of the metal used. We have already discussed the use of organocopper compounds, particularly *lithium dialkylcuprates,* R_2CuLi (Sections 1.11, 5.14), and *sodium acetylides,* R—C≡C—Na (Section 7.10), in *coupling reactions* with alkyl halides. The organocopper compounds react rapidly with hydroxylic solvents, but they do not usually attack the carbonyl groups of ketones or esters.

The two types of organometallic reagent that are most useful for addition to carbonyl groups are alkyllithium compounds (R—Li) and organomagnesium halides (e.g., R—Mg—Br). The latter compounds, better known as *Grignard reagents,* were discovered by Victor Grignard (France, 1871–1935). In 1912 he received the Nobel Prize in chemistry for his studies of these compounds.

The organolithium compounds are stronger bases than the corresponding Grignard reagents, and they sometimes undergo the side reaction of abstraction of an acidic hydrogen α to a carbonyl group (Section 4.7).

$$—C(=O)—C(H)< \xrightarrow{R—Li} —C(=O)—C^-< \longleftrightarrow —C(O^-)=C<$$

This side reaction occurs less frequently with Grignard reagents. The greatly enhanced acidity of a CH group adjacent to two carbonyl groups precludes the use of such dicarbonyl compounds with organometallic reagents. Only proton abstraction is observed; nucleophilic attack on the carbonyl groups does not occur.

$$-C(=O)-CH_2-C(=O)- \xrightarrow{R-M} -C(=O)-\overset{-}{C}H-C(=O)- \; M^+ + R-H$$

For similar reasons it is not usually possible to carry out nucleophilic attack on the carbonyl group of a polyfunctional compound that contains another relatively acidic group, especially an —OH group.

Preparation of Organometallic Reagents

Both alkyllithium and Grignard reagents are prepared by stirring the appropriate metal with a solution of an alkyl halide, using diethyl ether or another ether as solvent. The following equations summarize the preparation of these reagents:

$$R-Br \xrightarrow{Li} R-Li + LiBr$$

$$R-I \xrightarrow{Mg} R-MgI$$

The organometallic compounds are highly reactive and are not ordinarily isolated. However, they are relatively stable in solution, and solutions of some of the simpler derivatives (e.g., CH_3Li, CH_3MgBr, tBuLi) are sold commercially. These solutions can be used directly in reactions with carbonyl compounds.

Reactions of Alkyllithium and Grignard Reagents with Carbonyl Compounds

The reactions of alkyllithium and Grignard reagents with carbonyl compounds are analogous to the corresponding reactions of lithium aluminum hydride.* A single addition takes place with an aldehyde or ketone, and the resulting metal salt is converted to the alcohol during workup with water or some other acid.

$$>C=O \xrightarrow{R-Li} -C(OLi)(R)- \xrightarrow{H_2O} -C(OH)(R)-$$

$$>C=O \xrightarrow{R-MgBr} -C(OMgBr)(R)- \xrightarrow{H_2O} -C(OH)(R)-$$

*The reaction mechanisms illustrated here are somewhat oversimplified. For example, solutions of a Grignard reagent also contain other reactive species such as magnesium halides (which are relatively strong Lewis acids) and dialkylmagnesium compounds, which can also react with carbonyl compounds. In addition, there is some evidence that the overall reaction may proceed via transfer of an electron from the organometallic reagent to the carbonyl group to generate a pair of radicals, which couple in a subsequent step.

These reactions are illustrated by the following examples:

$$CH_3-\underset{}{\overset{CH_3}{\overset{|}{C}}}HBr \xrightarrow{Mg} CH_3-\overset{CH_3}{\overset{|}{C}}H-MgBr \xrightarrow[2.\ H_2O]{1.\ CH_3\overset{O}{\overset{\|}{C}}H} CH_3-\overset{CH_3}{\overset{|}{C}}H-\overset{OH}{\overset{|}{C}}H-CH_3 \quad 53\text{–}54\%$$

$$CH_3-\overset{O}{\overset{\|}{C}}-CH_3 \xrightarrow[2.\ H_2O]{1.\ CH_3CH_2CH_2CH_2MgBr} CH_3-\underset{CH_3}{\underset{|}{\overset{OH}{\overset{|}{C}}}}-CH_2CH_2CH_2CH_3 \quad 91\%$$

$$\text{cyclohexanone} \xrightarrow[2.\ H_2O]{1.\ (CH_3)_3C-Li} \text{1-}\textit{tert}\text{-butylcyclohexanol (OH, } C(CH_3)_3) \quad 53\%$$

$$\text{cyclohexyl–Cl} \xrightarrow{Mg} \text{cyclohexyl–MgCl} \xrightarrow[2.\ H_2O]{1.\ H_2C=O} \text{cyclohexyl–}CH_2OH \quad 64\text{–}69\%$$

In contrast to aldehydes and ketones, esters consume two equivalents of the organometallic reagent. After the first nucleophilic addition, an alkoxide ion is expelled to generate a new carbonyl compound, which undergoes a second reaction with the organometallic compound. The final product is an alcohol, formed by protonation of the metal salt during workup of the reaction.

$$\overset{O}{\overset{\|}{C}}-OCH_3 \xrightarrow{R-Li} -\underset{OCH_3}{\underset{|}{\overset{O-Li}{\overset{|}{C}}}}-R \longrightarrow \overset{O}{\overset{\|}{C}}-R \xrightarrow{R-Li} -\underset{R}{\underset{|}{\overset{O-Li}{\overset{|}{C}}}}-R \xrightarrow[\text{(workup)}]{H_2O} -\underset{R}{\underset{|}{\overset{OH}{\overset{|}{C}}}}-R$$

$$\overset{O}{\overset{\|}{C}}-OCH_3 \xrightarrow{R-MgBr} -\underset{OCH_3}{\underset{|}{\overset{O-MgBr}{\overset{|}{C}}}}-R \longrightarrow \overset{O}{\overset{\|}{C}}-R \xrightarrow{R-MgBr} -\underset{R}{\underset{|}{\overset{O-MgBr}{\overset{|}{C}}}}-R \xrightarrow[\text{(workup)}]{H_2O} -\underset{R}{\underset{|}{\overset{OH}{\overset{|}{C}}}}-R$$

Notice that two *identical* alkyl groups are introduced when a tertiary alcohol is formed by reaction of an ester with an alkyllithium or a Grignard reagent.

$$CH_3-(CH_2)_{14}-\overset{O}{\overset{\|}{C}}-OCH_3 \xrightarrow[2.\ H_2O]{1.\ C_2H_5MgBr} CH_3-(CH_2)_{14}-\underset{CH_2CH_3}{\underset{|}{\overset{OH}{\overset{|}{C}}}}-CH_2CH_3 \quad 96\%$$

$$CH_3CH_2CH_2CH_2Br \xrightarrow{Mg} CH_3CH_2CH_2CH_2MgBr \xrightarrow[2.\ H_2O]{1.\ H-\overset{O}{\overset{\|}{C}}-OC_2H_5} CH_3CH_2CH_2CH_2-\overset{OH}{\overset{|}{C}}H-CH_2CH_2CH_2CH_3$$

83–85%

In the preceding reactions we have sometimes shown the carbonyl compound as the starting material in the reaction, while in other cases we have depicted the alkyl halide as the reactant. Which way the reaction is drawn is simply a matter of convenience. These examples allow us to make a simple but important observation about all organometallic reactions with carbonyl compounds: The carbon atom of the product to which the OH is attached was originally the carbonyl carbon of the reactant. This piece of information can be very useful when you analyze a reaction or when you plan a synthesis (Section 11.10).

A careful comparison of the preceding examples will convince you that there is frequently more than one reaction which will yield a particular alcohol. For example, a tertiary alcohol with three different alkyl groups might be prepared by any of three pathways:

$$R{-}\overset{O}{\overset{\|}{C}}{-}R' \xrightarrow[2.\ H_2O]{1.\ R''MgBr} R{-}\underset{R'}{\underset{|}{\overset{OH}{\overset{|}{C}}}}{-}R''$$

$$R{-}\overset{O}{\overset{\|}{C}}{-}R'' \xrightarrow[2.\ H_2O]{1.\ R'MgBr} R{-}\underset{R'}{\underset{|}{\overset{OH}{\overset{|}{C}}}}{-}R''$$

$$R'{-}\overset{O}{\overset{\|}{C}}{-}R'' \xrightarrow[2.\ H_2O]{1.\ RMgBr} R{-}\underset{R'}{\underset{|}{\overset{OH}{\overset{|}{C}}}}{-}R''$$

The reactions of organometallic reagents with certain types of carbonyl compounds yield products that are highly characteristic of the reactants, and this can help you in analyzing reactions or planning a synthesis. As long as the carbonyl group of an ester has an alkyl substituent, reaction of the ester with an organometallic reagent will yield a tertiary alcohol, and at least two of the alkyl substituents will be the same. If the appropriate Grignard or alkyllithium reagent is used, all three alkyl groups could be equivalent. Reactions of organometallic reagents with carbonyl compounds having no alkyl substituents on the carbonyl carbon also afford characteristic products. You can see this in the following general equations that use formaldehyde ($H_2C{=}O$) and ethyl formate ($H{-}CO_2C_2H_5$); specific examples were shown previously in this section.

$$RMgI \xrightarrow[2.\ H_2O]{1.\ H_2C{=}O} R{-}CH_2OH \qquad \text{Primary alcohol}$$

$$RMgI \xrightarrow[2.\ H_2O]{1.\ H{-}\overset{O}{\overset{\|}{C}}{-}OCH_3} R{-}\overset{OH}{\overset{|}{CH}}{-}R \qquad \text{Symmetrical secondary alcohol}$$

Unsaturated alcohols can be synthesized from either unsaturated carbonyl compounds or unsaturated organometallic reagents. Alkenyl and aromatic Grignard and alkyllithium reagents can be prepared from the corresponding halides in the same way as their saturated counterparts, but acetylide derivatives are generated from the hydrocarbon by an acid-base reaction of a terminal alkyne with a base such as $NaNH_2$ (Section 7.8).

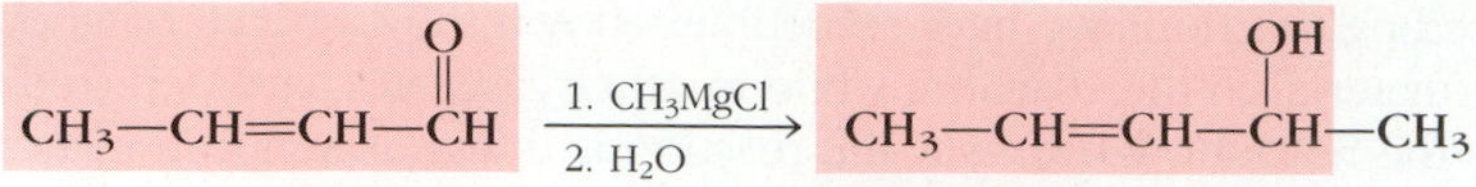

$$CH_3{-}CH{=}CH{-}\overset{O}{\overset{\|}{C}}H \xrightarrow[2.\ H_2O]{1.\ CH_3MgCl} CH_3{-}CH{=}CH{-}\overset{OH}{\overset{|}{C}}H{-}CH_3 \quad 81\text{–}86\%$$

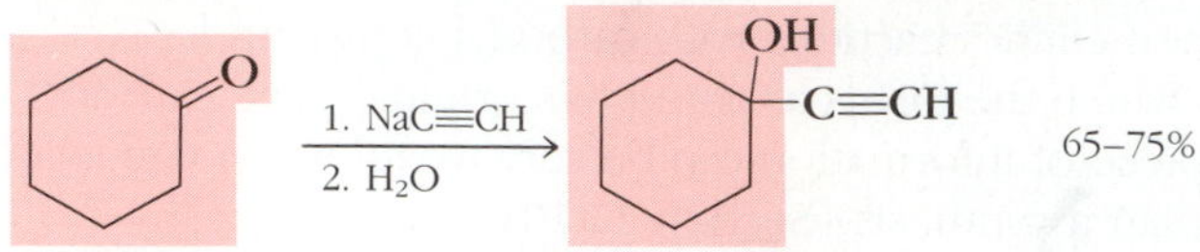

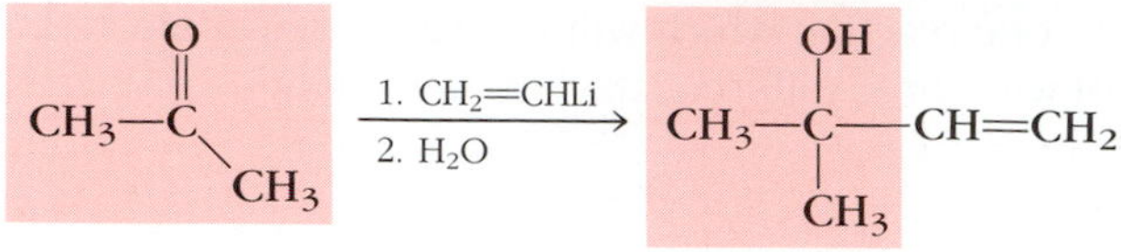

$$\text{cyclohexanone} \xrightarrow[2.\ H_2O]{1.\ NaC{\equiv}CH} \text{1-ethynylcyclohexanol (OH, } C{\equiv}CH) \quad 65\text{–}75\%$$

$$CH_3{-}\overset{O}{\overset{\|}{C}}{-}CH_3 \xrightarrow[2.\ H_2O]{1.\ CH_2{=}CHLi} CH_3{-}\underset{CH_3}{\underset{|}{\overset{OH}{\overset{|}{C}}}}{-}CH{=}CH_2 \quad 74\%$$

$$\xrightarrow[2.\ NH_4Cl,\ H_2O]{1.\ LiC{\equiv}CH} \text{HO, } C{\equiv}CH \quad 75\%$$

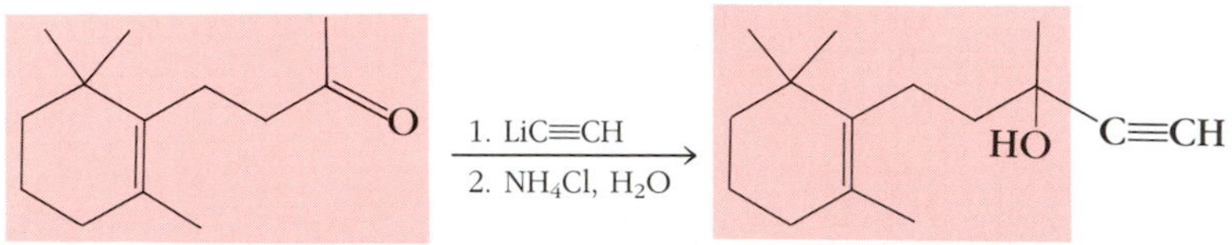

Difficulties are often encountered in attempts to synthesize polyfunctional compounds by using organometallic reagents. Carbon-carbon multiple bonds do not usually interfere with organometallic reactions, but this is not true for many other functional groups. Substituents such as —OH can act as proton donors and destroy the organometallic reagent in an acid-base reaction, so it is usually not possible to prepare an alkyllithium or Grignard reagent that has an —OH as a second functional group. The high reactivity of an organometallic center similarly precludes the presence of other reactive groups (halogen, carbonyl) in the organometallic reagent. In most cases this high reactivity also limits the functionality that can be present in the carbonyl compound. The organometallic reagent is both a nucleophile and a base, and undesired side reactions occur when the carbonyl compound has a second reactive site, such as hydroxyl, halogen, or a second carbonyl group.

One way that difunctional compounds can be prepared via organometallic reagents is the reaction of these agents with lactones. A lactone is a cyclic ester, so when the alkoxide is expelled from the carbonyl carbon during the course of the reaction, it still remains as part of the molecule. Consequently, a diol can be formed, as seen in the following example:

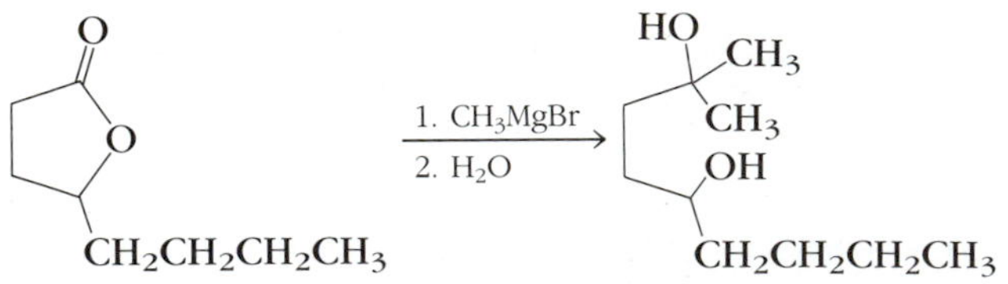

$$\text{lactone (}CH_2CH_2CH_2CH_3\text{)} \xrightarrow[2.\ H_2O]{1.\ CH_3MgBr} \text{HO, }CH_3\text{, }CH_3\text{, OH, }CH_2CH_2CH_2CH_3 \quad 57\%$$

EXERCISE 11.7

Draw the product expected from each of the following reactions:

(a) $\text{cyclopentanone} \xrightarrow[2.\ H_2O]{1.\ CH_3CH_2Li} ?$

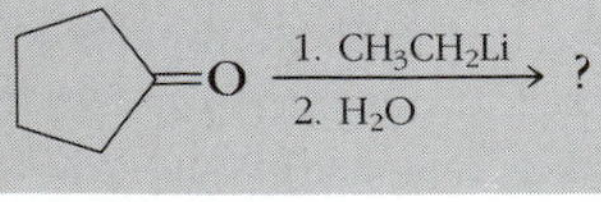

(b) $C_6H_5-\overset{O}{\overset{\|}{C}}-OCH_2CH_3 \xrightarrow[2.\ H_2O]{1.\ CH_3MgBr}$?

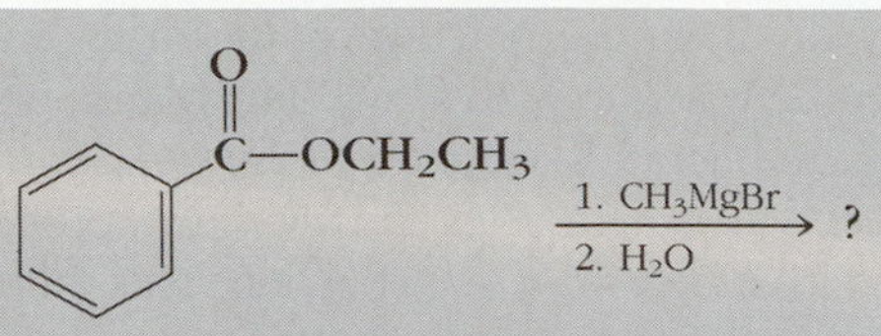

(c) $CH_3-\overset{CH_3}{\overset{|}{C}H}-CH_2-\overset{O}{\overset{\|}{C}H} \xrightarrow[2.\ H_2O]{1.\ C_6H_5MgBr}$?

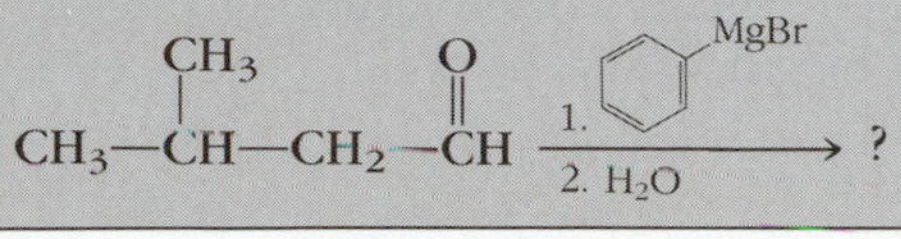

EXERCISE 11.8

What reagents would be needed to carry out each of the following conversions?

(a) cyclohexanone $\xrightarrow{?}$ 1-(CH(CH$_3$)$_2$)cyclohexanol (OH, CH$_3$, CH, CH$_3$)

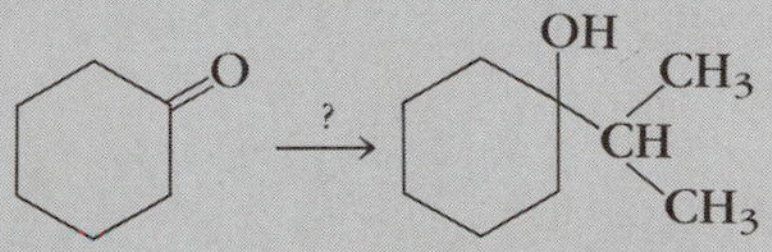

(b) $CH_3-\overset{O}{\overset{\|}{C}}-OCH_3 \xrightarrow{?} CH_3-\overset{OH}{\overset{|}{C}}$(cyclopentyl)$_2$

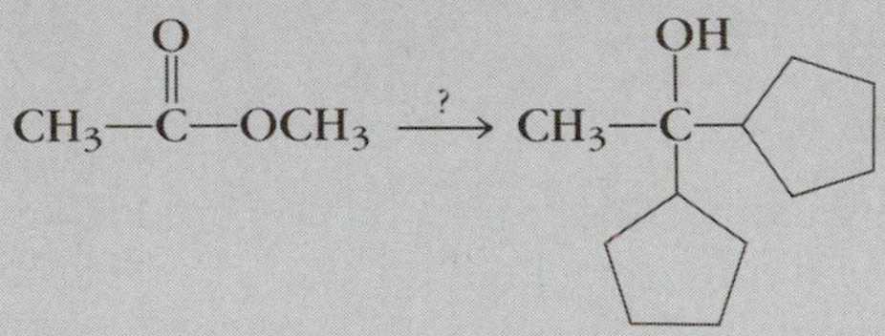

(c) $CH_3CH_2-\overset{CH_3}{\overset{|}{C}H}-CH_2I \xrightarrow{?} CH_3CH_2-\overset{CH_3}{\overset{|}{C}H}-CH_2CH_2OH$

EXERCISE 11.9

Draw the products of the reaction (after aqueous workup) of $CH_3CH_2CH_2Li$ with:

(a) Hexanal ($CH_3CH_2CH_2CH_2CH_2CHO$)
(b) Formaldehyde (CH_2O)
(c) Ethyl formate ($HCO_2CH_2CH_3$)

11.6 PREPARATION OF ALCOHOLS VIA REACTIONS OF ORGANOMETALLIC REAGENTS WITH EPOXIDES

We showed you previously (Section 3.4) that strained, three-membered rings are highly susceptible to ring-opening reactions, particularly in the case of epoxides (Section 5.12). Epoxides react readily with organometallic reagents, and this provides a useful method for synthesizing alcohols.

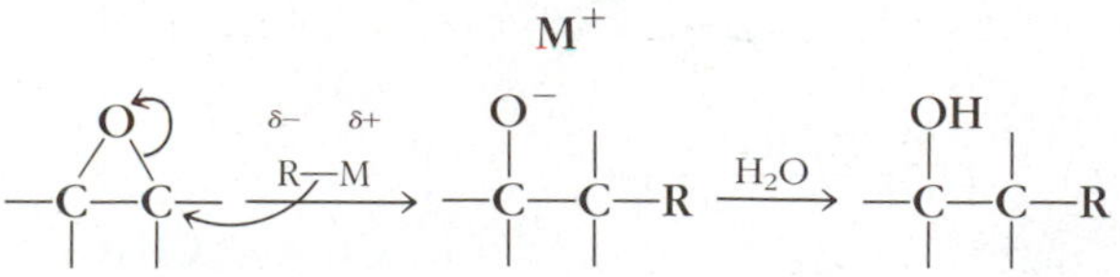

$$\text{epoxide } -\overset{|}{C}-\overset{|}{C}- \xrightarrow{\overset{\delta-}{R}-\overset{\delta+}{M}} -\overset{O^-\ M^+}{\overset{|}{C}}-\overset{|}{C}-R \xrightarrow{H_2O} -\overset{OH}{\overset{|}{C}}-\overset{|}{C}-R$$

This reaction differs in an important way from the reaction of an organometallic reagent with a carbonyl compound. In the reaction with a carbonyl compound, the alkyl group of the organometallic reagent becomes attached to the *same* carbon as the hydroxyl group. But in the reaction with an epoxide, the alkyl group becomes bonded to the carbon atom *adjacent* to that bearing the OH group. This difference can be very important when you plan a synthesis. The reaction gives good yields with ethylene oxide, and the chain length of the organometallic species is increased by two carbons.

$$\overset{O}{CH_2-CH_2} \xrightarrow[2.\ H_2O]{1.\ CH_3CH_2CH_2CH_2MgBr} CH_3CH_2CH_2CH_2-CH_2CH_2OH \quad 60\text{–}62\%$$

$$\overset{O}{CH_2-CH_2} \xrightarrow[2.\ H_2O]{1.\ CH_3C\equiv CLi} CH_3-C\equiv C-CH_2CH_2OH \quad 42\%$$

$$\overset{O}{CH_2-CH_2} \xrightarrow[2.\ H_2O]{1.\ C_6H_5MgCl} C_6H_5CH_2CH_2OH \quad 88\%$$

Lower yields often result if the organometallic reagent is prepared from a secondary or tertiary halide, apparently as a consequence of unfavorable steric interactions. The reaction of Grignard reagents with substituted epoxides is unreliable because the magnesium halide present in solution can act as a Lewis acid:

$$-\overset{|}{\underset{|}{C}}-\overset{:O:}{CH-CH_2} \xrightarrow{MgBr_2} -\overset{|}{\underset{|}{C}}-\overset{:\overset{+}{O}MgBr}{CH-CH_2} \rightleftharpoons -\overset{|}{\underset{|}{C}}-\overset{+}{C}H-\overset{:\ddot{O}MgBr}{\overset{|}{C}H_2}$$

This causes carbocation rearrangements and frequently leads to a mixture of products.

These problems are greatly alleviated in the case of alkyllithium and organocopper reagents. The latter, when available, are the reagents of choice.

$$\overset{O}{CH_2-CH-CH_2CH_3} \xrightarrow[2.\ H_2O]{1.\ (CH_3)_2CuLi} CH_3-CH_2-\overset{OH}{\overset{|}{C}H}-CH_2CH_3 \quad 88\%$$

$$C_6H_5\overset{O}{CH-CH_2} \xrightarrow[2.\ H_2O]{1.\ C_6H_5Li} C_6H_5\overset{OH}{\overset{|}{C}H}-CH_2C_6H_5 \quad 70\text{–}72\%$$

$$\text{cyclopentene oxide} \xrightarrow[2.\ H_2O,\ NH_4Cl]{1.\ Bu_2CuLi} \text{2-}(CH_2CH_2CH_2CH_3)\text{cyclopentanol (OH)} \quad 80\text{–}90\%$$

As you can see in the first two examples, the organometallic reagent attacks the less substituted (less sterically hindered) carbon atom of the epoxide ring.

The lower reactivity of the lithium organocuprates also permits the presence of other functional groups in the molecule containing the epoxide ring. In the following example reaction takes place selectively at the epoxide, and the keto group is unaffected.

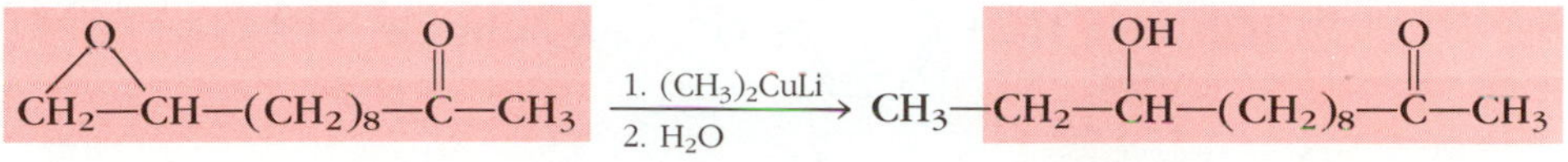

$$CH_2\overset{O}{—}CH—(CH_2)_8—\overset{O}{\overset{\|}{C}}—CH_3 \xrightarrow[2.\ H_2O]{1.\ (CH_3)_2CuLi} CH_3—CH_2—\overset{OH}{\overset{|}{C}H}—(CH_2)_8—\overset{O}{\overset{\|}{C}}—CH_3 \quad 68\%$$

Epoxides react with lithium aluminum hydride by a pathway analogous to that with an alkyllithium. As shown by the following example, the hydride reagent attacks the less hindered carbon atom to yield the more substituted alcohol.

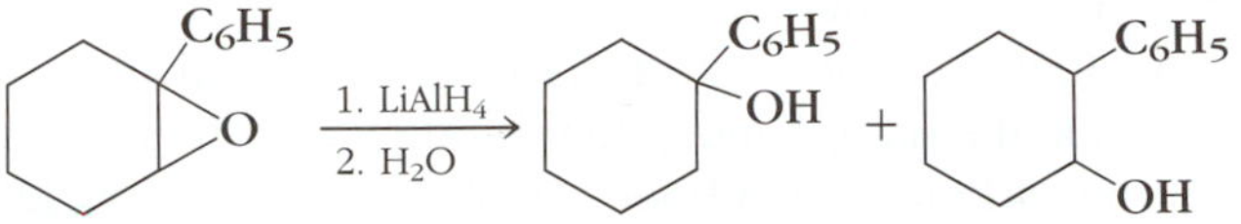

$$\text{1-phenylcyclohexene oxide } (C_6H_5) \xrightarrow[2.\ H_2O]{1.\ LiAlH_4} \text{1-phenylcyclohexanol } (C_6H_5, OH)\ 78\% + \text{2-phenylcyclohexanol } (C_6H_5, OH)\ 4\%$$

EXERCISE 11.10

Complete the following reactions:

(a) 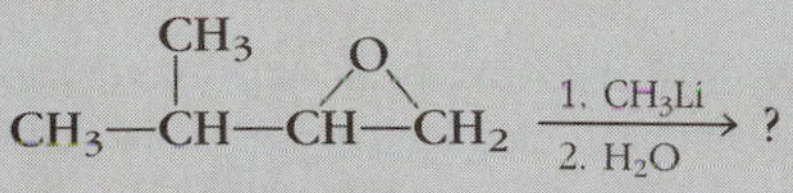 $CH_3—\overset{CH_3}{\overset{|}{C}H}—CH\overset{O}{—}CH_2 \xrightarrow[2.\ H_2O]{1.\ CH_3Li}$?

(b) $CH_3—\overset{CH_3}{\overset{|}{C}H}—MgBr \xrightarrow[2.\ H_2O]{1.\ CH_2\overset{O}{—}CH_2}$?

(c) cycloheptene oxide $\xrightarrow{?}$ 2-ethylcycloheptanol (OH, C_2H_5)

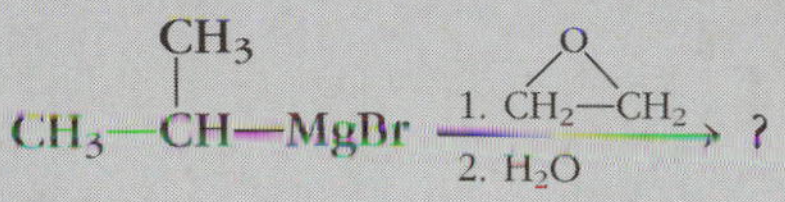

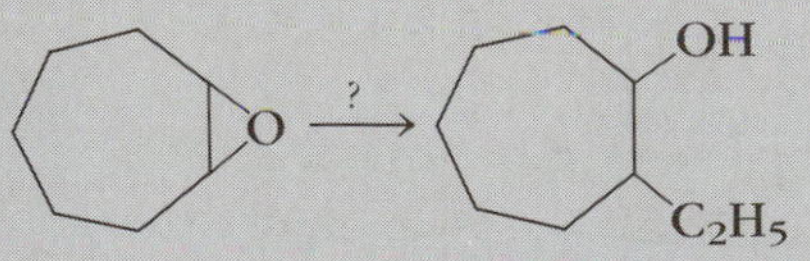

(d) ? $\xrightarrow{NaC{\equiv}CCH_3}$ $CH_3—\overset{OH}{\overset{|}{C}H}—CH_2—C{\equiv}C—CH_3$

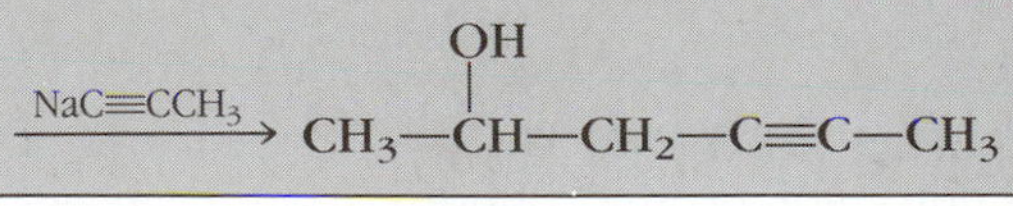

11.7 PHENOLS

Phenols are compounds having a hydroxyl group attached directly to an aromatic ring. The simplest example is phenol itself.

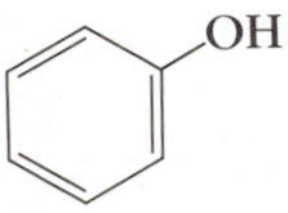

$$C_6H_5OH$$

Phenol

Phenols are similar in many ways to alcohols but show enough important differences to be considered a distinct functional class.

As we discussed in Section 10.9, phenols are stronger acids than water or alcohols, but they are much weaker acids than typical carboxylic acids.

$$CH_3OH \rightleftharpoons H^+ + CH_3O^- \qquad K_a = 10^{-16} \qquad \text{An alcohol}$$

$$C_6H_5\text{—}OH \rightleftharpoons H^+ + C_6H_5\text{—}O^- \qquad K_a = 10^{-10} \qquad \text{A phenol}$$

$$CH_3\text{—}\overset{\overset{\displaystyle O}{\|}}{C}\text{—}OH \rightleftharpoons H^+ + CH_3\text{—}\overset{\overset{\displaystyle O}{\|}}{C}\text{—}O^- \qquad K_a = 10^{-5} \qquad \text{A carboxylic acid}$$

The increased acidity of phenols relative to alcohols results from resonance stabilization of the *phenoxide ion* by the aromatic ring (Section 10.9).

The reactions of phenols involving only the OH group (formation of esters and ethers) are very similar to those of alcohols, and we will discuss them in the following section. On the other hand, phenols differ substantially from alcohols in the reactivity of the carbon-oxygen bond. Alcohols undergo a variety of reactions that result in loss or replacement of the hydroxyl group (Sections 5.13, 11.9), but these processes do not occur with phenols. Phenols also undergo characteristic substitution reactions on the aromatic ring, which will be discussed in Chapter 18, along with other aromatic substitution reactions. Phenols are prepared via reactions that are quite different from those used to prepare alcohols; methods for their synthesis are also found in Chapter 18.

Simple phenols are easily oxidized, but complex mixtures of products are often formed, and we will not discuss these reactions here. However, aromatic compounds having *two* hydroxyl groups (hydroquinones) can be cleanly oxidized to the corresponding *quinones*.

$$\underset{\text{Hydroquinone}}{HO\text{—}C_6H_4\text{—}OH} \underset{\text{reduction}}{\overset{\text{oxidation}}{\rightleftharpoons}} \underset{\text{Benzoquinone}}{O{=}C_6H_4{=}O}$$

Oxidation-reduction reactions of this type are important in several ways. In living organisms the chemical utilization of atmospheric oxygen proceeds by a sequence of several oxidation-reduction steps in which quinone oxidation-reductions play a key role. Hydroquinone itself has commercial importance as a major ingredient of *photographic developers*. The permanent image of a photograph results from the reduction to silver metal of silver halide crystals that have been exposed to light. Hydroquinone serves as an effective reducing agent in the photographic developing process.

$$HO\text{—}C_6H_4\text{—}OH + 2OH^- + 2Ag^+ \longrightarrow O{=}C_6H_4{=}O + 2H_2O + 2Ag^0$$

EXERCISE 11.11

Dihydroxybenzenes can be oxidized to quinones when the hydroxyl groups are *ortho* or *para* to each other, but not when they have a *meta* relationship. Use appropriate structural formulas to explain this.

11.8 FORMATION OF CARBON-OXYGEN BONDS: ALKYLATION AND ACYLATION OF OH GROUPS

One of the characteristic and important reactions of compounds containing the OH group is replacement of the bond to hydrogen by a bond to a carbon atom:

$$\text{—O—H} \longrightarrow \text{—O—}\overset{|}{\underset{|}{\text{C}}}\text{—}$$

A variety of methods are available to carry out such transformations, but we will restrict the present discussion to just a single method; displacement of a halo substituent from a carbon atom.

Alkylation of an alcohol can be carried out with an *alkyl halide* such as methyl iodide to produce an ether. Frequently the alcohol is first converted to its conjugate base, which is much more reactive.

$$\text{R—O—H} \xrightarrow{\text{base}} \text{R—O}^- \xrightarrow{CH_3\text{—I}} \text{R—O—}CH_3$$

Alkylation

$$\text{Ar—O—H} \xrightarrow{\text{base}} \text{Ar—O}^- \xrightarrow{CH_3\text{—I}} \text{Ar—O—}CH_3$$

Alkylation of the salt of an alcohol is known as the Williamson ether synthesis because the method was originally developed by the English chemist Alexander Williamson in the 1850s.

Acylation of an alcohol with an *acyl chloride* such as acetyl chloride (CH_3COCl) produces an *ester*.

$$\text{R—}\overset{\overset{\text{O}}{\|}}{\text{C}}\text{—} \qquad \text{R—}\overset{\overset{\text{O}}{\|}}{\text{C}}\text{—Cl} \qquad \text{R—}\overset{\overset{\text{O}}{\|}}{\text{C}}\text{—OR}$$

Acyl group — Acyl halide — Ester

$$\text{R—}\ddot{\underset{..}{\text{O}}}\text{—H} \xrightarrow{CH_3\text{—}\overset{\overset{\text{O}}{\|}}{\text{C}}\text{—Cl}} \text{R—}\ddot{\underset{..}{\text{O}}}\text{—}\overset{\overset{\text{O}}{\|}}{\text{C}}\text{—}CH_3$$

Acylation

$$\text{Ar—}\ddot{\underset{..}{\text{O}}}\text{—H} \xrightarrow{CH_3\text{—}\overset{\overset{\text{O}}{\|}}{\text{C}}\text{—Cl}} \text{Ar—}\ddot{\underset{..}{\text{O}}}\text{—}\overset{\overset{\text{O}}{\|}}{\text{C}}\text{—}CH_3$$

In these reactions the oxygen atom of the phenol or alcohol functions as a nucleophile. We will discuss the reaction mechanisms in subsequent chapters. For the present it is sufficient for you to recognize that the nonbonded electrons on the oxygen of the OH group (or of the conjugate base) can attack at the positive end of the polar carbon-halogen bond and displace a halide ion.

The following reactions illustrate the formation of ethers or esters by alkylation or acylation of alcohols and phenols. The alkylation reactions are generally restricted to methyl or primary alkyl halides, because secondary and tertiary halides yield elimination products (Section 5.13) when treated with alkoxide salts. The acylation of an alcohol is generally carried out in the presence of an amine such as pyridine to neutralize the HCl produced in the reaction.

$$CH_3CH_2CH_2CH_2OH \xrightarrow{Na} CH_3CH_2CH_2CH_2O^-Na^+ \xrightarrow{CH_3I} CH_3CH_2CH_2CH_2{-}O{-}CH_3 \quad 71\%$$

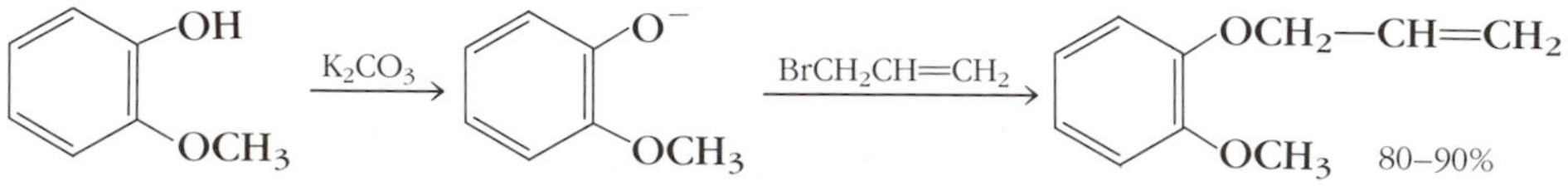

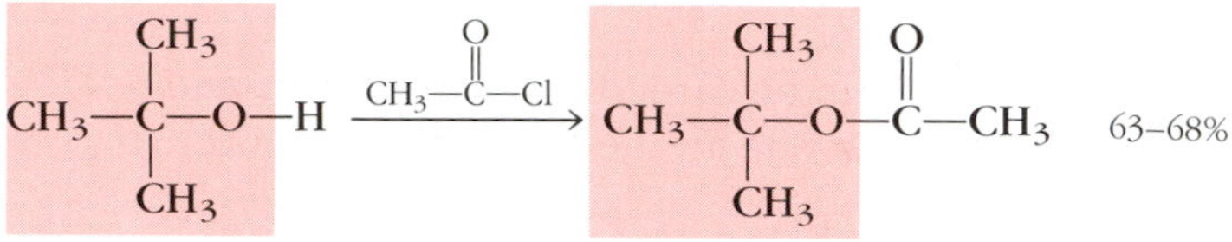

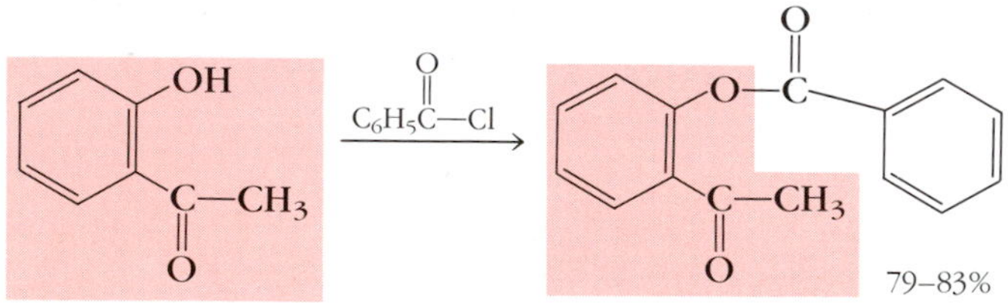

$$\text{2-methoxyphenol (OH, OCH}_3\text{)} \xrightarrow{K_2CO_3} \text{2-methoxyphenoxide (O}^-\text{, OCH}_3\text{)} \xrightarrow{BrCH_2CH{=}CH_2} \text{C}_6\text{H}_4(OCH_2{-}CH{=}CH_2)(OCH_3) \quad 80\text{–}90\%$$

$$(CH_3)_3C{-}O{-}H \xrightarrow{CH_3{-}\overset{O}{\overset{\|}{C}}{-}Cl} (CH_3)_3C{-}O{-}\overset{O}{\overset{\|}{C}}{-}CH_3 \quad 63\text{–}68\%$$

$$\text{2-HO-C}_6\text{H}_4{-}\overset{O}{\overset{\|}{C}}{-}CH_3 \xrightarrow{C_6H_5\overset{O}{\overset{\|}{C}}{-}Cl} \text{2-(C}_6\text{H}_5\overset{O}{\overset{\|}{C}}{-}O)\text{-C}_6\text{H}_4{-}\overset{O}{\overset{\|}{C}}{-}CH_3 \quad 79\text{–}83\%$$

EXERCISE 11.12

Suggest a method of synthesizing each of the following esters from an alcohol and an acyl chloride:

(a) $CH_3{-}\underset{}{\overset{CH_3}{\overset{|}{C}H}}{-}CH_2CH_2{-}O{-}\overset{O}{\overset{\|}{C}}{-}CH_2CH_3$

(b) $C_6H_5{-}CH_2{-}O{-}\overset{O}{\overset{\|}{C}}{-}CH_3$

(c) $\text{cyclopentyl}{-}O{-}\overset{O}{\overset{\|}{C}}{-}CH_2CH_3$

(d) $\text{cyclopentyl}{-}\overset{O}{\overset{\|}{C}}{-}O{-}CH_2CH_3$

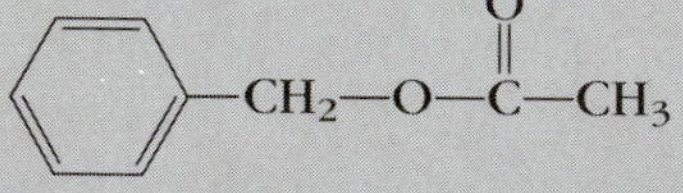

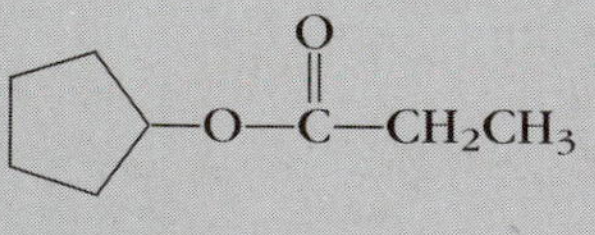

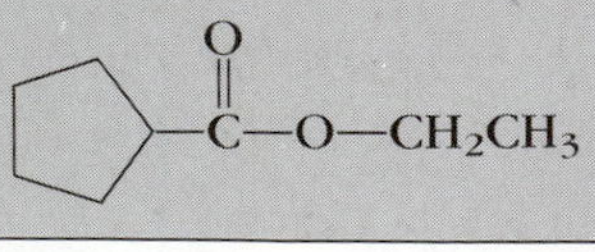

11.9 REPLACEMENT OF OH GROUPS: CONVERSION OF ALCOHOLS TO ALKYL HALIDES AND TOSYLATES

Alcohols play a key role in preparative organic chemistry. As we showed you in previous sections of this chapter, alcohols are the products of many important carbon-carbon bond-forming reactions of organometallic compounds. Alcohols are also valuable starting materials in many synthetic reactions. Oxidation of primary and secondary alcohols affords aldehydes and ketones, which can undergo reaction with organometallics. In addition, alcohols can be converted to halides, and these in turn can be used to prepare the organometallic reagents (Figure 11.1). For these reasons the replacement of OH by halogen is an important transformation in organic chemistry. In this section we will show you some specific examples of these reactions, and in Chapter 12 we will discuss the mechanisms by which they occur.

Simple treatment of the alcohol with concentrated acid has sometimes been used to prepare the corresponding halide.

$$CH_3{-}(CH_2)_{10}{-}CH_2OH \xrightarrow[H_2SO_4]{HBr} CH_3{-}(CH_2)_{10}{-}CH_2Br \qquad 91\%$$

However, the strongly acidic conditions can cause reaction with other functional groups or lead to carbocation rearrangements (Section 5.9).

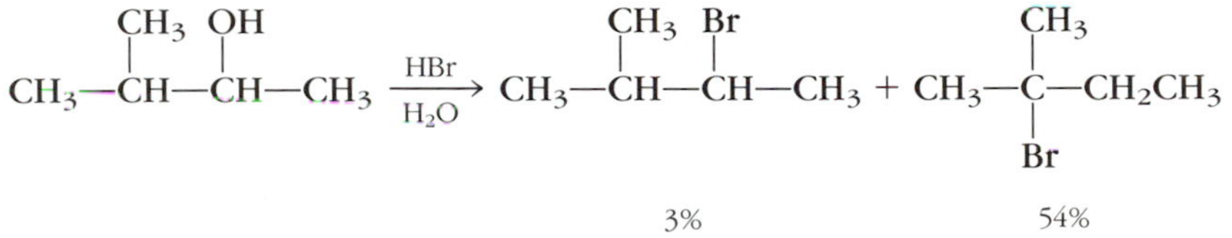

$$CH_3{-}\underset{|}{\overset{CH_3}{CH}}{-}\overset{OH}{CH}{-}CH_3 \xrightarrow[H_2O]{HBr} \underset{3\%}{CH_3{-}\overset{CH_3}{CH}{-}\overset{Br}{CH}{-}CH_3} + \underset{54\%}{CH_3{-}\overset{CH_3}{\underset{Br}{C}}{-}CH_2CH_3}$$

Such difficulties can usually be avoided by using inorganic halogen derivatives of phosphorus and sulfur. Phosphorus trichloride and phosphorus tribromide have

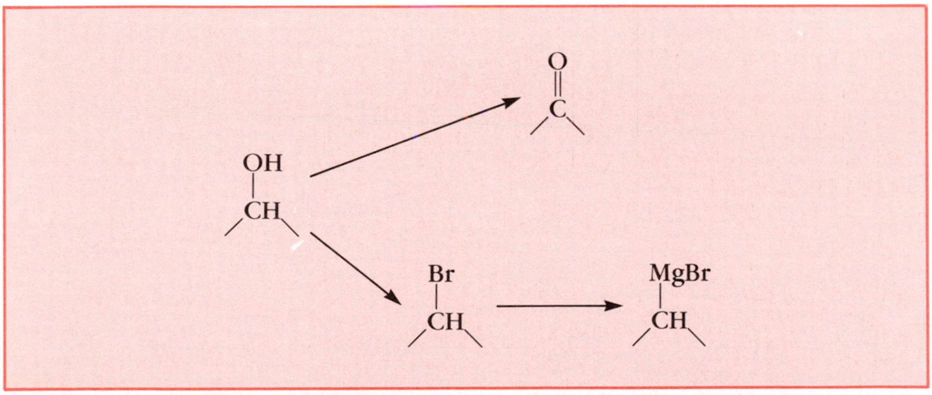

OH–CH< → >C=O

OH–CH< → Br–CH< → MgBr–CH<

Figure 11.1 Alcohols as Starting Materials in Organic Synthesis. Among the many reactions of alcohols are oxidation to carbonyl compounds and conversion (via alkyl halides) to organometallic reagents. Subsequent reactions between carbonyl compounds and organometallic reagents yield products with new carbon-carbon bonds.

been found to be particularly useful for conversion of alcohols to the chlorides and bromides, respectively. Chlorides have also been prepared by using a variety of other reagents such as phosphorus oxychloride ($POCl_3$), phosphorus pentachloride (PCl_5), and thionyl chloride ($SOCl_2$).

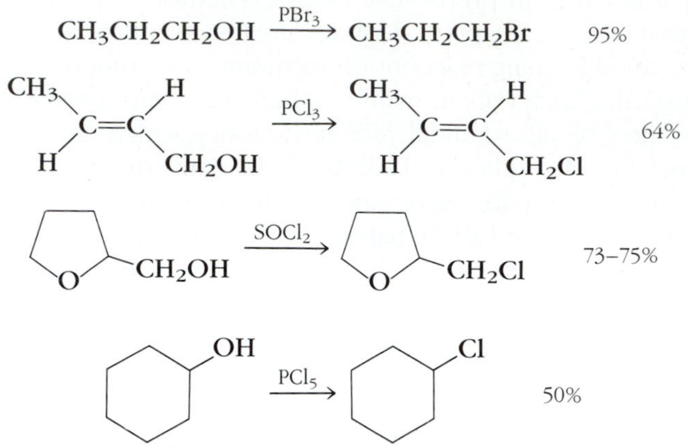

$$CH_3CH_2CH_2OH \xrightarrow{PBr_3} CH_3CH_2CH_2Br \quad 95\%$$

$$CH_3(H)C{=}C(H)CH_2OH \xrightarrow{PCl_3} CH_3(H)C{=}C(H)CH_2Cl \quad 64\%$$

$$\text{(tetrahydrofuran-2-yl)}CH_2OH \xrightarrow{SOCl_2} \text{(tetrahydrofuran-2-yl)}CH_2Cl \quad 73\text{–}75\%$$

$$\text{Cyclohexanol (OH)} \xrightarrow{PCl_5} \text{Chlorocyclohexane (Cl)} \quad 50\%$$

Alcohols can also be converted to *sulfonates,* which have many similarities to alkyl halides. Sulfonates are esters of sulfonic acids, of which two of the most common are methanesulfonic acid and *p*-toluenesulfonic acid.

$$CH_3{-}SO_2{-}OH \qquad CH_3{-}C_6H_4{-}SO_2{-}OH$$

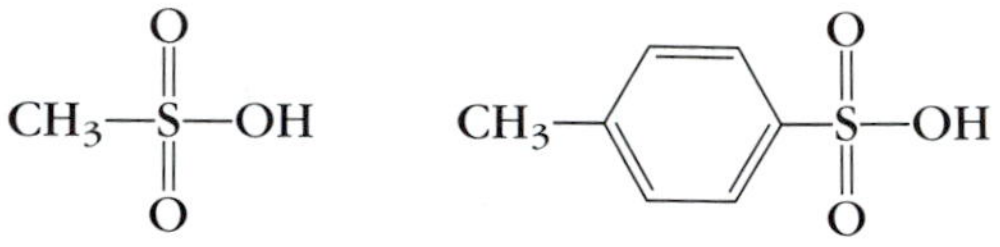

Methanesulfonic acid　　*p*-Toluenesulfonic acid

The acid chlorides of these compounds will react with alcohols to produce the corresponding sulfonates. The reactions are usually carried out in the presence of pyridine to neutralize the HCl produced. They are analogous to the acylation reactions we discussed in Section 11.8.

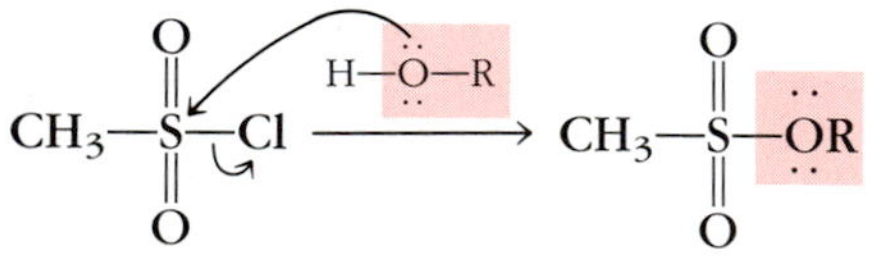

$$CH_3{-}SO_2{-}Cl \xrightarrow{H{-}\ddot{O}{-}R} CH_3{-}SO_2{-}\ddot{O}R$$

Methanesulfonyl chloride　　Methanesulfonate

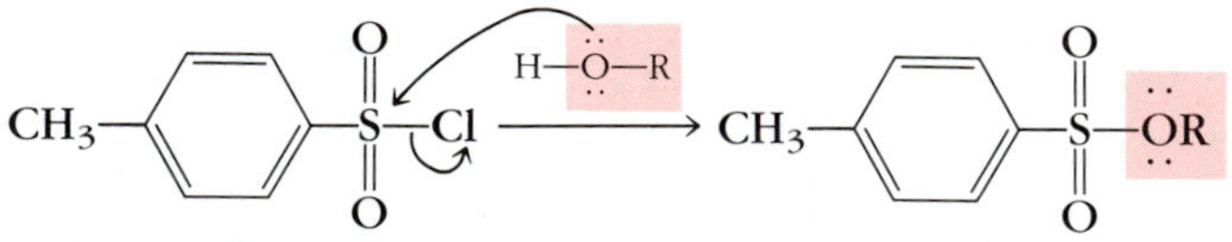

$$CH_3{-}C_6H_4{-}SO_2{-}Cl \xrightarrow{H{-}\ddot{O}{-}R} CH_3{-}C_6H_4{-}SO_2{-}\ddot{O}R$$

p-Toluenesulfonyl chloride (tosyl chloride, **TsCl**)　　*p*-Toluenesulfonate (a tosylate)

The *p*-toluenesulfonate esters are normally referred to as *tosylates,* and the tosyl group, $CH_3{-}C_6H_4{-}SO_2{-}$, is commonly abbreviated as −Ts. This allows us to conveniently rewrite the preceding reaction as:

$$\underset{\text{Alcohol}}{R-O-H} \xrightarrow[\text{pyridine}]{\text{TsCl}} \underset{\text{Tosylate}}{R-O-Ts}$$

Notice that the carbon-oxygen bond of the alcohol remains intact in this reaction. This can have significant stereochemical consequences, because the configuration of the carbon atom is unchanged. We will return to this important point in the next chapter. The following equations illustrate the preparation of sulfonate esters of alcohols and phenols:

$$CH_3-(CH_2)_{10}-CH_2OH \xrightarrow{\text{TsCl}} CH_3-(CH_2)_{10}-CH_2-OTs \quad 90\%$$

OH, NO2 (benzene ring) — CH_3-SO_2-Cl → $O-SO_2CH_3$, NO_2 (benzene ring) 89%

$$CH_3CH_2-\underset{|}{\overset{OH}{CH}}-CH_2CH_3 \xrightarrow{\text{TsCl}} CH_3CH_2-\overset{OTs}{CH}-CH_2CH_3 \quad 75\%$$

Sulfonic acids are strong acids, as you would expect from their structural similarity to sulfuric acid. The sulfonate ions are stabilized by resonance, so the sulfonates are good *leaving groups,* acting much like halides in elimination and nucleophilic substitution reactions. Since elimination reactions of sulfonates are carried out under basic conditions, this provides a useful way for carrying out the dehydration of an alcohol while avoiding strongly acidic conditions that could lead to carbocation rearrangements.

HO (steroid) — CH_3-SO_2-Cl → CH_3SO_2-O (steroid) — KOtBu → (alkene steroid) 65%

You must recognize one important distinction between sulfonates and halides: they behave similarly in elimination and substitution reactions, but the sulfonate esters *cannot* be used to prepare organometallic compounds.

EXERCISE 11.13

Complete the following reactions:

(a) 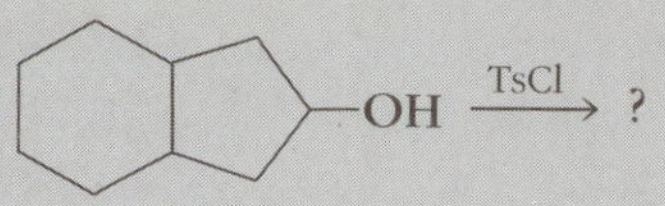 (bicyclic alcohol: cyclohexane ring fused to cyclopentane ring bearing OH) $\xrightarrow{TsCl}$?

(b) 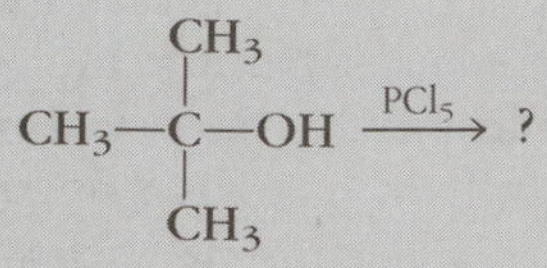 $CH_3-C(CH_3)_2-OH \xrightarrow{PCl_5}$?

(c) $(H)_2C{=}C(CH_2OH)(CH_2CH_3) \xrightarrow{CH_3-SO_2-Cl}$?

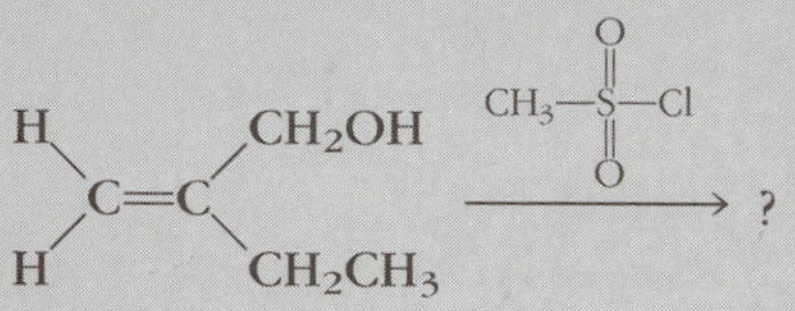

(d) 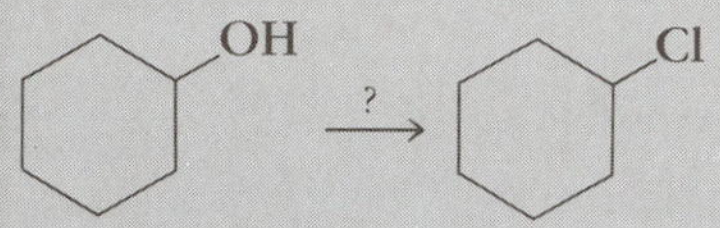 cyclohexanol $\xrightarrow{?}$ chlorocyclohexane

(e) 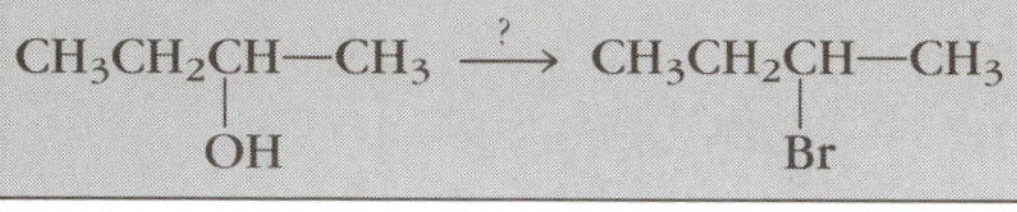 $CH_3CH_2CH(OH)-CH_3 \xrightarrow{?} CH_3CH_2CH(Br)-CH_3$

11.10 ALCOHOLS AND ORGANIC SYNTHESIS

Alcohols can play several important roles in organic synthesis:

1. The hydroxyl group is a common functional group, which is found in many compounds of commercial and biological importance. Consequently, the *final product* of an organic synthesis might be an alcohol.

2. Alcohols are readily converted to other functional classes (such as halides and carbonyl compounds), which can be used in carbon-carbon bond-forming reactions. Therefore alcohols can be very useful *starting materials* in organic synthesis.

3. In addition to the ready conversion of the hydroxyl group into a variety of other functional groups, alcohols are also produced as the *result* of carbon-carbon bond-forming reactions between organometallic reagents and carbonyl compounds or epoxides. This permits the use of alcohols as versatile **synthetic intermediates,*** i.e., compounds produced at an intermediate stage of an organic synthesis, which are converted to the final product in subsequent reactions.

*Note that *synthetic intermediates,* in contrast to *reactive intermediates* such as carbocations and free radicals, are stable, isolable compounds.

The examples we present in this section will illustrate some of the ways that alcohols can be used in synthetic organic chemistry. The approach to designing a synthetic plan is an extension of that developed in Sections 1.12, 5.14, and 7.10, but the versatility of alcohols as synthetic intermediates introduces many new possibilities.

As we discussed previously, the limitations that we will place on allowed starting materials for a synthesis must be somewhat arbitrary. In an introductory textbook we are more concerned with teaching you about organic chemistry than with developing the single best or least expensive method for preparing a compound. Therefore we will continue to employ such restrictions as "starting materials containing four or fewer carbon atoms" and ". . . containing no oxygen atoms."

Example 11.1 Suggest a synthesis of 2-methyl-3-pentanol from starting materials containing three or fewer carbon atoms.

Solution The key functional group is the hydroxyl group together with the carbon atom to which it is attached:

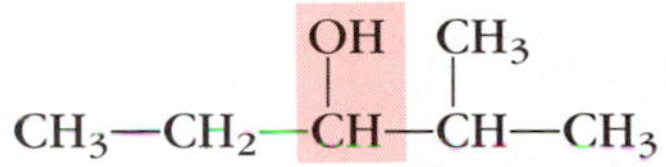

$$CH_3{-}CH_2{-}\underset{}{\overset{OH}{\overset{|}{CH}}}{-}\overset{CH_3}{\overset{|}{CH}}{-}CH_3$$

Formation of the alcohol could have taken place by simple functional group modification (reduction of a ketone, hydration of an alkene, etc.) or as a consequence of a carbon-carbon bond-forming reaction between an organometallic reagent and a carbonyl compound. The latter possibility is particularly attractive because the preparation of a six-carbon compound from three-carbon precursors clearly requires the formation of a new carbon-carbon bond. Simple division of the molecule into three-carbon fragments permits identification of potential precursors. The hydroxyl group is the key functional group, and it must originally have been a carbonyl group.

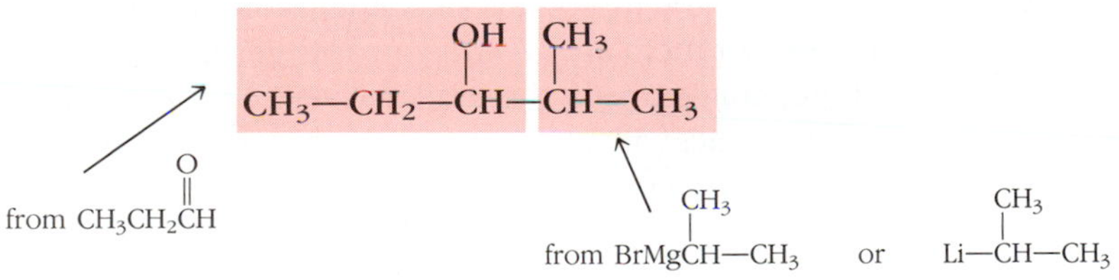

$$CH_3{-}CH_2{-}\overset{OH}{\overset{|}{CH}}{-}\overset{CH_3}{\overset{|}{CH}}{-}CH_3$$

from $CH_3CH_2\overset{O}{\overset{\|}{C}}H$ (gives $CH_3{-}CH_2{-}CH(OH)$ fragment)

from $BrMg\overset{CH_3}{\overset{|}{C}H}{-}CH_3$ or $Li{-}\overset{CH_3}{\overset{|}{CH}}{-}CH_3$ (gives $CH(CH_3){-}CH_3$ fragment)

We could propose an overall synthesis as follows:

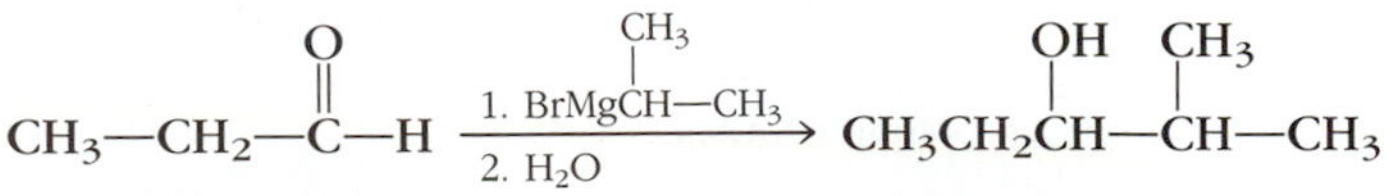

$$CH_3{-}CH_2{-}\overset{O}{\overset{\|}{C}}{-}H \xrightarrow[2.\ H_2O]{1.\ BrMg\overset{CH_3}{\overset{|}{C}H}{-}CH_3} CH_3CH_2\overset{OH}{\overset{|}{C}H}{-}\overset{CH_3}{\overset{|}{CH}}{-}CH_3$$

(Note that the Grignard reagent itself qualifies as an acceptable starting material in this case, so its preparation need not be shown.)

Example 11.2 Suggest a method of synthesis of 2-methyl-3-propyl-3-hexanol starting from compounds containing five or fewer carbon atoms and without using organocopper reagents.

Solution The key functional group is clearly the hydroxyl (together with the carbon atom to which it is attached):

$$CH_3—CH_2—CH_2—C(OH)(CH_2CH_2CH_3)—CH(CH_3)—CH_3$$

The limitation of 5 carbon atoms for allowed starting materials clearly requires carbon-carbon bond formation, because the target molecule contains 10 carbon atoms. The synthesis must almost certainly utilize the reaction of an organometallic compound with a carbonyl compound, and two alternatives are readily apparent:

$$CH_3CH_2CH_2—C(OH)(CH_2CH_2CH_3)—CH(CH_3)—CH_3$$

from $CH_3CH_2CH_2—C(=O)—CH_2CH_2CH_3$ and from $ClMg—CH(CH_3)—CH_3$

$$CH_3CH_2CH_2—C(OH)(CH_2CH_2CH_3)—CH(CH_3)—CH_3$$

from $CH_3CH_2CH_2MgCl$ and from $O=C(CH_2CH_2CH_2CH_3)—CH(CH_3)—CH_3$

In either case the carbonyl compound has more than 5 carbon atoms, and you might expect to synthesize it by some other carbon-carbon bond-forming reaction. (The sequence would presumably take place by reaction of a Grignard reagent with an aldehyde and subsequent oxidation of the resulting alcohol. We will leave this sequence for you as an exercise.) However, reexamination of the target molecule reveals that of the three alkyl substituents on the carbon atom bearing the hydroxyl group, two of them are *identical* propyl groups. In such cases a rather straightforward preparative method is available via an ester. If a 1-propyl Grignard reagent is allowed to react with an ester, two equivalents of the Grignard reagent will react to yield a tertiary alcohol.

$$CH_3CH_2CH_2—C(OH)(CH_2CH_2CH_3)—CH(CH_3)—CH_3$$

from $CH_3CH_2CH_2MgCl$ and from $CH_3O—C(=O)—CH(CH_3)—CH_3$

An alkylmagnesium bromide or iodide would work equally well, and propyllithium ($CH_3CH_2CH_2Li$) would also yield the desired product. A complete synthetic scheme could then be as follows:

$$CH_3-O-\overset{\overset{\displaystyle O}{\|}}{C}-\overset{\overset{\displaystyle CH_3}{|}}{CH}-CH_3 \xrightarrow[\text{2. } H_2O]{\text{1. } CH_3CH_2CH_2MgCl} CH_3CH_2CH_2-\underset{\underset{\displaystyle CH_2CH_2CH_3}{|}}{\overset{\overset{\displaystyle OH}{|}}{C}}-\overset{\overset{\displaystyle CH_3}{|}}{CH}-CH_3$$

Example 11.3 Suggest a method for synthesis of 2-cyclopentyl-2-heptanol from aldehydes and ketones containing five or fewer carbon atoms.

$$\text{cyclopentyl}-\underset{\underset{\displaystyle CH_3}{|}}{\overset{\overset{\displaystyle OH}{|}}{C}}-CH_2CH_2CH_2CH_2CH_3$$

Solution This example introduces two new degrees of difficulty. First, the allowed starting materials are restricted to aldehydes and ketones. This means that substantial functional group modification may be required in order to obtain any needed organometallic reagents. Second, the target molecule has a total of 12 carbon atoms, but each starting compound is limited to a maximum of only 5. At least *two* carbon-carbon bonds must be formed during the overall synthesis.

The key functional group of the target molecule is the alcohol group (—C—OH), and its three alkyl substituents are C_5, C_5, and C_1 groups. You do not yet know any methods for synthesizing five-membered rings, so the cyclopentyl group must be introduced as an intact C_5 subunit. Introduction of a C_5 group is also the most logical way to introduce the pentyl group, so the C_1 group originally must have been part of a two-carbon carbonyl compound:

$$\text{cyclopentyl}-\underset{\underset{\displaystyle CH_3}{|}}{\overset{\overset{\displaystyle OH}{|}}{C}}-CH_2CH_2CH_2CH_2CH_3$$

from cyclopentyl—MgBr from $CH_3-\overset{\overset{\displaystyle O}{\|}}{C}-$ from $BrMgCH_2CH_2CH_2CH_2CH_3$

Either of the C_5 alkyl groups could be introduced first, so two different synthetic schemes could be used to prepare the target molecule. In the following synthesis we have introduced the cyclopentyl group last. The other possible synthesis will be left for you as an exercise.

The last step of the synthesis is to be the reaction of a cyclopentyl Grignard reagent with a seven-carbon ketone.

$$\underset{\underset{\displaystyle CH_3}{|}}{\overset{O\diagdown\!\!\diagdown}{C}}-CH_2CH_2CH_2CH_2CH_3 \xrightarrow[\text{2. } H_2O]{\text{1. cyclopentyl}-MgBr} \text{cyclopentyl}-\underset{\underset{\displaystyle CH_3}{|}}{\overset{\overset{\displaystyle OH}{|}}{C}}-CH_2CH_2CH_2CH_2CH_3$$

But what is the origin of the seven-carbon ketone? A Grignard reaction was presumably necessary to join the C_2 and C_5 fragments, but that would yield an alcohol rather than a ketone. The answer to the dilemma is that the alcohol would indeed be produced first, but it could in turn be converted to the requisite ketone by *oxidation.*

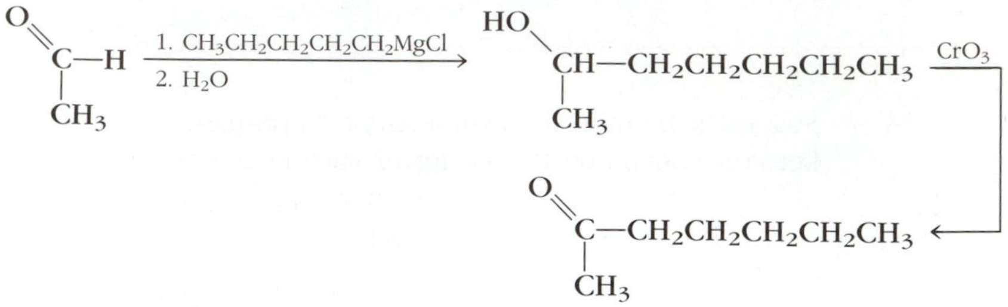

$$CH_3CH{=}O \xrightarrow[2.\ H_2O]{1.\ CH_3CH_2CH_2CH_2CH_2MgCl} CH_3CH(OH){-}CH_2CH_2CH_2CH_2CH_3 \xrightarrow{CrO_3} CH_3C({=}O){-}CH_2CH_2CH_2CH_2CH_3$$

The only remaining difficulty in the problem is the preparation of the Grignard reagents themselves. Only aldehydes and ketones were permitted as starting materials in this problem, so each of the Grignard reagents used here would have to be prepared by successive reduction of the appropriate carbonyl compound, conversion of the resulting alcohol to a halide, and reaction of the alkyl halide with magnesium. A variety of reagents are possible, and the following reactions simply illustrate one set of possibilities.

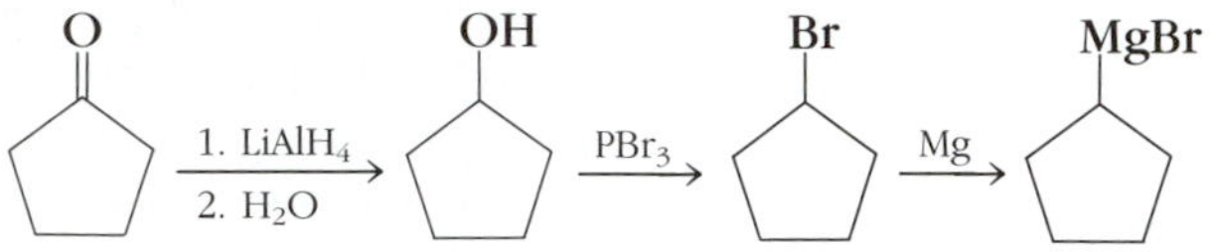

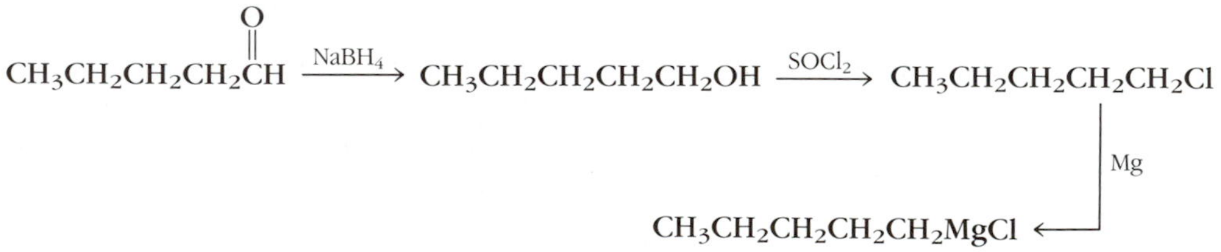

$$\text{cyclopentanone} \xrightarrow[2.\ H_2O]{1.\ LiAlH_4} \text{cyclopentanol (OH)} \xrightarrow{PBr_3} \text{cyclopentyl–Br} \xrightarrow{Mg} \text{cyclopentyl–MgBr}$$

$$CH_3CH_2CH_2CH_2CH{=}O \xrightarrow{NaBH_4} CH_3CH_2CH_2CH_2CH_2OH \xrightarrow{SOCl_2} CH_3CH_2CH_2CH_2CH_2Cl \xrightarrow{Mg} CH_3CH_2CH_2CH_2CH_2MgCl$$

Example 11.4 Suggest a method for synthesizing cyclopentylacetaldehyde from starting materials containing five or fewer carbon atoms.

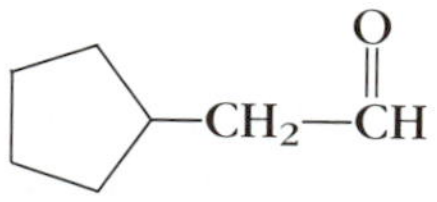

$$\text{cyclopentyl}{-}CH_2{-}CH{=}O$$

Solution The key functional group must be the carbonyl group, and one possible synthesis which comes to mind would employ the reaction of an appropriate organometallic reagent with formaldehyde ($H_2C{=}O$), followed by oxidation of the resulting alcohol. This would first require preparation of the six-carbon alcohol from C_5 and C_1 precursors, and we will leave it for you as an exercise. A more efficient plan is considered below.

Hydration of the corresponding alkyne (via hydroboration) would certainly yield the desired aldehyde, but the alkyne could not be prepared by alkylation of sodium acetylide. The corresponding cyclopentyl halide would be *secondary,* and you would expect elimination to predominate.

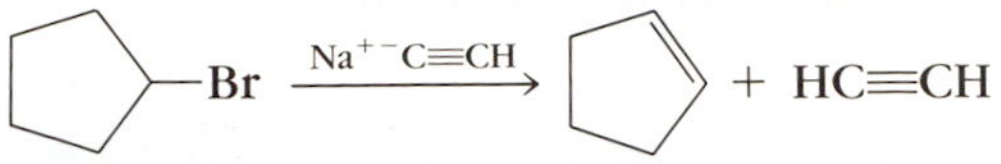

$$\text{cyclopentyl}-Br \xrightarrow{Na^{+}\,{}^{-}C\equiv CH} \text{cyclopentene} + HC\equiv CH$$

An aldehyde is readily formed by oxidation of a primary alcohol, so we can designate the corresponding alcohol as a new target molecule. This provides us with a new possibility for the key functional group, the —CH_2—OH, *together with* the adjacent carbon atom:

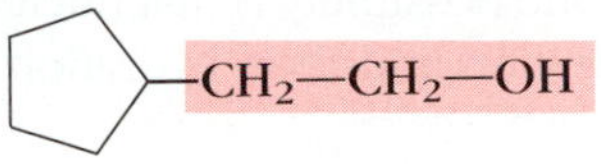

$$\text{cyclopentyl}-CH_2-CH_2-OH$$

The —CH_2—CH_2OH functionality results from attack of an organometallic reagent on ethylene oxide, and a very simple synthesis of the original aldehyde is therefore possible.

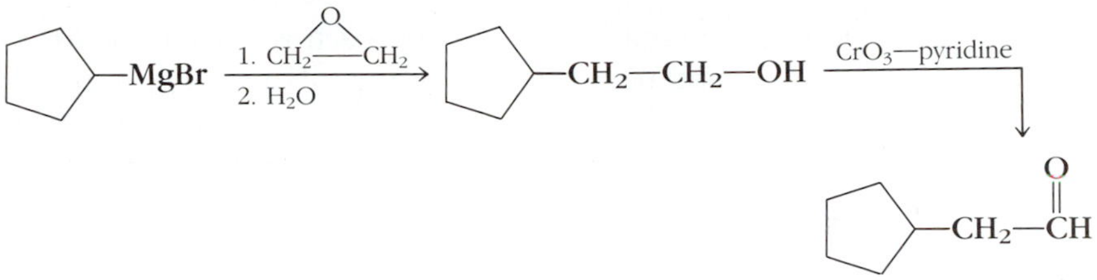

$$\text{cyclopentyl}-MgBr \xrightarrow[\text{2. } H_2O]{\text{1. } CH_2\!-\!CH_2 \text{ (O bridged)}} \text{cyclopentyl}-CH_2-CH_2-OH \xrightarrow{CrO_3\text{—pyridine}} \text{cyclopentyl}-CH_2-\overset{O}{\overset{\|}{C}}H$$

Another synthetic approach to this compound could be designed by using the coupling reactions of organocopper reagents with alkenyl halides (Section 5.14). Preparation of the appropriate alkene, followed by hydration and oxidation, would yield the required aldehyde. The complete description of this scheme is left for you as an exercise.

Example 11.5 Suggest a synthesis of 2-(2-cyclopentylethyl)cyclohexanol from compounds containing six or fewer carbon atoms.

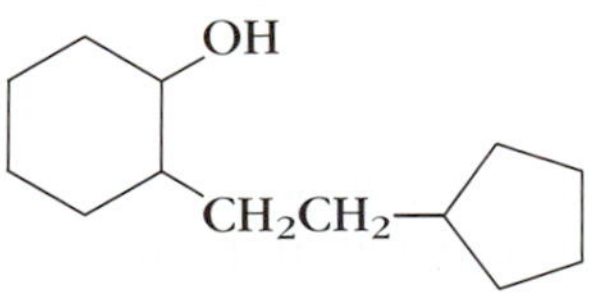

$$\text{cyclohexane ring with OH and } CH_2CH_2-\text{cyclopentyl on adjacent carbons}$$

Solution The target molecule has a total of 13 carbon atoms, so at least *two* carbon-carbon bonds must be formed during the synthesis. We have not shown you general ways to prepare six-membered rings, so it should be assumed that one of the starting materials was a cyclohexane derivative. The bond between the six-membered ring and the remaining seven-carbon fragment must therefore be formed during the synthesis. Evaluation of the key functional group (—C—OH) reveals that the C_7 alkyl substituent is attached to an *adjacent* carbon.

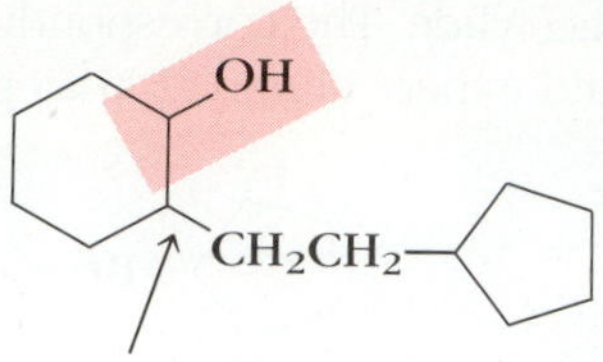

this bond must be formed during the synthesis

Two alternatives can be considered for such a situation in which the oxygen atom is on a carbon atom *adjacent* to the position at which a carbon-carbon bond is formed. One possibility is that nucleophilic attack at a carbonyl group was followed by dehydration and then hydration in the opposite orientation. For example,

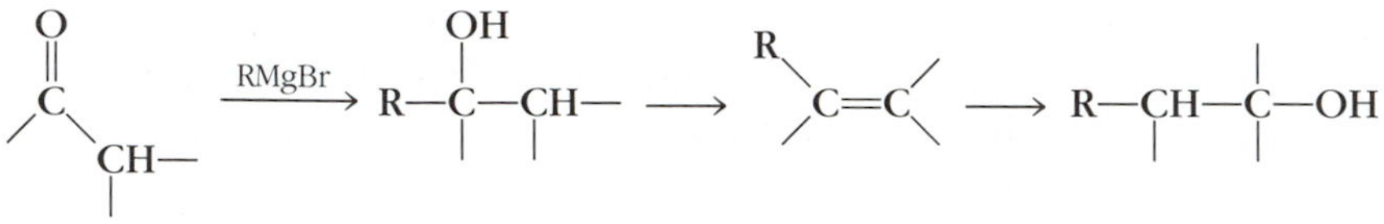

(Design of a synthesis by this approach is left for you as an exercise).

The second possibility is that the OH group was introduced *simultaneously,* with carbon-carbon bond formation by reaction of an organometallic compound with an epoxide (as we showed in Example 11.4). In the present case the last step would involve reaction with the epoxide of cyclohexene.

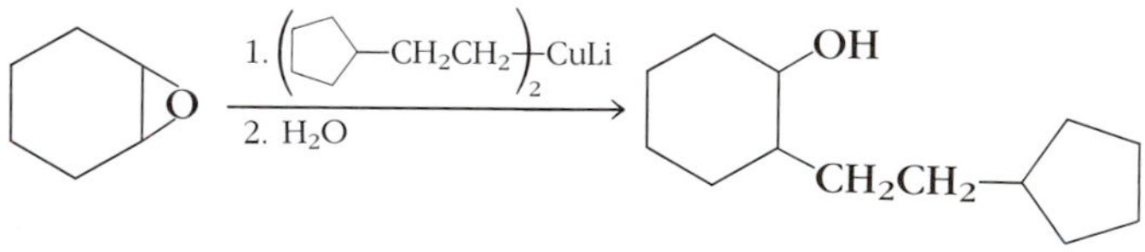

In reactions of substituted epoxides (i.e., those other than ethylene oxide itself) the organocopper derivatives are vastly superior reagents. The lithium dialkylcuprate needed for this reaction could be prepared from the seven-carbon alcohol via the following sequence:

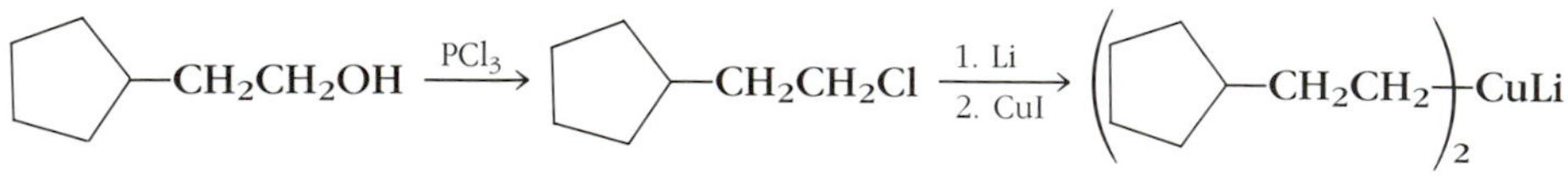

Preparation of the alcohol from starting materials containing six or fewer carbon atoms was illustrated in the previous example.

Example 11.6 Suggest a method for synthesizing 3-methyl-3-nonen-5-yne starting from compounds containing five or fewer carbon atoms.

Solution The key functional group might be either the double bond or the triple bond (or both together, since they are conjugated). The target molecule contains a total of 10 carbon atoms, and up to 5 can be used for a starting material, so the possibility of joining two C_5 fragments is certainly worth evaluating:

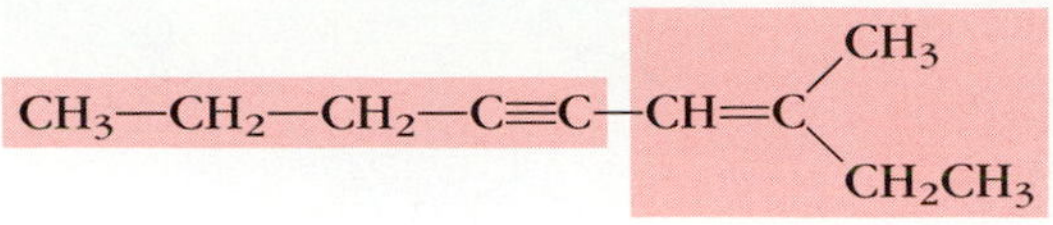

$$CH_3{-}CH_2{-}CH_2{-}C{\equiv}C{-}CH{=}C(CH_3)(CH_2CH_3)$$

One potential approach to the synthesis now becomes apparent: the C_5 unit containing the triple bond might have been introduced via the sodium salt of 1-pentyne. The direct formation of the target molecule would not be possible since displacement reactions work only with (primary) *alkyl* halides and not with *alkenyl* halides.

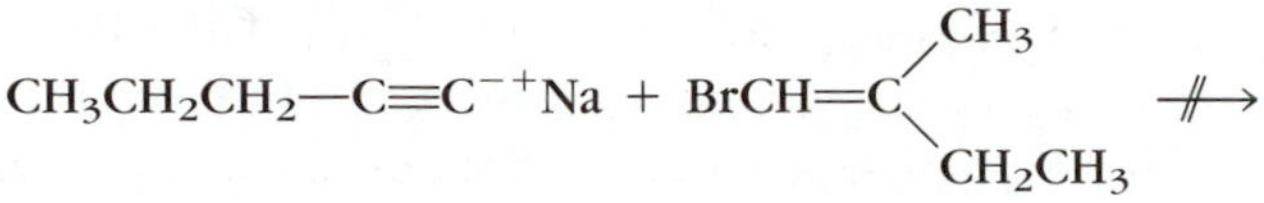

$$CH_3CH_2CH_2{-}C{\equiv}C^{-\,+}Na + BrCH{=}C(CH_3)(CH_2CH_3) \not\longrightarrow$$

On the other hand, reaction of the same sodium acetylide with the appropriate aldehyde would afford an alcohol that could subsequently be dehydrated to give the desired compound.

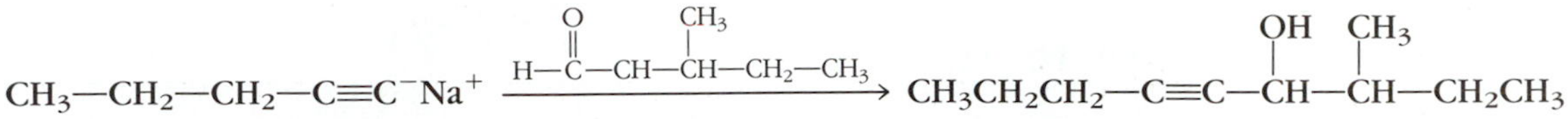

$$CH_3{-}CH_2{-}CH_2{-}C{\equiv}C^{-}Na^{+} \xrightarrow{H{-}C({=}O){-}CH{-}CH(CH_3){-}CH_2{-}CH_3} CH_3CH_2CH_2{-}C{\equiv}C{-}CH(OH){-}CH(CH_3){-}CH_2CH_3$$

While acid-catalyzed dehydration might afford the target molecule, it would be safer to avoid the possibility of side reactions that often occur under acidic conditions. As an alternative, the alcohol could first be converted to the tosylate and elimination could then be brought about by treatment with base.

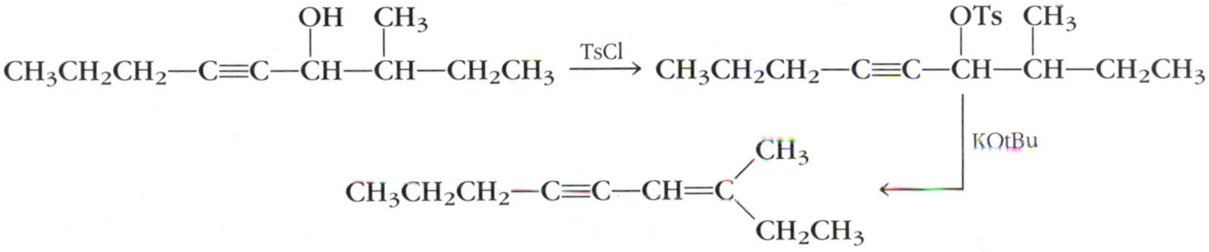

$$CH_3CH_2CH_2{-}C{\equiv}C{-}CH(OH){-}CH(CH_3){-}CH_2CH_3 \xrightarrow{TsCl} CH_3CH_2CH_2{-}C{\equiv}C{-}CH(OTs){-}CH(CH_3){-}CH_2CH_3$$

$$\xrightarrow{KOtBu} CH_3CH_2CH_2{-}C{\equiv}C{-}CH{=}C(CH_3)(CH_2CH_3)$$

(Elimination to form the conjugated double bond would occur quite easily in this reaction. If pyridine were used as a base in the tosylation, it is possible that elimination would occur immediately upon formation of the tosylate.)

Example 11.7 Suggest a method for synthesizing 1,2-epoxy-1-ethylcyclohexane using starting materials containing six or fewer carbon atoms.

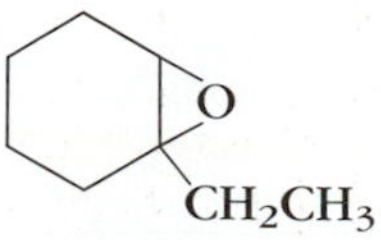

[Structure: cyclohexane ring with epoxide O bridging C1–C2; CH_2CH_3 on C1]

Solution The key functional group is the epoxide ring, but this would not be involved in any carbon-carbon bond-forming process. Instead, the reactive epoxide functionality must almost certainly be introduced at a later stage in the synthesis by epoxidation of the corresponding alkene:

$$\text{1-ethylcyclohexene} \xrightarrow{CH_3CO_3H} \text{1-ethyl-1,2-epoxycyclohexane}$$

The alkene could be synthesized in a variety of ways, and perhaps the simplest would be via a Grignard reaction with cyclohexanone, followed by acid-catalyzed dehydration of the resulting tertiary alcohol.

$$\text{cyclohexanone} \xrightarrow[2.\ H_2O]{1.\ CH_3CH_2MgI} \text{1-ethylcyclohexanol } (CH_2CH_3,\ OH) \xrightarrow{\text{conc. } H_2SO_4} \text{1-ethylcyclohexene } (CH_2CH_3)$$

In this reaction acid-catalyzed rearrangements would produce no difficulties, because the desired alkene is the most stable isomer. It is trisubstituted and both unsaturated carbons are part of the six-membered ring, a preferred arrangement for cyclohexyl derivatives.

EXERCISE 11.14

Complete the alternative synthetic proposals left as exercises in: (a) Example 11.2; (b) Example 11.3; (c) Example 11.4; (d) Example 11.5.

EXERCISE 11.15

Suggest methods of synthesizing the following compounds. Unless other restrictions are indicated, you may use any reagents you wish, but all the carbon atoms of the product must be derived from starting materials with the indicated number of carbon atoms.

(a) cyclopentyl–CH(OH)–CH_2CH_3 — From alcohols containing *five* or fewer carbon atoms

(b) cyclohexyl–CH_2–CH(OH)–cyclohexyl — From cyclohexanone and any other organic compounds containing *two* or fewer carbons atoms

(c) $CH_3{-}CH{=}C(CH_2CH_2CH_3){-}CH_2CH_2CH_3$ — From compounds containing *three* or fewer carbon atoms; no organocopper reagents may be used

11.11 INSECT PHEROMONES: A PRACTICAL EXAMPLE OF ORGANIC SYNTHESIS

While most animals do not communicate verbally, they are capable of other forms of communication. For lower animals such as insects this communication occurs by chemical means, wherein one animal secretes a messenger substance called a *pheromone,* which in turn elicits a specific response by another animal. This kind of communication is very primitive, but it can also be quite effective. For example, insect *sex attractants* are pheromones that attract insects of the opposite sex and elicit mating behavior, and they appear to be effective at the level of just a few *molecules.* These pheromones also tend to be very specific for each species.

Widespread interest currently exists in the possible use of pheromones for controlling insect populations in ways that are not harmful to the environment. For instance, a sex attractant could be used to lure insects of a destructive species to traps, where they could then be killed without the widespread introduction of chemicals that are harmful to other species. The insects themselves typically synthesize only minute quantities of pheromones, so any attempts to control an insect population with pheromones requires a larger-scale synthesis in the organic laboratory.

One pheromone that has been synthesized is the sex attractant of the Douglas fir tussock moth. The moth causes severe damage to fir trees in the western United States, and control of the insect could be quite beneficial. The pheromone has been identified as (*Z*)-6-henicosen-11-one (where the prefix "henicos-" designates a 21-carbon chain):

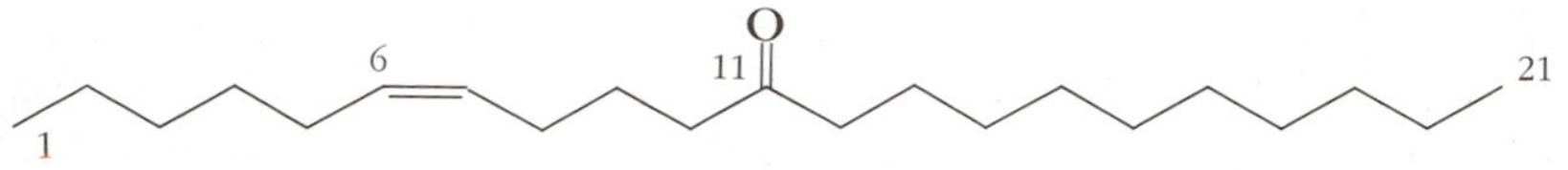

O
6 11 21
1

A synthesis of this pheromone, outlined in Figure 11.2, has been reported by workers at the Zoecon Corporation. The basic synthetic plan was to join C_{10} and C_{11} fragments via a Grignard reaction to give a C_{21} alcohol, which could then be modified by oxidation and reduction. (We have drawn the bonds to the C≡C unit as nonlinear for convenience only.)

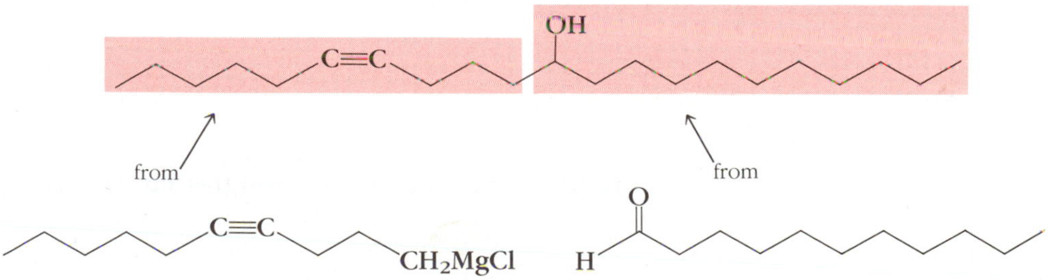

OH
C≡C
from from
O
C≡C
CH_2MgCl H

The C_{11} aldehyde is commercially available, but the C_{10} Grignard had to be synthesized from smaller fragments. The procedure selected for preparation of the corresponding chloride was alkylation of a sodium acetylide.

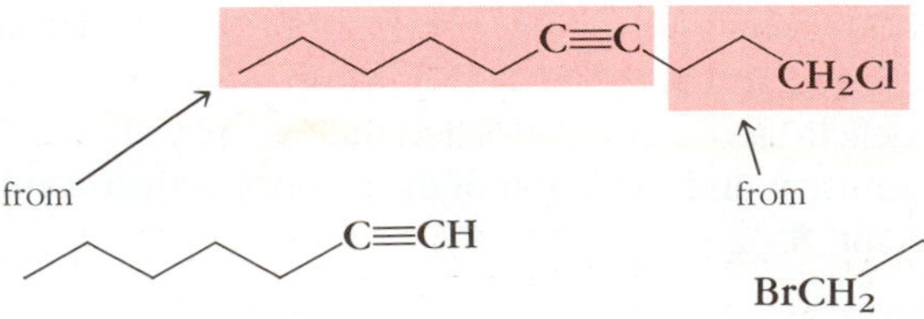

C≡C
CH_2Cl
from from
C≡CH
$BrCH_2$ CH_2Cl

Both 1-heptyne and 1-bromo-3-chloropropane are commercially available. The overall synthesis then proceeded as follows: Reaction with the lithium salt of

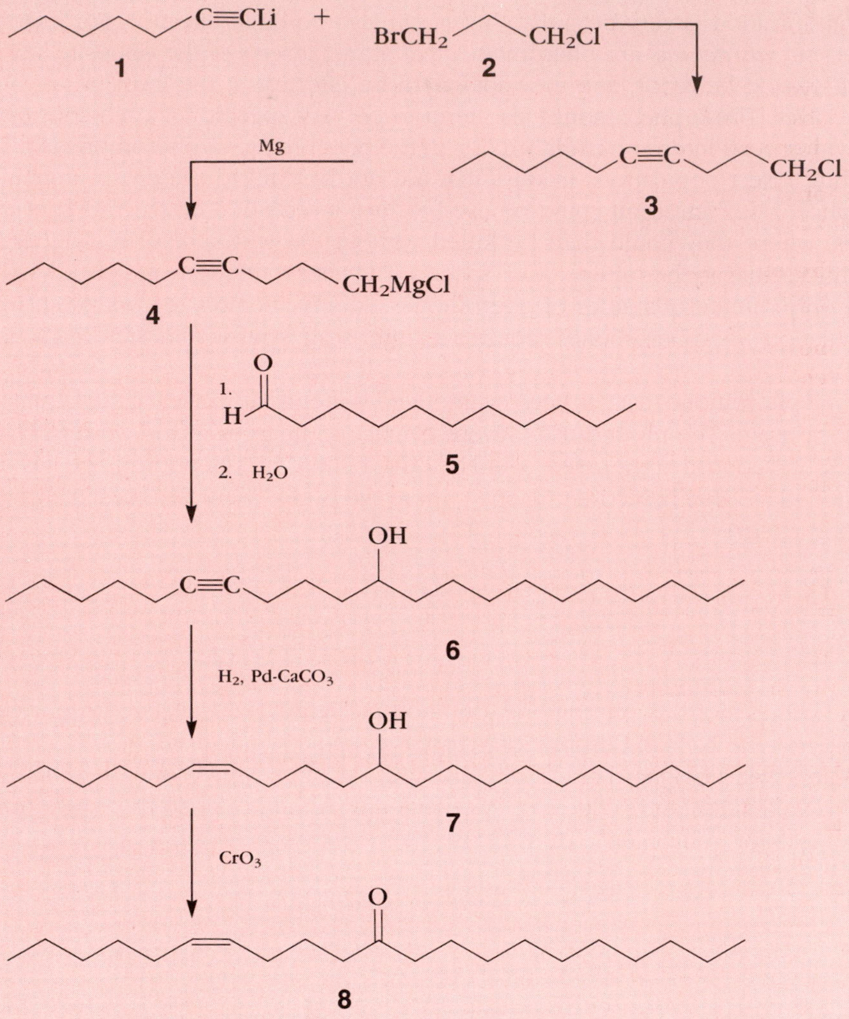

$C{\equiv}CLi$ + $BrCH_2$ CH_2Cl

1 2

$C{\equiv}C$ CH_2Cl

3

Mg

$C{\equiv}C$ CH_2MgCl

4

1. O H

2. H_2O

5

OH

$C{\equiv}C$

6

H_2, Pd-$CaCO_3$

OH

7

CrO_3

O

8

Figure 11.2 Synthesis of the Sex Attractant of the Douglas Fir Tussock Moth

1-heptyne (**1**) selectively displaced the more reactive *bromine* of the dihalopropane, **2** (Br^- is a better leaving group than Cl^-). The resulting 1-chloro-4-decyne (**3**) was converted to the corresponding Grignard reagent, **4**, by treatment with magnesium. Reaction with undecanal (**5**) afforded the hydroxy alkyne, **6**, which has all the carbon atoms of the final product but has the wrong oxidation state both at the triple bond and at the hydroxyl group. Selective reduction of the alkyne group using a deactivated catalyst afforded the hydroxy alkene, **7**, which has the correct *Z* stereochemistry, and oxidation of the alcohol with chromium trioxide yielded the sex attractant, **8**.

11.12 TERMS AND DEFINITIONS

Aldehyde. A carbonyl compound in which one of the two substituents on the carbonyl group is a hydrogen and the other is an alkyl (or aryl) group.

Carbonyl group. The C=O group; the functional group corresponding to a carbon-oxygen double bond.

Carboxylic acid. A carbonyl compound in which one of the substituents on the carbonyl group is an OH.

Chemoselectivity. Selective reaction at one of several different reactive centers in a molecule.

Ester. A carbonyl compound in which one of the substituents on the carbonyl group is alkoxy (or aryloxy) and the other substituent is alkyl, aryl, or hydrogen.

Ketone. A carbonyl compound in which both substituents on the carbonyl group are alkyl (or aryl) groups.

Organometallic compound. A reactive organic compound characterized by a carbon-metal bond. Organometallic reagents can be symbolized by the formula R—M, where M is a metal atom.

Pheromone. A chemical messenger secreted by an animal, that elicits a specific response in another animal (often of the opposite sex) of the same species.

Synthetic intermediate. A compound, produced at an intermediate stage of an organic synthesis, that is converted to the final product in subsequent reactions.

11.13 SUMMARY OF REACTIONS

The reactions of alcohols are summarized in Table 11.2, and various methods for the preparation of alcohols are summarized in Table 11.3.

TABLE 11.2 Reactions of Alcohols

Reaction	Comments
1. Acid-base reactions	Section 11.1
$R\text{—}OH \overset{H^+}{\rightleftharpoons} R\text{—}OH_2^+$ $R\text{—}OH \overset{\text{base}}{\rightleftharpoons} R\text{—}O^-$	Analogous to the reactions of water
2. Reaction with active metals	
$R\text{—}OH \xrightarrow{Na} R\text{—}O^-\ Na^+ + H_2$	Analogous to the reactions of water
3. Alkylation	Section 11.8
$R\text{—}\ddot{\underset{..}{O}}\text{—}H \xrightarrow{CH_3\text{—}X} R\text{—}\overset{+}{\underset{..}{O}}(CH_3)\text{—}H$ $R\text{—}O^- \xrightarrow{CH_3\text{—}X} R\text{—}O\text{—}CH_3$	Discussed more fully in Chapter 12

TABLE 11.2 Reactions of Alcohols (continued)

Reaction	Comments
4. Acylation	Section 11.8
$R{-}O{-}H \xrightarrow{Cl{-}C(=O){-}CH_3} R{-}O{-}C(=O){-}CH_3$	Discussed more fully in Chapter 16
5. Formation of sulfonate esters	Section 11.9
$R{-}O{-}H \xrightarrow{Cl{-}S(=O)_2{-}CH_3} R{-}O{-}S(=O)_2{-}CH_3$	The reaction is usually carried out in pyridine.
$R{-}O{-}H \xrightarrow{Cl{-}S(=O)_2{-}C_6H_4{-}CH_3} R{-}O{-}S(=O)_2{-}C_6H_4{-}CH_3$	
6. Oxidation of alcohols	Section 11.3
$-CH(OH)- \longrightarrow\ >C{=}O$	Industrial oxidations may use O_2 and a metal catalyst. Typical laboratory oxidizing agents are $KMnO_4$ and Cr(VI) reagents such as CrO_3, $Na_2Cr_2O_7$, H_2CrO_4, and CrO_3-pyridine.
$RCH_2{-}OH \xrightarrow[\text{agent}]{\text{oxidizing}} R{-}C(=O){-}OH$ $RCH_2{-}OH \xrightarrow{CrO_3\text{-pyridine}} R{-}C(=O){-}H$	Primary alcohols usually yield carboxylic acids. If the CrO_3-pyridine reagent is used, the oxidation stops at the aldehyde stage.
$R{-}CH(OH){-}R \xrightarrow[\text{agent}]{\text{oxidizing}} R{-}C(=O){-}R$	Secondary alcohols yield ketones (including oxidation with CrO_3-pyridine).
$R_3C{-}OH \xrightarrow[\text{agent}]{\text{oxidizing}}$ no reaction	Tertiary alcohols do not react under conditions used to oxidize primary and secondary alcohols.

TABLE 11.2 Reactions of Alcohols (continued)

Reaction	Comments
7. Conversion to halides (X = Cl, Br)	Section 11.9
R—OH ⟶ R—X	Sometimes conc. HX can be used, but rearrangements may result. Superior laboratory reagents are PCl_3, PCl_5, $POCl_3$, $SOCl_2$, and PBr_3.
8. Dehydration	Section 5.13
—CH—C—OH $\xrightarrow{\text{conc. } H_2SO_4}$ >C=C<	Acid-catalyzed rearrangements may result.
—CH—C—OH ⟶ —CH—C—Cl $\xrightarrow{\text{base}}$ >C=C< —CH—C—OH ⟶ —CH—C—OTs $\xrightarrow{\text{base}}$ >C=C<	Alternatively, the alcohol can first be converted to a halide or a sulfonate ester. Elimination can then be carried out under basic conditions (Section 11.9).

TABLE 11.3 Preparation of Alcohols

Reaction	Comments
Functional Group Modification, No Change in Carbon Skeleton	
1. Hydration of alkenes	
(a) Acid-catalyzed addition of H_2O	Section 5.8
>C=C< $\xrightarrow[H^+]{H_2O}$ —CH—C—OH	May involve acid-catalyzed rearrangements
(b) Hydroboration-oxidation	Section 5.11
>C=C< $\xrightarrow[\text{2. } H_2O_2, \text{ NaOH}]{\text{1. } BH_3}$ —CH—C—OH	No rearrangements. The OH group is introduced onto the less substituted carbon atom. A *syn* addition.

TABLE 11.3 Preparation of Alcohols (continued)

Reaction	Comments
Functional Group Modification, No Change in Carbon Skeleton	
(c) Oxymercuration-demercuration	Section 5.11
$\gt C{=}C\lt \xrightarrow[\text{2. NaBH}_4]{\text{1. Hg(OCCH}_3)_2\ (\text{C=O})} -\overset{\mid}{C}H-\underset{\mid}{\overset{\mid}{C}}-OH$	No rearrangements. The OH group is introduced onto the more substituted carbon atom.
2. Oxidation of alkenes: glycol formation	Section 5.12
$\gt C{=}C\lt \xrightarrow[\text{or alkaline KMnO}_4]{\text{OsO}_4,\ \text{NaClO}_3} HO-\underset{\mid}{\overset{\mid}{C}}-\underset{\mid}{\overset{\mid}{C}}-OH$	A *syn* addition
3. Allylic oxidation	Section 8.4
$\gt C{=}C\lt$ with $-\underset{\mid}{C}H-$ on the alkene carbon $\xrightarrow{\text{SeO}_2}$ $\gt C{=}C\lt$ with $\gt C-OH$ on the alkene carbon	
4. Reduction of carbonyl compounds	Section 11.4
(a) Catalytic hydrogenation	
$\gt C{=}O \xrightarrow[\text{catalyst}]{\text{H}_2} -\overset{OH}{\overset{\mid}{C}}H-$	Requires high pressures. Other multiple bonds may be reduced preferentially.
(b) Dissolving metal reactions	
$\gt C{=}O \xrightarrow[\text{CH}_3\text{CH}_2\text{OH}]{\text{Na}} -\overset{OH}{\overset{\mid}{C}}H-$	Most commonly done with sodium-ethanol but other metals and other proton sources can also be used.
$-\overset{O}{\overset{\Vert}{C}}-OR \xrightarrow[\text{CH}_3\text{CH}_2\text{OH}]{\text{Na}} -CH_2OH$	Aldehydes, ketones, and esters are all reduced to alcohols.

TABLE 11.3 Preparation of Alcohols (continued)

Reaction	Comments
Functional Group Modification, No Change in Carbon Skeleton	
(c) Metal hydride reductions	
$>C{=}O \xrightarrow[CH_3CH_2OH]{NaBH_4} -\overset{OH}{\underset{\vert}{C}}-H$	Sodium borohydride reduces aldehydes and ketones to alcohols. Carboxylic acids and esters are not reduced.
$>C{=}O \xrightarrow[2.\ H_2O]{1.\ LiAlH_4} -\overset{OH}{\underset{\vert}{C}}H$ $-\overset{O}{\overset{\Vert}{C}}-OR \xrightarrow[2.\ H_2O]{1.\ LiAlH_4} -CH_2OH$	Aldehydes, ketones, carboxylic acids, and esters are all reduced to alcohols with $LiAlH_4$.
5. Reduction of epoxides	Section 11.6
$-\overset{\vert}{C}\overset{O}{-}\overset{\vert}{C}- \xrightarrow[2.\ H_2]{1.\ LiAlH_4} H-\overset{\vert}{\underset{\vert}{C}}-\overset{\vert}{\underset{\vert}{C}}-OH$	
6. Hydrolysis of epoxides: glycol formation	Section 5.12
$-\overset{\vert}{C}\overset{O}{-}\overset{\vert}{C}- \xrightarrow[H^+]{H_2O} HO-\overset{\vert}{\underset{\vert}{C}}-\overset{\vert}{\underset{\vert}{C}}-OH$	
7. Hydrolysis of alkyl halides or tosylates (X = Cl, Br, OTs)	Discussed in Chapter 12
$R-X \xrightarrow{H_2O} R-OH$	
8. Hydrolysis of esters	Discussed in Section 16.6
$R-\overset{O}{\overset{\Vert}{C}}-OR' \xrightarrow[\text{acid or base}]{H_2O} R-\overset{O}{\overset{\Vert}{C}}OH + R'OH$	
9. Hydrolysis of ethers	Discussed in Section 15.3
$R-O-R' \xrightarrow[\text{acid}]{H_2O} R-OH + R'-OH$	

TABLE 11.3 Preparation of Alcohols (continued)

Reaction	Comments
Reactions Yielding Alcohols via Carbon-Carbon Bond Formation	
10. Reaction of organometallic reagents with carbonyl compounds	Section 11.5
(a) Aldehydes and ketones	
$>C{=}O \xrightarrow[2.\ H_2O]{1.\ RMgX\ \text{or}\ RLi} -\overset{OH}{\underset{\vert}{C}}-R$	The halogen (X) of the Grignard reagent can be Cl, Br, I.
$>C{=}O \xrightarrow[2.\ H_2O]{1.\ NaC{\equiv}CR\ \text{or}\ LiC{\equiv}Cr} -\overset{OH}{\underset{\vert}{C}}-C{\equiv}CR$	
(b) Esters	
$-\overset{O}{\overset{\Vert}{C}}-OCH_3 \xrightarrow[2.\ H_2O]{1.\ RMgX\ \text{or}\ RLi} -\overset{OH}{\underset{R}{C}}-R$	Two identical alkyl groups are introduced in the reaction of an ester.
11. Reaction of organometallic reagents with epoxides	Section 11.6
$-\underset{\vert}{C}\overset{O}{-}\underset{\vert}{C}- \xrightarrow[2.\ H_2O]{1.\ RMgX} -\overset{OH}{\underset{\vert}{C}}-\underset{\vert}{\overset{\vert}{C}}-R$	The alkyl group is introduced at the less substituted carbon atom of the epoxide. The halogen, X, of the Grignard reagent can be Cl, Br, I.
$-\underset{\vert}{C}\overset{O}{-}\underset{\vert}{C}- \xrightarrow[2.\ H_2O]{1.\ RLi\ \text{or}\ R_2CuLi} -\overset{OH}{\underset{\vert}{C}}-\underset{\vert}{\overset{\vert}{C}}-R$	Magnesium salts sometimes cause rearrangements with substituted epoxides. Lithium dialkylcuprates are superior reagents.
$-\underset{\vert}{C}\overset{O}{-}\underset{\vert}{C}- \xrightarrow[2.\ H_2O]{1.\ LiC{\equiv}C{-}R\ \text{or}\ NaC{\equiv}CR} -\overset{OH}{\underset{\vert}{C}}-\underset{\vert}{\overset{\vert}{C}}-C{\equiv}CR$	
12. Condensation reactions of carbonyl compounds	Chapter 14
$>C{=}O + -\overset{\vert}{C}H-\overset{O}{\overset{\Vert}{C}}- \xrightarrow[\text{or base}]{\text{acid}} -\overset{OH}{\underset{\vert}{C}}-\underset{\vert}{\overset{\vert}{C}}-\overset{O}{\overset{\Vert}{C}}-$	

11.14 PROBLEMS

11.1 For each of the structural formulas drawn below, give the correct IUPAC name:

(a) $CH_3-CH(CH_3)-CH_2-CH(OH)-CH(CH_3)-CH_3$

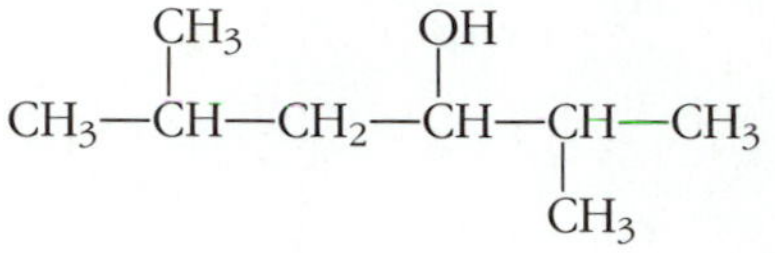

(b) $CH_3-CH_2-CH(OH)-CH_2-C(OH)(CH_3)-CH_2-CH_3$

(c) $CH_3-CH_2-CH(OH)-CH(CH_2-CH_2-CH_3)-CH_2-CH=CH_2$

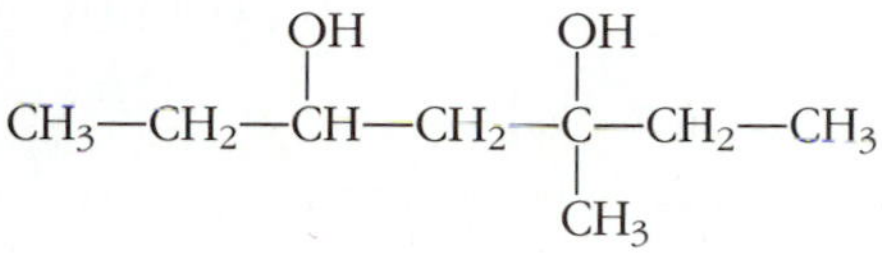

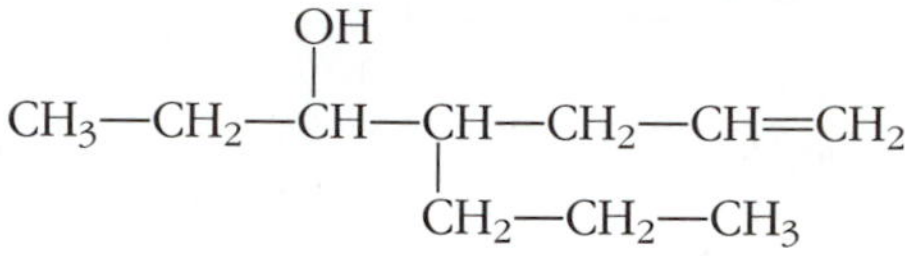

(d) $CH_3-CH_2-CH(Cl)-CH(CH_3)-CH_2-CH(OH)-CH_2Br$

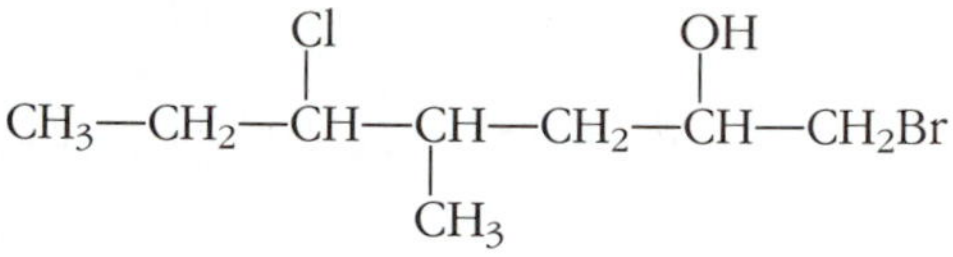

(e) cyclobutyl$-CH_2-CH_2-CH_2OH$

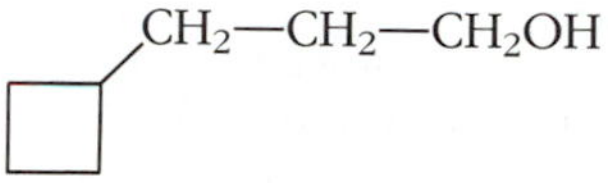

(f) cyclopentyl$-CH_2-CH(CH_3)-CH_2-CH(OH)-CH_3$

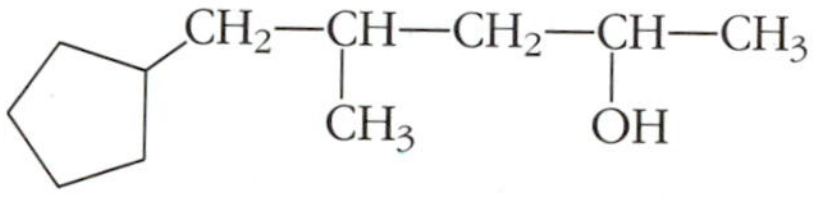

(g) $CH_3-CH(OH)-C(=CH_2)-CH_2-CH_3$

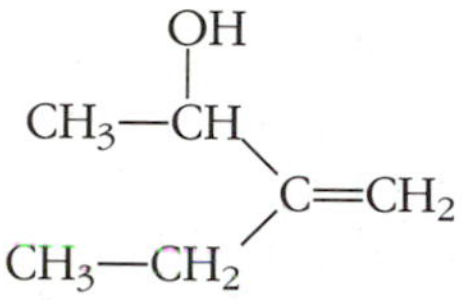

(h) cyclohexane ring with $C(CH_3)_3$ and OH on adjacent carbons

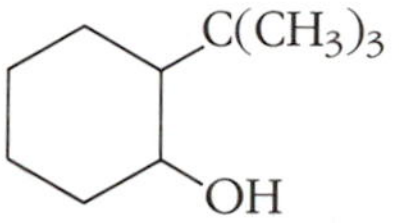

11.2 Draw the structure for each of the following compounds:
(a) 3-Methyl-2-hexanol
(b) 4,4-Dimethylcyclohexanol
(c) 2,2-Dichloro-3-isopropyl-5-methyl-1-octanol
(d) 2-Pentyllithium
(e) Lithium 3-methyl-3-pentoxide
(f) 3,3-Dimethyl-4-octen-1,2-diol
(g) 4-(2-Chloroethyl)cyclohexanol

11.3 Complete the following reactions:

(a) $\xrightarrow{PCl_5}$?

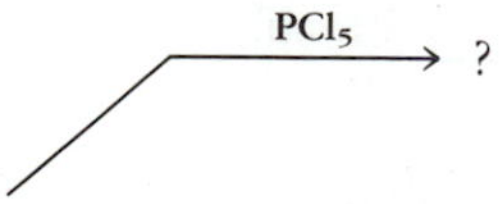

(b) $CH_3-CH_2-CH(CH_3)-CH_2OH \xrightarrow{KMnO_4}$?

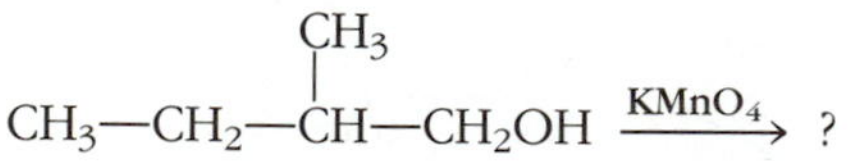

(c) $\xrightarrow{Na}$?

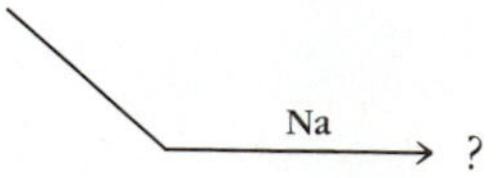

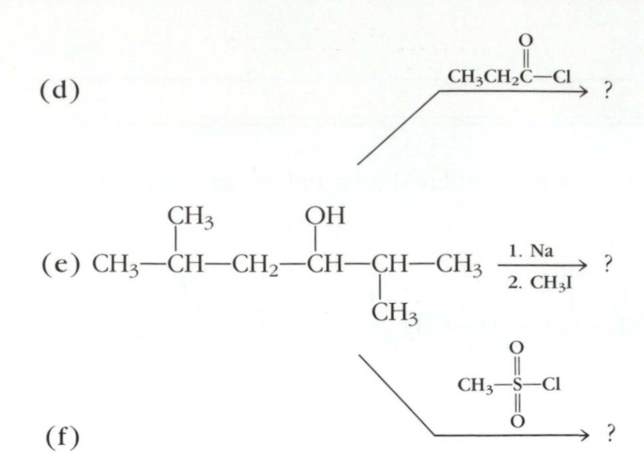

11.4 Complete the following reactions:

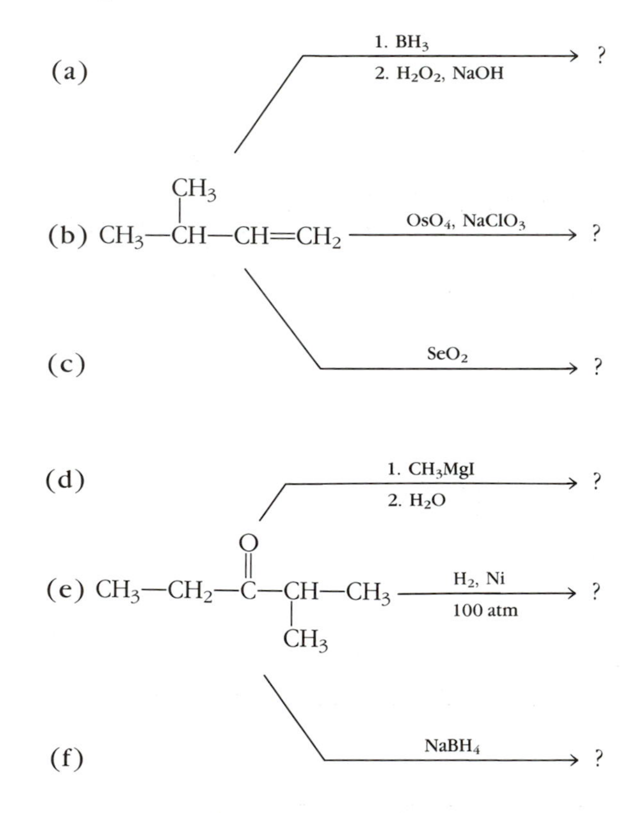

11.5 Complete the following reactions:

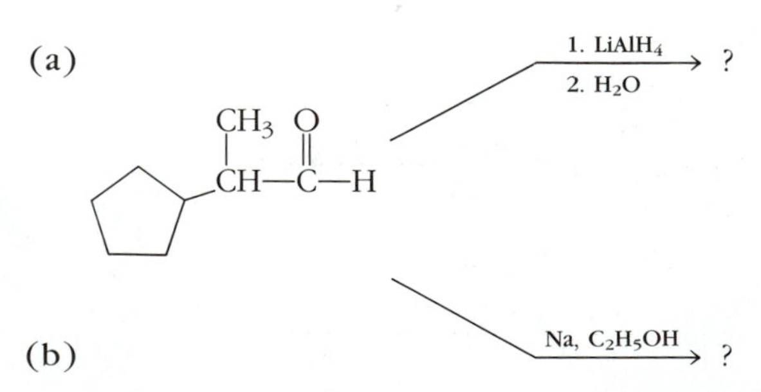

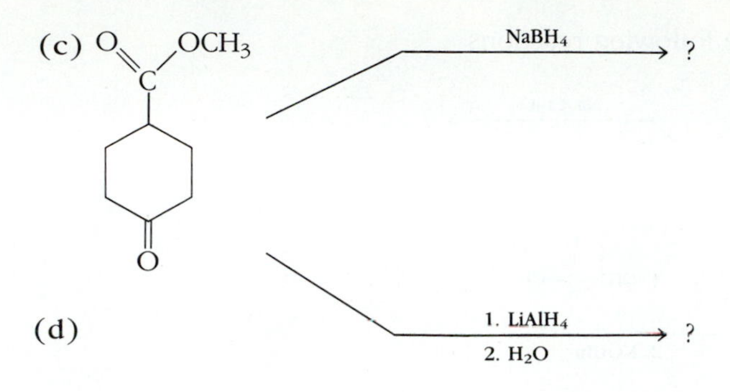

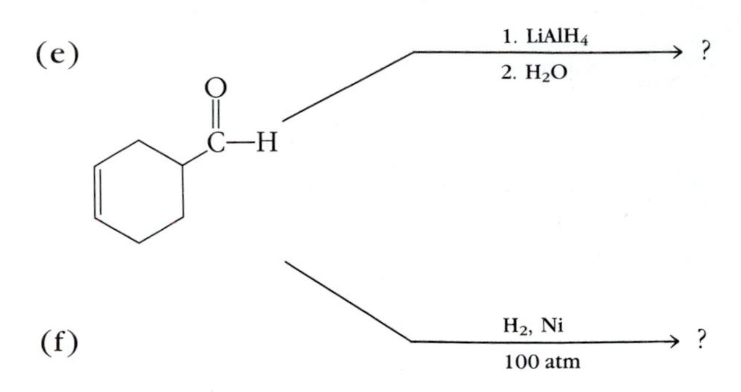

11.6 Complete the following reactions:

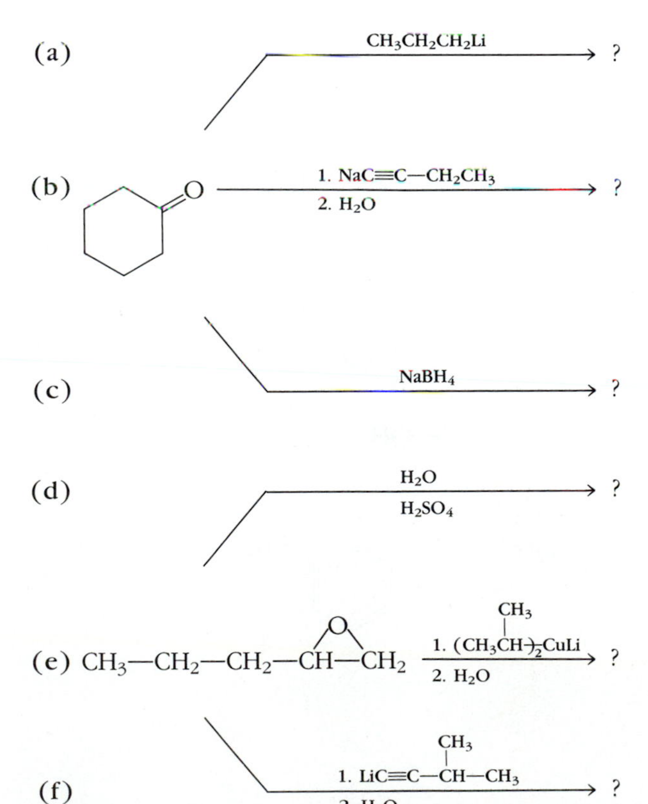

11.7 Complete the following reactions:

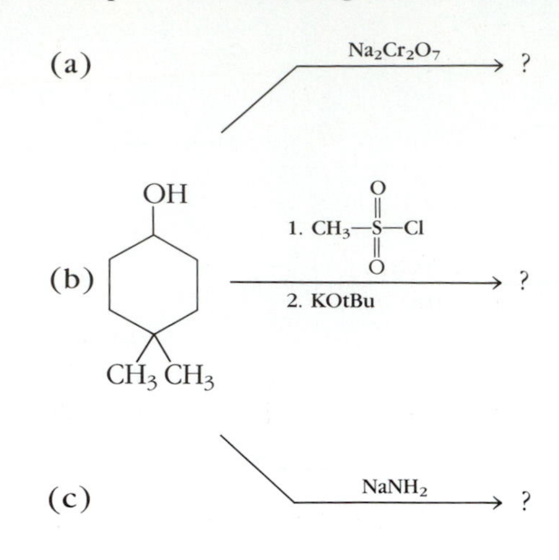

(d) $\xrightarrow[\text{2. } H_2O]{\text{1. } CH_3CH_2CH_2MgI}$?

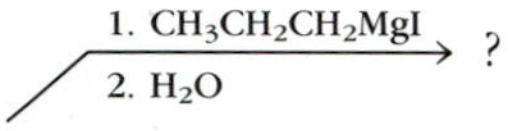

(e) cyclopentyl—C(=O)—O—C_2H_5 $\xrightarrow[\text{2. } H_2O]{\text{1. } LiAlH_4}$?

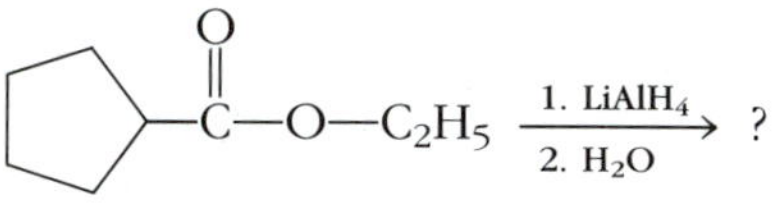

(f) $\xrightarrow[\text{2. } H_2O]{\text{1. } CH_3Li}$?

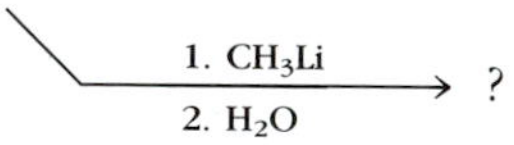

11.8 Suggest reagents for carrying out each of the following reactions:

(a) cyclopentanone $\xrightarrow{?}$ 1-cyclopentylcyclopentanol (OH)

(b) cyclopentanone $\xrightarrow{?}$ cyclopentanol (—OH)

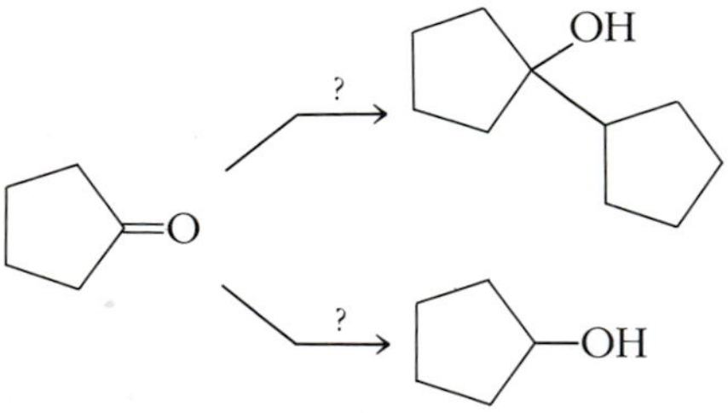

(c) C_6H_5—CH_2CH_2OH $\xrightarrow{?}$ C_6H_5—CH_2—C(=O)—H

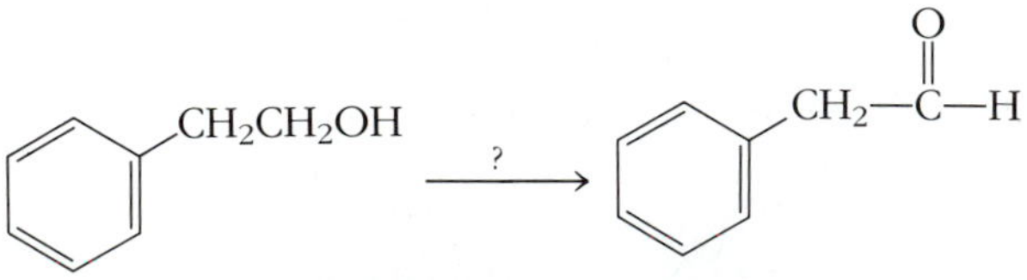

(d) $\xrightarrow{?}$ (cyclohex-3-enyl)CH_2OH

(cyclohex-3-enyl)$-\overset{\overset{O}{\|}}{C}-O-CH_3$

(e) $\xrightarrow{?}$ (cyclohexyl)$-\overset{\overset{O}{\|}}{C}-O-CH_3$

11.9 Write a detailed reaction mechanism for each of the reactions in Problem 11.8.

11.10 Suggest reagents for carrying out each of the following conversions. More than one step may be necessary in each case.

Starting material for (a)–(b): $CH_3-CH_2-CH(CH_3)-CH_2-\overset{\overset{O}{\|}}{C}-O-CH_3$

(a) $\xrightarrow{?}$ $CH_3CH_2-CH(CH_3)-CH_2-C(OH)(CH_2-CH_3)-CH_2-CH_3$

(b) $\xrightarrow{?}$ $CH_3CH_2-CH(CH_3)-CH_2-CH_2OH$

Starting material for (c)–(d): $CH_3-CH_2-CH=C(CH_3)_2$

(c) $\xrightarrow{?}$ $CH_3-CH_2-CH_2-C(CH_3)(OH)-CH_3$

(d) $\xrightarrow{?}$ $CH_3-CH_2-CH(OH)-CH(CH_3)-CH_3$

Starting material for (e)–(f): $CH_3-CH_2-C(CH_3)_2-CH_2-CH_2OH$

(e) $\xrightarrow{?}$ $CH_3-CH_2-C(CH_3)_2-CH=CH_2$

(f) $\xrightarrow{?}$ $CH_3-CH_2-C(CH_3)_2-CH_2-\overset{\overset{O}{\|}}{C}-OH$

11.11 Suggest methods for converting methylenecyclopentane into the following compounds. More than one step may be necessary in each case.

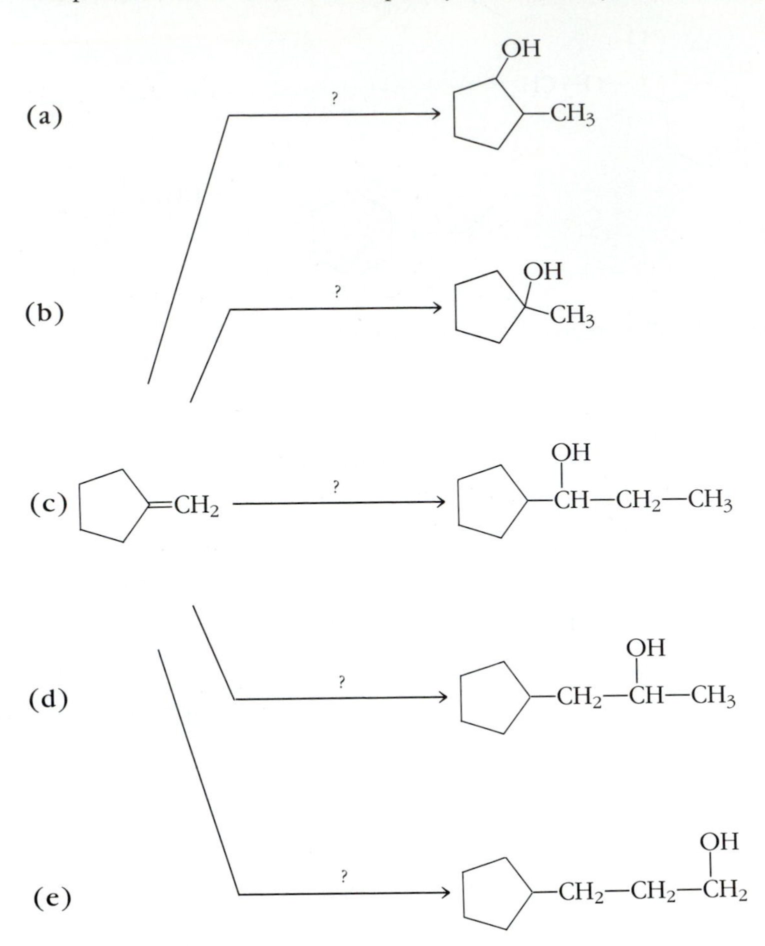

11.12 Write a detailed reaction mechanism for the reactions in Problem 11.11 (b) and (e).

11.13 Suggest methods for converting 2-ethyl-1-butanol into the following compounds. More than one step may be necessary in each case.

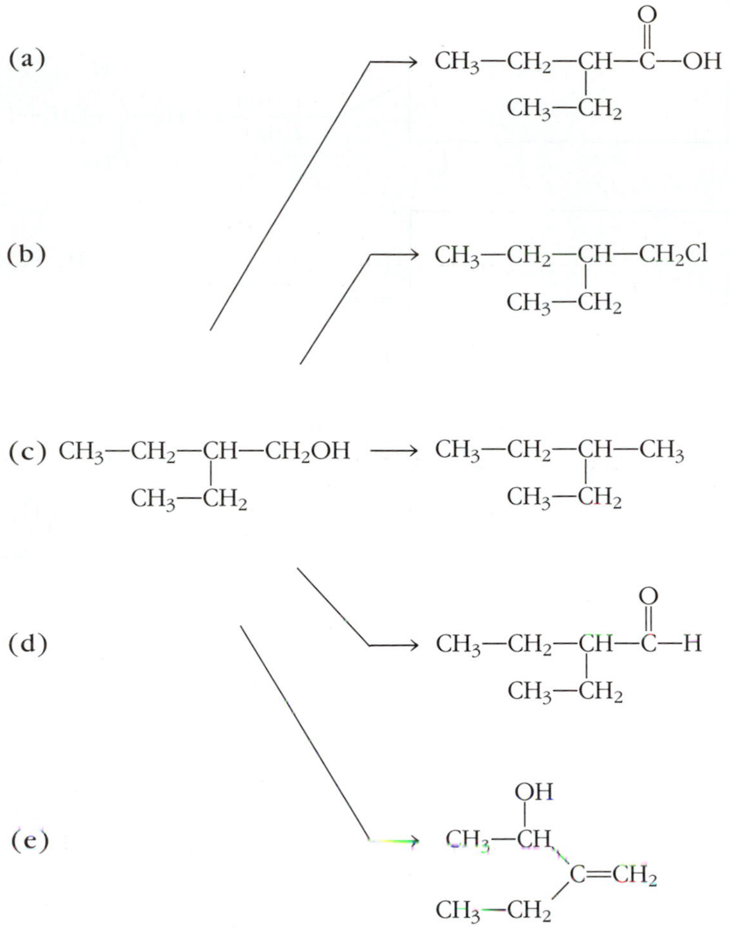

(a) → $CH_3-CH_2-CH(CH_2CH_3)-C(=O)-OH$

(b) → $CH_3-CH_2-CH(CH_2CH_3)-CH_2Cl$

(c) $CH_3-CH_2-CH(CH_2CH_3)-CH_2OH$ → $CH_3-CH_2-CH(CH_2CH_3)-CH_3$

(d) → $CH_3-CH_2-CH(CH_2CH_3)-C(=O)-H$

(e) → $CH_3-CH(OH)-C(=CH_2)-CH_2-CH_3$

11.14 Write a detailed reaction mechanism for the reactions in Problem 11.13 (c).

11.15 Deduce the structure of the starting material in each of the following reactions:

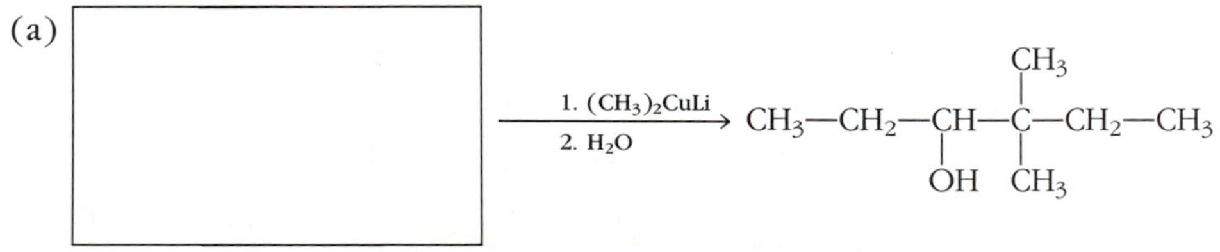

(a) [ ] $\xrightarrow[\text{2. } H_2O]{\text{1. } (CH_3)_2CuLi}$ $CH_3-CH_2-CH(OH)-C(CH_3)_2-CH_2-CH_3$

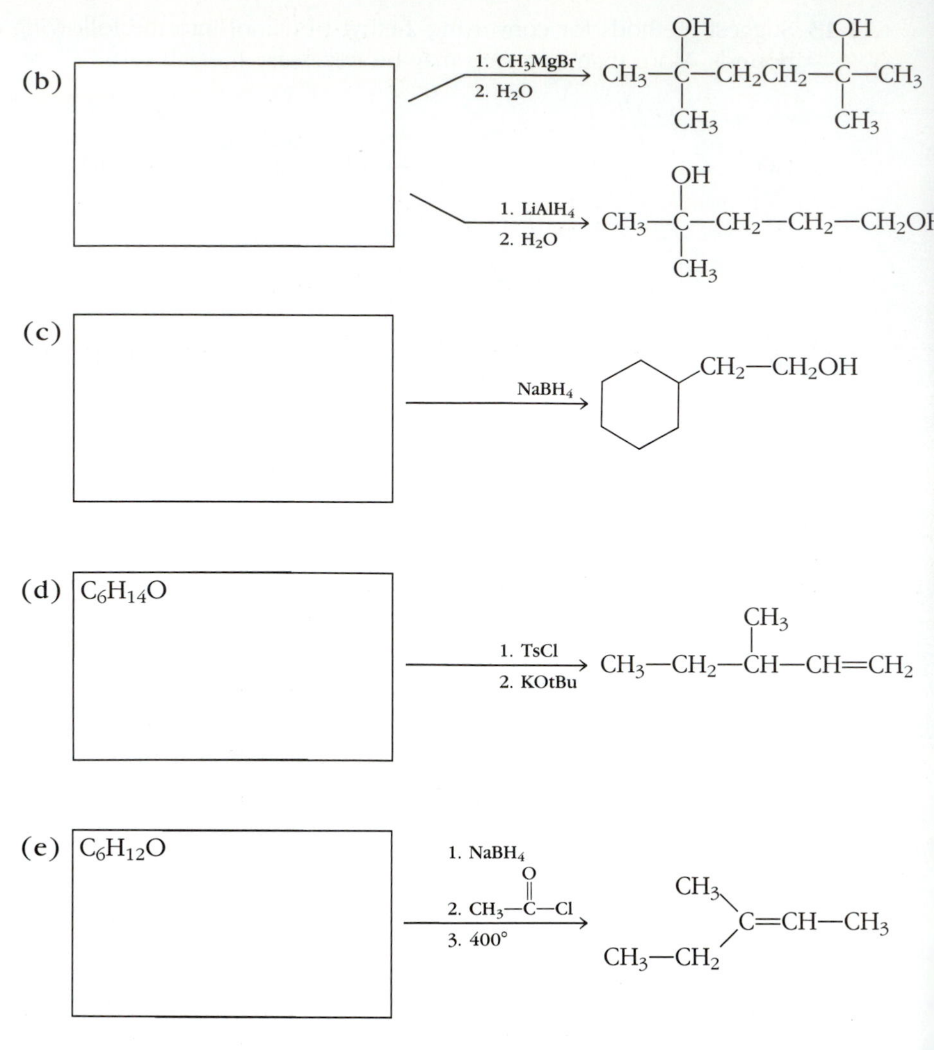

11.16 Suggest methods for carrying out the indicated conversions. More than one step may be necessary in each case.

(a) 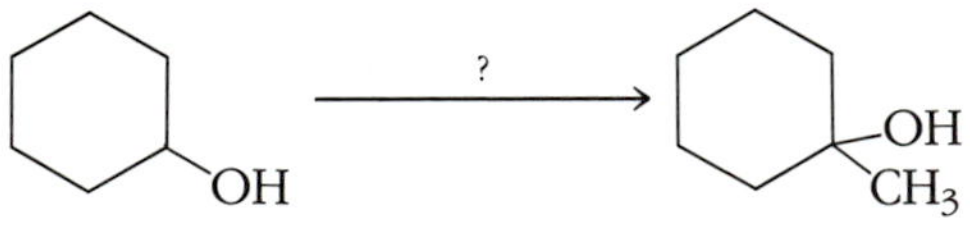 Cyclohexanol (OH) $\xrightarrow{?}$ 1-methylcyclohexanol (OH, CH_3)

(b) 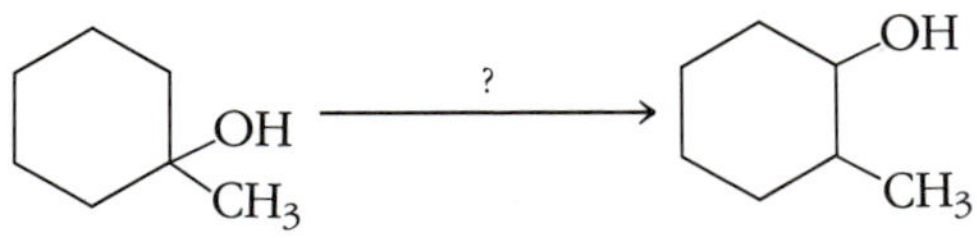 1-methylcyclohexanol (OH, CH_3) $\xrightarrow{?}$ 2-methylcyclohexanol (OH, CH_3)

(c) 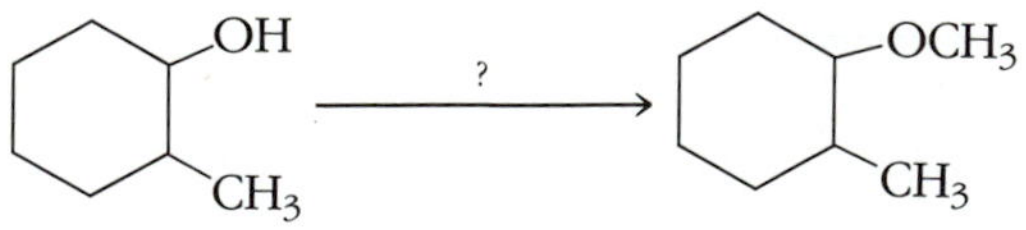 2-methylcyclohexanol (OH, CH_3) $\xrightarrow{?}$ 2-methylcyclohexyl methyl ether (OCH_3, CH_3)

(d) [2-methylcyclohexanol] $\xrightarrow{?}$ [cyclohexane with OH, C_2H_5 on one carbon and CH_3 on the adjacent carbon]

OH

CH_3

(e) [2-methylcyclohexanol] $\xrightarrow{?}$ [cyclohexane with C_2H_5 on one carbon and OH, CH_3 on the adjacent carbon]

11.17 **A–E** are all different compounds. Deduce the structures of **A–D** on the basis of the information provided.

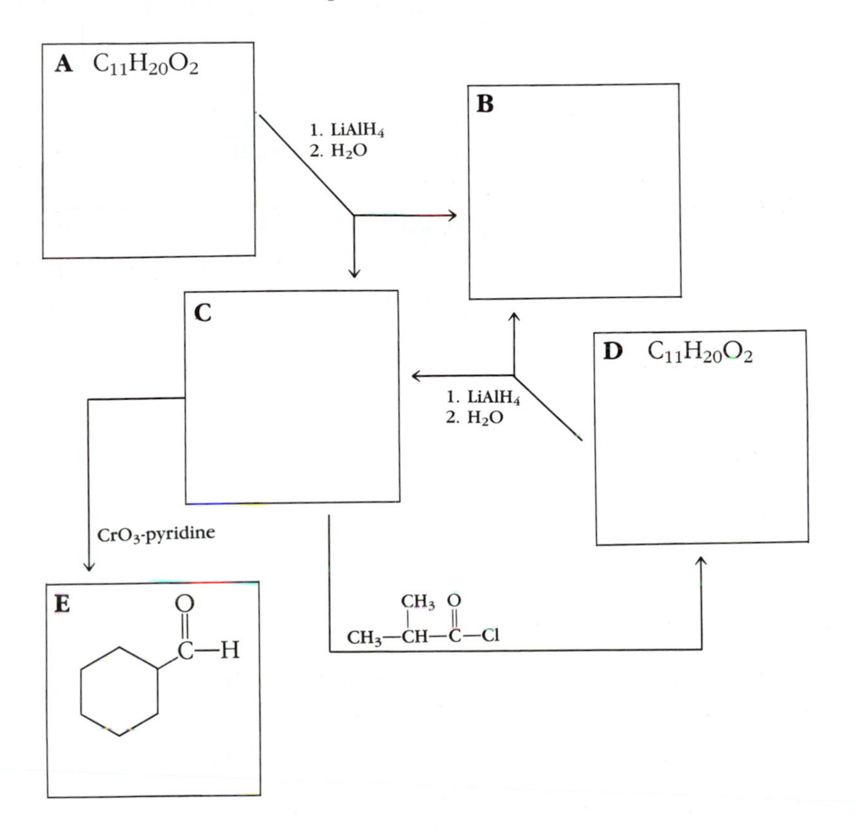

11.18 **A–F** are all different compounds. Deduce the structures of **A–E** on the basis of the information provided. Show stereochemistry whenever appropriate.

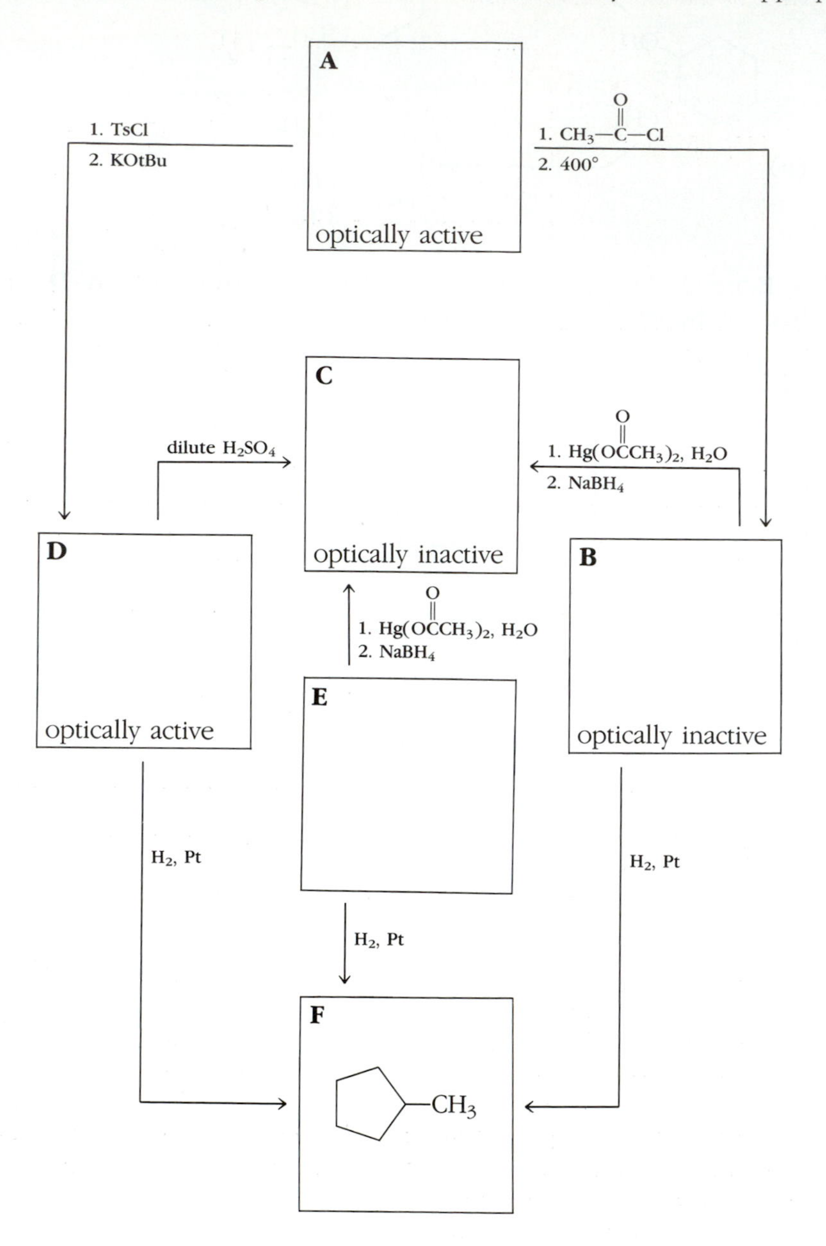

11.19 **A–G** are all different compounds. Deduce the structures of **A–E** on the basis of the information provided.

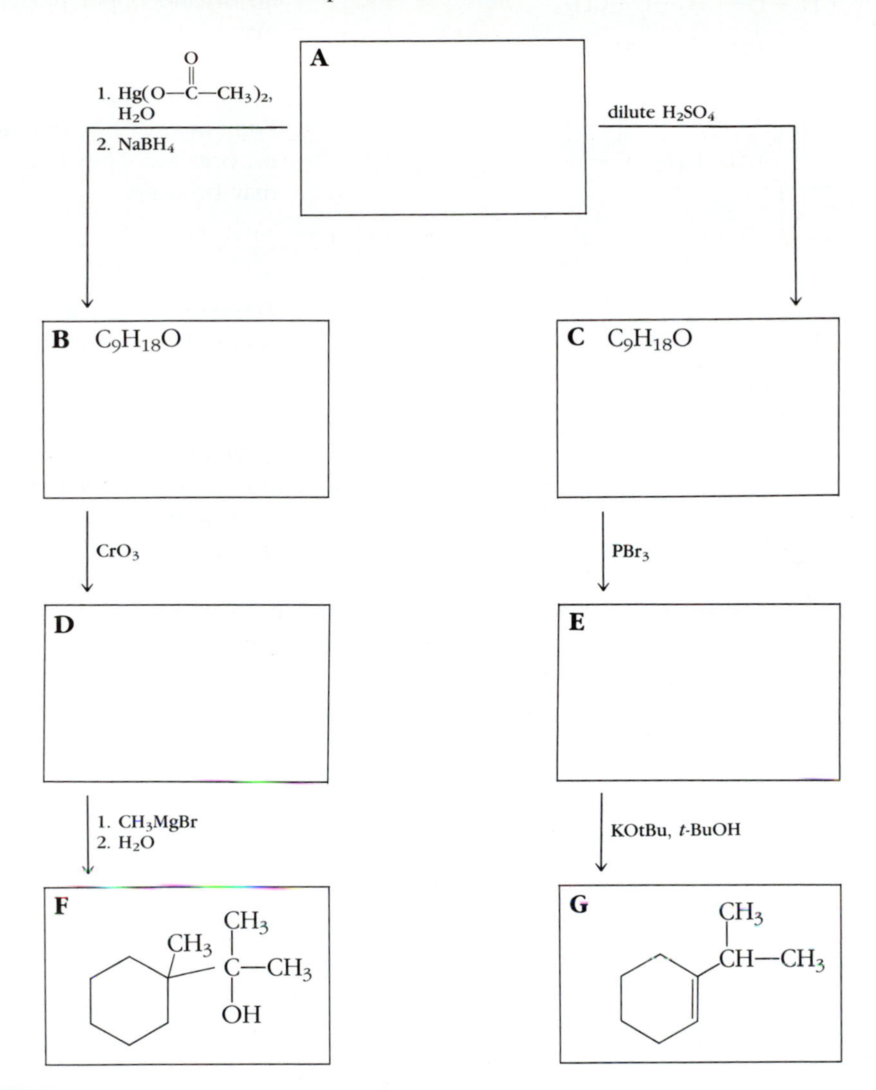

11.20 Suggest methods of synthesizing the following compounds. Unless other restrictions are indicated, you may use any reagents you wish, but all the carbon atoms of the product must be derived from starting materials with the indicated number of carbon atoms.

(a) $CH_3CH_2CH_2—C(OH)(CH_3)—CH_2CH_3$ *Three* or fewer carbon atoms

(b) $CH_3CH_2CH_2—CH(Cl)—CH_2—CH(CH_3)—CH_3$ *Three* or fewer carbon atoms; no organocopper reagents may be used

(c) 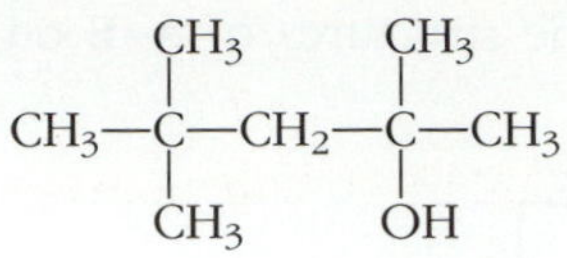
$$\begin{array}{c} \quad CH_3 \qquad\quad CH_3 \\ CH_3—C—CH_2—C—CH_3 \\ \quad CH_3 \qquad\quad OH \end{array}$$

Four or fewer carbon atoms; no organocopper reagents may be used

(d) 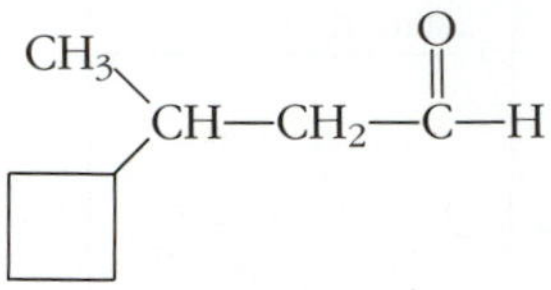 CH_3—CH(cyclobutyl)—CH_2—C(=O)—H
$$CH_3\diagdown CH—CH_2—\overset{O}{\overset{\|}{C}}—H$$
(with a cyclobutyl ring attached to the CH)

Four or fewer carbon atoms; no organocopper reagents may be used

(e) 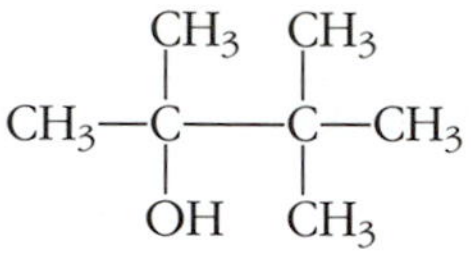
$$\begin{array}{c} \quad CH_3 \quad CH_3 \\ CH_3—C——C—CH_3 \\ \quad OH \quad\;\; CH_3 \end{array}$$

Three or fewer carbon atoms; no organocopper reagents may be used

(f) 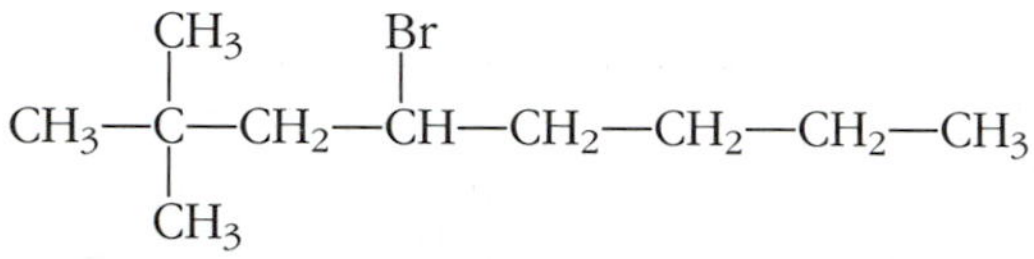
$$\begin{array}{l} \qquad\quad CH_3 \qquad\quad Br \\ CH_3—C—CH_2—CH—CH_2—CH_2—CH_2—CH_3 \\ \qquad\quad CH_3 \end{array}$$

Four or fewer carbon atoms; no organocopper reagents may be used

11.21 Suggest methods of synthesizing the following compounds. Unless other restrictions are indicated, you may use any reagents you wish, but all the carbon atoms of the product must be derived from starting materials with the indicated number of carbon atoms.

(a) 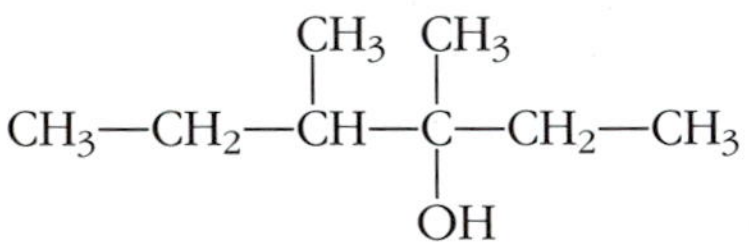
$$\begin{array}{c} \qquad\quad CH_3 \;\; CH_3 \\ CH_3—CH_2—CH—C—CH_2—CH_3 \\ \qquad\qquad\quad OH \end{array}$$

From *alcohols* containing *four* or fewer carbon atoms

(b) 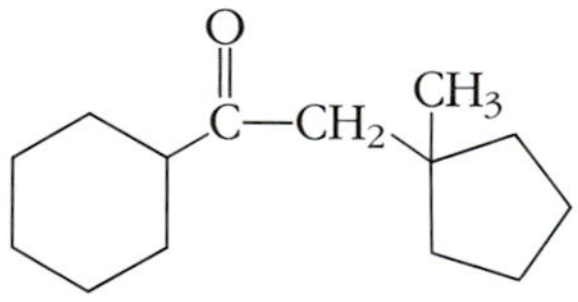 cyclohexyl—$\overset{O}{\overset{\|}{C}}$—$CH_2$—(1-methylcyclopentyl, with CH_3 on the ring carbon)

Six or fewer carbon atoms

(c) 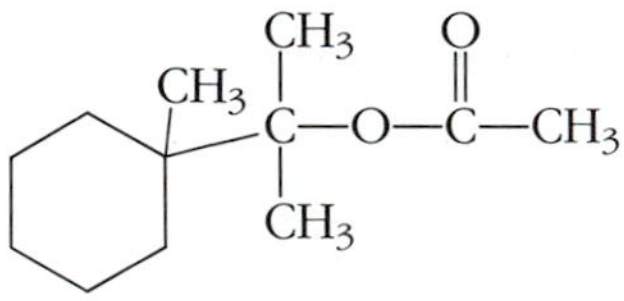 (1-methylcyclohexyl, with CH_3 on the ring carbon)—$\underset{CH_3}{\overset{CH_3}{C}}$—O—$\overset{O}{\overset{\|}{C}}$—$CH_3$

Six or fewer carbon atoms

(d) 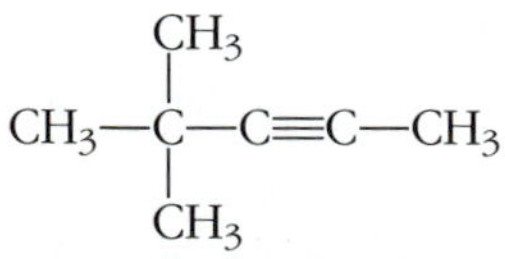
$$\begin{array}{c} CH_3 \\ CH_3—C—C\equiv C—CH_3 \\ CH_3 \end{array}$$

Four or fewer carbon atoms

(e) CO_2H, CH_3 (2-methylcyclohexanecarboxylic acid) — *Six* or fewer carbon atoms

(f) CH_3, $C(=O)—CH_3$ (1-acetyl-1-methylcyclopentane) — The carbon atoms of the five-membered ring must be derived from *cyclopentene.*

11.22 **A–H** are all different compounds. Deduce the structures of **A–G** on the basis of the information provided.

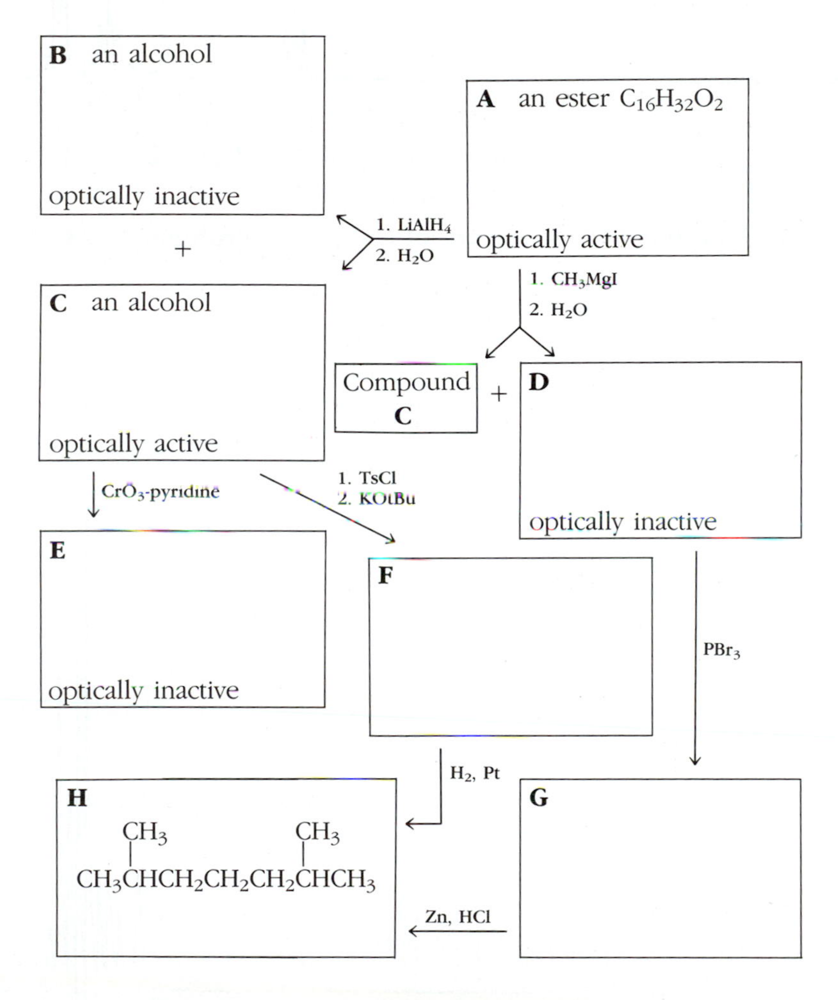

11.23 The reduction of an aldehyde with $NaBH_4$ yielded a product with the following proton NMR spectrum. Deduce the structure of both the aldehyde and the product.

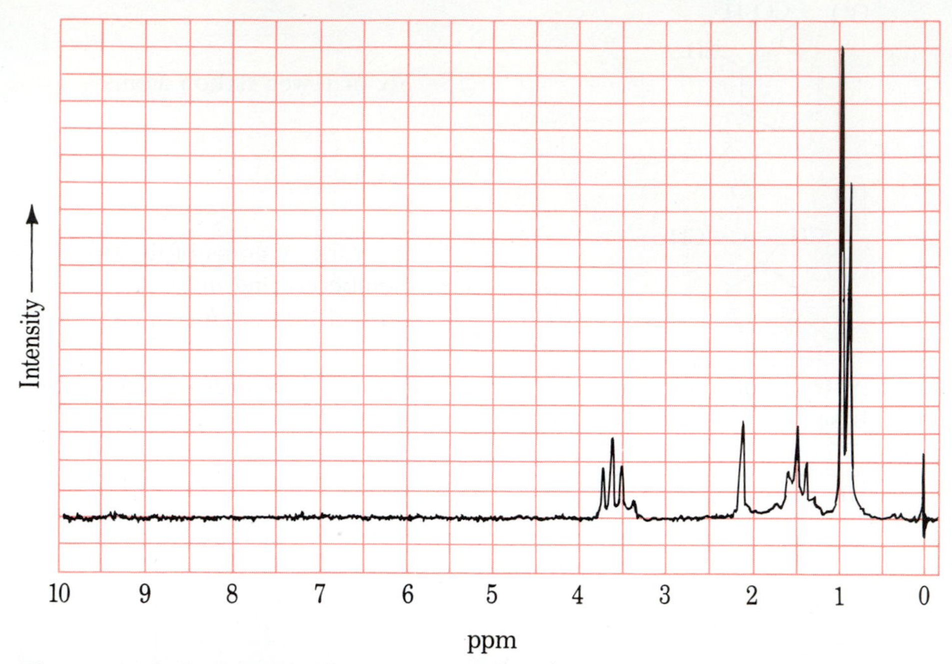

Figure 11.3 Proton NMR Spectrum of Unknown, Problem 11.23

11.24 An alcohol of uncertain structure was treated with acetyl chloride to form the corresponding ester. On the basis of their proton NMR spectra, deduce the structures of the alcohol and the ester.

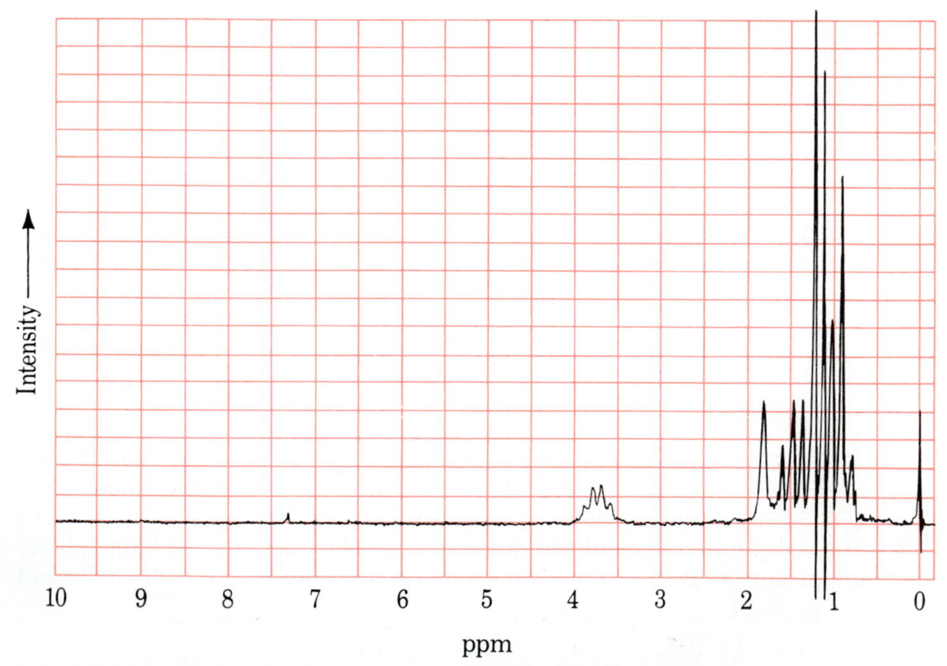

Figure 11.4 Proton NMR Spectrum of Alcohol, Problem 11.24

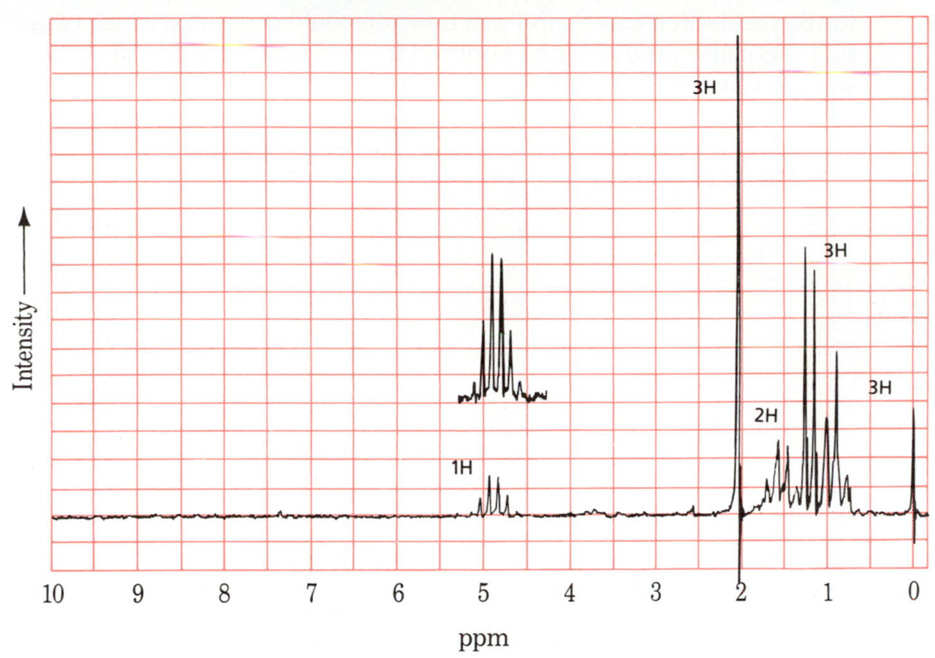

Figure 11.5 Proton NMR Spectrum of Ester, Problem 11.24

11.25 Complete each of the following reactions, and calculate the oxidation state (in both reactant and product) of each carbon atom that undergoes a change during the reaction. Which of these reactions are oxidation-reduction reactions?

(a) Propene + Br_2
(b) 2-Pentene + peracetic acid
(c) Propene + SeO_2
(d) Cyclohexene + H_2, Pt
(e) Cyclohexene + O_3, then Zn
(f) Propene + *N*-bromosuccinimide
(g) Propene + HBr

11.26 Oxidation of a primary alcohol with Cr(VI) reagents usually affords the corresponding carboxylic acid, but sometimes the aldehyde can be obtained in high yield. From your knowledge of physical properties and molecular structure (Section 4.9), explain why oxidation of a primary alcohol using aqueous $Na_2Cr_2O_7$ and H_2SO_4 is most likely to afford a high yield of the aldehyde when the aldehyde has a boiling point below 100°C.

11.27 Spectroscopy can often help you learn whether or not a reaction has followed the desired course. Indicate how you could determine whether the material isolated from the following reactions was the expected product or unreacted starting material. What specific changes in IR and NMR spectra would you look for?

(a) Reduction of 2-pentanal ($CH_3CH_2CH_2CH_2CHO$) with $NaBH_4$
(b) Alkylation of 2-pentanol with methyl iodide
(c) Reduction of ethyl benzoate ($C_6H_5CO_2CH_2CH_3$) with $LiAlH_4$

11.28 Spectroscopy can often help you learn whether or not a reaction has followed the desired course. Indicate how you could determine which of several alternative products was produced from the following reactions. What specific changes in IR and NMR spectra would you look for?

(a) Oxidation of 1-octanol to give either the aldehyde or carboxylic acid

(b) Oxymercuration of 4,4-dimethyl-2-hexene to yield either 4,4-dimethyl-2-hexanol or 4,4-dimethyl-3-hexanol

CHAPTER 12

NUCLEOPHILIC SUBSTITUTION

Nucleophilic substitution reactions have been intensely studied by chemists, and these investigations have greatly contributed to our understanding of how organic reactions occur. But the gains in chemical knowledge did not come easily. Sometimes groups of investigators placed different interpretations on similar results. At other times a stereochemical investigation and a study of reaction rates would lead to opposite conclusions. Only in the mid-1970s, after many years of continued (and sometimes angry) debate in the chemical literature, were most of the questions resolved.

In this chapter we will consider the various aspects of nucleophilic substitution reactions. We will present the reaction mechanisms by which nucleophilic substitution can occur, together with the structural and stereochemical consequences of each mechanism. Then we will show you how to use the reaction mechanism to predict the products that will be formed in a nucleophilic substitution reaction.

12.1 THE NATURE OF THE REACTANTS

Nucleophilic substitution is a type of reaction in which a *nucleophile* (e.g., Nuc^-) reacts with a **substrate** molecule (R—X) and replaces the **leaving group,** X.

$$Nuc^- + \overset{\delta+}{R}—\overset{\delta-}{X} \longrightarrow Nuc—R + X^-$$

Nucleophilic substitution reactions are highly important in organic chemistry, and you have already encountered examples such as the alkylation of sodium acetylides

$$R—C{\equiv}C^-Na^+ + CH_3—Br \longrightarrow R—C{\equiv}C—CH_3 + Br^-$$

and the conversion of alcohols to alkyl halides by reagents such as HBr:

$$R—OH \xrightarrow{\text{conc. HBr}} R—Br$$

Nucleophiles

What constitutes a nucleophile? Recall that a nucleophile is a species that seeks *positive charge.* Any chemical species with an available pair of electrons can function as nucleophile, so a nucleophile is essentially the same as a Lewis base. However, the term nucleophile is never applied to a species that attacks hydrogen in an acid-base reaction; it is most frequently used to describe a species that attacks a positively charged *carbon atom.* The carbon need not have a full ionic charge, and attack at the positive end of a polarized bond to carbon also describes the action of a nucleophile.

Many different anions can function as nucleophiles, including those derived from weak acids such as hydroxide ion (OH^-), alkoxide ions (RO^-), amide ion (NH_2^-) and its alkyl derivatives (RNH^-, R_2N^-), and the halide ions (F^-, Cl^-, Br^-, I^-). The following two equations illustrate nucleophilic substitution of an alkyl halide by fluoride ion and by lithium diethylamide ($LiN(CH_2CH_3)_2$):

$$CH_3(CH_2)_4CH_2—Cl \xrightarrow{KF} CH_3(CH_2)_4CH_2—F \quad 40\text{–}45\%$$

$$CH_3(CH_2)_6CH_2—Br \xrightarrow{LiN(CH_2CH_3)_2} CH_3(CH_2)_6CH_2—N(CH_2CH_3)_2 \quad 89\%$$

A negative charge is not a necessary requirement, and many neutral molecules having atoms with unshared pairs of electrons can also act as nucleophiles. These include the conjugate acids of the anionic nucleophiles that we just listed: H_2O, ROH, NH_3, RNH_2, and R_2NH. Nucleophilic substitution of an alkyl halide by a neutral nucleophile is illustrated by the reaction of ethyl bromide with an amine:

$$\text{3-}CH_3C_6H_4NH_2 \xrightarrow{C_2H_5Br} \text{3-}CH_3C_6H_4NHC_2H_5 \qquad 63\text{–}66\%$$

The reactivity of a species as a nucleophile usually reflects its reactivity as a base, and as you would expect, the anionic nucleophiles are better electron donors than their neutral conjugate acids. In other words, the anions are stronger nucleophiles. Neutral halogen compounds such as HF, HCl, HBr, and HI, which are very weak bases, are quite unreactive as nucleophiles.

When a neutral nucleophile also serves as the solvent, a nucleophilic substitution is called a **solvolysis** reaction (from the Greek *lysis,* cleavage by *solv*ent). *Hydrolysis* reactions are a specific subgroup. The reaction of triphenylmethyl chloride with ethanol is an example of a solvolysis reaction.

$$(C_6H_5)_3C{-}Cl + CH_3CH_2OH \longrightarrow (C_6H_5)_3C{-}OCH_2CH_3 \qquad 97\%$$

Substrates

What structural features are necessary for a molecule to undergo reaction with a nucleophile? Such compounds, which we will describe as the substrates in nucleophilic substitution reactions, share some common structural features. With few exceptions, the reactive center of the substrate cannot be part of a carbon-carbon multiple bond. (Nucleophilic attack at the carbon atom of carbon-oxygen and carbon-nitrogen multiple bonds occurs readily, and those reactions will be discussed in later chapters.) For the reactions that we will consider in this chapter, the first structural requirement for the substrate in a nucleophilic substitution reaction is that the reactive center must be a saturated (sp^3) carbon atom. The distinction between *nucleophilic substitution* and *electrophilic substitution* is a subtle one, since any species that reacts with a nucleophile is an electrophile. The classification of these substitution reactions is made according to the species that attacks the carbon atom undergoing substitution. Electrophilic substitution describes a reaction in which an unsaturated organic compound undergoes substitution after being attacked by an electrophile. In contrast, nucleophilic substitution occurs when the carbon atom undergoing substitution is attacked by a nucleophile.

The second structural requirement for the substrate in nucleophilic substitution is that one of the substituents at the reactive center must be a *good leaving group.* The leaving group is the substituent that is replaced in the substitution reaction. The nucleophile attacks the reactive center with a pair of electrons, and the leaving group departs with its bonding pair of electrons. In all four of the preceding examples, the leaving group was either a chloride ion or a bromide ion; in other words, a halo substituent departed with the bonding electrons to generate a halide ion.

A good leaving group affords a stable species when it leaves, that is, it generates an energetically favorable species that is relatively unreactive. Usually a leaving group, —X, is a neutral group in the reactant but becomes an anion when it leaves:

$$-\overset{|}{\underset{|}{C}}-X + Nuc^- \longrightarrow -\overset{|}{\underset{|}{C}}-Nuc + X^-$$

You can easily evaluate the stability of X^- from the strength of its conjugate acid, HX. If HX is a strong acid, then X^- is stable and —X is a good leaving group. Conversely, —X will not be a good leaving group if HX is a weak acid. Halides and sulfonates (Section 11.9) are the common leaving groups in which —X leaves as an anion, X^-.

Good Leaving Groups

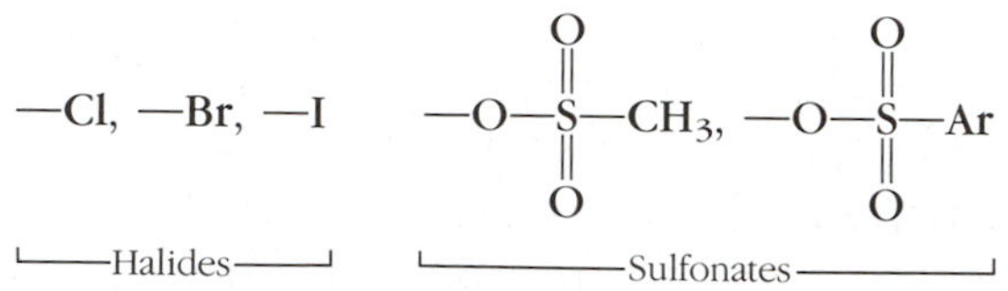

$$\underbrace{-Cl,\ -Br,\ -I}_{\text{Halides}} \qquad \underbrace{-O-\overset{O}{\overset{\|}{\underset{\underset{O}{\|}}{S}}}-CH_3,\ -O-\overset{O}{\overset{\|}{\underset{\underset{O}{\|}}{S}}}-Ar}_{\text{Sulfonates}}$$

In Chapter 11 we showed you a variety of substitution reactions of alcohols. Does this mean that —OH is a good leaving group? The answer is an emphatic *no.* For —OH to act as a leaving group it would have to depart as hydroxide ion, OH^-. This is the conjugate base of H_2O, which certainly is not a strong acid. How, then, do substitution reactions of alcohols occur? The hydroxyl group must *first* be converted into a good leaving group, for example, by protonation with acid:

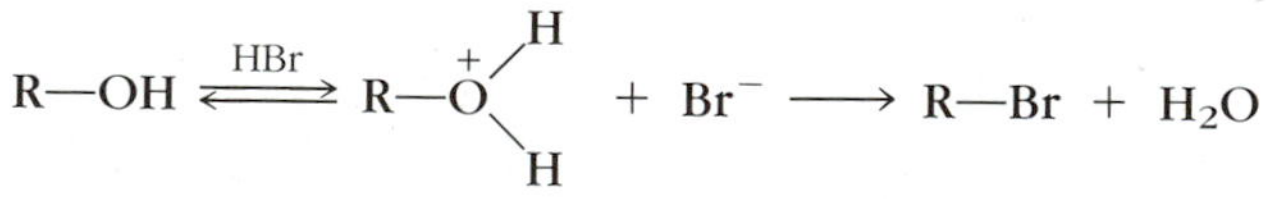

$$R-OH \underset{}{\overset{HBr}{\rightleftharpoons}} \underset{\text{Oxonium ion}}{R-\overset{+}{O}H_2} + Br^- \longrightarrow R-Br + H_2O$$

Protonation produces an **oxonium ion** (a species with a trivalent, positively charged oxygen atom—analogous to the hydronium ion, H_3O^+). Here the substituent is an $-OH_2^+$ group, which can leave with its bonding electrons to generate a stable, neutral water molecule. (The conjugate acid of H_2O is the strong acid, H_3O^+, and this again tells you that $-OH_2^+$ is a good leaving group.)

A variety of positively charged groups behave as good leaving groups. These groups usually contain trivalent oxygen (oxonium ions), trivalent sulfur (sulfonium ions), or tetravalent nitrogen (ammonium ions).

Good Leaving Groups

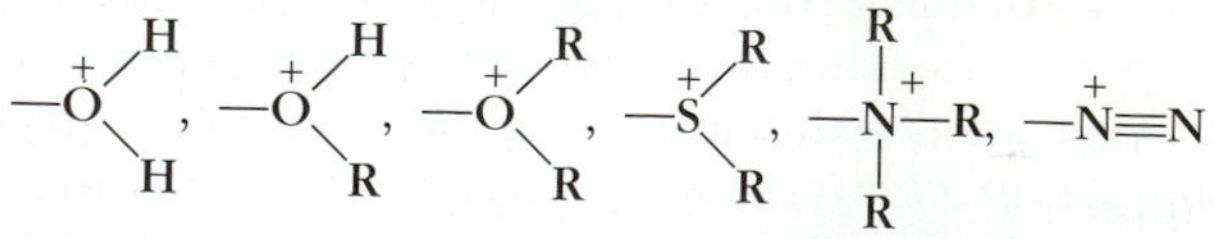

$$-\overset{+}{O}H_2,\quad -\overset{+}{O}HR,\quad -\overset{+}{O}R_2,\quad -\overset{+}{S}R_2,\quad -\overset{+}{N}R_3,\quad -\overset{+}{N}\equiv N$$

The last example, —N≡N$^+$, is the characteristic group of a *diazonium ion* (R—N≡N$^+$). When this group departs from the reactive center with its bonding pair of electrons, it forms molecular nitrogen, N_2. The very high stability of the nitrogen molecule causes —N≡N$^+$ to be an exceptionally good leaving group.

EXERCISE 12.1

For each of the following cases predict which species would be the stronger nucleophile (i.e., which would react more rapidly with CH_3Br).

(a) H_2O, OH^-
(b) OH^-, Cl^-
(c) NH_2^-, NH_3

EXERCISE 12.2

For each of the following pairs of structures predict which would have the better leaving group (i.e., which would react more rapidly in a nucleophilic substitution reaction).

(a) $CH_3—OH$, $CH_3—OH_2^+$ (b) $CH_3CH_2—\overset{+}{N}≡N$, $CH_3CH_2—Cl$

(c) $CH_3—C(CH_3)_2—NH_2$, $CH_3—C(CH_3)_2—\overset{+}{N}(CH_3)_3$

12.2 MECHANISMS OF NUCLEOPHILIC SUBSTITUTION

The mechanism of nucleophilic substitution reactions has long been a subject of great interest (as well as considerable controversy) among organic chemists. Interest in these reactions has been particularly intense because they often exhibit remarkable stereospecificity (Section 12.5). There is evidence for two different reaction mechanisms.

The first reaction mechanism involves collisions between the two species, nucleophile and substrate, that lead directly to the formation of product. The nucleophile attacks the partially positive carbon atom to which the electronegative leaving group is bonded. We will call these *direct displacement* reactions.

Direct Displacement Mechanism

$$Nuc^- + R—X \longrightarrow Nuc—R + X^-$$

Historically this type of reaction has been referred to as an S_N2 (*s*ubstitution, *n*ucleophilic, *bi*molecular) reaction, a term introduced by C. K. Ingold and his coworkers in England, who carried out pioneering studies on these reaction mechanisms. The term **bimolecular** refers to the dependence of the reaction rate (Section 12.3) upon the concentration of *both* the nucleophile and the substrate, because *two* molecules (both nucleophile and substrate) collide to form the products.

The second reaction mechanism involves an initial slow ionization to give a carbocation, which then undergoes a rapid reaction with a nucleophile to yield product.

Carbocation Mechanism

$$\mathrm{R{-}X} \xrightarrow[\text{slow}]{} \mathrm{R^+ + X^-} \xrightarrow[\text{fast}]{\mathrm{Nuc^-}} \mathrm{R{-}Nuc}$$

This type of reaction was designated by Ingold as an S_N1 (*s*ubstitution, *n*ucleophilic, *uni*molecular) reaction. The term **unimolecular** was introduced by Ingold to indicate the dependence of the reaction rate (Section 12.3) on the concentration of the substrate only, because the slow step, formation of a carbocation, involves only the substrate and not the nucleophile.

The key differences between the two mechanistic pathways for nucleophilic substitution are illustrated in Figure 12.1, which depicts a simplified view of the energetics for each mechanism. The direct displacement mechanism proceeds via a single transition state to yield the product directly. In contrast, the carbocation mechanism occurs in two steps. In the first step the leaving group departs to generate an intermediate carbocation. Then in a second step (with a much smaller energy barrier) the cation reacts with a nucleophile to form the product.

Effects of Substrate Structure on Mechanism

A vast amount of research on nucleophilic substitution has been carried out over the last 50 years. The results of individual experiments were often conflicting, but when all the data are considered together, a consistent picture emerges. What

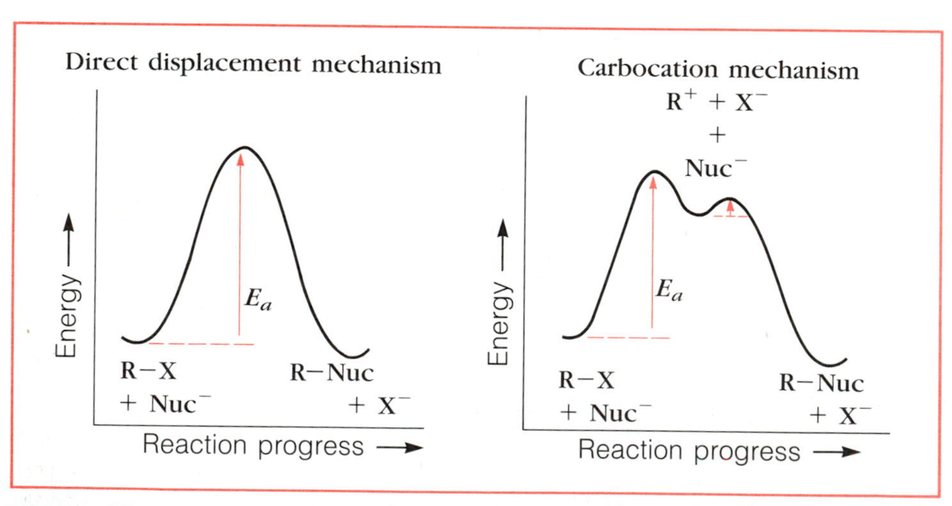

Figure 12.1 Energetics of Nucleophilic Substitution Reactions. In the direct displacement reaction the product is formed directly via a single transition state. For the carbocation mechanism there are two transition states: A carbocation intermediate is generated first, and this subsequently reacts with a nucleophile to form the product.

structural features of a compound will cause it to react preferentially by either a direct displacement or a carbocation mechanism? We will focus on two specific ways in which substrate structure can influence the mechanism of nucleophilic substitution: steric effects and electronic effects. *Steric effects* result from *non-bonded* interactions that arise in the course of the reaction. Normally we view these as repulsive effects that result when two atoms are forced into close proximity, and such effects are most commonly encountered in molecules containing branched, "bulky" alkyl groups. **Electronic effects** are the combination of inductive and resonance effects. These are usually not very important for neutral molecules, but they play a major role in determining the stability of an ionic species.

How do steric and electronic effects influence nucleophilic substitution reactions? *Direct displacement reactions are dominated by steric effects,* and bulky substituents at or near the reactive center inhibit attack by a nucleophile. We can therefore make the simple but accurate prediction that a direct displacement reaction should proceed most easily for methyl derivatives and then become progressively more difficult as the degree of substitution at the reactive center is increased. For a series of alkyl bromides the trend summarized in the box is expected.

Steric Effects on Direct Displacement Reactions

Direct displacement easiest .... Direct displacement most difficult

CH_3—Br . . . R—CH_2—Br . . . R—CH(R)—Br . ⋮ . R—C(R)(R)—Br

Steric interactions with an attacking nucleophile minimal ...... Steric interactions with an attacking nucleophile greatest

The steric interactions involved for attack on a tertiary alkyl derivative are sufficiently large that direct displacements do not occur. The vertical dotted line in the preceding comparison therefore defines the structural limitations for direct displacement:

Direct displacement reactions occur with methyl, primary, and secondary alkyl derivatives but not with tertiary alkyl derivatives.

Carbocation reactions are dominated by the electronic effects that determine cation stability, and steric effects play only a minor role. You are already aware of the increasing stabilities of cations from the least stable methyl cation to the increasingly stable primary, secondary, and tertiary cations (Section 2.7), but you must also consider *resonance stabilized* allylic and benzylic cations (Sections 8.3, 8.6). Both the degree of substitution (primary, secondary, etc.) *and* the possibility of resonance stabilization must be considered, so we cannot write one simple sequence for cation stability. In Figure 12.2 we have summarized the relative stabilities of representative cations that show variation in both these structural features. The relative stabilities of the cations are presented in terms of the ionization energy of the corresponding chlorides.

$$R\text{—}Cl \longrightarrow R^+ + Cl^-$$

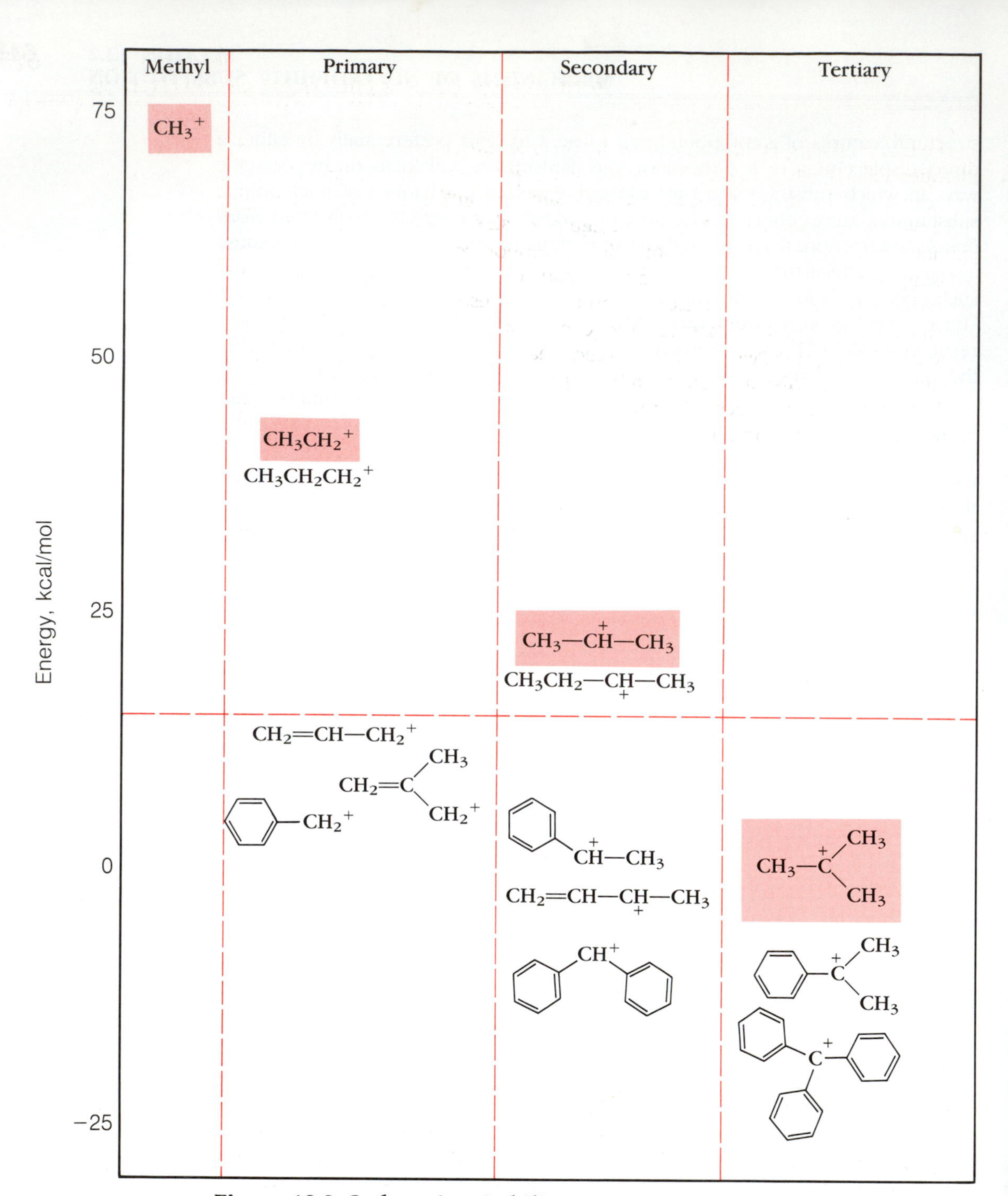

Figure 12.2 Carbocation Stability.
The stability of a carbocation is dependent on the substitution at the electron deficient center. Alkyl substitution stabilizes a cation, as seen by the trend:

$$CH_3^+ < CH_3CH_2^+ < (CH_3)_2CH^+ < (CH_3)_3C^+$$

An even greater stabilization results from an aryl or vinyl substituent, as can be seen by comparison of CH_3^+ with $CH_3CH{=}CH_2^+$ and with $C_6H_5CH_2^+$. (All the stabilities are relative to the *tert*-butyl cation, which is assigned an arbitrary energy of zero for purposes of comparison. Cations that lie above the dotted line are sufficiently unstable that they are not ordinarily formed in nucleophilic substitution reactions.)

The cations shown in Figure 12.2 are divided into two categories by the horizontal dotted line. Those cations in the lower half of the figure are relatively stable, and they can easily be formed during nucleophilic substitution reactions. However, those cations in the upper portion of the figure are all sufficiently unstable that they are not formed in ordinary nucleophilic substitution reactions.

Using steric and electronic effects, we can divide substrate molecules (alkyl derivatives) into three categories, and you can use these to predict the mechanism by which a nucleophilic substitution reaction will proceed. The first category is made up of simple methyl, primary, and secondary derivatives lacking any substituents that could stabilize a carbocation by resonance. Nucleophilic substitution proceeds by the direct displacement mechanism, because the energies of the corresponding cations are too high.

Compounds that Undergo Nucleophilic Substitution by the Direct Displacement Mechanism

$$CH_3\text{—}X \qquad R\text{—}CH_2\text{—}X \qquad R\text{—}\overset{\displaystyle R}{\overset{|}{C}H}\text{—}X$$

The second category consists of tertiary alkyl derivatives, which can ionize to moderately stable cations. Steric hindrance at a tertiary center precludes direct displacement.

Compounds that Undergo Nucleophilic Substitution by the Carbocation Mechanism

$$R\text{—}\underset{\underset{\displaystyle R}{|}}{\overset{\overset{\displaystyle R}{|}}{C}}\text{—}X \qquad \rangle C{=}\overset{|}{C}\text{—}\underset{\underset{\displaystyle R}{|}}{\overset{\overset{\displaystyle R}{|}}{C}}\text{—}X \qquad Ar\text{—}\underset{\underset{\displaystyle R}{|}}{\overset{\overset{\displaystyle R}{|}}{C}}\text{—}X$$

One or more alkenyl or aryl groups may be attached to the tertiary center.

The third category is made up of primary and secondary alkyl derivatives having one or two aryl or alkenyl groups attached directly to the reactive center. These compounds are not highly hindered, and the corresponding cations are resonance stabilized (Figure 12.2), so they can react by *either* the direct displacement *or* the carbocation mechanism. The reaction conditions will determine which pathway is actually followed.

Compounds that Undergo Nucleophilic Substitution by Either Direct Displacement or the Carbocation Mechanism

$$Ar\text{—}CH_2\text{—}X \qquad Ar\text{—}\overset{\displaystyle R}{\overset{|}{C}H}\text{—}X \qquad \rangle C{=}\overset{|}{C}\text{—}CH_2\text{—}X \qquad \rangle C{=}\overset{|}{C}\text{—}\overset{\displaystyle R}{\overset{|}{C}H}\text{—}X$$

The group designated as R could also be aryl or alkenyl

EXERCISE 12.3

For each of the following pairs of compounds, predict which would be more hindered toward direct displacement by a nucleophile.

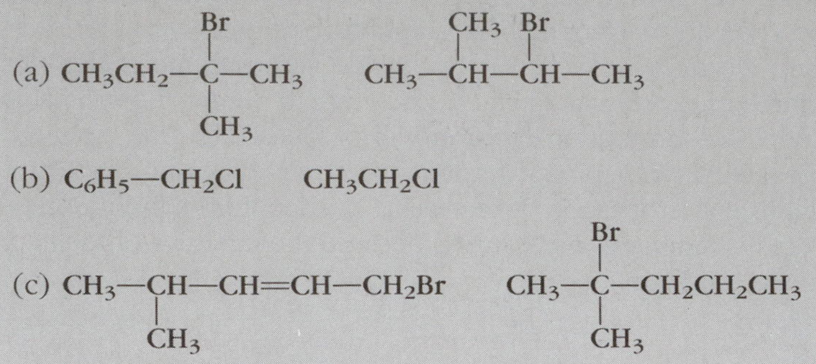

(a) $CH_3CH_2-\overset{Br}{\underset{CH_3}{C}}-CH_3$ $\quad$ $CH_3-\overset{CH_3}{CH}-\overset{Br}{CH}-CH_3$

(b) $C_6H_5-CH_2Cl$ $\quad$ CH_3CH_2Cl

(c) $CH_3-\underset{CH_3}{CH}-CH{=}CH-CH_2Br$ $\quad$ $CH_3-\overset{Br}{\underset{CH_3}{C}}-CH_2CH_2CH_3$

EXERCISE 12.4

For each of the following pairs of compounds, predict which would ionize more readily to a carbocation:

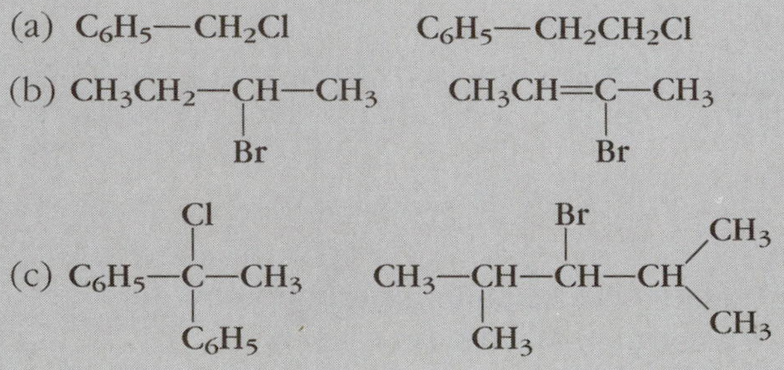

(a) $C_6H_5-CH_2Cl$ $\quad$ $C_6H_5-CH_2CH_2Cl$

(b) $CH_3CH_2-\underset{Br}{CH}-CH_3$ $\quad$ $CH_3CH{=}\underset{Br}{C}-CH_3$

(c) $C_6H_5-\overset{Cl}{\underset{C_6H_5}{C}}-CH_3$ $\quad$ $CH_3-\underset{CH_3}{CH}-\overset{Br}{CH}-CH(CH_3)_2$

EXERCISE 12.5

For each of the compounds in Exercises 12.3 and 12.4, predict whether a nucleophilic substitution reaction would proceed via:

(a) Direct displacement
(b) Carbocation formation
(c) Either mechanism, depending on reaction conditions

12.3 KINETICS OF NUCLEOPHILIC SUBSTITUTION REACTIONS

Mechanistic information about nucleophilic substitution has been obtained from many types of experiments, but studies of *rates* (kinetics) and *stereochemistry* of the reactions have been most productive. The stereochemical evidence is presented in Section 12.6, and we will present the kinetic evidence here.

The rate of a reaction is the *amount of product that is formed* (or of reactant that is consumed) *per unit time.* The amount of product formed is ordinarily expressed in concentration units of *moles per liter* (or molarity, *M*), and the usual time units are seconds, so a rate is expressed in units of moles per liter per second ($M\ s^{-1}$). Therefore you can see that a reaction rate indicates the change in *concentration* of a species as a function of time.

The rate for any particular step in a reaction is equal to the product of the concentrations of the reactants in that step times the **rate constant,** *k,* that is characteristic of that step. The rate constant reflects the activation energy for that step and is independent of the concentrations of the reacting species. For a single reaction step that is *bimolecular,* the following kinetic behavior will be observed:

Bimolecular Reaction Step

$$A + B \xrightarrow{k} C \qquad \text{Rate} = k[A][B] \tag{12.1}$$

The rate has units of $M\ s^{-1}$, and [A] and [B] both have units of *M,* so a bimolecular rate constant must have units of $M^{-1}\ s^{-1}$.

A reaction step that involves the reaction of a single species is described as *unimolecular,* and the following kinetic expression is observed:

Unimolecular Reaction Step

$$D \xrightarrow{k} E \qquad \text{Rate} = k[D] \tag{12.2}$$

The rate always has units of $M\ s^{-1}$ and [D] has units of *M,* so the units of a unimolecular rate constant are s^{-1}. For reactions that take place in solution a "unimolecular" step is often caused by collision with a solvent molecule, but this cannot be detected from the reaction kinetics.

Virtually all individual reaction steps fall into one of these two categories, although the overall kinetic behavior of a multistep process can be considerably more complex.

Kinetic Expectations

The kinetic behavior of a nucleophilic substitution reaction will depend markedly upon which mechanism is followed. At one extreme, a direct displacement reaction involves a single bimolecular step and necessarily occurs at a rate that is equal to the product of the rate constant times the concentrations of both nucleophile and substrate. The reaction is *first-order* with respect to the substrate (i.e., the exponent for its concentration is unity) and first-order with respect to the nucleophile, so it is *second-order* overall.

Direct Displacement (Bimolecular)

$$Nuc^- + R{-}X \xrightarrow{k} Nuc{-}R + X^- \qquad \text{Rate} = k[Nuc^-][RX] \tag{12.3}$$

At the other extreme the carbocation mechanism occurs by ionization of the substrate. If the carbocation were to react with a nucleophile as rapidly as it is formed, the rate of overall reaction would equal the rate of the initial unimolecular ionization. The reaction rate would then be equal to the product of the rate constant for ionization (k) times the concentration of substrate.

Unimolecular Ionization

$$\text{R—X} \xrightarrow[\text{slow}]{k} \text{R}^+ + \text{X}^- \xrightarrow[\text{fast}]{\text{Nuc}^-} \text{R—Nuc} \qquad \text{Rate} = k[\text{RX}] \tag{12.4}$$

The rate in this case depends only upon the *unimolecular* ionization step, so the overall reaction is described as *first-order;* it is *first-order* with respect to substrate and *zero-order* with respect to nucleophile.

Simple Kinetic Results

In the following discussion we will show you the results from actual kinetic studies of two specific compounds, methyl bromide and triphenylmethyl fluoride. These compounds illustrate the extremes for kinetic behavior, and you will see how these studies help to establish the mechanisms of nucleophilic substitution reactions.

The triphenylmethyl cation, $(C_6H_5)_3C^+$, is an exceptionally stable cation (Figure 12.2), because there is resonance stabilization by *three* aromatic rings. A triphenylmethyl halide is also tertiary (and therefore highly hindered to direct displacement), so it should react via a carbocation mechanism. When triphenylmethyl fluoride reacts with sodium hydroxide in an inert solvent, triphenylmethanol is formed.

$$(C_6H_5)_3C\text{—F} \xrightarrow{\text{NaOH}} \left[(C_6H_5)_3C^+ + F^- \right] \xrightarrow{^-\text{OH}} (C_6H_5)_3C\text{—OH}$$

Triphenylmethyl fluoride — Triphenylmethyl cation — Triphenylmethanol

The rate of this reaction was studied at varying concentrations of sodium hydroxide, and we have summarized the results in Table 12.1.

TABLE 12.1 Rate of Reaction of 0.008 *M* Triphenylmethyl Fluoride with Sodium Hydroxide

NaOH Concentration, *M*	Reaction Rate, $M\ s^{-1}$
0.0025	6.7×10^{-6}
0.0050	6.0×10^{-6}

A *twofold* increase in the concentration of nucleophile has almost *no effect* on the rate of reaction, the slight decrease in rate for the higher NaOH concentration resulting from small changes in reaction conditions. The data in Table 12.1 therefore show that the rate is independent of the concentration of nucleophile, as predicted by Equation 12.4.

In contrast to a triphenylmethyl derivative, a simple methyl halide should not undergo nucleophilic substitution via carbocation formation. The carbocation is highly unstable (Figure 12.2), so its rate of formation would be extremely slow. The reaction center is not hindered, however, and direct displacement could occur readily. The reaction of methyl bromide with pyridine in an inert solvent yields an ammonium salt, as shown in the following equation:

$$C_5H_5N: + CH_3\text{—}Br \longrightarrow C_5H_5\overset{+}{N}\text{—}CH_3 + Br^-$$

Pyridine

The rate of this reaction was measured at several different concentrations of pyridine (Table 12.2). As expected for a bimolecular reaction, the reaction rate is directly proportional to the concentration of nucleophile. When the concentration of pyridine is doubled and then tripled, the rate of reaction increases by factors of 2 and 3, respectively.

The last two entries in Table 12.2 also illustrate the dependence of rate on substrate concentration. The doubling of methyl bromide concentration (at the same concentration of nucleophile) also causes a twofold increase in reaction rate. The kinetics of this reaction are correctly described by Equation 12.3, as you should expect for a direct displacement reaction.

Ion Pairs as Reaction Intermediates

The methyl and triphenylmethyl halides that we just discussed represent the extremes of reactivity in nucleophilic substitution. For a very large number of other

TABLE 12.2 Rates of Reactions of Methyl Bromide with Pyridine

CH_3Br Concentration, M	Pyridine Concentration, M	Reaction Rate, $M\ s^{-1}$
0.12	0.10	5.6×10^{-6}
0.12	0.20	11.3×10^{-6}
0.12	0.30	17.0×10^{-6}
0.24	0.30	34.2×10^{-6}

compounds the kinetic behavior is intermediate between that illustrated in Tables 12.1 and 12.2. Frequently, the rate of reaction increases with (but is not directly proportional to) the concentration of nucleophile. One very important reason for such behavior in many cases is that free carbocations are not produced from compounds reacting via a carbocation mechanism. Instead, the negatively charged leaving group remains in the vicinity of the cation, and the resulting **ion pair,** $R^+ X^-$, exhibits chemical reactivity different from that of a completely free carbocation. For example, the leaving group, X^-, can function as a nucleophile to regenerate the starting material. In other words, *the initial ionization to form a carbocation is reversible.*

$$R{-}X \underset{k_{-1}}{\overset{k_1}{\rightleftharpoons}} [R^+ \; X^-] \xrightarrow[Nuc^-]{k_2} R{-}Nuc + X^-$$

The reversibility of ion pair formation (indicated by the forward and reverse rate constants, k_1 and k_{-1}, in the preceding equation) has profound effects on the kinetics of such a carbocation reaction, and the kinetic form of the carbocation mechanism is fairly complex. Depending on the relative values of the different rate constants and the concentrations of substrate and nucleophile, the *observed* kinetic behavior may be anywhere between the limits of first order and second order. For these reasons it is often difficult to determine the mechanism of a nucleophilic substitution reaction on the basis of reaction kinetics alone. Frequently, much more can be learned about a reaction mechanism by investigating the stereochemistry of a particular reaction (Sections 12.4, 12.5).

Rates of Solvolysis Reactions

Extensive studies have been carried out on nucleophilic substitution reactions in which the solvent is also the nucleophile. The overall rate of reaction can be related to three important factors: the reactivity of the nucleophile *(solvent nucleophilicity),* the ability of solvent to solvate ionic intermediates *(solvent polarity),* and *leaving group ability.* The two factors involving solvent will be discussed here, and leaving group effects will be considered later. For convenience we will use the abbreviation SOH as a general way to denote a hydroxylic solvent.

For nucleophilic substitution by a carbocation mechanism, ionization is usually the slow step, or **rate-limiting step** (Figure 12.1, Equation 12.4). Ionization to afford a carbocation is highly sensitive to the solvent polarity. A polar solvent in which ionic compounds are soluble facilitates ionization; reaction will occur more slowly in a less polar solvent. The *nucleophilicity* of the solvent has only a minor effect upon overall rate of solvolysis via a carbocation mechanism because nucleophilic attack by solvent occurs *after* the rate-limiting ionization step.

Solvolysis by the Carbocation Mechanism (SOH = Solvent)

$$R{-}X \overset{SOH}{\rightleftharpoons} [R^+X^-] \longrightarrow R{-}\overset{+}{O}(S)(H) + X^- \xrightarrow{\text{(proton transfer)}} R{-}OS + HX$$

Large effect of solvent polarity on rate
Small effect of solvent nucleophilicity on rate

The situation for solvolysis by a direct displacement reaction is quite different. The initially formed product is an oxonium ion:

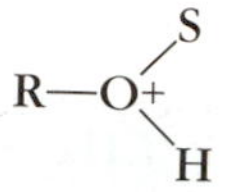

$$R{-}\overset{+}{O}(S)(H)$$

Oxonium ions are considerably more stable than carbocations, and only a small dependence of rate on solvent polarity is expected. Nucleophilic attack by solvent is rate-limiting (subsequent proton transfer is very rapid), so a large dependence of reaction rate on solvent nucleophilicity is expected.

Solvolysis by the Direct Displacement Mechanism (SOH = Solvent)

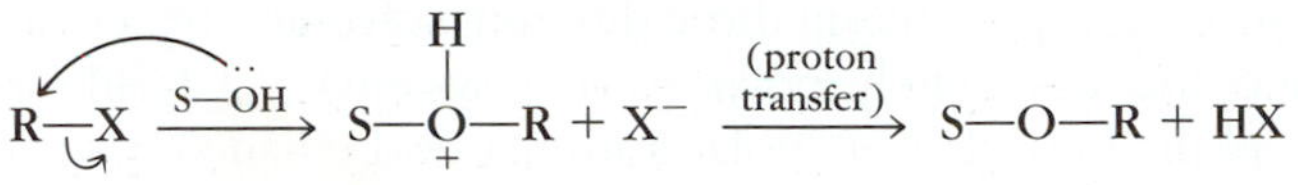

$$R{-}X \xrightarrow{S-OH} S{-}\overset{+}{O}(H){-}R + X^- \xrightarrow[\text{transfer})]{(\text{proton}} S{-}O{-}R + HX$$

Small effect of solvent polarity on rate
Large effect of solvent nucleophilicity on rate

What factors govern solvent polarity and solvent nucleophilicity? The implicit solvent for comparison is water, which is both polar and strongly nucleophilic relative to other neutral nucleophiles. Substitution of an organic group for one of the hydrogen atoms of water will result in a decrease in polarity, although the smallest decrease in polarity will be observed when the hydrogen atom is replaced by a strongly electron-withdrawing group. The examples in the following box illustrate trends for substituent effects on solvent polarity:

Relative Polarity of Some Hydroxylic Solvents

$$HOH > CF_3CH_2OH > CH_3\overset{O}{\overset{\|}{C}}OH > CH_3CH_2OH$$

Strongly polar — Weakly polar

Most organic solvents are also less nucleophilic than water, although replacement of a hydrogen atom of water with a simple alkyl group has very little effect on nucleophilicity. In contrast, the electron-withdrawing effect of an electronegative group makes the oxygen much less nucleophilic. The examples in the following box illustrate trends for substituent effects on solvent nucleophilicity:

Relative Nucleophilicity of Some Hydroxylic Solvents

$$HOH > CH_3CH_2OH > CH_3{-}\overset{\overset{\displaystyle O}{\|}}{C}{-}OH > CF_3CH_2OH$$

Strongly nucleophilic — Weakly nucleophilic

Frequently, *mixed solvents* are used in organic chemistry in order to obtain optimum combinations of reactivity and solubility properties. A mixed solvent system has properties which are an average of those of the individual components. For example, aqueous ethanol will dissolve many water-insoluble compounds, yet it is much more polar than ethanol alone. In Table 12.3 we have shown the rates of solvolysis for several compounds in three different solvents: 70% ethanol-30% H_2O (a nucleophilic and moderately polar solvent system), 2,2,2-trifluoroethanol (a weakly nucleophilic but relatively polar solvent), and ethanol (a strongly nucleophilic but weakly polar solvent). In each case we have shown the rate relative to that in 70% ethanol-30% H_2O for which a value of 1.0 is assigned.

The first compound in the table, methyl tosylate, reacts by a direct displacement mechanism, so you should expect solvent nucleophilicity to have the major effect on the rate. Indeed, the large reduction in nucleophilicity of trifluoroethanol results in a dramatic rate decrease, and the solvolysis rate of methyl tosylate in trifluoroethanol is only 0.005 of that in aqueous ethanol.

TABLE 12.3 Variation of Solvolysis Rate with Solvent*

Substrate	100% Ethanol (Nucleophilic, Nonpolar)	70% Ethanol/30% H_2O (Nucleophilic, Polar)	CF_3CH_2OH (Nonnucleophilic, Polar)
$CH_3{-}OTs$	0.2	1	0.005
$CH_3{-}CH(CH_3){-}OTs$	0.04	1	0.05
$CH_3{-}C(CH_3)_2{-}Cl$	0.003	1	4

*For each compound the rate is relative to that in 70% ethanol-30% water.

The third compound in Table 12.3, *tert*-butyl chloride, is a tertiary alkyl halide, which reacts via a carbocation. The rate depends predominantly on solvent ionizing power, and the change from aqueous ethanol to trifluoroethanol actually results in an *increase* in the rate of reaction. The rate of solvolysis in ethanol, which is only weakly polar, is lower than that in trifluoroethanol by a factor of more than 1000.

The data for methyl tosylate and *tert*-butyl chloride define the extremes in behavior. For compounds reacting by a direct displacement mechanism, the solvolysis rate depends mainly on the strength of the nucleophile, whereas the rate for a compound that reacts via a carbocation mechanism is governed primarily by solvent polarity. What happens for a compound intermediate in structure between the extremes of methyl tosylate and *tert*-butyl chloride? The answer is provided by the data for 2-propyl tosylate in Table 12.3. Its rate of reaction decreases upon changing from aqueous ethanol to the nucleophilic trifluoroethanol, but the change is smaller than that for methyl tosylate. Similarly, the effect of a decrease in solvent polarity is much less than that observed for *tert*-butyl chloride, but it is larger than that for methyl tosylate. For many years this intermediate kinetic behavior was interpreted to mean that secondary alkyl derivatives such as 2-propyl tosylate react by two competing pathways, that is, that direct displacement and ionization to a carbocation occur simultaneously. But recent stereochemical studies (Section 12.5) indicate that the reactions of secondary (as well as primary) alkyl derivatives are better described as proceeding by direct displacement alone.

Leaving Group Ability

Leaving group ability also plays an important role in the rate of solvolysis reactions and in nucleophilic substitution reactions in general. The pattern of reactivity is quite logical: the more stable the anion, X^-, the faster will be the reaction of R—X. An electrically *neutral* leaving group leads to even faster reaction. The "best" leaving groups are those which result in faster rates of reaction, and the relative leaving group ability correlates directly with the strength of the corresponding acid, HX. We have summarized the relative reactivities of a series of leaving groups in Table 12.4 by showing the relative rates of solvolysis of 1-phenyl-1-ethyl derivatives in acetone-water. The rates are shown as relative to that of the corresponding chloride, which therefore has a relative rate of unity.

Table 12.4 shows that sulfonate esters are much better leaving groups than halides. In fact, the sulfonates of tertiary alcohols are so reactive that it is very difficult to isolate and purify them. The same is true for allylic and benzylic derivatives. Consequently, the use of tosylates and other sulfonate esters is generally restricted to simple primary and secondary alkyl derivatives. The carboxylic acid esters (such as *p*-nitrobenzoate, abbreviated as OPNB) are quite poor leaving groups in comparison, and their use is restricted to reactive alkyl derivatives that afford fairly stable cations.

EXERCISE 12.6

Calculate the rate constants for the data in Tables 12.1 and 12.2. Are they constant?

TABLE 12.4 Relative Reactivities of Commonly Used Leaving Groups, —X

$$C_6H_5\text{—}CH(CH_3)\text{—}X \xrightarrow[\text{EtOH}]{H_2O} C_6H_5\text{—}CH(CH_3)\text{—}OH + H^+ + X^-$$

Leaving Group	Name	Common Abbreviation	Relative Rate (Compared with —Cl)
—O—S(=O)_2—CF_3	Trifluoromethanesulfonate (triflate)	—OTf	10^8
$\text{—O—S(=O)_2—C}_6\text{H}_4\text{—NO}_2$	*p*-Nitrobenzenesulfonate (nosylate)	—ONs	10^5
$\text{—O—S(=O)_2—C}_6\text{H}_4\text{—Br}$	*p*-Bromobenzenesulfonate (brosylate)	—OBs	10^5
$\text{—O—S(=O)_2—C}_6\text{H}_4\text{—CH}_3$	*p*-Toluenesulfonate (tosylate)	—OTs	4×10^4
$\text{—O—S(=O)_2—C}_6\text{H}_5$	Benzenesulfonate	$\text{—OSO}_2\text{C}_6\text{H}_5$	4×10^4
—O—S(=O)_2—CH_3	Methanesulfonate (mesylate)	—OMs	3×10^4
—I	Iodide		10^2
—Br	Bromide		10
—Cl	Chloride		1
$\text{—O—C(=O)—C}_6\text{H}_4\text{—NO}_2$	*p*-Nitrobenzoate	—OPNB	10^{-5}

12.4 STEREOCHEMICAL EXPECTATIONS IN NUCLEOPHILIC SUBSTITUTION

In the preceding section we showed that reaction kinetics do not always provide a clear distinction between the carbocation and direct displacement mechanisms for nucleophilic substitution. An alternative way to test the mechanism of a nucleophilic substitution reaction is to study the stereochemistry of the substitution process. Does the reaction proceed with *retention of configuration?*

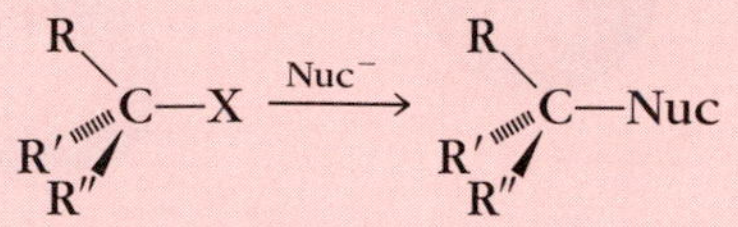

Retention of configuration:
Relative to the other three substituents the position of the nucleophile in the product is the same as that of the leaving group in the substrate.

Or does the reaction proceed with *inversion of configuration?*

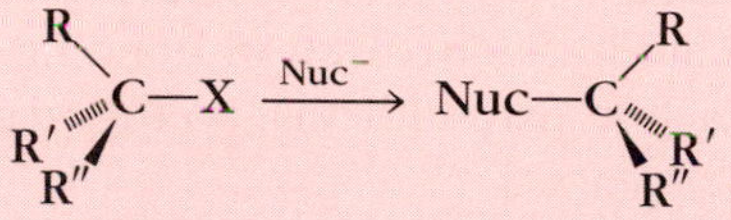

Inversion of configuration:
Relative to the other three substituents the position of the nucleophile in the product is *not* the same as that of the leaving group in the substrate.

Or as still another alternative, does the reaction proceed with a *mixture* of inversion and retention of configuration? These questions have been studied extensively, and we will evaluate some of the actual results in Section 12.5. First, however, we will discuss the stereochemical results that should be *expected* for each of the alternative mechanisms of nucleophilic substitution.

Carbocation Mechanism

What stereochemical results are expected for a nucleophilic substitution reaction that proceeds via an intermediate carbocation? We are concerned with the stereochemical relationship between the reactant and the *product,* so we must focus on the stereochemistry of the *product-forming step,* the reaction of the nucleophile with the carbocation:

$$R^+ \xrightarrow{Nuc^-} \text{product}$$

The key to the stereochemistry of this step is the interaction between orbitals of the nucleophile and the carbocation intermediate. The attacking nucleophile has a

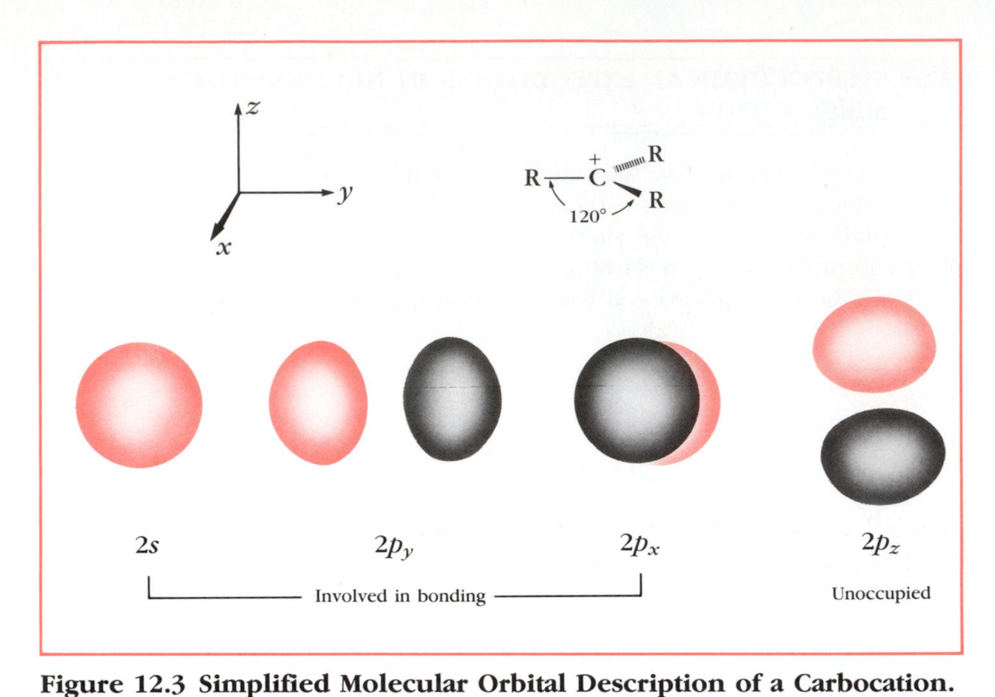

Figure 12.3 Simplified Molecular Orbital Description of a Carbocation. The central carbon atom has only three substituents, and the most stable arrangement is a *planar geometry,* with these three substituents lying in the same plane as the central carbon atom and separated by angles of 120°. Only three of the atomic orbitals of the central carbon atom are involved in bonding to the three substituents, so the p_z orbital (which is perpendicular to the plane of the carbocation) is "left over" and is *unoccupied.*

lone pair of electrons in an orbital that will be used in forming a bond to the electron-deficient carbon atom of the cation. The question to be evaluated is simply: What orbital of the carbocation will interact with the filled orbital of the nucleophile? The answer is equally simple: The filled orbital of the attacking nucleophile will interact with the lowest-energy "empty" orbital on the carbocation. This is the *lowest unoccupied molecular orbital,* or *LUMO*.

The positively charged carbon atom of the cation has only three substituents, and the optimum geometry is planar, with the three substituents separated by angles of 120° (Figure 12.3). Three atomic orbitals of the central carbon atom are hybridized to the three sp^2 orbitals that are involved in bonding to the three substituents. The third p orbital ($2p_z$) is "left over," and this is the lowest unoccupied molecular orbital in the carbocation.

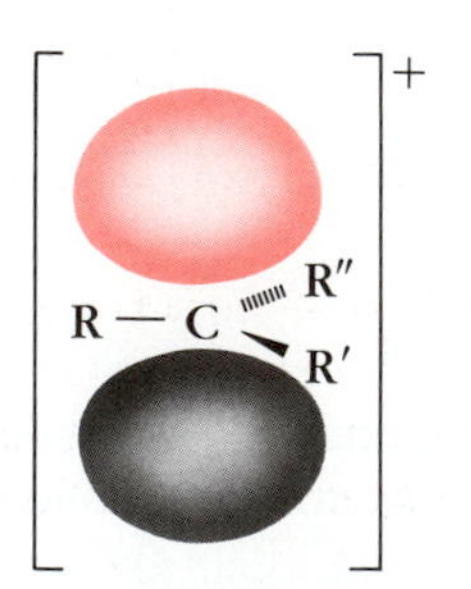

Lowest unoccupied molecular orbital (LUMO) of a carbocation

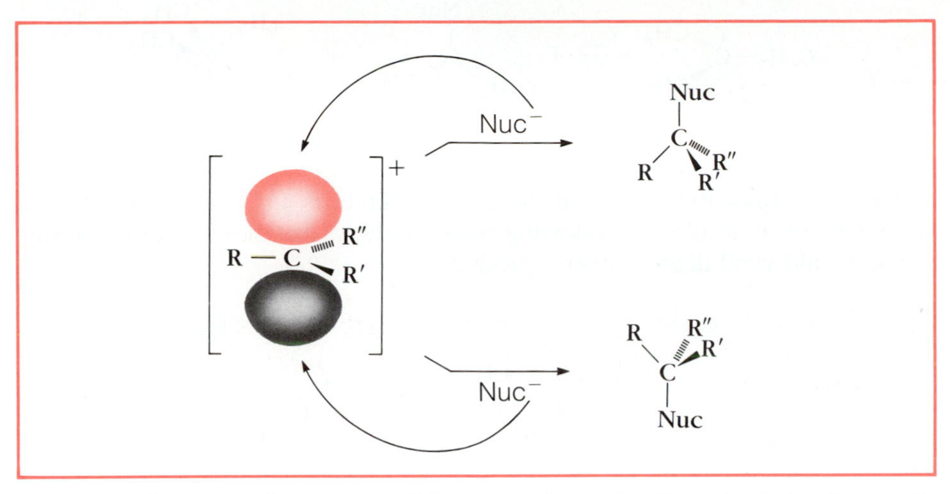

Figure 12.4 Stereochemistry of the Reaction of a Free Carbocation with a Nucleophile.
Attack by the nucleophile would occur at each lobe of the unoccupied *p* orbital of the cation.

The LUMO of a carbocation has two lobes, one above and one below the plane of the cationic center. Interaction with the occupied orbital of the nucleophile can occur with *either* of the two lobes (which differ only in phase), so you should expect product to be formed with *both* of the two possible configurations at the chiral center (Figure 12.4). This conclusion is also supported by steric arguments. The cationic center is flat, and the nucleophile can easily attack at either of its two faces.

Two important conclusions can be drawn from the reactions between a nucleophile and a carbocation that we have depicted in Figure 12.4. *First,* both stereoisomeric products are formed, so the overall *stereochemical result for nucleophilic substitution by a carbocation mechanism will be a mixture of products corresponding to both retention and inversion of configuration. Second,* the stereochemical integrity of the reactive center is destroyed when a free carbocation is formed. Regardless of the configuration of the reactive center in the starting material, the same carbocation would be formed. In other words, nucleophilic substitution by a carbocation mechanism is *not* a stereospecific process.

Several different stereochemical results could be obtained from nucleophilic attack on a carbocation. In the example shown in Figure 12.4 the reactive center becomes a chiral center in the products (and therefore was also a chiral center in the original reactant). As long as this is the only chiral center in the molecule, the two products would be enantiomers, and an optically active starting material would yield racemic product.

The stereochemical results will be different when two of the substituents on the reactive center are the same. In this case the central carbon is not chiral, so there is no difference between the products formed from attack at the two faces of the intermediate carbocation. We have illustrated this situation in the following equation:

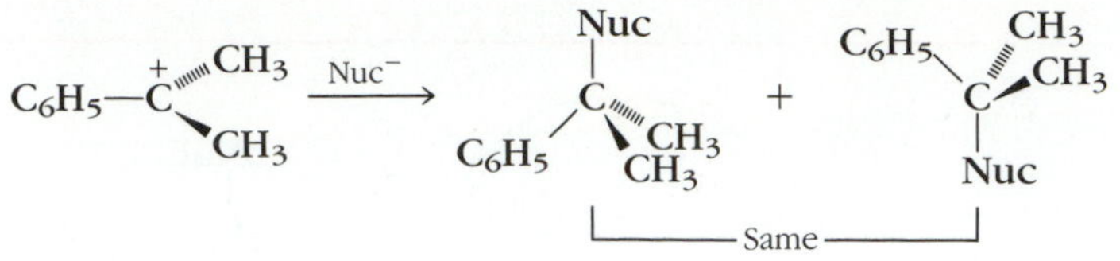

Still a third possibility arises if the reactive center is just one of several chiral centers in the molecule. The following equation shows that the carbocation in such a case would yield diastereomeric products.

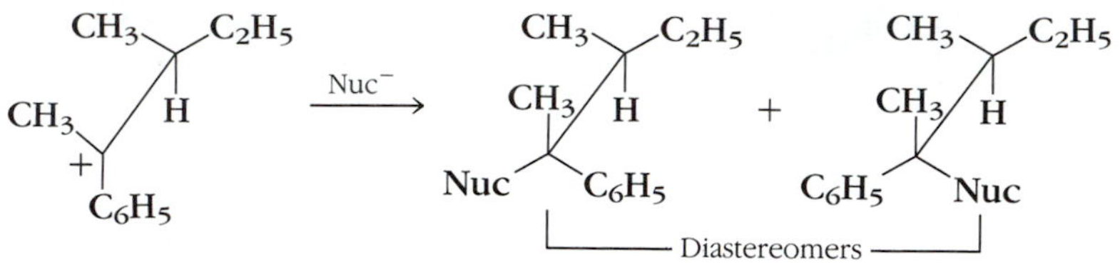

Diastereomers can also be formed in cyclic compounds, even when there are no chiral centers. (Recall that *cis-trans* isomers are diastereomers.)

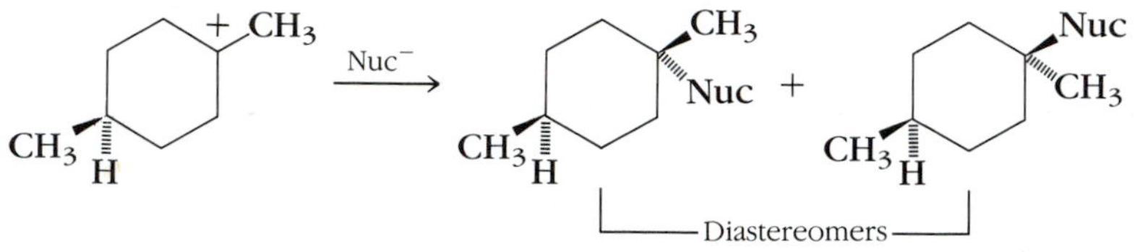

Whenever diastereomeric products are produced, you should expect unequal amounts of the two products. The transition states leading to the diastereomeric products must have different energies, and one product will be formed more rapidly than the other. Even when the products are enantiomeric, you should not expect a totally racemic product because the leaving group will usually have some influence upon the reaction of the cation. Only rarely is a free carbocation generated. You cannot ordinarily predict which stereoisomer will predominate, but a carbocation typically yields *comparable* (although not exactly equal) amounts of the two possible stereoisomers.

Direct Displacement Mechanism

What stereochemical results are expected for nucleophilic substitution reactions that proceed via the direct displacement mechanism? Once again, the stereochemical key is the interaction between orbitals of the two reacting species. The occupied orbital corresponding to the nonbonding electrons of the nucleophile will interact with the LUMO of the substrate. All the bonding molecular orbitals of a compound such as an alkyl halide are occupied, and the lowest *unoccupied* orbital is therefore *antibonding*. We have drawn this orbital in Figure 12.5 for methyl fluoride.

The interaction between this antibonding orbital and the occupied nonbonding orbital of the nucleophile will result in two *new* molecular orbitals in the transition state of the reaction (we will not attempt to draw these orbitals). Qualitatively, attack by a nucleophile will place electrons in the unoccupied orbital that is *antibonding* with respect to carbon and fluorine. This means that both the bonding and

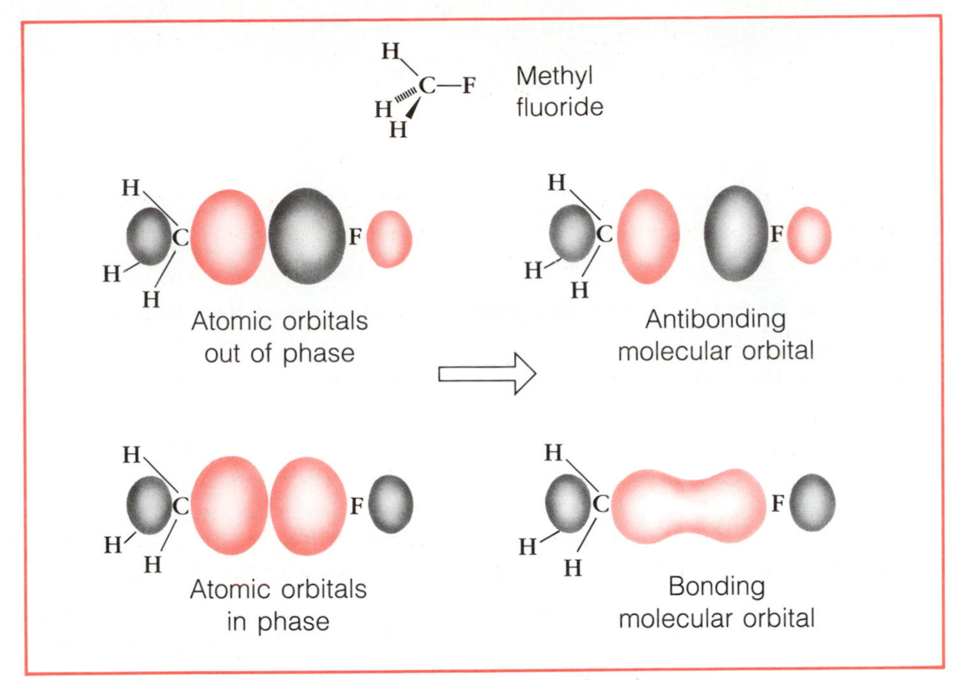

Figure 12.5 Sigma Bonding in Methyl Fluoride: A Simplified View of the Interaction between the *sp*^{3} Orbital of the Carbon Atom and the Corresponding Hybrid Orbital on Fluorine.
The interaction can be *in phase* (a bonding molecular orbital) or *out of phase* (an antibonding molecular orbital). The bonding orbital depicted in the lower part of the drawing is occupied by two electrons. The antibonding orbital shown at the top of the figure is the lowest unoccupied molecular orbital of methyl fluoride.

antibonding C—F sigma molecular orbitals will be populated at the transition state, so the bonding interaction between carbon and fluorine is essentially cancelled. As the reaction proceeds, orbital overlap generates a new bonding interaction between the carbon atom and the nucleophile. At the same time this causes "breaking" of the original bond to fluorine. The overall result is a one-step substitution reaction.

What is the preferred orientation for attack on the carbon atom by the nucleophile? The leaving group (in this case the departing fluoride ion) blocks the front, so the preferred orientation is for attack from the rear, or back side, of the carbon atom:

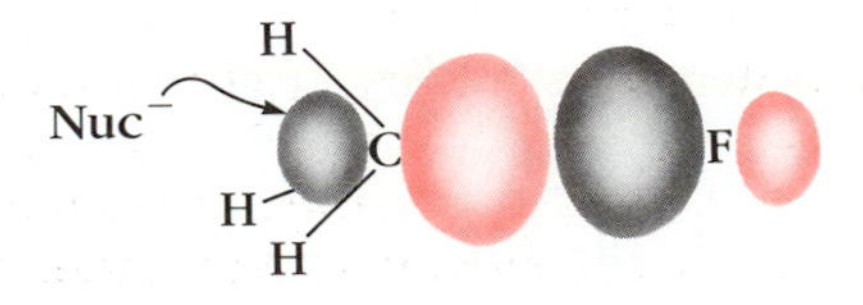

Back side attack

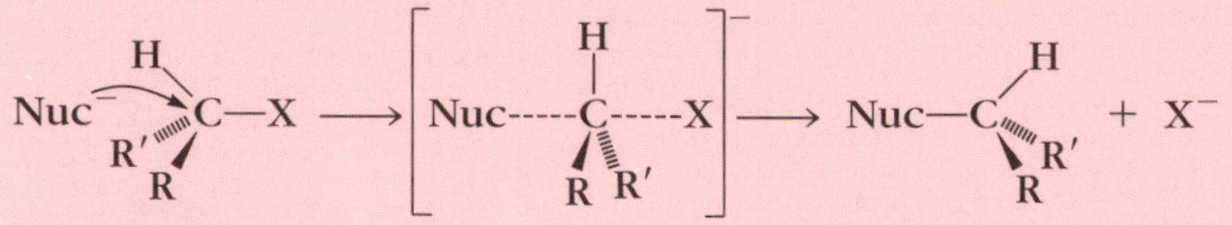

$$\mathrm{Nuc^-} + \mathrm{H(R)(R')C{-}X} \longrightarrow \left[\mathrm{Nuc{-}{-}{-}C(H)(R)(R'){-}{-}{-}X}\right]^- \longrightarrow \mathrm{Nuc{-}C(H)(R)(R')} + \mathrm{X^-}$$

Figure 12.6 Stereochemistry of a Direct Displacement Reaction. Backside attack leads to a pentavalent transition state and to inversion of configuration in the product.

Backside attack leads to a *pentavalent* transition state and then to a product with *inverted configuration*. This is illustrated in Figure 12.6 for the reaction of a secondary alkyl derivative in which the reaction center is also a chiral center.

The direct displacement reaction is therefore *stereospecific* because the configuration of the product is directly related to that of the reactant. In the following example we have illustrated the special case in which the attacking nucleophile is identical to the leaving group. A bromide ion acts as a nucleophile to attack a chiral alkyl bromide, so the leaving group is also a bromide ion.

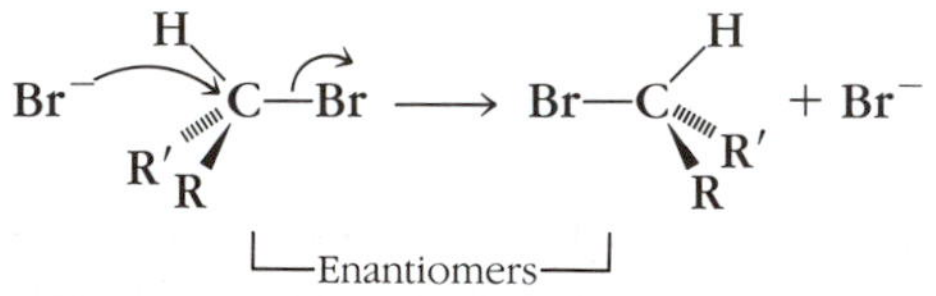

$$\mathrm{Br^-} + \mathrm{H(R)(R')C{-}Br} \longrightarrow \mathrm{Br{-}C(H)(R)(R')} + \mathrm{Br^-}$$

Enantiomers

The product and starting material in such a reaction are enantiomers, and this shows you the origin of the term **inversion of configuration,** which is used to describe the stereochemistry of direct displacement reactions.

Direct Displacement *vs* the Carbocation Mechanism

Most substrates can undergo nucleophilic substitution by only one of the two alternative mechanisms. But some compounds, typically primary or secondary alkyl derivatives that could yield resonance stabilized cations, are capable of reacting either by direct displacement or by ionization to a carbocation. The following equations show how the stereochemistry of a nucleophilic substitution reaction can be placed into one of three categories according to the structure of the substrate.

1. Stereochemistry: Primary or Secondary Alkyl

$\mathrm{R{-}CH_2{-}X}$ (1° Alkyl) or $\mathrm{R{-}CH(R){-}X}$ (2° Alkyl) $\longrightarrow$ Inversion of configuration (reaction occurs via direct displacement)

2. Stereochemistry: Tertiary Alkyl

$R_3C{-}X \longrightarrow$ Comparable amounts of inversion and retention of configuration (reaction occurs via carbocation)

3° Alkyl

3. Stereochemistry: Primary or Secondary Allylic or Benzylic

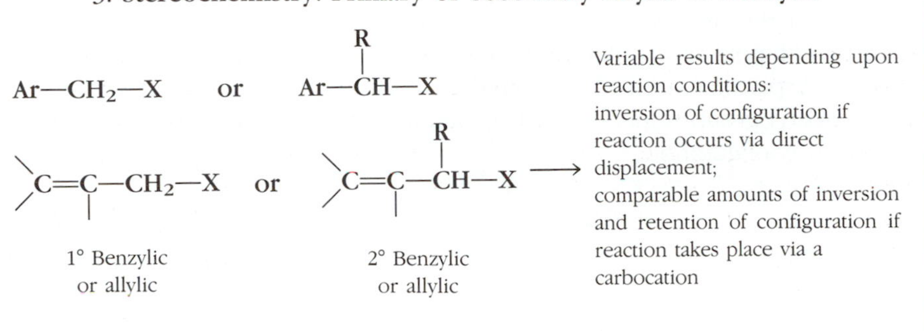

Ar—CH₂—X
or
Ar—CH(R)—X
C=C—CH₂—X
or
C=C—CH(R)—X
1° Benzylic or allylic
2° Benzylic or allylic
Variable results depending upon reaction conditions:
inversion of configuration if reaction occurs via direct displacement;
comparable amounts of inversion and retention of configuration if reaction takes place via a carbocation

EXERCISE 12.7

Draw the product or products expected for the following nucleophilic substitution reactions:

(a) $CH_3CH_2{-}C(Cl)(H)(CH_3) \xrightarrow{CH_3CH_2OH}$?

(b) $CH_3CH_2{-}C(Cl)(H)(CH_3) \xrightarrow{CF_3CH_2OH}$?

(c) $CH_3CH_2CH_2{-}C(CH_3)(Br)(CH_2CH_3) \xrightarrow{CH_3CH_2OH}$?

(d) $CH_3CH_2CH_2{-}C(CH_3)(Br)(CH_2CH_3) \xrightarrow{CF_3CH_2OH}$?

(e) $C_6H_5{-}C(CH_3)(CH_2CH_3)(Br) \xrightarrow{CH_3CH_2OH}$?

(f) $C_6H_5{-}C(CH_3)(CH_2CH_3)(Br) \xrightarrow{CF_3CH_2OH}$?

12.5 STEREOCHEMICAL RESULTS IN NUCLEOPHILIC SUBSTITUTION REACTIONS

At the end of the preceding section we made predictions for the stereochemical results of nucleophilic substitution reactions based on mechanistic categories. In this section we will present the actual results of a variety of nucleophilic substitution reactions and show you that these fully support our predictions. We will begin with *simple* (i.e., neither allylic nor benzylic) secondary alkyl derivatives and then proceed to the corresponding primary and methyl derivatives, followed by the tertiary analogs. Finally, we will show you the results for a group of compounds that could react by either of the two possible mechanisms, compounds that are primary or secondary but are *also* either allylic or benzylic.

Secondary Alkyl Derivatives

The stereochemistry of nucleophilic substitution at secondary carbon atoms has provided a great body of evidence about the mechanism of reaction. An early study of great importance was carried out by H. Phillips (England) in the early 1920s. He studied the solvolysis reaction of 2-octyl tosylate in ethanol, as summarized in Figure 12.7.

Optically active 2-octanol was converted to the ethyl ether by two different procedures. In the first synthesis 2-octanol was converted to its potassium salt by reaction with potassium metal, and the salt was allowed to react with ethyl bromide (a direct displacement on ethyl bromide). In this method the carbon-oxygen bond of the 2-octyl group remains intact throughout the sequence, so there can be no change in configuration.

$$CH_3{-}(CH_2)_5{-}\underset{}{\overset{CH_3}{\overset{|}{C}}}H{-}O{-}H \xrightarrow{K} CH_3{-}(CH_2)_5{-}\overset{CH_3}{\overset{|}{C}}H{-}O^-K^+$$

$$\xrightarrow{C_2H_5Br} CH_3{-}(CH_2)_5{-}\overset{CH_3}{\overset{|}{C}}H{-}O{-}C_2H_5$$

The second route involved conversion of the alcohol to the tosylate, and this also proceeds with retention of configuration because the carbon-oxygen bond remains intact. The tosylate was then solvolyzed in ethanol.

$$CH_3{-}(CH_2)_5{-}\overset{CH_3}{\overset{|}{C}}H{-}OH \xrightarrow{TsCl} CH_3{-}(CH_2)_5{-}\overset{CH_3}{\overset{|}{C}}H{-}OTs$$

$$\xrightarrow{C_2H_5OH} CH_3{-}(CH_2)_5{-}\overset{CH_3}{\overset{|}{C}}H{-}O{-}C_2H_5$$

The ethyl ether prepared by alkylation of 2-octanol and the tosylate (prepared by acylation of 2-octanol) both have the same relative configuration as the starting alcohol. But what of the ethyl ether that is produced by solvolysis of the tosylate? This material had an optical rotation that was equal in magnitude but *opposite in sign* to the ether formed by alkylation of 2-octanol. Therefore the displacement reaction of the secondary alkyl tosylate by ethanol must proceed with complete *inversion of configuration.*

A variety of other 2-octyl derivatives have been used to evaluate the stereochemistry of nucleophilic substitution reactions. Displacements both by neutral nucleophiles (solvolysis reactions) and by anions have been studied. Side reactions such as elimination (see Section 12.6) can lead to other products, but *substitution* products are invariably formed with complete inversion of configuration. The following

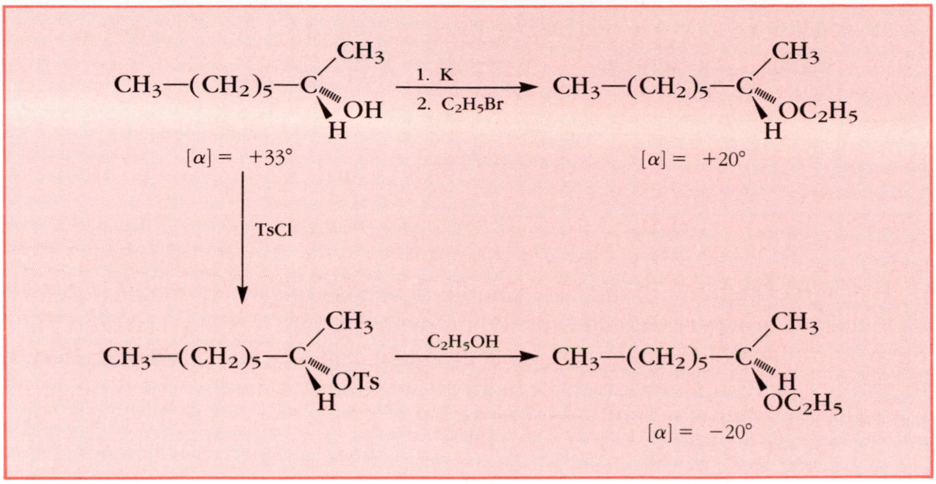

$$CH_3{-}(CH_2)_5{-}C(CH_3)(H)(OH)\ [\alpha] = +33° \xrightarrow[2.\ C_2H_5Br]{1.\ K} CH_3{-}(CH_2)_5{-}C(CH_3)(H)(OC_2H_5)\ [\alpha] = +20°$$

$$CH_3{-}(CH_2)_5{-}C(CH_3)(H)(OH) \xrightarrow{TsCl} CH_3{-}(CH_2)_5{-}C(CH_3)(H)(OTs) \xrightarrow{C_2H_5OH} CH_3{-}(CH_2)_5{-}C(CH_3)(H)(OC_2H_5)\ [\alpha] = -20°$$

Figure 12.7 Proof of Inversion of Configuration in Nucleophilic Substitution Reaction of 2-Octyl Tosylate.
Optically active 2-octanol was converted to the corresponding tosylate and to the ethyl ether by reactions that do not affect the configuration at the chiral center. The ether synthesized by alkylation of 2-octanol exhibited a positive rotation, but that formed by solvolysis of the tosylate in ethanol showed a rotation that was *negative* (although equal in magnitude to the other sample). The nucleophilic substitution reaction (solvolysis in ethanol) must have proceeded with complete *inversion of configuration.*

reactions illustrate this behavior. In contrast to the other chapters in this text, we will not show yields for many of the reactions in this section. The yields of these reactions often were not determined, because attention was focused on the structure and stereochemistry of the product. Abbreviations for the leaving groups have been defined previously (Table 12.4).

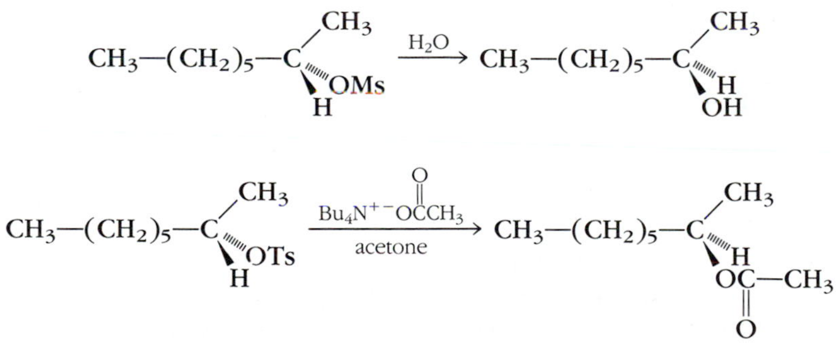

$$CH_3{-}(CH_2)_5{-}C(CH_3)(H)(OMs) \xrightarrow{H_2O} CH_3{-}(CH_2)_5{-}C(CH_3)(H)(OH)$$

$$CH_3{-}(CH_2)_5{-}C(CH_3)(H)(OTs) \xrightarrow[acetone]{Bu_4N^{+}\ {}^{-}OC(=O)CH_3} CH_3{-}(CH_2)_5{-}C(CH_3)(H)(OC(=O){-}CH_3)$$

The reaction of 2-octyl tosylate with tetrabutylammonium acetate in the second example was carried out in acetone as solvent. Acetone is quite nonnucleophilic, so the product is derived entirely from attack by the acetate ion. The tetrabutylammonium salt was employed because inorganic acetates such as sodium acetate are insoluble in acetone.

Secondary cyclic systems have also been studied, and the substitution products again are formed with complete inversion of configuration.

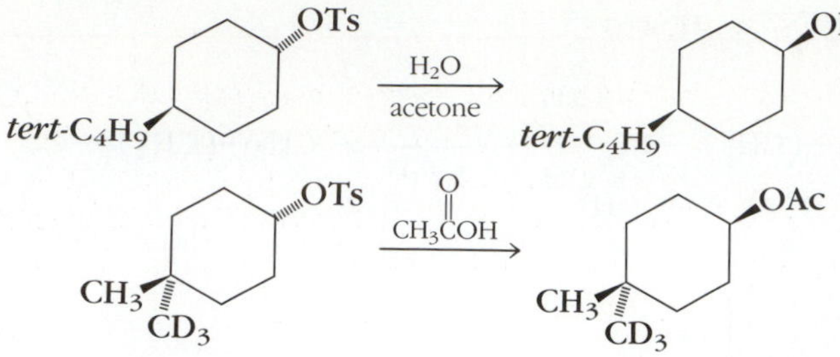

$$\text{cis-4-}tert\text{-}C_4H_9C_6H_{10}OTs \xrightarrow[\text{acetone}]{H_2O} \text{4-}tert\text{-}C_4H_9C_6H_{10}OH$$

$$CH_3(CD_3)C_6H_9OTs \xrightarrow{CH_3\overset{O}{\overset{\|}{C}}OH} CH_3(CD_3)C_6H_9OAc$$

The stereochemistry of the last example depends entirely on *isotopic labeling*. One methyl group is CH_3 and the other methyl group is CD_3. The steric and electronic effects of H and D are almost the same, and the only significant effect of the isotopic labeling is to serve as a marker for the stereochemistry by distinguishing between otherwise equivalent groups.

Primary Alkyl Derivatives

Under normal circumstances it is not possible to distinguish between inversion and retention of configuration at a primary carbon atom. The presence of two identical substituents (i.e., the two hydrogens of R—CH_2—X) requires that such a primary carbon atom be achiral, so only a single configuration is possible. However, the use of isotopic labeling changes the situation; thus replacement of one of the hydrogens of a CH_2 group with deuterium affords a chiral primary center.

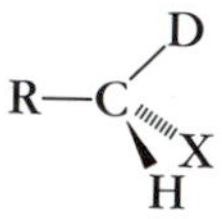

$$R\text{—}CHDX$$

Several such deuterium-substituted compounds have been investigated in order to learn the stereochemistry of nucleophilic substitution of simple primary alkyl derivatives. In each case the product of nucleophilic displacement exhibits *complete* inversion of configuration.

$$CH_3CH_2CH_2\text{—}CHD(ONs) \xrightarrow{H\text{—}\overset{O}{\overset{\|}{C}}OH} CH_3CH_2CH_2\text{—}CHD(O\text{—}\overset{}{C}H{=}O)$$

$$(CH_3)_3C\text{—}CHD(OTs) \xrightarrow{NaN_3} (CH_3)_3C\text{—}CHD(N_3)$$

Neopentyl group

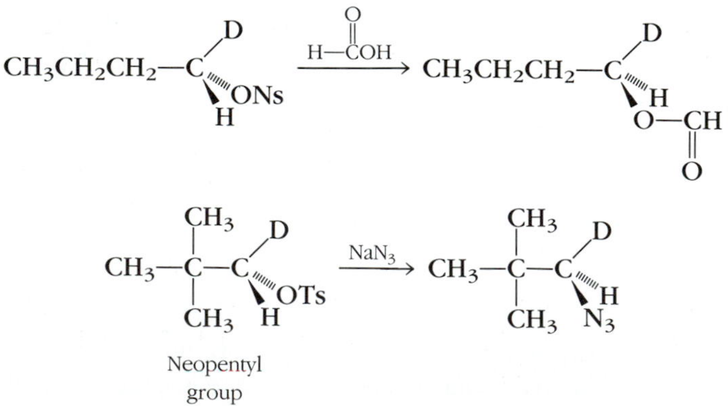

The second reaction would not give a very good yield of substitution product, because neopentyl systems tend to rearrange under the conditions of nucleophilic substitution reactions (Section 12.7). Nevertheless, the unrearranged product that is formed shows complete inversion of configuration.

Tertiary Alkyl Derivatives

The crowded steric environment of a tertiary center precludes direct displacement, so tertiary alkyl derivatives must undergo nucleophilic substitution by a carbocation mechanism. Very few optically active derivatives have been investigated, but the examples studied all react with substantial amounts of *both* inversion and retention of configuration.

$$CH_3CH(CH_3)-(CH_2)_3-C(CH_3)(Cl)(CH_2CH_3) \xrightarrow{CH_3CO_2H} CH_3CH(CH_3)-(CH_2)_3-C(CH_3)(CH_2CH_3)(O-C(=O)-CH_3)$$

> 67% inversion

$$+\ CH_3CH(CH_3)-(CH_2)_3-C(CH_3)(O-C(=O)-CH_3)(CH_2CH_3)$$

< 33% retention

$$C_6H_5-C(CH_3)(OPNB)(CH_2CH_3) \xrightarrow[CH_3CO_2K]{CH_3CO_2H} C_6H_5-C(CH_3)(O-C(=O)-CH_3)(CH_2CH_3) + C_6H_5-C(CH_3)(CH_2CH_3)(O-C(=O)-CH_3)$$

48% Retention 52% Inversion

The relative amounts of inversion and retention vary substantially in nucleophilic substitution reactions that proceed via carbocations; the results depend upon the reactivities of the nucleophile, the leaving group, and the cation. If the leaving group remains close to the cationic center, it will have different effects on the two pathways leading to retention and inversion. Greater dissociation of the leaving group from the cation would result in more complete racemization.

Primary and Secondary Allylic and Benzylic Derivatives

Primary and secondary allylic and benzylic compounds are capable of undergoing nucleophilic substitution by either of the two possible mechanistic pathways. In most instances a mixture of products corresponding to both inversion and retention of configuration is observed. This could be interpreted as reaction by competing pathways (direct displacement and carbocation) that operate simultaneously, but other evidence suggests that it is the result of reaction via an ion pair. Two important trends are observed for these derivatives: (1) strong nucleophiles tend to give increased amounts of inversion of configuration; and (2) more stable cations tend to result in more nearly equal amounts of inversion and retention. The following example of an allylic compound demonstrates that ionization to a carbocation can sometimes lead to complete racemization.

$$CH_3-CH{=}CH-C(CH_3)(OPNB)(H) \xrightarrow[acetone]{H_2O} CH_3-CH{=}CH-C(CH_3)(H)(OH) + CH_3-CH{=}CH-C(CH_3)(OH)(H)$$

50% Inversion 50% Retention

The effect of carbocation stability is seen in the following reactions of primary and secondary benzylic derivatives under the same conditions:

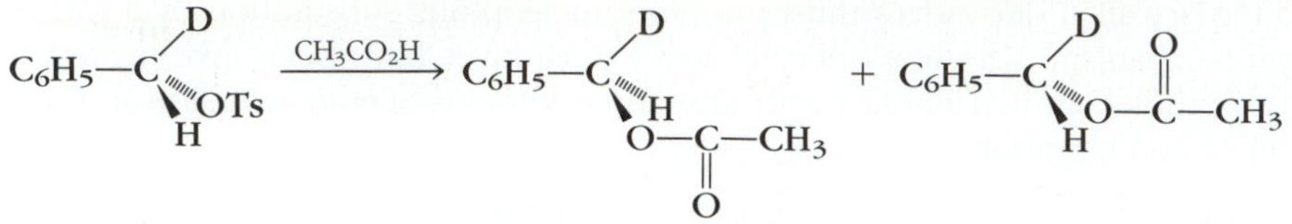

$$C_6H_5\text{—}C(D)(H)\text{OTs} \xrightarrow{CH_3CO_2H} C_6H_5\text{—}C(D)(H)\text{—O—}\overset{\overset{O}{\|}}{C}\text{—}CH_3 + C_6H_5\text{—}C(D)(H)\text{—O—}\overset{\overset{O}{\|}}{C}\text{—}CH_3$$

> 92% Inversion   < 9% Retention

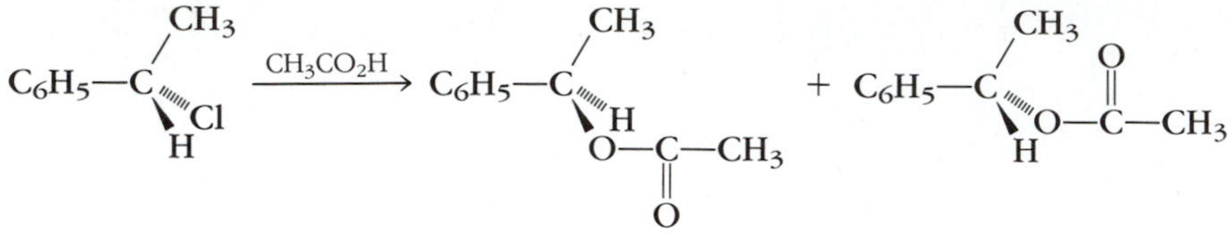

$$C_6H_5\text{—}C(CH_3)(H)\text{Cl} \xrightarrow{CH_3CO_2H} C_6H_5\text{—}C(CH_3)(H)\text{—O—}\overset{\overset{O}{\|}}{C}\text{—}CH_3 + C_6H_5\text{—}C(CH_3)(H)\text{—O—}\overset{\overset{O}{\|}}{C}\text{—}CH_3$$

56% Inversion   44% Retention

Ionization of the substrate in the second reaction would afford a carbocation that is secondary and benzylic, while that in the first reaction would be benzylic but primary. The greater stability of the *secondary* benzylic cation results in much more nearly equal amounts of retention and inversion.

The trend with respect to changing reaction conditions is illustrated by the following series of reactions of 1-chloro-1-phenylethane, in which the strength of the nucleophile increases steadily from trifluoroethanol to acetate ion:

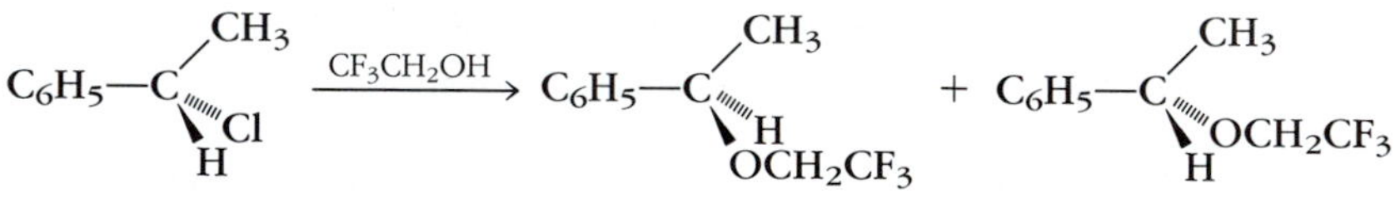

$$C_6H_5\text{—}C(CH_3)(H)\text{Cl} \xrightarrow{CF_3CH_2OH} C_6H_5\text{—}C(CH_3)(H)OCH_2CF_3 + C_6H_5\text{—}C(CH_3)(H)OCH_2CF_3$$

46% Inversion   54% Retention

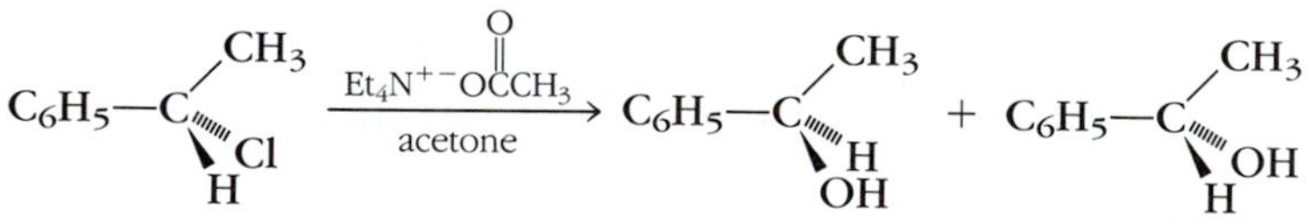

$$C_6H_5\text{—}C(CH_3)(H)\text{Cl} \xrightarrow[\text{acetone}]{Et_4N^{+}\,{}^{-}O\overset{\overset{O}{\|}}{C}CH_3} C_6H_5\text{—}C(CH_3)(H)OH + C_6H_5\text{—}C(CH_3)(H)OH$$

53% Inversion   47% Retention

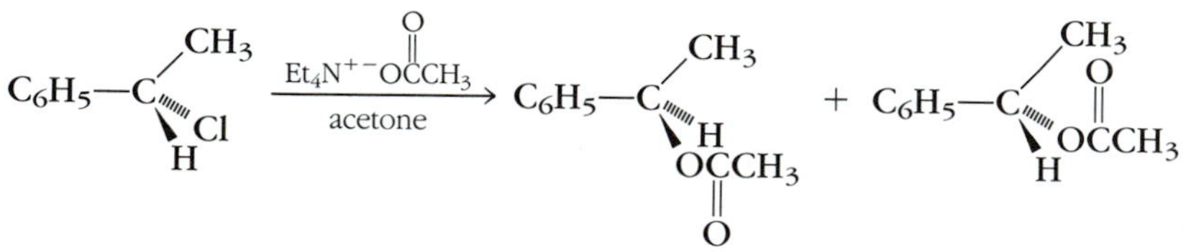

$$C_6H_5\text{—}C(CH_3)(H)\text{Cl} \xrightarrow[\text{acetone}]{Et_4N^{+}\,{}^{-}O\overset{\overset{O}{\|}}{C}CH_3} C_6H_5\text{—}C(CH_3)(H)O\overset{\overset{O}{\|}}{C}CH_3 + C_6H_5\text{—}C(CH_3)(H)O\overset{\overset{O}{\|}}{C}CH_3$$

82% Inversion   18% Retention

In the first reaction trifluoroethanol is a polar solvent but a very weak nucleophile, and the products of inversion and retention are formed in nearly equal amounts (as would be expected for a carbocation mechanism). As the strength of the nucleophile increases so does the relative amount of inversion of configuration, and this predominates in the last example.

The reactions that we have discussed in this section demonstrate that you can often predict the stereochemistry of nucleophilic substitution reactions with some certainty. Only in the case of primary and secondary benzylic or allylic derivatives do the results vary substantially with reaction conditions. For substrates of this type, a strong nucleophile (particularly one with a negative charge) favors inversion of configuration. Otherwise, you should expect comparable amounts of inversion and retention.

EXERCISE 12.8

Predict the stereochemistry of the substitution products that would be obtained from solvolysis in 2,2,2-trifluoroethanol of:

(a) (*R*)-2-Butyl brosylate
(b) *trans*-1-Chloro-3-methylcyclohexane
(c) (*S*)-2-Bromo-2-phenylbutane

EXERCISE 12.9

Predict the stereochemistry of the substitution products that would be obtained from (*R*)-2-chloro-2-phenylbutane in the following reactions:

(a) Solvolysis in ethanol
(b) Reaction with tetrabutylammonium acetate in acetone

12.6 ELIMINATION AS A SIDE REACTION IN NUCLEOPHILIC SUBSTITUTION

In Section 12.1 we explained that the fundamental structural requirement for a species to act as a nucleophile is a *pair of electrons* that can attack a positively charged (or partially positively charged) center. But this is the identical structural requirement for a species to act as a base. If the substrate has one or more hydrogen atoms located β to the leaving group, the nucleophile can also function as a base.

$$\text{Nuc}^- + \text{H—C—C—X} \longrightarrow \text{Nuc—H} + \text{C=C} + \text{X}^-$$

or

$$\text{Nuc}^- + \text{H—C—C}^+ \longrightarrow \text{Nuc—H} + \text{C=C}$$

In the preceding equations we have shown a nucleophile (or base) that is negatively charged, but you should keep in mind that only the pair of electrons (and not a negative charge) is required. Any species that can act as a nucleophile, including a neutral molecule such as water or an alcohol, can also act as a base. You should therefore expect that a nucleophilic substitution reaction will always be accompanied by a competing elimination reaction. (The converse is also true: substitution occurs as a side reaction in elimination reactions.)

This competition can be seen in the following examples. In Section 5.13 we showed that preparative elimination reactions of alkyl halides are usually carried out with a strong base. But even a solvolysis reaction can sometimes afford a substantial quantity of alkene, despite the absence of any strong base.

$$CH_3-\underset{CH_3}{\overset{CH_3}{C}}-Cl \xrightarrow[H_2O]{CH_3CH_2OH} CH_3-\underset{CH_3}{\overset{CH_3}{C}}-OC_2H_5 + CH_3-\underset{CH_3}{\overset{CH_3}{C}}-OH + CH_2{=}C(CH_3)_2$$

Substitution, 64% — Elimination, 36%

Alternatively, the reaction of an alkyl halide with strong base sometimes produces a substantial amount of substitution product in addition to the alkene formed by elimination.

$$CH_3-\overset{Br}{CH}-CH_3 \xrightarrow[C_2H_5OH]{NaOC_2H_5} CH_3-\overset{OC_2H_5}{CH}-CH_3 + CH_3-CH{=}CH_2$$

Substitution, 21% — Elimination, 79%

How can you anticipate whether substitution or elimination will predominate in these reactions? What factors govern the relative ease (and relative rates) of the elimination and substitution processes? Is it a function of the structure of the attacking nucleophile (i.e., attacking base)? Is it a function of the reaction conditions? Or is it a function of the structure of the substrate? In fact, all these factors can play a major role in determining the relative amounts of substitution *vs* elimination.

The product distribution in a reaction is often determined by only very small energy differences. For example, if a reaction yields two products in a 90%–10% distribution, the difference in activation energy for the two pathways is only about 1 kcal/mol. An energy difference of this magnitude is sufficiently small that even a minor change in reaction conditions can sometimes produce a large change in the product distribution. As long as the two products are not in equilibrium, the ratio in which they are formed must equal the ratio of the rate constants for the two pathways. The difference in activation energies, ΔE_a, is related to the respective rate constants, k_1 and k_2, by the following equation:

$$\Delta E_a \approx -2.3\ RT \log (k_1/k_2)$$

The relationships between rates, activation energies, and yields are illustrated in Figure 12.8.

Reaction conditions can play a significant role in determining the substitution-elimination ratio. For example, higher temperatures often increase the fraction of elimination. It is usually difficult to predict the effect of changes in reaction conditions such as solvent polarity, however. The only information that you can use reliably to evaluate substitution *vs* elimination comes from the structures of the substrate and of the nucleophile (or base).

Effect of Base Strength

The presence of a strong base favors elimination in the competition with nucleophilic substitution. The reaction of an alkyl halide or sulfonate with a strong base such as an alkoxide ion invariably leads to a greater proportion of elimination than does the corresponding reaction with a neutral nucleophile. Usually the alkene is

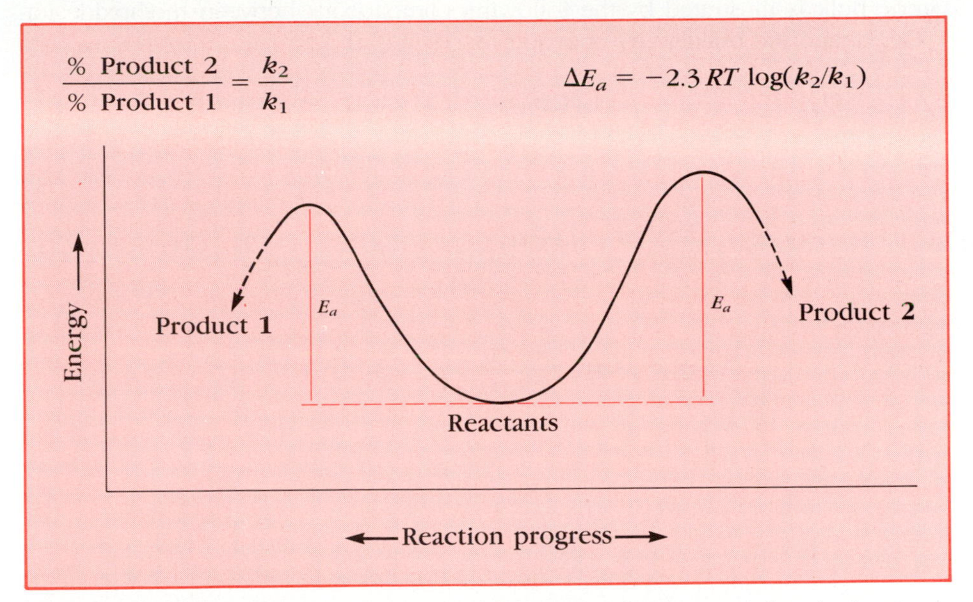

Figure 12.8 Dependence of the Product Distribution on the Activation Energies of Two Competing (Nonequilibrium) Pathways.
The relative amounts of the two products are directly proportional to the corresponding rate constants, and the ratio of the rate constants depends on the difference in activation energy for the two pathways. Reaction via the pathway with the smaller activation energy will occur more rapidly, and that product will predominate.

formed directly by attack of the base on the alkyl halide or sulfonate. By analogy to the S_N2 reaction, this is frequently described as E2 elimination. You can see the effect of a strong base in the following reactions of 2-octyl derivatives. (The percentages that we have shown in the examples in this section are product distributions rather than yields.)

$$CH_3{-}(CH_2)_5{-}\underset{}{CH}(OBs){-}CH_3 \xrightarrow{CH_3OH} CH_3{-}(CH_2)_5{-}CH(OCH_3){-}CH_3 \quad (\sim 100\%)$$

$$CH_3{-}(CH_2)_4{-}CH_2{-}CH(OSO_2C_6H_5){-}CH_3 \xrightarrow[CH_3OH]{NaOCH_3} CH_3{-}(CH_2)_4{-}CH_2{-}CH(OCH_3){-}CH_3 \quad 37\%$$

$$+\ CH_3{-}(CH_2)_4{-}CH{=}CH{-}CH_3 + CH_3{-}(CH_2)_4{-}CH_2CH{=}CH_2 \quad (63\%)$$

Steric Effects for Nucleophiles

Increased steric bulk of the attacking nucleophile favors elimination relative to substitution. A bulky nucleophile-base can easily abstract a proton from the periphery of the substrate, but nucleophilic attack at carbon is hindered. The effect of steric bulk is illustrated by the following comparisons between methoxide ion, $^{-}OCH_3$, and the bulkier *tert*-butoxide ion, $^{-}OC(CH_3)_3$:

$$CH_3(CH_2)_4CH_2CH_2OSO_2C_6H_5 \xrightarrow[CH_3OH]{NaOCH_3} \underset{\sim 100\%}{CH_3(CH_2)_4CH_2CH_2OCH_3} + \underset{\text{trace}}{CH_3(CH_2)_4{-}CH{=}CH_2}$$

$$CH_3(CH_2)_4CH_2CH_2OSO_2C_6H_5 \xrightarrow[tBuOH]{KOtBu} \underset{78\%}{CH_3(CH_2)_4CH_2CH_2OtBu} + \underset{22\%}{CH_3(CH_2)_4CH{=}CH_2}$$

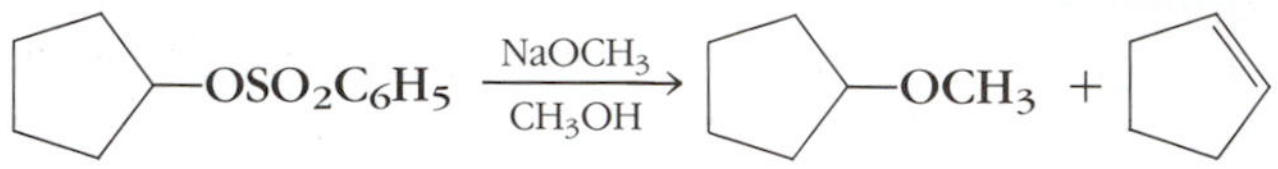

$$\text{cyclopentyl}{-}OSO_2C_6H_5 \xrightarrow[CH_3OH]{NaOCH_3} \underset{59\%}{\text{cyclopentyl}{-}OCH_3} + \underset{41\%}{\text{cyclopentene}}$$

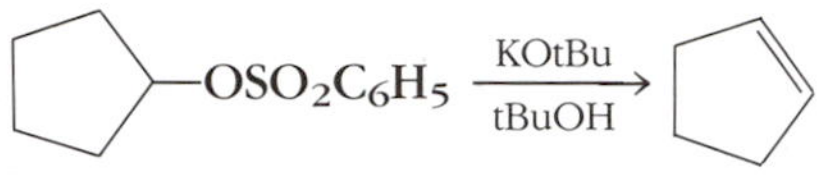

$$\text{cyclopentyl}{-}OSO_2C_6H_5 \xrightarrow[tBuOH]{KOtBu} \underset{100\%}{\text{cyclopentene}}$$

Structure of Substrate

The single most important criterion of reactivity is the degree of substitution at the reactive center. For similar alkyl derivatives *the ratio of elimination to substitution increases progressively from primary to secondary to tertiary alkyl.* You can see this pattern in the preceding examples, and it is clearly illustrated by the following reactions of alkyl bromides with sodium ethoxide:

$$CH_3CH_2Br \xrightarrow[C_2H_5OH]{NaOC_2H_5} \underset{99\%}{CH_3CH_2OC_2H_5} + \underset{1\%}{CH_2{=}CH_2}$$

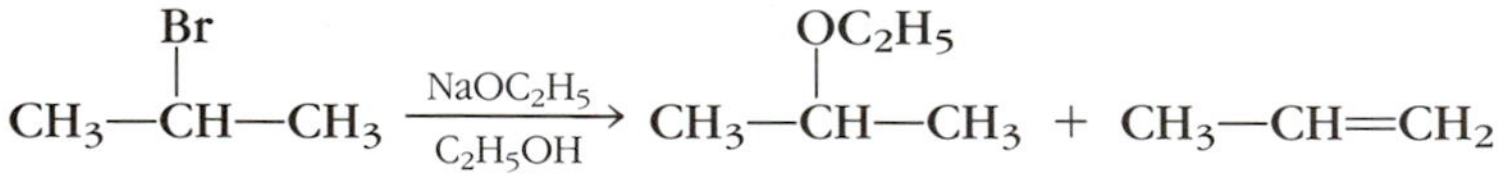

$$CH_3{-}CH(Br){-}CH_3 \xrightarrow[C_2H_5OH]{NaOC_2H_5} \underset{21\%}{CH_3{-}CH(OC_2H_5){-}CH_3} + \underset{79\%}{CH_3{-}CH{=}CH_2}$$

$$CH_3{-}C(CH_3)_2{-}Br \xrightarrow[C_2H_5OH]{NaOC_2H_5} CH_2{=}C(CH_3)_2 \qquad \sim 100\%$$

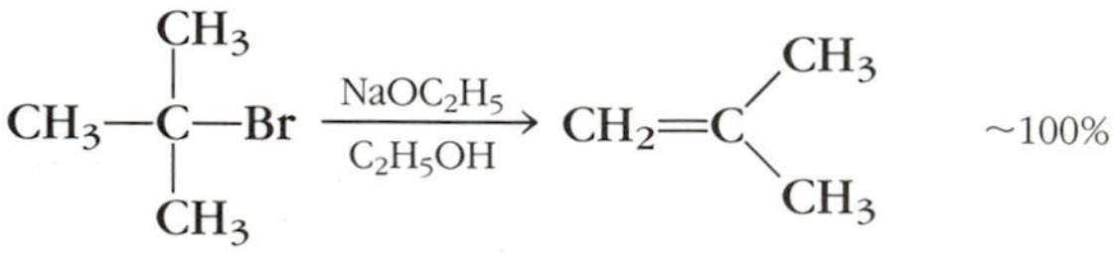

The overall amounts of elimination are less for reaction in a neutral medium, but you can still see the greater tendency for elimination with tertiary substrates. The formation of an alkene via a carbocation intermediate in the solvolysis of a tertiary alkyl halide is analogous to the S_N1 reaction, and elimination by this mechanism has often been described as E1 elimination.

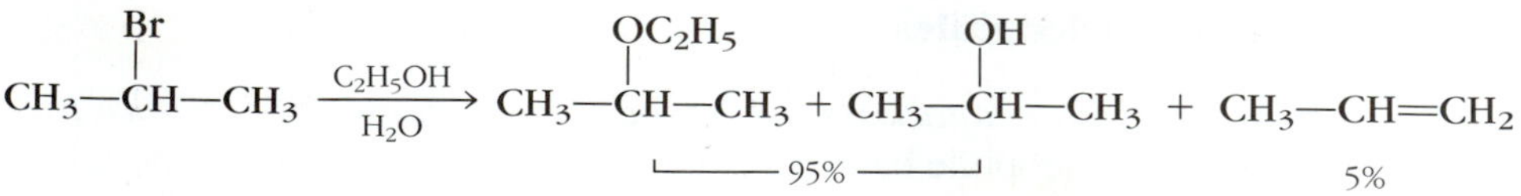

$$CH_3{-}CH(Br){-}CH_3 \xrightarrow[H_2O]{C_2H_5OH} \underbrace{CH_3{-}CH(OC_2H_5){-}CH_3 + CH_3{-}CH(OH){-}CH_3}_{95\%} + \underset{5\%}{CH_3{-}CH{=}CH_2}$$

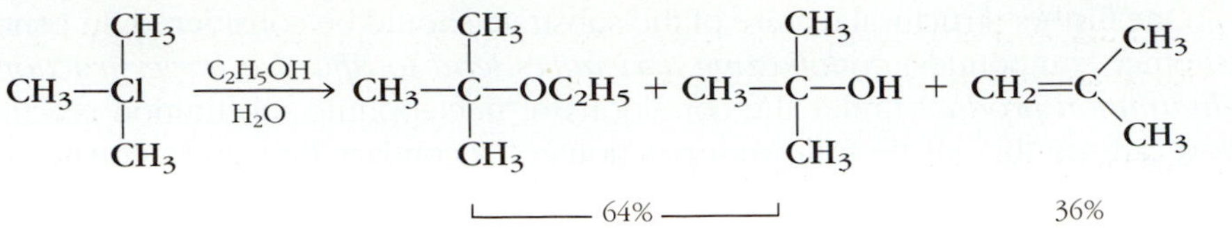

$$(CH_3)_3C{-}Cl \xrightarrow[H_2O]{C_2H_5OH} \underbrace{(CH_3)_3C{-}OC_2H_5 + (CH_3)_3C{-}OH}_{64\%} + \underbrace{CH_2{=}C(CH_3)_2}_{36\%}$$

This behavior can be ascribed to two factors. First, increased steric hindrance makes nucleophilic attack at carbon more difficult, while elimination is largely unaffected because it involves attack at hydrogens that are on the periphery of the molecule. Second, increased substitution at the reactive center allows the formation of a more stable alkene. The tertiary alkyl halide in the last example yields an alkene with a disubstituted double bond, while in the reaction which precedes it a secondary derivative yields an alkene with only one alkyl group on the double bond. These same factors also result in increased elimination for compounds with branching at the β position.

$$CH_3CH_2{-}CH_2Br \xrightarrow[C_2H_5OH]{NaOC_2H_5} \underset{91\%}{CH_3CH_2CH_2OC_2H_5} + \underset{9\%}{CH_3CH{=}CH_2}$$

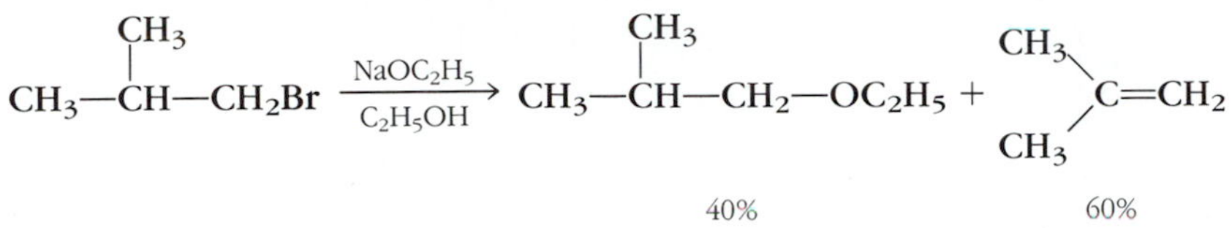

$$CH_3{-}CH(CH_3){-}CH_2Br \xrightarrow[C_2H_5OH]{NaOC_2H_5} \underset{40\%}{CH_3{-}CH(CH_3){-}CH_2{-}OC_2H_5} + \underset{60\%}{(CH_3)_2C{=}CH_2}$$

Other structural features of the substrate can also affect the relative yields of substitution *vs* elimination. In particular, the leaving group can influence the reaction, and *sulfonate esters afford a larger fraction of substitution product than do halides.* The reason for this behavior is not easily explained, but the following examples show that the effect can sometimes be substantial.

$$CH_3{-}(CH_2)_{15}{-}CH_2CH_2Br \xrightarrow{KOtBu} \underset{12\%}{CH_3{-}(CH_2)_{15}{-}CH_2CH_2OtBu} + \underset{88\%}{CH_3{-}(CH_2)_{15}{-}CH{=}CH_2}$$

$$CH_3{-}(CH_2)_{15}{-}CH_2CH_2OTs \xrightarrow{KOtBu} \underset{99\%}{CH_3{-}(CH_2)_{15}{-}CH_2CH_2OtBu} + \underset{1\%}{CH_3{-}(CH_2)_{15}{-}CH{=}CH_2}$$

The differences are usually much smaller, and they are sometimes obscured by other effects. For example, the double bond formed in each of the following reactions is conjugated with a phenyl group, and both the tosylate and the bromide yield alkene as the only product.

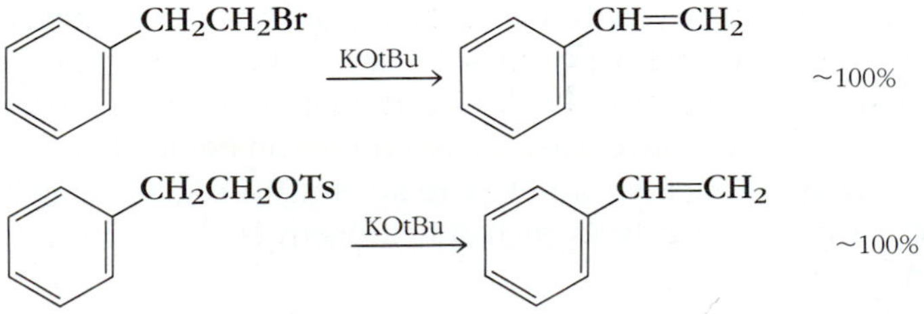

$$C_6H_5{-}CH_2CH_2Br \xrightarrow{KOtBu} C_6H_5{-}CH{=}CH_2 \quad \sim 100\%$$

$$C_6H_5{-}CH_2CH_2OTs \xrightarrow{KOtBu} C_6H_5{-}CH{=}CH_2 \quad \sim 100\%$$

One further structural feature of the substrate should be considered. In contrast to other compounds, *cyclohexane derivatives tend to afford a larger fraction of elimination product* under the conditions of nucleophilic substitution reactions. You can see this in the following reactions of secondary benzenesulfonates with sodium methoxide:

$$CH_3{-}(CH_2)_4{-}CH_2{-}\underset{\displaystyle|}{\overset{OSO_2C_6H_5}{CH}}{-}CH_3 \xrightarrow{NaOCH_3} CH_3{-}(CH_2)_4{-}CH_2{-}\overset{OCH_3}{CH}{-}CH_3 \quad 37\%$$

$$+ \; \underbrace{CH_3{-}(CH_2)_4{-}CH{=}CH{-}CH_3 + CH_3{-}(CH_2)_4{-}CH_2{-}CH{=}CH_2}_{63\%}$$

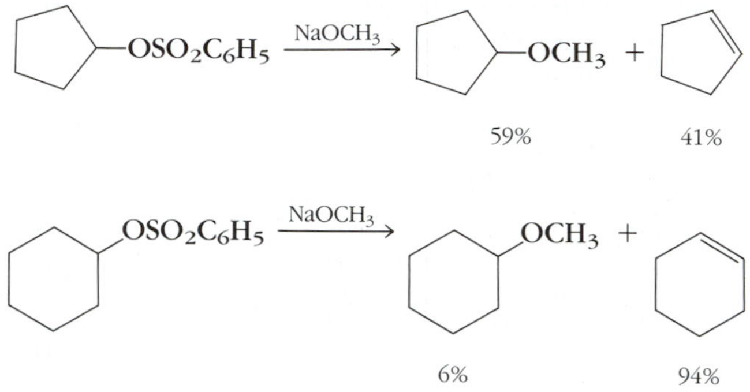

$$\text{cyclopentyl}{-}OSO_2C_6H_5 \xrightarrow{NaOCH_3} \text{cyclopentyl}{-}OCH_3 \; (59\%) + \text{cyclopentene} \; (41\%)$$

$$\text{cyclohexyl}{-}OSO_2C_6H_5 \xrightarrow{NaOCH_3} \text{cyclohexyl}{-}OCH_3 \; (6\%) + \text{cyclohexene} \; (94\%)$$

The enhanced elimination in cyclohexyl systems results from the favorable stability of cyclohexenes and from the ideal *anti* relationship for elimination when the leaving group is in an axial orientation (Section 6.9).

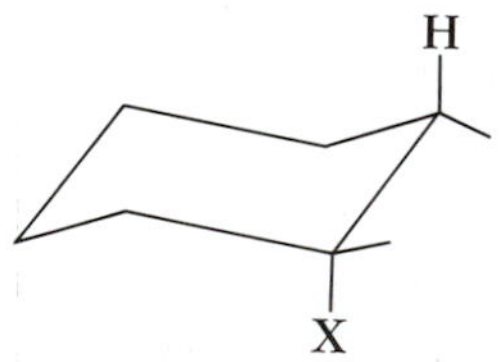

H

X

Nucleophilic Substitution: Prediction of Results

In the preceding discussion we listed criteria that allow you to evaluate the competition between elimination and substitution. We had previously discussed the factors that govern the competition between inversion and retention in the substitution process, so altogether you now have the capability for predicting the results of a reaction between an alkyl halide or sulfonate and a nucleophile-base.

There are many variables to be considered, so it is convenient to begin with the structure of the alkyl derivative. First, all substrates can be divided into two categories, depending upon whether or not there are any hydrogen atoms β to the reactive center. A substrate that lacks such β hydrogens is clearly incapable of direct elimination.

Substrate has β hydrogen	H—C—C—X	Direct elimination possible
Substrate has no β hydrogen	R—C—C—X	Direct elimination *not* possible

Second, all substrates can be divided into categories concerned with the mechanism of nucleophilic substitution.

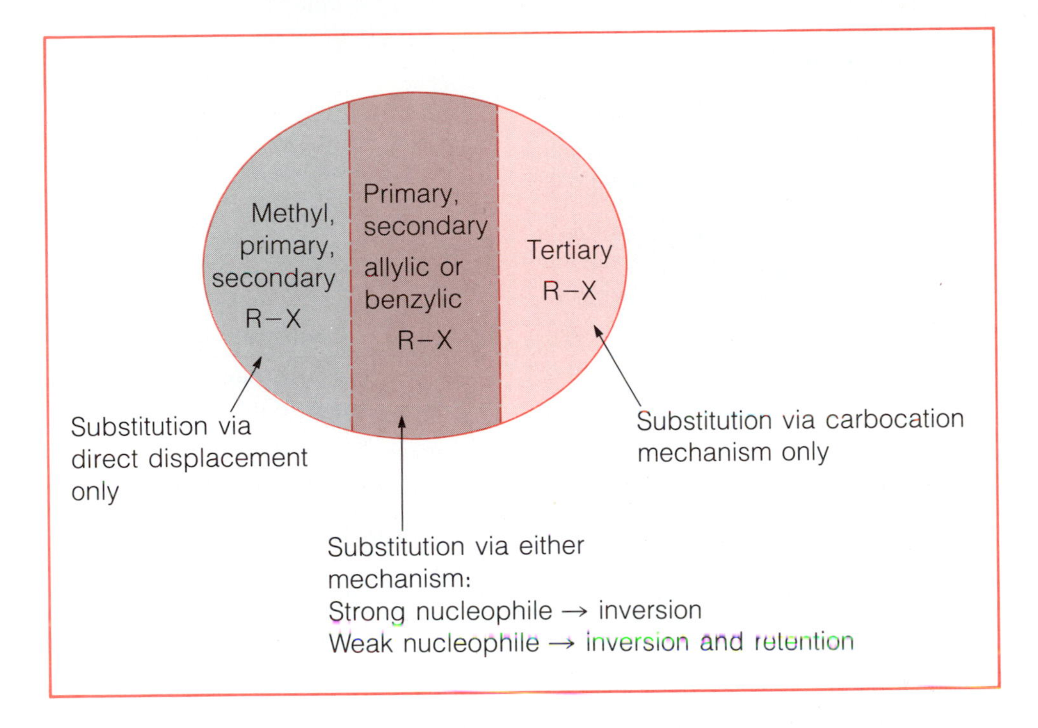

Third, the two preceding ways of categorizing substrates can be combined in order to evaluate other influences on the reaction, as shown in Figure 12.9. In all these descriptions we are considering reactions via ion pairs to fall in the classification of carbocation reactions.

Figure 12.9 is complex, but it summarizes the effects of a large number of variables. Three major categories result from the structure of the alkyl halide or sulfonate:

1. In the absence of any β hydrogens, direct elimination cannot occur, so the substitution reaction is favored. This corresponds to the shaded area of the figure, but it represents a relatively unusual situation. Some compounds of this type will react with rearrangement (Section 12.7).

2. Tertiary substrates with β hydrogens will invariably yield at least some some elimination product. If the nucleophile is also a strong base, elimination results exclusively.

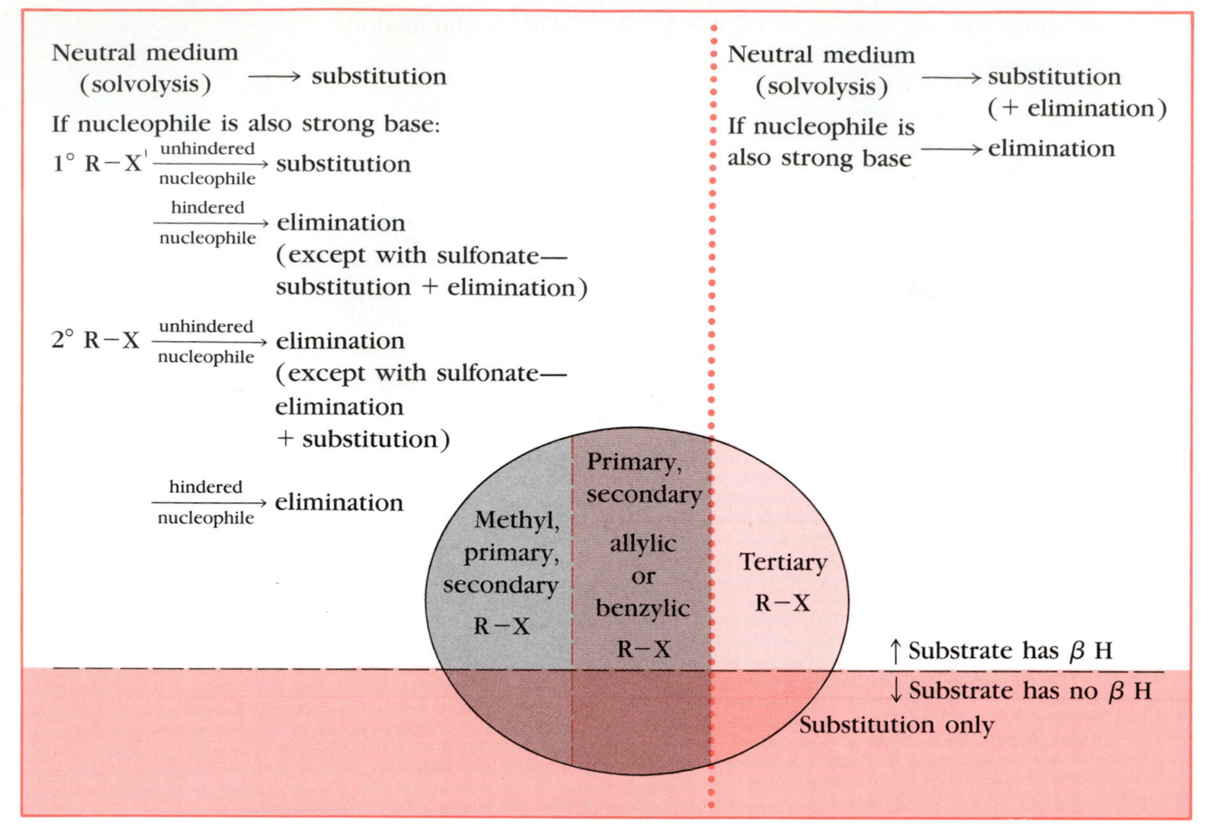

Figure 12.9 Substitution *vs* Elimination for the Reaction of an Alkyl Halide or Sulfonate with a Nucleophile.
None of the compounds in the lower part of the figure can give direct elimination. The remaining compounds are divided into two major classes by the heavy dotted line. Tertiary alkyl derivatives show one pattern of reactivity, while primary and secondary derivatives exhibit a different more complex pattern. The stereochemical consequences of the substitution reactions are described in the preceding discussion in the text. When two products are expected the major product is written first.

3. Methyl, primary, and secondary derivatives having β hydrogens comprise the largest and most complex group of substrates in Figure 12.9. They are all capable of undergoing both substitution and elimination, and the stereochemistry of the substitution reaction must be evaluated as well. If no strong base is present, nucleophilic substitution predominates for these compounds. When the nucleophile is also a strong base, the reactivity pattern must be evaluated in terms of primary *vs* secondary substrate, the steric bulk of the nucleophile, and the type of leaving group (halide vs sulfonate).

The following examples show you that it is indeed possible to predict the major product for a variety of substitution-elimination reactions.

Example 12.1 What are the expected products when 2-bromobutane undergoes solvolysis in aqueous ethanol?

$$CH_3CH_2\text{—}\underset{}{\overset{\displaystyle Br}{\overset{|}{C}H}}\text{—}CH_3 \xrightarrow[H_2O]{C_2H_5OH} ?$$

Solution This reaction of a secondary alkyl bromide falls into the upper left portion of Figure 12.9. There is no strong base, so *substitution* is the expected pathway. Both alcohol (from attack by H_2O) and ether (from attack by ethanol) would be formed. The actual results of this experiment showed the formation of a relatively small amount of alkene (7%) in addition to the substitution products, but our prediction of substitution as the major reaction is accurate.

$$CH_3CH_2\text{—}\overset{\displaystyle OH}{\overset{|}{C}H}\text{—}CH_3 + CH_3CH_2\text{—}\overset{\displaystyle OC_2H_5}{\overset{|}{C}H}\text{—}CH_3 + CH_3CH_2\text{—}CH{=}CH_2 + CH_3\text{—}CH{=}CH\text{—}CH_3$$

93% (substitution products) 7% (alkenes)

Example 12.2 What are the expected products in the reaction of the following bromoalkane with potassium *tert*-butoxide?

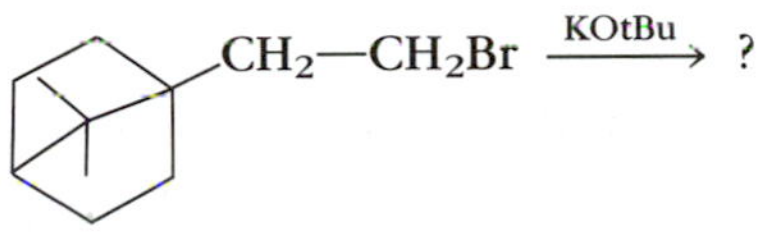

$$\text{(bicyclic)}\text{—}CH_2\text{—}CH_2Br \xrightarrow{KOtBu} ?$$

Solution This is another example which falls into the upper left portion of Figure 12.9. A strong base is present, so elimination and substitution are both likely. The reaction of a primary halide with a *hindered* base such as *tert*-butoxide ion is expected to yield mainly elimination product, and this is observed experimentally.

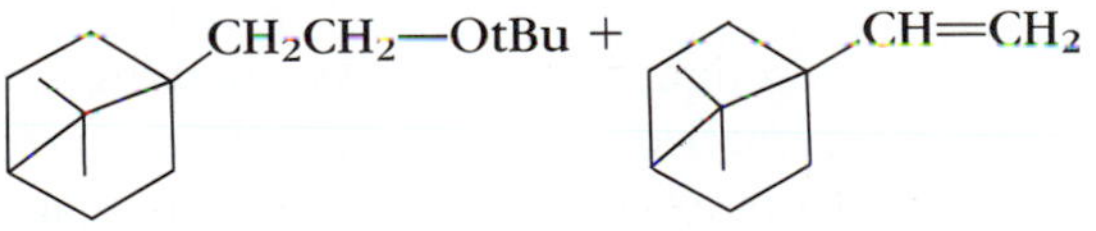

$$\text{(bicyclic)}\text{—}CH_2CH_2\text{—}OtBu + \text{(bicyclic)}\text{—}CH{=}CH_2$$

19% 81%

Example 12.3 What products would be expected when the following secondary benzylic *p*-nitrobenzoate is solvolyzed in aqueous acetone?

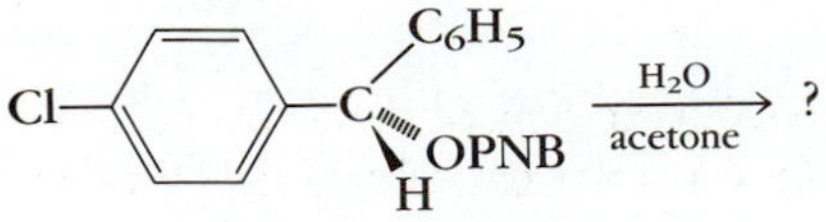

$$Cl\text{—}C_6H_4\text{—}C(C_6H_5)(H)\text{—}OPNB \xrightarrow[acetone]{H_2O} ?$$

Solution This reaction falls into the shaded area of Figure 12.9, because there are no hydrogen atoms β to the leaving group. The substrate could ionize to a resonance-stabilized cation, and the expected product is that of *substitution* with comparable amounts of both retention and inversion. The actual product was found to have a slight excess of retained configuration.

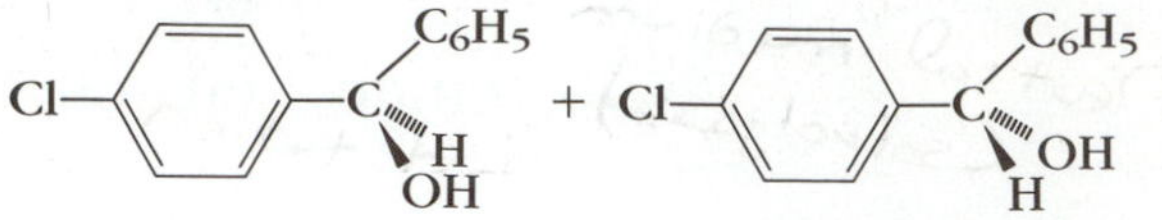

45% Inversion 55% Retention

Example 12.4 What products are expected in the reaction of the following tosylate with sodium methoxide?

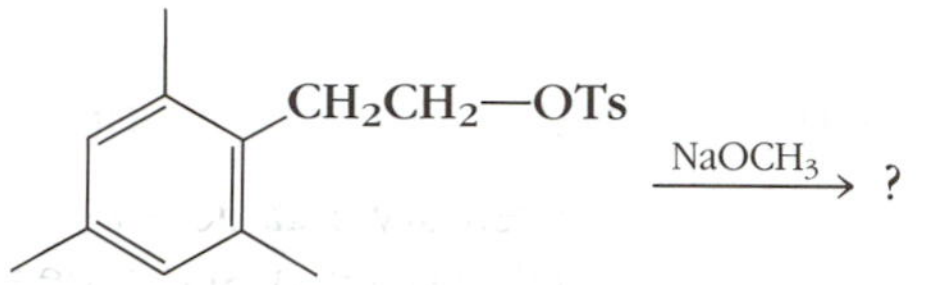

$CH_2CH_2—OTs \xrightarrow{NaOCH_3} ?$

Solution This reaction again falls into the upper left portion of Figure 12.9. A strong base is present, and a primary alkyl derivative could give either substitution or elimination, but the nucleophile is not hindered, so *substitution* should predominate. Indeed, the observed products consisted of the corresponding ether as the major product, with only a small amount of alkene.

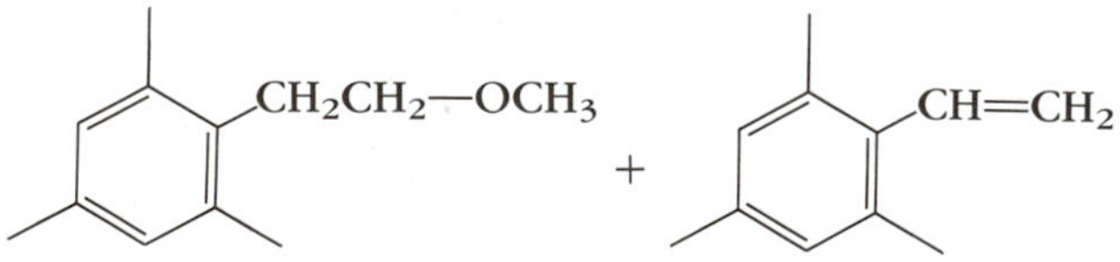

$CH_2CH_2—OCH_3$ + $CH{=}CH_2$

93% 7%

The preceding four examples show that even for fairly complex reactions the trends summarized in Figure 12.9 allow you to make a fairly accurate prediction of the course of a particular reaction. Elimination and substitution are invariably in competition with each other, and you can anticipate which will predominate for a given reaction. In addition, you can accurately suggest whether the substitution product will be formed with inversion of configuration or with a mixture of inversion and retention.

EXERCISE 12.10

(a) What products would be expected for Example 12.1 if the reaction were carried out with $NaOC_2H_5/C_2H_5OH$?
(b) What product (or products) would be expected for Example 12.1 if the leaving group were —OTs?

(c) What product (or products) would be expected for Example 12.2 if the reaction were carried out with $NaOCH_3$ instead of KOtBu?
(d) What product (or products) would be expected for Example 12.2 if the leaving group were —OTs instead of —Br?
(e) What product (or products) would be expected for Example 12.4 if the reaction were carried out with KOtBu instead of $NaOCH_3$?
(f) What product (or products) would be expected for Example 12.4 if the leaving group were —Br instead of —OTs?

12.7 NUCLEOPHILIC SUBSTITUTION: SOME SPECIAL CASES

In the preceding sections we have focused our discussion on the reactions pathways of nucleophilic substitution. In this section we will briefly look at several reactions that are *exceptions* to the typical pathways.

Rearrangements During Nucleophilic Substitution

During the early part of the twentieth century organic chemists began to investigate *reaction mechanisms.* The initial reaction of chemists to the possible intermediacy of carbocations in nucleophilic substitution reactions was highly skeptical, because this revolutionary idea was contrary to the interpretation that bonding in organic compounds was entirely covalent. Slowly, however, experimental data were accumulated in support of carbocation intermediates, and some of the most compelling evidence was provided by skeletal rearrangements of organic compounds. The first suggestions of carbocation intermediates came in 1922 from the work of Wagner and Meerwein in Germany, but only in the 1930s, when F. C. Whitmore (Pennsylvania State University) showed that carbocation intermediates could explain a large number of rearrangements, did these ideas finally achieve widespread acceptance. Rearrangements of carbocations via 1,2 alkyl shifts are frequently described as Wagner-Meerwein rearrangements.

The product of a carbocation rearrangement is usually a tertiary cation (or perhaps an even more stable allylic or benzylic cation). In Section 5.9 we indicated that rearrangement ordinarily generates a more stable cation, so this suggests that the rearrangement would begin with a primary or secondary cation. Yet we argued previously (Section 12.2) that primary and secondary cations are too unstable to be formed in nucleophilic substitution reactions. Considerable evidence suggests that the primary or secondary cation is not formed in these rearrangements, but that *ionization of the leaving group and rearrangement occur simultaneously,* as illustrated in the following equation:

$$R{-}\overset{\displaystyle R}{\underset{\displaystyle R}{C}}{-}CH_2{-}X \longrightarrow R{-}\overset{+}{\underset{\displaystyle R}{C}}{-}CH_2R$$

The most common rearrangements (Section 5.9) involve either a 1,2 hydride shift or a 1,2 alkyl shift to form a tertiary carbocation. In the latter case the carbon atom adjacent to the original reactive center must have been quaternary, as with the neopentyl (2,2-dimethyl-1-propyl) system:

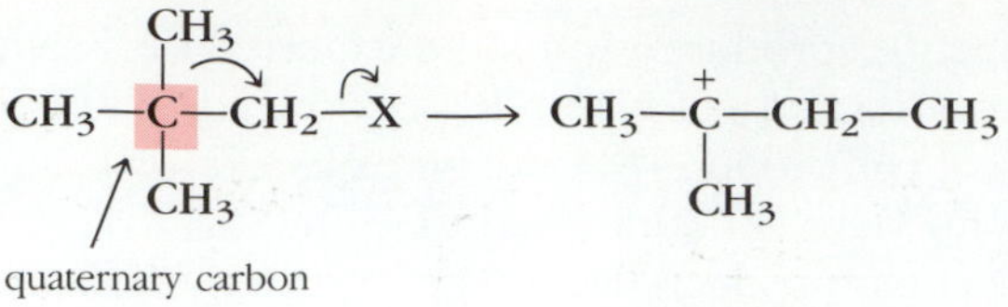

$$CH_3-C(CH_3)_2-CH_2-X \longrightarrow CH_3-\overset{+}{C}(CH_3)-CH_2-CH_3$$

quaternary carbon

The steric bulk of a quaternary carbon atom greatly hinders reaction at an *adjacent* carbon atom, so direct displacement reactions on neopentyl derivatives proceed extremely slowly. Only with a reactive (typically anionic) nucleophile does direct displacement compete with rearrangement. We have illustrated this in the reaction of the following tosylate with sodium cyanide, which affords unrearranged product with complete inversion of configuration.

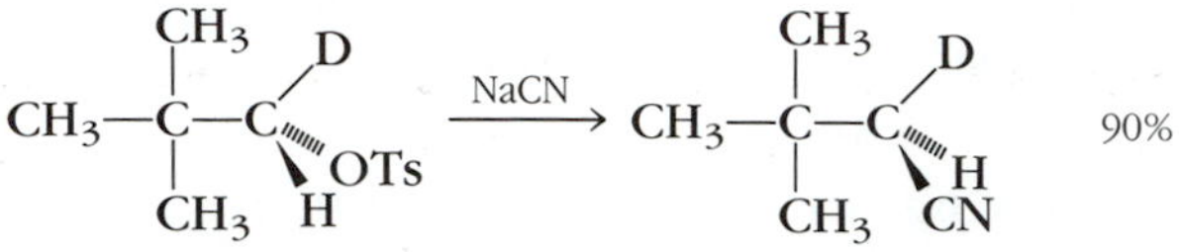

$$CH_3-C(CH_3)_2-CHD(OTs) \xrightarrow{NaCN} CH_3-C(CH_3)_2-CHD(CN) \quad 90\%$$

In contrast, solvolysis of the corresponding bromide yields rearranged products.

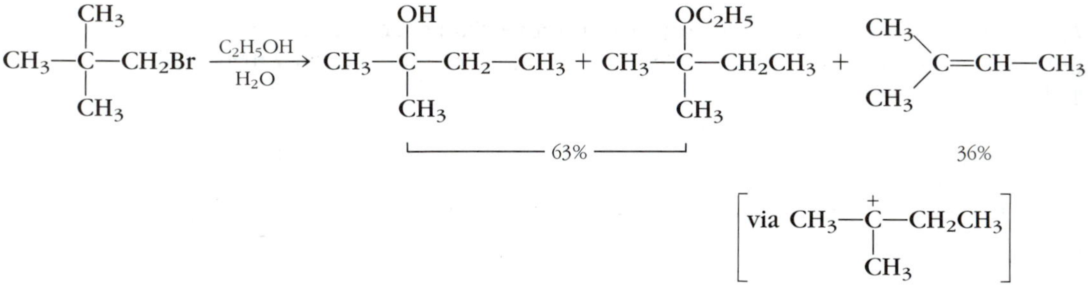

$$CH_3-C(CH_3)_2-CH_2Br \xrightarrow[H_2O]{C_2H_5OH} CH_3-C(OH)(CH_3)-CH_2-CH_3 + CH_3-C(OC_2H_5)(CH_3)-CH_2CH_3 + (CH_3)_2C{=}CH-CH_3$$

63% (alcohol + ether) 36%

$$\left[\text{via } CH_3-\overset{+}{C}(CH_3)-CH_2CH_3\right]$$

The nucleophiles (ethanol and water) are quite weak in the latter reaction, and rearrangement (or elimination) occurs more rapidly than direct displacement.

Rearrangement via hydride shift is illustrated by the following reactions:

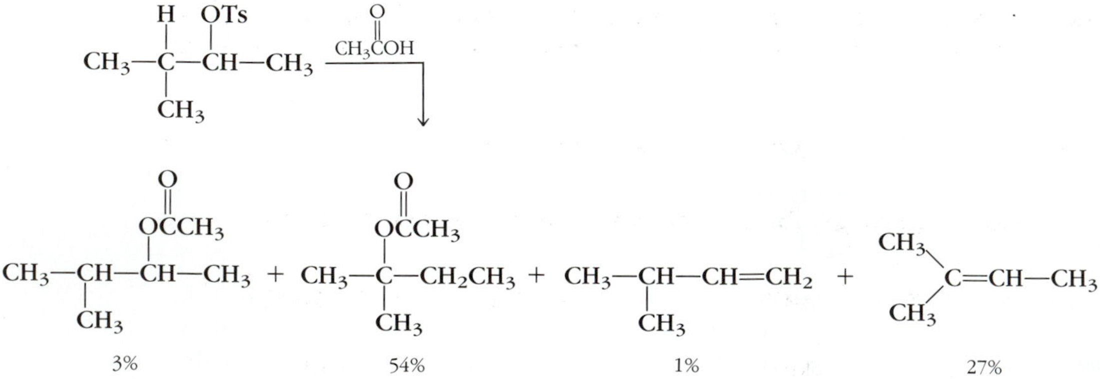

$$CH_3-CH(CH_3)-CH(OTs)-CH_3 \xrightarrow{CH_3COOH}$$

$$CH_3-CH(CH_3)-CH(OCOCH_3)-CH_3 + CH_3-C(OCOCH_3)(CH_3)-CH_2CH_3 + CH_3-CH(CH_3)-CH{=}CH_2 + (CH_3)_2C{=}CH-CH_3$$

3% 54% 1% 27%

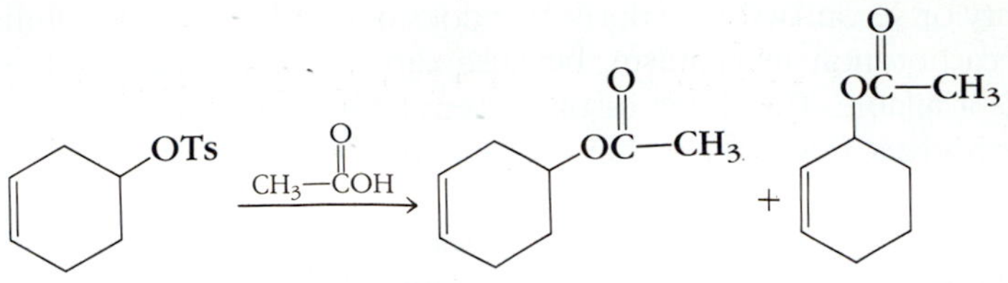

$$\text{cyclohex-2-enyl–OTs} \xrightarrow{CH_3\text{—}\overset{O}{\overset{\|}{C}}OH} \text{cyclohex-2-enyl–}O\overset{O}{\overset{\|}{C}}\text{—}CH_3 + \text{cyclohex-2-enyl–}O\overset{O}{\overset{\|}{C}}\text{—}CH_3$$

Unrearranged, 64% Rearranged, 30%

Neither of the last two examples involves a neopentyl-type substrate, so direct displacement can occur fairly easily, and the products of both direct displacement and rearrangement pathways are observed. Notice that in the second example the rearrangement leads to an *allylic* cation rather than to a tertiary cation.

Allylic Rearrangement

When an allylic derivative undergoes nucleophilic substitution via a carbocation, two products can result if the allylic cation is *unsymmetrical:*

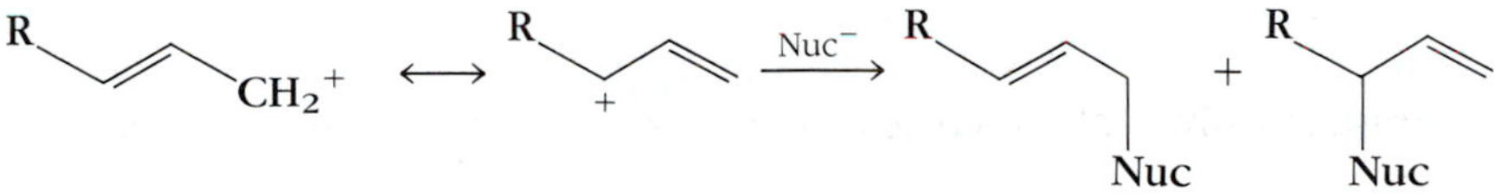

$$R\text{—}CH\text{=}CH\text{—}CH_2^+ \longleftrightarrow R\text{—}\overset{+}{C}H\text{—}CH\text{=}CH_2 \xrightarrow{Nuc^-} R\text{—}CH\text{=}CH\text{—}CH_2\text{—}Nuc + R\text{—}\underset{Nuc}{\underset{|}{C}H}\text{—}CH\text{=}CH_2$$

An allylic cation has *two* carbons that carry positive charge, and each of these can undergo attack by the nucleophile. This is illustrated by the alkaline hydrolysis of 1-chloro-2-butene.

$$CH_3\text{—}CH\text{=}CH\text{—}CH_2Cl \xrightarrow{H_2O} \underset{45\%}{CH_3\text{—}CH\text{=}CH\text{—}CH_2OH} + \underset{55\%}{CH_3\text{—}\overset{OH}{\overset{|}{C}H}\text{—}CH\text{=}CH_2}$$

Vinyl Cations

We stated in Section 12.1 that nucleophilic substitution is primarily restricted to systems in which the reactive center is a *saturated* carbon atom, that is, not part of a carbon-carbon multiple bond. In accord with this generalization, alkenyl halides are highly resistant to direct displacement reactions. On the other hand, alkenyl derivatives can undergo ionization to the corresponding alkenyl cations (commonly called *vinyl cations)* if a sufficiently reactive leaving group is employed.

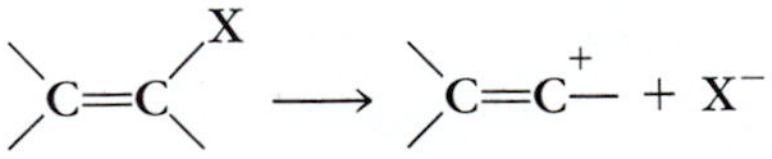

$$\rangle C\text{=}C\langle^{X} \longrightarrow \rangle C\text{=}\overset{+}{C}\text{—} + X^-$$

Vinyl cations are comparable in stability with their saturated counterparts:

$$\underset{1^\circ\ \text{Alkyl}}{R\text{—}CH_2^+} \simeq \underset{1^\circ\ \text{Alkenyl}}{\rangle C\text{=}\overset{+}{C}H} \qquad \underset{2^\circ\ \text{Alkyl}}{R\text{—}\overset{+}{C}H\text{—}\overset{|}{\underset{|}{C}}\text{—}} \simeq \underset{2^\circ\ \text{Alkenyl}}{R\text{—}\overset{+}{C}\text{=}C\langle}$$

A primary or secondary *alkyl* derivative does not undergo nucleophilic substitution by a carbocation mechanism, because direct displacement, a lower-energy pathway, is available. The direct displacement pathway is not of lower energy for alkenyl derivatives, however, so they are less reactive than their saturated counterparts. Only under unusual conditions, when a very good leaving group is used or when the cation is resonance-stabilized (for example, by an aryl group), does nucleophilic substitution occur via a vinyl cation. These possibilities are illustrated by the following two reactions:

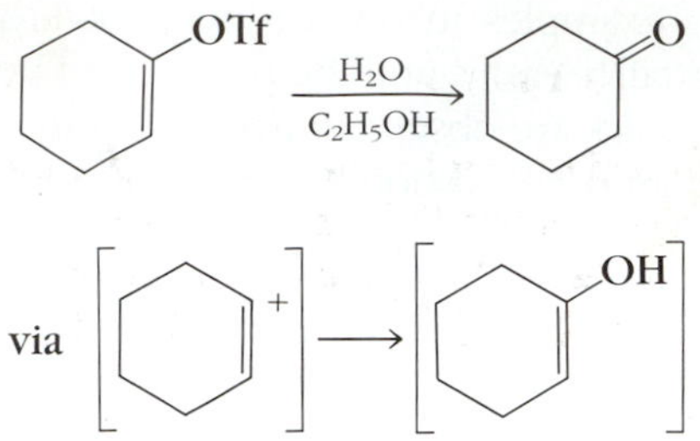

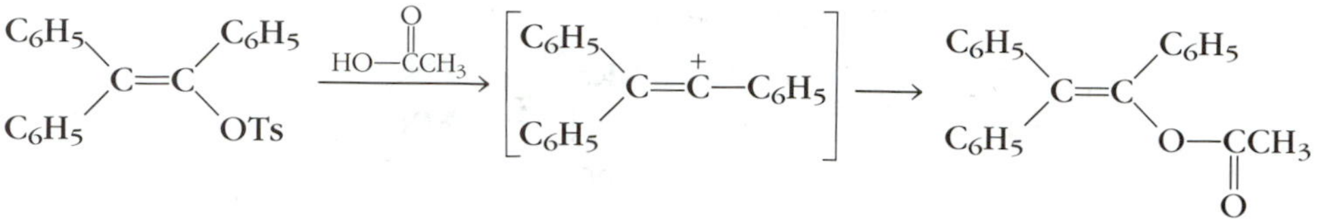

OTf
$\xrightarrow[C_2H_5OH]{H_2O}$
O

via [+] ⟶ [OH]

C_6H_5, C_6H_5, C=C, C_6H_5, OTs $\xrightarrow{HO-\overset{O}{\overset{\|}{C}}CH_3}$ [C_6H_5, C_6H_5, C=$\overset{+}{C}$—C_6H_5] ⟶ C_6H_5, C_6H_5, C=C, C_6H_5, O—$\underset{\underset{O}{\|}}{C}CH_3$

The leaving group in the first of these examples is a triflate, and if you refer back to Table 12.4, you will see that this is the most reactive leaving group available. (You may also wish to refer back to the enol → ketone interconversion that we introduced in Section 7.5 in our discussion of the hydration of alkynes.) The second example shows that a tosylate is an adequate leaving group when the vinyl cation is resonance-stabilized.

EXERCISE 12.11

For each of the reactions presented in this section write a complete mechanism, showing the formation and subsequent reaction of any carbocation that is formed.

EXERCISE 12.12

Draw the product or products expected from each of the following reactions:

(a) 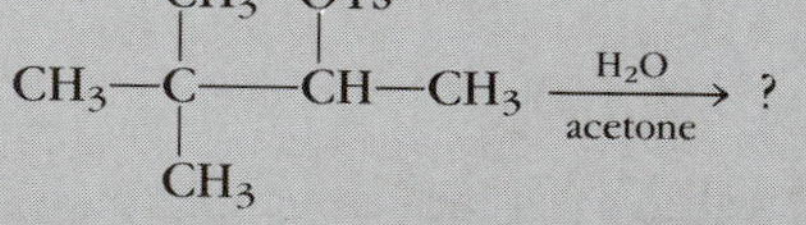 $CH_3-\overset{CH_3}{\underset{CH_3}{C}}-\overset{OTs}{CH}-CH_3 \xrightarrow[acetone]{H_2O}$?

(b) CH3, OTf $\xrightarrow{CH_3CO_2H}$?

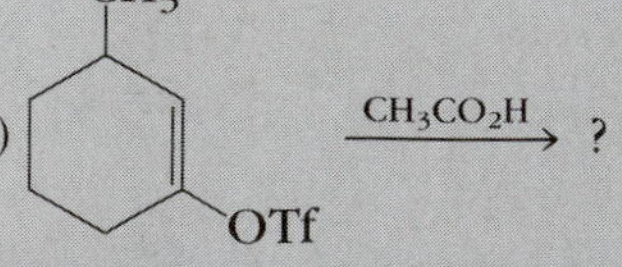

(c) Cl, CH3 $\xrightarrow{CH_3OH}$?

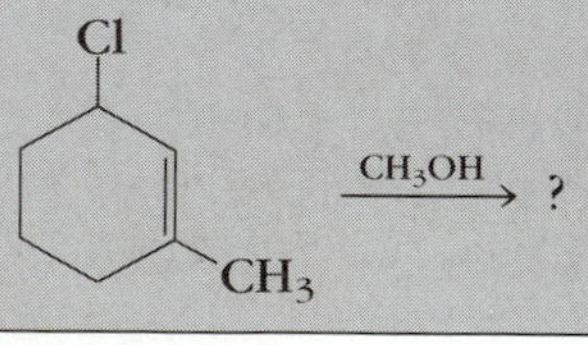

(d) 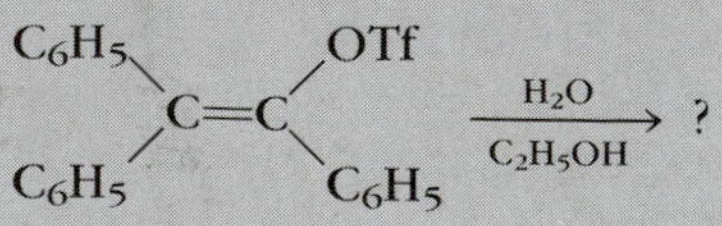 C_6H_5, C_6H_5, C=C, OTf, C_6H_5 $\xrightarrow[C_2H_5OH]{H_2O}$?

12.8 SYNTHETICALLY USEFUL NUCLEOPHILIC SUBSTITUTION REACTIONS

At the beginning of this chapter we stated that many important synthetic reactions involve nucleophilic substitution. In this section we will briefly discuss some of these reactions, with particular emphasis on structural requirements and stereochemical features.

Conversion of Alcohols to Halides

We have previously discussed the replacement of an OH group by a halo substituent (Section 11.9). The reagents used to carry out such transformations (i.e., HBr, HCl, PBr_3, PCl_3, PCl_5, $POCl_3$, $SOCl_2$) all produce an intermediate with a good leaving group. This intermediate undergoes nucleophilic substitution, as shown in the following examples:

$$\text{R—}\ddot{\underset{\cdot\cdot}{\text{O}}}\text{H} \xrightarrow{\text{HBr}} [\text{R—}\overset{+}{\underset{\cdot\cdot}{\text{O}}}\text{H}_2] \xrightarrow{\text{Br}^-} \text{R—Br}$$

$$\text{R—}\ddot{\underset{\cdot\cdot}{\text{O}}}\text{H} \xrightarrow{\text{PBr}_3} [\text{R—}\ddot{\underset{\cdot\cdot}{\text{O}}}\text{—P(Br)}_2] \xrightarrow{\text{Br}^-} \text{RBr}$$

$$\text{R—}\ddot{\underset{\cdot\cdot}{\text{O}}}\text{H} \xrightarrow{\text{SOCl}_2} [\text{R—}\ddot{\underset{\cdot\cdot}{\text{O}}}\text{—}\overset{\text{O}}{\overset{\|}{\ddot{\text{S}}}}\text{—Cl}] \xrightarrow{\text{Cl}^-} \text{R—Cl}$$

When an alcohol reacts with a hydrogen halide, the leaving group is water, generated by protonation of the hydroxyl group. The other reactions actually involve *two* nucleophilic substitutions: first, nucleophilic attack by the hydroxyl group on phosphorus or sulfur, and then attack by a halide ion on the carbon atom. The intermediates in these cases are esters of inorganic phosphorus and sulfur acids, and they resemble tosylates in their reactivity.

As you would expect, elimination can be an important side reaction in these nucleophilic substitution reactions. This can be seen from the product distribution obtained in reactions of 3-pentanol. Simple nucleophilic substitution would afford 3-bromopentane, but substantial amounts of the isomeric 2-bromopentane are also formed.

$$\text{CH}_3\text{CH}_2\text{—}\overset{\text{OH}}{\overset{|}{\text{C}}}\text{H—CH}_2\text{CH}_3 \xrightarrow[\text{H}_2\text{SO}_4]{\text{HBr}} \underset{60\%}{\text{CH}_3\text{CH}_2\text{—}\overset{\text{Br}}{\overset{|}{\text{C}}}\text{H—CH}_2\text{CH}_3} + \underset{40\%}{\text{CH}_3\text{CH}_2\text{CH}_2\text{—}\overset{\text{Br}}{\overset{|}{\text{C}}}\text{H—CH}_3}$$

$$\text{CH}_3\text{CH}_2\text{—}\overset{\text{OH}}{\overset{|}{\text{C}}}\text{H—CH}_2\text{CH}_3 \xrightarrow{\text{PBr}_3} \underset{73\%}{\text{CH}_3\text{CH}_2\overset{\text{Br}}{\overset{|}{\text{C}}}\text{H—CH}_2\text{CH}_3} + \underset{27\%}{\text{CH}_3\text{CH}_2\text{CH}_2\text{—}\overset{\text{Br}}{\overset{|}{\text{C}}}\text{H—CH}_3}$$

The formation of 2-bromopentane results from initial elimination to yield 2-pentene, which then undergoes addition of HBr to yield both 2- and 3-bromopentane.

$$CH_3CH_2\overset{OH}{\overset{|}{C}}H{-}CH_2CH_3 \longrightarrow CH_3CH_2{-}CH{=}CH{-}CH_3 \longrightarrow$$

$$CH_3CH_2\overset{Br}{\overset{|}{C}}H{-}CH_2CH_3 + CH_3CH_2CH_2{-}\overset{Br}{\overset{|}{C}}H{-}CH_3$$

The amount of "rearranged" 2-bromopentane is greater in the reaction with HBr, and for this reason the phosphorus or sulfur reagents are usually better for preparation of halides from alcohols.

The reactions of secondary (and primary) derivatives proceed by the direct displacement mechanism, and mainly inversion of configuration is observed.

$$CH_3CH_2{-}C(CH_3)(OH)(H) \xrightarrow{PBr_3} CH_3CH_2{-}C(CH_3)(H)(Br) + CH_3CH_2{-}C(CH_3)(Br)(H)$$

94% Inversion 6% Retention

$$CH_3CH_2{-}C(CH_3)(OH)(H) \xrightarrow{PCl_3} CH_3CH_2{-}C(CH_3)(H)(Cl)$$

100% Inversion

The formation of some product with *retention* of configuration merits comment. In the cases that have been investigated, the product with retained configuration is not the result of secondary carbocations. Instead, two other pathways have been found. First, *elimination* (an expected side reaction, as always) can be followed by addition of HCl or HBr to the resulting alkene. Addition can occur at either face of the alkene, so both configurations of the halide are produced.

$$>CH{-}C(CH_3)(OH)(H) \xrightarrow{\text{elimination}} >C{=}C(CH_3)(H) \xrightarrow{HX} >CH{-}C(CH_3)(H)(X) + >CH{-}C(CH_3)(X)(H)$$

Racemic

The second way that product can be formed with retention of configuration is by attack of halide ion on the initial product, which is formed with inversion. The result of two successive displacements, each with inversion of configuration, would be overall retention of configuration relative to the original alcohol.

$$R{-}C(CH_3)(OH)(H) \xrightarrow[\text{inversion}]{} R{-}C(CH_3)(H)(X) \xrightarrow[\text{inversion}]{X^-} R{-}C(CH_3)(X)(H)$$

Overall retention

Even in the case of a benzylic secondary derivative (where a carbocation mechanism would be a reasonable possibility), the reaction using a phosphorus reagent proceeds with predominant inversion of configuration.

C_6H_5—C(CH_3)(OH)(H) $\xrightarrow{POCl_3}$ C_6H_5—C(CH_3)(H)(Cl) + C_6H_5—C(CH_3)(Cl)(H)

93% Inversion 7% Retention

Displacement with Tetraalkylammonium Acetate

There are many times, either in a synthetic sequence or in a mechanistic study, that it would be useful to prepare an alcohol with the opposite configuration of the one already in hand. For example, in a study of substitution reactions of cyclobutyl derivatives, both *cis*- and *trans*-3-*tert*-butylcyclobutanol were needed. Reduction of the ketone (via attack of $LiAlH_4$ from the less hindered side of the ring) afforded the *cis* alcohol, which was then converted to the tosylate.

tert-C_4H_9 (cyclobutanone) $\xrightarrow{\text{1. } LiAlH_4 \text{ 2. } H_2O}$ *tert*-C_4H_9 (OH) $\xrightarrow{TsCl}$ *tert*-C_4H_9 (OTs)

Solvolysis of this *cis* tosylate in aqueous medium afforded rearranged (ring-opened) products rather than the desired *trans* alcohol. However, the *trans* alcohol could be formed by direct displacement with tetraethylammonium acetate in acetone.

tert-C_4H_9 (OTs) $\xrightarrow[\text{acetone}]{Et_4N^{+}\,{}^{-}O\text{—}C(=O)\text{—}CH_3}$ *tert*-C_4H_9 (O—C(=O)—CH_3) $\xrightarrow{\text{1. } LiAlH_4 \text{ 2. } H_2O}$ *tert*-C_4H_9 (OH)

Reduction of the resulting *trans* acetate with lithium aluminum hydride (cf. Section 11.4) then yielded *trans*-3-*tert*-butylcyclobutanol.

This reaction of a tosylate with a tetraalkylammonium acetate in acetone is an excellent method for causing inversion of configuration of an alcohol. Acetone is a nonpolar solvent, so carbocation formation (when a possibility) is retarded. Acetate ion is a fairly weak base, so elimination is minimized as well. The use of tetraalkylammonium salts simply increases the solubility of the salt relative to inorganic salts, which are usually quite insoluble in acetone. The following example of this reaction was found to proceed with 95–98% inversion of configuration, and the only other product isolated was a small amount of alkene.

TsO (steroid) $\xrightarrow[\text{acetone}]{(C_4H_9)_4N^{+}\,{}^{-}OAc}$ CH_3—C(=O)—O (steroid)

92%

Alkylation of Terminal Alkynes

The alkylation of the sodium or lithium salts of terminal alkynes is a useful method of carbon-carbon bond formation (Section 7.8). The terminal alkynes are very weak acids (K_a~10^{-25}), so their conjugate bases are strong bases, and elimination is a potentially serious side reaction. Indeed, with secondary (and of course tertiary) halides, elimination is the major pathway. This was the reason that we specified the requirement for a *primary* alkyl halide in Section 7.8. Sulfonate esters can also be used in the alkylation of alkynes, as shown by the following example:

$$C_6H_5{-}C{\equiv}C^-Na^+ \xrightarrow{CH_3CH_2OTs} C_6H_5{-}C{\equiv}C{-}CH_2CH_3 \quad 77\%$$

Reaction of Cyanide Ion with Alkyl Halides or Sulfonates

Another valuable reaction for forming carbon-carbon bonds is displacement with cyanide ion. The resulting cyano compounds can be converted to a variety of other compounds (Chapter 16), so they are valuable synthetic intermediates. HCN is moderately acidic (K_a~10^{-9}), so its conjugate base, cyanide ion, is not a strong base. Consequently, elimination is not as great a problem as it is for acetylide ions, and cyanide displacements can be carried out with both primary and secondary alkyl derivatives.

$$\text{(2,4,6-trimethylphenyl)}{-}CH_2Cl \xrightarrow{NaCN} \text{(2,4,6-trimethylphenyl)}{-}CH_2CN \quad 89\text{–}93\%$$

$$CH_3CH_2{-}CH(CH_3){-}Br \xrightarrow{NaCN} CH_3CH_2{-}CH(CH_3){-}CN \quad 90\%$$

Nucleophilic Attack on Epoxides

In previous chapters we have shown you several reactions that involve ring opening of epoxides. For example, hydrolysis of epoxides affords diols (Section 6.8), reduction with metal hydrides yields alcohols (cf. Section 11.6), and reaction with sodium or lithium acetylides leads to hydroxy alkynes (Section 11.6). When these reactions are viewed as nucleophilic substitutions some of their previously puzzling aspects become quite clear. Thus, nucleophilic attack usually takes place preferentially at the less substituted end of an epoxide because it is *less hindered.*

$$C_6H_5{-}\overset{O}{CH{-}CH_2} \xrightarrow[2.\ H_2O]{1.\ LiAlH_4} C_6H_5{-}CH(OH){-}CH_3 \quad 94\%$$

Similarly, the preference for direct displacement is illustrated by the *backside attack* that must occur in the following example (previously shown without stereochemistry in Section 11.6).

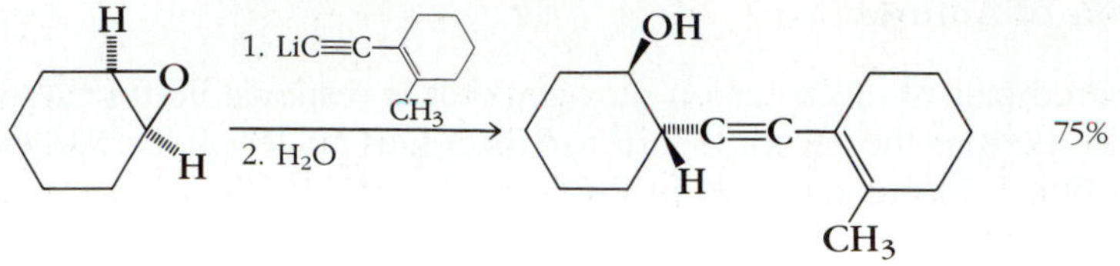

1. LiC≡C–(2-methylcyclohexenyl); 2. H_2O → trans-2-(2-methylcyclohex-1-enylethynyl)cyclohexanol, 75%

Alkylation of OH Groups

We discussed in Section 11.8 the alkylation of hydroxyl groups of alcohols and phenols to form ethers. This reaction can also be used in an *intramolecular* fashion to form epoxides. (Recall from Section 5.8 that halo alcohols are available from the alkenes.)

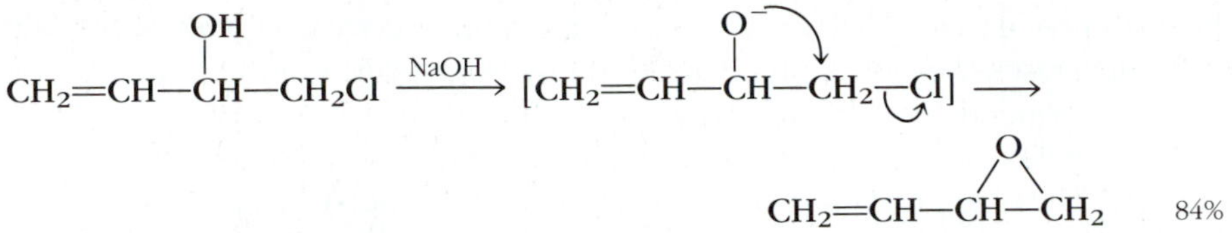

$$CH_2{=}CH{-}CH(OH){-}CH_2Cl \xrightarrow{NaOH} [CH_2{=}CH{-}CH(O^-){-}CH_2{-}Cl] \longrightarrow CH_2{=}CH{-}\underset{\text{epoxide}}{CH{-}CH_2} \quad 84\%$$

The OH group of carboxylic acids can be alkylated via direct displacement reactions to form esters. A useful way to carry out this conversion employs trimethyl- or triethyloxonium salts.

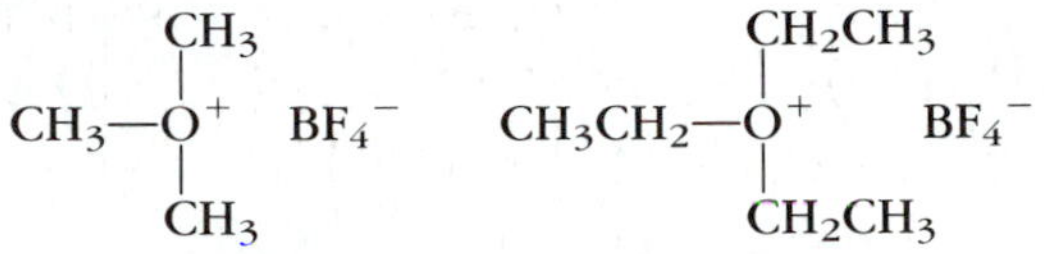

$(CH_3)_3O^+\ BF_4^-$ — Trimethyloxonium fluoroborate

$(CH_3CH_2)_3O^+\ BF_4^-$ — Triethyloxonium fluoroborate

These stable, crystalline salts are soluble in organic solvents and react rapidly with nucleophilic anions. This is illustrated by the following reaction.

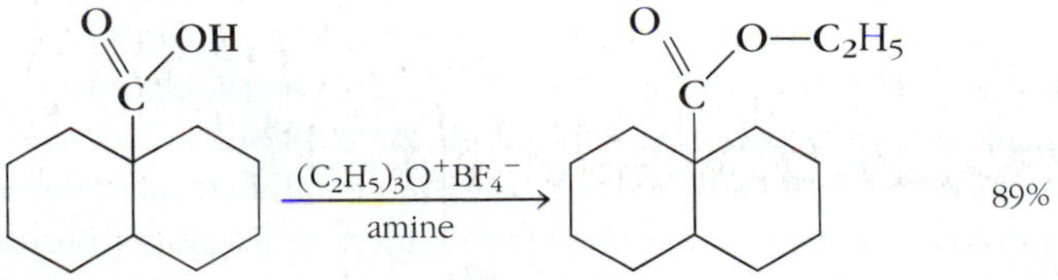

$$\text{decalin-COOH} \xrightarrow[\text{amine}]{(C_2H_5)_3O^+BF_4^-} \text{decalin-C(=O)O{-}}C_2H_5 \quad 89\%$$

The amine acts as a base, converting the carboxylic acid to the more nucleophilic carboxylate ion, which then attacks the oxonium salt.

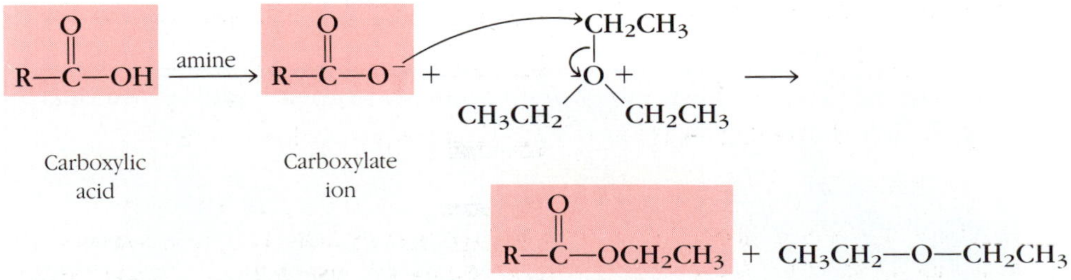

$$\underset{\text{Carboxylic acid}}{R{-}C(=O){-}OH} \xrightarrow{\text{amine}} \underset{\text{Carboxylate ion}}{R{-}C(=O){-}O^-} + (CH_3CH_2)_3O^+ \longrightarrow \underset{\text{Ester}}{R{-}C(=O){-}OCH_2CH_3} + CH_3CH_2{-}O{-}CH_2CH_3$$

Alkylation of Amines

The unshared pair of electrons on nitrogen that is responsible for the basicity of amines also permits these compounds to function as nucleophiles. We will discuss these reactions more extensively in Chapter 17, but the following equation illustrates the alkylation of an amine.

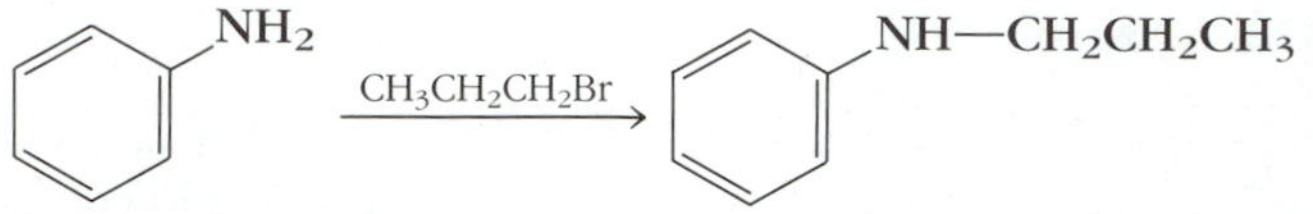

$$C_6H_5NH_2 \xrightarrow{CH_3CH_2CH_2Br} C_6H_5NH{-}CH_2CH_2CH_3 \quad 70\%$$

Alkylation of Active Methylene Compounds

The presence of a carbonyl group or of certain other electron-withdrawing groups greatly increases (Section 4.7) the acidity of hydrogen atoms at the adjacent position. It is frequently possible to *alkylate* the adjacent carbon by forming the conjugate base of the carbonyl compound and allowing it to react with an alkyl halide or sulfonate. These reactions are discussed in detail in Sections 14.4 and 21.6, but the following reaction shows a simple example.

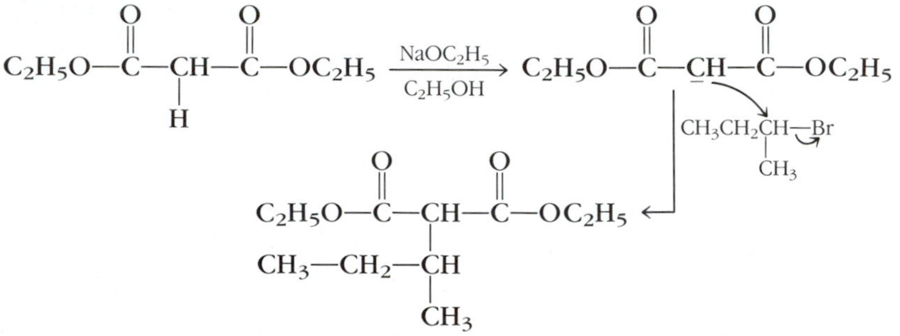

$$C_2H_5O{-}C(=O){-}CH_2{-}C(=O){-}OC_2H_5 \xrightarrow[C_2H_5OH]{NaOC_2H_5} C_2H_5O{-}C(=O){-}\underline{C}H^{-}{-}C(=O){-}OC_2H_5$$

$$\xrightarrow{CH_3CH_2CH(CH_3){-}Br} C_2H_5O{-}C(=O){-}CH(CH(CH_3)CH_2CH_3){-}C(=O){-}OC_2H_5 \quad 83\text{–}84\%$$

As you can see from the examples in this section, nucleophilic substitution reactions play many important roles in synthetic chemistry. In most cases the reactions involve *direct displacement* on an alkyl halide or sulfonate, and this has important consequences. First, these reactions are usually stereospecific, occurring with inversion of configuration. Second, elimination is always a potential problem, so the reactions are restricted to methyl, primary, and secondary alkyl derivatives. When the nucleophile is also a strong base, as in the case of salts of terminal alkynes, even secondary alkyl halides and sulfonates are excluded.

EXERCISE 12.13

Predict the product(s) expected from reaction of 1-bromopropane with each of the following reagents:

(a) $NaC{\equiv}CH$
(b) $Bu_4N^+\ CH_3CO_2^-$/acetone
(c) NaCN

EXERCISE 12.14

Predict the product (or products) expected from reaction of 2-butyl tosylate with each of the reagents in Exercise 12.13.

EXERCISE 12.15

Predict the product (or products) expected from each of the following reactions. Show the stereochemistry in each case.

(a) OTs, *tert*-C_4H_9 $\xrightarrow{Bu_4N^{+}\ ^{-}OC(=O)CH_3}$?

(b) $\xrightarrow{KCN}$?

(c) O $\xrightarrow{LiC{\equiv}CCH_3}$?

(d) $\xrightarrow[2.\ H_2O]{1.\ LiAlD_4}$?

12.9 NUCLEOPHILIC SUBSTITUTION REACTIONS OF BIOLOGICAL IMPORTANCE: BIOSYNTHESIS OF TERPENES AND STEROIDS

The chemical reactions that occur in plants and animals are in many ways very similar to those carried out in the laboratory. In this section we will discuss the biosynthesis of two important classes of compounds, steroids and terpenes, emphasizing the steps that involve nucleophilic substitution.

The term *biosynthesis* refers to the sequence of chemical reactions in a living organism that produces a particular compound, and there are several ways in which a biosynthesis differs from a laboratory synthesis. First, laboratory reactions often use sulfonate leaving groups, which are structurally related to *sulfuric* acid, while the usual leaving group in biological reactions is a derivative of *phosphoric* acid. Second, laboratory reactions are normally carried out in an organic solvent. In contrast, biological systems are inherently aqueous, although frequently an enzyme surrounds the molecules undergoing reaction and provides a "nonaqueous" environment for the reaction. Third, in biological systems an enzyme can bring reactants together in the appropriate orientation, and this can determine the stereochemistry of the reaction as well as enhance its rate.

Steroids comprise a class of compounds that serve several biological functions in animals. For example, the male and female sex hormones are members of this structural class. Steroids are characterized by a basic skeleton, consisting of three six-membered rings and one five-membered ring:

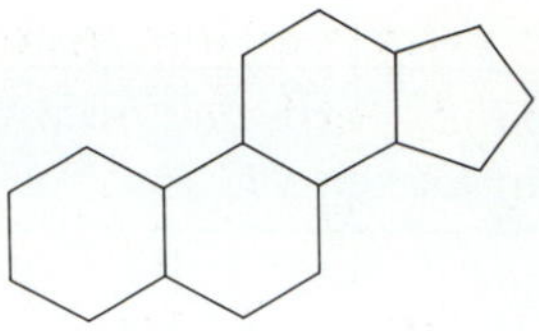

A variety of alkyl and oxygen substituents may also be present, as you will see in detail in Section 23.3.

Terpenes, which have the general molecular formula $(C_5H_8)_n$, comprise a class of hydrocarbons found in plants. Natural rubber (Section 8.5), for example, belongs to this class. A somewhat more general term, *terpenoids,* includes various alcohols, ketones, and other oxygen-containing derivatives of the actual terpenes. The biological function of terpenes in plants is not fully understood, but these compounds are used extensively in the flavor and fragrance industry. Our present discussion will be concerned with their biosynthesis, although we will return to other aspects of terpene chemistry in Section 23.2.

The tremendous interest in steroids for medical purposes beginning in the 1930s resulted in extensive studies on their biosynthesis by chemists in several countries, in particular K. Bloch (Harvard), J. W. Cornforth and G. Popjak (England), and F. Lynen (Germany). Their work led to the detailed description of the biosynthesis of steroids and of terpenes which is described in this section.

The Basic Building Block

The original hypotheses for the biosynthesis of terpenes were based on two observations. First, the carbon framework almost always could be described in terms of 2-methylbutyl subunits.

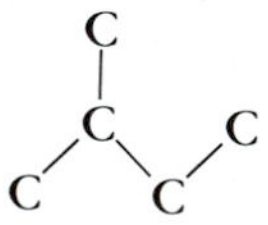

Second, their molecular formulas (which are multiples of C_5H_8) suggested *isoprene* as the basic building block.

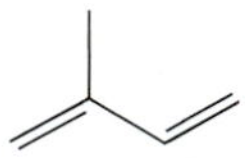

This original assumption was only partially correct, however. The actual substance involved in the biological reactions is *isopentenyl pyrophosphate* (the common name for the pyrophosphate ester of 3-methyl-3-buten-1-ol).

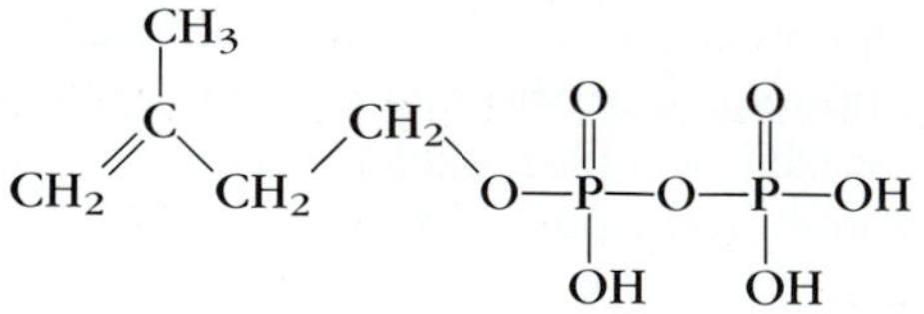

Isopentenyl pyrophosphate

The pyrophosphate group is a very good leaving group because pyrophosphoric acid (like phosphoric acid) is a strong acid.

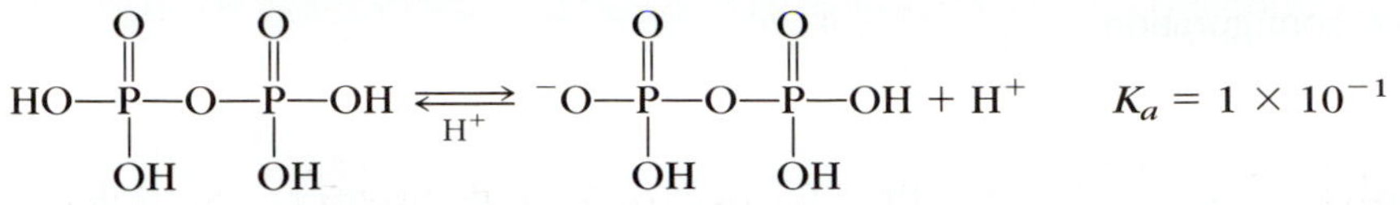

$$\mathrm{HO-P(=O)(OH)-O-P(=O)(OH)-OH} \underset{H^+}{\rightleftharpoons} \mathrm{{}^-O-P(=O)(OH)-O-P(=O)(OH)-OH} + H^+ \qquad K_a = 1 \times 10^{-1}$$

For convenience and clarity we will abbreviate the pyrophosphate group as —OPP. Isopentenyl pyrophosphate can then be drawn as:

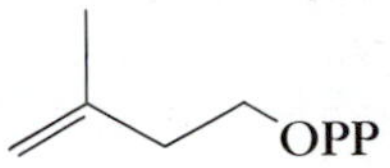

OPP

The first step in the biosynthesis of terpenes and steroids is isomerization of a molecule of isopentenyl pyrophosphate to *dimethylallyl pyrophosphate* (the common name for the pyrophosphate ester of 3-methyl-2-buten-1-ol). The isomerization occurs via protonation of the double bond to produce a tertiary cation followed by removal of a proton from the secondary carbon atom.

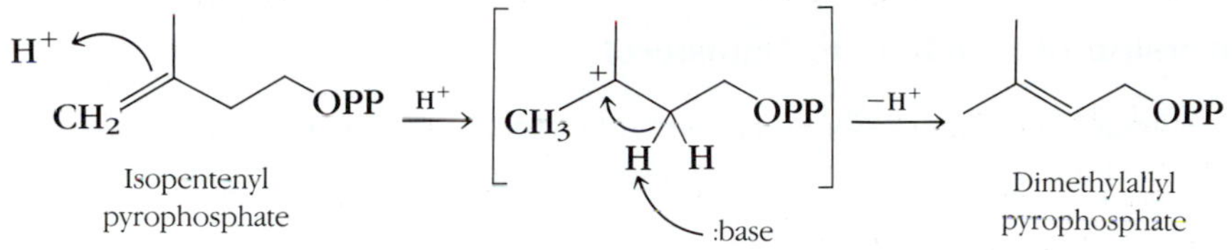

H^+ CH_2 OPP $\xrightarrow{H^+}$ [CH_3 + H H OPP :base] $\xrightarrow{-H^+}$ OPP

Isopentenyl pyrophosphate — Dimethylallyl pyrophosphate

Dimethylallyl pyrophosphate is a primary allylic derivative, which is capable of undergoing nucleophilic substitution either by a carbocation mechanism or by direct displacement. Stereochemical studies using isotopic hydrogen substitution have demonstrated that the biological reactions proceed with complete inversion of configuration. The precise role of the enzyme is not known for certain, but we will draw the reactions as direct displacement reactions.

The nucleophile that attacks dimethylallyl pyrophosphate is the electron pair of the π bond of isopentenyl pyrophosphate. This yields a tertiary carbocation, as shown in the following equation:

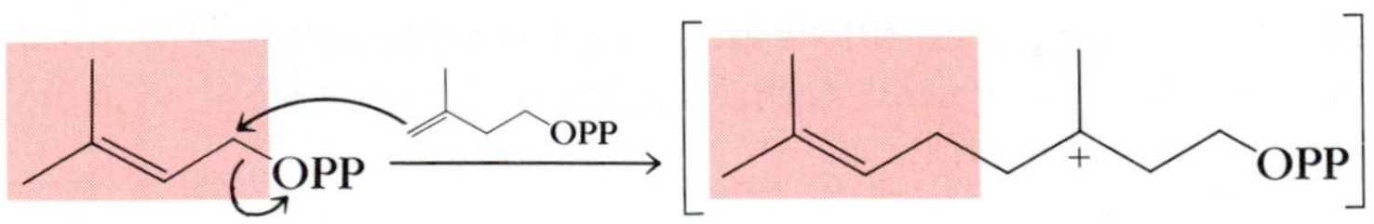

OPP OPP $\longrightarrow$ [+ OPP]

Transfer of a proton from the carbocation to some basic species yields a new allylic pyrophosphate, *geranyl pyrophosphate* (a C_{10} compound).

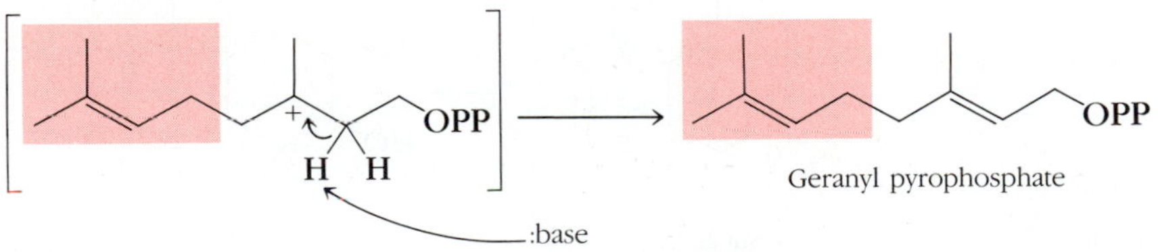

[+ H H OPP :base] $\longrightarrow$ OPP

Geranyl pyrophosphate

This nucleophilic substitution reaction is repeated with another molecule of isopentenyl pyrophosphate to yield *farnesyl pyrophosphate* (a C_{15} compound). As in the case of the geranyl derivative, the new double bond is formed specifically with the *E* configuration.

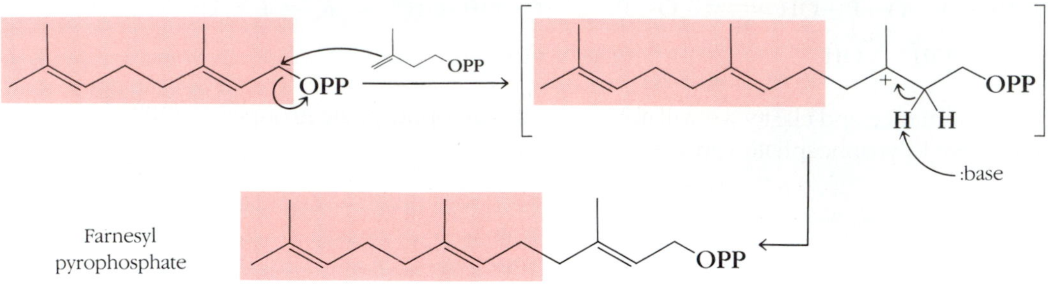

OPP OPP OPP H H :base

Farnesyl pyrophosphate OPP

Finally, two molecules of farnesyl pyrophosphate undergo a coupling reaction to yield squalene, a C_{30} compound that is the direct precursor of steroids and some terpenes. The formation of squalene is not a simple nucleophilic substitution and will not be considered here in detail. We have summarized the overall formation of squalene in Figure 12.10, and you can easily recognize the repeating isopentenyl subunit.

Conversion of Squalene to Lanosterol

The enzymatic conversion of squalene to lanosterol (a precursor to cholesterol) can seem quite surprising unless the structures are drawn appropriately. To emphasize the relationship between the two compounds, we will draw squalene in a different way. First, the terminal isopentenyl group will be abbreviated as R.

Squalene

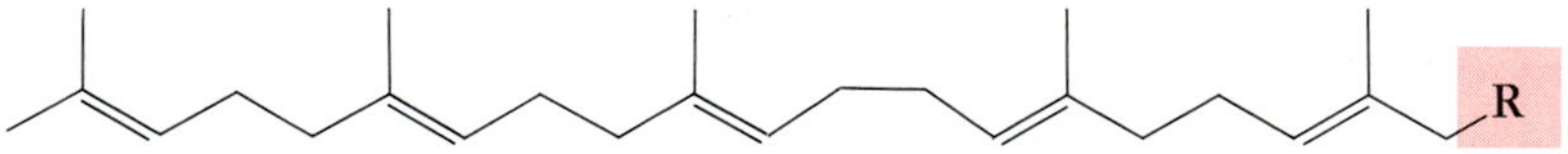

R

R = isopentenyl

Squalene can then be drawn in still another way to emphasize its structural similarity to lanosterol. The actual conversion is initiated by epoxidation of the double bond at the end of the chain.

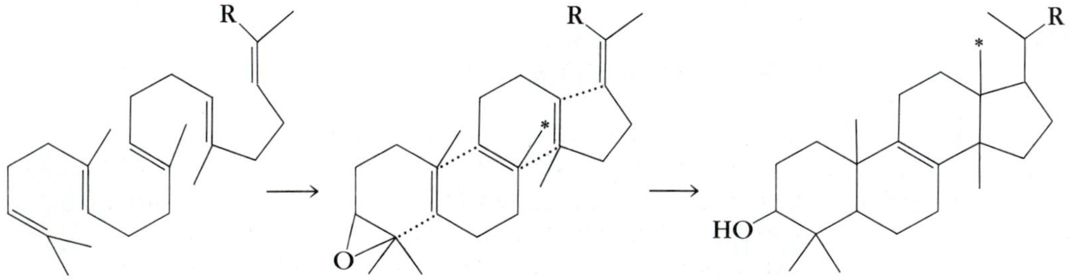

R R R HO O

Squalene → Squalene epoxide → Lanosterol

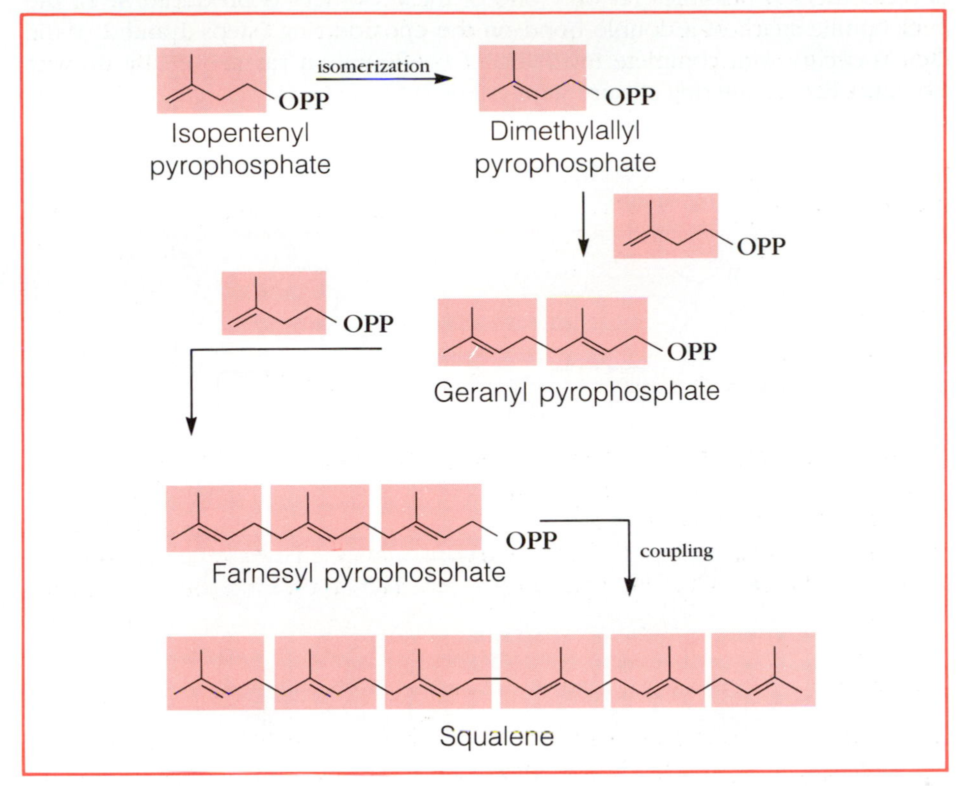

Figure 12.10 Isopentenyl Pyrophosphate as the Basic Building Block for Steroid Biosynthesis.
Isopentenyl pyrophosphate is initially isomerized to dimethylallyl pyrophosphate, and successive alkylations with isopentenyl pyrophosphate lead to geranyl and farnesyl pyrophosphate. Each of the alkylation steps can be viewed as a nucleophilic substitution reaction in which the attacking nucleophile is isopentenyl pyrophosphate. The nucleophile displaces pyrophosphate ion from dimethylallyl and geranyl pyrophosphate, respectively. Finally, two molecules of farnesyl pyrophosphate are coupled to yield squalene, which serves as the precursor of steroids and some terpenes.

These drawings emphasize the similarities of the main carbon chains of squalene and lanosterol. In fact the only difference, denoted by asterisks, is in the absence of a methyl group at one position and the presence of a methyl group at a second position. This difference arises through a series of carbocation rearrangements, which are outlined in Figure 12.11, along with the individual ring closure steps. These reactions show that there are no unusual reactions in the squalene-lanosterol conversion. The remarkable feature of the sequence is the occurrence of so many *normal* reactions in rapid succession.

Although the intermediates in Figure 12.11 are drawn as carbocations, they are certainly not *free* carbocations. Remember that this is an enzymatic process, and interactions with the enzyme (perhaps even formation of a covalent intermediate) may explain the otherwise surprising formation of a secondary cation in step 4 of Figure 12.11. It is not known precisely how the enzyme functions in this series of reactions, but all the steps are highly stereospecific. Lanosterol has seven chiral centers (although we have not shown the stereochemistry here), so a total of 128 stereoisomers is possible. Yet only one of these isomers is produced. Even the nucleophilic attack of a double bond on the epoxide ring (steps 1 and 2 of the figure) occurs with complete inversion of configuration (as proved by isotopic labeling). Each of the ring closure steps (steps 2–5) can be viewed as a nucleophilic attack on a cationic center, with a carbon-carbon double bond acting as a nucleophile.

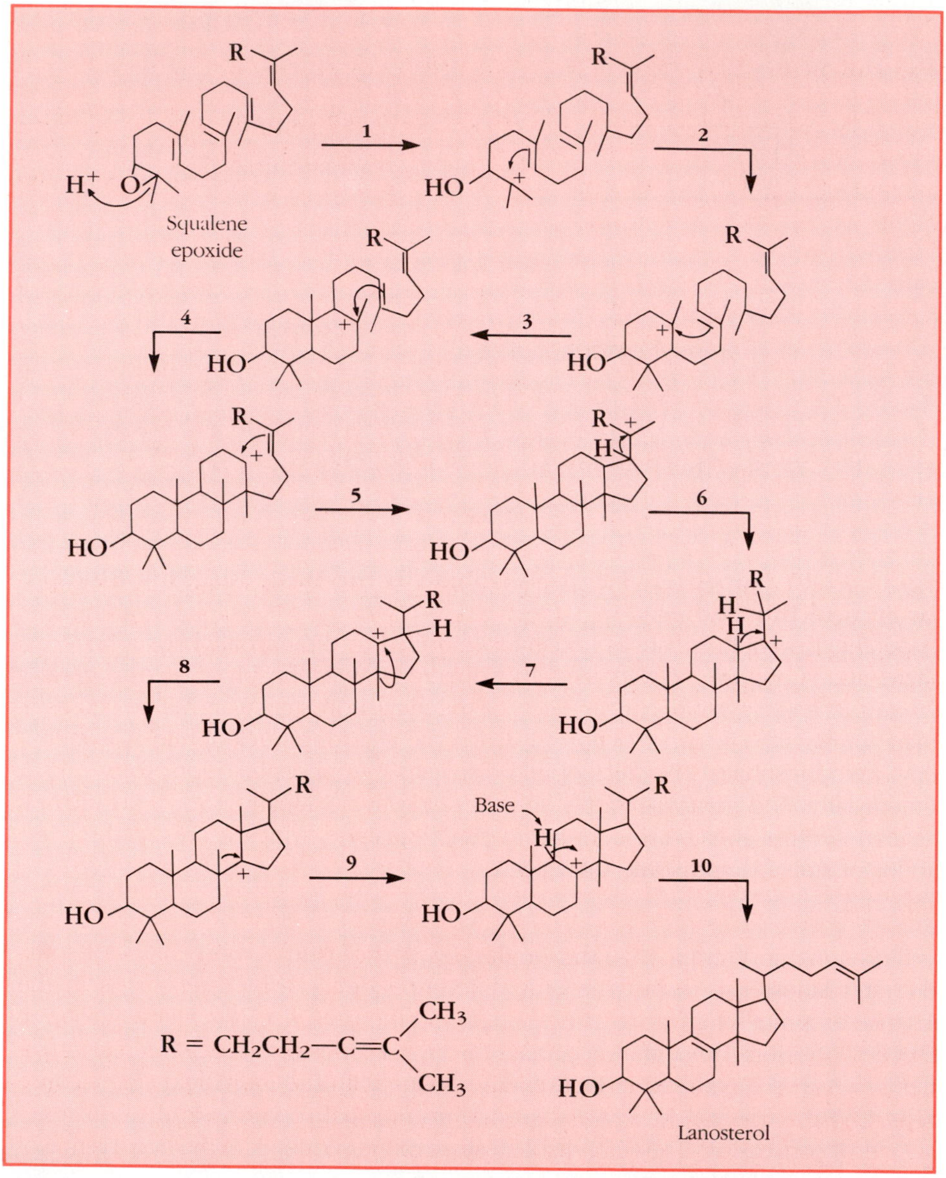

Squalene epoxide

$R = CH_2CH_2—C=C(CH_3)_2$

Lanosterol

Figure 12.11 Biosynthesis of Lanosterol from Squalene Epoxide. Steps 1 and 2 describe the nucleophilic attack by a double bond on an epoxide, while steps 2–5 involve nucleophilic attack by double bonds on carbocations (or ion pairs) to form the appropriate rings. The alkyl and hydride shifts of steps 6–9 yield the correct carbon skeleton, and loss of H^+ (step 10) affords lanosterol.

The rearrangement steps (steps 6–9) all involve tertiary cations, and the driving force for these rearrangements is not well defined. The interaction with the enzyme may cause each successive cation to be more stable, or the hydride and alkyl shifts may simply alleviate some unfavorable steric interactions. Finally, abstraction of a proton by "base" (presumably some group on the enzyme) in step 10 affords lanosterol. Lanosterol is further converted to cholesterol, the precursor of a wide variety of compounds having the basic steroid skeleton, but we will not consider the subsequent stages of steroid biosynthesis in detail.

The Constituents of Turpentine and the Biosynthesis of Terpenes

Turpentine is a familiar household solvent used for cleaning and for thinning paints. It is a by-product of the wood pulp and paper industry, and it is simply a mixture of volatile hydrocarbons from pine trees. Three major components are α-pinene, β-pinene, and limonene.

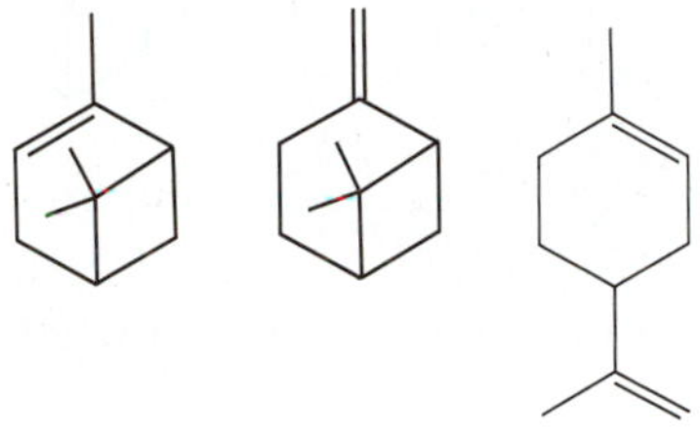

α-Pinene β-Pinene Limonene

These compounds all have the molecular formula $C_{10}H_{16}$, and this corresponds to two isopentenyl subunits. The biosynthesis proceeds via neryl pyrophosphate, which has the *Z* configuration of the double bond (in contrast to geranyl pyrophosphate, which is the *E* isomer).

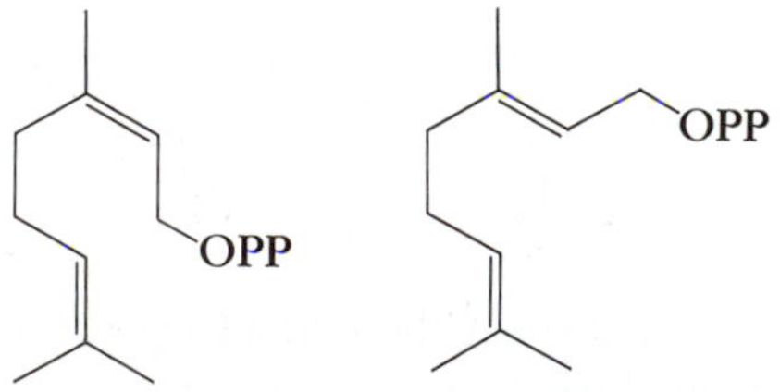

OPP OPP

Neryl pyrophosphate Geranyl pyrophosphate

All three of the major components of turpentine can be formed from neryl pyrophosphate, beginning with an initial nucleophilic displacement by a double bond to form a cyclic cation (again in the presence of an enzyme).

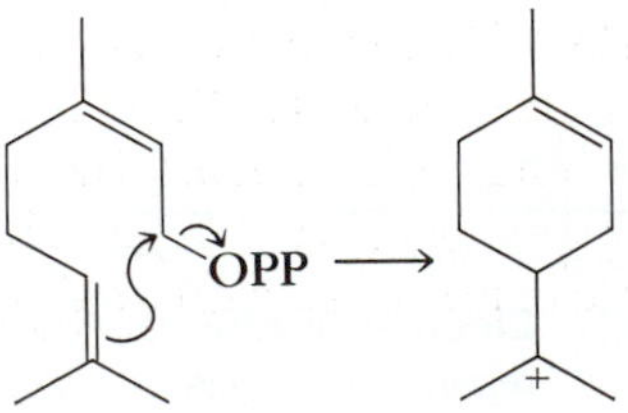

OPP ⟶ +

Abstraction of a proton from the resulting cation would produce limonene directly:

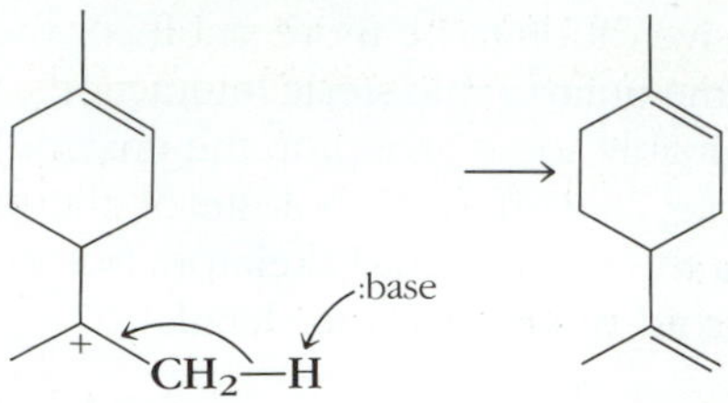

:base
CH_2—H
Limonene

Nucleophilic attack by the remaining double bond would yield the bicyclic carbon skeleton of the pinenes:

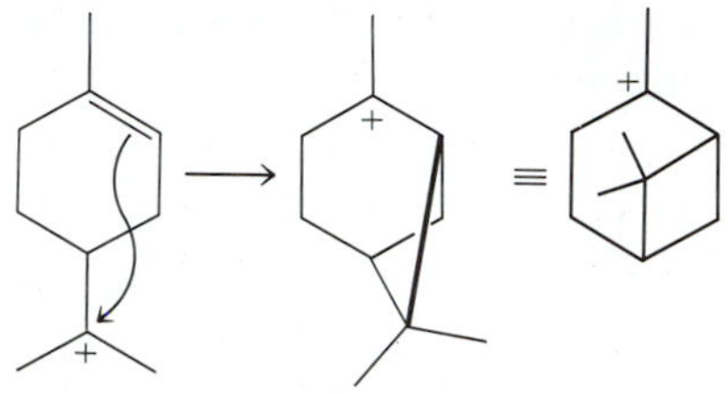

Removal of a proton from this cation could then occur in two ways to yield α-pinene or β-pinene.

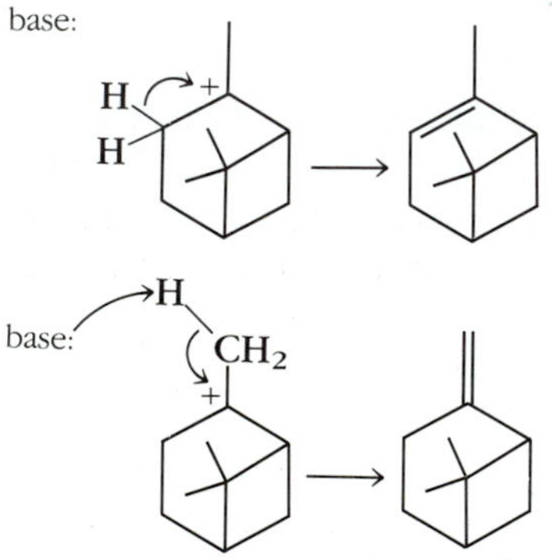

base:
H
H
α-Pinene

base:
H
CH_2
β-Pinene

The influence of the enzyme on these nucleophilic substitution reactions is illustrated by the following observation. Although the starting material (neryl pyrophosphate) is achiral, the products are not only chiral but optically active. The formation of just one of the two possible enantiomers can only be the result of interaction with the chiral enzyme. Curiously, the enantiomer that is obtained depends upon the species of the tree from which the turpentine is isolated. The pinenes obtained from trees in North America are the enantiomers of those obtained from European trees.

12.10 TERMS AND DEFINITIONS

Bimolecular. A term describing a reaction step in which two molecules collide.

Electronic effects. The combination of inductive and resonance effects.

Inversion of configuration. The result of a substitution reaction at a chiral center in which the new group does not occupy the same stereochemical site as the group that it replaces. (If the new and original groups were the same, reaction via inversion of configuration would produce the enantiomer of the reactant.)

Ion pair. A cation and anion that remain in close proximity as a result of electrostatic attraction between the two.

Leaving group. The substituent that is replaced in a nucleophilic substitution reaction. It leaves with its pair of bonding electrons.

Oxonium ion. A species such as the hydronium ion, H_3O^+, having a trivalent, positively charged oxygen atom.

Rate constant. A constant related to the activation energy of a reaction. The rate of a reaction is equal to the product of the rate constant and the appropriate concentration terms for the reactants.

Rate-limiting step. The slow step in a reaction; the step with the greatest activation energy.

Rate of reaction. The rate at which product is formed in a reaction (or the rate at which reactant is consumed). It is expressed as a change in concentration per unit time, typically moles per liter per second ($M\ s^{-1}$).

Solvolysis. A nucleophilic substitution reaction in which the solvent acts as the nucleophile.

Steroids. A biochemically important group of compounds with a ring skeleton having three six-membered rings and a five-membered ring. Examples are cholesterol, cortisone, and other hormones.

Substrate. A molecule that undergoes reaction, e.g., the compound R—X that undergoes nucleophilic substitution to yield a product in which a nucleophile has replaced the leaving group, X.

Terpenes. A class of hydrocarbons found in plants that have structures composed of C_5H_8 subunits.

Unimolecular. A term describing a reaction step that involves only a single molecule of reactant.

12.11 PROBLEMS

12.1 For each of the following pairs of structures, which would react more rapidly with CH_3OTs in a nucleophilic substitution reaction? Why?

(a) CH_3OH $\quad CH_3O^-$ (b) $CH_3—NH^-$ $\quad NH_3$ (c) H_2O $\quad NH_2^-$
(d) H_2O $\quad HI$ (e) Br^- $\quad CF_3CH_2OH$

12.2 For each of the following pairs of structures, which would undergo nucleophilic substitution more rapidly with a nucleophile such as CH_3OH? Why?

(a) $C_6H_5—CH_2OTs$ $\quad C_6H_5—CH_2OH$
(b) $CH_3—OSO_2C_6H_5$ $\quad CH_3—OSO_2CF_3$
(c) $CH_3CH_2—\overset{+}{O}H_2$ $\quad CH_3CH_2—NH_2$
(d) CH_3Br $\quad CH_3OH$
(e) $(CH_3)_2CH—CH_2—\overset{+}{S}(CH_3)_2$ $\quad (CH_3)_2CH—CH_2—OCH_3$
(f) $CH_2{=}CH—CH_2—OPNB$ $\quad CH_2{=}CH—CH_2—OBs$
(g) $CH_3—Cl$ $\quad CH_3—OSO_2C_6H_5$
(h) $CH_3—CH_2—CH(C_6H_5)—OPNB$ $\quad CH_3—CH_2—CH(C_6H_5)—Br$

12.3 For the solvolysis reaction

$$CH_3CH_2Br \xrightarrow{CH_3—\overset{O}{\overset{\|}{C}}—OH}$$

would the rate *increase* or *decrease* for each of the following changes? Why?

(a) Replace CH_3CH_2Br by $CH_3{-}C(CH_3)_2{-}CH_2Br$

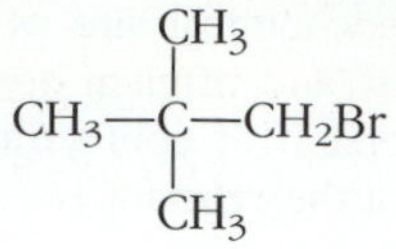

(b) Replace CH_3CH_2Br by CH_3CH_2OTs

(c) Replace $CH_3{-}\overset{O}{\overset{\|}{C}}{-}OH$ by H_2O

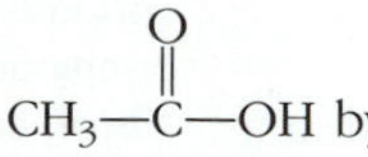

(d) Add $NaO{-}\overset{O}{\overset{\|}{C}}{-}CH_3$ to the mixture

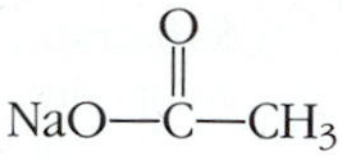

12.4 What product would be expected for the original reaction in Problem 12.3? What products would be expected for each set of conditions in (a)–(e)?

12.5 Write a detailed mechanism for the original reaction in Problem 12.3 and for each of the variations in (a)–(e).

12.6 For the solvolysis reaction

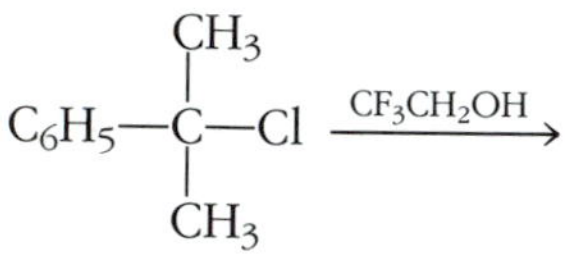

$$C_6H_5{-}C(CH_3)_2{-}Cl \xrightarrow{CF_3CH_2OH}$$

would the rate *increase* or *decrease* for each of the following changes? Why?

(a) Replace CF_3CH_2OH by $CH_3\overset{O}{\overset{\|}{C}}{-}OH$

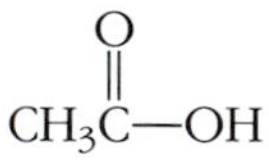

(b) Replace CF_3CH_2OH by $(CH_3)_3C{-}OH$

(c) Replace CF_3CH_2OH by CH_3CH_2OH

(d) Replace $C_6H_5{-}$ by $CH_3CH_2{-}$

(e) Replace —Cl by —OPNB

(f) Add CF_3CH_2ONa to the reaction mixture

12.7 What product would be expected for the original reaction in Problem 12.6? What products would be expected for each set of conditions in (a)–(f)?

12.8 Write a detailed mechanism for the original reaction in Problem 12.6 and for each of the variations in (a)–(f).

12.9 For each of the following pairs of compounds, which would undergo solvolysis more rapidly in CF_3CH_2OH? Why?

(a) 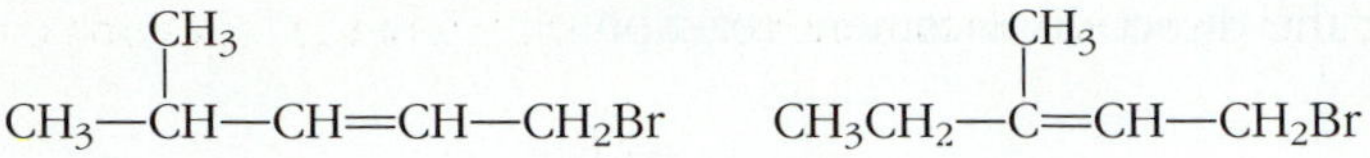 $CH_3—CH(CH_3)—CH=CH—CH_2Br$ $\quad$ $CH_3CH_2—C(CH_3)=CH—CH_2Br$

(b) $C_6H_5—CH_2OTs$ $\quad$ $C_6H_5—CH_2CH_2OTs$

(c) 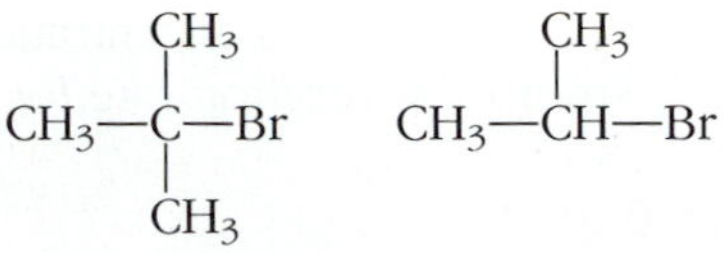 $CH_3—C(CH_3)_2—Br$ $\quad$ $CH_3—CH(CH_3)—Br$

(d) 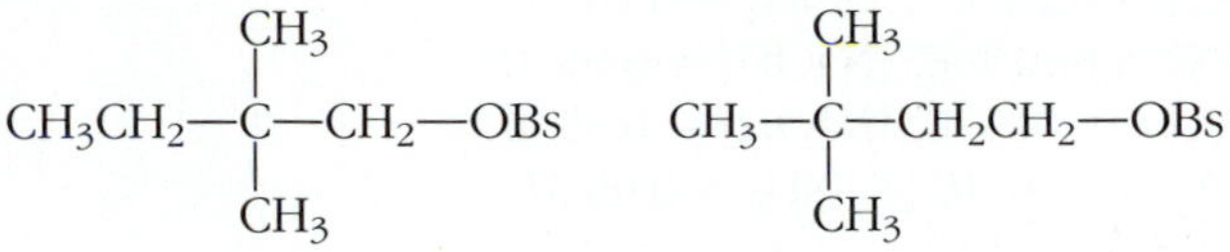 $CH_3CH_2—C(CH_3)_2—CH_2—OBs$ $\quad$ $CH_3—C(CH_3)_2—CH_2CH_2—OBs$

(e) 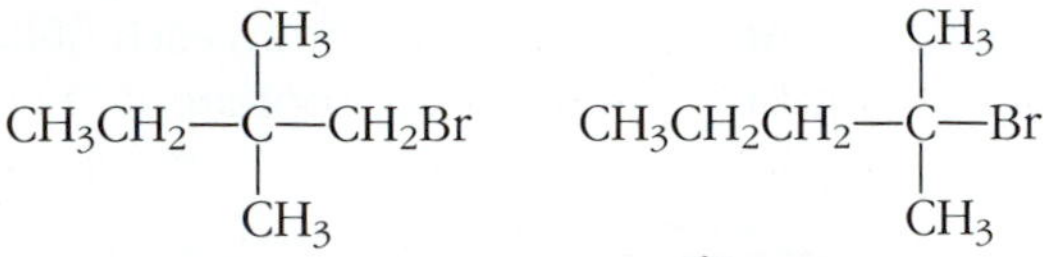 $CH_3CH_2—C(CH_3)_2—CH_2Br$ $\quad$ $CH_3CH_2CH_2—C(CH_3)_2—Br$

(f) 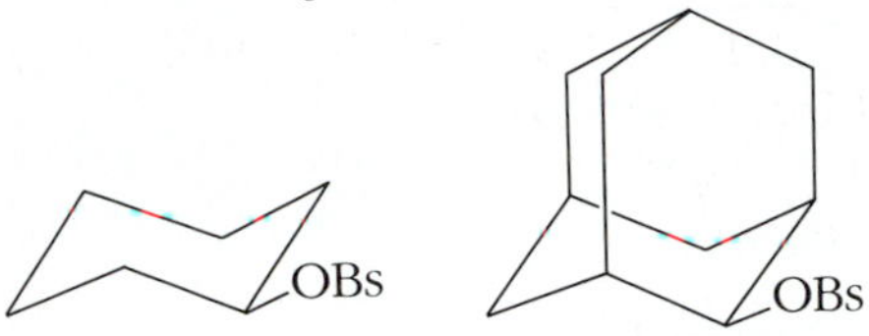 OBs $\quad$ OBs

12.10 Consider the data in the following table:

Rates of Reaction of R—X with Sodium Azide

[R—X], *M*	$[NaN_3]$, *M*	Reaction Rate, $M\ s^{-1}$
0.16	0.16	2.0×10^3
0.48	0.16	5.9×10^3
0.16	0.32	2.1×10^3
0.48	0.32	6.2×10^3

(a) Are these data more consistent for a direct displacement or a carbocation mechanism?

(b) Predict the rate if the concentrations were [R—X] = 0.80 *M* and $[NaN_3]$ = 0.16 *M*.

(c) Predict the rate if the concentrations were [R—X] = 0.16 *M* and $[NaN_3]$ = 0.68 *M*.

(d) Calculate the rate constant for each entry in the table. Is it constant, or does it vary with the reaction conditions?

12.11 For the direct displacement reaction

$$R{-}X \xrightarrow{NaOH} R{-}OH$$

when the RX concentration is 0.015 M and the OH^- concentration is 0.20 M, the rate constant is 0.033 $M^{-1}\ s^{-1}$. Predict the reaction rate for each of the following sets of conditions:
(a) [R—X] = 0.030 M, [NaOH] = 0.20 M
(b) [R—X] = 0.15 M, [NaOH] = 0.20 M
(c) [R—X] = 0.015 M, [NaOH] = 0.05 M
(d) [R—X] = 0.015 M, [NaOH] = 0.50 M
(e) [R—X] = 0.40 M, [NaOH] = 0.06 M

12.12 Draw the major product (or products) expected from each of the following reactions. Show stereochemistry whenever appropriate.

(a) 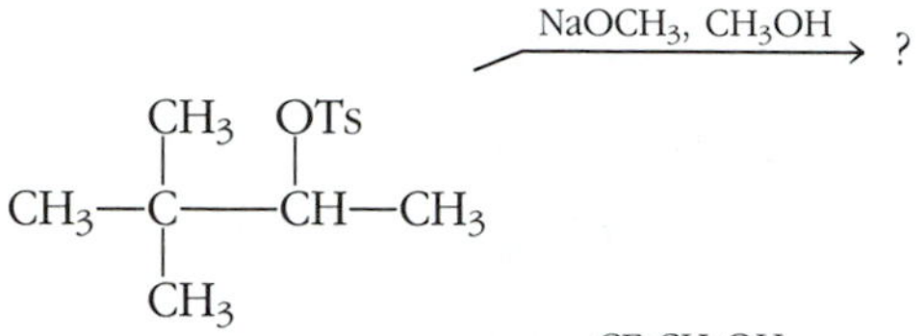 $CH_3{-}C(CH_3)_2{-}CH(OTs){-}CH_3 \xrightarrow{NaOCH_3,\ CH_3OH}$?

(b) [structure] $\xrightarrow{CF_3CH_2OH}$?

(c) [structure: cyclopentene ring with Br (wedge) and CH_3] $\xrightarrow{CF_3CH_2OH}$?

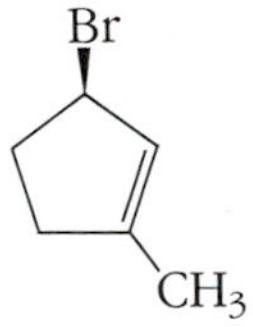

(d) 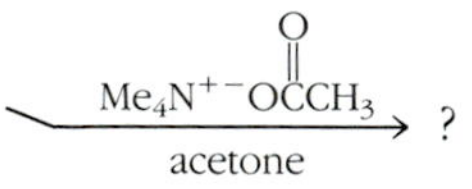 [structure] $\xrightarrow[\text{acetone}]{Me_4N^{+}\ {}^{-}OCCH_3\ (C{=}O)}$?

(e)  [structure: cyclopentane with OTs (hashed) and $(CH_3)_3C$ (wedge)] $\xrightarrow{CH_3OH}$?

(f) 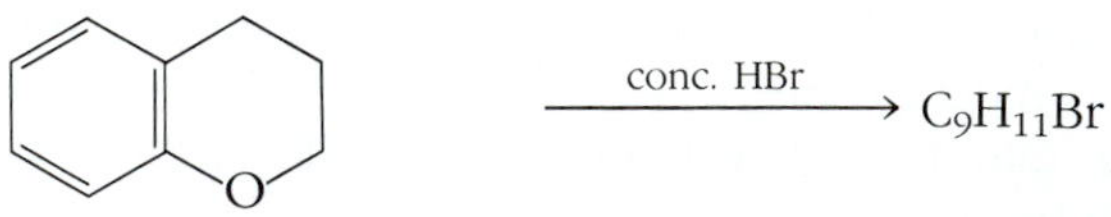 [structure: chromane (benzene fused to ring containing O)] $\xrightarrow{\text{conc. HBr}}$ $C_9H_{11}Br$

12.13 Draw the major product (or products) expected from each of the following reactions. Show stereochemistry whenever appropriate.

(a)  $\xrightarrow{CF_3CH_2OH}$?

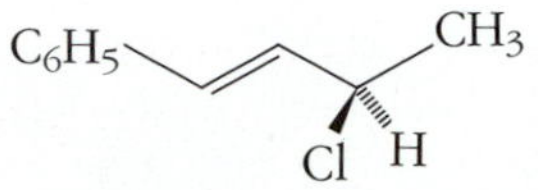

C_6H_5 CH_3 Cl H

(b)  $\xrightarrow{NaOEt}$?

(c) $\xrightarrow{CH_3OH}$?

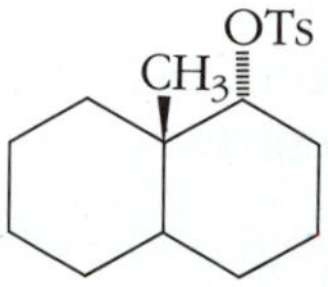

OTs CH_3

(d)  $\xrightarrow{KOtBu}$?

(e)  $\xrightarrow{KOtBu}$?

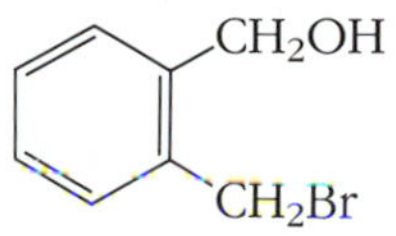

CH_2OH CH_2Br

(f)  $\xrightarrow{CH_3OH}$?

12.14 Draw the major product (or products) expected from each of the following reactions. Show stereochemistry whenever appropriate.

(a) 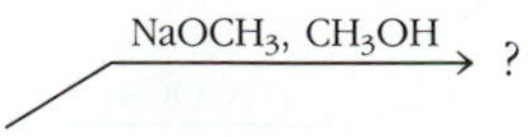 $\xrightarrow{NaOCH_3,\ CH_3OH}$?

(b) 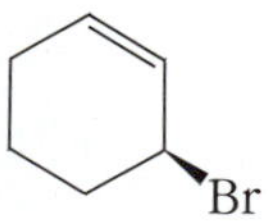 $\xrightarrow{CF_3CH_2OH}$?

Br

(c) 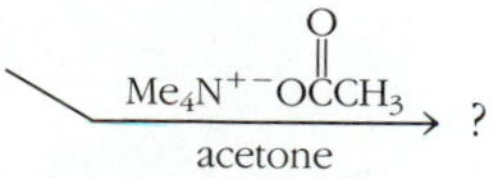 $\xrightarrow[\text{acetone}]{Me_4N^{+}\ {}^{-}OCCH_3}$?

O

(d)  $\xrightarrow{CH_3CH_2OH}$?

(e) 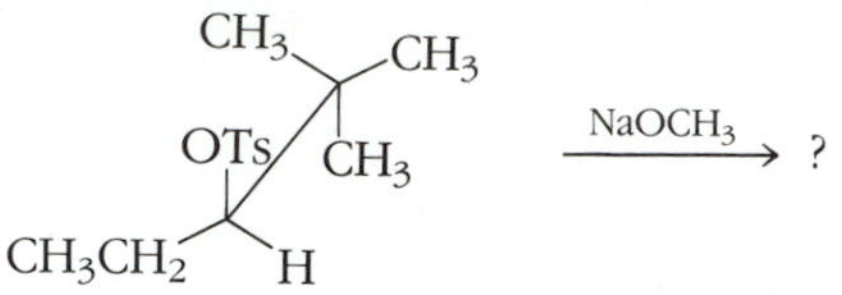 CH_3, CH_3, CH_3, OTs, CH_3CH_2, H $\xrightarrow{NaOCH_3}$?

(f) 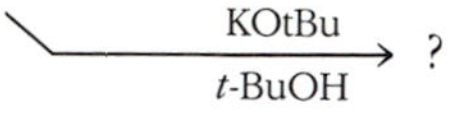 $\xrightarrow[t\text{-BuOH}]{KOtBu}$?

12.15 Draw the major product (or products) expected from each of the following reactions. Show stereochemistry whenever appropriate.

(a) 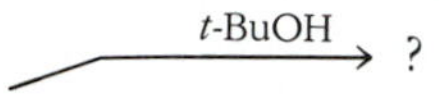 $\xrightarrow{t\text{-BuOH}}$?

(b) $CH_3—CH(OH)—CH_2CH_2CH_2—OTs$ $\xrightarrow[H_2O,\ acetone]{NaN_3}$?

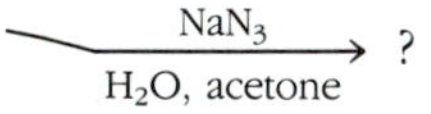

(c) 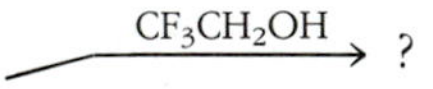 $\xrightarrow{CF_3CH_2OH}$?

(d) 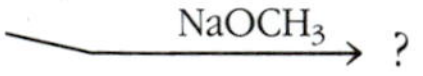 CH_3, Cl $\xrightarrow{NaOCH_3}$?

(e) 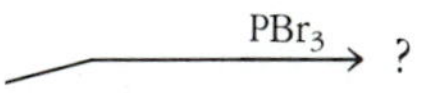 $\xrightarrow{PBr_3}$?

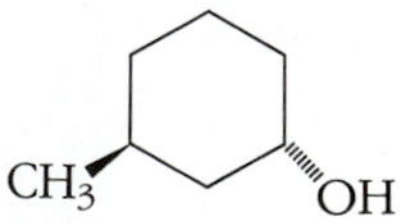

(f)  CH_3, OH $\xrightarrow{SOCl_2}$?

12.16 Draw the major product (or products) expected from each of the following reactions. Show stereochemistry whenever appropriate.

(a) $\xrightarrow{\text{NaCN}}$?

OTs

$(CH_3)_3C$

(b) $\xrightarrow[\text{acetone}]{Bu_4N^{+-}OAc}$?

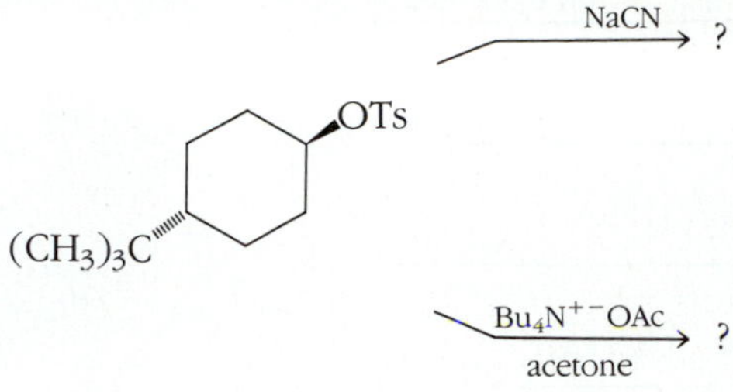

(c) $\xrightarrow{CF_3CH_2OH}$?

H_3C CH_3

OTs

(d) $\xrightarrow{CH_3CH_2OH}$?

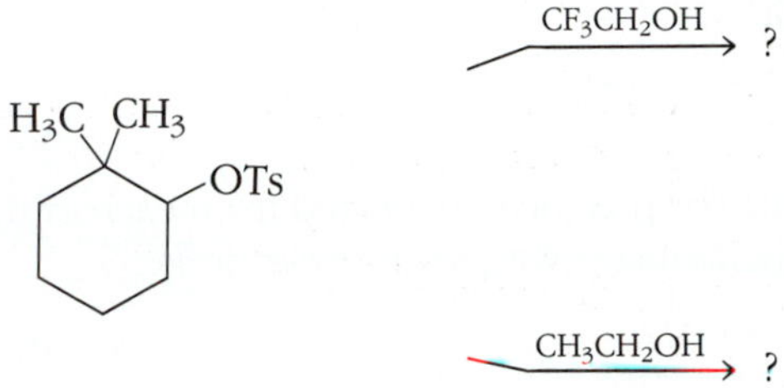

(e) $\xrightarrow[\text{acetone}]{H_2O}$?

OTf

(f) $\xrightarrow{CH_3CO_2H}$?

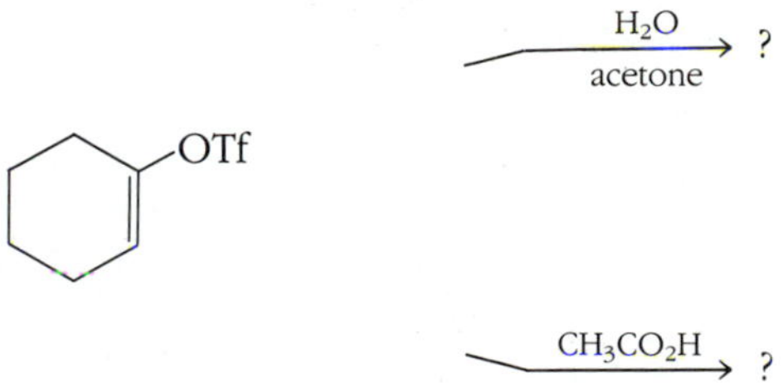

12.17 Draw the major product (or products) expected from each of the following reactions. Show stereochemistry whenever appropriate.

(a) $\xrightarrow{CF_3CH_2OH}$?

C_6H_5 CH_2CH_3

(b) C $\xrightarrow{\text{NaOEt}}$?

H Br

(c) $\xrightarrow{Et_4N^{+-}OAc}$?

(d) CH_3 $\xrightarrow{CF_3CH_2OH}$?

(e) $\xrightarrow{NaOCH_3}$?

Br

(f) $\xrightarrow[\text{acetone}]{H_2O}$?

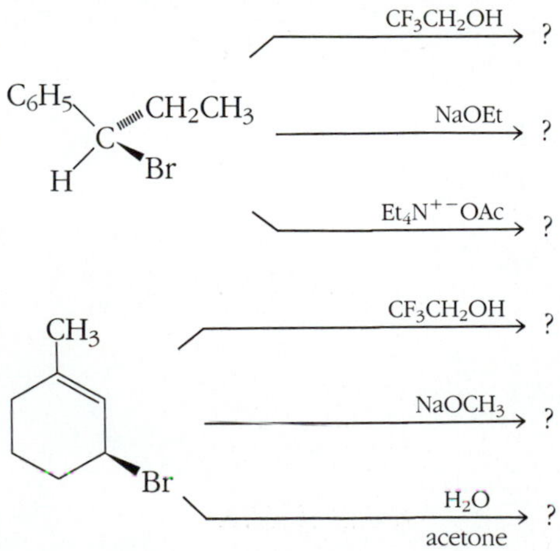

12.18 Draw the major product (or products) expected from each of the following reactions. Show stereochemistry whenever appropriate.

(a) $\xrightarrow[\text{acetone}]{Bu_4N^{+}\ {}^{-}OAc}$?

(b) $\xrightarrow{CF_3CH_2OH}$?

(c) $\xrightarrow{NaOCH_3}$?

CH_3, H, H, H, Cl, C_6H_5

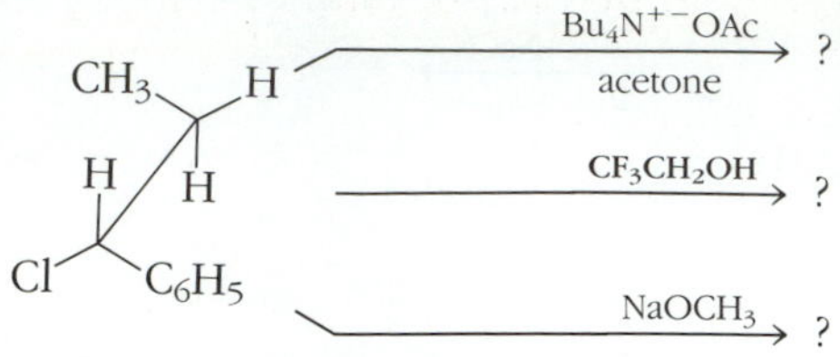

(d) $\xrightarrow{NaOCH_3,\ CH_3OH}$?

(e) $\xrightarrow[\text{acetone}]{Bu_4N^{+}\ {}^{-}OAc}$?

$(CH_3)_2C{=}CH{-}CH_2Br$

(f) $\xrightarrow{CF_3CH_2OH}$?

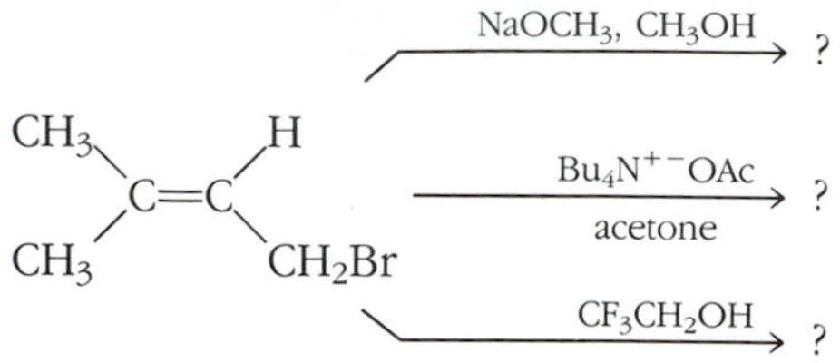

12.19 Draw the major product (or products) expected from each of the following reactions. Show stereochemistry whenever appropriate.

(a) $\xrightarrow{CH_3OH}$?

CH_3, Br

(b) $\xrightarrow{KOtBu}$?

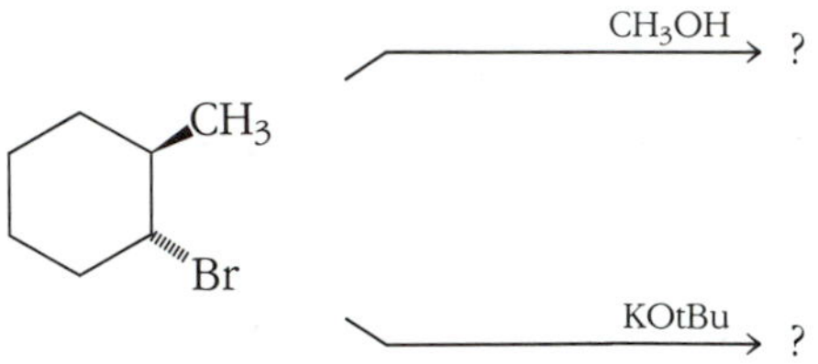

(c) CH_3, CH_2OTs $\xrightarrow{CH_3OH}$?

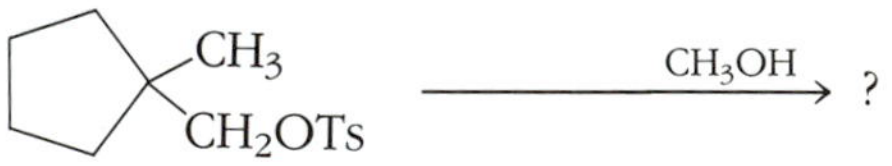

(d) H, CH_3 $\xrightarrow{H_2O/\text{acetone}}$?

(e) CH_3, Cl $\xrightarrow{NaOCH_3}$?

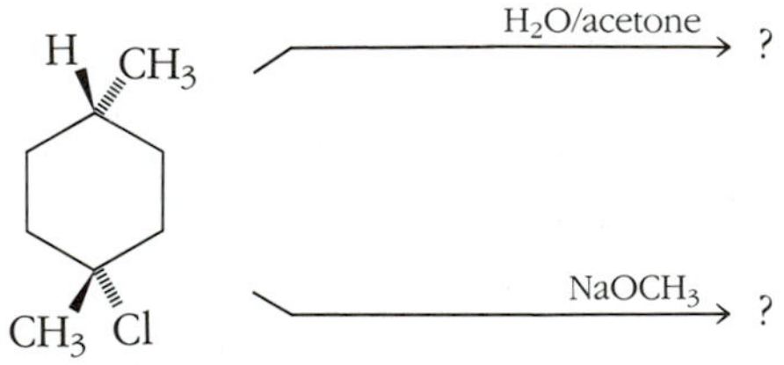

(f) CH_3, OTs $\xrightarrow{CF_3CH_2OH}$?

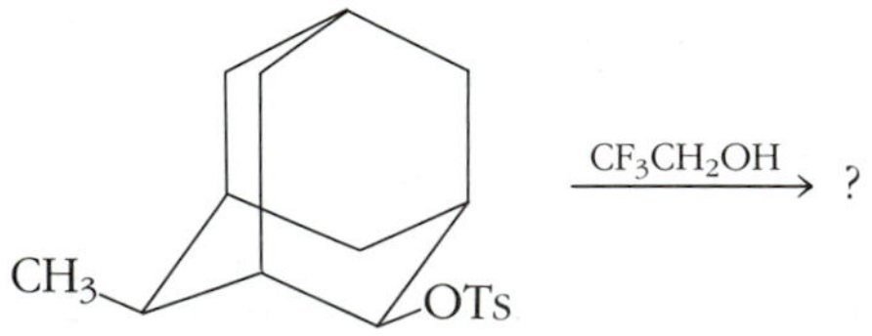

12.20 Draw the major product (or products) expected from each of the following reactions. Show stereochemistry whenever appropriate. If more than one product is expected, draw both major and minor products.

(a) Bromocyclohexane $\xrightarrow{CF_3CH_2OH}$?

Br

(b) Bromocyclohexane $\xrightarrow{NaOCH_3,\ CH_3OH}$?

(c) $\xrightarrow{CH_3CO_2H}$?

OBs

CH_3

(d) $\xrightarrow{NaOCH_3}$?

(e) $C_6H_5—CH_2Cl$ $\xrightarrow[tBuOH]{KOtBu}$?

(f) $C_6H_5—CH_2Cl$ $\xrightarrow{EtOH,\ H_2O}$?

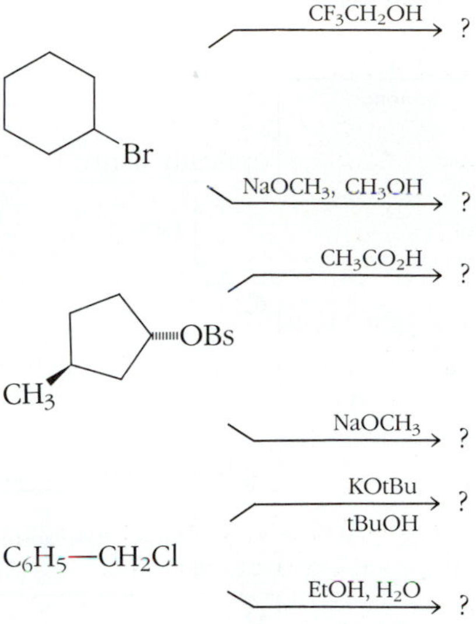

12.21 Suggest methods for carrying out the following conversions:

(a) Br → OH

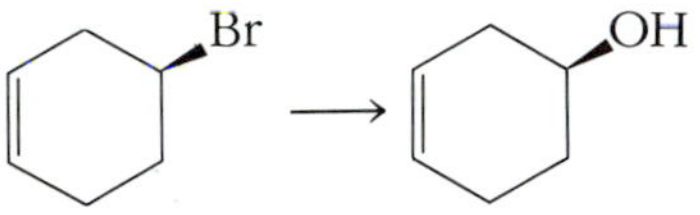

(b) H, CH_2CH_3, HO, H, CH_3, H → H, CH_2CH_3, CH_3, H, H, CN

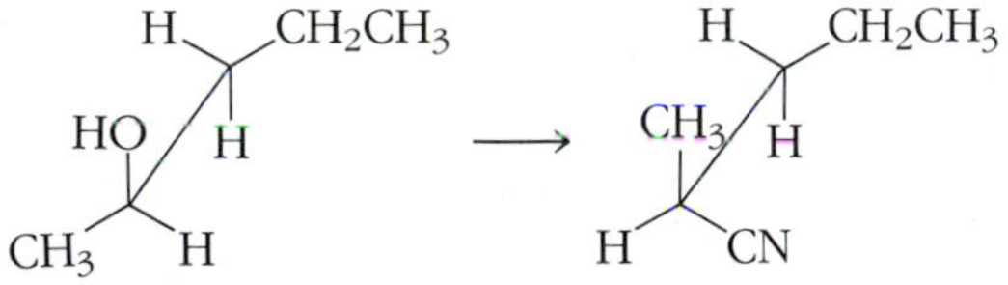

(c) $CH_3—C$(D)(OH)(H) → $CH_3—C$(H)(D)($C≡C—CH_3$)

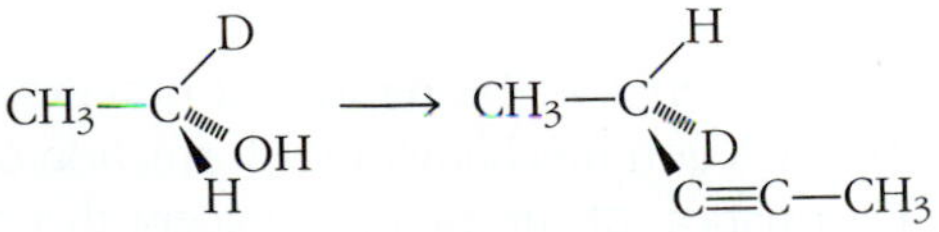

(d) CH_3, OH → CH_3, CN

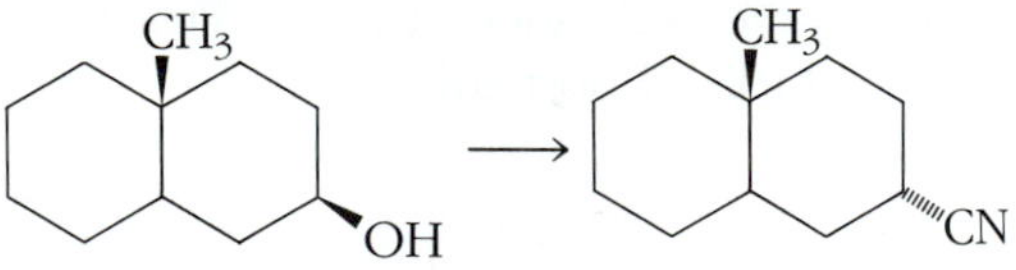

(e) (*R*)-2-Butanol ⟶ (*R*)-2-bromobutane

12.22 **A–H** are all different compounds. On the basis of the reactions shown, deduce the structures of compounds **A–F.** Show stereochemistry.

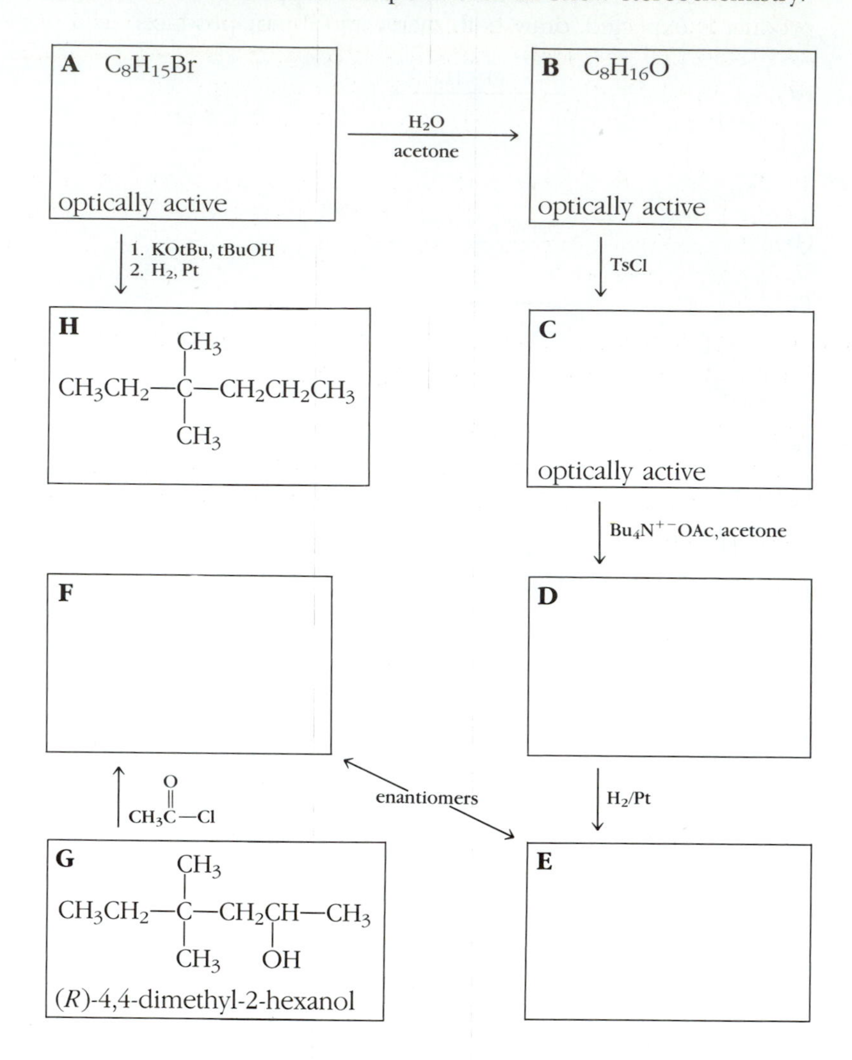

12.23 Neryl pyrophosphate can also serve as the biological precursor for terpenoids such as borneol. The reaction mechanism for formation of borneol is quite similar to that for formation of the pinenes except that a secondary "carbocation" is formed in the cyclization reaction. Propose a reasonable mechanism for the biosynthesis of borneol.

OH ≡ OH

Borneol

12.24 In Section 12.7 we pointed out that direct displacement reactions occur very slowly when the carbon adjacent to the reactive center is quaternary (i.e., with neopentyl systems). We showed that 2,2-dimethyl-1-propyl tosylate reacts (slowly) with sodium cyanide to give the product corresponding to direct displacement in excellent yield. The analogous direct displacement reaction of 3,3-dimethyl-2-butyl tosylate would fail. Why?

12.25 In Section 12.6 we showed that the reaction of 2-(2,4,6-trimethylphenyl)ethyl tosylate with sodium methoxide resulted in 93% substitution and 7% elimination. What results would be expected for 2-phenylethyl tosylate under these conditions?

12.26 Why are elimination reactions by the E1 mechanism not observed when an alkyl halide or tosylate is treated with strong base?

12.27 If compound A could react by one pathway (E_a = 13.2 kcal/mol) to give compound B or by a second pathway (E_a = 14.0 kcal/mol) to give compound C, what ratio of B to C would be expected? (Assume that B and C do not equilibrate.)

CHAPTER 13

ALDEHYDES AND KETONES

Aldehydes and ketones are the simplest classes of compounds that contain a carbonyl group. The carbonyl group of a ketone has two alkyl (or aryl) groups as its substituents, whereas the carbonyl group of an aldehyde has only one alkyl (or aryl) substituent, the other substituent being a hydrogen atom. (The compound in which both substituents are hydrogen is also classified as an aldehyde, formaldehyde, $H_2C{=}O$.)

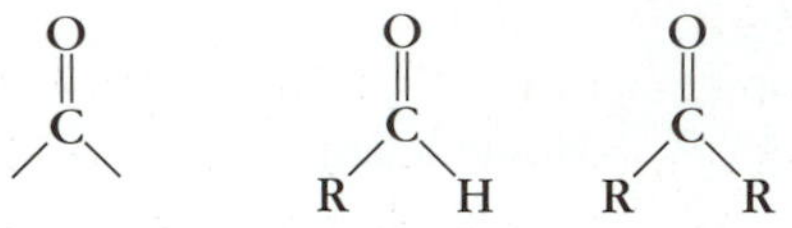

$$\underset{\text{Carbonyl group}}{>C{=}O} \qquad \underset{\text{Aldehyde}}{R{-}\overset{\overset{O}{\|}}{C}{-}H} \qquad \underset{\text{Ketone}}{R{-}\overset{\overset{O}{\|}}{C}{-}R}$$

The carbonyl group is found in a wide variety of both synthetic and naturally occurring compounds, and the reactions of the carbonyl group constitute a large part of both organic chemistry and biochemistry. For instance, the carbonyl group is present in compounds such as carbohydrates (aldehydes and ketone derivatives), amino acids (carboxylic acids), peptides and proteins (amides), and fats (esters). In this chapter we will examine the characteristic reactivity of the carbonyl group of aldehydes and ketones. This will give you the foundation for understanding the reactivity of the carbonyl group in other classes of compounds that we will cover in subsequent chapters.

Aldehydes and ketones play many important roles in industrial as well as laboratory organic chemistry. Aldehydes and ketones undergo many carbon-carbon bond-forming reactions, and they can also be transformed into other classes of compounds by functional group modification. Consequently, they are important starting materials and intermediates in synthetic organic chemistry. The lower-molecular-weight ketones also find considerable use as solvents in chemical processes, and other aldehydes and ketones are used in the flavor and fragrance industry (Section 13.2).

One of the most important ketones, acetone, is also the simplest; both substituents on the carbonyl group are methyl groups. Acetone is widely used as an industrial solvent and as a synthetic intermediate.

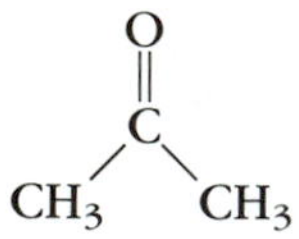

$$\underset{\text{Acetone}}{CH_3{-}\overset{\overset{O}{\|}}{C}{-}CH_3}$$

It is now made in other ways, but an important method for the preparation of acetone was developed in England during World War I by Chaim Weizmann (who is perhaps better known for his later role as the first president of the State of Israel). The fermentation of starch by this process produced large quantities of acetone. That development, and particularly the formation of large amounts of 1-butanol as a by-product, stimulated great advances in industrial chemistry, which was still in its infancy. We will present a more detailed discussion of acetone and its historical importance in Section 26.2.

13.1 NOMENCLATURE

Aldehydes and ketones are named by the same procedures that we previously introduced for hydrocarbons and alcohols.

Nomenclature of Aldehydes and Ketones

1. *Locate the principal carbon chain.* For a simple aldehyde or ketone this is the longest carbon chain in the molecule *which contains the C═O group.* The basic name is then formed by replacing the final *-e* in the name of the corresponding alkane by the suffix *-al* for an aldehyde or *-one* for a ketone.
   For a *polyfunctional* compound, the principal chain is selected according to the following priorities for the principal group:

   aldehyde > ketone > alcohol

2. *Name and number the substituents* on the principal chain.
   (a) The principal chain is numbered so that the lower of two possible numbers results for *the carbon bearing the carbonyl group.* This number is understood to be C-1 for an aldehyde, but for a ketone it is prefixed to the name of the principal chain with a hyphen. If the location of the carbonyl group does not distinguish between two different ways of numbering the principal chain, the substituents are evaluated so that the lower number results at the occasion of the first difference.
   (b) When there is more than one substituent, they are listed in alphabetical order. If several of the substituents are identical, this is indicated by the numerical prefixes *di-, tri-, tetra-, penta-, hexa-,* etc.
   (c) When a carbonyl group is not the principal group, the doubly bonded oxygen is named as a substituent by using the prefix *oxo-*. A one-carbon, aldehyde substituent can be named as a *formyl* group.

3. *Specify the stereochemistry* for any center in the molecule that could exist in two alternative configurations.

Monofunctional Compounds

The following C_3 compounds show the relationship between the names of alkanes and those of aldehydes or ketones with the same carbon skeleton:

$$CH_3{-}CH_2{-}CH_3 \qquad CH_3{-}CH_2{-}\overset{\overset{\displaystyle O}{\|}}{C}H \qquad CH_3{-}\overset{\overset{\displaystyle O}{\|}}{C}{-}CH_3$$

Propane — Propan*al* — 2-Propan*one*

The correct numbering of the principal chain is illustrated by the case of 5-methyl-3-hexanone.

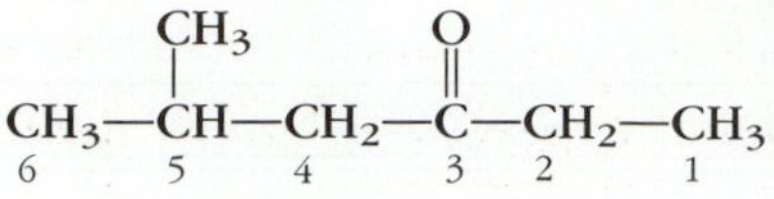

$$\underset{6}{CH_3}-\underset{5}{\overset{\overset{\displaystyle CH_3}{|}}{CH}}-\underset{4}{CH_2}-\underset{3}{\overset{\overset{\displaystyle O}{\|}}{C}}-\underset{2}{CH_2}-\underset{1}{CH_3}$$

5-methyl-3-hexanone
(not 2-methyl-4-hexanone)

In the case of an aldehyde the carbonyl carbon is always C-1 of the chain, so it need not be specified.

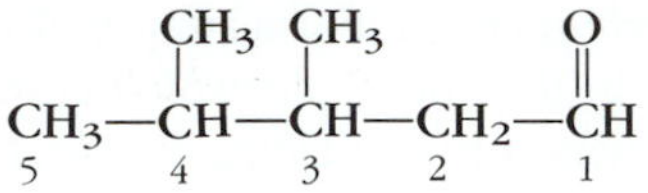

$$\underset{5}{CH_3}-\underset{4}{\overset{\overset{\displaystyle CH_3}{|}}{CH}}-\underset{3}{\overset{\overset{\displaystyle CH_3}{|}}{CH}}-\underset{2}{CH_2}-\underset{1}{\overset{\overset{\displaystyle O}{\|}}{CH}}$$

3,4-Dimethylpentanal

The following examples illustrate the naming of substituted ketones. Note that the carbonyl group is understood to be C-1 of a cyclic ketone.

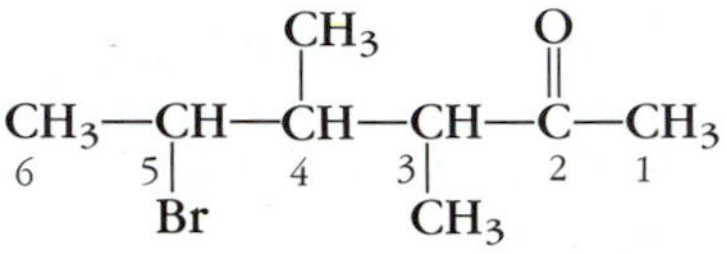

$$\underset{6}{CH_3}-\underset{\underset{\displaystyle Br}{5|}}{CH}-\underset{4}{\overset{\overset{\displaystyle CH_3}{|}}{CH}}-\underset{\underset{\displaystyle CH_3}{3|}}{CH}-\underset{2}{\overset{\overset{\displaystyle O}{\|}}{C}}-\underset{1}{CH_3}$$

5-Bromo-3,4-dimethyl-2-hexanone

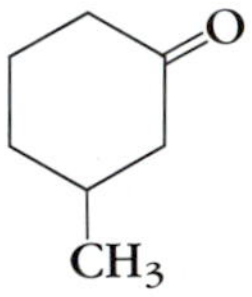

O
CH3

3-Methylcyclohexanone

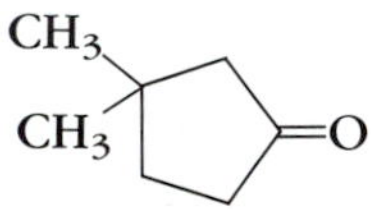

CH3
CH3
O

3,3-Dimethylcyclopentanone

Common Names and Alternative Nomenclature

Common names are not used frequently for ketones, with acetone (2-propanone) being an important exception. There are a number of aldehydes, however, for which you will often encounter common names (Table 13.1). In some cases the IUPAC name is virtually never used; such names are enclosed in parentheses in the table.

Phenyl ketones are generally named as *-phenones,* with the remainder of the name being derived from the corresponding carboxylic acid as illustrated by the following examples.

$C_6H_5-\overset{\overset{\displaystyle O}{\|}}{C}-CH_3$ *Aceto*phenone (HO$-\overset{\overset{\displaystyle O}{\|}}{C}-CH_3$ *Acet*ic acid)

$C_6H_5-\overset{\overset{\displaystyle O}{\|}}{C}-CH_2CH_3$ *Propio*phenone (HO$-\overset{\overset{\displaystyle O}{\|}}{C}-CH_2CH_3$ *Propion*ic acid)

$C_6H_5-\overset{\overset{\displaystyle O}{\|}}{C}-C_6H_5$ *Benzo*phenone (HO$-\overset{\overset{\displaystyle O}{\|}}{C}-C_6H_5$ *Benzo*ic acid)

TABLE 13.1 Common Names of Some Aldehydes

Aldehyde	Common Name	IUPAC Name
H—CHO	Formaldehyde	(Methanal)
CH_3—CHO	Acetaldehyde	(Ethanal)
CH_3CH_2—CHO	Propionaldehyde	Propanal
$CH_3CH_2CH_2$—CHO	Butyraldehyde	Butanal
CH_3—CH(CH_3)—CHO	Isobutyraldehyde	2-Methylpropanal
$CH_3CH_2CH_2CH_2CHO$	Valeraldehyde	Pentanal
CH_2=CH—CHO	Acrylaldehyde, acrolein	(Propenal)
C_6H_5—CHO	Benzaldehyde	(Benzenecarbaldehyde)

Ketones are also sometimes named by simply stating the two substituents on the carbonyl group in alphabetical order.

$CH_3CH_2—C(=O)—CH_3$ Ethyl methyl ketone (2-butanone)

$CH_3CH_2—C(=O)—CH_2CH_3$ Diethyl ketone (3-pentanone)

$C_6H_{11}—C(=O)—CH_2CH_3$ Cyclohexyl ethyl ketone

Aldehydes having a ring attached directly to the carbonyl group can be named as the corresponding *-carbaldehyde,* the —CHO group being treated as a principal chain with only a single carbon atom.

$C_6H_{11}—CHO$ Cyclohexanecarbaldehyde

Difunctional Compounds

When a compound contains more than one functional group, it is necessary to establish which is the *principal group.* For the few classes of compounds that we have introduced so far in this text, the following priorities are used:

aldehyde > ketone > alcohol

A ketoaldehyde would therefore be named as an aldehyde, but a ketoalcohol would be named as a ketone.

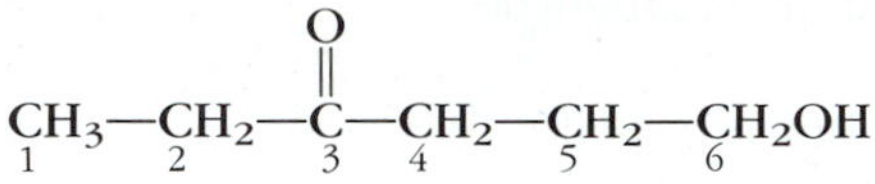

$$\underset{1}{CH_3}-\underset{2}{CH_2}-\underset{3}{C}(=O)-\underset{4}{CH_2}-\underset{5}{CH_2}-\underset{6}{CH_2OH}$$

6-Hydroxy-3-hexanone
(a keto alcohol)

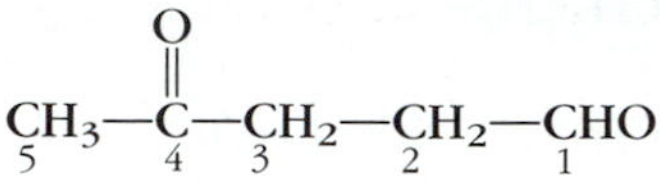

$$\underset{5}{CH_3}-\underset{4}{C}(=O)-\underset{3}{CH_2}-\underset{2}{CH_2}-\underset{1}{CHO}$$

4-Oxopentanal
(a keto aldehyde)

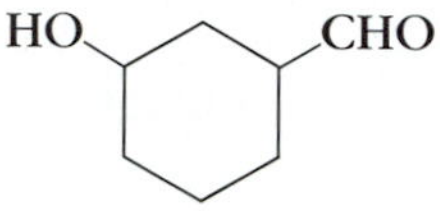

3-Hydroxycyclohexanecarbaldehyde

When the principal group occurs more than once in a molecule, the appropriate prefix (di-, tri-, etc.) is used, and the principal chain is selected to include as many of these groups as possible.

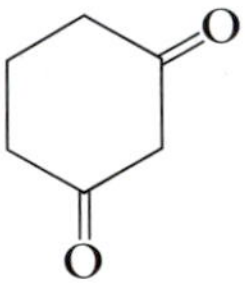

1,3-Cyclohexanedione

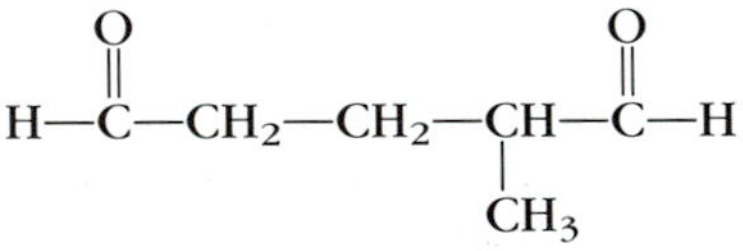

$$H-C(=O)-CH_2-CH_2-CH(CH_3)-C(=O)-H$$

2-Methylpentanedial

When an aldehyde group cannot be included in the principal chain, it can be named as a *formyl* substituent.

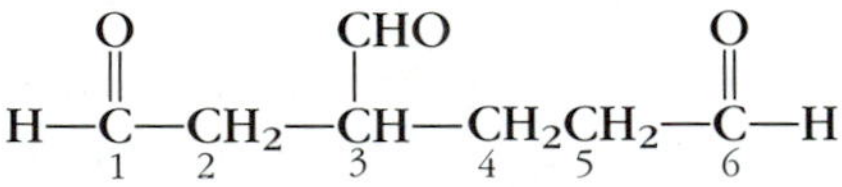

$$H-\underset{1}{C}(=O)-\underset{2}{CH_2}-\underset{3}{CH}(CHO)-\underset{4}{CH_2}\underset{5}{CH_2}-\underset{6}{C}(=O)-H$$

3-Formylhexandial

The double or triple bond is included in the principal chain whenever possible, and unsaturated ketones and aldehydes are named as *-enones* and *-enals,* respectively.

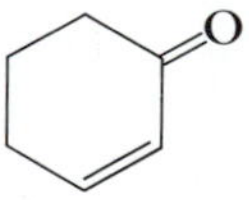

2-Cyclohexen-1-one

$$CH_3-CH{=}CH-CH_2CH_2-CHO$$

4-Hexenal

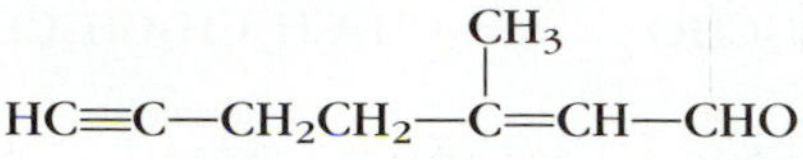

$$HC\equiv C-CH_2CH_2-\overset{\displaystyle CH_3}{\overset{|}{C}}=CH-CHO$$

3-Methyl-2-hepten-6-ynal

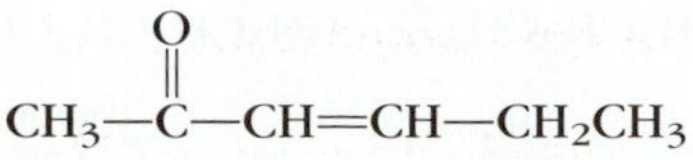

$$CH_3-\overset{\displaystyle O}{\overset{\|}{C}}-CH=CH-CH_2CH_3$$

3-Hexen-2-one

EXERCISE 13.1

Name each of the following compounds:

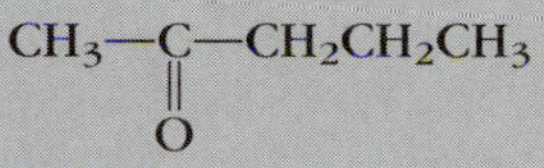

(a) 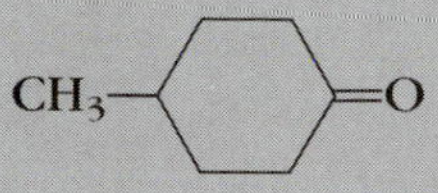 $CH_3-C_6H_{10}=O$ (4-methyl-substituted cyclohexane ring with =O)

(b) $CH_3-\underset{\underset{\displaystyle O}{\|}}{C}-CH_2CH_2CH_3$

(c) $CH_3-\overset{\displaystyle O}{\overset{\|}{C}}-CH_2CH_2-CHO$

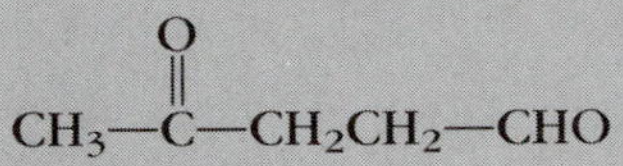

(d) 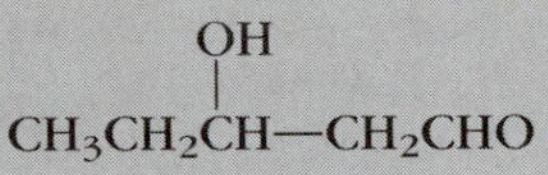 $CH_3CH_2\overset{\displaystyle OH}{\overset{|}{C}H}-CH_2CHO$

(e) 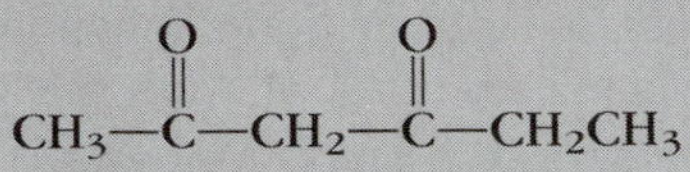 $CH_3-\overset{\displaystyle O}{\overset{\|}{C}}-CH_2-\overset{\displaystyle O}{\overset{\|}{C}}-CH_2CH_3$

EXERCISE 13.2

Draw each of the following compounds:

(a) 4-Methyl-2-heptanone
(b) 2-Hydroxypentanal
(c) 4-Oxocyclohexanecarbaldehyde
(d) Acetophenone
(e) Diisobutyl ketone

13.2 PHYSICAL PROPERTIES

The carbonyl group is a polar functional group

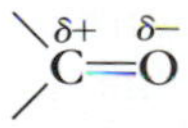

$$\gt\overset{\delta+}{C}=\overset{\delta-}{O}$$

and this is reflected in the physical properties of aldehydes and ketones (Section 4.9). For example, their boiling points are considerably *higher* than those of the analogous alkanes with similar molecular weights. On the other hand, the absence of hydrogen bonding leads to boiling points which are *lower* than those of the corresponding alcohols. Similarly, the melting points are intermediate between those of the nonpolar hydrocarbons and the more polar alcohols.

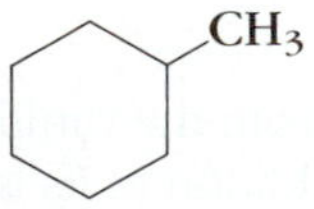

Methylcyclohexane	Cyclohexanone	Cyclohexanol
MW 98; bp 101°C; mp −126°C	MW 98; bp 156°C; mp −16°C	MW 100; bp 161°C; mp 25°C

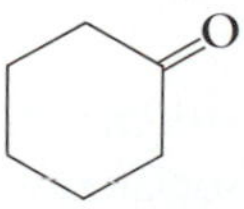

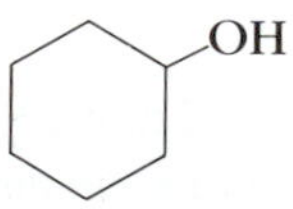

$CH_3CH_2CH_2CH_2CH_2CH_3$	$CH_3CH_2CH_2CH_2CH_2CHO$	$CH_3CH_2CH_2CH_2CH_2CH_2OH$
Heptane	Hexanal	1-Hexanol
MW 100; bp 98°C; mp −91°C	MW 100; bp 128°C; mp −56°C	MW 102; bp 158°C; mp −47°C

The *position* of the carbonyl group on a carbon chain has relatively little effect on physical properties, as you can see from the following series:

CHO O O O

Heptanal
bp 153°C; mp −43°C

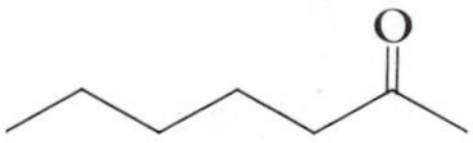

2-Heptanone
bp 151°C; mp −35°C

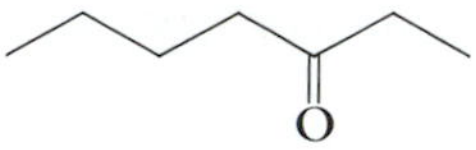

3-Heptanone
bp 147°C; mp −39°C

4-Heptanone
bp 144°C; mp −33°C

The lower-molecular-weight ketones, particularly acetone and 2-butanone, are widely used as solvents. Both are highly soluble in water, yet will dissolve a wide variety of organic compounds. They are also relatively volatile, so they are easily removed by distillation or evaporation. Other aldehydes and ketones are progressively less soluble in water as their molecular weights increase, and their boiling and melting points increase accordingly (Table 13.2).

Odor and Taste

The human perception of aldehydes and ketones by taste or smell is highly dependent upon chemical structure. Many of these compounds have pleasant fragrances or flavors, and they play important roles in the cosmetics and food industries. Perfumes and flavorings were used long before the development of synthetic organic chemistry, so many familiar flavors and scents are associated with naturally occurring compounds. Sometimes the flavor or aroma associated with a particular fruit or flower results from the combined effect of two or more constituents. As an example, the aroma of raspberries has been found to result from a mixture of at least 24 organic compounds present in varying amounts. In other cases a single constituent may be the major contributor to the taste or smell.

Several aldehydes are very common flavor-aroma ingredients. Benzaldehyde, the major constituent of oil of bitter almond (obtained through a distillation process), is the key ingredient of almond flavor, and cinnamaldehyde (the primary constituent of cinnamon oil) is responsible for the flavor and taste that you know as cinnamon.

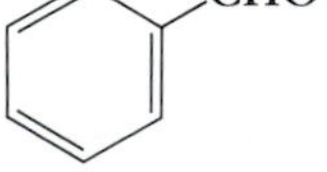

CHO CH=CH—CHO

Benzaldehyde 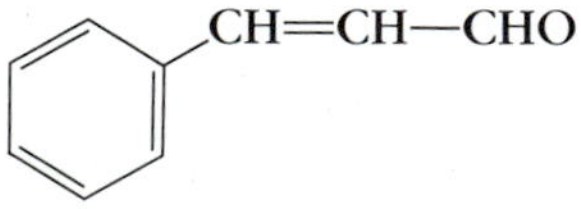 Cinnamaldehyde

A substituted benzaldehyde called *vanillin* can be extracted from the vanilla bean, and it exhibits the taste and odor associated with vanilla. In addition to its isolation from natural sources, vanillin is also synthesized commercially. A totally synthetic analog, *ethyl vanillin,* has an even more intense vanilla aroma than the naturally occurring compound and is used extensively in foods and cosmetics.

TABLE 13.2 Physical Properties of Some Common Aldehydes and Ketones

Number of Carbon Atoms	Structure	Name	BP, °C	MP, °C
1	HCHO	Formaldehyde	−21	−92
2	CH_3CHO	Acetaldehyde	21	−121
3	CH_3CH_2CHO	Propionaldehyde	49	−81
	$CH_3-\overset{O}{\overset{\|}{C}}-CH_3$	Acetone	56	−95
4	$CH_3CH_2CH_2CHO$	Butyraldehyde	76	−99
	$CH_3-\overset{CH_3}{\overset{\|}{CH}}-CHO$	Isobutyraldehyde	63	−66
	$CH_3-\overset{O}{\overset{\|}{C}}-CH_2CH_3$	2-Butanone	80	−86
5	$CH_3CH_2CH_2CH_2CHO$	Pentanal	103	−92
	(ring)=O	Cyclopentanone	131	−51
6	(ring)=O	Cyclohexanone	156	−16
7	C_6H_5-CHO	Benzaldehyde	178	−26
	$CH_3(CH_2)_5CHO$	Heptanal	153	−43
8	$CH_3-(CH_2)_5-\overset{O}{\overset{\|}{C}}-CH_3$	2-Octanone	173	−16
	(ring)$-\overset{O}{\overset{\|}{C}}-CH_3$	Acetophenone	202	20
	(ring)$-\overset{O}{\overset{\|}{C}}-CH_3$	Cyclohexyl methyl ketone	181	—

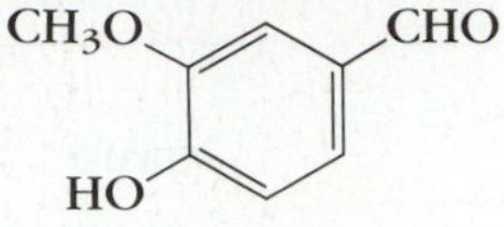

CH_3O CHO HO

Vanillin

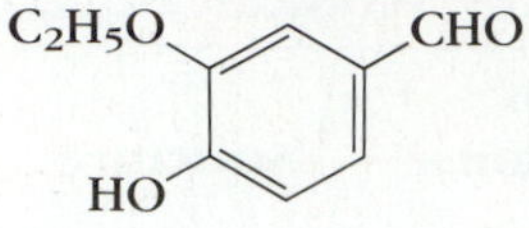

C_2H_5O CHO HO

Ethyl vanillin

Several naturally occurring, large-ring ketones exhibit a musk smell that has long been used in perfumery. The natural compounds can be obtained from animal sources but only in very small amounts. The scarcity of these compounds has led to substantial interest in their synthesis. Two examples are civetone and muskone; the latter is the primary ingredient of the natural substance obtained from the female musk deer (which is found in northern Asia and Tibet).

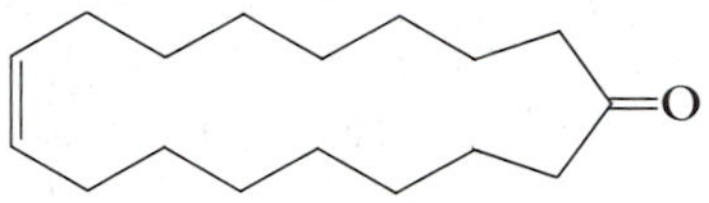

=O

Civetone

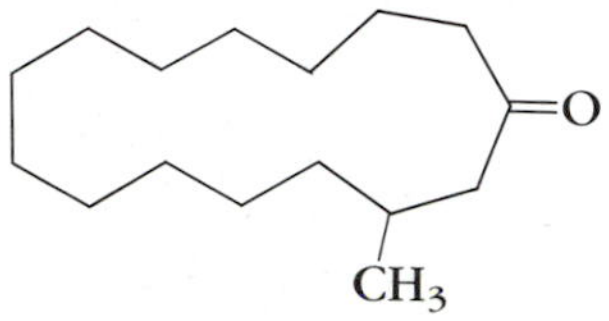

=O CH_3

Muskone

One of the most intriguing aspects of the perception of fragrance is its dependence not only upon basic chemical structure but even upon *chirality*. For example, the primary constituent responsible for the aroma of spearmint oil is the cyclic ketone *carvone*.

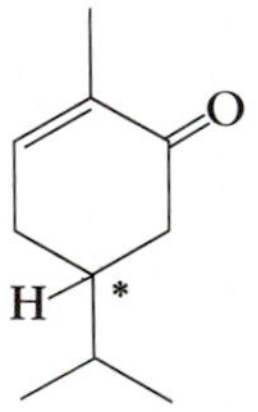

O H *

Carvone is also the principal ingredient responsible for the aroma of oil of caraway (the odor of caraway seeds, which may be more familiar to you as the smell of rye bread). The major ingredients of these two oils differ only in their configuration at the single chiral center (*) of carvone, yet this subtle structural change results in strikingly different odors.

13.3 PREPARATION OF ALDEHYDES AND KETONES VIA OXIDATION AND REDUCTION

Oxidation of Alcohols

We have previously discussed (Section 11.3) the various reactions that can be used to prepare aldehydes and ketones by oxidation, and we will only briefly review them here. Oxidation of secondary alcohols to ketones proceeds well with a variety of oxidizing agents, particularly Cr(VI) reagents. For example:

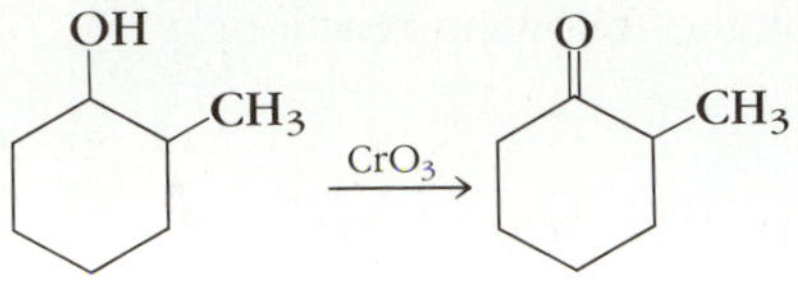

$$\text{2-methylcyclohexanol} \xrightarrow{CrO_3} \text{2-methylcyclohexanone} \quad 87\%$$

Care must be taken with primary alcohols, however, because oxidation can proceed past the aldehyde stage to yield a carboxylic acid. This problem is best solved by the use of CrO_3-pyridine reagents, as illustrated by the following reaction:

$$CH_3{-}(CH_2)_8{-}CH_2OH \xrightarrow{CrO_3\text{-pyridine}} CH_3{-}(CH_2)_8{-}CHO \quad 63\text{–}83\%$$

Aldehydes and ketones can be prepared from alkenes in a two-step process corresponding to addition of water (Section 5.11) followed by oxidation of the resulting alcohol by one of the procedures that we have just described.

Although not an oxidation, the *hydration* of alkynes (Section 7.5) can also be used to prepare aldehydes and ketones. Terminal alkynes can be converted to either the aldehyde or the ketone by an appropriate choice of reagents.

Oxidative Cleavage of Alkenes

Cleavage of alkenes into aldehydes and ketones via ozonolysis was previously discussed in Section 5.12. Reaction of an alkene with ozone generates an intermediate ozonide, which is not isolated but is treated with a reducing agent such as zinc metal.

$$CH_3{-}(CH_2)_7{-}CH{=}CH{-}(CH_2)_7CO_2CH_3 \xrightarrow[2.\ Zn,\ CH_3CO_2H]{1.\ O_3} CH_3{-}(CH_2)_7CHO\ (88\%) + H{-}\overset{O}{\overset{\|}{C}}{-}(CH_2)_7CO_2CH_3\ (77\%)$$

The intermediate ozonide can also be cleaved by a variety of other mild reducing agents, such as dimethyl sulfide (which is oxidized to dimethyl sulfoxide). This is illustrated by the ozonolysis of β-pinene.

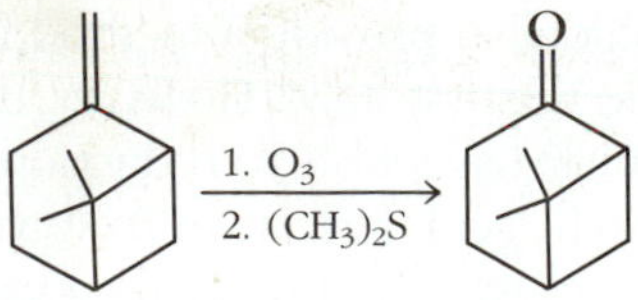

$$\beta\text{-pinene} \xrightarrow[2.\ (CH_3)_2S]{1.\ O_3} \text{ketone} \quad 70\%$$

Oxidative cleavage of alkenes can also be brought about by the periodate cleavage of 1,2-diols. The reaction of HIO_4 (or its salts) with a diol affords a cyclic intermediate that undergoes decomposition to yield carbonyl products and HIO_3 (or its salts).

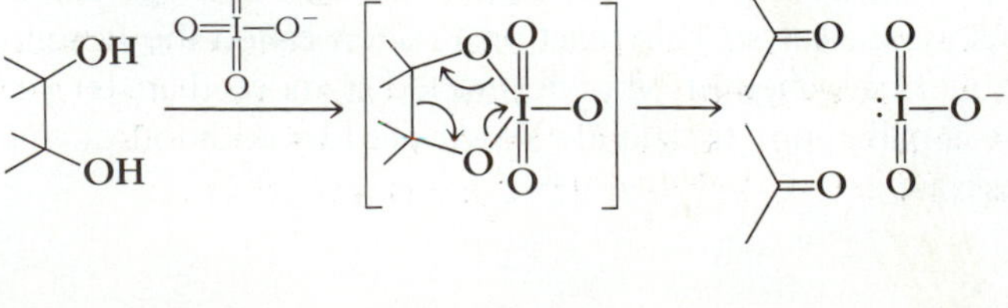

$$\text{diol (OH, OH)} \xrightarrow{O{=}I(O)(O){-}O^-} [\text{cyclic periodate ester}] \longrightarrow 2\ \text{acetone } (C{=}O) + :I(O)(O){-}O^-$$

This process is illustrated by the following reaction:

$$CH_3{-}(CH_2)_7{-}\overset{\displaystyle OH}{\overset{|}{C}H}{-}\overset{\displaystyle OH}{\overset{|}{C}H}{-}(CH_2)_7{-}CO_2H \xrightarrow{HIO_4} CH_3{-}(CH_2)_7{-}\overset{\displaystyle O}{\overset{\|}{C}}H + H\overset{\displaystyle O}{\overset{\|}{C}}{-}(CH_2)_7{-}CO_2H$$

87%

The 1,2-diols can be prepared by the oxidation of the corresponding alkenes with potassium permanganate or osmium tetroxide (Section 5.12). Alternatively, both reactions can be carried out together by using a set of reagents such as sodium periodate with a catalytic amount of osmium tetroxide or potassium permanganate.

$$\text{cyclohexene} \xrightarrow[NaIO_4]{OsO_4} \text{CHO / CHO} \quad 77\%$$

$$\text{=CH}_2 \xrightarrow[KMnO_4]{KIO_4} \text{=O} \quad 59\%$$

Note that the permanganate-periodate cleavage would not be appropriate for the preparation of aldehydes, because permanganate causes the further oxidation of aldehydes to carboxylic acids (Section 11.3, 13.11).

Reduction of Carboxylic Acid Derivatives

In Section 11.4 we showed that primary alcohols could be prepared by the reduction of carboxylic acids and esters, reactions that proceed via the aldehyde.

$$R{-}\overset{\displaystyle O}{\overset{\|}{C}}{-}X \xrightarrow{\text{reduction}} [R{-}\overset{\displaystyle O}{\overset{\|}{C}}{-}H] \xrightarrow{\text{reduction}} R{-}\overset{\displaystyle OH}{\overset{|}{C}}H_2$$

X = OH, OR, Cl

In principle this approach should also provide a method for preparing aldehydes by limiting the reduction only to the first stage. However, there is a major problem with this idea. Aldehydes are quite susceptible to reduction and are generally *more reactive* than the starting carboxylic acid derivative. To solve this problem chemists have modified the reaction in two ways: (1) by using a carboxylic acid derivative which is *more reactive* than the aldehyde; and (2) by utilizing a *less reactive* reducing reagent which will selectively attack the starting compound without causing the undesired reduction of the aldehyde product.

Acid chlorides (R—COCl) are the most reactive carboxylic acid derivatives, and they are the logical choice for carrying out a selective reduction to the aldehyde stage. Early approaches to the problem utilized catalytic hydrogenation with a deactivated, or poisoned, catalyst. This reaction is often called the Rosenmund reduction after the German chemist who developed it more than 60 years ago. The typical catalyst is palladium, frequently deactivated by addition of small quantities of a sulfur derivative.

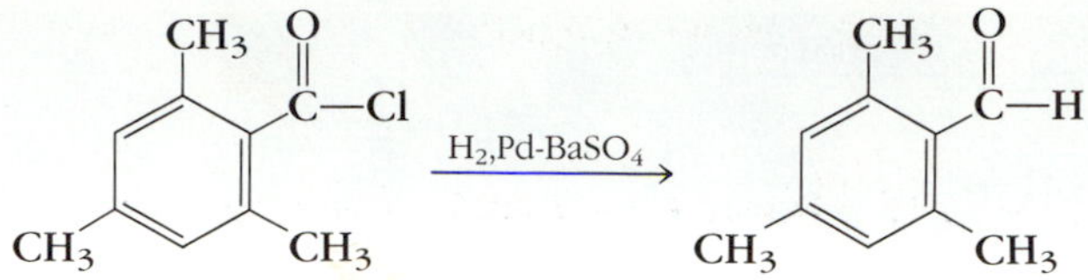

CH_3 O ‖ C—Cl (2,4,6-trimethylbenzoyl chloride) $\xrightarrow{H_2, Pd\text{-}BaSO_4}$ CH_3 O ‖ C—H (2,4,6-trimethylbenzaldehyde) 70–80%

O ‖ C—Cl (cyclohexanecarbonyl chloride) $\xrightarrow[Pd\text{-}C]{H_2}$ O ‖ C—H (cyclohexanecarbaldehyde) 78%

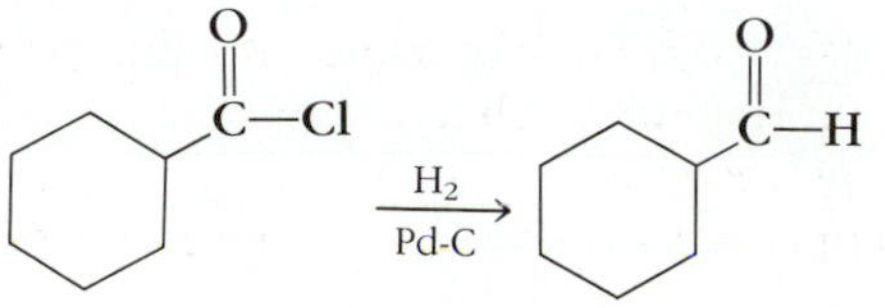

Another procedure was more recently developed by H. C. Brown (Purdue University), and this takes advantage of the reduced reactivity of lithium tri-*tert*-butoxyaluminum hydride. The reagent is prepared by the reaction of the much more reactive lithium aluminum hydride with *tert*-butyl alcohol.

$$LiAlH_4 \xrightarrow{t\text{-BuOH}} LiAl(OtBu)_3H + H_2$$

Three of the original four hydrogens are replaced by bulky *tert*-butoxy groups, and the resulting compound is a much less reactive, and more selective, reducing agent. Good yields of aldehydes can be obtained when this reagent is used for the reduction of acid chlorides.

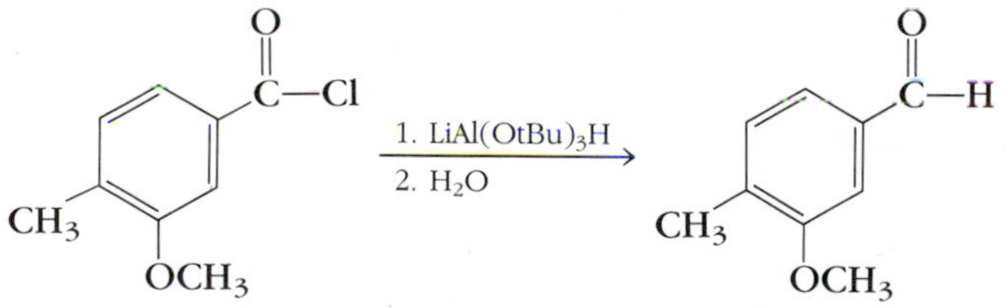

O ‖ C—Cl (CH_3, OCH_3 substituted benzoyl chloride) $\xrightarrow[2.\ H_2O]{1.\ LiAl(OtBu)_3H}$ O ‖ C—H (CH_3, OCH_3 substituted benzaldehyde) 62%

C_6H_5—CH=CH—C(=O)—Cl $\xrightarrow[2.\ H_2O]{1.\ LiAl(OtBu)_3H}$ C_6H_5—CH=CH—C(=O)—H 65%

$$CH_3\text{—}CH(CH_3)\text{—}C(=O)\text{—}Cl \xrightarrow[2.\ H_2O]{1.\ LiAl(OtBu)_3H} CH_3\text{—}CH(CH_3)\text{—}CH(=O) \quad 57\%$$

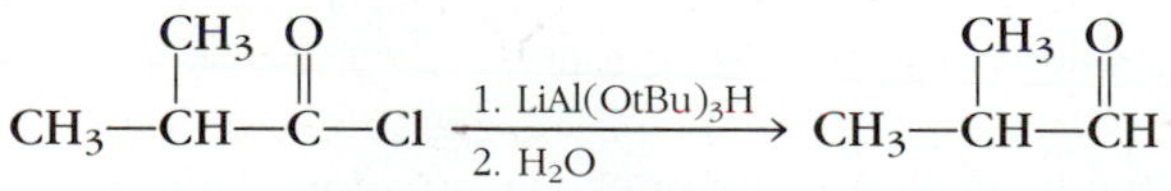

EXERCISE 13.3

Complete each of the following reactions:

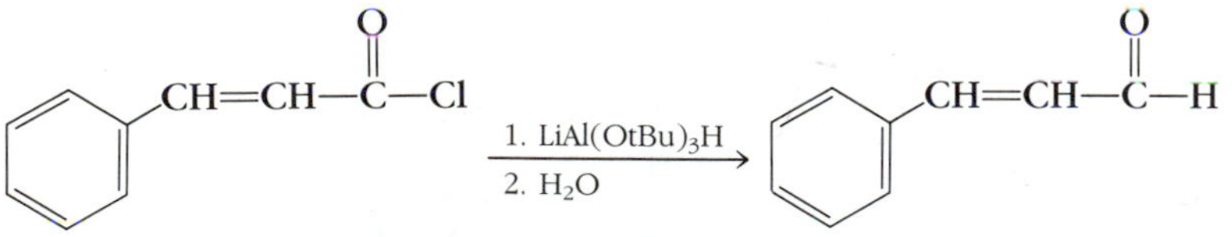

(a) 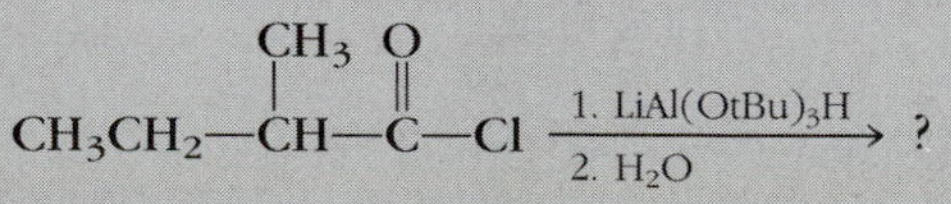 $CH_3CH_2\text{—}CH(CH_3)\text{—}C(=O)\text{—}Cl \xrightarrow[2.\ H_2O]{1.\ LiAl(OtBu)_3H} ?$

(b) 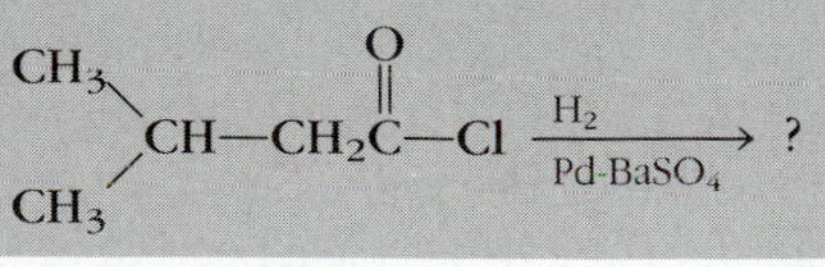 $(CH_3)_2CH\text{—}CH_2C(=O)\text{—}Cl \xrightarrow[Pd\text{-}BaSO_4]{H_2} ?$

(c) $CH_3CH_2CH_2OH \xrightarrow{CrO_3\text{-pyridine}} ?$

(d) $CH_3{-}CH{=}CH{-}CH_3 \xrightarrow[2.\ (CH_3)_2S]{1.\ O_3} ?$

(e) $CH_3CH_2{-}CH{=}C(CH_3)_2 \xrightarrow[OsO_4]{HIO_4} ?$

EXERCISE 13.4

Suggest reagents for carrying out each of the following conversions:

(a) 1-methylcyclohexene $\xrightarrow{?} CH_3{-}\overset{O}{\overset{\|}{C}}{-}(CH_2)_4{-}CHO$ (two different ways)

(b) $CH_3CH_2{-}\overset{CH_3}{\overset{|}{CH}}{-}CH_2CH_2OH \xrightarrow{?} CH_3CH_2{-}\overset{CH_3}{\overset{|}{CH}}{-}CH_2CHO$

(c) $CH_3CH_2{-}\overset{CH_3}{\overset{|}{CH}}{-}CH_2{-}\overset{O}{\overset{\|}{C}}{-}Cl \xrightarrow{?} CH_3CH_2{-}\overset{CH_3}{\overset{|}{CH}}{-}CH_2{-}CHO$ (two different ways)

13.4 PREPARATION OF KETONES VIA ORGANOMETALLIC REAGENTS

As we showed previously (Section 11.5), secondary alcohols can be prepared by the reaction of aldehydes with organometallic reagents. The secondary alcohols can in turn be oxidized to ketones, so this provides an *indirect* method for the preparation of ketones via organometallic reagents.

$$\underset{\text{Aldehyde}}{{-}\overset{O}{\overset{\|}{C}}{-}H} \xrightarrow[2.\ H_2O]{1.\ RMgX} \underset{2°\ \text{Alcohol}}{{-}\overset{OH}{\overset{|}{CH}}{-}R} \xrightarrow{CrO_3} \underset{\text{Ketone}}{{-}\overset{O}{\overset{\|}{C}}{-}R}$$

Direct methods for formation of ketones via organometallic reagents are also available, as depicted in the following equation:

$$-\overset{O}{\overset{\|}{C}}{-}X \xrightarrow{\overset{\delta-}{R}{-}\overset{\delta+}{M}} -\overset{O}{\overset{\|}{C}}{-}R + M^+X^-$$

These methods generally start from a carbonyl compound that has a leaving group as one of the two substituents of the carbonyl group. This leaving group (X in the preceding equation) is replaced by the alkyl group of the organometallic reagent. The reactant almost always has an alkyl group as the other substituent on the carbonyl group, so this procedure affords *ketones* and not aldehydes. A few specialized laboratory reactions for the preparation of aldehydes have been developed in recent years, but they will not be discussed here. An important industrial reaction, the *OXO process,* uses carbon monoxide as a one-carbon starting material for the synthesis of aldehydes, and we will return to this reaction in Section 26.1.

What specific types of carbonyl compounds are needed for reaction with an organometallic reagent? A carboxylic acid would usually be unsatisfactory, because the organometallic reagents are also bases and reaction would take place preferentially at the OH group.

$$—\overset{\overset{\displaystyle O}{\|}}{C}—OH \xrightarrow{R^-M^+} —\overset{\overset{\displaystyle O}{\|}}{C}—O^-\ M^+ + R—H$$

An ester presents other problems because organolithium and Grignard reagents undergo two successive carbonyl additions to yield a tertiary alcohol (Section 11.5). For example:

$$—\overset{\overset{\displaystyle O}{\|}}{C}—OCH_3 \xrightarrow{R—Li} —\overset{\overset{\displaystyle O^-}{|}}{\underset{\underset{\displaystyle OCH_3}{|}}{C}}—R \longrightarrow —\overset{\overset{\displaystyle O}{\|}}{C}—R \xrightarrow{R—Li} —\overset{\overset{\displaystyle O^-Li^+}{|}}{\underset{\underset{\displaystyle R}{|}}{C}}—R \xrightarrow[\text{(workup)}]{H_2O} —\overset{\overset{\displaystyle OH}{|}}{\underset{\underset{\displaystyle R}{|}}{C}}—R$$

These problems are quite similar to those we discussed for the reduction of carboxylic acid derivatives to aldehydes (Section 13.3). A highly reactive carboxylic acid derivative is needed, together with an organometallic reagent that has decreased reactivity.

One of the early solutions to this problem was the use of organo cadmium reagents, which are prepared from the corresponding Grignard reagents by reaction with cadmium chloride.

$$2\ RMgCl + CdCl_2 \longrightarrow R_2Cd + 2\ MgCl_2$$

Organocadmium compounds are less reactive than the Grignard reagents, and the ketone can be isolated in reasonable yield, as illustrated by the following reaction:

$$CH_3CH_2CH_2CH_2Br \xrightarrow[2.\ CdCl_2]{1.\ Mg} (CH_3CH_2CH_2CH_2)_2Cd \xrightarrow{ClCH_2—\overset{\overset{O}{\|}}{C}—Cl} ClCH_2—\overset{\overset{\displaystyle O}{\|}}{C}—CH_2CH_2CH_2CH_3$$

Several other procedures for preparation of ketones from organometallic reagents were developed subsequently, and the organocadmium method has been almost completely replaced by these newer techniques. One method that is quite useful for the preparation of ketones containing no other reactive functional groups is the reaction of an alkyllithium with a *carboxylic acid.*

$$R'—\overset{\overset{\displaystyle O}{\|}}{C}—OH \xrightarrow[2.\ H_2O\ \text{(workup)}]{1.\ RLi} R'—\overset{\overset{\displaystyle O}{\|}}{C}—R$$

Why does this reaction work, when an acid-base reaction should occur preferentially? The answer is that an acid-base reaction does occur *first,* but a subsequent reaction then takes place between the lithium salt of the carboxylic acid and a *second* equivalent of alkyllithium.

$$R'-\overset{\overset{\displaystyle O}{\|}}{C}-OH \xrightarrow{R-Li} [R'-\overset{\overset{\displaystyle O}{\|}}{C}-O^{-}Li^{+}] \xrightarrow[\text{2. } H_2O \text{ (workup)}]{\text{1. } R-Li} R'-\overset{\overset{\displaystyle O}{\|}}{C}-R$$

Nucleophilic attack of the organolithium reagent on the *negatively charged* carboxylate anion is somewhat surprising, but lithium ion is a relatively good Lewis acid, which bonds strongly to the oxygen of the carboxylate ion. The lithium salt is probably better drawn with a covalent bond between oxygen and lithium to indicate that the carboxyl group is essentially neutral. Reaction then leads to a dilithium salt, which is inert to further attack.

$$R'-\overset{\overset{\displaystyle O}{\|}}{C}-O-Li \xrightarrow{R-Li} R'-\underset{\underset{\displaystyle OLi}{|}}{\overset{\overset{\displaystyle OLi}{|}}{C}}-R \xrightarrow[\text{(workup)}]{H_2O} \left[R'-\underset{\underset{\displaystyle OH}{|}}{\overset{\overset{\displaystyle OH}{|}}{C}}-R\right] \longrightarrow R'-\overset{\overset{\displaystyle O}{\|}}{C}-R$$

Subsequent hydrolysis of this salt during workup leads to the ketone. (This proceeds via loss of water from the dihydroxy compound, discussed in Section 13.7). The following examples illustrate the use of this procedure for the preparation of ketones.

$$4\text{-}(CH_3)_3C\text{-cyclohexyl}-\overset{\overset{\displaystyle O}{\|}}{C}-OH \xrightarrow{CH_3Li} 4\text{-}(CH_3)_3C\text{-cyclohexyl}-\overset{\overset{\displaystyle O}{\|}}{C}-CH_3 \quad 91\%$$

$$C_6H_5-\overset{\overset{\displaystyle O}{\|}}{C}-OH \xrightarrow{t\text{-BuLi}} C_6H_5-\overset{\overset{\displaystyle O}{\|}}{C}-t\text{-Bu} \quad 66\text{–}67\%$$

Another useful procedure for the synthesis of ketones employs the reaction of a Grignard, or preferably an alkyllithium, reagent with a *nitrile* (a cyano compound). Addition of the organometallic reagent to the carbon-nitrogen triple bond produces the salt of an imine (a compound with a carbon-nitrogen double bond), and this is converted to the ketone by acidic hydrolysis during workup (see Section 13.7).

$$R'-C\equiv N \xrightarrow{R-Li} \left[R'-\overset{\overset{\displaystyle N-Li}{\|}}{C}-R\right] \xrightarrow[\text{workup}]{H_2O,\ HCl} R'-\overset{\overset{\displaystyle O}{\|}}{C}-R$$

The overall procedure is illustrated by the following reactions:

$$4\text{-}(CH_3)_3C\text{-}1\text{-}CH_3\text{-cyclohexyl}-C\equiv N \xrightarrow[\text{2. } H_2O,\ HCl]{\text{1. } CH_3Li} 4\text{-}(CH_3)_3C\text{-}1\text{-}CH_3\text{-cyclohexyl}-\overset{\overset{\displaystyle O}{\|}}{C}-CH_3 \quad 92\%$$

$$CH_3OCH_2C{\equiv}N \xrightarrow[2.\ H_2O,\ H_2SO_4]{1.\ C_6H_5Li} CH_3OCH_2{-}\overset{O}{\overset{\|}{C}}{-}C_6H_5 \qquad 71\text{–}78\%$$

One of the most valuable methods for preparing ketones via organometallic compounds is the reaction between a lithium dialkylcuprate and an acid chloride. The high selectivity of organocuprates permits the preparation of a wide variety of ketones, including those that have other reactive functional groups.

$$CH_3{-}\overset{O}{\overset{\|}{C}}{-}Cl \xrightarrow{(C_6H_5)_2CuLi} CH_3{-}\overset{O}{\overset{\|}{C}}{-}C_6H_5 \qquad 55\%$$

$$I{-}(CH_2)_{10}{-}\overset{O}{\overset{\|}{C}}{-}Cl \xrightarrow{(CH_3)_2CuLi} I{-}(CH_2)_{10}{-}\overset{O}{\overset{\|}{C}}{-}CH_3 \qquad 91\%$$

$$CH_3O{-}\overset{O}{\overset{\|}{C}}{-}CH_2CH_2{-}\overset{O}{\overset{\|}{C}}{-}Cl \xrightarrow{(CH_3CH_2CH_2CH_2)_2CuLi} CH_3O{-}\overset{O}{\overset{\|}{C}}{-}CH_2CH_2{-}\overset{O}{\overset{\|}{C}}{-}CH_2CH_2CH_2CH_3 \qquad 84\%$$

EXERCISE 13.5

Complete each of the following reactions:

(a) $CH_3{-}\overset{CH_3}{\overset{|}{C}H}{-}CH_2{-}\overset{O}{\overset{\|}{C}}{-}Cl \xrightarrow{(CH_3)_2Cd} ?$

(b) $CH_3{-}\overset{CH_3}{\overset{|}{C}H}{-}CH_2{-}\overset{O}{\overset{\|}{C}}{-}Cl \xrightarrow{CH_3CH_2Li} ?$

(c) $CH_3CH_2CH_2{-}\overset{O}{\overset{\|}{C}}{-}OH \xrightarrow{CH_3Li} ?$

(d) $CH_3CH_2CH_2{-}\overset{O}{\overset{\|}{C}}{-}Cl \xrightarrow{(CH_3CH_2)_2CuLi} ?$

EXERCISE 13.6

Suggest reagents for carrying out each of the following conversions:

(a) $CH_3{-}\overset{CH_3}{\overset{|}{C}H}{-}\overset{O}{\overset{\|}{C}}{-}Cl \xrightarrow{?} CH_3{-}\overset{CH_3}{\overset{|}{C}H}{-}\overset{O}{\overset{\|}{C}}{-}CH_2CH_2CH_3$ (two different ways)

(b) $CH_3CH_2{-}\overset{O}{\overset{\|}{C}}{-}OH \xrightarrow{?} CH_3CH_2{-}\overset{O}{\overset{\|}{C}}{-}CH_2CH_2CH_3$

13.5 REACTIVITY OF ALDEHYDES AND KETONES: THE POLARITY OF THE CARBONYL GROUP

In the first four sections of this chapter we have focused on the nomenclature and physical properties of aldehydes and ketones as well as on methods for their preparation. Now we will turn to the chemical reactions that aldehydes and ketones undergo. You can best understand the patterns of reactivity of these carbonyl compounds by first learning about the electronic structure of the carbon-oxygen double bond.

Structure of the Carbonyl Group

The structure of a carbonyl group is *planar:* the carbon and oxygen atoms and the two atoms bonded to the carbonyl carbon all lie in a single plane. In the simple case of formaldehyde all atoms of the molecule lie in a single plane.

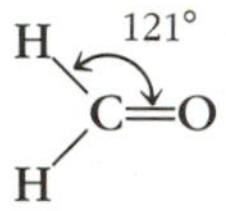

$$\underset{\text{H}}{\overset{\text{H}}{\diagdown}}\!\!\!\!\!\diagup \text{C}{=}\text{O} \quad (\text{H–C–O angle } 121^\circ)$$

Formaldehyde, a planar molecule

The carbonyl carbon is bonded to a total of three other atoms, which are separated by angles of approximately 120°, and we can conveniently describe it as having sp^2 hybridization. Three sp^2 hybrid orbitals are involved in formation of σ bonding molecular orbitals, and a p orbital remains for the π bond to oxygen. The carbon-oxygen double bond can therefore be described in terms of a σ and a π molecular orbital (Figure 13.1).

In contrast to the carbon atom, the carbonyl oxygen is bonded to only one atom, the carbonyl carbon. The σ bonding interaction (Figure 13.1) requires only a combination of 2s and a single 2p orbital on oxygen. Consequently, the oxygen should be sp hybridized, an sp hybrid orbital and a p orbital being utilized to form the σ and π molecular orbitals, respectively. This leaves the other sp orbital and the remaining p orbital on oxygen as nonbonding orbitals (Figure 13.2). Both of these are occupied by the unshared pairs of electrons of the neutral oxygen atom.

Polarity of the Carbonyl Group

The carbonyl group is polarized so that the oxygen atom has a partial negative charge and the carbonyl carbon atom has a partial positive charge.

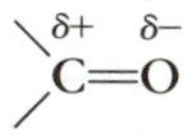

$$\diagdown\!\!\!\!\diagup \overset{\delta+}{\text{C}}{=}\overset{\delta-}{\text{O}}$$

This polarity is the consequence of several factors. As with any bond between carbon and a more electronegative element, the electrons of that bond are polarized toward the more electronegative element (Section 4.4).

$$\text{C} \,{+}\!\!\!\longrightarrow \text{O}$$

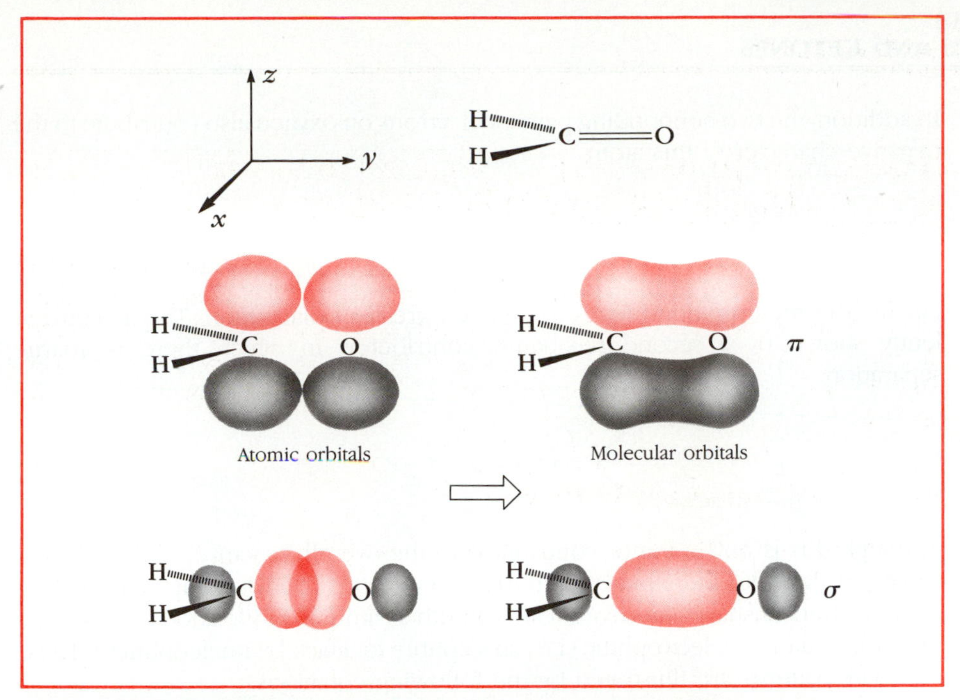

Figure 13.1 Molecular Orbital Description of the Carbon-Oxygen Double Bond of the Carbonyl Group of Formaldehyde.
(Only the bonding molecular orbitals are drawn.) In-phase combination of the $2p_z$ atomic orbitals leads to the bonding π molecular orbital. In-phase interaction of one of three sp^2 orbitals on carbon with an sp_y hybrid orbital on oxygen leads to the bonding σ orbital.

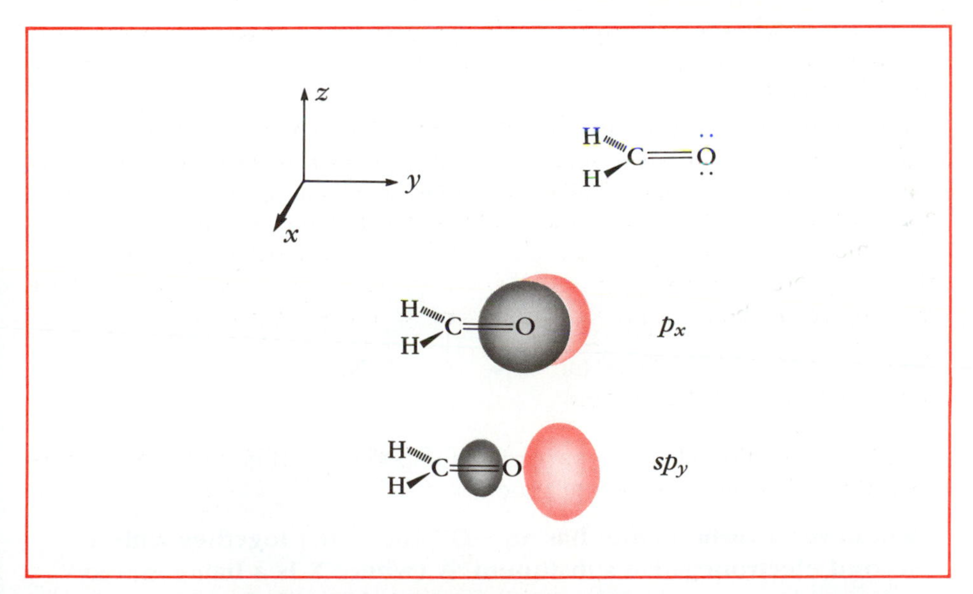

Figure 13.2 Molecular Orbital Description of the Lone Pair Orbitals on the Oxygen of a Carbonyl Group.
The bonding molecular orbitals drawn in Figure 13.1 (and of course the antibonding combinations as well) involve only the p_z and one of the two sp_y orbitals on oxygen. This leaves the p_x atomic orbital and the other sp_y hybrid orbital as nonbonding orbitals on oxygen. Both of these are doubly occupied in a normal carbonyl group.

In addition, the two nonbonding pairs of electrons on oxygen also contribute to the negative character of this atom.

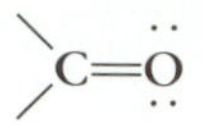

$$>C{=}\ddot{\underset{..}{O}}$$

Finally, the multiple bond leads to an even greater polarization. This is conveniently shown by a second resonance contributor, in which there is charge separation.

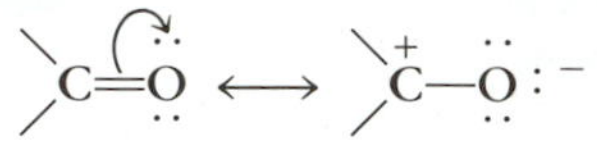

$$>C{=}\ddot{\underset{..}{O}} \longleftrightarrow >\overset{+}{C}{-}\ddot{\underset{..}{O}}{:}^{-}$$

Although this is only a minor contributor to the overall structure, it nevertheless serves to illustrate the characteristic reactivity of a carbonyl group. The oxygen atom, with its unshared electrons, can act both as a nucleophile and as a base, and the carbon atom is electrophilic (i.e., susceptible to attack by nucleophiles). These reactivity patterns are illustrated by the following equations:

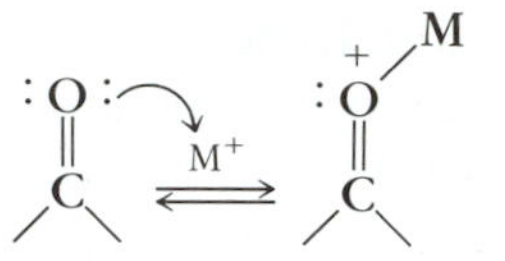

$$>C{=}\ddot{O}{:} \overset{M^+}{\rightleftharpoons} >C{=}\overset{+}{\ddot{O}}{-}M$$ Reaction of a carbonyl group *as* a nucleophile/base

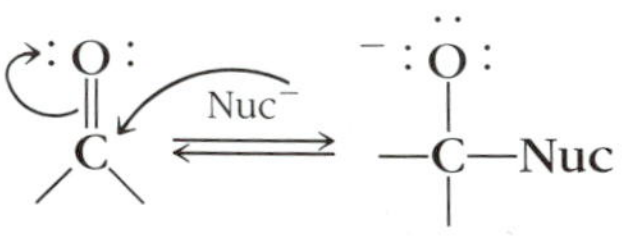

$$>C{=}\ddot{O}{:} \overset{Nuc^-}{\rightleftharpoons} {-}\underset{|}{\overset{\overset{-:\ddot{O}:}{|}}{C}}{-}\mathbf{Nuc}$$ Reaction of a carbonyl group *with* a nucleophile

In principle both these reactions should be reversible, but reversibility is not always observed in the latter process. We will present an extensive analysis of reversibility in nucleophilic attack on a carbonyl group in Sections 13.6 and 13.7, but the general requirement for reversibility is that Nuc$^-$ be able to act as a leaving group:

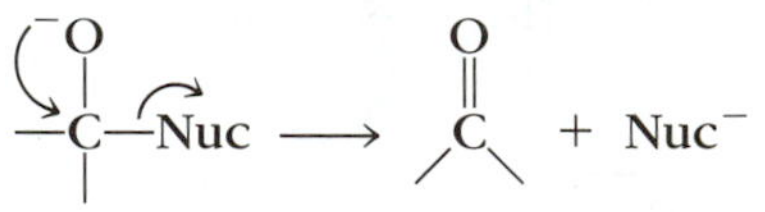

$${-}\underset{|}{\overset{\overset{^-O}{|}}{C}}{-}\mathbf{Nuc} \longrightarrow >C{=}O + \mathbf{Nuc}^-$$

This reverse reaction becomes increasingly important with good leaving groups and leads to the following generalization:

Whenever a carbon atom has an $-O^-$ (or $-OH$) together with a second electronegative substituent, X (where X is a halogen, oxygen, or nitrogen substituent), expulsion of X with formation of a carbonyl group occurs readily.

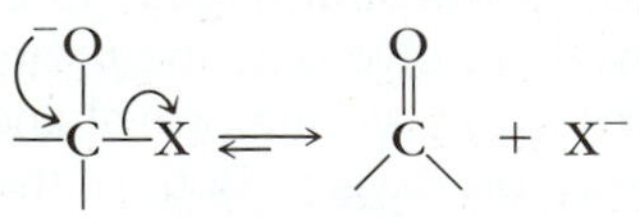

$${-}\underset{|}{\overset{\overset{^-O}{|}}{C}}{-}X \rightleftharpoons >C{=}O + X^-$$

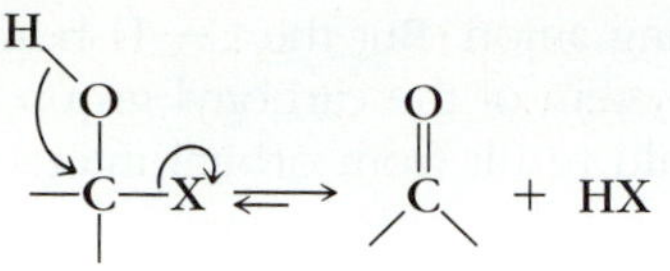

$$-\overset{|}{\underset{|}{C}}(OH)-X \rightleftharpoons >C{=}O + HX$$

You have previously encountered the expulsion of a leaving group from a carbon atom that also has an $-O^-$ substituent in the reactions of acid chlorides with metal hydrides (Section 13.3) and with organometallic reagents (Section 13.4). For example, in attack by "hydride ion":

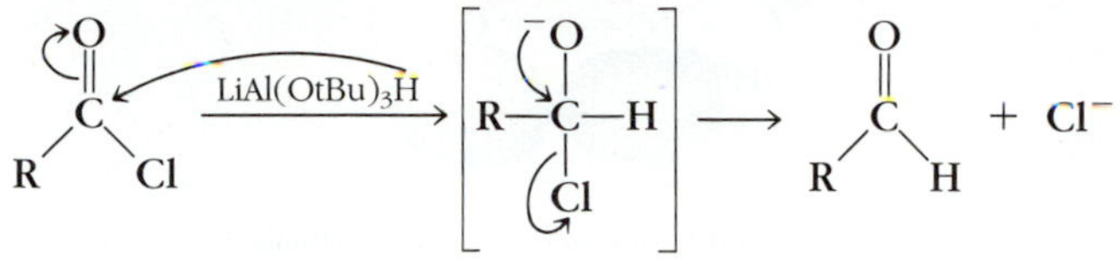

$$RC(=O)Cl \xrightarrow{LiAl(OtBu)_3H} [R-C(O^-)(Cl)-H] \longrightarrow RC(=O)H + Cl^-$$

The reactive center in the intermediate has both $-O^-$ and —Cl substituents, and expulsion of chloride ion leads to an aldehyde.

Acidity of α Hydrogens

The carbonyl group of aldehydes and ketones also exhibits a third type of reactivity: acidity of hydrogens on the α carbon. You can see the dramatic increase in acidity that results from the effect of an adjacent carbonyl group (Section 4.7) by comparing the following acid dissociation constants:

Cyclohexanone, K_a 10^{-17}

Cyclohexane, K_a 10^{-45}

This extremely large increase in acidity results from stabilization of the conjugate base by interactions with the π orbitals of the carbonyl group. This is conveniently illustrated by the two possible resonance contributors of the anion:

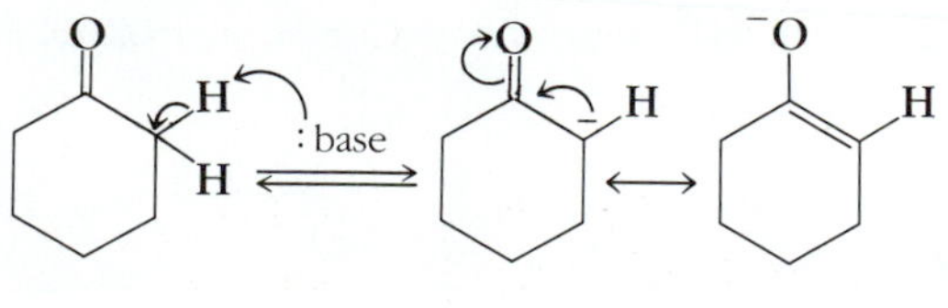

:base

The negative charge is not restricted to carbon but is delocalized over both the α carbon and the carbonyl oxygen. Although we will not consider it in detail, the molecular orbital description of this system corresponds very closely to that of the allyl anion (Section 8.3).

In contrast to the α hydrogens, the aldehyde hydrogen that is bonded directly to the carbonyl group does *not* exhibit such increased acidity. As you can see from the following drawing, the C—H bond for a hydrogen on an α carbon can be oriented parallel to the *p* orbitals on carbon and oxygen, and orbital interaction would stabilize the corresponding anion. But the C—H bond of an aldehydic C—H is perpendicular to the π system of the carbonyl group, and no stabilization of the corresponding anion could result from orbital interaction.

C—H bond is parallel to p orbitals of π system—*acidic*

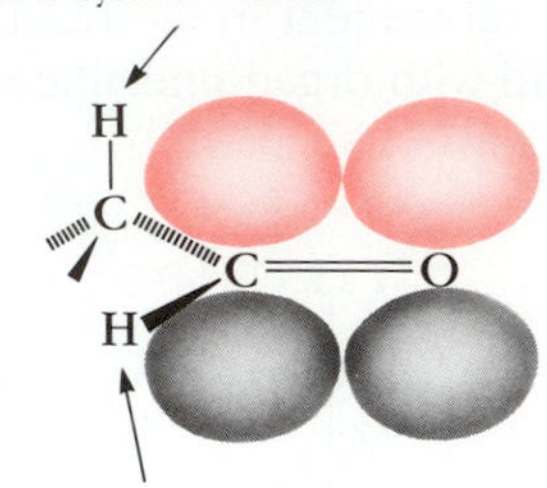

C—H bond is perpendicular to p orbitals of π system—*not acidic*

The acid dissociation constants for α hydrogens in typical aldehydes and ketones fall in the range of 10^{-17}–10^{-24}, so they are only slightly less acidic than alcohols ($K_a \sim 10^{-16}$–10^{-19}). This means that alkoxide ions are strong enough bases to generate small but significant concentrations of the conjugate bases of ketones and aldehydes.

$$-\overset{\displaystyle O}{\overset{\|}{C}}-\underset{\displaystyle H}{\underset{|}{\overset{|}{C}}}- \; \overset{^{-}OR}{\rightleftharpoons} \; -\overset{\displaystyle O}{\overset{\|}{C}}-\overset{-}{C}\langle \; \longleftrightarrow \; -\overset{\displaystyle O^{-}}{\overset{|}{C}}=C\langle$$

We will return to the importance of such acid-base reactions in Chapter 14, but for the present we have included it to give you a more complete description of the polarity of carbonyl compounds. A neutral aldehyde or ketone has partial positive charge ($\delta+$) on the carbonyl carbon and partial negative charge ($\delta-$) on the carbonyl oxygen. In addition an α carbon that has at least one hydrogen substituent can become negatively charged by reaction with a base.

EXERCISE 13.7

Where would each of the following species attack a carbonyl group, at carbon or at oxygen?

(a) H^+ (b) HO^- (c) Li in CH_3Li
(d) CH_3 in CH_3Li (e) Mg^{2+} (f) Br^-

EXERCISE 13.8

Which is (are) the most acidic hydrogen(s) in each of the following compounds?

(a) $CH_3CH_2-\overset{\displaystyle OH}{\overset{|}{C}H}-CH_2CH_3$ (b) $CH_3CH_2-\overset{\displaystyle O}{\overset{\|}{C}}-CH_2CH_3$

(c) $CH_3CH_2CH_2OH$ (d) CH_3CH_2CHO

13.6 IRREVERSIBLE ADDITION REACTIONS TO ALDEHYDES AND KETONES

Under what circumstances will nucleophilic addition to a carbonyl group be reversible?

$$\mathrm{C{=}O} \xrightarrow{\mathrm{Nuc^-}} -\mathrm{C(O^-)}-\mathrm{Nuc} \longrightarrow \mathrm{C{=}O} + \mathrm{Nuc^-}$$

In principle all carbonyl reductions could be reversible, but in practice many are not. The key requirement for a reversible reaction is that the nucleophile be a good leaving group. Conversely, if the nucleophile can not act as a leaving group, the reaction will be irreversible. The most common examples of irreversible carbonyl additions are the reactions of aldehydes and ketones with organometallic and metal hydride reagents. Expulsion of a free carbanion or a free hydride ion does not occur, because too much energy would be needed to form such high-energy intermediates.

$$\mathrm{C{=}O} \xrightarrow{\mathrm{RMgX}} -\mathrm{C(O^-)}-\mathrm{R} \not\longrightarrow \mathrm{C{=}O} + \mathrm{R^-}$$

$$\mathrm{C{=}O} \xrightarrow{\mathrm{LiAlH_4}} -\mathrm{C(O^-)}-\mathrm{H} \not\longrightarrow \mathrm{C{=}O} + \mathrm{H^-}$$

Even reversal of these reactions to regenerate the original metal derivative is unfavorable, because the forward reaction is usually quite exothermic and the energy requirements for the reverse reactions are too great (Figure 13.3). The greater stability of the products in these particular reactions reflects the much greater strength of carbon-carbon or carbon-hydrogen bonds in comparison with metal-carbon or metal-hydrogen bonds.

The following examples illustrate several such irreversible carbonyl addition reactions of aldehydes and ketones.

$$\mathrm{(C_6H_5)_2{-}CH{-}CH{=}O} \xrightarrow[\text{2. } \mathrm{H_2O}]{\text{1. } \mathrm{LiAlH_4}} \mathrm{(C_6H_5)_2{-}CH{-}CH_2OH} \quad 98\%$$

$$\text{2-adamantanone} \xrightarrow[\text{2. } \mathrm{H_2O}]{\text{1. } \mathrm{CH_3CH_2Li}} \text{2-adamantane with OH and } \mathrm{CH_2CH_3} \quad 83\%$$

$$\text{cyclohexyl}{-}\mathrm{CHO} \xrightarrow[\text{2. } \mathrm{H_2O}]{\text{1. } \mathrm{CH_2{=}CH{-}CH_2MgBr}} \text{cyclohexyl}{-}\mathrm{CH(OH){-}CH_2{-}CH{=}CH_2} \quad 72\%$$

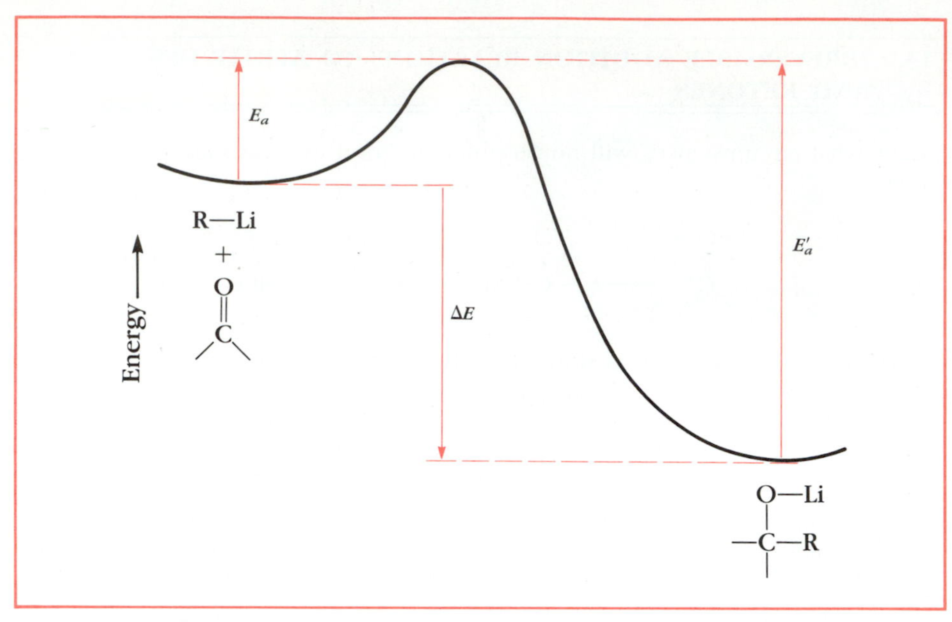

Figure 13.3 Reaction Coordinate for Attack of an Alkyllithium at the Carbonyl Group of an Aldehyde or Ketone.
The product is considerably more stable than the reactants (ΔE), and the activation energy of the forward reaction (E_a) is small. The activation energy for the reverse reaction (designated E'_a) is quite large, so the addition of an organometallic reagent to a carbonyl group is effectively irreversible.

13.7 REVERSIBLE ADDITION REACTIONS TO ALDEHYDES AND KETONES

In order for addition to a carbonyl group to be reversible, the nucleophile must also be able to act as a leaving group. The overall reaction can be summarized as the reversible addition of H and X.

$$\mathrm{>C{=}O} \underset{}{\overset{HX}{\rightleftharpoons}} \mathrm{-C(OH)-X}$$

The reaction may proceed under either acidic or basic conditions depending upon the specific example, and the nucleophile–leaving group can therefore be either HX or X^-. The reactions to be considered in this section are those in which HX is water (HOH), an alcohol (HOR), sodium bisulfite ($NaHSO_3$), hydrogen cyanide (HCN), or an amine (NH_2R).

Addition of Water

Ketones and aldehydes undergo a reversible reaction with water to yield a **geminal** (or *gem-*) diol. The term (derived from the Latin gemini, "twins") indicates that both substituents are attached to the same carbon atom.

$$H_2O + \underset{/\ \backslash}{\overset{\overset{O}{\|}}{C}} \overset{K_{hydration}}{\rightleftharpoons} -\overset{\overset{OH}{|}}{\underset{|}{C}}-OH$$

In most cases, particularly for ketones, the *gem*-diol is unstable relative to the carbonyl compound, and the equilibrium of the preceding equation lies to the left. However, some low-molecular-weight aldehydes undergo hydration to a substantial extent, and hydration is also favored by halogen substituents at the α position. The equilibrium constants for hydration of several carbonyl compounds are presented in Table 13.3.

The data in Table 13.3 show that there is a regular increase in hydration of aldehydes as the size of the alkyl substituent decreases, but even the smallest ketone (acetone) does not undergo substantial hydration. The last entry in the table (2,2,2-trichloroacetaldehyde) illustrates the effect of α-halo substituents. The hydrated form of this aldehyde is highly favored at equilibrium; it is a stable crystalline compound called *chloral hydrate* (well known in popular fiction as the active ingredient in "knockout drops").

The hydration of an aldehyde or ketone occurs under either acidic or basic conditions, and the reaction mechanism for each set of conditions is summarized in the following discussion.

Acidic Conditions

$$\underset{/\ \backslash}{\overset{\overset{:O:}{\|}}{C}} \overset{H^+}{\rightleftharpoons} \underset{/\ \backslash}{\overset{\overset{:\overset{+}{O}H}{\|}}{C}} \overset{H_2\ddot{O}:}{\rightleftharpoons} -\overset{\overset{:\ddot{O}H}{|}}{\underset{|}{C}}-\overset{+}{\ddot{O}}H_2 \overset{-H^+}{\rightleftharpoons} -\overset{\overset{:\ddot{O}H}{|}}{\underset{|}{C}}-\ddot{O}H$$

Under acidic conditions the first step of a carbonyl addition reaction is invariably *protonation of the carbonyl group*. This makes the carbonyl carbon much more susceptible to nucleophilic attack, as suggested by the two possible resonance

TABLE 13.3 Equilibrium Constants for Hydration of Aldehydes and Ketones

Aldehyde or Ketone	$K_{hydration}$
CH_2O	2,000
CH_3CHO	1
$(CH_3)_3C-CHO$	0.2
$CH_3-\overset{\overset{O}{\|}}{C}-CH_3$	0.002
Cl_3C-CHO	30,000

forms for the protonated carbonyl group:

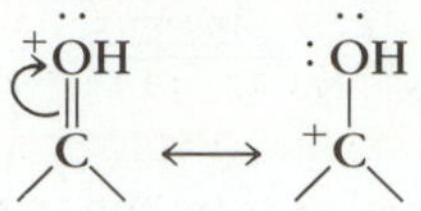

The initial protonation is followed by attack of the nucleophile (H_2O), and finally a proton transfer from the $-OH_2^+$ group (to H_2O or some other base) yields the *gem*-diol. The mechanism for the reverse process is exactly the opposite sequence:

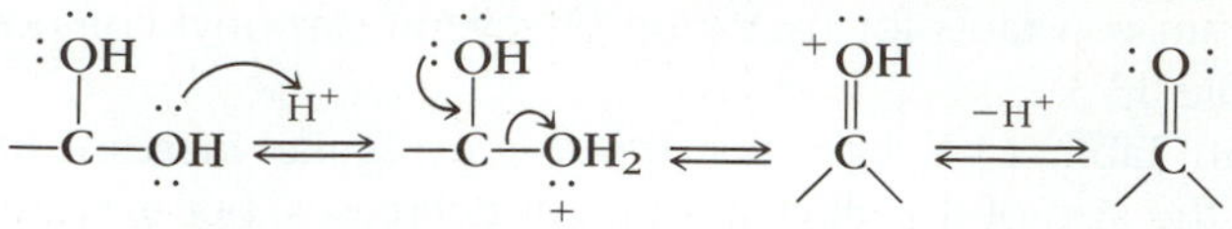

Protonation of one of the OH groups of the *gem*-diol is followed by loss of neutral water, and proton transfer then results in the neutral carbonyl group.

Basic Conditions

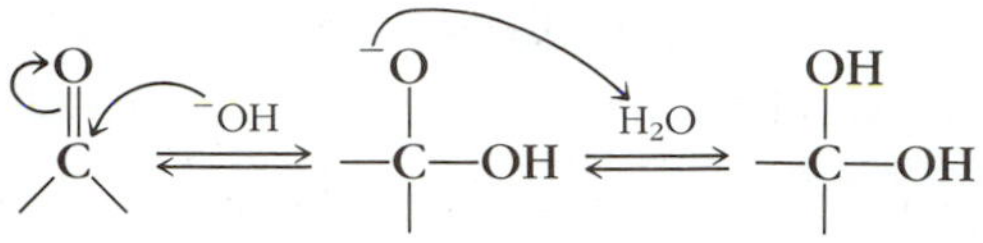

Under basic conditions the nucleophile is already deprotonated when it attacks the carbonyl group (i.e., the nucleophile is hydroxide ion rather than water). After nucleophilic attack the original carbonyl oxygen is protonated by proton transfer from solvent. The mechanism of the reverse process is again precisely the opposite: deprotonation of one of the OH groups of the *gem*-diol is followed by expulsion of a hydroxide ion to form the neutral carbonyl compound. (This reverse reaction illustrates one of the few situations in which hydroxide ion is observed as a leaving group.)

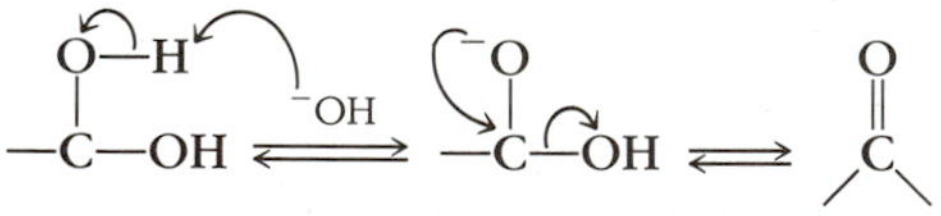

One of the most important aspects of the mechanisms for addition in acid and in base is their similarity; the major difference lies only in the sequence of the protonation and deprotonation steps. For example, in the acid-catalyzed addition the first step is protonation of the carbonyl oxygen, whereas this corresponds to the last step in the base-catalyzed addition.

Addition of Alcohols

The addition of an alcohol to an aldehyde or ketone affords a **hemiacetal** or **hemiketal,** respectively.

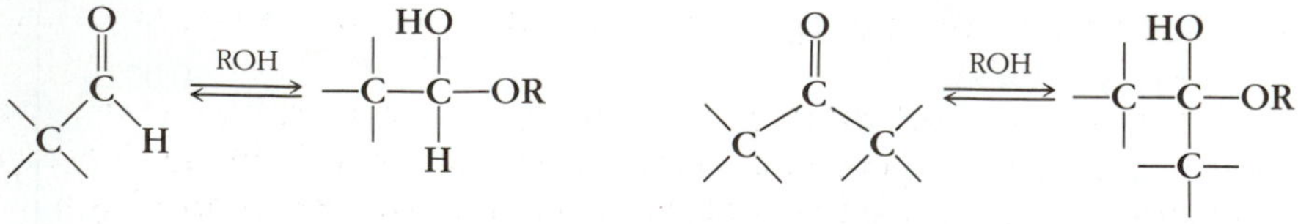

Aldehyde Hemiacetal Ketone Hemiketal

The reaction is analogous to the addition of water, and equilibrium usually favors the free aldehyde or ketone. Nevertheless, the hemiacetals and hemiketals are important intermediates in the formation of acetals and ketals (Section 15.5).

When the reaction is *intramolecular,* equilibrium is expected to favor the hemiacetal or hemiketal if a five- or six-membered ring results. Thus 5-hydroxypentanal exists predominantly in the hemiacetal form.

CHO, OH ⇌ OH, H, O

A similar situation is found for many sugars, which are polyhydroxy ketones and aldehydes (Section 19.2).

The mechanisms for addition of an alcohol to a ketone or aldehyde are essentially the same as those presented for the addition of water.

Acidic Conditions

$$\mathrm{>C{=}O} \overset{H^+}{\rightleftharpoons} \mathrm{>C{=}\overset{+}{O}H} \overset{R{-}\ddot{O}H}{\rightleftharpoons} \mathrm{-\overset{OH}{\underset{|}{C}}-\overset{+}{\underset{H}{O}}R} \overset{-H^+}{\rightleftharpoons} \mathrm{-\overset{OH}{\underset{|}{C}}-OR}$$

Basic Conditions

$$\mathrm{>C{=}O} \overset{^-OR}{\rightleftharpoons} \mathrm{-\overset{O^-}{\underset{|}{C}}-OR} \overset{H_2O}{\rightleftharpoons} \mathrm{-\overset{OH}{\underset{|}{C}}-OR}$$

Addition of Sodium Bisulfite

The reaction of a ketone or aldehyde with sodium bisulfite under weakly acidic conditions can lead to the bisulfite adduct.

$$\mathrm{NaHSO_3} + \mathrm{>C{=}O} \overset{K}{\rightleftharpoons} \mathrm{-\overset{OH}{\underset{|}{C}}-SO_3Na}$$

Ketone or aldehyde — Bisulfite adduct

These adducts can often be isolated as water-soluble, crystalline salts, and in such cases they provide an excellent method for separating an aldehyde or ketone from water-insoluble organic impurities. The addition is reversible, so the original aldehyde or ketone can easily be regenerated by acidification of the bisulfite adduct after it has been purified.

For aldehydes, equilibrium often favors the bisulfite adduct, but the free carbonyl compound usually predominates in the case of a ketone (Table 13.4). Even if the equilibrium constant is unfavorable, the adduct can often be prepared by using a large excess of sodium bisulfite.

TABLE 13.4 Equilibrium Constants for Addition of Sodium Bisulfite to Aldehydes and Ketones

Aldehyde or Ketone	*K*
CH_3CH_2CHO	100
C_6H_5CHO	7
$CH_3-C(=O)-CH_3$	3
$CH_3-C(=O)-CH_2CH_3$	0.5
$C_6H_5-C(=O)-C_6H_5$	~0

Bisulfite ion is the conjugate base of sulfurous acid, and it has nonbonded electrons on both oxygen and sulfur.

$$HO-\ddot{S}(=O)-OH \qquad HO-\ddot{S}(=O)-O^- \; Na^+$$

Sulfurous acid   Sodium bisulfite

As a result, both the oxygen and sulfur atoms are capable of acting as nucleophiles. The structure of the bisulfite adduct was debated for many years, but recent evidence has shown that it is the sulfur atom which attacks the carbonyl carbon:

$$>C=O \underset{}{\overset{H^+}{\rightleftharpoons}} >C=\overset{+}{O}H \xrightleftharpoons{:S(=O)(ONa)-O-H} -\overset{|}{\underset{|}{C}}(OH)-\overset{+}{S}(=O)(O-H)-ONa \overset{-H^+}{\rightleftharpoons} -\overset{|}{\underset{|}{C}}(OH)-S(=O)_2-ONa$$

This leads to the sodium salt of an α-hydroxy sulfonic acid (cf. Section 10.8).
The following example illustrates the preparation of bisulfite adducts:

$$Cl_3C-CH(=O) \xrightarrow[H_2O]{NaHSO_3} Cl_3C-CH(OH)-SO_3Na \qquad \sim 100\%$$

Addition of Hydrogen Cyanide

The addition of HCN to an aldehyde or ketone produces a *cyanohydrin*.

$$\underset{\text{Aldehyde or ketone}}{>C=O} \xrightleftharpoons{HC\equiv N} \underset{\text{Cyanohydrin}}{-\overset{OH}{\underset{|}{\overset{|}{C}}}-C\equiv N}$$

The equilibrium in this reaction is typically favorable for addition to both aldehydes and ketones. The cyanohydrins are relatively stable and can be isolated, although reversal occurs readily in the presence of base or at elevated temperatures.

Preparation of cyanohydrins by the addition of HCN is usually carried out in acid. HCN is a weak acid ($K_a \sim 10^{-10}$), but small amounts of cyanide ion are present, and the reaction proceeds by the following mechanism:

$$>C=O \xrightleftharpoons{H^+} >C=\overset{+}{O}H \xrightleftharpoons{^-C\equiv N} -\overset{OH}{\underset{|}{\overset{|}{C}}}-C\equiv N$$

The addition of hydrogen cyanide to acetone and to cycloheptanone illustrates the process.

$$CH_3-\overset{O}{\overset{\|}{C}}-CH_3 \xrightarrow[H_2SO_4]{NaCN} CH_3-\overset{OH}{\underset{CH_3}{\overset{|}{\underset{|}{C}}}}-C\equiv N \quad 77\text{–}78\%$$

$$\text{cycloheptanone}=O \xrightarrow[H_2SO_4]{KC\equiv N} \text{cycloheptane}\langle^{CN}_{OH} \quad 81\%$$

A sugar derivative of the cyanohydrin of benzaldehyde occurs naturally in bitter almonds. In this compound, called *amygdalin,* the hydroxyl hydrogen is replaced by a 12-carbon carbohydrate group consisting of two glucose residues. Hydrolysis of the oxygen-sugar linkage leads to the free cyanohydrin, and at high temperatures HCN is lost, yielding the aldehyde. This is the origin of benzaldehyde as the flavor ingredient in oil of bitter almond, which is obtained by distillation at relatively high temperatures.

$$\underset{\text{Amygdalin}}{C_6H_5-\overset{OC_{12}H_{21}O_{10}}{\overset{|}{C}H}-CN} \xrightarrow{\text{hydrolysis}} C_6H_5-\overset{OH}{\overset{|}{C}H}-CN \xrightarrow[-HCN]{\text{heat}} \underset{\text{Benzaldehyde}}{C_6H_5-\overset{O}{\overset{\|}{C}}-H}$$

Amygdalin is a relatively toxic substance, which is also found in apricot and peach pits. It has achieved some notoriety in recent years as the source of laetrile, a controversial drug sometimes advocated for anticancer therapy. Laetrile is produced by cleavage of one of the two glucose units, followed by a mild oxidation.

$$\underset{\text{Amygdalin}}{C_6H_5{-}\overset{\displaystyle OC_{12}H_{21}O_{10}}{\overset{|}{C}H}{-}CN} \xrightarrow[\text{and oxidation}]{\text{cleavage}} \underset{\text{Laetrile}}{C_6H_5{-}\overset{\displaystyle OC_6H_9O_6}{\overset{|}{C}H}{-}CN}$$

Although laetrile has reportedly been used as a treatment for cancer since 1845, no data have yet been published in the scientific literature which demonstrate that it actually has anticancer activity.

Addition of Ammonia Derivatives

The addition of ammonia derivatives to aldehydes and ketones occurs readily (typically with mild acid catalysis).

$$>C{=}O + R\ddot{N}H_2 \overset{H^+}{\rightleftharpoons} -\overset{OH}{\overset{|}{\underset{|}{C}}}-\ddot{N}HR$$

An amino alcohol

A new possibility arises in this situation, however. Not only could the resulting amino alcohol undergo the reverse reaction to regenerate the carbonyl compound, but it could also lose *water.*

$$-\overset{\ddot{O}H}{\overset{|}{\underset{|}{C}}}-\ddot{N}HR \overset{H^+}{\rightleftharpoons} -\overset{\overset{+}{O}H_2}{\overset{|}{\underset{|}{C}}}-\ddot{N}HR \rightleftharpoons >C{=}\overset{+}{N}HR \overset{-H^+}{\rightleftharpoons} >C{=}\ddot{N}{-}R$$

The net result of this sequence is replacement of the original carbon-oxygen double bond with a carbon-nitrogen double bond. This conversion plays a key role in a variety of organic reactions. Hydrazine (NH_2NH_2) and substituted hydrazines provide particularly important examples.

The reaction of *hydrazine* with an aldehyde or ketone leads to the formation of the corresponding *hydrazone.*

$$p\text{-}CH_3C_6H_4{-}\overset{O}{\overset{\|}{C}}{-}CH_3 \xrightarrow{NH_2NH_2} p\text{-}CH_3C_6H_4{-}\overset{N{-}NH_2}{\overset{\|}{C}}{-}CH_3$$

A hydrazone

Such hydrazones are formed as the first step in the Wolff-Kishner reduction (Section 13.11). The formation of arylhydrazones (from arylhydrazines) is a useful way of characterizing aldehydes and ketones. The arylhydrazones are usually highly crystalline derivatives with characteristic melting points.

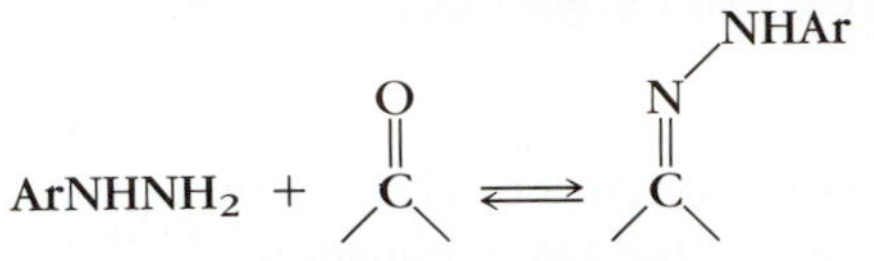

$$ArNHNH_2 + \text{>C=O} \rightleftharpoons \text{>C=N–NHAr} \quad \text{(precipitate)}$$

Arylhydrazone

EXERCISE 13.9

Draw the product of each of the following addition reactions.

(a) $CH_3CH_2\overset{O}{\overset{\|}{C}}H \xrightarrow{HCN} ?$

(b) $CH_3-\overset{O}{\overset{\|}{C}}-CH_2CH_3 \xrightarrow{CH_3OH} ?$

(c) $CH_3-\overset{O}{\overset{\|}{C}}-CH_2CH_3 \xrightarrow{NaHSO_3} ?$

(d) cyclohexanone $\xrightarrow{C_6H_5NHNH_2} ?$

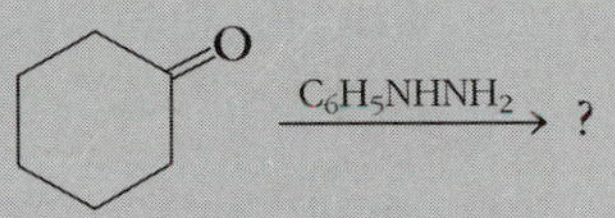

13.8 REACTIONS OF ALDEHYDES AND KETONES WITH YLIDES

The term **ylide** describes a neutral molecule having a negative charge on a carbon atom and a positive charge on an adjacent **heteroatom** such as phosphorus or sulfur (a heteroatom is any atom other than carbon in a chain or ring). Ylides undergo useful reactions with aldehydes and ketones.

$$(C_6H_5)_3\overset{+}{P}-CH_2^- \qquad (CH_3)_2\overset{+}{\underset{\cdot\cdot}{S}}-CH_2^-$$

Triphenylphosphonium methylide, a phosphorus ylide — Dimethylsulfonium methylide, a sulfur ylide

In the preceding drawings each individual atom has an octet of electrons. Consequently, a resonance structure having no charged atoms could be drawn only by allowing the heteroatom to have more than eight electrons in its valence shell. This is not unreasonable in the case of second-row elements such as phosphorus and sulfur, because these atoms have empty *d* orbitals of relatively low energy which are available for bonding.

$$(C_6H_5)_3\overset{+}{P}-CH_2^- \longleftrightarrow (C_6H_5)_3P{=}CH_2 \qquad (CH_3)_2\overset{+}{S}-CH_2^- \longleftrightarrow (CH_3)_2S{=}CH_2$$

Neutral resonance forms can therefore be drawn, but the ionic forms correctly predict the nucleophilic character of carbon in these ylides. Ylides are quite reactive, and they are ordinarily used as soon as they are generated in solution without any attempt being made to isolate them.

Phosphorus Ylides

The reaction of phosphorus ylides with aldehydes and ketones is called the Wittig reaction after Georg Wittig (Germany), who developed it. For this and related achievements Wittig was named a corecipient of the 1979 Nobel Prize in chemistry. The product of a Wittig reaction is an alkene, in which a double bond to carbon replaces the original double bond to oxygen. The reaction proceeds via a cyclic intermediate, as summarized by the following equation:

$$>C{=}O \xrightarrow{^{-}CH_2-\overset{+}{P}(C_6H_5)_3} >C(-O^{-})-CH_2-\overset{+}{P}(C_6H_5)_3 \longrightarrow >C(CH_2)(O)P(C_6H_5)_3 \longrightarrow >C{=}CH_2 + O{=}P(C_6H_5)_3$$

The phosphorus ylides are prepared by a two-step sequence, which begins by alkylating triphenylphosphine with an alkyl halide.

$$\underset{\text{Triphenylphosphine}}{(C_6H_5)_3P:} \xrightarrow{RCH_2-X} \underset{\text{A phosphonium salt}}{(C_6H_5)_3\overset{+}{P}-CH_2R + X^-}$$

This is a direct displacement reaction and works best with primary alkyl halides. Treatment of the resulting phosphonium salt with a strong base then affords the corresponding ylide.

$$\underset{\text{Phosphonium salt}}{(C_6H_5)_3-\overset{+}{P}-\underset{}{CH}(H)-R\ X^-} \xrightarrow{:Base} \underset{\text{Phosphorus ylide}}{(C_6H_5)_3\overset{+}{P}-\overset{-}{C}HR \longleftrightarrow (C_6H_5)_3P{=}CHR}$$

Some of the bases frequently used to prepare phosphorus ylides are potassium *tert*-butoxide, the sodium salt of dimethyl sulfoxide ($Na^{+}\,^{-}CH_2-SO-CH_3$), sodium hydride (NaH), and phenyllithium or butyllithium. In contrast to the hydrides of boron and aluminum, the hydrides of sodium and lithium are very strong bases but exhibit little reactivity as nucleophiles. The two organolithium reagents are already familiar to you as nucleophiles (Section 11.5, 13.4, and 13.6), but they react preferentially as bases with a phosphonium salt. For example

$$(C_6H_5)_3\overset{+}{P}-CH(H)-R \xrightarrow{C_6H_5-Li} (C_6H_5)_3P{=}CHR + C_6H_6$$

The following examples illustrate the use of the Wittig reaction with aldehydes and ketones:

$$(C_6H_5)_3\overset{+}{P}-CH_3\ I^- \xrightarrow{NaH} (C_6H_5)_3P{=}CH_2 \xrightarrow{\text{cyclohexanone}} \text{methylenecyclohexane } (=CH_2) \quad 86\%$$

$$(C_6H_5)_3\overset{+}{P}\text{—cyclopentyl}\ I^- \xrightarrow{C_4H_9Li} (C_6H_5)_3P\text{=cyclopentylidene} \xrightarrow{C_6H_5\text{—}CHO} C_6H_5\text{—}CH\text{=cyclopentylidene} \quad 65\%$$

$$(C_6H_5)_2C\text{=}O \xrightarrow{(C_6H_5)_3P\text{=}CHCH_3} (C_6H_5)_2C\text{=}C(H)CH_3 \quad 98\%$$

$$\text{2,2-dimethyl-1-tetralone} \xrightarrow{(C_6H_5)_3P\text{=}CH_2} \text{1-methylene-2,2-dimethyltetralin} \quad 83\%$$

$$\text{cyclohexanone} \xrightarrow{(C_6H_5)_3P\text{=}CHC_6H_5} \text{cyclohexylidene=}CHC_6H_5 \quad 60\%$$

$$C_6H_5CHO \xrightarrow{(C_6H_5)_3P\text{=cyclohexylidene}} \text{cyclohexylidene=}CHC_6H_5 \quad 70\%$$

The last two reactions show you that the same alkene can sometimes be prepared by two alternative pathways.

Sulfur Ylides

In contrast to the alkene-forming reactions of the phosphorus reagents, *sulfur ylides* react with aldehydes and ketones to yield *epoxides.* The reaction proceeds by nucleophilic attack on carbonyl carbon, followed by an intramolecular nucleophilic displacement reaction.

$$\text{>C=O} \xrightarrow{^-CH_2\text{—}\overset{+}{S}(CH_3)_2} \text{—}\underset{|}{\overset{^-O}{C}}\text{—}CH_2\text{—}\overset{+}{S}(CH_3)_2 \longrightarrow \text{>C(O)}CH_2 + (CH_3)_2S$$

The reactions of sulfur ylides have been extensively investigated by E. J. Corey (Harvard University) and B. M. Trost (University of Wisconsin). Substituted ylides such as $(CH_3)_2S\text{=}CHR$ have also been studied, but they sometimes undergo more complex reactions. Therefore we will restrict our discussion to the most common use of sulfur ylides, the introduction of a CH_2 group to form an epoxide, as shown in the preceding equation.

The sulfur ylides are also prepared by a two-step sequence: alkylation of a neutral sulfur compound followed by treatment of the resulting sulfonium salt with a strong base, such as the sodium salt of dimethyl sulfoxide (DMSO).

$$(CH_3)_2\ddot{S} \xrightarrow{CH_3\text{—}I} (CH_3)_2\overset{+}{S}\text{—}CH_2\text{—}H\ \ I^- \xrightarrow{Na^{+\,-}CH_2\overset{O}{\overset{\|}{S}}CH_3} (CH_3)_2\overset{+}{S}\text{—}CH_2^- \longleftrightarrow (CH_3)_2S\text{=}CH_2$$

The following examples illustrate the formation of epoxides by the reaction of aldehydes and ketones with sulfur ylides:

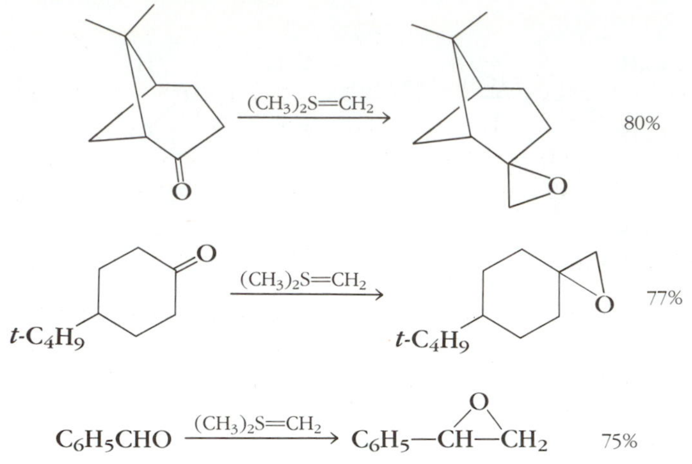

$\xrightarrow{(CH_3)_2S{=}CH_2}$ 80%

$t\text{-}C_4H_9$ $\xrightarrow{(CH_3)_2S{=}CH_2}$ $t\text{-}C_4H_9$ 77%

$$C_6H_5CHO \xrightarrow{(CH_3)_2S{=}CH_2} C_6H_5{-}\overset{O}{CH{-}CH_2} \quad 75\%$$

The preceding discussions of sulfur and phosphorus ylides show that these are valuable reagents for the preparation of epoxides and alkenes, respectively. In contrast to the methods that we discussed earlier in the text for preparing those classes of compounds, these new reactions proceed with carbon-carbon bond formation. Therefore the resulting alkenes or epoxides have a larger number of carbon atoms than the starting aldehydes or ketones from which they are prepared.

EXERCISE 13.10

Show how CH_3CH_2CHO could be converted to

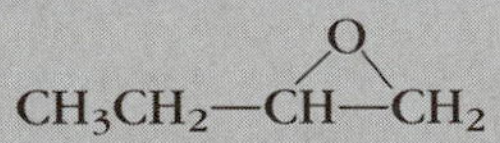

$$CH_3CH_2{-}\overset{O}{CH{-}CH_2}$$

(a) Using a sulfur ylide
(b) By a two-step sequence using a phosphorus ylide

EXERCISE 13.11

Complete the reaction of 3-pentanone with: (a) $(C_6H_5)_3P{=}CHCH_2CH_3$; (b) $(CH_3)_2S{=}CH_2$

$$CH_3CH_2{-}\overset{\overset{O}{\|}}{C}{-}CH_2CH_3 \longrightarrow ?$$

13.9 ENOLS AND ENOLATES

Enols were previously encountered as intermediates in the hydration of alkynes (Section 7.5) and in the reaction of vinyl triflates (Section 12.7). In both cases you saw that they normally undergo rapid rearrangement to isomeric aldehydes and ketones.

$$-C\equiv C- \xrightarrow{\text{hydration}} -CH=C(OH) \rightleftarrows -CH_2-C(=O)$$

enol

In this and subsequent chapters we will show that enols are important intermediates in many reactions of aldehydes and ketones. The conjugate bases of enols, called *enolate ions,* were also encountered previously (Section 4.7), and these too play an important role in the reaction of aldehydes and ketones.

Keto-Enol Equilibria

In the presence of either an acidic or basic catalyst, a rapid equilibrium can be established between an aldehyde or ketone and its isomeric enol form. This equilibrium is described as *keto-enol equilibrium.** For either an aldehyde or a ketone the carbonyl isomer is referred to as the *keto* form, and this usually predominates in the equilibrium mixture.

$$\underset{\text{Keto form}}{>C(=O)-CH<} \underset{}{\overset{\text{acid or base}}{\rightleftarrows}} \underset{\text{Enol form}}{>C(OH)=C<}$$

In the presence of base the interconversion between keto and enol forms occurs via the enolate ion. Abstraction of a proton from either the α carbon of the keto form or the OH group of the enol will produce the enolate ion. Two resonance forms can be drawn for the enolate ion, and these show that negative charge is localized both on the carbonyl oxygen and on the α carbon. Protonation of an enolate ion on oxygen affords the enol, and protonation on the α carbon generates the keto form. The base-catalyzed equilibrium is summarized by the equation in the following box in which SOH represents a hydroxylic solvent and SO^- is its conjugate base.

Keto-Enol Equilibrium: Base-Catalyzed

$$H-C-C(=O) \underset{SOH}{\overset{SO^-}{\rightleftarrows}} \left[>\overset{-}{C}-C(=O) \longleftrightarrow >C=C(O^-) \right] \underset{SO^-}{\overset{SOH}{\rightleftarrows}} >C=C(OH)$$

The keto-enol equilibrium can also be catalyzed by acid. As in other acid-catalyzed reactions of aldehydes and ketones, the first step is protonation of the carbonyl group. The positive charge in the resulting oxonium ion renders the α

*Rapid equilibrium between isomeric structures is sometimes known as *tautomerism,* and the isomeric compounds are called *tautomers.* The equilibrium between a carbonyl compound and its enol form is *keto-enol tautomerism.*

hydrogens much more acidic, and even a weak base such as neutral solvent can remove a proton to generate the enol.

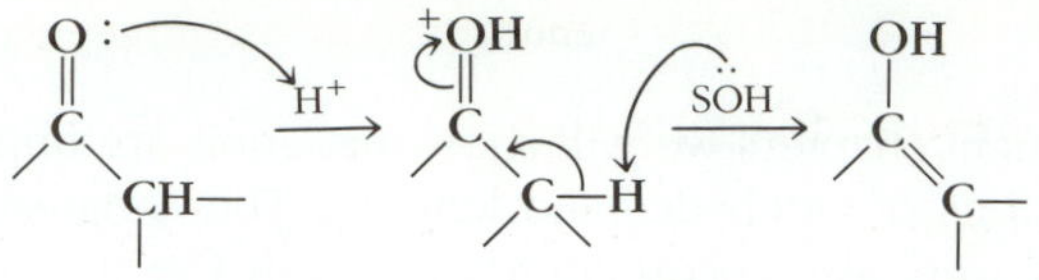

The conversion of an enol to the keto form proceeds via protonation on the α carbon to yield the same intermediate:

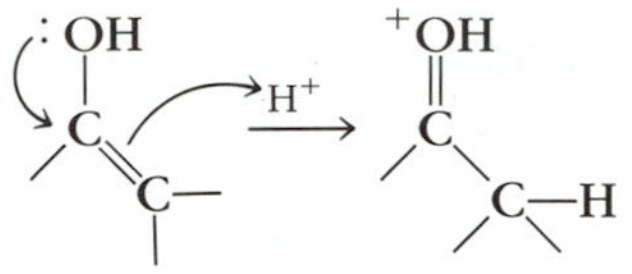

The acid-catalyzed equilibrium is summarized by the equation in the following box.

Keto-Enol Equilibrium: Acid-Catalyzed

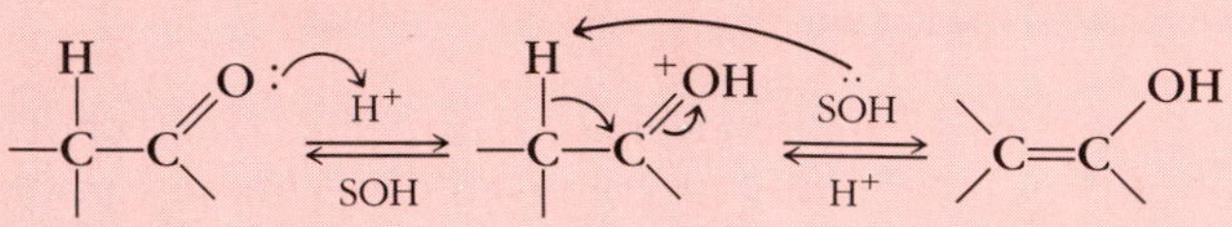

For most aldehydes and ketones the keto form is more stable by several kilocalories per mole. This is reflected by the small equilibrium constants for enol formation:

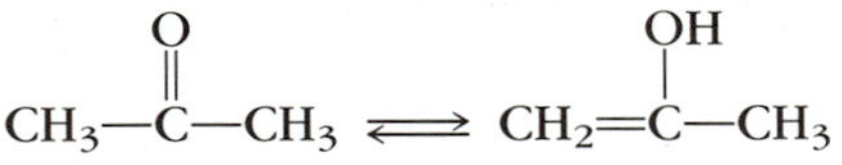

$$CH_3-\overset{O}{\overset{\|}{C}}-CH_3 \rightleftharpoons CH_2=\overset{OH}{\overset{|}{C}}-CH_3 \qquad K = 10^{-6}$$

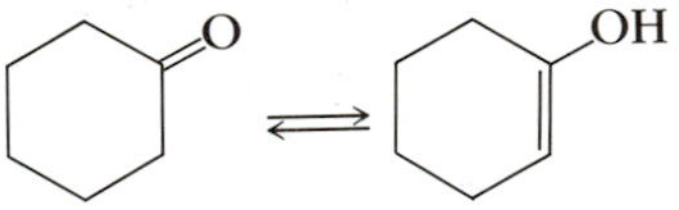

$K = 10^{-4}$

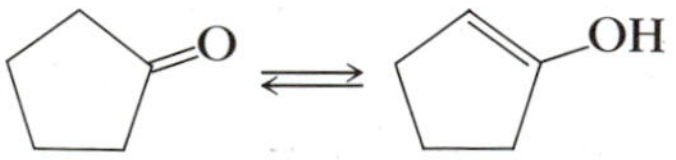

$K = 10^{-3}$

These three examples are typical of simple ketones, and in each case much less than 1% of the molecules exist in the enol form at any instant. Nevertheless the equilibrium is quite rapid in the presence of acid or base catalysts. Consequently, if the enol form undergoes some particular reaction, more can be formed almost as quickly as the existing enol is consumed. The same argument applies to enolate ions. Under equilibrium conditions enolates are often present in only very small amounts, but the rapid equilibrium permits reaction to proceed quite smoothly via the enolate ion. As soon as some enolate ion undergoes reaction, more is produced by way of the rapid acid-base equilibrium with a ketone.

β-Dicarbonyl Compounds

In contrast to simple ketones and aldehydes, some diketones and related dicarbonyl compounds exist to a substantial extent in the enol form. When the two carbonyl groups are separated by a single carbon atom (a *β-dicarbonyl compound*), the enol form is stabilized by a combination of effects: (1) production of a conjugated system

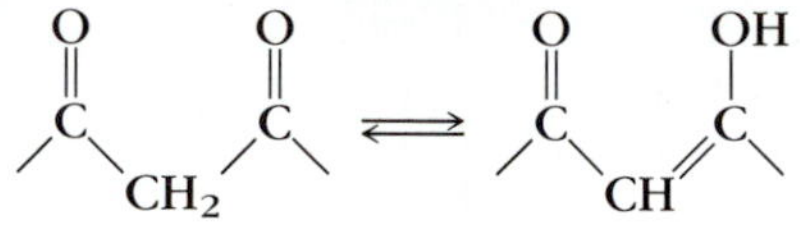

$$\text{—C(=O)—CH}_2\text{—C(=O)—} \rightleftharpoons \text{—C(=O)—CH=C(OH)—}$$

and (2) hydrogen bonding, particularly intramolecularly, via a six-membered ring

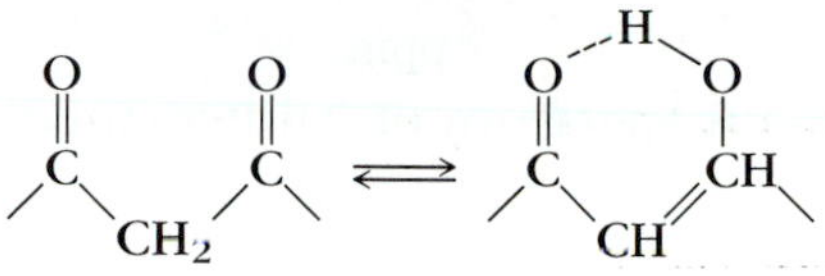

$$\text{—C(=O)—CH}_2\text{—C(=O)—} \rightleftharpoons \text{—C(=O}\cdots\text{H—O)—CH=CH—}$$

The percent of enol that is present for solutions of the following compounds illustrates the increased tendency of enolization of *β-dicarbonyl compounds*. Because hydrogen bonding to solvent is also important, the enol content depends markedly on solvent, and we have shown a range of values. Even the lower limits are much greater than the enol content of monocarbonyl compounds.

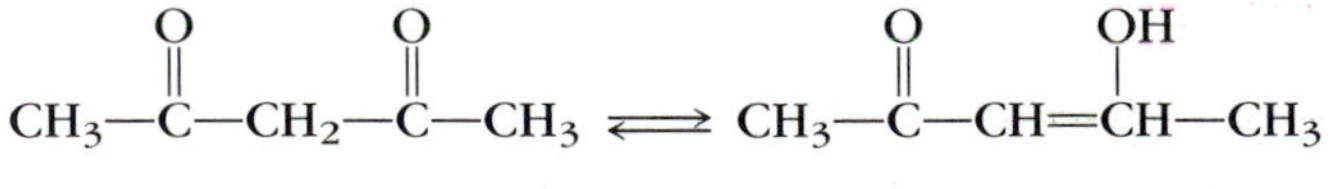

$$CH_3\text{—}\overset{\overset{O}{\|}}{C}\text{—}CH_2\text{—}\overset{\overset{O}{\|}}{C}\text{—}CH_3 \rightleftharpoons CH_3\text{—}\overset{\overset{O}{\|}}{C}\text{—}CH\text{=}\overset{\overset{OH}{|}}{C}H\text{—}CH_3 \qquad 16\text{–}92\%$$

$$CH_3\text{—}\overset{\overset{O}{\|}}{C}\text{—}\underset{\underset{CH_3}{|}}{C}H\text{—}\overset{\overset{O}{\|}}{C}\text{—}CH_3 \rightleftharpoons CH_3\text{—}\overset{\overset{O}{\|}}{C}\text{—}\underset{\underset{CH_3}{|}}{C}\text{=}\overset{\overset{OH}{|}}{C}\text{—}CH_3 \qquad 3\text{–}59\%$$

$$\text{5,5-dimethylcyclohexane-1,3-dione (O, O)} \rightleftharpoons \text{3-hydroxy-5,5-dimethylcyclohex-2-enone (O, OH)} \qquad 2\text{–}95\%$$

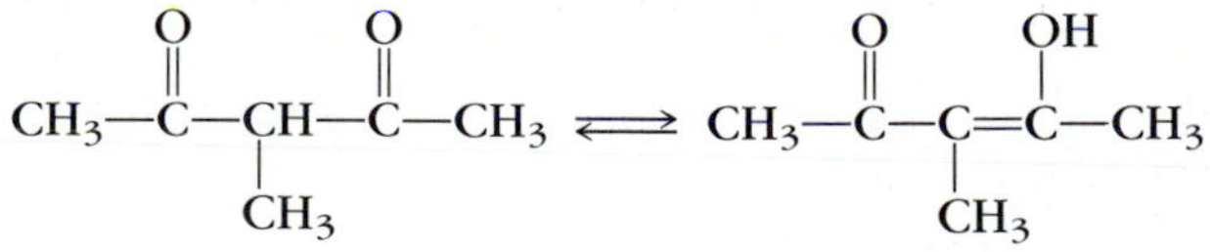

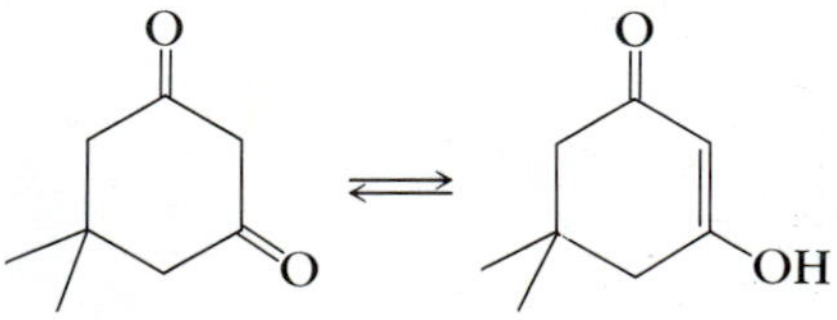

This increased tendency toward enolization with β-dicarbonyl compounds is clearly seen from the NMR spectrum of 2,4-pentanedione (Figure 13.4). In addition to the expected peaks for the keto form at 2.1 ppm (CH_3) and 3.6 ppm (CH_2), the larger peaks at 2.0 (CH_3) and 5.5 ppm (C=CH) demonstrate that the enol form predominates in this sample.

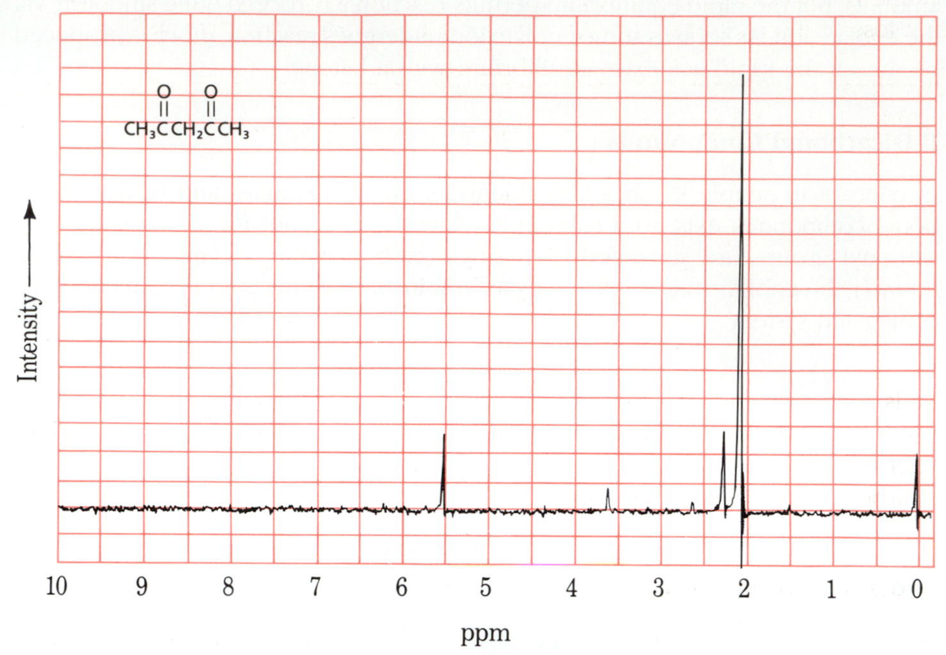

Figure 13.4 Proton NMR Spectrum of 2,4-Pentanedione.

Reactivity of Enols and Enolates

Enols and enolate ions both exhibit reactivity as nucleophiles. From the two resonance forms of an enolate ion it is apparent that carbon and oxygen can each be nucleophilic, but where will reaction actually occur? For both enols and enolate ions, most reactions occur preferentially at *carbon*. This is illustrated for an enolate ion by the following scheme, which shows reaction with an electrophile, symbolized as E^+.

Reaction of an Enolate Ion with an Electrophile

Either resonance form can be used to draw the enolate ion; remember, they are two drawings of the *same* ion. The choice of drawing has no effect on the reactivity of the enolate ion.

The reaction of a neutral enol closely resembles that of an enolate ion except that loss of the proton occurs *after* nucleophilic attack rather than before.

Reaction of an Enol with an Electrophile

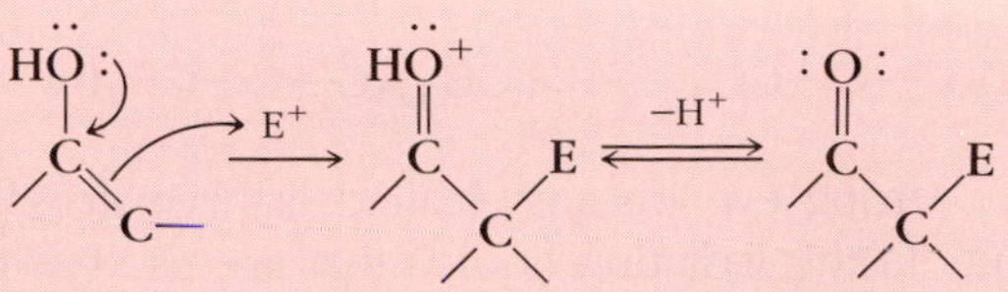

Enols and enolate ions are also capable of reaction as nucleophiles at oxygen, but this is not usually observed. Sometimes reaction at oxygen is reversible and at other times it is just slower. In any event most reactions occur preferentially at the α carbon, as depicted in the preceding equations.

Halogenation of Ketones

The reactions of enols and enolates as nucleophiles are illustrated by the halogenation of ketones. (Aldehydes tend to be *oxidized* under these conditions.) The reaction produces a ketone with a halogen on the carbon adjacent to the carbonyl group. Such compounds are called α-halo ketones. Under acidic conditions the reaction proceeds via the enol, as we have shown in the following mechanistic scheme for bromination:

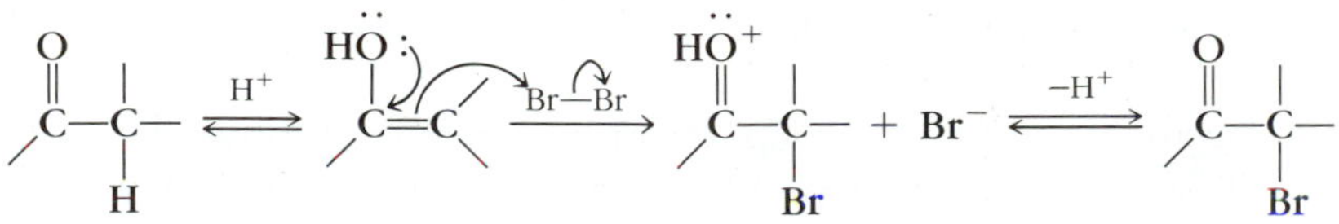

Halogenation under basic conditions proceeds via the corresponding enolate ion. The mechanism is almost the same as that for the enol, except that deprotonation occurs as the first step rather than the last.

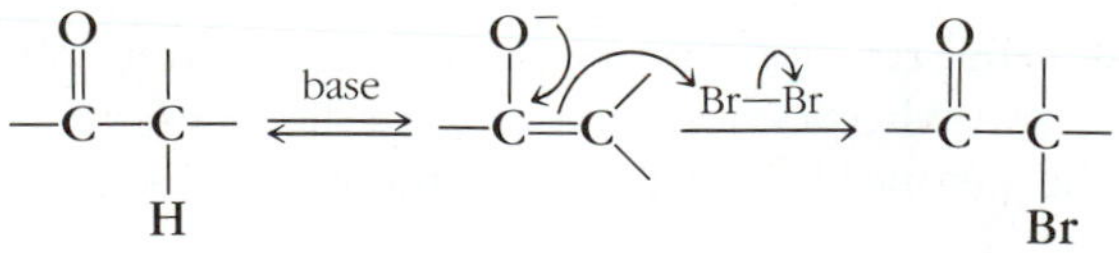

The acid-catalyzed halogenation can be controlled more easily, so it is commonly used for introduction of a single halogen atom. This is illustrated by the following reactions:

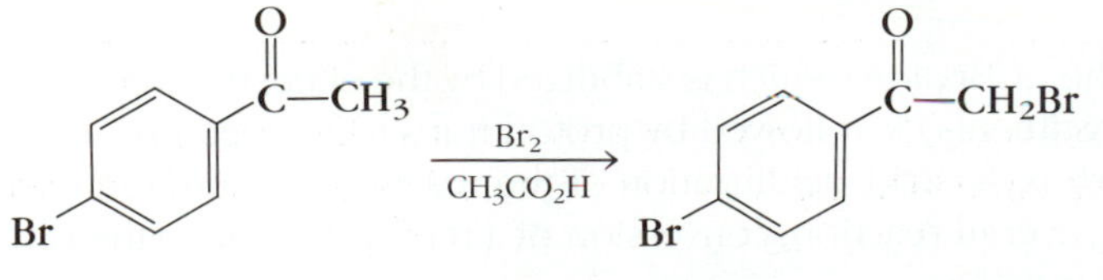

$$\text{4-Br-C}_6\text{H}_4\text{-C(=O)-CH}_3 \xrightarrow[\text{CH}_3\text{CO}_2\text{H}]{\text{Br}_2} \text{4-Br-C}_6\text{H}_4\text{-C(=O)-CH}_2\text{Br} \quad 69\text{–}72\%$$

$$\text{cyclopentanone} \xrightarrow[\text{CH}_3\text{CO}_2\text{H}]{\text{Br}_2} \text{2-bromocyclopentanone} \quad 78\%$$

$$(\text{CH}_3)_3\text{C}-\overset{\overset{\displaystyle\text{O}}{\|}}{\text{C}}-\text{CH}_3 \xrightarrow{\text{Br}_2} (\text{CH}_3)_3\text{C}-\overset{\overset{\displaystyle\text{O}}{\|}}{\text{C}}-\text{CH}_2\text{Br} \quad 63\%$$

α-Halo ketones are strong eye irritants (called lachrymators), and they are the principal ingredients in the irritating liquids that are used as tear gas.

Under basic conditions there is a strong tendency for the reaction to proceed past monohalogenation. This results from the inductive effect of the halogen, which further increases the acidity of other hydrogens on the α carbon. You can see this from the increased acid dissociation constant of 1-chloro-2-propanone relative to that of the parent ketone.

$$\text{CH}_3-\overset{\overset{\displaystyle\text{O}}{\|}}{\text{C}}-\text{CH}_2-\text{Cl} \qquad \text{CH}_3-\overset{\overset{\displaystyle\text{O}}{\|}}{\text{C}}-\text{CH}_3$$

$$K_a = 10^{-17} \qquad K_a = 10^{-20}$$

The halogenation of a *methyl ketone* in the presence of base produces unusual results. The initial reaction proceeds as you would expect, and an enolate is formed preferentially at the less substituted position. Reaction with halogen then generates the α-halo ketone, as we have shown for bromination in the following equation:

$$\text{R}-\overset{\overset{\displaystyle\text{O}}{\|}}{\text{C}}-\text{CH}_3 \overset{^-\text{OH}}{\rightleftharpoons} \text{R}-\overset{\overset{\displaystyle\text{O}^-}{|}}{\text{C}}=\text{CH}_2 \xrightarrow{\text{Br}-\text{Br}} \text{R}-\overset{\overset{\displaystyle\text{O}}{\|}}{\text{C}}-\text{CH}_2\text{Br}$$

The hydrogens of the CH_2Br group are more acidic than those of the original methyl ketone, and successive enolate formation and bromination occur at the same position to generate a trihalomethyl group.

$$\text{R}-\overset{\overset{\displaystyle\text{O}}{\|}}{\text{C}}-\text{CH}_2\text{Br} \xrightarrow[^-\text{OH}]{\text{Br}_2} \text{R}-\overset{\overset{\displaystyle\text{O}}{\|}}{\text{C}}-\text{CHBr}_2 \xrightarrow[^-\text{OH}]{\text{Br}_2} \text{R}-\overset{\overset{\displaystyle\text{O}}{\|}}{\text{C}}-\text{CBr}_3$$

At this point an unusual reaction occurs. As with any ketone, reversible attack on the carbonyl group by hydroxide ion can occur. Ordinarily, expulsion of hydroxide ion regenerates the ketone, but in this case there is a second potential leaving group, the $^-CBr_3$ group.

$$\text{R}-\overset{\overset{\displaystyle\text{O}}{\|}}{\text{C}}-\text{CBr}_3 \overset{^-\text{OH}}{\rightleftharpoons} \text{R}-\overset{\overset{\displaystyle ^-\text{O}}{|}}{\underset{\underset{\displaystyle\text{OH}}{|}}{\text{C}}}-\text{CBr}_3 \rightleftharpoons \left[\text{R}-\overset{\overset{\displaystyle\text{O}}{\|}}{\text{C}}-\text{OH} + {}^-\text{CBr}_3\right] \rightleftharpoons \text{R}-\overset{\overset{\displaystyle\text{O}}{\|}}{\text{C}}-\text{O}^- + \underset{\text{Bromoform}}{\text{CHBr}_3}$$

Expulsion of the $^-CBr_3$ ion (which is stabilized by the effect of three electronegative halogen substituents) is followed by proton transfer to yield bromoform and the salt of the carboxylic acid; acidification of the reaction mixture then produces the free acid. The overall reaction, conversion of a methyl ketone to the carboxylic acid and a trihalomethane, is called the *haloform reaction.* The reaction proceeds well with chlorine, bromine, or iodine to yield the corresponding haloform (i.e., chloroform, bromoform, or iodoform). The procedure is useful for identification of methyl ketones (Section 13.12), and the following examples show that it can also be a good preparative reaction.

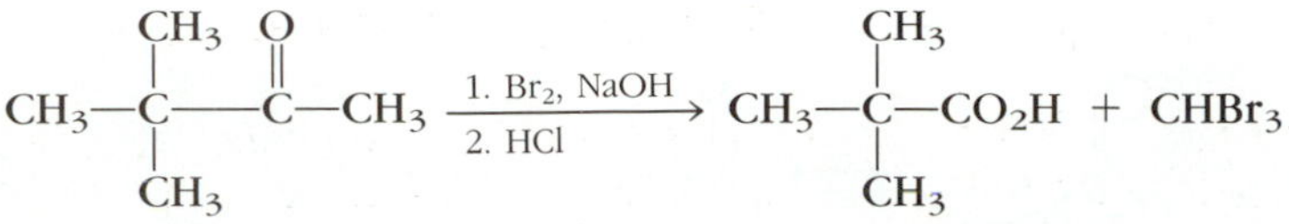

$$(CH_3)_3C{-}C(=O){-}CH_3 \xrightarrow[2.\ HCl]{1.\ Br_2,\ NaOH} (CH_3)_3C{-}CO_2H + CHBr_3$$

71–74%

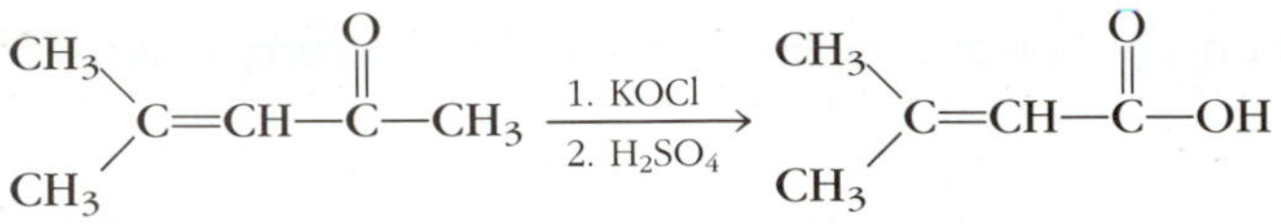

$$(CH_3)_2C{=}CH{-}C(=O){-}CH_3 \xrightarrow[2.\ H_2SO_4]{1.\ KOCl} (CH_3)_2C{=}CH{-}C(=O){-}OH \quad 49\text{–}53\%$$

EXERCISE 13.12

Complete each of the following reactions:

(a) 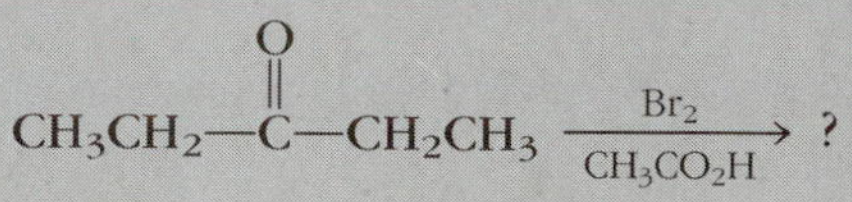 $CH_3CH_2{-}C(=O){-}CH_2CH_3 \xrightarrow[CH_3CO_2H]{Br_2} ?$

(b) $CH_3CH_2{-}C(=O){-}C(CH_3)_2{-}CH_3 \xrightarrow[CH_3CO_2H]{Cl_2} ?$

(c) $C_6H_5{-}C(=O){-}CH_3 \xrightarrow[NaOH]{I_2} ?$

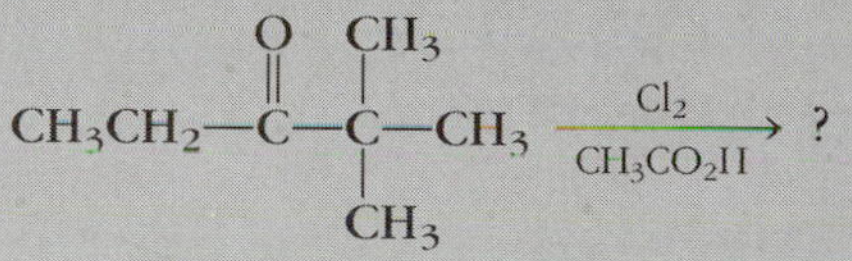

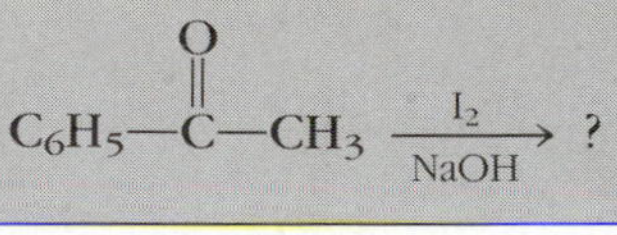

EXERCISE 13.13

Suggest reagents for carrying out each of the following conversions:

(a) 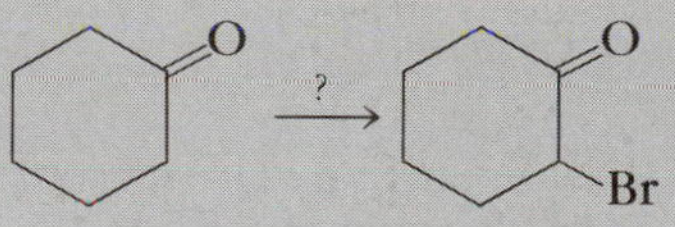 cyclohexanone $\xrightarrow{?}$ 2-bromocyclohexanone (ring with =O and Br)

(b) 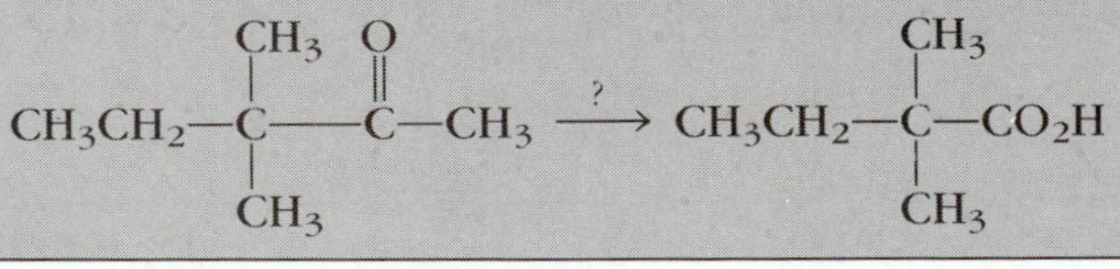 $CH_3CH_2{-}C(CH_3)_2{-}C(=O){-}CH_3 \xrightarrow{?} CH_3CH_2{-}C(CH_3)_2{-}CO_2H$

13.10 CONJUGATE ADDITION TO α,β-UNSATURATED ALDEHYDES AND KETONES

The polarity of a simple carbonyl group is indicated by the two resonance forms that can be drawn:

In the case of an α,β-unsaturated carbonyl compound, a third resonance contributor exists:

Thus both the carbonyl carbon and the unsaturated β carbon carry a partial positive charge.

As you would expect from these partial charges, nucleophilic attack can occur at either of two positions:

Carbonyl Addition

Conjugate Addition

Attack at the β position of a conjugated system, as shown in the latter equation, is called **conjugate addition,** and it is frequently observed in the reactions of α,β-unsaturated aldehydes and ketones.

Under basic conditions the intermediate that results from conjugate addition is an *enolate ion,* and subsequent protonation yields a saturated ketone (or aldehyde) derivative.

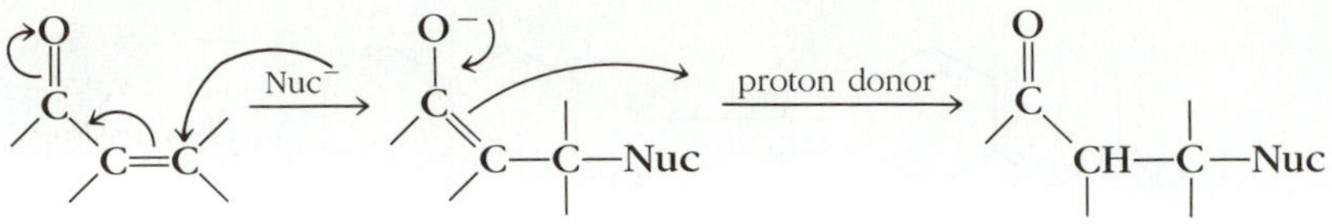

Conjugate addition also occurs under acidic conditions according to the following mechanism:

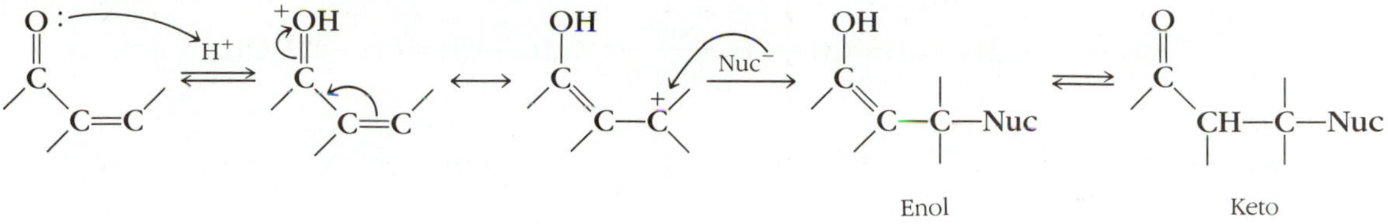

Enol Keto

Conjugate Addition vs Carbonyl Addition

What factors favor a particular mode of attack on an α,β-unsaturated ketone or aldehyde? Attack at carbonyl carbon usually occurs more rapidly than attack at the β position, but for ketones the product of conjugate addition tends to be more stable. Consequently, carbonyl addition is expected for *irreversible* reactions *(kinetic control).* In contrast, conjugate addition is usually observed for *reversible* additions to ketones *(thermodynamic control).*

Product of carbonyl addition ⇌ Product of conjugate addition

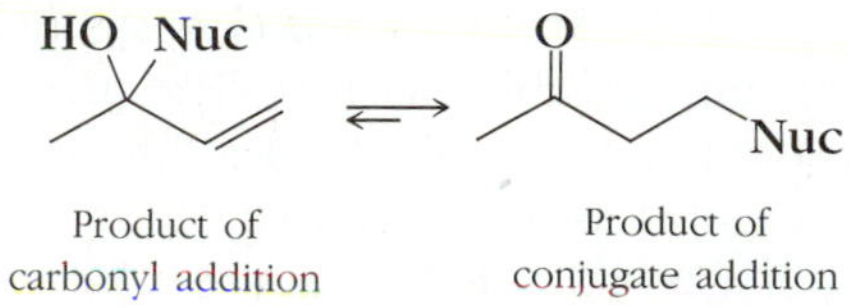

The dependence on kinetic and thermodynamic control is less for α,β-unsaturated aldehydes, which tend to yield the products of carbonyl addition in either situation.

The predominance of conjugate addition product in reversible reactions of ketones can be seen in the following additions of HCl and KCN:

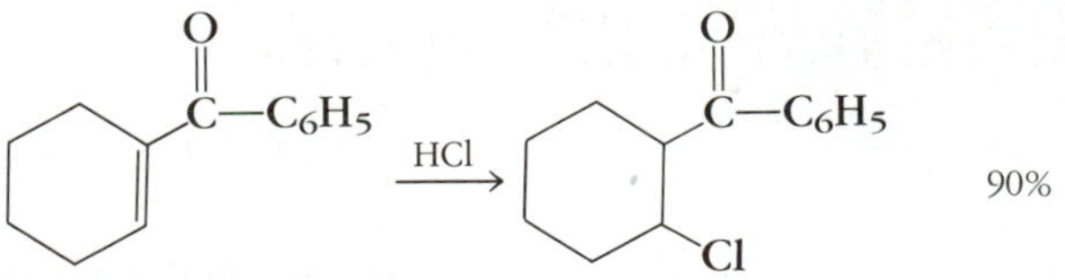

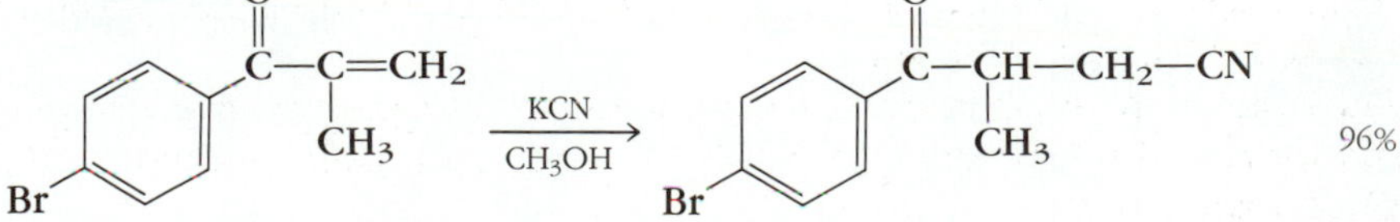

90%

96%

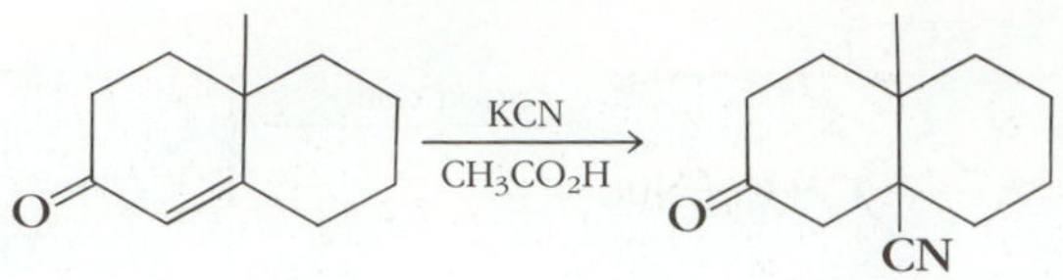

$\xrightarrow[CH_3CO_2H]{KCN}$ 76%

CN

The following irreversible reactions illustrate the tendency for lithium aluminum hydride, sodium borohydride, and Grignard reagents to attack at carbonyl carbon.

$$C_6H_5\text{—}CH\text{=}CH\text{—}\overset{O}{\overset{\|}{C}}H \xrightarrow[CH_3OH]{NaBH_4} C_6H_5\text{—}CH\text{=}CH\text{—}CH_2OH \quad 97\%$$

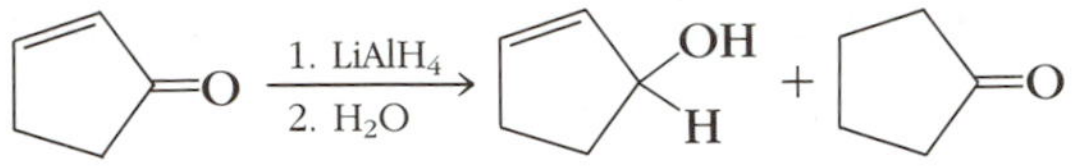

$\xrightarrow[2.\ H_2O]{1.\ LiAlH_4}$ OH H + O

85% 15%

$$CH_3\text{—}CH\text{=}CH\text{—}\overset{O}{\overset{\|}{C}}\text{—}CH_3 \xrightarrow[2.\ H_2O]{1.\ CH_3MgBr} CH_3\text{—}CH\text{=}CH\text{—}\underset{CH_3}{\overset{OH}{C}}\text{—}CH_3 + CH_3\text{—}\overset{CH_3}{CH}\text{—}CH_2\text{—}\overset{O}{\overset{\|}{C}}\text{—}CH_3$$

90% 3%

Organocopper reagents provide a striking contrast to the above generalizations. Their reactions are definitely not reversible, but they exhibit a strong preference for conjugate addition. This results from a reaction mechanism which is different from that shown at the beginning of this section. The mechanism is beyond the scope of the present section, and it is probably best for you to simply regard the reactivity of the copper reagents as an exception. The formation of products of conjugate addition for both aldehydes and ketones is illustrated by the following reactions of lithium dialkylcuprates:

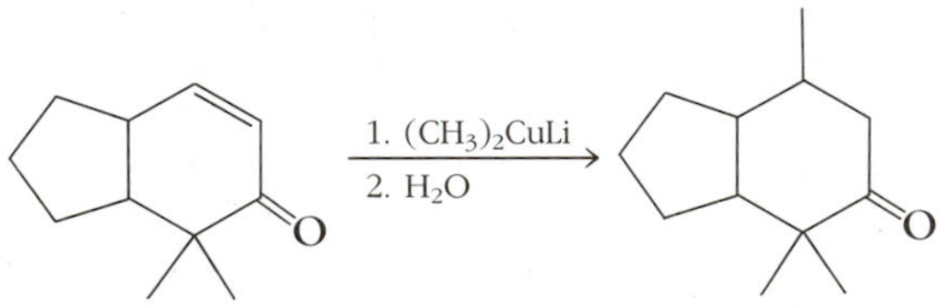

$\xrightarrow[2.\ H_2O]{1.\ (CH_3)_2CuLi}$ 84%

—CHO $\xrightarrow[2.\ H_2O]{1.\ [CH_3(CH_2)_6CH_2]_2CuLi}$ —CHO 88%

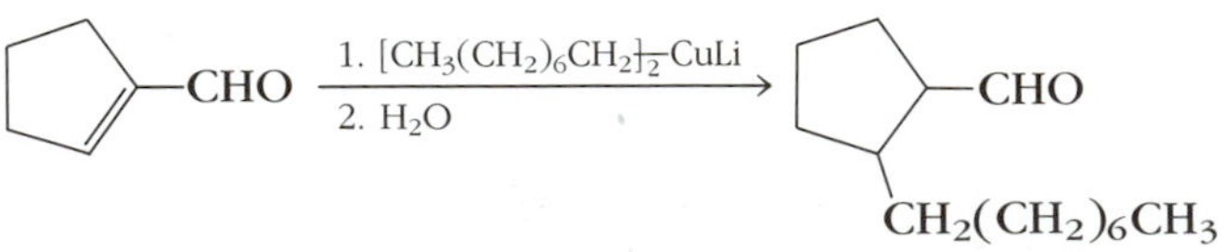

$CH_2(CH_2)_6CH_3$

EXERCISE 13.14

Complete the reaction of 4-hepten-3-one with each of the following reagents: (a) $LiAlH_4$; (b) HCN; (c) CH_3MgBr.

$$CH_3CH_2CH\text{=}CH\text{—}\overset{O}{\overset{\|}{C}}\text{—}CH_2CH_3 \longrightarrow ?$$

EXERCISE 13.15

Complete the reaction of 3-methyl-1-phenyl-2-buten-1-one with each of the following reagents: (a) $NaCN/CH_3CO_2H$; (b) $NaBH_4/CH_3OH$; (c) $(CH_3)_2CuLi$.

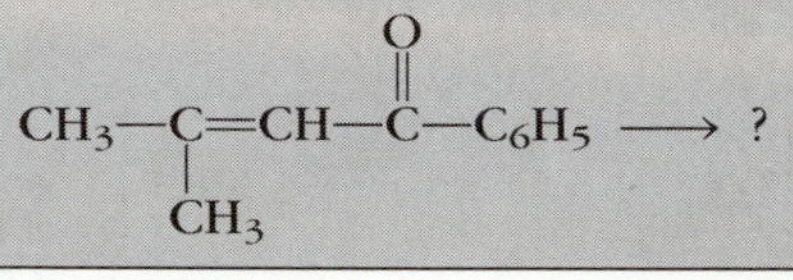

$$CH_3-C(CH_3)=CH-C(=O)-C_6H_5 \longrightarrow ?$$

13.11 OXIDATION AND REDUCTION OF ALDEHYDES AND KETONES

In addition to the reactions we have already discussed in Sections 11.3 and 11.4, a variety of other oxidation-reduction reactions can occur with aldehydes and ketones. Ketones can be oxidized at the reactive α position, and other reactions can cause oxidative carbon-carbon bond cleavage. Aldehydes and ketones can be reduced to alcohols as we described in Section 11.4, but with appropriate reagents the reduction can continue to the hydrocarbon stage. We will discuss these other possibilities in this section.

Oxidation of Aldehydes

Aldehydes are readily oxidized to the corresponding carboxylic acids by a variety of reagents. This was considered in Section 11.3 as one of the problems of preparing aldehydes via oxidation of primary alcohols, and CrO_3-pyridine reagents were used in those situations in which the aldehyde was the desired product. The following examples illustrate some of the reagents that can be used to oxidize aldehydes when the corresponding acids are the desired products.

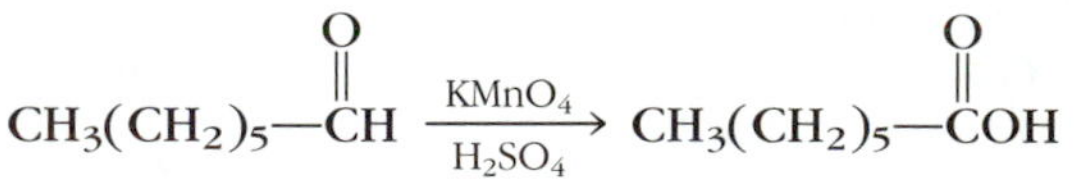

$$CH_3(CH_2)_5-\overset{O}{\overset{\|}{C}}H \xrightarrow[H_2SO_4]{KMnO_4} CH_3(CH_2)_5-\overset{O}{\overset{\|}{C}}OH \quad 76\text{–}78\%$$

$$\text{3-}tert\text{-}C_4H_9\text{-1-}C_6H_5\text{-cyclopentane-1-}\overset{O}{\overset{\|}{C}}H \quad H_2SO_4 \xrightarrow{CrO_3} \text{3-}tert\text{-}C_4H_9\text{-1-}C_6H_5\text{-cyclopentane-1-}\overset{O}{\overset{\|}{C}}OH \quad 85\%$$

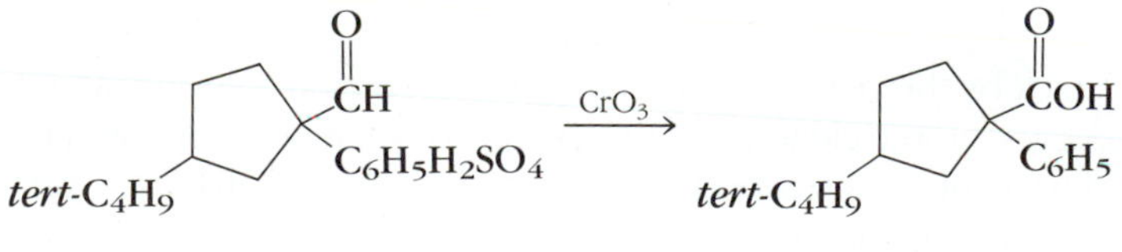

$$\text{cyclohex-3-ene-}\overset{O}{\overset{\|}{C}}H \xrightarrow{Ag_2O} \text{cyclohex-3-ene-}\overset{O}{\overset{\|}{C}}OH \quad 97\%$$

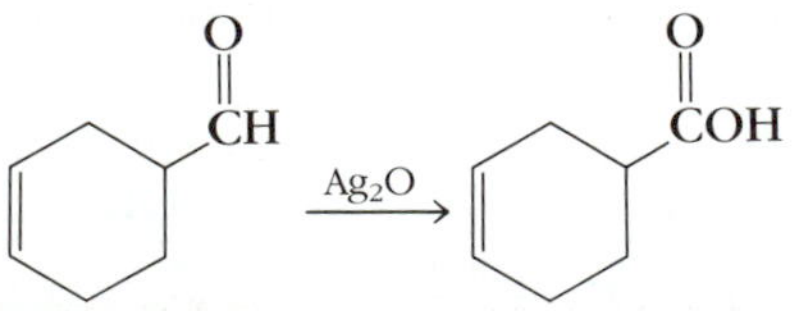

The last reaction, in which Ag(I) oxide is the oxidizing agent, illustrates that this mild reagent can selectively oxidize an aldehyde in the presence of a carbon-carbon double bond. Many other oxidizing agents would attack the alkene (Section 5.12). The redox reactions of Ag(I) salts are also used to detect the presence of an aldehyde group in a compound of unknown or uncertain structure (Section 13.12).

Oxidation of Ketones: the Baeyer-Villiger Reaction

In the late 1800s the German chemists A. Baeyer and V. Villiger found that treatment of a ketone with a peroxy acid afforded an ester.

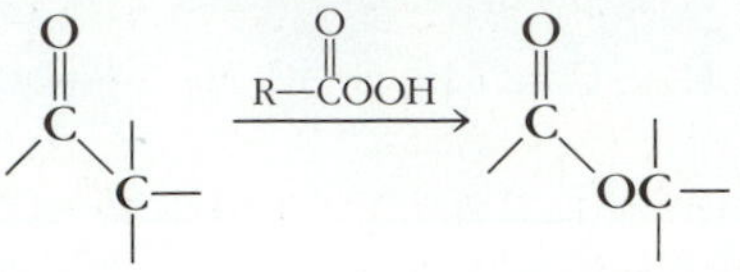

Ketone — R—COOH → Ester

This reaction has since been termed the *Baeyer-Villiger reaction.* A variety of peroxy acids have been utilized, and among the most useful are the following:

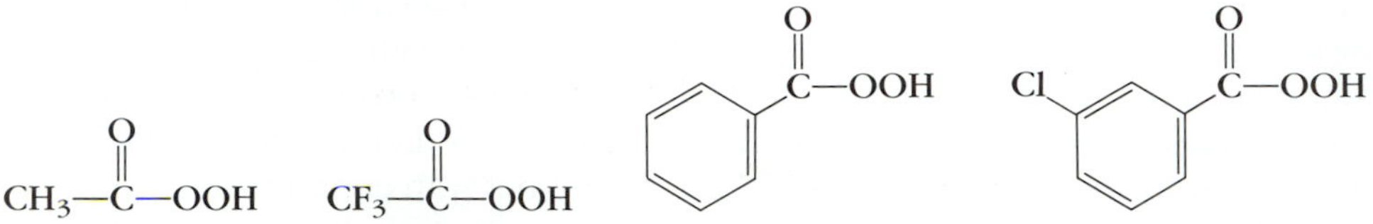

CH_3—C(O)—OOH Peracetic acid; CF_3—C(O)—OOH Peroxytrifluoro acetic acid; Perbenzoic acid; *m*-Chloroperbenzoic acid

The mechanism of the reaction is illustrated in the following equation with peracetic acid as the oxidizing agent:

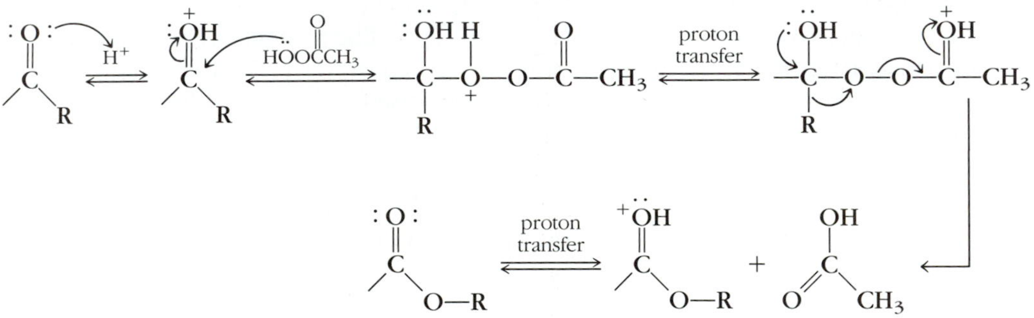

H^+; $HOOCCH_3$; proton transfer; proton transfer

Acid-catalyzed attack by the peroxy acid on the ketone carbonyl affords an intermediate which undergoes ionic cleavage of the oxygen-oxygen bond. At the same time there is a 1,2 alkyl shift from carbon to oxygen, and this affords the ester after proton transfer. Although the oxygen-oxygen cleavage and rearrangement steps appear to be concerted, the reaction can easily be understood by considering the two processes separately:

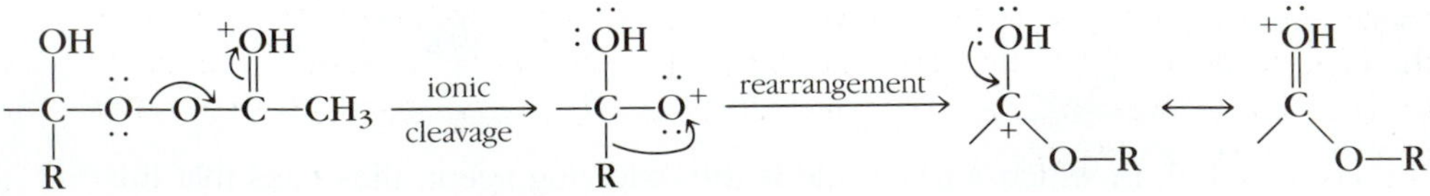

ionic cleavage; rearrangement

Such an ionic cleavage would produce an *electron-deficient oxygen* atom, and the 1,2 alkyl shift is precisely analogous to those observed in rearrangements to electron-deficient carbon (Section 5.9).

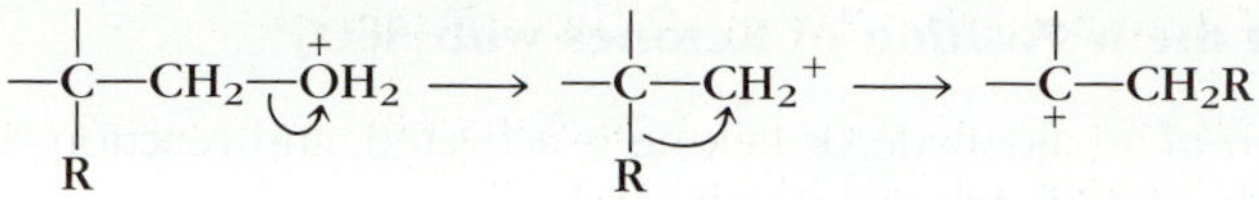

$$-\underset{R}{\overset{|}{C}}-CH_2-\overset{+}{O}H_2 \longrightarrow -\underset{R}{\overset{|}{C}}-CH_2^+ \longrightarrow -\underset{+}{\overset{|}{C}}-CH_2R$$

(As in the case of electron-deficient oxygen, a primary carbocation would be a highly unstable intermediate, and the two steps of the preceding equation are probably concerted.)

Baeyer-Villiger oxidation of a unsymmetrical ketone could yield two products, corresponding to migration of each of the two alkyl substituents on the original carbonyl group. Usually one group will rearrange preferentially according to the following **migratory aptitudes:**

Migratory Aptitudes

3° alkyl > aryl > 2° alkyl > 1° alkyl > methyl

These migratory preferences are seen in some of the examples of the Baeyer-Villiger reactions which follow.

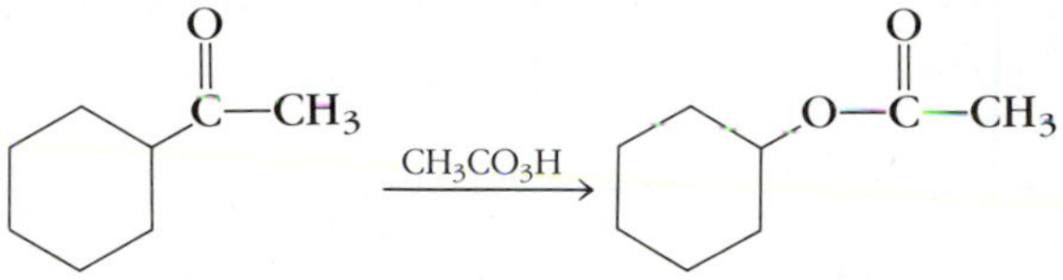

$$C_6H_{11}-\overset{O}{\overset{\|}{C}}-CH_3 \xrightarrow{CH_3CO_3H} C_6H_{11}-O-\overset{O}{\overset{\|}{C}}-CH_3 \quad 90\%$$

(2° > CH_3)

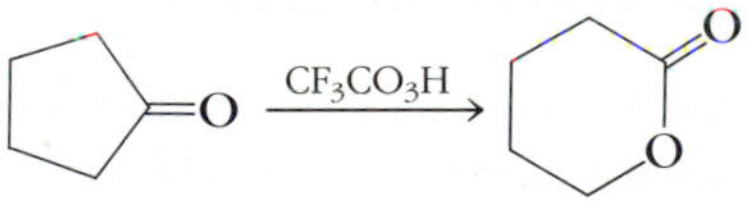

$$\text{cyclopentanone} \xrightarrow{CF_3CO_3H} \text{δ-valerolactone} \quad 81\%$$

$$\text{norcamphor} \xrightarrow{CH_3CO_3H} \text{bicyclic lactone} \quad 88\%$$

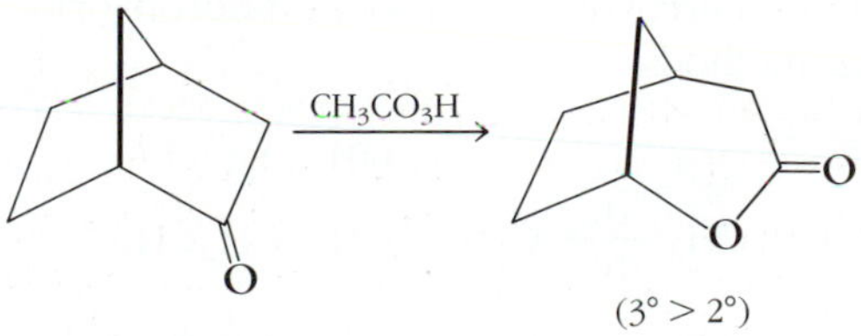

(3° > 2°)

Oxidative cleavage of ketones can also be carried out by vigorous oxidation with reagents such as CrO_3 or $KMnO_4$. These cleavage reactions are used infrequently in the laboratory, but related processes play an important role in industrial chemistry (Section 26.1). The oxidation of cyclohexanone to a dicarboxylic acid illustrates this type of oxidative cleavage.

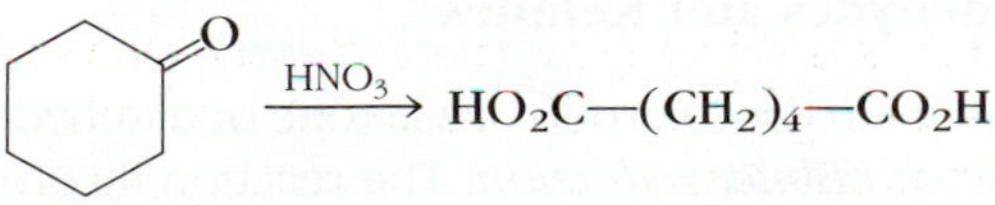

$$\text{cyclohexanone} \xrightarrow{HNO_3} HO_2C-(CH_2)_4-CO_2H \quad 72\%$$

Oxidation at the α Position of Ketones with SeO_2

The α position of an aldehyde or ketone is activated, and reaction with selenium dioxide affords an α-dicarbonyl compound.

$$-\overset{O}{\overset{\|}{C}}-CH_2- \xrightarrow{SeO_2} -\overset{O}{\overset{\|}{C}}-\overset{O}{\overset{\|}{C}}-$$

This reaction is closely related to allylic oxidation of alkenes (Section 8.4), but a carbonyl group (rather than an OH group) is introduced when an aldehyde or ketone is the reactant. This oxidation is illustrated by the following examples:

$$\text{cyclohexanone} \xrightarrow{SeO_2} \text{1,2-cyclohexanedione} \qquad 60\%$$

$$C_6H_5-\overset{O}{\overset{\|}{C}}-CH_3 \xrightarrow{SeO_2} C_6H_5-\overset{O}{\overset{\|}{C}}-\overset{O}{\overset{\|}{C}}H \qquad 69\text{–}72\%$$

$$\text{7-chlorobicyclo[2.2.1]heptan-2-one} \xrightarrow{SeO_2} \text{7-chlorobicyclo[2.2.1]heptane-2,3-dione} \qquad 58\%$$

Reduction of Aldehydes and Ketones to Alcohols

The reduction of aldehydes and ketones to primary and secondary alcohols, respectively, was discussed in detail in Section 11.4. The reduction can be accomplished using catalytic hydrogenation, dissolving metals, or complex metal hydrides such as lithium aluminum hydride and sodium borohydride. The following examples illustrate these methods:

$$C_6H_5-\overset{O}{\overset{\|}{C}}-CH_2CH_3 \xrightarrow[Pd]{H_2} C_6H_5-\overset{OH}{\overset{|}{C}H}-CH_2CH_3 \qquad 96\%$$

$$CH_3(CH_2)_7CH{=}CH(CH_2)_7CHO \xrightarrow[C_2H_5OH]{Na} CH_3(CH_2)_7CH{=}CH(CH_2)_7CH_2OH \qquad 49\text{–}51\%$$

$$(C_6H_5)_2CH-\overset{O}{\overset{\|}{C}}H \xrightarrow[2.\ H_2O]{1.\ LiAlH_4} (C_6H_5)_2CH-CH_2OH \qquad 98\%$$

Deoxygenation of Aldehydes and Ketones

A procedure for converting the C=O group of a ketone or aldehyde to a $-CH_2-$ group is known as the *Wolff-Kishner reduction.* The reaction involves treatment of the aldehyde or ketone with hydrazine and base, and it proceeds in two stages: formation of the hydrazone (Section 13.7) and second, its decomposition, with loss of molecular nitrogen, when heated at 150 to 250°C in a high-boiling solvent.

$$\mathrm{C{=}O} \xrightarrow[\text{KOH}]{NH_2{-}NH_2} \mathrm{C{=}N{-}NH_2} \longrightarrow CH_2 + N_2$$

The decomposition of the hydrazone proceeds via several proton transfer steps, as summarized for an aqueous system in the following equation:

$$\mathrm{C{=}N{-}NH{-}H} \underset{}{\overset{^-OH}{\rightleftharpoons}} \mathrm{{}^-C{-}N{=}NH} \overset{H_2O}{\rightleftharpoons} \mathrm{CH{-}N{=}N{-}H} \overset{^-OH}{\rightleftharpoons} \mathrm{CH{-}N{=}N^-} \xrightarrow{H_2O} CH_2 + :N{\equiv}N:$$

The deoxygenation of aldehydes and ketones via the Wolff-Kishner reduction is illustrated by the following examples. Ordinarily the hydrazone is not isolated but is decomposed directly to yield the deoxygenated product.

$$CH_3{-}(CH_2)_5{-}\overset{O}{\overset{\|}{C}}{-}CH_3 \xrightarrow[\substack{NaOCH_2CH_2OH \\ HOCH_2CH_2OH}]{NH_2NH_2} CH_3{-}(CH_2)_6{-}CH_3 \quad 75\%$$

$$p\text{-}CH_3C_6H_4{-}\overset{O}{\overset{\|}{C}}{-}CH_3 \xrightarrow{NH_2NH_2} \underset{88\%}{p\text{-}CH_3C_6H_4{-}\overset{N{-}NH_2}{\overset{\|}{C}}{-}CH_3} \xrightarrow{KOH} \underset{80\%}{p\text{-}CH_3C_6H_4{-}CH_2{-}CH_3}$$

$$CH_3{-}\overset{CH_3}{\overset{|}{C}H}{-}CH_2CH_2{-}\overset{O}{\overset{\|}{C}}{-}(CH_2)_3{-}\overset{O}{\overset{\|}{C}}OH \xrightarrow[\text{2. HCl (to neutralize)}]{\text{1. } NH_2NH_2/KOH} CH_3{-}\overset{CH_3}{\overset{|}{C}H}{-}(CH_2)_6{-}\overset{O}{\overset{\|}{C}}OH \quad 87\text{–}93\%$$

The last example shows that carboxyl groups of carboxylic acids are not affected by this procedure, because they are converted to the unreactive conjugate bases (until acidification during workup). Isolated double bonds are not affected by the reaction, but the reduction of α,β-unsaturated ketones by the Wolff-Kishner procedure frequently fails because the reaction with hydrazine generates a cyclic derivative. This cyclic compound does not lead to the desired product.

$$CH_3COCH{=}CH_2 \xrightarrow{NH_2NH_2} \text{(3-methyl-2-pyrazoline, N–N(H) ring)} \longrightarrow CH_3CH_2CH{=}CH_2$$

An alternative method for the deoxygenation of ketones and aldehydes is the *Clemmensen reduction.* This reaction involves treatment of the carbonyl compound with zinc amalgam and a mineral acid such as hydrochloric acid. For example:

$$CH_3{-}(CH_2)_5{-}\overset{\overset{\displaystyle O}{\|}}{C}{-}CH_3 \xrightarrow[HCl]{Zn\text{-}Hg} CH_3{-}(CH_2)_6{-}CH_3 \qquad 62\%$$

The use of such strongly acidic conditions imposes strict limitations on the use of the Clemmensen reduction with polyfunctional compounds, so it is used much less frequently than the Wolff-Kishner reduction.

EXERCISE 13.16

Complete the reaction of 3-pentanone with each of the following reagents: (a) $C_6H_5CO_3H$; (b) SeO_2; (c) NH_2NH_2/KOH; (d) N_aBH_4/CH_3OH.

$$CH_3CH_2{-}\overset{\overset{\displaystyle O}{\|}}{C}{-}CH_2CH_3 \longrightarrow ?$$

EXERCISE 13.17

Complete the reaction of phenylacetaldehyde with each of the following reagents: (a) CrO_3; (b) $LiAlH_4$, then H_2O; (c) Zn-Hg/HCl; (d) Ag_2O.

$$C_6H_5{-}CH_2{-}\overset{\overset{\displaystyle O}{\|}}{C}H \longrightarrow ?$$

13.12 CHEMICAL AND SPECTROSCOPIC CHARACTERIZATION OF ALDEHYDES AND KETONES

When a compound is isolated from a reaction mixture (or from any other source) how can a chemist determine its identity? What functional groups are present? What is its complete structure? Aldehydes and ketones exhibit a variety of physical and chemical properties that can be used to distinguish them from compounds belonging to other functional classes. Most of the spectroscopic and chemical behavior has been discussed previously in Chapter 9, but we will review it here with specific emphasis on the identification of aldehydes and ketones.

Spectroscopic Methods

NMR Spectroscopy In ^{13}C spectroscopy the carbonyl carbon of an aldehyde or ketone gives rise to a signal at 200–220 ppm that is highly characteristic of these functional groups (Table 9.2). In proton NMR an aldehydic hydrogen (—CHO) leads to an absorption between 9.4 and 11 ppm (Table 9.3). For both aldehydes and ketones the signals for the α hydrogens are shifted downfield by 1–2 ppm relative to the hydrocarbon analogs (Table 9.3).

IR Spectroscopy Aldehydes and ketones both exhibit a strong infrared absorption in the region of 1700 cm^{-1} (Table 9.6). With four- and five-membered-ring cyclic ketones, the absorption is shifted to higher wave number. The CHO group of aldehydes also gives rise to characteristic C—H absorptions; the peak near 2700 cm^{-1} is at a frequency that is outside the normal range for C—H stretching. You can see this absorption in Figures 9.41 and 9.45.

UV Spectroscopy The lowest-energy absorption of a saturated aldehyde or ketone is below 200 nm, which is outside the range of normal UV spectrophotometers, therefore aldehydes and ketones are said not to absorb in the UV region of the electromagnetic spectrum. However, α,β-unsaturated aldehydes and ketones show characteristic absorption maxima near 215 nm. The exact position of the peak is closely correlated with substitution pattern (Figure 9.15). Other unsaturated molecules also absorb in the UV (for example, conjugated dienes and aryl derivatives), so a peak near 215 nm must be interpreted with care.

Chemical Methods

Aldehydes, The Tollens Test The facile oxidation of aldehydes by Ag(I) salts (Section 13.11) is accompanied by the production of metallic silver. When the reaction is carefully carried out with a reagent prepared from $AgNO_3$ and aqueous ammonia, the silver metal forms as a shiny deposit on the inside of the reaction vessel.

$$\text{RCHO} \xrightarrow[\text{H}_2\text{O,NH}_3]{\text{Ag}^+} \text{R}\overset{\overset{\text{O}}{\|}}{\text{C}}\text{OH} + \text{Ag}^\circ$$

For this reason the reaction is often referred to as the *silver mirror test* for aldehydes.

Methyl Ketones: The Iodoform Test When a methyl ketone is halogenated under alkaline conditions, a cleavage reaction occurs, and the corresponding trihalomethane is formed (Section 13.9). When iodine is used as the halogen, the resulting trihalomethane is *iodoform,* a yellow solid melting at 119–121°C.

$$\text{R}-\overset{\overset{\text{O}}{\|}}{\text{C}}-\text{CH}_3 \xrightarrow[\text{NaOH}]{\text{I}_2} \text{R}-\overset{\overset{\text{O}}{\|}}{\text{C}}-\text{O}^-\text{Na}^+ + \underset{\text{Yellow precipitate}}{\text{CHI}_3}$$

The formation of a yellow precipitate of iodoform is characteristic of methyl ketones. Two other types of compounds also give positive iodoform tests: β-dicarbonyl compounds of the type

$$-\overset{\overset{\text{O}}{\|}}{\text{C}}-\text{CH}_2-\overset{\overset{\text{O}}{\|}}{\text{C}}-$$

that can undergo a double cleavage, and secondary alcohols

$$\text{R}-\overset{\overset{\text{OH}}{|}}{\text{C}}\text{H}-\text{CH}_3$$

that can be oxidized to methyl ketones under the reaction conditions.

Formation of 2,4-Dinitrophenylhydrazones We discussed the formation of hydrazones from hydrazines in Section 13.7, and this reaction is very useful in identifying aldehydes and ketones. One of the best reagents is 2,4-dinitrophenylhydrazine, which affords a 2,4-dinitrophenylhydrazone in reaction with an aldehyde or ketone. For example,

$$\mathrm{RCHO} \xrightarrow[\mathrm{H_2SO_4,\ C_2H_5OH}]{\mathrm{NH_2NH\text{-}C_6H_3(NO_2)_2}} \mathrm{RCH{=}N\text{-}NH\text{-}C_6H_3(NO_2)_2}$$

The 2,4-dinitrophenylhydrazones are usually insoluble in the reaction medium, so the formation of an orange or reddish precipitate is characteristic of an aldehyde or ketone.

EXERCISE 13.18

Answer the following questions about structures **1** through **8**.

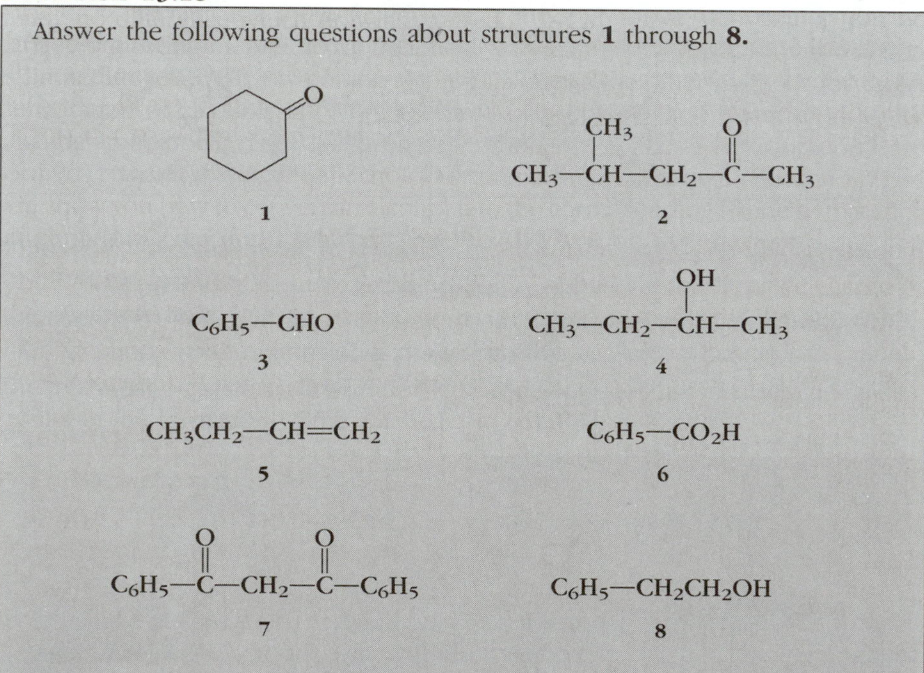

(a) Which would exhibit strong IR absorption near 1700 cm^{-1}?
(b) Which would show a pair of peaks in the IR in the 2700–2900 cm^{-1} range?
(c) Which would absorb in the UV between 200 and 400 nm?
(d) Which would show peaks near 200 ppm in the ^{13}C NMR spectrum?
(e) Which would show a peak in the proton NMR spectrum between 9 and 11 ppm?
(f) Which would show a positive Tollens test?
(g) Which would show a positive iodoform test?
(h) Which would show a positive test with 2,4-dinitrophenylhydrazine?

13.13 SELECTIVE REACTIONS OF UNSATURATED KETONES AND ALDEHYDES

When a compound contains more than one functional group, a common situation, an organic chemist must be able to predict which of the groups will react under a particular set of conditions. Will both functional groups react, or will reaction take place at just one of them? If so, at which one? In some cases the answer to such questions is straightforward because the two functional groups are very different. But when the functional groups exhibit similar reactivity, the problem is more difficult. We have already discussed these questions in Section 8.8 for compounds containing more than one carbon-carbon multiple bond. In this section we will consider molecules containing both carbon-carbon and carbon-oxygen double bonds, that is, unsaturated aldehydes and ketones.

The carbon-carbon and carbon-oxygen double bonds of α,β-unsaturated carbonyl compounds interact very strongly, so they tend not to react independently. Consequently, the selective reactions of such systems usually involve the alternatives of conjugate addition vs carbonyl addition. In Section 13.10 we indicated that most reversible reactions proceed via conjugate addition, whereas irreversible additions tend to occur by nucleophilic attack at carbonyl carbon. In this section we will concentrate on reactions of nonconjugated compounds, for which two functional groups react independently.

Polar Addition Reactions

The characteristic addition reaction of an alkene proceeds by electrophilic attack on the double bond to produce a carbocation. Aldehydes and ketones can react under the same conditions, but if the carbonyl addition is *reversible* (for example, reaction with HCl), then the net result may be selective reaction of the alkene double bond. In the following example addition of HBr occurs selectively to the isolated double bond rather than to the conjugated carbonyl system.

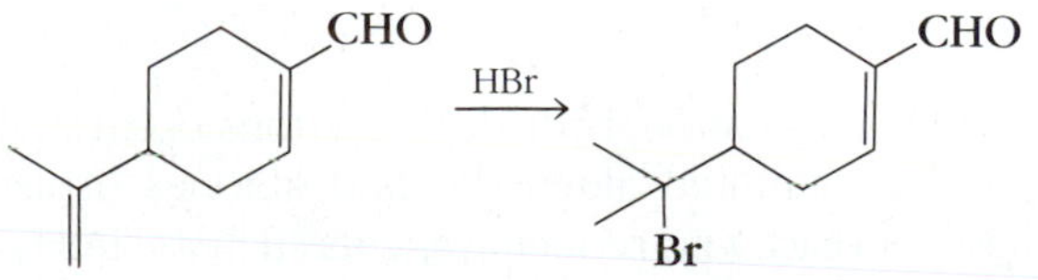

CHO —HBr→ CHO Br >41%

Reduction of Unsaturated Aldehydes and Ketones

Carbon-carbon and carbon-oxygen double bonds are both easily reduced, sometimes by the same reagents. Nevertheless, it is usually possible to reduce either of these functional groups selectively if you employ the proper reagents. Catalytic hydrogenation brings about preferential reduction of carbon-carbon double bonds, even in the case of α,β-unsaturated ketones.

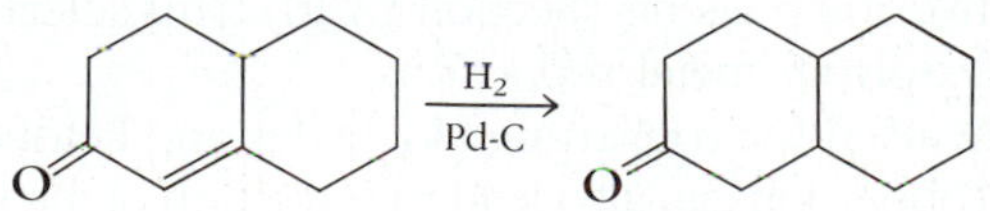

O —H_2, Pd-C→ O 80%

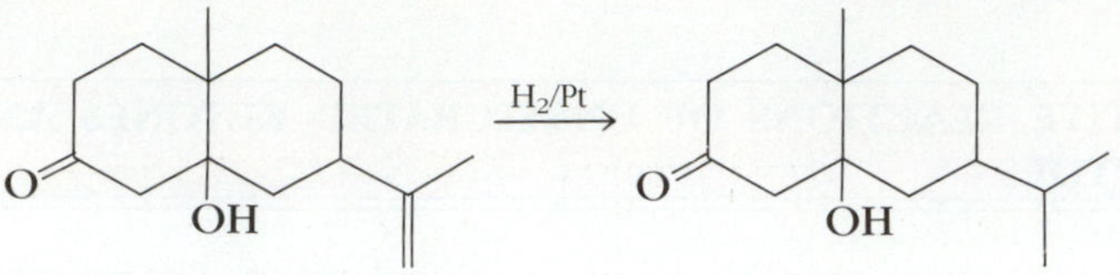

O OH $\xrightarrow{H_2/Pt}$ O OH 90%

Selective reduction of the carbon-carbon double bond of a conjugated carbonyl system can also be carried out by dissolving-metal reductions; the typical reagent is sodium or lithium metal in liquid ammonia.

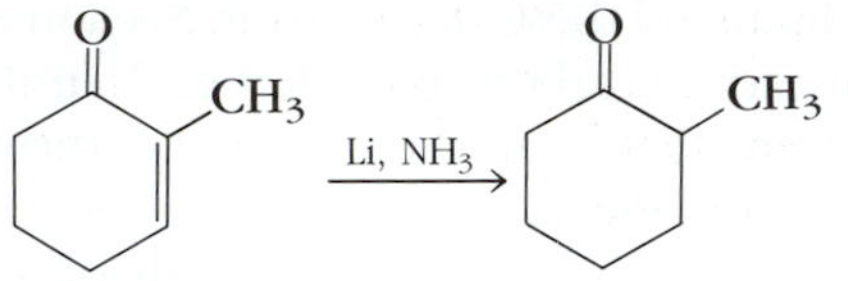

O CH_3 $\xrightarrow{\text{Li, } NH_3}$ O CH_3 50%

The initial reduction product is the enolate ion, which is not further reduced. Only during workup is the neutral ketone formed.

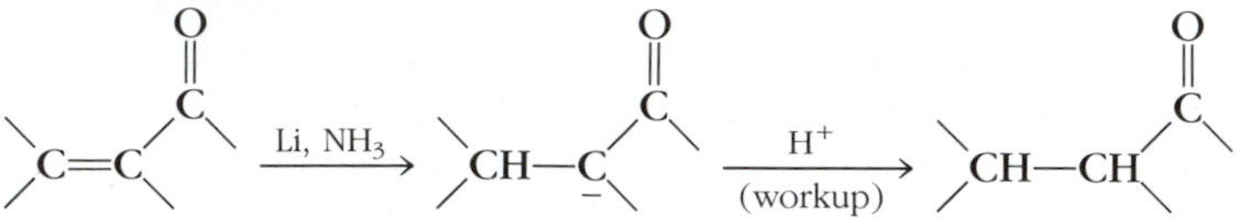

$$\text{C=C–C(=O)} \xrightarrow{\text{Li, NH}_3} \text{CH–}\overset{-}{\text{C}}\text{–C(=O)} \xrightarrow[\text{(workup)}]{H^+} \text{CH–CH–C(=O)}$$

Isolated double bonds are normally unaffected.

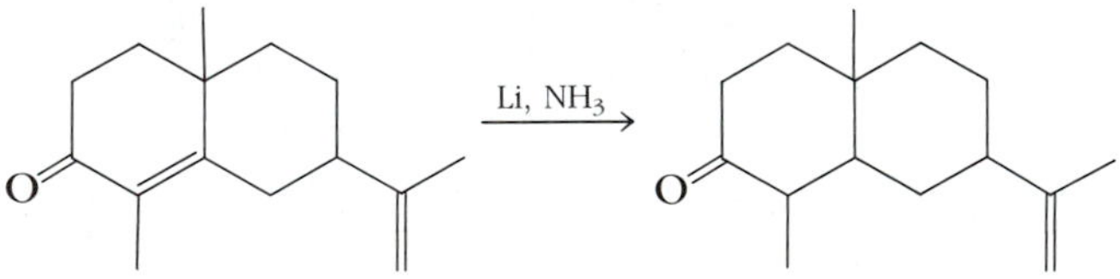

O $\xrightarrow{\text{Li, } NH_3}$ O 98%

In contrast to catalytic hydrogenation, hydride reductions occur at the carbonyl groups of nonconjugated unsaturated aldehydes and ketones. Isolated carbon-carbon double bonds do not react with reducing agents such as $LiAlH_4$ and $NaBH_4$.

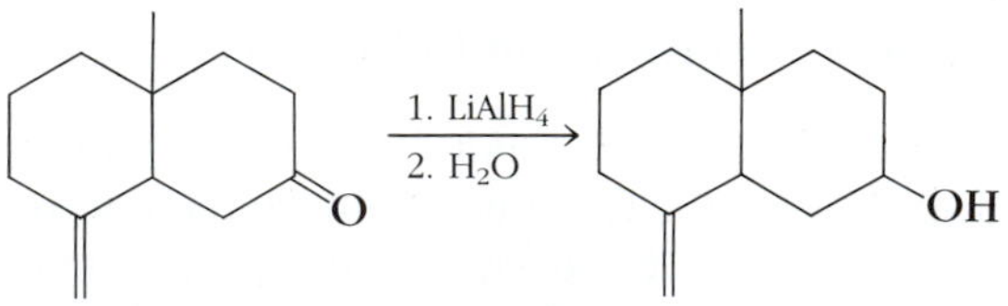

O $\xrightarrow[\text{2. } H_2O]{\text{1. } LiAlH_4}$ OH 83%

The carbonyl groups of α,β-unsaturated aldehydes and ketones are also reduced preferentially by these hydride reagents (Section 13.10). This selectivity is opposite to that obtained with dissolving metal reductions.

Although not typically used for converting aldehydes and ketones to the corresponding alcohols, hydroboration can also lead to reduction of the carbonyl group, as seen in the following example.

1. BH_3
2. H_2O_2/NaOH

OH
OH
70%

Oxidation of Unsaturated Aldehydes and Ketones

Reaction of a nonconjugated ketone with a peroxy acid tends to occur selectively at the carbon-carbon double bond. In other words, epoxidation occurs more rapidly than the Baeyer-Villiger reaction. The electron-withdrawing effect of a carbonyl group deactivates a conjugated double bond, so the isolated double bond undergoes selective epoxidation in the following example:

Cl CO$_3$H

Epoxidation of α,β-unsaturated ketones with peroxy acids does not occur smoothly. The carbon-carbon double bond is deactivated by the carbonyl group, and a mixture of epoxide and Baeyer-Villiger products is likely to be formed.

$$(CH_3)_2C{=}CH{-}\overset{O}{\overset{\|}{C}}{-}CH_3 \xrightarrow{CH_3CO_3H} (CH_3)_2\overset{O}{C{-}CH}{-}\overset{O}{\overset{\|}{C}}{-}CH_3 + (CH_3)_2\overset{O}{C{-}CH}{-}O{-}\overset{O}{\overset{\|}{C}}{-}CH_3$$

20% 80%

The major course for the preceding reaction was Baeyer-Villiger oxidation (followed by epoxidation of the product of that reaction).

Selective epoxidation of α,β-unsaturated ketones can be achieved by using hydrogen peroxide under alkaline conditions. This reaction does not affect nonconjugated double bonds, because it proceeds via nucleophilic attack of the hydroperoxide ion on the α carbon of the unsaturated carbonyl system.

H_2O_2, NaOH

70–72%

Even in the case of an unsaturated aldehyde this reaction can provide selective epoxidation of the conjugated double bond. The sensitive aldehyde group was not oxidized in the following example.

$$(CH_3)_2C{=}CH{-}CH_2CH_2{-}C(CH_3){=}CH{-}CHO \xrightarrow[NaOH]{H_2O_2} (CH_3)_2C{=}CH{-}CH_2CH_2{-}\overset{O}{C(CH_3){-}CH}{-}CHO \quad 80\%$$

Reactions of Organometallic Compounds and Ylides

As we stated previously in Section 13.10, reactions of lithium dialkylcuprates with conjugated ketones occur exclusively via conjugate addition. Most other organometallic reagents react preferentially at carbonyl carbon. Similarly, phosphorus and sulfur ylides of the types we discussed in Section 13.8 react selectively at carbonyl carbon rather than at the β position. None of these reagents undergo reaction with an isolated carbon-carbon double bond. This overall selectivity is illustrated by the following examples:

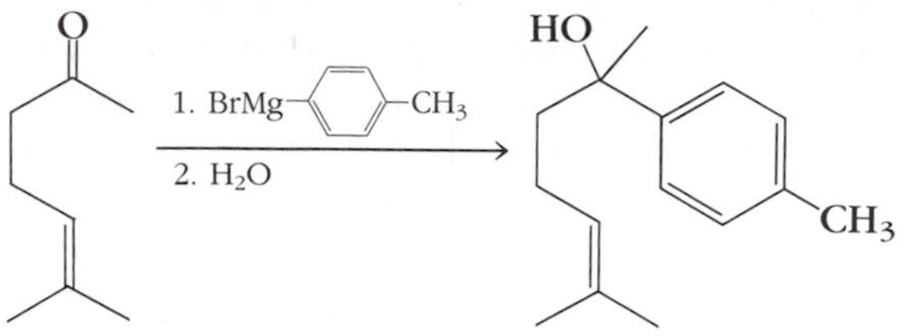

1. BrMg–C_6H_4–CH_3 2. H_2O → HO ... CH_3 48%

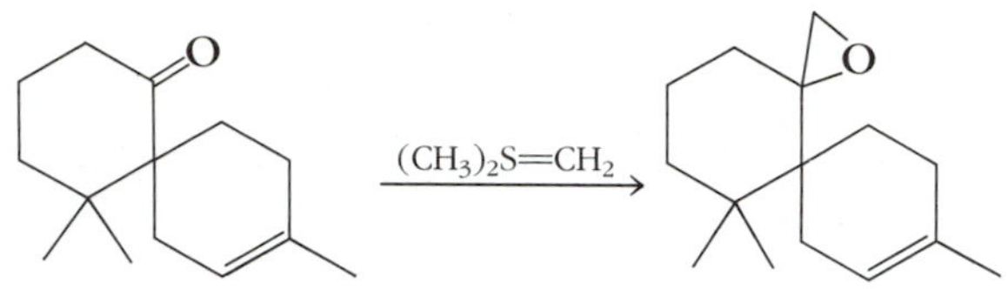

$(CH_3)_2S{=}CH_2$ → O 94%

$(C_6H_5)_3P{=}CH_2$ → 57%

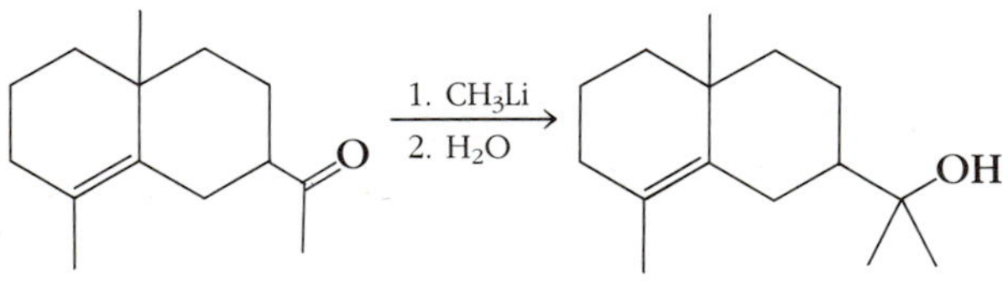

O 1. CH_3Li 2. H_2O → OH >40%

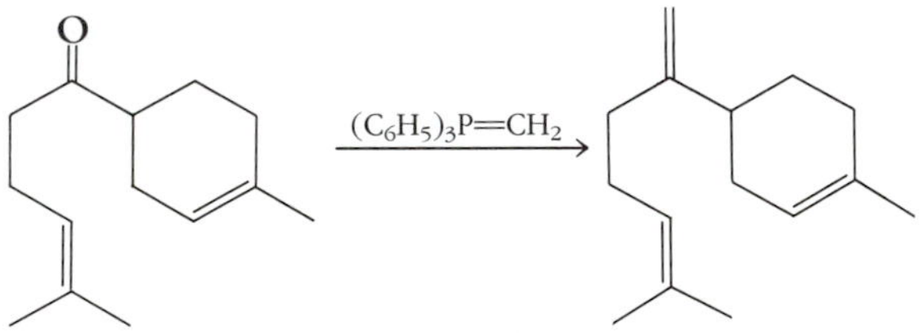

EXERCISE 13.19

Complete the reaction of 4-methylcyclohex-3-enone with each of the following reagents: (a) H_2/Pt; (b) CF_3CO_3H; (c) $LiAlH_4$, then H_2O; (d) $(CH_3)_2S{=}CH_2$.

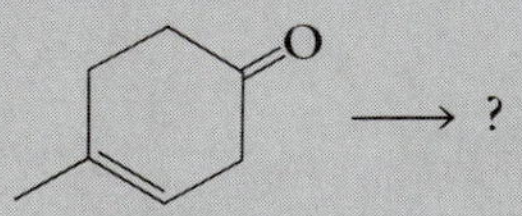

O → ?

EXERCISE 13.20

Complete the reaction of 1,6-octadien-3-one with each of the following reagents: (a) $NaBH_4/CH_3OH$; (b) $(C_6H_5)_3P{=}CH_2$; (c) CH_3Li; (d) $H_2O_2/NaOH$.

$$CH_3{-}CH{=}CH{-}CH_2CH_2{-}\overset{\overset{\displaystyle O}{\|}}{C}{-}CH{=}CH_2 \longrightarrow ?$$

13.14 TERMS AND DEFINITIONS

Conjugate addition. Addition to an α,β-unsaturated carbonyl system that is initiated by nucleophilic attack at the β position.

Geminal. Having two identical substituents on a single carbon atom, that is, a geminal diol (or *gem*-diol).

Hemiacetal. The product of the addition of an alcohol to the carbonyl group of an aldehyde, that is, a compound having a carbon on which the four substituents are an alkyl (or aryl) group, a hydrogen, an alkoxy group, and a hydroxyl group.

Hemiketal. The product of the addition of an alcohol to the carbonyl group of a ketone, that is, a compound having a carbon on which two of the substituents are alkyl (or aryl) groups and the remaining two are an alkoxy group and a hydroxyl group.

Heteroatom. An atom other than carbon in a chain or ring, that is, nitrogen, oxygen, sulfur, or phosphorus.

Migratory aptitude. The relative tendency of a substituent to undergo a 1,2 shift to an electron-deficient atom, that is, to migrate in the Baeyer-Villiger oxidation.

Ylide. A molecule having a negative charge on a carbon atom and a positive charge on an adjacent heteroatom such as phosphorus or sulfur; a double bond between the carbon and the heteroatom can be drawn only if more than eight electrons are considered to occupy the valence shell of the heteroatom.

13.15 SUMMARY OF REACTIONS

The reactions of aldehydes and ketones are summarized in Table 13.5, and various methods for their preparation are summarized in Table 13.6.

TABLE 13.5 Reactions of Aldehydes and Ketones

Reaction	Comments
1. Oxidation reactions	Section 13.11
(a) Oxidation of aldehydes $R{-}\overset{\overset{\displaystyle O}{\|}}{C}H \xrightarrow[\text{agent}]{\text{oxidizing}} R{-}\overset{\overset{\displaystyle O}{\|}}{C}OH$	A wide variety of oxidizing agents can be used: Cr(VI) and Mn(VII) are common. Reaction with Ag^+ to form Ag° is used as a test for aldehydes (Section 13.12).

TABLE 13.5 Reactions of Aldehydes and Ketones (continued)

Reaction	Comments
(b) Oxidative cleavage 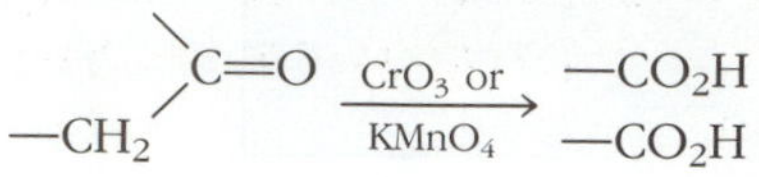 $\gt C{=}O$ with $-CH_2-$ $\xrightarrow[KMnO_4]{CrO_3\ \text{or}}$ $-CO_2H$ $-CO_2H$	Infrequently used as a laboratory method. Cyclic ketones yield dicarboxylic acids on vigorous oxidation.
(c) Baeyer-Villiger oxidation of ketones 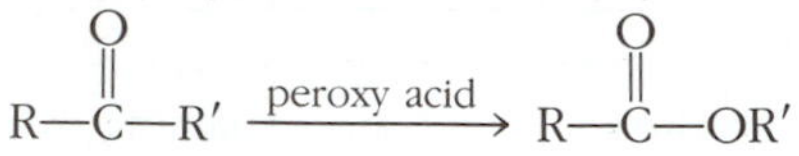 $R-\overset{O}{\overset{\parallel}{C}}-R' \xrightarrow{\text{peroxy acid}} R-\overset{O}{\overset{\parallel}{C}}-OR'$	A variety of peroxy acids can be used. The preference of alkyl group migration (R vs R′) is 3° > aryl > 2° > 1° > CH_3. Aldehydes are usually oxidized to the corresponding carboxylic acids.
(d) Epoxidation of *α,β*-unsaturated carbonyl compounds 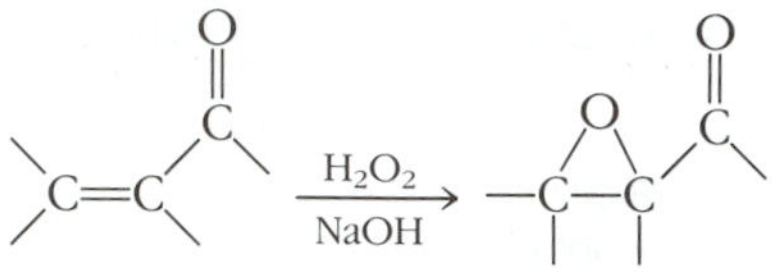 $\gt C{=}C\lt$ (with $-\overset{O}{\overset{\parallel}{C}}-$) $\xrightarrow[NaOH]{H_2O_2}$ $-C\overset{O}{-}C-$ (epoxide, with $-\overset{O}{\overset{\parallel}{C}}-$)	Isolated carbonyl groups and double bonds are unaffected by alkaline H_2O_2.
(e) Selenium dioxide oxidation 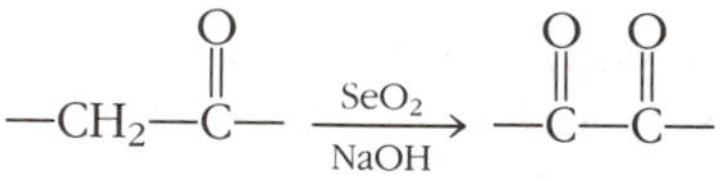 $-CH_2-\overset{O}{\overset{\parallel}{C}}- \xrightarrow[NaOH]{SeO_2} -\overset{O}{\overset{\parallel}{C}}-\overset{O}{\overset{\parallel}{C}}-$	Formation of *α*-dicarbonyl compounds.
2. Reduction of aldehydes and ketones	Sections 13.11, 11.4
(a) Hydride reduction 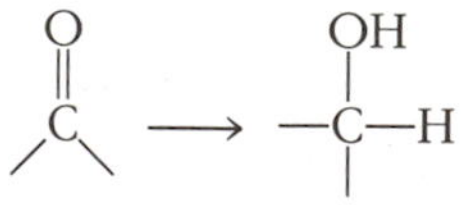 $\gt C{=}O \longrightarrow -\overset{OH}{\overset{\vert}{C}}-H$	Both $LiAlH_4$ and $NaBH_4$ work well. (Note that alkali metal hydrides such as LiH and NaH function as bases rather than as reducing agents.)
(b) Catalytic hydrogenation 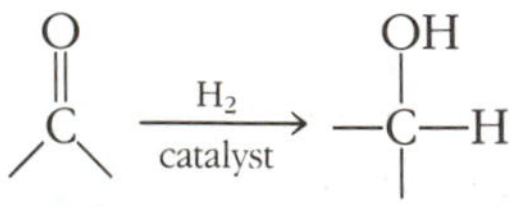 $\gt C{=}O \xrightarrow[\text{catalyst}]{H_2} -\overset{OH}{\overset{\vert}{C}}-H$	Carbon-carbon double bonds are reduced in preference to carbonyl groups.
(c) Dissolving metal reduction 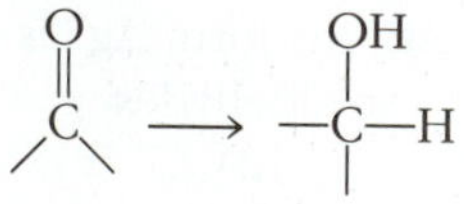 $\gt C{=}O \longrightarrow -\overset{OH}{\overset{\vert}{C}}-H$	Common reagents are sodium or lithium in ammonia. (Sodium-alcohol is sometimes used with saturated ketones or aldehydes.)

TABLE 13.5 Reactions of Aldehydes and Ketones (continued)

Reaction	Comments
$>C{=}C{-}C({=}O){-} \longrightarrow >CH{-}CH{-}C({=}O){-}$	α,β-Unsaturated aldehydes and ketones can be partially reduced to the saturated carbonyl compound (Section 13.13).
(d) Deoxygenation of aldehydes and ketones	
$>C{=}O \xrightarrow[KOH]{NH_2NH_2} -CH_2-$	The Wolff-Kishner reduction is generally superior to the Clemmensen reduction. Both procedures can yield other products in the reaction with α,β-unsaturated aldehydes or ketones.
$>C{=}O \xrightarrow[HCl]{Zn{-}Hg} -CH_2-$	
$>C{=}O \xrightarrow[2.\ H_2,\ Ni]{1.\ HS{-}CH_2CH_2{-}SH} -CH_2-$	Desulfurization (Section 15.5).
3. Reaction with carbon nucleophiles	Carbon-carbon bond-forming reactions.
(a) Organometallic reagents $>C{=}O \xrightarrow[2.\ H_2O]{1.\ RMgX\ or\ RLi} -C(OH)(R)-$	Sections 13.5 and 11.5 Carbonyl addition is favored with conjugated aldehydes and ketones.
(b) Lithium cuprates $>C{=}C{-}C({=}O){-} \xrightarrow[2.\ H_2O]{1.\ R_2CuLi} -C(R){-}CH{-}C({=}O){-}$	Section 13.10. Saturated aldehydes and ketones are unreactive; α,β-unsaturated compounds undergo conjugate addition.
(c) Phosphorus ylides $>C{=}O \xrightarrow{(C_6H_5)_3P{=}C<} >C{=}C<$	Section 13.8. Attack occurs preferentially at carbonyl carbon, even with α,β-unsaturated systems.
(d) Sulfur ylides $>C{=}O \xrightarrow{(CH_3)_2S{=}C<} -C{-}C-$ (epoxide, O bridging both carbons)	Section 13.8. Attack occurs preferentially at carbonyl carbon, even with α,β-unsaturated systems.

TABLE 13.5 Reactions of Aldehydes and Ketones (continued)

Reaction	Comments
(e) Hydrogen cyanide $\mathrm{>C{=}O \xrightleftharpoons[\text{acid or base}]{HCN} -C(OH)-CN}$	Section 13.7 The reaction is reversible, and conjugate addition is favored with α,β-unsaturated aldehydes and ketones.
4. Reaction with heteroatom nucleophiles	Section 13.7. Reversible reactions catalyzed by either acid or base.
(a) Water $\mathrm{>C{=}O \xrightleftharpoons{H_2O} -C(OH)-OH}$	Equilibrium usually favors the free carbonyl compound.
(b) Alcohols $\mathrm{>C{=}O \xrightleftharpoons{ROH} -C(OH)-OR}$	Hemiacetal or hemiketal formation. Equilibrium usually favors the free carbonyl unless a ring is formed.
$\mathrm{>C{=}O \xrightleftharpoons[H^+]{ROH} -C(OR)-OR}$	Ketal formation (Section 15.5).
(c) Hydrazine and primary amine derivatives $\mathrm{>C{=}O \xrightleftharpoons{R-NH_2} >C{=}N-R}$	The products may be intermediates in other reactions (such as the Wolff-Kishner reduction, Section 13.11). Hydrazones are useful for characterization of aldehydes and ketones (Section 13.12).
(d) Secondary amines $\mathrm{-C(=O)-CH< \xrightleftharpoons{R_2NH} -C(NR_2){=}C<}$	Enamine formation (Section 17.8).

TABLE 13.5 Reactions of Aldehydes and Ketones (continued)	
Reaction	**Comments**
(e) Thiols $\mathrm{>C{=}O \xrightleftharpoons{RSH} -C(SR)_2-}$	Preparation of thioketals (Section 15.5).
5. Halogenation	Section 13.9.
(a) Acid-catalyzed $\mathrm{-C({=}O)-CH< \xrightarrow[H^+]{X_2} -C({=}O)-CX<}$	Reaction via enol form. Can be used for *mono*halogenation.
(b) Base-catalyzed $\mathrm{-C({=}O)-CH< \xrightarrow[base]{X_2} -C({=}O)-CX<}$	Reaction via enolate ion. Typically more than one halogen is introduced.
$\mathrm{-C({=}O)-CH_3 \xrightarrow[^-OH]{X_2} -C({=}O)-O^- + HCX_3}$	Methyl ketones are cleaved (haloform reaction).
6. Alkylation of aldehydes and ketones	Section 14.3.
$\mathrm{-C({=}O)-CH< \xrightarrow[2.\ RX]{1.\ base} -C({=}O)-CR<}$	Reaction via enolate ion. Carbon-carbon bond formation.
7. Condensation reactions	Sections 14.1, 14.2
$\mathrm{-C({=}O)-CH_2- \xrightarrow[base\ or\ acid]{>C{=}O} -C({=}O)-C(-){=}C<}$	Reaction via the enol or enolate ion. Carbon-carbon bond formation.

TABLE 13.6 Preparation of Aldehydes and Ketones

Reaction	Comments
1. Oxidation of alcohols	Sections 13.3, 11.3.
$R{-}CH(OH){-}R \xrightarrow[\text{agent}]{\text{oxidizing}} R{-}C(=O){-}R$	A variety of oxidizing agents can be used to convert 2° alcohols to ketones: CrO_3, $KMnO_4$, $K_2Cr_2O_7$, etc.
$R{-}CH_2OH \xrightarrow{CrO_3\text{-pyridine}} R{-}CH{=}O$	Oxidation of 1° alcohols can continue past the aldehyde state to give carboxylic acids with other, more reactive oxidizing agents.
2. Hydration of alkynes	Section 7.5.
$-C{\equiv}C- \longrightarrow -CH_2{-}C(=O)-$	
3. Ozonolysis of alkenes	Sections 13.3 and 5.12.
$>C{=}C< \xrightarrow[\text{2. Zn}]{\text{1. } O_3} >C{=}O + O{=}C<$	The intermediate ozonide can also be reduced with $(CH_3)_2S$.
4. Periodate cleavage of 1,2-diols	Section 13.3.
$-C(OH){-}C(OH)- \xrightarrow{HIO_4} >C{=}O + O{=}C<$	
$>C{=}C< \xrightarrow[KMnO_4, KIO_4]{OsO_4, NaIO_4 \text{ or}} >C{=}O + O{=}C<$	Direct cleavage of an alkene via the diol without isolation.
5. Reduction of acid chlorides	Section 13.3, aldehyde synthesis.
$R{-}C(=O){-}Cl \xrightarrow[Pd\text{-}BaSO_4]{H_2} R{-}CH{=}O$	A deactivated catalyst is needed.
$R{-}C(=O){-}Cl \xrightarrow{LiAl(OtBu)_3H} R{-}CH{=}O$	With other hydride reducing agents such as $LiAlH_4$ or $NaBH_4$ the reduction proceeds further to give the 1° alcohol.

TABLE 13.6 Preparation of Aldehydes and Ketones (continued)

Reaction	Comments
6. Hydrolysis of blocked carbonyl groups	
$\mathrm{X(Y)C{<}} \xrightleftharpoons{H_2O} \mathrm{C{=}O}$	Reversal of carbonyl addition reactions (Section 13.7). X, Y = OR, OH, CN, etc.
$\mathrm{HO(X)C{<}} \xrightarrow{H_2O} \mathrm{C{=}O}$	Hydrolysis of acetals and ketals (Section 15.5).
$\mathrm{R{-}N{=}C(R){<}} \xrightleftharpoons{H_2O} \mathrm{C{=}O}$	Section 13.7.
7. Reaction of acid chlorides with organometallic reagents	Section 13.4, ketone synthesis.
$-\mathrm{C(=O)-Cl} \xrightarrow{R_2Cd} -\mathrm{C(=O)-R}$ $-\mathrm{C(=O)-Cl} \xrightarrow{R_2CuLi} -\mathrm{C(=O)-R}$	Organocadmium reagents have been largely replaced by organocuprates in modern organic chemistry. Other organometallic reagents react further to yield 3° alcohols.
8. Reaction of carboxylic acids with alkyllithium reagents	Section 13.4, ketone synthesis.
$-\mathrm{C(=O)-OH} \xrightarrow{R-Li} -\mathrm{C(=O)-R}$	Limited to alkyllithium reagents (primarily CH_3Li).
9. Reaction of nitriles with organometallic reagents	Section 13.4, ketone synthesis.
$\mathrm{R'-C{\equiv}N} \xrightarrow[2.\ H_2O]{1.\ R-M} \mathrm{R'-C(=O)-R}$	Alkyllithium reagents are preferable.
10. Friedel-Crafts acylation	Section 18.3.
$\mathrm{Ar-H} \xrightarrow[AlCl_3]{RC(=O)Cl} \mathrm{Ar-C(=O)-R}$	Synthesis of aromatic ketones.

TABLE 13.6 Preparation of Aldehydes and Ketones (continued)

Reaction	Comments
11. Acylation of enolate ions Synthesis of β-dicarbonyl compounds	Section 21.2.
$>C(=O)-\bar{C}< \xrightarrow{R-C(=O)-X} >C(=O)-C<-C(=O)-R$	Reaction via enol or enolate ion. Carbon-carbon bond formation.
12. Alkylation of aldehydes and ketones	Sections 14.3, 14.4.
$>C(=O)-CH< \xrightarrow[2.\ RX]{1.\ base} >C(=O)-C<-R$	Reaction via enolate ion. Carbon-carbon bond formation.
13. Condensation reactions	Sections 14.1, 14.2.
$>C(=O)-CH_2^- \xrightarrow[\text{base or acid}]{>C=O} >C(=O)-C=C<$	Reaction via enol or enolate ion. Carbon-carbon bond formation.

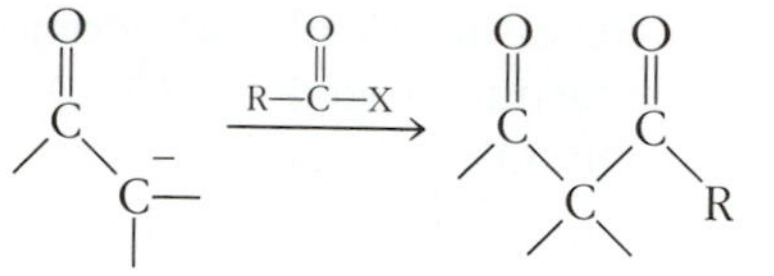

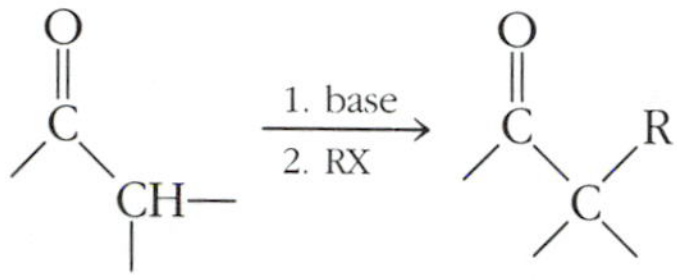

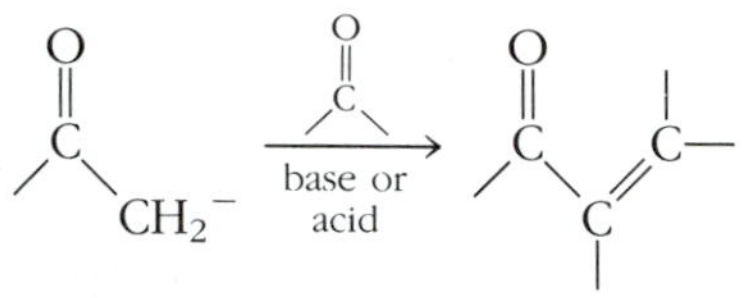

13.16 PROBLEMS

13.1 Name the following compounds:

(a) $CH_3-CH_2-CH_2-C(=O)-CH_2-CH_3$

(b) $CH_3-CH(CH_3)-CH=CH-CHO$

(c) 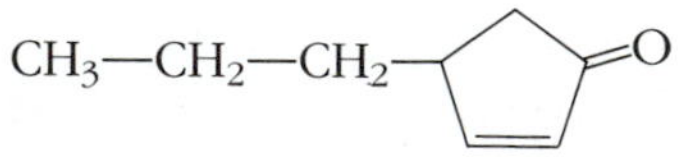 $CH_3-CH_2-CH_2-$(cyclopentenone)$=O$

(d) $(CH_3)_2C=CH-CH_2-CH_2-C(=O)-CH_2-CH_3$

(e) 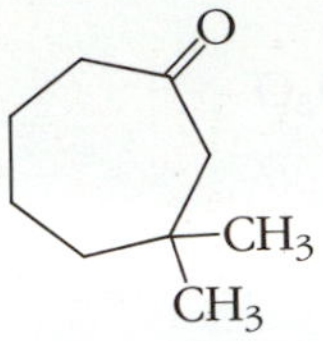 O, CH_3, CH_3

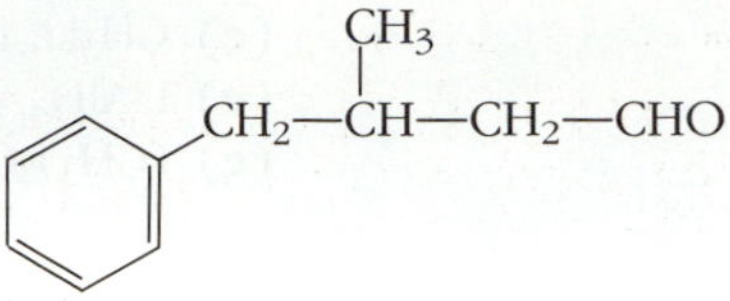

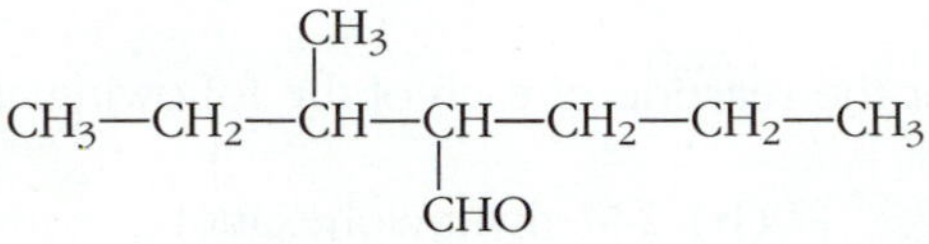

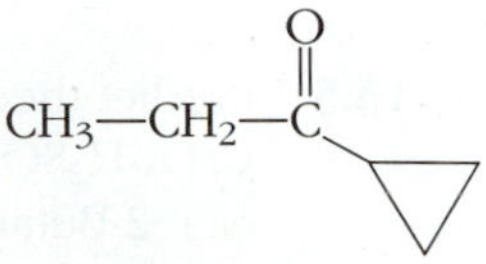

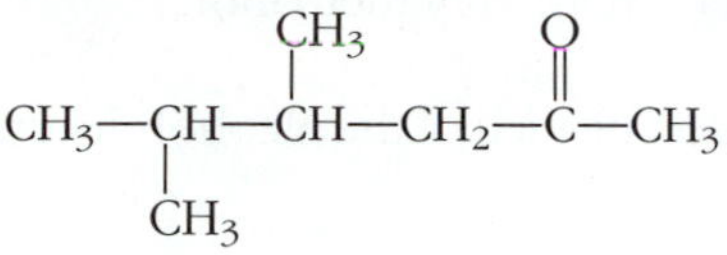

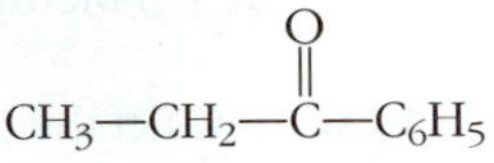

(f) CH_3; $CH_2—CH—CH_2—CHO$

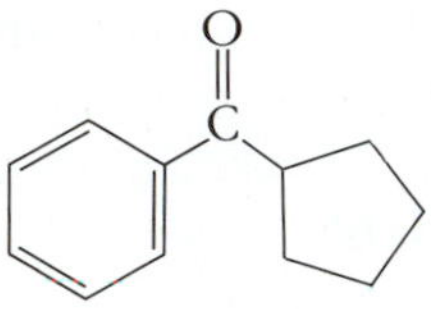

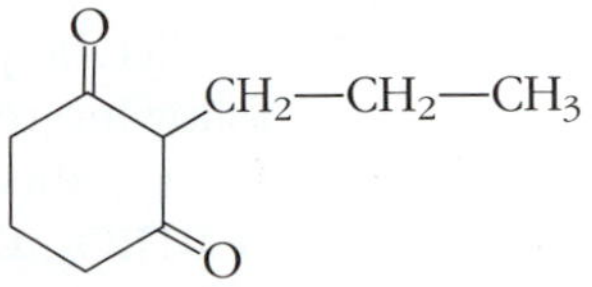

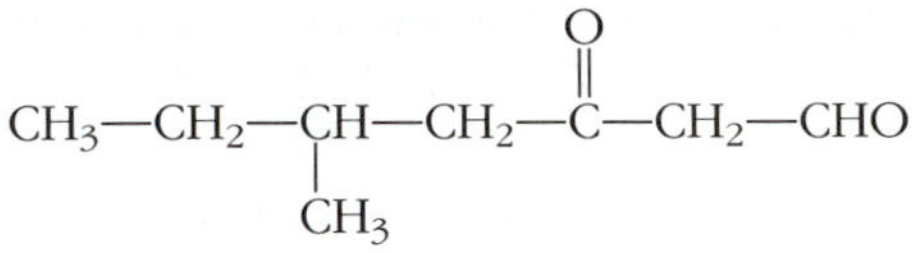

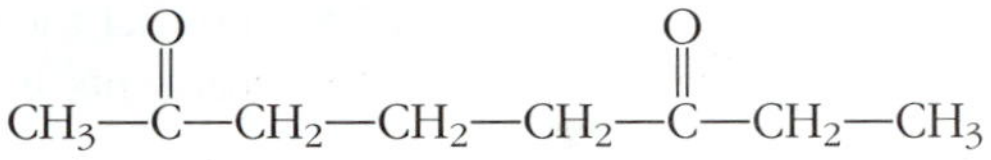

(g) $CH_3—CH_2—CH(CH_3)—CH(CHO)—CH_2—CH_2—CH_3$

(h) $CH_3—CH_2—C(=O)$—

(i) $CH_3—CH(CH_3)—CH(CH_3)—CH_2—C(=O)—CH_3$

(j) $CH_3—CH_2—C(=O)—C_6H_5$

(k) O, C

(l) O, $CH_2—CH_2—CH_3$, O

(m) $CH_3—CH_2—CH(CH_3)—CH_2—C(=O)—CH_2—CHO$

(n) $CH_3—C(=O)—CH_2—CH_2—CH_2—C(=O)—CH_2—CH_3$

(o) $CH_3—C(=O)—CH_2—CH_2—CH(CH_3)$—

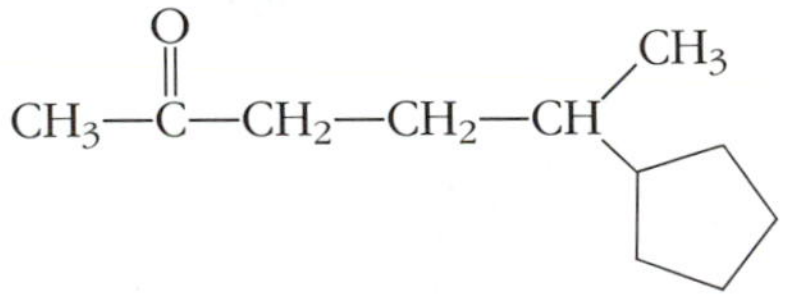

(p) 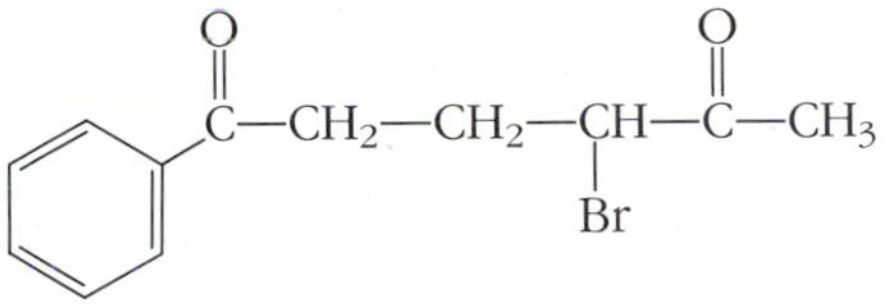 $—C(=O)—CH_2—CH_2—CH(Br)—C(=O)—CH_3$

13.2 Draw structures for the following compounds:

(a) 2-Methyl-3-heptanone
(b) 2-Chloro-5-methylcyclohexanone
(c) 4,4-Dimethylheptanal
(d) 3,4-Heptanedione
(e) Butyrophenone
(f) 2-Cyclobutyl-5,5-dimethyl-3-octanone
(g) 3,5-Dichloro-2-ethylcyclohex-2-en-1-one
(h) 6-Bromooct-3-yn-2-one
(i) 2,3,4-Trimethylhexanal
(j) 1-Cyclohexylpentane-1,2-dione
(k) 1-Hydroxy-4-methyl-3-heptanone
(l) 3-Chlorocyclopentanone
(m) 3-Oxocyclohexanecarboxaldehyde
(n) Cyclopropyl 2-methylcyclobutyl ketone
(o) 3,3,3-Trifluoro-1-phenyl-2-propanone

13.3 Draw the expected product for the reaction of benzoyl chloride (C_6H_5COCl) with each of the following reagents:

(a) H_2/Pd-$BaSO_4$ (b) $LiAl(OtBu)_3H$

(c) CH_3Li, then H_2O (d) $(CH_3)_2CuLi$
(e) $LiAlH_4$, then H_2O (f) CH_3MgBr, then H_2O
(g) $(CH_3)_2Cd$

13.4 Draw the expected reaction product for each of the reagents in Problem 13.3 with butyryl chloride ($CH_3CH_2CH_2COCl$).

13.5 Predict the product for the reaction of each of the following alcohols with CrO_3-H_2SO_4:
(a) 2-Butanol (b) 2-Methylcyclohexanol
(c) 3-Methyl-1-pentanol (d) Cyclopentylmethanol

13.6 Predict the product for the reaction of each of the alcohols listed in Problem 13.5 with CrO_3-pyridine.

13.7 Predict the product for the reaction of 2-methyl-2-pentene with each of the following reagents:
(a) O_3, then Zn, H_2O (b) HIO_4, OsO_4
(c) HIO_4, $KMnO_4$ (d) O_3, then $(CH_3)_2S$

13.8 Predict the product for the reaction of 1-ethylcyclohexene with each of the reagents in Problem 13.7.

13.9 Suggest methods for the preparation of 2-hexanone from each of the following starting materials. More than one reaction may be necessary in each case.
(a) Pentanal (b) Pentanoic acid
(c) 1-Hexene (d) 2,3-Dimethyl-2-heptene
(e) Acetylene (f) Acetyl chloride
(g) 1-Hexyne (h) 3,4-Dihydroxy-4-methyloctane

13.10 Suggest methods for the preparation of hexanal from each of the following starting materials. More than one reaction may be necessary in each case.
(a) 1-Heptene
(b) 1-Hexanol
(c) 1-Bromohexane
(d) Hexanoic acid
(e) Hexanoyl chloride ($CH_3CH_2CH_2CH_2CH_2COCl$)
(f) 2,3-Octanediol
(g) Acetylene

13.11 Which hydrogen in each of the following compounds should be most acidic?

(a) $CH_3—CH_2—CH_2—\overset{\overset{\large O}{\|}}{C}—CH_2—CH_2—CH_3$

(b) $CH_3—CH_2—\overset{\overset{\large O}{\|}}{C}—\overset{\overset{\large O}{\|}}{C}—CH_2—CH_3$

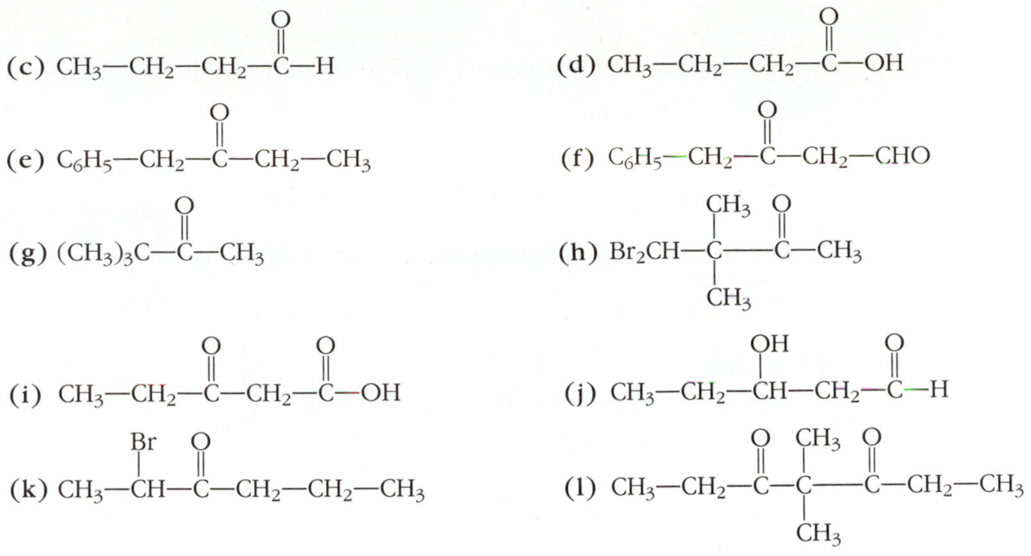

(c) $CH_3-CH_2-CH_2-\overset{O}{\overset{\|}{C}}-H$ (d) $CH_3-CH_2-CH_2-\overset{O}{\overset{\|}{C}}-OH$

(e) $C_6H_5-CH_2-\overset{O}{\overset{\|}{C}}-CH_2-CH_3$ (f) $C_6H_5-CH_2-\overset{O}{\overset{\|}{C}}-CH_2-CHO$

(g) $(CH_3)_3C-\overset{O}{\overset{\|}{C}}-CH_3$ (h) $Br_2CH-\underset{CH_3}{\overset{CH_3}{C}}-\overset{O}{\overset{\|}{C}}-CH_3$

(i) $CH_3-CH_2-\overset{O}{\overset{\|}{C}}-CH_2-\overset{O}{\overset{\|}{C}}-OH$ (j) $CH_3-CH_2-\overset{OH}{\overset{|}{CH}}-CH_2-\overset{O}{\overset{\|}{C}}-H$

(k) $CH_3-\overset{Br}{\overset{|}{CH}}-\overset{O}{\overset{\|}{C}}-CH_2-CH_2-CH_3$ (l) $CH_3-CH_2-\overset{O}{\overset{\|}{C}}-\underset{CH_3}{\overset{CH_3}{C}}-\overset{O}{\overset{\|}{C}}-CH_2-CH_3$

13.12 Predict the product for the reaction of 2-hexanone with each of the following reagents:

(a) Br_2, CH_3CO_2H
(b) CH_3CO_3H
(c) $LiAlH_4$, then H_2O
(d) C_2H_5MgCl, then H_2O
(e) $NH_2NHC_6H_5$, CH_3CO_2H
(f) Na, C_2H_5OH
(g) Zn-Hg, HCl
(h) C_6H_5Li, then H_2O
(i) Br_2, KOH
(j) NH_2NH_2, KOH
(k) $(C_6H_5)_3P{=}CH-CH_3$
(l) HCN, CH_3CO_2H
(m) $NaBH_4$, CH_3CH_2OH
(n) H_2, Pt
(o) $(CH_3)_2S{=}CH_2$
(p) CH_3Li, then H_2O
(q) $NaHSO_3$
(r) SeO_2

13.13 Predict the products for the reaction of 4-methylcyclohexanone with each of the reagents listed in Problem 13.12.

13.14 Predict the product for the reaction of hexanal with each of the following reagents:

(a) Ag_2O
(b) CH_3CO_3H
(c) $LiAlH_4$, then H_2O
(d) C_2H_5MgCl, then H_2O
(e) $NH_2NHC_6H_5$, CH_3CO_2H
(f) Na, C_2H_5OH
(g) Zn-Hg, HCl
(h) C_6H_5Li, then H_2O
(i) CrO_3
(j) NH_2NH_2, KOH
(k) $(C_6H_5)_3P{=}CH-CH_3$
(l) HCN, CH_3CO_2H
(m) $NaBH_4$, CH_3CH_2OH
(n) H_2, Pt
(o) $(CH_3)_2S{=}CH_2$
(p) CH_3Li, then H_2O
(q) $NaHSO_3$
(r) SeO_2

13.15 Predict the products for the reaction of cyclopentanecaraldehyde with each of the reagents listed in Problem 13.14.

13.16 Predict the product for Baeyer-Villiger oxidation of the following ketones with peroxytrifluoroacetic acid:

(a) 2-Methylcyclohexanone

(b) $CH_3-CH_2-CH_2-CH_2-C(=O)-CH_3$

(c) $CH_3-CH(CH_3)-C(=O)-CH_2-C_6H_5$

(d) $C_6H_5-C(=O)-CH_2-CH_3$

(e) 2,2,-Dimethylcyclopentanone

(f) $CH_3-CH_2-C(CH_3)_2-C(=O)-CH_3$

13.17 Predict the products for the reaction of propiophenone (ethyl phenyl ketone) with each of the following reagents:

(a) $NaBH_4$, CH_3OH
(b) NH_2NH_2, KOH
(c) HCN, H_2SO_4
(d) NH_2NH_2
(e) $(C_6H_5)_3P{=}CH-C_6H_5$
(f) C_2H_5Li, then H_2O
(g) $LiAlH_4$, then H_2O
(h) $(CH_3)_2S{=}CH_2$
(i) $C_6H_5CO_3H$
(j) I_2, NaOH

13.18 Write the mechanism for each of the reactions in Problem 13.17.

13.19 Predict the products for the reaction of 3-heptanone with each of the following reagents:

(a) $NaBH_4$, CH_3OH
(b) Zn-Hg, HCl
(c) HCN, H_2SO_4
(d) NH_2NH_2
(e) $(C_6H_5)_3P{=}C(CH_3)_2$
(f) C_2H_5Li, then H_2O
(g) $LiAlH_4$, then H_2O
(h) $(CH_3)_2S{=}CH_2$
(i) $C_6H_5CO_3H$
(j) Br_2, CH_3CO_2H
(k) SeO_2
(l) NH_2NH_2, KOH

13.20 Suggest methods for converting cyclohexylethanol into each of the indicated compounds. More than one reaction may be necessary in each case.

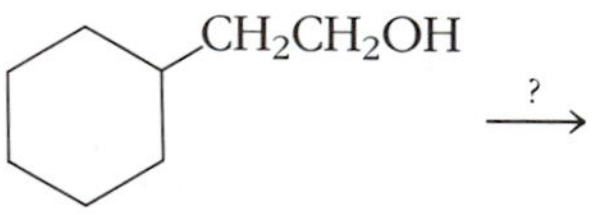

$C_6H_{11}-CH_2CH_2OH \xrightarrow{?}$

(a) 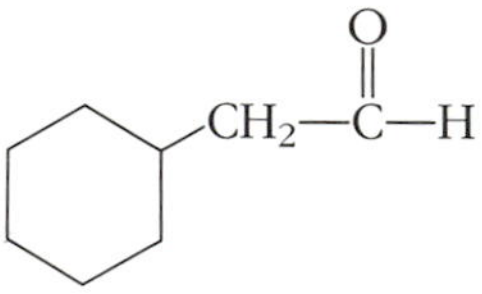 $C_6H_{11}-CH_2-C(=O)-H$

(b) $C_6H_{11}-C(=O)-H$

(c) $C_6H_{11}-CH_2-CH_2-C(=O)-H$

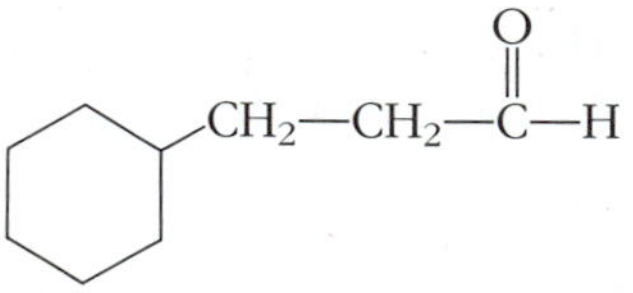

(d) 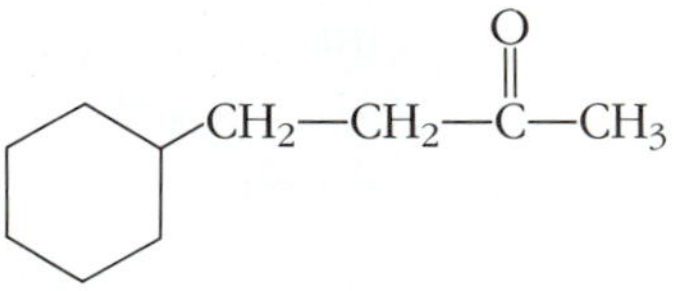 $C_6H_{11}-CH_2-CH_2-C(=O)-CH_3$

(e) 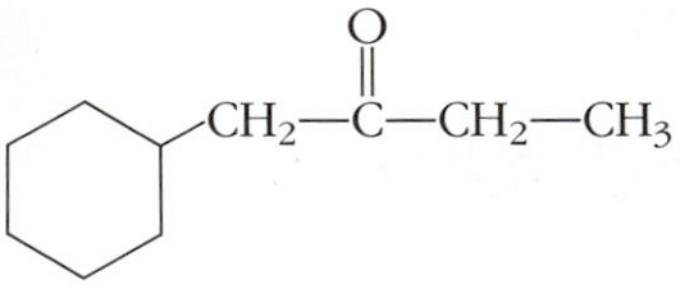 $C_6H_{11}-CH_2-C(=O)-CH_2-CH_3$

(f) 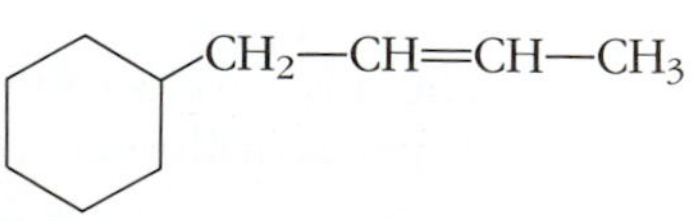 $C_6H_{11}-CH_2-CH{=}CH-CH_3$

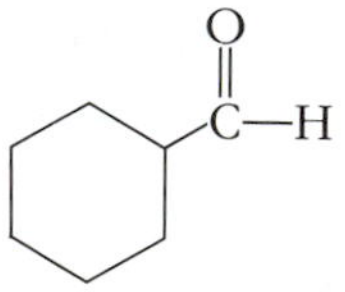

(g) 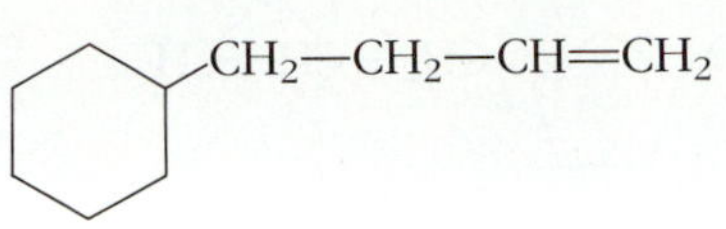 cyclohexyl–$CH_2—CH_2—CH=CH_2$ (h) 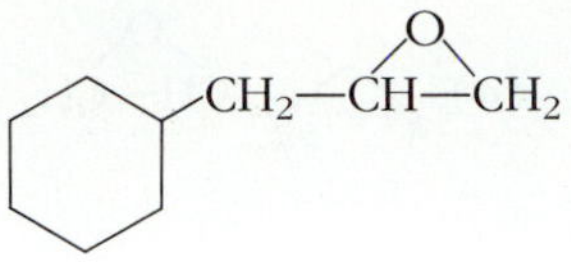 cyclohexyl–$CH_2—CH—CH_2$ (epoxide, O bridging CH and CH_2)

13.21 Write a complete mechanism for each of the reactions in Problem 13.20.

13.22 Suggest methods for converting cyclopentene into each of the indicated compounds. More than one reaction may be necessary.

(a) Cyclopentanone (b) 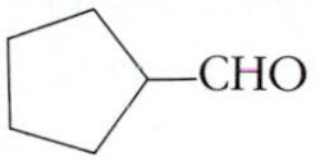 cyclopentyl–CHO (c) 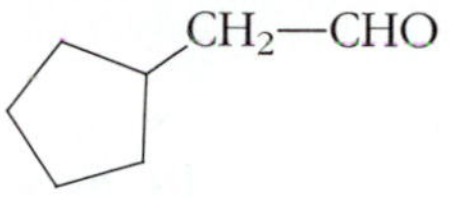 cyclopentyl–$CH_2—CHO$

(d) 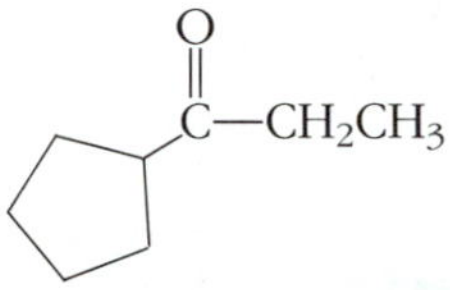 cyclopentyl–$\overset{O}{\overset{\|}{C}}—CH_2CH_3$ (e) 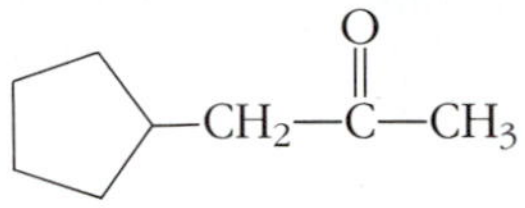 cyclopentyl–$CH_2—\overset{O}{\overset{\|}{C}}—CH_3$

13.23 Write a complete mechanism for each of the reactions in Problem 13.22.

13.24 Suggest methods for converting cyclohexanone into each of the indicated compounds. More than one reaction may be necessary.

(a) 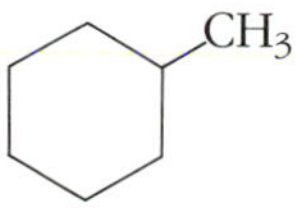 cyclohexyl–CH_3 (b) 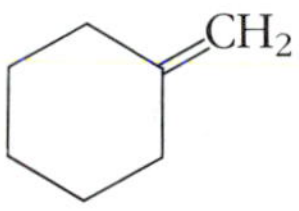 cyclohexylidene=CH_2

(c) 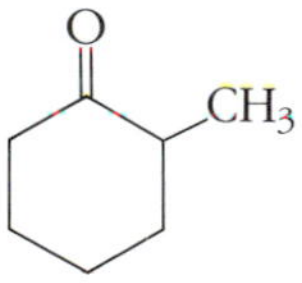 2-methylcyclohexanone (CH_3) (d) 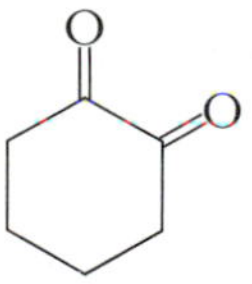 cyclohexane-1,2-dione (O, O)

(e) 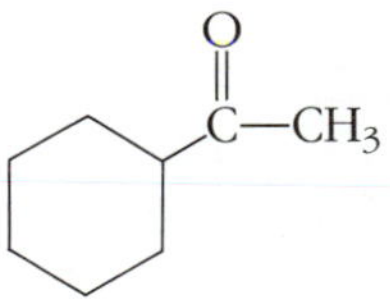 cyclohexyl–$\overset{O}{\overset{\|}{C}}—CH_3$ (f) 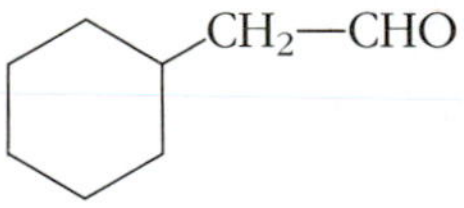 cyclohexyl–$CH_2—CHO$

13.25 Suggest methods for converting cyclopentylethylene into each of the indicated compounds. More than one reaction may be necessary.

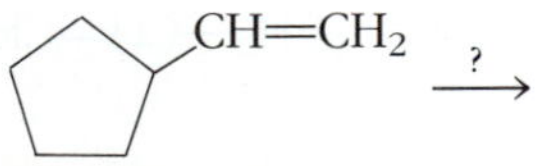

cyclopentyl–$CH=CH_2$ $\xrightarrow{?}$

(a) 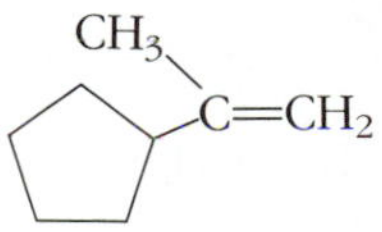 cyclopentyl–$C(CH_3)=CH_2$ (b) 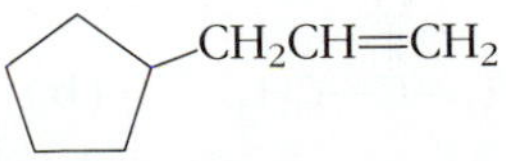 cyclopentyl–$CH_2CH=CH_2$

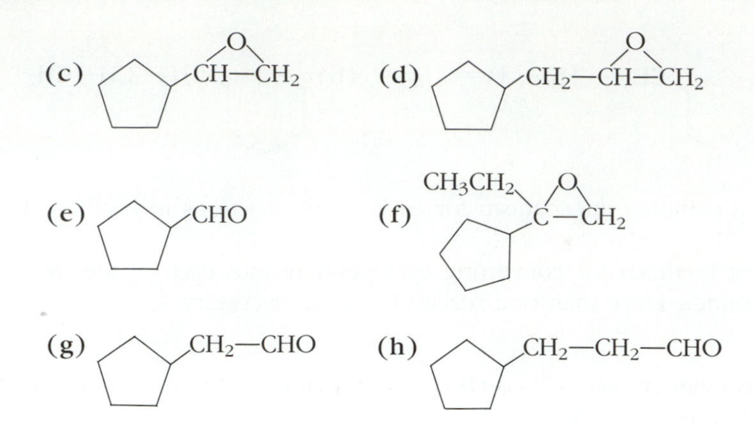

13.26 Suggest methods for converting 3-methylcyclohex-2-en-1-one into each of the indicated compounds. More than one reaction may be necessary.

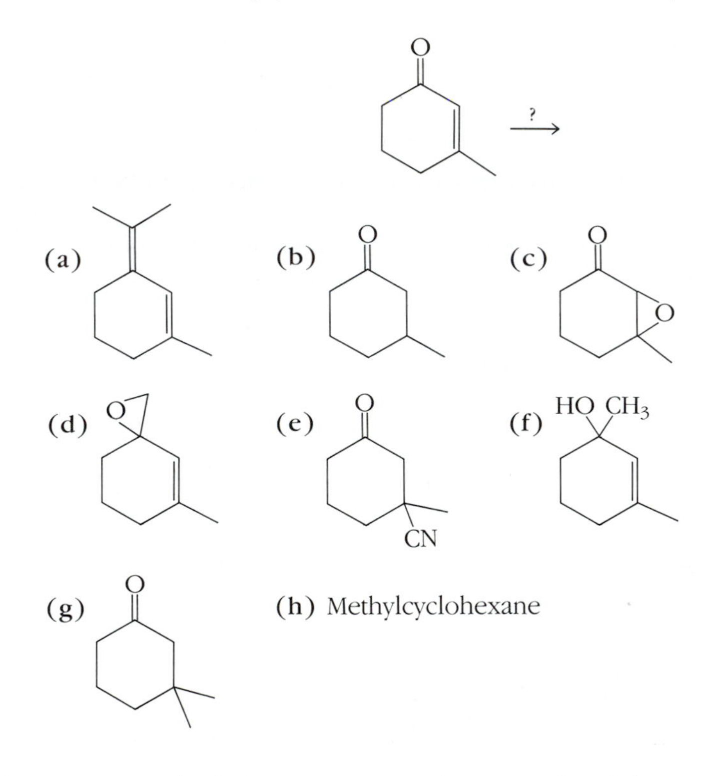

13.27 Suggest methods for converting 2-(2-propenyl)cyclohexanone into each of the indicated compounds. More than one reaction may be necessary.

O
$CH_2—CH{=}CH_2$ $\xrightarrow{?}$

(a) $CH_2—\overset{O}{\overset{\|}{C}}—CH_3$ (b) O $CH_2—CH_2—CH_3$

(c) CH_2—CH=CH_2

(d) O ‖ … CH_2—CH—CH_2 (O bridging)

(e) O–CH_2 … CH_2—CH=CH_2

13.28 **A–G** are all different compounds. From the information provided, deduce the structures of **A–F.**

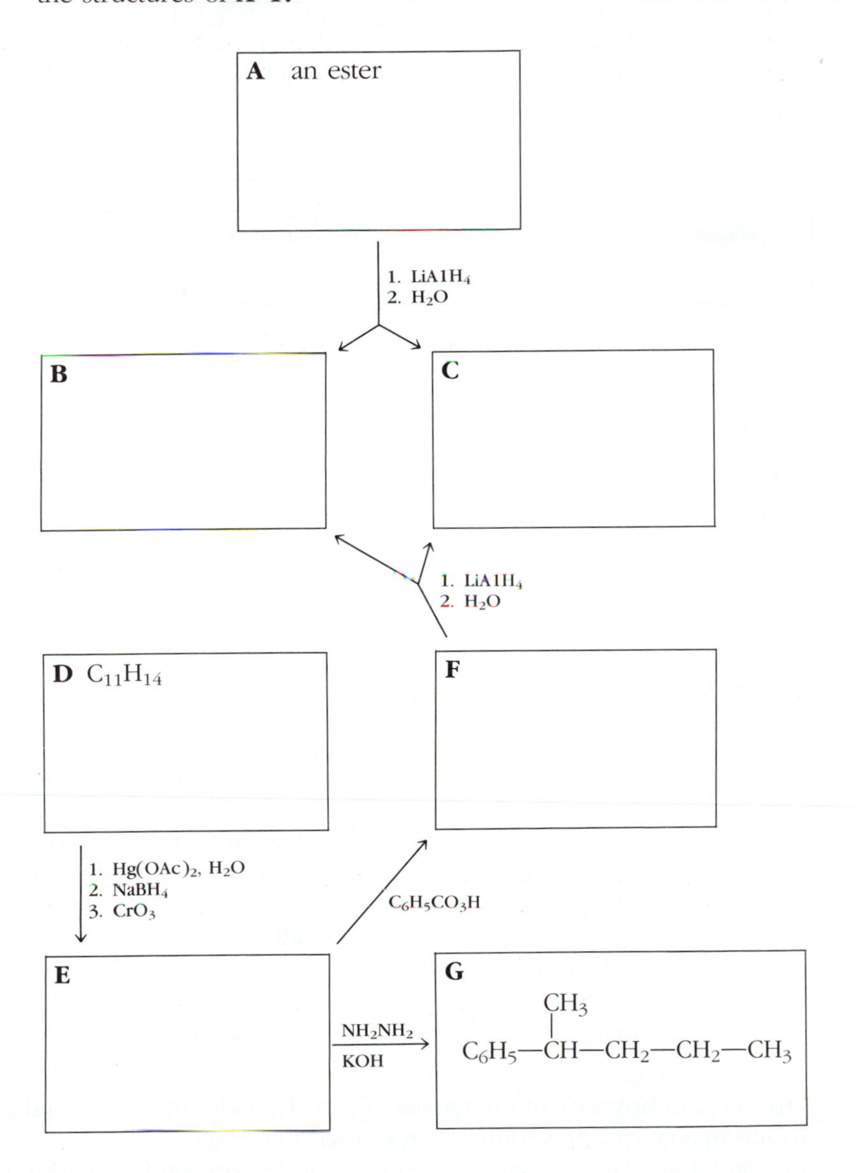

13.29 **A–G** are all different compounds. Compound **D** gives a positive Tollens test. On the basis of the information provided, deduce the structures of **A–F.**

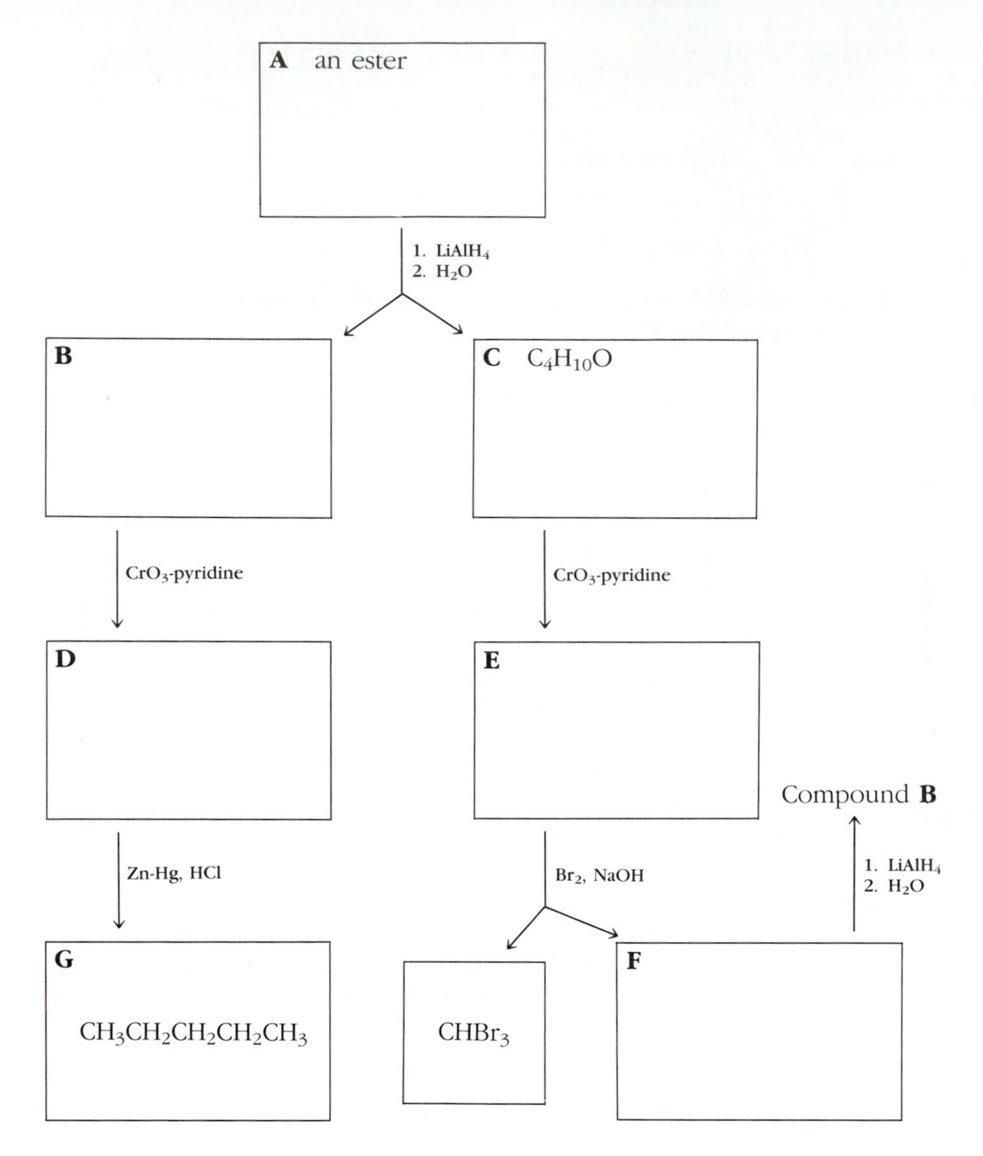

13.30 The mechanism drawn for the haloform reaction in Section 13.9 shows the formation of a halo-substituted carbanion, $^{-}CX_3$. Additional evidence for the formation of such carbanions is found in deuterium exchange studies. For example, heating dibromochloromethane with NaOD for 4 days at 105°C resulted in 43% deuterium incorporation.

$$CHBr_2Cl \xrightarrow[D_2O]{NaOD} [^{-}CBr_2Cl] \xrightarrow{D_2O} CDBr_2Cl$$

Treatment of bromodichloromethane, $CHBrCl_2$, under the same conditions results in only 16% deuterium incorporation. Other studies have shown that the rate of isotopic exchange is greater for iodoform and bromoform than for chloroform. What do these results suggest for the unusually high acidity of the trihalomethanes? Do the halogens stabilize the resulting carbanions only through inductive effects or also by interaction with empty *d* orbitals?

13.31 A compound of unknown structure gave a positive Tollens test. On the basis of this information and its NMR spectrum (Figure 13.5), deduce its structure.

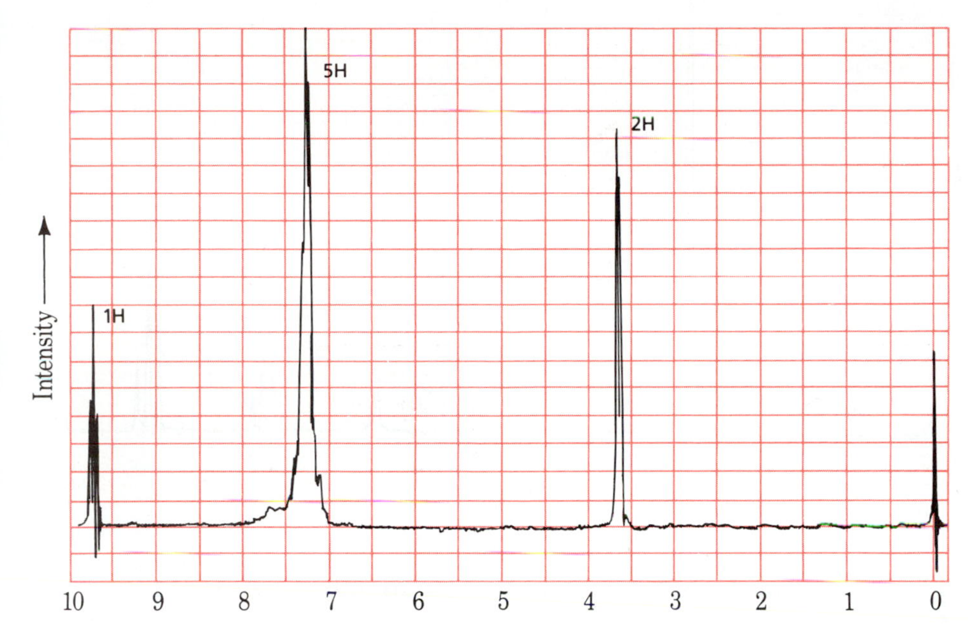

Figure 13.5 Proton NMR Spectrum of Unknown in Problem 13.31.

13.32 The following steroid derivative exhibits a peak at 1735 cm^{-1} in its IR spectrum (ester carbonyl), but there is no carbonyl peak corresponding to the aldehyde group. Suggest a reason for this unusual behavior.

CH_3
CHOH
O
CH
O
CH_3—C—O

13.33 Compound A (a hydrocarbon) was known to be an alkyne with a total of 10 carbon atoms, but its precise structure was not known. It gave no precipitate when treated either with cuprous chloride in aqueous ammonia or with alcoholic silver nitrate. Reaction with mercuric sulfate in aqueous sulfuric acid yielded two compounds, B and C. Both B and C exhibited IR absorptions near 1700 cm^{-1}, and each gave an orange precipitate with 2,4-dinitrophenylhydrazine reagent. Compound C also gave a positive iodoform test. The NMR spectrum of compound B is shown in Figure 13.6. On the basis of this information, deduce the structures of compounds A–C.

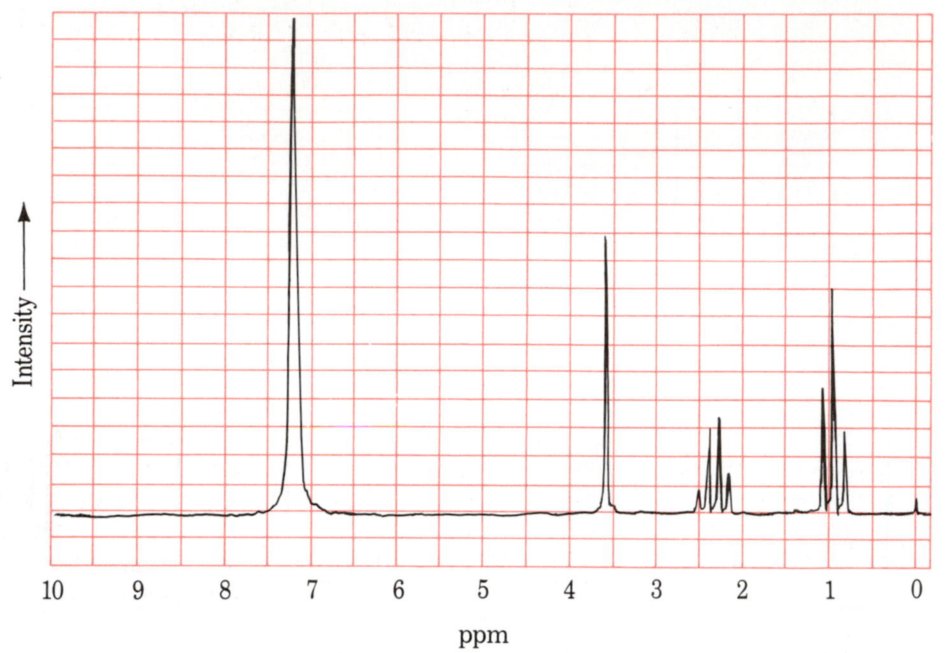

Figure 13.6 Proton NMR Spectrum 13.3 of Compound B (Problem 13.33).

13.34 When methyllithium was used to convert the following nitrile to the corresponding methyl ketone, complete inversion of configuration was observed at the carbon atom to which the cyano group was attached in the reactant. Explain this change in configuration.

CH_2 ... CN $\xrightarrow[\text{2. } H_2O,\ NH_4Cl]{\text{1. } CH_3Li}$ CH_2 ... H ... O ... CH_3 91%

13.35 For each of the reactions in Problem 13.17, suggest how you would use spectroscopic techniques (particularly NMR and IR) to determine whether the product isolated from the reaction is the expected product.

CHAPTER 14

CONDENSATION AND ALKYLATION REACTIONS OF ALDEHYDES AND KETONES

In Chapter 13 we showed that the reactivity of aldehydes and ketones could be divided into three categories:

1. Reaction at carbonyl oxygen (where M^+ is a metal or some other Lewis acid)

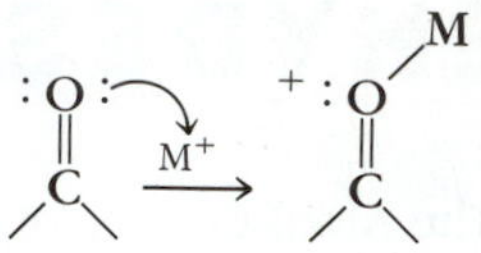

2. Attack at carbonyl carbon by a nucleophile

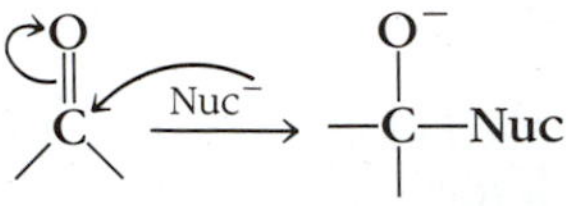

3. Reaction at the α carbon (via reaction of an electrophile with an enol or enolate ion)

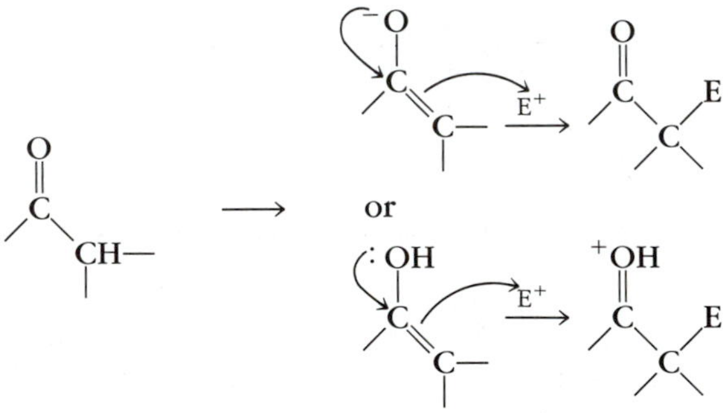

or

The third category, reaction at an α carbon, was discussed only for the reactions of enols and enolate ions with H^+ (or other proton donors) and with halogens. In this chapter we will focus on the reactions in which enols and enolate ions act as nucleophiles and attack *other carbon atoms.* Hence this chapter is concerned with *carbon-carbon bond forming reactions of enols and enolate ions.*

We will consider three basic classes of carbon-carbon bond-forming processes. The first reaction type involves nucleophilic attack of an enolate ion (or enol) on a second carbonyl group, a class of reactions that falls in the category of *aldol condensations.*

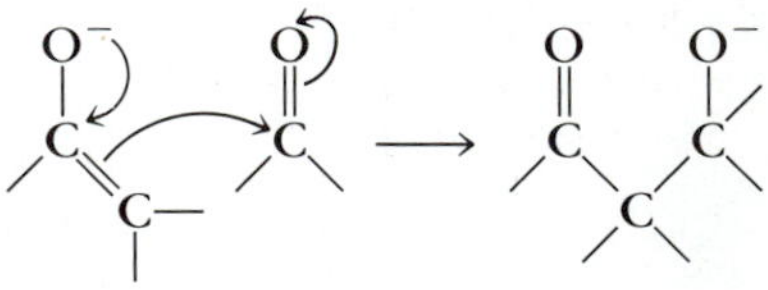

The second class of reactions involves nucleophilic attack of an enolate ion on an alkyl halide or sulfonate, a reaction described as *alkylation* of an aldehyde or ketone.

The third reaction type is that of *conjugate addition* of an enolate ion (or enol) to an α,β-unsaturated aldehyde or ketone. This third type of reaction is known as the *Michael reaction*.

All three types of reactions play very important roles in both synthetic and biological chemistry. In this chapter we will attempt to provide you with a basic understanding of how these processes take place, what products are to be expected, and what problems could arise during the course of these reactions.

14.1 THE ALDOL CONDENSATION

When treated with base, acetaldehyde undergoes reaction with itself to yield 3-hydroxybutanal.

$$CH_3-\overset{O}{\overset{\|}{C}}-H + CH_3-\overset{O}{\overset{\|}{C}}-H \xrightarrow{NaOH} CH_3-\overset{OH}{\overset{|}{C}H}-CH_2-\overset{O}{\overset{\|}{C}}-H \quad 50\%$$

The product of this reaction has both aldehyde and alcohol functional groups, and it was called by the common name *aldol* when the reaction was first discovered in the 1870s. Since then the term *aldol condensation* has been used to describe all such reactions of aldehydes and ketones which afford a β-hydroxy aldehyde or ketone (i.e., an *aldol* or *ketol*) or the corresponding dehydration product. The term *condensation* will be used in this book to describe a reaction of carbonyl compounds in which a carbon-carbon bond is formed between the α carbon of one molecule and the carbonyl carbon of the other. A generalized equation for the aldol condensation (and subsequent dehydration) is presented in Figure 14.1.

Base-Catalyzed Aldol Condensations

A wide variety of aldehydes and ketones will undergo self-condensation in the presence of a basic catalyst. Many different bases have been used, although NaOH is probably the most common. The following equations illustrate aldol condensations of a ketone and an aldehyde, respectively.

$$CH_3-\overset{O}{\overset{\|}{C}}-CH_3 + CH_3-\overset{O}{\overset{\|}{C}}-CH_3 \xrightarrow{Ba(OH)_2} CH_3-\underset{CH_3}{\underset{|}{\overset{OH}{\overset{|}{C}}}}-CH_2-\overset{O}{\overset{\|}{C}}-CH_3 \quad 71\%$$

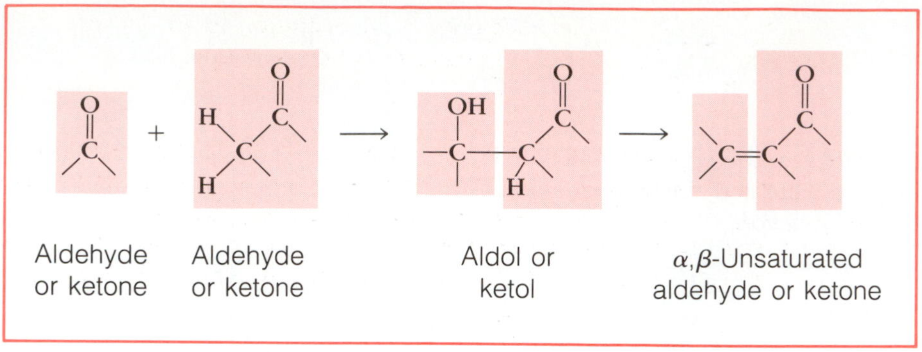

Aldehyde or ketone + Aldehyde or ketone ⟶ Aldol or ketol ⟶ α,β-Unsaturated aldehyde or ketone

Figure 14.1 The Aldol Condensation: a Generalized Equation.

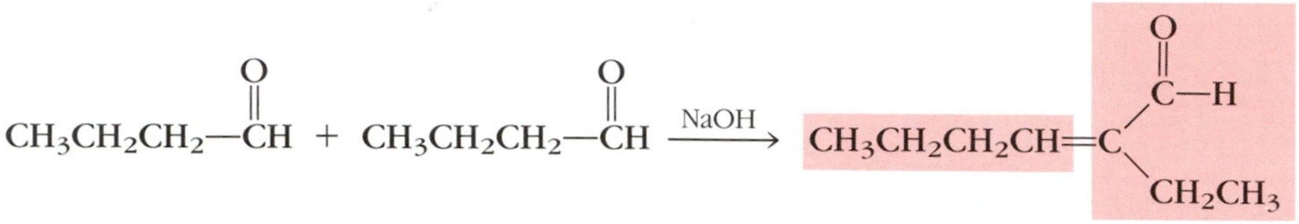

$$CH_3CH_2CH_2-\overset{\overset{\displaystyle O}{\|}}{C}H + CH_3CH_2CH_2-\overset{\overset{\displaystyle O}{\|}}{C}H \xrightarrow{NaOH} CH_3CH_2CH_2CH{=}C(CH_2CH_3)-\overset{\overset{\displaystyle O}{\|}}{C}-H \quad 86\%$$

Careful consideration of the preceding examples should lead to at least two questions: First, why did aldol condensations occur here but not in the examples we showed previously (Section 13.9) that involved treatment of aldehydes and ketones with base? Second, under what conditions does dehydration occur? The answer to the first question is straightforward. The aldol condensation is a bimolecular reaction between two molecules of the aldehyde or ketone, so its rate is highly dependent upon concentration. A relatively high concentration of aldehyde or ketone is ordinarily employed when carrying out an aldol condensation. For other base-catalyzed reactions, such as those we discussed in Chapter 13, the concentration of the carbonyl compound is kept low in order to minimize the formation of aldol products.

The second question is less easily answered, and no clear-cut rule can be given. Hydroxide ion is not a good leaving group, so dehydration under basic conditions would not normally be expected. Certainly, simple alcohols do not undergo dehydration in base. However, the OH group in an aldol product is β to a carbonyl group, and elimination leads to an α,β-unsaturated aldehyde or ketone in which the C=C and C=O double bonds are *conjugated.* The stabilizing effect of conjugation results in a lower activation energy for dehydration, and the unsaturated product is often obtained when vigorous reaction conditions are used: strong base, a long reaction time, and high temperature. Dehydration can also occur during workup if the reaction mixture is acidified or during distillation of a crude product.

Detailed reaction conditions are not normally specified in this text, and we will adopt the following oversimplified approach to predicting the products. *The normally expected product of an aldol condensation is an α,β-unsaturated aldehyde or ketone.* The one major exception to this generalization is the situation in which the β-hydroxy compound *cannot* undergo this dehydration reaction, that is, when there is no hydrogen on the α carbon of the β-hydroxy compound.

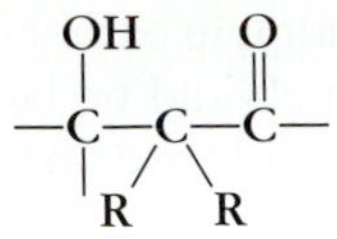

cannot undergo dehydration to an α,β-unsaturated carbonyl compound

Extensive studies of the aldol condensation have led to the reaction mechanism outlined in Figure 14.2. We have drawn the mechanism for the self-condensation of acetone, but it applies to all base-catalyzed aldol condensations. You should note that *all steps are reversible,* a very important feature of this mechanism. Nevertheless, the *overall* equilibrium usually favors the product, especially when dehydration occurs.

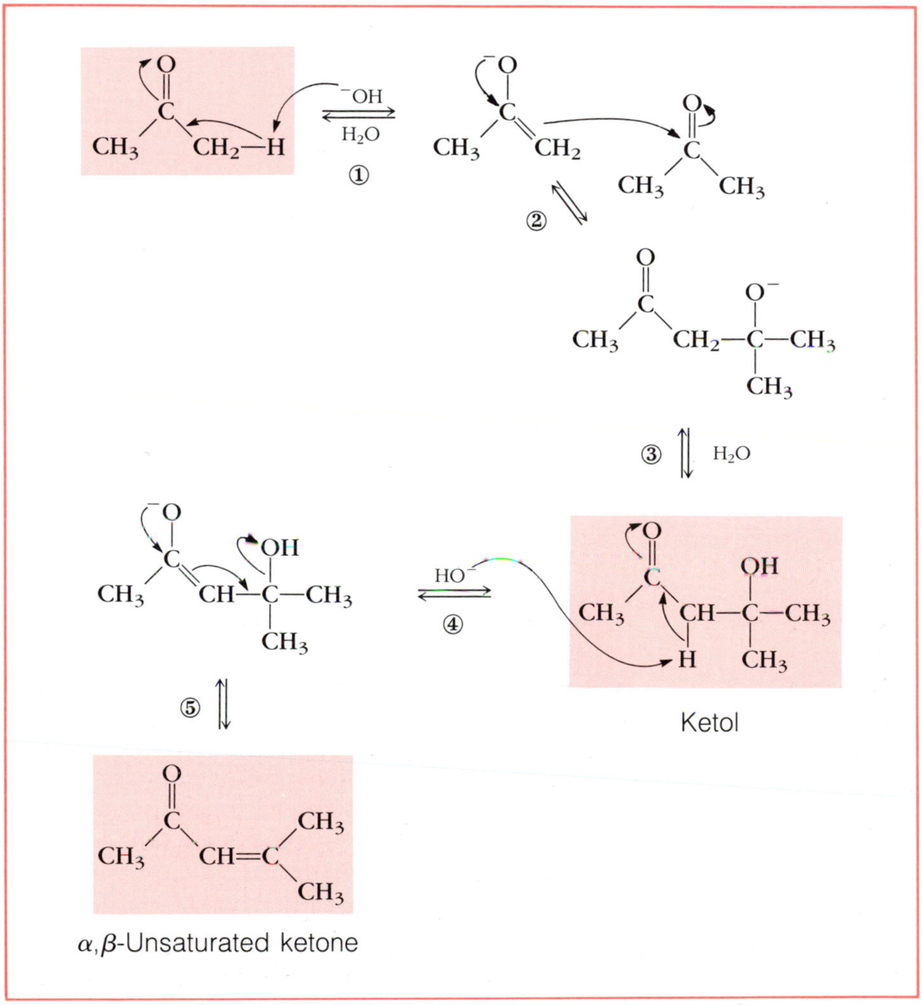

Ketol

α,β-Unsaturated ketone

Figure 14.2 Mechanism of the Base Catalyzed Aldol Condensation. The process is illustrated for the self-condensation of acetone with hydroxide ion as the base (and its conjugate acid, H_2O, as a proton source). The generally expected product is the α,β-unsaturated ketone, although the ketol can be isolated if mild reaction conditions are used.

We have not specifically discussed the aldol condensation in earlier chapters, but each of the individual steps in the reaction mechanism should be familiar to you (Sections 4.7 and 13.9). The reaction is initiated by attack of base on the ketone or aldehyde (step 1) to generate an enolate ion. The enolate ion then attacks the carbonyl group (step 2) to form a new carbon-carbon bond, and the resulting anion is protonated (step 3) to yield the ketol or aldol. Under vigorous conditions dehydration may occur by formation of the enolate ion (step 4) and subsequent expulsion of a hydroxide ion (step 5).

The *reversibility* of step 2 is in marked contrast to nearly all the other carbon-carbon bond-forming reactions that we have previously discussed. The reaction of a carbonyl compound with an enolate ion closely resembles the reaction with an organometallic reagent (Section 13.6).

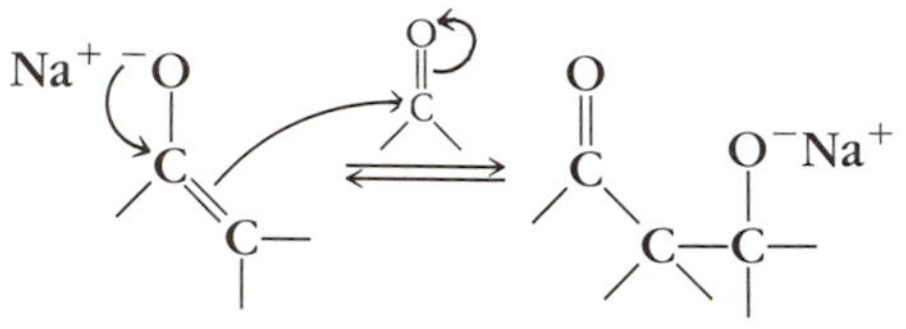

Enolate

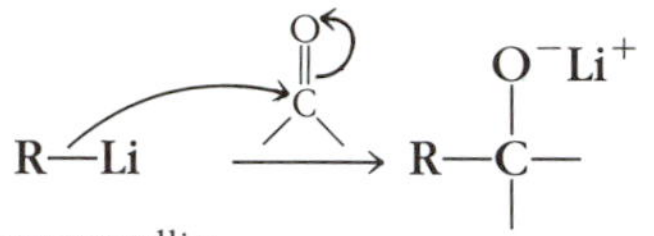

Organometallic reagent

Yet only the former reaction is reversible. Why? The answer is quite simple: An enol (or a ketone) is much more acidic than a hydrocarbon (Section 4.7), so the conjugate base is much more stable than that of a hydrocarbon. An enolate ion can act as a *leaving group*.

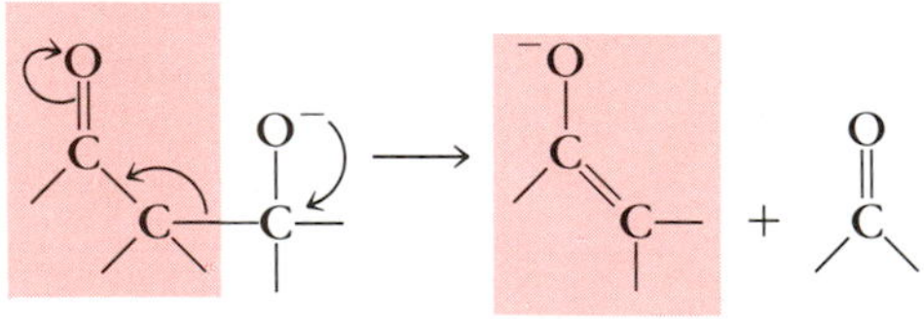

An enolate ion can act as a leaving group

We can also compare these processes by evaluating the overall energy changes. The reaction of an aldehyde or ketone with an organometallic reagent is much more exothermic than the corresponding reaction with an enolate ion. The activation energy for the reverse process is prohibitively high in the case of the organometallic reaction. You also encountered reversible carbon-carbon bond formation in the addition of HCN to the carbonyl group of aldehydes and ketones (Section 13.7). In that case the leaving group (^-CN) was again a relatively stable anion, and the equilibrium constants we showed for cyanohydrin formation clearly indicate that this is not a highly exothermic reaction.

Acid-Catalyzed Aldol Condensations

Although not used as frequently as basic conditions, acidic conditions can also be used to bring about an aldol condensation. For example, cyclohexanone undergoes the aldol condensation when treated with anhydrous hydrogen chloride. As you would expect, the β-hydroxy carbonyl compound undergoes dehydration under acidic conditions, but HCl can add to the resulting double bond. The unsaturated ketone is regenerated in good yield by treatment with base.

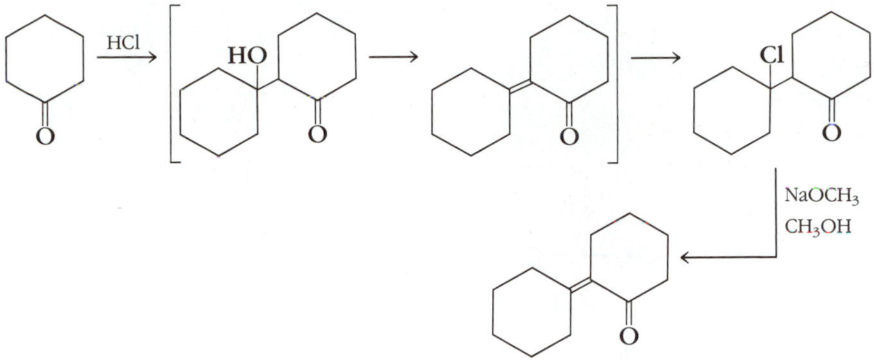

71%

88%

Studies of acid-catalyzed aldol condensations have led to the reaction mechanism depicted in Figure 14.3. The mechanism is quite similar to that of the base-catalyzed reaction (Figure 14.2), the major differences being in the sequence of the protonation and deprotonation steps. Once again the overall reaction is fully reversible, although dehydration favors the unsaturated carbonyl compound in the equilibrium.

Steps 1 and 2 are proton transfer reactions that produce the enol, and step 3 is the actual carbon-carbon bond forming step. The enol is a much weaker nucleophile than the enolate ion of Figure 14.2, but it attacks a highly reactive *protonated* ketone in step 3. This combination of a *less reactive* nucleophile and a *more reactive* electrophile permits the acid-catalyzed aldol condensation to proceed at a rate comparable with that of the base-catalyzed reaction. The ketol can be reversibly produced in step 4 (proton transfer), but the reaction ultimately proceeds via enolization (step 5), followed by protonation of the hydroxyl group in step 6 and elimination of a molecule of water (step 7). Finally, another proton transfer reaction (step 8) affords the unsaturated ketone.

Acid-catalyzed aldol condensations frequently yield complex mixtures of products. For example, the unsaturated ketone shown as the product in Figure 14.3 could undergo another aldol condensation, and it is sometimes difficult to stop the reaction sequence at the desired stage. Nevertheless, acid catalysis has at times been used successfully for aldol condensations.

EXERCISE 14.1

Draw the products expected for each of the following aldol condensation:

(a) $CH_3CH_2\overset{\overset{O}{\|}}{C}H \xrightarrow{KOH} ?$

(b) 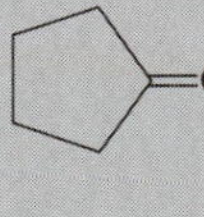 $=O \xrightarrow{HCl} ?$

(c) $CH_3{-}\underset{}{\overset{\overset{CH_3}{|}}{C}H}{-}CHO \xrightarrow{NaOH} ?$

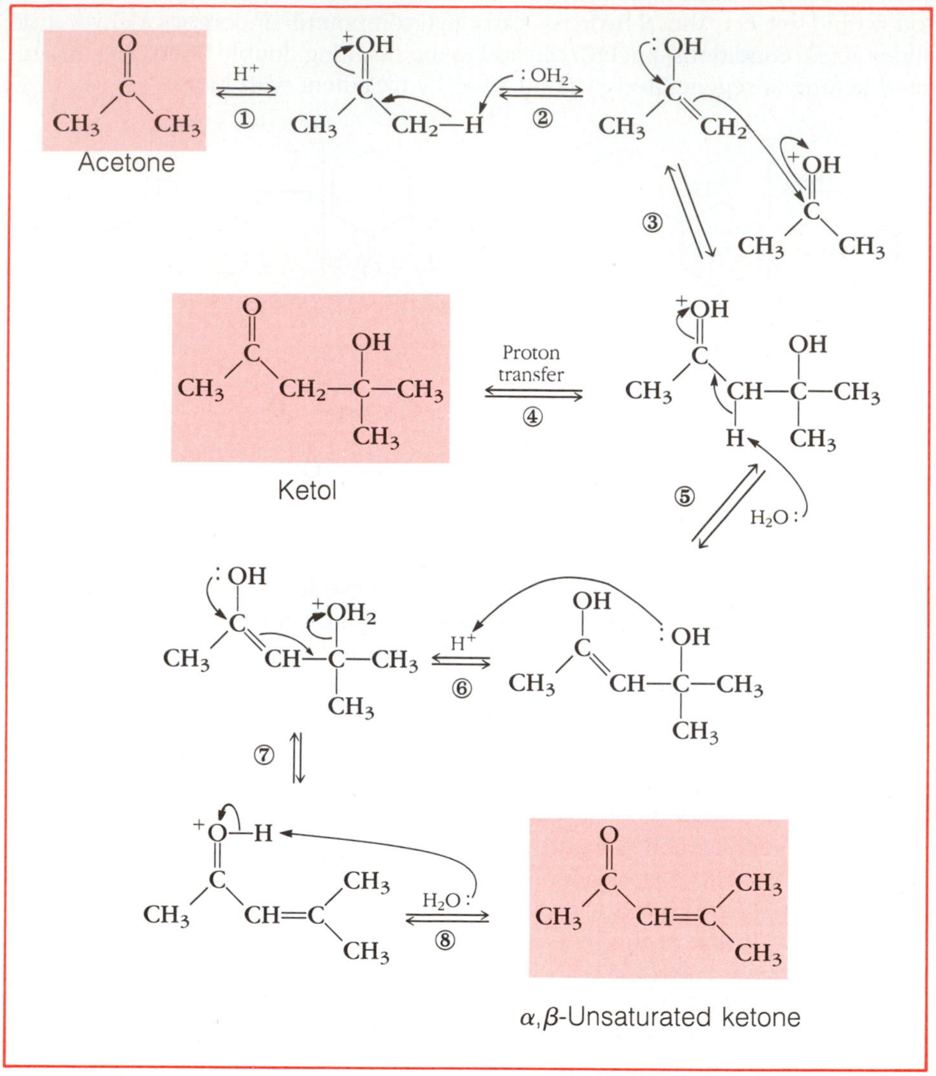

Acetone

Proton transfer

Ketol

α,β-Unsaturated ketone

Figure 14.3 Mechanism of the Acid-Catalyzed Aldol Condensation. The process is illustrated for the self-condensation of acetone with H^+/H_2O as the acid catalyst to produce the α,β-unsaturated carbonyl compound.

EXERCISE 14.2

Each of the following compounds resulted from an aldol condensation. Deduce the structure of the starting aldehyde or ketone in each case.

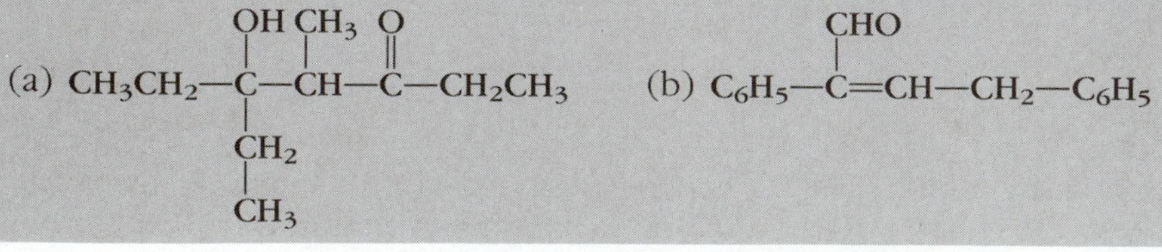

(a) $CH_3CH_2-C(OH)(CH_2CH_3)-CH(CH_3)-C(=O)-CH_2CH_3$

(b) $C_6H_5-C(CHO)=CH-CH_2-C_6H_5$

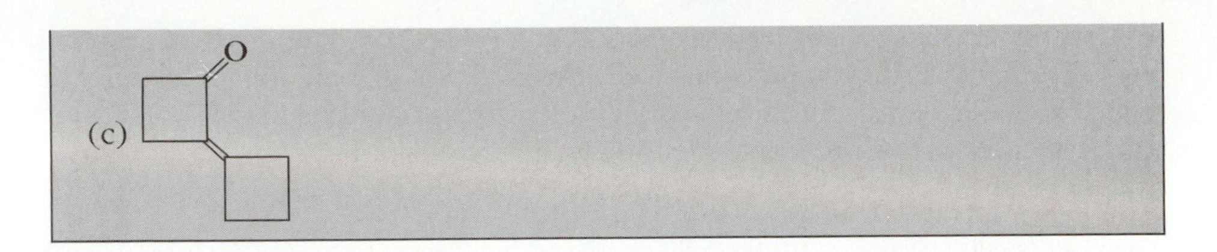

14.2 MIXED ALDOL CONDENSATIONS

The reactions that we discussed in the previous section all involved condensation between two molecules of the same compound (i.e., *self-condensation*). A *mixed aldol condensation* would involve attack of the enol or enolate of one compound on the carbonyl group of a second compound. What would happen if a mixture of two different carbonyl compounds was subjected to the conditions of the aldol condensation? Consider a simple example, the reaction of acetaldehyde and propanal in the presence of base.

$$CH_3\overset{O}{\overset{\|}{C}}H + CH_3CH_2\overset{O}{\overset{\|}{C}}H \xrightarrow{NaOH} ?$$

Enolate ions would be formed from each of the two aldehydes, and each of these could react with the two different aldehydes. Hence a total of four β-hydroxyaldehydes are possible. (An even larger number of products is possible if stereoisomers are considered.)

$$CH_3\overset{O}{\overset{\|}{C}}H \overset{^{-}OH}{\rightleftharpoons} CH_2{=}\overset{O^-}{\overset{|}{C}}H$$

$$CH_2{=}\overset{O^-}{\overset{|}{C}}H \xrightarrow{CH_3\overset{O}{\overset{\|}{C}}H} CH_3\overset{OH}{\overset{|}{C}}H{-}CH_2{-}\overset{O}{\overset{\|}{C}}H$$

$$CH_2{=}\overset{O^-}{\overset{|}{C}}H \xrightarrow{CH_3CH_2\overset{O}{\overset{\|}{C}}H} CH_3CH_2\overset{OH}{\overset{|}{C}}H{-}CH_2\overset{O}{\overset{\|}{C}}H$$

$$CH_3CH_2\overset{O}{\overset{\|}{C}}H \overset{^{-}OH}{\rightleftharpoons} CH_3CH{=}\overset{O^-}{\overset{|}{C}}H$$

$$CH_3CH{=}\overset{O^-}{\overset{|}{C}}H \xrightarrow{CH_3\overset{O}{\overset{\|}{C}}H} CH_3\overset{OH}{\overset{|}{C}}H{-}\underset{CH_3}{\underset{|}{C}}{-}\overset{O}{\overset{\|}{C}}H$$

$$CH_3CH{=}\overset{O^-}{\overset{|}{C}}H \xrightarrow{CH_3CH_2\overset{O}{\overset{\|}{C}}} CH_3CH_2\overset{OH}{\overset{|}{C}}H{-}\underset{CH_3}{\underset{|}{C}H}{-}\overset{O}{\overset{\|}{C}}H$$

In a condensation between two *ketones* the number of possible products can be even greater. For example, an aldol condensation between 2-butanone and

2-pentanone could initially lead to eight different β-hydroxy ketones (not counting stereoisomers). Each ketone could yield two different enolate ions, and the four possible enolate ions could react with either of the two neutral ketones. (Drawing the structures of the products is left for you as Exercise 14.3.)

$$CH_3CH_2\overset{\overset{O}{\|}}{C}CH_3 + CH_3CH_2CH_2\overset{\overset{O}{\|}}{C}CH_3 \xrightarrow{KOH} \textbf{eight possible aldol products}$$

Even the *self*-condensation of an unsymmetrical ketone can lead to two constitutionally isomeric products when two different enolate ions can be formed. The examples we have discussed here show that there are two major problems in mixed aldol condensations: *first* that nucleophilic attack can occur at two different carbonyl groups, and *second* that more than one enolate ion can be formed. By clearly defining these potential problems, a logical solution becomes possible:

Mixed Aldol Condensations

Mixed aldol condensations should be successful (that is, a single product should predominate) when two conditions are met: (1) One of the two carbonyl compounds is more susceptible to nucleophilic attack; and (2) only the other compound can form an enolate ion.

How can these conditions be satisfied? For the first condition (selective attack on one of the two carbonyl groups), we must analyze *steric effects.* In general, the carbonyl group of an aldehyde is less hindered than that of a ketone, so nucleophilic attack proceeds more easily at the aldehyde carbonyl. There will, of course, be variations within these categories, and the presence of substituents at the α position of a carbonyl group decreases its susceptibility to nucleophilic attack. Aryl groups are usually less bulky than substituted alkyl groups, so the most reactive aldehydes are those in which the substituent on the carbonyl group is an aromatic ring.

Steric Effects on Reactivity of Aldehydes and Ketones

$$Ar{-}\overset{\overset{O}{\|}}{C}{-}H \;\text{------}\; R{-}\overset{\overset{O}{\|}}{C}{-}H \;\text{------}\; R{-}\overset{\overset{O}{\|}}{C}{-}R$$

More susceptible toward nucleophilic attack — Less susceptible toward nucleophilic attack

The second problem (selective enolate ion formation) is easily avoided by using as one of the reactants a compound that cannot form an enolate ion. For example, aromatic aldehydes such as benzaldehyde have no α hydrogens and are frequently used in mixed aldol condensations. Aldehydes such as 2,2-dimethylpropanal, for which the α carbon is quaternary, are also unable to form enolate ions when treated with base, but the increased steric hindrance of the quaternary carbon would decrease the susceptibility to nucleophilic attack.

$$C_6H_5-\overset{O}{\overset{\|}{C}}-H \qquad CH_3-\underset{CH_3}{\overset{CH_3}{C}}-\overset{O}{\overset{\|}{C}}-H$$

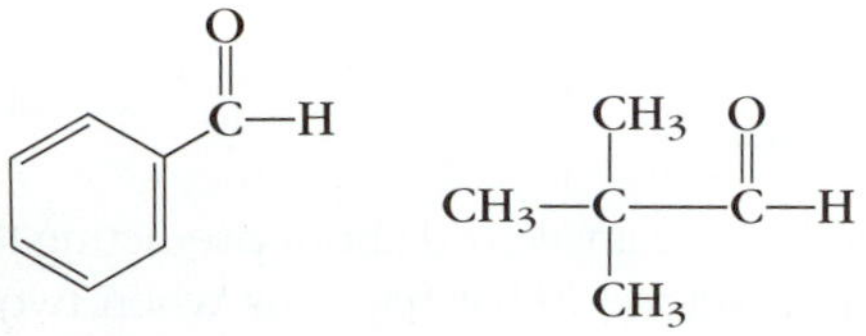

The optimum combination for a mixed aldol condensation is therefore an aromatic aldehyde (which cannot form an enolate ion but is highly susceptible to nucleophilic attack) and a ketone with α hydrogens (which can form an enolate ion but is less susceptible to nucleophilic attack). This is illustrated by the following reactions.

$$C_6H_5-\overset{O}{\overset{\|}{C}}H + CH_3-\overset{O}{\overset{\|}{C}}-C_6H_5 \xrightarrow{NaOH} C_6H_5-CH{=}CH-\overset{O}{\overset{\|}{C}}-C_6H_5 \qquad 85\%$$

$$C_6H_5-\overset{O}{\overset{\|}{C}}H + CH_3-\overset{O}{\overset{\|}{C}}-\underset{CH_3}{\overset{CH_3}{C}}-CH_3 \xrightarrow{NaOH} C_6H_5-CH{=}CH-\overset{O}{\overset{\|}{C}}-\underset{CH_3}{\overset{CH_3}{C}}-CH_3 \qquad 88\text{–}93\%$$

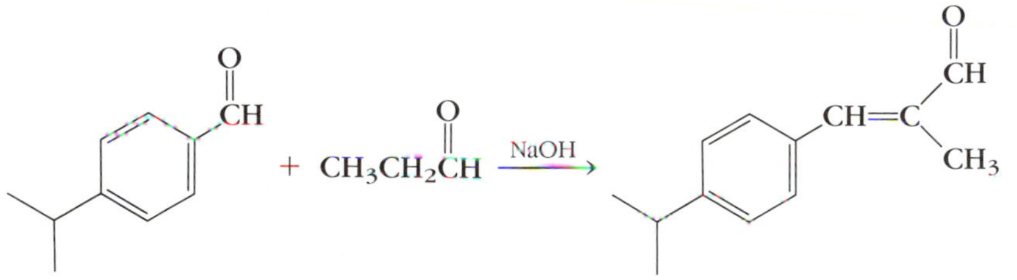

$$p\text{-}(CH_3)_2CH-C_6H_4-\overset{O}{\overset{\|}{C}}H + CH_3CH_2\overset{O}{\overset{\|}{C}}H \xrightarrow{NaOH} p\text{-}(CH_3)_2CH-C_6H_4-CH{=}C(CH_3)-\overset{O}{\overset{\|}{C}}H \qquad 80\%$$

The last of the preceding examples confirms that the carbonyl group of an aromatic aldehyde is selectively attacked even when the other carbonyl compound is also an aldehyde.

Because they are highly susceptible to nucleophilic attack, aldehydes can frequently be used in mixed condensations even if they have α hydrogens. A condensation between an aldehyde and a ketone usually proceeds by attack of the ketone enolate on the aldehyde carbonyl. The result is a β-hydroxy ketone (or an unsaturated ketone), as you can see from the following examples.

$$CH_3CH_2CH_2\overset{O}{\overset{\|}{C}}H + CH_3-\overset{O}{\overset{\|}{C}}-CH_3 \xrightarrow{NaOH} CH_3CH_2CH_2-\overset{OH}{\overset{|}{C}}H-CH_2-\overset{O}{\overset{\|}{C}}-CH_3 \qquad 84\%$$

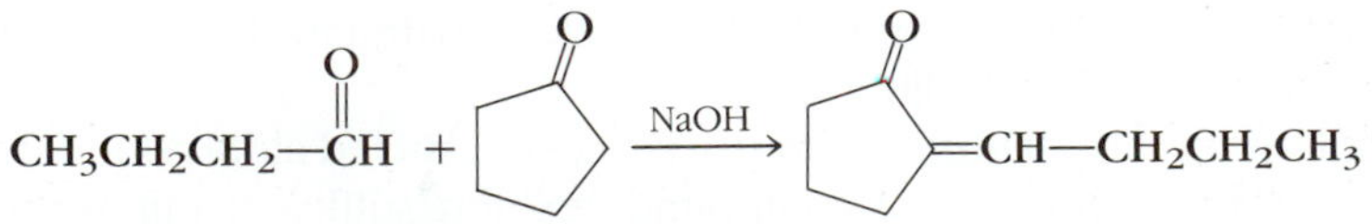

$$CH_3CH_2CH_2-\overset{O}{\overset{\|}{C}}H + \text{cyclopentanone} \xrightarrow{NaOH} \text{2-(butylidene)cyclopentanone } (\text{ring}{=}CH-CH_2CH_2CH_3) \qquad 46\%$$

Whenever the two carbonyl groups are of comparable reactivity, mixtures of products may result. This frequently occurs with mixed condensations between two aliphatic aldehydes or between two ketones. In order to provide you with working guidelines, we will consider such reactions to be unreliable.

Unsymmetrical Ketones

One final question needs to be answered about predicting the products of an aldol condensation involving a ketone. What happens when two different enolate ions are possible? 2-Butanone, for example, could yield enolate ions by removal of a proton from either C-1 or C-3.

$$\underset{1}{CH_3}-\underset{2}{\overset{O}{\overset{\|}{C}}}-\underset{3}{CH_2}\underset{4}{CH_3} \underset{\text{base}}{\rightleftharpoons} CH_2=\overset{O^-}{\overset{|}{C}}-CH_2CH_3$$

$$\underset{1}{CH_3}-\underset{2}{\overset{O}{\overset{\|}{C}}}-\underset{3}{CH_2}\underset{4}{CH_3} \underset{\text{base}}{\rightleftharpoons} CH_3-\overset{O^-}{\overset{|}{C}}=CH-CH_3$$

A definitive answer to the question is difficult, but some general trends have been observed. For *base*-catalyzed condensations enolate formation usually occurs at the *less substituted* α carbon. (Proton removal at the less hindered position is a typical result of kinetic control.) In contrast, *acid*-catalyzed reactions often proceed via enolization toward the *more substituted* α carbon. (This corresponds to the thermodynamically more stable enol.) These trends are illustrated in the following examples:

$$CH_3CH_2CH_2CH_2-\overset{O}{\overset{\|}{C}}-CH_3 \xrightarrow{Al(OtBu)_3} CH_3CH_2CH_2CH_2-\overset{O}{\overset{\|}{C}}-CH_2-\underset{CH_3}{\underset{|}{\overset{OH}{\overset{|}{C}}}}-CH_2CH_2CH_2CH_3 \quad 73\%$$

$$C_6H_5-\overset{O}{\overset{\|}{C}}H + CH_3-\overset{O}{\overset{\|}{C}}-CH_2CH_3 \xrightarrow{NaOH} C_6H_5-CH=CH-\overset{O}{\overset{\|}{C}}-CH_2CH_3 \quad 95\%$$

$$C_6H_5-\overset{O}{\overset{\|}{C}}H + CH_3-\overset{O}{\overset{\|}{C}}-CH_2-CH_3 \xrightarrow{HCl} C_6H_5-CH=C(-\overset{O}{\overset{\|}{C}}-CH_3)(CH_3) \quad 85\%$$

The use of $Al(OtBu)_3$ in the first of the preceding examples may seem unusual, but it is simply the aluminum salt of *tert*-butyl alcohol, one of many different bases that have been used for aldol condensations. The product is that expected for basic conditions, namely, that formed by reaction at the less substituted α carbon. The last two examples show how the same aldehyde-ketone pair affords different products under acidic and basic conditions.

As the examples in this section illustrate, a successful mixed aldol condensation is possible when the two carbonyl compounds exhibit sufficient differences in reactivity. The necessary differences are summarized in the following box.

Optimization of Mixed Aldol Condensations

1. *One* of the two carbonyl compounds must be more susceptible to nucleophilic attack. Usually this compound is an aldehyde with an unhindered carbonyl group.

2. The *second* compound must preferentially undergo enolate formation (or enol formation if acidic conditions are used). In many instances this preference is achieved because the *first* carbonyl compound has no α hydrogens (for example, if it is an aromatic aldehyde).

3. When the enolate is generated from an unsymmetrical ketone, reaction under the typical basic conditions usually proceeds by enolate formation at the *less* substituted α carbon of the ketone. (This orientation can sometimes be reversed if acidic conditions are employed.)

EXERCISE 14.3

Draw the eight possible constitutionally isomeric products of an aldol condensation between 2-butanone and 2-pentanone.

EXERCISE 14.4

Draw the products expected for each of the following aldol condensations:

(a) $C_6H_5{-}\overset{O}{\overset{\|}{C}}H + CH_3{-}\overset{O}{\overset{\|}{C}}{-}CH_3 \xrightarrow{KOH} ?$

(b) $C_6H_5{-}\overset{O}{\overset{\|}{C}}{-}CH_3 + CH_3{-}\overset{CH_3}{\overset{|}{C}}H{-}CHO \xrightarrow{NaOH} ?$

(c) $CH_3CH_2CH_2{-}\overset{O}{\overset{\|}{C}}H + C_6H_5{-}\overset{O}{\overset{\|}{C}}H \xrightarrow{KOH} ?$

EXERCISE 14.5

Each of the following compounds resulted from an aldol condensation. Draw the structures of the starting carbonyl compounds in each case.

(a) $C_6H_5{-}CH{=}CH{-}\overset{O}{\overset{\|}{C}}{-}CH_3$

(b) 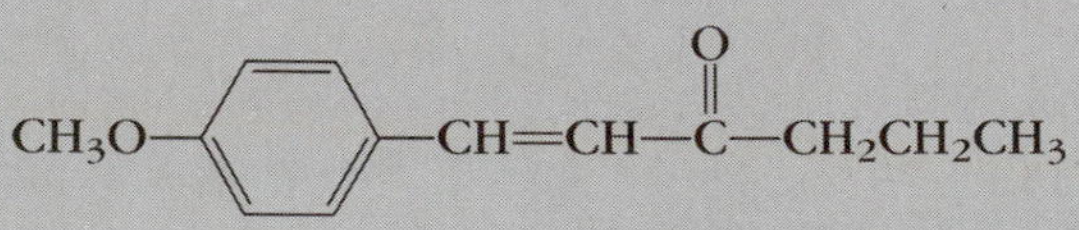 $CH_3O{-}C_6H_4{-}CH{=}CH{-}\overset{O}{\overset{\|}{C}}{-}CH_2CH_2CH_3$ (para-substituted benzene ring)

(c) 2-ethylidenecyclohexanone (cyclohexanone ring with C=O and adjacent =CH—CH$_3$)

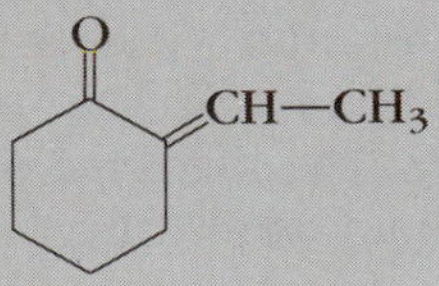

(d) $C_6H_5{-}CH{=}C(CHO)(CH_2CH_3)$

EXERCISE 14.6

Self-condensation of 2-butanone affords different products under acidic and basic conditions. Draw the product expected for each set of conditions.

14.3 ALKYLATION OF ENOLATE IONS

The examples that we have shown in the two preceding sections illustrate the usefulness of reactions involving nucleophilic attack by enolate ions on carbonyl groups. In this section we turn to a related reaction of enolate ions, nucleophilic attack on alkyl halides and sulfonates.

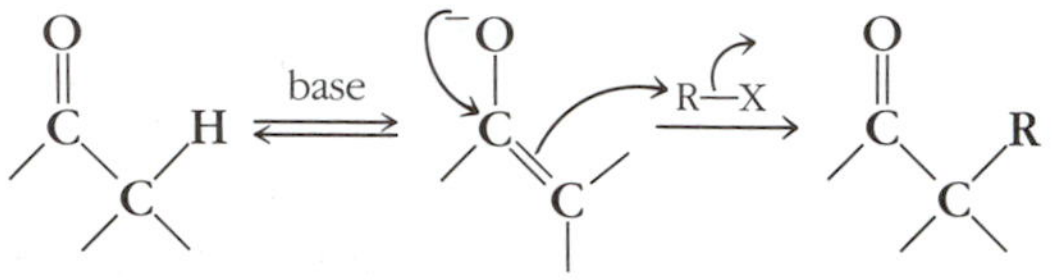

base

R—X

R

Such alkylation reactions could be very useful, but a number of potentially serious problems would first need to be circumvented:

1. The enolate ion is a strong base as well as a good nucleophile, and elimination of HX can compete with the desired displacement of X^-.

2. The enolate ion has two nucleophilic atoms (both carbon and oxygen), and alkylation of the oxygen atom (called *O*-alkylation) could occur.

3. The base used to generate the enolate ion can also act as a nucleophile and thereby destroy the alkyl halide or sulfonate.

4. The aldehyde or ketone could undergo a base-catalyzed aldol condensation.

5. If the product of the reaction still has α hydrogens, then polyalkylation could occur.

6. If an unsymmetrical ketone is used, two isomeric enolates could form, and a mixture of products might result.

Despite this formidable list of potential difficulties, alkylation of enolate ions is frequently a very useful reaction. In the following discussion we consider each of the problems in turn, along with ways in which they can be either avoided or minimized.

Elimination

Elimination to form a carbon-carbon double bond is almost always a competing side reaction in any nucleophilic substitution reaction. Direct displacement of a tertiary alkyl halide is not possible, and even secondary derivatives yield mainly elimination products (Figure 12.6). Consequently, if satisfactory yields are to be obtained in the alkylation of enolate ions, the alkyl halide (or sulfonate) must be restricted to either a methyl or primary alkyl derivative. The following reaction illustrates the alkylation of a ketone enolate with a primary alkyl halide.

$$p\text{-}CH_3OC_6H_4-CH_2-\overset{O}{\overset{\|}{C}}-C_6H_5 \xrightarrow[C_2H_5I]{NaOC_2H_5} p\text{-}CH_3OC_6H_4-\underset{C_2H_5}{\underset{|}{CH}}-\overset{O}{\overset{\|}{C}}-C_6H_5 \quad 80\%$$

O-Alkylation

The reaction of simple ketone enolates with alkyl halides usually occurs preferentially at *carbon,* so this problem is relatively unimportant for the present discussion. (In later chapters we will show you other reactions for which attack by enolate oxygen is a major pathway.)

Reaction of the Base with the Alkyl Halide

If the base used to generate the enolate ion is also a good nucleophile, then it can compete with the enolate ion in reaction with the alkylating agent. Manipulation of the reaction conditions (for example, use of excess alkylating agent) can help, but a more reliable solution can be obtained by selecting the appropriate base. The "ideal" base will exhibit two properties: first, it will be sterically hindered and therefore only weakly nucleophilic (for example, an alkoxide base such as potassium *tert*-butoxide); second, it will be a very strong base, which produces almost complete conversion to the enolate ion.

$$\overset{O}{\overset{\|}{C}}-CH- \;+\; \underset{\text{Strong base}}{{}^{-}B} \rightleftharpoons \overset{O^-}{\overset{|}{C}}=C- \;+\; HB$$

The very strong bases which have been used for this purpose include salts of ammonia, salts of amines, and sodium or lithium hydride. In contrast to the complex metal hydrides (i.e., $LiAlH_4$ and $NaBH_4$), the simple metal hydrides, LiH and NaH, do not normally act as nucleophiles with carbonyl compounds. Instead they function as bases, abstracting an α hydrogen to produce the metal salt of the enolate ion and a molecule of H_2. Metal salts of triphenylmethane have also been used as bases, and proton abstraction by the triphenylmethide ion yields the enolate and triphenylmethane. With these strong bases the conversion of the aldehyde or ketone to the enolate is usually carried out prior to the addition of alkylating agent to the reaction mixture.

$$\overset{O}{\overset{\|}{C}}-CH- \;+\; NaNH_2 \rightleftharpoons \overset{O^-Na^+}{\overset{|}{C}}=C- \;+\; NH_3$$

$$\overset{O}{\overset{\|}{C}}-CH- \;+\; LiH \longrightarrow \overset{O^-Li^+}{\overset{|}{C}}=C- \;+\; H_2$$

$$\text{O}=\text{C}<\text{CH}- \; + (C_6H_5)_3CK \rightleftharpoons \text{C}(\text{O}^-K^+)=\text{C}- \; + (C_6H_5)_3CH$$

The following examples illustrate the use of such bases to achieve alkylation of ketones:

KOtBu; Br—$C_{16}H_{33}$ → $C_{16}H_{33}$ 54%

1. $NaNH_2$; 2. Br—CH_2—CH=CH_2 → CH_2—CH=CH_2 54–62%

1. NaH; 2. CH_3I → 55%

Competing Aldol Condensations

This problem is particularly severe for aldehydes, but it can minimized by adjusting experimental conditions. An aldol condensation is a bimolecular reaction involving the carbonyl compound, so its rate can be reduced by decreasing the concentration of the carbonyl compound. By using one of the bases that causes essentially complete conversion to the enolate ion, the concentration of neutral aldehyde or ketone will be greatly reduced, and this will also decrease the rate of aldol condensation. The following reaction illustrates this approach.

CHO; 1. Ph_3CK; 2. CH_3I → CHO 91%

Polyalkylation

If an aldehyde or ketone has more than one α hydrogen, polyalkylation can occur. Reaction of an enolate ion with alkylating agent produces a neutral molecule.

This monoalkylated product could then undergo a proton transfer reaction with an unreacted enolate ion (or any other strong base that is present).

Reaction of the monoalkylated enolate ion could then yield dialkylated product.

The problem of polyalkylation is clearly illustrated by the methylation of 2-methylcyclohexanone (which has a total of three α hydrogens). Four products are formed, all in rather poor yield.

Several approaches have been developed to reduce the problem of polyalkylation. One of these, the use of an *activating group,* is discussed in Chapter 21. Another, much simpler, technique again depends on the base used in the reaction. The metal-oxygen bond in enolate salts is strongest when the cation is lithium. As a result, *proton transfer reactions occur more slowly with lithium enolates* than with sodium or potassium enolates. The proton transfer reactions of lithium enolates also occur more slowly than the alkylation reactions, and monoalkylation can often be achieved in good yield. Of course there are other occasions when polyalkylation is desired, and in such cases the use of sodium or potassium enolates (employing a base such as potassium *tert*-butoxide) together with excess alkylating agent gives favorable results.

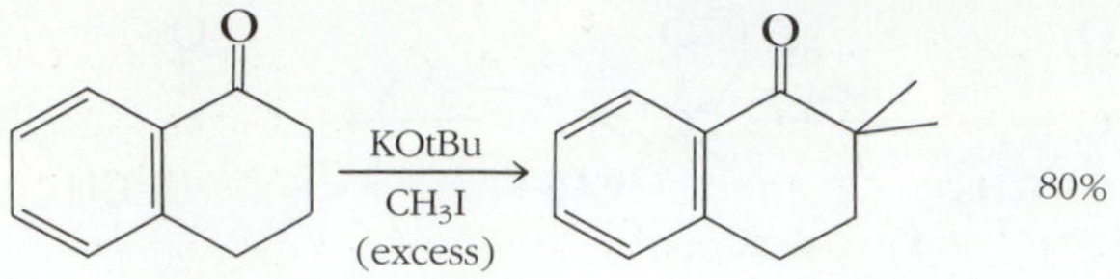

KOtBu; CH_3I (excess) — 80%

Formation of Isomeric Enolate Ions

An unsymmetrical ketone could yield two different enolate ions, and these in turn would afford isomeric alkylation products. In the preceding reaction of 2-methylcyclohexanone, both 2,2-dimethylcyclohexanone and 2,6-dimethylcyclohexanone were formed. This must have resulted from the formation of two different enolate ions.

$(C_6H_5)_3CK$

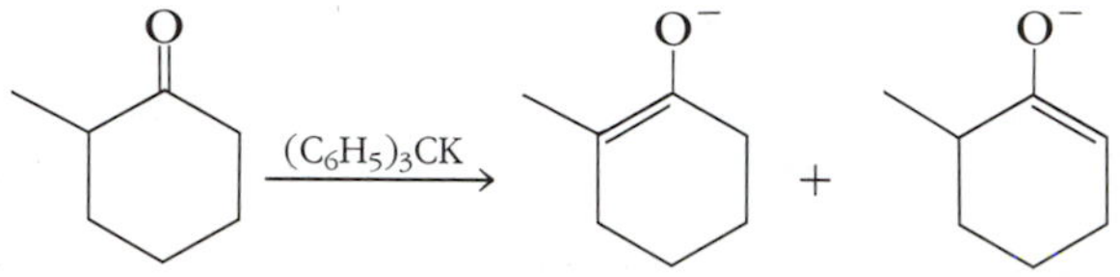

Is there a way by which just one enolate ion can be generated selectively? In fact, there are several ways to do this. These include the use of enamines (Section 17.8) and activating groups (Chapter 21), but a convenient alternative approach takes advantage of existing differences in reactivity of the two types of α hydrogens.

A *hindered* base preferentially abstracts a proton from the *less substituted* α carbon (kinetic control). In order to prevent subsequent equilibration of the enolate ions, the procedure works best when a lithium enolate is formed and the base used is strong enough to cause essentially complete conversion to the enolate ion. A reagent that meets both these criteria is lithium diisopropylamide, $LiN(iPr)_2$. This base is sterically hindered, so it is also a poor nucleophile.

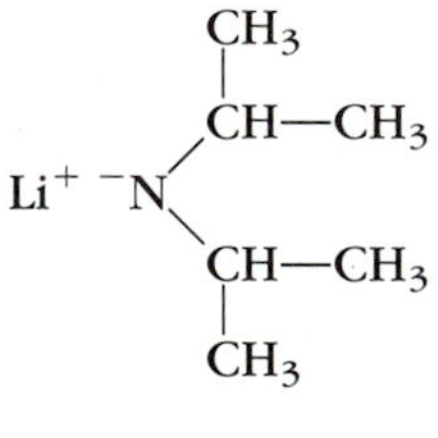

$Li^+ \ ^-N[CH(CH_3)_2]_2$

Lithium diisopropylamide
$LiN(iPr)_2$

Reaction of 2-methylcyclohexanone with lithium diisopropylamide proceeds almost entirely by abstraction of a proton from the less hindered C-6 position, and subsequent reaction with benzyl bromide affords mainly one compound.

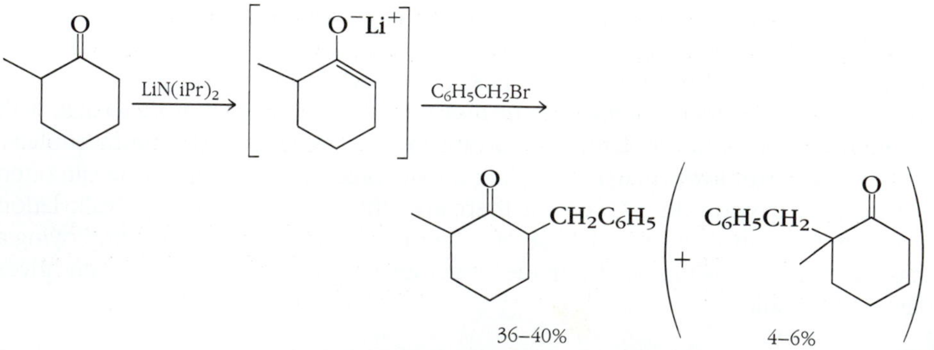

$LiN(iPr)_2$; O^-Li^+; $C_6H_5CH_2Br$

$CH_2C_6H_5$ — 36–40%

$C_6H_5CH_2$ — 4–6%

You should contrast this result with the mixture of four products obtained in the alkylation of the potassium enolate, which we described earlier in this section.

Sometimes alkylation of a specific enolate can be accomplished, not because it is formed more rapidly, but because it is *more stable* (thermodynamic control). In such cases equilibration of the enolate ions is needed, so bases that yield sodium or potassium enolates are used. These conditions often favor reaction at the more highly substituted carbon atom, particularly if the resulting enolate ion is conjugated with an aromatic ring or another double bond.

$$CH_3-\overset{\overset{O}{\|}}{C}-\underset{\underset{C_6H_5}{|}}{CH}-CH_2-\overset{\overset{CH_3}{|}}{C}=CH_2 \xrightarrow[CH_3I]{KOtBu} CH_3-\overset{\overset{O}{\|}}{C}-\underset{\underset{C_6H_5}{|}}{\overset{\overset{CH_3}{|}}{C}}-CH_2-\overset{\overset{CH_3}{|}}{C}=CH_2 \quad 80\%$$

$$\text{(6-methoxy-2-ethyl-1-tetralone; } CH_3O\text{, } C_2H_5\text{, O)} \xrightarrow[2.\ CH_2{=}CH{-}CH_2Br]{1.\ NaH} \text{(product with } CH_3O\text{, O, } C_2H_5\text{, } CH_2CH{=}CH_2\text{)} \quad \sim100\%$$

$$C_6H_5-CH_2-\overset{\overset{O}{\|}}{C}-CH_3 \xrightarrow[2.\ CH_3I]{1.\ (C_6H_5)_3CK} \underset{93\%}{C_6H_5-\underset{\underset{CH_3}{|}}{CH}-\overset{\overset{O}{\|}}{C}-CH_3} \left(+ \underset{<1\%}{C_6H_5-CH_2-\overset{\overset{O}{\|}}{C}-CH_2CH_3}\right)$$

In each of these examples the reaction proceeds via the more stable enolate, that in which the negative charge can be stabilized by the aromatic ring. In other situations, in which the two possible enolates are of comparable stability, a mixture of products would be expected. You already saw this in the alkylation of 2-methylcyclohexanone with $(C_6H_5)_3CK$ as the base.

Despite the difficulties that frequently arise, the alkylation of enolate ions can be a useful procedure for introducing substituents at the α position of an aldehyde or ketone. The limitations and requirements for this reaction can be summarized as in the following box.

Optimization of Alkylation of Aldehydes and Ketones

1. Only methyl or primary alkyl groups can be introduced successfully.

2. A sterically hindered base is needed. Lithium diisopropylamide usually abstracts a proton from the less hindered α position, resulting in alkylation at the less substituted α carbon.

3. Lithium enolates favor monoalkylation; bases that generate sodium or potassium enolates frequently result in alkylation at the more substituted α carbon and yield mixtures, including polyalkylated products.

EXERCISE 14.7

What product or products would be expected in each of the following alkylation reactions?

(a) $CH_3CH_2{-}\overset{O}{\overset{\|}{C}}{-}C_6H_5 \xrightarrow[\text{excess } CH_3I]{KOtBu} ?$

(b) $C_6H_5{-}CH_2{-}\overset{O}{\overset{\|}{C}}{-}CH_2CH_3 \xrightarrow[2.\ CH_2{=}CH{-}CH_2Br]{1.\ NaNH_2} ?$

(c) $CH_3CH_2{-}\overset{CH_3}{\overset{|}{C}H}{-}\overset{O}{\overset{\|}{C}}{-}CH_2CH_3 \xrightarrow[2.\ C_2H_5Br]{1.\ LiN(iPr)_2} ?$

EXERCISE 14.8

What by-product would be expected (in addition to the desired alkylation product) in each of the following reactions?

(a) $CH_3CH_2CH_2\overset{O}{\overset{\|}{C}}H \xrightarrow[C_2H_5Br]{KOtBu} ?$

(b) $CH_3CH_2\overset{O}{\overset{\|}{C}}{-}C_6H_5 \xrightarrow[CH_3I]{NaOC_2H_5} ?$

(c) $CH_3{-}\underset{CH_3}{\underset{|}{C}H}{-}\overset{O}{\overset{\|}{C}}{-}C_6H_5 \xrightarrow[(CH_3)_2CHBr]{KOtBu} ?$

(d) $CH_3CH_2{-}\overset{O}{\overset{\|}{C}}{-}CH_2CH_3 \xrightarrow[2.\ C_6H_5CH_2Br]{1.\ NaNH_2} ?$

14.4 ALKYLATION OF ENOLATE IONS GENERATED FROM α,β-UNSATURATED KETONES

Acidity of α,β-Unsaturated Carbonyl Compounds

The α hydrogen of an α,β-unsaturated ketone is not acidic. The C—H bond at the α position is perpendicular to the π system, so the resulting anion would not be stabilized by orbital interactions.

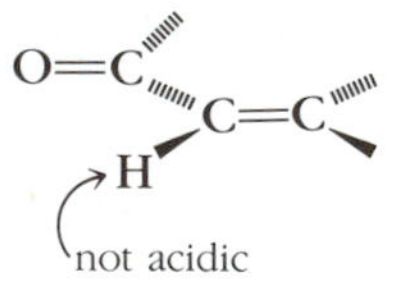

$$O{=}C{-}C(H){=}C$$

not acidic

Nevertheless, enolate ions can be formed from such compounds because the hydrogens on the γ carbon are acidic.

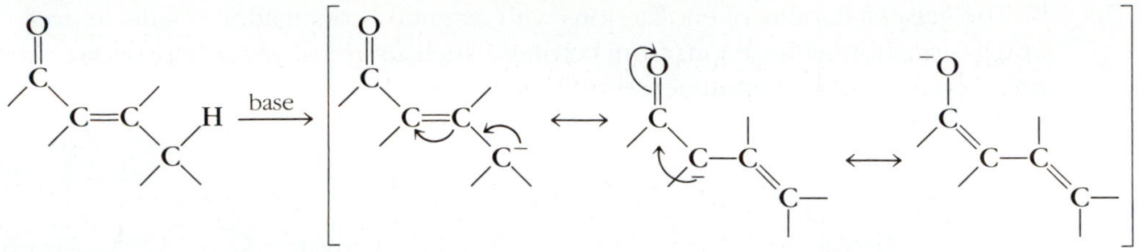

base

Abstraction of a proton from the γ position of an α,β-unsaturated ketone would lead to an enolate ion in which the charge is delocalized over a total of five atoms. Such *extended conjugation* results in greater stability of these enolate ions. As suggested by the three resonance contributors, such an enolate ion will exhibit reactivity as a base or nucleophile at both the α and γ carbons (as well as at oxygen). This is illustrated by the following reactions with a proton:

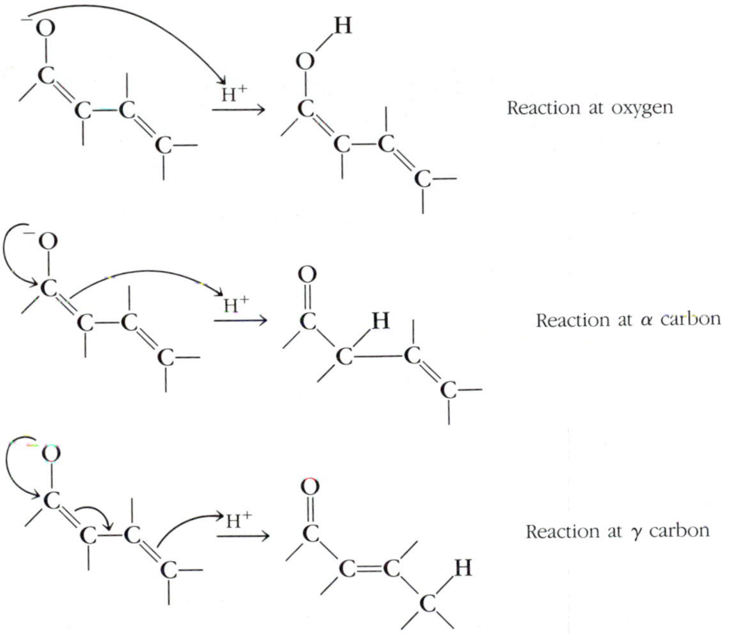

Reaction at oxygen

Reaction at α carbon

Reaction at γ carbon

Under conditions of *reversible* enolate ion formation, these pathways would result in equilibration of the α,β- and β,γ-unsaturated isomers. The conjugated isomer is generally more stable, so treatment with base provides a means of converting a β,γ-unsaturated ketone to the α,β-unsaturated isomer.

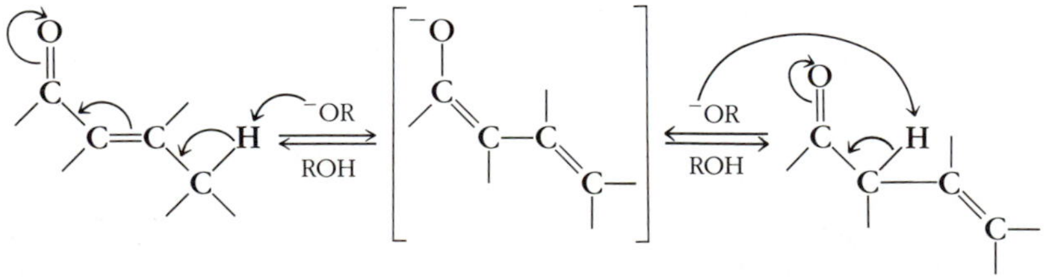

Equilibration of these isomers also occurs readily in the presence of acid, via formation of the enol. This facile interconversion provides the basis for many reactions of α,β- and β,γ-unsaturated aldehydes and ketones.

The greater stability of enolate ions with extended conjugation results in preferential formation of these ions from ketones which also have acidic hydrogen on the *other* α carbon (at least under equilibrating conditions).

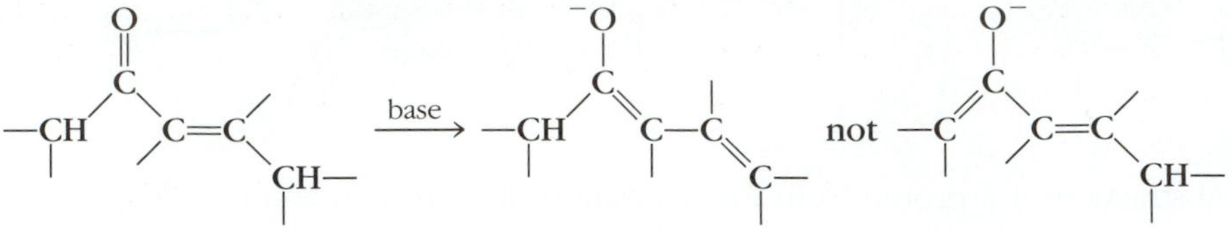

base

not

Alkylation of α,β-Unsaturated Ketones

Abstraction of a proton from the γ position of an aldehyde or ketone leads to an enolate ion with anionic character at both the α and γ carbons. Nevertheless, alkylation usually takes place at the α position.

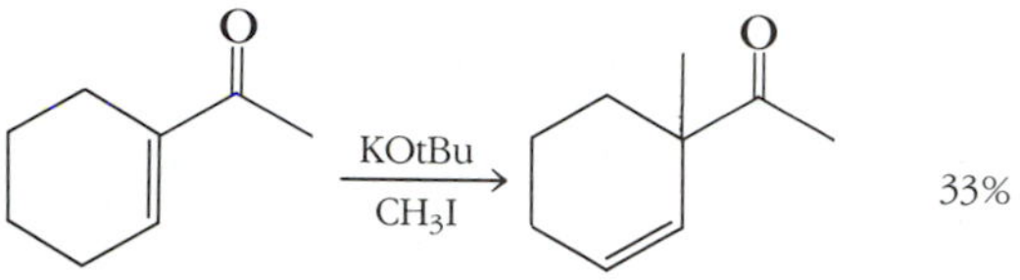

KOtBu, CH_3I — 33%

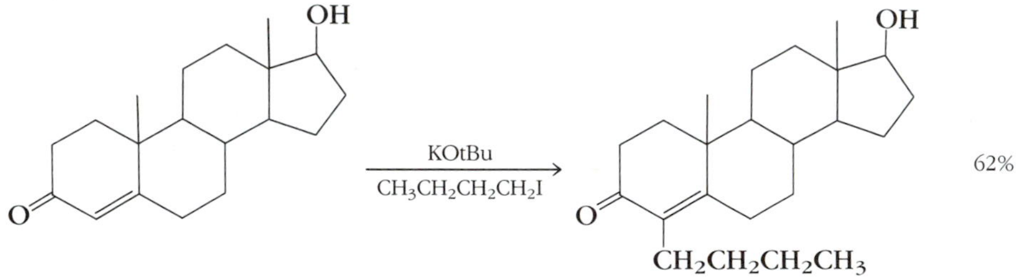

KOtBu, $CH_3CH_2CH_2CH_2I$ — $CH_2CH_2CH_2CH_3$ — 62%

Alkylation in the second reaction initially produced the β,γ-unsaturated ketone, but this was converted to the more stable, conjugated isomer in the presence of base.

In the presence of excess alkylating agent, a good yield of dialkylation product can be obtained (if there is a second acidic hydrogen).

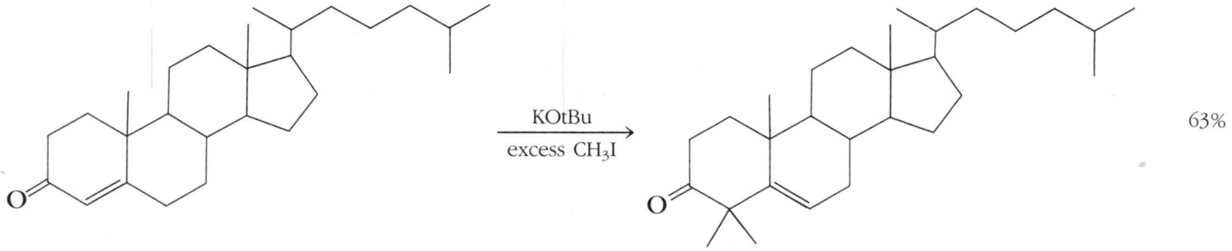

KOtBu, excess CH_3I — 63%

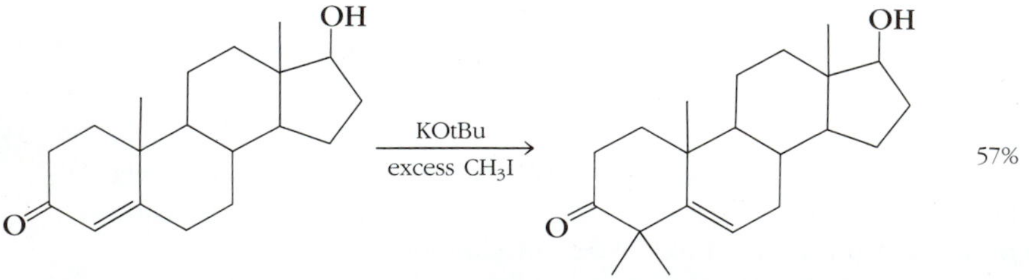

KOtBu, excess CH_3I — 57%

Generation of Enolate Ions by Reduction of α,β-Unsaturated Ketones

The dissolving metal reduction of an α,β-unsaturated carbonyl compound using Li/NH_3 (Section 13.13) takes place via successive electron transfer and proton transfer reactions. Initial transfer of an electron from the metal to the conjugated carbonyl system generates a *radical anion,* a species that has an unpaired electron and also bears a negative charge).

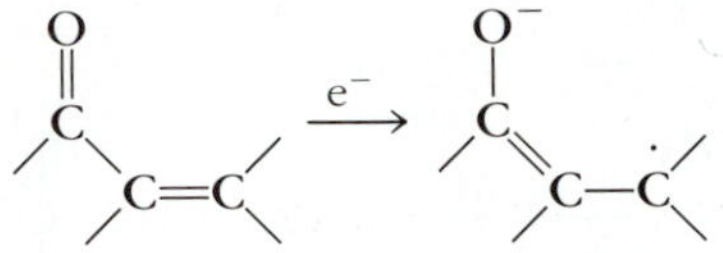

Other resonance forms could be drawn for the radical anion, but protonation of the oxygen atom, or coordination to a lithium cation as shown here, is followed by a second electron transfer to produce an allyl anion.

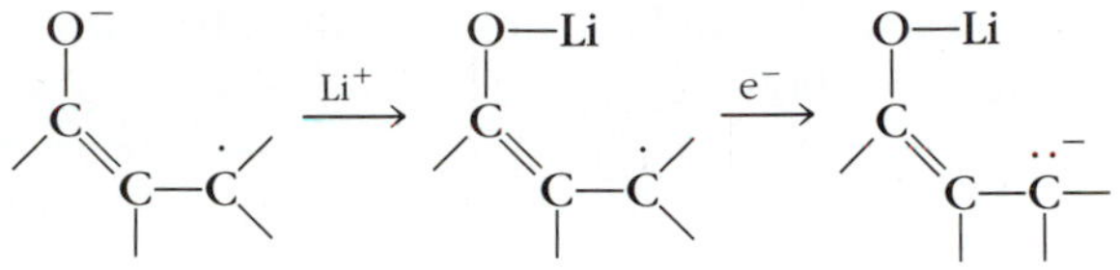

Protonation of the allyl anion by solvent then yields an enolate.

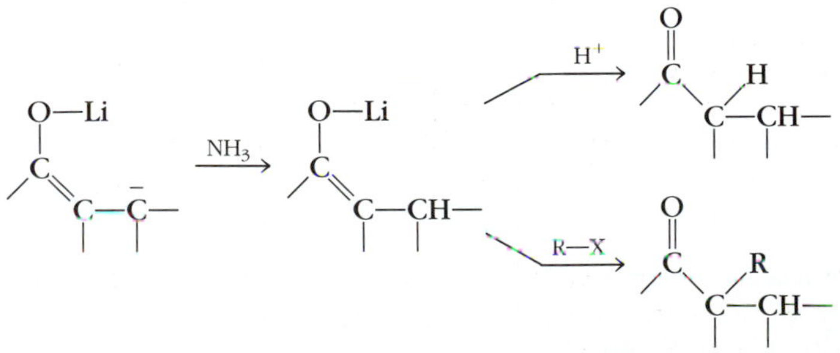

The enolate ion produced by lithium-ammonia reduction of an α,β-unsaturated ketone is stable under the reaction conditions. With normal reduction procedures it is protonated by acidification during workup to yield the saturated ketone. However, addition of an alkylating agent *prior to workup* permits alkylation of the enolate. The alkylation of enolates generated by reduction of unsaturated ketones is illustrated by the following examples:

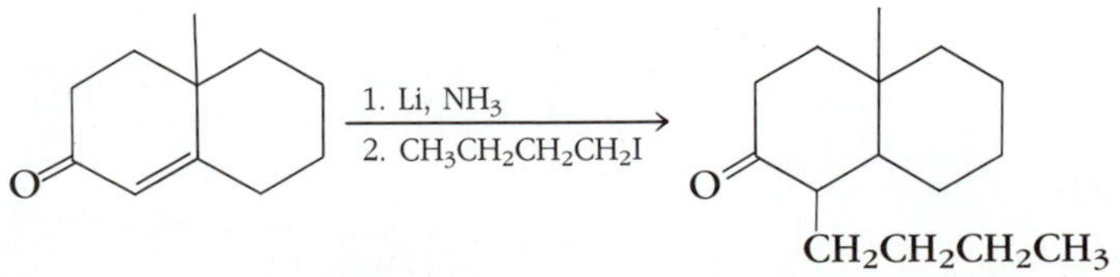

43%

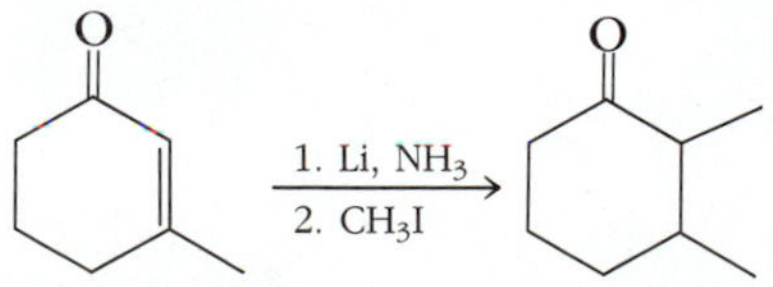

54%

$$\xrightarrow[\text{2. } CH_3I]{\text{1. Li, } NH_3} \qquad 40\%$$

$$C_6H_5\text{—}CH\text{=}C(CH_3)\text{—}C(\text{=}O)\text{—}CH_3 \xrightarrow[\text{2. } CH_3I]{\text{1. Li, } NH_3} C_6H_5\text{—}CH_2\text{—}C(CH_3)_2\text{—}C(\text{=}O)\text{—}CH_3 \qquad 57\text{–}67\%$$

Notice that the products of these reactions are *saturated ketones:* in addition to alkylation at the α position, the carbon-carbon double bond has been reduced. This type of reaction provides an excellent means of generating a specific lithium enolate, and alkylation tends to yield a single isomer, free from any product of alkylation at the other α carbon.

EXERCISE 14.9

What product would be expected in each of the following alkylation reactions?

$$CH_3\text{—}CH(CH_3)\text{—}CH\text{=}CH\text{—}C(\text{=}O)\text{—}CH_3$$

(a) With KOtBu, C_2H_5I

(b) With Li/NH_3, then C_2H_5I

(c) With KOtBu, $CH_2\text{=}CHCH_2Br$

(d) With Li/NH_3, then $CH_2CH\text{=}CH_2Br$

EXERCISE 14.10

What reagents are needed for conversion of

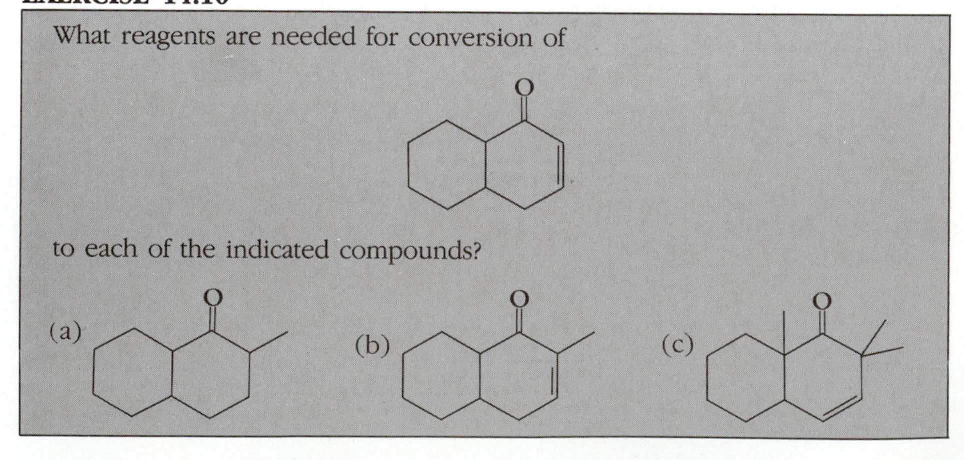

to each of the indicated compounds?

(a) (b) (c)

14.5 THE MICHAEL REACTION: CONJUGATE ADDITION OF ENOLATE IONS

The Michael reaction (named for A. Michael at Tufts University, who discovered it in the 1880s) proceeds by conjugate addition of an enolate to an α,β-unsaturated carbonyl compound to yield a 1,5-dicarbonyl compound:

$$\text{(enolate)} + \text{C=C–C=O} \xrightarrow{\text{base}} \text{O=C}^1\text{–C}^2\text{–C}^3\text{–CH}^4\text{–C}^5\text{=O}$$

The reaction is ordinarily carried out under basic conditions in a protic solvent (for example, with $NaOCH_3/CH_3OH$), and all the steps are fully reversible.

$$\text{O=C–C–H} \underset{}{\overset{CH_3O^-}{\rightleftharpoons}} \text{}^-\text{O–C=C} \rightleftharpoons \text{O=C–C–C–C=C–O}^- \rightleftharpoons \text{O=C–C–C–C(H)–C=O} \quad (H\text{—}OCH_3)$$

The last step of the reaction, protonation of the enolate product, regenerates the base which was consumed in the first step. Consequently, the base is strictly *catalytic,* and only small amounts are used. Indeed, the use of a full equivalent of base tends to promote the reverse reaction, called the *retrograde* Michael reaction:

$$\text{O=C–C–C–C(H)–C=O} \xrightarrow{^-OCH_3} \text{O=C–C–C–C=C–O}^- \longrightarrow \text{}^-\text{O–C=C–} + \text{–C=C–C=O}$$

The following examples show how 1,5-diketones can be prepared by the Michael reaction:

$$C_6H_5\text{—}\overset{O}{\overset{\|}{C}}\text{—}CH_2\text{—}C_6H_5 + CH_2\text{=}C(C_6H_5)\text{—}\overset{O}{\overset{\|}{C}}\text{—}C_6H_5 \xrightarrow[CH_3OH]{NaOCH_3} C_6H_5\text{—}\overset{O}{\overset{\|}{C}}\text{—}\underset{C_6H_5}{CH}\text{—}CH_2\text{—}\underset{C_6H_5}{CH}\text{—}\overset{O}{\overset{\|}{C}}\text{—}C_6H_5 \quad 99\%$$

$$C_6H_5\text{—}CH_2\text{—}\overset{O}{\overset{\|}{C}}\text{—}CH_2\text{—}C_6H_5 + C_6H_5\text{—}CH\text{=}CH\text{—}\overset{O}{\overset{\|}{C}}\text{—}C_6H_5 \xrightarrow[CH_3OH]{NaOCH_3} C_6H_5\text{—}CH_2\text{—}\overset{O}{\overset{\|}{C}}\text{—}\underset{C_6H_5}{CH}\text{—}\overset{C_6H_5}{CH}\text{—}CH_2\text{—}\overset{O}{\overset{\|}{C}}\text{—}C_6H_5 \quad 75\%$$

$$\text{2-methylcyclohexanone} + CH_2{=}CH{-}\overset{O}{\overset{\|}{C}}{-}C_6H_5 \xrightarrow[CH_3CH_2OH]{KOH} \text{2-methyl-2-}(CH_2CH_2{-}\overset{O}{\overset{\|}{C}}{-}C_6H_5)\text{cyclohexanone} \quad 64\%$$

Unsaturated aldehydes have a tendency to undergo self-condensation and other side reactions, but successful Michael reactions have been carried out:

$$CH_3{-}\overset{O}{\overset{\|}{C}}{-}CH_2CH_3 + CH_3{-}CH{=}CH{-}\overset{O}{\overset{\|}{C}}H \xrightarrow[CH_3OH]{NaOCH_3} CH_3{-}\overset{O}{\overset{\|}{C}}{-}\underset{CH_3}{\underset{|}{C}H}{-}\overset{CH_3}{\overset{|}{C}H}{-}CH_2{-}\overset{O}{\overset{\|}{C}}H \quad 60\%$$

$$C_6H_5{-}\overset{O}{\overset{\|}{C}}{-}CH_2{-}C_6H_5 + CH_2{=}CH{-}\overset{O}{\overset{\|}{C}}H \xrightarrow{NaOC_2H_5} C_6H_5{-}\overset{O}{\overset{\|}{C}}{-}\underset{C_6H_5}{\underset{|}{C}H}{-}CH_2{-}CH_2{-}\overset{O}{\overset{\|}{C}}H \quad 100\%$$

The last several reactions illustrate the tendency of ketones to react via the *more substituted* enolate ion under equilibrating conditions. We showed the same tendency for the *alkylation* of ketones when the enolate is formed reversibly (Section 14.3), and there are indeed many similarities between the alkylation of ketones and the Michael reaction. Sometimes β-halo carbonyl compounds are used in the Michael reaction, and this might appear to be a simple alkylation.

$$\overset{O}{\overset{\|}{C}}{-}\underset{|}{C}H{-} + {-}\overset{Cl}{\overset{|}{C}}{-}\underset{|}{C}H{-}\overset{O}{\overset{\|}{C}} \xrightarrow{\text{base}} \overset{O}{\overset{\|}{C}}{-}C{-}C{-}\underset{|}{C}H{-}\overset{O}{\overset{\|}{C}}$$

Actually, such reactions proceed by initial elimination of hydrogen halide to generate an unsaturated carbonyl compound, followed by a normal Michael reaction.

$${-}\overset{Cl}{\overset{|}{C}}{-}\underset{|}{C}H{-}\overset{O}{\overset{\|}{C}} \xrightarrow{\text{base}} {-}\overset{|}{C}{=}\underset{|}{C}{-}\overset{O}{\overset{\|}{C}}$$

The elimination of HCl (or other hydrogen halide) from a β-halo carbonyl compound proceeds very easily, because the α hydrogen is acidic and because a conjugated carbonyl system is formed. (Note that the use of a β-halo carbonyl compound as the reactant in a Michael reaction requires a full molar equivalent of base for the elimination reaction.)

The Michael reaction also shares many of the features of the alkylation of enolate ions and the aldol condensation. If an α,β-unsaturated aldehyde or ketone were to undergo *carbonyl addition* of an enolate, we would describe the process as an aldol condensation. The Michael reaction simply involves *conjugate addition* of the same enolate ion. As with other reversible additions (Section 13.10), conjugate addition is favored, and reactions of enolate ions with unsaturated carbonyl compounds usually yield the products of the Michael reaction rather than those of an aldol condensation.

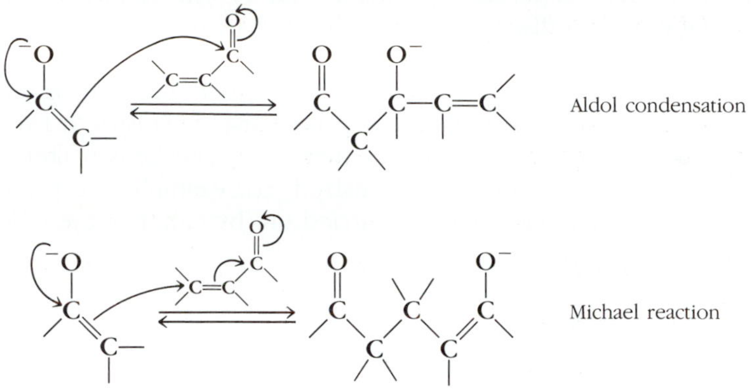

Aldol condensation

Michael reaction

Side Reactions

The Michael reaction involves two different carbonyl compounds in the presence of base, so many side reactions are possible. For example, the alcohol solvent could undergo conjugate addition (Section 13.10) to the unsaturated carbonyl compound. This is reversible, however, and causes no difficulties.

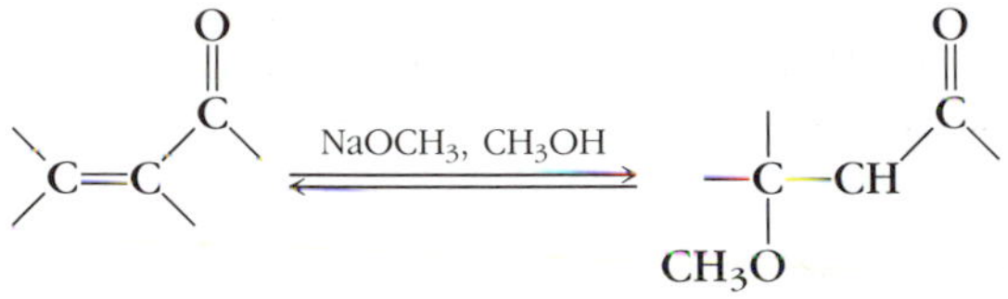

$NaOCH_3$, CH_3OH

CH_3O

Either or both of the two carbonyl reactants could undergo aldol condensations, as could the product of the Michael reaction. Moreover, the product is a dicarbonyl compound, and *intramolecular* aldol condensations occur readily. Fortunately, these intramolecular condensations often represent the desired course of reaction because they provide a means of forming *cyclic* compounds. We will present a detailed analysis of these cyclization reactions in Chapter 22.

Another side reaction that can occur is the retrograde Michael reaction. If there are hydrogen atoms on both C-2 and C-4 of the 1,5-dicarbonyl compound, then two different retrograde Michael reactions are possible:

base

base

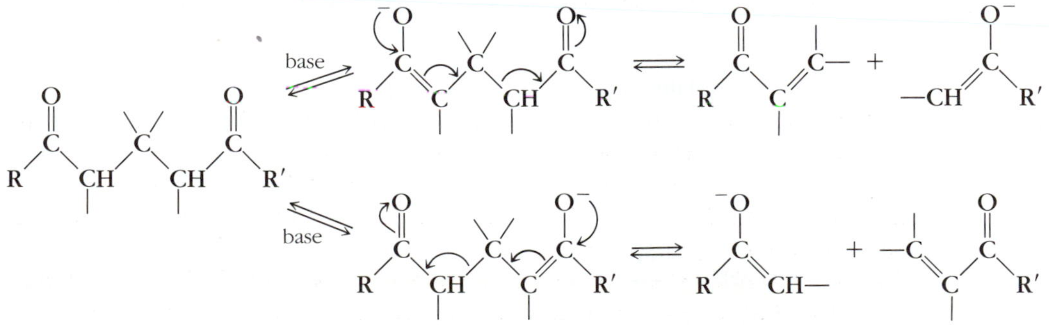

As we indicated previously, the retrograde Michael reaction does not usually take place as long as only a small *(catalytic)* quantity of base is used in the Michael reaction.

Most of these side reactions can be minimized by the careful selection of experimental conditions, although the "correct" conditions often can be determined only by carrying out several experiments in the laboratory. The important point for you to take from the present discussion is that the Michael reaction provides an excellent method for the preparation of 1,5-dicarbonyl compounds.

The possibility of two different pathways for a hypothetical retrograde Michael reaction has important consequences for preparative chemistry. If two pathways are possible for the reverse reaction, then there must also be two alternative ways in which the compound could be synthesized. For example, the preparation of 1,3-diphenyl-2,6-octanedione might be carried out by either of the following Michael reactions:

$$C_6H_5{-}CH_2{-}\overset{\overset{\displaystyle O}{\|}}{C}{-}CH_2{-}C_6H_5 + CH_2{=}CH{-}\overset{\overset{\displaystyle O}{\|}}{C}{-}CH_2CH_3 \xrightarrow{\text{base}} C_6H_5{-}CH_2{-}\overset{\overset{\displaystyle O}{\|}}{C}{-}\underset{\underset{\displaystyle C_6H_5}{|}}{CH}{-}CH_2CH_2{-}\overset{\overset{\displaystyle O}{\|}}{C}{-}CH_2CH_3$$

$$C_6H_5{-}CH_2{-}\overset{\overset{\displaystyle O}{\|}}{C}{-}\underset{\underset{\displaystyle C_6H_5}{|}}{C}{=}CH_2 + CH_3{-}\overset{\overset{\displaystyle O}{\|}}{C}{-}CH_2CH_3 \xrightarrow{\text{base}} C_6H_5{-}CH_2{-}\overset{\overset{\displaystyle O}{\|}}{C}{-}\underset{\underset{\displaystyle C_6H_5}{|}}{CH}{-}CH_2{-}CH_2{-}\overset{\overset{\displaystyle O}{\|}}{C}{-}CH_2CH_3$$

In this instance the first of the two pathways is superior for several reasons. Only one enolate ion can be formed from the symmetrical ketone, 1,3-diphenyl-2-propanone, and it is stabilized by the phenyl substituent.

$$C_6H_5{-}CH_2{-}\overset{\overset{\displaystyle O}{\|}}{C}{-}CH_2{-}C_6H_5 \xrightleftharpoons{\text{base}} C_6H_5{-}CH_2{-}\overset{\overset{\displaystyle ^{-}O}{|}}{C}{=}CH{-}C_6H_5$$

In contrast, two enolate ions could be formed from 2-butanone in the second pathway, and it is likely that reaction would occur via the "wrong" enolate ion (i.e., that which is more substituted).

$$CH_3{-}\overset{\overset{\displaystyle O}{\|}}{C}{-}CH_2CH_3 \xrightleftharpoons{\text{base}} CH_2{=}\overset{\overset{\displaystyle O^{-}}{|}}{C}{-}CH_2CH_3 \quad \text{or} \quad CH_3{-}\overset{\overset{\displaystyle O^{-}}{|}}{C}{=}CH{-}CH_3$$

Such an analysis of the alternatives frequently permits the selection of the superior pathway for preparing a 1,5-dicarbonyl compound via the Michael reaction.

EXERCISE 14.11

Predict the products for the following Michael reactions:

(a) 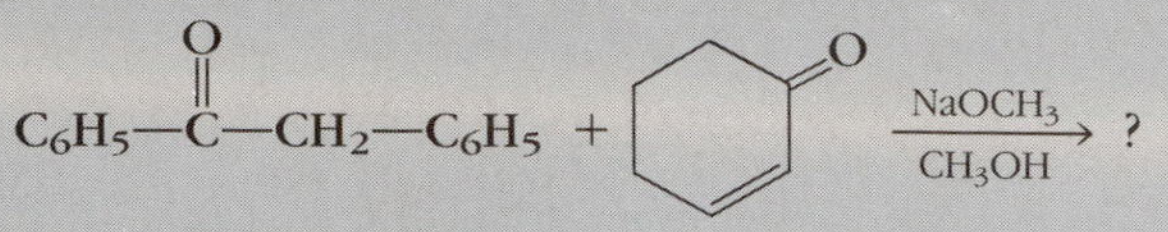 $C_6H_5-\overset{O}{\overset{\|}{C}}-CH_2-C_6H_5$ + (2-cyclohexenone) $\xrightarrow[CH_3OH]{NaOCH_3}$?

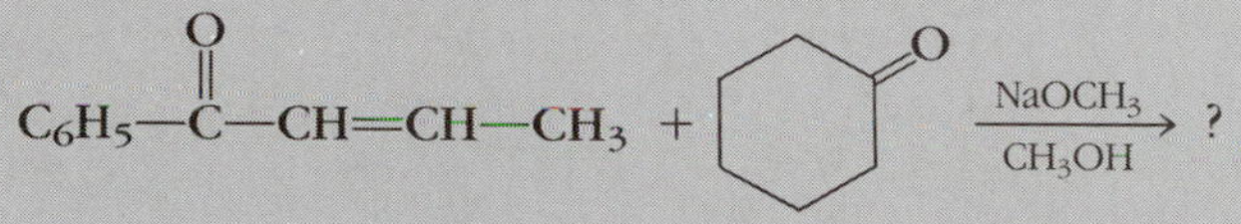

(b) $C_6H_5-\overset{O}{\overset{\|}{C}}-CH{=}CH-CH_3$ + (cyclohexanone) $\xrightarrow[CH_3OH]{NaOCH_3}$?

(c) $CH_3CH_2-\overset{O}{\overset{\|}{C}}-\underset{CH_3}{\underset{|}{CH}}-CH_3$ + (2-cyclopentenone) $\xrightarrow[CH_3OH]{NaOCH_3\ (>1\ equiv)}$?

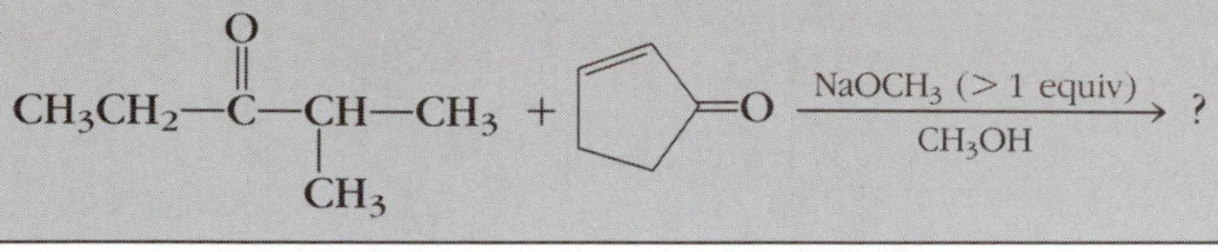

EXERCISE 14.12

What starting materials would yield the following 1,5-diketones via a Michael reaction?

(a) 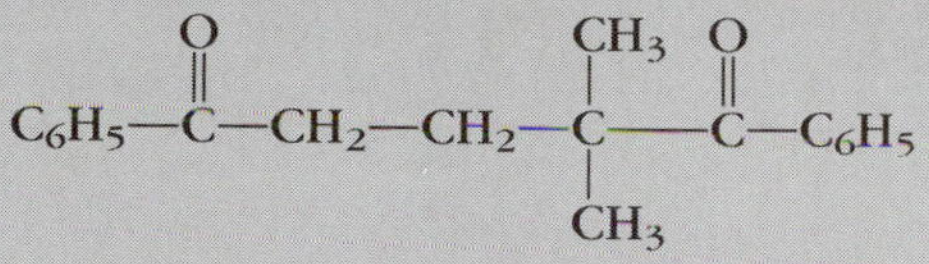 $C_6H_5-\overset{O}{\overset{\|}{C}}-CH_2-CH_2-\underset{CH_3}{\underset{|}{\overset{CH_3}{\overset{|}{C}}}}-\overset{O}{\overset{\|}{C}}-C_6H_5$

(b) 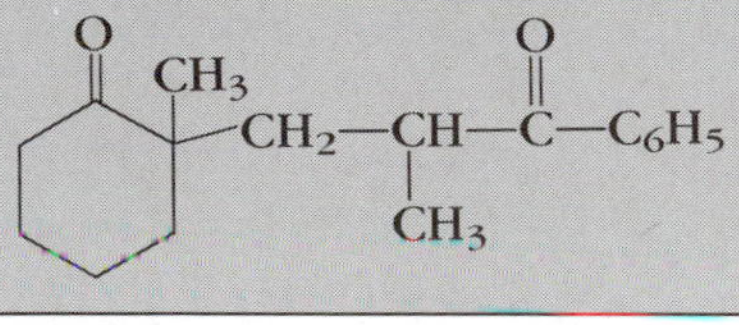 (2-methylcyclohexanone with CH_3 and $-CH_2-\underset{CH_3}{\underset{|}{CH}}-\overset{O}{\overset{\|}{C}}-C_6H_5$ on C-2)

14.6 CONDENSATION AND ALKYLATION REACTIONS IN ORGANIC SYNTHESIS

The reactions that we have discussed in this chapter play important roles in synthetic organic chemistry. The aldol condensation, enolate alkylation, and the Michael reaction all involve carbon-carbon bond formation. In addition, the product in each case contains functional groups that permit further reactions in a synthetic sequence.

In several ways the reactions are all interrelated, and they may be used in various combinations and sequences to carry out an organic synthesis. Several ways in which these reactions might be combined in a synthetic sequence are shown in Figure 14.4.

Of particular significance in Figure 14.4 is the relationship between the Michael reaction and the aldol condensation. The Michael reaction requires an α,β-unsaturated carbonyl compound as one of the reactants, and such unsaturated aldehydes and ketones can be prepared via the aldol condensation. Furthermore,

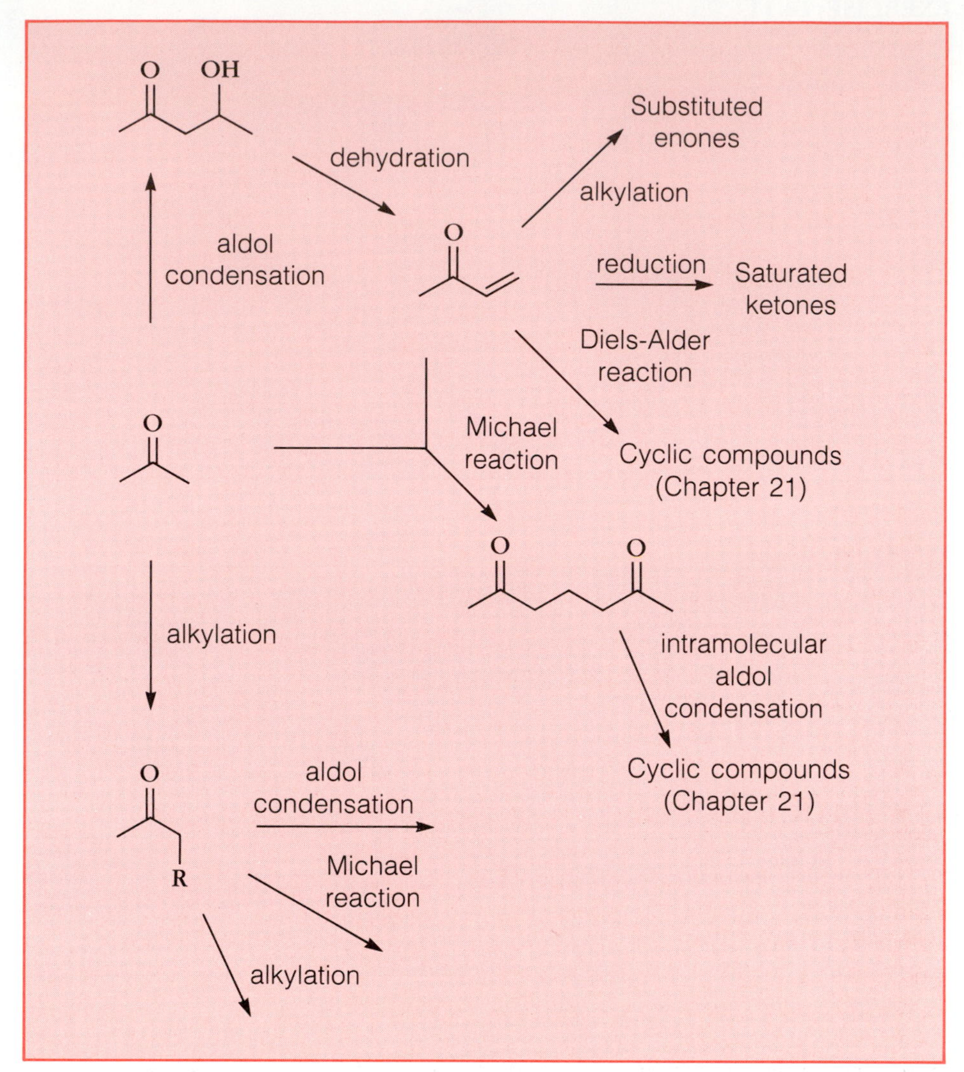

Figure 14.4 Some Synthetic Transformations Involving Carbon-Carbon Bond Formation via Enols and Enolate Ions of Ketones.
(Only interconversions of carbonyl compounds are shown because inclusion of various oxidation-reduction and organometallic reactions would make the scheme too complex).

the product of a Michael reaction is a 1,5-dicarbonyl compound, and these are frequently capable of undergoing intramolecular aldol condensations to form cyclic compounds.

We will not present a detailed discussion of these cyclization reactions in this chapter, but the following example illustrates the case of a Michael reaction followed by an intramolecular aldol condensation to produce the skeleton of a steroid (see Section 12.9).

CH_3O + NaNH₂ (Michael reaction) → [CH_3O] → (intramolecular aldol condensation) → [HO, CH_3O] → (dehydration) → CH_3O 55%

Several steps are involved (Michael reaction, aldol condensation, and dehydration). The tetracyclic unsaturated ketone is isolated directly from this reaction in 55% yield.

14.7 A BIOLOGICAL EXAMPLE OF THE ALDOL CONDENSATION

The need to carry out carbon-carbon bond-forming reactions is not limited to organic chemists; all living organisms must carry out such processes as part of their normal metabolism. We previously discussed the formation of carbon-carbon bonds via nucleophilic substitution in biological systems in Section 12.9, and here we will present an example of a biological aldol condensation.

Carbohydrates (Chapter 19) play a key role in the storage and regulation of energy in biological systems. The degradation of glucose releases energy upon conversion to smaller molecules (and ultimately to CO_2). *Glycolysis* is a sequence of reactions that converts glucose to lactate with concomitant storage of energy in the form of ATP (adenosine triphosphate). The enzyme *aldolase* catalyzes both the forward and the reverse reaction of one step in this metabolic pathway. The forward reaction involves the conversion of a C_6 compound, fructose-1,6-bisphosphate, into two C_3 compounds, glyceraldehyde-3-phosphate and dihydroxyacetone phosphate. The reverse reaction is an aldol condensation, and we have illustrated it in the following equation. (The phosphate group, $—OPO_3^{2-}$, is symbolized as —OP in this and subsequent equations for convenience.)

$$\underset{\text{Dihydroxyacetone phosphate}}{\underset{OP}{CH_2}-\overset{O}{\overset{\|}{C}}-\underset{OH}{CH_2}} + \underset{\text{Glyceraldehyde-3-phosphate}}{\overset{O}{\overset{\|}{HC}}-\underset{OH}{CH}-CH_2-OP} \rightleftharpoons \underset{\text{Fructose-1,6-bisphosphate}}{\underset{OP}{CH_2}-\overset{O}{\overset{\|}{C}}-\underset{OH}{CH}-\overset{OH}{CH}-\underset{OH}{CH}-CH_2OP}$$

Several features of this reaction are noteworthy. First, all the compounds involved are polyfunctional, and many side reactions would be expected if the reaction were attempted under laboratory conditions. The enzyme, however, is able to direct the reaction along the specific course shown here. Second, although no stereochemistry is depicted in the preceding equation, the condensation reaction generates two new chiral centers. Under the influence of the enzyme both these centers are formed stereospecifically, and the reaction yields a single stereoisomer. Third, the mechanism of the enzymatic reaction has been studied extensively, and it is known how the enzyme "holds" the ketone in the proper orientation for reaction to occur. Prior to the actual aldol condensation the ketone carbonyl reacts with an amino group on the enzyme, and a covalent bond is formed (see Section 13.7). Thus it is actually the nitrogen analog of the ketone (i.e., an imine) that undergoes the aldol condensation.

$$(PO)CH_2{-}C({=}O){-}CH_2OH \; + \; :NH_2{-}\text{enzyme} \rightleftharpoons (PO)CH_2{-}C({=}N{-}\text{enzyme}){-}CH_2OH$$

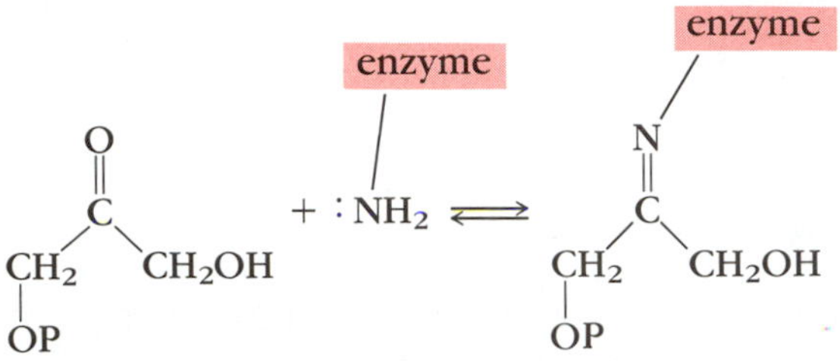

After the aldol condensation occurs, the nitrogen derivative is hydrolyzed to regenerate the carbonyl group. During the time that the ketone derivative is bound to the enzyme, other functional groups on the enzyme are appropriately situated to direct the reaction (by removal of the "correct" α hydrogen, etc.) In this way it is possible for enzymes to catalyze reactions that could not be effected by simply mixing the reagents in a flask.

14.8 SUMMARY OF REACTIONS

Table 14.1 presents a summary of the reactions discussed in this chapter. Many new and different bases have been introduced in the last two chapters, and we have summarized their properties and uses in Table 14.2.

TABLE 14.1 Reactions of Aldehydes and Ketones: Carbon-Carbon Bond Forming Reactions of Enols and Enolate Ions

Reaction	Comments
1. Aldol condensation 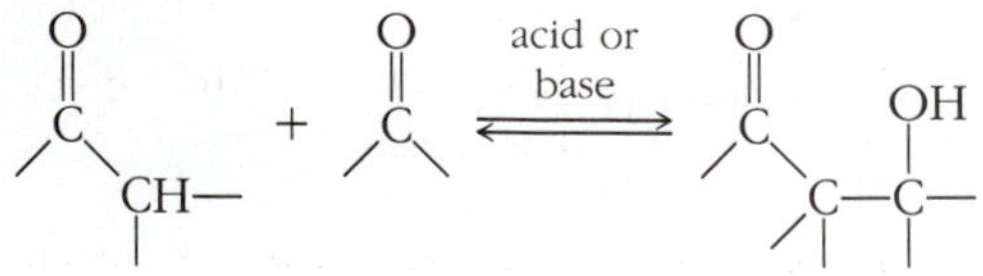 $-C({=}O){-}CH< \; + \; >C{=}O \xrightleftharpoons{\text{acid or base}} -C({=}O){-}C{-}C(OH)<$	Sections 14.1, 14.2 Base catalysis is more common. Typical bases are NaOH, $NaOCH_3$, and other salts of alcohols. The reaction is reversible.

TABLE 14.1 Reactions of Aldehydes and Ketones: Carbon-Carbon Bond Forming Reactions of Enols and Enolate Ions (continued)	
Reaction	**Comments**
$\text{>C(=O)CH}_2\text{—} + \text{>C=O} \longrightarrow \text{>C(=O)C=C<}$	Dehydration often occurs during the reaction if there is a second α hydrogen.
$\text{RC(=O)CH}_2\text{—} + \text{ArC(=O)H} \longrightarrow \text{RC(=O)C(—)=CH—Ar}$	Mixed aldol condensations work well when one compound is an aldehyde, especially an aromatic aldehyde. (The unhindered aldehyde carbonyl is highly susceptible to nucleophilic attack, and an aromatic aldehyde cannot form an enolate ion.)
$\text{RCH}_2\text{C(=O)CH}_3 + \text{ArC(=O)H} \xrightarrow{\text{base}} \text{RCH}_2\text{C(=O)CH=CHAr}$	Under basic conditions an unsymmetrical ketone usually reacts at the less substituted α carbon. (Reaction at the more substituted α carbon is sometimes found with acid catalysis.)
2. Alkylation of enolate ions: **saturated carbonyl compounds**	Sections 14.3, 14.4
$\text{>C(=O)CH(—)—} \xrightarrow[\text{2. RX}]{\text{1. base}} \text{>C(=O)C(R)<}$	A full equivalent of base is required. Side reactions (polyalkylation, aldol condensations, elimination) may occur. The alkyl halide (or sulfonate) must be methyl or primary. The base must be nonnucleophilic (KOtBu) or strong enough to cause complete conversion to the enolate ion [$NaNH_2$, LiH, NaH, $(C_6H_5)_3CK$, $LiN(iPr)_2$]. Lithium enolates lead to fewer side reactions.

TABLE 14.1 Reactions of Aldehydes and Ketones: Carbon-Carbon Bond Forming Reactions of Enols and Enolate Ions (continued)

Reaction	Comments
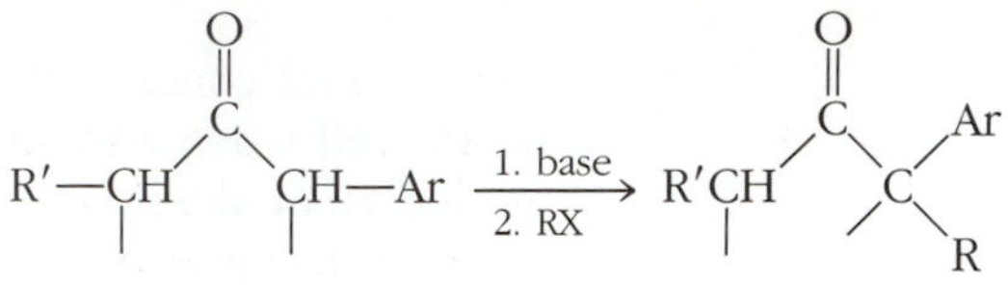 R′—CH—C(=O)—CH—Ar $\xrightarrow[2.\ RX]{1.\ base}$ R′CH—C(=O)—C(Ar)(R)	Formation of an enolate ion conjugated with an aryl substituent is favored.
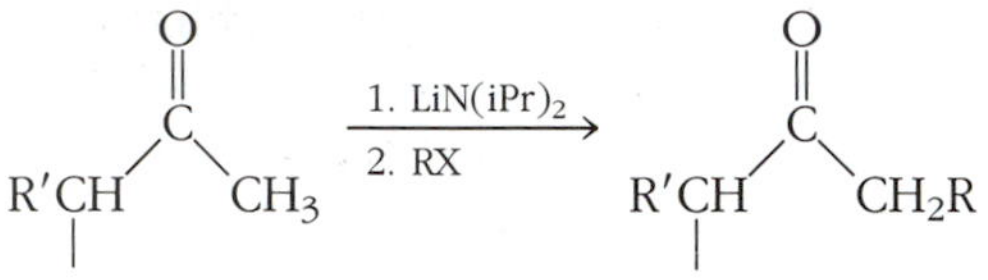 R′CH—C(=O)—CH_3 $\xrightarrow[2.\ RX]{1.\ LiN(iPr)_2}$ R′CH—C(=O)—CH_2R	Lithium diisopropylamide usually affords the less substituted enolate ion.
unsaturated carbonyl compounds	
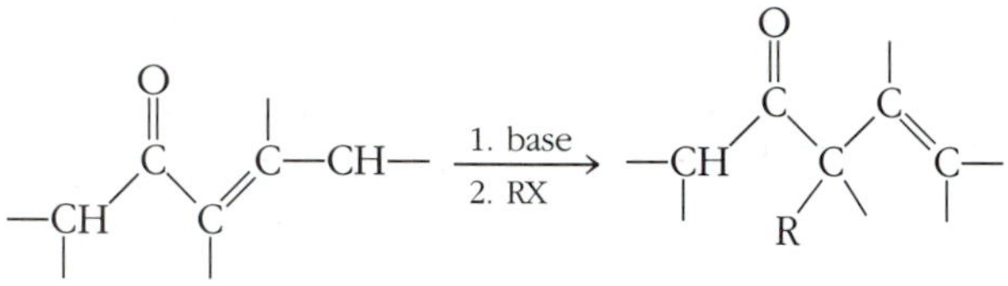 —CH—C(=O)—C=C—CH— $\xrightarrow[2.\ RX]{1.\ base}$ —CH—C(=O)—C(R)—C=C—	In reversible reactions α,β-unsaturated carbonyl compounds afford the more stable enolate ion (by removal of a γ hydrogen), but this alkylates at the α position.
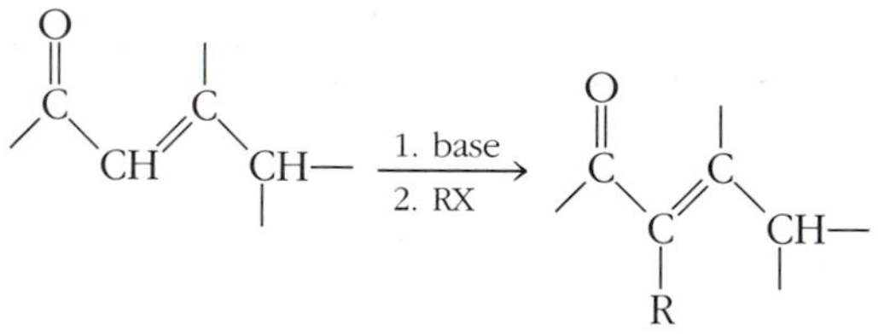 —C(=O)—CH=C—CH— $\xrightarrow[2.\ RX]{1.\ base}$ —C(=O)—C(R)=C—CH—	If the α carbon also has a hydrogen, the initial product usually rearranges to an α,β-unsaturated compound.
3. Reductive alkylation of an α,β-unsaturated carbonyl compound	Section 14.4
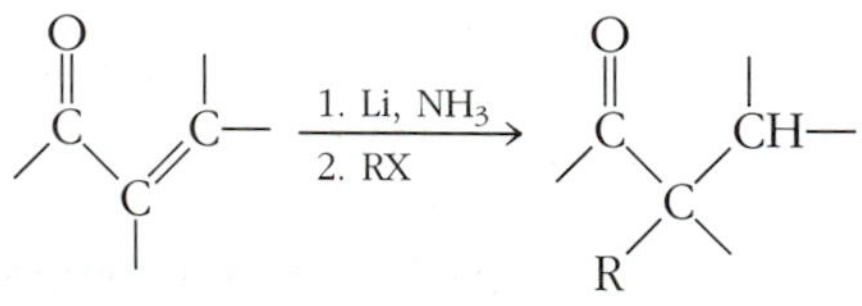 —C(=O)—C=C— $\xrightarrow[2.\ RX]{1.\ Li,\ NH_3}$ —C(=O)—C(R)—CH—	Reduction generates the enolate ion, which can then be alkylated. RX must be methyl or primary.

TABLE 14.1 Reactions of Aldehydes and Ketones: Carbon-Carbon Bond Forming Reactions of Enols and Enolate Ions (continued)

Reaction	Comments
4. Michael reaction	Section 14.5
$-C(=O)-C=C< \; + \; -CH-C(=O)- \;\underset{}{\overset{\text{base}}{\rightleftharpoons}}\; -C(=O)-CH-C-C-C(=O)-$	The reaction is reversible, but a catalytic quantity of base favors product formation. Typical bases are $NaOCH_3$, $NaOC_2H_5$. (Acid catalysis has been used on occasion.)
$-C(=O)-C=C< \; + \; CH_3-C(=O)-CH_2R \;\xrightarrow{\text{base}}\; -C(=O)-CH-C-CH_2-C(=O)-CH_2R$	Reaction via the less substituted enolate ion is typical.
$-C(=O)-C=C< \; + \; ArCH_2-C(=O)- \;\longrightarrow\; -C(=O)-CH-C-CH(Ar)-C(=O)-$	But aryl substitution favors enolate formation at that α carbon.

TABLE 14.2 Some Commonly Used Bases

Base	Conjugate Acid	K_a	Uses, Comments
NaOH, KOH, $R_4N^+OH^-$	H_2O	10^{-16}	Used to neutralize acidic reaction mixtures; catalysts in aldol condensations. Unsatisfactory if anhydrous or aprotic conditions are needed.
$NaOCH_3$, $NaOC_2H_5$	CH_3OH, C_2H_5OH	10^{-16}–10^{-18}	Used as catalysts in aldol condensations, Michael reactions. The alkoxide ions are weaker bases than enolate ions of aldehydes and ketones; only a small fraction of the carbonyl compound is converted to enolate at any instant.

TABLE 14.2 Some Commonly Used Bases (continued)

Base	Conjugate Acid	K_a	Uses, Comments
KOtBu	tBuOH	$\sim 10^{-18}$	Less nucleophilic than other alkoxide bases; can therefore be used in alkylation reactions of carbonyl compounds.
$(C_6H_5)_3CK$	$(C_6H_5)_3CH$	10^{-32}	More basic than enolate ions; produces essentially complete conversion of an aldehyde or ketone to the enolate ion.
$CH_3-S(=O)-CH_2^{-}Na^{+}$	$CH_3-S(=O)-CH_3$	10^{-35}	Prepared by reaction of DMSO with NaH. Often used to form ylides from the onium salts.
$NaNH_2$, $LiNH_2$	NH_3	10^{-35}	Often used with liquid ammonia as solvent. Converts terminal alkynes to the corresponding salts; nearly complete conversion of a ketone to the enolate.
$LiN(iPr)_2$	$HN(iPr)_2$	$\sim 10^{-36}$	Forms lithium enolate at the less substituted α carbon of a ketone. Sterically hindered, poor nucleophile. Excellent reagent for alkylation of a ketone.
LiH, NaH	H_2	(very small)	Powerful bases; evolution of H_2 allows complete conversion of ketone to enolate ion.
$LiAlH_4$	H_2	(very small)	Powerful base and powerful nucleophile. Converts OH groups to salts but acts as nucleophile at carbon. Rarely used as a base. (Note that $NaBH_4$ is a much weaker base, and it can be used in aqueous solution.)

TABLE 14.2 Some Commonly Used Bases (continued)

Base	Conjugate Acid	K_a	Uses, Comments
C_6H_5Li, *tert*-BuLi, $CH_3CH_2CH_2CH_2Li$	Benzene, isobutane, butane	(very small)	The organolithium reagents are powerful bases and powerful nucleophiles. Rarely used to generate enolates but are used to form ylides from onium salts.
RMgBr, etc.	R—H	(very small)	Grignard reagents are strongly basic and strongly nucleophilic. Almost always react as nucleophiles with carbonyl compounds but have been used as bases to convert terminal alkynes and NH groups to the corresponding salts.

14.9 PROBLEMS

14.1 Each of the following reactions involves carbon-carbon bond formation. Which can be described as aldol condensations? As alkylations of aldehydes or ketones? As Michael reactions?

(a) 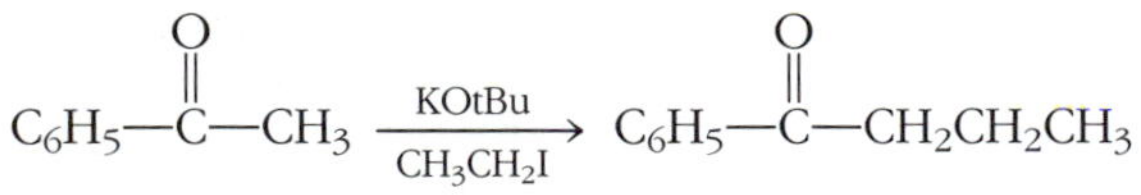 $C_6H_5\text{—}C(=O)\text{—}CH_3 \xrightarrow[CH_3CH_2I]{KOtBu} C_6H_5\text{—}C(=O)\text{—}CH_2CH_2CH_3$

(b) $C_6H_5\text{—}C(=O)\text{—}CH_3 \xrightarrow{NaOCH_3} C_6H_5\text{—}C(CH_3){=}CH\text{—}C(=O)\text{—}C_6H_5$

(c) $C_6H_5\text{—}C(=O)\text{—}CH_3 \xrightarrow[CH_2{=}CH\text{—}CHO]{NaOCH_3} C_6H_5\text{—}C(=O)\text{—}CH_2\text{—}CH_2\text{—}CH_2\text{—}C(=O)H$

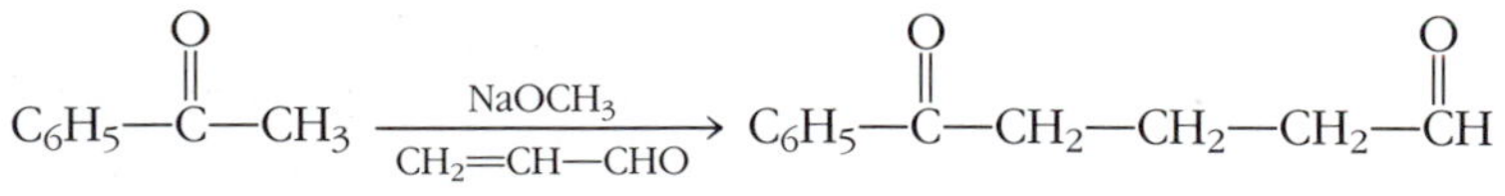

(d) 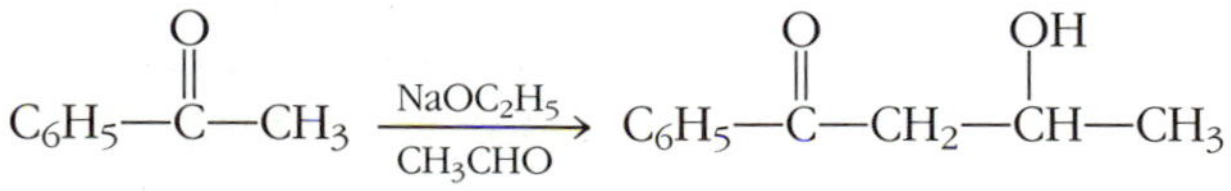 $C_6H_5\text{—}C(=O)\text{—}CH_3 \xrightarrow[CH_3CHO]{NaOC_2H_5} C_6H_5\text{—}C(=O)\text{—}CH_2\text{—}CH(OH)\text{—}CH_3$

(e) 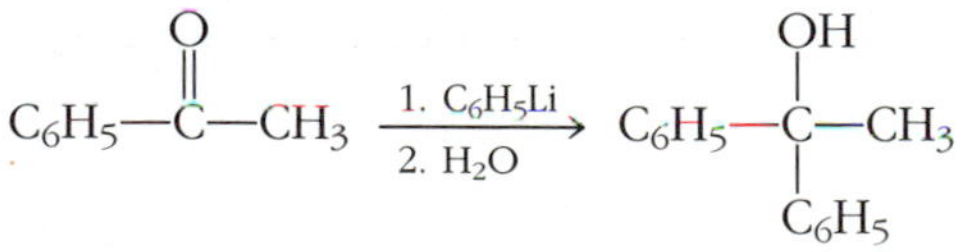 $C_6H_5\text{—}C(=O)\text{—}CH_3 \xrightarrow[2.\ H_2O]{1.\ C_6H_5Li} C_6H_5\text{—}C(OH)(C_6H_5)\text{—}CH_3$

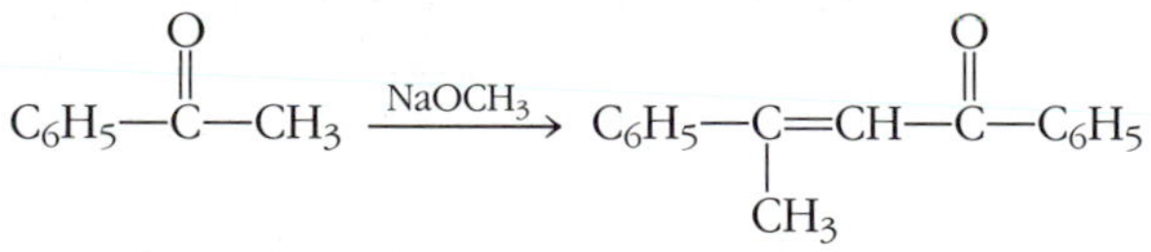

(f) $C_6H_5-\overset{O}{\overset{\|}{C}}-CH_3 \xrightarrow{(C_6H_5)_3P=CH_2} C_6H_5-\overset{CH_2}{\overset{\|}{C}}-CH_3$

(g) $C_6H_5-\overset{O}{\overset{\|}{C}}-CH_3 \xrightarrow[\text{2. } BrCH_2CH_2\overset{\|}{\underset{O}{C}}CH_3]{\text{1. KOtBu}} C_6H_5-\overset{O}{\overset{\|}{C}}-CH_2CH_2CH_2-\overset{O}{\overset{\|}{C}}-CH_3$

(h) $C_6H_5-CH=CH-\overset{O}{\overset{\|}{C}}-C_6H_5 \xrightarrow[CH_3\overset{O}{\overset{\|}{C}}CH_3]{NaOCH_3} CH_3-\overset{O}{\overset{\|}{C}}-CH_2-\underset{C_6H_5}{\underset{|}{CH}}-CH_2-\overset{O}{\overset{\|}{C}}-C_6H_5$

(i) $C_6H_5-CH=CH-\overset{O}{\overset{\|}{C}}-C_6H_5 \xrightarrow[\text{2. } H_2O]{\text{1. } (CH_3)_2CuLi} C_6H_5-\underset{CH_3}{\underset{|}{CH}}-CH_2-\overset{O}{\overset{\|}{C}}-C_6H_5$

14.2 Draw the product expected for each of the following aldol condensations:

(a) $CH_3CH_2-\overset{O}{\overset{\|}{C}}-CH_2CH_3 \xrightarrow{NaOH} ?$

(b) $CH_3CH_2CH_2\overset{O}{\overset{\|}{C}}H \xrightarrow{NaOCH_3} ?$

(c) cyclohexanone $\xrightarrow{KOtBu} ?$

(d) cyclopentane–CHO $\xrightarrow{KOH} ?$

(e) cyclobutanone $\xrightarrow{HCl} ?$

14.3 Draw the major product expected for each of the following aldol condensations:

(a) BR–C₆H₄–$\overset{O}{\overset{\|}{C}}H$ (para) $+ CH_3-\overset{O}{\overset{\|}{C}}-CH_3 \xrightarrow{NaOH} ?$

(b) CH_3–C₆H₄–$\overset{O}{\overset{\|}{C}}H$ (para) + cyclopentanone $\xrightarrow{KOC_2H_5} ?$

(c) $CH_3CH_2\overset{O}{\overset{\|}{C}}H$ + cyclohexanone $\xrightarrow{NaOCH_3} ?$

(d) $C_6H_5\text{—}\overset{O}{\overset{\|}{C}}H + CH_3\text{—}\overset{CH_3}{\overset{|}{C}}H\text{—}CH_2\text{—}\overset{O}{\overset{\|}{C}}\text{—}CH_3 \xrightarrow{NaOH} ?$

(e) CH_3, O (ring) + Cl, CHO (ring) $\xrightarrow{NaOCH_3}$?

(f) $CH_3CH_2CH_2CH_2\text{—}\overset{O}{\overset{\|}{C}}\text{—}CH_3 + C_6H_5\text{—}CHO \xrightarrow{HCl} ?$

(g) $CH_3CH_2CH_2\text{—}\overset{O}{\overset{\|}{C}}H + C_6H_5\text{—}CHO \xrightarrow{NaOH} ?$

14.4 Treatment of a mixture of 2-hexanone and 3-hexanone with base would yield isomeric aldol products. Draw the eight constitutionally isomeric α,β-unsaturated ketones that could be formed.

14.5 Each of the following compounds resulted from an aldol condensation. Draw the structures of the starting aldehydes and ketones.

(a) O (ring) $=CH\text{—}C_6H_5$

(b) $C_6H_5CH{=}CH\text{—}\overset{O}{\overset{\|}{C}}\text{—}CH_2CH_2CH_3$

(c) $CH_3\text{—}\overset{O}{\overset{\|}{C}}\text{—}CH_2\text{—}\overset{OH}{\overset{|}{C}}H\text{—}CH_3$

(d) CH_3O (ring) $CH{=}CH\text{—}\overset{O}{\overset{\|}{C}}$ (ring) Cl

(e) $(C_6H_5)_2C{=}C(CH_3)\text{—}\overset{O}{\overset{\|}{C}}\text{—}C_6H_5$

(f) $CH_3\text{—}\underset{CH_3}{\underset{|}{C}H}\text{—}CH_2(H)C{=}C(CH_3)\text{—}\overset{O}{\overset{\|}{C}}\text{—}CH_2CH_3$

(g) $CH_3CH_2\text{—}\overset{OH}{\overset{|}{C}}H\text{—}\underset{C_6H_5}{\underset{|}{C}H}\text{—}\overset{O}{\overset{\|}{C}}\text{—}CH_2\text{—}C_6H_5$

(h) $C_6H_5\text{—}\overset{O}{\overset{\|}{C}}\text{—}\underset{CH_2CH_3}{\underset{|}{C}H}\text{—}\overset{OH}{\overset{|}{\underset{C_6H_5}{\underset{|}{C}H}}}\text{—}CH_2CH_2CH_3$

14.6 What product or products would be expected in each of the following alkylation reactions?

(a) 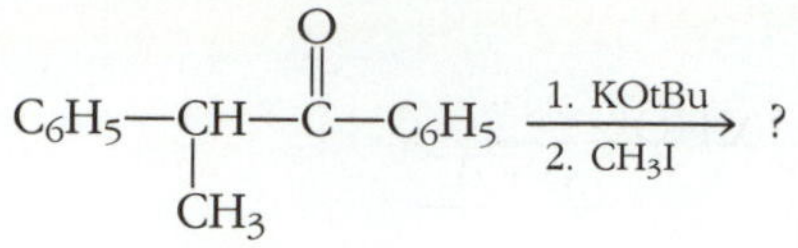 $C_6H_5-CH(CH_3)-C(=O)-C_6H_5 \xrightarrow[\text{2. } CH_3I]{\text{1. KOtBu}} ?$

(b) O, CH_3 $\xrightarrow[\text{2. } CH_3CH_2Br]{\text{1. } NaNH_2} ?$

(c) CH_3, O $\xrightarrow[\text{2. } CH_2{=}CH{-}CH_2Br]{\text{1. NaH}} ?$

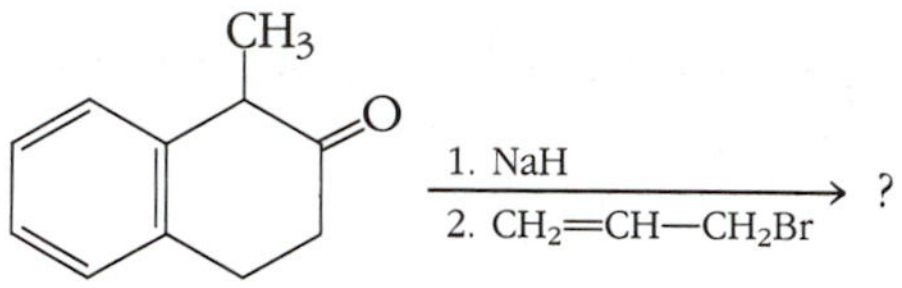

(d) 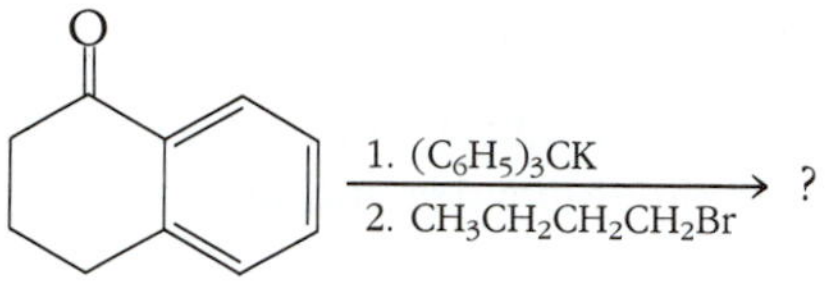 O $\xrightarrow[\text{2. } CH_3CH_2CH_2CH_2Br]{\text{1. } (C_6H_5)_3CK} ?$

(e) 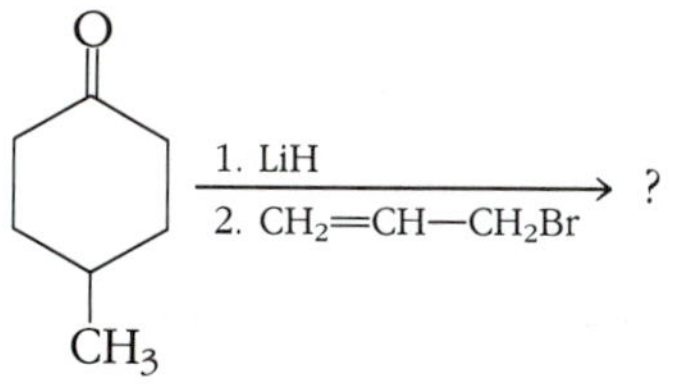 O, CH_3 $\xrightarrow[\text{2. } CH_2{=}CH{-}CH_2Br]{\text{1. LiH}} ?$

(f) 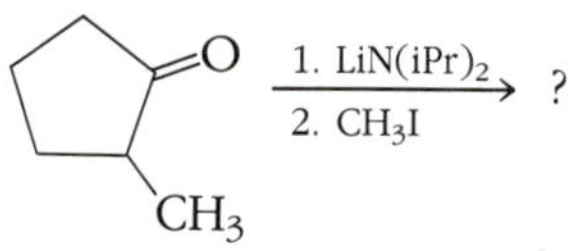 O, CH_3 $\xrightarrow[\text{2. } CH_3I]{\text{1. } LiN(iPr)_2} ?$

(g) 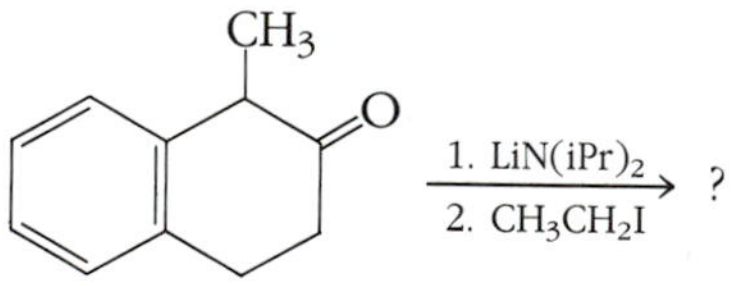 CH_3, O $\xrightarrow[\text{2. } CH_3CH_2I]{\text{1. } LiN(iPr)_2} ?$

(h) 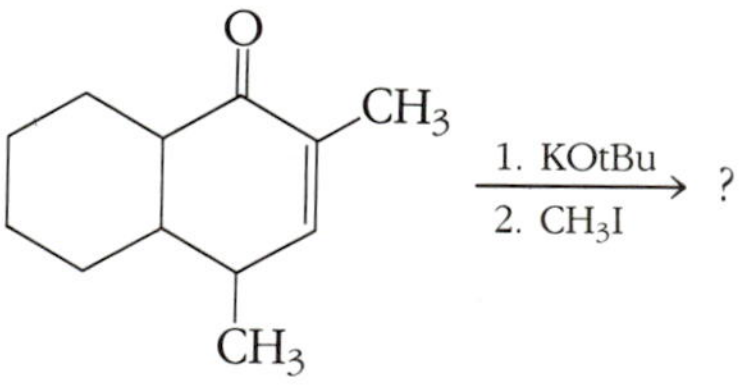 O, CH_3, CH_3 $\xrightarrow[\text{2. } CH_3I]{\text{1. KOtBu}} ?$

(i) 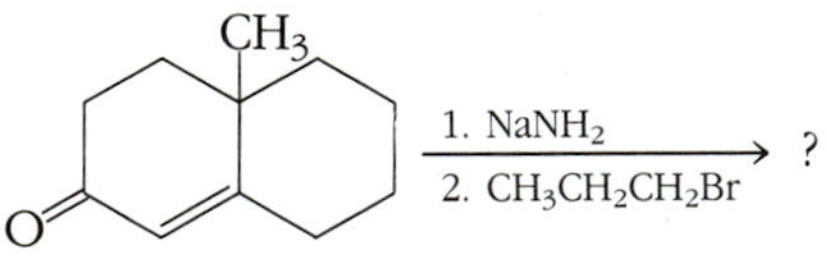 CH_3, O $\xrightarrow[\text{2. } CH_3CH_2CH_2Br]{\text{1. } NaNH_2} ?$

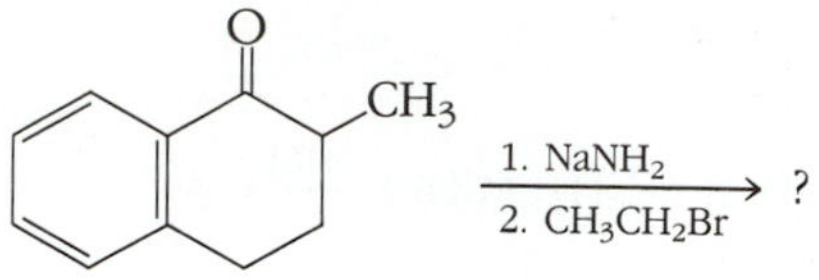

(j) 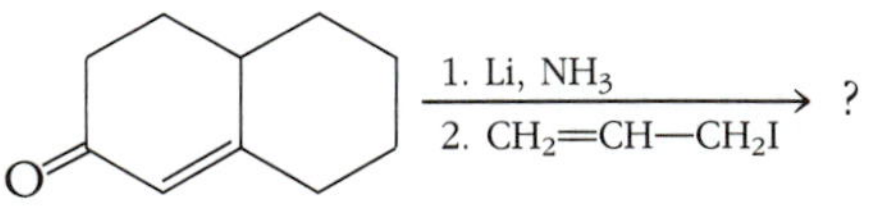 O $\xrightarrow[\text{2. } CH_2{=}CH{-}CH_2I]{\text{1. Li, } NH_3} ?$

14.7 Alkylation of the following ketone would be expected to take a different course depending upon the reaction conditions. Predict the major product in each case.

(a) $\xrightarrow{\text{KOtBu, excess } CH_2{=}CH{-}CH_2Br}$?

(b) $\xrightarrow[\text{2. } CH_2{=}CH{-}CH_2Br]{\text{1. Li(NiPr)}_2}$?

C_6H_5

(c) $\xrightarrow[\text{2. } CH_2{=}CH{-}CH_2Br]{\text{1. NaH}}$?

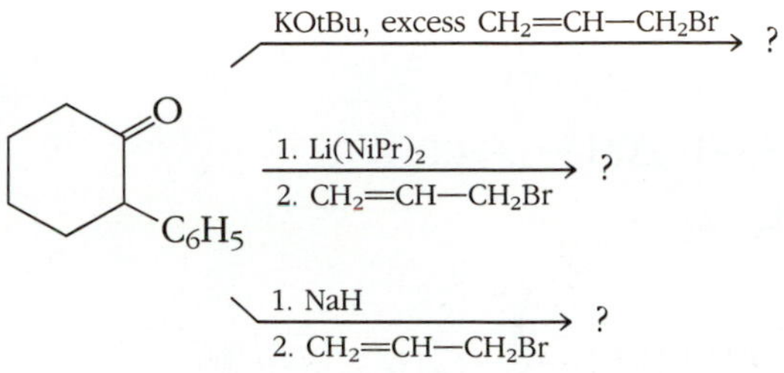

14.8 Alkylation of the following ketone would be expected to take a different course depending upon the reaction conditions. Predict the major product in each case.

(a) $\xrightarrow{\text{KOtBu, excess } CH_3I}$?

(b) $\xrightarrow[\text{2. } CH_3I]{\text{1. LiH}}$?

(c) $\xrightarrow[\text{2. } CH_3I]{\text{1. Li, } NH_3}$?

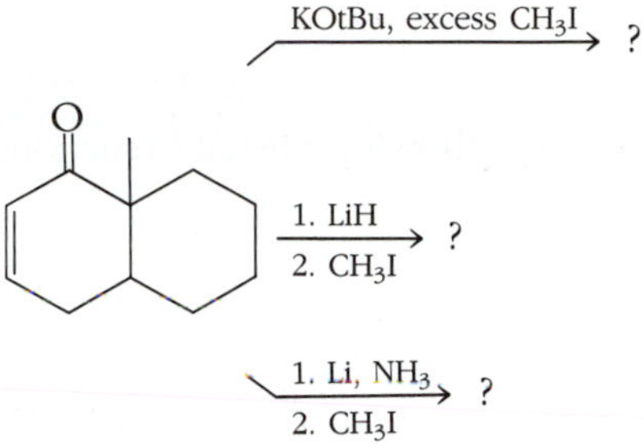

14.9 Alkylation of the following ketone would be expected to take a different course depending upon the reaction conditions. Predict the major product in each case.

(a) $\xrightarrow[\text{2. } C_2H_5Br]{\text{1. } (C_6H_5)_3CK}$?

(b) $\xrightarrow[\text{2. } C_2H_5Br]{\text{1. Li, } NH_3}$?

(c) $\xrightarrow[\text{2. } C_2H_5Br]{\text{1. KOtBu}}$?

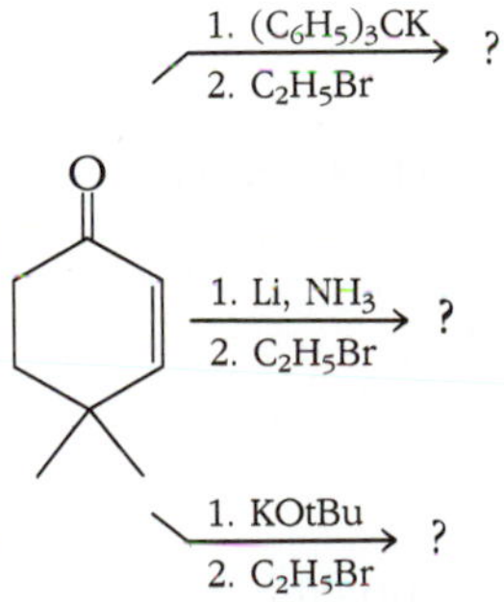

14.10 Each of the following diketones might be produced by two alternative Michael reactions. For each diketone draw the two alternative pairs of starting materials. Which set of starting materials would be expected to give the product in better yield? Why?

(a) $CH_3-C(=O)-CH(C_6H_5)-CH_2-CH_2-C(=O)-CH_2CH_3$

(b) $C_6H_5CH_2-C(=O)-CH_2CH_2CH(C_6H_5)-C(=O)-C_6H_5$

(c) $CH_3-C(CH_3)_2-C(=O)-CH_2CH_2CH_2-CH(=O)$

14.11 Predict the product for each of the following Michael reactions:

(a) 2-cyclohexen-1-one $+ CH_3CH_2-C(=O)-C_6H_5 \xrightarrow{NaOCH_3}$?

(b) cyclohexanone $+ CH_2=CH-C(=O)-C_6H_5 \xrightarrow{NaOCH_3}$?

(c) $CH_3-C(=O)-CH_2-C_6H_5 + C_6H_5-CH=CH-C(=O)-C_6H_5 \xrightarrow{NaOC_2H_5}$?

(d) 2-phenylcyclopentanone $+ CH_2=CH-CH(=O) \xrightarrow{NaOC_2H_5}$?

14.12 The base-catalyzed aldol condensation is completely reversible. Using the various intermediates drawn in Figure 14.2, write the mechanism for the retrograde aldol condensation in base.

14.13 The acid-catalyzed aldol condensation is completely reversible. Using the various intermediates drawn in Figure 14.3, write the mechanism for the retrograde aldol condensation in acid.

14.14 Suggest methods by which 4-phenyl-3-buten-2-one could be converted into each of the indicated compounds. More than one reaction may be necessary in each case.

$$C_6H_5{-}CH{=}CH{-}\overset{\overset{\displaystyle O}{\|}}{C}{-}CH_3$$

(a) $C_6H_5{-}CH_2CH_2{-}\overset{\overset{\displaystyle O}{\|}}{C}{-}CH_3$

(b) $C_6H_5{-}CH_2CH_2CH_2CH_3$

(c) $C_6H_5{-}CH{=}CH{-}CH{=}CH_2$

(d) $C_6H_5{-}CH{=}CH{-}\overset{\overset{\displaystyle OH}{|}}{C}H{-}CH_3$

(e) $C_6H_5{-}\overset{\overset{\displaystyle Br}{|}}{C}H{-}CH_2CH_2CH_3$

(f) $C_6H_5{-}\underset{\underset{\displaystyle CH_3}{|}}{C}H{-}CH_2{-}\overset{\overset{\displaystyle O}{\|}}{C}{-}CH_3$

(g) $C_6H_5{-}CH_2CH_2{-}\overset{\overset{\displaystyle O}{\|}}{C}{-}CH_2CH_3$

(h) $C_6H_5{-}CH_2CH_2{-}\overset{\overset{\displaystyle O}{\|}}{C}{-}OH$

(i) $C_6H_5{-}CH(CH(C_6H_5)C(=O)C_6H_5){-}CH_2{-}\overset{\overset{\displaystyle O}{\|}}{C}{-}CH_3$

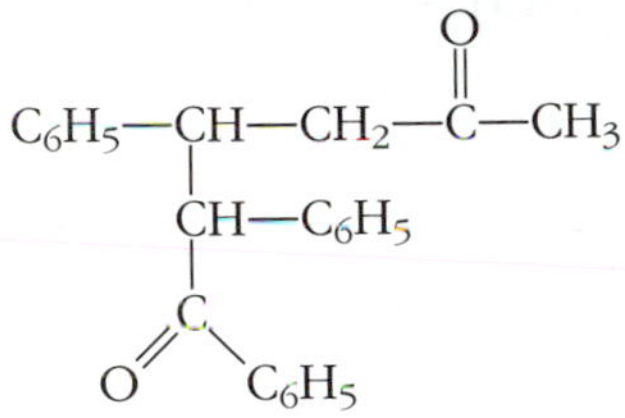

14.15 Suggest methods by which 2-cyclopenten-1-one could be converted into each of the following compounds. More than one reaction may be necessary in each case.

(a) O (b) OH (c)

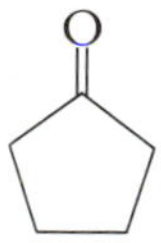

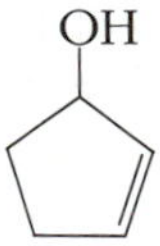

(d) 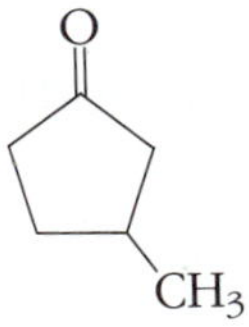 O CH$_3$ (e) O CH$_3$ (f) HO CH$_3$

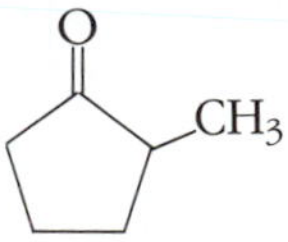

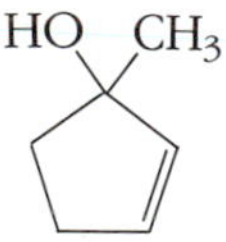

(g) HO CH$_3$ (h) CH$_3$ (i) CH$_2$ CH$_3$

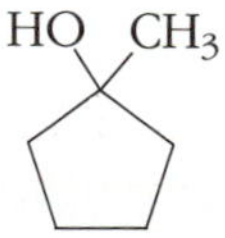

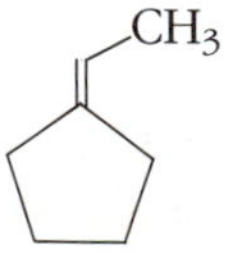

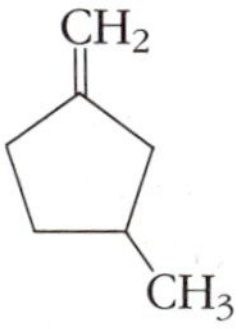

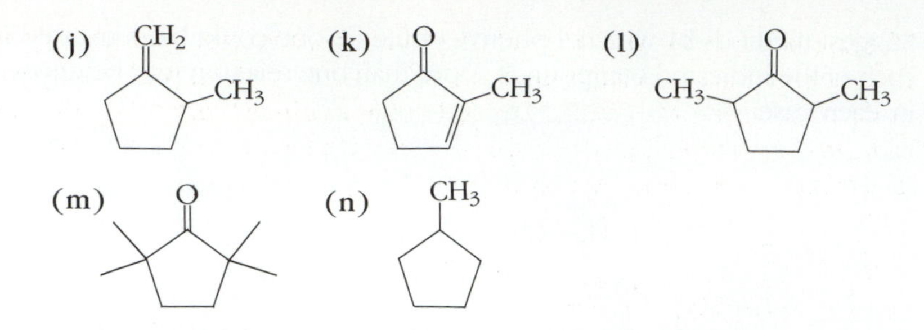

14.16 The base-catalyzed reaction between acetaldehyde and isobutyraldehyde is temperature-dependent. At 25°C the major product is 3-hydroxy-2,2-dimethylbutanal, but at 80°C the major product is 4-methyl-2-pentenal. Which is the product of kinetic control? What structural feature favors the formation of 4-methyl-2-pentenal at 80°C?

14.17 The choice of base would be important in any attempt to alkylate 3,3-dimethyl-2-butanone.

$$(CH_3)_3C{-}\overset{O}{\overset{\|}{C}}{-}CH_3 \xrightarrow[2.\ CH_3I]{1.\ base} (CH_3)_3C{-}\overset{O}{\overset{\|}{C}}{-}CH_2CH_3$$

Both potassium *tert*-butoxide and sodium amide should work reasonably well. In contrast, sodium methoxide would probably be unsatisfactory. Why?

14.18 In Section 14.4 the mechanism is shown for base-catalyzed equilibration of α,β- and β,γ-unsaturated carbonyl compounds. Draw the mechanism for interconversion of the isomers under acidic conditions.

14.19 Enolate ions generated from α,β-unsaturated ketones can take part in aldol condensations. The following reaction proceeded in 85% yield.

Compound A $\xrightarrow[C_6H_5CHO]{KOH}$ 5,5-dimethyl-3-(CH=CH—C_6H_5)-2-cyclohexenone

Deduce the structure of compound A and draw the enolate ion involved in the reaction.

14.20 The initially formed products of aldol condensations sometimes undergo further reaction to give a good yield of a single product, particularly when this involves ring formation. The following two reactions of acetone illustrate such reactions. Deduce the individual reaction steps and write the mechanisms for these reactions.

(a) 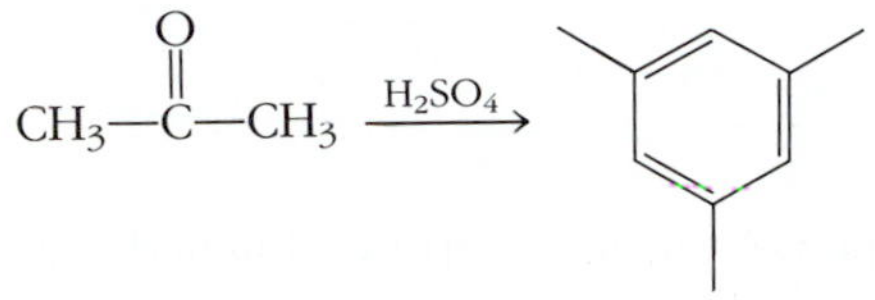

(b) 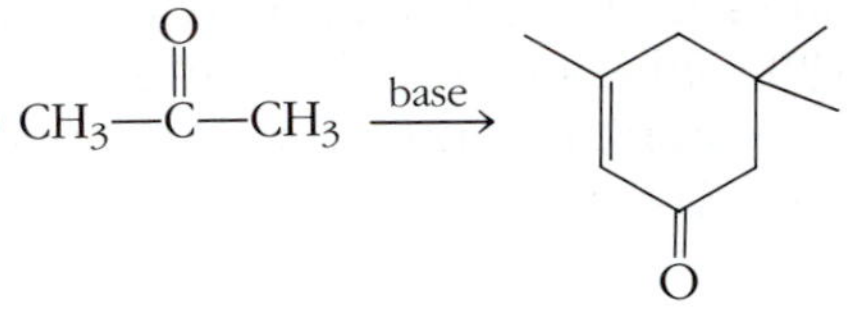

14.21 Treatment of the following α,β-unsaturated ketone with base resulted in the loss of the substituent on the five-membered ring. Draw a reaction mechanism which accounts for this.

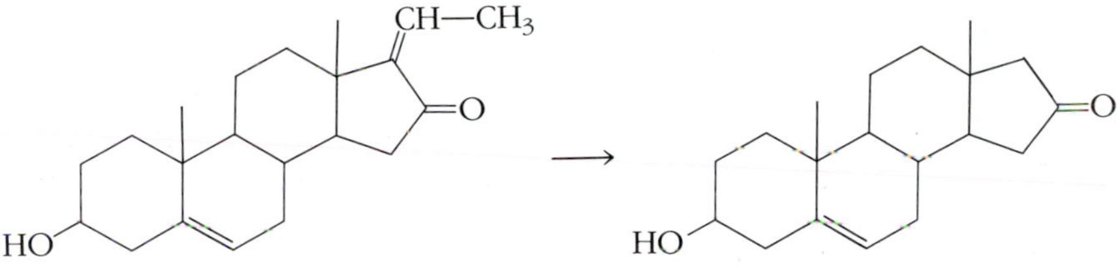

14.22 When 4-hydroxy-3-methyl-4-phenyl-2-butanone, an aldol product, is treated with acid, it undergoes the expected dehydration to give 3-methyl-4-phenyl-3-buten-2-one.

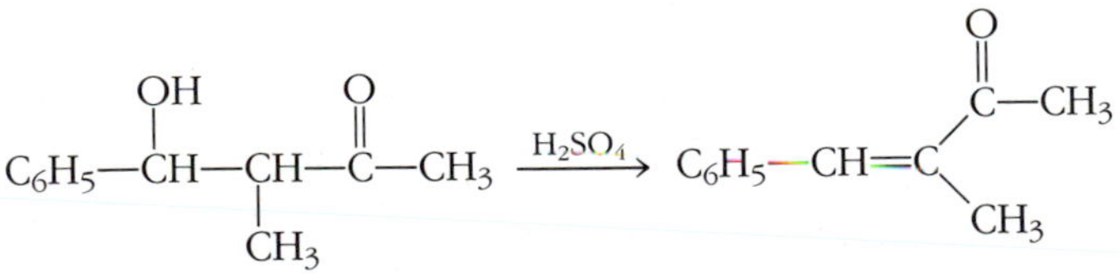

However, under basic conditions in the presence of excess 2-butanone the product is 5-phenyl-4-penten-3-one.

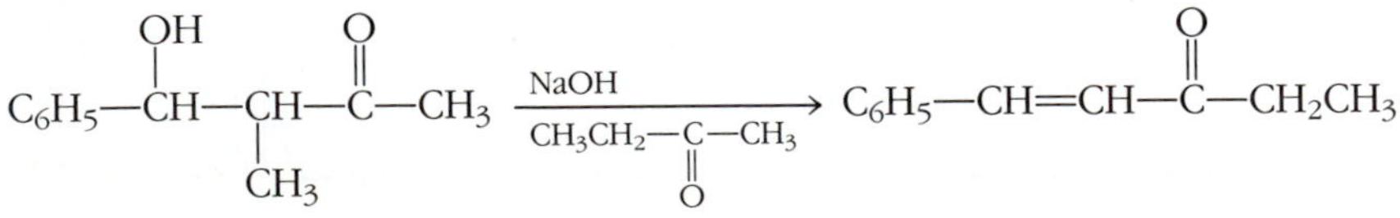

Suggest a mechanism which accounts for this product.

14.23 Alkylation of the following β,γ-unsaturated ketone was shown in Section 14.3 to give a monoalkylated product in 55% yield. In fact, this alkylation is highly sensitive to reaction conditions

1. base
2. CH_3I

and a very complex mixture of products can be obtained under different conditions. Draw the structures of the following products that might be expected in the alkylation with methyl iodide. (Disregard stereochemistry.)

(a) One C_{11} compound (not starting material)
(b) Four isomeric C_{12} compounds
(c) Four isomeric C_{13} compounds
(d) One C_{14} compound

Of these 10 possibilities, 8 have actually been isolated from the mixture of products.

14.24 When a ketone has two reactive α carbons, aldol condensation could occur at both positions. For example, the reaction of cyclohexanone in the presence of excess benzaldehyde and base leads to a compound with the molecular formula $C_{20}H_{18}O$. What is the structure of this compound?

14.25 Treatment of 2-butanone with potassium *tert*-butoxide and methyl iodide could lead to a variety of products, and the NMR spectra of three of these are shown below in Figures 14.5–14.7. Identify the compounds on the basis of their spectra.

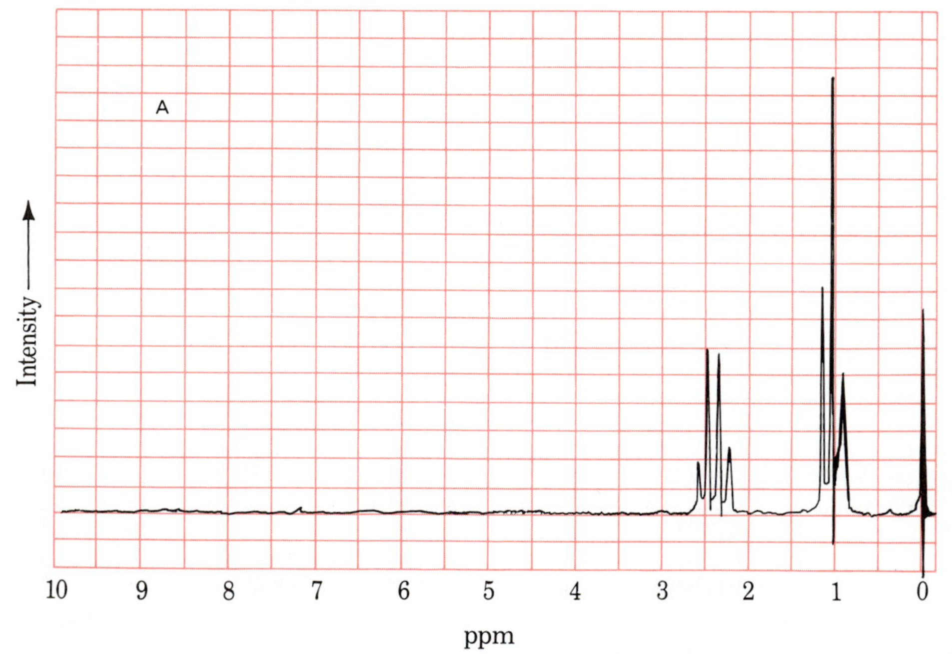

Figure 14.5 Proton NMR Spectrum of Compound A (Problem 14.25).

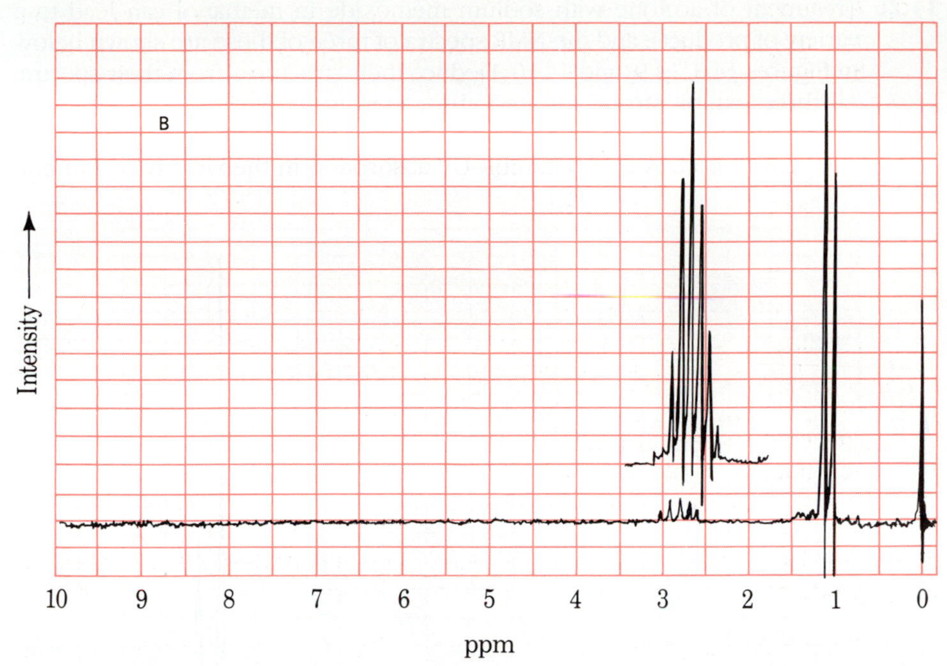

Figure 14.6 Proton NMR Spectrum of Compound B (Problem 14.25).

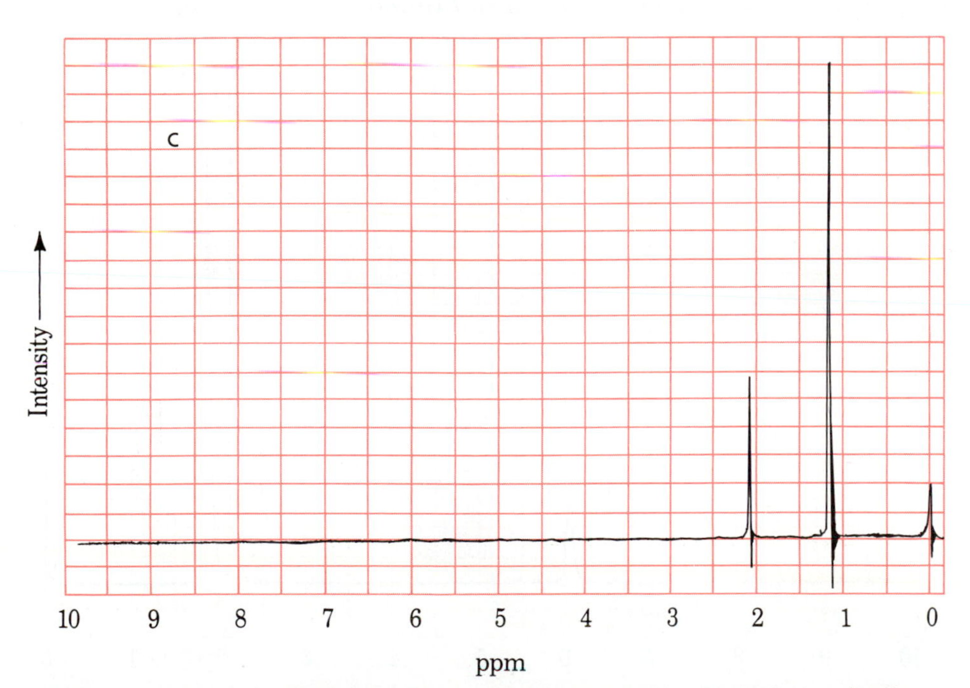

Figure 14.7 Proton NMR Spectrum of Compound C (Problem 14.25).

14.26 Treatment of acetone with sodium methoxide in methanol can lead to a variety of products, and the NMR spectra of three of these are shown below in Figures 14.8, 14.9, and 14.10. Deduce their structures from their spectra. All three exhibit strong infrared absorption in the vicinity of 1700 cm^{-1}. Compound **C** is the only one of the three which also exhibits a peak above 3200 cm^{-1}. Both **A** and **B** exhibit UV absorption in the vicinity of 240 nm.

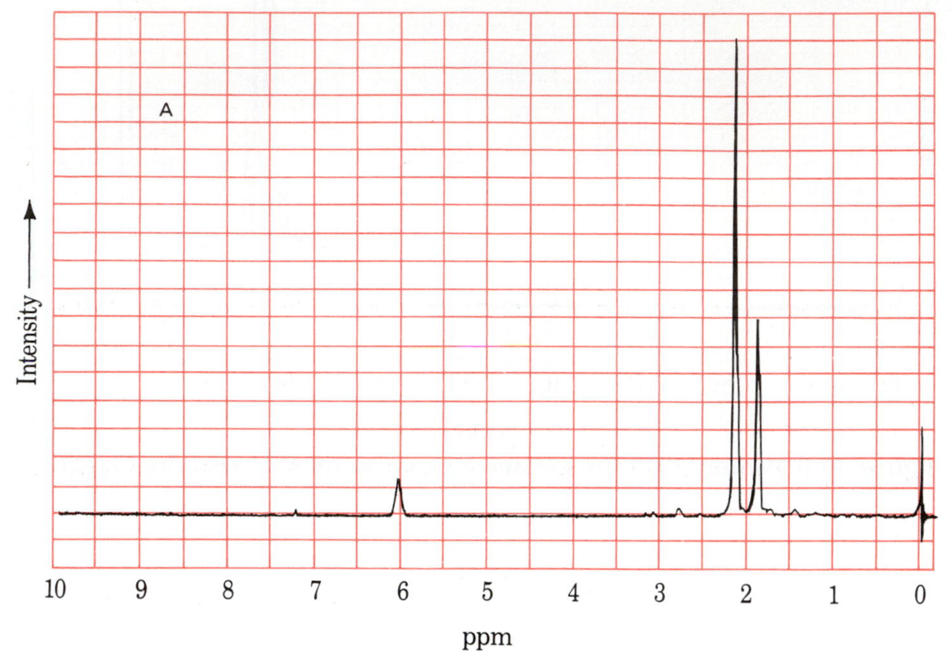

Figure 14.8 Proton NMR Spectrum of Compound A (Problem 14.26).

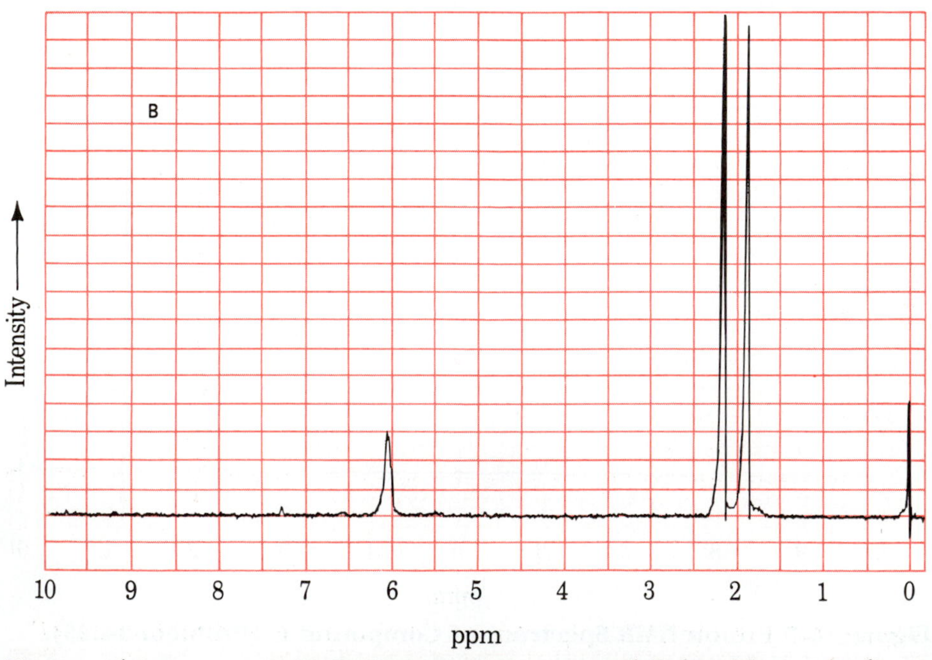

Figure 14.9 Proton NMR Spectrum of Compound B (Problem 14.26).

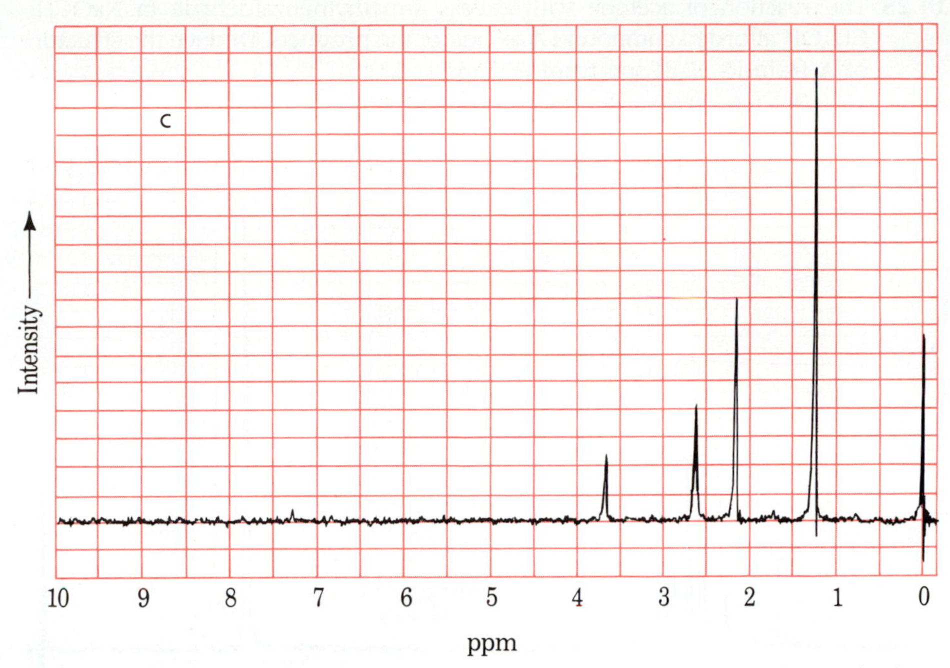

Figure 14.10 Proton NMR Spectrum of Compound C (Problem 14.26).

14.27 Compound **A** ($C_{10}H_{10}O$) was isolated as the product of an aldol condensation. Deduce the structure of **A** from its NMR spectrum (Figure 14.11).

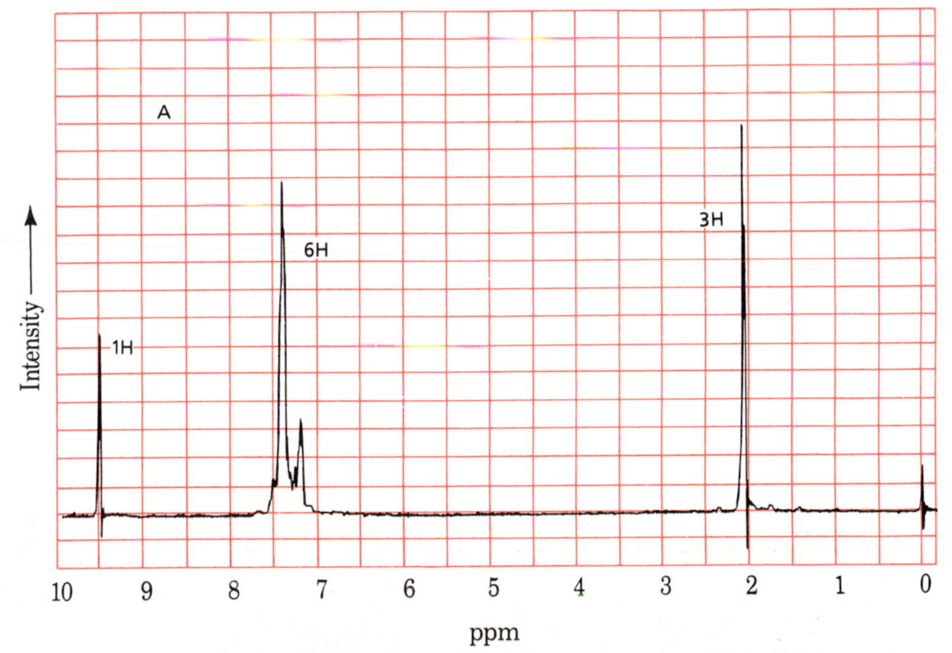

Figure 14.11 Proton NMR Spectrum of Compound A (Problem 14.27).

14.28 The reaction of acetone with excess 4-methylbenzaldehyde in $NaOCH_3$/CH_3OH afforded compound **A** as one of the products. Deduce the structure of **A** from its NMR spectrum (Figure 14.12).

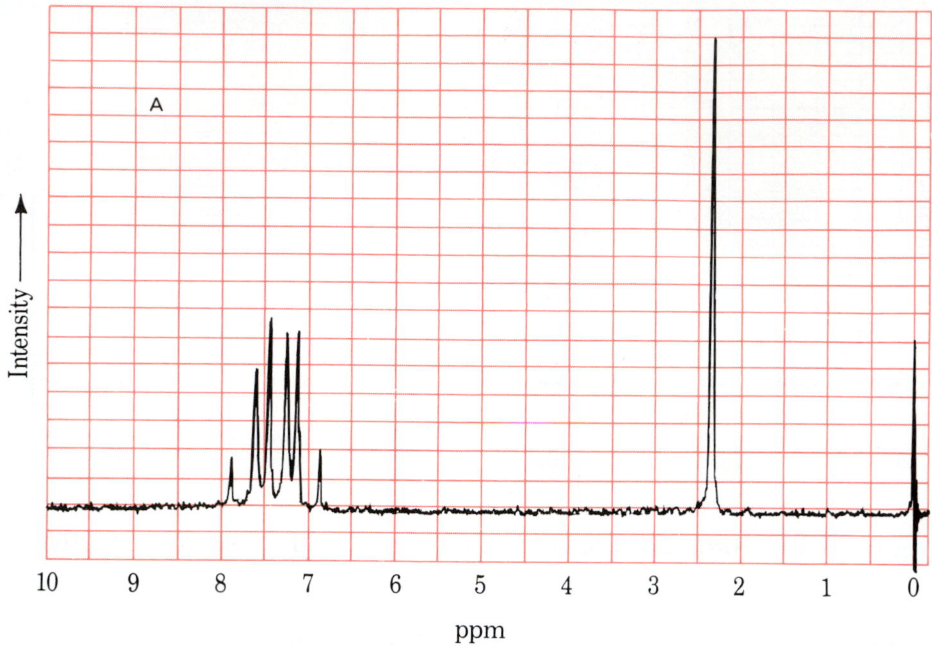

Figure 14.12 Proton NMR Spectrum of Compound A (Problem 14.28).

14.29 For each of the reactions in Problem 14.6, explain how you would employ NMR spectroscopy to verify that you had indeed obtained the expected product.

CHAPTER 15

ETHERS

In Chapter 11 we showed that alcohols can be considered derivatives of water. When *both* hydrogens of water are replaced by R groups, the resulting compound is an **ether.**

H—O—H	R—O—H	R—O—R
Water	An alcohol	An ether

The two R groups of an ether may be different, and they may be aryl or alkenyl as well as alkyl. The sequence of substitution from water to alcohols to ethers allows you to anticipate the chemical and physical properties of ethers. The absence of an OH group means that ethers will not exhibit the weak acidity shown by water and by alcohols. On the other hand, the ether oxygen atom still has two nonbonding pairs of electrons, so an ether can act as either a weak base or a nucleophile.

The best known member of this class of compounds is diethyl ether.

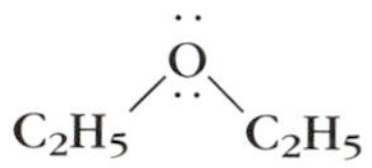

$$C_2H_5-\ddot{\underset{\cdot\cdot}{O}}-C_2H_5$$

Diethyl ether (sometimes called ethyl ether or just "ether") is an extremely useful solvent in chemical industry and in the laboratory, but it is better known to most people as an anesthetic. The use of diethyl ether as a general anesthetic was first publicly demonstrated at Massachusetts General Hospital (Boston) in the mid 1800s, but its potential was recognized as early as the sixteenth century, when the Swiss physician (and alchemist) Paracelsus wrote that " . . . it has an agreeable taste, so that even chickens take it gladly, and thereafter fall asleep for a long time, awakening unharmed . . ." Although it was the anesthetic of choice for many years, diethyl ether has largely been replaced by other compounds in modern medical procedures.

15.1 NOMENCLATURE

Acyclic ethers can be named by either of two methods. In the first method you simply list the two substituents on oxygen (as in *diethyl ether*), and in the second method you name an ether systematically as a substituted hydrocarbon. The first method is illustrated by the following examples. (Notice that the two groups are listed in alphabetical order.)

$CH_3-O-CH_2CH_3$

Ethyl methyl ether

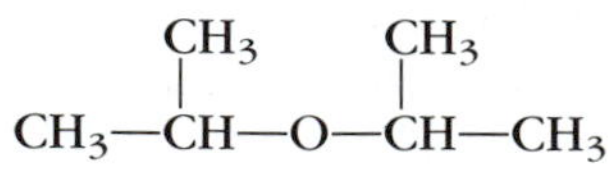

$CH_3-CH(CH_3)-O-CH(CH_3)-CH_3$

Diisopropyl ether

$C_6H_{11}-O-C_2H_5$

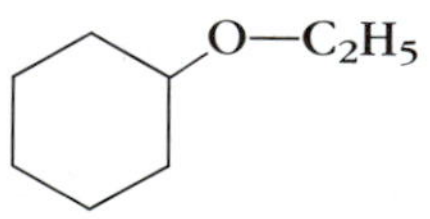

Cyclohexyl ethyl ether

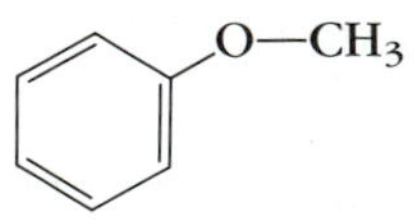

$C_6H_5-O-CH_3$

Methyl phenyl ether

Systematic Names

A systematic naming procedure is sometimes needed for complex structures and for polyfunctional ethers, and the rules are basically the same as those we presented in Section 13.1 for aldehydes and ketones.

Nomenclature of Ethers

1. *Locate the principal carbon chain.* This is the longest carbon chain in the molecule *which includes the principal group or has it as a substituent.* In *polyfunctional* compounds, the principal group is selected according to the following priorities:

Aldehyde > ketone > alcohol > ether

2. *Name and number the substituents* on the principal chain.
   (a) The principal chain is numbered in the direction that affords the lower number for *the carbon bearing the principal group.* This number is prefixed to the name of the principal chain with a hyphen. If the location of the principal group does not distinguish between two different ways of numbering, the substituents on the principal chain are evaluated so that the lower number results at the occasion of the first difference.
   (b) When there is more than one substituent, they are listed in alphabetical order. If several of the substituents are identical, this is indicated by the numerical prefixes *di-, tri-, tetra-, penta-, hexa-, etc.*
   (c) Ether substituents (—OR) are named as *alkoxy groups.* If the R group has four or fewer carbons and is unsubstituted, the "*-yl*" of the alkyl group, R, is replaced by *-oxy* in order to obtain the correct name, *alkoxy,* for the —OR group. Otherwise, the suffix *-oxy* is appended to the name of the alkyl group.

3. *Specify the stereochemistry* for any center in the molecule that could exist in two alternative configurations.

The following examples illustrate the use of this systematic nomenclature. The principal chain is highlighted in each case.

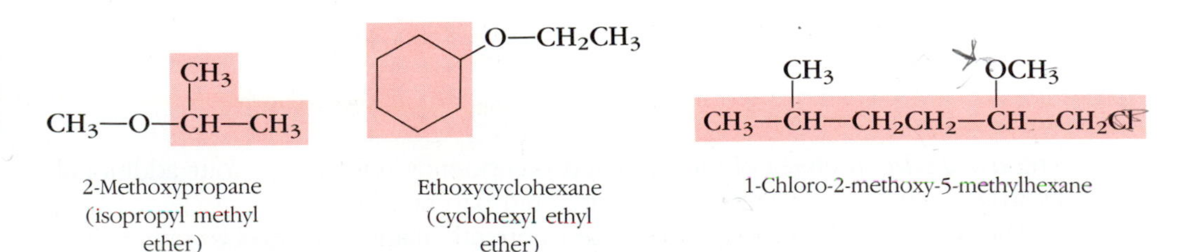

2-Methoxypropane (isopropyl methyl ether)

Ethoxycyclohexane (cyclohexyl ethyl ether)

1-Chloro-2-methoxy-5-methylhexane

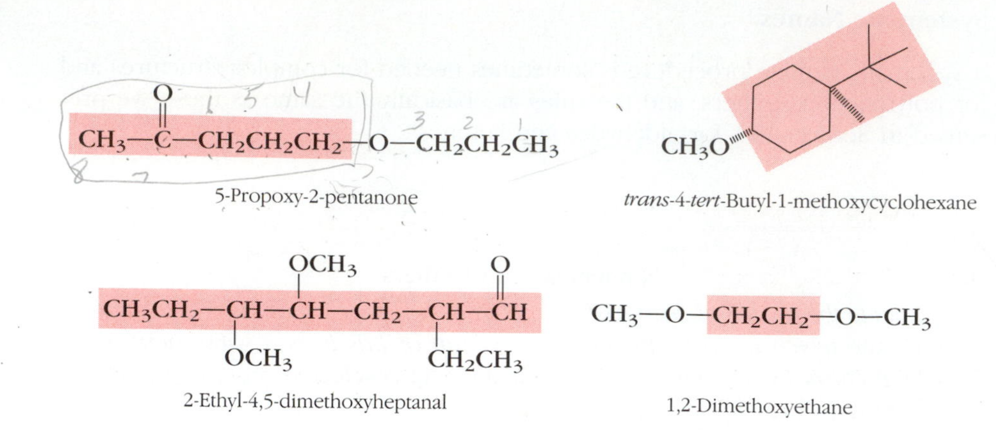

Cyclic Ethers

Cyclic ethers are named according to their ring size. The following parent compounds illustrate the most common ring sizes:

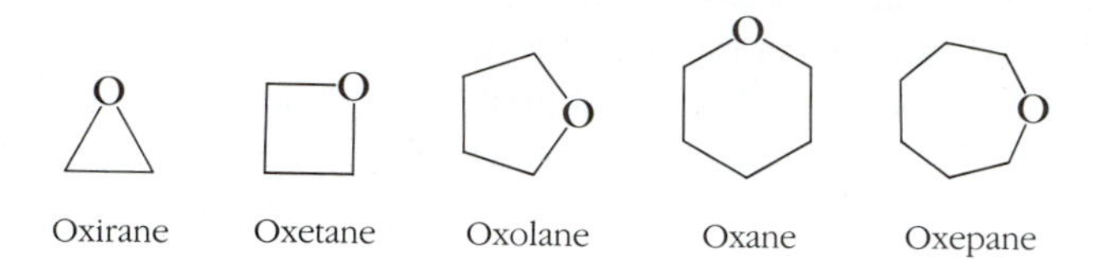

These are examples of *heterocyclic* compounds (from the Greek *heteros,* "other"), that is, cyclic compounds in which one or more of the atoms in the ring is not carbon. (We will discuss heterocyclic compounds more extensively in Chapter 24.) The *heteroatom* is always numbered as the 1 position in these systems, and other substituents on the ring are cited accordingly. For example:

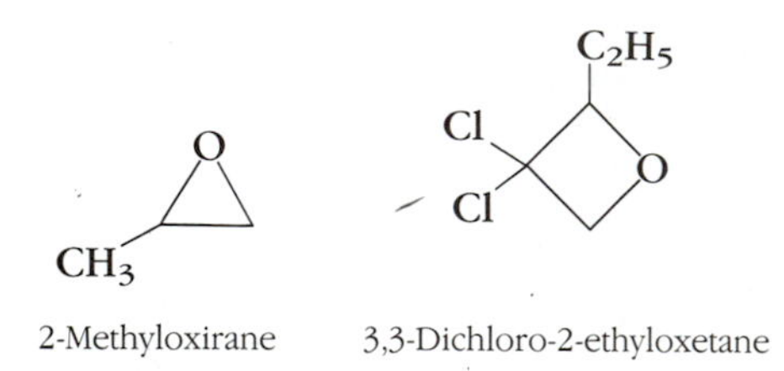

The five- and six- membered ring ethers also have common names that are used frequently.

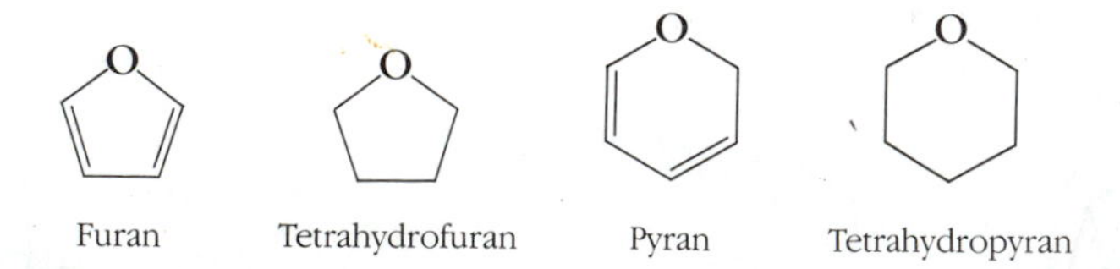

The *tetrahydro* prefixes of the saturated compounds refer to the four additional hydrogens that are absent in the unsaturated parent ethers.

The three-membered-ring ethers are frequently named as *epoxides* rather than as oxiranes. The oxygen is considered a substituent on two adjacent carbon atoms.

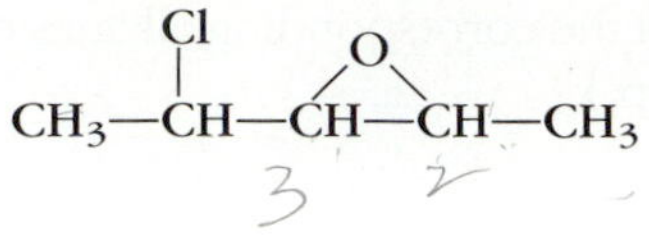

$$CH_3-\underset{\displaystyle |}{\overset{\displaystyle Cl}{CH}}-\overset{O}{CH-CH}-CH_3$$

4-Chloro-2,3-epoxypentane

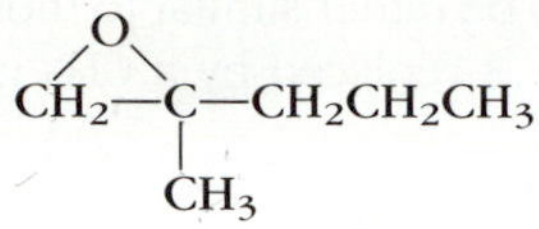

$$\overset{O}{CH_2-C}(CH_3)-CH_2CH_2CH_3$$

1,2-Epoxy-2-methylpentane

In general you will find that chemists name compounds by the easiest method available. When a common name can be used, it is often more convenient than a longer systematic name.

EXERCISE 15.1

Name the following compounds. Give alternative names if possible.

(a) 3-Chloro-3-methyl six-membered ring containing O (structure: ring with Cl and CH_3 on one carbon, O in ring)

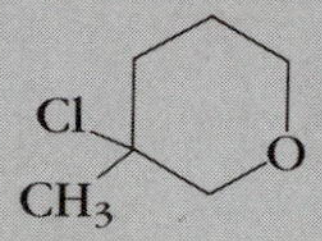

(b) $CH_3CH_2\overset{O}{CH-CH}-CH_3$

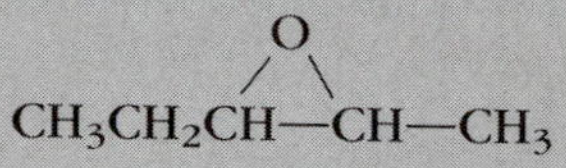

(c) $CH_3CH_2-O-CH_2CH_2-O-CH_2CH_3$

(d) $CH_3-\overset{O}{\overset{\|}{C}}-\overset{CH_3}{\overset{|}{CH}}-\underset{OCH_2CH_3}{\underset{|}{CH}}-CH_2Cl$

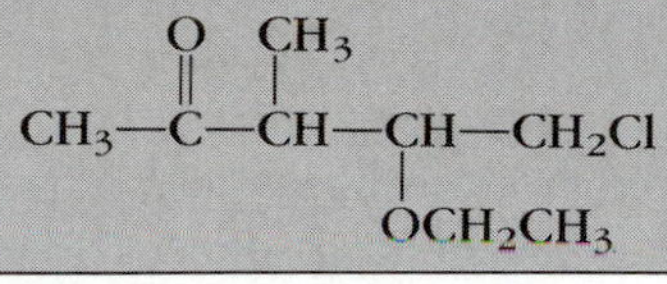

EXERCISE 15.2

Draw each of the following compounds:

(a) 4-Chloro-3-methoxyhexane
(b) 5-Chloro-2,2-dimethyloxane
(c) 2-Methyloxetane
(d) 3-Methyltetrahydrofuran
(e) 2-Isopropoxycyclohexanone

15.2 PHYSICAL PROPERTIES

The electronegative oxygen atom of an ether is the major influence on physical properties, particularly solubility. The nonbonded electrons on oxygen permit ethers to undergo hydrogen bonding with other compounds, for example with alcohols:

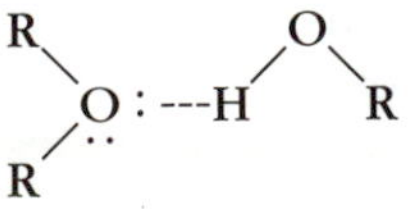

$$R_2\ddot{O}\text{:} \cdots H-O-R$$

Ether Alcohol

Hydrogen bonding between two molecules of the ether is not possible because there are no "acidic" hydrogens. For this reason the melting and boiling points of ethers tend to be rather similar to those of the corresponding alkanes (in which the ether oxygen is replaced by a CH_2 group.)

Heptane	Methyl pentyl ether	Butyl ethyl ether	Dipropyl ether
mp −91°C	—	−124°C	−122°C
bp 98°C	99–100°C	92°C	91°C

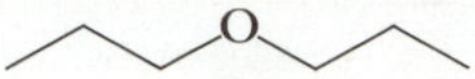

The preceding series of ethers also shows that melting and boiling points of these linear compounds do not show a large dependence on the location of the oxygen atom along the chain. The molecular shape (i.e., degree of branching) has a much greater influence. For example, the following branched ethers both have higher melting points and lower boiling points than their linear isomers.

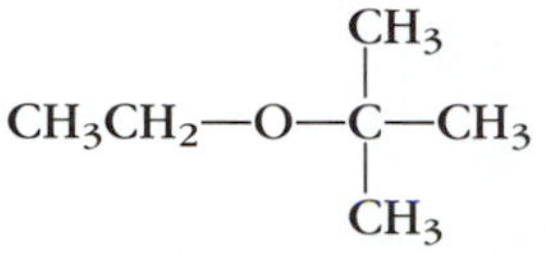

$CH_3CH_2—O—C(CH_3)_2—CH_3$ $\quad$ $CH_3—CH(CH_3)—O—CH(CH_3)—CH_3$

tert-Butyl ethyl ether (mp −94°C, bp 70°C) Diisopropyl ether (mp −86°C, bp 69°C)

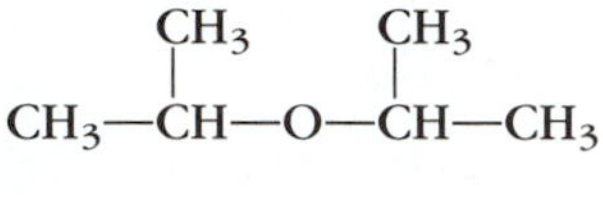

The absence of hydrogen bonding between ether molecules is also reflected by boiling points that are much lower than those of isomeric alcohols. This is illustrated by the following comparisons.

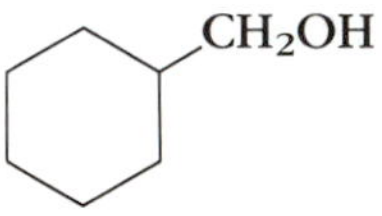

Cyclohexylmethanol ($C_6H_{11}CH_2OH$)	Cyclohexyl methyl ether ($C_6H_{11}OCH_3$)	1-Octanol $CH_3(CH_2)_6CH_2OH$	Dibutyl ether $CH_3CH_2CH_2CH_2OCH_2CH_2CH_2CH_3$
$C_7H_{14}O$	$C_7H_{14}O$	$C_8H_{18}O$	$C_8H_{18}O$
bp 184–186°C	133°C	194–195°C	142°C

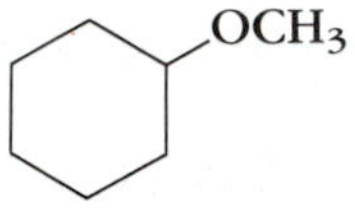

Ethers As Solvents

The nonbonded electrons of ethers can act as electron donors toward other molecules. Furthermore, ethers are relatively unreactive compounds, and these characteristics combine to make them excellent solvents for organic reactions. Both organic and inorganic compounds are relatively soluble in ethers. Ethers are only slightly polar, and van der Waals forces result in favorable interactions with nonpolar organic compounds. On the other hand, the oxygen atom of an ether can form hydrogen bonds to the OH groups of polar organic compounds (alcohols, carboxylic acids). Consequently, ethers tend to be excellent solvents for both polar and nonpolar organic compounds.

The ability of ethers to function as Lewis bases results in favorable interactions with metal ions.

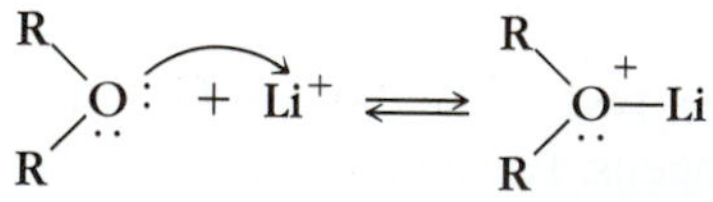

$$R_2\ddot{O}: + Li^+ \rightleftharpoons R_2\overset{+}{\ddot{O}}—Li$$

TABLE 15.1 Ethers Used as Solvents in Organic Reactions

Compound	IUPAC Name	Common Name(s)	bp, °C
$C_2H_5OC_2H_5$	Diethyl ether	Ether	35
(five-membered ring with O)	(Oxolane)	Tetrahydrofuran (THF)	64–65
$CH_3OCH_2CH_2OCH_3$	1,2-Dimethoxyethane (ethylene glycol dimethyl ether)	Glyme	85
(six-membered ring with two O, 1,4-positions)	1,4-Dioxane	Dioxane	101
$CH_3OCH_2CH_2OCH_2CH_2OCH_3$	Diethylene glycol dimethyl ether	Diglyme	161

Consequently, ethers are also excellent solvents for organic reactions involving reagents such as $LiAlH_4$ and various organometallic compounds such as alkyllithium and Grignard reagents.

Some of the most commonly used ethereal solvents are listed in Table 15.1. All but diethyl ether are also water-soluble, and this provides still another important use for these compounds, namely, as cosolvents for reactions that require aqueous conditions but involve water-insoluble organic compounds. For example, aqueous tetrahydrofuran (THF) is the standard solvent for oxidation of intermediate organoboranes in the hydroboration of alkenes (Section 5.11).

$$>C{=}C< \xrightarrow{BH_3} -\overset{H}{\underset{|}{\overset{|}{C}}}-\overset{\overset{|}{B}-}{\underset{|}{\overset{|}{C}}}- \xrightarrow[\text{NaOH, THF}]{H_2O_2} -\overset{H}{\underset{|}{\overset{|}{C}}}-\overset{OH}{\underset{|}{\overset{|}{C}}}-$$

15.3 REACTIONS OF ETHERS

Simple ethers are characterized by their relatively low reactivity. They lack the reactive functional groups (carbonyl and hydroxyl) that are present in more reactive oxygen-containing compounds. Nevertheless, ethers can be protonated by acids, and it is this acid-base equilibrium that gives you the key to understanding the reactivity of ethers.

$$R{-}\ddot{\underset{..}{O}}{-}R \overset{H^+}{\rightleftharpoons} R{-}\overset{+}{\underset{..}{O}}(H){-}R$$

Nucleophilic Displacement and Elimination

Under neutral or basic conditions, nucleophilic substitution or elimination reactions of an ether would require an *alkoxide ion* to function as a leaving group.

$$\text{Nuc}^- \;\; \text{R—}\ddot{\text{O}}\text{—R} \longrightarrow \text{Nuc—R} + {}^-\text{:}\ddot{\text{O}}\text{R}$$

$$\text{Base}^- + \text{H—}\overset{|}{\underset{|}{\text{C}}}\text{—}\overset{|}{\underset{|}{\text{C}}}\text{—}\ddot{\text{O}}\text{—R} \longrightarrow \text{>C=C<} + {}^-\text{:}\ddot{\text{O}}\text{R}$$

But alkoxide ions are poor leaving groups, and these processes do not normally occur. In fact we can state as a general rule that *ethers are unreactive under basic conditions.*

Acidic conditions present a quite different situation, because ether oxygen is readily protonated by strong acid. This means that a neutral alcohol molecule (ROH instead of its conjugate base, $^-$OR) could function as the leaving group.

$$\text{R—}\ddot{\text{O}}\text{—R} \xrightleftharpoons{\text{H}^+} \text{R—}\overset{+}{\ddot{\text{O}}}(\text{H})\text{—R} \xrightarrow{\text{Nuc}^-} \text{R—}\ddot{\text{O}}\text{—H} + \text{R—Nuc}$$

Recall (Chapter 12) that most strong nucleophiles are negatively charged. Under the acidic conditions necessary for reaction of ethers, these anionic nucleophiles would be protonated and rendered unreactive. As a result, even in acid there are only a few reactions that are typical of ethers; these are reactions in which the nucleophile is either neutral water, bromide ion, or iodide ion. Because HBr and HI are strong acids, bromide and iodide ions are among the few anionic nucleophiles that remain unprotonated under strongly acidic conditions. The conditions normally needed to carry out cleavage of ethers at a convenient rate are quite stringent: concentrated HBr or HI at elevated temperatures.

$$CH_3CH_2\text{—}\overset{CH_3}{\overset{|}{CH}}\text{—}(CH_2)_9CH_2OCH_3 \xrightarrow{HI} CH_3CH_2\text{—}\overset{CH_3}{\overset{|}{CH}}\text{—}(CH_2)_9CH_2I \quad 97\%$$

The mechanism of the preceding reaction merits discussion, because you would have expected the iodide ion to attack the protonated ether at the methyl group rather than at the more hindered CH_2 group. In fact this pathway does yield the iodide, as shown in the following equation, because the initially formed alcohol reacts further with HI (Sections 11.9, 12.8):

$$CH_3CH_2\text{—}\overset{CH_3}{\overset{|}{CH}}\text{—}(CH_2)_9\text{—}CH_2OCH_3 \xrightleftharpoons{H^+} CH_3CH_2\text{—}\overset{CH_3}{\overset{|}{CH}}\text{—}(CH_2)_9\text{—}CH_2\text{—}\overset{H}{\overset{|}{\underset{+}{O}}}\text{—}CH_3$$

(minor pathway, I^- attacks CH_2):

$$\longrightarrow CH_3CH_2\text{—}\overset{CH_3}{\overset{|}{CH}}\text{—}(CH_2)_9\text{—}CH_2\text{—}I$$

(major pathway, I^- attacks CH_3):

$$\longrightarrow CH_3CH_2\text{—}\overset{CH_3}{\overset{|}{CH}}\text{—}(CH_2)_9\text{—}CH_2\text{—}OH \xrightleftharpoons{H^+} CH_3CH_2\text{—}\overset{CH_3}{\overset{|}{CH}}\text{—}(CH_2)_9\text{—}CH_2\text{—}\overset{+}{O}H_2 \xrightarrow{I^-} CH_3CH_2\text{—}\overset{CH_3}{\overset{|}{CH}}\text{—}(CH_2)_9\text{—}CH_2\text{—}I$$

This type of cleavage reaction has frequently been used with aryl alkyl ethers. Nucleophilic attack on the protonated ether will occur only at the *saturated* carbon (Chapter 12), and this produces an alkyl halide along with the corresponding phenol.

$$\text{Ar—O—R} \xrightleftharpoons{\text{HBr (or HI)}} \text{Ar—}\overset{+}{\underset{\text{H}}{\text{O}}}\text{—R} \xrightarrow{\text{Br}^-} \text{ArOH} + \text{R—Br}$$

The reaction is illustrated by the following example with HBr:

$$o\text{-}C_6H_4(OCH_3)(OH) \xrightarrow{\text{48\% HBr}} o\text{-}C_6H_4(OH)_2 \quad (+CH_3Br) \quad 85\text{–}87\%$$

Acid-catalyzed cleavage of ethers does not always involve direct displacement (S_N2) reactions, but will also proceed smoothly if relatively stable carbocations can be formed. As an example, the cleavage of *tert*-butyl ethers proceeds by ionization of the protonated ether to the *tert*-butyl cation. The cation can then be attacked by a nucleophile (substitution) or it can transfer H^+ to some basic species (elimination).

$$\text{R—O—C(CH}_3)_3 \xrightleftharpoons{\text{HBr}} \text{R—}\overset{+}{\underset{\text{H}}{\text{O}}}\text{—C(CH}_3)_3 \rightleftharpoons \text{ROH} + (CH_3)_2\overset{+}{C}\text{—CH}_3 \rightarrow \text{Br—C(CH}_3)_3 \text{ or } (CH_3)_2C{=}CH_2$$

Tertiary ethers are cleaved more readily than the other alkyl ethers shown in the preceding reactions, and less stringent reaction conditions can be used. As a result the second alkyl group of the ether can often be isolated as the alcohol.

Epoxides provide an exception to the low reactivity of simple ethers. The strain of the three-membered ring is relieved upon breaking the ether linkage, and cleavage occurs even with fairly mild acid. Notice that *inversion of configuration* results from backside attack by the nucleophile.

$$\text{cycloheptene oxide} \xrightarrow{CH_3CO_2H} \textit{trans}\text{-2-(acetoxy)cycloheptanol } (\text{O—}\overset{\text{O}}{\overset{\|}{\text{C}}}\text{—CH}_3,\ \text{OH})$$

Epoxides are also the only simple ethers which can be cleaved by nucleophiles under basic or even neutral conditions.

$$CH_3\text{—}\overset{\text{O}}{\widehat{CH\text{—}CH_2}} \xrightarrow{\text{NaBr, }H_2O} CH_3\text{—}\underset{59\%}{\overset{\text{OH}}{CH}}\text{—}CH_2\text{—Br} \ (+\ CH_3\text{—}\underset{3\%}{\overset{\text{Br}}{CH}}\text{—}CH_2OH)$$

Notice that ring opening occurs preferentially via attack of the nucleophilic bromide ion at the less substituted carbon of the epoxide.

Oxidation of Ethers

Although ethers sometimes undergo reactions with common oxidizing agents, they are considerably less reactive than such functional groups as primary and secondary alcohols. Even the relatively reactive epoxide ring survives conditions used for the oxidation of alcohols.

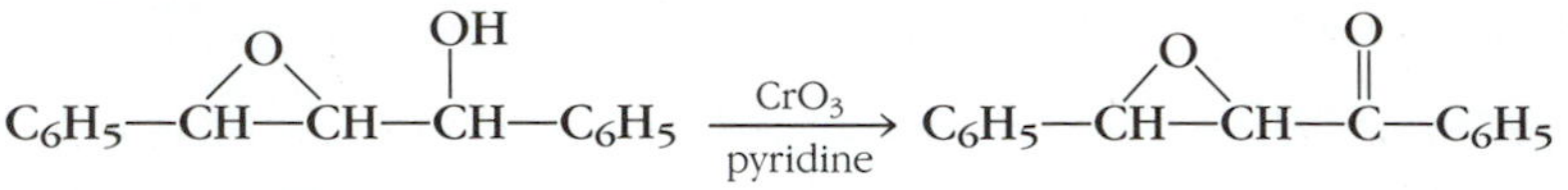

$$C_6H_5{-}\overset{O}{\widehat{CH{-}CH}}{-}\overset{OH}{\overset{|}{C}H}{-}C_6H_5 \xrightarrow[\text{pyridine}]{CrO_3} C_6H_5{-}\overset{O}{\widehat{CH{-}CH}}{-}\overset{O}{\overset{\|}{C}}{-}C_6H_5 \quad 65\%$$

In this case the CrO_3-pyridine reagent was used in order to avoid acidic conditions that might have opened the epoxide ring.

Although ethers can ordinarily be described as unreactive toward typical oxidizing agents, they do react very slowly with atmospheric oxygen. This reaction is of no synthetic value, but it constitutes a major safety hazard. The oxidation products are organic peroxides, which can decompose violently, and serious explosions have occurred when ethers have been handled improperly. Safe storage requires an appropriate metal container and the exclusion of air.

EXERCISE 15.3

What products would be expected in the following reactions?

(a) 3-methylanisole (benzene ring bearing OCH_3 and, meta to it, CH_3) $\xrightarrow{\text{conc. HI}}$

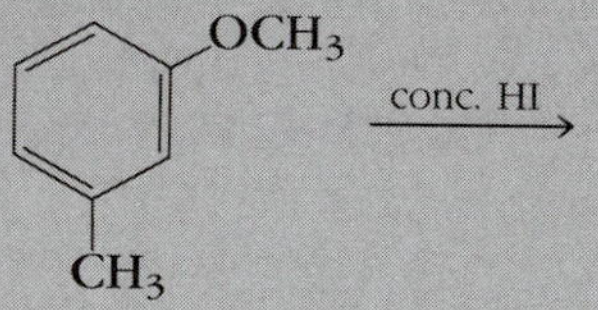

(b)
$$CH_3{-}\underset{CH_3}{\overset{CH_3}{\underset{|}{\overset{|}{C}}}}{-}O{-}\underset{CH_3}{\overset{CH_3}{\underset{|}{\overset{|}{C}}}}{-}CH_3 \xrightarrow{\text{conc. HBr}}$$

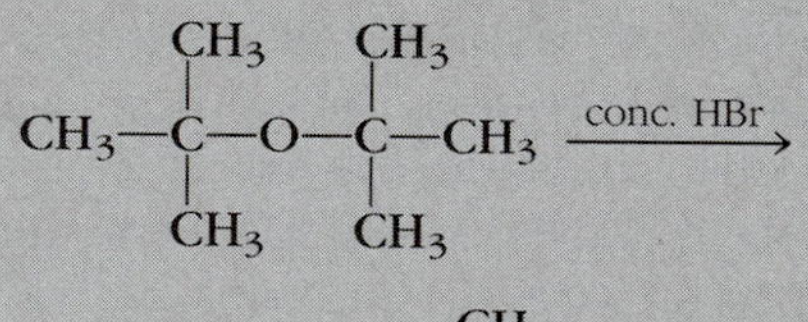

(c)
$$CH_3CH_2CH_2{-}O{-}\overset{CH_3}{\overset{|}{C}H}{-}CH_3 \xrightarrow{\text{conc. HBr}}$$

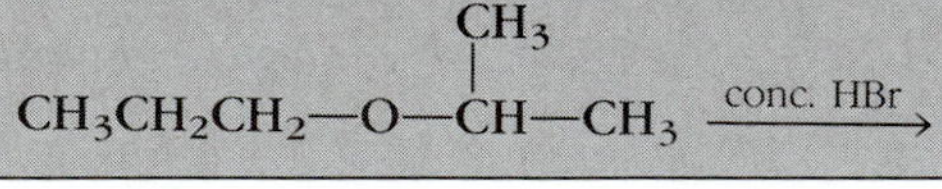

EXERCISE 15.4

Epoxides sometimes yield different products when ring opening is carried out under acidic conditions versus neutral or basic conditions. What product is expected to predominate in the reaction of the epoxide of 2-methylpropene with NaBr? A different product is expected for the reaction with HBr. What intermediate is attacked by bromide ion in the latter case?

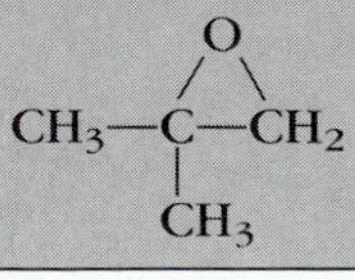

$$CH_3{-}\overset{O}{\widehat{\underset{CH_3}{\underset{|}{C}}{-}CH_2}} \qquad \text{1,2-Epoxy-2-methylpropane}$$

15.4 PREPARATION OF ETHERS

For the preparation of ethers we are concerned with the formation of *carbon-oxygen* bonds. This occurs by attack of some species with a nucleophilic oxygen atom on a second reactant. The methods summarized below either are nucleophilic substitution reactions (by direct displacement or by a carbocation mechanism) or else involve addition to a carbon-carbon double bond.

Acid-Catalyzed Condensation of Alcohols

When an alcohol is heated with a strong acid such as H_2SO_4, an ether can be formed. Protonation of the alcohol renders it susceptible to nucleophilic attack by a second molecule of alcohol, which generates the ether in its protonated form. Transfer of H^+ to another species then yields the ether.

$$\underset{\text{Alcohol}}{R{-}\ddot{O}H} \underset{}{\overset{H^+}{\rightleftharpoons}} \underset{\text{Protonated alcohol}}{R{-}\overset{+}{O}H_2} \overset{HOR}{\rightleftharpoons} \underset{\text{Protonated ether}}{R{-}\overset{H}{\overset{|}{O^+}}{-}R} \overset{ROH}{\rightleftharpoons} \underset{\text{Ether}}{R{-}\ddot{O}{-}R}$$

The reaction sequence is reversible, and the yield of ether varies widely for different starting alcohols. You will notice that the protonated alcohol is the same intermediate that is formed in acid-catalyzed dehydration of an alcohol (Section 5.13), so the reaction could easily follow several different pathways to give products. The best yields are often obtained when the product can be distilled from the reaction mixture. This method produces *symmetrical* ethers, because both alkyl groups are generated from a single alcohol. Its most important use is for the industrial synthesis of relatively simple ethers, although it sometimes works well under laboratory conditions.

$$CH_3CH_2CH_2CH_2CH_2{-}OH \xrightarrow{H_2SO_4} CH_3CH_2CH_2CH_2CH_2{-}O{-}CH_2CH_2CH_2CH_2CH_3 \quad 50\%$$

EXERCISE 15.5

Acids such as HCl are unsatisfactory for the conversion of alcohols to ethers. Why?

EXERCISE 15.6

Acid condensation of alcohols is unsatisfactory for preparing unsymmetrical ethers. What products would be expected if a mixture of 1-propanol and 1-butanol were heated with H_2SO_4?

Solvolysis of Alkyl Halides or Sulfonates

When an alkyl halide or sulfonate is dissolved in an alcohol and heated, nucleophilic substitution yields the corresponding ether. The alcohol is often used as a solvent, so this procedure has found its greatest use in preparing methyl and ethyl ethers (the corresponding alcohols of which are inexpensive). The solvolysis reaction occurs smoothly at a convenient rate when the alkyl halide can ionize to a fairly stable carbocation.

Alkylation of Alkoxide Ions

Ethers can often be prepared by alkylation of the salt of the alcohol with an alkyl halide or sulfonate.

$$RO:^- \xrightarrow{R'-X} R-\ddot{O}-R' + X^-$$

This is a direct displacement reaction. It is restricted to primary and secondary halides and sulfonates, because elimination occurs exclusively with tertiary derivatives (Section 12.6). Even secondary derivatives often give poor yields of the ether, and best results are obtained with primary and methyl halides or sulfonates.

The reaction of an alkoxide with an alkyl halide is frequently described as the Williamson synthesis of an ether (Section 11.8). Ordinarily, the alcohol is first converted to the alkoxide ion by reaction with sodium metal or by treatment with a strong base; the alkyl halide or sulfonate is added in a second step. Sometimes, however, the alcohol, base, and alkyl derivative are all mixed together at one time; the alcohol is often used as solvent. The following examples illustrate the preparation of ethers by this method.

$$CH_3-(CH_2)_5-OH \xrightarrow[2.\ CH_3CH_2I]{1.\ Na} CH_3-(CH_2)_5-O-CH_2CH_3 \quad 92\%$$

$$\text{cyclohexanol (C}_6\text{H}_{11}\text{OH)} \xrightarrow[2.\ CH_3-C_6H_4-CH_2Br]{1.\ NaNH_2} C_6H_{11}-O-CH_2-C_6H_4-CH_3 \quad 44\%$$

$$CH_3CH_2\underset{}{\overset{CH_3}{\overset{|}{C}H}}-CH_2-OSO_2C_6H_5 \xrightarrow[CH_3OH]{NaOCH_3} CH_3CH_2\overset{CH_3}{\overset{|}{C}H}-CH_2-O-CH_3 \quad 65\%$$

$$CH_3OH \xrightarrow[2.\ Cl-C_5H_9]{1.\ Na} CH_3O-C_5H_9 \text{ (cyclopentyl)} \quad 29\%$$

The last reaction illustrates the low yields which typically result when a secondary halide or sulfonate is used.

Addition to Alkenes

Simple ethers can sometimes be prepared by the acid-catalyzed addition of an alcohol to an alkene (Section 5.8). Phenols can sometimes be used as well, and this is illustrated in the following example:

$$C_6H_5OH \xrightarrow[H_2SO_4]{(CH_3)_2CH{=}CH_2} C_6H_5O-\underset{CH_3}{\underset{|}{\overset{CH_3}{\overset{|}{C}}}}-CH_3 \quad 61\%$$

The reaction takes place by initial protonation of the alkene to generate the *tert*-butyl cation, which then undergoes nucleophilic attack by the phenol.

$$(CH_3)_2C{=}CH_2 \xrightarrow{H^+} (CH_3)_2\overset{+}{C}{-}CH_3 \xrightarrow{C_6H_5\ddot{O}H} (CH_3)_2C(\overset{+}{H\ddot{O}}C_6H_5){-}CH_3 \xrightarrow[\text{transfer}]{\text{proton}} (CH_3)_2C(:\ddot{O}C_6H_5){-}CH_3$$

The strongly acidic conditions greatly restrict the use of the preceding method, and a more reliable procedure for the addition of an alcohol to an alkene in the laboratory is *oxymercuration* (Section 5.11). When water is used as solvent, the product of the reaction is an alcohol, but if the solvent is an alcohol, an ether is formed.

$$C_6H_5{-}CH{=}CH_2 \xrightarrow[\text{2. NaBH}_4]{\text{1. Hg(O}\overset{O}{\overset{\|}{C}}\text{CH}_3)_2,\ CH_3OH} C_6H_5{-}CH(OCH_3){-}CH_3 \quad 97\%$$

$$CH_3{-}C(CH_3)_2{-}CH{=}CH_2 \xrightarrow[\text{2. NaBH}_4]{\text{1. Hg(O}\overset{O}{\overset{\|}{C}}\text{CH}_3)_2,\ C_2H_5OH} CH_3{-}C(CH_3)_2{-}CH(OC_2H_5){-}CH_3 \quad 72\%$$

The results in these two examples agree with our previous suggestion that the reaction proceeds via development of positive charge (and subsequent nucleophilic attack) on the more substituted carbon atom of the double bond. Despite this orientation (which corresponds to a polar addition), no rearranged product was observed in the second example. Consequently, the oxymercuration method for preparing ethers provides a valuable alternative to reactions that proceed via carbocations.

Epoxides

In Section 11.6 we showed that these three-membered ring ethers are versatile intermediates for organic synthesis. Epoxides are usually prepared by oxidation of the corresponding alkene, but in some cases they can also be formed by the reaction of an aldehyde or ketone with a sulfur ylide (Section 13.8):

$$\text{cycloheptanone} \xrightarrow{(CH_3)_2S{=}CH_2} \text{spiro epoxide} \quad 97\%$$

Direct oxidation of alkenes to epoxides is normally carried out with a peroxycarboxylic acid (although alkaline H_2O_2 can be used to epoxidize α,β-unsaturated carbonyl compounds, Section 13.13). The most commonly used peroxyacids are:

$$CH_3{-}\overset{O}{\overset{\|}{C}}{-}OOH \qquad CF_3{-}\overset{O}{\overset{\|}{C}}{-}OOH \qquad C_6H_5{-}\overset{O}{\overset{\|}{C}}{-}OOH \qquad m\text{-}ClC_6H_4{-}\overset{O}{\overset{\|}{C}}{-}OOH$$

Peracetic acid (CH_3CO_3H)

Trifluoroperacetic acid (CF_3CO_3H)

Perbenzoic acid ($C_6H_5CO_3H$)

m-Chloroperbenzoic acid ($ClC_6H_4CO_3H$)

The following reactions illustrate the use of peracids for epoxidation:

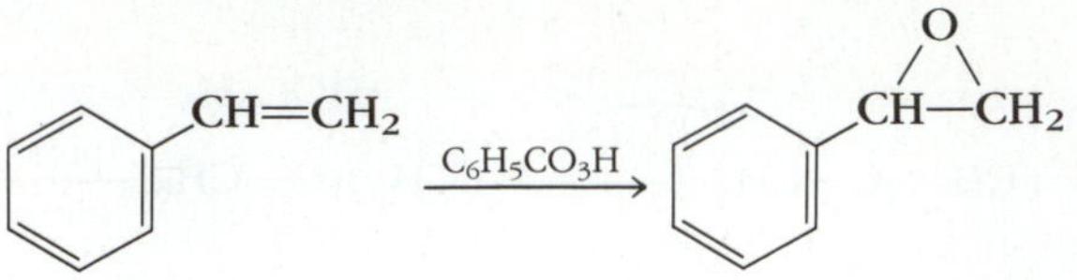

$$C_6H_5-CH{=}CH_2 \xrightarrow{C_6H_5CO_3H} C_6H_5-\overset{O}{CH-CH_2} \qquad 69\text{–}75\%$$

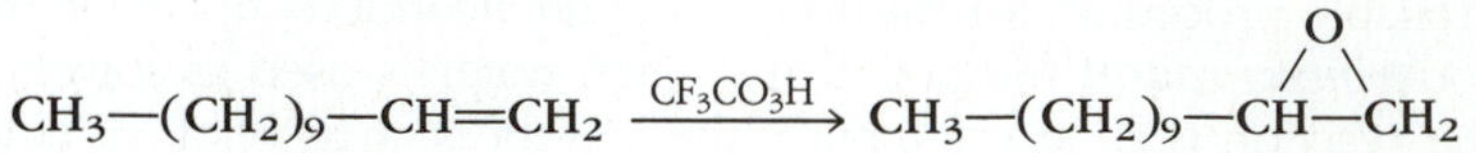

$$CH_3-(CH_2)_9-CH{=}CH_2 \xrightarrow{CF_3CO_3H} CH_3-(CH_2)_9-\overset{O}{CH-CH_2} \qquad 90\%$$

$$CH_3-(CH_2)_5-CH{=}CH_2 \xrightarrow{ClC_6H_4CO_3H} CH_3-(CH_2)_5-\overset{O}{CH-CH_2} \qquad 81\%$$

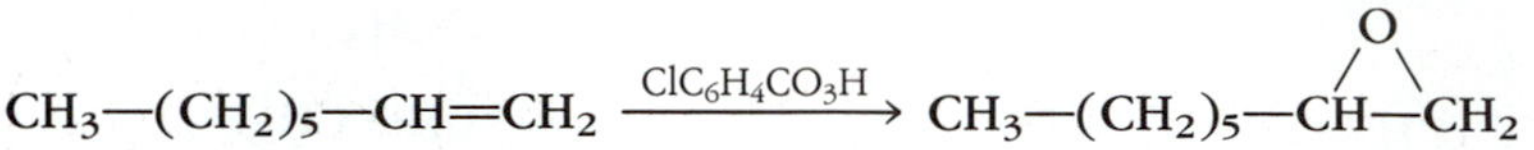

An alternative way of forming epoxides from alkenes involves initial formation of a halo alcohol, which subsequently cyclizes by an intramolecular nucleophilic displacement reaction.

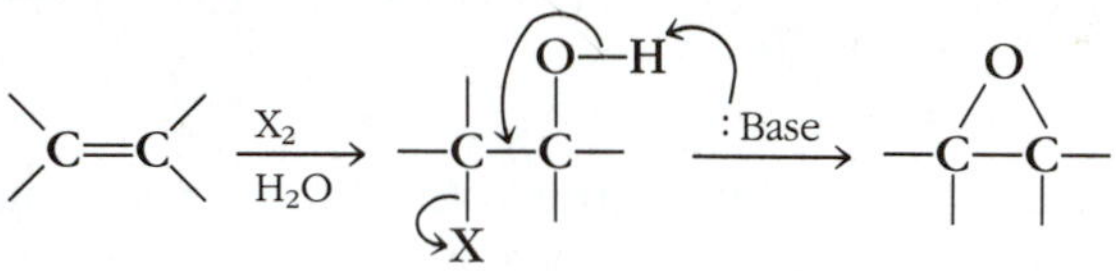

$$>C{=}C< \xrightarrow[H_2O]{X_2} -\underset{X}{C}-\overset{O-H}{C}- \xrightarrow{:Base} -\overset{O}{C-C}-$$

The halo alcohol can be prepared by the reaction of the alkene with Cl_2/H_2O, Br_2/H_2O or NBS/H_2O (Section 5.8). Cyclization is then carried out under mild alkaline conditions to avoid acid-catalyzed opening of the epoxide.

$$CH_2{=}CH-\overset{OH}{C}-CH_2-Cl \xrightarrow{NaOH} CH_2{=}CH-\overset{O}{CH-CH_2} \qquad 84\%$$

$$HO-CH_2CH_2CH_2CH_2-Cl \xrightarrow{H_2O} \text{tetrahydrofuran} \qquad 74\%$$

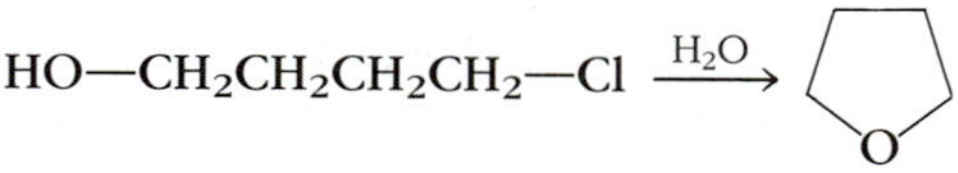

The second example shows that intramolecular nucleophilic substitution reactions can also be used to form cyclic ethers other than epoxides.

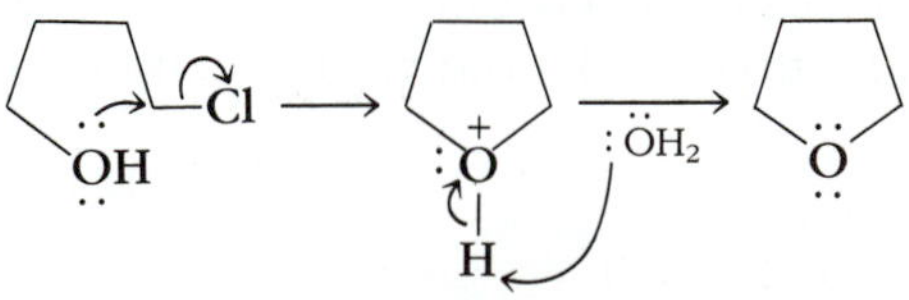

$$\text{HO}(CH_2)_4\text{Cl} \longrightarrow \text{cyclic oxonium } (\overset{+}{O}-H) \xrightarrow{:\ddot{O}H_2} \text{tetrahydrofuran}$$

In this case alkaline conditions were unnecessary because the five-membered ring is not readily cleaved.

EXERCISE 15.7

Draw the products expected for following reactions:

(a) $CH_3CH_2CH_2{-}Br \xrightarrow{NaOCH_2CH_3} ?$

(b) $CH_3{-}\underset{}{\overset{CH_3}{\overset{|}{C}H}}{-}CH_2OH \xrightarrow{H_2SO_4} ?$

(c) $CH_3CH_2CH{=}CH_2 \xrightarrow[2.\ NaBH_4]{1.\ Hg(O\overset{O}{\overset{\|}{C}}CH_3)_2,\ CH_3CH_2CH_2OH} ?$

EXERCISE 15.8

Suggest ways of making the following ethers. If several methods are applicable, show each of them.

(a) $CH_3{-}\underset{CH_3}{\underset{|}{\overset{CH_3}{\overset{|}{C}}}}{-}O{-}CH_3$ (b) $C_6H_5{-}O{-}\overset{CH_3}{\overset{|}{C}H}{-}CH_3$

(c) 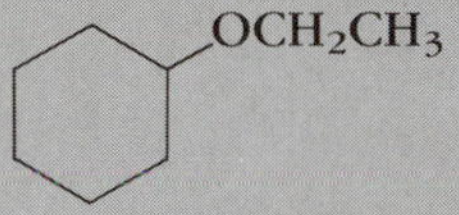 cyclohexane ring bearing OCH_2CH_3 (d) 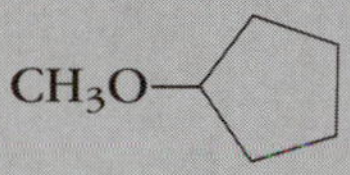 $CH_3O{-}$cyclopentyl

(e) $CH_3CH_2\overset{CH_3}{\overset{|}{C}H}{-}CH\overset{O}{—}CH_2$ (epoxide)

15.5 ACETALS AND KETALS

Acetals and **ketals** are compounds containing *two ether linkages to a single carbon atom,* and they are derived from aldehydes and ketones, respectively.

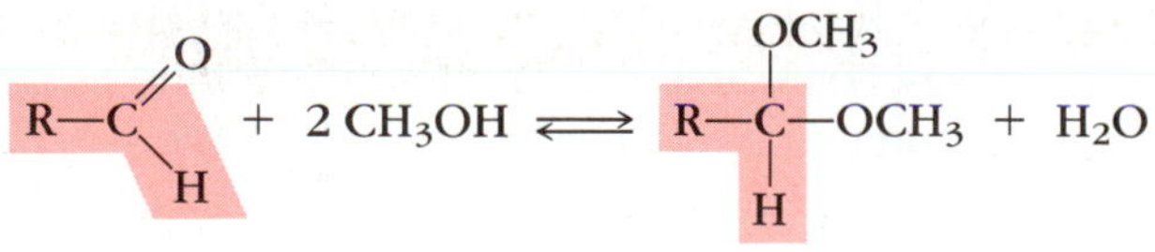

$$R{-}\overset{O}{\overset{\|}{C}}{-}H + 2\,CH_3OH \rightleftharpoons R{-}\underset{H}{\underset{|}{\overset{OCH_3}{\overset{|}{C}}}}{-}OCH_3 + H_2O$$

Aldehyde Acetal

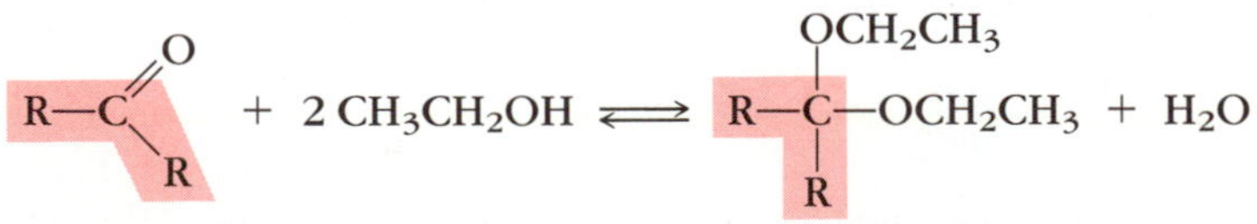

$$R{-}\overset{O}{\overset{\|}{C}}{-}R + 2\,CH_3CH_2OH \rightleftharpoons R{-}\underset{R}{\underset{|}{\overset{OCH_2CH_3}{\overset{|}{C}}}}{-}OCH_2CH_3 + H_2O$$

Ketone Ketal

Like simple ethers, acetals and ketals are unreactive under basic conditions. But in contrast to simple ethers, *acetals and ketals exhibit high reactivity under very mild acidic conditions.* They react cleanly with water in the presence of acid to generate the corresponding aldehyde or ketone. This pattern of reactivity has led to the use of these acetals and ketals as *protective groups* for polyfunctional compounds. They can be prepared from (and hydrolyzed back to) aldehydes and ketones under mild acid conditions that do not affect other functional groups.

Suppose you needed to oxidize the hydroxyl group of 7-hydroxyoctanal:

$$CH_3{-}CH(OH){-}(CH_2)_5{-}CH{=}O \longrightarrow CH_3{-}C({=}O){-}(CH_2)_5{-}CH{=}O$$

Oxidation of the alcohol presents no problem, but most oxidizing agents would also oxidize the aldehyde group to a carboxylic acid (Section 13.11).

$$CH_3{-}CH(OH){-}(CH_2)_5{-}CH{=}O \xrightarrow{\text{oxidation}} CH_3{-}C({=}O){-}(CH_2)_5{-}C({=}O){-}OH$$

However, the use of a **protecting group** (also known as a **blocking group** or **masking group**) would allow you to *temporarily modify* the reactive aldehyde until after you carried out the oxidation of the alcohol.

$$CH_3{-}CH(OH){-}(CH_2)_5{-}CH{=}O \xrightarrow{\text{protect aldehyde as acetal}} CH_3{-}CH(OH){-}(CH_2)_5{-}CH(OR){-}OR$$

$$CH_3{-}CH(OH){-}(CH_2)_5{-}CH(OR){-}OR \xrightarrow{\text{oxidize}} CH_3{-}C({=}O){-}(CH_2)_5{-}CH(OR){-}OR$$

$$CH_3{-}C({=}O){-}(CH_2)_5{-}CH(OR){-}OR \xrightarrow{\text{remove protecting group}} CH_3{-}C({=}O){-}(CH_2)_5{-}CH{=}O$$

We will return to the idea of protective groups and discuss them more fully in Section 15.7.

The facile hydrolysis of ketals under very mild acid catalysis has also been exploited for other purposes. For example, an automobile engine behaves irregularly when small amounts of water contaminate the fuel tank. There are several commercial products which alleviate this condition. They contain simple ketals such as 2,2-dimethoxypropane, which react with the water:

$$CH_3{-}C(OCH_3)(CH_3){-}OCH_3 + H_2O \xrightarrow{\text{traces of acid}} CH_3{-}C({=}O){-}CH_3 + 2\,CH_3OH$$

The two products of the preceding reaction, acetone and methanol, are both soluble in gasoline, and they both burn readily. The water is therefore removed, which allows the automobile engine to run smoothly.

Mechanism of Acetal and Ketal Formation

The mechanism of acetal and ketal formation is essentially the same for all aldehydes and ketones. We illustrate this general general mechanism for the reaction of a ketone with methanol.

$$R_2C{=}O \xrightarrow[\text{acid catalyst}]{CH_3OH} R_2C(OCH_3)_2$$

The first step (as with all acid-catalyzed reactions of aldehydes and ketones) is protonation of the carbonyl group. This is followed by nucleophilic attack of the alcohol on the protonated carbonyl, and proton transfer then affords the *hemiketal* (Section 13.7).

$$R_2C{=}O \overset{H^+}{\rightleftharpoons} R_2C{=}\overset{+}{O}H \overset{HOCH_3}{\rightleftharpoons} R_2C(OH){-}\overset{+}{O}(H)CH_3 \overset{HOCH_3}{\rightleftharpoons} R_2C(OH)(OCH_3)$$

Protonated ketone   Hemiketal

Protonation on the hydroxyl group of the hemiketal is followed by loss of water to produce a new oxonium ion, an *O*-alkylated ketone. (Notice the structural similarity of this intermediate to the protonated ketone in the preceding equation.)

$$R_2C(OH)(OCH_3) \overset{H^+}{\rightleftharpoons} R_2C(\overset{+}{O}H_2)(OCH_3) \rightleftharpoons R_2C{=}\overset{+}{O}CH_3$$

O-Alkylated ketone

Finally, nucleophilic attack by the alcohol on this *O*-alkylated ketone is followed by proton transfer to yield the ketal.

$$R_2C{=}\overset{+}{O}CH_3 \overset{HOCH_3}{\rightleftharpoons} R_2C(OCH_3){-}\overset{+}{O}(H)CH_3 \overset{HOCH_3}{\rightleftharpoons} R_2C(OCH_3)_2$$

Ketal

Several features of this reaction mechanism deserve special comment. First, all the cationic intermediates are *oxonium ions* rather than simple carbocations. These are structurally related to the hydronium ion, and the much greater stability of oxonium ions allows these reactions to proceed under very mild conditions (typically a trace amount of a strong acid). In contrast, the formation of simple ethers under acidic conditions requires either a direct displacement reaction or carbocation formation, and much more stringent reaction conditions are necessary. Second, the original carbonyl oxygen is lost as water. Both oxygen atoms of the ketal (or acetal) are derived from the alcohol used in the reaction. The carbon-oxygen bonds of the alcohol remain intact throughout the reaction, and only the carbon-oxygen bonds of the *carbonyl group* are readily cleaved. Third, the formation of acetals and ketals is completely reversible. Each of the individual steps is reversible, and the *same reaction mechanism* (in the opposite direction) describes the *hydrolysis* of acetals and ketals.

Preparation and Hydrolysis of Acetals and Ketals

Acetals and ketals are normally prepared by reaction of the aldehyde or ketone with an alcohol; a trace of a strong acid such as *p*-toluenesulfonic acid (TsOH) is added as a catalyst. In order to shift the equilibrium toward acetal or ketal formation, water is usually removed from the reaction mixture. Sometimes a drying agent is added to the reaction mixture, but more commonly benzene is used as the solvent, and water is removed by distillation. (Although the boiling point of benzene is 80°C, a mixture of benzene and water distills as a 10:1 benzene-water azeotrope at 69°C.) The following reactions illustrate the preparation of acetals and ketals:

$$m\text{-}O_2N\text{-}C_6H_4\text{-}CHO \xrightarrow[\text{HCl catalyst}]{CH_3OH} m\text{-}O_2N\text{-}C_6H_4\text{-}CH(OCH_3)_2 \qquad 76\text{–}85\%$$

$$CH_3CHO \xrightarrow[\text{TsOH}]{CH_3CH_2CH_2CH_2OH} CH_3\text{—}CH(\text{—}O\text{—}CH_2CH_2CH_2CH_3)\text{—}O\text{—}CH_2CH_2CH_2CH_3 \qquad 63\%$$

$$\text{cyclohexanone} \xrightarrow[\text{TsOH, } (CH_3)_2C(OCH_3)_2]{CH_2{=}CH\text{—}CH_2OH} \text{cyclohexane-1,1-}(O\text{—}CH_2\text{—}CH{=}CH_2)_2 \qquad 65\text{–}74\%$$

In the last example 2,2-dimethoxypropane is used to remove the water that is formed. (What are the other products of this reaction?)

The reaction of an aldehyde or ketone with a *diol* can produce a *cyclic* acetal or ketal.

$$CH_3CH_2CH_2CH{=}C(CHO)(CH_2CH_3) \xrightarrow[\text{TsOH, } C_6H_6]{HOCH_2CH_2OH} CH_3CH_2CH_2CH{=}C(CH_2CH_3)\text{—}CH(OCH_2CH_2O) \qquad 71\%$$

$$\text{cyclohexanone} \xrightarrow[\text{TsOH, } C_6H_6]{HOCH_2CH_2OH} \text{cyclohexanone ethylene ketal (spiro 1,3-dioxolane)} \qquad 75\text{–}85\%$$

Vinyl ethers

Vinyl ethers (also called *enol ethers*) are ethers in which an alkoxy group is a substituent on a carbon-carbon double bond:

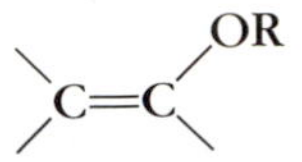

$$>C{=}C<^{OR}$$

Vinyl ether (enol ether)

The reactivity of vinyl ethers is very similar to that of acetals and ketals, because protonation of a vinyl ether occurs preferentially on carbon to give an oxonium ion.

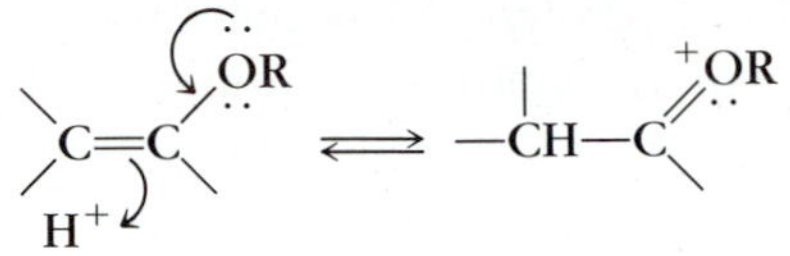

$$>C{=}C(OR)< + H^+ \rightleftharpoons -CH-C(={}^+OR)<$$

This is the same oxonium ion that is produced from the corresponding ketal (or acetal) in acid.

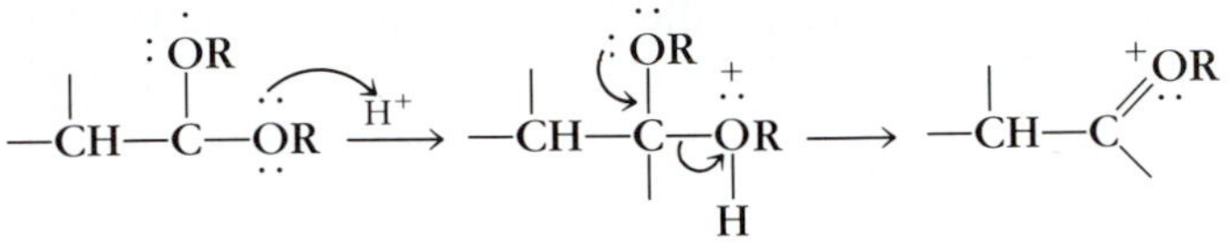

$$-CH-C(OR)-OR \xrightarrow{H^+} -CH-C(OR)-\overset{+}{O}R(H) \longrightarrow -CH-C(={}^+OR)<$$

The formation of a stable oxonium ion intermediate permits addition reactions of vinyl ethers to be carried out under very mild conditions. Reaction with an alcohol under mild acid catalysis results in addition of the alcohol to the double bond to produce an acetal or ketal.

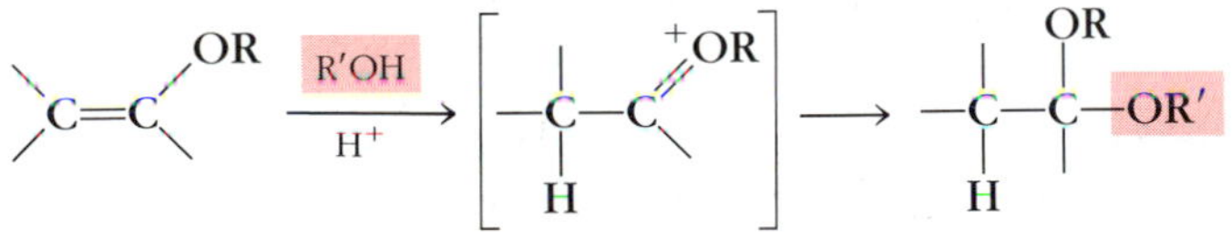

$$>C{=}C(OR)< \xrightarrow[H^+]{R'OH} \left[-C(H)-C(={}^+OR)<\right] \longrightarrow -C(H)-C(OR)-OR'$$

Dihydropyran is a particularly useful vinyl ether in organic synthesis. Alcohols add readily to the double bond under mildly acidic conditions to produce cyclic acetals called *tetrahydropyranyl ethers*. The acidic hydrogen of the alcohol is no longer present in the product, so the procedure is useful for protection of alcohol groups (see Section 15.7).

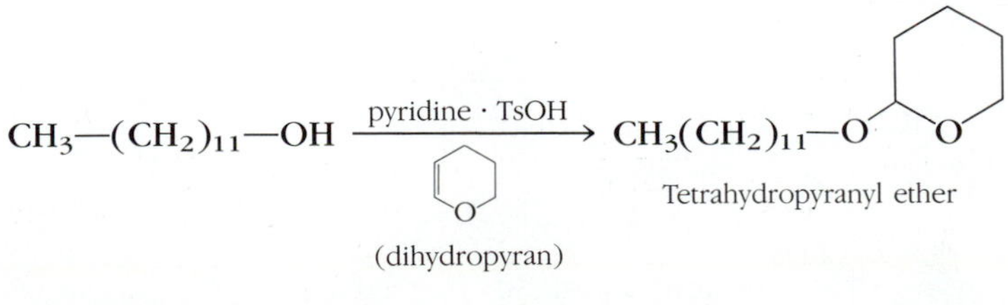

$$CH_3-(CH_2)_{11}-OH \xrightarrow[\text{(dihydropyran)}]{\text{pyridine} \cdot \text{TsOH}} CH_3(CH_2)_{11}-O-\text{(tetrahydropyran-2-yl)}$$

Tetrahydropyranyl ether

The acid catalyst used in this example is the ammonium salt formed by mixing pyridine and *p*-toluenesulfonic acid. It is a very mild acid that does not affect other functional groups.

Thioacetals and Thioketals

The use of thiols rather than alcohols allows aldehydes and ketones to be converted to their corresponding *thioacetals* and *thioketals*. One of the most common uses of this variation has employed ethanedithiol (HS—CH_2—CH_2—SH).

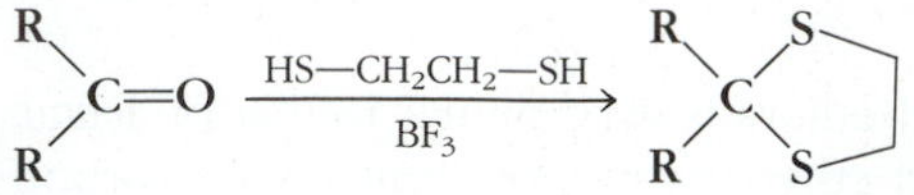

$$R_2C{=}O \xrightarrow[BF_3]{HS{-}CH_2CH_2{-}SH} R_2C(SCH_2CH_2S)$$

The conditions needed to hydrolyze thioacetals and thioketals back to the free carbonyl compounds are more vigorous than those needed to hydrolyze ordinary acetals and ketals. For this reason the sulfur derivatives are not usually hydrolyzed but are subjected to catalytic hydrogenation using a Raney nickel catalyst. This reaction causes *hydrogenolysis* of the carbon-sulfur bonds, and the net result is overall conversion of the original C=O group to CH_2. This provides an alternative to the Clemmensen and Wolf-Kishner reductions, which we discussed in Section 13.11. Preparation of a thioacetal or thioketal is typically carried out with BF_3 as the acid catalyst. The following sequence illustrates the use of a thioketal in the reduction of a ketone carbonyl.

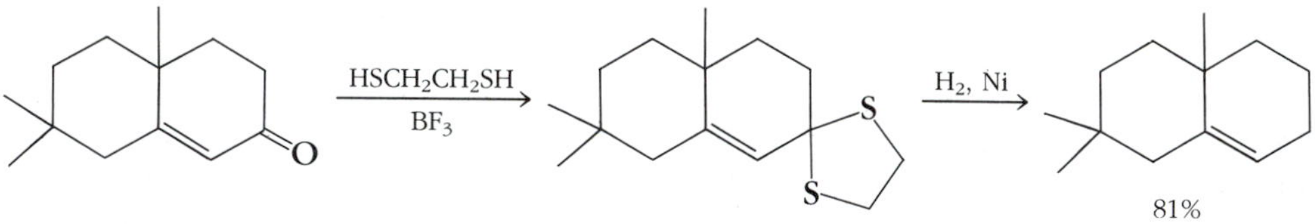

$$\xrightarrow[BF_3]{HSCH_2CH_2SH} \qquad \xrightarrow{H_2,\ Ni} \quad 81\%$$

Notice that the carbon-carbon double bond is not reduced under the mild conditions needed for hydrogenolysis of the carbon-sulfur bond.

EXERCISE 15.9

Draw the products expected for the following reactions:

(a) cyclopentanone $\xrightarrow[TsOH,\ (CH_3)_2C(OCH_3)_2]{CH_3OH}$?

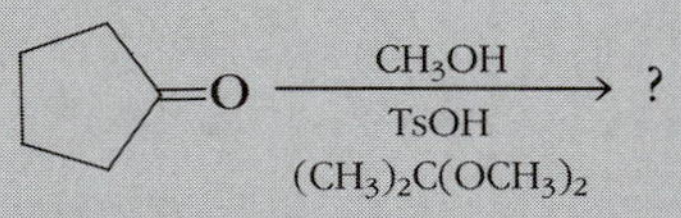

(b) $C_6H_5{-}CH(OCH_3){-}OCH_3 \xrightarrow{\text{dil. HCl}}$?

(c) 3-methylcyclohexanone (CH_3, O) $\xrightarrow{HS{-}CH_2CH_2{-}SH,\ BF_3}$?

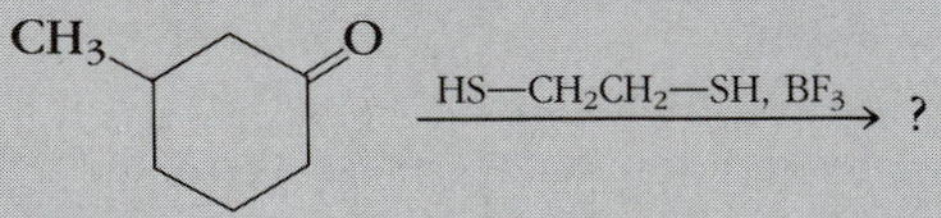

EXERCISE 15.10

Suggest a method for carrying out each of the following conversions:

(a) 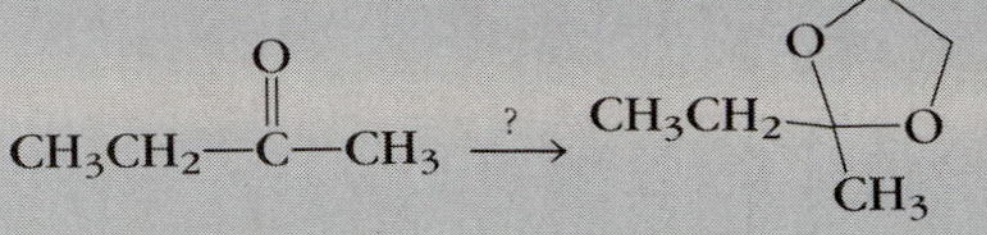 $CH_3CH_2-\overset{O}{\overset{\|}{C}}-CH_3 \xrightarrow{?}$ 2-ethyl-2-methyl-1,3-dioxolane (CH_3CH_2 and CH_3 on the ring carbon between the two O atoms)

(b) 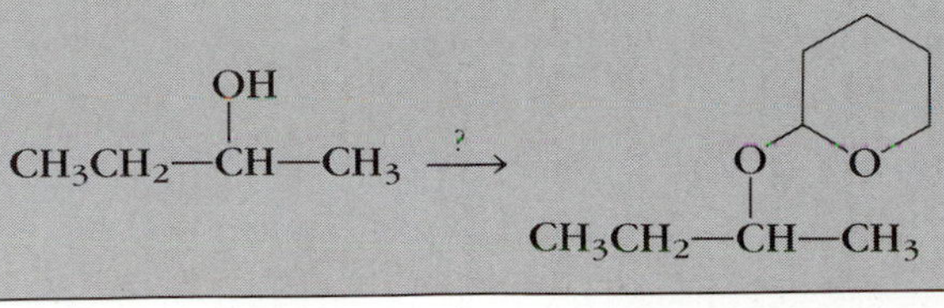 $CH_3CH_2-\overset{OH}{\overset{|}{C}H}-CH_3 \xrightarrow{?}$ 2-(sec-butoxy)tetrahydropyran ($CH_3CH_2-CH(O\text{-}THP)-CH_3$)

15.6 POLYETHERS

A variety of ethers have been synthesized that contain more than one ether linkage per molecule. Many of these are related to ethylene glycol and contain a repeating subunit, ($-OCH_2CH_2-$). In this category are the glymes, which are useful solvents (Table 15.1).

$$CH_3-OCH_2CH_2-OCH_3 \qquad CH_3-OCH_2CH_2-OCH_2CH_2-OCH_3 \qquad CH_3-(OCH_2CH_2)_3-OCH_3$$

Glyme — Diglyme — Triglyme

The higher molecular weight polyethylene glycols and their methyl ethers have a variety of commercial and industrial uses. These compounds range from high-boiling liquids to waxy solids as the number of repeating subunits, n, increases to values as high as 500 (Section 25.5).

$$H-(OCH_2CH_2)_n-OH \qquad CH_3-(OCH_2CH_2)_n-OCH_3$$

Polyethylene glycol — Polyethylene glycol dimethyl ether

These compounds are used in pharmaceutical and cosmetic ointments and as water-soluble lubricants. When one end of a polyethylene glycol is linked to a long-chain alkyl group, a useful *surfactant (surface active agent)* is produced. The polyglycol end is highly water-soluble, but the alkyl residue instead has an affinity for nonpolar organic materials. Such compounds are used as one of the components in commercial laundry detergents.

Crown Ethers

During the late 1960s important uses were discovered for another class of ethylene glycol polyethers, cyclic compounds called *crown ethers* (so named because of their three-dimensional crown shape). These compounds exhibit a high affinity for certain metal cations, because a complex is formed in which the metal resides within the crown and is simultaneously solvated by more than one oxygen of the polyether. This is illustrated for the case of 18-crown-6 (where 18 is the ring size and 6 is the number of oxygen atoms), a crown ether that binds strongly to potassium ions.

$$18\text{-crown-6} + K^+ \rightleftharpoons [K^+ \cdot 18\text{-crown-6}]$$

The coordination of metal ions by crown ethers has two important consequences for chemical reactivity. First, it permits inorganic salts to be dissolved in nonpolar organic solvents. This means that with the aid of crown ethers you could select a reaction solvent that is best for the organic compound (rather than for the inorganic reagent). Second, by physically separating the metal cation from the anion of the salt, the reactivity of the anion can be greatly enhanced. Together these effects make possible reactions that might otherwise be very difficult. The following examples illustrate the use of a crown ether to carry out nucleophilic substitution reactions with ionic reagents in nonpolar organic solvents.

$$CH_3(CH_2)_7Br \xrightarrow[\text{18-crown-6}]{\text{KF}} CH_3(CH_2)_7F \qquad 92\%$$

$$C_6H_5CH_2Cl \xrightarrow[\text{18-crown-6}]{\text{KCN}} C_6H_5CH_2CN \qquad 90\text{–}95\%$$

Crown ethers and related compounds offer great potential for specific binding to organic compounds as well as to metal ions, and the 1987 Nobel Prize in chemistry was awarded to three workers in the crown ether field: Donald J. Cram (University of California at Los Angeles), Charles J. Pedersen (du Pont Co.), and Jean-Marie Lehn (Université Louis Pasteur, France).

15.7 ACETALS AND KETALS AS PROTECTING GROUPS IN ORGANIC SYNTHESIS

In Chapter 8 we discussed some of the problems encountered in the reactions of difunctional compounds. In Section 8.8 we showed several examples of *selective* reactions, in which reaction occurs preferentially at one of two possible sites. Unfortunately, the desired site for reaction is often the less reactive group; in such cases the desired product can sometimes be produced by *protecting* or *blocking* the more reactive functional group. In this section we will summarize two such processes that illustrate the use of acetals and ketals in synthesis.

Protection of a Ketone as a Ketal

The sequence we describe here is part of a synthesis of the terpene *bullnesol,* a constituent of guaiacum wood. The oil obtained from the wood is used in perfumery. The synthesis was carried out at Northwestern University by J. A. Marshall and J. J. Partridge, who were faced with the problem of converting the keto ester **1** into the keto ether **6.**

1 → 6

The transformation requires reduction of the ester group, but ketone carbonyls are reduced more readily. Consequently, it was decided to first protect the ketone as a ketal, as shown in the following sequence (yields are shown in brackets):

1 —($HOCH_2CH_2OH$; TsOH, C_6H_6 [99%])→ 2 —(1. $LiAlH_4$ 2. H_2O [89%])→ 3 —(CH_3SO_2Cl; pyridine [94%])→ 4 —(C_6H_5OH, NaH [89%])→ 5 —(H_2O; HCl catalyst [90%])→ 6

Formation of the ketal, **2,** proceeded cleanly and did not affect the ester group. The ester group of **2** was then reduced to the primary alcohol (Section 11.4) with lithium aluminum hydride, and the methanesulfonate, **4,** was prepared by treatment of **3** with methanesulfonyl chloride (Section 11.9). Pyridine was used in this reaction to neutralize the HCl formed. Displacement of the methanesulfonate with sodium phenoxide (produced by the reaction between sodium hydride and phenol) yielded the ketal ether, **5.** Finally, the ketal was hydrolyzed with dilute hydrochloric acid to produce the desired keto ether, **6.** The overall yield of **6** from **1** was 66%, a very good yield for a five-step sequence.

Protection of an Alcohol as a Tetrahydropyranyl Ether

The female grape-berry moth, *Paralobesia viteana,* secretes a sex *pheromone* (Section 11.11), which attracts male moths. If large quantities of the pheromone were

available, it might be possible to control the insect population without resorting to toxic pesticides. Consequently, C. A. Henrick and B. A. Garcia of the Zoecon Corporation designed a synthesis of (*Z*)-9-dodecen-1-yl acetate **(11)**, the major constituent of the pheromone.

$$\underset{\mathbf{11}}{(Z)\text{-}CH_3CH_2CH{=}CH(CH_2)_8OC(=O)CH_3}$$

They intended to prepare the *Z* alkene by a *syn* addition of hydrogen to the alkyne, and this suggested that alkylation of a terminal alkyne could be used to build up the carbon chain.

$$CH_3CH_2{-}C{\equiv}C{-}(CH_2)_8{-}OC(=O)CH_3 \longleftarrow CH_3CH_2C{\equiv}CH + X{-}(CH_2)_8{-}OC(=O)CH_3$$

However, the strongly basic conditions needed to generate an acetylide ion are incompatible with an ester group. It would also be unsatisfactory to use the free alcohol because base would convert it to the nucleophilic alkoxide ion. The alcohol was therefore protected as the tetrahydropyranyl ether, and the acetate group was not introduced until the end of the synthesis.

$$\underset{\mathbf{7}}{Br{-}(CH_2)_8{-}OH} \xrightarrow[\text{trace conc. HCl}\ [80\%]]{\text{dihydropyran}} \underset{\mathbf{8}}{Br{-}(CH_2)_8{-}O{-}THP} \xrightarrow[NH_3\ [85\%]]{CH_3CH_2C{\equiv}C^-Li^+} \underset{\mathbf{9}}{CH_3CH_2{-}C{\equiv}C{-}(CH_2)_8{-}O{-}THP}$$

$$\underset{\mathbf{9}}{CH_3CH_2{-}C{\equiv}C{-}(CH_2)_8{-}O{-}THP} \xrightarrow[CH_3C(=O)Cl\ [85\%]]{CH_3C(=O)OH} \underset{\mathbf{10}}{CH_3CH_2{-}C{\equiv}C{-}(CH_2)_8{-}OC(=O)CH_3} \xrightarrow[\text{Pd—CaCO}_3\ [92\%]]{H_2} \underset{\mathbf{11}}{(Z)\text{-}CH_3CH_2CH{=}CH(CH_2)_8O{-}C(=O)CH_3}$$

The tetrahydropyranyl ether, **8,** was prepared by treatment of the bromo alcohol, **7,** with dihydropyran and an acid catalyst. This bromo ether was then used to alkylate the lithium salt of 1-butyne in liquid ammonia (Section 7.8). Acid-catalyzed cleavage of the acetal and acetylation (Section 11.8) of the alcohol were then carried out in a single step with acetic acid and acetyl chloride as reagents. Finally, the alkyne **10** was converted to the *Z*-alkene **11** by catalytic hydrogenation using a deactivated catalyst (Section 7.6). The yield for the entire synthesis of **11** from **7** proceeded in 53% overall yield, which corresponds to an average yield of 85% for each of the four steps in the sequence.

EXERCISE 15.11

(a) What product would be formed if the keto ester, **1** (p. 821), were reduced with $LiAlH_4$?
(b) What by-products might be produced if the alkylation reaction with sodium phenoxide were attempted with the free ketone rather than the protected derivative, **4**?

EXERCISE 15.12

What by-product would be expected if the lithium acetylide displacement were attempted with the unprotected bromide **7** (p. 822)?

15.8 TERMS AND DEFINITIONS

Acetal. A compound containing a saturated carbon atom with two ether linkages, one bond to hydrogen, and the remaining bond to carbon.

Blocking group. A group that is introduced *temporarily* during a synthesis to modify one reactive center and allow selective reaction at another site; also known as a *masking group* or *protecting group*.

Enol ether. An ether in which one of the two groups bonded to oxygen is an alkenyl group; also called a *vinyl ether*.

Ether. A compound containing two R groups bonded to an oxygen, where R can be alkyl, aryl, or alkenyl.

Heterocycle. A cyclic molecule in which one or more of the atoms of the ring is not carbon.

Ketal. A compound containing a saturated carbon atom with two ether linkages and two bonds to carbon atoms.

Masking group. A group that is introduced *temporarily* during a synthesis to modify one reactive center and allow selective reaction at another site; also known as a *blocking group* or *protecting group*.

Protecting group. A group that is introduced *temporarily* during a synthesis to modify one reactive center and allow selective reaction at another site; also known as a *blocking group* or *masking group*.

Vinyl ether. An ether in which one of the two groups bonded to oxygen is an alkenyl group; also called an *enol ether*.

15.9 SUMMARY OF REACTIONS

The reactions of ethers are summarized in Table 15.2 and various methods for preparation of ethers are summarized in Table 15.3.

TABLE 15.2 Reactions of Ethers (Section 15.3)

Reaction	Comments
1. Acid-catalyzed cleavage	Ethers are unreactive under basic conditions.
(a) Dialkyl ethers	
$R{-}O{-}R \xrightarrow{HX} RX$	Reaction requires conc. HBr or HI at high temperatures.
$R{-}O{-}tBu \xrightarrow{HX} ROH + tBu{-}X$	Epoxides and *tert*-butyl ethers react under less stringent conditions.
(b) Aryl alkyl ethers	
$Ar{-}O{-}R \xrightarrow{HX} ArOH + R{-}X$	Aryl alkyl ethers yield the phenol.
(c) Acetals and ketals	
$>C(OR)_2 \xrightarrow[H^+]{H_2O} >C{=}O + 2\,ROH$	Hydrolysis occurs under very mild acidic conditions.
(d) Thioacetals and thioketals	
$R_2C(SCH_2CH_2S) \xrightarrow{H_2,\ Ni} R_2CH_2$	Hydrogenolysis with Raney nickel yields the hydrocarbon.
2. Base-catalyzed cleavage	
$R{-}CH{-}CH_2$ (epoxide, O bridging) $\xrightarrow[H_2O]{NaOH} R{-}CH(OH){-}CH_2OH$	Epoxides are an exception; other ethers do not react under basic conditions. Other nucleophiles will also open an epoxide ring by attack at the less substituted carbon (Sections 11.6, 12.8).

TABLE 15.3 Preparation of Ethers

Reaction	Comments
1. Acid-catalyzed condensation of alcohols	Section 15.4.
$2\,ROH \xrightarrow{H^+} ROR$	Produces symmetrical ethers. Many side reactions; yields variable.

TABLE 15.3 Preparation of Ethers (continued)

Reaction	Comments
2. Solvolysis of alkyl halides or sulfonates	
$R{-}X + R'OH \longrightarrow ROR'$	Best results if RX can ionize to stable carbocation. Alcohol is typically methanol or ethanol.
3. Alkylation of alkoxide ions	
$RO^- + R'X \longrightarrow ROR'$	Limited to 1° and 2° halides and sulfonates (Williamson synthesis).
4. Addition to alkenes	
(a) Acid-catalyzed	
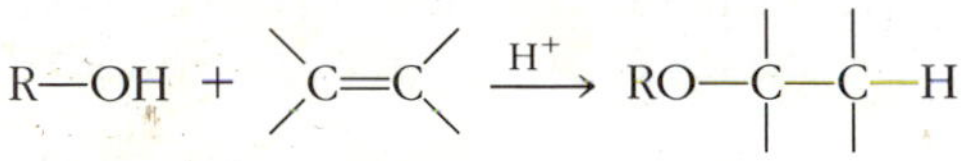 $R{-}OH + \diagdown C{=}C \diagup \xrightarrow{H^+} RO{-}\overset{\vert}{\underset{\vert}{C}}{-}\overset{\vert}{\underset{\vert}{C}}{-}H$	Acid conditions; rearrangements.
(b) Oxymercuration	
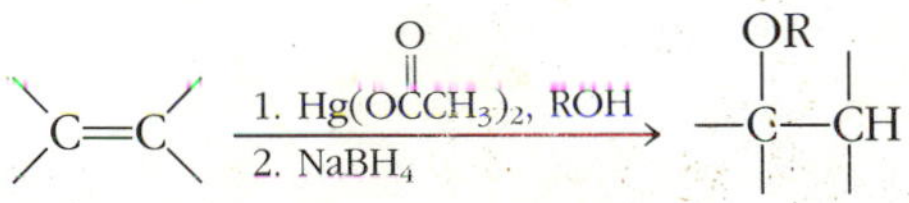 $\diagdown C{=}C \diagup \xrightarrow[\text{2. NaBH}_4]{\text{1. Hg(O}\overset{O}{\overset{\Vert}{C}}\text{CH}_3)_2\text{, ROH}} {-}\overset{OR}{\overset{\vert}{\underset{\vert}{C}}}{-}\overset{\vert}{\underset{\vert}{C}}H$	Nucleophilic attack by ROH occurs on the *more* substituted carbon atom. No rearrangements.
5. Epoxides	
(a) Epoxidation of alkenes	
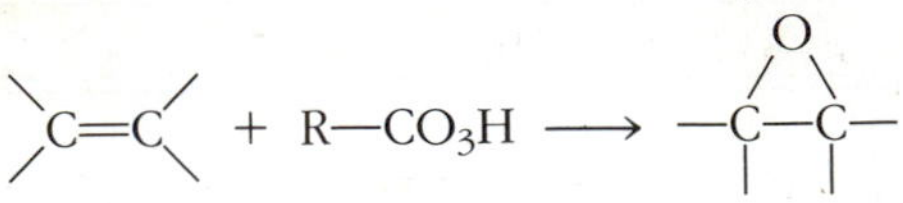 $\diagdown C{=}C \diagup + R{-}CO_3H \longrightarrow {-}\underset{\vert}{C}{-}\underset{\vert}{C}{-}$ (epoxide, O bridging the two C)	Most general method for forming epoxides.
(b) Via sulfur ylides	
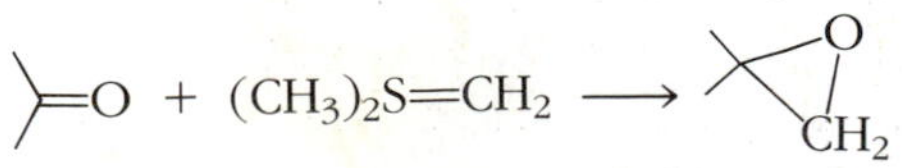 $\diagdown C{=}O + (CH_3)_2S{=}CH_2 \longrightarrow$ epoxide ($\diagdown C$ bridged by O and CH_2)	One additional carbon is introduced.
(c) Cyclization of halo alcohols	
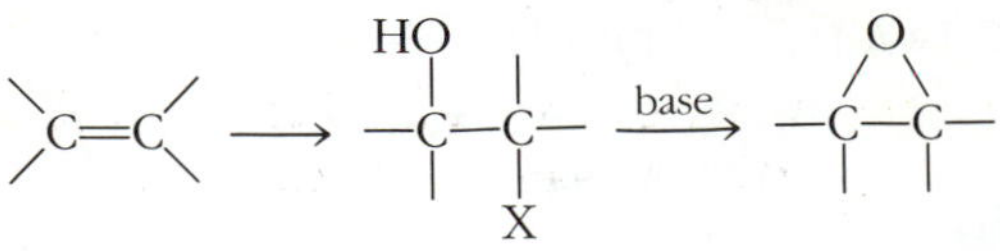 $\diagdown C{=}C \diagup \longrightarrow {-}\overset{HO}{\overset{\vert}{\underset{\vert}{C}}}{-}\overset{\vert}{\underset{X}{\underset{\vert}{C}}}{-} \xrightarrow{\text{base}} {-}\underset{\vert}{C}{-}\underset{\vert}{C}{-}$ (epoxide, O bridging the two C)	

TABLE 15.3 Preparation of Ethers (continued)

Reaction	Comments
6. Acetals and ketals	
(a) Protection of aldehydes and ketones	
$R'(R)C{=}O \underset{H^+}{\overset{ROH}{\rightleftharpoons}} R'{-}C(OR)(R){-}OR$ R = alkyl, R′ = alkyl or H	Reactions are reversible. Mild acid hydrolysis liberates aldehyde or ketone.
$R'(R)C{=}O \xrightarrow[C_6H_6,\ TsOH]{HOCH_2CH_2OH}$ cyclic acetal/ketal (R′, R on C in 1,3-dioxolane ring)	Ethylene glycol yields cyclic acetal or ketal.
$R'(R)C{=}O \xrightarrow[BF_3]{HS{-}CH_2CH_2{-}SH}$ cyclic thioacetal/thioketal (R′, R on C in 1,3-dithiolane ring)	Ethanedithiol yields cyclic thioacetal or thioketal. (Hydrogenolysis with Raney Ni/H_2 yields hydrocarbon).
(b) Protection of alcohols	
R—OH + dihydropyran $\xrightarrow{H^+}$ R—O–(tetrahydropyran-2-yl)	Tetrahydropyranyl ether: mild acid hydrolysis regenerates alcohol.

15.10 PROBLEMS

15.1 Name the following compounds. Give alternative names when possible.

(a) $CH_3CH_2CH(OH){-}CH_2OCH_2CH_3$

(b) $CH_3CH_2CH(CH_3){-}OCH_3$

(c) $CH_3CH_2CH{-}CHCH_2Cl$ (epoxide: O bridging the two CH carbons)

(d) $CH_3{-}C(=O){-}CH_2CH_2CH(CH_3){-}CH_2OCH_3$

(e) $CH_3CH_2CH(OCH_3){-}CH_2CH_2{-}CH(OCH_3){-}CH_2CH_2OCH_3$

(f) $CH_3{-}C(OCH_3)_2{-}CH_3$

(g) $CH_3CH_2{-}CH(CH_3){-}CH_2{-}O{-}CH(CH_3){-}CH_3$

15.2 Draw the following:
(a) 3-Methoxycyclohexanone
(b) 4-Chloro-3-propoxyoctane
(c) 2-Bromo-3,3-dimethyloxolane
(d) 2-Cyclohexyl-3,4-epoxyhexane
(e) 3,3-Dimethoxypentane
(f) 4-Methoxy-3,3-dimethylhexanol
(g) 2-Hexyloxyoctane

15.3 Which compound of each pair has the higher boiling point?

(a) $CH_3CH_2CH_2CH_2CH_2CH_2CH_2OCH_3$ and $CH_3C(CH_3)_2{-}O{-}C(CH_3)_2{-}CH_3$

(b) $CH_3CH_2CH_2{-}CH(OH){-}CH_2CH_2OCH_3$ and $CH_3CH_2CH(OCH_3){-}CH_2CH_2OCH_3$

(c) $CH_3CH_2CH_2{-}CH(OH){-}CH_2(OH)$ and $CH_3CH_2{-}O{-}CH_2{-}O{-}CH_2CH_3$

15.4 Predict the products of each of the following ethers with HBr:
(a) $CH_3{-}(CH_2)_4{-}CH_2OCH_3$
(b) $CH_3{-}(CH_2)_4{-}CH_2{-}O{-}C(CH_3)_3$
(c) $CH_3(CH_2)_4CH_2{-}O{-}C_6H_5$
(d) $CH_3(CH_2)_4CH_2OCH_2(CH_2)_4CH_3$

15.5 What products would be expected in the reaction of 1-methyl-1-phenyl-1,2-epoxypropane with:
(a) HBr
(b) NaCN
(c) Dilute HCl
(d) NaBr, H_2O

15.6 Suggest methods for preparing each of the following ethers. If more than one way would work, suggest several alternatives.

(a) $CH_3(CH_2)_4CH_2{-}O{-}CH_3$
(b) $C_6H_5CH_2{-}O{-}CH_2CH_2CH_3$
(c) $CH_3CH_2{-}O{-}CH(CH_3){-}CH_2CH_3$
(d) $C_6H_5{-}O{-}CH_2CH_2CH_3$
(e) 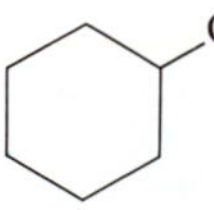 Cyclohexyl–OCH_2CH_3

15.7 Suggest stereospecific methods for preparing the following ethers from the appropriate alkene. If the structure shown is chiral, your synthesis should afford the racemic ether.

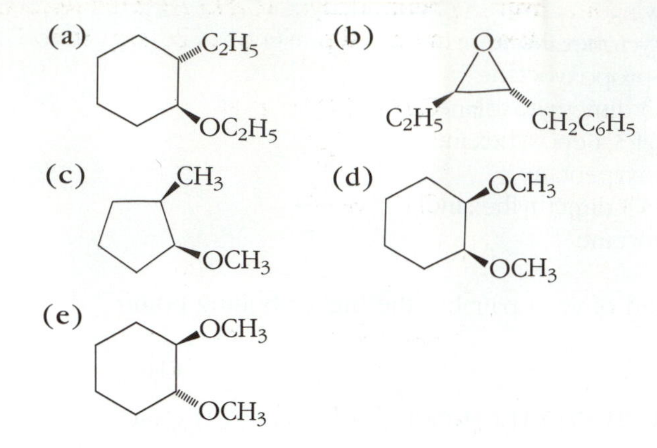

15.8 Suggest methods of preparing the following compounds from aldehydes and ketones:

(a) $CH_3CH_2CH(OCH_3)—OCH_3$

(b) $CH_3CH_2C(CH_3)_2—CH_2CH_2—C(OCH_3)_2—CH_3$

(c) 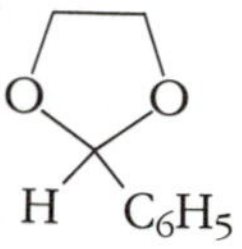 O, O, H, C_6H_5

(d) 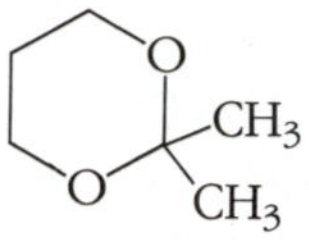 O, O, CH_3, CH_3

15.9 Write the mechanism for acid-catalyzed hydrolysis of 1,1-dimethoxycyclohexane to cyclohexanone.

15.10 Suggest methods for converting cyclohexanone into each of the following compounds. More than one step may be necessary in each case.

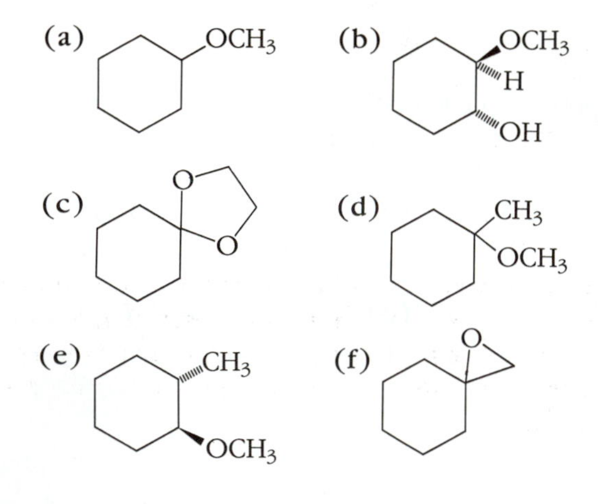

15.11 Suggest methods for converting benzaldehyde (C_6H_5CHO) into each of the following compounds. More than one step may be necessary in each case.

(a) $C_6H_5CH_2OCH_2CH_3$ (b) $C_6H_5CH_2OCH_2C_6H_5$

(c) 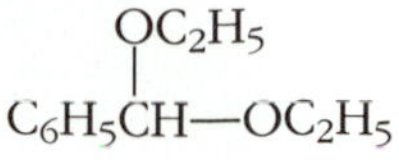 $C_6H_5CH(OC_2H_5)—OC_2H_5$ (d) 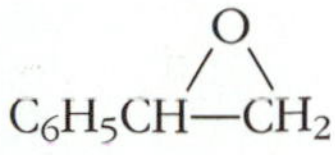 $C_6H_5CH—CH_2$ (epoxide, bridging O)

(e) 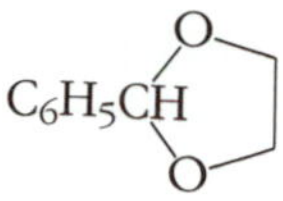 C_6H_5CH (cyclic acetal with two O in a five-membered ring) (f) 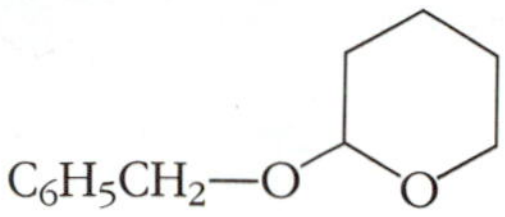 $C_6H_5CH_2—O$–(tetrahydropyran-2-yl)

(g) $C_6H_5CH_3$ (h) 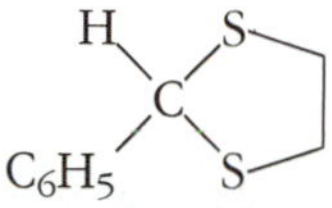 H and C_6H_5 on C of a 1,3-dithiolane ring (S, S)

15.12 Suggest methods for carrying out each of the following transformations. Each will require several steps.

(a) 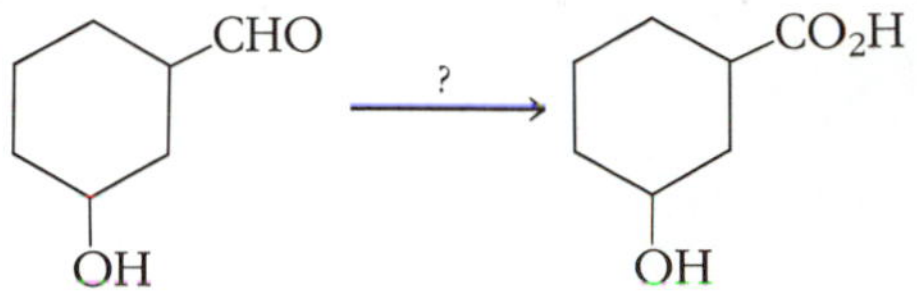 3-hydroxycyclohexane with CHO (OH, CHO) $\xrightarrow{?}$ (OH, CO_2H)

(b) 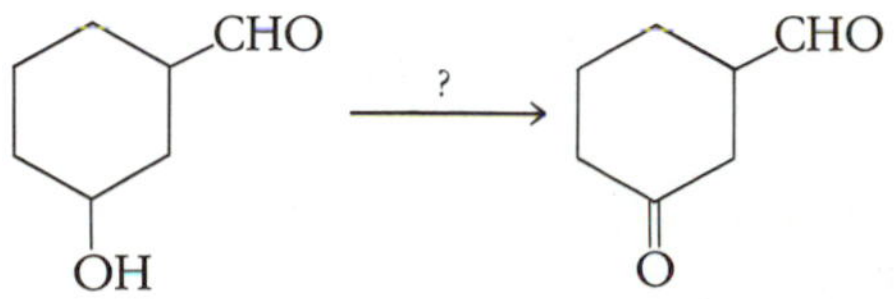 (OH, CHO) $\xrightarrow{?}$ (O, CHO)

(c) 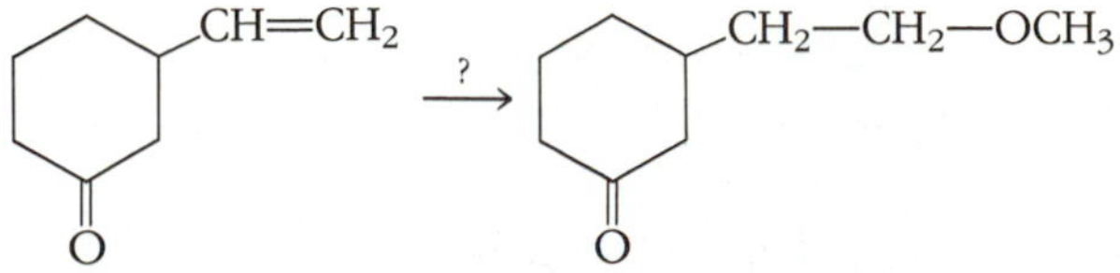 (O, $CH{=}CH_2$) $\xrightarrow{?}$ (O, $CH_2—CH_2—OCH_3$)

(d) 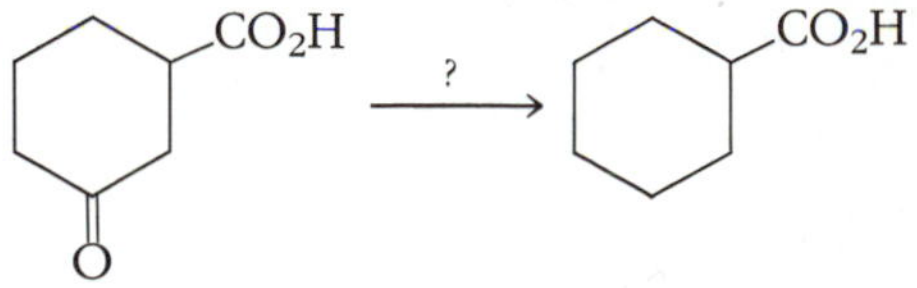 (O, CO_2H) $\xrightarrow{?}$ (CO_2H)

15.13 Spectroscopy can often help you learn whether or not a reaction has followed the desired course. Indicate how you could determine whether the material isolated from the last step of each of the conversions in Problem 15.10 had the desired structure. What specific changes in the IR and NMR spectra would you look for?

CHAPTER 16

CARBOXYLIC ACIDS AND THEIR DERIVATIVES

The chemistry of carboxylic acids and their derivatives affects our everyday lives to a greater extent than that of any other class of compounds we have discussed so far in this book. Many of the compounds you will encounter in this chapter are naturally occurring and play important roles in biochemical processes.

All carboxylic acid derivatives share the structural feature of a carbon atom with three bonds to heteroatoms such as oxygen and nitrogen. Most of these derivatives have a carbonyl group (two bonds to oxygen) with one additional heteroatom substituent. The carboxylic acids themselves are characterized by the **carboxyl group, —CO_2H.**

Carboxylic Acids and Carboxylic Acid Derivatives

$$\underset{\text{Carboxylic acid}}{R{-}\overset{O}{\overset{\|}{C}}{-}OH} \qquad \underset{\text{Ester}}{R{-}\overset{O}{\overset{\|}{C}}{-}OR} \qquad \underset{\text{Acid chloride}}{R{-}\overset{O}{\overset{\|}{C}}{-}Cl} \qquad \underset{\text{Amide}}{R{-}\overset{O}{\overset{\|}{C}}{-}N\langle}$$

$$\underset{\text{Nitrile}}{R{-}C{\equiv}N} \qquad \underset{\text{Acid anhydride}}{R{-}\overset{O}{\overset{\|}{C}}{-}O{-}\overset{O}{\overset{\|}{C}}{-}R} \qquad \underset{\text{Carboxylate salt}}{R{-}\overset{O}{\overset{\|}{C}}{-}O^{-}M^{+}}$$

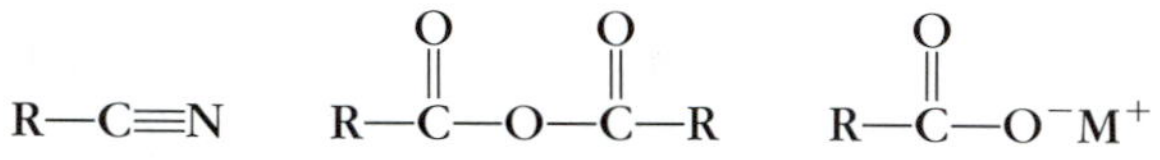

The three types of compounds that you will encounter most frequently are carboxylic acids, esters, and amides.

Carboxylic Acids Some **carboxylic acids** are already familiar to you. The low-molecular-weight carboxylic acids are moderately volatile, and their acidity leads to sharp, biting odors and tastes. Formic acid is found in ants and is a potent skin irritant, as you may have discovered if you have ever stepped in a nest of ants. Vinegar is merely a dilute solution of acetic acid in water.

$$\underset{\text{Formic acid}}{H{-}\overset{O}{\overset{\|}{C}}{-}OH} \qquad \underset{\text{Acetic acid}}{CH_3{-}\overset{O}{\overset{\|}{C}}{-}OH}$$

The next few carboxylic acids in the series have considerably less agreeable tastes and aromas. Some of these are formed in rancid butter (Latin *butyrum*, "butter"), and that is the origin of the name for butyric acid. The C_6, C_8 and C_{10} carboxylic acids were previously called by common names that reflect the occurrence of their derivatives in goat milk (Latin *caper*, "goat").

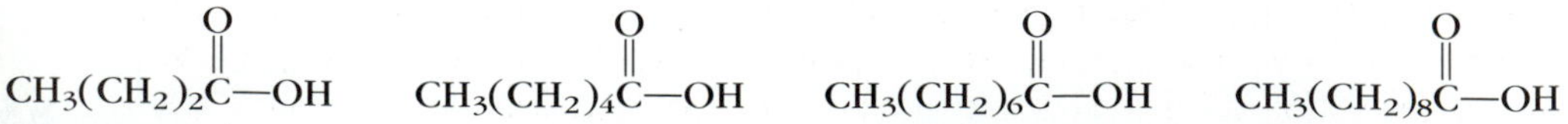

$$\underset{\text{Butyric acid}}{CH_3(CH_2)_2\overset{O}{\overset{\|}{C}}{-}OH} \qquad \underset{\text{Caproic acid}}{CH_3(CH_2)_4\overset{O}{\overset{\|}{C}}{-}OH} \qquad \underset{\text{Caprylic acid}}{CH_3(CH_2)_6\overset{O}{\overset{\|}{C}}{-}OH} \qquad \underset{\text{Capric acid}}{CH_3(CH_2)_8\overset{O}{\overset{\|}{C}}{-}OH}$$

Esters **Esters** of long-chain carboxylic acids are the primary constituents of fats. Among the most abundant fats in animal tissues are the triglycerides, which actually contain three ester linkages (the three substituents designated by R in the structure below may be different).

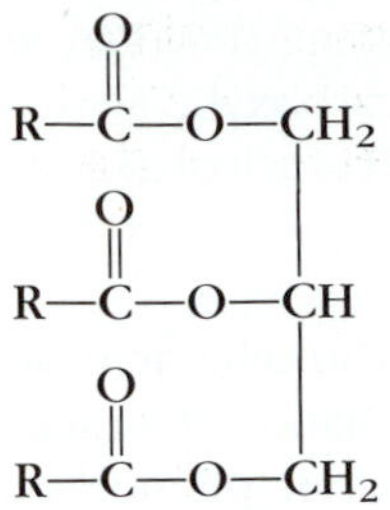

$$\begin{array}{l} R-C(=O)-O-CH_2 \\ \quad\quad\quad\quad\quad\quad | \\ R-C(=O)-O-CH \\ \quad\quad\quad\quad\quad\quad | \\ R-C(=O)-O-CH_2 \end{array}$$

Triglycerides

Alkaline hydrolysis of these compounds produces the salts of the carboxylic acids, which have long been used as soaps. The alkaline hydrolysis of esters in general is called *saponification* (from the Latin stem *sapon*, "soap").

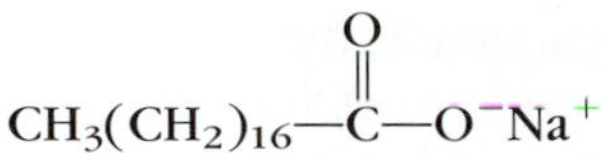

$$CH_3(CH_2)_{16}-C(=O)-O^-Na^+$$

Sodium stearate, a soap constituent

Esters often have rather pleasant, fruity aromas (in marked contrast to the corresponding acids). Many of the smells and flavors of your favorite foods result from complex mixtures of different esters.

Amides Simple **amides** are infrequently encountered outside the chemical laboratory, but polymeric amides are ubiquitous. The bond between nitrogen and the carbonyl group of an amide is the linkage that joins the individual amino acids of all enzymes and other proteins. (In biochemistry it is frequently referred to as the *peptide bond*.)

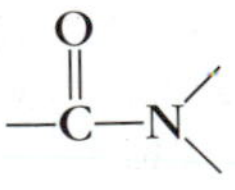

$$-C(=O)-N<$$

Amide linkage

The synthetic polymers known by the name *nylon* are also polyamides.

16.1 NOMENCLATURE

The nomenclature of carboxylic acid derivatives is based on that of the parent acids, so we shall first discuss the naming of carboxylic acids. Next we will present the nomenclature of their derivatives, and finally we will discuss ways of naming polyfunctional compounds. The basic rules for naming carboxylic acids are again an extension of the nomenclature we have introduced in preceding chapters, and you have now been exposed to almost all the major organic functional groups. The rules in the following box introduce you to the nomenclature of carboxylic acids and their derivatives, but they apply to all organic compounds.

Nomenclature of Carboxylic Acids

1. *Locate the principal carbon chain.* This is the longest carbon chain in the molecule *which includes the principal group or has it as a substituent.* The basic name is then formed by replacing the final *-e* of the corresponding alkane by the suffix *-oic acid* for a carboxylic acid.

In *polyfunctional* compounds, the principal group is selected according to the following priorities:

Carboxylic acid > carboxylic acid derivative (anhydride > ester > acyl halide > amide > nitrile) > ketone > aldehyde > alcohol or phenol > amine > ether

2. *Name and number the substituents* on the principal chain.
   (a) The principal chain is numbered in the direction that affords the lower number for *the carbon bearing the principal group. For a carboxylic acid derivative or an aldehyde the carbonyl carbon is understood to be C-1.* For other compounds, the number of the carbon bearing the principal group is prefixed to the name of the principal chain with a hyphen. If the location of the principal group does not distinguish between two different ways of numbering, the substituents on the principal chain are evaluated so that the lower number results at the occasion of the first difference.
   (b) When there is more than one substituent, they are listed in alphabetical order. If several of the substituents are identical, this is indicated by the numerical prefixes *di-, tri-, tetra-, penta-, hexa-,* etc.

3. *Specify the stereochemistry* for any center in the molecule that could exist in two alternative configurations.

Monofunctional Carboxylic Acids

The relationship between the name of an alkane and that of the corresponding carboxylic acid is shown by propane and propanoic acid.

$$CH_3{-}CH_2{-}CH_3 \qquad CH_3{-}CH_2{-}\overset{\overset{\displaystyle O}{\|}}{C}OH$$

Propane — Propan*oic acid*

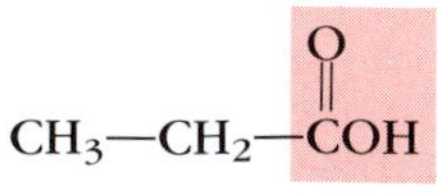

The following examples illustrate the implicit designation of the carboxyl carbon as C-1 of a carboxylic acid.

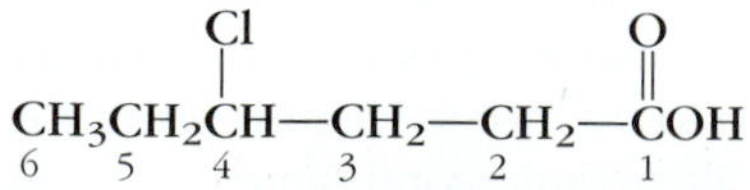

$$\underset{6\ \ 5}{CH_3CH_2}\underset{4}{\overset{\overset{\displaystyle Cl}{|}}{C}H}{-}\underset{3}{CH_2}{-}\underset{2}{CH_2}{-}\underset{1}{\overset{\overset{\displaystyle O}{\|}}{C}OH}$$

4-Chlorohexanoic acid

$$\underset{5}{CH_3}\underset{4}{\overset{\overset{\displaystyle CH_3}{|}}{C}H}{-}\underset{3}{CH_2}\underset{2}{\overset{\overset{\displaystyle Cl}{|}}{C}H}{-}\underset{1}{\overset{\overset{\displaystyle O}{\|}}{C}OH}$$

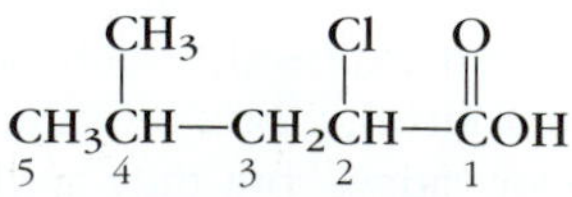

2-Chloro-4-methylpentanoic acid

When the carboxyl group is attached directly to a ring, the principal chain has only a single carbon. The ring is treated as a substituent of the parent "-carboxylic acid," and the numbering refers to the ring atoms (where C-1 is attached to the carboxyl group).

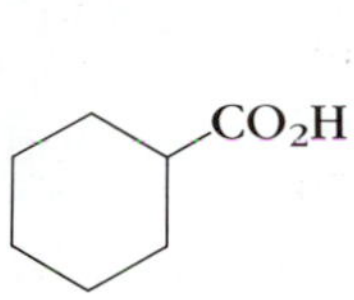

$$\text{C}_6\text{H}_{11}\text{–CO}_2\text{H}$$

Cyclohexanecarboxylic acid

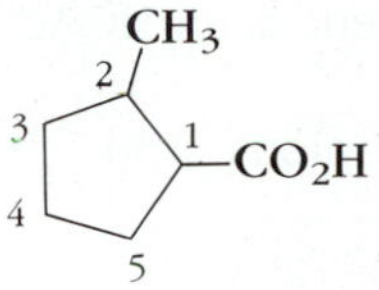

$$\text{2-CH}_3\text{–C}_5\text{H}_8\text{–1-CO}_2\text{H}$$

2-Methylcyclopentanecarboxylic acid

Common Names

Many carboxylic acids isolated from natural sources were assigned names before their structures were determined, and the use of these common names persists. Some of the most frequently encountered common names are listed in Table 16.1. The names *formic acid* and *acetic acid* are still used almost exclusively for the C_1 and C_2 acids, and the IUPAC rules for nomenclature still list the common names, *propionic acid* and *butyric acid,* as the normally preferred names of the C_3 and C_4 compounds. Nevertheless, systematic names are replacing the common names with increasing regularity.

Carboxylic Acid Derivatives

Acid anhydrides have names with the same stems as those of the parent acids, but *acid* is replaced by *anhydride.*

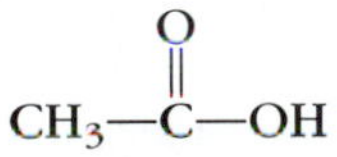

$$CH_3—\overset{O}{\overset{\|}{C}}—OH \qquad CH_3—\overset{O}{\overset{\|}{C}}—O—\overset{O}{\overset{\|}{C}}—CH_3$$

Acetic acid 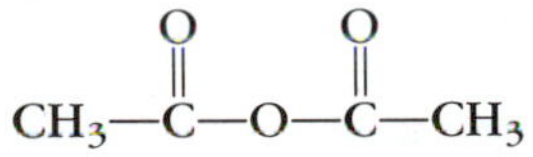 — Acetic anhydride

Acyl groups are the $R—\overset{O}{\overset{\|}{C}}—$ portions of the corresponding acids, $R—\overset{O}{\overset{\|}{C}}—OH$. These are named by replacing *-ic acid* with *-yl group.*

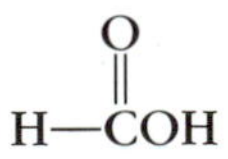

$$H—\overset{O}{\overset{\|}{C}}OH \qquad CH_3\overset{O}{\overset{\|}{C}}OH \qquad CH_3\overset{CH_3}{\overset{|}{C}}H—\overset{O}{\overset{\|}{C}}OH$$

Formic acid  — Acetic acid 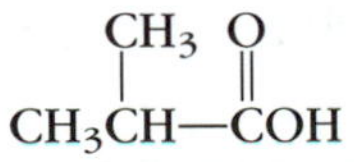 — 2-Methylpropanoic acid

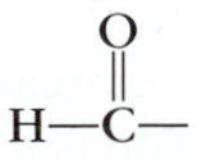

$$H—\overset{O}{\overset{\|}{C}}— \qquad CH_3\overset{O}{\overset{\|}{C}}— \qquad CH_3\overset{CH_3}{\overset{|}{C}}H—\overset{O}{\overset{\|}{C}}—$$

Formyl group 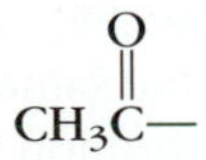 — Acetyl group 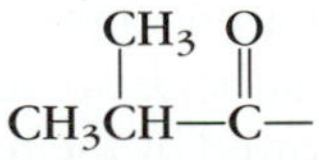 — 2-Methylpropanoyl group

Acid halides (or **acyl halides**) are named as the halides of the corresponding acyl groups.

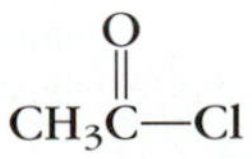

$$CH_3\overset{O}{\overset{\|}{C}}—Cl \qquad CH_3CH_2CH_2\overset{O}{\overset{\|}{C}}—Cl \qquad CH_3(CH_2)_4\overset{O}{\overset{\|}{C}}—Br$$

Acetyl chloride 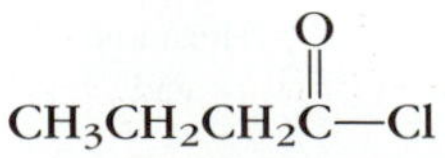 — Butanoyl chloride 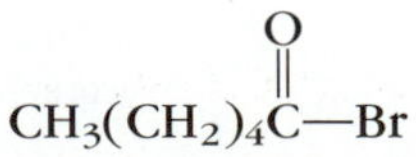 — Hexanoyl bromide

TABLE 16.1 Some Frequently Encountered Common Names of Carboxylic Acids

Acid	Common Name	Systematic Name*
$H{-}CO_2H$	Formic acid	(Methanoic acid)
$CH_3{-}CO_2H$	Acetic acid, HOAc	(Ethanoic acid)
$CH_3CH_2{-}CO_2H$	Propionic acid	(Propanoic acid)
$CH_3(CH_2)_2{-}CO_2H$	Butyric acid	(Butanoic acid)
$CH_3\overset{\overset{CH_3}{\vert}}{C}H{-}CO_2H$	Isobutyric acid	(2-Methylpropanoic acid)
$CH_3(CH_2)_3{-}CO_2H$	Valeric acid	(Pentanoic acid)
$CH_3\overset{\overset{CH_3}{\vert}}{C}H{-}CH_2{-}CO_2H$	Isovaleric acid	(2-Methylpropanoic acid)
$CH_3{-}\underset{\underset{CH_3}{\vert}}{\overset{\overset{CH_3}{\vert}}{C}}{-}CO_2H$	Pivalic acid	2,2-Dimethylpropanoic acid
$C_6H_5{-}CO_2H$	Benzoic acid	(Benzenecarboxylic acid)
$CH_3(CH_2)_{10}{-}CO_2H$	Lauric acid	Dodecanoic acid
$CH_3(CH_2)_{12}{-}CO_2H$	Myristic acid	Tetradecanoic acid
$CH_3(CH_2)_{14}{-}CO_2H$	Palmitic acid	Hexadecanoic acid
$CH_3(CH_2)_{16}{-}CO_2H$	Stearic acid	Octadecanoic acid

*Names shown in parentheses are used less frequently.

The abbreviation *Ac* is often used for the acetyl group, so acetyl chloride can be written as AcCl. Similarly, AcOH (or HOAc) and Ac_2O are common abbreviations for acetic acid and acetic anhydride, respectively.

Salts of carboxylic acids are named in the same way as inorganic salts: the name of the cation is followed by the name of the anion. The anion is named by changing *-ic acid* to *-ate*.

$HO{-}\overset{\overset{O}{\Vert}}{C}{-}CH_3$	$Na^{+}\,{}^{-}O\overset{\overset{O}{\Vert}}{C}CH_3$	$CH_3(CH_2)_4\overset{\overset{O}{\Vert}}{C}OH$	$CH_3(CH_2)_4\overset{\overset{O}{\Vert}}{C}O^{-}\,NH_4^{+}$
Acetic acid	Sodium acetate	Hexanoic acid	Ammonium hexanoate

Esters are named in the same way as salts, with the name of the group bonded to oxygen being used in place of the cation.

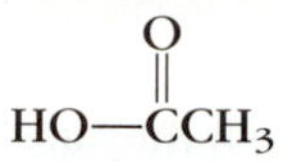

$$HO{-}\overset{O}{\overset{\|}{C}}CH_3 \qquad CH_3{-}O\overset{O}{\overset{\|}{C}}CH_3 \qquad CH_3{-}\overset{CH_3}{\overset{|}{C}H}{-}\overset{O}{\overset{\|}{C}}OH \qquad CH_3{-}\overset{CH_3}{\overset{|}{C}H}{-}\overset{O}{\overset{\|}{C}}{-}O{-}CH_2CH_3$$

Acetic acid 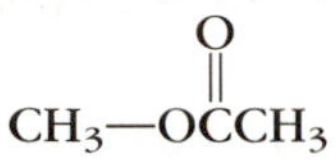 — Methyl acetate 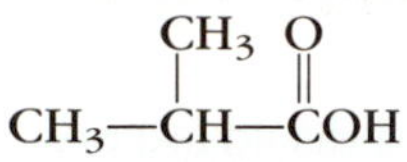 — 2-Methylpropanoic acid 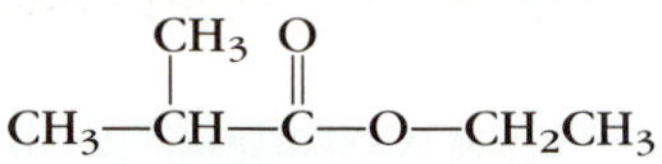 — Ethyl 2-methylpropanoate

The preceding esters are referred to as a methyl ester and an ethyl ester, respectively.

Amides in which the $—NH_2$ group is unsubstituted are named by changing *-oic acid* (or *-ic acid* for common names) to *-amide.*

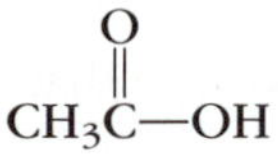

$$CH_3\overset{O}{\overset{\|}{C}}{-}OH \qquad CH_3\overset{O}{\overset{\|}{C}}{-}NH_2 \qquad CH_3(CH_2)_4\overset{O}{\overset{\|}{C}}OH \qquad CH_3(CH_2)_4\overset{O}{\overset{\|}{C}}{-}NH_2$$

Acetic acid 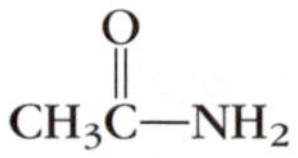 — Acetamide 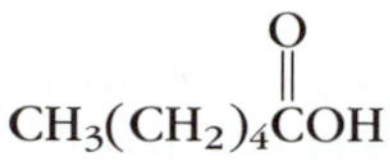 — Hexanoic acid 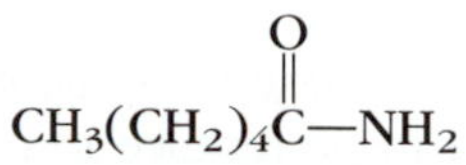 — Hexanamide

When one or both of the hydrogens of the $—NH_2$ group are replaced by an alkyl group, the compounds are named as *N*-substituted or *N,N*-disubstituted amides.

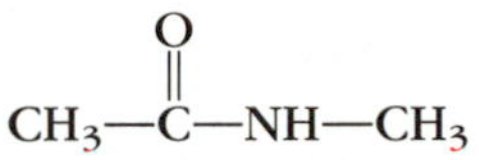

$$CH_3{-}\overset{O}{\overset{\|}{C}}{-}NH{-}CH_3 \qquad CH_3(CH_2)_4{-}\overset{O}{\overset{\|}{C}}{-}N(CH_2CH_3)_2$$

N-Methylacetamide 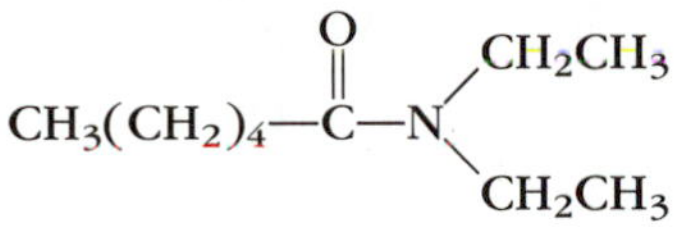 — *N,N*-Diethylhexanamide

Amides are classified as primary, secondary, or tertiary according to the total number of carbon atoms bonded to the nitrogen. $R{-}\overset{O}{\overset{\|}{C}}{-}NH_2$, for example, would be a primary amide.

Nitriles ($R—C{\equiv}N$) are named by several methods. When the corresponding acid ($R—CO_2H$) has a common name, the *-ic acid* (or *-oic acid*) is changed to *-onitrile.*

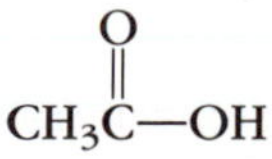

$$CH_3\overset{O}{\overset{\|}{C}}{-}OH \qquad CH_3{-}C{\equiv}N \qquad CH_3{-}\underset{CH_3}{\underset{|}{\overset{CH_3}{\overset{|}{C}}}}{-}CO_2H \qquad CH_3{-}\underset{CH_3}{\underset{|}{\overset{CH_3}{\overset{|}{C}}}}{-}C{\equiv}N$$

Acetic acid  — Acetonitrile 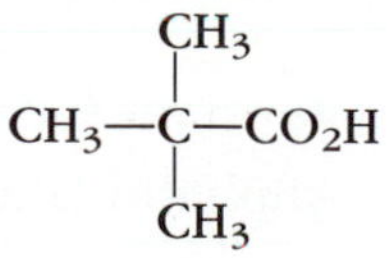 — Pivalic acid 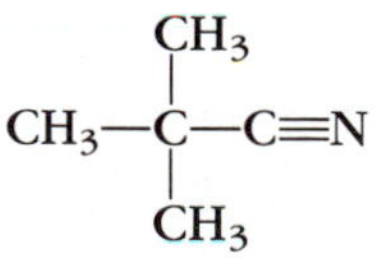 — Pivalonitrile

With systematic nomenclature the nitriles ($R—CN$) are named according to the parent alkanes ($R—CH_3$), and *-nitrile* is added to the name of the hydrocarbon.

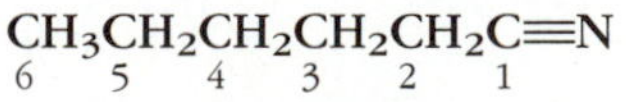

$$\underset{6}{CH_3}\underset{5}{CH_2}\underset{4}{CH_2}\underset{3}{CH_2}\underset{2}{CH_2}\underset{1}{C}{\equiv}N \qquad \underset{4}{CH_3}\underset{3}{\overset{CH_3}{\overset{|}{C}H}}{-}\underset{2}{CH_2}\underset{1}{C}{\equiv}N$$

Hexanenitrile 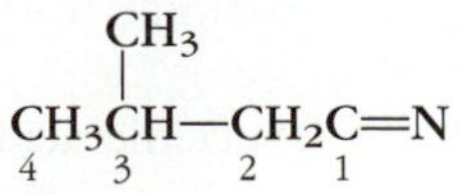 — 3-Methylbutanenitrile

Nitriles are on occasion named as *alkyl cyanides.* This would lead to a different parent name, because the carbon of the $—CN$ group would be part of the substituent, not part of the principal chain. Hexanenitrile, for example, could be called pentyl cyanide by this (less preferred) method.

Dicarboxylic Acids and Their Derivatives

If the two carboxyl groups of a diacid are on a single carbon chain (i.e., if they are not substituents on a ring), the principal chain has the two carboxyl carbons as its ends. The name is obtained by adding the suffix *-dioic acid* to the parent alkane. (As always, the chain is numbered in the direction that affords lower numbers for any substituents.)

$$\underset{1}{HO_2C}-\underset{2\ \ 3}{CH_2CH_2}-\underset{4}{CO_2H}$$

Butanedioic acid

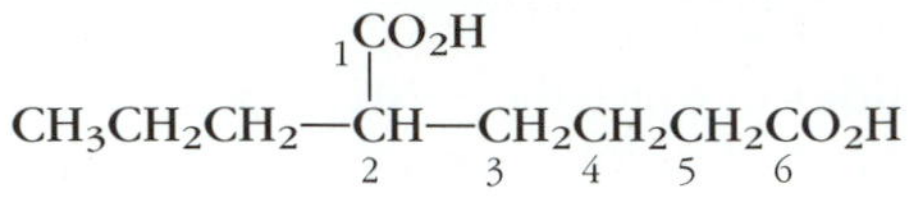

$$CH_3CH_2CH_2-\underset{2}{\overset{{}_1CO_2H}{\overset{|}{C}H}}-\underset{3\ \ 4\ \ 5\ \ 6}{CH_2CH_2CH_2CO_2H}$$

2-Propylhexanedioic acid

Several frequently encountered dicarboxylic acids have common names that are used in preference to the systematic names (Table 16.2).

Cyclic anhydrides of dicarboxylic acids are named by replacing *acid* with *anhydride.*

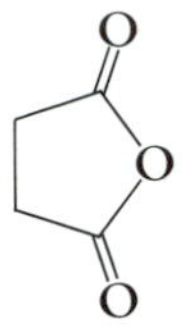

Succinic anhydride 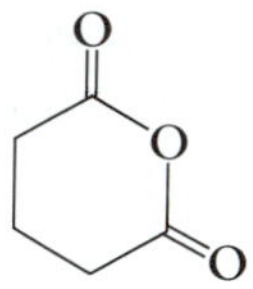   Glutaric anhydride

The analogous compounds in which both carbonyls are bonded to an NH group are called *imides,* and the *N*-bromo derivative of succinic acid is already familiar as a reagent for allylic bromination (Section 8.4).

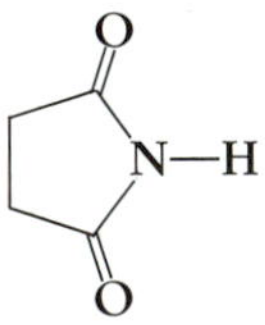

Succinimide 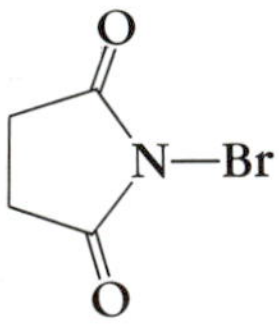   *N*-Bromosuccinimide

TABLE 16.2 Common Names of Some Aliphatic Dicarboxylic Acids*

Acid	Common Name	Systematic Name
HO_2C-CO_2H	Oxalic acid	(Ethanedioic acid)
$HO_2C-CH_2-CO_2H$	Malonic acid	(Propanedioic acid)
$HO_2C-CH_2CH_2-CO_2H$	Succinic acid	(Butanedioic acid)
$HO_2C-(CH_2)_3-CO_2H$	Glutaric acid	(Pentanedioic acid)
$HO_2C-(CH_2)_4-CO_2H$	Adipic acid	(Hexanedioic acid)

*Systematic names shown in parentheses are used less frequently.

Unsaturated Carboxylic Acids

Unsaturated carboxylic acids are named so that the principal chain includes the multiple bond.

$$\underset{6}{HC}\equiv\underset{5}{C}-\underset{4}{CH_2}\underset{3}{CH_2}\underset{2}{CH_2}\underset{1}{CO_2H}$$

5-Hexynoic acid

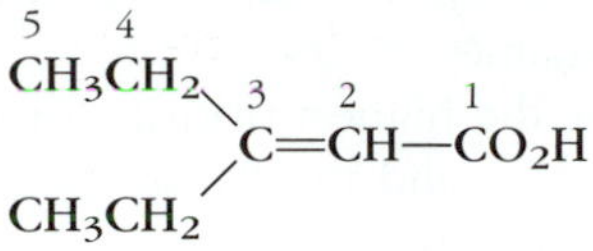

$$(\overset{5}{CH_3}\overset{4}{CH_2})(CH_3CH_2)\overset{3}{C}=\overset{2}{CH}-\overset{1}{CO_2H}$$

3-Ethyl-2-pentenoic acid

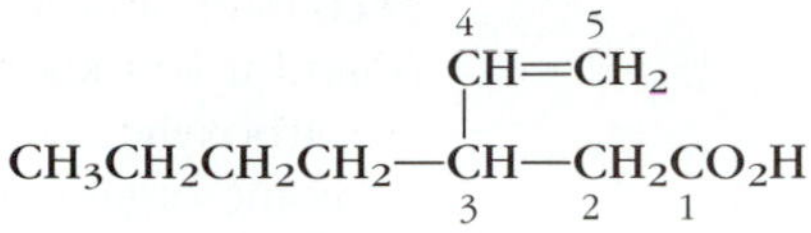

$$CH_3CH_2CH_2CH_2-\underset{3}{CH}(\overset{4}{CH}=\overset{5}{CH_2})-\underset{2}{CH_2}\underset{1}{CO_2H}$$

3-Butyl-4-pentenoic acid

A number of unsaturated acids and diacids also have common names that are frequently used in place of the systematic names, and these are listed in Table 16.3.

Polyfunctional Compounds

In most respects polyfunctional compounds present few difficulties. However, you must first establish which is the *principal functional group,* because this determines the *principal chain* and other functional groups are named as substituents. Previously you had only encountered a few functional groups, so the complete list

TABLE 16.3 Common Names of Some Unsaturated Carboxylic Acids and Diacids*

Acid	Common Name	Systematic Name
$CH_2=CH-CO_2H$	Acrylic acid	(Propenoic acid)
$HC\equiv C-CO_2H$	Propiolic acid	(Propynoic acid)
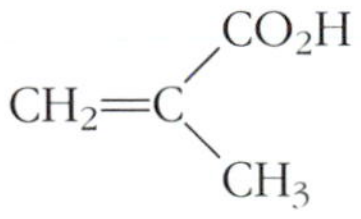 $CH_2=C(CO_2H)(CH_3)$	Methacrylic acid	(2-Methylpropenoic acid)
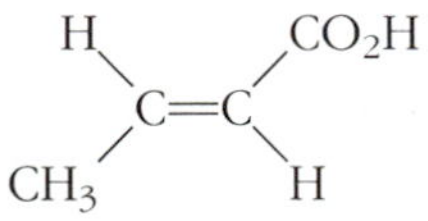 (H)(CH₃)C=C(CO₂H)(H) — $H(CH_3)C=C(CO_2H)H$	Crotonic acid	[(*E*)-2-Butenoic acid]
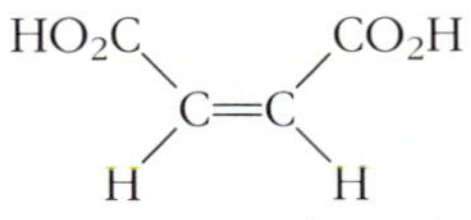 $(HO_2C)(H)C=C(CO_2H)(H)$	Maleic acid	[(*Z*)-2-Butenedioic acid]
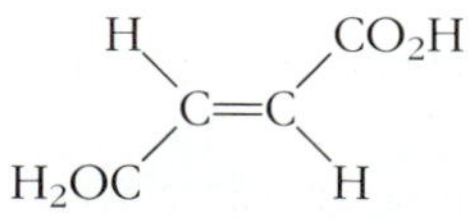 $(H)(H_2OC)C=C(CO_2H)(H)$	Fumaric acid	[(*E*)-2-Butenedioic acid]

*Systematic names shown in parentheses (or brackets) are used less frequently.

of priorities was deferred until this chapter. Even here we are only presenting the basics, and a more complete treatment of nomenclature should be consulted if you wish to name more complex structures. Nevertheless, our presentation of nomenclature will enable you to write the correct name for most of the structures you encounter, and it will allow you to draw the correct structural formula of a compound if you know its name.

Carboxylic acids have the highest priority for selecting the principal chain of a nonionic organic molecule, and the list in the following box shows the relative priority of other functional groups.

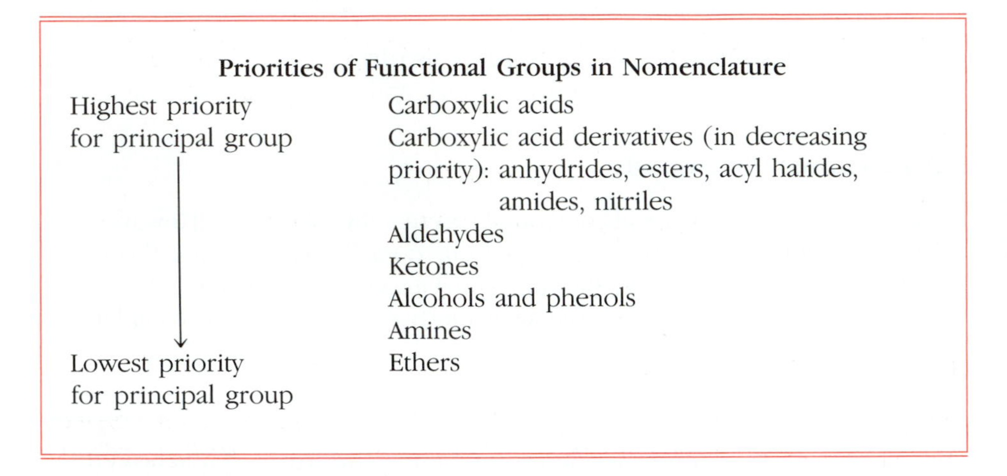

Priorities of Functional Groups in Nomenclature

Highest priority for principal group	Carboxylic acids
↓	Carboxylic acid derivatives (in decreasing priority): anhydrides, esters, acyl halides, amides, nitriles
	Aldehydes
	Ketones
	Alcohols and phenols
	Amines
Lowest priority for principal group	Ethers

After the principal group has been selected, other groups in a polyfunctional compound must be named as substituents. This can be complicated, and we will only present the basic ideas. The names for some of the more common groups are listed in the following box.

Names of Substituents

Substituent	Name
$—CO_2H$	Carboxy
$—CO_2R$	Alkoxycarbonyl
$—C{\equiv}N$	Cyano
—CHO	Formyl
=O	Oxo
—OH	Hydroxy
$—NH_2$	Amino
—OR	Alkoxy

The following examples are all carboxylic acids and are therefore named so that the principal chain includes the $—CO_2H$ group (or groups).

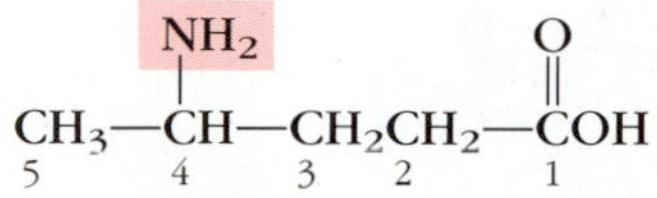

$$\underset{5}{CH_3}—\underset{4}{\overset{NH_2}{\overset{|}{CH}}}—\underset{3}{CH_2}\underset{2}{CH_2}—\underset{1}{\overset{O}{\overset{\|}{C}}OH}$$

4-*Amino*pentanoic acid

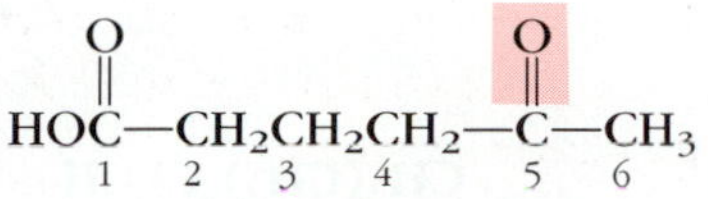

$$HO\underset{1}{\overset{O}{\overset{\|}{C}}}—\underset{2}{CH_2}\underset{3}{CH_2}\underset{4}{CH_2}—\underset{5}{\overset{O}{\overset{\|}{C}}}—\underset{6}{CH_3}$$

5-*Oxo*hexanoic acid

$$\underset{7}{HO_2C}—\underset{6}{CH_2}\underset{5}{CH_2}\underset{4}{CH_2}\underset{3}{CH_2}—\underset{2}{\overset{CN}{\overset{|}{CH}}}—\underset{1}{CO_2H}$$

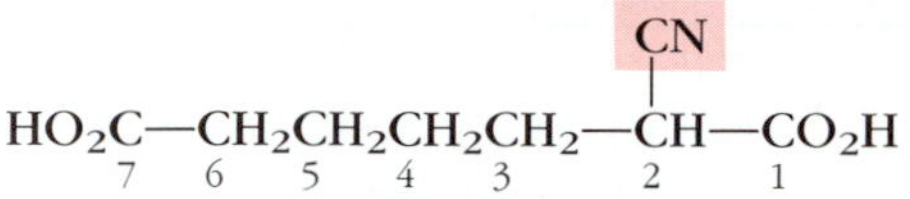

2-*Cyano*heptanedioic acid

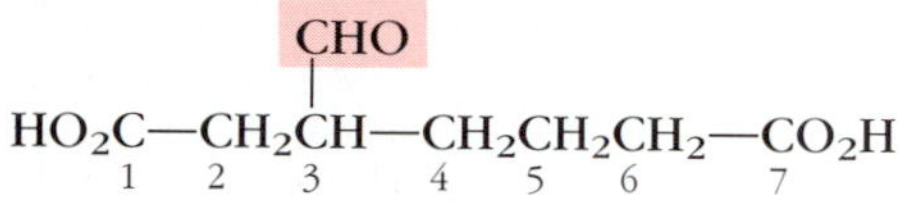

$$\underset{1}{HO_2C}—\underset{2}{CH_2}\underset{3}{\overset{CHO}{\overset{|}{CH}}}—\underset{4}{CH_2}\underset{5}{CH_2}\underset{6}{CH_2}—\underset{7}{CO_2H}$$

3-*Formyl*heptanedioic acid

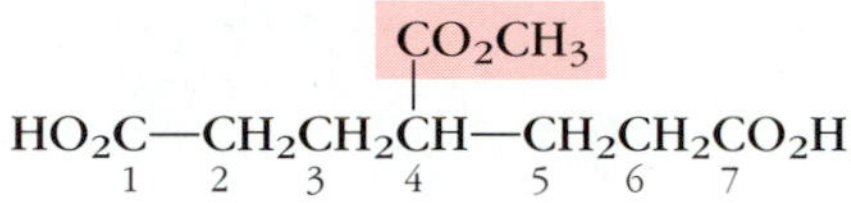

$$\underset{1}{HO_2C}—\underset{2}{CH_2}\underset{3}{CH_2}\underset{4}{\overset{CO_2CH_3}{\overset{|}{CH}}}—\underset{5}{CH_2}\underset{6}{CH_2}\underset{7}{CO_2H}$$

4-*Methoxycarbonyl*heptanedioic acid

Summary of Nomenclature

The discussion of nomenclature that we have presented in this section will allow you to name most aliphatic carboxylic acids. Table 16.4 summarizes the nomenclature of acids and their derivatives for three compounds: acetic acid (common name); hexanoic acid (systematic name); and cyclohexanecarboxylic acid (alternative systematic name).

EXERCISE 16.1

Name each of the following compounds:

(a) 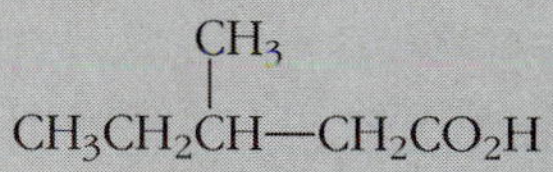 $CH_3CH_2\overset{CH_3}{\overset{|}{C}H}—CH_2CO_2H$

(b) $CH_3CH_2\underset{CO_2H}{\underset{|}{C}H}—CH_2CH_2CO_2H$

(c) $CH_3(CH_2)_7—\overset{O}{\overset{\|}{C}}—NH_2$

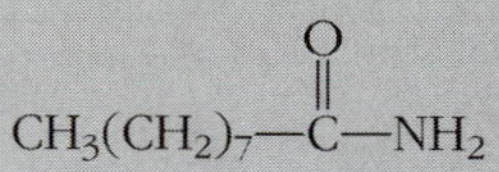

(d) 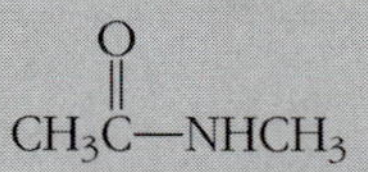 $CH_3\overset{O}{\overset{\|}{C}}—NHCH_3$

(e) 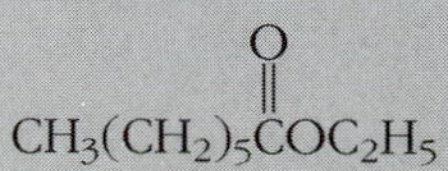 $CH_3(CH_2)_5\overset{O}{\overset{\|}{C}}OC_2H_5$

(f) 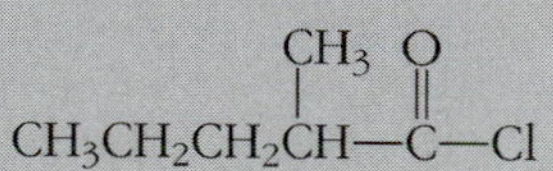 $CH_3CH_2CH_2\overset{CH_3}{\overset{|}{C}H}—\overset{O}{\overset{\|}{C}}—Cl$

EXERCISE 16.2

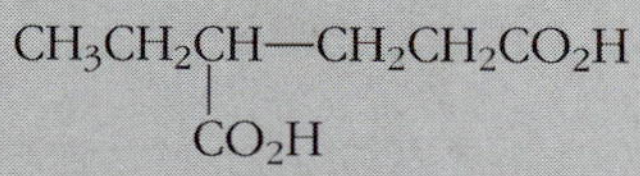

Name each of the following difunctional compounds:

(a) 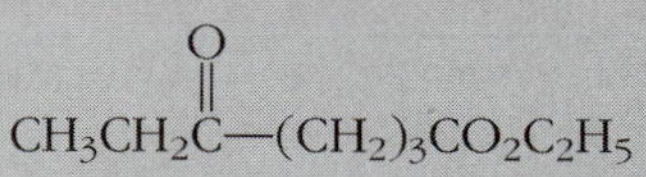 $CH_3CH_2\overset{O}{\overset{\|}{C}}—(CH_2)_3CO_2C_2H_5$

(b) 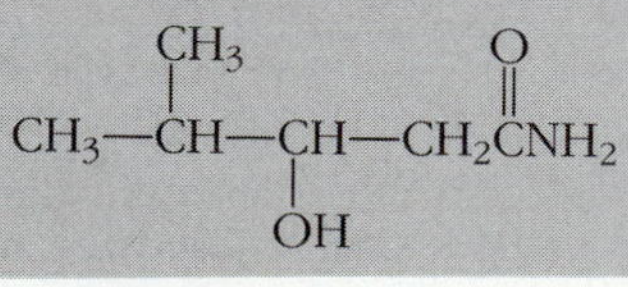 $CH_3—\overset{CH_3}{\overset{|}{C}H}—\underset{OH}{\underset{|}{C}H}—CH_2\overset{O}{\overset{\|}{C}}NH_2$

TABLE 16.4 Summary of Nomenclature for Carboxylic Acid Derivatives

Type of Derivative	$CH_3C(=O)OH$ Acetic Acid (Common Name)	$CH_3(CH_2)_4CO_2H$ Hexanoic Acid (Systematic Name)	$C_6H_{11}C(=O)OH$ (cyclohexane ring) Cyclohexanecarboxylic Acid (Alternative Systematic Name)
Anhydride	$CH_3C(=O)OC(=O)CH_3$ Acetic anhydride (Ac_2O)	$(CH_3(CH_2)_4C(=O))_2O$ Hexanoic anhydride	$C_6H_{11}C(=O)OC(=O)C_6H_{11}$ Cyclohexanecarboxylic anhydride
Acid chloride	$CH_3C(=O)Cl$ Acetyl chloride (AcCl)	$CH_3(CH_2)_4C(=O)Cl$ Hexanoyl chloride	$C_6H_{11}C(=O)Cl$ Cyclohexanecarbonyl chloride
Ethyl ester	$CH_3C(=O)OC_2H_5$ Ethyl acetate (EtOAc)	$CH_3(CH_2)_4C(=O)OC_2H_5$ Ethyl hexanoate	$C_6H_{11}C(=O)OC_2H_5$ Ethyl cyclohexanecarboxylate
Amide	$CH_3C(=O)NH_2$ Acetamide ($AcNH_2$)	$CH_3(CH_2)_4C(=O)NH_2$ Hexanamide	$C_6H_{11}C(=O)NH_2$ Cyclohexanecarboxamide
Nitrile	CH_3CN Acetonitrile	$CH_3(CH_2)_4CN$ Hexanenitrile	$C_6H_{11}CN$ Cyclohexanecarbonitrile

EXERCISE 16.2 (continued)

(c) $CH_3C(=O)CH_2CH_2C(=O)Cl$

(d) $CH_3CH_2CH(CH_2CH{=}CH_2)—CH_2CH_2C(=O)O^-Na^+$

16.2 PHYSICAL PROPERTIES

The carboxyl group and the corresponding $—C(=O)—X$ groups of carboxylic acid derivatives are highly polar, and this is reflected in the physical properties of these compounds. The lower-molecular-weight derivatives tend to exhibit high water solubility, and compounds such as formic acid, acetic acid, formamide, and acetamide are completely miscible with water. This water solubility results in large part from hydrogen bonding between water and the $-CO_2H$ or $-CONH_2$ groups, which can act as both donors and acceptors in hydrogen bonding.

```
 :O:                    :O:                          :O:---H—Ö—H
  ‖         ..           ‖   ..      ..               ‖
—C—OH---:OH₂           —C—NH---:OH₂                 —C—X
                             |
```

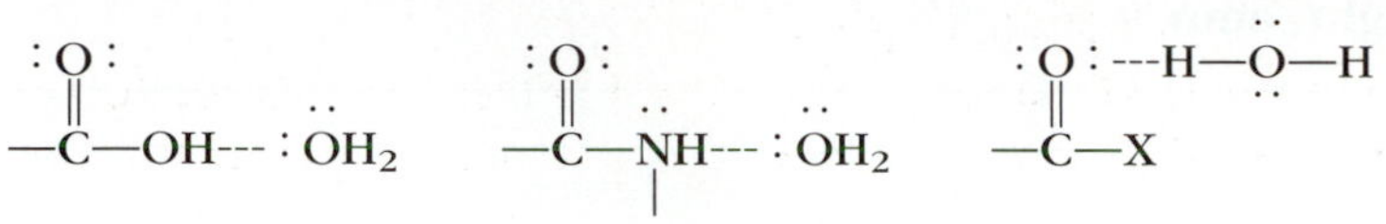

The remaining carboxylic acid derivatives have no hydrogens bound to oxygen or nitrogen and can only serve as acceptors in hydrogen bonding. Their decreased water solubilities are more similar to those of aldehydes and ketones.

As the chain length of carboxylic acids and their derivatives is increased, the compounds become increasingly nonpolar. This results in decreased water solubility, and the compounds with nine or more carbons are almost completely insoluble in water. The decrease in water solubility is accompanied by an increase in the solubility of these compounds in organic solvents.

Hydrogen bonding also plays an important role in other properties of carboxylic acid derivatives. Carboxylic acids, for example, can form *dimeric* structures with two hydrogen bonds.

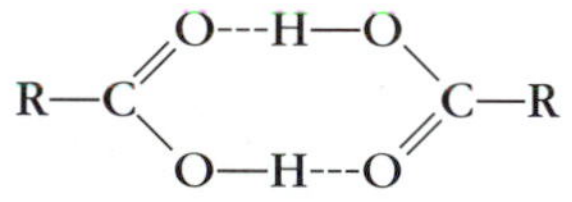

```
      O---H—O
     //       \
R—C             C—R
     \        //
      O—H---O
```

Strong intermolecular association such as this favors the crystalline state over the less ordered liquid state and results in high melting points for acids and amides. Some intermolecular association is still present in the liquid state, and this results in high boiling points for acids and amides as well. These patterns in physical properties are illustrated in Table 16.5, which summarizes the properties of a series of compounds containing the acetyl group.

Only two of the compounds in Table 16.5 have melting points above 0°C, namely, acetic acid and acetamide, which melt at 17°C and 82°C, respectively. These are the only two compounds in the table that are capable of intermolecular hydrogen bonding. This hydrogen bonding also results in boiling points that are much higher than those of other acetyl derivatives of similar molecular weight.

Boiling points tend to increase regularly with molecular weight, but there are three clear exceptions in Table 16.5. We can explain the first two (acetic acid and acetamide) on the basis of hydrogen bonding, but what about *N,N*-dimethylacetamide? Its boiling point of 165°C is even higher than that of acetic anhydride, which has a substantially greater molecular weight. The answer to this question comes from the very high polarity of the amide linkage, which leads to strong dipolar attractions between molecules in the liquid. We can illustrate this polarity with the following resonance structures:

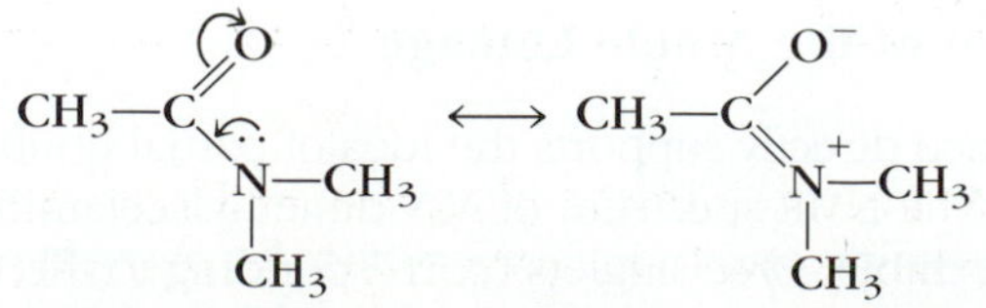

```
        ↷O                          O⁻
        //                          /
CH₃—C ↶..       ⟷     CH₃—C      +
        \                           \\
         N—CH₃                       N—CH₃
         |                           |
         CH₃                         CH₃
```

TABLE 16.5 Melting and Boiling Points of Acetic Acid and Related Acetyl Derivatives

Compound	Molecular Weight, amu	Melting Point, °C	Boiling Point, °C
$CH_3C(=O)H$	44	−121	21
$CH_3C(=O)CH_3$	58	−93	56
$CH_3C(=O)NH_2$	59	82	221
$CH_3C(=O)OH$	60	17	118
$CH_3C(=O)OCH_3$	74	−98	57
$CH_3C(=O)Cl$	78.5	−112	51
$CH_3C(=O)N(CH_3)_2$	87	−20	165
$CH_3C(=O)OC(=O)CH_3$	102	−73	140

In addition to explaining the high polarity of an amide group, the second resonance contributor also suggests that there should be some double bond character to the carbon-nitrogen linkage. This idea is supported by molecular orbital considerations because the *p* atomic orbitals of the nitrogen and the adjacent carbon and oxygen of the carbonyl interact in the same manner as the *p* orbitals of the allyl system (Section 8.3).

Restricted Rotation of the Amide Linkage

Experimental evidence directly supports the idea of partial double bond character of the amide bond. The NMR spectrum of *N,N*-dimethylacetamide (Figure 16.1) at room temperature exhibits *three* singlets (corresponding to three methyl groups),

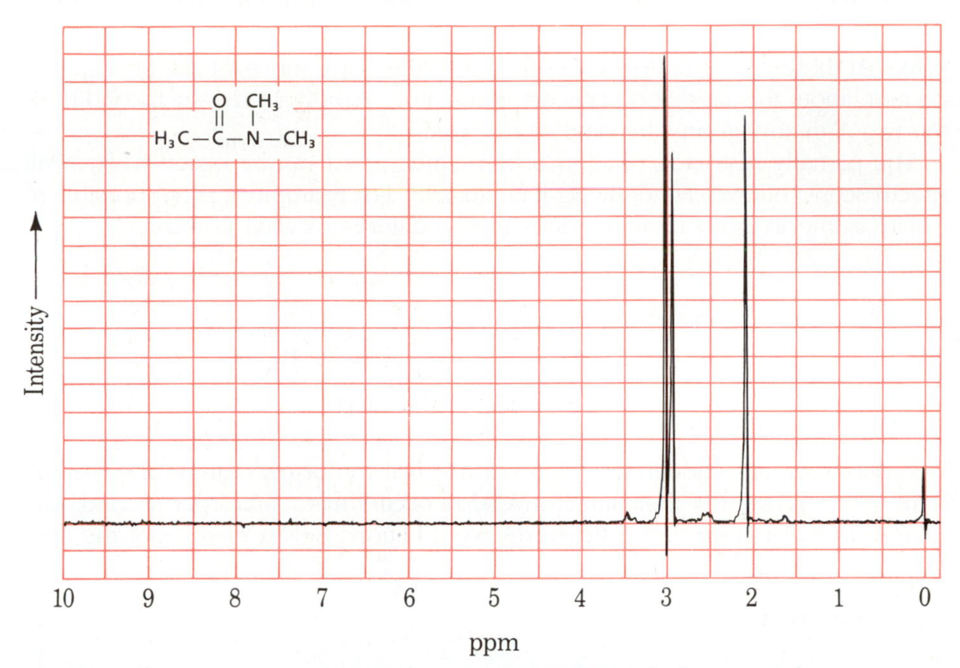

Figure 16.1 Proton NMR Spectrum of *N,N*-Dimethylacetamide.

and this means that the two *N*-methyl groups are *not equivalent.* We must therefore conclude that interchange of the two methyl groups via rotation about the nitrogen-carbonyl bond is slow (Figure 16.2).

At the transition state (Figure 16.2, center) the *p* orbital on nitrogen is perpendicular to the *p* orbitals of the carbonyl group, which eliminates the favorable orbital interaction of the optimum geometry. This rotation requires about 10–15 kcal/mol, which is substantially more than the barriers of 3–5 kcal/mol observed for rotation about single bonds in alkanes (Section 2.3). At higher temperatures the

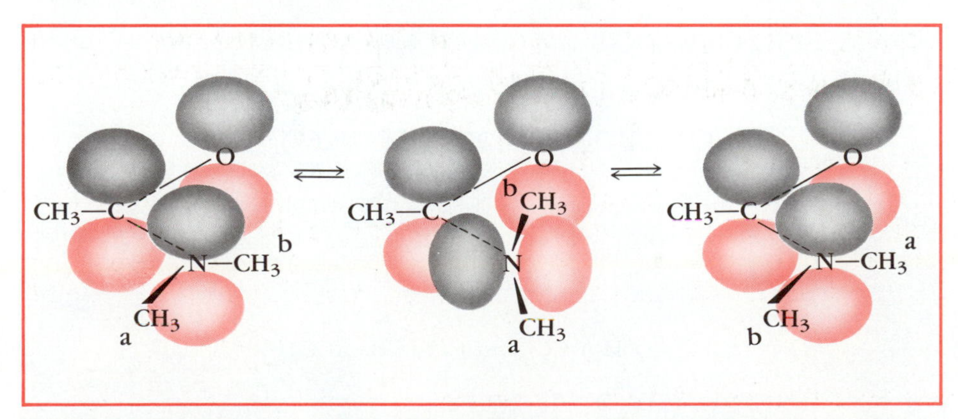

Figure 16.2 Molecular Orbital Description of Hindered Rotation About the Carbon-nitrogen Bond of an Amide.
Only the in phase atomic *p* orbital contributions are shown.

two *N*-methyl signals merge to a single resonance (which integrates for six protons), with a chemical shift midway between the values for the previously separated peaks. At these higher temperatures the molecules have more kinetic energy, and rotation about the amide bond occurs much more rapidly. Under such conditions the two *N*-methyl groups become *equivalent.*

The partially restricted rotation of the amide bond can be observed by NMR spectroscopy, but we cannot detect it chemically. For example, it is not possible to isolate isomeric forms of an amide with two different *N*-alkyl groups.

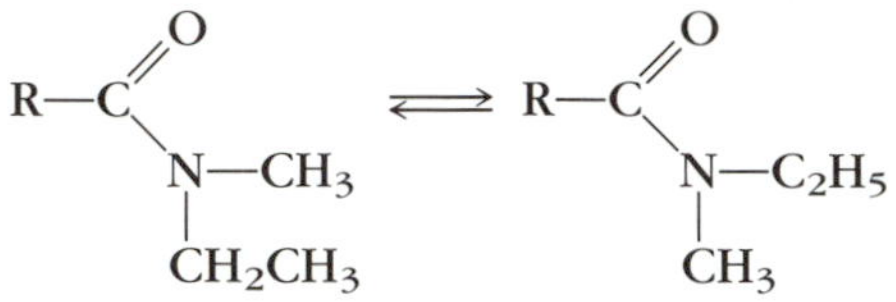

$$R{-}C({=}O){-}N(CH_3)(CH_2CH_3) \rightleftharpoons R{-}C({=}O){-}N(C_2H_5)(CH_3)$$

The rotation is slow enough that the two individual conformers can be observed by NMR spectroscopy, but their interconversion occurs many times per second. Any attempt to isolate one of the structures would afford only a mixture of the two.

EXERCISE 16.3

(a) For each of the following pairs of compounds, which of the two would be more soluble in water? Why?

$$CH_3CH_2\overset{O}{\overset{\|}{C}}NH_2 \text{ and } CH_3(CH_2)_7\overset{O}{\overset{\|}{C}}NH_2$$

$$CH_3(CH_2)_6\overset{O}{\overset{\|}{C}}OH \text{ and } H\overset{O}{\overset{\|}{C}}O(CH_2)_6CH_3$$

(b) For each of the following pairs of compounds, which of the two would be higher-boiling? Why?

$$CH_3(CH_2)_3\overset{O}{\overset{\|}{C}}OCH_3 \text{ and } CH_3(CH_2)_3\overset{O}{\overset{\|}{C}}{-}O{-}\overset{O}{\overset{\|}{C}}(CH_2)_3CH_3$$

$$CH_3(CH_2)_3\overset{O}{\overset{\|}{C}}OCH_3 \text{ and } CH_3(CH_2)_3\overset{O}{\overset{\|}{C}}OH$$

(c) For each of the following pairs of compounds, which of the two would be higher-melting? Why?

$$CH_3(CH_2)_3\overset{O}{\overset{\|}{C}}Cl \text{ and } CH_3(CH_2)_3\overset{O}{\overset{\|}{C}}OH$$

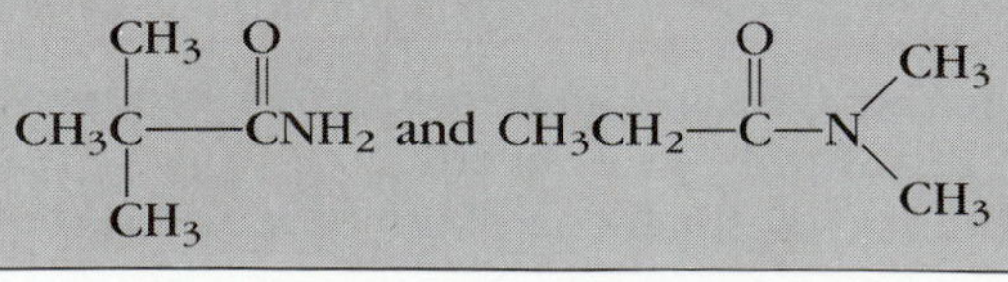

$$(CH_3)_3C{-}\overset{O}{\overset{\|}{C}}NH_2 \text{ and } CH_3CH_2{-}\overset{O}{\overset{\|}{C}}{-}N(CH_3)_2$$

16.3 REACTIVITY OF CARBOXYLIC ACIDS AND THEIR DERIVATIVES

Carboxylic acid derivatives, like aldehydes and ketones, undergo the characteristic reactions of the carbonyl group: attack of nucleophiles at the carbonyl carbon and acid-base reactions. We will first consider the acid-base reactions and then nucleophilic attack at the acyl group.

Carboxyl Derivatives as Bases

Carboxylic acids and their derivatives are all weak bases. They are somewhat weaker than but generally comparable in base strength with water. Protonation occurs at carbonyl oxygen to give an oxonium ion.

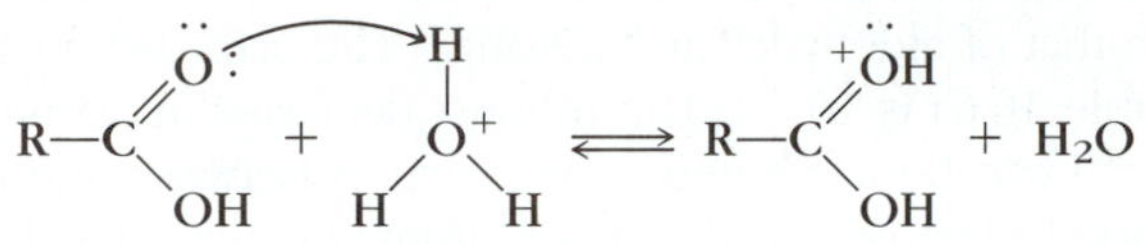

$$R{-}C(=O){-}OH + H_3O^+ \rightleftharpoons R{-}C(={}^+OH){-}OH + H_2O$$

You can readily understand the preference for protonation on carbonyl oxygen by looking at the three resonance contributors. In the case of the carboxylic acid itself, two equivalent oxonium ions can be drawn.

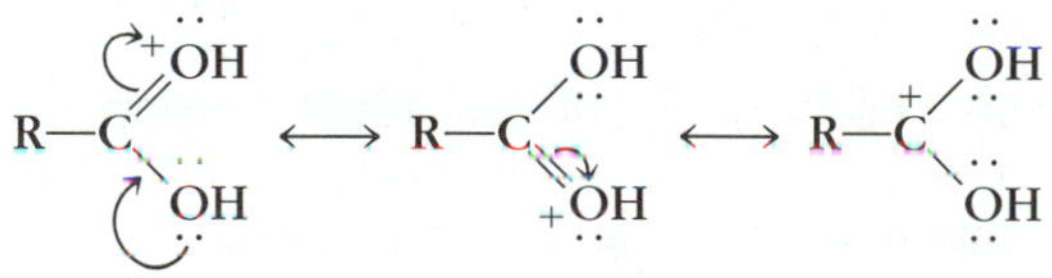

$$R{-}C(={}^+OH){-}OH \longleftrightarrow R{-}C(OH)={}^+OH \longleftrightarrow R{-}C^+(OH){-}OH$$

Protonation at the other oxygen would instead produce a less stable oxonium ion for which no other reasonable resonance structures could be drawn.

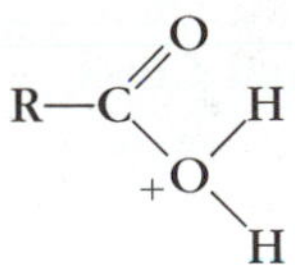

$$R{-}C(=O){-}{}^+OH_2$$

Protonation of other derivatives of carboxylic acids also occurs preferentially at carbonyl oxygen for the same reasons. The following structures show the ions produced by protonation of carbonyl oxygen in an acid chloride, ester, and amide, respectively.

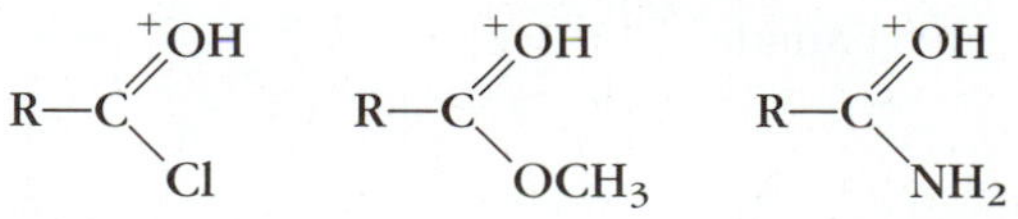

$$R{-}C(={}^+OH){-}Cl \qquad R{-}C(={}^+OH){-}OCH_3 \qquad R{-}C(={}^+OH){-}NH_2$$

All these ions are strong acids. The neutral carboxyl derivatives (the conjugate bases) are even weaker bases than water, so the protonated forms are stronger acids than H_3O^+.

Carboxyl Derivatives as Acids

If one of the atoms attached to a carbonyl group has hydrogen as a substituent, reaction with a base can produce the corresponding anion. We have discussed this extensively in Chapters 4, 13, and 14, and Table 4.3 provides a summary. We discussed substituent effects on the acidity of carboxylic acids in Section 4.8, and you may wish to review that material.

Most derivatives of carboxylic acids only act as acids by loss of a proton from *carbon*, as illustrated here for an ester:

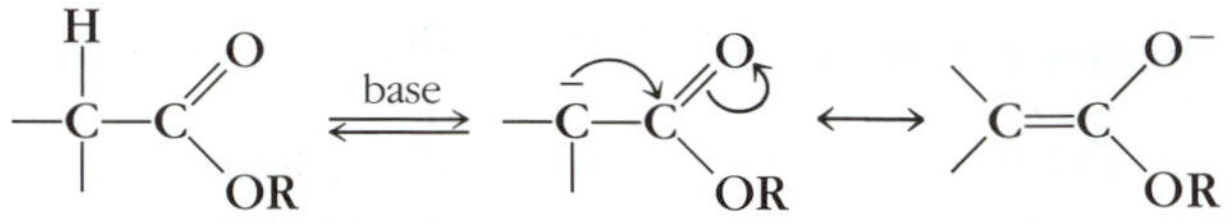

$$-\overset{\text{H}}{\underset{|}{\overset{|}{\text{C}}}}-\text{C}(=\text{O})\text{OR} \xrightleftharpoons{\text{base}} -\overset{-}{\underset{|}{\text{C}}}-\text{C}(=\text{O})\text{OR} \longleftrightarrow \text{>C=C}(\text{O}^-)\text{OR}$$

The resulting enolate ion resembles that produced from an aldehyde or ketone, and you should therefore expect the acidity of esters and related compounds to be comparable with that of aldehydes and ketones. The acid dissociation constant of ethyl acetate (Table 16.6) is 10^{-25}. The modest decrease in acidity relative to ketones ($K_a \sim 10^{-20}$) can be attributed to interaction between the carbonyl group and the nonbonded electrons of the heteroatom, as suggested by the following resonance forms:

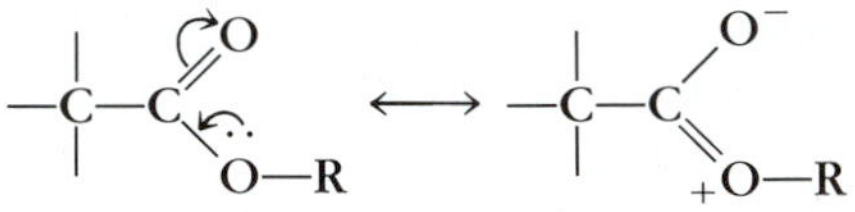

$$-\overset{|}{\underset{|}{\text{C}}}-\text{C}(=\text{O})-\ddot{\text{O}}-\text{R} \longleftrightarrow -\overset{|}{\underset{|}{\text{C}}}-\text{C}(\text{O}^-)={}^{+}\text{O}-\text{R}$$

This interaction decreases the ability of the carbonyl group to stabilize a negative charge on the α carbon, so enolate ion formation is inhibited.

The last two compounds in Table 16.6 are substantially more acidic; that is because the acidic hydrogen is bonded not to carbon but to nitrogen or oxygen.

TABLE 16.6 Acid Dissociation Constants for Some Carboxylic Acid Derivatives

Compound	Functional Class	K_a
$\mathbf{CH_3}-\overset{\text{O}}{\overset{\|}{\text{C}}}-OC_2H_5$	Ester	10^{-25}
$\mathbf{CH_3}-C\equiv N$	Nitrile	10^{-25}
$CH_3-\overset{\text{O}}{\overset{\|}{\text{C}}}-\mathbf{NH_2}$	Amide	10^{-17}
$CH_3-\overset{\text{O}}{\overset{\|}{\text{C}}}-\mathbf{OH}$	Carboxylic acid	10^{-5}

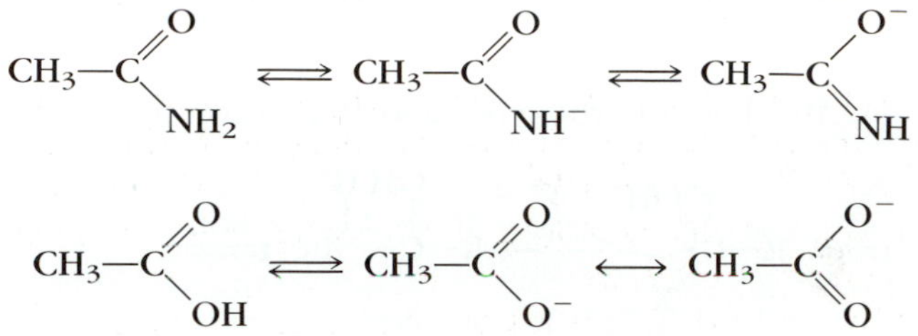

$$CH_3{-}C({=}O){-}NH_2 \rightleftharpoons CH_3{-}C({=}O){-}NH^- \rightleftharpoons CH_3{-}C(O^-){=}NH$$

$$CH_3{-}C({=}O){-}OH \rightleftharpoons CH_3{-}C({=}O){-}O^- \longleftrightarrow CH_3{-}C(O^-){=}O$$

Despite the 10^8 increase in acidity for the amide relative to an ester, it is still less acidic than water ($K_a = 10^{-16}$). The only monofunctional carboxyl derivatives that are more acidic than water are *carboxylic acids,* and they have acid dissociation constants of about 10^{-5}. This means that carboxylic acids are over 10^{10} times more acidic than water, so even relatively weak bases will cause complete conversion to the carboxylate ion. You will see shortly that this difference in acidity results in a pattern of reactivity for carboxylic acids that is somewhat different from that of other carboxyl derivatives.

Nucleophilic Attack at Carbonyl Carbon

Carboxylic acids and their derivatives all contain a carbonyl group. (Nitriles, of course, do not have a C=O group, but as we will show, the —C≡N group exhibits similar reactivity). In contrast to aldehydes and ketones (which undergo nucleophilic *addition*), the characteristic reaction of carboxyl derivatives with nucleophiles is *substitution.* This is a two-step process involving first addition of the nucleophile to the carbonyl group and then expulsion of another substituent from the carbonyl carbon.

Nucleophilic Attack at a Carboxylic Acid Derivative: Alkaline Conditions

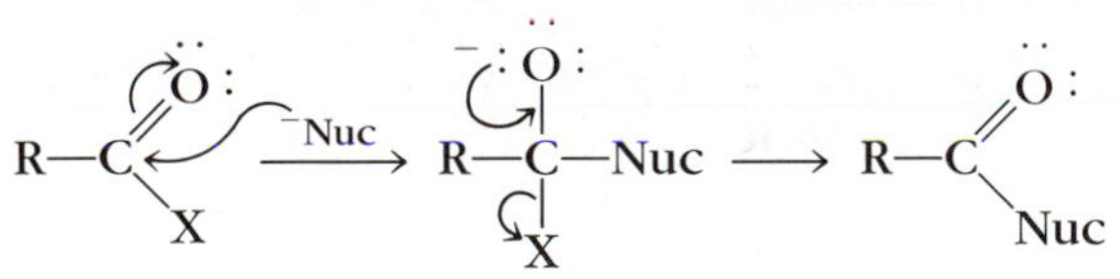

$$R{-}C({=}\ddot{O}{:}){-}X \xrightarrow{^-Nuc} R{-}C(X)(\ddot{O}{:}^-){-}Nuc \longrightarrow R{-}C(={\ddot{O}}{:}){-}Nuc$$

Why do we observe overall substitution rather than just addition? In contrast to the carbonyl groups of aldehydes and ketones, the carbonyl group of a carboxylic acid derivative has a substituent, —X (where —X is —Cl, —OR, —NH_2, etc.), that can act as a leaving group. As we discussed previously (Section 13.7), intermediates containing a tetrahedral carbon with two heteroatom substituents tend to expel one of them and form a carbonyl group because the carbonyl compound is more stable.

The mechanism for nucleophilic attack on a carboxyl group was shown above for alkaline conditions (anionic nucleophile, Nuc^-), but acidic conditions can often be used as well.

Nucleophilic Attack at a Carboxylic Acid Derivative: Acidic Conditions

$$\text{R—C(=O)X} \xrightleftharpoons{H^+} \text{R—C(=}\overset{+}{\text{O}}\text{H)X} \xrightarrow{\text{H—:Nuc}} \text{R—C(OH)(X)—}\overset{+}{\text{Nuc}}\text{H} \xrightleftharpoons{\text{SOH}} \text{R—C(OH)(X)—Nuc} \xrightleftharpoons{H^+} \text{R—C(OH)(Nuc)—}\overset{+}{\text{X}}\text{H} \rightleftharpoons \text{R—C(=}\overset{+}{\text{O}}\text{H)Nuc} \xrightleftharpoons{\text{SOH}} \text{R—C(=O)Nuc}$$

Except for a series of proton transfer reactions, the mechanism is very similar to that shown for alkaline conditions.

Nitriles initially undergo *addition* of a nucleophile to give an intermediate that is a nitrogen analog of a carboxylic acid derivative. This reaction also proceeds under either alkaline or acidic conditions.

$$\text{R—C}\equiv\text{N:} \xrightarrow{^-\text{Nuc}} \text{R—C(=N:}^-\text{)Nuc} \xrightleftharpoons{\text{SOH}} \text{R—C(=NH)Nuc} \qquad \text{Alkaline conditions}$$

$$\text{R—C}\equiv\text{N:} \xrightleftharpoons{H^+} \text{R—C}\equiv\overset{+}{\text{N}}\text{H} \xrightarrow{\text{H—:Nuc}} \text{R—C(=NH)—}\overset{+}{\text{Nuc}}\text{H} \xrightleftharpoons{\text{SOH}} \text{R—C(=NH)Nuc} \qquad \text{Acidic conditions}$$

The nitrogen analog shown as the product in the preceding equations often undergoes further reactions, and these will be discussed subsequently.

Relative Reactivities of Carboxylic Acid Derivatives

The reactivity of carboxylic acid derivatives, particularly under basic conditions, is governed largely by the leaving group ability of the substituent X.

$$\text{R—C(=O)X} + \text{Nuc}^- \longrightarrow \text{R—C(=O)Nuc} + \text{X}^-$$

Consequently, a very clear trend in reactivity toward nucleophilic attack is observed from the highly reactive acid halides and anhydrides down to the other, less reactive carboxylic acid derivatives.

Reactivity of Carboxylic Acid Derivatives

Most reactive toward nucleophilic attack — Least reactive toward nucleophilic attack

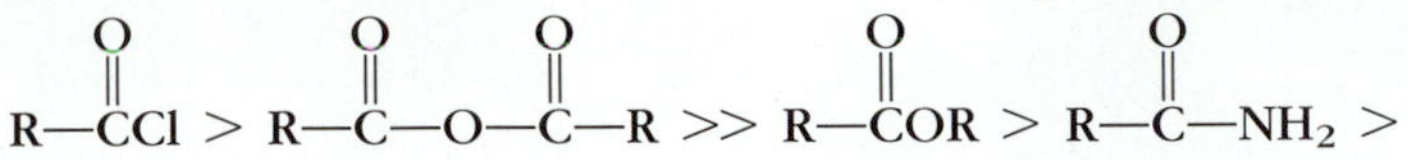

$$\underset{\text{Acid halide}}{\mathrm{R{-}\overset{O}{\overset{\|}{C}}Cl}} > \underset{\text{Anhydride}}{\mathrm{R{-}\overset{O}{\overset{\|}{C}}{-}O{-}\overset{O}{\overset{\|}{C}}{-}R}} \gg \underset{\text{Ester}}{\mathrm{R{-}\overset{O}{\overset{\|}{C}}OR}} > \underset{\text{Amide}}{\mathrm{R{-}\overset{O}{\overset{\|}{C}}{-}NH_2}} > \underset{\text{Nitrile}}{\mathrm{R{-}C{\equiv}N}} > \underset{\text{Carboxylic acid}}{\mathrm{R{-}\overset{O}{\overset{\|}{C}}{-}OH}}$$

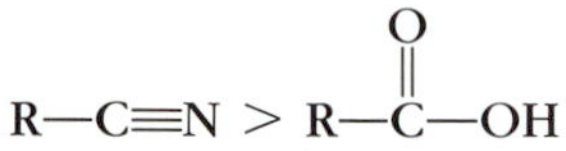

The order of leaving group ability among the first four entries has been established in previous chapters. The leaving groups for these four are $\mathrm{{-}Cl > {-}O\overset{O}{\overset{\|}{C}}{-}R > {-}OR > {-}NH_2}$, and their stabilities as anions are reflected by the corresponding strengths of their conjugate acids: $HCl > RCO_2H > ROH > NH_3$. What about the last two entries in the list? Nitriles are fairly similar to amides because attack by an anionic nucleophile places negative charge on nitrogen (even though there is no actual leaving group).

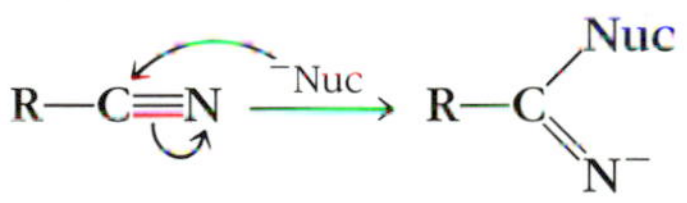

$$\mathrm{R{-}C{\equiv}N \xrightarrow{^{-}Nuc} R{-}C(Nuc){=}N^{-}}$$

Why are carboxylic acids the least reactive? If the leaving group were —OH, you would expect reactivities comparable with those of esters, where the leaving group is —OR. The answer to the question is that a different reaction ordinarily occurs first. Nucleophiles can also function as bases, and even a weak base will convert the acid to the corresponding carboxylate ion:

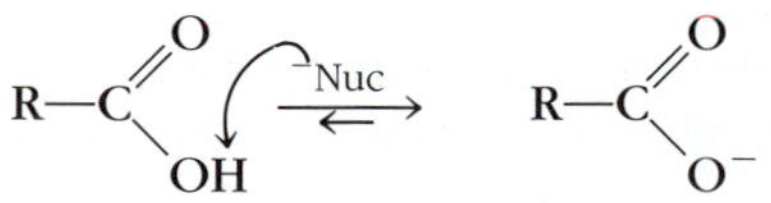

$$\mathrm{R{-}C({=}O)OH \underset{}{\overset{^{-}Nuc}{\rightleftarrows}} R{-}C({=}O)O^{-}}$$

Unreactive toward attack by nucleophiles

The carboxylate ion is very unreactive toward attack by nucleophiles because it already carries a negative charge. Consequently, the reactions of a carboxylic acid under neutral or alkaline conditions are acid-base reactions rather than substitution by nucleophilic attack at carbonyl carbon.

In order to carry out nucleophilic attack at the carbonyl carbon of a carboxylic acid it is necessary to use acidic conditions. This suppresses formation of the carboxylate ion, and the leaving group, —OH, departs in its neutral protonated form (i.e., as OH_2).

EXERCISE 16.4

Most carboxylic acid derivatives can act as either an acid or a base. For each of the following compounds draw the intermediates that would be generated from both (i) protonation and (ii) abstraction of a proton by a base.

(a) $CH_3CH_2\overset{O}{\overset{\|}{C}}Cl$ (b) $CH_3CH_2\overset{O}{\overset{\|}{C}}—NH_2$ (c) $CH_3CH_2\overset{O}{\overset{\|}{C}}OCH_2CH_3$

(d) $CH_3CH_2\overset{O}{\overset{\|}{C}}O\overset{O}{\overset{\|}{C}}CH_2CH_3$ (e) CH_3CH_2CN (f) $CH_3CH_2\overset{O}{\overset{\|}{C}}OH$

EXERCISE 16.5

Acetamide is 10^8 times more acidic than acetonitrile. How should *N,N*-dimethylacetamide compare with these two compounds?

16.4 OXIDATION-REDUCTION REACTIONS OF CARBOXYLIC ACIDS AND THEIR DERIVATIVES

Recall that carbon can exist in a variety of oxidation states (Section 11.3). For a carbon atom at the end of a carbon chain the highest oxidation state is +3 (carboxylic acid) and the lowest oxidation state is −3 (alkane). Reduction of a carboxylic acid to the alkane stage is rarely observed, so most of the oxidation-reduction reactions you will encounter in this chapter will be interconversions between compounds with oxidation states corresponding to primary alcohols, aldehydes, and carboxylic acids.

You can easily recognize these different oxidation states from the number of bonds to electronegative elements. For example, all three of the following compounds have the same oxidation state (+1 for the carbonyl carbon, *two* bonds to electronegative atoms, one bond to hydrogen):

$$R—\overset{O}{\overset{\|}{C}}H \qquad R—\overset{OCH_3}{\overset{|}{C}}H—OCH_3 \qquad R—\overset{NH}{\overset{\|}{C}}—H$$

Similarly, the following carboxylic acid derivatives each have *three* bonds to electronegative atoms, and they all have the same oxidation state (+3 for the carbonyl carbon).

$$R—\overset{O}{\overset{\|}{C}}OH \qquad R—\overset{O}{\overset{\|}{C}}—Cl \qquad R—\overset{O}{\overset{\|}{C}}—NH_2 \qquad R—\overset{NH}{\overset{\|}{C}}—NH_2 \qquad R—C{\equiv}N$$

Reduction of Carboxylic Acid Derivatives

Carboxylic acid derivatives are usually more difficult to reduce than aldehydes and ketones. Nevertheless, they will react with a variety of reducing agents, including H_2 with a metal catalyst, borane, sodium in liquid ammonia, and sodium in alcohol. The most widely applicable reagent for reducing carboxylic acid derivatives is *lithium aluminum hydride.* Sodium borohydride, in contrast, will not reduce most carboxylic acid derivatives. (Acid chlorides and anhydrides are sufficiently reactive, but they would react with the hydroxylic solvents normally used for $NaBH_4$ reductions.)

Carboxylic acids and esters are reduced by $LiAlH_4$ to primary alcohols (Section 11.4).

$$CH_3(CH_2)_{14}\overset{O}{\overset{\|}{C}}OC_2H_5 \xrightarrow[2.\ H_2O]{1.\ LiAlH_4} CH_3(CH_2)_{14}CH_2OH\ (+\ C_2H_5OH) \quad 98\%$$

Notice that *two* alcohols are produced in the reduction of an ester. Ordinarily the primary alcohol from reduction of the carbonyl group is the desired product, but sometimes the reaction is used to generate the alcohol from the *O*-alkyl portion of an ester.

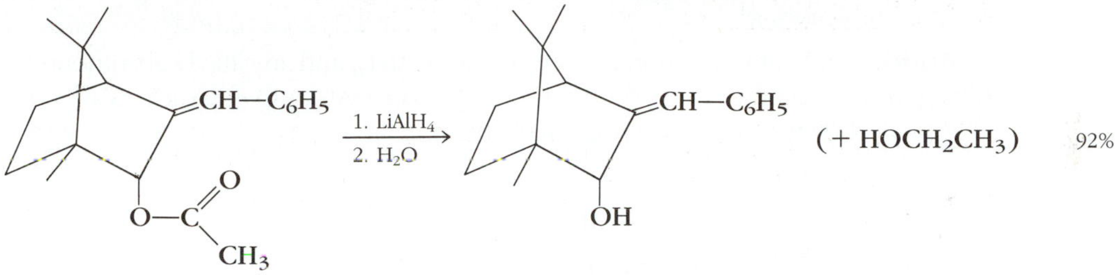

CH—C_6H_5 ; O—C(=O)—CH_3 $\xrightarrow[2.\ H_2O]{1.\ LiAlH_4}$ CH—C_6H_5 ; OH $(+\ HOCH_2CH_3)$ 92%

The reduction of carboxylic acids appears to contradict our previous arguments in this section that carboxylic acids should be unreactive toward nucleophiles that are also bases. Indeed, the first step in the reaction of a carboxylic acid with $LiAlH_4$ is an acid-base reaction to generate the lithium or aluminum salt of the acid (shown here as the aluminum salt).

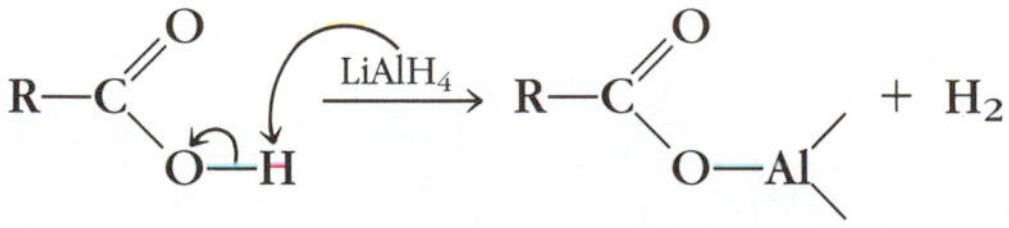

$$R{-}C(=O){-}O{-}H \xrightarrow{LiAlH_4} R{-}C(=O){-}O{-}Al\lt \quad +\ H_2$$

However, a metal cation is strongly coordinated to the negatively charged oxygen: Li^+ and Al^{3+} derivatives are both good Lewis acids. Consequently, this metal carboxylate is still susceptible to nucleophilic attack, and addition of a hydride generates an intermediate with two O—Al groups. One of these is expelled under the conditions of the reaction, and the resulting aldehyde is rapidly reduced to a primary alcohol. The expulsion of the O—Al group is assisted by complexation with a Lewis acid (Li^+ or an Al^{3+} species). Workup with H_2O yields the primary alcohol.

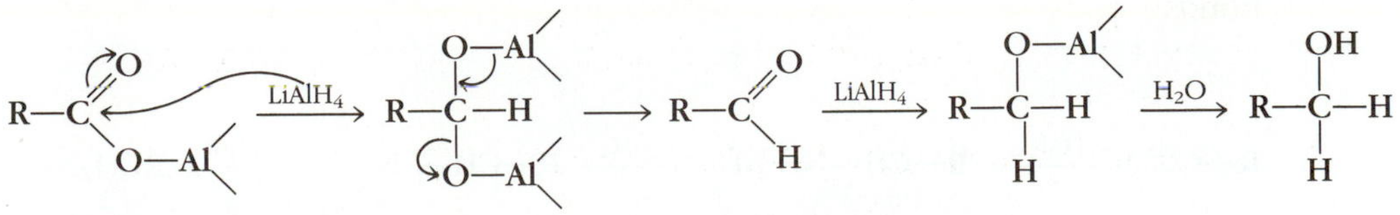

$$R{-}C(=O){-}O{-}Al\lt \xrightarrow{LiAlH_4} R{-}C(H)(O{-}Al\lt)_2 \longrightarrow R{-}C(=O){-}H \xrightarrow{LiAlH_4} R{-}CH_2{-}O{-}Al\lt \xrightarrow{H_2O} R{-}CH_2{-}OH$$

Acid chlorides and anhydrides are reduced to primary alcohols by $LiAlH_4$, but this reaction has little preparative value because the parent acid or one of its esters is usually more readily available as a starting material.

$$C_6H_5-\overset{O}{\overset{\|}{C}}-Cl \xrightarrow[2.\ H_2O]{1.\ LiAlH_4} C_6H_5CH_2OH \qquad 72\%$$

Recall, however, that the reduction of acid chlorides with the modified reagent $LiAl(OtBu)_3H$ affords *aldehydes* (Section 13.3).

$$ClOC-C_6H_4-COCl \xrightarrow{LiAl(OtBu)_3H} OHC-C_6H_4-CHO \qquad 77\%$$

Amides yield amines upon reduction with $LiAlH_4$, and any alkyl substituents on nitrogen are retained in the product. Once again O—Al (or O—Li) acts as a leaving group, as shown in the following mechanism.

$$R'-\overset{O}{\overset{\|}{C}}-\ddot{N}R_2 \xrightarrow{LiAlH_4} R'-\overset{O-Al}{\overset{|}{\underset{NR_2}{\underset{|}{C}}}}-H \longrightarrow R'-\overset{H}{\overset{|}{C}}=\overset{+}{N}R_2 \xrightarrow{LiAlH_4} R'-CH_2-\ddot{N}R_2$$

Notice that reactions of the carbon-nitrogen double bond parallel very closely those of the carbon-oxygen double bond. This is illustrated by the following examples:

$$C_6H_5OCH_2-C(=O)NH_2 \xrightarrow[2.\ H_2O]{1.\ LiAlH_4} C_6H_5OCH_2-CH_2NH_2 \qquad 80\%$$

$$C_6H_{11}-C(=O)N(CH_3)_2 \xrightarrow[2.\ H_2O]{1.\ LiAlH_4} C_6H_{11}-CH_2-N(CH_3)_2 \qquad 88\%$$

Nitriles are reduced to primary amines (RNH_2) by $LiAlH_4$. The reaction proceeds via two successive additions of hydride ion to carbon-nitrogen multiple bonds.

$$R-C\equiv N \xrightarrow{LiAlH_4} R-CH=N-Al\langle \xrightarrow{LiAlH_4} R-CH_2-N(Al\langle)_2 \xrightarrow{H_2O} RCH_2NH_2$$

Once again the strong interaction of the anion (this time on nitrogen) with a Lewis acid renders the intermediate susceptible to further nucleophilic attack, and good yields of amines are obtained.

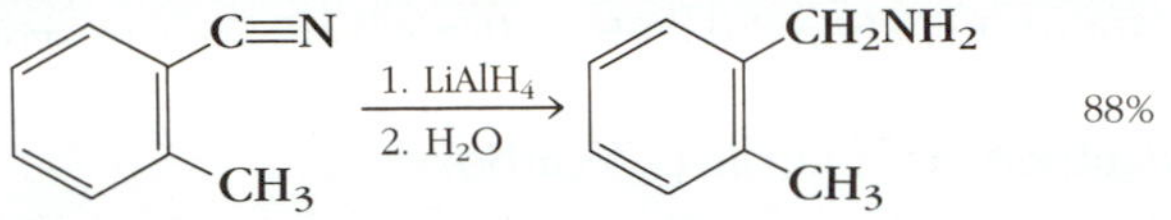

C≡N, CH_3 —1. $LiAlH_4$ 2. H_2O→ CH_2NH_2, CH_3 88%

Oxidation of Carboxylic Acid Derivatives

As we stated at the start of this section, carboxylic acids and their derivatives represent the highest oxidation state possible for a carbon atom that is part of a carbon chain. Consequently, further oxidation of these compounds is a relatively uncommon reaction. One important exception is the oxidation of formic acid derivatives. Formic acid is only a one-carbon compound, and (as suggested by its structural resemblance to an aldehyde) it can be easily oxidized. The product is carbon dioxide (formed via carbonic acid, H_2CO_3).

$$\mathrm{H{-}\overset{\overset{O}{\|}}{C}{-}OH \xrightarrow{\text{oxidation}} HO{-}\overset{\overset{O}{\|}}{C}{-}OH \rightleftharpoons H_2O + O{=}C{=}O}$$

The susceptibility of formic acid derivatives to oxidation is illustrated by the following reaction:

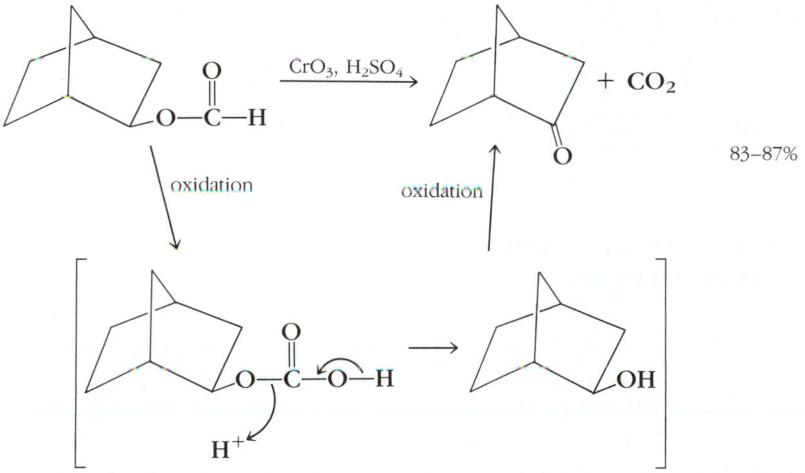

O—C(=O)—H —CrO_3, H_2SO_4→ O + CO_2 83–87%

oxidation oxidation

O—C(=O)—O—H, H^+ → OH

Oxidation of the formate ester of a secondary alcohol produces carbon dioxide and the ketone. Reaction occurs first at the formyl group to yield an ester of carbonic acid. Decomposition of this intermediate liberates carbon dioxide and the alcohol, which is further oxidized to the ketone.

The higher carboxylic acids and their derivatives can be oxidized only by cleavage of a carbon-carbon bond. Such oxidation of a carboxylic acid is called *oxidative decarboxylation* and can be carried out with transition metal compounds of lead or silver. We will not discuss these reactions in detail, but the following equation illustrates the use of lead tetraacetate for oxidative decarboxylation.

CO_2H —$Pb(OAc)_4$→ OAc 66%

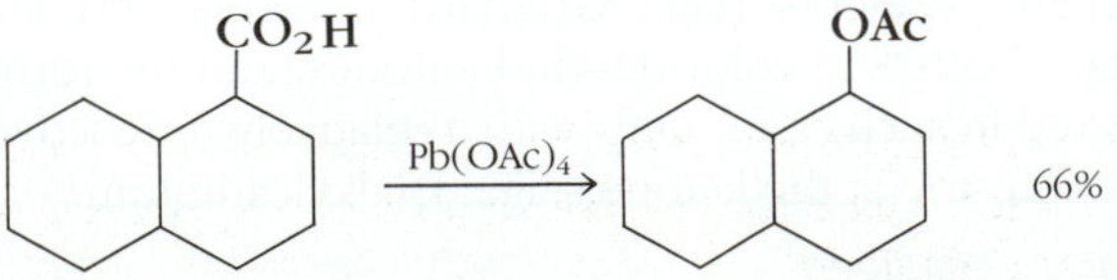

Such reactions are useful when the carboxylic acid is easier to synthesize than the corresponding alcohol of one less carbon. The alcohol could then be obtained (via the acetate) by subsequent oxidative decarboxylation.

EXERCISE 16.6

Write the products of reduction with $LiAlH_4$ for the following compounds:

(a) $CH_3CH_2\overset{O}{\overset{\|}{C}}OH$ (b) $CH_3CH_2\overset{O}{\overset{\|}{C}}Cl$ (c) $CH_3CH_2\overset{O}{\overset{\|}{C}}{-}O{-}\overset{O}{\overset{\|}{C}}CH_2CH_3$

(d) $CH_3CH_2\overset{O}{\overset{\|}{C}}OCH_3$ (e) $CH_3CH_2\overset{O}{\overset{\|}{C}}NH_2$ (f) $CH_3CH_2\overset{O}{\overset{\|}{C}}{-}NHCH_2CH_3$

(g) $CH_3CH_2\overset{O}{\overset{\|}{C}}{-}N(CH_3)_2$ (h) $CH_3CH_2C{\equiv}N$

EXERCISE 16.7

Write the products expected from the following oxidation reactions:

(a) $C_6H_5{-}CH_2CO_2H \xrightarrow{Pb(OAc)_4}$

(b) $C_6H_5{-}CH_2CH_2O\overset{O}{\overset{\|}{C}}H \xrightarrow{CrO_3}$

16.5 REACTION OF CARBOXYLIC ACIDS AND THEIR DERIVATIVES WITH ORGANOMETALLICS

We have previously presented many of the important reactions between organometallic reagents and carboxylic acid derivatives as methods of preparing alcohols (Section 11.5) and ketones (Section 13.4). In this section we will review those reactions as they apply to carboxylic acid derivatives.

Acid halides are the most reactive carboxylic acid derivatives, and they will react rapidly with a variety of organometallic reagents (step 1). The initial product is a ketone, and if the organometallic reagent is sufficiently reactive, attack by a second molecule of the organometallic compound (step 2) leads to the salt of a tertiary alcohol.

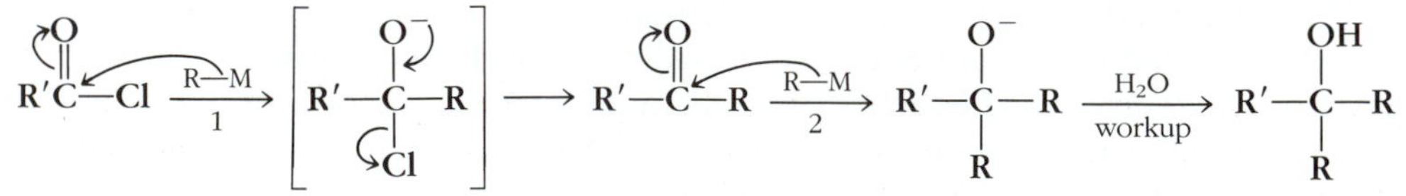

$$R'\overset{O}{\overset{\|}{C}}{-}Cl \xrightarrow[1]{R{-}M} \left[R'{-}\overset{O^-}{\underset{Cl}{\overset{|}{\underset{|}{C}}}}{-}R\right] \longrightarrow R'{-}\overset{O}{\overset{\|}{C}}{-}R \xrightarrow[2]{R{-}M} R'{-}\overset{O^-}{\underset{R}{\overset{|}{\underset{|}{C}}}}{-}R \xrightarrow[\text{workup}]{H_2O} R'{-}\overset{OH}{\underset{R}{\overset{|}{\underset{|}{C}}}}{-}R$$

Most organometallic reagents (e.g., Grignard reagents and alkyllithium compounds) will attack both acid chlorides and ketones, and the formation of tertiary alcohols is expected in such cases. Only with a relatively unreactive organometallic reagent will reaction stop at the ketone stage. Dialkylcadmium reagents have been used for this purpose, but lithium dialkylcuprates are the reagents of choice in modern organic chemistry (Section 13.4).

$$CH_3CH_2CH_2CH_2\overset{O}{\overset{\|}{C}}CH_2CH_2CH_2CH_2\overset{O}{\overset{\|}{C}}Cl \xrightarrow{(CH_3CH_2CH_2CH_2)_2CuLi} CH_3CH_2CH_2CH_2\overset{O}{\overset{\|}{C}}CH_2CH_2CH_2CH_2\overset{O}{\overset{\|}{C}}CH_2CH_2CH_2CH_3 \quad 83\%$$

Acid anhydrides exhibit much the same reactivity as acid halides. However, one half of the molecule acts as the leaving group in the first stage.

$$R{-}\overset{O}{\overset{\|}{C}}{-}O{-}\overset{O}{\overset{\|}{C}}{-}R \xrightarrow{R'{-}M} R{-}\overset{O}{\overset{\|}{C}}{-}O{-}\underset{R}{\overset{O^-}{C}}{-}R' \longrightarrow R'{-}\overset{O}{\overset{\|}{C}}{-}R \xrightarrow[\text{further}]{\text{reacts}} \quad + \quad R{-}\overset{O}{\overset{\|}{C}}{-}O^- \ \text{unreactive}$$

Of the two molecules of carboxylic acid required to form the anhydride, only one is converted to the desired product, and the other is discarded. This inefficiency causes acid chlorides to be preferable as synthetic reagents (although it can be a useful way of *differentiating* between two carboxyl groups in a cyclic anhydride of a dicarboxylic acid).

Esters react rapidly with alkyllithium and Grignard reagents to give alcohols, but unless the esters are α,β-unsaturated, they do not normally react with lithium dialkylcuprates. Remember that *two* alcohols are formed on reaction with a Grignard or alkyllithium reagent (Section 11.5).

$$R'{-}\overset{O}{\overset{\|}{C}}{-}OR'' \xrightarrow[2.\ H_2O]{1.\ RLi} R'{-}\underset{R}{\overset{OH}{C}}{-}R + HOR''$$

In most instances the tertiary alcohol ($R_2R'COH$) is the desired product of such a reaction.

$$CH_3CH_2CH_2\overset{O}{\overset{\|}{C}}OC_2H_5 \xrightarrow[2.\ H_2O,\ H_2SO_4]{1.\ CH_3MgI} CH_3CH_2CH_2\underset{CH_3}{\overset{OH}{C}}{-}CH_3 \ (+\ C_2H_5OH) \quad 88\%$$

But sometimes the other alcohol is the desired product. In other words, the process is used to cleave the ester and generate the corresponding alcohols without changing the carbon skeleton or stereochemistry.

$$CH_3{-}\overset{O}{\overset{\|}{C}}{-}O{-}(\text{cyclopropyl}){-}C_6H_5 \xrightarrow[2.\ H_2O,\ NH_4Cl]{1.\ CH_3Li} HO{-}(\text{cyclopropyl}){-}C_6H_5 \ (79\%) + CH_3{-}\underset{CH_3}{\overset{CH_3}{C}}{-}OH$$

The alcohol formed from the acyl portion of an ester is tertiary unless a *formate ester* is used. A formate ester leads to a *secondary* alcohol (with two identical substituents).

$$\text{H—}\overset{\text{O}}{\overset{\|}{\text{C}}}\text{—OC}_2\text{H}_5 \xrightarrow[\text{2. H}_2\text{O}]{\text{1. CH}_3\text{CH}_2\text{CH}_2\text{CH}_2\text{MgBr}} \text{CH}_3(\text{CH}_2)_3\text{—}\overset{\text{OH}}{\overset{|}{\text{CH}}}\text{—(CH}_2)_3\text{CH}_3 \quad (+\ \text{C}_2\text{H}_5\text{OH}) \qquad 85\%$$

Esters of *carbonic acid* react with three equivalents of the Grignard or alkyllithium reagent, and the resulting tertiary alcohol has *three* identical substituents.

$$\text{CH}_3\text{CH}_2\text{O}\overset{\text{O}}{\overset{\|}{\text{C}}}\text{OCH}_2\text{CH}_3 \xrightarrow[\text{2. H}_2\text{O}]{\text{1. CH}_3\text{CH}_2\text{CH}_2\text{MgBr}} \text{CH}_3\text{CH}_2\text{CH}_2\text{—}\underset{\text{CH}_2\text{CH}_2\text{CH}_3}{\underset{|}{\overset{\text{OH}}{\overset{|}{\text{C}}}}}\text{—CH}_2\text{CH}_2\text{CH}_3 \qquad 75\text{–}80\%$$

The first attack by the organometallic reagent displaces one of the two alkoxy groups of the dialkyl carbonate to generate another ester. The second stage produces a ketone, and finally the metal salt of the tertiary alcohol is formed.

$$\text{C}_2\text{H}_5\text{O}\overset{\text{O}}{\overset{\|}{\text{C}}}\text{OC}_2\text{H}_5 \xrightarrow[1]{\text{R—M}} \left[\text{R—}\overset{\text{O}}{\overset{\|}{\text{C}}}\text{OC}_2\text{H}_5\right] \xrightarrow[2]{\text{R—M}} \left[\text{R—}\overset{\text{O}}{\overset{\|}{\text{C}}}\text{—R}\right] \xrightarrow[3]{\text{R—M}} \text{R—}\underset{\text{R}}{\underset{|}{\overset{\text{O—M}}{\overset{|}{\text{C}}}}}\text{—R}$$

α,β-Unsaturated esters react with organocopper reagents in the same fashion as unsaturated ketones and aldehydes (Section 13.10). Conjugate addition takes place smoothly to generate an ester which is substituted at the β position.

$$\text{CH}_3(\text{CH}_2)_5\text{—CH=CH—}\overset{\text{O}}{\overset{\|}{\text{C}}}\text{OC}_2\text{H}_5 \xrightarrow[\text{2. H}_2\text{O}]{\text{1. (CH}_3)_2\text{CuLi}} \text{CH}_3(\text{CH}_2)_5\text{—}\overset{\text{CH}_3}{\overset{|}{\text{CH}}}\text{—CH}_2\text{—}\overset{\text{O}}{\overset{\|}{\text{C}}}\text{OC}_2\text{H}_5 \qquad 88\%$$

Amides are not ordinarily used as substrates in reactions with organometallic compounds, but they are attacked by these reagents. Attack at the carbonyl carbon (as with esters) is expected unless the amide nitrogen has a hydrogen as one of its substituents. In that case an acid-base reaction would produce the corresponding metal salt.

$$\text{—}\underset{|}{\overset{|}{\text{C}}}\text{—}\overset{\text{O}}{\overset{\|}{\text{C}}}\text{—N}\begin{smallmatrix}\text{H}\\ \\ \text{R}\end{smallmatrix} \xrightarrow{\text{R}'\text{—M}} \text{R}'\text{—H} + \text{—}\underset{|}{\overset{|}{\text{C}}}\text{—}\overset{\text{O}}{\overset{\|}{\text{C}}}\text{—N}^{-}\begin{smallmatrix}\text{M}^+\\ \\ \text{R}\end{smallmatrix}$$

In many cases such salts are insoluble and precipitate from the solution, so that no further reaction occurs.

Nitriles react readily with Grignard and alkyllithium reagents (although not with organocuprates). Only one equivalent of the organometallic reagent is consumed, and ketones can be prepared in good yield. The initial reaction generates a compound containing a carbon-nitrogen double bond, and this is hydrolyzed to the ketone during workup (see Section 13.7).

$$R'—C{\equiv}N \xrightarrow{R—M} R'—C(=N—M)—R \xrightarrow[\text{acid}]{H_2O} \left[R'—C(=NH)—R\right] \longrightarrow R'—C(=O)—R$$

The overall reaction is illustrated by the following equations. Ammonium chloride is used as a weak acid to neutralize the reaction mixtures in these two examples.

1-methyl-4-*t*-Bu-cyclohexanecarbonitrile $\xrightarrow[2.\ H_2O,\ NH_4Cl]{1.\ CH_3Li}$ 1-methyl-4-*t*-Bu-cyclohexyl methyl ketone 92%

2-C_6H_5-cyclopropanecarbonitrile $\xrightarrow[2.\ H_2O,\ NH_4Cl]{1.\ C_6H_5MgBr}$ 2-C_6H_5-cyclopropyl—C(=O)—C_6H_5 85%

Keep in mind that the high reactivity of alkyllithium and Grignard reagents precludes the use of nitriles that have any other reactive functional groups, such as —C=O, —NH, or —OH groups.

Carboxylic acids react rapidly with nearly all organometallic compounds in an acid-base reaction.

$$R—CO_2H + R'—M \longrightarrow R—CO_2^-M^+ + R'—H$$

In most instances the resulting carboxylate salt does not react further, but an important exception occurs with alkyllithium reagents (Section 13.4). The strong interaction between the carboxylate ion and a lithium cation makes the lithium carboxylate reactive to further attack by a powerful nucleophile. Such attack by a second alkyllithium produces a dilithium salt that does not react further until the reaction is quenched with water.

$$R—C(=O)—OH \xrightarrow{R'—Li} R—C(=O)—OLi \xrightarrow{R'—Li} \left[R—C(OLi)_2—R'\right] \xrightarrow{H_2O} R—C(OH)_2—R' \longrightarrow R—C(=O)—R'$$

As long as there are no other functional groups present that could react with the alkyllithium, this is a good method for preparing ketones from carboxylic acids.

2,2-$(C_6H_5)_2$-cyclopropane—CO_2H $\xrightarrow[2.\ H_2O,\ NH_4Cl]{1.\ C_6H_5Li}$ 2,2-$(C_6H_5)_2$-cyclopropyl—C(=O)—C_6H_5 73%

4-*t*-Bu-cyclohexylidene=CH—CO_2H $\xrightarrow[2.\ H_2O,\ NH_4Cl]{1.\ CH_3Li}$ 4-*t*-Bu-cyclohexylidene=CH—C(=O)—CH_3 96%

EXERCISE 16.8

What products would be expected from the reaction of the following carboxylic acid derivatives with methyllithium (followed by workup with mild aqueous acid)?

(a) $CH_3CH_2CH_2\overset{O}{\overset{\|}{C}}OH$ (b) $CH_3CH_2CH_2\overset{O}{\overset{\|}{C}}OCH_3$

(c) $CH_3CH_2CH_2\overset{O}{\overset{\|}{C}}Cl$ (d) $CH_3CH_2CH_2CN$

EXERCISE 16.9

What products would be expected from the reaction of the acid derivatives in Exercise 16.8 with methylmagnesium iodide (followed by workup with mild aqueous acid)? Which products are the same as those in the preceding question? Which are different?

EXERCISE 16.10

Which of the following compounds would be expected to react with lithium dimethylcuprate? Draw the products expected in those cases.

(a) $CH_3CH_2CH_2\overset{O}{\overset{\|}{C}}Cl$ (b) $CH_3CH_2CH_2\overset{O}{\overset{\|}{C}}OH$

(c) $CH_3CH{=}CH{-}\overset{O}{\overset{\|}{C}}OCH_3$ (d) $CH_3CH_2CH_2\overset{O}{\overset{\|}{C}}OCH_3$

16.6 INTERCONVERSION OF CARBOXYLIC ACID DERIVATIVES

Carboxylic acids and their derivatives are valuable starting materials for preparing a wide variety of organic compounds. In order to fully exploit their reactivity you must know how to interconvert them. In this section we will present the various reactions that can be used to convert each class of compound into the others. We will start with the acids themselves and then take up the various derivatives in turn.

Carboxylic acids

Carboxylic acids can be produced from any of their derivatives by *hydrolysis.* The most reactive of the carboxylic acid derivatives, acid chlorides, react rapidly with water even under neutral conditions.

$$R{-}\overset{O}{\overset{\|}{C}}{-}Cl \xrightarrow{:OH_2} R{-}\underset{Cl}{\underset{|}{\overset{^{-}O}{\overset{|}{C}}}}{-}\overset{+}{O}H_2 \xrightleftharpoons{\text{proton transfer}} R{-}\underset{Cl}{\underset{|}{\overset{OH}{\overset{|}{C}}}}{-}OH \longrightarrow R{-}\overset{^{+}OH}{\overset{\|}{C}}{-}OH \xrightleftharpoons{\text{proton transfer}} R{-}\overset{O}{\overset{\|}{C}}OH$$

Anhydrides also react with water under neutral conditions (although more slowly than acid chlorides.)

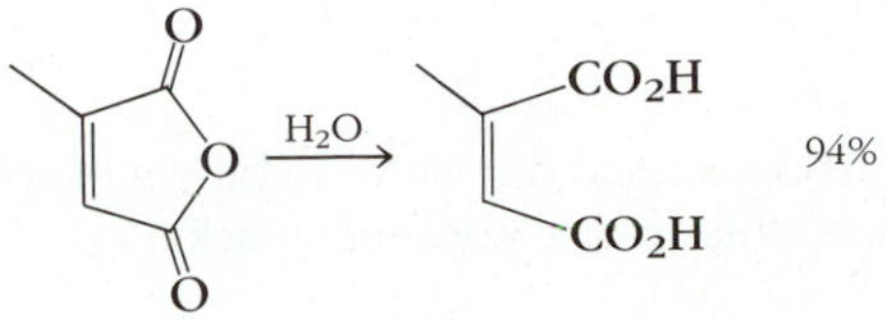

O O O $\xrightarrow{H_2O}$ CO_2H CO_2H 94%

The other carboxylic acid derivatives require either acid or base catalysis. The reaction mechanisms for hydrolysis of esters and amides are summarized in Figures 16.3 and 16.4 for acid and base catalysis, respectively. Most of the individual steps are reversible, but the overall hydrolysis is usually driven toward completion as a result of equilibria involving the products. The amine ($XH = NH_3$, NH_2R, etc.) produced in the acid hydrolysis of an amide will be protonated to give an ammonium ion (NH_4^+), which is not nucleophilic and therefore will not react with the acid.

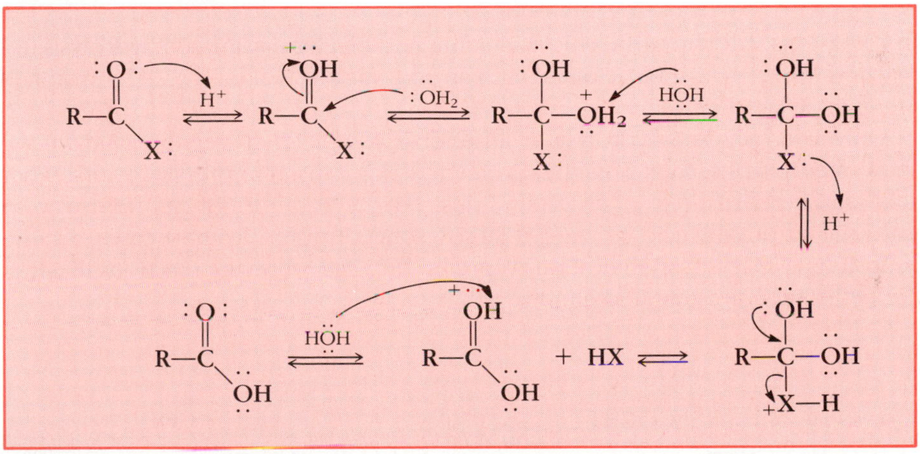

:O: R—C X: $\xrightleftharpoons{H^+}$ +OH R—C X: $\xrightleftharpoons{:OH_2}$:OH R—C—OH_2+ X: $\xrightleftharpoons{HOH}$:OH R—C—OH X: $\rightleftharpoons$ H^+

:OH R—C—OH +X—H $\rightleftharpoons$ +OH R—C OH + HX $\xrightleftharpoons{HOH}$:O: R—C OH

Figure 16.3 Mechanism for Acid-catalyzed Hydrolysis of Esters (—X is —OR′) and Amides (—X is —NH_2, —NR_2, etc.)

:O: R—C X $\xrightleftharpoons{^-:OH}$ $^-$:O: R—C—OH X HOH $\longrightarrow$:O: R—C O—H $\xrightleftharpoons{^-:OH}$:O: R—C :O:$^-$

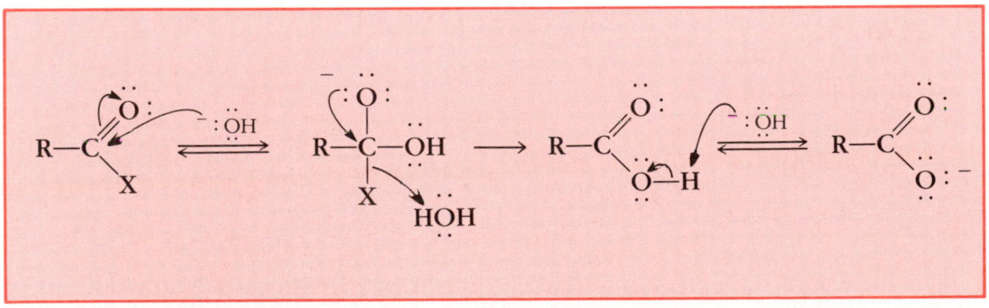

Figure 16.4 Mechanism for Base-catalyzed Hydrolysis of Esters (—X is —OR′) and Amides (—X is —NH_2, —NR_2, etc.)

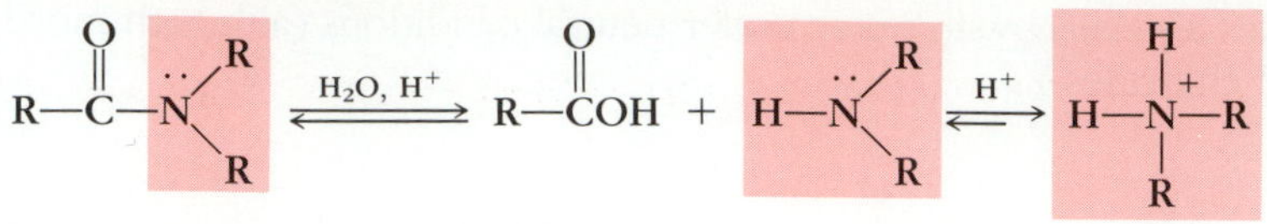

$$R\text{—}\overset{O}{\overset{\|}{C}}\text{—}\ddot{N}R_2 \xrightleftharpoons{H_2O,\ H^+} R\text{—}\overset{O}{\overset{\|}{C}}OH + H\text{—}\ddot{N}R_2 \xrightleftharpoons{H^+} H_2\overset{+}{N}R_2$$

Hydrolysis of any carboxylic acid derivative in base will generate the *carboxylate ion,* which is not susceptible to nucleophilic attack.

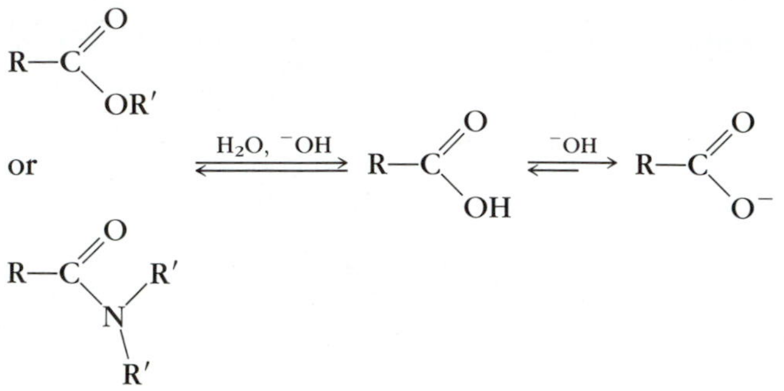

$$R\text{—}C(=O)OR' \text{ or } R\text{—}C(=O)NR'_2 \xrightleftharpoons{H_2O,\ ^-OH} R\text{—}C(=O)OH \xrightleftharpoons{^-OH} R\text{—}C(=O)O^-$$

In order to isolate the carboxylic acid at the end of a base-catalyzed hydrolysis, the reaction mixture must be acidified with an acid such as HCl or H_2SO_4 to protonate the carboxylate ion.

Only acid-catalyzed hydrolysis of an ester is completely reversible, and we will show later in this section how the reverse reaction is used to prepare esters. When acid-catalyzed *hydrolysis* of an ester is desired, the equilibrium can be driven toward the carboxylic acid by using a large excess of water. In most cases, alkaline hydrolysis of esters (saponification) is preferable. The following reactions illustrate the hydrolysis of esters and amides.

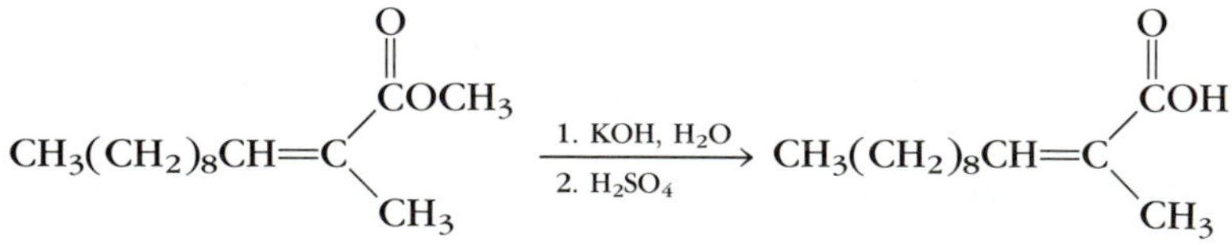

$$CH_3(CH_2)_8CH{=}C(CH_3)\overset{O}{\overset{\|}{C}}OCH_3 \xrightarrow[2.\ H_2SO_4]{1.\ KOH,\ H_2O} CH_3(CH_2)_8CH{=}C(CH_3)\overset{O}{\overset{\|}{C}}OH \qquad 68\text{–}83\%$$

$$BrCH_2\text{—}\overset{Br}{\overset{|}{C}H}\text{—}\overset{O}{\overset{\|}{C}}OCH_3 \xrightarrow[HBr]{H_2O} BrCH_2\overset{Br}{\overset{|}{C}H}\text{—}\overset{O}{\overset{\|}{C}}OH \qquad 72\%$$

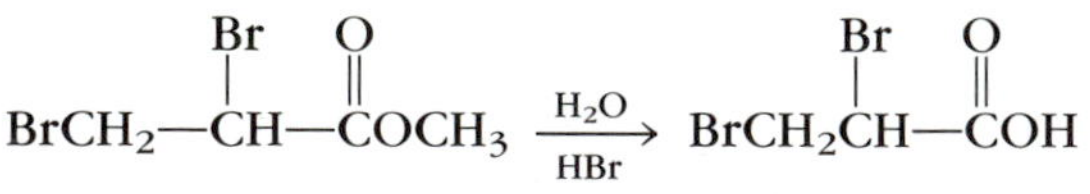

$$CH_3O\text{—}C_6H_4\text{—}CH_2\text{—}\overset{O}{\overset{\|}{C}}\text{—}NH_2 \xrightarrow[2.\ H_2SO_4]{1.\ KOH,\ H_2O} CH_3O\text{—}C_6H_4\text{—}CH_2\text{—}\overset{O}{\overset{\|}{C}}OH \qquad 85\%$$

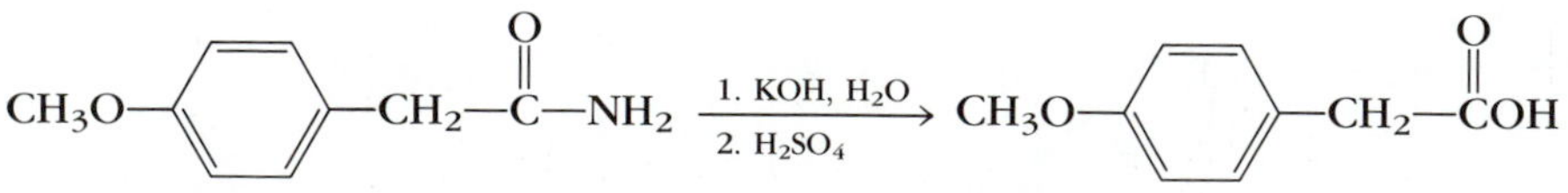

Which carbon-oxygen bond is broken during ester hydrolysis? You could imagine two possibilities: acyl cleavage and alkyl cleavage, in which the bond to be broken is between oxygen and the carbon of an acyl group or of an alkyl group, respectively. The following two equations show that very different reaction mechanisms would be required for the two processes.

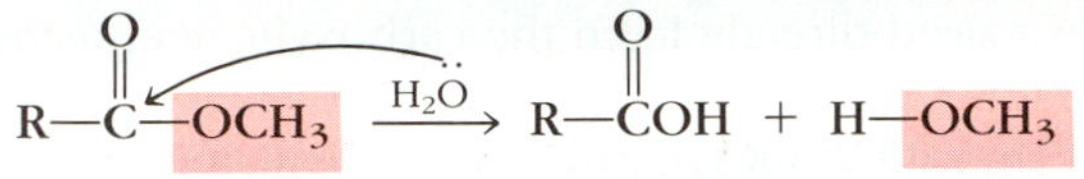

$$R{-}\overset{O}{\overset{\|}{C}}{-}OCH_3 \xrightarrow{H_2\ddot{O}} R{-}\overset{O}{\overset{\|}{C}}OH + H{-}OCH_3$$ Acyl cleavage

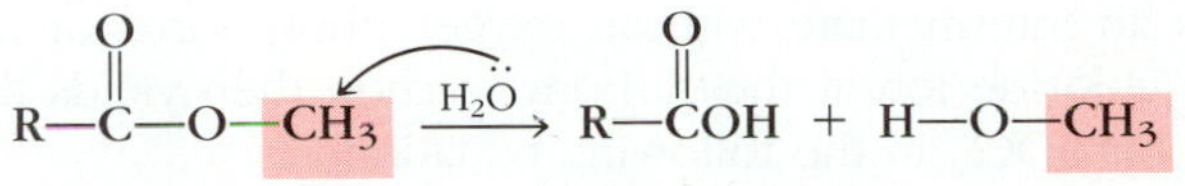

$$R{-}\overset{O}{\overset{\|}{C}}{-}O{-}CH_3 \xrightarrow{H_2\ddot{O}} R{-}\overset{O}{\overset{\|}{C}}OH + H{-}O{-}CH_3$$ Alkyl cleavage

The normal mechanism for ester hydrolysis (Figures 16.3 and 16.4) is that of acyl cleavage, in which the water or hydroxide ion attacks the *carbonyl group*. Alkyl cleavage, in contrast, would require a much more difficult nucleophilic substitution at saturated carbon. The hydrolysis of an ester via alkyl cleavage is unusual and occurs only with esters of highly acidic carboxylic acids such as CF_3CO_2H or with an ester that could ionize to produce a relatively stable carbocation (e.g., an allylic cation, Section 12.5).

The hydrolysis of nitriles begins with *addition* of water to the carbon-nitrogen triple bond, and the initially formed intermediate isomerizes to a primary amide. The amide is then hydrolyzed to the acid by the appropriate mechanism of Figure 16.3 or 16.4.

Acid Catalysis

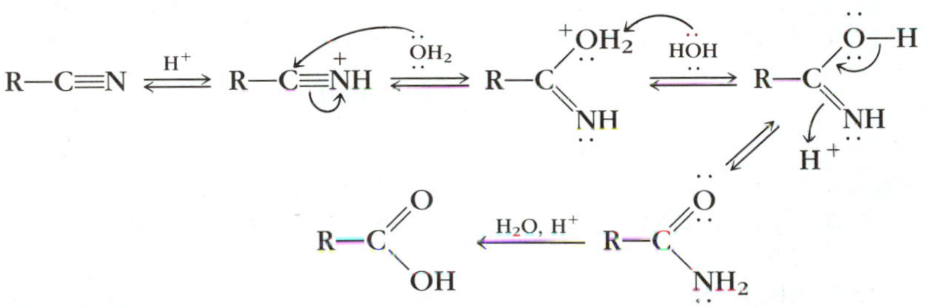

$$R{-}C{\equiv}N \overset{H^+}{\rightleftharpoons} R{-}C{\equiv}\overset{+}{N}H \overset{\ddot{O}H_2}{\rightleftharpoons} R{-}C(\overset{+}{O}H_2){=}NH \overset{H\ddot{O}H}{\rightleftharpoons} R{-}C(OH){=}NH \rightleftharpoons R{-}C(=O){-}NH_2 \xrightarrow{H_2O,\ H^+} R{-}C(=O){-}OH$$

Base Catalysis

$$R{-}C{\equiv}N \xrightarrow{^-OH} R{-}C(O{-}H){=}NH \xrightarrow{^-OH} R{-}C(=O){-}NH_2 \xrightarrow{H_2O,\ ^-OH} R{-}C(=O){-}O^-$$

H—OH H—OH

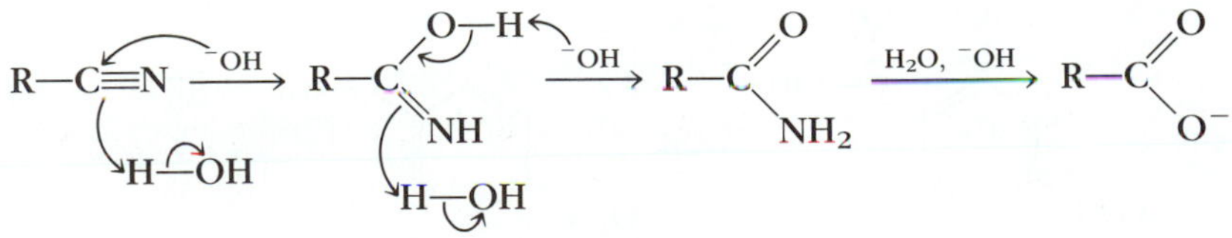

In some instances the hydrolysis of a nitrile can be stopped at the stage of the amide (we will present this reaction later in this section, together with other reactions that yield amides). The following reactions illustrate the hydrolysis of nitriles to carboxylic acids:

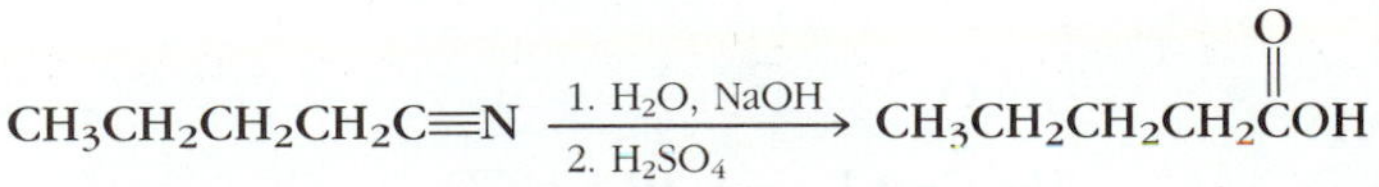

$$CH_3CH_2CH_2CH_2C{\equiv}N \xrightarrow[2.\ H_2SO_4]{1.\ H_2O,\ NaOH} CH_3CH_2CH_2CH_2\overset{O}{\overset{\|}{C}}OH \quad 71\%$$

$$p\text{-}O_2N{-}C_6H_4{-}CH_2C{\equiv}N \xrightarrow[H_2SO_4]{H_2O} p\text{-}O_2N{-}C_6H_4{-}CH_2{-}\overset{O}{\overset{\|}{C}}OH \quad 95\%$$

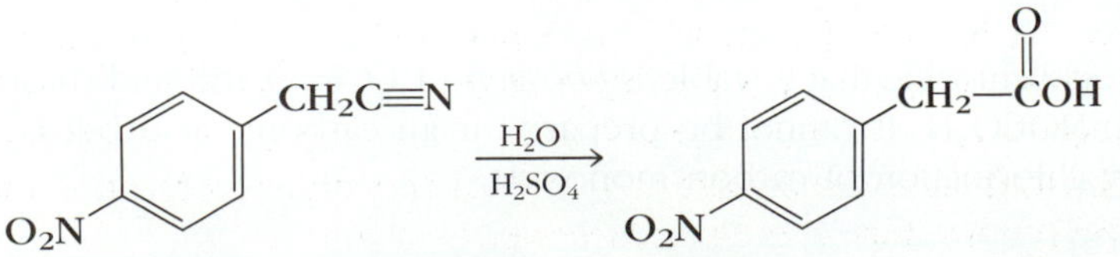

Acid chlorides are prepared directly from the carboxylic acid using some of the same reagents that you previously encountered for the conversion of alcohols to alkyl chlorides: thionyl chloride ($SOCl_2$), phosphorus pentachloride (PCl_5), and phosphorus trichloride (PCl_3). Initial reaction of the carboxylic acid with the inorganic halide generates an intermediate with an oxygen-phosphorus or oxygen-sulfur bond. Attack by chloride ion at the carbonyl carbon then yields the acid chloride, as illustrated for $SOCl_2$ in the following equation:

$$\mathrm{R{-}C({=}O){-}OH} + \mathrm{Cl{-}S({=}O){-}Cl} \longrightarrow \mathrm{R{-}C({=}O){-}\overset{+}{O}(H){-}S(O^-)(Cl){-}Cl} \longrightarrow \mathrm{R{-}C({=}O){-}\overset{+}{O}(H){-}S({=}O){-}Cl} \xrightarrow{\text{proton transfer}} \mathrm{R{-}C({=}O){-}O{-}S({=}O){-}Cl} \xrightarrow{Cl^-}$$

$$\mathrm{R{-}C({=}O){-}Cl} + SO_2 + Cl^- \longleftarrow \mathrm{R{-}C(O^-)(Cl){-}O{-}S({=}O){-}Cl}$$

Acyl bromides and fluorides are also known, but the chlorides are by far more common and we will not discuss the preparation of the others. The following examples illustrate the synthesis of acid chlorides:

$$\mathrm{CH_3{-}\overset{O}{\overset{\|}{C}}OH} \xrightarrow{PCl_3} \mathrm{CH_3{-}\overset{O}{\overset{\|}{C}}Cl} \qquad 62\%$$

$$\text{2,4,6-trimethylbenzoic acid } (\mathrm{Ar{-}C({=}O){-}OH}) \xrightarrow{SOCl_2} \mathrm{Ar{-}C({=}O){-}Cl} \qquad 90\text{–}97\%$$

$$O_2N{-}C_6H_4{-}\mathrm{C({=}O){-}OH} \xrightarrow{PCl_5} O_2N{-}C_6H_4{-}\mathrm{C({=}O){-}Cl} \qquad 90\text{–}96\%$$

As you will see in the subsequent discussion, acid chlorides are key synthetic intermediates in the preparation of many other carboxylic acid derivatives. One important exception is formic acid, the acid chloride of which is unstable; it decomposes to carbon monoxide and hydrogen chloride and so cannot be used as a laboratory reagent.

$$\mathrm{H{-}\overset{O}{\overset{\|}{C}}{-}Cl} \longrightarrow HCl + CO$$

A one-carbon acid chloride that is stable is *phosgene*, $Cl_2C{=}O$, the acid chloride of carbonic acid, $HOCO_2H$. It cannot be prepared from carbonic acid but is made industrially by chlorination of carbon monoxide.

$$Cl_2 + CO \longrightarrow Cl{-}\overset{\overset{\displaystyle O}{\|}}{C}{-}Cl$$

Phosgene is a useful laboratory reagent but is highly toxic. It was used as a chemical warfare agent during World War I.

Acid Anhydrides

Acid anhydrides of simple monocarboxylic acids are similar in reactivity to the corresponding acid chlorides. Acid chlorides are much easier to prepare, however, so only those anhydrides that are commercially available from industrial processes are used frequently. Acetic anhydride is the most important example, and it is prepared industrially by the reaction of acetic acid with *ketene*, $CH_2{=}C{=}O$ (a reactive compound generated by the decomposition of acetone at 700–750°C).

$$CH_3{-}C(OH){=}O{:} + O{=}C{=}CH_2 \longrightarrow \left[CH_3{-}\overset{\overset{\displaystyle H{-}O^+}{\|}}{C}{-}O{-}\overset{\overset{\displaystyle O^-}{|}}{C}{=}CH_2 \quad H^+ \right] \longrightarrow CH_3\overset{\overset{\displaystyle O}{\|}}{C}{-}O{-}\overset{\overset{\displaystyle O}{\|}}{C}CH_3$$

Other anhydrides can sometimes be prepared from acetic anhydride via an exchange reaction, as illustrated for the preparation of benzoic anhydride.

$$C_6H_5\overset{\overset{\displaystyle O}{\|}}{C}OH \xrightarrow{CH_3\overset{O}{\overset{\|}{C}}O\overset{O}{\overset{\|}{C}}CH_3} C_6H_5{-}\overset{\overset{\displaystyle O}{\|}}{C}{-}O{-}\overset{\overset{\displaystyle O}{\|}}{C}{-}C_6H_5 \qquad 72\text{–}74\%$$

This reaction involves two acyl transfers. Initial attack of benzoic acid on acetic anhydride generates a *mixed anhydride*, and attack by a second benzoic acid molecule produces benzoic anhydride.

$$C_6H_5{-}\overset{\overset{\displaystyle O}{\|}}{C}{-}OH + CH_3\overset{\overset{\displaystyle O}{\|}}{C}{-}O{-}\overset{\overset{\displaystyle O}{\|}}{C}CH_3 \longrightarrow C_6H_5{-}\overset{\overset{\displaystyle O}{\|}}{C}{-}O{-}\overset{\overset{\displaystyle O}{\|}}{C}{-}CH_3 + CH_3\overset{\overset{\displaystyle O}{\|}}{C}OH$$

Mixed anhydride

$$\xrightarrow{C_6H_5\overset{O}{\overset{\|}{C}}OH} C_6H_5{-}\overset{\overset{\displaystyle O}{\|}}{C}{-}O{-}\overset{\overset{\displaystyle O}{\|}}{C}{-}C_6H_5 + HO{-}\overset{\overset{\displaystyle O}{\|}}{C}CH_3$$

Mixed anhydrides can also result from the reaction of a carboxylic acid with an acid chloride. The attempted preparation of formic anhydride results in decomposition to give carbon monoxide, but mixed anhydrides of formic and other acids are known.

Esters

Esters are usually prepared from reactions of alcohols with acid chlorides, anhydrides, or carboxylic acids. Direct esterification of a carboxylic acid with an alcohol is an equilibrium reaction, and the mechanism is simply the reverse of that shown in Figure 16.3 for acid-catalyzed ester hydrolysis.

$$R'-C(=O)-OH \overset{H^+}{\rightleftharpoons} R'-C(=\overset{+}{O}H)-OH \overset{ROH}{\rightleftharpoons} R'-C(OH)_2-\overset{+}{O}(H)R \overset{OH_2}{\rightleftharpoons} R'-C(OH)_2-OR$$

$$\overset{H^+}{\rightleftharpoons} R'-C(OH)(\overset{+}{O}H_2)-OR \rightleftharpoons R'-C(=\overset{+}{O}H)-OR \overset{H_2O}{\rightleftharpoons} R'-C(=O)-OR$$

The reaction requires catalysis by a strong acid such as H_2SO_4, H_3PO_4, or TsOH, and this limits its use to molecules that do not contain acid-sensitive groups.

$$CH_3\overset{O}{\overset{\|}{C}}OH \xrightarrow[TsOH]{HOCH_2CH_2CH_2Cl} CH_3\overset{O}{\overset{\|}{C}}-OCH_2CH_2CH_2Cl \quad 93\text{–}95\%$$

$$HO\overset{O}{\overset{\|}{C}}-(CH_2)_4-\overset{O}{\overset{\|}{C}}OH \xrightarrow[H_2SO_4]{CH_3CH_2OH} CH_3CH_2O-\overset{O}{\overset{\|}{C}}(CH_2)_4\overset{O}{\overset{\|}{C}}-OCH_2CH_3 \quad 95\text{–}97\%$$

Often it is preferable to use an acid chloride (Section 11.8) or an anhydride for the preparation of an ester. However, the reaction with an acid chloride liberates HCl, and a trisubstituted amine is frequently added to prevent the reaction mixture from becoming too acidic.

$$CH_3-C(CH_3)_2-OH \xrightarrow[C_6H_5N(CH_3)_2]{CH_3-\overset{O}{\overset{\|}{C}}-Cl} CH_3-C(CH_3)_2-O-\overset{O}{\overset{\|}{C}}CH_3 \quad 62\%$$

$$\text{cyclopropyl}-\overset{O}{\overset{\|}{C}}-Cl \xrightarrow{CH_3CH_2OH} \text{cyclopropyl}-\overset{O}{\overset{\|}{C}}-OCH_2CH_3 \quad 54\%$$

$$C_6H_5-\overset{O}{\overset{\|}{C}}-\overset{OH}{\overset{|}{C}H}-C_6H_5 \xrightarrow{CH_3\overset{O}{\overset{\|}{C}}-O-\overset{O}{\overset{\|}{C}}CH_3} C_6H_5-\overset{O}{\overset{\|}{C}}-\overset{O-\overset{O}{\overset{\|}{C}}CH_3}{\overset{|}{C}H}-C_6H_5 \quad 86\text{–}90\%$$

Methyl and ethyl esters can be prepared by *O*-alkylation of carboxyl groups with reactive alkylating agents. These reactions proceed under mild conditions and are very useful with polyfunctional compounds. Diazomethane (CH_2N_2) gives excellent yields of methyl esters. Initial protonation of diazomethane by the carboxylic acid generates a methyl derivative with an extremely good leaving group, molecular nitrogen. Direct displacement then affords the methyl ester.

$$\mathrm{R{-}\overset{O}{\overset{\|}{C}}{-}O{-}H} + \mathrm{\overset{-}{C}H_2{-}\overset{+}{N}{\equiv}N} \longrightarrow \left[\mathrm{R{-}\overset{O}{\overset{\|}{C}}{-}O^-} + \mathrm{CH_3{-}\overset{+}{N}{\equiv}N}\right] \longrightarrow \mathrm{R{-}\overset{O}{\overset{\|}{C}}OCH_3} + N_2$$

Diazomethane has the disadvantages of being both highly toxic and explosive. These problems can be avoided by using a different alkylating agent. Trimethyloxonium and triethyloxonium fluoroborates ($Me_3O^+\ BF_4^-$ and $Et_3O^+\ BF_4^-$, *oxonium salts*) are stable, crystalline compounds that react rapidly with nucleophiles to transfer a methyl or ethyl group. The carboxylic acid is converted to the carboxylate ion by reaction with a sterically hindered (nonnucleophilic) amine, and this anion then attacks the oxonium ion to generate the ester (Section 12.8).

$$\mathrm{R{-}\overset{O}{\overset{\|}{C}}{-}O{-}H} \xrightarrow{\mathrm{Et\ddot{N}(iPr)_2}} \mathrm{R{-}\overset{O}{\overset{\|}{C}}{-}O^-} \xrightarrow{\mathrm{CH_3{-}\overset{+}{O}(CH_3)_2}} \mathrm{R{-}\overset{O}{\overset{\|}{C}}{-}O{-}CH_3} + \mathrm{CH_3OCH_3}$$

Notice that *both* oxygens of the ester are derived from the acid in these alkylation reactions. The methods other than *O*-alkylation for preparing esters involve acylation of an alcohol, and in esters so prepared only the carbonyl oxygen is derived from the acid.

(OH, O, CO_2H bicyclic keto acid) $\xrightarrow{CH_2N_2}$ (OH, O, CO_2CH_3 methyl ester) 88%

$$\mathrm{CH_3{-}\overset{O}{\overset{\|}{C}}{-}O{-}C_6H_4{-}\overset{O}{\overset{\|}{C}}{-}OH} \xrightarrow[\mathrm{EtN(iPr)_2}]{\mathrm{(CH_3CH_2)_2O^+{-}CH_2CH_3BF_4^-}} \mathrm{CH_3{-}\overset{O}{\overset{\|}{C}}{-}O{-}C_6H_4{-}\overset{O}{\overset{\|}{C}}{-}O{-}CH_2CH_3} \quad 85\text{–}95\%$$

One additional ester-producing reaction, *transesterification,* should be mentioned. When an ester is heated with an alcohol in the presence of either an acid or base catalyst, the alcohol can attack the ester carbonyl to give a different ester. For example:

$$\mathrm{R{-}\overset{O}{\overset{\|}{C}}{-}OCH_3} \underset{\text{acid or base catalyst}}{\overset{\mathrm{HOCH_2CH_3}}{\rightleftharpoons}} \mathrm{R{-}\overset{O}{\overset{\|}{C}}{-}OCH_2CH_3} + \mathrm{HOCH_3}$$

This is an equilibrium reaction, and the mechanism is very similar to that for hydrolysis under the appropriate conditions (Figure 16.3 or 16.4), the main difference being that ROH or RO^- replaces water or hydroxide ion as the attacking nucleophile. Transesterification is not often used preparatively because most esters can be prepared more easily by one of the other methods. However, it is sometimes observed with a polyfunctional compound as an undesired reaction pathway. Transesterification is illustrated by the following examples:

$$CH_2{=}CH{-}\overset{O}{\overset{\|}{C}}OCH_3 \xrightarrow[\text{TsOH}]{HOCH_2CH_2CH_2CH_3} CH_2{=}CH{-}\overset{O}{\overset{\|}{C}}{-}OCH_2CH_2CH_2CH_3 \quad 78\text{–}94\%$$

$$C_6H_5{-}\overset{O}{\overset{\|}{C}}{-}OCH_3 \xrightarrow[\text{KOtBu}]{(CH_3)_3COH} C_6H_5{-}\overset{O}{\overset{\|}{C}}{-}O{-}C(CH_3)_3 \quad 83\%$$

Amides

Amides are usually prepared by the reaction of ammonia or an amine with an acid chloride, anhydride, or ester. The mechanism is essentially the same in each case, involving nucleophilic attack of nitrogen at the carbonyl group, with displacement of X (where —X is —Cl, $-O\overset{O}{\overset{\|}{C}}-R$, or —OR′).

$$R{-}C(=O){-}X + HN\!< \longrightarrow R{-}C(O^-)(X){-}\overset{+}{N}H\!< \longrightarrow R{-}C(=O){-}\overset{+}{N}H\!< \underset{HN<}{\rightleftharpoons} R{-}C(=O){-}\ddot{N}\!<$$

As indicated in the preceding equation, the nitrogen atom of the amine must have at least one hydrogen that can be replaced.

$$CH_3{-}\underset{}{\overset{CH_3}{\overset{|}{C}H}}{-}\overset{O}{\overset{\|}{C}}{-}Cl \xrightarrow{\text{aqueous } NH_3} CH_3\overset{CH_3}{\overset{|}{C}H}{-}\overset{O}{\overset{\|}{C}}{-}NH_2 \quad 78\text{–}83\%$$

$$\text{2-}NO_2C_6H_4{-}NHCH_3 \xrightarrow{CH_3\overset{O}{\overset{\|}{C}}{-}O{-}\overset{O}{\overset{\|}{C}}CH_3} \text{2-}NO_2C_6H_4{-}N(CH_3){-}\overset{O}{\overset{\|}{C}}{-}CH_3 \quad 73\%$$

$$C_6H_{11}{-}NH_2 \xrightarrow{H\overset{O}{\overset{\|}{C}}{-}OC_2H_5} C_6H_{11}{-}NH{-}\overset{O}{\overset{\|}{C}}{-}H \quad 90\%$$

The partial hydrolysis of nitriles (in either acid or base) can also be used to prepare amides. For example, when phenylacetonitrile is heated with concentrated aqueous HCl under reflux,* the carboxylic acid is formed.

$$C_6H_5CH_2C{\equiv}N \xrightarrow[\text{reflux}]{12\,N\,\text{HCl}} C_6H_5CH_2\overset{\overset{\large O}{\|}}{C}OH \qquad 78\text{–}84\%$$

But when the reaction is carried out at a lower temperature, the amide can be isolated in good yield.

$$C_6H_5CH_2{-}C{\equiv}N \xrightarrow[40\text{–}50^\circ]{12\,N\,\text{HCl}} C_6H_5CH_2\overset{\overset{\large O}{\|}}{C}{-}NH_2 \qquad 82\text{–}86\%$$

Notice that the partial hydrolysis of a nitrile will always afford a *primary* amide, $R{-}CONH_2$.

Amides can also be prepared by the Ritter reaction. However, this reaction is mainly used for the synthesis of amines, and we will discuss it in Section 17.7.

Nitriles

Nitriles are prepared most easily by the dehydration of primary amides.

$$R{-}C(=O)NH_2 \xrightarrow{\text{dehydration}} R{-}C{\equiv}N$$

Phosphorus pentoxide (P_2O_5), phosphorus oxychloride ($POCl_3$), phosphorus pentachloride (PCl_5), and thionyl chloride ($SOCl_2$) have all been used for this purpose. The reagent first reacts at carbonyl oxygen (with displacement of a chloride ion from phosphorus or sulfur), as illustrated for $POCl_3$. Proton transfer is followed by an elimination to produce the nitrile.

$$R{-}C(=O){-}\ddot{N}H_2 \xrightarrow{O{=}PCl_3} R{-}C(OP(=O)Cl_2){=}\overset{+}{N}H_2 \xrightarrow[\text{transfer}]{\text{proton}} R{-}C(OP(=O)Cl_2){=}N{-}H \longrightarrow R{-}C{\equiv}N$$

The following reactions illustrate the conversion of primary amides to nitriles:

$$CH_3\overset{\overset{\large CH_3}{|}}{C}H{-}\overset{\overset{\large O}{\|}}{C}{-}NH_2 \xrightarrow{P_2O_5} CH_3\overset{\overset{\large CH_3}{|}}{C}H{-}C{\equiv}N \qquad 86\%$$

---

*Heating under reflux describes the experimental procedure in which a condenser is mounted on top of the reaction vessel. When the reaction mixture is heated to its boiling point, the vapors condense and the liquid falls back into the reaction vessel. This provides a constant reaction temperature with no loss of material to evaporation.

$$CH_3(CH_2)_3{-}\underset{|}{\overset{CH_3CH_2}{CH}}{-}\overset{O}{\overset{\|}{C}}{-}NH_2 \xrightarrow{SOCl_2} CH_3(CH_2)_3{-}\underset{|}{\overset{CH_3CH_2}{CH}}{-}C{\equiv}N \quad 86\text{–}94\%$$

$$N{\equiv}C{-}CH_2{-}\overset{O}{\overset{\|}{C}}NH_2 \xrightarrow{POCl_3} N{\equiv}C{-}CH_2{-}C{\equiv}N \quad 57\text{–}80\%$$

$$N{\equiv}C{-}CH_2{-}\overset{O}{\overset{\|}{C}}NH_2 \xrightarrow{PCl_5} N{\equiv}C{-}CH_2{-}C{\equiv}N \quad 67\text{–}80\%$$

Summary of Interconversions

The various interconversions of carboxylic acids and their derivatives that we have presented are summarized in Figure 16.5. Two features of this figure stand out: first, *all* the derivatives can be converted to carboxylic acids via hydrolysis, and second, the key intermediate for converting a carboxylic acid to other derivatives is often the acid chloride.

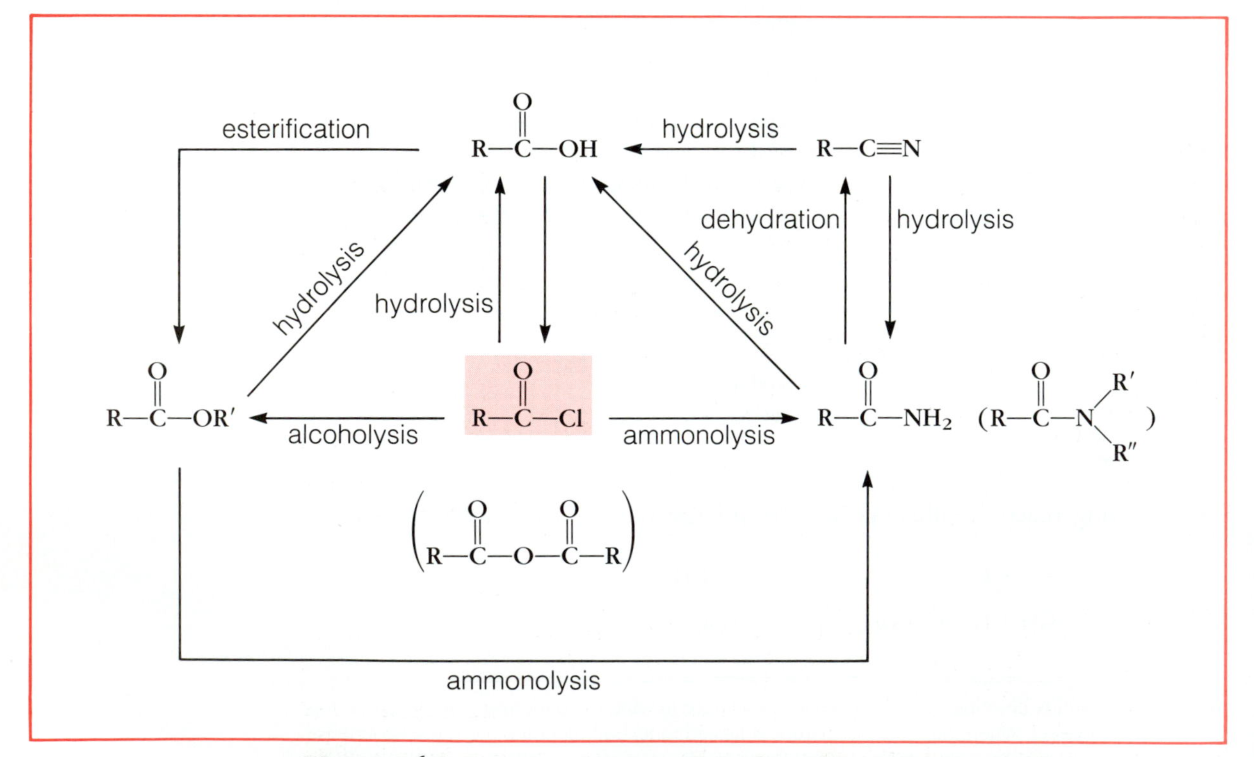

Figure 16.5 Interconversion of Carboxylic Acid Derivatives.

EXERCISE 16.11

Give the products expected from the reactions of

$$CH_3CH_2—\overset{\overset{\large O}{\|}}{C}OCH_2CH_3$$

with the following reagents:

(a) H_2O, KOH
(b) $HN(Et)_2$
(c) H_2O, H_2SO_4

EXERCISE 16.12

Using reagents (a)–(c) from Exercise 16.11, write the products expected for the reaction of

$$CH_3CH_2—\overset{\overset{\large O}{\|}}{C}—Cl$$

EXERCISE 16.13

Using reagents (a)–(c) from Exercise 16.11, write the products expected for the reaction of

$$CH_3CH_2\overset{\overset{\large O}{\|}}{C}O\overset{\overset{\large O}{\|}}{C}CH_2CH_3$$

EXERCISE 16.14

Suggest methods for converting propionic acid into each of the following compounds:

(a) $CH_3CH_2—\overset{\overset{\large O}{\|}}{C}—NH_2$ (b) $CH_3CH_2—\overset{\overset{\large O}{\|}}{C}Cl$

(c) $CH_3CH_2\overset{\overset{\large O}{\|}}{C}—OCH_3$ (d) $CH_3CH_2\overset{\overset{\large O}{\|}}{C}—OCH(CH_3)_2$

(e) $CH_3CH_2\overset{\overset{\large O}{\|}}{C}—N(CH_3)_2$ (f) $CH_3CH_2C{\equiv}N$

16.7 SYNTHESIS OF CARBOXYLIC ACIDS AND THEIR DERIVATIVES FROM COMPOUNDS OF OTHER FUNCTIONAL CLASSES

In the previous section we outlined methods for interconverting carboxylic acids and their derivatives. The relative ease with which these interconversions can be carried out usually means that synthesis of one of the derivatives would allow you to prepare any of the others. In this section we will consider some of the synthetic entries into the carboxylic acid class, particularly reactions involving formation or cleavage of carbon-carbon bonds. You have already encountered most of these reactions in earlier chapters, but here we have collected them in a single place.

Industrial Preparations

Most aliphatic and aromatic carboxylic acids are derived from petroleum and coal. Alkenes can be converted to carboxylic acids by a process called *hydrocarboxylation.* Reaction of the alkene with carbon monoxide in the presence of a nickel catalyst produces the acid with one additional carbon atom.

$$\text{>C=C<} \xrightarrow[\text{Ni catalyst, } H_2O]{CO} -\overset{H}{\underset{|}{\overset{|}{C}}}-\underset{|}{\overset{|}{C}}-\overset{O}{\overset{\|}{C}}-OH$$

Aromatic acids can be produced by oxidation of alkyl substituents. If a metal catalyst such as cobalt is employed, atmospheric oxygen can be used as the oxidant.

$$C_6H_5R \xrightarrow[\text{Co catalyst}]{O_2} C_6H_5CO_2H$$

This reaction succeeds because the aromatic ring is resistant to oxidation.

Low-molecular-weight aliphatic acids can also be produced by the oxidation of low-boiling petroleum distillates, again using atmospheric oxygen as the oxidizing agent.

$$C_nH_{2n+2} \xrightarrow[\substack{150\text{–}200°C \\ 30\text{–}90 \text{ atm}}]{O_2} CH_3\overset{O}{\overset{\|}{C}}OH \;\; (+\; H\overset{O}{\overset{\|}{C}}OH + CH_3CH_2\overset{O}{\overset{\|}{C}}OH)$$

Acetic acid is usually the major product of these reactions, but smaller amounts (10–20%) of formic and propionic acids may be obtained. Other carboxylic acids are prepared by oxidation of the corresponding alcohols and aldehydes.

The one-carbon acids and their derivatives deserve special comment. These are mainly produced from carbon monoxide. Chlorination of carbon monoxide generates phosgene (Section 16.6), and this in turn is used to prepare a wide variety of one-carbon acid derivatives, including substituted ureas, carbamates, and related polymers.

$$NH_2-\overset{O}{\overset{\|}{C}}-NH_2 \qquad NH_2-\overset{O}{\overset{\|}{C}}-OH \qquad NH_2-\overset{O}{\overset{\|}{C}}-OR$$

Urea — Carbamic acid — A carbamate

Urea itself, however, is manufactured from ammonia and carbon dioxide at high temperature and pressure.

Formic acid can be prepared from carbon monoxide by direct hydration in base or by addition of methanol followed by ammonolysis and hydrolysis. This sequence also produces other formic acid derivatives.

$$\mathrm{CO} \xrightarrow[\text{2. acidification}]{\text{1. NaOH}} \mathrm{H{-}\overset{\overset{\large O}{\|}}{C}{-}OH}$$

$$\mathrm{CO} \xrightarrow{\mathrm{CH_3OH},\ \text{high pressure}} \mathrm{H{-}\overset{\overset{\large O}{\|}}{C}{-}OCH_3} \xrightarrow{\mathrm{NH_3}} \mathrm{H{-}\overset{\overset{\large O}{\|}}{C}{-}NH_2} \xrightarrow{\mathrm{H_2O},\ \text{acid}} \mathrm{H{-}\overset{\overset{\large O}{\|}}{C}{-}OH}$$

Oxidation of Primary Alcohols and Aldehydes

Primary alcohols (Section 11.3) and aldehydes (Section 13.11) are readily oxidized to carboxylic acids with Cr(VI) reagents and with $KMnO_4$. The oxidation of aldehydes can also be carried out with Ag_2O. In addition to the examples we presented previously, the preparation of carboxylic acids by such oxidation is illustrated by the following reactions:

$$\text{(bicyclic)}{-}\mathrm{CH_2OH} \xrightarrow{\mathrm{KMnO_4}} \text{(bicyclic)}{-}\mathrm{CO_2H} \qquad 66\%$$

$$\text{(benzodioxole)}{-}\mathrm{CHO} \xrightarrow{\mathrm{KMnO_4}} \text{(benzodioxole)}{-}\mathrm{CO_2H} \qquad 78\text{–}84\%$$

$$\mathrm{CH_3O},\ \mathrm{HO}\text{–}\mathrm{C_6H_3}{-}\mathrm{CHO} \xrightarrow[\text{2. HCl}]{\text{1. } \mathrm{Ag_2O,\ NaOH}} \mathrm{CH_3O},\ \mathrm{HO}\text{–}\mathrm{C_6H_3}{-}\mathrm{CO_2H} \qquad 90\text{–}97\%$$

Oxidations Involving Carbon-Carbon Bond Cleavage

Alkenes with hydrogen substituents on the double bond can be cleaved to carboxylic acids by several methods.

$$\mathrm{R{-}CH{=}CH{-}R'} \xrightarrow{\text{oxidation}} \mathrm{RCO_2H + HO_2CR'}$$

Vigorous oxidation with permanganate or dichromate can be used, but this sometimes leads to overoxidation, with the formation of unwanted products. Ozonolysis also produces acids when an *oxidative* workup (in contrast to reductive workup with zinc metal) is utilized.

$$\mathrm{CH_3(CH_2)_5\overset{\overset{\large OH}{|}}{C}H{-}CH_2CH{=}CH{-}(CH_2)_7{-}\overset{\overset{\large O}{\|}}{C}OH} \xrightarrow[\text{2. } \mathrm{H_2SO_4}]{\text{1. } \mathrm{KMnO_4}} \mathrm{HO\overset{\overset{\large O}{\|}}{C}{-}(CH_2)_7{-}\overset{\overset{\large O}{\|}}{C}OH} \qquad 47\text{–}53\%$$

$$\text{cyclooctene} \xrightarrow[H_2O_2,\ NaOH]{O_3} HO_2C(CH_2)_6CO_2H \quad 63\%$$

Ketones can be converted to esters by the Baeyer-Villiger oxidation (Section 13.11). For example:

$$c\text{-}C_3H_5\text{—}\overset{O}{\overset{\|}{C}}\text{—}CH_3 \xrightarrow{CF_3CO_3H} c\text{-}C_3H_5\text{—}O\text{—}\overset{O}{\overset{\|}{C}}CH_3 \quad 53\%$$

Methyl ketones can often be cleaved by the haloform reaction (Section 13.9).

$$\text{2-naphthyl—}\overset{O}{\overset{\|}{C}}\text{—}CH_3 \xrightarrow[2.\ HCl]{1.\ NaOCl} \text{2-naphthyl—}CO_2H \quad 87\text{–}97\%$$

Aromatic Carboxylic Acids

Aromatic carboxylic acids can be prepared by oxidation of alkyl side chains with $KMnO_4$ or Cr(VI) reagents. The reaction appears to proceed via a benzylic free radical (Section 8.6), so the α carbon must have at least one hydrogen substituent.

$$Ar\text{—}\overset{R}{\overset{|}{C}H}\text{—}R \longrightarrow Ar\text{—}\dot{C}R_2 \longrightarrow Ar\text{—}\overset{O}{\overset{\|}{C}}OH$$

Although larger alkyl groups can be oxidized, methyl derivatives have been used most frequently.

$$\text{2-ClC}_6H_4CH_3 \xrightarrow[2.\ HCl]{1.\ KMnO_4} \text{2-ClC}_6H_4CO_2H \quad 76\text{–}78\%$$

$$\text{4-}O_2NC_6H_4CH_3 \xrightarrow[H_2SO_4]{Na_2Cr_2O_7} \text{4-}O_2NC_6H_4CO_2H \quad 82\text{–}86\%$$

The success of these reactions is a consequence of the stability of the aromatic ring (i.e., its resistance to oxidation), together with its ability to activate the α position of the side chain.

Synthesis of Carboxylic Acids and Their Derivatives by Carbon-Carbon Bond-Forming Processes

A variety of carbon-carbon bond-forming reactions yield carboxylic acids and their derivatives. Of particular importance are the alkylation and condensation reactions of enolate ions formed from esters and nitriles, and we will present a detailed discussion of these reactions in Chapter 21.

Three important laboratory syntheses of carboxylic acids proceed by introduction of a one-carbon fragment. In each case the one-carbon fragment becomes the carboxyl group (or its equivalent).

1. Alkyllithium and Grignard reagents (derived from the corresponding halides) react with carbon dioxide to give acids after neutralization with acid:

$$R{-}Mg{-}X \xrightarrow{O=C=O} R{-}C(=O){-}O{-}MgX \xrightarrow{H^+} R{-}C(=O){-}OH$$

In contrast to other organometallic reactions (Sections 11.5, 11.6, 13.4, 13,6) that yield alcohols or ketones, the initial product here is the *salt* of a carboxylic acid. It is therefore necessary to neutralize the reaction mixture with a strong acid in order to form the carboxylic acid.

$$CH_3CH_2CH(CH_3){-}Cl \xrightarrow[\text{2. } CO_2 \text{ 3. HCl}]{\text{1. Mg}} CH_3CH_2CH(CH_3){-}CO_2H \quad 76\text{–}86\%$$

$$CH_3{-}C(C_6H_5)_2{-}CH_2Cl \xrightarrow[\text{2. } CO_2 \text{ 3. HCl}]{\text{1. Li}} CH_3{-}C(C_6H_5)_2{-}CH_2{-}CO_2H \quad 59\%$$

2. Primary and secondary alkyl halides and sulfonates react with cyanide ion to yield nitriles (Section 12.8) by a direct displacement mechanism:

$$Na^+ \,^-CN + R{-}X \longrightarrow R{-}C{\equiv}N$$

$$C_6H_5CH_2Cl \xrightarrow{NaCN} C_6H_5CH_2{-}CN \quad 80\text{–}90\%$$

$$CH_3CH_2CH_2CH_2Br \xrightarrow{NaCN} CH_3CH_2CH_2CH_2{-}CN \quad 90\%$$

3. Carbon monoxide reacts with carbocations to give carboxylic acids (via acyl cations as intermediates):

$$R^+ \xrightarrow{:C{=}\ddot{O}} \left[R{-}\overset{+}{C}{=}\ddot{O} \longleftrightarrow R{-}C{\equiv}\overset{+}{O} \right] \xrightarrow{H_2\ddot{O}} R{-}C(=O)OH$$

This is known as the Koch-Haaf carboxylation reaction, named after the German chemists who discovered it in the 1950s. The carbon monoxide is normally generated by the dehydration of formic acid with concentrated sulfuric acid, and the same conditions generate the carbocation from an alcohol, alkene, or alkyl halide. In general, the reaction is restricted to carboxylation of tertiary carbocations.

$$CH_3{-}C(CH_3)_2{-}C(CH_3)_2{-}OH \xrightarrow[\text{conc. } H_2SO_4]{H{-}\overset{O}{\overset{\|}{C}}{-}OH} CH_3{-}C(CH_3)_2{-}C(CH_3)_2{-}\overset{O}{\overset{\|}{C}}OH \quad 88\%$$

$$\text{2-methylcyclohexanol} \xrightarrow[\text{conc. } H_2SO_4]{H{-}\overset{O}{\overset{\|}{C}}{-}OH} \text{1-methylcyclohexanecarboxylic acid (}CH_3\text{, }CO_2H\text{)} \quad 89\text{–}94\%$$

$$\left[\text{2-methylcyclohexyl cation (H shift)} \longrightarrow \text{1-methylcyclohexyl cation } (CH_3, +) \right]$$

As indicated by the second of the preceding examples, rearrangement (in this case by a hydride shift, see Section 5.9) to a tertiary cation should be expected if it is possible.

Each of the preceding methods offers certain advantages, but each has limitations. The carboxylation of organometallics will proceed not only with alkyl halides but also with alkenyl and aryl halides.

$$\text{2-bromo-1,3,5-trimethylbenzene (Br)} \xrightarrow[\text{2. } CO_2 \;\; \text{3. HCl}]{\text{1. Mg}} \text{2,4,6-trimethylbenzoic acid (}CO_2H\text{)} \quad 55\text{–}61\%$$

$$\text{1-bromocyclopentene (–Br)} \xrightarrow[\text{2. } CO_2 \;\; \text{3. HCl}]{\text{1. Mg}} \text{cyclopentene-1-}CO_2H \quad 63\%$$

Cyanide displacements in contrast are restricted to alkyl halides (i.e., to attack at *saturated* carbon). Moreover, direct displacement reactions will be accompanied by elimination as a side reaction, so the best results are obtained with primary halides and sulfonates (although secondary derivatives can sometimes be used). Yet cyanide displacements often succeed with compounds containing functional groups that would preclude formation of an organometallic reagent. You can see this in the following two examples, in which the reactants contain a second halogen and a hydroxyl group, respectively.

$$Cl{-}CH_2CH_2CH_2{-}Br \xrightarrow{NaCN} Cl{-}CH_2CH_2CH_2{-}CN \qquad 60\text{–}70\%$$

$$HO{-}CH_2CH_2{-}Br \xrightarrow{NaCN} HO{-}CH_2CH_2{-}CN \qquad 79\text{–}80\%$$

The Koch-Haaf carboxylation of carbocations is the most limited of the three methods because it requires a reactant that can be converted to a tertiary carbocation. The use of concentrated sulfuric acid as the reaction solvent can cause unwanted reactions of some aromatic rings (Chapter 18), so this method is further restricted to aliphatic compounds. Nevertheless, it complements the other two procedures, and one of the three methods will usually be satisfactory.

EXERCISE 16.15

Suggest methods of preparing the following two compounds using any starting materials you wish. If more than one of the methods in this section is possible, show all of them.

(a) $CH_3CH_2CO_2H$ (b) $(CH_3)_3CO_2H$

16.8 DIFUNCTIONAL CARBOXYLIC ACIDS AND CYCLIC CARBOXYLIC ACID DERIVATIVES

As we have previously discussed (particularly in Chapter 8), difunctional compounds can exhibit reactivity that is more complex than that of monofunctional compounds. Two different functional groups might both undergo reaction under a certain set of conditions, or they might react with each other via a cyclic intermediate or transition state. In this section we will first show you a way of introducing a second functional group into a carboxylic acid derivative, and then we will discuss some of the cyclic difunctional carboxylic acid derivatives that are frequently encountered.

α-Halocarboxylic Acids

Direct halogenation of carboxylic acids is generally unsatisfactory, because radical substitution is unselective (Section 2.6). On the other hand, *acid chlorides* can be halogenated at the α position in good yield, and the acid halide is hydrolyzed to the acid during aqueous workup.

$$-\overset{|}{\underset{H}{C}}-\overset{O}{\overset{\|}{C}}-Cl \xrightarrow{X_2} -\overset{|}{\underset{X}{C}}-\overset{O}{\overset{\|}{C}}-Cl \xrightarrow{H_2O} -\overset{|}{\underset{X}{C}}-\overset{O}{\overset{\|}{C}}OH$$

Bromine is usually employed, because it is a liquid at room temperature and is easier to handle than chlorine. The reaction has been suggested to proceed via addition of halogen to the enol of the acid chloride, but other evidence indicates that loss of HCl from the acid chloride sometimes produces a *ketene* intermediate, which then reacts further.

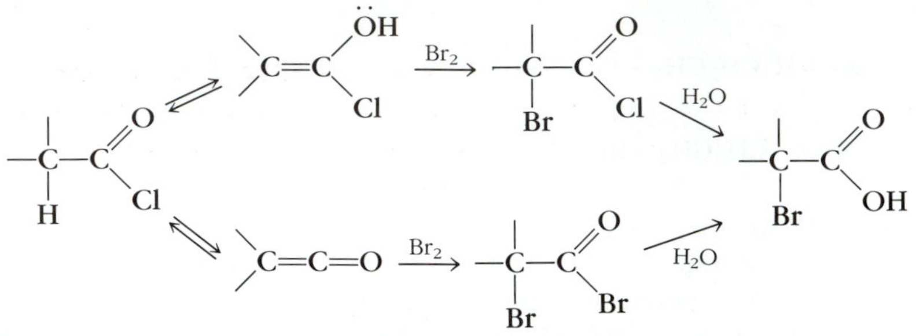

$$\text{H–C–C(=O)Cl} \rightleftharpoons \text{C=C(OH)Cl} \xrightarrow{Br_2} \text{–C(Br)–C(=O)Cl} \xrightarrow{H_2O} \text{–C(Br)–C(=O)OH}$$

$$\text{H–C–C(=O)Cl} \rightleftharpoons \text{C=C=O} \xrightarrow{Br_2} \text{–C(Br)–C(=O)Br} \xrightarrow{H_2O} \text{–C(Br)–C(=O)OH}$$

Such brominations can be carried out with either Br_2 or *N*-bromosuccinimide. Formation of the acid chloride and its subsequent halogenation can be carried out in the same reaction mixture by using PCl_3 or PBr_3 as one of the reagents. These procedures are illustrated by the following reactions:

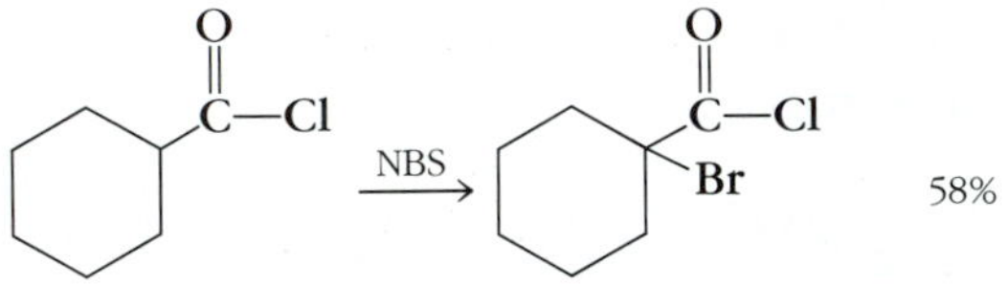

$$\text{C}_6\text{H}_{11}\text{–C(=O)–Cl} \xrightarrow{\text{NBS}} \text{C}_6\text{H}_{10}\text{(Br)–C(=O)–Cl} \quad 58\%$$

$$CH_3(CH_2)_3CH_2\overset{O}{\overset{\|}{C}}OH \xrightarrow[\text{2. } H_2O]{\text{1. } Br_2,\ PCl_3} CH_3(CH_2)_3\underset{Br}{\underset{|}{C}H}\text{—}\overset{O}{\overset{\|}{C}}OH \quad 83\text{–}89\%$$

$$HO\text{—}\overset{O}{\overset{\|}{C}}\text{—}CH_2CH_2CH_2CH_2\overset{O}{\overset{\|}{C}}OH \xrightarrow[\text{2. } Br_2 \text{ 3. } C_2H_5OH]{\text{1. } SOCl_2} C_2H_5O\text{—}\overset{O}{\overset{\|}{C}}\text{—}\underset{Br}{\underset{|}{C}H}\text{—}CH_2CH_2\text{—}\underset{Br}{\underset{|}{C}H}\text{—}\overset{O}{\overset{\|}{C}}\text{—}OC_2H_5 \quad 91\text{–}97\%$$

The last of these three examples shows that the intermediate acid chloride can be converted to an α-halo ester by reaction with an alcohol. Reaction with the alcohol occurs preferentially at the much more reactive acyl halide (rather than at the α position). α-Bromoacids have been used for a variety of synthetic purposes, including the preparation of the corresponding amino acids (Chapter 20). Dehydrohalogenation can produce α,β-unsaturated carboxylic acids which can then be used in Michael reactions (Chapter 21; cf. Section 14.5).

Dicarboxylic Acids

Dicarboxylic acids can form cyclic anhydrides readily if a five- or six-membered ring is produced. (Longer-chain diacids tend to form *inter*molecular anhydride linkages under the conditions for dehydration.)

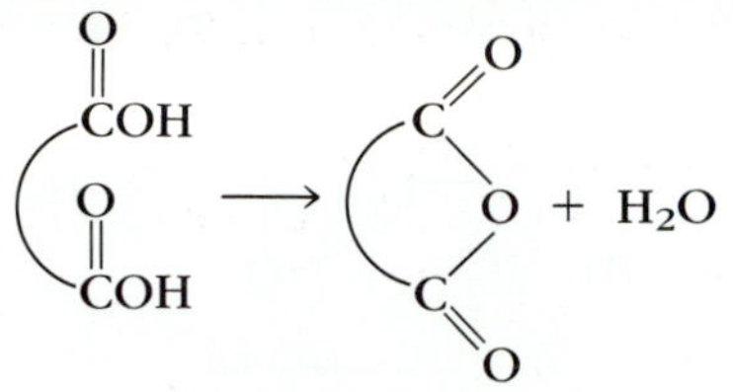

An extremely important feature of the reactivity of cyclic anhydrides is the formation of difunctional products corresponding to selective reaction at only one end of the diacid. Succinic anhydride, for example, reacts with methanol to produce the half-ester, methyl hydrogen succinate.

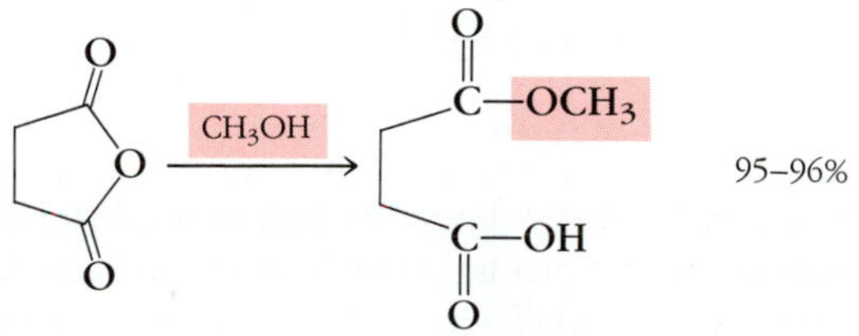

95–96%

Nucleophilic attack at either of the equivalent carbonyl groups of the anhydride results in displacement of the other carboxylate group.

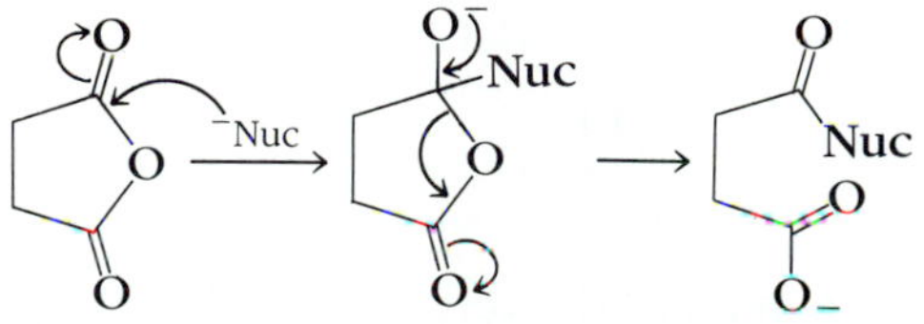

In the absence of acid catalysis this carboxylate ion (or neutral carboxyl group, depending on the pH) is unreactive to nucleophilic attack. As a result, cyclic anhydrides can be valuable synthetic reagents. Dicarboxylic acids can also form **imides,** in which two acyl groups are bonded to a single nitrogen atom. Succinic acid, for example, can be converted to succinimide, and this in turn is used to prepare *N*-bromosuccinimide.

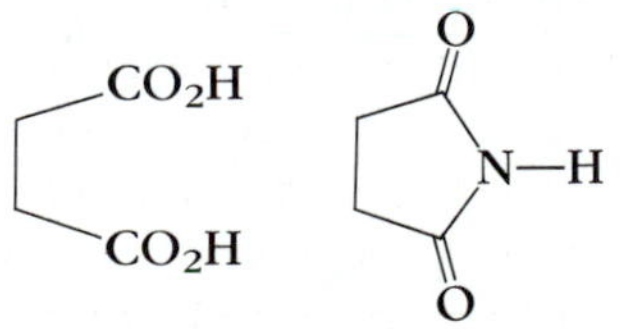

Succinic acid

Succinimide

Hydroxy Acids

Hydroxy acids can form cyclic esters, called **lactones,** by intramolecular esterification of the hydroxyl and carboxyl groups.

$$\text{HO-}\frown\text{-CO}_2\text{H} \longrightarrow \text{Lactone} + H_2O$$

Lactone

As in all the ring-forming reactions we have considered previously, such cyclization is usually observed only for the formation of five- and six-membered rings. (Other hydroxy acids esterify in normal, intermolecular fashion.)

$$CH_3CH(OH){-}CH_2CH_2COOH \xrightarrow{H^+} \text{(5-membered lactone)} \quad 94\%$$

In addition to their formation by direct acid-catalyzed esterification of hydroxy acids, lactones can sometimes be produced by heating hydroxy acids to moderately high temperatures without an acid catalyst. For example, addition of sodium acetylide to the keto acid shown in the following example generated a hydroxy acid, but distillation caused lactone formation to occur.

$$CH_3{-}\overset{O}{\overset{\|}{C}}{-}CH_2CH_2\overset{O}{\overset{\|}{C}}OH \xrightarrow[\text{2. } H_2SO_4 \text{ (neutralize)}]{\text{1. NaC}\equiv\text{CH}} HC{\equiv}C{-}\underset{OH}{\overset{CH_3}{C}}{-}CH_2CH_2\overset{O}{\overset{\|}{C}}OH \xrightarrow{\text{distillation}} HC{\equiv}C{-}(CH_3)\text{-lactone} \quad 80\%$$

Two equivalents of sodium acetylide were used in this reaction. The first reacted with the carboxylic acid to form the carboxylate salt, and the second equivalent attacked the ketone carbonyl.

When the equilibrium between lactone and hydroxy acid is unfavorable, cyclization can be effected with acetic anhydride. This generates the mixed anhydride, which in turn reacts with the hydroxyl group to form the lactone.

$$\text{(OH, O, CO}_2\text{H bicyclic)} \xrightarrow{CH_3{-}\overset{O}{\overset{\|}{C}}{-}O{-}\overset{O}{\overset{\|}{C}}{-}CH_3} \text{(lactone)} \quad 64\%$$

via

$$\left[\text{(:OH, O bicyclic)}{-}\overset{O}{\overset{\|}{C}}{-}O{-}\overset{O}{\overset{\|}{C}}{-}CH_3\right]$$

Lactones are also formed from the Baeyer-Villiger oxidation of cyclic ketones (Section 13.11). A variety of other reactions in which a carboxyl group interacts with a second reactive center in the molecule can also result in lactone formation, but we will not consider these in detail.

The reactivity of lactones is essentially the same as that of other esters. Hydrolysis, reduction, and reactions with organometallic compounds proceed in the same fashion. The only difference is that reaction of an acyclic ester typically produces *two* products (one each from the acid and alcohol portions), whereas in the corresponding reaction of a lactone both fragments will be part of the same molecule, a *difunctional* molecule. For example, a diol is produced by reduction of a lactone in the following reaction.

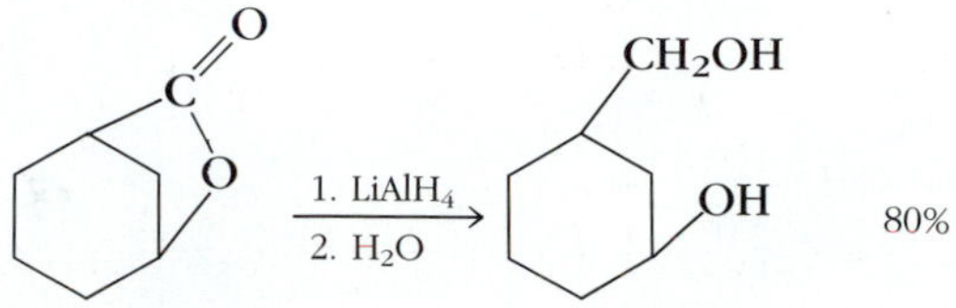

1. $LiAlH_4$ 2. H_2O → CH_2OH, OH 80%

Amino Acids

Amino acids can form cyclic amides called **lactams.** Once again, only five- and six-membered ring compounds are readily formed by cyclization. The reactivity of lactams is comparable with that of acyclic amides, and we will not consider it in detail.

An important group of lactams is that derived from β-amino carboxylic acids. These four-membered ring compounds comprise the medically useful *β-lactam antibiotics.* These are the penicillins and the cephalosporins (where R can be any of a variety of alkyl or aryl groups in the following structures).

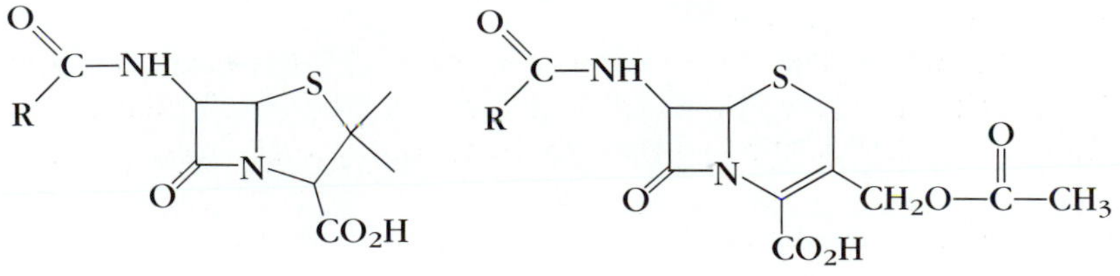

Penicillins

Cephalosporins

As you would expect for a strained four-membered ring, the β-lactam is very easily opened, and this high reactivity is responsible for the biological activity of the antibiotics. The β-lactam acylates certain amino groups of an enzyme involved in bacterial cell wall synthesis.

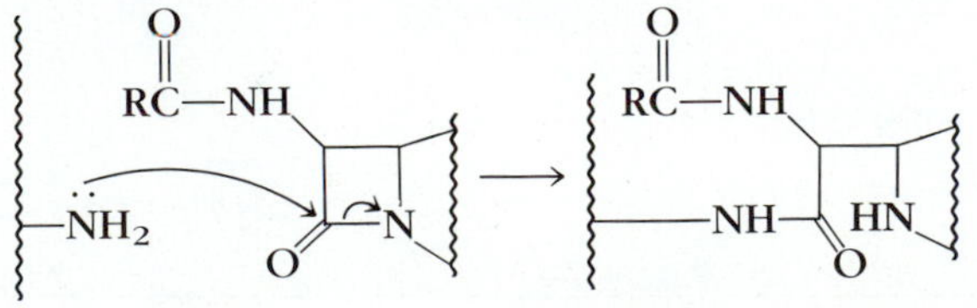

This inactivates the enzyme and is fatal to the bacteria. That enzyme is not found in humans (our cells do not have a cell wall structure), so bacterial infections can be treated with the β-lactam drugs. The drugs cause the specific (and fatal) interruption in metabolism only with the bacteria. See Section 24.5 for a more complete discussion of these drugs.

EXERCISE 16.16

What products are expected for the following reactions?

(a) $CH_3CH_2CH_2\overset{O}{\overset{\|}{C}}Cl \xrightarrow{NBS}$

(b) $CH_3CH_2CH_2\overset{O}{\overset{\|}{C}}OH \xrightarrow[\text{2. } Br_2 \quad \text{3. } H_2O]{\text{1. } SOCl_2}$

(c) $CH_3CH_2CH_2\overset{O}{\overset{\|}{C}}OH \xrightarrow[\text{2. } CH_3OH]{\text{1. } Br_2,\ PCl_3}$

EXERCISE 16.17

What products are expected for the reactions of

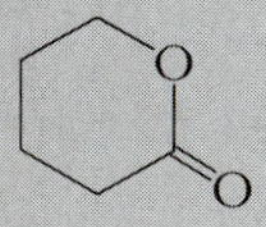

(δ-valerolactone: six-membered ring with ring O and C=O)

with the following reagents?

(a) $H_2O/NaOH$
(b) $LiAlH_4$, then H_2O
(c) CH_3MgBr, then H_2O

EXERCISE 16.18

What products are expected for the reactions of

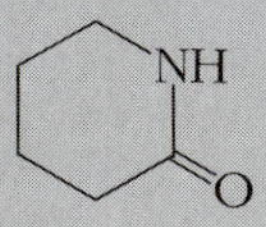

(δ-valerolactam: six-membered ring with NH and C=O)

with the following reagents?

(a) $LiAlH_4$, then H_2
(b) KOH/H_2O

16.9 CHEMICAL AND SPECTROSCOPIC CHARACTERIZATION OF CARBOXYLIC ACIDS AND THEIR DERIVATIVES

In this section we briefly discuss techniques that can be used to identify compounds of uncertain structure. The spectroscopic behavior of organic compounds was previously presented in Chapter 9, but we will focus specifically on carboxylic acid derivatives here. Similarly, the reactions of these compounds have been thoroughly discussed in the preceding sections, but here we will show how these reactions can be used to obtain structural information.

This is the first time we have considered structure determinations of nitrogen-containing compounds, and the previously stated (Section 7.11) procedure for calculating degrees of unsaturation must be modified. The general molecular formula of a saturated, acyclic nitrogen-containing molecule is $C_nH_{2n+2}(NH)_x$. In other words the total number of hydrogen atoms is two more than twice the number of carbons *plus* the number of nitrogen atoms. The number of degrees of unsaturation is equal to one-half the number of hydrogen atoms that are "missing" for the compound in question. As before, oxygens are ignored and halogens are counted as though they were hydrogen atoms.

The examples in Table 16.7 show how the number of degrees of unsaturation are correctly calculated for different compounds: a ring or double bond is equivalent to one degree of unsaturation and a triple bond corresponds to two degrees of unsaturation.

Spectroscopic Methods

NMR Spectroscopy All the carboxylic acid derivatives except nitriles, have a carbonyl group, and the carbonyl carbon gives rise to a characteristic ^{13}C NMR

TABLE 16.7 Degrees of Unsaturation for Some Nitrogen-Containing Compounds with *n* Carbon Atoms and *x* Nitrogen Atoms

Compound	*n*	*x*	$(2n + 2 + x)$	Number of H	Degrees of Unsaturation
CH_3NH_2	1	1	5	5	0
$CH_3—C(=O)NH_2$	2	1	7	5	1
$CH_3CH_2—C{\equiv}N$	3	1	9	5	2
cyclopentyl$—C(=O)—NHCH_3$	7	1	17	13	2

signal. The electronegative substituent on the carbonyl group results in substantial shifts relative to ketones and aldehydes, and a signal in the vicinity of 160–180 ppm is found for the carbonyl carbons of carboxylic acid derivatives. The cyano carbon of a nitrile is found near 120 ppm.

Proton NMR spectroscopy affords little information that is characteristic of carboxylic acid derivatives, although the carbonyl hydrogens of the acids themselves usually absorb far downfield, in the range of 11.5–12.5 ppm. Reference to Table 9.3 also shows that the hydrogen on the carbon bonded to the ester oxygen (—CH—O_2CR) absorbs about 0.5 ppm further downfield than the corresponding hydrogen of an alcohol (—CH—OH).

IR Spectroscopy Carboxylic acids exhibit a very broad and characteristic absorption peak over the range 2500–3500 cm^{-1}, and nitriles show a moderately strong peak (C≡N) in the vicinity of 2250 cm^{-1}, a region where most compounds show no peaks at all. All the remaining carboxylic acid derivatives exhibit a strong and characteristic peak in the carbonyl region of the spectrum (1650–1750 cm^{-1}). Substituent effects are summarized in Table 9.7.

UV Spectroscopy Carboxylic acid derivatives are similar to aldehydes and ketones, and there are no absorption peaks in normal UV spectra. α,β-Unsaturated derivatives, however, exhibit spectra that are similar to those for analogous unsaturated ketones (see Section 9.7).

Chemical Methods

Carboxylic acids can be detected by their acidity.

$$R-\overset{O}{\overset{\|}{C}}OH + NaOH \longrightarrow R-\overset{O}{\overset{\|}{C}}O^-Na^+ + H_2O$$

A small sample can be titrated by dropwise addition of dilute NaOH with phenolphthalein as an indicator. (Aqueous alcohol is usually an effective solvent if the compound does not dissolve in water.) As soon as an excess of NaOH is added, a sharp end point is seen. If the sample is a neutral compound, the end point will be observed upon addition of the first portion of base, but a carboxylic acid will react with a significant volume of base before the end point is reached. If the titration is carried out quantitatively with standardized base, the neutralization equivalent of the acid can be calculated. This is the weight of acid that reacts with one mole of base, and it is equal to the gram molecular weight for a monocarboxylic acid (one-half the gram molecular weight for a diacid.)

The acidity of carboxylic acids also leads to important solubility properties. The acids themselves (except for those with six or fewer carbons) are insoluble in water but soluble in organic solvents. The sodium salts of the acids, in contrast, are generally quite soluble in water but insoluble in organic solvents. This means that a carboxylic acid will dissolve in aqueous NaOH (because the acid is converted to the salt). Acidification of the alkaline solution with HCl will convert the salt back to the insoluble carboxylic acid.

$$\underset{\text{Water-insoluble}}{R-\overset{O}{\overset{\|}{C}}-OH} \xrightarrow{NaOH} \underset{\text{Water-soluble}}{R-\overset{O}{\overset{\|}{C}}-O^-Na^+} \xrightarrow{HCl} R-\overset{O}{\overset{\|}{C}}-OH$$

Acid chlorides and anhydrides react with hydroxylic solvents, so water solubility may be misleading. Their high reactivity permits them to be converted to the acid or to a derivative such as an amide, which can then be further characterized.

Esters do not react with aqueous sodium hydroxide in the nearly instantaneous manner of carboxylic acids, but they will hydrolyze slowly at elevated temperatures.

$$R{-}\overset{\overset{\displaystyle O}{\|}}{C}{-}OR' + NaOH \longrightarrow R{-}\overset{\overset{\displaystyle O}{\|}}{C}{-}O^-Na^+ + R'OH$$

This saponification reaction consumes hydroxide ion, producing the carboxylate salt, which is less basic. This decreases the pH, and the change can be detected with an indicator. It is also possible to carry out the hydrolysis quantitatively. The weight of ester that reacts with one mole of NaOH is the saponification equivalent, and it is equal to the gram molecular weight for the ester of a monocarboxylic acid.

Amides and **nitriles** are much less reactive than esters toward hydrolysis. Nevertheless, they can be hydrolyzed under strongly alkaline or strongly acidic conditions. Hydrolysis yields two products, the carboxylic acid and the amine (ammonia in the case of a nitrile or primary amide), which can be further characterized.

EXERCISE 16.19

Which of the following compounds would dissolve in aqueous base (dilute NaOH)?

(a) $CH_3CH_2\overset{\overset{\displaystyle O}{\|}}{C}OCH_2CH_3$ (b) $C_6H_5{-}\overset{\overset{\displaystyle O}{\|}}{C}OH$ (c) $Na^{+-}O\overset{\overset{\displaystyle O}{\|}}{C}CH_3$

(d) $Na^{+-}O\overset{\overset{\displaystyle O}{\|}}{C}{-}C_6H_5$ (e) $CH_3CH_2{-}\overset{\overset{\displaystyle O}{\|}}{C}OH$ (f) $CH_3(CH_2)_8\overset{\overset{\displaystyle O}{\|}}{C}{-}Cl$

EXERCISE 16.20

Which of the compounds in Exercise 16.19 would dissolve in aqueous acid?

EXERCISE 16.21

Compound **A** exhibits a carbonyl absorption in the infrared spectrum. **A** reacts slowly with aqueous NaOH to liberate ammonia and the salt of a carboxylic acid. Acidification of the alkaline solution results in a colorless precipitate, identified as 2,2-diphenylacetic acid. What is the structure of **A?**

16.10 TERMS AND DEFINITIONS

Acid halide. A carboxylic acid derivative in which the two substituents on the C=O group are alkyl (or aryl) and halo; an acyl halide.

Acyl group. The RCO group (i.e., C=O with an alkyl or aryl group as one of its substituents.

Acyl halide. An acid halide; a carboxylic acid derivative in which the two substituents on the carbonyl group are alkyl (or aryl) and halo.

Amide. A carboxylic acid derivative characterized by an acyl group bonded to an NH_2 (or NHR or NR_2) group.

Anhydride. A carboxylic acid derivative characterized by two acyl groups bonded to a single oxygen atom.

Imide. A cyclic derivative of a dicarboxylic acid in which the two acyl groups are bonded to a single NH (or NR) group.

Lactam. A cyclic amide derived from an amino-substituted carboxylic acid; the ring includes the carbonyl carbon and the nitrogen.

Lactone. A cyclic ester derived from a hydroxy-substituted carboxylic acid; the ring includes the carbonyl carbon and the ester oxygen.

Nitrile. A carboxylic acid derivative characterized by the presence of a cyano group (i.e., R—C≡N).

16.11 SUMMARY OF REACTIONS

The reactions of carboxylic acids and their derivatives are summarized in Table 16.8. Various methods for preparation of these compounds are summarized in Table 16.9.

TABLE 16.8 Reactions of Carboxylic Acids and Carboxylic Acid Derivatives

Reaction	Comments
1. Carboxylic acids	
(a) Reaction as bases	Section 16.3
$R{-}C(=O){-}OH \overset{H^+}{\rightleftharpoons} R{-}C(={}^{+}OH){-}OH$	Weaker bases than H_2O.
(b) Reaction as acids	Section 16.3
$RCO_2H \rightleftharpoons RCO_2^- + H^+$	Much stronger acids than H_2O. Anion is unreactive toward nucleophilic attack. Acidification regenerates carboxylic acid.
(c) Reduction	Section 16.4
$R{-}C(=O){-}OH \xrightarrow[2.\ H_2O]{1.\ LiAlH_4} RCH_2OH$	The most commonly used reducing agent is $LiAlH_4$.
(d) Oxidation	Section 16.4
$R{-}CO_2H \xrightarrow{Pb(OAc)_4} R{-}OAc + CO_2$	Unreactive (already in highest oxidation state), but oxidative decarboxylation can occur with certain transition metal salts.

TABLE 16.8 Reactions of Carboxylic Acids and Carboxylic Acid Derivatives (continued)

Reaction	Comments
(e) Reaction with organometallic compounds	Sections 13.5, 16.5
$R'\text{—}C(=O)\text{—}OH \xrightarrow{RLi} R'\text{—}C(=O)\text{—}R$	Only with alkyllithium reagent. Other organometallics just form salt of acid.
(f) Conversion to carboxylic acids	Not applicable.
(g) Conversion to acid chlorides	Section 16.6
$R\text{—}C(=O)\text{—}OH \xrightarrow{PCl_5} R\text{—}C(=O)\text{—}Cl$	$SOCl_2$ and PCl_3 also work well.
(h) Conversion to anhydrides	Section 16.6
$R\text{—}C(=O)\text{—}OH + Ac_2O \longrightarrow R\text{—}C(=O)\text{—}O\text{—}C(=O)\text{—}R$	Anhydrides are infrequently used unless commercially available. Can sometimes be prepared using acetic anhydride.
(i) Conversion to esters	Section 16.6
$R'\text{—}CO_2H \underset{}{\overset{ROH,\ H^+}{\rightleftharpoons}} R'CO_2\text{—}R$	Also via acid chloride (see entry 2i of this table). Equilibrium reaction.
$R\text{—}CO_2H \xrightarrow{CH_2N_2} RCO_2CH_3$	Section 11.8. Methyl esters with diazomethane; *O*-alkylation.
$R\text{—}CO_2H \xrightarrow[2.\ Et_3O^+BF_4^-]{1.\ (iPr)_2NEt} R\text{—}CO_2Et$	Section 11.8. Ethyl (and methyl) esters with oxonium salts; *O*-alkylation.
(j) Conversion to amides	Via acid chlorides, anhydrides, or esters. See entries 2j, 3j, and 4j of this table.
(k) Conversion to nitriles	Via amides; See entry 5k of this table.
2. Acid chlorides	
(a) Reaction as bases	Section 16.3
$R\text{—}C(=O)\text{—}Cl \overset{H^+}{\rightleftharpoons} R\text{—}C(=\overset{+}{O}H)\text{—}Cl$	Weaker bases than H_2O.

TABLE 16.8 Reactions of Carboxylic Acids and Carboxylic Acid Derivatives (continued)

Reaction	Comments
(b) Reaction as acids	Section 16.3
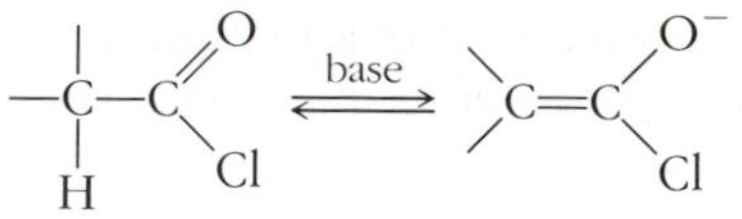 $-\overset{\vert}{\underset{\mathrm{H}}{\mathrm{C}}}-\mathrm{C}(=\mathrm{O})\mathrm{Cl} \underset{}{\overset{\text{base}}{\rightleftharpoons}} >\mathrm{C}=\mathrm{C}(\mathrm{O}^-)\mathrm{Cl}$	Enolate ion formation; comparable with ketones in acidity, but most bases will attack carbonyl.
(c) Reduction	Section 13.4
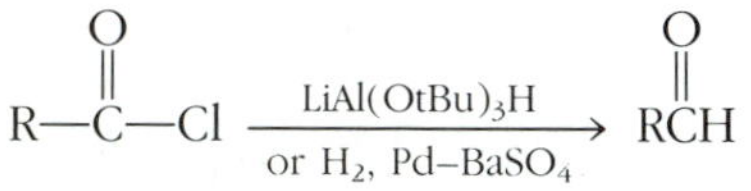 $\mathrm{R-\overset{O}{\overset{\Vert}{C}}-Cl} \xrightarrow[\text{or } H_2,\ Pd\text{–}BaSO_4]{LiAl(OtBu)_3H} \mathrm{R\overset{O}{\overset{\Vert}{C}}H}$	Selective reduction to aldehyde.
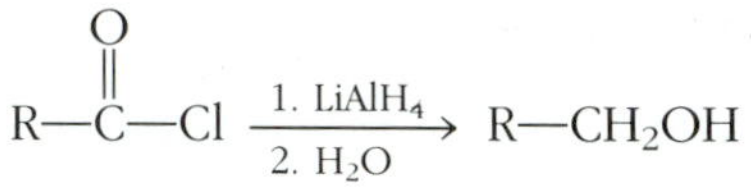 $\mathrm{R-\overset{O}{\overset{\Vert}{C}}-Cl} \xrightarrow[2.\ H_2O]{1.\ LiAlH_4} \mathrm{R-CH_2OH}$	Section 16.5 Other reducing agents will also yield primary alcohol.
(d) Oxidation	See entry 1d of this table.
(e) Reaction with organometallic reagents	Section 16.5
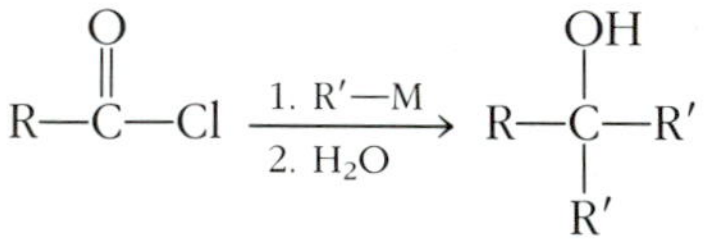 $\mathrm{R-\overset{O}{\overset{\Vert}{C}}-Cl} \xrightarrow[2.\ H_2O]{1.\ R'\text{—}M} \mathrm{R-\overset{OH}{\overset{\vert}{\underset{R'}{\underset{\vert}{C}}}}-R'}$	Highly reactive.
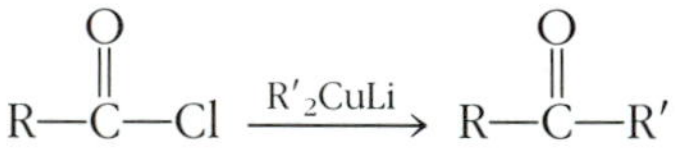 $\mathrm{R-\overset{O}{\overset{\Vert}{C}}-Cl} \xrightarrow{R'_2CuLi} \mathrm{R-\overset{O}{\overset{\Vert}{C}}-R'}$	Organocopper reagents are more selective than other organometallics.
(f) Conversion to carboxylic acids	Section 16.6
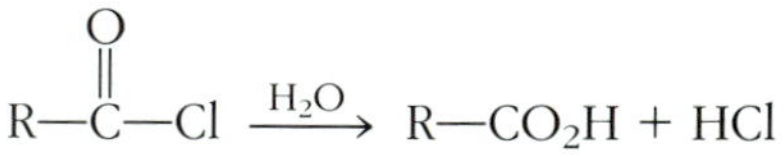 $\mathrm{R-\overset{O}{\overset{\Vert}{C}}-Cl} \xrightarrow{H_2O} \mathrm{R-CO_2H + HCl}$	Unintentional hydrolysis may occur if moisture is not excluded.
(g) Conversion to acid chlorides	Not applicable.
(h) Conversion to anhydrides	Section 16.6
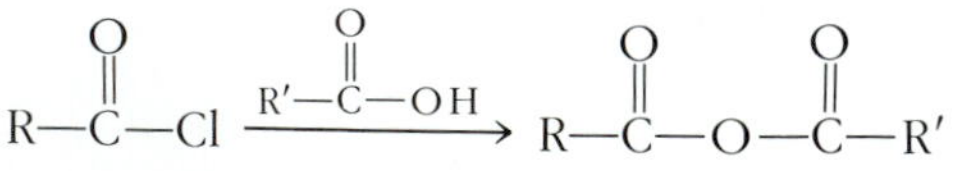 $\mathrm{R-\overset{O}{\overset{\Vert}{C}}-Cl} \xrightarrow{R'-\overset{O}{\overset{\Vert}{C}}-OH} \mathrm{R-\overset{O}{\overset{\Vert}{C}}-O-\overset{O}{\overset{\Vert}{C}}-R'}$	Mixed anhydride formation.
(i) Conversion to esters	Sections 11.8, 16.6
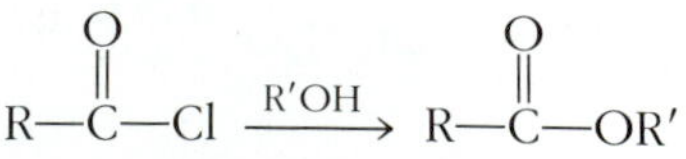 $\mathrm{R-\overset{O}{\overset{\Vert}{C}}-Cl} \xrightarrow{R'OH} \mathrm{R-\overset{O}{\overset{\Vert}{C}}-OR'}$	Acylation of alcohol.

TABLE 16.8 Reactions of Carboxylic Acids and Carboxylic Acid Derivatives (continued)

Reaction	Comments
(j) Conversion to amides	Section 16.6
$\mathrm{R{-}C({=}O){-}Cl \xrightarrow{HN<} R{-}C({=}O){-}N<}$	Amine must have at least one replaceable hydrogen.
(k) Conversion to nitriles	Via amides. See entry 5k in this table.
(l) α-Halogenation	Section 16.8
$\mathrm{{-}C(H){-}C({=}O){-}Cl \xrightarrow{NBS} {-}C(Br){-}C({=}O){-}Cl}$	Hydrolysis during workup affords the α-halo acid.
3. Acid anhydrides	
(a) Reaction as bases	Section 16.3
$\mathrm{R{-}C({=}O){-}O{-}C({=}O){-}R \underset{}{\overset{H^+}{\rightleftarrows}} R{-}C({=}O){-}O{-}C({=}{}^+OH){-}R}$	Weaker bases than H_2O.
(b) Reaction as acids	Section 16.3
$\mathrm{{-}C(H){-}C({=}O){-}O{-}C({=}O){-}R \overset{base}{\rightleftarrows} >C{=}C(O^-){-}O{-}C({=}O)R}$	Enolate ion formation; comparable with ketones in acidity. Most bases will attack carbonyl.
(c) Reduction	Section 16.4
$\mathrm{R{-}C({=}O){-}O{-}C({=}O){-}R \xrightarrow[2.\ H_2O]{1.\ LiAlH_4} R{-}CH_2OH}$	Not used preparatively.
(d) Oxidation	See entry 1d of this table.
(e) Reaction with organometallic reagents	Section 16.5
$\mathrm{R{-}C({=}O){-}O{-}C({=}O){-}R \xrightarrow[2.\ H_2O]{1.\ R'{-}M} R{-}C(OH)(R'){-}R' + R{-}CO_2^-}$	Highly reactive, not normally used preparatively. The two "halves" react differently when the R groups are not the same.

TABLE 16.8 Reactions of Carboxylic Acids and Carboxylic Acid Derivatives (continued)

Reaction	Comments
(f) Conversion to carboxylic acids	Section 16.6
$R{-}C({=}O){-}O{-}C({=}O){-}R \xrightarrow{H_2O} RCO_2H$	Unintentional hydrolysis may occur if moisture is not excluded.
(g) Conversion to acid chlorides	Not applicable.
(h) Conversion to anhydrides	Not applicable.
(i) Conversion to esters	Section 16.6
$R{-}C({=}O){-}O{-}C({=}O){-}R \xrightarrow{R'OH} R{-}C({=}O){-}OR' + RCO_2H$	Only one of the two acyl groups is converted to the ester.
(j) Conversion to amides	Section 16.6
$R{-}C({=}O){-}O{-}C({=}O){-}R \xrightarrow{HNR_2} R{-}C({=}O){-}NR_2 + RCO_2^-$	Only one of the two acyl groups is converted to the amide.
(k) Conversion to nitriles	Via amides; See entry 5k of this table.
4. Esters	
(a) Reaction as bases	Section 16.3
$R{-}C({=}O){-}OR' \overset{H^+}{\rightleftharpoons} R{-}C({=}\overset{+}{O}H){-}OR'$	Weaker bases than H_2O.
(b) Reaction as acids	Section 16.3
$H{-}\overset{\vert}{C}{-}CO_2R \overset{base}{\rightleftharpoons} {>}C{=}C(O^-)(OR)$	Enolate formation; comparable with ketones in acidity.
(c) Reduction	Section 16.4
$R{-}C({=}O){-}OR' \xrightarrow[2.\ H_2O]{1.\ LiAlH_4} RCH_2OH + R'OH$	Two alcohols are formed
(d) Oxidation	See entry 1d of this table.

TABLE 16.8 Reactions of Carboxylic Acids and Carboxylic Acid Derivatives (continued)

Reaction	Comments
(e) Reaction with organometallic reagents	Section 13.6
$R{-}\overset{O}{\overset{\Vert}{C}}{-}OR' \xrightarrow[2.\ H_2O]{1.\ R''{-}M} R{-}\underset{R''}{\overset{OH}{C}}{-}R'' + R'OH$	A tertiary alcohol is formed. with alkyllithium and Grignard reagents. A second alcohol is produced from the —OR′ portion of the ester.
(f) Conversion to carboxylic acids	Section 16.6. Hydrolysis
$R{-}\overset{O}{\overset{\Vert}{C}}{-}OR' \underset{}{\overset{H_2O,\ H^+}{\rightleftharpoons}} R{-}\overset{O}{\overset{\Vert}{C}}{-}OH + R'OH$	An equilibrium reaction under acidic conditions.
$R{-}\overset{O}{\overset{\Vert}{C}}{-}OR' \xrightarrow{H_2O,\ ^-OH} R{-}\overset{O}{\overset{\Vert}{C}}{-}O^- + R'OH$	Alkaline conditions (saponification); the salt of the acid is formed.
(g) Conversion to acid chlorides	Not applicable.
(h) Conversion to anhydrides	Not applicable.
(i) Conversion to esters	Section 16.6
$R{-}\overset{O}{\overset{\Vert}{C}}{-}OR' \overset{R''OH}{\rightleftharpoons} R{-}\overset{O}{\overset{\Vert}{C}}{-}OR'' + R'OH$	Transesterification occurs under either acidic or basic conditions.
(j) Conversion to amides	Section 16.6
$R{-}\overset{O}{\overset{\Vert}{C}}{-}OR' \xrightarrow{HN\langle} R{-}\overset{O}{\overset{\Vert}{C}}{-}N\langle + R'OH$	
(k) Conversion to nitriles	Via amides; see entry 5k of this table.
(l) Ester pyrolysis	Section 6.9
5. Amides	
(a) Reaction as bases	Section 16.3
$R{-}C({=}O){-}N\langle \overset{H^+}{\rightleftharpoons} R{-}C({=}\overset{+}{O}H){-}N\langle$	Comparable with H_2O as bases. Protonation can also occur on nitrogen.
(b) Reaction as acids	Section 16.3
$R{-}\overset{O}{\overset{\Vert}{C}}{-}NHR \overset{base}{\rightleftharpoons} R{-}\overset{O}{\overset{\Vert}{C}}{-}\overset{-}{N}{-}R$	Comparable with H_2O as acids.

TABLE 16.8 Reactions of Carboxylic Acids and Carboxylic Acid Derivatives (continued)

Reaction	Comments
$-\overset{\vert}{\underset{H}{C}}-\overset{O}{\overset{\Vert}{C}}-NR_2 \overset{base}{\rightleftharpoons} >C{=}C(O^-)(NR_2)$	Tertiary amides can only form enolate ions; comparable with ketones in acidity.
(c) Reduction	Section 16.4
$R-\overset{O}{\overset{\Vert}{C}}-N(R')(R'') \xrightarrow[2.\ H_2O]{1.\ LiAlH_4} R-CH_2-N(R')(R'')$	R′ and R″ may be alkyl, aryl, or hydrogen.
(d) Oxidation	See entry 1d of this table.
(e) Reaction with organometallic reagents	Section 16.5
$R-\overset{O}{\overset{\Vert}{C}}-NH_2 \xrightarrow{R-M} R-\overset{O}{\overset{\Vert}{C}}-\bar{N}H$ $R-\overset{O}{\overset{\Vert}{C}}-NH-R' \xrightarrow{R-M} R-\overset{O}{\overset{\Vert}{C}}-\bar{N}R'$	Not used preparatively. Primary and secondary amides undergo acid-base reaction. Tertiary amides can undergo nucleophilic attack at carbonyl group.
(f) Conversion to carboxylic acids	Section 16.6
$R-\overset{O}{\overset{\Vert}{C}}-N< \xrightarrow[H^+]{H_2O} R-CO_2H + >\overset{+}{N}H_2$	Hydrolysis. Acidic or basic conditions can be employed.
$R-\overset{O}{\overset{\Vert}{C}}-N< \xrightarrow[^-OH]{H_2O} R-CO_2^- + >NH$	Carboxylate salt is produced in base.
(g) Conversion to acid chlorides	Not applicable.
(h) Conversion to anhydrides	Not applicable.
(i) Conversion to esters	Not applicable.
(j) Conversion to amides	Not applicable.
(k) Conversion to nitriles	Section 16.6
$R-\overset{O}{\overset{\Vert}{C}}-NH_2 \xrightarrow{P_2O_5} R-C{\equiv}N$	Dehydration. Primary amides only. $SOCl_2$, $POCl_3$ and PCl_5 can also be used.

TABLE 16.8 Reactions of Carboxylic Acids and Carboxylic Acid Derivatives (continued)

Reaction	Comments
6. Nitriles	
(a) Reaction as bases	Section 16.3
$R{-}C{\equiv}N \overset{H^+}{\rightleftharpoons} R{-}C{\equiv}\overset{+}{N}H$	Much weaker bases than H_2O.
(b) Reaction as acids	Section 16.3
$-\underset{H}{\overset{\vert}{C}}-C{\equiv}N \overset{\text{base}}{\rightleftharpoons} \rangle C{=}C{=}N^-$	Enolate ion formation; comparable with ketones in acidity.
(c) Reduction	Section 16.4
$R{-}C{\equiv}N \xrightarrow[\text{2. } H_2O]{\text{1. } LiAlH_4} R{-}CH_2NH_2$	
(d) Oxidation	Not usually oxidized.
(e) Reaction with organometallic reagents	Section 16.5
$R{-}C{\equiv}N \xrightarrow[\text{2. } H_2O]{\text{1. } R'M} R{-}\overset{NH}{\overset{\Vert}{C}}{-}R' \longrightarrow R{-}\overset{O}{\overset{\Vert}{C}}{-}R'$	Grignard and alkyllithium reagents yield ketones.
(f) Conversion to carboxylic acids	Section 16.6
$R{-}C{\equiv}N \xrightarrow{H_2O,\ H^+} R{-}CO_2H$ $R{-}C{\equiv}N \xrightarrow{H_2O,\ ^-OH} R{-}CO_2^-$	Hydrolysis; acidic or basic conditions may be employed.
(g) Conversion to acid chlorides	Not applicable.
(h) Conversion to anhydrides	Not applicable.
(i) Conversion to esters	Not applicable.
(j) Conversion to amides	Section 16.6
$R{-}C{\equiv}N \xrightarrow[H^+ \text{ or } ^-OH]{H_2O} R{-}\overset{O}{\overset{\Vert}{C}}{-}NH_2$	Partial hydrolysis can be carried out in either acid or base under carefully controlled conditions.
(k) Conversion to nitriles	Not applicable.

TABLE 16.9 Preparation of Carboxylic Acids and Carboxylic Acid Derivatives

Reaction	Comments
1. Carboxylic acids	
(a) Oxidation of alcohols and aldehydes	Sections 11.3, 13.11, 16.7
$R{-}\overset{O}{\overset{\Vert}{C}}H \xrightarrow{\text{oxidation}} RCO_2H$	
$R{-}CH_2OH \xrightarrow{\text{oxidation}} RCO_2H$	
(b) Oxidation of alkenes	Section 16.7
$R{-}CH{=}CH{-}R' \xrightarrow{KMnO_4} RCO_2H + R'CO_2H$	Further oxidation may occur.
$R{-}CH{=}CH{-}R' \xrightarrow[\text{2. } H_2O_2,\text{ NaOH}]{\text{1. } O_3} RCO_2H + R'CO_2H$	Oxidative workup of ozonide.
(c) Oxidation of methyl ketones	Section 13.9
$R{-}\overset{O}{\overset{\Vert}{C}}{-}CH_3 \xrightarrow[\text{2. HCl}]{\text{1. NaOCl}} R{-}CO_2H + CHCl_3$	Haloform reaction (Cl_2, Br_2 or I_2 in aqueous base)
(d) Oxidation of alkyl substituents on aromatic rings	Section 16.7
$Ar{-}R \xrightarrow[Na_2Cr_2O_7]{KMnO_4\text{ or}} Ar{-}CO_2H$	
(e) Carboxylation of organometallic reagents	Section 16.7
$R{-}M \xrightarrow[\text{2. } H^+]{\text{1. } CO_2} R{-}CO_2H$	
(f) Koch-Haaf carboxylation of carbocations	Section 16.7
$R^+ \xrightarrow[H_2SO_4]{CO} R{-}CO_2H$	Must be moderately stable carbocation. Carbocation is generated by reactions of alcohol, alkene, or alkyl halide in H_2SO_4.
2. Esters	
(a) Oxidation of ketones	Sections 13.11, 16.7
$R{-}\overset{O}{\overset{\Vert}{C}}{-}R' \xrightarrow{CH_3CO_3H} R{-}\overset{O}{\overset{\Vert}{C}}{-}OR'$	Baeyer-Villiger oxidation.

TABLE 16.9 Preparation of Carboxylic Acids and Carboxylic Acid Derivatives (continued)

Reaction	Comments
(b) Condensation reactions of esters	Chapter 21
$-\underset{\vert}{\overset{\vert}{C}}H-CO_2R \xrightarrow[\text{base}]{R'-\overset{O}{\overset{\Vert}{C}}-X} R'-\overset{O}{\overset{\Vert}{C}}-\underset{\vert}{\overset{\vert}{C}}-CO_2R$	
(c) Alkylation of activated esters	Chapter 21
$-\overset{O}{\overset{\Vert}{C}}-CH_2-CO_2R \xrightarrow[\text{2. R'X}]{\text{1. base}} -\overset{O}{\overset{\Vert}{C}}-\underset{R'}{\underset{\vert}{C}}H-CO_2R$	
3. Nitriles	
(a) Cyanide displacement	Sections 12.8, 16.7
$R-X \xrightarrow{\text{NaCN}} R-CN$	RX must be primary or secondary halide or sulfonate.
(b) Alkylation of cyano esters	Chapter 21
$RO_2C-CH_2-CN \xrightarrow[\text{2. R'X}]{\text{1. base}} RO_2C-\underset{R'}{\underset{\vert}{C}}H-CN$	

16.12 PROBLEMS

16.1 Name each of the following, where R is H, CH_3, $CH(CH_3)_2$, $CH_2(CH_2)_3CH_3$, C_6H_5, and $CH_2C_6H_5$.

(a) $H-\overset{O}{\overset{\Vert}{C}}-OR$ (b) $CH_3-\overset{O}{\overset{\Vert}{C}}-OR$

(c) $CH_3-\overset{CH_3}{\overset{\vert}{C}}H-\overset{O}{\overset{\Vert}{C}}-OR$ (d) $CH_3(CH_2)_4CH_2-\overset{O}{\overset{\Vert}{C}}-OR$

(e) $ClCH_2-\overset{O}{\overset{\Vert}{C}}-OR$ (f) $C_6H_{11}-\overset{O}{\overset{\Vert}{C}}-OR$ (cyclohexyl)

16.2 Name each of the following, where X is OH, Cl, NH_2, $NHCH_3$, $N(CH_3)_2$, and CH_3—N—CH_2CH_3.

(a) H—C(=O)—X (b) CH_3—C(=O)—X (c) $CH_3CH_2CH_2$—C(=O)—X

(d) C_6H_5C(=O)—X (e) cyclopentyl—C(=O)—X

16.3 Draw each of the following compounds:

(a) *N*-Methylbutyramide
(b) *cis*-4-Hydroxycyclohexanecarboxylic acid
(c) 3,3-Dichlorooctanoyl chloride
(d) Methyl valerate
(e) *N,N*-Diisopropylbenzamide
(f) 4-Oxocyclohexanecarboxylic acid
(g) Cyclohexylacetic acid
(h) Cyclohexyl acetate
(i) Methyl isobutyrate
(j) 2-Methylbutyl formate
(k) Isobutyric anhydride
(l) Sodium pentanoate
(m) 4-Hydroxyoctanenitrile
(n) 3-Chloroadipic anhydride
(o) Benzonitrile
(p) Ammonium 2-chlorobutyrate

16.4 For each of the following pairs of compounds, which would be more soluble in water? Why?

(a) CH_3CN and $CH_3(CH_2)_4CN$
(b) HO_2C—$(CH_2)_4$—CO_2H and $CH_3(CH_2)_4CO_2H$
(c) CH_3O—C(=O)—$CH_2CH_2CH_2$—C(=O)—OCH_3 and HOC(=O)—CH(CH_3)—CH_2—CH(CH_3)—C(=O)—OH

16.5 For each of the pairs of compounds in Problem 16.4, which would have the higher boiling point? Why?

16.6 Draw the products of the reaction of each of the following compounds with aqueous sodium hydroxide:

(a) $CH_3CH_2CO_2CH_3$ (b) CH_3CH_2C(=O)NH_2

(c) CH_3CH(CH_3)—C(=O)—Cl (d) CH_3CH(CH_3)—CH_2CO_2H

(e) $CH_3(CH_2)_4CN$ (f) $C_6H_5CO_2C_2H_5$

16.7 For each of the compounds in Problem 16.6 that is not a carboxylic acid, show how it could be prepared from the corresponding acid.

16.8 Draw the products of the reaction of each of the compounds in Problem 16.6 with $LiAlH_4$ (followed by aqueous workup).

16.9 Draw the products of the reaction of each of the following compounds with CH_3CH_2OH/H_2SO_4:

(a) $C_6H_5CO_2CH_3$ (b) $CH_3(CH_2)_4CO_2H$
(c) $HO_2C—(CH_2)_3—CO_2H$ (d) $CH_3CH_2CO_2C_6H_5$

16.10 Draw the products of the reaction of each of the compounds in Problem 16.9 with butylamine.

16.11 Draw the products of the reaction of each of the following compounds with methyllithium (followed by aqueous workup):

(a) $C_6H_5CH_2CO_2CH_3$ (b) $CH_3CH(CH_3)—CH_2CO_2H$

(c) $CH_3CH_2CH_2—C(=O)—Cl$ (d) $(CH_3)_3C—CO_2CH_2CH_3$

(e) $CH_3(CH_2)_6—CN$

16.12 Draw the products of the reaction of each of the compounds in Problem 16.11 with ethylmagnesium bromide (followed by aqueous workup).

16.13 Draw the products of the reaction of each of the compounds in Problem 16.11 with $LiAlH_4$ (followed by aqueous workup).

16.14 For each of the compounds in Problem 16.11 that is not a carboxylic acid, show how it could be prepared from the corresponding acid.

16.15 Draw the products of the reaction of 3-methylbutanoic acid with each of the following compounds:

(a) CH_2N_2 (b) CH_3Li
(c) CH_3CH_2OH, H_2SO_4 (d) PCl_5
(e) $Pb(OAc)_4$ (f) NH_3
(g) $LiAlH_4$ followed by H_2O (h) $(CH_3)_3O^+BF_4^-$, $(iPr)_2NEt$

16.16 Draw the products of the reaction of benzoic acid with each of the reagents in Problem 16.15.

16.17 Draw the products of the reaction of $CH_3(CH_2)_4COCl$ with each of the following reagents:

(a) $LiAlH_4$ followed by water (b) 2-propanol
(c) Diethylamine (d) Water
(e) $LiAl(OtBu)_3H$ (f) C_6H_5MgBr
(g) NH_3, then P_2O_5 (h) CH_3Li
(i) $(CH_3)_2CuLi$

16.18 Draw the products of the reaction of $C_6H_5CH_2COCl$ with each of the reagents in Problem 16.17.

16.19 Draw the products of the reaction of acetic anhydride with each of the following reagents:
(a) 2-Butanol (b) Propylamine (c) H_2O
(d) CH_3CH_2Li (e) $HOCH_2CH_2CH_2OH$ (f) $LiAlH_4$, then H_2O

16.20 Draw the products of the reaction of benzoic anhydride with each of the reagents in Problem 16.19.

16.21 Draw the products of the reaction of butyl propionate with each of the following reagents:
(a) H_2O, H_2SO_4 (b) $NaOCH_3$, CH_3OH
(c) $LiAlH_4$, then H_2O (d) KOH, CH_3OH
(e) 2-Pentylamine (f) C_2H_5OH, H_2SO_4
(g) C_6H_5MgBr, then H_2O

16.22 Draw the products of the reaction of cyclohexyl acetate with each of the reagents in Problem 16.21.

16.23 Draw the products of the reaction of butyramide with each of the following reagents:
(a) $LiAlH_4$, then H_2O (b) KOH, alcohol
(c) CH_3MgBr (d) H_2SO_4, H_2O

16.24 Draw the products of the reaction of *N*-propylpropionamide with each of the reagents in Problem 16.23.

16.25 Suggest methods of converting each of the following alcohols (ROH) into carboxylic acids (RCO_2H). If more than one method would work, show the alternative ways.
(a) $CH_3(CH_2)_4CH_2OH$ (b) $CH_3CH_2{-}C(CH_3)_2{-}OH$
(c) Cyclohexanol (d) 1-Methylcyclohexanol
(e) 4-Hydroxycyclohexanol (f) $CH_3CH_2CH{=}CH{-}CH_2OH$

16.26 Suggest methods for preparing each of the following esters from the indicated starting material:
(a) Methyl 2-methylbutyrate from 2-methyl-1-butanol
(b) Ethyl propionate from 3-pentanone
(c) Methyl pivalate from 3,3-dimethyl-2-butanone
(d) *tert*-Butyl acetate from 3,3-dimethyl-2-butanone
(e) Methylhexanoate from hexanal

16.27 Draw the products of the reaction of succinic anhydride with each of the following reagents:
(a) NaOH, H_2O (b) $NaOCH_3$, CH_3OH (c) $LiAlH_4$, then H_2O
(d) Diethylamine (e) C_2H_5OH, H_2SO_4 (f) NBS

16.28 Complete the following reactions:

(a) 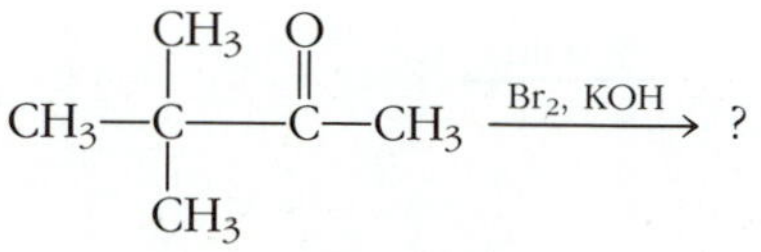 $CH_3-C(CH_3)_2-C(=O)-CH_3 \xrightarrow{Br_2,\ KOH} ?$

(b) (3-methylbenzaldehyde, $C(=O)-H$ and CH_3 on benzene ring) $\xrightarrow[\text{vigorous oxidation}]{KMnO_4} ?$

(c) $C_6H_5-CH_2-C(=O)-O-CH_2-CH_3 \xrightarrow{CH_3CH_2NH_2} ?$

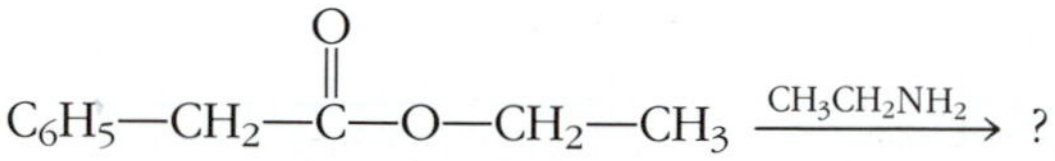

(d) 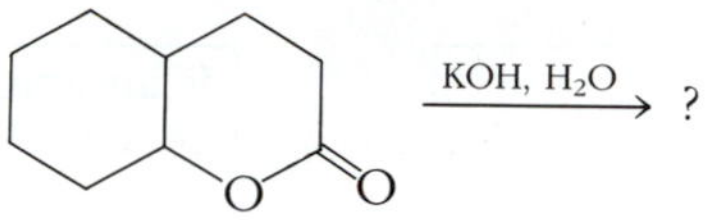 (bicyclic lactone: cyclohexane fused to ring containing O and C=O) $\xrightarrow{KOH,\ H_2O} ?$

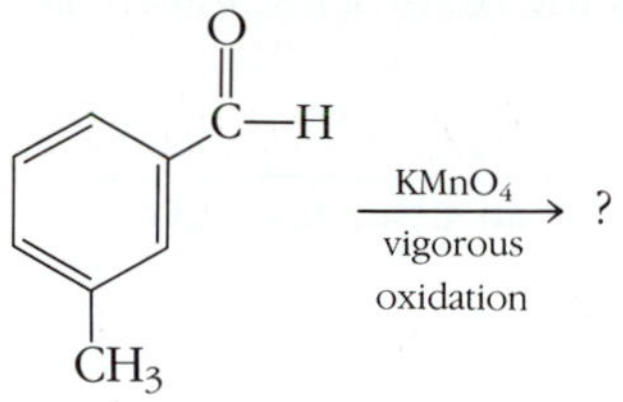

(e) 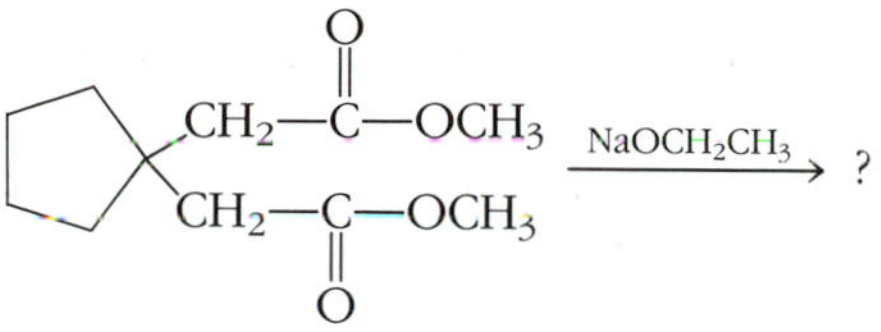 (cyclopentane bearing two $CH_2-C(=O)-OCH_3$ groups on the same carbon) $\xrightarrow{NaOCH_2CH_3} ?$

(f) 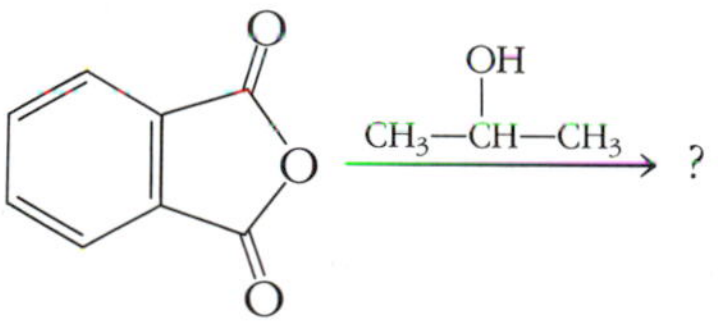 (phthalic anhydride) $\xrightarrow{CH_3-CH(OH)-CH_3} ?$

16.29 Complete the following reactions:

(a) 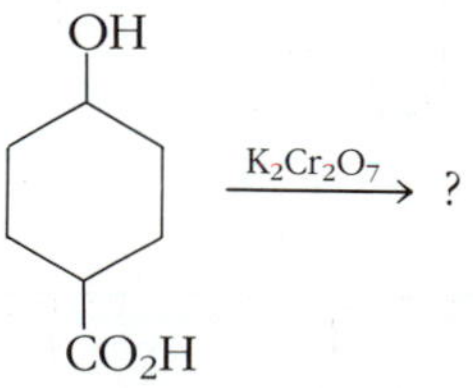 (cyclohexane with OH and CO_2H in 1,4 positions) $\xrightarrow{K_2Cr_2O_7} ?$

(b) 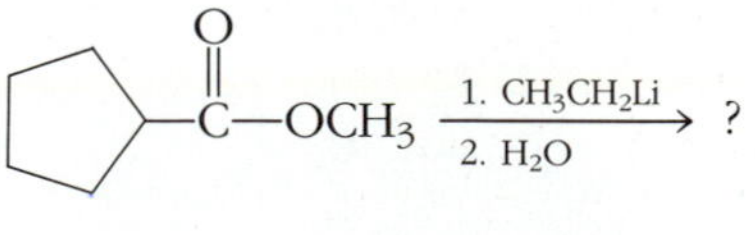 (cyclopentyl)$-C(=O)-OCH_3 \xrightarrow[2.\ H_2O]{1.\ CH_3CH_2Li} ?$

(c) (phenyl)$-C(=O)-NH-$(phenyl) $\xrightarrow[2.\ H_2O]{1.\ LiAlH_4} ?$

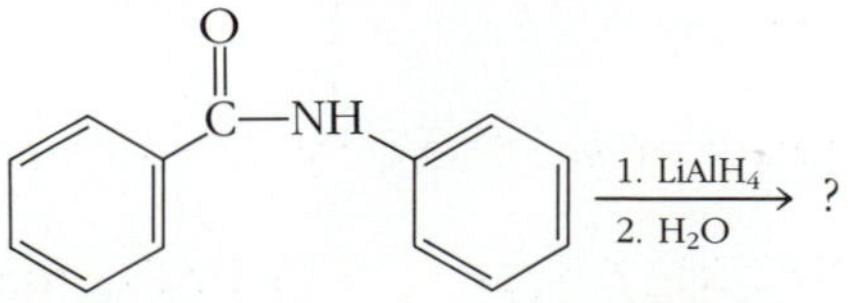

(d) $C_6H_5CH_2-\overset{O}{\overset{\|}{C}}-OCH_2CH_3 \xrightarrow[2.\ H_2O]{1.\ LiAlH_4} ?$

16.30 **A–I** are all different compounds. On the basis of the information provided, deduce their structures.

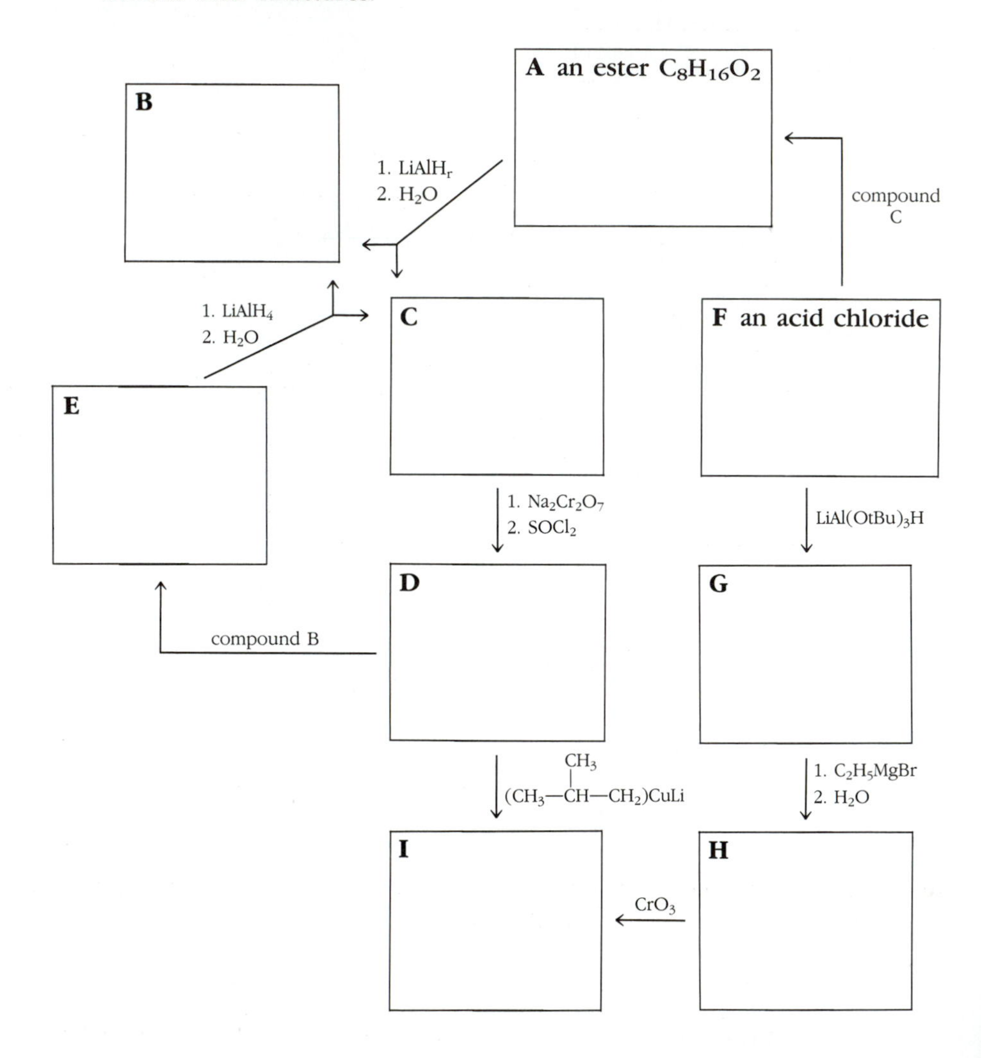

16.31 **A–G** are all different compounds. On the basis of the information provided, deduce their structures.

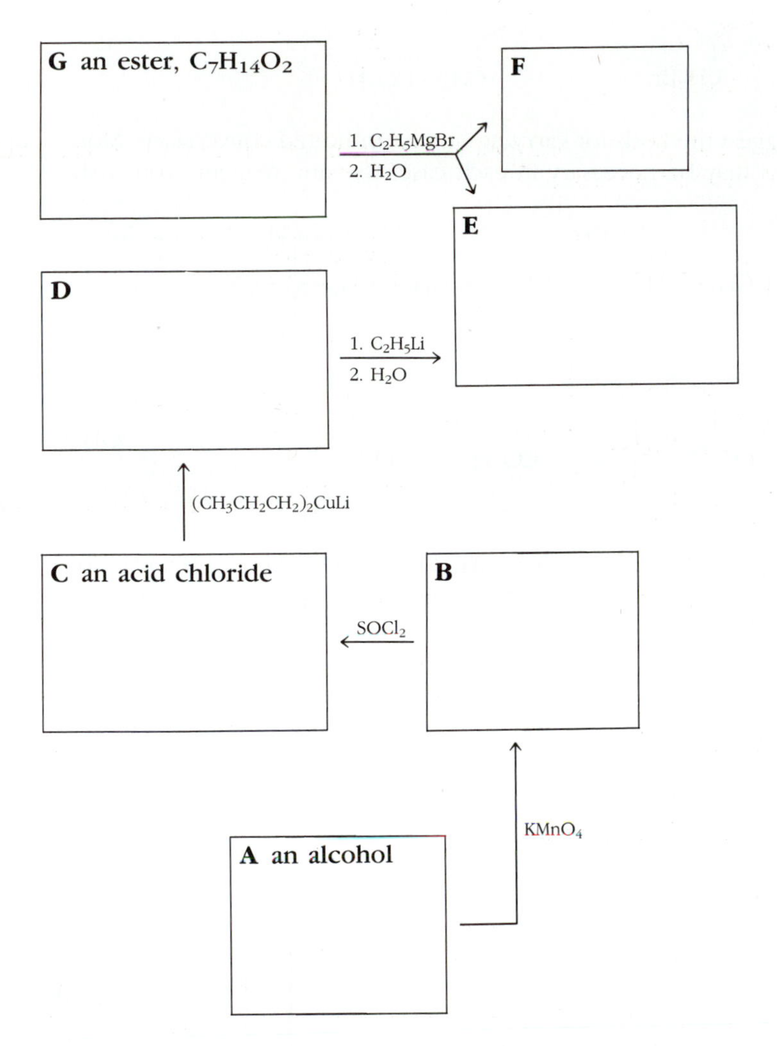

16.32 Suggest methods for carrying out the indicated conversions. More than one step may be necessary in each case. Use any reagents you wish.

(a) $CH_3-CH(CH_3)-CH_2-C(=O)-OH \longrightarrow CH_3-CH(CH_3)-CH_2-C(=O)-O-CH_2-CH_3$

(b) $CH_3-CH(CH_3)-CH_2-C(=O)-OH \longrightarrow CH_3-CH(CH_3)-CH_2-CH_2-O-C(=O)-CH_3$

(c) 2-bromotoluene (benzene ring with Br and CH_3 on adjacent carbons) $\longrightarrow$ 2-methylbenzoic acid (benzene ring with CO_2H and CH_3 on adjacent carbons)

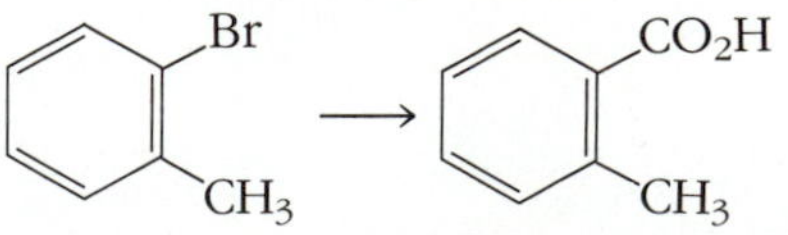

(d) m-(CH_2Br)$C_6H_4CO_2H$ ⟶ m-(CH_2—CO_2H)$C_6H_4CO_2H$

16.33 Suggest methods for carrying out the indicated conversions. More than one step may be necessary in each case. Use any reagents you wish.

(a) CH_3—CH_2—C(CH_3)$_2$—OH ⟶ CH_3—CH_2—C(CH_3)$_2$—CN

(b) O= (bicyclic ketone) —CO_2Et ⟶ O= (bicyclic ketone) —C(OH)(CH_3)$_2$

16.34 Saponification of ester **A** ($C_{12}H_{16}O_2$) afforded alcohol **B** and carboxylic acid **C.** On the basis of the proton NMR spectra of compounds **B** and **C,** deduce the structure of **A.**

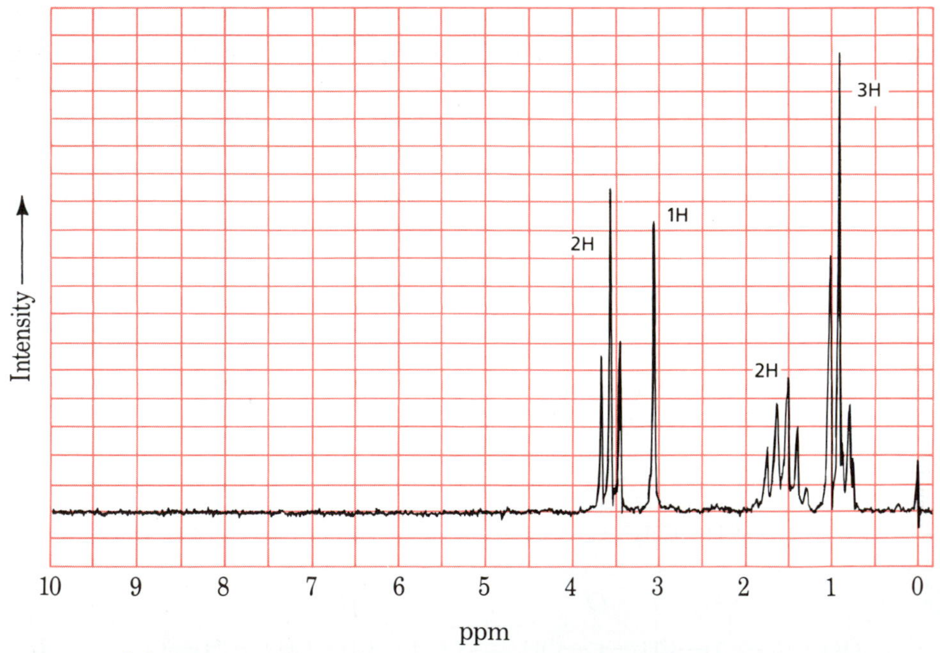

Figure 16.6 Proton NMR Spectrum of Compound B (Problem 16.34).

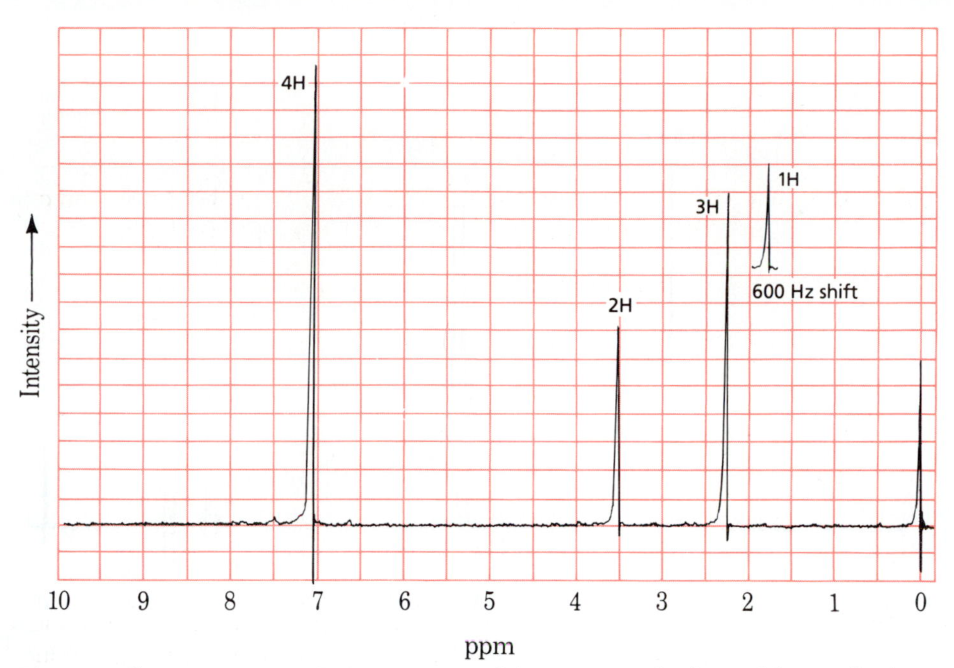

Figure 16.7 Proton NMR Spectrum of (Compound C) Problem 16.34.

16.35 An ester **A** ($C_{16}H_{16}O_2$) was reduced with lithium aluminum hydride to yield alcohols **B** and **C.** On the basis of the proton NMR spectra of **B** and **C,** deduce the structure of **A.**

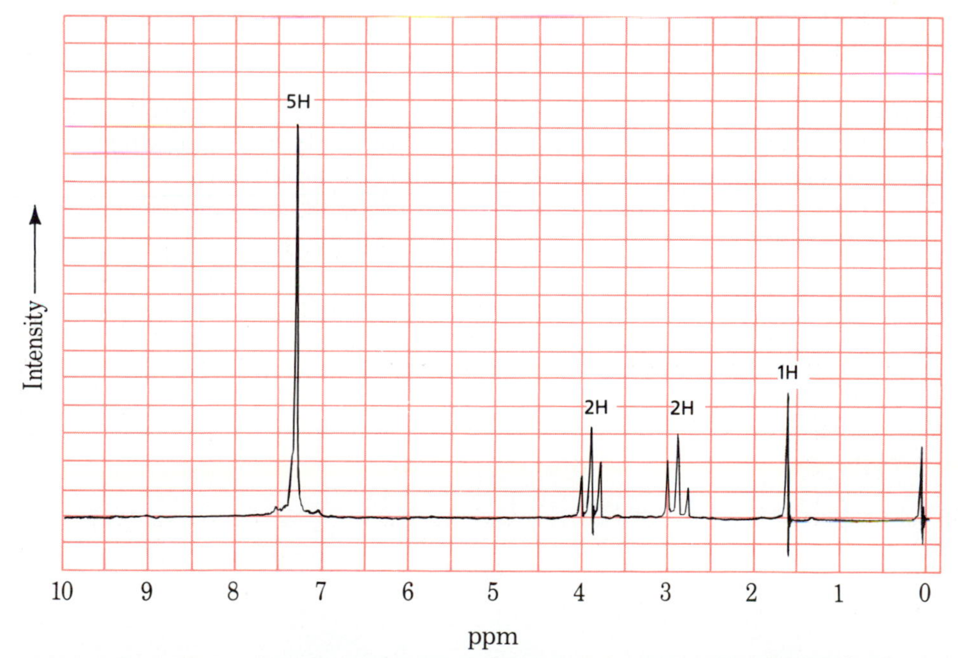

Figure 16.8 Proton NMR Spectrum of Compound B (Problem 16.35).

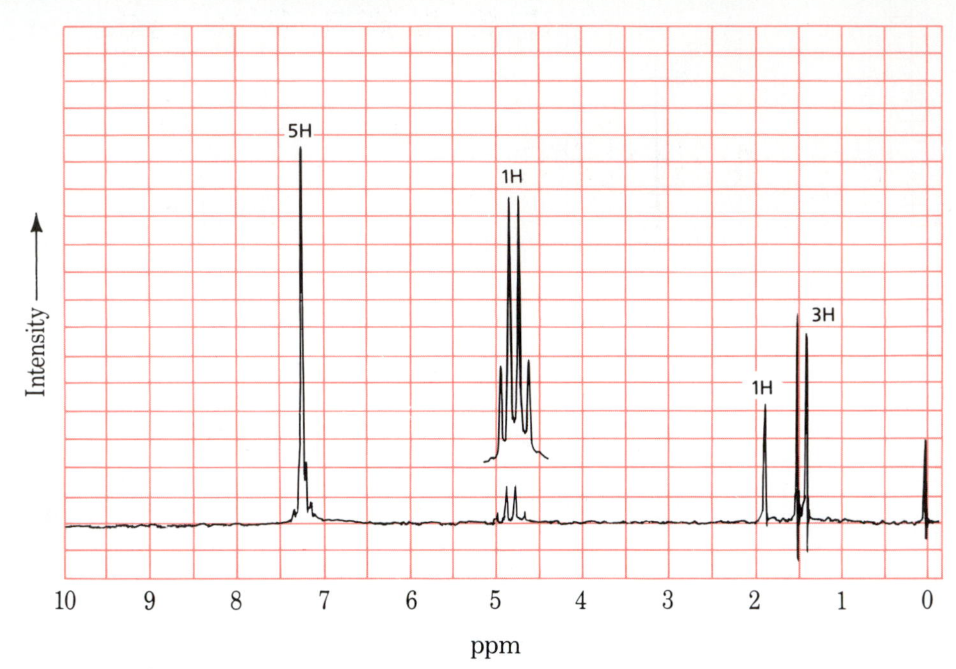

Figure 16.9 Proton NMR Spectrum of Compound C (Problem 16.35).

16.36 Spectroscopy can often help you learn whether or not a reaction has followed the desired course. Indicate how you could determine whether the material isolated from each of the reactions in Problem 16.15 had the expected structure. What specific changes in the IR and NMR spectra would you look for?

16.37 Spectroscopy can often help you learn whether or not a reaction has followed the desired course. Indicate how you could determine whether the material isolated from each of the reactions in Problem 16.17 had the expected structure. What specific changes in the IR and NMR spectra would you look for?

CHAPTER 17

AMINES

Amines are derivatives of ammonia in which one, two, or all three of the hydrogens of NH_3 have been replaced by alkyl or aryl groups. As you would expect for compounds that are structurally similar, the chemical reactivity of amines strongly resembles that of ammonia. Indeed, you can think of ammonia as being the simplest member of this functional class.

Amines have characteristic odors that are often reminiscent of ammonia, and at the same time they frequently have a "fishy" smell. In fact, it is the presence of volatile amines in their tissues that gives fish these characteristic odors. Two *di*amines that are released in decaying animal tissues have highly unpleasant odors, and those compounds have been aptly named:

$$H_2N—CH_2CH_2CH_2CH_2—NH_2 \qquad H_2N—CH_2CH_2CH_2CH_2CH_2—NH_2$$

Putrescine — Cadaverine

Amines play many important biological roles, and amine chemistry is vital to the functions of amino acids (along with enzymes and other proteins) and nucleic acids (including both DNA and RNA). These will be discussed in later chapters. Other amines are hormones; epinephrine, for example, helps regulate pulse rate and blood pressure.

$$\text{3,4-(HO)}_2C_6H_3—CH(OH)—CH_2—NH—CH_3$$

epinephrine (adrenaline)

A number of amines that are structurally related to epinephrine exhibit central nervous system (CNS) activity. Included in this group are a variety of valuable medicinal agents, as well as most of the psychoactive drugs that have been used illicitly in recent years (Section 23.4).

17.1 NOMENCLATURE

Monofunctional Amines

Amines are categorized as *primary, secondary,* or *tertiary* according to the number of carbons bonded to nitrogen.

$\ddot{N}H_3$	$R—\ddot{N}H_2$	$R—\ddot{N}H—R$	$R—\ddot{N}(R)—R$
Ammonia	Primary amine	Secondary amine	Tertiary amine

A primary amine containing no other functional groups is named as a substituted *-amine,* and the principal chain is the longest carbon chain that has the nitrogen as a substituent. Secondary and tertiary amines in which all alkyl (or aryl) substituents are the same utilize the prefix *di-* or *tri-*.

$CH_3CH_2{-}NH_2$ — Ethylamine

$CH_3CH_2{-}NH{-}CH_2CH_3$ — Diethylamine

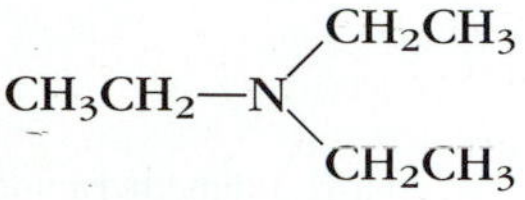

$CH_3CH_2{-}N(CH_2CH_3)_2$ — Triethylamine

If the groups on a secondary or tertiary amine are not the same, the compound is named as an *N*-alkyl derivative of a primary amine. This indicates that the alkyl group is a substituent on nitrogen and not a substituent on the principal chain.

$CH_3CH_2CH_2{-}NH{-}CH_3$ — *N*-Methylpropylamine

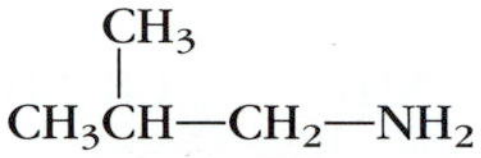

$CH_3CH(CH_3){-}CH_2{-}NH_2$ — 2-Methylpropylamine (isobutylamine)

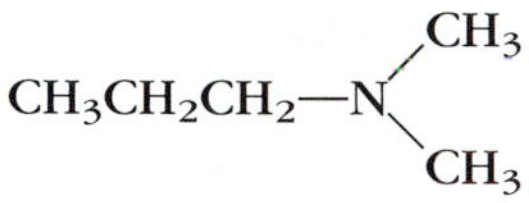

$CH_3CH_2CH_2{-}N(CH_3)_2$ — *N,N*-Dimethylpropylamine

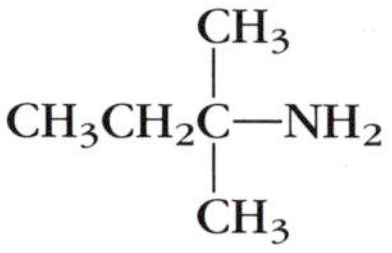

$CH_3CH_2C(CH_3)_2{-}NH_2$ — 1,1-Dimethylpropylamine

If the alkyl groups of a secondary or tertiary amine carry substituents, these alkyl groups can be differentiated with *primes.*

$\overset{2}{Cl}CH_2\overset{1}{C}H_2{-}NH{-}\overset{1'}{C}H_2\overset{2'}{C}H_2Cl$

2,2′-Dichlorodiethylamine

Aromatic amines frequently have common names. The simplest aromatic amine is called *aniline* (rather than phenylamine).

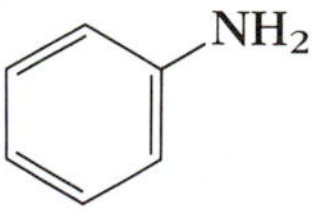

$C_6H_5{-}NH_2$

We have already presented the common names of other aromatic amines in Section 10.1.

Polyfunctional Amines

Amines rank very low in priority as the principal group in a polyfunctional compound (Section 16.1); only ethers rank lower among the commonly encountered functional groups. Consequently, most polyfunctional compounds must be named as *amino* derivatives of the parent compound.

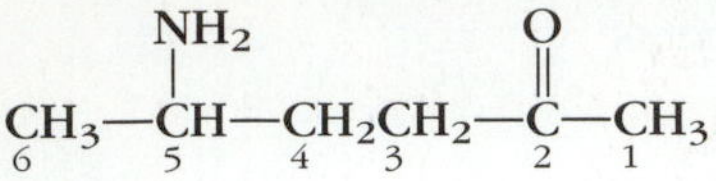

$$\underset{6}{CH_3}-\underset{5}{\overset{NH_2}{\overset{|}{C}H}}-\underset{4}{C}H_2\underset{3}{C}H_2-\underset{2}{\overset{O}{\overset{\|}{C}}}-\underset{1}{C}H_3$$

5-Amino-2-hexanone

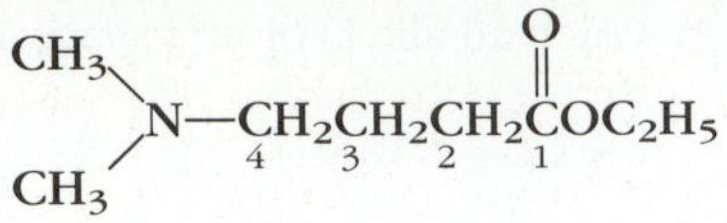

$$(CH_3)_2N-\underset{4}{C}H_2\underset{3}{C}H_2\underset{2}{C}H_2\underset{1}{\overset{O}{\overset{\|}{C}}}OC_2H_5$$

Ethyl 4-dimethylaminobutyrate

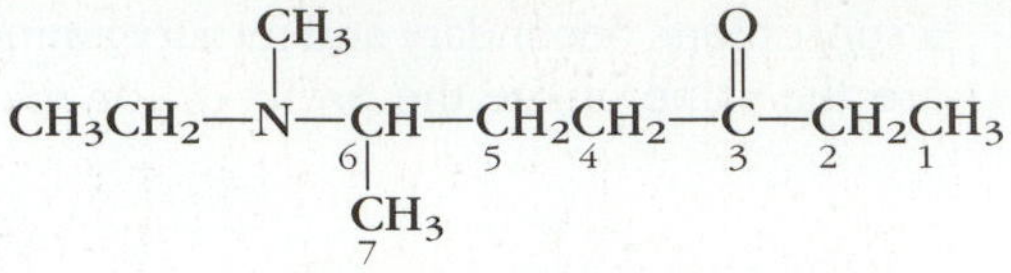

$$CH_3CH_2-\overset{CH_3}{\overset{|}{N}}-\underset{\underset{7}{CH_3}}{\underset{|}{\overset{6}{C}H}}-\underset{5}{C}H_2\underset{4}{C}H_2-\underset{3}{\overset{O}{\overset{\|}{C}}}-\underset{2}{C}H_2\underset{1}{C}H_3$$

6-(Ethylmethylamino)-3-heptanone

Ammonium Salts

Salts containing a tetravalent nitrogen are named as substituted ammonium salts, but in most other respects the names are comparable with those of the structurally related amines.

NH_3	CH_3NH_2	$CH_3CH_2-NH-CH_3$
Ammonia	Methylamine	*N*-Methylethylamine
$NH_4^+\ Cl^-$	$CH_3-NH_3^+\ Cl^-$	$CH_3CH_2-\overset{+}{N}H_2-CH_3\ Br^-$
Ammonium chloride	Methylammonium chloride	*N*-Methylethylammonium bromide

Ammonium salts are often named as products of a reaction between an amine and an acid. For example, $CH_3CH_2NH_3^+\ Cl^-$ (ethylammonium chloride) is sometimes called ethylamine hydrochloride (and is written as $CH_3CH_2NH_2 \cdot HCl$).

Quaternary ammonium compounds (four alkyl groups on nitrogen) are also named as substituted ammonium salts, and the alkyl groups are listed alphabetically.

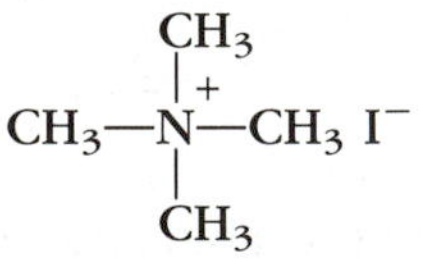

$$CH_3-\overset{CH_3}{\overset{|+}{\underset{\underset{CH_3}{|}}{N}}}-CH_3\ I^- \qquad CH_3CH_2-\overset{CH_3}{\overset{|+}{\underset{\underset{CH_3}{|}}{N}}}-CH_2CH_2CH_3\ {}^-OH$$

Tetramethylammonium iodide

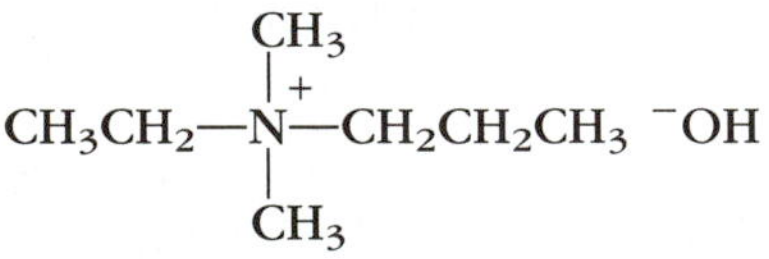

Ethyldimethylpropylammonium hydroxide

In contrast to amines, ammonium ions have the highest priority for a principal group, so they rarely need to be named as derivatives of other compounds. The list of priorities in the following box includes all the functional groups that we have discussed so far.

Priorities for Determining Principal Group

Onium ions > acids (carboxylic acid > sulfonic acid) > acid derivatives (anhydride > ester > acyl halide > amide > nitrile) > ketones > aldehydes > alcohols and phenols > amines > ethers > thioethers

EXERCISE 17.1

Name each of the following compounds:

(a) $CH_3CH_2CH(NH_2)—CH_2CH_3$

(b) $CH_3CH_2—NH—CH_2CH(CH_3)—CH_3$

(c) $CH_3—C(=O)—CH_2CH_2—CH(NHCH_3)—CH_2CH_3$

(d) $CH_3CH_2CH_2—\overset{+}{N}(CH_3)_2—CH_3\ Br^-$

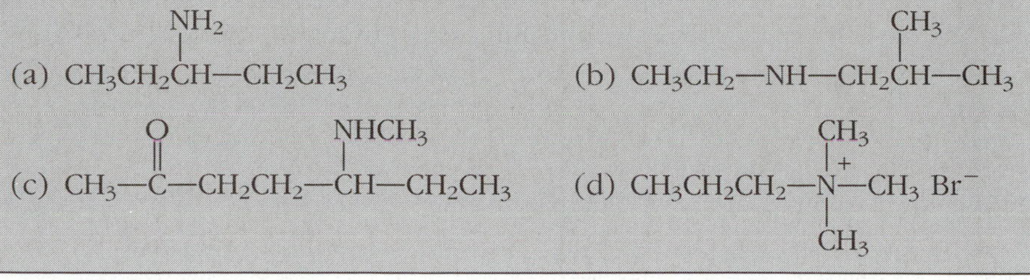

EXERCISE 17.2

Draw each of the following compounds:

(a) 2-Hexylamine
(b) *N,N*-Diethylisobutylamine
(c) Triisopropylammonium bromide
(d) 2-Amino-5-hydroxy-3-octanone

17.2 PHYSICAL PROPERTIES

Amines are moderately polar compounds but less so than structurally similar alcohols. Primary and secondary amines can act as both donors and acceptors in hydrogen bonding, and this leads to intermolecular association of the pure compounds.

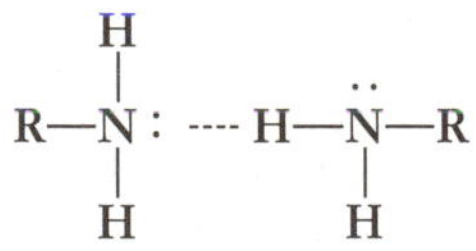

$$R—NH_2{:}\ \cdots\ H—\ddot{N}(H)—R$$

This interaction cannot occur with tertiary amines (there are no NH groups), but tertiary amines can still form hydrogen bonds to alcohols and other hydrogen bond donors.

The polarity and intermolecular association of primary amines is reflected by the boiling points of compounds with similar molecular weights (RCH_3, $R—NH_2$, and $R—OH$ all have the same molecular weight within 2 amu).

$CH_3CH_2CH_2CH_2—CH_3$	$CH_3CH_2CH_2CH_2—NH_2$	$CH_3CH_2CH_2CH_2—OH$
bp 36°C	bp 78°C	bp 117°C
$C_6H_5CH_2—CH_3$	$C_6H_5CH_2—NH_2$	$C_6H_5CH_2—OH$
bp 136°C	bp 185°C	bp 205°C

In each case the boiling point of the primary amine is substantially higher than that of the nonpolar alkane, but it is significantly lower than that of the corresponding alcohol.

For secondary amines the increase in boiling point above that of the related alkane is much smaller. This results in part from unfavorable steric interactions between the *N*-alkyl substituents when secondary amines undergo self-association. These steric effects weaken the intermolecular interactions favoring the liquid state. The oxygen analogs of secondary amines are ethers, which cannot self-associate via hydrogen bonding and have lower boiling points than the corresponding amines.

$CH_3(CH_2)_3—CH_2—(CH_2)_3CH_3$ bp 151°C

$CH_3(CH_2)_3—NH—(CH_2)_3CH_3$ bp 159°C

$CH_3(CH_2)_3—O—(CH_2)_3CH_3$ bp 142°C

$CH_3(CH_2)_3—CH_2—CH_2CH_3$ bp 98°C

$CH_3(CH_2)_3—NH—CH_2CH_3$ bp 108°C

$CH_3(CH_2)_3—O—CH_2CH_3$ bp 96°C

Self-association by hydrogen bonding is not possible for tertiary amines, so there is even less difference between the boiling points of these compounds and those of the structurally related alkanes. Sometimes the tertiary amine actually has a lower boiling point.

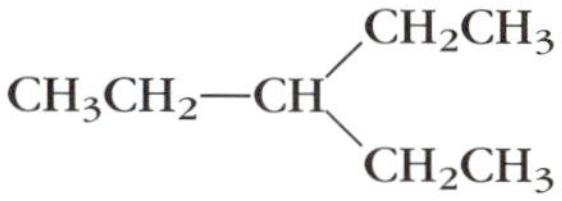

$CH_3CH_2—CH(CH_2CH_3)_2$ bp 94°C

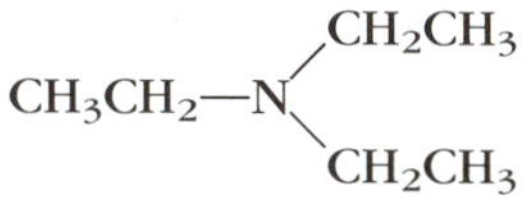

$CH_3CH_2—N(CH_2CH_3)_2$ bp 89°C

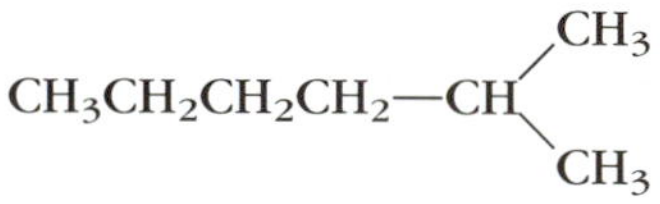

$CH_3CH_2CH_2CH_2—CH(CH_3)_2$ bp 90°C

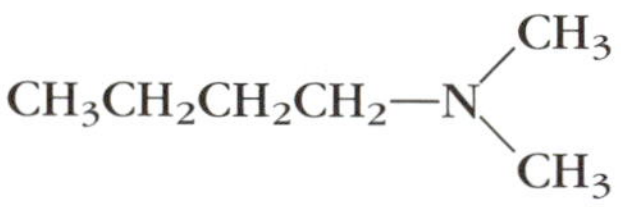

$CH_3CH_2CH_2CH_2—N(CH_3)_2$ bp 95°C

The hydrogen bonding ability of amines results in substantial water solubility for those compounds having a total of six or fewer carbon atoms. Although the higher-molecular-weight amines are not very soluble in water, their hydrogen bonding ability results in good solubility in protic solvents such as methanol and ethanol.

EXERCISE 17.3

Which compound of each pair would have a higher boiling point?

(a) 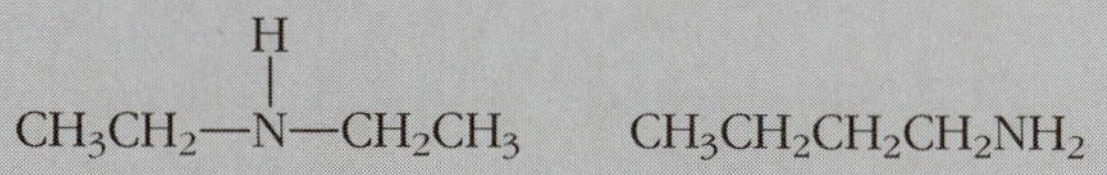 $CH_3CH_2—N(H)—CH_2CH_3$ $CH_3CH_2CH_2CH_2NH_2$

(b) 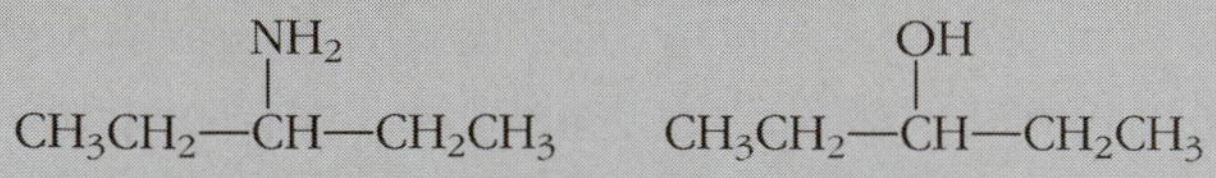 $CH_3CH_2—CH(NH_2)—CH_2CH_3$ $CH_3CH_2—CH(OH)—CH_2CH_3$

(c) 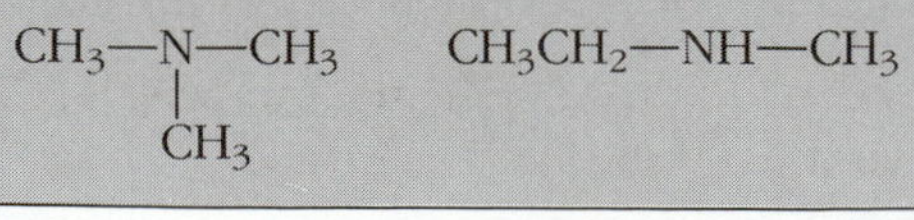 $CH_3—N(CH_3)—CH_3$ $CH_3CH_2—NH—CH_3$

EXERCISE 17.4

Which compound of each pair would be more soluble in H_2O?

(a) $(CH_3CH_2)_3N$ $(CH_3CH_2CH_2)_3N$

(b) $CH_3(CH_2)_3—NH—(CH_2)_3CH_3$ $CH_3(CH_2)_3—NH—(CH_2)_3—CH_2$ (with OH on the terminal CH_2)

17.3 MOLECULAR STRUCTURE OF AMINES

Because amines have only three substituents on the nitrogen atom, you might guess that they would have a planar arrangement around it. In fact amines are very nearly tetrahedral. While there are only three substituents, nitrogen also has a nonbonded pair of electrons. The most stable arrangement would be that in which the four electron pairs (three bonding, one nonbonding) are separated by angles of about 109.5°. The H—N—H angle in ammonia is 107° and the C—N—C angle of trimethylamine is 108°.

NH_3: 107° $N(CH_3)_3$: 108°

In a planar arrangement the three substituents would be separated by 120° (favorable), but each would be at only a 90° angle with the nonbonding orbital (unfavorable). These effects nearly counteract each other, and the planar geometry has an energy only about 6–10 kcal/mol above that of the optimum tetrahedral geometry. This means that amines have enough kinetic energy at room temperature to undergo rapid *inversion* from one tetrahedral geometry to another (Figure 17.1).

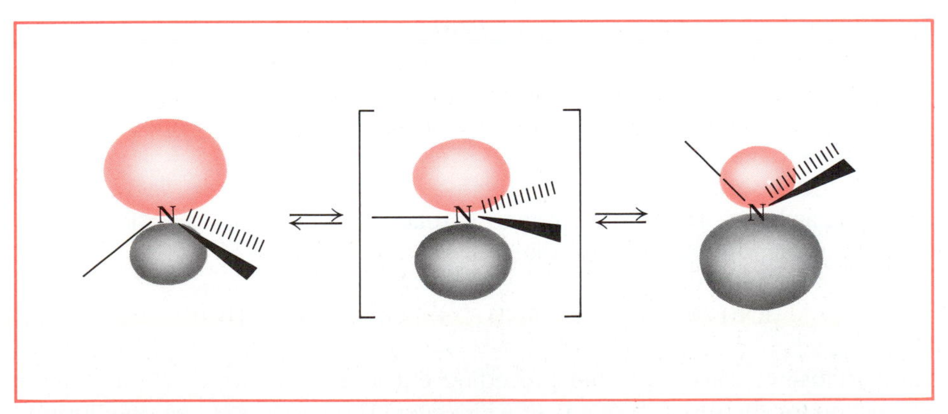

Figure 17.1 Conformational Inversion of Amines.
In the optimum geometry the amine is tetrahedral, and the nonbonding electrons occupy an orbital that is approximately sp^3 hybridized. The transition state for the inversion has the nitrogen and all three substituents in a single plane, with the nonbonding electrons occupying a pure p orbital.

Because this inversion is rapid, molecules that are chiral by virtue of three different substituents on nitrogen cannot be obtained in optically active form. Conformational inversion produces a 1:1 mixture of the two enantiomers. For example

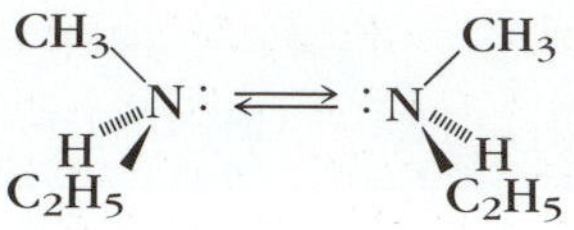

$$\mathrm{(CH_3)(H)(C_2H_5)N{:}} \rightleftharpoons \mathrm{{:}N(CH_3)(H)(C_2H_5)}$$

Of course, many optically active amines are known, but these are compounds whose optical activity results from the presence of one or more chiral *carbon* atoms.

In contrast to the free amines, quaternary ammonium salts have *four* substituents on nitrogen and are configurationally stable. Indeed, the following quaternary iodide was obtained in optically active form as early as 1899.

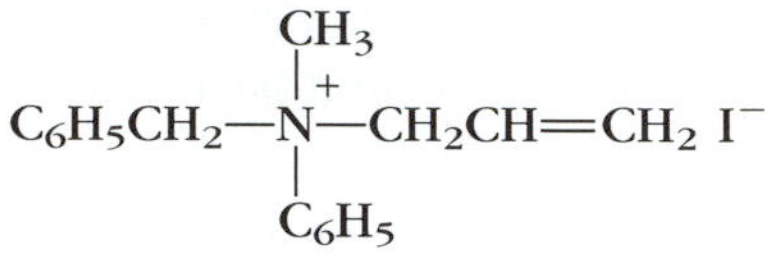

$$\mathrm{C_6H_5CH_2{-}\overset{+}{N}(CH_3)(C_6H_5){-}CH_2CH{=}CH_2\ I^-}$$

17.4 REACTIONS OF AMINES

Amines undergo a wide variety of reactions, but nearly all these result from a single structural feature, the nonbonding pair of electrons on nitrogen that reside in the highest occupied molecular orbital (HOMO).

Amines as Acids and Bases

Amines are ordinarily classified as *bases*, but primary and secondary amines (as well as ammonia) can also act as *acids*.

$$\mathrm{R{-}NH_2} \rightleftharpoons \mathrm{H^+ + RNH^-} \qquad K_a \sim 10^{-35}$$

The acid dissociation constant in the preceding equation shows that amines are extremely weak acids. Only in the presence of a very strong base will you observe the reactions of amines as acids.

In contrast to their capability of reacting as (extremely weak) acids, amines readily act as bases. Their base strength is relatively weak, but nevertheless they react to a significant extent with water.

$$\mathrm{RNH_2 + H_2O} \rightleftharpoons \mathrm{RNH_3{}^+ + {}^-OH} \qquad K_b \sim 10^{-4}$$

The equilibrium constant in the preceding equation is shown as K_b, a general designation for the reaction of a base with water. The corresponding equilibrium constant for hydroxide ion acting as a base is unity, and for salts of carboxylic acids the analogous equilibrium constants are about 10^{-9}. Therefore amines are weaker bases than hydroxide but are stronger bases than carboxylate ions.

Aromatic amines are less basic than aliphatic amines. For many years this was generally believed to result entirely from the orbital interaction between the nonbonding orbital on nitrogen and the π system of the aromatic ring. Protonation of the nitrogen can occur only at the expense of this favorable interaction, so this seemed to explain the lower basicity of aromatic amines. However, more recent studies have shown that relative basicities of various amines are quite different in solution and in the gas phase, and this means that the observed basicity of an amine results from a combination of both *solvation effects* and *electronic effects.* Nevertheless, inductive and resonance effects cause predictable changes in basicity, and this is illustrated by the three amines shown below. Cyclohexylamine is an ordinary aliphatic amine with a K_b of 5×10^{-4}, but the basicity of the aromatic amine, aniline, is a millionfold lower ($K_b = 5 \times 10^{-10}$). Substitution of the aromatic ring with a *p*-nitro group (for which there is a direct resonance interaction with the nonbonding electrons of the amino group) further decreases the basicity by a factor of 5000.

	Cyclohexylamine	Aniline	4-Nitroaniline
K_b of amine	5×10^{-4}	5×10^{-10}	1×10^{-13}
K_a of conjugate acid	2×10^{-11}	2×10^{-5}	1×10^{-1}

Table 17.1 summarizes the acid (and base) strengths of a series of important functional classes. Using this table you can directly compare the reactivity of amines as acids and bases with the corresponding reactivity of other organic compounds.

TABLE 17.1 Acid Dissociation Constants of Representative Classes of Organic Compounds*

	Acid	Approximate K_a of Acid	Conjugate Base	
Strongest acid	$ArNH_3^+$	10^{-4}	$ArNH_2$	Weakest base
	R—C(=O)OH	10^{-5}	R—C(=O)O$^-$	
	R—NH_3^+	10^{-10}	RNH_2	
	ArOH	10^{-11}	ArO^-	
	H_2O	10^{-16}	^-OH	
	ROH	10^{-17}	^-OR	
	R—C(=O)—CH_3	10^{-25}	R—C(=O)—CH_2^-	
	R—C≡CH	10^{-25}	R—C≡C$^-$	
	R—NH_2	10^{-35}	R—NH^-	
Weakest acid	R—CH_3	10^{-44}	R—CH_2^-	Strongest base

*See also Tables 4.3, 7.1, and 14.2.

Acylation of Amines

The nonbonding electrons on nitrogen allow an amine to function not only as a base but also as a nucleophile. A primary or secondary amine will react rapidly with an acid chloride or anhydride and more slowly with an ester to produce an amide.

$$R'-\overset{O}{\overset{\|}{C}}-X + H-\ddot{N}R_2 \longrightarrow R-\overset{O}{\overset{\|}{C}}-NR_2 + HX$$

These reactions are very useful for the preparation of amides. The reaction has been discussed more extensively in Section 16.7, so we will only show a single example here.

$$CH_3-\underset{CH_3}{\underset{|}{NH}} \xrightarrow{C_6H_5\overset{O}{\overset{\|}{C}}-Cl} CH_3-\underset{CH_3}{\underset{|}{N}}-\overset{O}{\overset{\|}{C}}-C_6H_5 \qquad 97\%$$

A tertiary amine is not acylated, because it lacks a replaceable hydrogen. The following equation shows the interaction of a tertiary amine with acetyl chloride. The amine can attack the carbonyl group and displace the chlorine, but the absence of a hydrogen substituent on nitrogen precludes proton transfer to generate the neutral amide. Consequently, the amine (or the corresponding ammonium salt) is regenerated when water is added during workup of the reaction.

$$R_3N: \xrightarrow{CH_3\overset{O}{\overset{\|}{C}}Cl} \left[R_3\overset{+}{N}-\underset{Cl}{\underset{|}{\overset{O^-}{\overset{|}{C}}}}-CH_3\right] \longrightarrow \left[R_3\overset{+}{N}-\overset{O}{\overset{\|}{C}}CH_3\right] \xrightarrow[\text{workup}]{H_2O} R_3N: \left(+ CH_3-\overset{O}{\overset{\|}{C}}-OH \right)$$

Primary and secondary amines react with sulfonyl chlorides to form *sulfonamides.* (Once again tertiary amines cannot give the corresponding product.)

$$H_2N-CH_2-\overset{CH_3}{\overset{|}{CH}}-CH_3 \xrightarrow{CH_3-C_6H_4-SO_2-Cl} CH_3-C_6H_4-\underset{O}{\underset{\|}{\overset{O}{\overset{\|}{S}}}}-NH-CH_2-\overset{CH_3}{\overset{|}{CH}}-CH_3 \qquad 98\%$$

The properties of the resulting sulfonamide make this reaction very useful for distinguishing among primary, secondary, and tertiary amines. We will discuss this in Section 17.9 (the Hinsberg test).

Sulfonamides occupy an important place in the history of medicinal chemistry. Prior to the 1930s there were but two or three specific organic compounds that had been found useful in the treatment of infectious disease, and none of them was applicable to the specific treatment of common bacterial infections. The discovery in 1936 that *sulfanilamide* inhibited the growth of bacterial cultures was a dramatic step forward in the development of modern medicine: this was the first *antibiotic.*

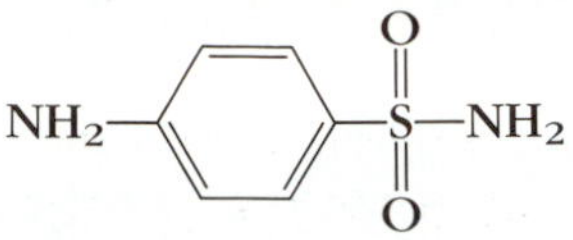

$$NH_2-C_6H_4-S(=O)_2-NH_2$$

Sulfanilamide

Subsequent research showed that sulfanilamide coordinates to a bacterial enzyme involved in the synthesis of *folic acid,* which is necessary for amino acid and nucleic acid metabolism (in humans as well as in bacteria). Sulfanilamide blocks the enzymatic step in which 4-aminobenzoic acid is incorporated.

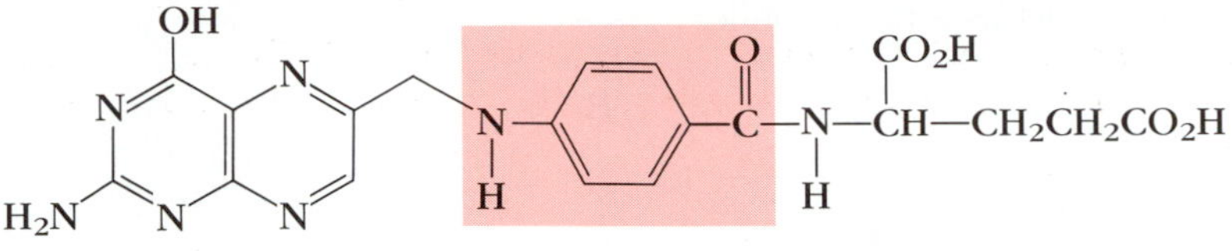

Folic acid

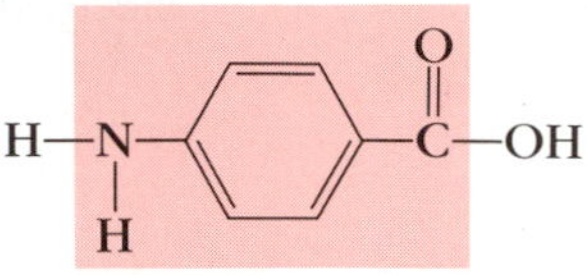

4-Aminobenzoic acid

The structural similarities between sulfanilamide and 4-aminobenzoic acid allow the former to bind to the enzyme in place of 4-aminobenzoic acid. The normal reaction cannot occur, and the synthesis of folic acid is therefore prevented. But why is this not equally harmful to the human patient? The answer is simple. The human body does not synthesize folic acid, and you must obtain it from dietary sources. As a result, the inhibition of folic acid biosynthesis is fatal to bacteria (which must synthesize their own) but not to humans.

The understanding of these principles led to the preparation of a number of derivatives of sulfanilamide that were also effective antibiotics, the *sulfa drugs.* During the 1940s the sulfa drugs helped to revolutionize medical treatment of bacterial diseases. For the first time it was possible to actively combat many bacterial infections, and diseases such as pneumonia were reduced from often fatal to merely serious illnesses.

Alkylation of Amines

Amines also react as nucleophiles with alkyl halides (and alkyl sulfonates).

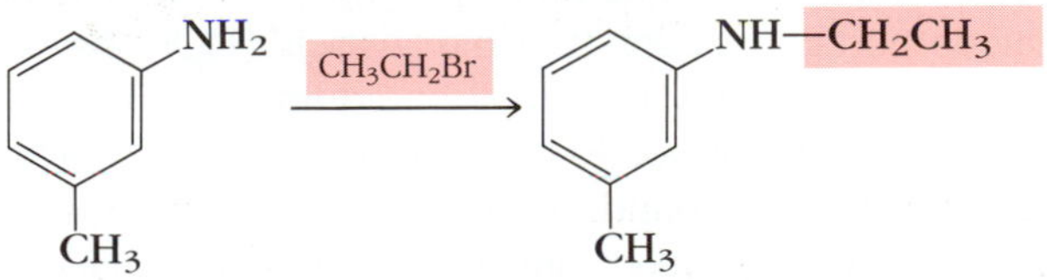

$$\text{3-}CH_3C_6H_4NH_2 \xrightarrow{CH_3CH_2Br} \text{3-}CH_3C_6H_4NH-CH_2CH_3 \quad 63\text{–}66\%$$

This process is sometimes highly useful for preparing substituted amines, but two important side reactions must be considered, *elimination* and *polyalkylation.* Amines are bases as well as nucleophiles, so they can also attack an alkyl halide (or sulfonate) by removal of a β hydrogen. The side reaction of elimination is relatively minor for primary halides, but it becomes increasingly important for secondary halides and occurs exclusively for tertiary halides. This is reflected by the high yield of alkene in the following reaction:

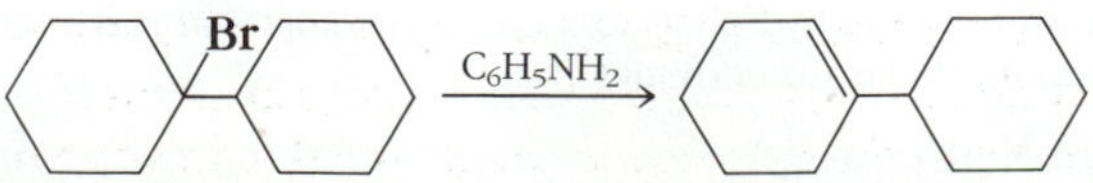

$$\xrightarrow{C_6H_5NH_2} \quad 90\%$$

Polyalkylation results from further reactions of the initial product, and it can continue to the stage of quaternization.

$$R'\text{—}\ddot{N}H_2 \xrightarrow{RX} R'\text{—}\ddot{N}HR \xrightarrow{RX} R'\text{—}\ddot{N}(R)\text{—}R \xrightarrow{RX} R'\text{—}\overset{+}{N}(R)_2\text{—}R \ X^-$$

The initial reaction generates an *ammonium salt,* and it might seem that this should prevent further reaction. The ammonium salt, after all, has no nonbonded electrons and cannot act as a nucleophile.

$$R'\text{—}\ddot{N}H_2 \xrightarrow{R\text{—}X} R'\text{—}\overset{+}{N}H_2\text{—}R \ X^-$$

The answer to this apparent dilemma is the very rapid proton transfer reaction (i.e., acid-base reaction) that takes place between the ammonium ion and unreacted amine.

$$R'\text{—}\overset{+}{N}H_2\text{—}R \underset{}{\overset{R\ddot{N}H_2}{\rightleftharpoons}} R'\text{—}\ddot{N}H\text{—}R + RNH_3^+$$

This acid-base reaction will always establish an equilibrium mixture of the various amines and their ammonium salts, so further alkylation can occur readily. The most common solution to the problem of polyalkylation has been to use an excess of the starting amine. The unalkylated amine is always present in higher concentration than any of the alkylation product, so it will preferentially react with the alkyl halide.

Reactions of Amines with Nitrous Acid

Amines undergo reaction with a variety of oxidizing agents, but the synthetic applications are limited. One important reaction uses nitrous acid (HO—N=O), a very mild oxidizing agent, which is usually prepared by adding acid to an aqueous solution of $NaNO_2$. The reaction proceeds by introduction of a nitroso group (—N=O) to give an *N*-nitroso compound. With secondary amines the *N*-nitroso compound ($R_2N\text{—}N{=}O$) can be isolated, but the reaction of tertiary amines is more complex.* The *N*-nitroso intermediate produced in the reaction of a primary amine loses water to generate a *diazonium ion:*

---

*The reaction of a tertiary aliphatic amine with nitrous acid generates an *N*-nitrosoammonium ion that loses HNO to form an iminium ion.

$$R_2N\text{—}CH\text{—} \xrightarrow{HONO} R_2\overset{+}{N}(NO)CH\text{—} \longrightarrow R_2\overset{+}{N}{=}C\langle + HNO$$

The iminium ion undergoes subsequent hydrolysis to a carbonyl compound and a secondary amine, which in turn is converted to the *N*-nitroso derivative.

$$R_2\overset{+}{N}{=}C\langle \xrightarrow[H_2O]{HONO} R_2N\text{—}NO + O{=}C\langle$$

$$\underset{\text{Primary amine}}{R{-}\ddot{N}H_2} \xrightarrow{\text{HONO}} \left[R{-}\overset{+}{N}H_2{-}N{=}O \right] \xrightarrow[H_2O]{\text{loss of}} \underset{\text{Diazonium ion}}{R{-}\overset{+}{N}{\equiv}N}$$

Diazonium ions exhibit remarkable reactivity. Molecular nitrogen (N≡N) is an exceptionally stable species, and as a consequence the —N_2^+ group of a diazonium ion is a very good leaving group. Aromatic diazonium ions are relatively stable species that can be isolated, and they undergo a variety of useful reactions, which we will discuss in Chapter 18. Aliphatic diazonium salts, in contrast, are much too reactive to be isolated. They are extremely susceptible to attack by nucleophiles (such as solvent), and they also have a strong tendency to lose nitrogen to form carbocations.

$$R{-}\overset{+}{N}{\equiv}N \xrightarrow{^-\text{Nuc}} \text{Nuc—R}$$

$$R{-}\overset{+}{N}{\equiv}N \longrightarrow N_2 + [R^+] \longrightarrow \textbf{substitution and elimination products}$$

The tendency to form carbocations often leads to mixtures of products, and substitution is invariably accompanied by elimination (Section 12.6).

$$\text{cyclohexyl—}NH_2 \xrightarrow[H_2O]{\text{HONO}} \underset{67\%}{\text{cyclohexyl—OH}} + \underset{22\%}{\text{cyclohexene}}$$

Carbocation formation often results in rearrangement to a more stable carbocation, and this has been exploited as a method for *ring enlargement:*

$$\text{cyclohexyl—}CH_2NH_2 \xrightarrow[H_2O]{\text{HONO}} \text{cycloheptyl—OH} \quad 35\text{–}40\%$$

$$\text{cyclopentyl—}CH_2NH_2 \xrightarrow[H_2O]{\text{HONO}} \text{cyclohexyl—OH} \quad 30\%$$

$$\left[\text{cyclopentyl—}CH_2{-}N_2^+ \longrightarrow \text{cyclohexyl cation } (+) \right]$$

The yields are typically rather low in this process, but the ability to generate a larger ring in a single step can sometimes offset that disadvantage.

EXERCISE 17.5

What products would you expect to be formed in the reaction of the following primary amines with nitrous acid?

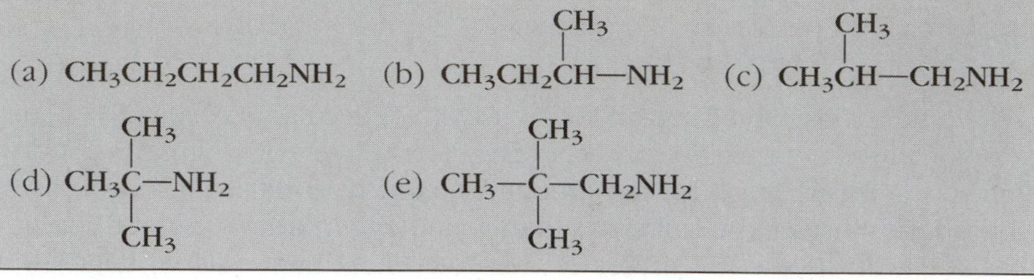

(a) $CH_3CH_2CH_2CH_2NH_2$ (b) $CH_3CH_2CH(CH_3)—NH_2$ (c) $CH_3CH(CH_3)—CH_2NH_2$

(d) $CH_3C(CH_3)_2—NH_2$ (e) $CH_3—C(CH_3)_2—CH_2NH_2$

17.5 ELIMINATION REACTIONS: CONVERSION OF AMINES TO ALKENES

In our previous discussions of elimination reactions we showed you ways in which alkenes could be prepared from alcohols by acid catalysis and from halides or sulfonates by the action of base (Section 5.13). We also discussed two pyrolysis reactions (of esters and amine oxides) in Section 6.9. With the exception of amine oxide pyrolysis, alcohols are usually the starting materials in these eliminations because they are readily converted into halides, sulfonates, and esters of carboxylic acids. In this section we will discuss two highly useful procedures that have been developed for carrying out elimination reactions starting with amines. One method is the Hofmann elimination, a reaction of quaternary ammonium salts that was developed by A.W. von Hofmann in Germany in the latter half of the 1800s. The second method is the pyrolysis of amine oxides developed by A.C. Cope at the Massachusetts Institute of Technology in the late 1940s.

Hofmann Elimination

The Hofmann elimination actually involves a sequence of reactions: the initial quaternization of an amine with a methylating agent such as methyl iodide or dimethyl sulfate, followed by the base-induced decomposition of the quaternary ammonium salt. A primary amine thus yields the alkene and trimethylamine as the final products.

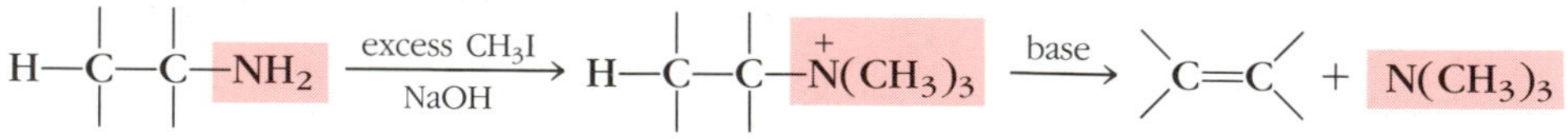

$$H—C—C—NH_2 \xrightarrow[\text{NaOH}]{\text{excess } CH_3I} H—C—C—\overset{+}{N}(CH_3)_3 \xrightarrow{\text{base}} >C=C< + N(CH_3)_3$$

The methylation reaction is carried out with an excess of methyl iodide. The initial product of alkylation is an ammonium salt, but the reaction is usually carried out in the presence of a base such as sodium hydroxide. The base liberates the free amine, which can then react further.

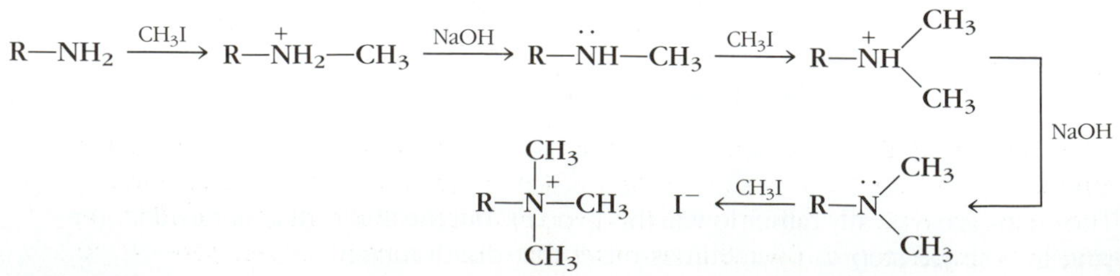

$$R—NH_2 \xrightarrow{CH_3I} R—\overset{+}{N}H_2—CH_3 \xrightarrow{\text{NaOH}} R—\ddot{N}H—CH_3 \xrightarrow{CH_3I} R—\overset{+}{N}H(CH_3)_2 \xrightarrow{\text{NaOH}} R—\ddot{N}(CH_3)_2 \xrightarrow{CH_3I} R—\overset{+}{N}(CH_3)_3 \; I^-$$

Some of the methyl iodide is destroyed by reaction with the sodium hydroxide, but an excess is employed, so this causes no difficulties.

The elimination step can be carried out with a base such as NaOH, KOH or NaOEt. Another procedure has sometimes been employed in which the quaternary ammonium iodide is treated with an aqueous suspension of silver oxide. The Ag_2O is in equilibrium with AgOH, and the following equation shows how this produces insoluble silver iodide and the quaternary ammonium *hydroxide*.

$$R{-}\overset{+}{N}(CH_3)_3\ I^- + AgOH \rightarrow AgI\downarrow + R{-}\overset{+}{N}(CH_3)_3\ {}^-OH$$

The hydroxide ion is a good base, so elimination can be brought about simply by vigorous heating (pyrolysis) of the quaternary ammonium hydroxide. The following reactions illustrate the Hofmann elimination:

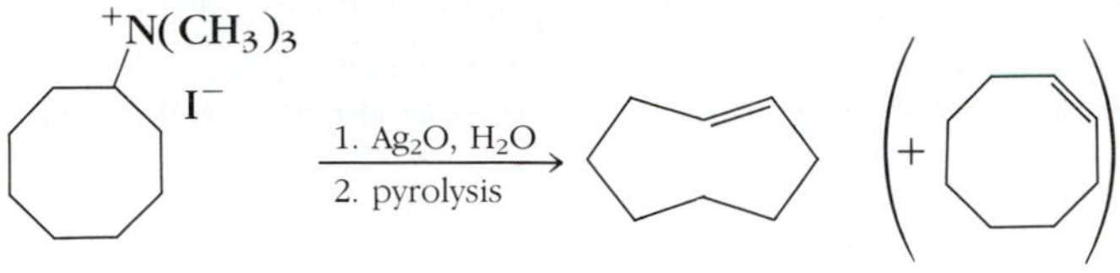

$$\text{cyclooctyl}{-}{}^+N(CH_3)_3\ I^- \xrightarrow[\text{2. pyrolysis}]{\text{1. } Ag_2O,\ H_2O} (E)\text{-cyclooctene} + (Z)\text{-cyclooctene}$$

45–46%, *E* 30–31%, *Z*

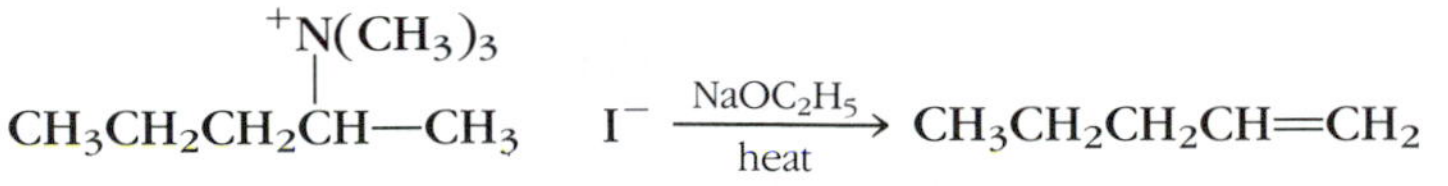

$$CH_3CH_2CH_2\underset{}{CH}({}^+N(CH_3)_3){-}CH_3\quad I^- \xrightarrow[\text{heat}]{NaOC_2H_5} CH_3CH_2CH_2CH{=}CH_2 \quad 63\%$$

$$CH_3CH_2{-}C(CH_3)_2{-}\overset{+}{N}(CH_3)_3\quad I^- \xrightarrow[\text{heat}]{KOC_2H_5} CH_3CH{=}C(CH_3)_2 + CH_3CH_2C(CH_3){=}CH_2$$

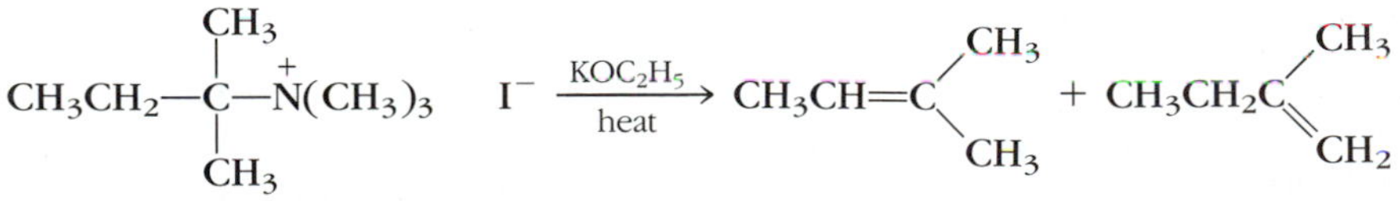

6% 78%

The first of these reactions shows the preparation of (*E*)-cyclooctene, the smallest *E* cycloalkene that is stable at room temperature (Section 5.6). The last two reactions illustrate an important feature of the Hofmann elimination: usually the *less substituted* alkene predominates. This is in contrast to the elimination reactions of halides, which we considered previously (Section 5.13). For example, the bromide corresponding to the last of the preceding examples affords the more highly substituted alkene in 59% yield (as compared with only 6% yield in the Hofmann elimination).

$$CH_3CH_2{-}C(CH_3)_2{-}Br \xrightarrow[C_2H_5OH]{NaOC_2H_5} CH_3CH{=}C(CH_3)_2 + CH_3CH_2C(CH_3){=}CH_2$$

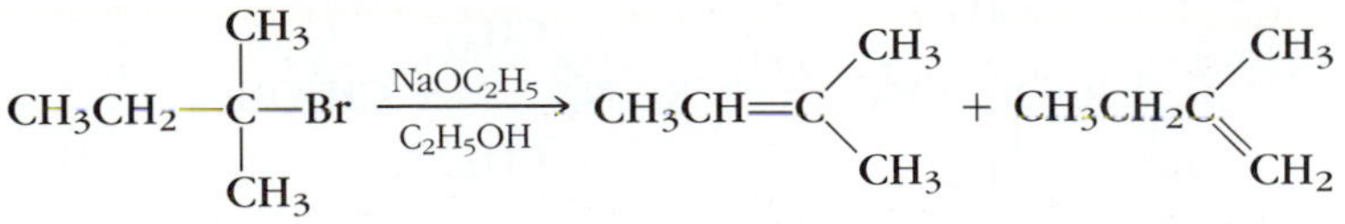

59% 36%

The preferential formation of the less substituted alkene appears to result from a combination of steric and electronic effects. With a quaternary ammonium salt the leaving group is very bulky, and this hinders approach of the base to secondary or tertiary hydrogen atoms. But primary hydrogens (i.e., methyl hydrogens) are on the exterior of the molecule, and base can attack them more easily. In addition to such steric factors, the positively charged onium ion exerts an electronic effect on the β hydrogens. This increases the acidity (and thus the ease of removal) of primary β hydrogens. Both effects favor formation of the less substituted alkene.

more accessible, more acidic → CH_3, CH_3; less accessible, less acidic → CH

Benzylic and allylic hydrogens are also more acidic than other alkyl hydrogens, and if elimination can produce a conjugated double bond, this pathway will predominate even if it yields a more highly substituted alkene.

The Hofmann elimination is an *anti elimination* (Section 6.9), as shown by the following example:

$$\text{H(C}_6\text{H}_5\text{)C}(\overset{+}{\text{N}}\text{(CH}_3\text{)}_3\text{)}-\text{C(H)(CH}_3\text{)(C}_6\text{H}_5\text{)} \xrightarrow[\text{C}_2\text{H}_5\text{OH}]{\text{NaOC}_2\text{H}_5} \text{H(C}_6\text{H}_5\text{)C}=\text{C(CH}_3\text{)(C}_6\text{H}_5\text{)} \quad 85\%$$

Two important side reactions are possible in the Hofmann elimination. First, if two different alkyl groups have β hydrogens, more than one alkene can be produced:

$$(CH_3CH_2CH_2)_2\overset{+}{N}(CH_2CH_2CH_2CH_3)_2 \rightarrow CH_3CH{=}CH_2 + CH_3CH_2CH_2{-}N(CH_2CH_2CH_2CH_3)_2 \quad 63\%$$

$$\rightarrow CH_3CH_2CH{=}CH_2 + (CH_3CH_2CH_2)_2N{-}CH_2CH_2CH_2CH_3 \quad 37\%$$

However, this is not usually a problem. When three of the substituents on nitrogen are methyl groups, elimination can only occur at the fourth alkyl substituent.

The second important side reaction is substitution. The base can behave as a nucleophile, and a neutral amine can act as the leaving group.

$$R{-}\overset{+}{N}(CH_3)_2{-}CH_3 \xrightarrow{^-OH} R{-}\ddot{N}(CH_3)_2 + CH_3OH$$

Such substitution usually occurs at methyl groups, but benzyl groups also react readily by this pathway. In the following example none of the alkyl groups has a β hydrogen, so substitution occurs exclusively to give the indicated distribution of products.

$$C_6H_5CH_2-\overset{+}{N}(CH_3)_3 \;\; {}^{-}OH \xrightarrow{\text{pyrolysis}} C_6H_5CH_2-N(CH_3)_2 + CH_3OH \quad 35\%$$

$$\phantom{C_6H_5CH_2-\overset{+}{N}(CH_3)_3 \;\; {}^{-}OH} \longrightarrow (CH_3)_2N-CH_3 + C_6H_5CH_2OH \quad 65\%$$

The Hofmann elimination was used extensively by Willstätter in his synthesis of cyclooctatetraene (Section 10.4), and we have summarized the synthesis in Figure 17.2.

Willstätter began with the naturally occuring amino ketone, pseudopelletierine, which was known to have an eight-membered ring with a nitrogen bridge. By first reducing the ketone and dehydrating the alcohol, he was able to produce a cyclooctene with a nitrogen bridge. This was followed by a sequence of methylations and Hofmann eliminations, which yielded cyclooctatriene. Introduction of the fourth double bond was carried out by first adding bromine to the triene. The resulting dibromo compound was used to alkylate dimethylamine, and a double Hofmann elimination finally yielded cyclooctatetraene.

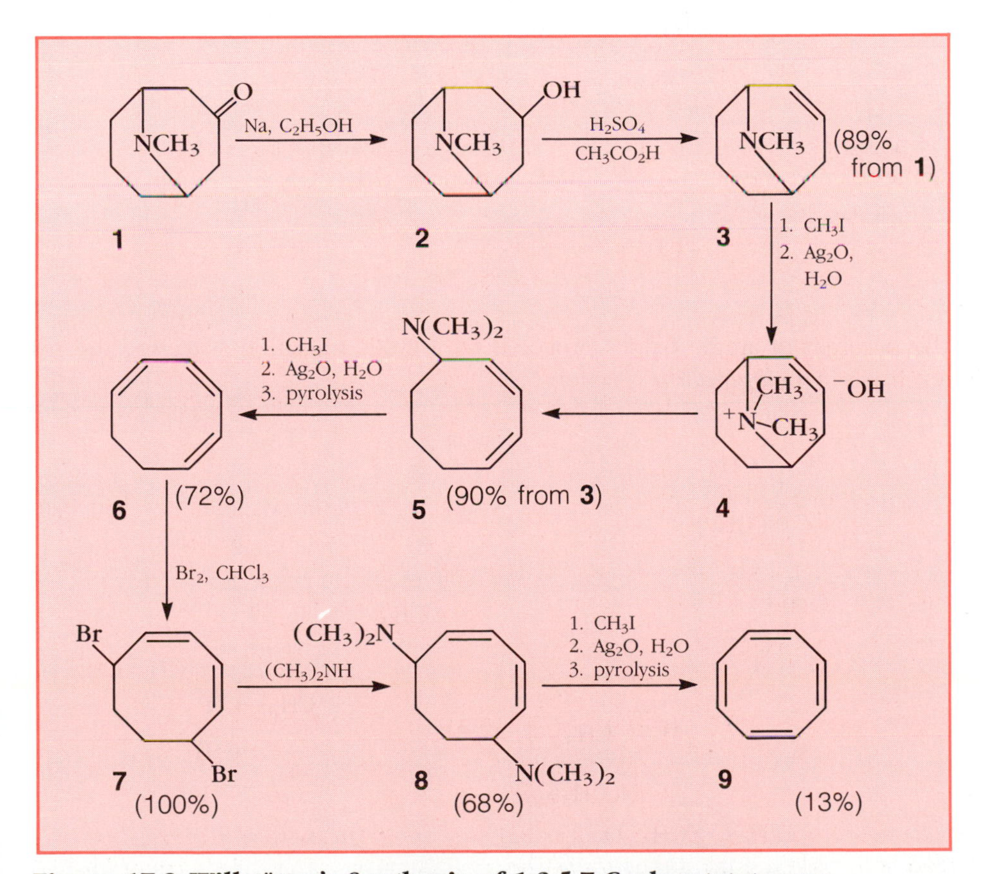

Figure 17.2 Willstätter's Synthesis of 1,3,5,7-Cyclooctatetraene.

Amine Oxide Pyrolysis

The pyrolysis of amine oxides is a two-step procedure that starts with a tertiary amine. The amine is first oxidized to the amine oxide, and then the amine oxide is pyrolyzed to yield an alkene and a hydroxylamine.

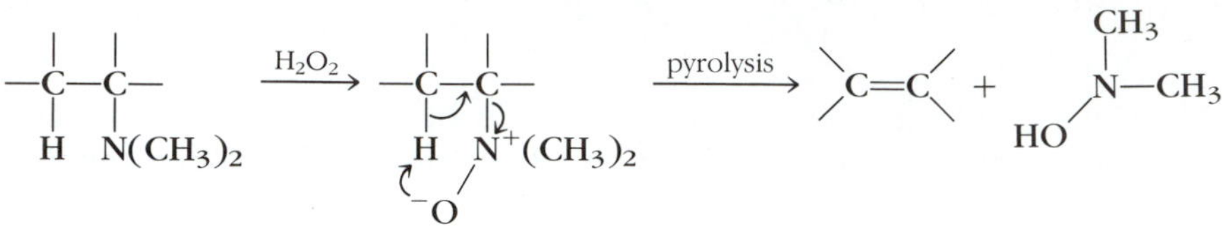

Amine oxides are readily prepared by the mild oxidation of tertiary amines with hydrogen peroxide or with peroxyacids. (Amine oxides of primary and secondary amines cannot be prepared; such structures would isomerize to the corresponding hydroxylamines, R—NH—OH or R_2N—OH.) The following examples illustrate the preparation of tertiary amine oxides.

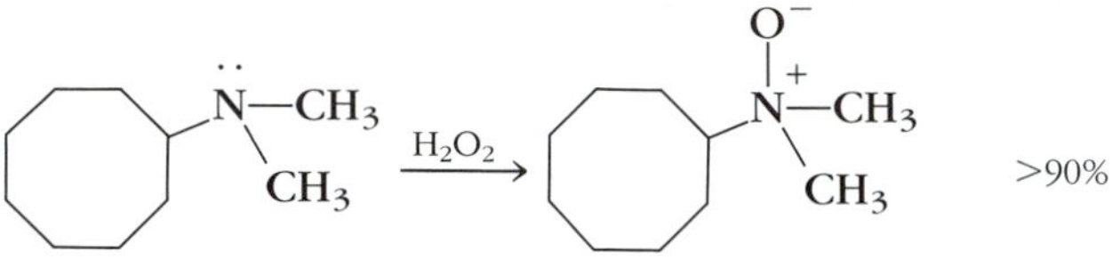

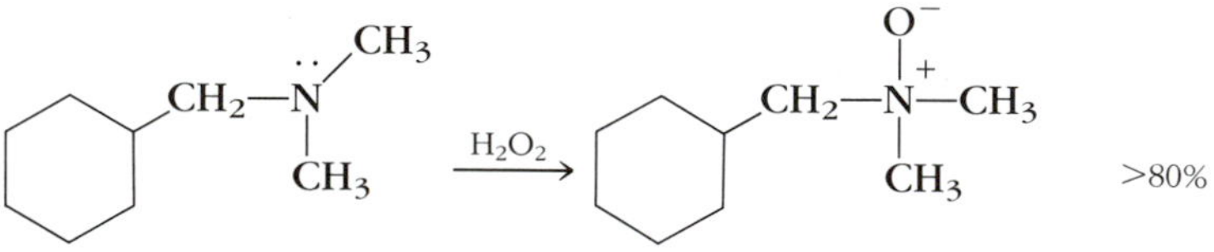

The pyrolysis of amine oxides proceeds by a cyclic transition state, and the net result is a *syn elimination* (Section 6.9).

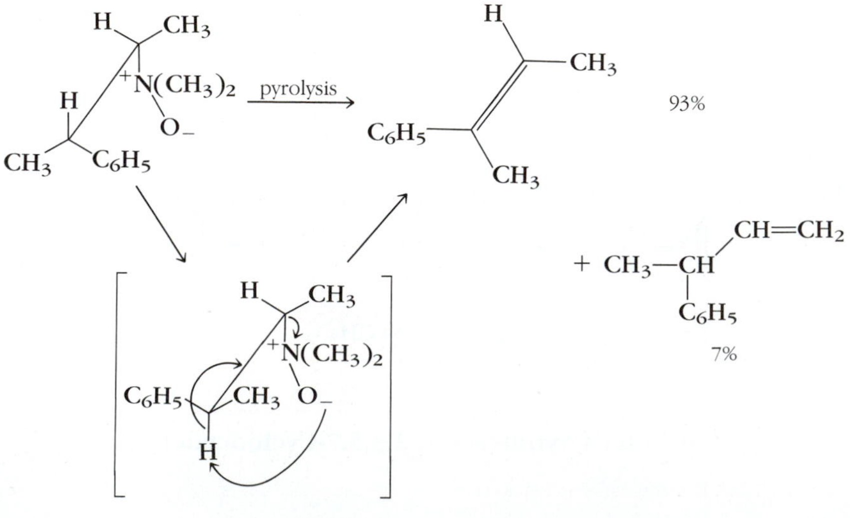

The direction of elimination tends to reflect the number of available β hydrogens, so the less substituted alkene usually predominates.

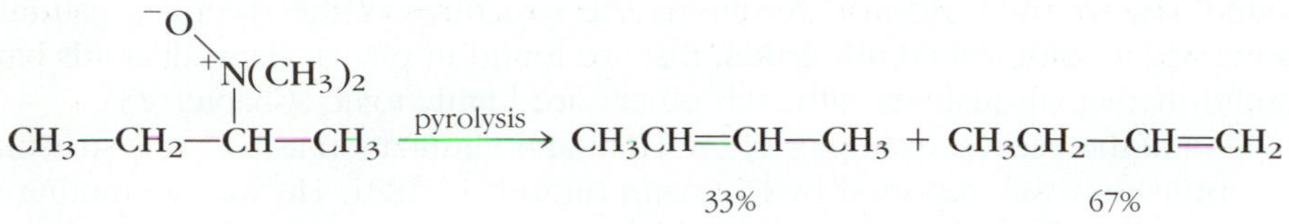

$$CH_3{-}CH_2{-}\underset{}{\overset{{}^{+}N(CH_3)_2{-}O^{-}}{CH}}{-}CH_3 \xrightarrow{\text{pyrolysis}} \underset{33\%}{CH_3CH{=}CH{-}CH_3} + \underset{67\%}{CH_3CH_2{-}CH{=}CH_2}$$

This statistical preference is sometimes outweighed by the greater stability of a product having a conjugated double bond. You can see this from the product mixture obtained in the pyrolysis of 3-phenyl-2-butylamine (the first of the two preceding examples). Despite a 3:1 statistical factor favoring removal of one of the methyl hydrogens, this pathway accounts for only 7% of the product.

Both the Hofmann elimination and amine oxide pyrolysis yield alkenes, but the stereochemical requirements of the two reactions can lead to quite different product distributions. These differences are illustrated by reaction of the following cyclohexane derivative under both sets of conditions:

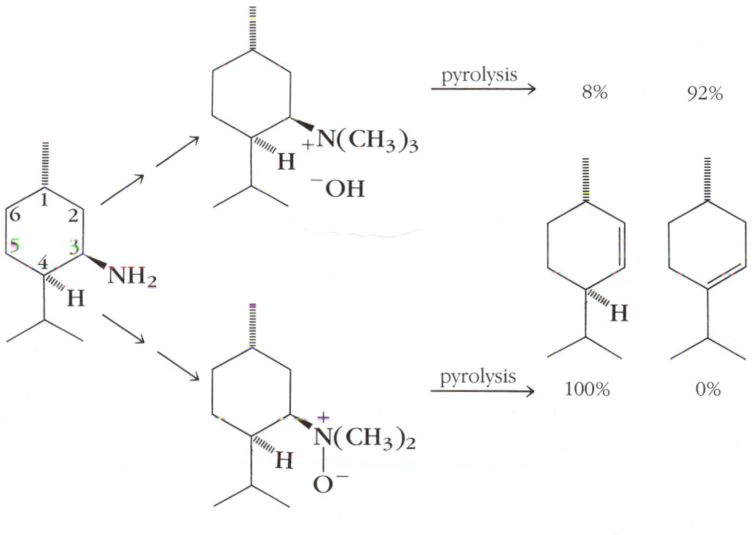

1 2 3 4 5 6 NH₂ H
$^{+}N(CH_3)_3$ H ^{-}OH — pyrolysis → 8% 92%
$\overset{+}{N}(CH_3)_2$, O^{-} H — pyrolysis → 100% 0%

Pyrolysis of the quaternary ammonium hydroxide yields a mixture of isomers, because both C-2 and C-4 have hydrogens *trans* to the leaving group. There is little difference in the steric accessibility of these hydrogens, and the more stable (more substituted) alkene predominates.

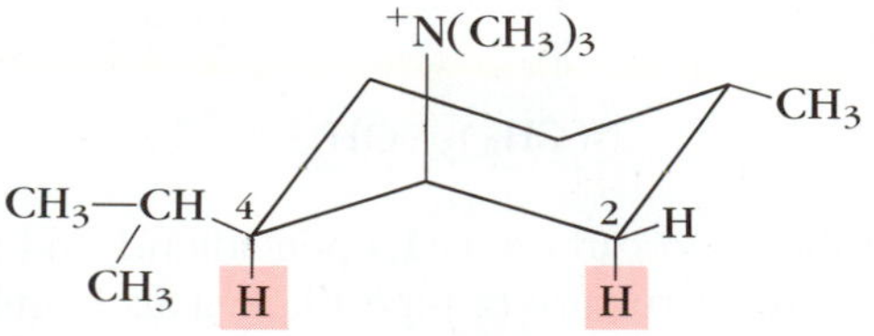

$^{+}N(CH_3)_3$ CH₃ $CH_3{-}CH$ 4 2 H CH_3 H H

Amine oxide pyrolysis requires a β hydrogen *cis* to the nitrogen, but the hydrogen on C-4 is *trans*. Consequently the elimination takes place exclusively by removal of the *cis* hydrogen on C-2.

Structure Determination of Alkaloids

The Hofmann elimination (along with amine oxide pyrolysis to a lesser extent) has found considerable use in determining the structures of the complex, naturally occurring amines, called **alkaloids,** that are found in plants. Many alkaloids have useful medicinal qualities, although others are highly toxic (Chapter 23).

One of the earliest examples of the Hofmann elimination as part of a structure determination was reported by Hofmann himself in 1881. He was attempting to determine the structure of piperidine, which had been obtained by hydrolysis of the naturally occurring amide *piperine.*

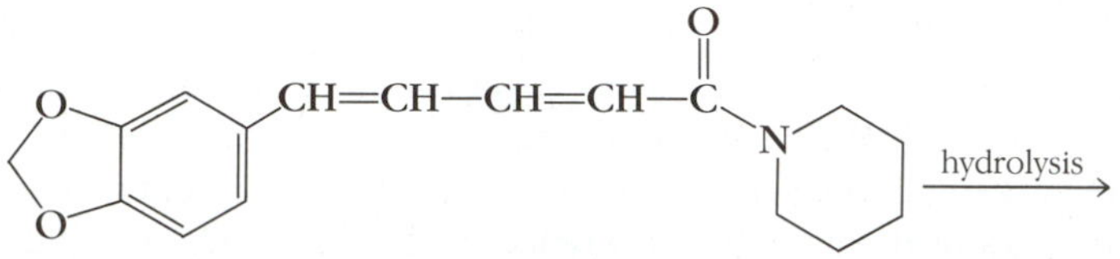

CH=CH—CH=CH—C(=O)—N (piperidine ring) —hydrolysis→

Piperine

HN (ring) + CH=CH—CH=CH—CO_2H

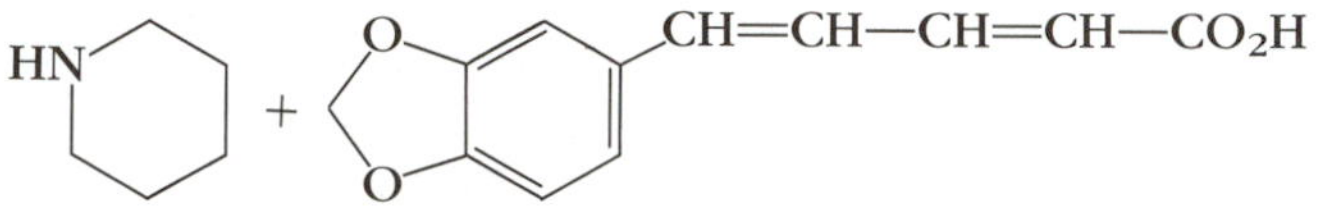

Piperidine   Piperic acid

Piperine can be isolated from black pepper (Latin, *Piper nigrum*) and it was named accordingly. The hydrolysis product, piperidine, had been shown to be an amine with the molecular formula $C_5H_{11}N$, but nothing else was known about its structure. Hofmann quaternized the amine with methyl iodide and subjected the quaternary hydroxide to pyrolysis. This reaction afforded only a single product (an amino alkene) rather than two fragments, so nitrogen must have been part of a ring.

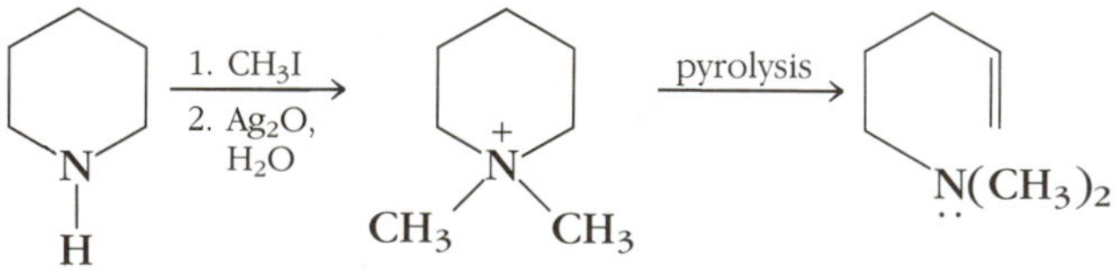

N—H —1. CH_3I; 2. Ag_2O, H_2O→ $\overset{+}{N}$ CH_3 CH_3 —pyrolysis→ $N(CH_3)_2$

The quaternization-elimination sequence was repeated, and this time two products were formed, trimethylamine and a hydrocarbon with the molecular formula C_5H_8.

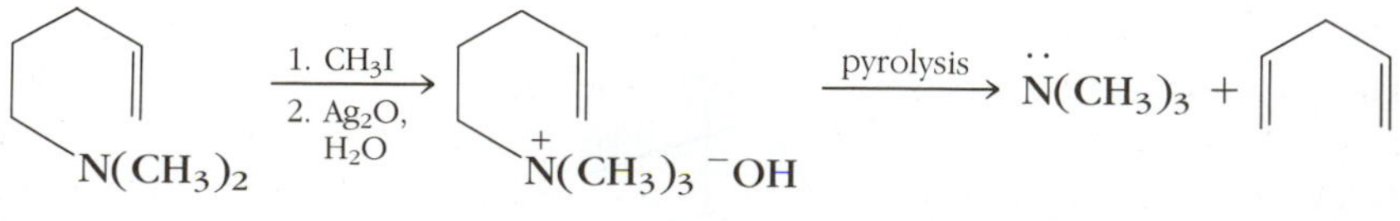

$N(CH_3)_2$ —1. CH_3I; 2. Ag_2O, H_2O→ $\overset{+}{N}(CH_3)_3\ ^{-}OH$ —pyrolysis→ $N(CH_3)_3$ +

Hofmann identified the hydrocarbon as 1,4-pentadiene, and this in turn allowed him to deduce the correct structure of piperidine: a six-membered ring with five CH_2 groups and one NH. There is a curious footnote to this story, because a bit of luck was involved on Hofmann's part. The 1,4-pentadiene produced in the second pyrolysis was isomerized to 1,3-pentadiene under the reaction conditions, but Hofmann did not realize this. By making the fortunate error of assigning the wrong structure to the diene (remember, this was in 1881), Hofmann arrived at the correct structure for piperidine.

EXERCISE 17.6

Draw the product (products) expected for each compound after quaternization with methyl iodide and subsequent base-induced elimination.

(a) $CH_3CH_2CH_2CH_2—N(CH_3)_2$

(b) $CH_3CH(CH_3)—CH(NHCH_3)—CH_3$

(c) $C_6H_5—CH_2—CH(N(CH_3)_2)—CH_3$

(d) 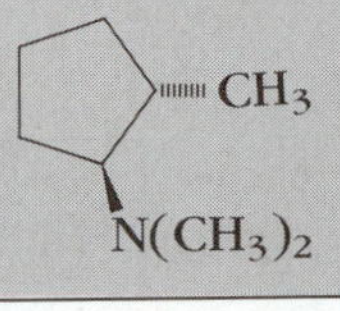 Cyclopentane bearing CH_3 (hashed bond) and $N(CH_3)_2$ (wedged bond) on adjacent carbons

EXERCISE 17.7

Draw the product (products) expected for each compound in Exercise 17.6 after conversion to amine oxide and subsequent pyrolysis.

17.6 PREPARATION OF AMINES BY REDUCTION OF NITROGEN-CONTAINING COMPOUNDS

In the next two sections we will present some of the most important ways of synthesizing amines. We will start in this section with the reduction of nitrogen-containing compounds, and in the following section we will consider procedures in which the amine is produced via formation of a new carbon-nitrogen bond.

Reduction of Nitro Compounds

A variety of reducing agents will convert a nitro group ($—NO_2$) to an amino group. This is a particularly important reaction in aromatic chemistry (Chapter 18), but aliphatic nitro compounds are also readily converted to amines.

$$R—NO_2 \xrightarrow{\text{reduction}} R—NH_2$$

Catalytic hydrogenation of nitro groups is an excellent reaction, and dissolving metal reductions in acid medium (zinc, iron, or tin and HCl) have also been used extensively. Metal hydrides are not used frequently; lithium aluminum hydride reduces nitro groups but sodium borohydride does not. Reductions with hydrogen sulfide and sodium sulfide provide a useful method for the reduction of aromatic nitro compounds to the corresponding amines. These reactions are illustrated by the following equations.

$$HOC(=O)—CH_2CH_2—CH(NO_2)—CH_3 \xrightarrow{H_2,\ Pt} HOC(=O)—CH_2CH_2—CH(NH_2)—CH_3 \quad 98\%$$

$$\text{2-methyl-1,4-dinitrobenzene (}CH_3\text{, }NO_2\text{, }NO_2\text{)} \xrightarrow[\text{2. NaOH (to neutralize)}]{\text{1. Fe, HCl}} \text{2-methyl-1,4-benzenediamine (}CH_3\text{, }NH_2\text{, }NH_2\text{)} \quad 74\%$$

$$CH_3CH_2\underset{|}{\overset{NO_2}{CH}}-CH_3 \xrightarrow[\text{2. } H_2O]{\text{1. } LiAlH_4} CH_3CH_2\overset{NH_2}{CH}-CH_3 \quad 85\%$$

$$\text{2,4-dinitrophenol (OH, }NO_2\text{, }NO_2\text{)} \xrightarrow{Na_2S} \text{2-amino-4-nitrophenol (OH, }NH_2\text{, }NO_2\text{)} \quad 58\text{–}61\%$$

The last example shows that sodium sulfide can be used selectively to reduce just one of two aromatic nitro groups. In an unsymmetrical dinitro compound you cannot readily predict which of the two nitro groups will be reduced preferentially, and mixtures of products often result. This reaction is best suited to the selective reduction of just one nitro group in a symmetrical dinitro compound.

Reduction of Amides and Nitriles

We discussed the reductions of amides and nitriles with lithium aluminum hydride, along with reduction of other carboxylic acid derivatives, in Section 16.5, but several features will be reviewed briefly here. Both amides and nitriles yield amines with a primary alkyl substituent (RCH_2—).

$$R-C\equiv N \xrightarrow[\text{2. } H_2O]{\text{1. } LiAlH_4} R-CH_2-NH_2$$

$$R-\overset{O}{\overset{\|}{C}}-N\langle \xrightarrow[\text{2. } H_2O]{\text{1. } LiAlH_4} R-CH_2-N\langle$$

Reduction of a nitrile always produces a primary amine (—NH_2), but *N*-substituted amides will afford correspondingly substituted amines (see Section 16.5 for additional examples).

$$CH_3(CH_2)_7-\overset{O}{\overset{\|}{C}}-NH-CH_3 \longrightarrow CH_3(CH_2)_7-CH_2-NH-CH_3 \quad 89\text{–}92\%$$

Reduction of Oximes

Oximes are prepared from ketones or aldehydes, and their reduction produces primary amines.

$$>C{=}O \xrightarrow{:NH_2OH} >C{=}N{-}OH \xrightarrow{\text{reduction}} >C(H)NH_2$$

Oxime

Several procedures can be used for this purpose, including dissolving metal reduction (e.g., sodium and alcohol), catalytic hydrogenation, and lithium aluminum hydride reduction. Each of these reagents not only reduces the carbon-nitrogen double bond but also brings about reductive cleavage of the nitrogen-oxygen bond.

$$CH_3{-}(CH_2)_5{-}CH{=}N{-}OH \xrightarrow[C_2H_5OH]{Na} CH_3{-}(CH_2)_5{-}CH_2{-}NH_2 \quad 60\text{–}73\%$$

$$CH_3CH_2{-}C({=}N{-}OH){-}CH_2CH_3 \xrightarrow{H_2,\ Ni} CH_3CH_2{-}CH(NH_2){-}CH_2CH_3 \quad 85\%$$

$$\text{2,2-}(C_6H_5)_2\text{cyclohexanone oxime } (C{=}N{-}OH) \xrightarrow[2.\ H_2O]{1.\ LiAlH_4} \text{2,2-}(C_6H_5)_2\text{cyclohexylamine } (NH_2) \quad 80\%$$

Notice that reduction of the oxime of an aldehyde (the first of the preceding reactions) affords the same amine that would have been obtained by reduction of the corresponding nitrile or primary amide.

EXERCISE 17.8

Show methods for preparing each of the following amines by reduction of a nitrogen-containing compound. When more than one method would work, show each of the alternatives.

(a) Benzylamine (b) 2-Propylamine
(c) Aniline (d) *N*-Methylpropylamine
(e) Cyclohexylamine (f) *N*-Methylcyclohexylamine

17.7 PREPARATION OF AMINES BY FORMATION OF CARBON-NITROGEN BONDS

Reductive Amination of Aldehydes and Ketones

Aldehydes and ketones react reversibly with ammonia and primary amines (Section 13.7) to form compounds with carbon-nitrogen double bonds, called **imines.** Secondary amines afford **iminium salts** under the same conditions. Equilibrium usually favors the aldehyde or ketone, but an imine or iminium salt can be reduced preferentially to form an amine. This process is called *reductive amination.*

$$\rangle C{=}O \xrightleftharpoons{R{-}NH_2} \rangle C{=}N{-}R \xrightarrow{\text{reduction}} \rangle C(NH{-}R)(H)$$

An imine

$$\rangle C{=}O \xrightleftharpoons{R_2NH} \rangle C{=}\overset{+}{N}R_2 \xrightarrow{\text{reduction}} {-}\underset{|}{\overset{H}{\overset{|}{C}}}{-}NR_2$$

An iminium ion

Reductive amination introduces an additional alkyl substituent on the nitrogen of the starting amine. When an aldehyde is used in the reaction, the new alkyl group is primary, and when a ketone is used, the alkyl group is secondary.

Catalytic hydrogenation has been used extensively for this reduction, but a new reagent gives superior results. *Sodium cyanoborohydride* ($NaBH_3CN$, in which one of the four hydrogens of the BH_4^- ion has been replaced by a cyanide) was developed in the late 1960s. This reducing agent, unlike $NaBH_4$, is stable in acid. Under acidic conditions the basic nitrogen of an imine is protonated, and this activates the carbon-nitrogen double bond toward reduction. Consequently, the imine is reduced in preference to the ketone, even though both are present in the equilibrium mixture. The following equation shows the change in reactivity toward a hydride reducing agent:

$$\rangle C{=}O \xrightleftharpoons{NH_3} \rangle C{=}NH \xrightleftharpoons{H^+} \rangle C{=}\overset{+}{N}H_2 \longleftrightarrow \rangle \overset{+}{C}{-}\ddot{N}H_2$$

Moderately reactive — Moderately reactive — Highly reactive

The following examples illustrate the preparation of amines by reductive amination:

$$C_6H_5{-}\overset{O}{\overset{\|}{C}}{-}CH_3 \xrightarrow[H_2,\ Ni]{NH_3} C_6H_5{-}\overset{NH_2}{\overset{|}{C}H}{-}CH_3 \qquad 44\text{–}52\%$$

$$3\text{-}CH_3C_6H_4{-}NH_2 \xrightarrow[H_2,\ Ni]{C_6H_5\overset{O}{\overset{\|}{C}}H} 3\text{-}CH_3C_6H_4{-}NH{-}CH_2C_6H_5 \qquad 89\text{–}94\%$$

$$C_6H_5{-}\overset{O}{\overset{\|}{C}}{-}CH_3 \xrightarrow[HCl,\ NaBH_3CN]{CH_3NH_2} C_6H_5{-}\overset{NHCH_3}{\overset{|}{C}H}{-}CH_3 \qquad 78\%$$

$$\text{cyclohexanone} \xrightarrow[\text{HCl, NaBH}_3\text{CN}]{\text{HN(CH}_3)_2} \text{cyclohexyl-N(CH}_3)_2 \quad 52\text{–}54\%$$

A very specialized type of reductive amination can be used for the introduction of methyl groups to generate a tertiary amine. Reaction of a primary amine with formaldehyde and formic acid generates an iminium ion, which is then reduced by *hydride transfer* from formate ion. (The imine is reduced to the amine, and formic acid is oxidized to carbon dioxide in the process.)

$$\text{R—NH}_2 \xrightleftharpoons{\text{O=CH}_2} \text{R—N=CH}_2 \xrightleftharpoons{\text{HCO}_2\text{H}} \text{R—}\overset{+}{\text{N}}\text{H=CH}_2 \xrightarrow{\text{H—C(=O)—O}^-} \text{R—NH—CH}_3$$

Reaction of the secondary amine with formaldehyde affords an iminium ion, and hydride transfer produces the tertiary amine.

$$\text{CH}_3\text{(R)NH} \xrightleftharpoons{\text{O=CH}_2} \text{CH}_3\text{(R)}\overset{+}{\text{N}}\text{=CH}_2 \xrightarrow{\text{H—C(=O)—O}^-} \text{CH}_3\text{(R)}\ddot{\text{N}}\text{—CH}_3$$

This introduction of two methyl groups converts a primary amine to the tertiary amine that is an appropriate starting material for oxidation and amine oxide pyrolysis (Section 17.5).

$$\text{C}_6\text{H}_5\text{CH}_2\text{CH}_2\text{—NH}_2 \xrightarrow[\text{HCO}_2\text{H}]{\text{HCHO}} \text{C}_6\text{H}_5\text{CH}_2\text{CH}_2\text{—N(CH}_3)_2 \quad 74\text{–}83\%$$

$$\text{NH}_3 \xrightarrow[\text{HCO}_2\text{H}]{\text{HCHO}} \text{CH}_3\text{—N(CH}_3)_2 \quad 85\text{–}90\%$$

The same result can be achieved by using a large excess of formaldehyde with sodium cyanoborohydride.

$$\text{3-NO}_2\text{C}_6\text{H}_4\text{NH}_2 \xrightarrow[\text{NaBH}_3\text{CN, CH}_3\text{CO}_2\text{H}]{\text{excess HCHO}} \text{3-NO}_2\text{C}_6\text{H}_4\text{N(CH}_3)_2 \quad 68\%$$

Alkylation of Ammonia and Amines

The reaction of ammonia or an amine with an alkyl halide or sulfonate introduces a new alkyl substituent on nitrogen. Two important side reactions can occur; elimination and polyalkylation (Section 17.4). Elimination, which converts the alkyl halide to an alkene, is minimized with primary alkyl halides, but tertiary halides yield alkenes exclusively. The importance of elimination as a side reaction for secondary derivatives is illustrated by the low yield obtained in the reaction of the following tosylate with ammonia:

$$\text{cis-4-}t\text{-Bu-cyclohexyl OTs} \xrightarrow{NH_3} \text{trans-4-}t\text{-Bu-cyclohexyl-}NH_2 \quad 37\%$$

Notice that this reaction proceeds with inversion of configuration, as expected for a direct displacement (Sections 12.4, 12.5).

Polyalkylation can be minimized by using a substantial excess of the amine, but this is impractical if the amine is expensive or difficult to make. On the other hand, this approach works well for the alkylation of ammonia and inexpensive amines.

$$Cl{-}CH_2{-}CH(OC_2H_5){-}OC_2H_5 \xrightarrow{NH_3} NH_2{-}CH_2{-}CH(OC_2H_5){-}OC_2H_5 \quad 71\text{–}74\%$$

$$C_6H_5NH_2 \xrightarrow{C_6H_5CH_2Cl} C_6H_5NH{-}CH_2C_6H_5 \quad 85\text{–}87\%$$

In the first of these examples the molar ratio of ammonia to halide was 72:1, and in the second reaction a 4:1 ratio of amine to halide was used. The alkylation of ammonia by this procedure has been used extensively for the synthesis of amino acids (Section 20.2).

Another approach to preventing polyalkylation utilizes a *blocking group* on nitrogen. The procedure was originally developed by S. Gabriel in 1887 while working in Hofmann's laboratory in Germany. The two carbonyl groups of phthalimide greatly increase the acidity of the proton on nitrogen ($K_a \sim 10^{-10}$), so the imide is almost completely converted to the anion by aqueous base. Alkylation produces the *N*-alkylphthalimide, and this neutral product does not react further with the alkylating agent. Subsequent hydrolysis yields the primary amine.

$$\text{phthalimide N–H} \underset{}{\overset{KOH}{\rightleftharpoons}} \text{phthalimide } N^-K^+ \xrightarrow{R{-}X} \text{phthalimide N–R} \xrightarrow{H_2O,\ KOH} C_6H_4(CO_2^-)_2 + R{-}NH_2$$

blocking group

The overall process requires several steps, but the blocking group totally prevents polyalkylation. The following example illustrates the alkylation of phthalimide:

$$\text{Phthalimide (N—H)} \xrightarrow[\text{2. } C_6H_5CH_2Cl]{\text{1. } K_2CO_3} \text{N—}CH_2C_6H_5 \text{ phthalimide} \quad 72\text{–}79\%$$

The Ritter Reaction: Alkylation of Nitriles with Carbocations

Nitriles are very weak bases, so even in concentrated sulfuric acid a significant fraction exists in the unprotonated form. Consequently, nitriles can be used for nucleophilic attack on carbocations in strongly acidic media. Under the same conditions amines are completely protonated, so they cannot behave as nucleophiles. The reaction with nitriles was developed by J.J. Ritter at New York University in the late 1940s, and it proceeds by way of a **nitrilium ion,** according to the following equation:

$$R\text{—}OH \xrightarrow[H_2SO_4]{\text{conc.}} R^+ \xrightarrow{:N\equiv C\text{—}R'} R\text{—}\overset{+}{N}\equiv C\text{—}R'$$

Nitrilium ion

When the reaction is worked up, the resulting nitrilium ion reacts readily with water to produce an amide.

$$R\text{—}\overset{+}{N}\equiv C\text{—}R' \xrightarrow{:OH_2} R\text{—}N{=}C(\overset{+}{O}H_2)R' \underset{\text{transfer}}{\overset{\text{proton}}{\rightleftharpoons}} R\text{—}\overset{+}{N}H{=}C(OH)R'$$

$$\rightleftharpoons R\text{—}NH\text{—}C(=\overset{+}{O}H)R' \underset{}{\overset{\text{proton transfer}}{\rightleftharpoons}} R\text{—}NH\text{—}\overset{O}{\overset{\|}{C}}\text{—}R'$$

Subsequent hydrolysis of the amide produces the primary amine. (Notice, however, that reduction with $LiAlH_4$ would yield a secondary amine.)

$$R\text{—}NH\text{—}\overset{O}{\overset{\|}{C}}\text{—}R' \xrightarrow[NaOH]{H_2O} R\text{—}NH_2$$

At first glance this seems to be no different than alkylation of ammonia with an alkyl halide. But the Ritter reaction must be used with alkyl derivatives that afford fairly stable carbocations in sulfuric acid, and these are usually *tertiary*. In contrast, the alkylation of amines with alkyl halides is limited to reactions with primary and secondary alkyl derivatives. Consequently, the two methods are complementary. In fact, the Ritter reaction provides the only general method for directly introducing a tertiary alkyl substituent on nitrogen.

Alkenes and other compounds, as well as alcohols, can be used as reactants in the Ritter reaction as long as they will react with concentrated sulfuric acid to form carbocations. The first of the following examples demonstrates that hydrogen cyanide reacts only at the nitrogen atom (because the carbon atom remains protonated in concentrated sulfuric acid).

$$\text{1-methylcyclohexanol} \xrightarrow[\text{2. } H_2O]{\text{1. } H{-}C{\equiv}N,\ \text{conc. } H_2SO_4} \text{1-methylcyclohexyl}{-}NH{-}\overset{O}{\overset{\|}{C}}H \xrightarrow{H_2O,\,KOH} \text{1-methylcyclohexyl}{-}NH_2 \quad 60\%$$

$$(CH_3)_2C{=}CH_2 \xrightarrow[\text{2. } H_2O]{\text{1. } CH_3{-}C{\equiv}N,\ \text{conc. } H_2SO_4} CH_3{-}\underset{CH_3}{\overset{CH_3}{C}}{-}NH{-}\overset{O}{\overset{\|}{C}}{-}CH_3 \quad 85\%$$

$$\text{spiro[4.5]decan-6-ol} \xrightarrow[\text{2. } H_2O]{\text{1. } CH_3{-}C{\equiv}N,\ \text{conc. } H_2SO_4} \text{4a-(acetylamino)decahydronaphthalene } (NH{-}\overset{O}{\overset{\|}{C}}{-}CH_3) \quad 90\%$$

via [spiro[4.5]dec-6-yl cation, $+$]

The last example serves as a reminder that a secondary cation will rearrange readily if the rearrangement leads to a more stable cation.

Amines via Hydroboration

In the usual hydroboration-oxidation reaction for the preparation of alcohols from alkenes (Section 5.11), the intermediate organoborane is oxidized with alkaline hydrogen peroxide.

$$R_3B \xrightarrow[\text{NaOH}]{\text{HOOH}} R{-}OH$$

If the organoborane is instead oxidized with chloramine (H_2NCl) or hydroxylamine-*O*-sulfonic acid ($H_2N{-}OSO_3H$), the amine can be prepared.

$$R_3B \xrightarrow[H_2N{-}OSO_3H]{H_2N{-}Cl \text{ or}} R{-}NH_2$$

As in the formation of alcohols from alkenes (Section 6.8) the reaction is stereospecific (overall *syn* addition), and nitrogen is introduced at the less substituted carbon of the original double bond.

$$\text{1-methylcyclopentene} \xrightarrow[\text{2. } H_2N\text{—}OSO_3N]{\text{1. } BH_3} \text{trans-2-methylcyclopentylamine} \quad 56\%$$

The oxidation has also been carried out with sodium hypochlorite in aqueous ammonia.

$$CH_3(CH_2)_7\text{—}CH{=}CH_2 \xrightarrow[\text{2. } NH_3,\ H_2O,\ NaOCl]{\text{1. } BH_3} CH_3\text{—}(CH_2)_7\text{—}CH_2CH_2NH_2 \quad 71\%$$

The hydroboration reaction does have limitations, because most carbonyl compounds are reduced by borane. Difficulties would therefore be encountered with many polyfunctional compounds.

Hofmann Degradation of Primary Amides

In 1881 Hofmann reported that a primary amide is degraded to the amine with one less carbon by the action of bromine in aqueous base.

$$R\text{—}\overset{O}{\overset{\|}{C}}\text{—}NH_2 \xrightarrow[KOH]{Br_2} R\text{—}NH_2$$

Despite the initial appearance of this reaction as an almost magical removal of the carbonyl group, the mechanism follows a series of logical steps. An initial acid-base reaction removes one of the hydrogens on the NH_2 group. (Recall that amides are comparable with water in acidity, Section 16.4.) The anion then reacts with bromine to produce the *N*-bromoamide.

$$RC(=O)NH_2 \underset{}{\overset{^-OH}{\rightleftharpoons}} RC(=O)\ddot{N}^-H \xrightarrow{Br\text{—}Br} RC(=O)N(Br)H$$

Base removes the remaining N—H hydrogen to give an unstable anion, which loses bromide ion, generating an intermediate with an electron-deficient nitrogen. This species has only six electrons in the valence shell of nitrogen, and like the analogous carbocations (Section 5.9), it is highly prone to rearrangement. (The mechanism of this reaction is discussed further in the Appendix of the Solutions Manual.)

$$RC(=O)N(Br)H \xrightarrow{^-OH} RC(=O)\ddot{N}^-\text{—}Br \longrightarrow RC(=O)\ddot{N}: + Br^-$$

A 1,2 shift of the R group with its bonding electrons generates what at first seems to be a species with both positive and negative charges, but when a second resonance form is drawn, you can see that it is actually a neutral compound, an *isocyanate*.

$$\text{R}-\overset{\overset{\text{O}}{\|}}{\text{C}}-\ddot{\text{N}}: \longrightarrow \overset{\overset{\text{O}}{\|}}{\text{C}^+}-\ddot{\text{N}}:^-(\text{R}) \longleftrightarrow \overset{\overset{\text{O}}{\|}}{\text{C}}=\ddot{\text{N}}-\text{R} \equiv \text{R}-\ddot{\text{N}}=\text{C}=\text{O}$$

Isocyanate

At this stage the carbonyl carbon has not yet "disappeared," but it is no longer bonded to the alkyl group. Finally, hydrolysis by the aqueous base liberates the free amine.

$$\text{R}-\text{N}=\text{C}=\text{O} \xrightarrow{^-\text{OH}} \text{R}-\underset{\text{H}}{\text{N}}-\overset{\text{OH}}{\text{C}}=\text{O} \underset{}{\overset{^-\text{OH}}{\rightleftharpoons}} \text{R}-\underset{\text{H}}{\text{N}}-\overset{\text{O}^-}{\text{C}}=\text{O} \longrightarrow \text{R}-\text{NH}_2 + \text{CO}_2$$

HO—H   HO—H

The following examples show you that amines can be prepared in good yield by this reaction. The limitations of the method are mainly in the availability of the appropriate carboxylic acid. Notice that a tertiary *N*-alkyl group can be introduced by the Hofmann degradation.

$$\text{CH}_3-\underset{\text{CH}_3}{\overset{\text{CH}_3}{\text{C}}}-\overset{\overset{\text{O}}{\|}}{\text{C}}-\text{NH}_2 \xrightarrow[\text{KOH, H}_2\text{O}]{\text{Br}_2} \text{CH}_3-\underset{\text{CH}_3}{\overset{\text{CH}_3}{\text{C}}}-\text{NH}_2 \quad 45\text{–}65\%$$

$$\text{CH}_3\underset{\text{CH}_3}{\overset{\text{CH}_3}{\text{C}}}-\text{CH}_2-\overset{\overset{\text{O}}{\|}}{\text{C}}-\text{NH}_2 \xrightarrow[\text{NaOH, H}_2\text{O}]{\text{Br}_2} \text{CH}_3\underset{\text{CH}_3}{\overset{\text{CH}_3}{\text{C}}}-\text{CH}_2-\text{NH}_2 \quad 94\%$$

$$3,4\text{-}(\text{CH}_3\text{O})_2\text{C}_6\text{H}_3-\overset{\overset{\text{O}}{\|}}{\text{C}}-\text{NH}_2 \xrightarrow[\text{NaOH, H}_2\text{O}]{\text{Cl}_2} 3,4\text{-}(\text{CH}_3\text{O})_2\text{C}_6\text{H}_3-\text{NH}_2 \quad 80\text{–}82\%$$

The last of these examples demonstrates that chlorine (actually, NaOCl) can be used in place of bromine in the reaction. The second example shows the formation of neopentylamine, a primary amine that cannot be prepared by alkylation of ammonia. (Recall that steric hindrance in neopentyl derivatives prevents most direct displacement reactions, Section 12.7.)

The rearrangement of an alkyl (or aryl) group to electron-deficient nitrogen is not limited to the Hofmann degradation, and you can find additional examples of this phenomenon in the Appendix of the Solutions Manual.

EXERCISE 17.9

Suggest methods for preparing the following primary amines. If more than one method will work, show all possibilities.

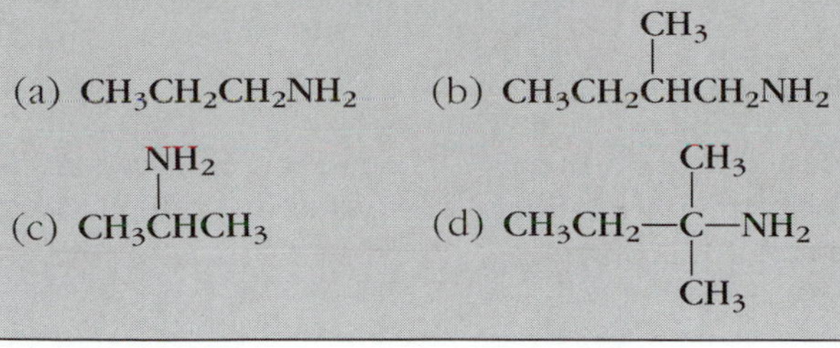

(a) $CH_3CH_2CH_2NH_2$ (b) $CH_3CH_2CH(CH_3)CH_2NH_2$

(c) $CH_3CH(NH_2)CH_3$ (d) $CH_3CH_2—C(CH_3)_2—NH_2$

17.8 ENAMINES

We have previously shown that the acid-catalyzed reaction of an aldehyde or ketone with a secondary amine can produce an iminium ion (Section 17.7).

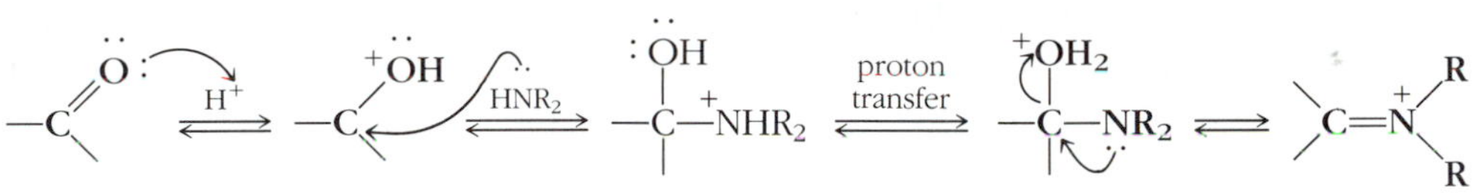

$$\text{—C(=O:)} \xrightleftharpoons{H^+} \text{—C=}\overset{+}{\text{O}}\text{H} \xrightleftharpoons{HNR_2} \text{—C(:OH)—}\overset{+}{\text{N}}\text{HR}_2 \xrightleftharpoons{\text{proton transfer}} \text{—C(}\overset{+}{\text{O}}\text{H}_2\text{)—NR}_2 \rightleftharpoons \text{C=}\overset{+}{\text{N}}\text{R}_2$$

When the original carbonyl compound has an α hydrogen, the iminium ion can lose a proton to generate an *enamine.*

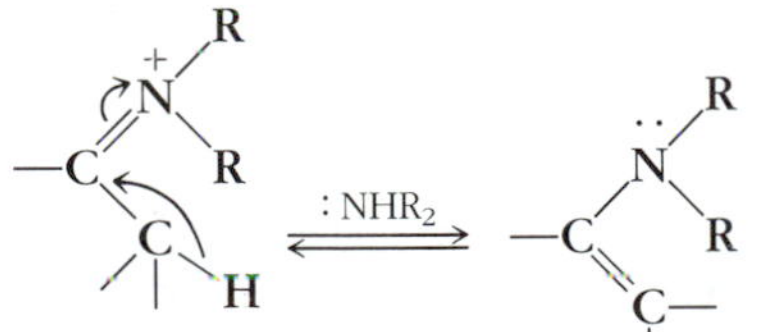

$$\text{Iminium ion} \xrightleftharpoons{:NHR_2} \text{Enamine}$$

Iminium ion — Enamine

Enamines are analogous to enols and enolates. Both nitrogen and the α carbon are nucleophilic, as suggested by the following resonance structures:

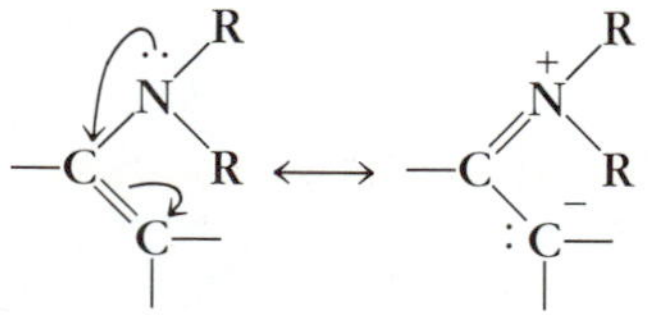

$$\text{—C(—}\ddot{\text{N}}\text{R}_2\text{)=C—} \longleftrightarrow \text{—C(=}\overset{+}{\text{N}}\text{R}_2\text{)—}\overset{-}{\ddot{\text{C}}}\text{—}$$

Molecular orbital theory also shows that both carbon and nitrogen are nucleophilic. The highest occupied molecular orbital (HOMO), which corresponds to the nonbonding electrons, is not localized on nitrogen but also has a contribution on the α carbon. Notice the similarity to the HOMO of the allyl anion (Section 8.2).

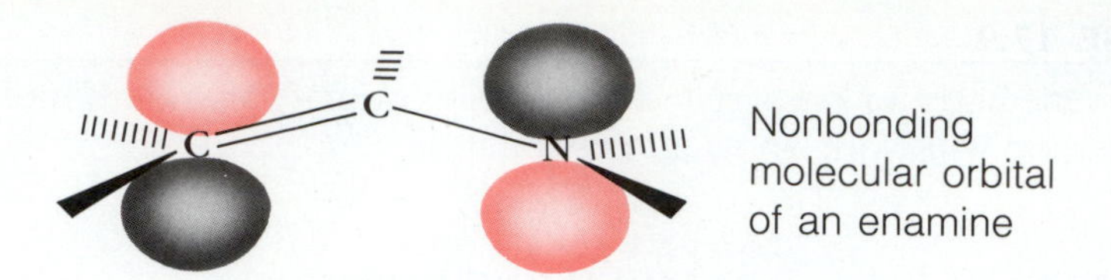

The nucleophilic character of the α carbon makes enamines useful reagents for organic synthesis. The secondary amines that have been used most frequently to prepare enamines are heterocyclic:

The following reactions illustrate the formation of enamines. Enamine formation is fully reversible, so water is typically removed from the reaction mixture by distillation as an azeotrope with benzene.

Some of the problems encountered in the alkylation of enolate ions (Section 13.4) are alleviated by the use of enamines.

There is no strong base present in the enamine reaction, so side reactions of the alkyl halide (elimination or nucleophilic attack by base) are greatly decreased. *N*-Alkylation is sometimes a problem, but this is minimized when the alkylating agent is a methyl, primary allylic, or benzylic derivative.

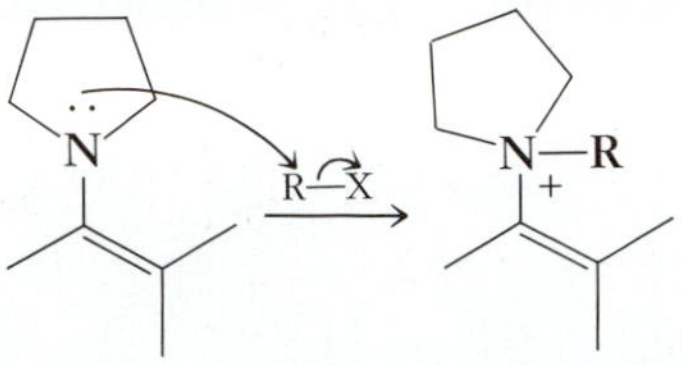

N R—X → N—R +

An important advantage of enamine alkylation is that unsymmetrical ketones can often be alkylated specifically at just one of the α carbons. This is because the less substituted enamine is usually more stable, and it is formed preferentially.

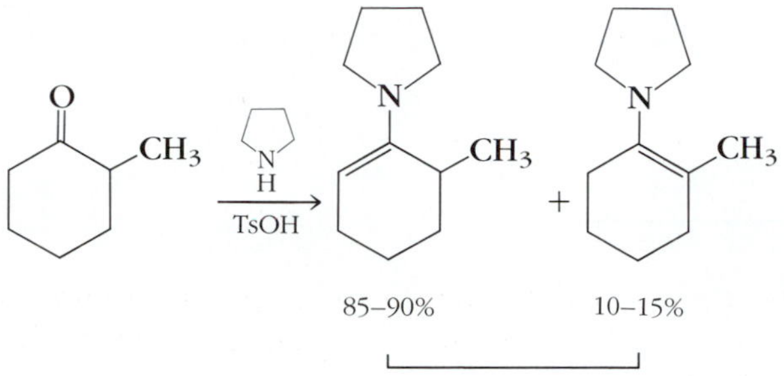

O CH$_3$ N H TsOH → N CH$_3$ + N CH$_3$ 53% yield

85–90% 10–15%

The orbital interactions in an enamine favor a structure in which the nitrogen atom, the two carbons of the double bond, and the substituents on these atoms all lie in a single plane. In the minor product this leads to an unfavorable steric interaction between the pyrrolidine ring and the methyl group. This interaction is not present in the major isomer, because the methyl group is attached to an sp^3 carbon and does not lie in the plane of the double bond.

The following examples illustrate the use of enamines for alkylating ketones:

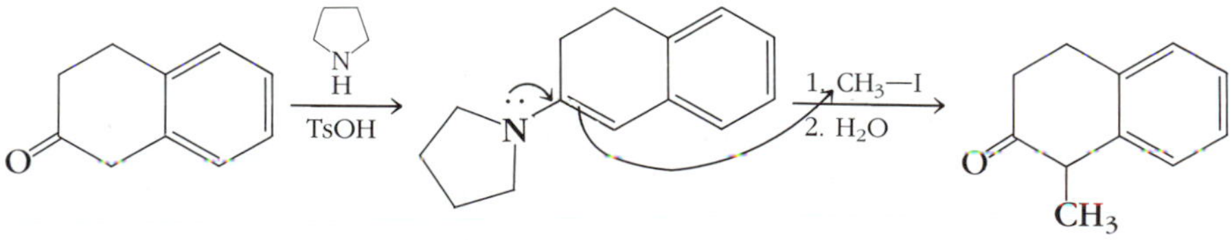

O N H TsOH → N 1. CH$_3$—I 2. H$_2$O → O CH$_3$

81%

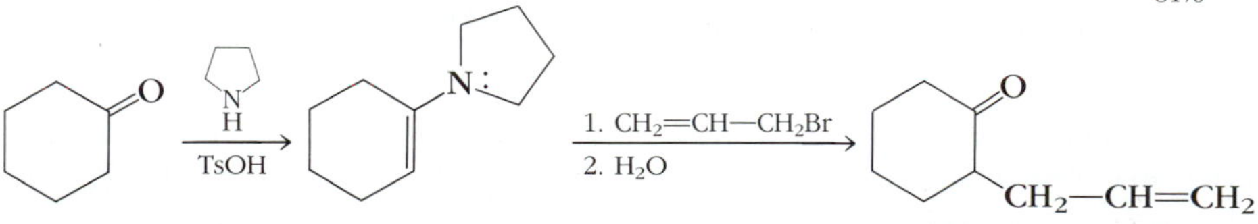

O N H TsOH → N: 1. CH$_2$=CH—CH$_2$Br 2. H$_2$O → O CH$_2$—CH=CH$_2$

Enamines also react with α,β-unsaturated carbonyl compounds by conjugate addition (Section 13.10), i.e., Michael addition (Section 14.5).

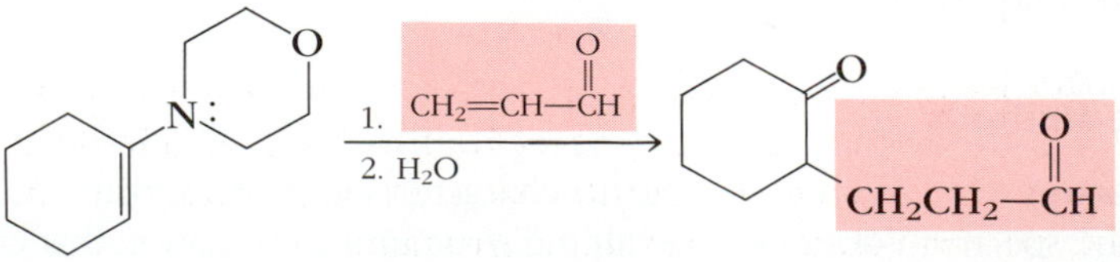

O N: 1. CH$_2$=CH—CH(=O) 2. H$_2$O → O CH$_2$CH$_2$—CH(=O) 60–65%

EXERCISE 17.10

Write the mechanism for the following enamine hydrolysis:

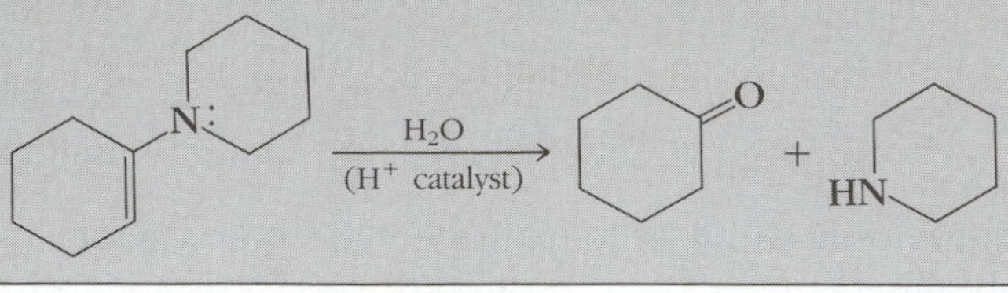

$$\text{1-(1-cyclohexenyl)piperidine} \xrightarrow[\text{(H}^+\text{ catalyst)}]{H_2O} \text{cyclohexanone} + \text{piperidine (HN)}$$

EXERCISE 17.11

Draw the products formed in each of the three steps for the following reactions:

(a) 2-methylcyclopentanone (CH_3, =O) $\xrightarrow{}$ 1. pyrrolidine (NH), TsOH; 2. $C_6H_5CH_2Br$; 3. H_2O

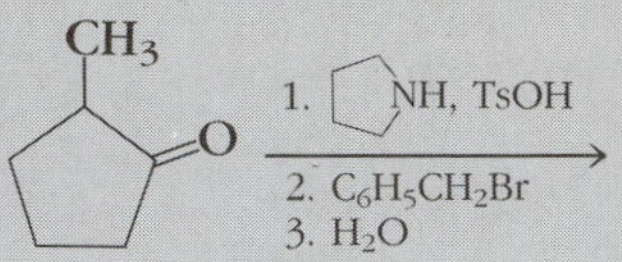

(b) 4-methylcyclohexanone (CH_3, =O) $\xrightarrow{}$ 1. morpholine (O, NH); 2. $CH_2{=}CH{-}CO_2C_2H_5$; 3. H_2O

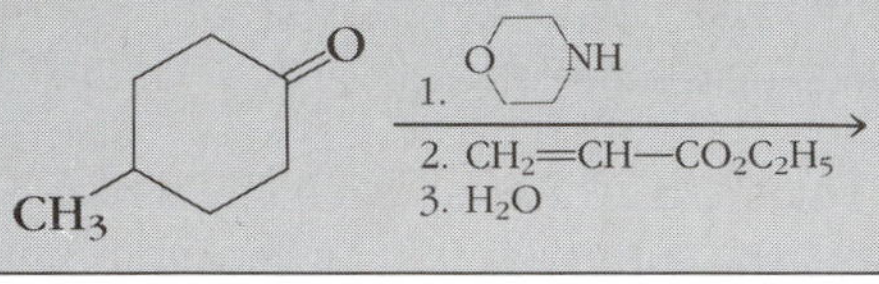

17.9 CHEMICAL AND SPECTROSCOPIC CHARACTERIZATION OF AMINES

The basicity of amines provides a simple method of distinguishing them from compounds of other functional classes. Even water-insoluble amines will normally dissolve in dilute HCl, because they are converted to ammonium salts.

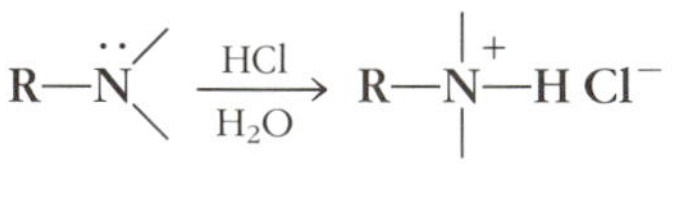

$$\underset{\text{Water-insoluble}}{R{-}\ddot{N}\!<} \xrightarrow[H_2O]{HCl} \underset{\text{Water-soluble}}{R{-}\overset{+}{N}{-}H\ Cl^-}$$

Amines can also be recognized by their ability to consume acid in a titration. Only when an excess of acid has been added will the pH of the solution decrease enough to cause a color change of an appropriate indicator.

The major problems in structure determination of amines are in identifying the substituents on nitrogen. Are the substituents alkyl or aryl? How many substituents are present, that is, is the amine primary, secondary, or tertiary? Both spectroscopic and chemical techniques are useful.

Spectroscopic Methods

NMR Spectroscopy Amines exhibit no characteristic peaks in the ^{13}C NMR spectrum, but the spectrum does provide all the structural information that can be obtained with other types of compounds. The substituent effect of an $-NH_2$ group is about +20 ppm, comparable with that for a —Cl substituent (see Table 9.2). Proton NMR spectroscopy also provides a powerful tool for structure determination of amines, although there are no specific peaks that are highly characteristic of amines. The substituent effect of an —NH_2 or —NR_2 group is +1.7 ppm, similar to that of other electronegative substituents (Table 9.3). The —NH signal of a primary or secondary amine usually falls in the range of 0.5–5 ppm, and it is also a very broad signal in many cases.

Many of the new, advanced NMR spectrometers can be used for ^{15}N spectroscopy, which provides direct information about the chemical environment of the nitrogen atom. Unfortunately, only 0.4% of all nitrogen atoms are ^{15}N (the major isotope is ^{14}N), so ^{15}N spectroscopy is experimentally difficult. The details of ^{15}N spectroscopy are beyond the scope of this book, and we will not discuss the topic in detail.

Infrared Spectroscopy The carbon-nitrogen bond of an amine leads to absorptions in the 1000–1400 cm^{-1} region, but other functional groups also give rise to peaks in this region. The N—H absorptions usually appear at 3300–3500 cm^{-1}, a slightly higher frequency than for alcohols (which give peaks in the 3200–3400 cm^{-1} region). The NH peaks of amines are usually less intense than the OH peaks of alcohols. The NH signal of ammonium salts is shifted to the region of 3000–3100 cm^{-1}.

UV-Visible Spectroscopy Amino groups do not absorb in the UV-visible region, so absorptions in this region will not be observed when the molecule also contains appropriately unsaturated groups. The UV spectrum of an aromatic amine will show strong pH dependence, because protonation of the nonbonding electron pair on nitrogen causes a substantial change in the orbital interaction between the nitrogen and the aromatic π system.

Chemical Methods

As we discussed earlier in this section, basic compounds are relatively easily identified as amines by solubility in aqueous acid or by titration with acid. Chemical methods can then be of great assistance in identifying the number and type of substituents on nitrogen.

Reaction With Nitrous Acid Primary, secondary, and tertiary amines all react with nitrous acid (Section 17.4), but *primary amines* give results that are most easily interpreted. Both aliphatic and aromatic primary amines yield diazonium salts, but the aliphatic diazonium ion undergoes rapid solvolysis with the evolution of nitrogen gas. This evolution of gas is a useful qualitative test for a primary aliphatic amine.

$$\underset{\text{Primary aliphatic amine}}{R{-}NH_2} \xrightarrow[\text{HCl, } H_2O]{\text{HONO}} [R{-}\overset{+}{N}{\equiv}N \quad Cl^-] \longrightarrow R{-}OH\ (+\ \text{alkene, etc.}) + \underset{\text{Evolution of gas}}{N_2}$$

$$\underset{\text{Primary aromatic amine}}{Ar{-}NH_2} \xrightarrow[\text{HCl, } H_2O]{\text{HONO}} \underset{\text{Diazonium salt}}{Ar{-}\overset{+}{N}{\equiv}N \quad Cl^-}$$

Aromatic diazonium salts are usually stable in solution (although they can decompose explosively when isolated as solids). A second test is carried out to detect an aromatic diazonium salt; this is normally done by mixing the nitrous acid solution with an alkaline solution of 2-naphthol. Formation of a brightly colored (red-orange) azo dye indicates the presence of an aromatic diazonium salt and shows that the original amine was both primary and aromatic.

$^{-}$OH

$:N\equiv \overset{+}{N}-Ar$

2-Naphthol

Azo dye

keto-enol equilibrium

During the early days of the organic chemical industry in the late 1800s, azo dyes were among the most widely used synthetic dyes. Variation of both aromatic groups (phenol and amine) allowed the synthesis of dyes with a wide range of colors and chemical properties. Modern organic chemistry has since provided other dyes with improved qualities, so the azo dyes are now less widely used.

The Hinsberg Test

A useful method for determining whether an amine is primary, secondary, or tertiary is based on its reaction with a sulfonyl chloride. An excess of either benzenesulfonyl chloride or *p*-toluenesulfonyl chloride is shaken with a mixture of the amine and aqueous base. After the reaction (if there is one) between amine and sulfonyl chloride is complete, the unreacted sulfonyl chloride hydrolyzes to produce the water-soluble salt of the sulfonic acid. At this stage the reaction mixture is examined. Several possibilities exist.

Tertiary amines do not react with the sulfonyl chloride, so no change occurs. Because a water-insoluble amine may be difficult to distinguish from an insoluble reaction product, the mixture is acidified. A tertiary amine will be converted to the soluble ammonium salt.

$$R_3N \xrightarrow[\text{NaOH, } H_2O]{ArSO_2Cl} \underset{\text{Usually insoluble}}{R_3N} \xrightarrow{HCl} \underset{\text{Soluble}}{R_3\overset{+}{N}H\ Cl^-}$$

(no reaction)

Secondary amines react with sulfonyl chlorides to give the corresponding sulfonamides, which are water-insoluble. In contrast to the results with a tertiary amine, this insoluble material (a neutral sulfonamide) will not dissolve in acid.

$$R_2NH \xrightarrow[NaOH,\ H_2O]{ArSO_2Cl} R_2N{-}S(=O)_2{-}Ar \xrightarrow[(no\ reaction)]{HCl} R_2N{-}S(=O)_2{-}Ar$$

Insoluble Insoluble

Primary amines also react to give a sulfonamide, but a primary sulfonamide still has one hydrogen on the nitrogen atom. The influence of the adjacent SO_2 group renders this NH acidic, and the primary sulfonamide reacts with base to form a water-soluble sodium salt. The resulting clear solution might be confused with the results for a water-soluble tertiary amine, but acidification regenerates the neutral, insoluble sulfonamide.

$$R{-}NH_2 \xrightarrow[NaOH,\ H_2O]{ArSO_2Cl} [R{-}NH{-}SO_2Ar] \longrightarrow R{-}\overset{-}{N}(Na^+){-}SO_2Ar \xrightarrow{HCl} RNH{-}SO_2Ar$$

Soluble Insoluble

This two-step procedure, involving initial reaction with a sulfonyl chloride in base and subsequent acidification of the mixture, gives results which clearly distinguish among primary, secondary, and tertiary amines.

Separation of Amines from Acidic and Neutral Compounds

Many organic reactions generate mixtures of compounds that must be separated in order to obtain the desired product. Sometimes unreacted starting material must be removed from the product, and other times mixtures of isomeric products must be separated. Such separations and purifications often require tedious distillation, crystallization, or chromatography. But there is one type of purification that can be done much more easily, namely, the separation of acidic, basic, and neutral components of a mixture. Reactions involving carboxylic acids or amines often produce such mixtures.

Consider a hypothetical hydrolysis reaction of the amide RCONHR that produces a mixture of the acid (RCO_2H), the amine (RNH_2), and some unreacted amide.

$$R{-}\overset{O}{\overset{\|}{C}}{-}NHR \xrightarrow{hydrolysis} R{-}\overset{O}{\overset{\|}{C}}OH + R{-}NH_2 + R{-}\overset{O}{\overset{\|}{C}}{-}NHR$$

acid base neutral (unreacted)

Separation of these can be accomplished by a series of extractions with a water-insoluble organic solvent (designated here as "organic").

At neutral pH each of the individual components would preferentially dissolve in the organic solvent rather than water, but this would not be true if the aqueous phase were either acidic or alkaline. If the aqueous phase were basic, the carboxylic acid would be converted to its water-soluble salt. Similarly, aqueous acid would convert the amine to its water-soluble salt. Acid-base reactions can therefore be used to change the solubilities of the components in our hypothetical reaction mixture. One possible sequence of extractions is shown in Figure 17.3.

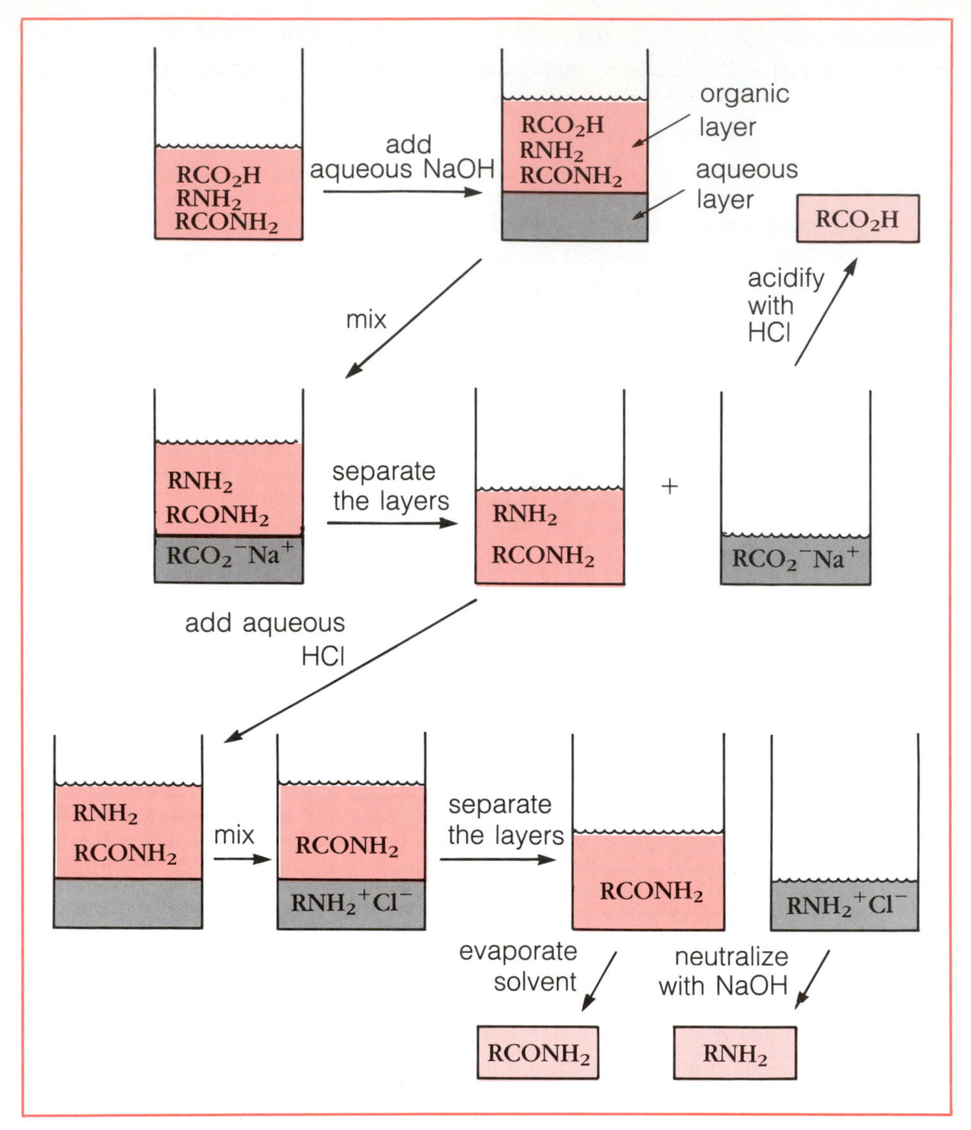

Figure 17.3 Separation of Acidic, Basic, and Neutral Components of an Organic Mixture by Extraction with Aqueous Acid and Base.
The aqueous phase is the lower or shaded layer in all cases.

Ordinarily, a *separatory funnel* would be used to carry out such extractions, but in Figure 17.3 we have drawn for convenience a vessel that resembles a simple beaker. The actual extraction takes place when the two phases are vigorously mixed, and the immiscible aqueous and organic layers can subsequently be separated. We have shown the organic phase as the upper layer here, but there are also organic solvents that are more dense than water.

The first step, extraction with aqueous base, removes the carboxylic acid from the organic mixture by converting it to the carboxylate salt, which dissolves in the aqueous layer. Acidification of the aqueous layer then regenerates the carboxylic acid free of the other organic compounds.

The second step, extraction with aqueous acid, then separates the remaining two components of the organic mixture. The amine is converted by the acid to an ammonium salt, which dissolves in the aqueous layer. Subsequent treatment of the aqueous layer with base regenerates the amine, and evaporation of the organic solvent affords the neutral amide.

EXERCISE 17.12

Which of the following compounds would evolve nitrogen in the reaction with nitrous acid?

(a) $CH_3CH_2NH_2$ (b) $CH_3CH_2—NH—CH_2CH_3$ (c) $CH_3—C(CH_3)_2—NH_2$

(d) $C_6H_5—NH_2$ (e) $C_6H_5—N(CH_3)_2$

EXERCISE 17.13

Which of the compounds in Problem 17.12 would give a precipitate in the reaction with $C_6H_5SO_2Cl$ and aqueous NaOH, followed by acidification?

EXERCISE 17.14

Suggest a method for removing unreacted octanoic acid from the mixture obtained by acid-catalyzed esterification with methanol.

EXERCISE 17.15

How could you isolate heptylamine from any neutral compounds that might be present in the mixture formed by the $LiAlH_4$ reduction of nitroheptane?

17.10 TERMS AND DEFINITIONS

Aliphatic. A term describing those organic compounds that are not aromatic.

Alkaloid. Any of a variety of naturally occurring physiologically active amines; usually obtained from plants.

Ammonium ion. A cation containing a tetrasubstituted, positively charged nitrogen atom (i.e., an alkyl- or aryl-substituted derivative of NH_4^+).

Enamine. A compound containing an amine that is directly bonded to a carbon-carbon double bond; a vinyl amine.

Imine. A compound containing a C=N double bond; the nitrogen analog of a ketone.

Iminium Ion. An ion with a $C=N^+$ group.

Nitrilium Ion. An ion with a —C≡N—R ion.

Quaternary ammonium ion. An ammonium ion in which all four substituents on the positively charged nitrogen are alkyl (or aryl) groups.

17.11 Summary of Reactions

The reactions of amines are summarized in Table 17.2, and the various methods for their preparation are summarized in Table 17.3.

TABLE 17.2 Reactions of Amines

Reaction	Comments
1. Reaction as base	Section 17.4
$R{-}\ddot{N}< + H^+ \rightleftharpoons R{-}\overset{+}{N}{-}H$	K_a of ammonium ion is approximately 10^{-5}.
2. Reaction as acid	Section 17.4
$R{-}\ddot{N}{-}H \rightleftharpoons H^+ + R{-}\ddot{N}:^-$	$K_a \sim 10^{-35}$. Amide ion is a strong base.
3. Acylation	Sections 16.6, 17.4
$R{-}\ddot{N}{-}H \xrightarrow{R'CCl\,(C{=}O)} R{-}\ddot{N}{-}\overset{O}{\overset{\Vert}{C}}{-}R'$	Formation of amides. Only primary and secondary amines react.
$R{-}\ddot{N}{-}H \xrightarrow{ArSO_2Cl} R{-}\ddot{N}{-}SO_2Ar$	Formation of sulfonamides.
4. Alkylation	Section 17.4
$R{-}\ddot{N}{-} \xrightarrow{R'X} R{-}\overset{+}{N}{-}R'$	Reaction with alkyl halide or sulfonate. Polyalkylation and elimination are important side reactions.
5. Reaction with nitrous acid	Section 17.4
$R{-}NH_2 \xrightarrow{HONO} R{-}\overset{+}{N}{\equiv}N \longrightarrow$ substitution and elimination	Primary aliphatic amines form unstable diazonium ions; secondary amines yield *N*-nitroso compounds; tertiary amines yield complex mixtures.
$Ar{-}NH_2 \xrightarrow{HONO} Ar{-}\overset{+}{N}{\equiv}N$	Primary aromatic amines form diazonium salts that are stable in solution (see Section 18.5).

TABLE 17.2 Reactions of Amines (continued)

Reaction	Comments
6. Hofmann elimination $-\overset{\mid}{\underset{H}{\overset{\mid}{C}}}-\overset{\mid}{\underset{\mid}{C}}-NH_2 \xrightarrow[CH_3I]{excess} -\overset{\mid}{\underset{H}{\overset{\mid}{C}}}-\overset{\mid}{\underset{\mid}{C}}-\overset{+}{N}(CH_3)_3$ $\downarrow$ base $>C=C< + N(CH_3)_3$	Section 17.5 Decomposition of quaternary ammonium salts. Less substituted alkene predominates. *Anti* elimination.
7. Amine oxide pyrolysis $-\overset{\mid}{\underset{H}{\overset{\mid}{C}}}-\overset{\mid}{\underset{\mid}{C}}-N(CH_3)_2 \xrightarrow{H_2O_2} -\overset{\mid}{\underset{H}{\overset{\mid}{C}}}-\overset{\mid}{\underset{\mid}{C}}-\underset{O^-}{\overset{+}{N}}(CH_3)_2$ $\downarrow$ pyrolysis $>C=C< + HON(CH_3)_2$	Section 17.5 *Syn* elimination.
8. Enamine formation $-C(=O)-\underset{\mid}{CH}- \xrightarrow{R_2NH} -C(NR_2)=\underset{\mid}{C}-$	Section 17.8 Enamines are useful for carrying out reaction at the α carbon of a ketone.

TABLE 17.3 Preparation of Amines

Reaction	Comments
1. Reduction of nitro compounds $R-NO_2 \xrightarrow{H_2,\ Pt} R-NH_2$ (or $Ar-NO_2$)	Section 17.6 Other reducing agents can be used: Zn, HCl; Sn, HCl; Fe, HCl; H_2S; $LiAlH_4$.

TABLE 17.3 Preparation of Amines (continued)	
Reaction	**Comments**
2. Reduction of amides	Sections 16.4, 17.6
$R{-}C({=}O){-}NR'R'' \xrightarrow[2.\ H_2O]{1.\ LiAlH_4} R{-}CH_2{-}NR'R''$	R' and R'' can be hydrogen, alkyl, or aryl.
3. Reduction of nitriles	Section 16.4, 17.6
$R{-}C{\equiv}N \xrightarrow[2.\ H_2O]{1.\ LiAlH_4} RCH_2NH_2$	Yields primary amines.
4. Reduction of oximes	Section 17.6
$>C{=}N{-}OH \xrightarrow[\text{or Na, alcohol}]{H_2,\ Ni} {-}\overset{\mid}{\underset{\mid}{C}}H{-}NH_2$	Yields primary amines. $LiAlH_4$ can also be used for reduction.
5. Reductive amination	Sections 13.7, 17.7
$>C{=}O \xrightleftharpoons{HNRR',\ H^+} >C{=}\overset{+}{N}RR' \xrightarrow{NaBH_3CN} {-}\overset{\mid}{\underset{\mid}{C}}H{-}NRR'$	R and R' can be hydrogen, alkyl, or aryl. Catalytic hydrogenation is sometimes used for reduction.
$R{-}NH_2 \xrightarrow{CH_2O,\ HCO_2H} R{-}N(CH_3)_2$	Special case of reductive amination with excess formaldehyde. Hydrogens are replaced by methyl groups to yield tertiary amines.
6. Alkylation of ammonia and amines	Section 17.7
$>NH \xrightarrow{R{-}X} >N{-}R$	Polyalkylation and elimination may occur.
7. Ritter reaction	Section 17.7
$R{-}OH \xrightarrow[R'CN]{conc.\ H_2SO_4} [R^+] \longrightarrow R{-}\overset{+}{N}{\equiv}C{-}R' \xrightarrow{H_2O} R{-}NH{-}C({=}O){-}R' \xrightarrow{\text{hydrolysis}} R{-}NH_2$	Alkylation of carbocations with nitriles. Nitrilium ion reacts with water to form amide. R^+ must be tertiary (or resonance-stabilized) cation.

TABLE 17.3 Preparation of Amines (continued)

Reaction	Comments
8. Hydroboration	Section 17.7
$>C{=}C< \xrightarrow{BH_3} -\overset{H}{\underset{\vert}{C}}-\overset{B-}{\underset{\vert}{C}}- \xrightarrow{H_2N-Cl \text{ or } H_2N-OSO_3H} -\overset{H}{\underset{\vert}{C}}-\overset{NH_2}{\underset{\vert}{C}}-$	*Syn* addition.
9. Hofmann degradation of primary amides	Section 17.7
$R-\overset{O}{\overset{\Vert}{C}}-NH_2 \xrightarrow{X_2,\ NaOH} R-NH_2$	X can be Cl, Br, or I.

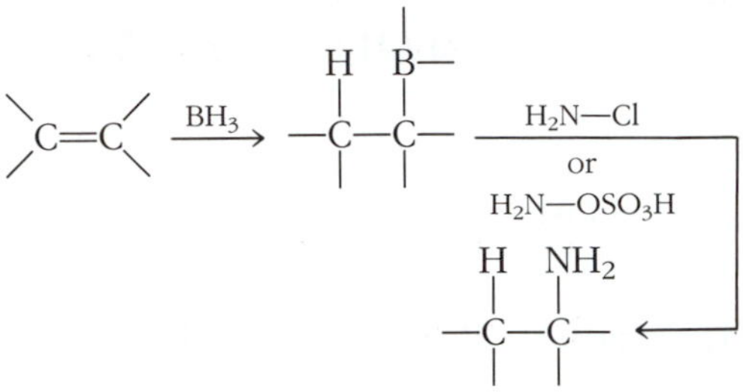

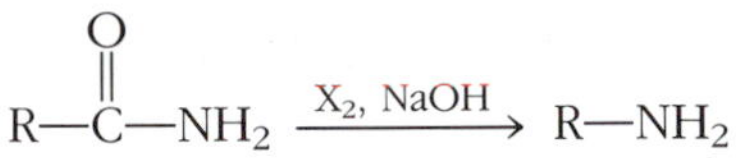

17.12 PROBLEMS

17.1 Name the following compounds:

(a) NEt_3

(b) $CH_3CH_2NHCH_3$

(c) $(CH_3)_3CNH_2$

(d) $CH_3CH_2\overset{CH_3}{\overset{\vert}{C}H}-NH-CH_2CH_3$

(e) $(CH_3CH_2CH_2)_4N^+\ Br^-$

(f) $CH_3CH_2\overset{NHCH_3}{\overset{\vert}{C}H}-CH_2CO_2CH_3$

(g) 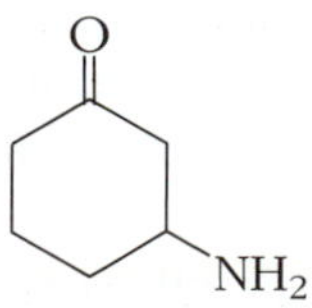 3-position NH_2 on cyclohexanone (ring with C=O; NH_2)

(h) 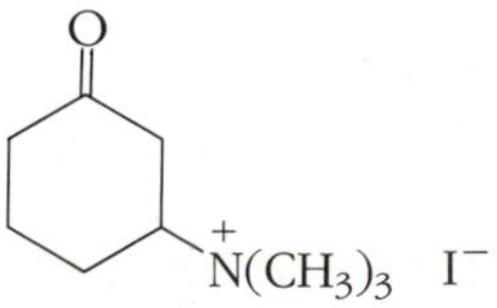 cyclohexanone ring with $\overset{+}{N}(CH_3)_3\ I^-$

(i) $CH_3CH_2-NH-CH_2CH_2OH$

17.2 Draw the following compounds:
(a) Isobutylammonium chloride
(b) Diethylamine
(c) 6-Aminohexanal
(d) *N,N*-Dibutylpentylamine

(e) 2,2-Dibutylpentylamine
(f) *N,N*-Dibutyl-2-pentylamine
(g) 1,4-Hexanediamine
(h) 3-Aminocyclopentylamine
(i) 2-(*N,N*-Dimethylamino)butyramide

17.3 For each of the following pairs, which of the two compounds would have a higher boiling point?

(a) $CH_3CH_2CH_2—N(CH_3)_2$ (N bearing two CH_3 groups) and $CH_3CH_2C(CH_3)_2—NH_2$ (C bearing CH_3 above and CH_3 below)

(b) $CH_3(CH_2)_5CH_2NH—CH_3$ and $CH_3(CH_2)_7CH_3$

(c) $CH_3CH_2—NH—CH_2CH_2—OH$ and $CH_3(CH_2)_4CH_2OH$

(d) $CH_3CH_2CH_2CH(NH_2)—CH_3$ (NH_2 on the CH) and $CH_3CH_2CH_2—NH—CH_2CH_3$

17.4 Draw the products expected when each of the following amines is treated with excess CH_3I and then heated with $NaOC_2H_5$ in ethanol:

(a) $CH_3CH_2CH(NH_2)CH_2CH_3$ (NH_2 on the central CH)

(b) *trans*-2-Methylcyclohexylamine

(c) $CH_3CH(NHCH_3)—CH_2—CH(CH_3)—CH_3$ ($NHCH_3$ below the first CH, CH_3 above the second CH)

(d) $CH_3CH_2—N(CH_3)—CH_2—CH(CH_3)—CH_2CH_2CH_3$ (CH_3 below N, CH_3 above CH)

(e) $CH_3—C(NH_2)(CH_3)—CH_2C_6H_5$ (NH_2 above and CH_3 below the C)

(f) *cis*-2-Ethylcyclopentylamine

(g) 1-Methylcycloheptylamine

(h) 2,2-Dimethylcyclopentylamine

17.5 Draw the products expected when each of the amines listed in Problem 17.4 is subjected to the following sequence: (1) HCO_2H, CH_2O; (2) H_2O_2; (3) pyrolysis.

17.6 Which of the amines in Problem 17.4 would afford a precipitate when treated with TsCl/aqueous NaOH? Draw the structures of the reaction products. Which of the others would afford a precipitate when the alkaline solution is acidified with HCl? Draw the structures of the reaction products.

17.7 Which of the amines in Problem 17.4 would react with nitrous acid with the evolution of a gas?

17.8 Which of the amines in Problem 17.4 would be acetylated by acetic anhydride? Draw the product in each such case.

17.9 Suggest methods for preparing each of the following amines from cyclohexanone:

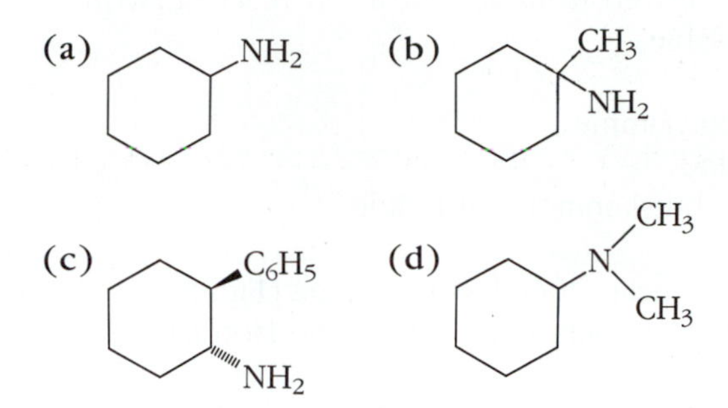

17.10 Write a detailed mechanism for each of the reactions that you wrote for Problem 17.9.

17.11 Suggest methods of converting benzoic acid into each of the following compounds:
(a) $C_6H_5CH_2NH_2$
(b) $C_6H_5CH_2CH_2NH_2$
(c) $C_6H_5CH_2NHC_2H_5$
(d) $C_6H_5{-}NH_2$
(e) $C_6H_5{-}\overset{\overset{CH_3}{|}}{C}H{-}NH_2$

17.12 Write a detailed mechanism for each of the reactions that you wrote for Problem 17.11.

17.13 Suggest methods for converting 1-pentanol into each of the following compounds:
(a) $CH_3(CH_2)_4{-}NH_2$
(b) $CH_3(CH_2)_3{-}NH_2$
(c) $CH_3(CH_2)_5{-}NH_2$

17.14 Many reactions of amines generate HCl (for example, the reaction of butylamine with acetyl chloride). The HCl can convert the amine to its nonnucleophilic ammonium salt, so a base is normally added to the reaction mixture.
(a) Which among CH_3CO_2Na, NaOH, $C_2H_5NH_2$, and $(C_2H_5)_3N$ are strong enough bases to deprotonate an ammonium salt?
(b) Which of these bases might cause undesired side reactions?
(c) Which base would be most satisfactory for general use?

17.15 Secondary amides such as *N*-methylbenzamide ($C_6H_5CONHCH_3$) do not dissolve in aqueous sodium hydroxide. Why is the corresponding sulfonamide ($C_6H_5{-}SO_2{-}NHCH_3$) soluble in aqueous base?

17.16 Suggest a method for converting butyric acid into each of the following compounds. Any carbon atoms in the product that do not originate from butyric acid must be derived from a one-carbon reactant. Write the most direct procedure available.
(a) Dibutylamine
(b) *N,N*-Dimethyl-1-butylamine
(c) *N*-Propylbutylamine
(d) *N,N,N*-Trimethyl-1-butylammonium iodide

17.17 Consider Wilstätter's synthesis of cyclooctatetraene (Figure 17.2). What factor seems to govern the orientation in each of the Hofmann eliminations?

17.18 Hofmann's structure determination of piperidine was based on the fortunate misassignment of the elimination product derived from 5-(*N,N*-dimethylamino)-1-pentene. If he had correctly assigned the structure of the diene as 1,3-pentadiene (without realizing that isomerization had occurred), what structure (or structures) would he have proposed for piperidine?

17.19 Complete the following reactions:

(a) $C_6H_5CH_2{-}NH{-}CH_2CH_2C_6H_5 \xrightarrow[\text{2. } Ag_2O,\ H_2O \quad \text{3. pyrolysis}]{\text{1. excess } CH_3I} ?$

(b) $CH_3{-}CH(CH_3){-}CH_2{-}CH_2{-}OH \xrightarrow{?} CH_3{-}CH(CH_3){-}CH_2{-}NH_2$

(c) $CH_3{-}CH(CH_3){-}CH_2{-}CH_2{-}OH \xrightarrow{?} CH_3{-}CH(CH_3){-}CH_2{-}CH_2{-}CH_2{-}NH_2$

(d) $CH_3{-}CH(CH_3){-}CH_2{-}CH_2{-}OH \xrightarrow{?} CH_3{-}CH(CH_3){-}CH_2{-}CH_2{-}NH_2$

(e) $CH_3{-}CH_2{-}C(=O){-}CH_3 \xrightarrow{?} CH_3{-}CH_2{-}CH(CH_3){-}NH{-}CH_2{-}CH_3$

17.20 **A–D** are all different compounds. On the basis of the information provided, deduce the structures of compounds **A–C**.

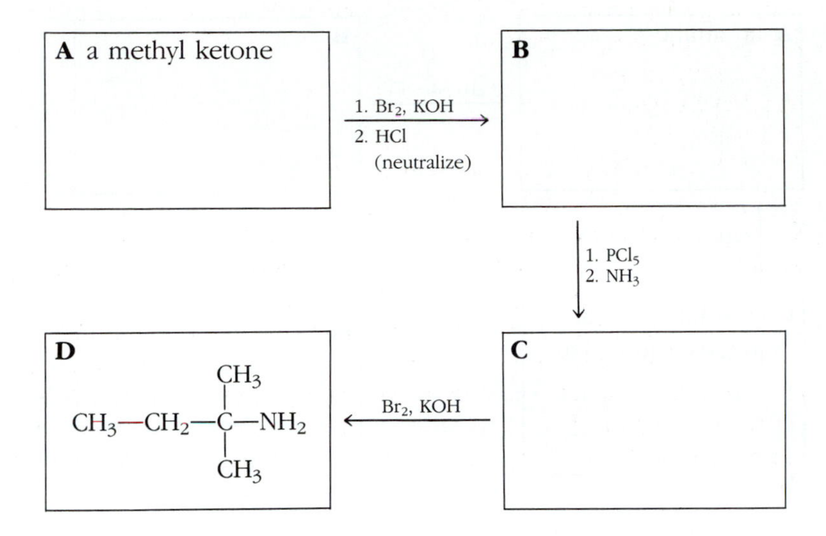

17.21 Write a detailed mechanism for each of the reactions in Problem 17.20.

17.22 **A–G** are all different compounds. On the basis of the information provided, deduce the structures of compounds **A–E.**

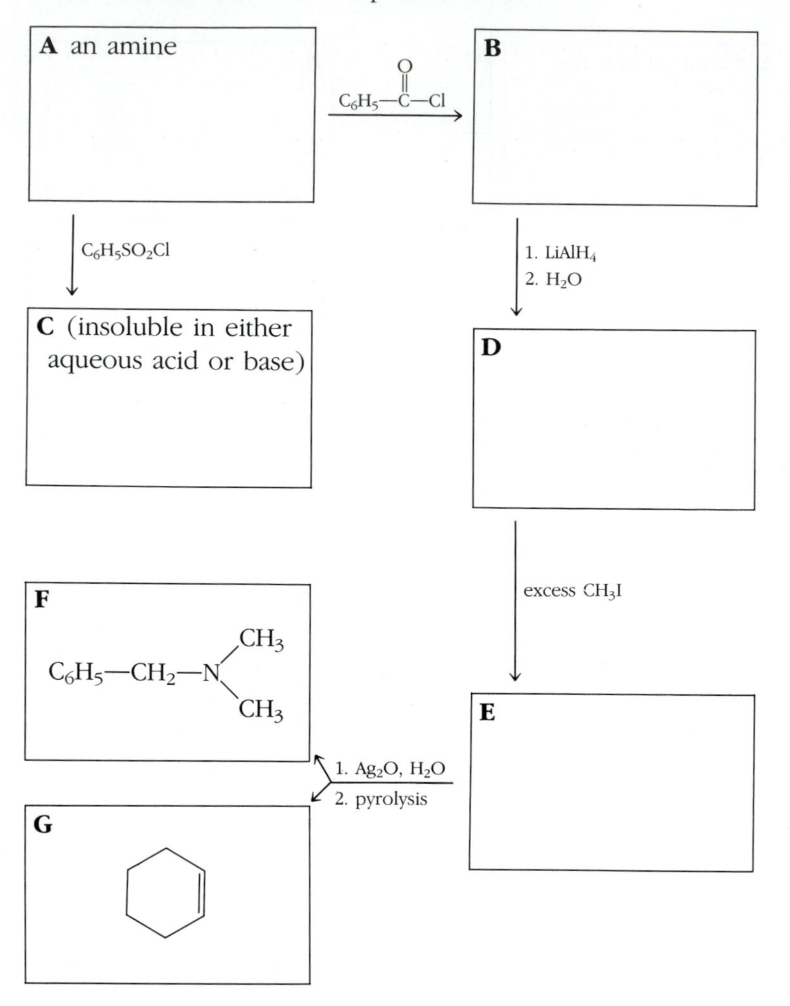

17.23 **A–C** are all different compounds. On the basis of the information provided, deduce their structures.

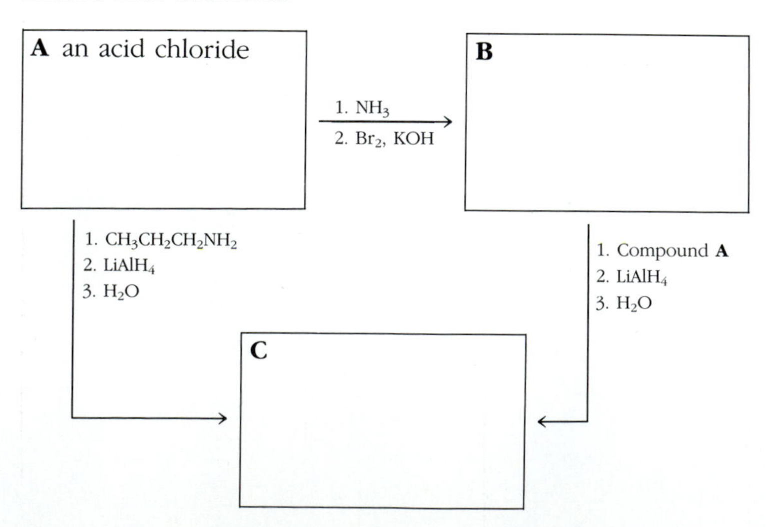

17.24 Draw the intermediate compounds formed for each of the reactants in Problem 17.23.

17.25 **A–H** are all different compounds. On the basis of the information provided, deduce the structures of compounds **A–F.**

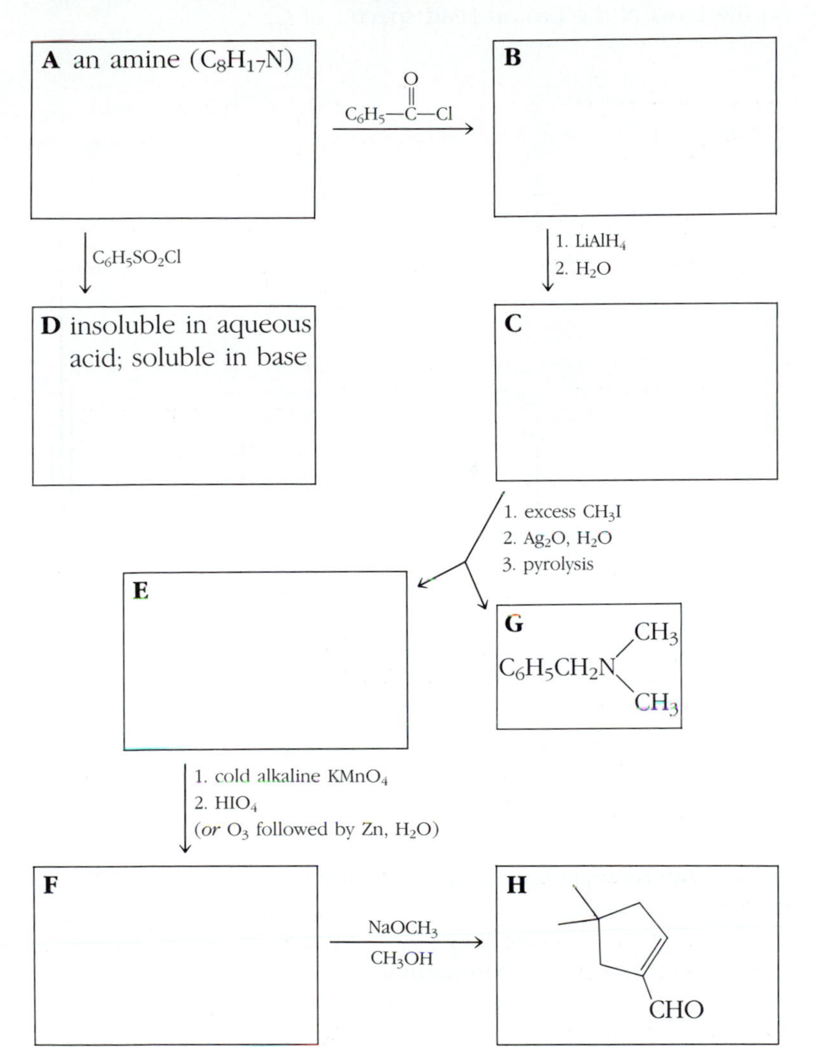

17.26 Compound **A** ($C_8H_{11}N$) is insoluble in water, but it dissolves readily in 1*N* HCl. The reaction of compound **A** with tosyl chloride yields compound **B,** which is insoluble in both aqueous acid and aqueous base. When compound **A** is treated with excess ethyl iodide and then subjected to a Hofmann elimination, compound **C** is formed. Deduce the structures of **A–C** on the basis of the proton NMR spectra of **C.**

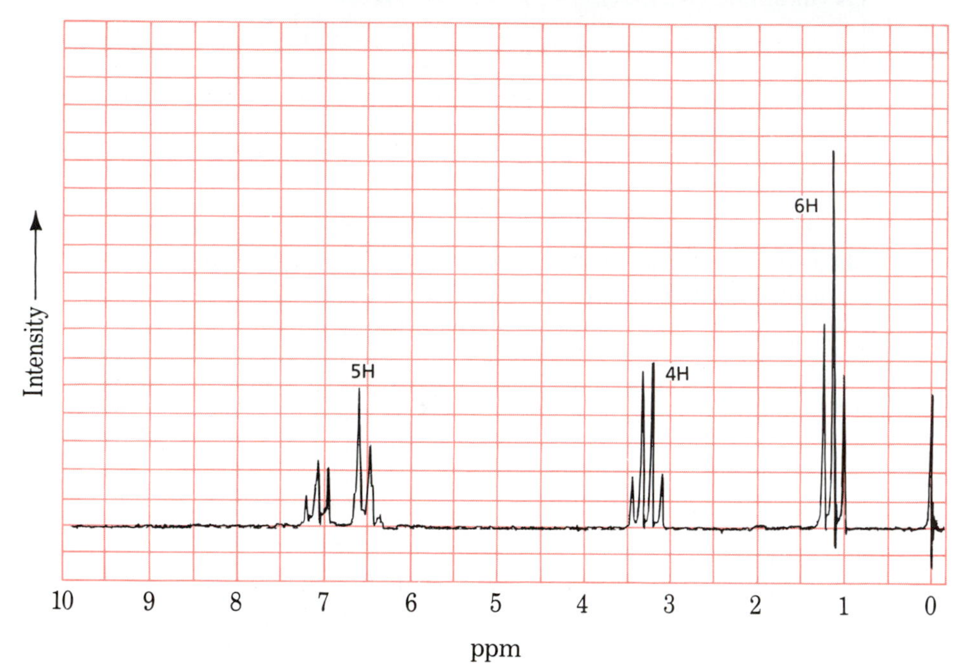

Figure 17.4 Proton NMR Spectrum of Compound C (Problem 17.26).

17.27 Explain the trends in the K_a values for the conjugate acids of the following amines:

Aniline	2×10^{-5}
3-Nitroaniline	3×10^{-3}
4-Nitroaniline	1×10^{-1}

CHAPTER 18

AROMATIC SUBSTITUTION REACTIONS

Aromatic compounds play many important roles in modern chemistry. Synthetic rubbers (elastomers) are used extensively in the tire industry, and *styrene* is a major starting compound for these materials. Synthetic fibers, such as polyesters of the *terephthalic acid* subunit, have helped to revolutionize textile and clothing manufacturing. Aromatic subunits are also found in a variety of other polymers. *Phenol,* for example, is the starting material for the class of plastics called phenolic resins.

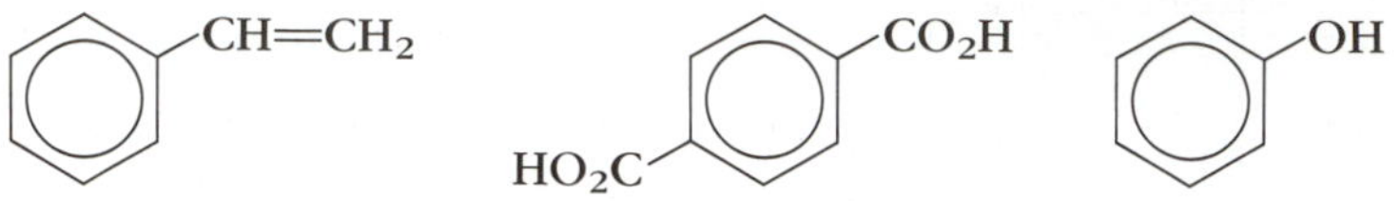

C_6H_5–CH=CH$_2$ HO$_2$C–C_6H_4–CO$_2$H C_6H_5–OH

Styrene   Terephthalic acid   Phenol

Most of the aromatic compounds that are important to the chemical industry can be synthesized from benzene, toluene, ethylbenzene, and the isomeric xylenes. These hydrocarbons were originally obtained from coal tar, but petroleum has replaced coal as the most economic source of simple aromatic hydrocarbons. Crude petroleum consists almost entirely of aliphatic hydrocarbons, but substantial quantities of aromatic compounds are generated in the catalytic reforming process. These simple aromatic hydrocarbons must then be converted to more complex molecules, and new functional groups are introduced via *substitution reactions of the aromatic ring*. In this chapter we will return to electrophilic substitution (Sections 10.7, 10.8). We will evaluate substituent effects on these reactions and show you how new carbon-carbon bonds can be formed via electrophilic substitution. We will also discuss other mechanistic pathways for substitution of the aromatic ring and show you some important methods for modifying the side chains of aromatic compounds.

18.1 THE MECHANISM OF ELECTROPHILIC SUBSTITUTION

We have already discussed the mechanism of electrophilic substitution reactions in Sections 10.7 and 10.8, but we will briefly review the basic process in general terms. An electrophile can interact with the π electrons of the aromatic ring to form an intermediate carbocation. This is shown in the following equation, where the electrophile is generalized as E^+.

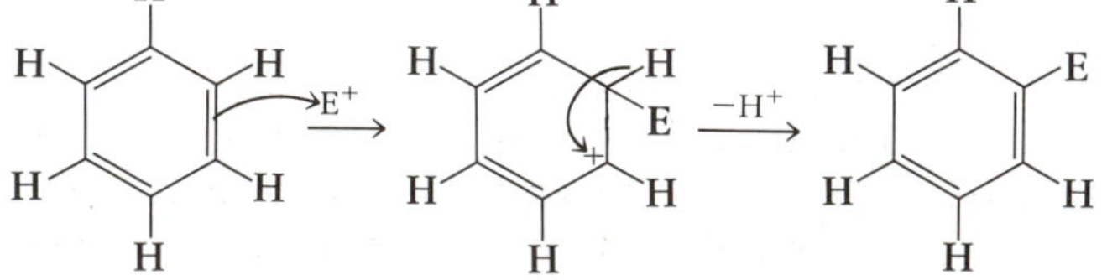

$$C_6H_6 \xrightarrow{E^+} [C_6H_6E]^+ \xrightarrow{-H^+} C_6H_5E$$

Proton transfer then generates a substituted benzene derivative in which E has replaced a hydrogen.

From your earlier reading (Section 10.9), you know that the positive charge will not be localized on a single carbon but will be delocalized around the ring. In the following section you will see how resonance theory can show which specific carbons of the ring carry the positive charge, and this will provide you with a powerful method for predicting the reactivities of aromatic compounds.

EXERCISE 18.1

Write a detailed mechanism for each of the following electrophilic substitution reactions of benzene: (a) nitration, (b) bromination, (c) sulfonation.

18.2 SUBSTITUENT EFFECTS IN ELECTROPHILIC SUBSTITUTION REACTIONS

When an electrophilic substitution reaction is carried out with a *substituted* benzene, two features of the reaction must be considered. First, what effect does the substituent have on reactivity, that is, on the *rate of reaction?* Does the substituted compound react more rapidly or more slowly than benzene itself under the same conditions? This is an important question, because reaction conditions such as temperature may have to be adjusted in order to compensate for changes in reactivity. Second, what effect does the substituent have on the relative reactivities of the various carbons of the benzene ring? We did not need to consider this in Section 10.8, because all six carbons of benzene are equivalent and reaction at any of the six yields the identical product. But reaction of a substituted benzene would yield different products depending upon which carbon atom is attacked by the electrophile. What will be the *orientation* of the newly introduced substituent with respect to the original substituent?

The nitration of toluene, for example, could produce the isomeric *ortho, meta,* and *para* isomers. Toluene has two positions that are *ortho* and two positions that are *meta* to the methyl group, but there is only a single *para* position. On a purely statistical basis, substitution of toluene should therefore afford 40% each of the *ortho* and *meta* products and only 20% of the *para* isomer. But the actual product distribution is quite different:

CH_3 CH_3 CH_3 CH_3

$\xrightarrow{NO_2^+\ BF_4^-}$ NO_2 + NO_2 + NO_2

65% 3% 32%

Product distribution

The *ortho* and *para* isomers are formed in the "expected" ratio of 2:1, but only a very small fraction of the total product is *m*-nitrotoluene.

Nitration of the fluorinated analog ($C_6H_5CF_3$) of toluene gives a quite different product distribution. The *ortho* and *para* products are again formed in a 2:1 ratio, but the *m*-nitro isomer constitutes a large majority of the total product.

CF_3 CF_3 CF_3 CF_3

$\xrightarrow{NO_2^+\ BF_4^-}$ NO_2 + NO_2 + NO_2

8% 88% 4%

Why do these substituents have such a dramatic influence on the product ratios? The influence of substituents in electrophilic substitution can be analyzed in terms of the same effects we have considered previously (Sections 4.7, 4.8, 10.9, 13.5) for other reactions: *inductive effects, resonance effects,* and *steric effects* on the intermediate carbocations.

Electron-Withdrawing Groups

Inductive Effects The inductive effect of a substituent is directly related to its electronegativity, and nearly all substituents other than simple alkyl groups exert inductive effects that are electron-withdrawing. Examples of such substituents are groups attached directly via an electronegative atom (nitrogen, oxygen, halogen), as well as groups that are attached via an unsaturated carbon atom (carbonyl, cyano). Even groups bonded via a saturated carbon atom can exert electron-withdrawing inductive effects when they are extensively substituted with electronegative atoms (e.g., the CF_3 group). Finally, some of the strongest electron-withdrawing groups are those which are positively charged (such as $—NH_3^+$).

Resonance Effects As we showed in Section 10.9, groups attached to an aromatic ring by an *unsaturated* atom can also exert an electron-withdrawing effect via *resonance.* Such substituents include nitro, carbonyl, and carboxyl groups; the resonance effect reinforces their electron-withdrawing inductive effects. Table 18.1 summarizes the different groups that exert electron-withdrawing effects. Also shown are the electron-donating groups that we will analyze in the following discussion.

Electron-Donating Groups

Very few substituents exert *inductive* effects that are electron-donating; of the commonly encountered groups, only alkyl substituents might be put into this category. There are, however, a number of substituents that can exert electron-donating *resonance effects.* Electron donation via resonance requires available electrons, and it is observed with two types of substituents: those attached via an atom with *nonbonding electrons* ($—\ddot{\underset{..}{O}}R$, $—\ddot{N}R_2$, $—\ddot{\underset{..}{Cl}}:$, etc.) and those for which the point of attachment is a carbon-carbon multiple bond with *π electrons* (vinyl and aryl groups). Do not confuse groups in the latter category with electron-withdrawing substituents containing a *polarized* multiple bond between carbon and a heteroatom.

Several types of substituents in Table 18.1 merit further comment. Substituted alkyl groups, for example, can exert either electron-withdrawing or electron-donating effects, depending upon the exact nature of the group. Simple alkyl groups are electron-donating, but this effect is reduced when the hydrogens are replaced by electronegative atoms. As a result, a trihalomethyl group is quite electron-withdrawing. Substituted alkyl groups (e.g., $—CH_2X$ or $—CHX_2$, where X is an electronegative atom or group) will also be less electron-donating than the corresponding unsubstituted alkyl group ($—CH_3$). Unfortunately, it is difficult to predict precisely where the changeover from electron-donating to electron-withdrawing will occur.

A second type of unusual behavior is illustrated by the last two types of substituents listed in Table 18.1, nonpolar multiple bonds and heteroatoms with nonbonding electrons. Each of these substituents exerts both *electron-withdrawing* induc-

TABLE 18.1 Examples of Typical Electron-Withdrawing and Electron-Donating Substituents

Type of Substituent	Electron-Withdrawing	Electron-Donating
Alkyl or substituted alkyl groups	$—CCl_3$, $—CF_3$, etc. [electron-withdrawing inductive effect]	$—CH_3$, $—CH_2R$, etc. [weak electron-donating inductive effect]
Cationic groups	$—\overset{+}{N}H_3$, $—\overset{+}{N}R_3$, etc. [strong electron-withdrawing inductive effect]	
Polarized multiple bonds	$—\overset{O}{\overset{\parallel}{C}}—$, $—C{\equiv}N$, $—\overset{+}{N}(=O)O^-$, etc. [electron-withdrawing by both inductive and resonance effects]	
Nonpolar (carbon-carbon) multiple bonds	$—CH{=}CH_2$, $—CH{=}CHR$, —(phenyl), etc. [weak electron-withdrawing inductive effect, electron-donating resonance effect]	
Heteroatom with nonbonding electrons	$—\ddot{\underset{..}{Cl}}:$, $—\ddot{\underset{..}{Br}}:$, $—\ddot{\underset{..}{O}}H$, $—\ddot{\underset{..}{O}}R$, $—\ddot{N}H_2$, $—\ddot{N}R_2$, etc. [electron-withdrawing inductive effect, electron-donating resonance effect]	

tive and *electron-donating* resonance effects. Which will win out? Ordinarily, the resonance effect predominates, and the net effect of hydroxyl, amino, and alkoxy groups is *electron-donating.* But for the halogens the two effects are comparable (as we illustrated previously in Table 10.5), and the electron-withdrawing inductive effect slightly outweighs the electron-donating resonance effect. The net effect of a halo substituent is therefore slightly *electron-withdrawing.* You can interpret this as a decreased ability of the halogens to form double bonds to carbon atoms according to the following resonance description:

$$\overset{+}{\rangle}C-\ddot{\underset{..}{Cl}}: \longleftrightarrow \rangle C=\ddot{\underset{..}{Cl}}^{+}$$

The ability of the halogens to stabilize an electron-deficient carbon by resonance decreases steadily along the series F, Cl, Br, I. Only in the case of fluorine do the nonbonded electrons occupy 2*p* orbitals; the *p* orbital on chlorine, for example, is a 3*p* orbital. The overlap between a carbon 2*p* orbital and a *p* orbital on the other halogens is much less effective. In contrast to the halogens, oxygen and nitrogen can readily form double bonds to carbon, and they have powerful electron-donating resonance effects. You will see further examples of these substituent effects later in this section.

Substituent Effects in Rate of Electrophilic Substitution

A study of the effect of substituents on the *rate* at which aromatic compounds undergo electrophilic substitution provides you with an excellent way to understand how these substituent effects operate. The rate of electrophilic substitution is usually governed by formation of the intermediate carbocation, and it is here that substituent effects on stability are greatest. (Of course rate measurements actually reflect the energy difference between reactant and *transition state*, but we can utilize the common approximation that the transition state closely resembles the actual carbocation.)

Any effect that is *electron-withdrawing* will destabilize the carbocation, the reaction will be energetically more difficult, and the rate will *decrease:*

Substituent Effects on Electrophilic Substitution

$$X-C_6H_5 \xrightarrow{E^+} \left[X-C_6H_5(H)(E)^{+}\right] \longrightarrow X-C_6H_4-E$$

An *electron-withdrawing* substituent *destabilizes* the carbocation and *decreases* the rate of reaction.

Electron-donating effects, in contrast, will stabilize the intermediate carbocation, the reaction will be energetically easier, and the rate will *increase.*

Substituent Effects on Electrophilic Substitution

X—C₆H₄—H $\xrightarrow{E^+}$ [X—(+)—H/E] ⟶ X—C₆H₄—E

An *electron-donating* substituent *stabilizes* the carbocation and *increases* the rate of reaction.

Figure 18.1 summarizes the substituent effects of a variety of groups on the rate of bromination of some benzene derivatives. This plot shows the relative electron-donating and -withdrawing abilities of various substituents. Moreover, the effects on reaction rate are enormous. Phenol does not react just slightly faster than benzene, it reacts more rapidly by a factor of 10^{12}. Similarly, nitrobenzene is less reactive than benzene by a factor of 10^9. You can place these huge differences in reactivity in perspective by comparing anisole ($—OCH_3$) and benzoic acid ($—CO_2H$). If a specific quantity of anisole were to react within a time period of a thousandth of a second, how long would it take for benzoic acid to react to the same extent? Remarkably, the answer is 3000 years! Clearly the different benzene derivatives exhibit very different reactivity toward electrophilic substitution. Under conditions in which a reactive benzene derivative, such as anisole, undergoes electrophilic substitution quite easily, another compound, such as nitrobenzene, may be totally unreactive.

Figure 18.1 can also be used to provide approximate numerical comparisons of substituent effects. As an example, you could estimate that the electron-*donating* effect of a methoxy group (relative rate for $C_6H_5OCH_3 = 10^9$) is about twice the electron-*withdrawing* effect of a carboxyl group (relative rate for $C_6H_5CO_2H = 10^{-4.7}$). You can therefore predict that a compound with both these substituents (i.e., one of the isomeric anisic acids) will be more reactive than benzene.

The shaded area in Figure 18.1 serves as a benchmark for comparing reactivities in electrophilic substitution reactions. All the compounds above this region are more reactive than benzene, and the substituents are described as *activating*. These are precisely the same substituents in Table 18.1 that were characterized as electron-*donating*. The halogens are the only electron-donating groups that do not result in enhanced reactivity, but we have already shown that their electron-donating resonance effect is counteracted by their electron-withdrawing inductive effect. The halobenzenes are slightly less reactive than benzene itself, but we have included them with benzene as a group of compounds that exhibit "normal" reactivity. In other words, they are neither highly activated nor highly deactivated. All the points below the shaded area of Figure 18.1 are substantially less reactive than benzene, and the substituents are called *deactivating*. The deactivating substituents are all groups that were classified in Table 18.1 as electron *withdrawing*. Hence we can summarize all these reactivity patterns by stating that electron-donating substit-

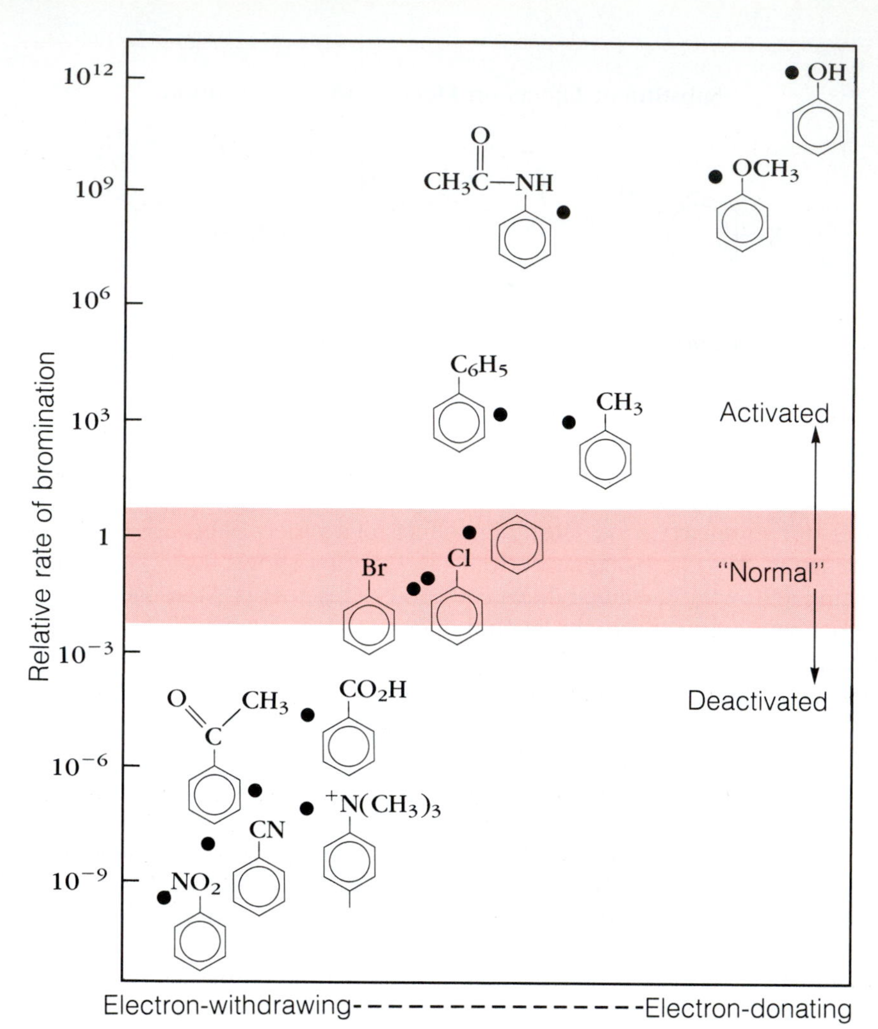

Figure 18.1 Substituent effects on the rate of electrophilic substitution. The compounds are ordered according to the electron-donating or electron-withdrawing abilities of their substituents.

uents will activate an aromatic ring toward electrophilic attack, while electron-withdrawing groups will have a deactivating effect.

Substituent Effects on Products of Electrophilic Substitution

The reactivity arguments we just presented now allow us to explain the different product distributions that we showed for the nitration of toluene and its trifluoromethyl analog. Toluene is activated toward electrophilic substitution by the electron-donating methyl group, but not all positions on the ring are activated equally. As we first stated in Section 10.7, the positive charge in the intermediate carbocation is delocalized, not to all atoms of the ring but only to those carbons that are *ortho* and *para* to the site of attack by the electrophile. This is further illustrated in Figure 18.2.

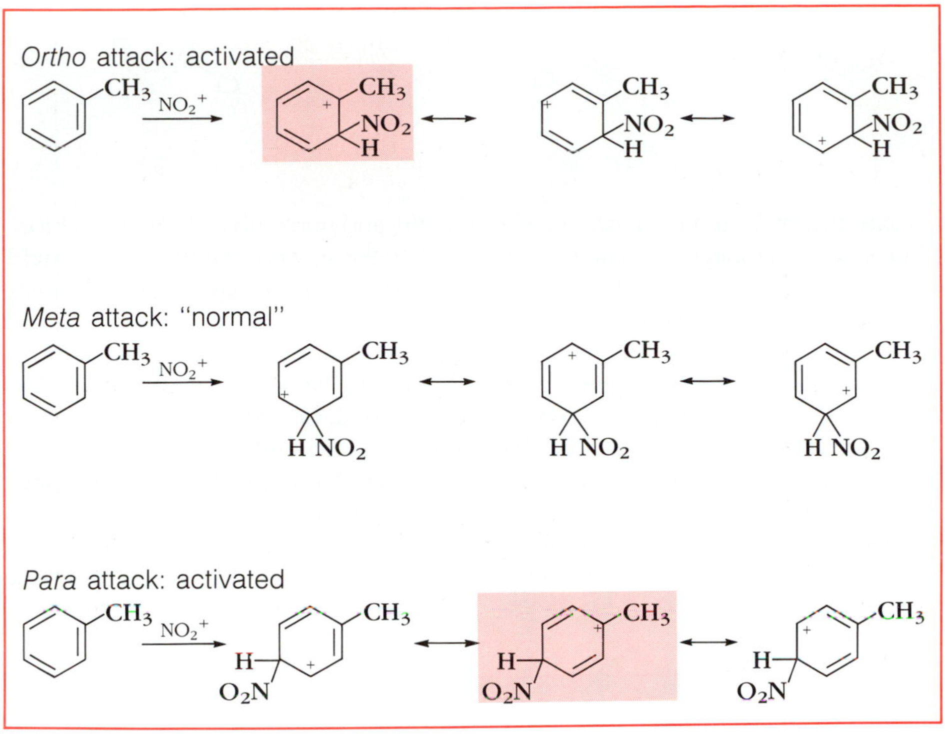

Ortho attack: activated

Meta attack: "normal"

Para attack: activated

Figure 18.2 Electrophilic attack on an activated benzene derivative. Only when attack occurs at the *ortho* or *para* positions can the electron-donating substituent directly stabilize the cation (shaded structures). Consequently, nitration occurs preferentially *ortho* and *para* to the activating group.

The equations in Figure 18.2 show that the influence of a substituent on a benzene ring is maximized when attack occurs at the *ortho* or *para* position, because the positive charge can reside on the same carbon atom as the substituent. For an activated benzene derivative stabilization of the cation lowers the energy of the transition states for attack at these positions, so the *ortho* and *para* products are formed at increased rates. Attack at the *meta* position produces a carbocation having no direct interaction with the electron-donating substituent. As a result, *meta* attack proceeds at a "normal" rate, but the increased rate of *ortho* and *para* attack leads to a predominance of those products.

Other electron-donating substituents also result in high proportions of *ortho* and *para* products, and this is illustrated by the following examples:

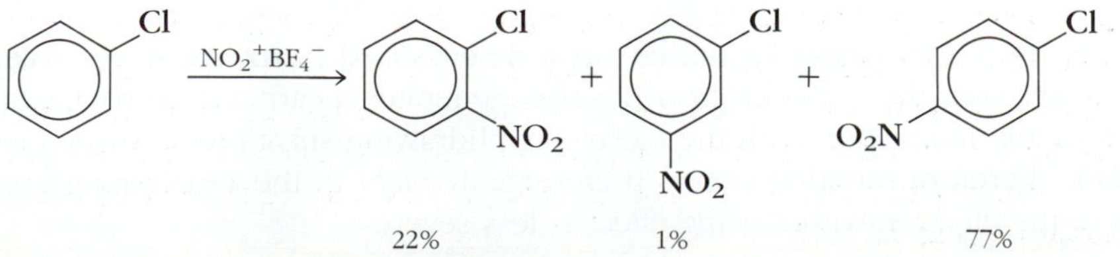

$NO_2^+BF_4^-$

22% 1% 77%

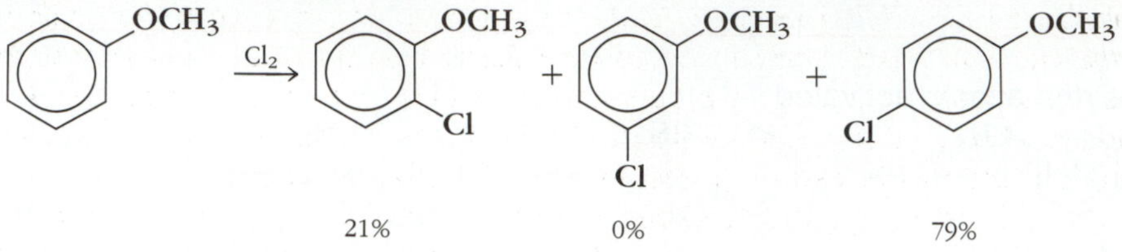

21% 0% 79%

Notice that there is a large preference for *ortho* and *para* substitution with chlorobenzene even though it is slightly deactivated. In the second example anisole yields no significant amount of the *meta* isomer, a result that is typical of such highly activated benzene derivatives.

Two other points are noteworthy in the preceding examples. First, the chlorination of anisole does not require a catalyst, and this behavior is also observed for other highly activated aromatic compounds. Second, both these examples show a ratio of *ortho* to *para* substitution that is much less than the statistical ratio of 2:1. Preferential *para* substitution is frequently observed, but it is difficult to anticipate with certainty when that will occur. To some extent it is a result of steric effects, which we will discuss later in this section.

The situation for deactivating substituents such as —CF_3 is quite the reverse, as shown by the equations in Figure 18.3. Electron-withdrawing inductive effects de-

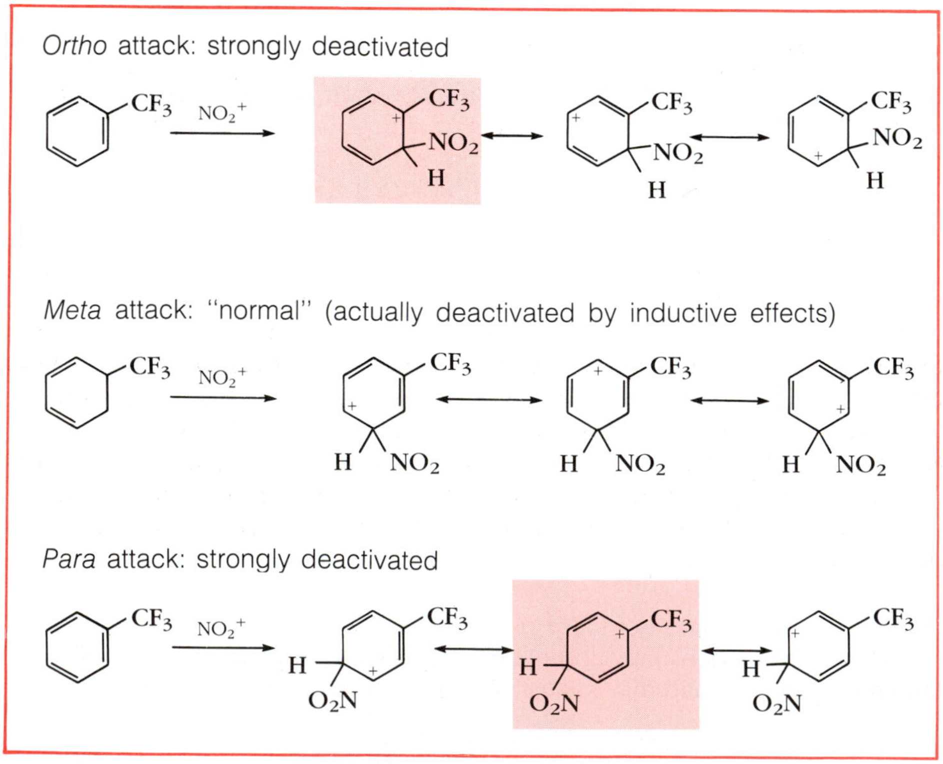

Ortho attack: strongly deactivated

Meta attack: "normal" (actually deactivated by inductive effects)

Para attack: strongly deactivated

Figure 18.3 Electrophilic attack on a deactivated benzene derivative. Attack at either the *ortho* or *para* position generates a carbocation having an unfavorable interaction with the electron-withdrawing substituent (shaded structures). Therefore nitration occurs preferentially *meta* to the deactivating group, where the electron-withdrawing effect is less severe.

stabilize a cation at any position in the ring, but the effect is particularly bad for *ortho* and *para* attack, because the positive charge resides on the same carbon that has the electron-withdrawing group. Consequently, the transition states for *ortho* and *para* substitution are more destabilized than the transition state for *meta* substitution, and this leads to a predominance of the *meta* product.

Other deactivating groups also result in preferential *meta* substitution, but some *ortho* and *para* products should usually be expected as well. This is illustrated by the following reactions of nitrobenzene and benzoic acid.

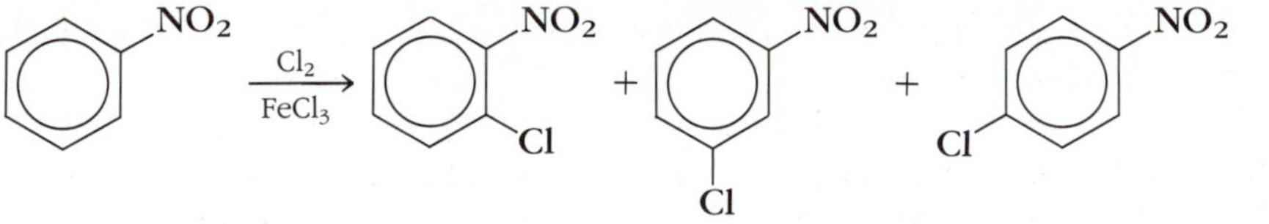

NO_2 $\xrightarrow[FeCl_3]{Cl_2}$ NO_2, Cl (1%) + NO_2, Cl (72%) + Cl, NO_2 (4%)

NO_2 $\xrightarrow{NO_2^+BF_4^-}$ NO_2, NO_2 (8%) + NO_2, NO_2 (90%) + O_2N, NO_2 (2%)

CO_2H $\xrightarrow{H_2SO_4/SO_3}$ CO_2H, SO_3H (0%) + CO_2H, SO_3H (86%) + HO_3S, CO_2H (14%)

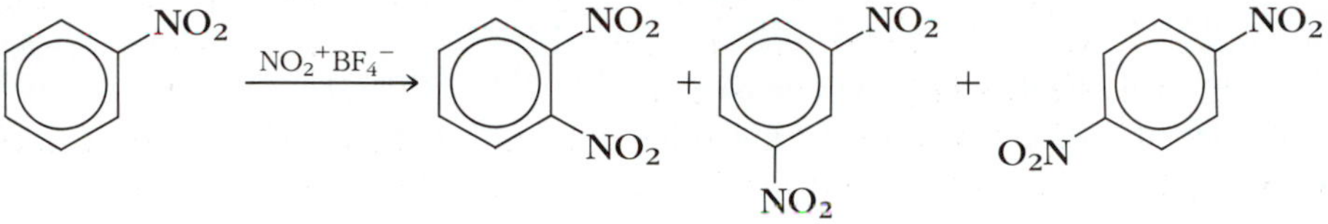

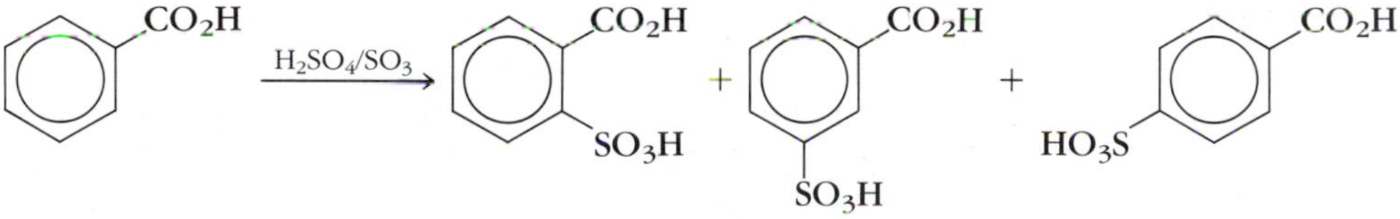

The mixtures of isomeric products that you would expect from an aromatic substitution reaction are not always observed when the reactions are actually carried out in the laboratory. Sometimes one isomer is formed to the almost complete exclusion of the others, but more commonly the minor constituents are simply removed during purification of the crude product. The following equations illustrate the common practice of reporting isolated yields of individual compounds after they have been separated from the crude reaction mixture. The isomers that are not shown may have been formed, but they were not isolated and purified. As you would expect, the *meta* isomer is the major product in reactions of deactivated benzenes.

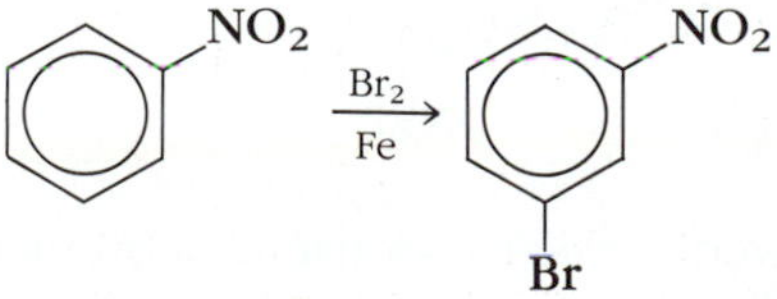

NO_2 $\xrightarrow[Fe]{Br_2}$ NO_2, Br 60–75%

CO_2CH_3 $\xrightarrow{HNO_3,\ H_2SO_4}$ CO_2CH_3, NO_2 81–85%

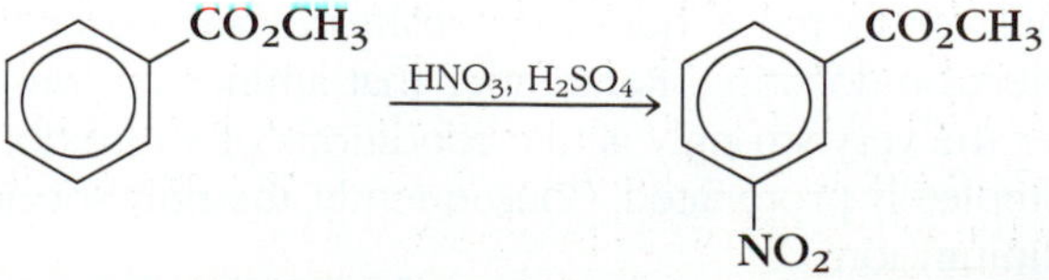

On the other hand, the *para* isomer usually predominates in the reaction of an activated benzene.

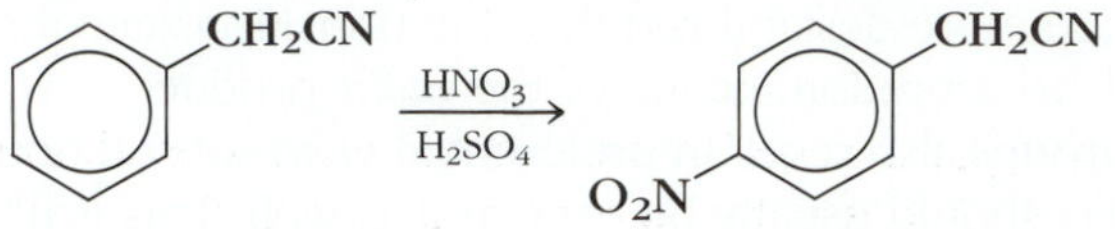

CH_2CN $\xrightarrow[H_2SO_4]{HNO_3}$ O_2N CH_2CN 50–54%

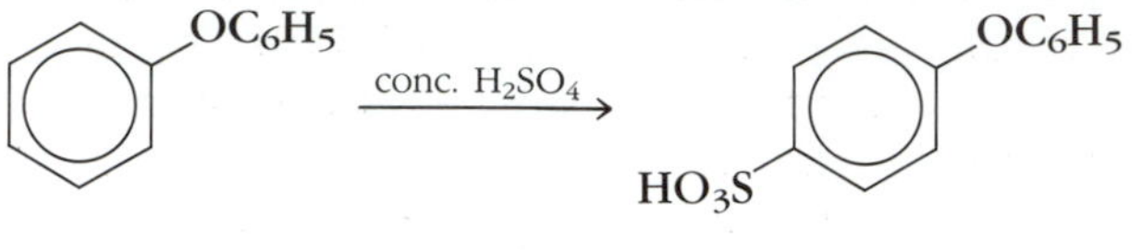

OC_6H_5 $\xrightarrow{\text{conc. } H_2SO_4}$ HO_3S OC_6H_5 93%

OH $\xrightarrow{Br_2}$ Br OH 80–84%

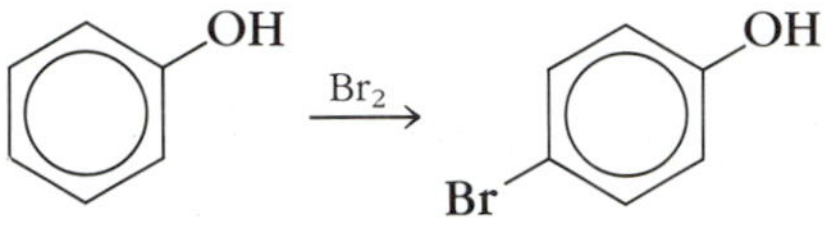

The powerful electron-donating effect of a phenolic hydroxyl group is illustrated by the last example, in which bromination proceeds readily in the absence of a catalyst.

Aniline Derivatives as Exceptions

The amino group is an even more powerful electron donor than the phenolic hydroxyl group, as is illustrated by the ease with which aniline undergoes bromination. No catalyst is needed and three bromine atoms are introduced despite the fact that each successive bromination decreases the reactivity of the ring. Notice that substitution occurs at the positions *ortho* and *para* to the amino group.

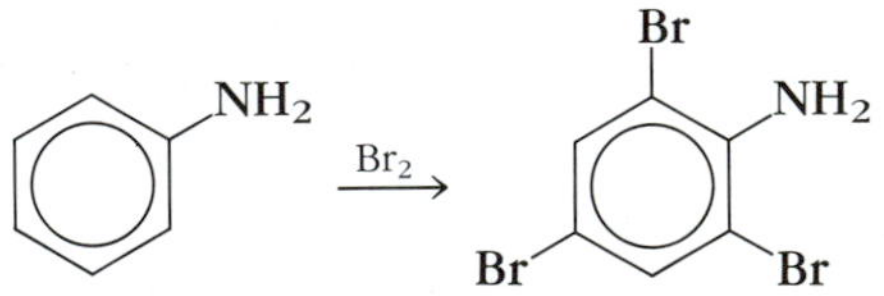

NH_2 $\xrightarrow{Br_2}$ Br, NH_2, Br, Br 89–95%

The powerful activating effect of the amino group might seem to be at odds with the following reaction of *N,N*-dimethylaniline:

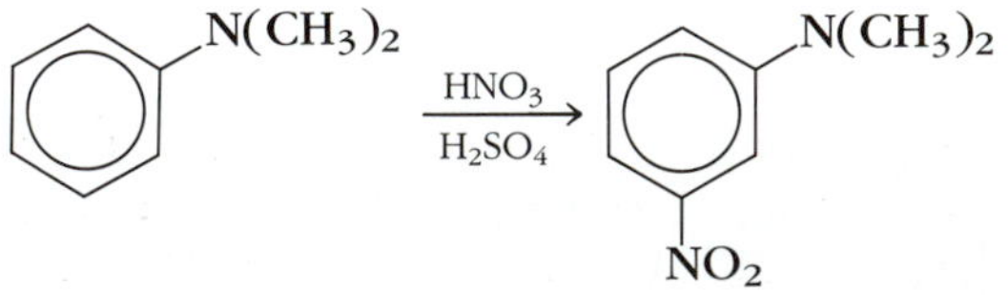

$N(CH_3)_2$ $\xrightarrow[H_2SO_4]{HNO_3}$ $N(CH_3)_2$, NO_2 56–63%

If amino groups are such powerful electron-donating substituents, why is only a single nitro group introduced? Furthermore, why does substitution occur at the *meta* position? The answer to these questions reaffirms our claim that amino groups are strongly electron-donating. Remember that amines are *bases,* so they react with acids. Under the very strongly acidic conditions of a nitration reaction, the amino group is completely protonated. Consequently, the only species present in solution is the anilinium ion.

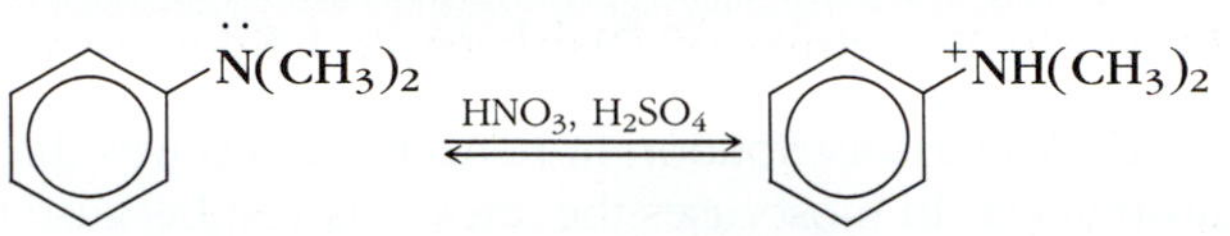

$$C_6H_5\ddot{N}(CH_3)_2 \xrightleftharpoons{HNO_3,\ H_2SO_4} C_6H_5\overset{+}{N}H(CH_3)_2$$

Therefore the species that actually undergoes nitration is highly deactivated, and substitution at the *meta* position is in full accord with all our previous discussions.

The ability of amino groups to act as powerful activating groups is observed only when strongly acidic conditions (and formation of the ammonium salts) are avoided. Many electrophilic substitution reactions require highly acidic conditions, however, so reactions of aniline and its derivatives are usually carried out with the aid of *blocking groups*. We will discuss this further in Sections 18.4 and 18.6.

Steric Effects in Electrophilic Substitution Reactions

Earlier in this section we argued that attack at the *ortho* position of a monosubstituted benzene is favored over reaction at the *para* position by a statistical factor of 2. This 2:1 ratio has been observed for nitration of both toluene and its CF_3 analog, but many other electrophilic substitution reactions give different results. Nitration of *tert*-butyl benzene, for example, affords the following isomer distribution:

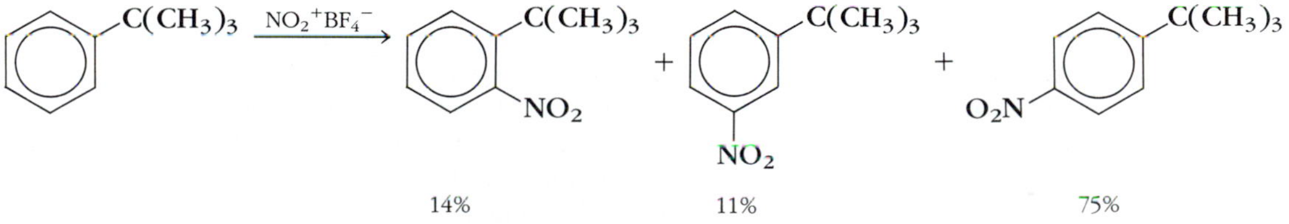

$$C_6H_5C(CH_3)_3 \xrightarrow{NO_2^+BF_4^-} o\text{-}O_2NC_6H_4C(CH_3)_3 \ (14\%) + m\text{-}O_2NC_6H_4C(CH_3)_3 \ (11\%) + p\text{-}O_2NC_6H_4C(CH_3)_3 \ (75\%)$$

The low *ortho/para* ratio can be explained on the basis of steric hindrance to *ortho* attack.

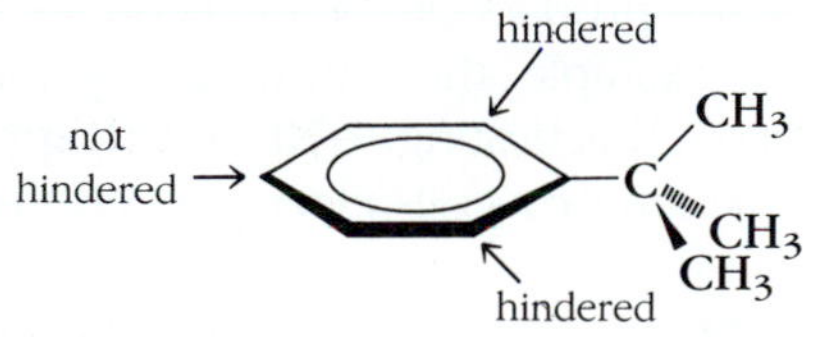

hindered
not hindered →
CH_3
C
CH_3
CH_3
hindered

The environment at the *para* position is "normal," but *ortho* substitution would generate unfavorable steric interactions between the bulky *tert*-butyl group and the incoming nitronium ion. As a result, the rate of reaction at the *ortho* positions is decreased and a lower *ortho/para* ratio is observed. The *ortho/para* ratio cannot be described strictly in terms of steric effects, because other structural features also contribute. As a general rule, activated benzene derivatives afford product mixtures in which the *para* isomer predominates, and you can see this in most of the preceding examples. Significant amounts of *ortho* product will sometimes be formed, however, and it is difficult to know when this will occur. For the purposes of answering problems in this book you should predict that both *ortho* and *para* products will be formed in the reaction of an activated benzene.

Substituent Effects in Reactions of Disubstituted Benzenes

Can you predict the product distribution in benzene derivatives that already have two or more substituents? In most cases the answer is yes, because the individual substituent effects are additive. As an example, bromobenzene reacts with a mixture of nitric and sulfuric acid under vigorous conditions to give a *dinitro* product. The first step is known to yield predominantly *p*-bromonitrobenzene, so the second nitro group is introduced *ortho* to the —Br and *meta* to the —NO_2 group. This is precisely the effect that you would expect for each substituent.

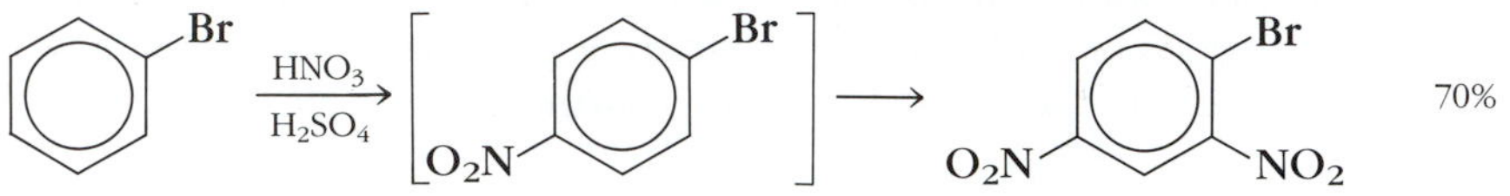

Br $\xrightarrow[H_2SO_4]{HNO_3}$ [O_2N … Br] ⟶ O_2N … Br, NO_2 70%

Similar cooperative effects operate in the nitration of salicylic acid. The —OH group is *ortho, para*–directing, and this favors attack at the 3 and 5 positions. These same positions are favored by the *meta*-directing carboxyl group.

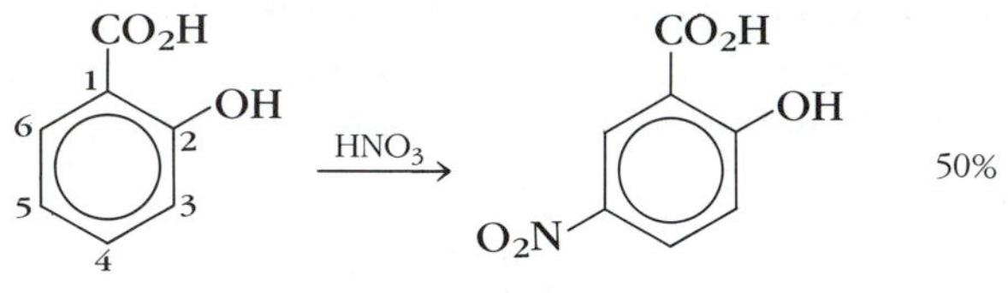

CO_2H, OH (1, 2, 3, 4, 5, 6) $\xrightarrow{HNO_3}$ CO_2H, OH, O_2N 50%

Salicyclic acid

The observed product (from reaction at C-5) reflects a preference for substitution *para* (rather than *ortho*) to the OH group. The mild reaction conditions (nitric acid without sulfuric acid) used in this reaction also demonstrate that salicylic acid is an *activated* benzene derivative. Reference to Figure 18.1 shows that a hydroxyl group increases the nitration rate of benzene by a factor of 10^{12}, whereas a carboxyl group in contrast reduces the rate by only a factor of 10^{-5}. Consequently, the two groups together are still quite strongly activating.

What if the effects of the two substituents operate in different directions? In such cases a prediction of the major product can be made on the basis of the substituent that is more activating. For example, the following reaction is dominated by the substituent effect of the strongly activating —$NHCOCH_3$ group. The *para* position is already occupied, so reaction proceeds cleanly to give *ortho* bromination.

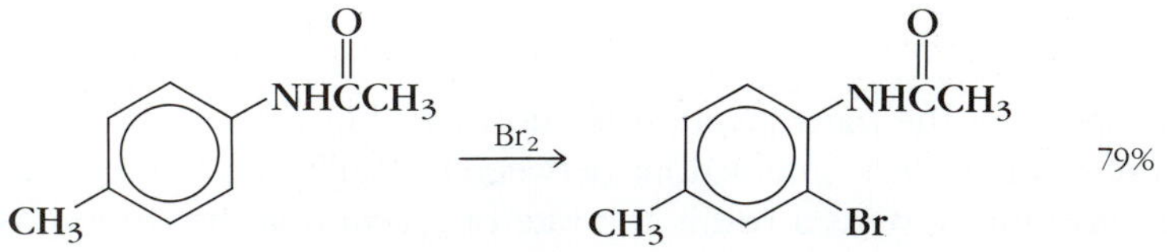

$NHCCH_3$ (C=O), CH_3 $\xrightarrow{Br_2}$ $NHCCH_3$ (C=O), CH_3, Br 79%

This reaction again shows you that a catalyst is not needed for the halogenation of a strongly activated aromatic ring.

Sulfonation and Equilibrium Effects

Most electrophilic substitution reactions afford product distributions that are relatively insensitive to temperature. Small changes in product ratios may be observed if the reaction is carried out at higher temperature, but no major changes would be expected. Sulfonation reactions provide a distinct exception to this pattern, as you can see from the reaction of toluene at 0° and at 100°C.

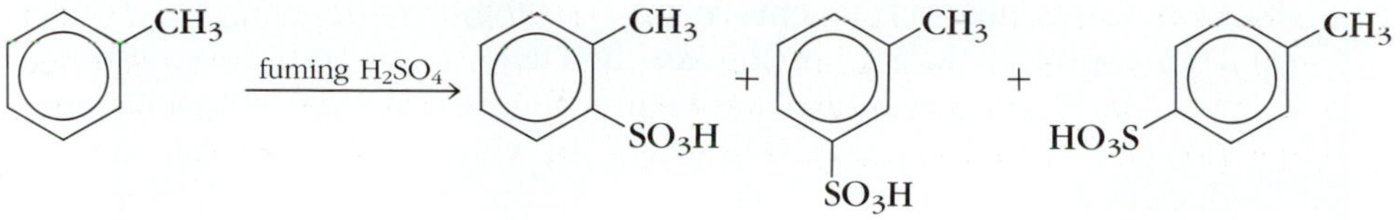

fuming H_2SO_4

	o-SO_3H	m-SO_3H	p-SO_3H
0°C	43%	4%	53%
100°C	13%	8%	79%

For most reactions the product ratios reflect the *rates* at which each product is formed (kinetic control). But this can only be true if the products are stable under the reaction conditions, and we have previously indicated (Section 10.8) that sulfonation is reversible. The behavior of toluene indicates that the reverse reaction is relatively slow at 0°C, so kinetic cotrol is observed. At elevated temperatures both forward and reverse reactions occur more rapidly, and an equilibrium is established.

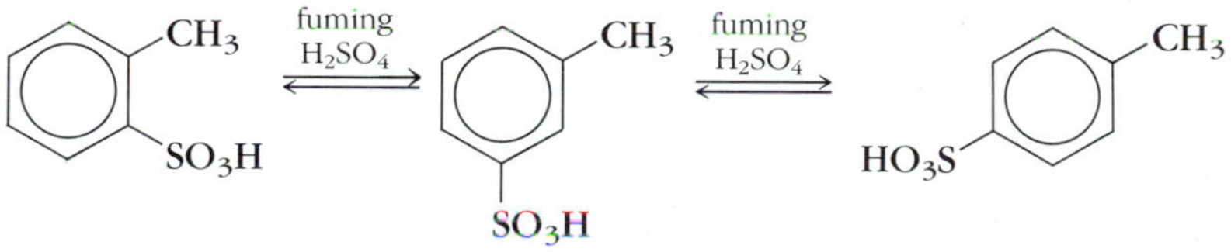

fuming H_2SO_4 ⇌ fuming H_2SO_4 ⇌

For this reason the relative amounts of *o*-, *m*-, and *p*-toluenesulfonic acid that are formed at 100°C reflect not their rates of formation but their relative stabilities. This is an example of thermodynamic control, and the predominance of the *para* product indicates that it is the most stable of the three isomeric sulfonic acids.

It is difficult to predict which of a series of isomeric reaction products would be most stable, so your predictions will be limited to kinetically controlled reactions. Sulfonation of nitrobenzene, as an example, affords the expected product of *meta* substitution in high yield.

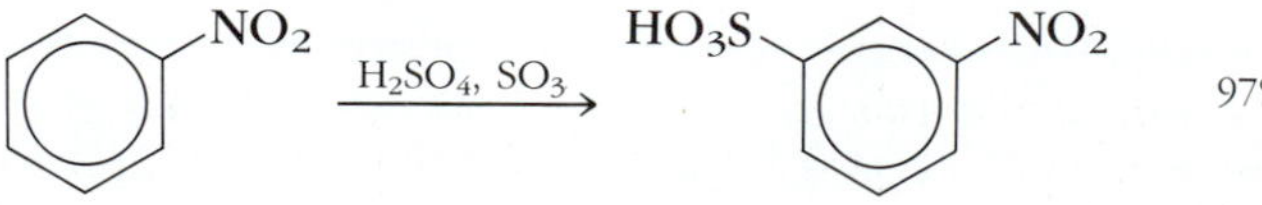

H_2SO_4, SO_3 → 97%

The following reaction also illustrates the dominance of the more powerful electron-donating group (—$NHCONH_2$) in governing the site of attack.

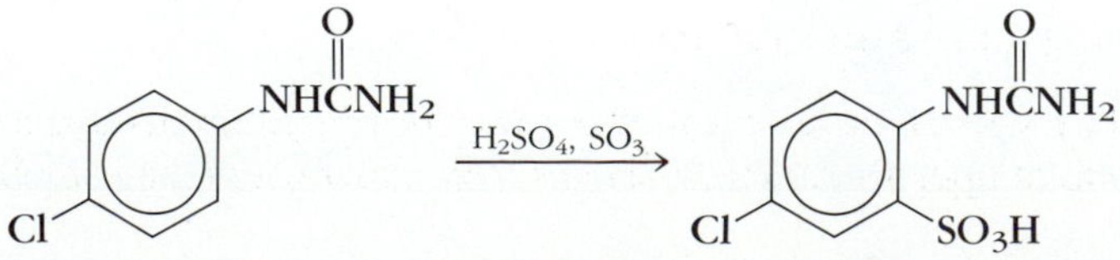

H_2SO_4, SO_3 → 75%

EXERCISE 18.2

(a) Classify each of the following as activating or deactivating substituents: $—OH$, $—SCH_3$, $—O_2C—CH_3$, $—NH_2$, $—NH_3^+$, $—Br$, $—CO_2CH_3$, $—CH_2CH_2CH_2CF_3$, $—CD_3$.
(b) Classify the same substituents as *o,p*-directing or *m*-directing.
(c) Four combinations are possible for the effects in (a) and (b): activating and *o,p*-directing; activating and *m*-directing; deactivating and *o,p*-directing; and deactivating and *m*-directing. In which of the four categories does each substituent fall? Two categories are empty (or nearly empty). Can you list other substituents that would fall into these two categories?

EXERCISE 18.3

Predict the product of monosubstitution for each of the following reactions:

(a) $C_6H_5CH_3 \xrightarrow{Br_2,\ FeBr_3}$?

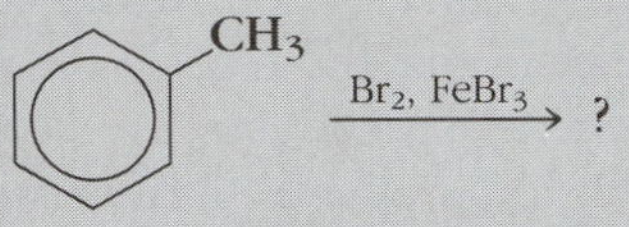

(b) $C_6H_5OCH_3 \xrightarrow{H_2SO_4,\ SO_3}$?

(c) $C_6H_5\overset{+}{N}(CH_3)_3 \xrightarrow{HNO_3,\ H_2SO_4}$?

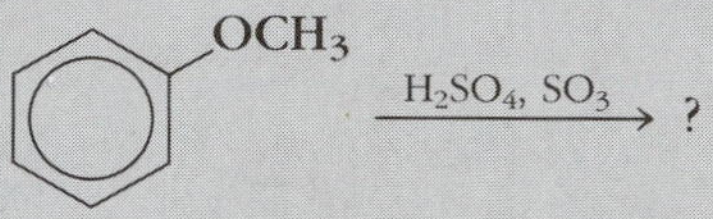

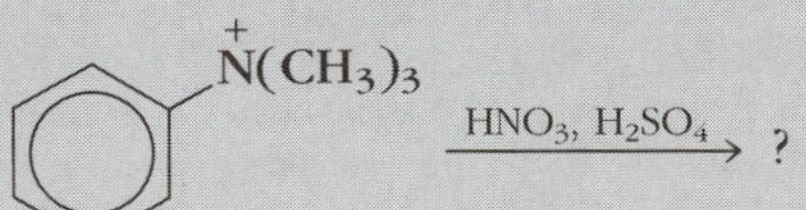

EXERCISE 18.4

Draw resonance forms for the intermediates produced by reaction of (a) $C_6H_5NO_2$ and (b) C_6H_5OH with dilute D_2SO_4 at all possible sites of reactivity. Which intermediates are most stable? Where will reaction occur preferentially?

EXERCISE 18.5

Using the quantitative changes in reactivity for the substituents shown in Figure 18.1, estimate whether each of the following would be more reactive than, less reactive than, or comparable in reactivity with benzene. What products would be expected from nitration of each?

(a) CO_2CH_3 / CH_3 (b) $NHCCH_3$ (C=O) / $^+N(CH_3)_3$ Cl^- (c) OH / Cl (d) Cl / NO_2

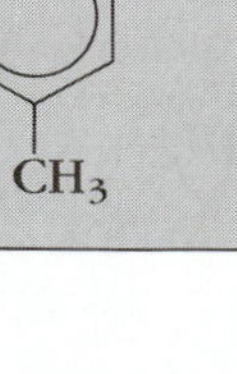

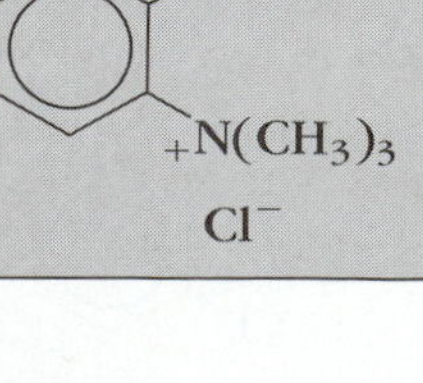

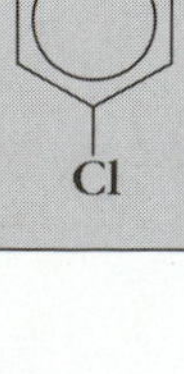

18.3 CARBON-CARBON BOND FORMATION BY ELECTROPHILIC SUBSTITUTION: THE FRIEDEL-CRAFTS REACTION

In 1877 the French chemist Charles Friedel and a visiting American colleague, James M. Crafts, discovered a new reaction. They reported that " . . . after mixing benzene with ethyl iodide and adding aluminum chloride, we observed the evolution of thick acid fumes containing hydrogen iodide and after treatment with water and distillation, we were able to isolate ethylbenzene."

$$C_2H_5I + C_6H_6 \xrightarrow{AlCl_3} C_6H_5C_2H_5 + HI$$

In addition to reactions of benzene with *alkyl* halides, they also observed its reactions with *acyl* halides to give ketones, as illustrated for benzoyl chloride in the following equation:

$$C_6H_5COCl + C_6H_6 \xrightarrow{AlCl_3} C_6H_5COC_6H_5 + HCl$$

Friedel and Crafts quickly recognized the importance of this reaction as a valuable new way of forming carbon-carbon bonds. The reaction of alkyl and acyl halides with aromatic compounds in the presence of aluminum chloride is now well known as the Friedel-Crafts reaction.

The alkylation reaction proceeds via electrophilic attack of a species that behaves as an alkyl cation. However, the aluminum chloride complex (rather than the free cation) is usually the actual electrophile, especially when a relatively unstable primary or secondary cation would be formed.

$$R\text{—}Cl \xrightarrow{AlCl_3} \overset{\delta+}{R}\text{---}Cl\text{---}\overset{\delta-}{AlCl_3} \quad (\xrightarrow{?} R^+)$$

Similarly, the electrophile in the acylation reaction can be either the aluminum chloride complex or the free cation, depending on the reaction conditions. Notice that the acyl cation is actually a fairly stable *oxonium* ion.

$$R\text{—}\overset{O}{\overset{\|}{C}}Cl \longrightarrow R\text{—}C(=\overset{\delta+}{O}\text{---}\overset{\delta-}{AlCl_3})Cl \rightleftharpoons R\text{—}C(=O)\overset{\delta+}{Cl}\text{---}\overset{\delta-}{AlCl_3} \quad (\xrightarrow{?} R\text{—}\overset{+}{C}{=}\ddot{O}: \longleftrightarrow R\text{—}C{\equiv}\overset{+}{O}:)$$

While these reactions are normally classified as electrophilic substitution of the aromatic ring, they could also be regarded as nucleophilic substitutions in which the aromatic ring acts as a nucleophile. As expected for such processes, only acyl and alkyl halides are reactive. Aryl halides and vinyl halides ordinarily do not react under the conditions of the Friedel-Crafts reaction.

After the initial studies of Friedel and Crafts it was found that many other reagents and catalysts could be used. Alkylation reactions can occur with cationic intermediates formed by protonation of alkenes or by reactions of alcohols or ethers with strong acid catalysts. Acylation reactions take place with anhydrides as well as with acid chlorides. Aluminum chloride is one of the best catalysts, but BF_3 and HF, as well as other acids, have also been used extensively. The following two reactions illustrate the use of alkenes in the alkylation reaction:

$$\text{benzene} + \text{cyclohexene} \xrightarrow{AlCl_3} \text{cyclohexylbenzene} \qquad 79\%$$

$$\text{benzene} + \text{cycloheptene} \xrightarrow{HF} \text{cycloheptylbenzene} \qquad 71\%$$

Limitations of Friedel-Crafts Alkylation

Friedel-Crafts *alkylation* has been developed extensively for industrial use, but it suffers several disadvantages that seriously limit its value as a laboratory method. One major problem is *polyalkylation;* the initial product is somewhat more reactive than the starting compound, and this favors further substitution. The severity of the problem is illustrated by the reaction of xylene with chloromethane:

$$C_6H_4(CH_3)_2 \xrightarrow[AlCl_3]{CH_3Cl} C_6H_3(CH_3)_3 + C_6H_2(CH_3)_4 + C_6H(CH_3)_5$$

Mixture of *o*, *m*, *p*-xylenes; 16%; 52%; 18%

Friedel-Crafts alkylation reactions also suffer from extensive *carbocation rearrangements.* For example, $AlCl_3$ converts 1-chloropropane to 2-chloropropane by a typical carbocation rearrangement (Section 5.9):

$$CH_3CH_2CH_2Cl \xrightarrow{AlCl_3} CH_3{-}CH(Cl){-}CH_3$$

Consequently, any attempt to alkylate an aromatic ring with 1-chloropropane leads to a mixture of the propyl and isopropyl products. Rearrangements are not limited to the alkyl substituent that is introduced; in the presence of a strong acid catalyst, an isomerization can occur which alters the location of substituents on the ring.

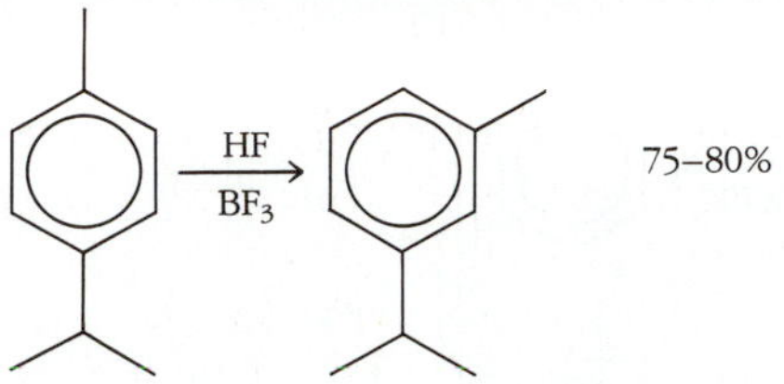

HF, BF_3 75–80%

Several mechanistic pathways can be written for the preceding rearrangement, but it probably occurs as a result of *ipso* protonation (i.e., protonation at one of the substituted carbons):

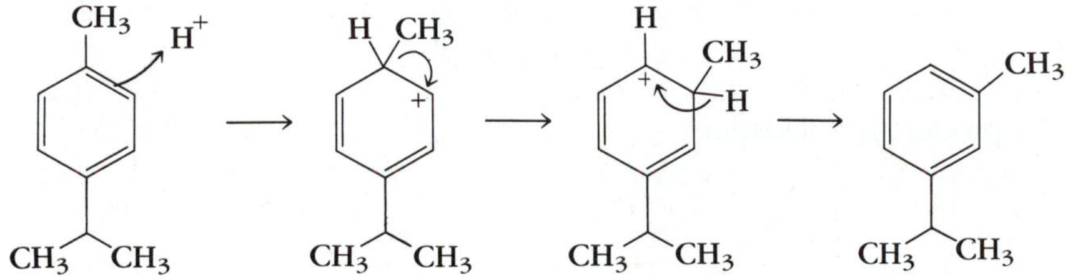

Protonation at the carbon bearing the methyl group, followed by a 1,2 methyl shift, would generate a new cation, and loss of a proton would yield the observed *meta* product. The yields of such rearrangement products are dependent upon reaction conditions such as temperature and reaction time, but it is usually quite difficult to effectively prevent their formation as by-products in Friedel-Crafts alkylations.

Friedel-Crafts Acylation

Many of the problems associated with Friedel-Crafts alkylation are not found in the acylation reaction. Because acylation introduces an electron-withdrawing group, the product is much less reactive than the starting material. This means that polyacylation is easily avoided, and a single substituent can be introduced. Isomerization of the electrophile is also avoided, because the relatively stable acyl cation is not prone to rearrangement.

Isomerization that changes the position of substituents on the ring is sometimes encountered, but Friedel-Crafts acylation is generally a reliable reaction that follows the same pattern of reactivity as other types of electrophilic substitution. The following examples illustrate the preparation of aromatic ketones by this method.

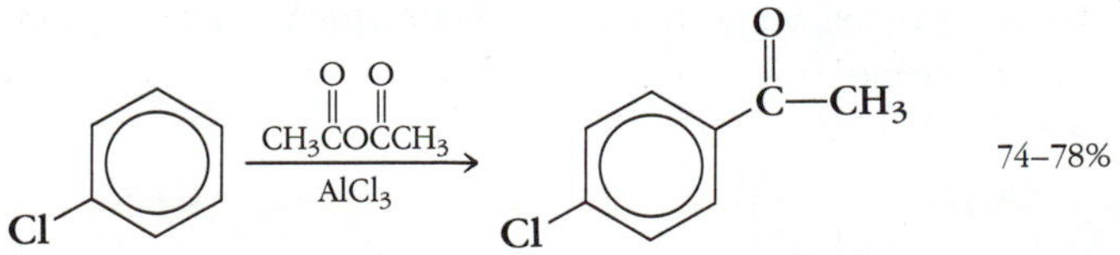

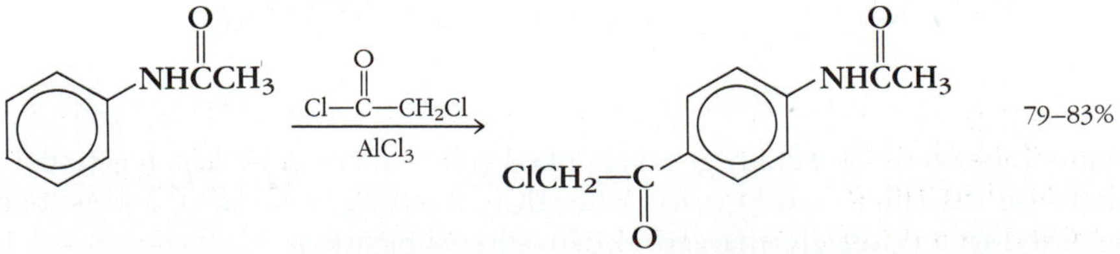

CH_3COCCH_3 (O O), $AlCl_3$ 74–78%

$Cl—C(=O)—CH_2Cl$, $AlCl_3$ 79–83%

$$\text{glutaric anhydride} \xrightarrow[AlCl_3]{C_6H_6} C_6H_5C(=O)CH_2CH_2CH_2CO_2H \quad 80\text{–}85\%$$

$$C_6H_5\text{—}CH_2\overset{O}{\overset{\|}{C}}OH \xrightarrow{PCl_3} C_6H_5CH_2\overset{O}{\overset{\|}{C}}Cl \xrightarrow[AlCl_3]{C_6H_6} C_6H_5\text{—}CH_2\overset{O}{\overset{\|}{C}}\text{—}C_6H_5 \quad 82\text{–}83\%$$

The first and third of the preceding reactions illustrate the use of anhydrides rather than acid chlorides. Only one of the acyl groups of an anhydride reacts (see Section 16.8), so the cyclic anhydride shown in the third example yields a keto acid. In contrast, when the acid chloride of a diacid is employed, both acyl groups react and the product is a diketone.

$$HO_2C(CH_2)_4CO_2H \xrightarrow{SOCl_2} ClC(=O)(CH_2)_4C(=O)Cl \xrightarrow[AlCl_3]{C_6H_6} C_6H_5C(=O)(CH_2)_4C(=O)C_6H_5 \quad 75\text{–}81\%$$

The only major limitation of the Friedel-Crafts acylation reaction is one of rate. When a deactivated aromatic compound is used, the rate of reaction may be quite slow and the yield of product correspondingly small. As a general rule you can expect that an acceptable yield of product will be obtained only in reactions of compounds that are at least as reactive as bromobenzene. This same limitation holds true for Friedel-Crafts alkylation.

Figure 18.1 provides a clear indication of the types of compounds that can and cannot be used in the Friedel-Crafts reaction. Any monosubstituted benzene with a deactivating group will be too unreactive. However, when a compound contains both activating and deactivating substituents, a Friedel-Crafts reaction may be possible. The net effect of several substituents can be estimated by combining the effects of the individual groups. For example, $\text{—}OCH_3$ is activating by about 10^9 and $\text{—}NO_2$ is deactivating by a factor of nearly 10^{-10}. The combined effect of the two groups should therefore be only slightly deactivating, and you would expect *o*-nitroanisole to react at a rate comparable with that of bromobenzene. In accord with this prediction, it reacts with phenylacetyl chloride to give a moderate yield of the corresponding ketone.

$$\text{o-}CH_3OC_6H_4NO_2 \xrightarrow[AlCl_3]{C_6H_5CH_2\overset{O}{\overset{\|}{C}}Cl} C_6H_5CH_2\text{—}\overset{O}{\overset{\|}{C}}\text{—}C_6H_3(OCH_3)(NO_2) \quad 50\%$$

Amino benzenes generally give very poor yields in Friedel-Crafts reactions. This is because the amino group reacts with $AlCl_3$, a strong Lewis acid, and is thereby converted to a positively charged, deactivating substituent.

Intramolecular Friedel-Crafts Acylation

Intramolecular Friedel-Crafts reactions often give high yields of cyclic products, and this provides a pathway by which six-membered rings can be built onto an aromatic nucleus. We have illustrated this in the following sequence of reactions:

C_6H_6, $AlCl_3$ → CO_2H 92–95%; Zn-Hg, HCl → CO_2H 82–89%; 1. $SOCl_2$ 2. $AlCl_3$ → 74–79%

The initial reaction with succinic anhydride yields a keto acid, and Clemmensen reduction affords 4-phenylbutyric acid. Conversion of this to the acid chloride with thionyl chloride, followed by an intramolecular Friedel-Crafts acylation, provides the bicyclic ketone in excellent yield.

Intramolecular processes that form five- or six-membered rings usually occur at faster rates than the corresponding intermolecular reactions (Section 8.7). This rate enhancement can partially offset the deactivating effect of an electron-withdrawing substituent, and an intramolecular Friedel-Crafts acylation can occur on a deactivated ring. For example, the acylation of toluene with phthalic anhydride generates a keto acid that can be cyclized without first reducing the keto group.

$C_6H_5CH_3$, $AlCl_3$ → CO_2H, CH_3 96%; fuming H_2SO_4 → CH_3 81–90%

This cyclization is so facile that it can be carried out with fuming sulfuric acid as catalyst rather than the more active aluminum chloride. Notice, however, that the electron-withdrawing carbonyl groups still prevent the intermolecular reaction of sulfonation of the aromatic rings by fuming sulfuric acid.

EXERCISE 18.6

(a) What monosubstitution products would be expected in the reaction of toluene and bromoethane in the presence of aluminum chloride? Which should be the major products?
(b) What by-products would be expected?
(c) What changes in product distribution might be expected with long reaction times?

EXERCISE 18.7

Predict the products formed by reaction of each of the following with propionyl chloride and $AlCl_3$:

(a) 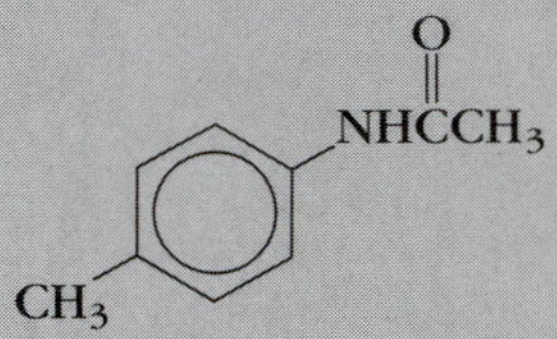 CH_3–C_6H_4–$NHCCH_3$ (O) (b) CH_3 (c) OH (d) 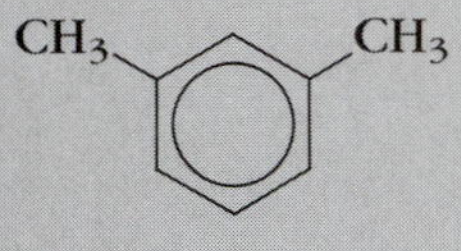 CH_3 CH_3

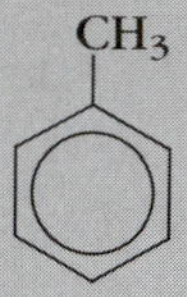

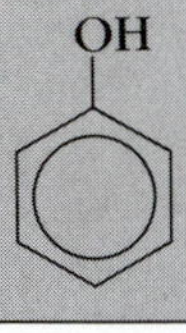

EXERCISE 18.8

(a) Predict whether each of the following compounds should be more or less reactive than benzene toward electrophilic substitution.
(b) For each compound predict whether or not it would undergo a Friedel-Crafts acylation reaction.

(i) 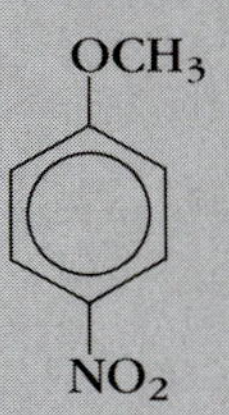 OCH_3, NO_2 (ii) 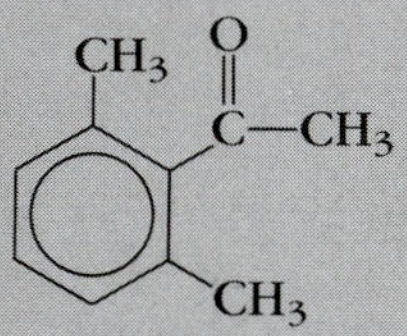 CH_3, O, $C—CH_3$, CH_3 (iii) 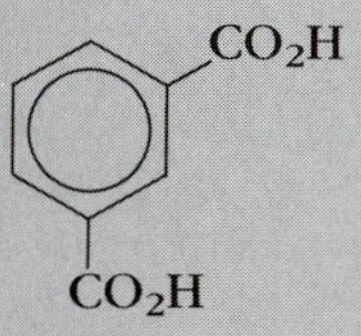 CO_2H, CO_2H (iv) 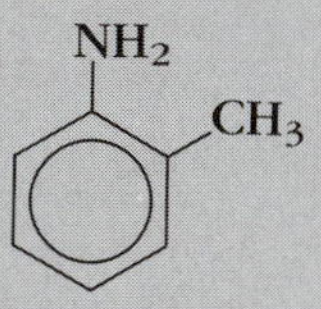 NH_2, CH_3

(v) 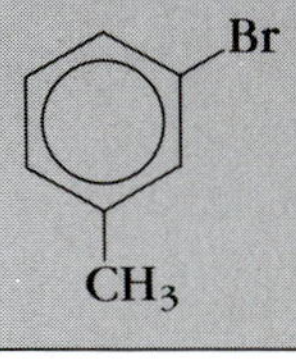 Br, CH_3 (vi) 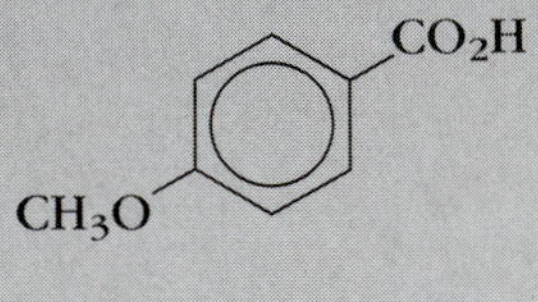 CO_2H, CH_3O (vii) 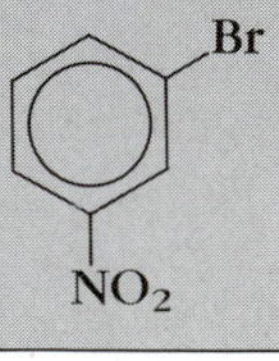 Br, NO_2

18.4 MODIFICATION OF SUBSTITUENTS ON AROMATIC RINGS

The reactions that we have introduced in this chapter can be combined with the reactions that you learned previously to prepare a wide variety of aromatic compounds by introducing new substituents on an aromatic ring. In this section we will show you how these substituents can be modified, thereby permitting the synthesis of an even wider range of aromatic compounds.

Halides

Unlike their aliphatic counterparts, aromatic halides do not undergo nucleophilic substitution and elimination reactions (see, however, Section 18.7). On the other hand, they can easily be converted to organometallic compounds as long as there are no other substituents that might interfere. The organometallic compounds can react with carbon dioxide to yield carboxylic acids (Section 16.7), with esters, aldehydes, and ketones to yield alcohols (Sections 11.5, 13.6, 16.5), with epoxides to form alcohols (Section 11.6), with acid chlorides to form ketones (Sections 13.4, 16.5), and with unsaturated carbonyl compounds to form products of conjugate addition (Section 13.10). This wide variety of carbon-carbon bond-forming reactions nicely complements the Friedel-Crafts reaction. Frequently only one of the two methods can be employed for the preparation of a particular compound (see Section 18.6).

Sulfonic Acids

Sulfonic acids can be converted to the corresponding sulfonyl chlorides with the same reagents that are used to prepare acid chlorides of carboxylic acids. The sulfonyl chlorides can then be used to prepare sulfonate esters (Section 11.9).

Aromatic sulfonation is reversible, so the —SO_3H group can be removed in a synthetically useful procedure. In the presence of strong *aqueous* acid the reverse reaction is favored, and good yields of desulfonated product can be obtained.

NH_2 / NO_2 / SO_3H — 1. H_2O, H_2SO_4; 2. NaOH (to neutralize) → NH_2 / NO_2 56%

The sulfonation-desulfonation procedure has sometimes been used as a blocking group technique. Introduction of an —SO_3H group at a particular position on the ring forces substitution to occur at some other site. Removal of the sulfonic acid residue then generates a product that would not have been favored in the absence of the blocking group.

A rather different reaction occurs when sulfonic acids are heated with molten KOH or NaOH at temperatures in the region of 250°–350°C. These extremely harsh conditions result in replacement of the —SO_3H group by an —OH. From the late 1800s through the first half of the twentieth century this was used as an industrial method for the preparation of phenol. The very harsh reaction conditions are not well suited for laboratory synthesis and most functional groups would not survive, but simple alkylbenzene sulfonic acids can be converted to the phenols, as shown by the following reactions:

SO_3H — 1. KOH, 250°C; 2. H_2SO_4 (neutralize) → OH 53%

SO_3H / CH_3 — 1. KOH, 300°C; 2. H_2SO_4 (neutralize) → OH / CH_3 63–72%

Alkylbenzenes

Alkyl groups are much less reactive than other functional groups, but we have previously discussed several reactions that permit modification of an alkyl substituent on an aromatic ring. *N*-Bromosuccinimide can be used to brominate a benzylic position (Section 8.6). This is a free radical reaction, and bromination of the ring, which occurs by an ionic process, does not take place under these conditions. Therefore bromination of either the ring or the side chain can be carried out selectively.

NBS; CH_2Br; CH_3; CH_3; CH_3; Br_2, $FeBr_3$; CH_3; Br; CH_3; 94–99%

Another reaction we presented earlier is the oxidation of an alkyl substituent to a carboxylic acid (Section 16.7). The vigorous conditions required for such oxidations will also result in the oxidation of many other functional groups, but the reaction can be carried out on halo- and nitro-substituted alkylbenzenes.

Phenols

The reactivity of phenols is very different than that of other hydroxylic compounds. In contrast to the hydroxyl groups of alcohols and carboxylic acids, the hydroxyl group of a phenol cannot be replaced by the action of reagents such as PCl_5 and PBr_3. Phenols are sensitive to oxidation (Section 11.7), but the preparative chemistry of phenolic hydroxyl groups is essentially limited to acid-base reactions and conversion to esters and ethers (Section 11.8). On the other hand, the powerful activating effect of the hydroxyl group allows phenols to undergo a remarkable variety of electrophilic substitution reactions under very mild conditions.

Nitro Compounds

Aromatic nitro compounds are among the most versatile synthetic intermediates, because the nitro group can be converted to a wide variety of other substituents. Although a nitro group will interfere with reduction and organometallic reactions, it can survive oxidizing conditions and reactions that require either strong acid or strong base. Reduction of nitro compounds yields primary amines (Section 17.6), and these can be alkylated or converted to amides. With primary *aromatic* amines, another important class of compounds can be prepared, diazonium salts ($Ar—\overset{+}{N}\equiv N\ X^-$). The preparation and reactions of diazonium salts are the subject of the next section of this chapter.

Amino Groups and Blocking Groups

An amino substituent on an aromatic ring can cause major problems: It is protonated by strong acid, and the resulting ammonium substituent is highly deactivating (and *meta*-directing).

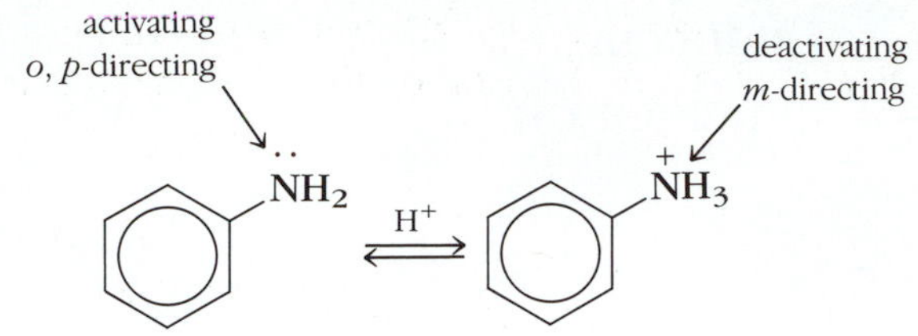

activating *o, p*-directing; deactivating *m*-directing

$$C_6H_5\ddot{N}H_2 \overset{H^+}{\rightleftharpoons} C_6H_5\overset{+}{N}H_3$$

On the other hand, a free amino group is very highly activating. As we showed in Section 18.2, aniline reacts with bromine in the absence of a catalyst to yield the tribromo derivative. These opposing effects ordinarily preclude direct introduction of a single substituent *ortho* or *para* to an amino group, and most electrophilic substitutions of amino compounds are carried out with the aid of blocking groups that prevent reaction at nitrogen.

The chemical reactivity of an amino group is most easily modified by *acylation*, particularly acetylation with acetyl chloride or acetic anhydride (Sections 16.6, 17.4).

$$Ar{-}NH_2 \xrightarrow{CH_3\overset{O}{\overset{\|}{C}}Cl} Ar{-}NH{-}\overset{O}{\overset{\|}{C}}{-}CH_3$$

The resulting amide is not basic, but the ring remains strongly activated (see Figure 18.1). The following reaction sequence illustrates the use of an acyl blocking group for the selective monosubstitution of 4-amino toluene. After first converting the amine to the amide, bromination proceeds smoothly (no catalyst is needed with this highly activated compound). Subsequent hydrolysis of the amide liberates the free amino group.

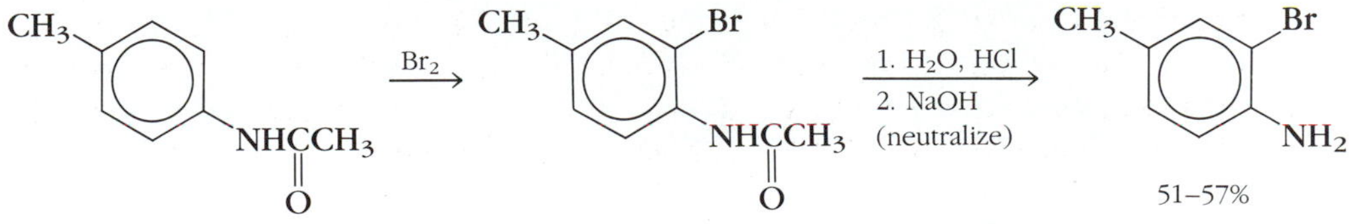

$$CH_3C_6H_4NH\overset{O}{\overset{\|}{C}}CH_3 \xrightarrow{Br_2} CH_3C_6H_3(Br)NH\overset{O}{\overset{\|}{C}}CH_3 \xrightarrow[\text{2. NaOH (neutralize)}]{\text{1. } H_2O,\ HCl} CH_3C_6H_3(Br)NH_2$$

51–57%

The use of a blocking group therefore permits the overall conversion of 4-aminotoluene to 3-bromo-4-aminotoluene.

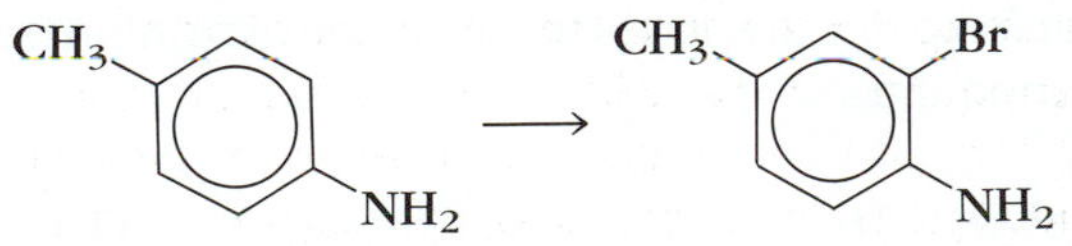

$$CH_3C_6H_4NH_2 \longrightarrow CH_3C_6H_3(Br)NH_2$$

Unless polysubstitution is desired (as in the preparation of 2,4,6-tribromoaniline, Section 18.2), an acyl blocking group is almost always employed for electrophilic substitution of aromatic amino compounds.

EXERCISE 18.9

Suggest reagents for carrying out the following conversions:

(a) $C_6H_5{-}Cl \longrightarrow C_6H_5CO_2H$

(b) $C_6H_5{-}Cl \longrightarrow C_6H_5\overset{O}{\overset{\|}{C}}CH_3$

(c) $C_6H_5Br \longrightarrow C_6H_5CH_2CH_2OH$

(d) $C_6H_5Br \longrightarrow C_6H_5{-}\underset{CH_3}{\underset{|}{C}}H{-}CH_2\overset{O}{\overset{\|}{C}}{-}CH_3$

(e) $C_6H_5SO_3H \longrightarrow C_6H_6$

(f) $C_6H_5{-}CH_3 \longrightarrow C_6H_5CO_2H$

(g) $C_6H_5{-}CH_3 \longrightarrow C_6H_5{-}CH_2Br$

(h) $C_6H_5{-}OH \longrightarrow C_6H_5{-}OCH_3$

(i) $C_6H_5{-}OH \longrightarrow C_6H_5O\overset{O}{\overset{\|}{C}}CH_3$

(j) $C_6H_5NO_2 \longrightarrow C_6H_5NH_2$

(k) $C_6H_5NO_2 \longrightarrow C_6H_5NH{-}\overset{O}{\overset{\|}{C}}{-}CH_3$

(l) $C_6H_5NO_2 \longrightarrow C_6H_5N(CH_3)_2$

EXERCISE 18.10

Suggest a method for carrying out the following conversion:

$$C_6H_5NH_2 \longrightarrow p\text{-}CH_3\overset{O}{\overset{\|}{C}}{-}C_6H_4{-}NH_2$$

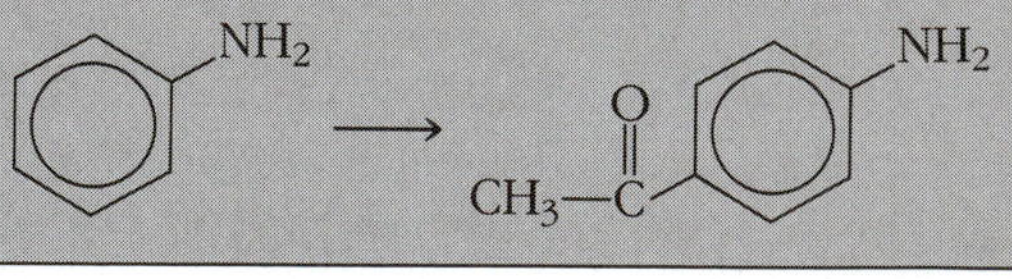

18.5 PREPARATION AND REACTIONS OF DIAZONIUM SALTS

As we have already shown in Section 17.4, aromatic amines react with nitrous acid to form *diazonium ions.* In some cases the aromatic diazonium salt can be isolated as a crystalline compound.

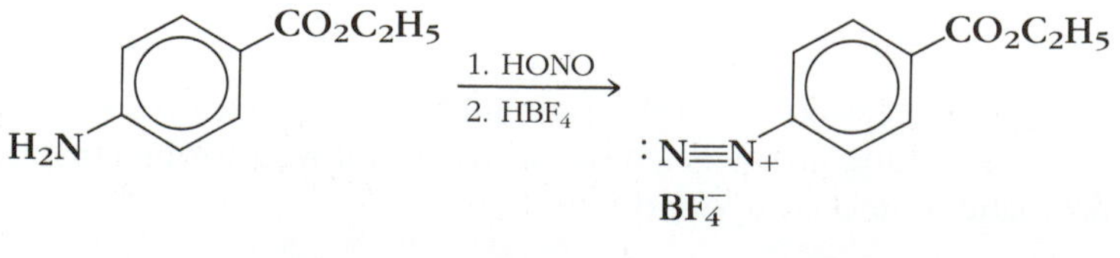

$$p\text{-}H_2N{-}C_6H_4{-}CO_2C_2H_5 \xrightarrow[2.\ HBF_4]{1.\ HONO} p\text{-}:N{\equiv}N^+{-}C_6H_4{-}CO_2C_2H_5\ BF_4^- \qquad 75\text{–}78\%$$

However, the potential for explosive decomposition of diazonium salts provides a strong argument against attempts at isolation. In most cases the solution of diazonium salt is used directly for the next step of a reaction sequence.

The great stability of molecular nitrogen makes it an excellent leaving group, but phenyl cations are relatively unstable, so carbocation formation does not occur readily in solution.

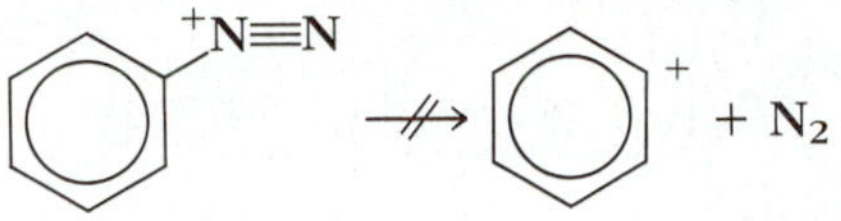

$$C_6H_5-\overset{+}{N}\equiv N \not\longrightarrow C_6H_5^+ + N_2$$

Neverthelesss, the N_2 group of a diazonium ion can be replaced by a variety of other substituents, particularly when a copper catalyst is employed. Procedures have been developed for replacement of $-N_2^+$ by —F, —Cl, —Br, —I, —CN, $-NO_2$, —OH, and —H. The last conversion ($Ar-N_2^+$ to Ar—H) extends the capability of blocking groups even further, because the entire amino group can be removed via diazotization and subsequent replacement of the diazonium group with a hydrogen atom. This allows the complete cycle to be carried out:

$$Ar-H \longrightarrow Ar-NO_2 \longrightarrow Ar-NH_2 \longrightarrow Ar-N_2^+ \longrightarrow Ar-H$$

The Sandmeyer Reaction

In 1884 T. Sandmeyer (Switzerland) found that the $-N_2^+$ group of a diazonium salt could be replaced by a variety of other substituents in a reaction with the appropriate cuprous salt. This type of substitution is often called the Sandmeyer reaction, and it proceeds via formation of an intermediate copper complex, which decomposes on mild heating.

Sandmeyer Reaction: X = Cl, Br, I, CN, NO₂

$$Ar-\overset{+}{N}\equiv N \xrightarrow{CuX} [\text{Cu complex}] \xrightarrow[\text{heating}]{\text{gentle}} Ar-X + N_2\uparrow$$

In a variation of this procedure copper powder is sometimes used in place of the copper salt. The anion, X^-, is then introduced as some other salt or as the corresponding acid, HX. The following reactions illustrate the Sandmeyer reaction:

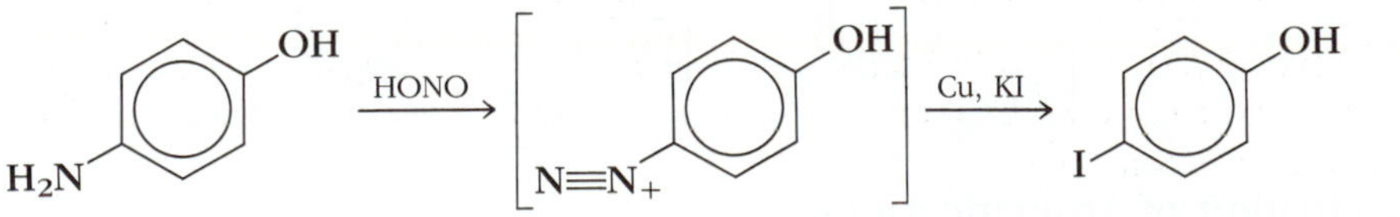

$$p\text{-}H_2N-C_6H_4-OH \xrightarrow{HONO} [p\text{-}\overset{+}{N}\equiv N-C_6H_4-OH] \xrightarrow{Cu,\ KI} p\text{-}I-C_6H_4-OH \quad 69\text{–}72\%$$

$$m\text{-}Cl-C_6H_4-NH_2 \xrightarrow{HONO} [m\text{-}Cl-C_6H_4-\overset{+}{N}\equiv N] \xrightarrow{CuBr} m\text{-}Cl-C_6H_4-Br \quad 91\text{–}94\%$$

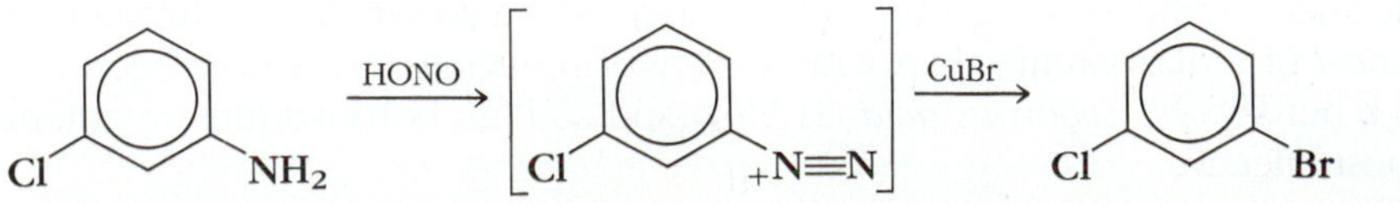

$$\text{o-}CH_3C_6H_4NH_2 \xrightarrow{HONO} [\text{o-}CH_3C_6H_4\overset{+}{N}\equiv N] \xrightarrow{CuCN} \text{o-}CH_3C_6H_4CN \quad 64\text{–}70\%$$

$$\text{4-nitro-1-naphthylamine } (NH_2, NO_2) \xrightarrow{HONO} [{}^{+}N_2 \text{ / } NO_2 \text{ naphthalene}] \xrightarrow{CuNO_2} \text{1,4-dinitronaphthalene } (NO_2, NO_2) \quad 52\text{–}60\%$$

Replacement by Fluoride

The use of a copper catalyst is not required for replacement of an aromatic diazonium group by —F. The fluoride is prepared by heating the dry BF_4^- salt (which is formed by adding HBF_4 to the initial solution of diazonium salt). This is illustrated by the preparation of fluorobenzene from aniline (as the hydrochloride salt).

$$C_6H_5\overset{+}{N}H_3\ Cl^- \xrightarrow[2.\ HBF_4]{1.\ HONO} C_6H_5\overset{+}{N}\equiv N\ BF_4^- \xrightarrow[\text{heating}]{\text{mild}} C_6H_5F$$

51–57%

Synthesis of Phenols

When a diazonium salt is refluxed in aqueous sulfuric acid, the corresponding phenol can be prepared in good yield.

$$CH_3, Br, NH_2 \text{ benzene} \xrightarrow{HONO} [CH_3, Br, N_2^+ \text{ benzene}] \xrightarrow[H_2SO_4]{H_2O} CH_3, Br, OH \text{ benzene} \quad 80\text{–}92\%$$

$$\text{m-}O_2NC_6H_4NH_2 \xrightarrow{HONO} [\text{m-}O_2NC_6H_4N_2^+] \xrightarrow[H_2SO_4]{H_2O} \text{m-}O_2NC_6H_4OH \quad 81\text{–}86\%$$

Deamination of Aromatic Amines

Replacement of the —N_2^+ group by hydrogen can be accomplished through the *reduction* of a diazonium salt. A wide variety of reducing agents have been employed, but *hypophosphorous acid* (H_3PO_2) and sodium borohydride are among the most effective.

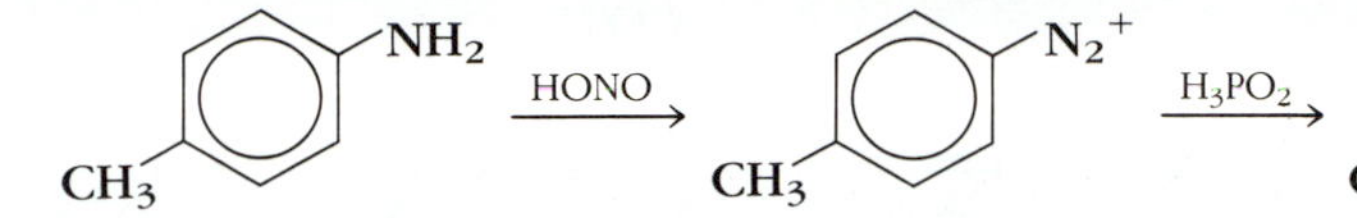

72%

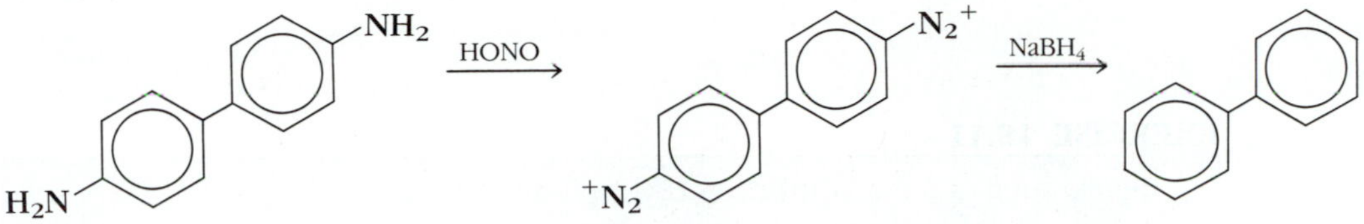

75%

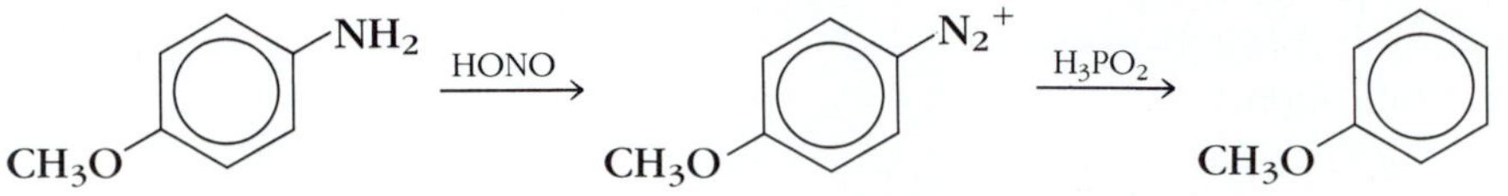

80%

Coupling Reactions: Formation of Azo Dyes

Aromatic diazonium salts react with phenols and aromatic amines to form products that contain the *azo* group, —N=N—. The azo compounds are usually very brightly colored, and the discovery of this reaction in the mid 1800s provided a great stimulus to the development of the synthetic dye industry. The following equation illustrates the coupling reaction between *N,N*-dimethylaniline and the diazonium ion formed from anthranilic acid.

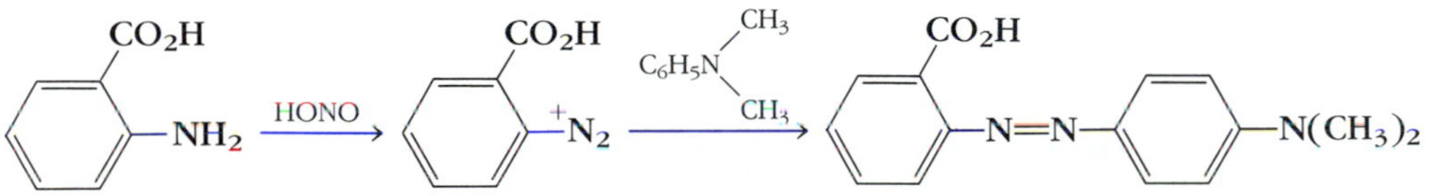

62–66%

The reaction proceeds via nucleophilic attack of the aniline derivative (activated at the *para* position by the amino group) on the diazonium group. Subsequent proton transfer generates the azo dye:

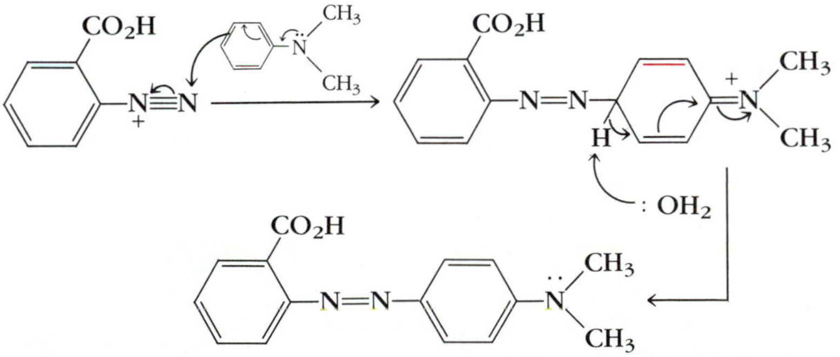

The azo dye shown here is known as *methyl red* and is frequently used as an indicator in acid-base titrations. In acidic medium ($pH < 5$) the compound exhibits a red color, but in neutral or alkaline solution it is yellow. The following equilibrium illustrates the pH sensitivity of methyl red.

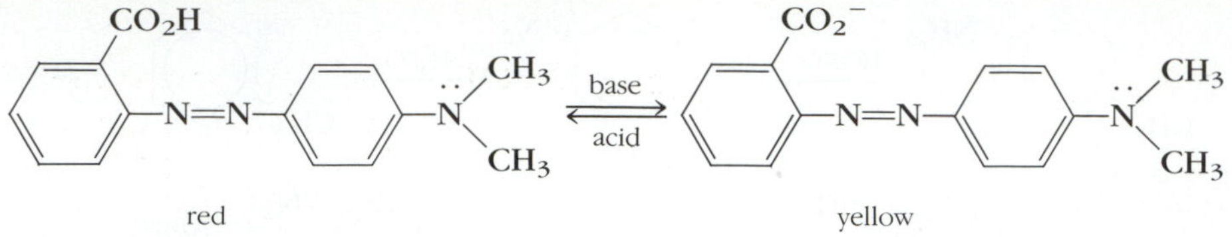

CO_2H —N=N— N(CH_3)$_2$ $\underset{\text{acid}}{\overset{\text{base}}{\rightleftharpoons}}$ CO_2^- —N=N— N(CH_3)$_2$

red yellow

EXERCISE 18.11

Suggest methods for converting *o*-chloroaniline to:

(a) *o*-Dichlorobenzene
(b) *o*-Chloronitrobenzene
(c) Chlorobenzene
(d) *o*-Chlorobenzoic acid

EXERCISE 18.12

Suggest methods for converting *p*-nitrotoluene to:

(a) *p*-Cresol (4-methylphenol)
(b) 4-Methylbenzonitrile
(c) *p*-Bromotoluene
(d) *p*-Fluorotoluene
(e) Toluene

18.6 SYNTHESIS OF AROMATIC COMPOUNDS

In the preceding sections of this chapter we have introduced a variety of electrophilic substitution reactions. In this section we will show you how these reactions can be combined (together with reactions from other chapters) to synthesize fairly complex aromatic compounds. The design of a synthetic plan has been considered previously in Sections 1.12, 5.14, 7.10, and 11.10. Our discussion here is a further extension of the ideas we presented earlier.

Example 18.1 Suggest a method of synthesizing *p*-bromoacetophenone starting from benzene.

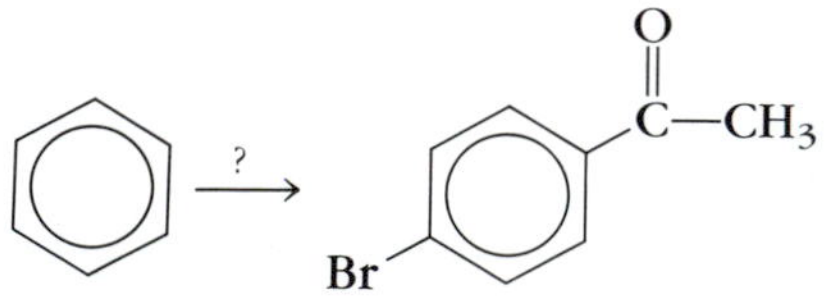

$\xrightarrow{?}$ Br, C($=$O)—CH_3

Solution Two substituents must be introduced on the aromatic ring. The major question to be answered is: Which of them is introduced first? If the first step in the synthesis were a Friedel-Crafts acylation, the *meta*-directing acyl group would cause bromination to occur at the wrong position.

$$C_6H_5COCH_3 \xrightarrow[FeBr_3]{Br_2} m\text{-}BrC_6H_4COCH_3$$

Consequently, it is the bromine (an *ortho, para*-directing group) that should be introduced first, and the following sequence would be satisfactory.

$$C_6H_6 \xrightarrow[FeBr_3]{Br_2} C_6H_5Br \xrightarrow[AlCl_3]{CH_3COCl} p\text{-}BrC_6H_4COCH_3$$

Some *o*-bromoacetophenone might be expected in the second step, but this could be removed during purification of the product.

Example 18.2 Propose a synthesis of 1-butylbenzene starting with compounds containing six or fewer carbon atoms.

$$C_6H_5CH_2CH_2CH_2CH_3$$

Solution The six-carbon limitation for the starting materials suggests that a four-carbon subunit must be joined to an intact benzene ring. One obvious possibility is a Friedel-Crafts alkylation, but this would yield a mixture of products resulting from carbocation rearrangements.

$$C_6H_6 \xrightarrow[AlCl_3]{Cl-CH_2CH_2CH_2CH_3} C_6H_5CH_2CH_2CH_2CH_3 + C_6H_5CH(CH_3)-CH_2CH_3$$

A much more reliable method would employ a Friedel-Crafts *acylation* to produce the desired carbon skeleton. The carbonyl group could then be reduced by one of several alternative methods (Section 13.11).

$$C_6H_6 \xrightarrow[AlCl_3]{Cl-C(=O)-CH_2CH_2CH_3} C_6H_5C(=O)-CH_2CH_2CH_3 \xrightarrow[HCl]{Zn-Hg} C_6H_5CH_2CH_2CH_2CH_3$$

The intermediate ketone could also be prepared by reaction of lithium diphenylcuprate with the acid chloride (Section 13.4).

$$C_6H_5Li \xrightarrow{CuI} (C_6H_5)_2CuLi \xrightarrow{Cl-\overset{O}{\overset{\|}{C}}CH_2CH_2CH_3} C_6H_5-\overset{O}{\overset{\|}{C}}-CH_2CH_2CH_3$$

In fact, the use of an organocopper reagent would permit the synthesis of butylbenzene directly via a coupling reaction (Section 1.11).

$$(C_6H_5)_2CuLi \xrightarrow{I-CH_2CH_2CH_2CH_3} C_6H_5-CH_2CH_2CH_2CH_3$$

The last two reactions should remind you that not all aromatic compounds must be prepared via electrophilic substitution.

Example 18.3 Devise a synthetic method for the preparation of *o*-nitroaniline from benzene. Recall, however, that the nitration of acetanilide (or other amides of aniline) affords the *para* product almost exclusively.

$$C_6H_6 \xrightarrow{?} o\text{-}(NH_2)(NO_2)C_6H_4$$

Solution The two substituents are *ortho* to each other, and this indicates that the electron-withdrawing nitro group must have been introduced second. But if substitution of an amino compound (protected as the amide) occurs at the *para* position, how can an ortho substituent be introduced?

$$C_6H_5NH\overset{O}{\overset{\|}{C}}CH_3 \xrightarrow{\text{substitution}} p\text{-}X\text{-}C_6H_4NH\overset{O}{\overset{\|}{C}}CH_3$$

This is a situation that calls for the use of a *blocking group*. Introduction of a sulfonic acid residue would block the position *para* to the protected nitrogen, but the *ortho* positions would still be reactive. The $-SO_3H$ and acetyl groups could be removed after the nitration reaction. The overall synthesis would start with the preparation of aniline via nitration of benzene and subsequent reduction of the $-NO_2$ group (Section 17.6). Substitution of an aromatic amine necessitates prior formation of the amide, so acetylation of aniline would be the third step.

$$C_6H_6 \xrightarrow[H_2SO_4]{HNO_3} C_6H_5NO_2 \xrightarrow{Sn,\ HCl} C_6H_5NH_2 \xrightarrow{CH_3\overset{O}{\overset{\|}{C}}Cl} C_6H_5NH\overset{O}{\overset{\|}{C}}CH_3$$

At this point sulfonation would introduce an $-SO_3H$ group in the *para* position (any *ortho* product could be removed during purification). In the subsequent nitration step both substituents would direct the reaction to a position *ortho* to the amide nitrogen. Finally, reaction with aqueous sulfuric acid would remove the sulfonic acid group, and these conditions would also cause hydrolysis of the *N*-acetyl group (Section 16.6).

$C_6H_5NHCOCH_3 \xrightarrow[H_2SO_4]{SO_3}$ p-$HO_3S{-}C_6H_4{-}NHCOCH_3 \xrightarrow[H_2SO_4]{HNO_3}$ $HO_3S{-}C_6H_3(NO_2){-}NHCOCH_3 \xrightarrow[\text{2. NaOH (neutralize)}]{\text{1. } H_2SO_4,\ H_2O}$ o-$NO_2{-}C_6H_4{-}NH_2$

Example 18.4 Suggest a method for synthesizing *m*-nitrotoluene starting from either benzene or toluene.

Benzene or toluene $\xrightarrow{?}$ m-$CH_3{-}C_6H_4{-}NO_2$

Solution Alkyl groups are activating, so nitration of toluene would yield only the products of *para* and *ortho* substitution in significant amounts. Consequently this one-step procedure would not work. What about the introduction of a methyl group into nitrobenzene (prepared from benzene by nitration)? While the nitro group is *meta*-directing, this is not a viable alternative either, because the Friedel-Crafts reaction fails with deactivated benzenes. This is a case in which the *para*-directing effect of the methyl can be used to *temporarily* introduce a strongly activating group (—X) that will in turn direct nitration to the correct position.

$C_6H_5CH_3 \longrightarrow$ p-$\ddot{X}{-}C_6H_4{-}CH_3 \longrightarrow$ $\ddot{X}{-}C_6H_3(NO_2){-}CH_3 \longrightarrow$ m-$CH_3{-}C_6H_4{-}NO_2$

Strongly activating group

The position *para* to substituent —X is blocked by the methyl group, so nitration must occur *ortho* to —X, i.e., *meta* to —CH_3. The desired activating group is —$NHCOCH_3$ (prepared by nitration, reduction, and acylation), and it can later be removed by hydrolysis to the amine, diazotization, and reduction. The entire process is outlined in the following equation:

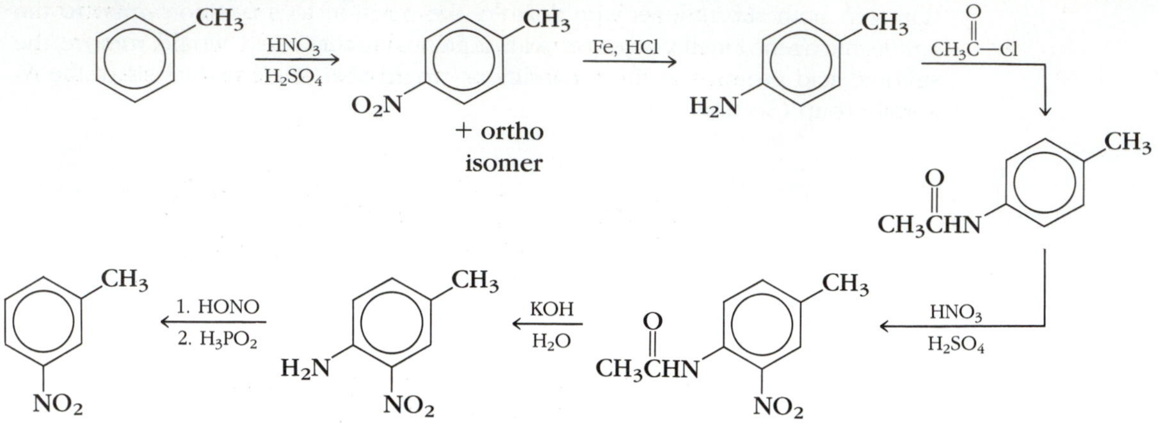

CH_3 $\xrightarrow[H_2SO_4]{HNO_3}$ O_2N CH_3 + ortho isomer $\xrightarrow{Fe, HCl}$ H_2N CH_3 $\xrightarrow{CH_3C(O){-}Cl}$ $CH_3C(O)HN$ CH_3 $\xrightarrow[H_2SO_4]{HNO_3}$ $CH_3C(O)HN$ CH_3 NO_2 $\xrightarrow[H_2O]{KOH}$ H_2N CH_3 NO_2 $\xrightarrow[2.\ H_3PO_2]{1.\ HONO}$ CH_3 NO_2

Example 18.5 How could 4-methoxy-4′-nitrobenzophenone be synthesized from benzene and/or toluene as the starting material?

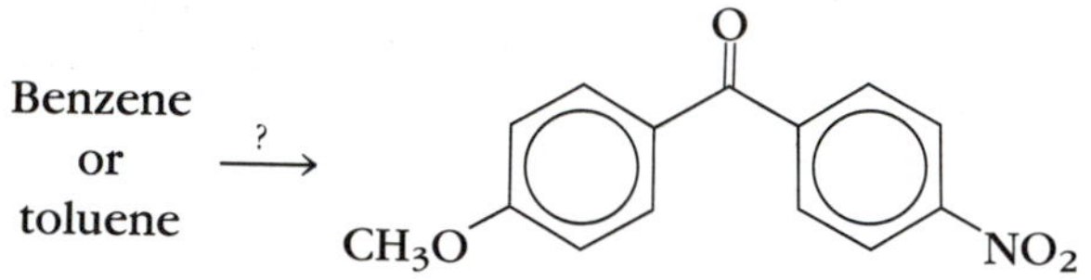

Benzene or toluene $\xrightarrow{?}$ CH_3O (O) NO_2

Solution The presence of an aromatic ketone group immediately suggests a Friedel-Crafts acylation, but which ring should be used as the acid chloride and which should be acylated? Or should the nitro and methoxy groups be introduced after the carbon skeleton has been formed? Consider structures **1**, **2**, and **3**:

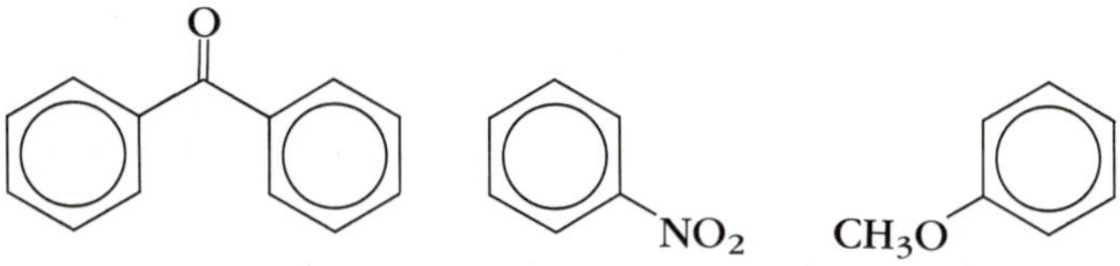

(O) — **1**; NO_2 — **2**; CH_3O — **3**

Both aromatic rings of **1** are deactivated, so any substitution reactions would occur at the *meta* position. Activating and blocking group techniques might be employed, but a more direct synthesis would be preferable. Friedel-Crafts acylation of **2** with *p*-methoxybenzoyl chloride would fail because **2** is a strongly deactivated compound. In contrast **3** would be acylated readily at the desired *para* position. The desired product could then be prepared by the following reaction:

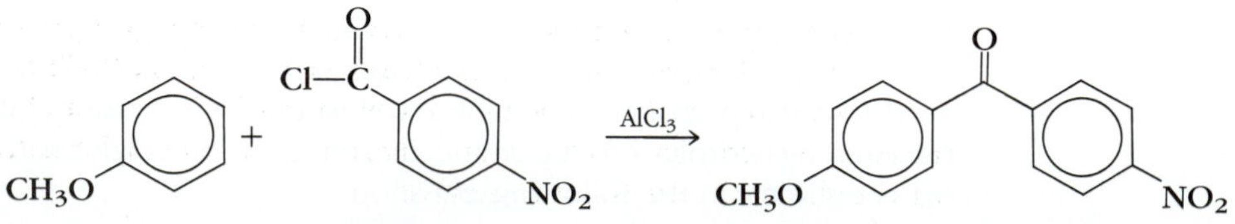

CH_3O + Cl—C(O) NO_2 $\xrightarrow{AlCl_3}$ CH_3O (O) NO_2

Further problems still remain with the synthesis of anisole and *p*-nitrobenzoyl chloride. The former could be prepared by alkylation of phenol (Section 11.8, 15.4), which in turn might be prepared from nitrobenzene via reduction and diazotization.

$$C_6H_6 \xrightarrow[H_2SO_4]{HNO_3} O_2N\text{–}C_6H_5 \xrightarrow{Sn,\ HCl} H_2N\text{–}C_6H_5 \xrightarrow[2.\ H_2O,\ H_2SO_4]{1.\ HONO} HO\text{–}C_6H_5 \xrightarrow[2.\ CH_3I]{1.\ NaNH_2} CH_3O\text{–}C_6H_5$$

Introduction of the two substituents of *p*-nitrobenzoic acid is more difficult, because both groups are *meta*-directing. One of these must have come from modification or replacement of another group that was *ortho, para*-directing. One reasonable alternative is that the carboxyl group was obtained by oxidation of a *methyl* group (Section 16.7). This would allow the synthesis of *p*-nitrobenzoyl chloride from toluene:

$$CH_3\text{–}C_6H_5 \xrightarrow[H_2SO_4]{HNO_3} CH_3\text{–}C_6H_4\text{–}NO_2 \xrightarrow{KMnO_4} HO_2C\text{–}C_6H_4\text{–}NO_2 \xrightarrow{PCl_5} Cl\text{–}\overset{O}{\overset{\|}{C}}\text{–}C_6H_4\text{–}NO_2$$

Example 18.6 Suggest a method of synthesizing 4-bromo-2-methylbenzoic acid from toluene:

$$\text{Toluene} \xrightarrow{?} \text{(4-bromo-2-methylbenzoic acid: } CH_3,\ CO_2H,\ Br\text{)}$$

Solution The allowed starting material has seven of the eight carbon atoms of the required skeleton. It is not difficult to introduce a substituent in the *ortho* position, but how can bromination be made to occur *meta* to the methyl group? Prior introduction of the carboxyl function would not help because it is *meta*-directing and the —Br must be *para* to it. This is again a situation in which it would be beneficial to first introduce a strongly activating group *ortho* to the methyl and then convert it to a carboxylic acid after bromination.

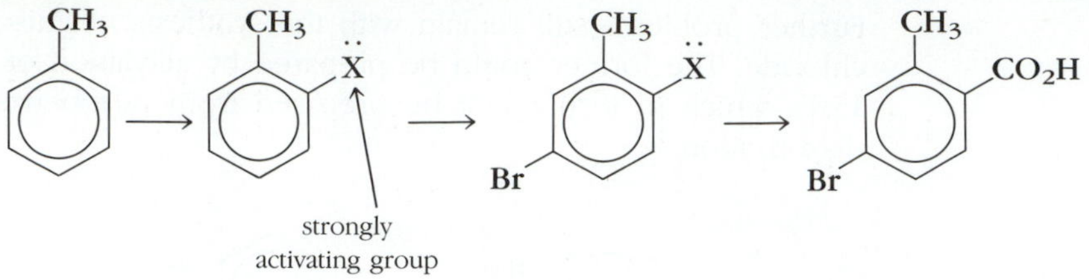

CH3 CH3 X CH3 X CH3 CO2H
Br Br
strongly
activating group

The key question now becomes: How can the activating group, —X, be converted to —CO_2H? Consider the several ways you have learned for the preparation of carboxylic acids (Section 16.7): (1) oxidation of an alkyl side chain; (2) oxidation of a methyl ketone; (3) carboxylation of an organometallic reagent; and (4) hydrolysis of a nitrile. Structures **1–4** illustrate possible intermediates for each of these methods, but not all are satisfactory.

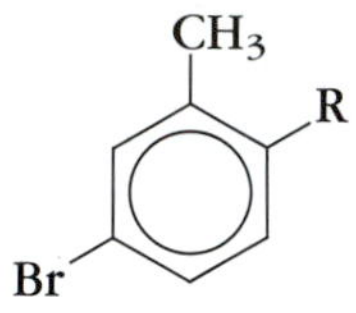

CH3 R Br **1**

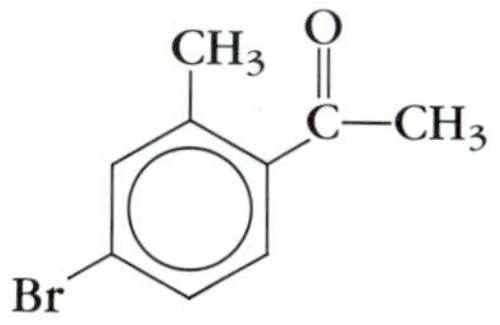

CH3 O C—CH3 Br **2**

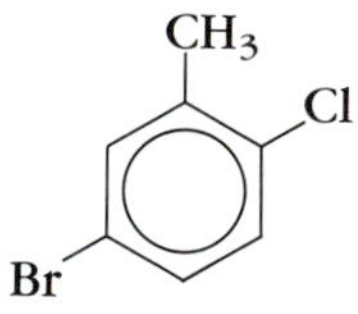

CH3 Cl Br **3**

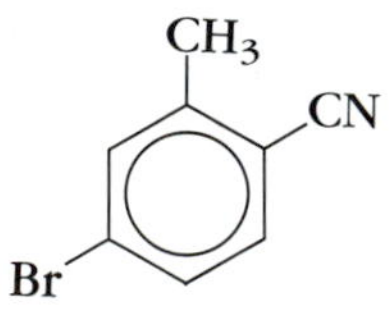

CH3 CN Br **4**

The first method would fail for several reasons. The —R group does not fit our requirement for a strongly activating substituent, and subsequent reaction with $KMnO_4$ would oxidize both the —R and —CH_3 groups. Moreover, introduction of the alkyl substituent by Friedel-Crafts alkylation would be accompanied by polysubstitution. The second method is no better, because the acyl group is *deactivating,* and introduction of bromine would occur at the wrong position.

The third method is again unsatisfactory. Organometallic reactions of polyfunctional compounds have many limitations, and in this case there are two halogen substituents that could react with lithium or magnesium. A complex mixture of products would be expected from this approach.

The fourth method appears to be quite reasonable. The cyano group could be generated by replacement of a diazonium group, and subsequent hydrolysis of the cyano group would result in the carboxylic acid. The diazonium group is formed from an —NH_2 group, so our requirement for a strongly activating substituent is fulfilled. A complete reaction sequence follows.

CH_3 $\xrightarrow[H_2SO_4]{HNO_3}$ CH_3, NO_2 (plus *para* isomer) $\xrightarrow[2.\ CH_3\overset{O}{\overset{\|}{C}}—Cl]{1.\ Fe,HCl}$ CH_3, $NH\overset{O}{\overset{\|}{C}}CH_3$ $\xrightarrow[Br_2]{FeBr_3}$ CH_3, $NH\overset{O}{\overset{\|}{C}}CH_3$, Br $\xrightarrow[2.\ HONO]{1.\ KOH,\ H_2O}$ CH_3, N_2^+, Br $\xrightarrow{CuCN}$ CH_3, CN, Br $\xrightarrow[H_2SO_4]{H_2O}$ CH_3, CO_2H, Br

Several of the examples we have presented in this section show that the preparation of substituted benzenes is not a trivial matter. On the other hand, now that you have learned some of the techniques for synthesis employing blocking groups and activating groups, you should be able to design reasonable synthetic pathways for a wide variety of compounds.

One final reaction illustrates a particularly useful possibility for introducing substituents in a *meta* orientation. The $—NO_2$ group is strongly *meta*-directing, so further nitration of nitrobenzene under vigorous conditions affords *m*-dinitrobenzene. The use of sodium sulfide as a reducing agent allows selective reduction of *just one* of the two nitro groups (Section 17.6).

NO_2, NO_2 $\xrightarrow{Na_2S}$ NO_2, NH_2 70–80%

The resulting *m*-nitroaniline is a highly versatile intermediate in aromatic synthesis. By using different combinations of protecting groups and diazotization reactions, either or both groups can be converted to a wide variety of aromatic compounds.

EXERCISE 18.13

We solved the problem posed in Example 18.4 on the basis of nitrating toluene in the *para* position. In actuality, *o*-nitrotoluene is the predominant isomer formed in that reaction. Suggest a scheme by which this could be used to prepare the same target molecule, *m*-nitrotoluene.

EXERCISE 18.14

In Example 18.5 we outlined a possible synthesis of *p*-nitrobenzoic acid, but other pathways are possible.

(a) Outline a scheme that proceeds via aniline, in which the amino group is eventually converted to a carboxyl group (via diazonium replacement).
(b) Outline still another scheme involving aniline, in which the carboxyl group is introduced via Friedel-Crafts acylation (and subsequent oxidation) and the nitro group is introduced by diazonium ion replacement.

18.7 AROMATIC SUBSTITUTION BY NUCLEOPHILIC AND ELIMINATION MECHANISMS

In Chapter 12 we argued that nucleophilic displacement reactions are limited to *alkyl* halides and sulfonates. Nevertheless, there are known examples in which *aryl* halides undergo nucleophilic substitution. In this section we will show you several of these reactions and describe how they proceed by mechanisms other than those that are observed for alkyl derivatives.

A simple aromatic halide such as chlorobenzene is unreactive toward base under normal laboratory conditions, but substitution does occur at very high temperatures. As an example, the hydrolysis of chlorobenzene was until recently a major industrial method for the preparation of phenol (the Dow process).*

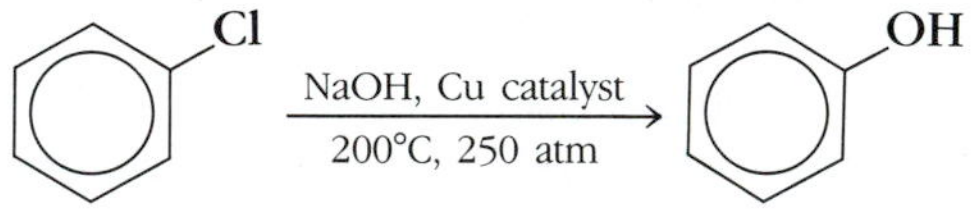

Cl — NaOH, Cu catalyst / 200°C, 250 atm → OH

Dow process for preparation of phenol

Certain nitro-substituted aromatic halides react under much milder conditions, however. Treatment of *p*-chloronitrobenzene with the potassium salt of 2,6-diisopropylphenol at 90°C yields the corresponding ether.

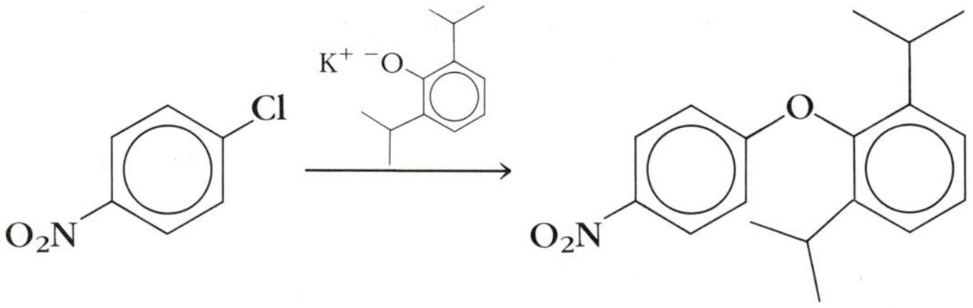

O_2N, Cl — K^+ ^-O → O_2N, O — 75%

The two preceding reactions take place by quite different mechanisms. Experiments with a variety of compounds have shown that nitro-substituted aromatic halides undergo nucleophilic substitution by a two-step process. The reaction proceeds *only* when the chlorine (or other leaving group) is *ortho* or *para* to a polar unsaturated substituent such as —NO_2, and the first step is an *addition* of the nucleophile. The following example illustrates the mechanism of this addition step for *o*-chloronitrobenzene:

*The modern industrial preparation of phenol is carried out by the oxidation of cumene. This is discussed in detail in Section 26.2.

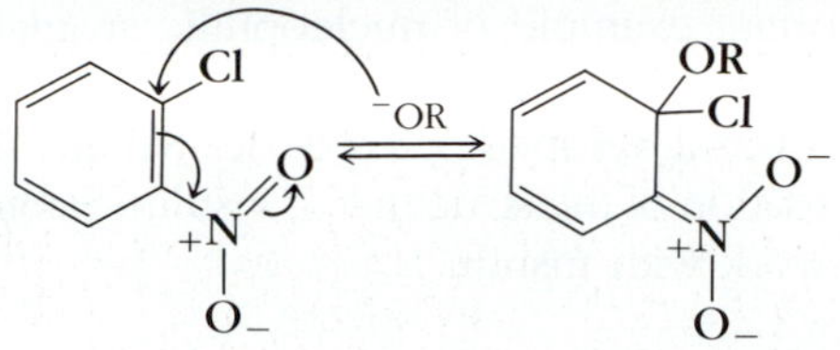

Cl, $^-$OR, O, +N, O$^-$ ⇌ OR, Cl, O$^-$, +N, O$^-$

This attack closely resembles conjugate addition to an unsaturated carbonyl group (Section 13.10) and the addition is reversible. But chloride is a better leaving group than $^-$OR so an overall substitution of —OR for —Cl occurs.

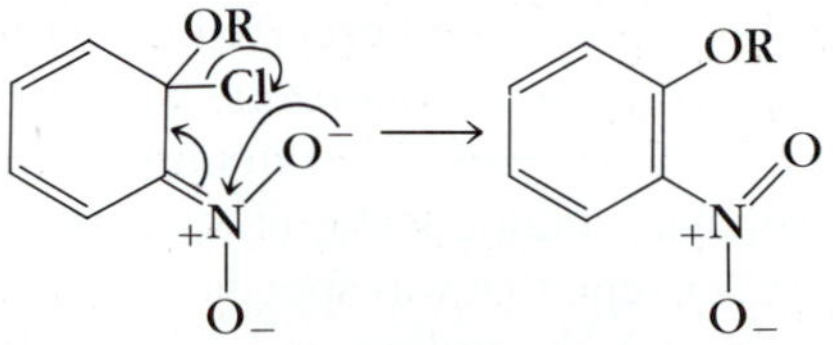

OR, Cl, O$^-$, +N, O$^-$ ⟶ OR, O, +N, O$^-$

Such addition-elimination reactions of activated aromatic halides are often called *nucleophilic aromatic substitutions.* Notice that in contrast to the situation with electrophilic aromatic substitution reactions, the nitro group acts as an *activating* substituent in these nucleophilic reactions. The attacking nucleophile can add to any carbon that is either *ortho* or *para* to a nitro group, but this may not be productive. For example, attack at the other carbon of *o*-chloronitrobenzene that is *ortho* to the nitro group would yield the following intermediate:

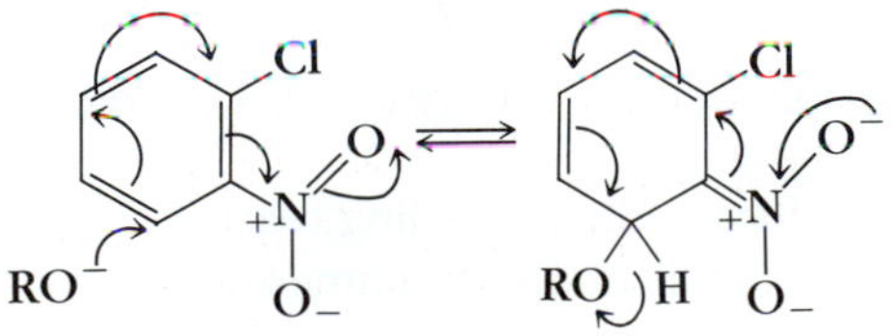

Cl, O, +N, O$^-$, RO$^-$ ⇌ Cl, O$^-$, +N, O$^-$, RO, H

This time —OR is the only leaving group on the carbon that has been attacked. Consequently, the addition is unproductive and the starting material is regenerated.

Nucleophilic aromatic substitution reactions have been used to prepare a variety of nitro-substituted aromatic compounds:

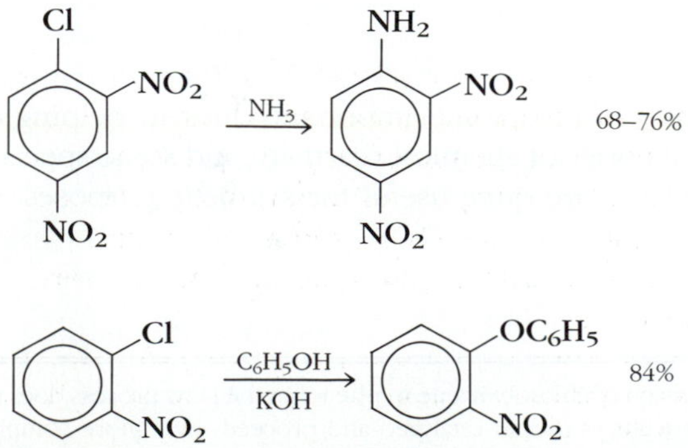

Cl, NO_2, NO_2 $\xrightarrow{NH_3}$ NH_2, NO_2, NO_2 68–76%

Cl, NO_2 $\xrightarrow[KOH]{C_6H_5OH}$ OC_6H_5, NO_2 84%

A particularly important example of nucleophilic aromatic substitution is the reaction of 2,4-dinitrofluorobenzene with the free amino groups of a protein (Section 20.6). This reagent was developed by Frederick Sanger (England), who used it extensively in his elucidation of the structure of insulin. Sanger received the Nobel Prize in 1958 for his work with insulin.

$$\text{2,4-}(O_2N)_2C_6H_3F \xrightarrow{H_2N{-}R} \text{2,4-}(O_2N)_2C_6H_3NH{-}R$$

The substitution reactions of unactivated aromatic halides proceed by a quite different mechanism. In 1953 John D. Roberts (California Institute of Technology) reported a study of the reaction of chlorobenzene with potassium amide in liquid ammonia. Replacement of —Cl by —NH_2 occurs under these conditions to yield aniline, but a strange result was obtained when radioactively labeled starting material was employed. Chlorobenzene that was specifically substituted with ^{14}C at the 1 position yielded a mixture of 1-^{14}C-aniline and 2-^{14}C-aniline. Clearly, the amino group was not specifically introduced at the carbon to which chlorine was originally bonded. Nor was the amino group specifically introduced at the adjacent position. Somehow reactivity had been conferred on *both* C-1 and C-2 of the ring. The evidence strongly indicated that an elimination reaction had occurred generating an intermediate with a highly strained triple bond.

$$\text{1-}^{14}\text{C-chlorobenzene} \xrightarrow{KNH_2} \text{benzyne} \xrightarrow{NH_3} \text{1-}^{14}\text{C-aniline} + \text{2-}^{14}\text{C-aniline}$$

The highly distorted triple bond is very reactive, and *addition* of ammonia occurs very rapidly.

The reactive intermediate in the preceding reaction is known as *benzyne,* and it undergoes addition of water, alcohols, or amines nearly as quickly as it is formed.*

$$p\text{-}CH_3C_6H_4Cl \xrightarrow[340°]{NaOH} [\text{methylbenzyne}] \longrightarrow p\text{-}CH_3C_6H_4OH\ (30\%) + m\text{-}CH_3C_6H_4OH\ (29\%)$$

Benzyne reactions provided some important clues in helping organic chemists develop modern theories of chemical reactivity, and some specialized reactions of benzyne intermediates are quite useful for synthetic purposes.

*Curiously, the conversion of chlorobenzene to phenol in the Dow process does not involve a benzyne intermediate. The reaction is copper-catalyzed and proceeds by a more complex mechanism.

EXERCISE 18.15

Write the detailed reaction mechanism for the conversion of *p*-nitrochlorobenzene to *p*-nitrophenol with sodium hydroxide.

EXERCISE 18.16

What product would be expected from the following reaction?

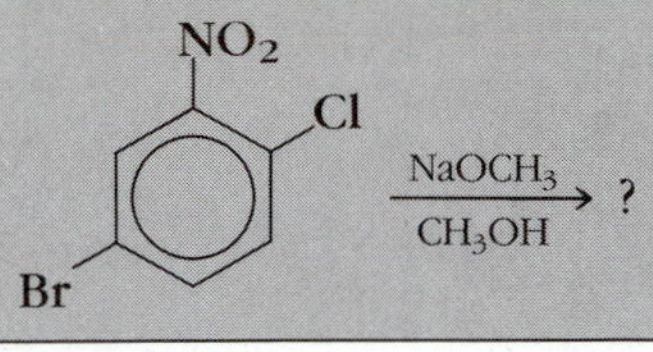

NO_2, Cl, Br (substituted benzene) $\xrightarrow[CH_3OH]{NaOCH_3}$?

EXERCISE 18.17

When *o*-chloroanisole was allowed to react with $NaNH_2$ in liquid NH_3, a 58% yield of *m*-anisidine was isolated. Write a detailed reaction mechanism that accounts for the formation of this product. What other product was probably formed (although not actually isolated)?

18.8 SUMMARY OF REACTIONS

Substitution reactions of benzene and benzene derivatives are summarized in Table 18.2, and some functional group interconversions of aromatic compounds are summarized in Table 18.3.

TABLE 18.2 Substitution Reactions of Benzene and Benzene Derivatives

Reaction	Comments
1. Hydrogen exchange	Section 10.7
$Ar{-}H \xrightarrow{^{*}H^{+}} Ar{-}H^{*}$	This reaction can only be observed by using isotopic labels.
2. Chlorination, bromination	Section 10.8
$Ar{-}H \xrightarrow[FeCl_3]{Cl_2} Ar{-}Cl$ $Ar{-}H \xrightarrow[FeBr_3]{Br_2} Ar{-}Br$	The ferric halide catalyst may not be necessary with highly reactive aromatic compounds.

TABLE 18.2 (continued)

Reaction	Comments
3. Nitration	Section 10.8
$Ar{-}H \xrightarrow[H_2SO_4]{HNO_3} Ar{-}NO_2$	
$Ar{-}H \xrightarrow{NO_2^+BF_4^-} Ar{-}NO_2$	
4. Sulfonation	Section 10.8, 18.2
$Ar{-}H \xrightarrow[H_2SO_4]{SO_3} Ar{-}SO_3H$	Sulfonation is reversible. Equilibrium mixtures of *ortho, meta,* and *para* isomers may be formed. Hydrolysis in strong acid replaces $—SO_3H$ by —H.
5. Friedel-Crafts alkylation	Section 18.3
$Ar{-}H \xrightarrow[AlCl_3]{R{-}X} Ar{-}R$	Carbocation rearrangements of R—X and of Ar—R are common. Polyalkylation is a major problem. Important for industrial synthesis; seldom used in laboratory synthesis. Deactivated compounds (Figure 18.1) do not react.
6. Friedel-Crafts acylation	Section 18.3
$Ar{-}H \xrightarrow[AlCl_3]{R{-}\overset{O}{\overset{\Vert}{C}}Cl} Ar{-}\overset{O}{\overset{\Vert}{C}}{-}R$	A versatile carbon-carbon bond-forming reaction for laboratory as well as industrial synthesis. Deactivated compounds (Figure 18.1) do not react.
$Ar{-}H \xrightarrow[AlCl_3]{R{-}\overset{O}{\overset{\Vert}{C}}{-}O{-}\overset{O}{\overset{\Vert}{C}}{-}R} Ar{-}\overset{O}{\overset{\Vert}{C}}{-}R$	Reduction of the acyl group provides a method for introducing an alkyl side chain.

TABLE 18.3 Functional Group Interconversions in Aromatic Compounds

Reaction	Comments
1. Sulfonic acids	Section 18.4
(a) In acid	
$Ar{-}SO_3H \xrightarrow[H_2SO_4]{H_2O} Ar{-}H$	Reverse of sulfonation: $-SO_3$ is replaced by H.
(b) In base	
$Ar{-}SO_3H \xrightarrow[250\text{–}300°C]{NaOH} Ar{-}OH$	Very stringent conditions; not generally useful for laboratory synthesis.
2. Halides (X═F, Cl, Br, I)	Section 18.7
(a) Reaction with strong base	
$Ar{-}X \xrightarrow[200\text{–}350°C]{NaOH} Ar{-}OH$ $Ar{-}X \xrightarrow[NH_3]{NaNH_2} Ar{-}NH_2$	Elimination-addition via benzyne intermediate. Requires stringent conditions; not generally useful for laboratory synthesis.
(b) Nucleophilic substitution of activated aromatic halides	
o-$NO_2C_6H_4X \xrightarrow{Nuc^-}$ o-$NO_2C_6H_4Nuc$	Limited to reaction of halides with an *o*- or *p*-NO_2 (or carbonyl) substituent. Nucleophile can be alkoxide ion or amino group.
(c) Formation of organometallic reagent	Sections 11.5, 13.4
$Ar{-}X \xrightarrow{M} Ar{-}M$	Useful intermediates for forming new carbon-carbon bonds. Other functional groups on the aromatic ring (e.g., OH, nitro, carbonyl) will interfere.

TABLE 18.3 (continued)

Reaction	Comments
3. Alkyl derivatives	
(a) Bromination	Sections 8.4, 18.4
$Ar—CH_2—R \xrightarrow{NBS} Ar—CH(Br)—R$	Benzylic bromination with NBS
(b) Oxidation	Sections 16.7, 18.4
$Ar—R \xrightarrow[Na_2Cr_2O_7]{KMnO_4 \text{ or}} Ar—CO_2H$	Other substituents may be oxidized also. (Halo and nitro groups are unaffected.)
4. Phenols	Sections 11.7, 11.8
(a) Alkylation	
$Ar—OH \xrightarrow[base]{RX} Ar—OR$	
(b) Acylation	
$Ar—OH \xrightarrow{R—\overset{O}{\overset{\Vert}{C}}Cl} Ar—O\overset{O}{\overset{\Vert}{C}}R$	
5. Amino derivatives	
(a) Acylation	Sections 16.6, 17.4, 18.3
$Ar—NH_2 \xrightarrow[\text{or } Ac_2O]{CH_3\overset{O}{\overset{\Vert}{C}}Cl} Ar—NH\overset{O}{\overset{\Vert}{C}}CH_3$	Useful as blocking group. Electrophilic substitution of amide is usually preferable to reaction of free amine. Secondary amines react similarly.
(b) Diazotization	Section 18.5
$Ar—NH_2 \xrightarrow{HONO} Ar—\overset{+}{N}≡N:$	Primary aromatic amines only. The diazonium salts are versatile synthetic intermediates.

TABLE 18.3 (continued)

Reaction	Comments
6. Diazonium salts	Section 18.5
(a) Replacement by —Cl, —Br, —I	
$Ar{-}N_2^+ \xrightarrow{CuCl} Ar{-}Cl$ $Ar{-}N_2^+ \xrightarrow{CuBr} Ar{-}Br$ $Ar{-}N_2^+ \xrightarrow{CuI} Ar{-}I$	Sandmeyer reaction. Copper powder plus a halide salt can be used in place of cuprous halide.
(b) Replacement by —F	
$Ar{-}N_2^+ \xrightarrow[\text{2. mild heating}]{\text{1. } HBF_4} Ar{-}F$	
(c) Replacement by —CN	
$Ar{-}N_2^+ \xrightarrow{CuCN} Ar{-}CN$	Sandmeyer reaction. Alternative method of carboxylic acid synthesis (cf. Section 16.7).
(d) Replacement by $—NO_2$	
$Ar{-}N_2^+ \xrightarrow{CuNO_2} Ar{-}NO_2$	Sandmeyer reaction. Permits sequence nitro → amino → nitro.
(e) Replacement by H	
$Ar{-}N_2^+ \xrightarrow[NaBH_4]{H_3PO_2 \text{ or}} Ar{-}H$	Reductive deamination: ($—NH_2 \rightarrow —N_2^+ \rightarrow H$). Permits use of nitro and amino groups as blocking or activating groups.
(f) Replacement by OH	
$Ar{-}N_2^+ \xrightarrow[H_2SO_4]{H_2O} Ar{-}OH$	Useful synthesis of phenols.
(g) Coupling with ArX ($X{=}NH_2$, OH)	
$Ar{-}N_2^+ \xrightarrow{C_6H_5{-}X} Ar{-}N{=}N{-}C_6H_4{-}X$	Preparation of azo dyes.

TABLE 18.3 (continued)

Reaction	Comments
7. Reduction of aromatic ring	
(a) Catalytic hydrogenation	Section 8.6
$\xrightarrow[\text{cat.}]{H_2}$	High pressures sometimes required.
(b) Birch reduction	Section 22.4
$\xrightarrow[NH_3]{Li}$	Hydrogens not usually added to positions with electron-donating groups. (Na or K can also be used.)

18.9 PROBLEMS

18.1 For each of the following substituents, predict whether it would be *o,p*-directing or *m*-directing. Would the corresponding monosubstituted benzene be more or less reactive than benzene in an electrophilic substitution reaction?

(a) $—CH_2CH_2CH_3$ (b) $—OC_6H_5$ (c) $—CF_3$ (d) $—\overset{O}{\overset{\|}{C}}—NH_2$

(e) $—NH—\overset{O}{\overset{\|}{C}}H$ (f) $—O\overset{O}{\overset{\|}{C}}CH_3$ (g) $—\overset{O}{\overset{\|}{C}}OCH_3$

18.2 Rank the following compounds in order of reactivity toward nitration with $NO_2^+ \ BF_4^-$.

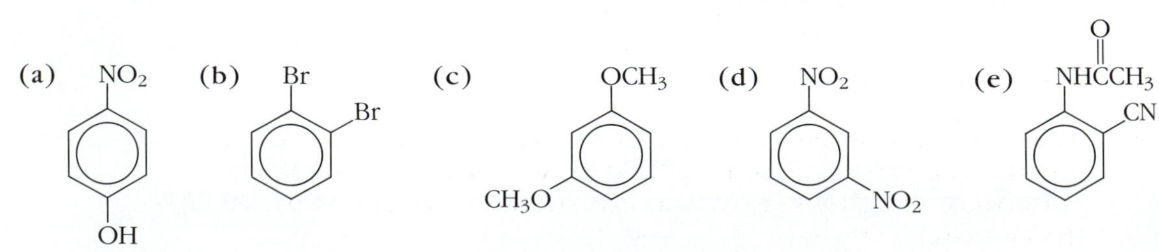

18.3 Draw the product of mononitration expected for each of the compounds in Problem 18.2.

18.4 Write a detailed reaction mechanism for the nitration of benzene. Draw all resonance forms for the intermediate cation.

18.5 Write a detailed reaction mechanism for the acylation of anisole with propionyl chloride and $AlCl_3$. Draw all resonance forms of the intermediate cations for *ortho, meta,* and *para* substitution. Indicate those resonance contributors that are particularly favorable or unfavorable in order to show why substitution occurs at the *ortho* and *para* positions.

18.6 Write a detailed reaction mechanism for the bromination of benzoic acid with bromine and $FeCl_3$. Draw all resonance forms of the intermediate cations for *ortho, meta,* and *para* substitution. Indicate those resonance contributors that are particularly favorable or unfavorable in order to show why substitution occurs at the *meta* position.

18.7 Depending upon the site of substitution, the cationic intermediate formed in electrophilic substitution of 4-methoxytoluene by bromine can sometimes be drawn as an oxonium ion. For the intermediate produced by reaction at each of the possible sites, draw all resonance forms. State whether or not there is a resonance form that is an oxonium ion.

18.8 Bromination of biphenyl is 1000 times faster than bromination of benzene, but fluorene reacts more than 1 million times more rapidly than benzene. Suggest an explanation.

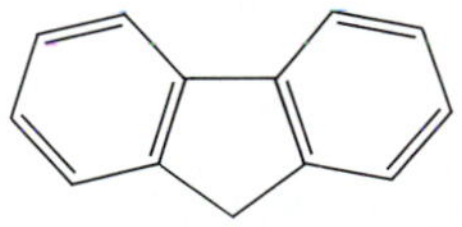

Fluorene 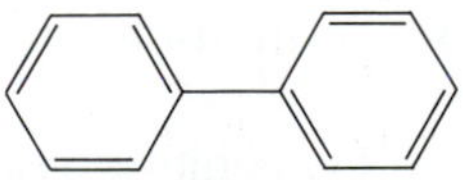 Biphenyl

18.9 The electron-donating abilities of methyl and isopropyl groups are comparable, yet *p*-cymene undergoes nitration to yield a product composed of more than 70% of the isomer in which the nitro group is *ortho* to the methyl group. Suggest an explanation.

NO_2

$\xrightarrow[H_2SO_4]{HNO_3}$

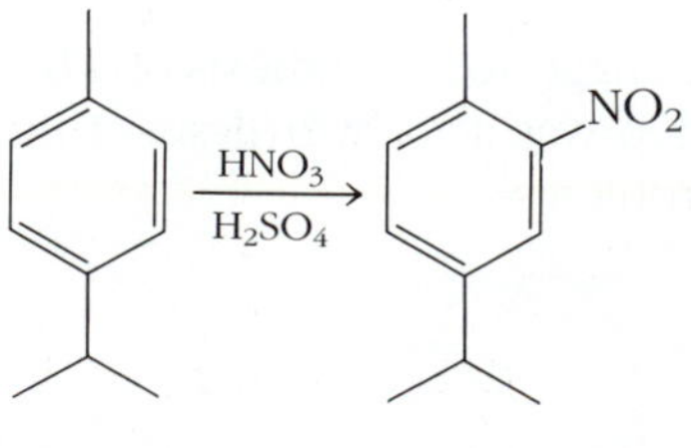

18.10 Dinitration of bromobenzene yields 2,4-dinitrobromobenzene. In Section 18.2 we illustrated this reaction as proceeding via *p*-nitrobromobenzene. What results would be expected if the initial nitration occurred *ortho* to the bromo substituent?

18.11 The coupling reactions of diazonium salts with phenols afford azo dyes in which reaction has occurred at the position *para* to the phenolic —OH. This characteristic reaction can be used as test for the presence of an aromatic diazonium ion (and hence for a primary aromatic amine via diazotization). Draw the structure of the product formed when aniline is treated first with nitrous acid and then with phenol.

18.12 Suggest a method for synthesizing diphenyl ether from benzene.

18.13 Suggest a method for preparing 2,4-dinitrofluorobenzene from benzene.

18.14 Attempted esterification of *o*-benzoylbenzoic acid via reaction with PCl_5 followed by treatment with ethanol yielded a brightly colored, neutral compound that absorbed in the carbonyl region of the infrared spectrum. However, no ethyl group was observed in the proton NMR. Deduce the structure of the product.

18.15 When diphenyl ether ($C_{12}H_{10}O$) is allowed to react with excess acetyl chloride in the presence of $AlCl_3$, a diacetyl product is formed in 60% yield. What is the most likely structure of this compound? Why?

18.16 When *p*-cymene (Prob. 18.9) reacts with acetyl chloride and aluminum chloride, the major product (50–55% yield) is a monoacetyl chloride. What is the most likely structure of this product? Why?

18.17 2-Methoxybiphenyl and 4-methoxybiphenyl behave differently in Friedel-Crafts acylations with acetyl chloride and $AlCl_3$. High yields are obtained, but in one case the acetyl and methoxy groups are on the same ring and in the other case the groups are on different rings. What are the structures of the products? Why do the isomeric reactants behave differently?

18.18 (a) 2,4-Dinitrobenzonitrile reacts with HNO_3/H_2SO_4 much more slowly than does chlorobenzene. The opposite order of reactivity is observed for reaction with KOH/CH_3OH. Explain this. What products are expected for each reaction?

(b) Benzonitrile reacts with HNO_3/H_2SO_4 much more slowly than does chlorobenzene. The opposite order of reactivity is observed for reaction with KOH/CH_3OH. Explain this. What products are expected for each reaction?

18.19 In Example 18.6 we suggested a synthesis of 4-bromo-2-methylbenzoic acid in which the last step was acid hydrolysis of the nitrile. Why might basic hydrolysis cause problems?

18.20 Suggest methods of synthesis for each of the following. All aromatic carbons must be derived from the indicated starting materials.

(a) *p*-Cymene

CH_3 from benzene and/or toluene

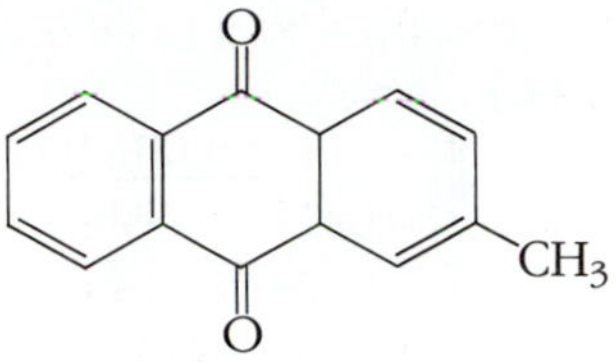

(b) 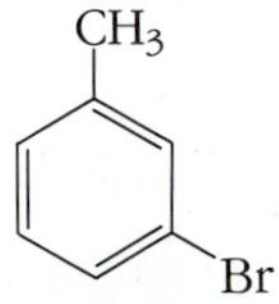 CH_3 Br from toluene

(c) 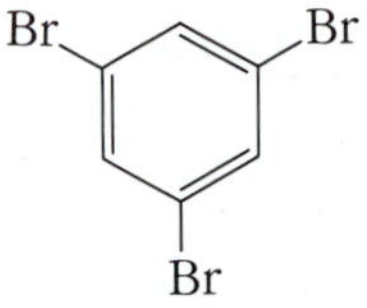 Br Br Br from benzene

(d)  NO_2 NO_2 from benzene

(e) 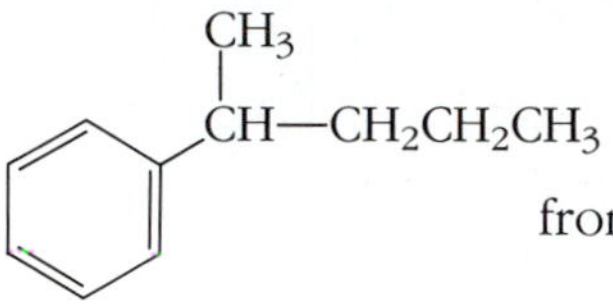 CH_3 CH—$CH_2CH_2CH_3$ from benzene

(f) 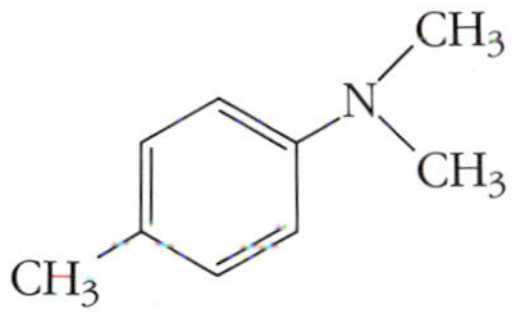 CH_3 N CH_3 CH_3 from toluene

18.21 Suggest methods for specifically converting *tert*-butylbenzene to:

(a) *p*-Bromo-*tert*-butylbenzene
(b) *m*-Bromo-*tert*-butylbenzene
(c) *o*-Bromo-*tert*-butylbenzene

18.22 Suggest methods for converting *p*-nitrotoluene to:

(a) *m*-Nitrotoluene
(b) *p*-Nitrobenzoic acid
(c) 3,4-Dinitrotoluene
(d) 2-Chloro-4-nitrotoluene

18.23 Complete the following reactions:

(a) 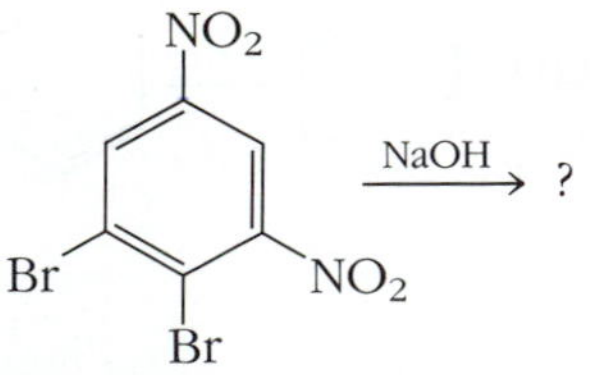 NO_2 Br Br NO_2 $\xrightarrow{NaOH}$?

(b) 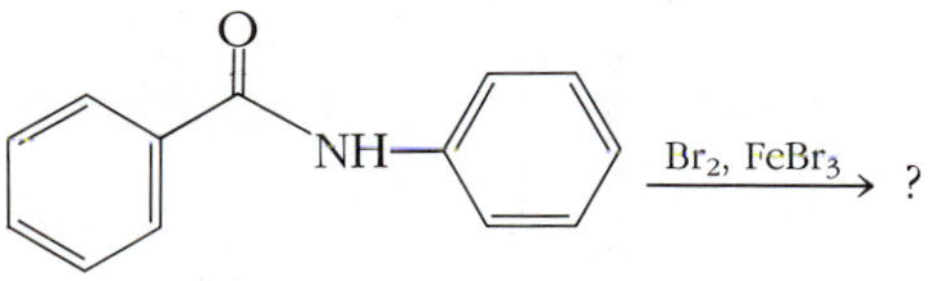 O NH— $\xrightarrow{Br_2,\ FeBr_3}$?

(c) $C_6H_5CH_2—C(=O)—OCH_2CH_3$ $\xrightarrow[2.\ H_2O]{1.\ LiAlH_4}$?

(d) $C_6H_5—C(=O)—CH_3$ $\xrightarrow{KOH,\ Br_2}$?

(e) C_6H_6 $\xrightarrow[AlCl_3]{CH_3CH_2CH_2Cl}$?

18.24 1,3,5-Trideuteriobenzene can be prepared in good yield from benzene in several steps. Outline a reasonable synthesis.

18.25 Suggest methods for carrying out the indicated conversions:

(a) benzene $\xrightarrow{?}$ NH_2 / CO_2H (meta-substituted benzene)

(b) H_2N–C_6H_5 $\xrightarrow{?}$ H_2N–C_6H_4–$CH_2CH_2CH_2CH(CH_3)—CH_3$

(c) naphthalene $\xrightarrow{?}$ 1-bromonaphthalene (Br)

(d) C_2H_5–C_6H_5 $\xrightarrow{?}$ C_2H_5–C_6H_4–SO_3H

(e) NO_2–C_6H_4–Cl $\xrightarrow{?}$ NO_2–C_6H_4–OCH_3

(f) CH_3–C_6H_4–NH_2 $\xrightarrow{?}$ CH_3–C_6H_4–Br

(g) benzene $\xrightarrow{?}$ C_6H_4(–$C(=O)—CH_3$)(–NO_2)

(h) benzene $\xrightarrow{?}$ C_6H_4(NO_2)(OH)

(i) $C_6H_5CH_2CH_2CH_2CH_2OH$ $\xrightarrow{?}$ tetralin

18.26 Suggest methods of synthesis for each of the following compounds. All aromatic carbons must be derived from the indicated starting materials.

(a) 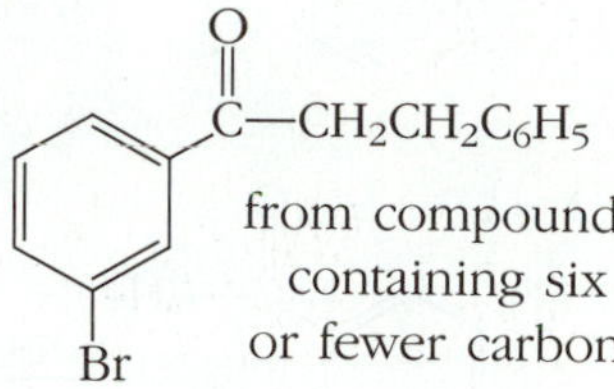 m-$BrC_6H_4C(=O)CH_2CH_2C_6H_5$ from compounds, containing six or fewer carbons

(b) 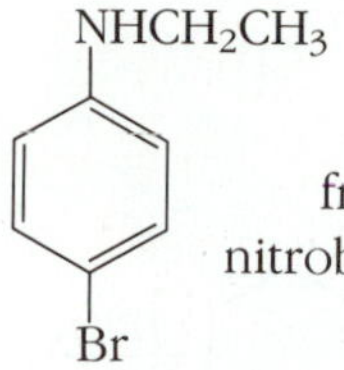 p-$BrC_6H_4NHCH_2CH_3$ from nitrobenzene

(c) 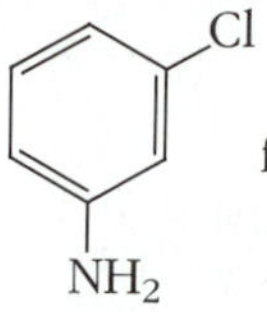 m-$ClC_6H_4NH_2$ from benzene

(d) 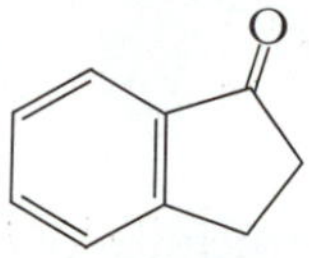 1-indanone from benzoic acid

(e) 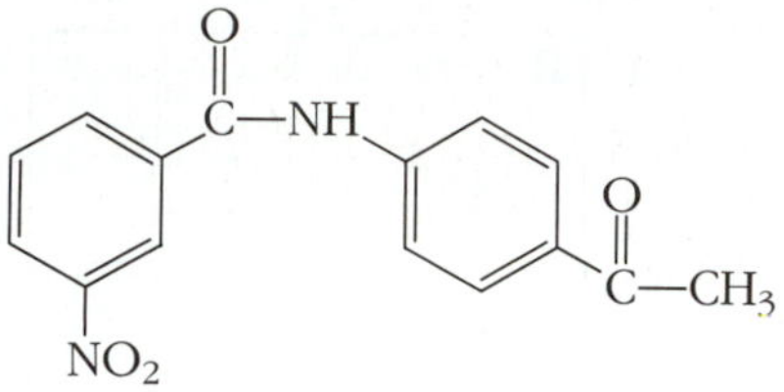 m-$O_2NC_6H_4C(=O)$–NH–C_6H_4-p-$C(=O)CH_3$ from compounds containing six or fewer carbons

(f) 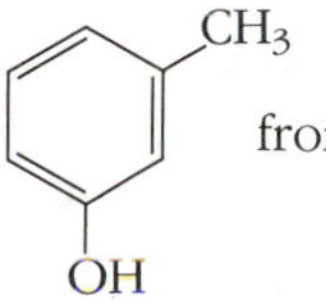 m-$CH_3C_6H_4OH$ from toluene

(g) 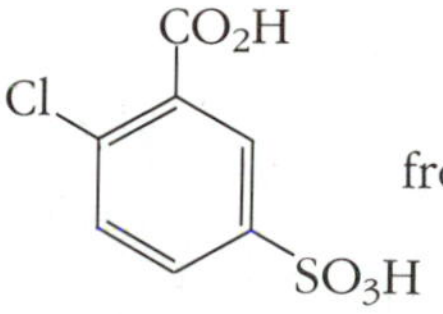 2-Cl, 5-SO_3H substituted $C_6H_3CO_2H$ from benzene

(h) 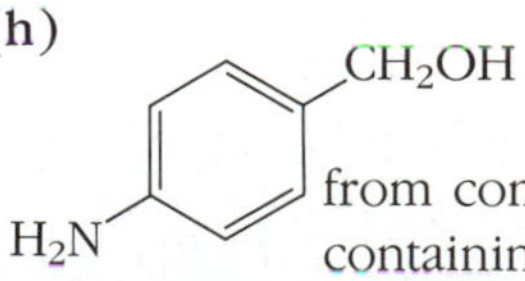 p-$H_2NC_6H_4CH_2OH$ from compounds containing six or fewer carbons

(i) 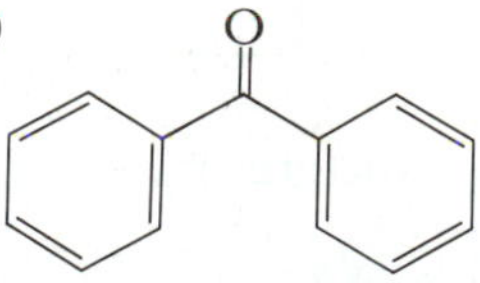 $C_6H_5C(=O)C_6H_5$ from compounds containing six or fewer carbons

18.27 Complete the following reactions:

(a) 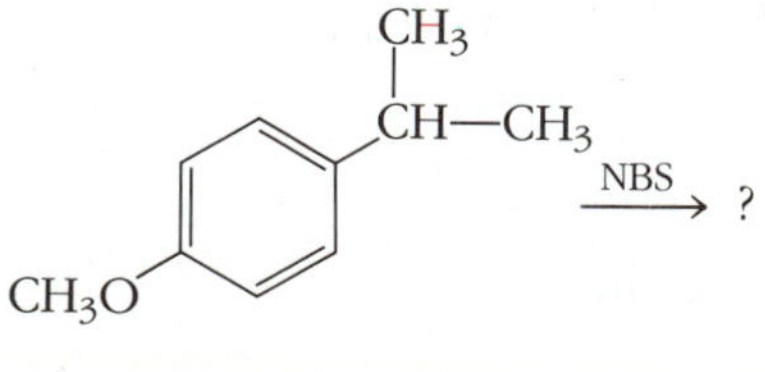 p-$CH_3OC_6H_4CH(CH_3)CH_3$ $\xrightarrow{NBS}$?

(b) 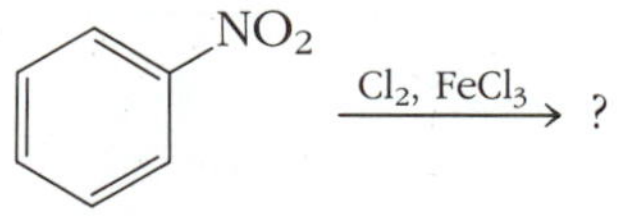 $C_6H_5NO_2$ $\xrightarrow{Cl_2,\ FeCl_3}$?

(c) $C_6H_5CO_2H$ $\xrightarrow[H_2SO_4]{HNO_3}$?

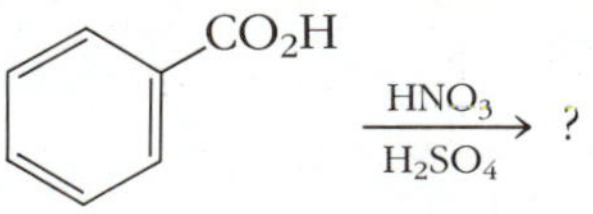

(d) $C_6H_5OCH_2C_6H_5$ $\xrightarrow[AlCl_3]{CH_3C(=O)-Cl}$?

(e) 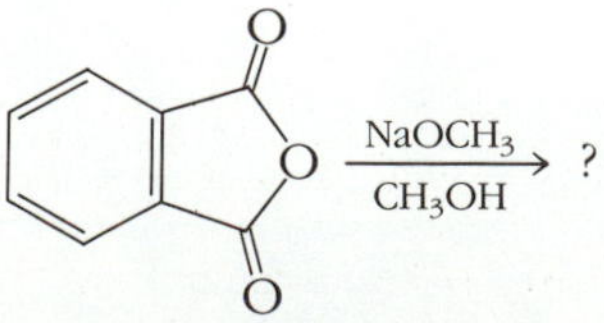 phthalic anhydride $\xrightarrow[CH_3OH]{NaOCH_3}$?

(f) 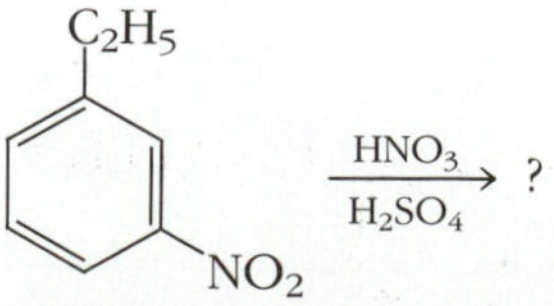 m-$C_2H_5C_6H_4NO_2$ $\xrightarrow[H_2SO_4]{HNO_3}$?

(g) OCH_3 / CH_3 $\xrightarrow{H_2SO_4,\ SO_3}$?

(h) $CH_2CH_2CO_2H$ $\xrightarrow{PCl_5}$?

(i) O $\xrightarrow[H_2SO_4]{HNO_3}$?

(j) O ‖ C—CH_3 / Br $\xrightarrow[KOH]{NH_2NH_2}$?

18.28 Deduce the structures of compounds **A** and **B**.

A $C_8H_8O_3$ $\xrightarrow[AlCl_3]{CH_3\overset{O}{\overset{\|}{C}}—Cl}$ **B** $\xrightarrow{NaOH,\ Br_2}$ **C** CH_3O / CO_2H / CO_2H

18.29 Deduce the structure of compound **A**.

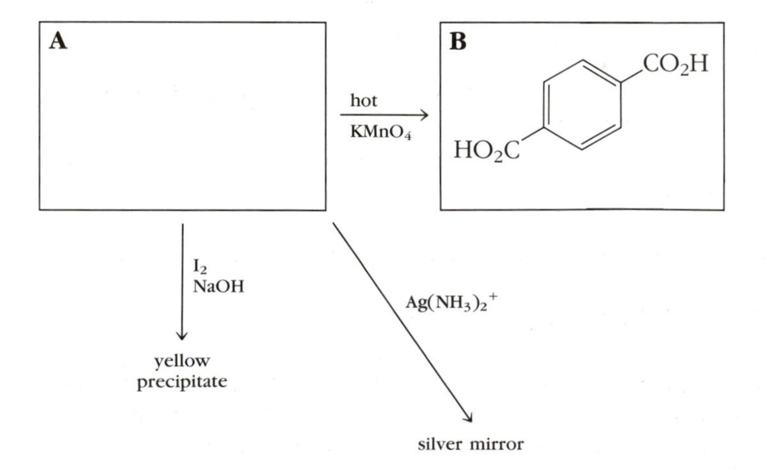

18.30 Suggest methods for carrying out each of the following conversions:

(a) C_2H_5 → Br, C_2H_5

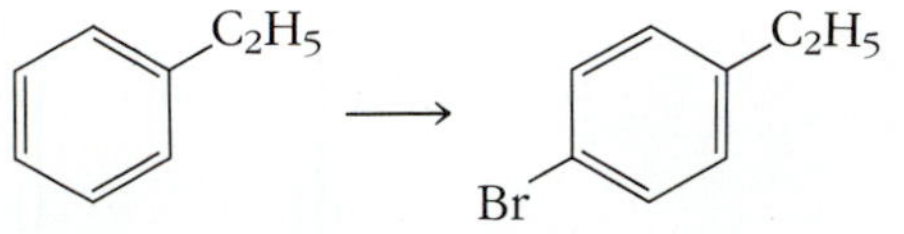

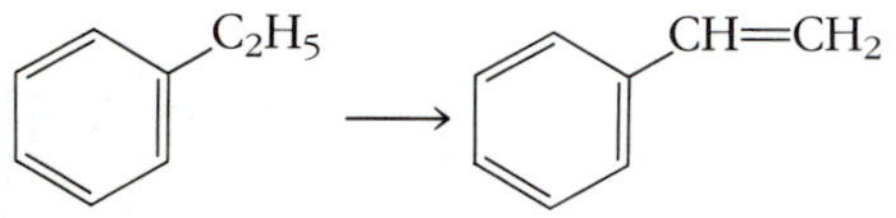

(b) C_2H_5 → $CH{=}CH_2$

(c) C_2H_5 → I, C_2H_5

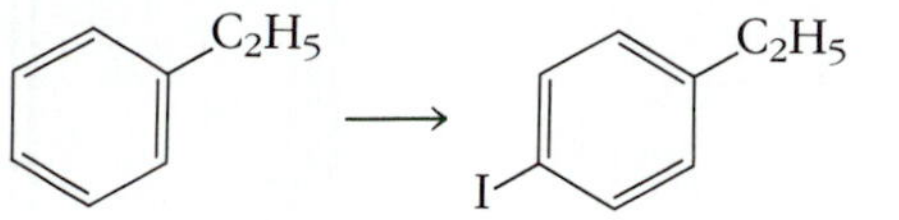

(d) C_2H_5 → NC, C_2H_5

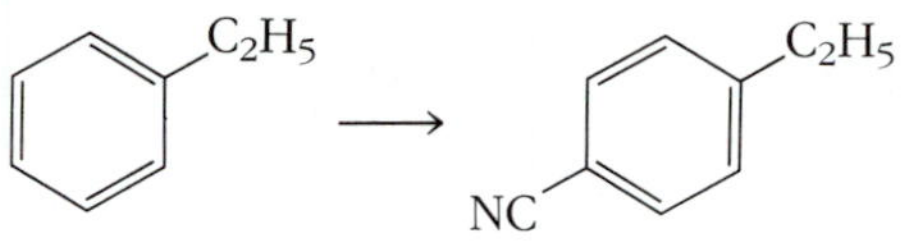

(e) C_2H_5 → CN

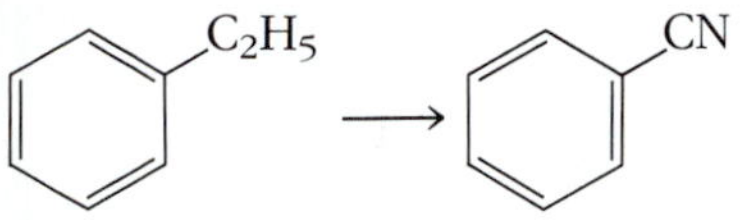

18.31 Deduce the structures of compounds **A–C.**

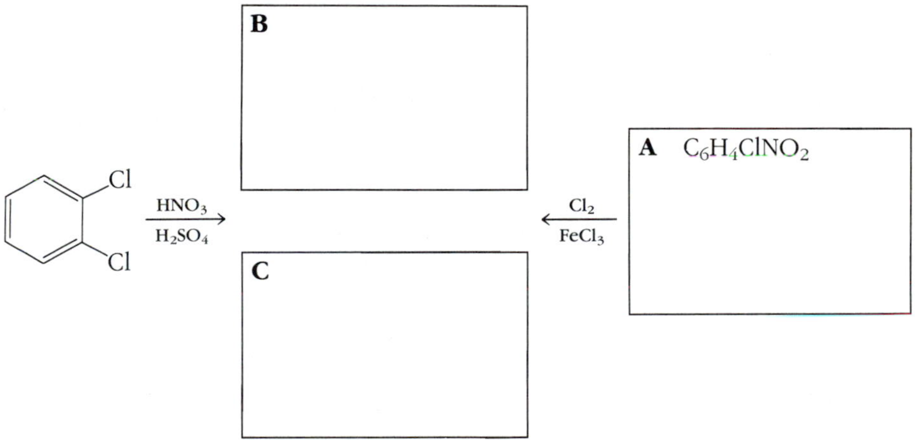

Cl, Cl $\xrightarrow[H_2SO_4]{HNO_3}$ **B** / **C** $\xleftarrow[FeCl_3]{Cl_2}$ **A** $C_6H_4ClNO_2$

18.32 In 1936 the synthesis of 3,4,5-[$^2H_3$]benzoic acid was reported. A Friedel-Crafts–type acylation was used to convert deuterated benzene (C_6D_6) to benzamide, and this was hydrolyzed to benzoic acid with aqueous acid. What isomeric compound is a more reasonable alternative to that which was originally claimed?

18.33 Suggest a synthesis of pentadeuteriobenzoic acid starting from C_6D_6.

18.34 How could C_6D_6 be prepared from C_6H_6?

18.35 A Friedel-Crafts reaction of benzene was carried out with 1-bromopropane and $AlCl_3$. Two of the products isolated from the reaction mixture were compounds **A** and **B,** and their proton NMR spectra are shown on page 1008. Deduce their structures.

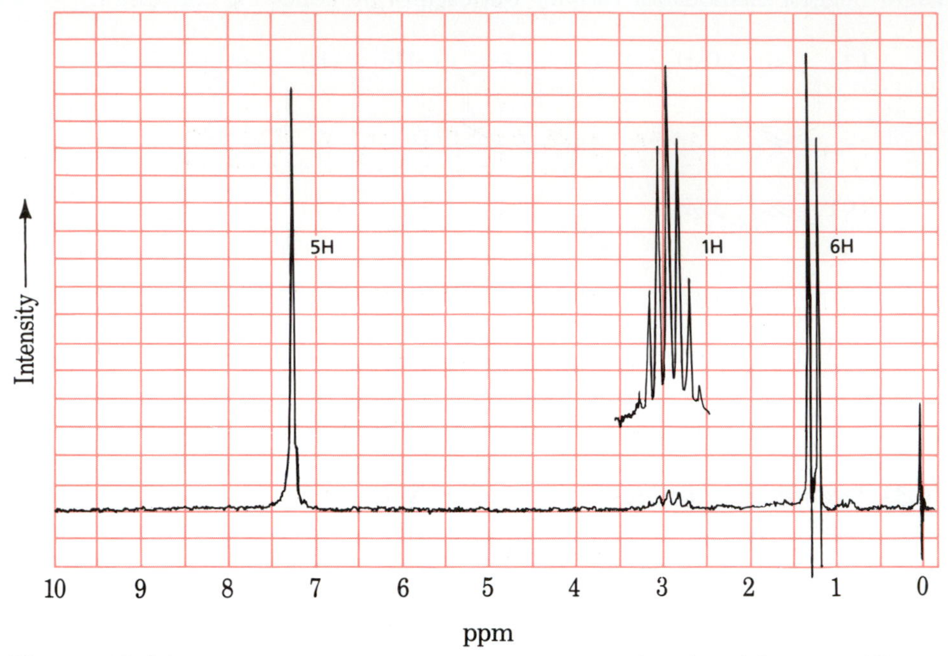

Figure 18.4 Proton NMR spectrum of Compound A (Problem 18.35).

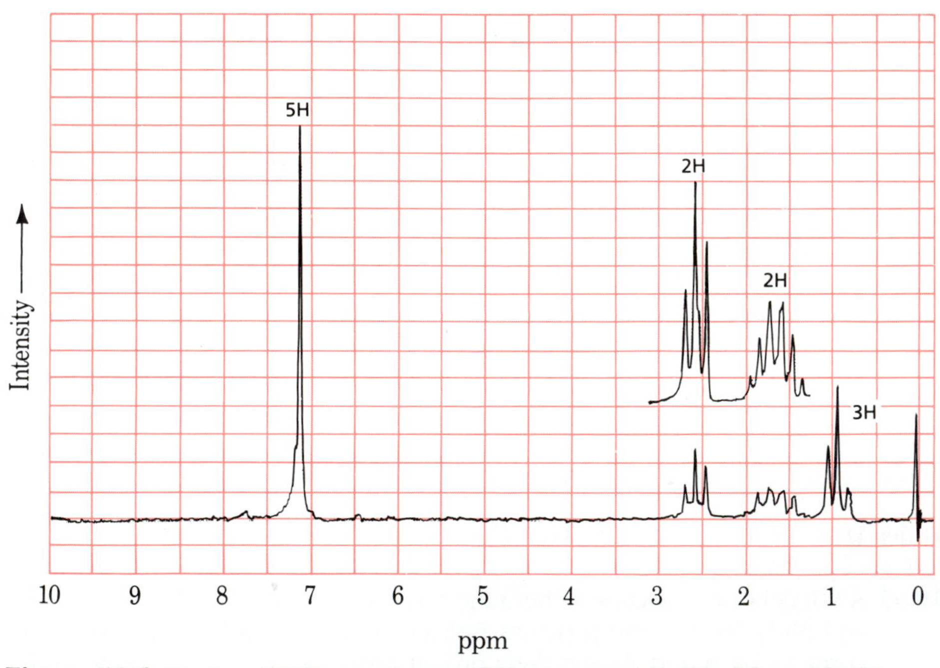

Figure 18.5 Proton NMR spectrum of Compound B (Problem 18.35).

18.36 For each of the reactions in Problem 18.27 describe how you could use spectroscopic techniques to verify that the product obtained actually has the expected structure. Indicate what specific peaks you would look for (or expect to be absent) in the NMR or IR spectrum.

CHAPTER 19

CARBOHYDRATES

Carbohydrates comprise the class of naturally occurring organic compounds that are more generally known as **sugars.** They play a basic role in the biochemistry of living organisms, particularly in the storage and production of energy. Photosynthesis in plants produces the sugar *glucose* from carbon dioxide converting solar energy to chemical energy. This chemical energy can be stored in an organism as a derivative of glucose, and oxidation of the glucose releases energy when it is needed. It is the oxidation of glucose that provides the energy your body requires to function normally.

Because of the extent to which carbohydrates are involved in every day life, some of the terms have developed more than one meaning. To a chemist the term "sugar" will indicate one of the many compounds belonging to the carbohydrate family. To the layman, however, "sugar" will almost certainly mean the specific substance, *sucrose* ($C_{12}H_{22}O_{11}$), that is obtained from sugar cane or sugar beets and is also called *table sugar.* Similarly, glucose is sometimes called *grape sugar* (from one of its commercial sources) or *dextrose* (which indicates the sign of its optical rotation).

The term *carbohydrate* was first used in the mid-1800s, when chemists recognized a pattern in the molecular formulas of many sugars. As seen for glucose and sucrose, the molecular formulas can be written as *hydrates of carbon.*

$$\text{Glucose} = C_6H_{12}O_6 = C_6 \cdot 6\ H_2O$$

$$\text{Sucrose} = C_{12}H_{22}O_{11} = C_{12} \cdot 11\ H_2O$$

Of course, we now know that the molecular structures are far more complex than suggested by this notation.

The most common naturally occurring carbohydrates are made up of subunits with a single five- or six-carbon chain. Sometimes these subunits are combined by ether linkages into more complex molecules, and carbohydrates are often classified according to the number of subunits. Glucose, for example, is a *monosaccharide,* while sucrose, with two six-carbon subunits (glucose and fructose), is a *disaccharide.* (In Section 19.9 we will discuss some important *poly*saccharides.) The term **saccharide** is yet another name for sugar, which can be traced back to the word for sugar in a variety of early languages (Latin, *saccharum;* Greek, *sakkharon;* Sanskrit, *sarkara;* Pali, *sakkhara*). In many places you will find that the terms carbohydrate, sugar, and saccharide are used almost interchangeably.

Despite its name, the artificial sweetener *saccharin* is not a carbohydrate at all. This compound, first synthesized in 1879 by Ira Remsen at Johns Hopkins University, is a sulfonamide analog of phthalimide.

Saccharin

Saccharin is more than 500 times sweeter than sucrose, and despite some questions about its potential as a carcinogen, it remains widely used as a sugar substitute. It does not enter the normal biochemical pathways for sugar metabolism, so it can be used as a sweetener by individuals with certain metabolic disorders, particularly diabetes, involving carbohydrates. Because saccharin is not metabolized by the human body, it is also used by individuals wishing to restrict the total caloric content of the food they eat. The use of saccharin in foods has decreased markedly in recent years, largely because of the introduction of another artificial sweetener, *aspartame* (an amino acid derivative, Chapter 20), which emerged as an important commercial product in 1985.

19.1 STRUCTURE AND NOMENCLATURE OF SIMPLE SUGARS

Simple monosaccharides have the molecular formula $(CH_2O)_n$. One of the carbons of the unbranched chain is a carbonyl, and each of the remaining carbons has a hydroxyl substituent. If the carbonyl group is at the end of the chain (i.e., if it is an aldehyde carbonyl), the sugar is called an **aldose,** and if it is a ketone carbonyl, the sugar is called a **ketose.**

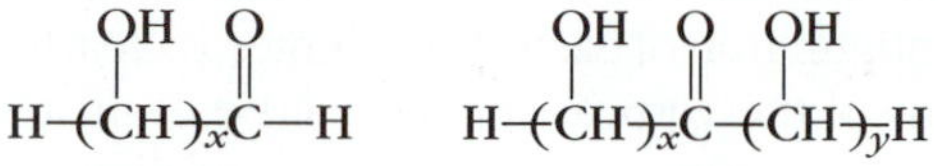

$$\mathrm{H{+}CH(OH){+}_x C(=O){-}H} \qquad \mathrm{H{+}CH(OH){+}_x C(=O){+}CH(OH){+}_y H}$$

Aldose — Ketose

Glucose is an aldose with six carbons; it is therefore called an aldohexose.

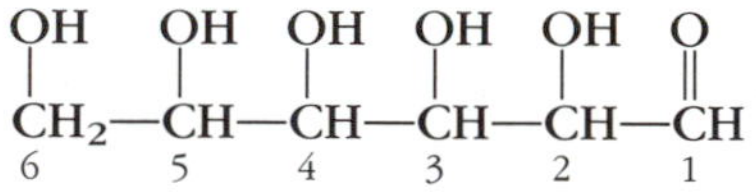

$$\mathrm{\underset{6}{CH_2(OH)}{-}\underset{5}{CH(OH)}{-}\underset{4}{CH(OH)}{-}\underset{3}{CH(OH)}{-}\underset{2}{CH(OH)}{-}\underset{1}{CH(=O)}}$$

Glucose, an aldohexose

Similarly, fructose is a ketohexose.

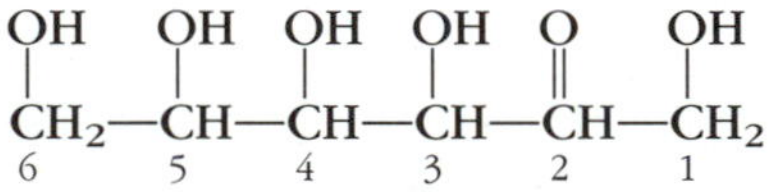

$$\mathrm{\underset{6}{CH_2(OH)}{-}\underset{5}{CH(OH)}{-}\underset{4}{CH(OH)}{-}\underset{3}{CH(OH)}{-}\underset{2}{C(=O)}{-}\underset{1}{CH_2(OH)}}$$

Fructose, a ketohexose

Structural formulas such as the preceding drawings of glucose and fructose are an awkward way to depict carbohydrates, and other types of structural formulas (particularly Fischer projections) are ordinarily used. The systematic names for carbohydrates are also cumbersome; for example that for glucose is 2,3,4,5,6-pentahydroxyhexanal. This problem becomes even worse when you recall that each of the chiral centers must also be designated as *R* or *S*. It is certainly easy to understand why the common name D-glucose is used in preference to (2*R*,3*S*,4*R*,5*R*)-2,3,4,5,6-pentahydroxyhexanal.

	CHO	
H	—	OH
HO	—	H
H	—	OH
H	—	OH
	CH₂OH	

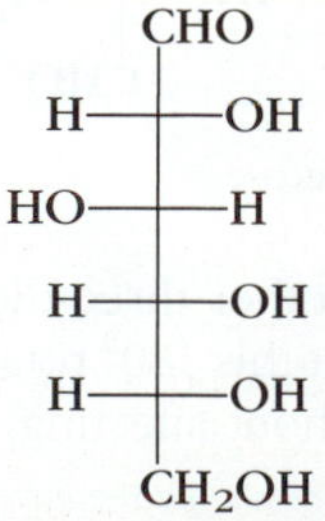

D-Glucose: (2*R*,3*S*,4*R*,5*R*)-2,3,4,5,6-Pentahydroxyhexanal

The aldohexoses have four chiral centers, so 16 stereoisomers (eight enantiomeric pairs) exist. A ketohexose has only three chiral centers, but there are still eight stereoisomers (four pairs of enantiomers). Each compound has a common name, and despite the effort required to remember the different structures, common names are used almost exclusively in carbohydrate chemistry.

Stereochemical Conventions for Carbohydrates

Before we present the common names of the more important monosaccharides, several stereochemical conventions used in sugar chemistry must be defined. Fischer projections are used extensively in carbohydrate chemistry, and we will briefly review the basic notation. (You may also wish to consult the earlier discussions in Section 6.4 and the appendix to Chapter 6 in the Student Solutions Manual). Fischer projections provide a very convenient and uniform way to denote the stereochemistry of the various structurally related sugars. They have the disadvantage as *projections* that they do not closely resemble the actual three-dimensional structures, but the configuration of each chiral center is clearly defined. In a Fischer projection the carbons of the vertical chain must be evaluated one at a time. For each of these carbons the configuration is defined as follows: If the carbon atom is taken to lie in the plane of the paper, the horizontal substituents are in front of the plane of the paper and the vertical substituents are behind the plane of the paper. This is illustrated for C-2 of D-glucose.

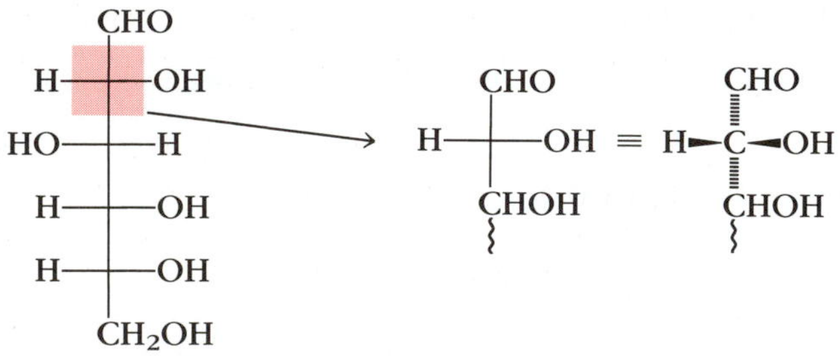

CHO
H—OH
HO—H
H—OH
H—OH
CH_2OH

CHO
H—OH ≡ H—C—OH
CHOH

The strict interpretation placed upon "vertical" and "horizontal" substituents of a Fischer projection precludes moving the drawing in any way other than rotating it 180° in the plane of the paper.

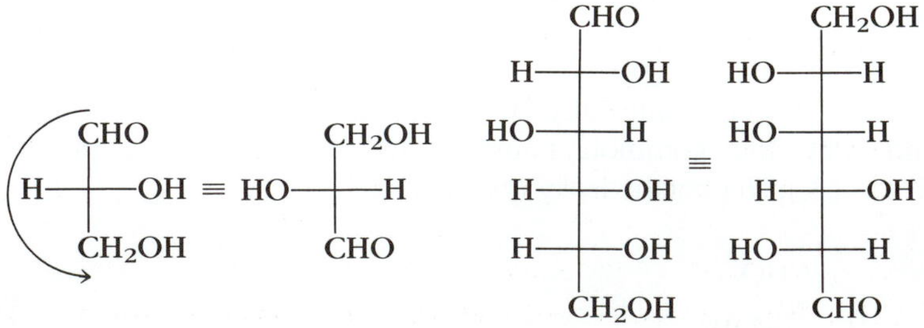

CHO
H—OH ≡ HO—H
CH_2OH / CHO

D-Glyceraldehyde

CHO: H—OH, HO—H, H—OH, H—OH, CH_2OH ≡ CH_2OH: HO—H, HO—H, H—OH, HO—H, CHO

D-Glucose

It is always possible to convert a Fischer projection into other three-dimensional representations that could be more easily manipulated, but this 180° rotation is the *only* way that a Fischer projection can be reoriented without affecting its stereochemical meaning.

You may occasionally encounter several different ways in which Fischer projections are drawn by other authors. Sometimes the individual carbon atoms are denoted by the letter C, but this is more tedious to draw and will not be used here. It was this type of drawing (Figure 19.1b) that was in fact originally introduced by Emil Fischer in 1891. Another drawing technique omits the hydrogens on chiral centers, showing only the hydroxyl groups (Figure 19.1c). In a fourth method only the bonds to the hydroxyl groups are shown and *none* of the horizontal substituents are drawn (Figure 19.1d).

The last two types of drawing shown in Figure 19.1 (i.e., 19.1c and 19.1d) employ very specific conventions, and great care must be used with them. Remember that in all other areas of organic chemistry we use a line to indicate bonds between carbon atoms, and such a bond extending from the principal chain would be used to indicate a methyl group.

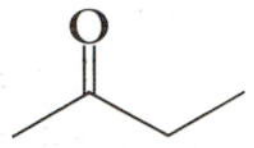

2-Butanone 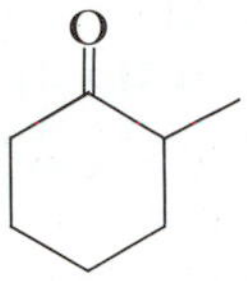 2-Methylcyclohexanol

Only in carbohydrate chemistry do we sometimes use a line to indicate other substituents such as hydrogen (Figure 19.1c) or hydroxyl groups (Figure 19.1d). Because the drawings are less cluttered, we will usually employ the convention shown in Figure 19.1c when we use Fischer projections to show the structures of carbohydrates in this chapter.

Diastereomeric sugars (with different physical and chemical properties) all have different names, but *enantiomeric* sugars have the same common name. Enantiomeric sugars are differentiated by the use of the prefix D or L, a notation that was first used with sugars in the late 1800s. In other areas of chemistry the use of D and

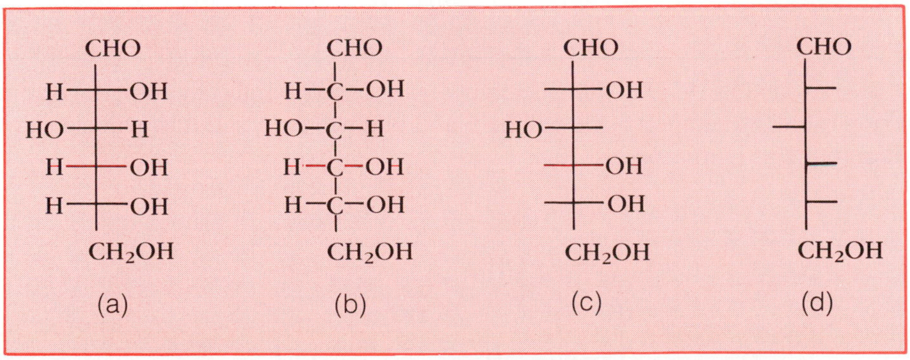

(a) (b) (c) (d)

Figure 19.1 D-Glucose, as drawn using several versions of the Fischer projection.
(a) The type that shows each substituent; (b) the convention originally used by Fischer in which the individual carbon atoms are drawn; (c) the type in which each C—H bond is drawn but the letter H is not; and (d) the variation in which each bond to OH is drawn (but the OH is not) and the C—H bonds are not shown.

L has been replaced by the *R/S* designation of configuration (Section 6.3), but the D/L notation has been retained by workers in the carbohydrate field.

The designation of a sugar as D or L specifically denotes only the configuration of the highest-numbered chiral center. This is always the hydroxyl end of an aldose, and ketoses are similarly numbered in the direction giving the lower possible number to the carbonyl carbon. To determine whether a particular structure belongs to the D or the L series of sugars, you must draw the Fischer projection with C-1 at the top and then examine the highest-numbered chiral center. When the hydroxyl substituent is on the right, the molecule is a D sugar, and when it is on the left, the compound is an L sugar. We have illustrated this with the following structures for the two possible configurations of glyceraldehyde:

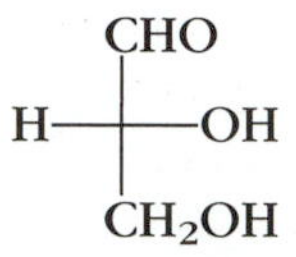

D-Glyceraldehyde 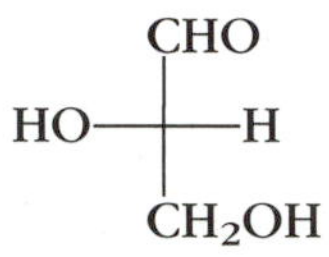 L-Glyceraldehyde

If two sugars have the same common name and differ only in the D or L designation, they are enantiomers. This means that they have the opposite configuration at *each* and *every* chiral center. Notice that the D/L designation is determined by the configuration at C-5 (the highest-numbered chiral center) for the following two aldohexoses:

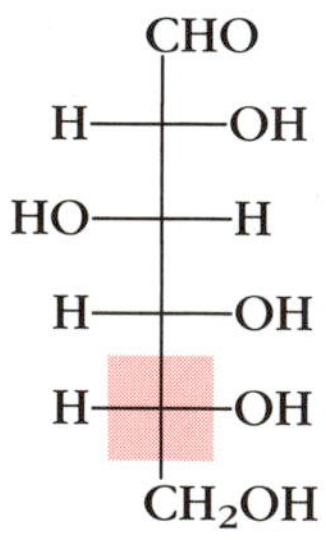

D-Glucose 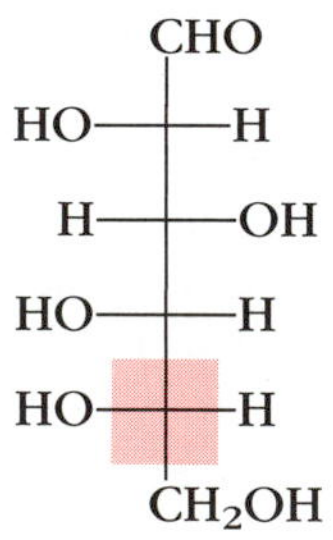 L-Glucose

Two sugars that differ at *only* the highest-numbered chiral center are not enantiomers but diastereomers. One will be D and one will be L, and they will also have different common names.

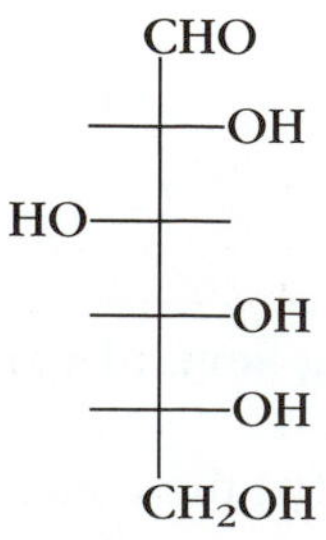

D-Glucose 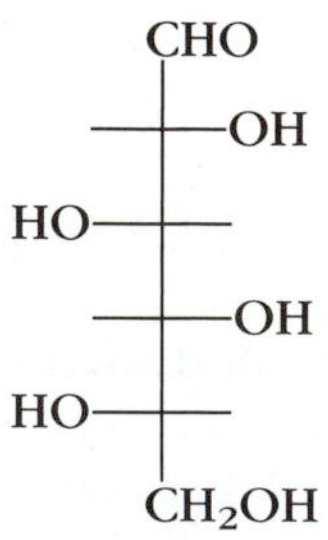 L-Idose

You must keep in mind that the D or L designation is an arbitrary device for nomenclature. Despite their similarity to *d* and *l* (which are sometimes used to indicate the sign of optical rotation), the D/L configuration does not correlate with the direction in which polarized light is rotated. D-Glucose and L-idose, for example, both exhibit a *positive* optical rotation.

The D Sugars

Naturally occurring sugars are optically active, and although both D and L sugars are well known, most have the D configuration. The reason that naturally occurring carbohydrates have the D configuration is a consequence of their biosynthesis from D-glyceraldehyde (Section 14.7).

The structures of *all* the D aldoses are drawn in Figure 19.2, which is organized in a way that emphasizes the structural relationships among the different sugars. The structures are arranged according to a hypothetical synthesis starting with D-glyceraldehyde, and each increase in the length of the carbon chain produces two new sugars that differ only in configuration at the "new" carbon atom. The two "new" aldoses are drawn below the "starting material" in each case.

An important kind of stereoisomeric relationship is often encountered in sugars. Two structures that differ in configuration at just one of several chiral centers are

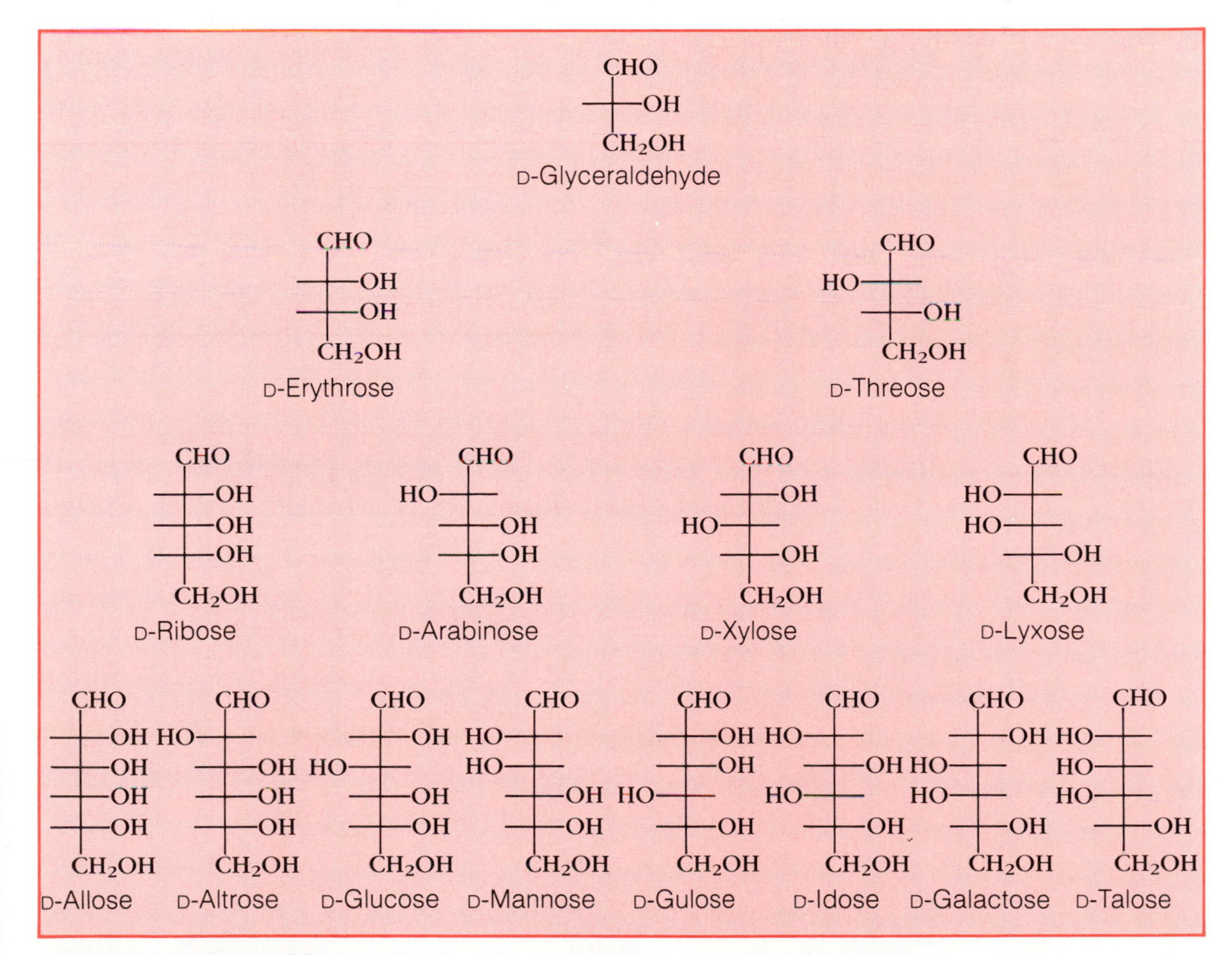

Figure 19.2 The D aldoses.

CH_2OH	CH_2OH	CH_2OH	CH_2OH
=O	=O	=O	=O
—OH	HO—	HO—	HO—
—OH	—OH	—OH	—OH
CH_2OH	CH_2OH	—OH	HO—
		CH_2OH	CH_2OH
D-Ribulose	D-Xylulose	D-Fructose	L-Sorbose

Figure 19.3 Some important naturally occurring ketoses.

epimers, and this provides a convenient way for specifying the stereochemical distinction between certain closely related diastereomers. D-Glucose and D-mannose, for example, are epimeric at C-2, whereas D-glucose and L-idose are C-5 epimers.

Of the various aldoses shown in Figure 19.2, only a limited number are isolated from natural sources. These include the aldohexoses D-glucose, D-mannose, and D-galactose, along with the aldopentoses D-ribose, L-arabinose, and D-xylose. The ketoses depicted in Figure 19.3 are also naturally occurring.

EXERCISE 19.1

Draw Fischer projections and write names for each of the L-aldopentoses.

EXERCISE 19.2

Draw and name the aldose that is epimeric with D-glucose at: (a) C-2; (b) C-3; (c) C-4; (d) C-5.

19.2 INTRAMOLECULAR INTERACTIONS IN SUGARS

With any polyfunctional compound the possibility exists for interactions between functional groups. Interaction between a hydroxyl group and the carbonyl group of an aldehyde or ketone can produce a cyclic hemiacetal or hemiketal (Section 13.7). The reaction is most important when a five- or six-membered ring is produced (cf. Section 8.7).

OH, O, C—H ⇌ O, OH, H

11% 89%

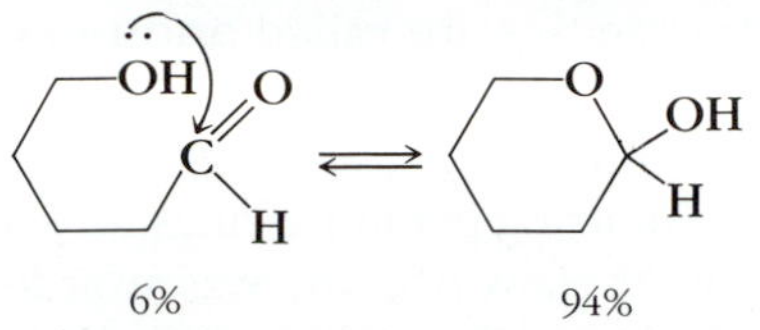

6% 94%

For three- and four-membered rings there is virtually none of the cyclic form at equilibrium, and for seven-membered and larger rings the open-chain hydroxy aldehyde predominates to the extent of at least 80–90%.

Cyclic Hemiacetals and Hemiketals of Sugars

The behavior of simple monohydroxy aldehydes and ketones is reflected by carbohydrates: both five- and six-membered-ring hemiacetals and hemiketals are formed. The names of these cyclic forms of sugars are based on the unsaturated cyclic ethers furan and pyran.

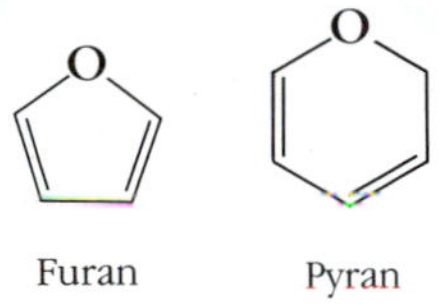

Furan Pyran

A five-membered-ring hemiacetal or hemiketal is a *furanose,* and the six-membered-ring analog is a *pyranose.*

In aqueous solution simple sugars exist almost entirely in the cyclic form, as mixtures of five- and six-membered-ring hemiacetals or hemiketals. The cyclic forms of sugars can be drawn in several ways, and we will use several examples to illustrate these drawings. Fischer projections can be used easily, as shown in the following equation for the equilibrium of the open-chain form of glucose and the two possible stereoisomeric pyranose forms. (We have not shown the furanose in this equation, but it would certainly be present in the equilibrium mixture.) Such a hemiacetal is named as a pyranose derivative of the original sugar, in this case as a D-*gluco*pyranose. (The three 90° angles in the bond between the C-5 oxygen and C-1 have no significance and should not be confused with other drawings of organic molecules in which such angles would indicate a carbon atom.)

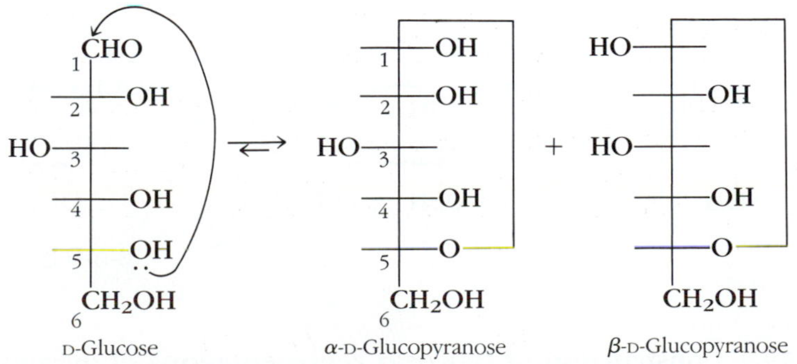

D-Glucose α-D-Glucopyranose β-D-Glucopyranose

Notice that this reaction yields two products because a new chiral center is generated at C-1. The carbonyl carbon is sometimes called the **anomeric carbon,** and the stereoisomeric glucopyranoses are called **anomers.** This is a special type of epimeric relationship, and the isomers are distinguished by the prefixes α and β, which are explained in the following discussion.

A more common way of drawing pyranose rings uses a *Haworth* representation, developed in England by W. N. Haworth, who was awarded the Nobel Prize in 1937 for his research on carbohydrates. The six-membered ring is depicted for convenience as planar, and vertical lines are used to denote the relative positions of the substituents. The plane of the ring is understood to be oriented perpendicular to the paper, and the oxygen atom is *always* placed at the right rear position with C-1 located at the extreme right of the structure. Hydrogens are often omitted for clarity.

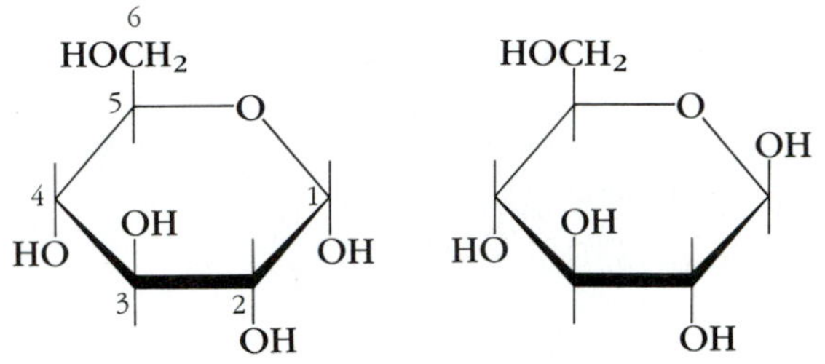

α-D-Glucopyranose $\qquad$ β-D-Glucopyranose

Figure 19.4 shows how you can convert a Fischer projection to a Haworth representation. The symbols α and β describe the orientation of the C-1 oxygen substituent. When a D sugar is drawn in a Haworth representation, a substituent on C-1 oriented below the ring is designated α and if it is above the ring it is β. (The *relative*

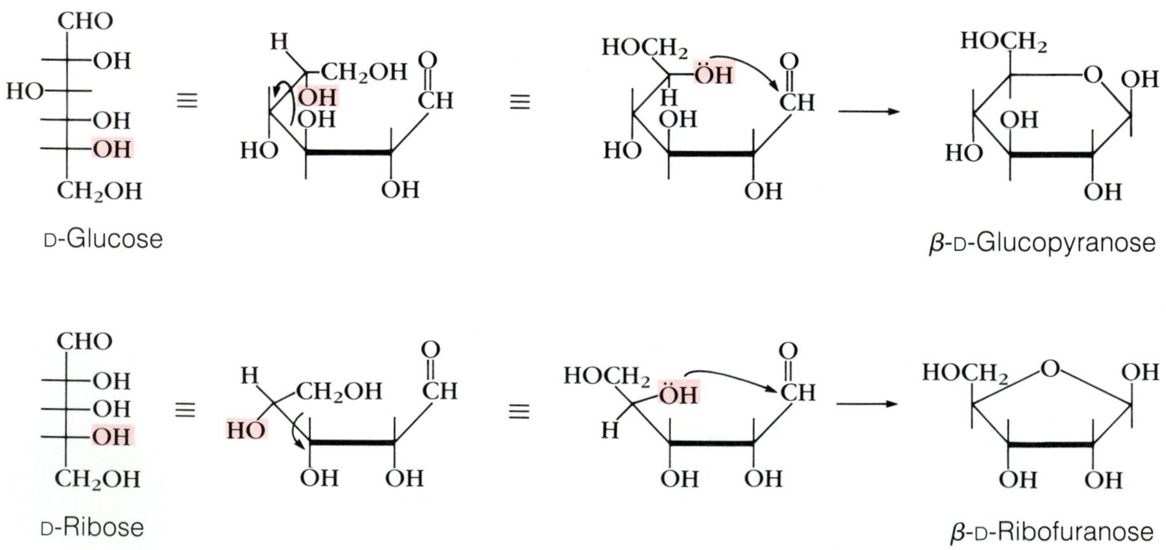

Figure 19.4 Construction of Haworth representations of pyranose and furanose ring systems starting from Fischer projections.
The β anomer has arbitrarily been drawn in each case.

positions of the groups are defined in the same manner for L sugars, and this means that the α-D and α-L forms of a sugar are enantiomeric.)

The α and β forms of glucopyranose have been isolated separately as crystalline solids. When dissolved in water they have different optical rotations, but over a period of time the rotation slowly changes until the same equilibrium value is observed from either starting form. This phenomenon of changing optical rotation with time is called **mutarotation,** and it is often observed when sugars isomerize in solution.

Furanose rings can also be drawn in several ways. As illustrated for D-ribose, Fischer projections can be used and two stereoisomeric furanoses can be formed.

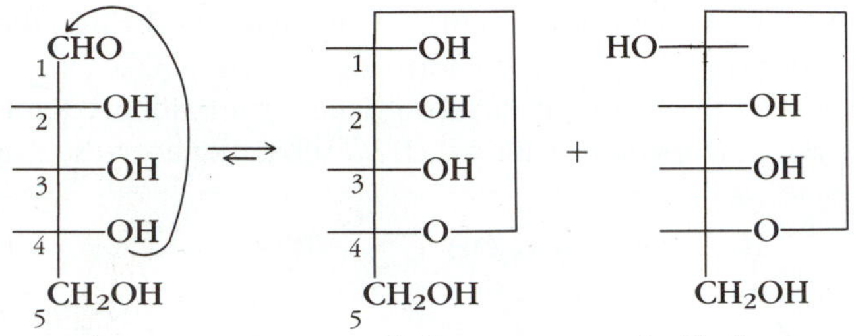

D-Ribose ⇌ α-D-Ribofuranose + β-D-Ribofuranose

Haworth representations are more common, however. In the Haworth representation of a furanose, the "planar" five-membered ring has an orientation perpendicular to the plane of the paper, with the ring oxygen farthest away from the viewer and C-1 at the right of the structure.

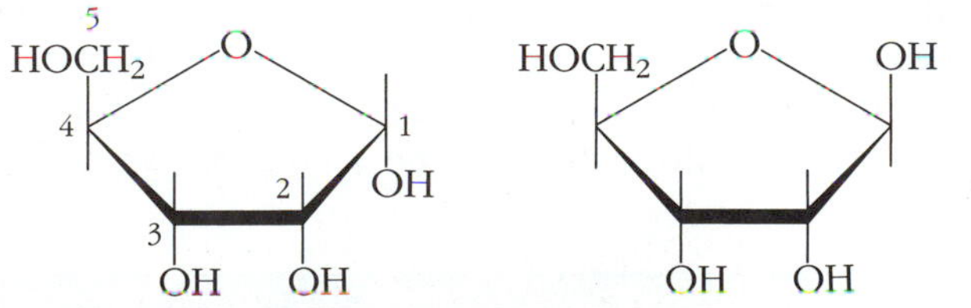

α-D-Ribofuranose β-D-Ribofuranose

The conversion of a Fischer projection to a Haworth representation of a furanose is also illustrated in Figure 19.4.

The preceding discussion has focused on aldoses, but the formation of cyclic derivatives is also observed with ketoses. D-Fructose, for example, can form two stereoisomeric hemiketals.

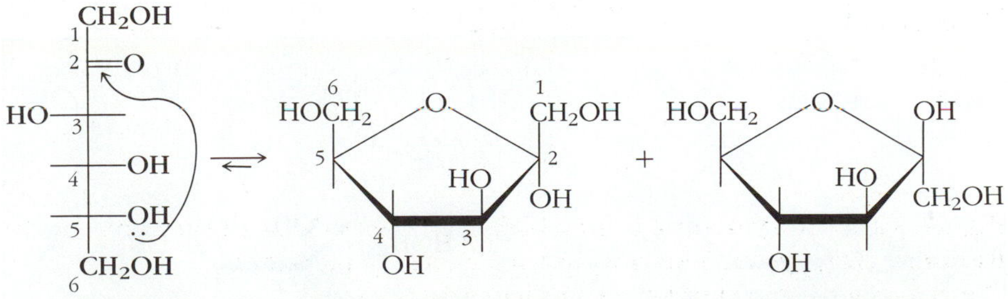

D-Fructose ⇌ α-D-Fructofuranose + β-D-Fructofuranose

Notice that the anomeric carbon (C-2 in the case of fructose) of a hemiketal has oxygen and *carbon* as its two substituents. Despite the tendency for both aldoses and ketoses to exist primarily in the respective hemiacetal or hemiketal forms, we will frequently draw them as the open-chain structures. This is done largely for convenience, and it greatly facilitates structural correlations using Figure 19.2. Nevertheless, you should always keep in mind the facile equilibrium between the open-chain and cyclic forms (with the open chain form present to only a small extent if a five- or six-membered ring is possible).

Chair Conformation of Pyranoses

Pyranose rings are commonly drawn with Haworth representations, but they are certainly not planar. As with other saturated six-membered rings, the most stable form of a pyranose ring is the chair conformation (Section 3.6). Of the two possible chairs, the one with the fewest axial substituents ordinarily predominates. This is illustrated for β-D-glucopyranose, for which *all* substituents on the ring are equatorial.

Despite the importance of the chair conformation in pyranose derivatives, we will typically use Haworth representations for several reasons. *First,* they are somewhat easier for you to work with. *Second,* as in previous chapters, we will not use conformational drawings (i.e., chairs) unless we are trying to illustrate conformational properties. *Third,* there are important cases in carbohydrate chemistry in which nonchair conformations are important, and it would be misleading to suggest that *only* the chair conformation can be found for pyranose derivatives.

EXERCISE 19.3

Draw Fischer projections and Haworth representations for each of the following:

(a) α-D-Galactopyranose
(b) β-D-Xylofuranose
(c) α-D-Allofuranose
(d) β-L-Ribofuranose
(e) β-D-Fructofuranose

EXERCISE 19.4

Name the following:

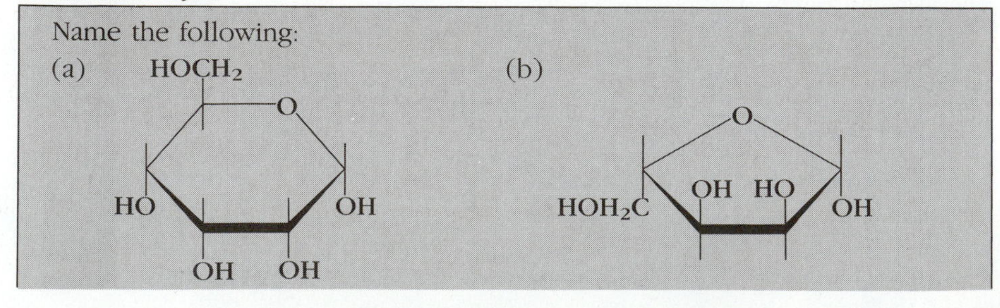

(c)

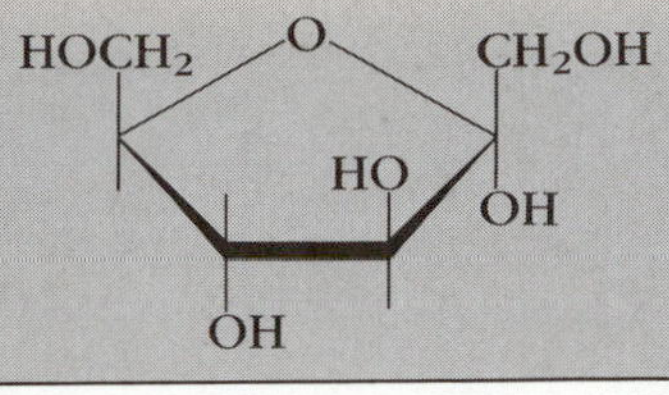

HOCH2 O CH2OH HO OH OH

EXERCISE 19.5

Draw the structures that you would expect to predominate for an aqueous solution of D-glucose at equilibrium. Show all stereochemistry.

19.3 ESTERS AND ETHERS OF CARBOHYDRATES

Cyclic Acetals and Ketals: Glycoside Formation

We have previously shown (Section 15.5) that the acid-catalyzed reaction of an aldehyde or ketone with an alcohol affords the corresponding acetal or ketal.

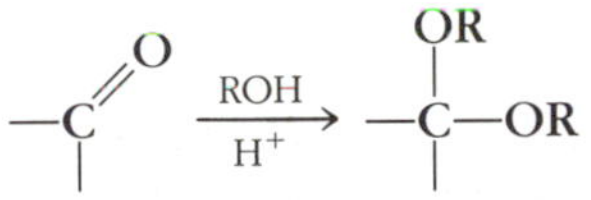

$$\text{—C(=O)—} \xrightarrow[\text{H}^+]{\text{ROH}} \text{—C(OR)—OR}$$

The cyclic hemiacetals and hemiketals are the predominant forms of most simple sugars in solution, and these are converted to cyclic acetals and ketals by the mechanism depicted in Figure 19.5. Each of the steps in this mechanism is fully reversible, but acid catalysis is required. Once the cyclic acetal or ketal has been isolated, it is a stable compound; in the absence of acid catalysts it is not in equilibrium with open-chain structures.

These cyclic acetals (or ketals) have one *intra*molecular ether linkage to the original carbonyl carbon, but the second ether linkage is formed *inter*molecularly with an alcohol such as methanol. These cyclic acetals and ketals of carbohydrates are called **glycosides.** Their names have two parts: the first word is the alkyl group corresponding to the alcohol used to make it, and the second part is the same as the name of the analogous hemiacetal (Section 19.1) except that the *-ose* ending of the hemiacetal (or hemiketal) is changed to *-oside.*

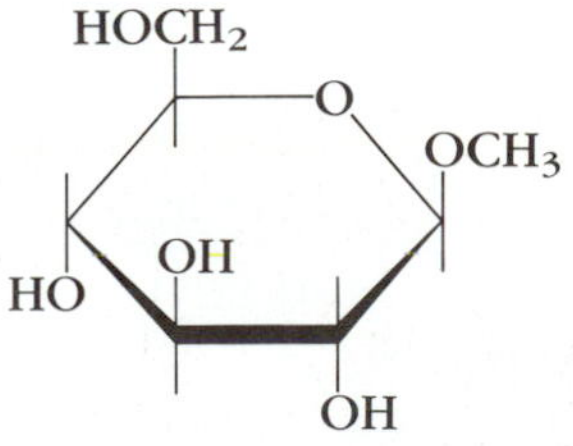

HOCH2 O OCH3 OH HO OH

Methyl β-D-glucopyranoside

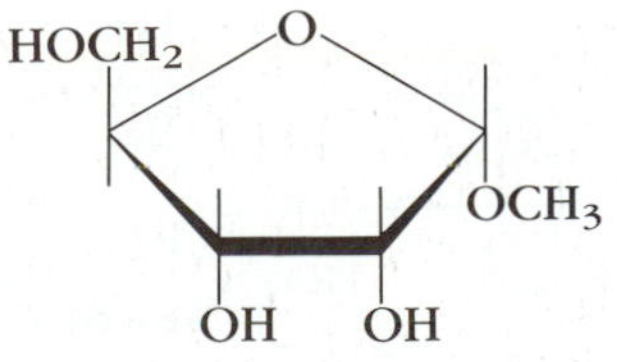

HOCH2 O OCH3 OH OH

Methyl α-D-ribofuranoside

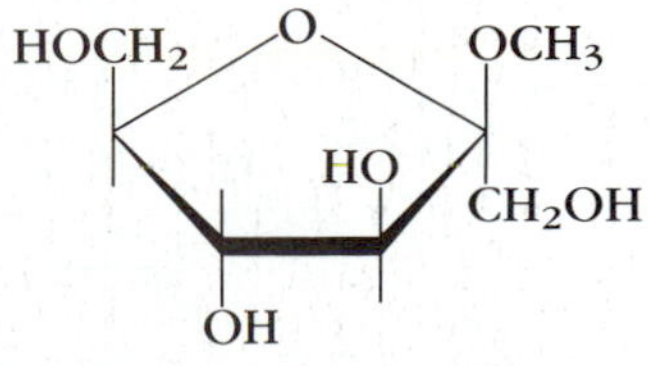

HOCH2 O OCH3 HO CH2OH OH

Methyl β-D-fructofuranoside

Notice that a *glucoside* is simply a glycoside derived from glucose.

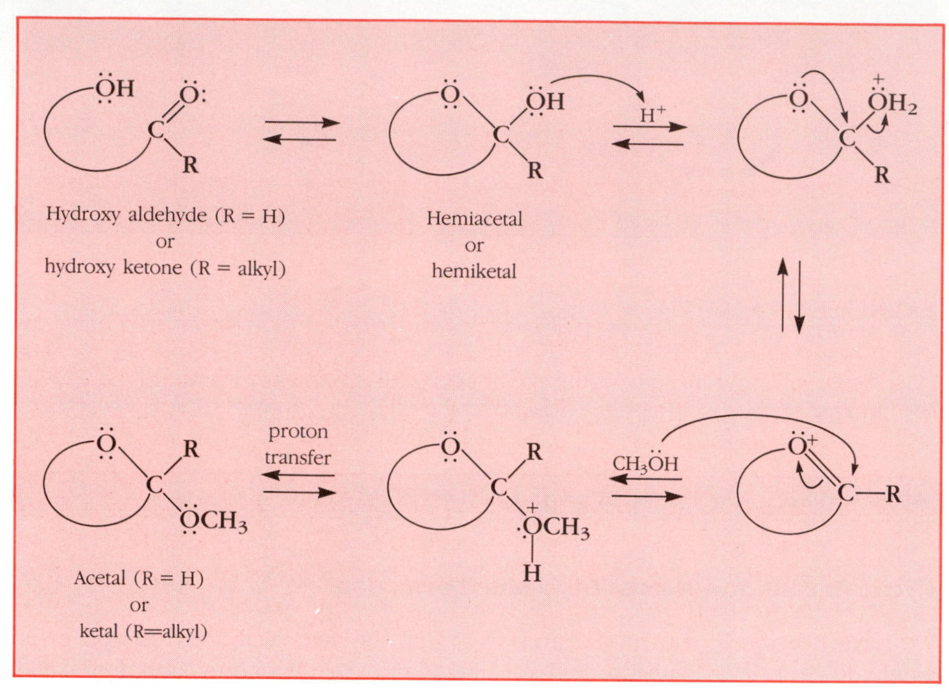

Figure 19.5 Mechanism of formation of a cyclic acetal or ketal from a hydroxy aldehyde or hydroxy ketone and methanol.

At this point we can start to illustrate our discussion with actual reactions of sugars. But first it is appropriate that we mention a general characteristic of these reactions: *Low yields of isolated product are typical in the reactions of carbohydrates.* There are always exceptions, of course, but several factors contribute to the low yields that are often observed. First, the products are often water-soluble and are difficult to separate from inorganic reagents. Second, many reactions of carbohydrates produce new chiral centers, and mixtures of diastereomers result. Third, the presence of a large number of functional groups in a sugar can result in side reactions, which directly reduce the yield of desired product. Despite these difficulties the starting compounds are often abundant and inexpensive, so low yields can be overlooked.

The following examples show the preparation and hydrolysis of methyl glycosides. As with other acetals and ketals (Section 15.7), both their formation and their hydrolysis occur readily under mildly acidic conditions.

CHO, OH, HO, OH, OH, CH_2OH — D-Glucose $\xrightarrow[\text{anhydrous HCl}]{CH_3OH}$ $HOCH_2$, O, HO, OH, OCH_3, OH — Methyl α-D-glucopyranoside 49%

D-Glucose

Methyl α-D-glucopyranoside

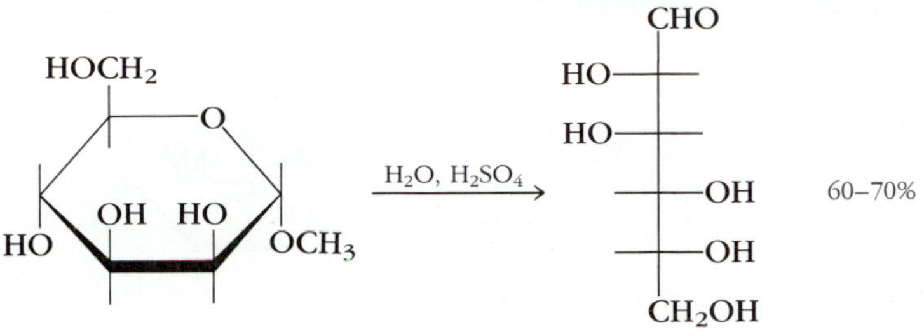

HOCH₂ ... $\xrightarrow{H_2O,\ H_2SO_4}$ ... 60–70%

Methyl α-D-mannopyranoside D-Mannose

The formation of the methyl pyranoside of glucose merits additional comment. Both the α and β isomers should be formed, and the latter should predominate because all substituents are equatorial in the chair conformation. How then can the α isomer be isolated in nearly 50% yield? The answer is that acetal formation is an equilibrium reaction and the α isomer is less soluble. When the reaction mixture was first cooled down, the α isomer crystallized from solution, but only in about 20% yield. Additional α glycoside was obtained by removing the crystals and heating the reaction solution to reestablish the equilibrium, but it took a total of four such cycles to reach a 49% yield.

Acetals and Ketals as Hydroxyl Protecting Groups

We have just shown that aldoses and ketoses will react with an alcohol in the presence of acid to form cyclic acetals and ketals. The resulting glycosides no longer exhibit the characteristic reactions of aldehydes and ketones because the carbonyl group has been *protected* (cf. Section 15.7). Only in the presence of acid will the acetal or ketal hydrolyze and permit normal carbonyl reactions to occur.

Acetals and ketals can also be used to protect the hydroxyl groups of sugars. Two of the OH groups of a carbohydrate can react with an aldehyde or ketone to form a cyclic acetal or ketal. This is an extremely important process in sugar chemistry, because it subsequently permits selective reaction of those hydroxyl groups which remain unprotected. The major problems lie in the difficulty of predicting where reaction will take place and whether one or two acetal or ketal linkages will be formed. In most cases complex equilibrium mixtures are produced. Unless the reaction has previously been reported, it may not be possible to predict the products in advance. The following equations illustrate these reactions:

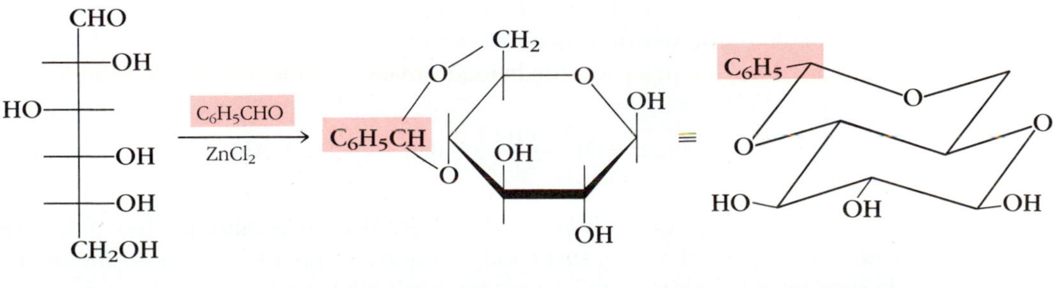

$\xrightarrow[ZnCl_2]{C_6H_5CHO}$ C_6H_5CH ... 42%

D-Glucose 42%

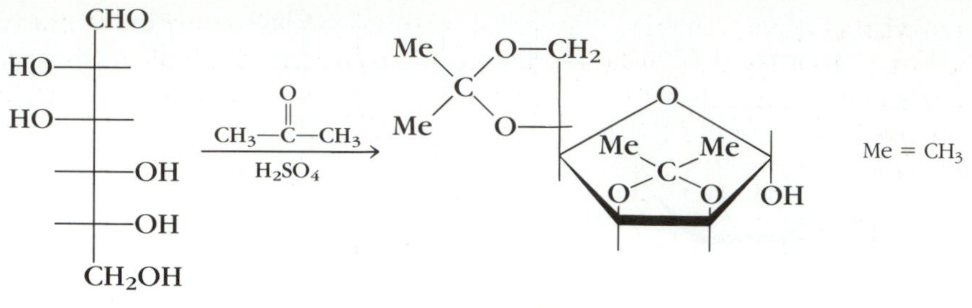

D-Mannose $\xrightarrow[H_2SO_4]{CH_3-\overset{O}{\overset{\|}{C}}-CH_3}$ 92%

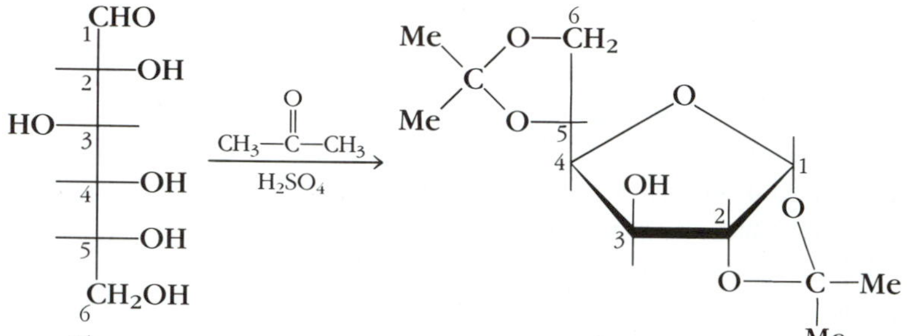

Me = CH_3

D-Glucose $\xrightarrow[H_2SO_4]{CH_3-\overset{O}{\overset{\|}{C}}-CH_3}$ 13%

In the first of these three reactions a Lewis acid ($ZnCl_2$) was employed, but H_2SO_4 was used in the other cases. The last example shows the unusual situation in which C-1 of the product is part of both a cyclic acetal and a cyclic ketal. You may be able to envision a variety of selective reactions that could be carried out on these protected compounds.

Alkylation of Hydroxyl Groups

The individual OH groups of carbohydrates undergo the same reactions as simple alcohols. Hydroxyl groups are nucleophilic, and they react with alkylating agents (Section 11.8) to form ethers. Methyl ethers have been used extensively in carbohydrate chemistry, and methyl groups can be introduced by several different procedures. Reaction of a carbohydrate with methyl iodide proceeds quite slowly unless the OH groups are first converted to metal salts by treatment with a strong base such as $NaNH_2$ or NaH.

$$\text{Sugar—OH} \xrightarrow[\text{2. } CH_3I]{\text{1. } NaNH_2} \text{sugar—}OCH_3$$

Alternatively, the reactivity of methyl iodide is greatly enhanced by the addition of silver oxide to the reaction mixture. Not only does this oxide act as a base, but the strong affinity of silver ion for halogen makes iodide a better leaving group.

$$\text{Sugar—OH} \xrightarrow[Ag_2O]{CH_3I} \text{sugar—}OCH_3 + AgI$$

However this method can only be used with glycosides, because the free aldose or ketose is readily oxidized by silver ion (Sections 13.11, 19.5). Because methyl iodide is relatively unreactive and cannot be used with the free sugars, a more powerful alkylating agent, dimethyl sulfate, has been used extensively by carbohydrate chemists. This is the dimethyl ester of sulfuric acid, and it readily alkylates sugars in the presence of aqueous base. (It also reacts with the aqueous base, so an excess of the reagent is employed.) The methylation of some carbohydrates is illustrated in the following equations:

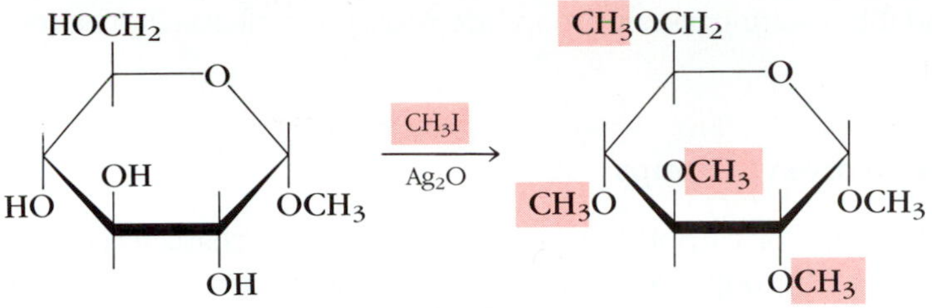

HOCH$_2$, HO, OH, OCH$_3$, OH $\xrightarrow[Ag_2O]{CH_3I}$ CH$_3$OCH$_2$, CH$_3$O, OCH$_3$, OCH$_3$, OCH$_3$ 97%

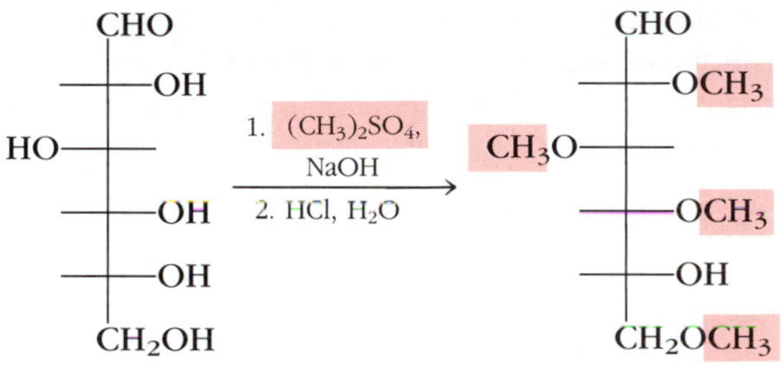

CHO, OH, HO, OH, OH, CH$_2$OH $\xrightarrow[2.\ HCl,\ H_2O]{1.\ (CH_3)_2SO_4,\ NaOH}$ CHO, OCH$_3$, CH$_3$O, OCH$_3$, OH, CH$_2$OCH$_3$ 46–55%

All the free hydroxyl groups are converted to the methyl ethers in these examples. The initial product of the second reaction is actually the methyl pyranoside, but treatment with hydrochloric acid hydrolyzes the acetal linkage. The four remaining *O*-methyl groups are normal ether linkages, however and are not affected by mild acid hydrolysis. The entire reaction is shown below.

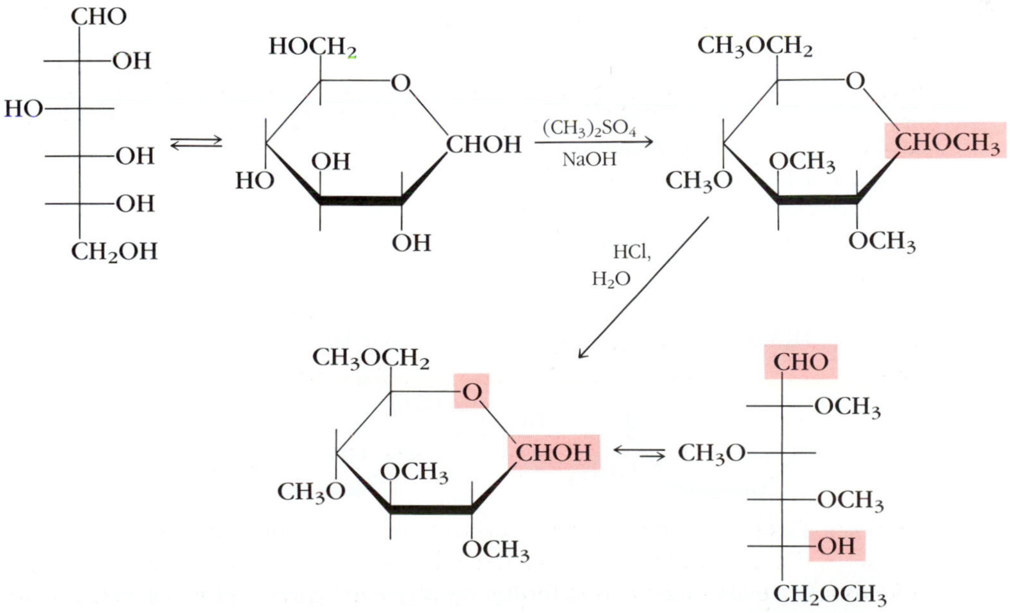

CHO, OH, HO, OH, OH, CH$_2$OH $\rightleftharpoons$ HOCH$_2$, HO, OH, CHOH, OH $\xrightarrow[NaOH]{(CH_3)_2SO_4}$ CH$_3$OCH$_2$, CH$_3$O, OCH$_3$, CHOCH$_3$, OCH$_3$

$\xrightarrow{HCl,\ H_2O}$ CH$_3$OCH$_2$, O, CH$_3$O, OCH$_3$, CHOH, OCH$_3$ $\rightleftharpoons$ CHO, OCH$_3$, CH$_3$O, OCH$_3$, OH, CH$_2$OCH$_3$

The preceding examples illustrate the great difference in reactivity between the OH group at the anomeric carbon of a furanose or pyranose and all the remaining hydroxyl groups. Ether formation (and cleavage) at the anomeric carbon is accomplished under very mild acid catalysis, and it is actually an *alkoxy* group that is introduced. It is quite easy to *selectively* introduce (or remove) an *O*-alkyl group at the anomeric carbon. Ether formation at all the other hydroxyl groups occurs via alkylation, and the resulting alkoxy groups are resistant to cleavage in either acid or base.

Acylation of Hydroxyl Groups

The hydroxyl groups of carbohydrates also undergo the characteristic acylation reactions of hydroxyl groups (Section 11.8). Esters can be prepared by using acid chlorides or anhydrides, and esters of sulfonic acids can be made by reaction with the appropriate sulfonyl chlorides. The reaction of D-glucose with acetic anhydride generates the cyclic pentaacetate. (Recall that D-glucose exists mainly as the cyclic hemiacetal, and it reacts accordingly.) In the following example the pentaacetate is not isolated but is converted to the 1-bromo derivative by reaction with hydrogen bromide. This last step emphasizes the greater reactivity of an oxygen substituent at C-1.

Ac_2O → [pentaacetate] → HBr → 80–87%

Ac = $CH_3C(=O)$

The treatment of such an acylated sugar with $NaOCH_3$ will regenerate the free OH groups via a transesterification reaction (Section 16.6).

The reaction of methyl α-D-galactopyranoside with tosyl chloride (in pyridine) points out another feature of the relative reactivity of different positions in a sugar. A less hindered, *primary* hydroxyl group reacts more rapidly than the other, secondary OH groups.

TsCl, pyridine → 22%

This tosylate also allows us to further emphasize the importance of intramolecular reactions in carbohydrate chemistry. When the toyslate is treated with sodium ethoxide in ethanol, the product is not that from nucleophilic displacement by ethoxide ion. Instead the ethoxide ion acts as a base to generate the anion of the C-3 hydroxyl group, and this anion displaces the tosylate intramolecularly, with formation of a five-membered ring.

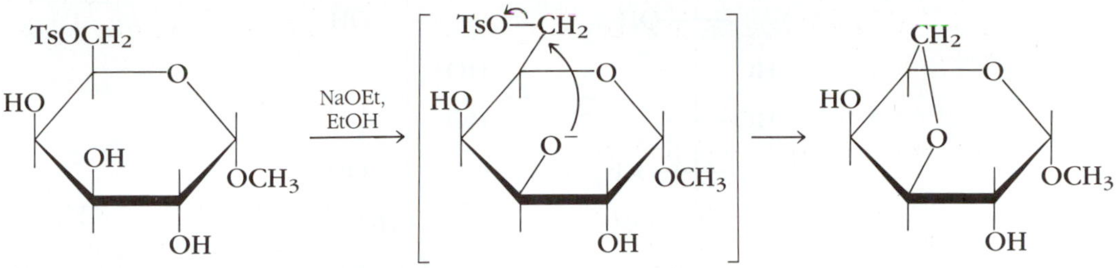

TsOCH₂ ... HO ... OH ... OCH₃ ... OH —NaOEt, EtOH→ [TsO—CH₂ ... HO ... O⁻ ... OCH₃ ... OH] → CH₂ ... HO ... O ... OCH₃ ... OH

EXERCISE 19.6

Write the reaction of D-galactose with each of the following reagents:

(a) CH_3OH, HCl
(b) $(CH_3)_2SO_4$, NaOH
(c) Ac_2O

EXERCISE 19.7

Write the reaction of methyl β-D-ribofuranoside with each of the following reagents:

(a) CH_3I, Ag_2O
(b) CH_3I, Ag_2O; followed by HCl, H_2O
(c) Ac_2O
(d) Tosyl chloride (one equivalent)

19.4 REDUCTION OF CARBOHYDRATES

Aldoses and ketoses can be reduced to polyhydroxy compounds called *alditols*. The reaction works better with aldoses, because a single product is formed. Reduction of a ketose generates a new chiral center, and two epimeric alditols are formed. The reduction can be carried out by several methods, including catalytic hydrogenation.

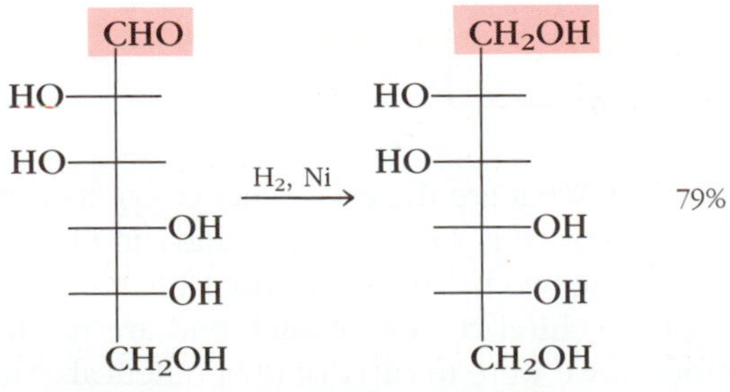

CHO; HO—; HO—; —OH; —OH; CH_2OH —H_2, Ni→ CH_2OH; HO—; HO—; —OH; —OH; CH_2OH 79%

D-Mannose D-Mannitol

Another reducing agent previously used for this purpose is sodium and alcohol, but it has largely been replaced by sodium borohydride in modern organic chemistry.

CHO | CH₂OH
—OH | —OH
HO— | HO—
HO— | HO—
—OH | —OH
CH_2OH | CH_2OH

$\xrightarrow[\text{pH 3–4}]{NaBH_4}$ 90%

D-Galactose → Galactitol

Aldoses exist almost totally as the hemiacetals, so it is necessary to employ acidic conditions to facilitate the equilibrium between cyclic and open-chain forms. Even though there is only a small amount of free aldose at any instant, more is formed as soon as some of it reacts. (Excess sodium borohydride must be used in this reduction because the reagent decomposes quite rapidly in acid.)

Alditols: Symmetry and its Consequences

Reduction of a monosaccharide to an alditol imparts a certain amount of symmetry to the molecule. This is certainly true if you consider only the constitution of the molecule because you will encounter precisely the same substituents starting from either end of the carbon chain. There may also be symmetry in a three-dimensional sense, and this can have very important consequences.

Each of the two alditols produced in the preceding reactions exhibits symmetry. D-Mannitol has *rotational symmetry,* and galactitol has *reflection symmetry.*

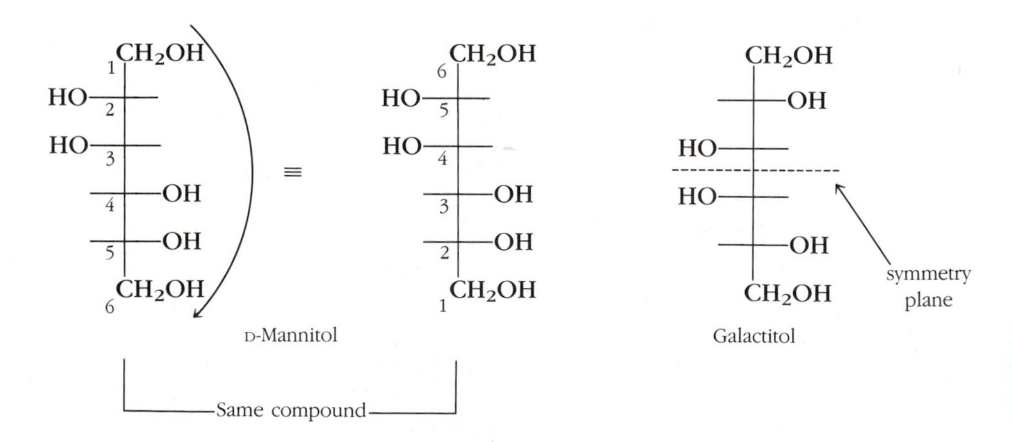

D-Mannitol

Galactitol

First consider galactitol. What are the consequences of its reflection symmetry? To begin with it is achiral (i.e., it is a *meso* compound) and is optically inactive (so no D or L prefix is used). The two halves of the molecule are mirror images of each other. This means that the chiral centers at each end are not the same, but have opposite configurations. If we were to carry out hypothetical oxidation reactions at the top and at the bottom of the structure, the two products would be enantiomers.

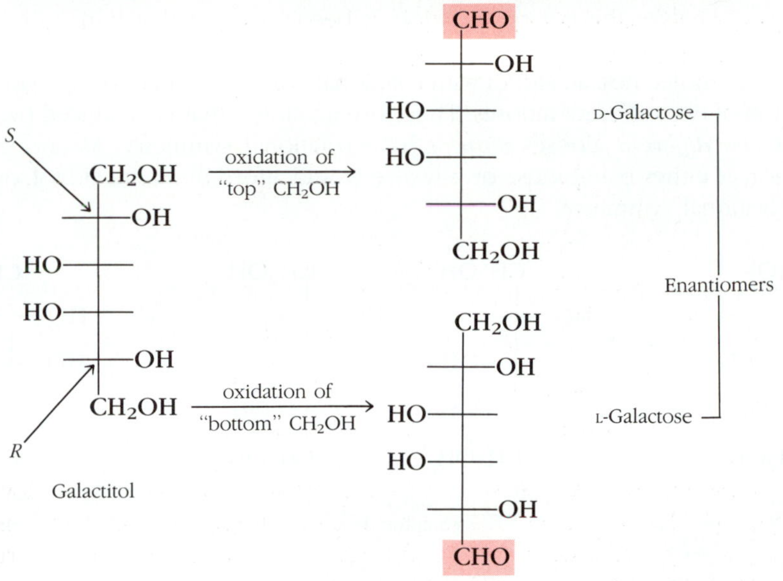

Galactitol

oxidation of "top" CH_2OH → D-Galactose

oxidation of "bottom" CH_2OH → L-Galactose

Enantiomers

Looking at this same reaction from the opposite direction, you can see that if two different aldoses yield the same *optically inactive* alditol, the starting aldoses must be enantiomers. Thus the presence of reflection symmetry (as detected experimentally by the lack of optical activity) provides structural information about both the alditol and the aldose or ketose from which it is derived.

Next consider mannitol. What are the consequences of its rotational symmetry? Rotational symmetry has no relationship to chirality, and D-mannitol is both chiral and optically active (the enantiomer of D-mannose would yield L-mannitol on reduction). In contrast to the case of galactitol, for which the two halves of the molecule are mirror images of each other, the rotational symmetry of D-mannitol requires that the two halves be completely *identical.* If we were to carry out a hypothetical oxidation reaction at either end of the molecule, we would obtain the identical product in each case.

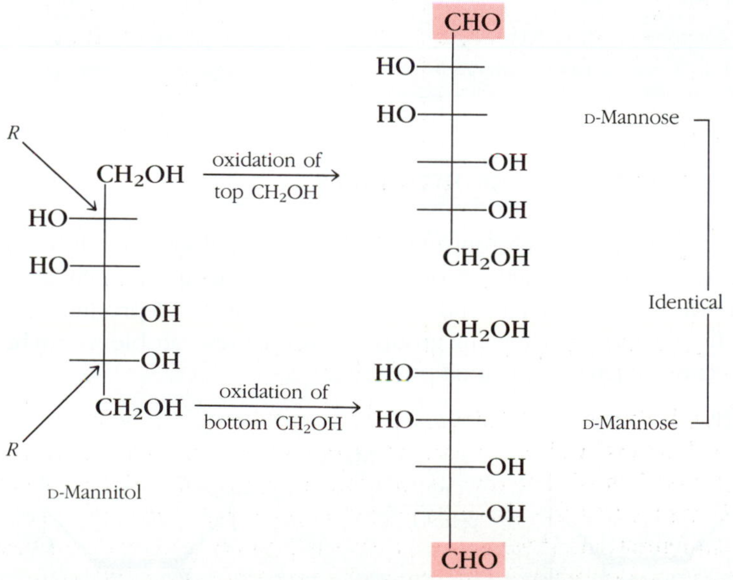

D-Mannitol

oxidation of top CH_2OH → D-Mannose

oxidation of bottom CH_2OH → D-Mannose

Identical

If you now analyze the reverse reaction, it becomes clear that D-mannose is the *only* aldose capable of yielding D-mannitol upon reduction. Extending this reasoning to other molecules, an alditol with rotational symmetry can never be formed by reduction of two different aldoses. Therefore an alditol that is produced by reduction of *two different* aldoses *cannot* have rotational symmetry. As an example, reduction of either D-arabinose or D-lyxose would afford the same alditol, one that lacks rotational symmetry.

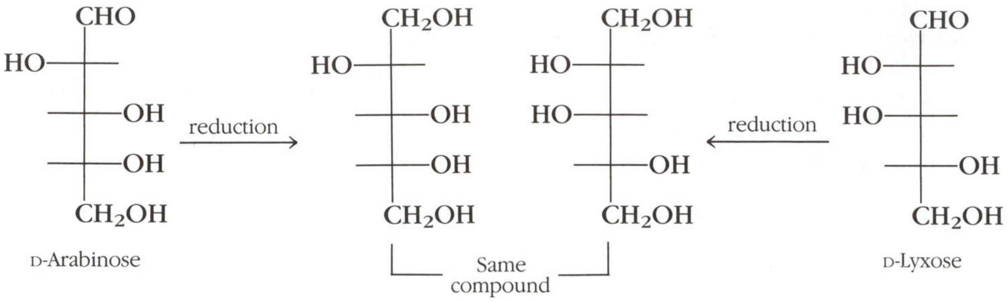

CHO / HO— / —OH / —OH / CH_2OH —reduction→ CH_2OH / HO— / —OH / —OH / CH_2OH ; CH_2OH / HO— / HO— / —OH / CH_2OH ←reduction— CHO / HO— / HO— / —OH / CH_2OH

D-Arabinose — Same compound — D-Lyxose

EXERCISE 19.8

Draw and name reduction products from treatment of the following sugars either with sodium in alcohol or with sodium borohydride:

(a) D-Arabinose
(b) D-Glucose
(c) D-Allose

EXERCISE 19.9

Which aldoses could yield ribitol upon reduction? (Consider both D and L aldoses.)

EXERCISE 19.10

(a) What two alditols would be formed by reduction of L-sorbose?
(b) What aldoses would also yield these alditols upon reduction?

19.5 OXIDATION OF CARBOHYDRATES

The presence of several different oxidizable functional groups can often lead to problems in oxidation reactions of carbohydrates. Reagents such as CrO_3 and $KMnO_4$ are not very selective, and extremely low yields of product are typically obtained. If appropriate protecting groups are used, these problems can be circumvented, making normal oxidations possible.

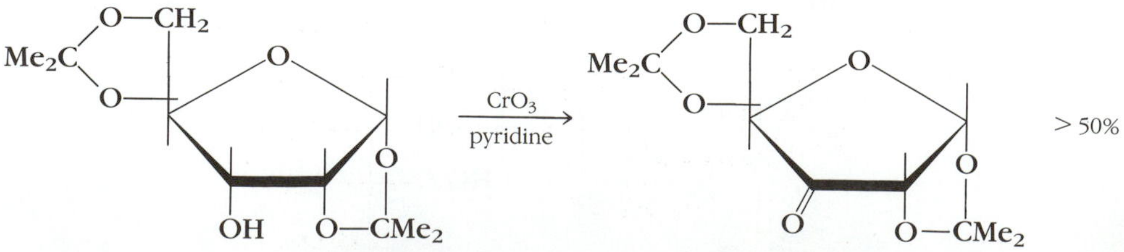

Me_2C, O—CH_2, O, OH, O—CMe_2 —CrO_3, pyridine→ Me_2C, O—CH_2, O, O, O—CMe_2 > 50%

In general, however, oxidations of carbohydrates are carried out with other reagents. Some of these find very little use outside the area of sugar chemistry, and we will only discuss the few that have been used most widely. One feature of these reactions that you will notice quickly is the ease of oxidation of the aldehyde group. Whether an aldose is in the open-chain or hemiacetal form, the preferential site for its oxidation is C-1.

Oxidation with Bromine Water: Aldonic Acids

The treatment of an aldose with bromine in aqueous solution results in oxidation of the aldehyde group to a carboxyl group. The resulting carboxylic acid is called an *aldonic acid.*

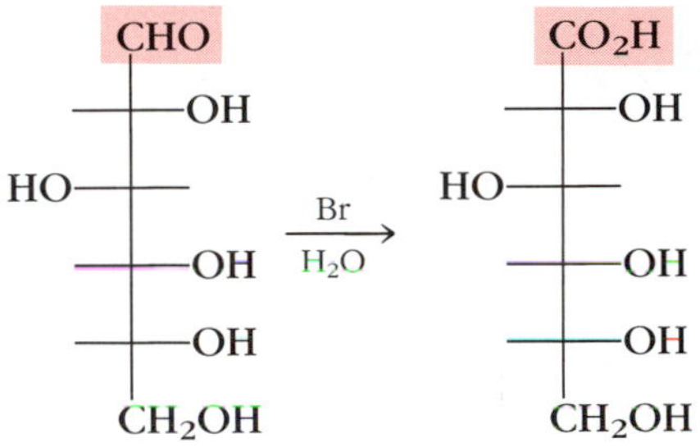

CHO, OH, HO, OH, OH, CH_2OH —Br, H_2O→ CO_2H, OH, HO, OH, OH, CH_2OH

D-Glucose D-Gluconic acid

96% (isolated as the calcium salt)

In most cases the reaction proceeds via reaction of bromine with the hemiacetal. This generates the hypobromite, which then loses HBr to yield a lactone. In acidic media the lactones are in equilibrium with the free aldonic acids, so the reaction conditions determine which compound is actually isolated.

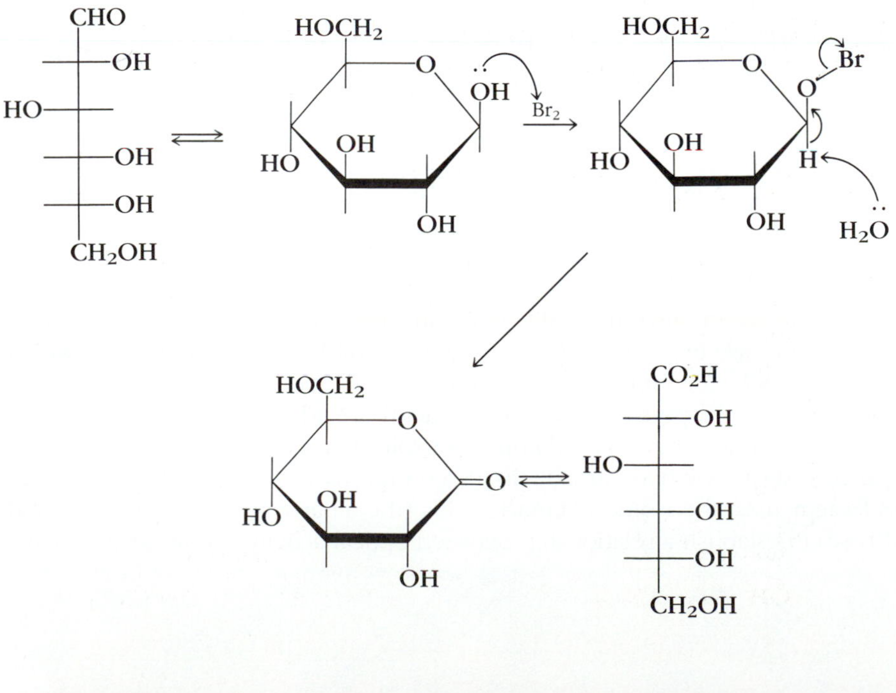

CHO, OH, HO, OH, OH, CH_2OH ⇌ $HOCH_2$, O, OH, OH, HO, OH —Br_2→ $HOCH_2$, O, O, Br, OH, HO, H, OH, H_2O

$HOCH_2$, O, =O, OH, HO, OH ⇌ CO_2H, OH, HO, OH, OH, CH_2OH

The reaction does not always proceed via the hemiacetal and lactone, however. There are no free OH groups in the tetraacetate of D-xylose, but it is oxidized to the tetraacetate of D-xylonic acid in good yield. In this case it is probably the hydrate of the aldehyde (i.e., a *gem*-diol, see Section 13.7) rather than a hemiacetal that reacts with bromine.

Oxidation with Nitric Acid: Aldaric Acids

Nitric acid, a potent oxidizing agent, oxidizes both termini of an aldose chain to yield the corresponding diacid. These diacids are called aldaric acids. (Recall that the *aldonic* acids are only *mono*carboxylic acids.) Reaction at C-1 is not surprising, because aldehydes are very easily oxidized. Selective reaction also occurs at the hydroxyl terminus of the carbon chain and reflects the preferential reaction of the less hindered primary hydroxyl groups.

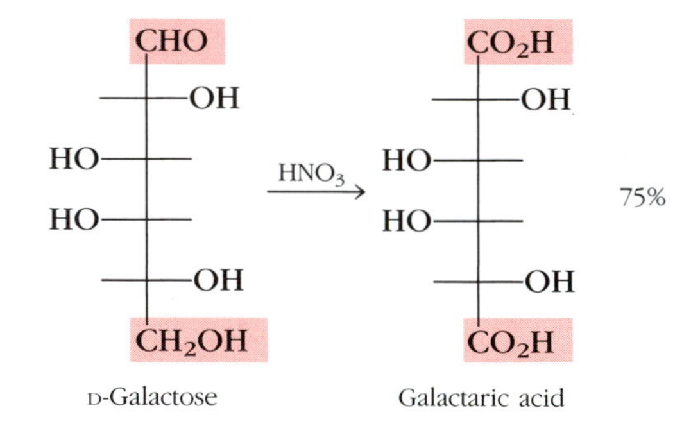

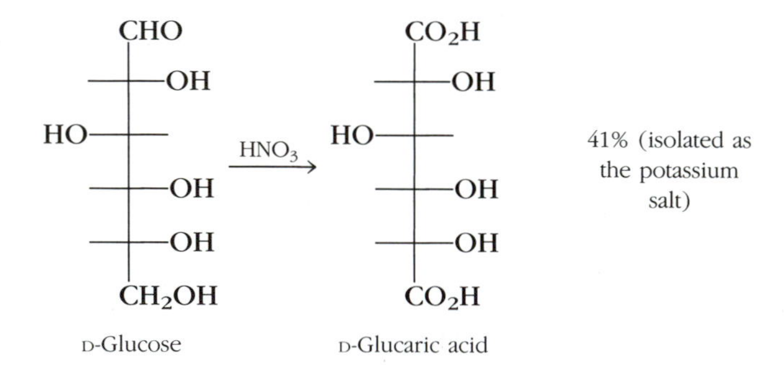

The aldaric acids are also in equilibrium with their lactones under acidic conditions. But there are two carboxyl groups and several hydroxyl groups with which they can interact, so complex mixtures often result.

The symmetry properties of the aldaric acids are identical to those of the alditols. As a result they have played important roles in establishing the structures of the various sugars. We have already discussed one such example in Section 6.6. In 1848 Pasteur used the isomeric tartaric acids (the aldaric acids from threose and erythrose) to establish a relationship between optical activity and molecular structure.

Oxidation with Silver and Copper Salts: Reducing Sugars

We have already established the ease with which aldehydes are oxidized, and oxidation with silver ion was discussed in Section 13.11 as a qualitative test for the aldehyde group. This oxidation was developed in the 1880s by the German chemist B. Tollens as a test for *reducing sugars.* Aldoses and ketoses will both reduce (i.e., be oxidized by) the Tollens reagent.

$$\begin{matrix} CHO \\ | \\ CH{-}OH \\ \wr \end{matrix} \quad \text{or} \quad \begin{matrix} CH_2OH \\ | \\ C{=}O \\ \wr \end{matrix} \xrightarrow{Ag(NH_3)_2{}^+OH^-} Ag^0 \text{ (precipitate)}$$

Only sugars containing a free aldehyde or free α-hydroxy keto groups will react, so these are called *reducing sugars.* Hemiacetals and hemiketals are in equilibrium with the free aldehydes and hydroxy ketones, so they are readily oxidized also. Glycosides (i.e., acetals and ketals), in contrast, are not reducing sugars.

A second reagent used to test for the aldehyde or α-hydroxy ketone group of reducing sugars is called *Fehling solution* (developed by H. Fehling in Germany during the 1800s). The reagent is prepared by mixing aqueous solutions of cupric sulfate and of the salt of tartaric acid. Reaction with a reducing sugar produces cuprous oxide as a red-orange precipitate.

$$\begin{matrix} CHO \\ | \\ CH{-}OH \\ \wr \end{matrix} \quad \text{or} \quad \begin{matrix} CH_2OH \\ | \\ C{=}O \\ \wr \end{matrix} \xrightarrow[\substack{\text{sodium tartrate} \\ H_2O}]{CuSO_4} Cu_2O \text{ (red precipitate)}$$

Oxidation of sugars with either Tollens reagent or Fehling solution affords complex mixtures of products. Consequently, these oxidations are not used preparatively but as diagnostic tools for structure determination. The Fehling reaction can also be used for quantitative determination of reducing sugars by determining the weight of cuprous oxide that is formed.

Oxidation with Phenylhydrazine: Osazone Formation

We have previously shown (Section 13.6, 13.12) that aldehydes and ketones react with hydrazines to form *hydrazones.* Sugars with free carbonyl groups (including hemiacetals and hemiketals) react with phenylhydrazine to form the corresponding phenylhydrazone.

$$\begin{matrix} CH{=}O \\ | \\ CH{-}OH \\ \wr \end{matrix} \xrightarrow{NH_2NHC_6H_5} \begin{matrix} CH{=}N{-}NHC_6H_5 \\ | \\ CH{-}OH \\ \wr \end{matrix}$$

If an excess of phenylhydrazine is used under acid conditions, the reaction proceeds further. Oxidation of an α-hydroxyl group takes place, and this is followed by formation of a bisphenylhydrazone, usually called a *phenylosazone,* or simply an **osazone.**

$$\begin{array}{l} CH{=}O \\ | \\ CH{-}OH \end{array} \xrightarrow[\text{}]{\text{excess } NH_2{-}NHC_6H_5} \left[\begin{array}{l} CH{=}N{-}NHC_6H_5 \\ | \\ CH{-}OH \end{array}\right] \longrightarrow \begin{array}{l} CH{=}N{-}NHC_6H_5 \\ | \\ C{=}N{-}NHC_6H_5 \end{array}$$

The formation of an osazone from an α-hydroxy aldehyde or ketone clearly involves an oxidation at the α position. The oxidizing agent is the hydrazine. The nitrogen-nitrogen single bond is fairly weak, and the N—N bond strength in phenylhydrazine (51 kcal/mol) is nearly the same as that for the O—O bond of hydrogen peroxide. The mechanism of the oxidation reaction involves formation of the enol form of the initially formed hydrazone.

$$\begin{array}{l} CH{=}N{-}NHC_6H_5 \\ | \\ CH{-}OH \end{array} \underset{}{\overset{H^+}{\rightleftharpoons}} \begin{array}{l} CH{=}\overset{+}{N}H{-}NHC_6H_5 \\ | \quad H \\ C \quad\quad :OH_2 \\ \quad OH \end{array} \rightleftharpoons \begin{array}{l} CH{-}NH{-}NHC_6H_5 \\ \| \\ C{-}OH \end{array}$$

The enol then decomposes, with cleavage of the N—N bond, and generates a keto imine. This can react with two more equivalents of phenylhydrazine to yield the osazone.

$$\begin{array}{l} CH{-}NH{-}NHC_6H_5 \\ \| \\ C{-}O{-}H \end{array} \overset{H^+}{\rightleftharpoons} \begin{array}{l} CH{-}NH{-}\overset{+}{N}H_2C_6H_5 \\ \| \\ C{-}O{-}H \end{array} \xrightarrow{:OH_2} \begin{array}{l} CH{=}NH \\ | \\ C{=}O \end{array} + C_6H_5NH_2$$

$$\begin{array}{l} CH{=}NH \\ | \\ C{=}O \end{array} \xrightarrow{NH_2{-}NHC_6H_5} \begin{array}{l} CH{=}N{-}NHC_6H_5 \\ | \\ C{=}N{-}NHC_6H_5 \end{array} + NH_3$$

The formation of phenylosazones is illustrated by the reaction of D-arabinose.

$$\begin{array}{c} CHO \\ HO{-}C{-}H \\ H{-}C{-}OH \\ H{-}C{-}OH \\ CH_2OH \end{array} \xrightarrow{NH_2{-}NHC_6H_5} \begin{array}{c} CH{=}N{-}NHC_6H_5 \\ C{=}N{-}NHC_6H_5 \\ H{-}C{-}OH \\ H{-}C{-}OH \\ CH_2OH \end{array} \quad 42\%$$

Although oxidation could in principle continue down the carbon chain, it ordinarily stops at the stage of the osazone. This results partly from hydrogen bonding, which stabilizes the osazone, and partly from the insolubility of the osazone, which causes it to precipitate from solution.

Ketoses also yield osazones. Although there are two different α-hydroxyl groups that could undergo oxidation, reaction occurs preferentially at the primary $-CH_2OH$ group. Thus D-fructose and D-glucose afford the same osazone. This oxidation destroys the chirality at C-2 of an aldose, so any two aldoses that are C-2 epimers yield the same osazone. Formation of the same osazone from different starting materials played an important role in elucidating the structures of carbohydrates.

Periodate Oxidations: C—C Bond Cleavage

We previously presented the periodate cleavage of 1,2-diols in Section 13.3. The reaction with HIO_4 or its salts is quite general, and cleavage will occur between any two carbon atoms bearing OH groups. Reaction proceeds via a cyclic intermediate.

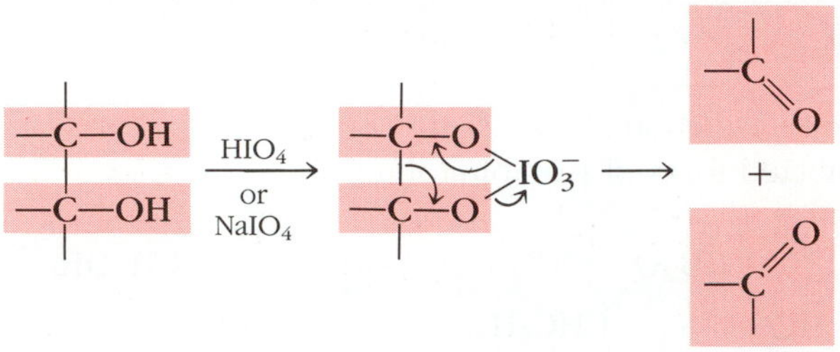

—C—OH
—C—OH
HIO_4 or $NaIO_4$ →
—C—O
—C—O
IO_3^- → —C=O + O=C—

Periodate oxidation has been one of the most important reactions used in the structure determination of carbohydrates. When appropriate protecting groups are used, it also provides an excellent preparative method based on degradation of one sugar to another of shorter chain length. Because the oxidation requires OH groups on adjacent carbons, cleavage will not occur if one of them is converted to an ether.

Cleavage also occurs readily between the carbonyl carbon and the α carbon of an α-hydroxy aldehyde or ketone. This can happen because the carbonyl form is in equilibrium with either a hemiacetal or the hydrated form.

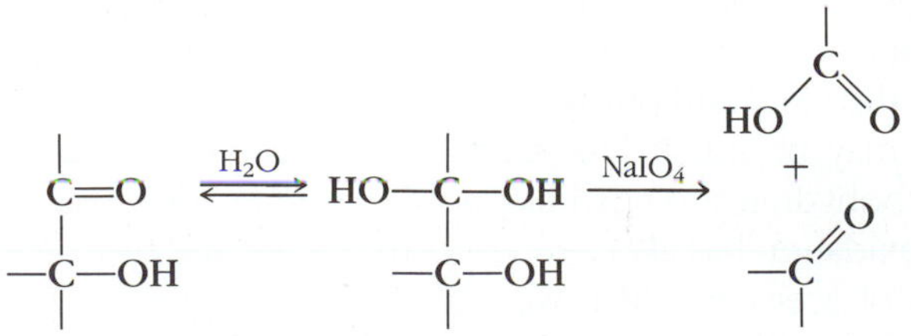

C=O
—C—OH
$\underset{}{\overset{H_2O}{\rightleftharpoons}}$
HO—C—OH
—C—OH
$\xrightarrow{NaIO_4}$
HO—C=O + —C=O

Carbon-hydrogen bonds are unaffected by this reaction, so the prediction of products is therefore relatively easy. The one-carbon fragments produced in periodate oxidations of sugars are formaldehyde, formic acid (or its salts), and carbon dioxide (or carbonate ion in alkaline solution). This is illustrated by the following hypothetical oxidation:

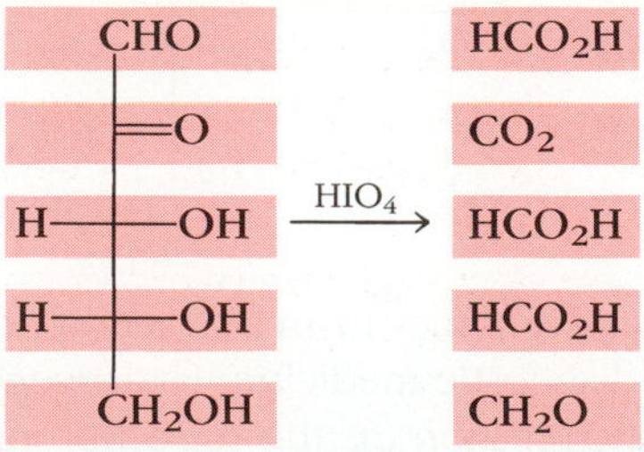

CHO
=O
H—OH
H—OH
CH_2OH
$\xrightarrow{HIO_4}$
HCO_2H
CO_2
HCO_2H
HCO_2H
CH_2O

Periodate oxidation has been used to prepare the four-carbon sugar D-erythrose from a D-glucose derivative:

CHO; H—C—OH; HO—C—H; H—C—O; H—C—OH; CH_2O (with O, O bridged by $CH—CH_3$) $\xrightarrow{NaIO_4}$ CHO; —C—O; —C—OH; CH_2O (with O, O bridged by $CHCH_3$) 70% $\xrightarrow[H_2SO_4]{H_2O}$ CHO; —C—OH; —C—OH; CH_2OH D-Erythrose

The epimeric four-carbon aldose, D-threose, can be prepared in a similar manner from a protected form of D-arabinitol.

CH_2O; HO—C—H; H—C—O (with O, O bridged by CHC_6H_5); H—C—OH; CH_2OH $\xrightarrow{NaIO_4}$ CH_2O; HO—C—; —C—O (with O, O bridged by CHC_6H_5); CHO 56% $\xrightarrow[CH_3CO_2H]{H_2O}$ CH_2OH; HO—C—; —C—OH; CHO $\equiv$ CHO; HO—C—; —C—OH; CH_2OH D-Threose 87%

We will present additional examples of periodate oxidations in Section 19.8 as part of the discussion of structure determination.

Microbial and Enzymatic Oxidations

The very high specificity of enzymatic reactions makes them an attractive alternative to the conventional laboratory processes. If an appropriate enzyme or microorganism is available, the yield and purity of product are often high, and the overall cost of the reaction may be much less. Certainly the use of microorganisms for the oxidation of carbohydrates to ethyl alcohol has been widely used for many centuries. Microbial oxidation has also been of commercial and historical importance in the preparation of acetone, which we will discuss in Section 26.2.

Molds of the *Aspergillus* and *Penicillium* genera oxidize D-glucose to D-gluconic acid in high yield, and the process has been employed commercially.

CHO; —C—OH; HO—C—; —C—OH; —C—OH; CH_2OH $\xrightarrow[O_2]{\textit{Aspergillus}}$ CO_2H; —C—OH; HO—C—; —C—OH; —C—OH; CH_2OH 90–99%

This oxidation is catalyzed by enzymes from a variety of sources, and this has given it unusual importance in diagnostic medicine. Hydrogen peroxide is formed as a by-product of the enzymatic reaction (as the reduction product of molecular oxygen). The quantity of H_2O_2 can be measured accurately by chemical methods, and this combination of enzymatic and chemical reactions has become the standard clinical method of analyzing for blood levels of D-glucose.

EXERCISE 19.11

Draw the products expected from reaction of D-glucose with:

(a) Br_2/H_2O
(b) HNO_3
(c) Excess phenylhydrazine
(d) HIO_4

EXERCISE 19.12

Which would yield a positive test with Tollens reagent (precipitate of silver metal) and/or with Fehling solution (precipitate of cuprous oxide)?

(a) CHO
—OH
—OH
CH_2OH

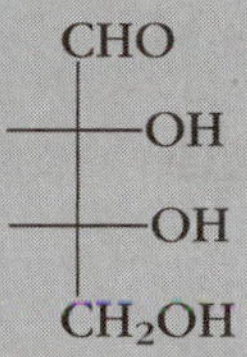

(b) CHO
—OCH_3
—OCH_3
CH_2OCH_3

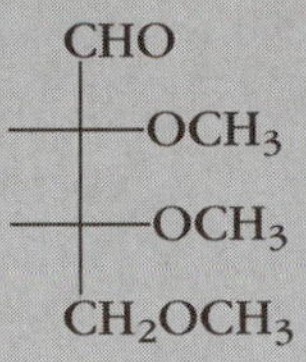

(c) $HOCH_2$ O OH OCH_3 OH

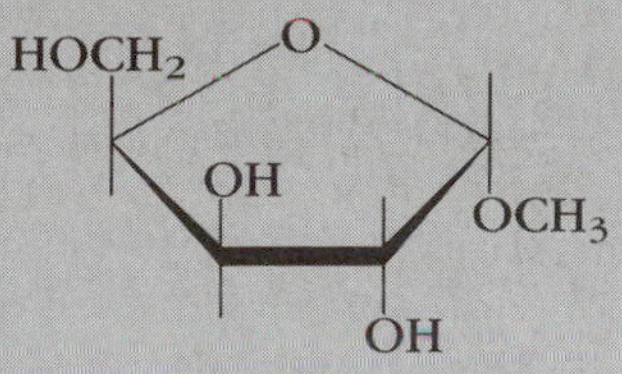

(d) $HOCH_2$ O OH HO OH OH

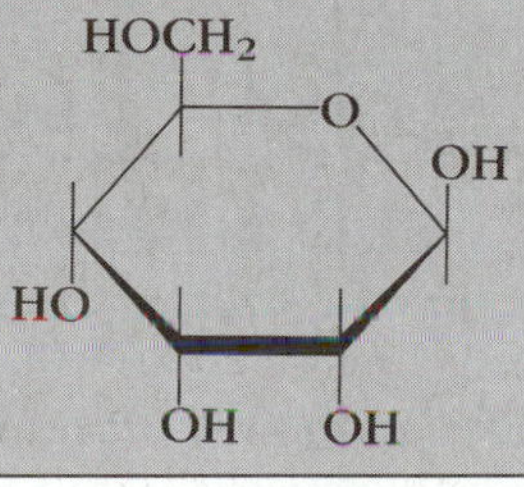

EXERCISE 19.13

Draw the products that would result from reaction of $NaIO_4$ with the compounds in Exercise 19.12.

19.6 REACTIONS THAT LENGTHEN OR SHORTEN THE CARBON CHAIN

Reactions that specifically lengthen or shorten the carbon chain of an aldose by a single carbon atom have been valuable for the synthesis of new sugar derivatives and for structure determination of others. Several methods are known for each of these processes, but we will only discuss two of them here: the Fischer-Kiliani chain-lengthening sequence and the Ruff degradation.

The Fischer-Kiliani Sequence: Chain Lengthening

In 1885 the German chemist H. Kiliani discovered that an aldose will form a cyanohydrin, which can be hydrolyzed to the corresponding carboxylic acid.

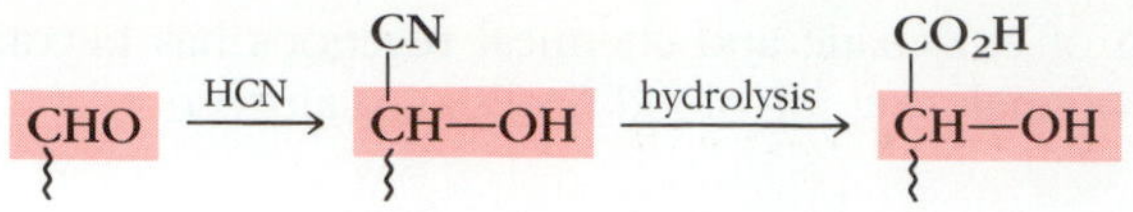

CHO $\xrightarrow{\text{HCN}}$ CN—CH—OH $\xrightarrow{\text{hydrolysis}}$ CO_2H—CH—OH

The product of this sequence is the aldonic acid of the sugar with one additional carbon atom. Actually, two epimeric aldonic acids are formed, because a new chiral center is generated in the reaction. For example, any of the pentoses shown in Figure 19.2 would yield a mixture of acids corresponding to the two isomeric hexoses located directly below the starting pentose in the figure.

The aldonic acid is in equilibrium with the lactone, and in 1889 Fischer discovered that the lactone could be reduced to the corresponding aldose by sodium amalgam. This reducing agent is milder than sodium metal alone, and Fischer was able to prepare D-mannose in 40% yield from the lactone of D-mannonic acid. (Portions of sulfuric acid are added to the reaction mixture periodically in this procedure to prevent the solution from becoming too alkaline.)

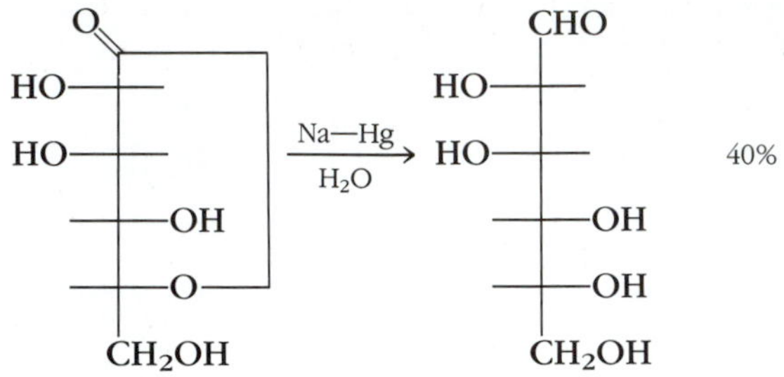

O, HO, HO, OH, O, CH_2OH $\xrightarrow[\text{H}_2\text{O}]{\text{Na—Hg}}$ CHO, HO, HO, OH, OH, CH_2OH 40%

The combination of Kiliani's cyanohydrin synthesis and Fischer's reduction permits an aldose to be converted to another aldose with one more carbon. Two diastereomers are expected, although they should not be formed in equal amounts. Moreover, solubility differences sometimes result in the isolation of just one of the two products. This is illustrated in the following sequence:

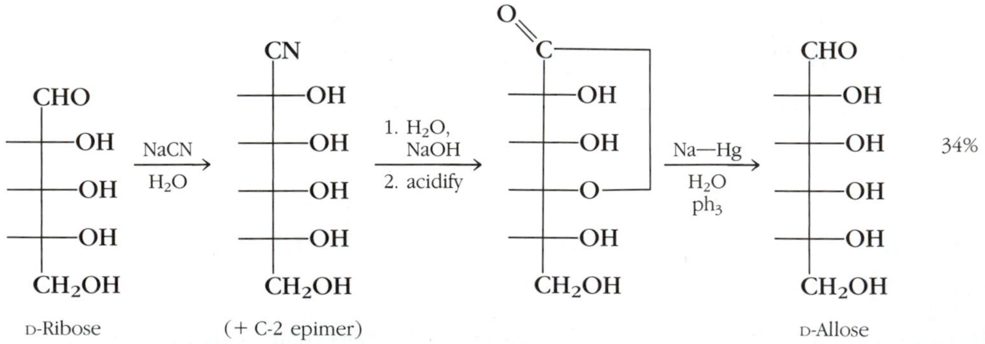

CHO, OH, OH, OH, CH_2OH (D-Ribose) $\xrightarrow[\text{H}_2\text{O}]{\text{NaCN}}$ CN, OH, OH, OH, OH, CH_2OH (+ C-2 epimer) $\xrightarrow[\text{2. acidify}]{\text{1. H}_2\text{O, NaOH}}$ O=C, OH, OH, O, OH, CH_2OH $\xrightarrow[\text{H}_2\text{O, ph}_3]{\text{Na—Hg}}$ CHO, OH, OH, OH, OH, CH_2OH (D-Allose) 34%

The Ruff Degradation: Chain Shortening

When an α-hydroxy carboxylic acid such as an aldonic acid is treated with hydrogen peroxide in the presence of a ferric salt, an *oxidative decarboxylation* occurs. This is a free radical process, initiated by the reaction of hydroperoxy radicals (HOO ·).

$$\begin{array}{l} CO_2H \\ | \\ CH{-}OH \\ \wr \end{array} \xrightarrow[Fe^{3+}]{H_2O_2} \begin{array}{c} CO_2 \\ + \\ CHO \\ \wr \end{array}$$

The discovery of this reaction by O. Ruff in Germany in the late 1890s was another important step in unraveling the complex chemistry of the carbohydrates. The Ruff degradation helped chemists carry out structural correlations between sugars with different chain lengths, and it is useful in synthesis when the higher molecular weight sugar is more readily available. A chiral center is destroyed in the reaction (C-2 of the starting aldose or aldonic acid), so two different sugars (epimeric at C-2) will yield the same aldose in this reaction. In terms of Figure 19.2, any of the aldoses shown could be formed from a Ruff degradation of *either* of the two aldoses drawn directly below it.

The overall sequence of the Ruff degradation proceeds in two steps: *first,* oxidation of the starting aldose to the aldonic acid with bromine water; and *second,* oxidative decarboxylation of the aldonic acid with hydrogen peroxide and ferric sulfate.

$$\begin{array}{l} CHO \\ | \\ CHOH \\ \wr \end{array} \xrightarrow[H_2O]{Br_2} \begin{array}{l} CO_2H \\ | \\ CHOH \\ \wr \end{array} \xrightarrow[Fe_2(SO_4)_3]{H_2O_2} \begin{array}{c} CO_2 \\ + \\ CHO \\ \wr \end{array}$$

We have previously shown examples of the oxidation of aldoses to aldonic acids (Section 19.5). Oxidative decarboxylation is illustrated by the degradation of D-galactonic acid to D-lyxose.

$$\begin{array}{c} CO_2H \\ \text{—OH} \\ \text{HO—} \\ \text{HO—} \\ \text{—OH} \\ CH_2OH \end{array} \xrightarrow[Fe_2(SO_4)_3]{H_2O_2} CO_2 + \begin{array}{c} CHO \\ \text{HO—} \\ \text{HO—} \\ \text{—OH} \\ CH_2OH \end{array} \quad 41\%$$

EXERCISE 19.14

Show how both D-threose and D-idose could be prepared from D-xylose.

EXERCISE 19.15

(a) What two aldohexoses will afford D-arabinose in the Ruff degradation?
(b) What two aldohexoses will be formed when the Fischer-Kiliani synthesis is applied to D-arabinose?

19.7 FISCHER'S DEDUCTION OF THE STRUCTURE OF GLUCOSE

The stereochemical complexity of relatively simple sugars can perplex even a modern organic chemist. But think how difficult it must have been for the early workers in sugar chemistry, who had no firm theory of structural chemistry to rely on. It was only in 1874 that van't Hoff and LeBel first proposed the idea of tetrahedral carbon, and their ideas were not readily accepted.

During the early 1880s Emil Fischer embarked on an imaginative and distinguished career in carbohydrate chemistry. He made the extremely important observation that the structural relationships of the isomeric sugars might be explained on the basis of the van't Hoff theory of tetrahedral carbon. Even more importantly, he realized that careful structural studies in the sugar series might allow a definitive evaluation of van't Hoff's theory. Fischer's work on carbohydrates during the period 1884–91 culminated in an elegant and unambiguous proof of the stereochemistry of glucose. His experiments and the structural relationships between sugars could be explained only on the basis of tetrahedral carbon, and the van't Hoff theory allowed Fischer to apply a single consistent interpretation to all the known reactions of sugars.

The foundations for modern stereochemistry were firmly established by Fischer's explanations of the sugars. The importance of this development is reflected by the first two Nobel Prizes that were awarded in chemistry: Van't Hoff received the first prize in 1901, and the 1902 prize was awarded to Fischer. In this section we summarize the arguments put forth by Fischer in his 1891 paper on the structure of glucose. Because of the tremendous historical significance of that paper, we will develop the proof using the same sequence of arguments originally set forth by Fischer.

1. The starting point for Fischer's analysis was a complete tabulation of all 16 possible aldohexose stereoisomers . These are drawn in Figure 19.6 in the same sequence as Fischer drew them. Because (+)-glucose was known to be an aldose with six carbons, it had to be one of these 16 structures.

2. Fischer recognized that only 10 stereoisomers existed for the aldaric acids, and he evaluated these first. The structures in Figure 19.6 are arranged so that for the diacids structures **11–16** are the same as **5–10** (they are merely rotated in the plane of the paper by 180°). Therefore the aldaric acid obtained from (+)-glucose must have one of the structures **1–10.**

3. Fischer found that the same aldaric acid was produced by oxidation of either (+)-glucose or a stereoisomeric aldose, (+)-gulose. Therefore the aldaric acid cannot have rotational symmetry (see Section 19.5). Structures **1–4,** which have rotational symmetry, can be excluded, so the aldaric acid produced from (+)-glucose must correspond to one of the six structures **5–10.**

4. The aldaric acid obtained from (+)-glucose is *optically active* and therefore does not have reflection symmetry. Structures **7** and **8** can thus be excluded, so only **5, 6, 9,** and **10** remain as possible structures for the aldaric acid obtained from (+)-glucose.

5. Fischer found that (+)-glucose and another aldose, (+)-mannose, yield the same osazone. These two sugars and their aldaric acids must therefore be

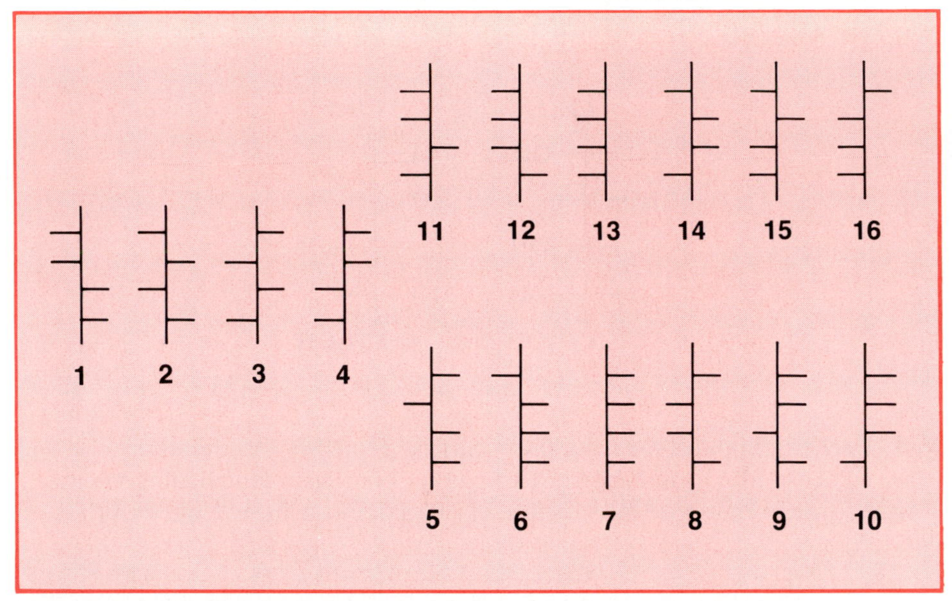

Figure 19.6 Fischer's tabulation of the 16 stereoisomeric forms of a carbon chain with four chiral centers.
Fischer arranged the structures so any two aldoses that would afford the same alditol or aldaric acid were paired together (one above the other). For such molecules the two ends of the structure have the same constitution, so **5–10** are the same as **11–16** drawn upside down, and only ten stereoisomers (**1–10**) are possible. Four aldoses (**1–4**) would yield aldaric acids with rotational symmetry, so none of these could be paired with a second aldose.

C-2 epimers. Fischer therefore argued that if the aldaric acid from (+)-glucose were either **6** or **10,** the diacid from (+)-mannose would have to be either **7** or **8** (the C-2 epimers of **6** and **10,** respectively). But **7** and **8** have reflection symmetry, while the aldaric acid prepared from (+)-mannose is optically active. Therefore neither **7** nor **8** could be the aldaric acid derived from mannose, so structures **6** and **10** can be ruled out for the aldaric acid prepared from (+)-glucose. Only **5** and **9** remain.

6. The two possible structures (**5** and **9**) for the aldaric acid are enantiomers. There was no way to distinguish between mirror image structures in 1891, so Fischer *arbitrarily* selected **5,** which we now call the D configuration. (It has since been shown that Fischer's arbitrary choice was correct!)

CO_2H
—OH
(+)-Glucose $\xrightarrow{HNO_3}$ HO—
—OH
—OH
CO_2H
5

7. If the aldaric acid from (+)-glucose has structure **5,** then (+)-glucose itself must be either **5** or structure **11** (which is directly above it in Figure 19.6).

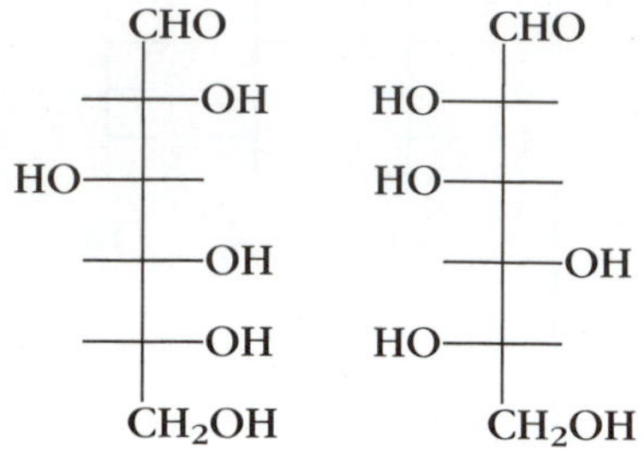

CHO
—OH
HO—
—OH
—OH
CH_2OH
5

CHO
HO—
HO—
—OH
HO—
CH_2OH
11

8. Fischer found that the naturally occurring pentose (+)-arabinose yielded (−)-glucose [the enantiomer of (+)-glucose] on chain lengthening via the cyanohydrin. Therefore the structure of (+)-arabinose must be either **A** or **B.**

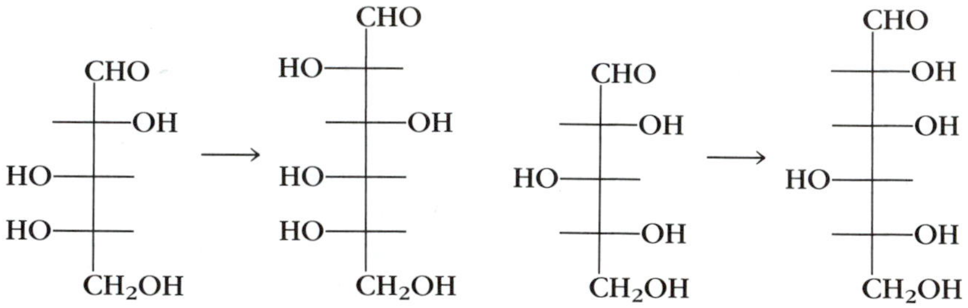

CHO
—OH
HO—
HO—
CH_2OH
A

⟶

CHO
HO—
—OH
HO—
HO—
CH_2OH
Enantiomer of **5**

CHO
—OH
HO—
—OH
CH_2OH
B

⟶

CHO
—OH
—OH
HO—
—OH
CH_2OH
Enantiomer of **11**

But the alditol and aldaric acid prepared from (+)-arabinose are *optically active,* which precludes structure **B.** On the basis of this evidence Fischer could state that (+)-arabinose has structure **A,** that (−)-glucose is the enantiomer of **5,** and that (+)-glucose has structure **5.**

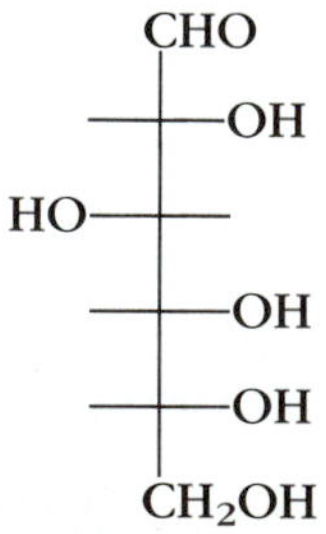

CHO
—OH
HO—
—OH
—OH
CH_2OH
5

(+)-Glucose

EXERCISE 19.16

(a) Referring to step 3 in Fischer's proof, what is the structure of (+)-glucose?
(b) Referring to step 5, what is the structure of (+)-mannose?

19.8 STRUCTURE DETERMINATION OF CARBOHYDRATES

A wide variety of carbohydrates are found in nature, and new compounds are still being isolated. How can the structures of these carbohydrates be determined? Physical methods such as NMR spectroscopy are certainly important, but their use with carbohydrates is beyond the scope of an introductory text. Consequently we will focus on some of the chemical methods which were employed prior to the development of modern spectroscopy and which are still quite useful today.

In addition to reactions and interconversions of the type discussed in the preceding section, *O*-methylation and periodate oxidation combine to provide substantial structural information. An aldose or ketose would be completely degraded to one-carbon fragments by periodate oxidation, but we showed in Section 19.5 that protected sugars are only partially degraded. This also holds for simple glycosides, which are cleaved only between carbons that have hydroxyl groups. Methyl β-D-ribofuranoside, for example, would undergo only a single cleavage to give a dialdehyde.

HOCH$_2$ O OCH$_3$ / OH OH $\xrightarrow{NaIO_4}$ HOCH$_2$ O OCH$_3$ / CH=O HC=O

The identical dialdehyde would be obtained by oxidation of methyl β-D-glucopyranoside.

HOCH$_2$ O OCH$_3$ HO OH OH $\xrightarrow{2\ NaIO_4}$ HOCH$_2$ O OCH$_3$ HC=O HC=O + HCO_2H

Clearly, the correlation of such oxidation products provides valuable structural information. The formation of the same dialdehyde in the preceding reactions requires both starting materials to have the D configuration and both to be β-glycosides.

Additional information is also available from the nature of the one-carbon fragments produced and from the quantity of periodate consumed. In the previous reaction of methyl β-D-ribofuranoside, the consumption of only a single mole of periodate and the absence of any one-carbon products both demonstrate that the reactant was a furanoside. Similarly, the six-membered ring of the glucose derivative is shown by the consumption of two moles of periodate to form the dialdehyde plus formic acid. (The methyl furanoside of glucose would have yielded formaldehyde rather than formic acid.)

The high reactivity of dialdehydes, analogous to that shown in the preceding reactions, has led to a modified procedure for structure determination of glycosides. The initial oxidation products, which still have an acetal linkage, are hydrolyzed and subsequently reduced with sodium borohydride. The resulting alditol fragments can then be identified. The dialdehyde described in the preceding reactions would yield glycerol and ethylene glycol by this procedure.

HOCH$_2$ O OCH$_3$ H CH CH H O O $\xrightarrow[HCl]{H_2O}$ [HOCH$_2$ OH H CHO + O CH CH O] $\xrightarrow{NaBH_4}$ H—C(CH$_2$OH)(CH$_2$OH)—OH + CH$_2$OH—CH$_2$OH

Notice that all stereochemical information would be lost in this sequence, and the constitution of the dialdehyde is not fully defined either. For example, the furanoside form of a ketopentose would also yield these two products.

H O CH$_2$OH H OCH$_3$ HO HO $\xrightarrow{NaIO_4}$ H O CH$_2$OH H CH CH OCH$_3$ O O $\xrightarrow[HCl]{H_2O}$ [H$_2$C(OH)CH=O + O=C(CH$_2$OH)CHO] $\xrightarrow{NaBH_4}$ CH$_2$OH—CH$_2$OH + H—C(CH$_2$OH)(CH$_2$OH)—OH

Additional information would therefore be required to completely identify the original sugar.

Glycosides that are *O*-alkylated at some positions will behave differently in periodate oxidations. For example, a glucopyranoside that is *O*-alkylated at the 3 position will be unreactive because no two adjacent carbons have free hydroxyl groups.

HOCH$_2$ O HO OR OCH$_3$ OH $\xrightarrow{NaIO_4}$ no reaction

The isomer that is *O*-alkylated at the 4 position would undergo cleavage between C-2 and C-3, but only one mole of periodate would be consumed and no one-carbon fragments would be formed.

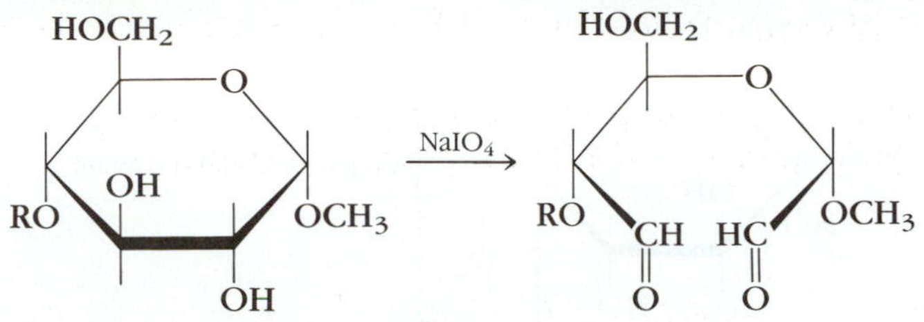

$HOCH_2$ … OH … RO … OCH_3 … OH $\xrightarrow{NaIO_4}$ $HOCH_2$ … RO … CH=O … HC=O … OCH_3

These last two examples illustrate behavior that has been extensively exploited in determining the structures of di- and polysaccharides (Section 19.9).

EXERCISE 19.17

Show the periodate oxidation products of each of the following:

(a) α-D-Ribofuranose
(b) methyl β-D-Galactopyranoside
(c) methyl β-D-Fructofuranoside

EXERCISE 19.18

Show the products that would be obtained from the compounds in Exercise 19.17 with the sequence $NaIO_4$ oxidation, hydrolysis, reduction.

EXERCISE 19.19

Methyl β-D-glucopyranoside consumes two moles of periodate and affords a dialdehyde plus formic acid. Show how this information precludes either a five- or seven-membered ring structure.

19.9 DISACCHARIDES AND POLYSACCHARIDES

Most naturally occurring sugars are not found as the free monosaccharides but as disaccharides and polysaccharides. In these complex sugars the individual subunits are joined by glycosidic linkages. Disaccharides and polysaccharides play important biological roles, ranging from energy storage and regulation (*storage polysaccharides*) to providing the structural support and integrity for plants and some invertebrates (*structural polysaccharides*). A detailed study of these compounds is beyond the scope of this text, but in this section we will provide a brief introduction to the structures and chemistry of some important disaccharides and polysaccharides. Additional discussion of polysaccharides will be found in Section 25.2.

Sucrose

Sucrose is the disaccharide that is well known as table sugar. It consists of α-D-glucopyranosyl and β-D-fructofuranosyl subunits joined by an ether linkage between the anomeric carbons of the two rings.

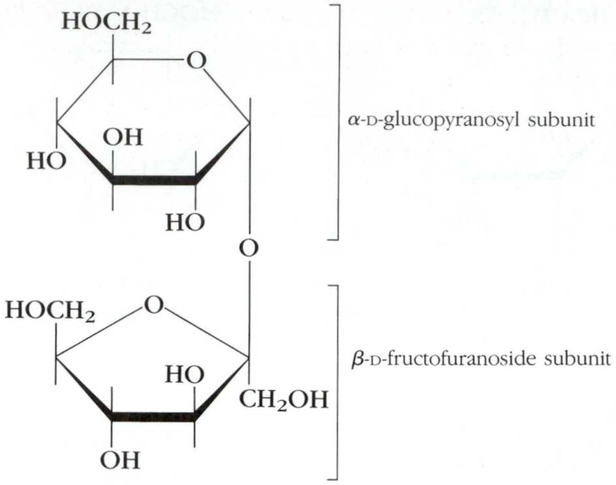

α-D-glucopyranosyl subunit

β-D-fructofuranoside subunit

Sucrose: α-D-glucopyranosyl β-D-fructofuranoside

The anomeric carbons of both subunits are involved in the glycosidic linkage, so sucrose is not a reducing sugar.

The glycosidic linkage is readily hydrolyzed in acid to yield a mixture of D-glucose and D-fructose. While sucrose exhibits a positive optical rotation, the hydrolysis product rotates polarized light in the opposite direction, and early workers called it *invert sugar*. The same cleavage of sucrose occurs by the action of the enzyme *invertase,* and the presence of this enzyme in bees is responsible for the conversion of sucrose to the mixture of D-glucose and D-fructose that constitutes honey. D-Glucose, sometimes called *dextrose* because of its positive optical rotation, exhibits a specific rotation of +74°, but D-fructose (also known as *levulose*) has a specific rotation of −106°. The 1:1 mixture of the two consequently exhibits a negative rotation.

Lactose

The milk of mammals contains approximately 4–8% of a disaccharide called *lactose* (milk sugar), and this provides the necessary carbohydrates for infant nutrition. Metabolism of lactose requires the enzyme *lactase,* and this is usually present at reduced levels in adult humans. For this reason, many adults have a limited tolerance of milk products in their diets, and ingestion of lactose can result in severe gastric disturbances. Lactose consists of D-glucose and D-galactose subunits and has the following structure:

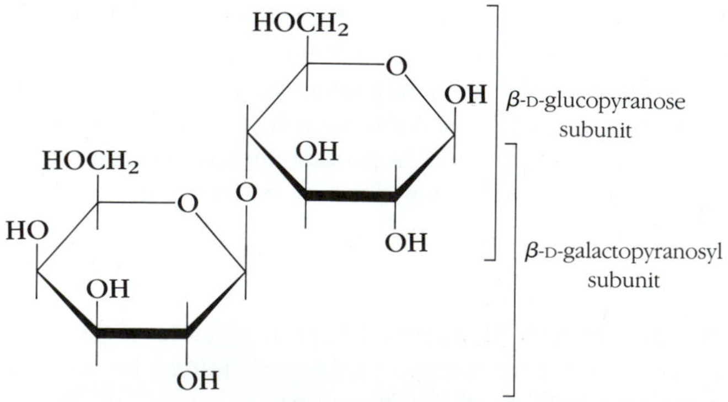

β-D-glucopyranose subunit

β-D-galactopyranosyl subunit

Lactose: 4-*O*-(β-D-galactopyranosyl)-β-D-glucopyranose

Lactose is a reducing sugar, because the glucose ring exists as the hemiacetal and is in equilibrium with the free aldehyde. (The α form of the glucopyranose ring is also present in the equilibrium mixture.)

Bromine water oxidation followed by acid hydrolysis shows that it is the glucose subunit (and not the galactose residue) of lactose that is present as the hemiacetal. This sequence yields gluconic acid, so it must be a glucose subunit that is the reducing sugar.

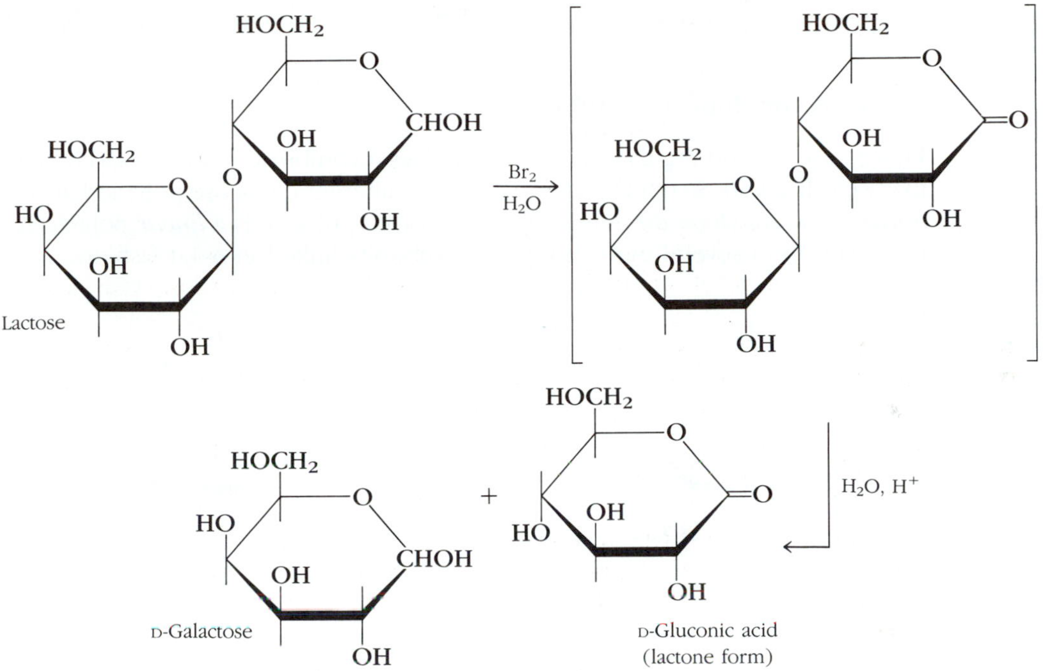

Lactose

$\xrightarrow[H_2O]{Br_2}$

$\downarrow$ H_2O, H^+

D-Galactose + D-Gluconic acid (lactone form)

Other Disaccharides

Sucrose and lactose are the only disaccharides that are commonly found in nature. Other disaccharides such as maltose and cellobiose (each of which has two glucose subunits) are obtained by the degradation of polysaccharides and will be discussed later in this section. The disaccharide *gentiobiose* also has two glucose subunits and is produced as a glycoside in several plants. It is the carbohydrate portion of amygdalin (Section 13.7), the precursor of the purported anticancer drug *laetrile*.

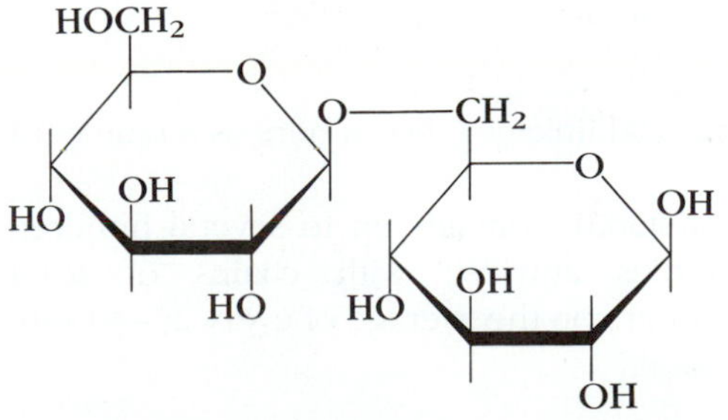

Gentiobiose

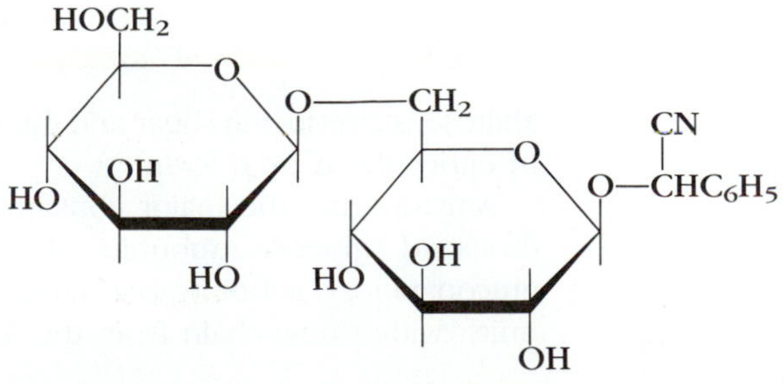

Amygdalin

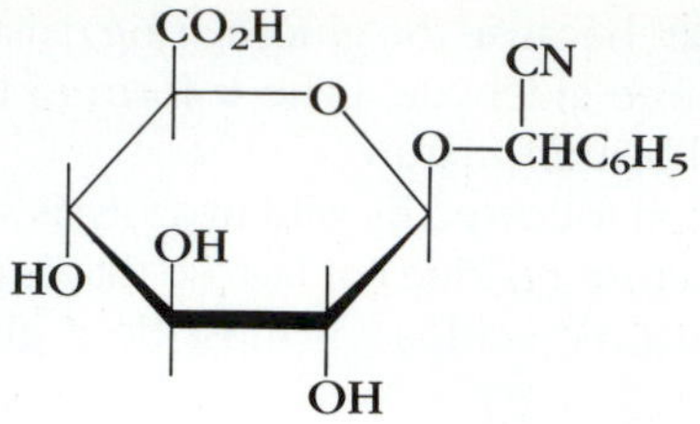

Laetrile

Amylose, Amylopectin, and Glycogen

Energy storage and regulation in plants utilizes a combination of polysaccharides that we call *starch.* Starch is composed of two different D-glucose polysaccharides, *amylose* and *amylopectin.* Amylose, the minor constituent, is a *linear* polysaccharide with up to several thousand glucose subunits linked in α-1,4 fashion.

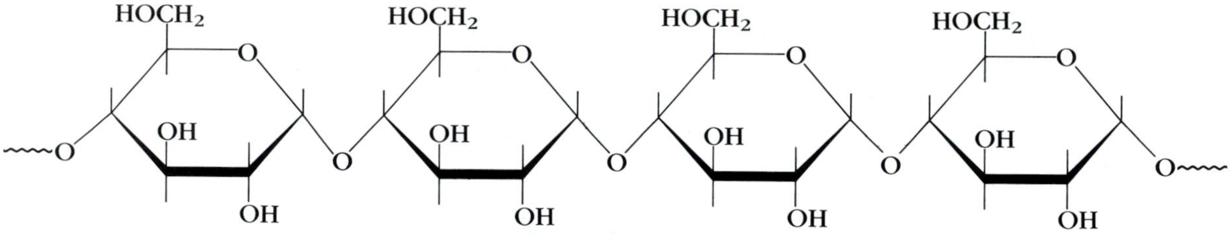

Amylose

Enzymatic degradation of amylose affords the disaccharide *maltose.*

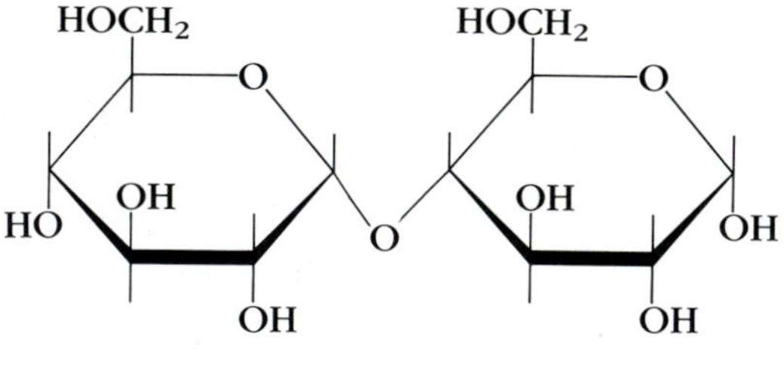

α-Maltose: 4-*O*-(α-glucopyranosyl)-α-D-glucopyranose

Maltose is a reducing sugar and the hemiacetal linkage (shown here as α) can exist in either the α or β form.

Amylopectin, the major constituent of starch, contains up to several hundred thousand glucose subunits. It resembles amylose with chains of α-1,4 glucopyranosyl subunits, but branching occurs on the average of every 25–30 subunits, with a new chain from the 6 position.

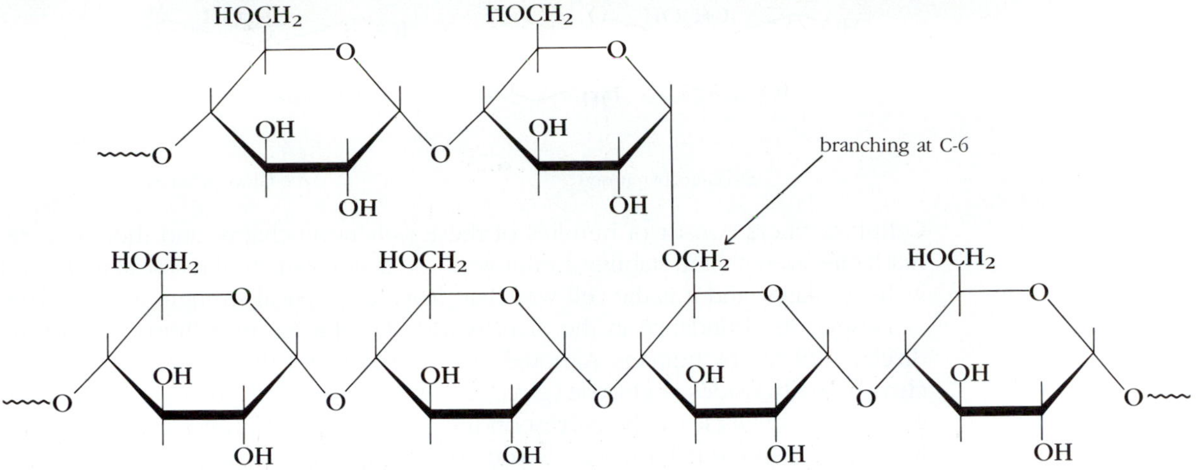

branching at C-6

Amylopectin

Glycogen is the main storage polysaccharide in animals. It is structurally similar to amylopectin, with a primary chain of α-1,4-glucopyranosyl subunits, but branching occurs much more frequently. There are only an average of 8–12 residues in the linear 1,4 chain before another chain originates as a branch at the 6 position of a glucose subunit. The rapid enzymatic degradation of glycogen to glucose permits the energy stored in the form of polysaccharides to be utilized by the organism as the need arises.

Cellulose

Cellulose is the main structural polysaccharide of plants. It is a highly regular polymer of D-glucose subunits linked in β-1,4 fashion.

Cellulose

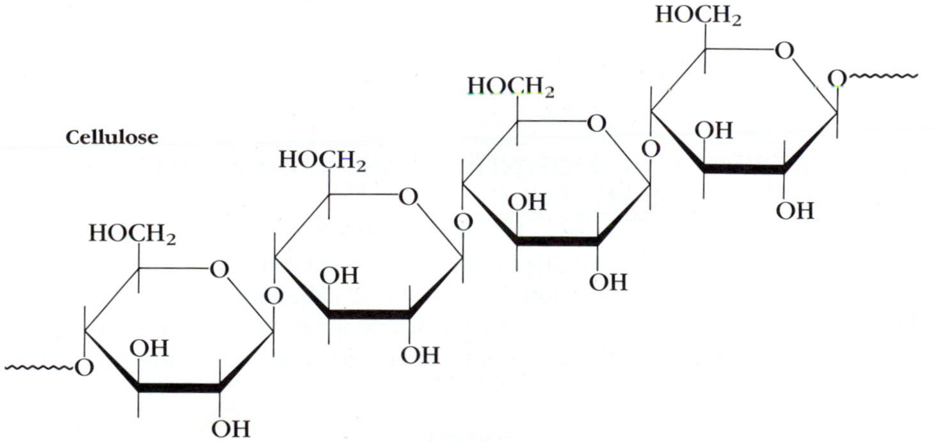

This β linkage is the only difference between cellulose and amylose, but it results in dramatically different physical properties. The individual subunits of cellulose can exist in the chair conformation with all substituents equatorial. Amylose, on the other hand, must either have the glucosyl substituent at the 1 position in an axial orientation or else exist in a nonchair conformation.

β-D-Glucopyranose α-D-Glucopyranose

Cellulose fibers consist of bundles of these polymeric chains, and they have remarkable strength and stability. Cellulose is the major structural material in the cell walls of plants, and it is the cell wall that provides physical strength for the plant.

Despite its abundance in the plant world, the glucose in cellulose cannot be directly utilized by humans. Although we have enzymes that are very efficient in cleaving α-glycosides of glucose (glycogen, amylose, amylopectin), we do not have enzymes that will cleave the corresponding β-1,4-linkage. Consequently, the cellulose present in the plant materials you consume passes unchanged through your gastrointestinal tract.

Partial hydrolysis of cellulose yields the disaccharide *cellobiose*.

β-Cellobiose: 4-*O*-(β-D-glucopyranosyl)-β-D-glucopyranose

Cellobiose is a reducing sugar, and the hemiacetal can exist as either the α or β anomer (where the latter is depicted in the structure shown here). The relationship between amylose and cellulose extends to the corresponding disaccharides. Maltose and cellobiose differ only by the stereochemistry of the glycosidic linkage.

19.10 ASCORBIC ACID: BIOSYNTHESIS AND COMMERCIAL SYNTHESIS OF VITAMIN C

The importance of fresh fruits in the human diet has been recognized for centuries, and their absence results in the nutritional disorder called scurvy. But only in 1917 did chemists identify the factor that is responsible for the beneficial properties of fresh fruit. The factor was called *vitamin C,* and its structure was established in 1933 as L-ascorbic acid.

L-Ascorbic acid (vitamin C)

Although the compound is clearly a lactone, it is named as an acid because the enediol form of the hydroxy ketone is rather acidic. The keto forms of this enediol are less stable than the enediol itself.

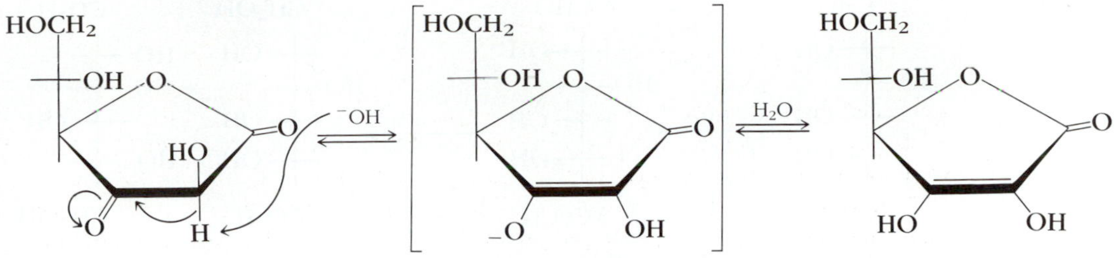

HOCH₂ … ^{-}OH … H_2O

Biosynthesis

Ascorbic acid is not produced by any of the biochemical pathways in humans. Nevertheless it is a necessary compound for human metabolism and is therefore classified as a vitamin, although its precise biochemical function is not yet understood. The biosynthesis of L-ascorbic acid in plants has been carefully studied and each individual step is known. These reactions are brought about by enzymes, but there is a close parallel to laboratory processes. The starting point for the biosynthesis is D-glucose, and this is at first a surprise. How is the L configuration of ascorbic acid produced from the D isomer of glucose? The answer is surprisingly simple. Ascorbic acid is designated as L because of the configuration at C-5, but this was originally C-2 of glucose.

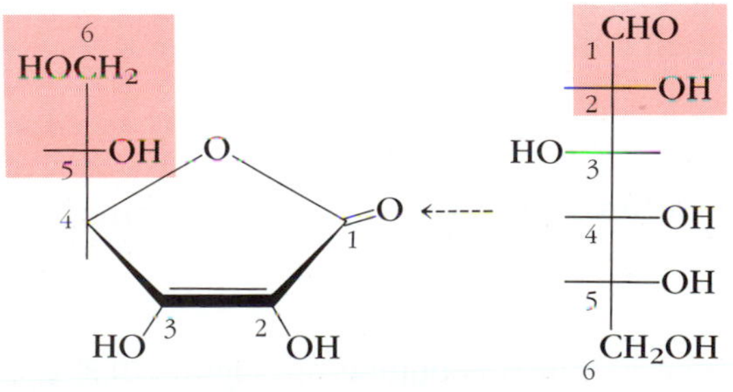

L-Ascorbic acid   D-Glucose

The entire biosynthesis is summarized in Figure 19.7.

The first steps in the biosynthesis involve a *chain end reversal,* in which the aldehyde end of glucose is reduced and the primary alcohol (C-6) is oxidized. Lactonization of the resulting aldonic acid is followed by specific oxidation of the position α to the carbonyl. The resulting keto lactone exists preferentially in the enol form, L-ascorbic acid.

Commercial Synthesis

A number of syntheses of L-ascorbic acid have been carried out, but the one we will discuss here was first accomplished in 1934 by T. Reichstein in Switzerland. (Reichstein received the Nobel Prize in 1950 for his work in organic chemistry.) Subsequent modifications have improved the yields, and this synthesis has remained an important and inexpensive method for the commercial production of

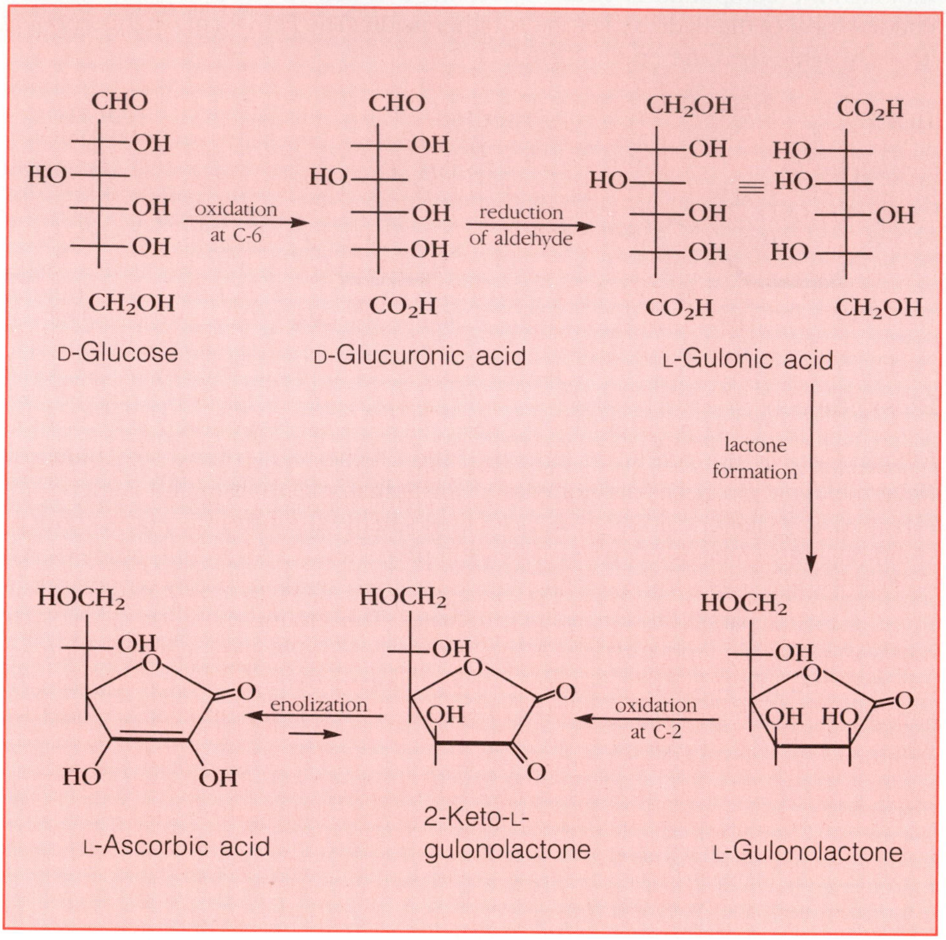

Figure 19.7 Biosynthesis of L-ascorbic acid in plants.

vitamin C. The synthesis, which is outlined in Figure 19.8, is very similar to the biosynthetic scheme shown in Figure 19.7.

The starting material in the commercial synthesis, as in the biosynthesis, is D-glucose (**1**). Reduction to the alditol **2** can be carried out with a variety of reagents, but catalytic hydrogenation was found to be most efficient when both yield and cost were considered. There are no direct chemical methods by which a polyfunctional compound such as **2** can be specifically oxidized at C-2, but enzymes do exhibit such specificity. Consequently the known capability of the bacterium *Acetobacter suboxydans* is employed to carry out the microbial oxidation in high yield. The resulting L-sorbose (**3a**), which is in equilibrium with the furanose form **3b,** is converted to its diketal **4** by reaction with acetone in the presence of sulfuric acid as catalyst. The primary alcohol of **4** can be oxidized to a carboxylic acid (**5**) in high yield, and this completes the chain end reversal. The original C-1 of glucose has been reduced to a primary alcohol (but is part of a ketal linkage in **5**), and the —CH_2OH terminus of glucose has been oxidized to a carboxyl group. Compound **5** is simply a protected derivative of 2-keto-L-gulonic acid, and the two ketal groups can be hydrolyzed by the action of aqueous acid. A variety of procedures have been

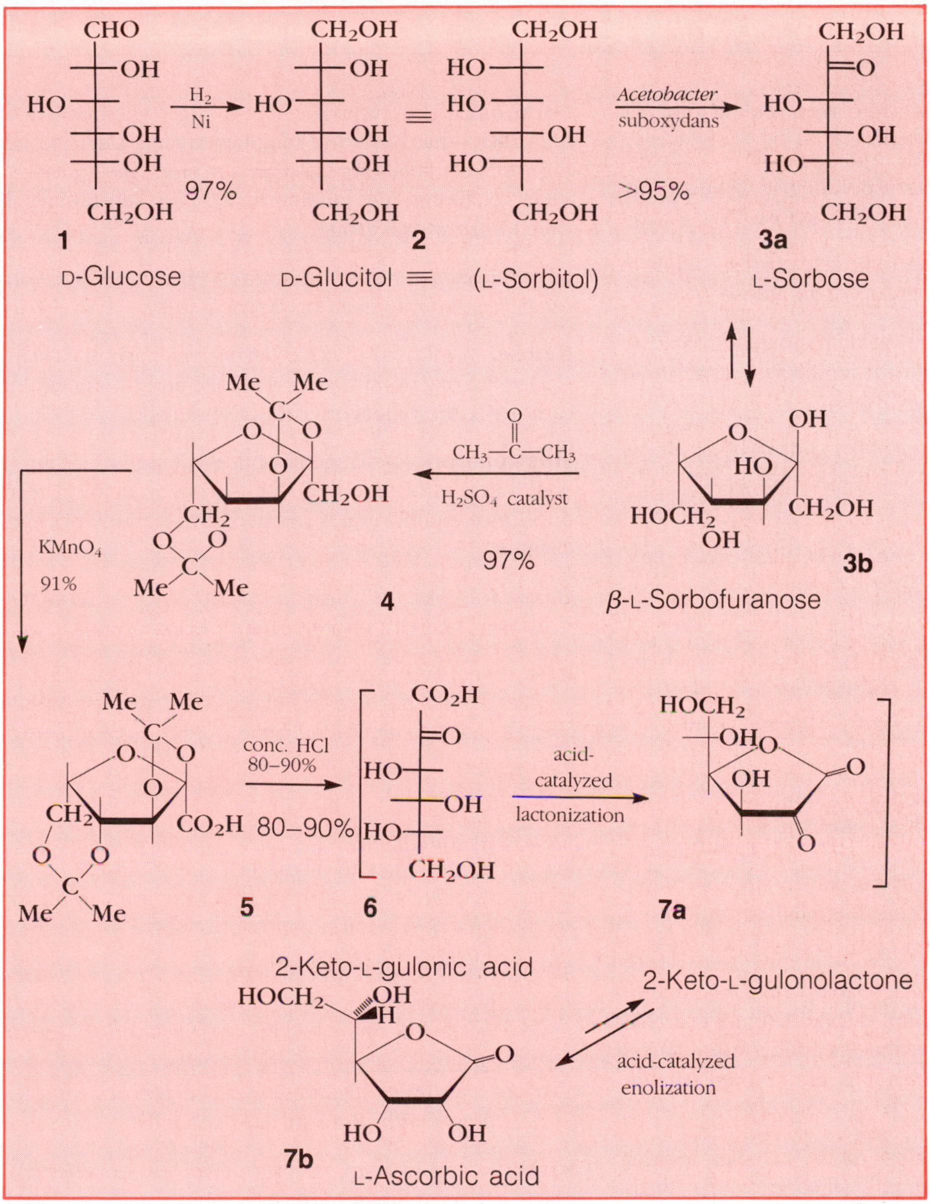

Figure 19.8 Commercial synthesis of ascorbic acid.
Yields shown here are optimal yields for each step, but the synthesis can be carried out in about 50% overall yield.

investigated, and treatment of a solution of **5** in an organic solvent such as benzene has been found to yield L-ascorbic acid (**7b**) directly. The initial hydrolysis of the ketal groups of **5** produces 2-keto-L-gulonic acid (**6**), which cyclizes to the lactone **7a** in the presence of acid. The keto lactone **7a** is in equilibrium with the enol form **7b,** which is rather insoluble in nonpolar organic solvents. Consequently, L-ascorbic acid crystallizes directly from the reaction mixture.

19.11 TERMS AND DEFINITIONS

Aldaric acid. A carboxylic acid corresponding to oxidation of both termini of an aldose.

Alditol. A polyalcohol corresponding to reduction of the carbonyl group of an aldose or ketose.

Aldonic acid. A monocarboxylic acid corresponding to oxidation of the aldehyde group of an aldose.

Aldose. A monosaccharide having the oxidation state of an aldehyde at C-1 and that of an alcohol at all other carbons.

Anomeric carbon. The carbonyl carbon of an aldose or ketose, which becomes a chiral center in the cyclic (acetal or ketal) forms.

Anomers. Stereoisomeric sugars that differ only in configuration at the anomeric carbon (a specific type of epimer).

Carbohydrate. A general term for compounds with the general formula $C_x \cdot (H_2O)_y$ (but also used to describe derivatives containing other atoms). The terms *sugar* and *saccharide* are synonyms.

α,β Configuration. A term describing the configuration of the anomeric carbon of a sugar in its cyclic form. When drawn in a Haworth representation, a D sugar has the α configuration if the oxygen substituent lies below the "plane" of the ring and the β configuration if it lies above. (The meaning of α and β is reversed for L sugars.)

D,L Configuration. A term describing the configuration of the highest-numbered chiral carbon of a monosaccharide. A simple D sugar has the *R* configuration at this chiral center. D-Glyceraldehyde is the reference compound.

Epimers. Stereoisomers which differ in configuration at only one of several chiral centers. All other chiral centers have the same configuration.

Furanose. The cyclic form of a sugar (hemiacetal or hemiketal) that has a five-membered ring. Corresponding acetals or ketals are called *furanosides;* the entire residue, excluding the anomeric oxygen substituent, is called a *furanosyl* group.

Glycoside. A general term for a cyclic acetal or ketal of a sugar (i.e., a furanoside or pyranoside).

Ketose. A monosaccharide having one carbon (usually C-2) at the oxidation state of a ketone and the remaining carbons at the oxidation state of an alcohol.

Mutarotation. A change in optical rotation of a sample over a period of time that results from a chemical reaction. Once equilibrium is reached, the rotation no longer changes.

Osazone. A bishydrazone (usually a bisphenylhydrazone) corresponding to an α-ketoaldose.

Pyranose. The cyclic form of a sugar (hemiacetal or hemiketal) that has a six-membered ring. Corresponding acetals or ketals are called *pyranosides;* the entire residue, excluding the anomeric oxygen substituent, is called a *pyranosyl* group.

Reducing sugar. A sugar (hemiacetal or hemiketal) that is in equilibrium with the free aldehyde or α-hydroxyketone and readily reacts with mild oxidizing agents such as silver ion and cupric ion.

Saccharide. Sugar; carbohydrate. The term *saccharide* is usually used with a prefix indicating the number of subunits joined by glycosidic linkages; a *monosaccharide* has only a single subunit, a *disaccharide* has 2, *oligosaccharides* have up to 10, and *polysaccharides* are complex carbohydrates with more than 10 subunits.

Sugar. Carbohydrate; a saccharide. In nonchemical usage the term *sugar* describes table sugar (sucrose).

19.12 SUMMARY OF REACTIONS

In Table 19.1 we summarize and review some of the important reactions that are used in carbohydrate chemistry.

TABLE 19.1 Reactions of Carbohydrates

Reaction	Comments
1. Intramolecular cyclization	Yields five- or six-membered rings preferentially.
(a) Hemiacetal-hemiketal formation	Section 19.2
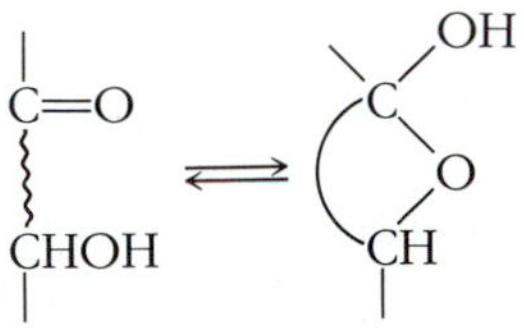 C=O ⋯ CHOH ⇌ cyclic (C–OH, O, CH)	Rapid equilibrium favoring cyclic form.
(b) Acetal-ketal formation	Section 19.3
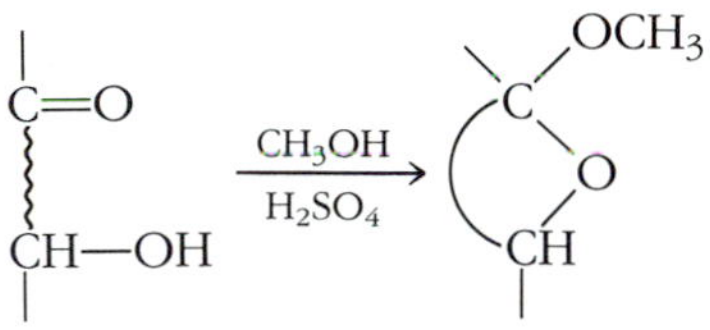 C=O ⋯ CH—OH $\xrightarrow[H_2SO_4]{CH_3OH}$ cyclic (C–OCH_3, O, CH)	Formation of methyl glycosides is most common. The glycoside is not a reducing sugar.
(c) Lactone formation	Section 19.5
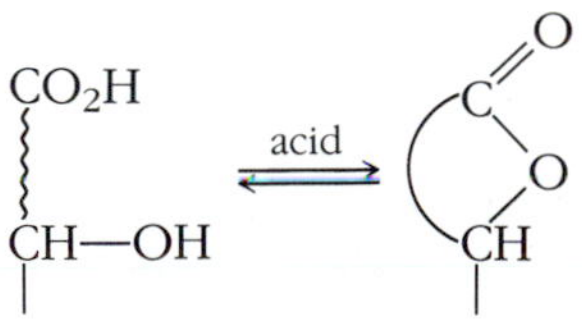 CO_2H ⋯ CH—OH $\overset{acid}{\rightleftharpoons}$ cyclic (C=O, O, CH)	Illustrated for an aldonic acid. Aldaric acids can yield products with two lactone rings. Ring opens to the salt of the hydroxy acid in base.
(d) Intramolecular *O*-alkylation	Section 19.3
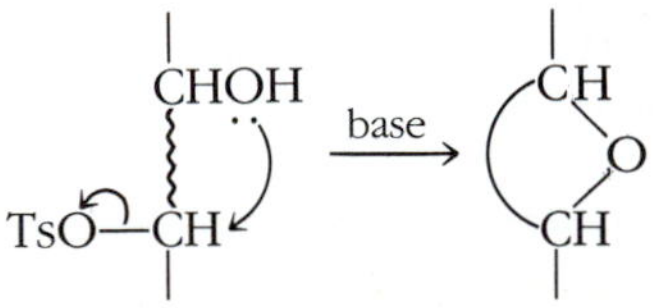 CHOH ⋯ TsO—CH $\xrightarrow{base}$ cyclic (CH, O, CH)	Three-membered rings (epoxides) can also be formed in this reaction.

TABLE 19.1 (continued)

Reaction	Comments
2. Protection of OH groups as acetals or ketals	Section 19.3
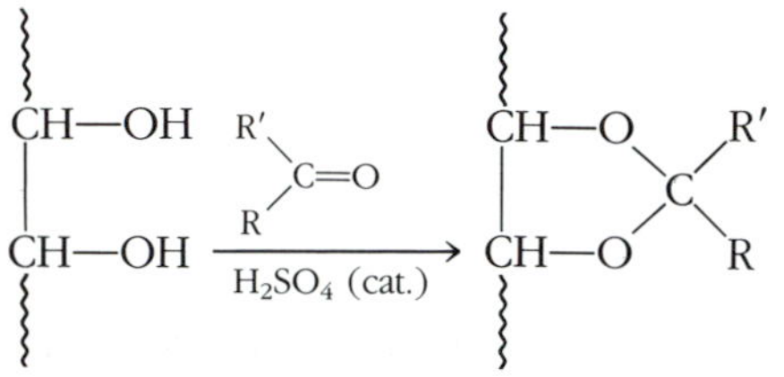 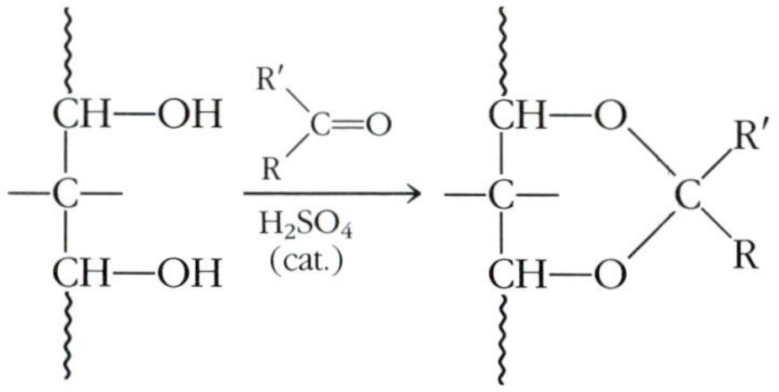 $\begin{matrix} \wr \\ CH{-}OH \\ \vert \\ CH{-}OH \\ \wr \end{matrix} \xrightarrow[H_2SO_4\ (cat.)]{R'(R)C{=}O} \begin{matrix} \wr \\ CH{-}O \\ \vert \\ CH{-}O \\ \wr \end{matrix}\!\!>\!C(R')(R)$	Five- or six-membered rings are formed. Common reagents include acetone ($R = R' = CH_3$), acetaldehyde ($R = CH_3$, $R' = H$), benzaldehyde ($R = C_6H_5$, $R' = H$).
$\begin{matrix} \wr \\ CH{-}OH \\ \vert \\ {-}C{-} \\ \vert \\ CH{-}OH \\ \wr \end{matrix} \xrightarrow[H_2SO_4\ (cat.)]{R'(R)C{=}O} \begin{matrix} \wr \\ CH{-}O \\ \vert \\ {-}C{-} \\ \vert \\ CH{-}O \\ \wr \end{matrix}\!\!>\!C(R')(R)$	
3. Methylation of hydroxyl groups	Section 19.3
$-\overset{\vert}{\underset{\vert}{C}}-OH \xrightarrow[\text{or } CH_3I,\ Ag_2O]{(CH_3)_2SO_4,\ NaOH} -\overset{\vert}{\underset{\vert}{C}}-OCH_3$	All free OH groups are alkylated. Silver salts cannot be used with reducing sugars.
4. Acylation of hydroxyl groups	Section 19.3
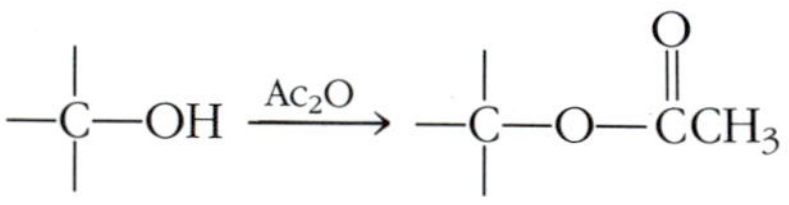 $-\overset{\vert}{\underset{\vert}{C}}-OH \xrightarrow{Ac_2O} -\overset{\vert}{\underset{\vert}{C}}-O-\overset{O}{\overset{\Vert}{C}}CH_3$	All free OH groups can be acylated. Acetic anhydride is a common reagent.
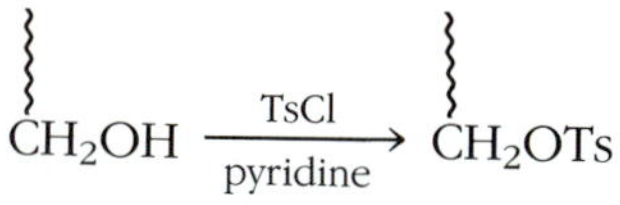 $\overset{\wr}{C}H_2OH \xrightarrow[\text{pyridine}]{TsCl} \overset{\wr}{C}H_2OTs$	Primary hydroxyl groups react preferentially.

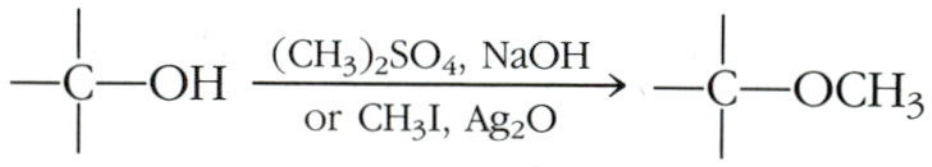

TABLE 19.1 (continued)

Reaction	Comments
5. Reduction	Section 19.4
(a) Alditol formation	
CHO–CHOH–⌇ or CH_2OH–C=O–⌇ $\xrightarrow{H_2\text{/cat. or Na/alcohol or } NaBH_4}$ CH_2OH–CHOH–⌇	The product from an aldose or ketose is an alditol.
(b) Lactone reduction	
Lactone (C=O, O, CH) $\xrightarrow[H_2O,\ pH\ 3]{Na\text{—}Hg}$ cyclic hemiacetal (CH–OH, O, CH)	With sodium amalgam the product is an aldose (sodium in alcohol would reduce the lactone to an alditol). Used in Fischer-Kiliani chain lengthening process.
6. Oxidation	Section 19.5
(a) Bromine water oxidation	
CHO–⌇ $\xrightarrow{Br_2,\ H_2O}$ CO_2H–⌇	An aldose yields an aldonic acid.
(b) Oxidation with nitric acid	
CHO–⌇–CH_2OH $\xrightarrow{HNO_3}$ CO_2H–⌇–CO_2H	An aldose yields an aldaric acid.

TABLE 19.1 (continued)

Reaction	Comments
(c) Oxidation with silver ion	
CHO (chain) or CH_2OH–C=O (chain) $\xrightarrow{Ag(NH_3)_2^+OH^-}$ Ag^0 (Precipitate)	Used for identification of free aldehyde or α-hydroxy ketone (Tollens test).
(d) Oxidation with cupric ion	
CHO (chain) or CH_2OH–C=O (chain) $\xrightarrow{CuSO_4}$ Cu_2O (Precipitate)	Used for identification (or quantitative analysis) of free aldehyde or α-hydroxy ketone. (Reagent is called Fehling solution.)
(e) Oxidation with phenylhydrazine	
CHO–CHOH (chain) or CH_2OH–C=O (chain) $\xrightarrow{NH_2NHC_6H_5}$ C=N—NHC_6H_5 / C=N—NHC_6H_5 (chain)	Osazone formation.
(f) Periodate oxidation	
CH=O / C=O / CHOH / CH_2OH $\xrightarrow{NaIO_4}$ HCO_2H / CO_2 / HCO_2H / CH_2O	Cleaves bonds between carbons bearing —OH or =O substituents. C—H bonds remain intact. No cleavage if OH groups are protected as ethers.
7. Chain lengthening	Section 19.6
CHO (chain) $\xrightarrow{HCN}$ CN–CH—OH (chain) $\xrightarrow{1.\ H_2O,\ ^-OH;\ 2.\ neutralize}$ CO_2H–CH—OH (chain)	Product is acid (or lactone) with one more carbon (two epimers are formed). Reduction of lactone with sodium amalgam completes the Fischer-Kiliani synthesis.

TABLE 19.1 (continued)

Reaction	Comments
8. Chain shortening	Section 19.6
CO_2H – CHOH (chain) $\xrightarrow[Fe^{3+}]{H_2O_2}$ CHO (chain) + CO_2	Ruff degradation. The aldonic acid is prepared by bromine water oxidation of the aldose.

19.13 PROBLEMS

19.1 Write the equation for the reaction of each of the following sugars with nitric acid:
(a) D-Mannose (b) β-D-Ribofuranose
(c) Lactose (d) D-Idose
(e) D-Threose (f) β-D-Gulopyranose

19.2. Write the equation for the reaction of each compound in Problem 19.1 with excess phenylhydrazine.

19.3 Write the equation for the reaction of each compound in Problem 19.1 with Br_2, H_2O.

19.4 Write the equation for the reaction of each product in Problem 19.3 with H_2O_2, ferric sulfate.

19.5 Write the equation for the reaction of each of the following with sodium borohydride:
(a) D-Fructose (b) β-D-Galactopyranose
(c) α-D-Xylofuranose (d) D-Ribulose
(e) L-Arabinose

19.6 Write the equation for the reaction of each compound in Problem 19.5 with methanol, HCl.

19.7 Which of the following compounds would give a positive test with Tollens' reagent?
(a) Maltose (b) D-Lyxose
(c) α-D-fructofuranose (d) Methyl β-D-ribofuranose
(e) Sucrose (f) L-Sorbose

19.8 Draw the products of periodate oxidation for all the compounds in Problem 19.7.

19.9 A student took a crystalline sample of α-D-glucopyranose, dissolved it in water, and measured its optical rotation. The specific rotation was +111°. Subsequently, the student's instructor checked the same solution and found the specific rotation to be +52°. Suggest a reasonable explanation. (Both the student's and instructor's measurements were correct.)

19.10 The same student returned to the laboratory and dissolved a sample of crystalline D-gulonic acid in dilute HCl. The solution exhibited a specific rotation of −7°, but it slowly changed. After a while the rotation was zero and the student was convinced that the sample had racemized. The next day the instructor checked the sample and found that the specific rotation was not zero but +12°. What had actually occurred?

19.11 Draw all pairs of aldohexoses that yield the same alditol on reduction with sodium borohydride.

19.12 Indicate which aldose would yield the same alditol as that formed by reduction of each of the following:
(a) L-sorbose (c) D-xylulose
(b) D-ribulose (d) D-fructose

19.13 Which D-aldohexoses would yield an optically inactive alditol as one of the products from Fischer-Kiliani chain lengthening followed by reduction?

19.14 Draw all possible C-5 aldaric acids and their aldose precursors.

19.15 Draw all possible C-6 aldaric acids and their aldose precursors.

19.16 Reduction of L-sorbose would yield two epimeric alditols. Which aldoses would yield these two?

19.17 Reduction of D-fructose would yield two epimeric alditols. Which aldoses would yield these two?

19.18 In 1891 Fischer oxidized xylose to an inactive diacid with HNO_3. He also converted (+)-xylose to (+)-gulose and (+)-gulose to (+)-glucaric acid (see Section 19.7). What could he conclude about the structure and stereochemistry of (−)-xylose?

19.19 On the basis of the information in Section 19.7, Fischer was able to assign the stereochemistry of (+)- and (−)-mannose. Show how this can be done.

19.20 Fischer's determination of the structure of glucose was carried to a successful conclusion despite some confusing and contradictory evidence. (a) Reduction of glucose with sodium in alcohol had previously been found to yield mannitol. Keeping in mind the reaction of sodium with ethanol, explain this observation. (b) Fischer's deduction of glucose stereochemistry is partly based on the conclusion that xylose yields an optically *inactive* aldaric acid. But it had previously been reported that xylose yields an optically *active* alditol with a specific rotation of 0.5°. why was Fischer able to disregard that report?

19.21 D-Glucose reacts with on equivalent of triphenylmethyl chloride (Section 12.1) to yield **A,** which contains a single triphenylmethyl group. Treatment of **A** with acetic anhydride affords **B** (which has four acetyl groups) in nearly 50% overall yield. The reaction of **B** with HBr produces a 55% yield of **C.** Compound **C** retains the four acetyl groups but has no triphenylmethyl group and does not react with Tollens' reagent. Suggest reasonable structures for **A, B,** and **C.**

19.22 When 100 g of commercial absorbent cotton is treated with acetic anhydride and sulfuric acid, about 65 g of a product **A** with the molecular formula $C_{12}H_{14}O_{11}(Ac)_8$ is obtained. Reaction of **A** with sodium methoxide in methanol provides a 94% yield of **B** ($C_{12}H_{22}O_{11}$). Acid hydrolysis of **B** affords D-glucose as the only product. What are the structures of **A** and **B**?

19.23 Reaction of lactose with dimethyl sulfate affords a 73% yield of an octamethyl derivative. What products would be expected at each stage if this heptamethyl derivative were subjected to acid hydrolysis followed by periodate oxidation?

19.24 Structural information about polysaccharides is provided by the reaction sequence consisting of: (1) periodate oxidation, (2) borohydride reduction, (3) complete methylation with dimethyl sulfate, and (4) acid hydrolysis.
(a) Draw the products that would be obtained from the terminal glucose subunit at the reducing end of an amylose chain.
(b) Draw the products that would be obtained from the terminal glucose subunit at the nonreducing end of an amylose chain.
(c) Draw the products that would be obtained from the internal glucose subunits of an amylose chain.

19.25 The reaction sequence of Problem 19.24 was applied to a sample of partially degraded amylose to yield, among other products, 1,3-di-*O*-methylglycerol and 1,4-di-*O*-methylerythritol in a ratio of 1:7. What was the *average* chain length of the sample?

19.26 The reaction sequence of Problem 19.24 was applied to a sample of amylopectin to yield, among other products, a 26:1 ratio of 1,4-di-*O*-methylerythritol and 1-*O*-methylerythritol. What was the average chain length between branch points in the sample?

19.27 A trisaccharide of D-glucopyranose subunits was subjected to borohydride reduction followed by periodate oxidation and reduction of the products by a second borohydride reduction. Acid hydrolysis then produced (per mole of trisaccharide):

HCO_2H (3 mol), CH_3OH (1 mol),

CHO
|
CH_2OH
2 mol

CH_2OH
—|—OH
CH_2OH
2 mol

CH_2OH
—|—OH
—|—OH
CH_2OH
1 mol

Assuming that all residues are α, what is the complete structure of the trisaccharide?

19.28 (a) Maltose, cellobiose, and gentiobiose (Section 19.9) are all disaccharides of D-glucose. Draw the products expected if each is subjected to methylation with dimethyl sulfate followed by hydrolysis with acid. (b) Assuming that the glucose residues were known to be pyranose rings but no other structural information was available, what would the products demonstrate about the structures of the three disaccharides?

19.29 Oxidation of aldohexose **A** yields the aldaric acid **B,** which is also formed by the oxidation of L-allose. The osazone **C** is formed from **A.** (a) On the basis of this information deduce the structures of compounds **A–C.** (b) Show the reagents needed to convert **A** to **B** and to convert **A** to **C.**

19.30 Complete the following reactions.

(a) D-Ribose $\xrightarrow[\text{2. Ba(OH)}_2\text{, H}_2\text{O}]{\text{1. HCN}}$? (two products)
3. lactonize
4. Na—Hg, H_2O, pH 3–5

(b) [furanose ring: $HOCH_2$ on C4, O in ring, OCH_3 on C1, OH on C2 and C3] $\xrightarrow{(CH_3)_2SO_4}$? $\xrightarrow[H_2SO_4]{H_2O}$?

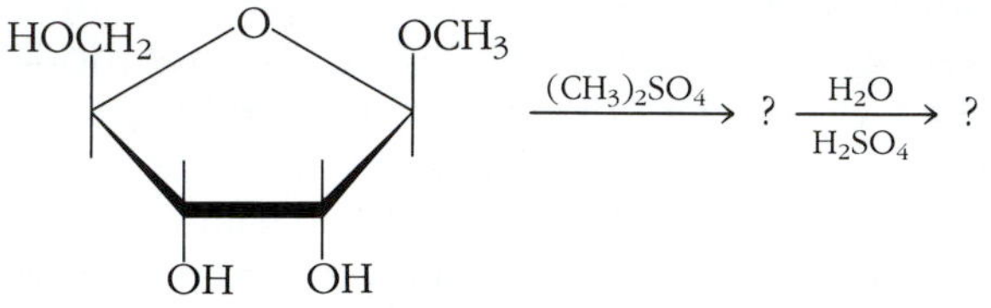

(c) ? $\xrightarrow{C_6H_5NHNH_2}$? $\xleftarrow{C_6H_5NHNH_2}$ D-Mannose

19.31 **A–H** are all different compounds and *all are optically active.* On the basis of the information provided, deduce the structures of **A–G.**

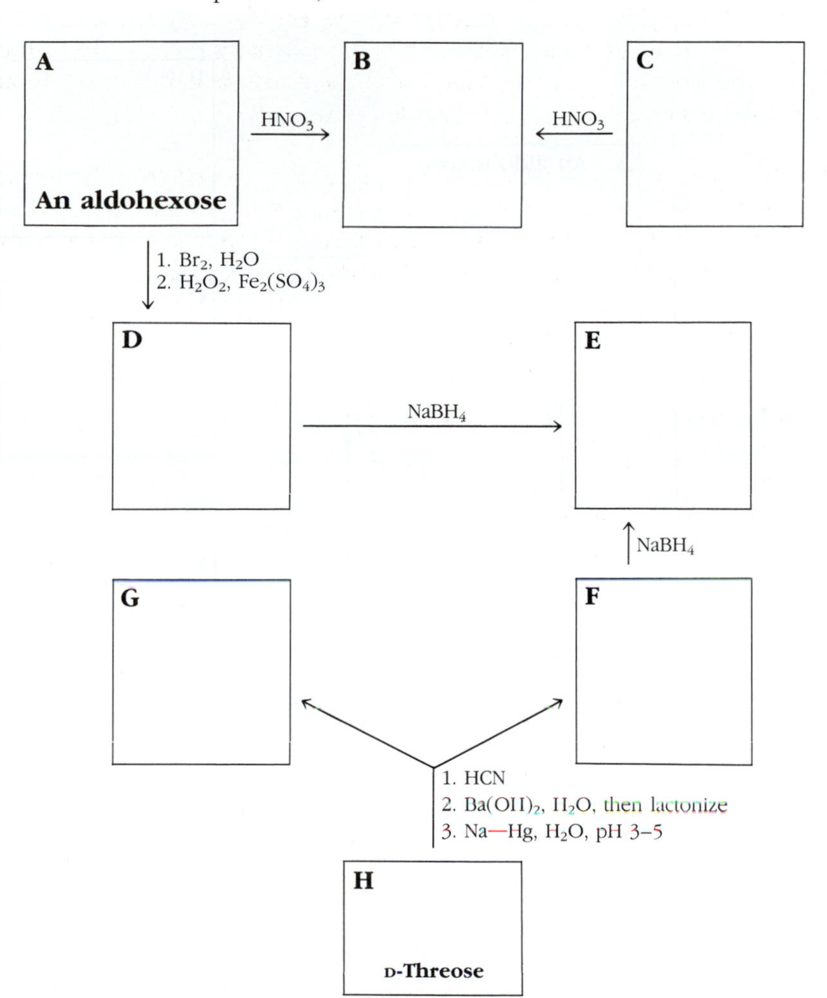

19.32 **A–H** are all different compounds and *all are optically active.* On the basis of the information provided, deduce the structures of **A–G.**

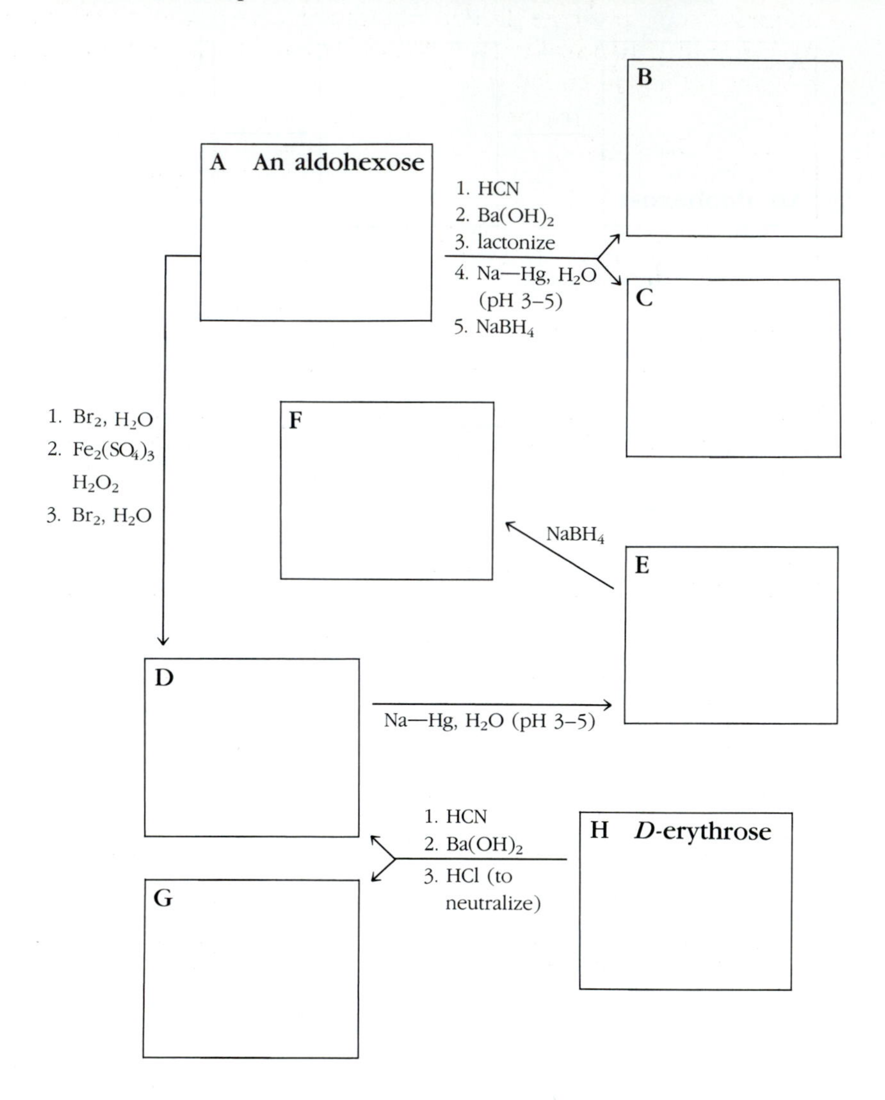

19.33 Compound **A** is a reducing sugar, $C_6H_{12}O_6$. It is not known whether it is an aldose or a ketose, and it is not known whether its cyclic acetal derivative is a furanose or a pyranose. Deduce the complete structures of compounds **A–F** from the information provided (**A–H** are all different compounds). Assume that any glycoside is α.

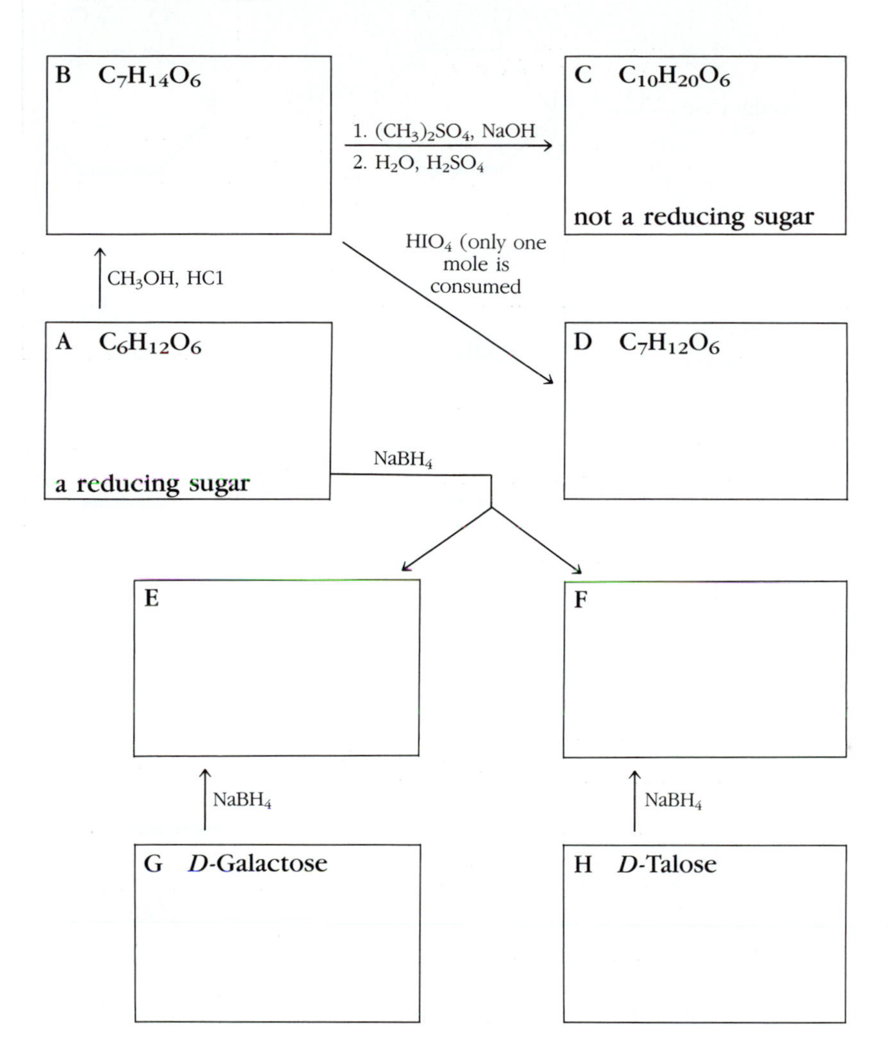

19.34 A synthesis of ascorbic acid is outlined below. Complete the missing portions.

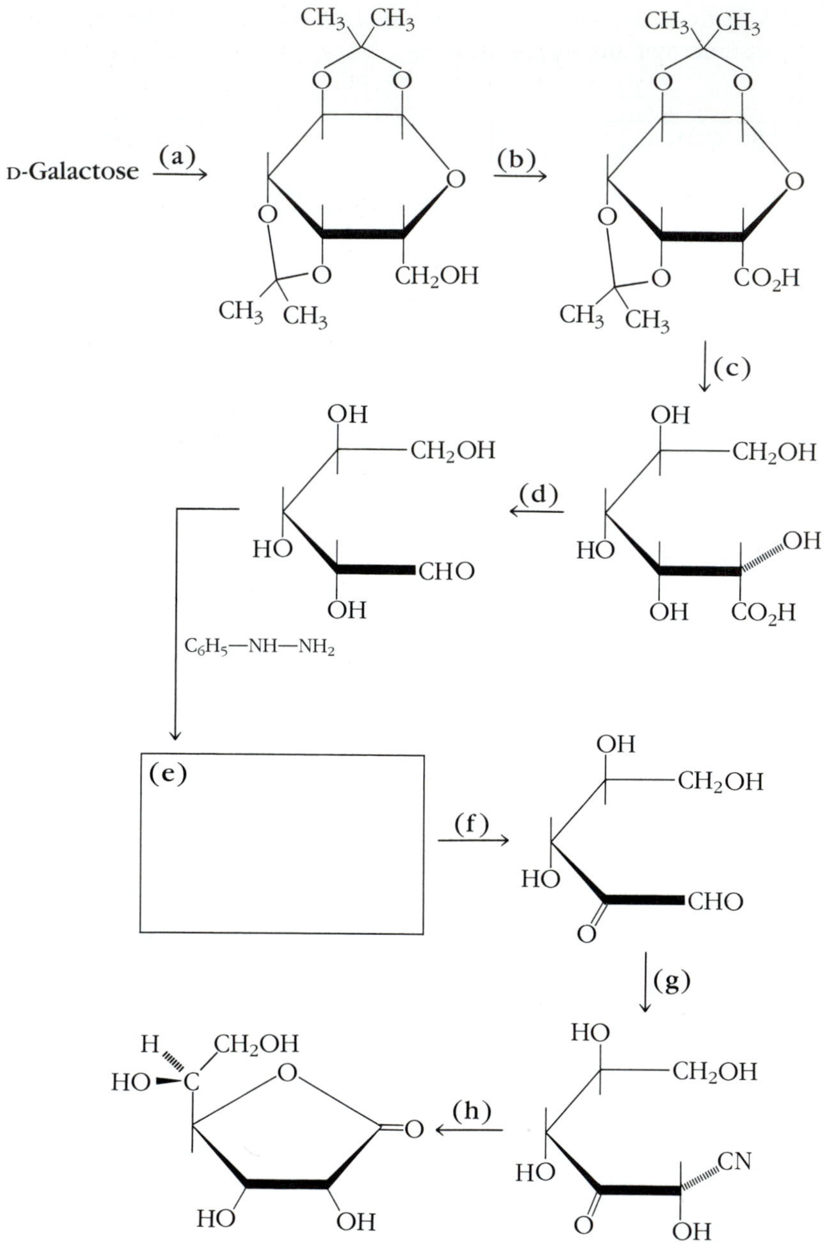

D-Galactose (a) →

CH_3 CH_3 O O O O CH_2OH CH_3 CH_3

(b) →

CH_3 CH_3 O O O O CO_2H CH_3 CH_3

↓ (c)

OH CH_2OH HO OH OH CO_2H

(d) ←

OH CH_2OH HO CHO OH

C_6H_5—NH—NH_2

(e)

(f) →

OH CH_2OH HO CHO O

↓ (g)

HO CH_2OH HO CN O OH

(h) ←

H CH_2OH HO C O O HO OH

CHAPTER 20

AMINO ACIDS AND PEPTIDES

Amino acids are the subunits of proteins; as such they are a major constituent of all living cells. The identification of proteins as polyamides made up of simple amino acids was achieved through the combined work of a variety of chemists (among them, Emil Fischer) around the turn of the century. Indeed, it was Fischer who proposed the name *peptide* for structures of the following type:

$$\sim\!\text{N}\!-\!\text{CH}\!-\!\overset{\overset{\displaystyle O}{\|}}{\text{C}}\!-\!\text{N}\!-\!\text{CH}\!-\!\overset{\overset{\displaystyle O}{\|}}{\text{C}}\!\sim$$

Other scientists at first found it difficult to accept the idea that proteins could have highly complex biological properties, such as their ability to catalyze biochemical reactions, if they were made up of such simple subunits. But ultimately the weight of evidence was overwhelming, and investigations continued into the details of structure and functions of these compounds.

Proteins play many roles in human biochemistry. Proteins of one class, enzymes, function as catalysts for various reactions; other proteins play structural roles, act as chemical messengers to regulate biochemical processes, or serve to transport various chemical substances from one place to another within the body. Smaller peptides (those with molecular weights less than 5000) are usually not classified as proteins, but they can serve some of these same functions. The ester of a very simple dipeptide has recently gained widespread use as an artificial sweetener. Aspartame (NutraSweet®), L-aspartyl-L-phenylalanine methyl ester, is 100–200 times sweeter than sucrose, and it is generally considered to be free of the bitter aftertaste of other artificial sweeteners.

$$NH_2\!-\!\underset{}{\overset{\overset{\displaystyle HO_2CCH_2}{|}}{CH}}\!-\!\underset{\underset{\displaystyle O}{\|}}{C}\!-\!NH\!-\!\overset{\overset{\displaystyle C_6H_5CH_2}{|}}{CH}\!-\!\underset{\underset{\displaystyle O}{\|}}{C}\!-\!OCH_3 \qquad \text{Aspartame}$$

The development of protein chemistry in this century has been truly remarkable. From a time when the structure of proteins was a complete mystery we have progressed to the stage at which scientists can seriously consider the chemical synthesis of large proteins. In this chapter we will present some of the fundamental chemistry of amino acids and peptides. We will consider the basic structural features of amino acids and their chemical consequences for procedures for joining amino acid subunits to produce peptides. We will then illustrate these concepts by presenting the synthesis of *oxytocin,* an important peptide hormone. Finally, we will evaluate the present status of peptide synthesis.

20.1 STRUCTURE AND CLASSIFICATION OF AMINO ACIDS

Hydrolysis of a typical protein produces significant quantities of up to 20 different amino acids. Other amino acids may be found to a lesser extent, but we will restrict our discussion to the 20 most common. All these amino acids share the structural features summarized by the following drawing:

H
C
—N CO_2H

In other words, they are all α-amino acids, the amino substituent being located on C-2 of the carboxylic acid chain. With the exception of proline (see Table 20.1) the amino group is primary, and the general structure can be drawn in the following way:

R H
C
NH_2 CO_2H $\equiv$ H_2N—(C)—H with CO_2H above and R below

The only differences among the various amino acid structures are found in the side chains, designated R. These structural differences are sufficient to produce the complex chemistry of amino acids and peptides.

The preceding drawings also indicate that C-2 is a chiral center (except in glycine, where R = H), and the naturally occurring compounds all have the same relative configuration. Because they all have the same stereochemistry as L-glyceraldehyde (cf. Sections 6.6 and 19.1), they are referred to as L-amino acids.

CHO / HO—(C)—H / CH_2OH

L-Glyceraldehyde

CO_2H / H_2N—(C)—H / R

L-Amino acid

Although C-2 of the L-amino acids always has the *S*-configuration (see Section 6.3), the older nomenclature has been retained by chemists working in this field.

The stereochemical homogeneity of amino acids has important consequences for protein structure. As illustrated for the following partial structure of a peptide, the substituents (designated R) are all oriented in the same direction relative to the amide linkages.

R_1 H R_2 H R_3 H R_4 H
C C C C
~NH C—NH C—NH C—NH C~
‖ ‖ ‖ ‖
O O O O

This regularity in stereochemistry permits proteins to adopt a coiled structure, designated as an α helix. This important feature of protein structure was first demonstrated by the x-ray crystallographic studies of Linus Pauling and Robert Corey at the California Institute of Technology in 1951. (Pauling was awarded the Nobel Prize for chemistry in 1954, and he subsequently received the Nobel Peace Prize in 1963.)

The structures of the 20 common amino acids are summarized in Table 20.1. We have also drawn in the table the ionized form that predominates for each compound under physiological conditions.

TABLE 20.1 The Common Amino Acids Found in Proteins $R\text{–}CH(NH_2)\text{–}CO_2H$

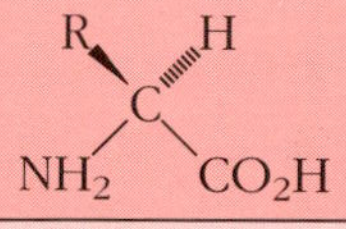

Structure	Name	Abbreviation	Ionic Form (pH ~ 7)	Classification (According to R Group)
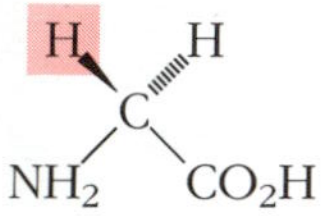 $H\text{–}CH(NH_2)CO_2H$	Glycine	Gly	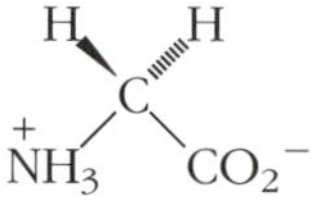 $H\text{–}CH(\overset{+}{N}H_3)CO_2^-$	Polar
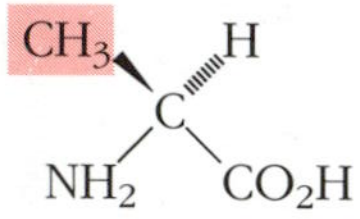 $CH_3\text{–}CH(NH_2)CO_2H$	Alanine	Ala	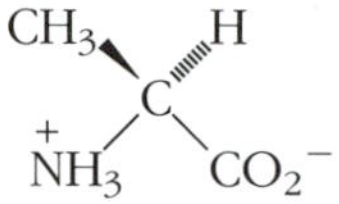 $CH_3\text{–}CH(\overset{+}{N}H_3)CO_2^-$	Nonpolar
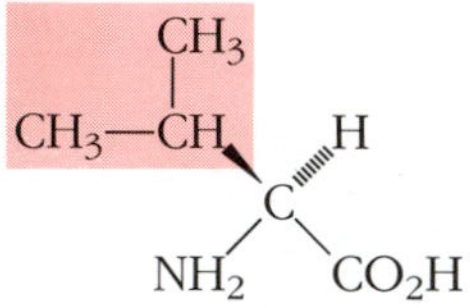 $(CH_3)_2CH\text{–}CH(NH_2)CO_2H$	Valine	Val	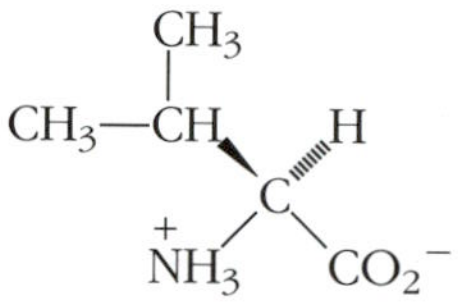 $(CH_3)_2CH\text{–}CH(\overset{+}{N}H_3)CO_2^-$	Nonpolar
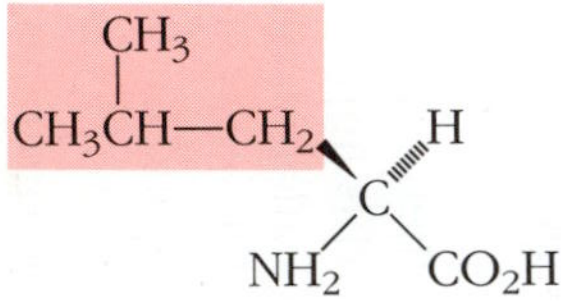 $(CH_3)_2CH\text{–}CH_2\text{–}CH(NH_2)CO_2H$	Leucine	Leu	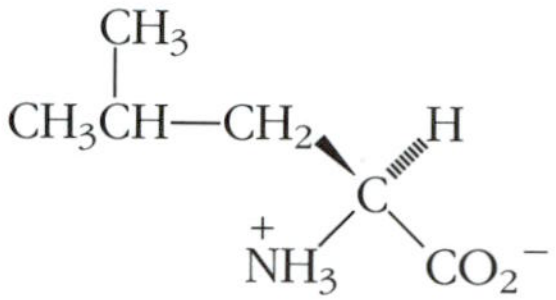 $(CH_3)_2CH\text{–}CH_2\text{–}CH(\overset{+}{N}H_3)CO_2^-$	Nonpolar
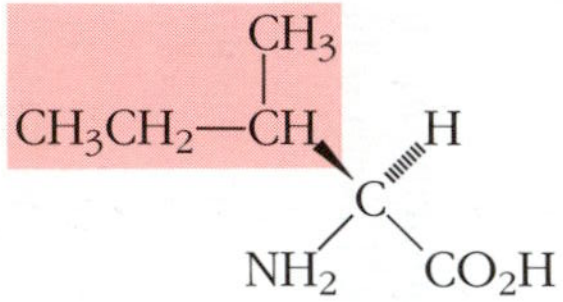 $CH_3CH_2\text{–}CH(CH_3)\text{–}CH(NH_2)CO_2H$	Isoleucine	Ile	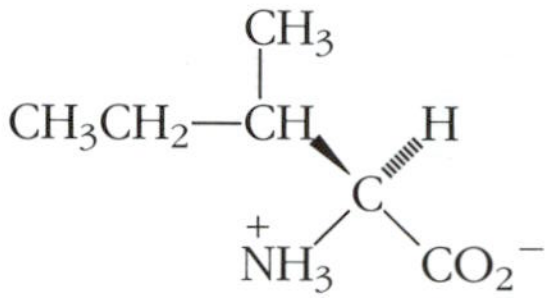 $CH_3CH_2\text{–}CH(CH_3)\text{–}CH(\overset{+}{N}H_3)CO_2^-$	Nonpolar

TABLE 20.1 (continued)

Structure	Name	Abbreviation	Ionic Form (pH ~ 7)	Classification (According to R Group)
H, C, N, CO_2H, H	Proline	Pro	H, C, $\overset{+}{N}$, CO_2^-, H, H	Nonpolar
$CH_3—S—CH_2CH_2$, H, C, NH_2, CO_2H	Methionine	Met	$CH_3—S—CH_2CH_2$, H, C, $\overset{+}{N}H_3$, CO_2^-	Nonpolar
CH_2, H, C, NH_2, CO_2H	Phenylalanine	Phe	CH_2, H, C, $\overset{+}{N}H_3$, CO_2^-	Aromatic nonpolar
CH_2, H, C, N, H, NH_2, CO_2H	Tryptophan	Trp	CH_2, H, C, N, H, $\overset{+}{N}H_3$, CO_2^-	Aromatic nonpolar
$HOCH_2$, H, C, NH_2, CO_2H	Serine	Ser	$HOCH_2$, H, C, $\overset{+}{N}H_3$, CO_2^-	Polar

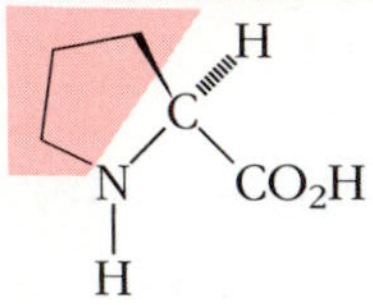

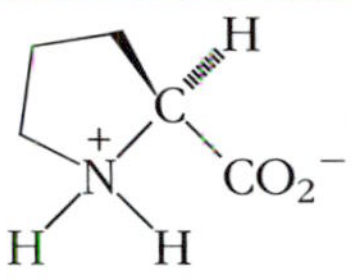

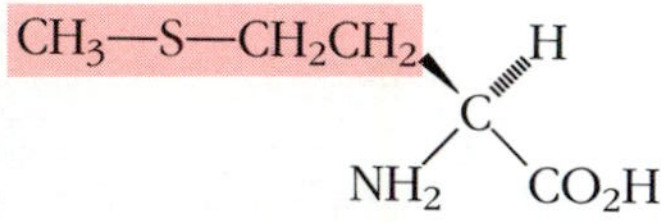

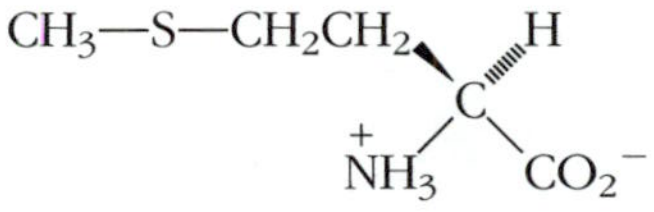

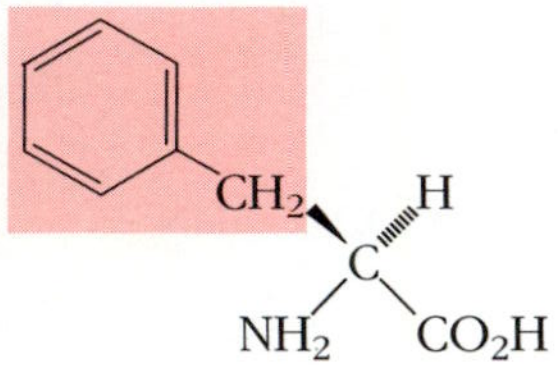

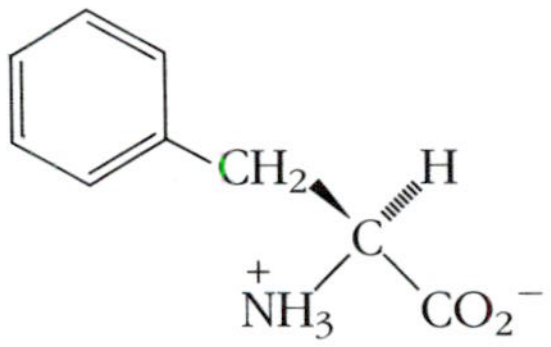

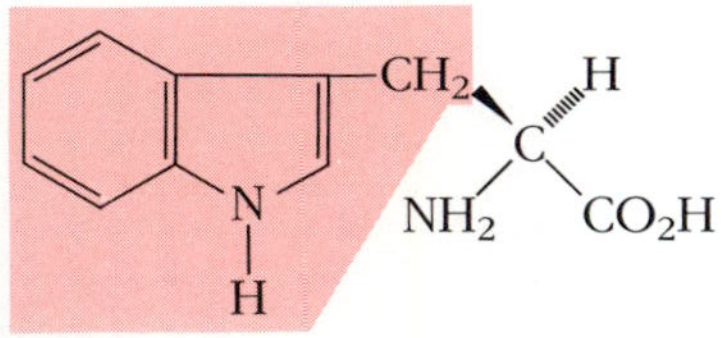

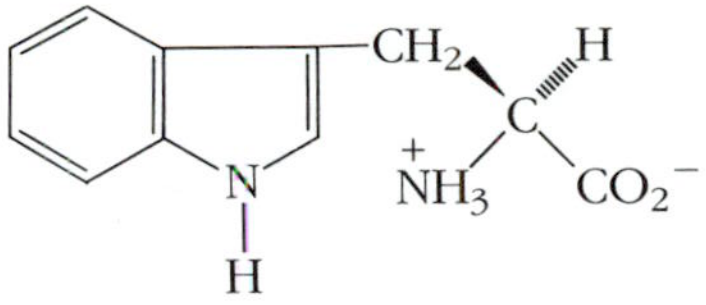

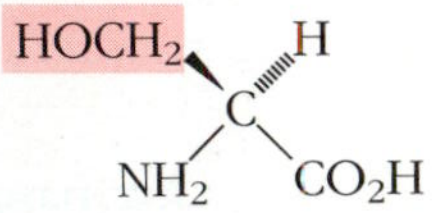

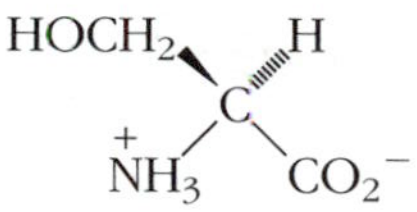

TABLE 20.1 (continued)

Structure	Name	Abbreviation	Ionic Form (pH ~ 7)	Classification (According to R Group)
$CH_3-CH(OH)-CH(NH_2)CO_2H$	Threonine	Thr	$CH_3-CH(OH)-CH(\overset{+}{N}H_3)CO_2^-$	Polar
$HS-CH_2-CH(NH_2)CO_2H$	Cysteine	Cys	$HS-CH_2-CH(\overset{+}{N}H_3)CO_2^-$	Polar
$HO-C_6H_4-CH_2-CH(NH_2)CO_2H$	Tyrosine	Tyr	$HO-C_6H_4-CH_2-CH(\overset{+}{N}H_3)CO_2^-$	Aromatic polar
$NH_2-C(=O)-CH_2-CH(NH_2)CO_2H$	Asparagine	Asn	$NH_2-C(=O)-CH_2-CH(\overset{+}{N}H_3)CO_2^-$	Polar
$NH_2-C(=O)-CH_2CH_2-CH(NH_2)CO_2H$	Glutamine	Gln	$NH_2-C(=O)-CH_2CH_2-CH(\overset{+}{N}H_3)CO_2^-$	Polar

TABLE 20.1 (continued)

Structure	Name	Abbreviation	Ionic Form (pH ~ 7)	Classification (According to R Group)
$HOC(=O)CH_2-C(H)(NH_2)CO_2H$	Aspartic acid	Asp	$^{-}OC(=O)CH_2-C(H)(\overset{+}{N}H_3)CO_2^{-}$	Acidic polar
$HOC(=O)CH_2CH_2-C(H)(NH_2)CO_2H$	Glutamic acid	Glu	$^{-}OC(=O)CH_2CH_2-C(H)(\overset{+}{N}H_3)CO_2^{-}$	Acidic polar
$NH_2-CH_2CH_2CH_2CH_2-C(H)(NH_2)CO_2H$	Lysine	Lys	$\overset{+}{N}H_3-CH_2CH_2CH_2CH_2-C(H)(\overset{+}{N}H_3)CO_2^{-}$	Basic polar
$NH_2-C(=NH)-NH-CH_2CH_2CH_2-C(H)(NH_2)CO_2H$	Arginine	Arg	$NH_2-C(=\overset{+}{N}H_2)-NH-CH_2CH_2CH_2-C(H)(\overset{+}{N}H_3)CO_2^{-}$	Basic polar
(imidazole ring: H—N, =N) $-CH_2-C(H)(NH_2)CO_2H$	Histidine	His	(imidazolium ring: H—N, $\overset{+}{N}$—H) $-CH_2-C(H)(\overset{+}{N}H_3)CO_2^{-}$	Basic polar

Before considering the chemical consequences of these structures, we will first make a few observations about the names of the amino acids. You probably will not see any immediate relationship between the structures in Table 20.1 and their names. *Glycine* has a sweet taste, and its name is derived from the same Greek word (*glykys,* "sweet") that affords the names of sugar derivatives such as glycogen. As with almost all the common amino acids, the *-ine* ending was used to indicate that the compound contains the functionality of an *amine. Alanine* was so named because it had been chemically converted to an *al*dehyde, and *valine* was named as a structural relative of iso*val*eric acid. The name of *leucine* comes from its white color (Greek, *leukos,* "white"), which was originally thought to be unusual, and the isomer of leucine was designated *isoleucine.*

Proline is an abbreviation of *pyrrolidine* and its name refers to its saturated five-membered heterocyclic ring. The name *methionine* is derived from the presence of both *methyl* and *thio* (sulfur) groups. *Phenylalanine* is also a name that describes the structure; this amino acid is simply a phenyl-substituted alanine. Another name that provides structural information is *threonine,* which has the same relative configuration as the four-carbon sugar D-*threose*:

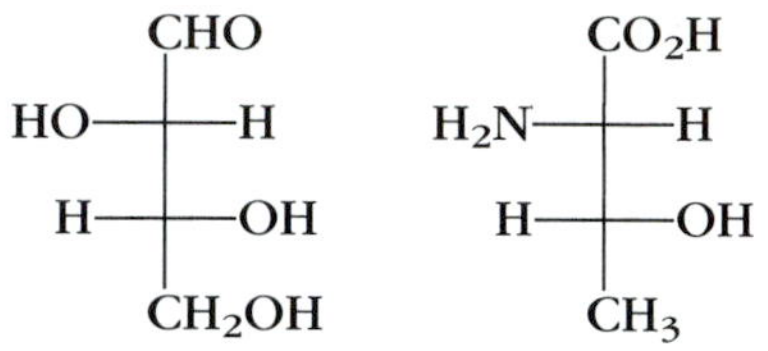

CHO
HO—+—H
H—+—OH
CH_2OH
D-Threose

CO_2H
H_2N—+—H
H—+—OH
CH_3
L-Threonine

Note that these D- and L- designations refer to different chiral centers in the two structures.

Some of the amino acids were named according to the materials from which they were originally isolated. In some cases the connection is straightforward, as with *asparagine* (from *asparagus*) and *glutamic* acid (from wheat *gluten*). On the other hand, *arginine* was first isolated as its silver salt (Latin, *argentum,* "silver") and the relationship is less obvious. *Cysteine* was so named because it is a component of urinary calculi (i.e., *cysts*), and *lysine* was first obtained by hydro*lysis* of a protein. Similarly, the names for *serine, tyrosine,* and *histidine* are indicative of their early isolation from silk (Latin *sericus,* "silken"), cheese (Greek, *tyros*), and tissue (Greek, *histion*). Finally, *tryptophan* was so named because of its isolation from a type of enzymatic hydrolysis of proteins called *tryptic* digestion.

The designations of nonpolar and polar in Table 20.1 are somewhat arbitrary, but they are meant to indicate the polarity of the side chains of the amino acids. The polarity of the entire molecule is also dependent upon the other functional groups that are present, the amino and carboxyl groups. All the amino acids are quite polar, being relatively soluble in water and insoluble in organic solvents. This solubility is typical of ionic compounds, as you would expect if the amino and carboxylic acid groups underwent the following reaction:

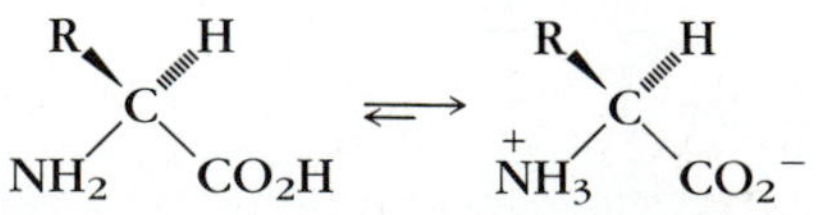

$$R(H)C(NH_2)(CO_2H) \rightleftharpoons R(H)C(\overset{+}{N}H_3)(CO_2^-)$$

This reaction does indeed take place, and the structures actually observed at neutral pH actually have two charged groups, a carboxylate anion and an ammonium cation. These internal salts are called *zwitterions* (German *zwitter,* "hybrid; hermaphrodite"). Despite the predominance of the zwitterionic form, for convenience we will nevertheless continue to draw many of the amino acids in this chapter with uncharged amino and carboxyl groups.

Several of the amino acids have a second acidic or basic substituent, and this has important consequences for the biological activity of peptides and proteins. Such functional groups can result in a local environment that is very different from the medium in which a protein is found, and this can aid an enzyme in its role as a chemical catalyst. Moreover, free carboxyl and amino groups can behave as specific acids, bases, or nucleophiles at the active site of an enzyme.

Our discussion in this section has centered on nomenclature of the individual amino acids, but you now have enough knowledge to name peptides as well. Whenever possible, we will draw peptides so that the free amino group (the *amino terminus*) is on the left. But regardless of the drawing, the name will always begin with the amino terminal residue. Each of the amino acid residues is then named in sequence, ending with the *carboxyl terminus.* Except for the last residue (the carboxylic acid) the other amino acid residues are each named as their acyl groups (in which the *-ine* ending of the amino acid is normally replaced by *-yl*). The following tripeptide is therefore named as L-cysteinyl-L-alanylglycine (or just cysteinylalanylglycine, if stereochemistry is ignored).

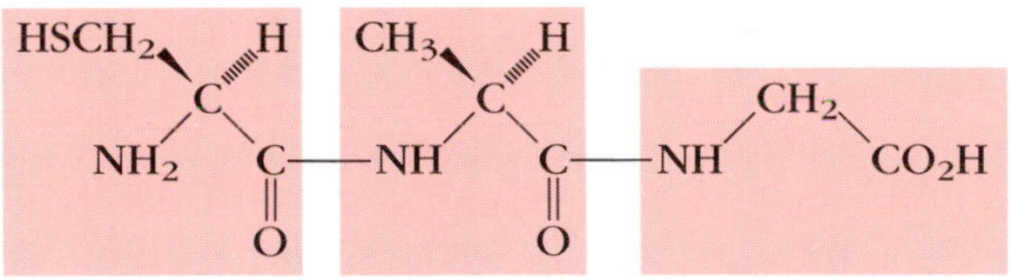

Cys Ala Gly

It can also be conveniently abbreviated as Cys-Ala-Gly (or preferably, to show the terminal groups, as NH_2—Cys—Ala—Gly—CO_2H). Notice that the amide linkages are not explicitly shown in this last type of representation. The subunits are nevertheless joined by such peptide bonds.

EXERCISE 20.1

Amino acids exist predominantly in the zwitterionic form at neutral pH, so there is clearly a difference in basicity between carboxylate ions and amino groups.

(a) Which is more basic? (Consider protonation of NH_2—CH(R)—CO_2^- to give the "neutral" amino acid; which group would be protonated?)
(b) Which is more acidic, a carboxyl group or an ammonium ion?

EXERCISE 20.2

Draw structures for all the amino acids that:

(a) Are positively charged at neutral pH
(b) Are negatively charged at neutral pH
(c) Have names describing (at least in part) their chemical structures
(d) Have names indicating their original isolation from a particular plant (or some other substance)

20.2 SYNTHESIS OF AMINO ACIDS

As chemists began to unravel the structural mysteries of amino acids and peptides, they found it necessary to synthesize samples of many amino acids. Although the common amino acids are readily available from hydrolysis of proteins, the preparation of isomeric (or otherwise structurally modified) amino acids remains important in several areas of research. In this section we will illustrate some of the ways by which amino acids can be synthesized. Note, however, that the syntheses we show here will all produce *racemic* products. Isolation of the L-amino acids would require resolution of the enantiomeric pairs.

In principle the synthesis of amino acids should be straightforward. We have previously covered synthetic methods for both carboxylic acids (Section 16.7) and amines (Sections 17.6 and 17.7). However, amino acids are *difunctional,* and this places some restrictions on the methods that can be used. We will discuss three of the most common procedures for amino acid synthesis. The first of these involves introduction of the amino group onto the appropriate carboxylic acid (i.e., the amino group is introduced after the side chain). The second proceeds by addition of cyanide to a carbonyl group, generating both the amino and "carboxyl" functions simultaneously; the side chain is already present as a substituent on the carbonyl compound. The third method begins with malonic acid, and the amino group and the side chain are introduced sequentially. These synthetic approaches are summarized in Figure 20.1.

All the sequences in Figure 20.1 represent only the final stages of a synthesis, and preparation of the indicated compounds can be carried out according to the methods described in previous chapters. The following synthesis of leucine illustrates the first method. 4-Methylpentanoic acid was treated with bromine and PCl_3 to generate the α-bromo acid. Only a catalytic quantity of PCl_3 was used, so the product of this reaction was the acid rather than the acid chloride. Reaction of the bromo acid with excess ammonia then yielded leucine.

CO_2H $\xrightarrow[PCl_3]{Br_2}$ Br CO_2H (63–66%) $\xrightarrow{NH_3}$ NH_2 CO_2H (43–45%)

Sometimes preparation of the bromo acid is more involved, as in the following synthesis of isoleucine, in which the carbon skeleton was first completed.

$CH_2(CO_2C_2H_5)_2$ $\xrightarrow[NaOC_2H_5]{\text{Br}}$ $CH(CO_2C_2H_5)_2$ $\xrightarrow[2.\ HCl]{1.\ KOH,\ H_2O}$ $CH(CO_2H)_2$ $\xrightarrow{Br_2}$ $C(Br)(CO_2H)_2$ $\xrightarrow[heat]{-CO_2}$ $CH(Br)CO_2H$ $\xrightarrow{NH_3}$ NH_2 CO_2H 34% overall yield

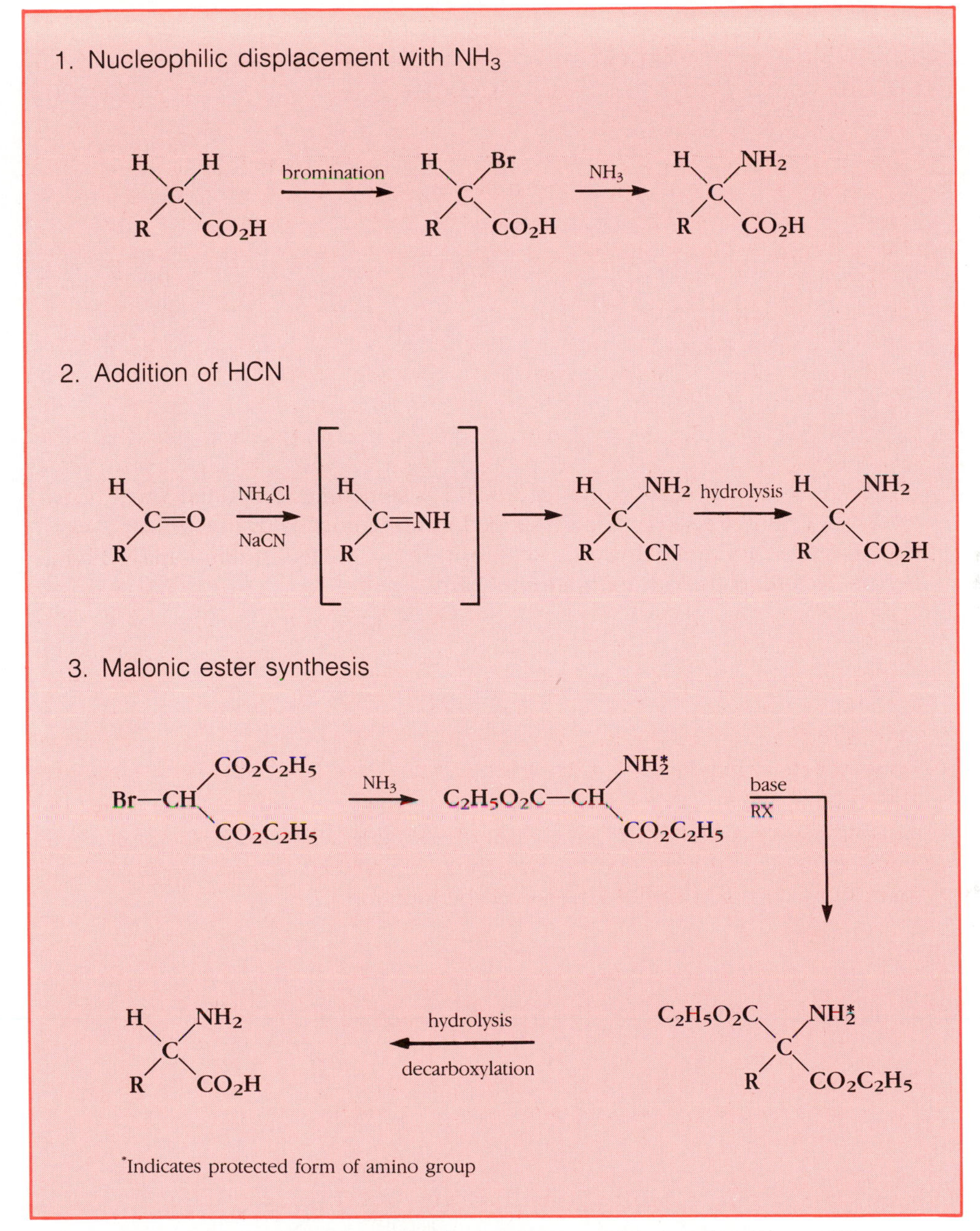

Figure 20.1 Some general methods for the synthesis of amino acids.

Bromination of the substituted malonic acid proceeds smoothly by an ionic mechanism because small but significant amounts of the enol form are present at equilibrium.

In a synthesis of serine the hydroxyl group was introduced (in a protected form) via oxymercuration in methanol. The methyl ether was subsequently cleaved by the action of HBr (Section 15.3).

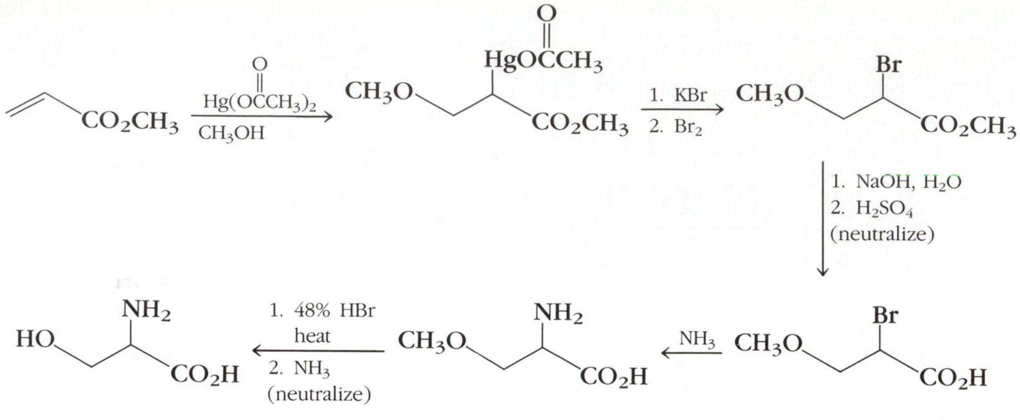

$$CH_2{=}CHCO_2CH_3 \xrightarrow[CH_3OH]{Hg(OCCH_3)_2} CH_3OCH_2CH(HgOCCH_3)CO_2CH_3 \xrightarrow[2.\ Br_2]{1.\ KBr} CH_3OCH_2CH(Br)CO_2CH_3$$

$$\xrightarrow{1.\ NaOH,\ H_2O;\ 2.\ H_2SO_4\ (neutralize)} CH_3OCH_2CH(Br)CO_2H \xrightarrow{NH_3} CH_3OCH_2CH(NH_2)CO_2H \xrightarrow{1.\ 48\%\ HBr,\ heat;\ 2.\ NH_3\ (neutralize)} HOCH_2CH(NH_2)CO_2H$$

30–40% overall yield

The second method shown in Figure 20.1 is sometimes called the *Strecker synthesis,* after Adolf Strecker, who developed it in the mid-1800s in Germany. When HCN is added to an aldehyde or ketone in the presence of ammonia, addition occurs preferentially with the *iminium ion.*

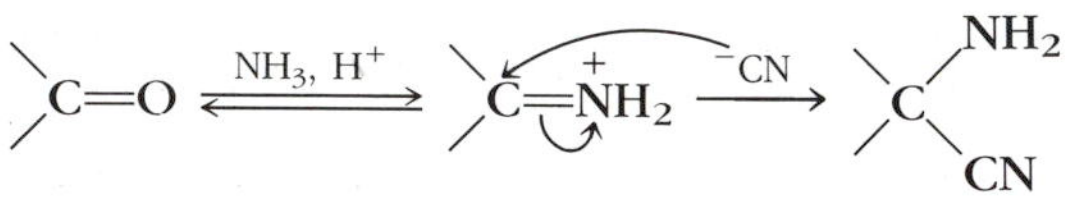

$$>C{=}O \xrightleftharpoons{NH_3,\ H^+} >C{=}\overset{+}{N}H_2 \xrightarrow{^-CN} >C(NH_2)(CN)$$

The resulting amino nitrile can then be hydrolyzed to the amino acid, as shown for the following synthesis of glycine. (Lead hydroxide was used to neutralize the ammonium salt in the last step, because the resulting lead salt was insoluble in the reaction mixture and could be removed by filtration.)

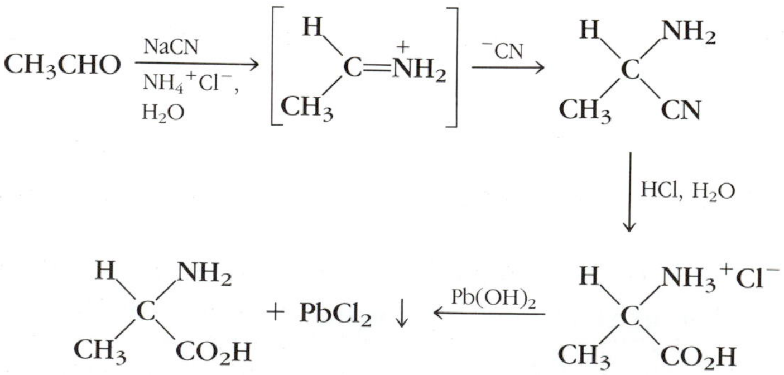

$$CH_3CHO \xrightarrow[NH_4^+Cl^-,\ H_2O]{NaCN} \left[\begin{matrix} H \\ CH_3 \end{matrix} > C{=}\overset{+}{N}H_2 \right] \xrightarrow{^-CN} H(CH_3)C(NH_2)(CN) \xrightarrow{HCl,\ H_2O} H(CH_3)C(NH_3^+Cl^-)(CO_2H) \xrightarrow{Pb(OH)_2} H(CH_3)C(NH_2)(CO_2H) + PbCl_2 \downarrow$$

52–60% overall

The third method for synthesis of amino acids begins with diethyl bromomalonate, and introduction of the amino nitrogen prior to introduction of the side chain requires the use of a blocking group on the nitrogen. Otherwise the alkylation step would generate substantial quantities of *N*-alkylated product. This is illustrated by the following synthesis of methionine.

$$Br{-}CH(CO_2C_2H_5)_2 \xrightarrow{\text{potassium phthalimide } (N^-K^+)} \text{phthalimido}{-}N{-}CH(CO_2C_2H_5)_2$$

82–85%

$$\xrightarrow[\text{2. } CH_3SCH_2CH_2{-}Cl]{\text{1. } NaOC_2H_5} \text{phthalimido}{-}N{-}C(CH_2CH_2SCH_3)(CO_2C_2H_5)_2$$

96–98%

$$\xrightarrow[\text{2. HCl (to neutralize)}]{\text{1. NaOH, } H_2O} CH_3SCH_2CH_2{-}CH(NH_2){-}CO_2H$$

59%

EXERCISE 20.3

Outline a synthesis of phenylalanine by (a) the first, (b) the second, and (c) the third of the three methods shown in Figure 20.1.

20.3 PEPTIDE SYNTHESIS: BLOCKING GROUPS FOR AMINO ACIDS

Amino acids are difunctional compounds, and this places limitations on the methods by which they can be converted to peptides. Suppose you wanted to link two amino acids to form a dipeptide:

$$NH_2{-}\underset{}{\overset{R}{CH}}{-}CO_2H + NH_2{-}\overset{R'}{CH}{-}CO_2H \Rightarrow NH_2{-}\overset{R}{CH}{-}\overset{O}{\overset{\|}{C}}{-}NH{-}\overset{R'}{CH}{-}CO_2H$$

The normal procedure for making an amide is to use an activated carboxyl derivative such as an acid chloride.

$$NH_2{-}\overset{R}{CH}{-}\overset{O}{\overset{\|}{C}}{-}X + NH_2{-}\overset{R'}{CH}{-}CO_2H \xrightarrow{?} NH_2{-}\overset{R}{CH}{-}\overset{O}{\overset{\|}{C}}{-}NH{-}\overset{R'}{CH}{-}CO_2H$$

Unfortunately, this approach would not work. In addition to the desired compound, you would observe products from reaction of the acid chloride with the amino group of a second acid chloride molecule. Even if the desired dipeptide were formed, it could react further to produce a tripeptide. Other difficulties would result if attempts were made to use a free amino acid as the second reactant, because the amino group would be in the protonated, zwitterionic form, in which the nitrogen has been rendered nonnucleophilic. These difficulties are further compounded with those amino acids having an additional carboxyl or amino group.

The only way around these problems is to employ a variety of protecting groups for both the amino and carboxyl groups. Other functional groups such as the thiol group of cysteine and the hydroxyl groups of serine, threonine, and tyrosine may also require protection during peptide synthesis. If this is done carefully, the desired bonds can be formed, and even complex peptides can be synthesized.

Protection of Amino Groups

The first widely used method for blocking amino groups in peptide synthesis involved formation of the *p*-toluenesulfonamide. Treatment of an amino acid with *p*-toluenesulfonyl chloride produces the *p*-toluenesulfonamide (Sections 17.4 and 17.9). Because HCl is generated, the reaction is ordinarily carried out in the presence of a base such as sodium hydroxide or a tertiary amine.

$$\text{>NH} \xrightarrow[\text{NaOH or } R_3N]{\text{TsCl}} \text{>N}-\overset{O}{\underset{O}{\overset{\|}{\underset{\|}{S}}}}-C_6H_4-CH_3$$

Protection of amino group by conversion to the *p*-toluenesulfonamide

The nitrogen of the *N*-tosyl derivative is no longer nucleophilic, and the protected amino acid behaves as a *monofunctional* carboxylic acid. The following examples illustrate the preparation of *N*-tosyl amino acids with valine and leucine, respectively.

$$CH_3CH(CH_3)-CH(NH_2)-CO_2H \xrightarrow[\text{2. HCl (to neutralize)}]{\text{1. TsCl, NaOH}} CH_3CH(CH_3)-CH(NH-Ts)-CO_2H \quad 59\%$$

$$CH_3CH(CH_3)-CH_2-CH(NH_2)-CO_2H \xrightarrow[\text{2. HCl (to neutralize)}]{\text{1. TsCl, } N(C_2H_5)_3} CH_3CH(CH_3)-CH_2-CH(NH-Ts)-CO_2H \quad 71\%$$

The second stage in using any blocking group is its removal following whatever chemical reaction is carried out. In the case of peptide synthesis the *N*-tosyl group must be removed without the cleavage of any amide linkages in the peptide, and this can be brought about by reduction with sodium in liquid ammonia. Electron transfer occurs preferentially at the aromatic sulfonyl group, producing the free amino compound (together with reduced sulfur compounds derived from the tosyl group). While *N*-tosyl groups can be removed under these strongly reducing conditions, they are highly resistant to conditions used to remove most other blocking groups, particularly hydrolysis in either acidic or alkaline media. Conversely, reduction with sodium in ammonia will also cleave other blocking groups that we will discuss subsequently, such as benzyloxycarbonyl and *S*-benzyl groups. The reaction cannot be employed when the carboxyl terminus of a peptide is present as an ester, because the carboxyl group would be reduced.

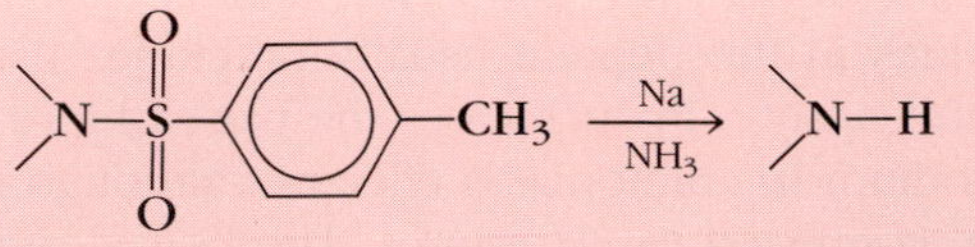

$$>N-SO_2-C_6H_4-CH_3 \xrightarrow[NH_3]{Na} >N-H$$

Removal of the *N*-tosyl protecting group

The reductive cleavage of *N*-tosyl groups is illustrated by the preparation of glutamylglycine.

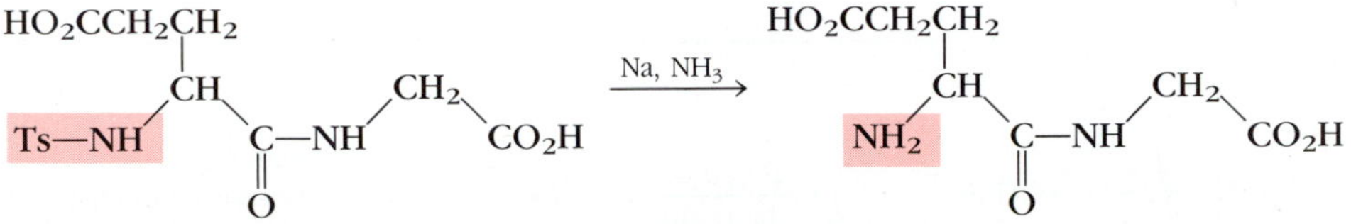

$$Ts-NH-CH(CH_2CH_2CO_2H)-C(=O)-NH-CH_2-CO_2H \xrightarrow{Na,\ NH_3} NH_2-CH(CH_2CH_2CO_2H)-C(=O)-NH-CH_2-CO_2H \quad 69\%$$

The most important amine-protecting group used in peptide synthesis is the benzyloxycarbonyl group.

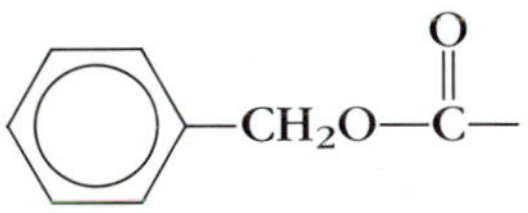

$$C_6H_5-CH_2O-C(=O)-$$

Benzyloxycarbonyl (Cbz) group

(This group was originally called "carbobenzoxy," which is the origin of the abbreviation Cbz). The protected derivative is actually a urethane (carbamate), since it is a half-ester and half-amide of carbonic acid. As in the case of the *N*-tosyl derivatives, the amide linkage renders the nitrogen nonnucleophilic. Benzyloxycarbonyl derivatives are prepared by reaction of benzyl chloroformate and the amine (or amino acid) in the presence of base:

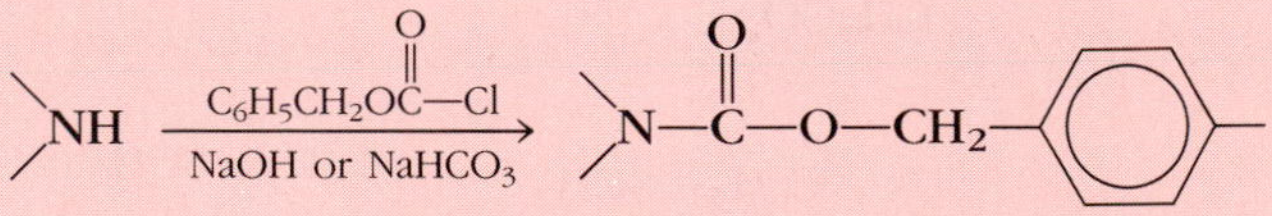

$$>NH \xrightarrow[NaOH\ or\ NaHCO_3]{C_6H_5CH_2OC(=O)-Cl} >N-C(=O)-O-CH_2-C_6H_5$$

Protection of amino group by formation of *N*-benzyloxycarbonyl derivative

The following reactions illustrate the formation of *N*-benzyloxycarbonyl derivatives of amino acids. Note that an amino group in the second example can be acylated selectively in the presence of an alcohol.

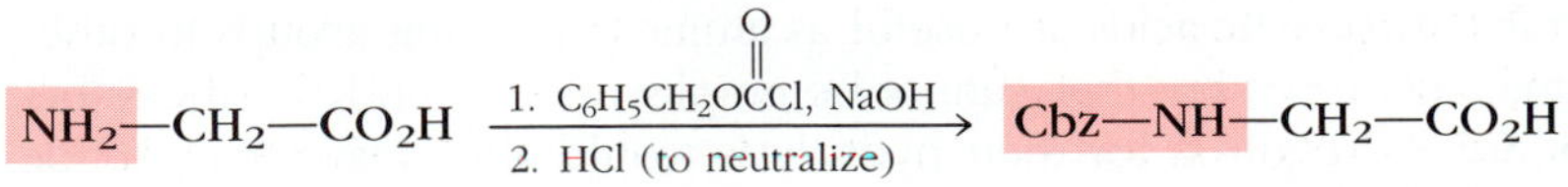

$$NH_2-CH_2-CO_2H \xrightarrow[2.\ HCl\ (to\ neutralize)]{1.\ C_6H_5CH_2OC(=O)Cl,\ NaOH} Cbz-NH-CH_2-CO_2H \quad 86\text{–}91\%$$

$$CH_3-CH(OH)-CH(NH_2)-CO_2H \xrightarrow[2.\ HCl\ (to\ neutralize)]{1.\ C_6H_5CH_2OC(=O)Cl,\ NaHCO_3} CH_3CH(OH)-CH(NH-Cbz)-CO_2H \quad 75\text{–}85\%$$

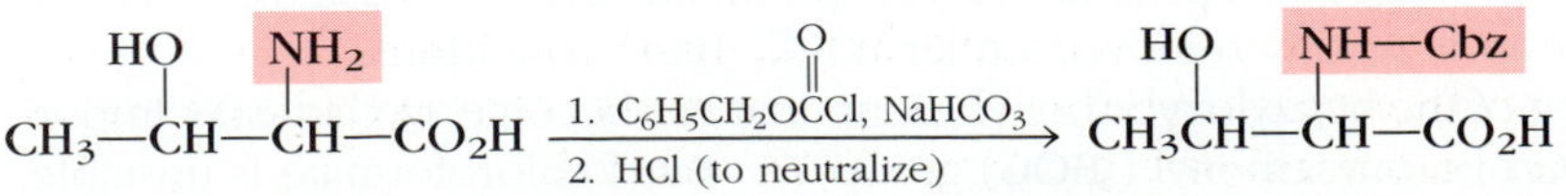

Although a benzyloxycarbonyl group can withstand the conditions needed for the coupling of two amino acids, it is very easily cleaved. Removal of the Cbz group can be carried out easily by mild catalytic hydrogenolysis (Section 8.6). The presence of sulfur-containing residues in the peptide usually precludes use of this method, because they poison the catalyst. In such cases the use of more strongly reducing conditions, such as sodium in ammonia, is effective, although disulfide bonds and *S*-benzyl groups are converted to SH groups. The benzyloxycarbonyl group is readily hydrolyzed by acid, and treatment with HBr is a common alternative for its removal. This blocking group is rather resistant to *alkaline* hydrolysis, however, and it is frequently possible to hydrolyze other ester groups in the molecule without cleavage of a Cbz protecting group.

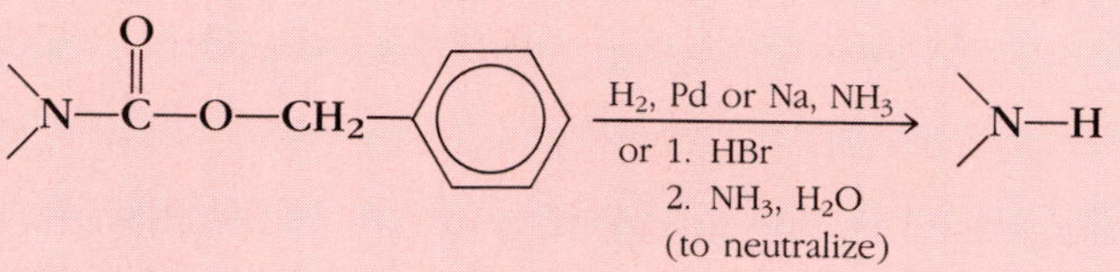

$$>N{-}\overset{O}{\overset{\|}{C}}{-}O{-}CH_2{-}C_6H_5 \xrightarrow[\text{or 1. HBr; 2. } NH_3,\ H_2O \text{ (to neutralize)}]{H_2,\ Pd \text{ or } Na,\ NH_3} >N{-}H$$

Removal of the benzyloxycarbonyl blocking group

The following reactions illustrate removal of the benzyloxycarbonyl group from peptides:

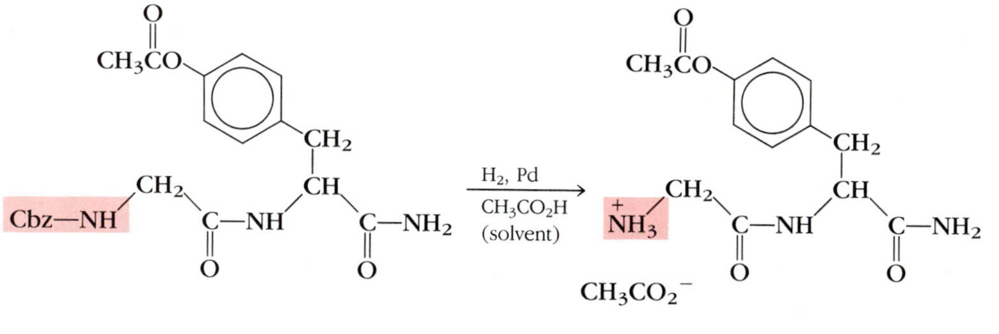

$$\text{Cbz—NH—}CH_2\text{—}\overset{O}{\overset{\|}{C}}\text{—NH—}CH(CH_2C_6H_4OCOCH_3)\text{—}\overset{O}{\overset{\|}{C}}\text{—}NH_2 \xrightarrow[CH_3CO_2H \text{ (solvent)}]{H_2,\ Pd} \overset{+}{N}H_3\text{—}CH_2\text{—}\overset{O}{\overset{\|}{C}}\text{—NH—}CH(CH_2C_6H_4OCOCH_3)\text{—}\overset{O}{\overset{\|}{C}}\text{—}NH_2 \quad CH_3CO_2^- \quad 92\%$$

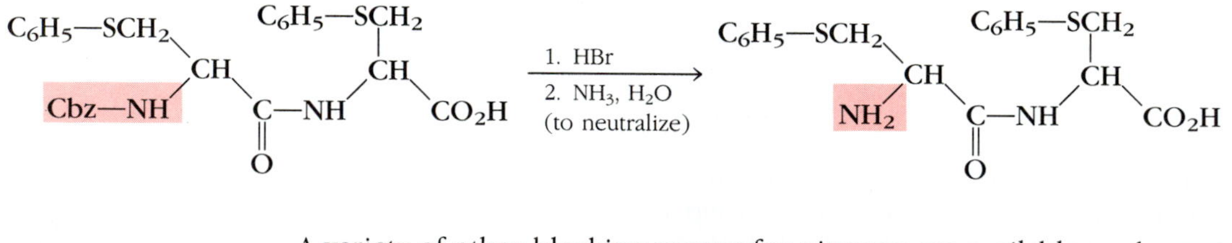

$$\text{Cbz—NH—}CH(CH_2S\text{—}C_6H_5)\text{—}\overset{O}{\overset{\|}{C}}\text{—NH—}CH(CH_2S\text{—}C_6H_5)\text{—}CO_2H \xrightarrow[\text{2. } NH_3,\ H_2O \text{ (to neutralize)}]{\text{1. HBr}} NH_2\text{—}CH(CH_2S\text{—}C_6H_5)\text{—}\overset{O}{\overset{\|}{C}}\text{—NH—}CH(CH_2S\text{—}C_6H_5)\text{—}CO_2H \quad 83\%$$

A variety of other blocking groups for nitrogen are available, and a number of ring-substituted benzyloxycarbonyl groups have been used in peptide synthesis. Amides of typical carboxylic acids are useful as amine protecting groups in other areas of organic chemistry, but they cannot be employed in peptide synthesis because the conditions required for their hydrolysis would also cleave the peptide bonds. Trifluoroacetic acid provides an exception, however, because the *N*-trifluoroacetyl group is readily removed under mildly basic conditions.

An analog of the benzyloxycarbonyl group that has become synthetically important is the *tert*-butoxycarbonyl (tBOC) group. *tert*-Butyl chloroformate is unstable, and the common reagent for introduction of the tBOC group is an acid *azide* rather than an acid chloride. This compound, *tert*-butyl azidoformate, is used despite its occasional tendency to decompose explosively.

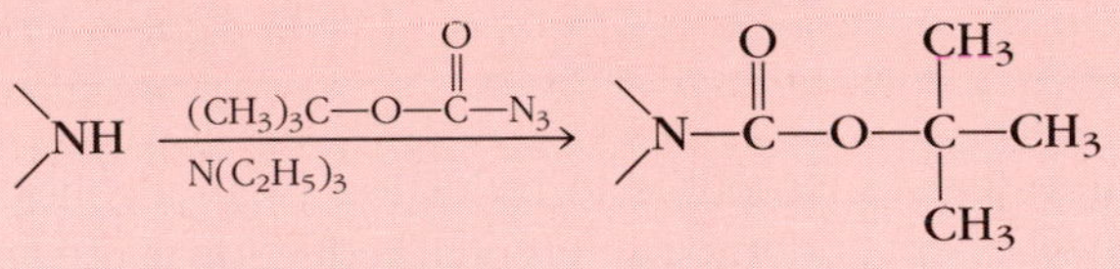

$$\text{>NH} \xrightarrow[\text{N(C}_2\text{H}_5)_3]{(\text{CH}_3)_3\text{C—O—}\overset{\text{O}}{\overset{\|}{\text{C}}}\text{—N}_3} \text{>N—}\overset{\text{O}}{\overset{\|}{\text{C}}}\text{—O—}\underset{\text{CH}_3}{\overset{\text{CH}_3}{\text{C}}}\text{—CH}_3$$

Protection of the amino group by formation of the *tert*-butyloxycarbonyl derivative

In contrast to the Cbz group, the tBOC group is stable to reducing conditions such as catalytic hydrogenolysis and sodium in ammonia, and this permits selective removal of a Cbz group in the presence of a tBOC group. Both groups are easily cleaved by acid.

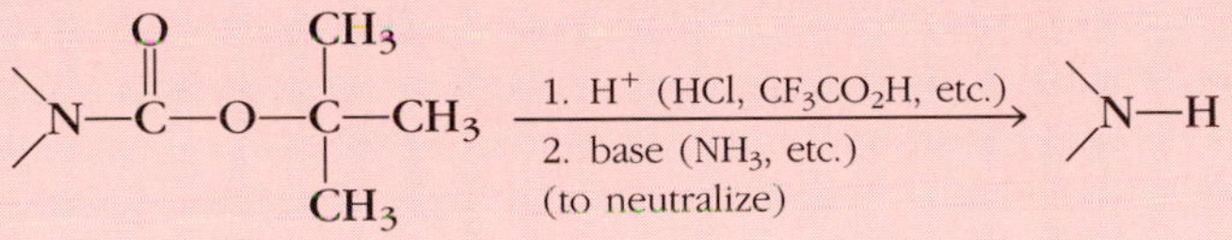

$$\text{>N—}\overset{\text{O}}{\overset{\|}{\text{C}}}\text{—O—}\underset{\text{CH}_3}{\overset{\text{CH}_3}{\text{C}}}\text{—CH}_3 \xrightarrow[\text{2. base (NH}_3\text{, etc.) (to neutralize)}]{\text{1. H}^+\text{ (HCl, CF}_3\text{CO}_2\text{H, etc.)}} \text{>N—H}$$

Removal of the *tert*-butyloxycarbonyl group

The following equations illustrate the introduction and removal of the *tert*-butoxycarbonyl group:

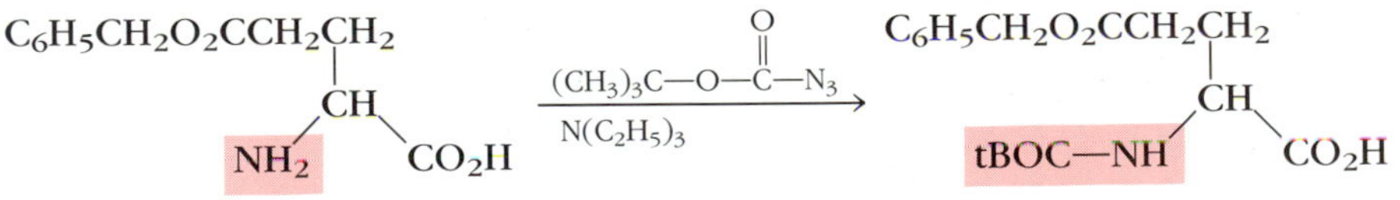

$$\text{C}_6\text{H}_5\text{CH}_2\text{O}_2\text{CCH}_2\text{CH}_2\text{CH(NH}_2\text{)CO}_2\text{H} \xrightarrow[\text{N(C}_2\text{H}_5)_3]{(\text{CH}_3)_3\text{C—O—}\overset{\text{O}}{\overset{\|}{\text{C}}}\text{—N}_3} \text{C}_6\text{H}_5\text{CH}_2\text{O}_2\text{CCH}_2\text{CH}_2\text{CH(tBOC—NH)CO}_2\text{H} \quad 92\%$$

$$\text{CH}_3\text{SCH}_2\text{CH}_2\text{CH(tBOC—NH)—}\overset{\text{O}}{\overset{\|}{\text{C}}}\text{—NH—CH}_2\text{—CO}_2\text{H} \xrightarrow[\text{2. NH}_3\text{, H}_2\text{O (to neutralize)}]{\text{1. HBr}} \text{CH}_3\text{SCH}_2\text{CH}_2\text{CH(NH}_2\text{)—}\overset{\text{O}}{\overset{\|}{\text{C}}}\text{—NH—CH}_2\text{—CO}_2\text{H} \quad 80\%$$

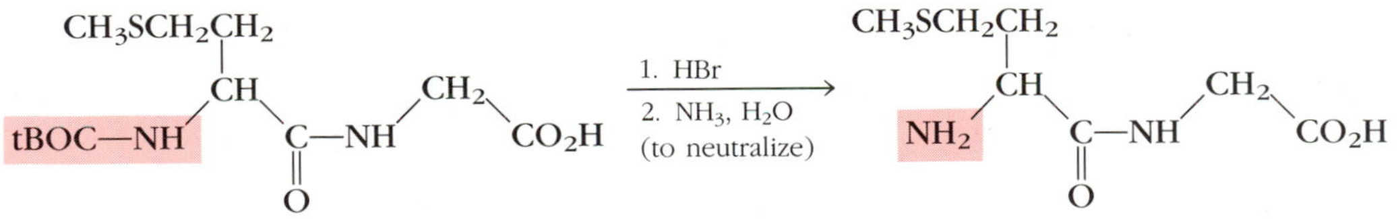

Protection of Carboxyl Groups

The most common carboxyl-protecting groups for peptide synthesis are esters. Methyl and ethyl esters have been used frequently; however, fairly vigorous alkaline hydrolysis is required for their removal, and this can cause side reactions at other sites in the peptide. This problem is alleviated by the use of phenyl or benzyl esters, which can be removed via catalytic hydrogenolysis. The esters are typically prepared by the acid-catalyzed reaction of an amino acid with the appropriate alcohol (Section 16.6).

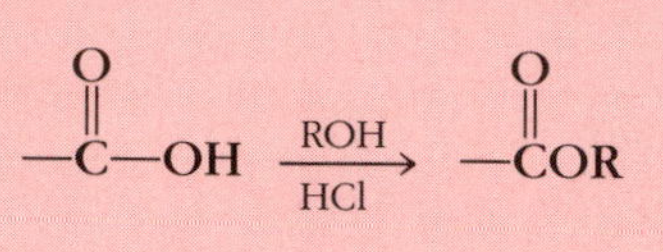

$$-\overset{\overset{\displaystyle O}{\|}}{C}-OH \xrightarrow[HCl]{ROH} -\overset{\overset{\displaystyle O}{\|}}{C}OR$$

Protection of the carboxyl group by esterification
$R = CH_3, C_2H_5, C_6H_5, C_6H_5CH_2$

The *carboxyl* group of an ester is susceptible to reduction to an alcohol with sodium in ammonia, but benzyl esters undergo preferential reduction at the *benzyl* group to liberate the free carboxyl group. Catalytic hydrogenolysis is a more common way of cleaving benzyl esters (although it cannot be employed with sulfur-containing peptides). Benzyl esters are also more susceptible to acid hydrolysis than other esters, and this provides yet another method for their cleavage.

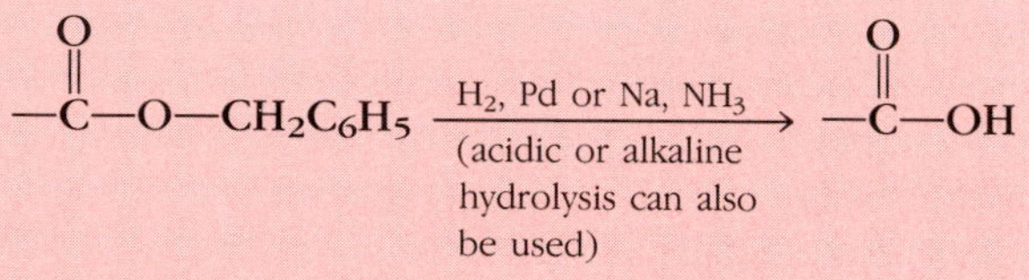

$$-\overset{\overset{\displaystyle O}{\|}}{C}-O-CH_2C_6H_5 \xrightarrow[\text{(acidic or alkaline hydrolysis can also be used)}]{H_2,\ Pd\ \text{or}\ Na,\ NH_3} -\overset{\overset{\displaystyle O}{\|}}{C}-OH$$

Cleavage of benzyl esters

Cleavage of benzyl esters is illustrated by the following reactions:

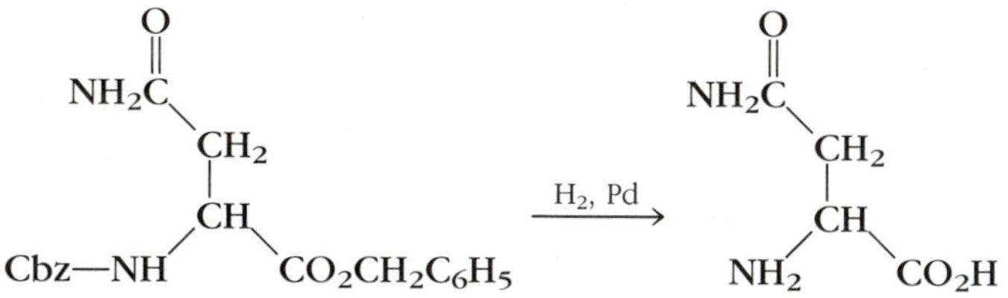

$$Cbz-NH-CH(CO_2CH_2C_6H_5)-CH_2-\overset{\overset{\displaystyle O}{\|}}{C}NH_2 \xrightarrow{H_2,\ Pd} NH_2-CH(CO_2H)-CH_2-\overset{\overset{\displaystyle O}{\|}}{C}NH_2 \quad 70\%$$

$$Cbz-NH-CH(\overset{\overset{\displaystyle O}{\|}}{C}-OCH_2C_6H_5)-CH_2-\overset{\overset{\displaystyle O}{\|}}{C}-NH-CH(CH_2-S-CH_2C_6H_5)-\overset{\overset{\displaystyle O}{\|}}{C}-NH-CH_2CO_2CH_3 \xrightarrow[\text{2. HCl (neutralize)}]{\text{1. NaOH}}$$

$$Cbz-NH-CH(CO_2H)-CH_2-\overset{\overset{\displaystyle O}{\|}}{C}-NH-CH(CH_2SCH_2C_6H_5)-\overset{\overset{\displaystyle O}{\|}}{C}-NHCH_2CO_2H \quad 90\%$$

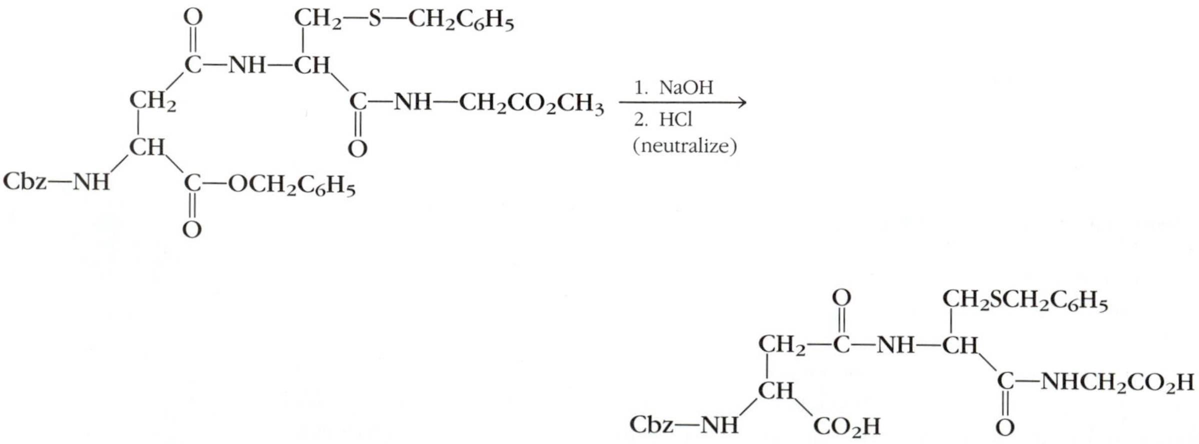

Notice that the Cbz group also undergoes hydrogenolysis in the first example. In the second reaction both the benzyl and methyl esters undergo alkaline hydrolysis but the Cbz group is unaffected.

Phenyl esters and substituted phenyl esters have been used extensively in peptide synthesis, because they are easily cleaved by nucleophilic attack at the carbonyl carbon. This reactivity simply reflects the leaving group ability. Phenols are stronger acids than alcohols, and phenoxy groups are therefore better leaving groups than alkoxy groups. In fact, some phenyl esters are sufficiently reactive that they can be used in coupling reactions during peptide synthesis (Section 20.4).

Protection of —SH Groups

Thiol groups (sulfhydryl groups) are relatively nucleophilic, and they are also very susceptible to oxidation in air (with formation of disulfide linkages). In order to avoid undesirable side reactions, the —SH groups of cysteine residues are often protected during peptide synthesis. The two most common ways of blocking the —SH group involve formation of the *S*-benzyl derivative and use of the disulfide dimer of cysteine (the dimeric amino acid is called *cystine*).

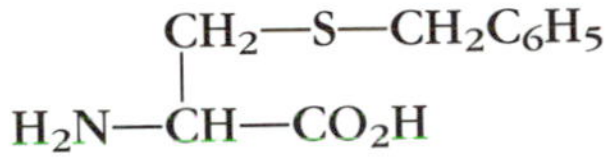

$$H_2N\text{—}CH(CH_2\text{—}S\text{—}CH_2C_6H_5)\text{—}CO_2H$$

S-Benzyl derivative of cysteine

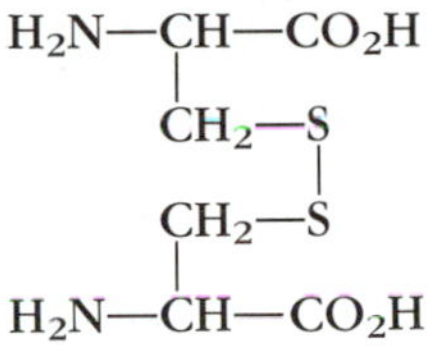

$$H_2N\text{—}CH(CO_2H)\text{—}CH_2\text{—}S\text{—}S\text{—}CH_2\text{—}CH(CO_2H)\text{—}NH_2$$

Cystine

In the latter case reactions are carried out at *both* amino groups or *both* carboxyl groups, and the final peptide is cleaved to produce two identical peptides with free —SH groups. An example of this strategy will be illustrated in Section 20.5.

The *S*-benzyl group is normally introduced by alkylation of the —SH group. The thiol group is sufficiently acidic ($K_a \sim 10^{-8}$) that it is readily deprotonated to the more nucleophilic —S$^-$ anion, so treatment of a thiol with base and benzyl chloride affords the corresponding *S*-benzyl derivative.

$$R\text{—}SH \xrightarrow[NaOH]{C_6H_5CH_2Cl} R\text{—}S\text{—}CH_2C_6H_5$$

Protection of the thiol group by formation of the *S*-benzyl derivative

Removal of the *S*-benzyl protecting group is usually accomplished by reduction with sodium in ammonia.

$$R\text{—}S\text{—}CH_2C_6H_5 \xrightarrow[\text{HBr in nonaqueous solvent}]{Na,\ NH_3\ \text{or}} R\text{—}SH$$

Removal of the *S*-benzyl protecting group

The following reactions illustrate the introduction and removal of the *S*-benzyl group as a thiol-protecting group for amino acids and peptides. Notice that the benzyl ester and *N*-benzyloxycarbonyl groups are also cleaved in the second example.

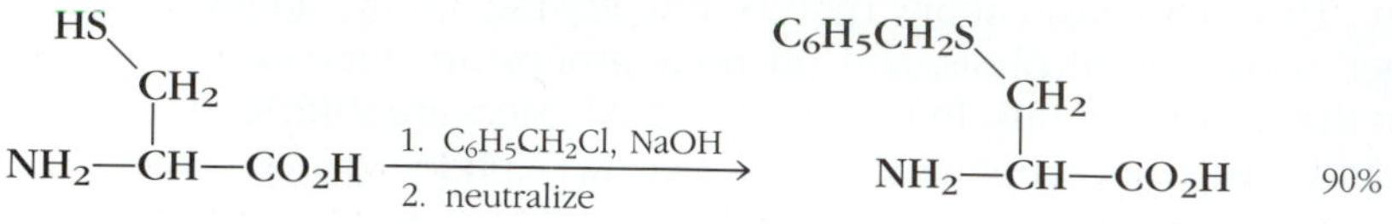

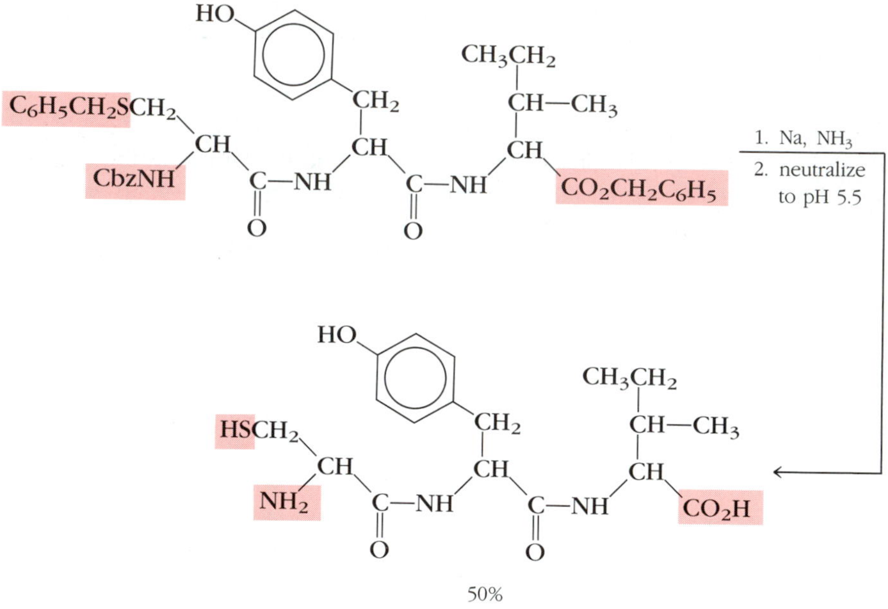

$$HSCH_2-CH(NH_2)-CO_2H \xrightarrow[\text{2. neutralize}]{\text{1. } C_6H_5CH_2Cl,\ NaOH} C_6H_5CH_2SCH_2-CH(NH_2)-CO_2H \quad 90\%$$

$$C_6H_5CH_2SCH_2-CH(CbzNH)-C(=O)-NH-CH(CH_2C_6H_4OH)-C(=O)-NH-CH(CH(CH_3)CH_2CH_3)-CO_2CH_2C_6H_5 \xrightarrow[\text{2. neutralize to pH 5.5}]{\text{1. Na, } NH_3}$$

$$HSCH_2-CH(NH_2)-C(=O)-NH-CH(CH_2C_6H_4OH)-C(=O)-NH-CH(CH(CH_3)CH_2CH_3)-CO_2H \quad 50\%$$

Protection of the Hydroxyl Group

The hydroxyl groups of amino acid residues present fewer problems in peptide synthesis than do the other functional groups we have just considered. This is because hydroxyl groups are less nucleophilic than amines, thiols, and carboxylate ions. The greater acidity of phenols means that aqueous base will produce significant amounts of the phenolate ions, so tyrosine residues are more apt to undergo side reactions than serine or threonine residues.

The methods commonly used for protection of OH groups in peptide synthesis involve formation of esters (Section 16.6) and silyl ethers (which we have not discussed). Tetrahydropyranyl ethers (Sections 15.7 and 23.3) are not generally employed in peptide synthesis because they are too easily hydrolyzed by mild acid. For protection of tyrosine residues, simple acetates are often satisfactory, and mild alkaline hydrolysis is sufficient for removal of the blocking group. Phenolate ions are good leaving groups, so the reaction proceeds more rapidly than in the case of alkyl esters. Since the OH group poses fewer problems in peptide synthesis than do the other functional groups, we will not discuss the protection of the OH group in detail here.

EXERCISE 20.4

Show how the amino group of phenylalanine could be protected as:

(a) the *N*-tosyl derivative;
(b) the *N*-benzyloxycarbonyl derivative; and
(c) the *N-tert*-butoxycarbonyl derivative.

EXERCISE 20.5

Show how the carboxyl group of phenylalanine could be protected as

(a) the methyl ester and
(b) the benzyl ester.

EXERCISE 20.6

Show how the thiol group of cysteine could be protected as

(a) the disulfide and
(b) the *S*-benzyl derivative.

EXERCISE 20.7

Show how the protecting group in each of the preceding questions could be removed.

20.4 PEPTIDE SYNTHESIS: COUPLING OF AMINO ACID RESIDUES

With a variety of blocking groups available, peptide synthesis is relatively straightforward. A dipeptide can be prepared by coupling two amino acids, each of which had been protected so that it would act as a monofunctional compound. This process is illustrated in Figure 20.2 for the use of benzyloxycarbonyl and benzyl esters as amino- and carboxyl-protecting groups, respectively. A fully protected dipeptide is the immediate result of the coupling reaction, but selective removal of the benzyloxycarbonyl group would produce a new "monofunctional" compound. Successive introduction of further amino acid residues would therefore allow synthesis of an extended polypeptide chain.

In our earlier discussions of the preparation of amides (Section 16.6) we were considering simple monofunctional compounds, and the reaction conditions were frequently more vigorous than would be acceptable for the synthesis of peptides. Successful formation of biologically active polypeptides not only requires high yield reactions but also must avoid any harsh acidic or basic conditions that could cause enolization of the carbonyl group. Formation of an enol or enolate would result in racemization of the amino acid residue, and the polypeptide would no longer possess the necessary stereochemical homogeneity. A variety of mild procedures have been developed for coupling amino acid residues, and we will illustrate several of the most important methods here.

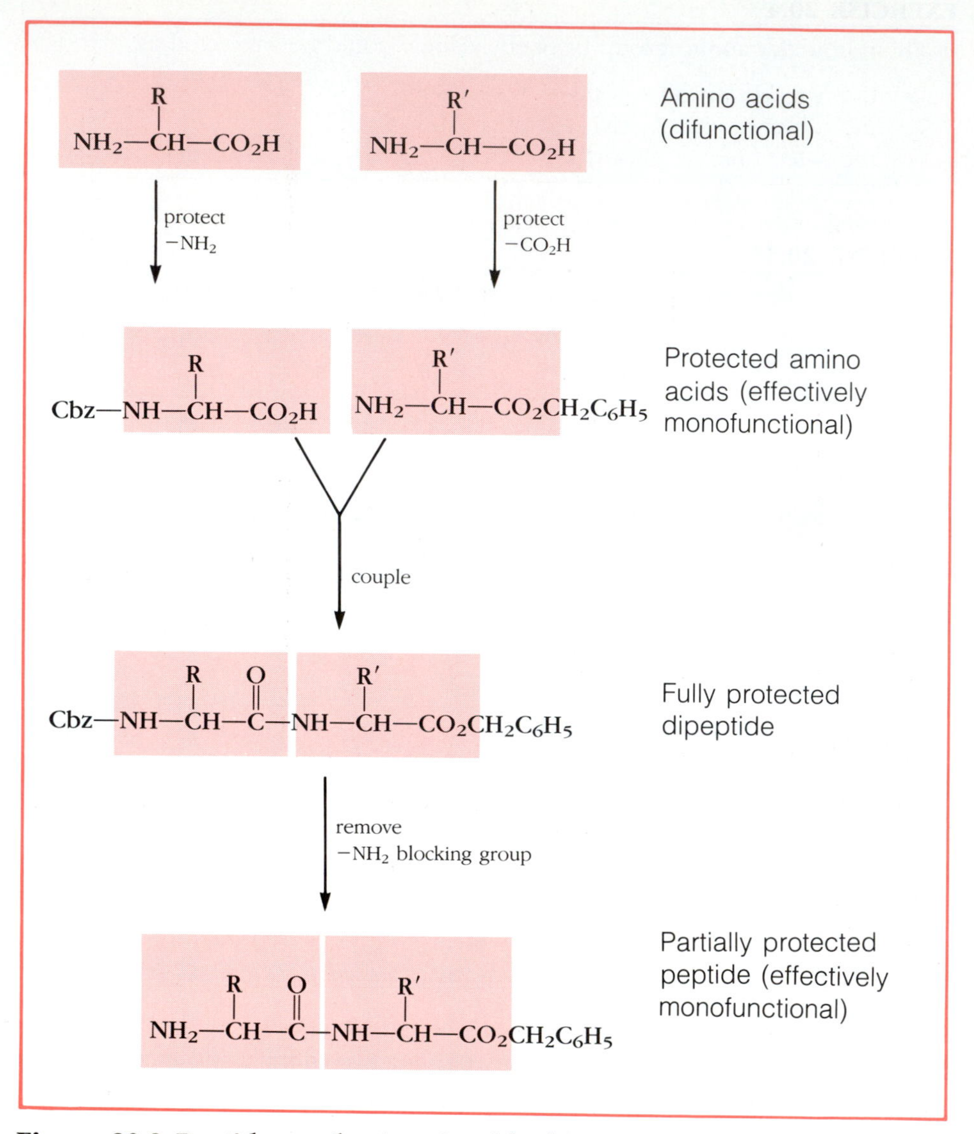

Figure 20.2 Peptide synthesis using blocking groups.
The individual amino acid residues are first converted to their amino-protected and carboxyl-protected derivatives, respectively. Coupling then produces a protected dipeptide. Finally, removal of the amino-protecting group generates a dipeptide, which could further react with another NH_2-protected amino acid to extend the peptide chain.

Acid Chloride Method

Some of the recent procedures for peptide synthesis are more satisfactory than the acid chloride procedure, but this method played an important role in the development of peptide synthesis. Reaction of an NH_2-protected amino acid with PCl_5 (or $SOCl_2$, etc.) produces an acid chloride which can react with a carboxyl-protected amino acid to yield a peptide. This is illustrated by the following reaction, which produces the disulfide derivative of cysteinylvaline.

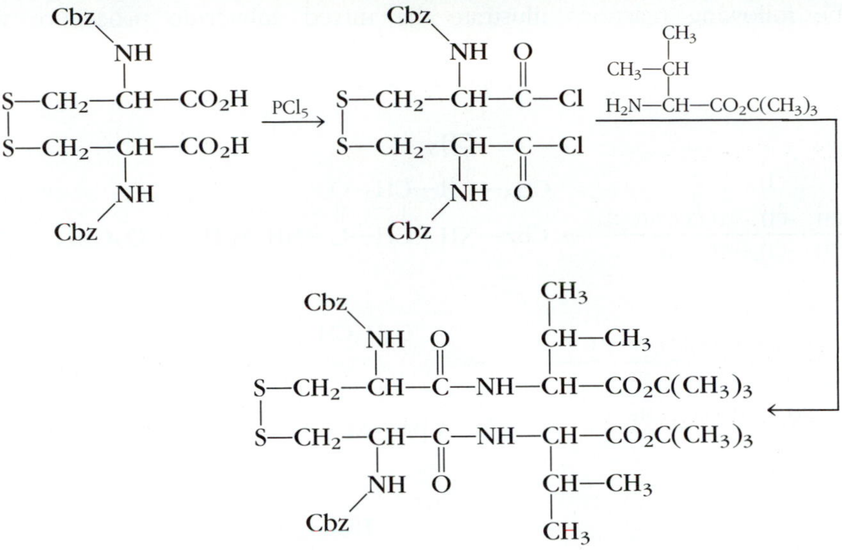

$$\text{(Cbz-NH)CH(CO}_2\text{H)-CH}_2\text{-S-S-CH}_2\text{-CH(NH-Cbz)CO}_2\text{H} \xrightarrow{PCl_5} \text{S-CH}_2\text{-CH(NH-Cbz)-C(=O)-Cl (dimer)} \xrightarrow{H_2N-CH(CH(CH_3)_2)-CO_2C(CH_3)_3}$$

$$\text{S-CH}_2\text{-CH(NH-Cbz)-C(=O)-NH-CH(CH(CH}_3)_2)\text{-CO}_2\text{C(CH}_3)_3 \text{ (disulfide dimer)}$$

77–82%

Notice that the disulfide linkage served as a protecting group for the sulfhydryl group of cysteine in this reaction.

Mixed Anhydride Method

The mixed anhydride procedure is closely related to the acid chloride method, but the conditions are milder and fewer side reactions result. An NH_2-protected amino acid is allowed to react with an acid chloride or anhydride to form a mixed anhydride (Section 16.6). For example:

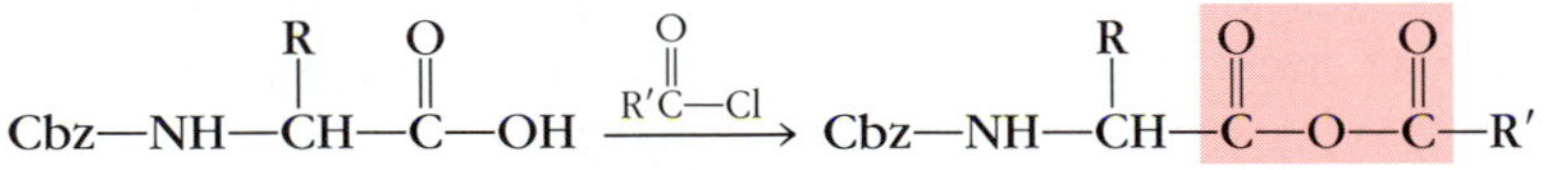

$$\text{Cbz—NH—CH(R)—C(=O)—OH} \xrightarrow{R'C(=O)—Cl} \text{Cbz—NH—CH(R)—C(=O)—O—C(=O)—R'}$$

If the mixed anhydride is suitably designed, nucleophilic attack by the amino group of a carboxyl-protected amino acid will occur selectively. There are two ways by which such selectivity can be obtained. First, if R′ is a bulky group, attack at the adjacent carbonyl group becomes sterically unfavorable. Second, if R′ is electron-withdrawing, then $R'—CO_2^-$ will be a better leaving group than the carboxylate anion of the amino acid.

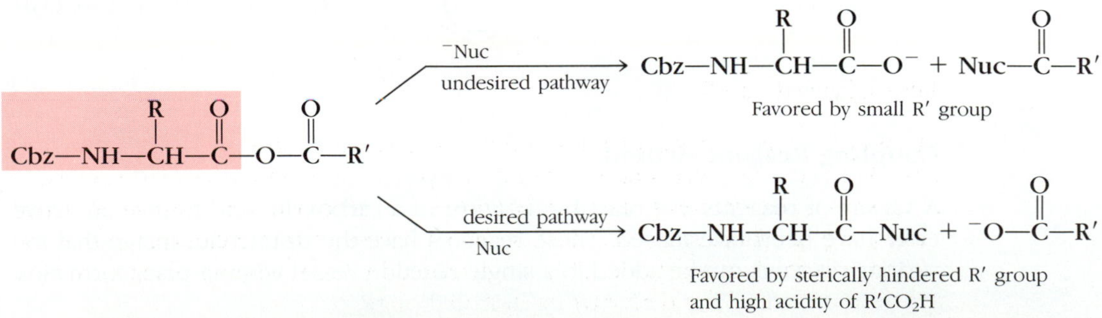

$$\text{Cbz—NH—CH(R)—C(=O)—O—C(=O)—R'} \xrightarrow[\text{undesired pathway}]{^-\text{Nuc}} \text{Cbz—NH—CH(R)—C(=O)—O}^- + \text{Nuc—C(=O)—R'}$$

Favored by small R′ group

$$\text{Cbz—NH—CH(R)—C(=O)—O—C(=O)—R'} \xrightarrow[^-\text{Nuc}]{\text{desired pathway}} \text{Cbz—NH—CH(R)—C(=O)—Nuc} + {}^-\text{O—C(=O)—R'}$$

Favored by sterically hindered R′ group and high acidity of $R'CO_2H$

The following reactions illustrate the mixed anhydride procedure with isovaleryl chloride, an alkyl chloroformate, and tetraethyl pyrophosphite (a phosphorous acid derivative) as the respective acylation reagents.

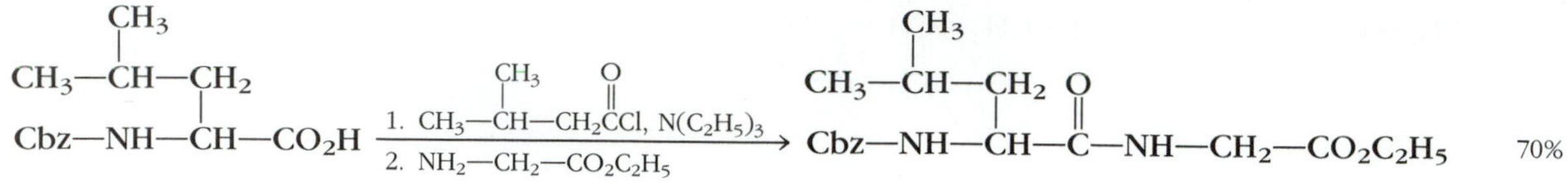

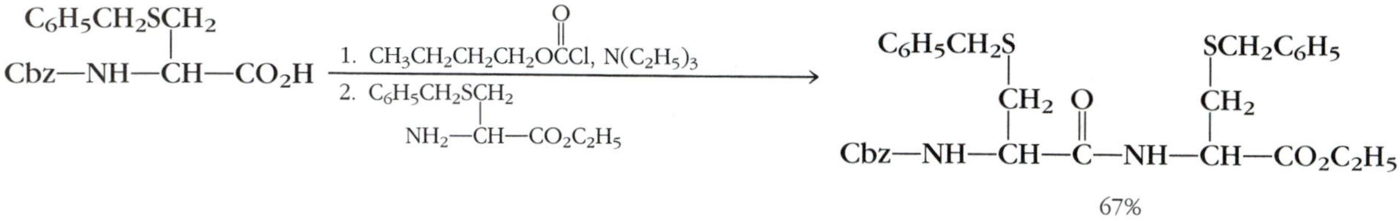

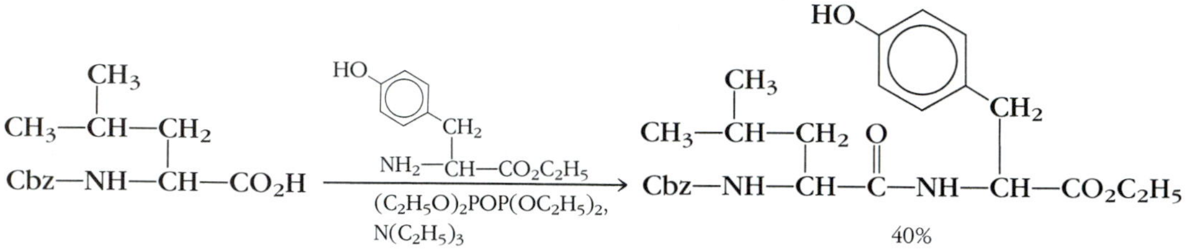

$$\text{Cbz—NH—CH(CH}_2\text{CH(CH}_3)_2\text{)—CO}_2\text{H} \xrightarrow[\text{2. NH}_2\text{—CH}_2\text{—CO}_2\text{C}_2\text{H}_5]{\text{1. CH}_3\text{—CH(CH}_3\text{)—CH}_2\text{C(=O)Cl, N(C}_2\text{H}_5)_3} \text{Cbz—NH—CH(CH}_2\text{CH(CH}_3)_2\text{)—C(=O)—NH—CH}_2\text{—CO}_2\text{C}_2\text{H}_5 \quad 70\%$$

$$\text{Cbz—NH—CH(CH}_2\text{SCH}_2\text{C}_6\text{H}_5\text{)—CO}_2\text{H} \xrightarrow[\text{2. NH}_2\text{—CH(CH}_2\text{SCH}_2\text{C}_6\text{H}_5\text{)—CO}_2\text{C}_2\text{H}_5]{\text{1. CH}_3\text{CH}_2\text{CH}_2\text{CH}_2\text{OC(=O)Cl, N(C}_2\text{H}_5)_3} \text{Cbz—NH—CH(CH}_2\text{SCH}_2\text{C}_6\text{H}_5\text{)—C(=O)—NH—CH(CH}_2\text{SCH}_2\text{C}_6\text{H}_5\text{)—CO}_2\text{C}_2\text{H}_5 \quad 67\%$$

$$\text{Cbz—NH—CH(CH}_2\text{CH(CH}_3)_2\text{)—CO}_2\text{H} \xrightarrow[(\text{C}_2\text{H}_5\text{O})_2\text{POP(OC}_2\text{H}_5)_2,\ \text{N(C}_2\text{H}_5)_3]{\text{NH}_2\text{—CH(CH}_2\text{C}_6\text{H}_4\text{OH)—CO}_2\text{C}_2\text{H}_5} \text{Cbz—NH—CH(CH}_2\text{CH(CH}_3)_2\text{)—C(=O)—NH—CH(CH}_2\text{C}_6\text{H}_4\text{OH)—CO}_2\text{C}_2\text{H}_5 \quad 40\%$$

Active Ester Method

Ordinarily, esters react with amines to form amides only under relatively vigorous conditions. But esters of phenols, particularly *p*-nitrophenol, are much more susceptible to nucleophilic attack. As a result they are useful intermediates in peptide synthesis. The following coupling of glycine *p*-nitrophenyl ester (protected as the phthalimide, cf. Section 17.7) and asparagine illustrates the method:

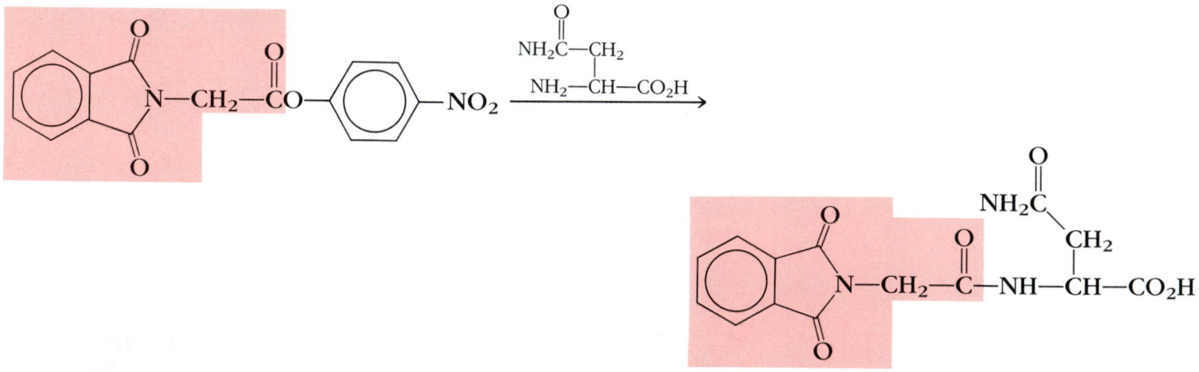

$$\text{PhthN—CH}_2\text{—CO—O—C}_6\text{H}_4\text{—NO}_2 \xrightarrow{\text{NH}_2\text{C(=O)—CH}_2\text{—CH(NH}_2\text{)—CO}_2\text{H}} \text{PhthN—CH}_2\text{—C(=O)—NH—CH(CH}_2\text{C(=O)NH}_2\text{)—CO}_2\text{H}$$

Coupling Reagent Method

A variety of reagents that react by *addition* of a carboxylic acid to give an active ester have been investigated. These reagents have the distinct advantage that the various reactants can be added to a single reaction vessel without prior formation of any intermediates. Two of the most important coupling reagents are dicyclohexylcarbodiimide (*DCC*), developed by J.C. Sheehan at the Massachussets Institute of Technology, and Woodward's *reagent K*, developed at Harvard by R.B. Woodward and R.A. Olofson (now at the Pennsylvania State University).

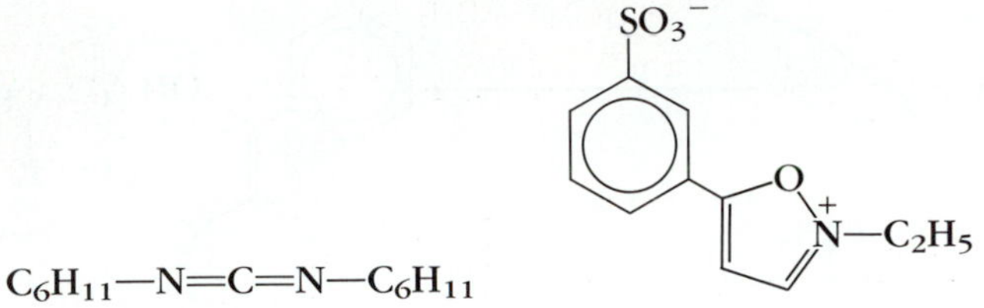

$$C_6H_{11}—N{=}C{=}N—C_6H_{11}$$

Dicyclohexylcarbodiimide (DCC)

Woodward's reagent K

The reaction of DCC with a carboxylic acid proceeds by addition of the carboxylic acid across one of the carbon-nitrogen double bonds to generate an active ester.

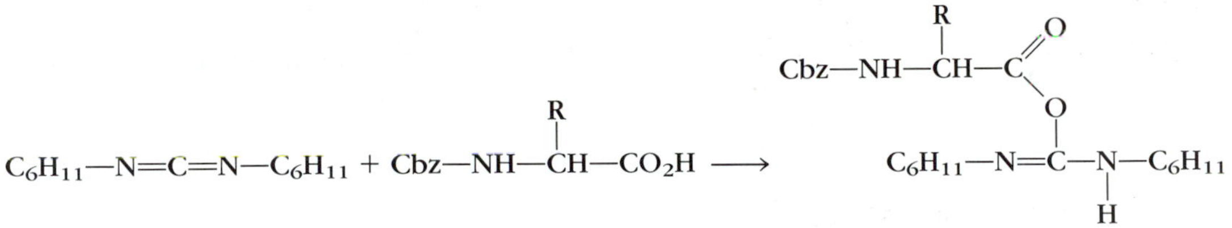

$$C_6H_{11}—N{=}C{=}N—C_6H_{11} + Cbz—NH—CH(R)—CO_2H \longrightarrow Cbz—NH—CH(R)—C({=}O)—O—C({=}N—C_6H_{11})—NH—C_6H_{11}$$

In this intermediate the carbonyl group of the amino acid is highly susceptible to nucleophilic attack, and dicyclohexylurea is formed as an insoluble by-product when reaction occurs.

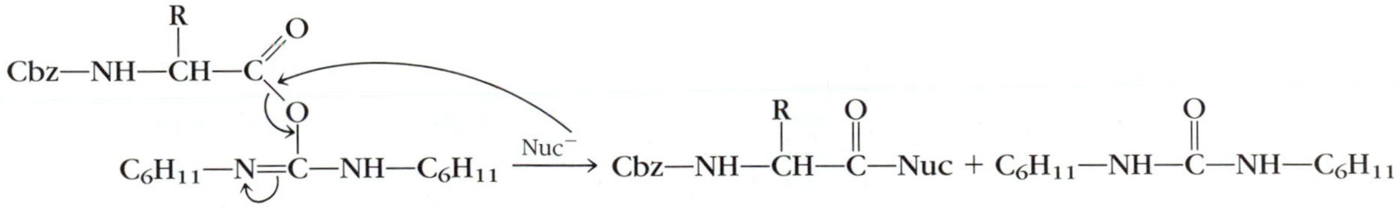

$$Cbz—NH—CH(R)—C({=}O)—O—C({=}N—C_6H_{11})—NH—C_6H_{11} \xrightarrow{Nuc^-} Cbz—NH—CH(R)—C({=}O)—Nuc + C_6H_{11}—NH—C({=}O)—NH—C_6H_{11}$$

The reaction with Woodward's reagent K follows a somewhat more complex sequence, starting with base-catalyzed cleavage of the five-membered heterocyclic ring.

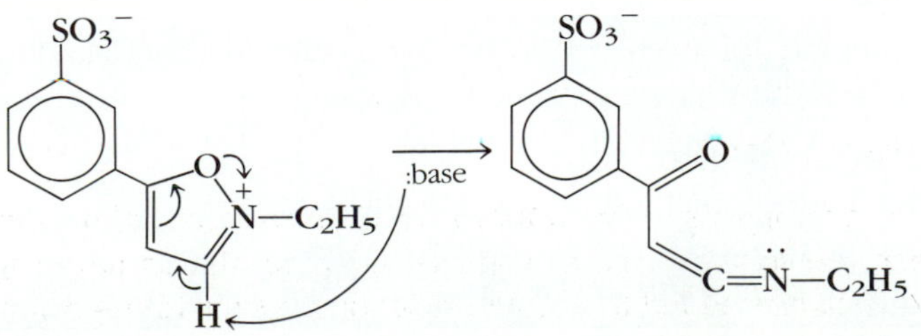

$$\xrightarrow{:base}$$

SO₃⁻ ... $N—C_2H_5$ ... $C{=}\ddot{N}—C_2H_5$

The carboxyl group of an amino acid adds to the carbon-nitrogen double bond generated in the first step, and this is followed by an intramolecular acyl transfer to form an active ester.

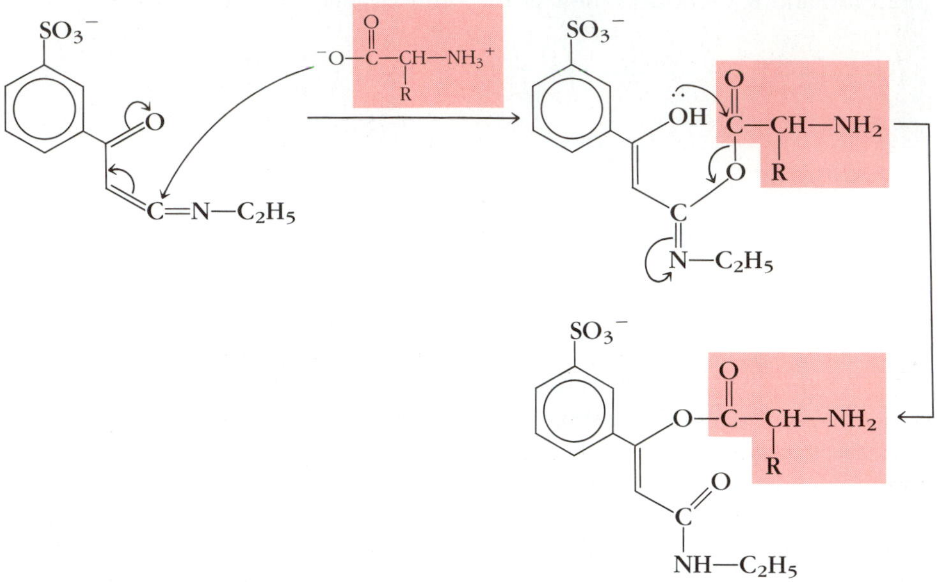

SO_3^- ... $^-O-C(=O)-CH(R)-NH_3^+$ ... $C{=}N-C_2H_5$ → SO_3^- ... OH $C(=O)-CH(R)-NH_2$ ... $N-C_2H_5$ → SO_3^- ... $O-C(=O)-CH(R)-NH_2$ ... $NH-C_2H_5$

Reaction of the active ester with the free amino group of an amino acid derivative then results in peptide bond formation.

The following equations illustrate the use of these coupling reagents in peptide synthesis:

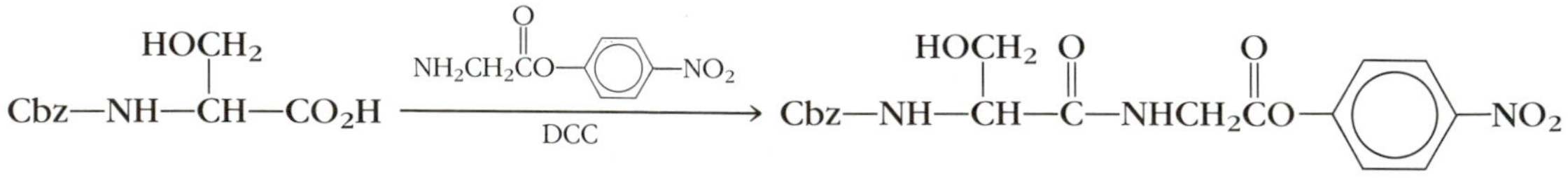

$$\text{Cbz—NH—CH(CH}_2\text{OH)—CO}_2\text{H} \xrightarrow[\text{DCC}]{\text{NH}_2\text{CH}_2\text{CO}_2\text{C}_6\text{H}_4\text{NO}_2} \text{Cbz—NH—CH(CH}_2\text{OH)—C(=O)—NHCH}_2\text{CO}_2\text{—C}_6\text{H}_4\text{—NO}_2$$

78–86%

$$\text{Cbz—NH—CH(CH}_2\text{C}_6\text{H}_5\text{)—CO}_2\text{H} \xrightarrow[\text{CH}_3\text{—CH(CH}_3\text{)—CH}_2\text{—CH(NH}_2\text{)—CO}_2\text{CH}_3]{\text{Woodward's reagent K, N(C}_2\text{H}_5)_3} \text{Cbz—NH—CH(CH}_2\text{C}_6\text{H}_5\text{)—C(=O)—NH—CH(CH}_2\text{CH(CH}_3)\text{CH}_3\text{)—CO}_2\text{CH}_3$$

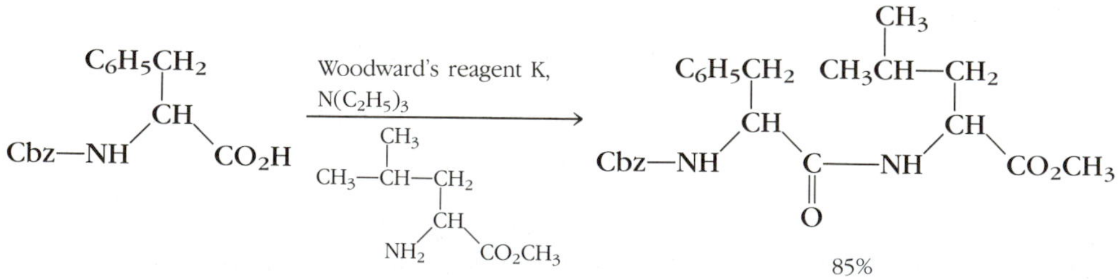

85%

EXERCISE 20.8

Suggest a method for preparing *N*-benzyloxycarbonylglycylalanine methyl ester by each of the following procedures, using the indicated reagent:

(a) Acid chloride method (PCl_5)
(b) Mixed anhydride method (isovaleryl chloride)
(c) Active ester method (via *p*-nitrophenyl ester)
(d) Coupling reagent method (DCC)

EXERCISE 20.9

Suggest a method for preparing *N*-tosylalanylglycine ethyl ester by using each of the following reagents:

(a) $SOCl_2$
(b) Tetraethyl pyrophosphite
(c) A *p*-nitrophenyl ester (i.e., using the active ester method)
(d) Woodward's reagent K

20.5 OXYTOCIN: SYNTHESIS OF A PEPTIDE HORMONE

During the early 1950s Vincent du Vigneaud and his coworkers at Cornell University Medical College carried out extensive studies on the hormone *oxytocin*. This hormone is secreted by the pituitary gland, and it plays very important roles in human physiology. Oxytocin stimulates smooth muscle contractions, particularly in the uterus, and it stimulates ejection of milk from the mammary gland in lactating females. It is employed medically to regulate these functions at the time of childbirth.

Du Vigneaud and his colleagues reported both the structure determination of oxytocin and the synthesis of the material within a single year, and this had a wide impact in several areas of science. The total synthesis of a peptide hormone (in this case with nine amino acid residues) indicated that it would also be possible to synthesize far more complex polypeptides. Moreover, the synthetic oxytocin was produced with full biological activity. This opened the door for medical use because the hormone occurs naturally in only miniscule quantities (for example, it is active at concentrations below 10^{-9} g/mL). In modern medical practice synthetic oxytocin (often called Pitocin® or Syntocinon®) is frequently administered to induce labor at the final stage of pregnancy.

The structure of oxytocin is complex even though it has only nine amino acid residues. Its molecular weight is approximately 1000, and the sulfur atoms of two cysteines are joined to form a ring involving six amino acids. The resulting 20-membered ring has a very high degree of conformational flexibility. The structure of oxytocin is written two different ways in Figure 20.3.

Although the complete molecular structure in Figure 20.3a shows all the bonding between atoms, the drawing in Figure 20.3b is much more concise. Moreover, the latter drawing also conveys the necessary chemical information as long as we specify the amino and carboxyl ends of the peptide chain. (Note that the carboxyl end of oxytocin occurs as the amide of ammonia.) Consequently, we will employ this more convenient representation for our discussions of oxytocin. In all cases the NH_2 and CO_2H groups (or their derivatives) will be drawn at the amino and carboxyl ends of the peptide chain.

As an illustration of the preparation of complex polypeptides, we will briefly present the synthesis of oxytocin as it was reported by du Vigneaud in 1953. The design of the synthesis was based in large part on his experimental observation that the disulfide bond of oxytocin could be regenerated by air oxidation after reduction to produce free SH groups.

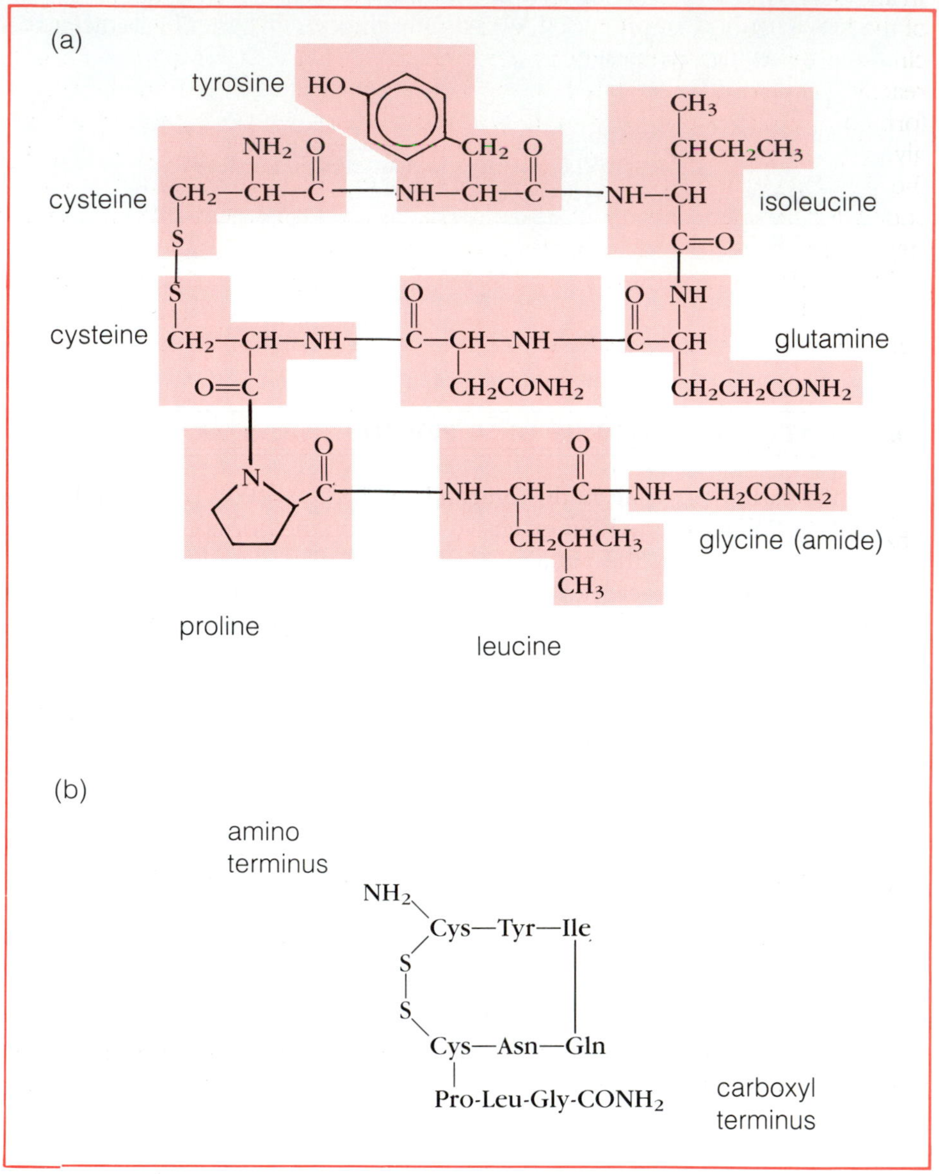

(a)

tyrosine HO—(benzene ring)—CH_2

cysteine CH_2—CH(NH_2)—C(=O)—NH—CH—C(=O)—NH—CH isoleucine

CH_3

$CHCH_2CH_3$

S

S

C=O

cysteine CH_2—CH—NH—C(=O)—CH—NH—C(=O)—CH(NH) glutamine

O=C

CH_2CONH_2

$CH_2CH_2CONH_2$

N (pyrrolidine ring)—C(=O)—NH—CH—C(=O)—NH—CH_2CONH_2

CH_2CHCH_3

CH_3

glycine (amide)

proline

leucine

(b)

amino
terminus

NH_2

Cys—Tyr—Ile

S

S

Cys—Asn—Gln

Pro-Leu-Gly-$CONH_2$

carboxyl
terminus

Figure 20.3 Oxytocin.
Drawn (a) to show the chemical structure and (b) to illustrate the amino acid sequence.

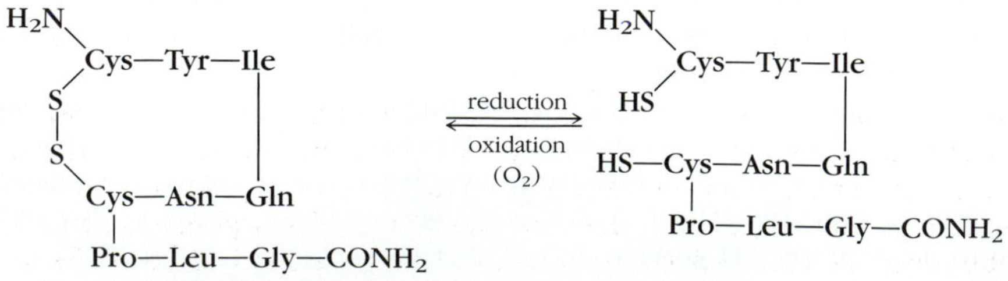

H_2N

Cys—Tyr—Ile

S

S

Cys—Asn—Gln

Pro—Leu—Gly—$CONH_2$

reduction ⇌ oxidation (O_2)

H_2N

Cys—Tyr—Ile

HS

HS—Cys—Asn—Gln

Pro—Leu—Gly—$CONH_2$

In addition it was found that benzylation of the SH groups and subsequent removal of the benzyl groups could be followed by air oxidation to give fully active oxytocin. As a result, the synthesis was designed to yield the nonapeptide with both reactive SH groups blocked; only as the final step would the 20-membered ring be formed. The plan was to begin at the carboxyl terminus of the peptide chain (i.e., glycine) and to introduce amino acid residues by coupling at the amino terminus of the growing peptide chain. In some instances a single amino acid group would be added, but at other times several amino acid residues could be added in a single step by employing the appropriate di- and tripeptides.

The synthesis began with the preparation of the ethyl ester of leucylglycine by the mixed anhydride method. The reactants were ethyl glycinate and *N*-benzyloxycarbonylleucine, so the protected dipeptide was formed.

$$\text{Cbz—NH—CH(CH}_2\text{CH(CH}_3\text{)—CH}_3\text{)—CO}_2\text{H} + \text{NH}_2\text{—CH}_2\text{—CO}_2\text{C}_2\text{H}_5 \xrightarrow{\text{CH}_3\text{CH(CH}_3\text{)—CH}_2\text{C(=O)Cl}} \text{Cbz—NH—CH(CH}_2\text{CH(CH}_3\text{)—CH}_3\text{)—C(=O)—NH—CH}_2\text{—CO}_2\text{C}_2\text{H}_5$$

60–70%

Using the shorthand notation, the reaction can be written in the following way:

$$\text{Cbz—NH—Leu—CO}_2\text{H} + \text{NH}_2\text{—Gly—CONH}_2 \xrightarrow{\text{CH}_3\text{CH(CH}_3\text{)—CH}_2\text{C(=O)Cl}} \text{Cbz—NH—Leu—Gly—CO}_2\text{C}_2\text{H}_5$$

The next step was introduction of the proline residue, and this was again accomplished by the mixed anhydride method after first removing the amino blocking group.

$$\text{Cbz—NH—Leu—Gly—CO}_2\text{C}_2\text{H}_5 \xrightarrow[\text{Pd—C}]{\text{H}_2} \text{NH}_2\text{—Leu—Gly—CO}_2\text{C}_2\text{H}_5$$

$$\xrightarrow{\text{CH}_3\text{CH(CH}_3\text{)—CH}_2\text{—C(=O)—Cl};\ \text{N-Cbz-proline (CO}_2\text{H)}} \text{Cbz—N—Pro—Leu—Gly—CO}_2\text{C}_2\text{H}_5$$

92%

Introduction of a cysteine residue was complicated by the fact that cysteine is a trifunctional amino acid. The sulfhydryl group was masked by employing the disulfide, and the amino groups were converted to the benzyloxycarbonyl derivatives. The bis(acid chloride) of this compound was then used to acylate two equivalents of the tripeptide (from which the *N*-terminal amino group had been liberated).

Cbz—N—Pro—Leu—Gly—$CO_2C_2H_5$ $\xrightarrow[Pd—C]{H_2}$ HN—Pro—Leu—Gly—$CO_2C_2H_5$

Cbz—NH—CH(C(=O)Cl)—S—S—CH(C(=O)Cl)—NH—Cbz

Cbz—NH—Cys(—S—S—)—Pro—Leu—Gly—$CO_2C_2H_5$ (dimer)

86%

Three changes were then made to the dimeric tetrapeptide: the ethyl esters of the glycine residues were converted to the necessary amides; the *N*-benzyloxycarbonyl groups were removed; and the disulfide linkage was cleaved (with the subsequent introduction of an *S*-benzyl protecting group). First the ethyl esters were hydrolyzed in base; subsequent reaction with sodium in liquid ammonia cleaved both the disulfide linkage and the benzyloxycarbonyl groups. This cleavage produced the monomeric tetrapeptide, and addition of benzyl chloride to the reaction mixture converted the sulfide ion to the corresponding *S*-benzyl ether. The glycine amide group was formed in two steps; esterification to generate the benzyl ester followed by reaction with ammonia yielded the corresponding amide.

Cbz—NH—Cys(—S—S—)—Pro—Leu—Gly—$CO_2C_2H_5$ (dimer) $\xrightarrow[2.\ HCl]{1.\ NaOH,\ H_2O}$ Cbz—NH—Cys(—S—S—)—Pro—Leu—Gly—CO_2H (dimer)

$\downarrow$ Na, NH_3

[NH_2—Cys($^-$S)—Pro—Leu—Gly—CO_2^-] $\xrightarrow[2.\ HCl\ (neutralize)]{1.\ C_6H_5CH_2Cl}$ NH_2—Cys($C_6H_5CH_2S$)—Pro—Leu—Gly—CO_2H

$\downarrow$ 1. $C_6H_5CH_2OH$, HCl; 2. NH_3

$C_6H_5CH_2S$—Cys—NH_2
|
Pro—Leu—Gly—$CONH_2$

72% overall yield

At this point the tetrapeptide was coupled with a tripeptide (Ile-Gln-Asn), and we will make a detour to show you how the tripeptide was prepared. The starting point for the tripeptide was the *N*-tosyl derivative of glutamic acid. This material was heated with phosphorus pentachloride to form an intermediate bis(acid chloride), which underwent cyclization.

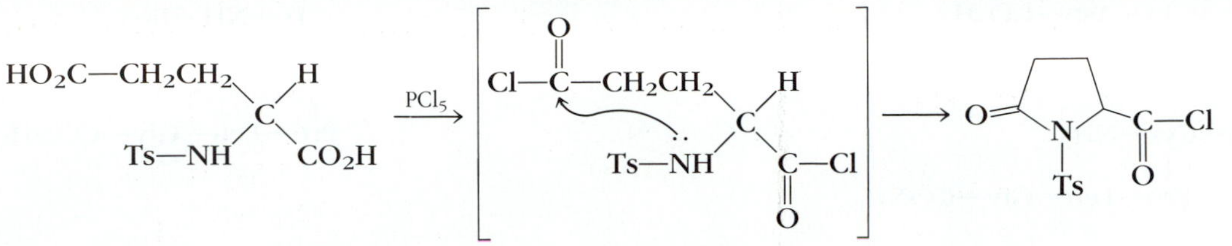

$$\mathrm{HO_2C{-}CH_2CH_2{-}CH(NH{-}Ts){-}CO_2H} \xrightarrow{PCl_5} \left[\mathrm{Cl{-}C(=O){-}CH_2CH_2{-}CH(NH{-}Ts){-}C(=O){-}Cl}\right] \longrightarrow \text{N-Ts-pyrrolidone-}\mathrm{C(=O){-}Cl}$$

90%

The resulting cyclic acid chloride was allowed to react with asparagine, and next the five-membered ring was opened by treatment with ammonia. Removal of the tosyl group then afforded the dipeptide glutaminylasparagine.

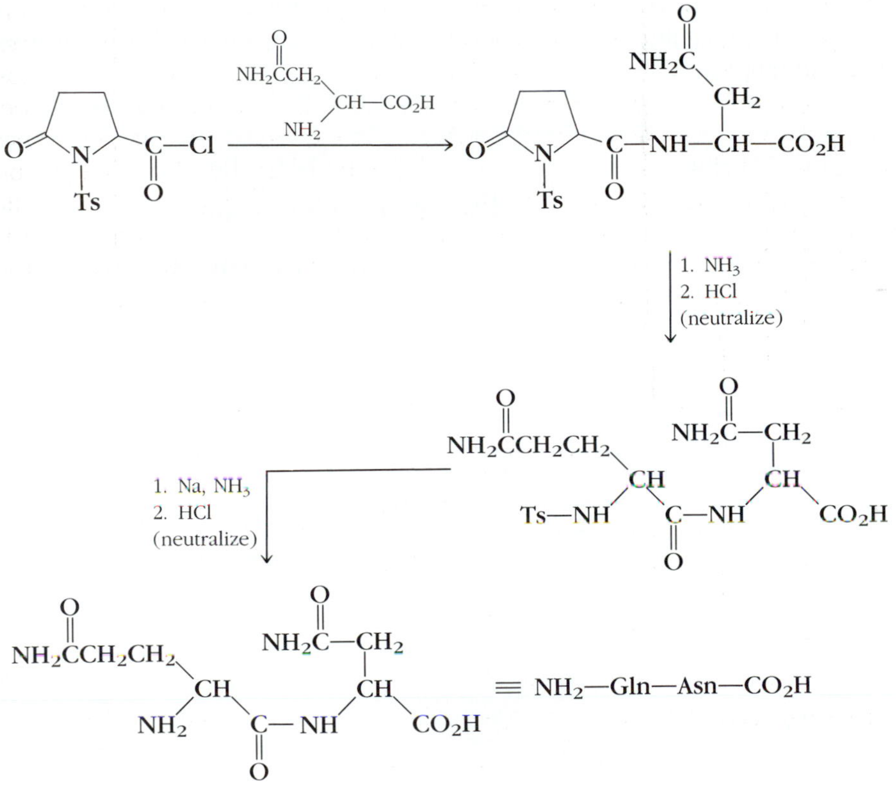

$$\text{N-Ts-pyrrolidone-}\mathrm{C(=O){-}Cl} \xrightarrow{\mathrm{NH_2C(=O)CH_2{-}CH(NH_2){-}CO_2H}} \text{N-Ts-pyrrolidone-}\mathrm{C(=O){-}NH{-}CH(CH_2C(=O)NH_2){-}CO_2H}$$

1. NH_3
2. HCl (neutralize)

$$\mathrm{NH_2C(=O)CH_2CH_2{-}CH(NH{-}Ts){-}C(=O){-}NH{-}CH(CH_2C(=O)NH_2){-}CO_2H}$$

1. Na, NH_3
2. HCl (neutralize)

$$\mathrm{NH_2C(=O)CH_2CH_2{-}CH(NH_2){-}C(=O){-}NH{-}CH(CH_2C(=O)NH_2){-}CO_2H} \equiv \mathrm{NH_2{-}Gln{-}Asn{-}CO_2H}$$

75% overall yield

The required tripeptide was then produced by the reaction of this dipeptide with the acid chloride of *N*-tosylisoleucine.

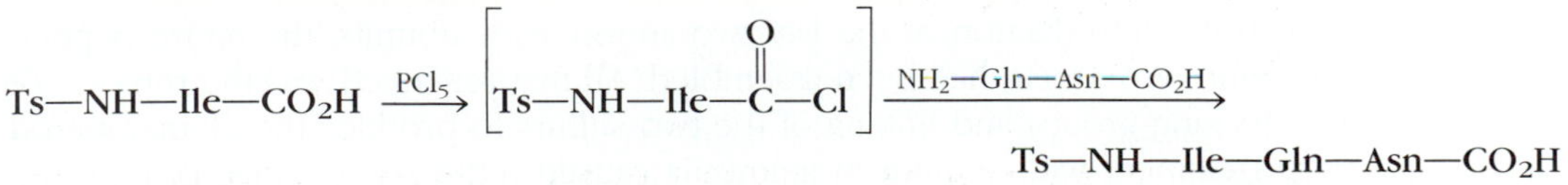

$$\mathrm{Ts{-}NH{-}Ile{-}CO_2H} \xrightarrow{PCl_5} \left[\mathrm{Ts{-}NH{-}Ile{-}C(=O){-}Cl}\right] \xrightarrow{\mathrm{NH_2{-}Gln{-}Asn{-}CO_2H}} \mathrm{Ts{-}NH{-}Ile{-}Gln{-}Asn{-}CO_2H}$$

58–62%

The stage was thus set for joining the tri- and tetrapeptides, and this was accomplished by the use of tetraethyl pyrophosphite.

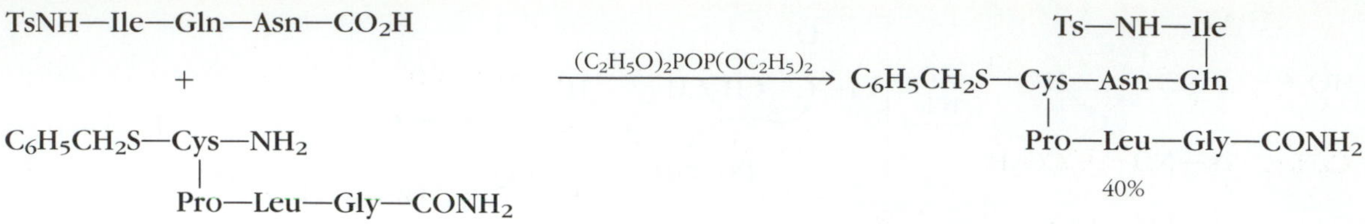

TsNH—Ile—Gln—Asn—CO_2H + $C_6H_5CH_2S$—Cys—NH_2 (Cys—Pro—Leu—Gly—$CONH_2$) $\xrightarrow{(C_2H_5O)_2POP(OC_2H_5)_2}$ $C_6H_5CH_2S$—Cys—Asn—Gln—Ile—NH—Ts (Cys—Pro—Leu—Gly—$CONH_2$)

40%

The remaining two amino acids of oxytocin were introduced as the appropriately protected derivative of cysteinyltyrosine, but it was first necessary to unblock the terminal amino group of the heptapeptide. Reductive cleavage of the *N*-tosyl group with sodium in ammonia was accompanied by loss of the *S*-benzyl group, and it was necessary to reintroduce this sulfhydryl protecting group before proceeding to the coupling reaction. For the coupling itself, tetraethyl pyrophosphite was again employed.

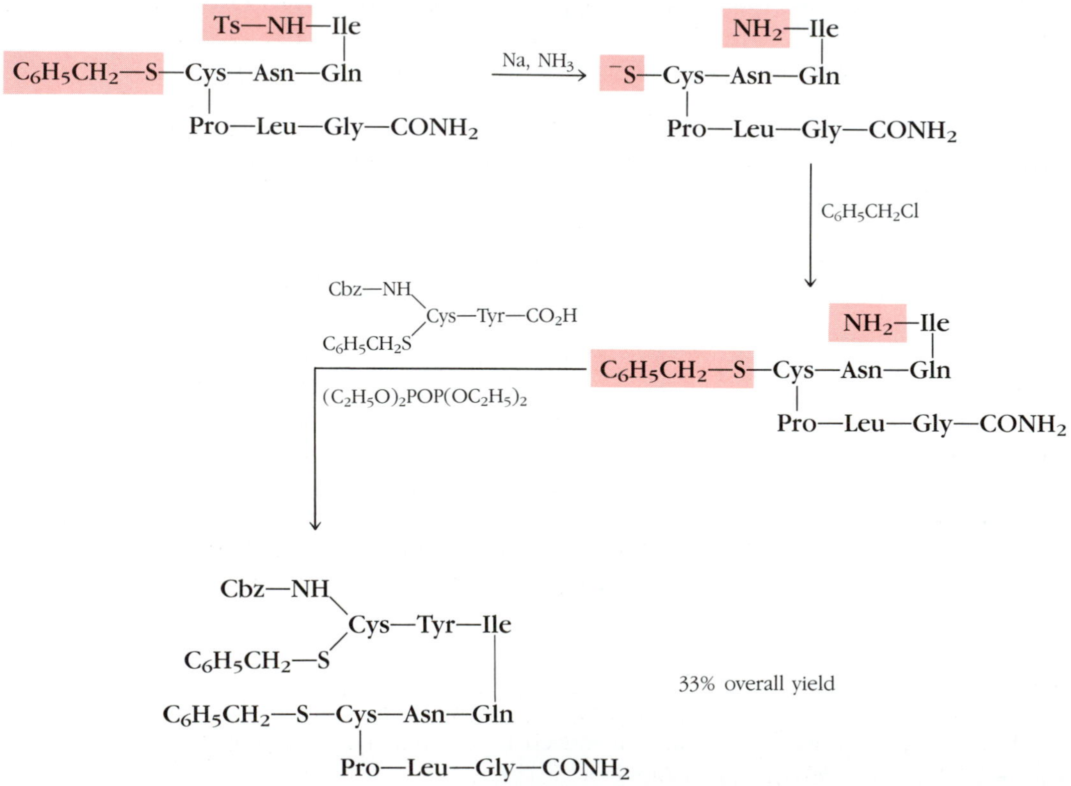

Ts—NH—Ile / $C_6H_5CH_2$—S—Cys—Asn—Gln / Pro—Leu—Gly—$CONH_2$ $\xrightarrow{Na,\ NH_3}$ NH_2—Ile / $^-$S—Cys—Asn—Gln / Pro—Leu—Gly—$CONH_2$

$\xrightarrow{C_6H_5CH_2Cl}$ NH_2—Ile / $C_6H_5CH_2$—S—Cys—Asn—Gln / Pro—Leu—Gly—$CONH_2$

+ Cbz—NH, $C_6H_5CH_2S$ —Cys—Tyr—CO_2H $\xrightarrow{(C_2H_5O)_2POP(OC_2H_5)_2}$

Cbz—NH, $C_6H_5CH_2$—S —Cys—Tyr—Ile / $C_6H_5CH_2$—S—Cys—Asn—Gln / Pro—Leu—Gly—$CONH_2$

33% overall yield

With the introduction of the last two amino acid subunits, the entire peptide backbone of oxytocin had been assembled. All that remained was the removal of the blocking groups and linking of the two sulfurs to produce the 20-membered ring. Treatment with sodium in ammonia caused reductive cleavage of both the *N*-benzyloxycarbonyl and *S*-benzyl groups. The crude product of the reduction was dissolved in water, and air was bubbled through the solution for several hours. This mild oxidation yielded synthetic oxytocin.

CbzNH
Cys—Tyr—Ile
$C_6H_5CH_2S$
$C_6H_5CH_2S$—Cys—Asn—Gln
Pro—Leu—Gly—$CONH_2$

1. Na, NH_3
2. NH_4Cl (neutralize)

NH_2
HS—Cys—Tyr—Ile
HS—Cys—Asn—Gln
Pro—Leu—Gly—$CONH_2$

O_2

NH_2
Cys—Tyr—Ile
S
S
Cys—Asn—Gln
Pro—Leu—Gly—$CONH_2$

Oxytocin,
20–30% overall yield for two steps

An important question still remained at that point, however. Would the synthetic material exhibit any significant biological activity? Did it posess the necessary three-dimensional structure for action as a hormone? If racemization of one or more of the amino acid residues had occurred at any point during the synthesis, nearly all activity might have been lost. To the investigators' relief, however, even the crude product exhibited biological activity that was about 20–30% of that for the natural hormone. After purification this relative activity increased to about 90%, a potency comparable with that of the material produced when the disulfide bridge of natural oxytocin was cleaved and then closed again.

The synthesis of oxytocin (summarized in Figure 20.4) was a major milestone in chemistry. For the first time it had become possible to synthesize a polypeptide with substantial medical value. Isolation of the hormone from natural sources (i.e., from other mammals) was economically impractical, but synthetic organic chemistry had provided an alternative. The door had been opened for the synthesis of even more complex peptides and perhaps even large proteins. A new era had been entered in chemistry, and du Vigneaud's contributions were recognized by the award of the Nobel Prize in chemistry in 1955. As you will see in the following section, there have been significant advances in the synthesis of polypeptides, but oxytocin remains one of the few synthetic peptides that is used in medicine.

EXERCISE 20.10

Isovaleryl chloride was used to form the peptide bond on two occasions in the synthesis of oxytocin (Cbz—NH—Leu—Gly—$CO_2C_2H_5$, Cbz—*N*—Pro—Leu—Gly—$CONH_2$). In each case draw the mixed anhydride that is formed as an intermediate.

EXERCISE 20.11

In the synthesis of oxytocin we described the formation of seven of the eight peptide bonds. For each of these bonds identify the method that was used to form it.

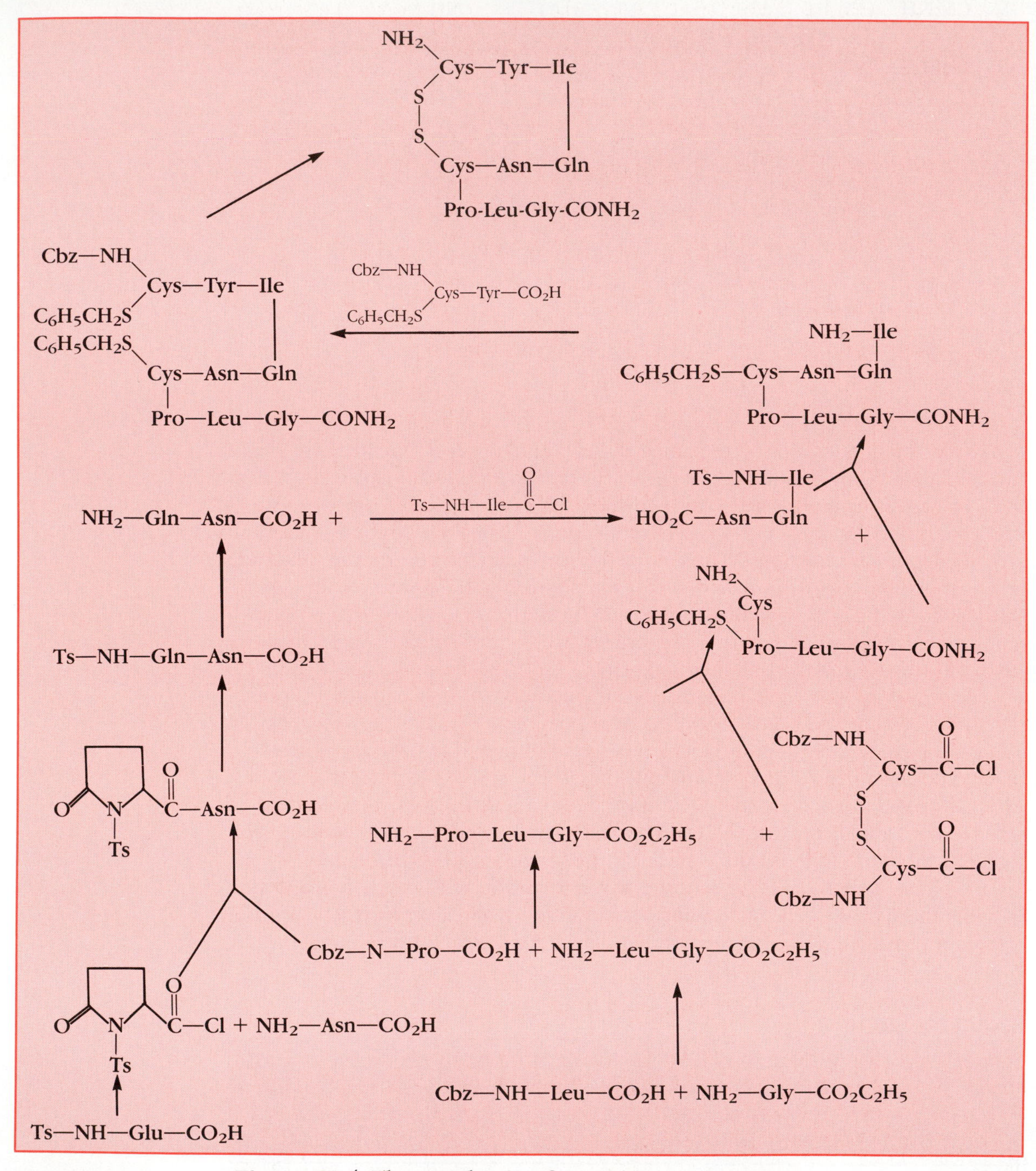

Figure 20.4 The synthesis of oxytocin—a summary.
(See text for details.)

EXERCISE 20.12

The starting point for the glutaminylasparagine dipeptide portion of oxytocin was *N*-tosylglutamic acid. Suggest a method for preparation of this compound from glutamic acid.

EXERCISE 20.13

One of the reagents used in the synthesis of oxytocin was *N*-tosylisoleucine. How could this compound be prepared from isoleucine?

20.6 MODERN DEVELOPMENTS IN PEPTIDE SYNTHESIS

The synthesis of oxytocin, accomplished in the early 1950s, was a major breakthrough in peptide chemistry. But what of the many other peptides and proteins that would be valuable in the realm of medicine? Unfortunately, the methods of peptide synthesis that we have outlined in the preceding sections are not applicable to many of these other situations. Why not? Because many of these other peptides and proteins are much larger molecules, and synthesis by those methods would be uneconomical. Even for a polypeptide of only 50 subunits (rather small in comparison with most proteins), a yield of 90% for each coupling reaction would afford the complete peptide chain in less than 1% overall yield. If the average yield of the individual steps were only 80%, the overall yield for the 50-subunit peptide would be reduced to 0.002%. Thus, commercial synthesis of large polypeptides by the procedures we have discussed in this chapter would not be satisfactory. The final yields would be extremely low, and only a small amount of material would be obtained from a large investment of time and effort. Futhermore, each step would demand purification of the intermediate, and the cost of both the chemicals and the personnel required for such a project would be prohibitive.

Insulin

Despite the difficulties we have just enumerated, ***insulin,*** a peptide hormone with 51 amino acid residues, has been synthesized by the classical methods that we have described in this chapter. Insulin is actually composed of two individual peptide chains (of 30 and 21 residues), that are joined by disulfide linkages, as shown in Figure 20.5.

Insulin occupies an important place in chemistry for several reasons. It was the first protein to be isolated in pure form (1926) and the first to have its complete amino acid sequence unraveled (1949). The total synthesis was achieved in 1963.

The primary function of insulin is the regulation of glucose concentration in the blood, and a deficiency of the hormone results in the condition known as *diabetes mellitus.* The hormone is produced in the pancreas, and it was discovered during the 1920s that insulin from one animal was frequently active in other species as well. We now know that insulin from a variety of mammals (dog, cow, pig, sheep, sperm whale) differ in only a few amino acid residues, and this does not significantly affect its activity in humans. As a result, relatively large quantities of insulin are available for medical use as a by-product of animals that are slaughtered for food.

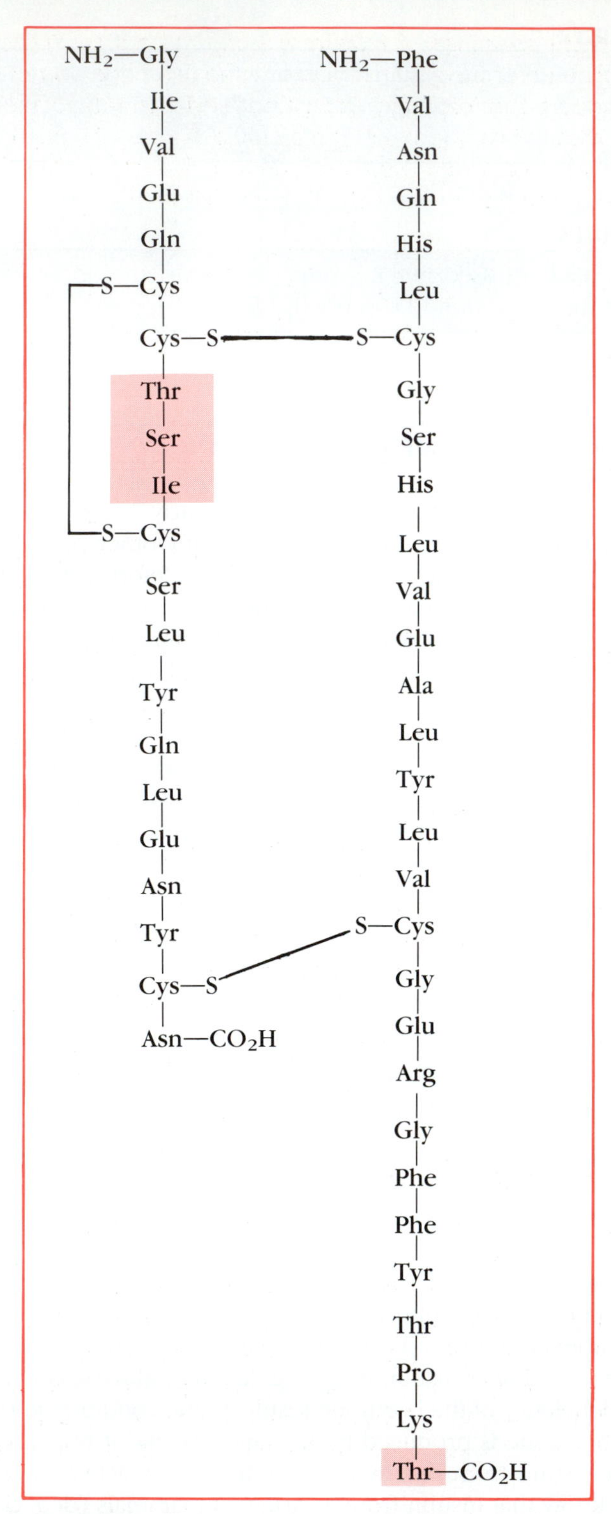

Figure 20.5 Human insulin.
The hormone is composed of two peptide chains joined by disulfide linkages. Other mammalian insulins have almost identical amino acid sequences. Sheep, bovine, and porcine insulin, for example, differ only in the residues indicated by the shaded boxes.

The amino acid sequence of insulin was worked out by Frederick Sanger in England (Cambridge University) in the late 1940s by using a technique known as *end group analysis.* Sanger found that 1-fluoro-2,4-dinitrobenzene (Section 18.7) reacts readily with the NH_2 group at the amino terminus of a peptide. Subsequent hydrolysis liberates the *N*-dinitrophenyl amino acid, and this identifies the particular amino acid at the end of the original peptide chain.

$$NH_2\text{—}CH(R)\text{—}C(=O)\text{—}NH\sim\sim \text{peptide} \xrightarrow{O_2N\text{—}C_6H_3(NO_2)\text{—}F} O_2N\text{—}C_6H_3(NO_2)\text{—}NH\text{—}CH(R)\text{—}C(=O)\text{—}NH\sim\sim \text{peptide}$$

$$\xrightarrow{\text{hydrolysis}} O_2N\text{—}C_6H_3(NO_2)\text{—}NH\text{—}CH(R)\text{—}CO_2H$$

By carrying out *partial* hydrolysis of a peptide and subjecting each of the smaller peptide fragments to such end group analysis, information is obtained about the amino acid sequence. The process is analogous to solving a puzzle, but we will not consider the specifics here. Sanger's analysis afforded the complete amino acid sequences of the two peptide chains of insulin, and he received the 1958 Nobel Prize in chemistry in recognition of this work.

In the mid 1960s three research groups (in West Germany, the USA, and the Peoples Republic of China) independently reported the total synthesis of insulin starting from the individual amino acids. This achievement represents another milestone in the development of peptide synthesis, but to some extent it also marks the end of the line for the classical methods. The yields in the synthesis of insulin were too low for development of a commercial process. For peptides with longer chains, the decrease in overall yield that results for each additional residue produces nearly insurmountable problems. Consequently, recent advances in protein synthesis have taken different directions.

Solid-Phase Peptide Synthesis

With the aim of alleviating some of the problems, R.B. Merrifield at the Rockefeller Institute in New York reported in 1963 a unique modification of the classical methods for peptide synthesis. He found that the carboxyl terminus of the peptide chain could be attached to an insoluble polymer (via a covalent bond). As each additional amino acid residue was coupled to the growing peptide chain, the polymer-bound product could be readily isolated by simple washings and filtrations. Loss of material from purification was therefore minimized. Moreover, much of the procedure is amenable to automation, so the individual chemist can become much more efficient.

Solid-phase synthesis of peptides is not without its limitations. For example, a peptide chain that fails to react in a particular coupling reaction will be carried on through subsequent stages of the synthesis. Nevertheless, it has made possible the synthesis of a variety of peptides that could not have been otherwise prepared. The sequence summarized in Figure 20.6 illustrates the technique for the synthesis of NH_2—Gly—Asp—Ser—Gly—CO_2H.

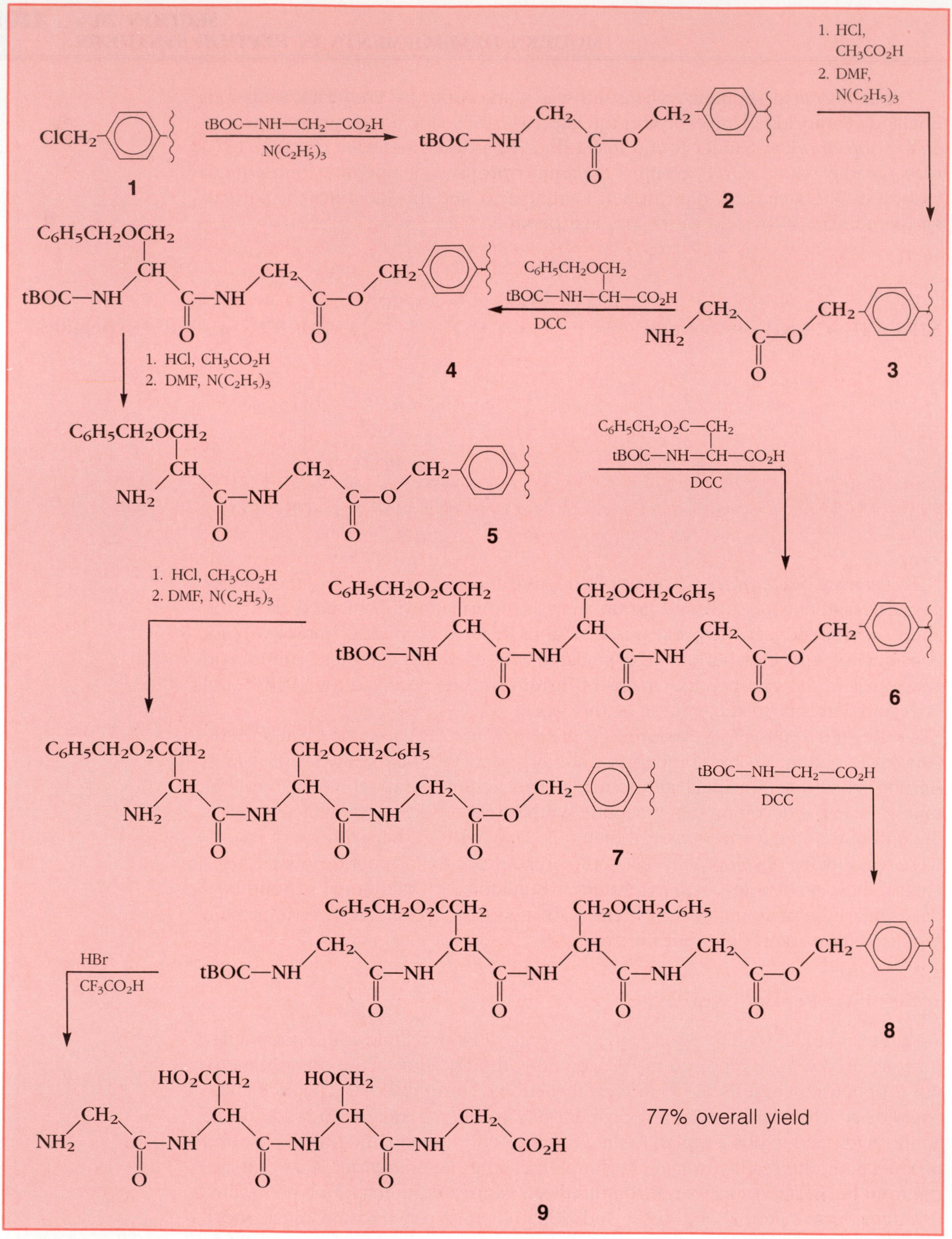

$ClCH_2-C_6H_4-$ (resin) **1**

$\xrightarrow[N(C_2H_5)_3]{tBOC-NH-CH_2-CO_2H}$

$tBOC-NH-CH_2-C(=O)-O-CH_2-C_6H_4-$ (resin) **2**

1. HCl, CH_3CO_2H
2. DMF, $N(C_2H_5)_3$

$NH_2-CH_2-C(=O)-O-CH_2-C_6H_4-$ (resin) **3**

$\xrightarrow[DCC]{C_6H_5CH_2OCH_2 \; | \; tBOC-NH-CH-CO_2H}$

$tBOC-NH-CH(CH_2OCH_2C_6H_5)-C(=O)-NH-CH_2-C(=O)-O-CH_2-C_6H_4-$ (resin) **4**

1. HCl, CH_3CO_2H
2. DMF, $N(C_2H_5)_3$

$NH_2-CH(CH_2OCH_2C_6H_5)-C(=O)-NH-CH_2-C(=O)-O-CH_2-C_6H_4-$ (resin) **5**

$\xrightarrow[DCC]{C_6H_5CH_2O_2C-CH_2 \; | \; tBOC-NH-CH-CO_2H}$

$tBOC-NH-CH(CH_2CO_2CH_2C_6H_5)-C(=O)-NH-CH(CH_2OCH_2C_6H_5)-C(=O)-NH-CH_2-C(=O)-O-CH_2-C_6H_4-$ (resin) **6**

1. HCl, CH_3CO_2H
2. DMF, $N(C_2H_5)_3$

$NH_2-CH(CH_2CO_2CH_2C_6H_5)-C(=O)-NH-CH(CH_2OCH_2C_6H_5)-C(=O)-NH-CH_2-C(=O)-O-CH_2-C_6H_4-$ (resin) **7**

$\xrightarrow[DCC]{tBOC-NH-CH_2-CO_2H}$

$tBOC-NH-CH_2-C(=O)-NH-CH(CH_2CO_2CH_2C_6H_5)-C(=O)-NH-CH(CH_2OCH_2C_6H_5)-C(=O)-NH-CH_2-C(=O)-O-CH_2-C_6H_4-$ (resin) **8**

$\xrightarrow[CF_3CO_2H]{HBr}$

$NH_2-CH_2-C(=O)-NH-CH(CH_2CO_2H)-C(=O)-NH-CH(CH_2OH)-C(=O)-NH-CH_2-CO_2H$ **9**

77% overall yield

Figure 20.6 Solid phase synthesis of NH_2—Gly—Asp—Ser—Gly—CO_2H.
Each amino acid residue is introduced by coupling the NH_2-blocked amino acid by use of dicyclohexylcarbodiimide. In the final step all benzyl groups are cleaved by treatment with relatively strong acid.

Notice that the individual steps and reagents employed in the solid phase technique are the same as those we presented earlier. The differences lie not in the chemistry but in the experimental conveniences that are provided. The first step is attachment of the carboxyl-terminal amino acid to the polymer. For the procedure summarized in Figure 20.6 this was done by a nucleophilic displacement of the carboxylate ion on the benzylic chloride of the polymer (**1** ⟶ **2**). Removal of the *tert*-butoxycarbonyl blocking group on the glycine nitrogen was effected by treatment with acid, and the product (**3**) was washed with dimethylformamide that contained triethylamine to neutralize any excess acid. Next, a serine residue was introduced by using the coupling reagent *N,N*-dicyclohexylcarbodiimide to form **4**. The removal of the tBOC group from nitrogen and coupling were repeated two more times to generate the polymer-bound tetrapeptide, **8**.

Notice that the hydroxyl group of the serine residue was protected as a benzyl ether. Similarly, the free carboxyl group of the aspartic acid was protected as a benzyl ester, and the linkage to the polymer was a substituted benzyl ester. The final step of the synthesis, treatment of the polymer-bound tetrapeptide with hydrogen bromide in trifluoroacetic acid, resulted in cleavage of all three benzyl groups as well as the *N*-tBOC group, and glycylaspartylserylglycine was produced in an overall yield of 77%. This corresponds to an average yield of 94% for each of the four coupling steps and is considerably better than the typical yields for coupling reactions in solution that we presented earlier in the chapter.

Solid phase peptide synthesis has had a major impact on modern peptide chemistry. It is now possible to prepare quickly, easily, and in substantial quantities peptides that previously would have required prohibitively large amounts of time and effort. Merrifield's contributions to this area were clearly recognized when he was awarded the 1984 Nobel Prize in chemistry.

Genetic Engineering

Even the advantages of solid-phase peptide synthesis may not be enough to allow that method to compete effectively with the modern techniques of biochemistry and molecular biology. It is now possible to take a protein and use recombinant DNA technology to produce the DNA (deoxyribonucleic acid) which codes for that protein. The same technology permits inclusion of the DNA into bacteria, and the bacterial culture will consequently produce the protein. Although the techniques are complex, they offer the hope of synthesizing large quantities of virtually any protein (assuming that a small amount of the protein is available at the start).

EXERCISE 20.14

Calculate the overall yield that would be obtained in a synthesis of the 30-subunit chain of insulin:

(a) For a solution-phase synthesis in which the individual coupling reactions proceed with an average yield of 75%.
(b) For a solid-phase synthesis in which the individual coupling reactions proceed with an average yield of 94%.

EXERCISE 20.15

Draw the structure of the dinitrophenyl derivative that would be produced by the reaction of 1-fluoro-2,4-dinitrobenzene with the tripeptide alanylglycylphenylalanine.

EXERCISE 20.16

Draw the amino acids (or amino acid derivatives) that would be produced upon hydrolysis of the reaction product in Problem 20.15.

20.7 SUMMARY OF BLOCKING GROUP REACTIONS

The following table summarizes the various blocking groups used in peptide synthesis that we have covered in this chapter. For each case we have listed the standard methods for both introduction and removal of the blocking group.

TABLE 20.2 Blocking Groups for Peptide Synthesis

Blocking Group (Abbreviation)	Reagents for Introduction	Reagents for Removal, Comments
Protection of $>N{-}H$		
p-Toluenesulfonyl (Ts) $>N{-}SO_2{-}C_6H_4{-}CH_3$	TsCl and base (NaOH or 3° amine)	Na, NH_3—strongly reducing conditions; will also cleave carbon-sulfur bonds, sulfur-sulfur bonds and benzyl groups; will reduce esters
Benzyloxycarbonyl (Cbz) $>N{-}C(=O){-}O{-}CH_2{-}C_6H_5$	Benzyl chloroformate ($C_6H_5CH_2{-}O{-}C(=O){-}Cl$) and base (NaOH or $NaHCO_3$)	H_2, Pd (not applicable for sulfur-containing peptides); HBr (in nonaqueous solvent); Na, NH_3
tert-Butoxycarbonyl (tBOC)co $>N{-}C(=O){-}O{-}C(CH_3)_3$	*tert*-Butyl azidoformate ($t\text{-}C_4H_9O{-}C(=O){-}N_3$) and a 3° amine, e.g., $N(C_2H_5)_3$	HBr (in nonaqueous solvent)
Trifluoroacetyl (TFA) $>N{-}C(=O){-}CF_3$	Trifluoroacetic anhydride $CF_3{-}C(=O){-}O{-}C(=O){-}CF_3$	aqueous NH_3; aqueous NaOH (dilute)

TABLE 20.2 Blocking Groups for Peptide Synthesis

Blocking Group (Abbreviation)	Reagents for Introduction	Reagents for Removal, Comments
Protection of $-C(=O)-O-H$		
Methyl or ethyl esters $-C(=O)-O-CH_3$, $-C(=O)-O-C_2H_5$	CH_3OH, C_2H_5OH, acid (cf. Section 15.7)	Hydrolysis in acid or base (some hydrolysis of peptide bonds may occur also)
Benzyl esters $-C(=O)-O-CH_2-C_6H_5$	$C_6H_5CH_2OH$, acid (cf., Section 15.7) or $C_6H_5CH_2Cl$, base (can also be used to form benzyl ethers for protection of OH groups)	H_2, Pd (not applicable to sulfur-containing peptides; will remove *N*-Cbz groups); Na, NH_3 (see comments under *N*-Ts); HBr in nonaqueous solvent; hydrolysis (acid or base catalysis—see comments under methyl or ethyl esters)
Protection of —S—H **Groups**		
Benzyl thioethers $-S-CH_2-C_6H_5$	$C_6H_5CH_2Cl$, NaOH	Na, NH_3 (see comments under *N*-Ts); HBr in nonaqueous solvent (will also cleave *O*-benzyl groups)
Disulfide bond formation $-S-S-$	Mild oxidation (e.g., O_2)	Na, NH_3; milder reducing agents such as $NaBH_4$ or $HOCH_2CH_2SH$ (mercaptoethanol) also cleave the disulfide bond.

20.8 PROBLEMS

20.1 (a) Draw the three-dimensional structure of the tripeptides L-alanyl-L-seryl-L-leucine (NH_2—Ala—Ser—Leu—CO_2H) and L-alanyl-D-seryl-L-leucine. (b) What is the stereochemical relationship between the two tripeptides?

20.2 Emil Fischer demonstrated that the naturally occurring amino acids serine and alanine have the same configuration. He did this by a series of chemical reactions that converted L-serine to L-alanine. Suggest a reaction sequence which would do this.

20.3 Suggest a method by which serine could be converted to cysteine. Would D-serine yield D- or L-cysteine?

20.4 Comparisons of molecular structures have helped chemists decipher ways in which different amino acids might be biochemically related. Show the structural similarities between the amino acids having side chains with:
(a) A single carbon
(b) two carbons
(c) three carbons

Suggest general ways in which the members of each group might be biochemically derived from each other or from a common precursor. (Do not be concerned with any specific biochemical reactions or with the chemical specificity that would be required with polyfunctional intermediates.)

20.5 For each of the acidic amino acids draw the actual species that would predominate under acidic (e.g., pH = 2) and basic (e.g., pH = 12) conditions. What other species would predominate at the appropriate intermediate pH values?

20.6 For each of the basic amino acids draw the actual species that would predominate under acidic (e.g., pH = 2) and basic (e.g., pH = 12) conditions. What other species would predominate at the appropriate intermediate pH values?

20.7 In Section 20.2 we illustrated a synthesis of serine via oxymercuration of methyl acrylate. (a) Show how threonine could be prepared by an analogous set of reactions. (b) Suggest reasons why the methyl ether was generated as an intermediate instead of the hydroxyl group being introduced directly. (c) Ordinarily, oxymercuration introduces an oxygen on the *more* substituted carbon of a double bond. Explain the apparent discrepancy in the reaction being considered here.

20.8 Outline a synthesis of valine by each of the three general methods depicted in Figure 20.1.

20.9 Outline a synthesis of isoleucine by each of the three general methods depicted in Figure 20.1.

20.10 Write a detailed mechanism for the complete reaction sequence for preparing the methyl ester of alanylglycine by each of the following methods. If an amino-blocking group is needed, use the Cbz group.

(a) The acid chloride method (using PCl_5)
(b) The mixed anhydride method (using ethyl chloroformate)
(c) The mixed anhydride method (using tetraethyl pyrophosphite)
(d) The active ester method (using *p*-nitrophenol)
(e) The coupling reagent method (using DCC)
(f) The coupling reagent method (using Woodward's reagent K)

20.11 In Section 20.3 we discussed the side reactions that could occur if the acid chloride of one amino acid (e.g., alanine) were allowed to react with a second amino acid such as leucine.

(a) Draw the dipeptide that would be formed from the desired reaction.
(b) Draw the product that would be produced if the acid chloride acylated the amino group of a second acid chloride.
(c) Draw the derivative that would be generated by cyclization of the product in (b).

20.12 A reaction between serine and the acid chloride of leucine could yield a variety of products and by-products. Draw those products containing either two or three amino acid residues that would be expected from this reaction.

20.13 A reaction between lysine and the acid chloride of alanine could yield a variety of products and by-products. Draw those products containing either two or three amino acid residues that would be expected from this reaction.

20.14 Write reactions for protection of the amino group of phenylalanine as: (a) the tosyl derivative; (b) the benzyloxycarbonyl derivative; and (c) the *tert*-butoxycarbonyl derivative.

20.15 Write reactions for regeneration of the free amino acid from each of the derivatives in Problem 20.14. If more than one set of reaction conditions would work, show all possibilities.

20.16 Write reactions for protection of the amino group of methionine as: (a) the tosyl derivative; (b) the benzyloxycarbonyl derivative; and (c) the *tert*-butoxycarbonyl derivative.

20.17 Write reactions for regeneration of the free amino acid from each of the derivatives in Problem 20.16. If more than one set of reaction conditions would work, show all possibilities.

20.18 The removal of an *N*-benzyloxycarbonyl group of alanine could be accomplished in a variety of ways. For each method suggest the fate of the Cbz group.

20.19 Explain why the *N*-trifluoroacetyl group is cleaved readily by aqueous base while other amide bonds in a peptide are stable to these conditions.

20.20 Explain why benzyl esters are more susceptible to acid hydrolysis than other esters.

20.21 Write reactions for protecting the carboxyl group of isoleucine as: (a) the benzyl ester, both by acid-catalyzed esterification and by alkylation of the carboxylate ion; and (b) the *p*-nitrophenyl ester.

20.22 Write reactions for regeneration of the free amino acid from each of the derivatives in Problem 20.21. If more than one set of conditions would work, show all possibilities.

20.23 Write reactions for protecting the thiol group of cysteine as: (a) the disulfide dimer and (b) the benzyl thioether.

20.24 Write reactions for regeneration of the free amino acid from each of the derivatives in Problem 20.23. If more than one set of conditions would work, show all possibilities.

20.25 Three examples of peptide synthesis via the mixed anhydride method are shown in Section 20.4. For each of these reactions draw the mixed anhydride that is formed and explain why subsequent nucleophilic attack occurs preferentially at the carbonyl carbon of the amino acid.

20.26 Outline a synthesis of leucylalanine (using all necessary blocking groups) by:
(a) The acid chloride method
(b) The mixed anhydride method
(c) The active ester method
(d) Coupling with dicyclohexylcarbodiimide
(e) Coupling with Woodward's reagent K
(f) The solid phase method

20.27 Answer Problem 20.26 for the synthesis of glycylcysteine.

20.28 Answer Problem 20.26 for the synthesis of tyrosylvaline.

20.29 Consider a synthesis of glycylvaline using Woodward's reagent K. Using appropriate structural drawings, show the fate of (a) reagent K itself and (b) each of the oxygen atoms of the reactants.

20.30 The protected dipeptide 4-benzyloxycarbonyl-*S*-benzylcysteinyltyrosine was used to introduce the last two amino acid residues in du Vigneaud's synthesis of oxytocin. Suggest a synthesis of this compound starting from the individual unprotected amino acids.

20.31 The formation of the tetrapeptide

NH_2
Cys—Pro—Leu—Gly—$CONH_2$
$C_6H_5CH_2S$

in the synthesis of oxytocin required cleavage of the dimeric species via reduction of the sulfur-sulfur bond. (a) Explain why the ethyl ester of the glycine residue was first hydrolyzed. (b) Does this reaction sequence (which involves amide formation via the benzyl ester) fit with our previous discussion (Section 16.3) of the relative reactivities of amides and esters? (c) Why could the glycine amide not have been prepared via the acid chloride rather than via the benzyl ester?

20.32 Based on the reactions used in the synthesis of oxytocin, which is more nucleophilic, an amino group or a sulfide anion? A sulfide anion or a carboxylate anion?

20.33 In du Vigneaud's synthesis of oxytocin, calculate the theoretical yield of oxytocin (MW = 1007) starting from 1.0 g of *N*-tosylglutamic acid. Using the % yields reported in Section 20.5, calculate the actual weight of oxytocin that would be obtained.

20.34 Consider a solid-phase synthesis of Ala—Leu—Ile—Phe—Ser—Gly. Suppose that in the step in which isoleucine is introduced into the growing peptide chain, only half the theoretical quantity of DCC were used by mistake. What major by-product (a peptide) would be obtained from this synthesis? What yields of this material might you expect for this impurity and for the desired hexapeptide (assuming an average yield of 90% for the individual coupling steps)?

20.35 A dipeptide was found to be hydrolyzed to phenylalanine and leucine, but it was not known which was the carboxyl terminus. The dipeptide reacted readily with 1-fluoro-2,4-dinitrobenzene, and subsequent hydrolysis yielded phenylalanine, together with the dinitrophenyl derivative of leucine. Write the complete reactions using the correct structure of the dipeptide.

20.36 A peptide with an experimental molecular weight of about 600 afforded only phenylalanine, leucine, and tyrosine upon complete hydrolysis. Partial hydrolysis also afforded several dipeptides: phenylalanyltyrosine, phenylalanylleucine, and leucylphenylalanine. When the original peptide was subjected to treatment with 1-fluoro-2,4-dinitrobenzene followed by hydrolysis, the dinitrophenyl derivative of phenylalanine could be isolated. Deduce the complete structure of the original peptide.

CHAPTER 21

CONDENSATION REACTIONS OF CARBOXYLIC ACID DERIVATIVES: β-DICARBONYL COMPOUNDS

Carbonyl groups greatly increase the acidity of hydrogens on an adjacent atom, a fact we have discussed in several previous chapters (Sections 4.7, 13.9, 16.3). The influence of *two* adjacent carbonyl groups has an even greater effect. The enolate ions formed from compounds such as β-diketones are much more stable than those from monoketones, and they can be prepared much more easily.

$$R{-}\overset{O}{\overset{\|}{C}}{-}CH_2{-}\overset{O}{\overset{\|}{C}}{-}R \xrightarrow{\text{base}} R{-}\overset{O}{\overset{\|}{C}}{-}\overset{-}{C}H{-}\overset{O}{\overset{\|}{C}}{-}R$$

Despite their increased stability, such enolate ions are still reactive nucleophiles. Consequently, they can be used reliably in many of the situations that result in serious problems with the more reactive enolates from monoketones (Section 14.2, 14.3).

In this chapter we will focus on the preparation and reactivity of β-dicarbonyl compounds, and we will show you how these are very useful for synthesizing organic compounds. We will also illustrate the role of such dicarbonyl compounds in the biosynthesis and metabolism of fatty acids.

21.1 ACIDITY OF β-DICARBONYL COMPOUNDS

Alkanes are extremely weak acids, but the presence of a carbonyl group brings about a dramatic increase in acidity; hydrogens α to the carbonyl group can be removed by moderately strong bases. When a CH or CH_2 group is bonded to *two* carbonyl groups, another large increase in acidity is observed, and such compounds are *stronger acids than water.*

$$CH_3{-}CH_3 \qquad CH_3{-}CH_2{-}\overset{O}{\overset{\|}{C}}CH_3 \qquad CH_3{-}\overset{O}{\overset{\|}{C}}{-}CH_2{-}\overset{O}{\overset{\|}{C}}{-}CH_3$$

$$K_a < 10^{-44} \qquad K_a \sim 10^{-20} \qquad K_a = 10^{-9}$$

This remarkable increase in acidity means that common bases such as sodium hydroxide and sodium alkoxides will bring about nearly complete conversion of a β-diketone to the corresponding enolate ion. Such enolate ions are stable species in aqueous or alcoholic solution, and they can often be isolated as crystalline salts.

$$CH_3{-}\overset{O}{\overset{\|}{C}}{-}CH_2{-}\overset{O}{\overset{\|}{C}}{-}CH_3 \overset{^{-}OH}{\rightleftharpoons} CH_3{-}\overset{O}{\overset{\|}{C}}{-}\overset{-}{C}H{-}\overset{O}{\overset{\|}{C}}{-}CH_3$$

The same bases produce only a small fraction of enolate ion with monocarbonyl ketones.

$$CH_3{-}\overset{O}{\overset{\|}{C}}{-}CH_3 \overset{^{-}OH}{\rightleftharpoons} CH_3{-}\overset{O}{\overset{\|}{C}}{-}CH_2^{-}$$

Why are the β-dicarbonyl compounds so much more acidic than simple ketones? We could explain it in terms of molecular orbital theory, but an even simpler answer is provided by drawing the resonance forms for the enolate ion. The negative charge is stabilized by interaction with *both* carbonyl groups.

$$CH_3C(=\ddot{O}:)\overset{..}{C}^{-}HC(=\ddot{O}:)CH_3 \longleftrightarrow CH_3C(-\ddot{O}:^{-})=CHC(=\ddot{O}:)CH_3 \longleftrightarrow CH_3C(=\ddot{O}:)CH=C(-\ddot{O}:^{-})CH_3$$

The high acidity of β-dicarbonyl compounds is not limited to diketones; ester carbonyl groups and cyano groups exert comparable stabilizing effects on enolate ions. Some representative β-dicarbonyl compounds are listed in Table 21.1, together with their K_a values. We have also included some appropriate monofunctional compounds for comparison.

Despite the substantial range in K_a values for the difunctional compounds listed in Table 21.1, all are considerably more acidic than water. In each case the sodium or potassium salt can be formed very easily. As you will see in subsequent parts of this chapter, these enolates are very useful as synthetic reagents.

TABLE 21.1 Acid Dissociation Constants for β-Dicarbonyl Compounds and Their Monofunctional Counterparts*

Compound	K_a	Name (Trivial Name)
$C_2H_5OC(=O)-CH_3$	10^{-25}	Ethyl acetate
$N\equiv C-CH_3$	10^{-25}	Acetonitrile
$CH_3-C(=O)-CH_3$	10^{-20}	Acetone
H_2O, ROH	10^{-16}–10^{-18}	Water, alcohols
$C_2H_5OC(=O)-CH_2-C(=O)OC_2H_5$	10^{-13}	Diethyl malonate
$CH_3-C(=O)-CH_2-C(=O)OC_2H_5$	10^{-11}	Ethyl 3-oxobutyrate (ethyl acetoacetate)
$N\equiv C-CH_2-C\equiv N$	10^{-11}	Malononitrile
$N\equiv C-CH_2-C(=O)OC_2H_5$	10^{-9}	Ethyl cyanoacetate (cyanoacetic ester)
$CH_3-C(=O)-CH_2-C(=O)-CH_3$	10^{-9}	2,4-Pentanedione (acetylacetone)

*Nitriles are considered "carbonyl" compounds for this table because of their similar reactivity.

EXERCISE 21.1

Draw all resonance forms for the enolate ion that would be formed from each of the compounds shown in Table 21.1.

EXERCISE 21.2

Indicate the most acidic hydrogen in each of the following compounds. State whether each compound would be more or less acidic than water.

(a) cyclohexane-1,3-dione (b) 3-oxocyclohexyl with CO_2CH_3 (c) $CH_2(CO_2H)_2$

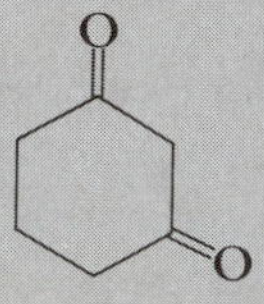

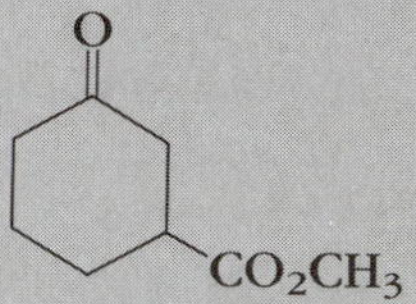

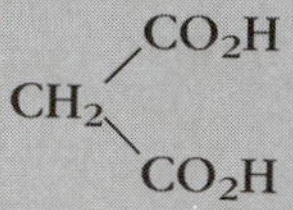

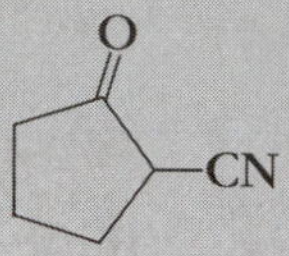

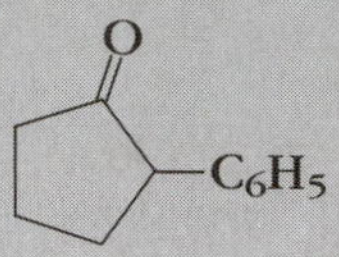

(d) 2-oxocyclopentyl with CN (e) 2-oxocyclopentyl with C_6H_5 (f) benzene ring with OH and CO_2CH_3

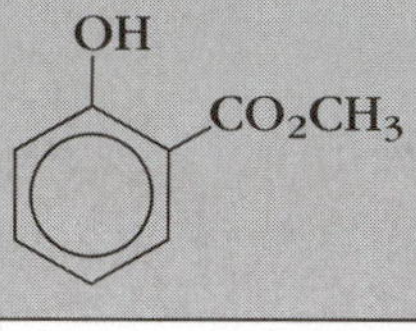

21.2 PREPARATION OF β-DICARBONYL COMPOUNDS: ACYLATION OF ENOLATE IONS

In earlier chapters we have shown various examples of nucleophilic attack on the carbonyl group of carboxylic acid derivatives. Amides, for example, are formed in the reactions of esters with amines (Section 16.7).

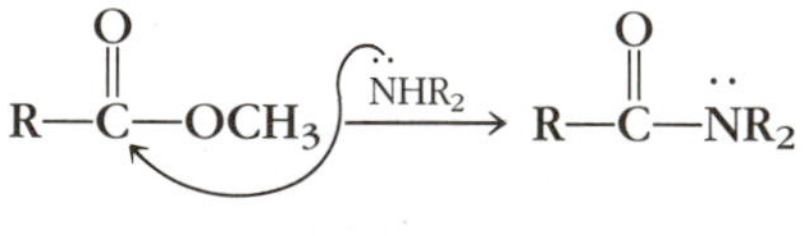

$$R{-}\overset{O}{\overset{\|}{C}}{-}OCH_3 \xrightarrow{\ddot{N}HR_2} R{-}\overset{O}{\overset{\|}{C}}{-}\ddot{N}R_2$$

Such acylation reactions are not limited to amines and the other nucleophiles involved in the interconversion of carboxylic acid derivatives. A particularly useful type of acylation reaction involves the attack of an enolate ion on a carboxylic acid derivative to yield a β-dicarbonyl compound.

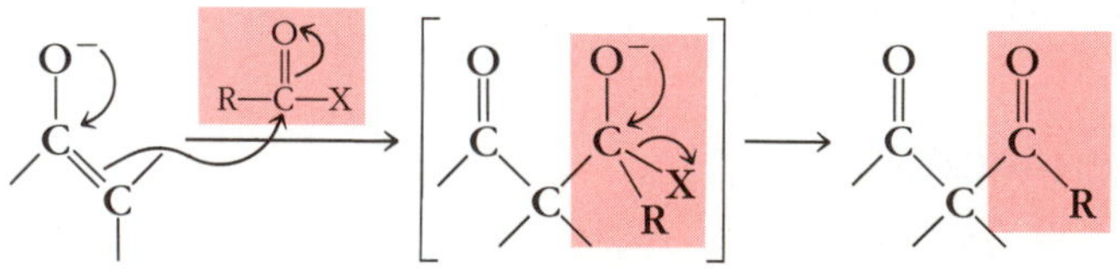

$$\text{enolate} + R{-}\overset{O}{\overset{\|}{C}}{-}X \longrightarrow \left[\text{tetrahedral intermediate } (O^-, X, R)\right] \longrightarrow \text{β-dicarbonyl compound } ({-}C(=O){-}C{-}C(=O)R)$$

β-Keto Esters

The acylation of an enolate ion is illustrated by the base-catalyzed condensation of ethyl acetate to produce ethyl acetoacetate. This specific reaction is sometimes called the acetoacetic ester condensation, and related ester condensation reactions are known as Claisen condensations (after the German chemist Ludwig Claisen, who studied them extensively in the late 1800s).

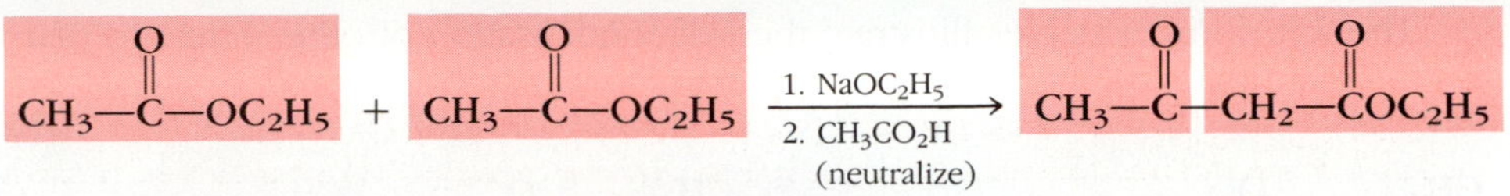

$$CH_3-\overset{O}{\overset{\|}{C}}-OC_2H_5 + CH_3-\overset{O}{\overset{\|}{C}}-OC_2H_5 \xrightarrow[\text{2. } CH_3CO_2H \text{ (neutralize)}]{\text{1. } NaOC_2H_5} CH_3-\overset{O}{\overset{\|}{C}}-CH_2-\overset{O}{\overset{\|}{C}}OC_2H_5 \quad 75\text{–}76\%$$

The mechanism of this reaction is illustrated in Figure 21.1. In the first step a small amount of enolate ion is generated by reaction of the starting ester with base. A second molecule of ester then undergoes nucleophilic attack by this enolate, and expulsion of ethoxide ion produces the β-keto ester. Every step in the reaction is reversible, and the overall equilibrium is favored by the last step, in which the β-keto ester is almost totally converted to the enolate. Only upon neutralization of the reaction mixture is this enolate fully converted to the neutral keto ester.

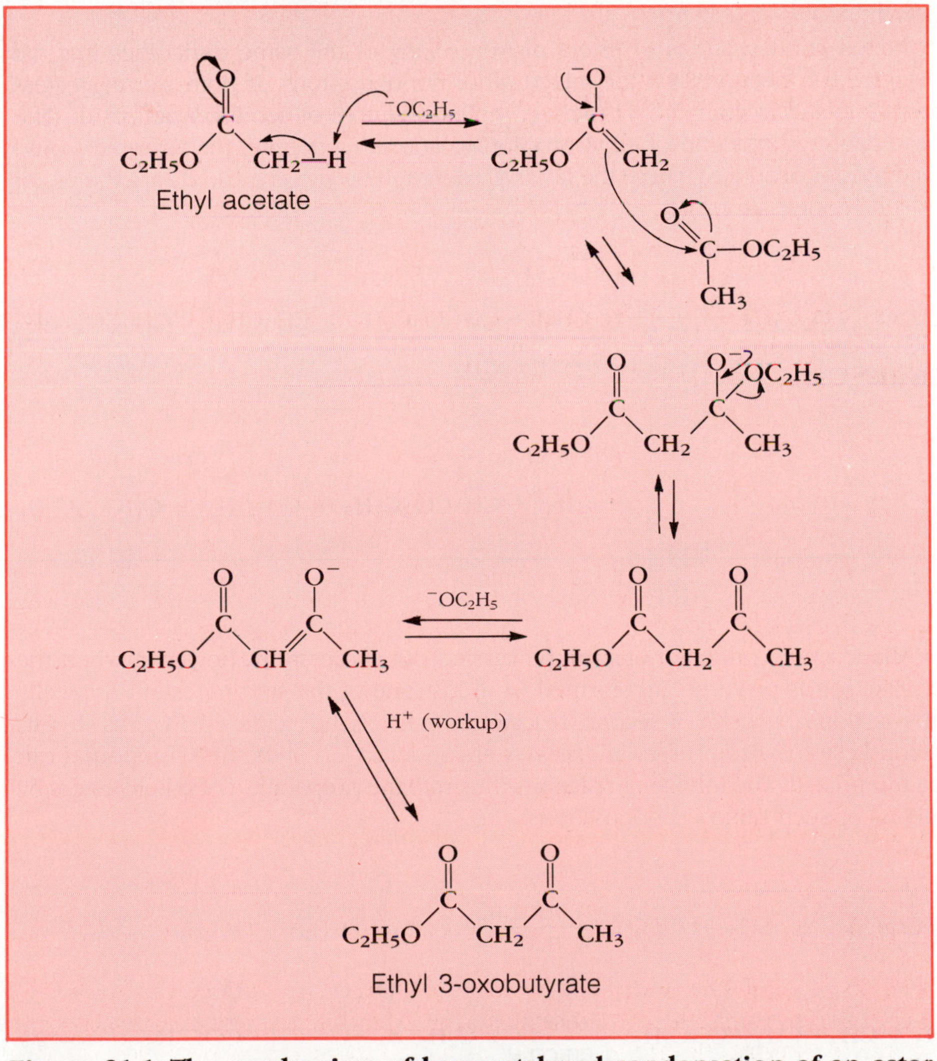

Ethyl acetate

$^-OC_2H_5$

$^-OC_2H_5$

H$^+$ (workup)

Ethyl 3-oxobutyrate

Figure 21.1 The mechanism of base-catalyzed condensation of an ester to produce a β-dicarbonyl compound.

The following examples illustrate the self-condensation of other esters to produce β-ketoesters:

$$\underset{\substack{|\\CH_3}}{}CH_3-\overset{CH_3}{\overset{|}{C}H}-CH_2-\overset{O}{\overset{\|}{C}}OC_2H_5 \xrightarrow[\text{2. } CH_3CO_2H]{\text{1. NaH}} CH_3-\overset{CH_3}{\overset{|}{C}H}-\underset{C_2H_5O_2C}{\underset{|}{C}H}-\overset{O}{\overset{\|}{C}}-CH_2-\overset{CH_3}{\overset{|}{C}H}-CH_3 \quad 52\%$$

$$CH_3CH_2CH_2CH_2\overset{O}{\overset{\|}{C}}OC_2H_5 \xrightarrow[\text{2. } CH_3CO_2H]{\text{1. NaOEt}} CH_3CH_2CH_2CH_2\overset{O}{\overset{\|}{C}}-\underset{CH_2CH_2CH_3}{\underset{|}{C}H}-CO_2C_2H_5 \quad 77\%$$

Notice that when an alkoxide base, ^{-}OR, is employed, it is usually the same alkoxy group as that in the starting ester. This circumvents problems with base-catalyzed transesterification. Transesterification can also be avoided by using potassium *tert*-butoxide, which is sterically hindered and therefore does not attack the ester carbonyl.

Mixed condensations of esters present many of the same difficulties that we discussed in Section 14.2 for mixed aldol condensations. If both starting esters form enolates at comparable rates, complex mixtures of products can result. The base-catalyzed reaction of ethyl propionate and ethyl acetate, for example, could lead to four products, where the last two (denoted below as **c** and **d**) are the result of self-condensation.

$$\begin{matrix} CH_3CH_2CO_2C_2H_5 \\ + \\ CH_3CO_2C_2H \end{matrix} \xrightarrow[\text{2. HCl}]{\text{1. } NaOC_2H_5} \underset{\mathbf{a}}{CH_3\overset{O}{\overset{\|}{C}}-\underset{CH_3}{\underset{|}{C}H}CO_2C_2H_5} + \underset{\mathbf{b}}{CH_3CH_2\overset{O}{\overset{\|}{C}}-CH_2CO_2C_2H_5}$$

$$+ \underset{\mathbf{c}}{CH_3\overset{O}{\overset{\|}{C}}-CH_2CO_2C_2H_5} + \underset{\mathbf{d}}{CH_3CH_2\overset{O}{\overset{\|}{C}}-\underset{CH_3}{\underset{|}{C}H}CO_2C_2H_5}$$

Mixed ester condensations can be carried out successfully, however, when the enolate ion is preferentially formed from just one of the starting esters. Typically this is done by using a second reactant which has no acidic hydrogens. If this second ester is also present in excess, self-condensation of the first compound can be minimized. The following reactions illustrate the preparation of β-keto esters by means of such mixed condensations.

$$C_6H_5\overset{O}{\overset{\|}{C}}OC_2H_5 + CH_3-\overset{O}{\overset{\|}{C}}OC_2H_5 \xrightarrow[\text{2. HCl}]{\text{1. } NaOC_2H_5} C_6H_5\overset{O}{\overset{\|}{C}}-CH_2-\overset{O}{\overset{\|}{C}}OC_2H_5 \quad 55\text{–}77\%$$

$$C_2H_5O\overset{O}{\overset{\|}{C}}-\overset{O}{\overset{\|}{C}}OC_2H_5 + CH_3-CH_2-\overset{O}{\overset{\|}{C}}OC_2H_5 \xrightarrow[\text{2. } CH_3CO_2H]{\text{1. } NaOC_2H_5} C_2H_5O\overset{O}{\overset{\|}{C}}-\overset{O}{\overset{\|}{C}}-\underset{CH_3}{\underset{|}{C}H}-\overset{O}{\overset{\|}{C}}OC_2H_5 \quad 60\text{–}70\%$$

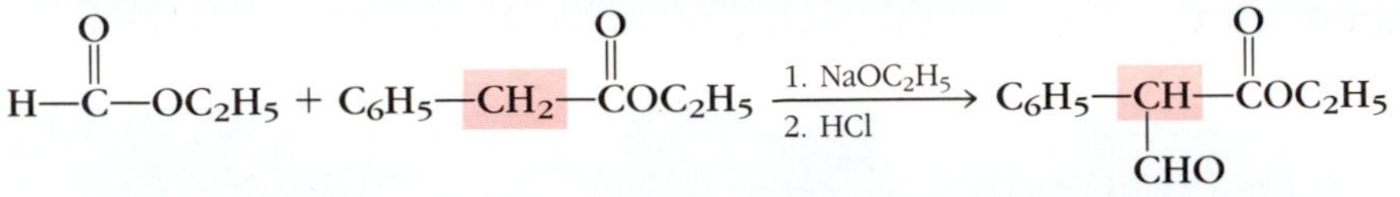

$$H\text{—}\overset{O}{\overset{\|}{C}}\text{—}OC_2H_5 + C_6H_5\text{—}CH_2\text{—}\overset{O}{\overset{\|}{C}}OC_2H_5 \xrightarrow[2.\ HCl]{1.\ NaOC_2H_5} C_6H_5\text{—}\underset{CHO}{\underset{|}{CH}}\text{—}\overset{O}{\overset{\|}{C}}OC_2H_5 \qquad 70\text{–}75\%$$

Acylation of nitriles can be accomplished under the same conditions used for ester condensations.

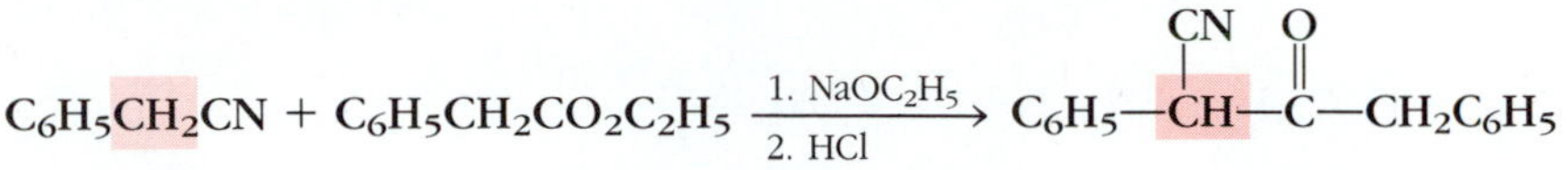

$$C_6H_5CH_2CN + C_6H_5CH_2CO_2C_2H_5 \xrightarrow[2.\ HCl]{1.\ NaOC_2H_5} C_6H_5\text{—}\overset{CN}{\overset{|}{CH}}\text{—}\overset{O}{\overset{\|}{C}}\text{—}CH_2C_6H_5 \qquad 85\%$$

Ketones also can be acylated by similar mixed condensations. The following reactions proceed by essentially the same mechanism as that shown in Figure 21.1 for the reaction of two molecules of ester. Strong bases such as sodium hydride are often employed to effect complete conversion of the ketone to its enolate, which reduces problems with competing aldol condensations (see Section 14.3).

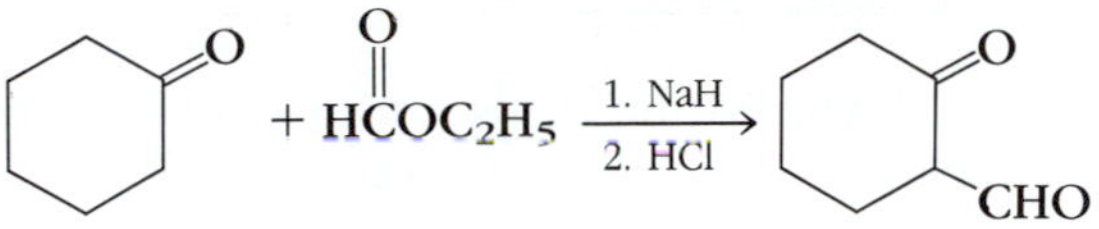

$$\text{cyclohexanone} + H\overset{O}{\overset{\|}{C}}OC_2H_5 \xrightarrow[2.\ HCl]{1.\ NaH} \text{2-formylcyclohexanone (CHO)} \qquad 70\text{–}74\%$$

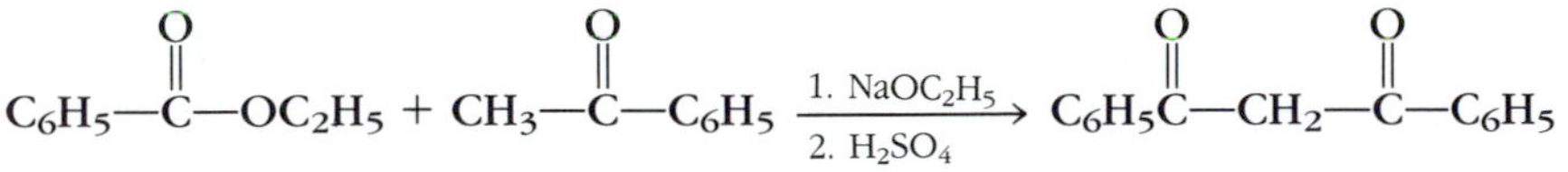

$$C_6H_5\text{—}\overset{O}{\overset{\|}{C}}\text{—}OC_2H_5 + CH_3\text{—}\overset{O}{\overset{\|}{C}}\text{—}C_6H_5 \xrightarrow[2.\ H_2SO_4]{1.\ NaOC_2H_5} C_6H_5\overset{O}{\overset{\|}{C}}\text{—}CH_2\text{—}\overset{O}{\overset{\|}{C}}\text{—}C_6H_5 \qquad 62\text{–}71\%$$

$$\text{cyclooctanone} + C_2H_5O\text{—}\overset{O}{\overset{\|}{C}}OC_2H_5 \xrightarrow[2.\ CH_3CO_2H]{1.\ NaH} \text{2-oxocyclooctane with } CO_2C_2H_5 \qquad 91\text{–}94\%$$

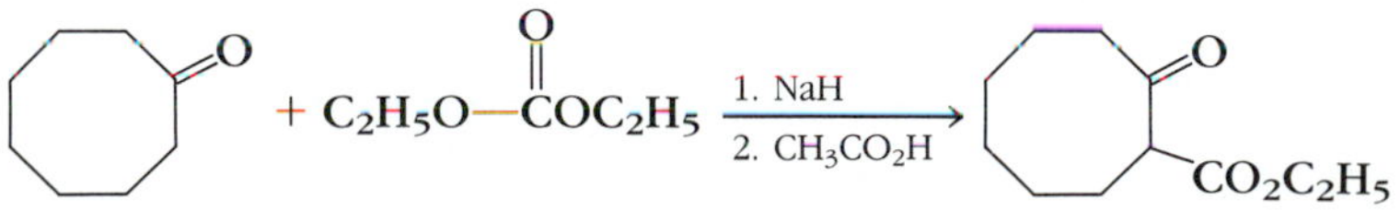

Notice that the acylation of cyclooctanone in the last of these reactions affords a β-keto ester rather than a β-diketone. In contrast to acylation with esters of other carboxylic acids, acylation with diethyl carbonate introduces an ester group (—$CO_2C_2H_5$) rather than an acyl group. A one-carbon unit can also be introduced with ethyl formate, as shown in the first example, where introduction of a formyl group (—CHO) produces a β-keto aldehyde.

EXERCISE 21.3

Self-condensations of ethyl esters are often carried out with sodium ethoxide in ethanol. What product would be expected if ethyl acetate were treated instead with sodium methoxide in methanol?

EXERCISE 21.4

Draw the complete reaction mechanism for the reaction of cyclooctanone with diethyl carbonate, using NaH as base.

EXERCISE 21.5

What products would be expected for each of the following reactions?

(a) $CH_3CH_2CH_2CO_2C_2H_5 \xrightarrow[2.\ HCl]{1.\ NaOC_2H_5}$

(b) $C_6H_5CO_2C_2H_5 + CH_3CO_2C_2H_5 \xrightarrow[2.\ HCl]{1.\ NaOC_2H_5}$

(c) $CH_3—CO_2C_2H_5 + CH_3CH(CH_3)—CO_2C_2H_5 \xrightarrow[2.\ HCl]{1.\ NaOC_2H_5}$

(d) $CH_3CH_2C(=O)C_6H_5 + C_2H_5O_2CCO_2C_2H_5 \xrightarrow[2.\ HCl]{1.\ NaH}$

21.3 ALDOL-TYPE CONDENSATIONS OF ESTERS

In the preceding section we discussed the *acylation* of enolate ions. Acylation results from nucleophilic attack on the carbonyl group of an ester (or another carboxylic acid derivative such as an acyl halide). What happens when nucleophilic attack instead takes place at the carbonyl group of an aldehyde or ketone? If the enolate is also derived from an aldehyde or ketone, the reaction would be classified as an aldol condensation.

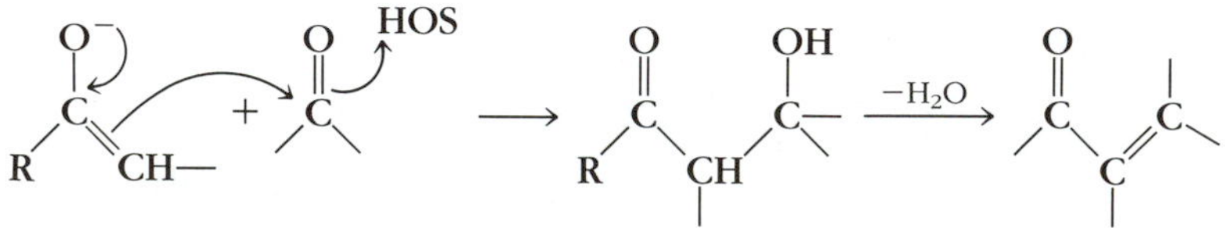

$$R—C(—O^-)=CH— + O=C< \xrightarrow{HOS} R—C(=O)—CH(—)—C(—OH)< \xrightarrow{-H_2O} —C(=O)—C(—)=C<$$

But the same kind of reaction could occur if the enolate were derived from an ester (i.e., if R in the preceding equation were replaced by alkoxy, OR).

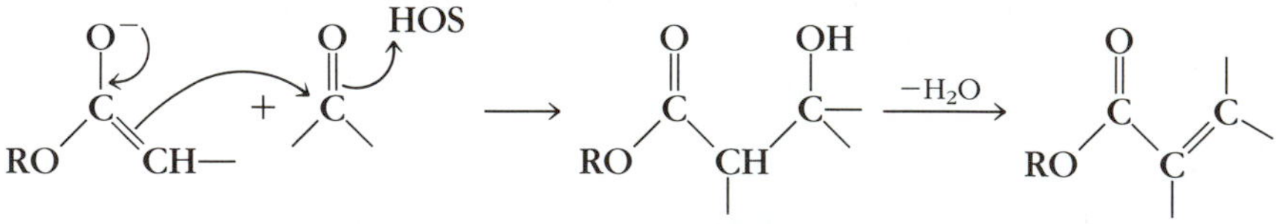

$$RO—C(—O^-)=CH— + O=C< \xrightarrow{HOS} RO—C(=O)—CH(—)—C(—OH)< \xrightarrow{-H_2O} RO—C(=O)—C(—)=C<$$

There are many potential difficulties with such mixed condensations, however, and we have considered these previously for mixed aldol condensations (Section 14.2) and mixed ester condensations (Section 21.2). In particular, there is the difficulty of preferentially forming the enolate of an ester in the presence of a ketone. Esters ($K_a \cong 10^{-25}$) are considerably less acidic than aldehydes or ketones ($K_a \cong 10^{-21}$), so these mixed condensations are often most successful with aldehydes and ketones that have no α hydrogens. In addition, each of the methods that we will illustrate here has some special feature that favors formation of the desired product.

One method for selectively generating the enolate of an ester was developed by Sergius Reformatsky (Russia) in the late 1800s. In a reaction that is related to both the Grignard reaction and the aldol condensation, an α-halo ester (Section 16.8) can be converted to its zinc enolate by the action of zinc metal.

$$\text{Zn:}\ \text{Br}{-}\text{C}{-}\text{C}(=\text{O})\text{OC}_2\text{H}_5 \longrightarrow \text{C}{=}\text{C}(\text{O}{-}\text{ZnBr})\text{OC}_2\text{H}_5$$

Treatment of a mixture of an α-halo ester and a ketone or aldehyde with zinc affords the corresponding β-hydroxy ester via an aldol-type condensation of the ester enolate with the aldehyde or ketone.

$$C_6H_5CHO + Br{-}CH_2{-}COOC_2H_5 \xrightarrow[\text{2. } H_2SO_4 \text{ (neutralize)}]{\text{1. Zn}} C_6H_5{-}CH(OH){-}CH_2{-}COOC_2H_5 \quad 61\text{–}64\%$$

$$C_6H_5{-}CO{-}CH_3 + CH_3{-}CH(Br){-}COOC_2H_5 \xrightarrow[\text{2. } H_2SO_4 \text{ (neutralize)}]{\text{1. Zn}} C_6H_5{-}C(OH)(CH_3){-}CH(CH_3){-}COOC_2H_5 \quad 75\text{–}81\%$$

The β-hydroxy esters are easily dehydrated to the unsaturated esters (cf. Section 14.1), so neutralization during workup is done with dilute H_2SO_4. An alternative method for preparing the unsaturated esters employs a modification of the Wittig reaction that we will discuss in Section 21.4.

Another aldol-type condensation of esters was discovered by Hans Stobbe in Poland in 1893. Condensation of *succinate* esters with aldehydes and ketones leads directly to an α,β-unsaturated product via an intermediate lactone. The formation of a five-membered ring and the subsequent elimination provide the driving force for this reaction. The alkoxide base corresponding to the alkoxy group of the succinate ester can be used, but potassium *tert*-butoxide and sodium hydride are often preferable.

$$C_2H_5O_2C{-}CH_2{-}CH_2{-}CO_2C_2H_5 + O{=}C(C_6H_5)_2 \xrightarrow[\text{2. HCl}]{\text{1. KOtBu}} HO_2C{-}CH_2{-}C(CO_2C_2H_5){=}C(C_6H_5)_2 \quad 90\text{–}94\%$$

Notice that one of the carboxylic ester groups has been converted to the acid in this reaction. Consideration of the mechanism shows that it is the "remote" carboxyl group that is hydrolyzed. Initial enolate formation and nucleophilic attack at the ketone proceed in the normal fashion.

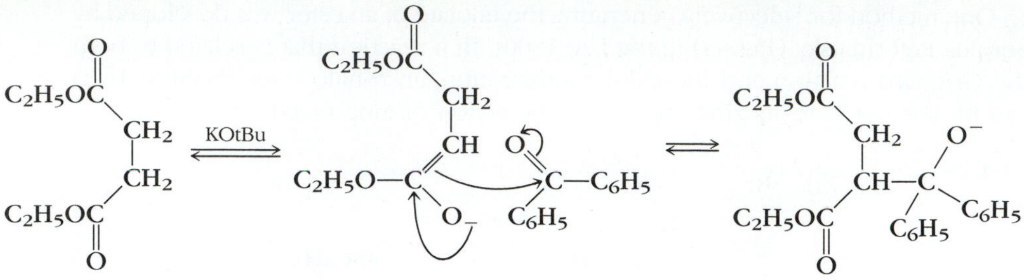

But the resulting oxygen anion can attack the remote carboxyl group in intramolecular fashion.

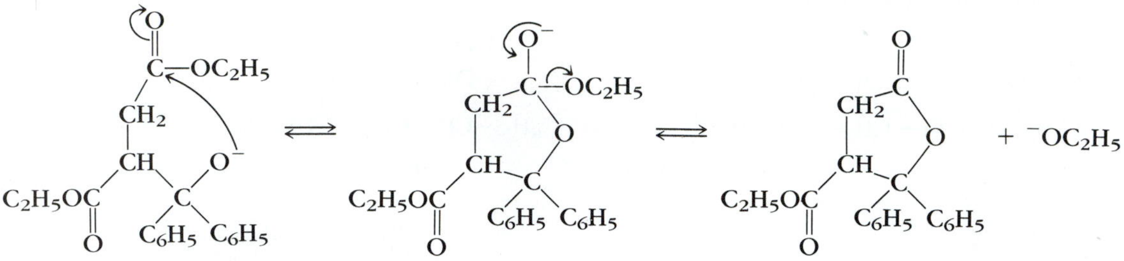

$$+\ ^{-}OC_2H_5$$

In the last intermediate, the carbon β to the ethoxycarbonyl has a good leaving group, and elimination occurs readily. Subsequent acidification affords the observed product.

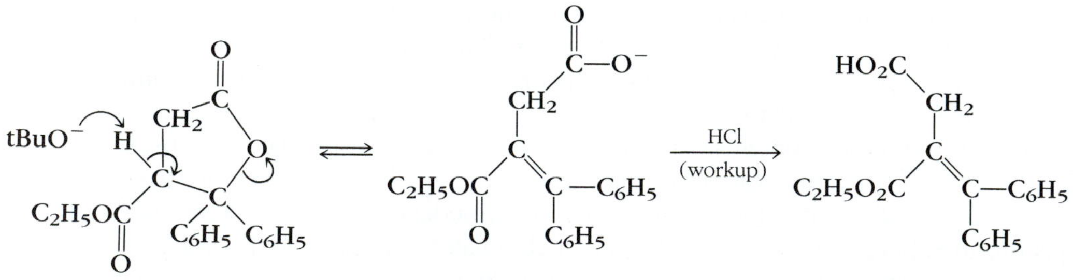

The following equation illustrates an additional example of the Stobbe condensation:

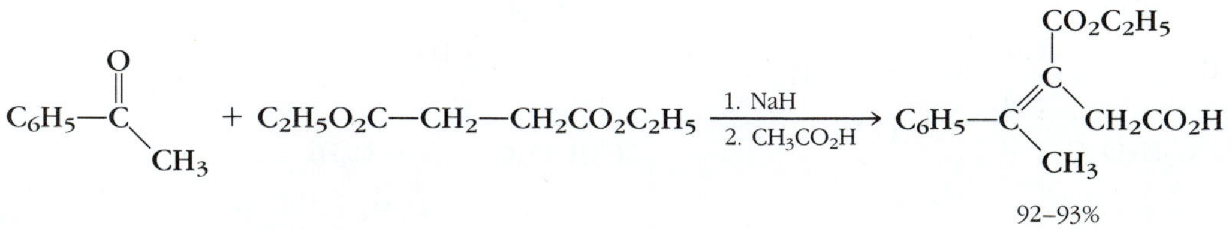

$$C_6H_5-\overset{O}{\overset{\|}{C}}-CH_3 + C_2H_5O_2C-CH_2-CH_2CO_2C_2H_5 \xrightarrow[2.\ CH_3CO_2H]{1.\ NaH} C_6H_5-C(CH_3)=C(CO_2C_2H_5)CH_2CO_2H$$

92–93%

(mixture of *E/Z* isomers)

A third kind of aldol-type condensation of carboxylic acid derivatives was developed by Sir William Perkin in England in the 1860s. This reaction is limited to condensations of aromatic aldehydes with *anhydrides* of carboxylic acids, and derivatives of cinnamic acid are produced. The base used in the reaction is the salt of the corresponding carboxylic acid, and rather high reaction temperatures of 150–180°C are needed with this weak base.

$$C_6H_5\overset{O}{\overset{\|}{C}}H + CH_3\overset{O}{\overset{\|}{C}}O\overset{O}{\overset{\|}{C}}CH_3 \xrightarrow[2.\ HCl]{1.\ KO_2CCH_3} C_6H_5{-}CH{=}CH{-}\overset{O}{\overset{\|}{C}}OH \quad 55\text{–}60\%$$

The driving force for the reaction is an *intramolecular* acylation that permits elimination to form the conjugated ester, as shown in the following equations. Initially, attack of the enolate on benzaldehyde generates the anion on the oxygen atom of the original benzaldehyde carbonyl group:

$$CH_3\overset{O}{\overset{\|}{C}}{-}O{-}\overset{O}{\overset{\|}{C}}CH_3 \underset{}{\overset{KO_2CCH_3}{\rightleftharpoons}} {}^{-}O{-}C({=}CH_2){-}O{-}\overset{O}{\overset{\|}{C}}{-}CH_3 + C_6H_5{-}CH{=}O \rightleftharpoons C_6H_5{-}CH(O^-){-}CH_2{-}\overset{O}{\overset{\|}{C}}{-}O{-}C({=}O){-}CH_3$$

The anion can attack the anhydride linkage in intramolecular fashion to form a six-membered-ring intermediate. Expulsion of a carboxylate anion then completes the acylation of the oxygen that was originally the carbonyl oxygen of benzaldehyde.

$$\text{(cyclic)}\ C_6H_5{-}CH(O^-){-}CH_2{-}C({=}O){-}O{-}C({=}O)CH_3 \rightleftharpoons C_6H_5{-}CH{-}CH_2{-}C({=}O){-}O{-}C(O^-)(CH_3){-}O\ \text{(ring)} \rightleftharpoons C_6H_5{-}CH(O{-}C({=}O)CH_3){-}CH_2{-}C({=}O)O^-$$

This oxygen is β to the carboxyl carbonyl and it is a good leaving group, so an elimination reaction occurs readily. In a final step acidification yields the neutral carboxylic acid.

$$CH_3CO_2^- + C_6H_5{-}CH(O\overset{O}{\overset{\|}{C}}CH_3){-}CH_2{-}\overset{O}{\overset{\|}{C}}O^- \rightleftharpoons C_6H_5{-}CH{=}CH{-}\overset{O}{\overset{\|}{C}}O^- \xrightarrow[\text{(workup)}]{HCl} C_6H_5{-}CH{=}CH{-}\overset{O}{\overset{\|}{C}}OH$$

Other examples of the Perkin condensation are shown in the following equations:

$$C_6H_5CHO \xrightarrow[2.\ HCl]{1.\ (CH_3CH_2\overset{O}{\overset{\|}{C}})_2O,\ CH_3CH_2CO_2Na} C_6H_5CH{=}C(CO_2H)(CH_3) \quad 60\text{–}70\%$$

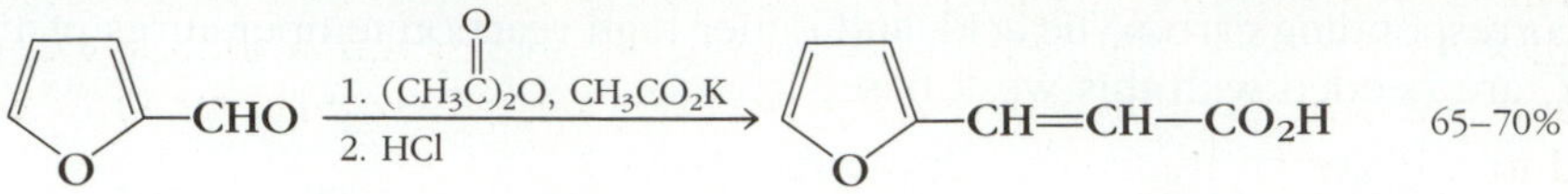

$$\text{(2-furyl)}-CHO \xrightarrow[\text{2. HCl}]{\text{1. }(CH_3\overset{O}{\overset{\|}{C}})_2O,\ CH_3CO_2K} \text{(2-furyl)}-CH{=}CH-CO_2H \qquad 65\text{–}70\%$$

EXERCISE 21.6

Predict the products in each of the following reactions:

(a) $CH_3O_2C-CH_2CH_2-CO_2CH_3 + CH_3CH_2CH_2CHO \xrightarrow[\text{2. HCl}]{\text{1. NaOCH}_3}$

(b) $BrCH_2CO_2C_2H_5 + C_6H_5\overset{O}{\overset{\|}{C}}CH_3 \xrightarrow[\text{2. HCl}]{\text{1. Zn}}$

(c) 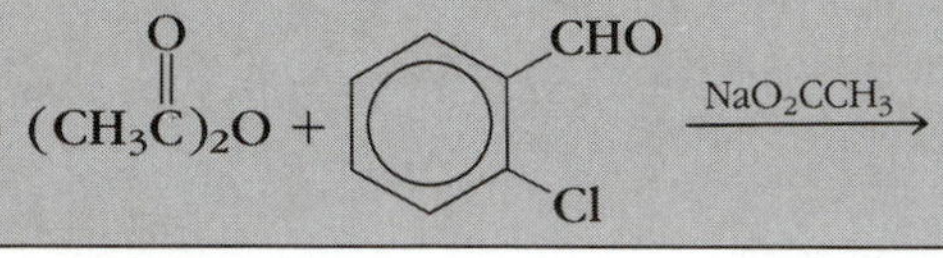 $(CH_3\overset{O}{\overset{\|}{C}})_2O$ + (2-chlorobenzaldehyde: benzene ring with CHO and Cl) $\xrightarrow{NaO_2CCH_3}$

21.4 ALDOL-TYPE CONDENSATIONS OF β-DICARBONYL COMPOUNDS

The great ease with which enols and enolate ions can be formed from β-dicarbonyl compounds makes them valuable and versatile synthetic reagents. In a mixed condensation with a monoketone or aldehyde, good yields of the corresponding aldol-type product can be obtained with little or no self-condensation of either reactant.

Knoevenagel Condensation

Some of the earliest studies of this type of reaction were carried out at the turn of the century by Emil Knoevenagel in Germany. He found that β-dicarbonyl compounds would condense with aldehydes and ketones in the presence of amines (including ammonia) or ammonium salts. These reactions are often called Knoevenagel condensations.

$$\begin{matrix}R'\\ \ \ \ \ C{=}O\\ R\end{matrix} + CH_2\begin{matrix}X\\ \\ Y\end{matrix} \xrightarrow[\text{ammonium salt}]{\text{amine or}} \begin{matrix}R'\\ \ \ \ \ C{=}C\\ R\end{matrix}\begin{matrix}X\\ \\ Y\end{matrix} + H_2O$$

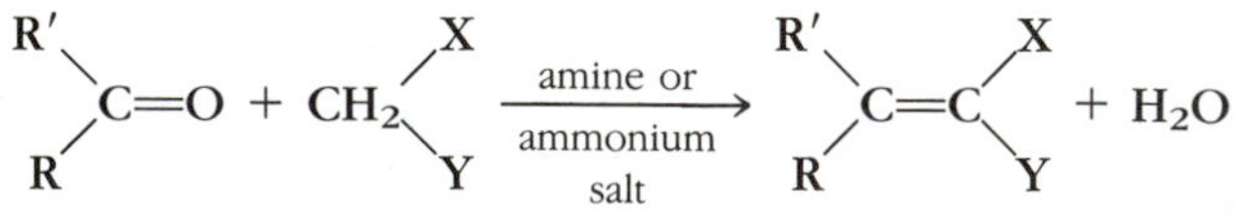

R,R′ = alkyl, aryl, H $\qquad$ X,Y = CO_2R, CO_2H, CN, $\overset{O}{\overset{\|}{C}}R$

The reaction mechanism is very similar to that of the aldol condensation except that reaction frequently proceeds via initial formation of an imine (or iminium salt).

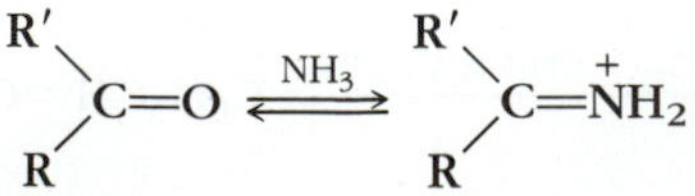

$$\begin{matrix}R'\\ \ \ \ \ C{=}O\\ R\end{matrix} \overset{NH_3}{\rightleftharpoons} \begin{matrix}R'\\ \ \ \ \ C{=}\overset{+}{N}H_2\\ R\end{matrix}$$

This imine is simply the nitrogen analog of a ketone, and it undergoes nucleophilic attack by an enol or enolate. Subsequent loss of ammonia affords the α,β-unsaturated compound. The following reactions show how this reaction can be used to prepare a variety of compounds.

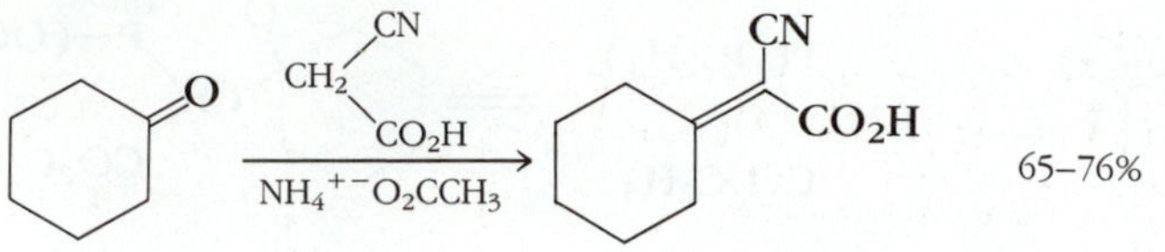

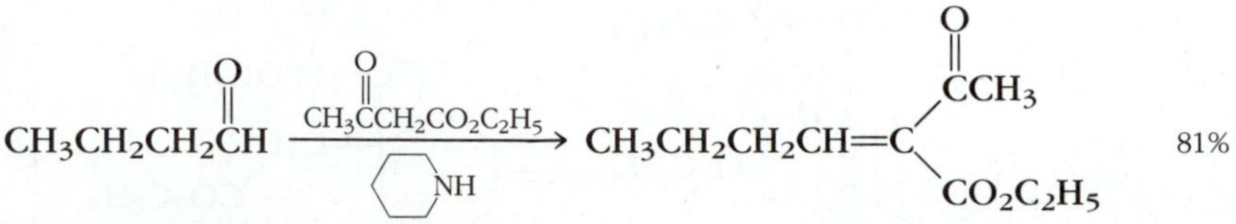

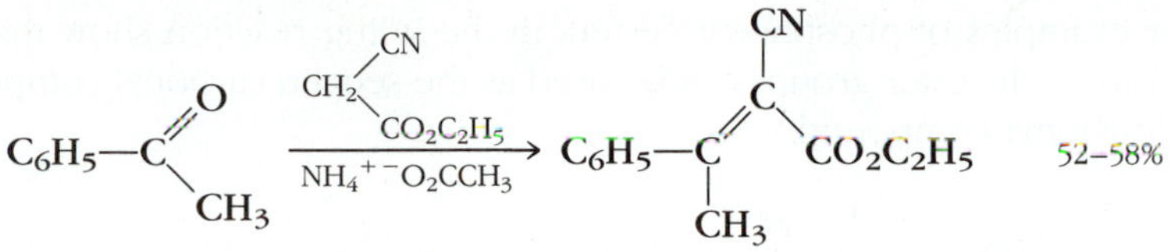

$$\text{cyclohexanone} \xrightarrow[NH_4^{+}\,^{-}O_2CCH_3]{CH_2(CN)CO_2H} \text{cyclohexylidene}{=}C(CN)CO_2H \quad 65\text{–}76\%$$

$$CH_3CH_2CH_2CHO \xrightarrow[\text{piperidine}]{CH_3COCH_2CO_2C_2H_5} CH_3CH_2CH_2CH{=}C(COCH_3)CO_2C_2H_5 \quad 81\%$$

$$C_6H_5{-}C({=}O){-}CH_3 \xrightarrow[NH_4^{+}\,^{-}O_2CCH_3]{CH_2(CN)CO_2C_2H_5} C_6H_5{-}C(CH_3){=}C(CN)CO_2C_2H_5 \quad 52\text{–}58\%$$

(*E*/*Z* mixture)

$$C_6H_5CHO \xrightarrow[\text{piperidinium } (NH_2^{+})\ ^{-}O_2CC_6H_5]{CH_2(CO_2C_2H_5)_2} C_6H_5CH{=}C(CO_2C_2H_5)_2 \quad 89\text{–}91\%$$

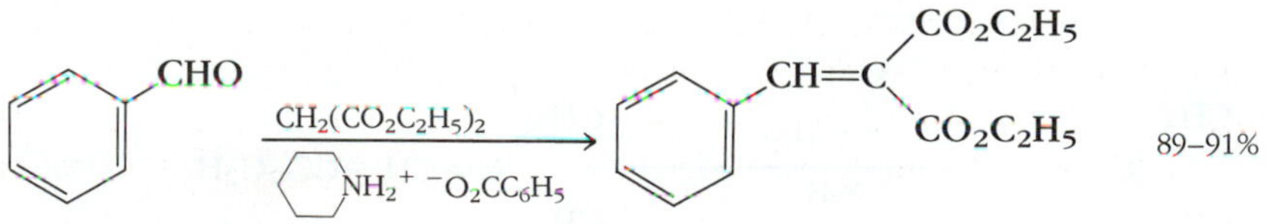

Modified Wittig Reaction

A closely related condensation that yields a monocarbonyl compound is a modification of the Wittig reaction (Section 13.8). A phosphonate ester is employed as one of the "carbonyl" groups of a β-dicarbonyl compound, and this is converted to its enolate by reaction with base.

$$(RO)_2\overset{O}{\overset{\|}{P}}{-}CH_2{-}\overset{O}{\overset{\|}{C}}OC_2H_5 \xrightarrow{\text{base}} (RO)_2\overset{O}{\overset{\|}{P}}{-}\overset{-}{C}H{-}\overset{O}{\overset{\|}{C}}OC_2H_5$$

In the ensuing reaction with an aldehyde or ketone, the phosphonyl group is lost and an α,β-unsaturated carbonyl compound is formed directly. This is illustrated by the reaction between cyclohexanone and triethylphosphonoacetate:

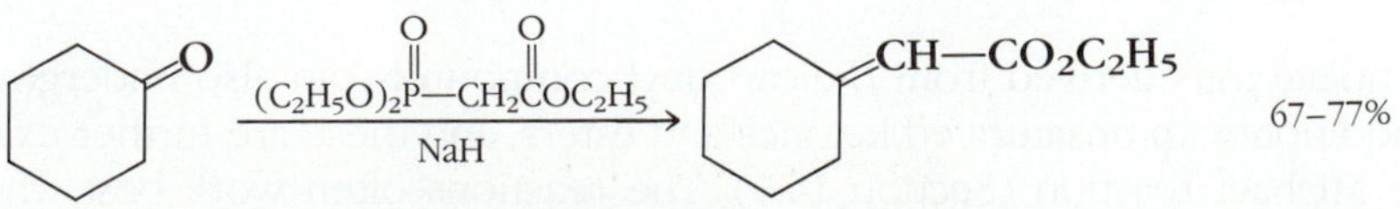

$$\text{cyclohexanone} \xrightarrow[NaH]{(C_2H_5O)_2P(O){-}CH_2C(O)OC_2H_5} \text{cyclohexylidene}{=}CH{-}CO_2C_2H_5 \quad 67\text{–}77\%$$

Attack of the phosphonate enolate at the carbonyl group produces an anion, which can decompose via four-membered ring formation and elimination of diethyl phosphate ion.

$$\text{cyclohexanone} + {}^{-}CH(CO_2C_2H_5)P(O)(OC_2H_5)_2 \rightleftharpoons \text{C}_6\text{H}_{10}(O^-)\text{–CH}(CO_2C_2H_5)\text{–P(O)(OC}_2H_5)_2$$

$$\rightleftharpoons \text{four-membered ring: } \text{C}_6\text{H}_{10}\text{–O–P(O}^-\text{)(OC}_2H_5)_2\text{–CH}(CO_2C_2H_5) \rightleftharpoons \text{C}_6\text{H}_{10}{=}CH\text{–}CO_2C_2H_5 + {}^{-}O\text{–}P(O)(OC_2H_5)_2$$

Other examples of phosphonate esters in the Wittig reaction show that cyano, acyl, or carboxylic ester groups can be used as the second carbonyl component of the β-dicarbonyl compound.

$$C_6H_5\text{–}CH{=}O \xrightarrow[\text{NaH}]{(C_2H_5O)_2P(O)\text{–}CH_2\text{–}CN} C_6H_5\text{–}CH{=}CH\text{–}CN \quad 66\%$$

$$(CH_3)_2C{=}O \xrightarrow[\text{NaH}]{(C_2H_5O)_2P(O)\text{–}CH_2CO_2C_2H_5} (CH_3)_2C{=}CH\text{–}CO_2C_2H_5 \quad 70\%$$

$$C_6H_5CHO \xrightarrow[\text{NaH}]{(C_2H_5O)_2P(O)\text{–}CH_2C(O)C_6H_5} C_6H_5CH{=}CH\text{–}C(O)\text{–}C_6H_5 \quad 61\%$$

The triphenylphosphonium salts normally used in Wittig reactions (Section 13.8) become less satisfactory when the α carbon carries a carbonyl substituent. The resulting ylide is less reactive, and reaction with aldehydes and ketones takes place very slowly.

$$(C_6H_5)_3\overset{+}{P}\text{–}CH_2\text{–}C(O)OR \xrightarrow{\text{base}} (C_6H_5)_3P{=}CH\text{–}C(O)OR$$

Relatively unreactive

Michael Additions

The enolate ions derived from β-dicarbonyl compounds can also undergo conjugate addition with unsaturated ketones and esters, and these are further examples of the Michael reaction (Section 14.5). The reactions often work best when the

β-dicarbonyl compound has *two* acidic hydrogens. This favors the overall equilibrium by formation of a stable enolate as the product.

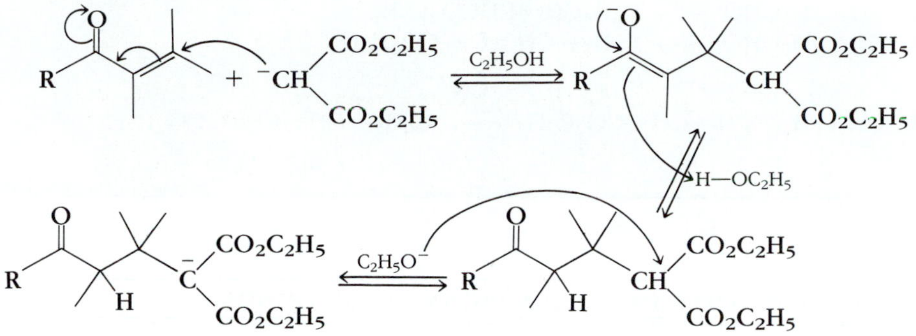

$CO_2C_2H_5$ R + ^-CH $CO_2C_2H_5$ $\xrightleftharpoons{C_2H_5OH}$ R CH $CO_2C_2H_5$ $CO_2C_2H_5$

H—OC_2H_5

O R H $\bar{C}$ $CO_2C_2H_5$ $CO_2C_2H_5$ $\xleftarrow{C_2H_5O^-}$ O R H CH $CO_2C_2H_5$ $CO_2C_2H_5$

The following examples illustrate Michael additions with β-dicarbonyl compounds.

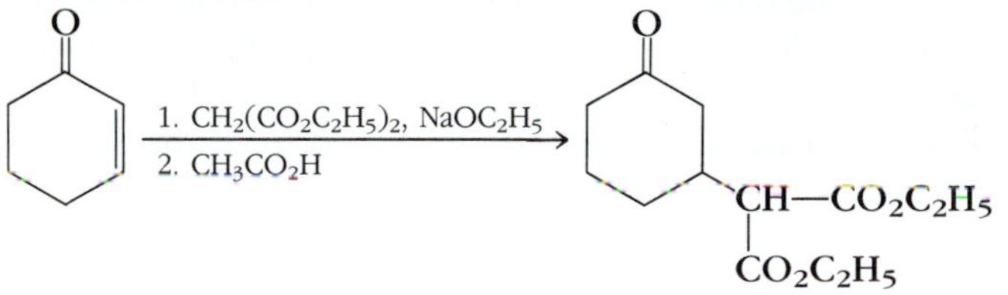

O $\xrightarrow[2.\ CH_3CO_2H]{1.\ CH_2(CO_2C_2H_5)_2,\ NaOC_2H_5}$ O CH—$CO_2C_2H_5$ $CO_2C_2H_5$ 90%

C_6H_5—CH=CH—CN $\xrightarrow[2.\ CH_3CO_2H]{1.\ CH_2(CO_2C_2H_5)_2,\ NaOC_2H_5}$ C_6H_5—CH($CH(CO_2C_2H_5)_2$)—CH_2—CN 83%

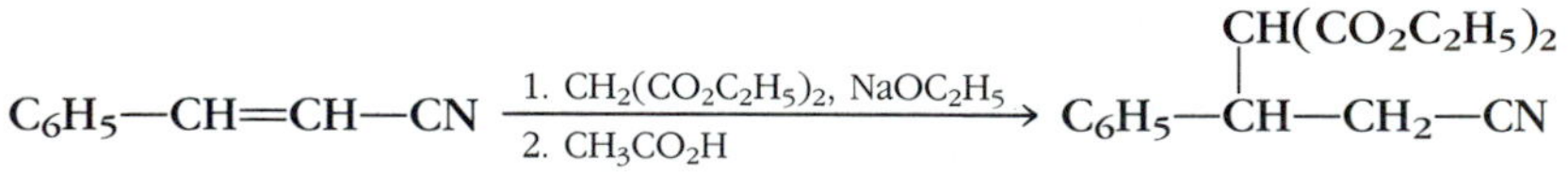

O $CO_2C_2H_5$ + $CO_2C_2H_5$ $\xrightarrow[2.\ CH_3CO_2H]{1.\ NaOC_2H_5}$ O $CO_2C_2H_5$ $CO_2C_2H_5$ 64%

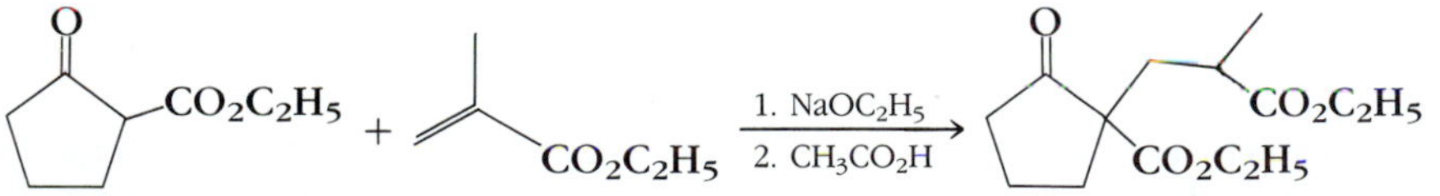

The products of these Michael additions are frequently capable of further reaction by way of *intramolecular* condensation, and we will return to this topic in the next chapter.

EXERCISE 21.7

Complete the following reactions:

(a) 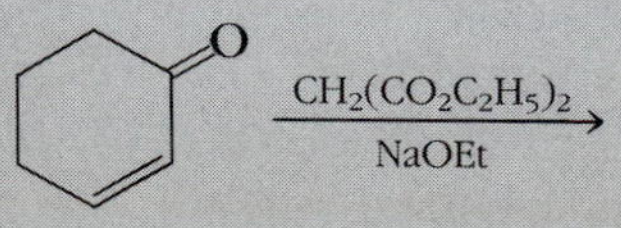 O $\xrightarrow[NaOEt]{CH_2(CO_2C_2H_5)_2}$

(b) 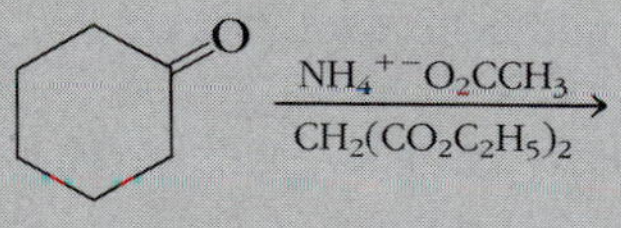 O $\xrightarrow[CH_2(CO_2C_2H_5)_2]{NH_4^{+\ -}O_2CCH_3}$

(c) 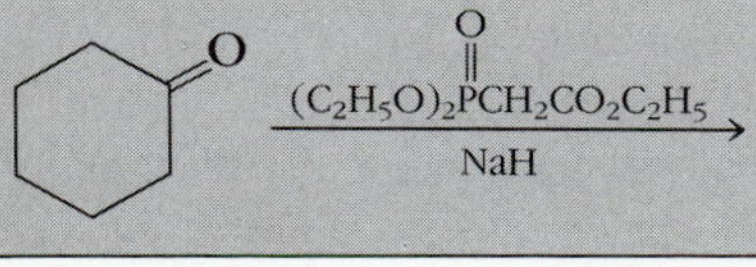 O $\xrightarrow[NaH]{(C_2H_5O)_2P(O)CH_2CO_2C_2H_5}$

EXERCISE 21.8

Suggest methods for carrying out the following conversions:

(a) $C_6H_5CHO \longrightarrow C_6H_5CH{=}CHCO_2C_2H_5$

(b) $C_6H_5CHO \longrightarrow C_6H_5{-}CH{=}C(CN){-}CO_2C_2H_5$

(c) $C_6H_5{-}CH{=}CH{-}CO_2C_2H_5 \longrightarrow C_6H_5{-}CH(CH(CN)_2){-}CH_2{-}CO_2C_2H_5$

21.5 ALKYLATION OF β-DICARBONYL COMPOUNDS

In Section 14.3 we discussed the many problems that are associated with attempts to alkylate enolate ions. Most of these difficulties are related to the fact that simple enolate ions are very strong bases, but the situation is quite different for enolates of β-dicarbonyl compounds. Remember that simple ketones have K_a values of about 10^{-21}, but β-dicarbonyl compounds are far more acidic, with a K_a in the range 10^{-9}–10^{-13}. This means that the enolate ions formed from the latter are much weaker bases.

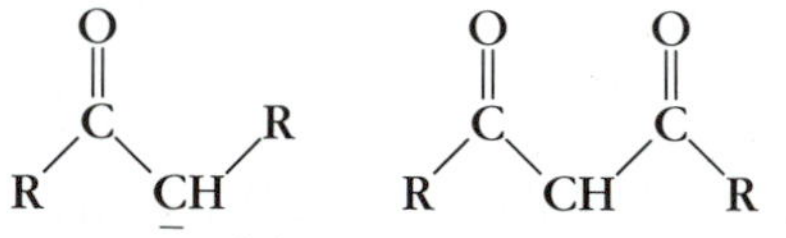

$$R{-}C(=O){-}\overset{-}{C}H{-}R \qquad\qquad R{-}C(=O){-}\overset{-}{C}H{-}C(=O){-}R$$

strong base — much weaker base

Because they are less basic, the enolates from β-dicarbonyl compounds tend not to cause base-catalyzed elimination or condensation reactions.

Another major problem in the alkylation of simple ketones is the formation of two different enolates from an unsymmetrical ketone. β-Dicarbonyl compounds, in contrast, will afford only the enolate that results from removal of a proton from the carbon atom *between* the two carbonyl groups. Other hydrogens in the molecule are less acidic by a factor of 10^7 or more.

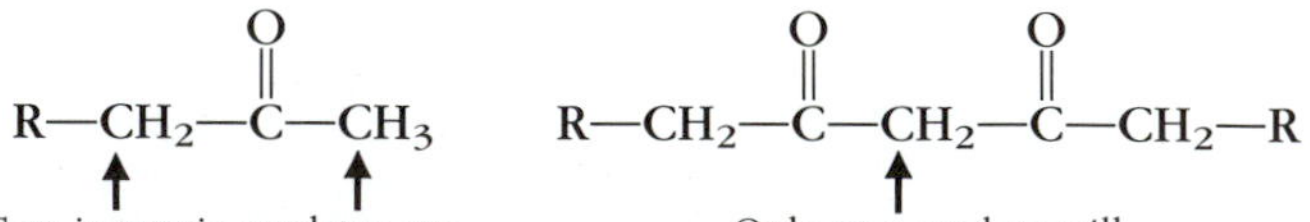

$$R{-}CH_2{-}C(=O){-}CH_3 \qquad\qquad R{-}CH_2{-}C(=O){-}CH_2{-}C(=O){-}CH_2{-}R$$

Two isomeric enolates can be formed — Only one enolate will be formed

The common procedure for alkylating a β-dicarbonyl compound employs initial formation of the enolate. This is usually done by reaction with sodium ethoxide (for ethyl esters) or bases of comparable strength, such as sodium hydroxide or even potassium carbonate. Then the alkyl halide is introduced, and the reaction mixture is heated until the reaction is complete.

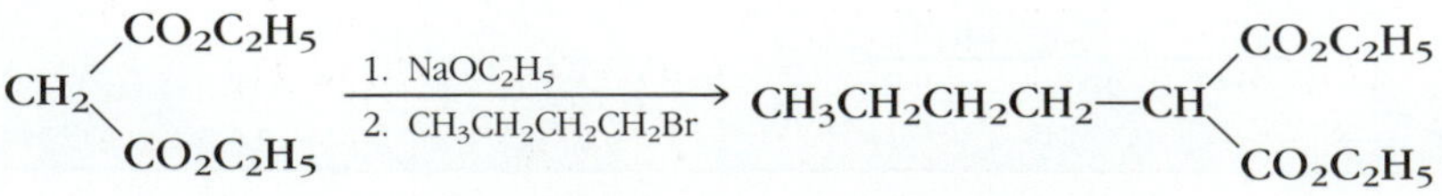

$$CH_2(CO_2C_2H_5)_2 \xrightarrow[\text{2. } CH_3CH_2CH_2CH_2Br]{\text{1. } NaOC_2H_5} CH_3CH_2CH_2CH_2{-}CH(CO_2C_2H_5)_2 \qquad 80\text{–}90\%$$

$$CH_2(COCH_3)(CO_2C_2H_5) \xrightarrow[2.\ CH_3CH_2CH_2CH_2Br]{1.\ NaOC_2H_5} CH_3CH_2CH_2CH_2{-}CH(COCH_3)(CO_2C_2H_5) \quad 69\text{–}72\%$$

$$CH_2(CO_2C_2H_5)_2 \xrightarrow[2.\ CH_3{-}C{\equiv}C{-}CH_2CH_2OTs]{1.\ NaOC_2H_5} CH_3{-}C{\equiv}C{-}CH_2CH_2{-}CH(CO_2C_2H_5)_2 \quad 69\%$$

In contrast to the alkylations of monoketones (Section 14.3), the alkylating agent is not limited to methyl and primary alkyl halides; β-dicarbonyl compounds can also be alkylated with secondary alkyl derivatives.

$$CH_2(CO_2C_2H_5)_2 \xrightarrow[2.\ CH_3CH_2CH(CH_3){-}Br]{1.\ NaOC_2H_5} CH_3CH_2CH(CH_3){-}CH(CO_2C_2H_5)_2 \quad 83\text{–}84\%$$

$$CH_2(CN)(CO_2C_2H_5) \xrightarrow[2.\ \text{3-bromocyclopentene}]{1.\ KOtBu} \text{(cyclopent-2-enyl)}{-}CH(CN)(CO_2C_2H_5) \quad > 48\%$$

Tertiary alkyl halides cannot be used, of course. As with all other attempts at direct displacement, tertiary compounds give exclusively elimination rather than substitution.

In most cases dialkylation can be prevented by using appropriate reaction conditions, and the monoalkylated product can be obtained in good yield. A second alkyl group can be introduced subsequently if desired.

$$CH_3CH_2CH(CH_3){-}CH(CO_2C_2H_5)_2 \xrightarrow[2.\ CH_2{=}CH{-}CH_2Br]{1.\ NaOC_2H_5} (CH_2{=}CHCH_2)(CH_3CH_2CH(CH_3))C(CO_2C_2H_5)_2 \quad 85\%$$

$$C_6H_5{-}C(CH_3)_2{-}CH(CO_2C_2H_5)_2 \xrightarrow[2.\ C_2H_5Br]{1.\ NaOC_2H_5} C_6H_5{-}C(CH_3)_2{-}C(C_2H_5)(CO_2C_2H_5){-}CO_2C_2H_5 \quad 74\%$$

Two identical alkyl groups can often be introduced in a single step if an excess of alkylating agent and excess base are used.

$$CH_2(CN)(CO_2C_2H_5) \xrightarrow[2.\ CH_3CH(I)CH_3\ (2\ equiv)]{1.\ excess\ NaOC_3H_7} (CH_3{-}CH(CH_3))_2C(CN)(CO_2C_2H_5) \quad 93\%$$

These reactions are not limited to the simple acylic examples we have just shown. Many of the most valuable uses involve complex cyclic and polycyclic systems, particularly β-keto esters and β-diketones.

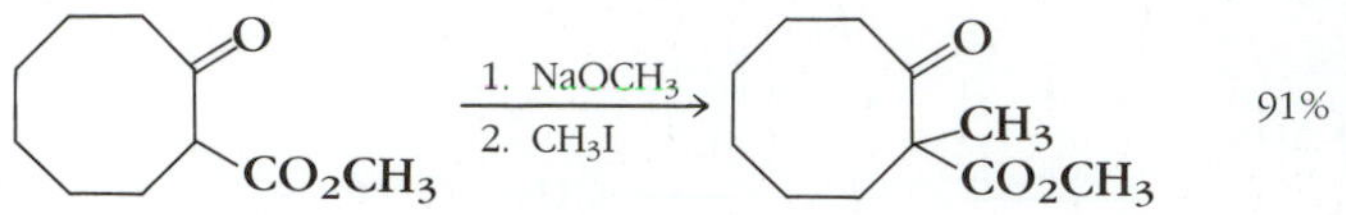

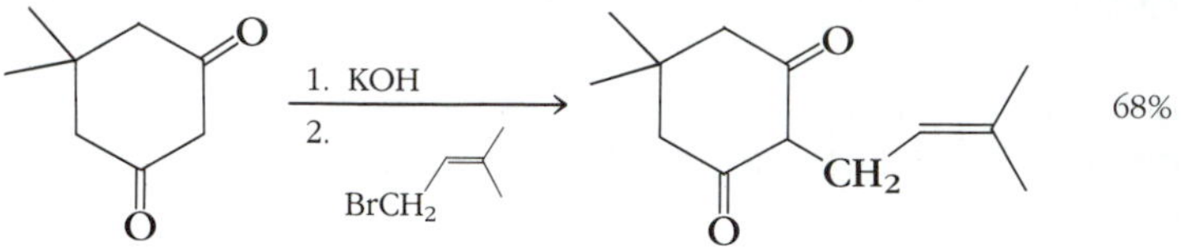

CO_2CH_3 —1. $NaOCH_3$; 2. CH_3I→ CH_3, CO_2CH_3 91%

—1. KOH; 2. $BrCH_2$→ CH_2 68%

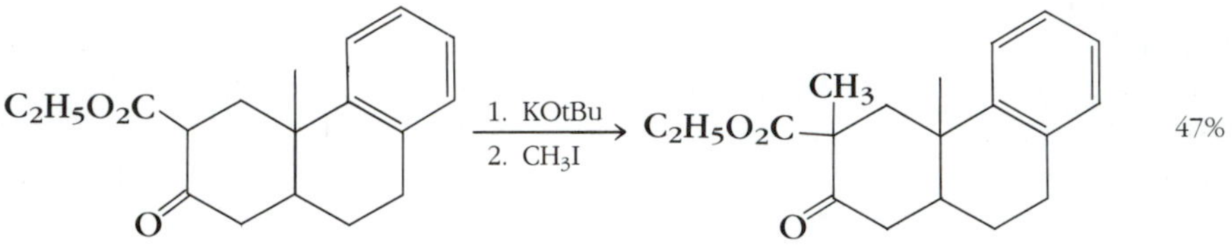

$C_2H_5O_2C$ —1. KOtBu; 2. CH_3I→ CH_3, $C_2H_5O_2C$ 47%

In the following sections we will return to these reactions and show you how the products of β-dicarbonyl alkylations can be further modified in a synthetic scheme.

EXERCISE 21.9

Complete the following reactions:

(a) $CH_2(CO_2C_2H_5)_2 \xrightarrow[2.\ C_2H_5I]{1.\ NaOH}$

(b) $CH_3\overset{O}{\overset{\|}{C}}CH_2CO_2C_2H_5 \xrightarrow[2.\ CH_3I\ (2\ equiv)]{1.\ excess\ NaOC_2H_5}$

(c) CO_2CH_3 —1. $NaOC_2H_5$; 2. $CH_2{=}C(CH_3){-}CH_2Br$→

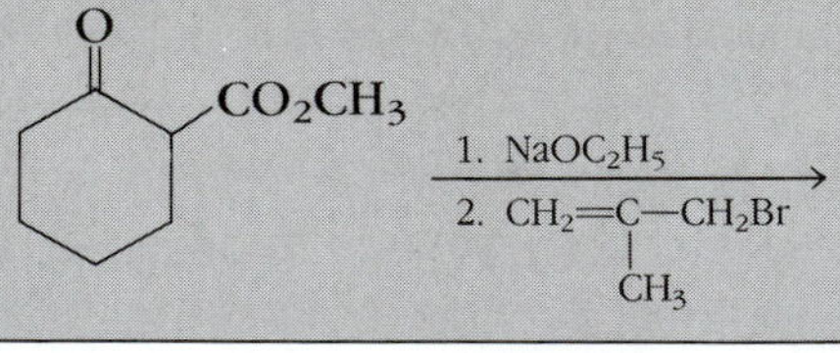

EXERCISE 21.10

Suggest methods for carrying out the following conversions:

(a) $CH_3{-}\overset{O}{\overset{\|}{C}}{-}CH_2{-}CO_2CH_3 \longrightarrow CH_3\overset{O}{\overset{\|}{C}}{-}C(CH_3)(CH_2CH_3){-}CO_2CH_3$

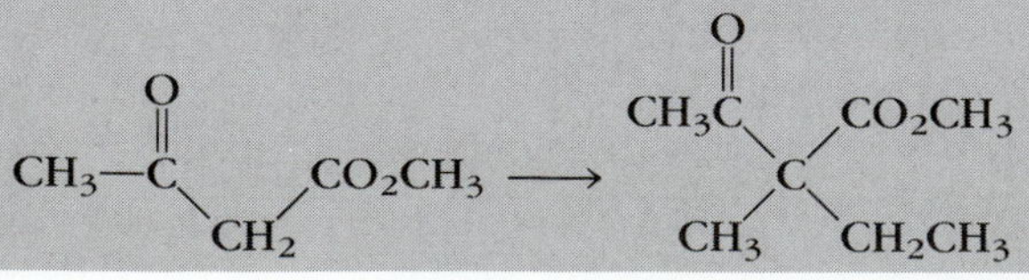

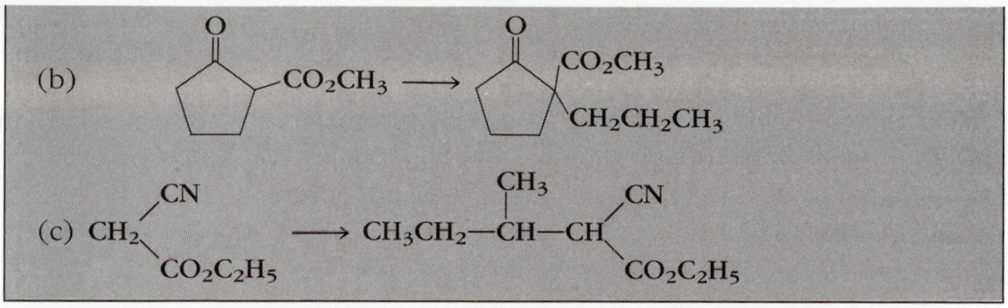

(b) 2-(CO_2CH_3)cyclopentanone ⟶ 2-(CO_2CH_3)-2-($CH_2CH_2CH_3$)cyclopentanone

(c) $CH_2(CN)(CO_2C_2H_5)$ ⟶ $CH_3CH_2{-}CH(CH_3){-}CH(CN)(CO_2C_2H_5)$

21.6 CLEAVAGE REACTIONS OF β-DICARBONYL COMPOUNDS

In the previous two sections we showed how β-dicarbonyl compounds can undergo condensation and alkylation reactions. The comparable reactions of monocarbonyl compounds are often accompanied by side reactions that give unwanted products, so the second carbonyl group affords important advantages. It is possible to introduce the second carbonyl group via an acylation reaction (Section 21.2). This allows alkylation (or aldol-type condensation) to proceed as desired, so the second carbonyl serves as an *activating group.* In a final stage the activating group can be removed in a reaction that causes carbon-carbon bond *cleavage.* The overall process, which is quite analogous to the use of a blocking group, is illustrated in Figure 21.2.

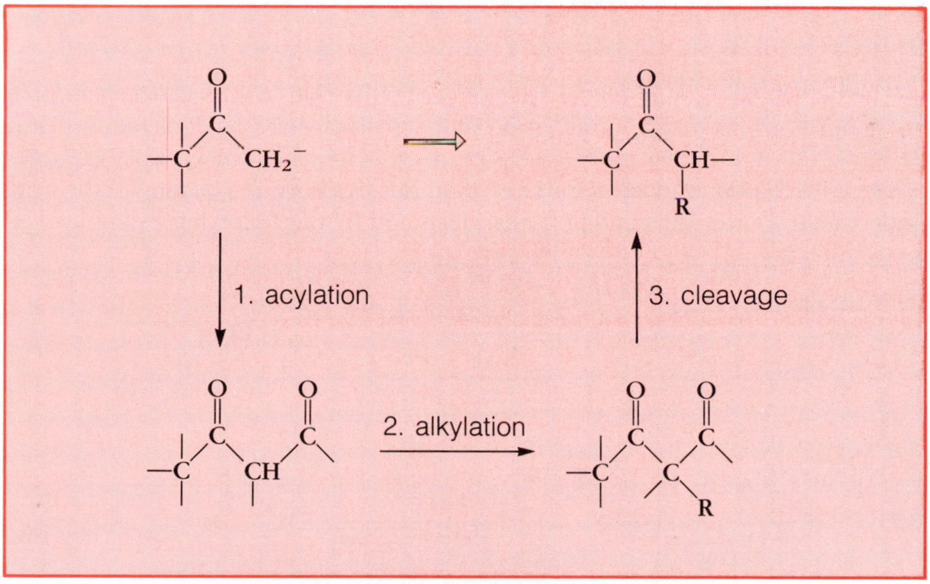

$-C-C(=O)-CH_2^-$ ⟹ $-C-C(=O)-CH(R)-$

1. acylation

2. alkylation

3. cleavage

Figure 21.2 The use of an activating group.
(1) In order to avoid the problems of alkylating a monocarbonyl compound, an activating group is introduced (*acylation*). (2) This produces a β-dicarbonyl compound, which is easily converted to its enolate in order to introduce the desired substituent (*alkylation*). (3) Finally, the alkylated derivative of the original monocarbonyl compound is obtained by removal of the activating group (*cleavage*).

There are two general pathways that can be used for cleavage of β-dicarbonyl compounds: *decarboxylation* and *retrocondensation* (i.e., reverse condensation). Decarboxylation is clearly limited to carboxylic acid derivatives, but retro condensations can occur with all types of β-dicarbonyl derivatives. In this section we will analyze both these processes.

Decarboxylation

In its most general sense the term *decarboxylation* describes a reaction in which carbon dioxide is cleaved from a carboxylic acid derivative. Usually it refers specifically to cleavage of a carbon-carbon bond.

$$R{-}\overset{\overset{O}{\|}}{C}{-}O{-}H \longrightarrow R{-}H + CO_2$$

We previously mentioned the oxidative decarboxylation of acids with lead tetraacetate (Section 16.4), but simple carboxylic acids do not normally lose CO_2 in the absence of such a reagent. When the carboxyl group is β to another carbonyl group, however, decarboxylation occurs readily under fairly mild conditions. For example, the hydrolysis of the following β-keto nitrile in aqueous acid at 100°C leads directly to the decarboxylated ketone:

$$C_6H_5{-}\overset{\overset{CN}{|}}{C}H{-}\overset{\overset{O}{\|}}{C}CH_3 \xrightarrow[H_2SO_4]{H_2O} C_6H_5{-}\overset{\overset{CO_2H}{|}}{C}H{-}\underset{\underset{O}{\|}}{C}{-}CH_3 \xrightarrow[CO_2]{\text{loss of}} C_6H_5{-}CH_2{-}\overset{\overset{O}{\|}}{C}{-}CH_3 \quad 77\text{–}86\%$$

The decarboxylation of β-keto acids (and of other acids with a carbonyl group at the β position) normally occurs via a cyclic transition state, as depicted in Figure 21.3.

Decarboxylation of β-keto acids and their analogs requires heating of the acids under neutral or acidic conditions, as illustrated by the following examples:

$$\text{cyclobutane-1,1-}(CO_2H)_2 \xrightarrow[(160\text{–}170°C)]{\text{mild heating}} \text{cyclobutane-}CO_2H + CO_2 \quad 86\text{–}89\%$$

$$\text{2-acetyl-}\gamma\text{-butyrolactone } (CH_3CO\text{–}) \xrightarrow[H_2O]{HCl} \left[CH_3COCH(CO_2H)CH_2CH_2Cl\right] \xrightarrow{-CO_2} CH_3COCH_2CH_2CH_2Cl \quad 79\text{–}90\%$$

$$CH_3CH_2\overset{\overset{CH_3}{|}}{C}H{-}CH(CO_2C_2H_5)_2 \xrightarrow{KOH,\ H_2O} CH_3CH_2\overset{\overset{CH_3}{|}}{C}H{-}CH(CO_2^-K^+)_2 \xrightarrow[H_2O]{H_2SO_4} CO_2 + CH_3CH_2\overset{\overset{CH_3}{|}}{C}H{-}CH_2CO_2H \quad 62\text{–}65\%$$

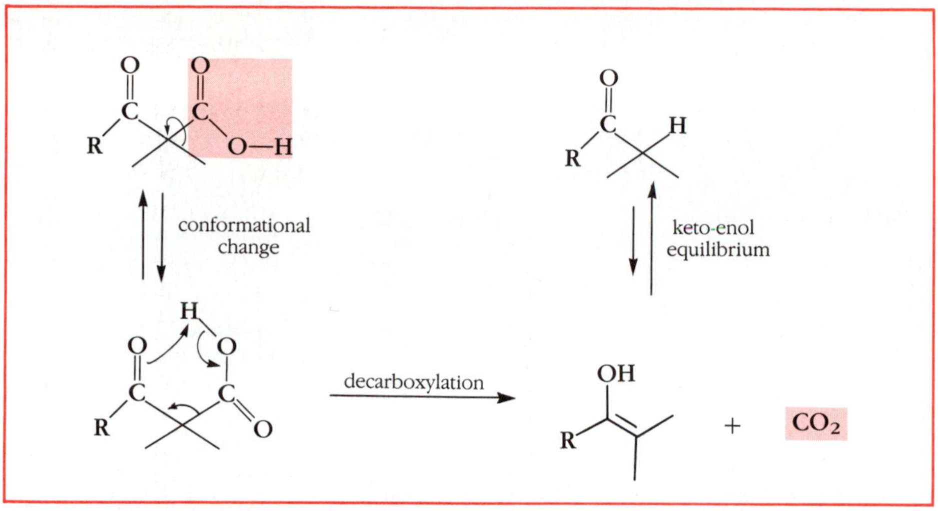

conformational change

keto-enol equilibrium

decarboxylation

$+ \; CO_2$

Figure 21.3 Decarboxylation of β-keto acids (R=alkyl, aryl) and malonic acid (R=OH) derivatives.
The cyclic elimination of carbon dioxide generates an enol, which isomerizes to the carbonyl compound.

Carboxylic acids lacking a carbonyl group in the β position will not undergo decarboxylation in this manner. This can be seen in the following equation, which shows hydrolysis of a keto triester accompanied by loss of only the carboxyl group that was β to the ketone. In other words, decarboxylation does not occur unless the carboxyl group and a second carbonyl are bonded to the *same* carbon atom.

$$C_2H_5O_2C{-}CH_2{-}CH(CO_2C_2H_5){-}C(=O){-}CO_2C_2H_5 \xrightarrow[H_2O]{HCl} HO_2C{-}CH_2{-}CH(CO_2H){-}C(=O){-}CO_2H \xrightarrow{-CO_2}$$

$$HO_2C{-}CH_2CH_2{-}C(=O){-}CO_2H \quad 92\text{–}93\%$$

Retrocondensation

The second process that leads to cleavage of β-dicarbonyl compounds is basically the reverse of the acylation reactions that we discussed in Section 21.2. These condensations are equilibrium reactions in the first place, so it is only a question of finding appropriate conditions to shift the equilibrium. Usually this is done by employing *aqueous* base, and the formation of a *carboxylate salt* makes the cleavage reaction essentially irreversible. The reaction mechanism for *retrocondensation* in aqueous base is shown in Figure 21.4.

Retrocondensation is illustrated in the following equations:

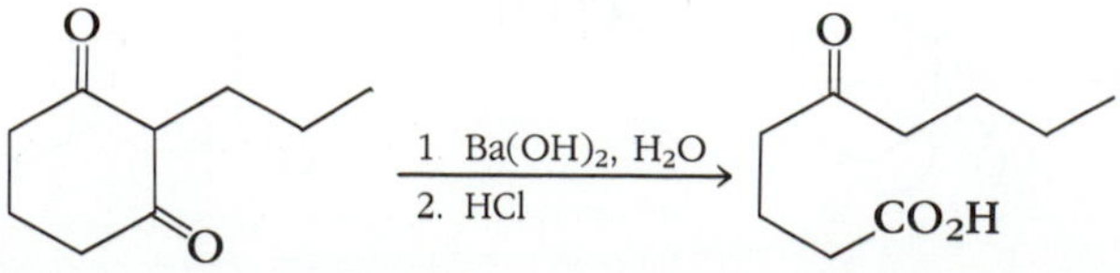

1. $Ba(OH)_2$, H_2O
2. HCl

CO_2H 81%

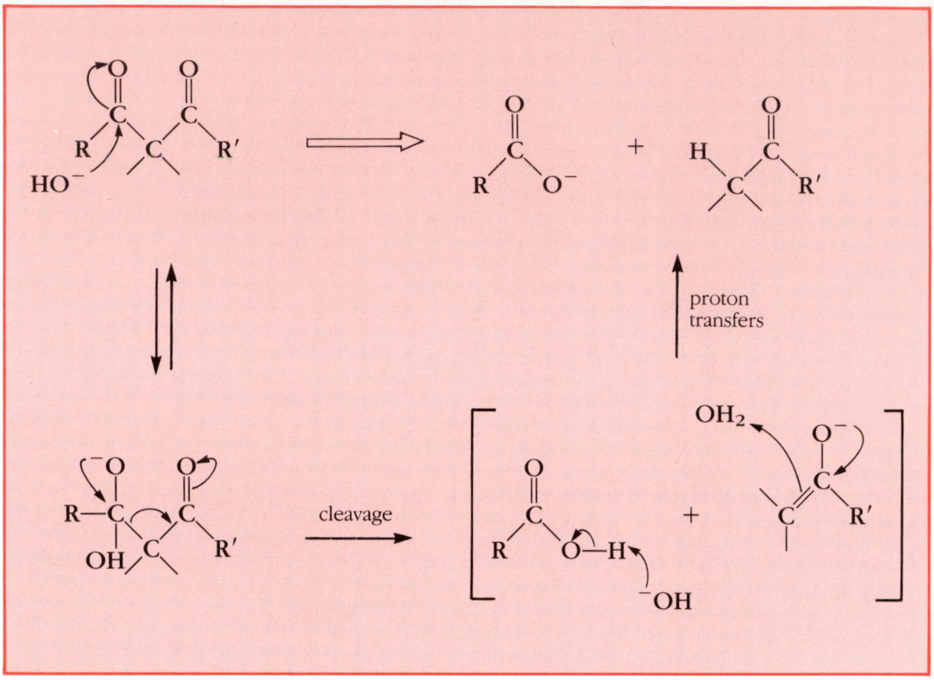

Figure 21.4 Cleavage of a β-dicarbonyl compound by retro-condensation.
The reaction is illustrated for aqueous base. Attack of base at the carbonyl group generates a tetrahedral intermediate, which expels an enolate ion as the leaving group in the cleavage step. Proton transfer reactions then afford the salt of a carboxylic acid and a neutral carbonyl compound. Note that reaction could occur at either carbonyl group even though only one of the two possibilities is depicted here.

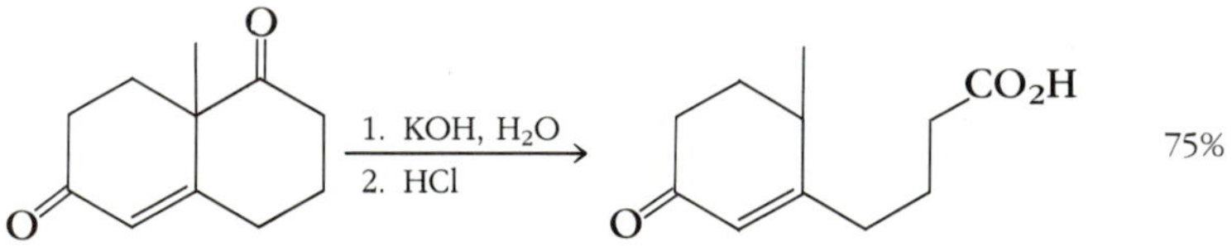

Notice that the second of these reactions proceeds by elimination of the enolate of an α,β-unsaturated ketone.

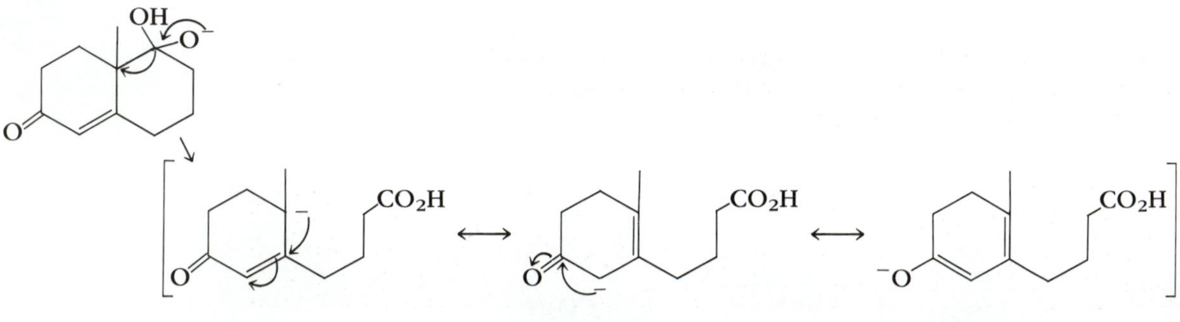

Retrocondensation is particularly facile when the enolate leaving group is stabilized by two carbonyl groups. Cleavage in the following reaction occurs under the mildly alkaline conditions of aqueous ammonia.

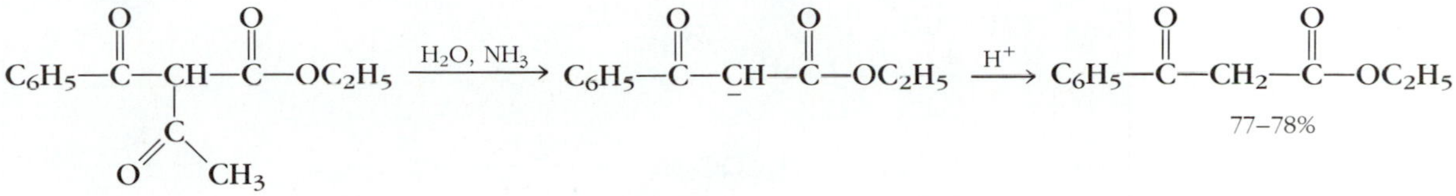

$$C_6H_5-\overset{O}{\overset{\|}{C}}-\underset{\underset{O \diagup\!\!\diagup \quad \diagdown CH_3}{\overset{|}{C}}}{CH}-\overset{O}{\overset{\|}{C}}-OC_2H_5 \xrightarrow{H_2O,\ NH_3} C_6H_5-\overset{O}{\overset{\|}{C}}-\underline{C}H-\overset{O}{\overset{\|}{C}}-OC_2H_5 \xrightarrow{H^+} C_6H_5-\overset{O}{\overset{\|}{C}}-CH_2-\overset{O}{\overset{\|}{C}}-OC_2H_5$$

77–78%

Under alkaline conditions the reaction of a β-keto ester can result in retrocondensation, as well as saponification of the ester group. 2-Ethoxycarbonylcyclohexanone affords the product from retrocondensation in good yield.

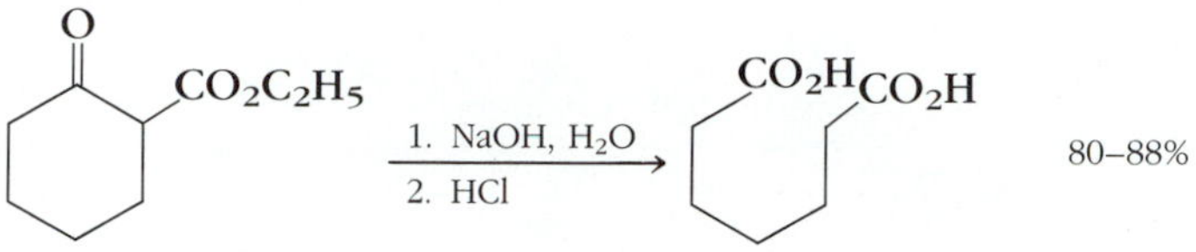

$$\text{2-ethoxycarbonylcyclohexanone } (CO_2C_2H_5) \xrightarrow[\text{2. HCl}]{\text{1. NaOH, } H_2O} \text{ring-opened product } (CO_2H,\ CO_2H) \quad 80\text{–}88\%$$

This reaction proceeds via attack of hydroxide ion at the ketone carbonyl and expulsion of an ester enolate.

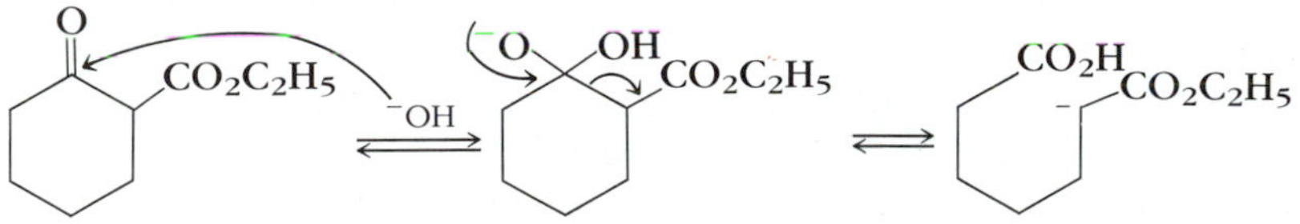

$$\text{O, } CO_2C_2H_5 \underset{}{\overset{^-OH}{\rightleftharpoons}} \text{ } ^-O\text{, OH, } CO_2C_2H_5 \rightleftharpoons CO_2H\text{, } ^-CO_2C_2H_5$$

Proton transfer reactions followed by ester saponification generate the salt of the diacid, which is protonated during workup.

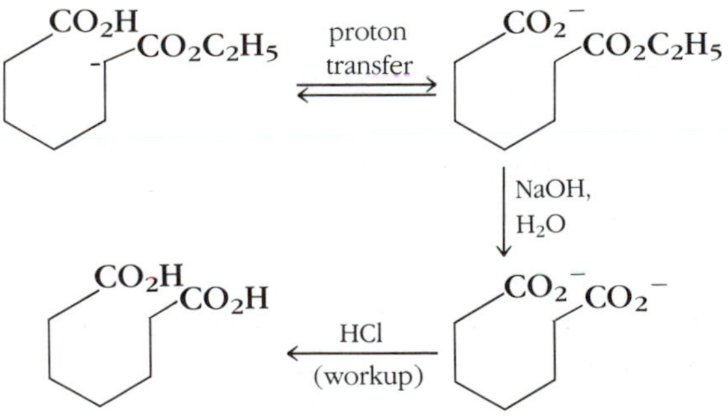

$$CO_2H\text{, } ^-CO_2C_2H_5 \overset{\text{proton transfer}}{\rightleftharpoons} CO_2^-\text{, } CO_2C_2H_5 \xrightarrow{NaOH,\ H_2O} CO_2^-\text{, } CO_2^- \xrightarrow[\text{(workup)}]{HCl} CO_2H\text{, } CO_2H$$

If ester hydrolysis followed by decarboxylation of a β-keto acid is the desired pathway, it is usually better to employ acidic conditions because alkaline conditions tend to favor retrocondensation.

EXERCISE 21.11

In Figure 21.4 reaction is shown for attack at just one of the two carbonyl groups. Draw the mechanism for the cleavage reaction that would result from attack at the other carbonyl group.

EXERCISE 21.12

Complete the following reactions:

(a) $CH_3{-}CH(CO_2C_2H_5)_2 \xrightarrow[\text{2. HCl}]{\text{1. KOH, } H_2O}$

(b) $C_6H_5CH_2CH_2\overset{O}{\overset{\|}{C}}CH_2\overset{O}{\overset{\|}{C}}OC_2H_5 \xrightarrow{H_2O,\ H_2SO_4}$

EXERCISE 21.13

Suggest methods of carrying out the conversion of

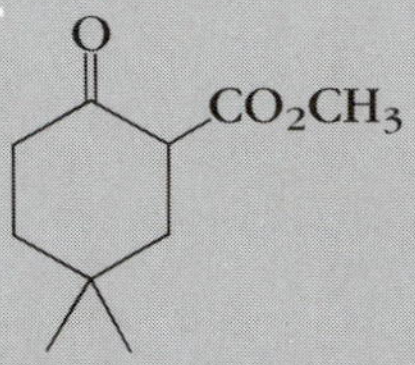

[2-(CO_2CH_3)-4,4-dimethylcyclohexanone structure]

to the indicated products.

(a) [4,4-dimethylcyclohexanone structure]

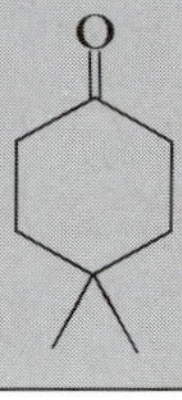

(b) HO_2C–$CH_2CH_2C(CH_3)_2CH_2CH_2$–$CO_2H$

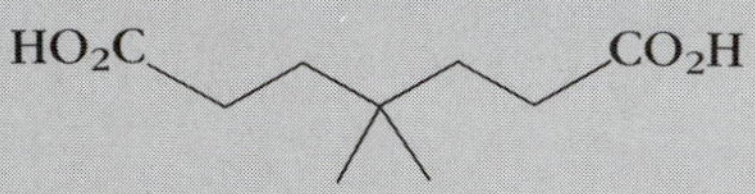

21.7 ORGANIC SYNTHESIS WITH β-DICARBONYL COMPOUNDS

In order to give you some feeling for how alkylations or condensations of β-dicarbonyl reactions can be put to use in synthetic chemistry, we will first present some specific examples of brief reaction sequences that actually have been carried out in the laboratory. Then we will turn to some hypothetical problems to show how you can design a reasonable synthetic strategy to prepare a desired compound. The yields in the following examples may at first seem to be quite poor, but we are showing overall yields for several steps. Even an average yield of 75% for each of three steps would produce an overall yield of 41%.

Example 21.1 Preparation of 2-heptanone from ethyl acetoacetate.

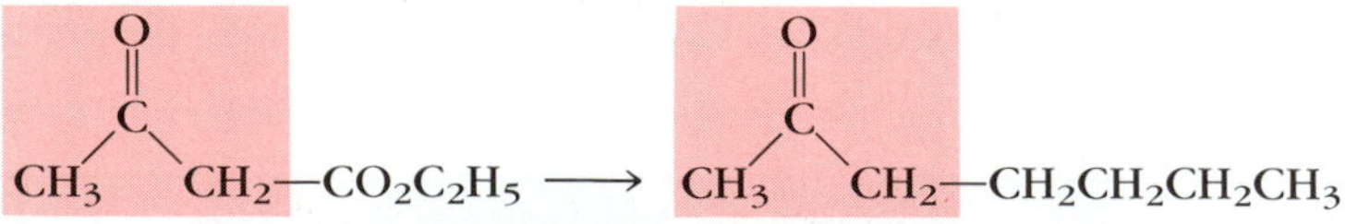

$$CH_3\overset{O}{\overset{\|}{C}}CH_2{-}CO_2C_2H_5 \longrightarrow CH_3\overset{O}{\overset{\|}{C}}CH_2{-}CH_2CH_2CH_2CH_3$$

Solution This conversion illustrates the use of an activating group ($-CO_2C_2H_5$) to alkylate a ketone. The commercially available starting material (ethyl acetoacetate) already contains the activating group, so alkylation followed by ester hydrolysis and decarboxylation affords the desired product. (Note that alkaline ester hydrolysis works well in this case.) The overall sequence employs *ethyl acetoacetate* to prepare an alkylated *acetone*.

$$CH_3-\overset{O}{\overset{\|}{C}}-CH_2-CO_2C_2H_5 \xrightarrow[2.\ CH_3CH_2CH_2CH_2Br]{1.\ NaOC_2H_5} CH_3\overset{O}{\overset{\|}{C}}-\overset{CO_2C_2H_5}{\overset{|}{C}H}-CH_2CH_2CH_2CH_3$$

$$\downarrow NaOH,\ H_2O$$

$$\underset{\text{36–44\% (overall yield for 3 steps)}}{CH_3\overset{O}{\overset{\|}{C}}CH_2CH_2CH_2CH_2CH_3} \xleftarrow[\text{heat } (-CO_2)]{H_2O,\ H_2SO_4} CH_3\overset{O}{\overset{\|}{C}}-\overset{CO_2^-Na^+}{\overset{|}{C}H}-CH_2CH_2CH_2CH_3$$

Example 21.2 Preparation of 3-methylpentanoic acid from diethyl malonate.

$$CH_2(CO_2C_2H_5)_2 \longrightarrow CH_3CH_2\overset{CH_3}{\overset{|}{C}H}-CH_2-CO_2H$$

Solution This synthesis illustrates the use of an activating group ($-CO_2C_2H_5$) to alkylate an ester. For preparing derivatives of *acetic acid* there is a commercially available reagent (*diethyl malonate*) that already contains the activating group. The necessary reaction sequence therefore involves alkylation followed by ester hydrolysis and decarboxylation.

$$CH_2(CO_2C_2H_5)_2 \xrightarrow[2.\ CH_3CH_2\underset{Br}{\underset{|}{C}H}CH_3]{1.\ NaOC_2H_5} CH_3CH_2\overset{CH_3}{\overset{|}{C}H}-CH(CO_2C_2H_5)_2 \xrightarrow[2.\ H_2SO_4,\ \text{heat } (-CO_2)]{1.\ KOH,\ H_2O} \underset{\text{50–53\% (overall yield)}}{CH_3CH_2\overset{CH_3}{\overset{|}{C}H}-CH_2CO_2H}$$

Example 21.3 Preparation of 1-phenyl-2-propanone from phenylacetonitrile and ethyl acetate.

$$C_6H_5CH_2-CN + C_2H_5O-\overset{O}{\overset{\|}{C}}-CH_3 \longrightarrow C_6H_5-CH_2-\overset{O}{\overset{\|}{C}}-CH_3$$

Solution This example shows how a ketone can be prepared via a mixed ester condensation (acylation) followed by hydrolysis and decarboxylation. That one of the "esters" is actually a nitrile in this case is only a minor change. The acylation reaction affords a β-keto nitrile instead of a β-keto ester.

$$C_6H_5CH_2CN + C_2H_5OC(=O)CH_3 \xrightarrow[\text{2. } CH_3CO_2H \text{ (neutralize)}]{\text{1. } NaOC_2H_5} C_6H_5-CH(CN)-C(=O)-CH_3 \xrightarrow[\text{heat}]{H_2O,\ H_2SO_4} C_6H_5-CH_2-C(=O)-CH_3$$

45–55% (overall yield)

In the first step the desired mixed condensation is favored by two factors. First, the phenyl substituent makes the nitrile considerably more acidic than the ester, and this results in preferential formation of the desired enolate. Second, nitriles are less susceptible to nucleophilic attack than are esters (Section 16.3), so this ensures that a mixed condensation will take place rather than self-condensation of the nitrile.

Example 21.4 Preparation of 2-oxoglutaric acid from diethyl succinate and diethyl oxalate.

$$CH_2(CO_2C_2H_5)-CH_2(CO_2C_2H_5) + C_2H_5O_2C-CO_2C_2H_5 \longrightarrow HO_2C-C(=O)-CH_2-CH_2-CO_2H$$

Solution The use of two diesters makes the reactions seem more complex at first, but in most respects this is nearly the same as the previous example. A mixed ester condensation results in acylation of one ester by a second ester, and the product is hydrolyzed and decarboxylated. The mixed ester condensation is favored because diethyl oxalate has no acidic hydrogens, so the keto ester is formed in excellent yield (86–91%). The hydrolysis and decarboxylation of this keto triester have already been discussed in Section 21.6, where we pointed out that only the single carboxyl group that is β to a carbonyl undergoes decarboxylation.

$$CH_2(CO_2C_2H_5)-CH_2(CO_2C_2H_5) + C_2H_5O_2C-CO_2C_2H_5 \xrightarrow[\text{2. HCl (neutralize)}]{\text{1. } NaOC_2H_5} O{=}C(CO_2C_2H_5)-CH(CO_2C_2H_5)-CH_2-CO_2C_2H_5 \xrightarrow[\text{heat}]{H_2O,\ HCl} \left[O{=}C(CO_2H)-CH(CO_2H)-CH_2-CO_2H\right] \xrightarrow{-CO_2} O{=}C(CO_2H)-CH_2-CH_2-CO_2H$$

63–76% overall yield

Example 21.5 Synthesis of 2-phenylsuccinic acid from benzaldehyde, diethyl malonate, and potassium cyanide.

$$C_6H_5CHO + CH_2(CO_2C_2H_5)_2 + KCN \longrightarrow C_6H_5{-}CH(CO_2H){-}CH_2{-}CO_2H$$

Solution The synthesis we will consider here is more complex than the preceding examples because it involves two different carbon-carbon bond-forming reactions. The first step is an aldol-type condensation between benzaldehyde and diethyl malonate catalyzed by an ammonium salt (Knoevenagel condensation). Next is the conjugate addition of cyanide to the unsaturated ester formed in the first step. Finally, hydrolysis and decarboxylation afford the desired product.

$$C_6H_5CHO + CH_2(CO_2C_2H_5)_2 \xrightarrow{\text{(piperidinium) } \overset{+}{N}H_2\,C_6H_5CO_2^-} C_6H_5CH{=}C(CO_2C_2H_5)_2$$

$$\xrightarrow{KCN,\ H_2O} C_6H_5{-}CH(CN){-}CH(CO_2C_2H_5)_2 \xrightarrow[\text{heat } (-CO_2)]{HCl,\ H_2O} C_6H_5{-}CH(CO_2H){-}CH_2{-}CO_2H$$

Example 21.6 Synthesis of 2-phenylglutaric anhydride from ethyl phenylcyanoacetate and acrylonitrile.

$$C_6H_5{-}CH(CN)(CO_2C_2H_5) + CH_2{=}CH{-}CN \longrightarrow \text{2-phenylglutaric anhydride (}C_6H_5\text{-substituted cyclic anhydride)}$$

Solution The first step in the sequence is a Michael reaction, in which the enolate of the cyanoacetate undergoes conjugate addition to the unsaturated nitrile. The product of this reaction is then hydrolyzed and decarboxylated, and the resulting diacid is converted to the anhydride by treatment with acetic anhydride (Section 16.6).

$$C_6H_5{-}CH(CO_2C_2H_5){-}CN + CH_2{=}CH{-}CN \xrightarrow[\text{2. HCl (neutralize)}]{\text{1. KOH}} C_6H_5{-}C(CH_2CH_2CN)(CO_2C_2H_5){-}CN$$

$$\xrightarrow{HCl,\ H_2O,\ heat} \left[C_6H_5{-}C(CO_2H)(CO_2H){-}CH_2CH_2CO_2H \right] \xrightarrow{(-CO_2)} C_6H_5{-}CH(CO_2H){-}CH_2CH_2CO_2H \xrightarrow{(CH_3CO)_2O} \text{cyclic anhydride (}C_6H_5\text{-substituted)}$$

57–71% overall yield

The last example, together with several others that we have presented in this section, helps to emphasize the idea that nitriles and esters can be used almost interchangeably in organic synthesis. Their α hydrogens have nearly the same acidity, and the only substantial difference between the two is the greater resistance of the cyano group to nucleophilic attack (Example 21.3).

Synthetic Planning

The six examples that we have presented in this section show ways in which β-dicarbonyl compounds can be used in synthetic chemistry. But how can you analyze a target molecule to determine what is the best (or at least a reasonable) method of synthesis? We will present several more examples that show the sort of reasoning process you need to employ, but first it is necessary to establish a few basic structural relationships between reactants and products.

As we have stated previously (Sections 1.12, 5.14, 7.10, and 11.10), synthetic planning requires you to deduce reasonable precursors for a given compound. This means that you must evaluate the various alternatives: what different reactions will afford a particular type of compound? We suggested previously that you first determine the *key functional groups,* and that is still the best procedure. Many of the reactions we are utilizing in this chapter yield *difunctional* compounds, so it is equally important for you to identify the relationship *between* the key functional groups. Some of these were depicted in Figure 14.4, and you should review that information. (No activating groups are shown in that figure, and you should consider the use of α-carboxyl groups to direct and improve those reactions.) Figure 21.5 summarizes the structural relationships among some different types of carbonyl compounds and the reactions that can be used to carry out the conversions. These patterns may help you to select a useful strategy for synthesis of such a difunctional compound.

1,3-Dicarbonyl compounds (β-dicarbonyl compounds):

via *acylation*

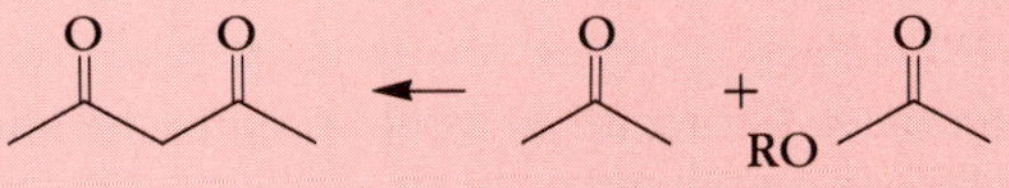

α,β-Unsaturated carbonyl compounds:

via *aldol-type condensation*

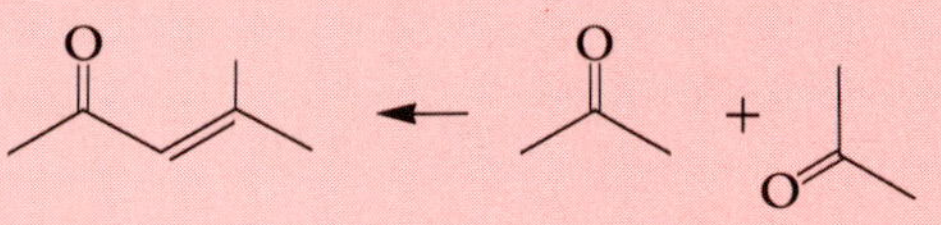

1,5-Dicarbonyl compounds:

via *Michael addition*

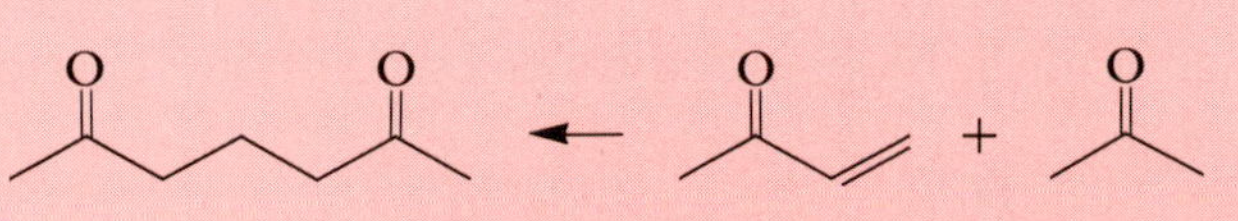

1,4-Dicarbonyl compounds:

via *conjugate addition of cyanide ion*

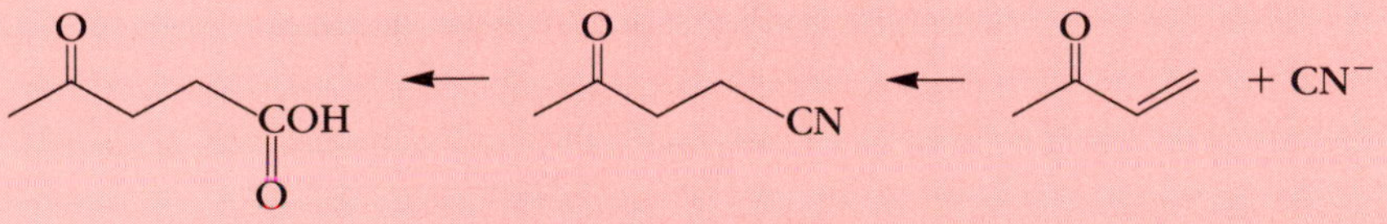

$+ CN^-$

Figure 21.5 Some difunctional carbonyl compounds and methods that can be used to prepare them.

Example 21.7 Suggest a method for synthesis of 2-heptanone starting from compounds containing four or fewer carbon atoms.

Solution Two pieces of information are available to design a synthetic procedure: the carbonyl group is the key functional group, and the allowed starting materials are limited to four carbon atoms. The target molecule contains only seven carbons, so one obvious possibility is the union of a three-carbon fragment with a fragment containing four carbons. Conceptually this would require formation of a carbon-carbon bond, as indicated by the following two possibilities:

A B

The first alternative (structure A) would require the joining of a C_4 group to the α carbon of a ketone, in other words, an alkylation. Of course, it would be preferable to use an activating group, so the necessary reactant would be ethyl acetoacetate.

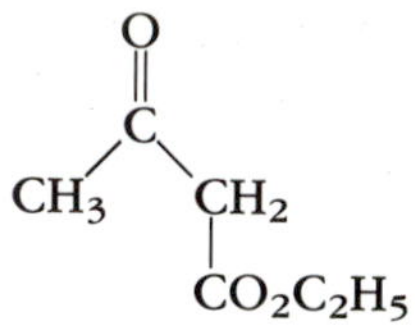

$$CH_3-\overset{\overset{O}{\|}}{C}-CH_2-CO_2C_2H_5$$

Ethyl acetoacetate is a four-carbon compound (as in earlier chapters, we will only use the longest *carbon chain* to meet the criterion for allowed starting materials) and is therefore an allowed starting material, although for the purpose of illustration it might be prepared either by acylation of acetone with diethyl carbonate

$$CH_3-\overset{\overset{O}{\|}}{C}-CH_3 + C_2H_5O\overset{\overset{O}{\|}}{C}OC_2H_5 \xrightarrow[\text{2. HCl (neutralize)}]{\text{1. NaH}} CH_3-\overset{\overset{O}{\|}}{C}-CH_2-\overset{\overset{O}{\|}}{C}OC_2H_5$$

or by self-condensation of ethyl acetate

$$CH_3CO_2C_2H_5 \xrightarrow[\text{2. HCl (neutralize)}]{\text{1. NaOC}_2\text{H}_5} CH_3-\overset{\overset{O}{\|}}{C}-CH_2-CO_2C_2H_5$$

The remaining C-4 group could be introduced by alkylation with 1-bromobutane, and subsequent hydrolysis and decarboxylation would yield 2-heptanone. In fact we have already shown this synthesis in detail in Example 21.1.

Alternative B could also be used to prepare 2-heptanone. This would require joining a three-carbon fragment to the β-carbon of a ketone, and this corresponds to conjugate addition. Therefore 2-heptanone could be prepared in a single step by reaction of 3-buten-2-one with lithium dipropylcuprate (see Section 13.10).

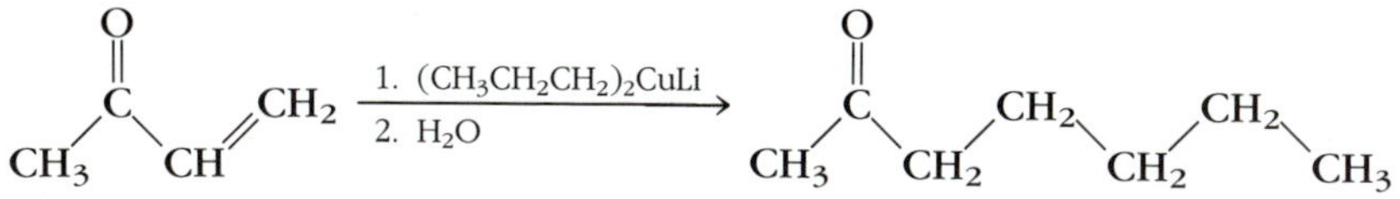

$$CH_3-\overset{\overset{O}{\|}}{C}-CH=CH_2 \xrightarrow[\text{2. H}_2\text{O}]{\text{1. (CH}_3\text{CH}_2\text{CH}_2)_2\text{CuLi}} CH_3-\overset{\overset{O}{\|}}{C}-CH_2-CH_2-CH_2-CH_2-CH_3$$

Example 21.8 Suggest a method for synthesis of 3-methylpentanoic acid from starting materials containing four or fewer carbon atoms.

$$CH_3CH_2\underset{}{\overset{\overset{CH_3}{|}}{C}H}-CH_2-CO_2H$$

Solution Once again the key functional group (CO_2H) and the four-carbon limitation suggest two alternative approaches.

$$\underbrace{CH_3CH_2{-}\underset{|}{\overset{CH_3}{C}}H{-}CH_2CO_2H}_{\textbf{A}} \quad \text{or} \quad \underbrace{CH_3CH_2\underset{|}{\overset{CH_3}{C}}H{-}CH_2CO_2H}_{\textbf{B}}$$

Approach A could be carried out via a conjugate addition analogous to that shown in the previous example. Note, however, that the free acid would destroy the organometallic reagent, so it would be necessary to use the corresponding ester (or nitrile) and later hydrolyze it to the acid.

$$CH_3{-}CH{=}CH{-}CO_2C_2H_5 \xrightarrow[2.\ H_2O]{1.\ (C_2H_5)_2CuLi} CH_3CH_2{-}\overset{CH_3}{\overset{|}{C}}H{-}CH_2{-}CO_2C_2H_5$$

$$\xrightarrow{H_2O,\ H_2SO_4} CH_3CH_2\overset{CH_3}{\overset{|}{C}}H{-}CH_2{-}CO_2H$$

The second alternative requires formation of a bond to the α position of a carboxylic acid, and this should immediately suggest a malonic ester synthesis. In fact, the synthesis of 3-methylpentanoic acid via *alkylation* of diethyl malonate has been shown in Example 21.2. Yet still another possibility remains: the malonic ester could be used in an aldol-type condensation. This would generate the required carbon skeleton, and reduction of the double bond could be followed by hydrolysis and decarboxylation.

$$(CH_3)(CH_3CH_2)C{=}O \xrightarrow[NH_4^+CH_3CO_2^-]{CH_2(CO_2C_2H_5)_2} (CH_3)(CH_3CH_2)C{=}C(CO_2C_2H_5)_2$$

$$\xrightarrow{H_2,\ Pt} CH_3CH_2\overset{CH_3}{\overset{|}{C}}H{-}CH(CO_2C_2H_5)_2$$

$$\xrightarrow[(-CO_2)]{1.\ NaOH,\ H_2O;\ 2.\ H_2SO_4,\ heat} CH_3CH_2\overset{CH_3}{\overset{|}{C}}H{-}CH_2CO_2H$$

Example 21.9 Suggest a method for synthesizing 3-methylglutaric anhydride starting from compounds containing four or fewer carbon atoms.

O
O
O

Solution The first step in answering this problem is to recognize that this is a derivative of the acyclic diacid, and that the diacid is a 1,5-dicarbonyl compound.

$$HO\overset{\overset{O}{\|}}{C}-CH_2-\overset{\overset{CH_3}{|}}{C}H-CH_2\overset{\overset{O}{\|}}{C}OH$$

The 1,5-dicarbonyl relationship strongly suggests a Michael addition (Figure 21.5), and such an addition could be carried out in only one way because the target molecule is symmetrical. It is therefore necessary to use the enolate of one acid derivative to attack a second α,β-unsaturated compound. As in other examples, a malonic ester is a superior reagent to a simple ester of acetic acid.

$$C_2H_5O_2C-CH{=}CH-CH_3 + CH_2(CO_2C_2H_5)_2 \xrightarrow[\text{2. HCl (neutralize)}]{\text{1. NaOC}_2\text{H}_5} C_2H_5O_2C-CH_2-\overset{\overset{CH_3}{|}}{C}H-CH(CO_2C_2H_5)_2$$

Hydrolysis and decarboxylation of the triester would generate the diacid, which could be converted to the cyclic anhydride by acetic anhydride.

$$C_2H_5O_2C-CH_2-\overset{\overset{CH_3}{|}}{C}H-CH(CO_2C_2H_5)_2 \xrightarrow[\text{2. H}_2\text{SO}_4\text{, heat } (-CO_2)]{\text{1. H}_2\text{O, KOH}} HO_2C-CH_2-\overset{\overset{CH_3}{|}}{C}H-CH_2CO_2H \xrightarrow{(CH_3CO)_2O} \text{(4-methyl cyclic anhydride)}$$

Example 21.10 Suggest a method for preparing 1,3-diphenyl-2-propanone from starting materials containing eight or fewer carbon atoms. You may not use any reagents that contain metallic elements other than sodium.

$$C_6H_5-CH_2-\overset{\overset{O}{\|}}{C}-CH_2-C_6H_5$$

Solution The limitation of metallic elements to sodium precludes the methods for ketone synthesis via organometallic reagents that we discussed in Section 13.4. Consequently we can turn our attention to synthetic reactions that involve enolate ions. The target molecule is a substituted acetone, and this may at first suggest an alkylation reaction. But only *alkyl* groups can be introduced by that process. The only other possibility is a condensation reaction, and that would afford a β-dicarbonyl compound. The target molecule has only a single carbonyl group, so we can now deduce that a second carbonyl group was removed via decarboxylation, and a new target molecule can be drawn.

$$C_6H_5CH_2-\overset{\overset{\large O}{\|}}{C}-\underset{\underset{\large CO_2H}{|}}{CH}-C_6H_5$$

Now two equivalent eight-carbon fragments ($C_6H_5—CH_2CO—$) can be recognized in the structure, and an ester condensation would produce the needed carbon skeleton. Subsequent hydrolysis and decarboxylation would complete the synthesis.

$$C_6H_5CH_2\overset{\overset{\large O}{\|}}{C}OC_2H_5 \xrightarrow[\text{2. HCl (neutralize)}]{\text{1. NaOC}_2\text{H}_5} C_6H_5CH_2\overset{\overset{\large O}{\|}}{C}-\underset{\underset{\large CO_2C_2H_5}{|}}{CH}-C_6H_5$$

$$\xrightarrow[\text{2. H}_2\text{SO}_4\text{, heat } (-\text{CO}_2)]{\text{1. H}_2\text{O, NaOH}} C_6H_5CH_2-\overset{\overset{\large O}{\|}}{C}-CH_2C_6H_5$$

EXERCISE 21.14

Suggest methods for synthesizing the following compounds from the indicated starting materials:

(a) $C_6H_5-CH_2-\overset{\overset{\large O}{\|}}{C}-\underset{\underset{\large CH_3}{|}}{CH}-CH_3$ from phenylacetonitrile and any other reagents you wish

(b) $CH_3CH_2\overset{\overset{\large O}{\|}}{C}-\underset{\underset{\large CH_3}{|}}{CH}-CH_2CH_3$ from ethyl propionate and ethyl iodide

(c) $CH_3CH_2CH_2-\underset{\underset{\large CH_3}{|}}{CH}-CO_2H$ from diethyl malonate and any other reagents you wish

(d) $H_2OC-CH_2-\underset{\underset{\large CH_3}{|}}{CH}-CH_2-CO_2H$ from ethyl cyanoacetate and ethyl crotonate (ethyl 2-butenoate)

21.8 BIOSYNTHESIS OF FATTY ACIDS: β-DICARBONYL COMPOUNDS IN BIOLOGICAL PROCESSES

Fatty acids are long-chain carboxylic acids that are normally found as esters in fatty tissues. Both plant and animal fats consist largely of triglycerides, that is, triesters of glycerol (1,2,3-propanetriol).

$$\begin{array}{l} CH_2\text{—O—}\overset{O}{\overset{\|}{C}}R \\ | \\ CH\text{—O—}\overset{O}{\overset{\|}{C}}R' \\ | \\ CH_2\text{—O—}\overset{O}{\overset{\|}{C}}R'' \end{array}$$

General formula for triglycerides—esters of glycerol with the fatty acids RCO_2H, $R'CO_2H$, and $R''CO_2H$. Frequently all three acyl groups are the same.

Saponification of these esters yields salts of the fatty acids, and these have long been used as soap (Chapter 16).

The most abundant fatty acids are C_{16} and C_{18} compounds such as palmitic, stearic, oleic, and linoleic acids.

C_{16}	$CH_3\text{—}(CH_2)_{14}\text{—}CO_2H$	Palmitic acid
C_{18}	$CH_3\text{—}(CH_2)_{16}\text{—}CO_2H$	Stearic acid
C_{18}	$CH_3\text{—}(CH_2)_7\text{—}CH{=}CH\text{—}(CH_2)_7\text{—}CO_2H$	Oleic acid
C_{18}	$CH_3\text{—}(CH_2)_4\text{—}CH{=}CH\text{—}CH_2\text{—}CH{=}CH\text{—}(CH_2)_7\text{—}CO_2H$	Linoleic acid

Triglycerides of the unsaturated fatty acids are predominant in many vegetable oils. Linseed oil contains a high concentration of esters of the unsaturated acid linoleic acid, and their polymerization is used to prepare the floor covering known as *linoleum.*

Nearly all the naturally occurring fatty acids contain an *even number* of carbon atoms, and this suggested to chemists that they were all derived from acetic acid as the basic subunit. Only recently was it shown that the actual "reagent" in fatty acid synthesis is a derivative of *malonic* acid rather than acetic acid. The actual biosynthesis is complicated, and it requires a system composed of seven different enzymes. But just as the pathway has evolved in a way that utilizes malonic acid for carbon-carbon bond formation, the other steps in the biosynthesis also proceed via reactions that are very similar to organic reactions that we have discussed in this text. In this section we will show you the biosynthesis of fatty acids from the perspective of organic chemistry. We will focus not on the enzymes involved but on the reactions that they catalyze.

Nearly all the intermediates in fatty acid biosynthesis are *thioesters,* that is, esters of thiols (R—SH). Thiols are more acidic and more nucleophilic than alcohols, but there is no evidence to show why this particular enzymatic pathway evolved. In order to simplify our discussion of the biosynthesis we will first define some of the biological "reagents" that will be encountered. Figure 21.6 shows two *coenzymes* (organic groups that function together with the enzyme in a reaction), *coenzyme A* and *NADPH* (the reduced form of *n*icotinamide *a*denine *d*inucleotide *p*hosphate). The —SH group of coenzyme A is readily acylated to give a thioester, and it acts as a carrier of acyl groups. NADPH functions as a reducing agent by transfer of a hydride.

A key step prior to the actual buildup of the fatty acid chain, is the carboxylation of the acetyl derivative of coenzyme A to form the malonyl derivative. (The enzymatic process takes place via the intermediacy of an *N*-carboxyl derivative of biotin,

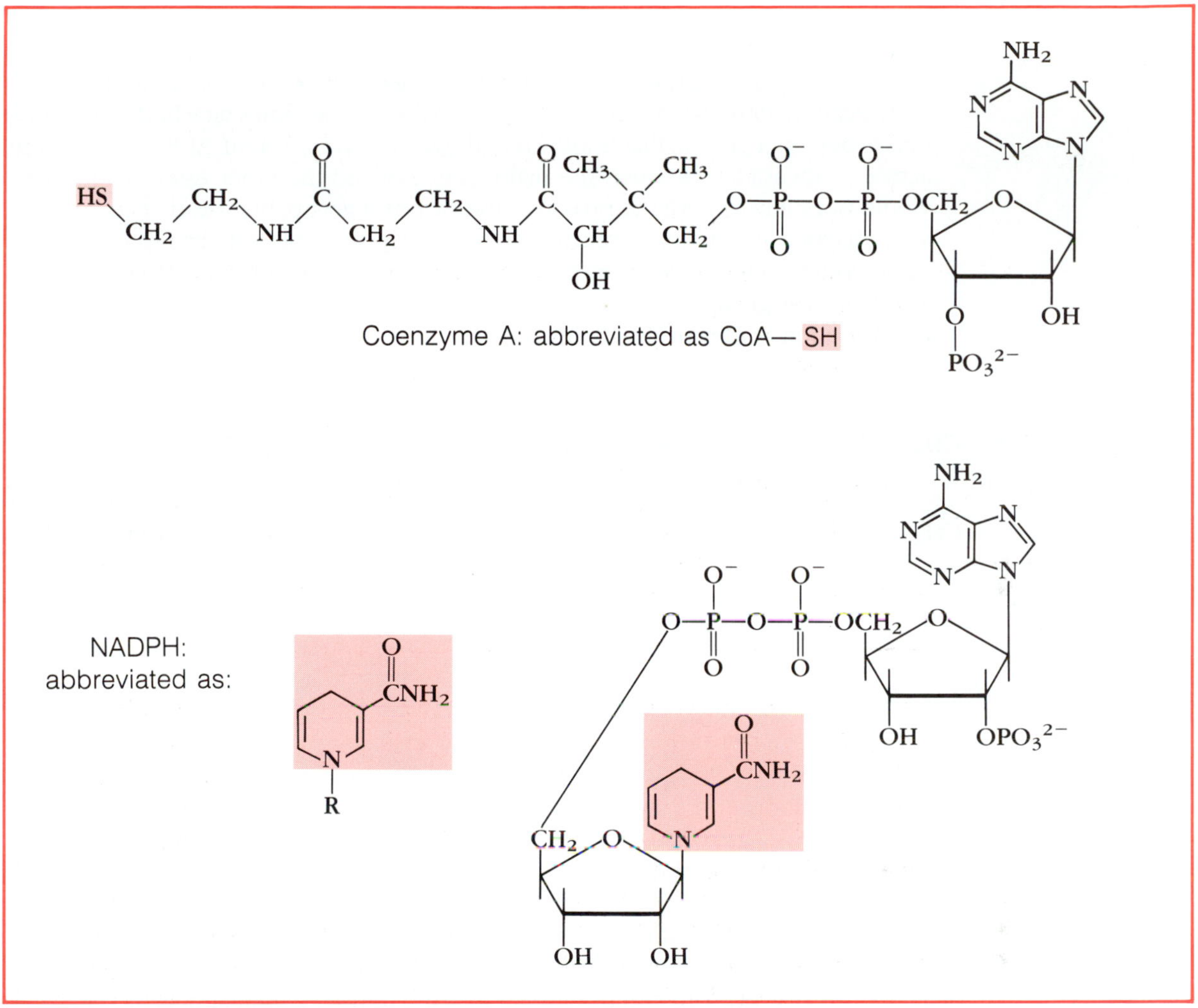

Figure 21.6 Complete and abbreviated structures for coenzyme A and NADPH, two important coenzymes in fatty acid biosynthesis.

but we will not show that here.) In a formal sense this involves a condensation reaction between an ester and carbon dioxide.

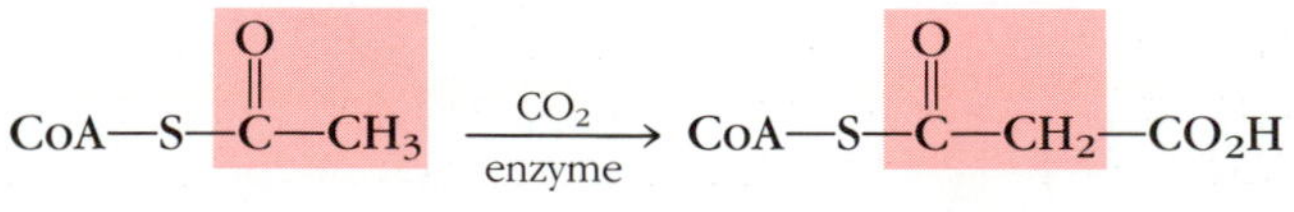

$$\text{CoA—S—}\overset{\text{O}}{\overset{\|}{\text{C}}}\text{—CH}_3 \xrightarrow[\text{enzyme}]{CO_2} \text{CoA—S—}\overset{\text{O}}{\overset{\|}{\text{C}}}\text{—CH}_2\text{—CO}_2\text{H}$$

Acetyl CoA — Malonyl CoA

These two reagents, acetyl coenzyme A and malonyl coenzyme A, provide all the carbon atoms for fatty acid biosynthesis. In fact, all the carbons can be traced back to acetic acid, but the free carboxyl group of malonyl CoA serves as an activating group that permits condensation reactions to occur under very mild conditions.

The remaining steps for building up the fatty acid skeleton are shown in Figure 21.7.

The first step is simply the acetylation of a particular thiol group in the enzyme system by a transesterification reaction with acetyl CoA. Next is a similar trans-esterification with malonyl CoA. (The malonyl group becomes attached to a sulfhydryl group on a protein that is known as the *a*cyl *c*arrier *p*rotein, ACP.) Step 3 is the actual condensation between the malonyl group and the other ester group. Base would normally remove the proton of the carboxyl group, but this does not necessarily happen in enzymatic reactions. A properly situated basic group on the enzyme could preferentially remove a proton from the CH_2 of the malonyl residue. The two acyl groups are held in close proximity in the enzyme system, so the condensation is effectively intramolecular:

$$-S-C(=O)-CH_3,\ -S-C(=O)-CH_2-CO_2H \longrightarrow -S-C(=O)-CH_3,\ -S-C(=O)-\bar{C}H-CO_2H \longrightarrow -S-C(O^-)(CH_3)-CH(CO_2H)-C(=O)-S-$$

$$\downarrow$$

$$-SH,\ -S-C(=O)-CH_2-C(=O)-CH_3 \longleftarrow -SH,\ -S-C(=O)-CH=C(OH)-CH_3 \xleftarrow{-CO_2} -SH,\ -S-C(=O)-CH(CO_2H)-C(=O)-CH_3$$

Decarboxylation of the initial product (step 4) is particularly easy here because there are *two* carbonyl groups β to the $-CO_2H$. At this stage of the biosynthesis the chain length has been increased by two carbons, but the chain still contains a keto group.

Steps 5–8 of Figure 21.7 result in the conversion of the ketone carbonyl of the β-ketoester to a CH_2 group. Initially (step 5), the carbonyl is reduced to a secondary alcohol by NADPH. This compound is a hydride transfer agent, and its biochemical reactivity is comparable with that of $NaBH_4$ in the laboratory.

$$\text{(1,4-dihydropyridine, R on N, } CONH_2\text{)} + >C=O \xrightarrow{H_2O} \text{(pyridinium, } {}^{+}N-R\text{, } CONH_2\text{)} + H-C(OH)<$$

Step 6 is a simple dehydration of the β-hydroxy ester, and step 7 is another reduction. The reducing agent is again NADPH, and you might argue that this is inconsistent with our previous statement that NADPH is analogous to sodium borohydride. After all, sodium borohydride reduces carbonyl groups, and step 7 is a carbon-carbon double bond reduction. But there is no inconsistency, because the double bond is conjugated with the ester carbonyl. Reduction proceeds via conjugate addi-

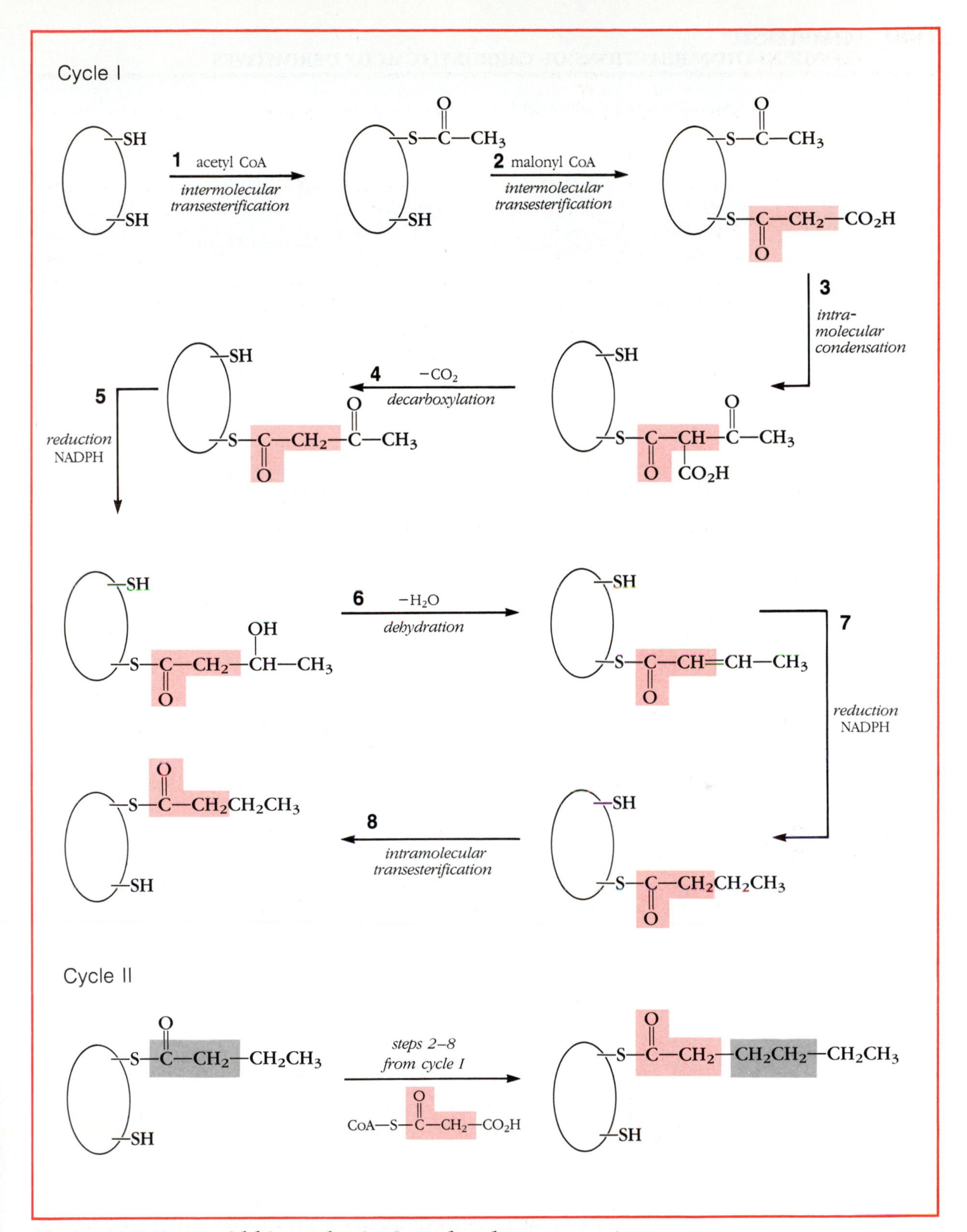

Figure 21.7 Fatty acid biosynthesis via malonyl coenzyme A. The individual steps are explained in the text. The different enzymes are represented as a single unit for convenience. Only the first two cycles are shown, but each cycle lengthens the carbon chain by two CH_2 groups (shaded portions of structures).

tion of hydride followed by conversion of the ester enol (or enolate) to the keto form.

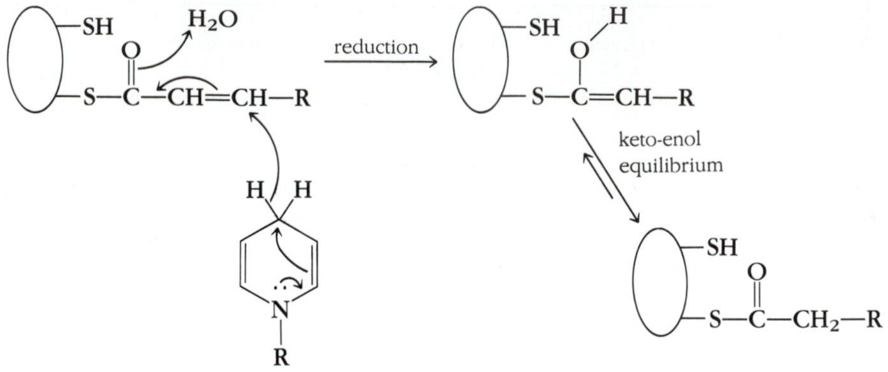

reduction

keto-enol equilibrium

The last step of the first cycle is a transesterification of the acyl group from the acyl carrier protein to the other thiol group. The reaction is shown here as proceeding by a base-catalyzed mechanism, which would require a basic group in the enzyme system to remove a proton from the —SH group. The resulting anion would then attack the acyl group that is held in close proximity.

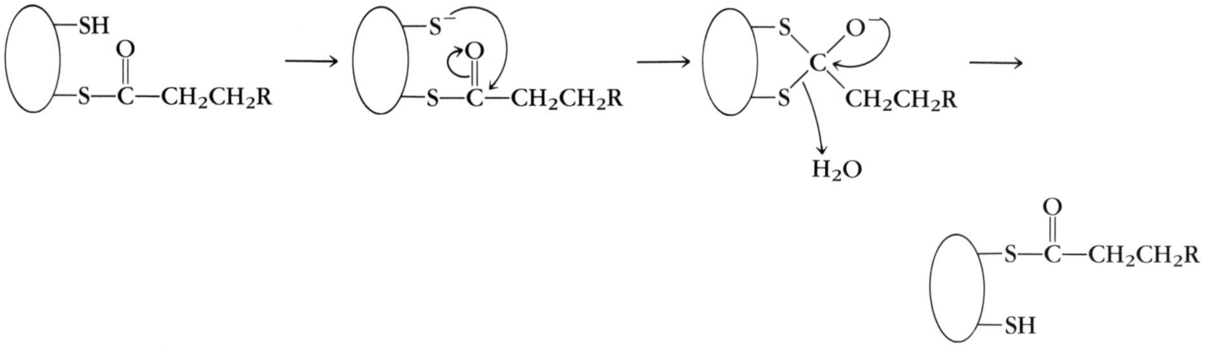

This last step frees the thiol group of the acyl carrier protein so that it can be acylated by another malonyl CoA to start the second cycle. Each cycle increases the chain length by two CH_2 units.

The unsaturated fatty acids are formed by the same series of reactions, and the double bonds are introduced at a later stage. For example, stearic acid is the biological precursor of oleic acid.

$$\underset{\text{Stearic acid}}{CH_3{-}(CH_2)_7{-}CH_2CH_2{-}(CH_2)_7{-}CO_2H} \xrightarrow[\text{oxidation}]{\text{biological}} \underset{\text{Oleic acid}}{CH_3{-}(CH_2)_7{-}CH{=}CH{-}(CH_2)_7{-}CO_2H}$$

The preceding reaction provides you with another example of the ways in which enzymes carry out reactions at specific individual sites even if other chemically similar groups are present in the substrate molecule. Clearly the enzyme must bind stearic acid in such a way that C-9 and C-10 are held in close proximity to the group that actually brings about the oxidation.

Fatty acid biosynthesis also reaffirms an important feature of biological reactions. Even though enzymes may permit selective reaction at a particular site, the reaction mechanism almost always corresponds to one of the simple organic reactions that we have presented in this text. For the same reasons that β-dicarbonyl compounds are useful synthetic intermediates in the laboratory, the biological system employs a malonic acid derivative for the carbon-carbon bond-forming step in fatty acid biosynthesis.

EXERCISE 21.15

Draw the structure of stearic acid. Indicate which carbon atoms are derived from malonyl CoA and which are derived from acetyl CoA without the intervention of malonyl CoA.

EXERCISE 21.16

Lauric acid, $C_{12}H_{24}O_2$, is produced in fatty acid biosynthesis. (a) From the mechanism of the biosynthesis deduce its structure. (b) Indicate which hydrogen atoms in the molecule are derived from NADPH.

21.9 SUMMARY OF REACTIONS

Some methods for the preparation of β-dicarbonyl compounds are summarized in Table 21.2. Table 21.3 shows aldol-type condensations of carboxylic acid derivatives, and Table 21.4 summarizes the alkylation and decarboxylation reactions of β-dicarbonyl compounds.

TABLE 21.2 Preparation of β-Dicarbonyl Compounds

Reaction	Comments
1. Ester condensation	Section 21.2 Produces β-ketoesters. Strong bases such as NaH can be used. The reactions are reversible.
(a) Self-condensation 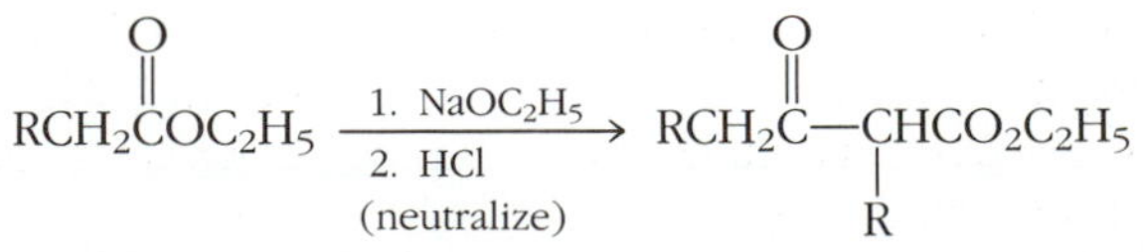 $RCH_2\overset{O}{\overset{\|}{C}}OC_2H_5 \xrightarrow[\text{2. HCl (neutralize)}]{\text{1. } NaOC_2H_5} RCH_2\overset{O}{\overset{\|}{C}}-\underset{R}{\underset{\|}{C}}HCO_2C_2H_5$	

TABLE 21.2 Preparation of β-Dicarbonyl Compounds (continued)

Reaction	Comments
(b) Mixed condensation	
$Ar-C(=O)-OC_2H_5 + RCH_2CO_2C_2H_5 \xrightarrow[\text{2. HCl (neutralize)}]{\text{1. NaOC}_2\text{H}_5} Ar-C(=O)-CH(R)CO_2C_2H_5$	Best results when one ester has no α hydrogens.
2. Acylation of ketones or nitriles	Section 21.2
$R-CO_2C_2H_5 + R'CH_2CN \xrightarrow[\text{2. HCl (neutralize)}]{\text{1. NaOC}_2\text{H}_5} RC(=O)-CH(R')CN$	Strong bases such as NaH are often used. Mixed condensation; preferential formation of just one enolate is needed. Yields β-keto nitriles or β-diketones.
$CH_3C(=O)CH_2- + RC(=O)OC_2H_5 \xrightarrow[\text{2. HCl (neutralize)}]{\text{1. NaOC}_2\text{H}_5} CH_3C(=O)CH(-)C(=O)R$	
$CH_3C(=O)CH_2- + C_2H_5OC(=O)OC_2H_5 \xrightarrow[\text{2. HCl (neutralize)}]{\text{1. NaH}} CH_3C(=O)CH(-)CO_2C_2H_5$	Acylation of a ketone with diethyl carbonate yields a β-keto ester.

TABLE 21.3 Aldol-Type Condensations of Carboxylic Acid Derivatives

Reaction	Comments
1. Reformatsky reaction	Section 21.3
$R_2C{=}O + BrCH(R')CO_2C_2H_5 \xrightarrow{\text{1. Zn; 2. HCl (neutralize)}} R_2C(OH){-}CH(R'){-}CO_2C_2H_5$	Yields β-hydroxy esters via zinc enolate.
2. Stobbe condensation	Section 21.3
$C_2H_5O_2C{-}CH_2{-}CH_2{-}CO_2C_2H_5 + O{=}CR_2 \xrightarrow{\text{1. NaH; 2. HCl (neutralize)}} R_2C{=}C(CH_2CO_2H)(CO_2C_2H_5)$	Condensation between succinic esters and ketones to give the unsaturated ester-acid. (Alkoxide salts can also be used as bases.)
3. Perkin condensation	Section 21.3
$ArCHO + (RCH_2\overset{O}{\overset{\Vert}{C}})_2O \longrightarrow ArCH{=}C(R){-}\overset{O}{\overset{\Vert}{C}}{-}OH$	Condensation between an anhydride and an aromatic aldehyde. (The corresponding carboxylate salt serves as the base.)
4. Knoevenagel condensation	Section 21.4
$R_2C{=}O + CH_2(X)(Y) \xrightarrow{NH_3} R_2C{=}C(X)(Y)$	Condensation of a β-dicarbonyl compound catalyzed by amine or ammonium salt. X and Y can be CN, COR, CO_2R, CO_2H.

TABLE 21.3 Aldol-Type Condensations of Carboxylic Acid Derivatives (continued)

Reaction	Comments
5. Modified Wittig reaction	Section 21.4
$R-C(=O)-R + (C_2H_5O)_2P(=O)-CH_2-X \xrightarrow{NaH} R-C(R)=CH-X$	Effectively a condensation of a β-dicarbonyl compound. The phosphonyl group is lost during the reaction. X can be CN, COR, CO_2R.
6. Michael addition	Section 21.4
$CH_3COCH=CH_2 + R-CH(X)(Y) \xrightarrow{base} CH_3COCH_2CH_2-C(R)(X)(Y)$	Conjugate addition of enolate ion from a β-dicarbonyl compound. (See also Section 14.5). X and Y can be CN, COR, CO_2R.

TABLE 21.4 Alkylation and Decarboxylation of β-Dicarbonyl Compounds

Reaction	Comments
1. Alkylation	
$-CH(X)(Y) \xrightarrow[2.\ RX]{1.\ NaOC_2H_5} R-C(X)(Y)$	RX is an alkyl halide or tosylate. R cannot be tertiary, and best results are obtained when R is primary or methyl. X and Y can be CN, COR, CO_2R.
2. Decarboxylation	Section 21.6
$-C(=O)-C(CO_2H) \longrightarrow -C(=O)-C(H)$	Removal of activating group after alkylation or condensation reaction. The carboxyl group must be β to the carbonyl group.
$-C(=O)-C(CN) \xrightarrow[H_2O]{H_2SO_4} -C(=O)-C(CO_2H) \xrightarrow{-CO_2} -C(=O)-C(H)$	Nitriles and esters must first be hydrolyzed to $-CO_2H$ (acid hydrolysis is preferred in order to avoid retrocondensation).

21.10 PROBLEMS

21.1 Indicate the most acidic hydrogen(s) in each of the following compounds:

(a) $CH_3CH_2\overset{O}{\overset{\|}{C}}CH_2CO_2H$

(b) $CH_3\underset{CN}{\underset{|}{C}H}-\overset{O}{\overset{\|}{C}}-CH_2C_6H_5$

(c) $C_6H_5-\overset{O}{\overset{\|}{C}}-\overset{O}{\overset{\|}{C}}-CH_3$

(d) $CH_3CH_2CH_2-\overset{O}{\overset{\|}{C}}-CH_2CO_2CH_3$

(e) $(C_2H_5O)_2\overset{O}{\overset{\|}{P}}-CH_2-\overset{O}{\overset{\|}{C}}-CH_2CH_2CO_2C_2H_5$

(f) $CH_2(CO_2CH_2CH_2OH)(CO_2CH_3)$

21.2 What products would be expected if ethyl butyrate were treated with each of the following?

(a) $NaOH$, H_2O (b) $NaOC_2H_5$ (c) HCl, H_2O
(d) NaH (e) NaH, $C_6H_5CO_2CH_3$
(f) $NaOC_2H_5$, $CH_3-CH(CH_3)-CO_2C_2H_5$; then HCl (to neutralize)
(g) $C_6H_5-CH_2CN$, NaH; then HCl (to neutralize)

21.3 Complete the following reactions:

(a) 2-methylcyclohexanone $+ C_2H_5OC(=O)-C(=O)OC_2H_5 \xrightarrow[2.\ HCl]{1.\ NaH}$?

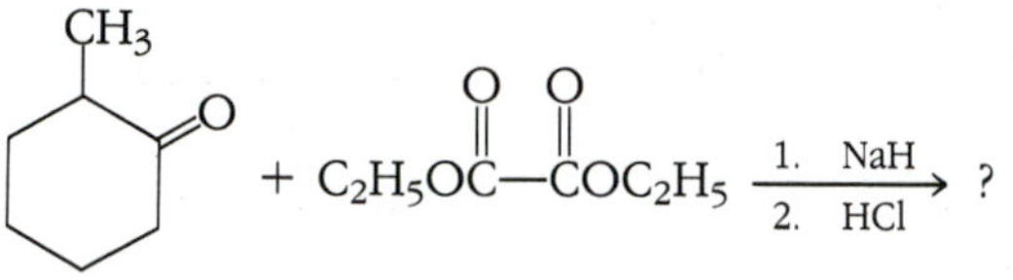

(b) cyclopentanone $+ HCO_2C_2H_5 \xrightarrow[2.\ HCl]{1.\ NaH}$?

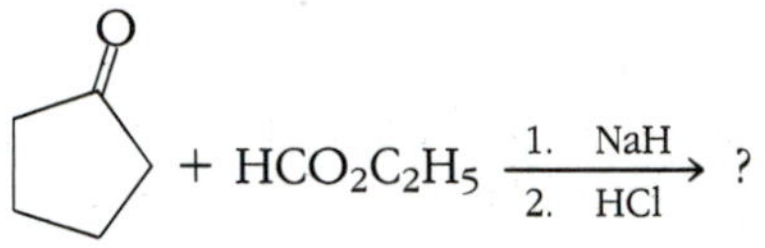

(c) cyclopentyl–$CH_2CO_2C_2H_5 \xrightarrow[2.\ HCl\ (neutralize)]{1.\ C_2H_5OC(=O)OC_2H_5,\ NaH}$?

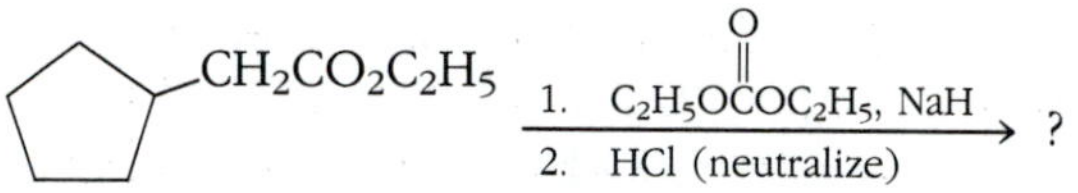

(d) $C_2H_5O_2CCH_2CH_2CO_2C_2H_5 + C_6H_5CHO \xrightarrow{NaOC_2H_5}$?

(e) cyclopentanone $\xrightarrow[CH_3CO_2^-Na^+]{(CH_3C(=O))_2O}$?

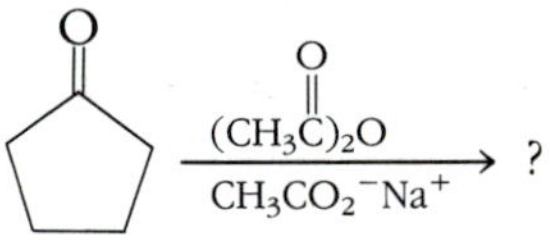

(f) $CH_3CH_2C(=O)CH_3 + BrCH_2CO_2C_2H_5 \xrightarrow[2.\ HCl]{1.\ Zn}$?

(g) $C_6H_5C(=O)CH_3 + C_6H_5CO_2CH_3 \xrightarrow[2.\ HCl]{1.\ NaOCH_3}$?

(h) $C_6H_5CH_2CN + CH_3CH_2CH_2CO_2CH_3 \xrightarrow[2.\ HCl]{1.\ NaH}$?

21.4 Suggest three different methods by which *p*-chlorobenzaldehyde could be converted to *p*-chlorocinnamic acid.

$$Cl-C_6H_4-CHO \longrightarrow Cl-C_6H_4-CH{=}CH-CO_2H$$

21.5 What products would be expected from treatment of acetophenone with the following reagents?

(a) $NC-CH_2-CO_2C_2H_5$, $NH_4^{+}\,^{-}O_2CCH_3$

(b) $C_2H_5O_2C-CH_2-P(=O)(OC_2H_5)_2$, NaH

(c) $BrCH_2CO_2C_2H_5$, Zn; then HCl (to neutralize)

(d) $(C_6H_5CH_2\overset{O}{\overset{\|}{C}})_2O$, $C_6H_5CH_2CO_2^-Na^+$

(e) $HCO_2C_2H_5$, NaH; then HCl (to neutralize)

21.6 What products would be expected from treatment of methyl cinnamate with the following reagents?
(a) NaCN, CH_3OH
(b) H_2O, H_2SO_4, heat
(c) $C_2H_5O_2C—CH_2—CN$, $NaOC_2H_5$; then H_2O, H_2SO_4, heat

21.7 What products would be formed on reaction of diethyl malonate with each of the following reagents?
(a) $NaOC_2H_5$; then $CH_3(CH_2)_6Br$
(b) Cyclohexanone, $NH_4^{+}\,{}^{-}O_2CCH_3$
(c) KOH, H_2O; then HCl (to neutralize); then heat
(d) $CH_3CH_2CH{=}CH—CO_2CH_3$, $NaOC_2H_5$; then HCl (to neutralize)

21.8 Compound **A** ($C_{13}H_{16}O_3$) was found to be a ketoester. When it was treated with concentrated NaOH, phenylacetic acid was isolated after acidification of the reaction mixture. However, when compound **A** was heated with aqueous sulfuric acid, the reaction product was 1-phenyl-2-butanone. Deduce the structure of compound **A.**

21.9 Suggest methods for carrying out the following conversions:

(a) cyclohexanone ⟶ 2-($CH_2CH_2CH_3$)cyclohexanone

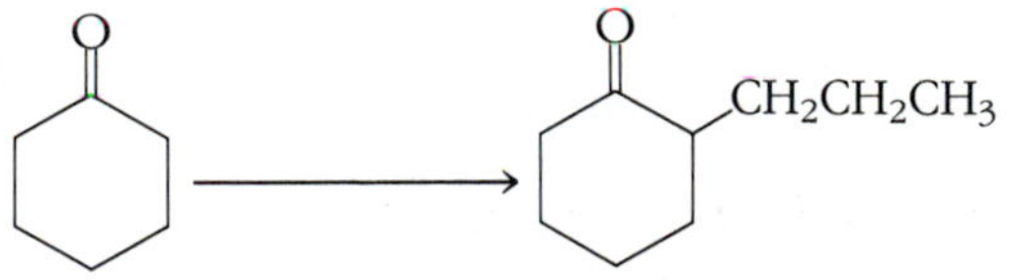

(b) cyclopentanone (=O) ⟶ cyclopentyl—CH_2CO_2H

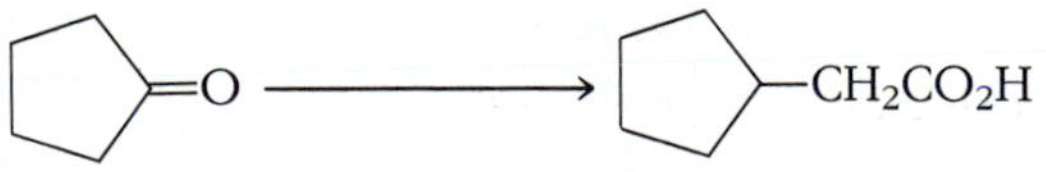

(c) cyclopentyl—CH_2Br ⟶ cyclopentyl—$CH_2CH_2—\overset{O}{\overset{\|}{C}}—CH_3$

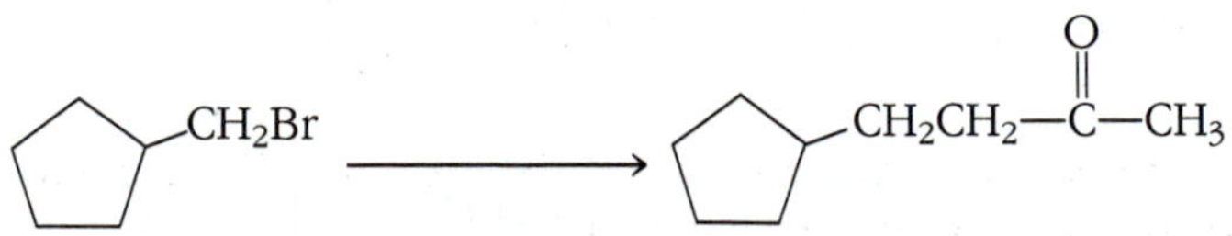

(d) cyclohexanone ⟶ 1-CH_3-1-(CH_2CO_2H)cyclohexane

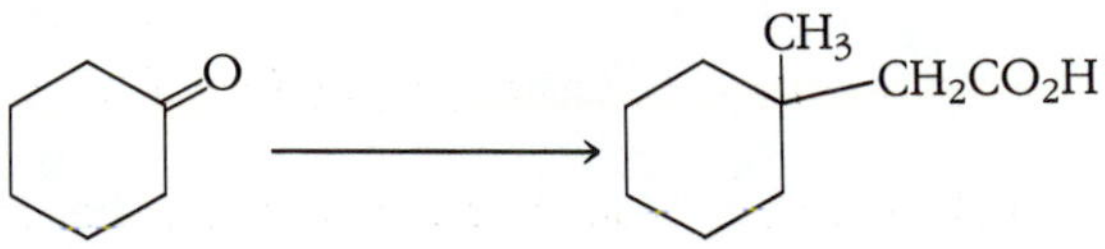

(e) $C_6H_5—CH_2Br \longrightarrow C_6H_5—\underset{CH_3}{\underset{|}{C}H}—CH_2CO_2CH_3$

(f) $C_6H_5CO_2H \longrightarrow C_6H_5—CH_2CH_2CO_2H$

21.10 Suggest methods of synthesis of the listed compounds from the indicated starting materials:

(a) 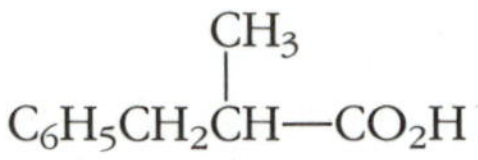 $C_6H_5CH_2CH(CH_3)—CO_2H$ — From diethyl malonate and any other reagents you wish

(b) 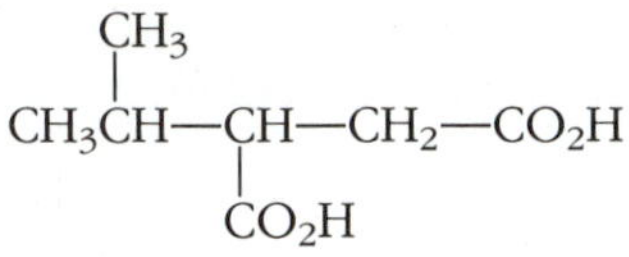 $CH_3CH(CH_3)—CH(CO_2H)—CH_2—CO_2H$ — From compounds containing four or fewer carbons

(c) 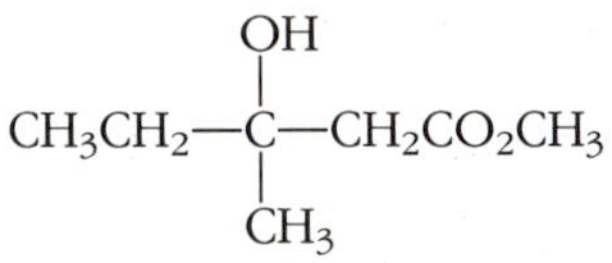 $CH_3CH_2—C(OH)(CH_3)—CH_2CO_2CH_3$ — From compounds containing four or fewer carbons

(d) 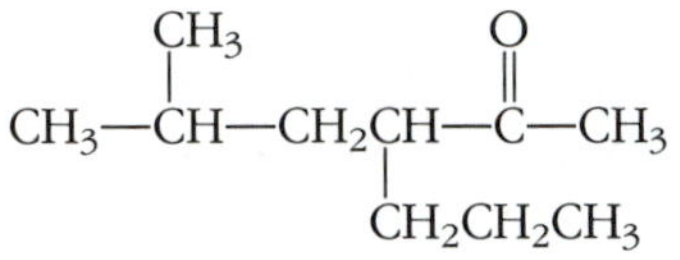 $CH_3—CH(CH_3)—CH_2CH(CH_2CH_2CH_3)—C(=O)—CH_3$ — From compounds containing four or fewer carbons

21.11 One of the common by-products of a Knoevenagel condensation contains *two* groups corresponding to the original β-dicarbonyl compound. For example, diethyl malonate can react with acetaldehyde to give a tetraester.

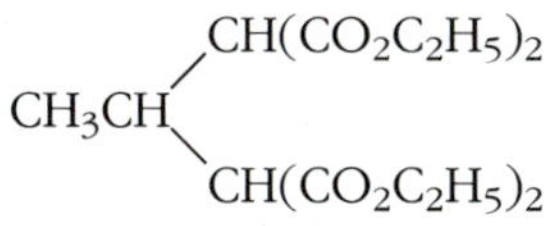

$$CH_3CH[CH(CO_2C_2H_5)_2]_2$$

Write a reasonable reaction mechanism that accounts for the formation of this compound.

21.12 The reaction of 2-furaldehyde with phenylacetic acid, acetic anhydride, and sodium acetate affords compound **A** as the major product. The reaction proceeds via formation of phenylacetic anhydride (or the mixed anhydride) in the reaction mixture by interaction of the acid with acetic anhydride (Section 16.6). Explain why **A** is formed in preference to **B.**

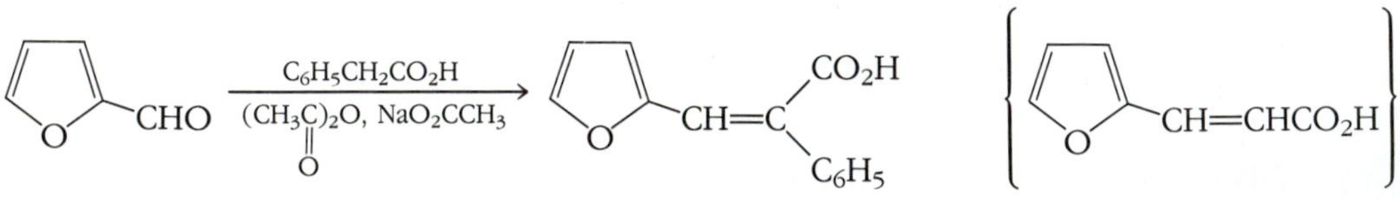

$$\text{2-furyl–CHO} \xrightarrow[(CH_3C(=O))_2O,\ NaO_2CCH_3]{C_6H_5CH_2CO_2H} \text{2-furyl–CH=C}(CO_2H)(C_6H_5)\ (\mathbf{A}) \qquad \{\text{2-furyl–CH=CHCO}_2\text{H}\}\ (\mathbf{B})$$

21.13 In Section 21.6 the reaction of aqueous ammonia with 2-ethoxycarbonyl-1-phenylbutane-1,3-dione was shown to yield ethyl 3-oxo-3-phenylpropionate via a retrocondensation. Draw the two other products which might also have been formed by retrocondensations. Suggest a reason why the pathway to the observed product is more favorable.

21.14 The reaction of ethyl benzoate with acetophenone in the presence of sodium ethoxide yields 1,3-diphenyl-1,3-propanedione upon acidification (Section 21.2). Write a detailed mechanism for this reaction.

21.15 In Section 21.2 we drew 2-formylcyclohexanone as the product of the base-catalyzed condensation reaction between cyclohexanone and ethyl formate. Frequently, the product of this reaction is drawn as 2-(hydroxymethylene)cyclohexanone. Why?

21.16 The reaction of ethyl phenylacetate with phenylacetonitrile in the presence of sodium ethoxide yields 2,4-diphenyl-3-oxobutyronitrile upon acidification (Section 21.2). Write a detailed mechanism for this reaction.

21.17 In Example 21.10 we suggested a synthesis of 1,3-diphenyl-2-propanone using a condensation reaction. Another approach to the carbon skeleton would employ alkyne chemistry. Outline a complete synthesis by that method.

21.18 Write a detailed mechanism for the Knoevenagel condensation between benzaldehyde and diethyl malonate using ammonium acetate as the catalyst. Assume an acid-catalyzed mechanism (that is, reaction via the enol rather than an enolate ion).

21.19 Suggest methods of synthesis for each of the following compounds. All carbon atoms of the product must be derived from the indicated starting materials in each case. Use any other reagents you wish.

(a) 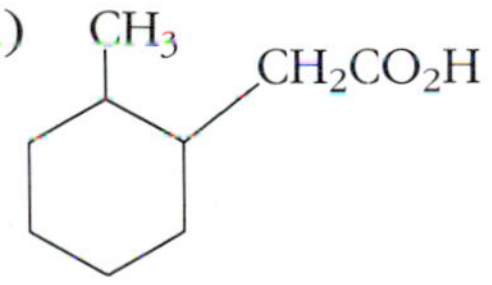 CH_3 CH_2CO_2H (cyclohexane ring) — From cyclohexanone and any other reagents you wish

(b) 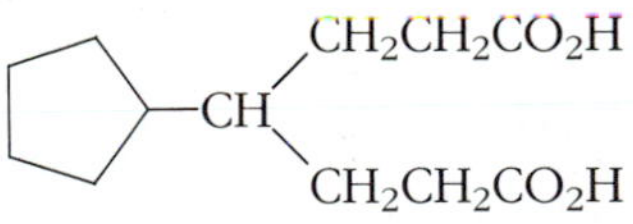 (cyclopentane ring)—$CH(CH_2CH_2CO_2H)_2$: $CH_2CH_2CO_2H$, CH, $CH_2CH_2CO_2H$ — From cyclopentanone and diethyl malonate

(c) 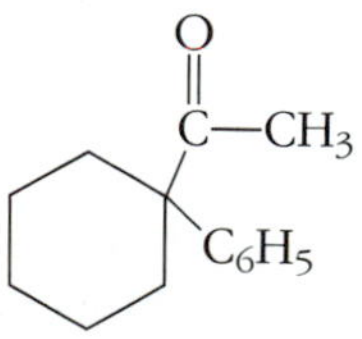 (cyclohexane ring) bearing O=C—CH_3 and C_6H_5 — From cyclohexanone and any other reagents you wish

(d) 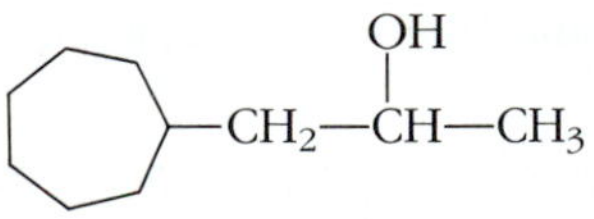 (cycloheptane ring)—CH_2—CH(OH)—CH_3 — From cycloheptanone and ethyl 3-oxobutyrate

21.20 **A–G** are all different compounds. On the basis of the information provided, deduce the structures of **A–F.**

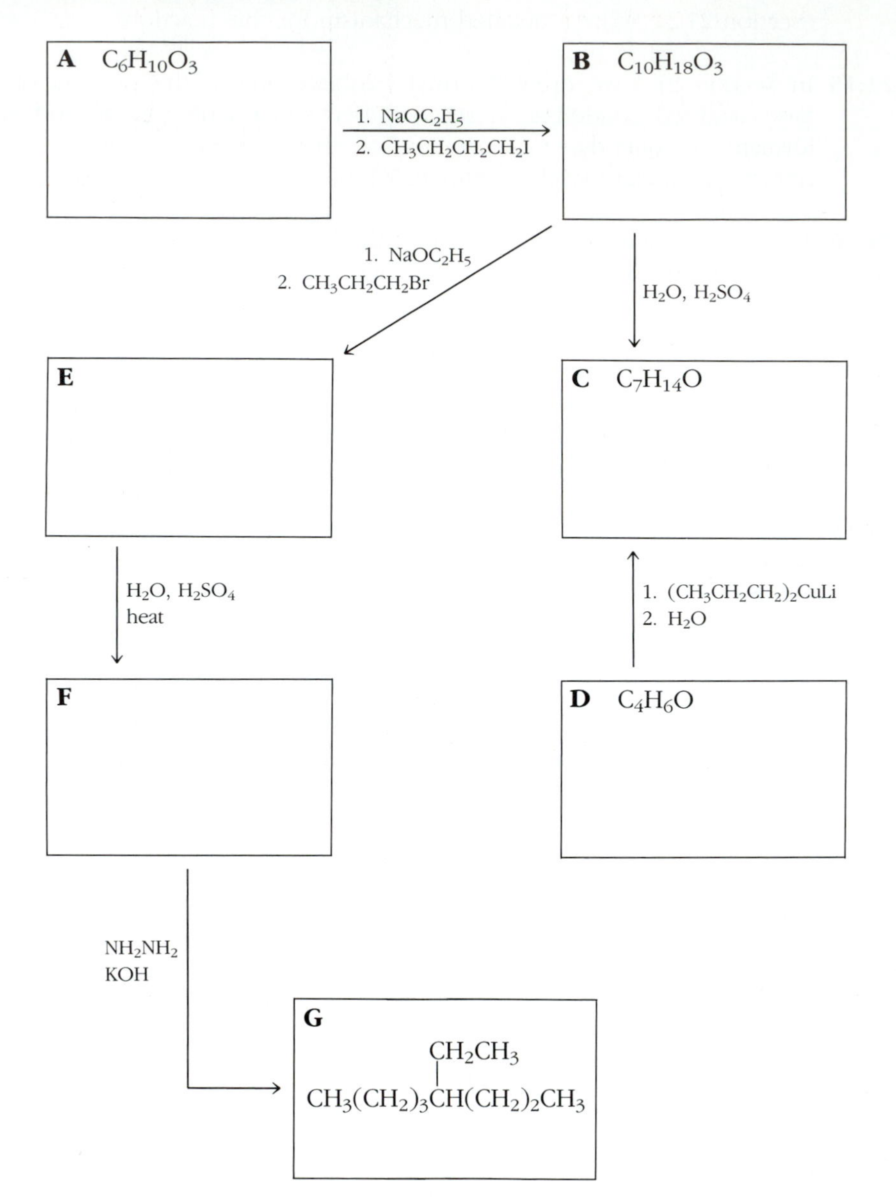

21.21 When 2-ethoxycarbonylcyclohexanone (**1**) is treated with sodium ethoxide followed by methyl iodide, the product isolated from the reaction is the 6-methyl derivative (**2**) rather than the 2-methyl derivative (**3**). Draw a reaction mechanism that explains this result. (Remember that ester condensations are reversible processes.)

O, $CO_2C_2H_5$ — **1**

CH_3, O, $CO_2C_2H_5$ — **2**

O, CH_3, $CO_2C_2H_5$ — **3**

21.22 When methyl benzoate and acetone are allowed to react in the presence of sodium amide, a product having the following NMR spectrum is formed in 63% yield. Deduce the structure of the compound.

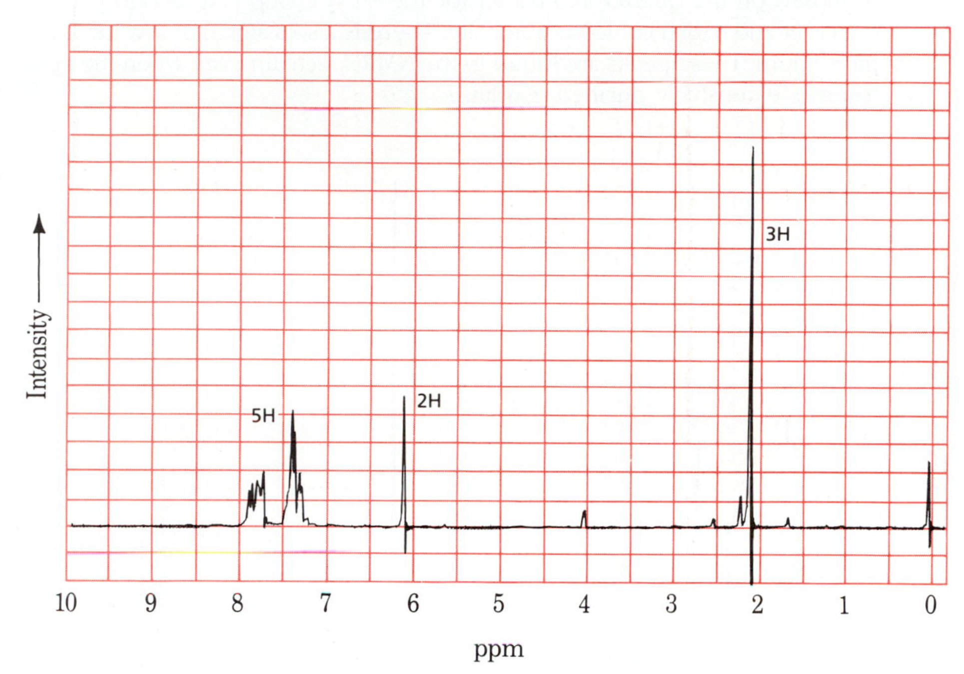

Figure 21.8 Proton NMR Spectrum of Unknown (Problem 21.22).

21.23 The NMR spectrum of ethyl 3-phenyl-3-oxopropionate (C_6H_5—CO—CH_2—$CO_2C_2H_5$) is shown below. At first glance you may find the spectrum to be in accord with the structure, but there are some discrepancies. Nearly superimposed on the quartet and triplet for the ethyl group is a second pattern (quartet and triplet) of lower intensity. There is also a signal of low intensity at 5.7 ppm. These peaks are found in the NMR spectrum even when the keto ester is extensively purified. Explain.

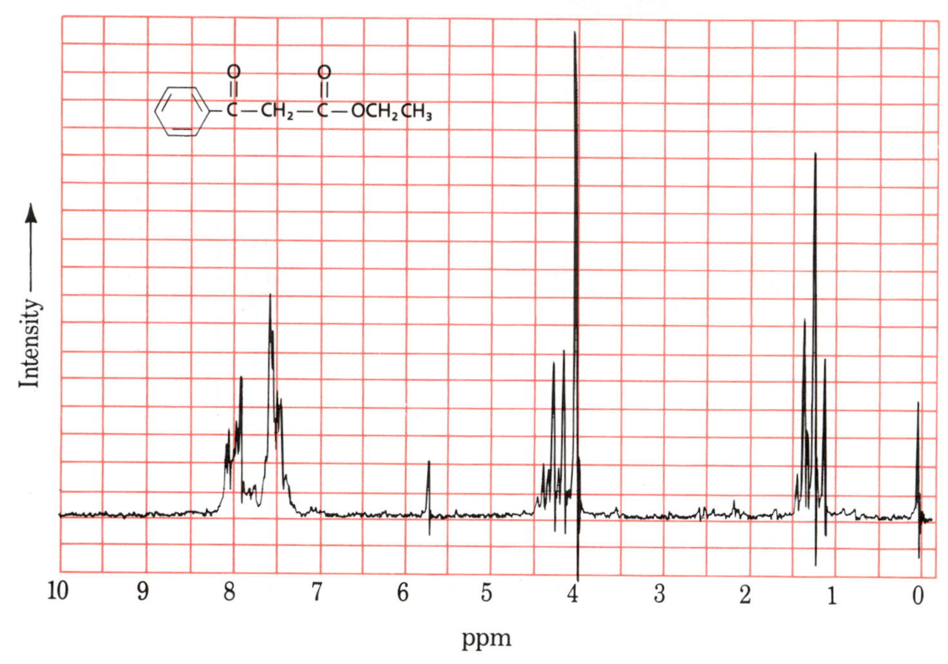

Figure 21.9 Proton NMR Spectrum of 3-phenyl-3-oxopropionate.

21.24 A condensation reaction was carried out between ethyl acetate and ethyl isobutyrate (ethyl 2-methylpropionate) by using sodium ethoxide as the base. One of the products isolated from the reaction mixture had the following NMR spectrum. Deduce its structure.

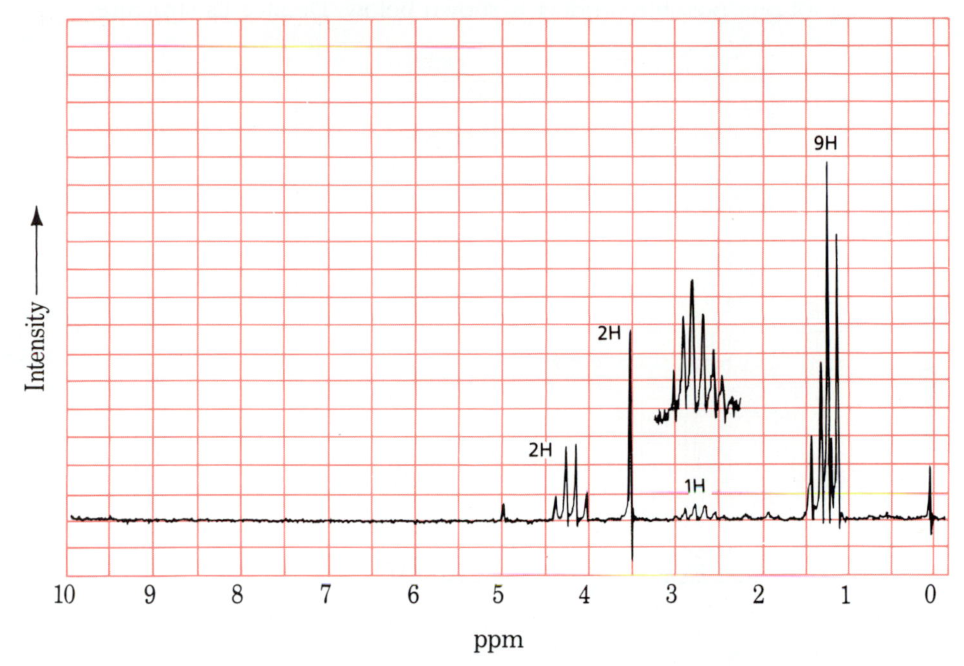

Figure 21.10 Proton NMR Spectrum of Unknown (Problem 21.24).

21.25 The alkylation of diethyl malonate (using sodium ethoxide to generate the enolate) with 1,3-dibromopropane can yield several pathways, depending upon the molar ratio of the reactants.

(a) What product is favored by a large ester/halide ratio?

(b) What product is favored by a low ester/halide ratio? The NMR spectrum of one possible product is shown below. Deduce its structure.

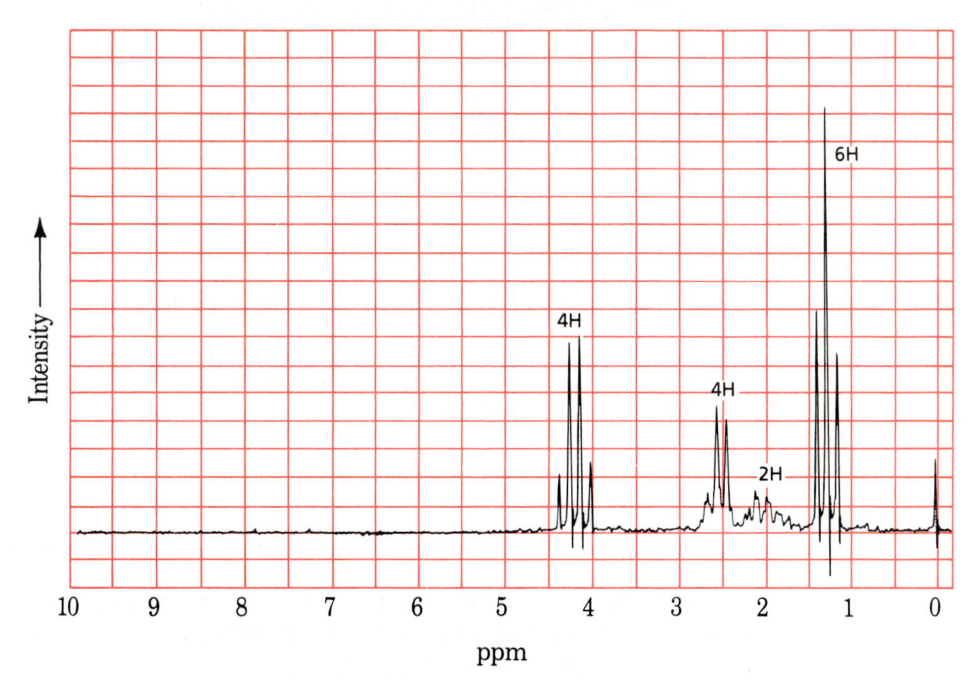

Figure 21.11 Proton NMR Spectrum of Unknown (Problem 21.25).

CHAPTER 22

ORGANIC SYNTHESIS USING POLYFUNCTIONAL COMPOUNDS

In the previous chapters of this text we have introduced the major functional classes of organic compounds and we have discussed the characteristic reactions of each class. We have presented many examples of the various reactions, but most of them involve compounds that contain only a single functional group. As a result we have so far provided you with only a very limited background on the behavior of compounds containing multiple functional groups. Yet such compounds and their reactions are extremely common and important in biological, medicinal, and industrial chemistry. We will now turn our attention to the reactions of polyfunctional compounds, with particular emphasis on the ways they can be used to synthesize complex molecules.

In many respects this discussion will be an extension of Chapter 8. In the former chapter our analysis was severely limited because you had encountered only a few of the functional groups. But now you have learned enough organic chemistry that we can consider a wide range of polyfunctional compounds. Our discussion will focus on two major aspects of compounds containing two or more functional groups: First, how can two groups be made to interact in a manner that results in *ring formation,* that is, how could you prepare cyclic compounds? Second, how can blocking groups and activating groups be employed to allow *selective* reactions at a specific site in a polyfunctional molecule?

22.1 ENERGETICS OF RING FORMATION

In Section 8.7 we briefly discussed the two factors that govern the ease of ring formation, probability and stability. We will return to this problem and focus on two questions that must be answered when evaluating potential cyclization reactions of polyfunctional compounds: (1) Will the reaction proceed inter- or intramolecularly? (2) If an intramolecular ring closure reaction does occur, what size ring will be formed?

First we will consider the ease with which rings of different size can be formed. The rates of intramolecular nucleophilic substitution have been measured for a series of halogenated amines, as depicted by the following equation:

$$Br—(CH_2)_n—NH_2 \longrightarrow (CH_2)_n\ NH + HBr$$

The reactivity varies dramatically with the number of intervening CH_2 groups. Table 22.1 summarizes the relative rates of cyclization for a series of compounds that yield rings containing 3 to 14 members. For convenience we have used the six-membered ring case as the reference, with a relative rate of unity.

The data in Table 22.1 clearly show you that five- and six-membered rings are formed much more easily (i.e., rapidly) than rings of other sizes. Only the reaction of the compound that yields a five-membered ring occurs more rapidly than the formation of the six-membered-ring reference compound. All the other compounds react more slowly, and the formation of rings with 10 or more atoms is at least 1 million times slower than the reference reaction. For other ring-forming reactions, the dependence of rate on ring size usually follows the same trend, although the exact relative rates usually show some dependence on the different structural features of the reactants. A second series of cyclization reactions is shown in Table 22.2.

TABLE 22.1 Relative Rates for Cyclization of a Series of Bromoamines:

$$Br\text{—}(CH_2)_n\text{—}NH_2 \longrightarrow \overbrace{(CH_2)_n\ NH}$$

Ring size $(n + 1)$	3	4	5	6	7	10	12	14
Relative rate	10^{-1}	10^{-3}	10^{2}	1	10^{-3}	10^{-10}	10^{-7}	10^{-6}

The reactions for which the relative rates are reported in Table 22.2 again proceed by intramolecular direct displacement, and the rate dependence on ring size is similar to that shown in Table 22.1. The relative cyclization rates for the larger rings are not quite as low as in the previous example, but they are still nearly a million times slower than that for six-membered ring formation.

What factors are responsible for these remarkable differences in reactivity? We have already briefly mentioned this phenomenon (Sections 3.2 and 8.7), but a more complete discussion is now in order. There are two main factors that govern the rate of an intramolecular reaction, *strain energy* and *probability*. These are related to the thermodynamic quantities *enthalpy* and *entropy*, but for this general discussion we will employ the former more general terms. As we illustrated previously in Figure 3.1, only three- and four-membered rings are strained to a substantial extent. But this strain is reflected in the transition state for cyclization, and reactions that form three- and four-membered rings are usually slower than those that form six-membered rings. Probability effects on ring formation are also highly dependent on ring size, and this factor makes cyclization increasingly difficult as ring size increases.

TABLE 22.2 Relative Rates of Cyclization of a Series of Bromoalkylphenols:

$$C_6H_4(\text{O—}(CH_2)_n\text{—Br})(\text{O—H}) \longrightarrow C_6H_4\text{(O—}(CH_2)_n\text{—O)}$$

Ring size $(n + 4)$	6	7	8	10	12	14
Relative rate	1	10^{-2}	10^{-4}	4×10^{-5}	2×10^{-5}	1×10^{-5}

The influence of probability can be visualized by drawing the conformation of the reactant that is necessary for cyclization. For example, 2-bromoethyl amine cyclizes via the conformation in which the amino and bromo substituents are *anti,* permitting *backside* displacement of bromide ion:

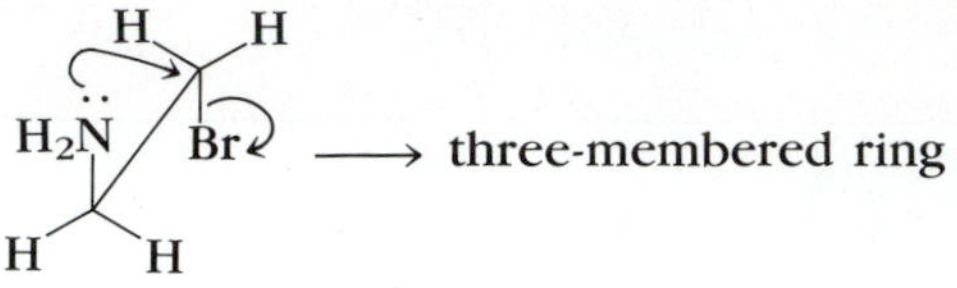

H H
H_2N Br ⟶ three-membered ring
H H

Rotation about the carbon-carbon bond leads to an equilibrium mixture of three conformers, and the reactive one is energetically favorable.

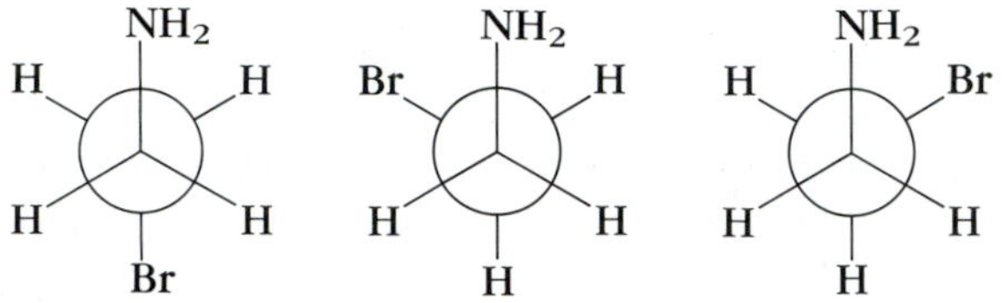

NH_2 H H H H Br

NH_2 Br H H H H

NH_2 H Br H H H

Reactive conformation

Cyclization to a three-membered ring therefore occurs readily, but the situation becomes quite different for the formation of larger rings. The conformation required for cyclization of 3-bromopropylamine is less favorable.

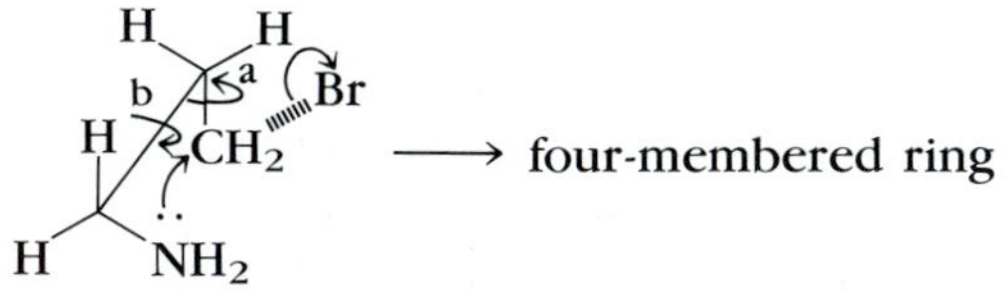

H H b a Br
H CH_2 ⟶ four-membered ring
H NH_2

Cyclization to a four-membered ring requires a specific conformation with regard to rotation about two different carbon-carbon bonds (arrows **a** and **b**). Moreover, this is one of the less preferred conformations, because the NH_2 and CH_2Br groups are *gauche* rather than *anti.* The molecule must react via this one specific geometry (of *nine* possible conformations generated by the rotations indicated with arrows **a** and **b**). For example, when the NH_2 and CH_2Br groups are in the more stable *anti* conformation, reaction cannot occur.

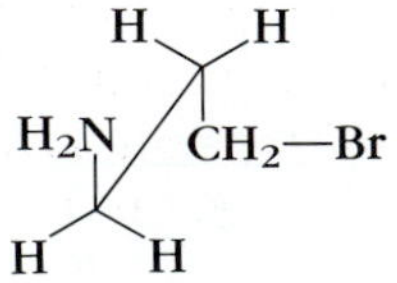

H H
H_2N CH_2—Br
H H

It should not surprise you, therefore, that cyclization to a four-membered ring proceeds more slowly than analogous formation of a three-membered ring.

The effects of ring size on rate of cyclization are depicted graphically in Figure 22.1, which illustrates the enthalpy (strain) and entropy (probability) contributions to the overall energy barrier for cyclization. Cyclization to a three-membered ring

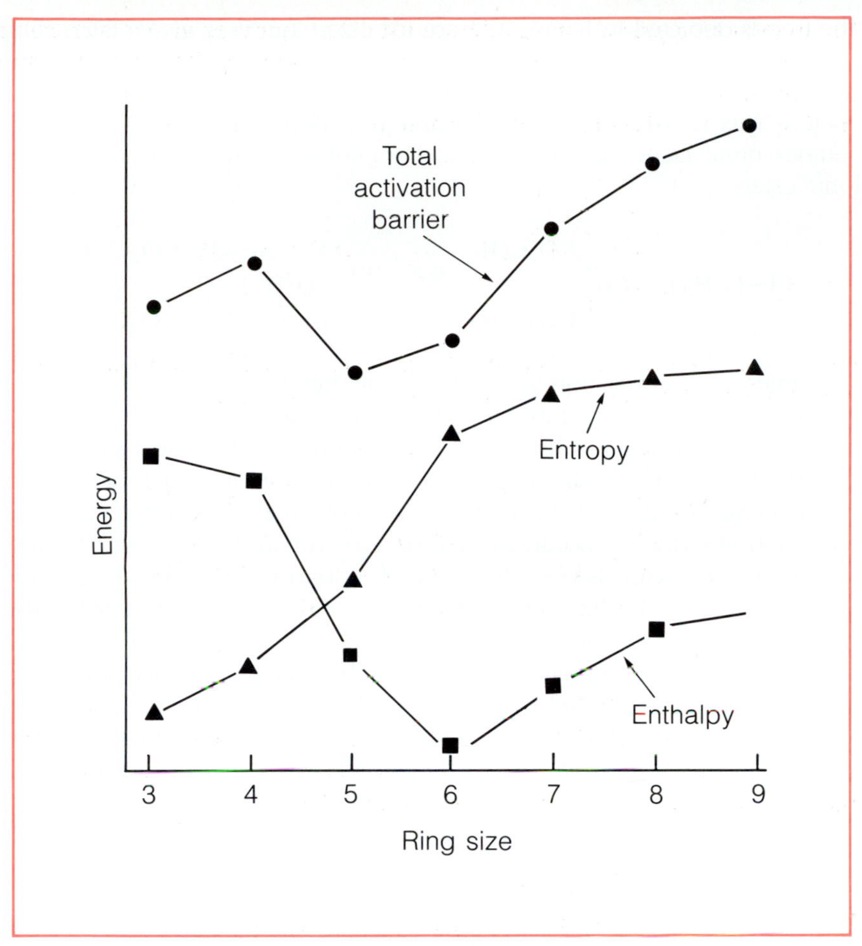

Figure 22.1 Idealized view of entropy (probability) and enthalpy (strain) contributions to the overall activation barrier in cyclization reactions.

is inhibited by a large strain energy, so the activation barrier is large despite the favorable probability effect. Formation of a four-membered ring is much less probable, so despite a slightly lower strain energy, the total barrier is greater.

The large decrease in strain energy for a five-membered ring more than compensates for the larger entropy requirement, and their combined effect results in a much lower activation energy. The same trend continues for cyclization to a six-membered ring, but now the change in probability is larger than the change in strain. Despite the fact that six-membered rings are the least strained, the activation barrier for their formation is slightly larger than that leading to the five-membered-ring analogs.

Proceeding to rings with more than six members, their formation has greater requirements for both enthalpy and entropy (although the effects of ring strain effectively disappear for rings with 13 or more atoms). The reason for the increasing entropy requirement is particularly easy to see. As the length of the carbon chain increases, it becomes less and less probable that the two ends of the chain will interact in the way needed for cyclization to occur.

The trends depicted in Figure 22.1 are idealized, but they give a fairly reliable picture of cyclization reactions. The contribution to the energy barrier from entropy is somewhat variable, because substituents can have dramatic effects on conformation. This is particularly true of three-membered ring formation, which is sometimes quite facile, as seen for the intramolecular alkylation of a series of malonic esters.

$$Cl{-}(CH_2)_n{-}CH(CO_2C_2H_5)_2 \xrightarrow{KOC(CH_3)_3} \overbrace{(CH_2)_n\ C}(CO_2C_2H_5)_2$$

Relative rates for this reaction are presented in Table 22.3.

In contrast to the data of Table 22.1, the fastest reaction in Table 22.3 is three-membered ring formation. Clearly, the favorable probability effect in this case is much more important than the strain of the cyclopropane ring that is being formed. The remaining compounds in Table 22.3 follow the expected pattern. Four-membered ring formation occurs much more slowly than cyclopropane formation, but a large rate increase is observed for cyclization to a cyclopentane derivative. Finally, a decrease in cyclization rate is again found on proceeding from five- to six-membered ring formation.

All the examples we have discussed here can be used to formulate a simple rule for predicting reactivity in cyclization reactions. Although there are occasional exceptions, the trend shown in the following box is usually observed for the relative rates of ring-forming reactions.

Cyclization Rate vs Ring Size

Cyclization fastest 5-ring $> \left\{\begin{matrix}\text{6-ring}\\ \text{3-ring}\end{matrix}\right\} > \left\{\begin{matrix}\text{4-ring}\\ \text{7-ring}\end{matrix}\right\} >>$ 8- and larger rings Cyclization slowest

TABLE 22.3 Relative Rates for Cyclization of a Series of Diethyl Chloroalkylmalonates:

$$Cl{-}(CH_2)_n{-}CH(CO_2C_2H_5)_2 \longrightarrow \overbrace{(CH_2)_n\ C}(CO_2C_2H_5)_2$$

Ring size ($n + 1$)	3	4	5	6
Relative rate	10^6	10^{-1}	10^3	1

For all practical purposes, only three-, five-, and six-membered rings can be formed by general methods. Rings of other sizes rings usually must be formed by specialized techniques.

Side Reactions: Ring Opening and Intermolecular Processes

Two additional features of cyclization reactions must be considered before we end our discussion. The first of these is ring opening, which often occurs in *reversible* reactions. Most of the condensation reactions we considered in Chapters 14 and 21 are reversible, and this can sometimes lead to unexpected products. Reversibility is particularly important in the case of strained rings, but the result can be predicted with ease: the strain of three- and four-membered rings favors the ring-opened products at equilibrium. Strained rings therefore cannot be prepared by reversible condensation reactions. In the case of larger rings, the exothermic formation of a carbon-carbon bond usually causes the cyclized product to be favored at equilibrium.

The second possible side reaction is *intermolecular* reaction. Almost any cyclization reaction is complicated by competing intermolecular processes.

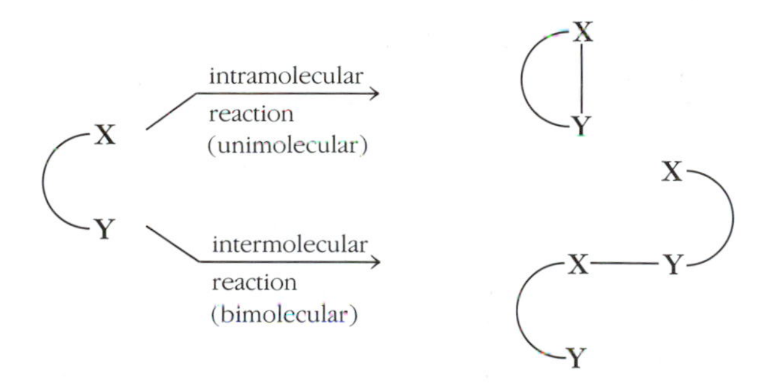

The course of the reaction depends upon the relative rates of the *intra-* and *inter*molecular pathways. Usually, it is only the formation of three-, five-, and six-membered rings that can compete effectively with a bimolecular process. Even the use of very dilute solutions (which decreases the rate of bimolecular reactions) is often inadequate when formation of larger rings is desired. This places additional restrictions on ring-forming reactions, and we are left with only five- and six-membered rings that can be prepared reliably via reversible reactions.

EXERCISE 22.1

An anhydride can sometimes be formed by heating a diacid. For which of the following diacids would you expect to obtain the cyclic anhydride in good yield?

(a) $CH_2(CO_2H)_2$ (b) $HO_2C{-}CH_2{-}CH_2{-}CO_2H$ (c) *cis*-$HO_2C{-}CH{=}CH{-}CO_2H$ (d) *trans*-$HO_2C{-}CH{=}CH{-}CO_2H$

22.2 RING CLOSURES VIA REACTIONS OF ENOLS AND ENOLATE IONS

Up to this point we have discussed very few ways for preparing cyclic compounds from acyclic precursors. Other than a few isolated examples, only the formation of cyclopropanes via addition of carbenes to alkenes (Section 5.7) has been presented. In this section we will turn to more general methods for generating rings. We will center our discussion on enols and enolate ions in which the presence of a second functional group allows an *intramolecular* reaction that produces a cyclic product. These cyclizations will be divided into three general areas: intramolecular *alkylation* of enolate ions, intramolecular *aldol-type* condensations, and intramolecular *acylation* of enolate ions.

Intramolecular Alkylation of Enolate Ions

Intramolecular alkylation provides one of the few general ways of closing rings with sizes that range from three to six members. Its success in forming strained three- and four-membered rings is a consequence of the *irreversibility* of the alkylation reaction; once formed, the product is usually stable to the reaction conditions. Frequently a dihalide is allowed to react with a β-dicarbonyl compound, and the intermediate difunctional compound is not isolated.

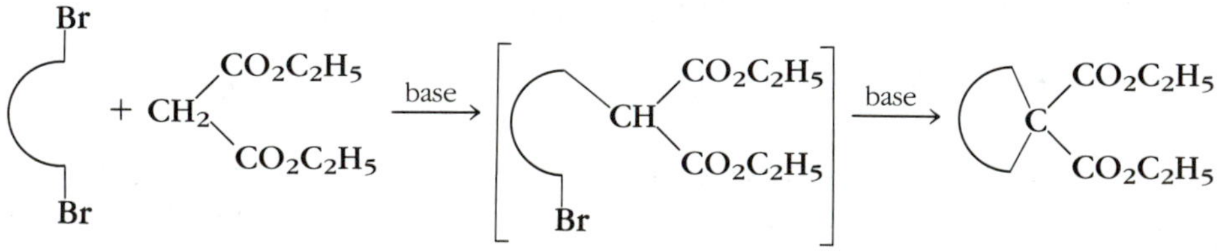

Br(chain)Br + $CH_2(CO_2C_2H_5)_2$ $\xrightarrow{\text{base}}$ [Br(chain)$CH(CO_2C_2H_5)_2$] $\xrightarrow{\text{base}}$ (ring)$C(CO_2C_2H_5)_2$

One of the major problems of this procedure is the *intermolecular* reaction of the difunctional intermediate. This often results in decreased yields, and it precludes the preparation of seven-membered or larger rings in nearly all cases. As illustrated by the following reaction, it may be necessary to separate the by-products in a purification step.

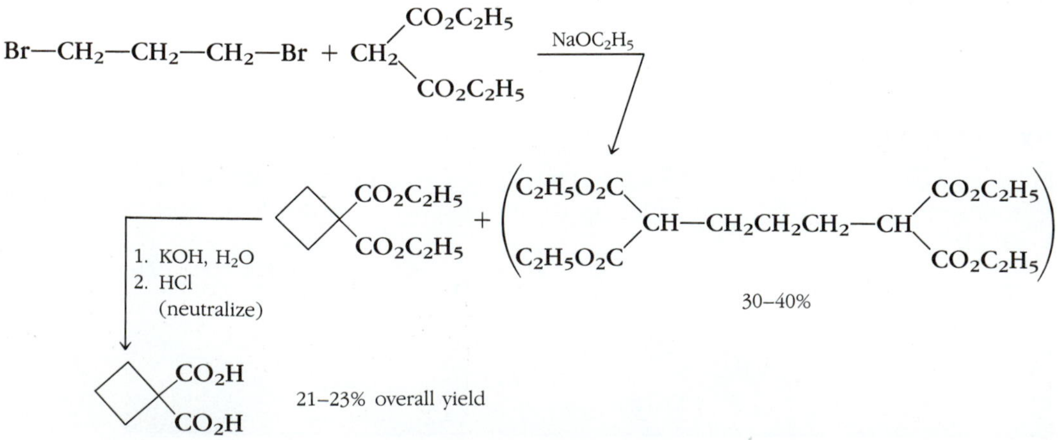

$Br{-}CH_2{-}CH_2{-}CH_2{-}Br + CH_2(CO_2C_2H_5)_2 \xrightarrow{NaOC_2H_5}$ (cyclobutane)$C(CO_2C_2H_5)_2$ + $(C_2H_5O_2C)_2CH{-}CH_2CH_2CH_2{-}CH(CO_2C_2H_5)_2$

30–40%

1. KOH, H_2O
2. HCl
(neutralize)

(cyclobutane)$C(CO_2H)_2$ 21–23% overall yield

(This reaction was the subject of Problem 21.25, and the NMR spectrum of the cyclic diester was shown).

The following examples show that the reactions of a dihalide with a β-dicarbonyl compound can be used to prepare three-, four-, and five-membered rings in excellent yield.

$$Br{-}CH_2{-}CH_2{-}Br + CH_2(CN)(CO_2C_2H_5) \xrightarrow{NaOC_2H_5} \text{(cyclopropane: } CH_2, CH_2, C(CN)(CO_2)\text{)} \quad 76\%$$

$$Br{-}CH_2CH_2CH_2{-}Cl + CH_2(CO_2C_2H_5)_2 \xrightarrow{NaOC_2H_5} \text{cyclobutane-}C(CO_2C_2H_5)_2 \quad 53\text{–}55\%$$

$$CH_3{-}CH(Br){-}CH_2{-}CH_2{-}CH(Br){-}CH_3 + CH_2(CN)(CO_2C_2H_5) \xrightarrow{NaOC_2H_5} \text{2,5-dimethylcyclopentane-}C(CN)(CO_2C_2H_5) \quad 81\%$$

$$\text{cyclohexene-4,5-}(CH_2OTs)_2 + CH_2(CO_2C_2H_5)_2 \xrightarrow{NaOC_2H_5} \text{bicyclic product (H, H)-}C(CO_2C_2H_5)_2 \quad 87\%$$

The last of these equations shows that tosylates can be employed in place of halides, and the third reaction is noteworthy as a successful example of the use of *secondary* halides.

The favorable probability effects of intramolecular reactions to form six-membered and smaller rings often permit the alkylation of *mono*carbonyl compounds, and this is illustrated by the following example:

$$\text{TsOCH}_2\text{, H (decalone)} \xrightarrow{CH_3{-}\overset{O}{\overset{\|}{S}}{-}CH_2^- \; Na^+} \left[\text{TsOCH}_2\text{, H (enolate, } O^-\text{)}\right] \longrightarrow \text{(tricyclic ketone, O)} \equiv \text{(tricyclic ketone, O)} \quad 90\%$$

Intramolecular Aldol Condensations

Addition of enols or enolate ions to aldehydes and ketones is a reversible process. This means that strained rings cannot be prepared by this method, so it is restricted to the formation of five- and six-membered rings. The procedure has been very widely used by synthetic chemists for the construction of cyclopentane and cyclohexane rings from dicarbonyl compounds. Use of 1,4- and 1,5-dicarbonyl compounds affords five- and six-membered rings, respectively. In each case a second pathway is possible, but the reaction is not observed because it would lead to a strained ring.

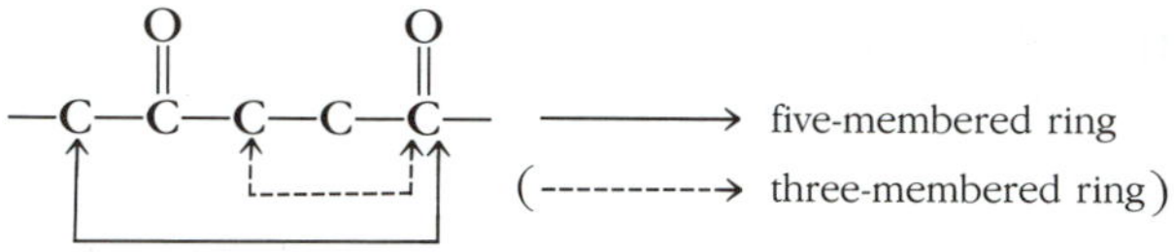

$$-C-\overset{O}{\overset{\|}{C}}-C-C-\overset{O}{\overset{\|}{C}}- \longrightarrow \text{five-membered ring}$$
$$(\dashrightarrow \text{three-membered ring})$$

1,4-Dicarbonyl
compound

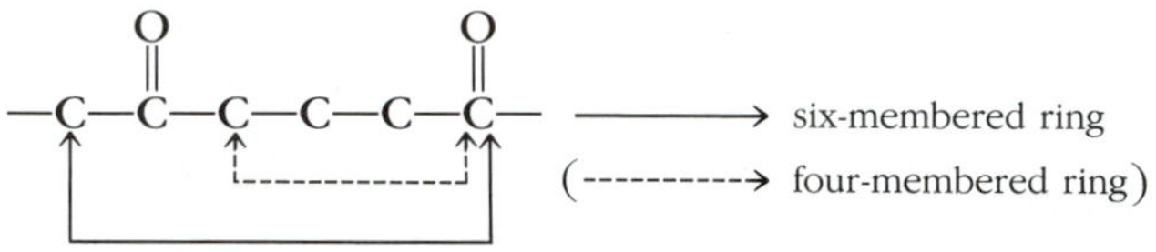

$$-C-\overset{O}{\overset{\|}{C}}-C-C-C-\overset{O}{\overset{\|}{C}}- \longrightarrow \text{six-membered ring}$$
$$(\dashrightarrow \text{four-membered ring})$$

1,5-Dicarbonyl
compound

Cyclization of 1,4- and 1,5-dicarbonyl compounds is illustrated by the following reactions:

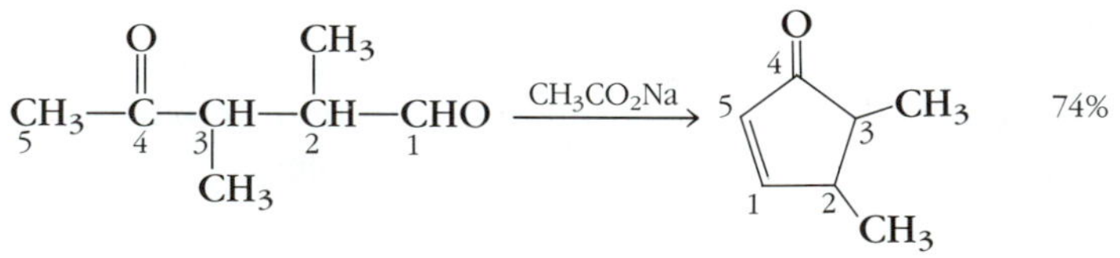

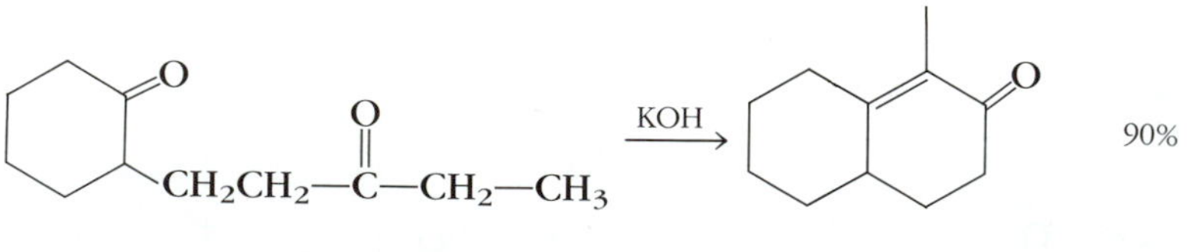

$$\underset{5}{CH_3}-\underset{4}{\overset{O}{\overset{\|}{C}}}-\underset{3}{\underset{CH_3}{CH}}-\underset{2}{\overset{CH_3}{CH}}-\underset{1}{CHO} \xrightarrow{CH_3CO_2Na} \text{2,3-dimethylcyclopent-4-en-1-one (numbered 1–5; } CH_3 \text{ at C-2 and C-3)} \quad 74\%$$

$$\text{2-}(CH_2CH_2-\overset{O}{\overset{\|}{C}}-CH_2-CH_3)\text{cyclohexanone} \xrightarrow{KOH} \text{bicyclic enone (methyl-substituted octalone)} \quad 90\%$$

As we will show in Section 22.3, cyclization of 1,5-dicarbonyl compounds (formed by Michael addition) provides a very important method for preparing six-membered rings.

Intramolecular condensation of 1,6- and 1,7-dicarbonyl compounds can also be used to form five- and six-membered rings, with an α carbon *between* the two carbonyls acting as the nucleophile. For these cyclizations the alternative pathways are not observed because they would lead to unfavorable seven- or eight-membered ring formation.

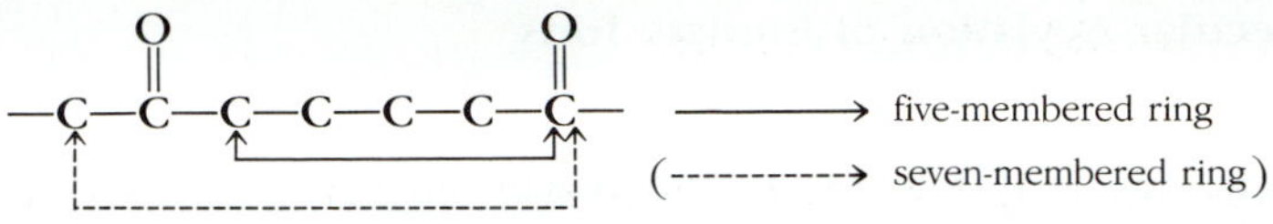

$-C-\overset{O}{\overset{\|}{C}}-C-C-C-C-\overset{O}{\overset{\|}{C}}-$ ⟶ five-membered ring
(- - - - - - → seven-membered ring)

1,6-Dicarbonyl
compound

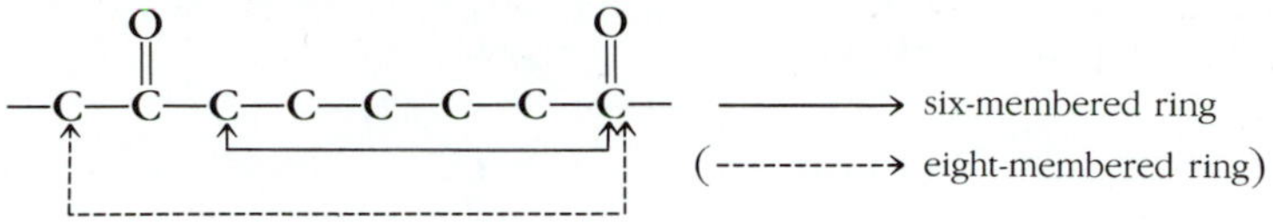

$-C-\overset{O}{\overset{\|}{C}}-C-C-C-C-C-\overset{O}{\overset{\|}{C}}-$ ⟶ six-membered ring
(- - - - - - → eight-membered ring)

1,7-Dicarbonyl
compound

The 1,4- and 1,5-dicarbonyl compounds are employed more frequently because they are easier to prepare (Section 21.7), but the 1,6- and 1,7-dicarbonyl compounds are quite satisfactory reactants when they are available. As in previous examples (Section 14.1), acidic or basic conditions can be used, but the latter are employed more frequently.

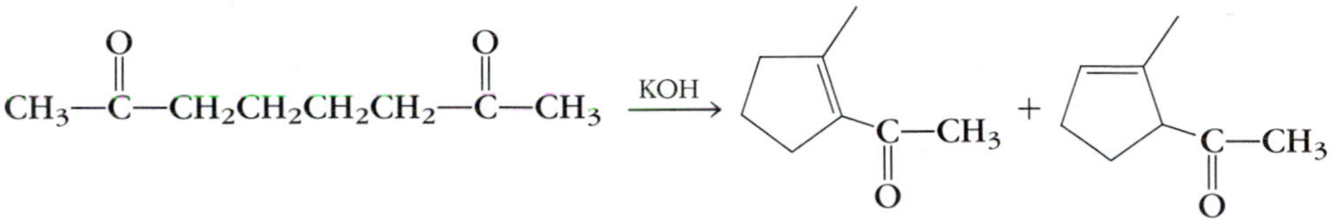

$$CH_3-\overset{O}{\overset{\|}{C}}-CH_2CH_2CH_2CH_2-\overset{O}{\overset{\|}{C}}-CH_3 \xrightarrow{KOH}$$ (cyclopentene product with $-\overset{}{C}(=O)-CH_3$) 83% + (isomeric cyclopentene product with $-C(=O)-CH_3$) 8%

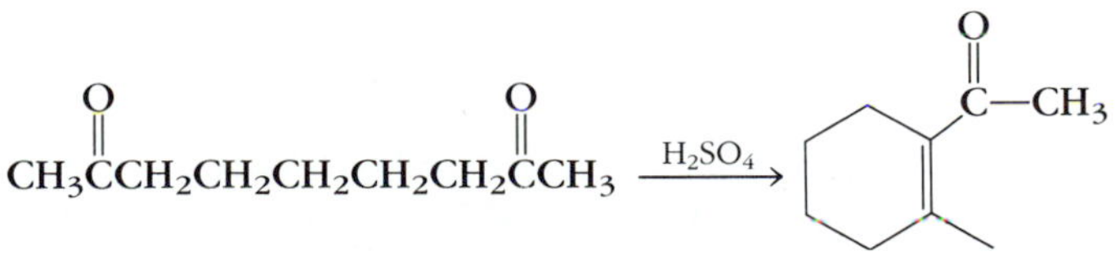

$$CH_3\overset{O}{\overset{\|}{C}}CH_2CH_2CH_2CH_2CH_2\overset{O}{\overset{\|}{C}}CH_3 \xrightarrow{H_2SO_4}$$ (cyclohexene product with $-C(=O)-CH_3$) 85%

$\xrightarrow{Na_2CO_3}$ 96%

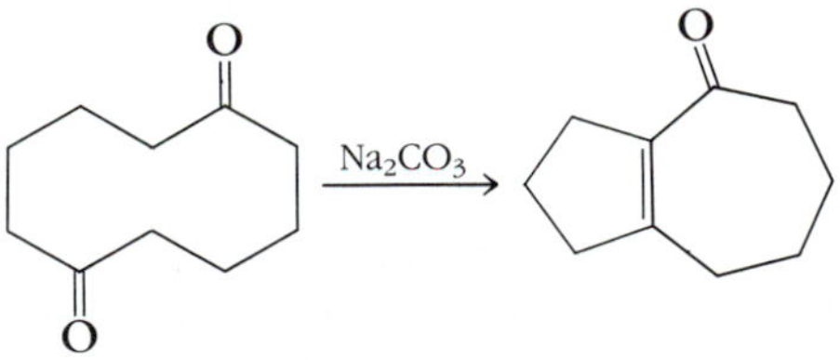

The preceding reactions show that five-membered rings can be formed from either 1,4- or 1,6-dicarbonyl compounds. Similarly, both 1,5- and 1,7-diketones yield six-membered rings. The generation of a seven-membered ring in the last example may be surprising, since we have argued that only five- and six-membered rings are formed in aldol condensations. But the favorable process in this case was indeed formation of a five-membered ring, and the seven-membered ring is only a consequence of the fact that both carbonyl groups were originally in the same ten-membered ring of the reactant.

Intramolecular Acylation of Enolate Ions

When an enolate ion is acylated by an ester, a β-keto carbonyl compound is produced (Section 21.2). If both the original reactive centers were in the same molecule, a cyclic product can result.

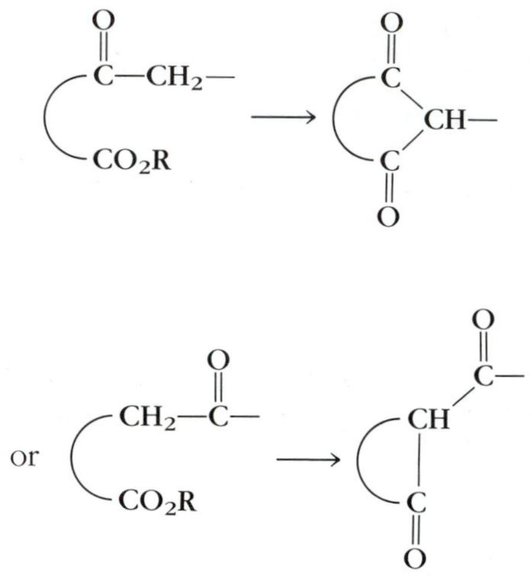

$$\text{(cyclic } \overset{O}{\overset{\|}{C}}\text{—}CH_2\text{—} \ldots CO_2R) \longrightarrow \text{(ring: } \overset{O}{\overset{\|}{C}}\text{—}CH\text{—}\underset{\underset{O}{\|}}{C}\text{)}$$

or

$$\text{(cyclic } CH_2\text{—}\overset{O}{\overset{\|}{C}}\text{—} \ldots CO_2R) \longrightarrow \text{(ring: } CH(\text{—}\overset{O}{\overset{\|}{C}}\text{—})\text{—}\underset{\underset{O}{\|}}{C}\text{)}$$

As in the case of aldol condensations, the reaction is reversible and is therefore limited to formation of five- and six-membered rings as a general procedure. The overall equilibrium is affected in the same ways that we discussed for intermolecular condensations in Section 21.2, and the reaction usually works best when the enolate ion is formed from a CH_2 group. This permits the β-dicarbonyl product to be converted to its enolate, thereby favoring the cyclized product at equilibrium.

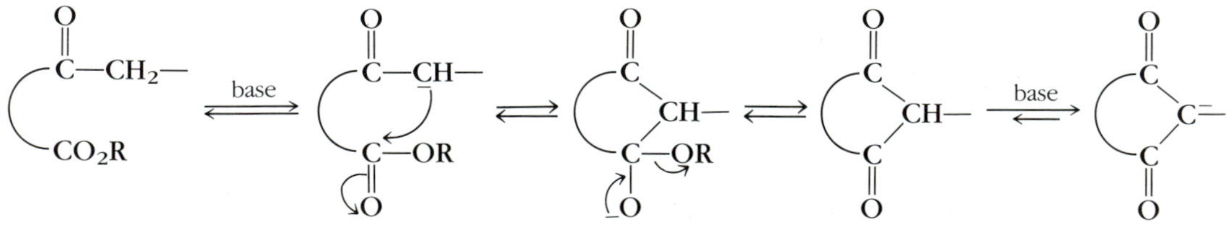

$$\text{C(=O)—CH}_2\text{— / CO}_2\text{R} \xrightleftharpoons{\text{base}} \text{C(=O)—CH}^-\text{— / C(=O)—OR} \rightleftharpoons \text{C(=O)—CH(—)—C(—O}^-\text{)—OR} \rightleftharpoons \text{C(=O)—CH(—)—C(=O)} \xrightleftharpoons{\text{base}} \text{C(=O)—C}^-\text{(—)—C(=O)}$$

Keto esters and diesters can both be readily cyclized, and other carboxylic acid derivatives are sometimes employed as well. The base-induced cyclization of a diester is often called a Dieckmann condensation (after the German chemist Walter Dieckmann, who discovered the reaction around the turn of the century). The following reactions illustrate a variety of intramolecular condensations that produce cyclic β-dicarbonyl compounds.

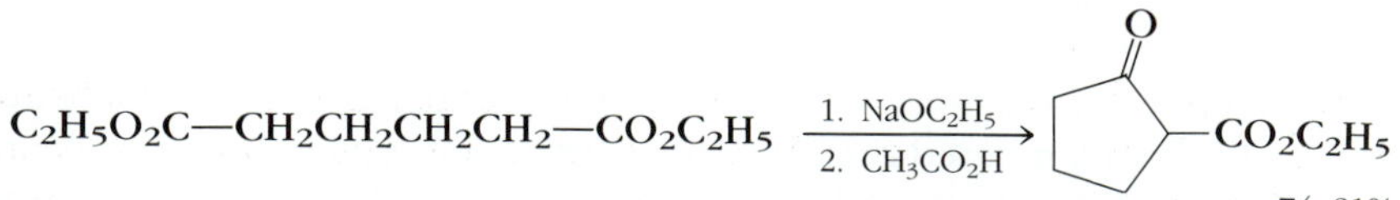

$$C_2H_5O_2C\text{—}CH_2CH_2CH_2CH_2\text{—}CO_2C_2H_5 \xrightarrow[\text{2. } CH_3CO_2H]{\text{1. } NaOC_2H_5} \text{2-oxocyclopentane—}CO_2C_2H_5$$

74–81%

$CH_2{-}CH_2{-}CO_2CH_3$ / $CH{-}CO_2CH_3$ / $CH_2{-}CH_2{-}CO_2CH_3$ $\xrightarrow[\text{2. } CH_3CO_2H]{\text{1. } NaOCH_3}$ (cyclohexanone with CO_2CH_3 at C-2 and C-4) 57–61%

$CH_3CH_2{-}\overset{O}{\overset{\|}{C}}{-}CH_2CH_2{-}CO_2C_2H_5$ $\xrightarrow[\text{2. HCl}]{\text{1. } NaOCH_3,\ CH_3OH}$ (2-methylcyclopentane-1,3-dione, O, O, CH_3) 70–71%

As you can see from the preceding examples, alkoxide and hydroxide bases are frequently employed, but stronger bases such as sodium hydride (or sodium amide) can be used also.

$CH_2CO_2C_2H_5$, $CH_2CO_2C_2H_5$ $\xrightarrow[\text{2. HCl}]{\text{1. NaH}}$ H, =O, H, $CO_2C_2H_5$ 83%

H $CH_2{-}\overset{O}{\overset{\|}{C}}{-}CH_3$, $CO_2C_2H_5$ $\xrightarrow[\text{2. HCl}]{\text{1. NaH}}$ H, O, O 83%

The following reaction proceeds via a cyclic intermediate even though the final product does not have an additional ring:

$\overset{O}{\overset{\|}{C}}{-}CH_3$, $O{-}\underset{O}{\underset{\|}{C}}{-}C_6H_5$ $\xrightarrow[\text{2. } CH_3CO_2H]{\text{1. KOH}}$ $\overset{O}{\overset{\|}{C}}{-}CH_2{-}\overset{O}{\overset{\|}{C}}{-}C_6H_5$, OH 80–85%

This reaction differs from the others, because the reactant is an ester of a hydroxy ketone rather than of a keto acid. The acyl group of the ester is *transferred* from oxygen to carbon.

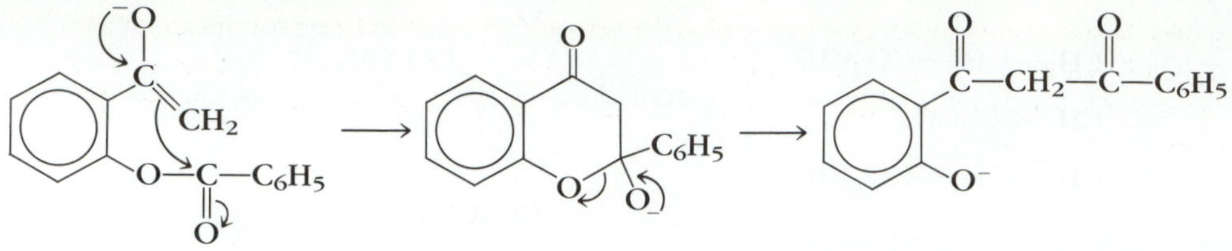

All the Dieckmann condensations in the preceding examples involved *symmetrical* diesters as starting materials. If two different enolates can be formed, diesters often yield mixtures of products. When only a single enolate is possible, as in the following example, the problem is avoided.

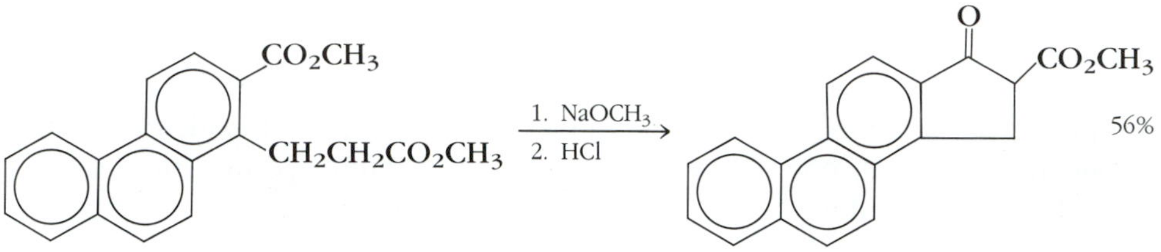

1. $NaOCH_3$
2. HCl

56%

Even when two enolates are possible, you should expect a single product if only one of the α carbons is a CH_2 group. This is very similar to the situation for intermolecular condensations that we discussed in Section 21.2, where the overall equilibrium is favored by forming the enolate of the β-dicarbonyl product. This is illustrated by the cyclization of the dimethyl ester of 2-phenylheptanedioic acid to give a single product.

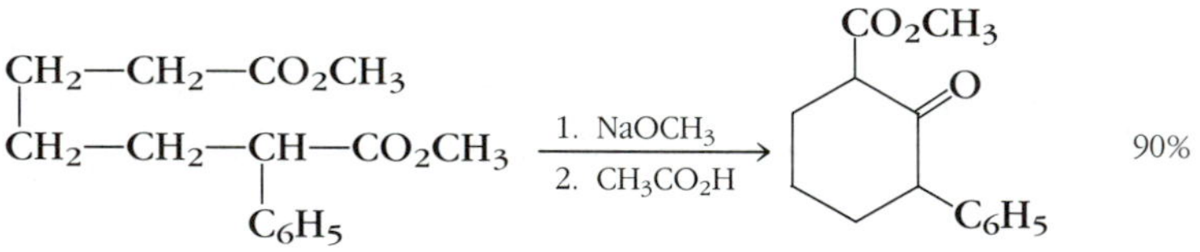

1. $NaOCH_3$
2. CH_3CO_2H

90%

You should draw the other product to convince yourself that it could not form a stable enolate (that is, one that is stabilized by two carbonyl groups).

The effect of enolate formation on the overall equilibrium is further emphasized by the following rearrangement:

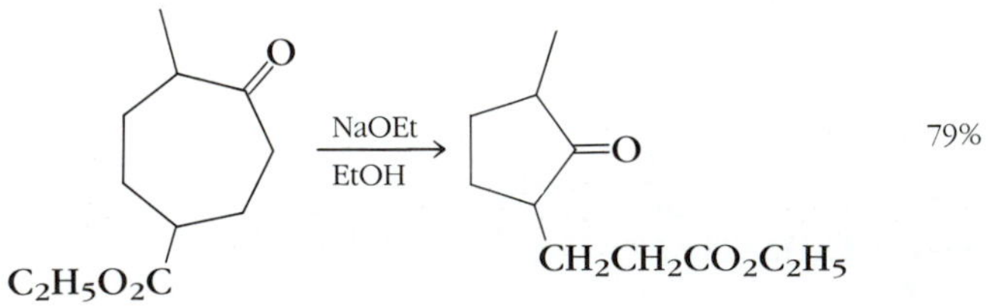

NaOEt
EtOH

79%

Despite the strange appearance of this example, the reaction pathway is precisely that which we have been discussing. Intramolecular attack of the ketone enolate on the ester yields a bicyclic diketone. (Note that reaction of the other enolate is unfavorable, because it would generate a four-membered ring.) The bicyclic diketone, having no hydrogen on the carbon between the two carbonyls, cannot be converted to a stable enolate ion and is therefore susceptible to retrocondensation. Attack by base at one ketone would regenerate the starting material, but reaction at the other keto group leads to the observed product, which is favored at equilibrium.

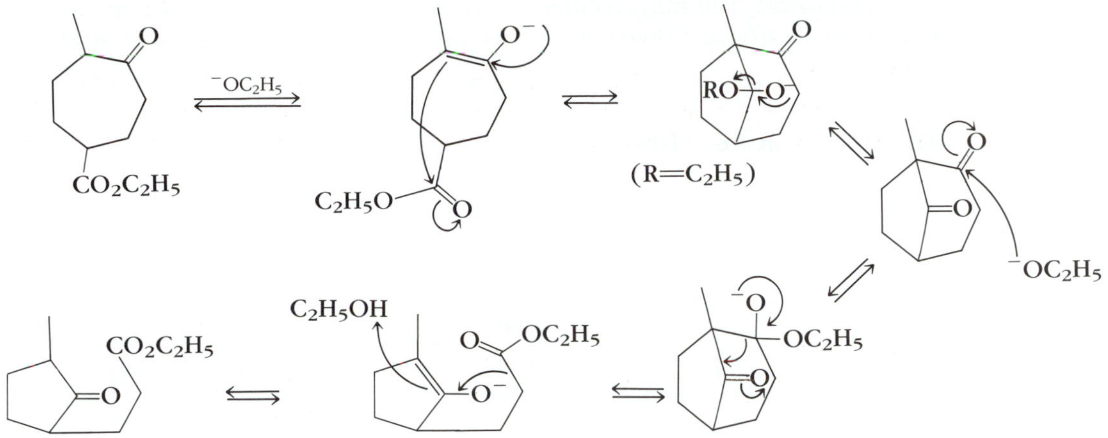

EXERCISE 22.2

For each of the intramolecular alkylations of an enolate ion shown in this section, draw the intermediate formed after the first alkylation step.

EXERCISE 22.3

For each of the intramolecular aldol condensations shown in this section, draw the β-hydroxy carbonyl compound formed prior to dehydration to the α,β-unsaturated product that we have written.

EXERCISE 22.4

For several of the intramolecular aldol condensations shown in this section, isomeric products could be formed via reaction of two different enolate ions. In each case draw the product that would be produced by reaction of the other enolate and explain why it is not the major product.

EXERCISE 22.5

For each of the intramolecular diester condensations shown in this section, draw the product formed prior to acidification of the reaction mixture.

22.3 GENERATION OF SIX-MEMBERED RINGS

Up to this point our discussion has been concerned with general methods for forming rings. With the exception of cyclopropanation reactions (Section 5.7), most cyclizations are not restricted to a particular size, although five- and six-membered rings are usually formed more easily than rings of other sizes (Section 22.1). Six-membered rings are found frequently in a wide variety of naturally occurring and medically useful compounds, and several important procedures have been developed which specifically produce six-membered rings. In this section we will present three such methods: the *Robinson annelation,* the *Diels-Alder reaction,* and the *Birch reduction.* The third method is not actually a cyclization reaction since it starts with an aromatic ring, but we have included it here because of its synthetic importance.

The Robinson Annelation

In Chapter 14 (Sections 14.5, 14.6) we showed that the Michael addition of an enolate to an α,β-unsaturated ketone generates a 1,5-dicarbonyl compound.

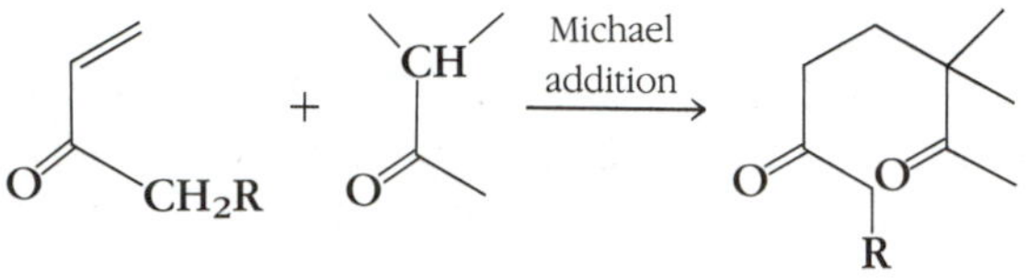

Michael addition

CH, CH_2R, O, R

As we have shown in the preceding sections, such a 1,5-diketone is capable of intramolecular aldol condensation to form a new six-membered ring.

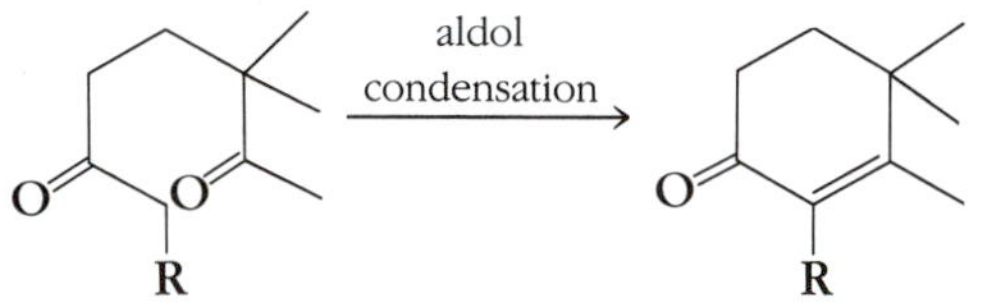

aldol condensation

O, R

The process of building up a new ring is called an *annelation* (from the Latin *annulus,* "ring"). The specific formation of a *cyclohexenone* by a Michael addition, followed by intramolecular aldol condensation, is often called the *Robinson annelation* (after the British chemist Sir Robert Robinson, who developed the technique in the 1930s). The overall process is summarized in the following equation:

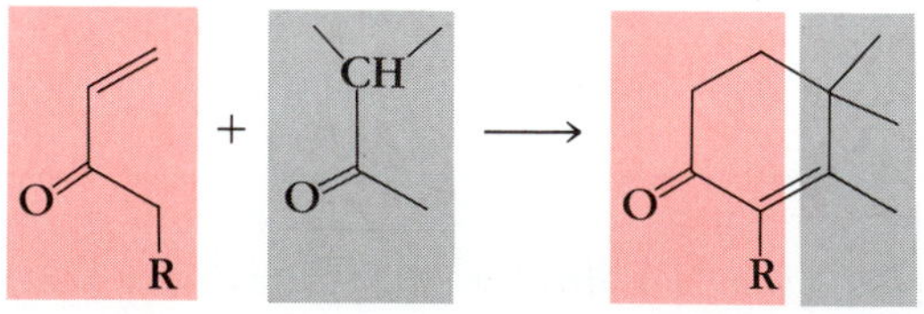

the Robinson annelation

There is considerable variation in the reaction conditions that have been used for Robinson annelations. Sometimes the aldol condensation occurs spontaneously under the conditions used to bring about the Michael addition. Other times the cyclization step requires the addition of another catalyst. For example, a secondary amine can cause the condensation to proceed via an enamine (see Section 17.8). Sometimes an enamine is used in the initial Michael addition as well.

A variation that is often employed in the first step is the use of a β-substituted ketone in place of an unsaturated ketone. The unsaturated ketones are fairly reactive and can decompose (via various condensations) when stored over long periods of time, but a β-haloketone is readily converted to the unsaturated ketone under the alkaline conditions of the Michael reaction.

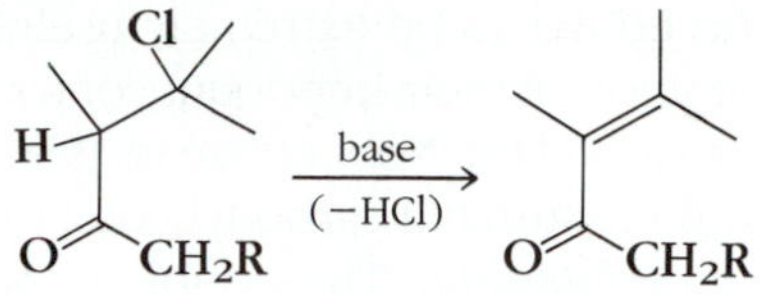

Cl H O CH$_2$R —base (−HCl)→ O CH$_2$R

In addition to β-halo ketones, β-amino ketones and the corresponding ammonium salts have been used extensively. In each case, rapid elimination to form the unsaturated ketone is followed by Michael addition and then an aldol condensation to close the six-membered ring.

The following equations illustrate a variety of Robinson annelations. Notice that a very common reagent is 3-buten-2-one (usually called *methyl vinyl ketone*), which is either used as the free ketone or generated in the reaction mixture from a β-substituted derivative of 2-butanone.

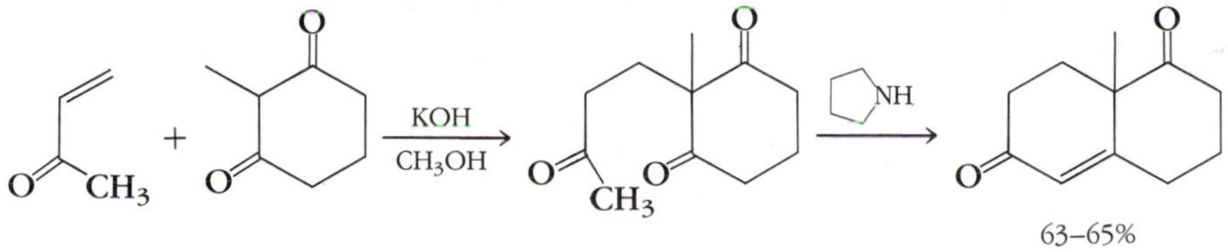

O CH$_3$ + O O —KOH, CH$_3$OH→ O CH$_3$ O O —NH→ O O 63–65%

$CO_2C_2H_5$ O + $CH_3\overset{+}{N}(C_2H_5)_2\ I^-$ O —NaOC$_2$H$_5$→ $CO_2C_2H_5$ O 70%

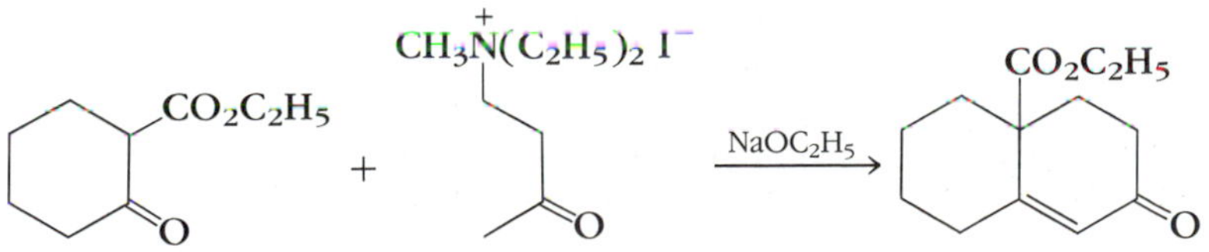

O + O —NaNH$_2$→ O 80%

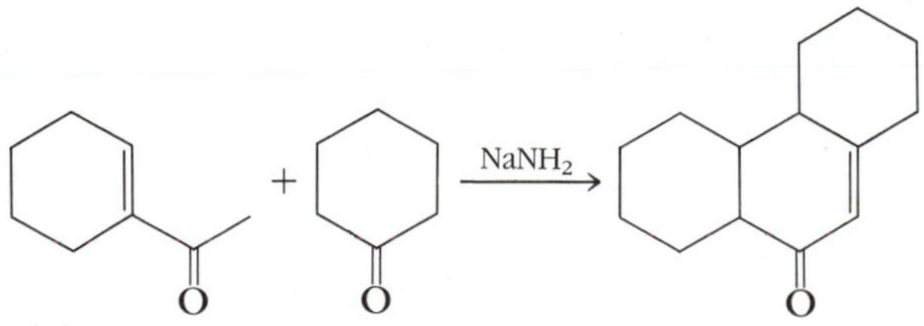

Cycloaddition: The Diels-Alder Reaction

Cycloaddition reactions are processes in which p or π orbitals of one molecule interact with an unsaturated molecule in such a way that two bonds are formed simultaneously to generate a cyclic product. You have already encountered one such process, the addition of a carbene or methylene to an alkene:

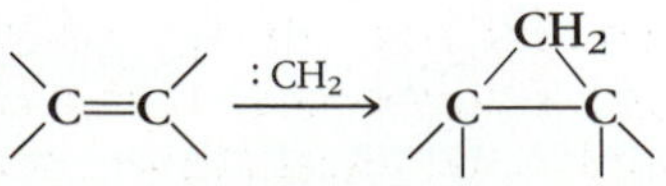

$$\mathrm{C{=}C} \xrightarrow{:CH_2} \mathrm{C{-}C\ (CH_2)}$$

The reaction mechanism for such cycloaddition reactions is difficult to describe. These are usually *concerted* reactions in which a single transition state intervenes between the reacting species and the cyclic product. In other words, no intermediates such as radicals or ions are formed in the cyclization process.

A particularly important cycloaddition reaction was developed in the 1920s by two German chemists, Otto Diels and Kurt Alder. The reaction, which generates a six-membered ring, has turned out to be extremely useful in the synthesis of organic compounds. In recognition of their important contribution to organic chemistry, Diels and Alder were awarded the Nobel Prize in 1950. The two reactants are a conjugated diene and an α,β-unsaturated carbonyl compound, and these undergo cycloaddition to produce a *cyclohexene*. The double bond of the cyclohexene is situated between the two carbon atoms that were originally the central carbons of the diene.

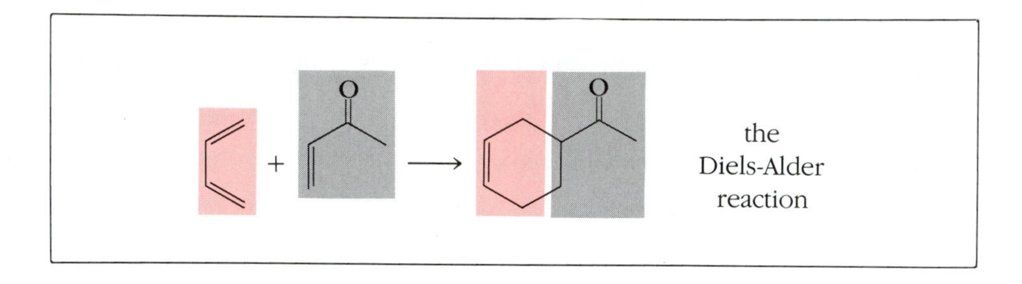

The reaction is a thermal addition and is usually carried out without a catalyst (although Lewis acids are occasionally useful). The diene reacts with an alkene, but the reaction is extremely slow unless the alkene has an electron-withdrawing group (such as a carbonyl substituent) in conjugation with the double bond. This substituted alkene is often called a *dienophile* because of its "affinity" for the diene. The electron-withdrawing group is abbreviated as X in the following generalized equation for the Diels-Alder reaction:

X → X

Diene Dieneophile

Even though the reaction is concerted, the "flow of electrons" indicated by the arrows in the preceding equation helps to keep track of the bonds that are formed.

The reaction results from interaction of the π molecular orbitals of the two reactants. It can conveniently be viewed as involving bond formation to each end of the alkene via the p atomic orbitals of the double bond and the p atomic orbitals at C-1 and C-4 of the diene unit. This leaves the two p orbitals on C-2 and C-3 of the original diene to interact as a double bond in the product.

X → X

The appearance of the double bond of the product between C-2 and C-3 of the original diene makes it relatively easy for you to identify the original diene and dienophile by inspection of the product.

The following examples illustrate the formation of six-membered rings by the Diels-Alder reaction:

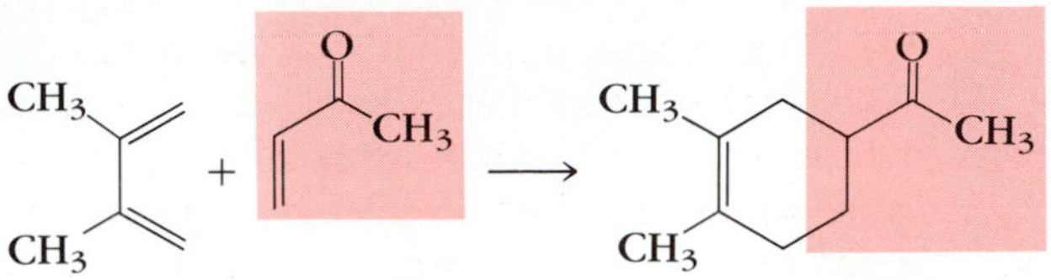

CH_3 CH_3 + CH_3 ⟶ CH_3 CH_3 CH_3 75–80%

+ H ⟶ H 100%

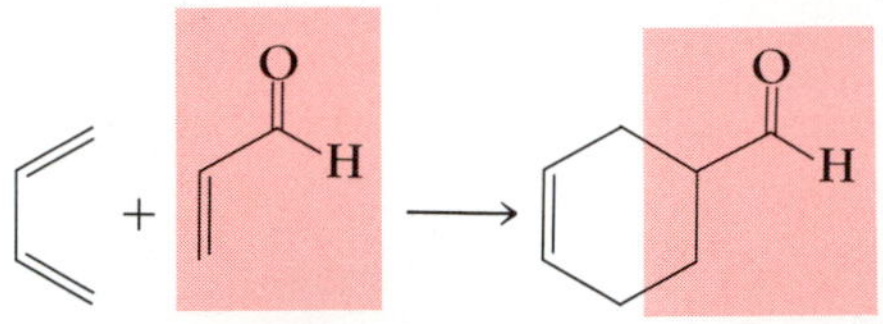

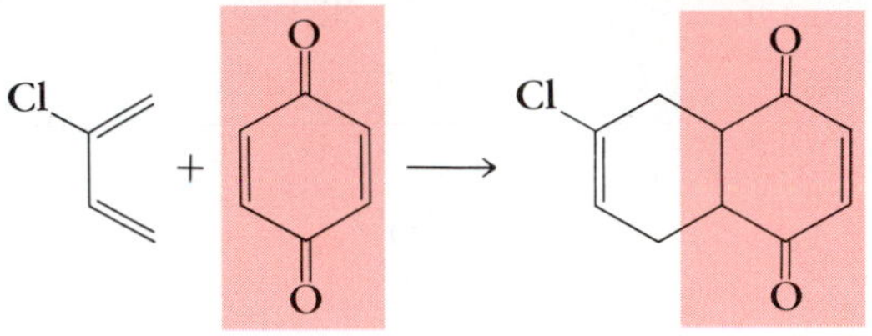

Cl + O O ⟶ Cl O O 95%

C_6H_5 C_6H_5 + $CO_2C_2H_5$ $CO_2C_2H_5$ ⟶ C_6H_5 C_6H_5 $CO_2C_2H_5$ $CO_2C_2H_5$ 90%

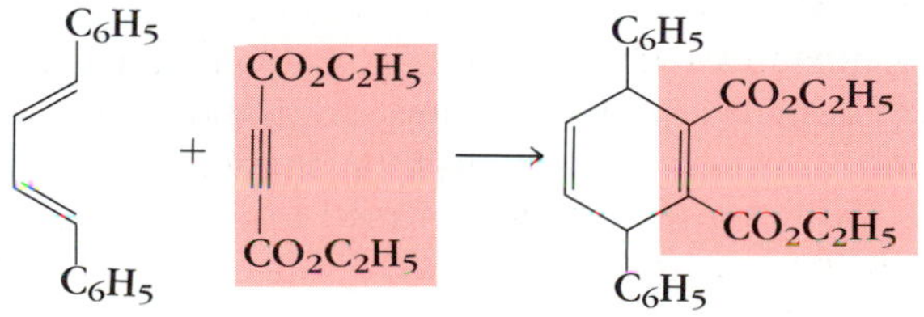

The last example shows that it is also possible to use an acetylenic dienophile. In this case the resulting six-membered ring has two double bonds on opposite sides of the ring.

The Diels-Alder reaction can be extremely useful in building up polycyclic compounds with six-membered rings; the product always contains one more (six-membered) ring than the total number of rings in the reactants. This is illustrated by the following examples.*

*A polycyclic compound contains multiple rings. Two rings of a polycyclic compound can be joined either as *bridged* rings or as *fused* rings.

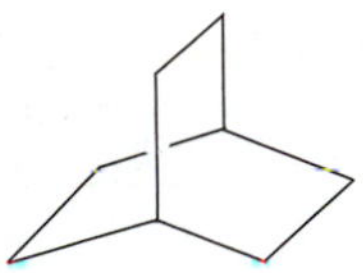

Bridged bicyclic compound

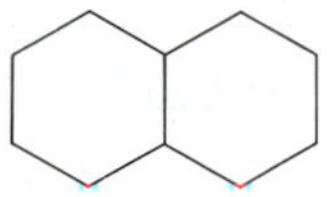

Fused ring system

The total number of rings is determined through a hypothetical process of ring opening by "cutting" bonds. The total number of bonds that must be cut to yield an acyclic structure is equal to the total

(footnote continued)

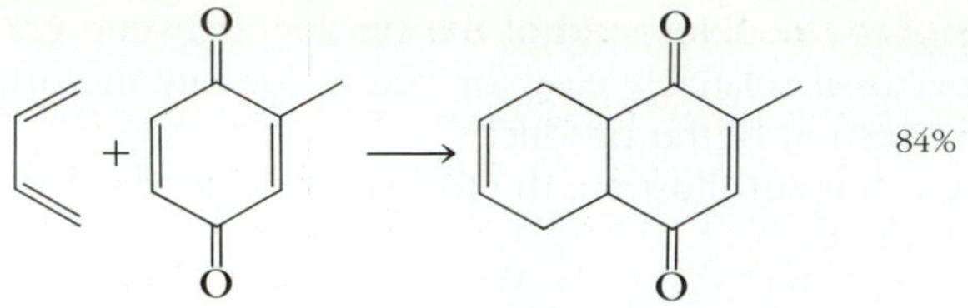

84%

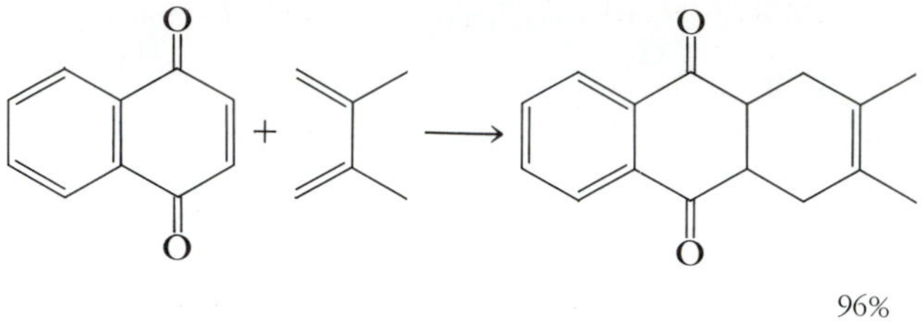

96%

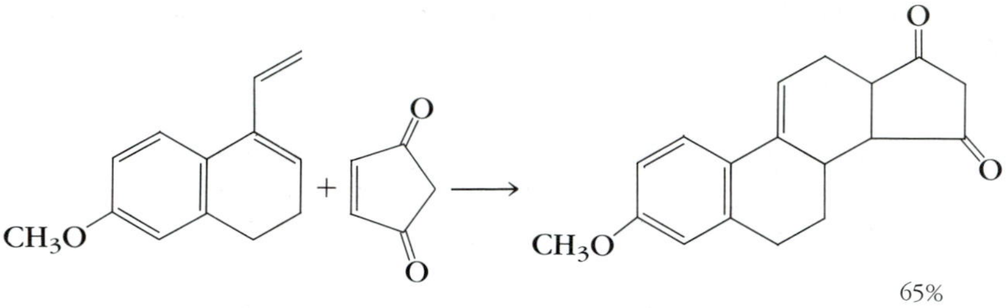

CH_3O CH_3O

65%

When two orientations are possible for the combination of diene and unsaturated carbonyl compound, mixtures of products are often obtained. The favored pathway is determined by a combination of steric and electronic effects, and it is often difficult to predict what the major product will be. For example, the two possible regioisomers in the following reaction are formed to equal extents:

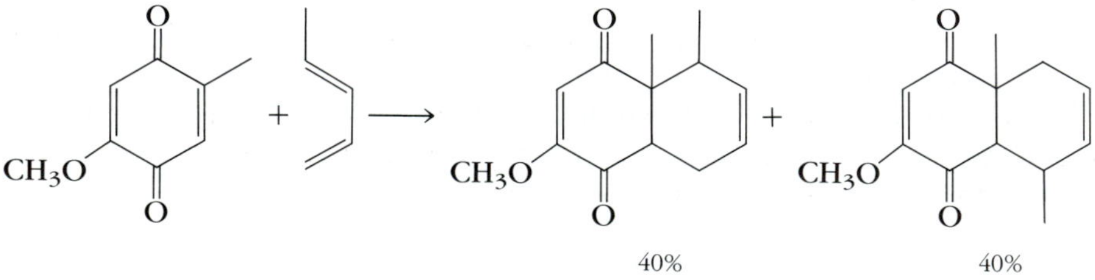

CH_3O CH_3O CH_3O

40% 40%

---

number of rings. This is illustrated for the two bicyclic compounds shown above by the following sequence of ring openings (our choice of which bonds to cut is arbitrary, but any sequence of ring openings would lead to the same conclusion).

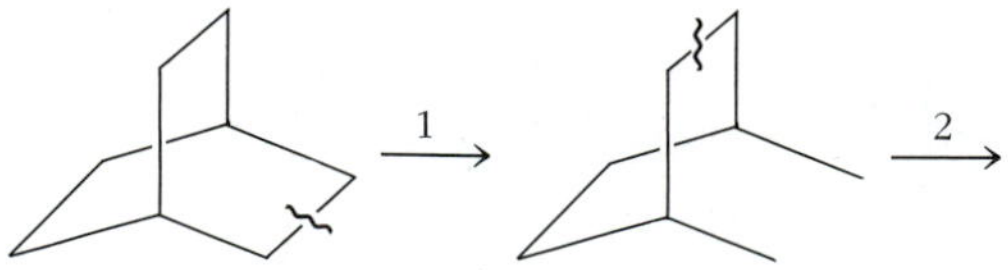

and

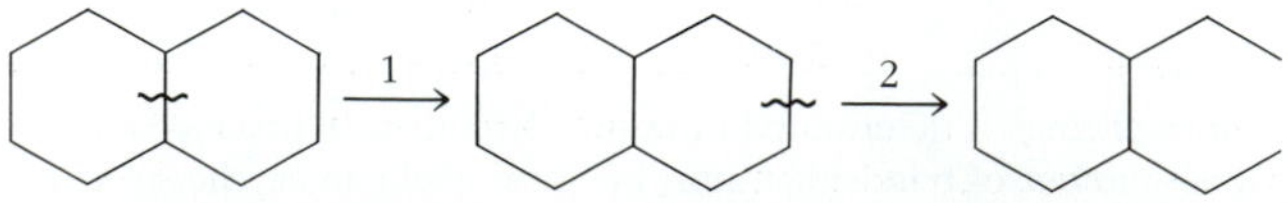

Stereochemistry of the Diels-Alder Reaction

One of the primary reasons that the Diels-Alder reaction has been so useful in organic synthesis is its high stereospecificity. This stereospecificity results both from restrictions on the *E*/*Z* configurations of the double bonds of the reactants and from conformational restrictions on the diene. The configurations of the double bonds are fixed, and the diene can react only in the *syn* conformation, where both C-1 and C-4 can interact with the dienophile simultaneously.

1
2
3
4

Anti conformation *Syn* conformation

Dienes that are restricted to an *anti* conformation will not undergo the Diels-Alder reaction.

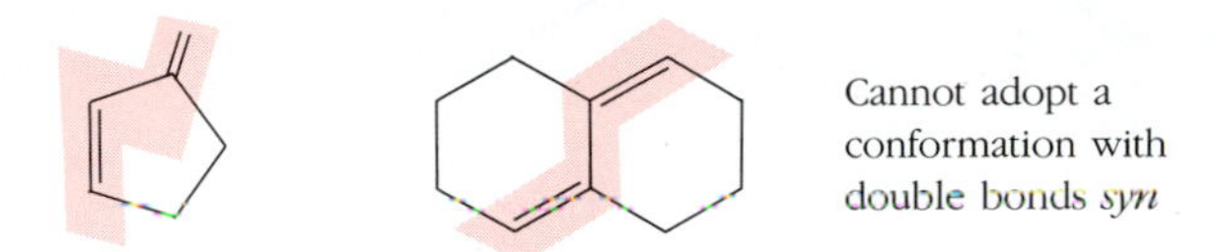

Figure 22.2 illustrates the geometry that the reactants must adopt in order to achieve the orbital interactions needed for cycloaddition to occur.* This orbital relationship determines the stereochemistry of the substituents on the newly formed ring. As is required for a stereospecific reaction, the stereochemistry of the product is correlated with the *E*/*Z* stereochemistry of the reactants.

A specific geometric relationship between the diene and dienophile is also favored in the transition state for the Diels-Alder reaction. This relationship places further limits on the stereochemistry of the substituents that were originally part of the dienophile relative to the substituents of the original diene. This preferred geometry is illustrated in Figure 22.3.

The following examples illustrate the stereospecificity of the Diels-Alder reaction. We have shown three-dimensional structures in parentheses to help you understand the stereochemistry of some of these reactions.

93–97%

*We will return to these molecular orbital requirements for the Diels-Alder reaction in Section 27.4 when we discuss *frontier orbital theory*.

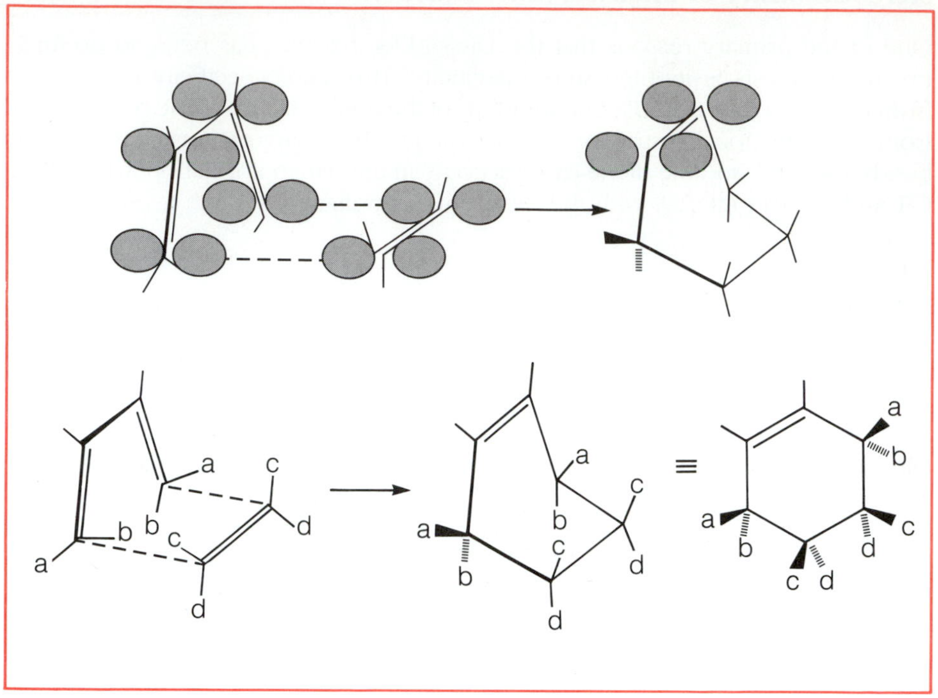

Figure 22.2 Stereochemistry of the Diels-Alder reaction.
The diene and dienophile interact so that the π molecular orbitals at the reaction centers are approximately parallel. (For convenience we have shown the orientation of the component *p* atomic orbitals rather than the actual π molecular orbitals.) This cycloaddition is a *syn* addition with respect to both reactants, and the original stereochemical relationships between substituents on each reactant are preserved.

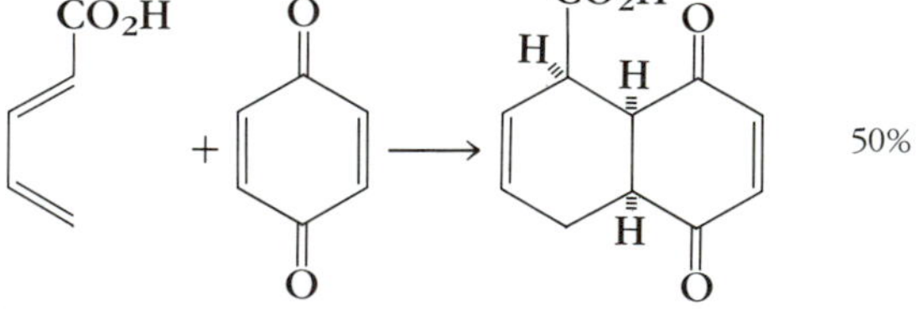

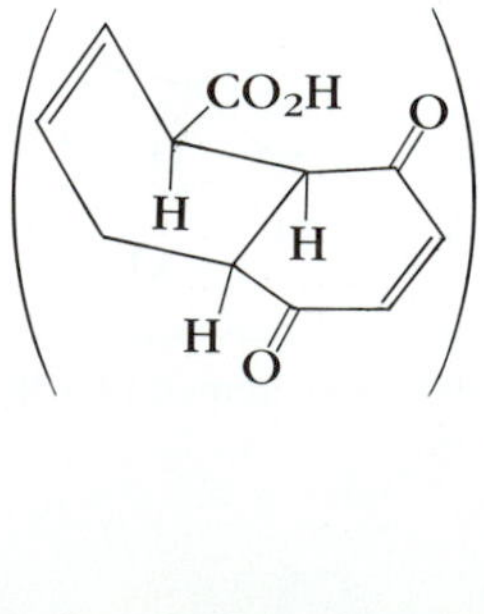

CO_2H O CO_2H O H H

\+ ⟶ 50%

H O O

CO_2H O H H H O

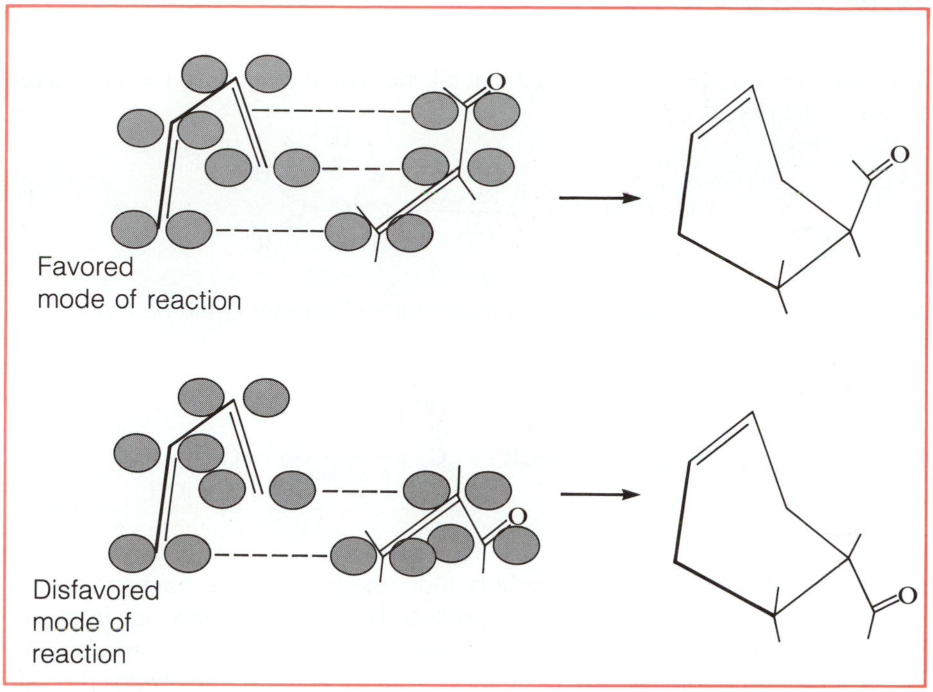

Favored
mode of reaction

Disfavored
mode of
reaction

Figure 22.3 Preferred orientation of the reactants in the Diels-Alder reaction.
The dienophile ordinarily has an electron-withdrawing substituent (such as the carbonyl group shown here) in conjugation with the double bond. The lower-energy reaction pathway is that with the optimum interaction between the π system of the diene and that of the substituent on the dienophile. (For convenience we have shown the orientation of the contributing *p* atomic orbitals rather than the actual π molecular orbitals.)

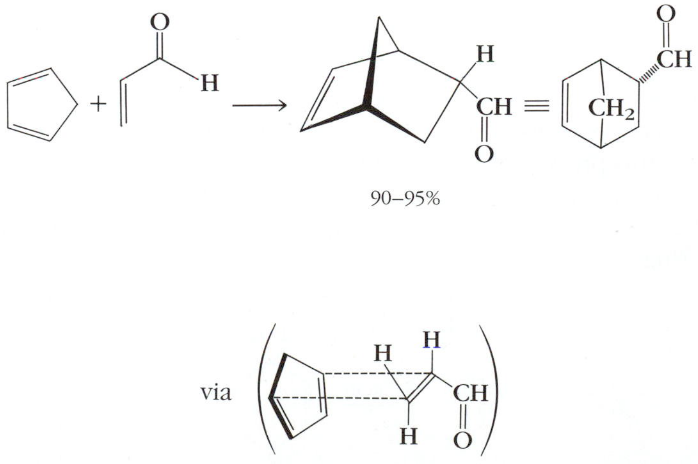

O H CH O CH CH_2 O

90–95%

via H H CH H O

The Birch Reduction

Solutions of sodium, lithium, or potassium metal in ammonia are powerful reducing agents. In Section 7.6 we showed that alkynes could be reduced to the corresponding alkenes

$$\text{R—C≡C—R} \xrightarrow[\text{NH}_3]{\text{Li, Na, or K}} \text{(R)(H)C=C(H)(R)}$$

and in Section 13.13 we showed that α,β-unsaturated ketones could be reduced to the saturated compounds.

$$\text{C=C—C(=O)} \xrightarrow[\text{NH}_3]{\text{Li, Na or K}} \left[\text{—CH—C=C—O}^-\right] \xrightarrow[\text{(workup)}]{\text{H}^+} \text{—CH—CH—C(=O)}$$

Saturated ketones are reduced to alcohols under these conditions, but enolate ions are not further reduced. The reactions proceed by transfer of electrons from the reducing agent to an unoccupied orbital (presumably the LUMO) of the substrate. This produces an anion that is subsequently protonated to give the observed product.

Aromatic rings are also reduced under these conditions, and this reaction is often called the Birch reduction, after the Australian chemist A. J. Birch, who studied it extensively. In a simplified view of the mechanism, addition of two electrons to benzene generates a dianion. We can draw the negative charges on carbons that have a *para* relationship, which corresponds to placing electrons in the lowest unoccupied molecular orbital of the aromatic ring.

$$\text{benzene} \xrightarrow{2\,e^-} \text{dianion (H, }^-\text{ at C-1 and C-4)}$$

Birch reductions are usually carried out in the presence of an alcohol, and the anions are rapidly protonated to yield a nonconjugated diene. Although a ring is not *formed* in this process, the reduction of an aromatic ring to a six-membered ring containing functional groups can be very useful. For this reason we have included the Birch reduction in this discussion of reactions that form six-membered rings.

$$\text{dianion} \xrightarrow{\text{ROH}} \text{1,4-cyclohexadiene (H H at C-3 and C-6)}$$

When an aromatic ring contains an electron withdrawing group X, the orbital interactions favor location of the anion (and subsequent protonation) on the substituted carbon.

X $\xrightarrow{2e^-}$ H, X $\xrightarrow{ROH}$ H, X, H, H

However, electron donating substituents (including alkyl groups) favor the other orientation, in which the negative charge is *not* on the substituted carbon.

R $\xrightarrow{2e^-}$ H, R, H $\rightarrow$ H H, R, H H

Thus benzoic acid and *o*-xylene are reduced to the products shown in the following equations:

CO_2H $\xrightarrow[C_2H_5OH]{Na,\ NH_3}$ H, CO_2H 89–95%

CH_3, CH_3 $\xrightarrow[C_2H_5OH]{Na,\ NH_3}$ CH_3, CH_3 77–92%

The synthetic potential of the Birch reduction of aromatic compounds is illustrated by the reduction of anisole:

OCH_3 $\xrightarrow[C_2H_5OH]{Li,\ NH_3}$ OCH_3 84%

The reaction product is a nonconjugated cyclohexadiene, and the methoxy group is located on one of the double bonds. This product is a *vinyl ether,* which can be easily hydrolyzed to a ketone (see Section 15.5). Reaction with aqueous hydrochloric acid converts this cyclohexadienyl ether to the unsaturated ketone. Very mild hydrolysis permits the nonconjugated ketone to be isolated, but it is readily isomerized to the α,β-unsaturated isomer.

OCH_3 $\xrightarrow[H_2O]{HCl}$ [O] $\rightleftarrows$ O 95%

As shown by the preceding example, the Birch reduction allows aromatic ethers to be converted to cyclohexenones, and these in turn are versatile synthetic intermediates for a wide variety of reactions.

EXERCISE 22.6

For each of the Robinson annelations shown in this section, draw the intermediate diketone that is formed by the initial Michael addition.

EXERCISE 22.7

Each of the following six-membered-ring compounds could be formed by one of the three methods discussed in this section. In each case deduce the reactants that would yield that compound.

(a) CO_2CH_3 CH_3O

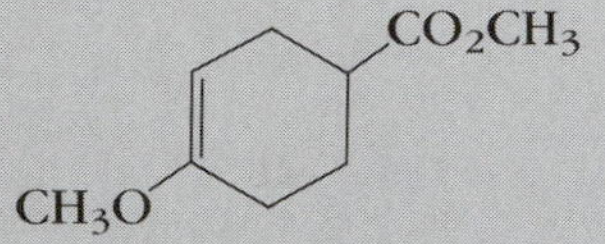

(b) CH_3 CH_3O

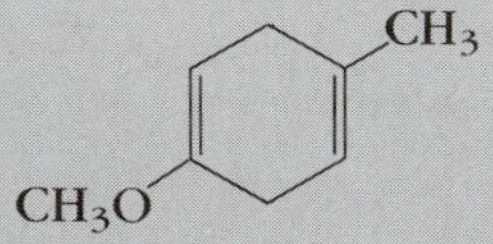

(c) CH_3 O CH_3

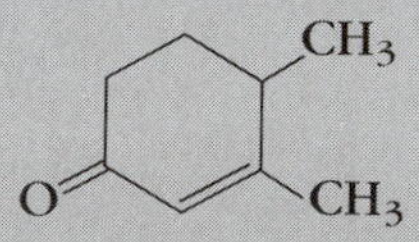

(d) O O

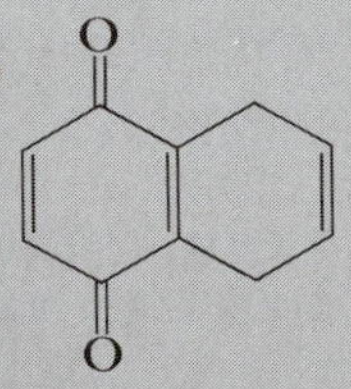

(e) O

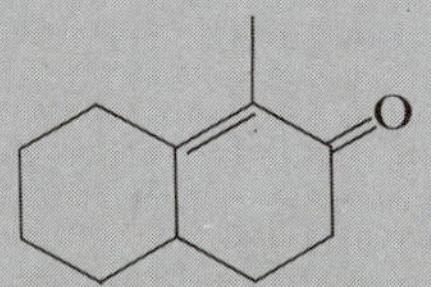

(f)

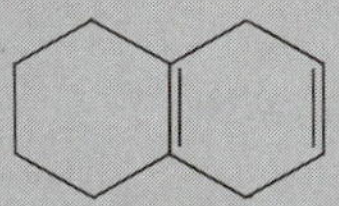

EXERCISE 22.8

Draw the product that would be expected in each of the following Diels-Alder reactions. (Disregard stereochemistry.)

(a) $CH_3{-}O{-}\overset{O}{\overset{\|}{C}}{-}C{\equiv}C{-}\overset{O}{\overset{\|}{C}}{-}O{-}CH_3$ + ⟶ ?

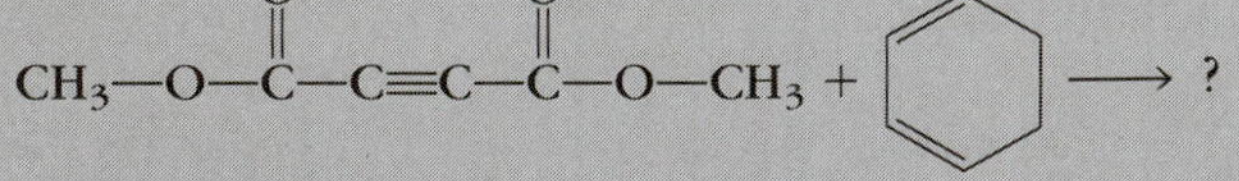

(b) $CH_2{=}\underset{CH_3}{\underset{|}{C}}{-}\overset{CH_3}{\overset{|}{C}}{=}CH_2 + CH_2{=}CH{-}CHO \longrightarrow$?

(c) $CH_2{=}CH{-}CH{=}CH_2$ + ⟶ ?

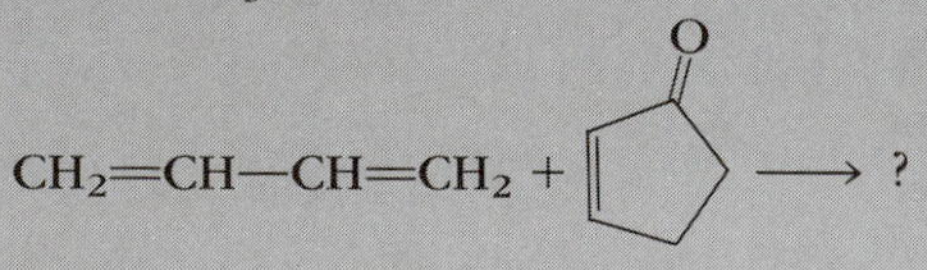

(d) + $CH_2{=}\underset{OCH_3}{\underset{|}{C}}{-}CH{=}CH_2 \longrightarrow$?

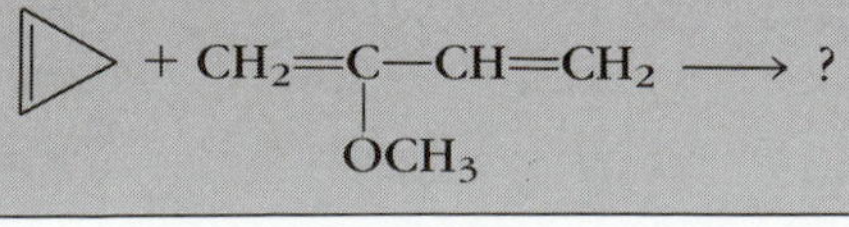

22.4 BLOCKING GROUPS AND ACTIVATING GROUPS: SOME EXAMPLES OF SYNTHETIC METHODOLOGY USING MASKED FUNCTIONALITY

When an organic chemist carries out a synthetic sequence that is aimed at preparing a polyfunctional compound, a series of obstacles may arise. The reactions used to build up one part of the molecule may cause unwanted transformations in other places. Sometimes these difficulties can be avoided by carrying out the reactions in a different sequence, so that the "sensitive" parts of the molecule are not introduced until a later stage of the synthesis. Frequently, however, it is necessary to use a blocking group to protect the sensitive functionality. We have discussed this problem previously (see sections 15.7, 20.3), but now we will return to it for more thorough consideration.

The heart of the problem is finding a way to carry out a reaction *selectively,* so that reaction occurs only at the desired site without affecting other sensitive groups. When a molecule contains two reactive sites, several approaches can be used to differentiate between them. One of these is the use of a *protective group* or *blocking group* to render one of the sites unreactive. A second approach utilizes an *activating group* to enhance reactivity at the desired site, and still another approach utilizes *latent functionality* to distinguish between reactive sites. This last approach is easily understood through an illustration. The $>C{=}CH_2$ group would resist typical carbonyl reactions (such as $LiAlH_4$ reduction or reaction with a Grignard reagent), but it could be converted to a $>C{=}O$ group by ozonolysis. Clearly, the original alkene functionality should not be described as a carbonyl group (or even as a protected carbonyl group), but its ready transformation by ozonolysis permits us to call it a *latent* carbonyl group. Sometimes the term *masked* is employed as a general term that includes both protected and latent functionality.

In this section we will present a series of examples that show you how these different methods can be employed in synthesis. Rarely is there only a single way to carry out a synthesis, and often it is not even possible to predict the best way with certainty. But we can outline a *reasonable* sequence for a proposed synthesis, and this frequently demands the use of masked functional groups so that selective reactions can be carried out.

Example 22.1 The conversion of keto ester **1** into the diketone **7** was carried out using both protecting groups and latent functionality.

O, $CO_2C_2H_5$ → O, O

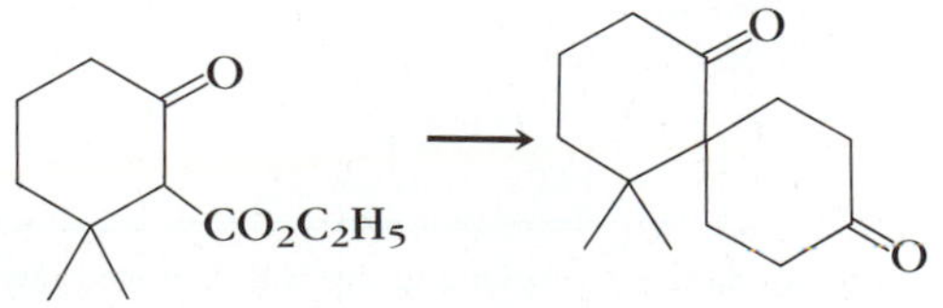

1 **7**

The synthetic plan involved formation of the new six-membered ring by a Diels-Alder reaction. This first required conversion of the keto ester to an unsaturated ketone, and it was necessary to protect the ketone carbonyl prior to reduction and dehydration.

$CO_2C_2H_5$ (1) $\xrightarrow[\text{TsOH, } C_6H_6 \text{ (70\%)}]{HOCH_2CH_2OH}$ $CO_2C_2H_5$ (2) $\xrightarrow[\text{2. } H_2O\text{, HCl (85\%)}]{\text{1. } LiAlH_4}$ CH_2OH (3)

(3) $\xrightarrow[\text{TsOH (72\%)}]{CH_3COCH_3}$ CH_2OH (4) $\xrightarrow[\text{(50\%)}]{\text{TsOH}}$ (5)

Lithium aluminum hydride reduction of ketal ester **2** yielded the ketal alcohol **3,** and the protecting group was removed under acidic conditions.* The β-hydroxy ketone **4** was readily dehydrated to **5** with *p*-toluenesulfonic acid, and a Diels-Alder reaction with 2-ethoxybutadiene gave **6** as the only product. This bicyclic material has the desired carbon skeleton, and the vinyl ether is merely a masked form of the ketone **7** (cf. Section 15.8). Mild acidic hydrolysis of **6** afforded the desired diketone in excellent yield.

(5) + OC_2H_5 $\xrightarrow{(20\%)}$ OC_2H_5 (6) $\xrightarrow[\text{(94\%)}]{H_2O\text{, HCl}}$ (7)

*In this reaction the blocking group was removed not by hydrolysis but by *ketal exchange:*

$$R_2C(OCH_2CH_2O) + (CH_3)_2C{=}O \overset{H^+}{\rightleftharpoons} R_2C{=}O + (CH_3)_2C(OCH_2CH_2O)$$

The use of excess acetone shifted the equilibrium to give the desired product.

Example 22.2 The following transformation involves no change in carbon skeleton, but it illustrates the idea of a methoxy substituted aromatic ring as a latent cyclohexanone; it also demonstrates the resistance of a vinyl ether (a protected ketone) to reduction by lithium in ammonia.

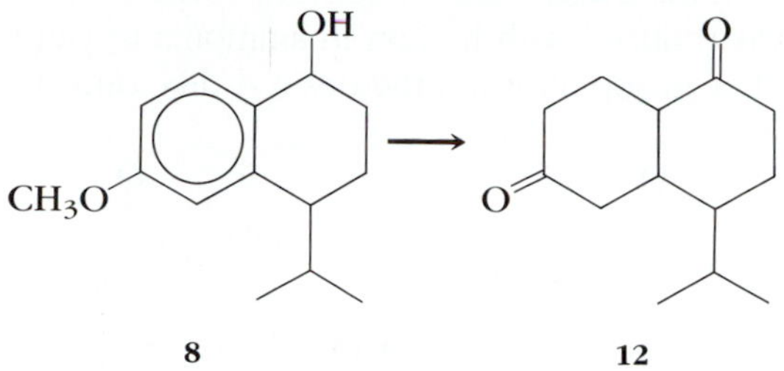

OH
CH_3O
O
O
8 **12**

When **8** was subjected to Birch reduction, the expected 1,4-cyclohexadiene was formed. Hydrogens were added (via protonation of intermediate carbanions) at the two unsubstituted positions of the aromatic ring that are *para* to each other. The next step, oxidation of the hydroxy group of **9** to the corresponding ketone, required mild, nonacidic conditions so that the vinyl ether would not be affected. The oxidizing agent used was one that we have not previously discussed, aluminum isopropoxide in acetone.

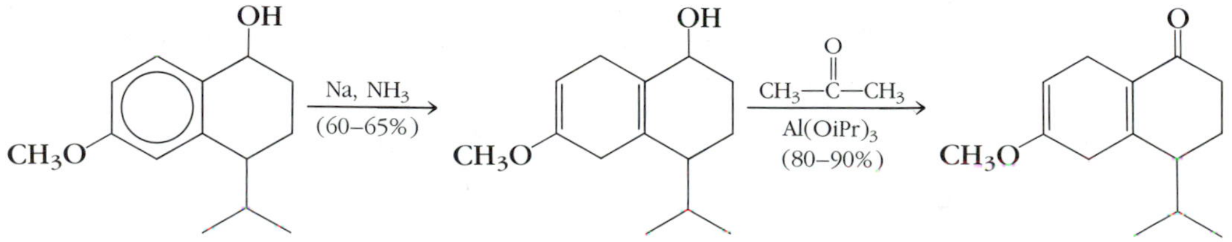

OH
CH_3O
Na, NH_3
(60–65%)
OH
CH_3O
$CH_3—C(=O)—CH_3$
$Al(OiPr)_3$
(80–90%)
O
CH_3O
8 **9** **10**

Oxidations with aluminum isopropoxide and acetone take place by a hydride transfer from the alcohol to acetone via a cyclic transition state. The aluminum cation serves as a template and holds the two reactants in the appropriate orientation.

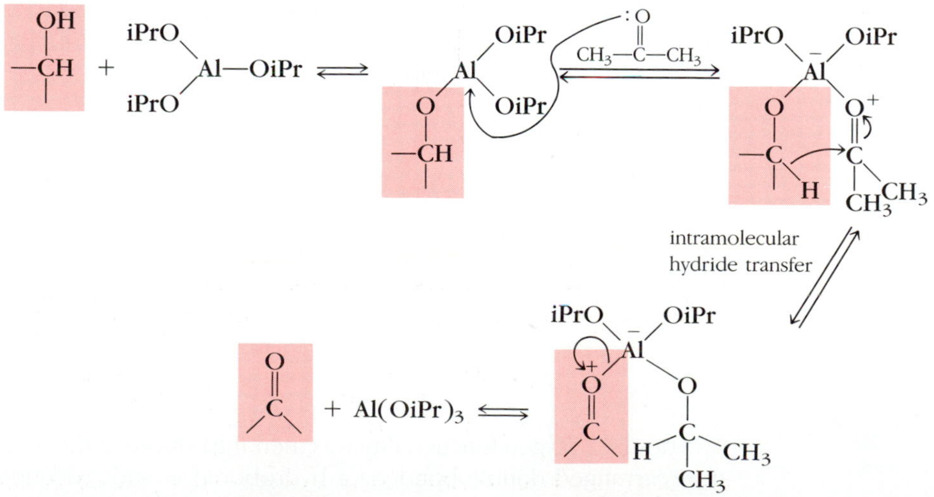

OH
—CH
+
iPrO
Al—OiPr
iPrO
OiPr
Al
O
—CH
OiPr
:O
$CH_3—C—CH_3$
iPrO
OiPr
Al
O
—C
H
O^+
C
CH_3
CH_3
intramolecular
hydride transfer
O
C
+ $Al(OiPr)_3$
iPrO
OiPr
Al
+
O
C
H
O
C
CH_3
CH_3

Intramolecular transfer of hydride from the alcohol to acetone effects simultaneous oxidation of the alcohol to a ketone and reduction of acetone to isopropyl alcohol (as the aluminum salt). The overall reaction is completely reversible, but the equilibrium is shifted toward the desired product by using a large excess of acetone. (This reaction of frequently called the Oppenauer oxidation, and the reverse reaction is known as the Meerwein-Ponndorf-Verley reduction.)

In the present case the ketone formed by oxidation is α,β-unsaturated, and this was reduced with lithium in ammonia to yield **11.** Finally, hydrolysis of the vinyl ether group afforded the desired diketone, **12.**

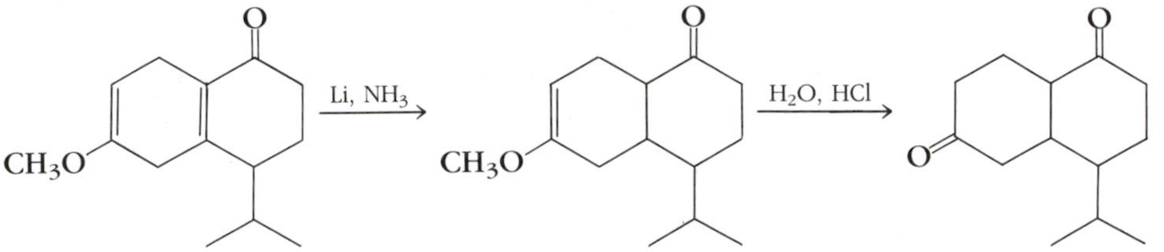

O, CH_3O — Li, NH_3 → O, CH_3O — H_2O, HCl → O, O — 30% (two steps)

10 **11** **12**

Example 22.3 The transformations we describe here were used to prepare the trifunctional compound **16.**

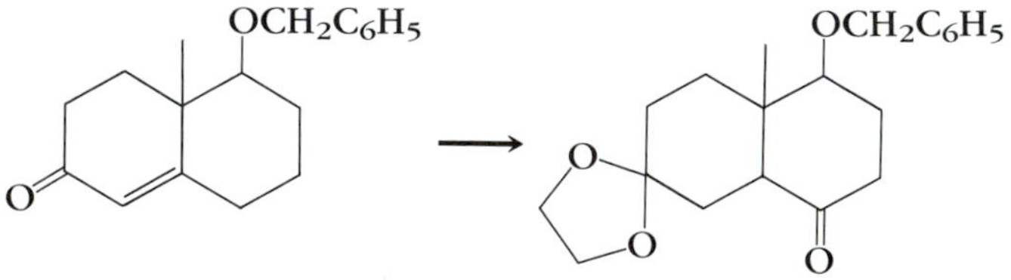

$OCH_2C_6H_5$, O → $OCH_2C_6H_5$, O, O, O

13 **16**

The first step in the sequence was protection of the ketone as the ethylene glycol ketal, and this was accompanied by isomerization of the double bond to the position that is β,γ to the original carbonyl. The *double bond isomerization is typical of ketal formation in this ring system* and should be expected to occur with structurally related compounds. (Unfavorable steric interactions are present when the double bond occupies its original position, but these are alleviated in the isomerized ketal.)

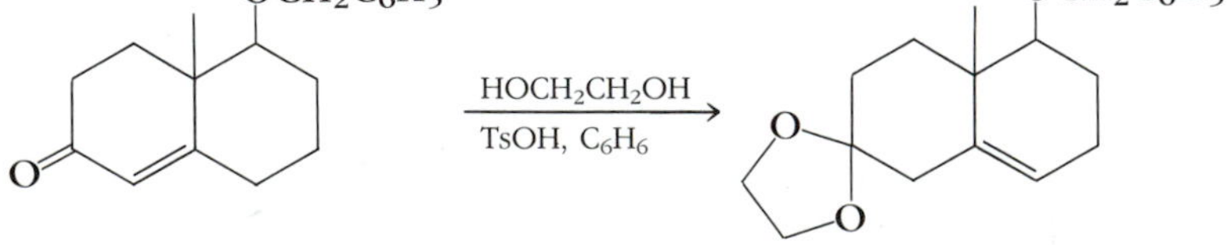

$OCH_2C_6H_5$, O — $HOCH_2CH_2OH$ / TsOH, C_6H_6 → $OCH_2C_6H_5$, O, O — 81%

13 **14**

The desired oxygen functionality was then introduced at the less substituted end of the rearranged double bond via a hydroboration and oxidation sequence.

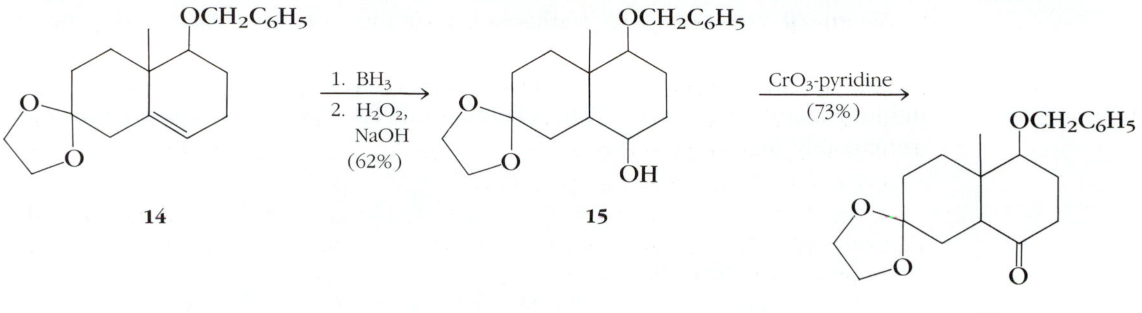

$OCH_2C_6H_5$ $\xrightarrow[\text{2. } H_2O_2\text{, NaOH (62\%)}]{\text{1. } BH_3}$ $OCH_2C_6H_5$, OH $\xrightarrow[(73\%)]{CrO_3\text{-pyridine}}$ $OCH_2C_6H_5$

14 **15** **16**

Notice that the ketal served two purposes in this sequence. It prevented reduction of the original carbonyl by borane, but more importantly, it caused isomerization of the double bond to activate a position that was originally unreactive.

Example 22.4 In order to prepare the ring system of **22** a carbocation rearrangement was employed. This required protection of one of the —OH groups of the starting diol (**17**) so that the other —OH could be converted to a reactive leaving group.

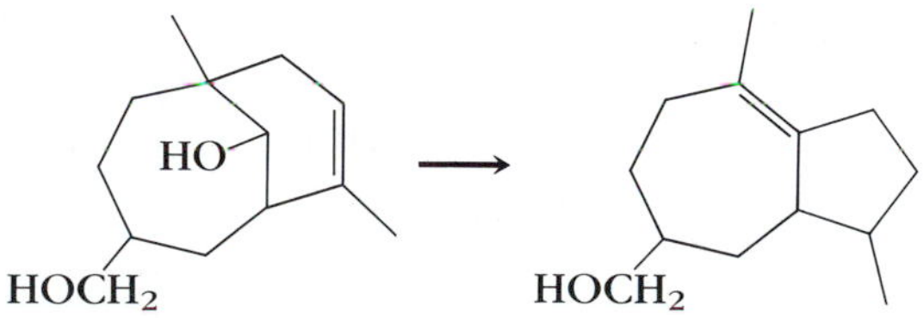

HO, $HOCH_2$ $\longrightarrow$ $HOCH_2$

17 **22**

The reaction of diol **17** with a single equivalent of acetic anhydride resulted in selective acylation of the primary alcohol. The more hindered secondary hydroxyl group was unaffected. After catalytic hydrogenation of the double bond, the secondary hydroxyl group was converted to the corresponding methanesulfonate (**20**).

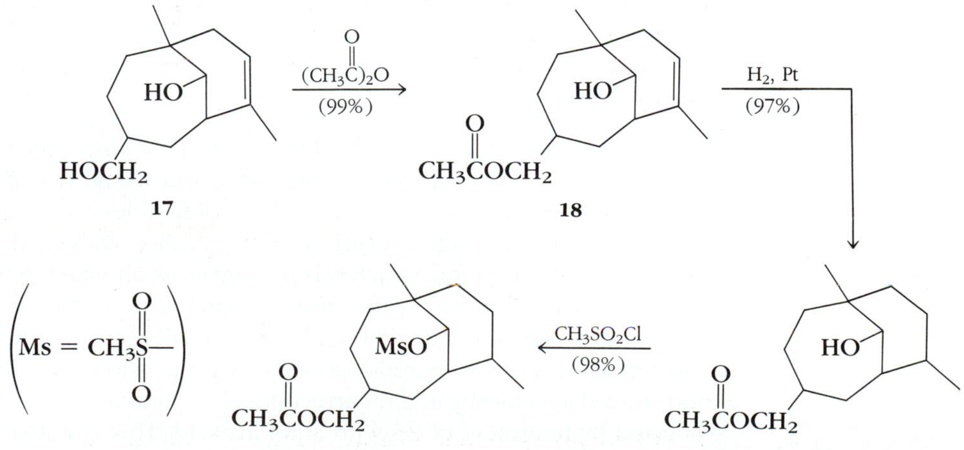

HO, $HOCH_2$ $\xrightarrow[(99\%)]{(CH_3\overset{O}{\overset{\|}{C}})_2O}$ HO, $CH_3\overset{O}{\overset{\|}{C}}OCH_2$ $\xrightarrow[(97\%)]{H_2\text{, Pt}}$

17 **18**

$\left(Ms = CH_3\overset{O}{\underset{O}{\overset{\|}{\underset{\|}{S}}}}—\right)$ MsO, $CH_3\overset{O}{\overset{\|}{C}}OCH_2$ $\xleftarrow[(98\%)]{CH_3SO_2Cl}$ HO, $CH_3\overset{O}{\overset{\|}{C}}OCH_2$

20 **19**

When **20** was heated in acetic acid, ionization of the methanesulfonate was accompanied by carbocation rearrangement to give a tertiary carbocation. Loss of a proton yielded the alkene acetate **21**, which was cleaved to the alcohol **22** by lithium aluminum hydride. The overall yield for this five-step sequence was a remarkably high 84%.

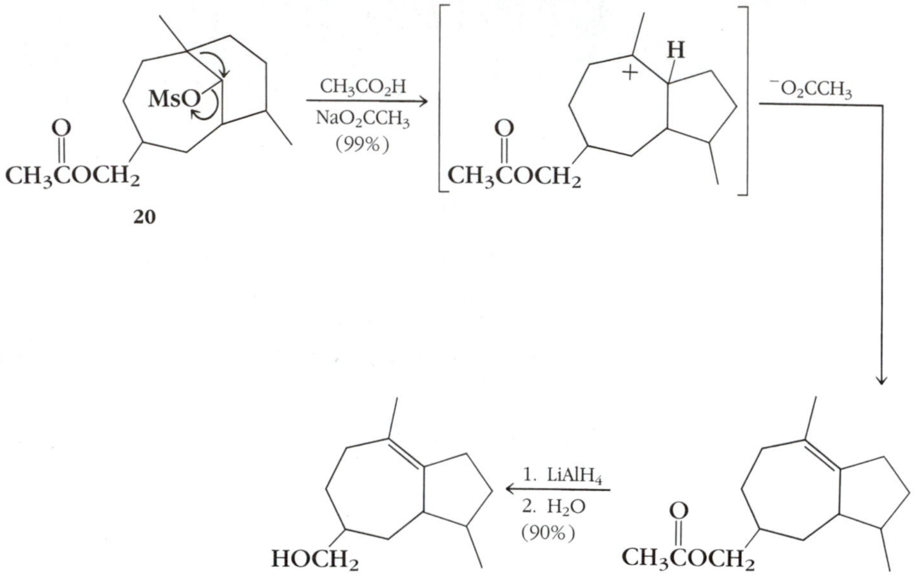

MsO
$CH_3\overset{O}{\overset{\|}{C}}OCH_2$
20

CH_3CO_2H
NaO_2CCH_3
(99%)

H
+
$CH_3\overset{O}{\overset{\|}{C}}OCH_2$

$^-O_2CCH_3$

$CH_3\overset{O}{\overset{\|}{C}}OCH_2$
21

1. $LiAlH_4$
2. H_2O
(90%)

$HOCH_2$
22

Example 22.5 The conversion of **23** to **26** required methylation of the tertiary carbon α to the carbonyl group.

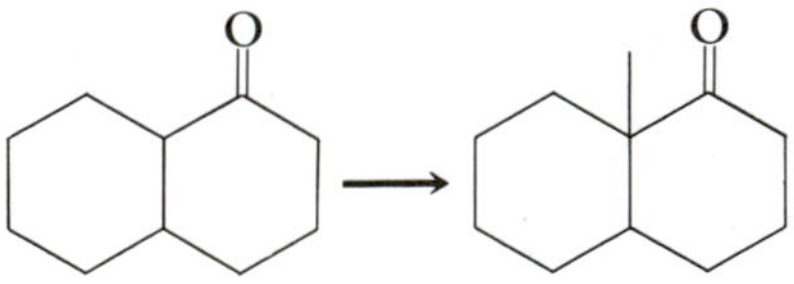

23 → **26**

Direct alkylation of **23** would lead to several problems (Section 14.3). Polyalkylation would be expected, and enolate formation would occur more rapidly at the less substituted α position. To avoid these difficulties a blocking group was introduced at the more reactive α carbon. Condensation with ethyl formate generated a β-dicarbonyl compound, which was present as its enolate ion in the basic medium. Reaction of this enolate with 2-iodopropane occurred via *O*-alkylation (Section 14.3) to give the vinyl ether **24.** Only one enolate ion could be formed from **24,** at the tertiary α carbon. Polyalkylation was not possible, so the methyl group was introduced specifically at the correct position. Removal of the blocking group was effected by treatment of **25** with aqueous acid. This resulted in hydrolysis of the vinyl ether followed by a retrocondensation, converting **25** to the ketone **26.**

23 → (1. $HCO_2C_2H_5$, $NaOCH_3$; 2. CH_3CHCH_3 with I) → 24 (=CH—OCH(CH_3)$_2$) → (1. KNH_2; 2. CH_3I) → 25 → (HCl, H_2O) → 26

43% overall yield

Example 22.6 The preparation of lactone **32** from ester **27** required the introduction of a four-carbon oxygen-containing fragment. You can see this by inspecting the corresponding hydroxy acid, **32a.**

$OCH_2C_6H_5$ — $CO_2C_2H_5$ **27** → **32** (C=O, O) ; **32a** (CO_2H, OH)

The strategy employed for this synthesis was to use methyl vinyl ketone in a Michael reaction. But this is a mixed condensation, and the strongly basic conditions needed to form an enolate of a monocarbonyl compound might have led to a variety of side reactions, such as self-condensation of the unsaturated ketone. Consequently, an *activating group* was employed. Treatment of the starting ester with diethyl carbonate and sodium hydride introduced a second —$CO_2C_2H_5$ group, and the resulting malonic ester reacted smoothly with methyl vinyl ketone.

27 ($OCH_2C_6H_5$, $CO_2C_2H_5$) → ($C_2H_5OCOC_2H_5$ (C=O), NaH) → **28** ($OCH_2C_6H_5$, H, $CO_2C_2H_5$, $CO_2C_2H_5$) → (CH_2=CH—C(=O)—CH_3, $KOC(CH_3)_3$) → **29** ($OCH_2C_6H_5$, $CO_2C_2H_5$, $CO_2C_2H_5$, O)

Sodium borohydride reduction of **29** gave the hydroxy diester **30,** and this was converted to lactone **31** by saponification of the ester groups followed by acidification. Finally, distillation of **31** resulted in decarboxylation (i.e., removal of the activating group) to give the target molecule **32.** The overall synthesis was carried out in 30% yield, which corresponds to an average yield of 79% for each of the five steps.

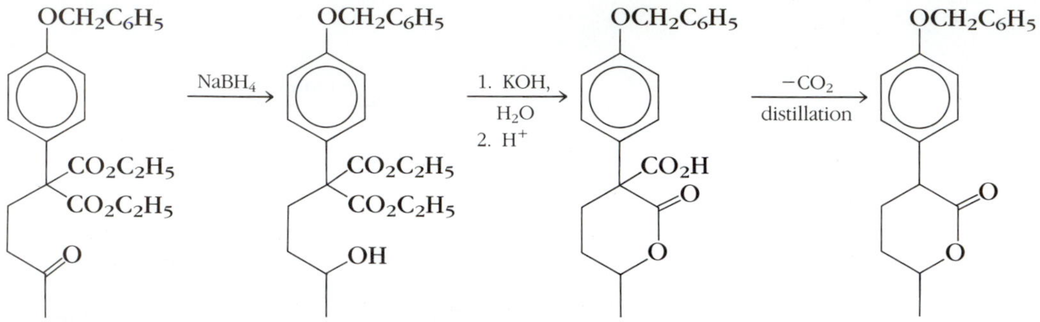

$OCH_2C_6H_5$ $CO_2C_2H_5$ $CO_2C_2H_5$ O — $NaBH_4$ → $OCH_2C_6H_5$ $CO_2C_2H_5$ $CO_2C_2H_5$ OH — 1. KOH, H_2O 2. H^+ → $OCH_2C_6H_5$ CO_2H O O — $-CO_2$ distillation → $OCH_2C_6H_5$ O O

29 **30** **31** **32**

Example 22.7 The oxidation of cholesterol (**33**) to the corresponding ketone, **36,** required protection of the double bond so that it would not isomerize to the more stable conjugated isomer.

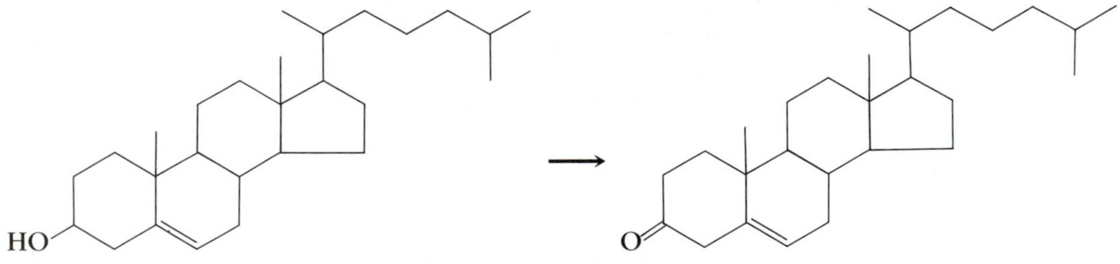

HO → O

33 **36**

Protection of the double bond was accomplished by converting it to the dibromide. Following oxidation of the alcohol, the double bond was regenerated in its original position by reduction of keto dibromide **35** with zinc. The final product was obtained in 71–72% yield.

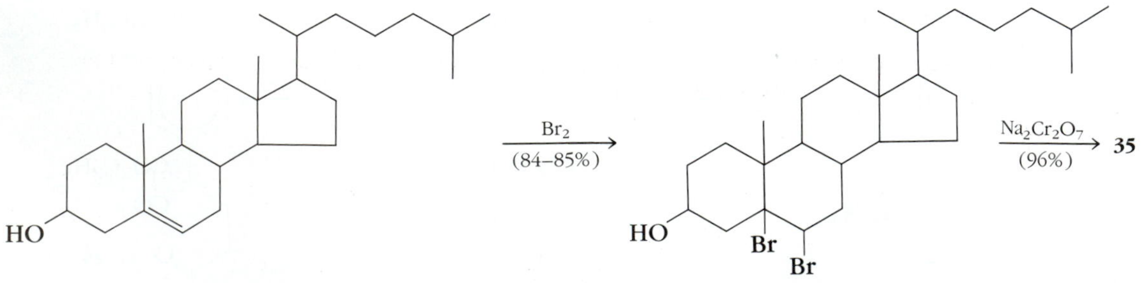

HO — Br_2 (84–85%) → HO Br Br — $Na_2Cr_2O_7$ (96%) → **35**

33 **34**

34 ⟶ [structure 35: 5,6-dibromo steroid ketone] $\xrightarrow[(88\%)]{Zn}$ [structure 36]

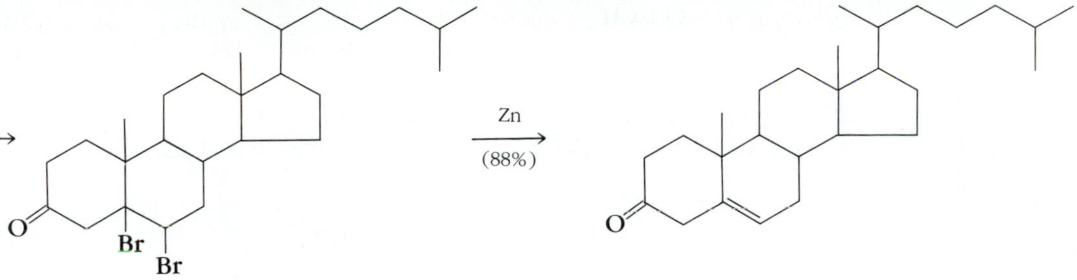

35 **36**

We originally discussed this type of reductive elimination with zinc in Section 5.13, and several closely related reductions have been developed. For example, reduction also occurs with an alkoxy bromide.

$$CH_2{=}CH{-}CH_2{-}CH(OC_2H_5){-}CH_2{-}Br \xrightarrow{Zn} CH_2{=}CH{-}CH_2{-}CH{=}CH_2 \quad 72\text{–}76\%$$

When a carbonyl compound with an α-bromo or α-acetoxy substituent is treated with zinc in acetic acid, the substituent is replaced by a hydrogen. This is again a reductive elimination, and it proceeds via the zinc enolate. The enolate is then converted to the keto form in the mildly acidic medium. Electron transfer must initially occur from zinc metal to the LUMO of the substrate, but we can summarize all these reactions by a common simplified mechanism.

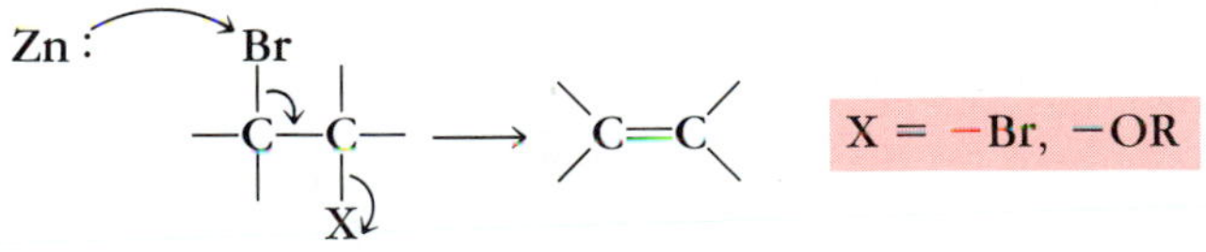

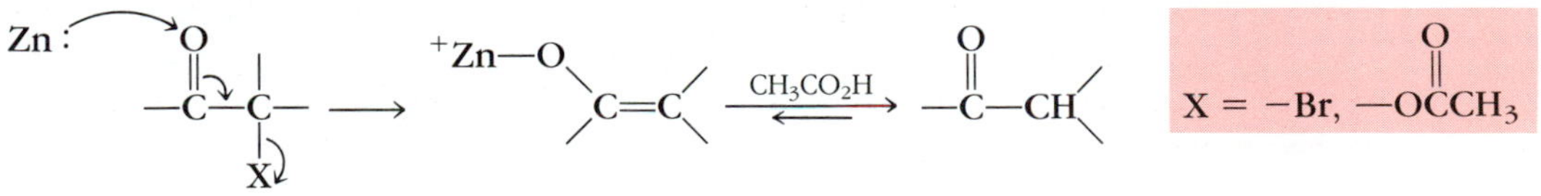

$$Zn{:}\ \ Br{-}C{-}C{-}X \longrightarrow C{=}C \qquad X = -Br,\ -OR$$

$$Zn{:}\ \ O{=}C{-}C{-}X \longrightarrow {}^{+}Zn{-}O{-}C{=}C \underset{}{\overset{CH_3CO_2H}{\rightleftharpoons}} -\overset{O}{\overset{\|}{C}}{-}CH \qquad X = -Br,\ -O\overset{O}{\overset{\|}{C}}CH_3$$

Such reductions have received widespread use in synthetic chemistry and are illustrated by the following equations:

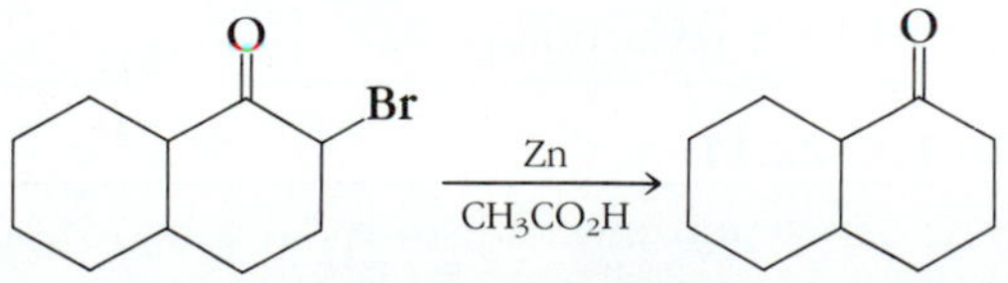

[2-bromo-1-decalone] $\xrightarrow[CH_3CO_2H]{Zn}$ [1-decalone] 91%

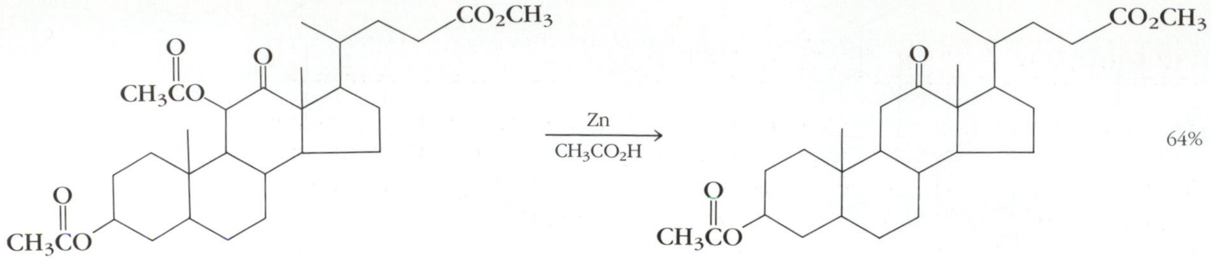

$\xrightarrow[CH_3CO_2H]{Zn}$ 64%

Notice that only the acetoxy group α to the ketone was reduced in the second of the preceding reactions. The two other ester groups were unaffected.

Example 22.8 Our final example illustrates how a carbon-carbon double bond can be used as a latent carbonyl group. Conversion of **37** to **39** is also an example of *ring contraction,* in which a cyclic compound is converted to a product in which the ring has one less carbon atom.

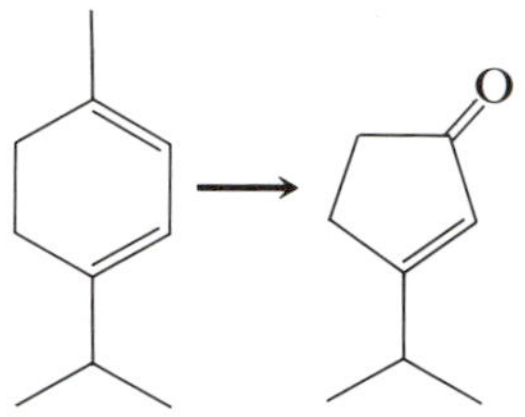

37 → **39**

The transformation was actually carried out in only two steps. Ozonolysis produced an acyclic 1,4-diketone (**38**). (The intermediate ozonide was reduced with iodide ion rather than with zinc metal in this instance.) Treatment of **38** with base effected an intramolecular aldol condensation to form **39.**

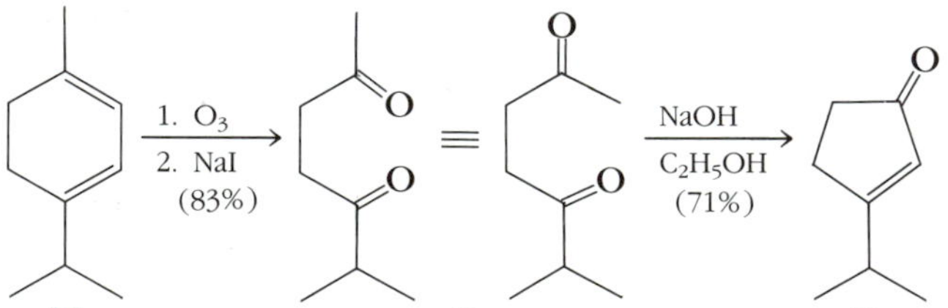

37 $\xrightarrow[\text{2. NaI (83\%)}]{\text{1. } O_3}$ **38** $\xrightarrow[C_2H_5OH \text{ (71\%)}]{NaOH}$ **39**

EXERCISE 22.9

Write a balanced equation for the conversion of compound **3** to compound **4** in Example 22.1.

EXERCISE 22.10

In Example 22.4, what other organic product is formed in the lithium aluminum hydride reduction of **21?**

EXERCISE 22.11

Draw a reaction mechanism for the hydrolysis of **25** to **26** Example 22.5. Show all products that are formed.

22.5 SUMMARY OF PROTECTING GROUPS

In this chapter we have shown you a variety of ways in which specific functional groups can be modified to change their reactivity. Depending on the particular instance, we might regard the modification as blocking, activating, or masking the original reactivity. We will use the term *protecting group* in a general sense that includes all of these effects. Table 22.4 summarizes the use of some common protecting groups.

TABLE 22.4 Some Common Protecting Groups

Protecting Group	Reagents for Introduction	Reagents for Removal; Comments
1. Protection of Aldehydes and Ketones		
(a) Acetals, ketals (acyclic)		Section 15.7
$>C{=}O \xrightarrow{ROH,\ H^+} >C(OR)_2$	Alcohol, acid	Dilute aqueous acid (typically, HCl or H_2SO_4)
(b) Acetals, ketals (cyclic)		Section 15.7
$>C{=}O \xrightarrow[TsOH,\ C_6H_6]{HOCH_2CH_2OH}$ cyclic ketal (C with O–CH₂CH₂–O ring)	Ethylene glycol, TsOH, benzene	Dilute aqueous acid (typically, HCl or H_2SO_4)
(c) Vinyl ethers	(Usually not prepared directly from aldehyde or ketone)	Section 15.7 Dilute aqueous acid (typically, HCl or H_2SO_4)
(d) Thioketals, thioacetals		Section 15.7
$>C{=}O \xrightarrow[BF_3]{HSCH_2CH_2SH}$ cyclic thioketal (C with S–CH₂CH₂–S ring)	Ethanedithiol, BF_3	Hydrolysis is slow; catalysis by silver or mercury salts is usually needed.
cyclic thioketal $\xrightarrow[H_2]{Ni} >CH_2$		Raney nickel cleaves the sulfurs by hydrogenolysis (to generate $—CH_2—$).

TABLE 22.4 (continued)

Protecting Group	Reagents for Introduction	Reagents for Removal; Comments
2. Protection of Carboxyl Groups		
(a) Methyl, ethyl esters		Section 16.6
	Methanol, ethanol (acid catalysis)	Hydrolysis in acid or base
(b) Benzyl esters	Benzyl alcohol (acid catalysis) or benzyl chloride, base	Section 16.6 Hydrogenolysis (Pd catalyst), HBr in nonaqueous solvent, or sodium, ammonia
(c) β-Bromo esters		cf. Section 22.4
$RCO_2H \xrightarrow[H^+]{HOCH_2CH_2Br} RCO_2CH_2CH_2Br$ $RCO_2CH_2CH_2Br \xrightarrow{Zn} RCO_2H$	β-Bromoethanol, acid (i.e., standard esterification methods)	Zinc, acetic acid (reductive elimination); trichloroethyl esters can also be used.

TABLE 22.4 (continued)

Protecting Group	Reagents for Introduction	Reagents for Removal; Comments
3. Protection of Amino Groups		
(a) Acetyl group		Section 16.6
$>N{-}H \xrightarrow[NEt_3]{Ac_2O \text{ or } AcCl} >N{-}Ac$ $>N{-}Ac \xrightarrow[^-OH \text{ or } H^+]{H_2O} >NH$	Acetic anhydride or acetyl chloride and a tertiary amine	Vigorous hydrolysis in acid or base
(b) Trifluoroacetyl group		Section 20.3
	Trifluoroacetic anhydride and a tertiary amine	Mild alkaline hydrolysis using dilute NaOH or aqueous ammonia

TABLE 22.4 (continued)

Protecting Group	Reagents for Introduction	Reagents for Removal; Comments
3. Protection of Amino Groups		
(c) *p*-Toluenesulfonyl group		Section 20.3
	p-Toluenesulfonyl chloride and a tertiary amine	Sodium in ammonia; esters, ketones, etc. are reduced under these conditions.
(d) Benzyloxycarbonyl group		Section 20.3
	Benzyl chloroformate and base (NaOH or $NaHCO_3$)	HBr in nonaqueous solvent, sodium in ammonia, or H_2/Pd
(e) *tert*-Butoxycarbonyl group (tBOC)		Section 20.3
	tert-Butyl azidoformate and a tertiary amine	HBr in nonaqueous solvent

TABLE 22.4 (continued)

Protecting Group	Reagents for Introduction	Reagents for Removal; Comments
4. Protection of the Hydroxyl Group		
(a) Esters, e.g., acetates		Sections 11.8, 16.6
$—O—\overset{\overset{\displaystyle O}{\|}}{C}—R$	Acetic anhydride or acetyl chloride and a tertiary amine	Hydrolysis (potassium carbonate); methanolic ammonia
(b) Benzyl ether		Sections 8.6, 11.8, 15.4
	Sodium amide or sodium hydride, then benzyl bromide	Hydrogenolysis with Pd catalyst

TABLE 22.4 (continued)

Protecting Group	Reagents for Introduction	Reagents for Removal; Comments
4. Protection of the Hydroxyl Group		
(c) Tetrahydropyranyl ether		Section 15.5
	Dihydropyran, *p*-toluenesulfonic acid	Mild acid hydrolysis (or alcoholysis), e.g., HOAc, H_2O, THF
(d) Silyl ethers		
	$(CH_3)_3SiCl$ and a tertiary amine	$Bu_4N^+ F^-$ (Primary hydroxyl groups are also regenerated by mild acid hydrolysis; secondary and tertiary derivatives react more slowly)
(e) Protection of diols: acetonide		Section 15.5
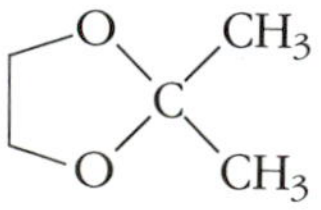 (cyclic –O–C(CH_3)(CH_3)–O– acetonide)	2,2-Dimethoxypropane and TsOH (ketal exchange, Section 22.4)	Dilute aqueous acid

22.6 PROBLEMS

22.1 Which of the following hypothetical reactions should work well? Which should fail? Explain your answers.

(a) 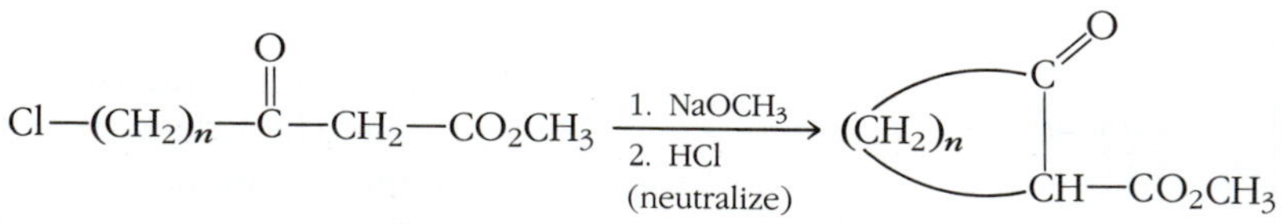 $Cl{-}(CH_2)_n{-}\overset{O}{\overset{\|}{C}}{-}CH_2{-}CO_2CH_3 \xrightarrow[\text{2. HCl (neutralize)}]{\text{1. NaOCH}_3}$ cyclic $(CH_2)_n$ ring containing $C{=}O$ and $CH{-}CO_2CH_3$

for $n = 1$; $n = 4$; $n = 7$

(b) 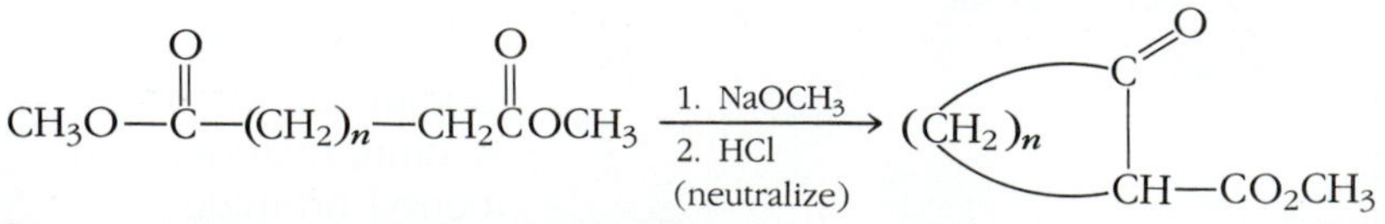 $CH_3O{-}\overset{O}{\overset{\|}{C}}{-}(CH_2)_n{-}CH_2\overset{O}{\overset{\|}{C}}OCH_3 \xrightarrow[\text{2. HCl (neutralize)}]{\text{1. NaOCH}_3}$ cyclic $(CH_2)_n$ ring containing $C{=}O$ and $CH{-}CO_2CH_3$

for $n = 1$; $n = 3$; $n = 5$

(c) 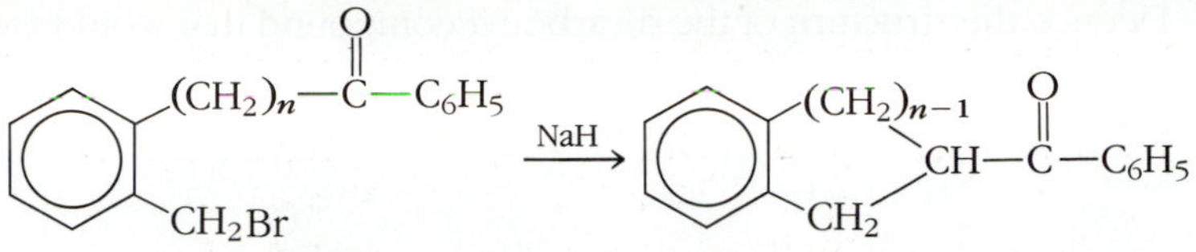 [ortho-disubstituted benzene with $(CH_2)_n-C(=O)-C_6H_5$ and CH_2Br] $\xrightarrow{NaH}$ [ortho-fused ring: benzene–$(CH_2)_{n-1}$–$CH(-C(=O)-C_6H_5)$–CH_2–]

for $n = 2$; $n = 3$; $n = 5$; $n = 6$

22.2 Complete each of the following reactions:

(a) $CH_3-\overset{O}{\overset{\|}{C}}-CH_2CH_2CH_2-\overset{O}{\overset{\|}{C}}-CH_3 \xrightarrow[CH_3OH]{KOH} ?$

(b) $C_6H_5-\overset{O}{\overset{\|}{C}}-CH_2CH_2CH_2-\overset{CH_3}{\overset{|}{C}H}-CHO \xrightarrow[C_2H_5OH]{NaOC_2H_5} ?$

(c) $C_6H_5-CH_2-\overset{O}{\overset{\|}{C}}-CH_2CH_2-\overset{CH_3}{\overset{|}{C}H}-CHO \xrightarrow[C_2H_5OH]{NaOC_2H_5} ?$

(d) $C_6H_5-CH_2CH_2-\overset{O}{\overset{\|}{C}}-CH_2-\overset{CH_3}{\overset{|}{C}H}-CHO \xrightarrow[C_2H_5OH]{NaOC_2H_5} ?$

(e) $CH_3-\overset{O}{\overset{\|}{C}}-CH_2CH_2-\overset{CH_3}{\overset{|}{C}H}-CH_2CH_2-\overset{O}{\overset{\|}{C}}-CH_3 \xrightarrow{KOH} ?$

22.3 Complete each of the following reactions:

(a) $C_6H_5-\overset{O}{\overset{\|}{C}}-CH_2CH_2CH_2CH_2-CO_2CH_3 \xrightarrow[\text{2. HCl (neutralize)}]{\text{1. }NaOCH_3} ?$

(b) $C_6H_5-CH_2-\overset{O}{\overset{\|}{C}}-CH_2CH_2CH_2-CO_2CH_3 \xrightarrow[\text{2. HCl (neutralize)}]{\text{1. }NaOCH_3} ?$

(c) $CH_3-\overset{CH_3}{\overset{|}{C}H}-CH_2-\overset{O}{\overset{\|}{C}}-CH_2CH_2-CO_2C_2H_5 \xrightarrow{NaH} ?$

(d) $CH_3O_2C-CH_2-\overset{CH_3}{\overset{|}{C}H}-CH_2-\overset{CH_3}{\overset{|}{C}H}-CO_2CH_3 \xrightarrow{NaOCH_3} ?$

(e) $C_2H_5O_2CCH_2CH_2-\overset{CO_2CH_3}{\overset{|}{C}H}-CH_2CH_2CH_2CO_2C_2H_5 \xrightarrow[\text{2. HCl (neutralize)}]{\text{1. NaH}} ?$

22.4 Deduce the structure of the dicarbonyl compound that would yield each of the following cyclic products upon treatment with sodium hydride (followed by neutralization with HCl during workup).

(a) 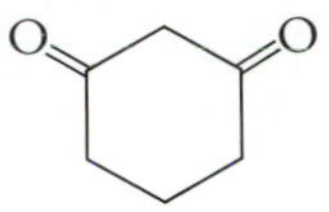 O O (b) 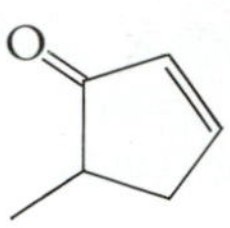 O (c) 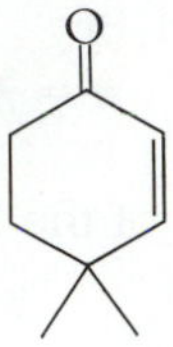 O

(d) 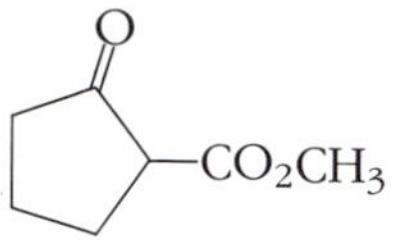 O $-CO_2CH_3$ (e) 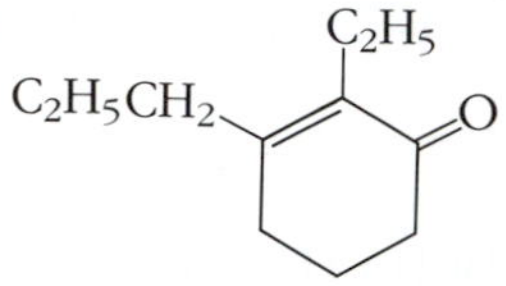 C_2H_5 $C_2H_5CH_2$ O (f) 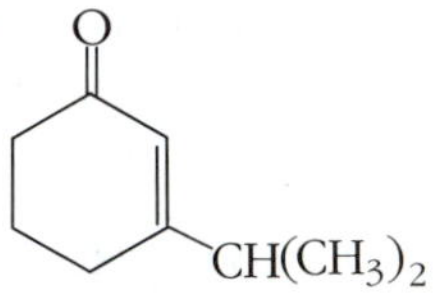 O $CH(CH_3)_2$

(g) 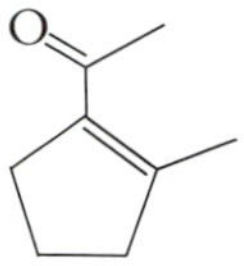 O (h) 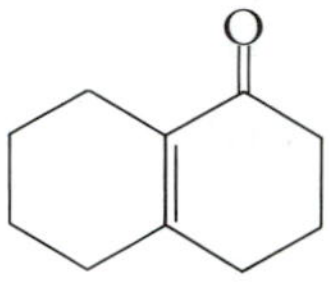 O

22.5 Complete each of the following reactions:

(a) $CH_3CH_2-\overset{O}{\overset{\|}{C}}-CH_2CH_3 + CH_3-\overset{O}{\overset{\|}{C}}CH{=}CH_2 \xrightarrow{NaOCH_3,\ CH_3OH} ?$

(b) $CH_3-CH{=}CH-CH{=}CH-CH_3 + CH_3CH{=}CH-\overset{O}{\overset{\|}{C}}-CH_3 \xrightarrow{heat} ?$

(c) 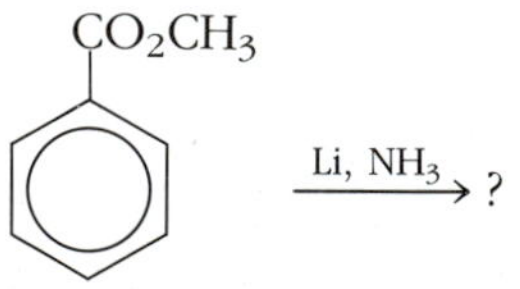 CO_2CH_3 $\xrightarrow{Li,\ NH_3} ?$

(d) 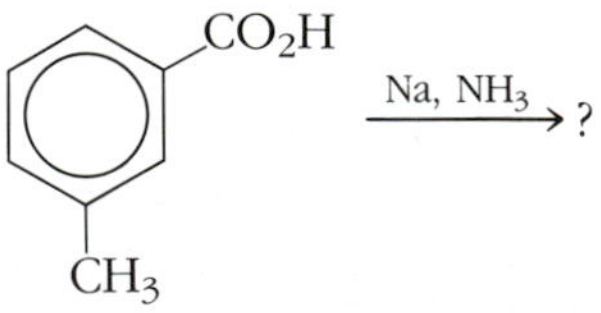 CO_2H CH_3 $\xrightarrow{Na,\ NH_3} ?$

(e) 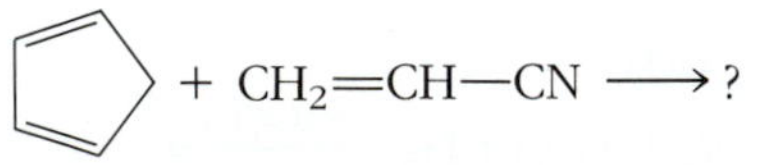 $+ CH_2{=}CH-CN \longrightarrow ?$

(f) 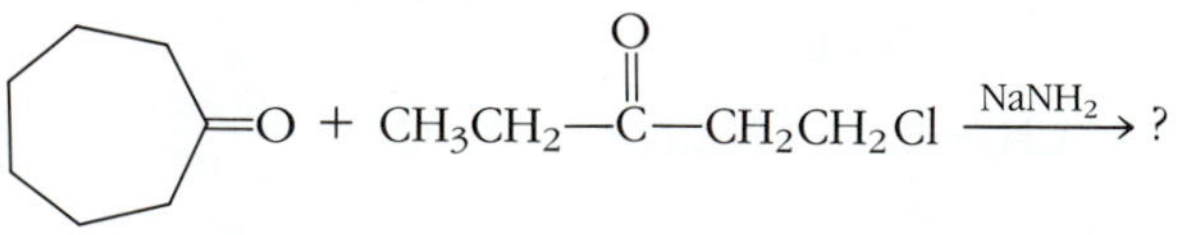 $=O + CH_3CH_2-\overset{O}{\overset{\|}{C}}-CH_2CH_2Cl \xrightarrow{NaNH_2} ?$

22.6 Deduce the acyclic starting materials and procedure that would afford each of the following six-membered-ring compounds:

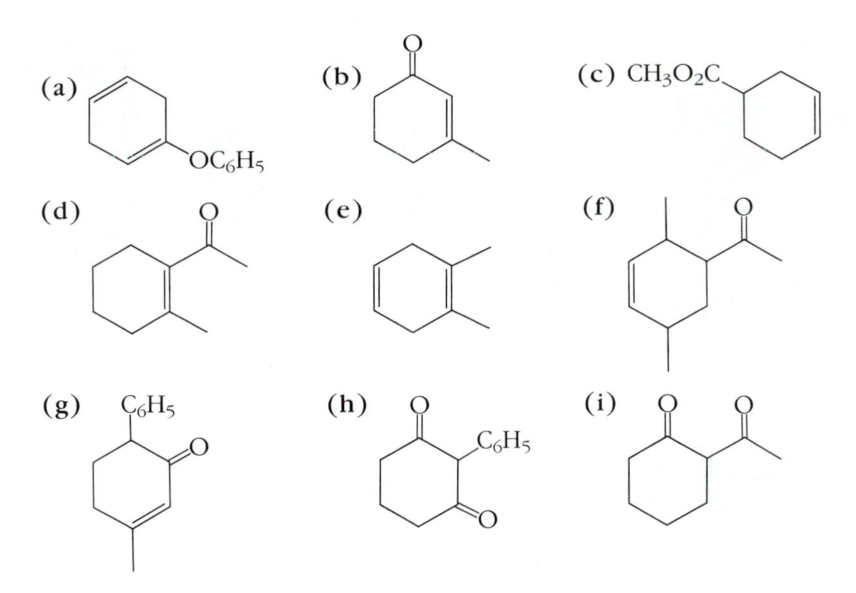

22.7 Sometimes more than one method can be used to prepare a particular compound. Show how the following unsaturated compound could be prepared by using: (a) the Diels-Alder reaction; (b) Robinson annelation; and (c) Birch reduction. These reactions may not yield the final product directly, however, so show any other reactions that may also be necessary.

22.8 For many of the Diels-Alder reactions in Section 22.3 we did not show the stereochemistry of the product. Draw three-dimensional representations of the preferred products for these reactions.

22.9 When 1 mol of the sodium salt of diethyl malonate was allowed to react with 1 mol of 1,2-dibromoethane, compound **A** was formed. When only 0.5 mol of the dibromide was used, the product was compound **B.** Deduce the structures of **A** and **B.**

22.10 Suggest methods for synthesizing the following compounds from starting materials containing no more than the indicated number of carbon atoms:

(a) six carbons

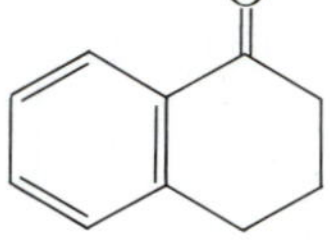

(b) CO_2CH_3 six carbons

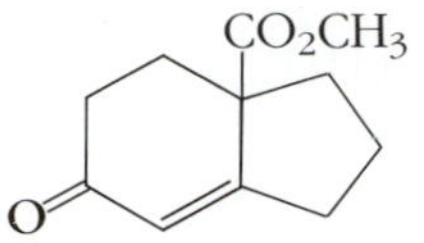

(c) CH_3O_2C CO_2CH_3 five carbons

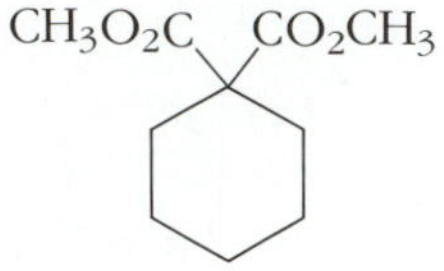

(d) C_6H_5 eight carbons

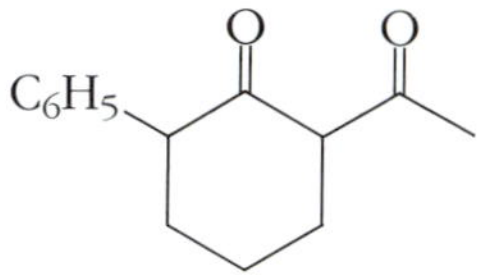

(e) CO_2CH_3 CO_2CH_3 five carbons

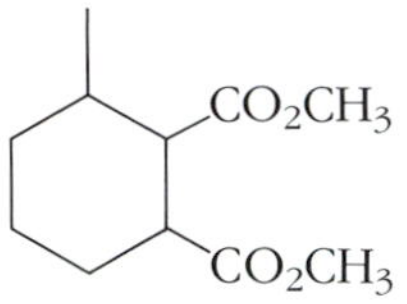

22.11 Write a reaction mechanism for the conversion of **13** to **14** in Example 22.3 that accounts for migration of the double bond from the α,β to the β,γ position.

22.12 In Example 22.3 an oxygen is protected as the benzyl ether. Suggest how this might be removed at the end of the sequence to generate the keto alcohol corresponding to **16** (more than one step might be necessary).

22.13 Complete the missing parts of the following synthetic sequence:

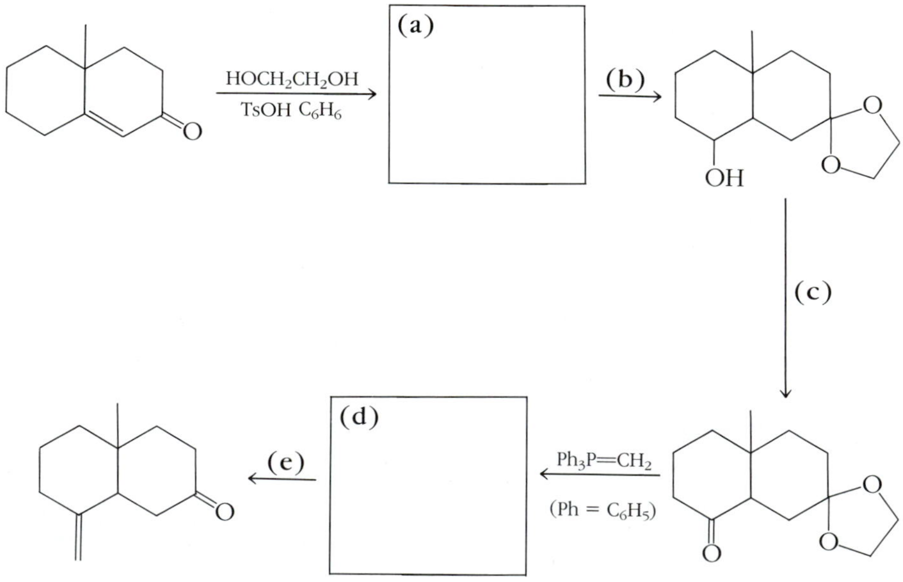

$HOCH_2CH_2OH$ / TsOH C_6H_6 → (a) → (b) → OH → (c) → $Ph_3P{=}CH_2$ ($Ph = C_6H_5$) → (d) → (e)

22.14 Complete the missing parts of the following synthetic sequence:

(a) (b)

C_2H_5O C_2H_5O

HCl, C_2H_5OH, H_2O

(c)

(d)

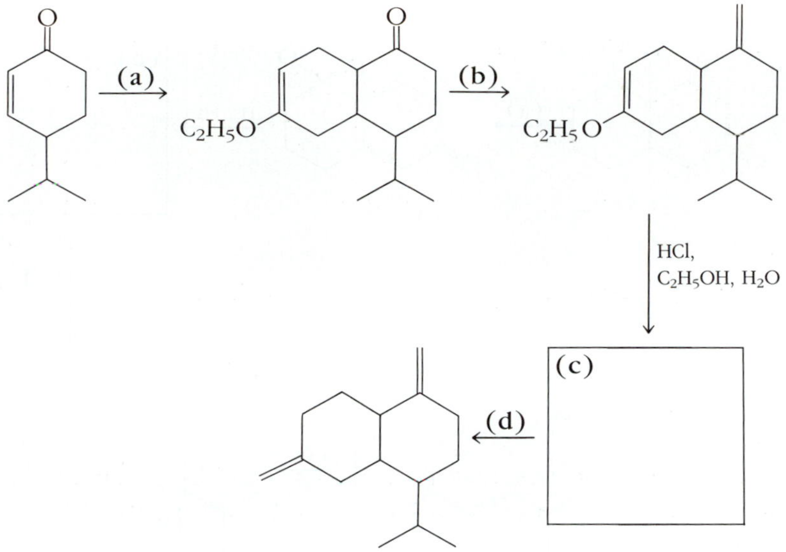

22.15 Complete the missing parts of the following synthetic sequence:

OH

(a)

O O

O O

1. $LiAlH_4$
2. H_2O

(b)

(c)

O O

(e) (d) NaCN

CO_2CH_3

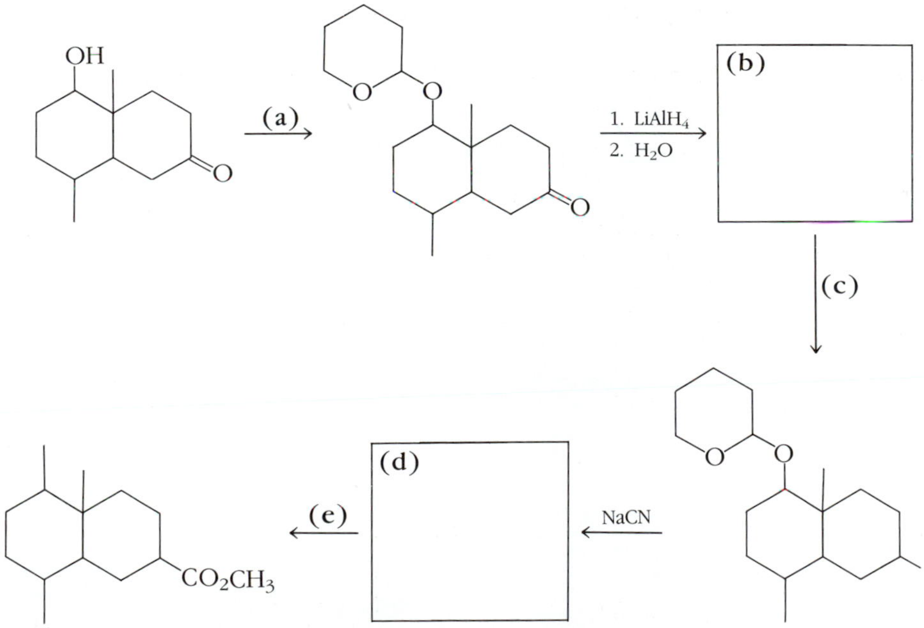

22.16 Complete the missing parts of the following synthetic sequence:

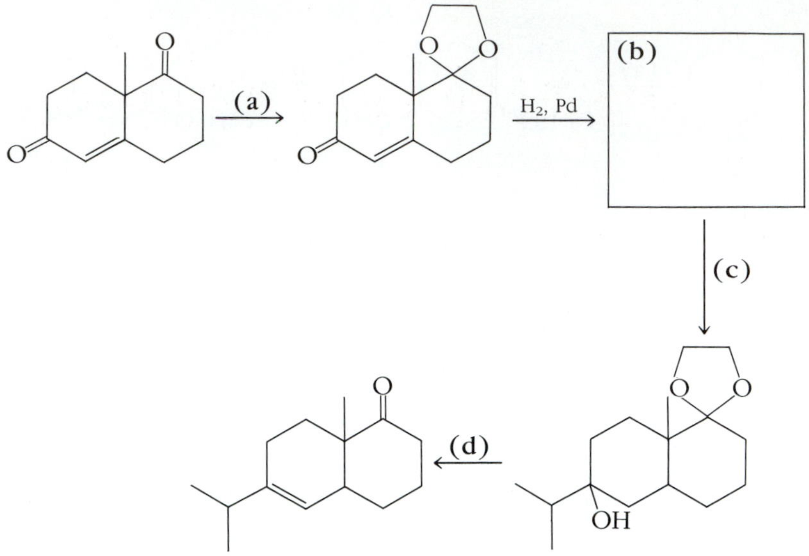

(a) H_2, Pd (b) (c) OH (d)

22.17 Suggest a method for carrying out the following conversion.

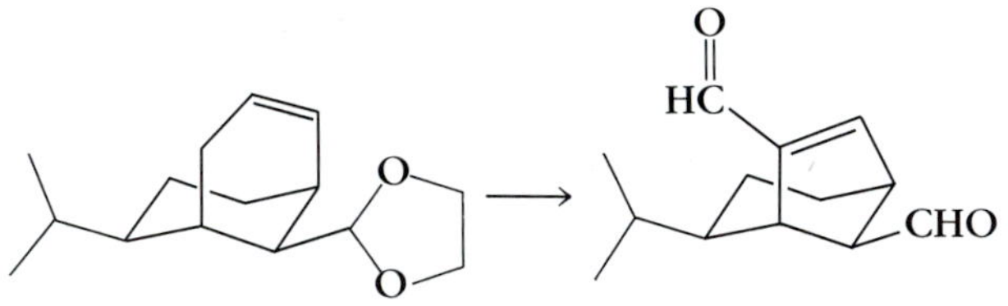

HC CHO

CHAPTER 23

NATURAL PRODUCTS

Organic chemistry has its origins in the studies of compounds that are synthesized by living organisms. Such compounds are called *natural products,* and they continue to be of great interest to organic chemists. If we are ever to gain a complete understanding of the chemistry of living organisms, a basic part of our knowledge will consist of the structure and reactivity of these naturally occurring substances.

The chemistry of natural products is a huge field, and we will only be able to provide a brief overview in this text. We have restricted our coverage to just a few classes of natural products: *terpenes, steroids, prostaglandins,* and *alkaloids.* All four classes of compounds can have a significant impact on humans. You will find additional discussions of natural products in several other chapters: for example, carbohydrates (Chapter 19, Section 25.2), peptides and proteins (Sections 20.5, 25.2); nucleic acids and nucleotide bases (Sections 24.4, 25.2); purine and pyrimidine biosynthesis (Section 24.4), insect pheromones (Sections 11.11, 26.3); the biosynthesis of terpenes and steroids (Section 12.9); vitamin C (Section 19.10); antibiotics (Section 24.5); and natural products with applications as pesticides (Section 26.3). Many categories of natural products are so structurally diverse that we cannot include them all in our brief discussion.

In each of the following sections we will first consider the structural and biological aspects of a particular class of natural products. We will then move on to illustrate some of the synthetic endeavors that have been carried out for specific compounds. Our reason for emphasizing the synthetic work is threefold. First, synthesis has been historically important as the ultimate proof of a proposed structure. Second, small modifications in a synthesis can be used to prepare *analogs* of natural products, and these can be of great value in medicine and in biochemical studies. Finally, we believe that these examples will provide you with an understanding of how practicing organic chemists actually use the many reactions you have learned in previous chapters.

23.1 TERPENES AND TERPENOIDS

Terpenes are hydrocarbons with structures built up from C_5H_8 subunits. They are found primarily in plants, although squalene (steroid biosynthesis, Section 12.9) is a notable exception, and they are produced by plants according to the biosynthetic sequence we outlined in Section 11.11. Oxidation of the terpene hydrocarbons often occurs in the plants to form modified compounds called *terpenoids,* which include a wide variety of alcohols, ketones, aldehydes, and ethers. The C_{10} and C_{15} terpenes have been most widely studied; these are *monoterpenes* and *sesquiterpenes,* respectively. The biological functions of most terpenes and terpenoids are not known. In some cases they seem to protect the plant from animals through a disagreeable taste or odor, but no general role has yet been established.

Many of the monoterpenes, sesquiterpenes, and their oxygenated derivatives are relatively volatile, and this has several important consequences. First, they are easily isolated by distillation from the nonvolatile plant materials, and therefore they were among the first natural products to be investigated by organic chemists. Second, they have distinctive, and sometimes quite pleasant, aromas. For this reason they have been used extensively in the flavor and fragrance industries. Figure 23.1 shows the structures of some terpenoids with aromas that may be familiar to you.

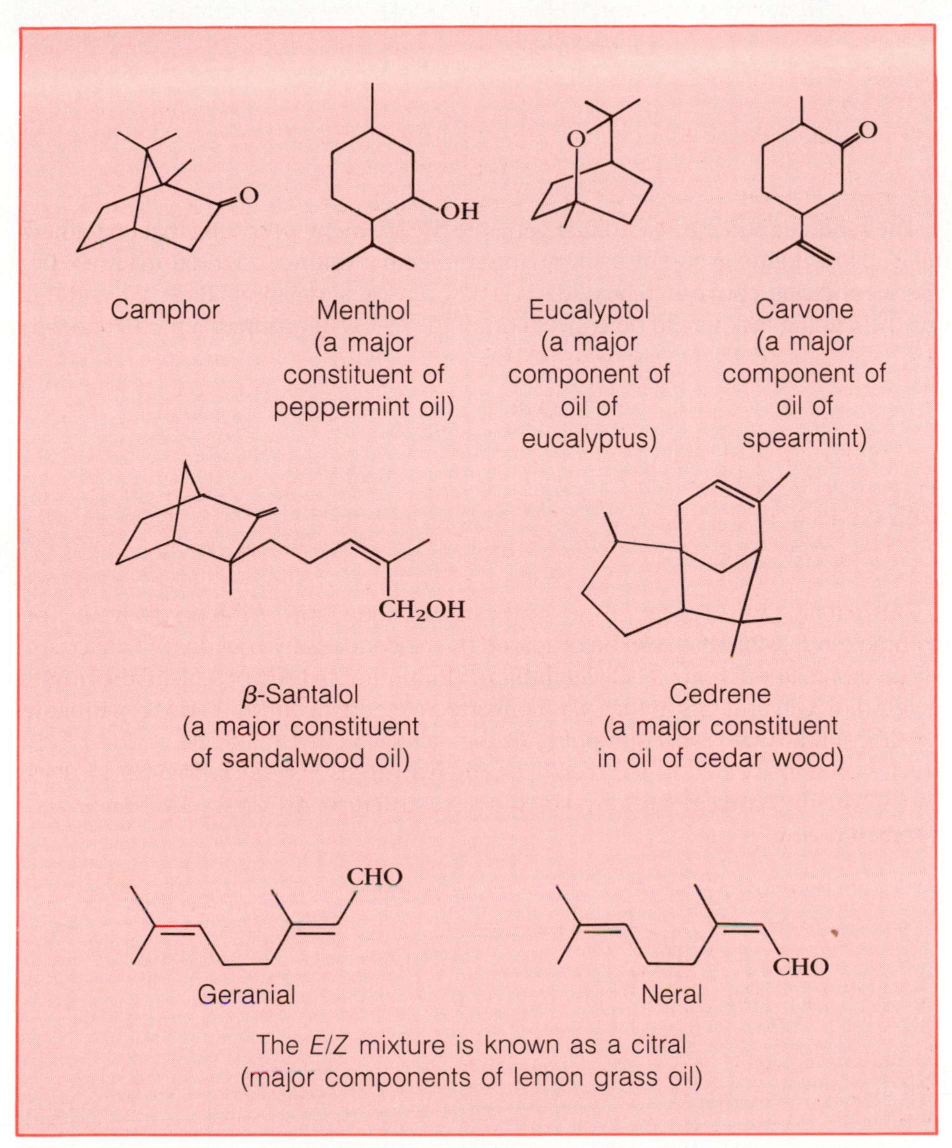

Figure 23.1 Some terpenes and terpenoids with characteristic odors.

Example 23.1 **The Synthesis of Isonootkatone**

Isonootkatone (**1-1**) is an important constituent of vetiver oil. The oil is isolated from vetiver grass, which grows in warm locations such as Java and Haiti. Vetiver oil has a pleasant aroma and is used in the manufacture of soaps and perfumes. The structure was originally misassigned, and it was the synthesis we describe here (by Marshall and coworkers at Northwestern University) that proved structure **1-1** to be correct.

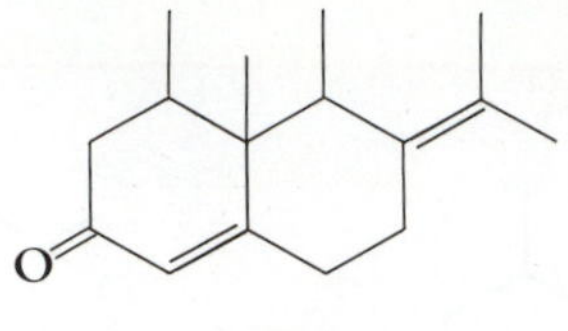

O

1–1

The synthetic strategy was to first generate the six-membered ring that contained the isopropylidene group by using an intramolecular ester condensation (**A** → **B**). The second ring could then be added via a Robinson annelation (**B** → **C**), and the only remaining task would be conversion of the —CO_2R group to a methyl group.

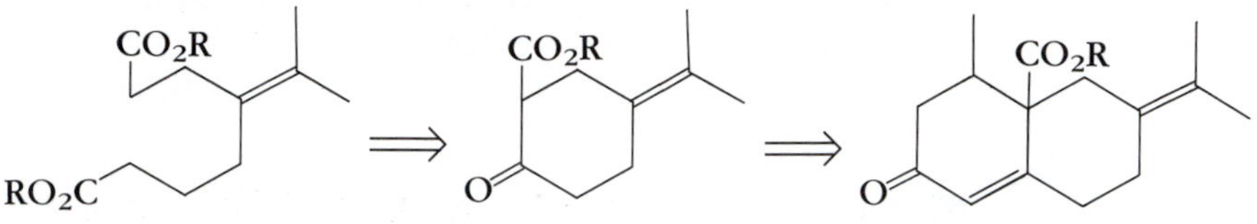

CO_2R RO_2C ⟹ CO_2R O ⟹ CO_2R O

A **B** **C**

The actual starting point of the synthesis was the isopropylidene derivative of malonic ester, **1-2,** which can be prepared by the condensation of diethyl malonate with acetone. Reduction of **1-2** with lithium aluminum hydride produced the unsaturated diol **1-3,** and this in turn was converted to the dibromide **1-4.**. Alkylation of the dibromide with two equivalents of the salt of diethyl malonate yielded **1-5,** which was transformed to the diacid **1-6** by hydrolysis and decarboxylation. This diacid had the correct carbon skeleton (cf. structure **A**) for cyclization a six-membered ring.

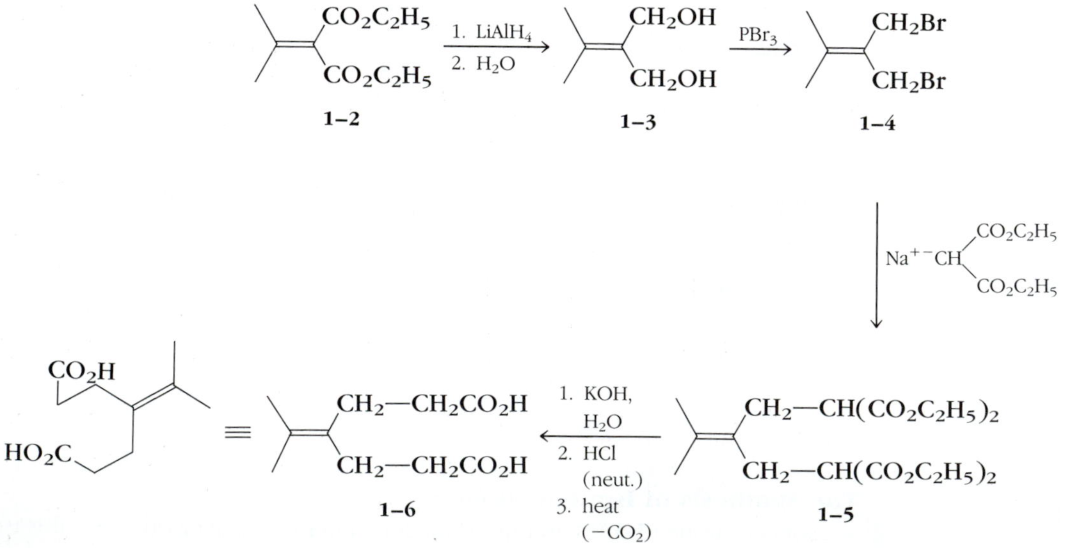

$CO_2C_2H_5$ / $CO_2C_2H_5$ (**1–2**) —1. $LiAlH_4$, 2. H_2O→ CH_2OH / CH_2OH (**1–3**) —PBr_3→ CH_2Br / CH_2Br (**1–4**)

↓ $Na^{+-}CH(CO_2C_2H_5)_2$

CO_2H / HO_2C ≡ $CH_2—CH_2CO_2H$ / $CH_2—CH_2CO_2H$ (**1–6**) ←1. KOH, H_2O 2. HCl (neut.) 3. heat ($-CO_2$)— $CH_2—CH(CO_2C_2H_5)_2$ / $CH_2—CH(CO_2C_2H_5)_2$ (**1–5**)

The diacid **1-6** was esterified, and treatment of the diester **1-7** with sodium hydride produced the cyclic keto ester **1-8.** Reaction of this keto ester with 3-penten-2-one and base resulted in a Robinson annelation (Section 22.3). The resulting bicyclic compound (**1-10**) has the required carbon skeleton corresponding to structure **C.**

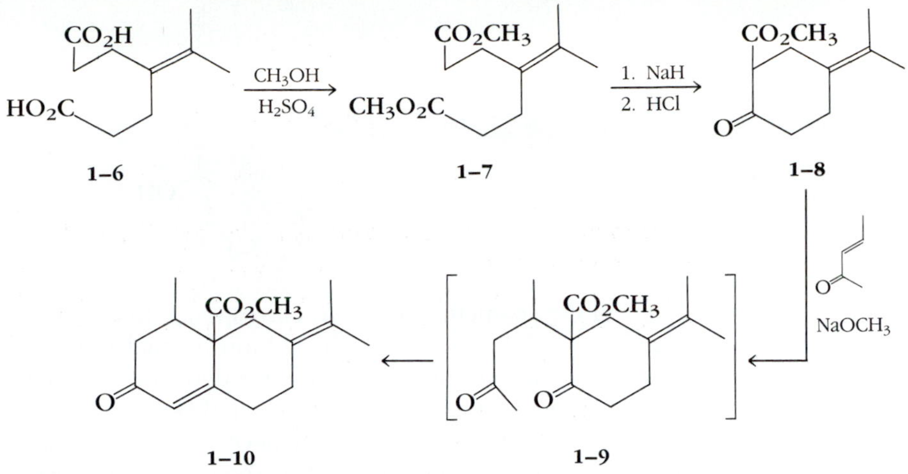

CO_2H HO_2C **1–6** $\xrightarrow[H_2SO_4]{CH_3OH}$ CO_2CH_3 CH_3O_2C **1–7** $\xrightarrow[2.\ HCl]{1.\ NaH}$ CO_2CH_3 O **1–8**

$NaOCH_3$ → [CO_2CH_3 O O **1–9**] → CO_2CH_3 O **1–10**

Conversion of the —CO_2CH_3 group of **1-10** to a methyl group required a reduction, but it was necessary to first protect the ketone carbonyl. Preparation of the ethylene glycol ketal (**1-11**) was accompanied by isomerization of the double bond, as you have seen in previous examples with this ring system (Section 22.4). Reduction with lithium aluminum hydride yielded the ketal alcohol **1-12,** which was next converted to the corresponding methanesulfonate. In a reaction analogous to metal reduction of alkyl halides (Section 1.11), the —OSO_2CH_3 group was replaced by —H upon reaction of **1-13** with lithium in ammonia. Finally, hydrolysis of the ketal afforded **1-1,** and this was found to be the same as naturally occurring isonootkatone (except that the synthetic material was racemic).

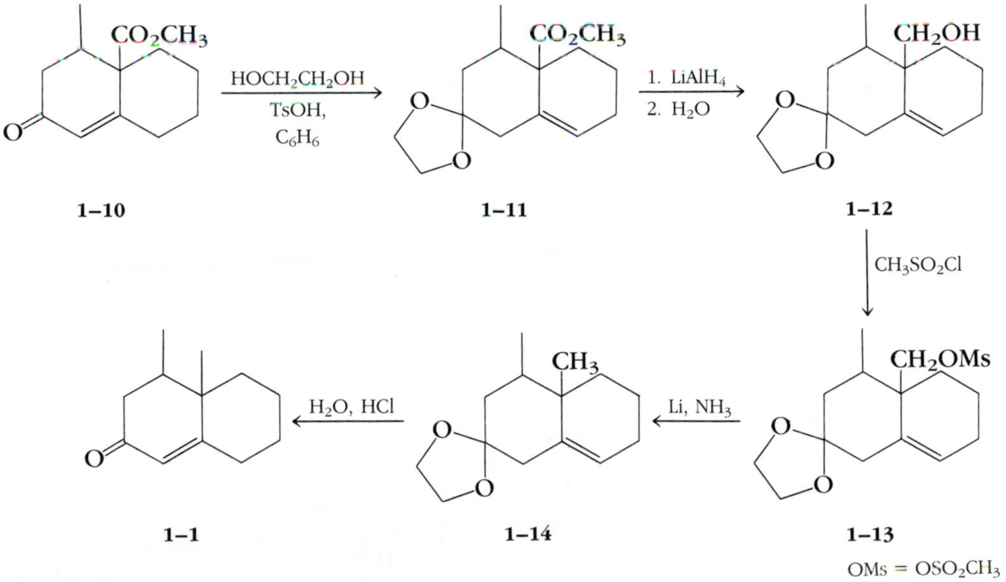

CO_2CH_3 O **1–10** $\xrightarrow[TsOH,\ C_6H_6]{HOCH_2CH_2OH}$ CO_2CH_3 O O **1–11** $\xrightarrow[2.\ H_2O]{1.\ LiAlH_4}$ CH_2OH O O **1–12**

CH_3SO_2Cl

O **1–1** $\xleftarrow{H_2O,\ HCl}$ CH_3 O O **1–14** $\xleftarrow{Li,\ NH_3}$ CH_2OMs O O **1–13**

OMs = OSO_2CH_3

Example 23.2 The Synthesis of Veticadinol

The synthesis of the bicyclic terpenoid veticadinol (**2-1**) was carried out by Vig and his coworkers in India.

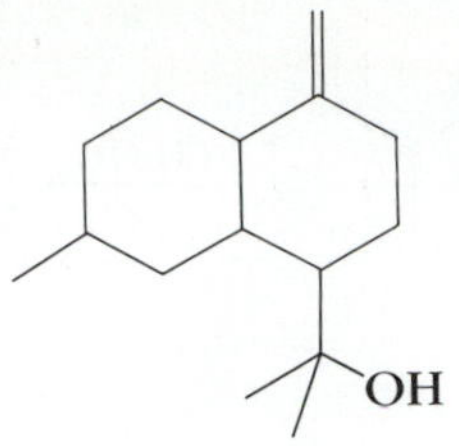

2–1

Neither of the six-membered rings corresponds to a cyclohexenone (i.e., an α,β-unsaturated cyclohexanone), so the Robinson annelation did not seem to offer the best approach toward building up the needed ring system. Instead it was decided to start with a cyclohexanone derivative and build up the second ring from fragments that would eventually allow the necessary substituents to be introduced. The general synthetic plan is illustrated by the following equation:

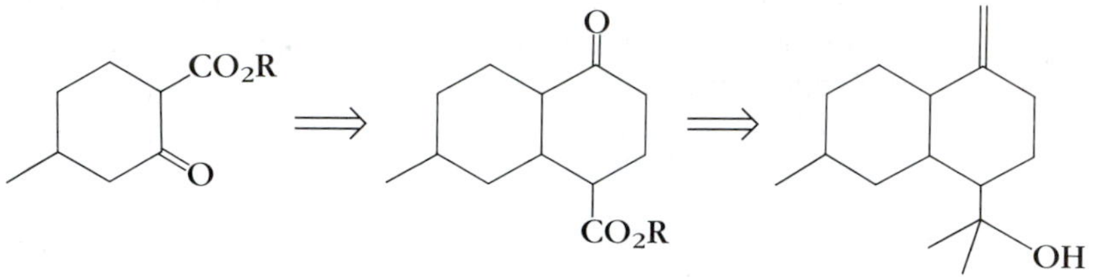

CO_2R O ⟹ O CO_2R ⟹ OH

A B 2–1

Once the ring structure of **B** had been created, the $=CH_2$ group could be introduced by a Wittig reaction, and the action of a Grignard on the ester group would generate the necessary tertiary alcohol.

The actual starting material for the synthesis was 2-ethoxycarbonyl-5-methylcyclohexanone (**2-2**), which underwent condensation with ethyl cyanoacetate to yield **2-3.** After reduction of the double bond by catalytic hydrogenation, a Michael addition with acrylonitrile was carried out to produce the dicyano diester **2-5.**

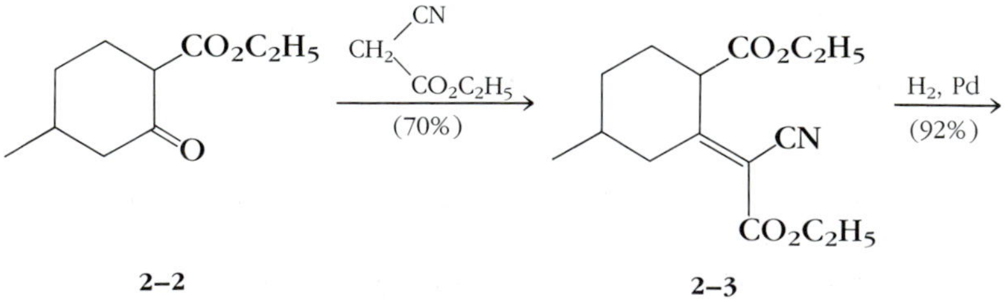

$CO_2C_2H_5$ O; $CH_2(CN)CO_2C_2H_5$ (70%) → $CO_2C_2H_5$ CN $CO_2C_2H_5$; H_2, Pd (92%) →

2–2 2–3

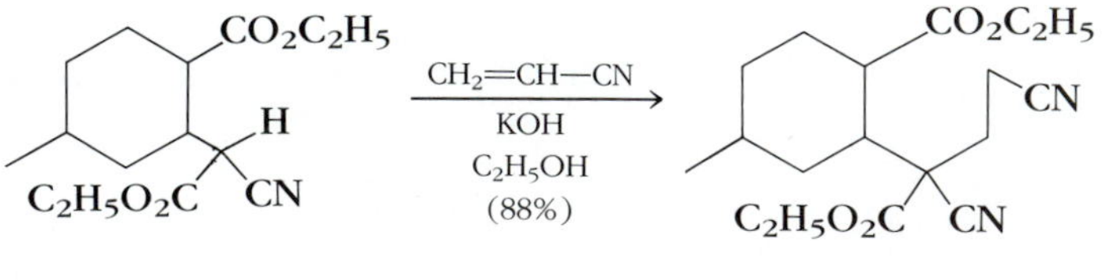

$CO_2C_2H_5$ H $C_2H_5O_2C$ CN; $CH_2{=}CH{-}CN$, KOH, C_2H_5OH (88%) → $CO_2C_2H_5$ CN $C_2H_5O_2C$ CN

2–4 2–5

The dicyano diester **2-5** was converted to the triester **2-8** by a sequence of hydrolysis, decarboxylation, and reesterification.

H_2O, HCl → 2-5 → [2-6] → ($-CO_2$) →

2-7 → C_2H_5OH, H_2SO_4 → 2-8 (64% yield for three steps)

Cyclization of triester **2-8** was accomplished by an intramolecular ester condensation, and the resulting β-keto ester (**2-9**) was hydrolyzed and decarboxylated. (Note that the other ester group was also hydrolyzed, so it was necessary to esterify it once again.)

2-8 → 1. $NaOC_2H_5$ 2. H_2O, H_2SO_4 → [2-9] → [2-10] → ($-CO_2$) → 2-11 → C_2H_5OH, H_2SO_4 → 2-12 (42% overall yield)

Final modification of **2-12** was accomplished according to the original synthetic plan. Reaction with a phosphorus ylide produced **2-13** and reaction of this ester with methylmagnesium iodide (followed by neutralization with ammonium chloride) yielded veticadinol.

2-12 → $(C_6H_5)_3P{=}CH_2$ → 2-13 → 1. CH_3MgI 2. NH_4Cl → 2-1

Example 23.3 The Synthesis of Chamigrene

Chamigrene is a bicyclic sesquiterpene with structure **3-1.**

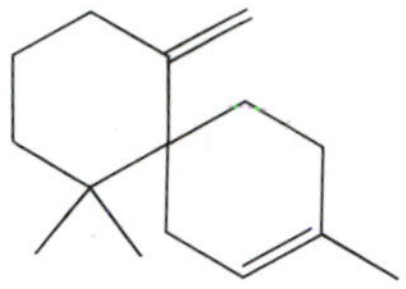

3–1

The ring structure is unusual, and chamigrene may represent a biogenetic link between other structural groups of sesquiterpenes. Carbocation rearrangements could interconvert ions **A, B,** and **C,** leading to chamigrene, thujopsene, and cuparene, respectively.

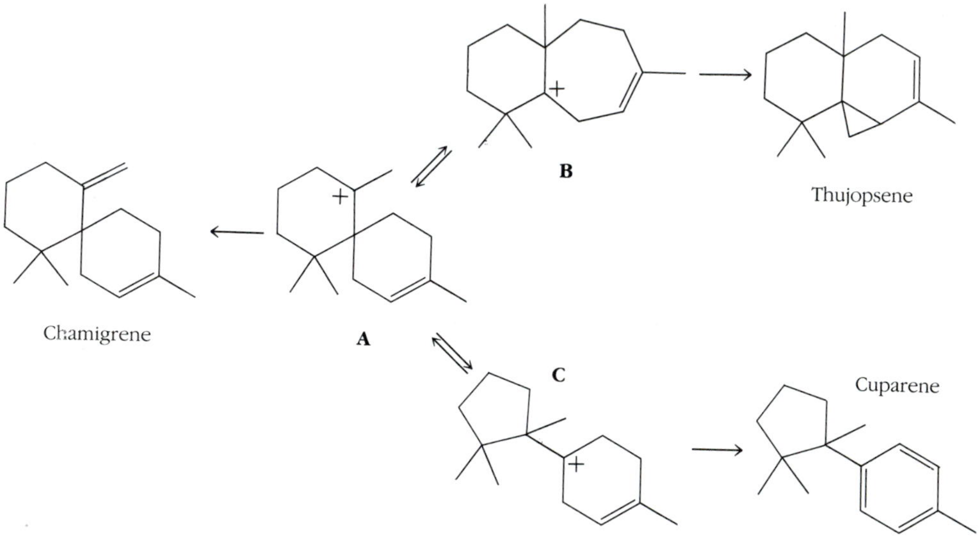

Chamigrene was synthesized by A. Yoshikoshi and coworkers in Japan. Their strategy was to start with a cyclohexanone derivative and generate the second six-membered ring via a Diels-Alder reaction. Dehydration of the β-hydroxyketone **3-2** yielded the unsaturated ketone **3-3.** The Diels-Alder reaction between **3-3** and 2-ethoxy-1,3-butadiene produced the bicyclic product **3-4,** which not only had the correct ring system but also contained oxygen substituents at those positions where further modification was necessary.

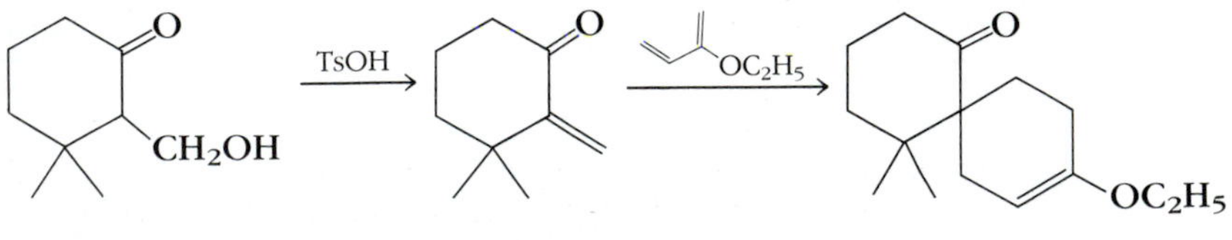

3–2 **3–3** **3–4**

Hydrolysis of the vinyl ether linkage in **3-4** produced the diketone **3-5,** in which the two carbonyl groups showed markedly different reactivity. The keto group of the original six-membered ring is sterically hindered, because the adjacent carbon is quaternary. Consequently, the reaction with methylmagnesium iodide took place exclusively at the other keto group to generate the keto alcohol **3-6.**

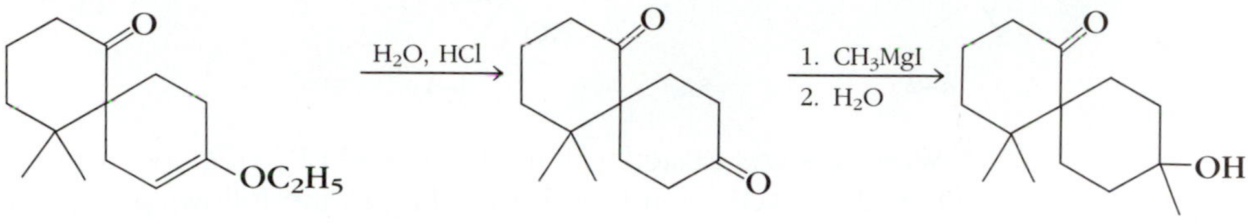

O OC_2H_5 $\xrightarrow{H_2O,\ HCl}$ O O $\xrightarrow[2.\ H_2O]{1.\ CH_3MgI}$ O OH

3–4 **3–5** **3–6**

Finally, dehydration of **3-6** produced the unsaturated ketone **3-7,** which was converted to chamigrene by reaction with a phosphorus ylide.

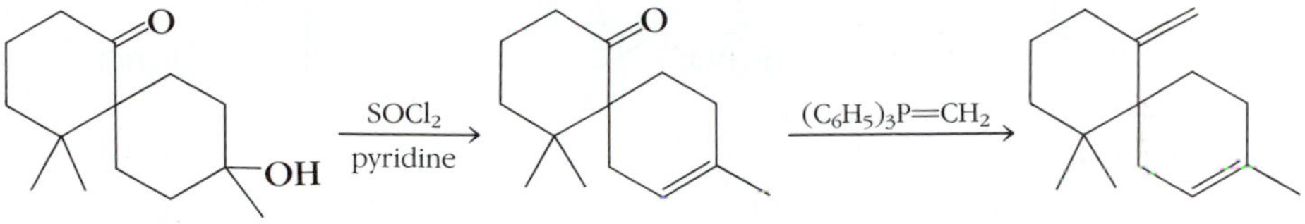

O OH $\xrightarrow[\text{pyridine}]{SOCl_2}$ O $\xrightarrow{(C_6H_5)_3P{=}CH_2}$

3–6 **3–7** **3–1**

Example 23.4 **Synthesis of Drimenol**

E. E. van Tamelen and his coworkers at Stanford University carried out a biomimetic synthesis of the bicyclic terpenoid drimenol (**4-1**).

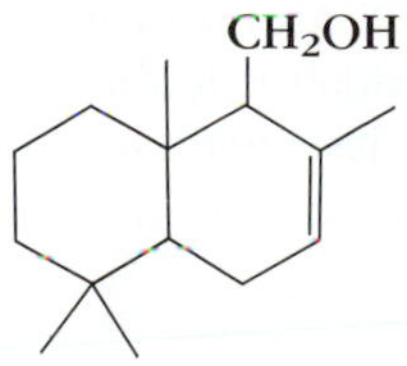

CH_2OH

4–1

The term *biomimetic* indicates that their laboratory synthesis was intended to *mimic* the actual biosynthetic pathway (see Section 12.9). Such biomimetic syntheses have been used to learn more about the biological processes, for example: Do stereochemical features of the natural products result from conformational preferences of the reactants, or are they induced by enzymes? Many chemists originally believed that acyclic precursors could be converted to polycyclic terpenoid structures only by the action of enzymes. But the following synthesis clearly demonstrates that such ring closures can also be effected by laboratory methods.

Van Tamelen and coworkers prepared the triene acetate **4-2,** which reacted with *N*-bromosuccinimide in aqueous medium (Section 5.8) to give **4-3.** This was converted to the epoxide **4-4** by mild treatment with base (intramolecular bromide displacement by the OH group).

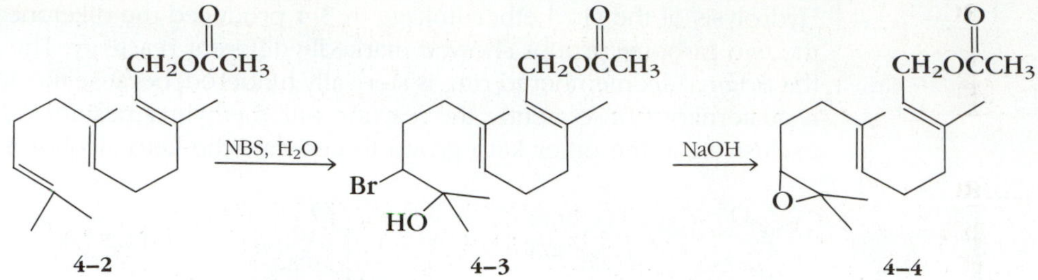

NBS, H_2O ; NaOH

4–2 4–3 4–4

When **4-4** was treated with the Lewis acid BF_3, ring opening of the epoxide was followed by cyclization to **4-5** according to the following sequence:

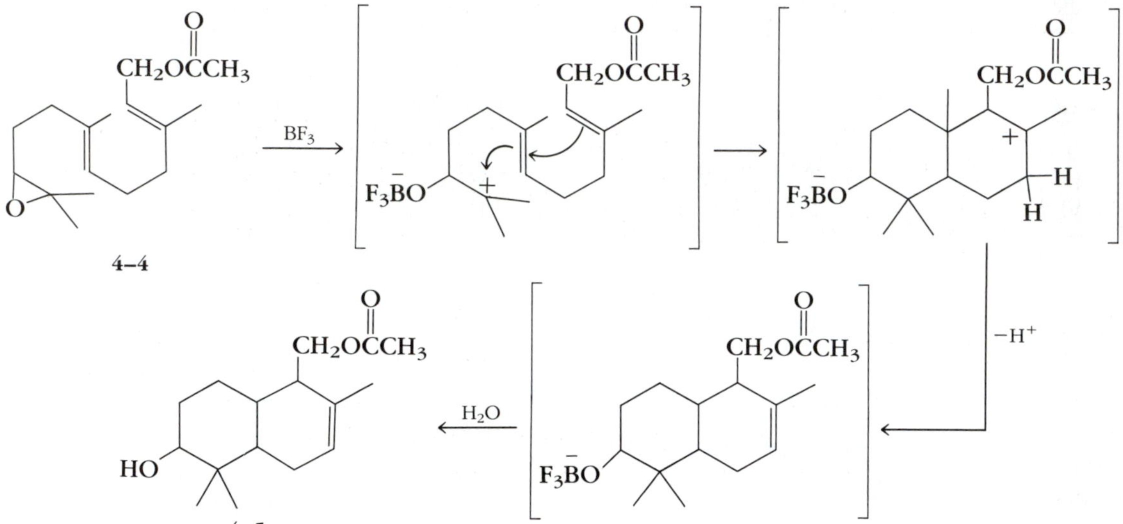

BF_3 ; $-H^+$; H_2O

4–4 4–5

The hydroxyl group of **4-5** was removed (cf. Section 15.5) in several steps, starting with oxidation to the ketone **4-6.** This was converted to the dithioketal **4-7,** and hydrogenolysis with a nickel catalyst gave **4-8.** Finally this was converted to drimenol by cleavage of the acetate with lithium aluminum hydride.

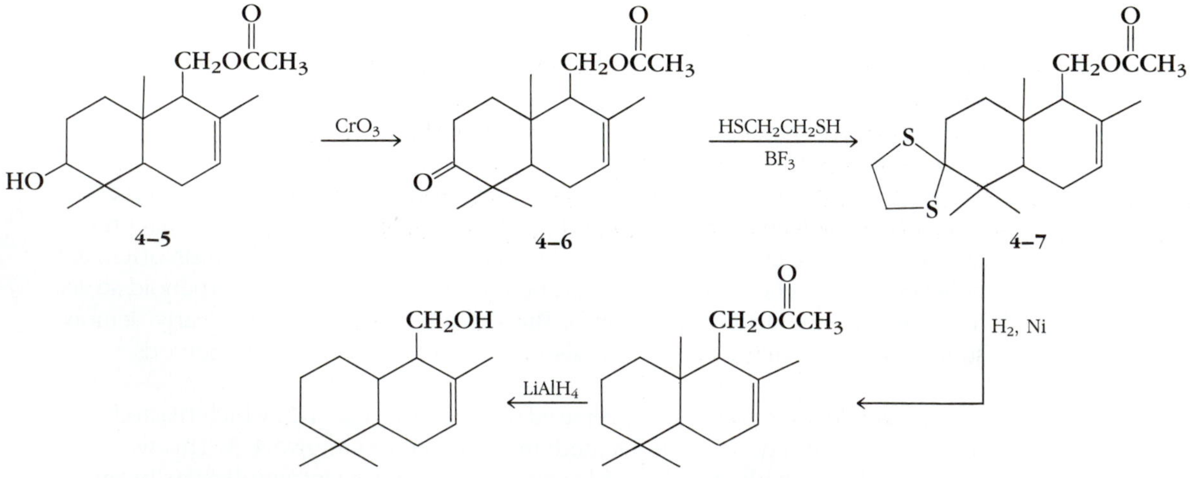

CrO_3 ; $HSCH_2CH_2SH$, BF_3 ; H_2, Ni ; $LiAlH_4$

4–5 4–6 4–7 4–8 4–1

EXERCISE 23.1

Show how compound **1-2** could be prepared from diethyl malonate and acetone.

EXERCISE 23.2

Which carboxyl group is lost in the decarboxylation of **2-6?** Why is only this carboxyl group lost?

EXERCISE 23.3

Show which bonds were formed in the Diels-Alder reaction leading to **3-4.**

EXERCISE 23.4

In the dehydration of **3-6** a proton could be removed from either of two CH_2 groups. Draw the products that would be formed by these two pathways. What is the relationship between them?

23.2 STEROIDS

Steroids are tetracyclic compounds that have a ring skeleton with three six-membered rings and a five-membered ring. The following structure illustrates both the numbering of the carbon atoms and the designation of the four rings by the letters **A-D.**

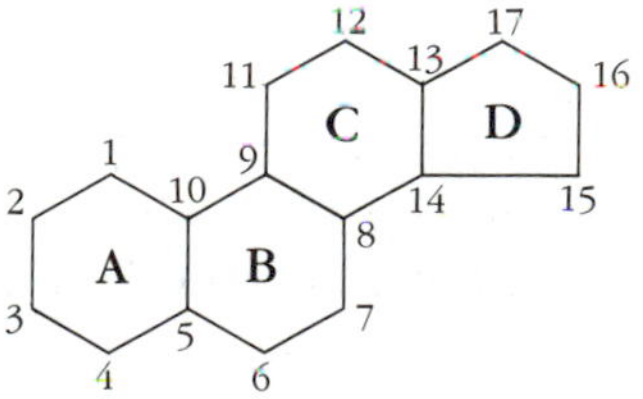

The steroid skeleton

Steroids serve a variety of important functions in human biochemistry. Their biosynthesis proceeds from isopentenyl pyrophosphate to give lanosterol and then cholesterol, as we have already described (Section 12.9). Cholesterol is the general precursor of other steroids and is present in large quantities, a total of about 150–250 g in a human adult. Most of this cholesterol is present as esters of C_{16} and C_{18} fatty acids. The following drawing shows the complete structure of cholesterol:

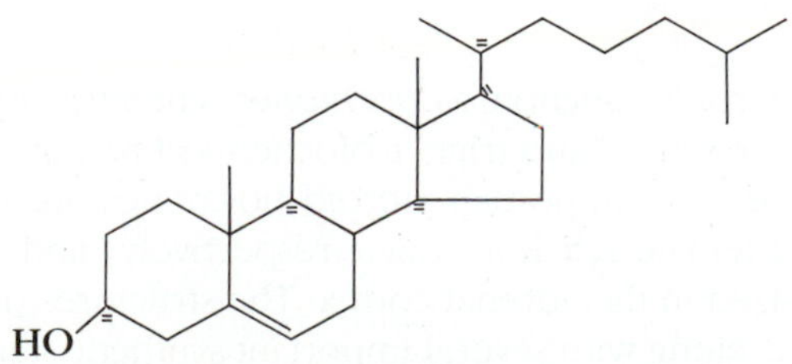

HO

Cholesterol

In this drawing of cholesterol and in many of the other drawings in this chapter, several important conventions will be employed. First, as we have done previously, we will use a short line segment to denote a methyl group. Methyl groups are found on C-10 and C-13 of cholesterol (as well as on the C-17 side chain). Second, the stereochemistry of a substituent will be shown only if it is below the "plane" of the ring (an orientation described as α in steroids, just as in carbohydrates). A β substituent will be drawn with a normal line, as shown for the C-10 and C-13 methyl groups of cholesterol. Third, if a hydrogen or a methyl group has the α orientation, we will only show it as a dashed line. We have used this convention previously with methyl groups, but now we will use the simple notation of *two dashes* to denote a hydrogen that is below the plane of the ring. These conventions permit chemists to draw highly complex molecules (such as cholesterol with its eight chiral centers) in such a way that the structure is unambiguous but still uncluttered.

Cholesterol is highly insoluble in aqueous media, and this insolubility is associated with several health problems. Gallstones, for example, are solid deposits formed in diseased gallbladders that consist mainly of crystalline cholesterol. Cholesterol also plays an important role in arteriosclerosis (hardening of the arteries), a condition resulting from the formation of plaque on the inner walls of arteries. The plaque again contains substantial quantities of cholesterol, which is why people are often advised to avoid foods, particularly animal fats, that are high in cholesterol content. Since cholesterol is biosynthesized in humans, it is not yet clear whether health problems are caused by excess dietary cholesterol or by impaired cholesterol metabolism.

Substantial quantities of cholesterol are metabolized in the liver to form *bile acids,* highly oxygenated steroids that help in the absorption of fats into the bloodstream from the intestine. An example is cholic acid. Reaction of cholic acid with the amino group of an amino acid yields the corresponding amide, which is the biologically active form. For example, the amide formed from glycine is glycocholic acid. One side of the molecule is very nonpolar, yet the other side has polar substituents that can interact strongly with water-soluble materials.

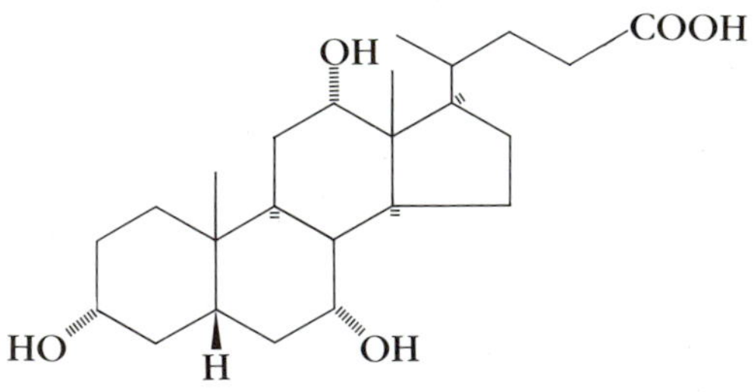

Cholic acid

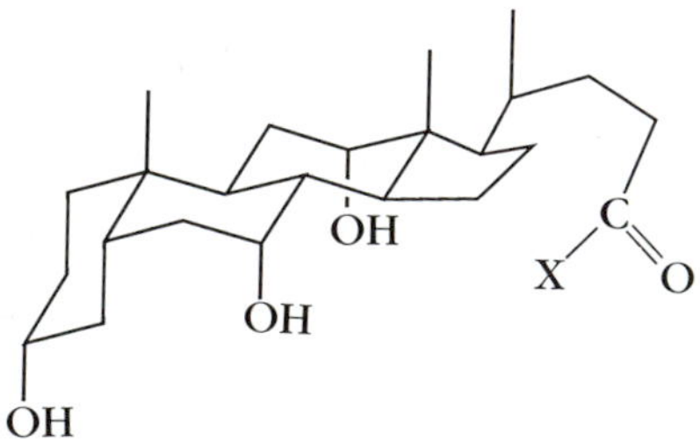

X = OH, cholic acid
X = NH—CH_2—CO_2H, glycocholic acid

The remaining steroids function as *hormones,* chemical messengers that are released into the bloodstream and trigger biochemical responses in other parts of the body. Among the most important steroid hormones are the *androgens* and *estrogens* (male and female sex hormones, respectively) and the wide variety of compounds synthesized in the adrenal cortex. The structures of some of these are shown in Figure 23.2, along with several important synthetic steroids (ethinyl estra-

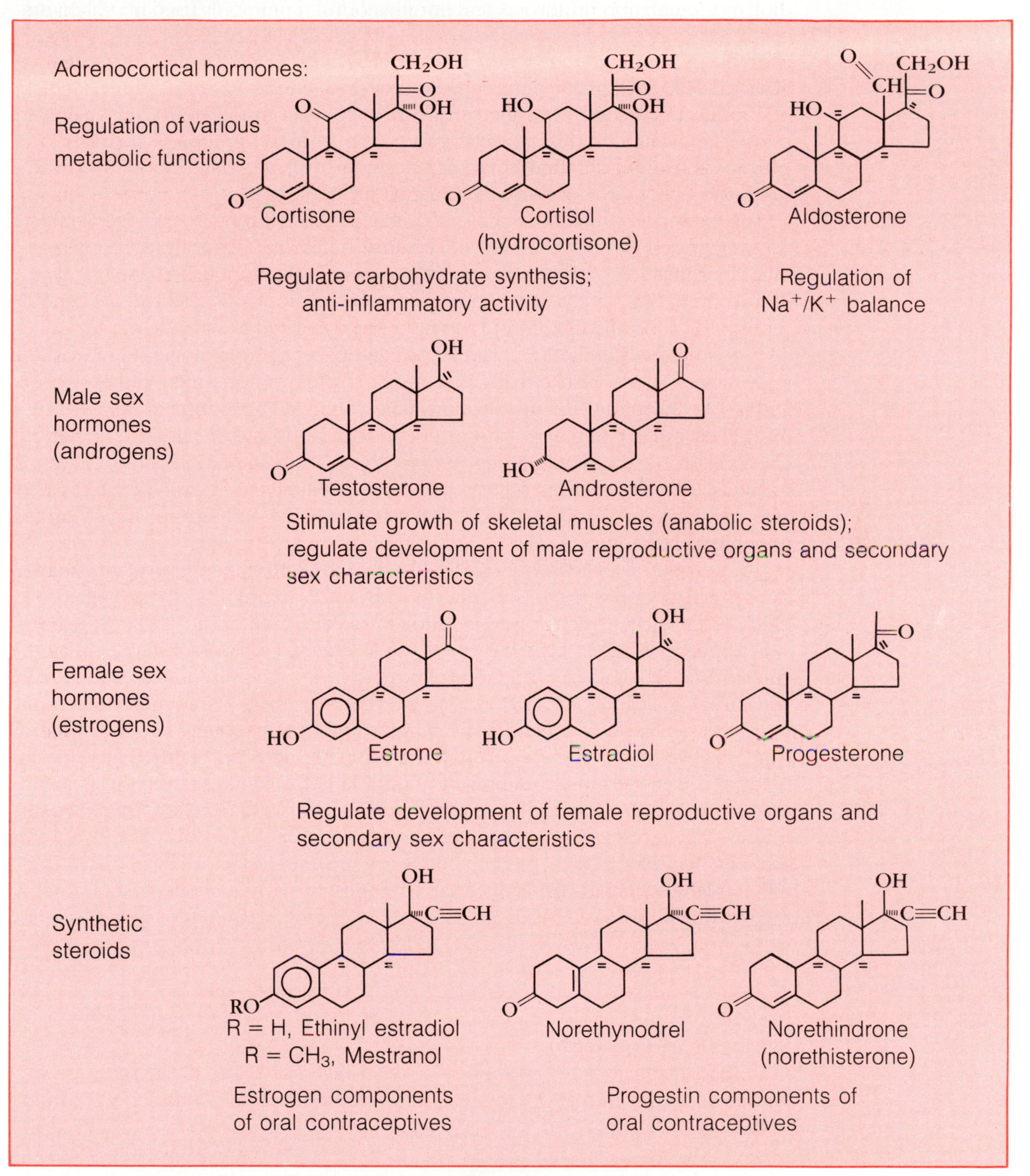

Figure 23.2 Examples of important steroids and some of their biological functions.

diol, mestranol, norethindrone, and norethynodrel) commonly used in oral contraceptives.

By inspecting Figure 23.2 you will notice that relatively small structural variations can lead to dramatic differences in biological effects. For example, the female sex hormone progesterone differs from the male sex hormone testosterone only in the D ring, which carries an acetyl group or a hydroxyl group, respectively. The carbon skeleton of another female sex hormone, estrone, differs from that of testosterone only by the absence of a methyl group at C-10.

The existence of chemical substances that produce strong physiological effects was recognized in the mid-nineteenth century, but isolation, purification, and structure determination of the steroid hormones were not accomplished until the twentieth century. The greatest advances occurred in the 1930s, following the 1929 report of estrone, the first steroid hormone to be isolated in a pure state. Why did it take so long for chemists to unravel a puzzle of such great biological and medical importance? There are two major reasons. First, the structures are highly complex, and it was difficult to find degradations that would yield products of known structure. Even if the complexity of the functional groups is disregarded, the stereochemical problems are staggering. As you can see from inspection of the structures in Figure 23.2, a compound such as aldosterone contains seven chiral centers. Even after the constitution was established, it was necessary to determine which of 128 possible stereoisomers corresponded to the correct structure.

The second major reason that it took so long to determine the structures of the steroid hormones is their presence in only trace amounts in living organisms. Much of the early work involved isolation of steroids from urine, and it was necessary to process some 15,000 liters of urine to obtain a total of 15 mg of androsterone in 1931. Higher concentrations of some steroids are present in animal tissues, and samples were available from slaughterhouses, but the amounts still are not large. The isolation of progesterone required more than 600 kg of ovarian tissue from 50,000 sows, and only 20 mg of the pure steroid was obtained. Clearly, any medical use of steroids was impractical until larger quantities could be obtained. This created a very strong demand on organic chemists to develop synthetic pathways to the steroids. The resulting efforts were highly successful, and most of the steroids that are used in modern medicine are the result of organic synthesis, either totally or in part (using materials from plants as synthetic precursors). In this section we will show you several historically important examples of steroid synthesis.

Example 23.5 Synthesis of Equilenin

W. Bachmann and his coworkers at the University of Michigan reported the synthesis of equilenin (**5-1**), in 1939; this was the first total synthesis of a natural steroid.

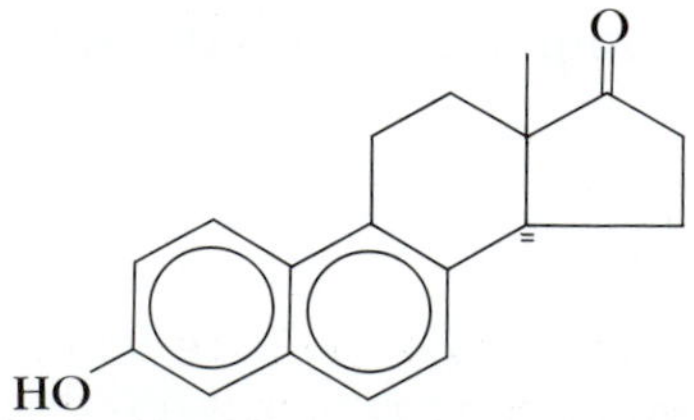

O

HO

5–1

Equilenin was a logical target for the first steroid synthesis because of its relative simplicity. There are only two functional groups, a phenolic hydroxyl and a ketone carbonyl. More importantly, both the A and B rings are aromatic, so there are only two chiral centers in the molecule. This means that nearly all stereochemical difficulties were circumvented by setting racemic equilenin as the initial synthetic goal. This material could then be resolved to prepare the naturally occurring enantiomer.

The synthetic strategy was based on conversion of a naphthalene derivative, **A,** to a tricyclic ketone with structure **B.** Introduction of a methyl group and formation of the five-membered ring would afford the required carbon skeleton **C.**

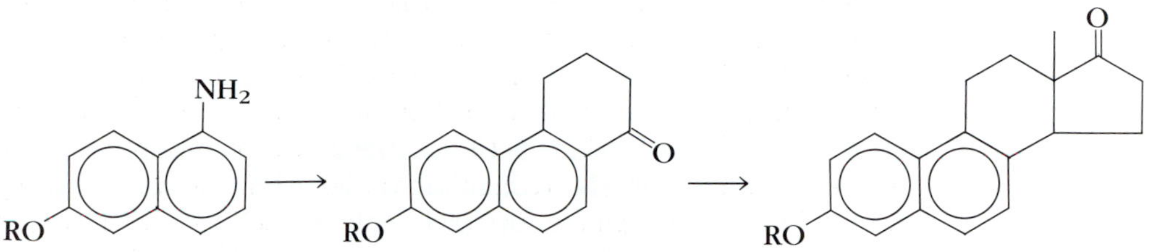

NH_2 RO RO O RO O

A **B** **C**

The actual starting material employed by Bachmann's group was 5-amino-2-naphthalenesulfonic acid (**5-2**), which was converted to the corresponding amino phenol (**5-3**) by heating it with potassium hydroxide. The amino group was protected as the amide so that the hydroxyl group could be methylated with dimethyl sulfate; subsequent removal of the blocking group yielded 6-methoxy-1-naphthylamine as the hydrochloride salt **5-6.**

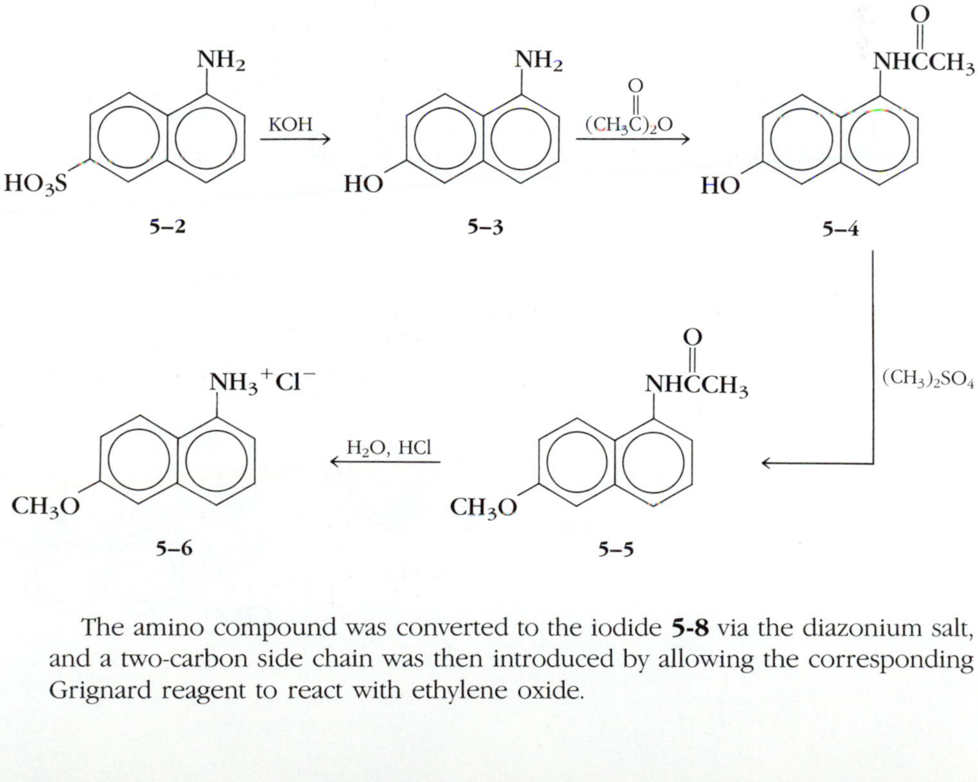

NH_2 HO_3S KOH NH_2 HO $(CH_3\overset{O}{\overset{\|}{C}})_2O$ $NH\overset{O}{\overset{\|}{C}}CH_3$ HO

5–2 **5–3** **5–4**

$(CH_3)_2SO_4$

$NH_3{}^+Cl^-$ CH_3O H_2O, HCl $NH\overset{O}{\overset{\|}{C}}CH_3$ CH_3O

Overall yield = 33%

5–6 **5–5**

The amino compound was converted to the iodide **5-8** via the diazonium salt, and a two-carbon side chain was then introduced by allowing the corresponding Grignard reagent to react with ethylene oxide.

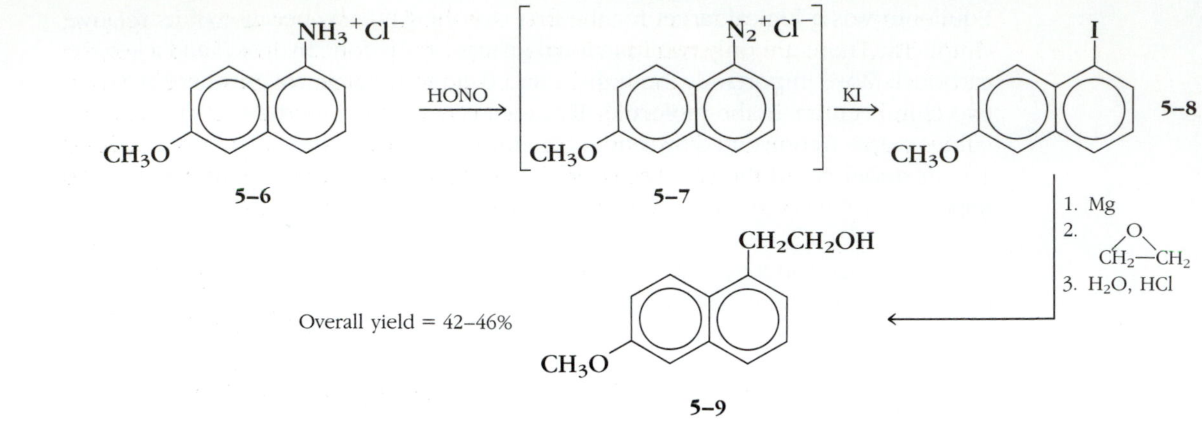

$NH_3^+Cl^-$ (5-6) → HONO → [$N_2^+Cl^-$ (5-7)] → KI → I (5-8)

5-8 → 1. Mg; 2. CH_2—CH_2 (O); 3. H_2O, HCl → CH_2CH_2OH (5-9)

Overall yield = 42–46%

In order to generate the C ring by an intramolecular Friedel-Crafts acylation, the side chain of **5-9** was first extended by two more carbon atoms using a malonic ester synthesis. The bromide **5-10** was prepared with phosphorus tribromide, and this was alkylated with the sodium salt of diethyl malonate. Hydrolysis and decarboxylation then yielded the carboxylic acid **5-13.**

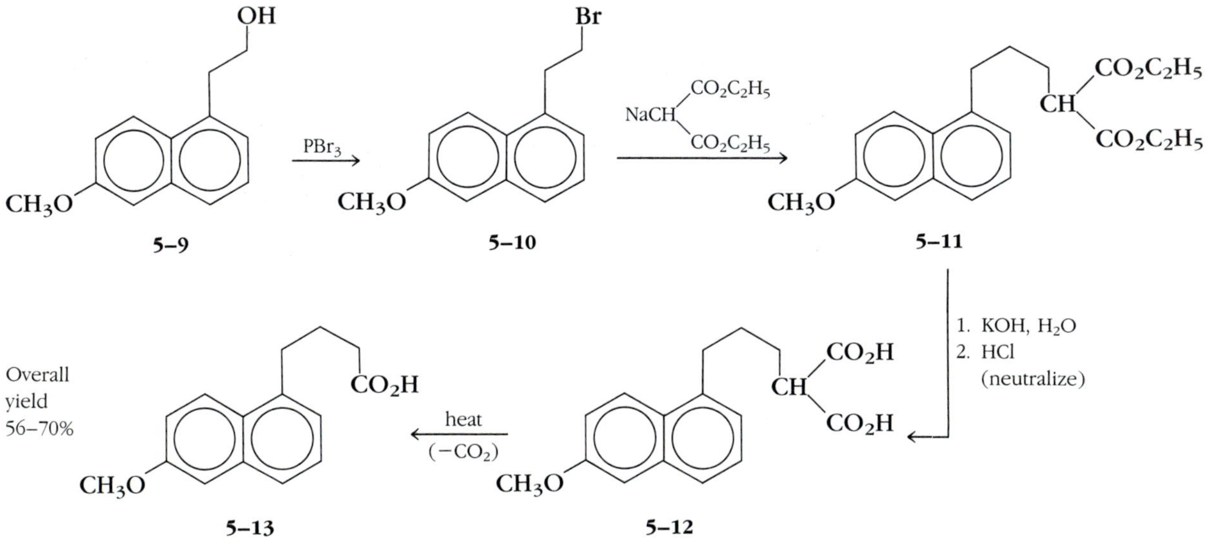

5-9 → PBr_3 → 5-10 → $NaCH(CO_2C_2H_5)_2$ → 5-11 → 1. KOH, H_2O; 2. HCl (neutralize) → 5-12 → heat ($-CO_2$) → 5-13

Overall yield 56–70%

Cyclization of **5-13** was effected by conversion to the acid chloride and subsequent treatment with the Lewis acid stannic chloride.

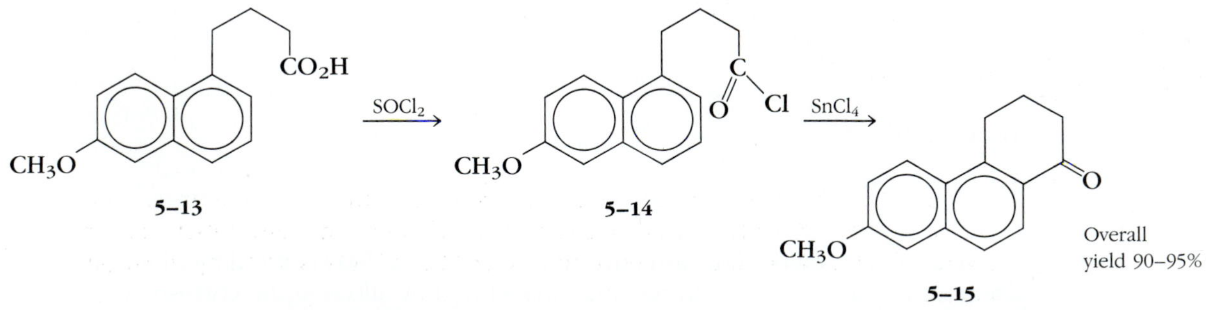

5-13 → $SOCl_2$ → 5-14 → $SnCl_4$ → 5-15

Overall yield 90–95%

A —CO_2CH_3 group was next introduced at the CH_2 adjacent to the carbonyl of **5-15.** This substituent served as an activating group for the introduction of the necessary methyl group, but it was also destined to become the carbonyl group of the final product. Instead of using dimethyl carbonate, dimethyl oxylate was employed, and carbon monoxide was extruded by pyrolysis. (This is a specialized reaction, which we will not discuss in detail.) Methylation of the β-keto ester **5-16** gave **5-17.** A Reformatsky reaction then yielded the diester **5-18.**

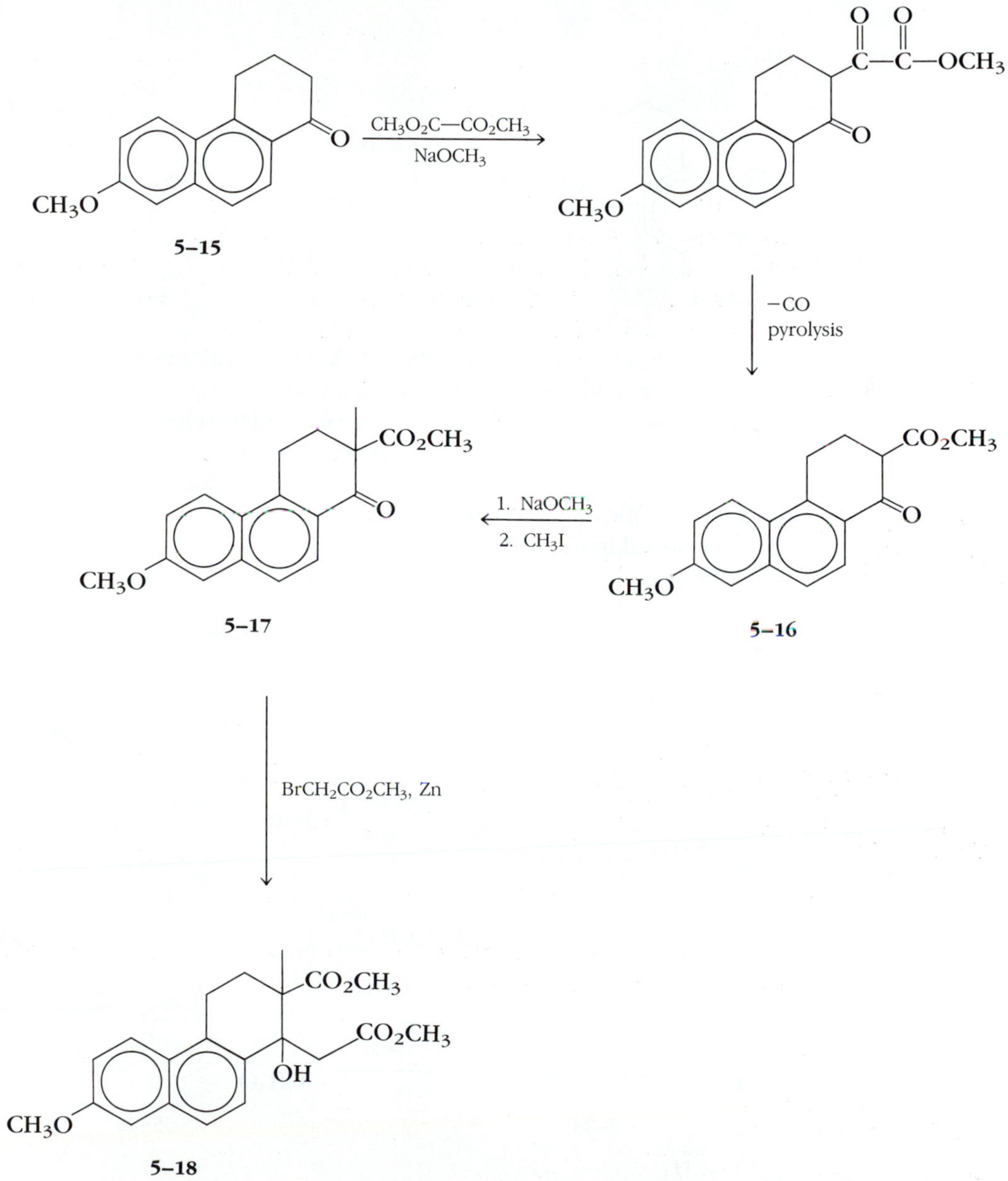

$CH_3O_2C—CO_2CH_3$, $NaOCH_3$

5–15

−CO
pyrolysis

1. $NaOCH_3$
2. CH_3I

5–17

5–16

$BrCH_2CO_2CH_3$, Zn

Overall yield 68–78%

5–18

The diester **5-18** was dehydrated and hydrolyzed to the salt of the diacid **5-20.** Reduction of **5-20** with sodium amalgam then afforded a mixture (about 1:1) of diastereomers, from which the desired isomer (**5-21a**) was isolated by recrystallization.

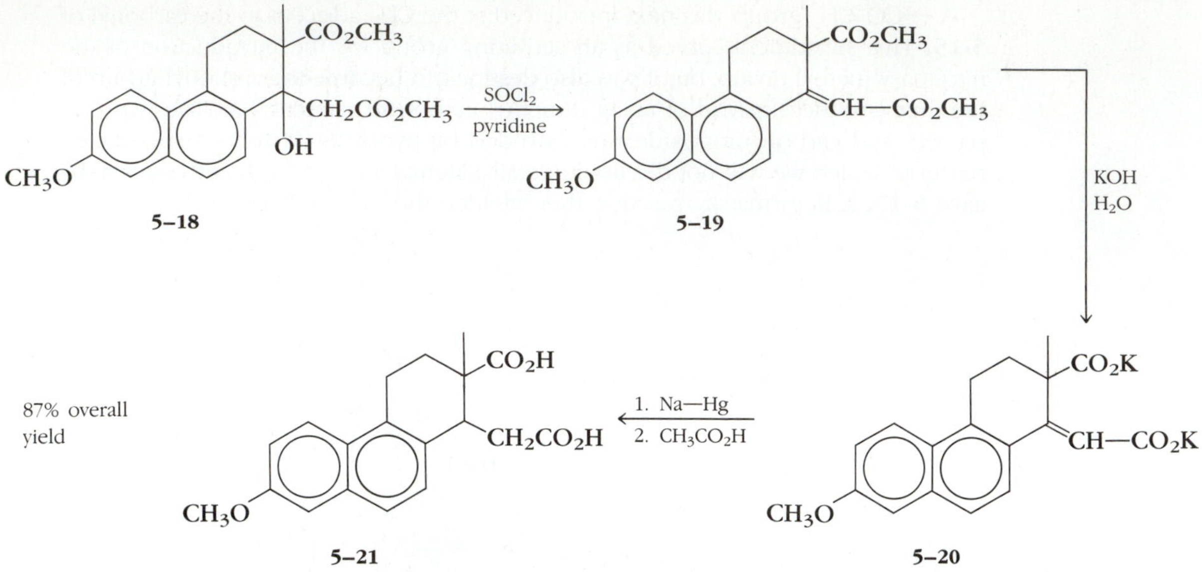

CO_2CH_3
$CH_2CO_2CH_3$
OH
CH_3O
5–18
$\xrightarrow[\text{pyridine}]{SOCl_2}$
CO_2CH_3
$CH—CO_2CH_3$
CH_3O
5–19
KOH
H_2O

CO_2K
$CH—CO_2K$
CH_3O
5–20
1. Na—Hg
2. CH_3CO_2H
CO_2H
CH_2CO_2H
CH_3O
5–21
87% overall yield

In order to close the five-membered ring by an intramolecular ester condensation, the acetic acid side chain had to be lengthened by one carbon. This was done by esterification of both carboxyl groups followed by selective hydrolysis of the less hindered one (to give **5-23**). Conversion of **5-23** to the acid chloride **5-24,** and reaction with diazomethane generated an intermediate, which underwent rearrangement to the diester **5-25.** (This procedure, known as the Arndt-Eistert synthesis, is discussed in the Appendix entitled "Molecular Rearrangements" in the Solutions Manual).

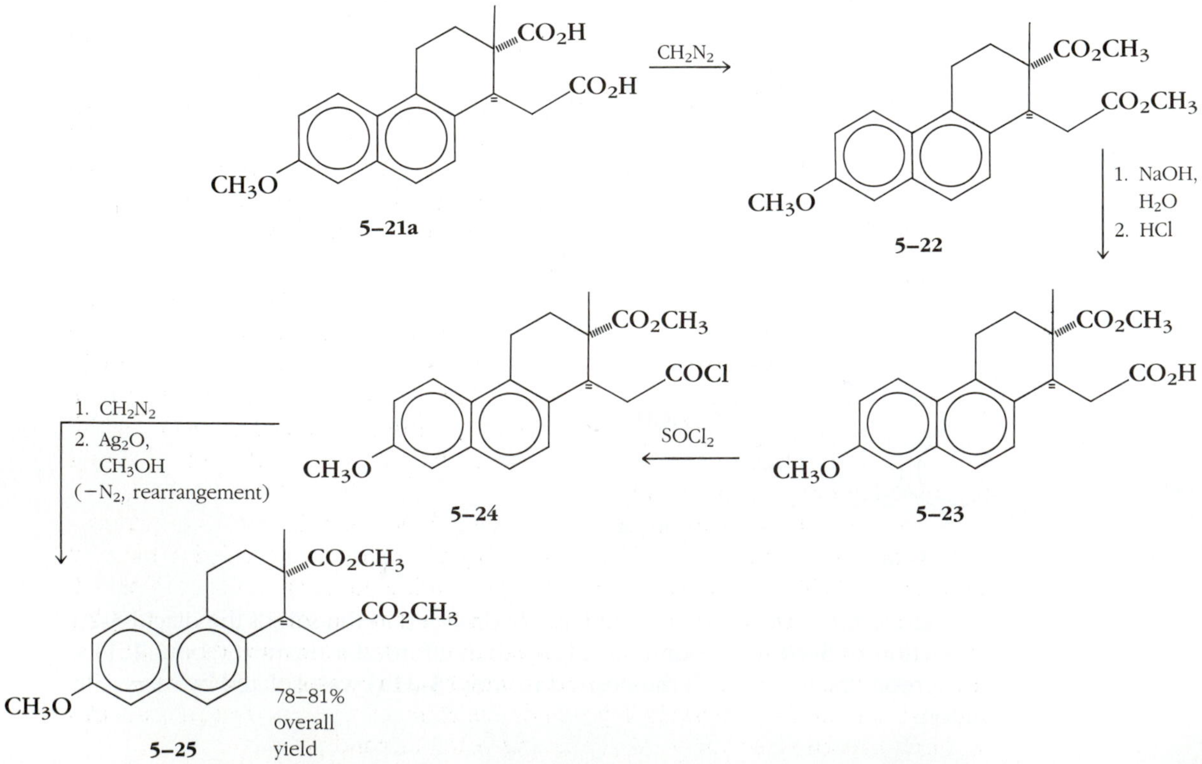

CO_2H
CO_2H
CH_3O
5–21a
$\xrightarrow{CH_2N_2}$
CO_2CH_3
CO_2CH_3
CH_3O
5–22
1. NaOH, H_2O
2. HCl

CO_2CH_3
CO_2H
CH_3O
5–23
$\xleftarrow{SOCl_2}$
CO_2CH_3
COCl
CH_3O
5–24
1. CH_2N_2
2. Ag_2O, CH_3OH
($-N_2$, rearrangement)

CO_2CH_3
CO_2CH_3
CH_3O
5–25
78–81% overall yield

The diester **5-25** possessed the requisite number of carbon atoms for cyclization, and treatment with sodium methoxide produced the tetracyclic keto ester **5-26.** Prolonged heating of this compound with concentrated hydrochloric acid not only caused hydrolysis and decarboxylation of the β-keto ester, but it also resulted in hydrolysis of the methyl ether to form racemic equilenin.

CO_2CH_3 CO_2CH_3 CH_3O **5–25** — 1. $NaOCH_3$ 2. CH_3CO_2H → O CO_2CH_3 CH_3O **5–26** — HCl, heat → O HO **5–1** 90–95% (two steps)

Example 23.6 Synthesis of Cortisone

By the early 1940s it was apparent that the steroid hormones of the adrenal cortex had enormous potential as medicinal agents. Unfortunately, research was severely hampered by the very limited availability of the compounds. For example, only about 200 mg of cortisone and 40 mg of hydrocortisone had been obtained from processing 500 kg of beef adrenal glands. In another set of experiments, the isolation of 1 g of cortisone required processing of the adrenal glands from nearly 200,000 sheep. In order to provide a greater supply of the hormones, a cooperative effort by several research laboratories in the United States was begun in 1942 under the sponsorship of the National Research Council. The goal of these efforts was the synthesis of substantial quantities of cortisone and other steroids. Several of these efforts were successful, particularly the approaches that employed the more abundant bile acids as starting materials. Many of the reactions developed at that time are still utilized to synthesize the steroids that are used in medical treatment, although steroidal precursors obtainable from some plants have allowed the development of very useful alternatives.

In this example we will present the total synthesis (i.e., from manmade starting materials) of cortisone (**6-1**) that was developed by L. H. Sarett and coworkers at Merck & Co. While the individual reactions should be familiar to you, the overall synthesis is complicated by the presence of several different functional groups. We have selected this example as a way of illustrating the use of protecting groups and latent functionality to synthesize complex polyfunctional organic molecules. The synthesis also demanded the generation of five chiral centers in cortisone, and this required careful selection of reactions and reaction conditions.

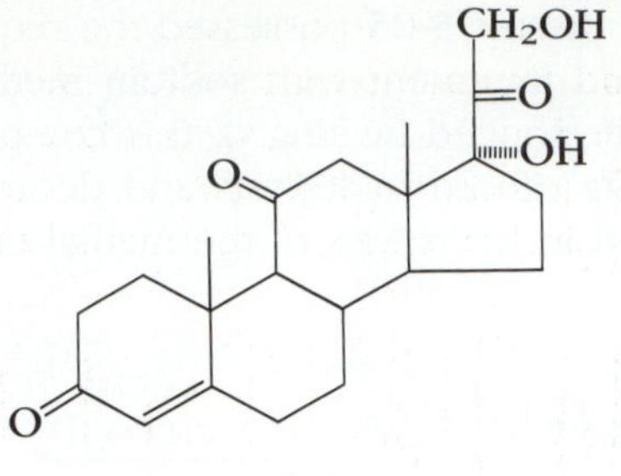

The synthetic strategy employed by the Merck group is outlined in Figure 23.3.

The strategy outlined in Figure 23.3 shows that a Diels-Alder reaction between an alkoxy-substituted diene (**I**) and a cyclohexenone derivative (**II**) would generate a bicyclic compound (**III**), which corresponds to the B and C rings of cortisone. A Robinson annelation would then form the A ring (**IV**). Formation of the five-membered ring would yield **V**, and modification of the various functional groups would lead to cortisone.

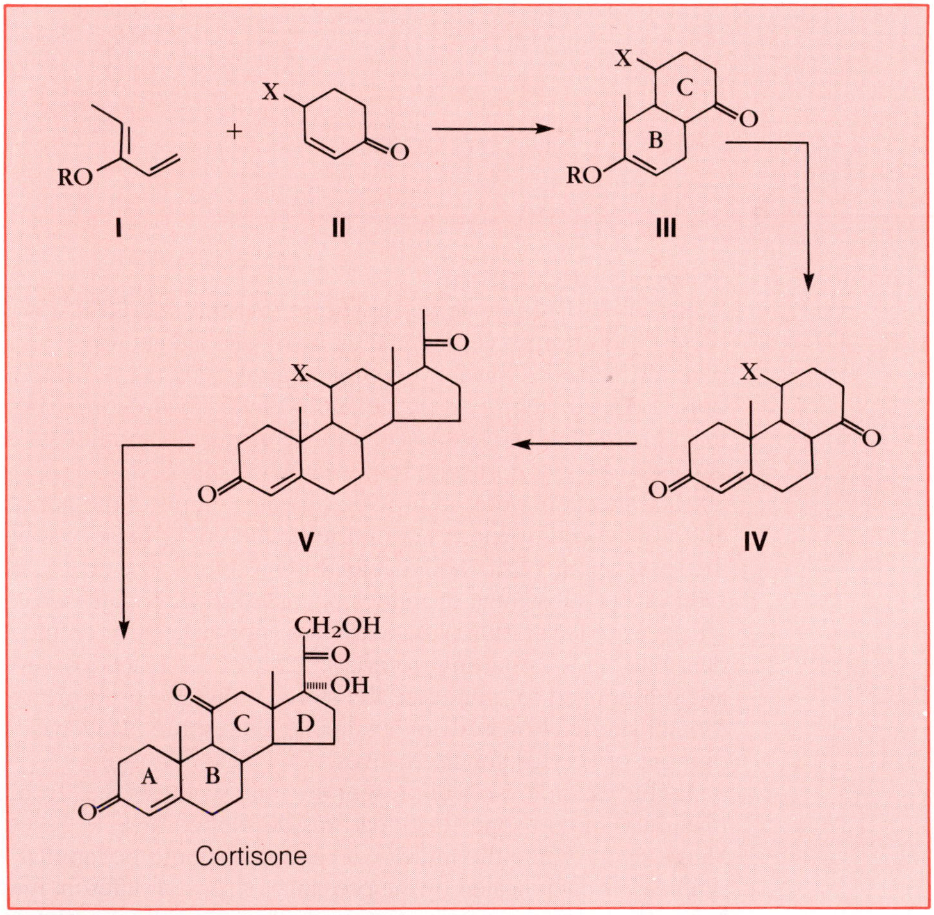

Figure 23.3 Outline of the synthetic strategy employed for the synthesis of cortisone.

The actual starting materials in the synthesis were (*Z*)-3-ethoxy-1,3-pentadiene (**6-2**) and benzoquinone (**6-3**). These reagents underwent a Diels-Alder reaction to give **6-4.** The vinyl ether was a latent carbonyl group, but before this functionality was unmasked, the other two keto groups were converted to alcohols. Catalytic hydrogenation reduced the double bond between these two keto groups, and $LiAlH_4$ reduction produced **6-5.** Hydrolysis of the vinyl ether and Robinson annelation of **6-6** with methyl vinyl ketone then yielded **6-7.**

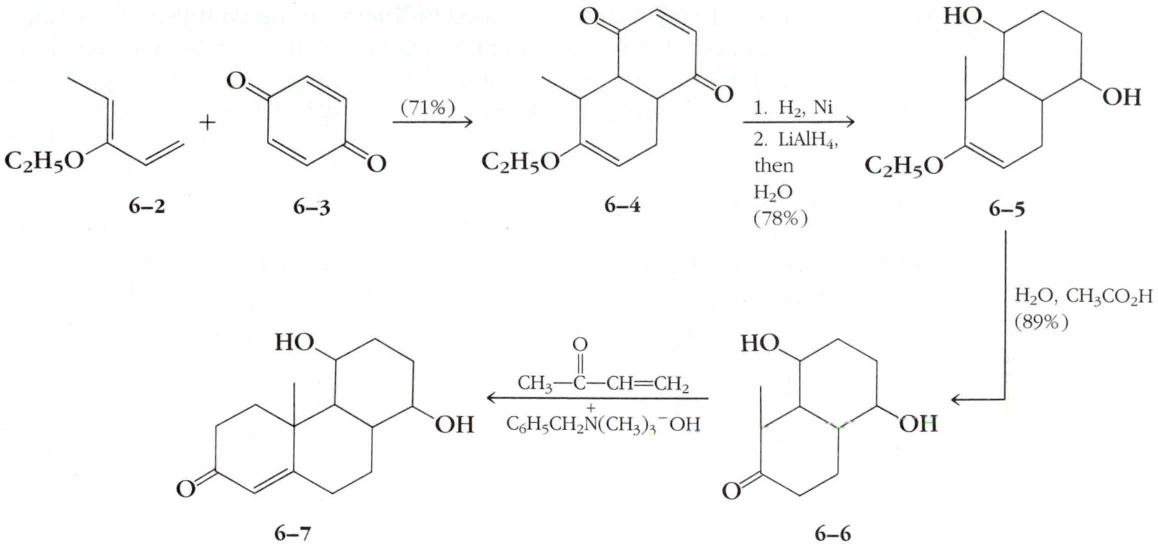

C_2H_5O | + | (71%) → | C_2H_5O | 1. H_2, Ni; 2. $LiAlH_4$, then H_2O (78%) → | HO, OH, C_2H_5O

6–2 6–3 6–4 6–5

H_2O, CH_3CO_2H (89%)

HO, OH | CH_3—C(=O)—CH=CH_2 / $C_6H_5CH_2\overset{+}{N}(CH_3)_3\,^{-}OH$ ← | HO, OH

6–7 6–6

Ketone **6-7** was next converted to a ketal so that other functional groups could be modified. Oxidation of the ketal with aluminum isopropoxide and cyclohexanone (cf. Example 22.2) resulted in oxidation of the less hindered hydroxyl group, and the resulting ketone was methylated with base and methyl iodide to generate **6-10.**

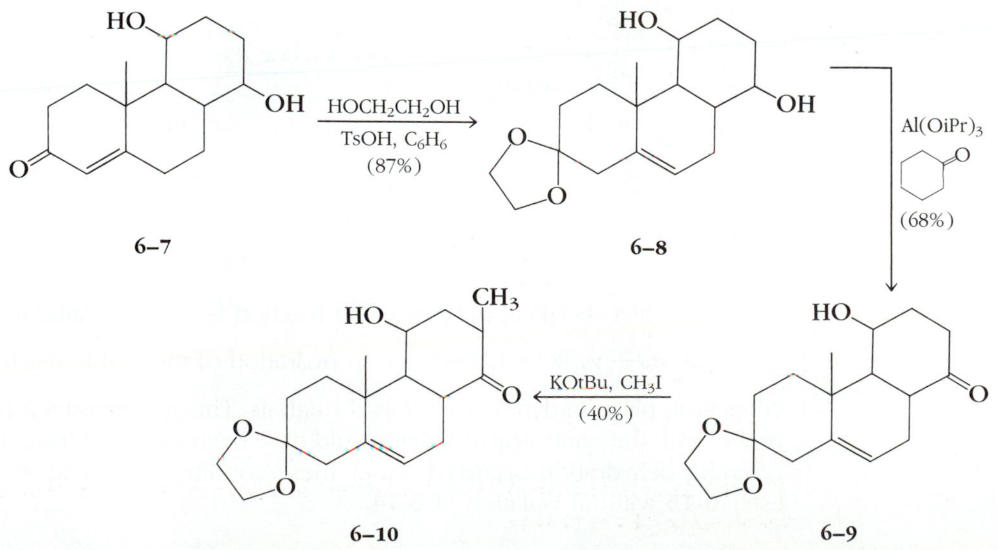

HO, OH | $HOCH_2CH_2OH$ / TsOH, C_6H_6 (87%) → | HO, OH

6–7 6–8

$Al(OiPr)_3$, cyclohexanone (68%)

HO, CH_3 | KOtBu, CH_3I (40%) ← | HO

6–10 6–9

Structure **6-10** corresponds to the tricyclic intermediate (IV) in Figure 23.3. The next phase of the synthesis was construction of the five-membered ring. Alkylation of the enolate of this ketone with 2-methyl-3-iodopropene yielded **6-11,** in which the $C{=}CH_2$ served as a latent carbonyl group. After several experiments it was found useful to oxidize the remaining hydroxyl group to a ketone before continuing to build up the carbon skeleton. The resulting oxidation of **6-11** to **6-12** is historically very important, because Sarett and his coworkers had to devise a way to carry out the oxidation without using acidic conditions that could hydrolyze the ketal group. It was to solve this and related problems in the synthesis of cortisone that they developed the CrO_3-pyridine reagent (Section 11.3). The resulting diketone (**6-12**) reacted preferentially at the less hindered carbonyl group when it was treated with the bromomagnesium salt of ethoxyacetylene.

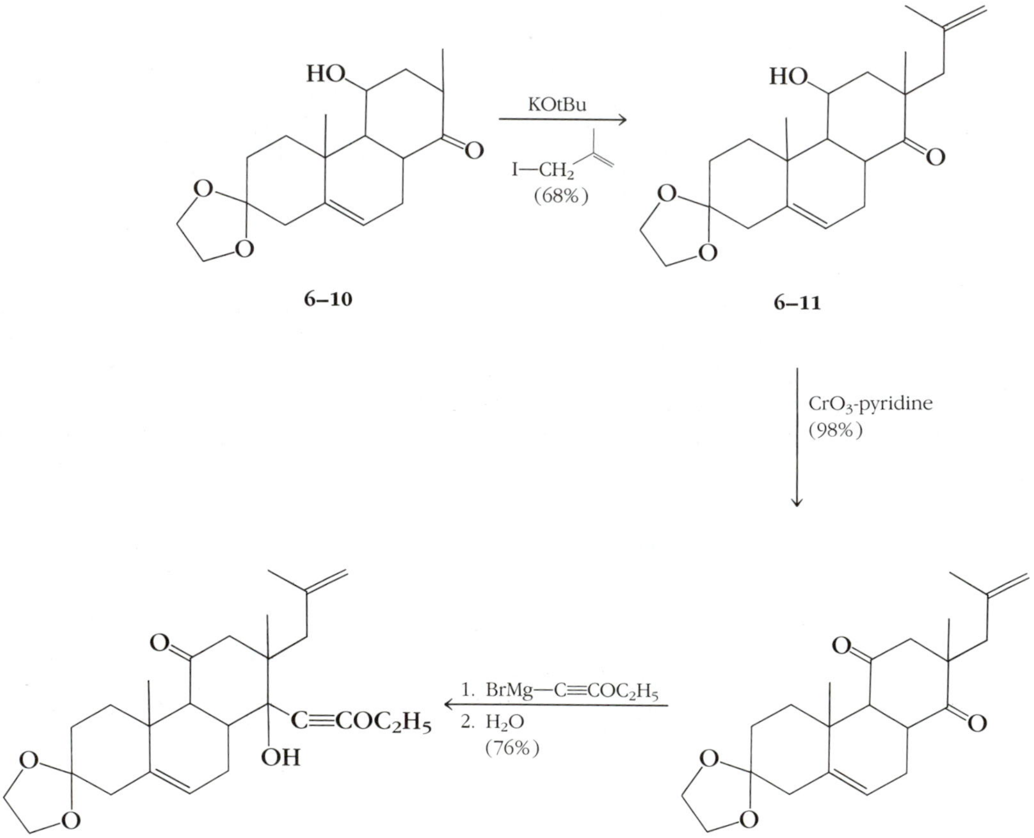

KOtBu, $I{-}CH_2C(CH_3){=}CH_2$ (68%): **6-10** → **6-11**

CrO_3-pyridine (98%): **6-11** → **6-12**

1. $BrMg{-}C{\equiv}COC_2H_5$ 2. H_2O (76%): **6-12** → **6-13** ($C{\equiv}COC_2H_5$, OH)

6-10 **6-11** **6-13** **6-12**

The product (**6-13**) of the preceding reaction is closely related to a vinyl ether ($RO{-}C{\equiv}C{-}$ vs $RO{-}CH{=}C\langle$), so hydration of the highly reactive acetylenic ether took place under very mild acid catalysis. This generated a β-hydroxy ester (**6-14**) with the same structure that would have been expected from a Reformatsky reaction; dehydration occurred under these conditions to yield the unsaturated ester **6-15** without isolation of **6-14.**

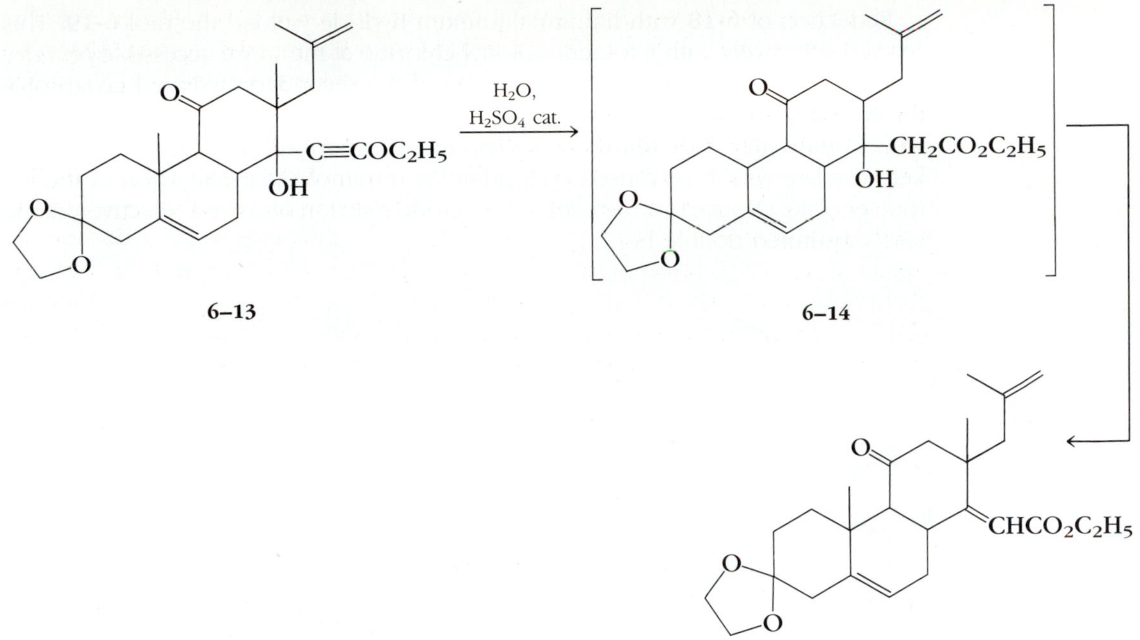

H_2O, H_2SO_4 cat.

$C{\equiv}COC_2H_5$ OH

6–13

$CH_2CO_2C_2H_5$ OH

6–14

$CHCO_2C_2H_5$

6–15

Compound **6-15** contained the necessary carbon atoms for forming the five-membered D ring, but several reduction and oxidation steps were necessary prior to cyclization. The unsaturated ester group could be reduced by potassium in ammonia, but better yields were obtained when the ester was first hydrolyzed to the free acid (**6-17**) and the keto group was reduced to an alcohol with sodium borohydride. Reduction of the unsaturated acid group then yielded **6-18.**

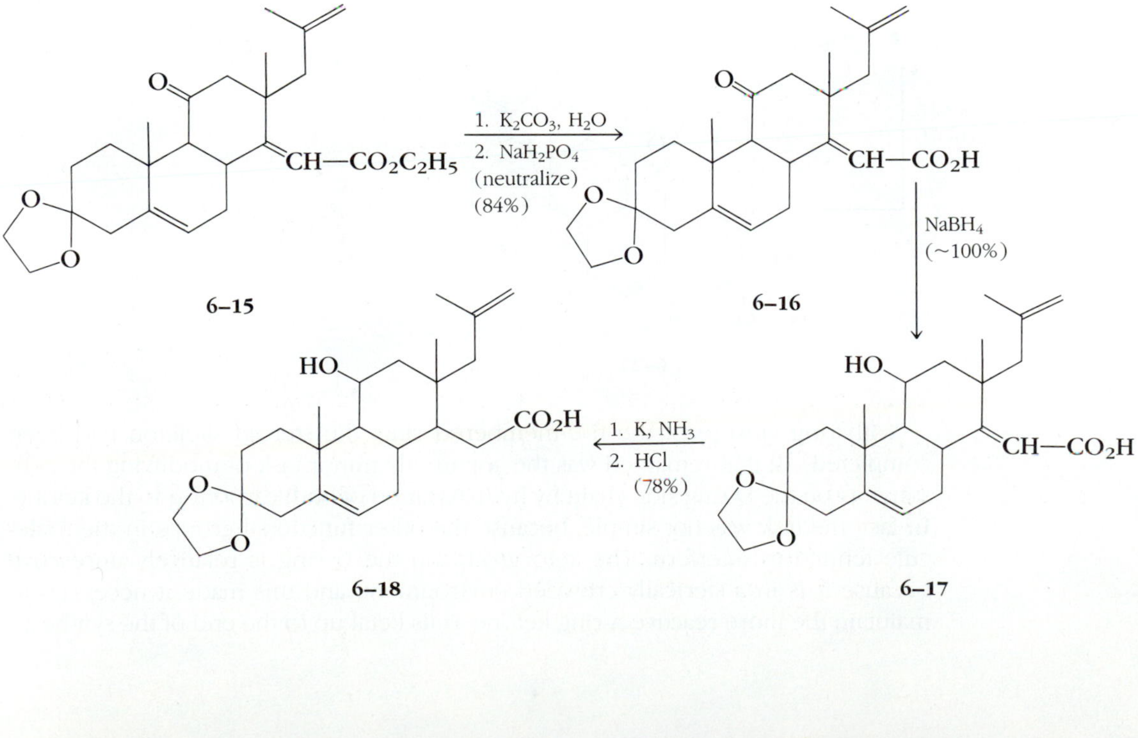

$CH{-}CO_2C_2H_5$

6–15

1. K_2CO_3, H_2O
2. NaH_2PO_4 (neutralize) (84%)

$CH{-}CO_2H$

6–16

$NaBH_4$ (~100%)

HO $CH{-}CO_2H$

6–17

1. K, NH_3
2. HCl (78%)

HO CO_2H

6–18

Reduction of **6-18** with lithium aluminum hydride yielded the diol **6-19.** This reacted selectively with *p*-toluenesulfonyl chloride at the more accessible *primary* hydroxyl group, and subsequent oxidation of the secondary hydroxyl group produced **6-20.** The latent carbonyl group was then liberated by formation of a diol with osmium tetroxide followed by cleavage with periodic acid. Treatment of the keto tosylate with base caused cyclization via intramolecular alkylation of the ketone enolate. (Notice that the osmium tetroxide reaction occurred selectively at the less substituted double bond.)

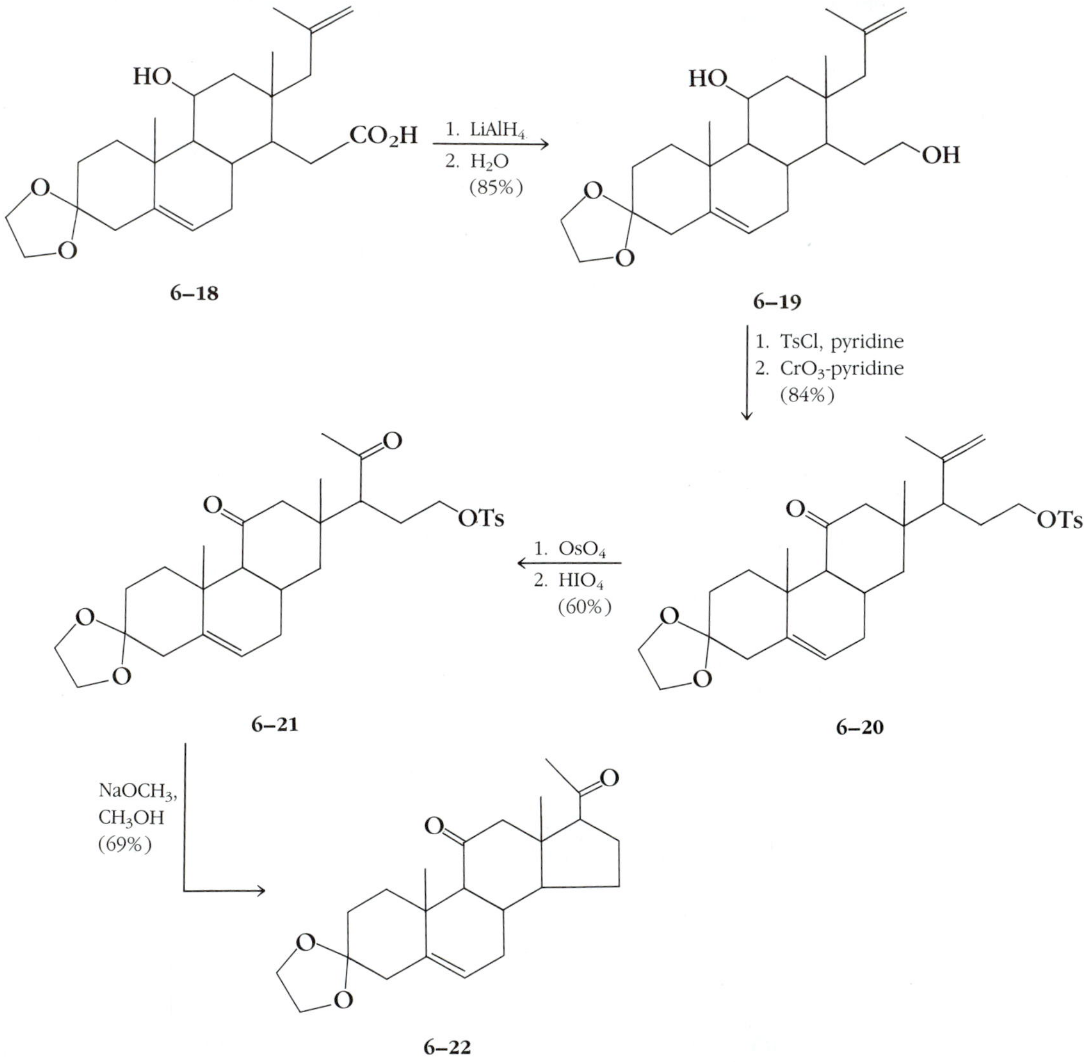

HO, CO_2H — 1. $LiAlH_4$ 2. H_2O (85%) → HO, OH

6–18 **6–19**

1. TsCl, pyridine
2. CrO_3-pyridine
(84%)

OTs ← 1. OsO_4 2. HIO_4 (60%) — OTs

6–21 **6–20**

$NaOCH_3$,
CH_3OH
(69%)

6–22

With ring closure of the five-membered ring, the steroid skeleton had been completed. All that remained was the apparently minor task of modifying the substituents on the D ring side chain by hydroxylation of both carbons α to the ketone. In fact, the task was not simple, because the other functional groups in the molecule tended to interfere. The keto group on the C ring is relatively unreactive because it is in a sterically crowded environment, and this made it necessary to maintain the more reactive A ring ketone as its ketal up to the end of the synthesis.

Introduction of the required oxygen substituents was carried out in two stages starting with the methyl carbon. Condensation of **6-22** with dimethyl oxalate generated the β-dicarbonyl compound **6-23.** This was easily halogenated to **6-24** with iodine and base, and the base also caused a retrocondensation of **6-24** to give the iodo ketone **6-25.** Displacement of the iodide by acetate ion then yielded the keto acetate **6-26.**

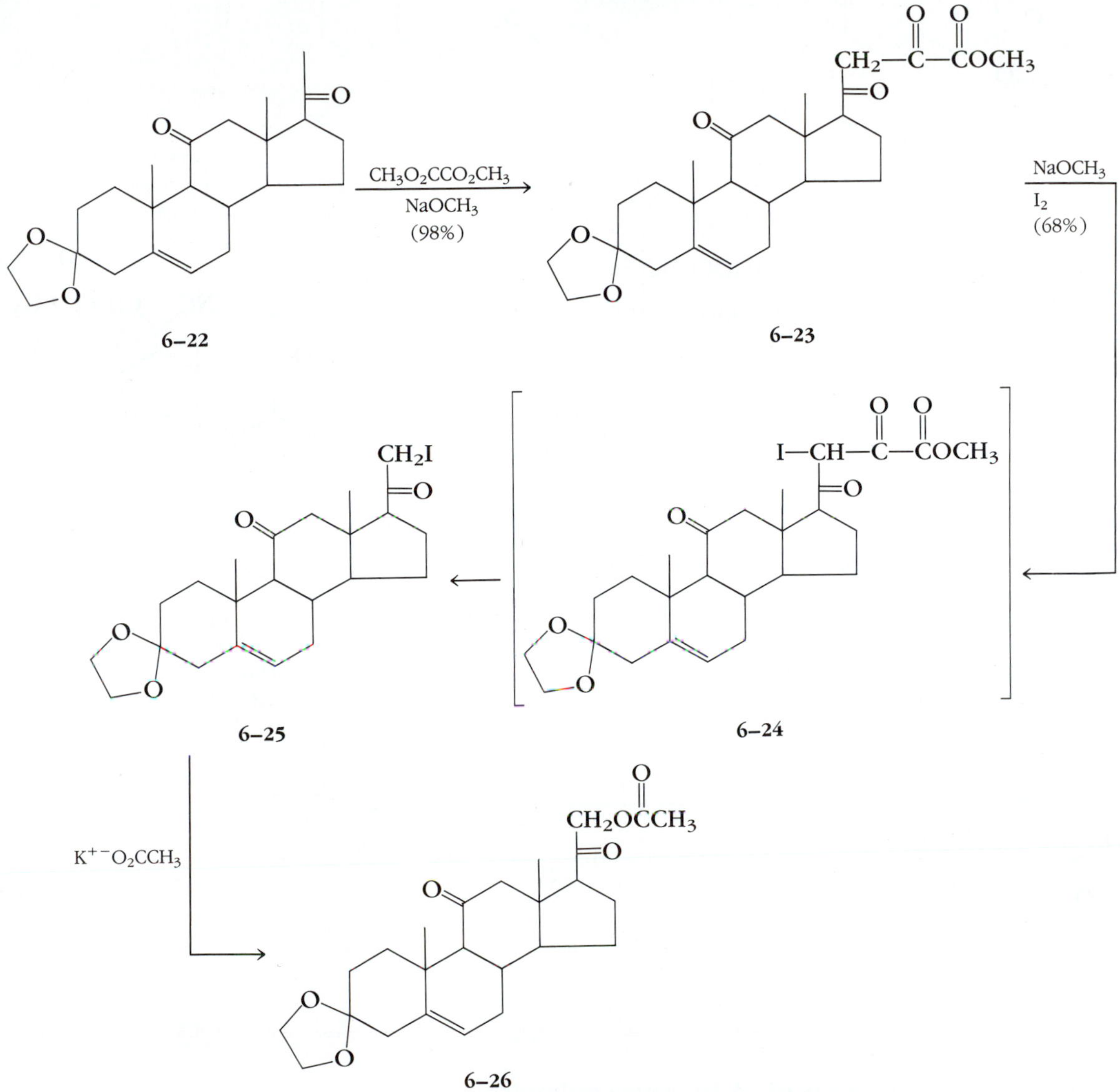

$CH_3O_2CCO_2CH_3$, $NaOCH_3$ (98%)

$NaOCH_3$, I_2 (68%)

$K^{+-}O_2CCH_3$

6–22 **6–23** **6–24** **6–25** **6–26**

The second oxygen was introduced by a novel method that employed formation of the cyanohydrin **6-27,** which was dehydrated to the unsaturated nitrile **6-28.** Reaction with potassium permanganate next generated the diol **6-29,** but this is also a cyanohydrin, and workup with base liberated the keto group to form **6-30.** Finally, cleavage of the ketal was carried out with mild acid. This was accompanied by isomerization of the double bond into conjugation with the ketone to yield **6-31,** the acetate of cortisone.

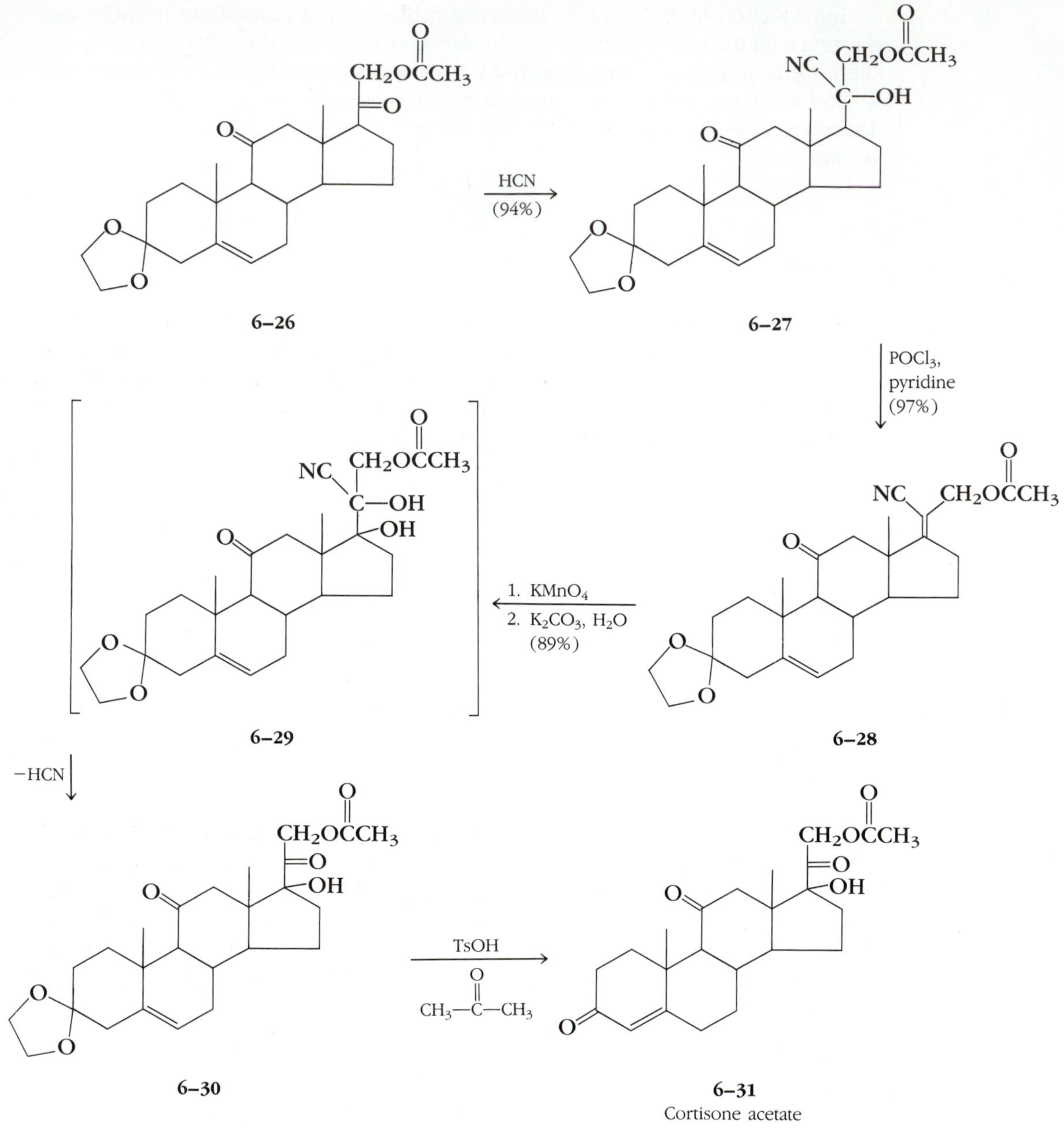

6–26

$\xrightarrow[\text{(94\%)}]{\text{HCN}}$

6–27

$POCl_3$, pyridine (97%)

6–28

1. $KMnO_4$
2. K_2CO_3, H_2O
(89%)

6–29

$-HCN$

6–30

TsOH, $CH_3{-}\overset{O}{\overset{\|}{C}}{-}CH_3$

6–31
Cortisone acetate

Example 23.7 **Synthesis of 25-Hydroxycholesterol**

The disease known as *rickets,* which results in faulty bone development, has historically been associated with conditions in which growing children were not exposed to adequate sunlight. In the early part of the twentieth century it was further established that an unknown substance, present in cod liver oil and other fish oils, could prevent the disease. The unknown factor was designated vitamin D_3, and its structure was later shown to be the steroid derivative **7-2.** When humans are exposed to sufficient sunlight, vitamin D_3 is produced by a photochemical reaction (Section 27.4) of 7-dehydrocholesterol (**7-1**), and no dietary intake of the vitamin is needed.

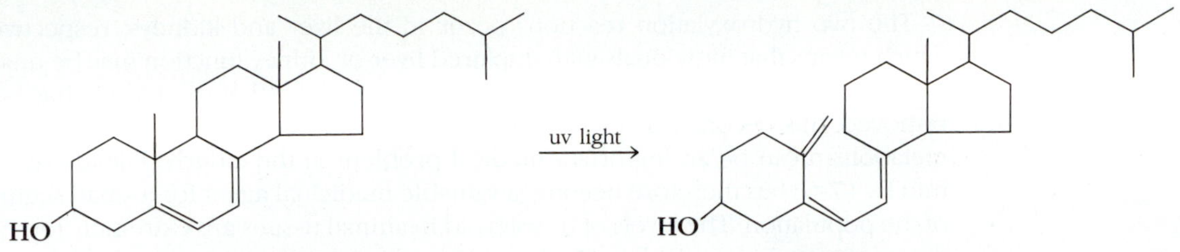

HO — uv light → HO

7-Dehydrocholesterol

7–1

Vitamin D_3 (cholicalciferol)

7–2

Another compound found to be effective in preventing rickets is vitamin D_2 (**7-4**), obtained by irradiation of the steroid ergosterol (**7-3**), which is present in significant quantities in yeast. The D_2 and D_3 series differ only by the presence of a methyl group and a double bond in the side chain of the former, and they react similarly in subsequent biochemical processes.

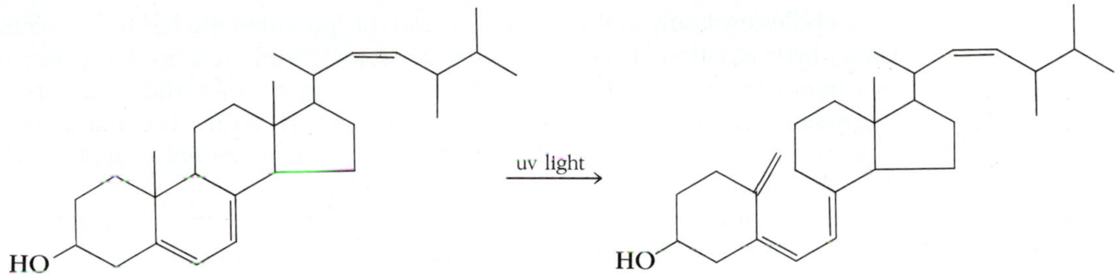

HO — uv light → HO

7–3

Ergosterol

7–4

Vitamin D_2 (calciferol)

During the 1960s work from several laboratories, and particularly that of H.F. DeLuca at the University of Wisconsin, demonstrated that further biochemical transformation of these compounds was necessary before they could play their necessary biochemical role in calcium regulation. By using radioactively labeled intermediates it was shown that vitamin D_3 undergoes two successive hydroxylation reactions at positions 1 and 25. Only then is the full biological activity observed.

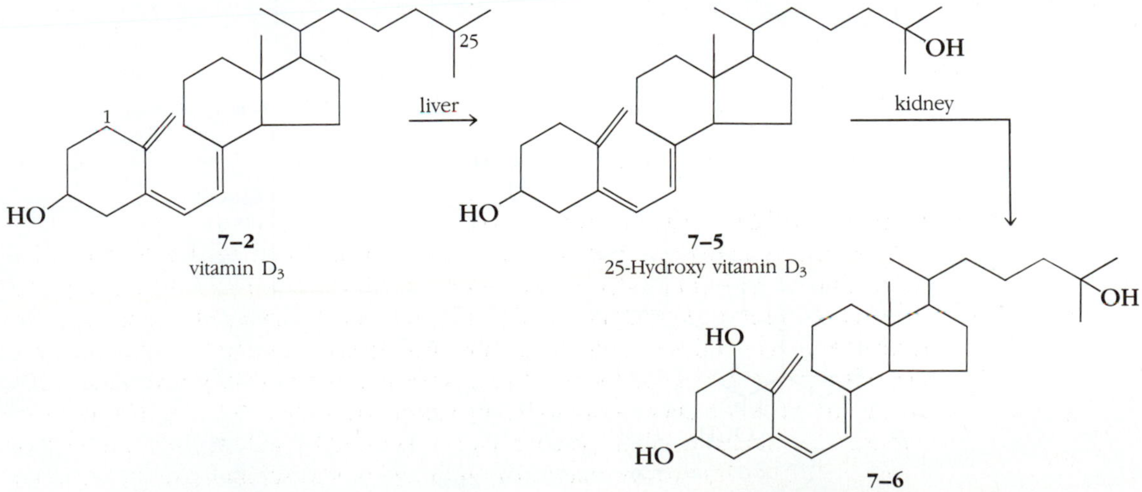

25 1 HO — liver → HO OH — kidney →

7–2

vitamin D_3

7–5

25-Hydroxy vitamin D_3

HO HO OH

7–6

1,25-Dihydroxy vitamin D_3

The two hydroxylation reactions occur in the liver and kidneys, respectively, which means that individuals with impaired liver or kidney function may be unable to utilize vitamin D_3 effectively. In extreme cases, in which the kidney has been removed, this becomes a very serious health problem. Irregularities in vitamin D metabolism can be an important medical problem in the elderly. Dihydroxy vitamin D_3 (**7-6**) has therefore become a valuable medicinal agent for a small segment of the population. The levels of this steroid in animal tissues are extremely low, and the compound is available only through synthesis.

In this example we will present a short but efficient synthesis of 25-hydroxycholesterol (**7-16**). This is an important synthetic intermediate, which has been further converted to 1,25-dihydroxyvitamin D_3 (by a sequence that is not shown here). The work was carried out by J. J. Partridge and his coworkers at Hoffmann-La Roche, Inc., and it makes possible the synthesis of the dihydroxy vitamin on a commercial scale.

The starting material for the synthesis was stigmasterol (**7-7**), a steroid that is available inexpensively and in large quantities from soybeans. Stigmasterol differs from cholesterol only in the side chain, and this provided the key to the synthesis. Direct hydroxylation of cholesterol at C-25 is not feasible (except for its formation as a minor by-product of cholesterol oxidation). In contrast, the double bond in stigmasterol allows modification of the side chain fairly easily. The first stage of the synthesis was protection of the B ring double bond. This was necessary because the branched alkyl substituents on the double bond of the side chain made it relatively unreactive, and direct cleavage of the side chain would have cleaved the B ring double bond as well. Conversion of stigmasterol to the tosylate **7-8**, followed by solvolysis in methanol, yielded the rearranged methoxy compound **7-9.** From reactions of similar compounds this rearrangement was expected, and it was also known that it could be reversed later in the synthesis.

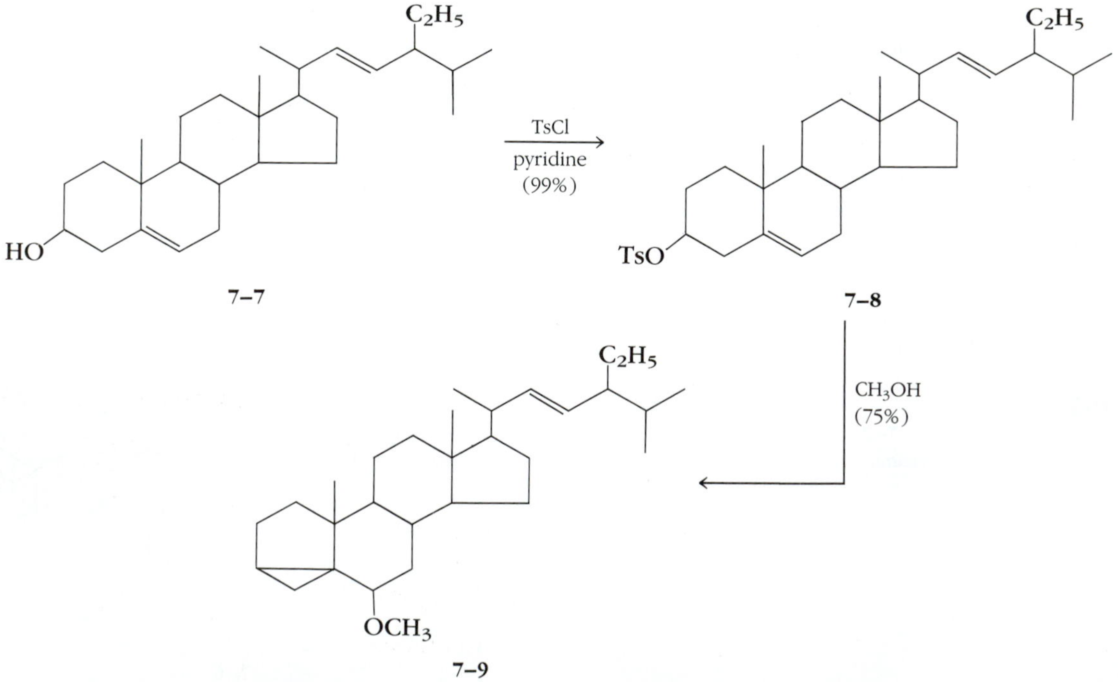

C_2H_5 HO **7–7** → TsCl, pyridine (99%) → C_2H_5 TsO **7–8**

7–8 → CH_3OH (75%) → C_2H_5 OCH_3 **7–9**

Only one double bond remains in **7-9,** and this allowed the unwanted portion of the side chain to be removed by ozonolysis. The intermediate ozonide was reduced with $NaAl(OCH_2CH_2OCH_3)_2H_2$, commonly known by the trade name Red-Al® (Aldrich Chemical Company). We have not discussed this reagent previously, but it is related to $LiAlH_4$ and $LiAl(OtBu)_3H$. This afforded the primary alcohol **7-10,** which was in turn converted to the corresponding tosylate, **7-11.**

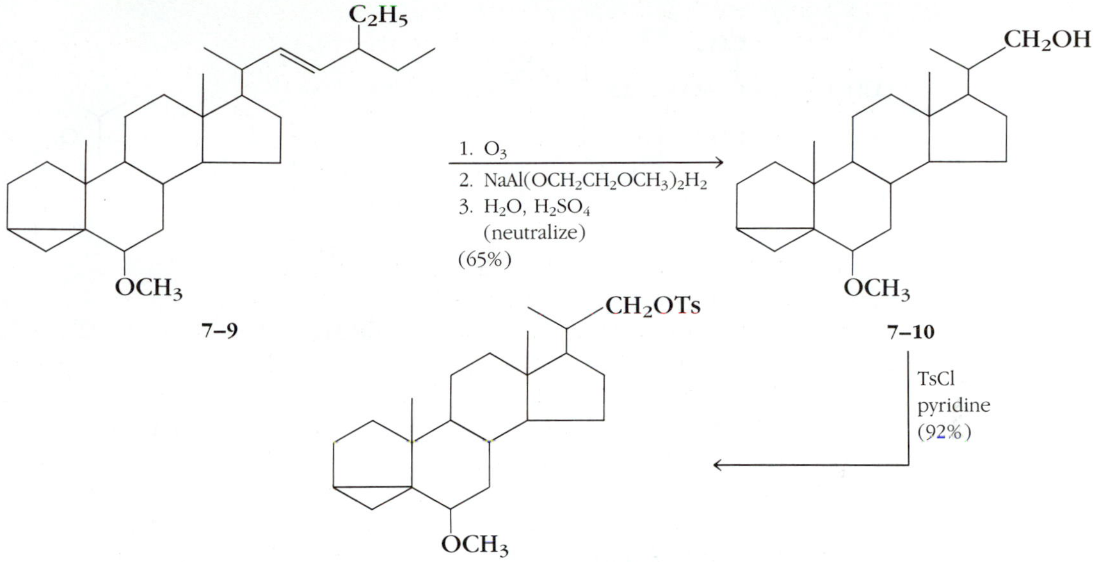

C_2H_5

CH_2OH

1. O_3
2. $NaAl(OCH_2CH_2OCH_3)_2H_2$
3. H_2O, H_2SO_4 (neutralize)
(65%)

OCH_3 **7–9**

OCH_3 **7–10**

TsCl
pyridine
(92%)

CH_2OTs

OCH_3 **7–11**

Reaction of tosylate **7-11** with the lithium acetylide **7-12** yielded the alkyne **7-13,** which had the required carbon skeleton for the side chain.

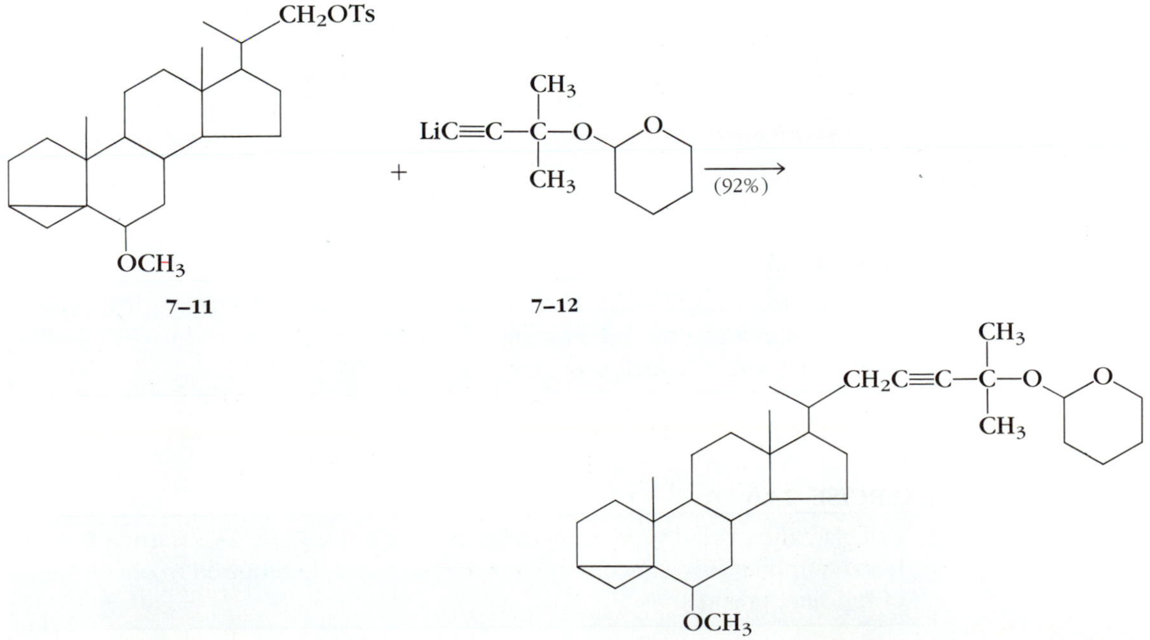

CH_2OTs

OCH_3 **7–11**

+ $LiC{\equiv}C{-}C(CH_3)_2{-}O{-}$(tetrahydropyranyl) **7–12** $\xrightarrow{(92\%)}$

$CH_2C{\equiv}C{-}C(CH_3)_2{-}O{-}$(tetrahydropyranyl)

OCH_3 **7–13**

The final stage of the synthesis consisted of reduction of the triple bond and removal of the two protecting groups. Catalytic hydrogenation reduced the alkyne smoothly, and treatment of **7-14** with aqueous acid caused hydrolysis of both the tetrahydropyranyl ether and the methyl ether at C-6. The methyl ether was activated by the adjacent cyclopropyl group, which underwent ring opening in the reaction to regenerate the original substitution pattern in the A and B rings. As a result, the hydrolysis produced the desired 25-hydroxycholesterol (**7-16**).

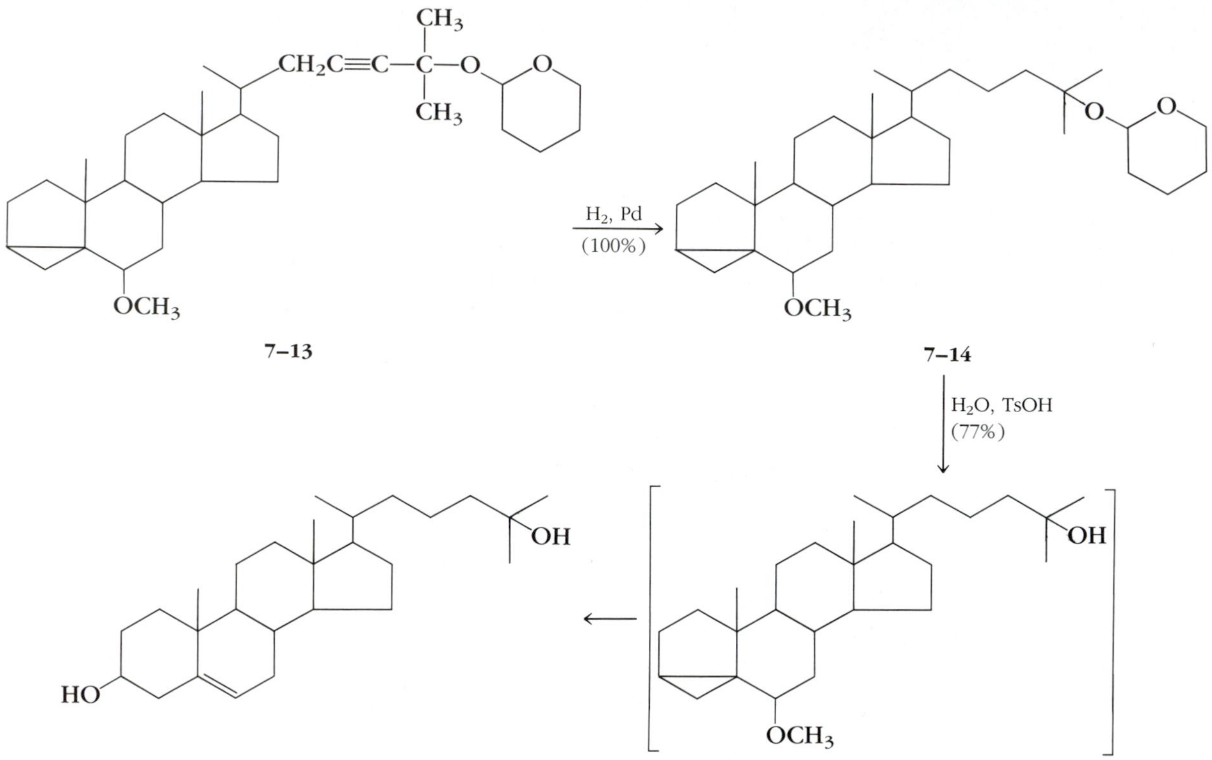

CH_3
$CH_2C{\equiv}C{-}C{-}O$ O
CH_3
OCH_3
7–13

H_2, Pd
(100%)

O O
OCH_3
7–14

H_2O, TsOH
(77%)

OH
HO
7–16

OH
OCH_3
7–15

EXERCISE 23.5

Calculate the overall yield of equilenin in Example 23.5 starting from the naphthalene derivative **5-2.** How much starting material would be required to obtain 1 g of the final product?

EXERCISE 23.6

Calculate the overall yield of cortisone acetate in Example 23.6 starting from benzoquinone. How much starting material would be required to obtain 1 g of the final product?

EXERCISE 23.7

In the oxidation of diol **6-8** to the keto alcohol **6-9,** identify the actual oxidizing agent (i.e., the species which is reduced). Write a balanced equation for the overall reaction.

EXERCISE 23.8

In the alkylation of **6-9** with methyl iodide and base, a by-product, believed to be a C_{19} compound, was formed in significant amounts. Suggest possible structures for this by-product.

EXERCISE 23.9

Write a detailed reaction mechanism for the acid-catalyzed conversion of **6-13** to **6-15.** First show the *addition* of water to produce **6-14** and then show the *elimination* of water to give **6-15.**

EXERCISE 23.10

In the hydrolysis of ester **6-15** to acid **6-16,** why were basic (rather than acidic) conditions used? What products would be expected from hydrolysis in acid?

EXERCISE 23.11

In the intramolecular alkylation of tosylate **6-21,** products might have been formed from alkylation of four different enolate ions, yet only **6-22** was observed. Draw the other three isomeric possibilities, and suggest reasons why they were not formed.

EXERCISE 23.12

(a) The retrocondensation of **6-24** in base gave **6-25,** which is the only observed product even though the β-diketone could be cleaved at either carbonyl. Draw the products of the other mode of cleavage. Which of the two carbonyls is more hindered? (α-Dicarbonyl compounds usually react with nucleophiles more rapidly than do monocarbonyl compounds; hence the use of an oxalyl substituent as an activating group.) (b) Retrocondensation occurs more easily with **6-24** than with most other β-diketones. Suggest an explanation. (What is the leaving group in the actual cleavage step?)

EXERCISE 23.13

The last step in the cortisone synthesis (**6-30** $\rightarrow$ **6-31**) is removal of the blocking group. This particular reaction is not a hydrolysis but a ketal *exchange.* Write the complete balanced equation.

EXERCISE 23.14

Solvolysis of **7-8** produced **7-9** via a carbocation rearrangement. Draw a reaction mechanism for this process. Another product was isolated in 24% yield; suggest a likely structure for this compound.

EXERCISE 23.15

Suggest a method for synthesizing the alkyne derivative **7-12** starting from compounds containing acetylene.

23.3 PROSTAGLANDINS

Prostaglandins are C-20 carboxylic acids that contain a cyclopentane ring, and they have the following general structure:

CO_2H

General formula of prostaglandins

Prostaglandins were first isolated in the 1930s from human seminal fluid and from vesicular glands of sheep. Their name resulted from the erroneous belief that these compounds were biosynthesized specifically in the prostate gland, but it has since been shown that they are produced in nearly all animal tissues.

The prostaglandins are intracellular regulators of hormone action, and they are intimately involved in many important biochemical processes. They aid in the regulation of gastric secretions and blood clotting and can induce relaxation of smooth muscle. These diverse effects have made prostaglandins and their analogs prime candidates as new useful medicinal agents. The ability to regulate gastric secretions, for example, suggests the possibility of using prostaglandins in the treatment of ulcers. It has also been found that aspirin inhibits prostaglandin biosynthesis, which indicates that the prostaglandins are involved in the action of pain relievers in general. The importance of prostaglandins in human biology was underscored by the award of the 1982 Nobel Prize in medicine and physiology to the Swedish chemists B. Sammuelsson and S. Bergstrom for their extensive work on the structure and biochemistry of these compounds.

The biosynthesis of the prostaglandins proceeds from the saturated fatty acid stearic acid. Lengthening of the chain by two carbons and introduction of four double bonds generates *arachidonic acid,* which undergoes oxidative cyclization to produce both prostaglandins and thromboxanes. (The latter are also involved in regulation of blood platelet aggregation.) The following equation shows just two of the many products formed in the biosynthesis.

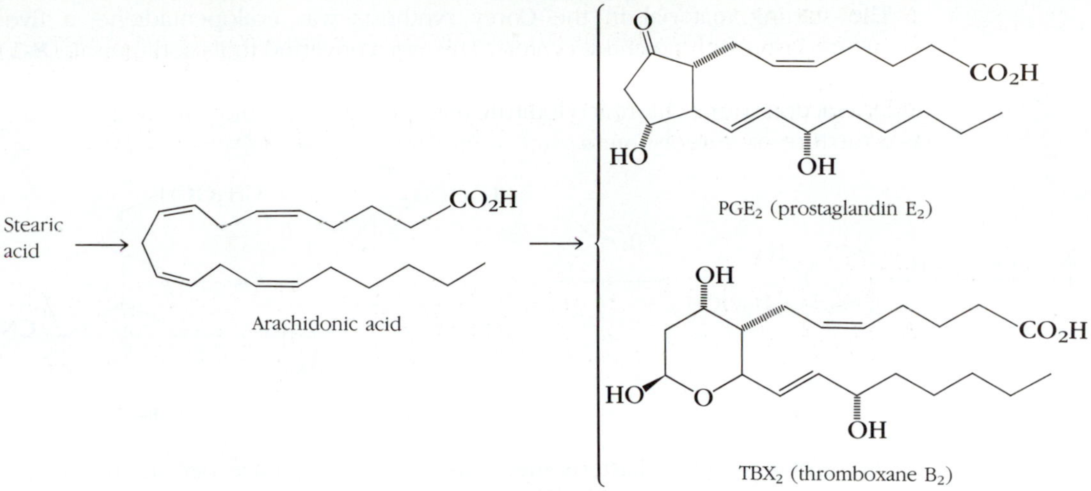

Stearic acid → Arachidonic acid → PGE_2 (prostaglandin E_2); TBX_2 (thromboxane B_2)

The nomenclature of the prostaglandins is somewhat complex. The letters A-I are used to indicate the substitution pattern on the five-membered ring, and the numerical subscripts indicate the total number of double bonds. In this section we will describe the synthesis of prostaglandins in the E and F series.

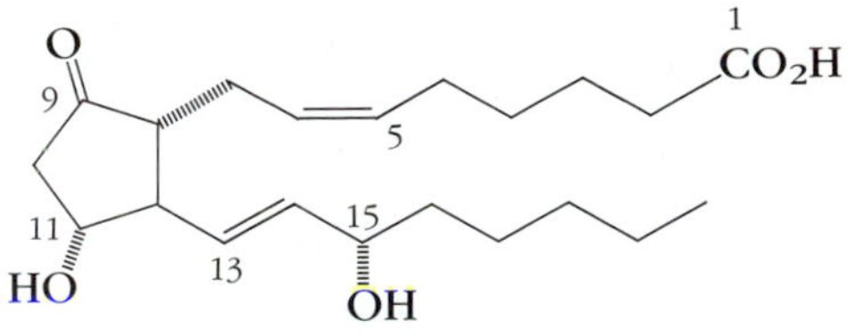

PGE_2 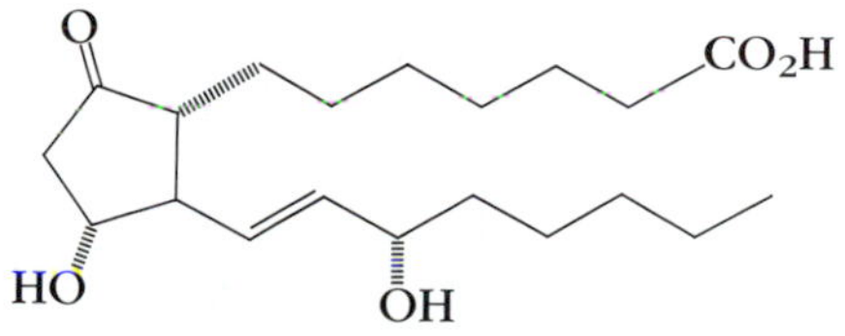 PGE_1

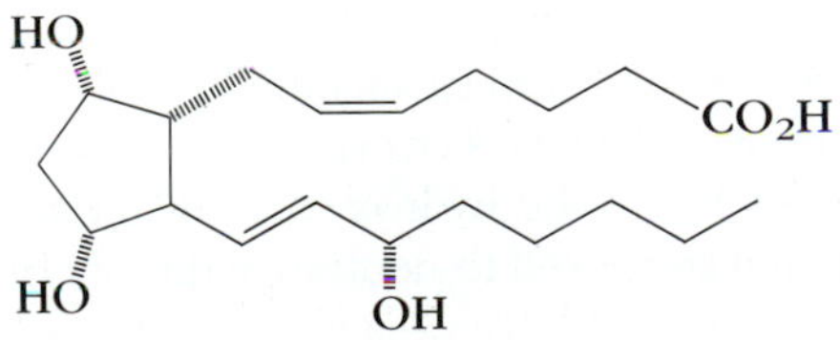

PGF_2 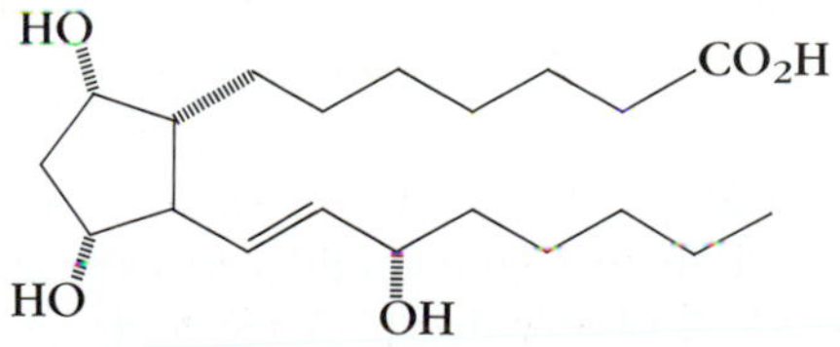 PGF_1

All these compounds have oxygen substituents at C-9, C-11, and C-15, but the E and F series differ by the oxidation state at C-9.

Example 23.8 **Synthesis of Prostaglandins**

Considerable synthetic work has been carried out in the prostaglandin area since the 1960s, and E.J. Corey at Harvard has been a leader in this field. The synthetic work that we will present here was carried out by his research group, and it is noteworthy for two reasons. First, the side chains are introduced in a way that would allow the preparation of analogs by closely related synthetic schemes. Second, stereochemical control was exercised throughout the sequence of reactions so that the desired stereoisomers were produced efficiently.

The starting material in the Corey synthesis was cyclopentadiene, a five-membered ring with no chiral centers. This was converted to its sodium salt (**8-1**) and alkylated with methyl chloromethyl ether to give the diene **8-2.** The Diels-Alder reaction with 2-chloroacrylonitrile then produced the bicyclic compound **8-3** as a mixture of stereoisomers.

Na^+ (8–1) $\xrightarrow{Cl—CH_2OCH_3}$ (8–2, H, CH_2OCH_3) $\xrightarrow{CH_2=C(Cl)CN}$ 8–3a (CH_3OCH_2, Cl, CN) + 8–3b (CH_3OCH_2, CN, Cl)

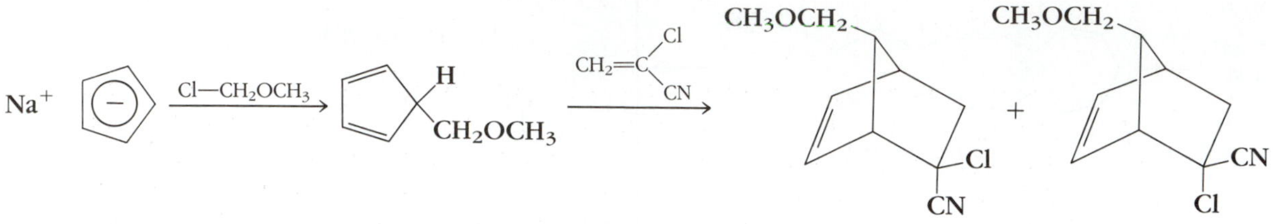

8–1 8–2 8–3a 8–3b

The formation of two diastereomers was of no consequence, because the stereochemistry at the carbon with the Cl and CN groups would be destroyed in the next reaction. More importantly, the other three chiral centers of **8-3** were formed with the desired configuration (although as a mixture of enantiomers). You can understand the formation of **8-3** with the correct stereochemistry by inspection of the transition state leading to **8-3a.** The optimum approach of the unsaturated nitrile is from the *less hindered* side of **8-2.**

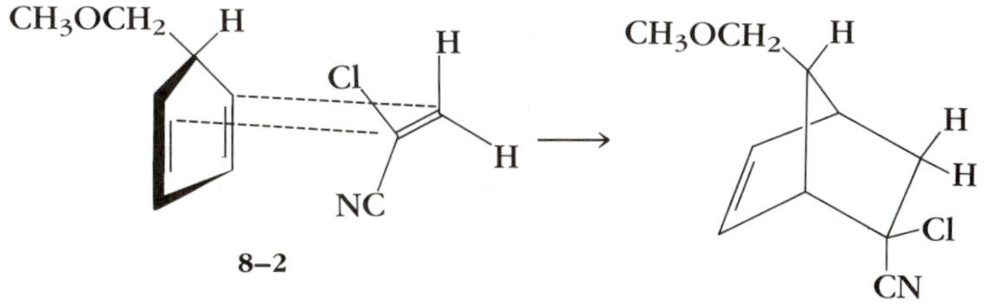

8–2 (CH_3OCH_2, H; Cl, H, H, NC) ⟶ 8–3a (CH_3OCH_2, H; H, H, Cl, CN)

When the mixture of stereoisomers of **8-3** was subjected to alkaline hydrolysis, the intermediate cyanohydrin afforded ketone **8-5.** This was converted to lactone **8-6** by Baeyer-Villiger oxidation, and hydrolysis led to the hydroxy acid **8-7.** (Notice the extremely mild acid, carbon dioxide in water, used to neutralize the salt of the acid-sensitive hydroxy acid).

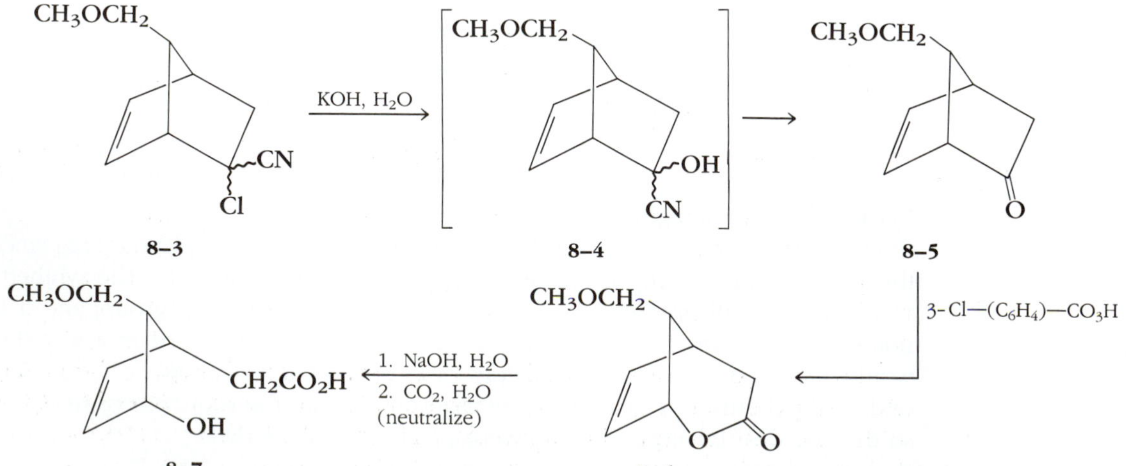

8–3 (CH_3OCH_2, CN, Cl) $\xrightarrow{KOH,\ H_2O}$ [8–4 (CH_3OCH_2, OH, CN)] ⟶ 8–5 (CH_3OCH_2, O)

8–5 $\xrightarrow{3\text{-}Cl—(C_6H_4)—CO_3H}$ 8–6 (CH_3OCH_2, O, O) $\xrightarrow{1.\ NaOH,\ H_2O;\ 2.\ CO_2,\ H_2O\ (neutralize)}$ 8–7 (CH_3OCH_2, CH_2CO_2H, OH)

8–3 8–4 8–5 8–7 8–6

At this stage an optically active amine was employed to *resolve* the hydroxy acid. When carboxylic acid **8-7** was allowed to react with (+)-ephedrine

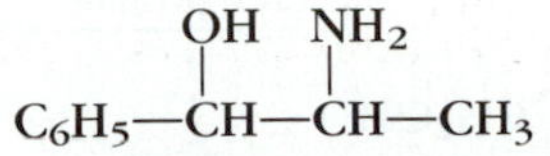

$$C_6H_5—\underset{}{\overset{OH}{\overset{|}{C}H}}—\overset{NH_2}{\overset{|}{C}H}—CH_3$$

the corresponding ammonium salt was formed. The acid **8-7** was racemic, so two diastereomeric salts were generated. The salt of the desired enantiomer (i.e., that with the stereochemistry that we have depicted in the drawings) is less soluble, and it precipitated from the solution. Conversion of this salt back to the hydroxy acid afforded optically active material for the remainder of the synthesis. All three chiral centers of **8-7** had the correct configuration at this point, and the next step was generation of the final chiral center. This was produced stereospecifically by forming an iodolactone (see Section 8.7). The new bond to oxygen was necessarily *cis* to the carboxylic acid side chain of **8-7.**

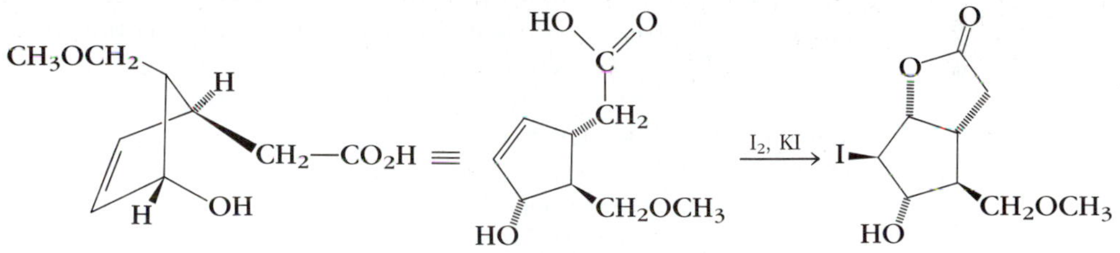

CH_3OCH_2 H $CH_2—CO_2H$ ≡ H OH HO O C CH_2 CH_2OCH_3 HO $\xrightarrow{I_2,\ KI}$ O O I CH_2OCH_3 HO

8–7 **8–8**

The unnecessary iodo substituent was removed by reduction with tributyltin hydride after first protecting the free hydroxyl group as the corresponding acetate. (Tributyltin hydride, a reagent we have not discussed previously, is used for selective reduction of carbon-halogen bonds.)

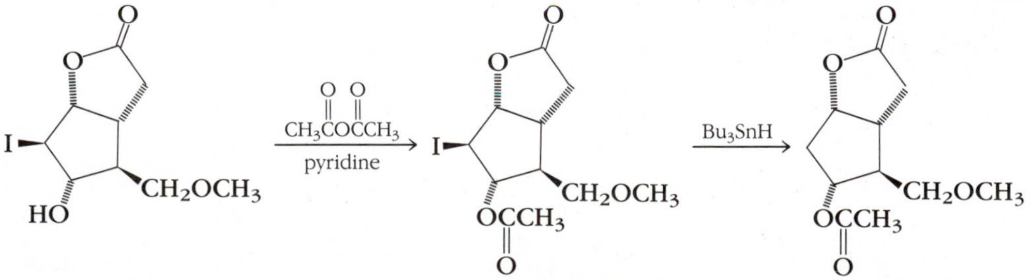

O O I CH_2OCH_3 HO $\xrightarrow[\text{pyridine}]{CH_3\overset{O}{\overset{||}{C}}O\overset{O}{\overset{||}{C}}CH_3}$ O O I CH_2OCH_3 $O\underset{O}{\underset{||}{C}}CH_3$ $\xrightarrow{Bu_3SnH}$ O O CH_2OCH_3 $O\underset{O}{\underset{||}{C}}CH_3$

8–8 **8–9** **8–10**

Elaboration of the side chains began with the selective cleavage of the methyl ether using the strong Lewis acid boron tribromide. The resulting alcohol **8-11** was oxidized to the corresponding aldehyde, and a modified Wittig reaction was carried out with the phosphonate derivative of 2-heptanone.

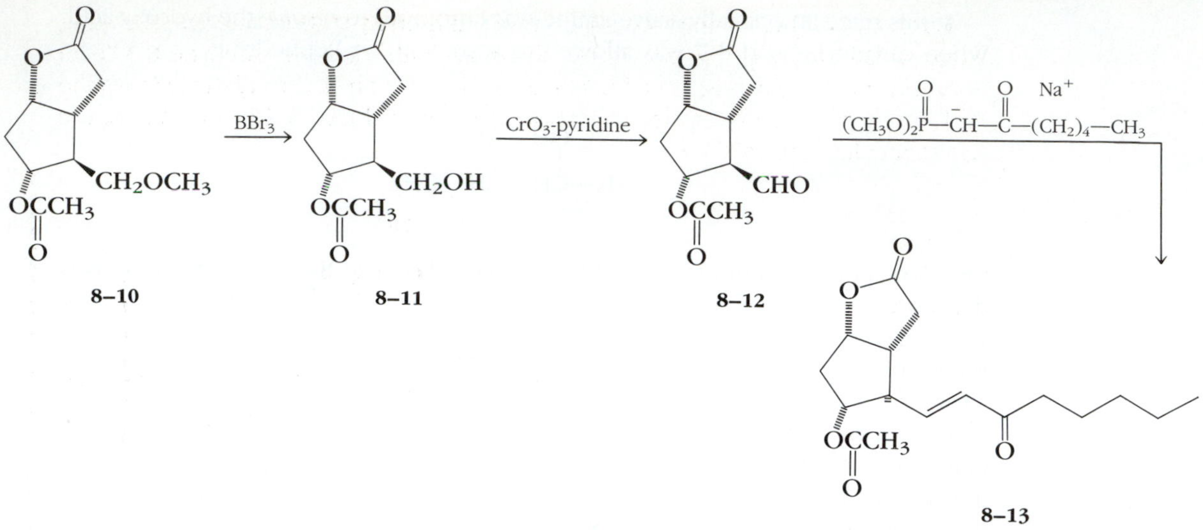

CH_2OCH_3 $OCCH_3$ — BBr_3 → CH_2OH $OCCH_3$ — CrO_3-pyridine → CHO $OCCH_3$ — $(CH_3O)_2P(=O)-\bar{C}H-C(=O)-(CH_2)_4-CH_3$ Na^+ →

8–10 8–11 8–12

$OCCH_3$

8–13

Prior to forming the other side chain it was necessary to introduce additional protecting groups. Borohydride reduction of **8-13** (using zinc rather than sodium borohydride) converted the keto group to a secondary alcohol. Two stereoisomers were formed in this reaction, and they were separated by chromatography. The "wrong" stereoisomer was recycled by oxidizing it back to **8-13.** Mild treatment of the desired stereoisomer, **8-14,** with base cleaved the acetate without opening the lactone ring, and both free hydroxyl groups were then converted to tetrahydropyranyl ethers (**8-16**).

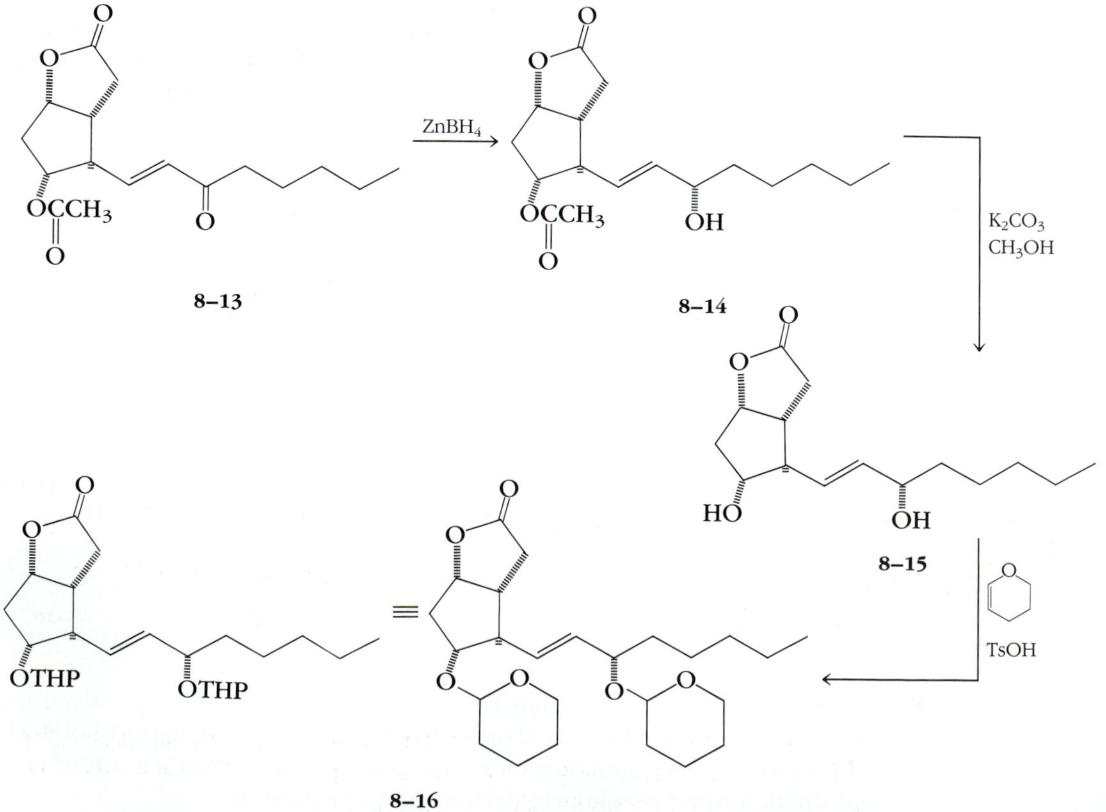

$OCCH_3$ — $ZnBH_4$ → $OCCH_3$ OH — K_2CO_3, CH_3OH →

8–13 8–14

HO OH

8–15

TsOH

OTHP OTHP ≡

8–16

Reduction of the lactone with diisobutylaluminum hydride (a selective reducing agent, which we have not previously discussed) afforded the aldehyde **8-17b** in equilibrium with the hemiacetal **8-17a.** A Wittig reaction of the aldehyde (with the ylide derived from 5-bromopentanoic acid) then produced **8-18** to complete the prostaglandin carbon skeleton.

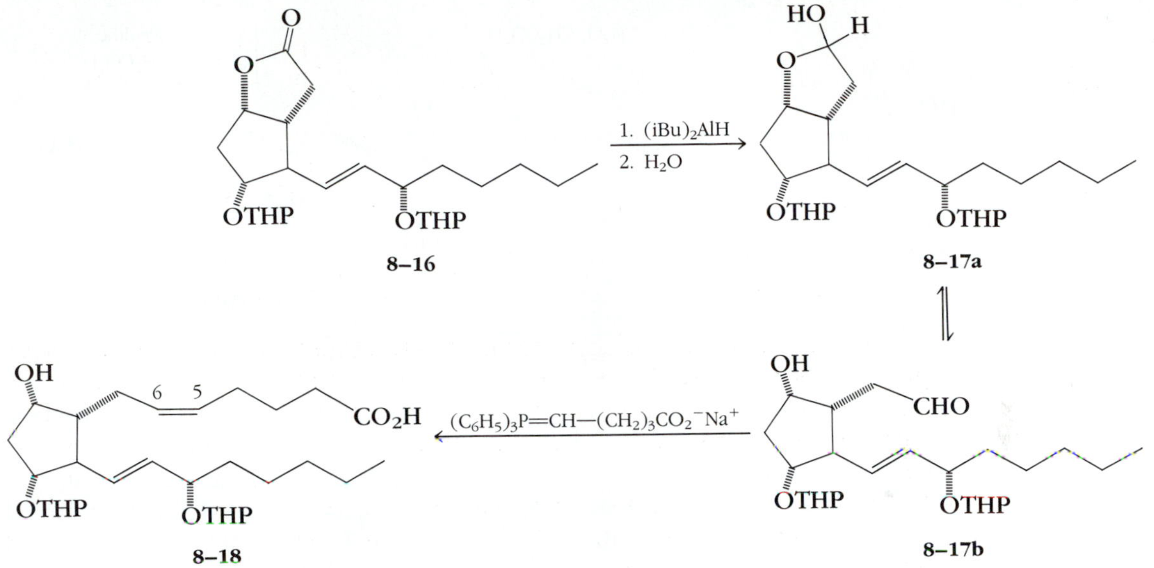

O
O
OTHP OTHP
8–16

1. $(iBu)_2AlH$
2. H_2O

HO H
O
OTHP OTHP
8–17a

⇌

OH
CHO
OTHP OTHP
8–17b

$(C_6H_5)_3P{=}CH{-}(CH_2)_3CO_2^-Na^+$

OH
6 5
CO_2H
OTHP OTHP
8–18

Compound **8-18** then served as the direct precursor of four prostaglandins. Mild hydrolysis of the tetrahydropyranyl ethers yielded prostaglandin F_2; if the free hydroxyl group was first oxidized, prostaglandin E_2 was obtained.

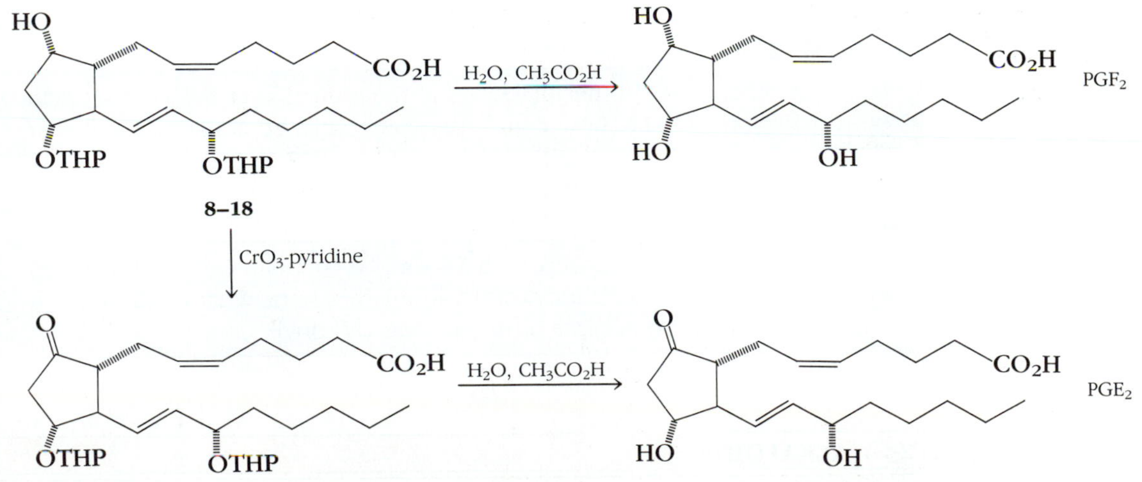

HO
CO_2H
OTHP OTHP
8–18

H_2O, CH_3CO_2H

HO
CO_2H
HO OH
PGF_2

CrO_3-pyridine

O
CO_2H
OTHP OTHP
8–19

H_2O, CH_3CO_2H

O
CO_2H
HO OH
PGE_2

Catalytic hydrogenation of **8-18** proceeded with selective reduction of the 5,6 double bond, affording **8-20,** which, when subjected to the same reactions used with **8-18,** yielded prostaglandins F_1 and E_1.

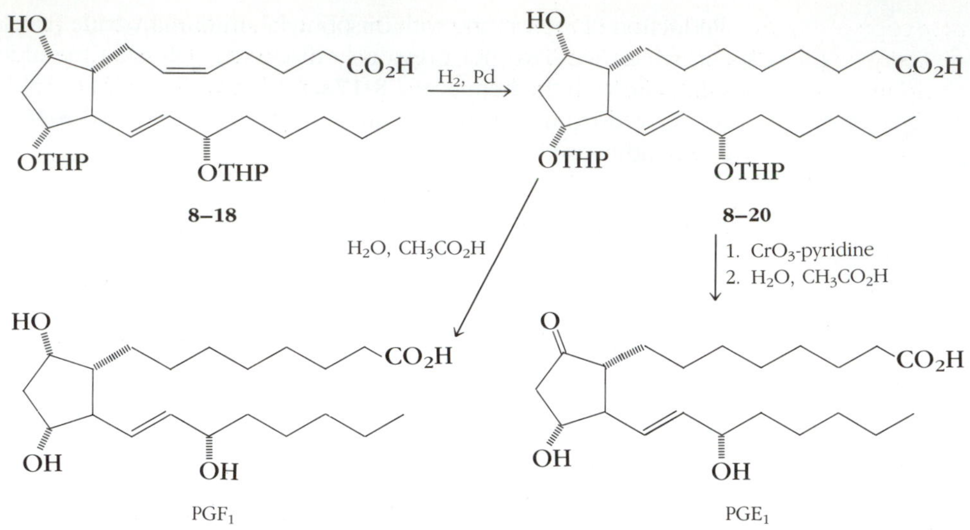

HO, CO_2H, OTHP, OTHP

8–18

H_2, Pd →

HO, CO_2H, OTHP, OTHP

8–20

H_2O, CH_3CO_2H

1. CrO_3-pyridine
2. H_2O, CH_3CO_2H

HO, CO_2H, OH, OH

PGF_1

O, CO_2H, OH, OH

PGE_1

EXERCISE 23.16

Using three-dimensional representations, write a reaction mechanism for formation of iodolactone **8-8.**

EXERCISE 23.17

How would the phosphonate Wittig reagent used to form **8-13** be prepared from 1-bromo-2-heptanone?

EXERCISE 23.18

Draw the other stereoisomer produced in the reduction of **8-13.** Other than by reoxidation to **8-13,** how could this compound be converted to **8-14?**

EXERCISE 23.19

Cleavage of the acetate group in **8-14** employed potassium carbonate in methanol. What was the other product in this reaction? What might occur if nucleophilic attack took place at the lactone carbonyl?

23.4 ALKALOIDS

Alkaloids are physiologically active amines that are usually obtained from plants. As in the case of terpenes, the biological function of alkaloids in plants is not well understood. Many alkaloids are extremely toxic, however, so they may have evolved as a defense mechanism that protects the plant against animal predators. Alkaloids exhibit variations in both structure and biological activity that far exceed the scope of this discussion, but several examples can give you an indication of this diversity. Figure 23.4 shows the structures of some well-known alkaloids; these compounds have physiological effects that can range from pleasant to fatal.

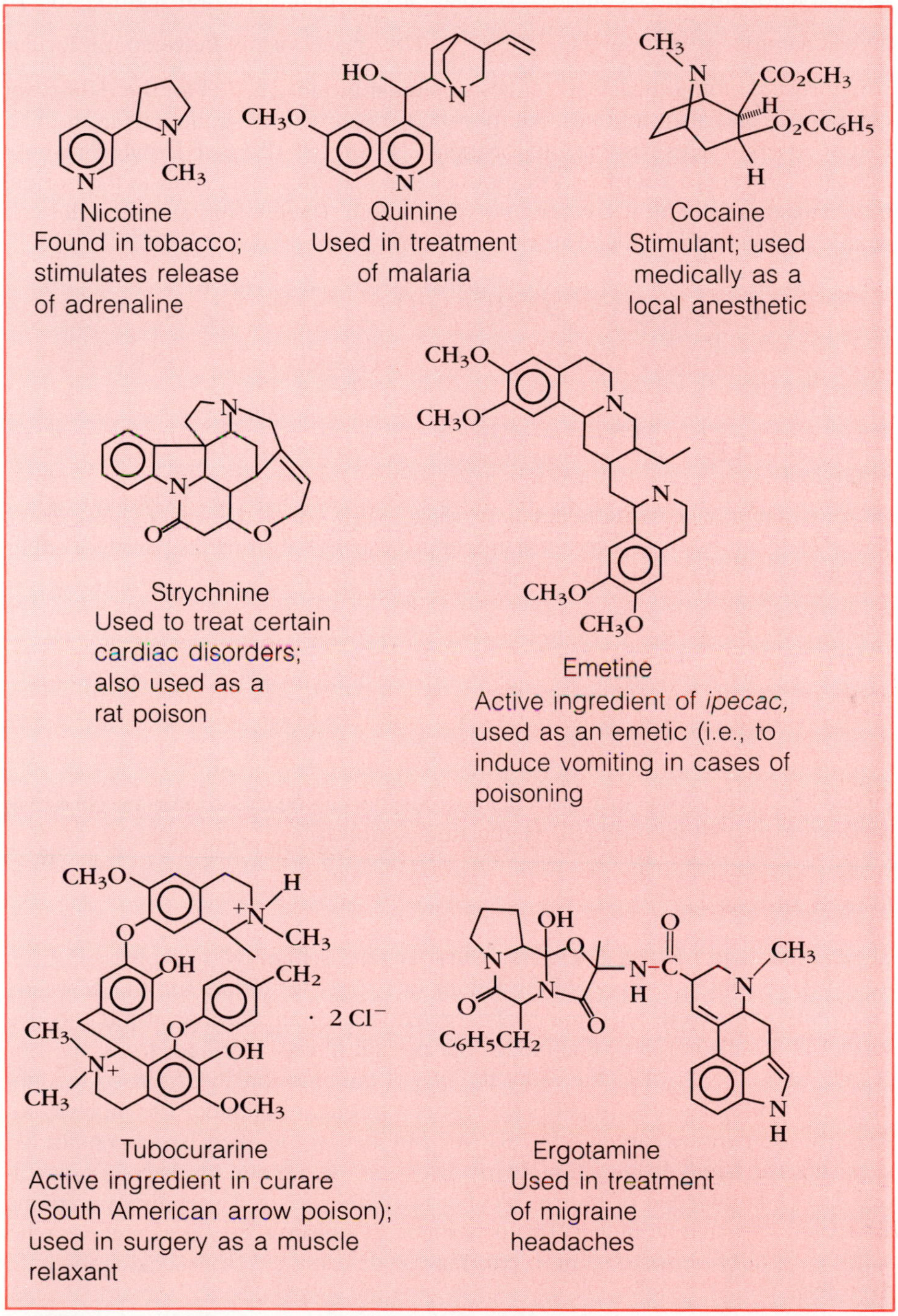

Nicotine
Found in tobacco; stimulates release of adrenaline

Quinine
Used in treatment of malaria

Cocaine
Stimulant; used medically as a local anesthetic

Strychnine
Used to treat certain cardiac disorders; also used as a rat poison

Emetine
Active ingredient of *ipecac,* used as an emetic (i.e., to induce vomiting in cases of poisoning

Tubocurarine
Active ingredient in curare (South American arrow poison); used in surgery as a muscle relaxant

Ergotamine
Used in treatment of migraine headaches

Figure 23.4 Some representative alkaloids.

Psychoactive Drugs

In the limited space of this chapter we will focus our attention on some alkaloids that exhibit central nervous system (CNS) activity. Many such compounds are also *psychoactive,* exhibiting effects on mood, emotions, or perception. Not all naturally occurring psychoactive drugs are alkaloids, but a great many are. Moreover, a very large fraction of these compounds share a common structural feature, the β-arylethylamine group, Ar—C—C—N<. This relationship between molecular structure and biological activity suggests a similar mode of action for these different compounds and also indicates that there may be structural similarities in the biological receptor sites responsible for their CNS effects. The β-arylethylamine subunit also appears in a series of compounds that play important roles in the normal biochemistry of the human central nervous system. These compounds are derived from the amino acids phenylalanine and tryptophan.

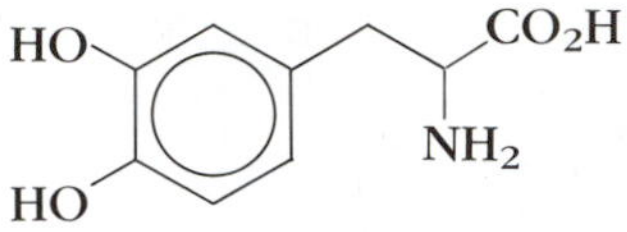

HO, HO, CO_2H, NH_2

DOPA
(dihydroxyphenylalanine)

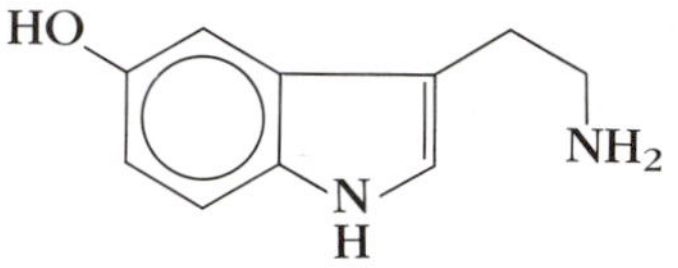

HO, N H, NH_2

Serotonin

(Several additional structures in this series are shown in at the top of Figure 23.5). The aryl groups in these β-arylethylamines are usually substituted derivatives either of benzene or of the bicyclic indole system.

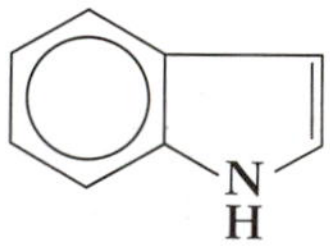

N H

Indole

Figure 23.5 shows some examples of alkaloids that can be classified as psychoactive and which also contain the β-arylethylamine subunit.

Mental Illness

The treatment of mental illness has a history that is filled with unpleasant, even incomprehensible examples of the failure of society to understand the problem. The tremendous medical advances of the twentieth century have provided effective treatment for many physical ailments, but mental illness continues to escape our search for a cure. Prior to the 1800s the fate of the emotionally disturbed was not likely to be pleasant, and imprisonment was a frequent outcome. The more fortunate were sometimes cared for by relatives, but they would likely have been hidden to protect the family from public shame. Even today, with our advances in psychology and in knowledge of human biochemistry, there is still sometimes a stigma associated with mental illness. This is particularly unfortunate, because those who suffer even minor emotional problems may be discouraged from seeking the help they need.

In the United States some progress in the care of the mentally ill was achieved during the mid-1800s. The pioneering efforts of Dorothea Dix led to increased

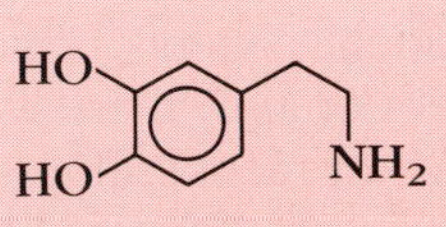

Dopamine

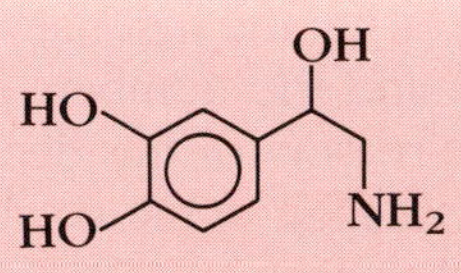

Noradrenalin (norepinephrine)

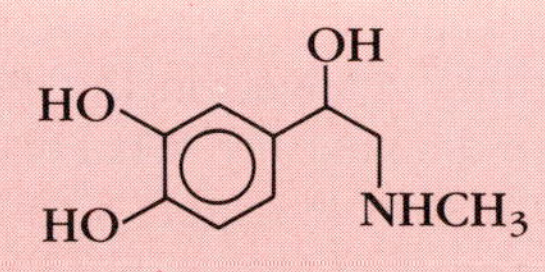

Adrenaline (epinephrine)

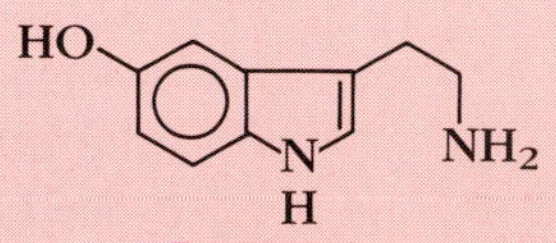

Serotonin

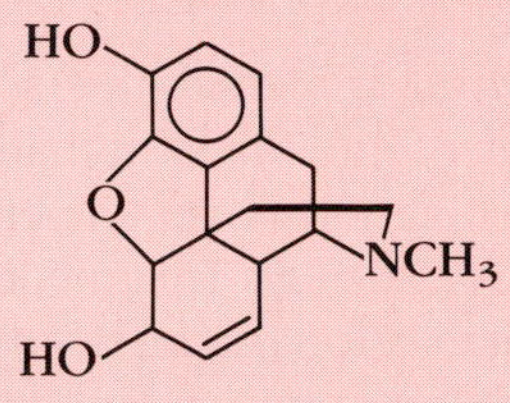

Morphine
Narcotic analgesic

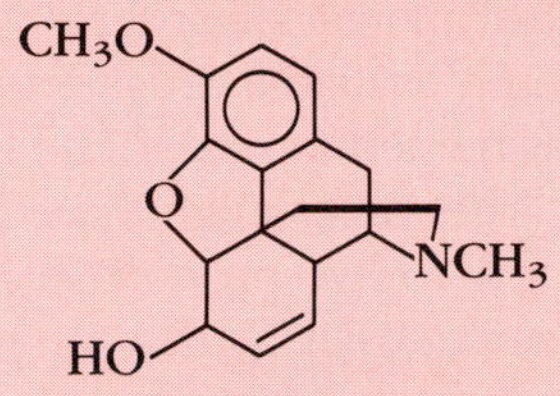

Codeine
Narcotic analgesic

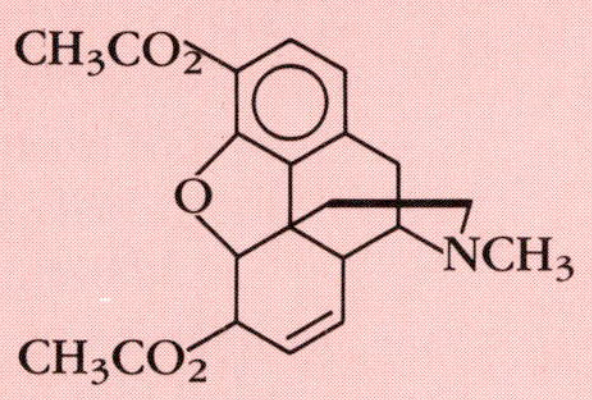

Heroin*
(diacetylmorphine)
Synthetic derivative of morphine; used medically as an analgesic outside the U.S.

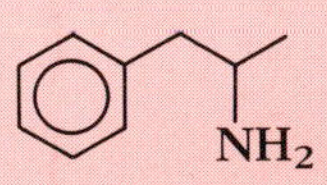

Amphetamine*
(+) is Dexedrine
(−) is Benzedrine
Used as appetite suppressant, stimulant

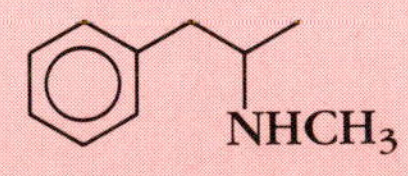

Methamphetamine*
Used as appetite suppressant, stimulant

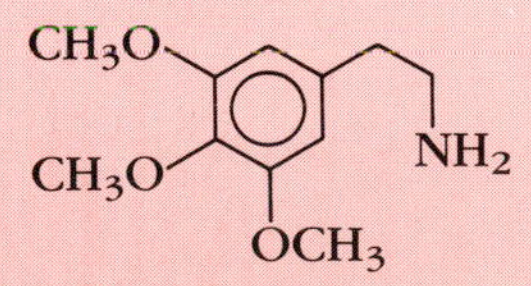

Mescaline
Hallucinogen; active ingredient in peyote

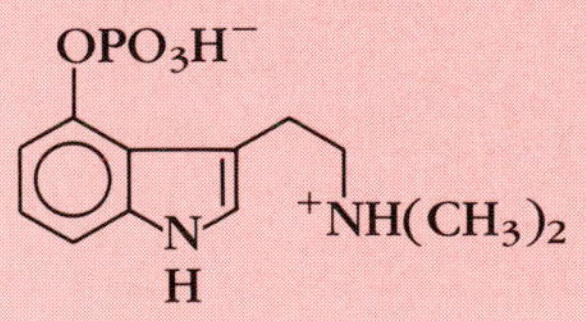

Psilocybin
Hallucinogen; active ingredient of the mushroom *Psilocybe Mexicana*

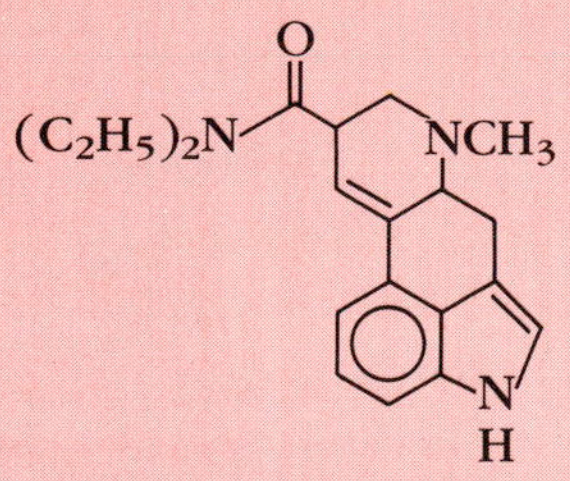

Lysergic acid diethylamide (LSD 25)*
Hallucinogen; synthetic derivative of ergot alkaloids

Figure 23.5 Some psychoactive alkaloids and synthetic (*) derivatives and analogs containing the β-arylethylamine subunit.

construction of mental hospitals, but severe overcrowding frequently permitted only custodial care. Figure 23.6 shows how hospitalization for mental illness continued to grow at a dramatic rate during the first half of the twentieth century. But something changed in the 1950s, and the number of resident patients began to decline dramatically. What was the reason for this decrease? Certainly there would be no reason to expect a sudden change in the occurrence of mental illness. Indeed, the decrease in hospital occupancy results not from a smaller number of patients but from a shorter duration of hospitalization. The total number of patients has continued to increase, but fewer are hospitalized at any one time. This change can be traced directly to the introduction in the 1950s of psychoactive drugs as an aid to the treatment of mental disorders.

None of the presently available drugs provides a *cure* for mental illness, but this type of treatment has helped to ameliorate the effects of the illness in many cases. Individuals who in earlier times might have required long-term or even permanent hospitalization are able to live productive lives in our society. Treatment can be obtained on an outpatient basis, so that time of hospitalization can be decreased or even eliminated. Individuals with less severe disturbances, minor emotional disorders for which hospitalization would not be considered, have also benefited greatly from psychoactive drugs. In many cases their symptoms can be almost completely alleviated with some of the drugs that are now available.

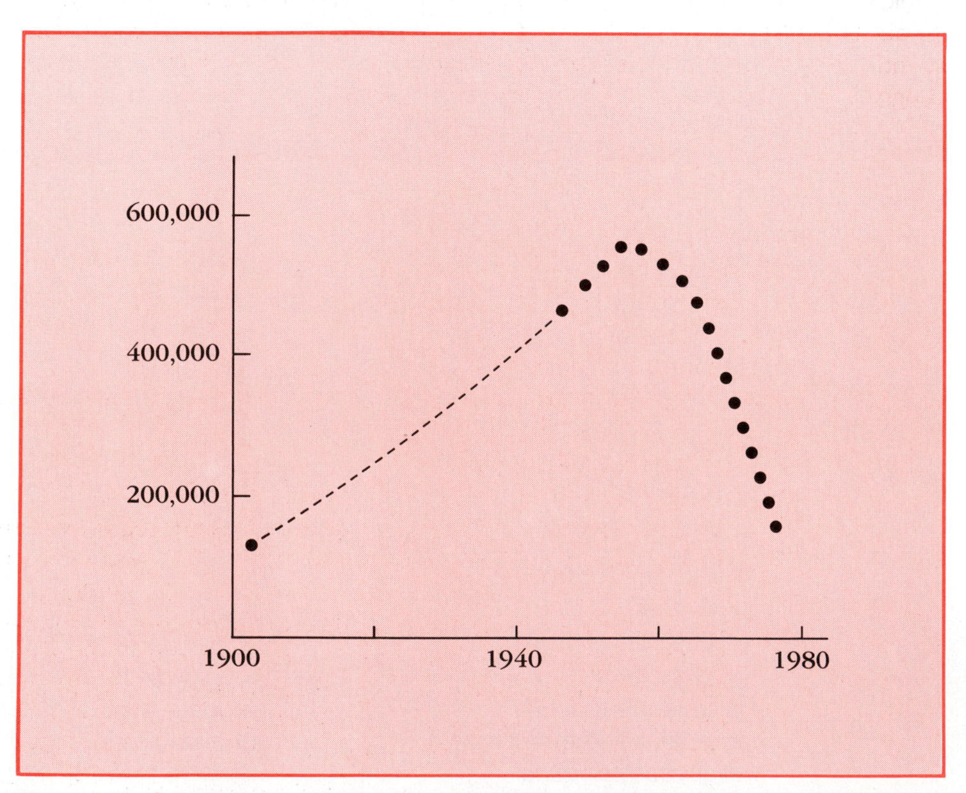

Figure 23.6 Resident patients in mental hospitals in the United States. The number of resident patients increased steadily through the first half of the twentieth century, but a dramatic decrease began in the mid-1950s, when psychoactive drugs were introduced.

Two of the first drugs used to treat mental illness were *chlorpromazine* and *reserpine*. The former is a totally synthetic material, but reserpine is an alkaloid isolated from the flowering plant *Rauwolfia,* which grows in India and Africa.

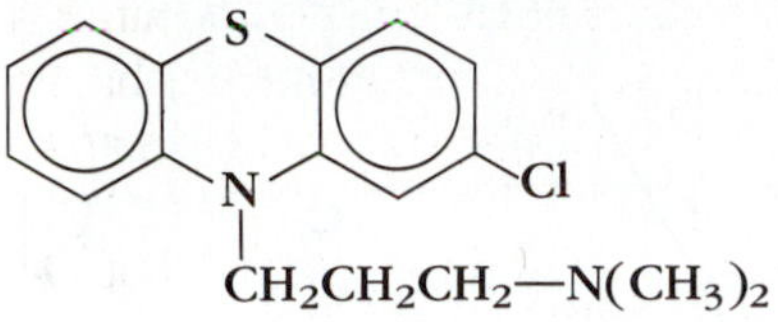

chlorpromazine

(Thorazine®): a phenothiazine derivative, used as a tranquilizer; also used as a sedative and as an antiemetic

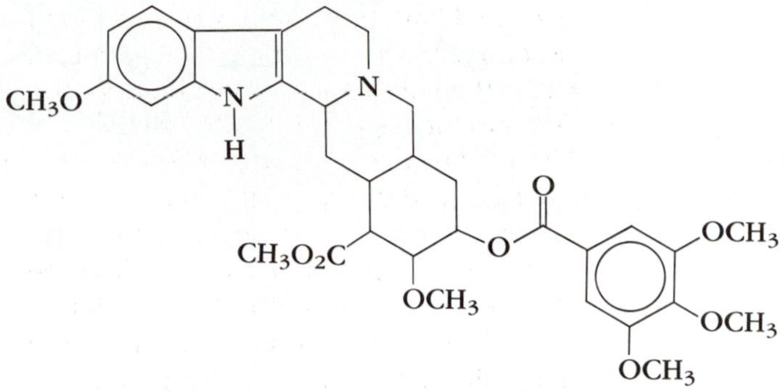

Reserpine: a tranquilizer

Many other psychoactive drugs have been developed for treating mental illness and minor emotional disorders. A complete discussion is beyond the scope of this book, but we would be remiss if we did not present the structure of the benzodiazepine derivative *diazepam*. Marketed under the name *Valium®* as an antianxiety agent, it has been until recently the most widely prescribed of all prescription drugs.

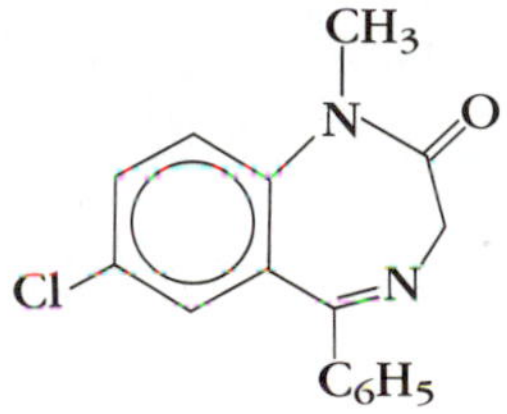

Diazepam

Example 23.9 Synthesis of Reserpine

The discovery of reserpine as a valuable therapeutic agent in the 1950s led to a period of intense chemical investigation. Despite the enormous structural complexity of the reserpine molecule, synthetic efforts were undertaken in order to prepare analogs as well as the natural alkaloid. The synthesis we show here was carried out by R. B. Woodward and his collaborators at Harvard University, and it is noteworthy for two reasons. First, the synthesis (after relatively minor modification) was so well designed that synthetic reserpine could be produced at a cost competitive with extraction of the natural material from the *Rauwolfia* plant. Second, the techniques developed by Woodward for controlling stereochemistry in this synthesis represent a milestone in organic chemistry. When Woodward received by Nobel Prize in chemistry in 1965, the award was considered by many to merely confirm the view that he was the most outstanding organic chemist of our century. The synthesis was designed to follow the general plan outlined in Figure 23.7.

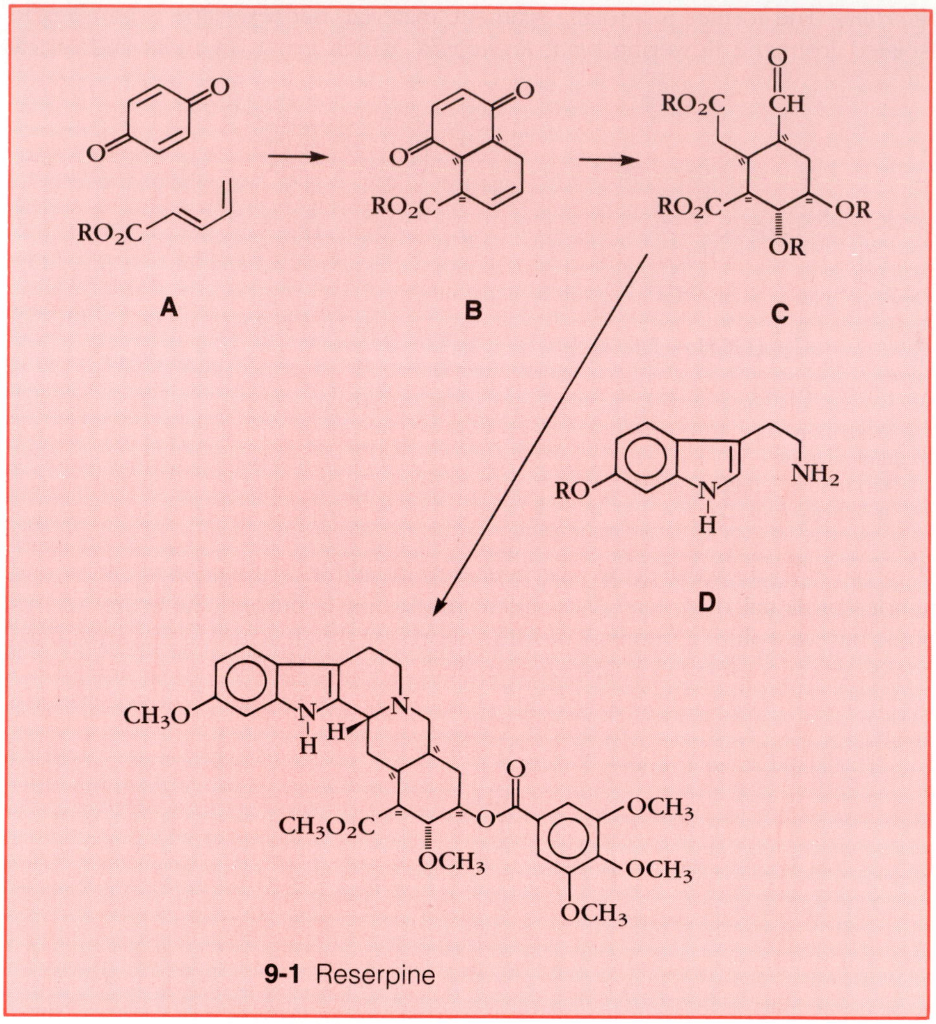

9-1 Reserpine

Figure 23.7 General plan for the synthesis of reserpine (9-1).

The portion of the molecule designated as fragment **C** in Figure 23.7, contains the greatest stereochemical complexity (five adjacent chiral centers); this structure was the first synthetic target. The plan was to generate the six-membered ring with a Diels-Alder reaction as a template. The β-arylethylamine (**D**) would then be introduced at a later stage to complete the synthesis.

The first reaction of the synthesis generated three of the six chiral centers that would eventually be needed. The Diels-Alder reaction is highly stereospecific, so the observed stereochemistry of adduct **9-4** was anticipated from the *Z* and *E* configurations of the double bonds in **9-2** and **9-3,** respectively (the stereochemistry of this specific reaction was discussed previously in Section 22.3). Of course, both enantiomers of **9-4** were produced, but we have only shown the one corresponding to natural reserpine.

The double bond in the lower ring of **9-4** provided the opportunity for introducing the necessary oxygen substituents via epoxidation. It was known, however, that the wrong stereochemistry would be obtained if direct epoxidation and ring opening were attempted. Nucleophilic attack on epoxide **E** would lead to **F,** with the attacking nucleophile and the OH group occupying positions on the six-membered ring that are *trans* and diaxial. The desired stereochemistry has the nucleophile *trans* to the carboxyl group, but in **F** the relationship is *cis.*

In order to obtain the correct stereochemistry, compound **9-4** was modified so that it could be forced into a less favorable conformation. When **9-4** was esterified and then reduced with aluminum isopropoxide and isopropyl alcohol (Example 22.2), both keto groups were reduced and the five-membered-ring lactone **9-6** was formed. When this lactone was treated with bromine, the cationic intermediate underwent intramolecular attack by the free OH group to produce the bridged ether **9-7.**

The following three-dimensional representations show how the last reaction proceeds. The lactone ring locks the molecule into a conformation that holds the free OH group above the double bond of the lower ring. Cyclization to a five-membered-ring ether then occurs readily once the double bond is attacked by bromine.

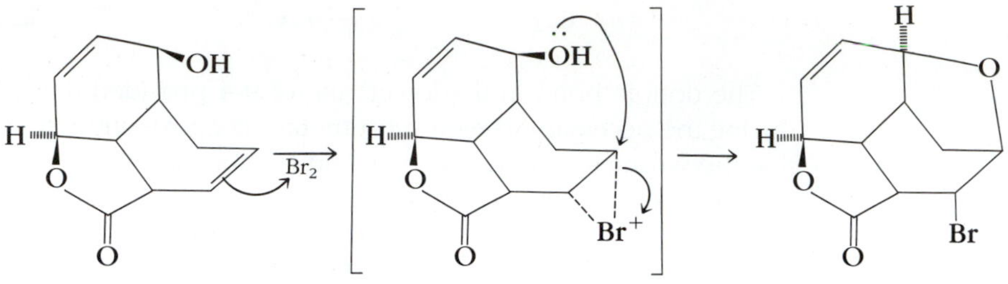

9–6 9–7

Treatment of **9-7** with sodium methoxide next produced the corresponding methyl ether with *retention* of configuration. Note, however, that this was *not* a nucleophilic displacement. Since **9-7** is a β-bromocarbonyl compound, the base rapidly caused elimination to generate an unsaturated lactone (**9-8**), which then underwent conjugate addition of methanol. The overall retention of stereochemistry was a simple consequence of attack from the less hindered side.

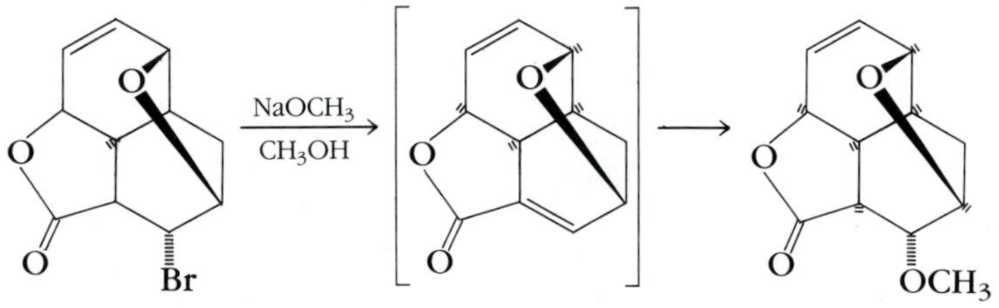

9–7 9–8 9–9

The formation of methyl ether **9-9** completed the first phase of the synthesis because the lower six-membered ring contained all five chiral centers with the correct configurations. The next phase was cleavage of the upper six-membered ring, together with removal of an "extra" carbon atom, to generate the key intermediate, **C.** This process was begun by reaction of **9-9** with aqueous *N*-bromosuccinimide to add Br and OH to the double bond. Tremendous structural simplification resulted when **9-10** was first oxidized with chromium trioxide and next reduced with zinc in acetic acid to yield the bicyclic product **9-12.**

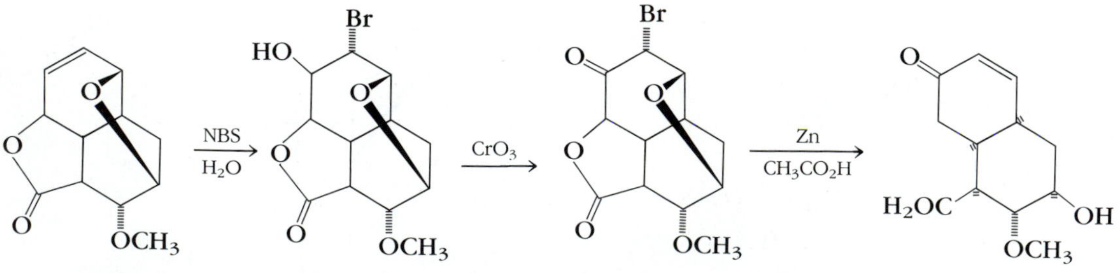

9–9 9–10 9–11 9–12

The zinc reduction of **9-11** is an example of the reductive eliminations that we discussed in Example 22.7, and two such eliminations actually occur. The following equation summarizes the overall process:

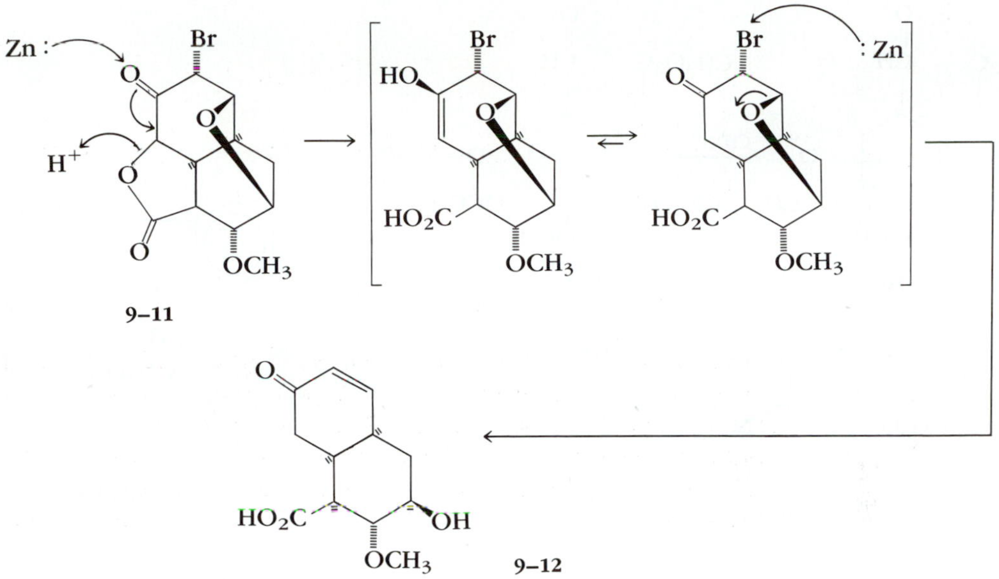

9–11 → [⇌] → **9–12**

After esterification of the carboxyl group of **9-12** and protection of the alcohol as an acetate, the upper six-membered ring was cleaved. The acetate ester **9-14** was oxidized with osmium tetroxide, and the resulting diol (**9-15**) was treated with periodic acid. This generated **9-16** via cleavage of the bonds between the two OH groups and between the ketone carbonyl and the adjacent OH groups.

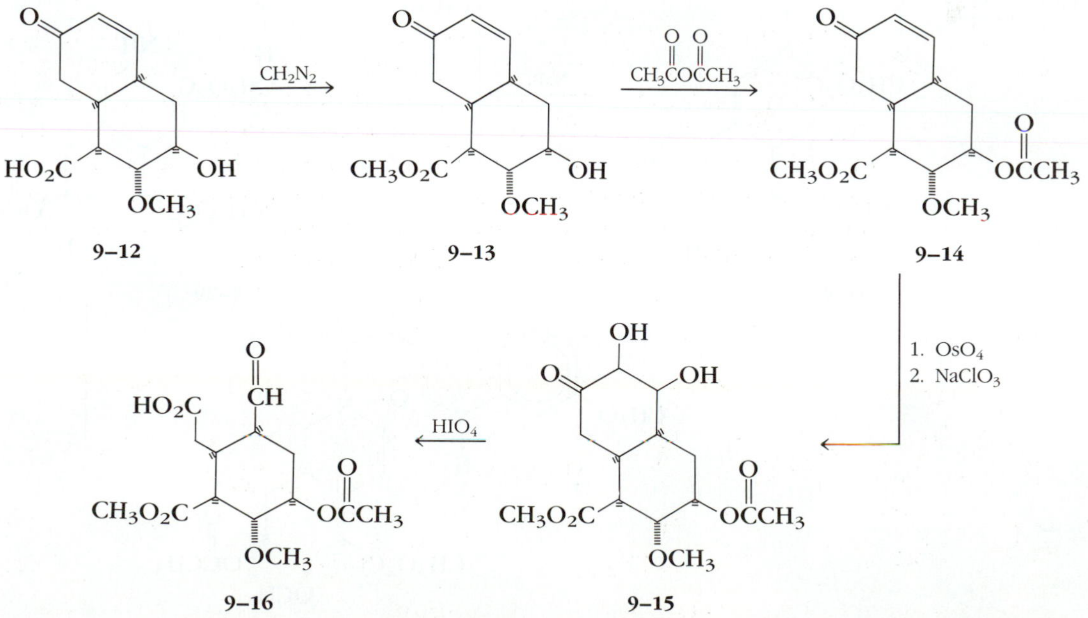

9–12 $\xrightarrow{CH_2N_2}$ **9–13** $\xrightarrow{CH_3COCCH_3}$ **9–14** $\xrightarrow{1.\ OsO_4;\ 2.\ NaClO_3}$ **9–15** $\xrightarrow{HIO_4}$ **9–16**

With polyfunctional compound **9-16** in hand, the second phase of the synthesis had been completed. This intermediate was esterified with diazomethane, and reaction of the resulting ester **9-17** with the indole derivative **9-18** produced the imine **9-19,** which contains all the carbon atoms of the reserpine skeleton.

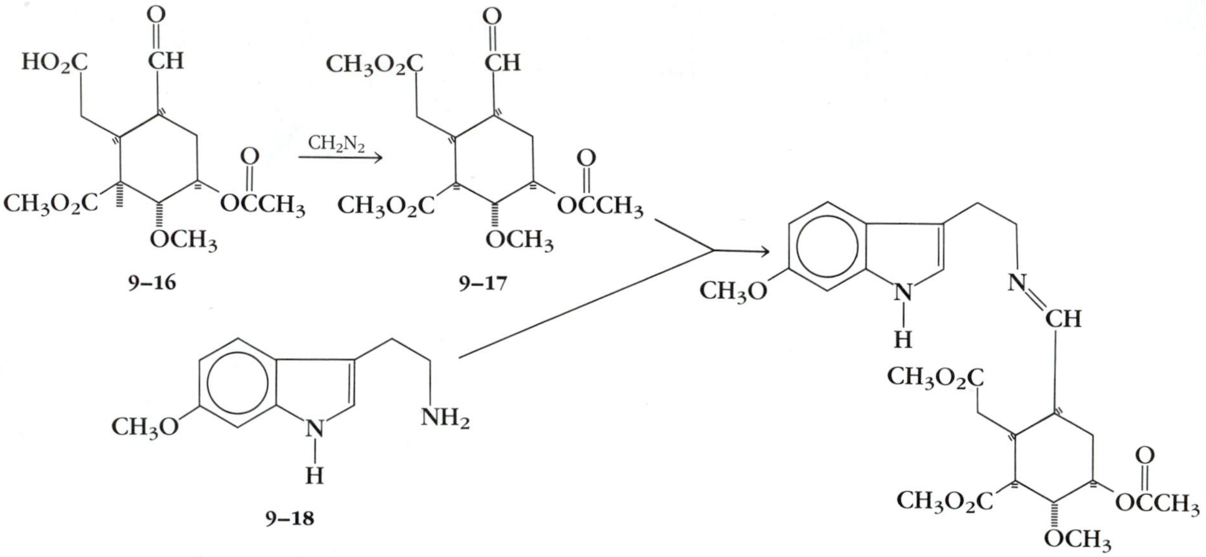

CH_2N_2

9–16 **9–17** **9–18** **9–19**

When the carbon-nitrogen double bond of **9-19** was reduced with sodium borohydride and the product was gently heated, intramolecular amide formation took place, yielding **9-21.**

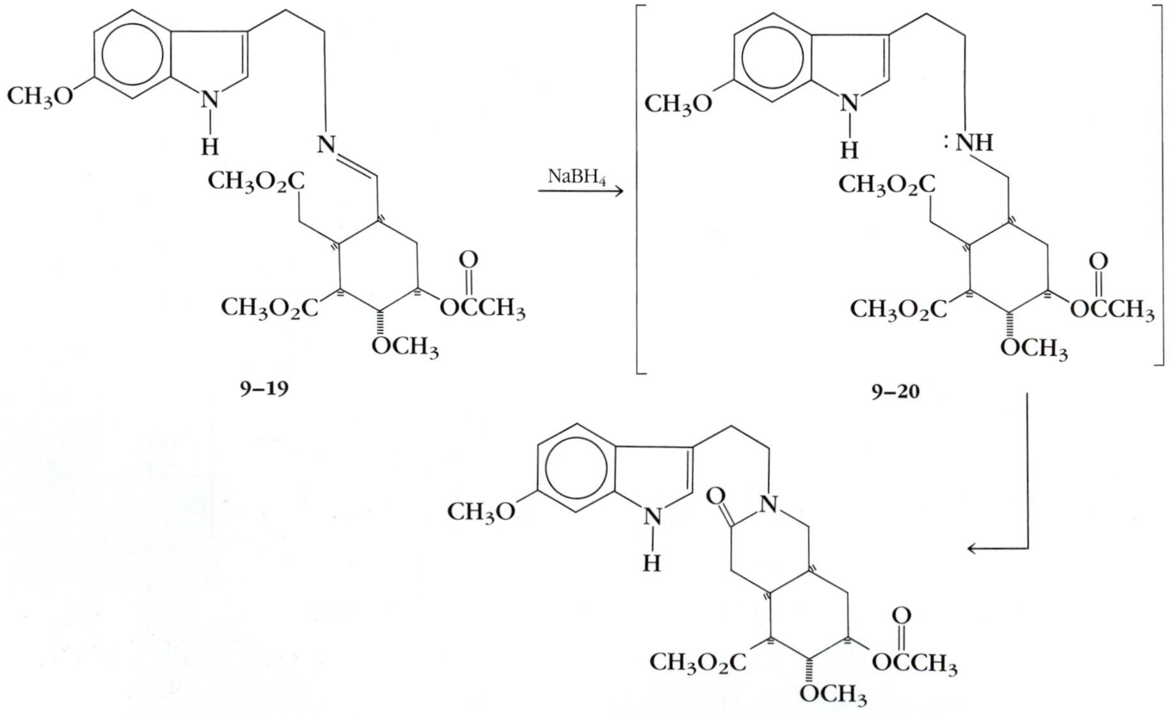

$NaBH_4$

9–19 **9–20** **9–21**

Lactam **9-21** was next subjected to an intramolecular Friedel-Crafts–type acylation with phosphorus oxychloride as the catalyst. The cyclization proceeded by the following mechanism (where —Ⓟ is used to indicate a phosphonyl group, such as —$POCl_2$):

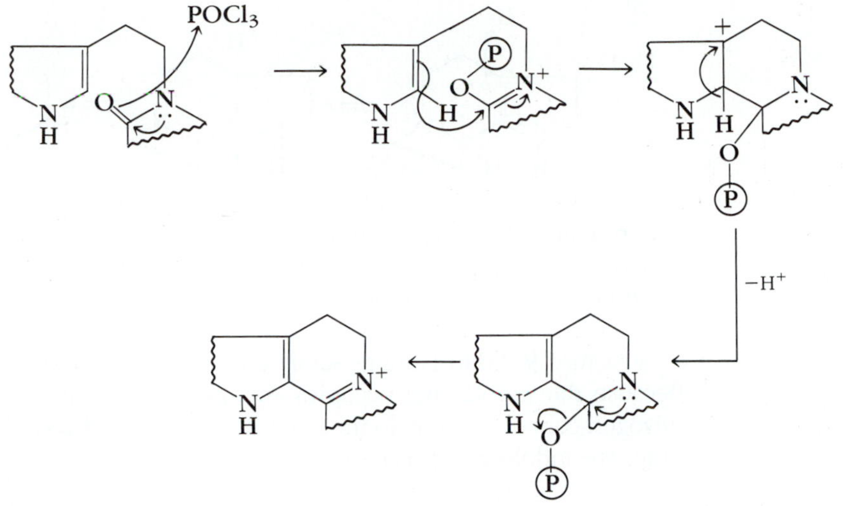

$POCl_3$

$-H^+$

When the resulting iminium salt (**9-22**) was reduced with sodium borohydride, it was hoped that the product would be identical with racemic reserpine except for the acetate (in place of a trimethoxybenzoate) on the bottom ring. Regrettably, there was also another difference: the last reduction had created the final chiral center (C-3) with the *wrong stereochemistry*.

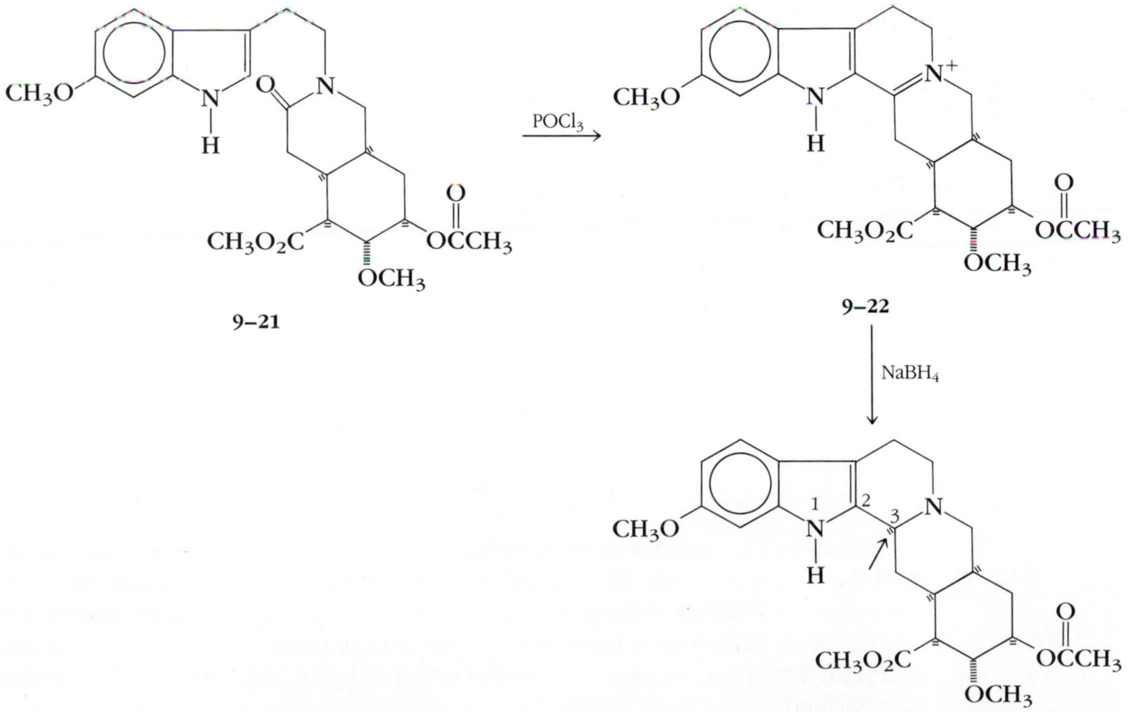

CH_3O CH_3O_2C OCH_3 $OCCH_3$

9–21

$POCl_3$

9–22

$NaBH_4$

9–23iso

The formation of **9-23iso** was a disappointment but not a disaster, because it was known that epimerization at C-3 (see structure **9-23iso**) occurs in acid. The isomerization occurs through reversible loss of the hydrogen at C-3 by a mechanism such as the following:

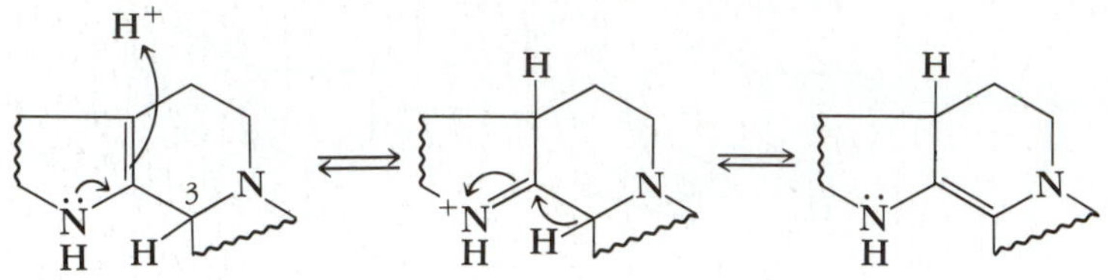

H^+ N H H 3 N ⇌ H $^+$N H H N ⇌ H N H N

Unfortunately the equilibrium favors **9-23iso** rather than the stereoisomer corresponding to reserpine. But Woodward was not without his resources, and once again he turned to the device of locking a molecule in an unfavorable conformation.

Structure **9-23iso** is more stable than the isomer corresponding to reserpine because the indole ring occupies an *equatorial* position on the six-membered nitrogen-containing ring. In the epimer (designated **9-23**) corresponding to reserpine, the indole group is *axial.*

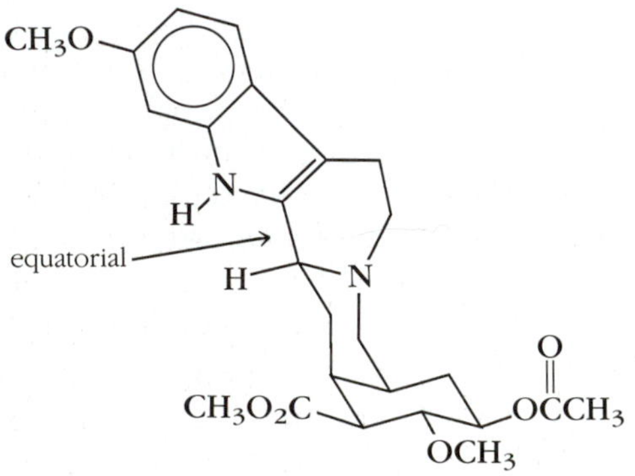

CH_3O N H equatorial H N CH_3O_2C O $OCCH_3$ OCH_3

9–23iso

H CH_3O N H N axial CH_3O_2C O $OCCH_3$ OCH_3

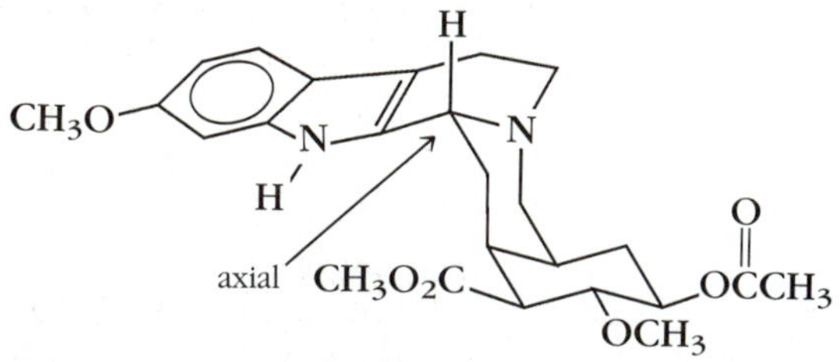

9–23

In order to reverse the relative stabilities of the C-3 epimers, Woodward modified the substituents on the bottom ring of **9-23iso** to produce a lactone. Ester hydrolysis of **9-23iso** with aqueous potassium hydroxide followed by lactone formation using dehydrating agent dicyclohexylcarbodiimide (Section 20.4) produced **9-25iso.** When this lactone was treated with mild acid, the "correct" C-3 epimer was formed.

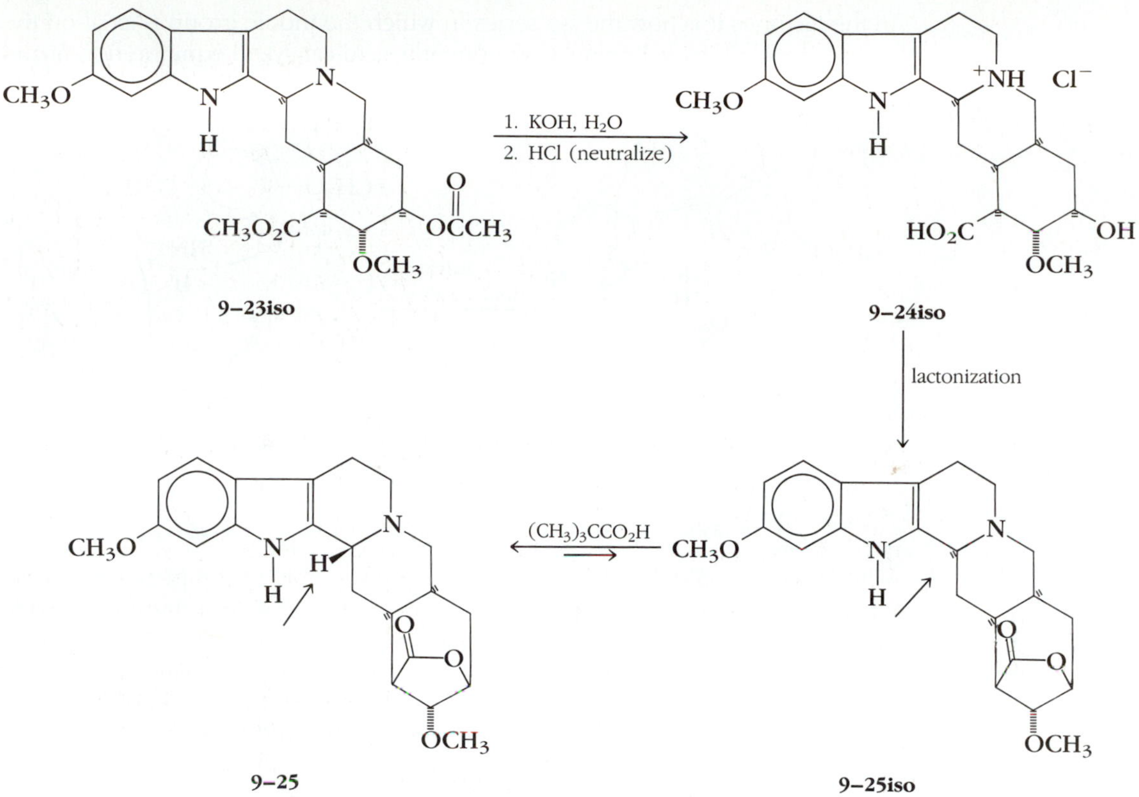

CH_3O N H N CH_3O_2C OCH_3 O $OCCH_3$

1. KOH, H_2O
2. HCl (neutralize)

CH_3O N H ^{+}NH Cl^- HO_2C OCH_3 OH

9–23iso **9–24iso**

lactonization

CH_3O N H H N O O OCH_3

$(CH_3)_3CCO_2H$

CH_3O N H N O O OCH_3

9–25 **9–25iso**

Why does the lactone ring cause a change in the preferred configuration at C-3? Inspection of the conformational drawing of **9-24iso** shows that both the —CO_2CH_3 and —O_2CCH_3 groups are equatorial. In order to form the lactone, the molecule must adopt a second (less stable) conformation in which these groups are both axial. (The new conformation is drawn from a slightly different perspective in order to best show its structural features.)

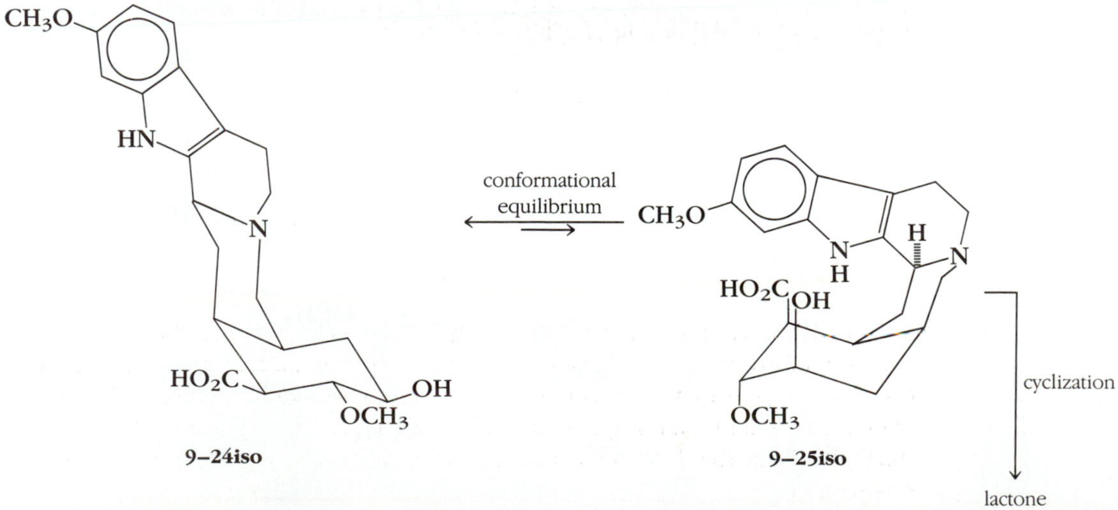

CH_3O HN N HO_2C OH OCH_3

conformational equilibrium

CH_3O N H H N HO_2C OH OCH_3

9–24iso **9–25iso**

cyclization

lactone

In the lactones it is now the iso series in which the indole group is axial on the six-membered heterocyclic ring. Consequently, acid-catalyzed equilibration favors the stereoisomer corresponding to reserpine.

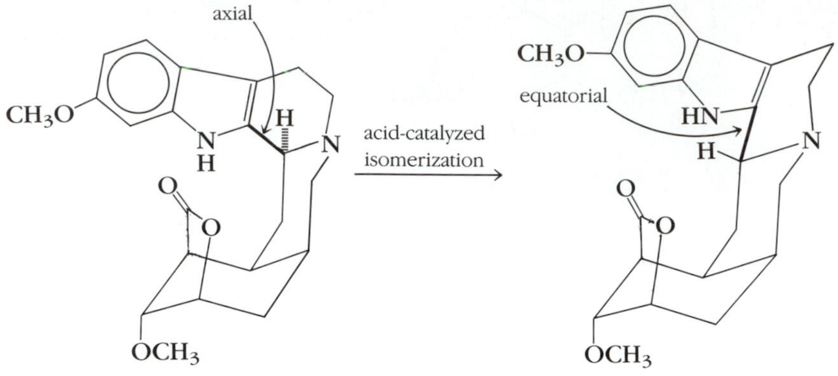

axial

CH_3O N H H N O O OCH_3

acid-catalyzed isomerization →

CH_3O equatorial HN H N O O OCH_3

9–25iso **9–25**

With lactone **9-25** in hand, Woodward's group was able to complete the synthesis of reserpine in only two steps. Cleavage of the lactone with sodium methoxide in methanol afforded the ester-alcohol **9-26,** and acylation with the acid chloride of 3,4,5-trimethoxybenzoic acid yielded racemic reserpine (**9-1**). Finally, the racemic material was resolved by using a sulfonic acid derived from (+)-camphor to give (−)-reserpine that was chemically indistinguishable from the natural alkaloid. Reserpine prepared by the synthesis we have just shown (with only minor modification) has been a valuable and helpful medication for a great many individuals over the last 30 years.

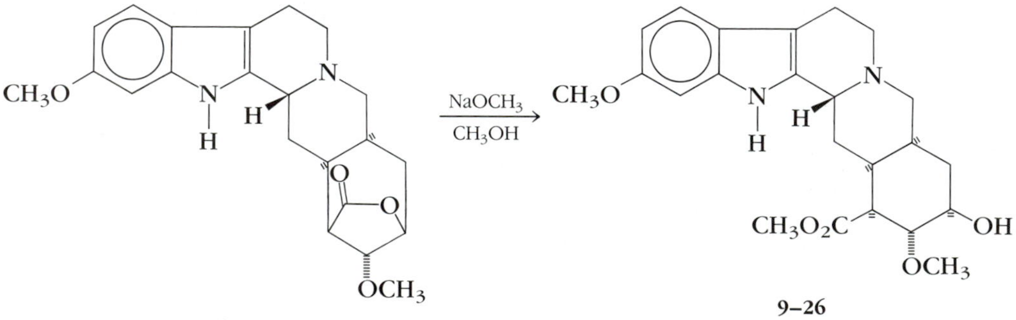

CH_3O N H H N O O OCH_3

9–25

$NaOCH_3$ / CH_3OH →

CH_3O N H H N CH_3O_2C OH OCH_3

9–26

O C Cl OCH_3 OCH_3 OCH_3

CH_3O N H H N CH_3O_2C OCH_3 O O OCH_3 OCH_3 OCH_3

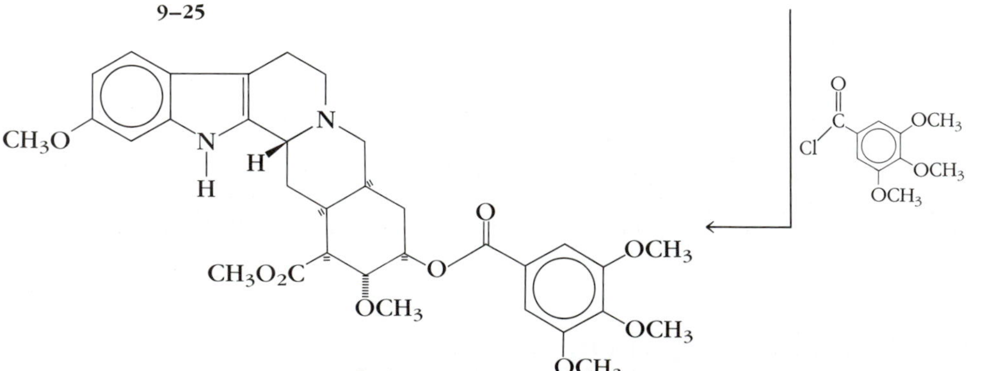

9–1

EXERCISE 23.20

(a) Explain why epoxidation of **9-4** occurs at the double bond of the lower ring rather than at the ring that was originally the quinone. (b) Using three-dimensional drawings, explain why **9-4** is preferentially epoxidized from the "bottom" (i.e., to give the stereochemistry shown in structure **E**).

EXERCISE 23.21

Write a balanced equation for the reduction of **9-5.** What species is actually oxidized?

EXERCISE 23.22

Reduction of the two carbonyl groups of **9-5** gave the possibility of forming two different lactones (one of which is **9-6**). (a) Draw the other possibility. (b) The actual product had a carbonyl peak in the infrared spectrum at approximately 1760 cm^{-1}. Does this support structure **9-6** or the other possibility?

EXERCISE 23.23

Draw the structure of **9-8** in a three-dimensional representation. Why does attack of methoxide ion lead to the stereochemistry shown for **9-9?**

EXERCISE 23.24

The periodate oxidation of **9-15** leads to **9-16.** What is the fate of the missing carbon atom? How many moles of periodate are consumed by each mole of **9-15?**

EXERCISE 23.25

Compound **9-20** has *two* secondary amino groups and *three* ester groups, yet cyclization leads to **9-21** only. (a) Why does only one of the amino groups react? (b) Draw the two products that would be formed by attack of the reactive amino group at the other two esters. (c) Explain why the products from (b) are not formed.

EXERCISE 23.26

Explain the conformational preference of **9-24iso.** How many axial substituents (total) can you find for each conformer?

23.5 PROBLEMS

23.1 Complete the missing portions of the following synthesis of thujopsene, a terpene isolated from wood oil of the Japanese Hiba tree.

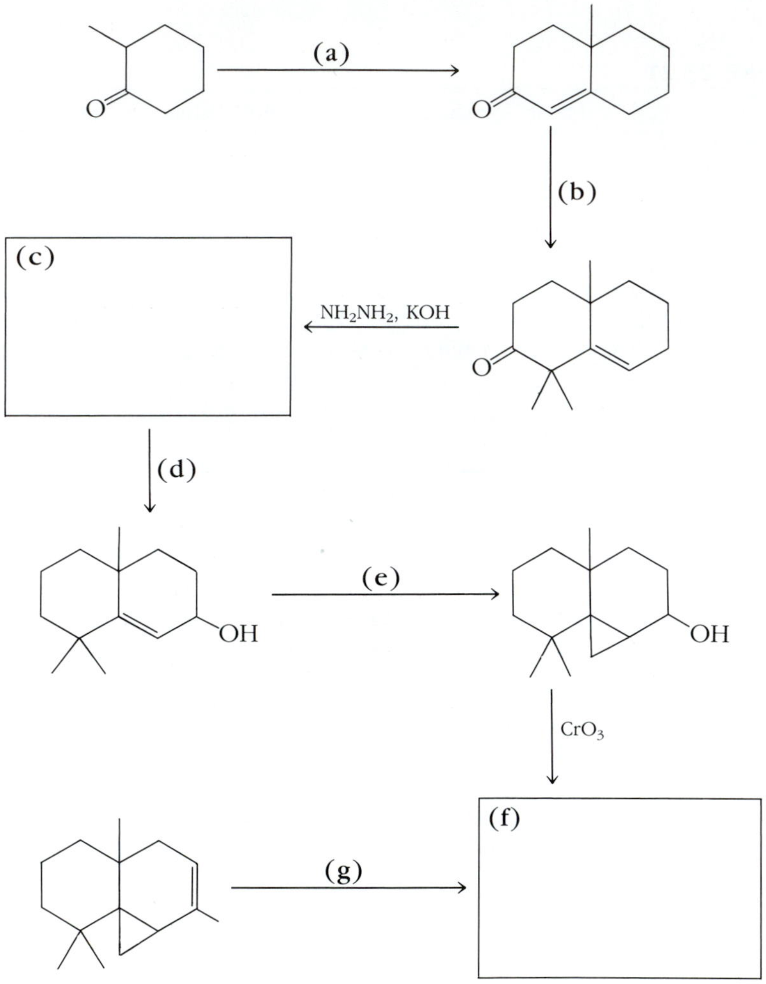

(a)

(b)

(c)

NH_2NH_2, KOH

(d)

OH

(e)

OH

CrO_3

(f)

(g)

23.2 Complete the missing portions of the following synthesis of the bicyclic terpenoid nootkatone.

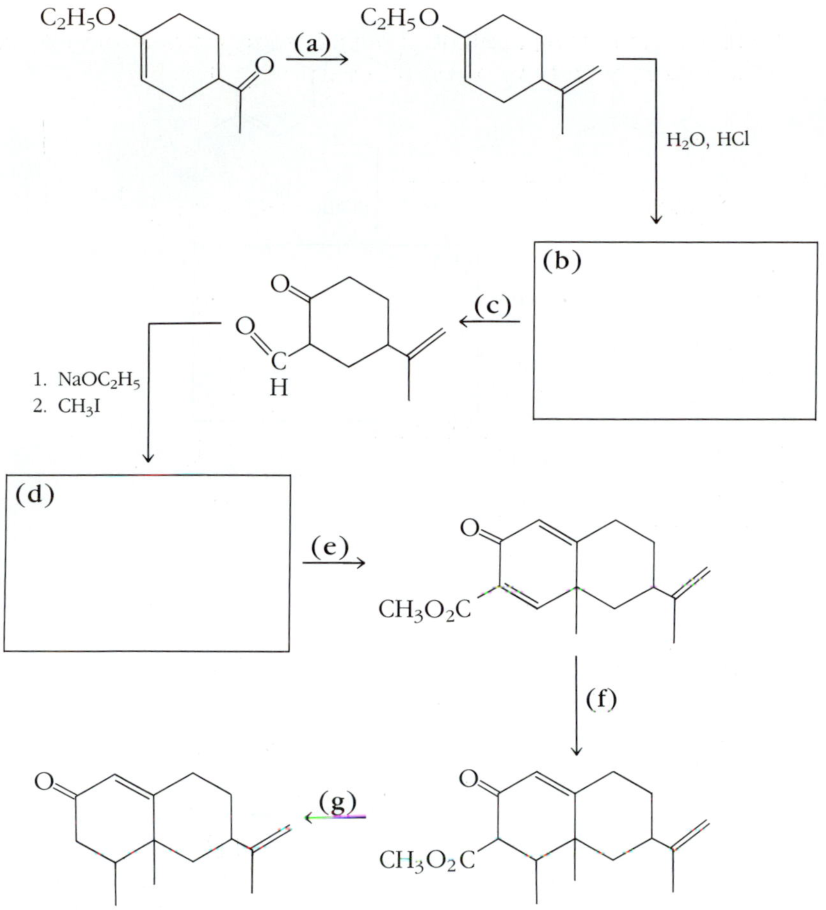

C_2H_5O (a) C_2H_5O

H_2O, HCl

(b)

(c)

O C H

1. $NaOC_2H_5$
2. CH_3I

(d)

(e)

O

CH_3O_2C

(f)

O (g) O

CH_3O_2C

23.3 Complete the missing portions of the following synthesis of β-bisabolene, a terpene that is formed by cyclization of farnesyl pyrophosphate.

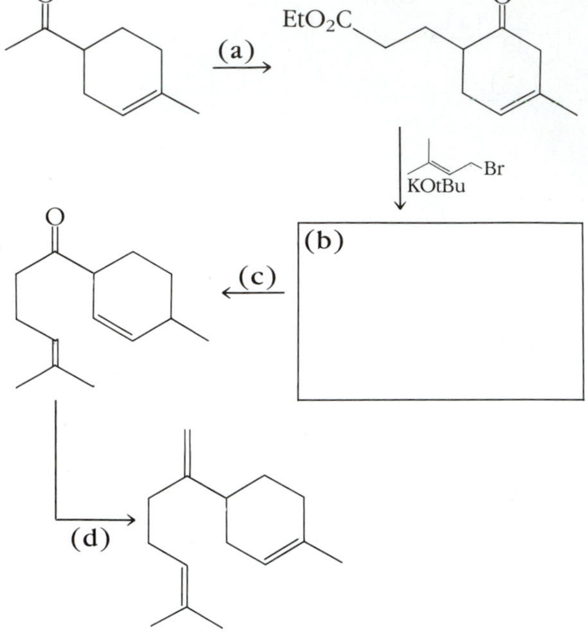

(a)

EtO_2C

Br
KOtBu

(b)

(c)

(d)

23.4 Complete the missing portions of the following synthesis of clovene, a rearrangement product of the terpene caryophyllene.

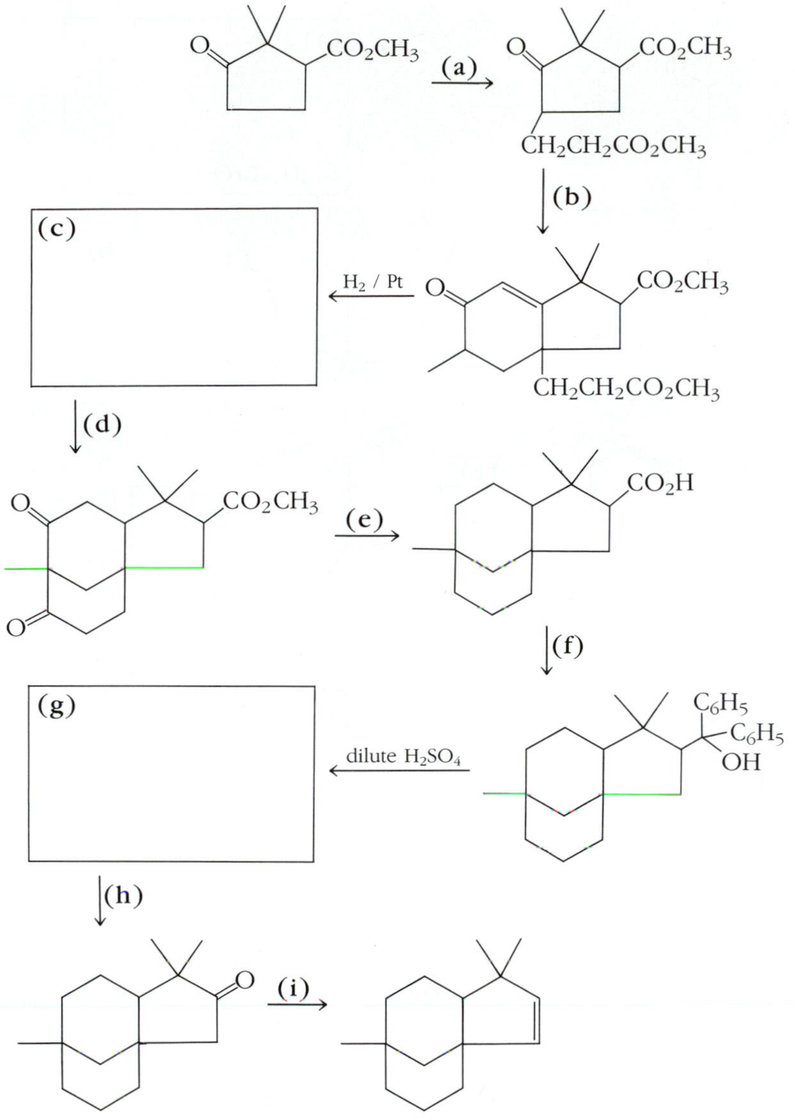

CO_2CH_3 (a) → CO_2CH_3 $CH_2CH_2CO_2CH_3$

(b)

(c) ← H_2 / Pt O CO_2CH_3 $CH_2CH_2CO_2CH_3$

(d)

O CO_2CH_3 O (e) → CO_2H

(f)

(g) ← dilute H_2SO_4 C_6H_5 C_6H_5 OH

(h)

O (i) →

clovene

23.5 Complete the missing portions of the following synthesis of aromadendrene, a tricyclic terpene that contains a cyclopropane ring.

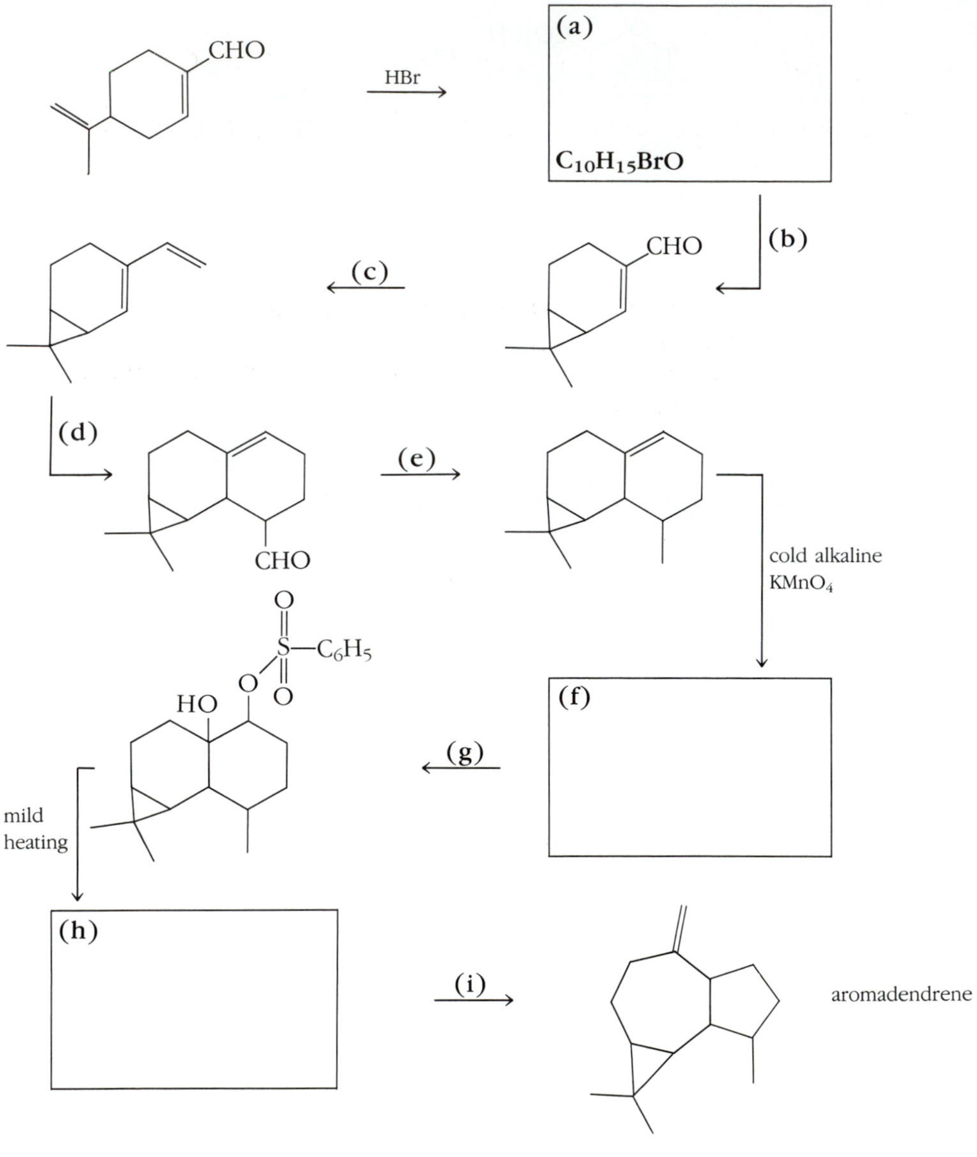

CHO

HBr

(a)

$C_{10}H_{15}BrO$

(b)

CHO

(c)

(d)

CHO

(e)

cold alkaline $KMnO_4$

O

S—C_6H_5

O

O

HO

(f)

(g)

mild heating

(h)

(i)

aromadendrene

23.6 β-Eudesmol (structure **C**) is a bicyclic terpenoid. Its synthesis was accomplished by starting with compound **A**, which was converted to the key intermediate **B.**

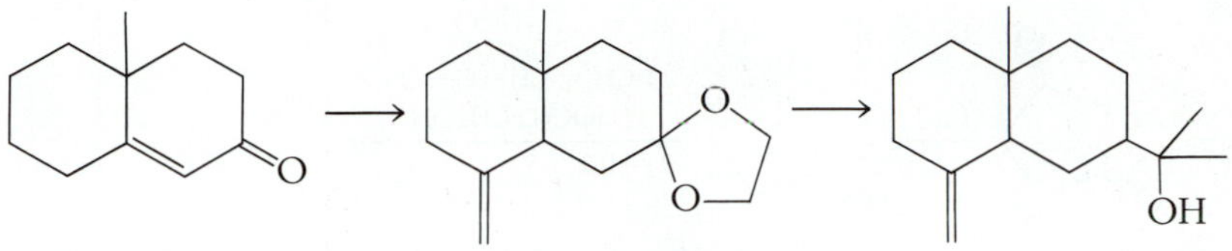

O O O OH

A **B** **C**

Suggest a series of reactions that would effect the appropriate conversion of (a) **A** to **B** and (b) **B** to **C.**

You should draw the structures of compounds that are intermediates in your sequence of reactions.

23.7 Complete the missing portions of the following synthesis of the steroid derivative retroprogesterone.

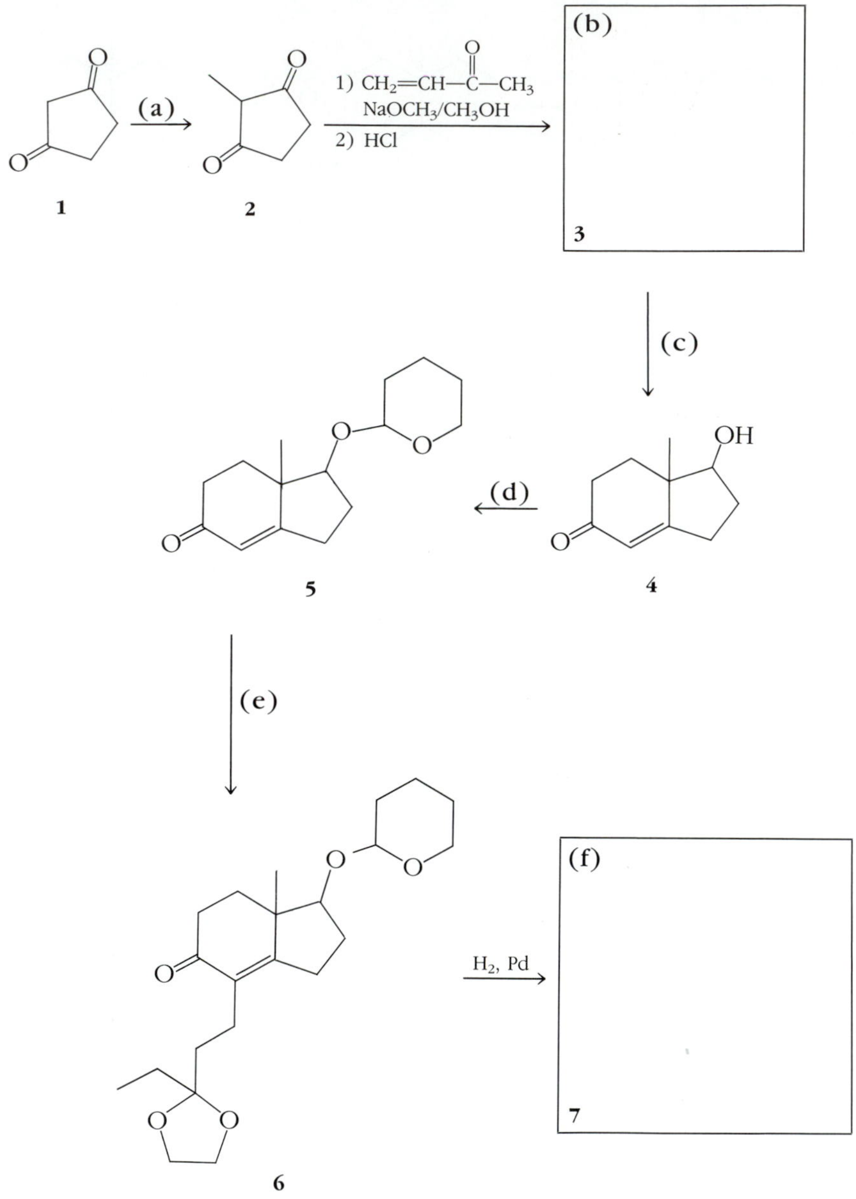

(a)

1

2

1) $CH_2{=}CH{-}\overset{O}{\overset{\|}{C}}{-}CH_3$
NaOCH$_3$/CH$_3$OH
2) HCl

(b)

3

(c)

OH

(d)

5

4

(e)

6

H_2, Pd

(f)

7

Problem 23.7, continued

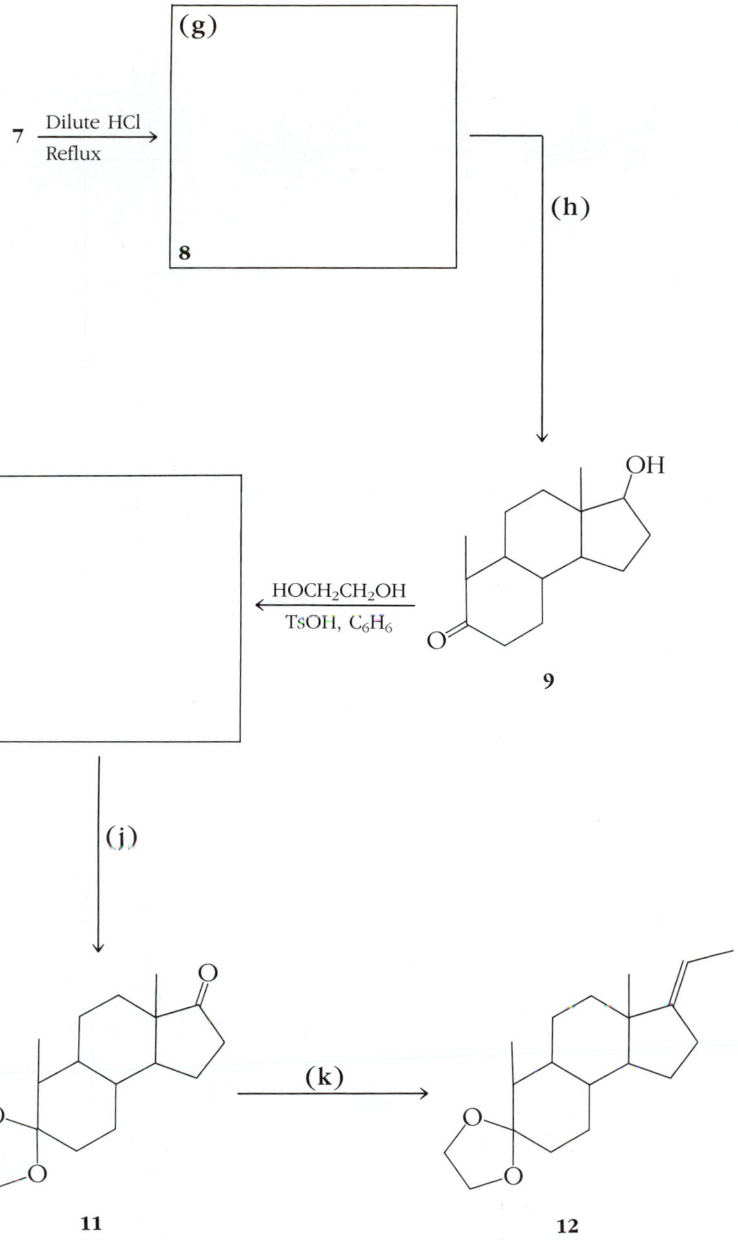

(g)

7 $\xrightarrow[\text{Reflux}]{\text{Dilute HCl}}$ 8

(h)

OH

O

9

(i)

$\xleftarrow[\text{TsOH, } C_6H_6]{HOCH_2CH_2OH}$

10

(j)

O

O O

11

(k)

O O

12

Problem 23.7, continued

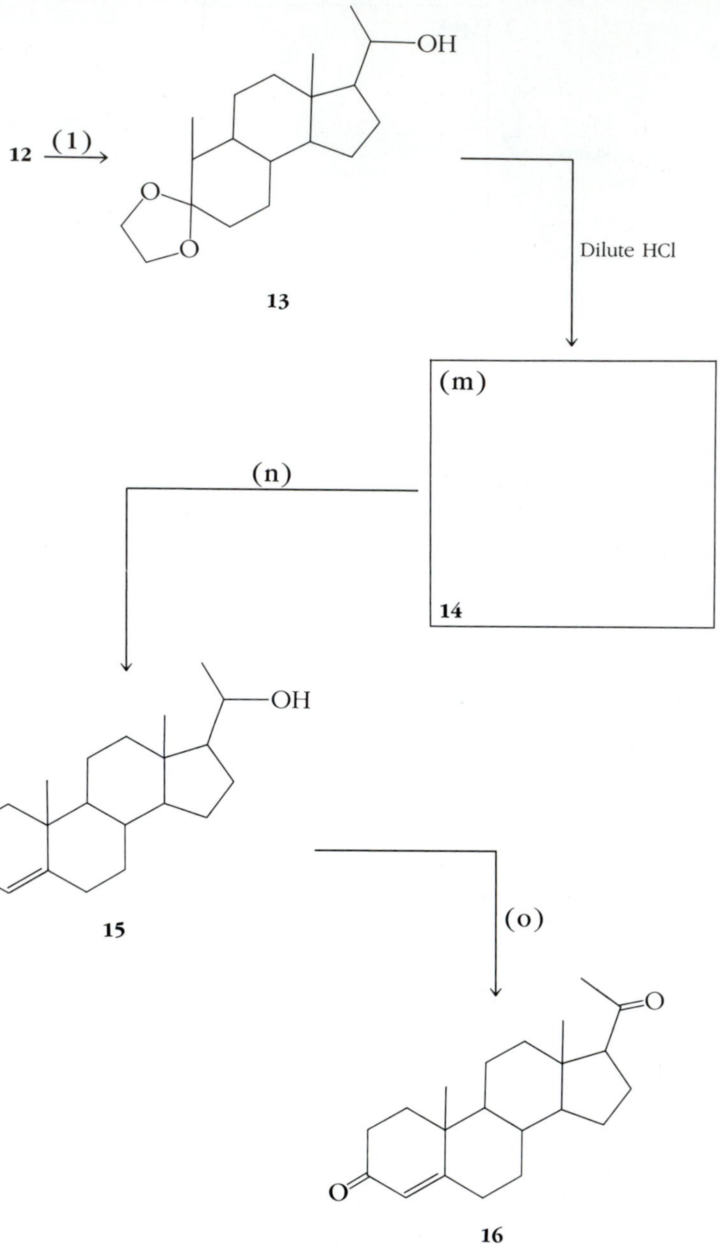

12 —(l)→ 13

OH

O

O

13

Dilute HCl

(m)

14

(n)

OH

O

15

(o)

O

O

16

23.8 Complete the missing portions of the following synthesis of estrone methyl ether:

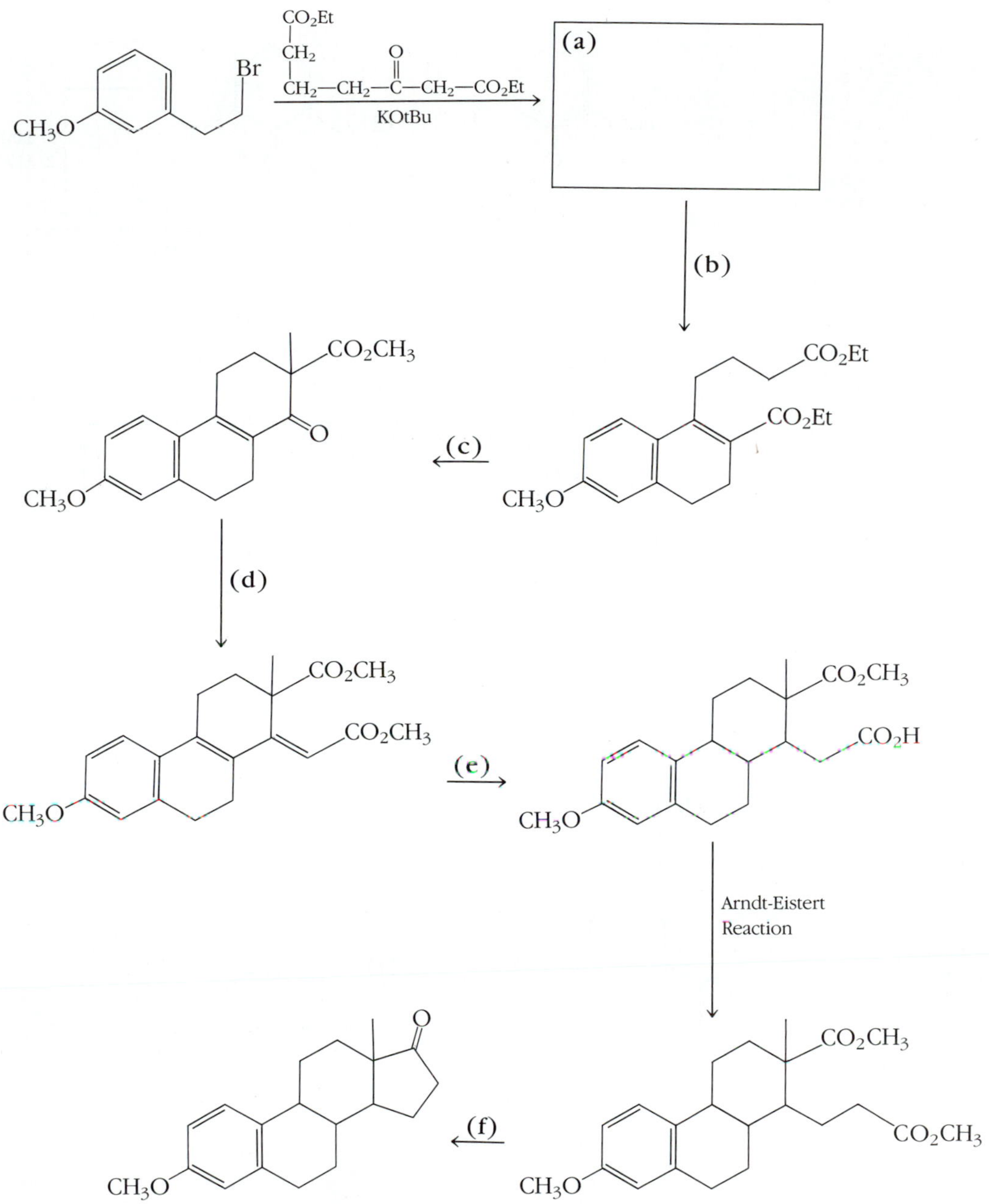

CH$_3$O Br

CO$_2$Et | CH$_2$ | CH$_2$—CH$_2$—C(=O)—CH$_2$—CO$_2$Et

KOtBu

(a)

(b)

CO$_2$Et

CO$_2$Et

CH$_3$O

(c)

CO$_2$CH$_3$

O

CH$_3$O

(d)

CO$_2$CH$_3$

CO$_2$CH$_3$

CH$_3$O

(e)

CO$_2$CH$_3$

CO$_2$H

CH$_3$O

Arndt-Eistert Reaction

CO$_2$CH$_3$

CO$_2$CH$_3$

CH$_3$O

(f)

O

CH$_3$O

23.9 Complete the missing portions of the following synthesis of cholesterol:

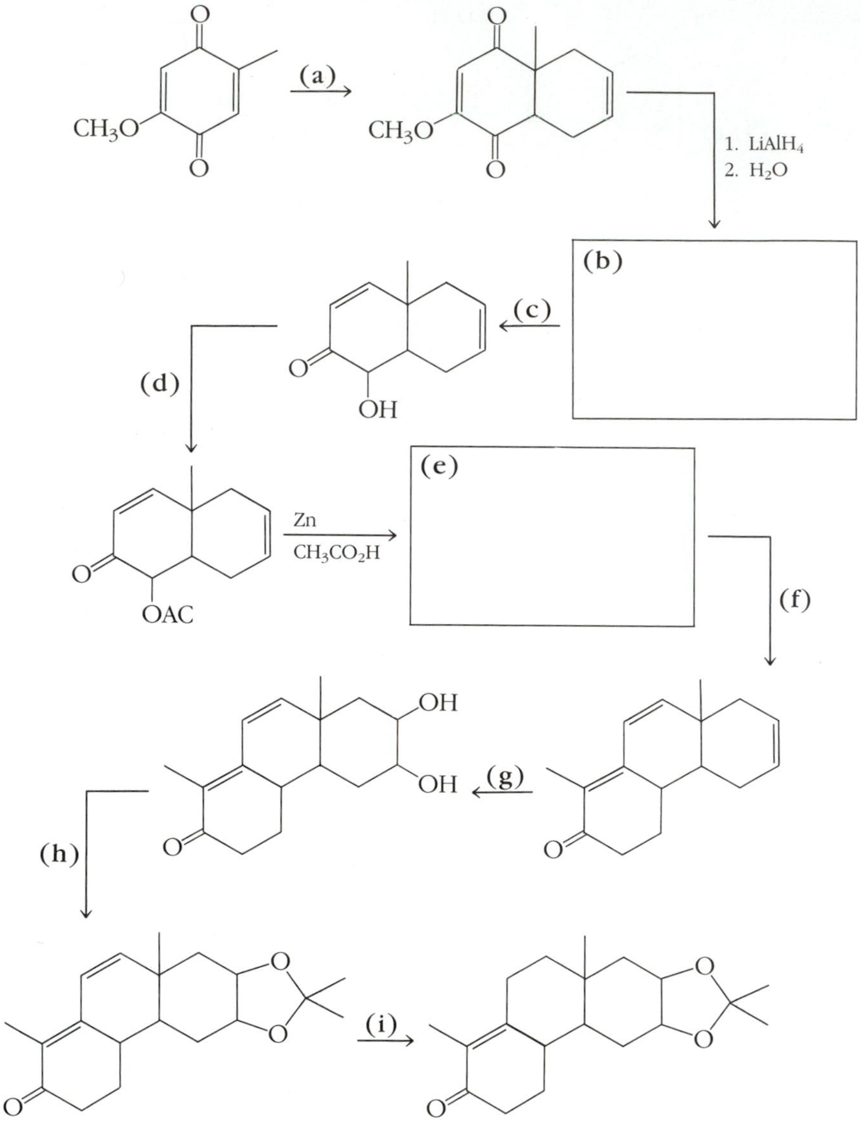

CH_3O (a) CH_3O

1. $LiAlH_4$
2. H_2O

(b)

(c)

(d)

OH

Zn
CH_3CO_2H

(e)

OAC

(f)

OH

OH (g)

(h)

O

O

O

O

(i)

Problem 23.9, continued

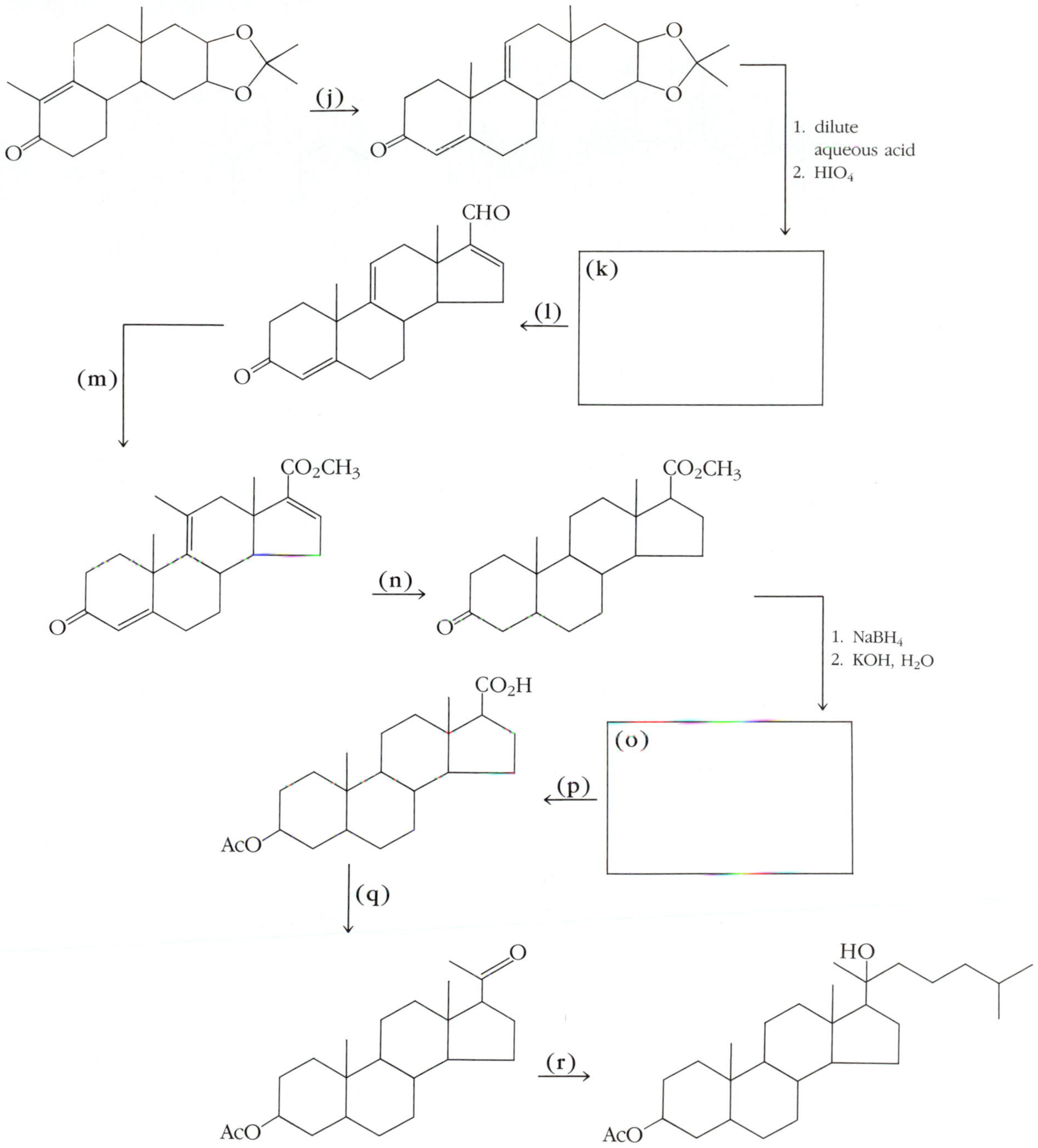

(j)

1. dilute aqueous acid
2. HIO_4

CHO

(k)

(l)

(m)

CO_2CH_3

(n)

CO_2CH_3

1. $NaBH_4$
2. KOH, H_2O

CO_2H

(o)

(p)

AcO

(q)

O

(r)

HO

AcO

AcO

Problem 23.9, continued

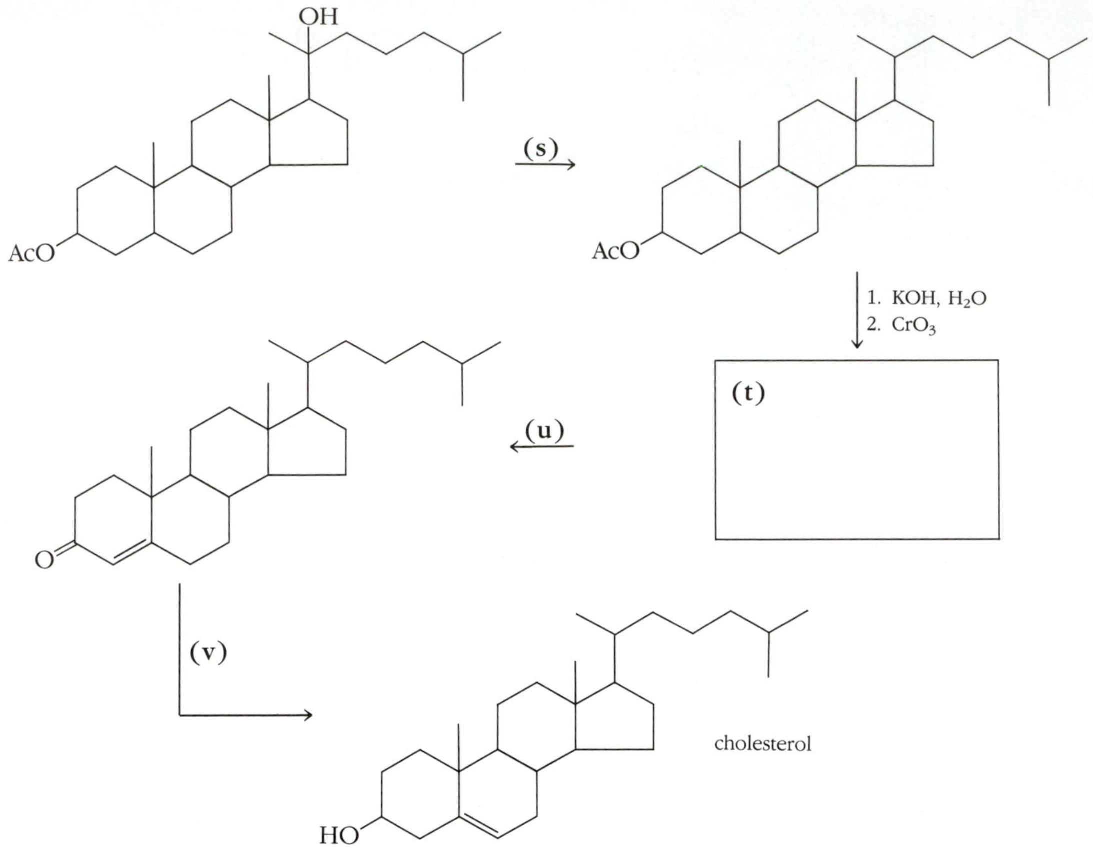

OH

AcO

(s)

AcO

1. KOH, H_2O
2. CrO_3

(t)

(u)

O

(v)

HO

cholesterol

23.10 (a) 6-Methoxytetralone (**1**) is a versatile intermediate in steroid synthesis, as illustrated by its conversion to either **2** or **3**. Complete the reactions for preparation of **2** and **3**.

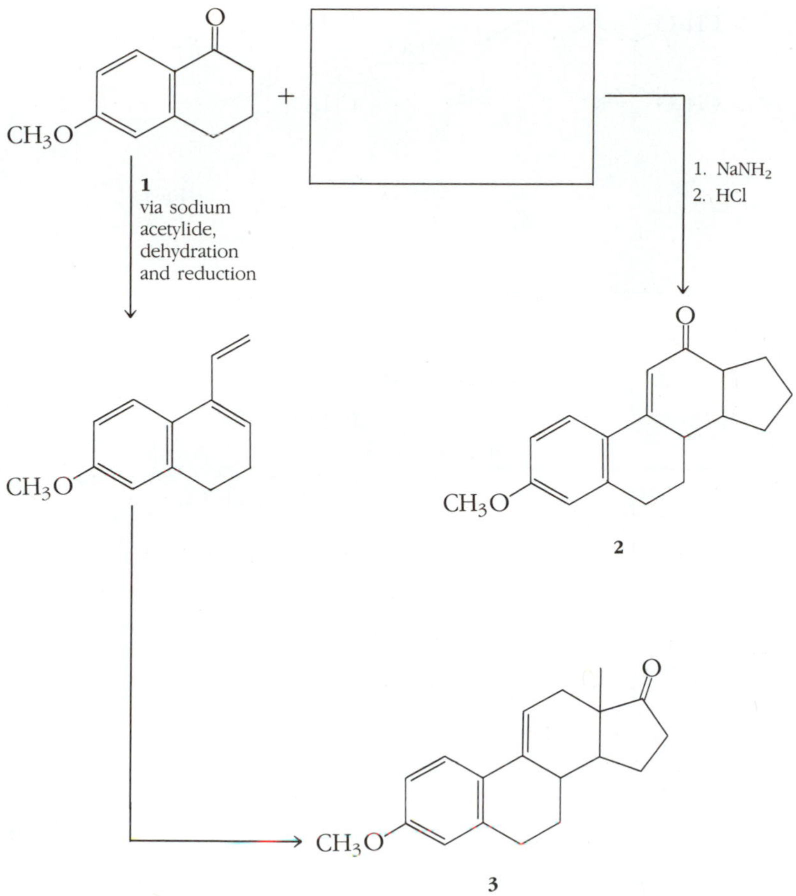

CH_3O **1** + [ ]

1. $NaNH_2$
2. HCl

1 via sodium acetylide, dehydration and reduction

CH_3O

CH_3O **2**

CH_3O **3**

(b) Suggest a method for preparing **1** starting from compounds that contain seven or fewer carbon atoms.

23.11 Complete the missing portions of the following synthesis of protoemetine, a compound that has been used as an intermediate in the synthesis of the alkaloid emetine.

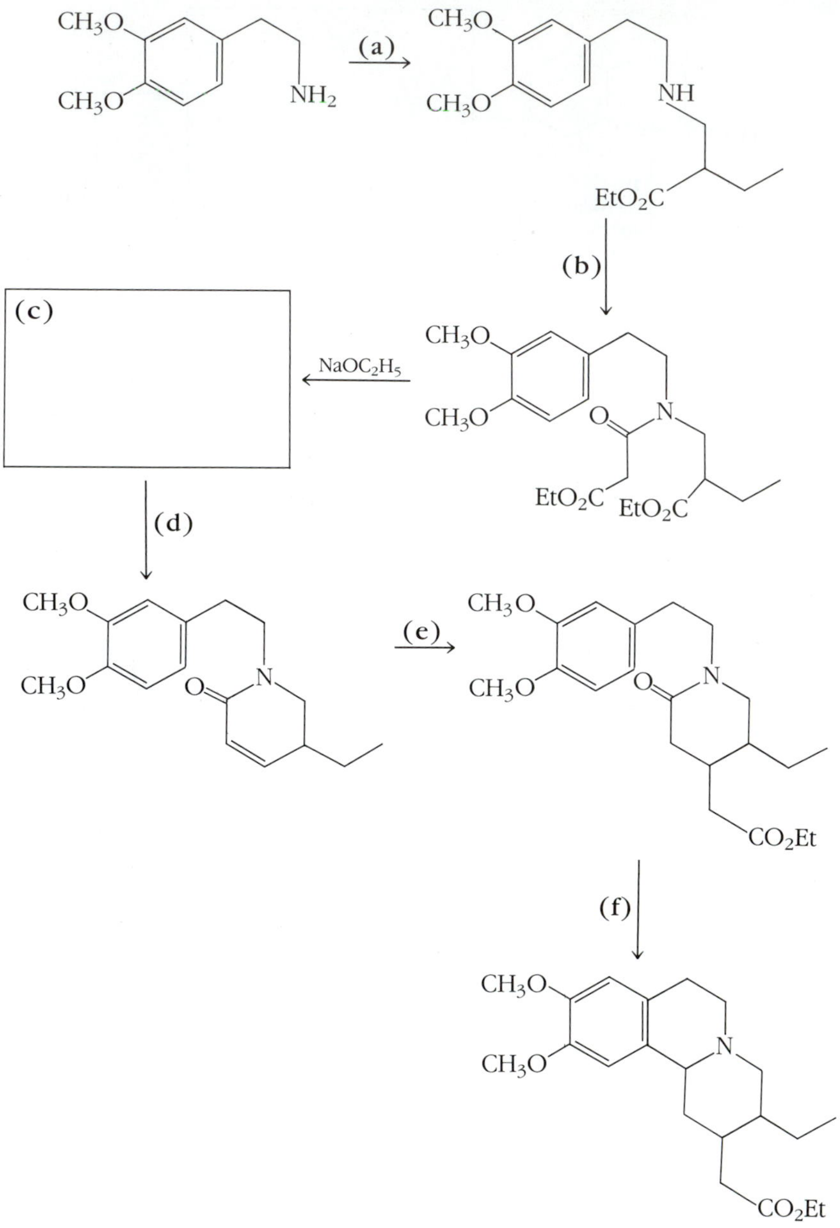

CH_3O
CH_3O
NH_2
(a)
CH_3O
CH_3O
NH
EtO_2C
(b)
(c)
$NaOC_2H_5$
CH_3O
CH_3O
O
N
EtO_2C
EtO_2C
(d)
CH_3O
CH_3O
O
N
(e)
CH_3O
CH_3O
O
N
CO_2Et
(f)
CH_3O
CH_3O
N
CO_2Et

23.12 Complete the missing portions of the following synthesis of the indole alkaloid yohimbine:

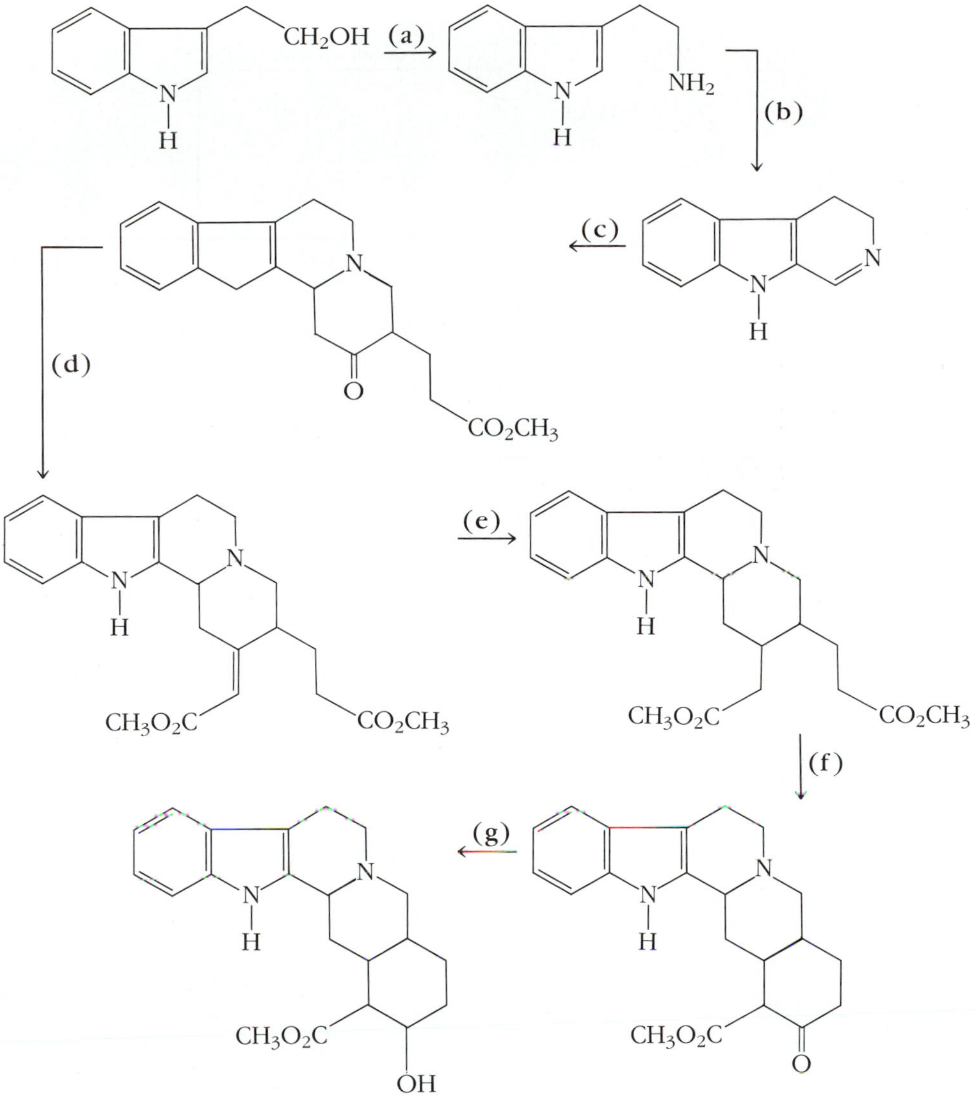

23.13 Complete the missing portions of the following synthesis of the alkaloid mesembrine:

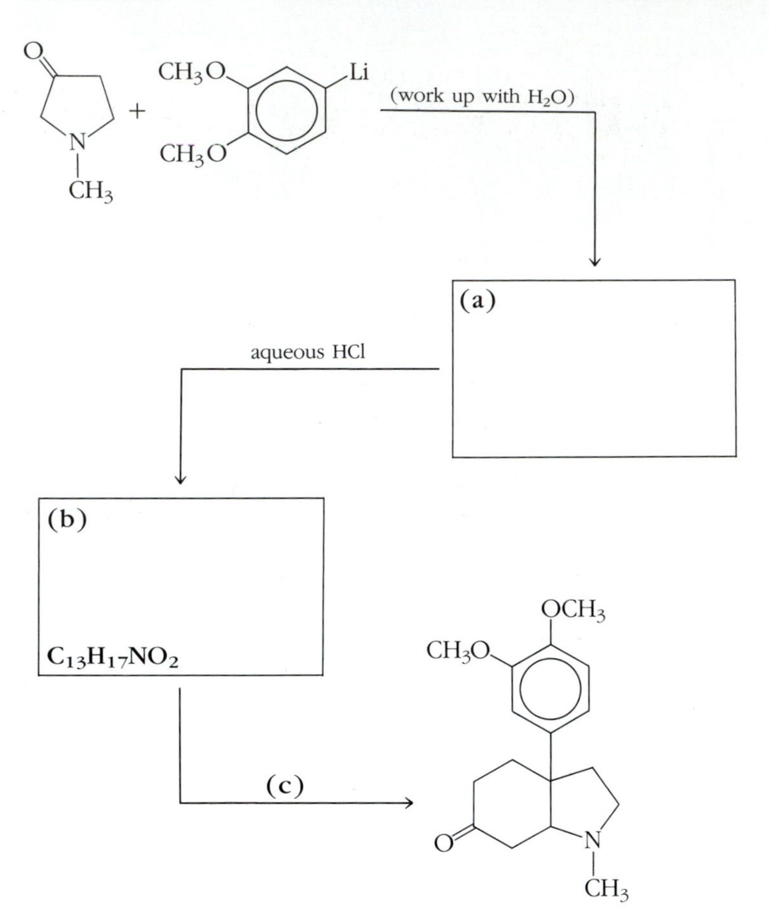

CHAPTER 24

HETEROCYCLIC COMPOUNDS

Heterocyclic compounds comprise a wide array of organic molecules that play important biochemical and economic roles. From some of the most common amino acids to the constituents of nucleic acids, these compounds pervade all areas of biochemistry. Many important pharmaceuticals are heterocycles, and the economic impact of some of these has been quite large. An example is *Valium* (diazepam), an antianxiety agent, which until recently was the most widely prescribed drug on the market. A second example is 3′-deoxy-3′-azidothymidine (AZT), the first drug approved for treatment of acquired immune deficiency syndrome (AIDS).

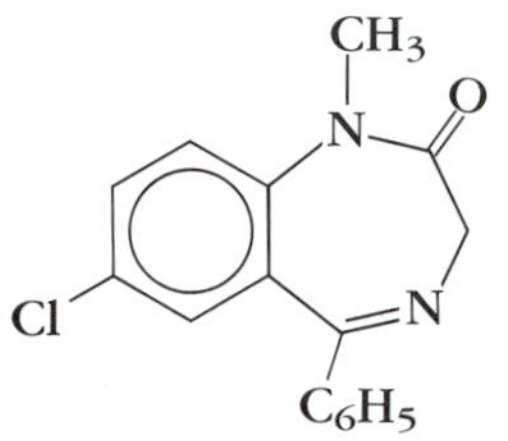

Valium

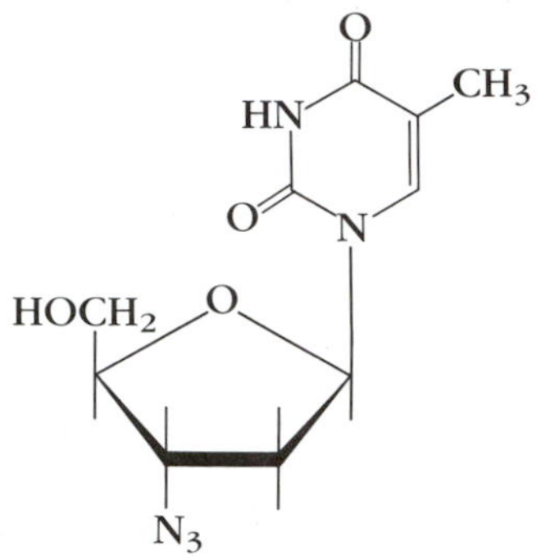

AZT

What constitutes a heterocycle? Quite simply, the name describes a cyclic compound for which the ring atoms include more than one element. Often a ring is made up only of carbon atoms (i.e., the compound is a *carbocycle*), but replacement of one or more carbons by other atoms produces a *heterocycle*. As you will see, the two most important heteroatoms are nitrogen and oxygen, but heterocycles can also be formed from a wide variety of other elements (including sulfur, phosphorus, and boron). You have already encountered a substantial number of heterocyclic compounds in earlier chapters. These include ethylene oxide and other epoxides (Section 6.8 and later chapters); aromatic heterocycles such as pyridine (Section 10.6); cyclic ethers, including acetals and ketals (Chapter 15); a variety of carbohydrates (Chapter 19); alkaloids (Section 23.4); and several amino acids (Chapter 20).

In this chapter we will provide a brief overview of some of the more important types of heterocycles and their characteristic reactions. We will then present a short discussion of some biologically significant heterocycles, with particular emphasis on either their biosynthesis or their mechanism of action.

24.1 MONOCYCLIC HETEROCYCLIC COMPOUNDS

Reactivity of Aromatic Heterocycles

In Section 10.6 we argued that replacement of one of the carbon atoms of benzene with a heteroatom would lead to another aromatic compound. Pyridine, for example, is an unsaturated heterocycle that is usually described as aromatic. But the replacement of a carbon by a nitrogen means that pyridine must exhibit at least some differences from benzene in its chemical behavior.

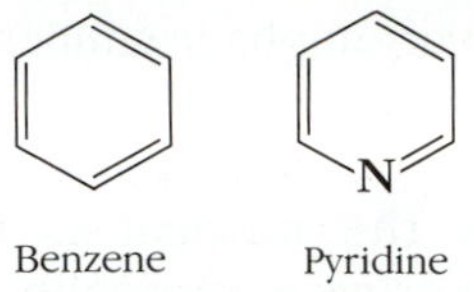

Benzene Pyridine

What types of behavior might we expect for a heterocyclic compound such as pyridine? Using X to denote the heteroatom, we can consider three general modes of reaction.

1. *Reaction at the heteroatom:* The nonbonding electrons of X can coordinate to an H^+ (or to any Lewis acid or electrophile):

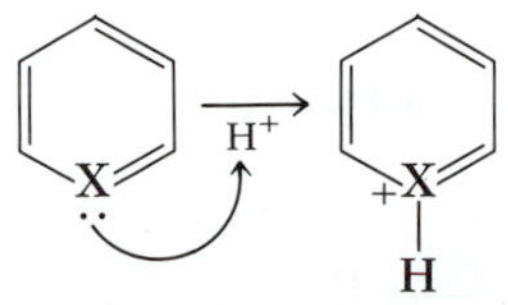

2. *Reaction of the aromatic π system:* Electrophilic substitution would occur in the same manner as for benzene derivatives. (We will discuss subsequently the preferred site for electrophilic substitution.)

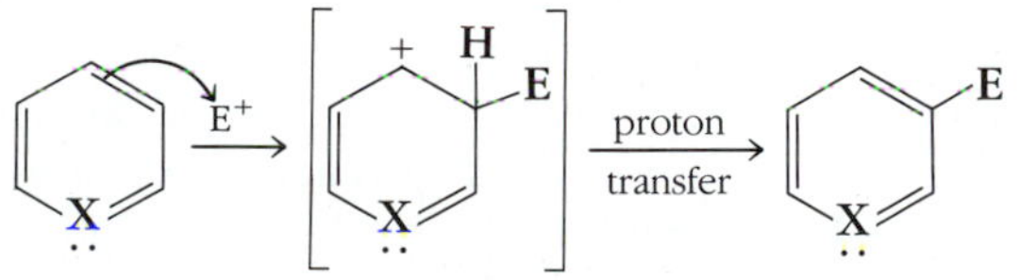

3. *Nucleophilic attack at a C—X "double bond":* The ring is aromatic, so this is not an isolated double bond, but we can use a single resonance form to show the reaction:

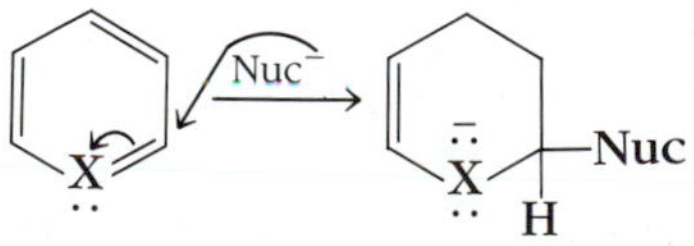

Such nucleophilic attack on a ring carbon can lead to a variety of products, including those formed by ring opening or by replacement of another substituent on the ring. (The latter process also occurs in some benzene derivatives; see Section 18.7.)

The result obtained in any specific reaction depends upon the heteroatom X, the other substituents on the ring, and the reaction conditions. In the following discussion we will illustrate how these factors can control the reactions of aromatic heterocycles.

Pyridine

Pyridine exhibits the characteristic basicity of an amine although it is a somewhat weaker base than most aliphatic amines. The K_a of its conjugate acid is 10^{-5}, whereas conjugate acids of tertiary aliphatic amines have K_a values of about 10^{-10}. The decreased basicity of pyridine results from differences in solvation between the neutral and charged forms; in the gas phase the basicity of pyridine is quite similar to that of other amines. The principal site for reaction of pyridine and its derivatives with either an acid or an electrophile is the *nitrogen atom.*

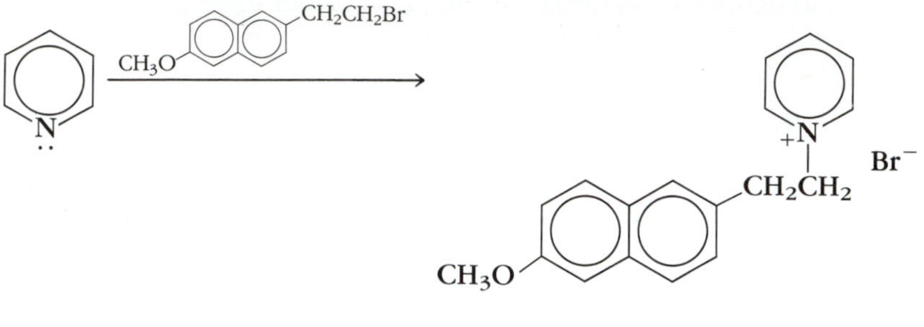

CH_2CH_2Br CH_3O +N Br^- CH_2CH_2 CH_3O ~100%

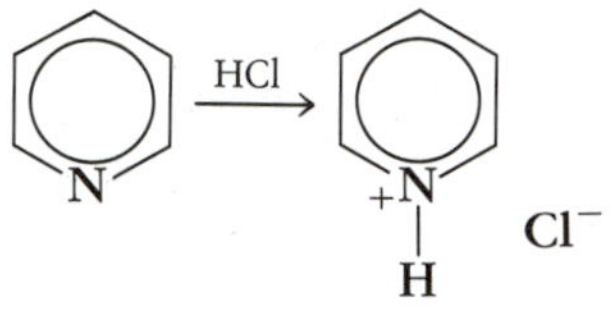

N HCl +N H Cl^- 98–100%

In each of the preceding reactions the pyridinium salt can be isolated as a crystalline solid. These are thermally stable compounds, although the protonated and alkylated pyridines will react very readily with a base or other nucleophile.

Because pyridine reacts preferentially at nitrogen, electrophilic substitution of the ring cannot normally be accomplished directly. The situation is very similar to that described for aniline in Section 18.2. The acidic conditions (protic acid or Lewis acid) result in nearly complete conversion to an unreactive pyridinium salt.

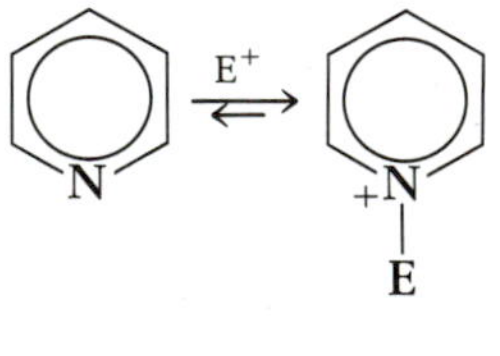

N E^+ +N E (unreactive to further electrophilic attack)

This low reactivity is illustrated by the drastic conditions needed for sulfonation.

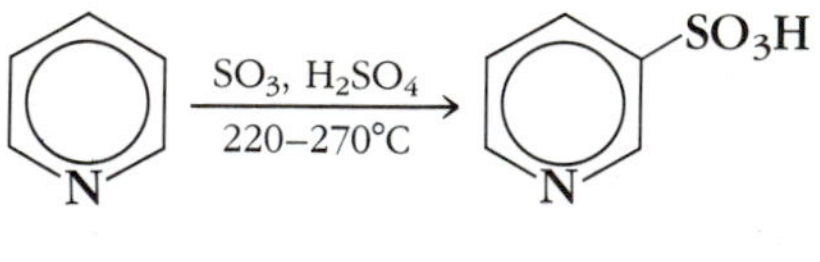

N SO_3, H_2SO_4 220–270°C SO_3H N 70%

This example also shows that electrophilic substitution of a pyridine occurs preferentially at C-3 or C-5 (note that the nitrogen is position 1). The reason for this preference is quite simple: electrophilic attack at C-4, C-2, or C-6 of a pyridine will produce an intermediate that has (partial) positive charge on the electronegative nitrogen atom. In contrast, attack at C-3 or C-5 does not result in such an unfavorable intermediate. (We will leave it as an exercise for you to draw the appropriate resonance forms.)

Because the pyridines themselves are so resistant to electrophilic substitution, many synthetic efforts have taken advantage of the greater reactivity of the *N*-oxides. These are readily prepared by reaction of the pyridine with hydrogen peroxide or a peroxy acid (Section 17.4).

$$\text{N} \xrightarrow{CH_3CO_3H} {}^{+}\text{N}-\text{O}^{-} \qquad 78\text{–}83\%$$

Other oxidizing agents do not usually affect the pyridine ring, as illustrated by the following side chain oxidation:

$$\text{N}\;\; CH_3 \xrightarrow[\text{2. HCl}]{\text{1. KMnO}_4} {}^{+}\text{N}(\text{H})\;\; CO_2H \;\; Cl^{-} \qquad 50\text{–}51\%$$

Although the nitrogen of the *N*-oxide has a *formal* positive charge, the electron distribution is quite different from that in pyridinium salts. The oxygen exerts a directive effect analogous to that found with phenols, and substitution at C-4 is observed.

$$CH_3,\; {}^{+}\text{N}-\text{O}^{-} \xrightarrow[H_2SO_4]{HNO_3} NO_2,\; CH_3,\; {}^{+}\text{N}-\text{O}^{-} \qquad 70\text{–}73\%$$

Nucleophilic attack at the carbon-nitrogen "double bond" of a pyridine occurs only with highly reactive reagents such as organolithium compounds and other very strong bases.

$$\text{N} \xrightarrow{C_6H_5Li} \left[\text{N}(\text{Li})\;\; \text{H}\;\; -C_6H_5\right] \longrightarrow \text{N}\;\; C_6H_5 \qquad 49\%$$

Notice that *addition* of phenyllithium in the preceding equation was followed by a loss of LiH to give an aromatic product.

The C—N "double bond" also increases the acidity of benzylic-type hydrogens. A hydrogen can be removed from the α carbon of an alkyl-substituted pyridine with strong base, and the resulting salt is a useful nucleophile.

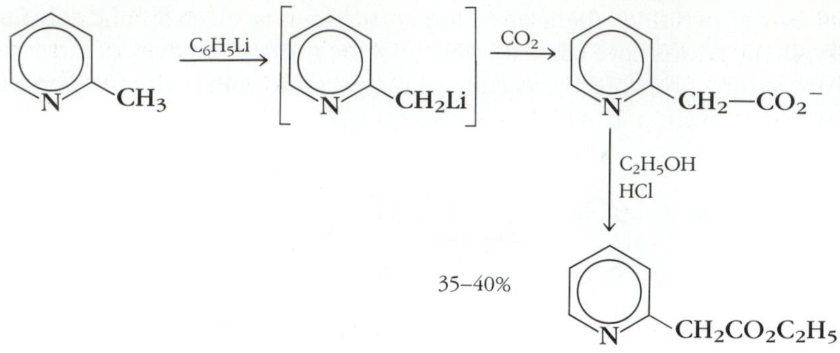

C_6H_5Li → CH_2Li → CO_2 → $CH_2—CO_2^-$

C_2H_5OH
HCl

35–40% $CH_2CO_2C_2H_5$

Nucleophilic addition occurs much more readily with pyridine *N*-oxides. Under acidic conditions, the oxygen of the *N*-oxide is lost in an elimination reaction to produce a ring-substituted pyridine.

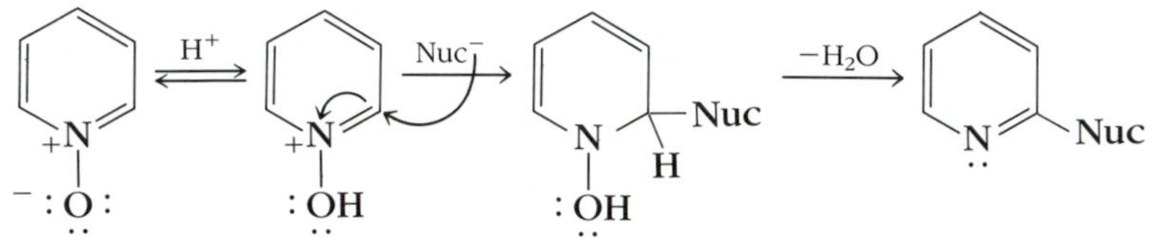

H^+ Nuc$^-$ Nuc H :OH $-H_2O$ Nuc

This process is illustrated by the reaction of pyridine *N*-oxide to give a mixture of 2- and 4-chloropyridine.

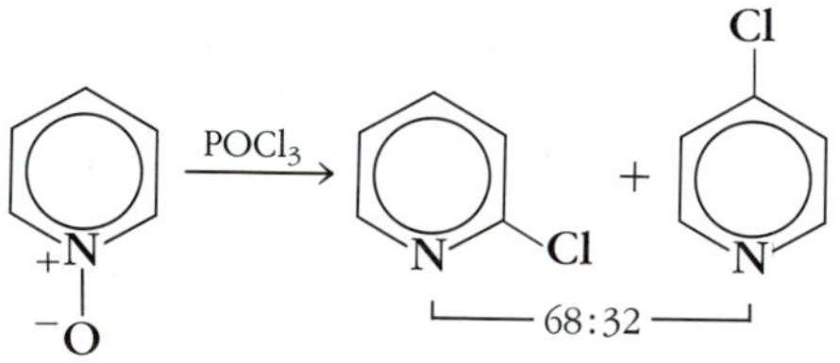

$POCl_3$ Cl + Cl 63%

68:32

Nucleophilic displacement of halides and NO_2 groups is a convenient method for modifying a pyridine derivative. The reaction proceeds via initial *addition* of the nucleophile to the pyridine ring, so it is restricted to attack at the 2, 4, and 6 positions. The reaction proceeds much more easily with the *N*-oxides than with the free pyridines.

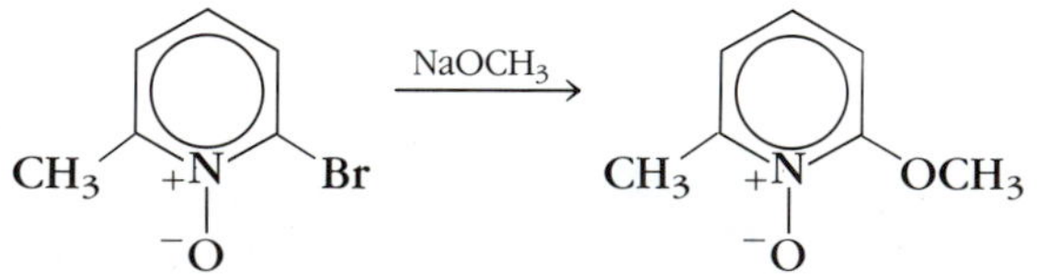

CH_3 Br $NaOCH_3$ → CH_3 OCH_3 67%

NO_2 CH_3 HCl → Cl CH_3 70%

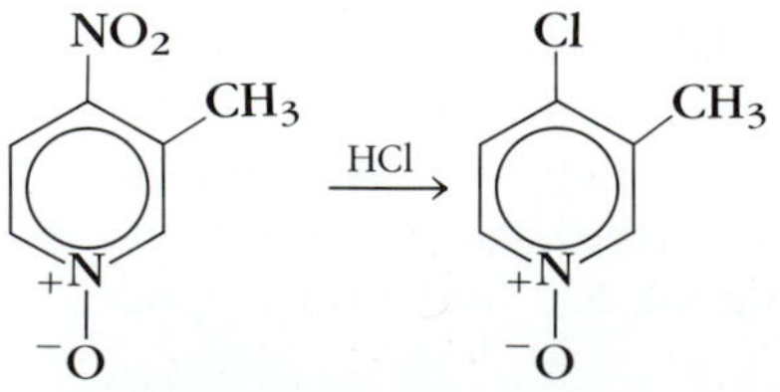

The synthesis of pyridine rings from acyclic precursors has been carried out in a variety of ways, and we will only show you a few examples here. One method involves condensation of ammonia or its derivatives with 1,5-dicarbonyl compounds, as illustrated by the following equation:

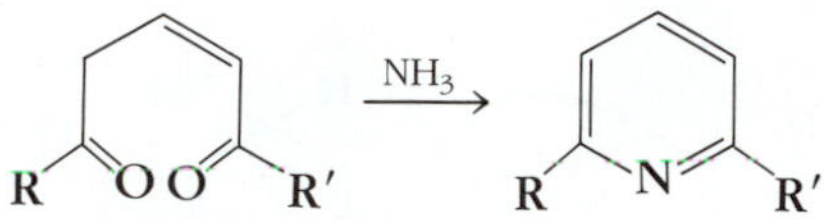

NH_3

R O O R′ → R N R′

This reaction occurs stepwise by the following mechanism:

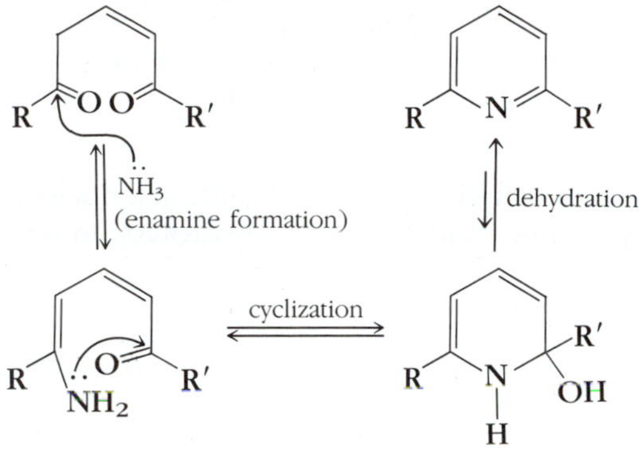

R O O R′

NH_3
(enamine formation)

R NH_2 O R′

cyclization

R N R′ OH
H

dehydration

R N R′

This method is particularly useful when a pyrylium salt is used in place of the dicarbonyl compound, and we will show you specific examples subsequently.

Among the derivatives of pyridine that play important biological roles are compounds in which the ring contains a second nitrogen. These are called *diazines,* and three isomeric types exist:

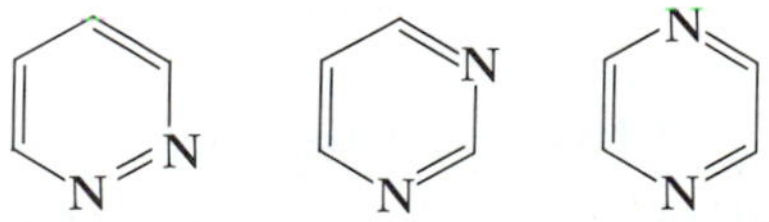

Pyridazine Pyrimidine Pyrazine

In most respects the diazines exhibit reactivity similar to that of pyridine. We will not consider the reactions of diazines in detail, but some of the chemistry of pyrimidine and pyrazine derivatives will be presented in Section 24.4.

Pyrylium Salts

Pyrylium salts are structurally similar to pyridine, but they are quite different in reactivity. If nonnucleophilic anions such as ClO_4^- are employed, these oxonium salts can be isolated as crystalline solids.

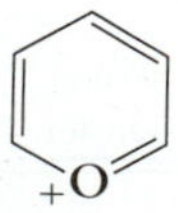

+O

Pyrylium ion

Pyrylium salts exhibit some properties that allow them to be classified as aromatic, yet as oxonium salts they are also highly susceptible to nucleophilic attack. Nucleophilic attack at the 2, 4, and 6 positions is the predominant mode of reaction for pyrylium ions. (Note that these are the same positions that are susceptible to nucleophilic attack in pyridine.)

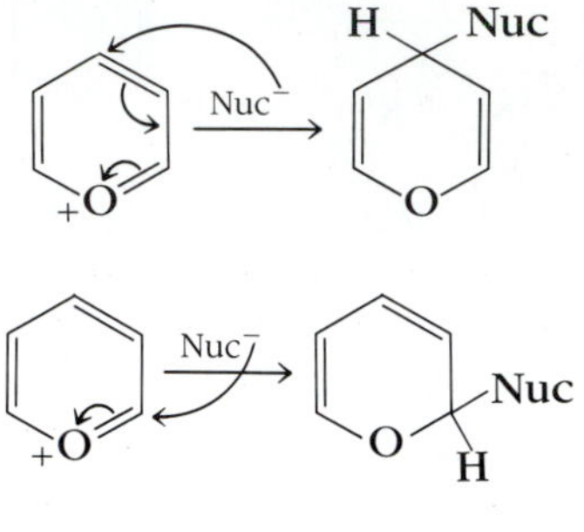

H Nuc
Nuc⁻
+O O

Nuc⁻
+O O Nuc
H

When the attacking species is an oxygen or nitrogen nucleophile, reversible ring opening frequently occurs, as illustrated for hydrolysis in the following equation:

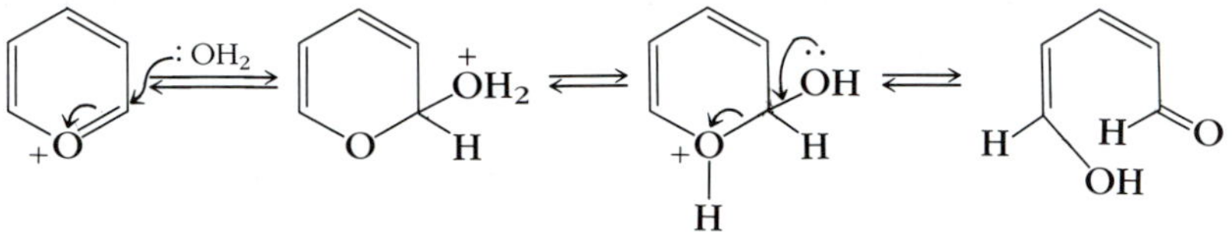

: OH_2
+O O $\overset{+}{O}H_2$ H +O OH H H H O
H OH

Reaction with ammonia and primary amines affords the corresponding pyridine via ring opening followed by reclosure to nitrogen.

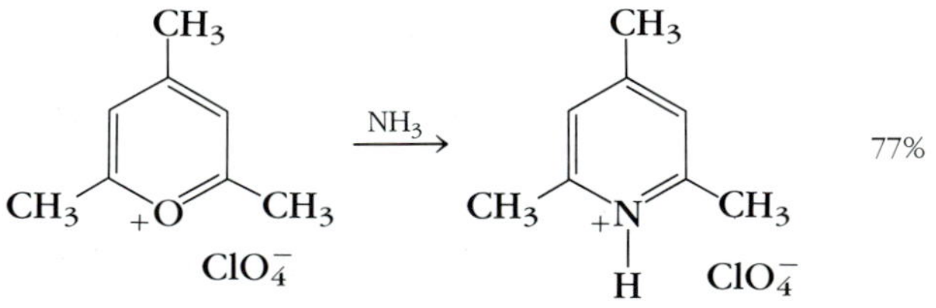

CH_3 CH_3
NH_3
77%
CH_3 +O CH_3 CH_3 +N CH_3
ClO_4^- H ClO_4^-

The reaction can be explained by the following mechanism:

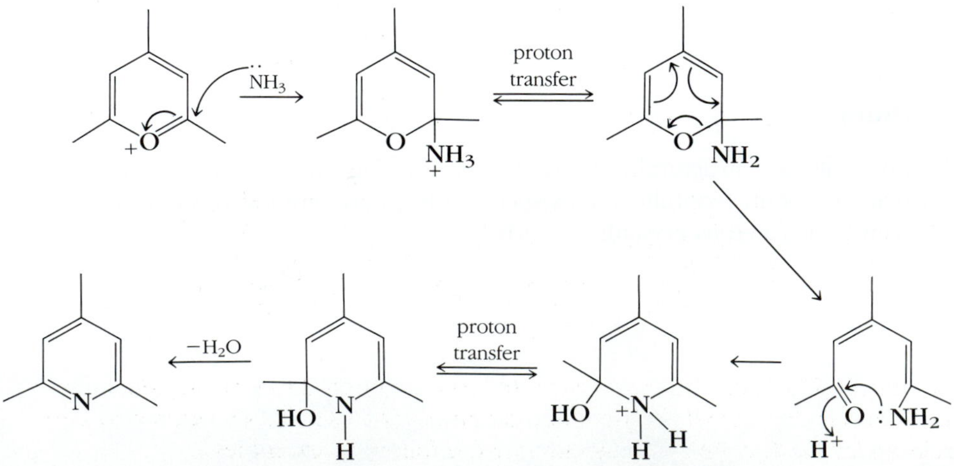

proton transfer
NH_3
+O O $\overset{+}{N}H_3$ O NH_2

proton transfer
$-H_2O$
N HO N HO +N O : NH_2
H H H H^+

Pyrrole, Furan, and Thiophene

The five-membered ring heterocycles pyrrole, furan, and thiophene are all six-π-electron systems, and they exhibit aromatic properties (Section 10.6).

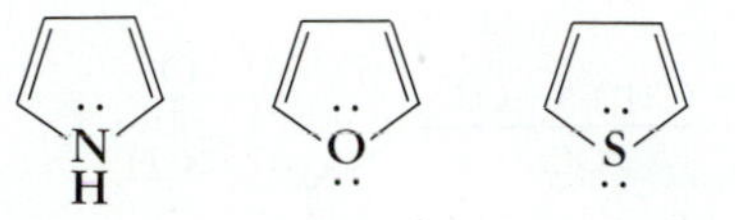

Pyrrole Furan Thiophene

Because the "nonbonding" electrons of nitrogen are actually involved in the aromatic π system, the basicity of pyrrole is extremely low, and it behaves as a neutral compound in aqueous solution. Similarly, there is little tendency for reaction with electrophiles to occur at nitrogen. Oxygen and sulfur compounds are inherently less basic than amines, so all three of these aromatic heterocycles behave as neutral compounds.

The primary mode of reaction observed for these five-membered ring heterocycles with an electrophile is aromatic substitution (Chapter 18). Their relative reactivities parallel those of the corresponding benzene derivatives, although the heterocycles are generally more reactive.

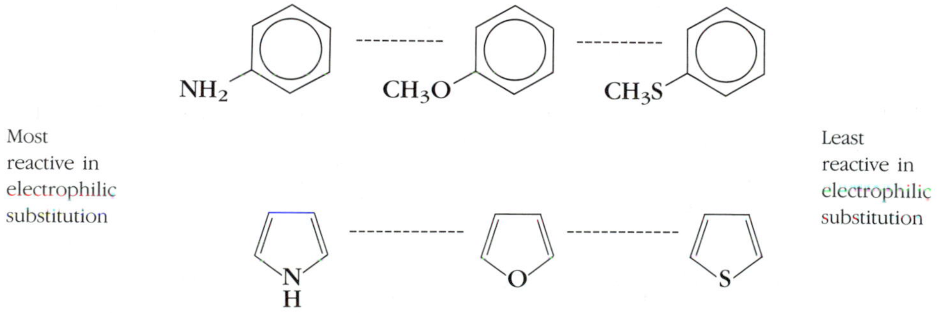

NH_2 ---------- CH_3O ---------- CH_3S

Most reactive in electrophilic substitution

Least reactive in electrophilic substitution

The directive effect of the heteroatom favors substitution at the 2 position (the heteroatom always defines the 1 position). You can easily understand this preference by evaluating the resonance forms of the cationic intermediates:

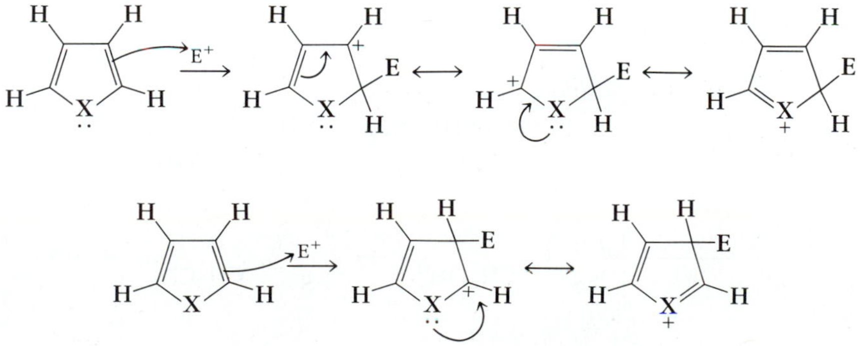

Both intermediates can be drawn as onium ions, but only attack at C-2 generates a cation that is also "allylic." The latter cation is more stable, so reaction occurs preferentially at C-2. This is illustrated by the following examples:

Thiophene $\xrightarrow[H_3PO_4]{CH_3COCCH_3 \text{ (O, O)}}$ 2-acetylthiophene (CCH_3, O) 74–79%

Pyrrole (N–H) $\xrightarrow[POCl_3]{(CH_3)_2N—CH \text{ (=O)}}$ pyrrole-2-carbaldehyde (CH, =O) 78–79%

The reactivity of these heterocycles is fairly high, and strong Lewis acid catalysts such as $AlCl_3$ are usually not required. Introduction of a formyl group with dimethylformamide and $POCl_3$ is a widely used technique in heterocyclic synthesis. The reaction proceeds by the following two-step sequence:

$$(1)\ HC(=O)—\ddot{N}R_2 \xrightarrow{POCl_3} HC(OPCl_2{=}O)=\overset{+}{N}R_2$$

(2) heterocycle (X) + $CH(OPCl_2{=}O)=\overset{+}{N}R_2$ $\longrightarrow$ σ-complex ($+X$, H, $CH(O—PCl_2{=}O)NR_2$) $\xrightarrow{:base}$ 2-substituted heterocycle–$CH(O—PCl_2{=}O)\ddot{N}R_2$ $\longrightarrow$ heterocycle–$CH=\overset{+}{N}R_2$ $\xrightarrow[H_2O]{workup}$ heterocycle–$CH(=O)$

Reactions with iminium ions ($>C=\overset{+}{N}R_2$) or unsaturated oxonium ions ($>C=\overset{+}{O}R$ or $>C=\overset{+}{O}H$) are similar to the preceding reaction and are also used extensively.

2-methylfuran (CH_3, O) $\xrightarrow[\text{[from } CH_2=O \text{ and } NH(CH_3)_2]]{CH_2=\overset{+}{N}(CH_3)_2}$ CH_3–furan–$CH_2—N(CH_3)_2$ 69–76%

Thiophene $\xrightarrow[HCl]{CH_2=O}$ [2-thienyl–CH_2OH] $\longrightarrow$ 2-thienyl–CH_2Cl 40–41%

Substituents on the heterocyclic ring can affect the preferred site of reaction by a combination of steric and electronic effects. When both C-2 and C-5 are already substituted, reaction occurs at C-3 or C-4.

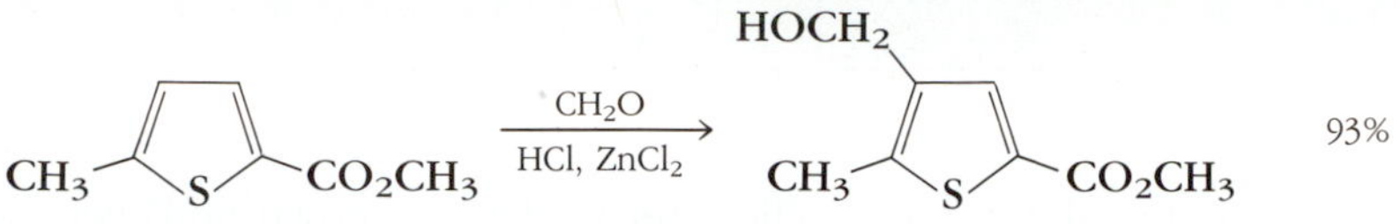

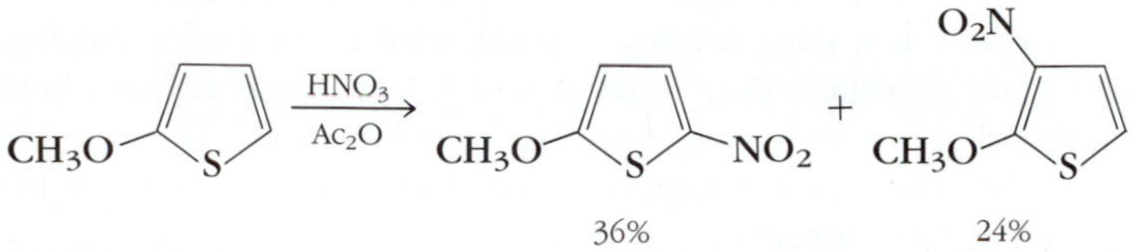

$$CH_3\text{-(thiophene)-}CO_2CH_3 \xrightarrow[\text{HCl, ZnCl}_2]{CH_2O} \text{3-}(HOCH_2)\text{-}CH_3\text{-(thiophene)-}CO_2CH_3 \quad 93\%$$

$$CH_3O\text{-(thiophene)} \xrightarrow[Ac_2O]{HNO_3} CH_3O\text{-(thiophene)-}NO_2 \;(36\%) + CH_3O\text{-(3-}O_2N\text{-thiophene)} \;(24\%)$$

The five-membered-ring aromatic heterocycles can be synthesized by a variety of pathways, but one of the most general methods employs 1,4-dicarbonyl compounds. This is illustrated by the following equation:

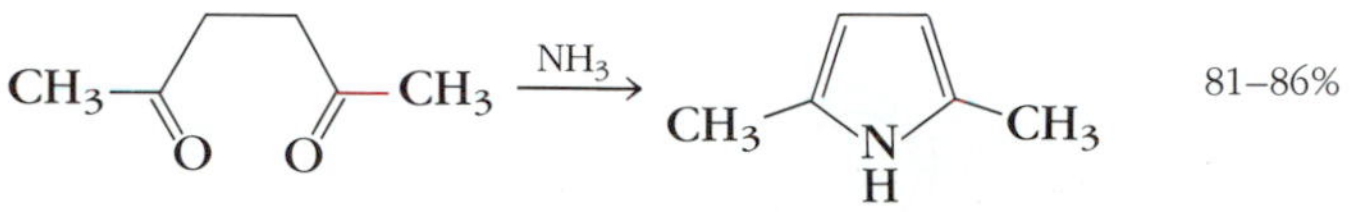

$$CH_3COCH_2CH_2COCH_3 \xrightarrow{NH_3} \text{2,5-dimethylpyrrole } (CH_3\text{-(pyrrole N-H)-}CH_3) \quad 81\text{–}86\%$$

The aromatic stability of pyrroles, furans, and thiophenes allows a wide variety of reactions to be carried out on their substituents. The facile oxidation of pyrroles is a notable exception to this pattern, however. In contrast to pyridines, which are stable to oxidations with $KMnO_4$ and Cr(VI) reagents, pyrroles react readily with such oxidizing agents to give intractable mixtures of products.

EXERCISE 24.1

Write the product expected from the reaction of pyridine with: (a) HCl; (b) $C_6H_5CH_2Br$; (c) C_6H_5Li; (d) $C_6H_5CO_2H$.

EXERCISE 24.2

Write the product expected from the reactions of pyrrole with: (a) Br_2; (b) HNO_3; (c) CH_2O, HCl.

EXERCISE 24.3

Write the product expected from the reactions of furan with the reagents listed in Exercise 24.2.

EXERCISE 24.4

Write the product expected from the reactions of thiophene with the reagents listed in Exercise 24.2.

24.2 POLYCYCLIC AROMATIC HETEROCYCLES

A wide variety of bicyclic and polycyclic heterocycles are known, and even a survey of the different types would be beyond the scope of this text. Instead we will briefly discuss a few examples, emphasizing several systems of biological significance. Some of these are bicyclic compounds in which both rings contain heteroatoms. Other compounds have only one ring that is heterocyclic, the remaining rings being carbocyclic. The latter compounds have chemical properties that closely resemble those of the monocyclic analogs.

Indoles

Indole is a *benzopyrrole,* a frequently encountered structural unit in biological systems.

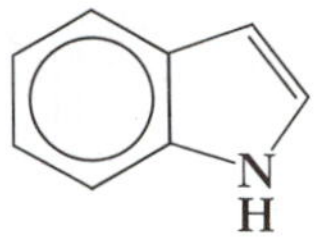

N H

Indole

The indole nucleus is found in many alkaloids and psychoactive compounds (Section 23.4). Particularly important are those with an aminoethyl side chain derived from the amino acid tryptophan.

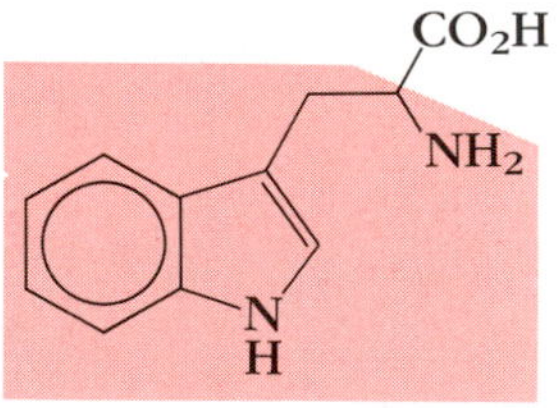

CO_2H NH_2 N H

Tryptophan
(the β-indolethylamine subunit found in many psychoactive compounds is highlighted)

Like the pyrrole nitrogen, the indole nitrogen is not appreciably basic. In contrast to pyrrole, however, indole normally undergoes electrophilic substitution at the 3 position. In this way the reactivity of indole is very similar to that of an enamine (Section 17.8).

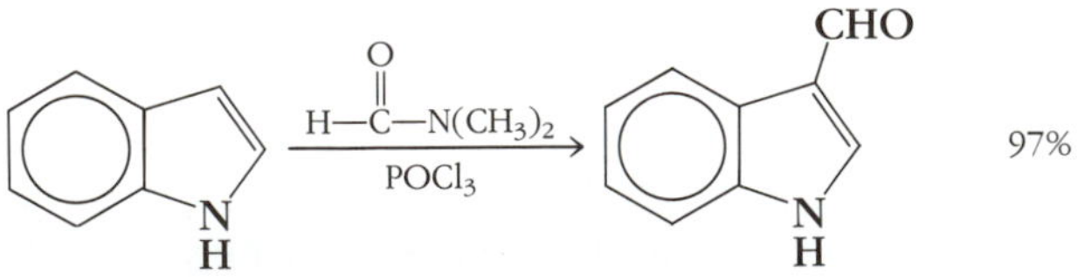

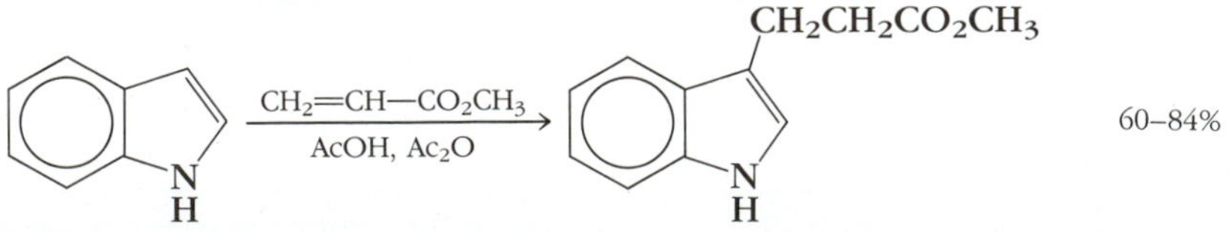

N H $\xrightarrow[POCl_3]{H-C(=O)-N(CH_3)_2}$ (CHO) N H 97%

N H $\xrightarrow[AcOH,\ Ac_2O]{CH_2{=}CH-CO_2CH_3}$ ($CH_2CH_2CO_2CH_3$) N H 60–84%

Notice that these reactions occur on the pyrrole ring rather than the benzene ring. This reemphasizes our earlier statement that the five-membered-ring aromatic heterocycles undergo electrophilic substitution more easily than do the benzene analogs.

Substitution at the 2 position of indoles does occur readily in *intramolecular* reactions, particularly when it generates a six-membered ring.

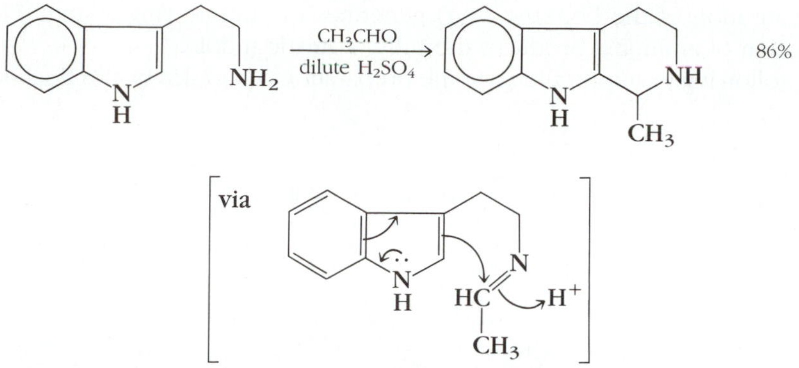

CH_3CHO
dilute H_2SO_4
86%

via

This type of cyclization was encountered previously in the synthesis of reserpine (Section 23.4).

The indole system can be synthesized in a variety of ways, but one of the most general methods is the *Fischer indole synthesis,* first reported by Emil Fischer in the 1880s. In this process phenylhydrazine (or one of its derivatives) is heated with an aldehyde or ketone in the presence of a Lewis acid such as $ZnCl_2$ (Figure 24.1).

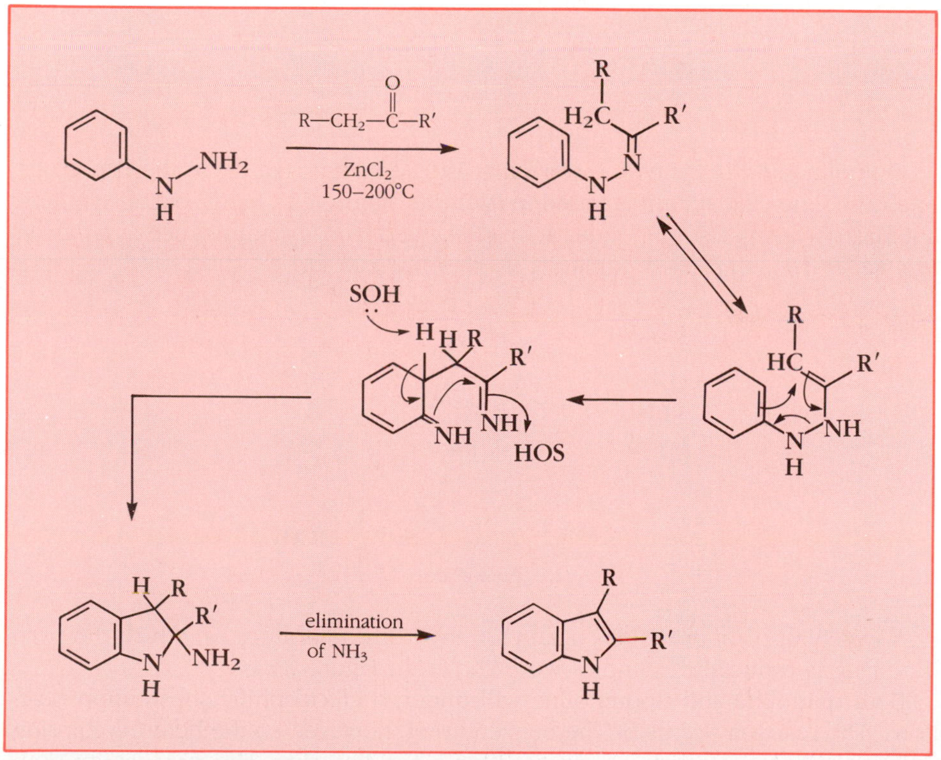

$R—CH_2—C(=O)—R'$
$ZnCl_2$
150–200°C

elimination of NH_3

Figure 24.1 The Fischer indole synthesis.

Condensation of the NH_2 group with the carbonyl compound affords the phenylhydrazone, which is in equilibrium with the enamine analog. High temperatures are employed, and the enamine undergoes an electrocyclic reaction (Section 27.4). This cleaves the nitrogen-nitrogen bond of the original hydrazine. Nucleophilic attack of one nitrogen on the other carbon-nitrogen double bond (together with rearomatization of the benzene ring) generates the indole ring system. Finally, elimination of ammonia produces the fully aromatic indole.

The following examples illustrate the preparation of indoles by the Fischer synthesis:

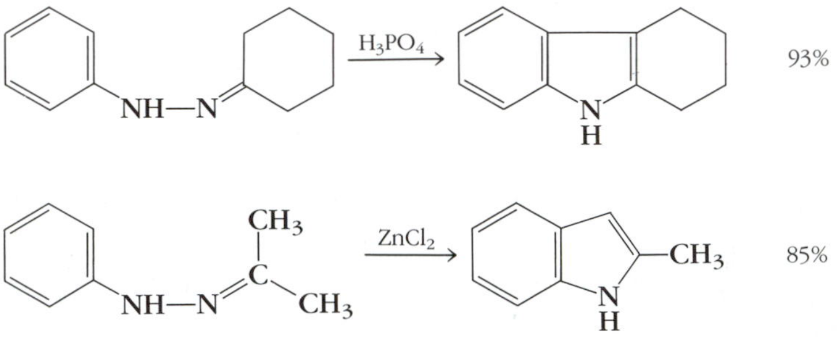

NH—N $\xrightarrow{H_3PO_4}$ N H 93%

CH_3 NH—N=C—CH_3 $\xrightarrow{ZnCl_2}$ —CH_3 N H 85%

Quinolines and Isoquinolines

Quinoline and isoquinoline are *benzopyridines,* and their reactivity resembles not only that of pyridine itself but also that of substituted benzenes.

N
Quinoline

N
Isoquinoline

Quinoline and isoquinoline were first isolated from coal tar in the late 1800s. The quinoline ring system is found in *quinine* (Figure 23.4) and several synthetic antimalarial agents. Isoquinolines are represented by compounds such as *papaverine,* which has been used medically as an antispasmodic.

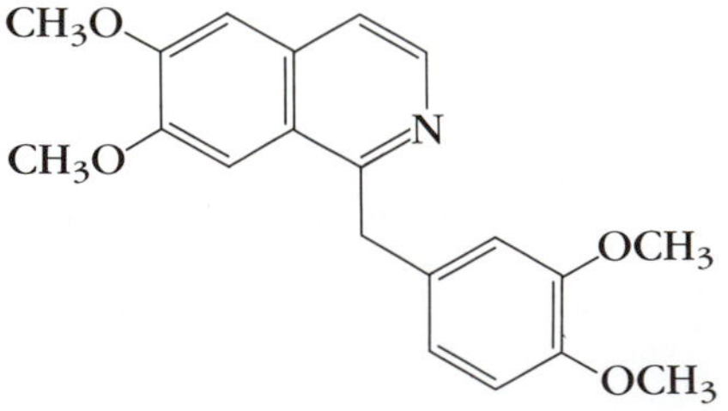

CH_3O CH_3O N OCH_3 OCH_3

Papaverine

A variety of plant alkaloids contain tetrahydroisoquinolines (in which the pyridine ring is reduced); examples were shown in Figure 23.4.

Both quinoline and isoquinoline will undergo electrophilic substitution reactions. The reaction occurs on the benzene ring, however, so the heterocyclic ring remains rather unreactive, as expected for a pyridine ring. The reactivity of both compounds under acidic conditions is comparable with that of the anilinium ion, $C_6H_5NH_3^+$

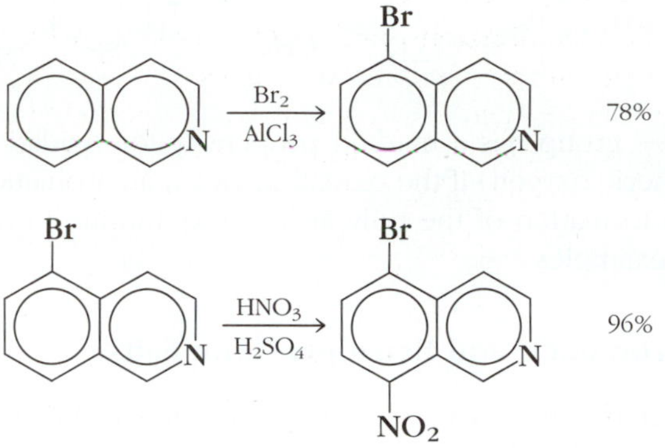

Br
Br_2 / $AlCl_3$
N N
78%

Br Br
HNO_3 / H_2SO_4
N N
NO_2
96%

The most versatile methods for synthesizing quinolines and isoquinolines involve formation of the heterocyclic ring via condensation of an amino-substituted benzene with an appropriate carbonyl or dicarbonyl compound.

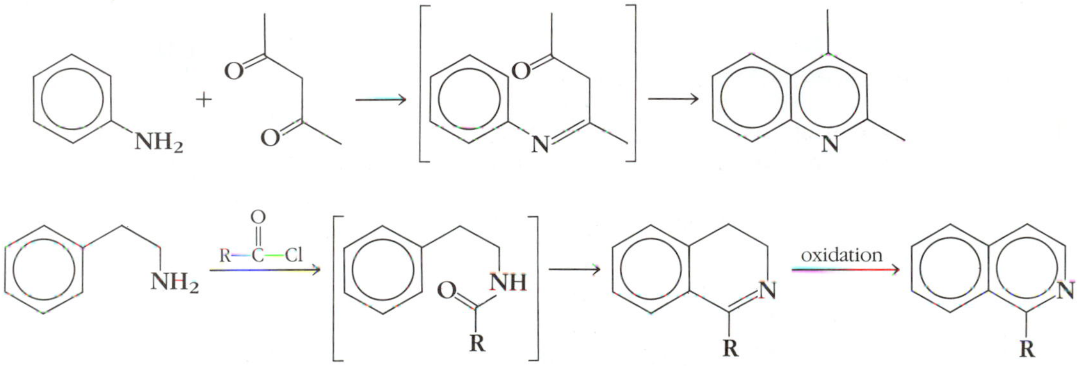

NH_2 + O O → [O N] → N

NH_2 $\xrightarrow{R-C(=O)-Cl}$ [O NH R] → N R $\xrightarrow{\text{oxidation}}$ N R

The second of these two pathways initially affords a dihydroisoquinoline, which must be oxidized to the fully aromatic heterocycle. The dihydroisoquinolines are useful intermediates for the synthesis of tetrahydroisoquinolines, as we illustrated previously in the synthesis of reserpine (Section 23.4). The following equations show further examples of the synthesis of quinoline and isoquinoline derivatives.

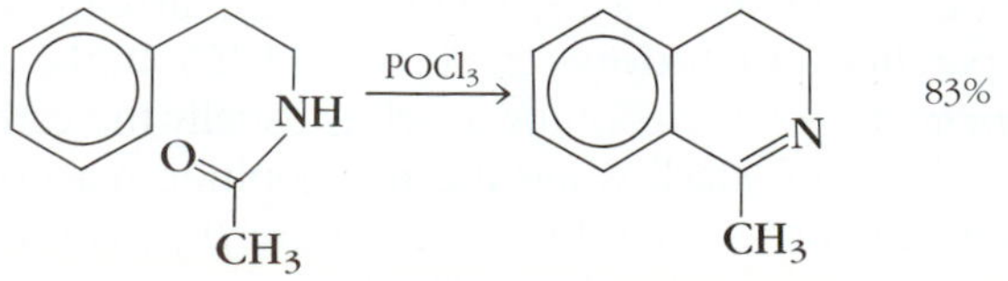

NH O CH_3 $\xrightarrow{POCl_3}$ N CH_3
83%

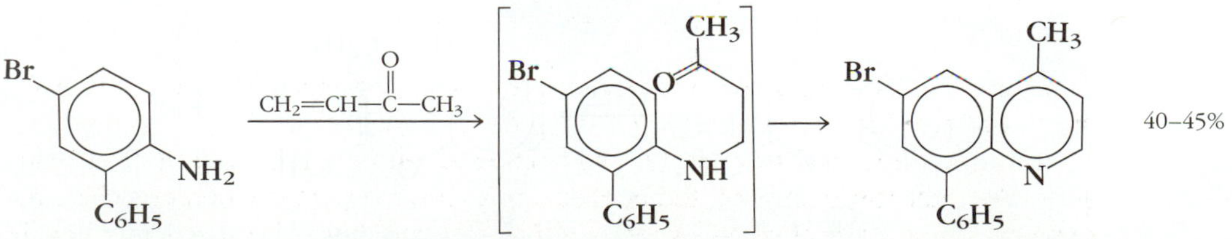

Br NH_2 C_6H_5 $\xrightarrow{CH_2=CH-C(=O)-CH_3}$ [Br CH_3 O NH C_6H_5] → Br CH_3 N C_6H_5
40–45%

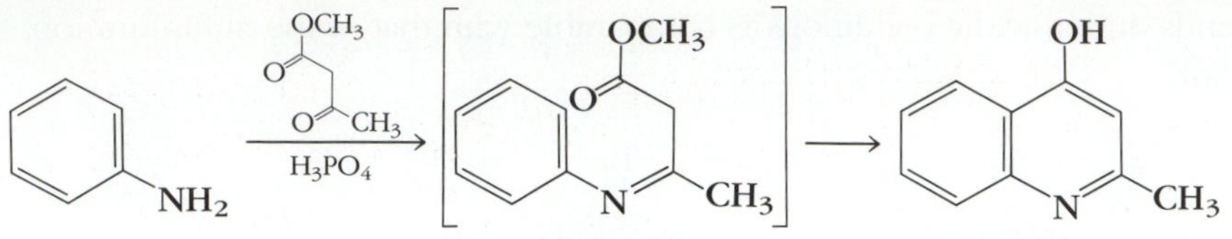

$\xrightarrow{H_3PO_4}$ 70%

The —CH—NH— group has a marked propensity for oxidation to —C=N— (even by atmospheric oxygen) if the oxidation yields an aromatic product. This is illustrated by the formation of the fully aromatic quinoline in the second of the three preceding examples.

Coumarins, Chromones, and Benzopyrylium Salts

A variety of important heteroaromatic compounds are naphthalene derivatives in which C-1 has been replaced by oxygen. These are related to quinoline in the same way that pyran derivatives are related to pyridine. The 1-benzopyrylium ion is found in many naturally occurring compounds; the conjugate bases of the 2- and 4-hydroxy derivatives form classes of compounds known as *coumarins* and *chromones,* respectively.

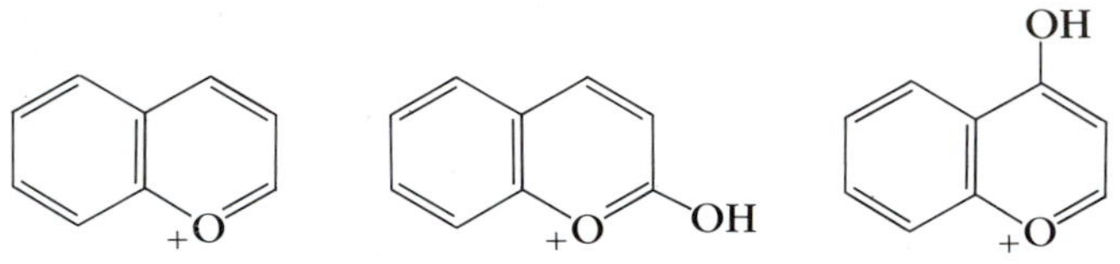

1-Benzopyrylium ion

2-Hydroxy-1-benzopyrylium ion

4-Hydroxy-1-benzopyrylium ion

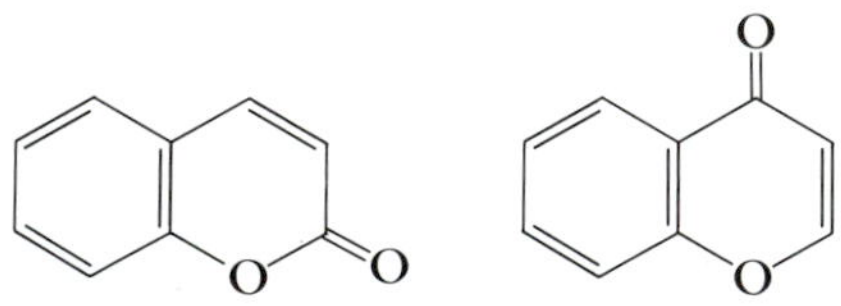

Coumarin

Chromone

Nucleophilic attack on these compounds can occur at either C-2 or C-4, and it is frequently difficult to predict which will predominate in a specific reaction. All these compounds can be viewed as α, β-unsaturated carbonyl compounds, and there are strong parallels in reactivity (cf. Section 13.10). In the absence of steric hindrance the *kinetic* site of nucleophilic attack is usually the carbonyl carbon, so this is the expected site of attack when the nucleophilic reaction is *irreversible. Reversible* additions, in contrast, tend to occur at the β position. These reactivity patterns are illustrated in the following examples:

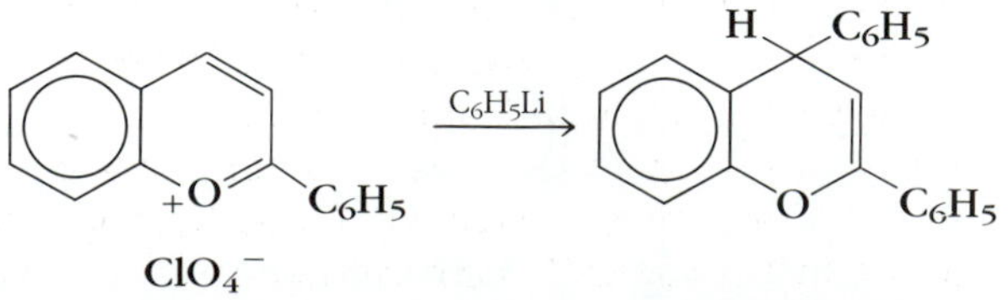

ClO_4^- $\xrightarrow{C_6H_5Li}$ 86%

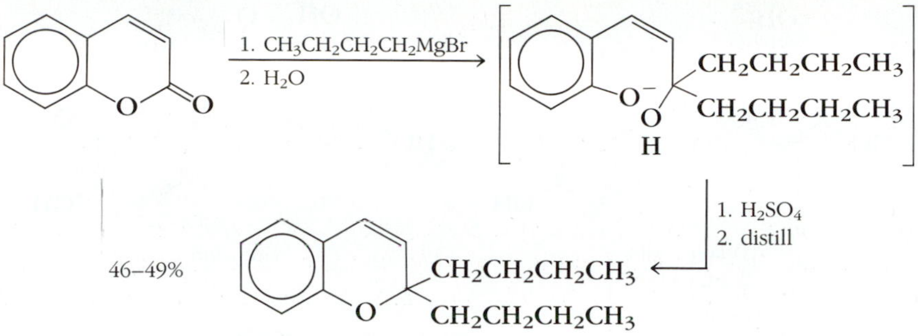

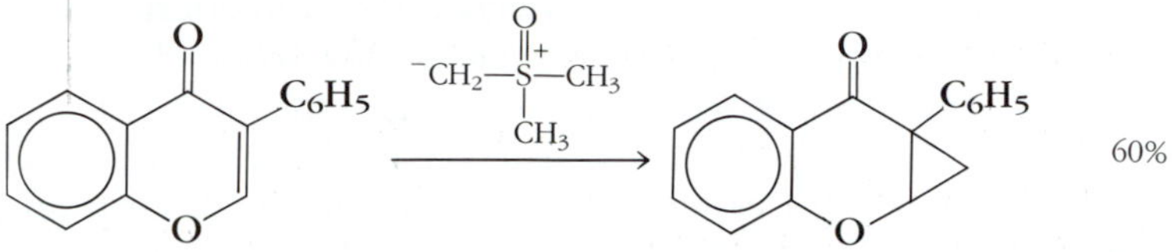

1. $CH_3CH_2CH_2CH_2MgBr$
2. H_2O

$CH_2CH_2CH_2CH_3$
$CH_2CH_2CH_2CH_3$

1. H_2SO_4
2. distill

46–49%

$CH_2CH_2CH_2CH_3$
$CH_2CH_2CH_2CH_3$

C_6H_5

$^{-}CH_2$—$\overset{+}{S}$(=O)—CH_3, CH_3

C_6H_5

60%

The last reaction involves *reversible* addition of a sulfur ylide (cf. Section 13.8) at the β position of the unsaturated carbonyl system.

Despite the many similarities between their reactivities and those of pyrylium salts, 1-benzopyrylium salts do not afford quinolines upon reaction with amines or ammonia.

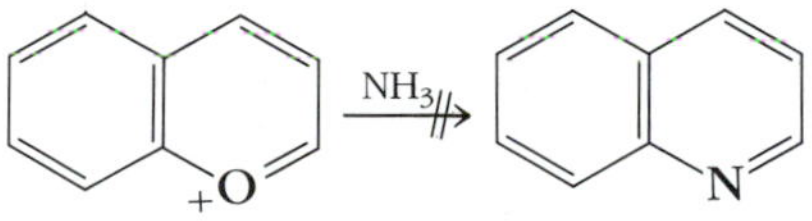

NH_3

You can see the reason for this difference by evaluating the ring-opened intermediate that would be formed:

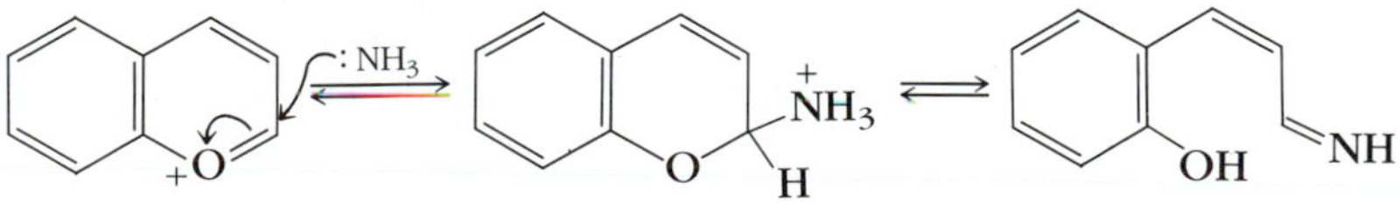

: NH_3

$\overset{+}{N}H_3$

OH NH

The only way that nitrogen could replace the original oxygen would be by nucleophilic attack on the *keto* form of the phenol:

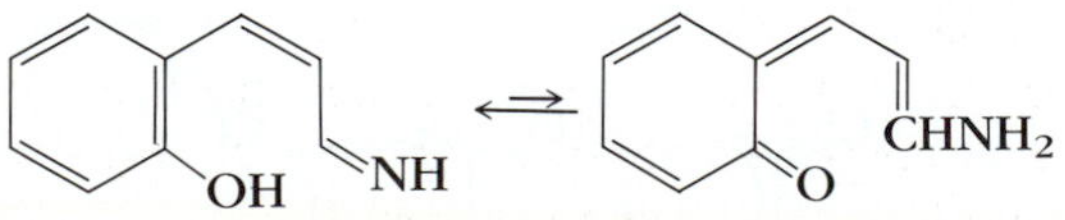

OH NH

CHNH₂

When ring opening occurs with pyrylium salts, the keto form is favored, but for the 1-benzopyrylium salts the enol form (a phenol) is highly favored. As a consequence, replacement of oxygen by nitrogen is not observed.

Two important classes of plant pigments are based on the benzopyrylium nucleus: the anthocyanins, which are hydroxylated 2-phenyl-1-benzopyrylium salts, and the flavones, which are hydroxylated 2-phenylchromones.

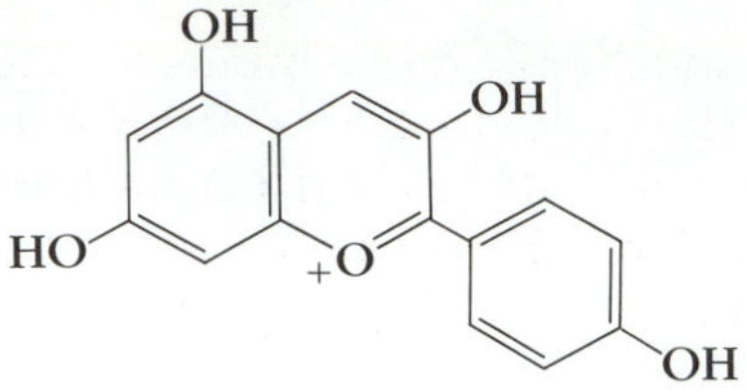

Pelargonidin
(an anthocyanin)

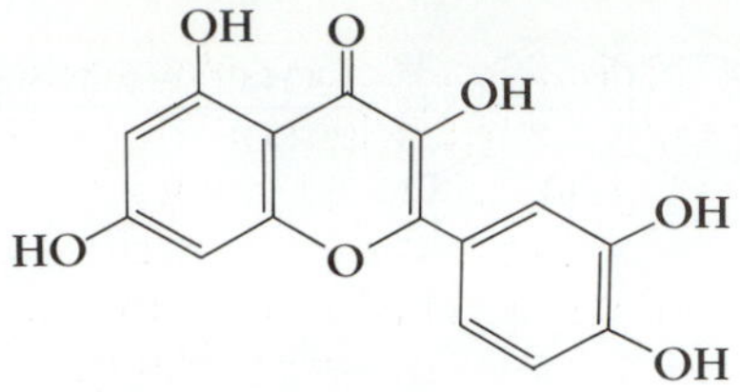

Guercitin
(a flavone)

Coumarins have found significant medical and commercial use because of their biological effects. Some coumarins act as *anticoagulants,* interfering with vitamin K biochemistry and disrupting the normal pathways for blood clotting. *Warfarin* (developed at the University of Wisconsin and named for the Wisconsin Alumni Research Foundation) is a potent anticoagulant that has been used extensively as a rodenticide. Ingestion of this compound by rats results in death from internal hemorrhaging.

Warfarin

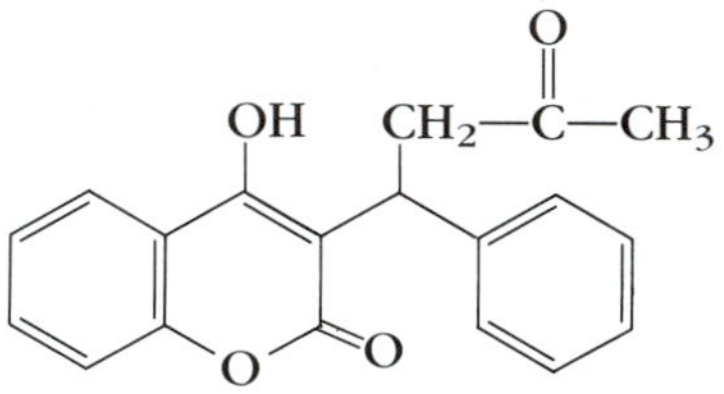

Dioxins

The dioxins are unsaturated six-membered-ring heterocycles with two oxygens in a 1,4 relationship. The best-known dioxins are tricyclic analogs of anthracene.

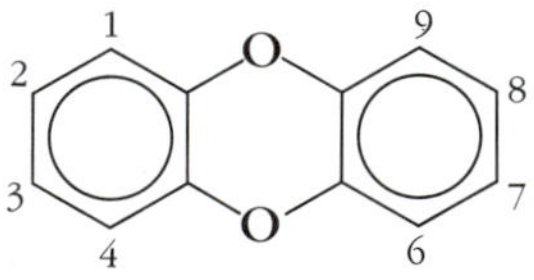

Dioxin
(correctly called
dibenzo-*p*-dioxin

In some respects these represent only a minor class of organic compounds, but their role in environmental problems has become extremely important in recent years. The term *dioxin* is used carelessly, particularly by the news media, and the specific compound that has become well-known as an environmental hazard is 2,3,7,8,-tetrachlorodibenzo-*p*-dioxin.

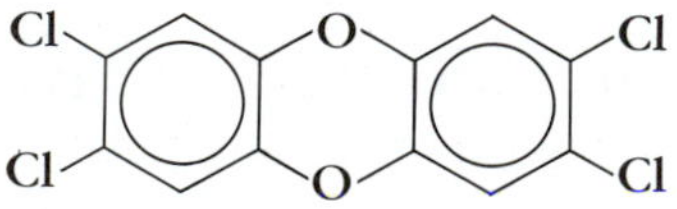

This compound can have a variety of adverse effects on human health. There is strong evidence that it is carcinogenic (i.e., cancer-causing) and teratogenic (i.e., capable of inducing birth defects in the offspring of individuals exposed during pregnancy).

The problem of dioxin toxicity was first widely recognized during the Vietnam War, when the herbicide 2,4,5-T was used extensively as a defoliant (as a component of "agent orange"). During the commercial synthesis of 2,4,5-T small amounts of 2,3,7,8-tetrachlorodibenzo-*p*-dioxin can be formed, and use of the contaminated herbicide resulted in the problem of human exposure to "dioxin." The legal consequences of these events are still being debated.

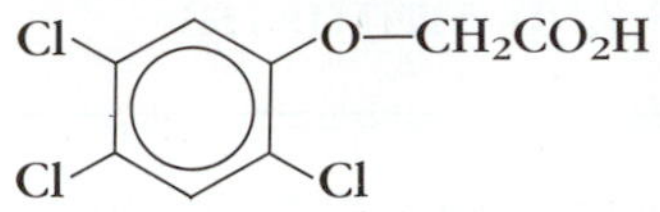

Cl, Cl, Cl; $O{-}CH_2CO_2H$

2,4,5-T
(2,4,5-trichlorophenoxyacetic acid)

Formation of dioxins from halogenated phenols probably occurs via nucleophilic displacement of halide by oxygen, as summarized by the following equation:

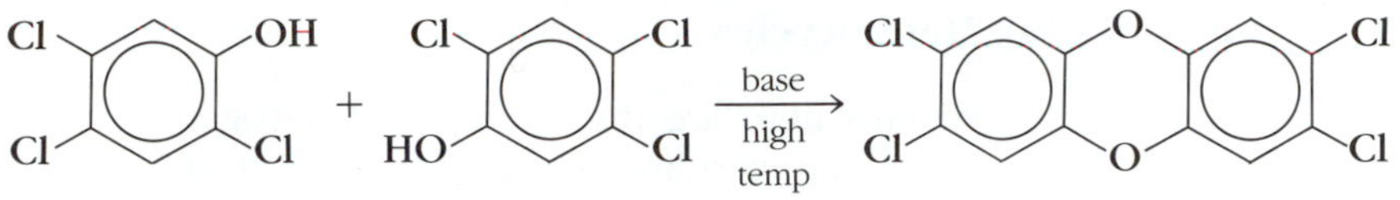

Cl, Cl, OH, Cl + Cl, HO, Cl, Cl $\xrightarrow[\text{high temp}]{\text{base}}$ Cl, Cl, O, O, Cl, Cl

Now that the health hazards have been recognized, commercial synthesis of the herbicide 2,4,5-T has been modified to avoid formation of dangerous levels of dioxins.

There are some indications that formation of dioxins can also occur during combustion of halogenated aromatic compounds. Partial oxidation would generate halogenated phenolic intermediates, and these could cyclize to the dioxins under the high temperatures of the combustion process. The dioxins exert their toxic effects at very low concentration levels, so their possible formation as by-products in such reactions poses serious problems in the disposal of certain organic waste materials.

EXERCISE 24.5

Show the relationship of coumarins and chromones to 1-benzopyrylium salts by drawing the conjugate acids of coumarin and chromone. Use appropriate resonance forms.

EXERCISE 24.6

Deduce the starting materials needed for the synthesis of the following compound via a Fischer indole synthesis:

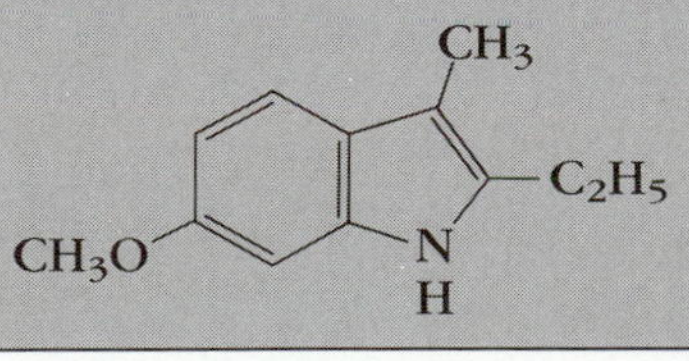

CH_3, C_2H_5, CH_3O, N, H

EXERCISE 24.7

Quinoline is less reactive than isoquinoline in nitration with HNO_3 by about a factor of 10. Draw intermediates for each reaction that can explain this reactivity difference.

24.3 NONAROMATIC HETEROCYCLES AND SYNTHETIC APPLICATIONS

Enormous numbers of heterocyclic compounds, both aromatic and nonaromatic, have been synthesized and studied. We will therefore limit our discussion in this section to a brief overview of some of the compounds that play important roles in synthetic organic chemistry.

Three-Membered Ring Heterocycles

The two most important systems here are the oxygen and nitrogen derivatives known as epoxide (or oxirane) and aziridine derivatives, respectively.

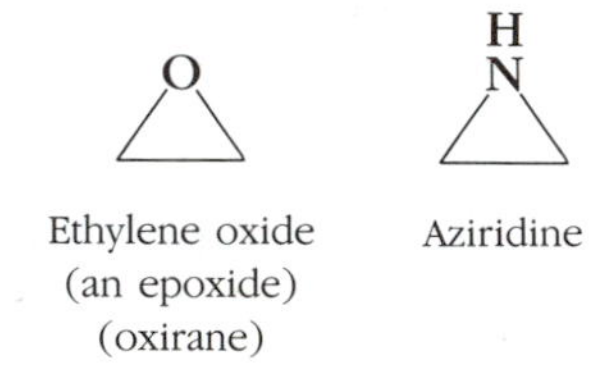

Both these classes of compounds undergo nucleophilic ring opening according to the following equation:

H^+ X Nuc$^-$ ⟶ Nuc XH   X = O, NH

We have discussed such reactions of epoxides extensively in earlier chapters (Sections 5.12, 6.8, 11.6, 12.8, 13.8 and 15.3).

Epoxides are generally prepared by peracid oxidation of the corresponding alkene (Sections 6.8 and 15.4), by base-catalyzed cyclization of a halo alcohol (Section 15.4), or by reaction of a ketone or aldehyde with a sulfur ylide (Section 13.8).

C=C —RCO_3H→
HO—C—C—X —base→ —C—C— (O bridged epoxide)
—C(=O)— —$(CH_3)_2S{=}CH_2$→

Aziridines are generally prepared either by cyclization of a β-substituted amine or by addition of a nitrene (see Solutions Manual Appendix, Molecular Rearrangements) to an alkene, although the latter is usually restricted to strained double bonds. This is illustrated by the following examples:

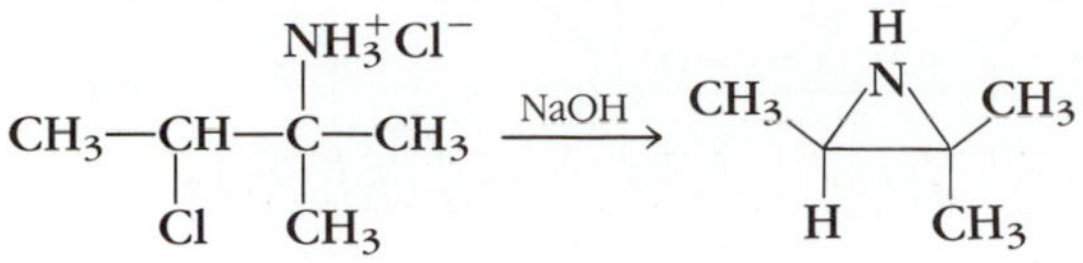

$$CH_3-CH(Cl)-C(NH_3^+Cl^-)(CH_3)-CH_3 \xrightarrow{NaOH} \text{2,2,3-trimethylaziridine} \quad 73\%$$

$$\text{norbornene} \xrightarrow{C_6H_5C(=O)N_3} \text{aziridine}-N-C(=O)-C_6H_5 \quad 87\%$$

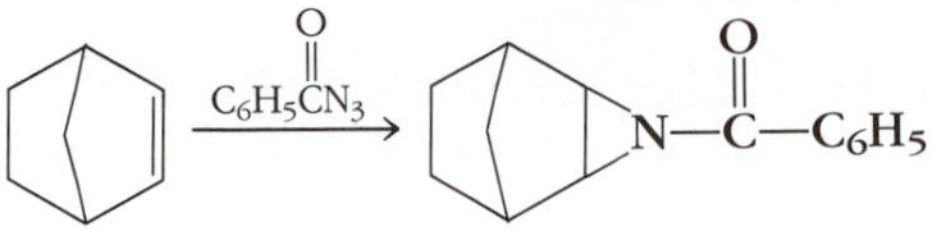

The latter reaction may actually proceed via cycloaddition of the azide followed by extrusion of molecular nitrogen, rather than by direct addition of the free nitrene. The net result, however, is the same.

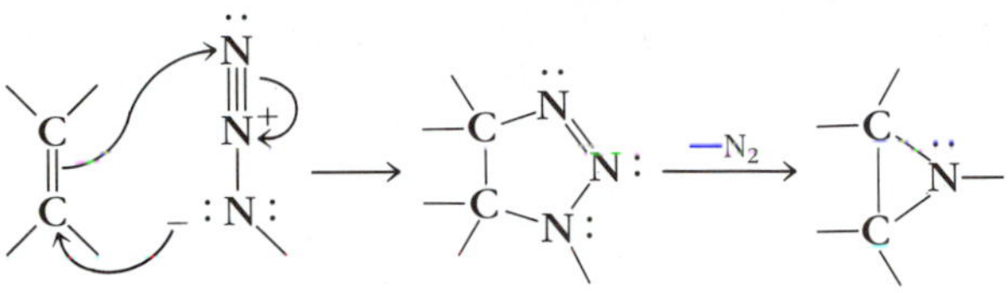

$$\xrightarrow{-N_2}$$

Four-Membered-Ring Heterocycles

The four-membered-ring heterocycles, azetidine and oxetane are infrequently encountered and therefore are not used extensively in organic synthesis.

Azetidine   Oxetane

One oxetane derivative that does have synthetic importance is *ketene dimer,* which can be converted to the highly reactive monomer by pyrolysis.

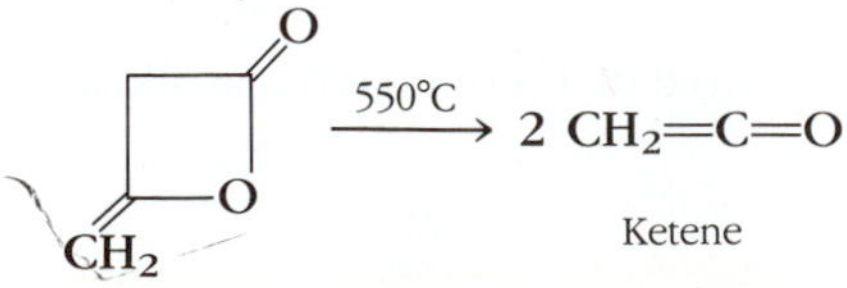

$$\text{Ketene dimer} \xrightarrow{550°C} 2\ CH_2{=}C{=}O \quad 46\text{–}55\%$$

Ketene dimer   Ketene

Ketenes can in turn be employed as useful reagents. They react rapidly with alcohols (with only a trace of acid catalyst) to form esters, and they undergo cycloaddition with alkenes to produce cyclobutanones. These reactions are illustrated in the following examples:

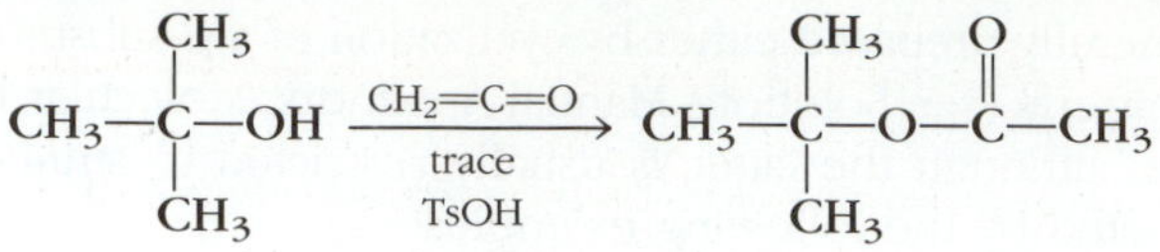

$$CH_3-\underset{CH_3}{\overset{CH_3}{C}}-OH \xrightarrow[\text{trace TsOH}]{CH_2=C=O} CH_3-\underset{CH_3}{\overset{CH_3}{C}}-O-\overset{O}{\overset{\|}{C}}-CH_3 \quad 86\text{–}89\%$$

$$\xrightarrow{(CH_3)_2C=C=O} \quad 77\%$$

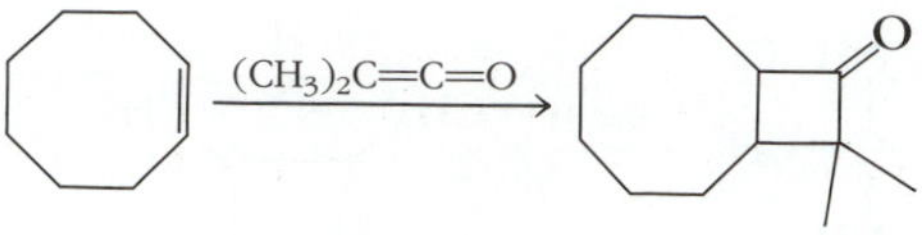

Five-Membered Ring Heterocycles

Various saturated or partially reduced derivatives of pyrrole, furan, and thiophene have synthetic importance. Some of these are used extensively as solvents in organic chemistry:

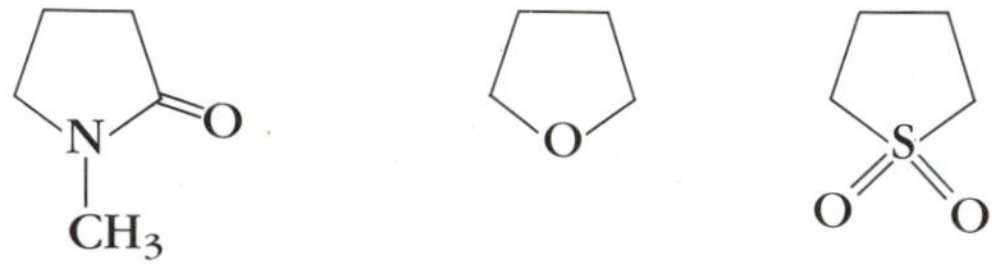

N-Methylpyrrolidone — Tetrahydrofuran (THF) — Sulfolane

The dioxolanes have been widely used as protecting groups for aldehyde and ketones (Sections 15.6 and 22.4).

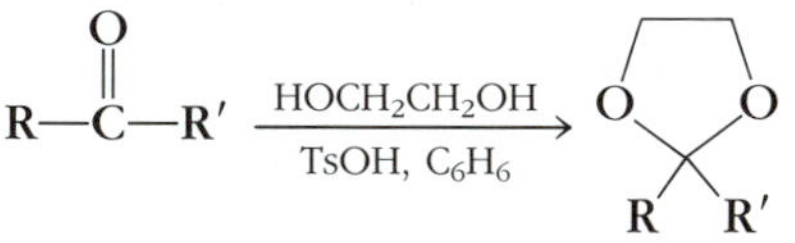

$$R-\overset{O}{\overset{\|}{C}}-R' \xrightarrow[\text{TsOH, } C_6H_6]{HOCH_2CH_2OH}$$

Pyrrolidine has frequently been employed for the preparation of enamines (Section 17.8).

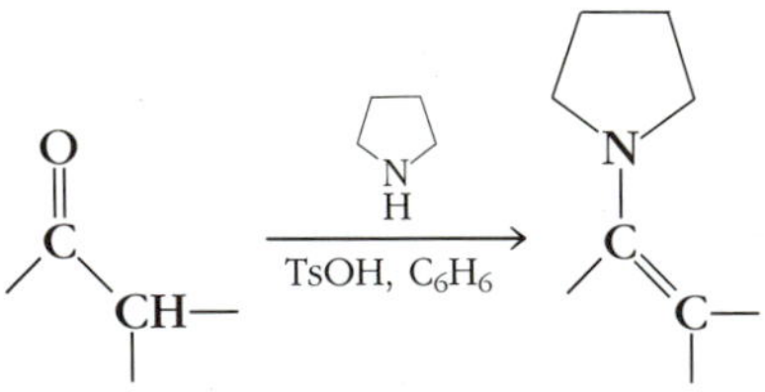

$$\xrightarrow[\text{TsOH, } C_6H_6]{\text{pyrrolidine}}$$

3-Sulfolene is a convenient source of 1,3-butadiene, as illustrated by the following Diels-Alder reaction:

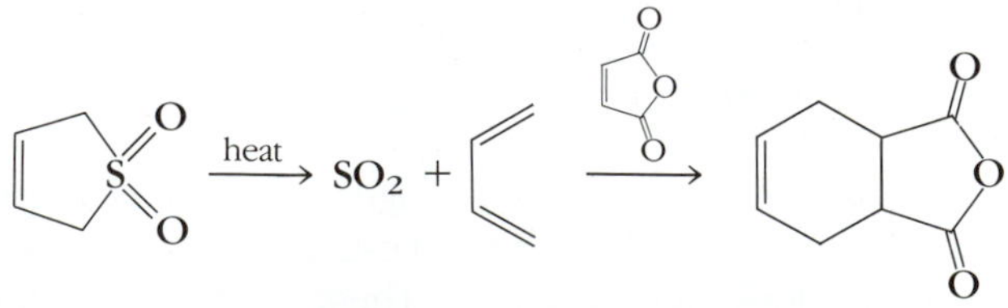

$$\xrightarrow{\text{heat}} SO_2 + \qquad 82\text{–}90\%$$

The preparation of five-membered-ring heterocycles can be carried out in a variety of ways. Sometimes the corresponding aromatic heterocycle can be reduced directly. Another useful method involves cyclization of 1,4-disubstituted alkanes via intramolecular nucleophilic displacement reactions.

$$\text{2-furylmethanol } (CH_2OH) \xrightarrow{H_2,\ Ni} \text{tetrahydrofurfuryl alcohol } (CH_2OH) \qquad 85\%$$

$$CH_3CH(OH)CH_2CH_2CH_2OH \xrightarrow[H_2SO_4]{H_3PO_4} \text{2-methyltetrahydrofuran } (CH_3) \qquad 80\%$$

$$CH_3CHClCH_2CH_2CH_2Cl \xrightarrow[K_2CO_3]{CH_3CH_2CH_2CH_2NH_2} \text{2-methyl-N-}(CH_2CH_2CH_2CH_3)\text{pyrrolidine} \qquad 43\%$$

Acyclic amines can be converted to pyrrolidines in a free radical cyclization reaction.

$$(CH_3CH_2CH_2CH_2)_2NH \xrightarrow[2.\ NaOH]{1.\ Cl_2} (CH_3CH_2CH_2CH_2)_2N\text{—}Cl \xrightarrow[\substack{2.\ NaOH \\ (neutralize)}]{1.\ H_2SO_4} \text{N-}(CH_2CH_2CH_2CH_3)\text{pyrrolidine} \qquad 70\text{–}80\%$$

Initial reaction of the amine with chlorine (or hypohalite) generates an *N*-haloamine (or the corresponding *N*-haloammonium ion), and this undergoes either thermal or photolytic cleavage.

$$R\text{—}NH\text{—}(CH_2)_3CH_3 \xrightarrow{Cl_2} R(H)\overset{+}{N}(Cl)\text{—}(CH_2)_3CH_3 \longrightarrow R(H)\overset{\cdot +}{N}\text{—}(CH_2)_3CH_3 + Cl\cdot$$

The intermediate cation-radical extracts a hydrogen via a six-membered-ring transition state, and the resulting carbon radical combines with the chlorine atom formed in the preceding step.

$$R(H)\overset{\cdot +}{N}\text{—}(CH_2)_3\text{—}CH_2\text{—}H \longrightarrow R(H)_2\overset{+}{N}\text{—}(CH_2)_3\text{—}\dot{C}H_2 \xrightarrow{Cl\cdot} R\overset{+}{N}H_2\text{—}(CH_2)_3\text{—}CH_2Cl$$

Finally, treatment with base generates the free amine, which can cyclize by displacement of chlorine.

R, $^{+}NH_2$, CH_2Cl —base→ R, NH, CH_2—Cl → ^{+}N(R)(H) —base→ N—R

Five-membered-ring heterocycles containing both nitrogen and oxygen have been exploited synthetically by A. I. Meyers and coworkers (Colorado State University) for the protection and activation of carboxyl groups. A carboxylic acid will react with a *β*-amino alcohol to form an *oxazoline*.

Br, C—OH, O —(H_2N, HO)→ Br, N, O > 90%

The "carbonyl carbon" of an oxazoline is resistant to attack by many organometallic and hydride reagents, but acidic hydrolysis will regenerate the carboxylic acid.

C_6H_5—C(=O), N, O —(H_2O, H_2SO_4)→ C_6H_5—C(=O), CO_2H > 90%

The *α* hydrogens are sufficiently acidic that the conjugate base can be prepared and alkylated. This is illustrated by the following example:

$CH_3CH_2CH_2C$(=O)OH —(H_2N, HO)→ $CH_3CH_2CH_2$—(N, O)

↓ C_4H_9Li

[CH_3CH_2CH(Li)—(N, O)] —(1. CH_3O, CH_3O, CHO; 2. H_2O)→ CH_3CH_2—CH—(N, O), CH—OH, CH_3O, CH_3O

94%

Six-Membered-Ring Heterocycles

Heterocyclic analogs of cyclohexane and cyclohexene are employed in a variety of ways for organic synthesis. Two of the nitrogen-containing compounds, piperidine and morpholine

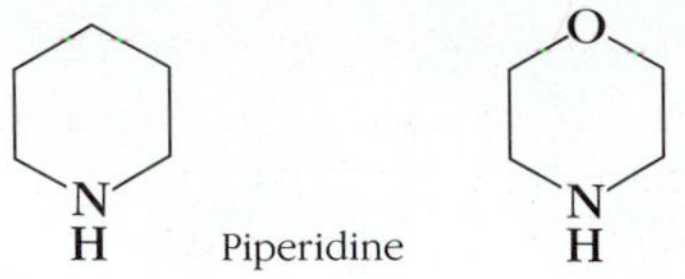

Piperidine Morpholine

have been used extensively for the preparation of enamines (Section 17.8).

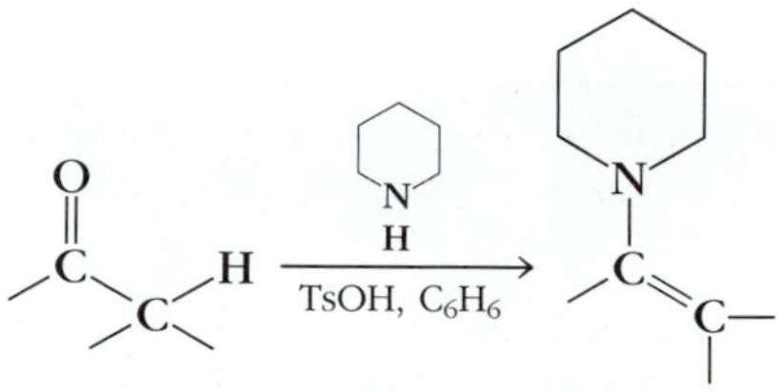

TsOH, C_6H_6

1,4-Dioxane has been used as a solvent in many organic reactions, particularly when a partially aqueous medium is required. Dioxane has excellent properties as a solvent for many organic compounds, and it is also miscible with water. Consequently, aqueous dioxane can frequently provide the desirable properties of both aqueous and organic solvents.

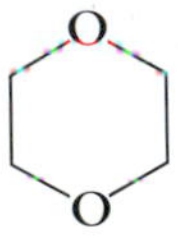

1,4-Dioxane

Acetals and ketals with six-membered rings are sometimes employed as protecting groups (Section 15.5).

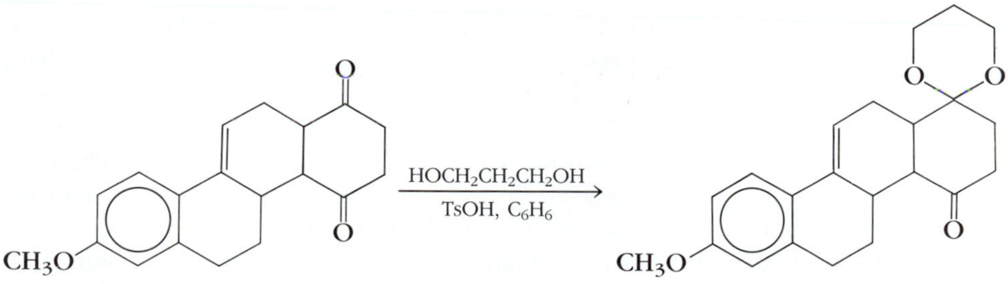

CH_3O $\xrightarrow[\text{TsOH, } C_6H_6]{HOCH_2CH_2CH_2OH}$ CH_3O 61%

In general, the five-membered ring analogs (dioxolanes) are more useful synthetic intermediates.

Pyrans are also important protecting groups for hydroxyl groups (Section 15.6).

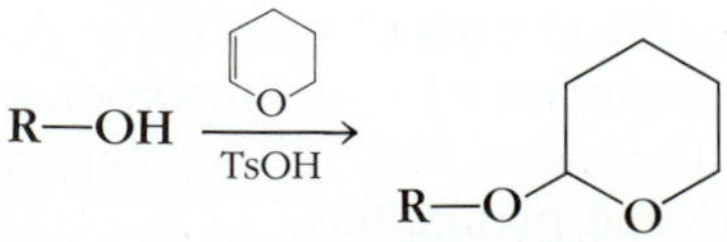

R—OH $\xrightarrow{\text{TsOH}}$ R—O

EXERCISE 24.8

Suggest how each of the following could be prepared readily from a nonaromatic heterocyclic compound:

(a) $CH_3OCH_2CH_2OH$ (b) $CH_3CH_2OCH_2CH_2—NH_2$ (c) $CH_2=C=O$

(d) 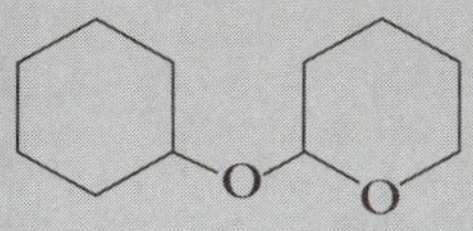 (e)

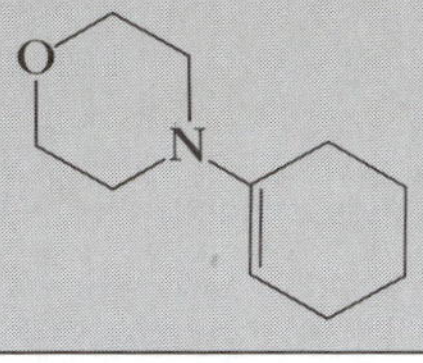

EXERCISE 24.9

Show how each of the following could be prepared from nonheterocyclic compounds.

(a) 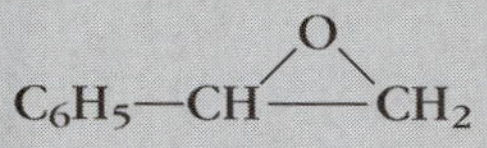 $C_6H_5—CH—CH_2$ (epoxide, O bridging) (b) $C_6H_5—CH—CH_2$ (aziridine, N–H bridging)

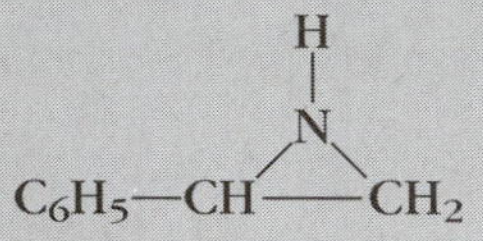

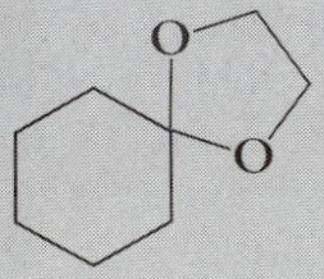

(c) (d) 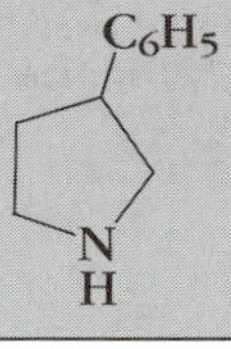 C_6H_5 (e) 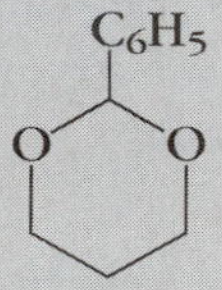 C_6H_5

24.4 PURINES AND PYRIMIDINES: BIOSYNTHESIS OF NUCLEIC ACID BUILDING BLOCKS

Of the vast number of heterocyclic ring systems that can be formed biochemically, we have chosen the purines and pyrimidines to illustrate biosynthesis. We have done this for two reasons. First, the purines and pyrimidines play critical roles in biochemistry, not the least of which is to serve as the basis of the genetic code (see Section 25.2). Second, these reactions will show you once again that the chemical reactions occurring in living cells are highly similar to those carried out in a flask by an organic chemist, although enzymatic catalysis permits much milder conditions for the biochemical reactions. Figure 24.2 shows the purines (adenine and guanine) and the pyrimidines (cytosine, uracil, and thymine) that are the major building blocks of nucleic acids. Also shown in the figure are the corresponding ribosyl and deoxyribosyl derivatives (nucleosides) and their monophosphates (nucleotides).

The carbon atoms of the purines and pyrimidine ring systems are derived from several different amino acids and from C_1 units corresponding to formic acid and carbon dioxide. The nitrogen atoms are either incorporated directly as ammonia or as part of an amino acid. These structural relationships are summarized in Figure 24.3 for the major purines and pyrimidines.

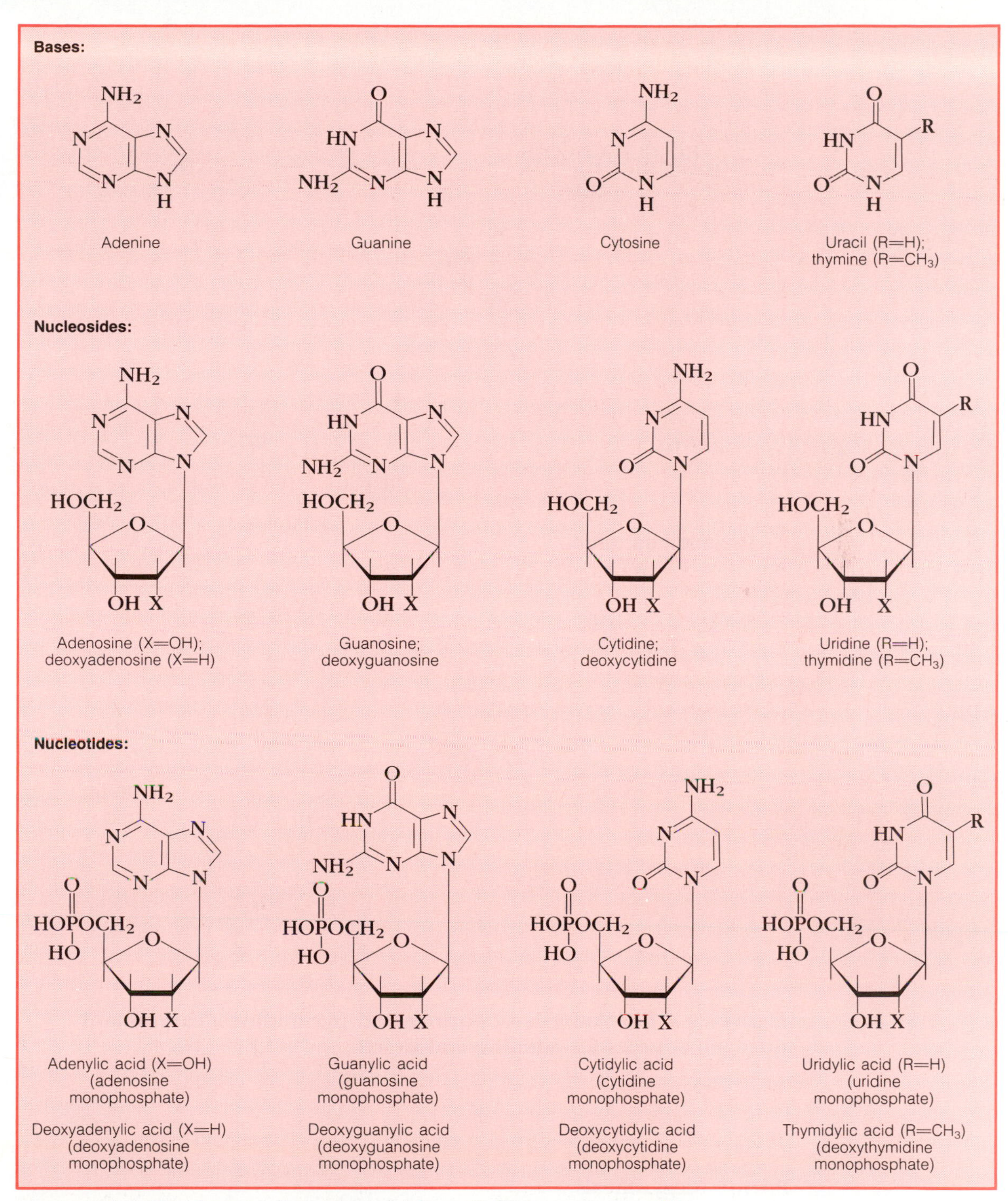

Figure 24.2 The bases of nucleic acids and their corresponding nucleosides and nucleotides.

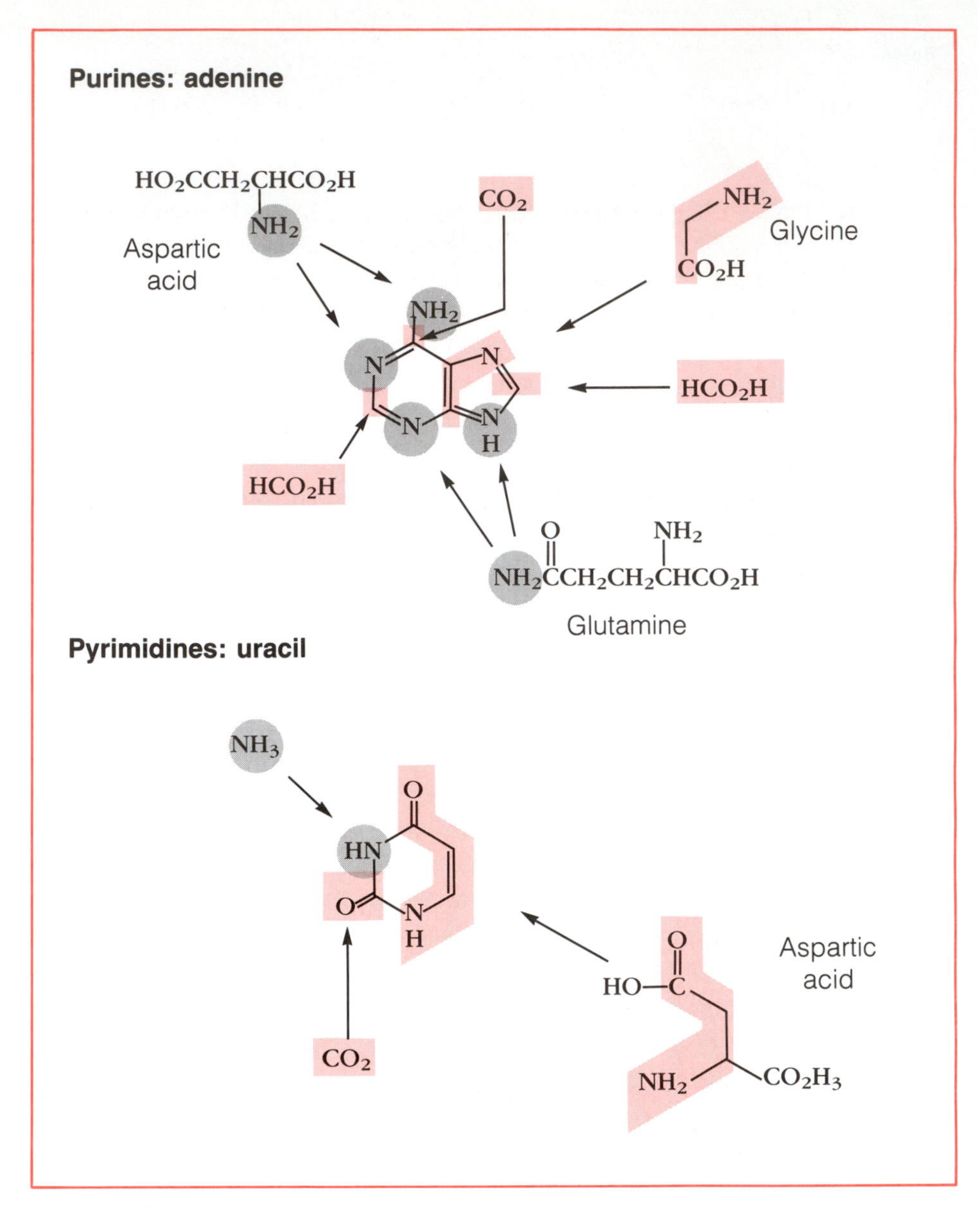

Figure 24.3 Biosynthesis of purines and pyrimidines: carbon and nitrogen sources for adenine and uracil.

Purine Biosynthesis

A key feature of purine biosynthesis is the origination of the pathway with ribose-5-phosphate; the sugar residue is present from the start. The C-1 hydroxyl group is activated by enzymatic phosphorylation, much as a laboratory chemist might convert an —OH to a tosylate.

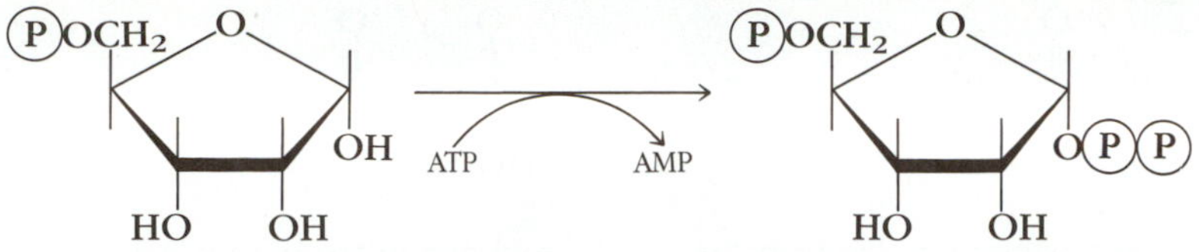

Two aspects of this equation require further discussion. First, in this chapter as in previous discussions (Section 12.9) we will employ —O—P, —O—P—P, and —O—P—P—P as convenient abbreviations for mono-, di-, and triphosphates. Second, there is a curved offshoot from the arrow between reactant and products. This is a device frequently employed in biochemistry to show the chemical fate of a reagent. In the present case adenosine triphosphate (ATP) is converted to adenosine monophosphate (AMP), the diphosphate group having been transferred to the ribose. The technique allows us to focus on the modification of the sugar, but at the same time we can keep track of changes in other reagents.

The diphosphate group at C-1 of the ribose is a good leaving group, and nucleophilic substitution by the amide nitrogen of glutamine (followed by hydrolysis of the amide linkage) produces the corresponding amino sugar.

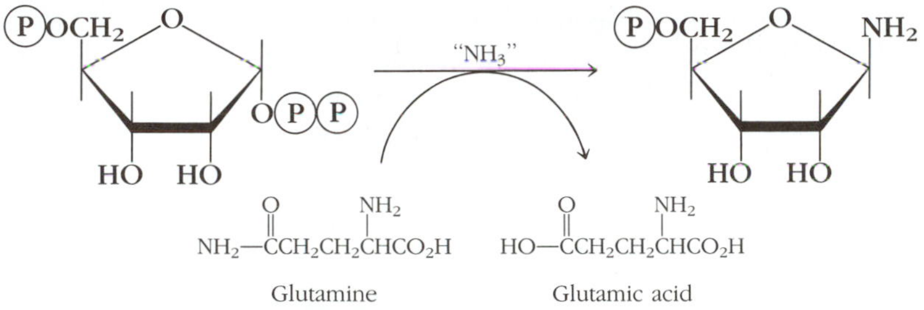

Glutamine Glutamic acid

Loss of pyrophosphate from the reactant would generate a relatively stable cation (an oxonium ion), but the observed stereochemistry is that of complete inversion. This reflects the ability of an enzyme to hold the reacting groups in a specific relative orientation.

For those parts of the biosynthesis following the introduction of the NH_2 group as a substituent on the sugar ring, we can simplify subsequent structural formulas by using *ribose-P* to represent the β-D-ribosyl-5-phosphate group. For example, the next step in purine biosynthesis, formation of an amide with the amino acid, glycine, can be represented in the following way:

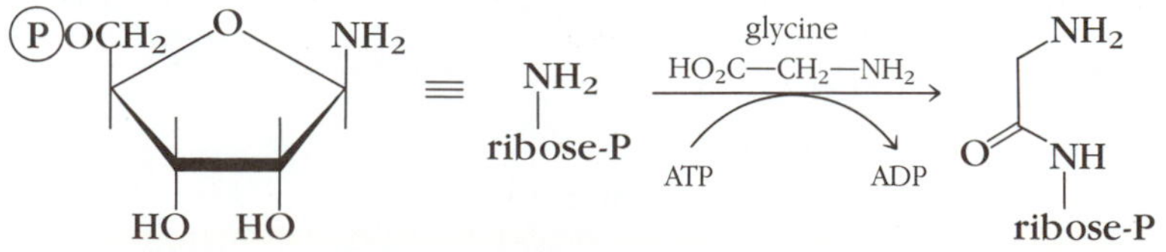

The requirement for ATP in this reaction should remind you that amines and carboxylic acids do not react directly to form amides at normal temperatures. In the laboratory a chemist would first convert the acid to the corresponding acid chloride, but the enzyme system instead converts glycine to a mixed anhydride of phosphoric acid, which is susceptible to nucleophilic attack at the carbonyl carbon.

$$HO-\overset{O}{\overset{\|}{C}}-CH_2NH_2 + ATP \longrightarrow HO-\underset{HO}{\underset{|}{\overset{O}{\overset{\|}{P}}}}-O-\overset{O}{\overset{\|}{C}}-CH_2NH_2 + ADP$$

The next two steps involve transfer of a formyl group from a derivative of tetrahydrofolic acid (the details of which we will not consider) and introduction of a second nitrogen from glutamine.

formyl tetrahydrofolic acid; glutamine "NH_3"; ATP → ADP; ribose-P

Once again, you see a reaction that would not occur directly under laboratory conditions, and the *driving force* is provided by ATP. The amide carbonyl is activated to nucleophilic attack by phosphorylation of the carbonyl oxygen (much like the reaction between $POCl_3$ and dimethyformamide that we showed in Section 24.1), and this results in subsequent conversion of the original C=O to C=NH.

In still another ATP-driven reaction, activation of the formyl group results in closure of the imidazole ring of the purine system. This compound can act as an enamine (Section 17.8), and it reacts with CO_2 to give the carboxylated intermediate.

ATP → ADP; CO_2; HO_2C; H_2N; ribose-P

Formation of an amide with aspartic acid (via a mixed anhydride) is followed by an enzymatic elimination reaction, which yields fumaric acid and the simple amide.

HO_2C; H_2N; ribose-P; CO_2H–CH_2–$CHNH_2$–CO_2H; ATP → ADP; CO_2H, H—CH, HO_2C—CH—NHC(=O); elimination; HO_2C–CH=CH–CO_2H + H_2N–C(=O) imidazole, H_2N, ribose-P

The final carbon atom of the purine skeleton is introduced as another formyl group, and ring closure is followed by dehydration to yield inosinic acid.

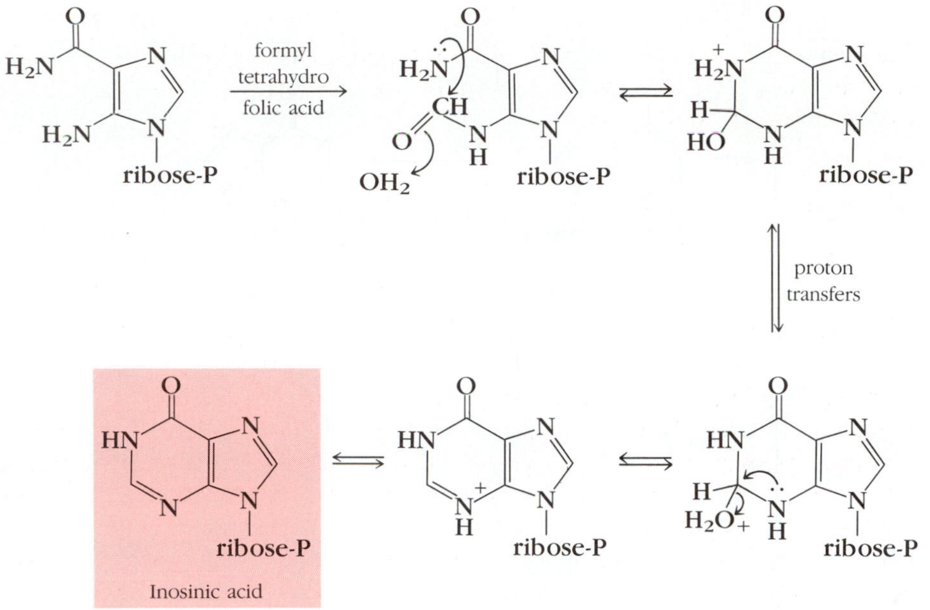

formyl tetrahydro folic acid

ribose-P

proton transfers

Inosinic acid

The purine nucleotides of major importance in nucleic acid biochemistry are all derived from inosinic acid, as outlined in Figure 24.4. The two stages of purine biosynthesis therefore are: (1) the buildup of the heterocyclic base on the ribosyl-5-phosphate group; and (2) subsequent modifications such as conversion to adenylic acid and guanylic acid. Reduction to the deoxynucleotides occurs only after the purine synthesis is completed.

Pyrimidine Biosynthesis

A major difference in the biosynthesis of pyrimidine from that of the purines is the formation of the heterocyclic ring *prior to* introduction of the sugar residue. The initial reaction in the sequence results in the formation of carbamyl phosphate from carbon dioxide, ammonia, and ATP.

$$NH_3 + CO_2 + ATP \xrightarrow{\text{enzyme}} NH_2{-}\overset{O}{\overset{\|}{C}}{-}O{-}\overset{O}{\overset{\|}{P}}O_2H_2 + ADP$$

Carbamoyl phosphate

In the next step the activated carbonyl group of carbamoyl phosphate is attacked by the NH_2 group of the amino acid aspartic acid.

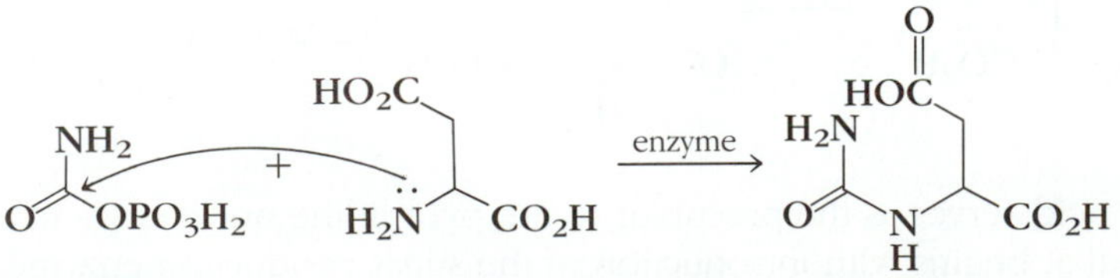

enzyme

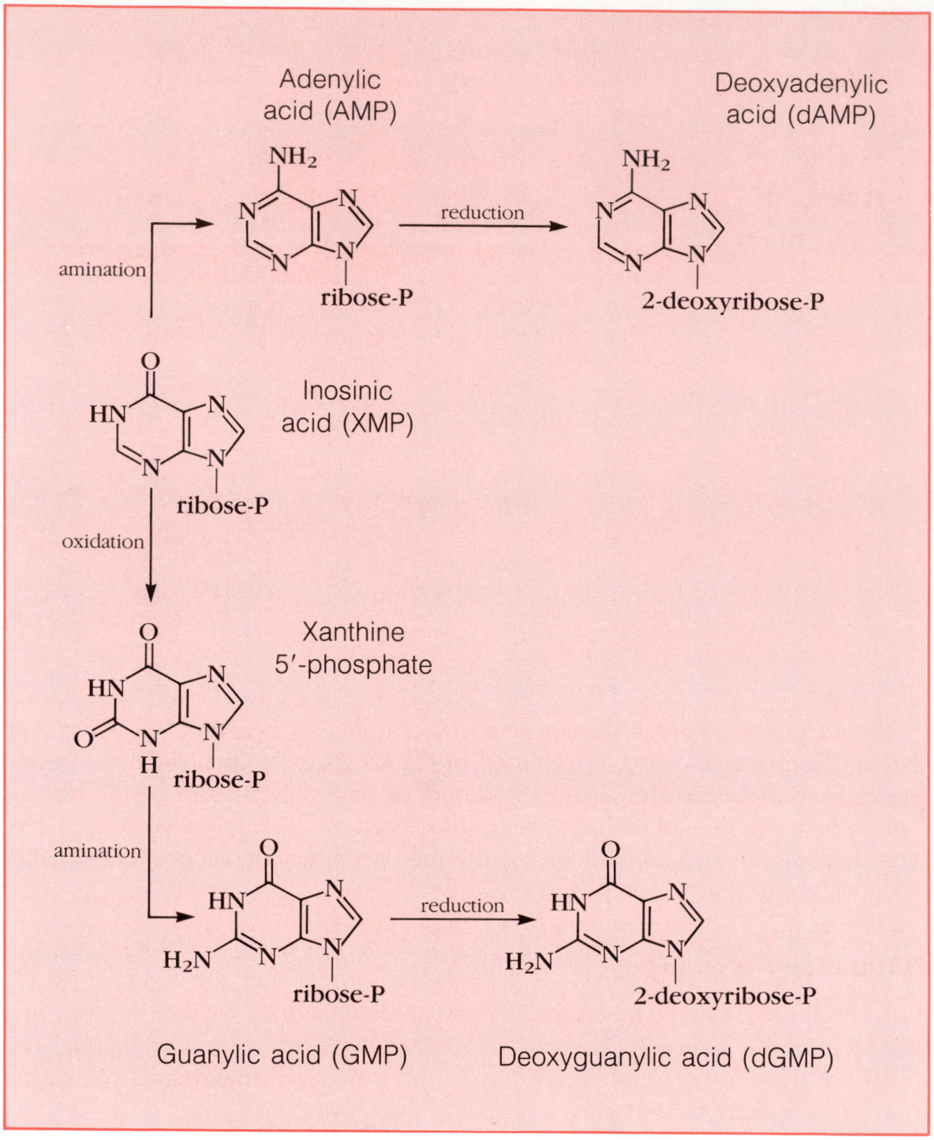

Adenylic acid (AMP)

NH_2

N N N N

ribose-P

reduction

Deoxyadenylic acid (dAMP)

NH_2

N N N N

2-deoxyribose-P

amination

O

HN N N N

ribose-P

Inosinic acid (XMP)

oxidation

O

HN N O N N H

ribose-P

Xanthine 5′-phosphate

amination

O

HN N H_2N N N

ribose-P

reduction

O

HN N H_2N N N

2-deoxyribose-P

Guanylic acid (GMP)

Deoxyguanylic acid (dGMP)

Figure 24.4 Inosinic acid as the precursor of purine nucleotides and deoxynucleotides.

The resulting intermediate undergoes enzymatic ring closure with loss of water, and oxidation produces a pyrimidine, orotic acid.

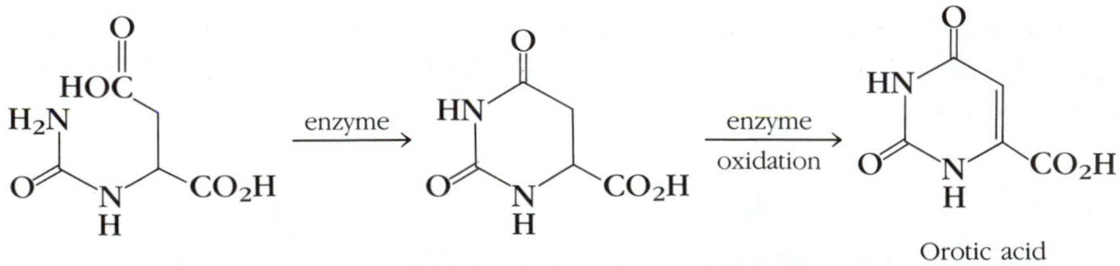

O HOC H_2N O N H CO_2H —enzyme→ O HN O N H CO_2H —enzyme, oxidation→ O HN O N H CO_2H

Orotic acid

Orotic acid serves as the precursor to the pyrimidine nucleotides in a series of reactions that begins with introduction of the sugar residue. An enzyme-catalyzed reaction with the pyrophosphate derivative of ribose-5-phosphate proceeds with complete inversion of configuration at C-1 of the ribose (as is also observed in the initial step of purine biosynthesis).

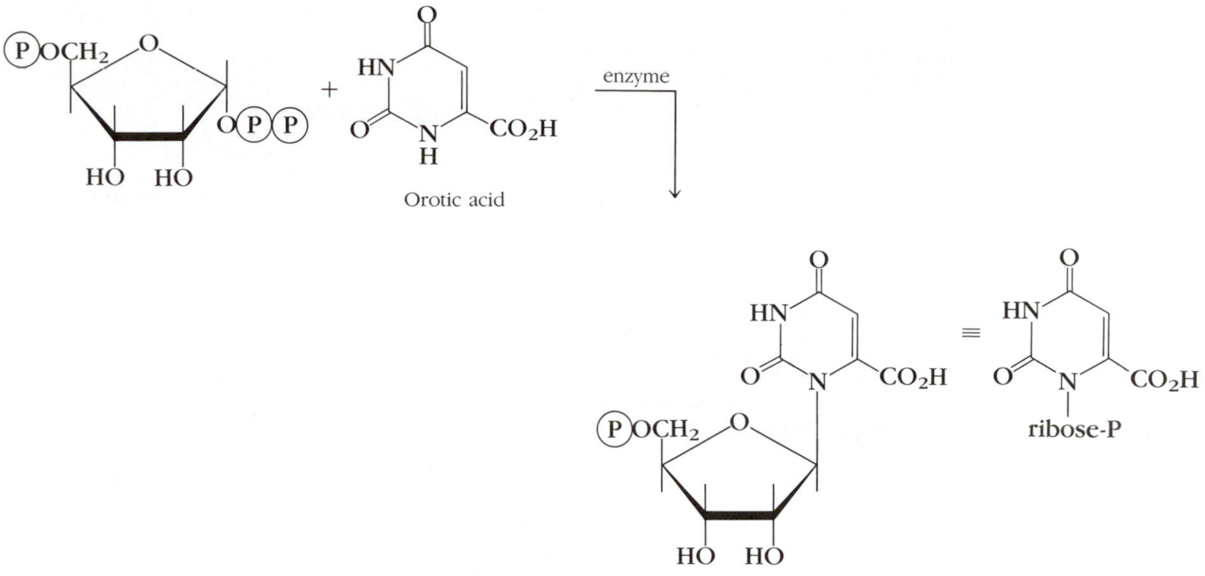

Orotic acid

enzyme

Orotidine-5-phosphate

Further modification of orotidine-5-phosphate produces the other pyrimidine nucleotides, as illustrated in Figure 24.5.

EXERCISE 24.10

Draw structures similar to those in Figure 24.3 to show the origin of the individual carbon and nitrogen atoms for each of the purines and pyrimidines discussed in this section.

27.5 HETEROCYCLES AS MEDICINAL AGENTS

A vast array of heterocyclic compounds have been investigated as pharmaceuticals, and many of these have enjoyed great medical success. We showed you several examples in our discussion of alkaloids in Section 23.4, but space precludes a broad survey of the topic. Instead we will discuss three specific classes of heterocyclic compounds that are medically important: the *barbiturates,* the *sulfa drugs,* and the *β-lactam antibiotics.*

The Barbiturates

During the mid-1800s chemists were attempting to deduce the structure of the naturally occurring purine uric acid. The German chemist Baeyer isolated a key degradation product, which he named (for a friend, Barbara) *barbituric acid.*

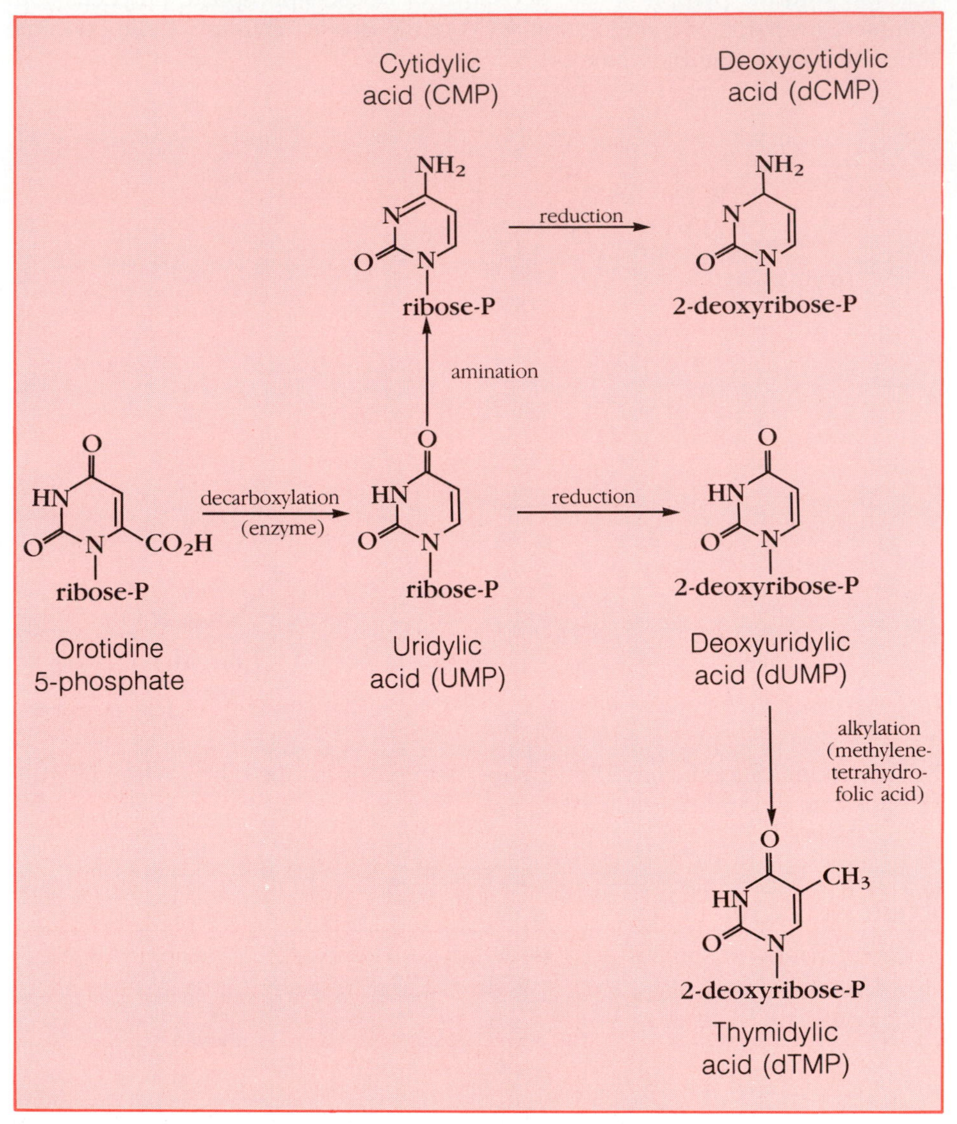

Figure 24.5 Biosynthesis of pyrimidine nucleotides.
The pathways have been oversimplified by abbreviation of all the sugar residues as ribose-P or 2-deoxyribose-P. The enzymatic reactions involve some of the di- and triphosphates as well.

degradation

Uric acid

Barbituric acid

The study of uric acid in turn led to investigations of barbituric acid and its derivatives.

Barbituric acid was first synthesized directly from urea and malonic acid (by reaction with PCl_3), but the base-catalyzed condensation of urea with diethyl malonate was later found to be a more general procedure.

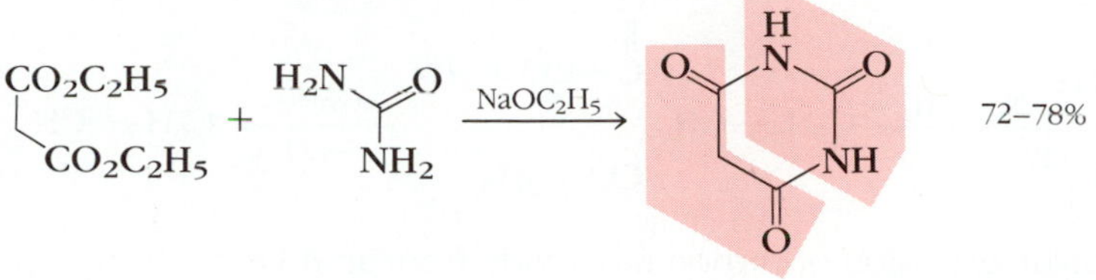

$$CH_2(CO_2C_2H_5)_2 + H_2NC(=O)NH_2 \xrightarrow{NaOC_2H_5} \text{barbituric acid} \quad 72–78\%$$

This synthetic approach to barbituric acid derivatives has been employed extensively, and its use by Emil Fischer led to the first commercial drug, *Veronal (barbital),* in 1903.

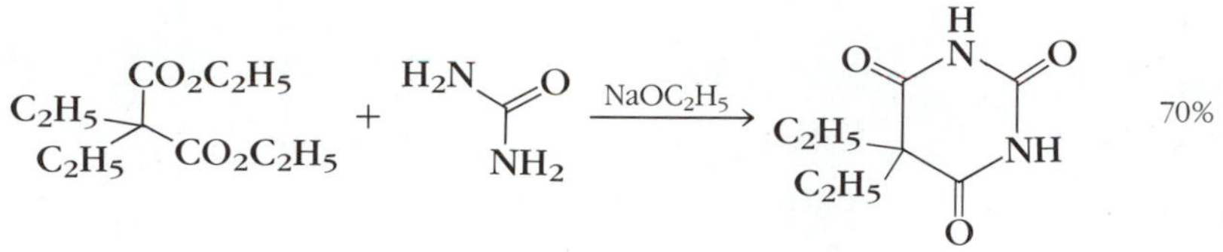

$$(C_2H_5)_2C(CO_2C_2H_5)_2 + H_2NC(=O)NH_2 \xrightarrow{NaOC_2H_5} \text{5,5-diethylbarbituric acid} \quad 70\%$$

The barbiturates as a class act as sedatives, but there is substantial variation of activity among the different derivatives. They have long been used as sleeping pills and are sometimes classified as *sedative hypnotics.* They have been prescribed less frequently by physicians in recent years because dangerous side effects have been recognized. A major difficulty with most of these drugs is their poor *therapeutic ratio* (the ratio of the toxic dose to the optimal dosage). In addition, many barbiturates are addictive and they also tend to exhibit pronounced adverse interactions with other drugs, particularly with CNS (central nervous system) depressants. The interactions with alcohol and with tranquilizers can be fatal and have resulted in many deaths. The barbiturates are properly used only under close medical supervision because of the high potential for accidental death or suicide in depressed patients.

Despite these disadvantages, the clinical need for safe and effective sedatives stimulated the synthesis of a wide variety of barbituric acid derivatives. Among the compounds that have been used commercially are *pentobarbital* (sometimes sold under the trade name *Nembutal, phenobarbital,* and *secobarbital* (sold as *Seconal*).

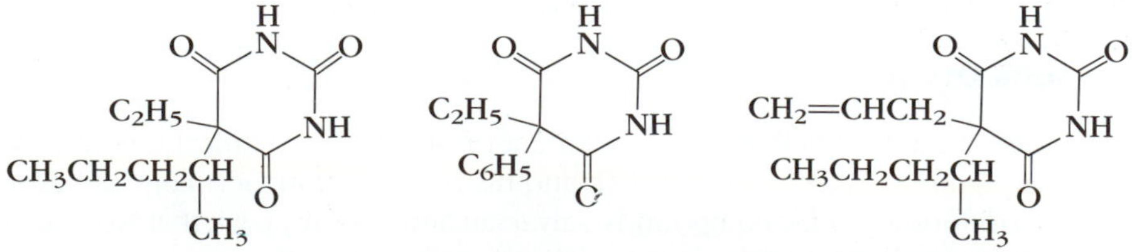

Pentobarbital: 5-ethyl-5-($CH_3CH_2CH_2CH(CH_3)$)barbituric acid

Phenobarbital: 5-ethyl-5-phenyl (C_2H_5, C_6H_5) barbituric acid

Secobarbital: 5-($CH_2{=}CHCH_2$)-5-($CH_3CH_2CH_2CH(CH_3)$)barbituric acid

Pentobarbital Phenobarbital Secobarbital

It is not always possible to start with diethyl malonate, but these barbituric acids can all be synthesized by condensation of the appropriately substituted malonic ester with urea. The use of a substituted malonic ester is illustrated by the following synthesis of phenobarbital. In the first step a condensation between ethyl phenylacetate and diethyl oxalate affords the required phenyl-substituted malonic ester.

$$C_2H_5O{-}C(=O){-}C(=O){-}OC_2H_5 + C_6H_5CH_2CO_2C_2H_5 \xrightarrow{NaOC_2H_5} C_6H_5{-}CH(C(=O){-}CO_2C_2H_5)(CO_2C_2H_5) \xrightarrow[-CO]{\text{pyrolysis}} C_6H_5{-}CH(CO_2C_2H_5)_2$$

The pyrolytic extrusion of carbon monoxide from an α-keto ester is a reaction we have not considered in detail, although we showed you an example previously in Section 23.2 (the conversion of **5-15** to **5-16**).

Alkylation of the phenylmalonate with ethyl iodide generates the appropriately disubstituted malonic ester, and phenobarbital is produced by condensation with urea.

$$H(C_6H_5)C(CO_2C_2H_5)_2 \xrightarrow[2.\ C_2H_5I]{1.\ NaOC_2H_5} (C_2H_5)(C_6H_5)C(CO_2C_2H_5)_2 \xrightarrow[NaOC_2H_5]{NH_2{-}C(=O){-}NH_2} \text{phenobarbital}$$

87%

One of the medical uses for which barbiturates have been employed is the management of convulsive disorders. Recently it has been found that *O*-alkylation of the salt of phenobarbital produces a compound that may be quite effective for this medical application.

$$\text{phenobarbital} \xrightarrow[2.\ Cl{-}CH_2OCH_3]{1.\ KOtBu} \text{4,6-bis}(CH_3OCH_2O)\text{-5-}C_2H_5\text{-5-}C_6H_5\text{-pyrimidin-2(5H)-one analog}$$

64%

The *O*-alkylated compound has reduced hypnotic activity, so its dangerous side effects are alleviated, and at the same time its anticonvulsant activity is greater than that of the other barbiturates. Unfortunately, the relationship between this small structural change and the resulting changes in biological activity is not yet fully understood.

The Sulfa Drugs

At the start of the twentieth century treatment of disease by chemical methods was extremely limited. By 1912 Paul Ehrlich and his coworkers in Germany had developed two different arsenic compounds, salvarsan and neosalvarsan, that were active against *Treponema palladium,* the causative agent of syphilis.

$$Cl^-\ H_3N^+,\ HO{-}C_6H_3{-}As{=}As{-}C_6H_3{-}OH,\ NH_3^+\ Cl^-$$

Salvarsan

$$HO{-}C_6H_4{-}As{=}As{-}C_6H_3(NH_2)(NH{-}CH_2{-}SO_3^-\ Na^+){-}OH$$

Neosalvarsan

Neosalvarsan remained the standard treatment for syphilis until the 1940s, when effective antibiotics were developed.

A few other chemical substances were discovered that were effective against certain pathogenic microorganisms, but these had very limited applications. Not until 1932, with the development of the sulfa drugs, did chemical agents become available that were effective against a fairly wide range of bacterial infections. The sulfa drugs are still used medically, and most are structurally related to *sulfanilamide* (*p*-aminobenzenesulfonamide). Their effectiveness is related to the enzymatic processes in bacteria that produce folic acid and related compounds such as the tetrahydrofolic acid derivatives necessary for purine biosynthesis (Section 24.4). Sulfanilamide is structurally similar to *p*-aminobenzoic acid, and it can block the enzyme site at which the latter is normally incorporated into the folic acid skeleton.

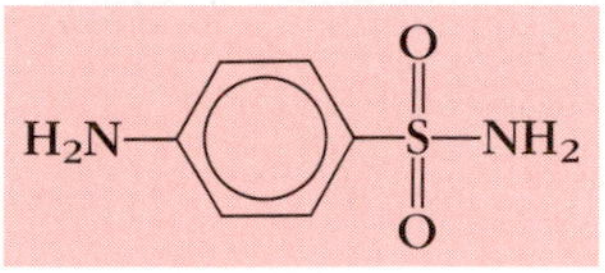

Sulfanilamide

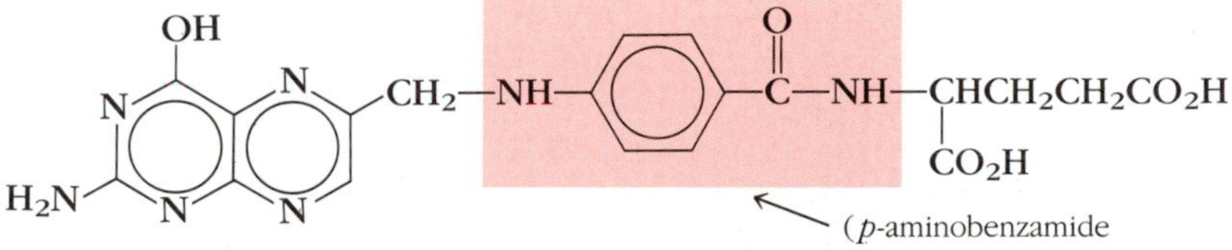

Folic acid

(*p*-aminobenzamide subunit)

Sulfanilamide therefore inhibits folic acid biosynthesis, and this in turn restricts the synthesis of nucleic acid. This chain of events prevents multiplication of the bacteria, but why does it not also harm the human patient? The answer is surprisingly simple: humans (and other animals) do not have the enzyme systems for biosynthesis of folic acid, so there is no enzyme for the drug to inhibit. We require folic acid as part of our diets; in other words it is a vitamin, which we obtain from plant sources.

Despite the development of many other effective antibiotics, sulfa drugs are still widely used in modern medicine. The combination of sulfamethoxazole and trimethoprim (a structurally unrelated pyrimidine derivative) is frequently prescribed for certain common infections. This combination is marketed under trade names such as *Septra* and *Bactrim,* and you can readily see from its structural formula the relationship of sulfamethoxazole to sulfanilamide.

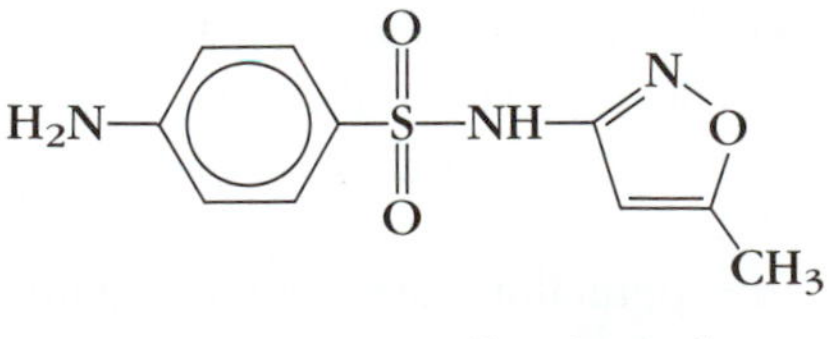

Sulfamethoxazole

The *β*-Lactam Antibiotics

Discovery of the remarkable specificity of the sulfa drugs in interfering with the biochemistry of infectious bacteria but not of the host animal represented a major advance in modern medicine. The next step was the discovery of naturally occurring antibiotics, organic compounds with antibacterial activity that are biosynthesized by fungi or other microorganisms. These compounds apparently play a defensive role in the organisms that manufacture them, inhibiting the growth of other, competing organisms. This phenomenon was discovered by Alexander Fleming in 1928 when a bacterial culture was allowed to remain unattended for several days. He found that a mold (later identified as *Penicillium notatum*) had grown in part of the culture medium. In the region where the mold was growing the bacterial growth had terminated, and he correctly interpreted this to mean that the mold had produced a substance that was toxic to the bacteria. Fleming gave this unknown material the name *penicillin.*

Although Fleming recognized that penicillin was not toxic to animals, the medical applications were not pursued actively for almost 10 years. By 1942 a yellow powder had been isolated from the mold and used to successfully treat meningitis. During the 1940s the efforts of a large number of chemists around the world showed that *penicillin* was not a single compound but was instead a mixture of structurally related compounds. All of these compounds shared a basic heterocyclic skeleton, differing only in the nature of the side chain (abbreviated as R in the following structure).

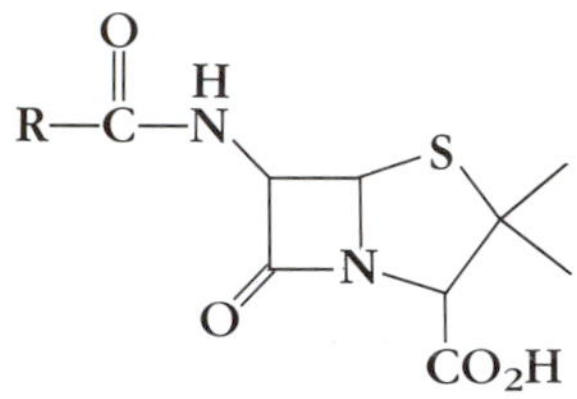

Penicillin antibiotics
(general structure)

The mode of action of the penicillins involves interference with the biosynthesis of *cell walls* in bacteria. The cell wall of a bacterium plays a major structural role, and survival of the organism requires an intact cell wall. The penicillins inhibit an enzyme that causes cross-linking (i.e., formation of chemical bonds between subunits on different polymer chains) of the polysaccharide responsible for the physical strength of the cell wall. The polysaccharide has amino acid substituents, and the cross-linking occurs by formation of amide bonds to the amino acid groups. The active site of this enzyme reacts with the strained amide linkage of the β-lactam antibiotic and becomes inactivated. Bacterial growth is thereby inhibited, because cell division produces deficient organisms, which do not have intact cell walls and cannot survive.

Like the sulfa drugs, the penicillins are specific against bacteria, in this case because the cell wall structure is not present in animals. Human cells, for example, are surrounded by *cell membranes,* but these are not made up of polysaccharides. Consequently, our biochemistry is not directly affected by penicillins (although allergic reactions are fairly common).

The compound known as *penicillin G* was of major importance in early use of penicillin antibiotics, and it was found that it could be prepared in large quantities by fermentation of *Penicillium notatum* in a culture medium enriched in phenylacetic acid. Further modifications in the medical properties of the penicillins were still needed, however. The improvements became possible through an understanding of the chemistry and biochemistry of these compounds.

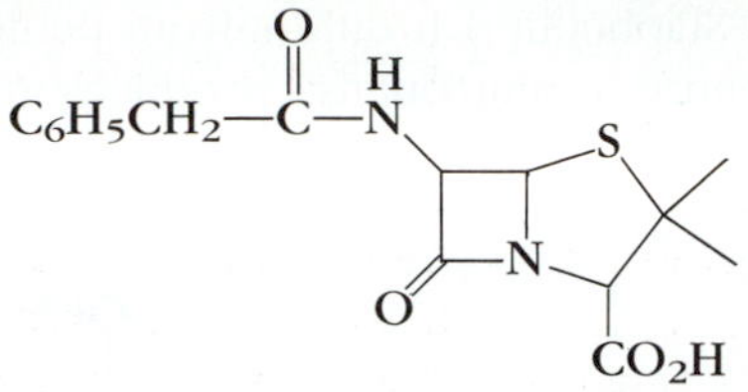

Penicillin G
(benzyl penicillin)

As you should expect, the bicyclic ring system of the penicillins is highly strained, particularly the four-membered-ring portion. This four-membered ring lactam is known as a β-lactam, and that description is used for the entire class of related antibiotics. (We have already discussed these briefly in Section 16.8.) The β-lactam group is readily cleaved by hydrolysis with either acid or base catalysis. Consequently, the high acidity of gastric juices (and the alkaline environment of the duodenum, the beginning of the small intestine) severely limited oral administration of the penicillins. Moreover, some strains of bacteria developed enzymes (called β-lactamases) that catalyze hydrolysis of the β-lactam.

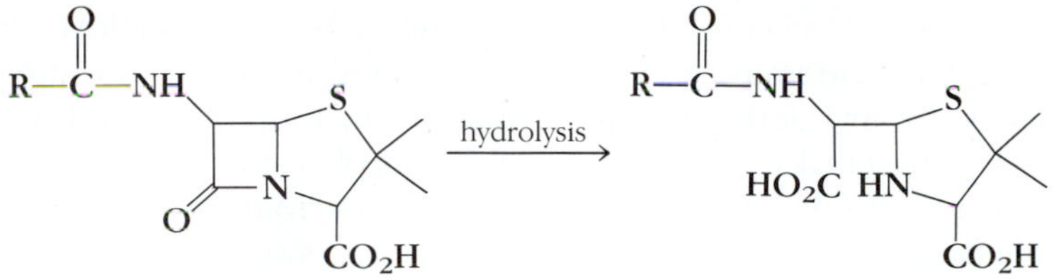

hydrolysis

Despite their highly successful use in treating certain infectious diseases, the limitations of the early penicillins prompted an intensive search for analogs that would have good antibiotic activity but would also be more stable to hydrolysis, whether acid-catalyzed, base-catalyzed, or enzymatic. Total synthesis of these complex molecules is impractical on a large scale, but when *Penicillium notatum* is cultured in an appropriate growth medium, 6-aminopenicillic acid can be isolated in large quantities. This material can in turn be modified, either by simple acylation of the 7-amino group or by more complex series of reactions. Such procedures have led to a number of useful antibiotics.

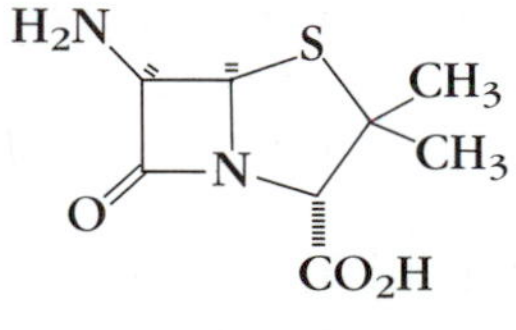

6-Aminopenicillic acid
(the key subunit for the
penicillin antibiotics)

One successful approach to preparation of a penicillin derivative that is more resistant to hydrolysis employed the logic of basic organic chemistry. The rationale was to use a hindered carboxylic acid so that subsequent degradation by bacterial β-lactamases would be more difficult. The use of 2,6-dimethoxybenzoic acid as the precursor of the carboxyl portion of the molecule produced such a compound, *methicillin* (marketed as Staphcillin.) It differs from penicillin G only by the two methoxy groups on the benzene ring, but this produces much greater resistance to bacterial inactivation.

$$\text{2,6-(OCH}_3)_2\text{C}_6\text{H}_3\text{—C(=O)—Cl} + \text{6-aminopenicillanic acid (H}_2\text{N, S, N, O, CO}_2\text{H)} \xrightarrow[\text{2. HCl; 3. NaHCO}_3]{\text{1. Et}_3\text{N}} \text{2,6-(OCH}_3)_2\text{C}_6\text{H}_3\text{—C(=O)—NH—(penam, CO}_2\text{H)}$$

This compound was prepared by reaction of the substituted benzoyl chloride with 6-aminopenicillic acid in the presence of triethylamine. The reaction mixture was extracted with hydrochloric acid to remove any unreacted amine, and the resulting penicillin derivative was converted to its sodium salt for isolation.

Another modification of penicillin G that has led to improved oral activity (i.e., better stability at nonneutral pH) is the use of an amino acid rather than a simple carboxylic acid to form the side chain. The use of *two polyfunctional* compounds, an amino acid and 6-aminopenicillic acid, presents new problems, and the techniques used in peptide synthesis (Sections 20.3 and 20.4) were employed to generate the amide. First, the amino group of α-aminophenylacetic acid was protected as the benzyloxycarbonyl derivative, and a mixed anhydride of the carboxyl group was formed using ethyl chloroformate. Acylation of 6-aminopenicillin with this mixed anhydride afforded the 4-benzyloxycarbonyl derivative of the desired product, and the blocking group was removed by hydrogenolysis.

$$\text{C}_6\text{H}_5\text{—CH(NH}_2\text{)—CO}_2\text{H} \xrightarrow{\text{C}_6\text{H}_5\text{CH}_2\text{OC(=O)Cl}} \text{C}_6\text{H}_5\text{—CH(NH—Cbz)—C(=O)—OH} \quad 100\%$$

$$\xrightarrow[\text{N(C}_2\text{H}_5)_3]{\text{Cl—C(=O)—OC}_2\text{H}_5} \text{C}_6\text{H}_5\text{—CH(NH—Cbz)—C(=O)—O—C(=O)—OC}_2\text{H}_5$$

$$\xrightarrow[\text{NaHCO}_3]{\text{6-aminopenicillanic acid}} \text{C}_6\text{H}_5\text{—CH(NH—Cbz)—C(=O)—NH—(penam, CO}_2\text{H)} \quad 45\%\ \text{overall (2 steps)}$$

$$\xrightarrow[\text{2. HCl}]{\text{1. H}_2\text{, Pd}} \text{C}_6\text{H}_5\text{—CH(NH}_2\text{)—C(=O)—NH—(penam, CO}_2\text{H)} \quad 40\%$$

Ampicillin

18% overall

The resulting penicillin derivative, *ampicillin,* has been widely and successfully used for treatment of a variety of bacterial infections by oral administration.

Another widely used penicillin derivative, *amoxicillin* (also sold as *Amoxil*), is closely related to ampicillin but has the amino and amide nitrogens of the side chain tied up as a heterocyclic ring analogous to a ketal. Amoxicillin is stable to the acidic conditions of the stomach, and it is widely used for oral administration. The acetone "ketal" undergoes slow hydrolysis in the intestine, and the active antibiotic is actually the *p*-hydroxy derivative of ampicillin.

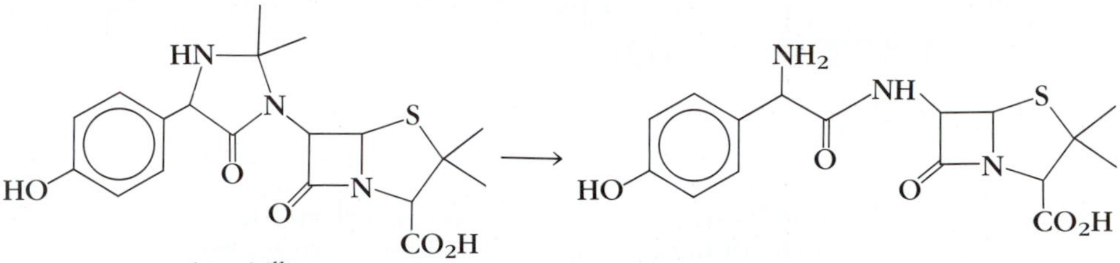

Amoxicillin

Amoxicillin can be prepared by a sequence analogous to that shown previously for ampicillin. The final step, reaction with acetone, is catalyzed by triethylamine.

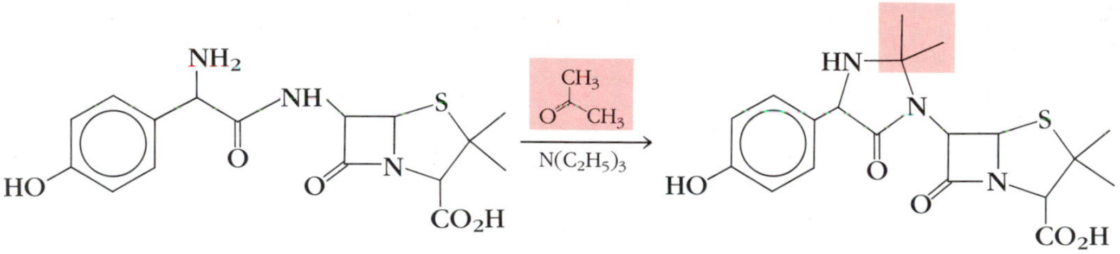

$N(C_2H_5)_3$

The idea that penicillins play a defensive role for *Penicillium notatum* suggested that antibiotics might be obtained from other fungi as well. Furthermore, the most likely place to find antibiotics that are effective against pathogenic organisms might be in places where fungi are growing in the midst of such pathogens. This line of reasoning led to a continuing research effort, and in the 1950s a new organism, *Cephalosporium,* was isolated from sewage effluent on the island of Sardinia, off the coast of Italy. This organism produces a mixture of antibiotics that are structurally related to the penicillins. The parent bicyclic system, 7-aminocephalosporanic acid, differs from the corresponding penicillic acid in that the sulfur atom is connected to a carbon atom corresponding to one of the methyl groups of the penicillins to form a six- rather than a five-membered ring. The *cephalosporins* exhibit greater stability toward acid and toward β-lactamase activity than do the penicillins.

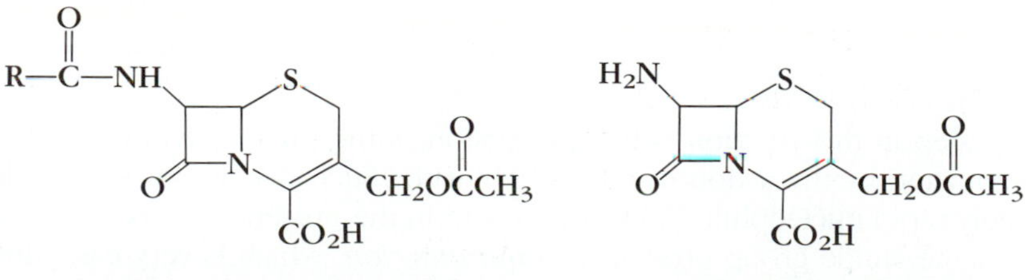

Cephalosporins (general structure)

7-Aminocephalosporanic acid

As with the penicillins, a wide variety of cephalosporins have been synthesized in an effort to obtain compounds with good antibiotic activity and minimal side effects. Among the cephalosporins used clinically are *cephalexin* and *cefadroxil* (analogs of ampicillin and amoxicillin), and *cefaclor* (marketed as Ceclor).

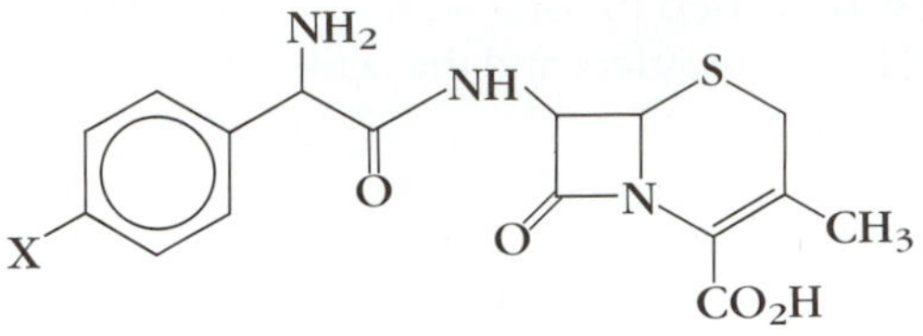

Cephalexin, X = H
cefadroxil, X = OH

Cefaclor

Notice that the allylic acetoxy group of the natural cephalosporins is not necessary for the antibiotic activity of these compounds. All three are semisynthetic derivatives prepared by modification of the naturally occurring material.

In contrast to the case of the penicillins, it was not possible to obtain the bicyclic nucleus, 7-aminocephalosporanic acid, directly from a fungal culture. Consequently, a chemical removal of the amino acid side chain of *cephalosporin C,* the natural material, had to be developed. But how could this amide linkage be cleaved in the presence of the much less stable β-lactam?

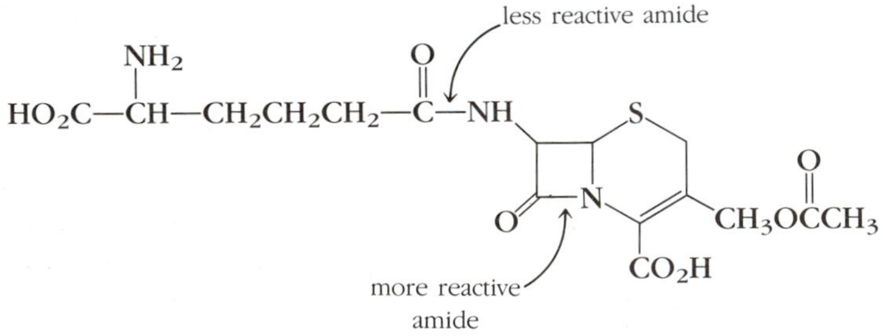

less reactive amide

more reactive amide

Cephalosporin C

The answer was to employ a highly selective reaction using nitrosyl chloride in formic acid:

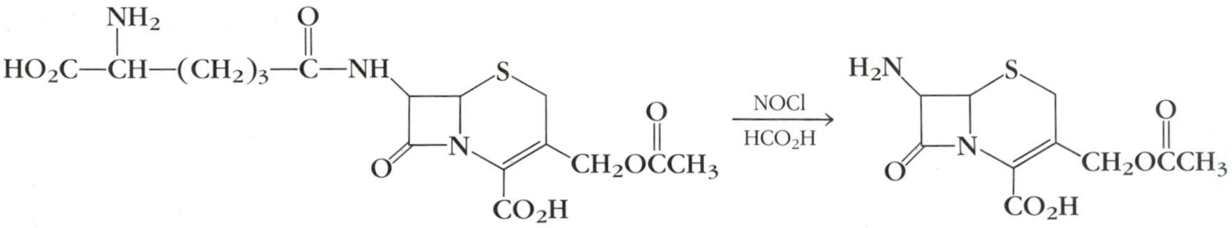

$\xrightarrow[HCO_2H]{NOCl}$

40%

The key step in this reaction is the conversion of the primary amino group to a diazonium ion by the action of nitrosyl chloride. Such diazonium ions undergo extremely rapid nucleophilic substitution, and in the present case *intramolecular* attack by the amide group produces an *iminium ion,* which is very easily hydrolyzed by water. The mechanism is illustrated in the following sequence:

H_2N CO_2H
CH
O
C—NH
S
N
O
O
CH_2OCCH_3
CO_2H

NOCl
HCO_2H

$N{\equiv}N^+$ CO_2H
O
C—NH
O

CO_2H
O
NH
$+OH_2$
O

H_2O :

CO_2H
O
$\overset{+}{N}$
H
O

proton transfers

CO_2H
O
$\overset{+}{N}H_2$
HO :
O

H_2N
S
N
O
O
CH_2OCCH_3
CO_2H

Conversion of 7-aminocephalosporanic acid to an antibiotic such as *cephalexin* again required the techniques of peptide synthesis (Sections 20.3 and 20.4). The amide side chain was introduced by a mixed anhydride reaction, in which the amino group of the amino acid was protected as the *tert*-butoxycarbonyl derivative.

O
$(CH_3)_3C—O—C$
NH O
$C_6H_5—CH—C—OH$

1. $N(C_2H_5)_3$
2. $(CH_3)_2CHCH_2OC—Cl$
O

O
$(CH_3)_3C—O—C$
NH O O CH_3
$C_6H_5—CH—C—O—COCH_2CHCH_3$

1. H_2N S O N O CH_2OCCH_3 CO_2H
2. HCl
(neutralize)

O
$(CH_3)_3C—O—C$ H
N O
$C_6H_5—CHCNH$
S
N
O
O
CH_2OCCH_3
CO_2H

74%

The remaining modifications were carried out by first cleaving the allylic acetoxy group via hydrogenolysis and finally removing the *tert*-butoxycarbonyl protecting group in acid.

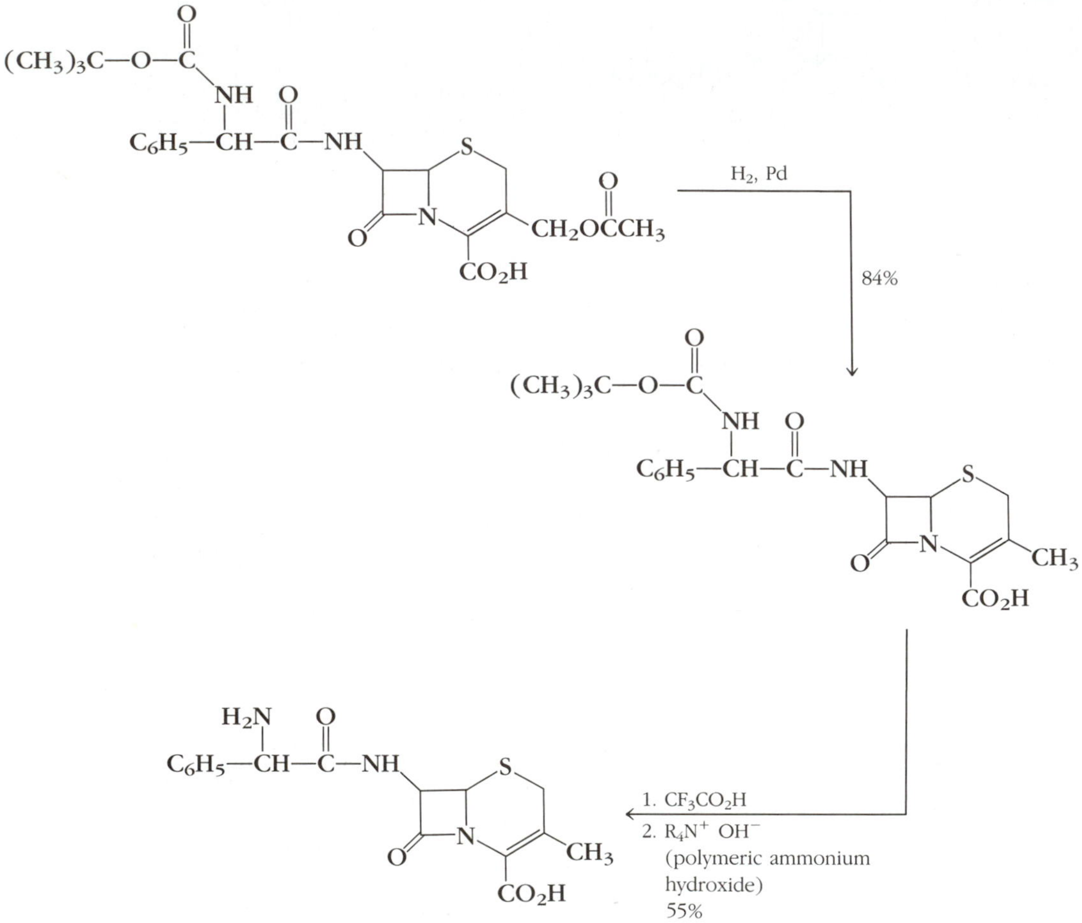

H_2, Pd

84%

1. CF_3CO_2H
2. $R_4N^+\ OH^-$
(polymeric ammonium hydroxide)
55%

The β-lactam antibiotics have proved highly valuable in modern medicine, and chemists continue to search for new compounds, both synthetic and naturally occurring, that have even better therapeutic properties. Among these are the cephamycins (obtained from *Streptomyces* cultures) and the *penems,* synthetic hybrids of the penicillins and cephalosporins.

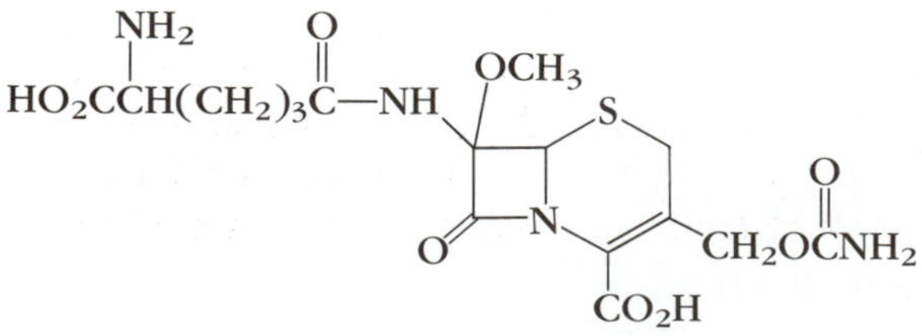

Cephamycin C

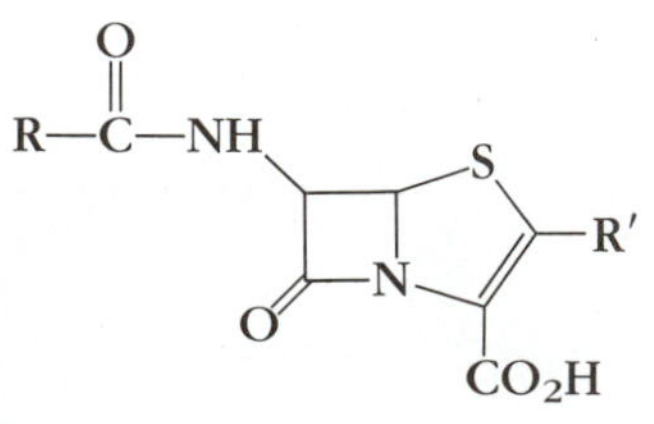

The penems

Chemists have not yet fully unraveled the relationship between molecular structure and biological activity, but enough is understood about the β-lactam antibiotics that new, medically valuable drugs continue to be developed.

EXERCISE 24.11

Why are uric acid and barbituric acid called "acids," when neither has a carboxyl group? Use appropriate structures to support your answer.

EXERCISE 24.12

Outline a complete synthesis of veronal (barbital), starting from malonic acid.

EXERCISE 24.13

Compare the structures of 6-aminopenicillic acid and 7-aminocephalosporanic acid. Show which carbon atoms of the former correspond to those of the latter.

EXERCISE 24.14

Removal of the side chain of cephalosporin C is effected with nitrosyl chloride because this results in selective hydrolysis of just one amide linkage. Draw the products that would be expected for simple acid or base hydrolysis of cephalosporin C.

24.6 PROBLEMS

24.1 When pyridine is treated with acetyl chloride, a crystalline adduct is formed. This material has the molecular formula C_7H_8ClNO, and it reacts rapidly with water to form pyridine, acetic acid, and HCl. Suggest a reasonable structure for the crystalline solid.

24.2 The product obtained when pyridine undergoes aromatic sulfonation was drawn as a sulfonic acid in Section 24.1, but this is an oversimplification. Draw the structure that more properly describes the structure of this difunctional compound.

24.3 Draw the resonance forms for the intermediate formed by attack of an electrophile (E^+) at (a) C-2, (b) C-3, and (c) C-4 of pyridine. Indicate any resonance structures which would indicate that an intermediate is either stabilized or destabilized relative to the other two intermediates. What does this analysis predict for the preferred site of electrophilic substitution reactions of pyridine?

24.4 Write reaction mechanisms for nitration of pyridine *N*-oxide at (a) C-2, (b) C-3, and (c) C-4. Indicate any resonance structures which would indicate that an intermediate is either stabilized or destabilized relative to the other two intermediates. What does this analysis predict for the preferred site of electrophilic substitution reactions of pyridine *N*-oxide?

24.5 Write a detailed mechanism for the reaction of pyridine *N*-oxide with $POCl_3$. Use appropriate resonance forms to explain why chlorine substitution occurs only at C-2 and C-4.

24.6 Write a detailed mechanism for the reaction of sodium methoxide with (a) 2-chloropyridine and (b) the corresponding *N*-oxide. Draw appropriate resonance forms to explain why the *N*-oxide reacts more rapidly.

24.7 Draw appropriate resonance forms to explain why sodium methoxide reacts readily with 2- or 4-chloropyridine and the corresponding *N*-oxides but not with the 3-chloro isomers.

24.8 Unless the positions are already occupied, electrophilic substitution in thiophenes (as well as furans and pyrroles) usually occurs at C-2 and C-5. Use appropriate resonance forms to explain why nitration of 2-methoxythiophene (Section 24.1) leads to substitution at C-3 and C-5 instead of just C-5.

24.9 Complete the reactions of:

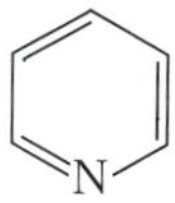

N
with **(a)** HBr, **(b)** C_2H_5I, **(c)** C_6H_5Li

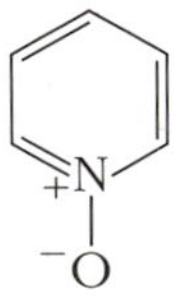

+N
−O
with **(d)** C_6H_5Li, **(e)** HNO_3, H_2SO_4, **(f)** PBr_3

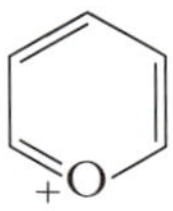

+O
ClO_4^-
with **(g)** CH_3Li, **(h)** $CH_3CH_2NH_2$, **(i)** $NaBH_4$

24.10 Complete the reactions of:

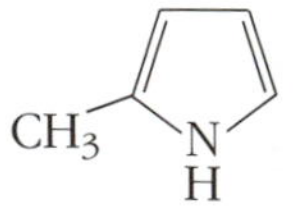

CH_3 N H
with **(a)** $(CH_3)_2NH$, CH_2O, HCl, **(b)** $CH_3\overset{O}{\overset{\|}{C}}—N(CH_3)_2$, $POCl_3$

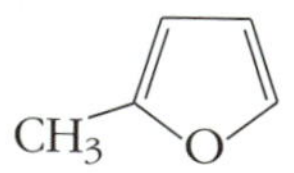

CH_3 O
with **(c)** $CH_3\overset{O}{\overset{\|}{C}}O\overset{O}{\overset{\|}{C}}CH_3$, **(d)** CH_2O, HCl

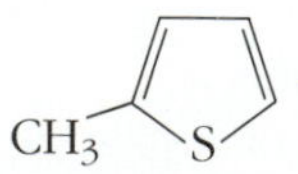

CH_3 S
with **(e)** $CH_3\overset{O}{\overset{\|}{C}}Cl$, $ZnCl_2$, **(f)** $CH_3\overset{O}{\overset{\|}{C}}N(CH_3)_2$, $POCl_3$, **(g)** Br_2, **(h)** HNO_3, H_2SO_4

24.11 Complete the reaction of 2-methylpyridine with each of the following: (a) C_6H_5Li then D_2O; (b) CH_3CO_3H; (c) CH_3I.

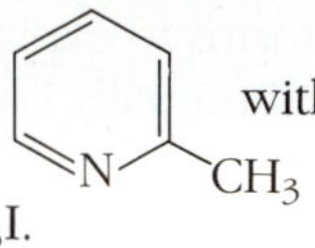

24.12 In many respects the differences in reactivity between pyridine and pyridine *N*-oxide are similar to those between benzene and phenol (or the corresponding phenoxide ion). Draw the resonance structures for the intermediate produced by attack of NO_2^+ at the 4 position of (a) pyridine *N*-oxide and (b) the conjugate base of phenol. Are the nonbonding electrons of oxygen able to interact favorably with the electron-deficient carbon in each case?

24.13 Write the reaction mechanism for conversion of 4-nitropyridine *N*-oxide to 4-methoxypyridine *N*-oxide by the action of sodium methoxide.

24.14 Treatment of pyrylium salts with nucleophiles such as ammonia often leads to new ring structures via ring opening and subsequent reclosure. Explain the following reaction with sodium hydroxide.

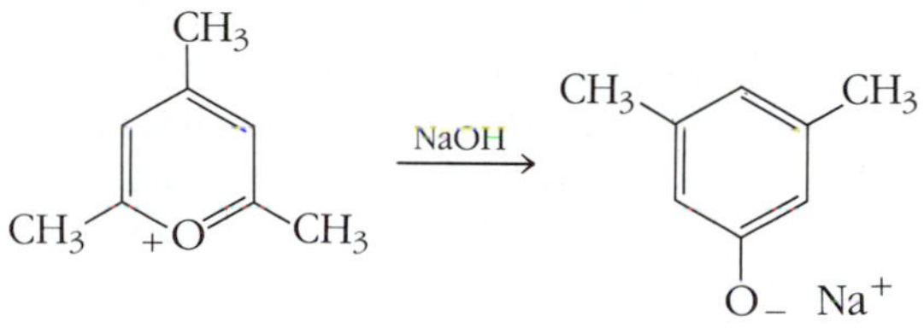

2,4,6-trimethylpyrylium (CH_3, CH_3, CH_3, $+O$) $\xrightarrow{NaOH}$ 3,5-dimethylphenoxide (CH_3, CH_3, O^- Na^+)

24.15 Write a reaction mechanism for the conversion of 2,4,6-trimethylpyrylium perchlorate to 2,4,6-trimethylpyridinium perchlorate when the former is treated with ammonia (Section 24.1).

24.16 Electrophilic substitution of the following disubstituted thiophene could conceivably occur at either C-3 or C-4. The observed reaction with formaldehyde (Section 24.1) results in a 93% yield of C-4 substitution product. Draw the two alternative intermediates produced by reaction with an electrophile (E^+). Explain by means of appropriate resonance structures why attack at C-4 is preferred.

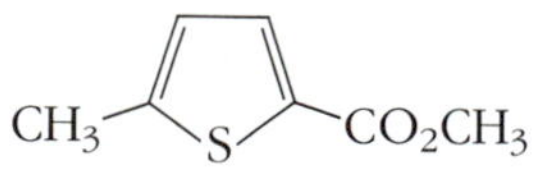

CH_3–thiophene(S)–CO_2CH_3

24.17 Draw resonance forms showing the location of positive charge in the intermediate formed by attack on indole by an electrophile (E^+) at the 2 and 3 positions of the pyrrole ring.

24.18 Draw resonance forms of quinoline showing the location of positive charge in the intermediate formed by attack of an electrophile at the four different positions of the benzene ring. Assuming that the nitrogen is not protonated under the reaction conditions, would you expect quinoline to react more or less rapidly than aniline? Explain.

24.19 Use appropriate resonance forms to explain why reaction of an electrophile (E^+) with indole occurs preferentially on the heterocyclic ring, while both quinoline and isoquinoline react preferentially on the carbocyclic aromatic ring.

24.20 The reaction of isoquinoline with bromine and aluminum chloride results in a high yield of a single bromo product. Subsequent nitration of the bromoisoquinoline leads to a nitro derivative in high yield.

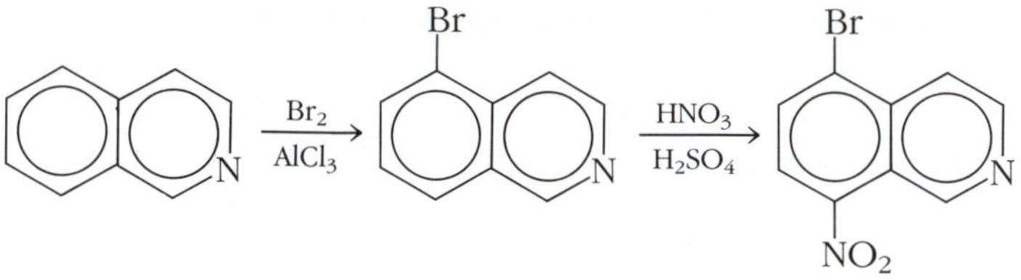

Br Br
$\xrightarrow[AlCl_3]{Br_2}$ $\xrightarrow[H_2SO_4]{HNO_3}$
N N N
NO_2

(a) Use appropriate resonance forms to explain why each of these products is formed. (b) On the basis of the product formed, which species actually undergoes bromination, free isoquinoline or its aluminum complex?

24.21 The colors of the anthocyanins are highly dependent on pH and the local environment in a plant. (a) Draw structures for the products of deprotonation of each of the different OH groups of pelargonidin (Section 24.2). Show in each case only the structure that has no charges, if possible. The electronic differences between these structures would be sufficient to account for significant variation in colors. (b) At what positions (if any) on the anthocyanin nucleus would hydroxyl substituents not be directly conjugated with the positively charged pyrylium group?

24.22 Complete each of the following reactions:

(a) N H $\xrightarrow[HCl]{CH_2O}$?

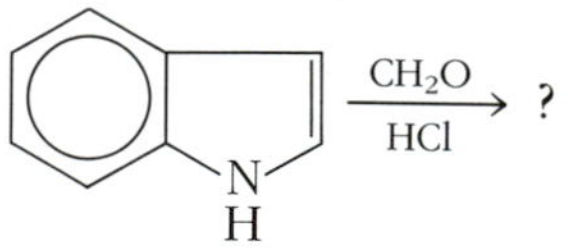

(b) N CH_3O $\xrightarrow[H_2SO_4]{HNO_3}$?

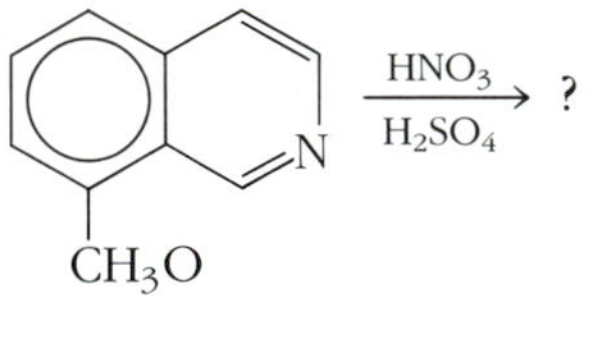

(c) N $\xrightarrow[H_2SO_4]{HNO_3}$?

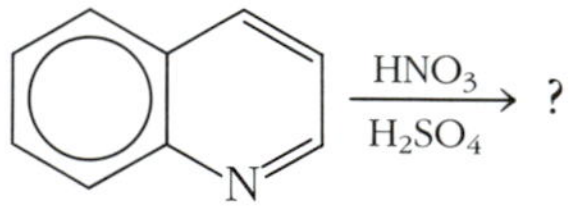

(d) O O $\xrightarrow{NaCN}$?

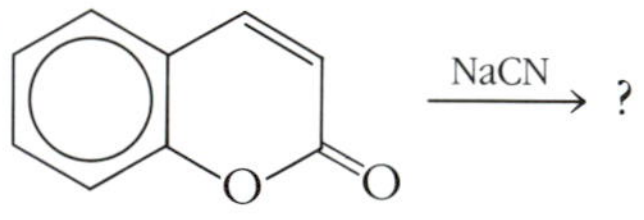

(e) O O $\xrightarrow{NaBH_4}$?

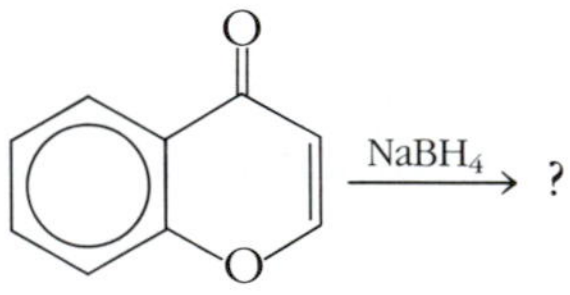

(f) $^+$O $\xrightarrow{CH_3MgBr}$?

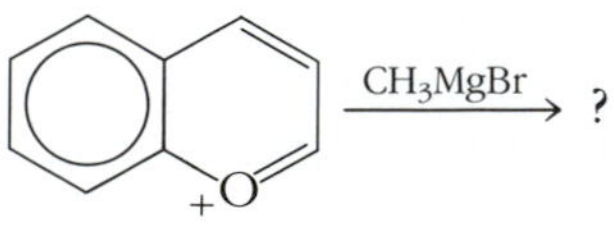

24.23 Show how each of the following could be prepared from starting materials that contain no heterocyclic ring.

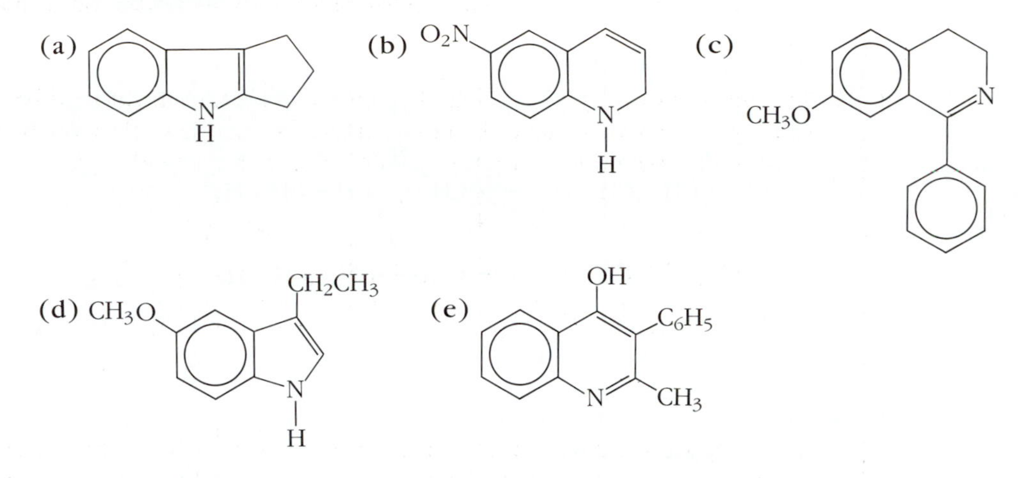

24.24 Suggest an explanation of why the five-membered-ring aromatic heterocycles pyrrole and furan tend to undergo electrophilic substitution more rapidly than benzene, while the six-membered-ring heterocycles (e.g., pyridine) are less reactive than benzene.

24.25 The following questions refer to compounds involved in the synthesis of reserpine, which was discussed in Section 23.4. (a) Suggest how the aromatic amine (**9-18**) could be prepared starting from an appropriately substituted phenylhydrazine. (b) Using appropriate resonance forms, suggest whether the methoxy substituent on the benzene ring aids or inhibits cyclization of the lactam (**9-21**) to the pentacyclic material (**9-22**).

24.26 Nitro and halogen substituents can be displaced from pyridine *N*-oxides by reaction with nucleophiles. What products would be expected from reaction of the following isomers with $NaOCH_3$? Which would react more rapidly? Use appropriate resonance forms to explain your answer.

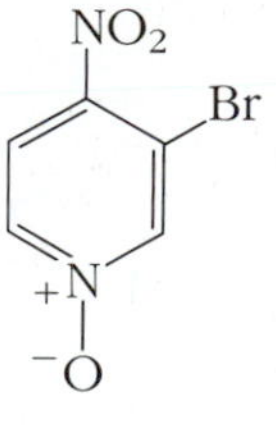

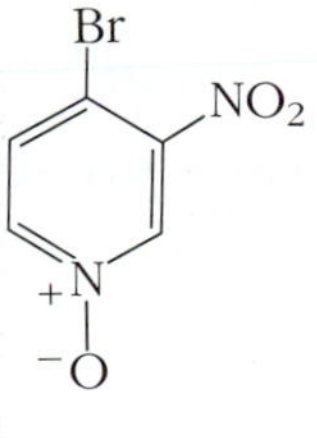

NO_2
Br
^{+}N
^{-}O

Br
NO_2
^{+}N
^{-}O

24.27 Suggest methods for carrying out each of the following conversions:

(a) → NCC_6H_5 (O above the C, double-bonded)

(b) $CH_2—CH—CH_2CH_3$ (O bridging the two ring carbons) ⟶ $CH_2—CH—CH_2CH_3$ (N–H bridging the two ring carbons)

(c) $CH_2{=}CH—CH_2CH_3$ ⟶ $CH—CH—CH_2CH_3$ (O bridging the two ring carbons)

(d) CH_3CH_2CHO ⟶ $CH—CH—CH_2CH_3$ (O bridging the two ring carbons)

24.28 Suggest how the following synthesis could be carried out by using an oxazoline intermediate:

N, CH_3 ⟶ N, CO_2H, CHC_6H_5

24.29 Suggest methods for carrying out each of the following conversions:

(a) CH_2OH, O ⟶ CHO, O

(b) $CH_3—CH(OH)—CH_2—CH(CH_3)—CH_2OH$ ⟶ CH_3, O, CH_3

(c) $CH_2{=}CH—CH_2—CH(Br)—CH_2CH_3$ ⟶ N–H, CH_2CH_3

(d) $CH_3CH_2CH_2CH_2CH_2CH_2—NHCH_3$ ⟶ N–CH_3, CH_2CH_3

24.30 Heterocyclic rings are often involved in the use of activating and protecting groups. Illustrate the use of heterocyclic derivatives for the following:
(a) Protection of an aldehyde
(b) Activation of a ketone
(c) Protection of a carboxylic acid
(d) Activation of a carboxylic acid
(e) Protection of a simple alcohol
(f) Protection of a diol

24.31 Heterocyclic compounds sometimes act as precursors or intermediates in cycloaddition reactions. Suggest ways in which this type of reaction could be used to prepare each of the following:

(a) NC_6H_5 from

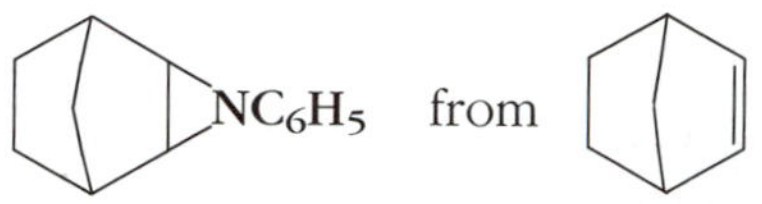

(b) CN from CN

(c) O from

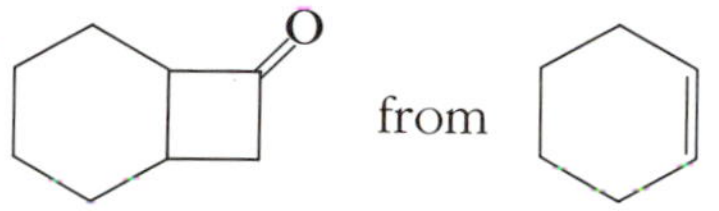

24.32 Suggest methods for the synthesis of (a) secobarbital and (b) pentobarbital from diethyl malonate.

24.33 Why can diethyl malonate not be used as the starting material for the synthesis of phenobarbital?

24.34 The formation from phenobarbital of the anticonvulsant described in Section 24.5 requires two *O*-alkylations. Draw the intermediates involved in each alkylation step.

24.35 (a) Which are the most acidic hydrogens in barbituric acid? (b) Explain the following reaction:

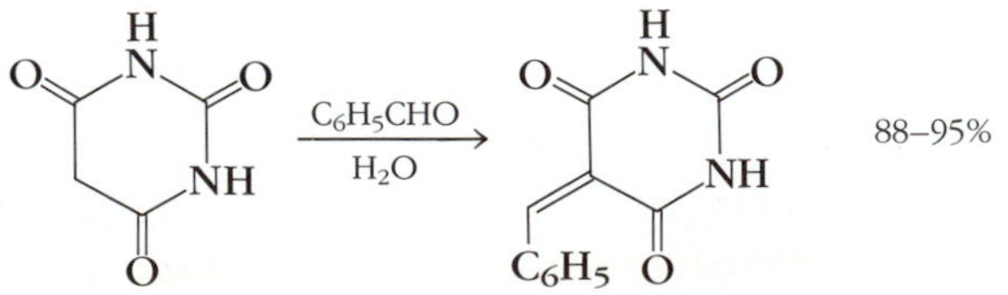

$$\xrightarrow[H_2O]{C_6H_5CHO}$$ C_6H_5 88–95%

24.36 Suggest a reason why the β-lactamase susceptibility of a penicillin derivative is lowered by using a bulky acyl group to acylate the 6-amino group, even though the β-lactamase attacks the other amide group (i.e., that of the four-membered ring).

24.37 The side chain of 7-aminocephalosporanic acid is best removed by the action of nitrosyl chloride. When aqueous nitrous acid is used to form the diazonium salt, *two* moles of N_2 are generated and little or no 7-aminocephalosporanic acid is obtained. Explain this.

24.38 The reaction of *N*-chloroamines with acid ordinarily yields pyrrolidines. Explain why a six-membered ring is produced in the following reaction:

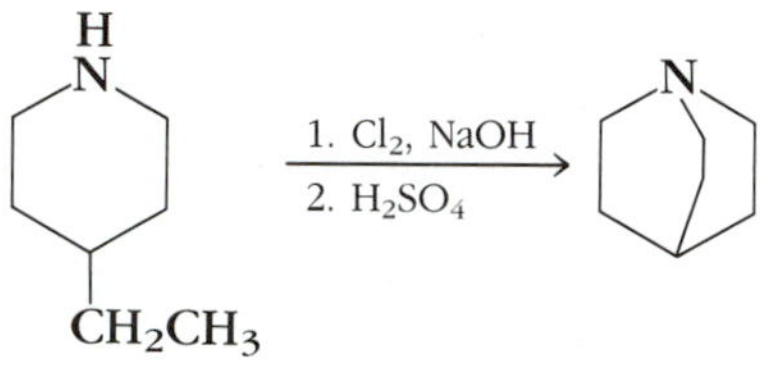

H
N
CH_2CH_3
1. Cl_2, NaOH
2. H_2SO_4
N

CHAPTER 25

POLYMERS

Polymers are everywhere: in the chair on which you are sitting, the pages of this book, your muscles, your genes, and perhaps the jeans that you are wearing. All these materials and substances are made up primarily of large molecules called *polymers.* These large molecules, frequently described as *macromolecules,* are characterized by a substantial number of repeating subunits. There are no precise definitions for the size of a polymer, but typical molecular weights fall in the range of 10^4–10^7 amu, and there may be as many as 100,000 subunits.

In addition to the macromolecules that carry out major biochemical functions (proteins, nucleic acids, etc.), other biopolymers have been adapted to human use for many centuries. Two of the major uses have been for structural materials and clothing. Cellulose, for example, is a polymer of glucose (Section 19.9). It is the major constituent in cell walls of plants and provides the structural strength of wood and of fibers such as cotton. Other naturally occurring fibers such as silk and wool are proteins, consisting of long polymer chains derived from amino acids.

Considering the great importance of naturally occurring polymers for our shelter and clothing, you should not be surprised that considerable scientific effort has been expended in studying them. This has led not only to chemical modification of biopolymers but also to the preparation of a wide variety of *synthetic polymers.* Indeed, the success of polymer research over the last 50 years has provided a wide variety of synthetic polymers that are used in nearly all phases of modern life. Whether it is the vinyl upholstery of your car, the plastic knobs of your television set, the polyester shirt you may be wearing, or the pen with which you are writing, you are never very far from at least one of the many synthetic polymers that are used in our society.

In this chapter we will present a brief introduction to the chemistry of polymers. We will discuss many of the polymers that are used commercially, and we will analyze the relationships between their molecular structures and physical properties. Finally, we will show you how different structural features and physical properties lead to specific uses for various polymers.

25.1 POLYMER STRUCTURE

The single most recognizable feature of polymer structure is the repeating occurrence of one or more basic subunits. The actual compounds with the subunit structure that react to yield polymers are called *monomers.* The polymer is named by affixing the name of the monomer to the prefix *poly.** For example, polystyrene has a long carbon backbone, in which you can easily identify the individual styrene subunits.

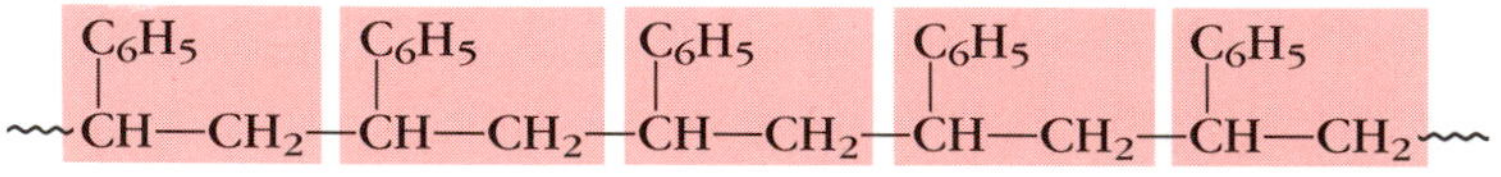

$$\sim\!\underset{\displaystyle |}{\overset{C_6H_5}{}}\!\!CH-CH_2-\overset{C_6H_5}{\underset{|}{CH}}-CH_2-\overset{C_6H_5}{\underset{|}{CH}}-CH_2-\overset{C_6H_5}{\underset{|}{CH}}-CH_2-\overset{C_6H_5}{\underset{|}{CH}}-CH_2\!\sim \qquad \overset{C_6H_5}{\underset{|}{CH}}=CH_2$$

Polystyrene 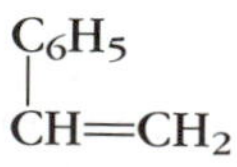 — Styrene

*The name of the monomeric unit is often enclosed in parentheses to avoid ambiguity when naming a polymer. This is particularly important when the name of the monomer consists of more than one word.

It is not practical to draw the entire structure of a molecule that may have thousands of repeating subunits, so chemists have developed more convenient ways of denoting polymer structures. The most common way is to enclose the subunit in parentheses and indicate the repeating nature by the subscript *n*:

$$\left(\overset{\displaystyle C_6H_5}{\underset{}{\overset{|}{CH}}}-CH_2\right)_n$$

Polystyrene

But this drawing should leave you with two important questions: (1) What atoms or groups are found at the *ends* of the polymer chain? (2) What is the value of *n*? The answers to these two questions will help you understand the unique aspects of polymer chemistry.

The value of *n*, frequently called the *degree of polymerization,* can vary tremendously depending upon the way a polymer is prepared. Typical commercial polymers consist of 100–1000 subunits, and molecular weights fall in the range of 10^5–10^6 amu. However, we can only consider *average values* for the degree of polymerization and molecular weight, because individual polymer molecules are not all the same. With "ordinary" (i.e., nonpolymeric) compounds of low molecular weight, purification affords a sample in which essentially all molecules have the same chemical formula. This is not true for polymers. With the exception of certain biological macromolecules, a characteristic feature of polymeric compounds is the presence of individual molecules that exhibit a substantial variation in molecular weight because they have different numbers of subunits.

If even a "pure" polymer consists of individual molecules that have different degrees of polymerization, how can a polymer behave as a well-defined substance with reliable and reproducible physical and chemical properties? The answer to this question is also the answer to our previous question about the nature of the end groups. Because of their very high molecular weights, the physical and chemical properties of macromolecules are determined almost entirely by the large number of repeating subunits. Their physical and chemical properties are nearly independent of the precise number of subunits and the nature of the end groups. The drawings in Figure 25.1 show this schematically, and inspection of the drawings may help you visualize the relative unimportance of changes in the end groups or of variation in the degree of polymerization.

Structural Types of Polymers

The useful properties of polymeric materials depend not only upon the structures of the individual subunits but also upon the overall structure of the macromolecule. Before discussing these relationships, we must first establish some of the common terms and definitions that are used to describe polymer structures.

In the absence of any chain branching, the structure is that of a *linear polymer,* while a *branched polymer* has polymeric chains that extend from the main chain. These can be described schematically in the following way:

(a) XAAAAAAAAAAAAAAAAAAAAAAAAAAAAAAAAAAAAAAAAAAAAAAAAAAAAAAAAAAAAAAAAAAAAAA
A
A
A
XAAAAAAAAAAAAAAAAAAAAAAAAAAAAAAAAAAAAAAAAAAAAAAAAAAAAAAAAAAAAAAAAAAAAAA

(b) XAAAAAAAAAAAAAAAAAAAAAAAAAAAAAAAAAAAAAAAAAAAAAAAAAAAAAAAAAAAAAAAAAAAA
A
A
A
XAAAAAAAAAAAAAAAAAAAAAAAAAAAAAAAAAAAAAAAAAAAAAAAAAAAAAAAAAAAAAAAAAAAA

(c) YAAAAAAAAAAAAAAAAAAAAAAAAAAAAAAAAAAAAAAAAAAAAAAAAAAAAAAAAAAAAAAAAAAAAAA
A
A
A
YAAAAAAAAAAAAAAAAAAAAAAAAAAAAAAAAAAAAAAAAAAAAAAAAAAAAAAAAAAAAAAAAAAAAAA

Figure 25.1 Physical and chemical properties of polymers are largely independent of precise molecular weight and end groups.
The overall molecular properties are a function of the large number of repeating —A— subunits, and the drawings suggest that there should be little effect on molecular properties from small changes in degree of polymerization [(*a*), $n = 143$, (*b*) $n = 139$] or from changes in end groups [(*a*) end groups = X, (*c*) end groups = Y].

```
                                   A—A—A~~
                                   |
                                   A
                                   |
                           ~~A—A—A—A—A—A—A—A~~
                                         |
                                         A
                                         |
~~A—A—A—A—A—A—A~~              ~~A—A—A—A

   Linear polymer                  Branched polymer
```

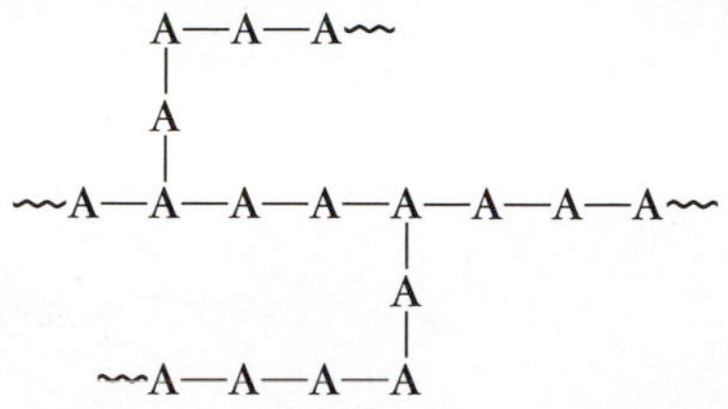

Sometimes a polymeric material can be caused to undergo reactions which join the different chains, and a *cross-linked* or *network polymer* results.

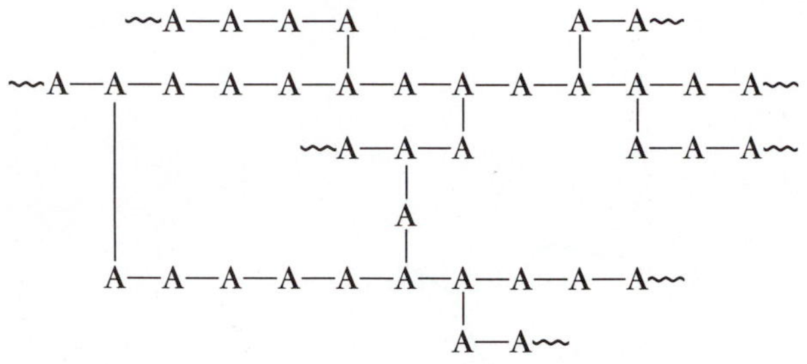

```
     ~~A—A—A—A                A—A~~
              |               |
~~A—A—A—A—A—A—A—A—A—A—A—A—A~~
   |                |         |
   |         ~~A—A—A          A—A—A~~      Cross-linked polymer
   |             |                          (network polymer)
   |             A
   |             |
   A—A—A—A—A—A—A—A—A—A~~
                   |
                   A—A~~
```

These structural types are depicted in another way in Figure 25.2.

What is the relationship between polymer structure (i.e., branching and cross-linking) and physical properties? Consider first the effects of cross-linking. When the various polymer chains are joined in this way via *covalent bonds,* the substance becomes rigid, brittle, and insoluble. The cross-links between the main polymer chains result in huge molecules, and the material can be deformed or dissolved only if covalent bonds are broken. Such polymers are sometimes called *duromers,* and they retain their shape even at elevated temperatures.

Polymers are much less rigid in the absence of cross-linking and frequently they can be deformed (i.e., molded) at elevated temperatures. Such materials are called *thermoplastics* (or sometimes just plastics). They are also more soluble in organic solvents than cross-linked polymers. Linear polymers (perhaps with some branching) are particularly useful as *fibers.* These materials are quite flexible, but they also exhibit great strength. In order to break such a fiber it is necessary to overcome a very large number of intermolecular forces between the long segments of individual molecules that interact. Hence both covalent bonds and large numbers of intermolecular attractions contribute to the strength of polymeric materials. Linear polymers with limited branching often exhibit the phenomenon of *elasticity,* the ability to return to their original shape after being deformed.

Copolymers

All the polymer structures we have considered to this point have had a single repeating subunit, and these are called *homopolymers.* But many macromolecules contain two or more different subunits, and such macromolecules are called *heteropolymers.* When more than one monomer is used to prepare a synthetic polymer, the resulting heteropolymer is called a *copolymer.* Copolymers formed from a reaction mixture of two monomers can yield either *random* or *alternating* copolymers, depending on the nature of the reactants.

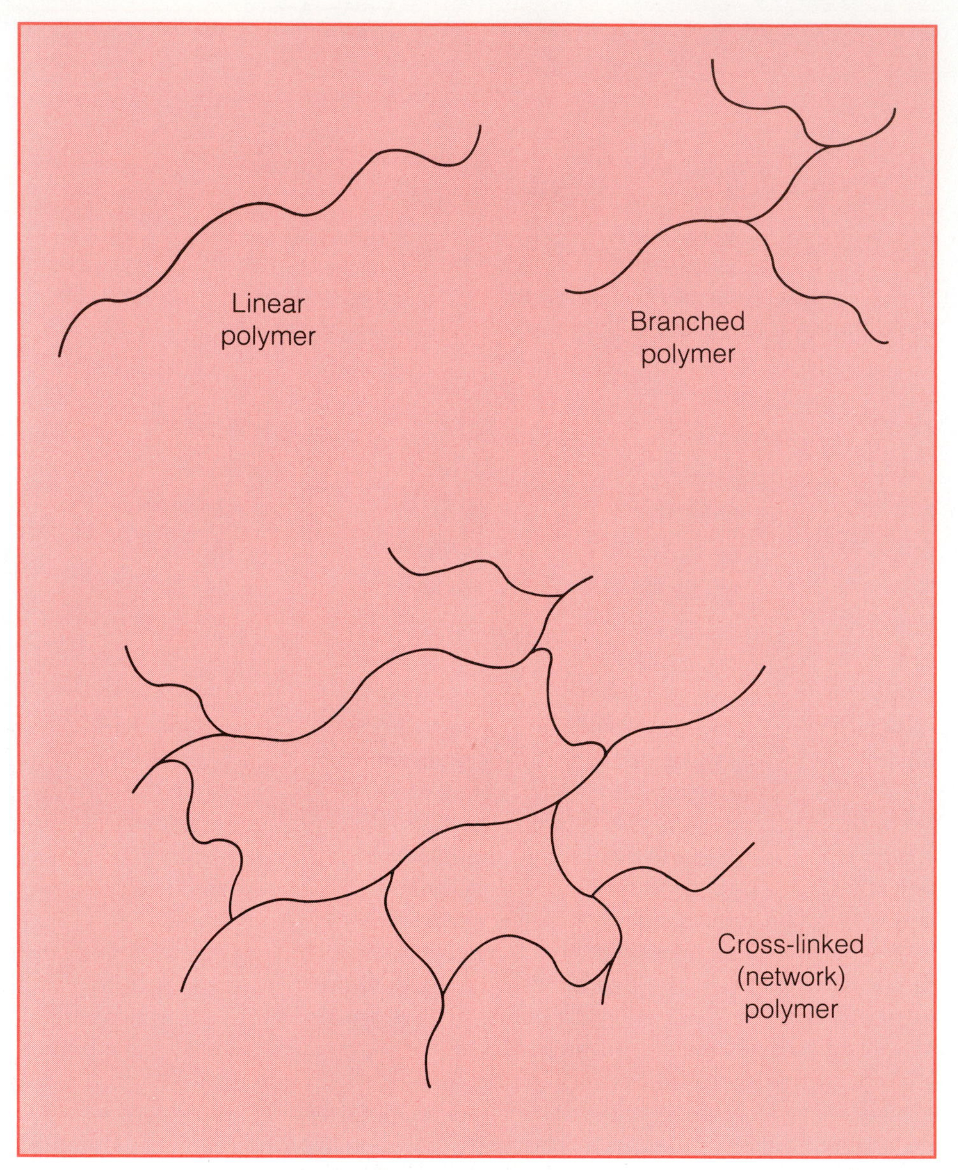

Figure 25.2 Structural types of polymers.

—A—B—A—B—A—B—A—B—A—B—A—B—A—

Alternating copolymer

—A—B—A—B—B—A—A—B—A—A—B—B—A—A—A—

Random copolymer

When short chains (each made up of a single monomer) are joined, a *block copolymer* is formed, in which blocks of each monomer occur regularly.

—A—A—A—A— + —B—B—B—B— ⟶

—A—A—A—A—B—B—B—B—A—A—A—A—B—B—B—B—

Block copolymer

A copolymer can also be produced by introduction of new side chains onto an existing homopolymer to form a *graft copolymer.*

—A—A—A—A—A—A—A—A—A—A—A—A—A— $\xrightarrow{\text{—B—B—B—B—}}$

Homopolymer

```
     B—B—B—
     |
     B
     |
—A—A—A—A—A—A—A—A—A—A—A—A—A—
                    |
                    B
                    |
                    B—B—B—B—
```

Graft copolymer

EXERCISE 25.1

Some of the most important commercial polymers are linear homopolymers. Examples are polyethylene and poly(vinyl chloride). For each of these draw a portion of the polymeric chain that contains five subunits.

EXERCISE 25.2

Draw a short portion (five subunits) of the structure corresponding to an alternating copolymer of isobutylene and vinyl chloride.

25.2 BIOPOLYMERS: STRUCTURE AND FUNCTION

Polymer science is a development of the twentieth century, but it would be a great mistake to conclude that polymers were unimportant in earlier times. By the mid-1800s chemical modification of naturally occurring macromolecules had become industrially important, but even that was not new. For many centuries people have employed naturally occurring polymeric materials for clothing and shelter, and chemical modification of these materials was standard practice. Animal hides (largely consisting of protein) were preseved by chemical treatment (tanning), and natural fibers (cotton, wool, linen, and silk) were subjected to various chemical treatments, including dyeing to produce colored fabrics. But the importance of macromolecules far precedes any recorded history, because proteins, nucleic acids, and polysaccharides are all essential to life as we know it. All living organisms use these polymers and depend on them for their existence.

Rubber

Natural rubber is a polymeric hydrocarbon that is obtained from plants found primarily in South America and Asia. It is a polymer of isoprene (2-methyl-1,3-butadiene), but it undoubtedly is formed from isopentenyl pyrophosphate by the normal pathways for terpene biosynthesis (Section 12.9).

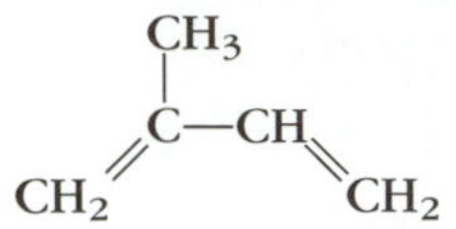

$$CH_2{=}C(CH_3){-}CH{=}CH_2$$

Isoprene

The double bonds in natural rubber all have the *Z* configuration, which is an important structural feature.

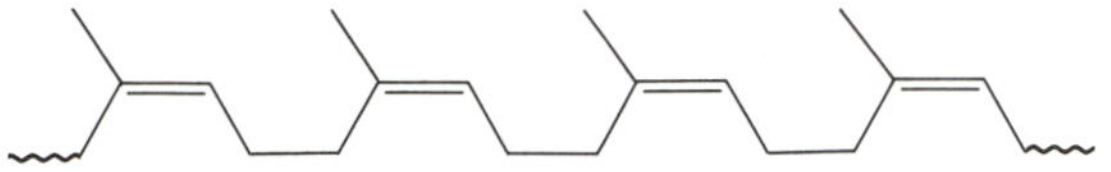

(*Z*)-polyisoprene
(natural rubber)

The stereoisomeric polymer, gutta percha, is obtained from a group of related tropical plants.

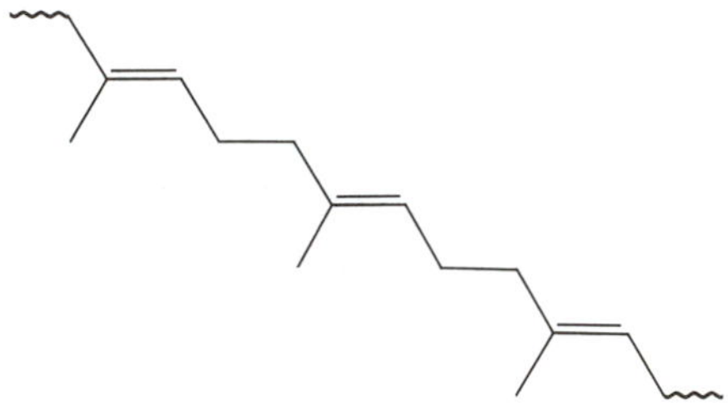

(*E*)-polyisoprene
(gutta percha)

The two stereoisomeric forms have very different physical properties, even though they have similar molecular weights (in the range of 500,000 amu). The relatively small change in molecular structure results in quite different intermolecular interactions. Gutta percha is a much harder, more brittle material with greatly reduced elasticity. One commercial use for this material has been as the hard covering on golf balls. Natural rubber, on the other hand, is quite soft and flexible; a familiar example of its use is the rubber tubing found in laboratories. Some tropical plants produce a mixture of the two polyisoprene stereoisomers, such as the material called *chicle,* which has long been used to make chewing gum.

Polysaccharides

We have already discussed some aspects of polysaccharides in Section 19.9, so our treatment here will be brief. Polymers of glucose are particularly important for energy storage in living systems, and their metabolism can provide the necessary chemical energy for cellular activity during periods when the external supply of nutrients is limited. These glucose polymers, known as *storage polysaccharides,* can be viewed as derivatives of the linear polymer amylose.

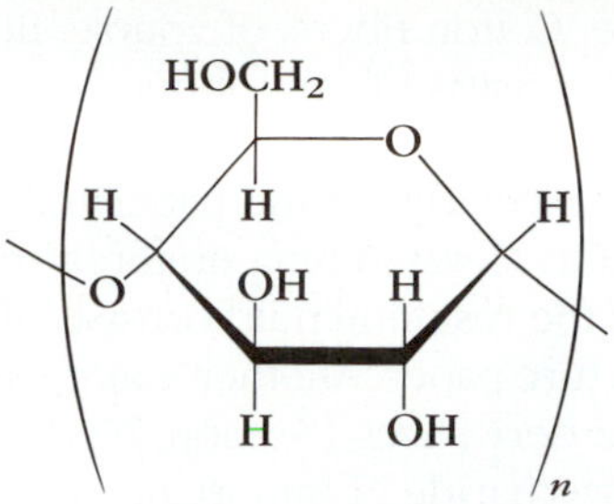

Amylose

The occurrence of branching (via additional chains extending from the C-6 position of glucose subunits) results in amylopectin, which is the major storage polysaccharide in plants. The mixture of amylose and amylopectin is known as starch. The primary storage polysaccharide in animals is glycogen, which is similar in structure to amylopectin but is more highly branched.

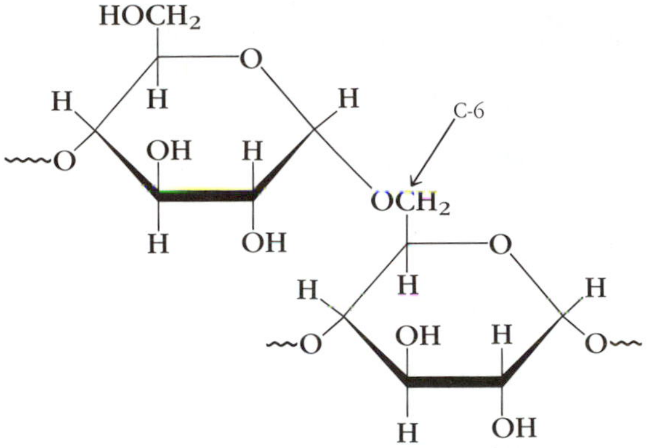

Branching structure of amylopectin and glycogen

Cellulose is the main *structural polysaccharide* of plants and is again a linear polymer of glucose. It differs from amylose in the stereochemistry at C-1, which is β rather than α.

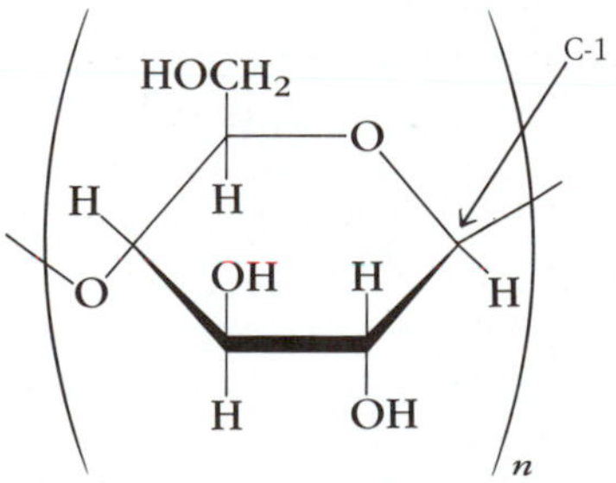

Cellulose

Recall from Section 19.2 that the β-D form of glucopyranose can exist in the chair conformation with all substituents equatorial. The intermolecular forces between polymer chains of cellulose are particularly strong. The chains fit together in ways that facilitate hydrogen bonding, and the resulting material has great physical strength combined with significant flexibility.

For commercial purposes the major sources of cellulose are cotton and wood. Cotton fibers consist almost entirely of cellulose, so very little chemical treatment is needed to obtain pure cellulose. Cotton fibers, of course, find a major use in the textile industry. Although wood consists of only about 40 percent cellulose (after removal of water), the availability of large quantities of wood have made it an attractive source of cellulose, particularly for the paper industry. Chemical digestion of wood with acid or alkali at elevated temperatures and pressures removes most of the other materials, and the resulting pulp consists of relatively pure cellulose that can be used to manufacture paper. Another major use of this pulp is in the preparation of various cellulose derivatives (Section 25.3).

An additional structural polysaccharide of importance is *chitin,* which forms the hard exoskeletons of insects and crustacea. It is structurally related to cellulose, but the monomer is an *amino sugar* in which an *N*-acetyl group is found in place of the C-2 hydroxyl group of glucose. The great physical strength of this polymer is well-known to anyone who has enjoyed a lobster dinner.

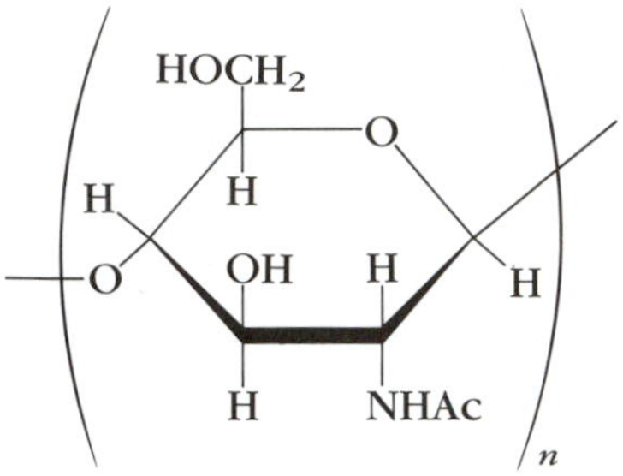

Chitin

$Ac = \overset{O}{\overset{\|}{C}}—CH_3$

Still another polysaccharide associated with food is *carrageenin,* which is obtained from algal seaweeds. Carrageenin is actually a mixture of several different polymers with a D-galactose subunit. A major component has alternating α-D-galactopyranose-4-sulfate and 3,6-anhydro-α-D-galactopyranose subunits.

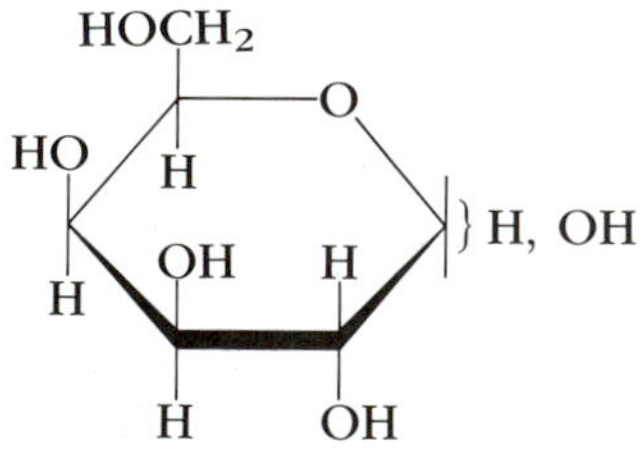

D-Galactopyranose

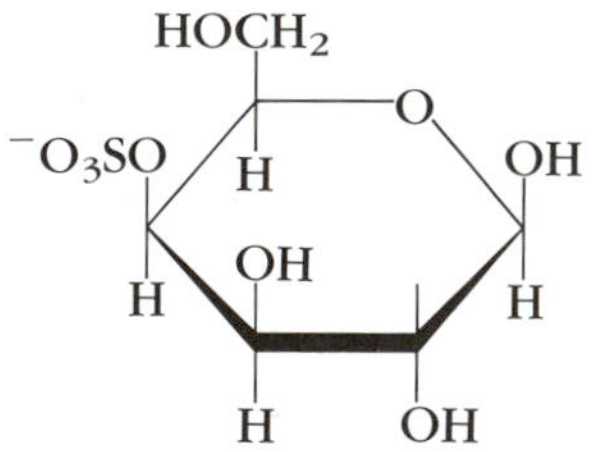

β-D-Galactopyranose-4-sulfate

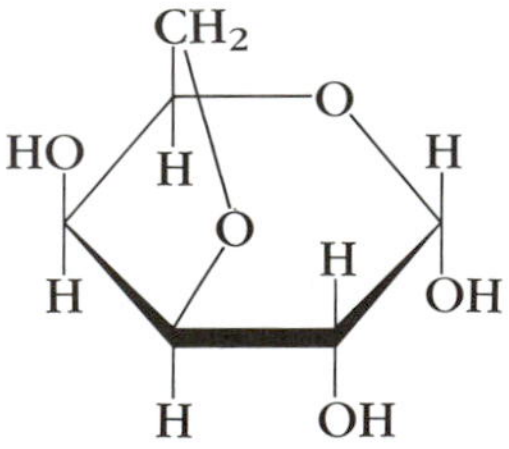

3,6-Anhydro-α-D-galactopyranose

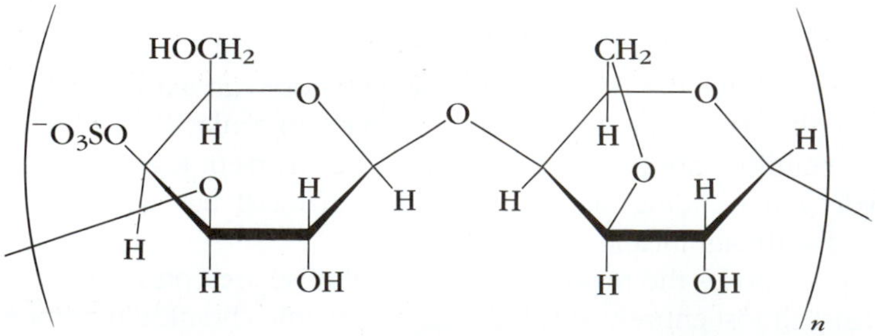

κ-Carrageenan

Carrageenin is used commercially in many dairy products, particularly in less expensive brands of ice cream and other dessert products, to provide a smooth, creamy texture. Look at the label the next time you eat or drink such a commercial dairy product.

Proteins

Proteins are polymers based primarily on α-amino acid subunits, where R represents various side chains.

$$H_2N{-}\underset{\displaystyle R}{\underset{|}{CH}}{-}CO_2H$$

Generalized structure of an amino acid

The amino acid subunits combine to form a polyamide, and a typical sequence of subunits is illustrated in the following partial structure:

$$\sim NH{-}\underset{R_1}{\underset{|}{CH}}{-}\overset{O}{\overset{\|}{C}}{-}NH{-}\underset{R_2}{\underset{|}{CH}}{-}\overset{O}{\overset{\|}{C}}{-}NH{-}\underset{R_3}{\underset{|}{CH}}{-}\overset{O}{\overset{\|}{C}}{-}NH{-}\underset{R_4}{\underset{|}{CH}}{-}\overset{O}{\overset{\|}{C}}\sim$$

Such polyamides built up from natural amino acids are also known as *polypeptides* or sometimes just *peptides*.

We have already discussed peptides and amino acids in Chapter 20, so only a few highlights will be given here. Certainly one of the most important aspects of proteins is that they are found in all living organisms. *Enzymes* are proteins, and they catalyze and regulate virtually all biochemical reactions. Proteins also form the fibers of muscle tissue in animals, playing a structural role in tissues such as collagen (connective tissue) and keratin (hair, nails). Many of these proteins have significant numbers of cross-links between the polypeptide chains, and this imparts great physical strength to them.

There are many intermolecular hydrogen bonds between the polymer chains of proteins, and these combine with the flexibility of linear polymers to produce the desirable qualities of a fiber. Wool and silk are two naturally occurring fibers that are made up of proteins, and these have been used to make clothing for many centuries. Even with the development of synthetic polymers, silk and wool retain a position of importance in the world economy.

In contrast to all synthetic polymers and the other naturally occurring polymers that we have considered up to this point, proteins exhibit a structural feature that is unique: they are *structurally homogeneous.* In other words, for a particular protein each molecule is the same. The sequence of amino acids is the same, and the molecular weight is the same. Even for proteins having more than 1000 subunits and made up of 20 different amino acids, any two molecules will have the identical number and sequence of amino acids. It is this structural homogeneity that allows enzymes and other proteins to carry out their biochemical roles with extremely high specificity.

Nucleic Acids

Nucleic acids are linear polymers of phosphorylated sugars. Their name is derived from two sources: they comprise the genetic material in the *nucleus* of a cell, and the presence of the phosphate groups makes them *acidic.* There are two types of nucleic acids, ribonucleic acid (RNA) and deoxyribonucleic acid (DNA). The sugar residues in these polymers are D-ribose and 2-deoxy-D-ribose, respectively, and they exist exclusively in the furanose form.

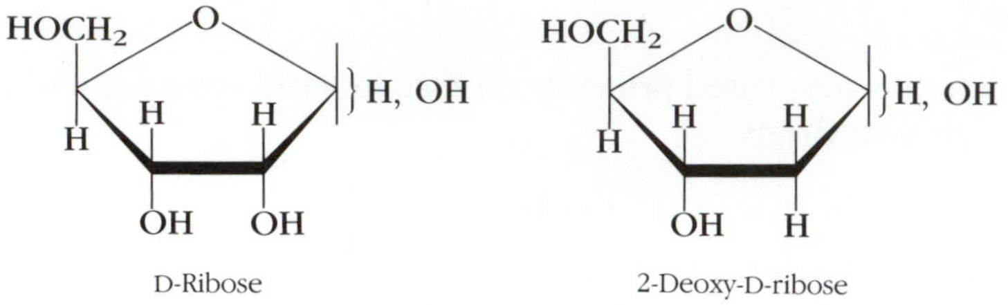

D-Ribose

2-Deoxy-D-ribose

Considering only the polymer backbone, we can draw RNA and DNA as shown in Figure 25.3; those drawings can be further simplified to the following representations.

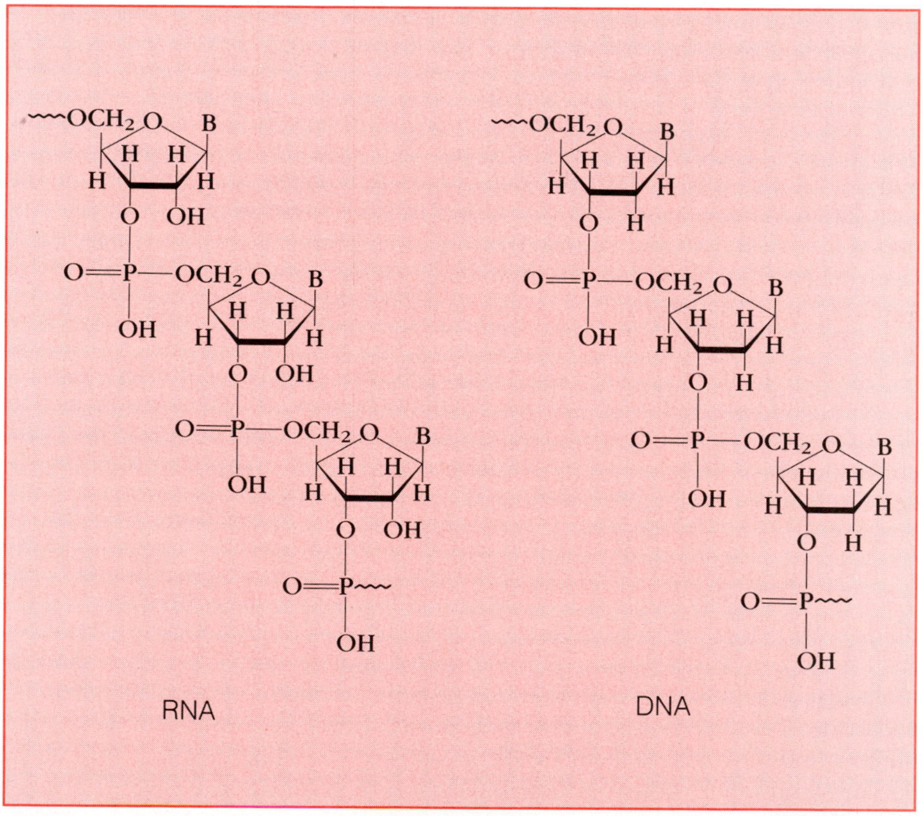

RNA

DNA

Figure 25.3 The polymer backbone of nucleic acids.
Ribonucleic acid (RNA) and deoxyribonucleic acid (DNA) differ only by the presence or absence of the hydroxyl group on C-2 of the ribose ring. The phosphate groups are shown in their protonated form for convenience, and the purine and pyrimidine bases are designated by the letter B.

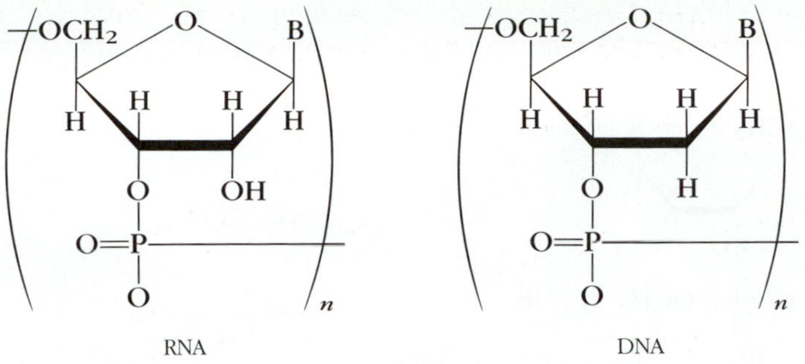

RNA DNA

The preceding drawings are very different from the typical representations found in biology and biochemistry textbooks, and there is good reason for this. From a biological or biochemical standpoint, the identities of the various purine and pyrimidine bases along the polymer backbone define the function of the nucleic acid. For that reason, a typical representation focuses on the various bases, and the carbohydrate portions are likely to be abbreviated or even omitted. In contrast, the drawings in Figure 25.3 are intended to illustrate not the biological function but the chemical structure of the polymer chain. This is a series of ribose (or deoxyribose) residues joined by phosphate ester linkages, and the heterocyclic bases, denoted by the letter B, are merely substituents on the polymer chain.

Four bases are found in DNA, and four are found in RNA. When coordinated to a ribose or a deoxyribose they are called *nucleosides,* and the derivatives in which the sugar groups are phosphorylated are called *nucleotides.* The structures of the bases (together with their nucleoside and nucleotide derivatives) were summarized in Figure 24.2.

The formation of nucleic acids in biological systems takes place via *triphosphate* derivatives of the nucleosides. Pyrophosphate ($H_3P_2O_7^-$ or one of its more ionized forms) is a good leaving group, so the triphosphate grouping is quite susceptible to nucleophilic attack. This is illustrated by the following equation, which shows such nucleophilic attack on adenosine triphosphate (ATP).

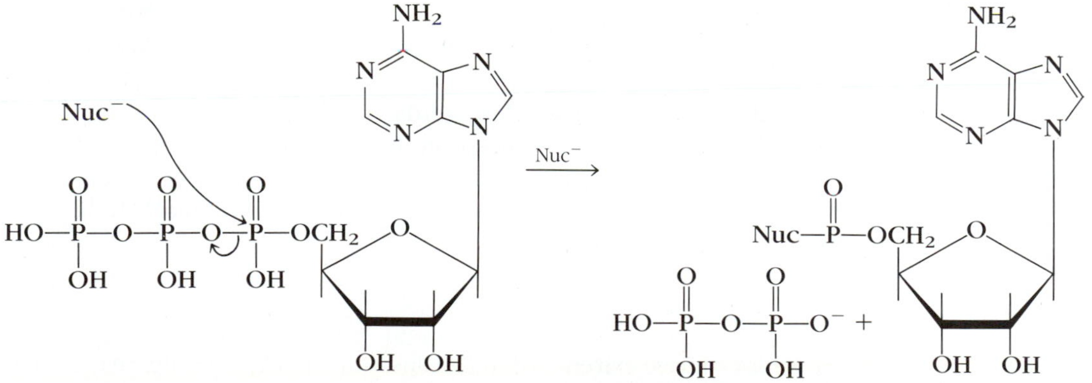

$$\xrightarrow{\text{Nuc}^-}$$

In the formation of long chain nucleic acids these reactions are enzyme-mediated, but the general mechanism is the same. The free 3′-hydroxyl group (where 3′ indicates the 3 position of the ribose rather than the base) at the end of the existing polymer chain acts as a nucleophile to attack the triphosphate group of a simple nucleoside triphosphate, and the polymer chain is lengthened by one subunit. This process is illustrated in Figure 25.4.

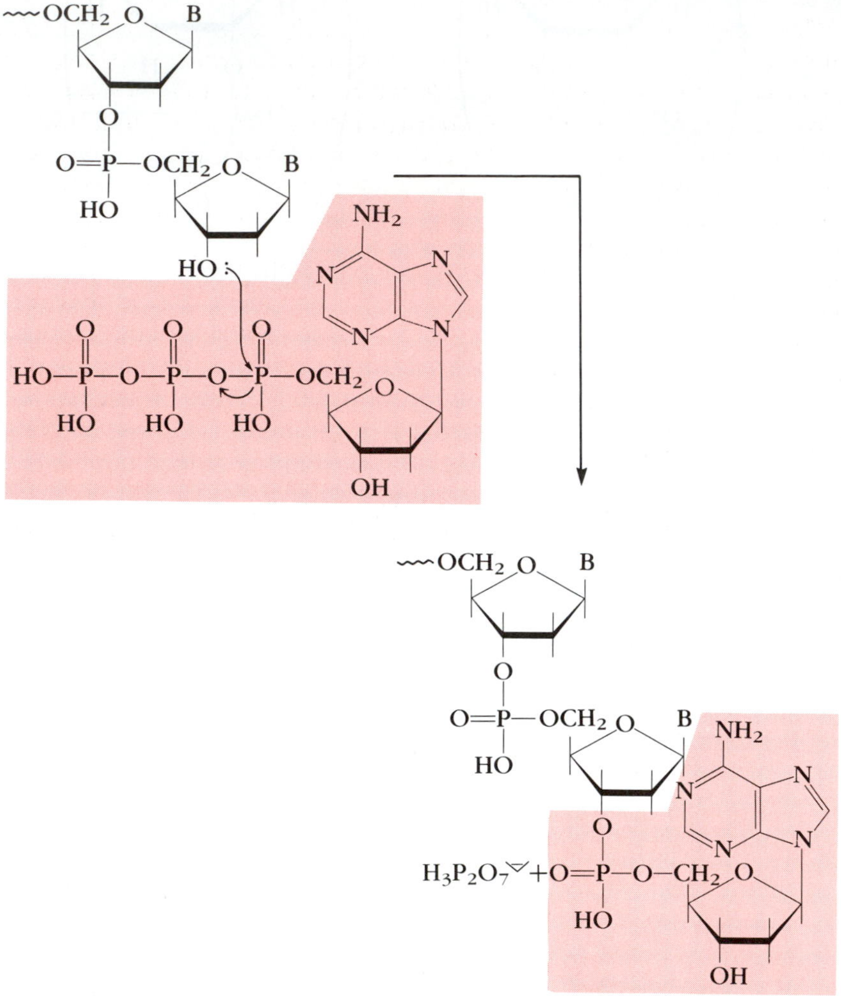

Figure 25.4 Biosynthesis of nucleic acids.
Nucleophilic displacement of pyrophosphate from a nucleoside triphosphate increases the chain length by one subunit. The reaction shown here illustrates DNA synthesis, in which the 3′-hydroxyl group attacks the triphosphate linkage of a deoxyadenosine triphosphate.

Nucleic acids can have extremely long chains; some DNA molecules may contain as many as *1 million* subunits. This in turn results in a molecular weight of more than 10^9 amu, and the length of an individual molecule can be *more than one millimeter*! These are gigantic molecules. Despite their enormous size, DNA molecules are structurally homogeneous. A particular nucleic acid molecule with hundreds of thousands of subunits can be synthesized many times over by an organism, and each time the new molecule will have the identical number and sequence of nucleotides. The capability of living systems to carry out such remarkable syntheses provides the chemical foundation for life as we know it. Whether you consider an entire organism or merely the products of cell division, nucleic acids carry all the genetic information that allows each new generation to carry on in the same manner as the preceding generation.

What structural features of nucleic acids result in their remarkable ability to maintain structural homogeneity and to pass it on by synthesis of new, identical molecules in the process called *replication*? This question fascinated and perplexed scientists for many years after it was first recognized that nucleic acids somehow encoded all the genetic information of living organisms. The answer resulted in the award of the 1962 Nobel Prize in medicine and physiology to James D. Watson of the United States and the British scientists Francis H. C. Crick and Maurice H. F. Wilkins. Watson and Crick were working together in the latter's laboratory in Cambridge; using the available data (including x-ray crystallographic data obtained by Wilkins and Rosalind Franklin), they published their interpretation of DNA structure in 1953. It had been well established that DNA contains equal numbers of adenine and thymine residues; similarly, guanine and cytosine residues are present in equal numbers. From their analysis of x-ray crystallographic data on DNA, Watson and Crick concluded that there were strong hydrogen bonding interactions between pairs of bases. When paired appropriately, these bases seem to be ideally suited for such interactions; such base pairs are called *complementary*. The complementary base pairs and their hydrogen bonding interactions are shown in Figure 25.5.

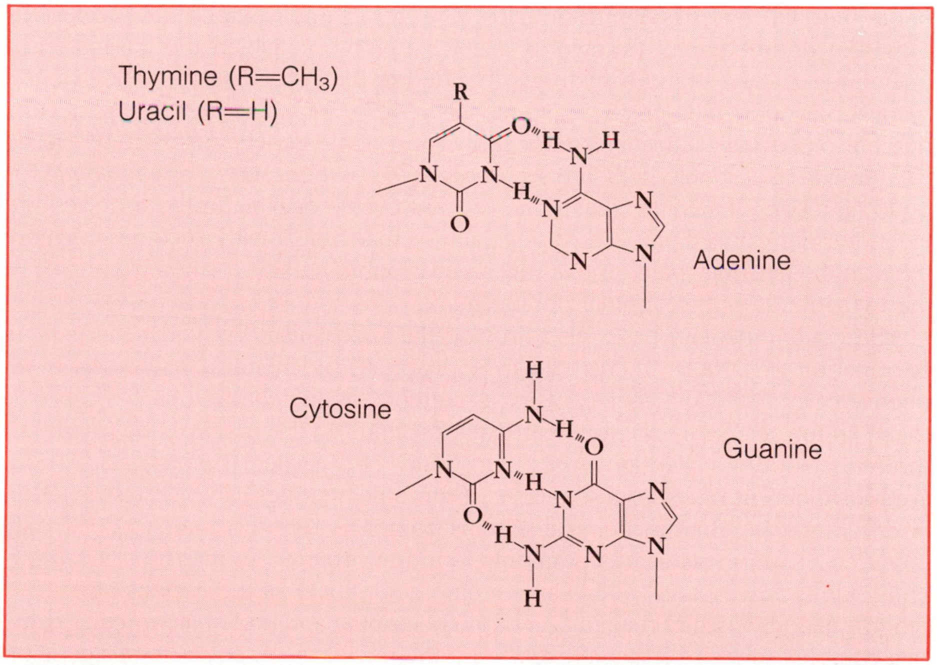

Thymine ($R{=}CH_3$)
Uracil ($R{=}H$)

Adenine

Cytosine

Guanine

Figure 25.5 Hydrogen bonding between complementary bases of nucleic acids.
Adenine always pairs with either thymine (DNA) or uracil (RNA), and guanine always pairs with cytosine. Other combinations do not afford the favorable hydrogen bonding shown here.

One structural feature of the nucleotide bases that we have so far neglected is the possibility of keto-enol equilibria. In contrast to simple phenols, these heterocyclic systems do not exhibit a large preference for the enol form. Thymine, for example, could exist as any of the following isomers:

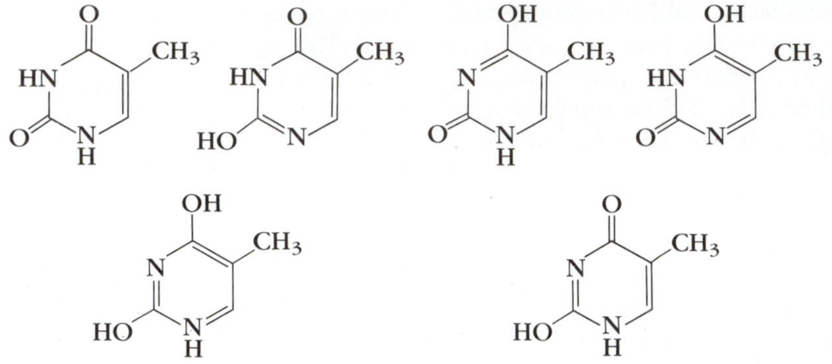

That the first of the preceding structures would lead to favorable hydrogen bonding with adenine was not initially recognized. The lack of a clearly defined preferred structure of the various bases posed a substantial impediment to learning the structure and function of nucleic acids. But once Watson and Crick had proposed the specific interactions shown in Figure 25.5, rapid progress became possible.

We now know that complementary base pairing is responsible for a wide range of the special properties of nucleic acids. DNA, for example, almost always exists as a twisted double strand (known as a *double helix*), in which the base of each nucleotide on one strand is hydrogen-bonded to the complementary base on the second polynucleotide chain. This not only helps to stabilize the DNA molecule, but it also provides a mechanism for replication of the genetic material. When the two strands of a double helix are separated, each of them serves as a *template* for building a complementary chain. The process occurs enzymatically, and each nucleotide is introduced onto a growing chain by matching it with its complementary base on the existing chain. In this way a new double helix is formed from each of the two strands of the original double helix.

The complementary bases also provide the mechanism by which the genetic information of DNA is interpreted in appropriate biochemical processes. A sequence of DNA called a *gene* contains a sequence of nucleotides that corresponds to the amino acid sequence for a particular protein.* In the *genetic code* a series of three nucleotides, called a *codon,* corresponds to a particular amino acid. There are four different bases, so a sequence of three nucleotides can be constructed in 4^3 different ways. But biochemical systems do not require that many codons; only 20 different amino acids are commonly found in proteins. In many instances several codons correspond to the same amino acid. Some of the codons also serve other functions, such as recognition of the beginning and end of a particular gene. In a complex series of enzymatic reactions, the nucleic acid sequence of a gene on a DNA molecule is converted to a complementary sequence of RNA in a process

*The DNA of higher organisms sometimes contains "extra" blocks of nucleotides that do not match a specific amino acid sequence. These blocks of nucleotides are called *introns,* and their function is not yet known.

called *transcription.* This RNA molecule is then *translated* into the appropriate protein. The protein is synthesized, one amino acid at a time, according to the series of codons on the RNA. Despite the apparent complexity of these processes, it is clear that nucleic acids have evolved into extremely useful polymer molecules that can store and accurately transmit highly complex chemical information.

EXERCISE 25.3

Which of the biopolymers discussed in this section are: (a) Linear? (b) Branched? (c) Cross-linked?

EXERCISE 25.4

(a) Which of the biopolymers discussed in this section have randomly distributed molecular weights? (b) Which are structurally homogeneous?

EXERCISE 25.5

Which of the biopolymers discussed in this section are heteropolymers (i.e., have polymer chains not made up entirely of carbon atoms)?

25.3 SEMISYNTHETIC POLYMERS

The earliest examples of useful polymers produced by chemical reactions involved modification of naturally occurring polymers. For such derivatives of naturally occurring macromolecules we will use the term *semisynthetic polymers.* The first important example was provided by the rubber industry. Natural rubber is quite soft and has poor durability for many purposes. In the 1840s Charles Goodyear in the United States found that a much more rigid and durable material was obtained when natural rubber was heated with sulfur. This resulted in the formation of many cross-links, each consisting of one or more sulfur atoms between the polymer chains, and it produced a polymer with much greater mechanical strength and resistance to wear. This material was called *vulcanized rubber,* and it achieved great economic importance with the subsequent development of the automobile and tire industries.

The most important raw material for semisynthetic polymers is cellulose. It was first subjected to chemical modification in the 1840s by treatment with nitric acid. The resulting *nitrocellulose,* known as guncotton, was extensively used as an explosive. In this material each of the three hydroxyl groups in nearly all the glucose subunits of cellulose have been converted to nitrate esters.

NO_2OCH_2, O, H, O, ONO_2, H, ONO_2, n

Nitrocellulose

It was found in the mid-1800s that a mixture of nitrocellulose and camphor produced a moldable, clear plastic material called *celluloid.* This found a wide variety of uses and was the primary material for photographic film for many years, despite its high flammability.

An important procedure involving chemical modification of cellulose is the *viscose process,* in which cellulose (usually from wood pulp) is converted to the xanthate derivative. Reaction of some (but not all) of the hydroxyl groups with carbon disulfide under alkaline conditions converts the —OH to a xanthate derivative (—O—CS_2^-, a sulfur analog of a carbonate):

$$\text{Cellulose—OH} \xrightarrow[\text{NaOH}]{CS_2} \text{cellulose—O—}\overset{\overset{\displaystyle S}{\|}}{C}\text{—}S^-Na^+$$

As you would expect from its similarity to a carboxyl group, the —CS_2H group is acidic, so the modified cellulose is soluble in the alkaline aqueous medium. The resulting viscous liquid can be extruded into aqueous acid as a filament. The xanthate groups are hydrolyzed under these conditions, and the resulting fibers of regenerated cellulose are known as *rayon.* Rayon is an important fiber in the textile and tire industries. When the xanthate solution is extruded through a slit into aqueous acid, a film (rather than a fiber) of regenerated cellulose is produced. This clear, transparent film is known as *cellophane* and is widely used in commercial packaging.

A variety of cellulose derivatives are used in modern technology. The flammable and unstable cellulose nitrate has been largely replaced by other cellulose esters such as cellulose acetate in the manufacture of photographic film. Cellulose acetate is also the fiber material in filter cigarettes, and fibers of cellulose acetate are used to make a fabric known by trade names such as Arnel (Celanese Co.). Many cellulose ethers have also found widespread commercial use. For example, carboxymethylcellulose (as the polysodium salt) can be prepared from wood pulp under alkaline conditions.

HOCH$_2$ ... O ... H ... O ... OH ... H ... OH ... n $\xrightarrow[\text{NaOH}]{\text{Cl—CH}_2\text{CO}_2\text{H}}$ ROCH$_2$ ... O ... H ... O ... OR ... H ... OR ... n R = $CH_2CO_2^-$ Na^+

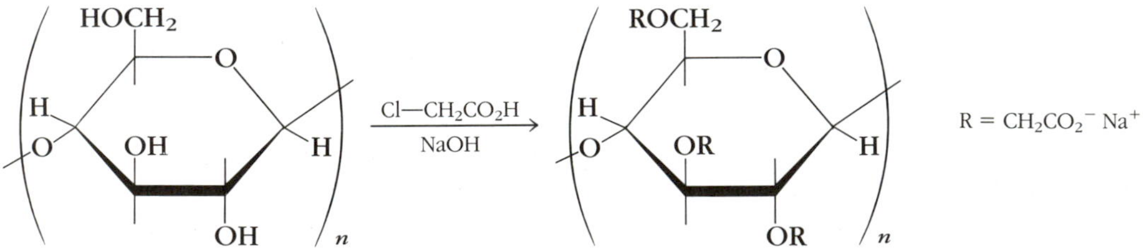

Cellulose   Carboxymethylcellulose

Not all the hydroxyl groups are alkylated in this process, but the preceding equation illustrates the reaction. Carboxymethylcellulose is employed in diverse ways: as an ingredient in laundry detergents; as an additive in foods, pharmaceuticals, and cosmetics; and as an emulsifier in water-base paints. Because they are nontoxic, a number of different ethers of cellulose (particularly methylcellulose) are employed as emulsifiers or thickening agents in commercial food products.

Cross-Linking

We will discuss two important examples of the modification of natural polymers by cross-linking. In each case changes in the shape of the material result. The first example involves textile fabrics. Treatment of cotton or a cotton-synthetic blend with appropriate difunctional organic compounds results in a modified fabric with wrinkle-resistant properties known as a *permanent press* fabric. The process is illustrated schematically in the following equation:

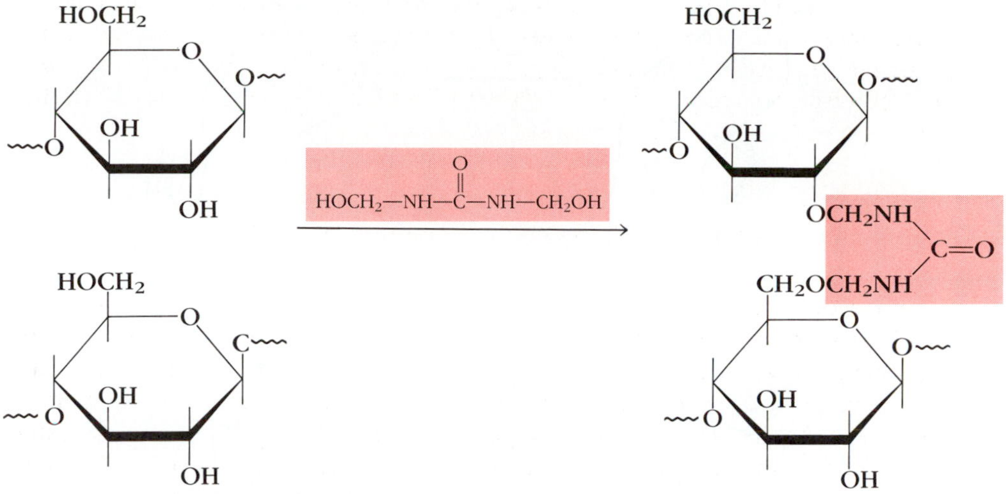

The cross-links impart additional strength to the fabric and help it to resist physical deformation. Although the fabric may be twisted or folded, the cross-links favor a return to the original, unwrinkled shape.

A related phenomenon is found in the second example, the *permanent wave* process for hair. The natural shape of a person's hair (curly or straight) is dependent on the structure of the constituent proteins. One common feature of the proteins in hair is a large number of intramolecular disulfide (—S—S—) linkages between amino acid residues along the polypeptide chain. The first step in the permanent wave process involves cleavage of these disulfide bridges (abbreviated as R—S—S—R) with a reducing agent such as thioglycolic acid ($HS{-}CH_2CO_2H$):

$$R{-}S{-}S{-}R + 2\,HSCH_2CO_2H \longrightarrow 2\,R{-}SH + \begin{array}{l} S{-}CH_2CO_2H \\ | \\ S{-}CH_2CO_2H \end{array}$$

Cleavage of the disulfide linkages of the protein makes the polypeptide chain more flexible, but the hair becomes limp and less resilient. Therefore another step is needed. After setting the strands of hair in the desired shape or curl, a mild oxidation (for example, with sodium bromate or even the oxygen in air) is used to regenerate disulfide bridges.

$$R{-}\overset{\displaystyle H}{\overset{|}{S}} + \overset{\displaystyle H}{\overset{|}{S}}{-}R \xrightarrow{\text{oxidation}} R{-}S{-}S{-}R$$

Setting the strands of hair causes changes in the molecular conformation of the proteins, and many of the groups originally joined as disulfide linkages will no longer be in close proximity. As a result, new disulfides can be formed as *cross-links* between different polypeptide chains. These cross-links help to hold the strands of hair in their new shape, hence the "permanent" wave. The overall process is illustrated schematically in Figure 25.6.

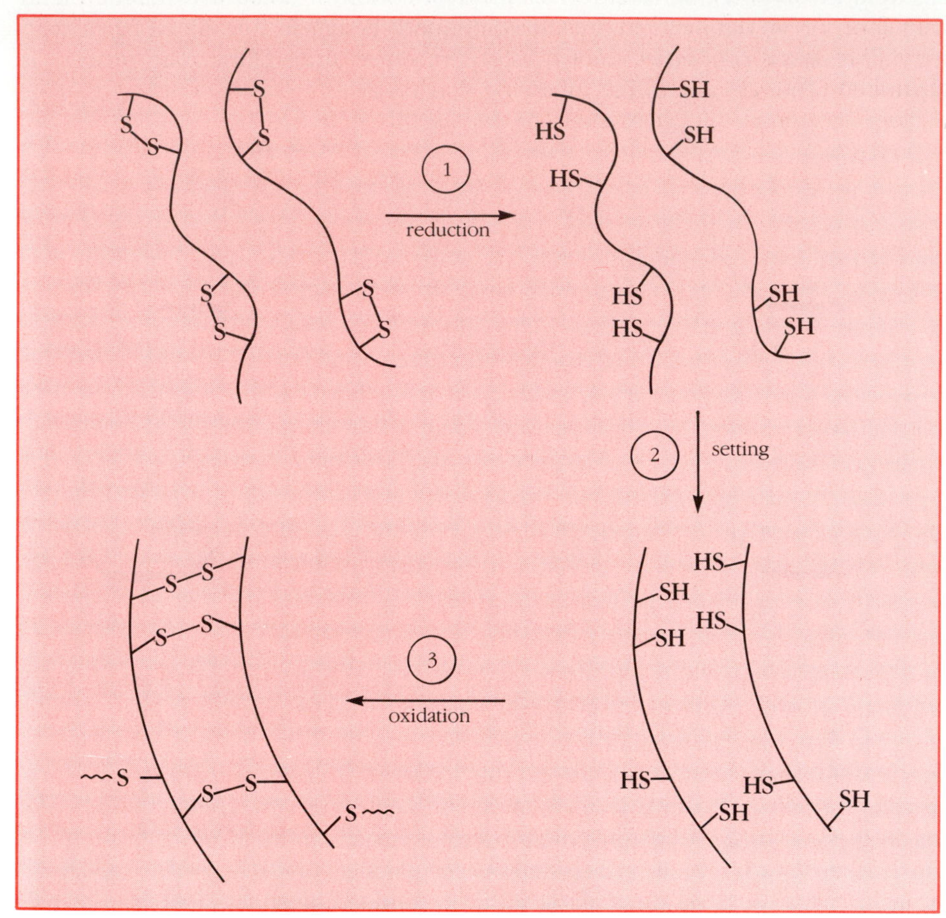

Figure 25.6 Schematic representation of the permanent wave process. In the first step the intramolecular disulfide bridges are cleaved by reduction. In the second step the hair is set into a new shape, and this causes conformational changes in the polypeptide chains. In the third step an oxidation regenerates disulfide linkages, but many of these are intermolecular. The net result is therefore a cross-linking of the proteins, which holds the hair in the new shape.

25.4 SYNTHETIC POLYMERS: PREPARATION

Historical Developments

The earliest synthetic polymers were probably not recorded, and even the first recorded examples were not recognized as having possible value. Many organic reactions yield varying amounts of polymeric material as an unwanted by-product, and this is usually discarded. When did polymer synthesis become an intentional practice of organic chemists? And how did polymer chemistry evolve into the multibillion dollar business that it is today?

The first commercially important, synthetic polymer was Bakelite (Union Carbide Co.), a material developed by L. H. Bakeland in the United States during the early years of the twentieth century. This polymer results from the reaction of phenol and formaldehyde under alkaline conditions in what is essentially a series of aldol condensations and Michael additions:

OH $H_2C{=}O$ H_2O OH_2 O CH_2OH H ^-OH

HO CH_2OH O CH_2 ^-O CH_2—OH O^-

^-O O HO HO ⟶ ⟶ polymer

Because the *para* position as well as both *ortho* positions of the phenol are reactive, a highly cross-linked, network polymer can be formed, which has the following structural features.

OH OH OH OH OH OH OH OH OH

Bakelite is very hard and rigid and cannot be molded once it is formed. Consequently, the reactants are mixed in a container of the appropriate shape, and polymerization produces a product in its final form and shape when it is removed from the mold. Bakelite has high electrical resistance and was used extensively in the developing radio and electronics industry during the first half of the 1900s.

Subsequent development of polymer synthesis was greatly assisted by the careful structural, mechanistic, physical, and theoretical studies that were carried out by a variety of scientists. Among these were Herman Staudinger in Germany, Herman Mark (first in Germany and later at the Polytechnic Institute of Brooklyn), and Wallace H. Carothers at du Pont. Staudinger's contributions earned him the Nobel Prize in 1953, and Carothers earned a place in history with his discovery of nylon in 1935. The importance of developments in polymer science was further underscored by the 1963 Nobel Prize in chemistry, which was jointly awarded to Karl Ziegler (Germany) and Giulio Natta (Italy) for their independent studies of alkene polymerization.

Mechanisms of Polymerization

Free radical polymerization is the most important method for commercial synthesis of polymers with a carbon chain. It is an example of a *chain reaction* that involves three steps: first, *initiation* (to generate a reactive intermediate); second, *propagation* (the actual series of product-forming steps); and third, *termination* (a step, often accidental, that breaks the chain of propagation steps). These are the same three steps that we discussed for the free-radical halogenation of alkanes in Section 2.6. The key to the overall chain reaction is that each reaction of a free radical in a propagation step produces a *new free radical* as the product. The polymerization can continue for a very large number of steps, and very large molecules can be formed. The process is illustrated here for the formation of polystyrene.

Initiation: Thermal decomposition of a peroxide or azo compound generates a free radical (RO· or R·):

$$\mathrm{R{-}O{-}O{-}R} \longrightarrow 2\ \mathrm{R{-}O\cdot}$$
$$\mathrm{R{-}N{=}N{-}R} \longrightarrow 2\ \mathrm{R\cdot} + \mathrm{N_2}$$

and one of these radicals can then add to the double bond of styrene:

$$\mathrm{CH(C_6H_5){=}CH_2} \xrightarrow{\mathrm{RO\cdot}} \cdot\mathrm{CH(C_6H_5){-}CH_2{-}OR}$$

Propagation: The benzylic radical produced in the initiation step can add to the double bond of another styrene monomer:

$$\mathrm{CH(C_6H_5){=}CH_2} \xrightarrow{\cdot\mathrm{CH(C_6H_5){-}CH_2{-}OR}} \cdot\mathrm{CH(C_6H_5){-}CH_2{-}CH(C_6H_5){-}CH_2{-}OR}$$

Continuation of this process builds up the number of subunits in the growing polymer chain:

$$\cdot\mathrm{CH(C_6H_5){-}CH_2}\!\left(\mathrm{CH(C_6H_5){-}CH_2}\right)_n\!\mathrm{OR}$$

Termination: Growth of the polymer is finally ended by a reaction between two radicals. This step is fairly improbable because free radicals are present in very low concentrations. As a result, long polymer chains can form before termination occurs. The reaction of two radicals can involve either coupling or hydrogen transfer.

$$\sim CH_2-\underset{}{\overset{C_6H_5}{\overset{|}{C}H}}\cdot + \cdot\overset{C_6H_5}{\overset{|}{C}H}-CH_2\sim \xrightarrow{\text{coupling}} \sim CH_2-\overset{H_5C_6}{\overset{|}{C}H}-\overset{C_6H_5}{\overset{|}{C}H}-CH_2\sim$$

$$\sim CH_2-\overset{C_6H_5}{\overset{|}{C}H}\cdot + \cdot\overset{C_6H_5}{\overset{|}{C}H}-\underset{H}{\underset{|}{C}H}\sim \xrightarrow[\text{transfer}]{\text{hydrogen}} \sim CH_2-\overset{C_6H_5}{\overset{|}{C}H_2} + \overset{C_6H_5}{\overset{|}{C}H}=CH\sim$$

Both these reactions destroy the free radical, so the chain reaction (for each of the two molecules involved) is terminated. Notice that modification of the repeating subunit occurs *only* at the growing end of the chain. As we discussed previously, the nature of the end groups is unimportant for a long chain polymer.

Cationic polymerization is another commercially significant method for the synthesis of polymers. It is of particular importance in the polymerization of isobutylene, and this is illustrated here.

Initiation: Cationic polymerization is typically initiated with electrophilic Friedel-Crafts catalysts such as BF_3 and $AlCl_3$. For convenience we will show the electrophile as an H^+ (from hydrogen chloride), which protonates the double bond of isobutylene.

$$(CH_3)_2C=CH_2 \xrightarrow{H^+} (CH_3)_2\overset{+}{C}-CH_2-H$$

Propagation: The tertiary carbocation formed in the initiation step can add to the double bond of another isobutylene.

$$(CH_3)_2C=CH_2 + (CH_3)_2\overset{+}{C}-CH_2-H \longrightarrow (CH_3)_2\overset{+}{C}-CH_2-\underset{CH_3}{\underset{|}{\overset{CH_3}{\overset{|}{C}}}}-CH_2-H$$

The addition can be repeated many times to build up the polymer chain.

$$(CH_3)_2\overset{+}{C}-CH_2\left(\underset{CH_3}{\underset{|}{\overset{CH_3}{\overset{|}{C}}}}-CH_2\right)_n-H$$

Termination: Termination occurs primarily by removal of an H^+ from the carbocation or by addition of some nucleophile (such as a chloride ion) to the cationic center.

$$(CH_3)_2\overset{+}{C}-CH(H)-\left(C(CH_3)_2-CH_2\right)_n-H \xrightarrow{\text{deprotonation}} (CH_3)_2C{=}CH-\left(C(CH_3)_2-CH_2\right)_n-H + HCl$$

$$Cl^-$$

$$(CH_3)_2\overset{+}{C}-CH_2-\left(C(CH_3)_2-CH_2\right)_n-H \xrightarrow[\text{addition}]{\text{nucleophilic}} CH_3-C(CH_3)(Cl)-CH_2-\left(C(CH_3)_2-CH_2\right)_n-H$$

$$Cl^-$$

Notice, however, that these termination steps would usually be reversible. The HCl generated in the first of the preceding reactions might react with the double bond, and the alkyl chloride formed in the second reaction could ionize to regenerate the carbocation.

Anionic polymerization has only limited commercial significance, but it has been extensively studied in the laboratory. It is largely restricted to systems in which the anion is stabilized by a strong electron-withdrawing substituent. We will illustrate the anionic polymerization of acrylonitrile.

Initiation: Various organometallic compounds and strong bases have been used to initiate anionic polymerization, one of which is sodium amide.

$$CH(CN){=}CH_2 \xrightarrow{^-NH_2\ Na^+} {}^-CH(CN)-CH_2-NH_2$$

Although drawn in the preceding equation as a carbanion, the negative charge is also delocalized onto the cyano group.

Propagation: A series of Michael additions occurs

$$CH(CN){=}CH_2 + {}^-CH(CN)-CH_2-NH_2 \longrightarrow {}^-CH(CN)-CH_2-CH(CN)-CH_2-NH_2$$

and this generates a polymer chain.

$${}^-CH(CN)-CH_2-\left(CH(CN)-CH_2\right)_n-NH_2$$

Termination: A variety of termination steps are possible, but the most likely one is any process that quenches the anion by proton transfer. For example

$${}^-CH(CN)-CH_2-\left(CH(CN)-CH_2\right)_n-NH_2 \xrightarrow{NH_3} CH_2(CN)-CH-\left(CH(CN)-CH_2\right)_n-H$$

Transition metal–catalyzed polymerization was discovered during the 1950s by Karl Ziegler (Germany), who found that complex catalyst mixtures such as $TiCl_4$—$Al(C_2H_5)_3$ were very effective in the polymerization of alkenes. These catalysts were also studied by Giulio Natta (Italy), and as we indicated previously, Ziegler and Natta shared the 1963 Nobel Prize in chemistry for this work. The polymerization of alkenes (particularly ethylene and propene) by transition metal catalysts has since become an important commercial process. One of the most important features of this type of polymerization is the stereochemistry of the resulting polymer. All the chiral centers of polypropylene, for example, tend to have the *same* configuration (all *R* or all *S*). A polymer with this type of stereochemical homogeneity is called *isotactic*. In contrast, a polymer in which there is a random distribution of *R* and *S* centers is called *atactic*. (A less common type of polymer stereochemistry involves alternating *R* and *S* centers, and polymers with this property are called *syndiotactic*.)

A detailed analysis of transition metal bonding is beyond the scope of this text, but the general aspects of polymerization by transition metal catalysts can still be explained fairly easily. In the following equations we will use the abbreviation M to designate the transition metal, and we will only show the metal substituents that are involved in growth of the polymer chain. For the case of the $TiCl_4$—$Al(C_2H_5)_3$ system, the reaction appears to originate by formation of a titanium species with at least one ethyl group. The transition metal is a Lewis acid, so it can coordinate to an alkene.

$$CH_3CH_2\text{—}M \xrightarrow{CH_2=CH_2} CH_3CH_2\text{—}\underset{\vdots}{M}\,(CH_2=CH_2)$$

The complex is formed by donation of π electron density from the alkene to the metal, and the carbons of the double bond become electron-deficient. This is followed by rearrangement of the original alkyl substituent from the metal center to one of the carbons of the double bond. This is easier to see when a polar resonance form of the alkene complex is drawn.

$$\underset{CH_3CH_2\text{—}M}{CH_2=CH_2} \longrightarrow \underset{CH_3CH_2\text{—}M^-}{{}^+CH_2\text{—}CH_2} \longrightarrow \underset{M}{CH_3CH_2\text{—}CH_2\text{—}CH_2}$$

The net result of the rearrangement is *insertion* of a —CH_2CH_2— subunit between the transition metal and the original alkyl substituent. Repetition of this insertion reaction produces a polymer.

$$CH_3CH_2\text{—}(CH_2CH_2)_n\text{—}M \xrightarrow{CH_2=CH_2} CH_3CH_2\text{—}(CH_2CH_2)_{n+1}\text{—}M$$

Condensation Polymers

All the polymerization mechanisms in the preceding discussion involved *addition* of the monomer to a reactive species. A second general type of polymerization occurs via a series of stepwise reactions. Many of these reactions produce water as a by-product, and the general term *condensation polymerization* has been used to describe them. The formation of Bakelite from phenol and formaldehyde is an example of this type of polymerization.

In most condensation polymers the polymer chain is not just a carbon chain but also contains amide or ester linkages. Formation of such polymers frequently involves high-temperature condensation of the appropriate carboxylic acids and amines or alcohols. Such reactions of a free carboxylic acid do not readily occur under typical laboratory conditions, and very high temperatures are required. This is illustrated by the following equation for the preparation of nylon 66, one of several common nylons.

$$HO_2C{-}CH_2CH_2CH_2CH_2{-}CO_2H + NH_2{-}(CH_2)_6{-}NH_2 \xrightarrow[16\ atm]{270\text{–}280°C} {+}\overset{O}{\overset{\|}{C}}{-}CH_2CH_2CH_2CH_2{-}\overset{O}{\overset{\|}{C}}{-}NH{-}CH_2CH_2CH_2CH_2CH_2CH_2{-}NH{+}_n + H_2O\uparrow$$

Nylon 66

Notice that condensation polymers must be formed from *difunctional monomers.* Nylon 66 is prepared from two monomers, a diacid and a diamine, and its name ("nylon six six") is derived from the number of carbons in each monomer.

Another common nylon is nylon 6, which is formed from a single monomer that has both amino and carboxylic acid functionality. Actually, it is the lactam rather than the amino acid that is used in the industrial synthesis (although small amounts of water are present to generate some of the ring-opened monomer).

$$\text{(caprolactam)} \xrightarrow{200°C} {+}\overset{O}{\overset{\|}{C}}{-}CH_2CH_2CH_2CH_2CH_2{-}NH{+}_n$$

In the preceding reactions we illustrated the formation of polyamides; similar condensations are used in the commercial synthesis of polyesters. In some cases esters or acid chlorides are used in place of the free acids; the by-products in such cases are alcohols and HCl, respectively. These reactions proceed under much milder conditions, but the lower cost of the carboxylic acids generally favors their direct condensation in commercial applications.

Copolymers

As synthetic polymers have become increasingly important in our society during the last 50 years, scientists have continually been searching for ways to produce new materials with better properties. Hardness, flexibility, impact resistance, elasticity, degradation resistance, and solvent resistance are just some of the features that might be important for a commercial polymer. Sometimes the physical properties of a polymer are unsatisfactory for a particular purpose, and chemists have found ways to alter these properties. Sometimes the solubility of a polymer (and its ease of processing) can be improved by the addition of small amounts of an appropriate high-boiling solvent. Such additives are called *plasticizers,* and they are used extensively in the formulation of synthetic rubbers. A widely used plasticizer is dioctyl phthalate.

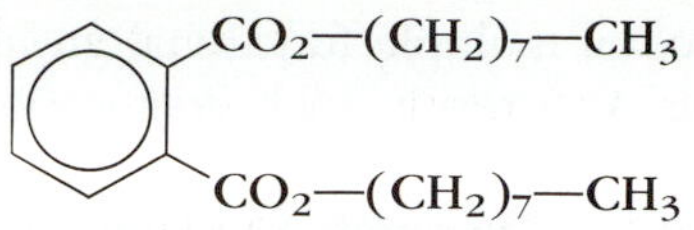

$$C_6H_4(CO_2{-}(CH_2)_7{-}CH_3)_2$$

What if two desirable properties correspond to different polymers? Can anything be done in such situations? One approach is to *blend* two polymers, each of which has one of the desired properties; there are many cases in which polymer blends have commercial applications. But the blending process is severely limited. Polymers usually are not highly soluble in each other, so even if molten polymers are mixed, the resulting material may not be homogeneous. Instead it may consist of a mixture of small particles, each of which is made up of just one of the polymers. In order to achieve a better mixing of two polymers, a second approach has been developed, in which the different types of polymer chain are incorporated into a single molecule. In other words, the goal is synthesis of a *copolymer*.

Copolymerization of a mixture of two monomers has gained extensive use in industrial polymer synthesis. One of the most important synthetic polymers in modern technology is styrene-butadiene rubber (SBR), a synthetic rubber that is used in the tire industry, in electrical insulation, and in many other rubberized materials. The number and distribution of styrene subunits depends on the relative amount of each monomer used in the polymerization, so no simple formula can be drawn. The following structure shows what part of the chain might be like.

$$\sim CH_2{-}CH{=}CH{-}CH_2{-}CH(C_6H_5){-}CH_2{-}(CH_2{-}CH{=}CH{-}CH_2)_3{-}CH(C_6H_5){-}CH_2 \sim$$

SBR rubber

Another important application of such copolymerization is the addition of a monomer that is chlorinated or brominated. The presence of halogenated residues greatly reduces the flammability of the resulting polymer. Such fire resistance is particularly important in the manufacture of fibers for the textile and clothing industry.

Graft copolymers have a main polymer chain with a single repeating subunit, but they also have branches that are built up from a different monomer. There are two general ways of forming graft copolymers, and side chains are introduced onto an existing polymer in both cases. In the first method the side chains are introduced as preformed polymers, and in the second method a polymerization reaction occurs which involves reactive sites on the original chain.

The first method is illustrated by the following scheme, in which hydroxyl groups of one polymer react with carboxyl groups of a second polymer to form ester linkages. Only short segments of the chains are shown.

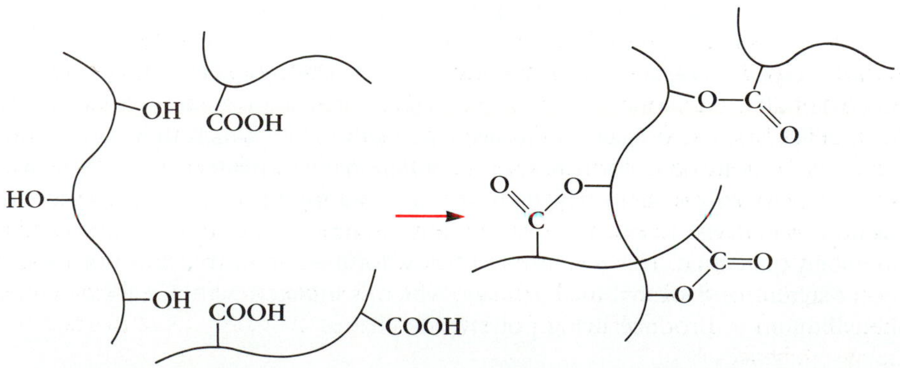

OH COOH HO OH COOH COOH → O—C=O, O=C—O, C=O—O

Notice that the presence of multiple functional groups on the polymer chains will result in cross-linking. As a result, such graft copolymers are apt to be very strong but also quite rigid.

The other method for producing a graft copolymer is illustrated by the polymerization of styrene in the presence of poly(butadiene). Poly(butadiene) is highly unsaturated, and the free radicals formed by the initiation of styrene polymerization could interact with a poly(butadiene) chain either by addition to a double bond or by hydrogen abstraction to form an allylic radical.

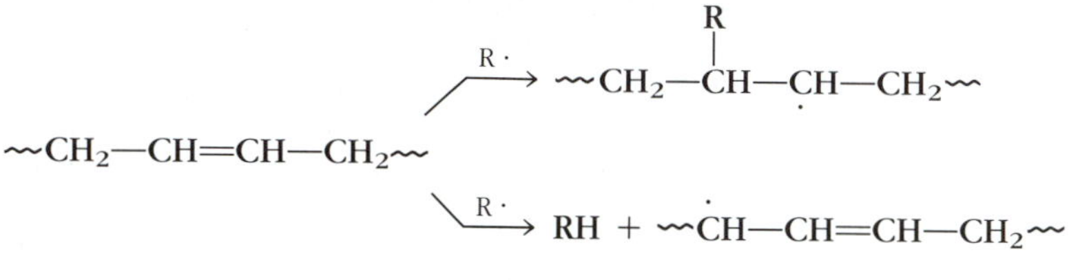

$$\sim CH_2-CH=CH-CH_2\sim \xrightarrow{R\cdot} \sim CH_2-\underset{}{\overset{R}{\overset{|}{C}H}}-\dot{C}H-CH_2\sim$$

$$\sim CH_2-CH=CH-CH_2\sim \xrightarrow{R\cdot} RH + \sim\dot{C}H-CH=CH-CH_2\sim$$

In either case a radical center is formed on the poly(butadiene) chain, and this can enter into polymerization with a styrene monomer to generate a polystyrene side chain on the main poly(butadiene) chain. For example:

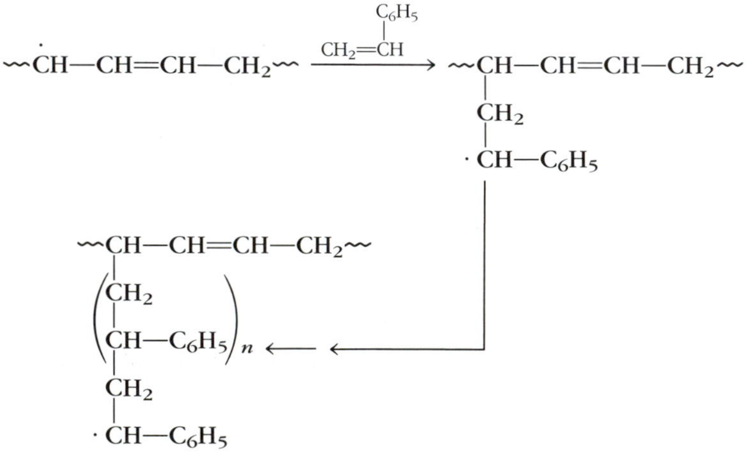

$$\sim\dot{C}H-CH=CH-CH_2\sim \xrightarrow{CH_2=CH(C_6H_5)} \sim CH(CH_2-\dot{C}H-C_6H_5)-CH=CH-CH_2\sim$$

$$\longrightarrow \sim CH\left(\left(CH_2-CH(C_6H_5)\right)_n-CH_2-\dot{C}H-C_6H_5\right)-CH=CH-CH_2\sim$$

Possibilities exist for cross-linking either by coupling of two radical side chains or by reaction of the radical side chain with another poly(butadiene) chain. The material prepared in this way is a very strong polymer, known as high-impact polystyrene.

Block copolymers are linear polymers that consist of alternating segments (*blocks*) of different monomers. As a group these are the most important commercial copolymers. Once again, there are two methods by which they can be prepared. The first involves formation of a *living polymer*. This term describes chain reaction polymerization in which the reactive intermediate is stable under the reaction conditions. Termination is therefore unimportant until the reaction is intentionally quenched. This is illustrated below for the anionic block copolymerization of styrene and methyl methacrylate, which is initiated with a reagent such as phenyllithium to produce living polystyrene chains.

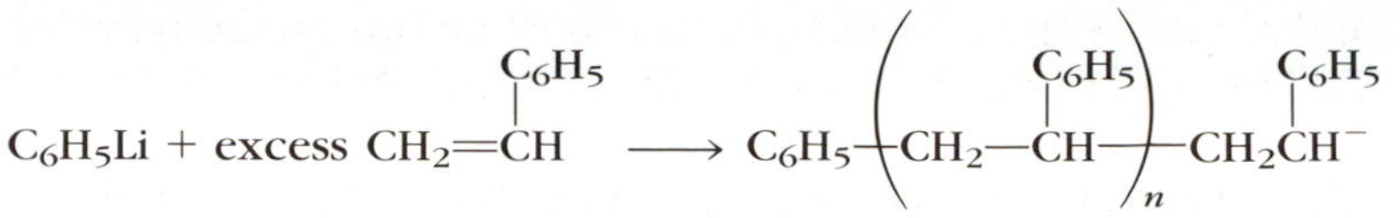

$$C_6H_5Li + \text{excess } CH_2{=}\underset{|}{\overset{C_6H_5}{CH}} \longrightarrow C_6H_5\text{-}\left(CH_2\text{—}\overset{C_6H_5}{\underset{}{CH}}\right)_n\text{-}CH_2\overset{C_6H_5}{CH^-}$$

When all the styrene monomer has been consumed, methyl methacrylate is added, and a new polymerization takes place.

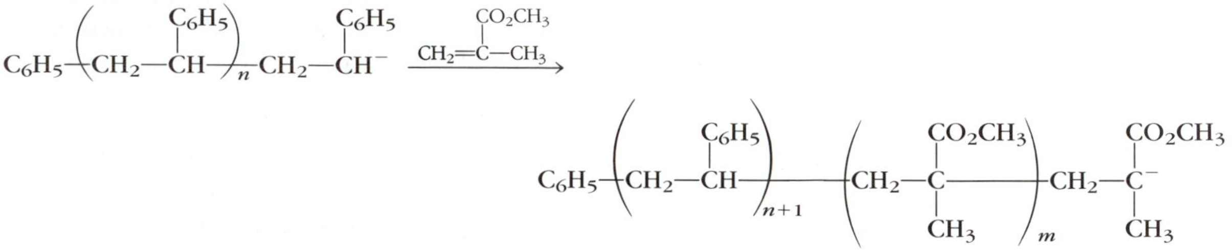

$$C_6H_5\text{-}\left(CH_2\text{—}\overset{C_6H_5}{CH}\right)_n\text{-}CH_2\text{—}\overset{C_6H_5}{CH^-} \xrightarrow{CH_2{=}\overset{CO_2CH_3}{C}\text{—}CH_3}$$

$$C_6H_5\text{-}\left(CH_2\text{—}\overset{C_6H_5}{CH}\right)_{n+1}\text{-}\left(CH_2\text{—}\overset{CO_2CH_3}{\underset{CH_3}{C}}\right)_m\text{-}CH_2\text{—}\overset{CO_2CH_3}{\underset{CH_3}{C^-}}$$

Quenching of the living polymer (e.g., with acid) produces the final copolymer.

The second method for preparing block copolymers is one in which preformed polymer chains of two different types are joined together. This procedure is very similar to the first method shown for graft copolymers, but the only reactive functional groups are at the *ends* of the chains. Such a block copolymer could be produced from one reactant with —OH as end groups and another with —CO_2H at the chain termini.

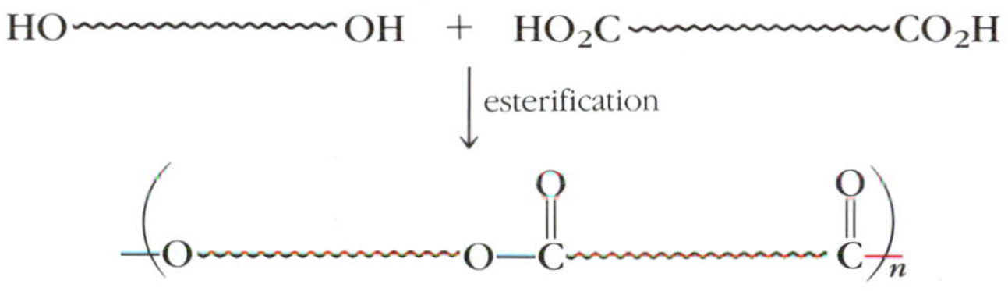

$$HO\sim\sim\sim OH + HO_2C\sim\sim\sim CO_2H \xrightarrow{\text{esterification}} \left(O\sim\sim\sim O\text{—}\overset{O}{\overset{\|}{C}}\sim\sim\sim\overset{O}{\overset{\|}{C}}\right)_n$$

The various methods we have discussed here can be used to prepare an enormous number of polymers and copolymers. It would not be possible to evaluate all of them, but some of the more important examples are discussed in the following section.

EXERCISE 25.6

Ethylene oxide can be polymerized in a chain reaction that can be initiated by either acid or base. (a) Draw the structure of the polymer. (b) Draw the mechanism for the polymerization initiated by H_2SO_4. (c) Draw the mechanism for the polymerization initiated by $NaOCH_3$.

EXERCISE 25.7

Using the monomers shown below, suggest examples of (a) linear homopolymers; (b) linear (random) copolymers; (c) block copolymers. Suggest how you could make each.

(i) $CH_2{=}CH_2$ (ii) $CH_2{=}CH\text{—}CH_3$

(iii) $HO_2CCH_2CH_2CH_2NH_2$ (iv) $HO_2CCH_2CH_2CH_2CH_2NH_2$

25.5 SYNTHETIC POLYMERS: APPLICATIONS

In the preceding sections of this chapter we have introduced you to some important types of macromolecules and the reactions by which they are formed. We have also emphasized the importance of polymers in nearly all aspects of our daily lives, but only in the case of biopolymers have we discussed the specific roles of the various macromolecules. In this section we will briefly discuss a number of commercially important polymers so that you will better understand the structure and function of these commonly encountered materials.

Polymeric Hydrocarbons

Polyethylene and *polypropylene* are saturated hydrocarbons that are often prepared by transition metal catalysis, although free radical polymerization is of major importance for commercial preparation of polyethylene.

$$-\!\!\left(CH_2CH_2\right)\!\!-_n \qquad -\!\!\left(\underset{}{\overset{CH_3}{\overset{|}{C}H}}-CH_2\right)\!\!-_n$$

Polyethylene　　Polypropylene

Both these polymers are used extensively in the manufacture of molded housewares (for example, plastic pails and buckets) and toys. Both melt above 100°C, so they are not affected by boiling water. These two polymers, particularly polypropylene, find some use in textiles. Polypropylene (under the trade name *Herculon*) is used in indoor-outdoor carpeting and in upholstery material. Polyethylene can be prepared as a thin film, which is sold for household use (an example is *Glad Wrap*, manufactured by Union Carbide).

Polystyrene has unsaturated substituents (i.e., the phenyl groups), but its main polymer chain is a saturated hydrocarbon.

$$-\!\!\left(\overset{C_6H_5}{\overset{|}{C}H}-CH_2\right)\!\!-_n \qquad \text{polystyrene}$$

Polystyrene is a fairly hard, brittle solid that is used to make many molded housewares and toys. When manufactured so that gas bubbles form in the molten polymer, it cools to a solid foam. This material (*Styrofoam,* Dow Chemical Co.) is used as an insulating agent and as a packing material.

Poly(vinyl chloride) is not a hydrocarbon, but we have included it here for convenience because the monomer, chloroethylene (vinyl chloride), is a simple monosubstituted alkene.

$$-\!\!\left(\overset{Cl}{\overset{|}{C}H}-CH_2\right)\!\!-_n \qquad \text{poly(vinyl chloride)}$$

The polymer, often called PVC, is used for pipes in household plumbing, for phonograph records, for the soles and heels of shoes, and for insulation of electrical wire. Formulation with plasticizers yields a softer polymer that is used to manufacture coated fabrics such as *Naugahyde* (U.S. Rubber Co.). When vinyl chloride is polymerized together with approximately 20% of vinylidene chloride

($CH_2{=}CCl_2$), the resulting copolymer can be processed as a film. The copolymer manufactured by the Dow Chemical Co. is called *Saran,* and the film, sold as *Saran Wrap,* has many kitchen and household uses.

Halocarbon Polymers

Poly(tetrafluoroethylene), or PTFE, is a polymer with high thermal stability and very low solubility. It can be machined to different shapes, is an excellent electrical insulator, is highly resistant to most chemicals, and has a low coefficient of friction. The material made by du Pont has the trade name *Teflon.*

$$\left(CF_2 - CF_2 \right)_n \qquad \text{Poly(tetrafluoroethylene) (PTFE)}$$

The physical properties and chemical stability of PTFE have resulted in its widespread use in the electrical and chemical industries as parts (or coatings of parts) that come into contact with corrosive chemicals. It has also been used with increasing frequency for laboratory apparatus, and thin layers of the polymer are used to create nonstick surfaces for cooking utensils.

Poly(chlorotrifluoroethylene) is similar in many ways to the fully fluorinated polymer. It is more flexible and transparent, but its resistance to organic solvents and its electrical insulating properties are not as good. It is sold under several trade names, including *Kel-F* (3M Corp.), and it is used to manufacture such products as tubing, gaskets, and electrical wire insulation.

Elastomers

Elastomers are materials with elastic properties, and some of them are commercially important as *synthetic rubber.* An example that we discussed in Section 25.4 is styrene-butadiene rubber (SBR), which played an important role during the second World War, when sources of natural rubber became unavailable.

Polychloroprene is a structural analog of natural rubber in which —Cl has replaced $-CH_3$ as the substituent on the monomeric diene.

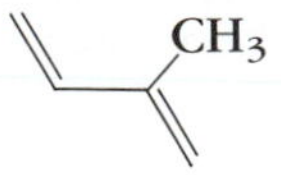

CH_3 Cl

Isoprene 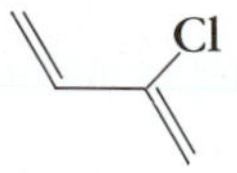 Chloroprene

The polymer is a synthetic rubber with excellent chemical resistance. It is used in industrial applications (hoses and belts), to make heels for shoes, and for coated fabrics. The material marketed by du Pont is known as *Neoprene.*

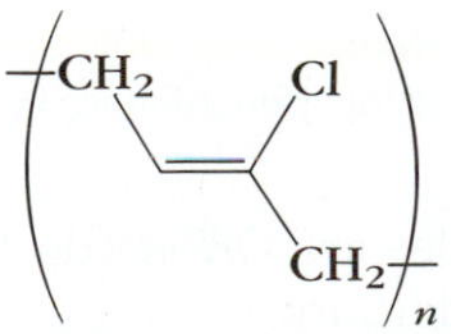

$$\left(CH_2-CH{=}C(Cl)-CH_2 \right)_n \qquad \text{Polychloroprene}$$

Butyl rubber is a copolymer of isobutylene (2-methylpropene) and small amounts (1–3%) of isoprene. Since there are few isoprene subunits, we can use the following approximate representation for the polymer:

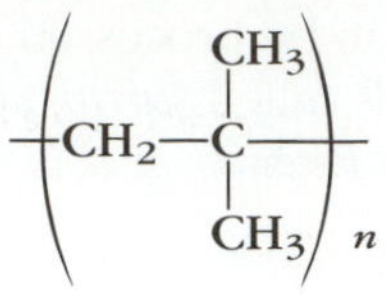

$$\left(CH_2 - \underset{CH_3}{\overset{CH_3}{C}} \right)_n$$

Butyl rubber

Butyl rubber exhibits very low permeability to gases, and it is used extensively for the manufacture of inner tubes in the tire industry.

Silicone rubber has a skeletal backbone of alternating silicon and oxygen atoms, so it is quite different from the carbon-chain polymers of the other types of synthetic rubber.

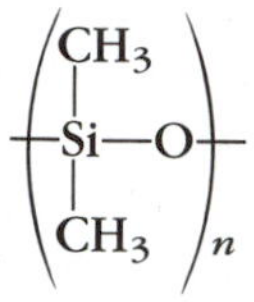

$$\left(\underset{CH_3}{\overset{CH_3}{Si}} - O \right)_n$$

Silicone rubber

Silicone rubber retains its elastic properties over a very wide temperature range, and it is used for gaskets in combustion engines. Chemists frequently use a silicone rubber disk as the septum through which samples are injected into a gas chromatograph.

Acrylics

The reactive double bond of acrylic acid and its derivatives has permitted the synthesis of a wide range of acrylic polymers. The two most important monomers are methyl methacrylate and acrylonitrile.

$$CH_2{=}\overset{CH_3}{C}{-}CO_2CH_3 \qquad CH_2{=}CH{-}CN$$

Methyl methacrylate Acrylonitrile

Polyacrylonitrile is widely used as a fiber in the textile industry. Commercial formulations are often copolymers in which small quantities of a second monomer such as methyl methacrylate or vinyl acetate are employed to improve the qualities of the polymer. Such copolymers, as well as the homopolymer, are known as *acrylics.*

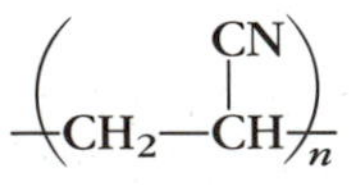

$$\left(CH_2 - \overset{CN}{CH} \right)_n$$

Polyacrylonitrile (acrylic)

Examples of acrylic fibers used in textiles are *Orlon* (du Pont), *Acrilan* (Chemstrand Corp.), and *Dynel* (Union Carbide Corp.)

Poly(methyl methacrylate) is a colorless, optically transparent material that is prepared in the form of sheets, rods, and tubes. It is familiar to many people as *Lucite* (du Pont) or *Plexiglas* (Rohm & Haas).

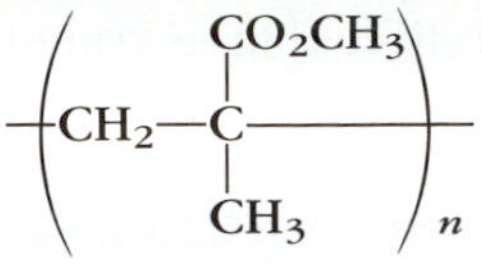

$$\left(CH_2{-}\overset{\overset{\displaystyle CO_2CH_3}{|}}{\underset{\underset{\displaystyle CH_3}{|}}{C}} \right)_n$$

Poly(methyl methacrylate)

Poly(methyl methacrylate) is used as a substitute for glass, particularly in the automotive industry, because of its optical clarity.

Polyamides

The development of *nylon* by Carothers at du Pont in the 1930s afforded the first completely synthetic fiber, and nylon made its commercial appearance in the form of women's stockings in 1940. The polymer used was nylon 66, which is still widely used in the textile industry and for manufacturing rope and tire cord. It is also molded into specific shapes for use as gears and bearings, because it has good resistance to abrasion. The original trademark of du Pont has lapsed, and the term *nylon* is often used almost interchangeably with *polyamide.*

A variety of nylons are used commercially, including nylon 6, which is marketed under the trade names *Perlon* (GAF, Inc.) and *Capran* (Allied Chemical Corp.) Nylon 6 is also used for tire cord and as a fabric, although the latter use is more important in Europe. Nylon 610 and nylon 11 are commercially important in Europe, particularly for textiles in the case of nylon 11.

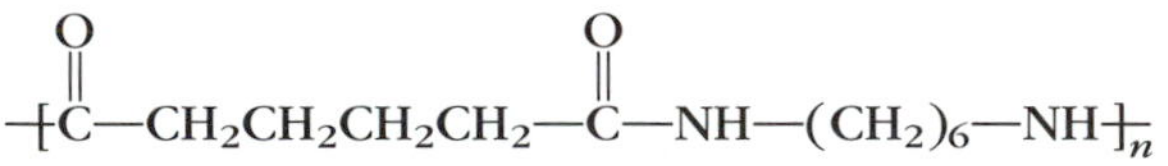

$$\left[\overset{O}{\overset{\|}{C}}{-}CH_2CH_2CH_2CH_2{-}\overset{O}{\overset{\|}{C}}{-}NH{-}(CH_2)_6{-}NH \right]_n \quad \text{Nylon 66}$$

$$\left[\overset{O}{\overset{\|}{C}}{-}CH_2CH_2CH_2CH_2{-}\overset{O}{\overset{\|}{C}}{-}NH{-}(CH_2)_{10}{-}NH \right]_n \quad \text{Nylon 610}$$

$$\left[\overset{O}{\overset{\|}{C}}{-}(CH_2)_5{-}NH \right]_n \quad \text{Nylon 6}$$

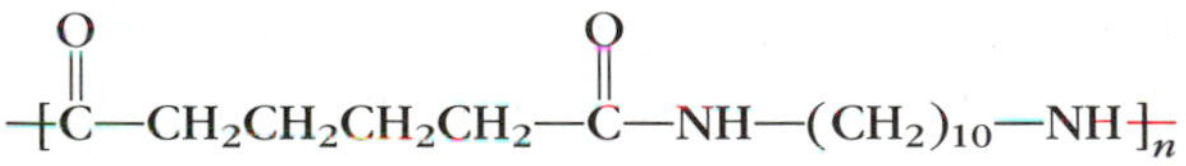

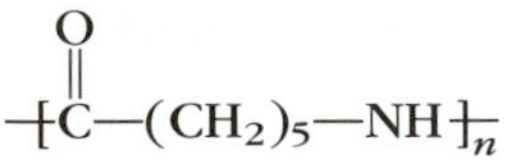

$$\left[\overset{O}{\overset{\|}{C}}{-}(CH_2)_{10}{-}NH \right]_n \quad \text{Nylon 11}$$

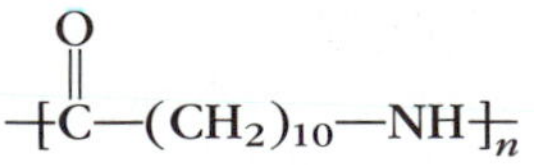

Polyesters

Although polymers such as poly(methyl methacrylate) contain ester groups in the repeating subunit, they are not called polyesters. Poly(methyl methacrylate) has a backbone containing only carbon atoms, whereas in *polyesters* the repeating carboxyl groups are part of the polymer chain.

Poly(ethylene terephthalate), or *PETP,* is the most important commercial polyester. It is derived from terephthalic acid and ethylene glycol.

$$HO_2C{-}C_6H_4{-}CO_2H \qquad HOCH_2CH_2OH \qquad \left(\overset{O}{\overset{\|}{C}}{-}C_6H_4{-}\overset{O}{\overset{\|}{C}}OCH_2CH_2O \right)_n$$

Terephthalic acid — Ethylene glycol — Poly(ethylene terephthalate)

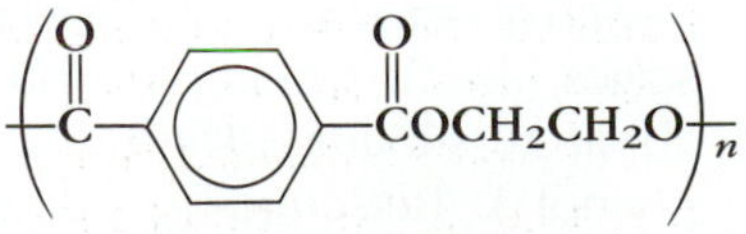

PETP has great physical strength, which is reflected in its commercial applications. It is used as a textile fiber, in tire cord, and in the form of a film for recording tape and motion picture film. Some of its well-known trade names are *Dacron* (du Pont), *Fortrel* (Fiber Industries, Inc.), *Kodel* (Eastman Kodak), and *Mylar* (du Pont).

Polycarbonates

Polycarbonates are actually polyesters in which the diacid is the one-carbon compound carbonic acid (H_2CO_3). H_2CO_3 decomposes to CO_2 and H_2O, so direct esterification to form a polymer is not possible. Instead, the corresponding acid chloride, phosgene, or a monomeric carbonate ester is allowed to react with an appropriate diol. The major commercial polymer is derived from 2,2-bis(4-hydroxyphenyl)propane, known as *bisphenol A.*

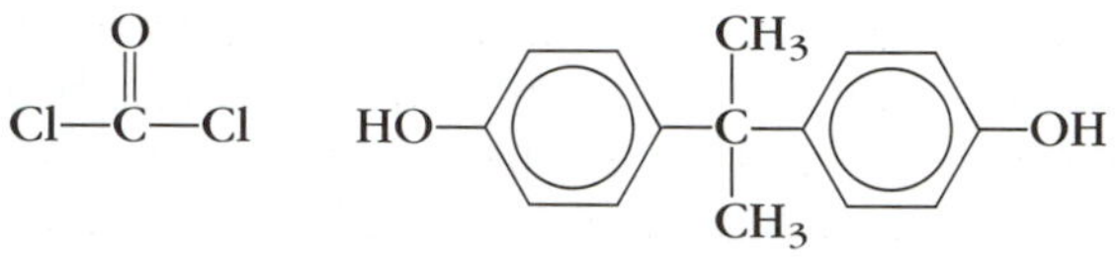

$$Cl-\overset{\overset{O}{\|}}{C}-Cl \qquad HO-C_6H_4-\underset{\underset{CH_3}{|}}{\overset{\overset{CH_3}{|}}{C}}-C_6H_4-OH$$

Phosgene Bisphenol A

$$\left(O-C_6H_4-\underset{\underset{CH_3}{|}}{\overset{\overset{CH_3}{|}}{C}}-C_6H_4-O-\overset{\overset{O}{\|}}{C} \right)_n$$

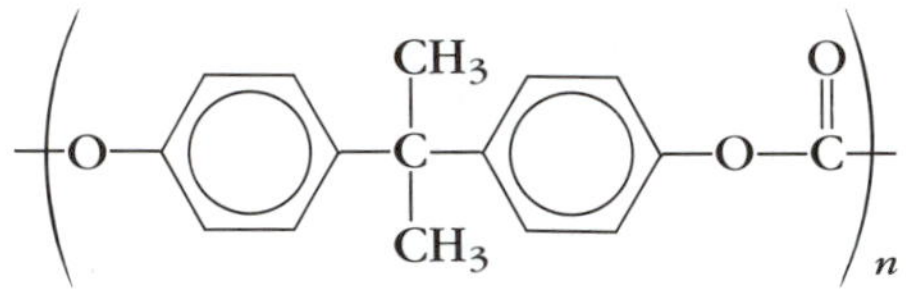

Polycarbonate

Polycarbonate is an optically transparent material that has very high impact resistance. It is particularly useful for laboratory safety shields and is used for the face covers of astronauts' space suits. The polymer, made by the General Electric Company, is known as *Lexan.* Its impact resistance has also made it a good material for manufacturing toys, and it is sometimes used for molecular models.

Polyethers

Several structural types of polyethers are known, but the polymers of greatest commercial importance are derived from epoxides. The strained three-membered ring is readily cleaved by nucleophilic attack under either acidic or basic conditions, and polymerization occurs readily.

Poly(ethylene glycol) can be obtained by polymerization of ethylene oxide, and the resulting polymers are water-soluble. They are used as thickening agents and stabilizers in adhesives and in cosmetic and pharmaceutical formulations.

$$\left(CH_2-CH_2-O \right)_n$$

The material known as *Carbowax* (Union Carbide) is familiar to chemists as a stationary phase for gas chromatography.

Epoxy resins are actually a family of polymers. They are usually prepared in two stages, the first being formation of a *prepolymer,* a copolymer containing only about 10 subunits. The most common monomers are epichlorohydrin and bisphenol A. These are both difunctional compounds, and they produce a polymer with an epoxide group at each terminus.

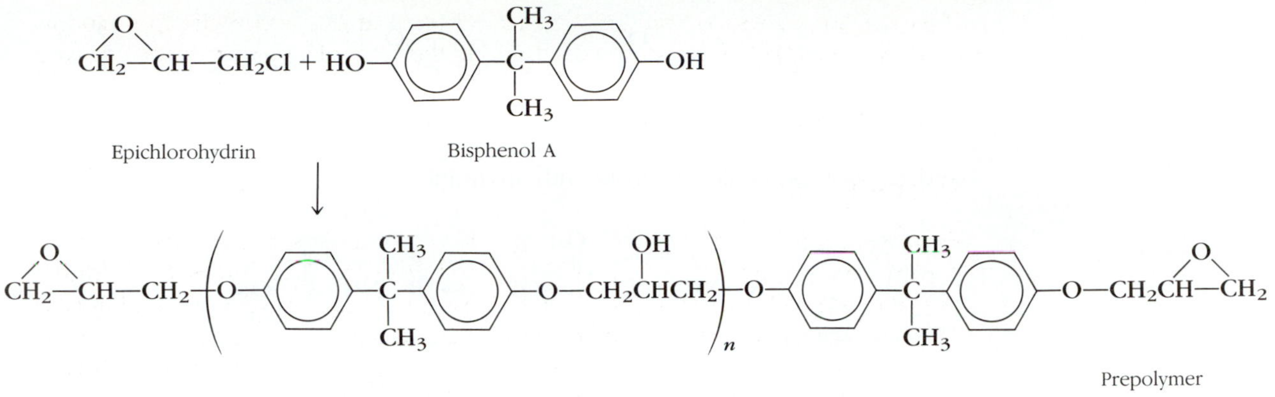

Epichlorohydrin + Bisphenol A $\longrightarrow$ Prepolymer

The prepolymer is mixed with a cross-linking agent (curing agent) to produce the final epoxy resin. The use of curing agents such as diamines results in highly cross-linked polymers because each nitrogen can react with two different epoxide groups, as illustrated by the following equation. Additional branching (not shown) can also occur by reaction of the hydroxyl groups on the polymer chain with epoxide residues.

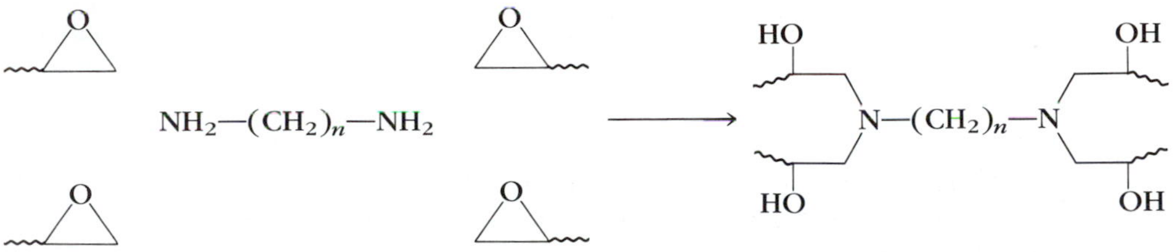

$$NH_2-(CH_2)_n-NH_2 \longrightarrow N-(CH_2)_n-N$$

The epoxy resins are extremely tough and durable. They are employed in laminating applications, in the manufacture of molded materials, and as surface coatings. They are used extensively to make *fiber glass* materials, in which the resin is reinforced with glass fibers. Epoxy resin adhesives are available for both commercial and household use. The two components of the adhesive are the prepolymer and the curing agent.

Polyurethanes

Urethanes (carbamates) are derivatives of carbonic acid having both ester and amide characteristics.

$$R-NH-\overset{O}{\overset{\|}{C}}-O-R$$ A urethane (carbamate)

While a carbamic acid (RNH—CO_2H) is unstable and can slowly lose carbon dioxide, the corresponding carbamates (RNH—CO_2R) are esters and are quite stable. The major pathways by which the urethanes are formed are nucleophilic attack on an isocyanate

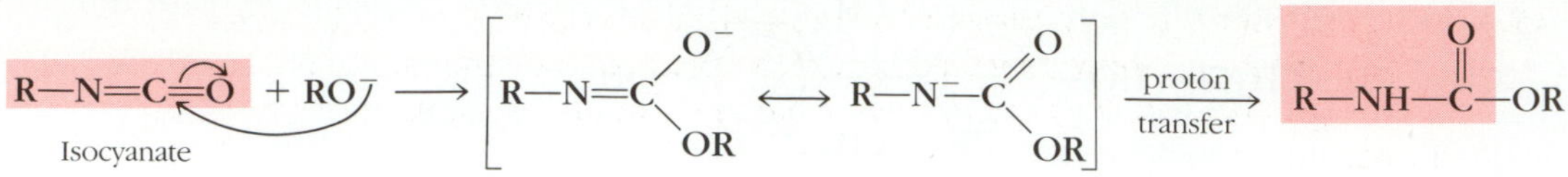

$$R{-}N{=}C{=}O + RO^- \longrightarrow \left[R{-}N{=}C(O^-)(OR) \longleftrightarrow R{-}\bar{N}{-}C({=}O)(OR) \right] \xrightarrow[\text{transfer}]{\text{proton}} R{-}NH{-}C({=}O){-}OR$$

Isocyanate

and reaction of a chloroformate with an amine

$$R{-}\ddot{N}H_2 + Cl{-}C({=}O){-}OR \longrightarrow R{-}NH{-}C({=}O){-}OR$$

Chloroformate

Polyurethanes can therefore be prepared by reaction between a diisocyanate and a diol:

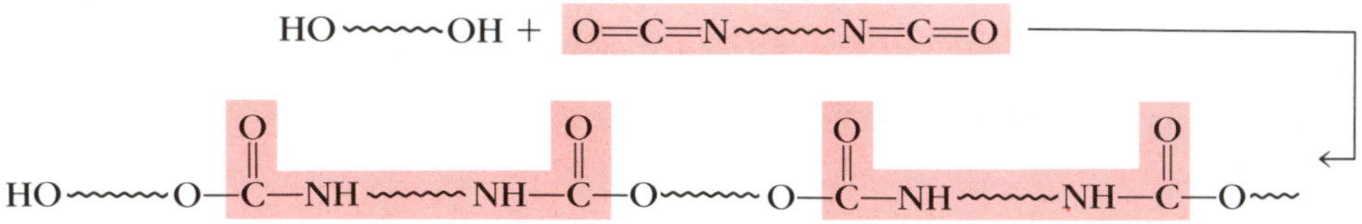

$$HO{\sim\sim}OH + O{=}C{=}N{\sim\sim}N{=}C{=}O \longrightarrow$$

$$HO{\sim\sim}O{-}C({=}O){-}NH{\sim\sim}NH{-}C({=}O){-}O{\sim\sim}O{-}C({=}O){-}NH{\sim\sim}NH{-}C({=}O){-}O{\sim}$$

Polyurethane

Polyurethanes have a variety of commercial uses. When produced as a foam, the polymer is an excellent material for thermal insulation. Other polyurethanes are used as elastomers, as coatings, and as textile fibers. An important example of the latter are the *spandex* fibers; these are elastic fibers that are used to make stretch fabrics such as *Lycra* (du Pont). Spandex is actually a complex copolymer in which one of the "monomers" (the diol) is itself a low-molecular-weight polymer such as poly(butanediol)

$$H{\left(O{-}CH_2CH_2CH_2CH_2\right)_n}OH$$

25.6 PROBLEMS

25.1 Explain why the end groups of a macromolecule have little effect on its chemical and physical properties.

25.2 For each of the following polymers, determine the average degree of polymerization (i.e., number of subunits) for a sample that has an average molecular weight of 200,000 amu.

(a) Polyethylene
(b) Poly(vinyl chloride)
(c) Nylon 6
(d) Nylon 11
(e) Cellulose
(f) Poly(tetrafluoroethylene)
(g) Poly(ethylene terephthalate)

25.3 Draw a short segment (four subunits) of each of the polymers in Problem 25.2.

25.4 When ethylene is polymerized by free radical initiators, the product has some branching. Show how this branching can arise when the radical end of a growing chain abstracts a hydrogen from the "middle" of another polymer molecule.

25.5 Contrast the structures of DNA and RNA. Suggest a reason why RNA is more easily degraded by hydrolysis.

25.6 The end groups of a polymer are often determined by the initiator of a chain reaction. What end groups are introduced when the polymerization of ethylene is initiated by (a) benzoyl peroxide and (b) azobisisobutyronitrile?

$$C_6H_5\overset{\overset{O}{\|}}{C}O O\overset{\overset{O}{\|}}{C}C_6H_5$$

Benzoyl peroxide

$$CH_3-\underset{\underset{CN}{|}}{\overset{\overset{CH_3}{|}}{C}}-N{=}N-\underset{\underset{CN}{|}}{\overset{\overset{CH_3}{|}}{C}}-CH_3$$

Azobisisobutyronitrile

25.7 Why are only high-boiling solvents used as plasticizers?

25.8 Write mechanisms that illustrate the polymerization of ethylene oxide (a) by acid (use H^+ for simplicity) and (b) by $NaOCH_3$.

25.9 A derivative of poly(methyl methacrylate) is used in plastic lenses for eyeglasses. The derivative, poly(cyclohexyl methacrylate), is prepared from poly(methyl methacrylate). Suggest a reasonable procedure for the preparation of poly(cyclohexyl methacrylate).

25.10 Suggest a reasonable set of conditions for the preparation of poly(ethylene terephthalate).

25.11 Show the reaction mechanism by which polycarbonate could be produced from bisphenol A and dimethyl carbonate.

25.12 Why is poly(ethylene glycol) water-soluble?

25.13 In the formation of the prepolymer for epoxy resins, epoxy groups are always formed at the termini even though nucleophilic attack on epichlorohydrin usually occurs at the epoxy group rather than at the $—CH_2Cl$ group. Suggest an explanation.

25.14 What cyclic ether is a likely source for the poly(butanediol) used to make Spandex polyurethane?

25.15 Draw the structure of a short segment (eight subunits) of polypropylene that is:
(a) Isotactic
(b) Syndiotactic
(c) Atactic

25.16 Which of the following polymers would be capable of showing stereochemical features (i.e., isotactic or atactic)?

(a) Polyethylene
(b) Polystyrene
(c) Teflon
(d) PVC
(e) Poly(methyl methacrylate)
(f) Nylon 66

25.17 For each of the polymers in the preceding problem that could show such stereochemistry, draw a short (five-subunit) segment of the isotactic structure. Indicate whether the configuration of each chiral center is *R* or *S*.

CHAPTER 26

INDUSTRIAL ORGANIC CHEMISTRY

Industrial organic chemistry pervades our everyday lives, and our contact with industrial organic chemicals is both constant and inescapable. The clothing you wear, from your underwear to your shirt, is likely to be made in part from synthetic fibers; only by wearing garments made exclusively of cotton, wool, or silk would you be able to avoid the synthetic materials. Your shoes, whether just the heels or perhaps the entire soles, are almost certainly made from a synthetic elastomer rather than from natural rubber. The story continues far beyond your clothing. Your immediate environment is filled with synthetic organic materials: plastics, fibers, dyes, printing inks, paint, carpeting, draperies, etc. The list is endless.

How did this come about? Less than 100 years ago synthetic materials were virtually nonexistent, and now they are everywhere. In this chapter we will introduce you to some of the key aspects of industrial chemistry. We will show how its evolution has been inextricably linked to both historical and economic developments. The story is not a simple one, but by selecting a few relatively narrow areas for exploration we hope to provide you with an understanding of the chemical industry and its interrelationships with other aspects of the modern world.

26.1 THE PETROCHEMICAL INDUSTRY: ITS ORIGINS AND ITS FUTURE

Petroleum

Chemical energy has always been a primary concern to humans, starting from the discovery that fire could be used to provide warmth, cook food, and provide light. As civilization developed, wood and coal were satisfactory fuels for generating heat, but candles provided the only convenient way to provide light when the sun was down. Whale oil afforded an improvement, and then by the mid 1800s a satisfactory illuminating oil had become available from coal.

During the middle of the nineteenth century a small number of Americans were trying to think of a use for the black, oily substance that was sometimes found in springs or water supplies. Samuel Kier of Pittsburgh took advantage of its reputation as a home remedy to market "Petroleum, or Rock Oil . . . a pure, unadulterated article, without any chemical change," as a medicinal agent. He claimed it was effective in treating a variety of diseases, ". . . giving increased and renewed energy to all the organs of life." While we may think less of those ideas today, it is more difficult to ignore the importance of Kier's subsequent preparation of "carbon oil" by distilling it from the crude petroleum. Within a relatively few years whale oil as well as coal oil had been superseded, and kerosene lanterns were providing reliable illumination in the absence of sunlight for the first time in human history.

The next steps, of course, were development of oil well drilling and petroleum refining. Experimentation showed that petroleum is actually a mixture of many hydrocarbons with different boiling points, and these could be separated by distillation. The fraction of interest, kerosene, was a mixture of hydrocarbons boiling in the region of 160–250°C. the lower-boiling materials were too flammable for use in illuminating lamps, and they were unwanted by-products. The early kerosene industry was faced with the problem of how to dispose of the "useless" fraction known as gasoline!

At that time, when kerosene was the major usable product obtained from petroleum, thermal cracking (Section 1.11) of the high-boiling residue was discovered.

Heating of the residue to temperatures above 600°C resulted in the decomposition of long-chain hydrocarbons and the production of additional kerosene. Whereas only about 10–15% of the crude oil could be recovered as kerosene by distillation alone, thermal cracking increased that yield substantially.

Gasoline

Unquestionably, the major development in petroleum chemistry was the automobile. As the automotive industry blossomed in the early part of the twentieth century, gasoline changed from its status as an unwanted by-product to the most important fraction in petroleum. Economic pressures increased for the petroleum industry to produce more and more gasoline, and this in turn produced chemical breakthroughs. Careful studies of the cracking process showed that silica-alumina catalysts permitted effective cracking at lower temperatures, and the overall yield of gasoline from crude petroleum was greatly increased.

A highly simplified view of petroleum refining is presented in Figure 26.1. This

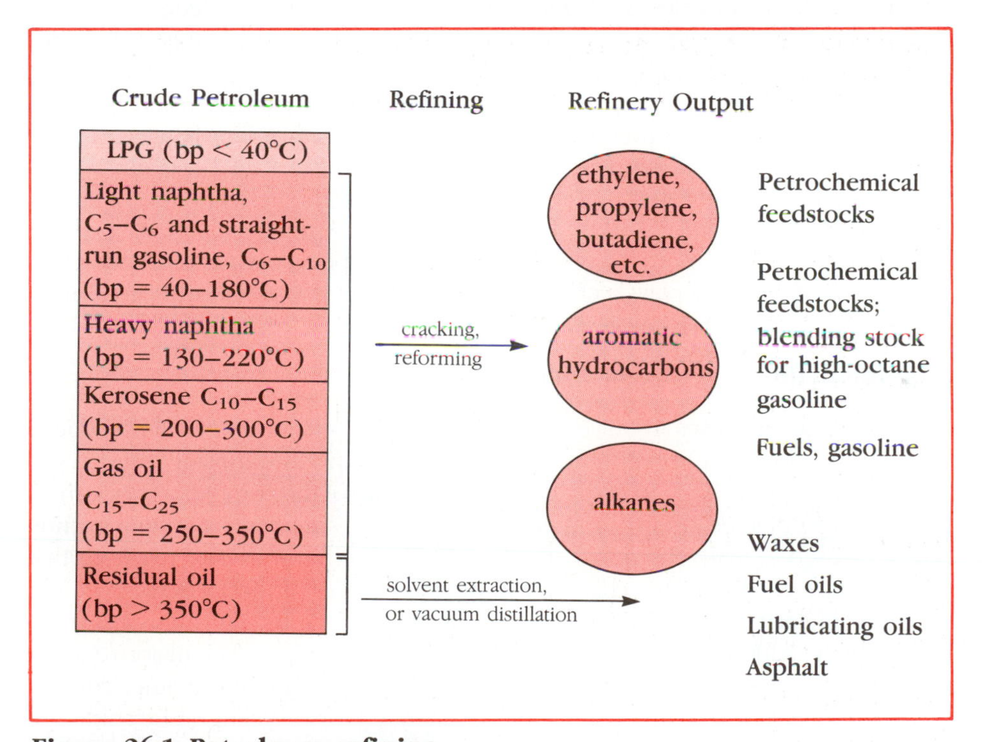

Figure 26.1 Petroleum refining.
Crude petroleum is distilled to produce a series of fractions ranging from *liquefied petroleum gas* (LPG) to *gas oil,* leaving behind the nonvolatile *residual oil.* The approximate composition of the crude oil is given by the size of each box in the figure, but this varies tremendously with the source of the petroleum. The boiling ranges specified are also approximate, since the exact ranges depend on the subsequent processing to be carried out. Some of the major products of the refining process are listed in the ovals at the right of the figure. Relative quantities of these products can be varied by the refiner to suit economic needs.

figure is intended to summarize refining in general rather than to describe a specific plan, because the processes can be varied to suit specific needs. For example, what fraction of the crude distillate will be subjected to cracking and reforming? If the major need were gasoline and kerosene (jet fuel), then it would not be logical to subject the low-boiling fractions to cracking; instead, gas oil would be the logical feedstock. On the other hand, if production of ethylene and propylene were the primary goal, you would surely include kerosene and the naphthas in the feedstock for cracking.

Petrochemicals

One of the curious developments in petroleum chemistry was the production of aromatic compounds (benzene, toluene, xylenes) during reforming with platinum catalysts. This development spelled the end of coal as the major natural resource for the chemical industry. By the 1930s it had become possible to obtain all the basic chemicals, both aliphatic and aromatic, from a single source—that source was petroleum. Whether genius or folly, the entire world economy and its growing chemical industry was developing a heavy dependence on petroleum.

The automobile was no longer the only factor in petroleum chemistry. Polymer chemistry was also developing rapidly in the first half of this century. Scientists had dreams of synthetic materials that could replace natural rubber, wood, silk and other fabrics, and perhaps even metals for structural uses. This idea had its origin in the apparently unlimited supply of petroleum as a source of the necessary monomeric compounds. After World War II a dramatic rise of the petrochemical industry occurred. Although fuel products (gasoline, oils for heating and energy production) still account for approximately 80 percent of all petroleum consumption, the petrochemical industry plays a critical role in both the U.S. and world economies.

Monomers for the Polymer Industry

Organics find many uses in our culture, but of all the organic chemicals produced industrially, more than three-fourths of the total quantity is ultimately used in the manufacture of polymers. In order to show you some of the complex relationships between the polymer industry and petroleum, we will simply consider the starting materials used to make the polymers discussed in Section 25.5. The reactions that we present throughout this chapter will differ in three important ways from the reactions in earlier chapters. *First,* we will not list any yields. For industrial processes a high yield of a specific product is not always necessary. The major requirement is one of economics, and a low yield may be acceptable if the other components in the product mixture can either be separated for recycling or be sold for a profit. *Second,* many of the reactions involve catalytic processes, and they are run at very high temperatures. Typically these processes are not suitable for laboratory use, and often they will work only with the specific compound for which they were developed. *Third,* the reagents employed for industrial reactions are typically inexpensive. As an example, molecular oxygen is a very common reagent for oxidations (often catalytic). Why would a chemical company wish to purchase an expensive oxidizing agent when an unlimited supply of air is available at no cost?

1. *Polyethylene and polypropylene* are made from ethylene and propylene, *primary petrochemicals* that are obtained directly from petroleum refining.

$$\text{Petroleum} \xrightarrow{\text{cracking}} CH_2{=}CH_2 + CH_3{-}CH{=}CH_2 + \text{other compounds}$$

These two organic compounds rank first and second among primary petrochemicals in total quantity produced in the United States; approximately 14×10^9 kg of ethylene and 7×10^9 kg of propylene were produced in 1984. Almost half of the total ethylene and 25 percent of the propylene are converted to polyethylene and polypropylene, respectively. The remaining supplies are converted to other organic compounds.

2. *Polystyrene* is made from styrene, and this in turn is prepared by catalytic dehydrogenation of ethylbenzene.

$$C_6H_5CH_2CH_3 \xrightarrow[630°]{\text{ZnO or other metal oxide catalyst}} C_6H_5CH{=}CH_2 + H_2$$

The ethylbenzene is first manufactured by Friedel-Crafts alkylation of benzene with ethylene.

$$C_6H_6 \xrightarrow[\text{AlCl}_3 \text{ or } H_3PO_4,\ 95°C]{CH_2{=}CH_2} C_6H_5CH_2CH_3$$

Benzene is also a primary petrochemical, produced in the catalytic reforming process. It ranks third among primary petrochemicals in the United States, with production of approximately 5×10^9 kg in 1984. About half the total benzene is converted to ethylbenzene.

3. *Poly(vinyl chloride)* is prepared by polymerization of vinyl chloride, which is made from ethylene by addition of Cl_2 followed by elimination of HCl. The first reaction occurs readily at low temperatures in dichloroethane as the solvent, and pyrolysis of the 1,2-dichloroethane (often with pumice as a catalyst) yields vinyl chloride.

$$CH_2{=}CH_2 \xrightarrow{Cl_2} Cl{-}CH_2{-}CH_2{-}Cl \xrightarrow[\text{pumice}]{500°C} CH_2{=}CHCl + HCl$$

Other processes for manufacture of vinyl chloride (e.g., addition of HCl to acetylene) have also been used, but the low cost of ethylene from petroleum has made it the preferred starting material. Almost all the vinyl chloride that is manufactured is in turn used for poly(vinyl chloride) synthesis, and this use accounts for approximately 15 percent of the total ethylene production in the United States.

4. *Poly(tetrafluoroethylene)*, more commonly known as Teflon, requires tetrafluoroethylene as the starting material. Because fluorine is so reactive, the direct fluorination of a C_1 or C_2 monomer with F_2 is impractical. Instead, chloroform is first treated with hydrogen fluoride to produce chlorodifluoromethane. The chloroform is made (along with other chlorinated methanes, Section 1.10) by chlorination of methane, another primary petrochemical that is obtained from natural gas and petroleum cracking.

$$CHCl_3 + 2\ HF \rightarrow CHClF_2 + 2\ HCl$$

Cracking of the chlorofluoromethane generates difluoromethylene ($:CF_2$) as a reactive intermediate, and this dimerizes to form tetrafluoroethylene.

$$CHClF_2 \xrightarrow[\text{Pt}]{600°C} CF_2{=}CF_2$$

A related sequence of reactions is used to manufacture the monomer chlorotrifluoroethylene. Hexachloroethane can be obtained by the chlorination of ethane, and reaction with hydrogen fluoride affords a mixture of products, including 1,1,2-trichlorotrifluoroethane. Reductive elimination of chlorine with zinc then yields the necessary monomer for production of poly(chlorotrifluoroethylene).

$$CCl_2F{-}CClF_2 \xrightarrow[C_2H_5OH]{Zn} (Cl)(F)C{=}C(F)(F)$$

5. *Butadiene* and its simple derivatives, isoprene and chloroprene, along with styrene are the most important monomers in the elastomer industry. Together they account for more than 90 percent of all synthetic rubber produced worldwide. Butadiene itself is obtained as one of the cracking products in petroleum refining and ranks fifth among the primary petrochemicals, with U.S. production at 1×10^9 kg in 1984.

$$\text{Petroleum} \xrightarrow{\text{cracking}} CH_2{=}CH{-}CH{=}CH_2$$

A sequence involving addition of chlorine and elimination of HCl is used to produce chloroprene.

$$CH_2{=}CH{-}CH{=}CH_2 \xrightarrow[\text{2. elimination of HCl}]{\text{1. addition of } Cl_2} CH_2{=}C(Cl){-}CH{=}CH_2$$

The process is complex, and isomerization of the initially formed mixture of dichlorobutenes is employed to increase the yield of chloroprene.

Isoprene is synthesized by catalytic dehydrogenation of the 2-methylbutenes that are produced in petroleum refining.

$$\text{Petroleum} \xrightarrow{\text{cracking}} \begin{matrix} CH_2{=}C(CH_3){-}CH_2CH_3 \\ + \\ CH_3{-}C(CH_3){=}CHCH_3 \end{matrix} \longrightarrow CH_2{=}C(CH_3){-}CH{=}CH_2 + H_2$$

6. *Acrylonitrile and methyl methacrylate* are the monomers for acrylic polymers, but their origins are quite different. Acrylonitrile has been produced by

several routes, but of current importance is that from propylene. In the presence of complex catalysts (bismuth or uranium salts), propylene reacts with ammonia and oxygen to yield acrylonitrile directly.

$$CH_2{=}CH{-}CH_3 \xrightarrow[\text{catalyst}]{NH_3,\ O_2} CH_2{=}CH{-}C{\equiv}N + H_2O$$

Methyl methacrylate is made from acetone (Section 26.2) by formation of the cyanohydrin, followed by acid-catalyzed dehydration with accompanying hydrolysis to the amide, and subsequent reaction with methanol.

$$CH_3{-}\overset{\overset{\displaystyle O}{\|}}{C}{-}CH_3 \xrightarrow[\text{NaOH}]{\text{HCN}} CH_3{-}\underset{\underset{\displaystyle OH}{|}}{\overset{\overset{\displaystyle CH_3}{|}}{C}}{-}CN \xrightarrow{98\%\ H_2SO_4} CH_2{=}\overset{\overset{\displaystyle CH_3}{|}}{C}{-}\overset{\overset{\displaystyle O}{\|}}{C}{-}NH_2$$

$$\xrightarrow[H_2SO_4]{CH_3OH} CH_2{=}\overset{\overset{\displaystyle CH_3}{|}}{C}{-}CO_2CH_3$$

The methanol employed in the last step is another important petrochemical, and it is obtained from both petroleum and natural gas. Methane is converted to carbon monoxide and hydrogen (a mixture known as *synthesis gas*) by high-temperature reaction with water.

$$CH_4 + H_2O \xrightarrow[\text{nickel oxide catalyst}]{700°C} CO + 3\,H_2$$

Subsequent partial reduction of the carbon monoxide yields methanol.

$$CO + 2\,H_2 \xrightarrow[\text{Cr—Zn catalyst}]{400°C} CH_3OH$$

7. *Polyamides* are prepared from a variety of monomers, but some of the most common are six-carbon compounds. Not surprisingly, these can be traced back to petroleum: specifically to either benzene or cyclohexane, which are obtained from catalytic reforming. A major pathway for the industrial synthesis of the six-carbon monomers relies on the conversion of benzene to phenol (see Section 26.2) and subsequent hydrogenation to cyclohexanone. Cyclohexanone can be converted to caprolactam (the monomer for nylon 6) by a Beckmann rearrangement (Solutions Manual Appendix, Molecular Rearrangements) or it can be cleaved to adipic acid by oxidation with nitric acid.

$$\text{phenol} \xrightarrow[\text{catalyst}]{H_2} \text{cyclohexanone} \begin{cases} \xrightarrow[H_2SO_4]{NH_2OH} \text{caprolactam (N—H)} \\ \xrightarrow[\text{catalysts}]{HNO_3} HO_2C(CH_2)_4CO_2H \end{cases}$$

Adipic acid is one of the monomers for nylon 66, and its conversion to the nitrile (via the amide) followed by reduction leads to the other monomer, hexamethylenediamine.

$$HO_2C(CH_2)_4CO_2H \xrightarrow[\text{2. pyrolysis}]{\text{1. } NH_2} H_2NOC(CH_2)_4CONH_2 \xrightarrow{\text{dehydration}} NC(CH_2)_4CN \xrightarrow[\text{catalyst}]{H_2} H_2NCH_2(CH_2)_4CH_2NH_2$$

8. *Poly(ethylene terephthalate),* better known as polyester, is made from terephthalic acid and ethylene glycol. Terephthalic acid is manufactured by oxidation of *p*-xylene, another primary petrochemical. *p*-Xylene ranks fourth among the primary petrochemicals in the amount produced, with a U.S. output of 2×10^9 kg in 1984. Almost all the xylene manufactured is converted to terephthalic acid or its dimethyl ester.

$$p\text{-}CH_3C_6H_4CH_3 \xrightarrow[\text{Co-Mn catalyst}]{O_2,\ 200°C} p\text{-}HO_2CC_6H_4CO_2H$$

Ethylene glycol is produced by hydrolysis of ethylene oxide, which in turn is prepared by catalytic oxidation of ethylene.

$$CH_2{=}CH_2 \xrightarrow[\text{Ag, 280°C}]{O_2} \underset{\text{(epoxide, O bridging)}}{CH_2\text{—}CH_2} \xrightarrow[\text{1\% } H_2SO_4,\ 60°C]{H_2O} HO\text{—}CH_2\text{—}CH_2\text{—}OH$$

Ethylene oxide also serves as the monomer for poly(ethylene glycol).

9. *Polycarbonates* of different structures are known, but the major commercial product is a polymer made from bisphenol A and phosgene. Bisphenol A is prepared by the following acid-catalyzed reaction of phenol with acetone. (The industrial preparation of both phenol and acetone is discussed in Section 26.2.)

$$CH_3{-}\overset{\overset{\displaystyle O}{\|}}{C}{-}CH_3 + 2\ C_6H_5OH \xrightarrow{\text{acid}} HO{-}C_6H_4{-}\overset{\overset{\displaystyle CH_3}{|}}{\underset{\underset{\displaystyle CH_3}{|}}{C}}{-}C_6H_4{-}OH$$

Phosgene is manufactured by the reaction of carbon monoxide with chlorine.

$$CO + Cl_2 \longrightarrow Cl_2C{=}O$$

10. *Epoxy resins* are cross-linked polymers derived from bisphenol A, epichlorohydrin, and diamines. The production of bisphenol A has just been discussed, and epichlorohydrin is manufactured from propylene by chlorination, followed by addition of HOCl (Section 5.8) and subsequent cyclization with base (Section 15.4).

$$CH_2{=}CH{-}CH_3 \xrightarrow[400\text{–}500°C]{Cl_2} CH_2{=}CH{-}CH_2Cl \xrightarrow[H_2O]{Cl_2}$$

$$\overset{\overset{\displaystyle OH}{|}}{CH}{-}\underset{\underset{\displaystyle Cl}{|}}{CH}{-}CH_2Cl \xrightarrow{Ca(OH)_2} \overset{\quad O}{CH_2{-}CH}{-}CH_2Cl$$

A variety of different amines are used in the manufacture of epoxy resins. Although we will not discuss their preparation, ethylene serves as a feedstock for some of the more important diamines.

As you can see from the preceding discussion, a vast array of complex organic molecules is derived from petroleum. The polymers discussed in this section are derived almost entirely from petrochemicals, and can in fact be linked to a rather small number of primary petrochemicals. The relationships are summarized in Figure 26.2 for most of the polymers discussed in this section.

All the polymers shown in Figure 26.2 can be traced back to eight primary petrochemicals: ethylene, propylene, butadiene, benzene, *p*-xylene, ethane, methane, and the 2-methylbutenes. The figure not only shows how they are structurally related but should also help to provide you with an understanding of the economic interrelationships. A change in demand for any of the polymers or any of the intermediates could easily affect the supply (and therefore the cost) of many other petrochemicals.

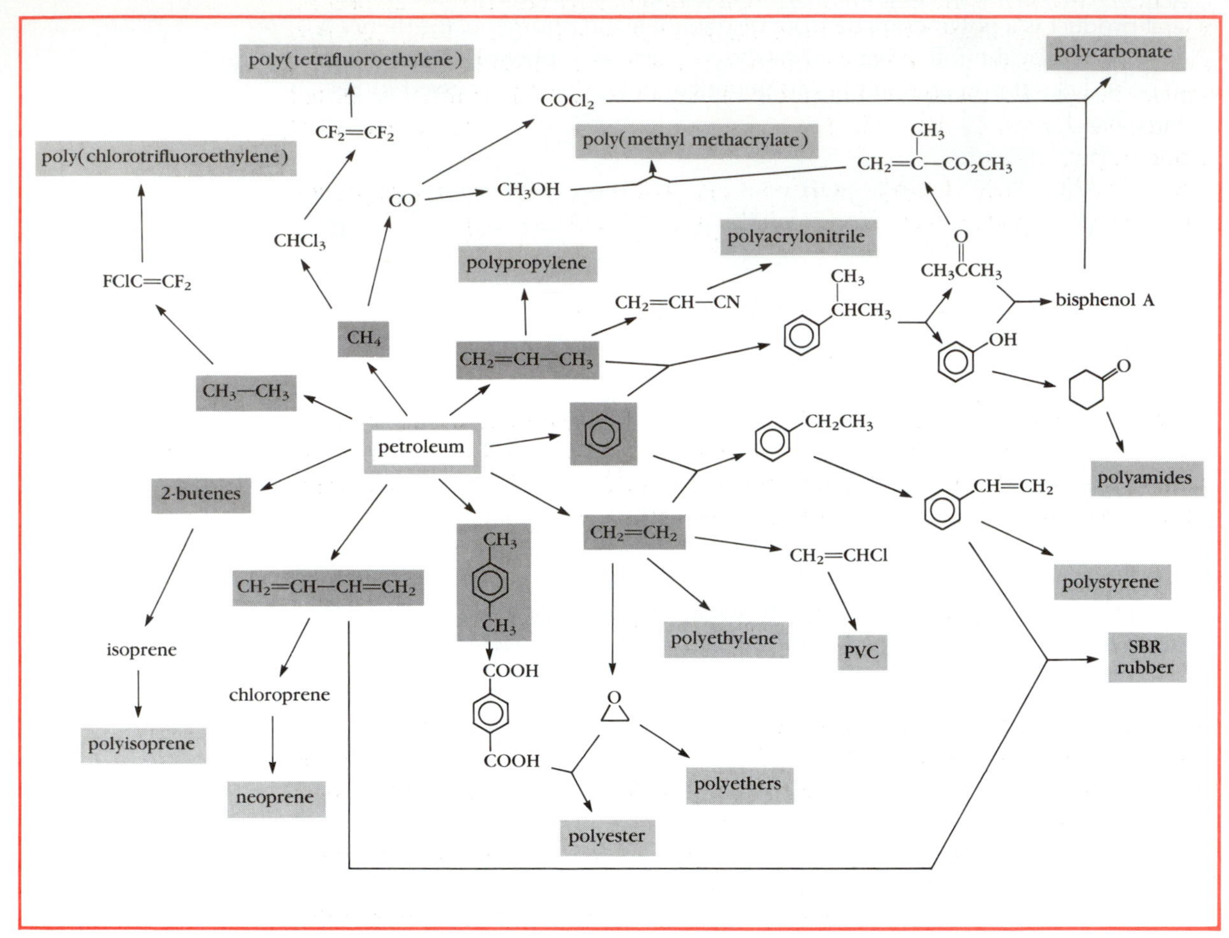

Figure 26.2 Relationships between petroleum, primary petrochemicals and polymers.

Coal and One-Carbon Compounds

Since the oil embargo of 1973 energy resources have been a continual source of worldwide concern. Governments as well as the general public have become increasingly aware that fossil fuels are available in only limited amounts, and at some point the supplies will be exhausted. The available data indicate that natural gas and petroleum supplies will be depleted first, and this has spurred considerable effort to obtain alternative energy sources.

But energy is not the only problem. Even if safe and efficient energy supplies can be developed with atomic energy, we must still face the problem of organic chemicals. Gasoline and synthetic polymers are an integral part of modern life, and depletion of petroleum supplies will eventually require an alternative source of the hydrocarbons used as feedstocks. Where will we find such an alternative source? A likely possibility is *coal,* in large part because it is believed that the amount of coal remaining in worldwide deposits will be relatively large even when petroleum supplies become exhausted.

Actually, this is not the first time such a situation has arisen. Prior to the Second World War, in anticipation that its petroleum supplies would be cut off, Germany undertook extensive investigations of coal as a source of organic compounds. A similar situation has faced South Africa in recent years, and this politically isolated country has chosen to use its large coal supplies as much as possible instead of imported petroleum.

Since the late 1800s it has been known that coal reacts with water at high temperatures to produce a flammable gas mixture, known as *water gas*.

$$C_{(coal)} \xrightarrow[1000°C]{H_2O} CO + CO_2 + H_2$$

At high temperatures carbon monoxide (rather than CO_2) predominates, so it can be obtained in large quantities from coal. During the 1920s the German chemists Franz Fischer and Hans Tropsch developed a catalytic process for converting water gas to a mixture of hydrocarbons resembling petroleum.

$$n\,CO + 2n\,H_2 \xrightarrow[\text{Fe catalyst}]{250-300°C} \sim\!(CH_2)_n\!\sim + n\,H_2O$$

The Fischer-Tropsch process was discontinued in Germany at the end of World War II, but South Africa continues to employ it as a source of hydrocarbons. The crude mixture of products can contain more than 70% of C_5—C_{11} alkanes, so it is a good feedstock for gasoline production.

Another important industrial process that employs carbon monoxide is the *hydroformylation of alkenes* to produce aldehydes. This process, formerly known as the OXO reaction, is used extensively with propylene to generate mixtures of the C_4 aldehydes.

$$CH_3{-}CH{=}CH_2 \xrightarrow[\text{cobalt catalyst}]{CO} CH_3{-}\underset{}{\overset{CHO}{\overset{|}{C}H}}{-}CH_3 + CH_3CH_2CH_2CHO$$

The aldehydes can in turn be subjected to a variety of reactions, and this process serves as a major source of butanols by reduction of the aldehydes.

The major feedstock for the industrial synthesis of low-molecular-weight compounds is ethylene (although propylene is also used to a lesser extent). At the present time our chemical industry is almost entirely based on this *two-carbon* feedstock, but a shift toward *one-carbon* feedstocks appears to be inevitable if we are to utilize our coal resources. An example of this trend is illustrated by a decision of Tennessee Eastman in 1980 to produce acetic anhydride from coal.

As a subsidiary of Eastman Kodak, Tennessee Eastman's interest in acetic anhydride is not surprising. Cellulose acetate is the polymeric support for photographic films, and it is prepared by acylation of cellulose with acetic anhydride. As petroleum prices have increased, one-carbon compounds from coal have become economically more attractive. Earlier in this section we described the synthesis of methanol from carbon monoxide, and the recent development of catalytic *carbonylation* reactions (i.e., insertion of CO) have made it possible to proceed all the way from coal to acetic anhydride with only one-carbon feedstocks.

$$\text{coal} \xrightarrow{H_2O} CO + H_2 \xrightarrow{\text{catalyst}} CH_3OH$$

$$2\,CH_3OH + CO \xrightarrow{\text{Rh catalyst}} CH_3{-}\overset{\overset{\displaystyle O}{\|}}{C}{-}OCH_3 + H_2O$$

$$CH_3{-}\overset{\overset{\displaystyle O}{\|}}{C}{-}OCH_3 + CO \xrightarrow{\text{Rh catalyst}} CH_3{-}\overset{\overset{\displaystyle O}{\|}}{C}{-}O{-}\overset{\overset{\displaystyle O}{\|}}{C}{-}CH_3$$

Although the trend is a new one, you should expect to see in the coming years an increasing number of reports about the industrial synthesis of organic compounds from coal.

EXERCISE 26.1

List the polymers described in this section that can be traced back entirely or in part to ethylene as the original precursor.

26.2 ACETONE: A CASE HISTORY

At the beginning of this chapter we stated that industrial chemistry and its development have been closely intertwined with history and economics. Nowhere has this been more true than with acetone, and in this section we will present a brief history of its manufacture.

Acetone has been known for several hundred years, and prior to the early 1900s it was prepared by the pyrolysis of calcium acetate:

$$Ca(O\overset{\overset{\displaystyle O}{\|}}{C}CH_3)_2 \xrightarrow{400°C} CaCO_3 + CH_3\overset{\overset{\displaystyle O}{\|}}{C}CH_3$$

The push for an alternative synthesis of acetone can be traced directly to the military needs of the British in the First World War. For 500 years the explosive material used for military purposes had been *black powder,* a mixture of potassium nitrate, sulfur, and charcoal. At the turn of the century the British began using *cordite,* in which the explosive components were nitroglycerin and nitrocellulose. Acetone was needed as a solvent for the manufacture of cordite, but the available supplies were limited. Anticipating the military needs of World War I, the British government was anxious to find a new source of acetone.

During the same general period a group of chemists at the University of Manchester had been searching for a way to make synthetic rubber. They hoped to use butadiene as a reactant, but they lacked a source of four-carbon starting materials. Led by Chaim Weizmann, a Russian-born immigrant, these chemists turned to the work reported 50 years earlier by Louis Pasteur. Pasteur had found that some fermentation processes yield 1-butanol as a product. The Weizmann group began investigating different fermentations, and by 1912 they had cultured a new bacterium, *Clostridium acetobutylicum,* which produced 1-butanol from carbohydrates.

Weizmann's work with fermentation took a much different course than he had originally planned. The fermentation did not generate just one product but yielded ethanol (12%) and acetone (32%) along with the 1-butanol (56%). The total conversion of carbohydrate to these three compounds was about 30%, so large quantities could be prepared. When the British government learned of the laboratory process, it immediately asked for Weizmann's assistance in producing acetone for the war effort. In only a short time a number of private distilleries were converted from whiskey production to the manufacture of acetone, and butanol had become merely a by-product.

Weizmann's work with acetone earned him the gratitude of the British government. It has been reported that the Balfour Declaration, in which the British gave support to the Zionist movement, was in part a repayment of their debt to Weizmann's contribution. Weizmann subsequently became the first president of the state of Israel when it was founded in 1948.

After World War I the need for acetone declined, and the Weizmann fermentation process appeared to have little future. But this was a time of change, and the U.S. automotive industry was growing rapidly. One of the problems facing automobile manufacturers was the exterior finish on their cars. The traditional varnishing techniques were slow, and the finish did not bind well to metal surfaces. New nitrocellulose lacquers were being developed, but faster drying solvents were needed. The ideal solvent had to be volatile but not too volatile, and it turned out that 1-butanol, along with a simple derivative, 1-butyl acetate, had the necessary properties. The economic importance of the Weizmann process was thereby maintained, but the principal purpose had shifted back to the original goal of butanol production. Bacterial cultures were improved even further, and a 3:1 ratio of 1-butanol to acetone was achieved. For a number of years automobiles, 1-butanol, and acetone were tied together by a strong economic bond.

Nothing lasts forever though, and the emerging petrochemical industry made fermentation uneconomical. New solvents replaced butanol in the automotive industry, and acetone could be produced from propylene via hydration to 2-propanol.

$$CH_3{-}CH{=}CH_2 \xrightarrow{85\%\ H_2SO_4} CH_3{-}\overset{\overset{\displaystyle OSO_3H}{|}}{C}H{-}CH_3 \xrightarrow{H_2O} CH_3{-}\overset{\overset{\displaystyle OH}{|}}{C}H{-}CH_3$$

This hydration afforded a convenient way of isolating a useful material from the gaseous products of petroleum cracking. By letting the gases rise up through an absorption tower as sulfuric acid was trickling down, propylene could be selectively converted to 2-propanol while ethylene was unreactive under these conditions.

Several processes were developed for catalytic conversion of 2-propanol to acetone, including both oxidation and dehydrogenation.

$$CH_3{-}\underset{}{\overset{OH}{\overset{|}{C}}H}{-}CH_3 + O_2 \xrightarrow[500\text{–}600°C]{\text{Cu catalyst}} CH_3{-}\overset{O}{\overset{\|}{C}}{-}CH_3 + H_2O$$

$$CH_3{-}\overset{OH}{\overset{|}{C}}H{-}CH_3 \xrightarrow[400\text{–}500°C]{\text{ZnO catalyst}} CH_3{-}\overset{O}{\overset{\|}{C}}{-}CH_3 + H_2$$

The importance of acetone was not restricted to its use as a solvent in the munitions industry. It also aided the development of the rayon industry, where the availability of high-purity acetone was a major factor. Growth of the petrochemical industry created other new uses, and acetone is now employed primarily as a chemical intermediate and only secondarily as a solvent. Some of its major uses are in the polymer industry, where it is a key intermediate in the manufacture of methyl methacrylate and bisphenol A.

In the 1950s a new process was developed for the production of acetone starting from benzene and propylene. Friedel-Crafts alkylation with a catalyst such as aluminum chloride yields cumene.

$$C_6H_6 + CH_3{-}CH{=}CH_2 \xrightarrow{AlCl_3} C_6H_5{-}CH(CH_3)_2$$

The tertiary (benzylic) position of cumene is highly reactive, and a hydroperoxide is readily generated by reaction with the molecular oxygen in air.

$$C_6H_5{-}CH(CH_3)_2 \xrightarrow[\text{pH } 8.5\text{–}10.5]{O_2,\ 130°C} C_6H_5{-}C(CH_3)_2{-}OOH$$

Cumene hydroperoxide decomposes under acidic conditions to a mixture of acetone and phenol.

$$C_6H_5{-}C(CH_3)_2{-}OOH \xrightarrow[45\text{–}65°C]{10\%\ H_2SO_4} C_6H_5{-}OH + CH_3{-}\overset{O}{\overset{\|}{C}}{-}CH_3$$

The reaction occurs via a migration to electron-deficient oxygen, a type of rearrangement that is discussed more fully in the Solutions Manual Appendix, Molecular Rearrangements. The oxygen cation may not be a discrete intermediate, but it is conveniently represented in that way.

$$C_6H_5-\underset{CH_3}{\overset{CH_3}{C}}-O-\ddot{O}-H \underset{}{\overset{H^+}{\rightleftharpoons}} C_6H_5-\underset{CH_3}{\overset{CH_3}{C}}-\ddot{O}-OH_2^+ \rightleftharpoons C_6H_5-\underset{CH_3}{\overset{CH_3}{C}}-\ddot{O}^+ \rightarrow {}^+\underset{CH_3}{\overset{CH_3}{C}}-\ddot{O}-C_6H_5$$

$$\rightarrow (CH_3)_2C=\overset{+}{O}-C_6H_5 \overset{H_2O}{\rightleftharpoons} H_2\overset{+}{O}-\underset{CH_3}{\overset{CH_3}{C}}-O-C_6H_5 \overset{\text{proton transfer}}{\rightleftharpoons} HO-\underset{CH_3}{\overset{CH_3}{C}}-\overset{+}{O}(H)C_6H_5$$

$$\rightleftharpoons H\overset{+}{O}=C(CH_3)_2 + HO-C_6H_5 \overset{\text{proton transfer}}{\rightleftharpoons} O=C(CH_3)_2 + HOC_6H_5$$

A major economic advantage of this new method for synthesizing acetone was the simultaneous production of phenol. Bisphenol A (Section 25.5) is a key intermediate in polymer synthesis, particularly in polycarbonate and epoxy resin production, and its industrial synthesis requires both phenol and acetone. Consequently, the cumene hydroperoxide process made it possible for an individual company to produce bisphenol A directly from a single feedstock, cumene, without having to purchase either acetone or phenol.

EXERCISE 26.2

Of the various methods discussed here for production of acetone, which might become economically attractive as petroleum resources become depleted? Explain your answer.

26.3 PESTICIDES

Since the earliest days of human civilization, insects have been a bane to mankind. The bee may pollinate plants and it may make honey, but its sting is never a good thing for the recipient. Insects, along with certain other members of both the animal and plant kingdoms, have long been regarded as pests, and considerable human effort has been aimed at their destruction. The control of pests was not an easy problem to solve, and early efforts using flyswatters and scarecrows were not efficient. Only with the advent of chemical methods did humans begin to make serious progress in their fight against the lower organisms. In this section we will briefly examine the history and chemistry of pesticides, particularly those which are employed in modern agriculture.

First-Generation Pesticides

For many centuries the battle against pests was largely a personal one, but the Industrial Revolution, with its accompanying shift to large-scale agriculture, made it a problem of major economic significance. The earliest known pesticides were natural products obtained from plants, and some of them are still in use today. Among these are nicotine, pyrethrum, and rotenone.

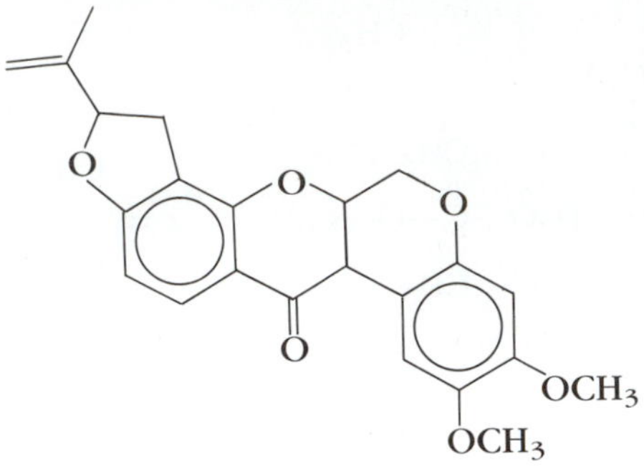

Rotenone

In the first half of the twentieth century nicotine was used extensively to combat aphids, mites and ticks, but it has been replaced by other pesticides since World War II. It does still find limited applications in home gardening, and it is the primary ingredient in *Black Leaf 40.*

Pyrethrum is a mixture of several compounds that are structurally related to pyrethrin I. Natural pyrethrum is obtained from chrysanthemum flowers, and synthetic analogs are widely used in modern pesticides.

Rotenone, obtained from the roots of certain tropical trees, has been used as a fish toxin for hundreds of years. The fish killed by rotenone can still be eaten, and the use of rotenone as a means of gathering fish for food is still used in some undeveloped countries. Rotenone was first used as an insecticide in the mid 1800s, and it is still used for home gardening.

Other first-generation pesticides were largely inorganic materials. In 1864 it was found that Paris green, a copper acetate/copper arsenite complex, was effective against the Colorado potato beetle. Other arsenic compounds were also used as insecticides, and lead arsenate was introduced against the gypsy moth in 1892.

Herbicides have also played important roles in modern agriculture, and a recent estimate indicated that perhaps 50 percent of all American crops would be lost in the absence of herbicides. Ashes and inorganic salts have long been known to kill plants, but the first *selective* herbicidal chemical was discovered in 1896. At that time copper salts were found to act selectively against broadleafs in the presence of cereals. Hence these copper salts could be used to kill weeds without harming a crop of grain.

The discoveries of potent insecticides and herbicides were accompanied by new chemical methods for killing various other pests. But all the chemical methods suffered from a common disadvantage: toxicity was not restricted to the target organism. In particular, many of the compounds used were found to be highly toxic to humans.

Second-Generation Pesticides

As the modern chemical industry began growing in the twentieth century, synthetic compounds offered new possibilities, and a different approach to pesticides began to emerge. No longer was it sufficient merely to control the pest. Instead, people began to think in terms of complete eradication (at least on a local scale).

The first synthetic organic compounds to exhibit activity as pesticides were phenols, and *DNOC* (dinitro-*o*-cresol) was the first synthetic material to be patented as a selective herbicide. It was highly effective against weeds in cereal crops and quickly replaced the more dangerous inorganic herbicides.

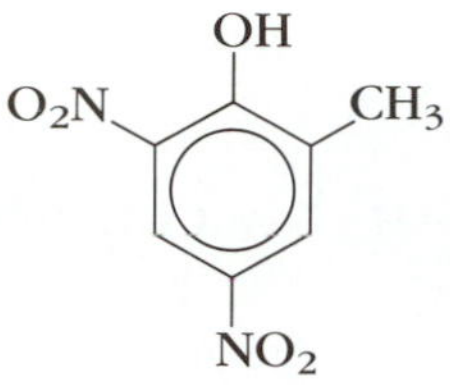

DNOC

The herbicidal action of 2,4-dichlorophenoxyacetic acid (2,4-D) was discovered in the 1940s, and it soon became a major agricultural chemical. (You will recall from Section 24.2 that 3,4,8,9-tetrachloro-*p*-dibenzodioxin can be formed as a by-product of 2,4-D synthesis; potential dioxin contamination has been a continuing environmental concern.)

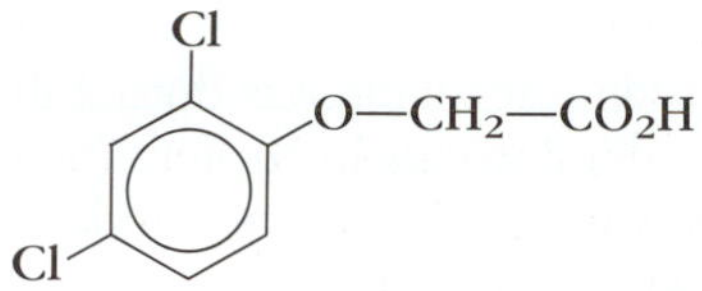

2,4-D

Highly selective in its action, 2,4-D also displays activity at remarkably low concentrations. Application of only several hundred grams per acre is adequate for control of broadleaf plants in cereal crops.

Many other organic herbicides have been developed over the last 40 years, including a variety of heterocycles and chlorinated heterocycles. Among these are picloram, atrazine, and paraquat.

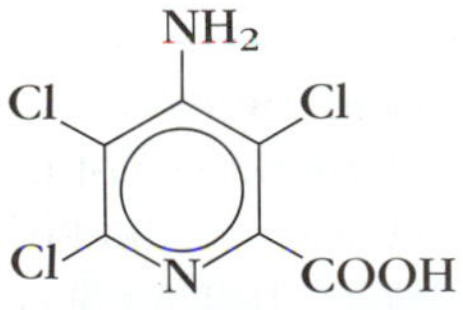

Picloram

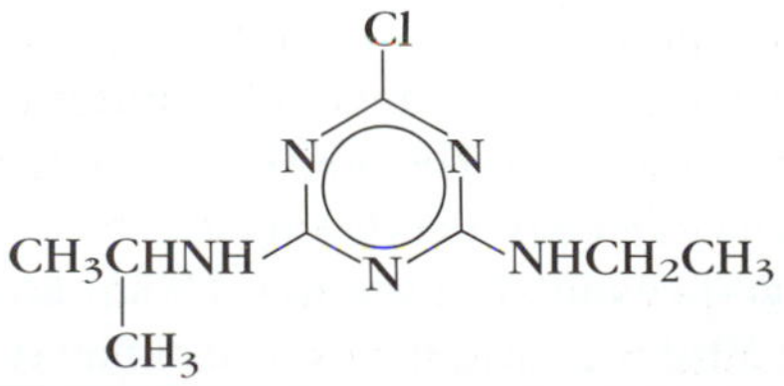

Atrazine

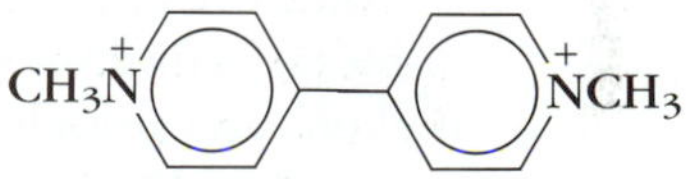

Paraquat

DDT: A Second-Generation Insecticide. The chlorinated hydrocarbon 1,1,1-trichloro-2,2-bis(*p*-chlorophenyl)ethane was synthesized in the 1800s, but only in 1939 did Paul Müller of the Swiss firm Geigy discover its insecticidal properties. Better known as DDT, an abbreviation for *d*ichloro*d*iphenyl*t*richloroethane, this became the first of the second-generation insecticides. In contrast to earlier insecticides, it was found to be much less dangerous to humans and other animals despite its effectiveness against insects. As a result of ecological and health concerns, DDT use has been largely discontinued, but its early success was outstanding.

DDT could be manufactured inexpensively, which is one of the reasons that it was used so extensively. DDT synthesis occurs by a Friedel-Crafts reaction of chlorobenzene with chloral; the starting materials are in turn prepared from acetaldehyde and benzene, respectively. Because acetaldehyde is made from ethylene by oxidation (using O_2 and a complex catalyst system of copper and palladium chlorides), both of the organic starting materials for DDT synthesis can be traced back to primary petrochemicals.

$$CH_2{=}CH_2 \xrightarrow[\text{CuCl}_2,\ \text{PdCl}_2 \ 100^\circ\text{C}]{O_2} CH_3{-}\overset{\overset{\displaystyle O}{\|}}{C}{-}H \xrightarrow{Cl_2} Cl_3C{-}\overset{\overset{\displaystyle O}{\|}}{C}{-}H$$

$$C_6H_6 \xrightarrow[\text{FeCl}_3]{Cl_2} C_6H_5Cl$$

$$2\,C_6H_5Cl + Cl_3C{-}CHO \xrightarrow{H_2SO_4} Cl_3C{-}CH(C_6H_4Cl)_2 \quad \text{DDT}$$

The first major use of DDT as an insecticide was during the Second World War, when an outbreak of typhus hit Allied troops in Naples. The disease is transmitted by body lice, and the outbreak was quickly brought under control by the use of DDT. This success spurred large-scale employment of the remarkable new compound against the insects that transmit other diseases. Exceptional results were obtained in reducing the worldwide incidence of malaria, yellow fever, and plague. The beneficial results of DDT were so impressive that its discoverer, Paul Müller, was awarded the 1948 Nobel Prize in medicine and physiology.

The effectiveness of DDT against the mosquito population that transmits malaria is dramatically illustrated by medical statistics from India. Prior to the use of DDT there were approximately 100 million cases of malaria each year, which was reduced to less than 200,000 in the mid-1960s. Fatalities decreased from almost 1 million to less than 2000 annually.

Similar results were achieved in Sri Lanka (the large island off the subcontinent of India, formerly known as Ceylon). The annual incidence of malaria was above 3 million cases prior to the use of DDT, decreasing to fewer than 100 cases during the early 1960s. But something changed in the mid-1960s, and in 1968 a half million cases were reported. By 1970 the incidence was estimated at 2 million cases annually! What was the explanation? The use of DDT had been discontinued.

Why would an agent as effective as DDT in controlling harmful insects be discontinued? The answer is complex, and there have been extensive debates as to whether or not the decision was the correct one. One thing is clear, however: DDT was overused to an extent that harmful effects on the environment could be observed. National and world leaders simply had not been prepared to think of pollution on the scale that became possible after World War II. Previously, chemical contamination almost always had been highly localized, and natural detoxification appeared to be quite effective. But when worldwide use of DDT reached annual levels of 4×10^9 kg in 1963, this was far more than any natural degradation system could handle. Moreover, DDT turned out to be highly persistent in the environment.

By the early 1960s a number of insect populations had developed resistance to DDT, apparently because they had an enzyme capable of converting DDT to the less toxic dichlorodiphenyldichloroethylene (DDE).

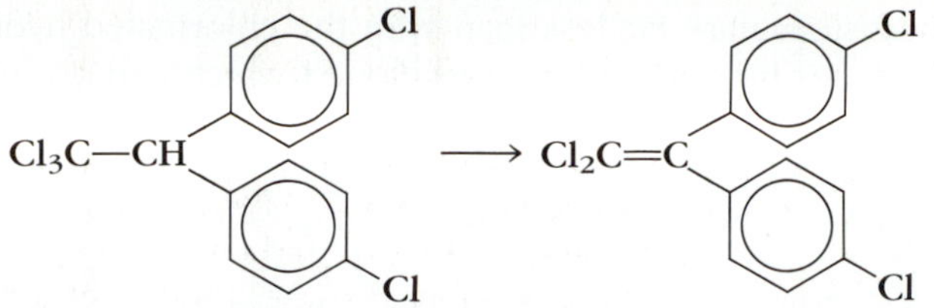

$$Cl_3C{-}CH(C_6H_4Cl)_2 \longrightarrow Cl_2C{=}C(C_6H_4Cl)_2$$

DDE

As the number of examples of DDT resistance was increasing, another, more important development occurred. Scientific evidence was mounting that DDT was accumulating in the fatty tissues of animals, and the concentrations increased along the food chain. This spurred concerns about possible long-term effects on human health. DDT was identified as an agent which interfered with certain enzymatic processes in birds, resulting in weakened eggshells and increased mortality rates. A combination of concerns about human health and environmental quality led to a complete ban on the use of DDT in the United States after 1972, and worldwide use has also been greatly (although not totally) curtailed.

Other Second-Generation Insecticides. Even when the use of DDT was in its infancy, many other pesticides were being introduced. The success of DDT stimulated the manufacture of a variety of chlorinated hydrocarbons that were also potent insecticides. Among these are chlordane, aldrin, mirex, and toxaphene.

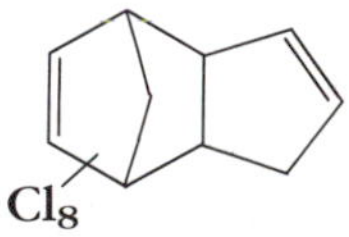

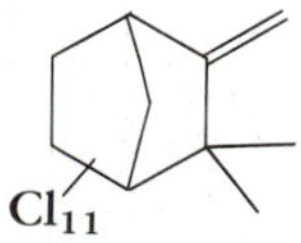

Cl_8 Chlordane | Cl_6 Aldrin | Cl_{12} Mirex | Cl_{11} Toxaphene

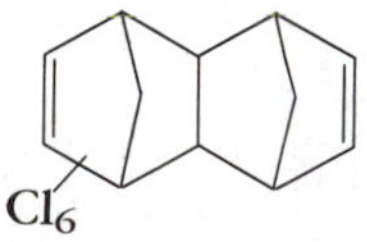

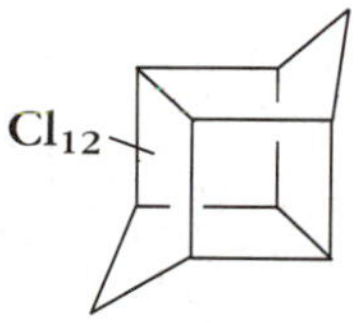

Environmental and health concerns, as well as the availability of alternative insecticides, have resulted in greatly decreased usage of these chlorinated hydrocarbons in recent years.

Research by the German chemist Gerhard Schrader in the late 1930s led to the development of a series of organophosphate compounds with strong insecticidal properties. These compounds function by blocking the action of acetylchlolinesterase, thus disrupting the nervous system. The earliest insecticide to result from these studies was schradan. The organophosphorus compounds also exhibit high toxicity in mammals, and research into this area during World War II led to the development of the "nerve gases" tabun and sarin.

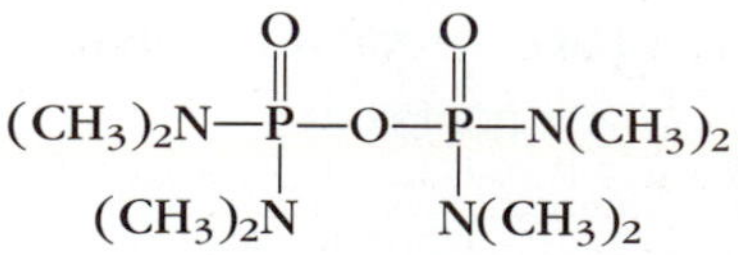

$$[(CH_3)_2N]_2P(=O){-}O{-}P(=O)[N(CH_3)_2]_2$$

Schradan

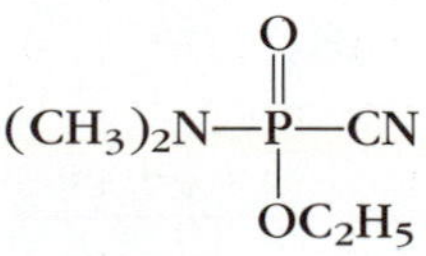

$$(CH_3)_2N{-}P(=O)(OC_2H_5){-}CN$$

Tabun

$$CH_3{-}CH(CH_3){-}O{-}P(=O)(CH_3){-}F$$

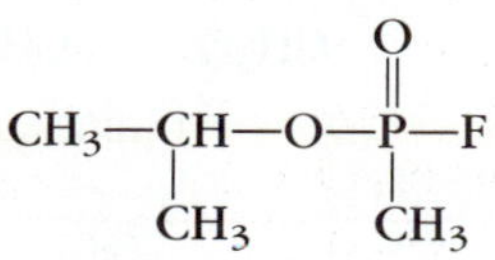

Sarin

Schrader synthesized parathion in 1944, and this is still a widely used insecticide today. As do all the organophosphorus insecticides, parathion hydrolyzes in water, so problems of persistence are far less than with the chlorinated hydrocarbons. Other organophosphates have since been developed which exhibit much lower mammalian toxicity, so they can be used more safely. These include malathion and fenitrothion.

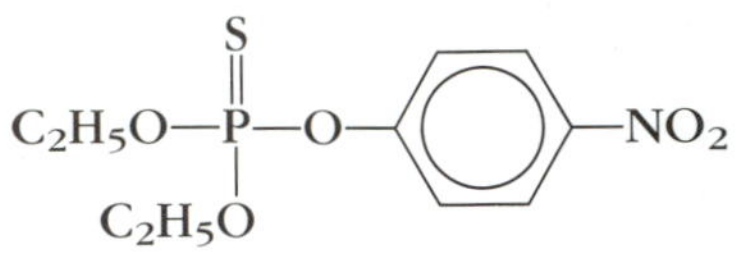

$(C_2H_5O)_2P(=S)\text{—}O\text{—}C_6H_4\text{—}NO_2$

Parathion

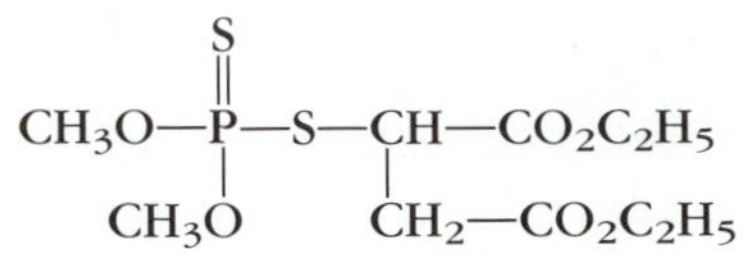

$(CH_3O)_2P(=S)\text{—}S\text{—}CH(CO_2C_2H_5)\text{—}CH_2\text{—}CO_2C_2H_5$

Malathion

$(CH_3O)_2P(=S)\text{—}O\text{—}C_6H_3(CH_3)\text{—}NO_2$

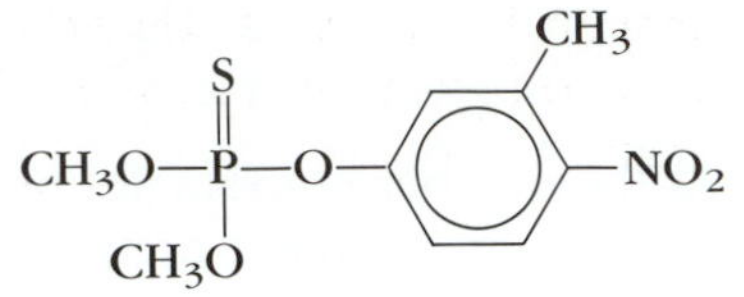

Fenitrothion

Although the relationship between structure and toxicity is not well understood in general, the selective toxicity of malathion to insects is a result of biochemical differences between insects and mammals. As shown in Figure 26.3, mammals have enzymes, called carboxyesterases, which convert malathion to a less toxic derivative by hydrolysis of one of the carboxylic ester groups. Insects not only have much lower concentrations of carboxyesterases than do mammals, but they also have higher concentrations of oxidases, which convert malathion to an even more toxic substance, malaoxon.

Some of the newer organophosphate insecticides have sufficiently low mammalian toxicity that they can be used as *systemics.* An example is cythioate (Proban), which is administered to the animal and enters its bloodstream. Concentration

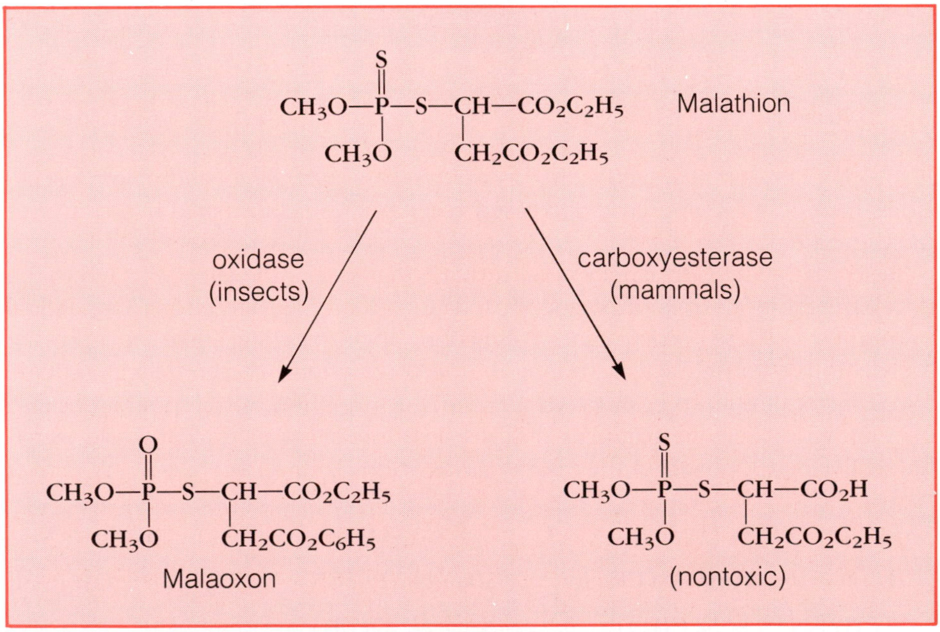

$(CH_3O)_2P(=S)\text{—}S\text{—}CH(CO_2C_2H_5)CH_2CO_2C_2H_5$ Malathion

oxidase (insects) → $(CH_3O)_2P(=O)\text{—}S\text{—}CH(CO_2C_2H_5)CH_2CO_2C_6H_5$ Malaoxon

carboxyesterase (mammals) → $(CH_3O)_2P(=S)\text{—}S\text{—}CH(CO_2H)CH_2CO_2C_2H_5$ (nontoxic)

Figure 26.3 Selective toxicity of malathion.
In mammalian tissues the higher concentrations of carboxyesterases convert malathion to nontoxic compounds. In contrast, other enzymes in insects convert malathion to malaoxon, which is a more potent acetylchlolinesterase inhibitor by a factor of at least 10^3.

levels in the blood are high enough to kill insects such as ticks without harmful effects to the host animal.

Carbamate insecticides form another subgroup of modern pesticides that act on the nervous system by blocking cholinesterase action. These compounds all share the carbamate (urethane) group, —N—CO_2—. Many of them are fairly toxic to mammals, but the carbamate group is hydrolyzed by water, so the danger of exposure to humans is decreased. Carbaryl and carbofuran are examples of the carbamates.

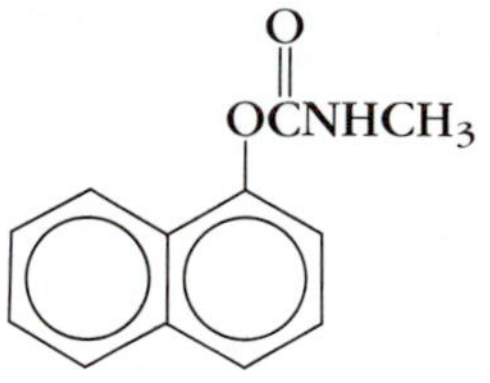

Carbaryl

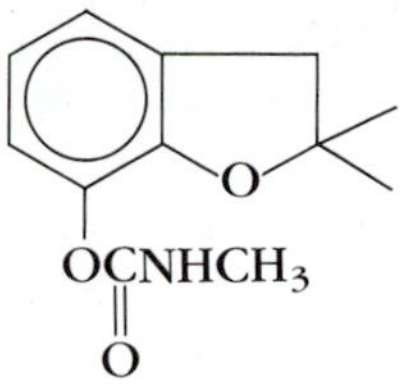

Carbofuran

The most recently introduced group of synthetic insecticides is the pyrethroids, analogs of the natural pyrethrins, which can be prepared economically. Two examples are fenvalerate (Pydrin) and permethrin (Ambush, Pounce). The latter exhibits unusually low toxicity in mammals, so it has become a popular ingredient in household insecticides.

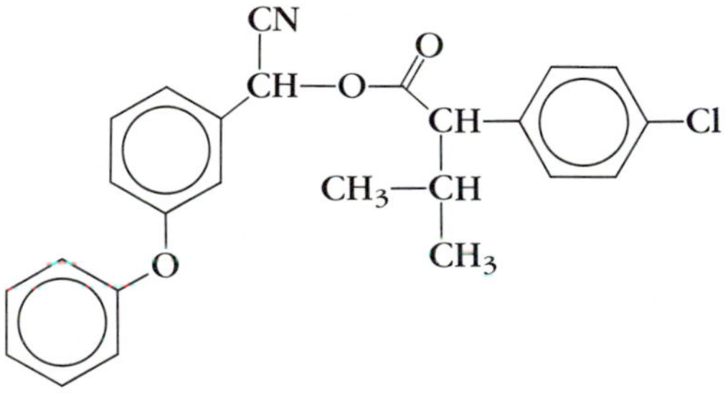

Fenvalerate

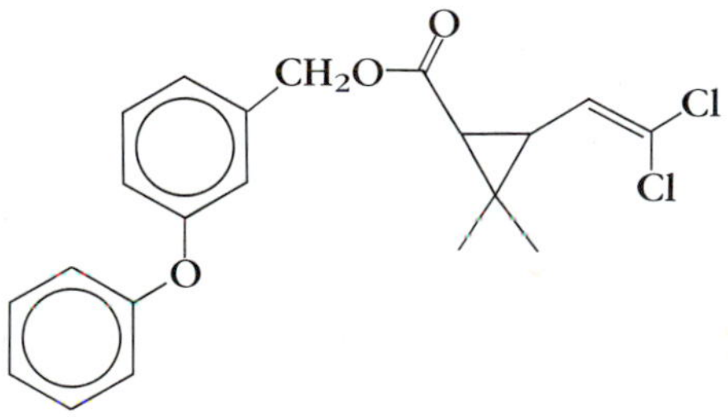

Permethrin

Pesticide Toxicities

A comparison of pesticide toxicities is provided in Table 26.1. The table shows acute LD_{50} values (the dosage that is lethal to 50 percent of the sample animals) for rats; the lower the LD_{50}, the greater the toxicity. The data are expressed in milligrams per kilogram which is the dose of pesticide in milligrams needed for each kilogram of body weight of the test animal. Direct extrapolation to insect toxicity is not always valid, because the mode of application (e.g., dermal contact vs. ingestion) or biological action may differ.

Data are not available for humans, but Table 26.1 provides an indication of toxicity in mammals. Typical body weight for adult humans is in the 50–100 kg range, and crude estimates of a lethal dose (in milligrams) can be obtained by multiplying the LD_{50} by 100. Of course, this does not mean that a dose less than the estimated LD_{50} would be safe; even sublethal doses can cause very bad health effects. Clearly, pesticides are designed to kill living organisms. Whatever their

TABLE 26.1 Acute Oral Toxicities of Pesticides in Rats

Compound[a]	LD_{50} (mg/kg)[b]
Aldrin	39
Atrazine[c]	3,080
Carbaryl	307
Carbofuran	8
Chlordane	283
2,4-Dichlorophenoxyacetic acid[c]	375
DDT	200[d]
DNOC[c]	30
Fenitrothion	250
Fenvalerate	451
Malathion	885
Methoprene[e]	34,600
Mirex	235
Nicotine	50
Paraquat[c]	150
Parathion	3
Permethrin	2,000
Pyrethrin I	200
Rotenone	60
Schradan	5
Toxaphene	40

[a] Insecticides unless otherwise noted. [b] G. W. Ware, "The Pesticide Book," W. H. Freeman & Co., San Francisco, 1978, except as noted. [c] Herbicide. [d] F. L. McEwen and G. R. Stephenson, "The Uses and Significance of Pesticides in the Environment," Wiley, New York, 1979, p. 169. [e] Insect hormone analog.

mode of action, it should not surprise you that they will show some toxicity to organisms other than those that they are specifically intended to kill.

Third-Generation Insecticides

The ideal pesticide would be one that acts specifically against the target organism, does not persist in the environment, is nontoxic to other organisms, and can be produced inexpensively. While this may seem to be an impossible combination, we are indeed at the threshold of an era in which such pesticides will be readily available.

Extensive research on the chemistry and biochemistry of insects during the 1960s showed that the life cycle of insects is controlled by a group of several hormones. These include the *molting hormone, ecdysone* (which triggers molting several times prior to maturation as an adult), and *juvenile hormone,* which induces molting to produce another larval stage rather than an adult insect. Because the molting is unique to insects and arthropods, these hormones should have little or no biological effect on higher animals.

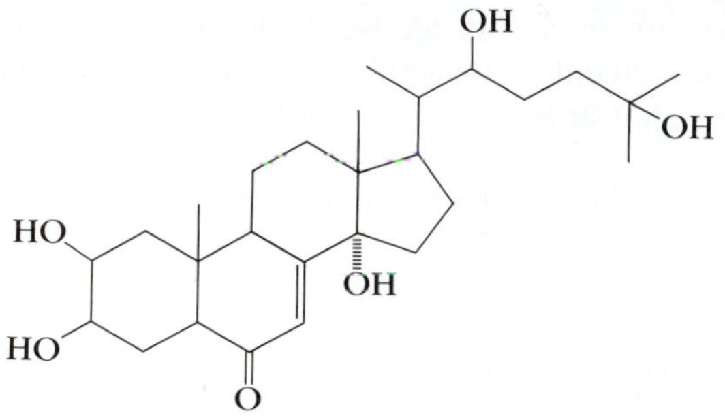

OH OH HO OH HO O

Ecdysone

Juvenile hormone is not a single compound but a group of compounds that exhibit structural variations for different insects. Moreover, these compounds typically are too structurally complex to be attractive targets for large-scale synthesis. But *analogs* offer great potential. *Methoprene* is an example of a juvenile hormone mimic that is effective against a variety of insects.

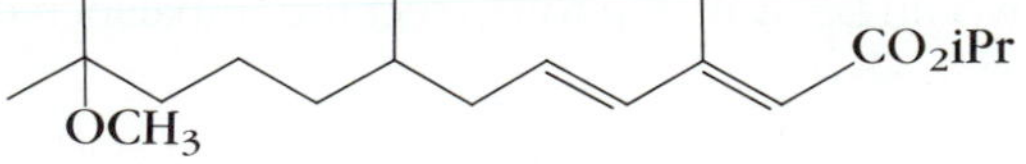

CO_2iPr OCH_3

Methoprene

The successful use of juvenile hormone analogs requires application of the substance at a time when the insect is ready to undergo metamorphosis into an adult. At such times the presence of juvenile hormone can cause molting into a form that cannot survive. Because mosquito breeding cycles are related to rainfall, the appropriate time for methoprene application can be predicted. Methoprene has been used with excellent results in mosquito control, and its exceptionally low mammalian toxicity (Table 26.1) further emphasizes the exciting potential of this new kind of pesticide.

Still another recent development was the discovery of insect *pheromones* (Section 11.11). These chemical messengers exhibit high specificity for individual species, and as a result they offer the greatest hope for controlling specific insect populations without disturbing other aspects of the environment. The extensive use of insect sex attractants is likely to be expensive, because the specific pheromone (or at least an analog) would have to be synthesized for each species. In addition, the pheromone is not directly harmful to the insects, and a mechanism for killing the insects would be necessary once they had been attracted to the site of pheromone application. Despite the problems that must be overcome in their development, insect hormones and pheromones show great promise as agents for the control of insects in the future.

EXERCISE 26.3

Draw the structures of the most toxic and least toxic insecticides listed in Table 26.1. In what categories (e.g., halogenated hydrocarbons) do they belong?

26.4 PROBLEMS

26.1 Cyclic compounds are important feedstocks for the polymer industry. If petroleum refining suddenly failed to yield any cyclic compounds, which of the polymers discussed in Section 26.1 would become unavailable?

26.2 List all the polymers discussed in Section 26.1 that include carbon atoms derived from:

(a) Methane (b) Ethane
(c) Ethylene (d) Propylene
(e) Butadiene (f) 2-Methylbutene
(g) Benzene (h) *p*-Xylene

26.3 Assume that you are operating a chemical manufacturing business that synthesizes bisphenol A from cumene.

(a) What reagents (other than catalysts) would you need to use?
(b) An excess of one compound will be produced during the manufacturing process. Write the necessary balanced equations to show which compound this is.
(c) What options would be available to correct the imbalance discussed in (b)?

26.4 Acetone is converted into a variety of commercially important compounds. Suggest how each of the following compounds could be made from acetone by industrial procedures. Use inexpensive reagents and metal catalysts whenever possible. (It is not necessary to specify a specific catalyst, but the process should be analogous to one that is discussed in this chapter.)

(a) Diacetone alcohol, $(CH_3)_2C(OH){-}CH_2{-}C(=O){-}CH_3$

(b) Mesityl oxide, $(CH_3)_2C{=}CH{-}C(=O){-}CH_3$

(c) Methyl isobutyl ketone
(d) Methyl isobutyl carbinol (4-methyl-2-pentanol)
(e) Hexylene glycol (4-methyl-2,4-pentanediol)
(f) Isophorone (3,5,5-trimethylcyclohex-2-en-1-one)

26.5 Much of the modern production of acetone is by the cumene hydroperoxide route. Since cumene must first be prepared from benzene and propylene, what advantages are offered over direct production of acetone from propylene via 2-propanol?

26.6 In 1945 James Crafts showed that the 4,6-dinitrophenols had herbicide activity that varied with the nature of the alkyl group at the 2 position. This led to the development of *dinoseb* (2,4-dinitro-2-*sec*-butylphenol) as the most commonly used phenolic herbicide in the United States. From your knowledge of organic chemistry (and also knowing the identity of the chemist who developed it: see Chapter 18) suggest a reasonable industrial synthesis for dinoseb starting from phenol.

26.7 Esters of phthalic acid are important as plasticizers in the polymer industry. Suggest a logical petroleum-based feedstock and a catalytic method for converting it to phthalic acid. At high temperatures, would you expect to isolate the diacid or one of its derivatives?

26.8 Phenol is produced commercially by several routes in addition to the cumene hydroperoxide method. One of these is the Dow process (Section 18.7), which begins with conversion of benzene to chlorobenzene. Consider this process (and likely impurities in the reactants and products), and suggest how it could be utilized for the synthesis of herbicides such as 2,4-D and 2,4,5-T.

26.9 Bisphenol A and DDT are both prepared by the acid-catalyzed reaction of a substituted benzene with a simple carbonyl compound. Compare the reaction mechanisms for the processes used to manufacture these two compounds.

26.10 The organophosphorus pesticides all decompose in the environment by reaction with water to give relatively nontoxic products. Draw the hydrolysis products of: (a) Parathion; (b) Malathion; (c) Fenitrothion.

26.11 The carbamate pesticides all decompose in the environment by reaction with water to give relatively nontoxic products. Draw the hydrolysis products of: (a) Carbaryl (b) Carbofuran.

26.12 In contrast to catalytic hydrogenation, catalytic dehydrogenation is very limited as a synthetic method, yet two examples of such dehydrogenation were presented in this chapter as important industrial syntheses.

(a) Identify these two processes.

(b) What is there about these two reactions that allows catalytic dehydrogenation to give a single product in high yield?

CHAPTER 27

MOLECULAR ORBITAL THEORY

Throughout this book we have employed qualitative descriptions of molecular orbitals to help explain the preferred *structures* of organic molecules. In several instances (for example, allylic intermediates in Section 8.3) we have also shown how simple drawings of molecular orbitals can help you to understand the *reactions* of organic molecules. These simple drawings of molecular orbitals, constructed from atomic orbitals, are an outgrowth of quantum mechanics, but they do not require the use of mathematics. Consequently, qualitative molecular orbital theory (MO theory) provides a simple yet powerful tool for analyzing structure and reactivity in organic chemistry.

The importance of molecular orbital theory in modern organic chemistry is reflected by the award of the Nobel Prize in chemistry for achievement in this area on two different occasions in the last 20 years. In 1966 Robert S. Mulliken at the University of Chicago received the Nobel Prize for his many contributions to the development of MO theory. The 1981 prize was jointly awarded to two scientists, Roald Hoffmann of Cornell University and Kenichi Fukui of Japan, for their work on *frontier orbital theory.* In this chapter we will present some basic principles that will allow you to draw molecular orbitals for many organic molecules. We will then show how these drawings can be employed to make useful predictions about chemical reactivity and molecular structure.

27.1 INTERACTION OF ATOMIC ORBITALS TO FORM MOLECULAR ORBITALS

Whether we use a full quantum mechanical calculation or merely draw an approximate picture, molecular orbitals are generated by combining appropriate *atomic orbitals* of the individual atoms in a molecule. In this text we are concerned only with the qualitative aspects, but it is still necessary to establish a reliable procedure by which the atomic orbitals can be combined. The eight rules that we discuss in this section provide a framework for drawing molecular orbitals and for determining their relative energies. Some of this material was covered in previous chapters, but it is important to set down the entire analysis in one place. The first four rules are concerned with how atomic orbitals interact and what the resulting molecular orbitals look like. The last four rules deal with the relative energies of the molecular orbitals.

1. **Interaction of *n* atomic orbitals produces *n* molecular orbitals.** In other words, the total number of orbitals is constant no matter how you analyze a particular species. (This is also true for the generation of hybrid atomic orbitals from pure *s* and *p* atomic orbitals; see Section 1.7.) As a result you can always determine the total number of molecular orbitals (bonding, nonbonding, and antibonding) of a molecular species if you know its molecular formula. The total number of molecular orbitals must equal the sum of the number of atomic orbitals for each of the constituent atoms. For atoms of first-row elements (hydrogen and helium) only a single atomic orbital, the 1*s* orbital, needs to be considered; other orbitals would be so much higher in energy that they can normally be neglected. Atoms of the second-row elements (lithium, beryllium, boron, carbon, nitrogen, oxygen, fluorine, and neon) each have five atomic orbitals, 1*s*, 2*s*, and three 2*p* orbitals. For the discussion in this chapter we will be concerned only with the four *valence-*

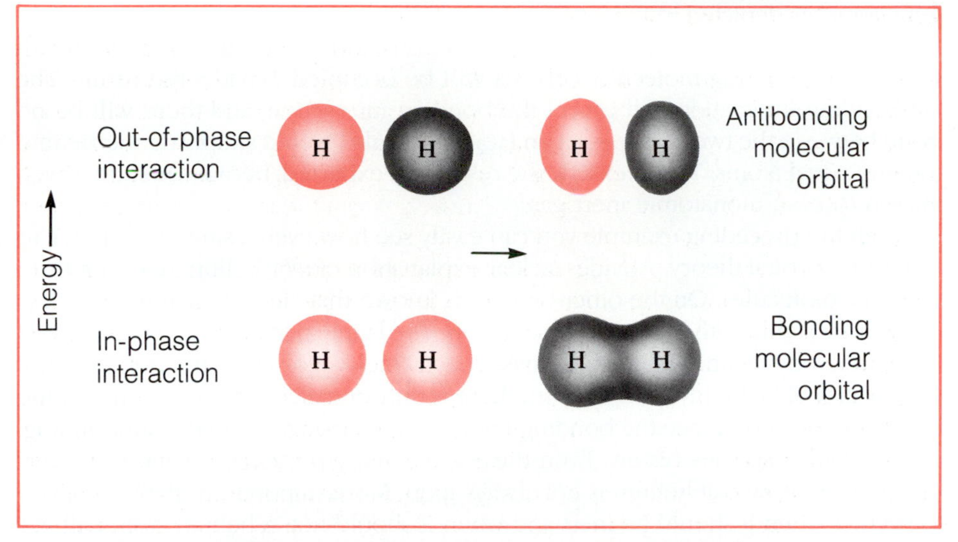

Figure 27.1 Molecular orbitals of the hydrogen molecule.
The 1*s* atomic orbitals can interact in either an in-phase or out-of-phase manner. The former leads to a bonding molecular orbital, while the latter leads to an antibonding molecular orbital. The two electrons of the hydrogen molecule reside in the lower-energy bonding molecular orbital.

level atomic orbitals of these second-row elements, namely, the 2*s* and the three 2*p* atomic orbitals.

In analyzing structure and reactivity with MO theory, you must first describe the various molecular orbitals. Only after this is finished do you finally "add" the appropriate number of electrons. Orbitals are normally filled with two electrons each, starting with the lowest-energy orbitals. The final result therefore describes the electronic structure of a species in terms of the occupied molecular orbitals.

2. **Two orbitals can interact in either a bonding or an antibonding manner.** When the spatially overlapping regions of two orbitals have the *same* phase, the interaction is *bonding*. Conversely, when the overlapping regions have the *opposite* phase, the interaction is *antibonding*. These two situations are shown in Figure 27.1, which illustrates the interactions of the 1*s* orbitals of two hydrogen atoms to form the bonding and antibonding orbitals of the hydrogen molecule.*

Ordinarily, only the lower-energy bonding molecular orbitals are occupied by electrons. Antibonding orbitals are involved in reactions when they interact with molecular orbitals of a second reagent (Sections 27.3, 27.4), but they are usually unoccupied in an isolated molecular species. What would happen if the antibonding orbitals were also occupied? We can answer this question by considering the interaction between two helium atoms to form He_2. Helium has only a 1*s* orbital of

*For additional illustrations of how atomic orbitals interact to form molecular orbitals, see Figures 5.3, 5.4, 7.2, 7.3, 7.4, 7.6, 8.1, 8.3, 8.5, 10.8, 10.9, 10.10, 12.2, and 12.4.

accessible energy, so the orbital interaction will be the same as that for two hydrogen atoms, as depicted in Figure 27.1. But a helium atom has two electrons, so our He_2 "molecule" will have a total of *four* electrons. This means that both the bonding and antibonding molecular orbitals will be occupied. What is the result? The antibonding interaction will cancel the bonding interaction, and there will be no bond between the two atoms; such an He_2 "molecule" would simply dissociate into the individual atoms. This is exactly the result you expected, because helium is well known to be a monatomic inert gas.

From the preceding example you can easily see how even a simple approach to molecular orbital theory provides a clear explanation of *why* helium does not form diatomic molecules. On the other hand, it is known that He_2^+, which has one less electron, is stable with regard to dissociation. The latter species has been generated and studied by techniques such as mass spectrometry. For He_2^+ the bonding molecular orbital is doubly occupied, but there is only one electron in the antibonding molecular orbital. Hence the bonding interaction is stronger than the antibonding, and a "stable" species results. Even these elementary principles of molecular orbital theory show that helium is not always inert. More importantly, they allow you to predict when it should be inert and when it should not. A helium atom will not react with a second helium atom to form He_2, but it will react with an He^+ ion to yield He_2^+.

There is a second consequence of the interaction of atomic orbitals to form one bonding and one antibonding molecular orbital: it allows us to ignore interactions for the orbitals corresponding to *core electrons.* For all atoms beyond hydrogen in the periodic table, the 1*s* orbital is always doubly occupied. Consequently, any interaction between two different 1*s* orbitals will yield a doubly occupied bonding orbital and a doubly occupied antibonding orbital. The effects cancel, so to a first approximation there is no net interaction between the 1*s* orbitals of different atoms. This is why we can restrict our analysis to just the orbitals of the *valence level,* and only in the case of hydrogen (or helium) is the valence level a 1*s* orbital. For other atoms you will be concerned only with the 2*s* and 2*p* orbitals.

3. **Only orbitals of the same symmetry will interact.** As shown in Figure 27.2, an orbital can be either *symmetric* or *antisymmetric* with respect to a plane that bisects the orbital. Such a bisecting plane produces two halves that by definition are symmetric with regard to *shape.* The phases of the two halves can be either the *same* or *opposite,* and the orbital is described as symmetric or antisymmetric, respectively. An *s* atomic orbital is *symmetric* no matter how it is bisected, because the two halves are identical in both shape and phase. A *p* orbital is also symmetric with respect to a plane that bisects both lobes of the orbital, but it is *antisymmetric* with respect to a plane that lies between the two lobes. For the latter plane the two halves are mirror images with respect to shape, but they are opposite in phase.

The interaction of two orbitals of the *same* symmetry is illustrated in Figure 27.1 for the hydrogen molecule, and other such interactions have been shown in previous chapters. The interaction between two *p* orbitals, for example, was shown in Figure 5.4, which illustrates the π molecular orbitals of ethylene. But what happens when we attempt to consider the interaction of two orbitals of *different* symmetry? Consider the interaction between the 1*s* orbital of the hydrogen atom of a C—H group and the *p* orbital (on carbon) that is perpendicular to the C—H bond. A

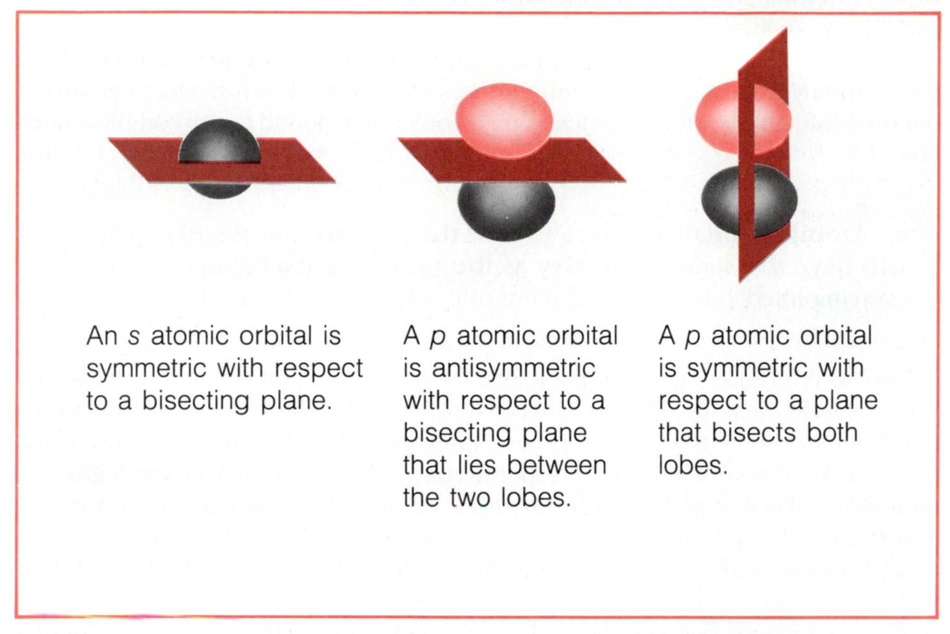
An *s* atomic orbital is symmetric with respect to a bisecting plane.
A *p* atomic orbital is antisymmetric with respect to a bisecting plane that lies between the two lobes.
A *p* atomic orbital is symmetric with respect to a plane that bisects both lobes.

Figure 27.2 Symmetry properties of *s* and *p* atomic orbitals.

C—H bond will always have two *local* perpendicular symmetry planes that describe the symmetry of the C—H subunit, even if the entire molecule does not possess any symmetry. One of the symmetry planes is the plane of the paper on which the C—H is drawn, and both orbitals are symmetric with regard to it. The second local symmetry plane is a horizontal plane that is perpendicular to the plane of the paper (symbolized by a dashed line in the following drawing).

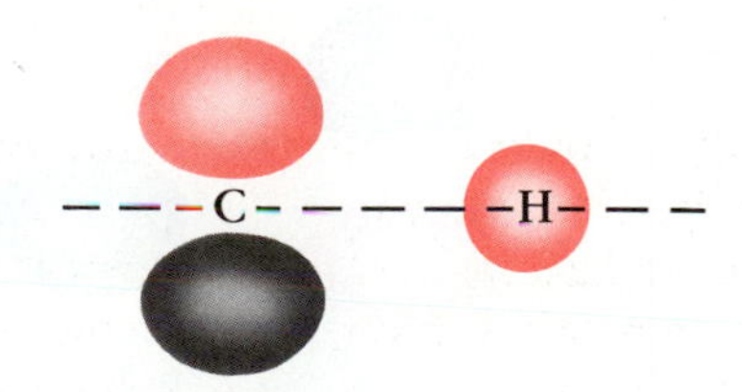
C
H

With respect to the horizontal plane of local symmetry, the *s* orbital of the hydrogen is *symmetric* but the *p* orbital on carbon is *antisymmetric*. Consequently, the bonding interaction between the hydrogen 1*s* orbital and the upper lobe of the *p* orbital is exactly cancelled by the antibonding interaction with the bottom lobe of the *p* orbital. You can therefore conclude that there is no *net interaction* between two orbitals unless they have the same symmetry.

Interaction of an *s* orbital and a *p* orbital is not always precluded by symmetry, however. For the *p* orbital on carbon that is directed *toward* a hydrogen, the local symmetry is the same for both orbitals.

C H

For this subunit any plane that includes the dotted line is a local symmetry plane, and for any such plane both atomic orbitals are symmetric. There is a net interaction, and we have depicted an in-phase, bonding interaction. We could also have drawn the hydrogen 1*s* orbital with the opposite phase, which would generate an out-of-phase, nonbonding interaction that would be included in any complete analysis.

4. Atomic orbitals interact so that the resulting molecular orbitals will have the same symmetry as the molecule itself. This statement is oversimplified, but it is a convenient rule if we place it in the proper context.

Take as reference planes the plane of the paper together with the horizontal and vertical planes that are perpendicular to the paper. Then orient the structure (if possible) so that one or more of these three planes are symmetry planes. If the molecule itself is symmetric with respect to one of these planes, then any legitimate molecular orbital must be either *symmetric* or *antisymmetric* with respect to that same plane.*

We alluded to this principle in our discussion of allylic intermediates in Section 8.3, and now we can complete our explanation. The lowest-energy MO in the allyl system (π_1) has no out-of-phase interactions, but the next MO (π_2) must have one such interaction. The following combinations of atomic orbitals all have a single out-of-phase interaction, but the first two are not legitimate molecular orbitals for the allyl group.

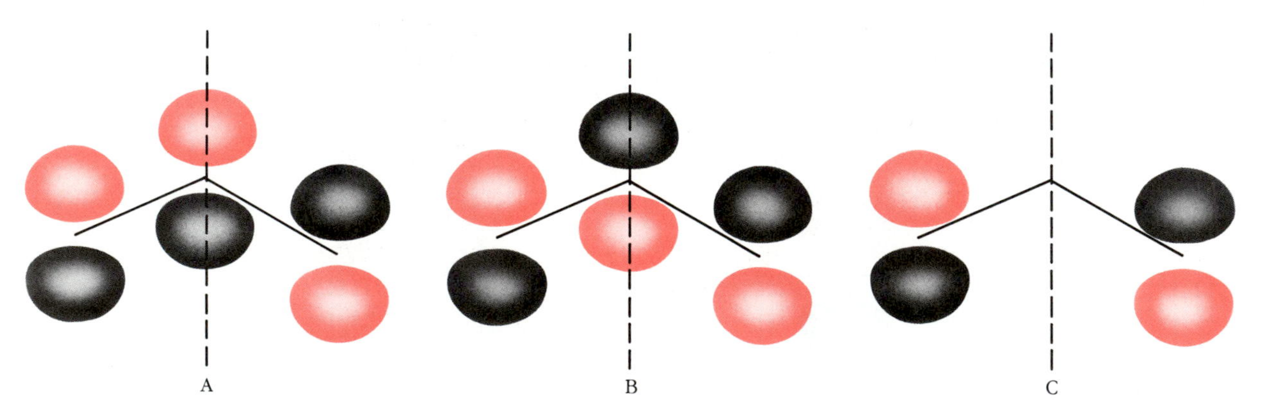

As we have drawn it here, the allyl group itself lies in the horizontal plane that is perpendicular to the paper; the vertical plane perpendicular to the paper is a symmetry plane (dotted line). The first two possibilities (**A** and **B**) are *neither* symmetric *nor* antisymmetric with respect to this symmetry plane, so they cannot be legitimate molecular orbitals.† Only when there is no contribution from the *p*

*Similarly, if the *x*, *y*, or *z* axis (formed by the intersection of these planes) is a C_2 axis for the molecule, then any legitimate molecular orbital must be either symmetric or antisymmetric with respect to that C_2 axis. (Recall that a C_2 axis means that rotation by 180° about the axis generates an identical structure.)

†For example, in **A** the *p* orbitals on C-1 and C-3 are *antisymmetric* with respect to the symmetry plane (dotted line), but the *p* orbital on C-2 is *symmetric* about that plane. When all three *p* orbitals are taken together, the result is *neither* symmetric nor antisymmetric.

atomic orbital of the central carbon (possibility **C**) is the resulting molecular orbital antisymmetric with respect to the symmetry plane. We might also consider omitting the *p* orbital at one of the ends of the allyl system. The resulting *p* orbital contributions would then resemble those of an individual π bond. But this would not afford an acceptable MO for the allyl system, because it does not have the necessary symmetry, even with respect to *shape*.

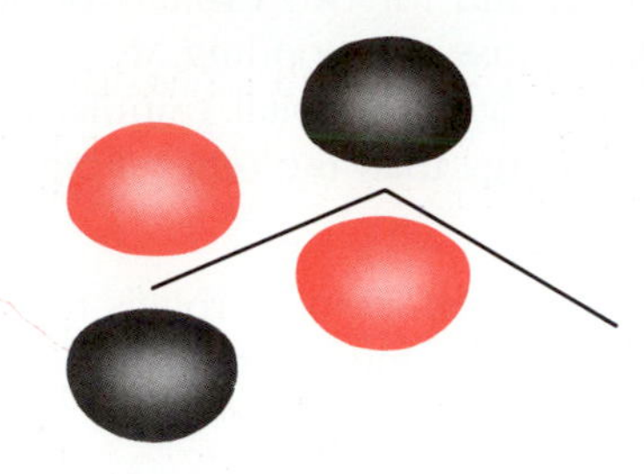

Similar symmetry arguments can be used in constructing the correct molecular orbitals of other symmetrical molecules such as benzene (Figure 10.8) and butadiene (Figure 8.5).

5. Orbitals interact most strongly when they overlap physically to the greatest extent. Remember that the pictures we draw for orbitals are rather crude attempts to represent very complex mathematical functions. The aim of the drawings is to show the general *shape* of an orbital. As we draw it, an orbital is an envelope that encompasses much (but not all) of the region in which an electron in that orbital would be found. A "better" description of the orbitals might include larger regions of space, but the corresponding drawings would be confusing. Thus the circles we used to represent atomic orbitals of the hydrogen atoms in Figure 27.1 do not intersect, but the orbitals themselves do overlap in space.

When two atoms are far apart in space, there is less overlap of their atomic orbitals. In fact, you can generally assume that the orbital interaction is negligible for atoms that are separated by a distance of twice a normal bond length. This is the basis for describing π_2 of the allyl group (Figure 8.1) as nonbonding rather than antibonding.

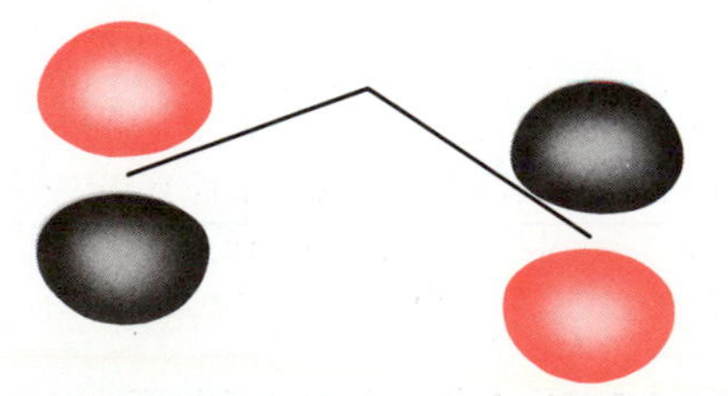

The orbitals are out of phase, but they are physically far apart, so the overlap is so small that the net interaction is negligible.

6. Orbitals interact most strongly when they have similar energies. The interaction of two orbitals produces a bonding and an antibonding combination. When the two interacting orbitals have the same (or similar) energy, a large stabilization is seen for the bonding combination. Similarly, a large destabilization is seen for the antibonding combination. The bonding molecular orbital is usually occupied, so this corresponds to a *strong bond.* When the two interacting orbitals have very different energies, the stabilization and destabilization of the resulting molecular orbitals are much less. This is illustrated in Figure 27.3, which shows the interactions for two orbitals of the same energy (Figure 27.3a) and for two orbitals of quite different energy (Figure 27.3b). In the latter case the bonding MO is only slightly lower in energy than the lower-energy atomic orbital. (Similarly, the antibonding MO is only slightly higher in energy than the other atomic orbital.)

This lessened interaction between atomic orbitals that are far apart in energy corresponds to a *weak bond* when the bonding MO is occupied by a pair of electrons.

7. When two orbitals interact, the increase in energy for the antibonding combination is slightly greater than the stabilization of the bonding combination. Antibonding orbitals are not normally occupied, so you do not usually need to worry about this. However, it can be important for photochemical processes, in which absorption of a photon generates an *excited state* species with an electron occupying an antibonding orbital. It also furthers our understanding of why helium and the other inert gases are monatomic. The net result of fully occupying both the bonding and antibonding molecular orbitals in the diatomic species is not merely negligible but actually slightly repulsive.

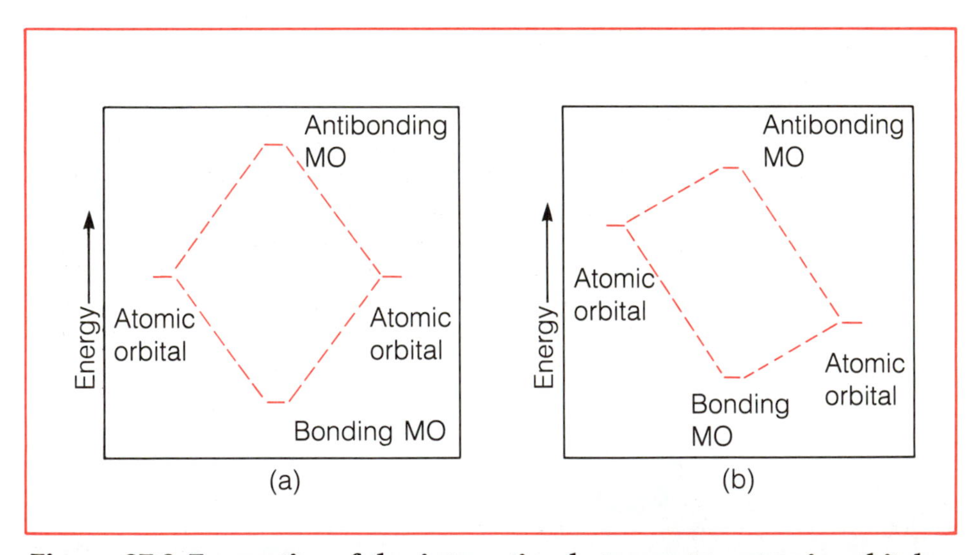

Figure 27.3 Energetics of the interaction between two atomic orbitals to produce two molecular orbitals.
(a) When the interacting orbitals are close in energy, the resulting bonding MO exhibits a large decrease in energy (and the antibonding MO shows a corresponding increase in energy). (b) When the interacting atomic orbitals are not close in energy, the resulting bonding MO shows only a small stabilization (and the antibonding MO shows only a small increase in energy).

8. The energies of molecular orbitals increase with the number of nodal planes. A *nodal plane* is a plane that separates two regions of an orbital having opposite phase. The term *node* derives from mathematical usage and corresponds to the location at which the two lobes of opposite phase intersect. There is zero probability that an electron in the orbital will be found at this location. For example, any *p* atomic orbital or π molecular orbital must have a nodal plane. For the following drawing of a *p* orbital (oriented perpendicular to the plane of the paper) the paper corresponds to a nodal plane.

For any *p* orbital the nodal plane is the symmetry plane about which the orbital is antisymmetric (see Figure 27.2). This local symmetry element is present in all π orbitals and we will call it *π symmetry*. Not all nodal planes need be symmetry planes, however. For the molecular orbitals depicted in Figure 27.4 only the plane bisecting all four atomic *p* orbitals (which is present as a nodal plane in each of the molecular orbitals) is also a symmetry plane.

Figure 27.4 illustrates the relationship between orbital energy and the number of nodal planes in the orbital, and we have used it to show the ordering of the π molecular orbitals for a conjugated diene. The drawings in Figure 27.4 show only the contributing *p* atomic orbitals, but you can refer to Figure 8.5 to see the molecular orbitals. With the exception of the plane that is common to all the individual *p* orbitals (shown only for π_1), each of the nodal planes corresponds to an antibonding interaction between adjacent atoms. This is why an increasing number of nodal planes results in increasing energy for the molecular orbitals; the molecular orbitals become increasingly antibonding.

The correlation between the number of nodes in an orbital and the energy of the orbital is also consistent with the relative energies of π and σ orbitals. You can see this by looking at the bonding σ and π orbitals that correspond to the double bond of ethylene (Figures 5.3 and 5.4). The following drawings show the atomic orbital contributions to these molecular orbitals. You will recognize that the sp^2 hybrid orbitals that combine to produce the σ molecular orbital overlap to a greater extent than do the *p* orbitals that generate the π molecular orbital. This is because the orbitals that produce the σ molecular orbital point toward each other, while the *p* orbitals that form the π molecular orbital do not.

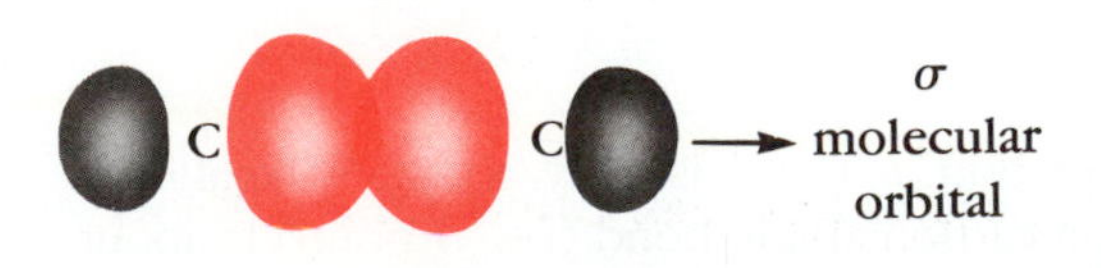

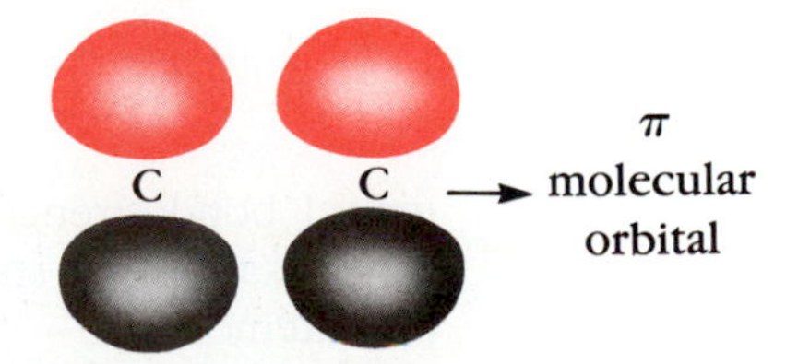

The greater overlap of the hybrid atomic orbitals leading to this σ molecular orbital results in more bonding character and a lower orbital energy. The difference in orbital energies agrees well with experimental observations that the bond strengths of π bonds are less than those of σ bonds between the same atoms. A

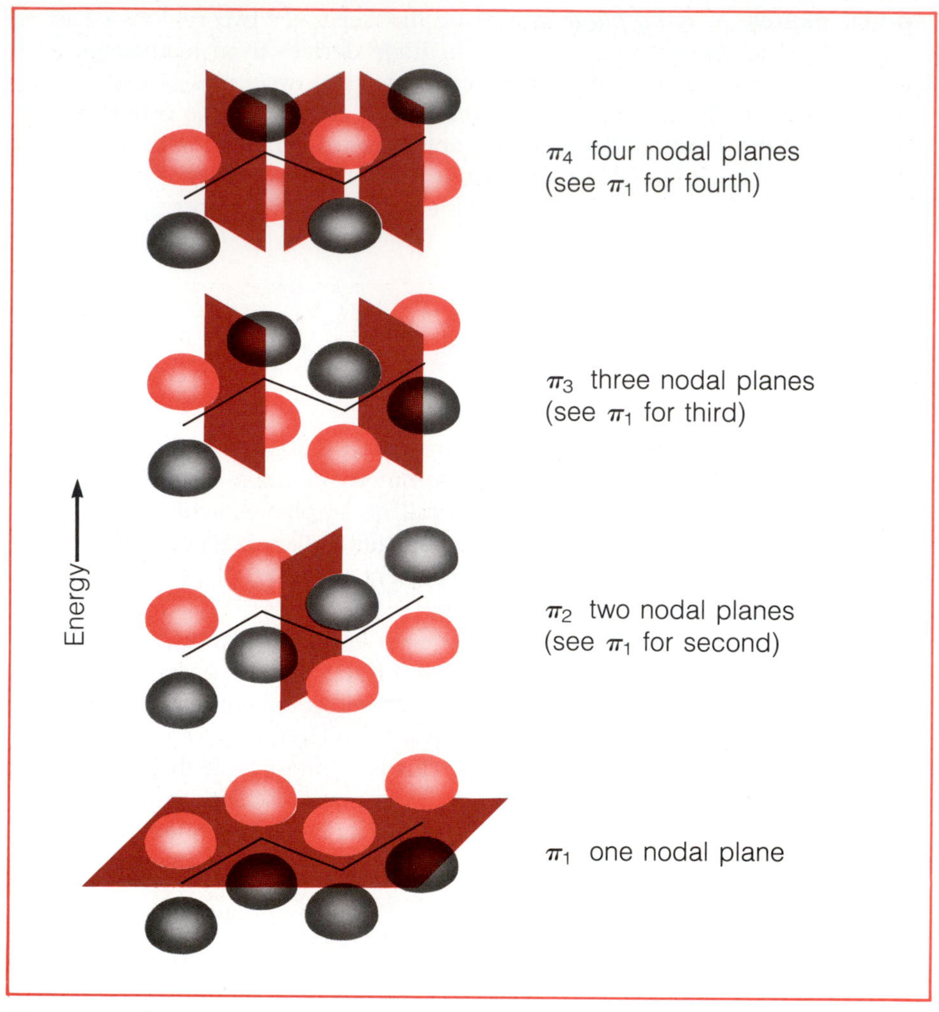

Figure 27.4 Atomic *p* orbital contributions to the four π molecular orbitals of a conjugated diene.
The nodal planes are shown for each MO but only for π_1 have we drawn the nodal plane that is common to all four MOs, a horizontal plane perpendicular to the plane of the paper that includes the nodal plane of each of the contributing atomic *p* orbitals).

typical bond strength for a carbon-carbon π bond is only about 50 kcal/mol, whereas the bond strength of a carbon-carbon single bond (i.e., σ bond) is about 80 kcal/mol.

The eight rules discussed in this section provide the framework for using qualitative molecular orbital theory. In the following sections we will show you how this information can be used to analyze questions about molecular structure and reactivity.

EXERCISE 27.1

How many valence-level atomic orbitals are available for formation of molecular orbitals in each of the following molecules? How many total molecular orbitals are there in each case?

(a) CO_2 (b) CH_4 (c) CF_4 (d) NH_3
(e) H_2O (f) HCN (g) C_6H_6 (h) $C_{15}H_{30}$

EXERCISE 27.2

Which of the following combinations of atomic orbitals would lead to legitimate molecular orbitals? For those combinations that would not, explain why.

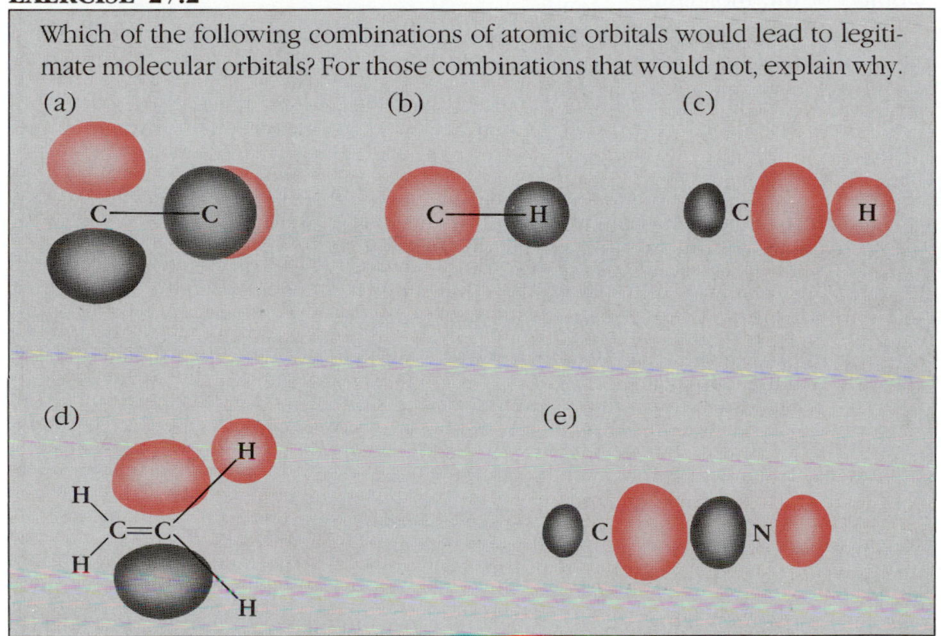

27.2 MOLECULAR GEOMETRY

Discussions of molecular geometry are often based on repulsion of electron pairs, and we have also used such arguments in this book. For example, the water molecule is said to be nonlinear because the oxygen has four electron pairs in its valence shell, two bonding pairs and two nonbonding pairs. On this basis it is similar to methane (in which all four electron pairs are bonding electrons), so an angle close to the tetrahedral angle in CH_4 is expected. Indeed, the H—O—H angle of 105° is not far from the 109.5° angle found in ethane.

Sometimes these arguments are of limited value, however. For example, methylene ($:CH_2$) has been observed in both *singlet* (no unpaired electrons) and *triplet* (two unpaired electrons) states. You might be tempted to describe the singlet as a species with sp^2 hybridization, so that the three electron pairs (two bonding, one nonbonding) would all be in the same plane. You would then expect the H—C—H angle to be about 120°. The *triplet,* in contrast, might be expected to have sp^3 hybridization, with two paris of bonding electrons and two nonbonding orbitals each occupied by a single electron. You would then expect the H—C—H angle of triplet methylene to be about 109°. In fact the opposite trend is observed. The singlet has a geometry that is more highly bent (102°) than the triplet (138°).

If electron pair repulsion arguments do not always work, what alternatives are available to explain (or predict) molecular geometries? A very useful procedure employs qualitative molecular orbital theory in a way suggested by B.M. Gimarc at the University of South Carolina. You can first construct a set of molecular orbitals for a highly symmetrical structure and next determine how the energy of each MO is affected by distortion to a less symmetrical structure. By knowing which orbitals are occupied, you can estimate the effect of the distortion on the total energy (i.e., stability) of the molecule.

For the analysis of methylene we will first construct molecular orbitals for a linear structure, using only the valence level 2*s* and 2*p* atomic orbitals of carbon. For each combination we will then determine whether the energy would increase or decrease upon bending the H—C—H away from linearity. The actual analysis will be done for the more general case of H—X—H (where X can be any of the second-row elements from lithium to neon), and it is summarized in Figure 27.5.

The lowest-energy MO (ϕ_1) will result from interaction of the *s* atomic orbital on each atom. Distortion from linearity does not significantly affect overlap between H and X orbitals, and it produces a small in-phase interaction between the *s* orbitals of the two hydrogens. Consequently, this molecular orbital will decrease in energy when bending occurs. The effect is small, however, because the hydrogens are relatively distant from each other.

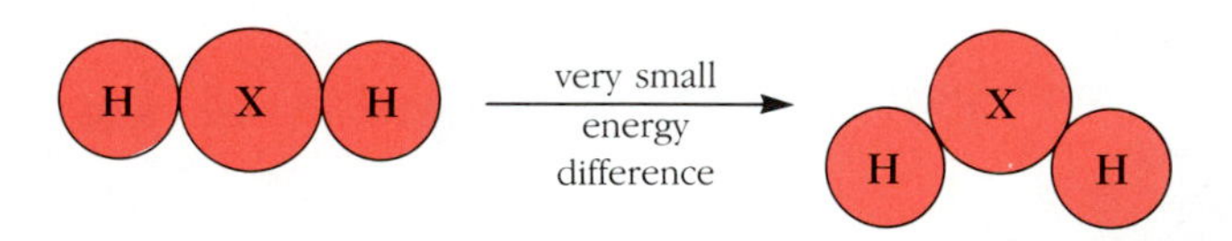

The next higher MO (ϕ_2) results from σ interaction of the hydrogen 1*s* orbitals with the *p* orbital of the central atom. This MO will increase in energy with nonlinear distortion for two reasons: first, an out-of-phase interaction will develop between the hydrogen 1*s* orbitals; and second, the overlap of the hydrogen orbital with the *p* orbital will decrease as the hydrogen moves away from the axis along which the *p* orbital is directed. These effects are again relatively small, but they are larger than for the previous MO.

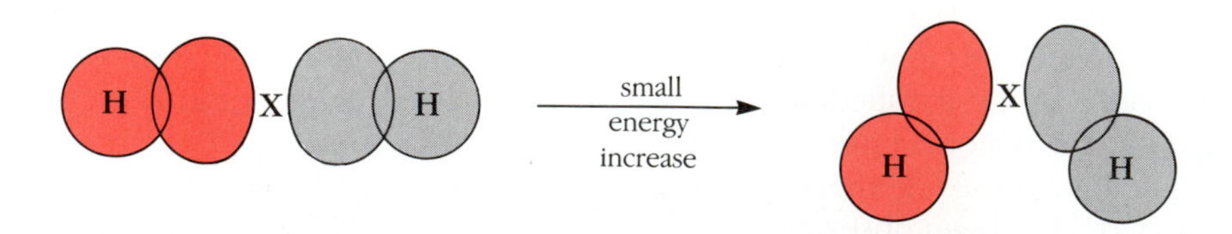

Next consider the *p* orbitals on X that are oriented perpendicular to the H—X—H axis; these can be designated ϕ_3 and ϕ_4. The *p* orbital that is perpendicular to the plane of the paper will always have the wrong symmetry for interaction with the hydrogens no matter what the H—X—H angle. Consequently, its energy will be unaffected by the H—X—H angle.

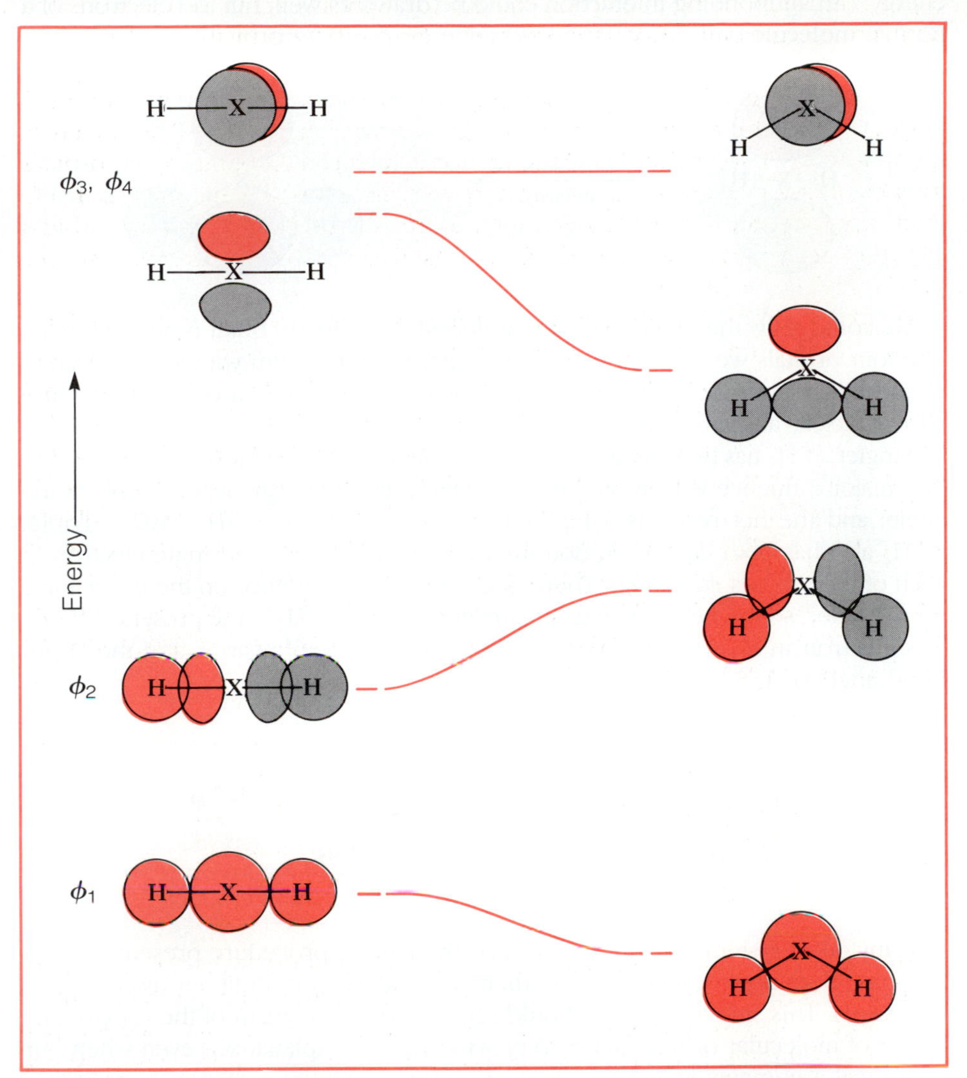

Figure 27.5 Qualitative MO analysis of the geometry of H—X—H. When only ϕ_1 is occupied, a nonlinear geometry (right) is favored. When ϕ_2 is also doubly occupied, a linear array (left) is favored. Additional electrons will occupy ϕ_3, resulting in a strong preference for nonlinear geometry.

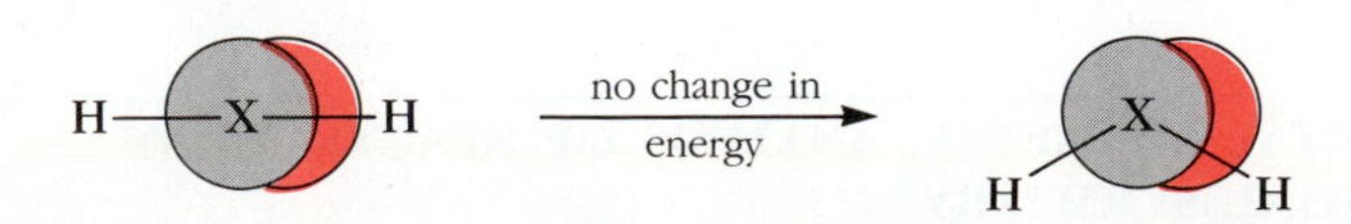

The other p orbital is oriented in a vertical direction and lies in the plane of the paper. While it has the wrong symmetry for interaction when H—X—H is linear, bending allows both 1s orbitals of the hydrogens to interact with one lobe of the p orbital. This results in a strong bonding interaction, so the MO will decrease in energy. (An antibonding interaction could be drawn as well, but the electrons of a normal molecule will occupy the lower-energy, *bonding* orbital.)

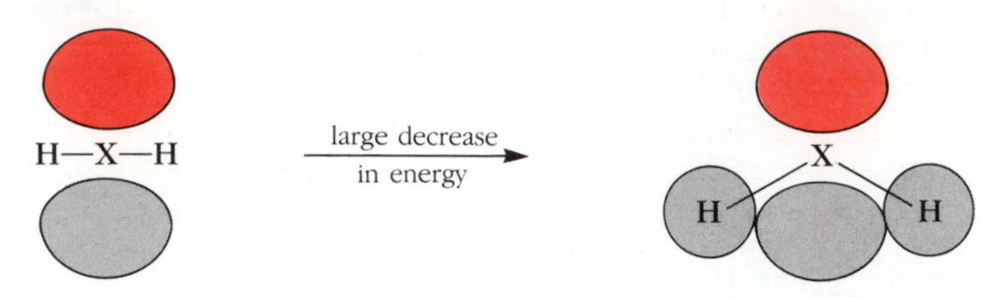

We could draw the two remaining (antibonding) orbitals, but it is not necessary. The four orbitals we have shown are sufficient for any system with a total of up to eight valence electrons, and they are all shown together in Figure 27.5. This figure allows a clear answer regarding the structure of methylene.

Singlet $:CH_2$ has two electrons in each of the three lowest MOs (ϕ_1, ϕ_2, and ϕ_3). The major influence is from ϕ_3 because it has a large energy dependence on the angle, and the net result is a highly bent structure (H—C—H = 102°). Triplet $:CH_2$ also has both ϕ_1 and ϕ_2 doubly occupied, but there is a single electron in each of the orbitals ϕ_3 and ϕ_4. There is no energy dependence on the bond angle for the latter, so the net change from singlet to triplet $:CH_2$ is the presence of one less electron in ϕ_3. This reduces the preference for nonlinearity, and the larger bond angle of 138° is observed.

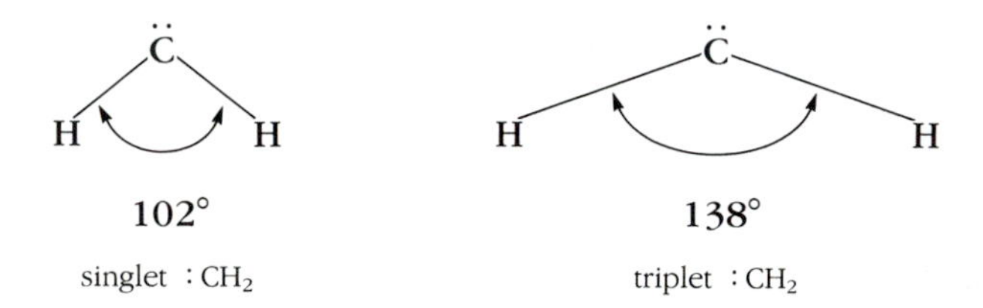

Many other systems can be analyzed by the type of procedure presented here, but space does not permit us to show them. Instead we must limit our discussion to H—X—H. This single example should remind you once again of the remarkable power of molecular orbital theory to provide correct explanations even when our traditional arguments fail.

EXERCISE 27.3

Using the molecular orbital arguments summarized in Figure 27.5, suggest whether each of the following should be linear or bent: (a) BeH_2; (b) $^+BH_2$; (c) NH_2^-; (d) H_2O.

27.3 ORBITAL SYMMETRY: ANALYSIS OF REACTIONS BY ORBITAL SYMMETRY

The information we have presented up to this point shows how molecular orbitals can be drawn for various molecules. Moreover, in the preceding section we showed how MO theory can be used to deduce the optimum structure of a molecule. But how can you use MO theory to study reactions? Assuming that you know the structures of the reactants and of the possible products, you would like to determine which of several alternative products will be obtained. In other words, which reaction pathway is energetically most favorable?

To fully assess the energetics of two or more alternative pathways you would really want to determine the relative energies of the transition states. But how can you evaluate the energy of these exact structures? A solution to this problem was put forth in the 1960s by R.B. Woodward and R. Hoffmann (both then at Harvard; Hoffmann is now at Cornell). They proposed that in a *concerted* reaction (Section 5.7), that is, a one-step process with a single transition state and no intermediate, the molecular orbitals of the reactant and product would show a correlation according to symmetry. If the occupied molecular orbitals of the reactant showed a symmetry correlation with high-energy product molecular orbitals, then that pathway would proceed via a high-energy transition state. Conversely, if the occupied molecular orbitals of the reactant showed a symmetry correlation with low-energy product orbitals, then a lower-energy transition state would afford the preferred reaction path. This requirement for a symmetry correlation of product and reactant MOs has been described as the *conservation of orbital symmetry.*

Figure 27.6 shows *correlation diagrams* for two hypothetical reactions. In reaction A the lowest-energy molecular orbital (ϕ_1) is occupied by two electrons, and it is the only occupied MO. This orbital is *antisymmetric* (designated **A**) with regard to some unspecified symmetry plane, and it correlates with the product MO (ϕ_2'). In other words, the orbital energy of the occupied MO would show a large increase as the reaction progressed, and this would lead to a high activation energy. The situation is quite different for reaction B, because here the occupied MO of the reactant (ϕ_1) correlates with a low-energy product orbital (ϕ_1'). Both of these are

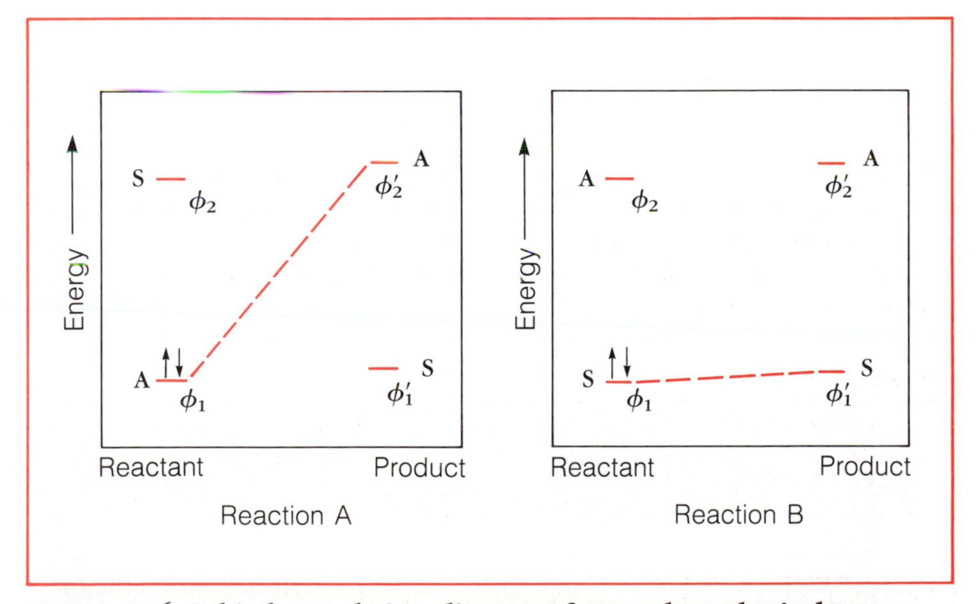

Figure 27.6 Orbital correlation diagrams for two hypothetical reactions.
In Reaction A the doubly occupied reactant orbital ϕ_1 (antisymmetric, **A**) correlates with the higher-energy product orbital ϕ_2', so a high activation barrier is expected. In reaction B the doubly occupied reactant orbital ϕ_1 (symmetric, **S**) correlates with product orbital ϕ_1', and a much lower activation barrier is expected.

symmetric (designated **S**) with respect to our unspecified symmetry plane. There is no large increase in energy of the occupied orbital, so a much lower activation energy would be expected. (There will of course be contributions to the activation barrier in addition to those which might result from orbital symmetry restrictions.)

The following discussion will show how orbital symmetry can be used to understand a specific reaction, the dimerization of ethylene to form cyclobutane. This is a *cycloaddition* related to the Diels-Alder reaction (Section 22.3).

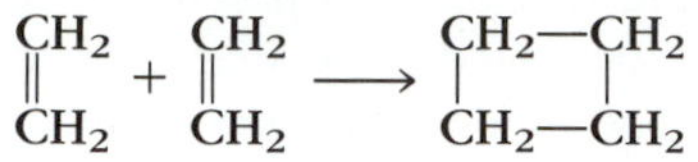

$$CH_2{=}CH_2 + CH_2{=}CH_2 \longrightarrow \text{cyclo-}(CH_2)_4$$

Remember, we are restricting the use of orbital symmetry arguments to concerted (one-step) processes, in which no reactive intermediates are formed. Both new carbon-carbon bonds must be formed simultaneously.

The first step in the analysis must be to determine what symmetry is maintained by the *atoms* throughout the reaction. Only then will it be possible to classify the orbitals. We will analyze the pathway in which two ethylene molecules approach each other so that their molecular planes are parallel, one above the other, as shown in Figure 27.7.

Three symmetry planes are maintained during the entire course of the reaction, two of which are illustrated by the following drawings of the transition state.

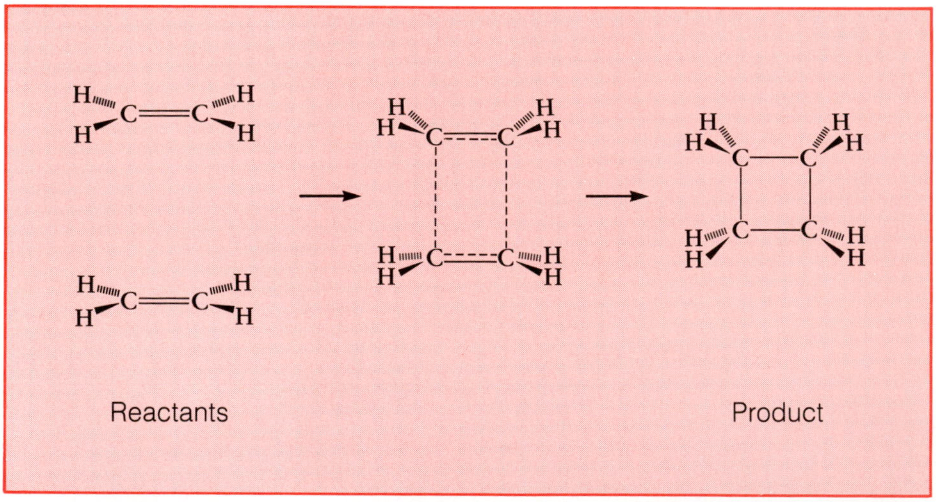

Reactants

Product

Figure 27.7 Geometry for cycloaddition of two ethylenes molecules to form cyclobutane.
The carbon atoms of the ethylenes lie in the plane of the paper, and the molecular plane of each ethylene is perpendicular to the page. For the pathway under consideration, three symmetry planes are maintained throughout the reaction: (1) a horizontal plane perpendicular to the page which is equidistant from the two ethylene units; (2) a vertical plane perpendicular to the page, which bisects the C—C bond of each subunit; and (3) the plane of the paper.

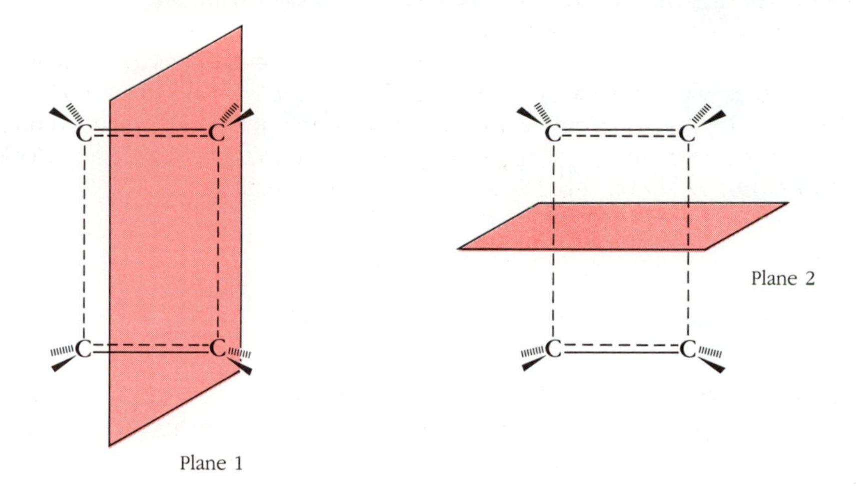

The third plane of symmetry is the plane of the paper. It turns out that only the first two planes are important for our analysis: we are only concerned with *symmetry elements that bisect bonds which are broken or formed.* Thus plane 1 bisects the two double bonds that are broken in the reaction, and plane 2 bisects the two single bonds that are formed.

The next step in the analysis is to draw all pertinent molecular orbitals. Once again, we need to focus on only those MO's corresponding to bonds that are broken or formed. Consequently, we will look at only the π_{cc} orbitals of the ethylenes and the σ_{cc} orbitals of cyclobutane that are oriented in a vertical direction. There is a small complication that results from the symmetry of the system: Drawings such as the following do not represent legitimate orbitals, because they lack symmetry with respect to one of the molecular symmetry planes.

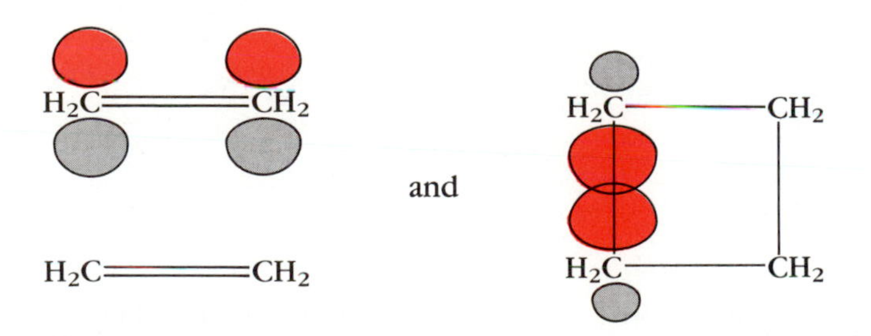

The first drawing (which shows the atomic orbital contributions to just one of the ethylenes) is neither symmetric nor antisymmetric with respect to plane 2. Similarly, the second drawing (which shows just one localized σ_{cc} orbital) is neither symmetric nor antisymmetric with respect to plane 1.

To resolve the problem that all orbitals must be either symmetric or antisymmetric with respect to both planes, we must employ *combinations* of the localized orbitals. The two possible combinations of the bonding π orbitals for the ethylenes have in-phase $(\pi_1 - \pi_2)$ and out-of-phase $(\pi_1 + \pi_2)$ interactions, respectively. (The shaded lobes have the *same* orientations in the $\pi_1 + \pi_2$ combination; they have *opposite* orientations in the $\pi_1 - \pi_2$ combination).

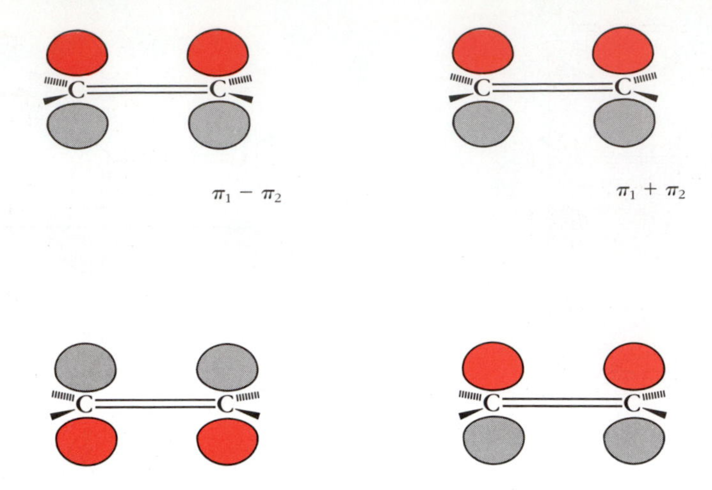

$\pi_1 - \pi_2$ $\pi_1 + \pi_2$

In a similar way we can draw two combinations for the bonding sigma MOs of cyclobutane. The combination with both contributions having the same phase is designated $\sigma_1 + \sigma_2$, and the other is $\sigma_1 - \sigma_2$.

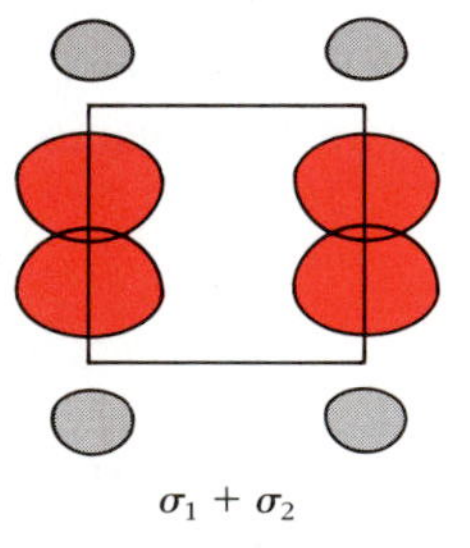

$\sigma_1 + \sigma_2$

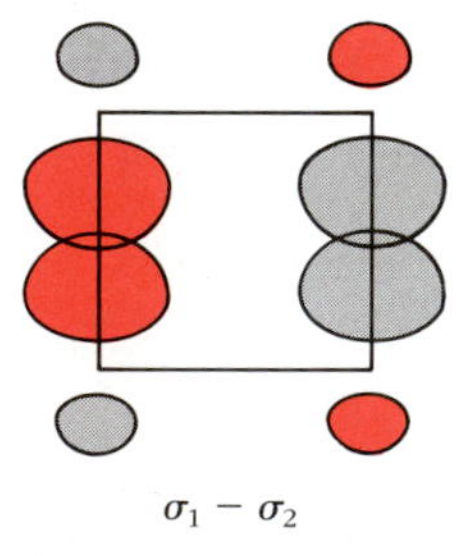

$\sigma_1 - \sigma_2$

Analogous combinations can be drawn for the antibonding π MO's of the ethylenes and for the antibonding σ MO's of cyclobutane. The complete set of orbitals is shown in Figure 27.8. Notice that the relative energy of an orbital for either reactant or product simply corresponds to the number of nodal planes. The lowest bonding σ orbital of cyclobutane is shown at lower energy than the bonding π orbitals of the ethylenes because of the greater overlap in σ interactions. (Similarly, the highest antibonding combinations for cyclobutane are drawn at higher energy than the ethylene π^* orbitals.)

Each of the MO's in Figure 27.8 is labeled as symmetric and/or antisymmetric according to the vertical and horizontal symmetry planes. Thus $\pi_1 + \pi_2$ is symmetric (**S**) with respect to the vertical plane but antisymmetric (**A**) with respect to the horizontal plane.

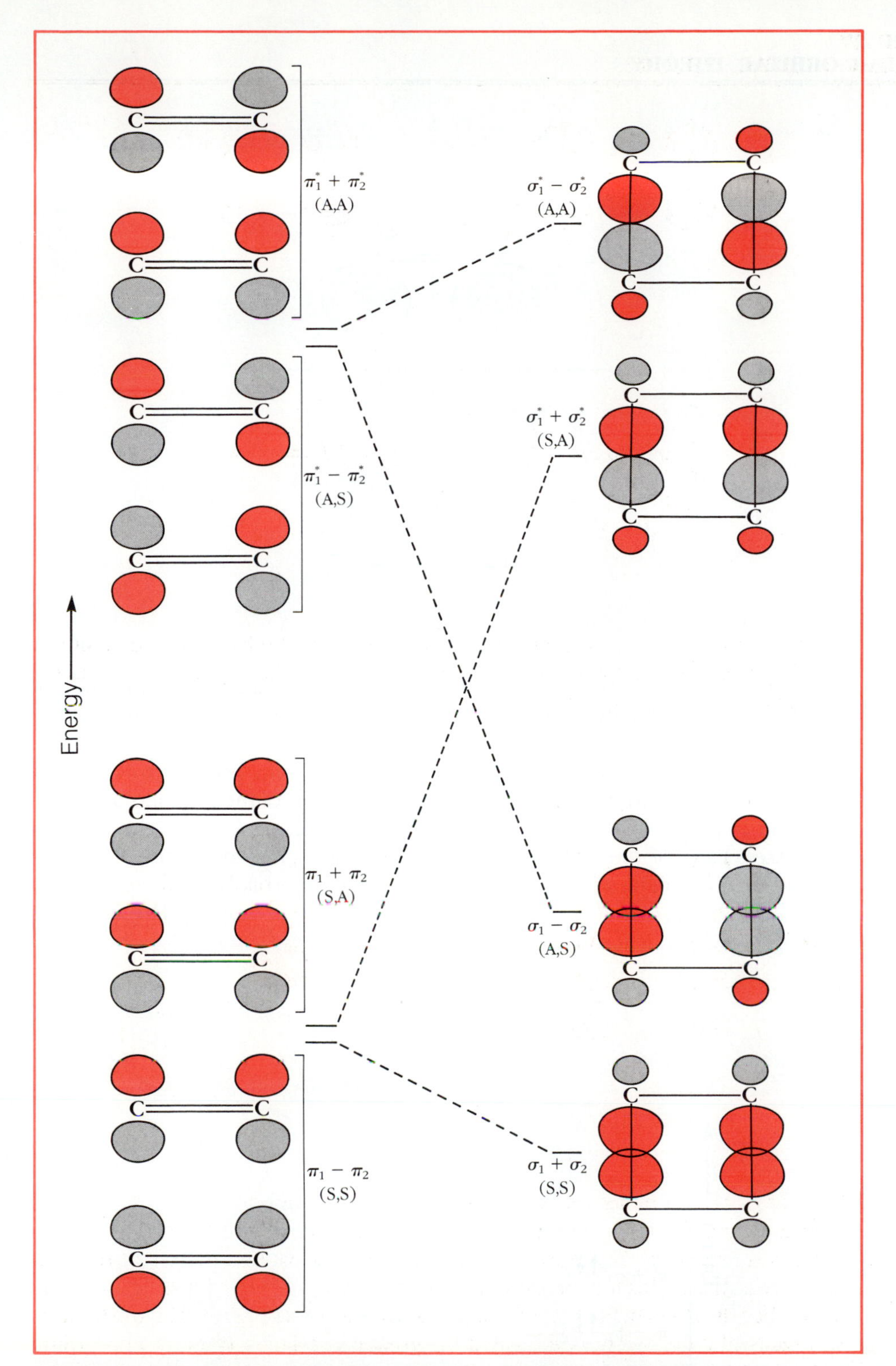

Figure 27.8 Orbital correlation diagram for the concerted cycloaddition of two ethylene molecules to form cyclobutadiene. The geometry is that described in Figure 27.7. Symmetry designations are with respect to the vertical and horizontal symmetry planes (in that order). To a first approximation the orbitals of the free ethylenes do not interact, so the in-phase and out-of-phase combinations are shown at the same energy. Product and reactant orbitals are shown for convenience as having approximately the same energies.

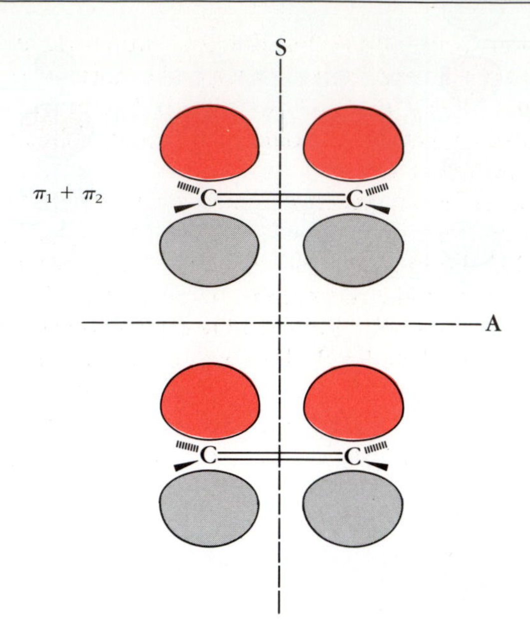

The overall symmetry label for this orbital is therefore designated *SA*. Inspection of Figure 27.8 shows that this correlates with an antibonding MO ($\sigma_1^* + \sigma_2^*$) of the product.

Now that we have drawn a complete correlation diagram, how can this be used to analyze the reaction? Simply determine which orbitals of the reactants are occupied, and evaluate what happens to the energies of these orbitals as the reaction progresses. Each of the isolated ethylene molecules has two electrons in its π system, so there will be a total of four electrons in the π orbitals of the reactants. In other words, the two lowest MOs should each be doubly occupied. The following drawing shows the features of interest.

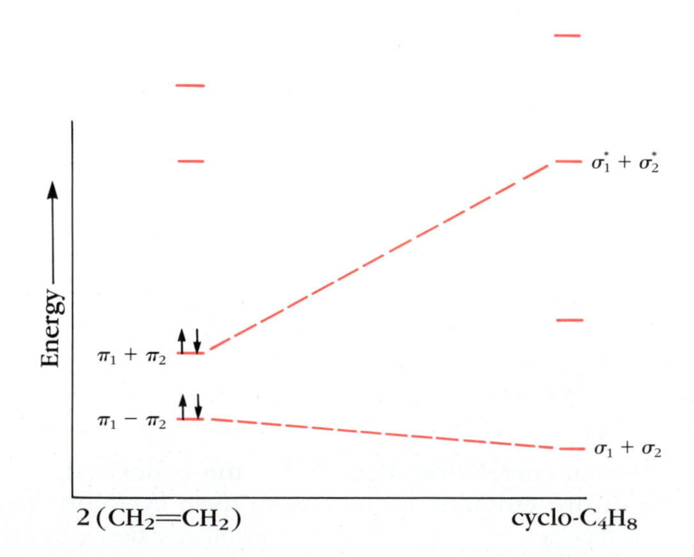

The occupied $\pi_1 + \pi_2$ MO of the reactant correlates with an *antibonding* MO ($\sigma_1^* + \sigma_2^*$) of the product, which means that the total orbital energy will be very high at the transition state of the concerted reaction. We can therefore reach the following conclusion: Concerted cycloaddition of two ethylene molecules is unfavorable and should not be expected to occur under normal conditions. This conclusion is supported by experimental observation. In contrast to the facile Diels-Alder reaction of alkenes with dienes, dimerization to cyclobutanes is not observed when alkenes are heated alone.

Why have we expended all this effort to analyze a reaction that does not occur? The answer is that the reaction is not observed under "normal" conditions. When energy is supplied in the form of heat, no *thermal* dimerization is observed, but cycloaddition does occur under *photochemical* conditions. The difference between thermal and photochemical behavior is readily explained using Figure 27.8. When the alkene mixture is irradiated with the appropriate wavelength of UV light (see Section 9.7), an electron is promoted to a higher-energy MO to generate an *excited state.* If an excited singlet state of ethylene (designated as $CH_2{=}CH_2^*$) and *ground state* (i.e., "normal") ethylene were to interact, this would correspond to the following drawing.

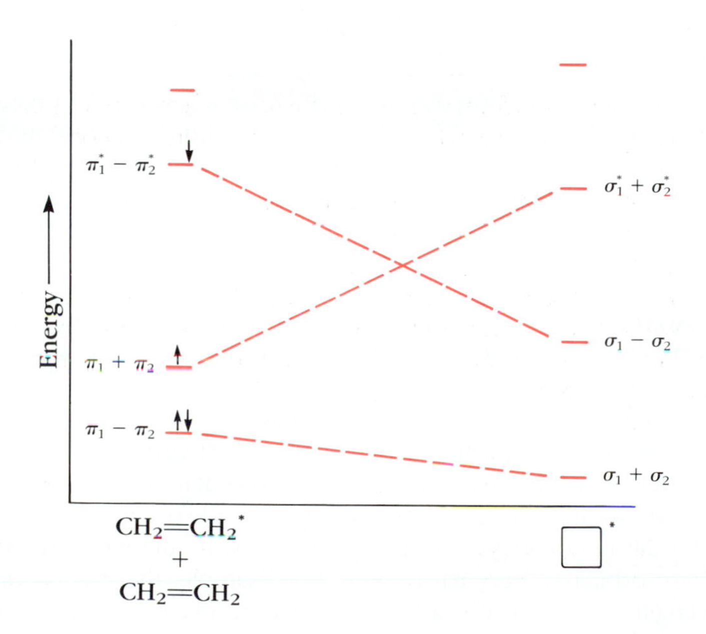

For this photochemical process the increase in energy of the $(\pi_1 + \pi_2)$ MO is compensated by a corresponding energy decrease for $(\pi^*_1 - \pi^*_2)$, where both orbitals are singly occupied. Consequently, at the transition state there should be no large increase in the total energy of the occupied orbitals.

Our analysis suggests that photochemical cycloaddition of alkenes should be an energetically reasonable process, and experimental results confirm the suggestion. The overall reaction proceeds in several steps, starting with photoexcitation of one of the alkenes. Next follows a concerted cycloaddition with a second (ground state) alkene. This cycloaddition generates an excited state of the cyclobutane (with one electron in the $\sigma_1^* + \sigma_2^*$ orbital), and the last step is relaxation to the ground state via thermal transfer of the excess energy. The following reaction illustrates the photochemical cycloaddition of alkenes to generate cyclobutanes:

C_6H_5(H)C=C(H)(C_6H_5) + $(CH_3)_2C=C(CH_3)_2$ $\xrightarrow{h\nu}$ cyclobutane product (95%)

It would not have been possible to analyze this specific reaction by the identical procedure used for ethylene dimerization because there is less symmetry. Nevertheless, the predictions we made for ethylene work quite well for other nonpolar alkenes: Thermal cycloaddition is unfavorable and is not observed, while photochemical cycloaddition to form cyclobutanes occurs readily.

Many other reactions can be studied using a full correlation of orbital symmetries, but we will limit our discussion to the single case just presented. In the next section we will show you a related (but easier) method for analyzing concerted reactions.

EXERCISE 27.4

Orbital symmetry correlation diagrams can be used to evaluate a reaction in both directions. Use Figure 27.8 to answer the following question: Should cyclobutane undergo a concerted fragmentation to two ethylenes upon heating?

27.4 FRONTIER ORBITAL THEORY: CONCERTED THERMAL AND PHOTOCHEMICAL REACTIONS

In several earlier discussions in this text we have considered the importance of interactions involving two specific types of orbital, the highest occupied molecular orbital (HOMO) and the lowest unoccupied molecular orbital (LUMO). In any molecule these two orbitals define a border between the occupied and unoccupied MO's, and for this reason they are called *frontier orbitals.* In many cases, the importance of these orbitals is very easily seen. For example, there is no difficulty in understanding the reactivity of a carbocation in terms of a vacant orbital on carbon, the LUMO. Similarly, there is no difficulty in interpreting the reactivity of a nucleophile such as $:NH_3$ in terms of the nonbonding electrons, which occupy the HOMO.

A great many chemical reactions can be analyzed by focusing specifically on the interactions between the HOMO of one reactant and the LUMO of the other. By this method we explained the stereochemistry of nucleophilic substitution in Section 12.4. When two species undergo a reaction, there are many orbital interactions, but the HOMO-LUMO interaction is often the most important in determining the energetics of a process.

When *conservation of orbital symmetry* was put forth as a general principle for concerted reactions by Woodward and Hoffmann in 1965, the study of orbital correlation diagrams suggested a remarkable simplification: *The energetics of a concerted pathway can be analyzed by simply evaluating the HOMO-LUMO interaction.* In the following discussion we will show you how frontier orbital theory can be used to analyze a variety of concerted reactions. As in previous sections we will use the atomic orbital contributions to depict the MOs. All these analyses are based on reactions of molecules with substantial symmetry, and from a rigorous mathematical viewpoint the arguments apply only to those specific cases. Nevertheless, the patterns of reactivity that we will analyze for symmetrical molecules are also observed for a wide variety of structurally similar (but less symmetrical) compounds.

Cycloaddition

In contrast to the detailed analysis of the preceding section, we can evaluate the cycloaddition of two alkenes merely by drawing the HOMO-LUMO interaction. Both reactant molecules are the same, so the interaction is between the π orbital (HOMO) of one alkene molecule and the π^* orbital (LUMO) of the other.

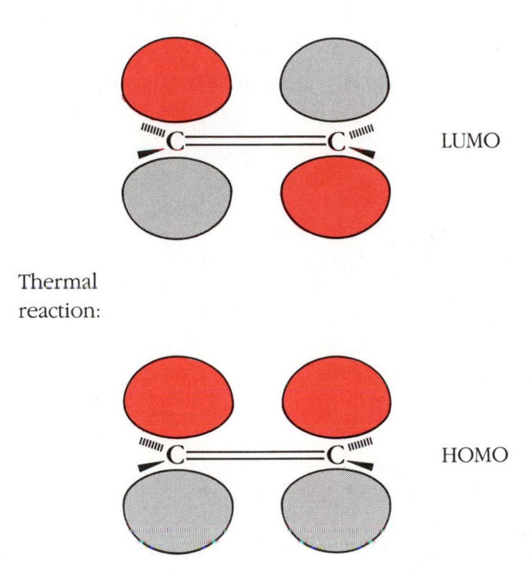

There is no net interaction because the two interacting orbitals have different symmetry. Hence the reaction is not favorable, and this agrees both with our previous analysis and with experimental results. The opposite prediction is obtained for the photochemical reaction. We now use the π^* orbital as the HOMO of the excited ethylene (π and π^* each have one electron), and π^* is also the LUMO of the other (ground state) ethylene molecule. This time the two interacting orbitals have the same symmetry, so a favorable (bonding) interaction can be obtained.

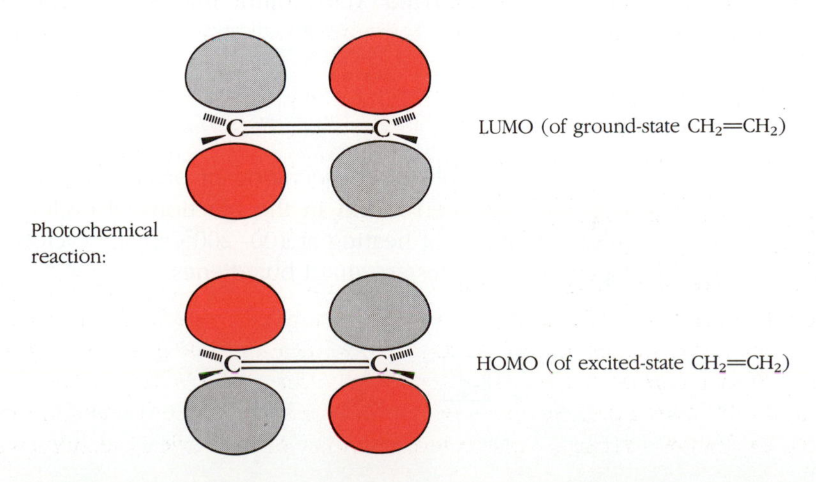

In accord both with our earlier analysis and with experimental results, the photochemical cycloaddition of two alkene molecules is predicted to be an energetically favorable reaction.

The Diels-Alder reaction (Section 22.3) can also be explained by frontier orbital theory. Although Diels-Alder reactions usually involve substituted derivatives, we will employ butadiene and ethylene to analyze the reaction. In this case the two reactants are different, so which HOMO and which LUMO should be used? The alkenes employed in Diels-Alder reactions typically have an electron-withdrawing substituent (e.g., —CN, —CO_2R), so they can be considered electron-deficient. A reasonable choice would therefore be to look at the LUMO of the alkene and its interaction with the HOMO of the diene. (Actually, either HOMO-LUMO combination will yield exactly the same final prediction.) The highest occupied MO of a diene (π_2 of Figure 8.5) has a single out-of-phase interaction between adjacent atomic p orbitals, and the LUMO of the alkene is simply the antibonding orbital π^*. The following drawing shows the geometry corresponding to the experimentally observed results that we previously illustrated in Figure 22.2

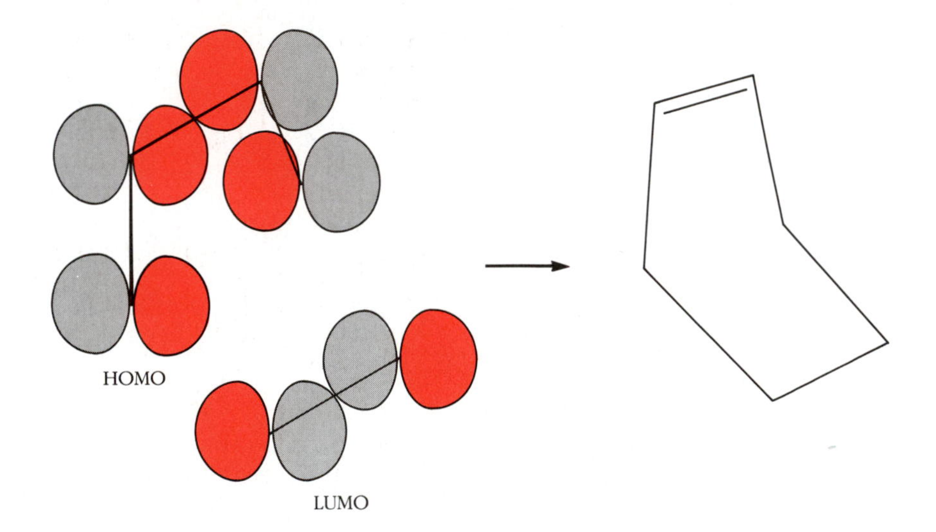

The HOMO and LUMO orbitals both have the same symmetry: They are antisymmetric with respect to the plane of the paper, and this is the symmetry plane that is maintained throughout the course of the reaction. This drawing therefore explains why the Diels-Alder reaction is a concerted cycloaddition that occurs between two molecules in their ground states.

Electrocyclic Reactions

Prior to the 1960s, when the principle of conservation of orbital symmetry was developed, a curious puzzle had been found in the reactions of cyclobutenes. Under the relatively mild conditions of heating at 100–200°C many cyclobutenes would undergo ring opening to the less strained butadienes.

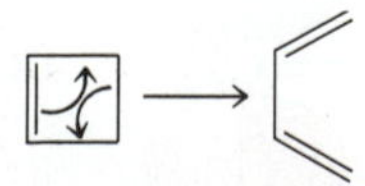

For a variety of different cyclobutenes the ring opening was found to be highly stereospecific. For example, *cis*-3,4-dichlorocyclobutene affords exclusively (1*E*, 4*Z*)-dichloro-1,3-butadiene.

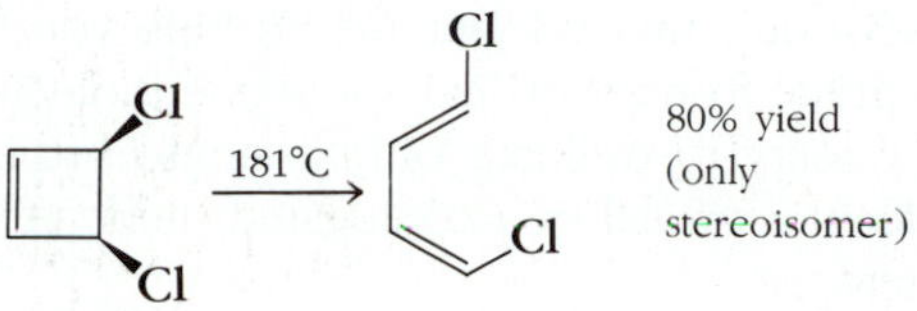

181°C

80% yield (only stereoisomer)

Analogous stereochemical results were found for *cis*-3,4-dimethylcyclobutene, while the *trans* isomer was found to yield only the *E,E* isomer of the diene.

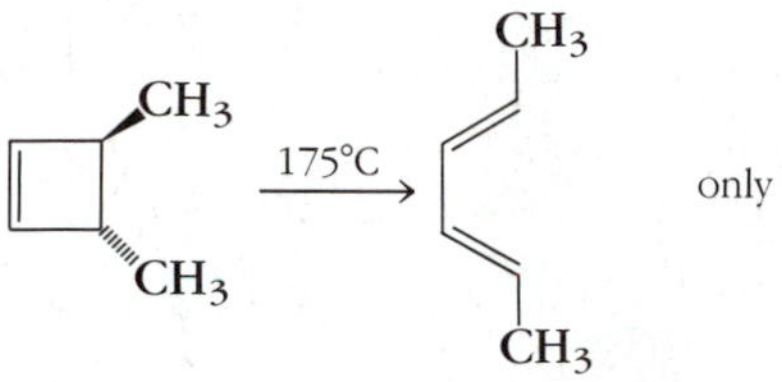

175°C

only

Orbital symmetry theory permitted the stereochemical puzzle to be unraveled, and now chemists can predict the outcome of a wide variety of related reactions. Once again, either an orbital correlation diagram or a frontier orbital analysis would provide a satisfactory explanation, but we will show the latter because of its simplicity. A further convenience arises from the fact that the same orbital symmetry arguments apply to both the forward and reverse directions of a reaction. In this case it is easier to visualize the orbital interactions of the diene, so we will analyze the diene → cyclobutene process even though the reverse is observed under thermal conditions.

There are two alternative modes by which the diene and the cyclobutene could be interconverted. Both ends could twist in the same direction (called *conrotatory*):

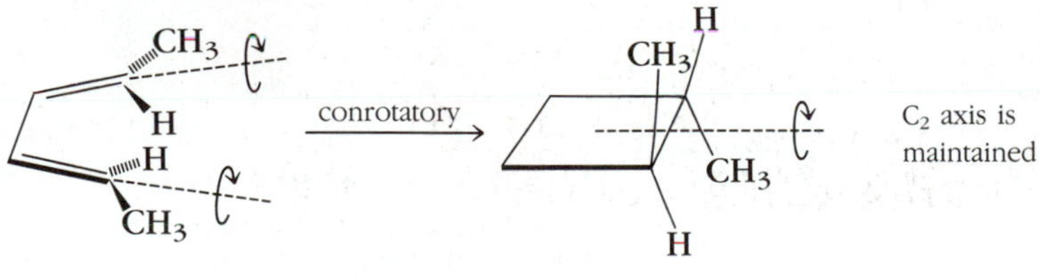

conrotatory

C_2 axis is maintained

Notice that the only molecular symmetry element maintained throughout the conrotatory process is a C_2 axis.

In the other mode the termini could rotate in opposite directions (called *disrotatory*):

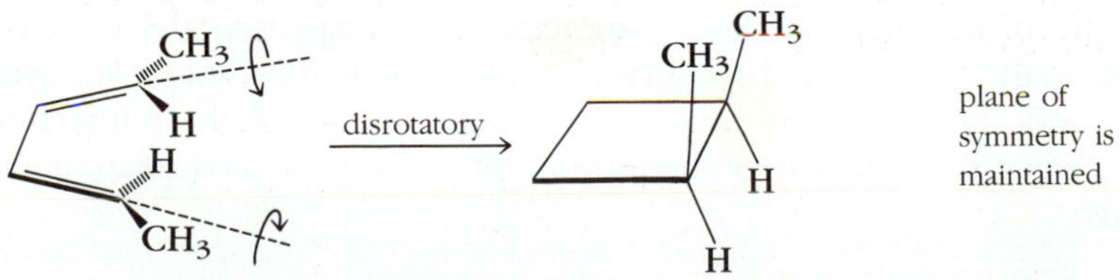

disrotatory

plane of symmetry is maintained

Here a symmetry plane (the plane of the paper) is maintained throughout the entire process.

Clearly, it is the conrotatory pathway that occurs in the thermal reactions for which we have shown experimental results. The disrotatory pathway predicts the wrong diene from *cis*-3,4-dimethylcyclobutene. To help you understand why, we will analyze the appropriate frontier orbital. Only one molecule is involved in the reaction, and it is necessary to evaluate only a single orbital, the HOMO. The highest occupied molecular orbital of a conjugated diene (Figure 8.5) is π_2, and this is the orbital of interest.

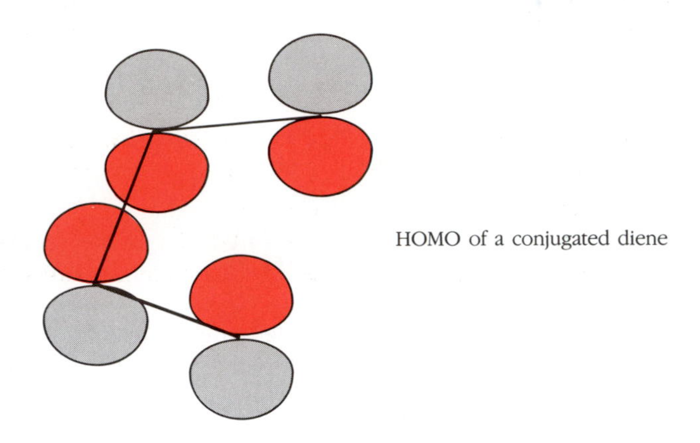
HOMO of a conjugated diene

In the conrotatory process a favorable in-phase interaction develops between the *p* orbitals at the termini.

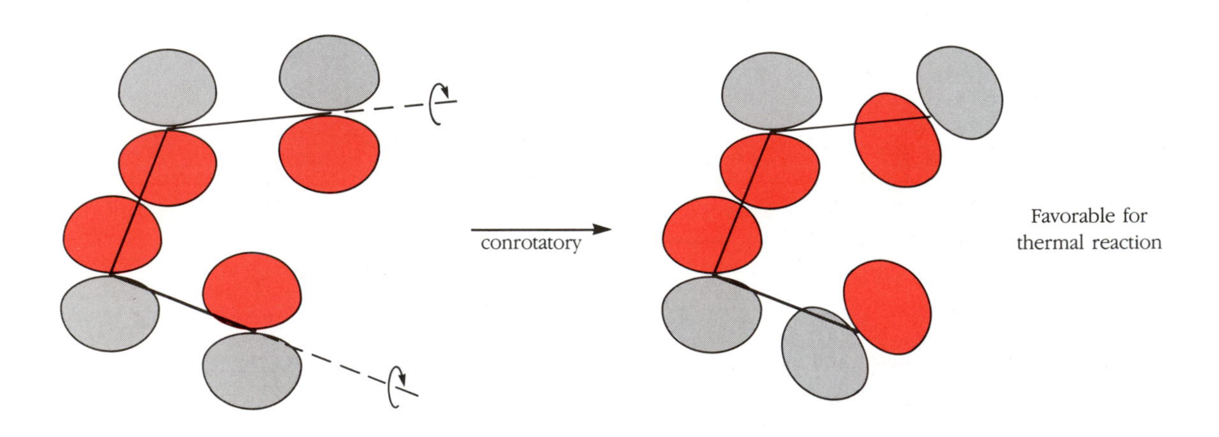

Favorable for thermal reaction

The HOMO of the diene remains symmetric with respect to the C_2 axis of the molecule, so the conrotatory process is also favorable in terms of orbital symmetry.

What about the disrotatory mode? An unfavorable out-of-phase interaction develops between the *p* orbitals at the termini when the two ends rotate in opposite directions.

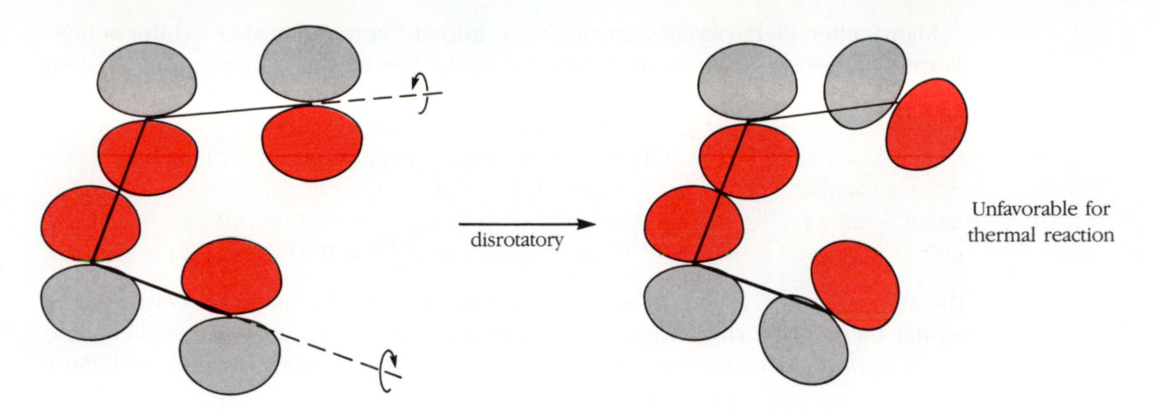

The HOMO of the starting diene is antisymmetric with respect to the molecular symmetry plane, and if this orbital symmetry were maintained, a full σ antibonding interaction would develop between the termini. Clearly the disrotatory process is not favored for a concerted thermal reaction.

The photochemical process is quite different because the lowest excited state has an electron in π_3 of the diene, and it is this orbital which governs the reaction.

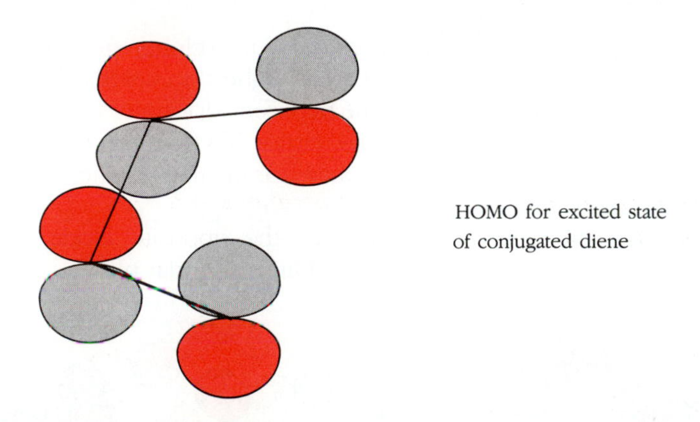

The symmetry of π_3 is different from that of the ground state HOMO, and the opposite stereochemistry is observed. The disrotatory mode is observed experimentally for the photochemical reaction. A second major difference in the photochemical process is the *direction* of the reaction. A conjugated diene can be selectively excited with radiation of about 225 nm (see Section 9.7), and the resulting photochemical reaction will cause ring *closure* to form a cyclobutene. The product contains only a single, isolated double bond, so it does not absorb radiation of that wavelength. Therefore ring *opening* does not occur under photochemical conditions. This is illustrated by the following example.

UV light → 60%

The ring closes photochemically by the disrotatory mode, which you can recognize from the *cis* relationship of the two hydrogens on the four-membered ring.

Many other electrocyclic reactions are known, and these also exhibit a high degree of stereospecificity. For larger systems the ring-closed product is often favored in thermal processes, as illustrated by the following reaction:

CH_3 CH_3 $\xrightarrow{132°C}$ CH_3 CH_3 ~100%

The stereochemistry of this thermal reaction can again be analyzed with frontier orbital theory. For ring closure of a conjugated triene to a cyclohexadiene the favored mode is disrotatory, as seen by the following drawing of the HOMO.

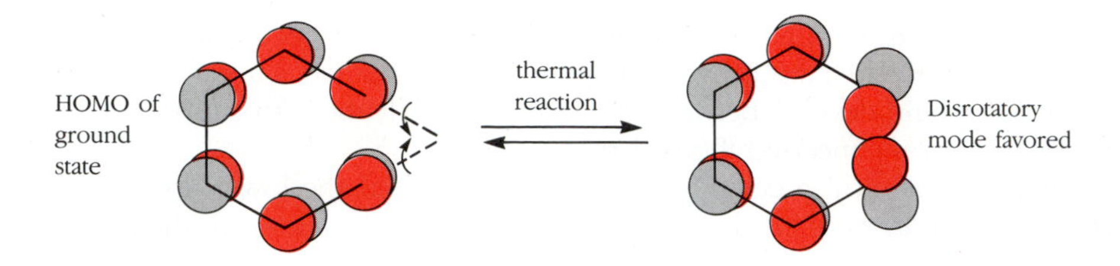

The change from the preferred conrotatory mode in cyclobutene formation results from the different symmetry of the HOMO. The highest occupied orbital of a conjugated diene is π_2, with a single nodal plane between adjacent *p* orbitals; the highest occupied orbital of a conjugated triene, in contrast, has two such nodal planes. As a result, the *p* orbitals of the termini are in phase for the triene, whereas they are out of phase for the diene HOMO.

The photochemical reaction proceeds by the opposite mode because the HOMO of the lowest excited state has an additional nodal plane.

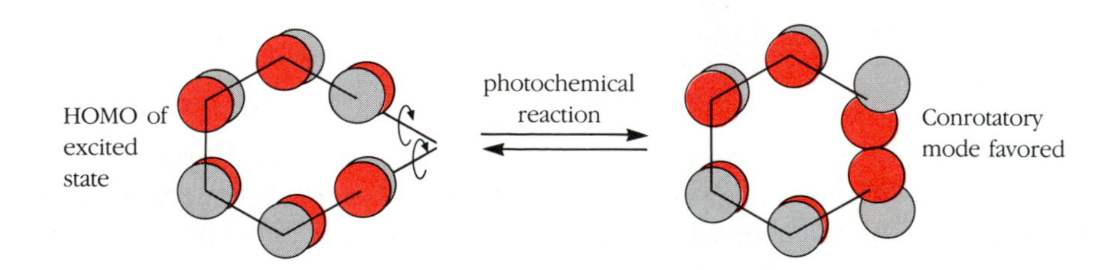

Other electrocyclic processes that form or cleave rings of different sizes are also well known. The analysis of the frontier orbitals is carried out in the same way that we have shown here for the diene and triene cases. Some of these will be left for you as problems in this text.

Sigmatropic Rearrangements

In many reactions of unsaturated systems, isomerizations have been observed in which a σ-bonded substituent migrates to a carbon that was originally part of the π system. These processes are called *sigmatropic* rearrangements. The following equation illustrates the migration of a hydrogen by such a reaction.

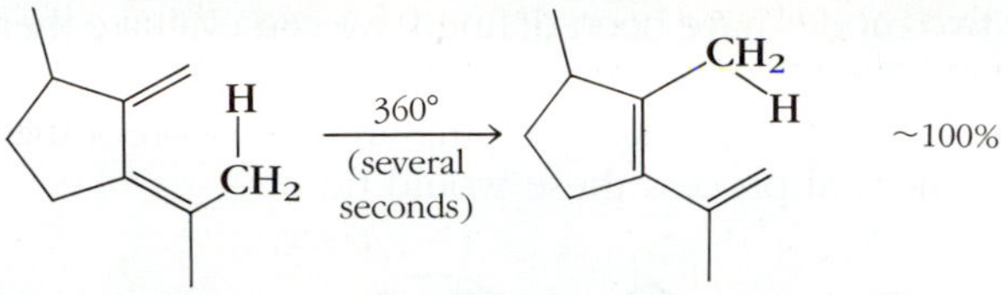

360° (several seconds) ~100%

To analyze these reactions we will begin by considering a 1,3 allylic shift of a hydrogen atom.

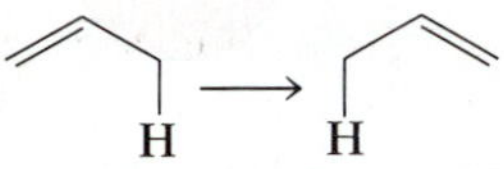

This type of reaction is different from those we have considered previously, because there is an inherent *lack* of symmetry during the reaction as the hydrogen moves from one end of the system to the other. Symmetry suddenly arises at the transition state, when the hydrogen is equidistant from the two ends of the π system.

As we discussed previously, it is difficult to analyze such transition states because we know little about their structure. Woodward and Hoffmann suggested an elegant and simple solution to this problem in 1965. By considering a *different reaction* that leads to the *same transition state,* frontier orbital theory could be used to determine the optimal pathway. For the present case the combination of a hydrogen radical and an allyl radical serves this purpose.

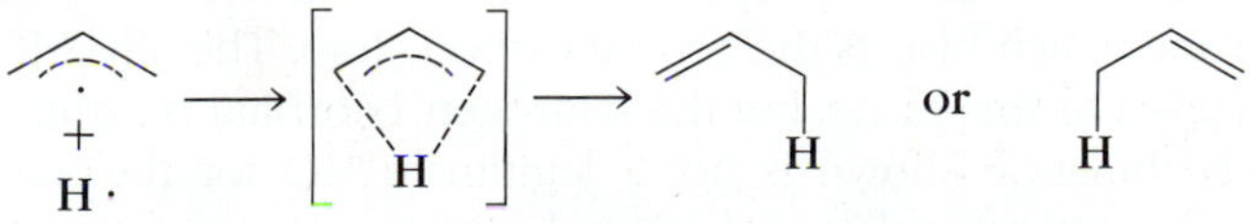

Two alternative modes must be considered for a concerted sigmatropic reaction: the *suprafacial* mode, in which the migrating group remains on the same face of the π system

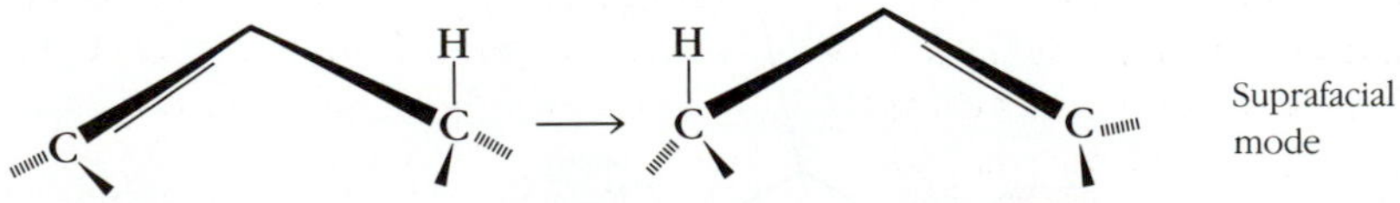

Suprafacial mode

and the *antarafacial* mode, in which the migrating group moves between the top and bottom faces of the π system.

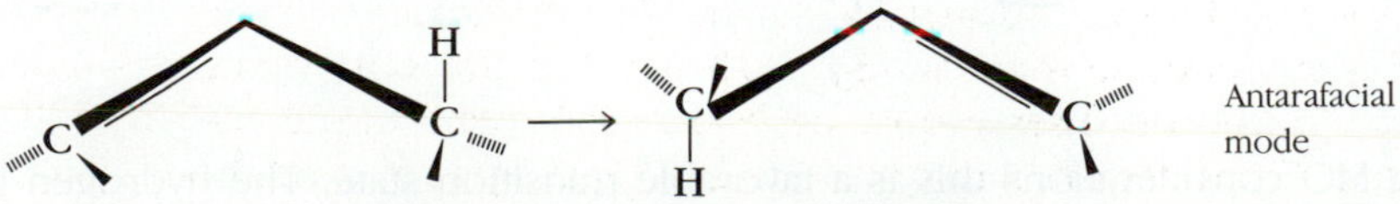

Antarafacial mode

Now that the two modes have been defined, we can evaluate them according to the frontier orbitals. The MO's of interest are those on each species that are singly occupied (that is, that which contains the unpaired electron of the free radical in each case). For a thermal process these would be

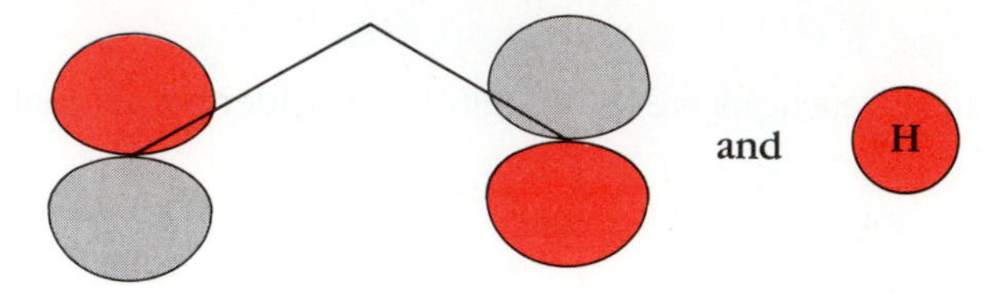

Clearly, the suprafacial mode corresponds to a high-energy transition state because the interaction of the hydrogen orbital must be antibonding with respect to one of the *p* orbitals of the allyl MO.

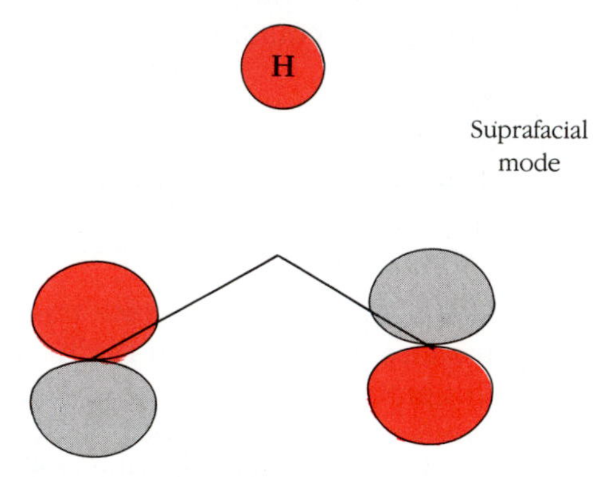

Another way of looking at the suprafacial mode is to consider that the combination of orbitals shown here lacks the proper symmetry. The molecular symmetry is that of a vertical plane which bisects the central carbon atom. The allyl MO is antisymmetric with respect to this plane, but the hydrogen 1*s* orbital is symmetric. Consequently the combination shown is not a legitimate MO for the transition state.

What about the antarafacial mode? Here the hydrogen must move from above the plane of the allyl group to below it, and at the transition state it must lie in this plane. Distortion of the carbon *p* orbitals to increase overlap would then lead to the following orbital representation:

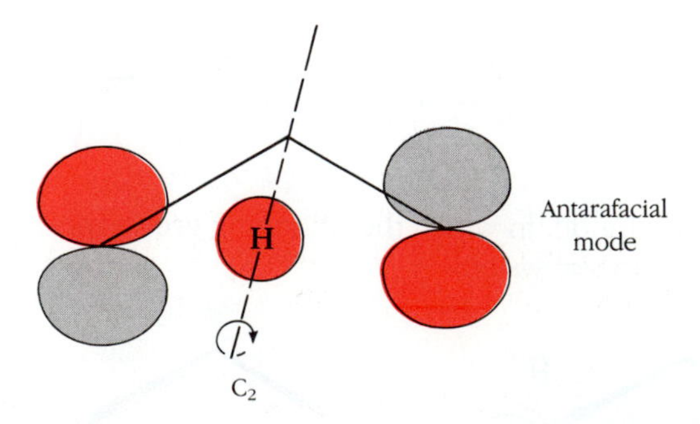

From MO considerations this is a favorable transition state. The hydrogen has a bonding interaction with both carbons, and both the hydrogen and allyl orbital

components are symmetric with respect to the molecular C_2 axis. But such a shift is *physically* difficult, because the hydrogen is initially situated in the wrong orientation for interaction with the bottom lobe of the carbon p orbital.

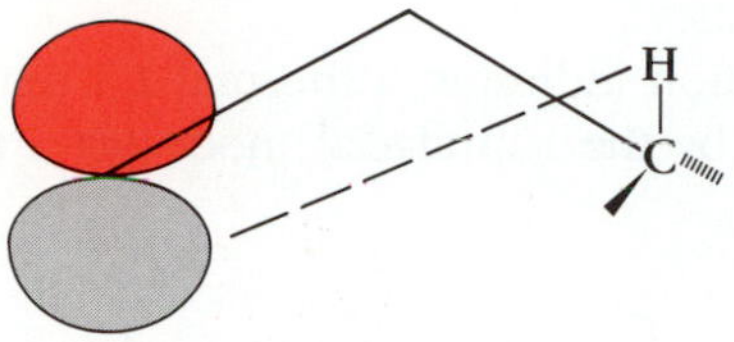

H
C

Consequently, thermal 1,3 shifts of allylic hydrogens are not expected by any mode, and they are not observed experimentally.

Once again the situation is reversed for the photochemical reactions. The excited state of an allyl radical would involve the highest allyl MO, which has the correct symmetry for a suprafacial migration.

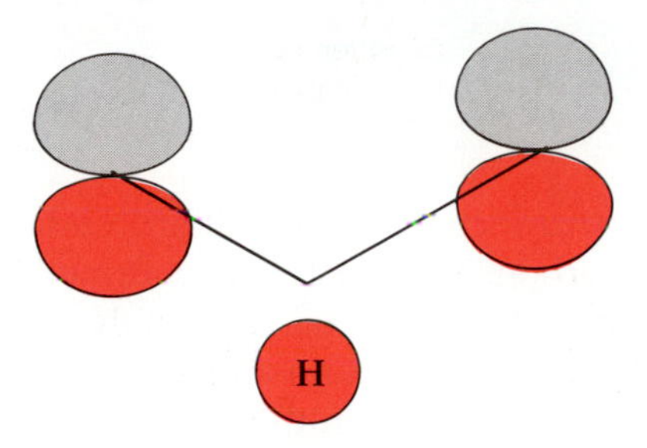

As you might expect from the preceding analysis, photochemical 1,3 hydrogen shifts are well documented in allylic systems.

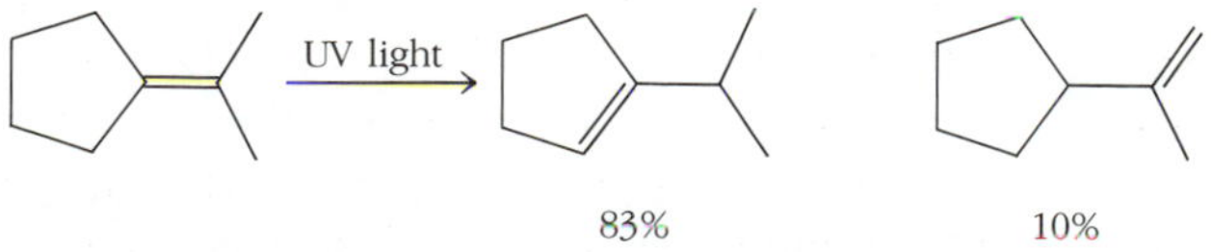

UV light

83% 10%

Other sigmatropic migrations of hydrogen are well documented, but we will leave the MO analysis for you as an exercise. Instead we will turn to sigmatropic reactions of carbon.

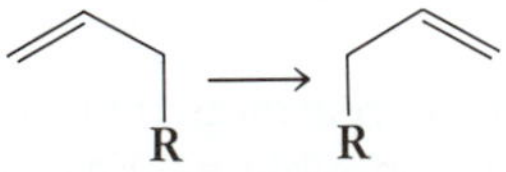

R R

The approach to analyzing the transition state is the same as we used for hydrogen migration, namely, analysis of a reaction in which the same transition state is reached by combination of two free radicals.

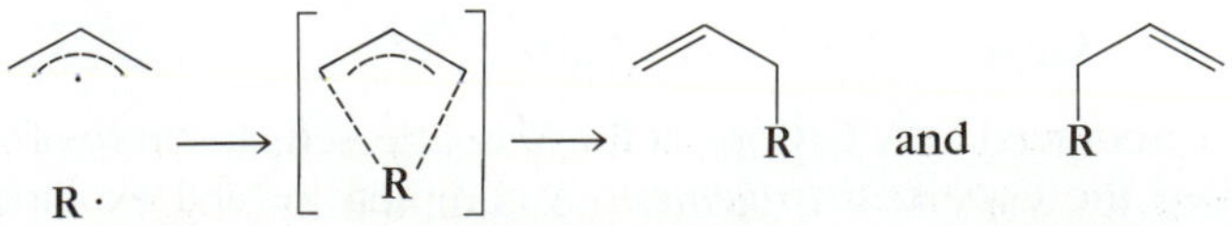

R· R R and R

There is an important difference for carbon migrations, however. While the orbital of interest for hydrogen was a 1*s* orbital, the unpaired electron of a carbon radical should occupy an orbital with substantial *p* character. The symmetry of a *p* orbital is different from that of an *s* orbital, so you should expect different results for carbon rearrangements.

As the following equation indicates, a thermal 1,3 rearrangement of a carbon substituent can proceed by the suprafacial mode via a transition state with the correct orbital symmetry.

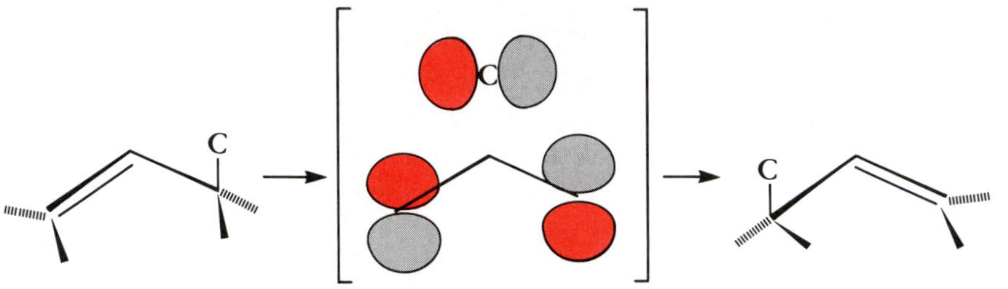

C

C

C

Transition state for suprafacial mode

This drawing also implies that such reactions will proceed with *inversion of configuration* at the migrating carbon. This has been confirmed by experimental observation. In the following example, allylic migration of the carbon bearing the deuterium occurs with inversion. The deuterium and acetoxy groups are *trans* to each other in the reactants, but they are *cis* in the product.

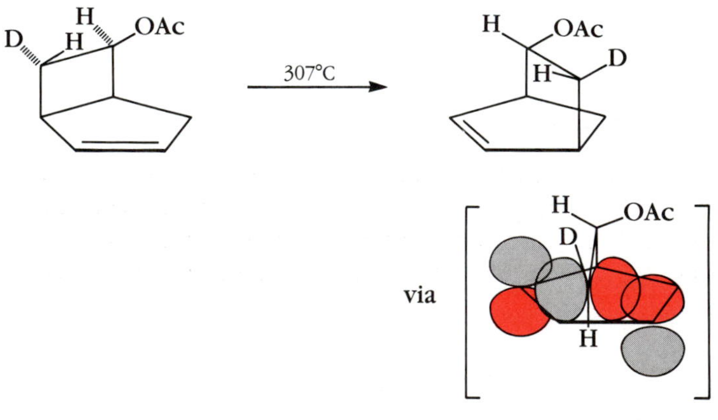

D H H OAc

307°C

H OAc H D

via

H OAc D H

Still another type of sigmatropic rearrangement has been found useful for synthetic purposes. This reaction is a double allylic rearrangement of a 1,5-diene to form another 1,5-diene. The reaction is reversible, but the formation of substituted double bonds favors the product.

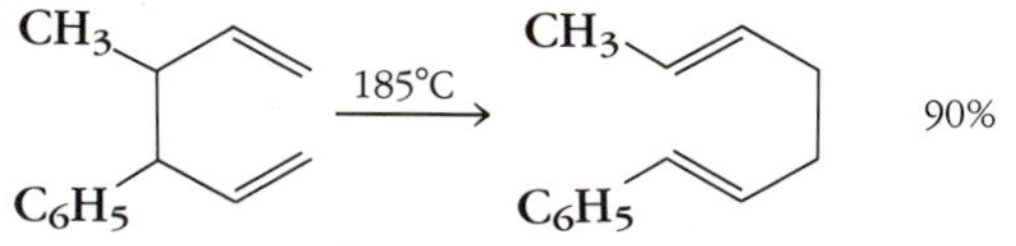

CH_3 C_6H_5 185°C CH_3 C_6H_5 90%

The reaction, discovered by A. C. Cope at the Massachusetts Institute of Technology, is usually called the *Cope rearrangement.* A complete orbital explanation of the process is again difficult because inadequate symmetry is maintained during the reaction. But the same transition state could be generated by interaction of two allylic radicals, and that permits a frontier orbital description. The MOs of interest are the singly occupied MO (π_2) of each allyl fragment. Interaction via a chairlike transition state has the proper symmetry and in-phase orbital interactions.

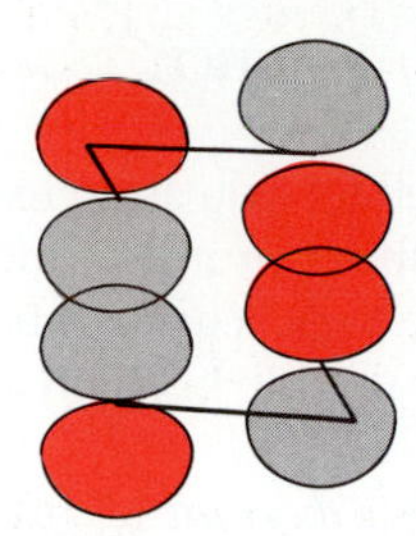

The overall reaction can then be drawn in the following way:

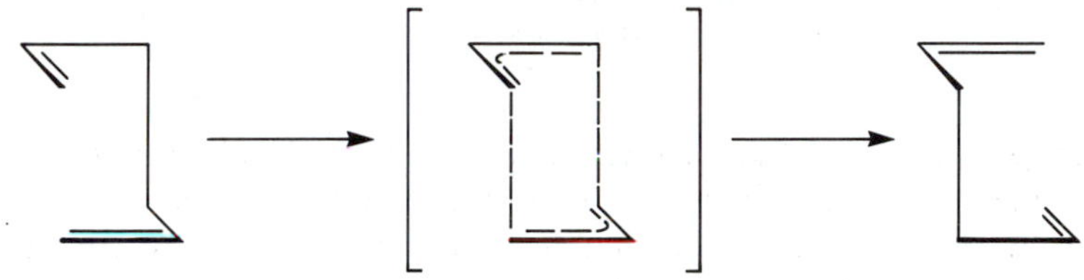

We will present additional examples of this type of sigmatropic rearrangement in Section 28.4.

Formation of Vitamin D_3

In Example 23.7 we discussed the importance of sunlight in vitamin D formation. Sunlight causes the conversion of 7-dehydrocholesterol or ergosterol to the corresponding vitamin D product, and we will now return to this reaction. Armed with a new understanding of orbital symmetry, you will be able to see what a remarkable transformation this is.

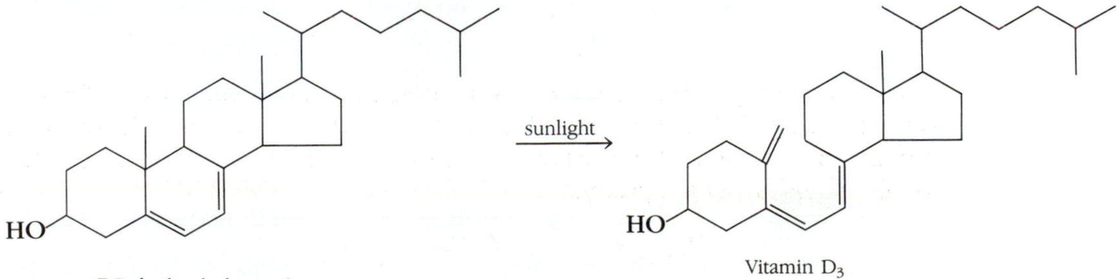

HO — sunlight → HO

7-Dehydrocholesterol

Vitamin D_3

The overall reaction proceeds in two steps, the first being an electrocyclic ring opening to give a conjugated triene called *previtamin D_3*.

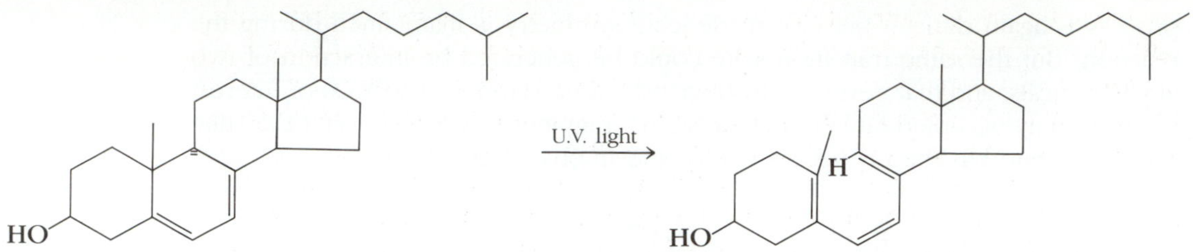

HO U.V. light HO H

An orbital symmetry analysis (left for you as an exercise) will show that the photochemical reaction proceeds by the conrotatory mode. Thermal ring opening is not observed, because it would have to proceed by the disrotatory mode, which would generate a highly strained *trans* double bond in a six-membered ring of the product.

Previtamin D_3 next undergoes a *thermal* 1,7 hydrogen shift, and extension of the arguments we have presented shows that this should proceed by the *antarafacial* mode. This hydrogen migration then yields vitamin D_3.

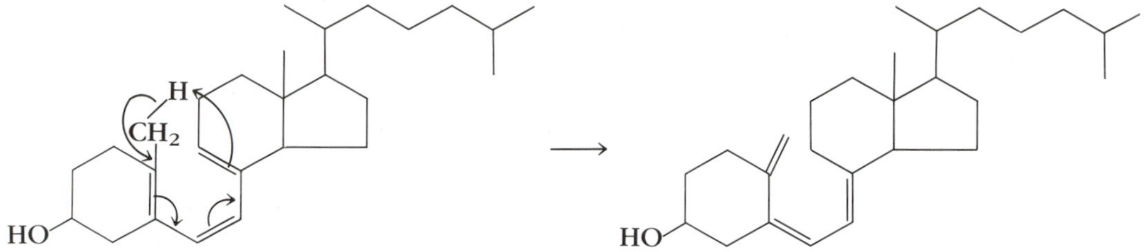

H CH_2 HO → HO

Previtamin D_3 Vitamin D_3

EXERCISE 27.5

The photochemical cyclization of 7-dehydrocholesterol to previtamin D_3 is a conrotatory process.

(a) Draw the HOMO of the triene product to show why this is so.
(b) Draw the product that would result from a *thermal,* disrotatory ring opening. Why is this not observed?

EXERCISE 27.6

Draw the frontier orbitals corresponding to the 1,7 hydrogen shift observed in the thermal reaction of previtamin D_3. Show why the antarafacial mode is favored.

27.5 GROUP ORBITALS

Up until this point we have restricted most of our discussion to interactions between orbitals on two or three adjacent atoms. Only for conjugated π systems have we analyzed molecular orbitals with contributions from more than three atomic orbitals. Just as the p orbitals from several atoms can all interact to form π molecular orbitals, so can σ molecular orbitals be formed by contributions from several atoms. Because sigma bonding involves interactions between hybrid atomic orbitals, the appropriate contributions from the various *s* and *p* atomic orbitals can be difficult to determine. Fortunately, there is a convenient device that makes the process much easier. We can first define a set of *group orbitals* that are centered on CH, CH_2, and CH_3 groups. The appropriate molecular orbitals can then be generated by combining group orbitals according to the rules that we set down in Section 27.1. For convenience, we will only draw the atomic orbital contributions rather than attempt to depict the overlap and draw the actual group orbitals.

A note of caution: The orbitals drawn in this section will seem to be quite different from the traditional representations in which carbon is shown as being exactly *sp*, sp^2, or sp^3 hybridized. In fact, hybridization to *sp*, sp^2, or sp^3 is not a property of the orbitals but is only a way that chemists sometimes attempt describe them. The end result, the molecular orbitals, will be the same regardless of how we first choose to describe the atomic orbitals.

Before proceeding with our analysis, we will first review some definitions and terms:

Atomic orbitals are the orbitals on individual atoms that we combine to generate other orbitals; they may be pure *s* and *p* or they may be combinations of *s* and *p* orbitals. The latter are frequently called *hybrid atomic orbitals,* but they are atomic orbitals nonetheless. The distinguishing feature of an atomic orbital is that it consists of orbital contributions from only a single atom.

Group orbitals provide a way of looking at the orbital interactions in common structural subunits. By first analyzing all the interactions between atomic orbitals of a CH_3 group, for example, we can generate a set of CH_3 *group orbitals.* These are merely combinations of the various atomic orbitals of the CH_3 group (in accord with appropriate symmetry restrictions). The use of group orbitals makes it easier to see how the CH_3 group will interact with the rest of the molecule.

Molecular orbitals provide a description of the entire molecule. They are constructed by appropriate combinations of atomic orbitals (or group orbitals) according to the rules enumerated in Section 27.1. Some molecular orbitals may include contributions from all atoms in the molecule, while others might involve only a single atom. In either case they describe the orbital interactions (or lack thereof) between the various atoms of the molecule.

CH Group Orbitals

In order to describe a C—H bond we must allow an interaction between the 1*s* orbital of the hydrogen and a hybrid orbital on carbon. The optimum interaction will result from overlap of the hydrogen 1*s* orbital with a hybrid orbital on carbon made up by combining the 2*s* orbital and the 2*p* orbital that is oriented toward the hydrogen. We can then draw hybrid orbitals for the carbon, as shown in Figure 27.9. Note that we have not indicated a 50:50 contribution of *s* and *p* to the hybrid orbitals; the method of group orbitals is an approximate method, and the exact contributions of *s* and *p* to the hybrid orbital are unimportant.

The two hybrid atomic orbitals are *sp* combinations, and one has its large lobe oriented toward the hydrogen. This orbital will be involved in forming the C—H bonding and antibonding orbitals.

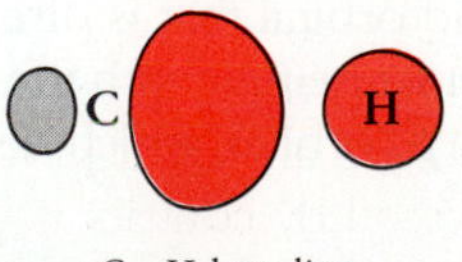

C—H bonding

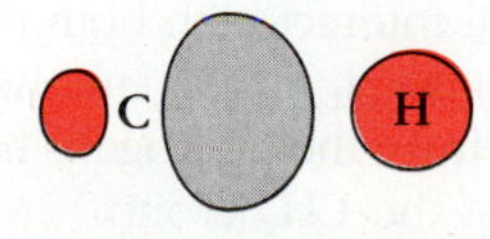

C—H antibonding

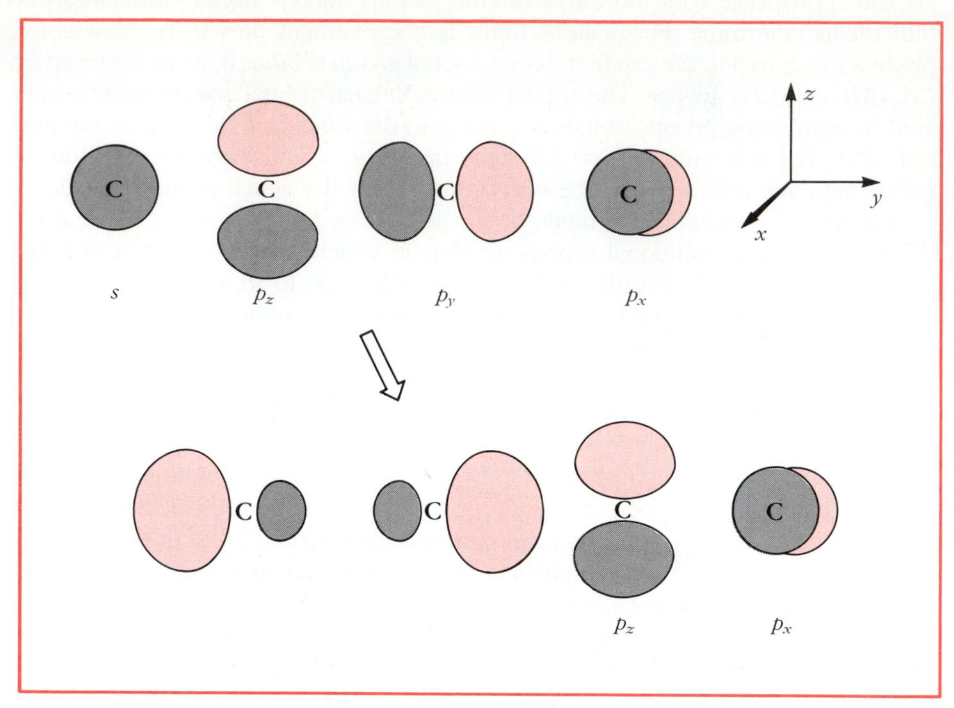

Figure 27.9 Hybrid atomic orbitals on carbon for the generation of CH group orbitals.
(For convenience the p orbitals are labeled according to their directionality along the x, y, or z axis.) The hybrid orbitals shown here are generated from combinations of the 2s and 2p_y orbitals.

The remaining p orbitals (p_x and p_z) have the wrong symmetry for inter action with the hydrogen. The large lobe of the other *sp* hybrid orbital points *away* from the hydrogen and will be used (together with the p_x and p_z orbitals) for bonding of the CH fragment to the remainder of the molecule. The five group orbitals for the CH group are shown in Figure 27.10.

CH_2 Group Orbitals

The hybrid atomic orbitals that we used for the CH group (Figure 27.9) will also be satisfactory for the CH_2 group. In order to take advantage of the local symmetry of the CH_2 group, it is drawn in an orientation such that its twofold symmetry axis corresponds to the y axis of our coordinate system. The hydrogen 1s orbitals of the CH_2 group will interact with both the *sp* hybrid orbital that is directed toward the hydrogens and the p_z orbital, but they will interact neither with the *sp* hybrid that is directed away from the hydrogens nor with the p_x orbital, which lies perpendicular to the plane of the CH_2 group.

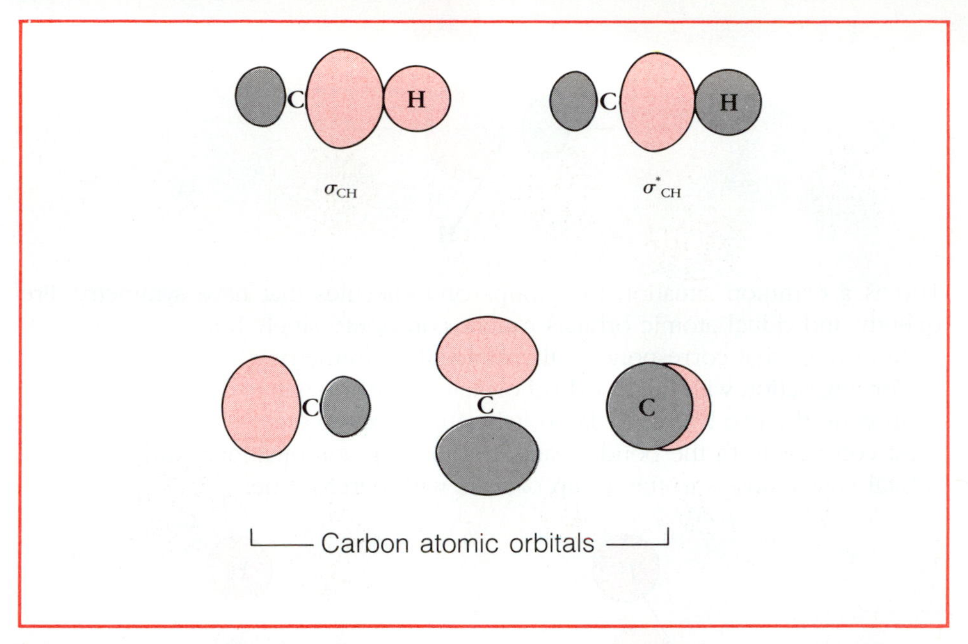

Figure 27.10 Group orbitals for C—H (shown as the atomic orbital contributions).
The four atomic orbitals from carbon and the single atomic orbital of hydrogen yield five group orbitals. One of these is bonding with respect to carbon and hydrogen, and another is antibonding. The three remaining orbitals are hybrid atomic orbitals on carbon, which will be used for bonding of the C—H group to the rest of the molecule.

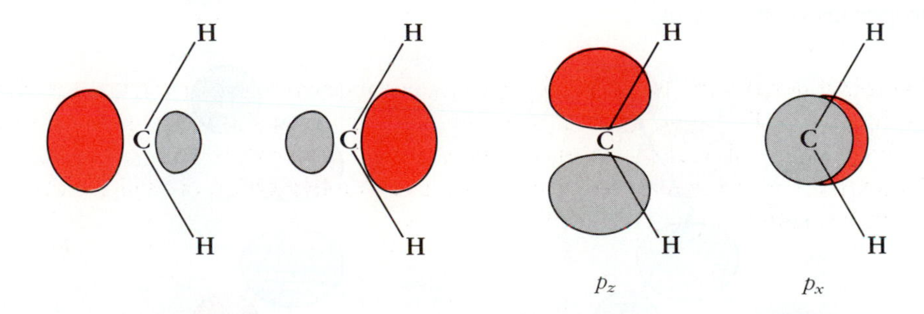

The first of these hybrid atomic orbitals does not interact significantly with the hydrogens because its large lobe is directed away from them. The p_x orbital cannot interact with the hydrogen 1*s* orbitals because they do not have the same symmetry. (The 1*s* orbitals of hydrogen are symmetric with respect to the plane of the paper, while the carbon p_x orbital is antisymmetric).

The remaining two carbon atomic orbitals can interact with the hydrogen 1*s* orbitals, but we must use the orbitals from *both* hydrogens. The interaction with a single hydrogen does not produce a legitimate MO because it lacks the symmetry of the CH_2 group (dotted line).

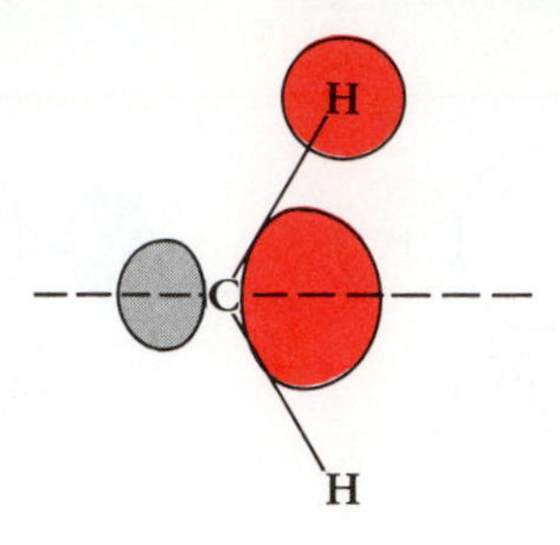

This is a common situation for groups or molecules that have symmetry. Frequently, individual atomic orbitals cannot contribute singly but must interact in combinations that correspond to the molecular symmetry.

For interaction with the *sp* hybrid orbital on carbon that is directed toward the hydrogens, the two hydrogen 1*s* orbitals must have the same phase. As always, we must consider both the bonding and antibonding combinations, and the atomic orbital contributions to the group orbitals will therefore be:

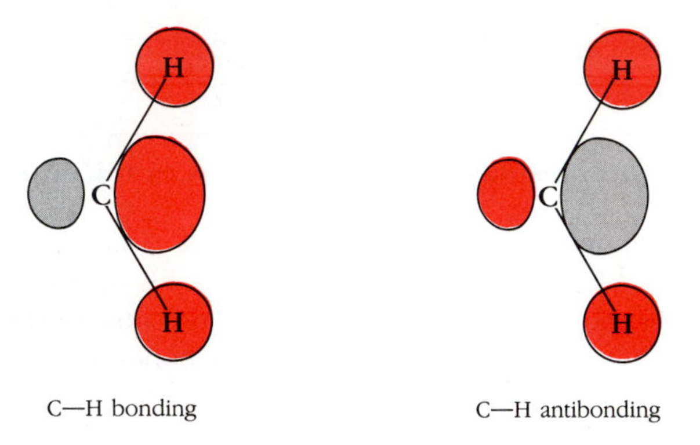

Interaction of the hydrogen 1*s* orbitals with the carbon p_z orbital requires that the phase of the two hydrogen orbitals be different, and the resulting group orbitals are antisymmetric about a nodal plane. In other words they have the symmetry properties of π *orbitals.*

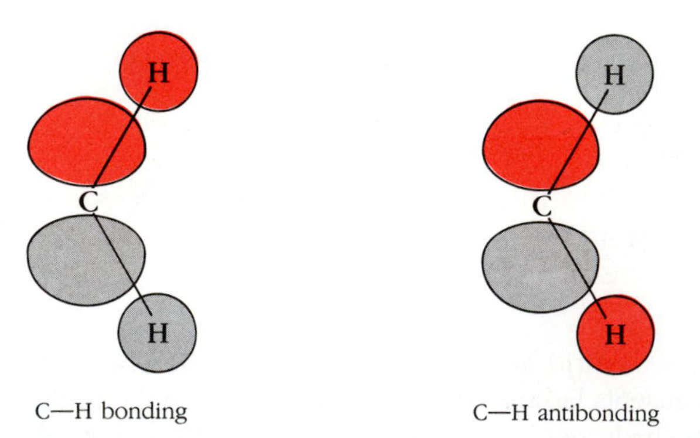

The six group orbitals for a CH_2 group are drawn in Figure 27.11.

CH_3 Group Orbitals

To describe the localized orbitals of a methyl group, the carbon hybrid atomic orbitals shown in Figure 27.9 will once again suffice. In order to take advantage of the local symmetry of the CH_3 group, we have drawn it so that its threefold symme-

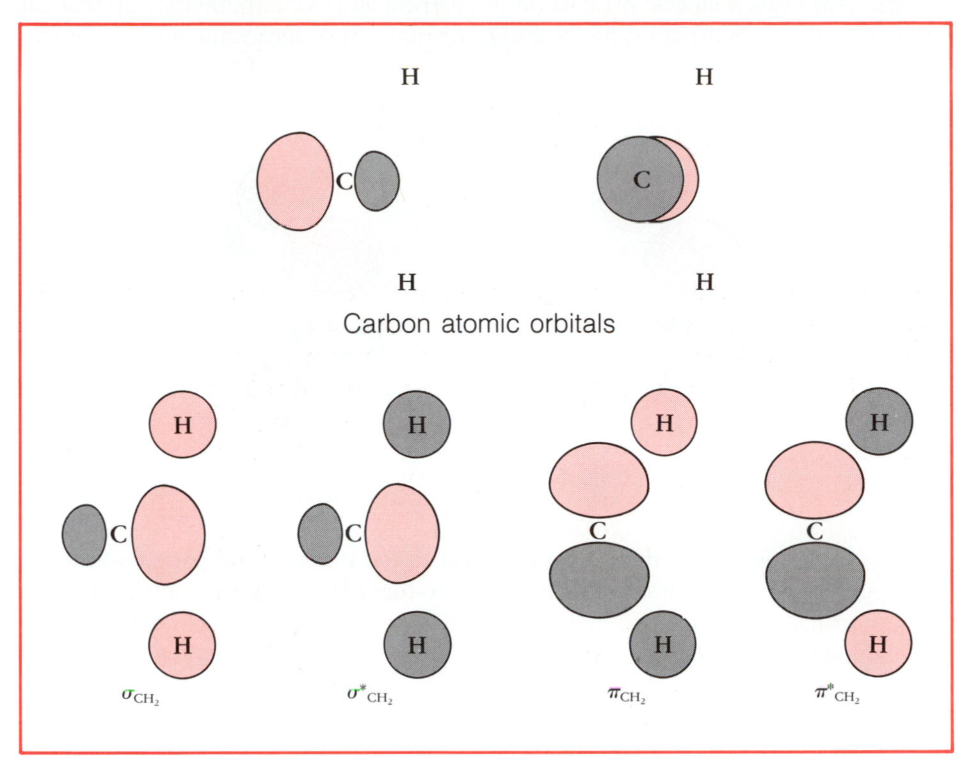

Figure 27.11 Group orbitals for CH_2 (shown as the atomic orbital contributions).
The four atomic orbitals of carbon plus the two hydrogen 1*s* orbitals generate six group orbitals. Two of these are bonding between carbon and hydrogen (σ_{CH_2} and π_{CH_2}), while two are antibonding between these nuclei ($\sigma^*_{CH_2}$ and $\pi^*_{CH_2}$). The two remaining orbitals are carbon atomic orbitals that will be used for bonding of the CH_2 group to the rest of the molecule. Because σ interactions are stronger, σ_{CH_2} is lower in energy than π_{CH_2}. (Similarly, $\sigma^*_{CH_2}$ is higher in energy than $\pi^*_{CH_2}$.)

try axis corresponds to the *y* axis of our coordinate system. As with CH and CH_2 groups, the *sp* hybrid that points toward the three hydrogens interacts with the hydrogen 1*s* orbitals to form bonding and antibonding σ combinations. (Notice that all three 1*s* orbitals of the hydrogens must have the same phase in order to maintain the symmetry of the CH_3 group.)

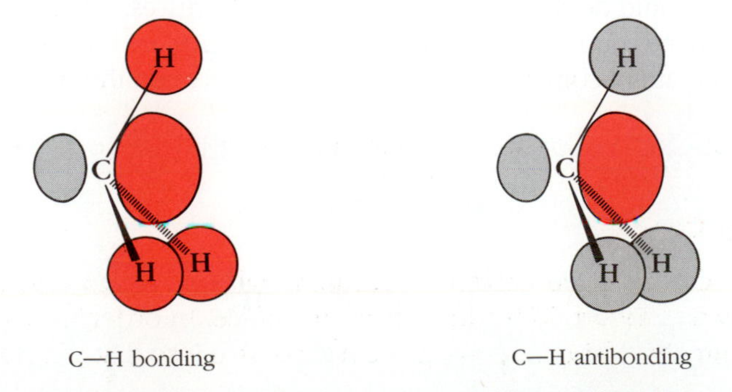

There are two p atomic orbitals on the carbon atom of a methyl group that can interact with the hydrogens, the p_z and p_x orbitals. In order to maintain the π-type symmetry, interaction with p_z requires that the 1s orbital of the "top" hydrogen have the opposite phase from the two other hydrogen orbitals.

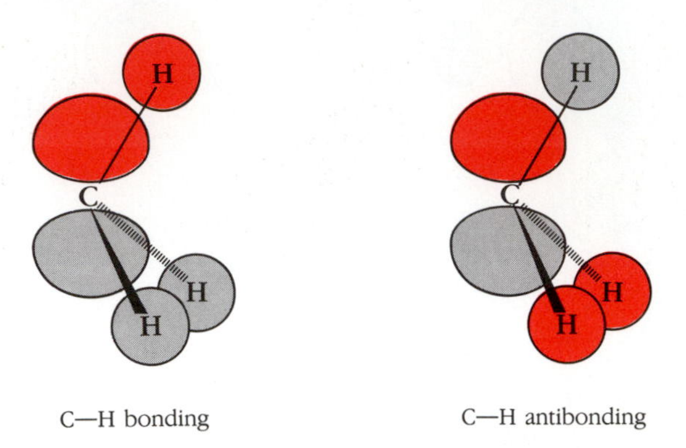

C—H bonding C—H antibonding

The p_x orbital is antisymmetric with respect to the plane of the paper, so the top hydrogen (which is symmetric with respect to that plane) cannot contribute to the group orbital. The other two hydrogen orbitals must have opposite phase in order to maintain π symmetry.

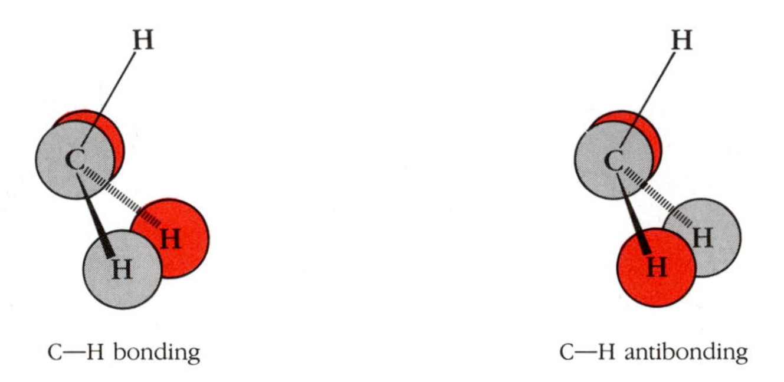

C—H bonding C—H antibonding

The combination of four atomic orbitals of carbon together with one atomic orbital from each of the hydrogens yields the seven group orbitals shown in Figure 27.12.

Molecular Orbitals from Group Orbitals: Ethane

It is beyond the scope of this chapter to construct molecular orbitals for a large variety of molecules. In fact, we will present just a single example, although this will show how you could do the same thing for other structures. The molecule we have selected is ethane, and we have chosen to analyze it in the *eclipsed* geometry because this creates a convenient symmetry plane between the two methyl groups.

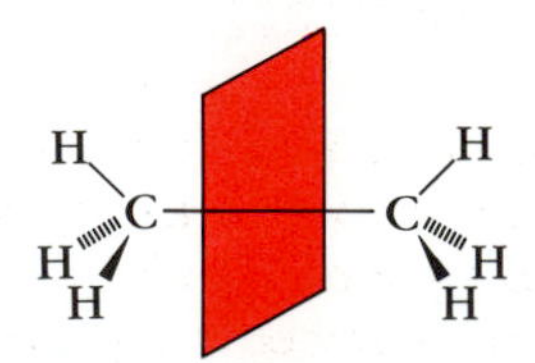

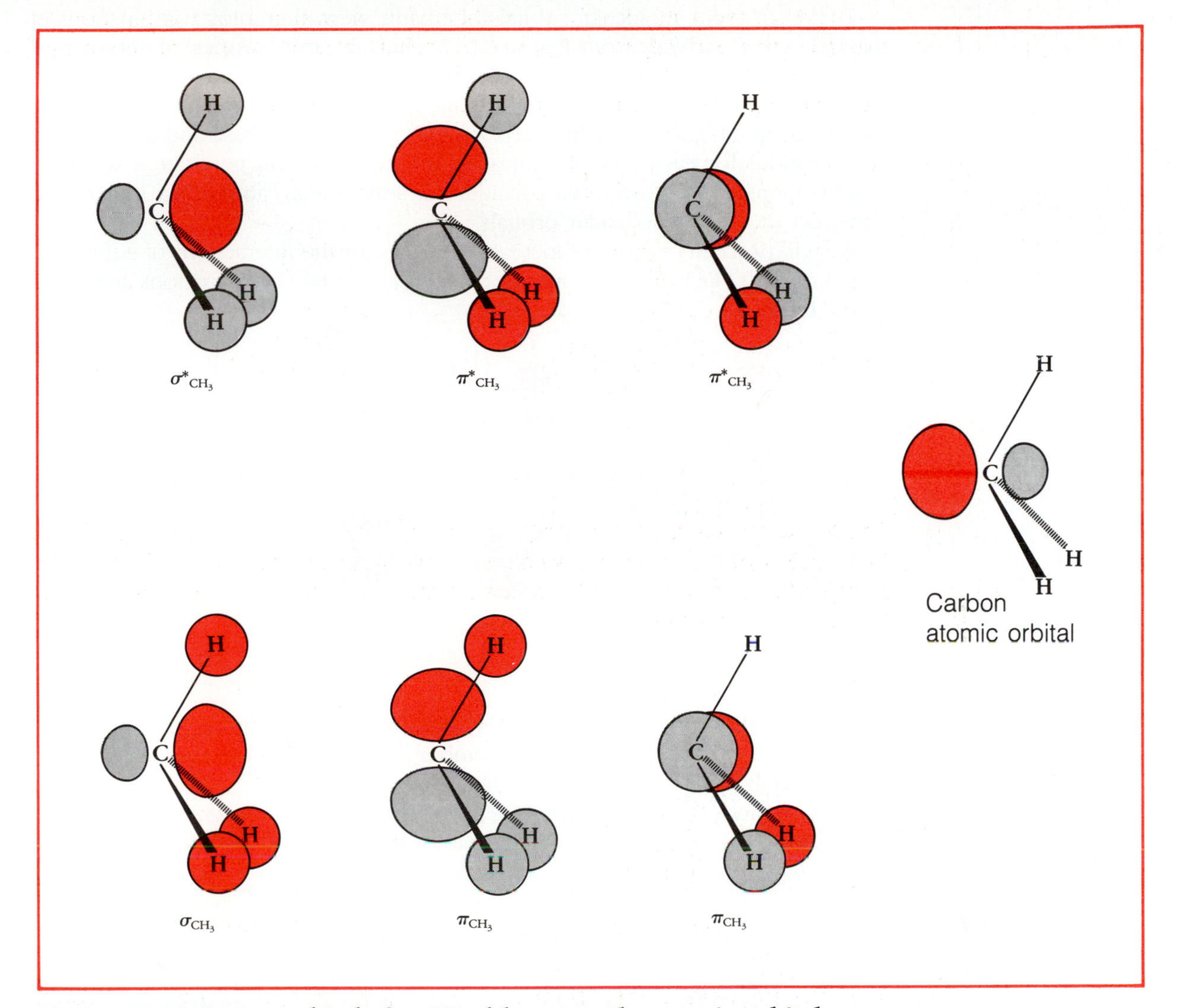

Figure 27.12 Group orbitals for CH_3 (shown as the atomic orbital contributions).
The four carbon atomic orbitals plus the three hydrogen 1*s* orbitals generate seven group orbitals. Three of these are bonding between carbon and hydrogen (σ_{CH_3} and both π_{CH_3}'s) while three more are antibonding between carbon and hydrogen ($\sigma^*_{CH_3}$ and both $\pi^*_{CH_3}$'s). The remaining carbon atomic orbital will be used for bonding to the rest of the molecule. The ordering of the orbitals in energy is:

$$\sigma_{CH_3} < \pi_{CH_3} < \text{atomic orbital} < \sigma^*_{CH_3} < \pi^*_{CH_3}$$

The two π_{CH_3} orbitals are degenerate (that is, they have the same energy), and the same is true of the two $\pi^*_{CH_3}$ orbitals.

The eclipsed ethane molecule therefore has two symmetry planes: the plane of the paper and a vertical plane that is perpendicular to the paper and bisects the C—C bond. In accord with rule 4 of Section 27.1, all molecular orbitals of the molecule must be either symmetric or antisymmetric with respect to each of these planes.

To construct the molecular orbitals of ethane, we will employ combinations of the CH_3 *group orbitals* from Figure 27.12, choosing the "proper" combinations according to the rules elaborated in Section 27.1. The overall procedure is quite straightforward. We will start with the full set of seven group orbitals for each methyl group of ethane. The first step will be to take the in-phase and out-of-phase combinations for each pair of equivalent orbitals. This will generate a set of 14 combinations, and we will then consider ways that these can interact in order to construct the actual molecular orbitals.

We will proceed in order of increasing energy for the interactions of equivalent group orbitals (Figure 27.12), starting with σ_{CH_3}. The two combinations are designated **1** and **2**:

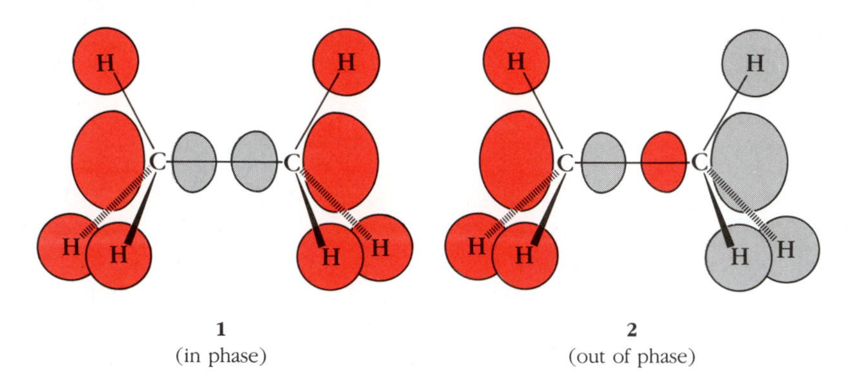

1
(in phase)

2
(out of phase)

These two orbitals are similar, but they differ by a small bonding or antibonding interaction between the carbons. Remember, however, that the main property of these orbitals is that they are C—H bonding.

Next consider the π_{CH_3} orbitals. The group orbitals have different symmetries, so each can interact only with its counterpart at the other end of the molecule. The four component π_{CH_3} group orbitals will therefore lead to combinations **3–6**.

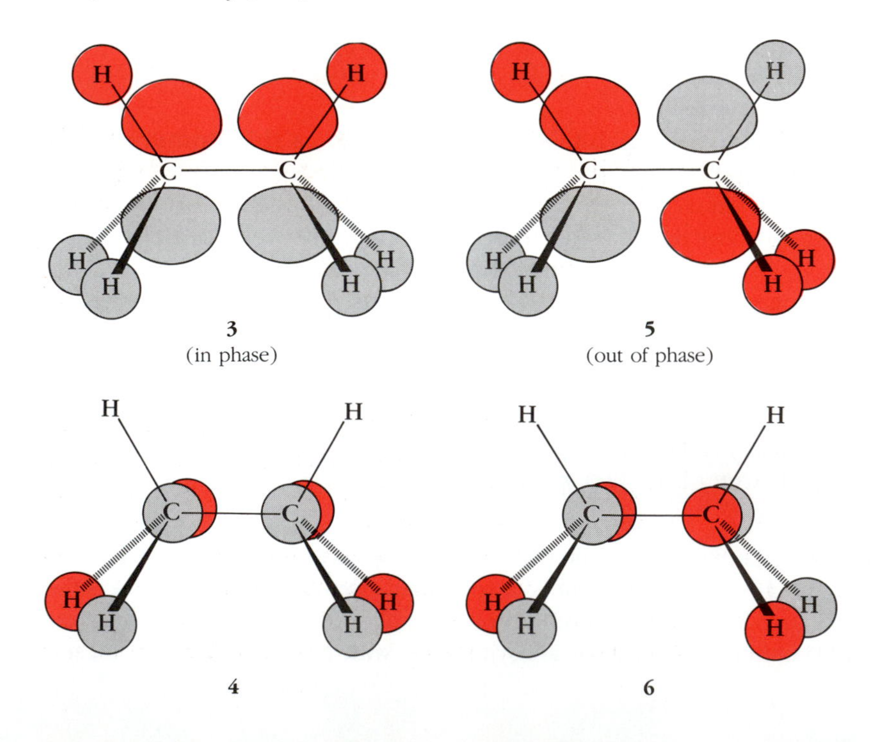

3
(in phase)

5
(out of phase)

4

6

Notice that combinations **3** and **4** are also bonding with regard to the carbon atoms, whereas **5** and **6** are C—C antibonding. Nevertheless, all four are strongly C—H bonding.

The next group orbital is the *sp* hybrid carbon atomic orbital. In-phase and out-of-phase interactions of this orbital for the two CH_3 groups yield a C—C bonding and a C—C antibonding combination.

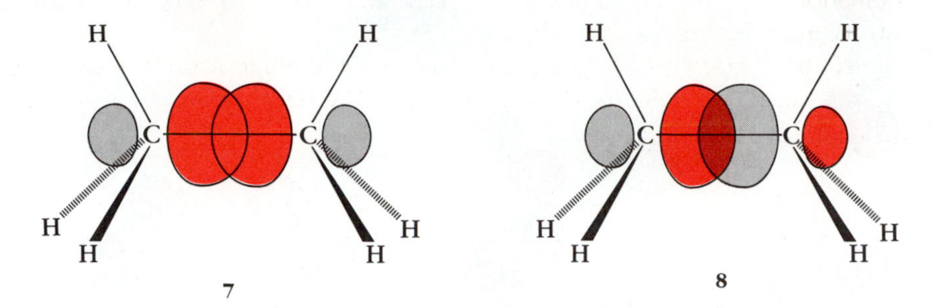

The three remaining CH group orbitals (one $\sigma^*_{CH_3}$ and two $\pi^*_{CH_3}$) for each methyl will combine to generate six more combinations. The resulting atomic orbital contributions to the molecular orbitals are shown in Figure 27.13.

The final step in constructing the molecular orbitals is to consider interaction between any of the combinations that have the same symmetry. To draw the *bonding* molecular orbitals we will employ combinations of orbitals **1** through **7**. (The seven antibonding combinations are of much higher energy and do not interact significantly with the low-energy orbitals even if they have the correct symmetry.) The only two combinations with the same symmetry are **1** and **7** (symmetric with respect to both molecular symmetry planes). Their interaction leads to the following (designated **1** + **7** and **1** − **7** for the in- and out-of-phase combinations, respectively).

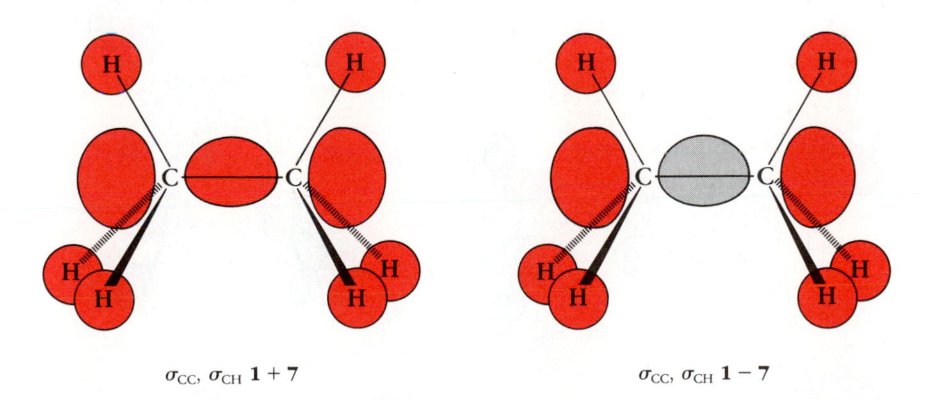

Each of these is bonding with respect to both C—C and C—H interactions, but the presence of two nodal planes makes the second orbital somewhat higher in energy. This results in a change in the relative energies of the orbitals we have been considering. The lowest-energy molecular orbital is that corresponding to combination **1** + **7** (labeled as ϕ_1 in Figure 27.13), and this is followed by combination **2** (ϕ_2). The next two orbitals are the in-phase π combinations, **3** and **4** (ϕ_3 and ϕ_4, respectively), and these are followed by combination **1** − **7** (ϕ_5). The two highest occupied molecular orbitals are the out-of-phase π combinations **5** and **6** (ϕ_6 and ϕ_7).

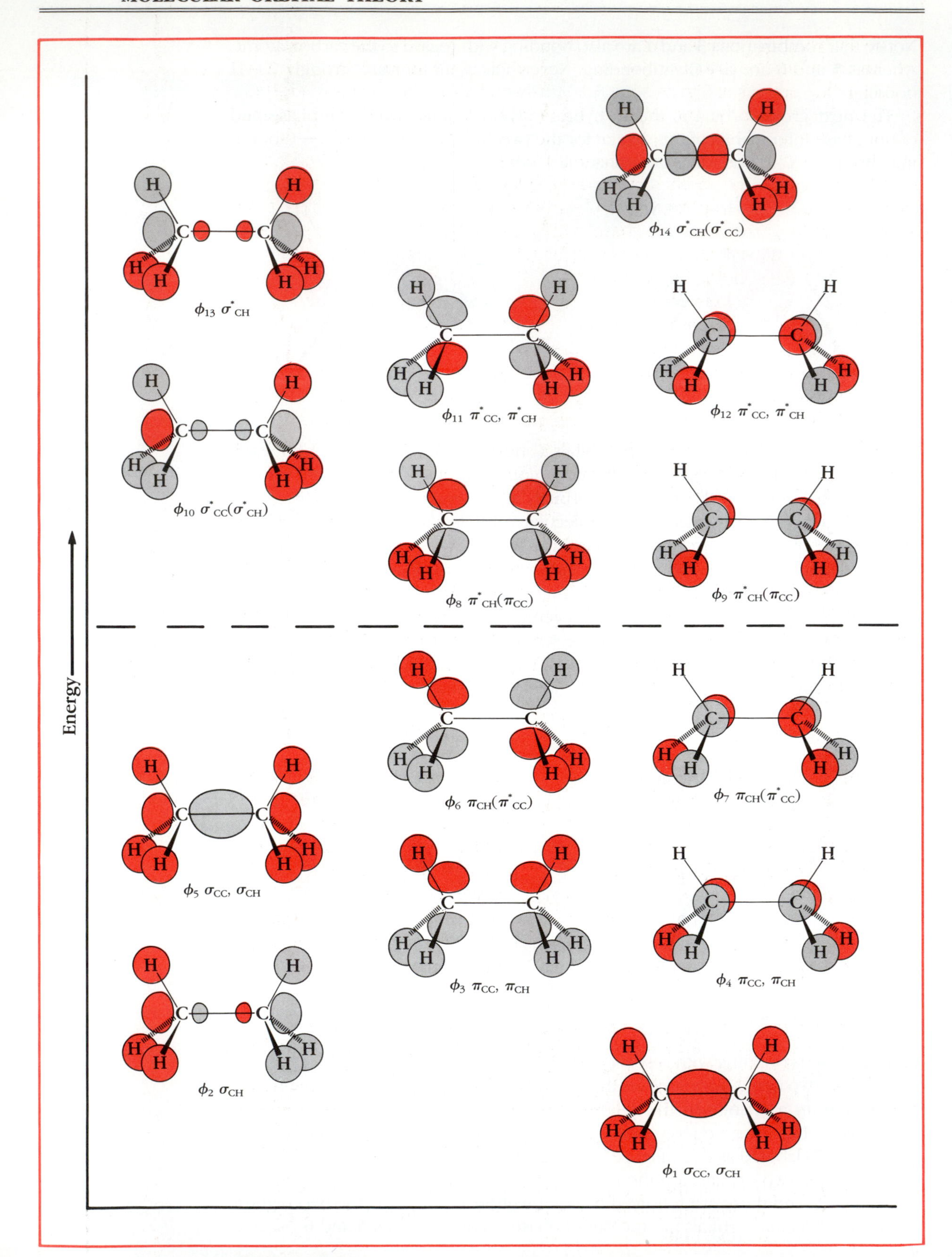
Energy
$\phi_{14}\ \sigma^*_{CH}(\sigma^*_{CC})$
$\phi_{13}\ \sigma^*_{CH}$
$\phi_{11}\ \pi^*_{CC},\ \pi^*_{CH}$
$\phi_{12}\ \pi^*_{CC},\ \pi^*_{CH}$
$\phi_{10}\ \sigma^*_{CC}(\sigma^*_{CH})$
$\phi_8\ \pi^*_{CH}(\pi_{CC})$
$\phi_9\ \pi^*_{CH}(\pi_{CC})$
$\phi_6\ \pi_{CH}(\pi^*_{CC})$
$\phi_7\ \pi_{CH}(\pi^*_{CC})$
$\phi_5\ \sigma_{CC},\ \sigma_{CH}$
$\phi_3\ \pi_{CC},\ \pi_{CH}$
$\phi_4\ \pi_{CC},\ \pi_{CH}$
$\phi_2\ \sigma_{CH}$
$\phi_1\ \sigma_{CC},\ \sigma_{CH}$

Figure 27.13

The antibonding orbitals can be constructed similarly, and Figure 27.13 shows the atomic orbital contributions to all 14 molecular orbitals of eclipsed ethane.

Each of the orbitals in Figure 27.13 is labeled according to the major interaction, C—C or C—H bonding, and antibonding orbitals are designated by an asterisk. Ethane has a total of 14 valence-level electrons (four for each carbon and one for each hydrogen), so the lowest seven orbitals will be doubly occupied. This corresponds precisely to the seven *bonding* molecular orbitals. The lowest-energy orbital (ϕ_1) is bonding with regard to all carbons and hydrogens. As we consider higher-energy orbitals, there are some antibonding interactions, but these are all minor contributors to the bonding molecular orbitals. For example, the π_{CH_3} orbitals, ϕ_3 and ϕ_4, are bonding for both C—C and C—H interactions. The only out-of-phase interactions are between hydrogens on the same methyl group, and we do not consider such hydrogens to be bonded to each other in any event. Even for ϕ_6 and ϕ_7, where there is an antibonding interaction between the carbons, the primary interaction is between carbon and hydrogen, and this is *bonding*.

We have not attempted to indicate the magnitudes of the contributions for the various atomic orbitals to a given MO in Figure 27.13, because these could be obtained only by sophisticated quantum mechanical calculations. The results of such a calculation indicate, for example, that there is a *larger* contribution of the hydrogen 1*s* orbitals to ϕ_6 than to ϕ_3. Conversely, the contribution of carbon atomic orbitals is *less* to ϕ_6 than to ϕ_3. Hence ϕ_3 is primarily a CC orbital, and ϕ_6 is mainly a CH orbital.

Only when we consider the antibonding orbitals do the out-of-phase interactions begin to predominate. Thus ϕ_8 and ϕ_9 are primarily CH *antibonding,* even though the CC interactions are in phase. It is difficult to assess the relative importance of such opposing interactions in the absence of computations, but you could order most of the molecular orbitals of Figures 27.13 by counting the number of nodal planes. Difficulties will arise only for cases in which two orbitals have the same number of nodes; the different energies of the contributing atomic orbitals require the energies of these MOs to depend on more than just the number of nodal planes.

The molecular orbitals of many other structures can be drawn by procedures comparable with those we have illustrated here for ethane. Clearly, the total number of orbitals increases substantially as the number of atoms increases. But it is usually not necessary to draw all the molecular orbitals. If you are interested in qualitatively describing the molecular orbitals at a reactive center, it is usually sufficient to analyze a simple compound that contains the same functional group. The orbital description of a nitrile, for example, could be provided by drawing all the molecular orbitals of HCN. The major features of these orbitals will be the same as those of a larger nitrile, and the effort needed to draw them will be far less.

EXERCISE 27.7

Draw the molecular orbitals (by showing atomic orbital contributions) of: (a) formaldehyde; (b) ethylene. Designate each MO as bonding, nonbonding, or antibonding. Indicate those orbitals that are occupied.

Figure 27.13 Atomic orbital contributions to the molecular orbitals of ethane.

The orbitals are drawn for the eclipsed geometry of ethane, and they are shown in order of increasing energy (with the dotted line indicating the nonbonding level). See page 1440.

27.6 PROBLEMS

27.1 For each of the following molecules, how many valence-level atomic orbitals are available for formation of molecular orbitals? How many total molecular orbitals are there in each case? How many are bonding? Nonbonding? Antibonding? (a) $CH_2{=}CH_2$; (b) CH_3NH_2; (c) $CH_3OH_2^+$; (d) BH_3; (e) CH_3^+; (f) $CH_2{=}CH{-}CH{=}CH_2$; (g) $CH_2{=}C{=}CH_2$.

27.2 Consider the molecular orbitals for molecular nitrogen (N_2). (*Hint:* The sigma bonding between the nitrogens must involve an orbital that is a hybrid of *s* and *p;* analyze the problem on the basis that each nitrogen has two *sp* atomic orbitals and two *p* atomic orbitals).

(a) Sketch the four atomic orbitals for one of the nitrogens.

(b) By allowing appropriate interaction between the atomic orbitals, sketch all the molecular orbitals of nitrogen. (For convenience draw the MOs as the atomic orbital contributions.)

(c) Order the MOs in energy.

(d) Identify each MO as bonding, antibonding, or nonbonding.

(e) Which orbitals are occupied?

27.3 Answer Problem 27.2 for oxygen (O_2). Explain how these molecular orbitals account for the fact that oxygen is a diradical (that is, it has two unpaired electrons in its ground state).

27.4 Answer Problem 27.2 for HCN.

27.5 Which of the following combinations of atomic orbitals would represent legitimate molecular orbitals? For those which would not, indicate why.

(a) Water

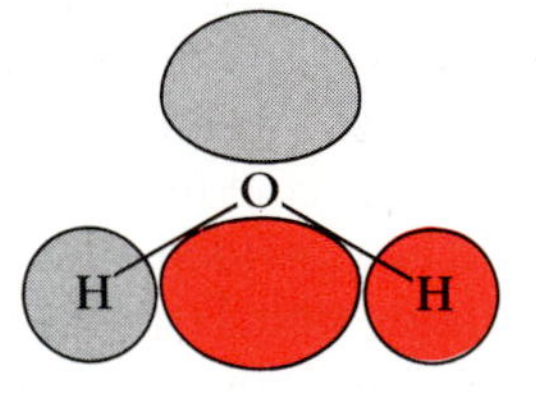

(b) Methyl cation

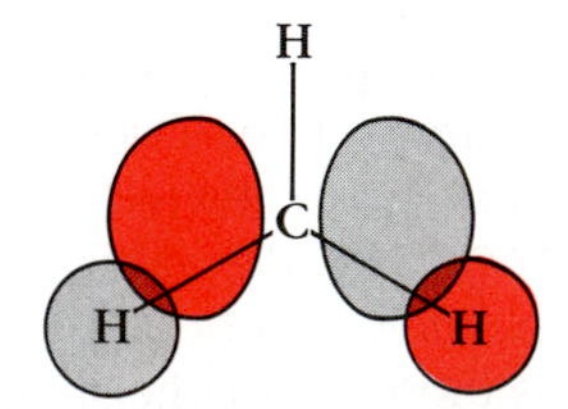

(c) Carbon dioxide

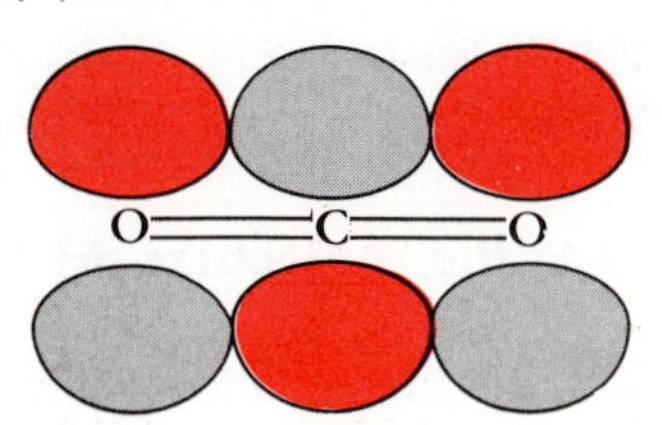

(d) Water

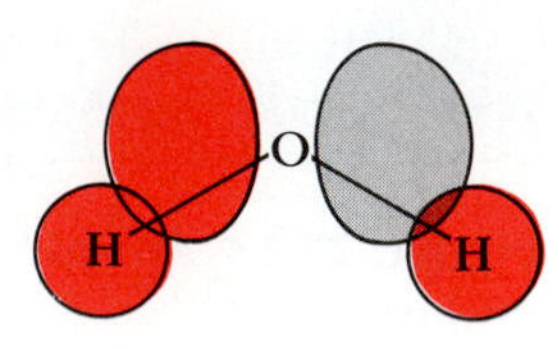

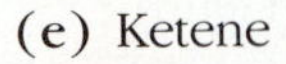

(e) Ketene

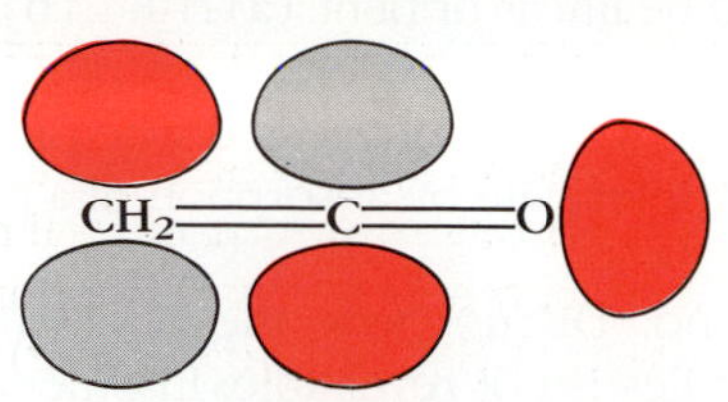

(f) Methyl radical

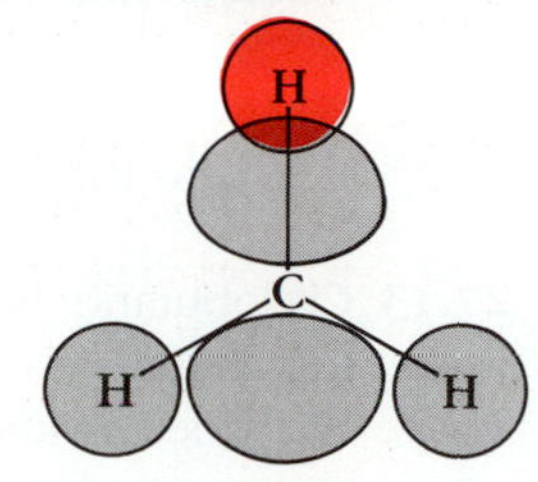

(g) Cyclobutane

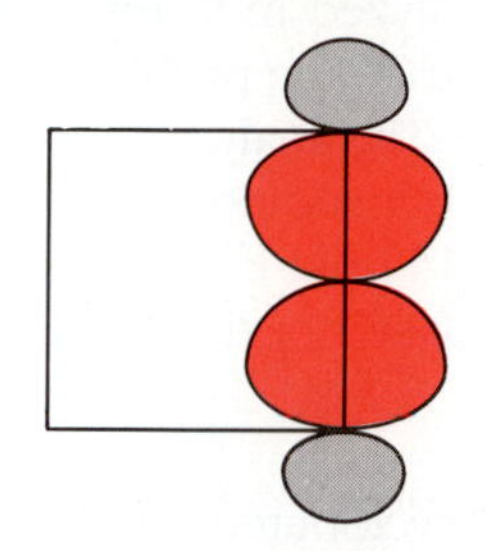

(h) Water

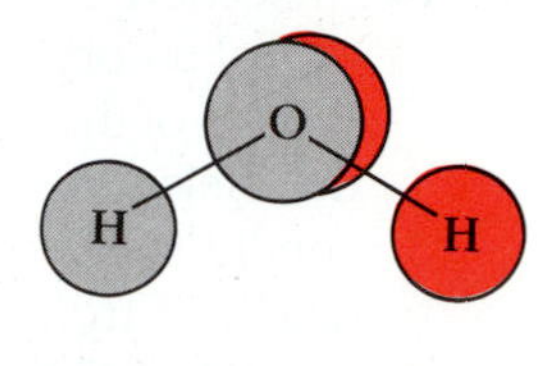

27.6 Draw the complete set of molecular orbitals for acetylene. Use atomic orbital contributions to represent the MOs, and use the group orbital method to generate the MOs.

27.7 In Section 27.2 we stated that hybridized atomic orbitals on carbon would not have the proper symmetry for a linear H—C—H fragment. What symmetry is needed? Show how the *sp* hybrid orbitals used in Figure 27.9 lack this symmetry. Show how the 2*s* and 2*p* atomic orbitals of carbon satisfy this symmetry restriction.

27.8 The H—O—H angle of water is 105°. Using qualitative MO analysis as presented in Section 27.2, predict whether the H—O—H angle would be larger or smaller for the lowest-energy excited state of water. Note that you will have to analyze the two antibonding molecular orbitals (which we did not describe for H—X—H).

27.9 Consider the molecular orbitals for the diatomic molecules of the first row of the periodic table (N_2, O_2, and F_2). Note that to a first approximation the MO's are the same for all three; they differ in only in the number of orbitals that are occupied. Explain why we usually draw the nitrogen, oxygen, and fluorine molecules with triple, double, and single bonds, respectively.

27.10 Diimide (HN═NH) is a reactive compound that acts as a reducing agent by transferring two hydrogens to a carbon-carbon double bond and generating molecular nitrogen. Although diimide is closely related to acetylene, it is not linear. Using qualitative MO theory, show why the nonlinear geometry is favored for diimide. (*Hint:* Construct the molecular orbitals for a hypothetical, linear diimide and then evaluate the effects on orbital energies for distortion from linearity. Note that the MO's for linear diimide are the same as those for acetylene).

27.11 Using the molecular orbital arguments summarized in Figure 27.5, suggest whether each of the following should be linear or bent: (a) H_2F^+; (b) NH_2^+.

27.12 What change in the geometry of BH_2^+ would be expected when it is converted to the lowest-energy excited triplet state by absorption of a photon?

27.13 Cyclobutane is a strained compound. Discuss why the photochemical cycloaddition of two ethylene molecules is not reversible. In other words, why does irradiation of cyclobutane with 180 nm light not result in cleavage to two ethylene molecules?

27.14 In Section 27.4 the Diels-Alder reaction was analyzed in terms of frontier orbital theory. The frontier orbitals used to illustrate the problem were the HOMO of the diene and the LUMO of the alkene. The analysis indicated that the Diels-Alder reaction should be a concerted cycloaddition under thermal conditions. Show that the same prediction should hold if the other combination of frontier orbitals were used, i.e., the LUMO of the diene and the HOMO of the alkene.

27.15 In this chapter we showed that ring opening of cyclobutenes to butadienes is a concerted thermal reaction in which the ring-opened compounds are energetically favored. "Dewar benzene" contains *two* strained cyclobutene rings but is nevertheless quite resistant to ring opening when it is heated. Why?

Dewar benzene

27.16 The cyclopropyl cation can undergo a concerted ring opening to form the allyl cation. This process is so facile that the cyclopropyl cation cannot be observed, and ionization of cyclopropyl derivatives leads directly to allyl cations. Using frontier orbital theory, predict the stereochemistry of the allyl cation that would result from ionization of *cis,cis*-3-chloro-1,2-dimethylcyclopropane.

27.17 When chemists were attempting to learn whether or not cyclodecapentaene should be aromatic (Section 10.6), many attempts were made to prepare this compound. One synthetic strategy was based on thermal ring opening of dihydronaphthalene.

The hydrogens at the ring junction could be either *cis* or *trans*, and the two stereoisomers would be expected to yield products that were also stereoisomers. Draw the two reactions, showing all stereochemical features of the reactants and products. (The original syntheses were unsuccessful because the dihydronaphthalenes are more stable and are favored at equilibrium.)

27.18 Thermal cyclization of 2,4,6,8-decatetraene appears to be a concerted reaction, which produces 7,8-dimethyl-1,3,5-cyclooctatriene. Using frontier orbital theory, deduce the stereochemistry of the starting tetraene that would lead to the cyclic product having the methyls *cis* and *trans*, respectively.

27.19 Cyclopentadienes undergo facile thermal rearrangements in which a substituent appears to migrate around the ring.

R H → R H H + R H H

In fact this results from hydrogen migrations, and these appear to be concerted 1,5 shifts. Using frontier orbital theory, deduce the stereochemistry (suprafacial or antarafacial) of the reaction.

INDEX

The references given in bold refer to first-time entries in the text which are defined in the Terms and Definitions sections or which are defined on that page.